SIXTH EDITION

Genetics

From Genes to Genomes

Leland H. Hartwell
FRED HUTCHISON CANCER CENTER

Michael L. Goldberg
CORNELL UNIVERSITY

Janice A. Fischer
UNIVERSITY OF TEXAS AT AUSTIN

Leroy Hood
THE INSTITUTE FOR SYSTEMS BIOLOGY

Mc Graw Hill Education

GENETICS: FROM GENES TO GENOMES, SIXTH EDITION

Published by McGraw-Hill Education, 2 Penn Plaza, New York, NY 10121.

This book is printed on acid-free paper.

1 2 3 4 5 6 7 8 9 LWI 21 20 19 18 17

ISBN 978–1–259–70090–3
MHID 1–259–70090–9

Portfolio Manager: *Justin K. Wyatt, PhD*
Lead Product Developer: *Fran Simon*
Senior Marketing Manager: *Kelly Brown*
Senior Content Project Manager: *Vicki Krug*
Lead Content Project Manager: *Christina Nelson*
Buyer: *Jennifer Pickel*
Lead Designer: *David Hash*
Lead Content Licensing Specialist: *Lori Hancock*
Content Licensing Specialist: *Lori Slattery*
Cover Image: © *Sydney F. Vize*
Design Element: *Front and Back Matter DNA helix header, © MedicalRF.com*
Compositor: *Aptara, Inc*
Printer: *LSC Communications*

Library of Congress Cataloging-in-Publication Data

Names: Hartwell, Leland, author.
Title: Genetics : from genes to genomes / Leland Hartwell, Michael Goldberg, Janice Fischer, Lee Hood, Charles F. Aquadro, Bruce Bejcek.
Description: Sixth edition. | New York, NY : McGraw-Hill Education, [2018] | Includes index.
Identifiers: LCCN 2017017175| ISBN 9781259700903 (alk. paper) | ISBN 1259700909 (alk. paper)
Subjects: LCSH: Genetics.
Classification: LCC QH430 .G458 2018 | DDC 576.5—dc23 LC record available at https://lccn.loc.gov/2017017175

mheducation.com/highered

About the Authors

Dr. Leland Hartwell is President and Director of Seattle's Fred Hutchinson Cancer Research Center and Professor of Genome Sciences at the University of Washington.

Dr. Hartwell's primary research contributions were in identifying genes that control cell division in yeast, including those necessary for the division process as well as those necessary for the fidelity of genome reproduction. Subsequently, many of these same genes have been found to control cell division in humans and often to be the site of alteration in cancer cells.

Dr. Hartwell is a member of the National Academy of Sciences and has received the Albert Lasker Basic Medical Research Award, the Gairdner Foundation International Award, the Genetics Society Medal, and the 2001 Nobel Prize in Physiology or Medicine.

Dr. Michael Goldberg is a Professor at Cornell University, where he teaches introductory genetics and human genetics. He was an undergraduate at Yale University and received his Ph.D. in biochemistry from Stanford University. Dr. Goldberg performed postdoctoral research at the Biozentrum of the University of Basel (Switzerland) and at Harvard University, and he received an NIH Fogarty Senior International Fellowship for study at Imperial College (England) and fellowships from the Fondazione Cenci Bolognetti for sabbatical work at the University of Rome (Italy). His current research uses the tools of *Drosophila* genetics and the biochemical analysis of frog egg cell extracts to investigate the mechanisms that ensure proper cell cycle progression and chromosome segregation during mitosis and meiosis.

Dr. Janice Fischer is a Professor at The University of Texas at Austin, where she is an award-winning teacher of genetics and Director of the Biology Instructional Office. She received her Ph.D. in biochemistry and molecular biology from Harvard University, and did postdoctoral research at The University of California at Berkeley and The Whitehead Institute at MIT. In her research, Dr. Fischer used *Drosophila* first to determine how tissue-specific transcription works, and then to examine the roles of ubiquitin and endocytosis in cell signaling during development.

Dr. Lee Hood received an M.D. from the Johns Hopkins Medical School and a Ph.D. in biochemistry from the California Institute of Technology. His research interests include immunology, cancer biology, development, and the development of biological instrumentation (for example, the protein sequencer and the automated fluorescent DNA sequencer). His early research played a key role in unraveling the mysteries of antibody diversity. More recently he has pioneered systems approaches to biology and medicine.

Dr. Hood has taught molecular evolution, immunology, molecular biology, genomics and biochemistry and has co-authored textbooks in biochemistry, molecular biology, and immunology, as well as *The Code of Codes*—a monograph about the Human Genome Project. He was one of the first advocates for the Human Genome Project and directed one of the federal genome centers that sequenced the human genome. Dr. Hood is currently the president (and co-founder) of the cross-disciplinary Institute for Systems Biology in Seattle, Washington.

Dr. Hood has received a variety of awards, including the Albert Lasker Award for Medical Research (1987), the Distinguished Service Award from the National Association of Teachers (1998) and the Lemelson/MIT Award for Invention (2003). He is the 2002 recipient of the Kyoto Prize in Advanced Biotechnology—an award recognizing his pioneering work in developing the protein and DNA synthesizers and sequencers that provide the technical foundation of modern biology. He is deeply involved in K–12 science education. His hobbies include running, mountain climbing, and reading.

Brief Contents

Contents

PART II

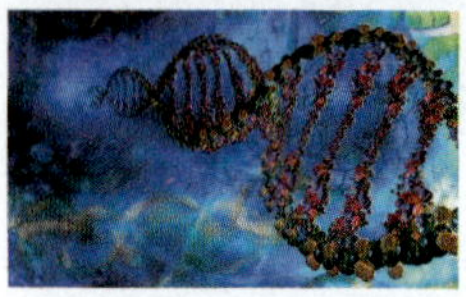

PART III

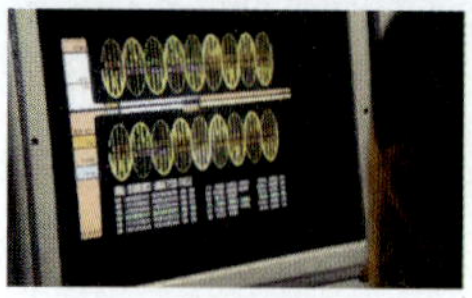

PART IV

© SPL/Science Source

© Michael Goldberg, Cornell University, Ithaca, NY

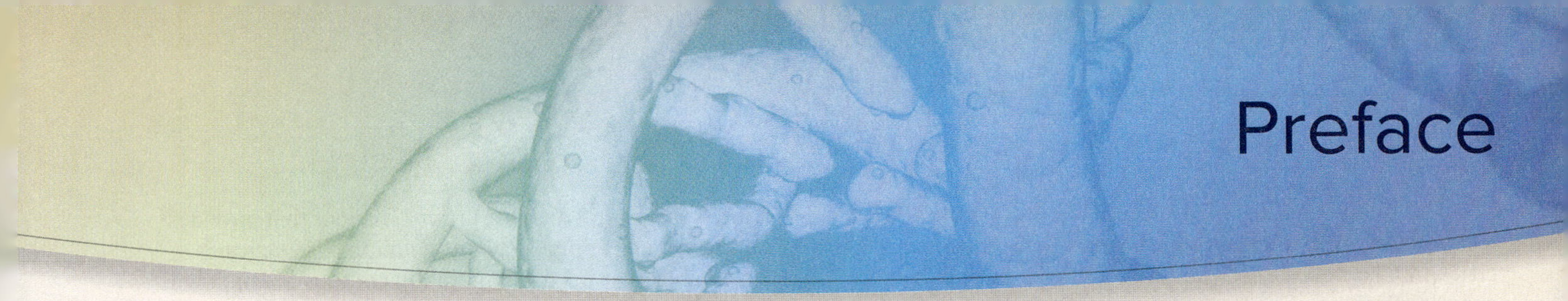

Preface

A Note from the Authors

The science of genetics is less than 150 years old, but its accomplishments within that short time have been astonishing. Gregor Mendel first described genes as abstract units of inheritance in 1865; his work was ignored and then rediscovered in 1900. Thomas Hunt Morgan and his students provided experimental verification of the idea that genes reside within chromosomes during the years 1910–1920. By 1944, Oswald Avery and his coworkers had established that genes are made of DNA. James Watson and Francis Crick published their pathbreaking structure of DNA in 1953. Remarkably, less than 50 years later (in 2001), an international consortium of investigators deciphered the sequence of the 3 billion nucleotides in the human genome. Twentieth century genetics made it possible to identify individual genes and to understand a great deal about their functions.

Today, scientists are able to access the enormous amounts of genetic data generated by the sequencing of many organisms' genomes. Analysis of these data will result in a deeper understanding of the complex molecular interactions within and among vast networks of genes, proteins, and other molecules that help bring organisms to life. Finding new methods and tools for analyzing these data will be a significant part of genetics in the twenty-first century.

Our sixth edition of *Genetics: From Genes to Genomes* emphasizes both the core concepts of genetics and the cutting-edge discoveries, modern tools, and analytic methods that will keep the science of genetics moving forward.

The authors of the sixth edition have worked together in revising every chapter in an effort not only to provide the most up-to-date information, but also to provide continuity and the clearest possible explanations of difficult concepts in one voice.

Our Focus—An Integrated Approach

Genetics: From Genes to Genomes represents a new approach to an undergraduate course in genetics. It reflects the way we, the authors, currently view the molecular basis of life.

We integrate:

- **Formal genetics:** the rules by which genes are transmitted.
- **Molecular genetics:** the structure of DNA and how it directs the structure of proteins.
- **Digital analysis and genomics:** recent technologies that allow a comprehensive analysis of the entire gene set and its expression in an organism.
- **Human genetics:** how genes contribute to health and diseases, including cancer.
- **The unity of life-forms:** the synthesis of information from many different organisms into coherent models.
- **Molecular evolution:** the molecular mechanisms by which biological systems, whole organisms, and populations have evolved and diverged.

The strength of this integrated approach is that students who complete the book will have a strong command of genetics as it is practiced today by both academic and corporate researchers. These scientists are rapidly changing our understanding of living organisms, including ourselves. Ultimately, this vital research may create the ability to replace or correct detrimental genes—those "inborn errors of metabolism," as researcher Archibald Garrod called them in 1923, as well as the later genetic alterations that lead to the many forms of cancer.

The Genetic Way of Thinking

Modern genetics is a molecular-level science, but an understanding of its origins and the discovery of its principles is a necessary context. To encourage a genetic way of thinking, we begin the book by reviewing Mendel's principles and the chromosomal basis of inheritance. From the outset, however, we aim to integrate organism-level genetics with fundamental molecular mechanisms.

Chapter 1 presents the foundation of this integration by summarizing the main biological themes we explore. In Chapter 2, we tie Mendel's studies of pea trait inheritance to the actions of enzymes that determine whether a pea is round or wrinkled, yellow or green, etc. In the same chapter, we point to the relatedness of the patterns of heredity in all organisms. Chapters 3–5 cover extensions to Mendel, the chromosome theory of inheritance, and the fundamentals of gene linkage and mapping. Starting in Chapter 6, we focus on the physical characteristics of DNA, on mutations, and on how DNA encodes, copies, and transmits biological information.

Beginning in Chapter 9, we move into the digital revolution in DNA analysis with a look at modern genetics techniques, including gene cloning, PCR, microarrays, and high-throughput genome sequencing. We explore how bioinformatics, an emergent analytical tool, can aid in discovery of genome features. This section concludes in Chapter 11 with case studies leading to the discovery of human disease genes.

The understanding of molecular and computer-based techniques carries into our discussion of chromosome specifics in Chapters 12–15, and also informs our analysis of gene regulation in Chapters 16 and 17. Chapter 18 describes the most recent technology that scientists can use to manipulate genomes at will – for research and practical purposes including gene therapy. Chapter 19 describes the use of genetic tools at the molecular level to uncover the complex interactions of eukaryotic development. In Chapter 20, we explain how our understanding of genetics and the development of molecular genetic technologies is enabling us to comprehend cancer and in some cases to cure it.

Chapters 21 and 22 cover population genetics, with a view of how molecular tools have provided information on species relatedness and on genome changes at the molecular level over time. In addition, we explain how bioinformatics can be combined with population genetics to understand inheritance of complex traits and to trace human ancestry.

Throughout our book, we present the scientific reasoning of some of the ingenious researchers of the field—from Mendel, to Watson and Crick, to the collaborators on the Human Genome Project. We hope student readers will see that genetics is not simply a set of data and facts, but also a human endeavor that relies on contributions from exceptional individuals.

Student-Friendly Features

As digital components of the text become more and more crucial, we are very excited that Janice Fischer, a textbook author, is taking on a dual role as Digital Editor! Janice will ensure the important consistency between text and digital.

We have taken great pains to help the student make the leap to a deeper understanding of genetics. Numerous features of this book were developed with that goal in mind.

- ***One Voice Genetics:*** *Genes to Genomes* has a friendly, engaging reading style that helps students master the concepts throughout this book. The writing style provides the student with the focus and continuity required to make the book successful in the classroom.
- ***Visualizing Genetics*** The highly specialized art program developed for this book integrates photos and line art in a manner that provides the most engaging visual presentation of genetics available. Our Feature Figure illustrations break down complex processes into step-by-step illustrations that lead to greater student understanding. All illustrations are rendered with a consistent color theme—for example, all presentations of phosphate groups are the same color, as are all presentations of mRNA.
- ***Accessibility*** Our intention is to bring cutting-edge content to the student level. A number of more complex illustrations are revised and segmented to help the student follow the process. Legends have been streamlined to highlight only the most important ideas, and throughout the book, topics and examples have been chosen to focus on the most critical information.
- ***Problem Solving*** Developing strong problem-solving skills is vital for every genetics student. The authors have carefully created problem sets at the end of each chapter that allow students to improve upon their problem-solving ability.
- ***Solved Problems*** These cover topical material with complete answers provide insight into the step-by-step process of problem solving.
- ***Review Problems*** More than 700 questions involving a variety of levels of difficulty that develop excellent problem-solving skills. The problems are organized by chapter section and in order of increasing difficulty within each section for ease of use by instructors and students. The companion online *Study Guide and Solutions Manual,* completely revised for the 6th edition by Michael Goldberg and Janice Fischer, provides detailed analysis of strategies to solve all of the end-of-chapter problems.

SOLVED PROBLEMS

I. The following figure shows a screen shot from the UCSC Genome Browser, focusing on a region of the human genome encoding a gene called *MFAP3L.* (*Note: hg38* refers to version 38 of the human genome RefSeq.) If you do not remember how the browser represents the genome, refer to the key at the bottom of Fig. 10.3.

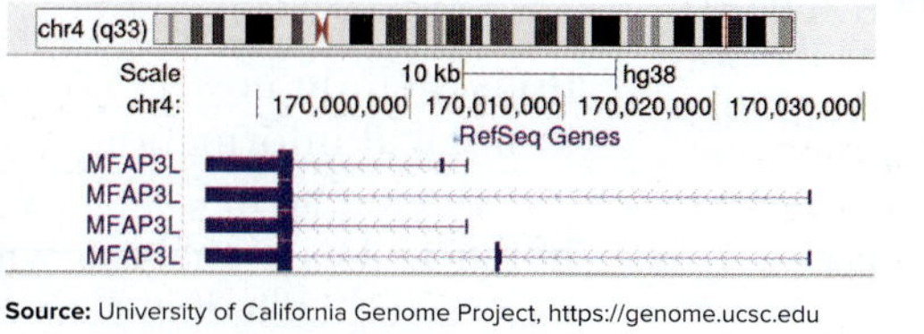

Source: University of California Genome Project, https://genome.ucsc.edu

a. Describe in approximate terms the genomic location of *MFAP3L.*

b. Is the gene transcribed in the direction from the centromere-to-telomere or from the telomere-to-centromere?

c. How many alternative splice forms of *MFAP3L* mRNA are indicated by the data?

b. The arrows within the introns of the gene show that the direction of transcription is from the telomere of *4q* toward the centromere of chromosome 4.

c. The data indicate four alternatively spliced forms of the mRNA. In the following parts, we list these as A to D from top to bottom.

d. The data suggest two promoters. One is roughly at position 170,037,000 and allows the transcription of a primary RNA alternatively spliced to produce mRNAs B and D. The other is roughly at position 170,013,000 and leads to the transcription of a primary RNA alternatively spliced to generate mRNAs A and C.

e. The data indicate that the *MFAP3* gene can encode two different but closely related proteins. mRNAs A, B, and C all encode the same protein; mRNA D a slightly larger protein that includes at its N terminus additional amino acids not found in the other protein. Otherwise these two proteins appear to be the same. The ORF that encodes the A B C protein form is about 880 bp long (a rough estimate); this corre-

Changes in the 6th Edition: A Chapter-by-Chapter Summary

The sixth edition has been revised and modernized significantly as compared with the fifth edition. We scrutinized the entire text and clarified the language wherever possible. In total, we created more than 50 new Figures and Tables, and revised more than 100 in addition. We also wrote more than 125 new end-of-chapter problems, and revised many other problems for clarity. The entire *Solutions Manual and Study Guide* was corrected and revised for clarity. We added several new Fast Forward, Genetics and Society, and Tools of Genetics Boxes on modern topics. Chapter 9 in the 5th edition was split into two separate chapters in the 6th edition: Chapter 9 (Digital Analysis of DNA) and Chapter 10 (Genome Annotation).

Along with the numerous text changes, the authors have also spent a great deal of time updating the test bank and question bank content to align more closely to the text. There will also be new video tutorials for difficult concepts in every chapter!

Every chapter of the sixth edition was improved significantly from the fifth edition. The most important changes in the sixth edition are summarized below:

Chapter 3 Extensions to Mendel's Laws
- Relationship between epistasis and complementation explained more clearly.
- Discussion of two-gene versus multifactorial inheritance now separated for clarity.
- Comprehensive Example about dog coat colors expanded to include molecular explanations for the various gene activities.

Chapter 4 The Chromosome Theory of Inheritance
- Figures and text altered to clarify that each chromatid has a centromere.
- New Fast Forward Box: *Visualizing X Chromosome Inactivation in Transgenic Mice*

Chapter 5 Linkage, Recombination, and the Mapping of Genes on Chromosomes
- New Fast Forward Box: *Mapping the Crossovers that Generate Individual Human Sperm*

Chapter 6 DNA Structure, Replication, and Recombination
- Improvements to diagrams of DSB repair model of recombination.
- New section about site-specific recombination.

Chapter 7 Anatomy and Function of Gene: Dissection Through Mutation
- Reorganized and clarified material to separate the discussion of DNA sequence alteration mechanisms from DNA repair mechanisms.

Chapter 9 Digital Analysis of DNA
- Improved depiction of plasmid cloning vectors.
- Renovated explanation of paired-end whole-genome shotgun sequencing.

Chapter 10 Genome Annotation
- Improved depiction of alternative RNA splicing.
- New illustration of consensus amino acid sequences in proteins.
- New material on the evolution of *de novo* genes.

Chapter 12 The Eukaryotic Chromosome
- New material on synthetic yeast chromosomes.

Chapter 15 Organellar Inheritance
- New Fast Forward Box about the Mitchondrial Eve concept.

Chapter 17 Gene Regulation in Eukaryotes
- New Tools of Genetics Box: *The Gal4/UAS$_G$ Binary Gene Expression System*
- New part of Epigenetics section: *Can Environmentally Acquired Traits Be Inherited?*
- New part of Regulation After Transcription section: Trans-*acting Proteins Regulate Translation*

Chapter 18 Manipulating the Genomes of Eukaryotes
- New part of Targeted Mutagenesis section: *CRISPR/Cas9 Allows Targeted Gene Editing in Any Organism*
- New Tools of Genetics Box: *How Bacteria Vaccinate Themselves Against Viral Infections with CRISPR/Cas9*
- New Genetics and Society Box: *Should We Alter the Genomes of Human Germ Lines?*

Chapter 19 The Genetic Analysis of Development
- Comprehensive Example of *Drosophila* body patterning revised to clarify that homeotic genes function within parasegments, and to clarify the concept of a morphogen.

Chapter 20 The Genetics of Cancer
- Clarified the fact that mutation drives cancer progression.
- Improved explanation of driver and passenger mutations.
- Increased coverage of tumor genome sequencing and the heterogeneity of mutations in different individuals with cancers in the same organ.

Chapter 22 The Genetics of Complex Traits
- Revised the section on heritability to clarify: lines of correlation and correlation coefficients; how to use different kinds of human twin studies to estimate the heritability of complex quantitative traits and complex discrete traits.
- New explanation of how to use the chi-square test for independence for GWAS.

Robust Analytics and Reporting

©Hero Images/Getty Images

- Connect Insight® generates easy-to-read reports on individual students, the class as a whole, and on specific assignments.
- The Connect Insight dashboard delivers data on performance, study behavior, and effort. Instructors can quickly identify students who struggle and focus on material that the class has yet to master.
- Connect automatically grades assignments and quizzes, providing easy-to-read reports on individual and class performance.

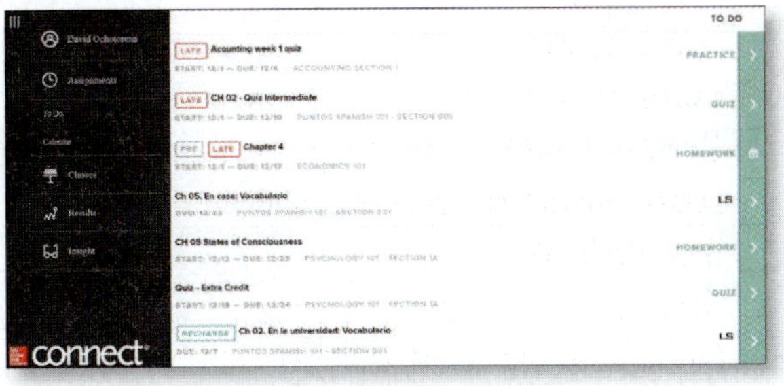

Impact on Final Course Grade Distribution

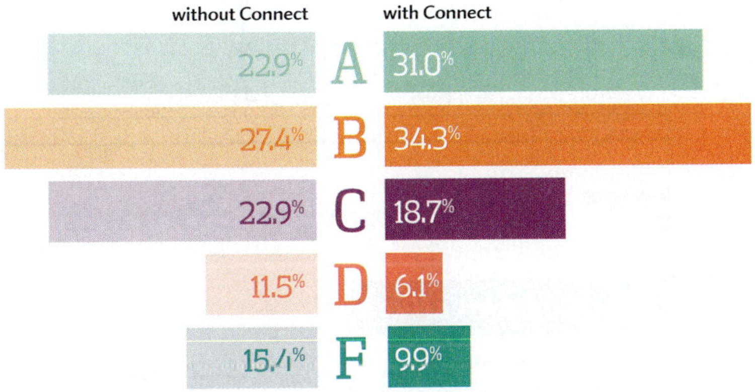

More students earn **As** and **Bs** when they use **Connect**.

Trusted Service and Support

- Connect integrates with your LMS to provide single sign-on and automatic syncing of grades. Integration with Blackboard®, D2L®, and Canvas also provides automatic syncing of the course calendar and assignment-level linking.
- Connect offers comprehensive service, support, and training throughout every phase of your implementation.
- If you're looking for some guidance on how to use Connect, or want to learn tips and tricks from super users, you can find tutorials as you work. Our Digital Faculty Consultants and Student Ambassadors offer insight into how to achieve the results you want with Connect.

www.mheducation.com/connect

Guided Tour

Integrating Genetic Concepts

Genetics: From Genes to Genomes takes an integrated approach in its presentation of genetics, thereby giving students a strong command of genetics as it is practiced today by academic and corporate researchers. Principles are related throughout the text in examples, essays, case histories, and connections sections to make sure students fully understand the relationships between topics.

Chapter Outline

Every chapter opens with a brief outline of the chapter contents.

chapter outline

- 18.1 Creating Transgenic Organisms
- 18.2 Uses of Transgenic Organisms
- 18.3 Targeted Mutagenesis
- 18.4 Human Gene Therapy

Learning Objectives

Learning Objectives appear before each section, and are carefully written to clearly outline expectations.

18.2 Uses of Transgenic Organisms

learning objectives

1. Describe how transgenes can clarify which gene causes a mutant phenotype.
2. Summarize the use of transgene reporter constructs in gene expression studies.
3. Discuss examples of how transgenic organisms serve to produce proteins needed for human health.
4. List examples of GM organisms and discuss the pros and cons of their production.
5. Explain the use of transgenic animals to model gain-of-function genetic diseases in humans.

essential concepts

- A wild-type transgene can be inserted into an embryo homozygous for a recessive mutant allele. If the normal phenotype is restored, then the transgene identifies the gene that was mutated.
- The creation of reporter constructs allows easy detection of when and in which tissues a gene is turned on or turned off in eukaryotes.
- Transgenic organisms produce medically important human proteins including insulin, blood clotting factors, and erythropoietin; transgenic crop plants can potentially make ingestible vaccines.
- GM soybeans are resistant to the weed killer glyphosate. Many crops, such as corn, soybean, canola, and cotton have been genetically modified to express Bt protein which discourages insect predation.
- Adding a transgene that carries a disease-causing, gain-of-function allele to a nonhuman animal model allows researchers to observe disease progression and to test possible therapeutic interventions.

Essential Concepts

After each section, the most relevant points of content are now provided in concise, bulleted statements to reinforce crucial concepts and learning objectives for students.

WHAT'S NEXT

Manipulation of the genome is the basis for many of the experimental strategies we will describe in Chapter 19, where we discuss how genetic analysis has been a crucial tool in unraveling the biochemical pathways of development—the process by which a single-celled zygote becomes a complex multicellular organism. Transgenic technology is key to cloning the genes identified in mutant screens that are crucial for regulating development, and also to manipulating these genes in order to understand their precise functions in the organism.

What's Next

Each chapter closes with a What's Next section that serves as a bridge between the topics in the chapter just completed to those in the upcoming chapter or chapters. This spirals the learning and builds connections for students.

New! Exciting Revised Content

Every chapter of the sixth edition has been revised and modernized significantly as compared with the fifth edition. More than 50 new Figures and Tables were created, and more than 100 were revised. More than 125 new end-of-chapter problems were written, and many more revised for clarity. The entire Solutions Manual and Study Guide was updated, corrected, and revised by Michael Goldberg and Janice Fischer. Several new Fast Forward, Genetics and Society, and Tools of Genetics Boxes covering modern topics were created. For breadth and clarity, Chapter 9 in the 5th edition was split into two separate chapters in the 6th edition: Chapter 9 (Digital Analysis of DNA) and Chapter 10 (Genome Annotation).

Figure 6.30 One site-specific recombination mechanism. The Cre and Flp enzymes discussed in the text function as shown. The *red* and *blue* target DNA sequences are identical to each other but are represented in different colors for clarity. These targets are embedded in different DNA molecules (*black* and *gray dots*). The subunits of the recombinase tetramer are *yellow ovals;* this enzyme catalyzes all steps of the reaction. *Black triangles* are sites where recombinase cleaves single-stranded DNA. Note that resolution of the Holliday junction intermediate involves cleavage of the *blue* and *red* DNA strands that were not cleaved initially.

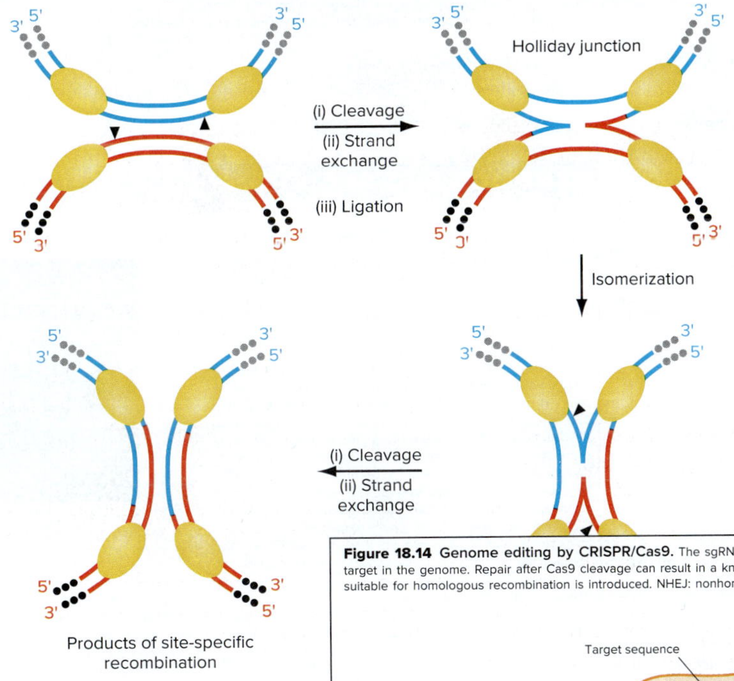

Figure 18.14 Genome editing by CRISPR/Cas9. The sgRNA sequence is designed to bring the Cas9 endonuclease to a specific target in the genome. Repair after Cas9 cleavage can result in a knockout or a knockin, depending on whether or not a DNA fragment suitable for homologous recombination is introduced. NHEJ: nonhomologous end-joining.

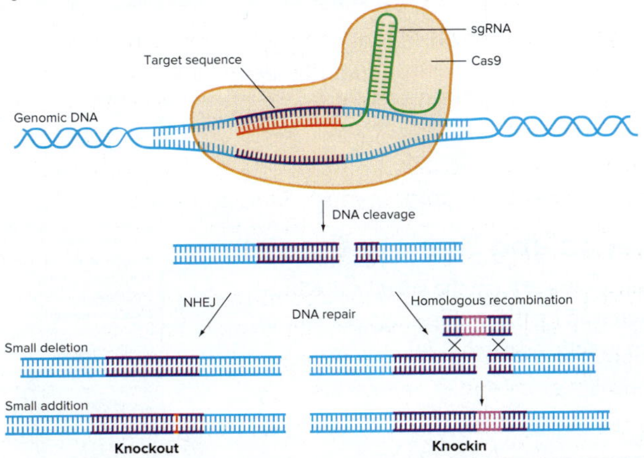

FAST FORWARD

Visualizing X Chromosome Inactivation in Transgenic Mice

Scientists have recently used molecular techniques and transgenic technology (similar to that described in the earlier Fast Forward Box *Transgenic Mice Prove That SRY Is the Maleness Factor*) to visualize the pattern of X chromosome inactivation in mice. The researchers generated XX mice containing two different *transgenes* (in this case, genes from a different species). One of these transgenes was a jellyfish gene that specifies green fluorescent protein (GFP); the other was a gene from red coral that makes red fluorescent protein (RFP) **(Fig. A)**.

In the XX mice, the GFP gene is located on the X chromosome from the mother, and the RFP gene resides on the X chromosome from the father. Clonal patches of cells are either green or red depending on which X chromosome was turned into a Barr body in the original cell that established the patch **(Fig. B)**.

Different XX mice display different green and red patchwork patterns, providing a clear demonstration of the random nature of X chromosome inactivation. The patchwork patterns reflect the cellular memory of which X chromosome was inactivated in the founder cell for each clonal patch. Geneticists currently use these transgenic mice to decipher the genetic details of how cells "remember" which X to inactivate after each cell division.

Figure A **Cells of transgenic mice glow either green or red in response to X chromosome inactivation.** The mouse carries a green (GFP) transgene inserted in the maternal X chromosome (X^M), and a red (RFP) transgene in the paternal X chromosome (X^P). Cells in which X^P is inactivated (*top*) glow green; cells glow red (*bottom*) when X^M is inactivated.

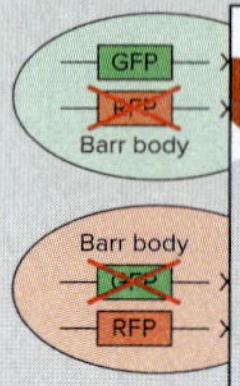

Figure B **Heart cells of a transgenic mouse reveal a clonal patchwork of X inactivation.** Patches of red or green cells represent cellular descendants of the founders that randomly inactivated one of their X chromosomes.
© Hao Wu and Jeremy Nathans, Molecular Biology and Genetics, Neuroscience, and HHMI, Johns Hopkins Medical School.

Fast Forward

This feature is one of the methods used to integrate the Mendelian principles introduced early in the content with the molecular content that will follow.

TOOLS OF GENETICS

How Bacteria Vaccinate Themselves Against Viral Infections with CRISPR/Cas9

Researchers discovered clustered sequence repeats (CRISPRs) in bacterial genomes as early as 1987. When in 2005 some of these sequences were found to originate from bacteriophage genomes, several astute scientists speculated that CRISPRs might mediate a viral immunity system in bacteria. These ideas were largely ignored for several more years until the mechanism of resistance became clarified. And finally, in 2012–2013, the so-called *CRISPR craze* reached its full bloom when researchers including Feng Zhang, Jennifer Doudna, and Emmanuelle Charpentier developed methods to adapt this viral immunity system to engineer genomes in bacterial cells and in eukaryotic organisms.

At the *CRISPR* locus of bacterial genomes, short direct repeats are interrupted at regular intervals by unique spacer sequences (**Fig. A**). The spacer sequences are fragments of bacteriophage genomes captured by the host cell and integrated into the host genome by the action of two bacterially encoded Cas proteins (Cas1 and Cas2). The repeats within the *CRISPR* arrays are added by these endonucleolytic enzymes during the capture and integration process.

Viral immunity results from steps that begin with transcription of the *CRISPR* array into long RNA molecules called *pre-crRNAs* that are processed into short (24–48 nt) so-called *CRISPR RNAs* (*crRNAs*). In the bacterial species

Figure A The *CRISPR/Cas9* locus vaccinates bacteria against viruses.

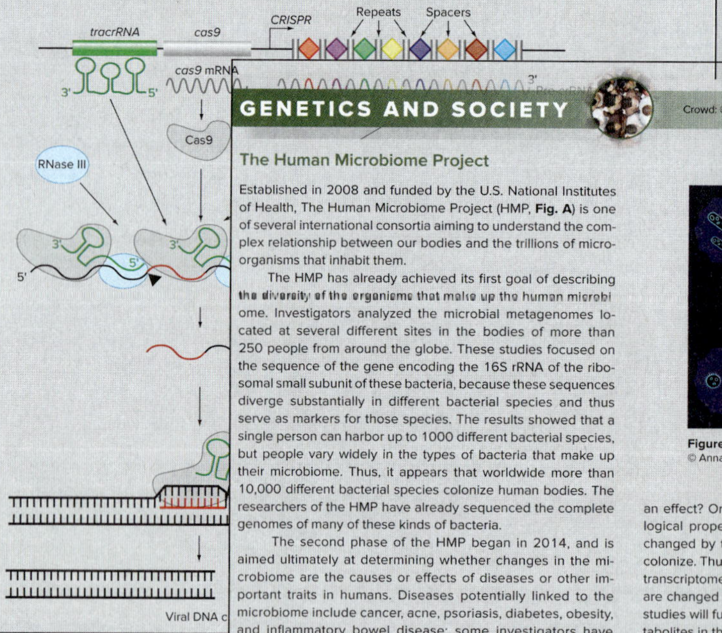

Tools of Genetics Essays

Current readings explain various techniques and tools used by geneticists, including examples of applications in biology and medicine.

GENETICS AND SOCIETY

Crowd: © Image Source/Getty Images RF

The Human Microbiome Project

Established in 2008 and funded by the U.S. National Institutes of Health, The Human Microbiome Project (HMP, **Fig. A**) is one of several international consortia aiming to understand the complex relationship between our bodies and the trillions of microorganisms that inhabit them.

The HMP has already achieved its first goal of describing the diversity of the organisms that make up the human microbiome. Investigators analyzed the microbial metagenomes located at several different sites in the bodies of more than 250 people from around the globe. These studies focused on the sequence of the gene encoding the 16S rRNA of the ribosomal small subunit of these bacteria, because these sequences diverge substantially in different bacterial species and thus serve as markers for those species. The results showed that a single person can harbor up to 1000 different bacterial species, but people vary widely in the types of bacteria that make up their microbiome. Thus, it appears that worldwide more than 10,000 different bacterial species colonize human bodies. The researchers of the HMP have already sequenced the complete genomes of many of these kinds of bacteria.

The second phase of the HMP began in 2014, and is aimed ultimately at determining whether changes in the microbiome are the causes or effects of diseases or other important traits in humans. Diseases potentially linked to the microbiome include cancer, acne, psoriasis, diabetes, obesity, and inflammatory bowel disease; some investigators have suggested that the composition of microbiomes could also influence the mental health of their hosts. The first step in these studies will be to establish whether statistical correlations exist between specific kinds of microbial communities and disease states. As one example, one HMP phase II project currently underway is an analysis of vaginal host cells and microbes during pregnancy. Approximately 2000 pregnant women will be studied and their birth outcomes recorded. The goal of this project is to determine if changes in the microbiome correlate with premature birth or other complications of pregnancy.

Of course, the existence of any correlations found between microbiomes and disease does not prove cause or effect. But even if bacteria correlated with a disease state do not cause the disease, the existence of the correlation could be useful as a way to diagnose certain conditions. Nevertheless, the most ex-

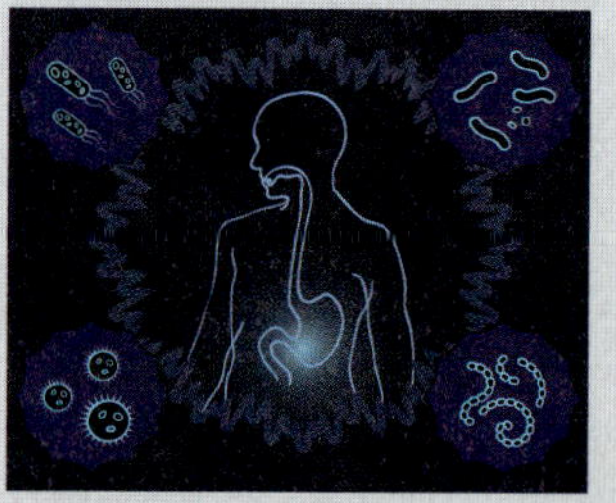

Figure A
© Anna Smirnova/Alamy RF

an effect? One method is to investigate in detail how the biological properties of the microbiome and the host might be changed by the interactions of bacteria and the humans they colonize. Thus, scientists will characterize whether and how the transcriptomes and proteomes of the bacteria and human cells are changed by bacterial colonization of human organs. These studies will further delve into *metabolomics* (characterizing metabolites in the human bloodstream).

A second and even more powerful method for establishing the cause and effect of microbiome changes is the use of *germ-free mice* raised in sterile environments. Surprisingly, germ-free mice can survive although they are not normal: they have altered immune systems, poor skin, and they need to eat more calories than do normal mice to maintain a normal body weight. Researchers can populate germ-free mice with a single bacterial species or a complex microbial community, and thus determine how microbiomes influence physiological states. Problem 8 at the end of this chapter will allow you to explore this approach by discussing an experiment recently performed with germ-free mice that asks if the microbiome plays a causal role in obesity.

If microbial communities indeed contribute to disease states in humans, then future treatments might aim to alter resident microbiomes. Thus, the flip side of the HMP is to in-

Genetics and Society Essays

Dramatic essays explore the social and ethical issues created by the multiple applications of modern genetic research.

Visualizing Genetics

Full-color illustrations and photographs bring the printed word to life. These visual reinforcements support and further clarify the topics discussed throughout the text.

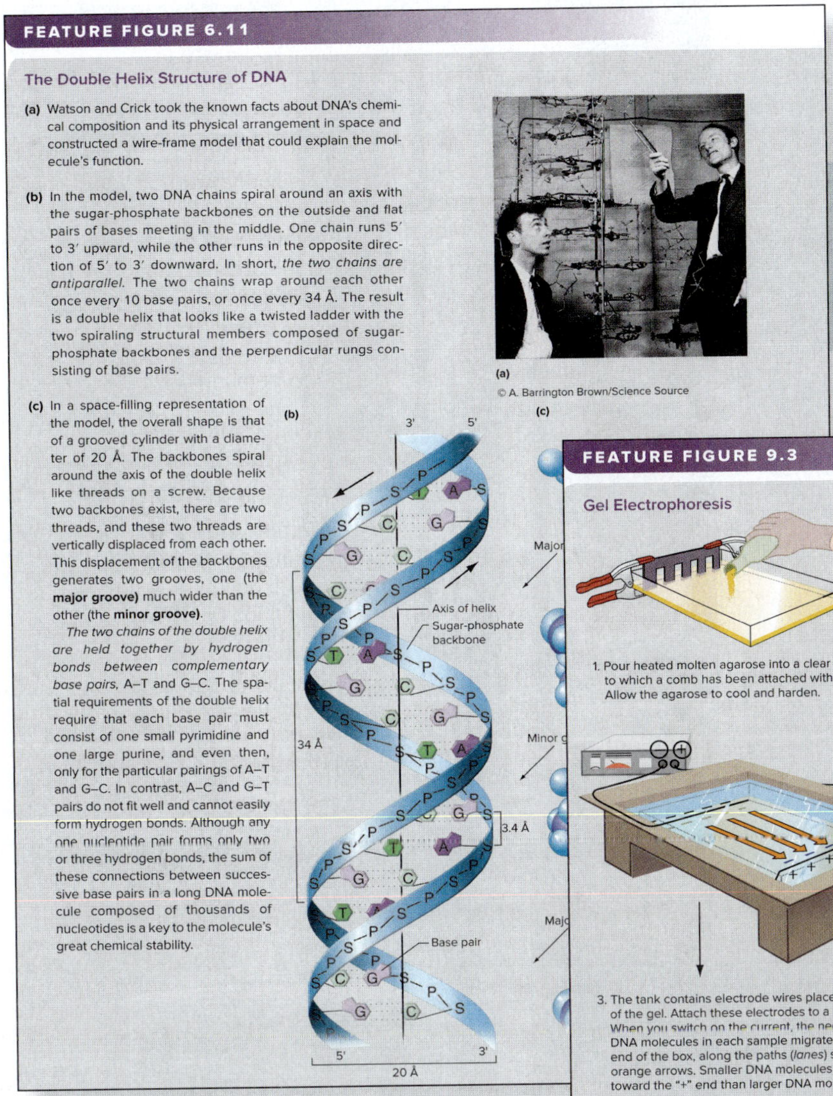

FEATURE FIGURE 6.11

The Double Helix Structure of DNA

(a) Watson and Crick took the known facts about DNA's chemical composition and its physical arrangement in space and constructed a wire-frame model that could explain the molecule's function.

(b) In the model, two DNA chains spiral around an axis with the sugar-phosphate backbones on the outside and flat pairs of bases meeting in the middle. One chain runs 5′ to 3′ upward, while the other runs in the opposite direction of 5′ to 3′ downward. In short, *the two chains are antiparallel.* The two chains wrap around each other once every 10 base pairs, or once every 34 Å. The result is a double helix that looks like a twisted ladder with the two spiraling structural members composed of sugar-phosphate backbones and the perpendicular rungs consisting of base pairs.

(c) In a space-filling representation of the model, the overall shape is that of a grooved cylinder with a diameter of 20 Å. The backbones spiral around the axis of the double helix like threads on a screw. Because two backbones exist, there are two threads, and these two threads are vertically displaced from each other. This displacement of the backbones generates two grooves, one (the **major groove**) much wider than the other (the **minor groove**).

The two chains of the double helix are held together by hydrogen bonds between complementary base pairs, A–T and G–C. The spatial requirements of the double helix require that each base pair must consist of one small pyrimidine and one large purine, and even then, only for the particular pairings of A–T and G–C. In contrast, A–C and G–T pairs do not fit well and cannot easily form hydrogen bonds. Although any one nucleotide pair forms only two or three hydrogen bonds, the sum of these connections between successive base pairs in a long DNA molecule composed of thousands of nucleotides is a key to the molecule's great chemical stability.

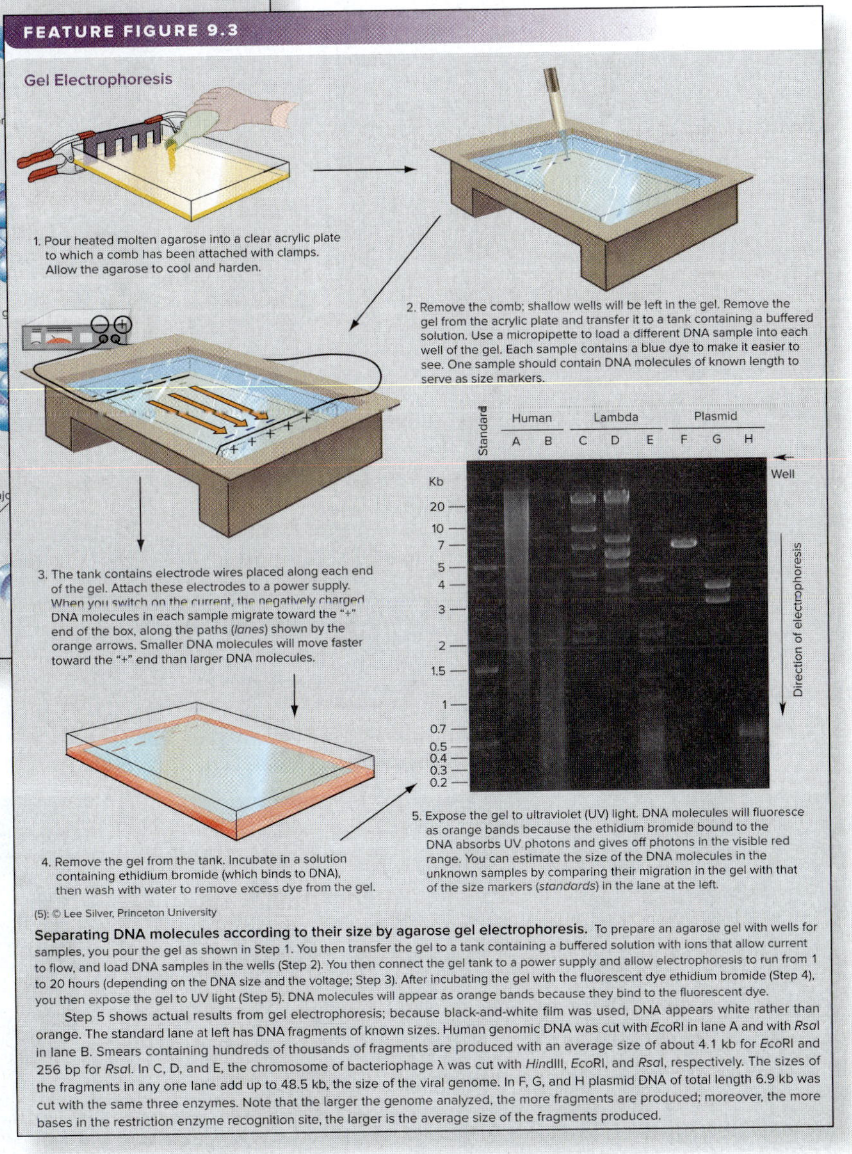

FEATURE FIGURE 9.3

Gel Electrophoresis

1. Pour heated molten agarose into a clear acrylic plate to which a comb has been attached with clamps. Allow the agarose to cool and harden.

2. Remove the comb; shallow wells will be left in the gel. Remove the gel from the acrylic plate and transfer it to a tank containing a buffered solution. Use a micropipette to load a different DNA sample into each well of the gel. Each sample contains a blue dye to make it easier to see. One sample should contain DNA molecules of known length to serve as size markers.

3. The tank contains electrode wires placed along each end of the gel. Attach these electrodes to a power supply. When you switch on the current, the negatively charged DNA molecules in each sample migrate toward the "+" end of the box, along the paths (*lanes*) shown by the orange arrows. Smaller DNA molecules will move faster toward the "+" end than larger DNA molecules.

4. Remove the gel from the tank. Incubate in a solution containing ethidium bromide (which binds to DNA), then wash with water to remove excess dye from the gel.

5. Expose the gel to ultraviolet (UV) light. DNA molecules will fluoresce as orange bands because the ethidium bromide bound to the DNA absorbs UV photons and gives off photons in the visible red range. You can estimate the size of the DNA molecules in the unknown samples by comparing their migration in the gel with that of the size markers (*standards*) in the lane at the left.

(5): © Lee Silver, Princeton University

Separating DNA molecules according to their size by agarose gel electrophoresis. To prepare an agarose gel with wells for samples, you pour the gel as shown in Step 1. You then transfer the gel to a tank containing a buffered solution with ions that allow current to flow, and load DNA samples in the wells (Step 2). You then connect the gel tank to a power supply and allow electrophoresis to run from 1 to 20 hours (depending on the DNA size and the voltage; Step 3). After incubating the gel with the fluorescent dye ethidium bromide (Step 4), you then expose the gel to UV light (Step 5). DNA molecules will appear as orange bands because they bind to the fluorescent dye.

Step 5 shows actual results from gel electrophoresis; because black-and-white film was used, DNA appears white rather than orange. The standard lane at left has DNA fragments of known sizes. Human genomic DNA was cut with *Eco*RI in lane A and with *Rsa*I in lane B. Smears containing hundreds of thousands of fragments are produced with an average size of about 4.1 kb for *Eco*RI and 256 bp for *Rsa*I. In C, D, and E, the chromosome of bacteriophage λ was cut with *Hin*dIII, *Eco*RI, and *Rsa*I, respectively. The sizes of the fragments in any one lane add up to 48.5 kb, the size of the viral genome. In F, G, and H plasmid DNA of total length 6.9 kb was cut with the same three enzymes. Note that the larger the genome analyzed, the more fragments are produced; moreover, the more bases in the restriction enzyme recognition site, the larger is the average size of the fragments produced.

Feature Figures

Special multipage spreads integrate line art, photos, and text to summarize in detail important genetic concepts.

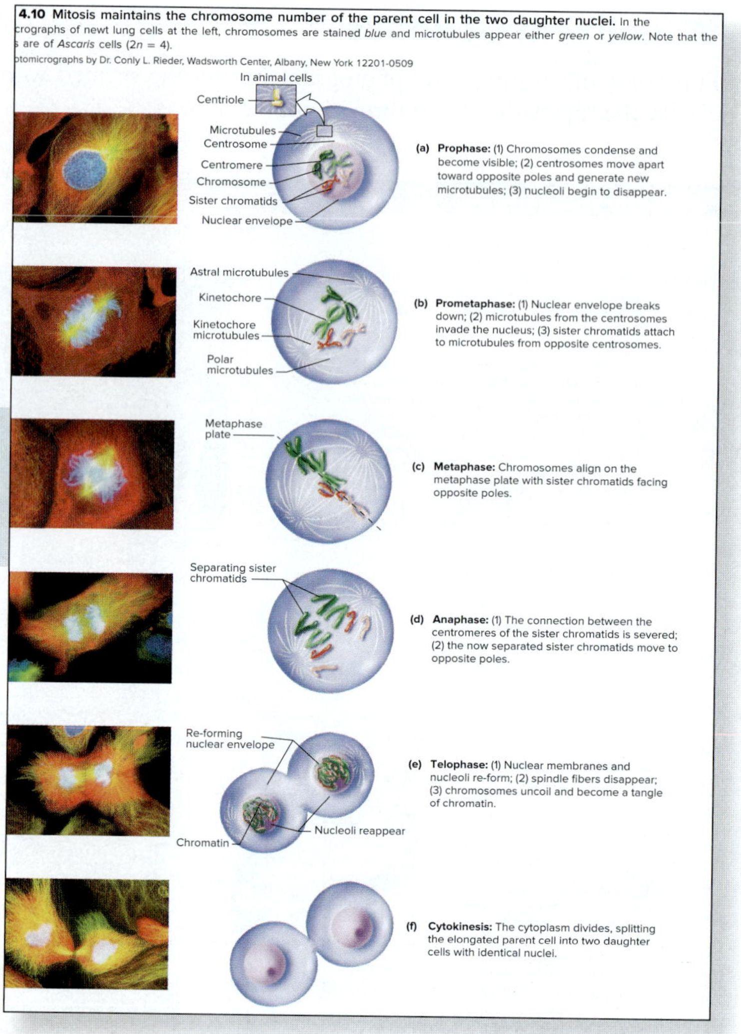

Process Figures

Step-by-step descriptions allow the student to walk through a compact summary of important details.

Micrographs

Stunning micrographs bring the genetics world to life.

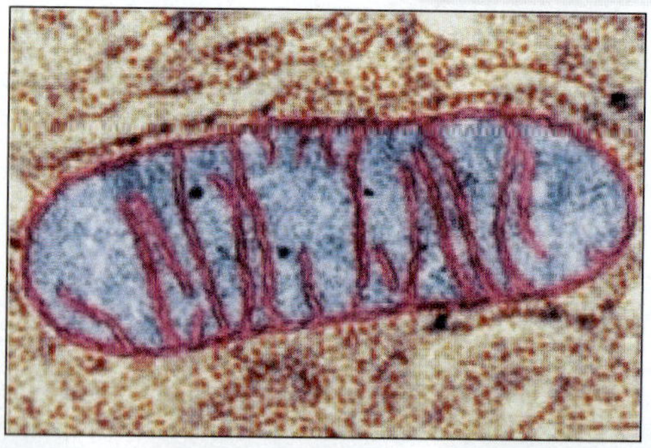

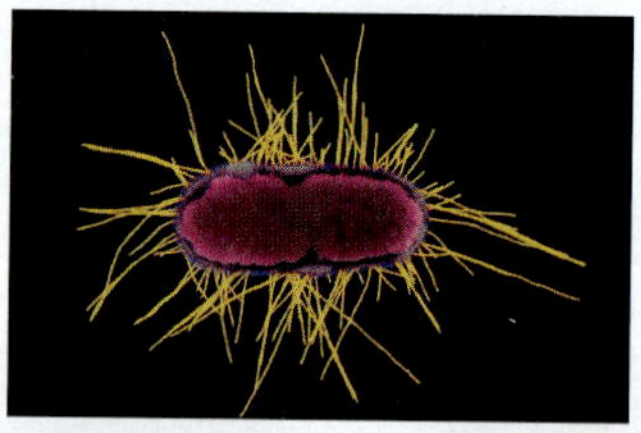

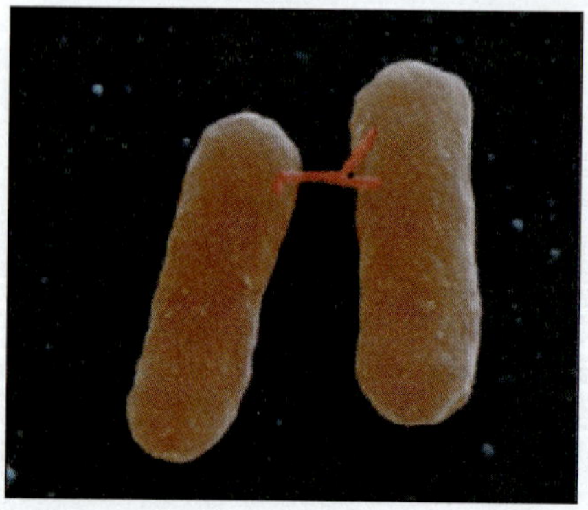

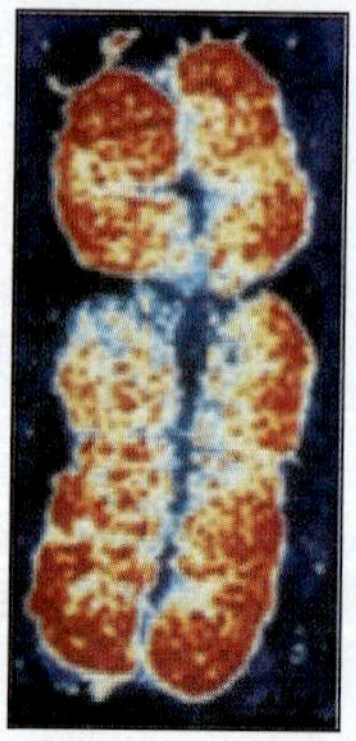

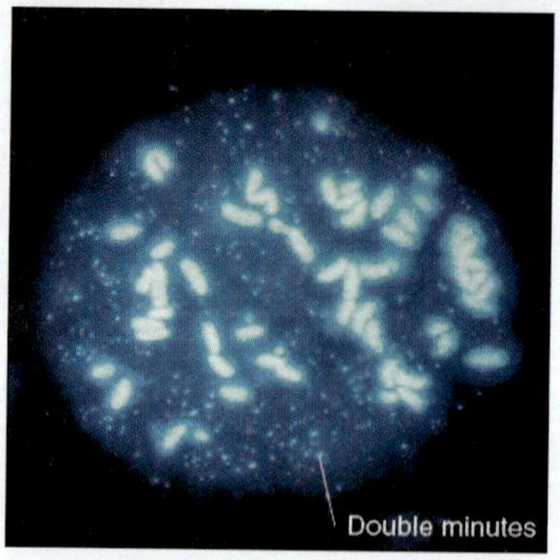

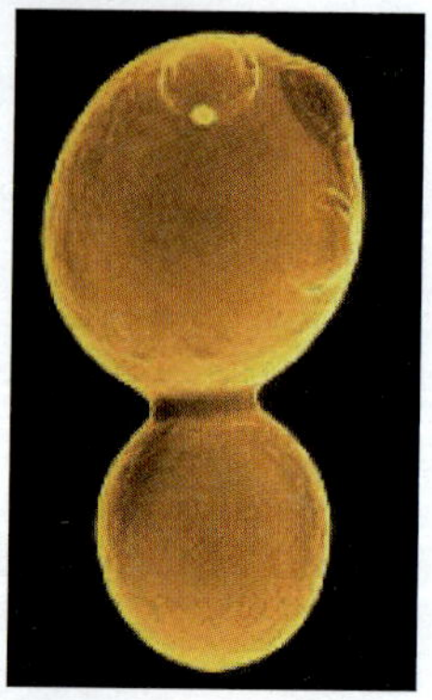

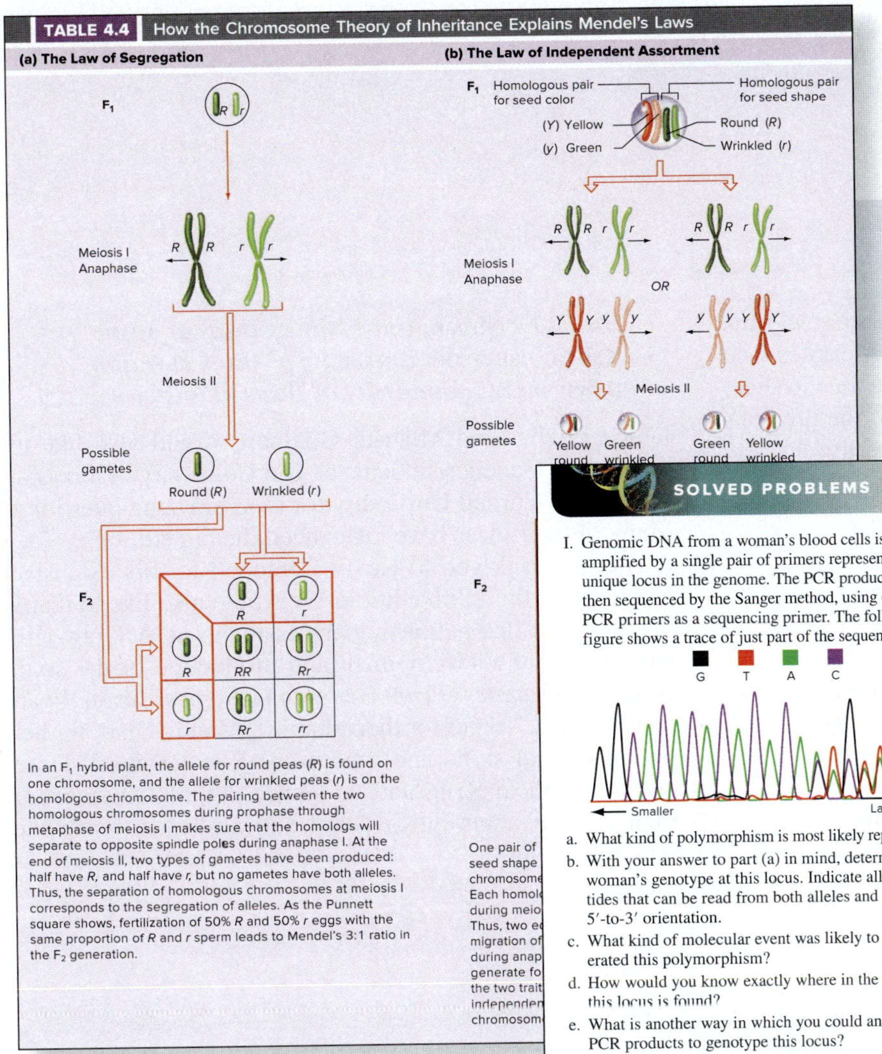
TABLE 4.4 How the Chromosome Theory of Inheritance Explains Mendel's Laws

(a) The Law of Segregation

In an F_1 hybrid plant, the allele for round peas (*R*) is found on one chromosome, and the allele for wrinkled peas (*r*) is on the homologous chromosome. The pairing between the two homologous chromosomes during prophase through metaphase of meiosis I makes sure that the homologs will separate to opposite spindle poles during anaphase I. At the end of meiosis II, two types of gametes have been produced: half have *R*, and half have *r*, but no gametes have both alleles. Thus, the separation of homologous chromosomes at meiosis I corresponds to the segregation of alleles. As the Punnett square shows, fertilization of 50% *R* and 50% *r* eggs with the same proportion of *R* and *r* sperm leads to Mendel's 3:1 ratio in the F_2 generation.

(b) The Law of Independent Assortment

One pair of seed shape chromosome... Each homol... during meio... Thus, two e... migration of... during anap... generate fo... the two trait... independen... chromosom...

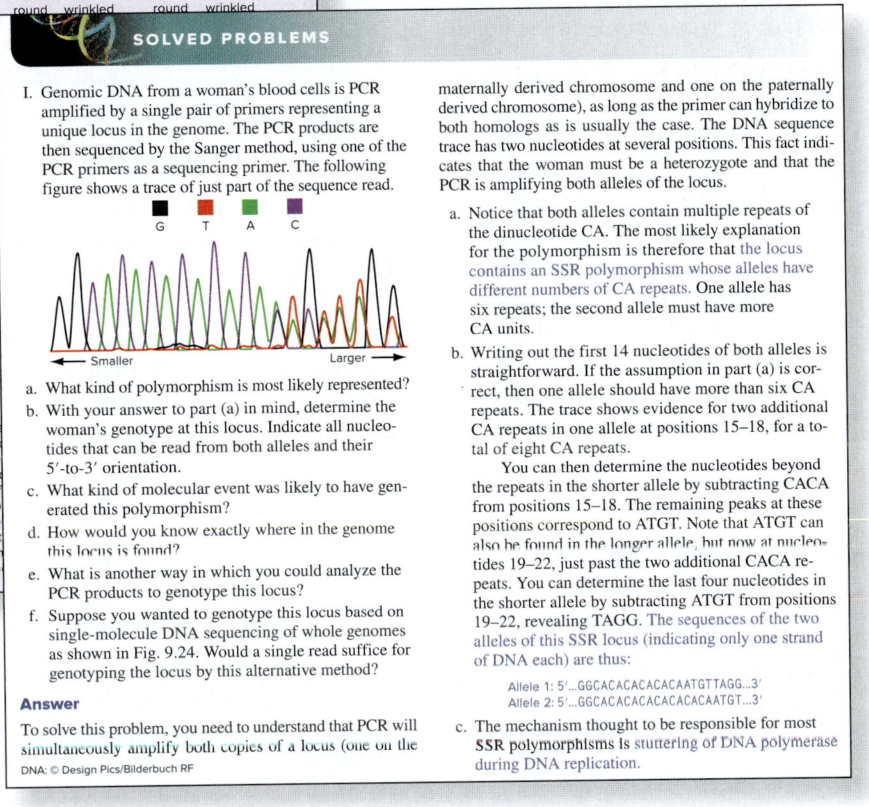
SOLVED PROBLEMS

I. Genomic DNA from a woman's blood cells is PCR amplified by a single pair of primers representing a unique locus in the genome. The PCR products are then sequenced by the Sanger method, using one of the PCR primers as a sequencing primer. The following figure shows a trace of just part of the sequence read.

a. What kind of polymorphism is most likely represented?
b. With your answer to part (a) in mind, determine the woman's genotype at this locus. Indicate all nucleotides that can be read from both alleles and their 5′-to-3′ orientation.
c. What kind of molecular event was likely to have generated this polymorphism?
d. How would you know exactly where in the genome this locus is found?
e. What is another way in which you could analyze the PCR products to genotype this locus?
f. Suppose you wanted to genotype this locus based on single-molecule DNA sequencing of whole genomes as shown in Fig. 9.24. Would a single read suffice for genotyping the locus by this alternative method?

Answer

To solve this problem, you need to understand that PCR will simultaneously amplify both copies of a locus (one on the maternally derived chromosome and one on the paternally derived chromosome), as long as the primer can hybridize to both homologs as is usually the case. The DNA sequence trace has two nucleotides at several positions. This fact indicates that the woman must be a heterozygote and that the PCR is amplifying both alleles of the locus.

a. Notice that both alleles contain multiple repeats of the dinucleotide CA. The most likely explanation for the polymorphism is therefore that the locus contains an SSR polymorphism whose alleles have different numbers of CA repeats. One allele has six repeats; the second allele must have more CA units.
b. Writing out the first 14 nucleotides of both alleles is straightforward. If the assumption in part (a) is correct, then one allele should have more than six CA repeats. The trace shows evidence for two additional CA repeats in one allele at positions 15–18, for a total of eight CA repeats.
You can then determine the nucleotides beyond the repeats in the shorter allele by subtracting CACA from positions 15–18. The remaining peaks at these positions correspond to ATGT. Note that ATGT can also be found in the longer allele, but now at nucleotides 19–22, just past the two additional CACA repeats. You can determine the last four nucleotides in the shorter allele by subtracting ATGT from positions 19–22, revealing TAGG. The sequences of the two alleles of this SSR locus (indicating only one strand of DNA each) are thus:
Allele 1: 5′...GGCACACACACACAATGTTAGG...3′
Allele 2: 5′...GGCACACACACACACACAATGT...3′
c. The mechanism thought to be responsible for most SSR polymorphisms is stuttering of DNA polymerase during DNA replication.

DNA: © Design Pics/Bilderbuch RF

Comparative Figures

Comparison illustrations lay out the basic differences of often confusing principles.

Solving Genetics Problems

The best way for students to assess and increase their understanding of genetics is to practice through problems. Found at the end of each chapter, problem sets assist students in evaluating their grasp of key concepts and allow them to apply what they have learned to real-life issues.

Review Problems

Problems are organized by chapter section and in order of increasing difficulty to help students develop strong problem-solving skills. The answers to select problems can be found in the back of this text.

Solved Problems

Solved problems offer step-by-step guidance needed to understand the problem-solving process.

Acknowledgements

The creation of a project of this scope is never solely the work of the authors. We are grateful to our colleagues who answered our numerous questions, or took the time to share with us their suggestions for improvement of the previous edition. Their willingness to share their expertise and expectations was a tremendous help to us.

- Charles Aquadro, *Cornell University*
- Daniel Barbash, *Cornell University*
- Johann Eberhart, *University of Texas at Austin*
- Tom Fox, *Cornell University*
- Kathy Gardner, *University of Pittsburgh*
- Larry Gilbert, *University of Texas at Austin*
- Nancy Hollingsworth, *Stony Brook University*
- Mark Kirkpatrick, *University of Texas at Austin*
- Alan Lloyd, *University of Texas at Austin*
- Paul Macdonald, *University of Texas at Austin*
- Kyle Miller, *University of Texas at Austin*
- Debra Nero, *Cornell University*
- Howard Ochman, *University of Texas at Austin*
- Kristin Patterson, *University of Texas at Austin*
- Inder Saxena, *University of Texas at Austin*

Janice Fischer and Michael Goldberg would also like to thank their Genetics students at The University of Texas at Austin and Cornell University for their amazing questions. Many of their ideas have influenced the 6th edition.

A special thank-you to Kevin Campbell for his extensive feedback on this sixth edition. We would also like to thank the highly skilled publishing professionals at McGraw-Hill who guided the development and production of the sixth edition of ***Genetics: From Genes to Genomes:*** Justin Wyatt and Michelle Vogler for their support; Mandy Clark for her organizational skills and tireless work to tie up all loose ends; and Vicki Krug and the entire production team for their careful attention to detail and ability to move the schedule along.

chapter 1

Genetics: The Study of Biological Information

Information can be stored in many ways, including the patterns of letters and words in books and the sequence of nucleotides in DNA molecules.
© James Strachan/Getty Images

chapter outline

GENETICS, THE SCIENCE of heredity, is at its core the study of biological information. All living organisms—from single-celled bacteria and protozoa to multicellular plants and animals—must store and use vast quantities of information to develop, survive, and reproduce in their environments (**Fig. 1.1**). Geneticists examine how organisms use biological information during their lifetimes and pass it on to their progeny.

This book introduces you to the field of genetics as currently practiced in the early twenty-first century. Several broad themes recur throughout this presentation. First, we know that biological information is encoded in *DNA*, and that the *proteins* responsible for an organism's many functions are built from this code. Second, we have found that all living forms are related at the molecular level. With the aid of high-speed computers and other technologies, we can now study *genomes* at the level of DNA sequence. These new methods have revealed that genomes have a modular construction that has allowed rapid evolution of complexity. Finally, our focus in this book is on human genetics and the application of genetic discoveries to human problems.

Figure 1.1 The biological information in DNA generates an enormous diversity of living organisms.
(a): © Kwangshin Kim/Science Source; (b): © Frank & Joyce Burek/Getty Images RF; (c): © Carl D. Walsh/Getty Images RF; (d): © Brand X Pictures/PunchStock RF; (e): © H. Wiesenhofer/PhotoLink/Getty Images RF; (f): © Ingram Publishing RF; (g): Source: Carey James Balboa. https://en.wikipedia.org/wiki/File:Red_eyed_tree_frog_edit2.jpg; (h): © Digital Vision RF

(a) Bacteria **(b) Clown fish** **(c) Lion** **(d) Oak tree**

(e) Poppies **(f) Hummingbird** **(g) Red-eyed tree frog** **(h) Humans**

1.1 DNA: Life's Fundamental Information Molecule

learning objectives

1. Relate the structure of DNA to its function.
2. Differentiate between a chromosome, DNA, a gene, a base pair, and a protein.

The process of **evolution**—the change in traits of groups of organisms over time—has taken close to 4 billion years to generate the amazing mechanisms for storing, replicating, expressing, and diversifying biological information seen in organisms now inhabiting the earth. The linear **DNA** molecule stores biological information in units known as **nucleotides.** Within each DNA molecule, the sequence of the four letters of the DNA alphabet—G, C, A, and T—specify which proteins an organism will make as well as when and where protein synthesis will occur. The letters refer to the **bases**—guanine, cytosine, adenine, and thymine—that are components of the nucleotide building blocks of DNA. The DNA molecule itself is a double strand of nucleotides carrying complementary G–C or A–T base pairs (**Fig. 1.2**). These **complementary base pairs** bind together through hydrogen bonds. The molecular **complementarity** of double-stranded DNA is its most important property and the key to understanding how DNA functions.

Figure 1.2 Complementary base pairs are a key feature of the DNA molecule. A single strand of DNA is composed of nucleotide subunits each consisting of a deoxyribose sugar (*white pentagons*), a phosphate (*yellow circles*), and one of four nitrogenous bases—adenine, thymine, cytosine, or guanine (designated as *lavender* or *green As*, *Ts*, *Cs*, or *Gs*). Hydrogen bonds (*dotted lines*) enable A to associate tightly with T, and C to associate tightly with G. Thus the two strands are complementary to each other. The arrows labeled *5′* to *3′* show that the strands have opposite orientations.

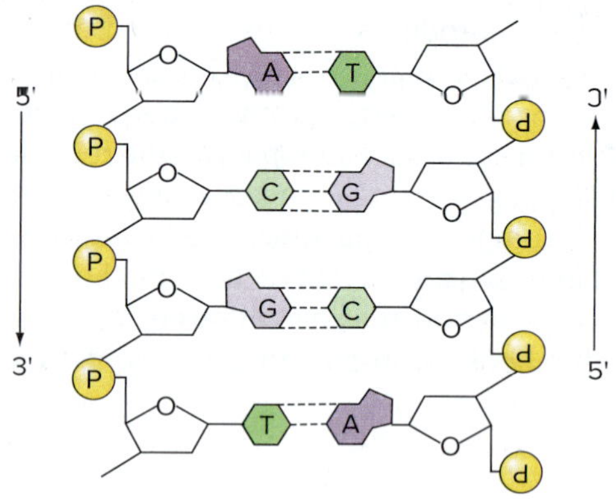

Although the DNA molecule is three-dimensional, most of its information is one-dimensional and digital. The information is one-dimensional because it is encoded as a specific sequence of letters along the length of the molecule. It is digital because each unit of information—one of the four letters of the DNA alphabet—is discrete. Because genetic information is digital, it can be stored as readily in a computer memory as in a DNA molecule. Indeed, the

Figure 1.3 A human chromosome. Each chromosome contains hundreds to thousands of genes.

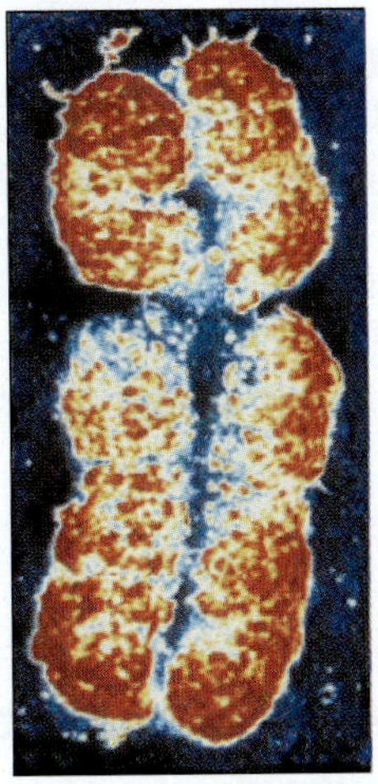

combined power of DNA sequencers, computers, and DNA synthesizers makes it possible to store, interpret, replicate, and transmit genetic information electronically from one place to another anywhere on the planet.

The DNA regions that encode proteins are called **genes.** Just as the limited number of letters in a written alphabet places no restrictions on the stories one can tell, so too the limited number of letters in the genetic code alphabet places no restrictions on the kinds of proteins and thus the kinds of organisms genetic information can define.

Within the cells of an organism, DNA molecules carrying the genes are assembled into **chromosomes:** organized structures containing DNA and proteins that package and manage the storage, duplication, expression, and evolution of DNA (**Fig. 1.3**). The DNA within the entire collection of chromosomes in each cell of an organism is its **genome.** Human cells, for example, contain 24 distinct kinds of chromosomes carrying approximately 3×10^9 base pairs and roughly 27,000 genes. The amount of information that can be encoded in this size genome is equivalent to 6 million pages of text containing 250 words per page, with each letter corresponding to one *base pair*.

To appreciate the long journey from a finite amount of genetic information easily storable on a computer disk to the production of a human being, we next must examine proteins, the primary molecules that determine how complex systems of cells, tissues, and organisms function.

essential concepts

- *DNA*, a double-stranded macromolecule composed of four nucleotides, is the repository of genetic information.
- DNA is organized into *chromosomes* (of 24 different types in humans) that collectively constitute an organism's *genome*.
- The human genome contains about 27,000 genes, most of which encode *proteins*.

1.2 Proteins: The Functional Molecules of Life Processes

learning objectives

1. Compare the chemical structures of DNA and proteins.
2. Differentiate between the functions of DNA and the functions of proteins.

Although no single characteristic distinguishes living organisms from inanimate matter, you would have little trouble deciding which entities in a group of objects are alive. Over time, these living organisms, governed by the laws of physics and chemistry as well as a genetic program, would be able to reproduce themselves. Most of the organisms would also have an elaborate and complicated structure that would change over time—sometimes drastically, as when an insect larva metamorphoses into an adult. Yet another characteristic of life is the ability to move. Animals swim, fly, walk, or run, while plants grow toward or away from light. Still another characteristic is the capacity to adapt selectively to the environment. Finally, a key characteristic of living organisms is the ability to use sources of energy and matter to grow—that is, the ability to convert foreign material into their own body parts. The chemical and physical reactions that carry out these conversions are known as *metabolism.*

Most properties of living organisms arise ultimately from the class of molecules known as **proteins**—large polymers composed of hundreds to thousands of **amino acid** subunits strung together in long chains. Each chain folds into a specific three-dimensional conformation dictated by the sequence of its amino acids (**Fig. 1.4**). Most proteins are composed of 20 different amino acids. The information in the DNA of genes dictates, via a *genetic code,* the order of amino acids in a protein molecule.

You can think of proteins as constructed from a set of 20 different kinds of snap beads distinguished by color and shape. If you were to arrange the beads in any order, make strings of a thousand beads each, and then fold or twist the chains into shapes dictated by the order of their beads, you would be able to make a nearly infinite number of different three-dimensional shapes. The astonishing diversity of three-dimensional protein structures generates the extraordinary diversity of protein functions that is the basis

Figure 1.4 Proteins are polymers of amino acids that fold in three dimensions. **(a)** Structural formulas for two amino acids: alanine and tyrosine. All amino acids have a basic amino group (—NH_2; *green*) at one end and an acidic carboxyl group (—COOH; *blue*) at the other. The specific side chain (*red*) determines the amino acid's chemical properties. **(b)** The amino acid sequences of two different human proteins: the ß chain of hemoglobin (*green*), and the enzyme lactate dehydrogenase (*purple*). **(c)** The different amino acid sequences of these proteins dictate different three-dimensional shapes. The specific sequence of amino acids in a chain determines the precise three-dimensional shape of the protein.

(a)

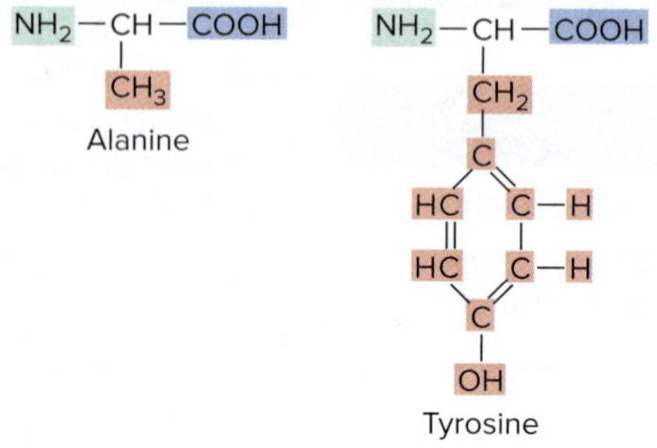

(b)

Hemoglobin β chain

MVHLTPEEKSAVTALWGKVNVDEVGEALGRLLVVYPWTQRLFESFGDLFTPDAVMGNPKVKAHG
KKVLGAFSDGPAHLDNLKGTFATLSELHCDKLHVDPENFRLLGNVLVCVLAHHFGKEFTPPVQAA
YQKVVAGVANALAHKYH

Lactate dehydrogenase

MATIKSELIKNFAEEEAIHHNKISIVGTGSVGVACAISILLKGLSDELVLVDVDEGKLKGETMDL
QHGSPFMKMPNIVSSKDYLVTANSNLVIITAGARQKKGETRLDLVQRNVSIFKLMIPNITQYSPH
CKLLIVTNPVDILTYVAWKLSGFPKNRVIGSGCNLDSARFRYFIGQRLGIHSESCHGLILGEHGD
SSVPVWSGVNIAGVPLKDLNPDIGTDKDPEQWENVHKKVISSGYEMVKMKGYTSWGISLSVADLT
ESILKNLRRVHPVSTLSKGLYGINEDIFLSVPCILGENGITDLIKVKLTLEEEACLQKSAETLWEIQKELKL

A = Ala = alanine	G = Gly = glycine	M = Met = methionine	S = Ser = serine
C = Cys = cysteine	H = His = histidine	N = Asn = asparagine	T = Thr = threonine
D = Asp = aspartic acid	I = Ile = isoleucine	P = Pro = proline	V = Val = valine
E = Glu = glutamic acid	K = Lys = lysine	Q = Gln = glutamine	W = Trp = tryptophan
F = Phe = phenylalanine	L = Leu = leucine	R = Arg = arginine	Y = Tyr = tyrosine

(c)

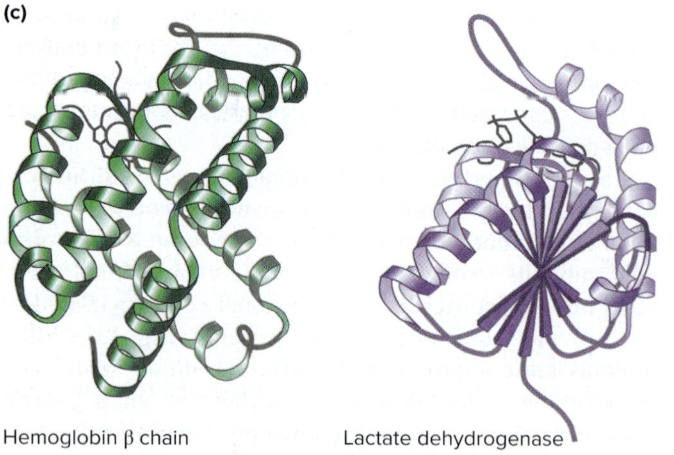

of each organism's complex and adaptive behavior (Fig. 1.4b and c). The structure and shape of the hemoglobin protein, for example, allow it to transport oxygen in the bloodstream and release it to the tissues. In contrast, lactate dehydrogenase is an enzyme that converts lactate to pyruvate, an important step in producing cellular energy. Most of the properties associated with life emerge from the constellation of protein molecules that an organism synthesizes according to instructions contained in its DNA.

essential concepts

- *Proteins* are responsible for most biological functions of cells and organisms.
- A protein is a macromolecule consisting of *amino acids* linked in a linear sequence.
- The sequences of amino acids in proteins are encoded by *genes* within the DNA.

1.3 Molecular Similarities of All Life-Forms

learning objective

1. Summarize the molecular evidence for the common origin of living organisms.

The evolution of biological information is a fascinating story spanning the 4 billion years of earth's history. Many biologists think that **RNA** was the first information-processing molecule to appear. Very similar to DNA, RNA molecules are also composed of four subunits: the bases G, C, A, and U (for uracil, which replaces the T of DNA). Like DNA, RNA has the capacity to store, replicate, and express information; like proteins, RNA can fold in three dimensions to produce molecules capable of catalyzing the chemistry of life. In fact, you will learn that the ultimate function of some genes is to encode RNA molecules instead of proteins. RNA molecules, however, are intrinsically unstable. Thus, it is probable that the more stable DNA took over the linear information storage and replication functions of RNA, while proteins, with their far greater capacity for diversity, preempted in large part the functions derived from RNA's three-dimensional folding. With this division of labor, RNA became primarily an intermediary in converting the information in DNA into the sequence of amino acids in protein (**Fig. 1.5a**). The separation that placed information storage in DNA and biological function mainly in proteins was so successful that all known organisms alive today descend from the first organisms that happened upon this molecular specialization.

The evidence for the common origin of all living forms is present in their DNA sequences. All living organisms use essentially the same **genetic code** in which various triplet groupings of the four letters of the DNA and RNA alphabets encode the 20 letters of the amino acid alphabet (**Fig. 1.5b**).

Figure 1.5 RNA is an intermediary in the conversion of DNA information into protein via the genetic code. (a) The linear bases of DNA are copied through molecular complementarity into the linear bases of RNA. The bases of RNA are read three at a time (that is, as triplets) to encode the amino acid subunits of proteins. **(b)** The genetic code dictionary specifies the relationship between RNA triplets and the amino acid subunits of proteins.

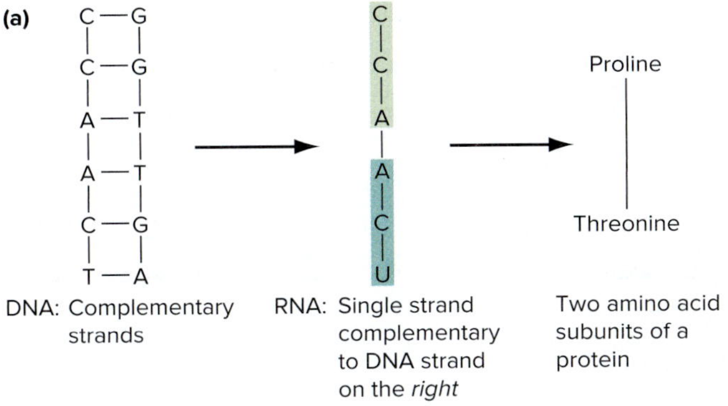

(b) Second letter (columns); First letter (rows); Third letter (U, C, A, G within each row)

First letter	U	C	A	G	Third letter
U	UUU Phe	UCU Ser	UAU Tyr	UGU Cys	U
	UUC Phe	UCC Ser	UAC Tyr	UGC Cys	C
	UUA Leu	UCA Ser	UAA Stop	UGA Stop	A
	UUG Leu	UCG Ser	UAG Stop	UGG Trp	G
C	CUU Leu	CCU Pro	CAU His	CGU Arg	U
	CUC Leu	CCC Pro	CAC His	CGC Arg	C
	CUA Leu	CCA Pro	CAA Gln	CGA Arg	A
	CUG Leu	CCG Pro	CAG Gln	CGG Arg	G
A	AUU Ile	ACU Thr	AAU Asn	AGU Ser	U
	AUC Ile	ACC Thr	AAC Asn	AGC Ser	C
	AUA Ile	ACA Thr	AAA Lys	AGA Arg	A
	AUG Met	ACG Thr	AAG Lys	AGG Arg	G
G	GUU Val	GCU Ala	GAU Asp	GGU Gly	U
	GUC Val	GCC Ala	GAC Asp	GGC Gly	C
	GUA Val	GCA Ala	GAA Glu	GGA Gly	A
	GUG Val	GCG Ala	GAG Glu	GGG Gly	G

The relatedness of all living organisms is also evident from comparisons of genes with similar functions in very different organisms. A striking similarity exists between the genes for many corresponding proteins in bacteria, yeast, plants, worms, flies, mice, and humans. For example, most of the amino acids in the cytochrome c proteins of diverse species are identical to each other (**Fig. 1.6**), indicating that these proteins all derived from a common ancestral protein. It is also important to note that some amino acids in these various cytochrome c proteins are different. The reason is that different **mutations,** that is, changes in nucleotide pairs, can occur when genes are passed from one generation of an organism to the next. The accumulation of these mutations in genomes is the main driving force of evolution.

Figure 1.6 Comparisons of gene products in different species provide evidence for the relatedness of living organisms. This chart shows the amino acid sequence for equivalent portions of the cytochrome c protein in six species: *Saccharomyces cerevisiae* (yeast), *Arabidopsis thaliana* (a weedlike flowering plant), *Caenorhabditis elegans* (a nematode), *Drosophila melanogaster* (the fruit fly), *Mus musculus* (the house mouse), and *Homo sapiens* (humans). Consult Fig. 1.4b for the key to amino acid names.

S. cerevisiae	GPNLHGIFGRHSGQVKGYSYTDANINKNVKW
A. thaliana	GPELHGLFGRKTGSVAGYSYTDANKQKGIEW
C. elegans	GPTLHGVIGRTSGTVSGFDYSAANKNKGVVW
D. melanogaster	GPNLHGLIGRKTGQAAGFAYTDANKAKGITW
M. musculus	GPNLHGLFGRKTGQAAGFSYTDANKNKGITW
H. sapiens	GPNLHGLFGRKTGQAPGYSYTAANKNKGIIW
	** ***.** .* *. *. ** *.. *

S. cerevisiae	DEDSMSEYLTNPKKYIPGTKMAFAGLKKEKDR
A. thaliana	KDDTLFEYLENPKKYIPGTKMAFGGLKKPKDR
C. elegans	TKETLFEYLLNPKKYIPGTKMVFAGLKKADER
D. melanogaster	NEDTLFEYLENPKKYIPGTKMIFAGLKKPNER
M. musculus	GEDTLMEYLENPKKYIPGTKMIFAGIKKKGER
H. sapiens	GEDTLMEYLENPKKYIPGTKMIFVGIKKKEER
	... *** *********** * *.** .*

* Indicates identical and . indicates similar

Despite the occurrence of mutations that alter DNA and thus protein sequences, it is often possible to place a gene from one organism into the genome of a very different organism and see it function normally in the new environment. Human genes that help regulate cell division, for example, can replace related genes in yeast and enable the yeast cells to function normally.

One of the most striking examples of relatedness at this level of biological information was uncovered in studies of eye development. Both insects and vertebrates (including humans) have eyes, but they are of very different types (**Fig. 1.7**). Biologists had long assumed that the evolution of eyes occurred independently, and in many evolution textbooks, eyes were used as an example of *convergent evolution,* in which structurally unrelated but functionally analogous organs emerge in different species as a result of natural selection. Studies of a gene called *Pax6* have turned this view upside down.

Mutations in the *Pax6* gene lead to a failure of eye development in both people and mice, and molecular studies have suggested that *Pax6* might play a central role in the initiation of eye development in all vertebrates. Remarkably, when the human *Pax6* gene is expressed in cells along the surface of the fruit fly body, it induces numerous little eyes to develop there. It turns out that fruit flies also have a gene specifying

Figure 1.7 The eyes of insects and humans have a common ancestor. (a) A fly eye and **(b)** a human eye.
(a): © Science Source; (b): © Nick Koudis/Getty Images RF

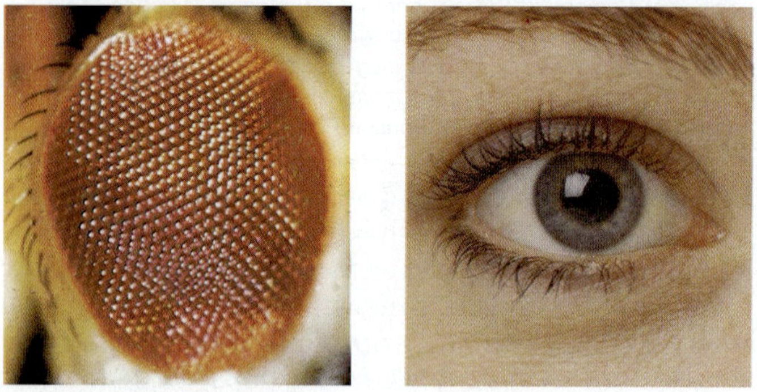

(a) (b)

Figure 1.8 How genes arise by duplication and divergence. Ancestral gene *A* contains exons (*green, red,* and *purple*) separated by introns in *blue*. Gene *A* is duplicated to create two copies that are originally identical, but mutations in either or both (*other colors*) cause the copies to diverge. Additional rounds of duplication and divergence create a family of related genes.

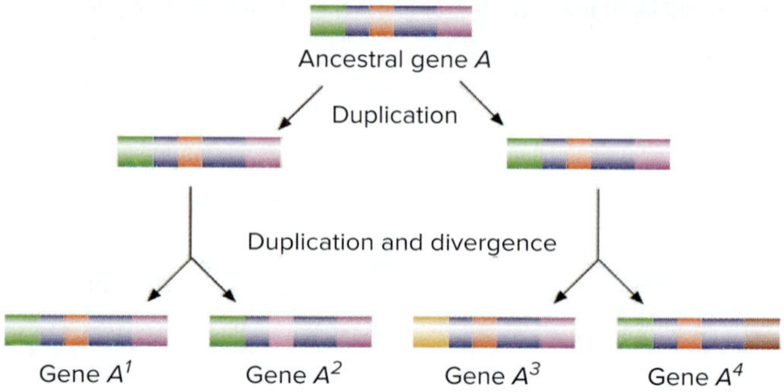

a protein whose amino acid sequence is distantly but clearly related to that of the protein specified by human *Pax6*; and furthermore, certain mutations in the fly gene result in animals lacking eyes. Taken together, these results demonstrate that during 600 million years of *divergent evolution,* an ancestral gene that served as the main control switch for initiating eye development accumulated different mutations in the lineages leading to people and fruit flies, but the gene still serves the same function in both species.

The usefulness of the relatedness and unity at all levels of biological information cannot be overstated. It means that in many cases, the experimental manipulation of organisms known as *model organisms* can shed light on gene functions in humans. If genes similar to human genes function in simple model organisms such as fruit flies or bacteria, scientists can determine gene function and regulation in these experimentally manipulable organisms and bring these insights to an understanding of the human organism.

essential concepts

- Living organisms exhibit marked similarities at the molecular level in the ways they use DNA and RNA to make proteins.
- Certain genes have persisted throughout the evolution of widely divergent species.

1.4 The Modular Construction of Genomes

learning objectives

1. Describe mechanisms by which new genes could arise.
2. Explain how regulation of gene expression can alter gene function.

We have seen that roughly 27,000 genes direct human growth and development. How did such complexity arise? Recent technical advances have enabled researchers to complete structural analyses of the entire genome of many organisms. The information obtained reveals that **gene families** have arisen by duplication of a primordial gene; after duplication, mutations may cause the two copies to diverge from each other (**Fig. 1.8**). In both humans and chimpanzees, for example, four different genes produce different rhodopsin proteins that are expressed in photoreceptors of distinct retinal cells. Each of these proteins functions in a slightly different way so that four kinds of retinal cells respond to light of different wavelengths and intensities, resulting in color vision. The four rhodopsin genes arose from a single primordial gene by several duplications followed by slight divergences in structure.

Duplication followed by divergence underlies the evolution of new genes with new functions. This principle appears to have been built into the genome structure of all multicellular organisms. The protein-coding region of most genes is subdivided into as many as 10 or more small pieces (called *exons*), separated by DNA that does not code for protein (called *introns*) as shown in Fig. 1.8. This modular construction facilitates the rearrangement of different modules from different genes to create new combinations during evolution. It is likely that this process of modular reassortment facilitated the rapid diversification of living forms about 570 million years ago (see Fig. 1.8).

The tremendous advantage of the duplication and divergence of existing pieces of genetic information is evident in the history of life's evolution (**Table 1.1**). *Prokaryotic* cells such as bacteria, which do not have a membrane-bounded nucleus, evolved about 3.7 billion years ago; *eukaryotic* cells such as algae, which have a membrane-bounded nucleus, emerged around 2 billion years ago; and multicellular eukaryotic organisms appeared 700–600 million years ago. Then, about 570 million years

TABLE 1.1 Fossil Evidence for Some Major Stages in the Evolution of Life

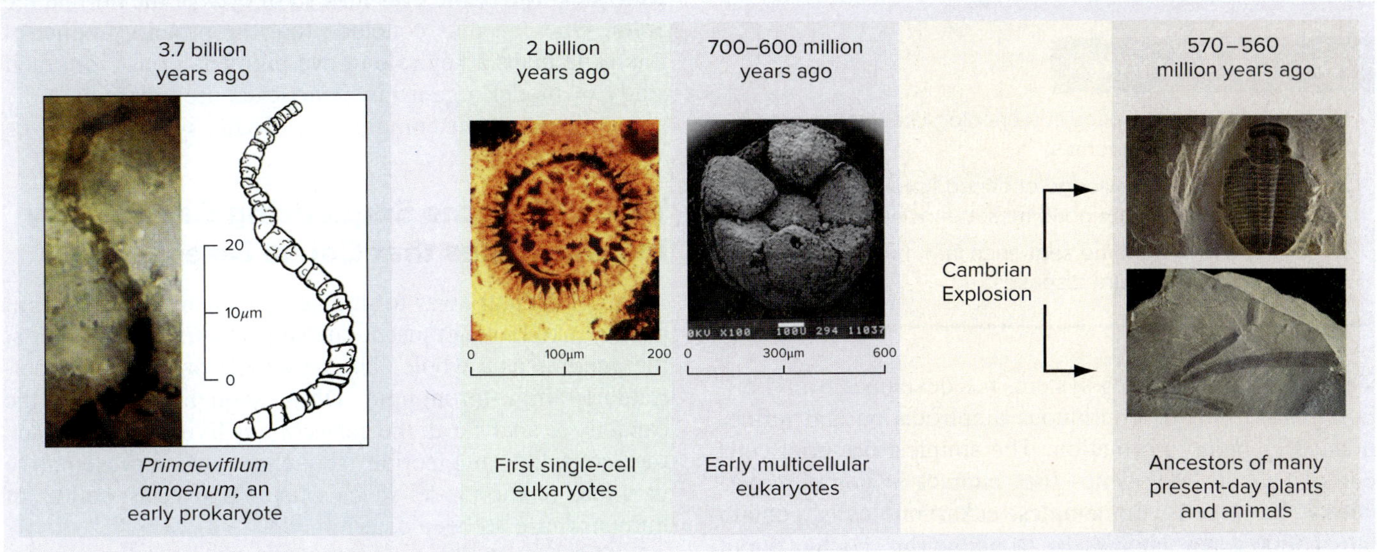

(prokaryote): © J.W. Schopf; (eukaryotes): © Prof. Andrew Knoll; (trilobite): © Brand X Pictures/PunchStock RF; (sponge): © Alan Sirulnikoff/Science Source

ago, within the relatively short evolutionary time of roughly 20–50 million years known as the *Cambrian explosion*, the multicellular life-forms diverged into a bewildering array of organisms, including primitive vertebrates.

Figure 1.9 Two-winged and four-winged flies. Geneticists converted a contemporary normal two-winged fly to a four-winged insect resembling the fly's evolutionary antecedent. They accomplished this by mutating a key element in the fly's regulatory network. Note the club-shaped halteres (*arrows*) behind the wings of the fly at the top.

(both): © Edward Lewis, California Institute of Technology

A fascinating question is: How could the multicellular forms achieve such enormous diversity in only 20–50 million years? The answer lies, in part, in the hierarchic organization of the information encoded in chromosomes. Exons are arranged into genes; genes duplicate and diverge to generate gene families; and gene families sometimes rapidly expand to **gene superfamilies** containing hundreds of related genes. In both mouse and human adults, for example, the immune system is encoded by a gene superfamily composed of hundreds of closely related but slightly divergent genes. With the emergence of each successively larger informational unit, evolution gains the ability to duplicate increasingly complex informational modules through single genetic events.

Probably even more important for the evolution of complexity is the rapid change of regulatory networks that specify how genes behave (that is, when, where, and to what degree they are expressed) during development. For example, the two-winged fly evolved from a four-winged ancestor not because of changes in gene-encoded structural proteins, but rather because of a rewiring of the regulatory network, which converted one pair of wings into two small balancing organs known as *halteres* (**Fig. 1.9**).

essential concepts

- *Gene duplication* followed by the *divergence* of copies is one explanation for how new functions evolve.
- The *reshuffling of exons* in eukaryotes provides another mechanism for the rapid diversification of genomes.
- Changes in DNA that affect *gene regulation*—where, when, and to what degree genes are expressed—also generate evolutionary change.

1.5 Modern Genetic Techniques

learning objectives

1. Explain how advances in technology have accelerated the analysis of genomes.
2. Compare the knowledge obtained from genetic dissection and from genome sequencing.
3. Discuss how genome sequence information can be used to treat or cure diseases.

The complexity of living systems has developed over 4 billion years from the continuous amplification and refinement of genetic information. The simplest bacterial cells contain about 1000 genes that interact in complex networks. Yeast cells, the simplest eukaryotic cells, contain about 5800 genes. Nematodes (roundworms) contain about 20,000 genes, and fruit flies contain roughly 13,000 genes. Humans have approximately 27,000 genes; surprisingly, the flowering plant *Arabidopsis* has as many and the zebrafish *D. rerio* has even more (**Fig. 1.10**). Each of these organisms has provided valuable insights into aspects of biology that are conserved among all organisms as well as other phenomena that are species-specific.

Genetic Dissection of Model Organisms Reveals the Working of Biological Processes

Model organisms including bacteria, yeast, nematodes, fruit flies, *Arabidopsis*, zebrafish, and mice are extremely valuable to researchers, who can use these organisms to analyze the complexity of a genome piece by piece. The logic used in *genetic dissection* is quite simple: inactivate a gene in a model organism and observe the consequences. For example, the loss of a gene for visual pigment produces fruit flies with white eyes instead of eyes of the normal red color. One can thus conclude that the protein product of this gene plays a key role in eye pigmentation. From their study of model organisms, geneticists are amassing a detailed picture of the complexity of living systems.

Whole-Genome Sequencing Can Identify Mutant Genes that Cause Disease

A complementary way to study an organism's genetic complexity is to look not just at one gene at a time, but rather at the genome as a whole. The new tools of **genomics,** particularly high-throughput DNA sequencers, have the capacity to analyze all the genes of any living thing. In fact, the complete nucleotide sequences of representative genomes of the model species listed above, as well as of humans, have all been determined.

The first draft of the human genome sequence announced by the Human Genome Project in 2001 cost $3 billion and took over 10 years to produce. Since then, rapid advances in genome sequencing technology have made it possible in 2016 to determine the genome sequence of an individual in just a few days for about $1000. Alongside the advances in DNA sequencing technology have been the development of computer algorithms to analyze the sequence data and the establishment of online databases that catalog the differences in individual genome sequences.

No example better illustrates the power of genome sequencing technology than its use in the identification of gene mutations that cause human genetic diseases. For diseases that result from mutation of a single gene, the gene responsible often may be identified by determining the genome sequence of just a few people or sometimes even that of a single individual.

In the case shown in **Fig. 1.11a,** geneticists analyzed whole genome sequences to find a gene mutation underlying a rare brain malformation disease called *microcephaly*. The

Figure 1.10 Seven model organisms whose genomes were sequenced as part of the Human Genome Project. The chart indicates genome size in millions of base pairs, or megabases (Mb). The bottom row shows the approximate number of genes for each organism. (*E. coli*): © David M. Phillips/Science Source; (*S. cerevisiae*): © CMSP/Getty Images; (*C. elegans*): © Sinclair Stammers/Science Source; (*A. thaliana*): Source: Courtesy USDA/Peggy Greb, photographer; (*D. melanogaster*): © Hemis.fr/SuperStock; (*D. rerio*): © A Hartl/Blickwinkel/agefotostock; (*M. musculus*): © imageBROKER/SuperStock RF

Organism	*E. coli*	*S. cerevisiae*	*C. elegans*	*A. thaliana*	*D. melanogaster*	*D. rerio*	*M. musculus*
Genome size	4.6 Mb	12 Mb	100 Mb	125 Mb	130 Mb	1500 Mb	2700 Mb
Number of protein-coding genes (approximate)	4300	5800	20,000	27,000	13,000	36,000	25,000

Figure 1.11 A causal gene for microcephaly identified by genome sequencing. **(a)** Magnetic resonance images of normal and microcephalic brains. **(b)** Sequence analysis of normal and mutant copies of the *WDR62* gene. The mutation is a deletion of the four nucleotides TGCC (*green*) that causes a major change in the amino acid sequence of the protein product of the gene. The letters above each triplet sequence identify the encoded amino acid. **(c)** Five different mutations in the *WDR62* gene in five different families are shown. Four of the mutations affect the identity of a single amino acid in the protein encoded by the gene. For example, W224S means that the 224th amino acid is normally W (tryptophan) but is changed to S (serine) by the mutation. The *arrow* indicates the position of the TGCC deletion mutation shown in (b).

(a): Source: Images produced by the Yale University School of Medicine. M. Bakircioglu, et al., "The Essential Role of Centrosomal NDE1 in Human Cerebral Cortex Neurogenesis," *The American Journal of Human Genetics,* 88(5): 523–535, Fig. 2C, 13 May 2011. Copyright © Elsevier Inc. http://www.cell.com/action/showImagesData?pii=S0002-9297%2811%2900135-2. CC-BY

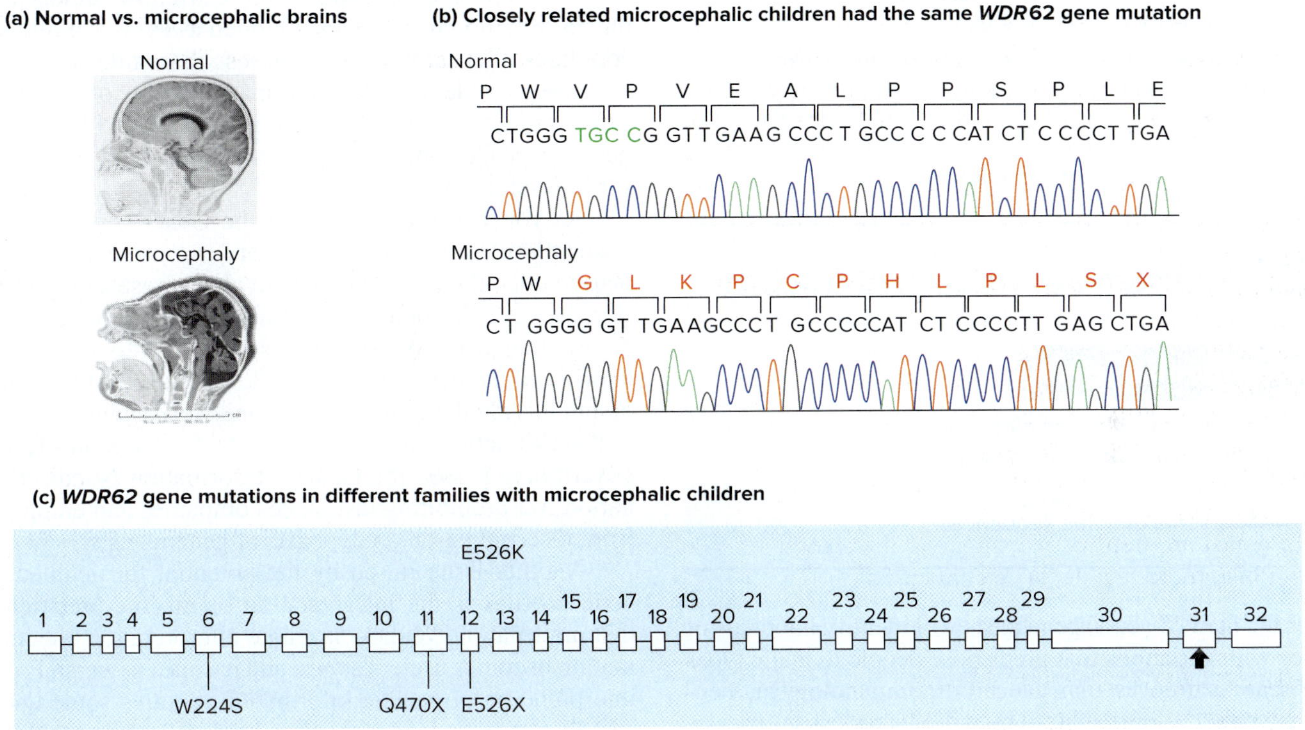

inheritance pattern observed for microcephaly indicated that it is a so-called *recessive* disease, meaning that diseased people inherit two mutant gene copies, one from each normal parent. The parents have one normal gene copy and one mutant copy, explaining why the parents' brains do not have this malformation. Sequencing and analysis of the genomes of two children with microcephaly from the same family identified a single rare gene mutation present in both siblings—a deletion of four base pairs within a gene called *WDR62* (**Fig. 1.11b**). Each parent was found to have one normal copy of *WDR62* and one copy with the four base pair deletion (Fig. 1.11b). Subsequently, the researchers found that different families with microcephalic children harbored different mutations in the same gene (**Fig. 1.11c**), thus confirming *WDR62* as a causative gene for microcephaly.

Gene Therapy May Help Cure Genetic Diseases

By enabling rapid disease gene discovery, the sequencing of whole genomes is revolutionizing medicine. Knowledge of disease genes can inform parents whether their children might suffer from a severely debilitating disease like microcephaly, allowing the parents to consider ways in which to avoid or to prepare for such an outcome. Moreover, identification of a disease gene provides information about the protein encoded by that gene that can sometimes guide the design of effective therapeutics to treat the disease. This strategy has not yet been useful in the case of microcephaly, but it has already been of tremendous value in developing drugs to combat other genetic diseases, including certain kinds of cancer.

Dramatic progress within the last few years offers hope that medical scientists may eventually be able to treat genetic diseases by modifying the genomes of the *somatic* (body) *cells* affected by the disease syndrome. One method under development is called **gene therapy;** here, scientists introduce normal copies of genes into human cells, where they can be expressed and compensate for their mutant, nonfunctional counterparts in the genome. An alternative, very recent gene therapy approach is *genome editing,* in which researchers change the base-pair sequence of a mutant gene to that of its normal counterpart. Gene therapy and genome editing have been used in model organisms such as mice to restore proper gene function and sometimes reverse the disease process, but the application of these techniques to human conditions is still only in very early stages.

essential concepts

- Scientists analyze mutations in model organisms to understand the molecular basis of many biological processes.
- Automated sequencing and computer analysis have made possible the rapid determination of DNA sequences in genomes, allowing researchers to pinpoint genes responsible for genetic diseases.
- Knowledge of disease genes helps parents make informed reproductive decisions, allows pharmaceutical companies to design effective drugs, and in the future will enable medical researchers to manipulate somatic cell genomes to reverse disease effects.

1.6 Human Genetics and Society

learning objectives

1. Describe the types of information that can be obtained from an individual's genome sequence.
2. Discuss the social issues that arise from the availability of personal genome sequences.

Over the next 25 years, geneticists will identify hundreds of genes with variations that predispose people to many types of disease: cardiovascular, cancerous, immunological, mental, metabolic. Some mutations will always cause disease, as in the microcephaly example just discussed; others will only predispose to disease. For example, a change in a specific single DNA base pair in the *β-globin* gene will nearly always cause sickle-cell anemia, a painful, life-threatening condition that leads to severe anemia. By contrast, a mutation in the *breast cancer 1 (BRCA1)* gene increases the risk of breast cancer to between 40% and 80% in a woman carrying one copy of the mutation. This conditional state arises because the *BRCA1* gene interacts with environmental factors that affect the probability of activating the cancerous condition, and because various forms of other genes can modify the effect of the *BRCA1* gene mutation.

Physicians may be able to use DNA diagnostics—a collection of techniques for characterizing genes—to analyze an individual's DNA for genes that predispose to some diseases. With this genetic profile, doctors may be able to write out a predictive health history based on probabilities for some medical conditions.

Many Social Issues Need to Be Addressed

Although biological information is similar to other types of information from a strictly technical point of view, it is as different as can be in its meaning and impact on individual human beings and on human society as a whole. The difference lies in the personal nature of the unique genetic profile carried by each person from birth. Within the genome of each individual is complex information that provides greater or lower susceptibility or resistance to many diseases, as well as greater or lesser potential for the expression of many physiological and neurological attributes that distinguish people from each other. Until now, almost all of this information has remained hidden away. But if research continues at its present pace, in less than a decade it will become possible to understand many aspects of a person's genetic make-up, and with this information will come the power to make some limited predictions about future possibilities and risks.

As you will see in many of the Genetics and Society boxes throughout this book, society can use genetic information not only to help people but also to restrict their lives (for example, by denying insurance or employment). We believe that just as our society respects an individual's right to privacy in other realms, it should also respect the privacy of an individual's genetic profile and work against all types of discrimination. Indeed, in 2008 the United States federal government passed the Genetic Information Nondiscrimination Act prohibiting insurance companies and employers from discrimination on the basis of genetic tests.

Another issue raised by the potential for detailed genetic profiles is the interpretation or misinterpretation of that information. Without accurate interpretation, the information becomes useless at best and harmful at worst. Proper interpretation of genetic information requires some understanding of statistical concepts such as risk and probability. As an example, women found to have the *BRCA1* mutation described above need to weigh the possible risks and benefits of preventive treatments such as mastectomy against the statistical probability that they would suffer otherwise from breast cancer. To help people understand these concepts, widespread education in this area will be essential.

To many people, the most frightening potential of the new genetics is the development of gene therapy technologies that can alter or add to the genes present within the *germ line* (reproductive cell precursors) of human embryos. In contrast with the previously described use of such *genetic engineering* to treat diseases in non-inheritable somatic cells, germ-line cells that have been manipulated in these ways can be transmitted between generations and thus have the potential to influence the evolution of our species.

Some people caution that developing the power to alter our own genomes is a step we should not take, arguing that if genetic information and technology are misused, as they certainly have been in the past, the consequences could be horrific. Attempts to use genetic information for social purposes were prevalent in the early twentieth century, leading to enforced sterilization of individuals thought to be inferior, to laws that prohibited interracial marriage, and to laws prohibiting immigration of certain ethnic groups. The basis of these actions was not scientific and has been thoroughly discredited.

Others agree that we must not repeat the mistakes of the past, but they warn that if the new technologies could help children and adults lead healthier, happier lives, we cannot outlaw their application. Most agree that the biological revolution we are living through will have a greater impact on human society than any technological revolution of the past and that education and public debate are the key to preparing for the consequences of this revolution.

The focus on human genetics in this book looks forward into the new era of biology and genetic analysis. These new possibilities raise serious moral and ethical issues that will demand wisdom and humility. It is in the hope of educating young people for the moral and ethical challenges awaiting the next generation that we write this book.

essential concepts

- Genome sequences may identify not only genes that always cause disease but also genes that predispose an individual to a disease.
- As a society, we must ensure that genetic knowledge is properly interpreted and that individuals' privacy is protected.

WHAT'S NEXT

Genetics, the study of biological information, is also the study of the DNA and RNA molecules that store, replicate, transmit, express, and evolve information for the construction of proteins. At the molecular level, all living things are closely related, and as a result, observations of model organisms as different as yeast and mice can provide insights into general biological principles as well as human biology.

Remarkably, more than 75 years before the discovery of DNA, Gregor Mendel, an Augustinian monk, delineated the basic laws of gene transmission with no knowledge of the molecular basis of heredity. He accomplished this by following simple traits, such as flower or seed color, through several generations of the pea plant (*Pisum sativum*). We now know that his findings apply to all sexually reproducing organisms. Chapter 2 describes Mendel's studies and insights, which became the foundation of the field of genetics.

PROBLEMS

Vocabulary

1. Choose the phrase from the right column that best fits the term in the left column.

a. complementarity	1. a linear polymer of amino acids that folds into a particular shape
b. nucleotide	2. part of a gene that does not contain protein coding information
c. chromosomes	3. a polymer of nucleotides that is an intermediary in the synthesis of proteins from instructions in DNA
d. protein	4. G–C and A–T base pairing in DNA through hydrogen bonds
e. genome	5. alteration of DNA sequence
f. gene	6. part of a gene that can contain protein coding information
g. uracil	7. DNA/protein structures that contain genes
h. exon	8. DNA information for a single function, such as production of a protein
i. intron	9. the entirety of an organism's hereditary information
j. DNA	10. a double-stranded polymer of nucleotides that stores the inherited blueprint of an organism
k. RNA	11. subunit of the DNA macromolecule
l. mutation	12. the only one of the four bases in RNA that is not in DNA

Section 1.1

2. If one strand of a DNA molecule has the base sequence 5′-AGCATTAAGCT-3′, what is the base sequence of the other, complementary strand?
3. The size of one copy of the human genome is approximately 3 billion base pairs, and it contains about 27,000 genes organized into 23 chromosomes.
   a. Human chromosomes vary in size. What would you predict is the size of the average chromosome?
   b. Assuming that genes are spread evenly among chromosomes, how many genes does an average human chromosome contain?
   c. About half of the DNA in chromosomes contains genes. How large (in base pairs) is an average human gene?

Section 1.2

4. Indicate whether each of the following words or phrases applies to proteins, DNA, or both.
   a. a macromolecule composed of a string of subunits
   b. double-stranded
   c. four different subunits
   d. 20 different subunits
   e. composed of amino acids
   f. composed of nucleotides
   g. contains a code used to generate other macromolecules
   h. performs chemical reactions
5. a. How many different DNA strands composed of 100 nucleotides could possibly exist?
   b. How many different proteins composed of 100 amino acids could possibly exist?

Section 1.3

6. RNA shares with proteins the ability to fold into complex three-dimensional shapes. As a result, RNA molecules can, like protein molecules, catalyze biochemical reactions (that is, both kinds of molecules can act as *enzymes*, or biological catalysts). These statements are not true of DNA. Why can some RNA molecules act as enzymes whereas DNA molecules cannot? (*Hint:* Most RNA molecules consist of a single strand of nucleotides while most DNA molecules are double helixes made of two strands of nucleotides.)
7. The human protein lactate dehydrogenase shown in Fig. 1.4 has 332 amino acids. What is the smallest possible combined size of the parts of the gene that specify this protein using the genetic code?
8. a. Are the triplets in the genetic code table shown in Fig. 1.5b written as DNA or RNA?
   b. Two amino acids are each specified only by a single triplet. Identify these two amino acids and the corresponding triplets.
   c. If you know the sequence of amino acids in a protein, what does the genetic code table allow you to infer about the sequence of base pairs in the gene that specifies that protein?
9. Why do scientists think that all forms of life on earth have a common origin?
10. Why would a geneticist study a yeast cell or a fruit fly or a mouse in order to understand human genes and human biology?
11. How can a scientist tell if a protein present in bacteria and a fruit fly protein have a common origin? How can a scientist determine whether a protein with a common origin in bacteria and a fruit fly function in a common pathway?
12. Figure 1.6 shows the amino acid sequences of parts of the cytochrome c proteins from several different organisms. Some of these amino acids are highlighted in *dark orange,* some in *light orange,* and some are not highlighted at all. Which of these three classes of amino acids is likely to be most important for the biochemical function of cytochrome c proteins?

Section 1.4

13. Why do scientists think that new genes arise by duplication of an original gene and divergence by mutation?
14. Explain how the exon/intron structure of genes contributes to the generation of new gene functions during evolution.
15. Mutations in genes that change their pattern of expression (the time and cell type in which the gene product is produced) are thought to be a major factor in the evolution of different organisms. Would you expect the same protein to work in the same way (for example, to perform the same kind of enzymatic reaction) in two different types of cells (for example, cells in the retina of the eye and muscle cells)? Is it possible that the same protein might function in different biochemical pathways in eye cells and muscle cells even if the protein's basic mechanism always remains the same?

Section 1.5

16. A single zebrafish gene function was inactivated completely by mutation, and a zebrafish with this mutation had none of its normal horizontal stripes. For each of the following statements, indicate whether the

statement is certainly true, certainly untrue, or if there is insufficient information to decide.

a. The normal gene function is required for the viability of the zebrafish.

b. The normal gene function is required for the formation of stripes.

c. The normal gene function is required to make the pigment deposited in the stripes.

d. The gene is required in zebrafish only for stripe formation.

17. Different mutations in the *WDR62* gene that inactivate gene function were found in the genomes of many different people with microcephaly. This information provided strong support for the idea that the *WDR62* gene mutation causes microcephaly.

a. The human genome sequence identified *WDR62* as one of the approximately 27,000 genes in the human genome. What information about the function of *WDR62* do you think was learned originally from the DNA sequence of the normal human genome?

b. What additional information was provided by identification of *WDR62* as the microcephaly disease gene?

c. The mouse genome contains a gene similar to human *WDR62*. Experiments in mice have shown that the mouse *WDR62* gene is expressed in the brain. Technology is now available that allows scientists to generate mice in which the two normal copies of the *WDR62* gene are replaced with mutant copies of the gene that are nonfunctional. Why would a scientist want to generate such *WDR62* mutant mice?

18. Researchers have successfully used gene therapy to ameliorate some human genetic diseases by adding a normal gene copy to cells whose genomes originally had only nonfunctional mutant copies of that gene. For example, a form of blindness due to the lack of a single protein called RPE65 has been reversed by introduction of a normal *RPE65* gene to cells of the retina of adults.

a. The success of this gene therapy approach provides us with clues about the role of the RPE65 protein in the retina. Do you think that RPE65 is needed for the proper development of the human eye?

b. Can you see a potential difficulty in applying this gene therapy approach for diseases like microcephaly?

Section 1.6

19. By the time this book is published, it will likely be possible for you to obtain the sequence of your genome at nominal cost. Do you want this information? Explain the factors that affected your decision. (You may not be able to answer this question until you are finished reading this book.)

chapter 2

Mendel's Principles of Heredity

Although Mendel's laws can predict the probability that an individual will have a certain genetic makeup, the chance meeting of particular male and female gametes determines an individual's actual genetic fate.

chapter outline

A QUICK GLANCE at an extended family portrait is likely to reveal children who resemble one parent or the other, or who look like a combination of the two (**Fig. 2.1**). Some children, however, look unlike any of the assembled relatives and more like a great, great grandparent. What causes the similarities and differences of appearance and the skipping of generations?

The answers lie in our **genes,** the basic units of biological information, and in **heredity,** the way genes transmit physiological, anatomical, and behavioral traits from parents to offspring. Each of us starts out as a single fertilized egg cell that develops, by division and differentiation, into a mature adult made up of 10^{14} (a hundred trillion) specialized cells capable of carrying out all the body's functions and controlling our outward appearance. Genes, passed from one generation to the next, underlie the formation of every heritable trait. Such traits are as diverse as the presence of a cleft in your chin, the tendency to lose hair as you age, your hair, eye, and skin color, and even your susceptibility to certain cancers. All such traits run in families in predictable patterns that impose some possibilities and exclude others.

Genetics, the science of heredity, pursues a precise explanation of the biological structures and mechanisms that determine inheritance. In some instances, the relationship between gene and trait is remarkably simple. A single change in a single gene, for example, results in sickle-cell anemia, a disease in which the hemoglobin molecule found in red blood cells is defective. In other instances, the correlations between genes and traits are bewilderingly complex. An example is the genetic basis of facial features, in which many genes determine a large number of molecules that interact to generate the combination we recognize as a friend's face.

Gregor Mendel (1822–1884; **Fig. 2.2**), a stocky, bespectacled Augustinian monk and expert plant breeder, discovered the basic principles of genetics in the mid-nineteenth century. He published his findings in 1866, just seven years after Darwin's *On the Origin of Species* appeared in print. Mendel lived and worked in Brünn, Austria (now Brno in the Czech Republic), where he examined the inheritance of clear-cut alternative traits in pea plants, such as purple versus white flowers or yellow versus green seeds. In so

Figure 2.1 A family portrait. The extended family shown here includes members of four generations.
© Bruce Ayres/Getty Images

Figure 2.2 Gregor Mendel. Photographed around 1862 holding one of his experimental plants.
© Science Source

Figure 2.3 Like begets like and unlike. A Labrador retriever with her litter of pups.
© Saudjie Cross Siino/Weathertop Labradors

doing, he inferred genetic laws that allowed him to make verifiable predictions about which traits would appear, disappear, and then reappear, and in which generations.

Mendel's laws are based on the hypothesis that observable traits are determined by independent units of inheritance not visible to the naked eye. We now call these units *genes*. The concept of the gene continues to change as research deepens and refines our understanding. Today, a gene is recognized as a region of DNA that encodes a specific protein or a particular type of RNA. In the beginning, however, it was an abstraction—an imagined particle with no physical features, the function of which was to control a visible trait by an unknown mechanism.

We begin our study of genetics with a detailed look at what Mendel's laws are and how they were discovered. In subsequent chapters, we discuss logical extensions to these laws and describe how Mendel's successors grounded the abstract concept of hereditary units (genes) in an actual biological molecule (DNA).

Four general themes emerge from our detailed discussion of Mendel's work. The first is that variation, as expressed in alternative forms of a trait, is widespread in nature. This genetic diversity provides the raw material for the continuously evolving variety of life we see around us. Second, observable variation is essential for following genes from one generation to the next. Third, variation is not distributed solely by chance; rather, it is inherited according to genetic laws that explain why like begets both like and unlike. Dogs beget other dogs—but hundreds of breeds of dogs are known. Even within a breed, such as Labrador retrievers, genetic variation exists: Two black dogs could have a litter of black, chocolate (brown), and yellow puppies (**Fig. 2.3**). Mendel's insights help explain why this is so. Fourth, the laws Mendel discovered about heredity apply equally well to all sexually reproducing organisms, from protozoans to peas to dogs to people.

2.1 The Puzzle of Inheritance

learning objectives

1. Relate how Mendel's experimental approach is similar to the process of modern scientific inquiry.
2. Describe how Mendel cross-fertilized and self-fertilized pea plants.
3. Explain the importance of Mendel's inclusion of reciprocal crosses within his controlled breeding program of pea plants.
4. Predict the type of progeny produced by Mendel's crosses between pure-breeding plants with discrete, antagonistic traits, such as purple versus white flowers.

Several steps lead to an understanding of genetic phenomena: the careful observation over time of groups of organisms, such as human families, herds of cattle, or fields of corn or tomatoes; the rigorous analysis of systematically recorded information gleaned from these observations; and the development of a theoretical framework that can explain the origins and relationships of these phenomena. In the mid-nineteenth century, Gregor Mendel became the first person to combine data collection, analysis, and theory in a successful pursuit of the true basis of heredity. For many thousands of years before that, the only genetic practice was the selective breeding of domesticated plants and animals, with no guarantee of what a particular mating would produce.

Artificial Selection Was the First Applied Genetic Technique

A rudimentary use of genetics was the driving force behind a key transition in human civilization, allowing hunters and gatherers to settle in villages and survive as shepherds and farmers. Even before recorded history, people practiced applied genetics as they domesticated plants and animals for their own uses. From a large litter of semitamed wolves, for example, they sent the savage and the misbehaving to the stew pot while sparing the alert sentries and friendly companions for longer life and eventual mating. As a result of this **artificial selection**—purposeful control over mating by choice of parents for the next generation—the domestic dog (*Canis lupus familiaris*) slowly arose from ancestral wolves (*Canis lupus*). The oldest bones identified indisputably as dog (and not wolf) are a skull excavated from a 20,000-year-old Alaskan settlement. Many millennia of evolution guided by artificial selection have produced massive Great Danes and minuscule Chihuahuas as well as hundreds of other modern breeds of dog. By 10,000 years ago, people had begun to use this same kind of genetic manipulation to develop economically valuable herds of reindeer, sheep, goats, pigs, and cattle that produced life-sustaining meat, milk, hides, and wools.

Farmers also carried out artificial selection of plants, storing seed from the hardiest and tastiest individuals for the next planting, eventually obtaining strains that grew better, produced more, and were easier to cultivate and harvest. In this way, scrawny, weed-like plants gradually, with human guidance, turned into rice, wheat, barley, lentils, and dates in Asia; corn, squash, tomatoes, potatoes, and peppers in the Americas; yams, peanuts, and gourds in Africa. Later, plant breeders recognized male and female organs in plants and carried out artificial pollination. An Assyrian frieze carved in the ninth century B.C., pictured in **Fig. 2.4,** is the oldest known visual record of this kind of genetic experiment. It depicts priests brushing the flowers of female date palms with selected male pollen. By this method of artificial selection, early practical geneticists produced several hundred varieties of dates, each differing in specific observable qualities, such as the fruit's size, color, or taste. A 1929 botanical survey of three oases in Egypt turned up 400 varieties of date-bearing palms, twentieth-century evidence of the natural and artificially generated variation among these trees.

Figure 2.4 The earliest known record of applied genetics. In this 2800-year-old Assyrian relief from the Northwest Palace of Assurnasirpal II (883–859 B.C.), priests wearing bird masks artificially pollinate flowers of female date palms.

Desirable Traits Sometimes Disappear and Reappear

In 1822, the year of Mendel's birth, what people in Austria understood about the basic principles of heredity was not much different from what the people of ancient Assyria had known. By the nineteenth century, plant and animal breeders had created many strains in which offspring often carried a prized parental trait. Using such strains, breeders could produce plants or animals with desired characteristics for food and fiber, but they could not always predict why a valued trait would sometimes disappear and then reappear in only some offspring.

For example, selective breeding practices had resulted in valuable flocks of merino sheep producing large quantities of soft, fine wool, but at the 1837 annual meeting of the Moravian Sheep Breeders Society, one breeder's dilemma epitomized the state of the art. He possessed an outstanding ram that would be priceless "if its advantages are inherited by its offspring," but "if they are not inherited, then it is worth no more than the cost of wool, meat, and skin." Which would it be? According to the meeting's recorded minutes, current breeding practices offered no definite answers. In his concluding remarks at this sheep-breeders meeting, the Abbot Cyril Napp pointed to a possible way out. He proposed that breeders could improve their ability to predict what traits would appear in the offspring by finding the answers to three basic questions: What is inherited? How is it inherited? What is the role of chance in heredity?

This quandary is where matters stood in 1843 when 21-year-old Gregor Mendel entered the monastery in Brünn, presided over by the same Abbot Napp. Although Mendel was a monk trained in theology, he was not a rank amateur in science. The province of Moravia, in which Brünn was located, was a center of learning and scientific activity. Mendel was able to acquire a copy of Darwin's *On the Origin of Species* shortly after it was translated into German in 1863. Abbot Napp, recognizing Mendel's intellectual abilities, sent him to the University of Vienna—all expenses paid—where he prescribed his own course of study. Mendel's choices were an unusual mix: physics, mathematics, chemistry, botany, paleontology, and plant physiology. Christian Doppler, discoverer of the Doppler effect, was one of his teachers. The cross-pollination of ideas from several disciplines would play a significant role in Mendel's discoveries. One year after he returned to Brünn, he began his series of seminal genetic experiments. **Figure 2.5** shows where Mendel worked and the microscope he used.

Figure 2.5 Mendel's garden and microscope. (a) Gregor Mendel's garden was part of his monastery's property in Brünn. **(b)** Mendel used this microscope to examine plant reproductive organs and to pursue his interests in natural history.

(a): © Biophoto Associates/Science Source; (b): © James King-Holmes/Science Source

(a)

(b)

Mendel Devised a New Experimental Approach

Before Mendel, many misconceptions clouded people's thinking about heredity. Two of the prevailing errors were particularly misleading. The first was that one parent contributes most to an offspring's inherited features; Nicolaas Hartsoeker, one of the earliest microscopists, contended in 1694 that it was the male, by way of a fully formed *homunculus* inside the sperm (**Fig. 2.6**). Another deceptive notion was the concept of *blended inheritance*, the idea that parental traits become mixed and forever changed in the offspring, as when blue and yellow pigments merge to green on a painter's palette. The theory of blending may have grown out of a natural tendency for parents to see a combination of their own traits in their offspring. While blending could account for children who look like a combination of their parents, it could not explain obvious differences between biological brothers and sisters nor the persistence of variation within extended families.

The experiments Mendel devised would lay these myths to rest by providing precise, verifiable answers to the three questions Abbot Napp had raised almost 15 years earlier: What is inherited? How is it inherited? What is the

Figure 2.6 The homunculus: A misconception. Well into the nineteenth century, many prominent microscopists believed they saw a fully formed, miniature fetus crouched within the head of a sperm.
© Klaus Guldbrandsen/SPL/Science Source

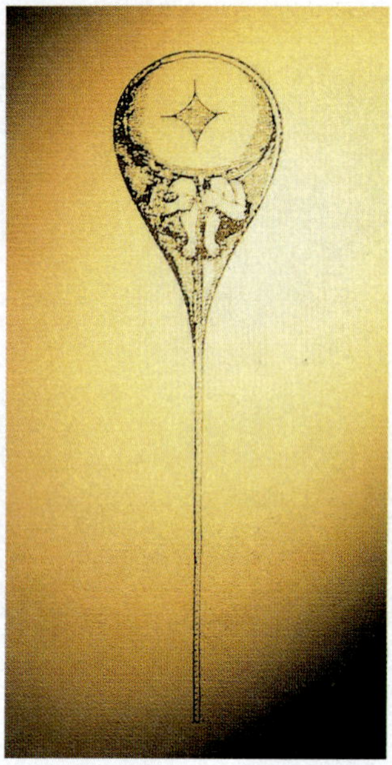

role of chance in heredity? A key component of Mendel's breakthrough was the way he set up his experiments.

What did Mendel do differently from those who preceded him? First, he chose the garden pea *(Pisum sativum)* as his experimental organism (**Figs. 2.7a** and **b**). Peas grew well in Brünn, and with male and female organs in the same flower, they were normally self-fertilizing. In **self-fertilization** (or **selfing**), both egg and pollen come from the same plant. The particular anatomy of pea flowers, however, makes it easy to prevent self-fertilization and instead to **cross-fertilize** (or *cross*) two individuals by brushing pollen from one plant onto a female organ of another plant, as illustrated in **Fig. 2.7c.** Peas offered yet another advantage. For each successive generation, Mendel could obtain large numbers of individuals within a relatively short growing season. By comparison, if he had worked with sheep, each mating would have generated only a few offspring and the time between generations would have been several years.

Second, Mendel examined the inheritance of clear-cut alternative forms of particular traits—purple versus white flowers, yellow versus green peas. Using such either-or traits, he could distinguish and trace unambiguously the transmission of one or the other observed characteristic, because no intermediate forms existed. (The opposite of these so-called **discrete traits** are **continuous traits,** such as height and skin color in humans. Continuous traits show many intermediate forms.)

Third, Mendel collected and perpetuated lines of peas that bred true. Matings within such **pure-breeding** (or **true-breeding) lines** produce offspring carrying specific parental traits that remain constant from generation to generation. These lines are also called *inbred* because they have been mated only to each other for many generations. Mendel observed his pure-breeding lines for up to eight generations. Plants with white flowers always produced offspring with white flowers; plants with purple flowers produced only offspring with purple flowers. Mendel called constant but mutually exclusive, alternative traits,

Figure 2.7 Mendel's experimental organism: The garden pea. **(a)** Pea plants with white flowers. **(b)** Pollen is produced in the anthers. Mature pollen lands on the stigma, which is connected to the ovary (which becomes the pea pod). After landing, the pollen grows a tube that extends through the stigma to one of the ovules (immature seeds), allowing fertilization to take place. **(c)** To prevent self-fertilization, breeders remove the anthers from the female parents (here, the white flower) before the plant produces mature pollen. Pollen is then transferred with a paintbrush from the anthers of the male parent (here, the purple flower) to the stigma of the female parent. Each fertilized ovule becomes an individual pea (mature seed) that can grow into a new pea plant. All of the peas produced from one flower are encased in the same pea pod, but these peas form from different pollen grains and ovules.
(a): © Andrea Jones Images/Alamy

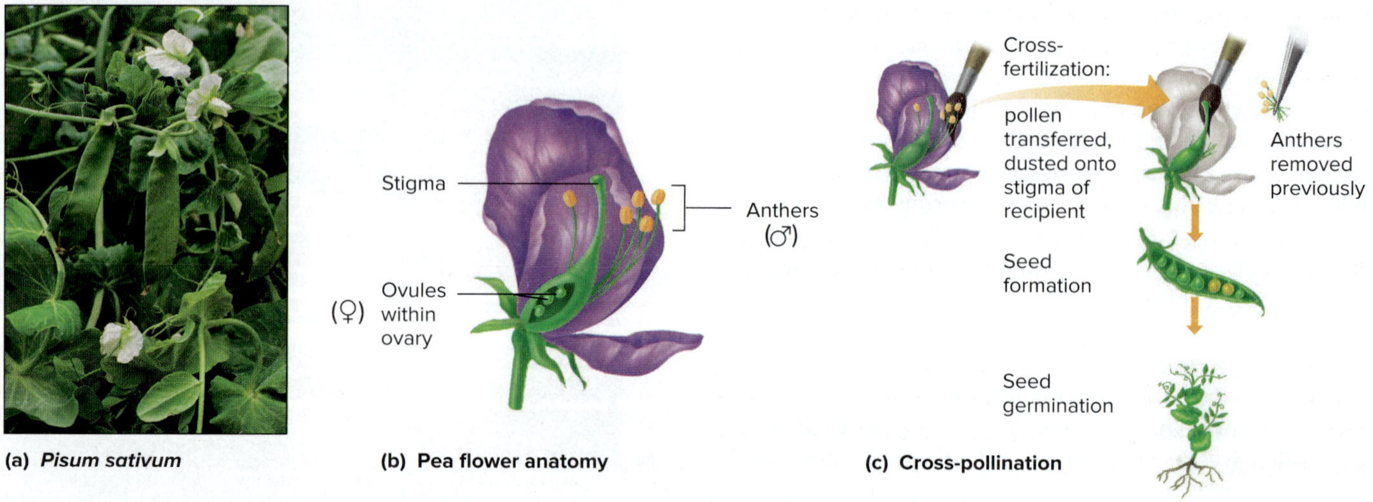

such as purple versus white flowers or yellow versus green seeds *antagonistic pairs*, and he settled on seven such pairs for his study (**Fig. 2.8**). In his experiments, Mendel not only perpetuated pure-breeding stocks for each member of a pair, but he also cross-fertilized pairs of plants to produce **hybrids,** offspring of genetically dissimilar parents, for each pair of antagonistic traits. Figure 2.8 shows the appearance of the hybrids he studied.

Fourth, being an expert plant breeder, Mendel carefully controlled his matings, going to great lengths to ensure that the progeny he observed really resulted from the specific fertilizations he intended. Thus Mendel painstakingly prevented the intrusion of any foreign pollen and assured self- or cross-pollination as the experiment demanded. Not only did this allow him to carry out controlled breedings of selected traits, but he could also make **reciprocal crosses.** In such crosses, he reversed the traits of the male and female parents, thus controlling whether a particular trait was transmitted via the egg cell within the ovule or via a sperm cell within the pollen. For example, he could use pollen from a purple flower to fertilize the eggs of a white flower and also use pollen from a white flower to fertilize the eggs of a purple flower. Because the progeny of these reciprocal crosses were similar, Mendel demonstrated that the two parents contribute equally to inheritance. "It is immaterial to the form of the hybrid," he wrote, "which of the parental types was the seed or pollen plant."

Fifth, Mendel worked with large numbers of plants, counted all offspring, subjected his findings to numerical analysis, and then compared his results with predictions based on his models. He was the first person to study inheritance in this manner, and no doubt his background in physics and mathematics contributed to this quantitative approach. Mendel's careful numerical analysis revealed patterns of transmission that reflected basic laws of heredity.

Finally, Mendel was a brilliant practical experimentalist. When comparing tall and short plants, for example, he made sure that the short ones were out of the shade of the tall ones so their growth would not be stunted. Eventually he focused on certain traits of the pea seeds themselves, such as their color or shape, rather than on traits of the plants arising from the seeds. In this way, he could observe many more individuals from the limited space of the monastery garden, and he could evaluate the results of a cross in a single growing season.

In short, Mendel purposely set up a simplified black-and-white experimental system and then figured out how it worked. He did not look at the vast number of variables that determine the development of a prize ram nor at the origin of differences between species. Rather, he looked at discrete traits that came in two mutually exclusive forms and asked questions that could be answered by observation and computation.

Figure 2.8 The mating of parents with antagonistic traits produces hybrids. Note that each of the hybrids for the seven antagonistic traits studied by Mendel resembles only one of the parents. The parental trait that shows up in the hybrid is known as the *dominant trait.*

essential concepts

- People practiced *artificial selection* of crop plants and domestic animals for thousands of years without understanding how heredity works.
- Mendel established *pure-breeding* lines of peas in which a specific characteristic would remain constant from one generation to the next.
- When Mendel crossed pure-breeding lines with alternative traits, the *hybrid* progeny always had the characteristics of one parent.
- In Mendel's experiments, the hybrid progeny produced by reciprocal cross-fertilizations had the same characteristics; it did not matter which parent was male and which was female.

2.2 Genetic Analysis According to Mendel

learning objectives

1. Explain Mendel's law of segregation and how it predicts the 3:1 dominant-to-recessive phenotypic ratio among the F_2 generation of a monohybrid cross.
2. Distinguish between a monohybrid cross and a testcross.
3. Explain Mendel's law of independent assortment and how the 9:3:3:1 phenotypic ratio among the F_2 of a dihybrid cross provides evidence for this law.
4. Interpret phenotypic ratios of progeny to infer how particular traits are inherited.
5. Predict the genotypic and phenotypic ratios among progeny of complex multihybrid crosses using simple rules of probability.
6. Cite the most common molecular explanations for dominant and recessive alleles.

In early 1865 at the age of 43, Gregor Mendel presented a paper entitled *Experiments on Plant Hybrids* before the Natural Science Society of Brünn. Despite its modest heading, this was a scientific paper of uncommon clarity and simplicity that summarized a decade of original observations and experiments. In it Mendel describes in detail the transmission of visible characteristics in pea plants, defines unseen but logically deduced units (genes) that determine when and how often these visible traits appear, and analyzes the behavior of genes in simple mathematical terms to reveal previously unsuspected principles of heredity.

Published the following year, the paper would eventually become the cornerstone of modern genetics. Its stated purpose was to see whether there is a "generally applicable law governing the formation and development of hybrids." Let us examine its insights.

Monohybrid Crosses Reveal the Law of Segregation

Once Mendel had isolated pure-breeding lines for several sets of characteristics, he carried out a series of matings between individuals that differed in only one trait, such as seed color or stem length. In each cross, one parent carries one form of the trait, and the other parent carries an alternative form of the same trait. **Figure 2.9** illustrates one such mating. Early in the spring of 1854, for example, Mendel planted pure-breeding green peas and pure-breeding yellow peas and allowed them to grow into the **parental (P) generation.** Later that spring when the plants had flowered, he dusted the female stigma of green-pea plant flowers with pollen from yellow-pea plants. He also performed the reciprocal cross, dusting yellow-pea plant stigmas with green-pea pollen. In the fall, when he collected and separately analyzed the progeny peas of these reciprocal crosses, he found that in both cases, the peas were all yellow.

These yellow peas, progeny of the P generation, were the beginning of what we now call the **first filial (F_1) generation.** To learn whether the green trait had disappeared entirely or remained intact but hidden in these F_1 yellow peas, Mendel planted them to obtain mature F_1 plants that he allowed to self-fertilize. Such experiments involving hybrids for a single trait are often called **monohybrid crosses.** He then harvested and counted the peas of the resulting **second filial (F_2) generation,** progeny of the F_1 generation. Among the progeny of one series of F_1 self-fertilizations, there were 6022 yellow and 2001 green F_2 peas, an almost

Figure 2.9 Analyzing a monohybrid cross. Cross-pollination of pure-breeding parental plants produces F_1 hybrids, all of which resemble one of the parents. Self-pollination of F_1 plants gives rise to an F_2 generation with a 3:1 ratio of individuals resembling the two original parental types. For simplicity, we do not show the plants that produce the peas or that grow from the planted peas.

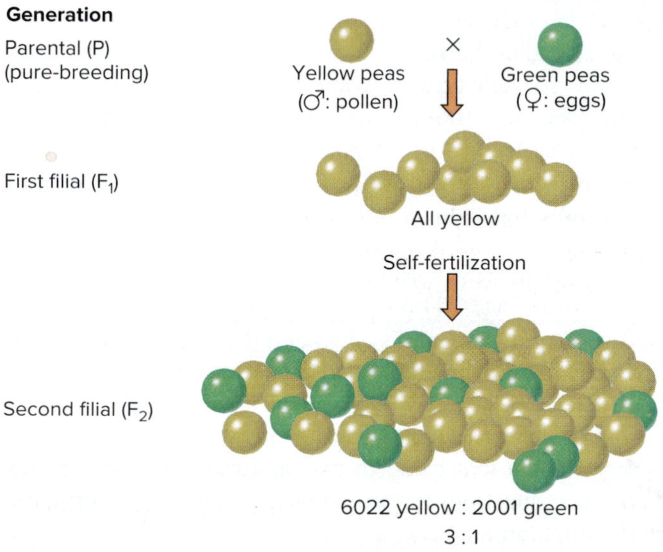

perfect ratio of 3 yellow : 1 green. F_1 plants derived from the reciprocal of the original cross produced a similar ratio of yellow to green F_2 progeny.

Reappearance of the recessive trait

The presence of green peas in the F_2 generation was irrefutable evidence that blending had not occurred. If it had, the information necessary to make green peas would have been lost irretrievably in the F_1 hybrids. Instead, the information remained intact and was able to direct the formation of 2001 green peas actually harvested from the second filial generation. These green peas were indistinguishable from their green grandparents.

Mendel concluded that two types of yellow peas must exist: those that breed true like the yellow peas of the P generation, and those that can yield some green offspring like the yellow F_1 hybrids. This second type somehow contains latent information for green peas. He called the trait that appeared in all the F_1 hybrids—in this case, yellow seeds—**dominant** (see Fig. 2.8) and the antagonistic green-pea trait that remained hidden in the F_1 hybrids but reappeared in the F_2 generation **recessive.** But how did he explain the 3:1 ratio of yellow to green F_2 peas?

Genes: Discrete units of inheritance

To account for his observations, Mendel proposed that for each trait, every plant carries two copies of a unit of inheritance, receiving one from its maternal parent and the other from the paternal parent. Today, we call these units of inheritance **genes.** Each unit determines the appearance of a specific characteristic. The pea plants in Mendel's collection had two copies of a gene for seed color, two copies of another for seed shape, two copies of a third for stem length, and so forth.

Mendel further proposed that each gene comes in alternative forms, and combinations of these alternative forms determine the contrasting characteristics he was studying. Today we call the alternative forms of a single gene **alleles.** The gene for pea color, for example, has yellow and green alleles; the gene for pea shape has round and wrinkled alleles. In Mendel's monohybrid crosses, one allele of each gene was dominant, the other recessive. In the P generation, one parent carried two dominant alleles for the trait under consideration; the other parent, two recessive alleles. The F_1 generation hybrids carried one dominant and one recessive allele for the trait. Individuals having two different alleles for a single trait are **monohybrids.**

The law of segregation

If a plant has two copies of every gene, how does it pass only one copy of each to its progeny? And how do the offspring then end up with two copies of these same genes, one from each parent? Mendel drew on his background in plant physiology and answered these questions in terms of the two biological mechanisms behind reproduction: gamete formation and the random union of gametes at fertilization.

Gametes are the specialized cells—eggs within the ovules of the female parent and sperm cells within the pollen grains—that carry genes between generations. Mendel imagined that during the formation of eggs and sperm, the two copies of each gene in the parent separate (or *segregate*) so that each gamete receives only one allele for each trait (**Fig. 2.10a**). Thus, each egg and each sperm receives only one allele for pea color (either yellow or green).

At fertilization, a sperm with one or the other allele unites at random with an egg carrying one or the other allele, restoring the two copies of the gene for each trait in the fertilized egg, or **zygote** (**Fig. 2.10b**). If the sperm carries yellow and the egg green, the result will be a hybrid yellow pea like the F_1 monohybrids that resulted when pure-breeding parents of opposite types mated. If the yellow-carrying sperm unites with a yellow-carrying egg, the result will be a yellow pea that grows into a pure-breeding plant like those of the P generation that produced only yellow peas.

Figure 2.10 The law of segregation. (a) The two identical alleles of pure-breeding plants separate (segregate) during gamete formation. As a result, each sperm or egg carries only one of each pair of parental alleles. **(b)** Cross-pollination and fertilization between pure-breeding parents with antagonistic traits result in F_1 hybrid zygotes with two different alleles. For the seed color gene, a *Yy* hybrid zygote will develop into a yellow pea.

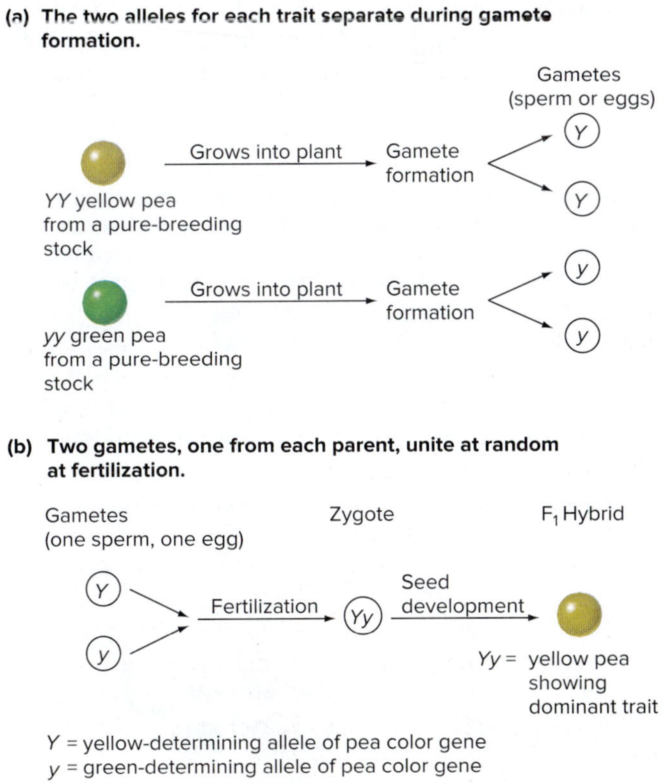

And finally, if sperm carrying the allele for green peas fertilizes a green-carrying egg, the progeny will be a pure-breeding green pea.

Mendel's **law of segregation** encapsulates this general principle of heredity: *The two alleles for each trait separate (segregate) during gamete formation, and then unite at random, one from each parent, at fertilization.* Throughout this book, the term **segregation** refers to such *equal segregation* in which one allele, and only one allele, of each gene goes to each gamete. Note that the law of segregation makes a clear distinction between the **somatic cells** (*body cells*) of an organism, which have two copies of each gene, and the **gametes,** which bear only a single copy of each gene.

The Punnett square

Figure 2.11 shows a simple way of visualizing the results of the segregation and random union of alleles during gamete formation and fertilization. Mendel invented a system of symbols that allowed him to analyze all his crosses in the same way. He designated dominant alleles with a capital *A*, *B*, or *C* and recessive ones with a lowercase *a*, *b*, or *c*. Modern geneticists have adopted this convention for naming genes in peas and many other organisms, but they often choose a symbol with some reference to the trait in question—a *Y* for yellow or an *R* for round. Throughout this book, we present gene symbols in italics. In Fig. 2.11, we denote the dominant yellow allele with a capital *Y* and the recessive green allele with a lowercase *y*. The pure-breeding plants of the parental generation are either *YY* (yellow peas) or *yy* (green peas). The *YY* parent can produce only *Y* gametes, the *yy* parent only *y* gametes. You can see in Fig. 2.11 why every cross between *YY* and *yy* produces exactly the same result—a *Yy* hybrid—no matter which parent (male or female) contributes which particular allele.

Figure 2.11 The Punnett square: Visual summary of a cross. This Punnett square illustrates the combinations that can arise when an F_1 hybrid undergoes gamete formation and self-fertilization. The F_2 generation should have a 3:1 ratio of yellow to green peas.

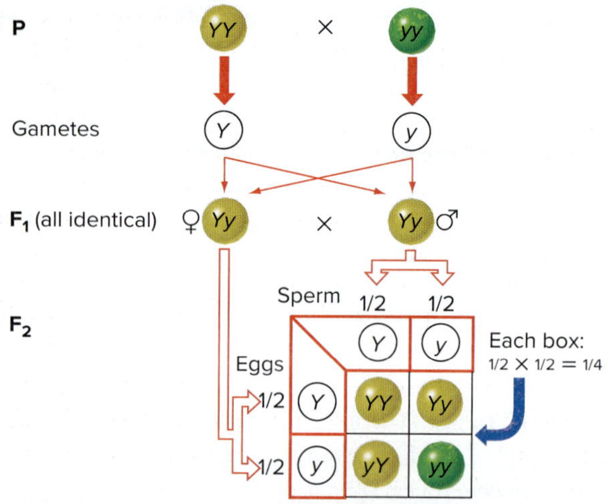

Next, to visualize what happens when the *Yy* hybrids self-fertilize, we set up a **Punnett square** (named after the British mathematician Reginald Punnett, who introduced it in 1906; Fig. 2.11). The square provides a simple and convenient method for tracking the kinds of gametes produced as well as all the possible combinations that might occur at fertilization. As the Punnett square shows in the first column and the first row, each hybrid produces two kinds of gametes, *Y* and *y*, in a ratio of 1:1. Thus, half the sperm and half the eggs carry *Y*, the other half of each gamete type carries *y*.

Each box in the Punnett square in Fig. 2.11 containing a colored pea represents one possible fertilization event. At fertilization, 1/4 of the progeny will be *YY*, 1/4 *Yy*, 1/4 *yY*, and 1/4 *yy*. Because the gametic source of an allele (egg or sperm) for the traits Mendel studied had no influence on the allele's effect, *Yy* and *yY* are equivalent. This means that 1/2 of the progeny are yellow *Yy* hybrids, 1/4 *YY* true-breeding yellows, and 1/4 true-breeding *yy* greens. The diagram illustrates how the segregation of alleles during gamete formation and the random union of egg and sperm at fertilization can produce the 3:1 ratio of yellow to green that Mendel observed in the F_2 generation.

Mendel's Results Reflect Basic Rules of Probability

Though you may not have realized it, the Punnett square illustrates two simple rules of probability—the *product rule* and the *sum rule*—that are central to the analysis of genetic crosses. These rules predict the likelihood that a particular combination of events will occur.

The product rule

The **product rule** states that the probability of two or more *independent events* occurring together is the *product* of the probabilities that each event will occur by itself. With independent events:

Probability of event 1 *and* event 2 =

Probability of event 1 × probability of event 2.

Consecutive coin tosses are obviously independent events; a heads in one toss neither increases nor decreases the probability of a heads in the next toss. If you toss two coins at the same time, the results are also independent events. A heads for one coin neither increases nor decreases the probability of a heads for the other coin. Thus, the probability of a given combination is the product of their independent probabilities. For example, the probability that both coins will turn up heads is:

$$1/2 \times 1/2 = 1/4.$$

Similarly, the formation of egg and sperm are independent events; in a hybrid plant, the probability is 1/2 that a given gamete will carry *Y* and 1/2 that it will carry *y*. Because fertilization happens at random, the probability that a particular combination of maternal and paternal alleles will occur simultaneously in the same zygote is the product of the independent probabilities of these alleles being packaged in egg and sperm. Thus, to find the chance of a *Y* egg (formed as the result of one event) uniting with a *Y* sperm (the result of an independent event), you simply multiply $1/2 \times 1/2$ to get 1/4. This is the same fraction of *YY* progeny seen in the Punnett square of Fig. 2.11, which demonstrates that the Punnett square is simply another way of depicting the product rule. It is important to realize that each box in the Punnett square represents an equally likely outcome of the cross (an equally likely fertilization event) *only because* each of the two types of sperm and eggs (*Y* and *y*) are produced at equal frequencies.

The sum rule

While we can describe the moment of random fertilization as the simultaneous occurrence of two independent events, we can also say that two different fertilization events are mutually exclusive. For instance, if *Y* combines with *Y*, it cannot also combine with *y* in the same zygote. A second rule of probability, the **sum rule,** states that the probability of either of two such *mutually exclusive events* occurring is the *sum* of their individual probabilities. With mutually exclusive events:

Probability of event 1 *or* event 2 =

Probability of event 1 + probability of event 2.

To find the likelihood that an offspring of a *Yy* hybrid self-fertilization will be a hybrid like the parents, you add 1/4 (the probability of maternal *Y* uniting with paternal *y*) and 1/4 (the probability of the mutually exclusive event where paternal *Y* unites with maternal *y*) to get 1/2, again the same result as in the Punnett square.

In another use of the sum rule, you could predict the ratio of yellow to green F_2 progeny. The fraction of F_2 peas that will be yellow is the sum of 1/4 (the event producing *YY*) plus 1/4 (the mutually exclusive event generating *Yy*) plus 1/4 (the mutually exclusive event producing *yY*) to get 3/4. The remaining 1/4 of the F_2 progeny will be green. So the yellow-to-green ratio is 3/4 to 1/4, or more simply, 3:1.

Further Crosses Verify the Law of Segregation

The law of segregation was a hypothesis that explained the data from simple crosses involving monohybrid peas, but Mendel needed to perform additional experiments to check its validity. Mendel's hypothesis, summarized in Fig. 2.11, made the testable prediction that the F_2 should have two kinds of yellow peas (*YY* and *Yy*) but only one kind of green pea (*yy*). In addition, his hypothesis predicted that the *YY* and *Yy* yellow peas in the F_2 should be present in a ratio of 1*YY* : 2*Yy*.

Figure 2.12 Yellow F_2 peas are of two types: Pure breeding and hybrid. The distribution of a pair of contrasting alleles (*Y* and *y*) after two generations of self-fertilization. The homozygous individuals of each generation breed true, whereas the hybrids do not.

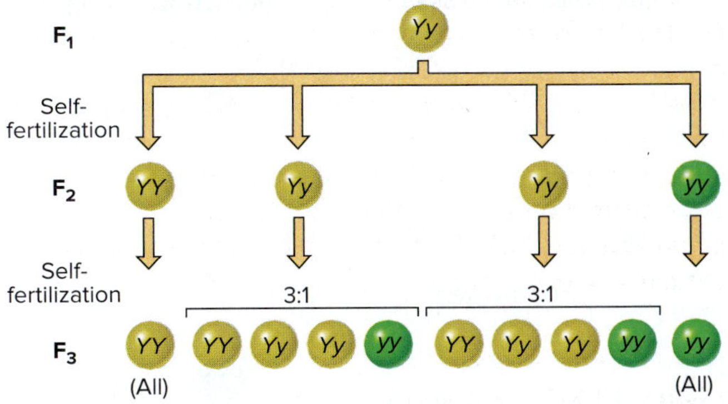

To verify these expectations, Mendel allowed self-fertilization of all the plants in the F_2 generation and counted the types of F_3 progeny (**Fig. 2.12**). He found that the plants that developed from F_2 green peas all produced only green peas in the F_3, and when the resulting F_3 plants self-fertilized, the next generation (the F_4) also produced green peas (*not shown*). This is what we (and Mendel) would expect of pure-breeding *yy* lines carrying two copies of the recessive allele. The yellow peas were a different story. When Mendel allowed 518 F_2 plants that developed from yellow peas to self-fertilize, he observed that 166, roughly 1/3 of the total, were pure-breeding yellow through several generations, but the other 352 (2/3 of the total yellow F_2 plants) were hybrids because they gave rise to yellow and green F_3 peas in a ratio of 3:1. Therefore, as Mendel's theory anticipated, the ratio of *YY* to *Yy* among the 518 F_2 yellow pea plants was indeed 1:2.

It took Mendel years to conduct such rigorous experiments on seven pairs of pea traits, but in the end, he was able to conclude that the segregation of dominant and recessive alleles during gamete formation and their random union at fertilization could indeed explain the 3:1 ratios he observed whenever he allowed hybrids to self-fertilize. His results, however, raised yet another question, one of some importance to future plant and animal breeders. Plants showing a dominant trait, such as yellow peas, can be either pure-breeding (*YY*) or hybrid (*Yy*). How can you distinguish one from the other? For self-fertilizing plants, the answer is to observe the appearance of the next generation. But how would you distinguish pure-breeding from hybrid individuals in species that do not self-fertilize?

Figure 2.13 Genotype versus phenotype in homozygotes and heterozygotes. The relationship between genotype and phenotype with a pair of contrasting alleles where one allele (*Y*) shows complete dominance over the other (*y*).

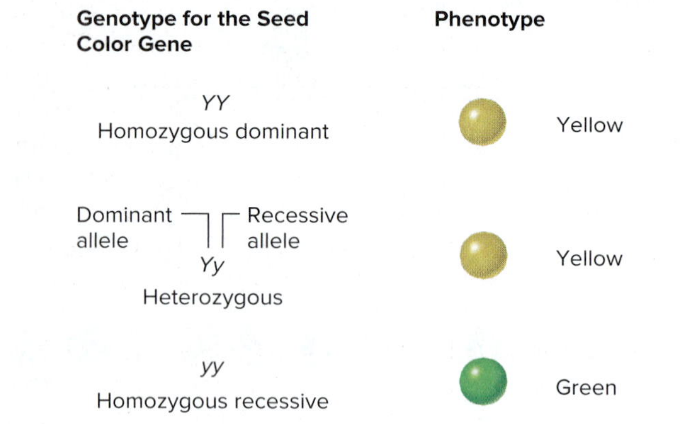

Figure 2.14 How a testcross reveals genotype. An individual of unknown genotype, but dominant phenotype, is crossed with a homozygous recessive. If the unknown genotype is homozygous, all progeny will exhibit the dominant phenotype (*cross A*). If the unknown genotype is heterozygous, half the progeny will exhibit the dominant trait, half the recessive trait (*cross B*).

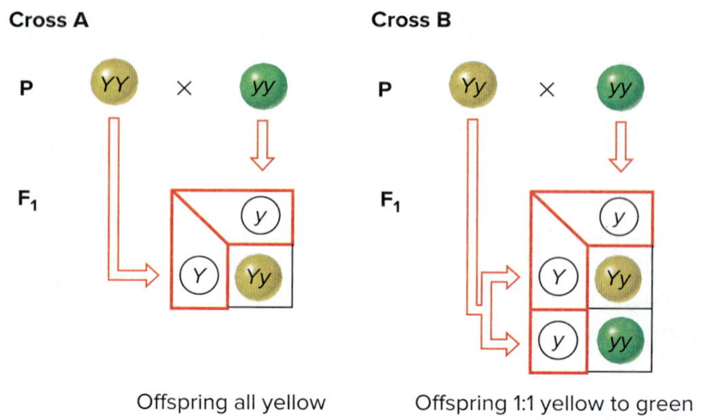

Testcrosses: A way to establish genotype

Before describing Mendel's answer, we need to define a few more terms. An observable characteristic, such as yellow or green pea seeds, is a **phenotype,** while the actual pair of alleles present in an individual is its **genotype.** A *YY* or a *yy* genotype is called **homozygous,** because the two copies of the gene that determine the particular trait in question are the same. In contrast, a genotype with two different alleles for a trait is **heterozygous;** in other words, it is a hybrid for that trait (**Fig. 2.13**). An individual with a homozygous genotype is a **homozygote;** one with a heterozygous genotype is a **heterozygote.**

Note that the phenotype of a heterozygote (that is, of a hybrid) defines which allele is dominant: Because *Yy* peas are yellow, the yellow allele *Y* is dominant to the *y* allele for green. If you know the genotype and the dominance relation of the alleles, you can accurately predict the phenotype. The reverse is not true, however, because some phenotypes can derive from more than one genotype. For example, the phenotype of yellow peas can result from either the *YY* or the *Yy* genotype.

With these distinctions in mind, we can look at the method Mendel devised for deciphering the unknown genotype. We'll call it *Y*–, responsible for a dominant phenotype; the dash represents the unknown second allele, either *Y* or *y*. This method, called the **testcross,** is a mating in which an individual showing the dominant phenotype, for instance, a *Y*– plant grown from a yellow pea, is crossed with an individual expressing the recessive phenotype, in this case a *yy* plant grown from a green pea. As the Punnett squares in **Fig. 2.14** illustrate, if the dominant phenotype in question derives from a homozygous *YY* genotype, all the offspring of the testcross will show the dominant yellow phenotype. But if the dominant parent of unknown genotype is a heterozygous hybrid (*Yy*), half of the progeny are expected to be yellow peas, and the other half green. In this way, the testcross establishes the genotype behind a dominant phenotype, resolving any uncertainty.

As we mentioned earlier, Mendel deliberately simplified the problem of heredity, focusing on traits that come in only two forms. He was able to replicate his basic monohybrid findings with corn, beans, and four-o'clocks (plants with tubular, white or bright red flowers). As it turns out, his concept of the gene and his law of segregation can be generalized to almost all sexually reproducing organisms.

Dihybrid Crosses Reveal the Law of Independent Assortment

Having determined from monohybrid crosses that genes are inherited according to the law of segregation, Mendel turned his attention to the simultaneous inheritance of two or more apparently unrelated traits in peas. He asked how two pairs of alleles would segregate in a **dihybrid** individual, that is, in a plant that is heterozygous for two genes at the same time.

To construct such a dihybrid, Mendel mated true-breeding plants grown from yellow round peas (*YY RR*) with true-breeding plants grown from green wrinkled peas (*yy rr*). From this cross he obtained a dihybrid F_1 generation (*Yy Rr*) showing the two dominant phenotypes, yellow and round (**Fig. 2.15**). He then allowed these F_1 dihybrids to self-fertilize to produce the F_2 generation. Mendel could not predict the outcome of this mating. Would all the F_2 progeny be **parental types** that looked like either the original yellow round parent or the green wrinkled parent? Or would some new combinations of phenotypes occur that were not seen in the parental lines, such as yellow wrinkled or green round peas? New phenotypic combinations like these are called **recombinant types.**

Figure 2.15 A dihybrid cross produces parental types and recombinant types. In this dihybrid cross, pure-breeding parents (P) produce a genetically uniform generation of F_1 dihybrids. Self-pollination or cross-pollination of the F_1 plants yields the characteristic F_2 phenotypic ratio of 9:3:3:1.

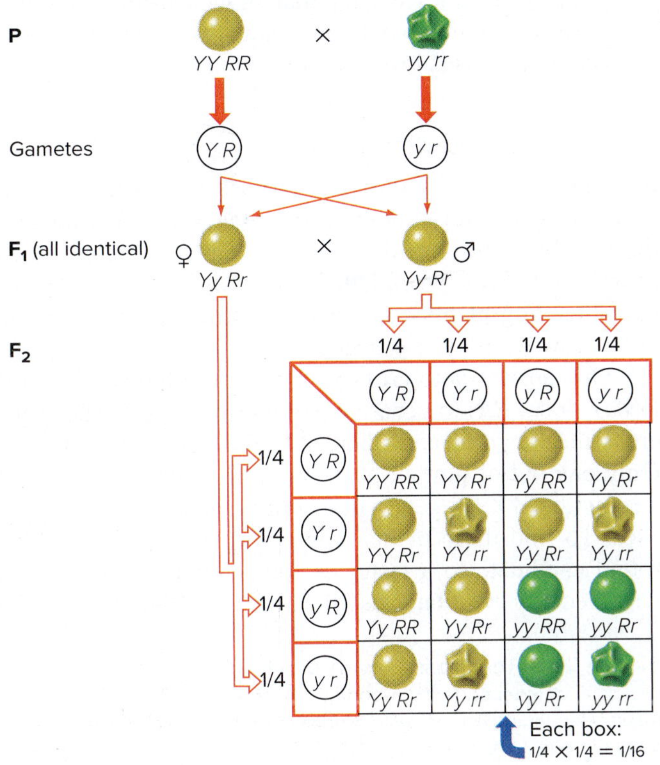

Type	Genotype	Phenotype	Number	Phenotypic Ratio
Parental	*Y– R–*	yellow round	315	9/16
Recombinant	*yy R–*	green round	108	3/16
Recombinant	*Y– rr*	yellow wrinkled	101	3/16
Parental	*yy rr*	green wrinkled	32	1/16

Ratio of yellow (dominant) to green (recessive) = 12:4 or 3:1

Ratio of round (dominant) to wrinkled (recessive) = 12:4 or 3:1

When Mendel counted the F_2 generation of one experiment, he found 315 yellow round peas, 108 green round, 101 yellow wrinkled, and 32 green wrinkled. Both yellow wrinkled and green round recombinant phenotypes did, in fact, appear, providing evidence that some shuffling of the alleles of different genes had taken place.

The law of independent assortment

From the observed ratios, Mendel inferred the biological mechanism of that shuffling—the **independent assortment** of gene pairs during gamete formation. Because the genes for pea color and for pea shape assort independently, the allele for pea shape in a gamete carrying *Y* could with equal likelihood be either *R* or *r*. Thus, the presence of a particular allele of one gene, say, the dominant *Y* for pea color, provides no information whatsoever about the allele of the second gene. Each dihybrid of the F_1 generation can therefore make four kinds of gametes: *Y R, Y r, y R*, and *y r*. In a large number of gametes, the four kinds will appear in an almost perfect ratio of 1:1:1:1, or put another way, roughly 1/4 of the eggs and 1/4 of the sperm will contain each of the four possible combinations of alleles. That "the different kinds of germinal cells [eggs or sperm] of a hybrid are produced on the average in equal numbers" was yet another one of Mendel's incisive insights.

At fertilization then, in a mating of dihybrids, 4 different kinds of eggs can each combine with any 1 of 4 different kinds of sperm, producing a total of 16 possible zygotes. Once again, a Punnett square is a convenient way to visualize the process (Fig. 2.15). Using the same kind of logic previously applied to the Punnett square for monohybrid crosses (review Fig. 2.11), each of the 16 boxes with colored peas in the Punnett square for the dihybrid cross in Fig. 2.15 represents an equally likely fertilization event. Again, each box is an equally likely outcome *only because* each of the different gamete types is produced at equal frequency in each parent. Therefore, using the product rule, the frequency of the progeny type in each box is $1/4 \times 1/4 = 1/16$.

If you look at the square in Fig. 2.15, you will see that some of the 16 potential allelic combinations are identical. In fact, only nine different genotypes exist—*YY RR, YY Rr, Yy RR, Yy Rr, yy RR, yy Rr, YY rr, Yy rr*, and *yy rr*—because the source of the alleles (egg or sperm) does not make any difference. If you look at the combinations of traits determined by the nine genotypes, you will see only four phenotypes—yellow round, green round, yellow wrinkled, and green wrinkled—in a ratio of 9:3:3:1. If, however, you look only at pea color or only at pea shape, you can see that each trait is inherited in the 3:1 ratio predicted by Mendel's law of segregation. In the Punnett square, there are 12 yellow for every 4 green and 12 round for every 4 wrinkled. In other words, the ratio of each dominant trait (yellow or round) to its antagonistic recessive trait (green or wrinkled) is 12:4, or 3:1. This means that the inheritance of the gene for pea color is unaffected by the inheritance of the gene for pea shape, and *vice versa*.

The preceding analysis became the basis of Mendel's second general genetic principle, the **law of independent assortment:** *During gamete formation, different pairs of alleles segregate independently of each other* (**Fig. 2.16**). The independence of their segregation and the subsequent random union of gametes at fertilization determine the phenotypes observed. Using the product rule for assessing the probability of independent events, you can see mathematically how the 9:3:3:1 phenotypic ratio observed in a dihybrid cross derives from two separate 3:1 phenotypic ratios. If the two sets of

Figure 2.16 The law of independent assortment. In a dihybrid cross, each pair of alleles assorts independently during gamete formation. In the gametes, *Y* is equally likely to be found with *R* or *r* (that is, *Y R* = *Y r*); the same is true for *y* (that is, *y R* = *y r*). As a result, all four possible types of gametes (*Y R*, *Y r*, *y R*, and *y r*) are produced in equal frequency among a large population.

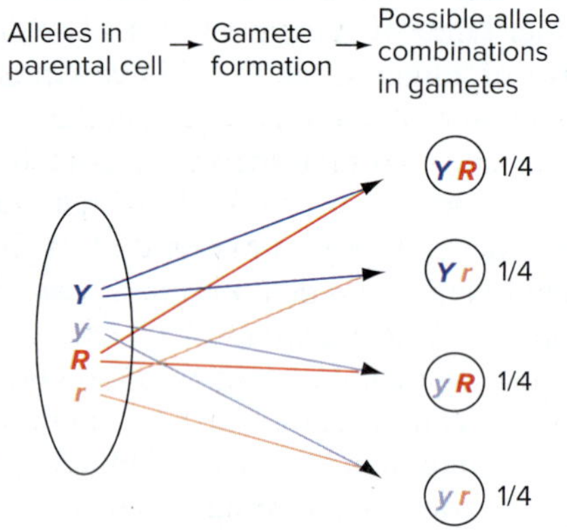

alleles assort independently, the yellow-to-green ratio in the F_2 generation will be 3/4 : 1/4, and likewise, the round-to-wrinkled ratio will be 3/4 : 1/4. To find the probability that two independent events such as yellow and round will occur simultaneously in the same plant, you multiply as follows:

Probability of yellow round = 3/4 × 3/4 = 9/16

Probability of green round = 1/4 × 3/4 = 3/16

Probability of yellow wrinkled = 3/4 × 1/4 = 3/16

Probability of green wrinkled = 1/4 × 1/4 = 1/16

Thus, in a population of F_2 plants, there will be a 9:3:3:1 phenotypic ratio of yellow round to green round to yellow wrinkled to green wrinkled.

Branched-line diagrams

A convenient way to keep track of the probabilities of each potential outcome in a genetic cross is to construct a **branched-line diagram (Fig. 2.17)**, which shows all the possible genotypes or phenotypes for each gene in a sequence of columns. In Fig. 2.17, the first column shows the two possible pea color phenotypes; the second column demonstrates that each pea color can occur with either of two pea shapes. Again, the 9:3:3:1 ratio of phenotypes is apparent. You will see later that branched-line diagrams are more convenient than Punnett squares for predicting the outcomes of crosses involving more than two genes.

Figure 2.17 Following crosses with branched-line diagrams. A branched-line diagram, which uses a series of columns to track every gene in a cross, provides an organized overview of all possible outcomes. This branched-line diagram of a dihybrid cross generates the same phenotypic ratios as the Punnett square in Fig. 2.15, showing that the two methods are equivalent.

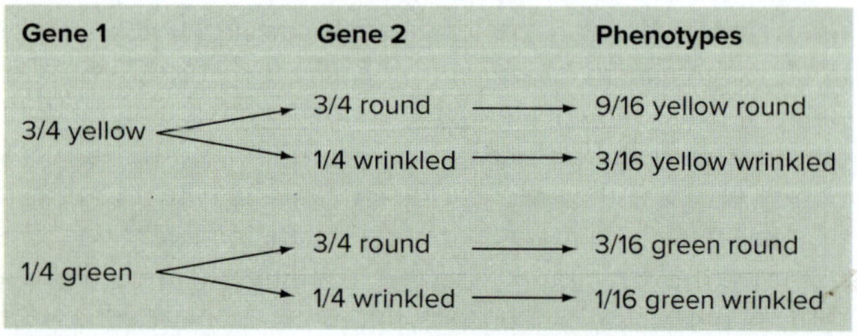

Testcrosses with dihybrids

An understanding of dihybrid crosses has many applications. Suppose, for example, that you work for a nursery that has three pure-breeding strains: yellow wrinkled, green round, and green wrinkled. Your assignment is to grow pure-breeding plants guaranteed to produce yellow round peas. How would you proceed?

One answer is to cross your two pure-breeding strains (*YY rr* × *yy RR*) to generate a dihybrid (*Yy Rr*). Then self-cross the dihybrid and plant only the yellow round peas. Only one out of nine of such progeny—those grown from peas with a *YY RR* genotype—will be appropriate for your uses. To find these plants, you could subject each yellow round candidate to a testcross for genotype with a green wrinkled (*yy rr*) plant, as illustrated in **Fig. 2.18**. If the

Figure 2.18 Testcrosses with dihybrids. Testcrosses involving two pairs of independently assorting alleles yield different, predictable results depending on the tested individual's genotype for the two genes in question.

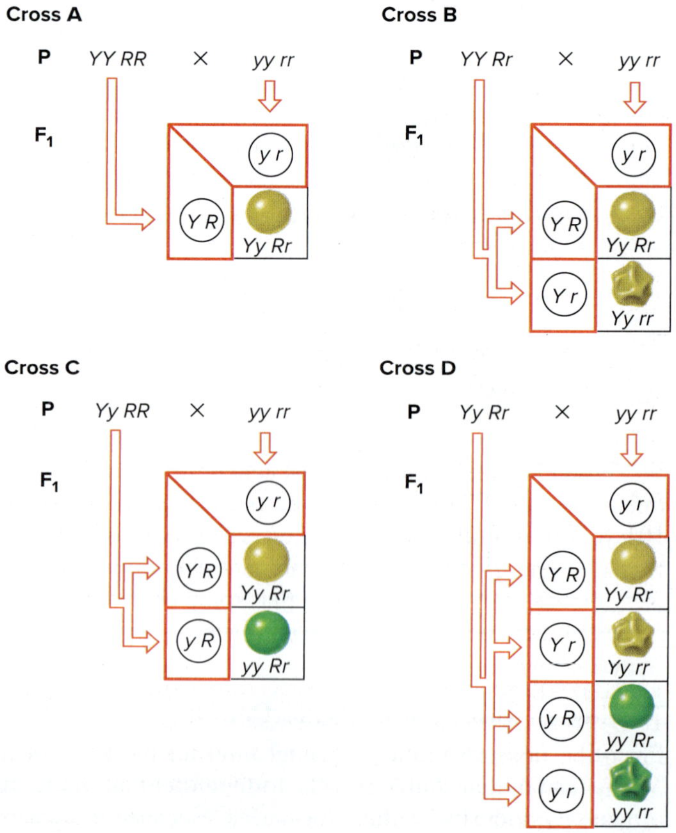

testcross yields all yellow round offspring (testcross A), you can sell your test plant, because you know it is homozygous for both pea color and pea shape. If your testcross yields 1/2 yellow round and 1/2 yellow wrinkled (testcross B), or 1/2 yellow round and 1/2 green round (testcross C), you know that the candidate plant in question is genetically homozygous for one trait and heterozygous for the other and must therefore be discarded. Finally, if the testcross yields 1/4 yellow round, 1/4 yellow wrinkled, 1/4 green round, and 1/4 green wrinkled (testcross D), you know that the plant is a heterozygote for both the pea color and the pea shape genes.

Geneticists Use Mendel's Laws to Calculate Probabilities and Make Predictions

Mendel performed several sets of dihybrid crosses and also carried out **multihybrid crosses:** matings between the F_1 progeny of pure-breeding parents that differed in three or more unrelated traits. In all of these experiments, he observed numbers and ratios very close to what he expected on the basis of his two general biological principles: The alleles of a gene segregate during the formation of egg or sperm, and the alleles of different genes assort independently of each other. Mendel's laws of inheritance, in conjunction with the mathematical rules of probability, provide geneticists with powerful tools for predicting and interpreting the results of genetic crosses. But as with all tools, they have their limitations. We examine here both the power and the limitations of Mendelian analysis.

First, the power: Using simple Mendelian analysis, it is possible to make accurate predictions about the offspring of extremely complex crosses. Suppose you want to predict the occurrence of one specific genotype in a cross involving several independently assorting genes. For example, if hybrids that are heterozygous for four traits are allowed to self-fertilize—*Aa Bb Cc Dd* × *Aa Bb Cc Dd*—what proportion of their progeny will have the genotype *AA bb Cc Dd?* You could set up a Punnett square to answer the question. Because for each trait there are two different alleles, the number of different eggs or sperm is found by raising 2 to the power of the number of differing traits (2^n, where n is the number of traits). By this calculation, each hybrid parent in this cross with 4 traits would make $2^4 = 16$ different kinds of gametes. The Punnett square depicting such a cross would thus contain 256 boxes (16 × 16).

Setting up such a square may be fine if you live in a monastery with a bit of time on your hands, but not if you're taking a 1-hour exam. It would be much simpler to analyze the problem by breaking down the multihybrid cross into four independently assorting monohybrid crosses. Remember that the genotypic ratios of each monohybrid cross are 1 homozygote for the dominant allele, to 2 heterozygotes, to 1 homozygote for the recessive allele = 1/4 : 2/4 : 1/4. Thus, you can find the probability of *AA bb Cc Dd* by multiplying the probability of each independent event: *AA* (1/4 of the progeny produced by *Aa* × *Aa*); *bb* (1/4); *Cc* (2/4); *Dd* (2/4):

$$1/4 \times 1/4 \times 2/4 \times 2/4 = 4/256 = 1/64.$$

The Punnett square approach would provide the same answer, but it would require much more time.

If instead of a specific genotype, you want to predict the probability of a certain phenotype, you can again use the product rule as long as you know the phenotypic ratios produced by each pair of alleles in the cross. For example, if in the multihybrid cross of *Aa Bb Cc Dd* × *Aa Bb Cc Dd*, you want to know how many offspring will show the dominant A trait (genotype *AA* or *Aa* = 1/4 + 2/4, or 3/4), the recessive b trait (genotype *bb* = 1/4), the dominant C trait (genotype *CC* or *Cc* = 3/4), and the dominant D trait (genotype *DD* or *Dd* = 3/4), you simply multiply:

$$3/4 \times 1/4 \times 3/4 \times 3/4 = 27/256.$$

In this way, the rules of probability make it possible to predict the outcome of very complex crosses.

You can see from these examples that particular problems in genetics are amenable to particular modes of analysis. As a rule of thumb, Punnett squares are excellent for visualizing simple crosses involving a few genes, but they become unwieldy in the dissection of more complicated matings. Direct calculations of probabilities, such as those in the two preceding problems, are useful when you want to know the chances of one or a few outcomes of complex crosses. If, however, you want to know all the outcomes of a multihybrid cross, a branched-line diagram is the best way to go as it will keep track of all the possibilities in an organized fashion.

Now, the limitations of Mendelian analysis: Like Mendel, if you were to breed pea plants or corn or any other organism, you would most likely observe some deviation from the ratios you expected in each generation. What can account for such variation? One element is chance, as witnessed in the common coin toss experiment. With each throw, the probability of the coin coming up heads is equal to the likelihood it will come up tails. But if you toss a coin 10 times, you may get 30% (3) heads and 70% (7) tails, or *vice versa.* If you toss it 100 times, you are more likely to get a result closer to the expected 50% heads and 50% tails. The larger the number of trials, the lower the probability that chance significantly skews the data. The statistical benefit is one reason Mendel worked with large numbers of pea plants.

Mendel's laws, in fact, have great predictive power for populations of organisms, but they do not tell us what will happen in any one individual. With a garden full of self-fertilizing monohybrid pea plants, for example, you can expect that 3/4 of the F_2 progeny will show the dominant phenotype and 1/4 the recessive, but you cannot predict the phenotype of any particular F_2 plant. In Chapter 5, we discuss mathematical methods for assessing whether the

Figure 2.19 The science of genetics begins with the rediscovery of Mendel. Working independently near the beginning of the twentieth century, Correns, de Vries, and von Tschermak each came to the same conclusions as those Mendel summarized in his laws.
(a, c): © SPL/Science Source; (b): © INTERFOTO/Alamy; (d): © ullstein bild/Getty Images

(a) Gregor Mendel **(b)** Carl Correns **(c)** Hugo de Vries **(d)** Erich von Tschermak

chance variation observed in a sample of individuals within a population is compatible with a genetic hypothesis.

Mendel's Genius Was Unappreciated Before 1900

Mendel's insights into the workings of heredity were a breakthrough of monumental proportions. By counting and analyzing data from hundreds of pea plant crosses, he inferred the existence of genes—independent units that determine the observable patterns of inheritance for particular traits. His work explained the reappearance of hidden traits, disproved the idea of blended inheritance, and showed that mother and father make an equal genetic contribution to the next generation. The model of heredity that he formulated was so specific that he could test predictions based on it by observation and experiment.

With the exception of Abbot Napp, none of Mendel's contemporaries appreciated the importance of his research. Mendel did not teach at a prestigious university and was not well known outside Brünn. Even in Brünn, members of the Natural Science Society were disappointed when he presented *Experiments on Plant Hybrids* to them. They wanted to view and discuss intriguing mutants and lovely flowers, so they did not appreciate his numerical analyses. Mendel, it seems, was far ahead of his time. Sadly, despite written requests from Mendel that others try to replicate his studies, no one repeated his experiments. Several citations of his paper between 1866 and 1900 referred to his expertise as a plant breeder but made no mention of his laws. Moreover, at the time Mendel presented his work, no one had yet seen the structures within cells, the *chromosomes,* that actually carry the genes. That would happen only in the next few decades (as described in Chapter 4). If scientists had been able to see these structures, they might have more readily accepted Mendel's ideas, because the chromosomes are actual physical structures that behave exactly as Mendel predicted.

Mendel's work might have had an important influence on early debates about evolution if it had been more widely appreciated. Charles Darwin (1809–1882), who was unfamiliar with Mendel's work, was plagued in his later years by criticism that his explanations for the persistence of variation in organisms were insufficient. Darwin considered such variation a cornerstone of his theory of evolution, maintaining that natural selection would favor particular variants in a given population in a given environment. If the selected combinations of variant traits were passed on to subsequent generations, this transmission of variation would propel evolution. He could not, however, say how that transmission might occur. Had Darwin been aware of Mendel's ideas, he might not have been backed into such an uncomfortable corner.

For 34 years, Mendel's laws lay dormant—untested, unconfirmed, and unapplied. Then in 1900, 16 years after Mendel's death, Carl Correns, Hugo de Vries, and Erich von Tschermak independently rediscovered and acknowledged his work (**Fig. 2.19**). The scientific community had finally caught up with Mendel. Within a decade, investigators had coined many of the modern terms we have been using: phenotype, genotype, homozygote, heterozygote, gene, and genetics, the label given to the twentieth-century science of heredity. Mendel's paper provided the new discipline's foundation. His principles and analytic techniques endure today, guiding geneticists and evolutionary biologists in their studies of genetic variation.

The Influence of Molecules on Phenotype Determines Whether Alleles are Dominant or Recessive

We now know that genes specify the proteins (and RNAs) that cells produce and that dictate cellular structure and function. Recently, two genes were identified that are likely to correspond to Mendel's genes for seed shape and seed

Figure 2.20 Molecular explanations of Mendel's pea shape and pea color genes. The *R* allele of the pea shape gene specifies the enzyme Sbe1, which converts unbranched starch (amylose) to branched starch (amylopectin). The *r* allele does not produce Sbe1. The buildup of unbranched starch in *rr* peas ultimately causes seed wrinkling. The *Y* allele of the pea color gene specifies the enzyme Sgr, which functions in a pathway to break down chlorophyll during pea maturation, resulting in yellow peas. The *y* allele does not produce Sgr. Chlorophyll is not broken down in *yy* peas, and they remain green.

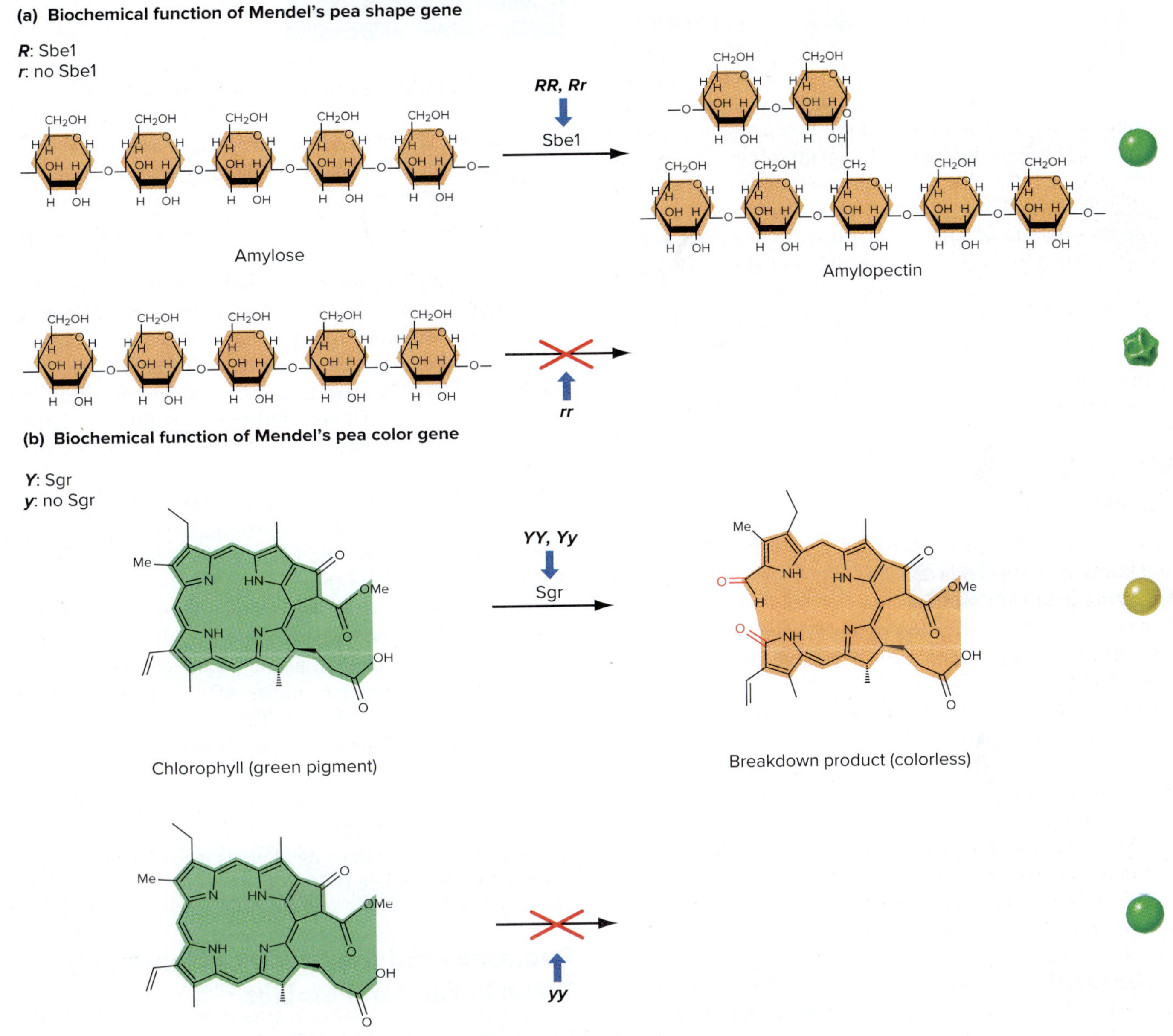

color. The pea shape gene specifies an enzyme known as Sbe1 (for Starch-branching enzyme 1). Sbe1 catalyzes the conversion of amylose, an unbranched linear molecule of starch, to amylopectin, a starch molecule composed of several branching chains (**Fig. 2.20a**). The dominant *R* allele of the pea shape gene determines a normal, functioning Sbe1 enzyme. In contrast, the recessive allele *r* specifies no Sbe1 enzyme. As a result, *RR* homozygotes contain a high proportion of branched starch molecules, which allows the peas to maintain a rounded shape. In homozygous recessive *rr* peas, sucrose builds up because less of it is converted into starch. The excess sucrose modifies the osmotic pressure, causing water to enter the young seeds. As the seeds mature, they lose the water, shrink, and wrinkle. The single dominant allele in *Rr* heterozygotes apparently specifies enough of the normal Sbe1 enzyme to prevent wrinkling.

The pea color gene determines an enzyme called Sgr (for Stay green). Sgr performs one step in the pathway leading to the breakdown of the green pigment chlorophyll, a process that occurs naturally in peas as they mature (**Fig. 2.20b**). The dominant *Y* allele specifies Sgr, and the recessive *y* allele does not. Homozygous *YY* or heterozygous *Yy* peas are yellow because they each have enough Sgr to break down all the chlorophyll. Homozygous *yy* peas stay green because they lack the Sgr enzyme, and the chlorophyll remains.

Two general principles emerge from these molecular discoveries. First, a specific gene determines a specific protein (in these cases an enzyme). The activity of the protein may affect the phenotype of the pea plant in any number of ways, depending on the biochemical pathway in which it functions. Second, a pattern can be seen in both of these examples: The dominant allele determines a normally functioning protein, while the recessive allele does not specify a functional protein. You will see in Chapter 3 that, although it is certainly not always the case, the molecular explanation described here is the most common reason why one allele is dominant to another (recessive) allele. Genes likely to be those Mendel described for stem length and flower color have also been identified recently. In both cases, the dominant allele encodes a normally functioning protein, and the recessive allele specifies either no protein or a less functional version of the normal protein.

essential concepts

- Discrete units called *genes* control the appearance of inherited traits; genes come in alternative forms called *alleles.*
- A sexually reproducing organism's body cells contain two *alleles* for every gene. These alleles may be the same (in a *homozygote*) or different (in a *heterozygote*).
- *Genotype* refers to the alleles an individual possesses; *phenotype* refers to the traits the individual exhibits.
- The *dominant* allele controls the phenotype of a trait in heterozygotes; the other allele in the heterozygote is *recessive.* In monohybrid crosses, the dominant and recessive phenotypes will appear in the progeny in a ratio of 3:1.
- Alleles segregate during the formation of gametes, which thus contain only one allele of each gene. Male and female gametes unite at random at fertilization. These two processes correspond to Mendel's *law of segregation.*
- The segregation of alleles of any one gene is independent of the segregation of the alleles of other genes. This principle is Mendel's *law of independent assortment.* According to this law, crosses between *Aa Bb* dihybrids will generate progeny with a phenotypic ratio of 9 (*A– B–*) : 3 (*A– bb*) : 3 (*aa B–*) : 1 (*aa bb*).
- Most often, the dominant allele of a gene specifies a functional product (a protein), while the recessive allele determines either a less functional or nonfunctional version of the protein, or no protein at all.

2.3 Mendelian Inheritance in Humans

learning objectives

1. Analyze human pedigrees to determine whether a genetic disease exhibits recessive or dominant inheritance.
2. Explain why Huntington disease is inherited as a dominant allele while cystic fibrosis is caused by a recessive allele.

Although many human traits clearly run in families, most do not show a simple Mendelian pattern of inheritance. Suppose, for example, that you have brown eyes, but both your parents' eyes appear to be blue. Because blue is normally considered recessive to brown, does this mean that you are adopted or that your father isn't really your father? Not necessarily, because eye color is influenced by more than one gene.

Like eye color, most common and obvious human phenotypes arise from the interaction of many genes. In contrast, single-gene traits in people usually involve an abnormality that is disabling or life-threatening. Examples are the progressive neurological damage of Huntington disease and the clogged lungs and potential respiratory failure of cystic fibrosis. A defective allele of a single gene gives rise to Huntington disease; defective alleles of a different gene are responsible for cystic fibrosis. **Table 2.1** lists some of the roughly 6000 such single-gene, or Mendelian, traits known in humans as of 2016. As you will see, the allele that causes Huntington disease is dominant and the normal (nondisease) allele of this gene is recessive. The opposite is true for cystic fibrosis—the disease-causing allele is recessive and the normal (nondisease) allele is dominant.

Pedigrees Aid the Study of Hereditary Traits in Human Families

Determining a genetic defect's pattern of transmission is not always an easy task because people make slippery genetic subjects. Their generation time is long, and the families they produce are relatively small, which makes statistical analysis difficult. Humans do not base their choice of mates on purely genetic considerations. Thus, no pure-breeding lines exist and no controlled matings are possible. Furthermore, people rarely produce a true F_2 generation (like the one in which Mendel observed the 3:1 ratios from which he derived his rules) because brothers and sisters almost never mate.

Geneticists circumvent these difficulties by working with a large number of families or with several generations

TABLE 2.1 Some of the Most Common Single-Gene Traits in Humans

Disease	Effect	Incidence of Disease
Caused by a Recessive Allele		
Thalassemia (chromosome 16 or 11)	Reduced amounts of hemoglobin; anemia, bone and spleen enlargement	1/10 in parts of Italy
Sickle-cell anemia (chromosome 11)	Abnormal hemoglobin; sickle-shaped red cells, anemia, blocked circulation; increased resistance to malaria	1/625 African-Americans
Cystic fibrosis (chromosome 7)	Defective cell membrane protein; excessive mucus production; digestive and respiratory failure	1/2000 Caucasians
Tay-Sachs disease (chromosome 15)	Missing enzyme; buildup of fatty deposit in brain that disrupts mental development	1/3000 Eastern European Jews
Phenylketonuria (PKU) (chromosome 12)	Missing enzyme; mental deficiency	1/10,000 Caucasians
Caused by a Dominant Allele		
Hypercholesterolemia (chromosome 19)	Missing protein that removes cholesterol from the blood; heart attack by age 50	1/122 French Canadians
Huntington disease (chromosome 4)	Abnormal Huntingtin protein; progressive mental and neurological damage; neurologic disorders by ages 40–70	1/25,000 Caucasians

of a very large family. In this way, scientists can study the large numbers of genetically related individuals needed to establish the inheritance patterns of specific traits. A family history, known as a **pedigree,** is an orderly diagram of a family's relevant genetic features, extending back to at least both sets of grandparents and preferably through as many additional generations as possible. From systematic pedigree analysis in the light of Mendel's laws, geneticists can tell if a trait is determined by alternative alleles of a single gene and whether a single-gene trait is dominant or recessive. Because Mendel's principles are so simple and straightforward, a little logic can go a long way in explaining how traits are inherited in humans.

Figure 2.21 shows how to interpret a family pedigree diagram. Squares (□) represent males, circles (○) are females, diamonds (◊) indicate that the sex is unspecified. Family members affected by the trait in question are indicated by a filled-in symbol (for example, ■). A single horizontal line connecting a male and a female (□–○) represents a mating; a double connecting line (□=○) designates a **consanguineous mating,** that is, a mating between relatives; and a horizontal line above a series of symbols (○ □ ○) indicates the children of the same parents (a *sibship*) arranged and numbered from left to right in order of their birth. Roman numerals to the left or right of the diagram indicate the generations.

To reach a conclusion about the mode of inheritance of a family trait, human geneticists must use a pedigree that supplies sufficient information. For example, researchers could not determine whether the allele causing the disease depicted at the bottom of Fig. 2.21 is dominant or recessive solely on the basis of the simple pedigree shown. The data are consistent with both possibilities. If the trait is dominant, then the father and the affected son are heterozygotes, while the mother and the unaffected son are homozygotes for the recessive normal allele. If instead the trait is recessive, the father and affected son are homozygotes for the recessive disease-causing allele, while the mother and the unaffected son are heterozygotes.

Several kinds of additional information could help resolve this uncertainty. Human geneticists would particularly want to know the frequency at which the trait in question is found in the population from which the family came. *If the trait is rare in the population, then the allele giving rise to the trait should also be rare, and the most likely hypothesis*

Figure 2.21 Symbols used in pedigree analysis. In the simple pedigree at the bottom, I-1 is the father, I-2 is the mother, and II-1 and II-2 are their sons. The father and the first son are both affected by the disease trait.

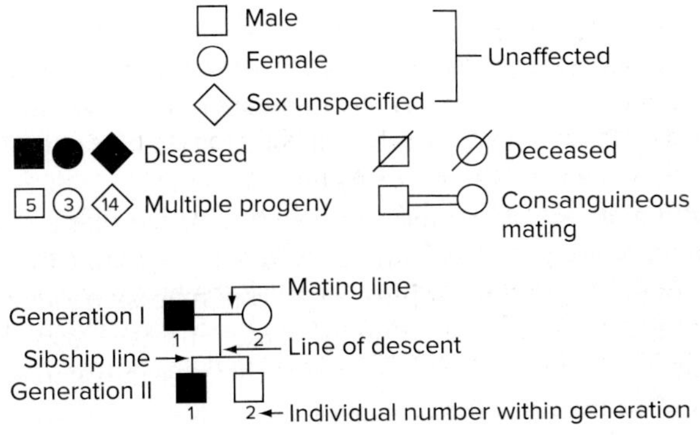

would require that the fewest genetically unrelated people carry the allele. Only the father in Fig. 2.21 would need to have a dominant disease-causing allele, but both parents would need to carry a recessive disease-causing allele (the father two copies and the mother one). However, even the information that the trait is rare does not allow us to draw the firm conclusion that it is inherited in a dominant fashion. The pedigree in the figure is so limited that we cannot be sure the two parents are themselves unrelated. As we discuss later in more detail, related parents might have both received the same rare recessive allele from their common ancestor. This example illustrates why human geneticists try to collect family histories that cover several generations.

We now look at more extensive pedigrees for the dominant trait of Huntington disease and for the recessive condition of cystic fibrosis. The patterns by which these traits appear in the pedigrees provide important clues that can indicate modes of inheritance and allow geneticists to assign genotypes to family members.

Figure 2.22 Huntington disease: A rare dominant trait. All individuals represented by filled-in symbols are heterozygotes (except I-1, who could instead have been homozygous for the dominant *HD* disease allele); all individuals represented by open symbols are homozygotes for the recessive HD^+ normal allele. Among the 14 children of the consanguineous mating, DNA testing shows that some are *HD HD,* some are *HD* HD^+, and some are HD^+ HD^+. The diamond designation masks personal details to protect confidentiality.

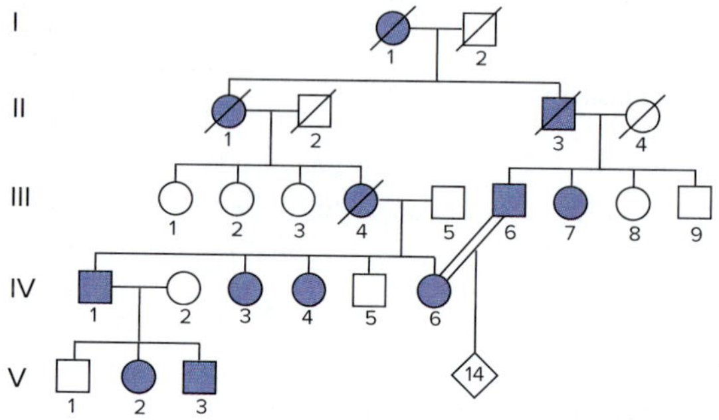

A Vertical Pattern of Inheritance Indicates a Rare Dominant Trait

Huntington disease is named for George Huntington, the New York physician who first described its course. This illness usually shows up in middle age and slowly destroys its victims both mentally and physically. Symptoms include intellectual deterioration, severe depression, and jerky, irregular movements, all caused by the progressive death of nerve cells. If one parent develops the symptoms, his or her children usually have a 50% probability of suffering from the disease, provided they live to adulthood. Because symptoms are not present at birth and manifest themselves only later in life, Huntington disease is known as a **late-onset genetic trait.**

How would you proceed in assigning genotypes to the individuals in the Huntington disease pedigree depicted in **Fig. 2.22**? First, you would need to find out if the disease-producing allele is dominant or recessive. Several clues suggest that Huntington disease is transmitted by a dominant allele of a single gene. Everyone who develops the disease has at least one parent who shows the trait, and in several generations, approximately half of the offspring are affected. The pattern of affected individuals is thus *vertical:* If you trace back through the ancestors of any affected individual, you would see at least one affected person in each generation, giving a continuous line of family members with the disease. When a disease is rare in the population as a whole, a vertical pattern is strong evidence that a dominant allele causes the trait; the alternative would require that many unrelated people carry a rare recessive allele. (A recessive trait that is extremely common might also show up in every generation; we examine this possibility in Problem 40 at the end of this chapter.)

In tracking a dominant allele through a pedigree, you can view every mating between an affected and an unaffected partner as analogous to a testcross. If some of the offspring do not have Huntington disease, you know the parent showing the trait is a heterozygote. As an exercise, you should check your own genotype assignments against the answers in the caption to Fig. 2.22.

Notice also in the legend to Fig. 2.22 that human geneticists use different symbols than Mendel's for alleles of genes. In human genotypes, all alleles are written in uppercase. If the allele specifies a normally functioning gene product, the allele symbol has a superscript +. Alleles that specify no gene product or abnormal gene products sometimes have no superscript at all, as in the Fig. 2.22 legend, but in other cases they have a superscript other than + that signifies a particular type of abnormal allele. (See the Appendix *Guidelines for Gene Nomenclature* for further discussion of genetic notation.)

Like Mendel's pea genes, the gene that causes Huntington disease has been identified and studied at the molecular level. In fact, in 1988 this was the first human disease gene identified molecularly using methods that will be described in Chapter 11. The protein product of the Huntington disease gene, called Huntingtin or Htt, is needed for the proper physiology of nerve cells, but the protein's precise role in these cells is not yet understood. The dominant disease allele (*HD*) specifies a defective Htt protein that over time damages nerve cells (**Fig. 2.23**).

The disease allele is dominant to the normal allele because the presence of the normal Htt protein in heterozygotes does not prevent the abnormal protein from damaging the cells. It is important to note that this explanation for the Huntington disease allele is only one of many different

Figure 2.23 Why the allele for Huntington disease is dominant. People who are *HD HD* or *HD HD*$^+$ exhibit Huntington disease because the *HD* allele produces an abnormal Htt protein that damages nerve cells. Homozygotes for the normal allele (*HD*$^+$ *HD*$^+$) produce only normal Htt protein and do not have the disease. The disease allele (*HD*) is dominant because even when the normal protein is present—in *HD HD*$^+$ heterozygotes—the abnormal protein damages nerve cells. *HD HD* homozygosity is possible because the abnormal Htt protein retains some function of the normal protein.

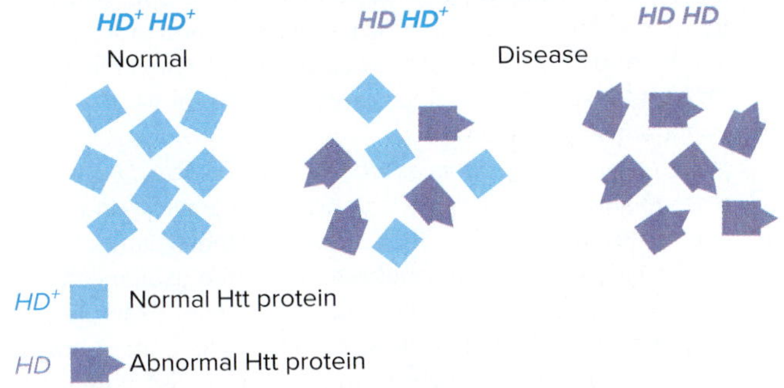

molecular mechanisms that may result in a disease allele that is dominant to the normal allele of a particular gene.

No effective treatment yet exists for Huntington disease, and because of its late onset, there was until the 1980s no way for children of a Huntington parent to know before middle age—usually until well after their own childbearing years—whether they carried the Huntington disease allele (*HD*). Most people with the disease allele are *HD HD*$^+$ heterozygotes, so their children would have a 50% probability of inheriting *HD* and, before they are diagnosed, a 25% probability of passing the defective allele on to one of their children.

In the mid-1980s, with new knowledge of the gene, molecular geneticists developed a DNA test that determines whether an individual carries the *HD* allele. (This test will be explained in detail in Chapter 11.) Because of the lack of effective treatment for the disease, some young adults whose parents died of Huntington disease prefer not to be tested so that they will not learn their own fate prematurely. However, other at-risk individuals employ the test for the *HD* allele to guide their decisions about having children. If someone whose parent had Huntington disease does not have *HD*, he or she has no chance of developing the disease or of transmitting it to offspring. If the test shows the presence of *HD*, the at-risk person and his or her partner might choose to conceive a child using *in vitro* fertilization (IVF) technology (described in Chapter 11) that allows for genotyping of early-stage embryos. Using IVF, only embryos lacking the *HD* disease allele would be introduced into the mother's womb.

The Genetics and Society Box *Developing Guidelines for Genetic Screening* discusses significant social and ethical issues raised by information obtained from family pedigrees and molecular tests.

A Horizontal Pattern of Inheritance Indicates a Rare Recessive Trait

Unlike Huntington disease, most confirmed single-gene diseases in humans are caused by recessive alleles. One reason is that, with the exception of late-onset traits, deleterious dominant alleles are unlikely to be transmitted to the next generation. For example, if people affected with Huntington disease all died by the age of 10, the disease would disappear from the population. In contrast, individuals can carry one allele for a recessive disease without ever being affected by any symptoms.

Figure 2.24 shows three pedigrees for cystic fibrosis, the most commonly inherited recessive disease among Caucasian children in the United States. A double dose of the recessive *CF* allele (meaning the absence of a *CF*$^+$ allele) causes a fatal disorder in which the lungs, pancreas, and other organs become clogged with a thick, viscous mucus that can interfere with breathing and digestion. One in every 2000 white Americans is born with cystic fibrosis, and only 10% of them survive into their 30s.

Note two salient features of the cystic fibrosis pedigrees. First, the family pattern of people showing the trait is often *horizontal:* The parents, grandparents, and great-grandparents of children born with cystic fibrosis do not themselves manifest the disease, while several brothers and sisters in a single generation may. A horizontal pedigree pattern is a strong indication that the trait is recessive. The unaffected parents are heterozygous **carriers:** They bear a dominant

Figure 2.24 Cystic fibrosis: A recessive condition. In **(a)**, the two affected individuals (VI-4 and VII-1) are *CF CF*; that is, homozygotes for the recessive disease allele. Their unaffected parents must be carriers, so V-1, V-2, VI-1, and VI-2 must all be *CF CF*$^+$. Individuals II-2, II-3, III-2, III-4, IV-2, and IV-4 are probably also carriers. We cannot determine which of the founders (I-1 or I-2) was a carrier, so we designate their genotypes as *CF*$^+$–. Because the *CF* allele is relatively rare, it is likely that II-1, II-4, III-1, III-3, IV-1, and IV-3 are *CF*$^+$*CF*$^+$ homozygotes. The genotype of the remaining unaffected people (VI-3, VI-5, and VII-2) is uncertain (*CF*$^+$–). **(b** and **c)** These two families demonstrate horizontal patterns of inheritance. Without further information, the unaffected children in each pedigree must be regarded as having a *CF*$^+$– genotype.

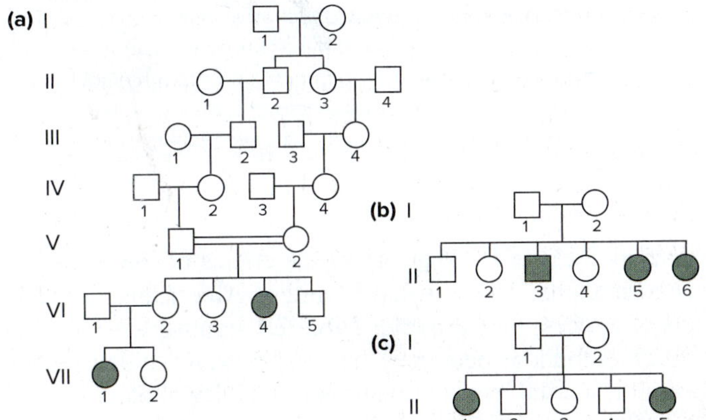

GENETICS AND SOCIETY

Crowd: © Image Source/Getty Images RF

Developing Guidelines for Genetic Screening

In the early 1970s, the United States launched a national screening program for carriers of sickle-cell anemia, a recessive genetic disease that afflicts roughly 1 in 600 African-Americans. The disease is caused by a particular allele, called $Hb\beta^S$, of the β-globin gene; the dominant normal allele is $Hb\beta^A$. The protein determined by the β-globin gene is one component of the oxygen-carrying hemoglobin molecule. $Hb\beta^S$ $Hb\beta^S$ homozygotes have sickle-shaped red blood cells; these patients suffer a decrease in oxygen supply, tire easily, and often develop heart failure from stress on the circulatory system.

The national screening program for sickle-cell anemia was based on a simple test of hemoglobin mobility: Normal and *sickling* hemoglobins move at different rates in a gel. People who participated in the screening program and found they were carriers could use the test results to make informed reproductive decisions. A healthy man, for example, who learned he was a carrier (that is, that he was a $Hb\beta^S$ $Hb\beta^A$ heterozygote) would not have to worry about having an affected child if his mate was a noncarrier.

The original sickle-cell screening program, based on detection of the abnormal hemoglobin protein, was not an unqualified success, largely because of insufficient educational follow-through. Many who learned they were carriers mistakenly thought they had the disease. Moreover, employers and insurance companies that obtained access to the information denied jobs or health insurance to some heterozygotes for no acceptable reason. Problems of public relations and education thus made a reliable screening test into a source of dissent and alienation.

Today, at-risk families may be screened for a growing number of genetic disorders, thanks to the ability to evaluate genotypes directly. The need to establish guidelines for genetic screening thus becomes more and more pressing. Several related questions reveal the complexity of the issue.

- *Why carry out genetic screening at all?* The first reason for screening is to obtain information that will benefit individuals. For example, if you learn at an early age that you have a genetic predisposition to heart disease, you can change your lifestyle to improve your chances of staying healthy. You can also use the results from genetic screening to make informed reproductive decisions.
- The second reason for genetic screening, which often conflicts with the first, is to benefit groups within society. Insurance companies and employers, for example, would like to know who is at risk for various genetic conditions.
- *Should I be screened if a test is available?* For most inherited diseases, no cures presently exist. The psychological burden of anticipating a fatal late-onset disease for which there is no treatment could be devastating, and therefore some people might decide not to be tested. Others may object to testing for religious reasons, or because of confidentiality concerns.
- *If a screening program is established, who should be tested?* The answer depends on what the test is trying to accomplish as well as on its expense. Ultimately, the cost of a procedure must be weighed against the usefulness of the data it provides. In the United States, for example, only one-tenth as many African-Americans as Caucasians are affected by cystic fibrosis, and Asians almost never exhibit the disease. Should all racial groups be tested for cystic fibrosis, or only Caucasians?
- *Should private employers and insurance companies be allowed to test their clients and employees?* Some employers advocate genetic screening to reduce the incidence of occupational disease, arguing that they can use genetic test results to make sure employees are not assigned to environments that might cause them harm. Critics of this position say that screening violates workers' rights, including the right to privacy, increases racial and ethnic discrimination in the workplace, and provides insurers with an excuse to deny coverage. In 2008, President George W. Bush signed into law the Genetic Information Nondiscrimination Act, which prohibits insurance companies and employers in the United States from discriminating on the basis of information derived from genetic tests.
- *Finally, how should people be educated about the meaning of test results?* In one small-community screening program, people identified as carriers of the recessive, life-threatening blood disorder known as β-thalassemia were ostracized; as a result, carriers ended up marrying one another, only making medical matters worse. By contrast, in Ferrara, Italy, where 30 new cases of β-thalassemia had been reported every year, extensive screening combined with education was so successful that the 1980s passed with only a few new cases of the disease.

Given all of these considerations, what kind of guidelines would you like to see established to ensure that genetic screening reaches the right people at the right time, and that information gained from such screening is used for the right purposes?

normal allele (CF^+) that masks the effects of the recessive abnormal one. An estimated 12 million Americans are carriers of a recessive *CF* allele. **Table 2.2** summarizes some of the clues found in pedigrees that can help you decide whether a trait is caused by a dominant or a recessive allele.

The second salient feature of the cystic fibrosis pedigrees is that many of the couples who produce afflicted children are blood relatives; that is, their mating is consanguineous (as indicated by the double line). In Fig. 2.24a, the consanguineous mating in generation V is between third cousins. Of course, children with cystic fibrosis can also have unrelated carrier parents, but because relatives share genes, their offspring have a much greater than average chance of receiving two copies of a rare allele. Whether or

TABLE 2.2 How to Recognize Dominant and Recessive Traits in Pedigrees

Dominant Traits	
1.	Affected children always have at least one affected parent.
2.	Dominant traits show a *vertical pattern* of inheritance: The trait shows up in every generation.
3.	Two affected parents can produce unaffected children, if both parents are heterozygotes.
Recessive Traits	
1.	Affected individuals can be the children of two unaffected carriers, particularly as a result of consanguineous matings.
2.	All the children of two affected parents should be affected.
3.	*Rare* recessive traits show a *horizontal pattern* of inheritance: The trait first appears among several members of one generation and is not seen in earlier generations.
4.	Recessive traits may show a vertical pattern of inheritance if the trait is extremely common in the population.

not they are related, carrier parents are both heterozygotes. Thus among their offspring, the proportion of unaffected to affected children is expected to be 3:1. To look at it another way, the chances are that 1 out of 4 children of two heterozygous carriers will be homozygous cystic fibrosis sufferers.

You can gauge your understanding of this inheritance pattern by assigning a genotype to each person in Fig. 2.24 and then checking your answers against the caption. Note that for several individuals, such as the generation I individuals in part (a) of the figure, it is impossible to assign a full genotype. We know that one of these people must be the carrier who supplied the original *CF* allele, but we do not know if it was the male or the female. As with an ambiguous dominant phenotype in peas, the unknown second allele is indicated by a dash (–).

In Fig. 2.24a, a mating between the unrelated carriers VI-1 and VI-2 produced a child with cystic fibrosis. How likely is such a marriage between unrelated carriers for a recessive genetic condition? The answer depends on the gene in question and the particular population into which a person is born. As Table 2.1 shows, the incidence of genetic diseases (and thus the frequency of their carriers) varies markedly among populations. Such variation reflects the distinct genetic histories of different groups. The area of genetics that analyzes differences among groups of individuals is called *population genetics*, a subject we cover in detail in Chapter 21. Notice that in Fig. 2.24a, several unrelated, unaffected people, such as II-1 and II-4, married into the family under consideration. Although it is highly probable that these individuals are homozygotes for the normal allele of the gene (CF^+CF^+), a small chance (whose magnitude depends on the population) exists that any one of them could be a carrier of the disease.

Genetic researchers identified the cystic fibrosis gene in 1989, soon after the Huntington disease gene was identified. The normal, dominant CF^+ allele makes a protein called cystic fibrosis transmembrane conductance regulator (CFTR). CFTR protein forms a channel in the cell membranes that controls the flow of chloride ions through lung cells. Recessive *CF* disease alleles either produce no CFTR or produce nonfunctional or less functional versions of the protein (**Fig. 2.25**). Because of osmosis, water flows into lung cells without CFTR, while a thick, dehydrated

Figure 2.25 Why the allele for cystic fibrosis is recessive. The CFTR protein regulates the passage of chloride ions (*green spheres*) through the cell membrane. People who are homozygous for a cystic fibrosis disease allele (*CF CF*) have the disease because recessive disease alleles either specify no CFTR protein as shown, or encode abnormal CFTR proteins that do not function at all or do not function as well as the normal protein (*not shown*). Disease alleles (*CF*) are recessive because *CF CF*$^+$ heterozygotes produce CFTR from the normal (*CF*$^+$) allele, and this amount of CFTR is sufficient for normal lung function.

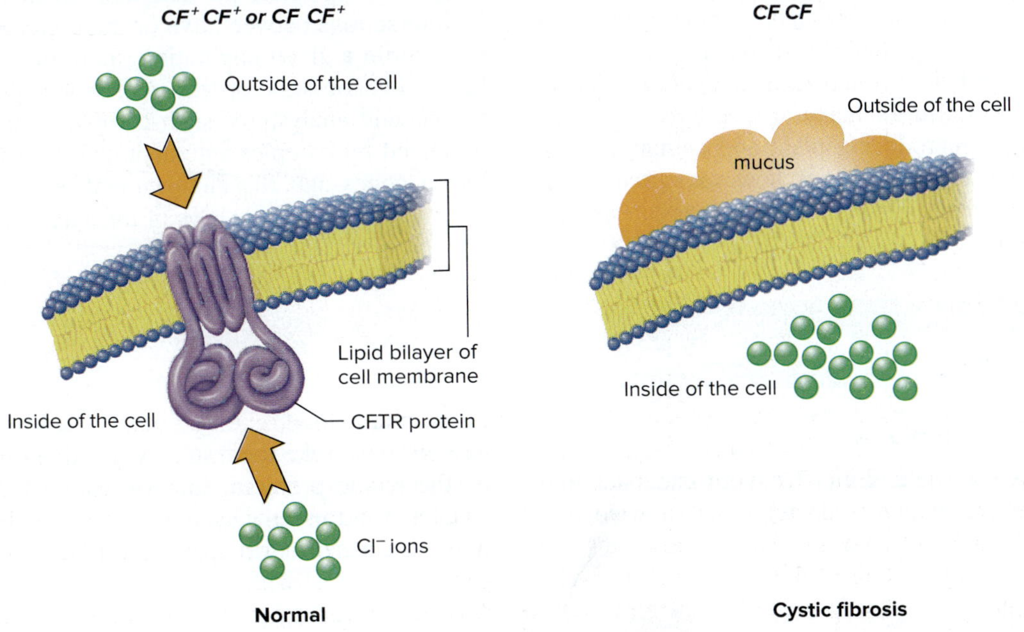

mucus builds up outside the cells. Thus, *CF CF* homozygotes have no functional CFTR (or not enough of this protein) and exhibit cystic fibrosis. Gene therapy—insertion of a normal CF^+ gene into lung cells of patients—has been tried to ameliorate the disease's debilitating symptoms, but so far without success.

Despite the failure to date of gene therapy, identification of the gene responsible for cystic fibrosis has very recently led to effective treatments for the disease in patients with particular mutant alleles. For example, in 2015 the U.S. Food and Drug Administration approved a drug cocktail called Orkambi® that helps the particular defective form of CFTR specified by one of these alleles to function properly. Varied approaches to the treatment of cystic fibrosis and other inherited diseases will be discussed later in the book.

essential concepts

- In a *vertical* pattern of transmission, a trait that appears in an affected individual also appears in at least one parent, one of the affected parent's parents, and so on. If a trait is rare, a pedigree with a vertical pattern usually indicates that the disease-causing allele is dominant.
- In a *horizontal* pattern of transmission, a trait that appears in an affected individual may not appear in any ancestors, but it may appear in some of the person's siblings. A pedigree with a horizontal pattern usually indicates a rare recessive disease-causing allele. Affected individuals are often products of consanguineous mating.
- Various kinds of biochemical events may explain why some disease alleles are dominant. In the case of Huntington disease, the disease-causing *HD* allele specifies an abnormal, deleterious version of the protein produced by the normal, recessive allele.
- Recessive disease alleles, like the *CF* alleles that cause cystic fibrosis, usually specify either no protein or less-functional versions of the protein that the normal, dominant allele produces.

WHAT'S NEXT

Mendel answered the three basic questions about heredity as follows: To *What is inherited?* he replied, "alleles of genes." To *How is it inherited?* he responded, "according to the principles of segregation and independent assortment." And to *What is the role of chance in heredity?* he said, "for each individual, inheritance is determined by chance, but within a population, this chance operates in a context of strictly defined probabilities."

Within a decade of the 1900 rediscovery of Mendel's work, numerous breeding studies had shown that Mendel's laws hold true not only for seven pairs of antagonistic characteristics in peas, but also for many diverse traits in a wide variety of sexually reproducing plant and animal species. Some of these same breeding studies, however, raised a challenge to the new genetics. For certain traits in certain species, the studies uncovered unanticipated phenotypic ratios, or the results included F_1 and F_2 progeny with novel phenotypes that resembled those of neither pure-breeding parent.

These phenomena could not be explained by Mendel's hypothesis that for each gene, two alternative alleles, one completely dominant, the other recessive, determine a single trait. We now know that most common traits, including skin color, eye color, and height in humans, are determined by interactions between two or more genes. We also know that within a given population, more than two alleles may be present for some of those genes. Chapter 3 shows how the genetic analysis of such complex traits, that is, traits produced by complex interactions between genes and between genes and the environment, extended rather than contradicted Mendel's laws of inheritance.

SOLVED PROBLEMS

Solving Genetics Problems

The best way to evaluate and increase your understanding of the material in the chapter is to apply your knowledge in solving genetics problems. Genetics word problems are like puzzles. Take them in slowly—don't be overwhelmed by the whole problem. Identify useful facts given in the problem, and use the facts to deduce additional information. Use genetic principles and logic to work toward the

DNA: © Design Pics/Bilderbuch RF

solutions. The more problems you do, the easier they become. In solving problems, you will not only solidify your understanding of genetic concepts, but you will also develop basic analytical skills that are applicable in many disciplines.

Note that some of the problems at the end of each chapter are designed to introduce supplementary but important concepts that expand on the information in the text. You can nonetheless answer such problems using logical inferences from your reading.

Solving genetics problems requires much more than simply plugging numbers into formulas. Each problem is unique and requires thoughtful evaluation of the information given and the question being asked. The following are general guidelines you can follow in approaching these word problems:

a. Read through the problem once to get some sense of the concepts involved.

b. Go back through the problem, noting all the information supplied to you. For example, genotypes or phenotypes of offspring or parents may be given to you or implied in the problem. Represent the known information in a symbolic format—assign symbols for alleles; use these symbols to indicate genotypes; make a diagram of the crosses including genotypes and phenotypes given or implied. Be sure that you do not assign different letters of the alphabet to two alleles of the same gene, as this can cause confusion. Also, be careful to discriminate clearly between the upper- and lowercases of letters, such as *C*(*c*) or *S*(*s*).

c. Now, reassess the question and work toward the solution using the information given. Make sure you answer the question being asked!

d. When you finish the problem, check to see that the answer makes sense. You can often check solutions by working backwards; that is, see if you can reconstruct the data from your answer.

e. After you have completed a question and checked your answer, spend a minute to think about which major concepts were involved in the solution. This is a critical step for improving your understanding of genetics.

For each chapter, the logic involved in solving two or three types of problems is described in detail.

I. In cats, white patches are caused by the dominant allele *P*, while *pp* individuals are solid-colored. Short hair is caused by a dominant allele *S*, while *ss* cats have long hair. A long-haired cat with patches whose mother was solid-colored and short-haired mates with a short-haired, solid-colored cat whose mother was long-haired and solid-colored. What kinds of kittens can arise from this mating, and in what proportions?

Answer

The solution to this problem requires an understanding of dominance/recessiveness, gamete formation, and the independent assortment of alleles of two genes in a cross.

First make a representation of the known information:

Mothers:	solid, short-haired		solid, long-haired
Cross:	cat 1 patches, long-haired	×	cat 2 solid, short-haired

What genotypes can you assign? Any cat showing a recessive phenotype must be homozygous for the recessive allele. Therefore the long-haired cats are *ss;* solid cats are *pp*. Cat 1 is long-haired, so it must be homozygous for the recessive allele (*ss*). This cat has the dominant phenotype of patches and could be either *PP* or *Pp,* but because the mother was *pp* and could only contribute a *p* allele in her gametes, cat 1 must be *Pp*. Cat 1's full genotype is *Pp ss*. Similarly, cat 2 is solid-colored, so it must be homozygous for the recessive allele (*pp*). Because this cat is short-haired, it could have either the *SS* or *Ss* genotype. Its mother was long-haired (*ss*) and could only contribute an *s* allele in her gamete, so cat 2 must be heterozygous *Ss*. The full genotype is *pp Ss*.

The cross is therefore between *Pp ss* (cat 1) and *pp Ss* (cat 2). To determine the types of kittens, first establish the types of gametes that can be produced by each cat and then set up a Punnett square to determine the genotypes of the offspring. Cat 1 (*Pp ss*) produces *Ps* and *ps* gametes in equal proportions. Cat 2 (*pp Ss*) produces *pS* and *ps* gametes in equal proportions. Four types of kittens can result from this mating with equal probability: *Pp Ss* (patches, short-haired), *Pp ss* (patches, long-haired), *pp Ss* (solid, short-haired), and *pp ss* (solid, long-haired).

Cat 2 \ Cat 1	*P s*	*p s*
p S	*Pp Ss*	*pp Ss*
p s	*Pp ss*	*pp ss*

The following table demonstrates that you could also work through this problem using the product rule of probability instead of a Punnett square. The principles are the same: Gametes produced in equal amounts by either parent are combined at random.

Cat 1 gamete		**Cat 2 gamete**	**Progeny**
1/2 *P s*	×	1/2 *p S*	1/4 *Pp Ss* patches, short-haired
1/2 *P s*	×	1/2 *p s*	1/4 *Pp ss* patches, long-haired
1/2 *P s*	×	1/2 *p S*	1/4 *pp Ss* solid-colored, short-haired
1/2 *P s*	×	1/2 *p s*	1/4 *pp ss* solid-colored, long-haired

II. In tomatoes, red fruit is dominant to yellow fruit, and purple stems are dominant to green stems. The progeny from one mating consisted of 305 red fruit, purple stem plants; 328 red fruit, green stem plants; 110 yellow fruit, purple stem plants; and 97 yellow fruit, green stem plants. What were the genotypes of the parents in this cross?

Answer

This problem requires an understanding of independent assortment in a dihybrid cross as well as the ratios predicted from monohybrid crosses.

Designate the alleles:

R = red, r = yellow

P = purple stems, p = green stems

In genetics problems, the ratios of offspring can indicate the genotype of parents. You will usually need to total the number of progeny and approximate the ratio of offspring in each of the different classes. For this problem, in which the inheritance of two traits is given, consider each trait independently. For red fruit, there are 305 + 328 = 633 red-fruited plants out of a total of 840 plants. This value (633/840) is close to 3/4. About 1/4 of the plants have yellow fruit (110 + 97 = 207/840). From Mendel's work, you know that a 3:1 phenotypic ratio results from crosses between plants that are hybrid (heterozygous) for one gene. Therefore, the genotype for fruit color of each parent must have been *Rr*.

For stem color, 305 + 110 or 415/840 plants had purple stems. About half had purple stems, and the other half (328 + 97) had green stems. A 1:1 phenotypic ratio occurs when a heterozygote is mated to a homozygous recessive (as in a testcross). The parents' genotypes must have been *Pp* and *pp* for stem color.

The complete genotype of the parent plants in this cross was *Rr Pp* × *Rr pp*.

III. Tay-Sachs is a recessive lethal disease in which there is neurological deterioration early in life. This disease is rare in the population overall but is found at relatively high frequency in Ashkenazi Jews from Eastern Europe. A woman whose maternal uncle had the disease is trying to determine the probability that she and her husband could have an affected child. Her father does not come from a high-risk population. Her husband's sister died of the disease at an early age.

a. Draw the pedigree of the individuals described. Include the genotypes where possible.

b. Determine the probability that the couple's first child will be affected.

Answer

This problem requires an understanding of dominance/recessiveness and probability. First diagram the pedigree, and then assign as many genotypes as possible using the following allele designations:

T = normal allele; *t* = Tay-Sachs allele

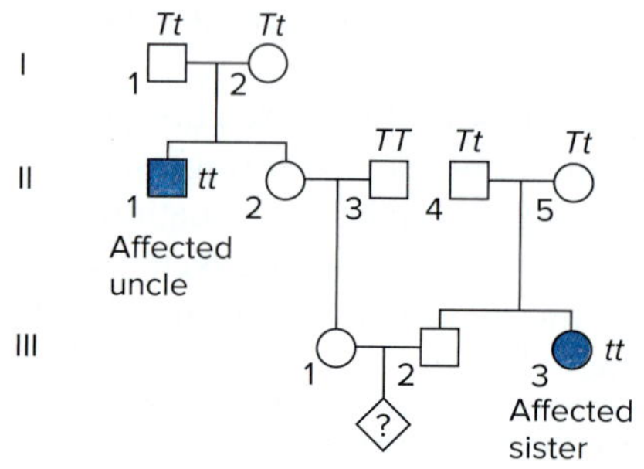

The genotypes of the two affected individuals, the woman's uncle (II-1) and the husband's sister (III-3), are *tt*. Because the uncle was affected, both of his parents must have been heterozygous.

Similarly, as the husband's sister (III-3) is affected, both of her parents (II-4 and II-5) must be heterozygotes. Finally, because individual II-3 is not from a high-risk population, the most likely assumption is that he is *TT*.

You next need to determine the chance that a child of III-1 and III-2 (that is, individual IV-1) would have Tay-Sachs (*tt*). For that to be possible, both III-1 and III-2 must be *Tt* given that neither is *tt*. For III-1 to be *Tt*, II-2 must be *Tt*. Calculating the chance that II-2 is *Tt* is a bit tricky. At first, it appears that the chance is 1/2 that the daughter of two heterozygous (*Tt*) parents would be *Tt*: the expected progeny ratio is 1 *TT* : 2 *Tt* : 1 *tt*. However, in this case you have additional information to consider: II-2 is unaffected and thus the genotype *tt* is ruled out. That leaves 1 *TT* : 2 *Tt*, or a 2/3 chance that II-2 is *Tt*. If so, the chance that II-2 would transmit the *t* allele to III-1 is 1/2. Thus, the probability that III-1 is *Tt* is 2/3 × 1/2 = 1/3. This fact implies that II-2 could be either *TT* (probability = 1/3) or *Tt* (probability = 2/3). If II-2 is *Tt,* the chance that she would transmit the *t* allele to III-1 is 1/2. Thus, the probability that III-1 is *Tt* is 2/3 × 1/2 = 1/3.

What is the chance that III-2 is *Tt*? Both of his parents are heterozygous, and he is unaffected. Thus, using similar logic, the likelihood that III-2 is *Tt* is 2/3.

The probability that both III-1 and III-2 are *Tt* is 1/3 × 2/3 = 2/9. The chance that the child of two *Tt* parents would be *tt* is 1/4. Thus, the overall likelihood that IV-1, the child of III-1 and III-2, would have Tay-Sachs is 2/9 × 1/4 = 1/18.

PROBLEMS

Vocabulary

1. For each of the terms in the left column, choose the best matching phrase in the right column.

a. phenotype	1. having two identical alleles of a given gene
b. alleles	2. the allele expressed in the phenotype of the heterozygote
c. independent assortment	3. alternate forms of a gene
d. gametes	4. observable characteristic
e. gene	5. a cross between individuals both heterozygous for two genes
f. segregation	6. alleles of one gene separate into gametes randomly with respect to alleles of other genes
g. heterozygote	7. reproductive cells containing only one copy of each gene
h. dominant	8. the allele that does not contribute to the phenotype of the heterozygote
i. F_1	9. the cross of an individual of ambiguous genotype with a homozygous recessive individual
j. testcross	10. an individual with two different alleles of a gene
k. genotype	11. the heritable entity that determines a characteristic
l. recessive	12. the alleles an individual has
m. dihybrid cross	13. the separation of the two alleles of a gene into different gametes
n. homozygote	14. offspring of the P generation

Section 2.1

2. During the millennia in which selective breeding was practiced, why did breeders fail to uncover the principle that traits are governed by discrete units of inheritance (that is, by genes)?

3. Describe the characteristics of the garden pea that made it a good organism for Mendel's analysis of the basic principles of inheritance. Evaluate how easy or difficult it would be to make a similar study of inheritance in humans by considering the same attributes you described for the pea.

Section 2.2

4. An albino corn snake is crossed with a normal-colored corn snake. The offspring are all normal-colored. When these first-generation progeny snakes are crossed among themselves, they produce 32 normal-colored snakes and 10 albino snakes.
   a. How do you know that only a single gene is responsible for the color differences between these snakes?
   b. Which of these phenotypes is controlled by the dominant allele?
   c. A normal-colored female snake is involved in a testcross. This cross produces 10 normal-colored and 11 albino offspring. What are the genotypes of the parents and the offspring?

5. Two short-haired cats mate and produce six short-haired and two long-haired kittens. What does this information suggest about how hair length is inherited?

6. Piebald spotting is a condition found in humans in which there are patches of skin that lack pigmentation. The condition results from the inability of pigment-producing cells to migrate properly during development. Two adults with piebald spotting have one child who has this trait and a second child with normal skin pigmentation.
   a. Is the piebald spotting trait dominant or recessive? What information led you to this answer?
   b. What are the genotypes of the parents?

7. As a *Drosophila* research geneticist, you keep stocks of flies of specific genotypes. You have a fly that has normal wings (dominant phenotype). Flies with short wings are homozygous for a recessive allele of the wing-length gene. You need to know if this fly with normal wings is pure-breeding or heterozygous for the wing-length trait. What cross would you do to determine the genotype, and what results would you expect for each possible genotype?

8. A mutant cucumber plant has flowers that fail to open when mature. Crosses can be done with this plant by manually opening and pollinating the flowers with pollen from another plant. When closed × open crosses were done, all the F_1 progeny were open. The F_2 plants were 145 open and 59 closed. A cross of closed × F_1 gave 81 open and 77 closed. How is the closed trait inherited? What evidence led you to your conclusion?

9. In a particular population of mice, certain individuals display a phenotype called *short tail,* which is inherited as a dominant trait. Some individuals display a recessive trait called *dilute,* which affects coat color. Which of these traits would be easier to eliminate from the population by selective breeding? Why?

10. In humans, a dimple in the chin is a dominant characteristic controlled by a single gene.
    a. A man who does not have a chin dimple has children with a woman with a chin dimple whose

mother lacked the dimple. What proportion of their children would be expected to have a chin dimple?

b. A man with a chin dimple and a woman who lacks the dimple produce a child who lacks a dimple. What is the man's genotype?

c. A man with a chin dimple and a nondimpled woman produce eight children, all having the chin dimple. Can you be certain of the man's genotype? Why or why not? What genotype is more likely, and why?

11. Some inbred strains of the weedy plant *Arabidopsis thaliana* flower early in the growing season, but other strains flower at later times. Four different *Arabdiposis* plants (1–4) were crossed, and the resulting progeny were tabulated as follows:

Mating	Progeny
1 × 2	77 late : 81 early
1 × 3	134 late
1 × 4	93 late : 32 early
2 × 3	111 late
2 × 4	65 late : 61 early
3 × 4	126 late

a. Explain the genetic basis for the difference in flowering time. How do you know that among this group of plants, the flowering time trait is influenced by the action of a single gene? Which allele is dominant and which recessive?

b. Ascribe genotypes to the four plants.

c. What kinds of progeny would you expect if you allowed plants 1–4 to self-fertilize, and in what ratios?

12. Among Native Americans, two types of earwax (cerumen) are seen, dry and sticky. A geneticist studied the inheritance of this trait by observing the types of offspring produced by different kinds of matings. He observed the following numbers:

		Offspring	
Parents	Number of mating pairs	Sticky	Dry
Sticky × sticky	10	32	6
Sticky × dry	8	21	9
Dry × dry	12	0	42

a. How is earwax type inherited?

b. Why are no 3:1 or 1:1 ratios present in the data shown in the chart?

13. Imagine you have just purchased a black stallion of unknown genotype. You mate him to a red mare, and she delivers twin foals, one red and one black. Can you tell from these results how color is inherited, assuming that alternative alleles of a single gene are involved? What crosses could you do to determine how color is inherited?

14. If you roll a die (singular of dice), what is the probability you will roll: (a) a 6? (b) an even number? (c) a number divisible by 3? (d) If you roll a pair of dice, what is the probability that you will roll two 6s? (e) an even number on one and an odd number on the other? (f) matching numbers? (g) two numbers both over 4?

15. In a standard deck of playing cards, four suits exist (red suits = hearts and diamonds, black suits = spades and clubs). Each suit has 13 cards: Ace (A), 2, 3, 4, 5, 6, 7, 8, 9, 10, and the face cards Jack (J), Queen (Q), and King (K). In a single draw, what is the probability that you will draw a face card? A red card? A red face card?

16. How many genetically different eggs could be formed by women with the following genotypes?

a. *Aa bb CC DD*

b. *AA Bb Cc dd*

c. *Aa Bb cc Dd*

d. *Aa Bb Cc Dd*

17. What is the probability of producing a child that will phenotypically resemble either one of the two parents in the following four crosses? How many phenotypically different kinds of progeny could potentially result from each of the four crosses?

a. *Aa Bb Cc Dd* × *aa bb cc dd*

b. *aa bb cc dd* × *AA BB CC DD*

c. *Aa Bb Cc Dd* × *Aa Bb Cc Dd*

d. *aa bb cc dd* × *aa bb cc dd*

18. A mouse sperm of genotype *a B C D E* fertilizes an egg of genotype *a b c D e*. What are all the possibilities for the genotypes of (a) the zygote and (b) a sperm or egg produced by the mouse that develops from this fertilization?

19. Your friend is pregnant with triplets. She thinks that it is equally likely that she will be the mother of 3 sons, 3 daughters, 2 sons and 1 daughter, or 1 son and 2 daughters. Is she correct? Explain. (Assume that each of the triplets is from a separate fertilization, and that boys and girls are equally likely.)

20. Galactosemia is a recessive human disease that is treatable by restricting lactose and glucose in the diet. Susan Smithers and her husband are both heterozygous for the galactosemia gene.

a. Susan is pregnant with twins. If she has fraternal (nonidentical) twins, what is the probability both of the twins will be girls who have galactosemia?

b. If the twins are identical, what is the probability that both will be girls and have galactosemia?

For parts (c–g), assume that none of the children is a twin.

c. If Susan and her husband have four children, what is the probability that none of the four will have galactosemia?

d. If the couple has four children, what is the probability that at least one child will have galactosemia?

e. If the couple has four children, what is the probability that the first two will have galactosemia and the second two will not?

f. If the couple has three children, what is the probability that two of the children will have galactosemia and one will not, regardless of order?

g. If the couple has four children with galactosemia, what is the probability that their next child will have galactosemia?

21. Albinism is a condition in which pigmentation is lacking. In humans, the result is white hair, nonpigmented skin, and pink eyes. The trait in humans is caused by a recessive allele. Two normal parents have an albino child. What are the parents' genotypes? What is the probability that the next child will be albino?

22. A cross between two pea plants, both of which grew from yellow round seeds, gave the following numbers of seeds: 156 yellow round and 54 yellow wrinkled. What are the genotypes of the parent plants? (Yellow and round are dominant traits.)

23. A third-grader decided to breed guinea pigs for her school science project. She went to a pet store and bought a male with smooth black fur and a female with rough white fur. She wanted to study the inheritance of those features and was sorry to see that the first litter of eight contained only rough black animals. To her disappointment, the second litter from those same parents contained seven rough black animals. Soon the first litter had begun to produce F_2 offspring, and they showed a variety of coat types. Before long, the child had 125 F_2 guinea pigs. Eight of them had smooth white coats, 25 had smooth black coats, 23 were rough and white, and 69 were rough and black.

a. How are the coat color and texture characteristics inherited? What evidence supports your conclusions?

b. What phenotypes and proportions of offspring should the girl expect if she mates one of the smooth white F_2 females to an F_1 male?

24. The self-fertilization of an F_1 pea plant produced from a parent plant homozygous for yellow and wrinkled seeds and a parent homozygous for green and round seeds resulted in a pod containing seven F_2 peas. (Yellow and round are dominant.) What is the probability that all seven peas in the pod are yellow and round?

25. The achoo syndrome (sneezing in response to bright light) and trembling chin (triggered by anxiety) are both dominant traits in humans.

a. What is the probability that the first child of parents who are heterozygous for both the achoo gene and trembling chin will have achoo syndrome but lack the trembling chin?

b. What is the probability that the first child will have neither achoo syndrome nor trembling chin?

26. A pea plant from a pure-breeding strain that is tall, has green pods, and has purple flowers that are terminal is crossed to a plant from a pure-breeding strain that is dwarf, has yellow pods, and has white flowers that are axial. The F_1 plants are all tall and have purple axial flowers as well as green pods.

a. What phenotypes do you expect to see in the F_2?

b. What phenotypes and ratios would you predict in the progeny from crossing an F_1 plant to the dwarf parent?

27. The following table shows the results of different matings between jimsonweed plants that had either purple or white flowers and spiny or smooth pods. Determine the dominant allele for the two traits and indicate the genotypes of the parents for each of the crosses.

Parents	Offspring			
	Purple Spiny	White Spiny	Purple Smooth	White Smooth
a. purple spiny × purple spiny	94	32	28	11
b. purple spiny × purple smooth	40	0	38	0
c. purple spiny × white spiny	34	30	0	0
d. purple spiny × white spiny	89	92	31	27
e. purple smooth × purple smooth	0	0	36	11
f. white spiny × white spiny	0	45	0	16

28. A pea plant heterozygous for plant height, pod shape, and flower color was selfed. The progeny consisted of 272 tall, inflated pods, purple flowers; 92 tall, inflated, white flowers; 88 tall, flat pods, purple; 93 dwarf, inflated, purple; 35 tall, flat, white; 31 dwarf, inflated, white; 29 dwarf, flat, purple; 11 dwarf, flat, white. Which alleles are dominant in this cross?

29. In the fruit fly *Drosophila melanogaster,* the following genes and mutations are known:

Wing size: recessive allele for tiny wings *t;* dominant allele for normal wings *T.*

Eye shape: recessive allele for narrow eyes *n;* dominant allele for normal (oval) eyes *N.*

For each of the four following crosses, give the genotypes of each of the parents.

	Male			Female		
	Wings	**Eyes**		**Wings**	**Eyes**	**Offspring**
1	tiny	oval	×	tiny	oval	78 tiny wings, oval eyes 24 tiny wings, narrow eyes
2	normal	narrow	×	tiny	oval	45 normal wings, oval eyes 40 normal wings, narrow eyes 38 tiny wings, oval eyes 44 tiny wings, narrow eyes
3	normal	narrow	×	normal	oval	35 normal wings, oval eyes 29 normal wings, narrow eyes 10 tiny wings, oval eyes 11 tiny wings, narrow eyes
4	normal	narrow	×	normal	oval	62 normal wings, oval eyes 19 tiny wings, oval eyes

30. Based on the information you discovered in the previous problem, answer the following:
    a. A female fruit fly with genotype *Tt nn* is mated to a male of genotype *Tt Nn*. What is the probability that any one of their offspring will have normal phenotypes for both characters?
    b. What phenotypes would you expect among the offspring of this cross? If you obtained 200 progeny, how many of each phenotypic class would you expect?
31. Considering the yellow and green pea color phenotypes studied by Gregor Mendel:
    a. What is the biochemical function of the protein that is specified by the gene responsible for the pea color phenotype?
    b. *A null allele* of a gene is an allele that does not specify any of the biochemical function that the gene normally provides. Of the two alleles *Y* and *y*, which is more likely to be a null allele?
    c. In terms of the underlying biochemistry, why is the *Y* allele dominant to the *y* allele?
    d. Why are peas that are *yy* homozygotes green?
    e. The amount of the protein specified by a gene is roughly proportional to the number of functional copies of the gene carried by a cell or individual. What do the phenotypes of *YY* homozygotes, *Yy* heterozygotes, and *yy* homozygotes tell us about the amount of the Sgr enzyme (the product of the pea color gene) needed to produce a yellow color?
    f. The Sgr enzyme is not needed for the survival of a pea plant, but the genomes of organisms contain many so-called *essential genes* needed for an individual's survival. For such genes, heterozygotes for the normal allele and the null allele survive, but individuals homozygous for the null allele die soon after the male and female gametes, each with a null allele, come together at fertilization. In light of your answer to part (e), what does this fact tell you about the advantage to an organism of having two copies of their genes?
    g. Do you think that a single pea pod could contain peas with different phenotypes? Explain.
    h. Do you think that a pea pod could be of one color (say, green) while the peas within the pod could be of a different color (say, yellow)? Explain.
32. What would have been the outcome (the genotypic and phenotypic ratios) in the F_2 of Mendel's dihybrid cross shown in Fig. 2.15 if the alleles of the pea color gene (*Y*, *y*) and the pea shape gene (*R*, *r*) did not assort independently and instead the alleles inherited from a parent always stayed together as a unit?
33. Recall that Mendel obtained pure-breeding plants with either long or short stems and that hybrids had long stems (Fig. 2.8). Monohybrid crosses produced an F_2 generation with a 3:1 ratio of long stems to short stems, indicating that this difference in stem length is governed by a single gene. The gene that likely controlled this trait in Mendel's plants has been discovered, and it specifies an enzyme called G3βH, which catalyzes the reaction shown in the accompanying figure. The product of the reaction, gibberellin, is a growth hormone that makes plants grow tall. What is the most likely hypothesis to explain the difference between the dominant allele (*L*) and the recessive allele (*l*)?

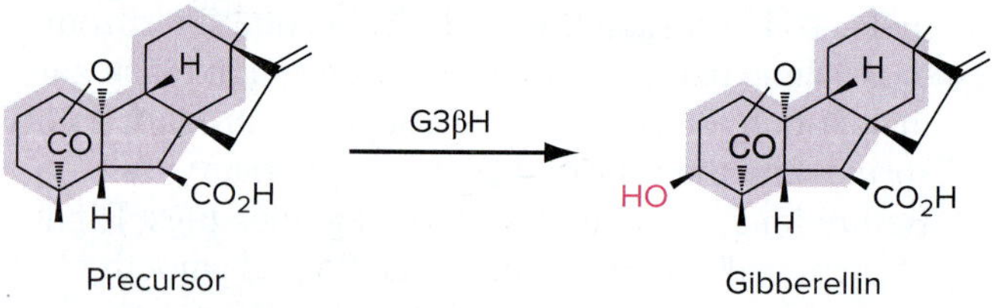

34. The gene that likely controlled flower color (purple or white) in Mendel's pea plants has also been identified. The flower color gene specifies a protein

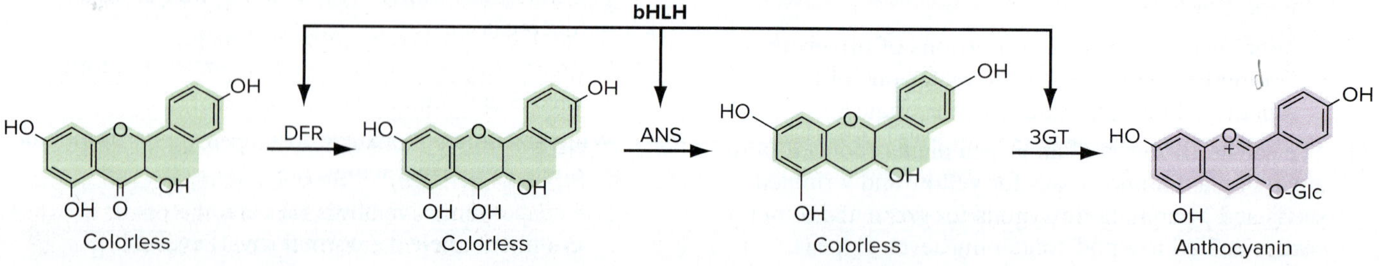

called bHLH required by cells to make three different enzymes (DFR, ANS, and 3GT) that function in the pathway shown in the accompanying figure, leading to synthesis of the purple pigment anthocyanin.

a. What is the most likely explanation for the difference between the dominant allele (*P*) and the recessive allele (*p*) of the gene responsible for these flower colors?

b. Given the biochemical pathway shown, could a different gene have been the one governing Mendel's flower colors?

Section 2.3

35. For each of the following human pedigrees, indicate whether the inheritance pattern is recessive or dominant. What feature(s) of the pedigree did you use to determine the mode of inheritance? Give the genotypes of affected individuals and of individuals who carry the disease allele but are not affected.

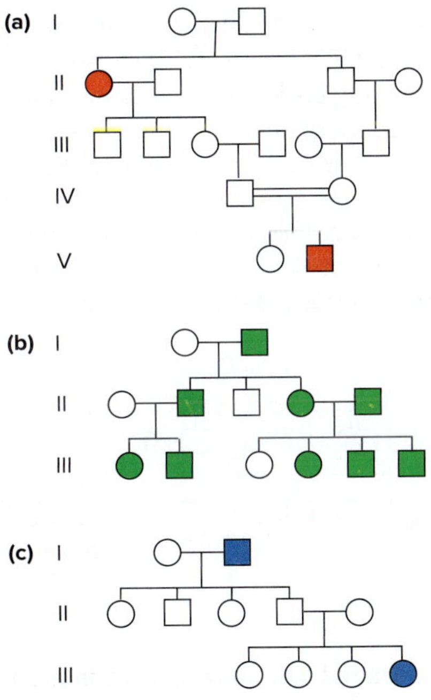

36. Consider the pedigree that follows for cutis laxa, a connective tissue disorder in which the skin hangs in loose folds.

a. Assuming that the trait is rare, what is the apparent mode of inheritance?

b. What is the probability that individual II-2 is a carrier?

c. What is the probability that individual II-3 is a carrier?

d. What is the probability that individual III-1 is affected by the disease?

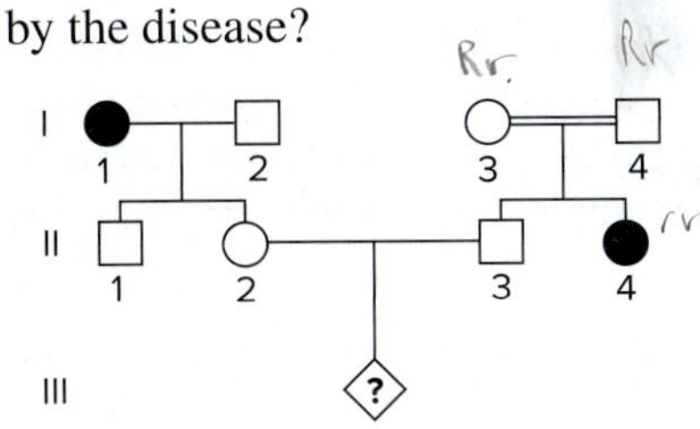

37. A young couple went to see a genetic counselor because each had a sibling with cystic fibrosis. (Cystic fibrosis is a recessive disease, and neither member of the couple nor any of their four parents is affected.)

a. What is the probability that the female of this couple is a carrier?

b. What are the chances that their child will have cystic fibrosis?

c. What is the probability that their child will be a carrier of the cystic fibrosis disease allele?

38. Huntington disease is a rare fatal, degenerative neurological disease in which individuals start to show symptoms in their 40s. It is caused by a dominant allele. Joe, a man in his 20s, just learned that his father has Huntington disease.

a. What is the probability that Joe will also develop the disease?

b. Joe and his new wife have been eager to start a family. What is the probability that their first child will eventually develop the disease?

39. Is the disease shown in the following pedigree caused by a dominant or a recessive allele? Why? Based on this limited pedigree, do you think the disease allele is rare or common in the population? Why?

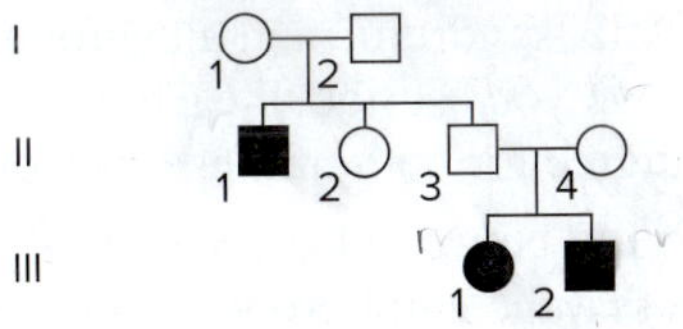

40. Figure 2.22 shows the inheritance of Huntington disease in a family from a small village near Lake Maracaibo in Venezuela. The village was founded by a small number of immigrants, and generations of their descendants have remained concentrated in this isolated location. The allele for Huntington disease has remained unusually prevalent there.

a. Why could you not conclude definitively that the disease is the result of a dominant or a recessive allele solely by looking at this pedigree?

b. Is there any information you could glean from the family's history that might imply the disease is due to a dominant rather than a recessive allele?

41. Consider the cystic fibrosis pedigree in Figure 2.24a.
    a. Assuming that one of the individuals in generation I was a carrier, and that no one from outside the family was a carrier, what was the probability that any single child of the consanguineous couple in generation V would have cystic fibrosis? (Assume that none of their children is born yet, so you don't know that VI-4 has the disease.)
    b. Assuming that one of the individuals in generation I was a carrier and that 1/1000 people in the population is a carrier, and knowing that VI-4 has the disease, how likely was it that VII-1 would be affected?

42. The common grandfather of two first cousins has hereditary hemochromatosis, a recessive condition causing an abnormal buildup of iron in the body. Neither of the cousins has the disease nor do any of their relatives.
    a. If the first cousins had a child, what is the chance that the child would have hemochromatosis? Assume that the unrelated, unaffected parents of the cousins are not carriers.
    b. How would your calculation change if you knew that 1 out of every 10 unaffected people in the population (including the unrelated parents of these cousins) was a carrier for hemochromatosis?

43. People with nail-patella syndrome have poorly developed or absent kneecaps and nails. Individuals with alkaptonuria have arthritis as well as urine that darkens when exposed to air. Both nail-patella syndrome and alkaptonuria are rare phenotypes. In the following pedigree, vertical red lines indicate individuals with nail-patella syndrome, while horizontal green lines denote individuals with alkaptonuria.
    a. What are the most likely modes of inheritance of nail-patella syndrome and alkaptonuria? What genotypes can you ascribe to each of the individuals in the pedigree for both of these phenotypes?
    b. In a mating between IV-2 and IV-5, what is the chance that the child produced would have both nail-patella syndrome and alkaptonuria? Nail-patella syndrome alone? Alkaptonuria alone? Neither defect?

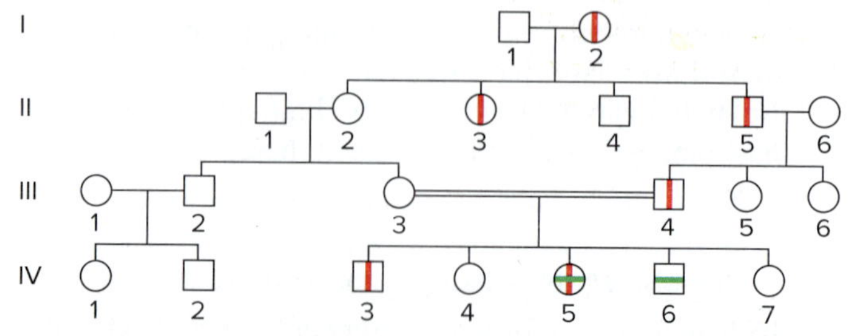

44. Midphalangeal hair (hair on top of the middle segment of the fingers) is a common phenotype caused by a dominant allele *M*. Homozygotes for the recessive allele (*mm*) lack hair on the middle segment of their fingers. Among 1000 families in which both parents had midphalangeal hair, 1853 children showed the trait while 209 children did not. Explain this result.

45. A man with Huntington disease (he is heterozygous $HD\ HD^{+}$) and a normal woman have two children.
    a. What is the probability that only the second child has the disease?
    b. What is the probability that only one of the children has the disease?
    c. What is the probability that none of the children has the disease?
    d. Answer (a) through (c) assuming that the couple had 10 children.
    e. What is the probability that 4 of the 10 children in the family in (d) have the disease?

46. Explain why disease alleles for cystic fibrosis (*CF*) are recessive to the normal alleles (CF^{+}), yet the disease alleles responsible for Huntington disease (*HD*) are dominant to the normal alleles (HD^{+}).

47. The following pedigree shows the inheritance of red hair in a family in Scotland. Red hair is caused by homozygosity for a recessive allele of a gene called *MC1R*. Although worldwide red hair is the rarest of human hair colors, red hair is not uncommon in Scotland. In fact, 40% of Scots without red hair are nonetheless carriers of the red hair allele.

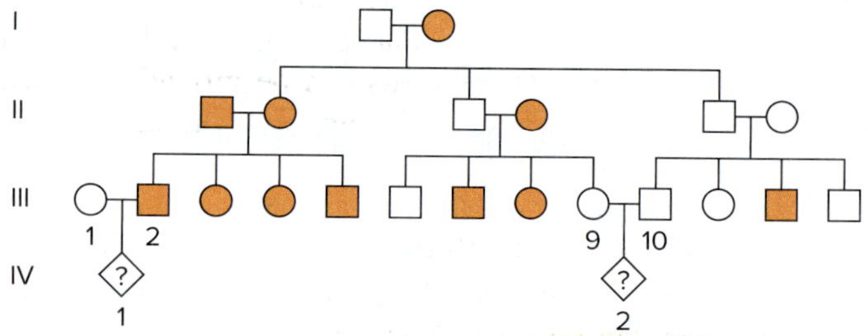

    a. Why does red hair show a horizontal inheritance pattern in this particular pedigree even though the trait is caused by a recessive allele?
    b. Assuming that individual III-2 has a child with the Scottish woman shown (III-1) who is not a close relative, what is the probability that this child (IV-1) will have red hair?
    c. What is the probability that the child of first cousins III-9 and III-10 (IV-2), will have red hair?

chapter 3

Extensions to Mendel's Laws

In this array of green, brown, and red lentils, some of the seeds have speckled patterns, while others are clear.

chapter outline

UNLIKE THE PEA traits that Mendel examined, most human characteristics do not fall neatly into just two opposing phenotypic categories. These *complex traits,* such as skin and hair color, height, athletic ability, and many others, seem to defy Mendelian analysis. The same can be said of traits expressed by many of the world's food crops: Their size, shape, succulence, and nutrient content vary over a wide range of values.

Lentils (*Lens culinaris*) provide a graphic illustration of this variation. A type of legume, lentils are grown in many parts of the world as a rich source of both protein and carbohydrate. The mature plants set fruit in the form of diminutive pods that contain two small seeds. These seeds can be ground into meal or used in soups, salads, and stews. Lentils come in an intriguing array of colors and patterns (**Fig. 3.1**), and commercial growers always seek to produce combinations to suit the cuisines of different cultures. But crosses between pure-breeding lines of lentils result in some startling surprises. A cross between pure-breeding tan and pure-breeding gray parents, for example, yields an all-brown F_1 generation. When these hybrids self-pollinate, the F_2 plants produce not only tan, gray, and brown lentils, but also green.

Beginning in the first decade of the twentieth century, geneticists subjected many kinds of plants and animals to controlled breeding tests, using Mendel's 3:1 phenotypic ratio as a guideline. If the traits under analysis behaved as predicted by Mendel's laws, then they were assumed to be determined by a single gene with alternative dominant and recessive alleles. Many traits, however, did not behave in this way. For some, no definitive dominance and recessiveness could be observed, or more than two alleles could be found in a particular cross (Fig. 3.1). Other traits turned out to be determined by two genes. Yet other traits were *multifactorial*, that is, determined by several different genes, or by the interaction of genes with the environment. The seed color of lentils is a multifactorial trait because color is controlled by multiple genes.

Because traits can arise from an intricate network of interactions, they do not always generate straightforward Mendelian phenotypic ratios. Nonetheless, simple extensions of Mendel's hypotheses can clarify the relationship between genotype and

Figure 3.1 Some phenotypic variation poses a challenge to Mendelian analysis. Lentils show complex speckling patterns that are controlled by a gene that has more than two alleles.

phenotype, allowing explanation of the observed deviations without challenging Mendel's basic laws.

One general theme stands out from these breeding studies: To make sense of the enormous phenotypic variation of the living world, geneticists usually try to limit the number of variables under investigation at any one time. Mendel did this by using pure-breeding, inbred strains of peas that differed from each other by one or a few traits, so that the action of single genes could be detected. Similarly, twentieth-century geneticists used inbred populations of fruit flies, mice, and other experimental organisms to study specific traits. Of course, geneticists cannot study people in this way. Human populations are typically far from inbred, and researchers cannot ethically perform breeding experiments on people. As a result, the genetic basis of much human variation remained a mystery. The advent of molecular biology in the 1970s provided new tools that geneticists now use to unravel the genetics of complex human traits, as will be described in later chapters.

3.1 Extensions to Mendel for Single-Gene Inheritance

learning objectives

1. Categorize allele interactions as completely dominant, incompletely dominant, or codominant.
2. Recognize progeny ratios that imply the existence of recessive lethal alleles.
3. Predict from the results of crosses whether a gene is polymorphic or monomorphic in a population.

William Bateson was an early interpreter and defender of Mendel. Bateson, who coined the terms *genetics*, *allelomorph* (later shortened to *allele*), *homozygote*, and *heterozygote*, entreated the audience at a 1908 lecture: "Treasure your exceptions! … Keep them always uncovered and in sight. Exceptions are like the rough brickwork of a growing building which tells that there is more to come and shows where the next construction is to be." Consistent exceptions to simple Mendelian ratios revealed unexpected patterns of single-gene inheritance. By distilling the significance of these patterns, Bateson and other early geneticists extended the scope of Mendelian analysis and obtained a deeper understanding of the relationship between genotype and phenotype. We now look at the major extensions to Mendelian analysis elucidated over the last century.

Dominance Is Not Always Complete

A consistent working definition of dominance and recessiveness depends on the F_1 hybrids that arise from a mating between two pure-breeding lines. If a hybrid is identical to one parent for the trait under consideration, the allele carried by that parent is deemed dominant to the allele carried by the parent whose trait is not expressed in the hybrid. If, for example, a mating between a pure-breeding white line and a pure-breeding blue line produces F_1 hybrids that are white, the white allele of the gene for color is dominant to the blue allele. If the F_1 hybrids are blue, the blue allele is dominant to the white one (**Fig. 3.2**).

Mendel described and relied on complete dominance in sorting out his ratios and laws, but it is not the only kind of dominance he observed. Figure 3.2 diagrams two situations in which neither allele of a gene is completely dominant. As the figure shows, crosses between true-breeding strains can produce hybrids with phenotypes that differ from both parents. We now explain how these phenotypes arise.

Figure 3.2 Different dominance relationships. The phenotype of the heterozygote defines the dominance relationship between two alleles of the same gene (here, A^1 and A^2). Dominance is complete when the hybrid resembles one of the two pure-breeding parents. Dominance is incomplete when the hybrid resembles neither parent; its novel phenotype is usually intermediate. Codominance occurs when the hybrid shows traits from both pure-breeding parents.

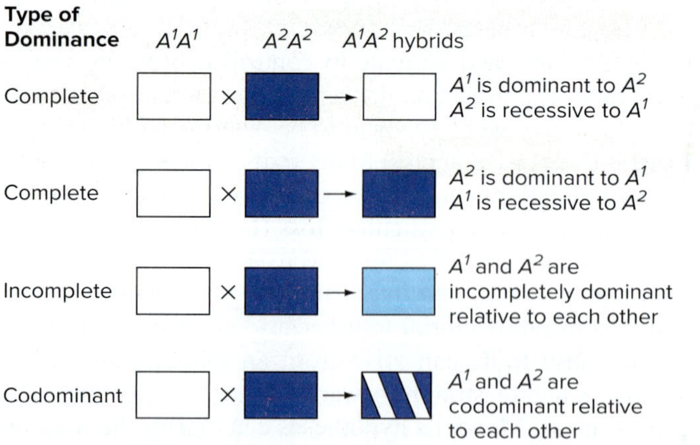

Incomplete dominance: The F_1 hybrid resembles neither pure-breeding parent

A cross between pure late-blooming and pure early-blooming pea plants results in an F_1 generation that blooms in between the two extremes. This is just one of many examples of **incomplete dominance,** in which the hybrid does not resemble either pure-breeding parent. F_1 hybrids that differ from both parents often express a phenotype that is intermediate between those of the pure-breeding parents. Thus, with incomplete dominance, neither parental allele is dominant or recessive to the other; both contribute to the F_1 phenotype. Mendel observed plants that bloomed midway between two extremes when he cultivated various types of pure-breeding peas for his hybridization studies, but he did not pursue the implications. Blooming time was not one of the seven characteristics he chose to analyze in detail, almost certainly because in peas, the time of bloom was not as clear-cut as seed shape or flower color.

In many plant species, flower color serves as a striking example of incomplete dominance. With the floret clusters of snapdragons, for instance, a cross between pure-breeding red-flowered parents and pure-breeding white yields hybrids with pink blossoms, as if a painter had mixed red and white pigments to get pink (**Fig. 3.3a**). If allowed to self-pollinate, the F_1 pink-blooming plants produce F_2 progeny bearing red, pink, and white flowers in a ratio of 1:2:1 (**Fig. 3.3b**). This is the familiar *genotypic ratio* of an ordinary single-gene F_1 self-cross. What is new is that because the heterozygotes look unlike either homozygote, the *phenotypic ratios* are an exact reflection of the genotypic ratios.

The simplest biochemical explanation for this type of incomplete dominance is that each allele of the gene under analysis specifies an alternative form of a protein molecule with an enzymatic role in red pigment production. The white allele (A^2) does not give rise to a functional enzyme, but the red allele (A^1) does. Thus, in snapdragons, two red alleles per cell (A^1A^1) produce a double dose of a red-producing enzyme, which generates enough pigment to make the flowers look fully red. In the heterozygote (A^1A^2), one copy of the red allele per cell results in only enough pigment to make the flowers look pink. In the homozygote for the white allele (A^2A^2), where there is no functional enzyme and thus no red pigment, the flowers appear white.

Codominance: The F_1 hybrid exhibits traits of both parents

A cross between pure-breeding spotted lentils and pure-breeding dotted lentils produces heterozygotes that are both spotted and dotted (**Fig. 3.4a**). These F_1 hybrids illustrate a second significant departure from complete dominance. The progeny look like both parents, which means that neither the spotted nor the dotted allele is dominant or recessive to the other. Because both traits show up equally in the heterozygote's phenotype, the alleles are termed **codominant.** Self-pollination of the spotted/dotted F_1 generation generates F_2 progeny in the ratio of 1 spotted : 2 spotted/dotted : 1 dotted. The Mendelian 1:2:1 ratio among these F_2 progeny establishes that the spotted and dotted traits are determined by alternative alleles of a single gene. Once again, because the heterozygotes can be distinguished from both homozygotes, the phenotypic and genotypic ratios coincide.

In humans, some of the complex membrane-anchored molecules that distinguish different types of red blood cells exhibit codominance. For example, one gene (I) with alleles I^A and I^B controls the presence of a sugar polymer

Figure 3.3 Pink flowers are the result of incomplete dominance. (a) Color differences in these snapdragons reflect the activity of one pair of alleles. **(b)** The F_1 hybrids from a cross of pure-breeding red and white strains of snapdragons have pink blossoms. Flower colors in the F_2 appear in the ratio of 1 red : 2 pink : 1 white. This ratio signifies that the alleles of a single gene determine these three colors.

a: © Henry Hemming/Getty Images RF

(a) *Antirrhinum majus* (snapdragons)

(b) A Punnett square for incomplete dominance

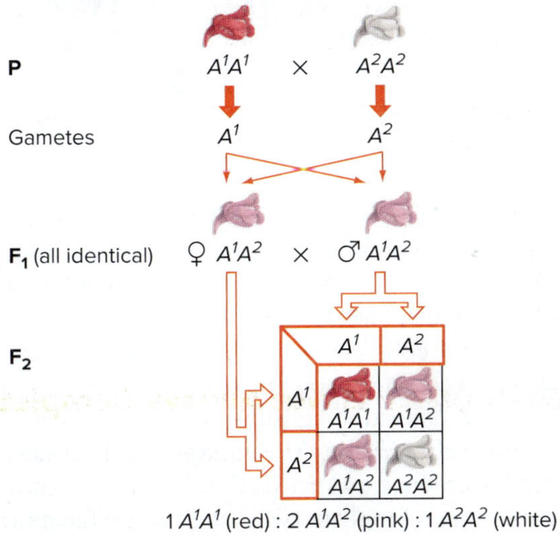

Figure 3.4 In codominance, F_1 hybrids display the traits of both parents. (a) A cross between pure-breeding spotted lentils and pure-breeding dotted lentils produces heterozygotes that are both spotted and dotted. Each genotype has its own corresponding phenotype, so the F_2 ratio is 1:2:1. **(b)** The I^A and I^B blood group alleles are codominant because the red blood cells of an I^AI^B heterozygote have both kinds of sugars at their surface.

(a) Codominant lentil coat patterns

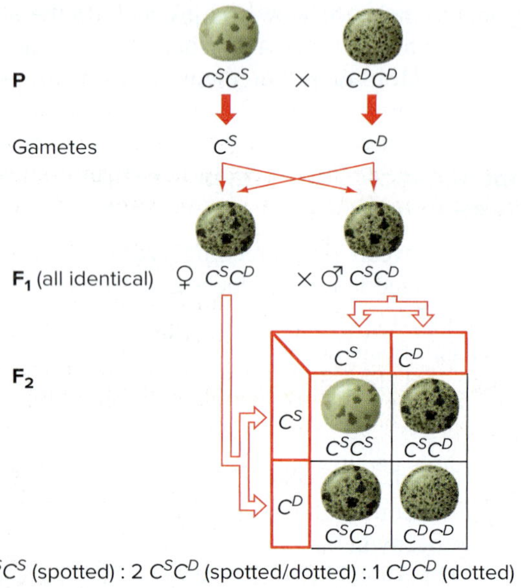

(b) Codominant blood group alleles

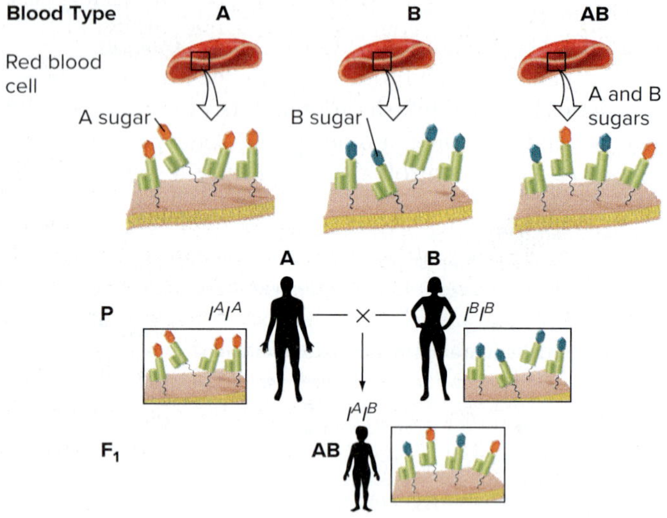

that protrudes from the red blood cell membrane. Each of the alternative alleles encodes a slightly different form of an enzyme that causes production of a slightly different form of the complex sugar. In heterozygous individuals, the red blood cells carry both the I^A-determined and the I^B-determined sugars on their surface, whereas the cells of homozygous individuals display the products of either I^A or I^B alone (**Fig. 3.4b**). As this example illustrates, when both alleles produce a functional gene product, they are usually codominant for phenotypes analyzed at the molecular level.

Figure 3.2 summarizes the differences between complete dominance, incomplete dominance, and codominance for phenotypes reflected in color variations. Determinations of dominance relationships depend on the phenotype that appears in the F_1 generation. With complete dominance, F_1 progeny look like one of the true-breeding parents. Complete dominance, as we saw in Chapter 2, results in a 3:1 ratio of phenotypes in the F_2. With incomplete dominance, hybrids resemble neither of the parents and thus display neither pure-breeding trait. With codominance, the phenotypes of both pure-breeding lines show up simultaneously in the F_1 hybrid. Both incomplete dominance and codominance yield 1:2:1 F_2 ratios.

Mendel's law of segregation still holds

The dominance relations of a gene's alleles do not affect the alleles' transmission. Whether two alternative alleles of a single gene show complete dominance, incomplete dominance, or codominance depends on the kinds of proteins determined by the alleles and the biochemical function of those proteins in the cell. These phenotypic dominance relations, however, have no bearing on the segregation of the alleles during gamete formation.

As Mendel proposed, cells still carry two copies of each gene, and these copies—a pair of either similar or dissimilar alleles—segregate during gamete formation. Fertilization then restores two alleles to each cell without reference to whether the alleles are the same or different. Variations in dominance relations thus do not detract from Mendel's laws of segregation. Rather, they reflect differences in the way gene products control the production of phenotypes, adding a level of complexity to the tasks of interpreting the visible results of gene transmission and inferring genotype from phenotype.

A Gene May Have More Than Two Alleles

Mendel analyzed *either-or* traits controlled by genes with two alternative alleles, but for many traits, more than two alternatives exist. Here, we look at three such traits: human ABO blood types, lentil seed coat patterns, and human histocompatibility antigens.

ABO blood types

If a person with blood type A mates with a person with blood type B, it is possible in some cases for the couple to have a child that is neither A nor B nor AB, but a fourth blood type called O. The reason? The gene for the ABO blood types has three alleles: I^A, I^B, and *i* (**Fig. 3.5a**). Allele I^A gives rise to blood type A by specifying an enzyme that adds sugar A, I^B results in blood type B by specifying an enzyme that adds sugar B; *i* does not produce a functional sugar-adding enzyme. Alleles I^A and I^B are both dominant to *i,* and blood type O is therefore a result of homozygosity for allele *i.*

Figure 3.5 ABO blood types are determined by three alleles of one gene. (a) Six genotypes produce the four blood group phenotypes. **(b)** Blood serum contains antibodies against foreign red blood cell molecules. **(c)** If a recipient's serum has antibodies against the sugars on a donor's red blood cells, the blood types of recipient and donor are incompatible, and coagulation of red blood cells will occur during transfusions. In this table, a plus (+) indicates compatibility, and a minus (–) indicates incompatibility. Antibodies in the donor's blood usually do not cause problems because the amount of transfused antibody is small.

(a)

Genotypes	Corresponding Phenotypes: Type(s) of Molecule on Cell
I^AI^A I^Ai	A
I^BI^B I^Bi	B
I^AI^B	AB
ii	O

(b)

Blood Type	Antibodies in Serum
A	Antibodies against B
B	Antibodies against A
AB	No antibodies against A or B
O	Antibodies against A and B

(c)

Blood Type of Recipient	Donor Blood Type (Red Cells) A	B	AB	O
A	+	–	–	+
B	–	+	–	+
AB	+	+	+	+
O	–	–	–	+

Note in Fig. 3.5a that the A phenotype can arise from two genotypes, I^AI^A or I^Ai. The same is true for the B blood type, which can be produced by I^BI^B or I^Bi. But a combination of the two alleles I^AI^B generates blood type AB.

We can draw several conclusions from these observations. First, as already stated, a given gene may have more than two alleles, or *multiple alleles*; in our example, the series of alleles is denoted I^A, I^B, and i.

Second, although the ABO blood group gene has three alleles, each person carries only two of the alternatives—I^AI^A, I^BI^B, I^AI^B, I^Ai, I^Bi, or ii. Thus six possible ABO genotypes exist. Because each individual carries no more than two alleles for each gene, no matter how many alleles are in a series, Mendel's law of segregation remains intact, because in a sexually reproducing organism, the two alleles of a gene separate during gamete formation.

Third, an allele is not inherently dominant or recessive; its dominance or recessiveness is always relative to a second allele. In other words, dominance relations are unique to a pair of alleles. In our example, I^A is completely dominant to i, but it is codominant with I^B. Given these dominance relations, the six genotypes possible with I^A, I^B, and i generate four different phenotypes: blood groups A, B, AB, and O. With this background, you can understand how a type A and a type B parent could produce a type O child: The parents must be I^Ai and I^Bi heterozygotes, and the child receives an i allele from each parent.

An understanding of the genetics of the ABO system has had profound medical and legal repercussions. Matching ABO blood types is a prerequisite of successful blood transfusions, because people make antibodies to foreign blood cell molecules. A person whose cells carry only A molecules, for example, produces anti-B antibodies; B people manufacture anti-A antibodies; AB individuals make neither type of antibody; and O individuals produce both anti-A and anti-B antibodies (**Fig. 3.5b**). These antibodies cause coagulation of cells displaying the foreign molecules (**Fig. 3.5c**). As a result, people with blood type O have historically been known as *universal donors* because their red blood cells carry no surface molecules that will stimulate an antibody attack in a transfusion recipient. In contrast, people with blood type AB are considered *universal recipients,* because they make neither anti-A nor anti-B antibodies, which, if present, would target the surface molecules of incoming blood cells.

Information about ABO blood types can also be used as legal evidence in court, to exclude the possibility of paternity or criminal guilt. In a paternity suit, for example, if the mother is type A and her child is type B, logic dictates that the I^B allele must have come from the father, whose genotype may be I^AI^B, I^BI^B, or I^Bi. In 1944, the actress Joan Barry (phenotype A) sued Charlie Chaplin (phenotype O) for support of a child (phenotype B) whom she claimed he fathered. The scientific evidence indicated that Chaplin could not have been the father, since he was apparently ii and did not carry an I^B allele. This evidence was admissible in court, but the jury was not convinced, and Chaplin had to pay. Today, the molecular genotyping of DNA (*DNA fingerprinting,* see Chapter 11) provides a powerful tool to help establish paternity, guilt, or innocence, but juries still often find it difficult to evaluate such evidence.

Lentil seed coat patterns

Lentils offer another example of multiple alleles. A gene for seed coat pattern has five alleles: spotted, dotted, clear (pattern absent), and two types of marbled. Reciprocal crosses between pairs of pure-breeding lines of all patterns (marbled-1 × marbled-2, marbled-1 × spotted, marbled-2 × spotted, and so forth) have clarified the dominance relations of all possible pairs of the alleles to reveal a **dominance series** in which alleles are listed in order from most dominant to most recessive. For example, crosses of marbled-1 with marbled-2, or of marbled-1 with spotted or dotted or clear, produce the marbled-1 phenotype in the F_1 generation and a ratio of three marbled-1 to one of any of the other phenotypes in the F_2. These results indicate that the marbled-1 allele is completely dominant to each of the other four alleles.

Analogous crosses with the remaining four phenotypes reveal the dominance series shown in **Fig. 3.6.** Recall that dominance relations are meaningful only when comparing two alleles: An allele, such as marbled-2, can be recessive to a second allele (marbled-1) but dominant to a third and fourth (dotted and clear). The fact that all tested pairings of lentil seed coat pattern alleles yielded a 3:1 ratio in the F_2 generation (except for spotted × dotted, which yielded the 1:2:1 phenotypic ratio reflective of codominance) indicates that these lentil seed coat patterns are determined by different alleles of the same gene.

Figure 3.6 How to establish the dominance relations between multiple alleles. Pure-breeding lentils with different seed coat patterns are crossed in pairs, and the F_1 progeny are self-fertilized to produce an F_2 generation. The 3:1 or 1:2:1 F_2 monohybrid ratios from all of these crosses indicate that different alleles of a single gene determine all the traits. The phenotypes of the F_1 hybrids establish the dominance relationships (*bottom*). Spotted and dotted alleles are codominant, but each is recessive to the marbled alleles and is dominant to clear.

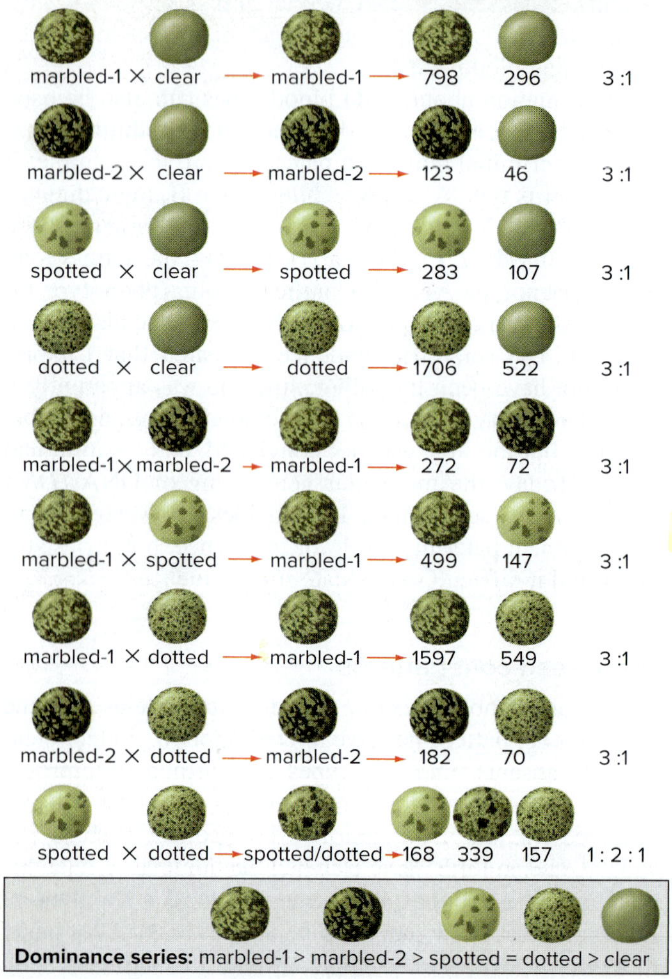

Parental Generation Parental seed coat pattern in cross Parent 1 × Parent 2	F_1 Generation F_1 phenotype	F_2 Generation Total F_2 frequencies and phenotypes	Apparent phenotypic ratio
marbled-1 × clear	marbled-1	798 296	3:1
marbled-2 × clear	marbled-2	123 46	3:1
spotted × clear	spotted	283 107	3:1
dotted × clear	dotted	1706 522	3:1
marbled-1 × marbled-2	marbled-1	272 72	3:1
marbled-1 × spotted	marbled-1	499 147	3:1
marbled-1 × dotted	marbled-1	1597 549	3:1
marbled-2 × dotted	marbled-2	182 70	3:1
spotted × dotted	spotted/dotted	168 339 157	1:2:1

Dominance series: marbled-1 > marbled-2 > spotted = dotted > clear

Histocompatibility in humans

In some multiple allelic series, each allele is codominant with every other allele, and every distinct genotype therefore produces a distinct phenotype. This happens particularly with traits defined at the molecular level. An extreme example is the group of three major genes that encode a family of related cell surface molecules in humans and other mammals known as *histocompatibility antigens.* Carried by all of the body's cells except the red blood cells and sperm, histocompatibility antigens play a crucial role in facilitating a proper immune response that destroys intruders (viral or bacterial, for example) while leaving the body's own tissues intact. Because each of the three major histocompatibility genes (called *HLA-A, HLA-B,* and *HLA-C* in humans) has between 400 and 1200 alleles, the number of possible allelic combinations in an individual creates a powerful potential for the phenotypic variation of cell surface molecules. Other than identical (that is, *monozygotic*) twins, no two people are likely to carry the same array of histocompatibility antigens on the surfaces of their cells.

The extreme variation in these proteins has important medical consequences, because people can make antibodies to *non-self* histocompatibility antigens different from their own. These antibodies can lead to rejection of transplanted organs. Doctors thus attempt to match as closely as possible the histocompatibility antigen types of transplant donors and recipients. Family members usually make the best organ donors, as the closer the genetic relationship between two people, the more likely they are to share *HLA* alleles.

Mutations Are the Source of New Alleles

How do the multiple alleles of an allelic series arise? The answer is that chance alterations of the genetic material, known as **mutations,** arise spontaneously in nature. Once they occur in gamete-producing cells, they are inherited faithfully. Mutations that have phenotypic consequences can be counted, and such counting reveals that they occur at low frequency. The frequency of gametes carrying a new mutation in a particular gene varies anywhere from 1 in 10,000 to 1 in 1,000,000. This range exists because different genes have different mutation rates.

Mutations make it possible to follow gene transmission. If, for example, a mutation specifies an alteration in an enzyme that normally produces yellow so that it now makes green, the new phenotype (green) will make it possible to recognize the new mutant allele. In fact, it takes at least two different alleles, that is, some form of variation, to "see" the transmission of a gene. Thus, in segregation studies, geneticists can analyze only genes with variants; they have no way of following a gene that comes in only one form. If all peas were yellow, Mendel would not have been able to decipher the transmission patterns of the gene for the seed color trait. We discuss mutations in greater detail in Chapter 7.

Allele frequencies and monomorphic genes

Because each organism carries two copies of every gene, you can calculate the number of copies of a gene in a given population by multiplying the number of individuals by 2. Each allele of the gene accounts for a percentage of the total number of gene copies, and that percentage is known as the **allele frequency.** The most common alleles in a population are usually called the **wild-type alleles,** often designated by a superscript plus sign ($^+$). An allele is considered wild-type if it is present in the population at a frequency greater than 1%. A rare allele in the same population is considered a **mutant allele.** (Note that the definitions of *wild-type* versus *mutant alleles* are not static. A newly induced mutation generates a mutant allele whose frequency can increase and over time, the allele can become wild-type.)

In mice, for example, one of the main genes determining coat color is the *agouti* gene. The wild-type allele (A) produces fur with each hair having yellow and black bands that blend together from a distance to give the appearance of dark gray, or agouti. Researchers have identified in the laboratory 14 distinguishable mutant alleles for the *agouti* gene. One of these (a^t) is recessive to the wild type and gives rise to a black coat on the back and a yellow coat on the belly; another (a) is also recessive to A and produces a pure black coat (**Fig. 3.7**). In nature, wild-type agoutis (AA) survive to reproduce, while very few black-backed or pure black mutants (a^ta^t or aa) do so because their dark coat makes it hard for them to evade the eyes of predators. As a result, A is present at a frequency of much more than 99% and is thus the only wild-type allele in mice for the *agouti* gene. A gene with only one common, wild-type allele is **monomorphic.**

Allele frequencies and polymorphic genes

In contrast, some genes have more than one common allele, which makes them **polymorphic.** For example, in the ABO blood type system, all three alleles—I^A, I^B, and i—have appreciable frequencies in most human populations. Although all three of these alleles can be considered to be wild-type, geneticists instead usually refer to the high-frequency alleles of a polymorphic gene as **common variants.** Certain rare genes are so polymorphic that hundreds of allelic variants can be found in populations. We have already discussed the case of the *HLA* histocompatibility genes in humans, which encode cell surface proteins that help the immune system deal with pathogenic invaders such as bacteria and viruses. Some scientists think that evolution favors the emergence of new *HLA* gene alleles to ensure that no single pathogen among the many to which we are exposed in the environment could destroy the entire human population. That is, at least a few individuals with particular *HLA* gene alleles would be protected from any given pathogen.

Figure 3.7 The mouse *agouti* gene: One wild-type allele, many mutant alleles. (a) Black-backed, yellow-bellied (*top left*); black (*top right*); and agouti (*bottom*) mice. **(b)** Genotypes and corresponding phenotypes for alleles of the *agouti* gene. **(c)** Crosses between pure-breeding lines reveal a dominance series. Interbreeding of the F_1 hybrids (*not shown*) yields 3:1 phenotypic ratios of F_2 progeny, indicating that A, a^t, and a are in fact alleles of one gene.

(a) *Mus musculus* (house mouse) coat colors

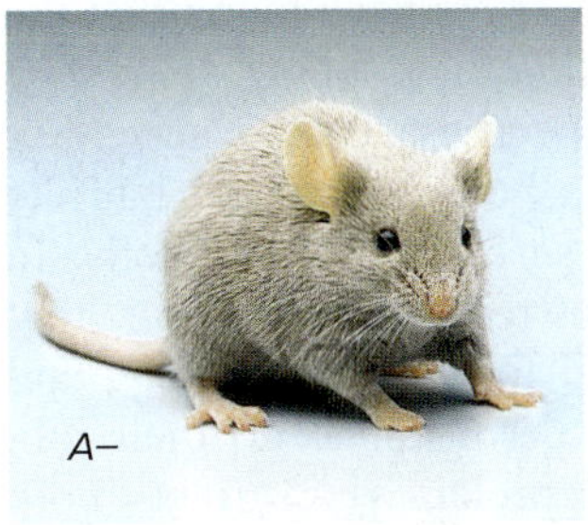

(b) Alleles of the *agouti* gene

Genotype	Phenotype
$A–$	agouti
a^ta^t	black/yellow
aa	black
a^ta	black/yellow

(c) Evidence for a dominance series

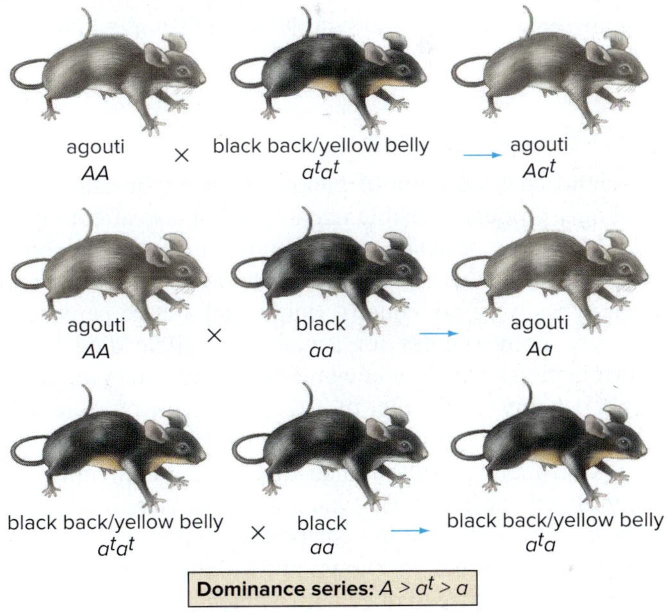

One Gene May Contribute to Several Characteristics

Mendel derived his laws from studies in which one gene determined one trait, but, always the careful observer, he himself noted possible departures. In listing the traits selected for his pea experiments, Mendel remarked that specific seed coat colors are always associated with specific flower colors.

The phenomenon of a single gene determining a number of distinct and seemingly unrelated characteristics is known as **pleiotropy.** Because geneticists now know that each gene determines a specific protein (or RNA) and that each gene product can have a cascade of effects on an organism, we can understand how pleiotropy arises. Among the Maori people of New Zealand, for example, many men develop respiratory issues and are also sterile. These men are said to exhibit a **syndrome**—a group of problems that are usually seen together. Researchers have found that the fault lies with the recessive allele of a single gene. The gene's normal dominant allele specifies a protein necessary for the action of cilia and flagella, both of which are hairlike structures extending from the surfaces of some cells. In men who are homozygous for the recessive allele, cilia that normally clear the airways fail to work effectively, and flagella that normally propel sperm fail to do their job. Thus, one gene determines a protein that affects both respiratory function and reproduction. Because most proteins act in a variety of tissues and influence multiple biochemical processes, mutations in almost any gene may have pleiotropic effects.

Recessive lethal alleles

A significant variation of pleiotropy occurs in alleles that not only produce a visible phenotype but also affect viability. Mendel assumed that all genotypes are equally viable—that is, they have the same likelihood of survival. If all genotypes were not equally viable, and a large percentage of, say, homozygotes for a particular allele died before germination or birth, you would not be able to count them. This lethality would alter the 1:2:1 genotypic ratios and the 3:1 phenotypic ratios predicted for the F_2 generation.

Consider the inheritance of coat color in mice. As mentioned earlier, wild-type agouti (*AA*) animals have black hairs with a yellow stripe that appear dark gray to the eye. One of the 14 known mutant alleles of the *agouti* gene, A^Y, gives rise to mice with a much lighter, almost yellow color. When pure-breeding *AA* mice are mated to yellow mice, the offspring always emerge in a 1:1 ratio of the two coat colors (**Fig. 3.8a**). From this result, we can draw three conclusions: (1) All yellow mice must carry the wild-type *A* allele even though they do not express the agouti phenotype; (2) yellow is therefore dominant to agouti; and (3) all yellow mice are A^YA heterozygotes.

Figure 3.8 A^y **is a pleiotropic and recessive lethal allele.** **(a)** A cross between inbred agouti mice and yellow mice yields a 1:1 ratio of yellow to agouti progeny. The yellow mice are therefore A^yA heterozygotes, and for the trait of coat color, A^y (for yellow) is dominant to *A* (for agouti). (Note we assume that if they could survive, A^YA^Y mice would have the same coat color as A^YA mice.) **(b)** Yellow mice do not breed true. In a yellow × yellow cross, the 2:1 ratio of yellow to agouti progeny indicates that the A^y allele is a recessive lethal.

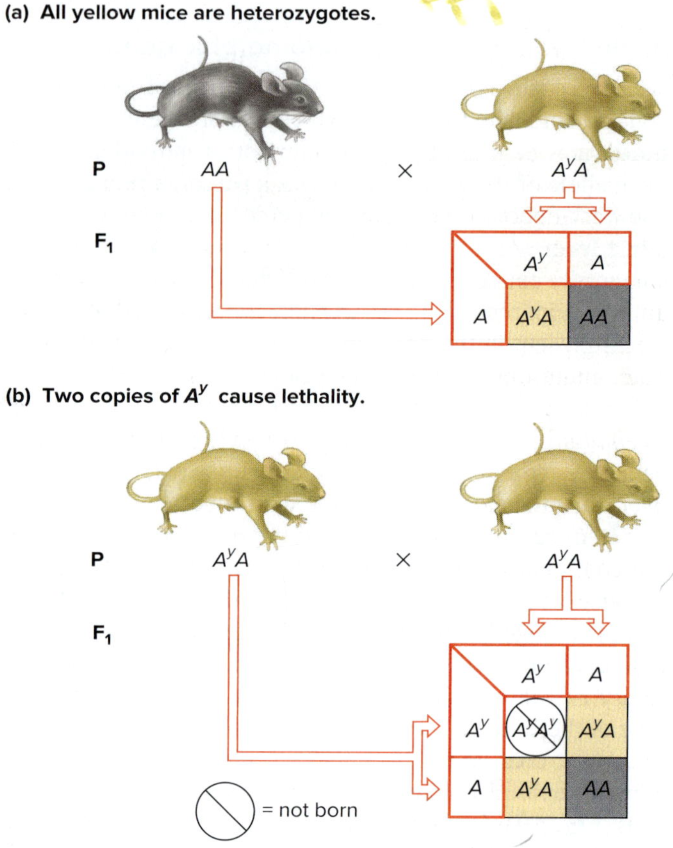

Note again that dominance and recessiveness are defined in the context of each pair of alleles. Even though, as previously mentioned, agouti (*A*) is dominant to the a^t and *a* mutations for black coat color, it can still be recessive to the yellow coat color allele. The yellow mice in the preceding cross are A^YA heterozygotes, and the agoutis, *AA* homozygotes. So far, no surprises. But a mating of yellow to yellow produces a skewed phenotypic ratio of two yellow mice to one agouti (**Fig. 3.8b**). Among these progeny, matings between agouti mice show that the agoutis are all pure-breeding and therefore *AA* homozygotes as expected. However, no pure-breeding yellow mice appear among the progeny. When the yellow mice are mated to each other, they unfailingly produce 2/3 yellow and 1/3 agouti offspring, a ratio of 2:1, so the yellow mice must be heterozygotes (A^YA). In short, one can never obtain pure-breeding yellow mice (A^YA^Y).

How can we explain this phenomenon? The Punnett square in Fig. 3.8b suggests an answer. Two copies of the

A^y allele prove fatal to the animal carrying them, whereas one copy of the allele produces a yellow coat. This means that the A^y allele affects two different traits: It is dominant to A in the determination of coat color, but it is recessive to A in the production of lethality. An allele, such as A^y, that negatively affects the survival of a homozygote is known as a **recessive lethal allele.** Note that the same two alleles (A^y and A) can display different dominance relationships when looked at from the point of view of different phenotypes; we return later to this important point.

Because the A^y allele is dominant for yellow coat color, it is easy to detect carriers of this particular recessive lethal allele in mice. Such is not the case, however, for the vast majority of recessive lethal mutations, as these usually do not simultaneously show a visible dominant phenotype for some other trait. Lethal mutations can arise in many different genes, and as a result, most animals, including humans, carry some recessive lethal mutations. Such mutations usually remain hidden, except in rare cases of homozygosity, which in people are often caused by consanguineous matings (that is, matings between close relatives). If a mutation produces an allele that prevents production of a crucial molecule, homozygous individuals will not make any of the vital molecule and will not survive. Heterozygotes, by contrast, with only one copy of the deleterious mutation and one wild-type allele, can produce 50% of the wild-type amount of the normal molecule; this is usually sufficient to sustain normal cellular processes such that life goes on.

Delayed lethality

In the preceding discussion, we have described recessive alleles that result in the death of homozygotes prenatally; that is, *in utero.* With some mutations, however, homozygotes may survive beyond birth and die later from the deleterious consequences of the genetic defect. An example is seen in human infants with Tay-Sachs disease. The seemingly normal newborns remain healthy for five to six months but then develop blindness, paralysis, mental impairment, and other symptoms of a deteriorating nervous system; the disease usually proves fatal by the age of six. Tay-Sachs disease results from the absence of an active lysosomal enzyme called hexosaminidase A, leading to the accumulation of a toxic waste product inside nerve cells. The approximate incidence of Tay-Sachs among live births is 1/35,000 worldwide, but it is 1/3000 among Jewish people of Eastern European descent. Reliable tests that detect carriers, in combination with genetic counseling and educational programs, have all but eliminated the disease in the United States.

Recessive alleles causing prenatal or early childhood lethality can be passed on to subsequent generations only by heterozygous carriers because affected homozygotes die before they can mate. However, for late-onset diseases causing death in adults, homozygous patients can pass on the lethal allele before they become debilitated. An example is provided by the degenerative disease Friedreich ataxia: Some homozygotes first display symptoms of ataxia (loss of muscle coordination) at age 30–35 and die about five years later from heart failure.

Dominant alleles causing late-onset lethality can also be transmitted to subsequent generations; Figure 2.22 illustrates this fact for the inheritance of Huntington disease. By contrast, if the lethality caused by a dominant allele occurs instead during fetal development or early childhood, the allele will not be passed on, so all dominant early lethal mutant alleles must be new mutations.

Table 3.1 summarizes Mendel's basic assumptions about dominance, the number and viability of one gene's alleles, and the effects of each gene on phenotype, and then compares these assumptions with the extensions contributed by his twentieth-century successors. Through carefully controlled monohybrid crosses, these later geneticists analyzed the transmission patterns of the alleles of single genes, challenging and then confirming the law of segregation.

TABLE 3.1 For Traits Determined by One Gene: Alterations of the 3:1 Monohybrid Ratio

What Mendel Described	Extension	Extension's Effect on Heterozygous Phenotype	Extension's Effect on Ratios Resulting from an $F_1 \times F_1$ Cross
Complete dominance	Incomplete dominance Codominance	Unlike either homozygote	Phenotypes coincide with genotypes in a ratio of 1:2:1
Two alleles	Multiple alleles	Multiplicity of phenotypes	A series of 3:1 or 1:2:1 ratios
All alleles are equally viable	Recessive lethal alleles	Heterozygotes survive but may have visible phenotypes	2:1 instead of 3:1
One gene determines one trait	Pleiotropy: One gene influences several traits	Several traits affected in different ways, depending on dominance relations	Different ratios, depending on dominance relations for each affected trait

A Comprehensive Example: Sickle-Cell Disease Illustrates Many Extensions to Mendel's View of Single-Gene Inheritance

Sickle-cell disease is the result of a faulty hemoglobin molecule. Hemoglobin is composed of two types of polypeptide chains, alpha (α)-globin and beta (β)-globin, each specified by a different gene: *Hbα* for *α*-globin and *Hbβ* for β-globin. Normal red blood cells are packed full of millions upon millions of hemoglobin molecules, each of which picks up oxygen in the lungs and transports it to all the body's tissues.

Multiple alleles

The β-globin gene has a normal wild-type allele ($Hb\beta^A$) that gives rise to fully functional β-globin, as well as close to 400 mutant alleles that have been identified so far. Some of these mutant alleles result in the production of hemoglobin that carries oxygen only inefficiently. Other mutant alleles prevent the production of β-globin, causing a hemolytic (blood-destroying) disease called *β-thalassemia*. Here, we discuss the most common mutant allele of the β-globin gene, $Hb\beta^S$, which specifies an abnormal polypeptide that causes sickling of red blood cells (**Fig. 3.9a**).

Pleiotropy

The $Hb\beta^S$ allele of the β-globin gene affects more than one trait (**Fig. 3.9b**). Hemoglobin molecules in the red blood cells of homozygous $Hb\beta^S$ $Hb\beta^S$ individuals undergo an aberrant transformation after releasing their oxygen. Instead of remaining soluble in the cytoplasm, they aggregate to form long fibers that deform the red blood cell from a normal biconcave disk to a sickle shape (see Fig. 3.9a). The deformed cells clog small blood vessels, reducing oxygen flow to the tissues and giving rise to muscle cramps, shortness of breath, and fatigue. The sickled cells are also fragile and easily broken. Consumption of fragmented cells by phagocytic white blood cells leads to a low red blood cell count, a condition called *anemia.*

On the positive side, $Hb\beta^S$ $Hb\beta^S$ homozygotes are resistant to malaria because the organism that causes the disease, *Plasmodium falciparum,* can multiply rapidly in normal red blood cells but cannot do so in cells that sickle. Infection by *P. falciparum* causes sickle-shaped cells to break down before the malaria organism has a chance to multiply.

Figure 3.9 Pleiotropy of sickle-cell anemia: Dominance relations vary with the phenotype under consideration. (a) A normal red blood cell (*top*) is easy to distinguish from the sickled cell in the scanning electron micrograph at the *bottom*. **(b)** Different levels of analysis identify various phenotypes. Dominance relationships between the $Hb\beta^S$ and $Hb\beta^A$ alleles of the *Hbβ* gene vary with the phenotype and sometimes even change with the environment.

a (top): © BSIP/Newscom; a (bottom): Source: Janice Haney Carr/CDC

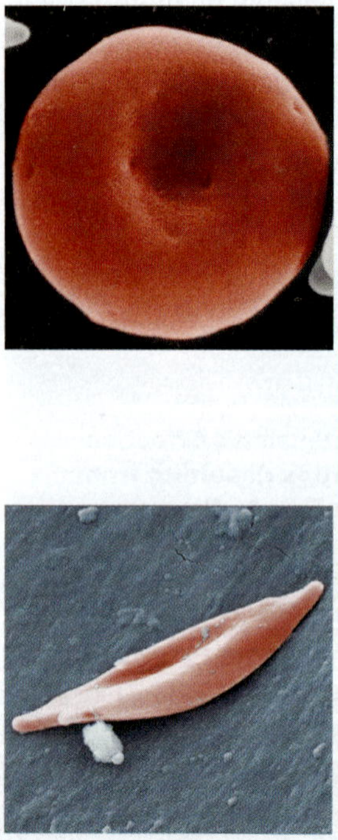

(a)

Phenotypes at Different Levels of Analysis	Normal $Hb\beta^A$ $Hb\beta^A$	Carrier $Hb\beta^A$ $Hb\beta^S$	Diseased $Hb\beta^S$ $Hb\beta^S$	Dominance Relations at Each Level of Analysis
Red blood cell shape at sea level	Normal	Normal	Sickled cells present	$Hb\beta^A$ is dominant $Hb\beta^S$ is recessive
Red blood cell concentration at sea level	Normal	Normal	Lower	
β-globin polypeptide production	A protein	A and S proteins	S proteins	$Hb\beta^A$ and $Hb\beta^S$ are codominant
Red blood cell shape at high altitudes	Normal	Sickled cells present	Severe sickling	
Red blood cell concentration at high altitudes	Normal	Lower	Very low, anemia	$Hb\beta^A$ and $Hb\beta^S$ show incomplete dominance
Susceptibility to malaria	Normal susceptibility	Resistant	Resistant	$Hb\beta^S$ is dominant $Hb\beta^A$ is recessive

(b)

Recessive lethality

People who are homozygous for the recessive $Hb\beta^S$ allele often develop heart failure because of stress on the circulatory system. Many sickle-cell sufferers die in childhood, adolescence, or early adulthood.

Different dominance relations

Comparisons of heterozygous carriers of the sickle-cell allele—individuals whose cells contain one $Hb\beta^A$ and one $Hb\beta^S$ allele—with homozygous $Hb\beta^A\ Hb\beta^A$ (normal) and homozygous $Hb\beta^S\ Hb\beta^S$ (diseased) individuals make it possible to distinguish different dominance relationships for different phenotypic aspects of sickle-cell anemia (Fig. 3.9b).

At the molecular level—the production of β-globin proteins—both alleles are expressed such that $Hb\beta^A$ and $Hb\beta^S$ are codominant. At the cellular level, in their effect on red blood cell shape, the $Hb\beta^A$ and $Hb\beta^S$ alleles show either complete dominance or codominance depending on altitude. Under normal oxygen conditions, the great majority of a heterozygote's red blood cells have the normal biconcave shape ($Hb\beta^A$ is dominant to $Hb\beta^S$). When oxygen levels drop, however, sickling occurs in some $Hb\beta^A\ Hb\beta^S$ cells ($Hb\beta^A$ and $Hb\beta^S$ are codominant). During World War II, soldiers who were heterozygous carriers and who were airlifted in transport planes to cross the Pacific experienced sickling crises for this reason.

Considering the trait of resistance to malaria, the $Hb\beta^S$ allele is dominant to the $Hb\beta^A$ allele. The reason is that infected $Hb\beta^A\ Hb\beta^S$ cells are resistant to malaria because they break down before the malarial organism has a chance to reproduce, just like the $Hb\beta^S\ Hb\beta^S$ cells described previously. But luckily for the heterozygote, for the phenotypes of anemia or death, $Hb\beta^S$ is recessive to $Hb\beta^A$. A corollary of this observation is that in its effect on general health under normal environmental conditions and its effect on red blood cell count, the $Hb\beta^A$ allele is dominant to $Hb\beta^S$.

Thus, for the β-globin gene, as for other genes, dominance and recessiveness are not an inherent quality of alleles in isolation; rather, they are specific to each pair of alleles and to the level of physiology at which the phenotype is examined. When discussing dominance relationships, it is therefore essential to define the particular phenotype under analysis.

The complicated dominance relationships between the $Hb\beta^A$ and $Hb\beta^S$ alleles help explain the puzzling observation that the normally deleterious allele $Hb\beta^S$ is widespread in certain populations. In areas where malaria is endemic, heterozygotes are better able to survive and pass on their genes than are either type of homozygote. $Hb\beta^S\ Hb\beta^S$ individuals often die of sickle-cell disease, while those with the genotype $Hb\beta^A\ Hb\beta^A$ often die of malaria. Heterozygotes, however, are relatively immune to both conditions, so high frequencies of both alleles persist in tropical environments where malaria is found. We explore this phenomenon in more quantitative detail in Chapter 21 on population genetics.

essential concepts

- Two alleles of a single gene may exhibit *complete dominance*, in which heterozygotes resemble the homozygous dominant parent; *incomplete dominance*, in which heterozygotes have an intermediate phenotype; and *codominance,* in which heterozygotes display aspects of each homozygous phenotype.
- New alleles of a gene arise by mutation. Alleles with a frequency greater than 1% in a population are *wild-type;* alleles that are less frequent are *mutant.*
- When two or more wild-type alleles (*common variants*) exist for a gene, the gene is *polymorphic;* a gene with only one wild-type allele is *monomorphic.*
- In *pleiotropy,* one gene contributes to multiple traits. The dominance relationship between any two alleles can vary depending on the trait.
- Homozygotes for a *recessive lethal allele* that fails to provide an essential function will die. If a recessive lethal allele has dominant effects on a visible trait, two-thirds of the surviving progeny of a cross between heterozygotes will display this trait.

3.2 Extensions to Mendel for Two-Gene Inheritance

learning objectives

1. Conclude from the results of crosses whether a single gene or two genes control a trait.
2. Infer from the results of crosses the existence of interactions between alleles of different genes including: additivity, epistasis, redundancy, and complementation.

Two genes can interact in several ways to determine a single trait, such as the color of a flower, a seed coat, a chicken's feathers, a dog's fur, or the shape of a plant's leaves. In a dihybrid cross like Mendel's, each type of interaction produces its own signature of phenotypic ratios. In the following examples, the alternate alleles of each of the two genes are completely dominant (such as *A* and *B*) and recessive (*a* and *b*). For simplicity, we sometimes refer to a gene name using the symbol for the dominant allele, for example, gene *A*. In addition, we refer to the protein product of allele *A* as protein A (no italics), and when appropriate, that of allele *a* as protein a (no italics).

Additive Interactions Between Two Genes Controlling a Single Trait Can Produce Novel Phenotypes

In the chapter opening, we described a mating of tan and gray lentils that produced a uniformly brown F_1 generation and then an F_2 generation containing brown, tan, gray, and green lentil seeds. An understanding of how this can happen emerges from experimental results demonstrating that the ratio of the four F_2 colors is 9 brown : 3 tan : 3 gray : 1 green (**Fig. 3.10a**). Recall from Chapter 2 that this is the same ratio Mendel observed in his analysis of the F_2 generations from dihybrid crosses following two independently assorting genes. In Mendel's studies, each of the four classes consisted of plants that expressed a combination of two unrelated traits. With lentils, however, we are looking at a single trait—seed color. The simplest explanation for the parallel ratios is that a combination of genotypes at two independently assorting genes interacts *additively* to produce the phenotype of seed color in lentils.

Results obtained from self-crosses with the various types of F_2 lentil plants support a two-gene explanation. Self-crosses of F_2 green individuals show that they are pure-breeding, producing an F_3 generation that is entirely green. Tan individuals generate either all tan offspring, or a mixture of tan offspring and green offspring. Grays similarly produce either all gray, or gray and green. Self-crosses of brown F_2 individuals can have four possible outcomes: all brown, brown plus tan, brown plus gray, or all four colors (**Fig. 3.10b**). The two-gene hypothesis explains why:

- only one green genotype exists: pure-breeding *aa bb,* but
- two types of tans exist: pure-breeding *AA bb* as well as tan- and green-producing *Aa bb,* and
- two types of grays exist: pure-breeding *aa BB* and gray- and green-producing *aa Bb,* yet
- four types of browns exist: true-breeding *AA BB,* brown- and tan-producing *AA Bb,* brown- and gray-producing *Aa BB,* and *Aa Bb* dihybrids that give rise to plants producing lentils of all four colors.

Figure 3.10 How two genes interact to produce seed colors in lentils. (a) In a cross of pure-breeding tan and gray lentils, all the F_1 hybrids are brown, but four different phenotypes appear among the F_2 progeny. The 9:3:3:1 ratio of F_2 phenotypes suggests that seed coat color is determined by two independently assorting genes. **(b)** Expected results of selfing individual F_2 plants of the indicated phenotypes to produce an F_3 generation, if seed coat color results from the interaction of two genes. The third column shows the proportion of the F_2 population that would be expected to produce the observed F_3 phenotypes. **(c)** Other two-generation crosses involving pure-breeding parental lines also support the two-gene hypothesis. In this table, the F_1 hybrid generation has been omitted.

(a) A dihybrid cross with lentil coat colors

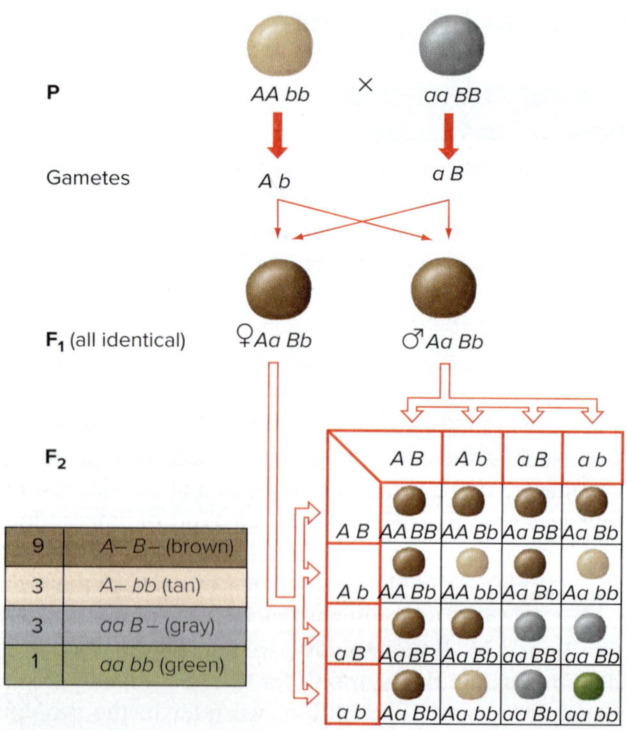

(b) Self-pollination of the F_2 to produce an F_3

Phenotypes of F_2 Individuals	Observed F_3 Phenotypes	Expected Proportion of F_2 Population*
Green	Green	1/16
Tan	Tan	1/16
Tan	Tan, green	2/16
Gray	Gray, green	2/16
Gray	Gray	1/16
Brown	Brown	1/16
Brown	Brown, tan	2/16
Brown	Brown, gray	2/16
Brown	Brown, gray, tan, green	4/16

*This 1:1: 2 : 2 :1:1: 2 : 2 : 4 F_2 genotypic ratio corresponds to a 9 brown :3 tan : 3 gray :1 green F_2 phenotypic ratio.

(c) Sorting out the dominance relations by select crosses

Seed Coat Color of Pure-Breeding Parents	F_2 Phenotypes and Frequencies	Ratio
Tan × green	231 tan, 85 green	3:1
Gray × green	2586 gray, 867 green	3:1
Brown × gray	964 brown, 312 gray	3:1
Brown × tan	255 brown, 76 tan	3:1
Brown × green	57 brown, 18 gray, 13 tan, 4 green	9:3:3:1

In short, for the two genes that determine seed color, both dominant alleles must be present to yield brown (*A– B–*); the dominant allele of one gene produces tan (*A– bb*); the dominant allele of the other specifies gray (*aa B–*); and the complete absence of dominant alleles (that is, the double recessive) yields green (*aa bb*). Thus, the four color phenotypes arise from four **genotypic classes,** with each class defined in terms of the presence or absence of the dominant alleles of two genes: (1) both present (*A– B–*), (2) one present (*A– bb*), (3) the other present (*aa B–*), and (4) neither present (*aa bb*). Note that the *A–* notation means that the second allele of this gene can be either *A* or *a,* while *B–* denotes a second allele of either *B* or *b*. Note also that only with a two-gene system in which the dominance and recessiveness of alleles at both genes is complete can the nine different genotypes of the F_2 generation be categorized into the four phenotypic classes described. With incomplete dominance or co-dominance, the F_2 genotypes could not be grouped together in this simple way, as they would give rise to more than four phenotypes.

Further crosses between plants carrying lentils of different colors confirmed the two-gene hypothesis (**Fig. 3.10c**). Thus, the 9:3:3:1 phenotypic ratio of brown to tan to gray to green in an F_2 descended from pure-breeding tan and pure-breeding gray lentils tells us not only that dominant and recessive alleles of two genes assort independently and interact to produce the seed color, but also that each genotypic class (*A– B–, A– bb, aa B–,* and *aa bb*) determines a particular phenotype.

How can we explain the 9:3:3:1 phenotypic ratio in terms of the action of the protein products of these two genes? We cannot answer this question definitively because the genes controlling lentil seed color have not been identified at the molecular level, and the biochemical pathways in which they function are not known. However, information available about the mechanisms of seed color inheritance in other plant species allows us to formulate a plausible model for this system in lentils (**Fig. 3.11**). The model illustrates an important implication of the 9:3:3:1 ratio: The two independently assorting genes controlling the same trait probably function additively in independent biochemical pathways, so that in this case, tan (the product of one pathway) + gray (the product of the other pathway) = brown.

Figure 3.11 A biochemical model for the inheritance of lentil seed colors. The seed has an opaque outer layer (the seed coat) and an inner layer (the cotyledon). The green chlorophyll in the cotyledon is not visible if the seed coat is colored. Allele *A* encodes enzyme A. Allele *a* does not produce this enzyme. Allele *B* of a second gene encodes a different enzyme; *b* produces none of this enzyme. Seeds appear brown if the tan and gray pigments are both present. In the absence of both enzymes (*aa bb*), the seed coat is nonpigmented, so the green chlorophyll in the cotyledon will show through. The 9:3:3:1 ratio implies that the *A* and *B* genes operate in independent biochemical pathways.

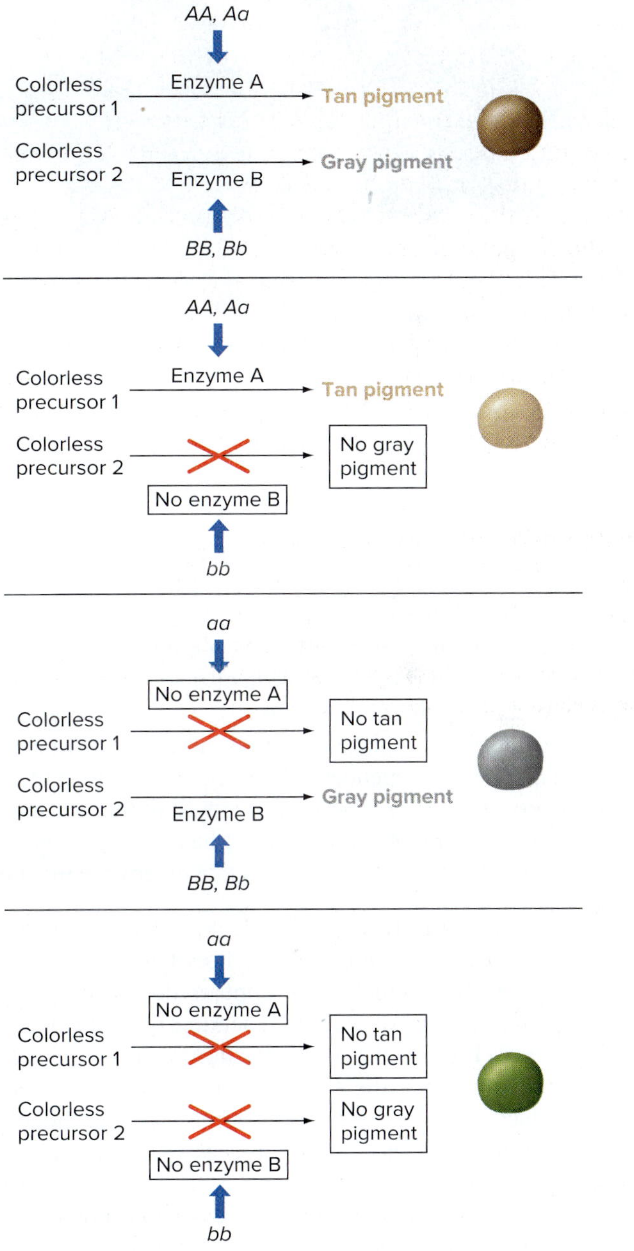

Epistasis: Alleles of One Gene Can Mask the Phenotypic Effects of Alleles of Another Gene

Sometimes, when two genes control a single trait, the four Mendelian genotypic classes produce fewer than four observable phenotypes because one gene masks the phenotypic effects of another. A gene interaction in which an allele at one gene hides the effects of alleles at another gene is known as **epistasis;** the allele that is doing the masking is *epistatic* to the gene that is being masked (the *hypostatic* gene).

Figure 3.12 Recessive epistasis determines coat color in Labrador retrievers. (a) Labrador retriever colors. **(b)** Yellow Labrador retrievers are homozygous for the recessive *e* allele, which masks the effects of the *B* or *b* alleles of a second coat color gene. In *E–* dogs, a *B–* genotype produces black and a *bb* genotype produces brown.

a: © Vanessa Grossemy/Alamy

(a) Chocolate, yellow, and black Labrador retrievers

(b) A dihybrid cross showing recessive epistasis

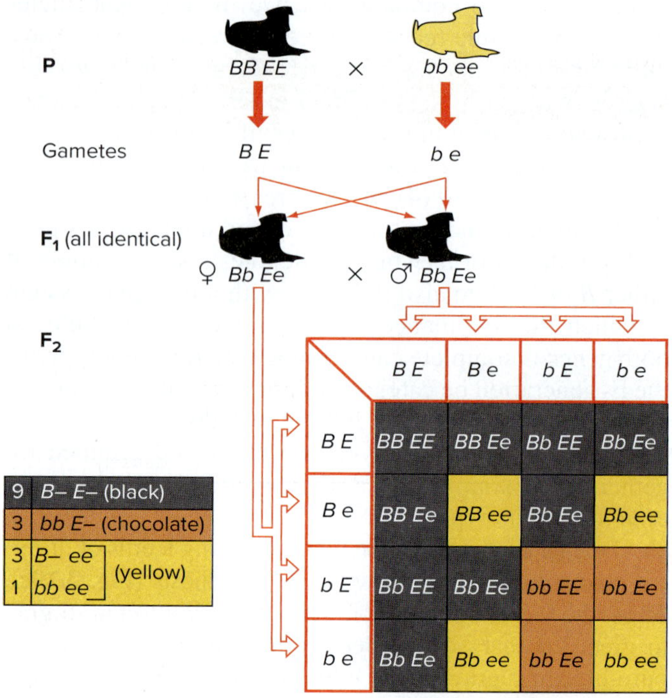

Recessive epistasis

We present here three examples of **recessive epistasis,** where homozygosity for a recessive allele of one gene hides the effect of a second gene. In other words, when an individual is homozygous for the epistatic recessive allele of the first gene, the phenotype is independent of the alleles present at the second (hypostatic) gene. The final example in this section describes a surprising phenomenon in which recessive epistasis is reciprocal between the two genes that determine the trait.

Yellow Labrador retrievers The sleek, short-haired coat of Labrador retrievers can be black, chocolate brown, or yellow (**Fig. 3.12a**). Which color appears depends on the allelic combinations of two independently assorting coat color genes (**Fig. 3.12b**). When the dominant *E* allele of the first gene is present, the *B* allele of the second gene determines black, and the recessive *bb* homozygote is chocolate. However, a double dose of the recessive allele (*ee*) hides the effect of any combination of the black or chocolate alleles to yield yellow. Thus, the recessive *ee* homozygous genotype is epistatic to any allelic combination at the second, hypostatic gene, *B*.

Let's look at the phenomenon in greater detail. Crosses between pure-breeding black retrievers (*BB EE*) and one type of pure-breeding yellow retriever (*bb ee*) create an F_1 generation of dihybrid black retrievers (*Bb Ee*). Crosses between these F_1 dihybrids produce an F_2 generation with nine black dogs (*B– E–*) for every three brown (*bb E–*) and four yellow (*– – ee*) (Fig. 3.12b). Note that only three phenotypic classes exist because the two genotypic classes without a dominant *E* allele—the three *B– ee* and the one *bb ee*—combine to produce a yellow phenotype. The telltale ratio of recessive epistasis in the F_2 generation is thus 9:3:4, with the 4 representing a combination of 3 (*B– ee*) + 1 (*bb ee*). Because the *ee* genotype masks the influence of the other gene for coat color, you cannot tell by looking at a yellow Labrador whether its genotype at the *B* locus is *B–* (black) or *bb* (chocolate).

Scientists understand with some precision the biochemical pathways in which different alleles of the *B* and *E* genes operate (**Fig. 3.13**). All coat color in dogs comes from two pigments synthesized from a common precursor: a dark pigment called eumelanin and a light pigment called pheomelanin. When Labrador retrievers have at least one copy of the *E* allele, the resultant protein E ensures that the animals will make only eumelanin and no pheomelanin. The protein specified by the *B* allele is required for eumelanin synthesis and its deposition in the hair, while the protein made by the *b* allele is less efficient. As a result, chocolate *E– bb* dogs have less eumelanin in their hairs than black dogs with at least one *B* allele (*E– B–*). But in the absence of the E protein (in *ee* dogs), only pheomelanin is synthesized, and so the dogs appear yellow. It is easy to

Figure 3.13 A biochemical explanation for coat color in Labrador retrievers. Protein E activates an enzyme that generates eumelanin from a colorless precursor. When protein E is present, only eumelanin is produced. Protein B deposits eumelanin densely so that the hair is black. Pigment is deposited less densely without protein B, producing brown hair (chocolate). In the absence of protein E, no eumelanin is produced and instead, pheomelanin (yellow pigment) is synthesized. Homozygous *ee* dogs are always yellow regardless of the gene *B* genotype. The reason is that protein B affects eumelanin only, but these *ee* animals have no eumelanin.

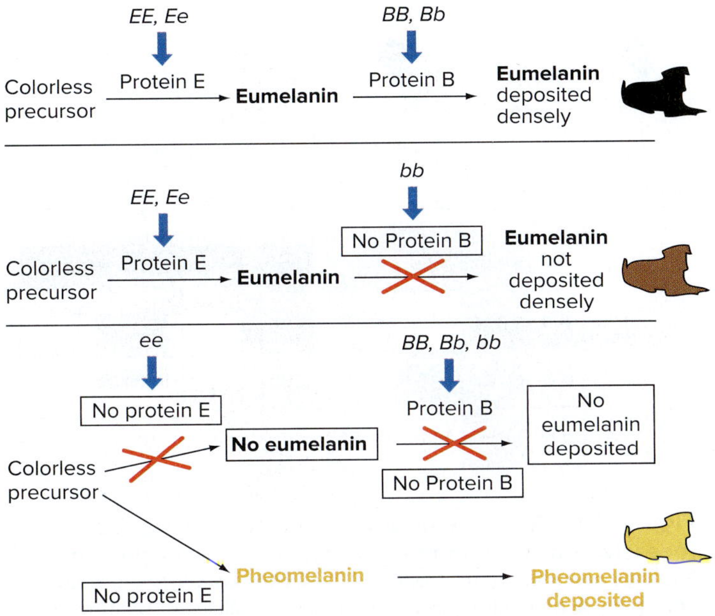

see in Fig. 3.13 why *ee* is epistatic to both alleles of gene *B*: In *ee* dogs, no eumelanin is present, so the dogs are yellow regardless of whether they are *B*– or *bb*.

The Bombay phenotype in humans An understanding of recessive epistasis made it possible to resolve an intriguing puzzle in human genetics. In rare instances, two parents who appear to have blood type O, and thus would be predicted to be genotype *ii*, produce a child who is either blood type A (genotype I^Ai) or blood type B (genotype I^Bi). This phenomenon occurs because an extremely rare trait, called the Bombay phenotype after its discovery in Bombay, India, superficially resembles blood type O. As **Fig. 3.14a** shows, the Bombay phenotype actually arises from homozygosity for a mutant recessive allele (*hh*) of a second gene that masks the effects of any ABO alleles that might be present.

Here's how it works at the molecular level (Fig. 3.14a). In the construction of the red blood cell surface molecules that determine blood type, type A individuals make an enzyme that adds polysaccharide A onto a sugar polymer known as substance H; type B individuals make an altered form of the enzyme that adds polysaccharide B onto the sugar polymer H; and type O individuals make neither A-adding nor B-adding enzyme and thus have an exposed substance H in the membranes of their red blood cells. All people of A, B, or O phenotype carry at least one dominant wild-type *H* allele for the second gene and thus produce some substance H. In contrast, the rare Bombay-phenotype individuals, with

Figure 3.14 Recessive epistasis in humans causes a rare blood type. (a) Homozygosity for the *h* Bombay allele is epistatic to the *I* gene determining ABO blood types. *hh* individuals fail to produce substance H, which is needed for the addition of A or B sugars at the surface of red blood cells. **(b)** Because *h* is epistatic to *I*, rare individuals may appear to have blood type O despite having an I^A or I^B allele. When the masked *I* allele is expressed in their *Hh* progeny, these people may be surprised by their child's blood type.

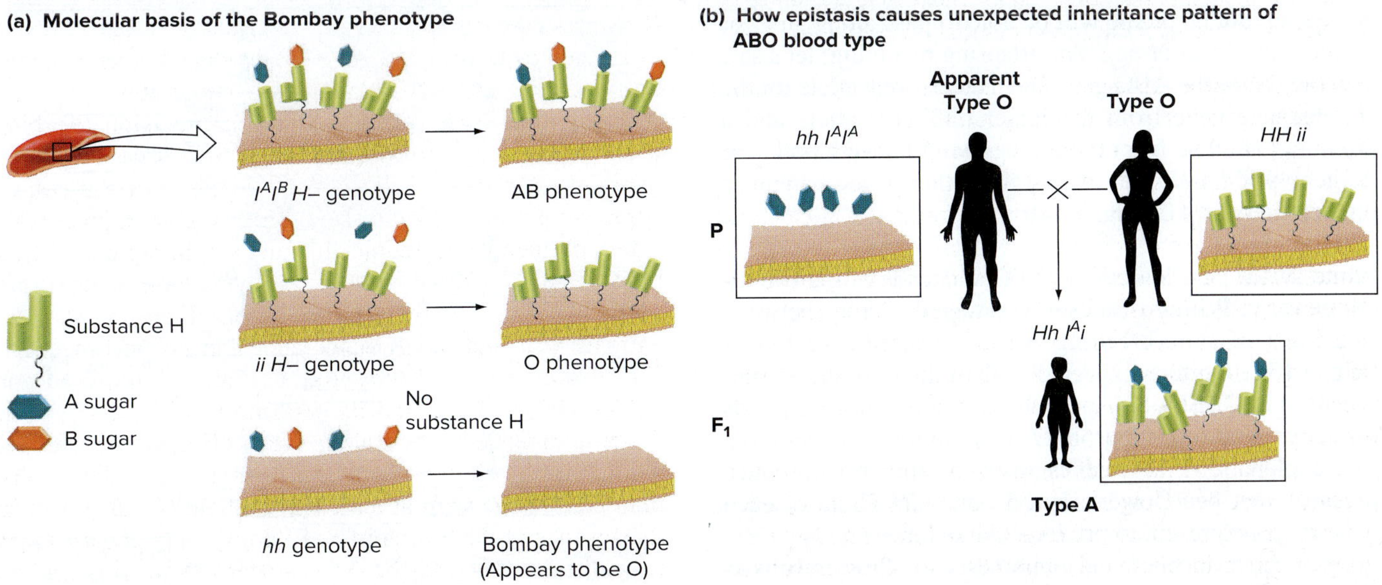

Figure 3.15 Dominant alleles of two genes needed for purple color in sweet peas. (a) White and purple sweet pea flowers. **(b)** The 9:7 ratio of purple to white F_2 plants indicates that at least one dominant allele of each gene is necessary for the development of purple color.

(a) ***Lathyrus odoratus*** **(sweet peas)**

(b) A dihybrid cross showing reciprocal recessive epistasis

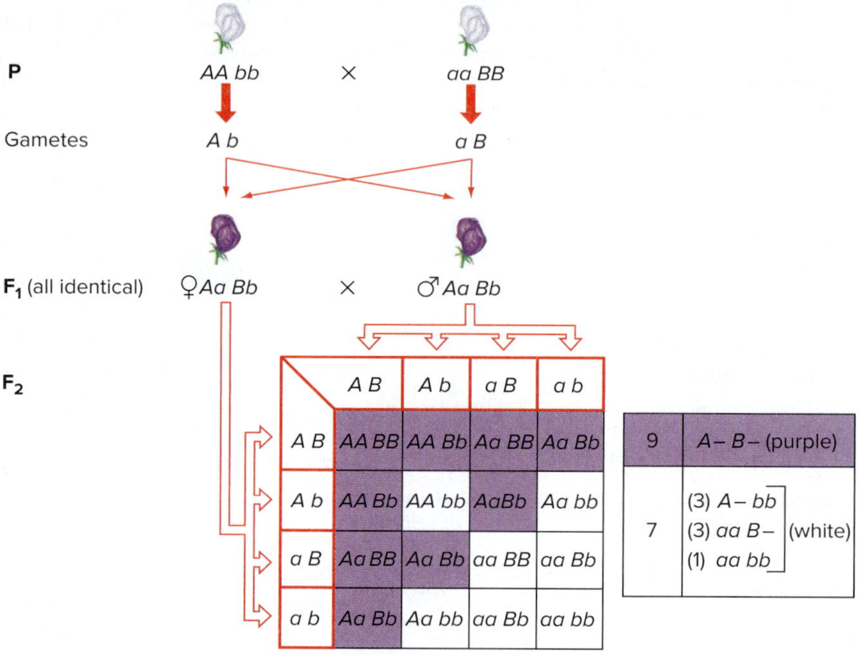

genotype *hh* for the second gene, do not make substance H at all. Thus, even if these people make an enzyme that would add A or B to this polysaccharide base, they have nothing to add it onto; as a result, Bombay-phenotype individuals appear to be type O. For this reason, homozygosity for the recessive *h* allele of the H-substance gene masks the effects of the *ABO* gene, making the *hh* genotype epistatic to any combination of I^A, I^B, and *i* alleles (except for *ii*.).

A person who carries I^A, I^B, or both I^A and I^B but is also an *hh* homozygote for the H-substance gene may appear to be type O, but he or she will be able to pass along an I^A or I^B allele in sperm or egg. An offspring receiving, let's say, an I^A allele for the ABO gene and a recessive *h* allele for the H-substance gene from the father plus an *i* allele and a dominant *H* allele from the mother would have blood type A (genotype I^Ai *Hh*), even though neither of the parents is phenotype A or AB (**Fig. 3.14b**).

White sweet pea flowers In the first decade of the twentieth century, William Bateson conducted a cross between two lines of pure-breeding white-flowered sweet peas (**Fig. 3.15a**). Quite unexpectedly, all of the F_1 progeny were purple (**Fig. 3.15b**). Self-pollination of these novel hybrids produced a ratio of 9 purple: 7 white in the F_2 generation. The explanation? Two genes work in tandem to produce purple sweet pea flowers, and a dominant allele of each gene must be present to produce that color.

A simple biochemical hypothesis for these results is shown in **Fig. 3.16.** Because it takes two enzymes catalyzing

Figure 3.16 A biochemical explanation for reciprocal recessive epistasis in the generation of sweet pea color. Enzymes specified by the dominant alleles of two genes are both necessary to produce pigment. The recessive alleles of both genes specify no enzymes. In *aa* homozygotes, no intermediate precursor 2 is generated, so even if enzyme B is available, it cannot produce purple pigment.

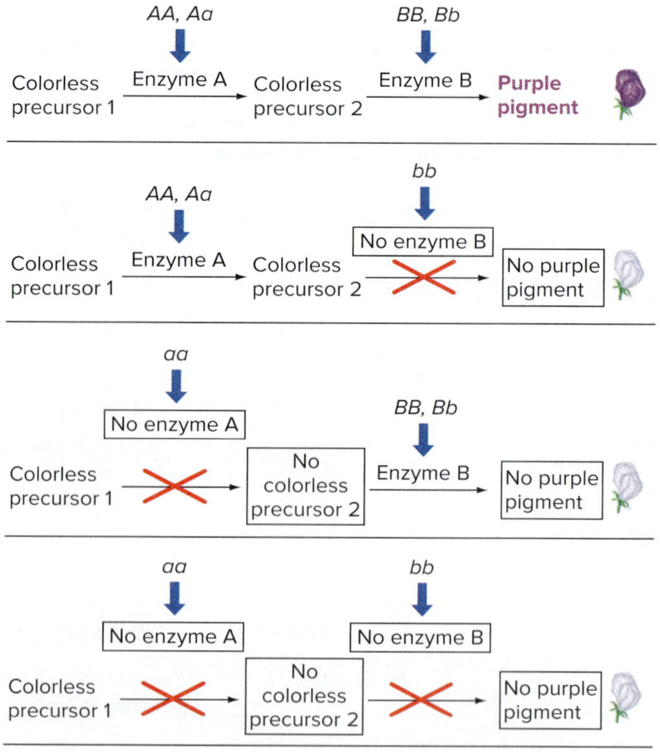

sequential biochemical reactions to change a colorless precursor into a purple pigment, only the *A– B–* genotypic class, which produces active forms of both required enzymes, can generate colored flowers. The other three genotypic classes (*A– bb, aa B–,* and *aa bb*) become grouped together with respect to phenotype because they do not specify functional forms of one or the other requisite enzyme and thus give rise to no color, which is the same as white. It is easy to see how the 7 part of the 9:7 ratio encompasses the 3:3:1 of the 9:3:3:1 F_2 ratio.

The 9:7 ratio is the phenotypic signature of this type of **reciprocal recessive epistasis** in which the dominant alleles of two genes acting together (*A– B–*) produce color or some other trait, while the other three genotypic classes (*A– bb, aa B–,* and *aa bb*) do not (see **Fig. 3.15b**). Given that the phenotype associated with either allele *A* or allele *B* is purple, then we can say that *aa* is epistatic to *B*, and *bb* is epistatic to *A*. If the sweet peas are either *aa* or *bb*, their flowers will be white regardless of whether or not they have a dominant allele of the other gene.

Dominant epistasis

Epistasis can also be caused by a dominant allele. Depending on the details of the biochemical pathway involved, dominant epistasis can result in either of two different phenotypic ratios.

Squash fruit color In summer squash, two genes influence the color of the fruit (**Fig. 3.17a**). With one gene, the dominant allele (*A–*) determines yellow, while homozygotes for the recessive allele (*aa*) are green. A second gene's dominant allele (*B–*) produces white, while *bb* fruit may be either yellow or green, depending on the genotype of the first gene. In the interaction between these two genes, the presence of *B* hides the effects of either *A–* or *aa,* producing white fruit, and *B–* is thus epistatic to any genotype of the *A* gene. The recessive *b* allele has no effect on fruit color determined by gene *A*. Epistasis in which the dominant allele of one gene hides the effects of another gene is called **dominant epistasis.** In a cross between white F_1 dihybrids (*Aa Bb*), the F_2 phenotypic ratio is 12 white : 3 yellow : 1 green (Fig. 3.17a). The 12 includes two genotypic classes: 9 *A– B–* and 3 *aa B–*.

Figure 3.17 Dominant epistasis may result in a 12:3:1 phenotypic ratio. (a) In summer squash, the dominant *B* allele causes white color and is sufficient to mask the effects of any combination of *A* and *a* alleles. As a result, yellow (*A–*) or green (*aa*) color is expressed only in *bb* individuals. **(b)** The *A* allele encodes enzyme A, while the *a* allele specifies no enzyme. Therefore, yellow pigment is present in *A–* squash and green pigment in *aa* squash. Deposition of either pigment depends on protein b encoded by allele *b*, the normal (wild-type) allele of a second gene. However, the mutant dominant allele *B* encodes an abnormal version B of this protein that prevents pigment deposition, even when the normal protein b is present. Therefore, in order to be colored, the squash must have protein b but not protein B (genotype *bb*).

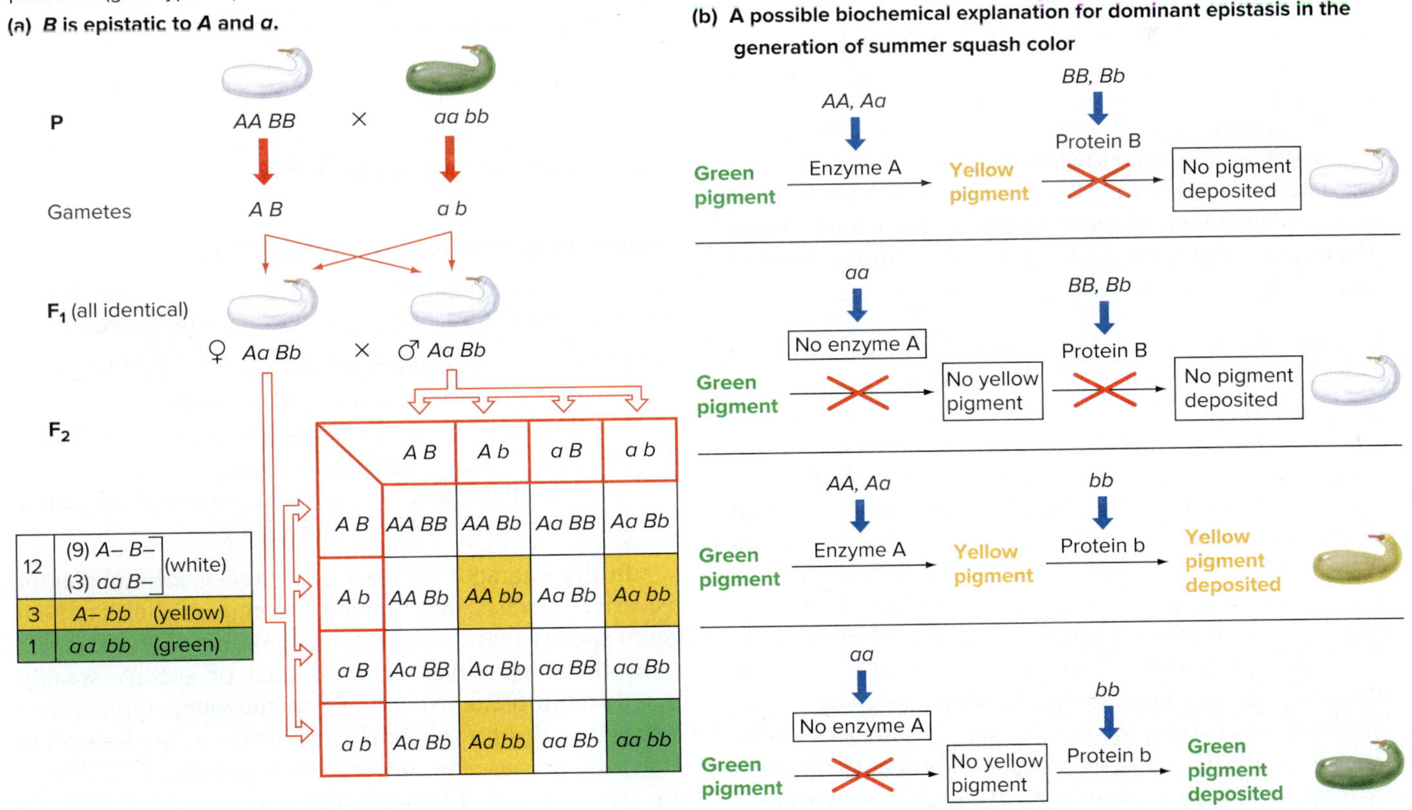

Figure 3.18 Dominant epistasis may also result in a 13:3 phenotypic ratio. (a) In the F_2 generation resulting from a dihybrid cross between white leghorn and white wyandotte chickens, the ratio of white birds to birds with color is 13:3. This ratio emerges because at least one copy of *A* and the absence of *B* is needed to produce color. **(b)** Enzyme A, encoded by allele *A*, is needed to synthesize pigment. Allele *a* encodes no enzyme. Pigment deposition in the feathers depends on protein b encoded by allele *b*, the normal (wild-type) allele of a second gene. The mutant dominant allele *B*, however, encodes an abnormal version of the protein that prevents pigment deposition, even when the normal protein b is present.

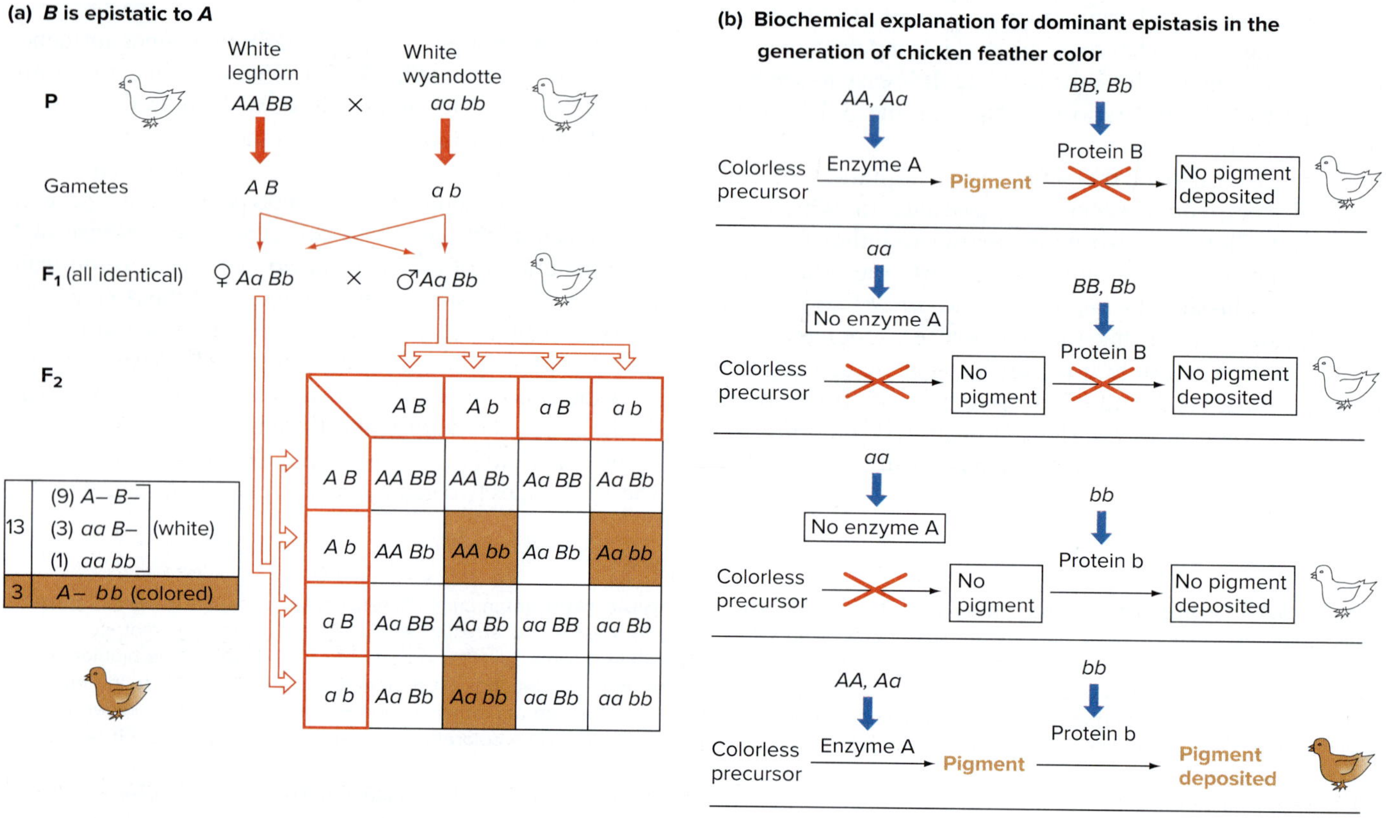

The squash genes *A* and *B* have not been identified at the molecular level, and the biochemical pathway in which they interact is unknown. However, based on knowledge of similar phenomena in other plants, a likely biochemical pathway underlying the 12:3:1 phenotypic ratio is shown in **Fig. 3.17b.**

Chicken feather color A variant ratio indicating dominant epistasis is seen in the feather color of certain chickens (**Fig. 3.18a**). White leghorns have a doubly dominant *AA BB* genotype for feather color; white wyandottes are homozygous recessive for both genes (*aa bb*). A cross between these two pure-breeding white strains produces an all-white dihybrid (*Aa Bb*) F_1 generation, but birds with color in their feathers appear in the F_2, and the ratio of white to colorful is 13:3 (Fig. 3.18a). We can explain this ratio by assuming a kind of dominant epistasis in which *B* is epistatic to *A;* the *A* allele produces color only in the absence of *B;* and the *a, B,* and *b* alleles produce no color. The interaction is characterized by a 13:3 ratio because the 9 *A– B–*, 3 *aa B–*, and 1 *aa bb* genotypic classes combine to produce only one phenotype: white. The biochemical pathway known to underly the 13:3 ratio for chicken feather color is shown in **Fig. 3.18b.**

Important points regarding epistasis

Several important points emerge from the examples of recessive and dominant epistasis we have discussed:

- Epistasis is an interaction between alleles of different genes, not between alleles of the same gene.
- In dihybrid crosses, the F_2 phenotypic ratios resulting from epistasis depend on the functions of the specific alleles and the particular biochemical pathways in which the genes participate.

In the Labrador retriever and sweet pea examples of recessive epistasis, the completely dominant alleles of both genes specify normally functional protein, while the recessive alleles are either nonfunctional or specify weakly functional protein. Nevertheless, the phenotypic ratios among the F_2 of a dihybrid cross differ in the Labradors and peas because the underlying biochemical pathways are not identical. Likewise, the two dominant epistasis

Figure 3.19 Redundant genes result in a 15:1 phenotypic ratio. (a) Normal maize leaves (*AA BB*) and a leaf lacking both dominant alleles *A* and *B* (*aa bb*). **(b)** In maize, either dominant allele *A* or *B* is sufficient for normal leaf development. Only the absence of both dominant alleles (*aa bb*) results in malformed, thin leaves. The result is a 15:1 ratio in the dihybrid cross F_2.

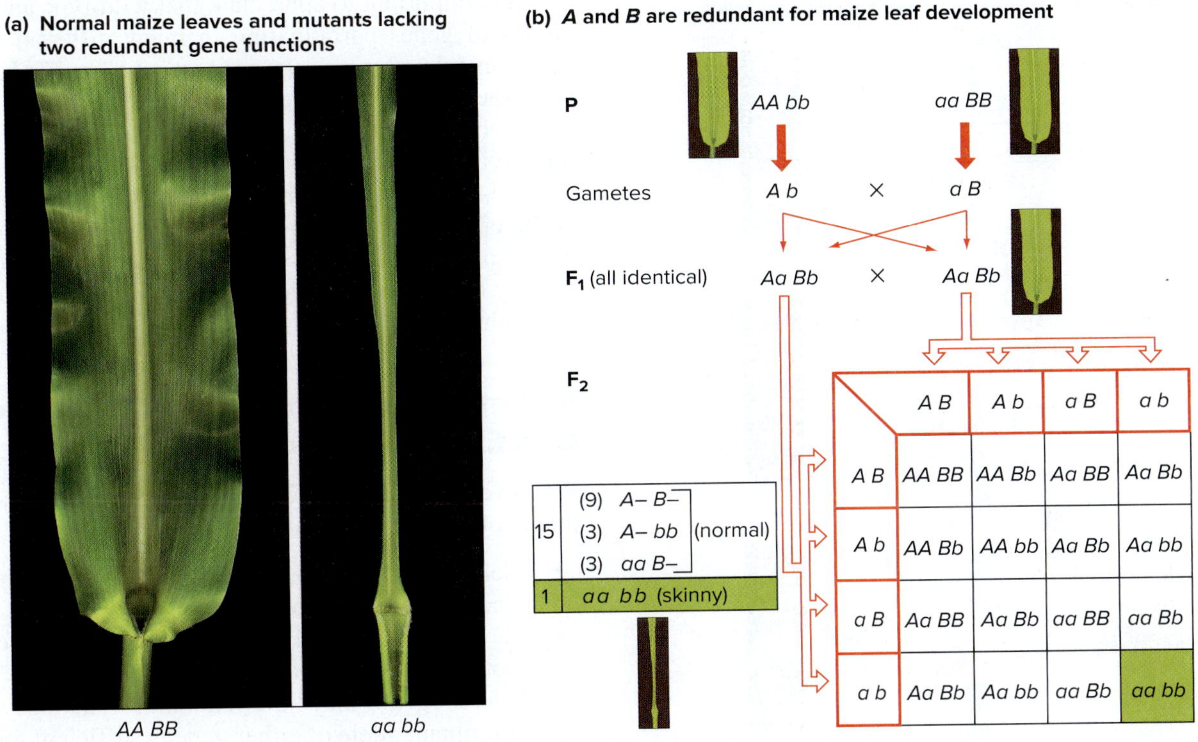

	A B	A b	a B	a b
A B	AA BB	AA Bb	Aa BB	Aa Bb
A b	AA Bb	AA bb	Aa Bb	Aa bb
a B	Aa BB	Aa Bb	aa BB	aa Bb
a b	Aa Bb	Aa bb	aa Bb	aa bb

examples (squash and chicken colors) yield different F_2 phenotypic ratios because of differences in the biochemical pathways involved.

- Recessive epistasis usually indicates that the dominant alleles of the two genes function in the same pathway to achieve a common outcome. In the Labrador retriever example, *B* and *E* both function to generate black hairs.
- Dominant epistasis usually indicates that the dominant alleles of the two genes have antagonistic functions. Both in the cases of squash and chicken color, the dominant allele of gene *B* prevents deposition of a pigment whose synthesis depends on the dominant allele of gene *A*.

Redundancy: One or More Genes in a Pathway Are Superfluous

In maize, two genes, *A* and *B*, control leaf development. Normal broad leaves develop as long as the plant has either a dominant *A* allele or a dominant *B* allele (*A– B–*, *A– bb*, or *aa B–*). However, the leaves of plants that have neither dominant allele (*aa bb*) are skinny because they contain too few cells (**Fig. 3.19a**). Given that leaves are malformed only in the absence of both *A* and *B* (*aa bb*), the F_2 phenotypic ratio signifying **redundant gene action** is 15:1 (**Fig. 3.19b**).

The proteins (A and B) encoded by the dominant alleles act in parallel, redundant pathways that recruit precursor cells to become part of the leaf (**Fig. 3.20**). That is, if either pathway functions, the leaves will develop their normal broad shape.

Often, as in this case, redundant genes specify nearly identical proteins that perform the same function. Why does the organism have two genes that do the same thing? One answer is that redundant genes often arise by chance evolutionary processes that duplicate genes, as will be explained in Chapter 10.

Summary: A Variety of Different Biochemical Pathways Can Produce Any Given Altered Mendelian Ratio

So far we have seen that when two independently assorting genes interact to determine a trait, the 9:3:3:1 ratio of the four Mendelian genotypic classes in the F_2 generation can produce a variety of phenotypic ratios, depending on the nature of the gene interactions. The result may be four, three, or two phenotypes, composed of different combinations of

Figure 3.20 A biochemical explanation for redundant gene action. The dominant alleles *A* and *B* specify proteins that function in independent pathways to instruct cells to become part of the leaf. The recessive alleles *a* and *b* specify no proteins. Because either pathway is sufficient, only plants that lack both dominant alleles have thin leaves.

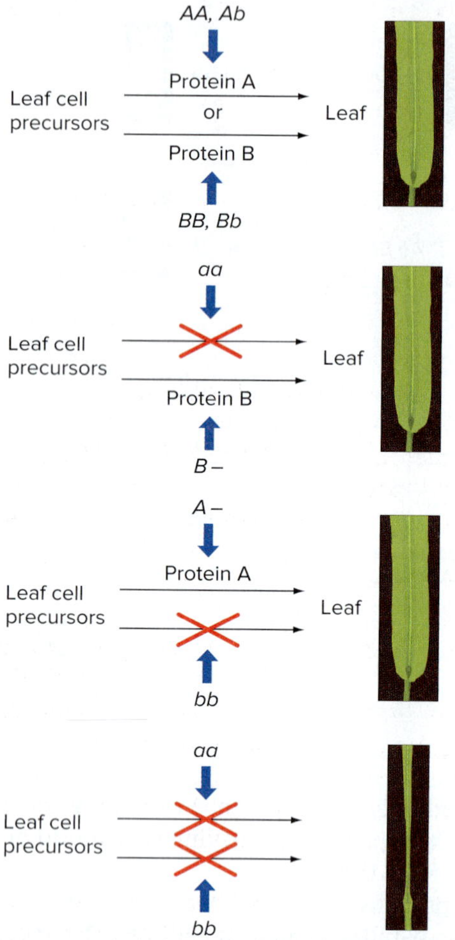

the four genotypic classes. **Table 3.2** summarizes some of the possibilities, correlating the phenotypic ratios with the genetic phenomena they reflect.

It is important to appreciate that wild-type and mutant alleles of genes participating in many different types of biochemical pathways may produce any specific F_2 phenotypic ratio shown in Table 3.2, such as 9:7 or 12:3:1. Thus, if you observe a certain ratio in a cross, you cannot infer the underlying pathway, although you can exclude some possibilities. On the other hand, as you will see in the problems at the end of this chapter, if you know the pathway's biochemistry, you can predict accurately the phenotypic ratios among the progeny of a cross involving the genes that determine the trait.

Incomplete Dominance or Codominance Can Expand Phenotypic Variation

We have identified to this point several variations on the theme of two-gene inheritance:

- alleles of different genes can interact *additively* to generate novel phenotypes;
- one gene's alleles can mask the effects of alleles at another gene (*epistasis*);
- different genes may have *redundant* functions so that a dominant allele of either gene is sufficient for the production of a particular normal phenotype.

All but the first of these interactions between different genes resulted in the merging of two or more of Mendel's four genotypic classes into one phenotypic class. For example, when genes are redundant, *A– B–*, *A– bb*, and *aa B–* have the same phenotype. In examining each of these

TABLE 3.2 Summary of Two-Gene Interactions

Gene Interaction	Example	F_2 Genotypic Ratios from an F_1 Dihybrid Cross: *A– B–*	*A– bb*	*aa B–*	*aa bb*	F_2 Phenotypic Ratio
Additive: Four distinct F_2 phenotypes	Lentil: seed coat color (see Fig. 3.10a)	9	3	3	1	9:3:3:1
Recessive epistasis: When homozygous, recessive allele of one gene masks both alleles of another gene	Labrador retriever: coat color (see Fig. 3.12b)	9	3	3	1	9:3:4
Reciprocal recessive epistasis: When homozygous, recessive allele of each gene masks the dominant allele of the other gene	Sweet pea: flower color (see Fig. 3.15b)	9	3	3	1	9:7
Dominant epistasis I: Dominant allele of one gene hides effects of both alleles of the other gene	Summer squash: color (see Fig. 3.17a)	9	3	3	1	12:3:1
Dominant epistasis II: Dominant allele of one gene hides effects of dominant allele of other gene	Chicken feathers: color (see Fig. 3.18a)	9	3	3	1	13:3
Redundancy: Only one dominant allele of either of two genes is necessary to produce phenotype	Maize: leaf development (see Fig. 3.19b)	9	3	3	1	15:1

categories, for the sake of simplicity, we have looked at examples in which one allele of each gene in a pair showed complete dominance over the other. But for any type of gene interaction, the alleles of one or both genes may exhibit incomplete dominance or codominance, and these possibilities increase the potential for phenotypic diversity. For example, **Fig. 3.21** shows how incomplete dominance at both genes in a dihybrid cross results not in a collapse of several genotypic classes into one but rather an expansion—each of the nine genotypes in the dihybrid cross F_2 corresponds to a different phenotype.

A simple biochemical explanation for the phenotypes in Fig. 3.21 is similar to that for incomplete dominance in Fig. 3.3b, where the amount of red pigment produced was proportional to the amount of an enzyme. The difference here is that purple pigmentation requires the action of two enzymes, A and B, and one is more efficient than the other, resulting in one gene (in this case, the *A* gene) contributing more to the purple phenotype than the other gene.

Figure 3.21 With incomplete dominance, the interaction of two genes can produce nine different phenotypes for a single trait. In this example, two genes produce purple pigments. Alleles A^1 and A^2 of the first gene exhibit incomplete dominance, as do alleles B^1 and B^2 of the second gene. The two alleles of each gene can generate three different phenotypes, so double heterozygotes can produce nine (3 × 3) different colors in a ratio of 1:2:2:1:4:1:2:2:1.

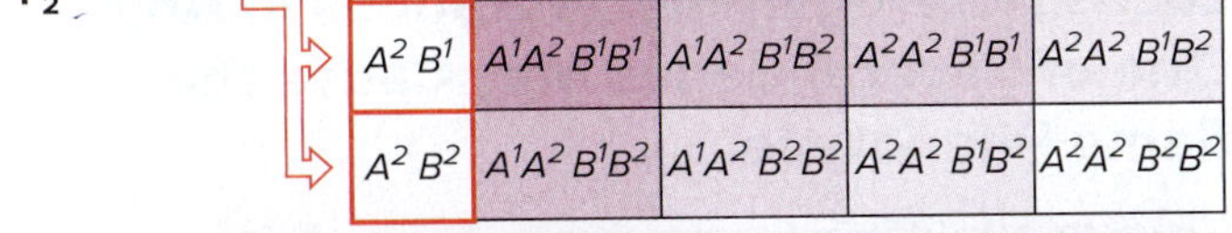

1	$A^1A^1\ B^1B^1$	purple shade 9
2	$A^1A^1\ B^1B^2$	purple shade 8
2	$A^1A^2\ B^1B^1$	purple shade 7
1	$A^1A^1\ B^2B^2$	purple shade 6
4	$A^1A^2\ B^1B^2$	purple shade 5
1	$A^2A^2\ B^1B^1$	purple shade 4
2	$A^1A^2\ B^2B^2$	purple shade 3
2	$A^2A^2\ B^1B^2$	purple shade 2
1	$A^2A^2\ B^2B^2$	purple shade 1 (white)

Although the possibilities for variation are manifold, none of the observed departures from Mendelian phenotypic ratios contradicts Mendel's genetic laws of segregation and independent assortment. The alleles of each gene still segregate as he proposed. Interactions between the alleles of many genes simply make it harder to unravel the complex relation of genotype to phenotype.

Breeding Studies Help Geneticists Determine Whether One or Two Genes Determine a Trait

How do geneticists know whether a particular trait is caused by the alleles of one gene or by two genes interacting in one of a number of possible ways? Breeding tests can usually resolve the issue. Phenotypic ratios diagnostic of a particular mode of inheritance (for instance, the 9:7 or 13:3 ratios indicating that two genes are interacting) can provide the first clues and suggest hypotheses. Further breeding studies can then show which hypothesis is correct.

As an example, a mating of one strain of pure-breeding white albino mice with pure-breeding brown results in black hybrids; and a cross between the black F_1 hybrids produces 90 black, 30 brown, and 40 albino offspring. What is the genetic constitution of these phenotypes? We could assume that we are seeing the 9:3:4 ratio of recessive epistasis and hypothesize that two interacting genes (call them *B* and *C*) control color. In this model, each gene has completely dominant and recessive alleles, and the homozygous recessive of one gene is epistatic to both alleles of the other gene (**Fig. 3.22a**). This idea makes sense, but it is not the only hypothesis consistent with the data.

We might also explain the data—160 progeny in a ratio of 90:30:40—by the activity of one gene (**Fig. 3.22b**). According to this one-gene hypothesis, albinos would be homozygotes for one allele (B^1B^1), brown mice would be homozygotes for a second allele (B^2B^2), and black mice would be heterozygotes (B^1B^2) that have their own novel phenotype because B^1 and B^2 are incompletely dominant. Under this system, a mating of black (B^1B^2) to black (B^1B^2) would be expected to produce 1 B^2B^2 brown : 2 B^1B^1 black : 1 B^1B^1 albino, or 40 brown : 80 black : 40 albino. Is it possible that the 30 brown, 90 black, and 40 albino mice actually counted were obtained from the inheritance of a single gene? Intuitively, the answer is yes because the ratios 40:80:40 and 30:90:40 do not seem that different. We know that if we flip a coin 100 times, it doesn't always come up 50 heads : 50 tails; sometimes it's 60:40 just by chance. So, how can we decide between the two-gene and the one-gene model?

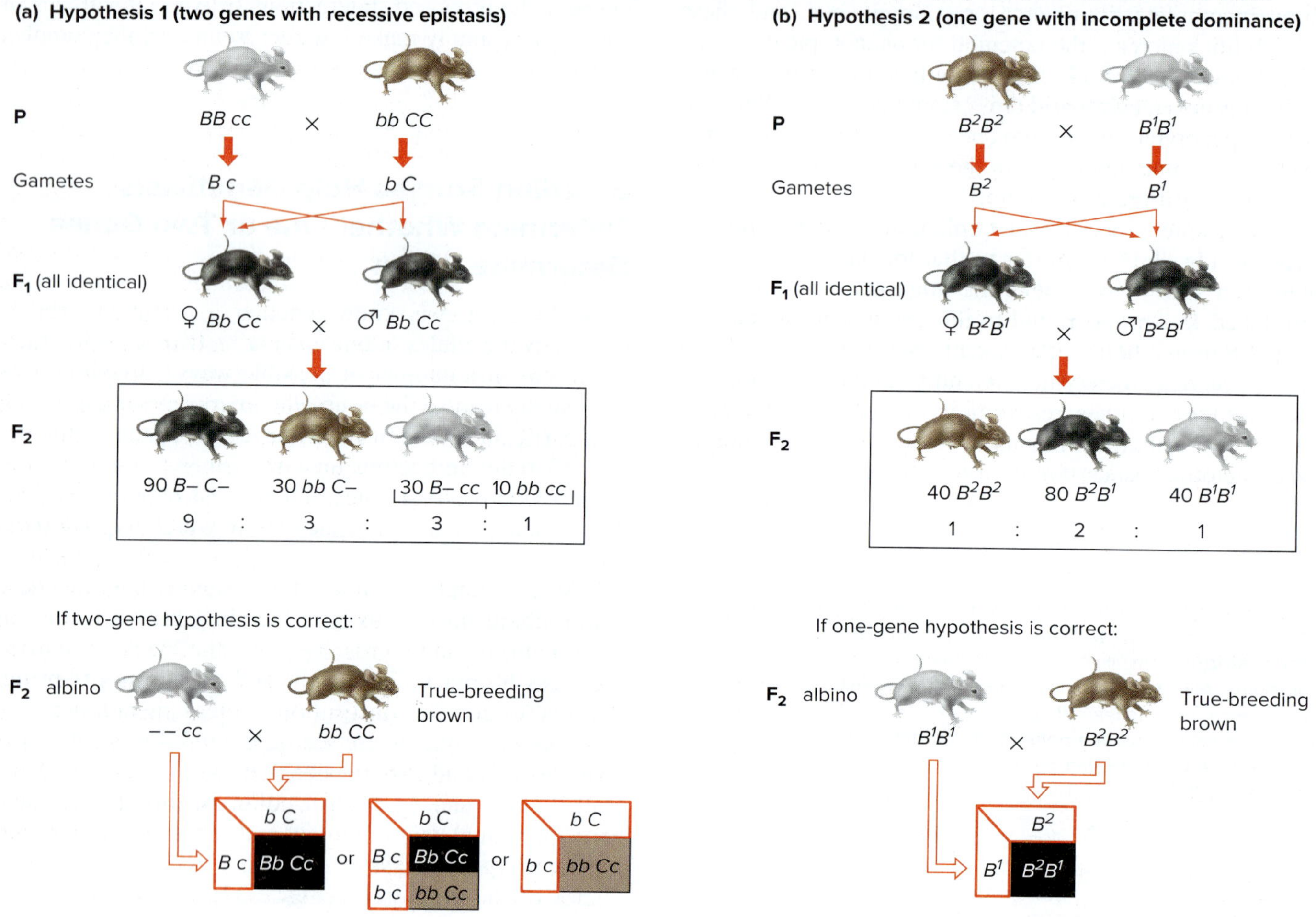

Figure 3.22 Specific breeding tests can help decide between hypotheses. Either of two models could explain the results of a cross tracking coat color in mice. **(a)** In one hypothesis, two genes interact with recessive epistasis to produce a 9:3:4 ratio. **(b)** In the other hypothesis, a single gene with incomplete dominance between the alleles generates the observed results. One way to decide between these models is to cross each of several albino F_2 mice with true-breeding brown mice. The two-gene model predicts several different outcomes depending on the – – *cc* albino's genotype at the *B* gene. The one-gene model predicts that all progeny of all the crosses will be black.

The answer is that we can use other types of crosses to verify or refute the hypotheses. For instance, if the one-gene hypothesis were correct, a mating of F_2 albinos with pure-breeding brown mice similar to those of the parental generation would produce all black heterozygotes [brown (*BB*) × albino (*bb*) = all black (*Bb*)] (Fig. 3.22b). But if the two-gene hypothesis is correct, with recessive mutations at an albino gene (called *C*) epistatic to all expression from the *B* gene, different matings of pure-breeding brown (*bb CC*) with the F_2 albinos (– – *cc*) will give different results—all progeny are black; half are black and half brown; all are brown—depending on the albino's genotype at the *B* gene (see Fig. 3.22a). In fact, when the experiment is actually performed, the diversity of results confirms the two-gene hypothesis.

Locus Heterogeneity: Mutations in Any One of Several Genes May Cause the Same Phenotype

Close to 50 different genes have mutant alleles that can cause deafness in humans. Many genes generate the developmental pathway that brings about hearing, and a loss of function in any part of the pathway, for instance, in one small bone of the middle ear, can result in deafness. In other words, it takes a dominant wild-type allele at each of these 50 genes to produce normal hearing. Thus, deafness is a **heterogeneous trait:** A mutation at any one of a number of genes can give rise to the same phenotype. We saw earlier (Fig. 3.15b) that whiteness of sweet pea flowers is also a heterogeneous trait; *AA bb* and *aa BB* flowers, each homozygous for recessive, nonfunctional alleles of different genes, were both white.

Evidence for locus heterogeneity in human pedigrees

Careful examination of many family pedigrees can reveal whether **locus heterogeneity**—a property of a trait where mutations in any one of two or more genes results in the same mutant phenotype—explains the inheritance pattern of a trait. In the case of deafness, for example, whether a particular nonhearing man and a particular nonhearing woman carry mutations in the same gene or different ones can be determined if they have children together. If they have only children who can hear, the parents most likely carry mutations at two different genes, and the children carry one normal, wild-type allele for both of those genes (**Fig. 3.23a**). By contrast, if all of their children are deaf, it is likely that both parents are homozygous for a mutation in the same gene, and all of their children are also homozygous for this same mutation (**Fig. 3.23b**).

Complementation and complementation tests

The method outlined in Fig. 3.23 for discovering whether a particular phenotype arises from mutations in the same or separate genes is a naturally occurring version of an experimental genetic tool called the **complementation test.** Simply put, when what appears to be an identical *recessive* phenotype arises in two separate breeding lines, geneticists want to know whether mutations in the same gene are responsible for the phenotype in both lines. They answer this question by setting up a mating between affected individuals from the two lines. If offspring receiving the two mutations—one from each parent—express the wild-type phenotype, **complementation** has occurred. The observation of complementation means that the original mutations affected two different genes, and for both genes, the normal allele from one parent can provide what the mutant allele of the same gene from the other parent cannot. Note that a finding of complementation implies that the trait in question must be heterogeneous.

You previously saw an example of complementation in Fig. 3.15b. There, the white parental plants were homozygous for nonfunctional alleles of different genes required for purple pigment synthesis. The F_1 were purple because the gamete of each parent provided the wild-type allele that the other lacked. The pedigree for deafness in Fig. 3.23a, in which all the children of two deaf parents had normal hearing, provides another example of this same phenomenon. By contrast, if all the offspring of affected parents express the mutant phenotype, no complementation has occurred. Each offspring received two recessive mutant alleles—one from each parent—of the same gene (Fig. 3.23b). A lack of complementation does not exclude the possibility that a trait could be heterogeneous, but instead it simply indicates that the parents involved in the particular cross had mutant alleles of the same gene.

You can quiz your understanding of the related concepts of locus heterogeneity and complementation by considering a form of albinism known as ocular-cutaneous albinism (OCA). People with this inherited condition have little or no pigment in their skin, hair, and eyes (**Fig. 3.24a**). The horizontal inheritance pattern seen in **Fig. 3.24b** suggests that OCA is determined by the recessive allele of one gene, with albino family members being homozygotes for that allele. But a 1952 paper on albinism reported a family in which two albino parents produced three normally pigmented children (**Fig. 3.24c**). How would you explain this phenomenon?

The answer is that albinism is another example of locus heterogeneity: Mutant alleles at any one of several different genes can cause the condition. The reported mating was, in effect, an inadvertent complementation test. The complementation observed showed that one albino parent was homozygous for an OCA-causing mutation in gene *A*, while the other albino parent was homozygous for an OCA-causing mutation in a different gene, *B*.

Figure 3.23 **Locus heterogeneity in humans: Mutations in many genes can cause deafness.** **(a)** Two deaf parents can have hearing offspring if the mother and father are homozygous for recessive mutations in different genes. **(b)** Two deaf parents with mutations in the same gene may produce all deaf children.

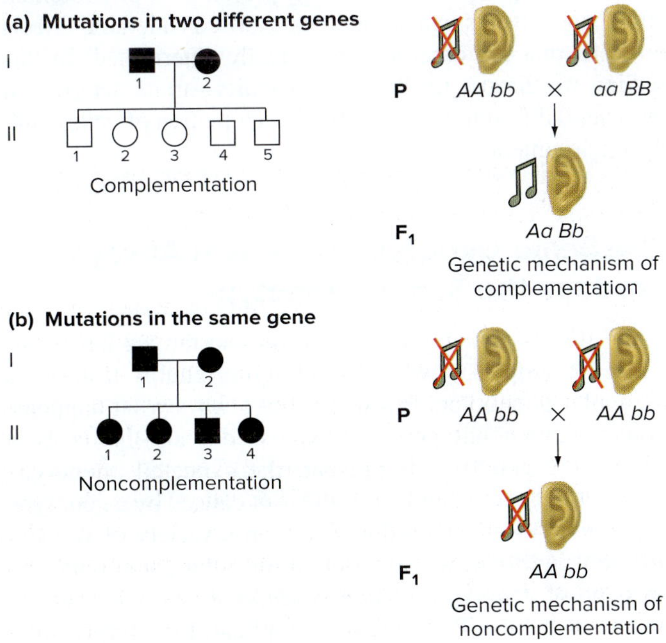

Figure 3.24 Family pedigrees help unravel the genetic basis of ocular-cutaneous albinism (OCA). **(a)** An albino Nigerian girl and her sister celebrating the conclusion of the All Africa games. **(b)** A pedigree following the inheritance of OCA in an inbred family indicates that the trait is recessive. **(c)** A family in which two albino parents have nonalbino children demonstrates that homozygosity for a recessive allele of either of two genes can cause OCA.

a: © Radu Sigheti/Reuters

(a) Ocular-cutaneous albinism (OCA)

(b) OCA is recessive

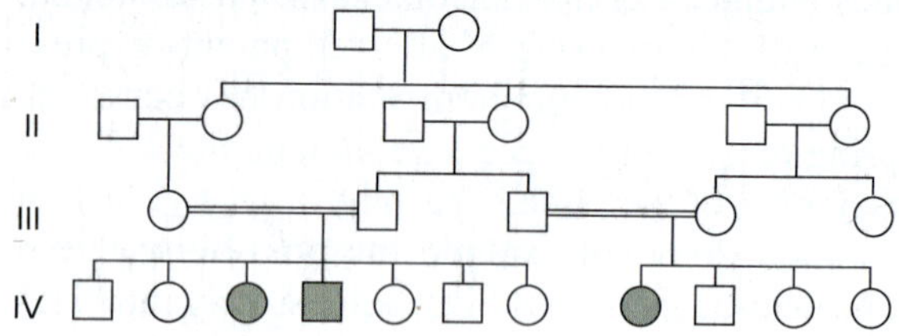

(c) Complementation for albinism

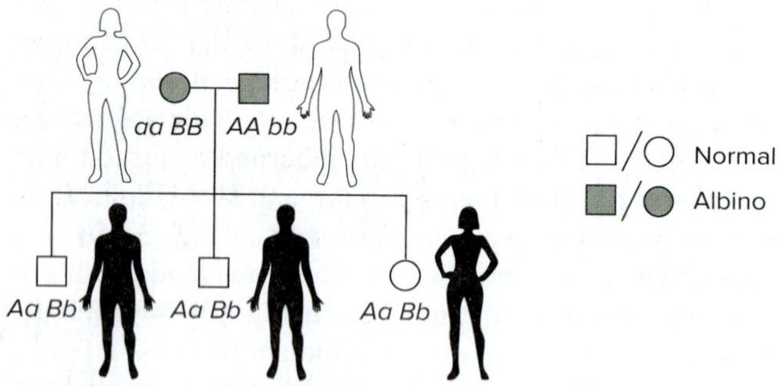

essential concepts

- Two genes may interact to affect a single trait; these interactions may be detected by ratios that can be predicted from Mendelian principles.
- Retention of the 9:3:3:1 phenotypic ratio usually indicates that two genes function in independent pathways and their alleles interact *additively*.
- In *epistasis*, an allele at one gene can hide traits otherwise caused by alleles at another gene.
- When genes display *redundancy* for a trait, one dominant and normally functioning allele of either gene is sufficient to generate the normal phenotype.
- Many traits exhibit *locus heterogeneity*, in which homozygosity for mutations at any one of several genes can produce the same mutant phenotype.
- *Complementation* occurs in the progeny of pure-breeding parents with the same mutant phenotype if the parents are homozygous for recessive, nonfunctional alleles of different genes whose products function in a common pathway.

3.3 Extensions to Mendel for Multifactorial Inheritance

learning objectives

1. Discuss the factors that can cause different individuals with the same genotype to be phenotypically dissimilar.
2. Explain how Mendelian genetics is compatible with the fact that many traits, such as human height and skin colors, exhibit continuous variation.

The inheritance of many traits appears to be more complex than can be explained by the participation of only one or two genes in patterns compatible with straightforward Mendelian principles. Of course, one reason for this complexity is that more than two genes can influence certain traits. But a second reason is that genes are not the only players: The environment and chance events can sometimes exert considerable effects on traits that are otherwise genetically determined. In this section we discuss **multifactorial traits**—traits determined by several different genes, or by the interaction of genes with the environment.

The Same Genotype Does Not Always Produce the Same Phenotype

In our discussion of gene interactions so far, we have considered examples in which a genotype reliably fashions a particular phenotype. But this is not always what happens. Sometimes a genotype is not expressed at all; that is, even though the genotype is present, the expected phenotype does not appear. Other times, the trait caused by a genotype is expressed to varying degrees or in a variety of ways in different individuals. Factors that alter the phenotypic expression of genotype include *modifier genes,* the environment, and chance. These factors complicate the interpretation of breeding experiments.

Penetrance and expressivity

Retinoblastoma, the most malignant form of eye cancer, arises from a dominant mutation in one gene, but only about 75% of people who carry the mutant allele develop the disease. Geneticists use the term **penetrance** to describe the proportion of individuals with a particular genotype who show the expected phenotype. Penetrance can be *complete* (100%), as in the traits that Mendel studied, or *incomplete,* as in retinoblastoma. For retinoblastoma, the penetrance is ~75%.

In some people with retinoblastoma, only one eye is affected, while in other individuals with the phenotype, both eyes are diseased. **Expressivity** refers to the degree or intensity with which a particular genotype is expressed in a phenotype. Expressivity can be *variable,* as in retinoblastoma (one or both eyes affected), or *unvarying,* as in pea color (all *yy* peas are green). As we will see, the incomplete penetrance and variable expressivity of retinoblastoma are mainly the result of chance, but in other cases, it is other genes and/or the environment that cause variations in phenotype. **Figure 3.25** summarizes in graphic form the differences between complete penetrance, incomplete penetrance, variable expressivity, and unvarying expressivity.

Modifier genes

Not all genes that influence the appearance of a trait contribute equally to the phenotype. Major genes have a large influence, while **modifier genes** have a more subtle, secondary effect. Modifier genes alter the phenotypes produced by the alleles of other genes. No formal distinction exists between major and modifier genes. Rather, a continuum exists between the two, and the cutoff is arbitrary. Scientists sometimes call the set of unknown modifier genes that influence the action of known genes the **genetic background.**

Modifier genes influence the length of a mouse's tail. The mutant *T* allele of the tail-length gene causes a shortening of the normally long wild-type tail. But not all mice carrying the *T* mutation have the same length tail. A comparison of several inbred lines points to modifier genes as the cause of this variable expressivity. In one inbred line, mice carrying the *T* mutation have tails that are approximately 75% as long as normal tails; in another inbred line, the tails are 50% normal length; and in a third line, the tails are only 10% as long as wild-type tails. Because all members of each inbred line grow the same length tail, no matter what the environment (for example, diet, cage temperature, or bedding), geneticists conclude it is genes and not the environment or chance that determine the length of a mutant mouse's tail. Different inbred lines most likely carry different alleles of the modifier genes that determine exactly how short the tail will be when the *T* mutation is present; that is, these lines have different genetic backgrounds.

Figure 3.25 Phenotypes may show variations in penetrance and expressivity. A genotype is completely penetrant when all individuals with that genotype have the same phenotype (*green*). Some genotypes are incompletely penetrant—some individuals with the same genotype show the phenotype and others do not. Genotypes may also show variable expressivity, meaning that individuals with the same genotype may show the trait but to different degrees.

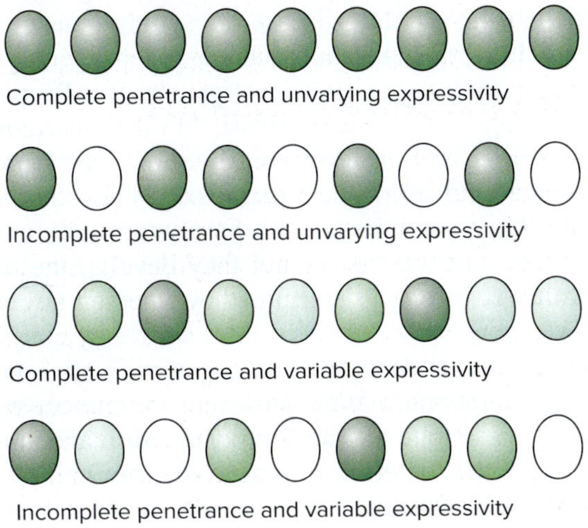

Environmental effects on phenotype

Temperature is one element of the environment that can have a visible effect on phenotype. For example, temperature influences the unique coat color pattern of Siamese cats (**Fig. 3.26**). These domestic felines are homozygous for one of the multiple alleles of a gene that encodes an enzyme catalyzing the production of the dark pigment melanin. The form of the enzyme generated by the variant *Siamese* allele does not function at the cat's normal core body temperature. It becomes active only at the lower temperatures found in the cat's extremities, where it promotes the production of melanin, which darkens the animal's ears, nose, paws, and tail. The enzyme is thus *temperature sensitive.* Under the normal environmental conditions in temperate climates, the Siamese phenotype does not vary much in expressivity from one cat to another, but one can imagine the expression of a very different phenotype—no dark extremities—in equatorial deserts, where the ambient temperature is at or above normal body temperature.

Temperature can also affect survivability. In one type of experimentally bred fruit fly (*Drosophila melanogaster*), some individuals develop and multiply normally at temperatures between 18°C and 29°C; but if the thermometer climbs beyond that cutoff for a short time, they become reversibly paralyzed; and if the temperature remains high for more than a few hours, they die. These insects carry a temperature-sensitive allele of the *shibire*

Figure 3.26 In Siamese cats, temperature affects coat color. **(a)** A Siamese cat. **(b)** Melanin is produced only in the cooler extremities. This phenomenon occurs because Siamese cats are homozygous for a mutation that renders an enzyme involved in melanin synthesis temperature sensitive. The mutant enzyme is active at lower temperatures but inactive at higher temperatures.

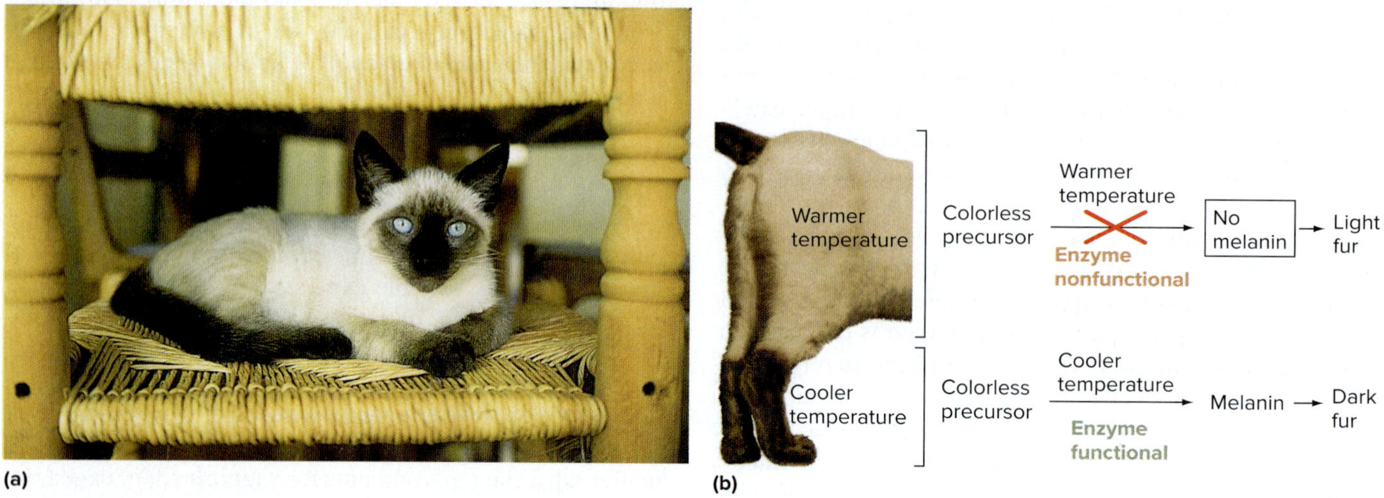

gene, which specifies a protein essential for nerve cell transmission. This type of allele is known as a **conditional lethal** because it is lethal only under certain conditions. The range of temperatures under which the insects remain viable are **permissive conditions;** the lethal temperatures above that are **restrictive conditions.** Thus, at one temperature, the allele gives rise to a phenotype that is indistinguishable from the wild type, while at another temperature, the same allele generates a mutant phenotype (in this case, lethality). Flies with the wild-type *shibire* allele are viable even at the higher temperatures. The fact that some mutations are lethal only under certain conditions clearly illustrates that the environment can affect the penetrance of a phenotype.

Even in genetically normal individuals, exposure to chemicals or other environmental agents can have phenotypic consequences that are similar to those caused by mutant alleles of specific genes. A change in phenotype arising in such a way is known as a **phenocopy.** By definition, phenocopies are not heritable because they do not result from a change in a gene. In humans, ingestion of the sedative thalidomide by pregnant women in the early 1960s produced a phenocopy of a rare dominant trait called *phocomelia.* By disrupting limb development in otherwise normal fetuses, the drug mimicked the effect of the phocomelia-causing mutation. When this problem became evident, thalidomide was withdrawn from the market.

Some types of environmental change may have a positive effect on an organism's survivability. In the following example, a straightforward application of medical science artificially reduces the penetrance of a mutant phenotype. Children born with the recessive trait known as *phenylketonuria,* or *PKU,* will develop a range of neurological problems, including convulsive seizures and mental impairment, unless they are put on a special diet. Homozygosity for the mutant PKU allele eliminates the activity of a gene encoding the enzyme phenylalanine hydroxylase. This enzyme normally converts the amino acid phenylalanine to the amino acid tyrosine. Absence of the enzyme causes a buildup of phenylalanine, and this buildup results in neurological problems. Today, a reliable blood test can detect the condition in newborns. Once a baby with PKU is identified, a protective diet that excludes phenylalanine is prescribed. The diet must also provide enough calories to prevent the infant's body from breaking down its own proteins, thereby releasing the damaging amino acid from within. Such dietary therapy—a simple change in the environment—now enables many PKU infants to develop into healthy adults.

Finally, two of the top killer diseases in the United States—cardiovascular disease and lung cancer—also illustrate how the environment can alter phenotype by influencing both expressivity and penetrance. People may inherit a propensity to heart disease, but the environmental factors of diet and exercise contribute to the occurrence (penetrance) and seriousness (expressivity) of their condition. Similarly, some people are born genetically prone to lung cancer, but whether or not they develop the disease (penetrance) is strongly determined by whether they choose to smoke.

Thus, various aspects of an organism's environment, including temperature, diet, and exercise, interact with its genotype to generate the phenotype, the ultimate combination of traits that determines what a plant or animal looks like and how it behaves.

The effects of random events on penetrance and expressivity

Whether a carrier of the retinoblastoma mutation described earlier develops the phenotype, and whether the cancer affects one or both eyes, depend on additional genetic events that occur at random. To produce retinoblastoma, these events must alter the second allele of the gene in specific body cells. Examples of random events that can trigger the onset of the disease include cosmic rays (to which humans are constantly exposed) that alter the genetic material in retinal cells or mistakes made during cell division in the retina. Chance events provide the second *hit*—a mutation in the second copy of the retinoblastoma gene—necessary to turn a normal retinal cell into a cancerous one. The phenotype of retinoblastoma thus results from a specific heritable mutation in a specific gene, but the incomplete penetrance and variable expressivity of the disease depend on random genetic events that affect the other allele in certain cells. The relationship between genotype and phenotype as it applies to cancer will be discussed fully in Chapter 20.

By contributing to incomplete penetrance and variable expressivity, modifier genes, the environment, and chance give rise to phenotypic variation. The probability of penetrance and the level of expressivity cannot be derived from the original Mendelian principles of segregation and independent assortment; they are established empirically by observation and counting.

Mendelian Principles Can Also Explain Continuous Variation

In Mendel's experiments, height in pea plants was determined by two segregating alleles of one gene (in the wild, plant height is determined by many genes, but in Mendel's inbred populations, the alleles of all but one of these genes were invariant). The phenotypes that resulted from these alternative alleles were clear-cut, either short or tall, and pea plant height was therefore known as a **discontinuous trait** (or **discrete trait**). In contrast, because human populations are not inbred, height in people is determined by alleles of many different genes whose interaction with each other and the environment produces continuous phenotypic variation; height in humans is thus an example of a **continuous trait** (or **quantitative trait**). Within human populations, individual heights vary over a range of values that when charted on a graph produce a bell curve (**Fig. 3.27a**). In fact, many human traits, including height, weight, and skin color, show continuous variation, rather than the clear-cut alternatives analyzed by Mendel.

Continuous traits often appear to blend and unblend. Think for a moment of skin color. Children of marriages between people of African and Northern European descent, for example, often seem to be a blend of their parents' skin colors. Progeny of these F_1 individuals produce offspring displaying a wide range of skin pigmentation; a few may be as light as the original Northern European parent, a few as dark as the original African parent, but most will fall in a range between the two (**Fig. 3.27b**). For such reasons, early human geneticists were slow to accept Mendelian analysis. Because they were working with *outbred populations* (populations in which individuals differ in alleles of many genes), these scientists found very few examples of either-or Mendelian traits in normal, healthy people.

By 1930, however, studies of corn and tobacco demonstrated conclusively that it is possible to provide a Mendelian explanation of continuous variation by simply increasing the number of genes contributing to a phenotype. The more genes, the more phenotypic classes, and the more classes, the more the variation appears continuous.

As a hypothetical example, consider a series of genes (*A*, *B*, *C*, . . .) all affecting the height of pole beans. For

Figure 3.27 Continuous traits in humans. **(a)** Women runners at the start of a 5th Avenue mile race in New York City demonstrate that height is a trait showing continuous variation. **(b)** The skin color of most F_1 offspring is usually between the parental extremes, while the F_2 generation exhibits a broader distribution of continuous variation.

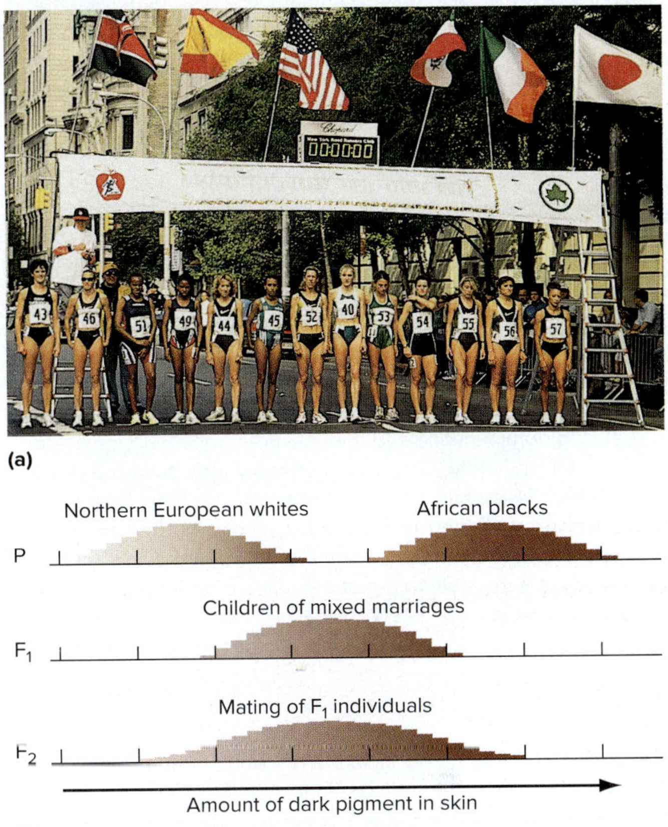

(a) (b)

GENETICS AND SOCIETY

Crowd: © Image Source/Getty Images RF

Disease Prevention Versus the Right to Privacy

In one of the most extensive human pedigrees ever assembled, a team of researchers traced a familial pattern of blindness back through five centuries of related individuals to its origin in a couple who died in a small town in northwestern France in 1495. More than 30,000 French men and women alive today descended from that one fifteenth-century couple, and within this direct lineage reside close to half of all reported French cases of hereditary juvenile glaucoma. The massive genealogical tree for the trait (when posted on the office wall, it was over 100 feet long) showed that the genetic defect follows a simple Mendelian pattern of transmission determined by the dominant allele of a single gene (**Fig. A**). The pedigree also showed that the dominant genetic defect displays incomplete penetrance. Not all people receiving the dominant allele become blind; these sighted carriers may unknowingly pass the blindness-causing dominant allele to their children.

Unfortunately, people do not know they have the disease until their vision starts to deteriorate. By that time, their optic fibers have sustained irreversible damage, and blindness is all but inevitable. Surprisingly, the existence of medical therapies that make it possible to arrest the nerve deterioration created a quandary in the late 1980s. Because effective treatment has to begin before symptoms of impending blindness show up, information in the pedigree could have helped doctors pinpoint people who are at risk, even if neither of their parents is blind. The researchers who compiled the massive family history therefore wanted to give physicians the names of at-risk individuals living in their area, so that doctors could monitor them and recommend treatment if needed. However, a long-standing French law protecting personal privacy forbids public circulation of the names in genetic pedigrees. The French government agency interpreting this law maintained that if the names in the glaucoma pedigree were made public, potential carriers of the disease might suffer discrimination in hiring or insurance.

France thus faced a serious ethical dilemma: On the one hand, giving out names could save perhaps thousands of people from blindness; on the other hand, laws designed to protect personal privacy precluded the dissemination of specific names. The solution adopted by the French government at the time was a massive educational program to alert the general public to the problem so that concerned families could seek medical advice. This approach addressed the legal issues but was only partially helpful in dealing with the medical problem, because many affected individuals escaped detection.

Figure A Pedigree showing the transmission of juvenile glaucoma. A small part of the genealogic tree: The vertical transmission pattern over seven generations shows that a dominant allele of a single gene causes juvenile glaucoma. The lack of glaucoma in V-2 followed by its reappearance in VI-2 reveals that the trait is incompletely penetrant.

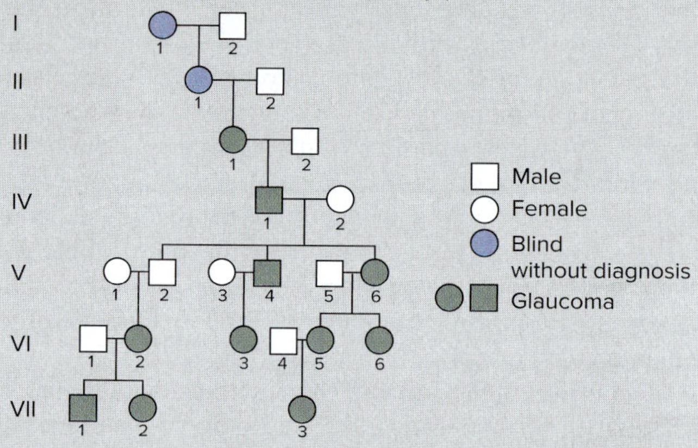

By 1997, molecular geneticists had identified the gene whose dominant mutant allele causes juvenile glaucoma. This gene specifies a protein called myocilin whose normal function in the eye is at present unknown. The mutant allele encodes a form of myocilin that folds incorrectly and then accumulates abnormally in the tiny canals through which eye fluid normally drains into the bloodstream. Misfolded myocilin blocks the outflow of excess vitreous humor, and the resulting increased pressure within the eye (glaucoma) eventually damages the optic nerve, leading to blindness.

Knowledge of the specific disease-causing mutations in the *myocilin* gene has more recently led to the development of diagnostic tests based on the direct analysis of genotype. (We describe methods for direct genotype analysis in Chapter 11.) These DNA-based tests can not only identify individuals at risk, but they can also improve disease management. Detection of the mutant allele before the optic nerve is permanently damaged allows for timely treatment. If these tests become sufficiently inexpensive in the future, they could resolve France's ethical dilemma. Doctors could routinely administer the tests to all newborns and immediately identify nearly all potentially affected children; private information in a pedigree would thus not be needed.

each gene, two alleles exist; a *0* allele that contributes nothing to height and a *1* allele that increases the height of a plant by one unit. All alleles exhibit incomplete dominance relative to alternative alleles of the same gene. The phenotypes determined by all these genes are additive. What would be the result of a two-generation cross between pure-breeding plants carrying only *0* alleles at each height gene and pure-breeding plants carrying only *1* alleles at each height gene? If only one gene were responsible for height, and if environmental effects could be discounted, the F_2 population would be distributed among three classes: homozygous A^0A^0 plants with 0 height (they

Figure 3.28 A Mendelian explanation of continuous variation. The more genes or alleles, the more possible phenotypic classes, and the greater the similarity to continuous variation. In these examples, several pairs of incompletely dominant alleles have additive effects. Percentages shown at the *bottom* denote frequencies of each genotype expressed as fractions of the total population.

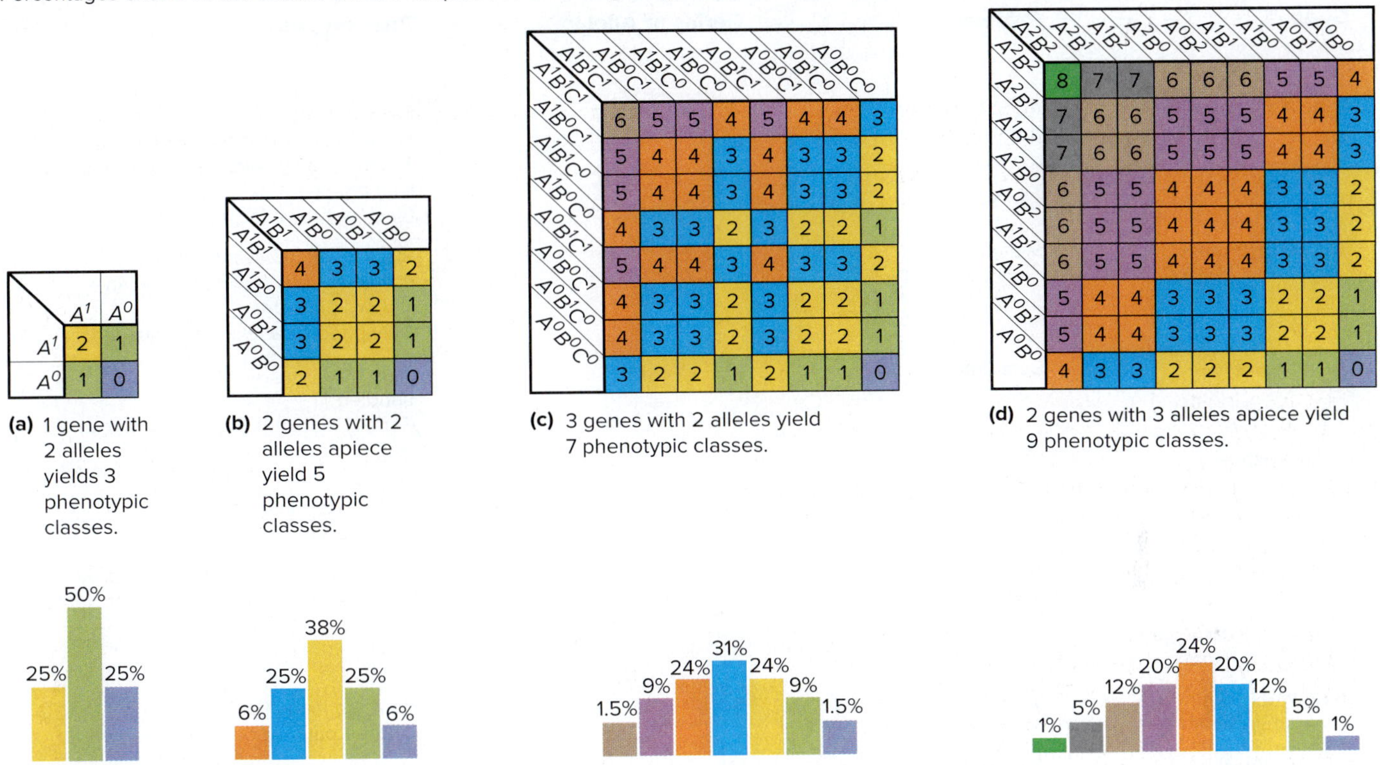

(a) 1 gene with 2 alleles yields 3 phenotypic classes.

(b) 2 genes with 2 alleles apiece yield 5 phenotypic classes.

(c) 3 genes with 2 alleles yield 7 phenotypic classes.

(d) 2 genes with 3 alleles apiece yield 9 phenotypic classes.

lie prostrate on the ground); heterozygous A^0A^1 plants with a height of 1; and homozygous A^1A^1 plants with a height of 2 (**Fig. 3.28a**). This distribution of heights over three phenotypic classes does not make a continuous curve. But for two genes, five phenotypic classes will appear in the F_2 generation (**Fig. 3.28b**); for three genes, seven classes (**Fig. 3.28c**); and for four genes, nine classes (*not shown*).

The distributions produced by three and four genes thus begin to approach continuous variation, and if we add a small contribution from environmental variation, a smoother curve will appear. After all, we would expect bean plants to grow better in good soil, with ample sunlight and water. The environmental component effectively converts the stepped bar graph to a continuous curve by producing some variation in expressivity within each genotypic class. Moreover, additional variation might arise from more than two alleles at some genes (**Fig. 3.28d**), unequal contribution to the phenotype by the various genes involved (review Fig. 3.21), interactions with modifier genes, and chance. Thus, from what we now know about the relation between genotype and phenotype, it is possible to see how only a handful of genes that behave according to known Mendelian principles can easily generate continuous variation.

Continuous (or quantitative) traits vary over a range of values and can usually be measured: the length of a tobacco flower in millimeters, the amount of milk produced by a cow per day in liters, or the height of a person in meters. Continuous traits are usually **polygenic**—controlled by multiple genes—and show the additive effects of a large number of alleles, which creates an enormous potential for variation within a population. Differences in the environments encountered by particular individuals contribute even more variation. We discuss quantitative, multifactorial traits in detail in Chapter 22.

A Comprehensive Example: Multiple Alleles of Several Genes Determine Dog Coat Color

Domestic dogs have a wide variety of coat colors and markings. The *E* and *B* genes described previously in Labrador retrievers (recall Fig. 3.13) are only two of at least 12 genes that control dog coat color and pattern. The roles of seven of these genes—*E*, *B*, *A*, *K*, *D*, *S*, and *M*—are the best understood. **Table 3.3** lists the proteins these genes specify and the nature of various alleles found in domesticated dogs.

TABLE 3.3 Some of the Genes Affecting Domestic Dog Coat Color and Pattern

Gene	Protein	Dominance Series of Alleles		Phenotypes
Pigment-type switch genes				
Gene *A Agouti*	Agouti signaling protein (ASIP)	$A^y > a^w > a^t > a$	A^y a^w a^t a	Fawn (lots of light pigment on hair) Agouti (light stripe on dark hair) Tan belly (only hairs on belly have some light pigment) Black or brown (no light stripe on hairs)
Gene *E Extension*	Melanocortin receptor (MC1R)	$E^m > E > e$	E^m E e	Black mask on fawn or brindle Eumelanin (dark) and pheomelanin (yellow) pigments Only pheomelanin (cream, tan, red)
Gene *K Kurokami*	Beta-defensin	$K^b > k^{br} > k^y$	K^b k^{br} k^y	Solid color Brindled Gene *A* markings expressed normally
Dilution genes				
Gene *D Dilute*	Melanophilin (MLPH)	$D > d$	D d	Colors not dilute Colors dilute
Gene *B Brown*	Tyrosine-related protein (TYRP1)	$B > b$	B b	Black: eumelanin deposited densely Brown: eumelanin deposited less densely
Pigment cell development and survival genes				
Gene *S Spotting*	Micropthalmia-associated transcription factor (MITF)	$S > s^p$	S s^p	No white markings Colored patches on white background
Gene *M Merle*	Premelanosome protein (PMEL)	$M^1 = M^2$	M^1 M^2	Coat color diluted (homozygote has various health problems) Normal color

Genes *A, E,* and *K* control the switch from eumelanin to pheomelanin production

Skin cells called *melanocytes* make the pigments deposited in each dog hair. Melanocytes can produce either a dark pigment (eumelanin), or a light pigment (pheomelanin). The MC1R protein, specified by the *E* gene, spans the cell membrane and acts a switch that determines which pigment a melanocyte produces (**Fig. 3.29a**). Melanocytes produce eumelanin only when MC1R is switched on; pheomelanin is produced when the switch is turned off. By binding to MC1R at the cell surface, two proteins made by nearby skin cells control the MC1R switch. Binding of ASIP (specified by gene *A*) turns the switch off, but when Beta-defensin (specified by gene *K*) successfully outcompetes ASIP for MC1R binding, the switch flips on. The different alleles of the *E, A,* and *K* genes found in different dog breeds result in various colors and pigment distribution patterns on each hair and over the dog's body as a whole.

As described earlier, Labrador retrievers have two different gene *E* alleles; *E* specifies MC1R while *e* is nonfunctional. All Labrador retrievers are homozygous for the *a* allele of gene *A,* which specifies an inactive ASIP protein, and for the K^b allele of gene *K*, whose product is a functional Beta-defensin. Labradors that are *E–* produce eumelanin because MC1R is present and is switched on by the K^b Beta-defensin protein (**Fig. 3.29b**). These *E–* dogs are either black or chocolate depending on their gene *B* allele (as will be explained). In contrast, the melanocytes of *ee* Labradors have no MC1R, so the switch cannot be turned on; pheomelanin is made by default, and the dogs are yellow. (The particular shade of yellow in *ee* dogs, which varies from cream to red, is controlled by other genes not yet identified at the molecular level.)

Other dog breeds can have various alleles of genes *A*, *E* and *K*. Four different alleles of the *A* gene form a dominance series (Table 3.3). As just described, the *a* allele makes a nonfunctional protein. The proteins made by the other three alleles direct the pigment switch with different efficiencies or in different parts of the dog's body. Although the *A* gene in dogs is the same as the *A* gene in mice (review Fig. 3.7), the A^Y allele behaves differently. In contrast with mice, dogs may be homozygous A^YA^Y; they are an overall light brown color called fawn because the hairs contain a

Figure 3.29 Genes *E, A,* and *K* control the switch between light and dark pigment synthesis in melanocytes. MCR1 is specified by gene *E*. **(a)** At *left,* Beta-defensin (specified by gene *K*) out-competes ASIP for MC1R binding and switches MC1R on; melanocytes produce eumelanin (dark pigment). At *right,* melanocytes produce pheomelanin (light pigment) when ASIP (specified by gene *A*) out-competes Beta-defensin for MC1R binding. When bound by ASIP, MCR1 does not signal. **(b)** At *left,* the version of Beta-defensin made by the Labrador's K^b alleles always binds MC1R because it out-competes the version of ASIP made by their *a* alleles. As a result, eumelanin is synthesized and the dog is either black or chocolate (depending on its alleles of gene *B*). At *right,* the melanocytes of yellow Labrador retrievers lack MC1R (their *ee* alleles are nonfunctional), and pheomelanin is made by default.

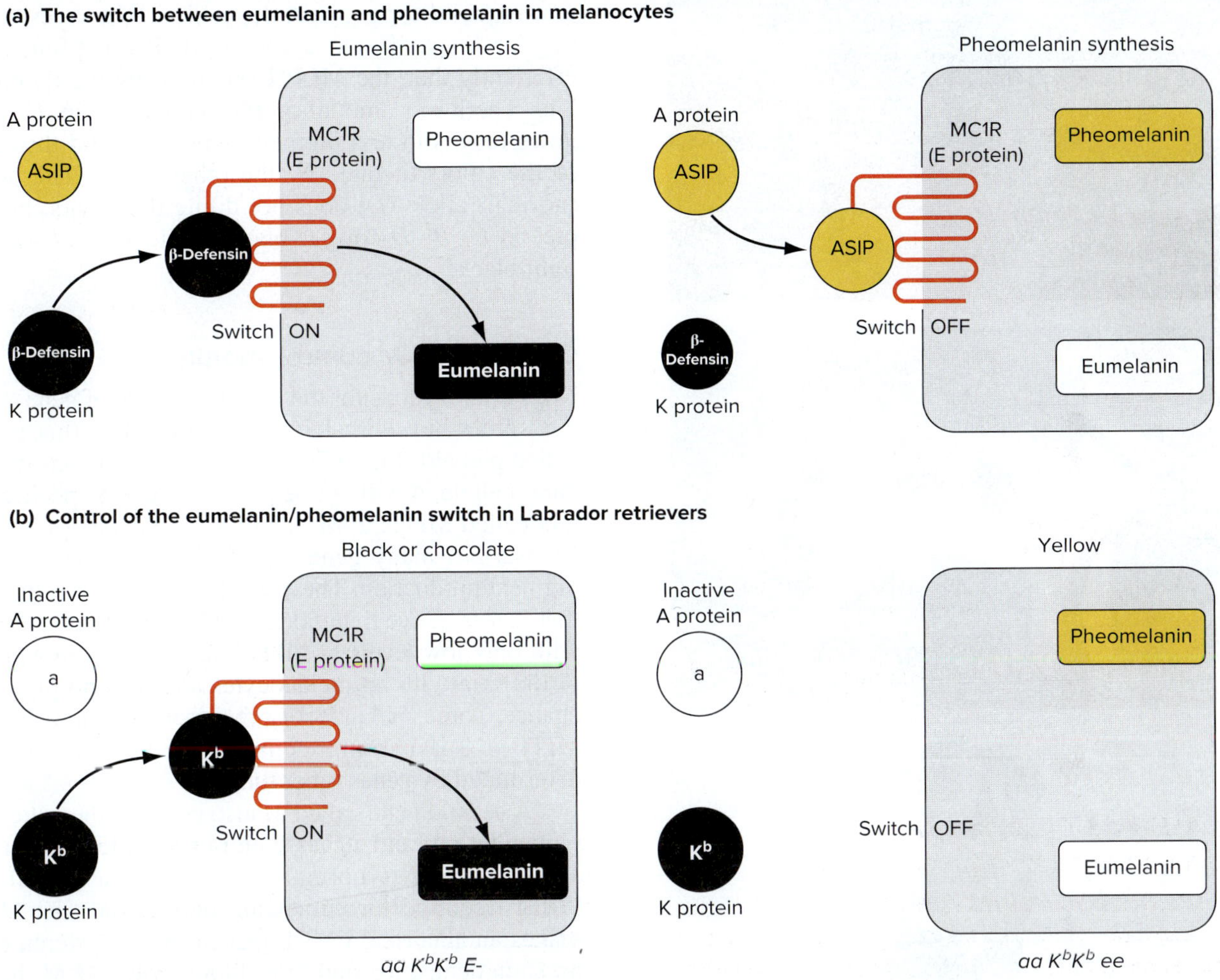

lot of pheomelanin in addition to eumelanin. The a^w allele (like the *A* allele in mice) gives the dogs an overall gray (agouti) color. In an agouti dog (or mouse), the hairs are mainly black with a single yellow stripe. In both species, the a^t allele results in lighter hair on the belly and solid dark hairs on the back.

The *ee* genotype is epistatic to A^Y, a^w, and a^t because these *A* gene phenotypes require the hair to have some dark pigment. Some dogs breeds have an allele of gene *E* called E^m that is dominant to both *E* and *e* and directs the formation of a dark mask (**Fig. 3.30**). The E^m allele specifies a version of MC1R thought to cause more eumelanin production than normal, and the melanocytes around the muzzle are most sensitive to the effects of this increased eumelanin.

Three alleles of the *K* gene determine whether or not the *A* gene markings are visible, and we consider two of them here. The K^b allele present in Labradors specifies a version of Beta-defensin that is made in all melanocytes and can out-compete any of the different ASIP proteins made by gene *A* alleles. Thus in a K^b– dog, eumelanin is made all of the time, regardless of whether or not ASIP is present. If an *E*– dog is also K^b– (like black or chocolate Labrador retrievers), the dog is solidly dark regardless of its *A* gene alleles: K^b is epistatic to all alleles of gene *A*. In contrast, the version of Beta-defensin made by the k^y allele

Figure 3.30 Dog color pattern is a polygenic trait. The major alleles determining three kinds of dog coat color patterns. (top): © Tierfotoagentur/Alamy; (middle): © Vanessa Grossemy/Alamy; (bottom): © Martin Rogers/Getty Images

allows ASIP to inhibit MC1R sometimes, permitting pheomelanin production. As a result, $k^y k^y$ homozygotes allow expression of the fawn, agouti, or tan traits associated with the A^Y, a^w, or a^t alleles of the *A* gene.

Genes *B* and *D* control deposition of all pigments

Gene *B* specifies TYRP1, a multifunctional protein required for eumelanin synthesis and for depositing pigment in melanocytes. The *B* allele makes fully functional TYRP1, and the *b* allele specifies a less active version. As described earlier, *E– B–* Labrador retrievers are black because eumelanin is densely deposited, and their *E– bb* counterparts are brown because less eumelanin is produced and it is deposited less densely. Because the role of TYRP1 depends on the presence of MC1R (made by gene *E*), *ee* is epistatic to both *B* and *b* (recall Figs. 3.12 and 3.13).

Gene *D* specifies Melanophilin (MLPH), another protein required for pigment deposition. The recessive allele of gene *D* specifies a version of MLPH that functions less efficiently than the MLPH specified by the dominant allele. The lower amount of MLPH activity in *dd* homozygotes results in less pigment deposition and thus dilution of the colors dictated by the other genes. The dominant (normal) allele (*D*) does not dilute the colors. For example, an *E– B– D–* dog is black, while an *E– B– dd* dog is light black.

Genes *S* and *M* control spotting

Dogs homozygous for the recessive allele of gene *S* (that is, $s^P s^P$) are white with large spots of color; this pattern is called piebald (Fig. 3.30). As long as a dog has one dominant *S* allele, it will not be piebald. Gene *S* specifies a protein called MITF, a *transcription factor* needed to express (transcribe) many genes specifying enzymes needed for pigment production. The s^P allele makes a version of MITF that is less active than normal. Melanocyte precursor cells with only low levels of MITF die, resulting in white areas of the skin with no melanocytes and thus no pigment. By chance, some melanocyte precursor cells have sufficient MITF to survive, producing colored spots; the color is determined by genes other than *S*.

A second gene called *M* also controls the patterning of pigmentation, and it has codominant alleles M^1 and M^2. The M^2 allele is normal—it specifies a protein called PMEL required for eumelanin deposition. The M^1 allele makes an abnormal PMEL protein that interferes with eumelanin deposition and thus dilutes color. M^1M^2 heterozygotes, called *merle* dogs, have patches of diluted color (the M^1 phenotype) and patches of normal color (the M^2 phenotype) (Fig. 3.30). Breeders would never mate two merle dogs because the *M* gene is pleiotropic. The M^1 allele has recessive deleterious effects: So-called double merle dogs (M^1M^1) usually have serious health problems, including defects in hearing and vision. The eye and ear problems are due to the death of retinal and ear pigment cells caused by the abnormal PMEL protein.

This example of coat color in dogs gives some idea of the potential for variation from just half of the genes known to affect coat color. Amazingly, this is just the tip of the iceberg. When you realize that both dogs and humans carry roughly 27,000 genes, the number of interactions that connect the various alleles of these genes in the expression of phenotype is in the millions, if not the billions. The potential for variation and diversity among individuals is staggering indeed.

essential concepts

- In *incomplete penetrance,* a phenotype is expressed in fewer than 100% of individuals having the same genotype. In *variable expressivity,* a phenotype is expressed at different levels among individuals with the same genotype.
- A *continuous trait* can have any value of expression between two extremes. Most traits of this type are *polygenic,* that is, determined by the interactions of multiple genes.
- The environment and random events can interact with genes to influence the expression of many so-called *multifactorial traits.*

WHAT'S NEXT

Part of Mendel's genius was to look at the genetic basis of variation through a very narrow window, focusing his first glimpse of the mechanisms of inheritance on simple yet fundamental phenomena. Mendel worked on just a handful of traits in inbred populations of one species. For each trait, he manipulated one gene with one completely dominant and one recessive allele that determined two distinguishable, or discontinuous, phenotypes. Both the dominant and recessive alleles showed complete penetrance and negligible differences in expressivity.

In the first few decades of the twentieth century, many biologists questioned the general applicability of Mendelian analysis, for it seemed to shed little light on the complex inheritance patterns of most plant and animal traits or on the mechanisms producing continuous variation. Simple embellishments, however, clarified the genetic basis of continuous variation and provided explanations for other apparent exceptions to Mendelian analysis as described in this chapter. Each embellishment extends the range of Mendelian analysis and deepens our understanding of the genetic basis of variation. And no matter how broad the view, Mendel's basic conclusions, embodied in his first law of segregation, remain valid.

But what about Mendel's second law that genes assort independently? As it turns out, its application is not as universal as that of the law of segregation. Many genes do assort independently, but some do not; rather, the two genes appear to be linked and transmitted together from generation to generation. An understanding of this fact emerged from studies that located Mendel's hereditary units, the genes, in specific cellular organelles, the chromosomes. In describing how researchers deduced that genes travel with chromosomes, Chapter 4 establishes the physical basis of inheritance, including the segregation of alleles, and clarifies why some genes assort independently while others do not.

SOLVED PROBLEMS

I. Imagine you purchased an albino mouse (genotype *cc*) in a pet store. The *c* allele is epistatic to other coat color genes. How would you go about determining the genotype of this mouse at the brown locus? (In pigmented mice, *BB* and *Bb* are black, *bb* is brown.)

Answer

This problem requires an understanding of gene interactions, specifically epistasis. You have been placed in the role of experimenter and need to design crosses that will answer the question. To determine the alleles of the *B* gene present, you need to eliminate the blocking action of the *cc* genotype. Because only the recessive *c* allele is epistatic, when a *C* allele is present, no epistasis will occur. To introduce a *C* allele during the mating, the test mouse you mate to your albino can have the genotype *CC* or *Cc*. (If the mouse is *Cc,* half the progeny will be albino and will not contribute useful information, but the non-albinos from this cross would be informative.) What alleles of the *B* gene should the test mouse carry? To make this decision, work through the expected results using each of the possible genotypes.

Test mouse genotype		Albino mouse	Expected non-albino progeny
BB	×	*BB*	all black
	×	*Bb*	all black
	×	*bb*	all black
Bb	×	*BB*	all black
	×	*Bb*	3/4 black, 1/4 brown
	×	*bb*	1/2 black, 1/2 brown
bb	×	*BB*	all black
	×	*Bb*	1/2 black, 1/2 brown
	×	*bb*	all brown

From these hypothetical crosses, you can see that a test mouse with either the *Bb* or *bb* genotype would yield distinct outcomes for each of the three possible albino mouse genotypes. However, a *bb* test mouse would be more

DNA: © Design Pics/Bilderbuch RF

useful and less ambiguous. First, it is easier to identify a mouse with the *bb* genotype because a brown mouse must have this homozygous recessive genotype. Second, the results are completely different for each of the three possible genotypes when you use the *bb* test mouse. (In contrast, a *Bb* test mouse would yield both black and brown progeny whether the albino mouse was *Bb* or *bb;* the only distinguishing feature is the ratio.) To determine the full genotype of the albino mouse, you should cross it to a brown mouse (which could be *CC bb* or *Cc bb*).

II. In a particular kind of ornamental flower, the wild-type flower color is deep purple, and the plants are true-breeding. In one true-breeding mutant stock, the flowers have a reduced pigmentation, resulting in a lavender color. In a different true-breeding mutant stock, the flowers have no pigmentation and are thus white. When a lavender-flowered plant from the first mutant stock was crossed to a white-flowered plant from the second mutant stock, all the F_1 plants had purple flowers. The F_1 plants were then allowed to self-fertilize to produce an F_2 generation. The 277 F_2 plants were 157 purple : 71 white : 49 lavender.

a. Explain how flower color is inherited. Is this trait controlled by the alleles of a single gene?

b. What kinds of progeny would be produced if lavender F_2 plants were allowed to self-fertilize?

Answer

a. Are any modes of single-gene inheritance compatible with the data? The observations that the F_1 plants look different from either of their parents and that the F_2 generation is composed of plants with three different phenotypes exclude complete dominance. The ratio of the three phenotypes in the F_2 plants has some resemblance to the 1:2:1 ratio expected from codominance or incomplete dominance, but the results would then imply that purple plants must be heterozygotes. This conflicts with the information provided that purple plants are true-breeding.

Consider now the possibility that two genes are involved. From a cross between plants heterozygous for two genes (*W* and *P*), the F_2 generation would contain a 9:3:3:1 ratio of the genotypes *W– P–*, *W– pp*, *ww P–*, and *ww pp* (where the dash indicates that the allele could be either a dominant or a recessive form). Would any combinations of the 9:3:3:1 ratio be close to that seen in the F_2 generation in this example? The numbers appear to fit best with a 9:4:3 ratio. What hypothesis would support combining two of the classes (3 + 1)? If *w* is epistatic to the *P* gene, then the *ww P–* and *ww pp* genotypic classes would have the same white phenotype. With this explanation, 1/3 of the F_2 lavender plants would be *WW pp*, and the remaining 2/3 would be *Ww pp*.

b. Upon self-fertilization, *WW pp* plants would produce only lavender (*WW pp*) progeny, while *Ww pp* plants would produce a 3:1 ratio of lavender (*W– pp*) and white (*ww pp*) progeny.

III. Huntington disease is a rare dominant condition in humans that results in a slow but inexorable deterioration of the nervous system. The disease shows what might be called age-dependent penetrance, which is to say that the probability that a person with the Huntington genotype will express the phenotype varies with age. Assume that 50% of those inheriting the *HD* allele will express the symptoms by age 40. Susan is a 35-year-old woman whose father has Huntington disease. She currently shows no symptoms. What is the probability that Susan will show symptoms in five years?

Answer

This problem involves probability and penetrance. Two conditions are necessary for Susan to show symptoms of the disease. A 1/2 (50%) chance exists that she inherited the mutant allele from her father, and if she does inherit the disease allele, a 1/2 (50%) chance exists that she will express the phenotype by age 40. Because these are independent events, the probability is the product of the individual probabilities, or 1/4.

PROBLEMS

Vocabulary

1. For each of the terms in the left column, choose the best matching phrase in the right column.

a. epistasis	1. one gene affecting more than one phenotype
b. modifier genes	2. the alleles of one gene mask the effects of alleles of another gene
c. conditional lethal	3. both parental phenotypes are expressed in the F_1 hybrids
d. permissive condition	4. a heritable change in a gene
e. reduced penetrance	5. genes whose alleles alter phenotypes produced by the action of other genes
f. multifactorial trait	6. less than 100% of the individuals possessing a particular genotype express it in their phenotype
g. incomplete dominance	7. environmental condition that allows conditional lethals to live

h. codominance	8. a trait produced by the interaction of alleles of at least two genes or from interactions between gene and environment
i. mutation	9. individuals with the same genotype have related phenotypes that vary in intensity
j. pleiotropy	10. a genotype that is lethal in some situations (for example, high temperature) but viable in others
k. variable expressivity	11. the heterozygote resembles neither homozygote

Section 3.1

2. In four-o'clocks, the allele for red flowers is incompletely dominant to the allele for white flowers, so heterozygotes have pink flowers. What ratios of flower colors would you expect among the offspring of the following crosses: (a) pink × pink, (b) white × pink, (c) red × red, (d) red × pink, (e) white × white, and (f) red × white? If you specifically wanted to produce pink flowers, which of these crosses would be most efficient?

3. The *Aa* heterozygous snapdragons in Fig. 3.3 are pink, while *AA* homozygotes are red. However, Mendel's *Pp* heterozygous pea flowers were every bit as purple as those of *PP* homozygotes (Fig. 2.8). Assuming that the *A* allele and the *P* allele specify functional enzymes, and the *a* and *p* alleles specify no protein at all, explain why the alleles of gene *A* and the alleles of gene *P* interact so differently.

4. Recall from Chapter 2 (Fig. 2.20) that Mendel's *R* gene specifies an enzyme called Sbe1 that forms branched starches. The dominant allele (*R*) makes protein, and the recessive allele (*r*) is nonfunctional. When considering the phenotype of round or wrinkled peas, *R* is completely dominant to *r: RR* and *Rr* peas are both equally round and *rr* peas are wrinkled. Imagine that the phenotype described is instead the average number of Sbe1 protein molecules in a pea. How would you describe the dominance relation between *R* and *r* in this case?

5. In the fruit fly *Drosophila melanogaster,* very dark (ebony) body color is determined by the *e* allele. The e^+ allele produces the normal wild-type, honey-colored body. In heterozygotes for the two alleles (but not in e^+e^+ homozygotes), a dark marking called the trident can be seen on the thorax, but otherwise the body is honey-colored. The e^+ snd *e* alleles are thus considered to be incompletely dominant.

   a. When female e^+e^+ flies are crossed to male e^+e flies, what is the probability that progeny will have the dark trident marking?

   b. Animals with the trident marking mate among themselves. Of 300 progeny, how many would be expected to have a trident, how many ebony bodies, and how many honey-colored bodies?

6. A cross between two plants that both have yellow flowers produces 80 offspring plants, of which 38 have yellow flowers, 22 have red flowers, and 20 have white flowers. If one assumes that this variation in color is due to inheritance at a single locus, what is the genotype associated with each flower color, and how can you describe the inheritance of flower color?

7. In radishes, color and shape are each controlled by a single locus with two incompletely dominant alleles. Color may be red (*RR*), purple (*Rr*), or white (*rr*) and shape can be long (*LL*), oval (*Ll*), or round (*ll*). What phenotypic classes and proportions would you expect among the offspring of a cross between two plants heterozygous at both loci?

8. A wild legume with white flowers and long pods is crossed to one with purple flowers and short pods. The F_1 offspring are allowed to self-fertilize, and the F_2 generation has 301 long purple, 99 short purple, 612 long pink, 195 short pink, 295 long white, and 98 short white. How are these traits being inherited?

9. Assuming no involvement of the Bombay phenotype (in case you've already read ahead to Section 3.2):

   a. If a girl has blood type O, what could be the genotypes and corresponding phenotypes of her parents?

   b. If a girl has blood type B and her mother has blood type A, what genotype(s) and corresponding phenotype(s) could the other parent have?

   c. If a girl has blood type AB and her mother is also AB, what are the genotype(s) and corresponding phenotype(s) of any male who could *not* be the girl's father?

10. Several genes in humans in addition to the ABO gene (*I*) give rise to recognizable antigens on the surface of red blood cells. The *MN* and *Rh* genes are two examples. The *Rh* locus can contain either a positive or a negative allele, with positive being dominant to negative. *M* and *N* are codominant alleles of the *MN* gene. The following chart shows several mothers and their children. For each mother-child pair, choose the father of the child from among the males in the right column, assuming one child per male.

	Mother	Child	Males
a.	O M Rh(pos)	B MN Rh(neg)	O M Rh(neg)
b.	B MN Rh(neg)	O N Rh(neg)	A M Rh(pos)
c.	O M Rh(pos)	A M Rh(neg)	O MN Rh(pos)
d.	AB N Rh(neg)	B MN Rh(neg)	B MN Rh(pos)

11. Alleles of the gene that determines seed coat patterns in lentils can be organized in a dominance series: marbled > spotted = dotted (codominant alleles) > clear. A lentil plant homozygous for the marbled seed coat pattern allele was crossed to one homozygous for the spotted pattern allele. In another cross, a homozygous dotted lentil plant was crossed to one homozygous for clear. An F_1 plant from the first cross was then mated to an F_1 plant from the second cross.
    a. What phenotypes in what proportions are expected from this mating between the two F_1 types?
    b. What are the expected phenotypes of the F_1 plants from the two original parental crosses?
12. One of your fellow students tells you that there is no way to know that the spotted and dotted patterns on the lentils in Fig. 3.4a are due to codominant alleles (C^S and C^D) of a single gene *C*. He claims that spotting could be controlled by gene *S*, with a completely dominant allele *S* that directs spotting and a recessive allele *s* that directs no spots. Likewise, he claims that dotting could be controlled by a separate gene *D*, with a completely dominant allele *D* that directs dotting and a recessive allele *d* that directs no dots. Is he correct, or does the information in Fig. 3.4a argue against this idea? Explain.
13. In a population of rabbits, you find three different coat color phenotypes: chinchilla (C), himalaya (H), and albino (A). To understand the inheritance of coat colors, you cross individual rabbits with each other and note the results in the following table.

Cross number	Parental phenotypes	Phenotypes of progeny
1	H × H	3/4 H : 1/4 A
2	H × A	1/2 H : 1/2 A
3	C × C	3/4 C : 1/4 H
4	C × H	all C
5	C × C	3/4 C : 1/4 A
6	H × A	all H
7	C × A	1/2 C : 1/2 A
8	A × A	all A
9	C × H	1/2 C : 1/2 H
10	C × H	1/2 C : 1/4 H : 1/4 A

    a. What can you conclude about the inheritance of coat color in this population of rabbits?
    b. Ascribe genotypes to the parents in each of the 10 crosses.
    c. What kinds of progeny would you expect, and in what proportions, if you crossed the chinchilla parents in crosses 9 and 10?
14. In clover plants, the pattern on the leaves is determined by a single gene with multiple alleles that are related in a dominance series. The gene is not pleiotropic. Seven different alleles of this gene are known; an allele that determines the absence of a pattern is recessive to the other six alleles, each of which produces a different pattern. All heterozygous combinations of alleles show complete dominance.
    a. How many different kinds of leaf patterns (including the absence of a pattern) are possible in a population of clover plants in which all seven alleles are represented?
    b. What is the largest number of different genotypes that could be associated with any one phenotype? Is there any phenotype that could be represented by only a single genotype?
    c. In a particular field, you find that the large majority of clover plants lack a pattern on their leaves, even though you can identify a few plants representative of all possible pattern types. Explain this finding.
15. Fruit flies with one allele for curly wings (*Cy*) and one allele for normal wings (Cy^+) have curly wings. When two curly-winged flies were crossed, 203 curly-winged and 98 normal-winged flies were obtained. In fact, all crosses between curly-winged flies produce nearly the same curly : normal ratio among the progeny.
    a. What is the approximate phenotypic ratio in these offspring?
    b. Suggest an explanation for these data.
    c. If a curly-winged fly was mated to a normal-winged fly, how many flies of each type would you expect among 180 total offspring?
16. In certain plant species such as tomatoes and petunias, a highly polymorphic *incompatibility* gene *S* with more than 100 known alleles prevents self-fertilization and promotes outbreeding. In this form of incompatibility, a plant cannot accept sperm carrying an allele identical to either of its own incompatibility alleles. If, for example, pollen carrying sperm with allele S^1 of the incompatibility gene lands onto the stigma (a female organ) of a plant that also carries the S^1 allele, the sperm cannot fertilize any eggs in that plant. (This phenomenon occurs because the pollen grain on the stigma cannot grow a pollen tube to allow the sperm to unite with the egg.)

    For the following crosses, indicate whether any progeny would be produced, and if so, list all possible genotypes of these progeny.
    a. ♂ $S^1 S^2$ × ♀ $S^1 S^2$
    b. ♂ $S^1 S^2$ × ♀ $S^2 S^3$
    c. ♂ $S^1 S^2$ × ♀ $S^3 S^4$
    d. Explain how this mechanism of incompatibility would prevent plant self-fertilization.

e. How does this incompatibility system ensure that all plants will be heterozygotes for different alleles of the *S* gene?

f. How do you know that peas are not governed by this incompatibility mechanism?

g. Explain why evolution would favor the emergence of new incompatibility alleles, making the gene increasingly polymorphic in populations of tomatoes or petunias.

17. In a species of tropical fish, a colorful orange and black variety called montezuma occurs. When two montezumas are crossed, 2/3 of the progeny are montezuma and 1/3 are the wild-type, dark grayish green color. Montezuma is a single-gene trait, and montezuma fish are never true-breeding.

 a. Explain the inheritance pattern seen here and show how your explanation accounts for the phenotypic ratios given.

 b. In this same species, the morphology of the dorsal fin is altered from normal to ruffled by homozygosity for a recessive allele designated *f*. What progeny would you expect to obtain, and in what proportions, from the cross of a montezuma fish homozygous for normal fins to a green, ruffled fish?

 c. What phenotypic ratios of progeny would be expected from the crossing of two of the montezuma progeny from part (b)?

18. People heterozygous for normal and nonfunctional mutant alleles of the *SMARCAD1* gene have a condition known as *adermatoglyphia:* they have no fingerprints. Sometimes adermatoglyphia is called *immigration delay disease* because people lacking fingerprints have trouble obtaining passports.

 Homozygotes for nonfunctional mutant alleles do not exist—they are never born. Describe the dominance relation between the mutant and wild-type alleles of *SMARCAD1*.

19. Using old Fugate family Bibles and the Perry County, Kentucky historical record, a hematologist in the 1960s constructed the pedigree of the *Blue People of Troublesome Creek*. Many members of the Fugate family had blue skin, a rare but harmless condition known as *methemoglobinemia;* other people in the pedigree had blue lips and fingertips but their skin was otherwise normal. The blue color is due to lack of function of the enzyme NADH diaphorase, which repairs hemoglobin damaged by oxidation. Unrepaired hemoglobin accumulates as blue pigment.

© James Devaney/Getty Images

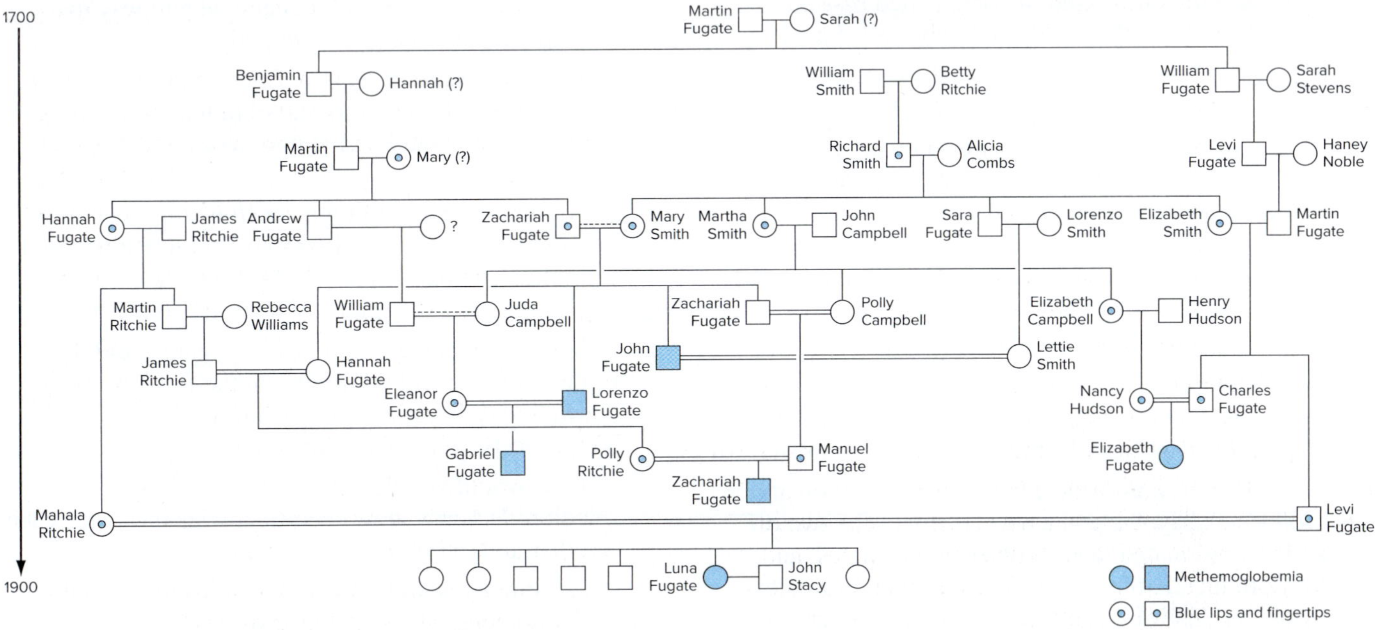

a. Based on the pedigree, describe the dominance relation between the wild-type and mutant alleles of the gene for NADH diaphorase.

b. The pedigree indicates certain people who were known to have only blue lips and fingertips. However, the historical record is incomplete. Which other people in the diagram must have had this phenotype? Explain any ambiguities that exist. (In case you've already read Section 3.3, assume that the blue lip and fingertip phenotype is fully penetrant.)

c. Two of the matings in the pedigree are shown as possibly consanguineous, as indicated by a dotted horizontal line above a solid horizontal line. The reason for the uncertainty is that the historical record does not say whether or not Mary [Mary (?)], the wife of the Martin Fugate at the top left of the diagram, was a Ritchie or a Smith or was instead unrelated to either family. Explain why a geneticist would think that Mary is likely a Ritchie or a Smith.

d. All of the Blue People (people with methemoglobemia) in the pedigree are Fugates, yet the blue mutation did not originate in the Fugate family. Which person or people introduced the mutant NADH diaphorase allele(s) into the Fugate family?

Section 3.2

20. A rooster with a particular comb morphology called *walnut* was crossed to a hen with a type of comb morphology known as *single*. The F_1 progeny all had walnut combs. When F_1 males and females were crossed to each other, 93 walnut and 11 single combs were seen among the F_2 progeny, but there were also 29 birds with a new kind of comb called *rose* and 32 birds with another new comb type called *pea*.

    a. Explain how comb morphology is inherited.

    b. What progeny would result from crossing a homozygous rose-combed hen with a homozygous pea-combed rooster? What phenotypes and ratios would be seen in the F_2 progeny?

    c. A particular walnut rooster was crossed to a pea hen, and the progeny consisted of 12 walnut, 11 pea, 3 rose, and 4 single chickens. What are the likely genotypes of the parents?

    d. A different walnut rooster was crossed to a rose hen, and all the progeny were walnut. What are the possible genotypes of the parents?

21. A black mare was crossed to a chestnut stallion and produced a bay son and a bay daughter. The two offspring were mated to each other several times, and they produced offspring of four different coat colors: black, bay, chestnut, and liver. Crossing a liver grandson back to the black mare gave a black foal, and crossing a liver granddaughter back to the chestnut stallion gave a chestnut foal. Explain how coat color is being inherited in these horses.

22. Filled-in symbols in the pedigree that follows designate individuals who are deaf.

    a. Study the pedigree and explain how deafness is being inherited.

    b. What is the genotype of the individuals in generation V? Why are they not deaf?

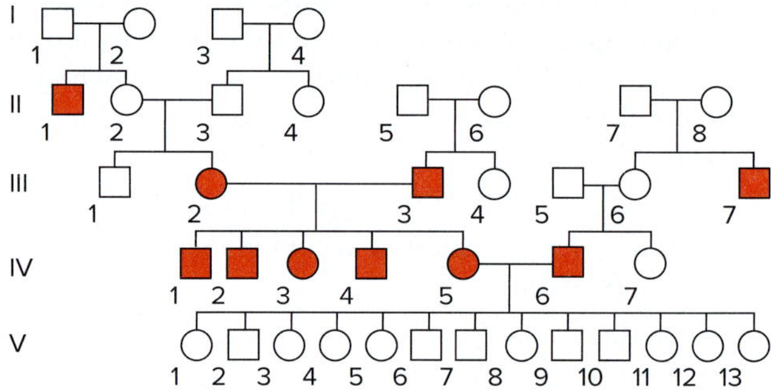

23. You perform a cross between two true-breeding strains of zucchini. One has green fruit and the other has yellow fruit. The F_1 plants are all green, but when these are crossed, the F_2 plants consist of 9 green : 7 yellow.

    a. Explain this result. What were the genotypes of the two parental strains?

    b. Indicate the phenotypes, with frequencies, of the progeny of a testcross of the F_1 plants.

    c. Describe the epistasis interactions observed.

    d. Suppose that the dominant alleles specify functional enzymes, and the recessive alleles are nonfunctional. Propose a biochemical pathway that could explain the gene interactions.

    e. Is it possible to cross two different pure-breeding yellow zucchini strains and obtain all green progeny? What would be the genotypes of the parents and progeny?

    f. Assuming that wild-type zucchini are green, how would you describe the phenomenon that occurred in the F_1 of part (e)?

24. Two true-breeding white strains of the plant *Illegitimati noncarborundum* were mated, and the F_1 progeny were all white. When the F_1 plants were allowed to self-fertilize, 126 white-flowered and 33 purple-flowered F_2 plants grew.

    a. How would you describe inheritance of flower color? Describe how specific alleles influence each other and therefore affect phenotype.

    b. A white F_2 plant is allowed to self-fertilize. Of the progeny, 3/4 are white-flowered, and 1/4 are purple-flowered. What is the genotype of the white F_2 plant?

c. A purple F_2 plant is allowed to self-fertilize. Of the progeny, 3/4 are purple-flowered, and 1/4 are white-flowered. What is the genotype of the purple F_2 plant?

d. Two white F_2 plants are crossed with each other. Of the progeny, 1/2 are purple-flowered, and 1/2 are white-flowered. What are the genotypes of the two white F_2 plants?

25. Suppose the intermediate called *Colorless precursor 2* in the pathway shown in Fig. 3.16 was blue instead of colorless.

 a. What would be the phenotypic ratio of the F_2? (Blue color is distinct from purple.)

 b. Describe the type of genetic interaction that corresponds to this new phenotypic ratio.

26. Explain the difference between epistasis and dominance. How many loci are involved in each case?

27. The dominant allele *H* reduces the number of body bristles in fruit flies, giving rise to a *hairless* phenotype. In the homozygous condition, *H* is lethal. The dominant allele *S* has no effect on bristle number except in the presence of *H*, in which case a single *S* allele suppresses the hairless phenotype, thus restoring the bristles. However, *S* is also lethal in homozygotes.

 a. What ratio of flies with normal bristles to hairless individuals would we find in the live progeny of a cross between two normal flies both carrying the *H* allele in the suppressed condition?

 b. When the hairless progeny of the previous cross are crossed with one of the parental normal flies from part (a) (meaning a fly that carries *H* in the suppressed condition), what phenotypic ratio would you expect to find among their live progeny?

28. *Secretors* (genotypes *SS* and *Ss*) secrete their A and B blood group antigens into their saliva and other body fluids, while *nonsecretors* (*ss*) do not. What would be the apparent phenotypic blood group proportions among the offspring of an I^AI^B *Ss* woman and an I^AI^A *Ss* man if typing was done using saliva?

29. Normally, wild violets have yellow petals with dark brown markings and erect stems. Imagine you discover a plant with white petals, no markings, and prostrate stems. What experiment could you perform to determine whether the non-wild-type phenotypes are due to several different mutant genes or to the pleiotropic effects of alleles at a single locus? Explain how your experiment would settle the question.

30. A woman who is blood type B has a child whose blood type is A; her husband is blood type O. Despite his wife's claims of innocence, the irate father claims that the child is not his. Do you think that the wife is necessarily guilty of adultery? Explain.

31. The following table shows the responses of blood samples from the individuals in the pedigree to anti-A and anti-B sera. A plus (+) in the anti-A row indicates that the red blood cells of that individual were clumped by anti-A serum and therefore the individual made A antigens, and a minus (−) indicates no clumping. The same notation is used to describe the test for the B antigens.

	I-1	I-2	I-3	I-4	II-1	II-2	II-3	III-1	III-2
anti-A	+	+	−	+	−	−	+	+	−
anti-B	+	−	+	+	−	−	+	−	−

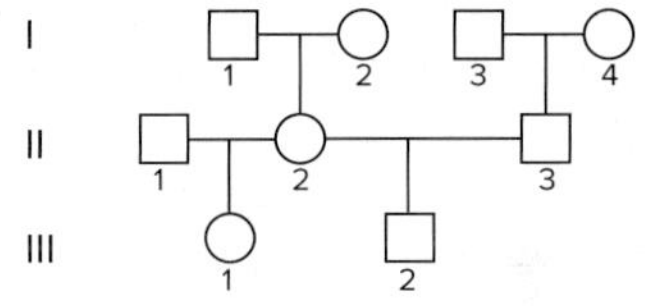

 a. Deduce the blood type of each individual from the data in the table.

 b. Assign genotypes for the blood groups as accurately as you can from these data, explaining the pattern of inheritance shown in the pedigree. Assume that all genetic relationships are as presented in the pedigree (that is, there are no cases of false paternity).

32. Three different pure-breeding strains of corn that produce ears with white kernels were crossed to each other. In each case, the F_1 plants were all red, while both red and white kernels were observed in the F_2 generation in a 9:7 ratio. These results are summarized in the following table.

	F_1	F_2
white-1 × white-2	red	9 red : 7 white
white-1 × white-3	red	9 red : 7 white
white-2 × white-3	red	9 red : 7 white

 a. How many genes are involved in determining kernel color in these three strains?

 b. Define your symbols and show the genotypes for the pure-breeding strains white-1, white-2, and white-3.

 c. Diagram the cross between white-1 and white-2, showing the genotypes and phenotypes of the F_1 and F_2 progeny. Explain the observed 9:7 ratio.

33. In mice, the A^Y allele of the *agouti* gene is a recessive lethal allele, but it is dominant for yellow coat color. What phenotypes and ratios of offspring would you expect from the cross of a mouse heterozygous at the *agouti* gene (genotype A^YA) and also at the *albino* gene (*Cc*) to an albino mouse (*cc*) heterozygous at the *agouti* gene (A^YA)?

34. A student whose hobby was fishing pulled a very unusual carp out of Cayuga Lake: It had no scales on its body. She decided to investigate whether this strange nude phenotype had a genetic basis. She therefore obtained some inbred carp that were pure-breeding for the wild-type scale phenotype (body covered with scales in a regular pattern) and crossed them with her nude fish. To her surprise, the F_1 progeny consisted of a 1:1 ratio of wild-type fish and fish with a single linear row of scales on each side.

   a. Can a single gene with two alleles account for this result? Why or why not?

   b. To follow up on the first cross, the student allowed the linear fish from the F_1 generation to mate with each other. The progeny of this cross consisted of fish with four phenotypes: linear, wild type, nude, and scattered (the latter had a few scales scattered irregularly on the body). The ratio of these phenotypes was 6:3:2:1, respectively. How many genes appear to be involved in determining these phenotypes?

   c. In parallel, the student allowed the phenotypically wild-type fish from the F_1 generation to mate with each other and observed, among their progeny, wild-type and scattered carp in a ratio of 3:1. How many genes with how many alleles appear to determine the difference between wild-type and scattered carp?

   d. The student confirmed the conclusions of (c) by crossing those scattered carp with her pure-breeding wild-type stock. Diagram the genotypes and phenotypes of the parental, F_1, and F_2 generations for this cross and indicate the ratios observed.

   e. The student attempted to generate a true-breeding nude stock of fish by inbreeding. However, she found that this was impossible. Every time she crossed two nude fish, she found nude and scattered fish in the progeny, in a 2:1 ratio. (The scattered fish from these crosses bred true.) Diagram the phenotypes and genotypes of this gene in a nude × nude cross and explain the altered Mendelian ratio.

   f. The student now felt she could explain all of her results. Diagram the genotypes in the linear × linear cross performed by the student in (b). Show the genotypes of the four phenotypes observed among the progeny and explain the 6:3:2:1 ratio.

35. Suppose that blue flower color in a plant species is controlled by two genes, *A* and *B*. The dominant alleles *A* and *B* specify proteins that function in the pathways shown below. The A and B proteins are both required to make blue pigment from a colorless precursor. A and B proteins also independently inhibit the production of blue pigment from a different colorless precursor; that is, the presence of either protein A or protein B is sufficient to prevent blue pigment production from precursor 2. The recessive mutant alleles *a* and *b* specify no protein. Two different pure-breeding mutant strains with white flowers were crossed and complementation was observed so that all the F_1 were blue.

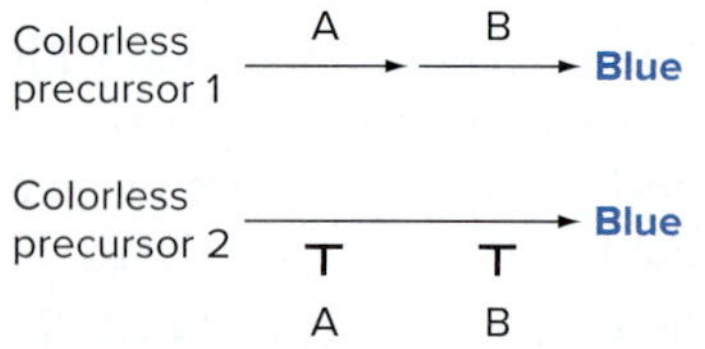

   a. What are the genotypes of each white mutant strain and the F_1?

   b. If the F_1 are selfed, what would be the phenotypic ratio of the F_2?

36. This problem examines possible biochemical explanations for variations of Mendel's 9:3:3:1 ratio. Except where indicated, compounds 1, 2, 3, and 4 have different colors, as do mixtures of these compounds. A and B are enzymes that catalyze the indicated steps of the pathway. Alleles *A* and *B* specify functional enzymes A and B, respectively; these are completely dominant to alleles *a* and *b*, which do not specify any of the corresponding enzyme. If functional enzyme is present, assume that the compound to the left of the arrow is converted completely to the compound to the right of the arrow. For each pathway, what phenotypic ratios would you expect among the progeny of a dihybrid cross of the form *Aa Bb* × *Aa Bb*?

   a. Independent pathways

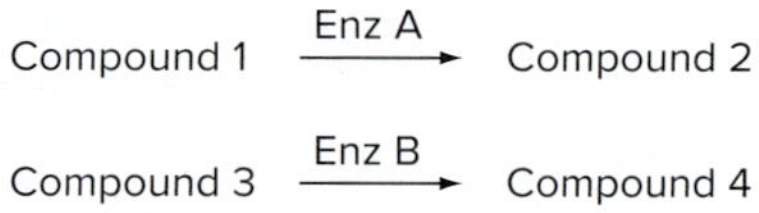

   b. Redundant pathways

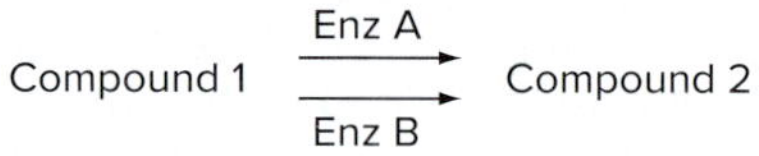

   c. Sequential pathway

   d. Enzymes A and B both needed to catalyze the reaction indicated.

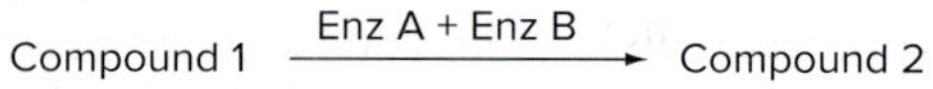

   e. Branched pathways (assume enough of compound 1 for both pathways)

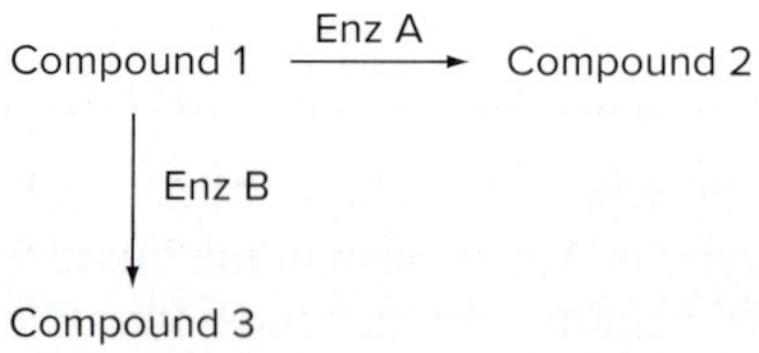

f. Now consider independent pathways as in (a), but the presence of compound 2 masks the colors due to all other compounds.

g. Next consider the sequential pathway shown in (c), but compounds 1 and 2 are the same color.

h. Finally, examine the pathway that follows. Here, compounds 1 and 2 have different colors. The protein encoded by A prevents the conversion of compound 1 to compound 2. The protein encoded by B prevents protein A from functioning.

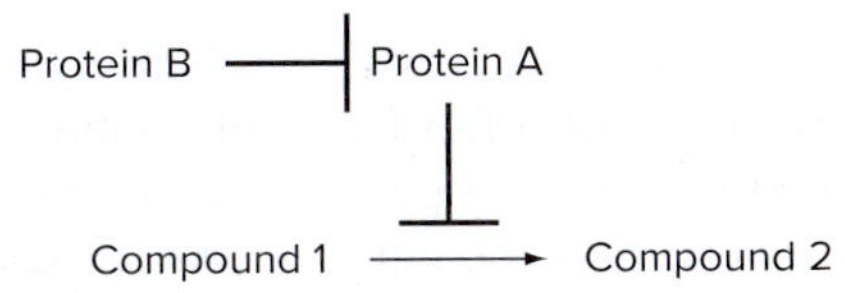

37. Considering your answers to Problem 36, does the existence of a particular variation of a 9:3:3:1 ratio among the F_2 progeny allow you to infer the operation of a specific biochemical mechanism responsible for these phenotypes? Inversely, if you know a biochemical mechanism of gene interaction, can you predict the ratios of the phenotypes you would see among the F_2 progeny?

Section 3.3

38. You picked up two mice (one female and one male) that had escaped from experimental cages in the animal facility. One mouse is yellow in color, and the other is brown agouti. (Agouti hairs have bands of yellow, while non agouti hairs are solid-colored.) You know that this mouse colony has animals with different alleles at only three coat color genes: the agouti (A) or non-agouti (a) or yellow (A^Y) alleles of the A gene ($A^Y > A > a$; A^Y is a recessive lethal), the black (B) or brown (b) alleles of the B gene ($B > b$), and the albino (c) or non-albino (C) alleles of the C gene ($C > c$; cc is epistatic to all other phenotypes). However, you don't know which alleles of these genes are actually present in each of the animals that you've captured. To determine the genotypes, you breed the two escaped mice together. The first litter has only three pups. One is albino, one is brown (non-agouti), and the third is black agouti.

a. What alleles of the *A, B,* and *C* genes are present in the two mice you caught?

b. After raising several litters from these two parents, you have many offspring. How many different coat color phenotypes (in total) do you expect to see expressed in the population of offspring? What are the phenotypes and corresponding genotypes?

39. Figure 3.21 and Fig. 3.28b both show traits that are determined by two genes, each of which has two incompletely dominant alleles. But in Fig. 3.21 the gene interaction produces nine different phenotypes, while the situation depicted in Fig. 3.28b shows only five possible phenotypic classes. How can you explain this difference in the amount of phenotypic variation?

40. Three genes in fruit flies affect a particular trait, and one dominant allele of *each* gene is necessary to get a wild-type phenotype.

a. What phenotypic ratios would you predict among the progeny if you crossed triply heterozygous flies?

b. You cross a particular wild-type male in succession with three tester strains. In the cross with one tester strain (*AA bb cc*), only 1/4 of the progeny are wild type. In the crosses involving the other two tester strains (*aa BB cc* and *aa bb CC*), half of the progeny are wild type. What is the genotype of the wild-type male?

41. The garden flower *Salpiglossis sinuata* (painted tongue) comes in many different colors. Several crosses are made between true-breeding parental strains to produce F_1 plants, which are in turn self-fertilized to produce F_2 progeny.

Parents	F_1 phenotypes	F_2 phenotypes
red × blue	all red	102 red, 33 blue
lavender × blue	all lavender	149 lavender, 51 blue
lavender × red	all bronze	84 bronze, 43 red, 41 lavender
red × yellow	all red	133 red, 58 yellow, 43 blue
yellow × blue	all lavender	183 lavender, 81 yellow, 59 blue

a. State a hypothesis explaining the inheritance of flower color in painted tongues.

b. Assign genotypes to the parents, F_1 progeny, and F_2 progeny for all five crosses.

c. In a cross between true-breeding yellow and true-breeding lavender plants, all of the F_1 progeny are bronze. If you used these F_1 plants to produce an F_2 generation, what phenotypes in what ratios would you expect? Are there any genotypes that might produce a phenotype that you cannot predict from earlier experiments, and if so, how might this alter the phenotypic ratios among the F_2 progeny?

42. In foxgloves, three different petal phenotypes exist: white with red spots (WR), dark red (DR), and light red (LR). Two different kinds of true-breeding WR strains (WR-1 and WR-2) can be distinguished by two-generation intercrosses with true-breeding DR and LR strains:

			F_2		
	Parental	F_1	WR	LR	DR
1	WR-1 × LR	all WR	480	39	119
2	WR-1 × DR	all WR	99	0	32
3	DR × LR	all DR	0	43	132
4	WR-2 × LR	all WR	193	64	0
5	WR-2 × DR	all WR	286	24	74

a. What can you conclude about the inheritance of the petal phenotypes in foxgloves?

b. Ascribe genotypes to the four true-breeding parental strains (WR-1, WR-2, DR, and LR).

c. A WR plant from the F_2 generation of cross 1 is now crossed with an LR plant. Of 500 total progeny from this cross, there were 253 WR, 124 DR, and 123 LR plants. What are the genotypes of the parents in this WR × LR mating?

43. In a culture of fruit flies, matings between any two flies with hairy wings (wings abnormally containing additional small hairs along their edges) always produce both hairy-winged and normal-winged flies in a 2:1 ratio. You now take hairy-winged flies from this culture and cross them with four types of normal-winged flies; the results for each cross are shown in the following table. Assuming that only two possible alleles of the hairy-winged gene exist (one for hairy wings and one for normal wings), what can you say about the genotypes of the four types of normal-winged flies?

	Progeny obtained from cross with hairy-winged flies	
Type of normal-winged flies	**Fraction with normal wings**	**Fraction with hairy wings**
1	1/2	1/2
2	1	0
3	3/4	1/4
4	2/3	1/3

44. As shown in the picture that follows, flowers of the plant *Arabidopsis thaliana* (mustard weed) normally contain four different types of organs: sepals (leaves), petals, anthers (male sex organs), and carpels (female sex organs). The mutant strain shown in the picture at the right has abnormal flower morphology—the flower is made up entirely of sepals! Three genes (called *SEP1*, *SEP2*, and *SEP3*) function redundantly in the pathway for generating petals, anthers, and carpels. For normal flower morphology, the plant requires only one dominant, normally functioning allele of any one of these genes: *SEP1* (*A*) or *SEP2* (*B*) or *SEP3* (*C*). Recessive mutant alleles of these genes (*a*, *b*, or *c*) specify no protein.

a. What is the genotype of the mutant plant below?

b. In a trihybrid cross of the type *AA bb cc* × *aa BB CC*, where all of the F_1 are *Aa Bb Cc*, what is the expected fraction of normal plants among the F_2 progeny?

c. Suggest a model to explain how the *Arabidopsis thaliana* genome came to acquire three redundant genes.

© Sandra Biewers, www.sysflo.edu

45. A couple wants to know the probability that their expected child will suffer from split-hand deformity, which affects the prospective father, who is indicated by an arrow in the pedigree shown. (The arrow means that he is the *proband*—the person in the family who first brought the disorder to the attention of medical professionals.) This trait, shown in the following photo, is rare in the population, and the prospective parents are not related to each other.

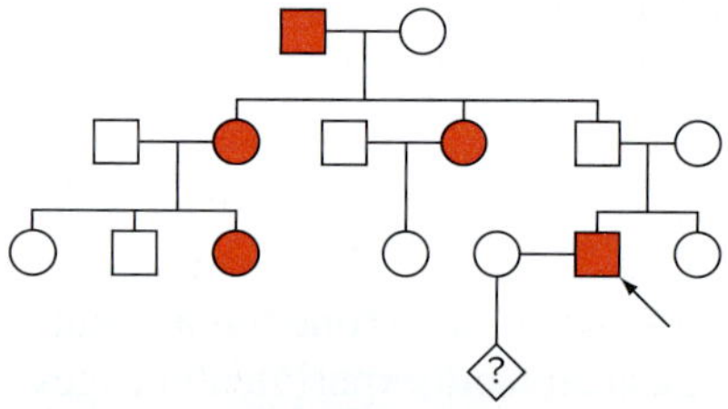

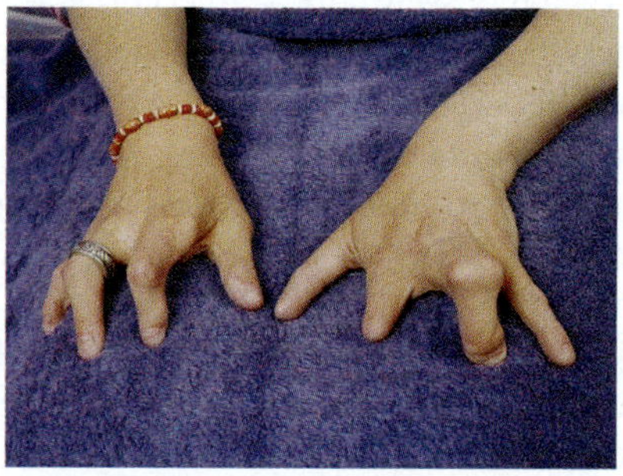
© Maria Platt-Evans/Science Source

a. What is the mode of inheritance of this trait?

b. What is the penetrance of this trait [that is, the ratio between the number of individuals in the pedigree who display the trait (numerator), and the number of individuals you know from the pedigree *must* have the trait-determining genotype regardless of whether they have the trait or not (denominator)]?

c. Using your answer to part (b) above, what would you tell the parents about the numerical likelihood their expected child will have split-hand deformity?

d. Why is it possible that the likelihood the child will be affected is actually less than the number you just answered in part (c)? You should specify the lowest numerical likelihood that could possibly be consistent with the data.

46. This problem illustrates why classical geneticists in the days before DNA analysis usually needed to work with traits showing complete penetrance. Consider the sweet peas shown in Fig. 3.15, where the *A– B–* genotypic class normally produces purple flowers and all other genotypic classes have white flowers.
    a. If the parental generation is *AA bb* × *aa BB,* what phenotypic ratio do you expect in the F_2 generation, assuming complete penetrance?
    b. Suppose now that only 75% of *A– B–* individuals have purple flowers (that is, the penetrance of this trait is 75%). What phenotypic ratio do you now expect among the F_2 plants?
    c. In doing these types of crosses, what kinds of results (other than an unexpected F_2 ratio) might suggest that penetrance of the purple phenotype is incomplete?

47. Spherocytosis is an inherited blood disease in which the erythrocytes (red blood cells) are spherical instead of biconcave. This condition can be inherited in a dominant fashion, with *ANK1* (the nonfunctional mutant allele) dominant to *ANK1*$^+$. In people with spherocytosis, the spleen recognizes the spherical red blood cells as defective and removes them from the bloodstream, leading to anemia. The spleen in different people removes the spherical erythrocytes with different efficiencies. Some people with spherical erythrocytes suffer severe anemia and some mild anemia, yet others have spleens that function so poorly no symptoms of anemia exist at all. When 2400 people with the genotype *ANK1 ANK1*$^+$ were examined, it was found that all of them had spherical erythrocytes, 2250 had anemia of varying severity, and 150 had no symptoms. (Assume that *ANK1 ANK1* homozygotes do not exist.)
    a. Does this description of people with spherocytosis represent incomplete penetrance, variable expressivity, or both? Explain your answer. Can you derive any values from the numerical data to measure penetrance or expressivity?
    b. Suggest a treatment for spherocytosis and describe how the incomplete penetrance and/or variable expressivity of the condition might affect this treatment.

48. Familial hypercholesterolemia (FH) is an inherited trait in humans that results in higher-than-normal serum cholesterol levels [measured in milligrams of cholesterol per deciliter of blood (mg/dl)]. People with serum cholesterol levels that are roughly twice normal have a 25 times higher frequency of heart attacks than unaffected individuals. People with serum cholesterol levels three or more times higher than normal have severely blocked arteries and almost always die before they reach the age of 20. The following pedigrees show the occurrence of FH in four Japanese families:
    a. What is the most likely mode of inheritance of FH based on these data? Do any individuals in any of these pedigrees not fit your hypothesis? What special conditions might account for such individuals?
    b. Why do individuals in the same phenotypic class (unfilled, yellow, or red symbols) show such variation in their levels of serum cholesterol?

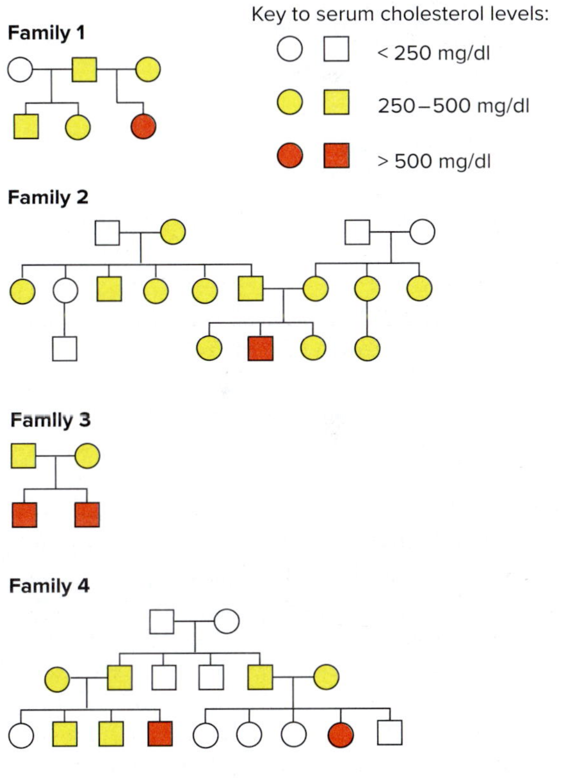

49. You have come into contact with two unrelated patients who express what you think is a rare phenotype—a dark spot on the bottom of the foot. According to a medical source, this phenotype is seen in 1 in every 100,000 people in the population. The two patients give their family histories to you, and you generate the pedigrees that follow.
    a. Given that this trait is rare, do you think the inheritance is dominant or recessive? Are there any special conditions that appear to apply to the inheritance?
    b. Which nonexpressing members of these families must carry the mutant allele?

c. If this trait is instead quite common in the population, what alternative explanation would you propose for the inheritance?

d. Based on this new explanation in (c), which non-expressing members of these families must have the genotype normally causing the trait?

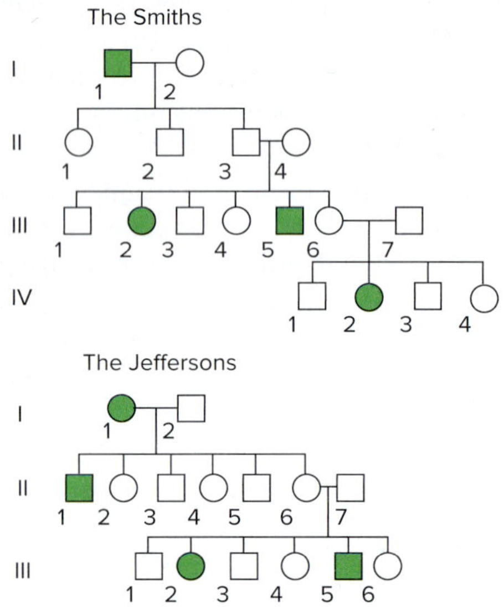

50. Polycystic kidney disease is a dominant trait that causes the growth of numerous cysts in the kidneys. The condition eventually leads to kidney failure. A child with polycystic kidney disease is born to a couple, neither of whom shows the disease. What possibilities might explain this outcome?

51. Identical (monozygotic) twins have similar, but not identical, fingerprints. Given that all the alleles of all the genes of identical twins are the same, explain how this outcome is possible.

© McGraw-Hill Education/Gary He

© The Print Collector/Getty Images

52. Using each of the seven coat color genes discussed in the text (listed in Table 3.3), propose a possible genotype for each of the three Labrador retrievers in Fig. 3.12a. Keep in mind that the Labrador retrievers are pure-breeding for uniformly colored coats without spots or eye masks. Explain any ambiguities in your genotype assignments.

chapter 4

The Chromosome Theory of Inheritance

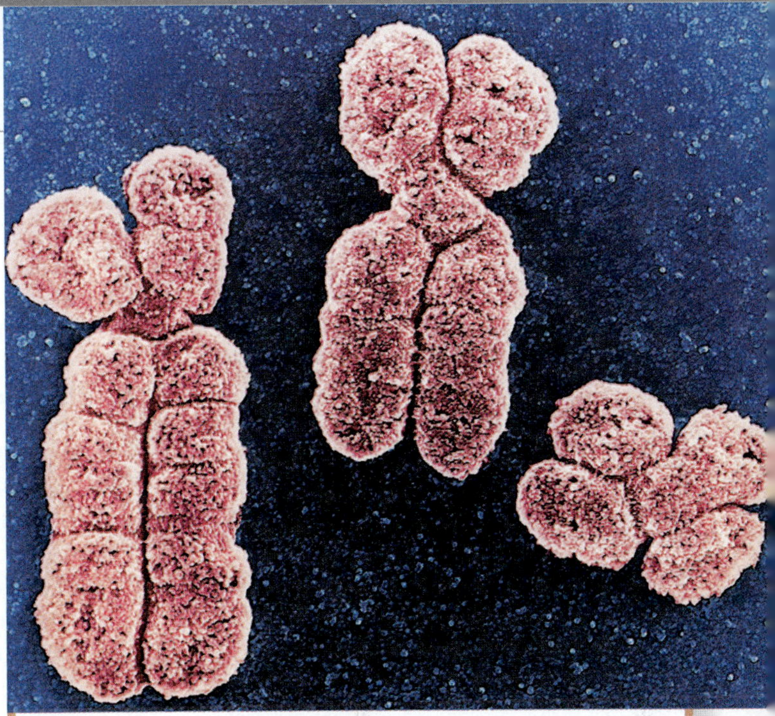

Each of these three human chromosomes carries hundreds of genes.

chapter outline

IN THE SPHERICAL, membrane-bound nuclei of plant and animal cells prepared for viewing under the microscope, chromosomes appear as brightly colored, threadlike bodies. The nuclei of normal human cells carry 23 pairs of chromosomes for a total of 46. Noticeable differences in size and shape exist among the 23 pairs, but within each pair, the two chromosomes appear to match exactly. (The only exceptions are the male's sex chromosomes, designated X and Y, which constitute an unmatched pair.)

Down syndrome was the first human genetic disorder attributable not to a gene mutation but to an abnormal number of chromosomes. Children born with Down syndrome have 47 chromosomes in each somatic cell nucleus because they carry three, instead of the normal pair, of a very small chromosome referred to as number 21. The aberrant genotype, called *trisomy 21*, gives rise to an abnormal phenotype, including a wide skull that is flatter than normal at the back, an unusually large tongue, learning disabilities caused by the abnormal development of the hippocampus and other parts of the brain, and a propensity to respiratory infections as well as heart disorders, rapid aging, and leukemia (**Fig. 4.1**).

How can one extra copy of a chromosome that is itself of normal size and shape cause such wide-ranging phenotypic effects? The answer has two parts. First and foremost, chromosomes are the cellular structures responsible for transmitting genetic information. In this chapter, we describe how geneticists concluded that chromosomes are the carriers of genes, an idea that became known as the **chromosome theory of inheritance.** The second part of the answer is that proper development depends not just on what type of genetic material is present, but also on how much of it there is. Thus the mechanisms governing gene transmission during cell division must vigilantly maintain each cell's chromosome number.

Cell division proceeds through the precise chromosome-parceling mechanisms of *mitosis* (for somatic, or body cells) and *meiosis* (for gametes—eggs and sperm). When

Figure 4.1 Down syndrome: One extra chromosome 21 has widespread phenotypic consequences. Trisomy 21 usually causes changes in physical appearance as well as in the potential for learning. Many children with Down syndrome, such as the fifth grader at the center of the photograph, can participate fully in regular activities.

the machinery does not function properly, errors in chromosome distribution can have dire repercussions on the individual's health and survival. Down syndrome, for example, is the result of a failure of chromosome segregation during meiosis. The meiotic error gives rise to an egg or sperm carrying an extra chromosome 21 which, if incorporated in the zygote at fertilization, is passed on via mitosis to every cell of the developing embryo. Trisomy—three copies of a chromosome instead of two—can occur with other chromosomes as well, but in nearly all of these cases, the condition is prenatally lethal and results in a miscarriage.

Two themes emerge in our discussion of meiosis and mitosis. First, direct microscopic observations of chromosomes during gamete formation led early twentieth-century investigators to recognize that chromosome movements parallel the behavior of Mendel's genes, so chromosomes are likely to carry the genetic material. This chromosome theory of inheritance was proposed in 1902 and was confirmed in the following 15 years through elegant experiments performed mainly using the fruit fly *Drosophila melanogaster*. Second, the chromosome theory transformed the concept of a gene from an abstract particle to a physical reality—part of a chromosome that could be seen and manipulated.

4.1 Chromosomes: The Carriers of Genes

learning objectives

1. Differentiate among somatic cells, gametes, and zygotes with regard to the number and origin of their chromosomes.
2. Distinguish between homologous and nonhomologous chromosomes.
3. List the differences between sister chromatids and nonsister chromatids.

One of the first questions asked at the birth of an infant—is it a boy or a girl?—acknowledges that male and female normally are mutually exclusive characteristics like the yellow versus green of Mendel's peas. What's more, among humans and most other sexually reproducing species, a roughly 1:1 ratio exists between the two sexes. Both males and females produce cells specialized for reproduction—sperm or eggs—that serve as a physical link to the next generation. In bridging the gap between generations, these **gametes** must each contribute half of the genetic material for making a normal, healthy son or daughter. Whatever part of the gamete carries this material, its structure and function must be able to account for the either-or aspect of sex determination as well as the generally observed 1:1 ratio of males to females. These two features of sex determination were among the earliest clues to the cellular basis of heredity.

Genes Reside in the Nucleus

The nature of the specific link between sex and reproduction remained a mystery until Anton van Leeuwenhoek, one of the earliest and most astute of microscopists, discovered in 1667 that semen contains spermatozoa (literally *sperm animals*). He imagined that these microscopic creatures might enter the egg and somehow achieve fertilization, but it was not possible to confirm this hypothesis for another 200 years.

Then, during a 20-year period starting in 1854 (about the same time Gregor Mendel was beginning his pea experiments), microscopists studying fertilization in frogs and sea urchins observed the union of male and female gametes and recorded the details of the process in a series of drawings. These drawings, as well as later micrographs (photographs taken through a microscope), clearly show that egg and sperm nuclei are the only elements contributed equally by maternal and paternal gametes. This observation implies that something in the nucleus contains the hereditary material. In humans, the nuclei of the gametes are less than 2 millionths of a meter in diameter. It is indeed remarkable that the genetic link between generations is packaged within such an exceedingly small space.

Genes Reside in Chromosomes

Further investigations, some dependent on technical innovations in microscopy, suggested that yet smaller, discrete structures within the nucleus are the repository of genetic information. In the 1880s, for example, a newly discovered combination of organic and inorganic dyes revealed the existence of the long, brightly staining, threadlike bodies within the nucleus that we call **chromosomes** (literally *colored bodies*). It was now possible to follow the movement of chromosomes during different kinds of cell division.

In embryonic cells, the chromosomal threads split lengthwise in two just before cell division, and each of the two newly forming daughter cells receives one-half of every split thread. The kind of nuclear division followed by cell division that results in two daughter cells containing the same number and type of chromosomes as the original parent cell is called **mitosis** (from the Greek *mitos* meaning *thread* and *-osis* meaning *formation* or *increase*).

In the cells that give rise to male and female gametes, the chromosomes composing each pair become segregated, so that the resulting gametes receive only one chromosome from each chromosome pair. The kind of nuclear division that generates egg or sperm cells containing half the number of chromosomes found in other cells within the same organism is called **meiosis** (from the Greek word for *diminution*).

Fertilization: The union of haploid gametes to produce diploid zygotes

In the first decade of the twentieth century, cytologists—scientists who use the microscope to study cell structure—showed that the chromosomes in a fertilized egg actually consist of two matching sets, one contributed by the maternal gamete, the other by the paternal gamete. The corresponding maternal and paternal chromosomes appear alike in size and shape, forming pairs (with one exception—the *sex chromosomes*—which we discuss in a later section).

Gametes and other cells that carry only a single set of chromosomes are called **haploid** (from the Greek word for *single*). Zygotes and other cells carrying two matching sets are **diploid** (from the Greek word for *double*). The number of chromosomes in a normal haploid cell is designated by the shorthand symbol ***n***. The number of chromosomes in a normal diploid cell is then **2*n***. **Figure 4.2** shows diploid cells as well as the haploid gametes that arise from them in *Drosophila,* where $2n = 8$ and $n = 4$. In humans, $2n = 46$; $n = 23$.

You can see how the halving of chromosome number during meiosis and gamete formation, followed by the union of two gametes' chromosomes at fertilization, normally allows a constant $2n$ number of chromosomes to be maintained from generation to generation in all individuals of a species. The chromosomes of every pair must segregate from each other during meiosis so that the haploid gametes will each have one complete set of chromosomes. After fertilization forms the zygote, the process of mitosis then ensures that all the somatic cells of the developing individual have identical diploid chromosome sets.

Species variations in the number and shape of chromosomes

Scientists analyze the chromosomal makeup of a cell when the chromosomes are most visible—at a specific moment in the cell cycle of growth and division, just before the nucleus divides. At this point, known as *metaphase* (described in detail later), individual chromosomes have duplicated and condensed from thin threads into compact rodlike structures. Each chromosome now consists of two identical halves known as **sister chromatids (Fig. 4.3)**.

The specific location at which sister chromatids are attached to each other is called the **centromere**. Each sister chromatid has its own centromere (Fig. 4.3), but in the duplicated chromosome, the two sister centromeres are pulled together so tightly that they form a *constriction* within which they cannot be resolved from each other, even in images obtained in the scanning electron microscope (see the picture at the beginning of the chapter).

Figure 4.2 Diploid versus haploid: 2*n* versus *n*. Fruit fly somatic cells are diploid: They carry a maternal and paternal copy of each chromosome. Meiosis generates haploid gametes with only one copy of each chromosome. In *Drosophila,* diploid cells have eight chromosomes ($2n = 8$), while gametes have four chromosomes ($n = 4$). Note that the chromosomes in this diagram are pictured before their replication. The X and Y chromosomes determine the sex of the individual.

Drosophila melanogaster

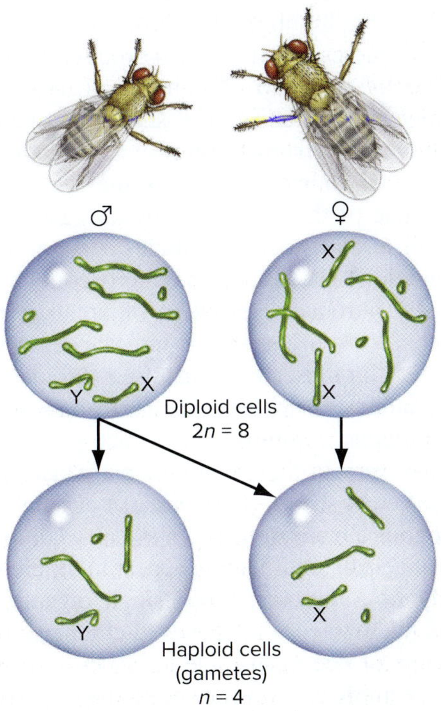

Figure 4.3 Metaphase chromosomes can be classified by centromere position. Before cell division, each chromosome replicates into two sister chromatids connected at their centromeres. In highly condensed metaphase chromosomes, the centromeres can appear near the middle (a *metacentric chromosome*), very near an end (an *acrocentric chromosome*), or anywhere in between. In a diploid cell, one homologous chromosome in each pair is from the mother and the other from the father.

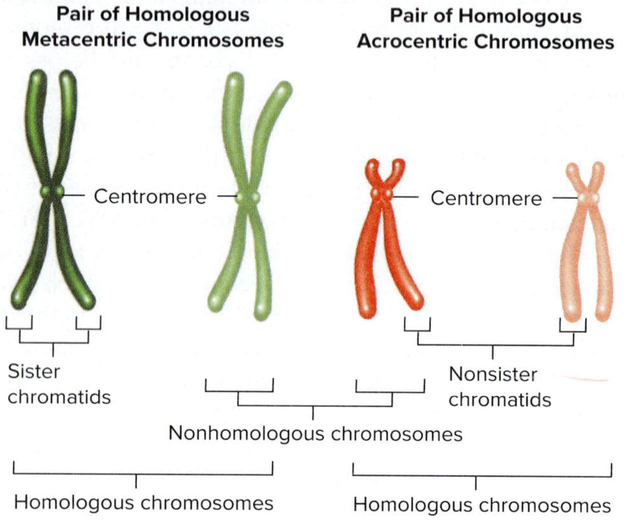

Geneticists often describe chromosomes according to the location of the centromere (Fig. 4.3). In **metacentric** chromosomes, the centromere is more or less in the middle; in **acrocentric** chromosomes, the centromere is very close to one end. Chromosomes thus always have two *arms* separated by a centromere, but the relative sizes of the two arms can vary in different chromosomes.

Cells in metaphase can be fixed and stained with one of several dyes that highlight the chromosomes and accentuate the centromeres. The dyes also produce characteristic banding patterns made up of lighter and darker regions. Chromosomes that match in size, shape, and banding are called **homologous chromosomes,** or **homologs.** The two homologs of each pair contain the same set of genes, although for some of those genes, they may carry different alleles. The differences between alleles occur at the molecular level and don't show up in the microscope.

Figure 4.3 introduces a system of notation employed throughout this book, using color to indicate degrees of relatedness between chromosomes. Thus, sister chromatids, which are identical duplicates, appear in the same shade of the same color. Homologous chromosomes, which carry the same genes but may vary in the identity of particular alleles, are pictured in different shades (light or dark) of the same color. **Nonhomologous chromosomes,** which carry completely unrelated sets of genetic information, appear in different colors.

To study the chromosomes of a single organism, geneticists arrange micrographs of the stained chromosomes in homologous pairs of decreasing size to produce a **karyotype.** Karyotype assembly can now be speeded and automated by

Figure 4.4 Karyotype of a human male. Photos of metaphase human chromosomes are paired and arranged in order of decreasing size. In a normal human male karyotype, 22 pairs of autosomes are present, as well as an X and a Y ($2n = 46$). Homologous chromosomes share the same characteristic pattern of dark and light bands.

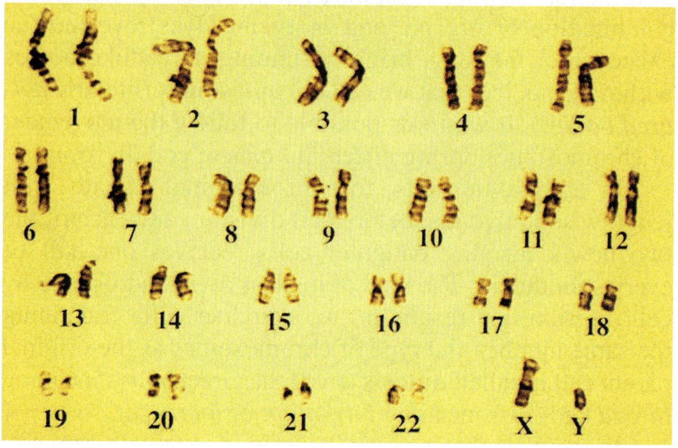

computerized image analysis. **Figure 4.4** shows the karyotype of a human male, with 46 chromosomes arranged in 22 matching pairs of chromosomes and one nonmatching pair. The 44 chromosomes in matching pairs are known as **autosomes.** The two unmatched chromosomes in this male karyotype are called *sex chromosomes* because they determine the sex of the individual. (We discuss sex chromosomes in more detail in subsequent sections.)

Modern methods of DNA analysis can reveal differences between the maternally and paternally derived chromosomes of a homologous pair and can thus track the origin of the extra chromosome 21 that causes Down syndrome in individual patients. In 80% of cases, the third chromosome 21 comes from the egg; in 20%, from the sperm. The Genetics and Society box entitled *Prenatal Genetic Diagnosis* describes how physicians use karyotype analysis and a technique called *amniocentesis* to diagnose Down syndrome prenatally, roughly three months after a fetus is conceived.

Through thousands of karyotypes on normal individuals, cytologists have verified that the cells of each species carry a distinctive diploid number of chromosomes. For example, Mendel's peas contain 14 chromosomes in 7 pairs in each diploid cell, the fruit fly *Drosophila melanogaster* carries 8 chromosomes (4 pairs), macaroni wheat has 28 (14 pairs), giant sequoia trees 22 (11 pairs), goldfish 94 (47 pairs), dogs 78 (39 pairs), and people 46 (23 pairs). Differences in the size, shape, and number of chromosomes reflect differences in the assembled genetic material that determines what each species looks like and how it functions. As these figures show, the number of chromosomes does not always correlate with the size or complexity of the organism.

In the next section, you will see that the discovery that chromosomes carry information about an individual's sex led to the realization that chromosomes carry the genes that determine all traits.

GENETICS AND SOCIETY

Prenatal Genetic Diagnosis

With new technologies for observing chromosomes and the DNA in genes, modern geneticists can define an individual's genotype directly. Doctors can use this basic strategy to diagnose, before birth, whether or not a baby will be born with a genetic condition.

The methods first developed for prenatal diagnosis were to obtain fetal cells whose DNA and chromosomes could be analyzed for genotype. The most frequently used method for acquiring these cells is **amniocentesis** (**Fig. A**). To carry out this procedure, a doctor inserts a needle through a pregnant woman's abdominal wall into the amniotic sac in which the fetus is growing; this procedure is performed about 16 weeks after the woman's last menstrual period. By using ultrasound imaging to guide the location of the needle, the doctor then withdraws some of the amniotic fluid in which the fetus is suspended into a syringe. This fluid contains living cells called *amniocytes* that were shed by the fetus. When placed in a culture medium, these fetal cells undergo several rounds of mitosis and increase in number. Once enough fetal cells are available, clinicians look at the chromosomes and genes in those cells. In later chapters, we describe techniques that allow the direct examination of the DNA constituting particular disease genes.

Amniocentesis also allows the diagnosis of Down syndrome through the analysis of chromosomes by karyotyping. Because the risk of Down syndrome increases rapidly with the age of the mother, more than half the pregnant women in North America who are over the age of 35 currently undergo amniocentesis. Although the goal of this karyotyping is usually to learn whether the fetus is trisomic for chromosome 21, many other abnormalities in chromosome number or shape may show up when the karyotype is examined.

More recently, scientists have been able to analyze the genotype of fetuses from the mother's blood, bypassing the need to obtain fetal cells. This procedure is made possible because the mother's blood contains cell-free fetal DNA. Fetal cells leak into the mother's bloodstream and then break down, releasing their DNA. Modern DNA sequencing techniques allow geneticists not only to genotype this material for particular disease-associated alleles, but even to determine the fetus's entire genome sequence. The analysis of fetal DNA obtained from the mother's blood is still experimental, but it likely will replace amniocentesis in the near future because drawing blood from the mother is inexpensive and noninvasive. The normal risk of miscarriage at 16 weeks' gestation is about 2–3%, and amniocentesis increases that risk by about 0.5% (about 1 in 200 precedures). In contrast, analyzing cell-free DNA from the mother's blood cannot harm the fetus.

Figure A Obtaining fetal cells by amniocentesis. A physician guides the insertion of the needle into the amniotic sac (aided by ultrasound imaging) and extracts amniotic fluid containing fetal cells into the syringe.

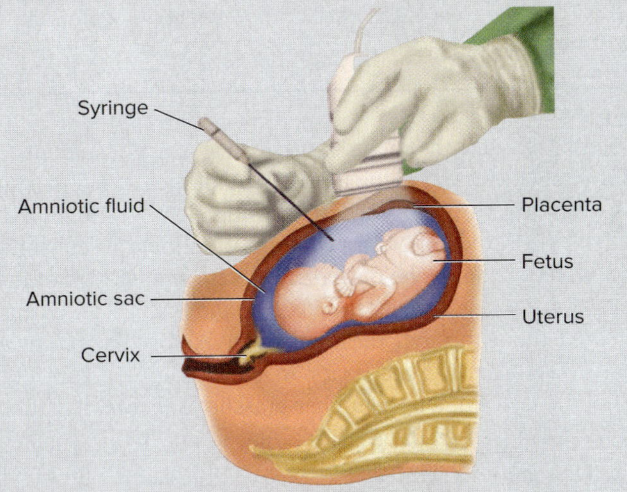

The availability of amniocentesis and **cell-free fetal DNA analysis** for prenatal diagnosis is intimately entwined with the personal and societal issue of abortion. The large majority of amniocentesis procedures are performed with the understanding that a fetus whose genotype indicates a genetic disorder, such as Down syndrome, will be aborted. Some prospective parents who are opposed to abortion still elect to undergo amniocentesis so that they can better prepare for an affected child, but this is rare.

The ethical and political aspects of the abortion debate influence many of the practical questions underlying prenatal diagnosis. For example, parents must decide which genetic conditions would be sufficiently severe that they would be willing to abort the fetus. From the economic point of view, society must decide who should pay for prenatal diagnosis procedures. In current practice, the risks and costs of amniocentesis generally restrict its application to women over age 35 or to mothers whose fetuses are at high risk for a testable genetic condition because of family history. The personal and societal equations determining the frequency of prenatal testing may, however, need to be overhauled in the not-too-distant future because technological advances such as the analysis of cell-free fetal DNA will minimize the costs and risks.

essential concepts

- *Chromosomes* are cellular structures specialized for the storage and transmission of genetic material.
- *Genes* are located on chromosomes and travel with them during cell division and gamete formation.
- *Somatic cells* carry a precise number of *homologous pairs* of chromosomes, which is characteristic of the species.
- In *diploid* organisms, one *homolog* of a pair is of maternal origin, and the other paternal.

4.2 Sex Chromosomes and Sex Determination

learning objectives

1. Predict the sex of humans with different complements of X and Y chromosomes.
2. Describe the basis of sex reversal in humans.
3. Compare the means of sex determination in different organisms.

Walter S. Sutton, a young American graduate student at Columbia University in the first decade of the twentieth century, was one of the earliest cytologists to realize that particular chromosomes carry the information for determining sex. In one study, he obtained cells from the testes of the great lubber grasshopper (*Brachystola magna;* **Fig. 4.5**) and followed them through the meiotic divisions that produce sperm. He observed that prior to meiosis, precursor cells within the testes of a great lubber grasshopper contain a total of 24 chromosomes. Of these, 22 are found in 11 matched pairs and are thus autosomes. The remaining two chromosomes are unmatched. He called the larger of these the X chromosome and the smaller the Y chromosome.

After meiosis, the sperm produced within these testes are of two equally prevalent types: one-half have a set of 11 autosomes plus an X chromosome, while the other half have a set of 11 autosomes plus a Y. By comparison, all of the eggs produced by females of the species carry an 11-plus-X set of chromosomes like the set found in the first class of sperm. When a sperm with an X chromosome fertilizes an egg, an XX female grasshopper results; when a Y-containing sperm fuses with an egg, an XY male develops. Sutton concluded that the X and Y chromosomes determine sex.

Figure 4.5 The great lubber grasshopper. In this mating pair, the smaller male is astride the female.

Figure 4.6 The X and Y chromosomes determine sex in humans. **(a)** This colorized micrograph shows the human X chromosome on the *left* and the human Y on the *right.* **(b)** Children can receive only an X chromosome from their mother, but they can inherit either an X or a Y from their father.

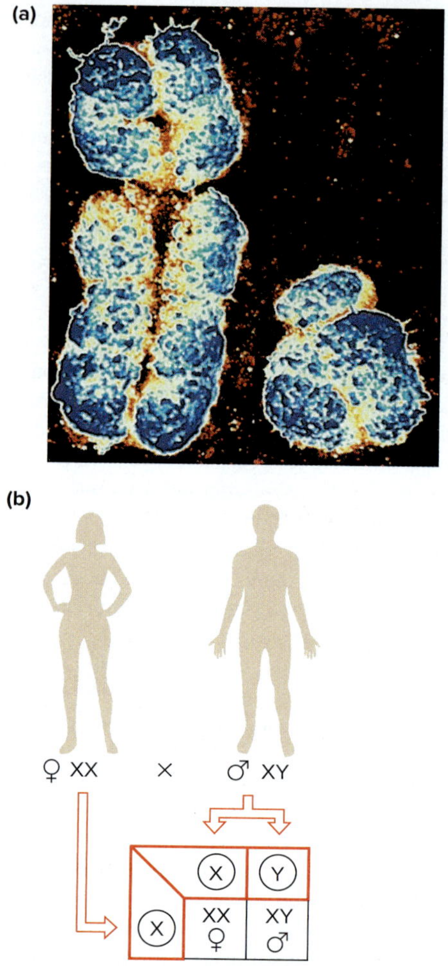

Several researchers studying other organisms soon verified that in many sexually reproducing species, two distinct chromosomes—known as the **sex chromosomes**—provide the basis of sex determination. One sex carries two copies of the same chromosome (a matching pair), while the other sex has one of each type of sex chromosome (an unmatched pair). The cells of normal human females, for example, contain 23 pairs of chromosomes. The two chromosomes of each pair, including the sex-determining X chromosomes, appear to be identical in size and shape. In males, however, one unmatched pair of chromosomes is present: the larger of these is the X; the smaller, the Y (Fig. 4.4 and **Fig. 4.6a**). Apart from this difference in sex chromosomes, the two sexes are not distinguishable at any other pair of chromosomes. Thus, geneticists can designate women as XX and men as XY and represent sexual reproduction as a simple cross between XX and XY.

If sex is an inherited trait determined by a pair of sex chromosomes that separate to different cells during gamete formation, then an XX × XY cross could account for both the mutual exclusion of sexes and the near 1:1 ratio of males to females, which are hallmark features of sex determination (**Fig. 4.6b**). And if chromosomes carry information defining the two contrasting sex phenotypes, we can easily infer that chromosomes also carry genetic information specifying other characteristics as well.

In Humans, the *SRY* Gene Determines Maleness

You have just seen that humans and other mammals have a pair of sex chromosomes that are identical in the XX female but different in the XY male. Several studies have shown that in humans, it is the presence or absence of the Y that actually makes the difference; that is, any person carrying a Y chromosome will look like a male. For example, rare humans with two X and one Y chromosome (XXY) are males displaying certain abnormalities collectively called *Klinefelter syndrome*. Klinefelter males are typically tall, thin, and sterile, and they sometimes show mental retardation. That these individuals are males shows that two X chromosomes are insufficient for female development in the presence of a Y.

In contrast, humans carrying an X and no second sex chromosome (XO) are females with *Turner syndrome*. Turner females are usually sterile, lack secondary sexual characteristics such as pubic hair, are of short stature, and have folds of skin between their necks and shoulders (webbed necks). Even though these individuals have only one X chromosome, they develop as females because they have no Y chromosome.

In 1990, researchers discovered that it is not the entire Y chromosome, but rather a single Y-chromosome-specific gene called ***SRY*** (*sex determining region of Y*) that is the primary determinant of maleness. The evidence implicating *SRY* came from so-called **sex reversal:** the existence of XX males and XY females (**Fig. 4.7**). In many sex-reversed XX males, one of the two X chromosomes carries a portion of the Y chromosome. Although in different XX males, different portions of the Y chromosome are found on the X, one particular gene—*SRY*—is always present. Sex-reversed XY females, in contrast, always have a Y chromosome lacking a functional *SRY* gene; the portion of the Y chromosome containing *SRY* is either replaced by a portion of the X chromosome, or the Y contains a nonfunctional mutant copy of *SRY* (Fig. 4.7). Later experiments with mice confirmed that *SRY* indeed determines maleness. These experiments are described in the Fast Forward Box *Transgenic Mice Prove That* SRY *Is the Maleness Factor.*

SRY is one of about 110 protein-coding genes on the Y chromosome. The two ends of the Y chromosome are called the **pseudoautosomal regions (PARs)** because homologous DNA sequences are present at the ends of the X chromosome (**Fig. 4.8**). The two PARs (PAR1 and PAR2) together contain about 30 genes, copies of which are found on both the X and Y chromosomes.

Most of the Y chromosome, however, is called the male-specific region (MSY) (Fig. 4.8); the functions of only some of the genes in the MSY are understood. The MSY includes four Y-specific (and therefore male-specific) genes: *SRY* and three genes required for spermatogenesis. The name MSY is somewhat misleading because eight of the genes in the MSY also exist on the X chromosome, but unlike the PAR genes, they are not grouped together in one region of either the X or Y. These eight MSY genes affect the functions of cells and tissues all over the body. In fact, several of these MSY genes shared with X are essential for male viability because without the Y-linked copies, the single gene copies on the X chromosome do not supply sufficient protein. (Females normally express both alleles of the X-linked copies of these eight genes, as these genes escape a phenomenon described later in this chapter.)

Figure 4.7 Sex reversal. Sex-reversed XX males have a part of the Y including the *SRY* gene on one of their X chromosomes. Sex-reversed XY females lack *SRY* on their Y chromosome either because it has been replaced by part of the X chromosome or because it has been inactivated by mutation.

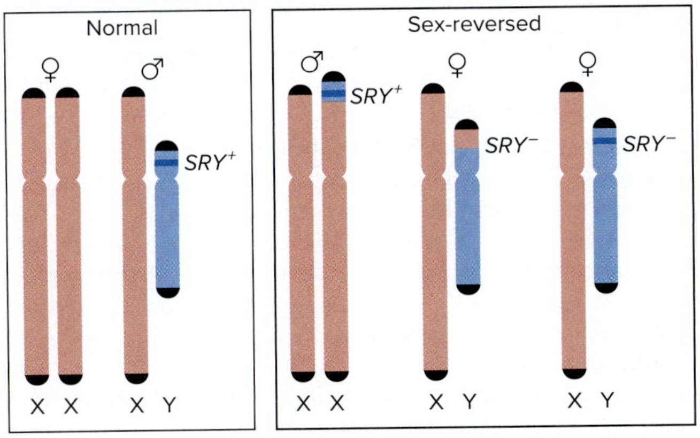

Figure 4.8 Human sex chromosomes have both shared and unique genes. PAR1 and PAR2 (*black*) are homologous regions of the X and Y chromosomes that together contain about 30 genes. The MSY region contains genes needed for maleness itself (*SRY*), genes for male fertility, and essential genes shared with the X required for male viability because their X-linked counterparts alone do not produce enough protein.

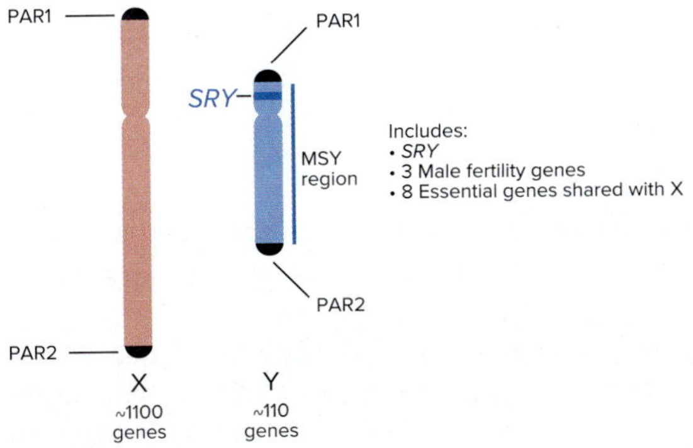

FAST FORWARD

Sprinters: © Robert Michael/Corbis RF

Transgenic Mice Prove That *SRY* Is the Maleness Factor

Genes similar to human *SRY* have been identified on the Y chromosomes of nearly all mammalian species. In 1991, researchers used mouse transgenic technology to show definitively that the *SRY* gene is the crucial determinant of maleness. A *transgenic mouse* is one whose genome contains copies of a gene that came from another individual—or even from another species. Such genes are called *transgenes*. One focus of genetic engineering is technology for the manipulation and insertion of transgenes.

To determine if *SRY* is sufficient to determine maleness, researchers wanted to introduce copies of the mouse *SRY* gene into the genome of chromosomally female (XX) mice. If *SRY* is the crucial determinant of maleness, then XX mice containing an *SRY* transgene would nevertheless be male.

First, the scientists isolated the DNA of the mouse *SRY* gene using cloning technology to be discussed in later chapters. Next, using a method called *pronuclear injection*, transgenic mice were generated that contained the *SRY* gene on one of their autosomes. To perform pronuclear injection, researchers collected many fertilized mouse eggs from mated females and injected the sperm or egg nucleus (called a *pronucleus* when in the zygote) with hundreds of copies of the *SRY* gene DNA (**Fig. A**). Enzymes in the pronucleus integrated the DNA into random locations in the genome (Fig. A).

After the injected zygotes matured into early embryos, they were implanted into surrogate mothers. When the mice were born, cells were taken from their tails and tested for the presence of the *SRY* transgene using molecular biology techniques.

Figure B shows at the right a transgenic mouse (*transformed* with *SRY*) obtained in this study. Although it is chromosomally XX, it is phenotypically male. This result demonstrates conclusively that the *SRY* gene alone is sufficient to determine maleness.

Figure A Using pronuclear injection to generate mice transgenic for the SRY gene.

(1): © Brigid Hogan, Howard Hughes Medical Institute, Vanderbilt University; (2): © Charles River Laboratories

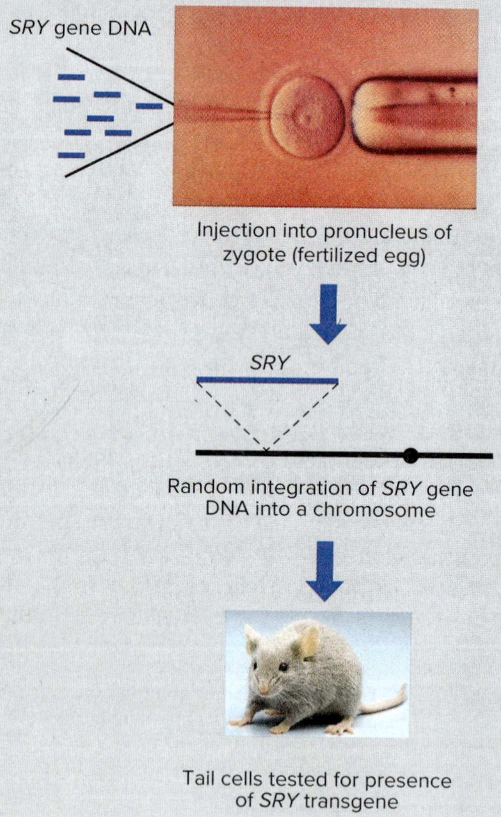

Figure B An XX mouse transformed with *SRY* is phenotypically male. Both the transformed XX mouse at the *right* and its normal XY littermate at the *left* have normal male genitalia. Arrows point to the penis.

© Medical Research Council/Science Source

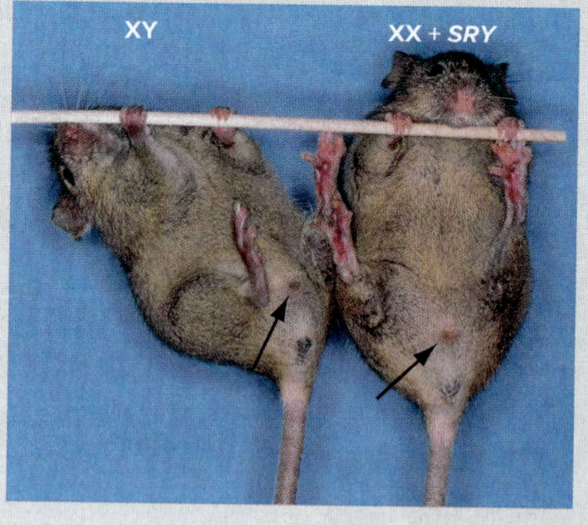

The X chromosome contains about 1100 genes, most of which have nothing to do with sex; they specify proteins needed by both males and females.

Why does having an *SRY* gene mean that you will be male and not having *SRY* mean that you will be female? Approximately six weeks after fertilization, SRY protein activates testes development in XY (or sex-reversed XX) embryos. The embryonic testes secrete hormones that trigger the development of male sex organs and prevent the formation of female sex organs. In the absence of SRY protein, an ovary develops instead of a testis, and other female sex organs develop by default.

TABLE 4.1 Sex Determination in Fruit Flies and Humans

	Complement of Sex Chromosomes						
	XXX	**XX**	**XXY**	**XO**	**XY**	**XYY**	**OY**
Drosophila	Dies	Normal female	Normal female	Sterile male	Normal male	Normal male	Dies
Humans	Nearly normal female	Normal female	Klinefelter male (sterile); tall, thin	Turner female (sterile); short, webbed neck	Normal male	Normal or nearly normal male	Dies

Humans can tolerate extra X chromosomes (e.g., XXX) better than can *Drosophila* because in humans all but one X chromosome becomes a Barr body, as discussed later in this chapter. Complete absence of an X chromosome is lethal to both fruit flies and humans. Additional Y chromosomes have little effect in either species. Although the Y chromosome in *Drosophila* does not determine whether a fly looks like a male, it is necessary for male fertility; XO flies are thus sterile males.

Species Vary Enormously in Sex Determining Mechanisms

Other species show variations on this XX versus XY chromosomal strategy of sex determination. In fruit flies, for example, although normal females are XX and normal males XY (see Fig. 4.2), it is ultimately the number of X chromosomes (and not the presence or absence of the Y) that determines sex. The different responses of humans and *Drosophila* to the same unusual complements of sex chromosomes (**Table 4.1**) reveal that the mechanisms for sex determination differ in flies and humans. XXY flies are female because they have two X chromosomes, but XXY humans are male because they have a Y. Conversely, because they have one X chromosome, XO flies are male, while XO humans are female because they lack a Y.

The XX = female / XY = male strategy of sex determination is by no means universal. In some species of moths, for example, the females are XX, but the males are XO. In *C. elegans* (one species of nematode), males are similarly XO, but XX individuals are not females; they are instead self-fertilizing hermaphrodites that produce both eggs and sperm. In birds and butterflies, males have the matching sex chromosomes, while females have an unmatched set; in such species, geneticists represent the sex chromosomes as ZZ in the male and ZW in the female. The sex having two different sex chromosomes is termed the **heterogametic sex** because it gives rise to two different types of gametes; conversely, the sex with two similar sex chromosomes is the **homogametic sex.** The gametes of the heterogametic sex would contain either X or Y in the case of male humans, and either Z or W in the case of female birds; the gametes of the homogametic sex would contain only an X (humans) or only a Z (birds). Yet other variations include the complicated sex-determination mechanisms of bees and wasps, in which females are diploid and males haploid, and the systems of certain fish, in which sex is determined by changes in the environment, such as fluctuations in temperature. **Table 4.2** summarizes some of the astonishing variety in the ways that different species have solved the problem of assigning sex to individuals.

In spite of these many differences between species, early researchers concluded that chromosomes can carry the genetic information specifying sexual characteristics—and probably many other traits as well. Sutton and other early adherents of the chromosome theory realized that the perpetuation of life itself therefore depends on the proper distribution of chromosomes during cell division. In the next sections, you will see that the behavior of chromosomes during mitosis and meiosis is exactly that expected of cellular structures carrying genes.

TABLE 4.2 Mechanisms of Sex Determination

	♀	♂
Humans and *Drosophila*	XX	XY
Moths and *C. elegans*	XX (hermaphrodites in *C. elegans*)	XO
Birds and Butterflies	ZW	ZZ
Bees and Wasps	Diploid	Haploid
Lizards and Alligators	Cool temperature	Warm temperature
Tortoises and Turtles	Warm temperature	Cool temperature
Anemone Fish	Older adults	Young adults

In the species in the *top three rows*, sex is determined by sex chromosomes. The species in the *bottom four rows* have identical chromosomes in the two sexes, and sex is determined instead by environmental or other factors. Anemone fish (*bottom row*) undergo a sex change from male to female as they age.

essential concepts

- Many sexually reproducing organisms have *sex chromosomes* that are sex-specific and that determine sex.
- In humans, male sex determination is triggered by a Y-linked gene called *SRY*; female sex determination occurs in XX embryos by default.
- Mechanisms of sex determination vary remarkably; in some species sex is determined by environmental factors rather than by specific chromosomes.

4.3 Mitosis: Cell Division That Preserves Chromosome Number

learning objectives

1. Describe the key chromosome behaviors during mitosis.
2. Diagram the forces and structures that dictate chromosomal movement during mitosis.

The fertilized human egg is a single diploid cell that preserves its genetic identity unchanged through more than 100 generations of cells as it divides again and again to produce a full-term infant ready to be born. As the newborn infant develops into a toddler, a teenager, and an adult, yet more cell divisions fuel continued growth and maturation. Mitosis, the nuclear division that apportions chromosomes in equal fashion to two daughter cells, is the cellular mechanism that preserves genetic information through all these generations of cells. In this section, we take a close look at how the nuclear division of mitosis fits into the overall scheme of cell growth and division.

If you were to peer through a microscope and follow the history of one cell through time, you would see that for much of your observation, the chromosomes resemble a mass of extremely fine tangled string—called **chromatin**—surrounded by the **nuclear envelope.** Each convoluted thread of chromatin is composed mainly of DNA (which carries the genetic information) and protein (which serves as a scaffold for packaging and managing that information, as described in Chapter 12). You would also be able to distinguish one or two darker areas of chromatin called *nucleoli* (singular, **nucleolus,** literally *small nucleus*); nucleoli play a key role in the manufacture of ribosomes, organelles that function in protein synthesis. During the period between cell divisions, the chromatin-laden nucleus houses a great deal of invisible activity necessary for the growth and survival of the cell. One particularly important part of this activity is the accurate duplication of all the chromosomal material.

With continued vigilance, you would observe a dramatic change in the nuclear landscape during one very short period in the cell's life history: The chromatin condenses into discrete threads, and then each chromosome compacts even further into the twin rods clamped together at their centromeres that can be identified in karyotype analysis (review Fig. 4.3). Each rod in a duo is called a **chromatid;** as described earlier, it is an exact duplicate of the other sister chromatid to which it is connected. Continued observation would reveal the doubled chromosomes beginning to jostle around inside the cell, eventually lining up at the cell's midplane. At this point, the sister chromatids of each chromosome separate to opposite poles of the now elongating cell, where they become identical sets of chromosomes. Each of the two identical sets eventually ends up enclosed in a separate nucleus in a separate cell. The two cells, known as *daughter cells,* are thus genetically identical.

The repeating pattern of cell growth (an increase in size) followed by division (the splitting of one cell into two) is called the **cell cycle** (**Fig. 4.9**). Only a small part of the cell cycle is spent in division (or **M phase**); the period between divisions is called **interphase.**

During Interphase, Cells Grow and Replicate Their Chromosomes

Interphase consists of three parts: gap 1 (**G_1**), synthesis (**S**), and gap 2 (**G_2**) (Fig. 4.9). G_1 lasts from the birth of a new cell to the onset of chromosome replication; for the genetic material, it is a period when the chromosomes are neither duplicating nor dividing. During this time, the cell achieves most of its growth by using the information from its genes to make and assemble the materials it needs to function normally. G_1 varies in length more than any other phase of the cell cycle. In rapidly dividing cells of the human embryo, for example, G_1 is as short as a few hours. In contrast, mature brain cells

Figure 4.9 The cell cycle: An alternation between interphase and mitosis. (a) Chromosomes replicate to form sister chromatids during synthesis (S phase); the sister chromatids segregate to daughter cells during mitosis (M phase). The gaps between the S and M phases, during which most cell growth takes place, are called the G_1 and G_2 phases. In multicellular organisms, some terminally differentiated cells stop dividing and arrest in a G_0 stage. **(b)** Interphase consists of the G_1, S, and G_2 phases together.

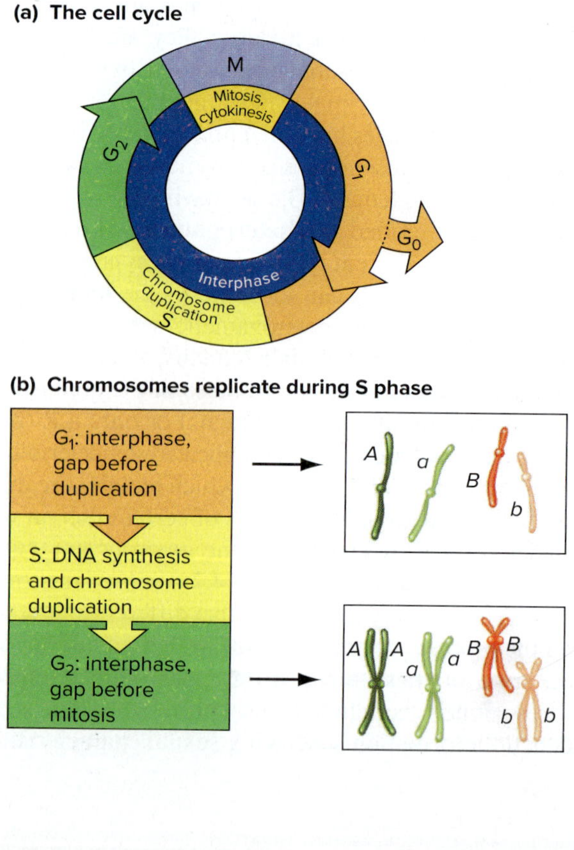

become arrested in a resting form of G_1 known as G_0 and do not normally divide again during a person's lifetime.

Synthesis (S) is the time when the cell duplicates its genetic material by synthesizing DNA. During duplication, each chromosome doubles to produce identical sister chromatids that will become visible when the chromosomes condense at the beginning of mitosis. The two sister chromatids remain joined to each other at their centromeres. (Note that this joined structure is considered a single chromosome as long as the connection between sister chromatids is maintained.) The replication of chromosomes during S phase is crucial; the genetic material must be copied exactly so that both daughter cells receive identical sets of chromosomes.

Gap 2 (G_2) is the interval between chromosome duplication and the beginning of mitosis. During this time, the cell may grow (usually less than during G_1); it also synthesizes proteins that are essential to the subsequent steps of mitosis itself.

In addition, during interphase an array of fine microtubules crucial for many biochemical processes becomes visible outside the nucleus. The microtubules radiate out into the cytoplasm from a single organizing center known as the **centrosome,** usually located near the nuclear envelope. In animal cells, the discernible core of each centrosome is a pair of small, darkly staining bodies called **centrioles** (**Fig. 4.10a**); the microtubule-organizing center of plants does not contain centrioles. During the S and G_2 stages of interphase, the centrosomes replicate, producing two centrosomes that remain in extremely close proximity.

During Mitosis, Sister Chromatids Separate and Two Daughter Nuclei Form

Although the rigorously choreographed events of nuclear and cellular division occur as a dynamic and continuous process, scientists traditionally analyze the process in separate stages marked by visible cytological events. The artist's sketches in Fig. 4.10 illustrate these stages in the nematode *Ascaris,* whose diploid cells contain only four chromosomes (two pairs of homologous chromosomes).

Prophase: Chromosomes condense (Fig. 4.10a)

During all of interphase, the cell nucleus remains intact, and the chromosomes are indistinguishable aggregates of chromatin. At **prophase** (from the Greek *pro-* meaning *before*), the gradual emergence, or **condensation,** of individual chromosomes from the undifferentiated mass of chromatin marks the beginning of mitosis. Each condensing chromosome has already been duplicated during interphase and thus consists of sister chromatids attached at their centromeres. At this stage in *Ascaris* cells, therefore, four chromosomes exist with a total of eight chromatids.

The progressive appearance of an array of individual chromosomes is a truly impressive event. Interphase DNA molecules as long as 3–4 cm condense into discrete chromosomes whose length is measured in microns (millionths of a meter). This process is equivalent to compacting a 200 m length of thin string (as long as two football fields) into a cylinder 8 mm long and 1 mm wide.

Another visible change in chromatin also takes place during prophase: The darkly staining nucleoli begin to break down and disappear. As a result, the manufacture of ribosomes ceases, providing one indication that general cellular metabolism shuts down so that the cell can focus its energy on chromosome movements and cellular division.

Several important events that characterize prophase occur outside the nucleus in the cytoplasm. The centrosomes, which replicated during interphase, now move apart and become clearly distinguishable as two separate entities in the light microscope. At the same time, the interphase scaffolding of long, stable microtubules disappears and is replaced by a set of dynamic microtubules that rapidly grow from and shrink back toward their centrosomal organizing centers. The centrosomes continue to move apart, migrating around the nuclear envelope toward opposite ends of the nucleus, apparently propelled by forces exerted between interdigitated microtubules extending from the two centrosomes.

Prometaphase: The spindle forms (Fig. 4.10b)

Prometaphase (*before middle stage*) begins with the breakdown of the nuclear envelope, which allows microtubules extending from the two centrosomes to invade the nucleus. Chromosomes attach to these microtubules through the **kinetochore,** a structure in the centromere region of each chromatid that is specialized for conveyance. Each kinetochore contains proteins that act as molecular motors, enabling the chromosome to slide along the microtubule. When the kinetochore of a chromatid originally contacts a microtubule at prometaphase, the kinetochore-based motor moves the entire chromosome toward the centrosome from which that microtubule radiates. Microtubules growing from the two centrosomes capture chromosomes by connecting first to the kinetochore of one of the two sister chromatids, chosen at random. As a result, it is sometimes possible to observe groups of chromosomes congregating in the vicinity of each centrosome. In this early part of prometaphase, for each chromosome, one chromatid's kinetochore is attached to a microtubule, but the sister chromatid's kinetochore remains unattached.

During prometaphase, three different types of microtubule fibers together form the **mitotic spindle.** All of these microtubule classes originate from the centrosomes, which function as the two *poles* of the spindle apparatus. Microtubules that extend between a centrosome and the kinetochore of a chromatid are called **kinetochore microtubules,** or *centromeric fibers*. Microtubules from each centrosome that are directed toward the middle of the cell are **polar microtubules;** polar microtubules originating in opposite centrosomes interdigitate near the cell's equator. Finally, short

Figure 4.10 Mitosis maintains the chromosome number of the parent cell in the two daughter nuclei. In the photomicrographs of newt lung cells at the left, chromosomes are stained *blue* and microtubules appear either *green* or *yellow*. Note that the drawings are of *Ascaris* cells ($2n = 4$).

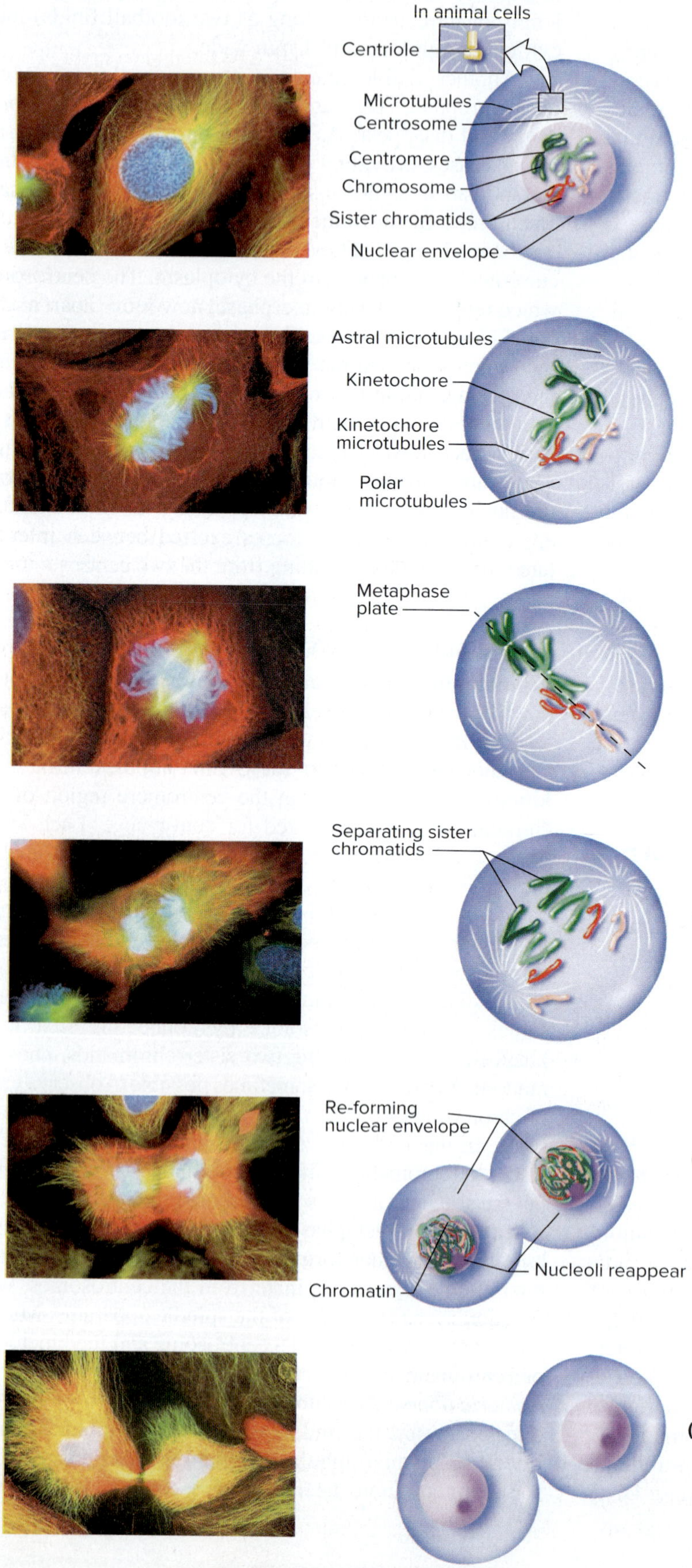

(a) Prophase: (1) Chromosomes condense and become visible; (2) centrosomes move apart toward opposite poles and generate new microtubules; (3) nucleoli begin to disappear.

(b) Prometaphase: (1) Nuclear envelope breaks down; (2) microtubules from the centrosomes invade the nucleus; (3) sister chromatids attach to microtubules from opposite centrosomes.

(c) Metaphase: Chromosomes align on the metaphase plate with sister chromatids facing opposite poles.

(d) Anaphase: (1) The connection between the centromeres of the sister chromatids is severed; (2) the now separated sister chromatids move to opposite poles.

(e) Telophase: (1) Nuclear membranes and nucleoli re-form; (2) spindle fibers disappear; (3) chromosomes uncoil and become a tangle of chromatin.

(f) Cytokinesis: The cytoplasm divides, splitting the elongated parent cell into two daughter cells with identical nuclei.

astral microtubules extend out from the centrosome toward the cell's periphery.

Soon before the end of prometaphase, the kinetochore of each chromosome's previously unattached sister chromatid now associates with microtubules extending from the opposite centrosome. This event orients each chromosome such that one sister chromatid faces one pole of the cell and the other faces the opposite pole. Experimental manipulation has shown that if both kinetochores become attached to microtubules from the same pole, the configuration is unstable; one of the kinetochores will detach repeatedly from the spindle until it associates with microtubules from the other pole. The attachment of sister chromatids to opposite spindle poles is the only stable arrangement.

Metaphase: Chromosomes align at the cell's equator (Fig. 4.10c)

During **metaphase** (*middle stage*), the connection of sister chromatids to opposite spindle poles sets in motion a series of jostling movements that cause the chromosomes to move toward an imaginary equator halfway between the two poles. The imaginary midline is called the **metaphase plate.** When the chromosomes are aligned along it, the forces pulling sister chromatids toward opposite poles are in a balanced equilibrium maintained by tension across the chromosomes. Tension results from the fact that the sister chromatids are pulled in opposite directions while they are still connected to each other by the tight cohesion of their centromeres. Tension compensates for any chance movement away from the metaphase plate by restoring the chromosome to its position equidistant between the poles.

Anaphase: Sister chromatids move to opposite spindle poles (Fig. 4.10d)

The nearly simultaneous severing of the centromeric connections between the sister chromatids of all chromosomes indicates that **anaphase** (from the Greek *ana-* meaning *up* as in *up toward the poles*) is underway. The separation of sister chromatids allows each chromatid to be pulled toward the spindle pole to which it is linked by kinetochore microtubules; as the chromatid moves toward the pole, its kinetochore microtubules shorten. Because the arms of the chromatids lag behind the kinetochores, metacentric chromatids have a characteristic V shape during anaphase. The attachment of sister chromatids to microtubules emanating from opposite spindle poles means that the genetic information migrating toward one pole is exactly the same as its counterpart moving toward the opposite pole.

Telophase: Identical sets of chromosomes are enclosed in two nuclei (Fig. 4.10e)

The final transformation of chromosomes and the nucleus during mitosis happens at **telophase** (from the Greek *telo-* meaning *end*). Telophase is like a rewind of prophase. The spindle fibers begin to disperse; a nuclear envelope forms around the group of chromatids at each pole; and one or more nucleoli reappear. The former chromatids now function as independent chromosomes, which decondense (uncoil) and dissolve into a tangled mass of chromatin. Mitosis, the division of one nucleus into two identical nuclei, is over.

Cytokinesis: The cytoplasm divides (Fig. 4.10f)

In the final stage of cell division, the daughter nuclei emerging at the end of telophase are packaged into two separate daughter cells. This final stage of division is called **cytokinesis** (literally *cell movement*). During cytokinesis, the elongated parent cell separates into two smaller independent daughter cells with identical nuclei. Cytokinesis usually begins during anaphase, but it is not completed until after telophase.

The mechanism by which cells accomplish cytokinesis differs in animals and plants. In animal cells, cytoplasmic division depends on a **contractile ring** that pinches the cell into two approximately equal halves, similar to the way the pulling of a string closes the opening of a bag of marbles (**Fig. 4.11a**). Intriguingly, some types of molecules that form the contractile ring also participate in the mechanism responsible for muscle contraction. In plants, whose cells are surrounded by a rigid cell wall, a membrane-enclosed disk, known as the *cell plate,* forms inside the cell near the equator and then grows rapidly outward, thereby dividing the cell in two (**Fig. 4.11b**).

During cytokinesis, a large number of important organelles and other cellular components, including ribosomes, mitochondria, membranous structures such as Golgi

Figure 4.11 Cytokinesis: The cytoplasm divides, producing two daughter cells. (a) In this dividing frog zygote, the contractile ring at the cell's periphery has contracted to form a *cleavage furrow* that will eventually pinch the cell in two. **(b)** In this dividing onion root cell, a cell plate that began forming near the equator of the cell expands to the periphery, separating the two daughter cells.

a: © Don W. Fawcett/Science Source; b: © McGraw-Hill Education/Al Telser

(a) Cytokinesis in an animal cell

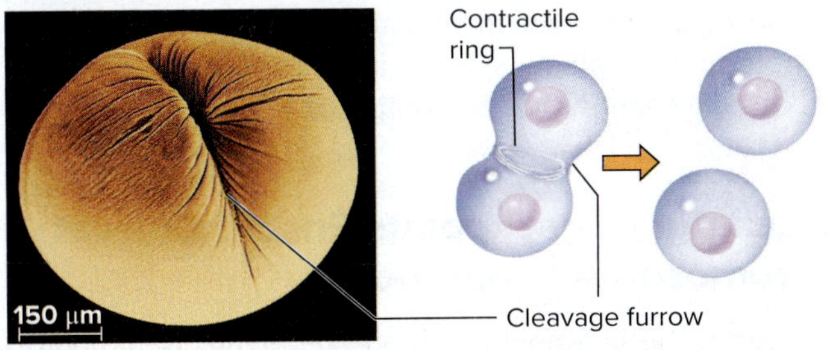

(b) Cytokinesis in a plant cell

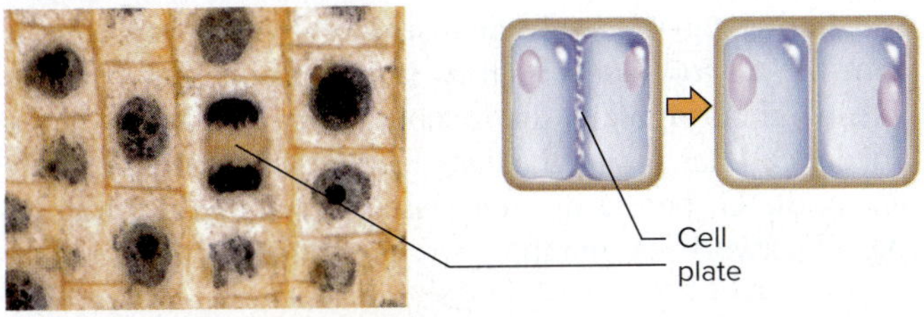

Figure 4.12 If cytokinesis does not follow mitosis, one cell may contain many nuclei. In fertilized *Drosophila* eggs, 13 rounds of mitosis take place without cytokinesis. The result is a single-celled syncytial embryo that contains several thousand nuclei. The photograph shows part of an embryo with dividing nuclei; chromosomes are in *red,* and spindle fibers are in *green.* Nuclei at the *upper left* are in metaphase, while nuclei toward the *bottom right* are progressively later in anaphase. Membranes eventually grow around these nuclei, dividing the embryo into cells.

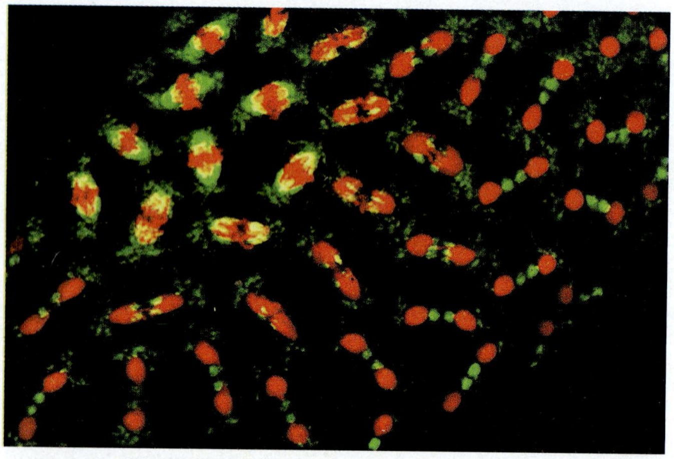

bodies, and (in plants) chloroplasts, must be parceled out to the emerging daughter cells. The mechanism accomplishing this task does not appear to predetermine which organelle is destined for which daughter cell. Instead, because most cells contain many copies of these cytoplasmic structures, each new cell is bound to receive at least a few representatives of each component. This original complement of structures is enough to sustain the cell until synthetic activity can repopulate the cytoplasm with organelles.

Sometimes cytoplasmic division does not immediately follow nuclear division, and the result is a cell containing more than one nucleus. An animal cell with two or more nuclei is known as a **syncytium.** The early embryos of fruit flies are multinucleated syncytia (**Fig. 4.12**), as are the precursors of spermatozoa in humans and many other animals. A multinucleate plant tissue is called a **coenocyte;** coconut milk is a nutrient-rich food composed of coenocytes.

Regulatory Checkpoints Ensure Correct Chromosome Separation

The cell cycle is a complex sequence of precisely coordinated events. In higher organisms, a cell's "decision" to divide depends on both intrinsic factors, such as conditions within the cell that register a sufficient size for division, and signals from the environment, such as hormonal cues or contacts with neighboring cells that encourage or restrain division. Once a cell has initiated events leading to division, usually during the G_1 period of interphase, everything else follows like clockwork. A number of **checkpoints**—moments at

Figure 4.13 Checkpoints help regulate the cell cycle. Cellular checkpoints (*red wedges*) ensure that important events in the cell cycle occur in the proper sequence. At each checkpoint, the cell determines whether prior events have been completed before it can proceed to the next step of the cell cycle. (For simplicity, we show only two chromosomes per cell.)

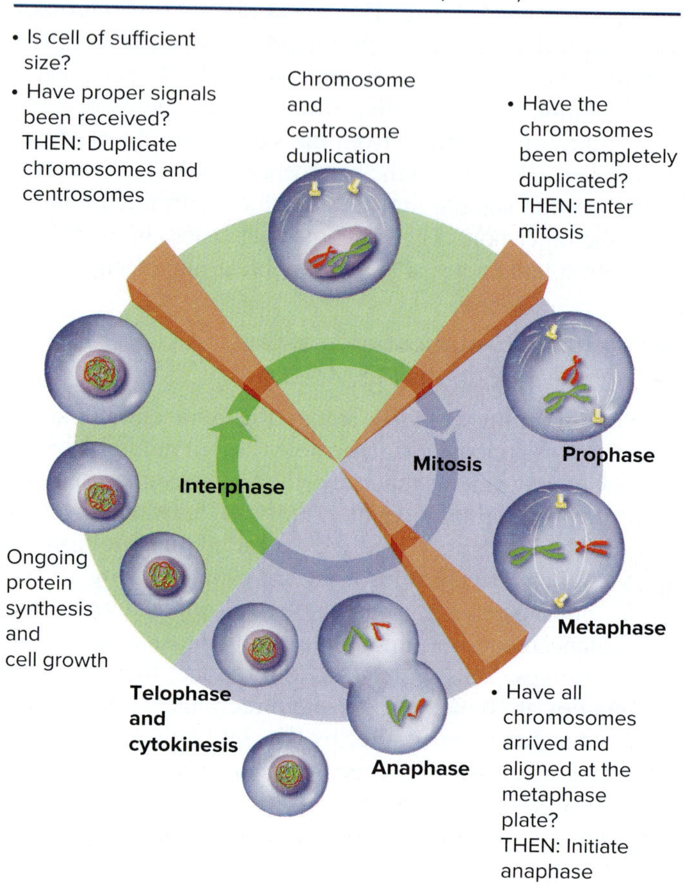

which the cell evaluates the results of previous steps—allow the sequential coordination of cell-cycle events (**Fig. 4.13**). For example, the enzymes operating in one type of checkpoint monitor DNA replication to ensure that cells do not begin mitosis until all the chromosomes have been completely copied. If this checkpoint did not exist, at least one of the daughter cells would lose DNA every cell cycle.

In a second illustration of the molecular basis of checkpoints, even a single kinetochore that has not attached to spindle fibers generates a molecular signal that prevents the sister chromatids of all chromosomes from separating at their centromeres. This signal makes the beginning of anaphase dependent on the prior proper alignment of all the chromosomes at metaphase. As a result of this cell-cycle checkpoint, each daughter cell reliably receives the right number of chromosomes.

Breakdown of the mitotic machinery can produce division mistakes that have crucial consequences for the cell. Improper chromosome segregation, for example, can cause serious malfunction or even the death of daughter cells. Gene mutations that disrupt mitotic structures, such as the spindle, kinetochores, or centrosomes, are one source of improper segregation. Other problems occur in cells where the normal restraints on cell division, such as checkpoints, have broken down. Such cells may divide uncontrollably, leading to a tumor. We present the details of cell-cycle regulation, checkpoint controls, and cancer formation in Chapter 20.

essential concepts

- Through *mitosis,* diploid cells produce identical diploid progeny cells.
- At *metaphase,* the *sister chromatids* are being pulled at their kinetochores toward opposite spindle poles; these poleward forces are balanced because the chromatids are connected at their centromeres.
- At the beginning of *anaphase,* the connections between sister centromeres are severed so sister chromatids separate and move to opposite spindle poles.
- Cell cycle *checkpoints* help ensure correct duplication and separation of chromosomes.

4.4 Meiosis: Cell Divisions That Halve Chromosome Number

learning objectives

1. Describe the key chromosome behaviors during meiosis that lead to haploid gametes.
2. Compare chromosome behaviors during mitosis and meiosis.
3. Explain how the independent alignment of homologs, and also crossing-over during the first meiotic division, each contribute to the genetic diversity of gametes.

During the many rounds of cell division within an embryo, most cells either grow and divide via the mitotic cell cycle just described, or they stop growing and become arrested in G_0. These mitotically dividing and G_0-arrested cells are the so-called **somatic cells** whose descendants continue to make up the vast majority of each organism's tissues throughout the lifetime of the individual. Early in the embryonic development of animals, however, a group of cells is set aside for a different fate. These are the **germ cells:** cells destined for a specialized role in the production of gametes. Germ cells arise later in plants, during floral development instead of during embryogenesis. The germ cells become incorporated in the reproductive organs—ovaries and testes in animals, ovaries and anthers in flowering plants—where they ultimately undergo meiosis, the special two-part cell division that produces gametes (eggs and sperm) containing half the number of chromosomes other body cells have.

Figure 4.14 An overview of meiosis: The chromosomes replicate once, while the nuclei divide twice. In this figure, all four chromatids of each chromosome pair are shown in the same shade of the same color. Note that the chromosomes duplicate before meiosis I, but they do not duplicate between meiosis I and meiosis II.

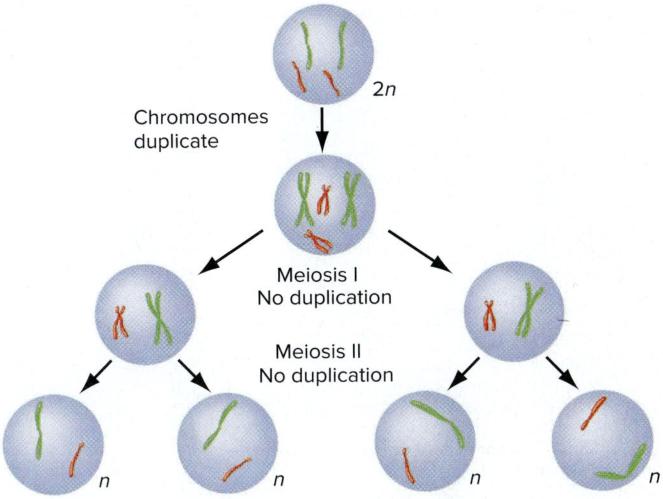

The union of haploid gametes at fertilization yields diploid offspring that carry the combined genetic heritage of two parents. Sexual reproduction therefore requires the alternation of haploid and diploid generations of cells. If gametes were diploid rather than haploid, the number of chromosomes would double in each successive generation. In humans, for example, the children would have 92 chromosomes per cell, the grandchildren 184, and so on. Meiosis prevents this lethal, exponential accumulation of chromosomes.

In Meiosis, the Chromosomes Replicate Once but the Nucleus Divides Twice

Unlike mitosis, meiosis consists of two successive nuclear divisions, logically named *division I of meiosis* and *division II of meiosis*, or simply **meiosis I** and **meiosis II.** With each round, the cell passes through a prophase, metaphase, anaphase, and telophase. In meiosis I, the parent nucleus divides to form two daughter nuclei; in meiosis II, each of the two daughter nuclei divides, resulting in four nuclei (**Fig. 4.14**). These four nuclei—the final products of meiosis—become partitioned in four separate daughter cells because cytokinesis occurs after both rounds of division. The chromosomes duplicate at the start of meiosis I, but they do not duplicate in meiosis II, which explains why the gametes contain half the number of chromosomes found in somatic cells. A close look at each round of meiotic division reveals the mechanisms by which each gamete comes to receive one full haploid set of chromosomes.

During Meiosis I, Homologs Pair, Exchange Parts, and Then Segregate

The events of meiosis I are unique among nuclear divisions (**Fig. 4.15**, meiosis I). The process begins with the replication

FEATURE FIGURE 4.15

Meiosis: One Diploid Cell Produces Four Haploid Cells

Meiosis I: A reductional division

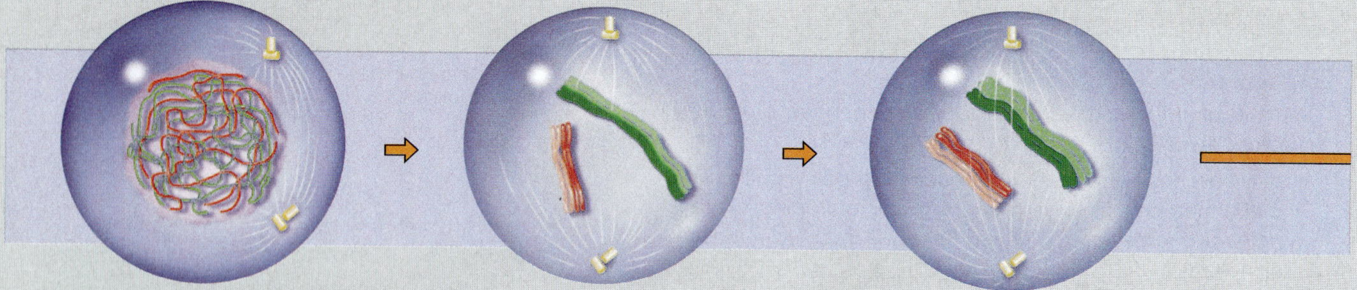

Prophase I: Leptotene
1. Chromosomes thicken and become visible, but the chromatids remain invisible.
2. Centrosomes begin to move toward opposite poles.

Prophase I: Zygotene
1. Homologous chromosomes enter *synapsis*.
2. The *synaptonemal complex* forms.

Prophase I: Pachytene
1. Synapsis is complete.
2. *Crossing-over*, genetic exchange between nonsister chromatids of a homologous pair, occurs.

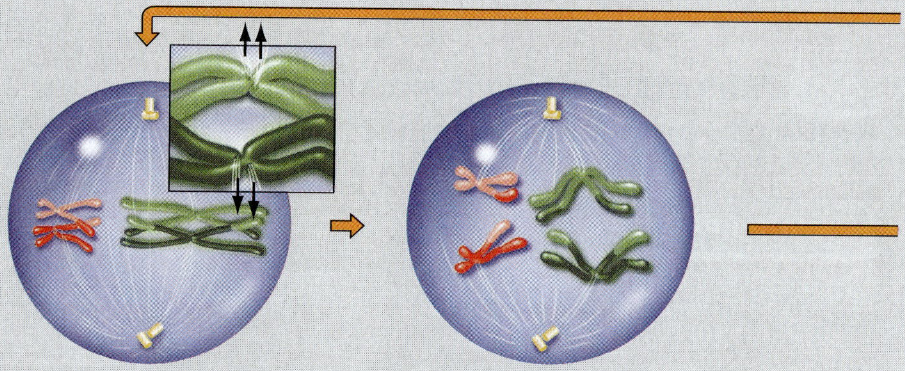

Metaphase I
1. Tetrads line up along the *metaphase plate*.
2. Each chromosome of a homologous pair attaches to fibers from opposite poles.
3. Sister chromatids attach to fibers from the same pole.

Anaphase I
1. Sister centromeres remain connected to each other.
2. The chiasmata dissolve.
3. Homologous chromosomes move to opposite poles.

Meiosis II: An equational division

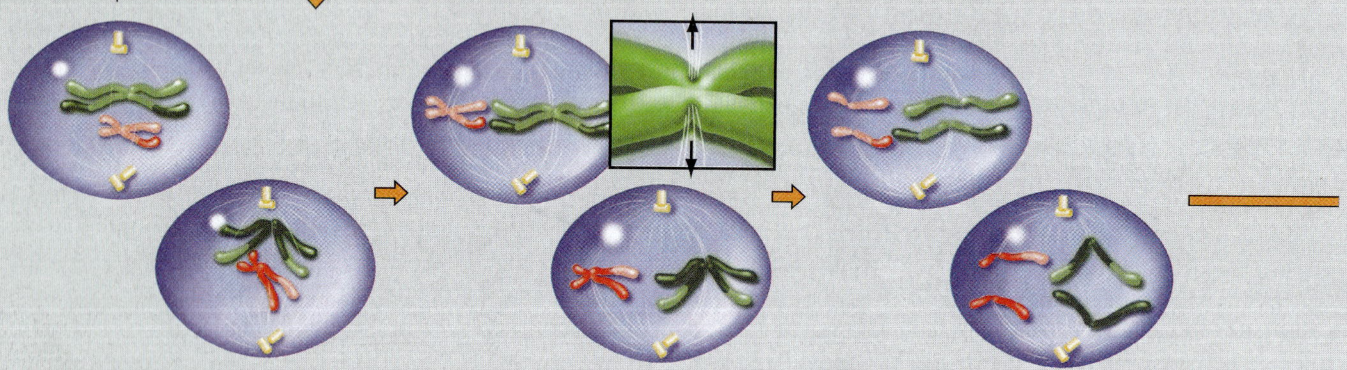

Prophase II
1. Chromosomes condense.
2. Centrioles move toward the poles.
3. The nuclear envelope breaks down at the end of prophase II (*not shown*).

Metaphase II
1. Chromosomes align at the metaphase plate.
2. Sister chromatids attach to spindle fibers from opposite poles.

Anaphase II
1. Sister centromeres detach from each other, allowing sister chromatids to move to opposite poles.

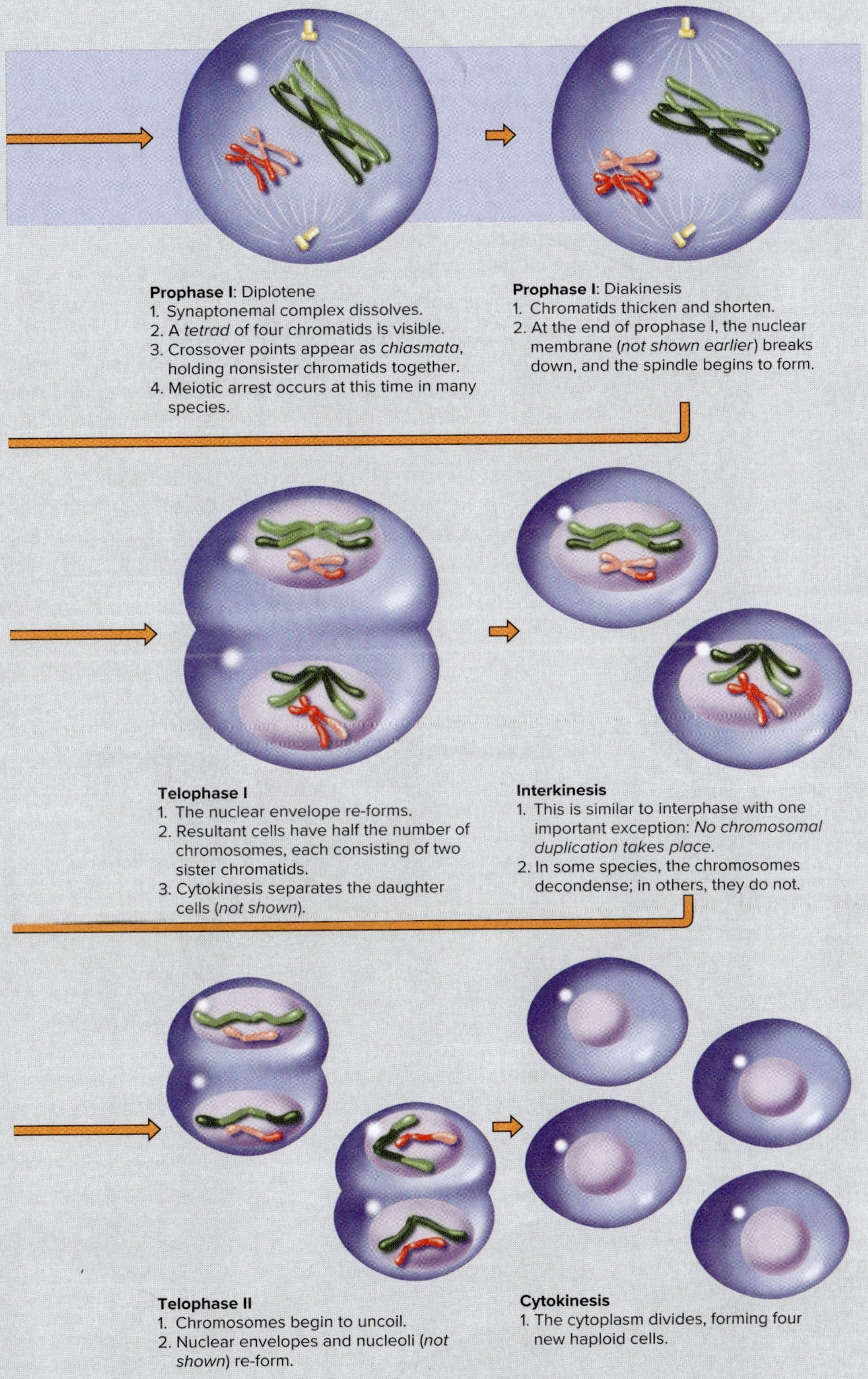

Figure 4.15 To aid visualization of the chromosomes, the figure is simplified in two ways: (1) The nuclear envelope is not shown during prophase of either meiotic division. (2) The chromosomes are shown as fully condensed at zygotene; in reality, full condensation is not achieved until diakinesis.

of chromosomes, after which each one consists of two sister chromatids. A key to understanding meiosis I is the observation that the centromeres of these sister chromatids remain connected throughout the entire division, rather than separating from each other as in mitosis.

As meiosis I proceeds, homologous chromosomes align across the cellular equator to form a coupling that ensures proper chromosome segregation later in the division. Moreover, during the time homologous chromosomes face each other across the equator, the maternal and paternal chromosomes of each homologous pair may exchange parts, creating new combinations of alleles at different genes along the chromosomes. Afterward, the two homologous chromosomes, each still consisting of two sister chromatids connected at their centromeres, are pulled to opposite poles of the spindle. As a result, it is homologous chromosomes (rather than sister chromatids as in mitosis) that segregate into different daughter cells at the conclusion of the first meiotic division. With this overview in mind, let us take a closer look at the specific events of meiosis I, remembering that we analyze a dynamic, flowing sequence of cellular events by breaking it down somewhat arbitrarily into the easily pictured, traditional phases.

Prophase I: Homologs condense and pair, and crossing-over occurs

Among the crucial events of **prophase I** are the condensation of chromatin, the pairing of homologous chromosomes, and the reciprocal exchange of genetic information between these paired homologs. Figure 4.15 shows a generalized view of prophase I; however, research suggests that the exact sequence of events may vary in different species. These complicated processes can take many days, months, or even years to complete. For example, in the female germ cells of several species, including humans, meiosis is suspended at prophase I for many years until ovulation (as will be discussed further in Section 4.5).

Leptotene (from the Greek for *thin* and *delicate*) is the first definable substage of prophase I, the time when the long, thin chromosomes begin to thicken (see **Fig. 4.16a** for a more detailed view). Each chromosome has already duplicated prior to prophase I (as in mitosis) and thus consists of two sister chromatids affixed at their centromeres. At this point, however, these sister chromatids are so tightly bound together that they are not yet visible as separate entities.

Zygotene (from the Greek for *conjugation*) begins as each chromosome seeks out its homologous partner and the matching chromosomes become zipped together in a

Figure 4.16 Prophase I of meiosis at very high magnification.

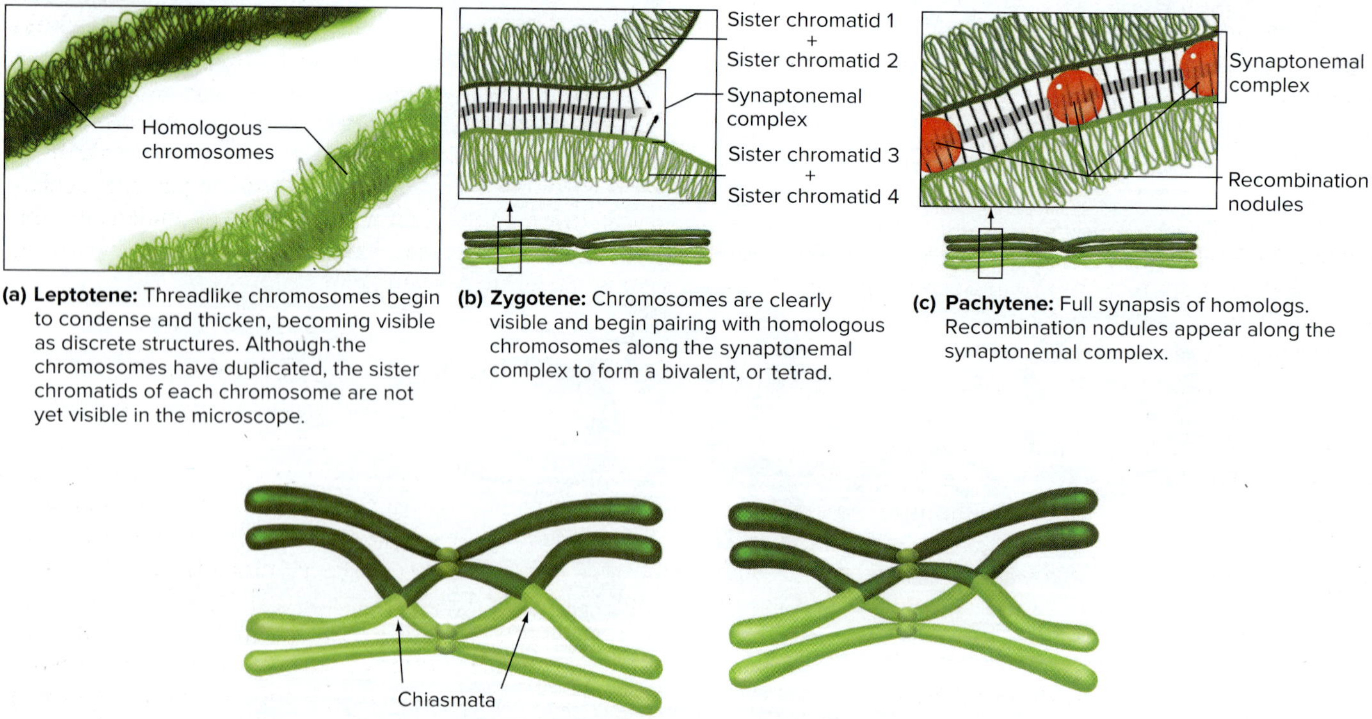

(a) Leptotene: Threadlike chromosomes begin to condense and thicken, becoming visible as discrete structures. Although the chromosomes have duplicated, the sister chromatids of each chromosome are not yet visible in the microscope.

(b) Zygotene: Chromosomes are clearly visible and begin pairing with homologous chromosomes along the synaptonemal complex to form a bivalent, or tetrad.

(c) Pachytene: Full synapsis of homologs. Recombination nodules appear along the synaptonemal complex.

(d) Diplotene: Bivalent pulls apart slightly, but homologous chromosomes remain connected due to recombination at crossover sites (chiasmata).

(e) Diakinesis: Further condensation of the bivalent.

process known as **synapsis.** The "zipper" itself is an elaborate protein structure called the **synaptonemal complex** that aligns the homologs with remarkable precision, juxtaposing the corresponding genetic regions of the chromosome pair (**Fig. 4.16b**).

Pachytene (from the Greek for *thick* or *fat*) begins at the completion of synapsis when homologous chromosomes are united along their length. Each synapsed chromosome pair is known as a **bivalent** (because it encompasses two chromosomes), or a **tetrad** (because it contains four chromatids). On one side of the bivalent is a maternally derived chromosome, on the other side a paternally derived one. Because X and Y chromosomes are not identical, they do not synapse completely. However, the pseudoautosomal regions previously shown in Fig. 4.8 provide small stretches of similarity between the X and the Y chromosomes that allow them to pair with each other during meiosis I in males.

During pachytene, structures called **recombination nodules** begin to appear along the synaptonemal complex, and an exchange of parts between nonsister (that is, between maternal and paternal) chromatids occurs at these nodules (see **Fig. 4.16c** for details). Such an exchange is known as **crossing-over;** it results in the **recombination** of genetic material. As a result of crossing-over, chromatids may no longer be of purely maternal or paternal origin; however, no genetic information is gained or lost, so all chromatids retain their original size.

Diplotene (from the Greek for *twofold* or *double*) is signaled by the gradual dissolution of the synaptonemal zipper complex and a slight separation of regions of the homologous chromosomes (see **Fig. 4.16d**). The aligned homologous chromosomes of each bivalent nonetheless remain very tightly merged at intervals along their length called **chiasmata** (singular, *chiasma*), which represent the sites where crossing-over occurred.

Diakinesis (from the Greek for *double movement*) is accompanied by further condensation of the chromatids. Because of this chromatid thickening and shortening, it can now clearly be seen that each tetrad consists of four separate chromatids, or viewed in another way, that the two homologous chromosomes of a bivalent are each composed of two sister chromatids held together at their centromeres (see **Fig. 4.16e**). Nonsister chromatids that have undergone crossing-over remain closely associated at chiasmata. The end of diakinesis is analogous to the prometaphase of mitosis: The nuclear envelope breaks down, and the microtubules of the spindle apparatus begin to form.

Metaphase I: Paired homologs attach to spindle fibers from opposite poles

During mitosis, each sister chromatid has a kinetochore that becomes attached to microtubules emanating from opposite spindle poles. During meiosis I, the situation is different. The kinetochores of sister chromatids fuse, so that each chromosome contains only a single functional kinetochore. During **metaphase I** (see Fig. 4.15, meiosis I), it is the kinetochores of homologous chromosomes that attach to microtubules from opposite spindle poles. As a result, in chromosomes aligned at the metaphase plate, the kinetochores of maternally and paternally derived chromosomes are subject to pulling forces from opposite spindle poles, balanced by the physical connections between homologs at chiasmata. Each bivalent's alignment and hookup is independent of that of every other bivalent, so the chromosomes facing each pole are a random mix of maternal and paternal origin.

Anaphase I: Homologs move to opposite spindle poles

At the onset of **anaphase I,** the chiasmata joining homologous chromosomes dissolve, which allows the maternal and paternal homologs to begin to move toward opposite spindle poles (see Fig. 4.15, meiosis I). Note that in anaphase of the first meiotic division, the sister centromeres do not separate as they do in mitosis. Thus, from each homologous pair, one chromosome consisting of two sister chromatids joined at their centromeres segregates to each spindle pole.

Recombination through crossing-over plays an important role in the proper segregation of homologous chromosomes during the first meiotic division. The chiasmata hold the homologs together and thus ensure that their kinetochores remain attached to opposite spindle poles throughout metaphase. When recombination does not occur within a bivalent, mistakes in hookup and conveyance may cause homologous chromosomes to move to the same pole, instead of segregating to opposite poles. In some organisms, however, proper segregation of nonrecombinant chromosomes nonetheless occurs through other pairing mechanisms. Investigators do not yet completely understand the nature of these processes, and they are currently evaluating several models to explain them.

Telophase I: Nuclear envelopes re-form

The telophase of the first meiotic division, or **telophase I,** takes place when nuclear membranes begin to form around the chromosomes that have moved to the poles. Each of the incipient daughter nuclei contains one-half the number of chromosomes in the original parent nucleus, but each chromosome consists of two sister chromatids joined at their centromeres (see Fig. 4.15, meiosis I). Because the number of chromosomes is reduced to one-half the normal diploid number, meiosis I is often called a **reductional division.**

In most species, cytokinesis follows telophase I, with daughter nuclei becoming enclosed in separate daughter cells. A short interphase then ensues. During this time, the chromosomes usually decondense, in which case they must recondense during the prophase of the subsequent second meiotic division. In some species, however, the chromosomes simply stay condensed. Most importantly, no S phase exists during the interphase between meiosis I and meiosis II; that is, the chromosomes do not replicate during meiotic interphase. The relatively brief interphase between meiosis I and meiosis II is known as **interkinesis.**

During Meiosis II, Sister Chromatids Separate to Produce Haploid Gametes

The second meiotic division (meiosis II) proceeds in a fashion very similar to that of mitosis, but because the number of chromosomes in each dividing nucleus has already been reduced by half, the resulting daughter cells are haploid. The same process occurs in each of the two daughter cells generated by meiosis I, producing four haploid cells at the end of this second meiotic round (see Fig. 4.15, meiosis II).

Prophase II: The chromosomes condense

If the chromosomes decondensed during the preceding interphase, they recondense during **prophase II.** At the end of prophase II, the nuclear envelope breaks down, and the spindle apparatus re-forms.

Metaphase II: Chromosomes align at the metaphase plate

The kinetochores of sister chromatids attach to microtubule fibers emanating from opposite poles of the spindle apparatus, just as in mitotic metaphase. Nonetheless, two significant features of **metaphase II** distinguish it from mitosis. First, the number of chromosomes is one-half that in mitotic metaphase of the same species. Second, in most chromosomes, the two sister chromatids are no longer strictly identical because of the recombination through crossing-over that occurred during meiosis I. The sister chromatids still contain the same genes, but they may carry different combinations of alleles.

Anaphase II: Sister chromatids move to opposite spindle poles

Just as in mitosis, severing of the connection between sister centromeres allows the sister chromatids to move toward opposite spindle poles during **anaphase II.**

Telophase II: Nuclear membranes re-form, and cytokinesis follows

Membranes form around each of four daughter nuclei in **telophase II,** and cytokinesis places each nucleus in a separate cell. The result is four haploid gametes. Note that at the end of meiosis II, each daughter cell (that is, each gamete) has the same number of chromosomes as the parental cell present at the beginning of this division. For this reason, meiosis II is termed an **equational division.**

Mistakes in Meiosis Produce Defective Gametes

Segregational errors during either meiotic division can lead to aberrations, such as trisomies, in the next generation. If, for example, the homologs of a chromosome pair do not segregate during meiosis I (a mistake known as **nondisjunction**), they may travel together to the same pole and eventually become part of the same gamete. Such an error may at fertilization result in any one of a large variety of possible trisomies. Most autosomal trisomies in humans, as we already mentioned, are lethal *in utero;* one exception is trisomy 21, the genetic basis of Down syndrome. Like trisomy 21, extra sex chromosomes may also be viable but cause a variety of mental and physical abnormalities, such as those seen in Klinefelter syndrome (see Table 4.1).

Meiosis Contributes to Genetic Diversity

The wider the assortment of different gene combinations among members of a species, the greater the chance that at least some individuals will carry combinations of alleles that allow survival in a changing environment. Two aspects of meiosis contribute to genetic diversity in a population. First, because only chance governs which paternal or maternal homologs migrate to the two poles during the first meiotic division, different gametes carry a different mix of maternal and paternal chromosomes. **Figure 4.17a** shows how different patterns of homolog migration produce different mixes of parental chromosomes in the gametes. The amount of potential variation generated by this random independent assortment increases with the number of chromosomes. In *Ascaris,* for example, where $n = 2$ (the chromosome complement shown in Fig. 4.17a), the random assortment of homologs could produce only 2^2, or four types of gametes. In a human being, however, where $n = 23$, this mechanism alone could generate 2^{23}, or more than 8 million genetically different kinds of gametes.

A second feature of meiosis, the reshuffling of genetic information through crossing-over during prophase I, ensures an even greater amount of genetic diversity in gametes. Because crossing-over recombines maternally and paternally derived genes, each chromosome in each different gamete could consist of different combinations of maternal and paternal alleles (**Fig. 4.17b**).

Of course, sexual reproduction adds yet another means of producing genetic diversity. At fertilization, any one of a vast number of genetically diverse sperm can fertilize an egg with its own distinctive genetic constitution. It is thus not very surprising that, with the exception of identical twins, the 6 billion people in the world are each genetically unique.

Figure 4.17 How meiosis contributes to genetic diversity. (a) The variation resulting from the independent assortment of nonhomologous chromosomes increases with the number of chromosomes in the genome. **(b)** Crossing-over between homologous chromosomes ensures that each gamete is unique.

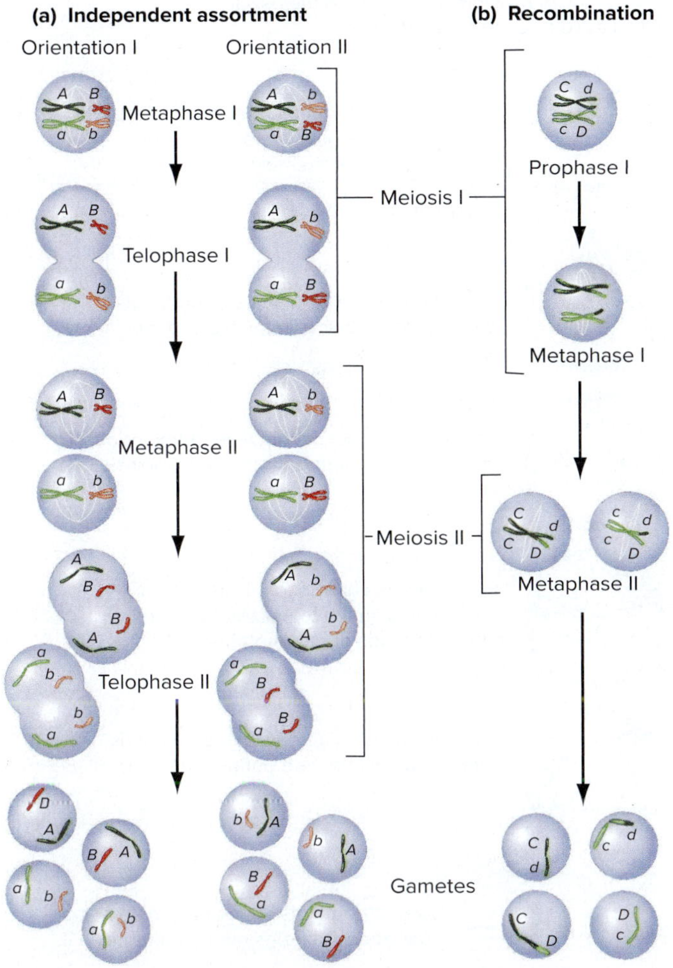

Mitosis and Meiosis: A Comparison

Mitosis occurs in all types of eukaryotic cells (that is, cells with a membrane-bounded nucleus) and is a conservative mechanism that preserves the genetic *status quo.* Mitosis followed by cytokinesis produces growth by increasing the number of cells. It also promotes the continual replacement of roots, stems, and leaves in plants and the regeneration of blood cells, intestinal tissues, and skin in animals.

Meiosis, on the other hand, occurs only in sexually reproducing organisms, in just a few specialized germ cells within the reproductive organs that produce haploid gametes. It is not a conservative mechanism; rather, the extensive combinatorial changes arising from meiosis are one source of the genetic variation that fuels evolution. **Table 4.3** illustrates the significant contrasts between the two mechanisms of cell division.

essential concepts

- In *meiosis,* chromosomes replicate once (before meiosis I), but the nucleus divides twice (meiosis I and II).
- During *metaphase I,* homologous chromosomes connect to opposite spindle poles. The independent alignment of each pair of homologs ensures the independent assortment of genes carried on different chromosomes.
- *Crossing-over* during the first meiotic division maintains the connection between homologous chromosomes until *anaphase I* and contributes to the genetic diversity of gametes.
- Sister chromatids separate from each other during *meiosis II* so that gametes have only one copy of each chromosome.
- *Fertilization*—the union of egg and sperm—restores the diploid number of chromosomes (2*n*) to the zygote.
- Errors during meiosis may produce gametes with missing or extra chromosomes, which often is lethal to offspring.

4.5 Gametogenesis

learning objectives

1. Compare the processes of oogenesis and spermatogenesis in humans.
2. Distinguish between the sex chromosome complements of human female and male germ-line cells at different stages of gametogenesis.

In all sexually reproducing animals, the embryonic germ cells (collectively known as the **germ line**) undergo a series of mitotic divisions that yield a collection of specialized diploid cells, which subsequently divide by meiosis to produce haploid cells. As with other biological processes, many variations on this general pattern have been observed. In some species, the haploid cells resulting from meiosis are the gametes themselves, while in other species, those cells must undergo a specific plan of differentiation to fulfill that function. Moreover, in certain organisms, the four haploid products of a single meiosis do not all become gametes. Gamete formation, or **gametogenesis,** thus gives rise to haploid gametes marked not only by the events of meiosis *per se* but also by cellular events that precede and follow meiosis. Here we illustrate gametogenesis with a description of egg and sperm formation in humans. The details of gamete formation in several other organisms appear throughout the book in discussions of specific experimental studies.

TABLE 4.3 Comparing Mitosis and Meiosis

Mitosis	Meiosis
Occurs in somatic cells and germ-line precursor cells Haploid and diploid cells can undergo mitosis One round of division	Occurs in germ cells as part of the sexual cycle Two rounds of division, meiosis I and meiosis II Only diploid cells undergo meiosis
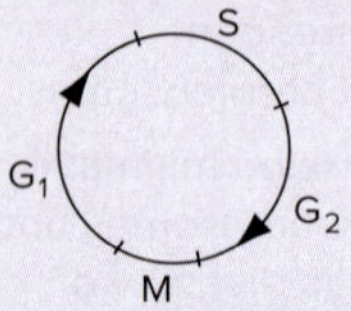 Mitosis is preceded by S phase (chromosome duplication).	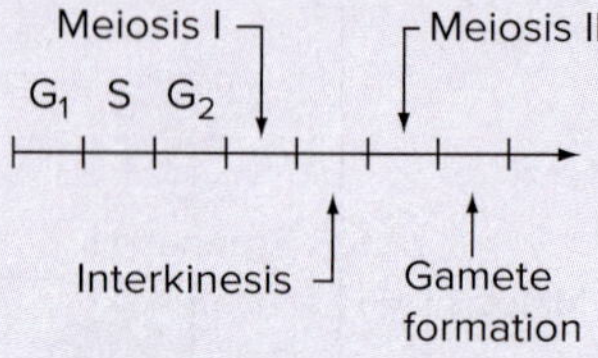 Chromosomes duplicate prior to meiosis I but not before meiosis II.
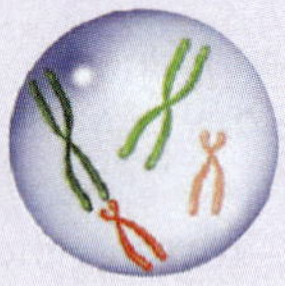 Homologous chromosomes do not pair.	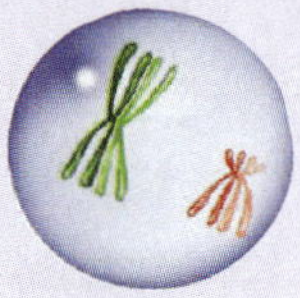 During prophase of meiosis I, homologous chromosomes pair (synapse) along their length.
Genetic exchange between homologous chromosomes is very rare.	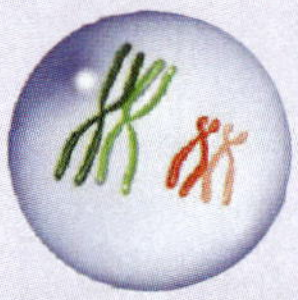 Crossing-over occurs between homologous chromosomes during prophase of meiosis I.
 Sister chromatids attach to spindle fibers from opposite poles during metaphase.	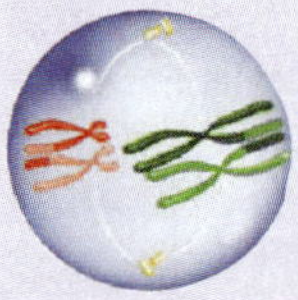 Homologous chromosomes (not sister chromatids) attach to spindle fibers from opposite poles during metaphase I.
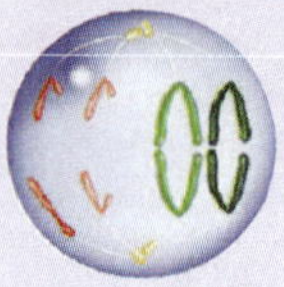 The centromeres of the sister chromatids separate at the beginning of anaphase.	 The centromeres of the sister chromatids remain tightly attached during meiosis I.
	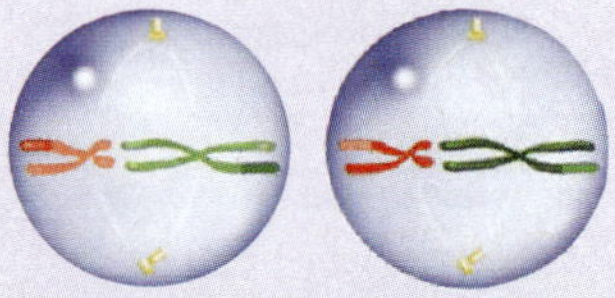 Sister chromatids attach to spindle fibers from opposite poles during metaphase II.
	 The centromeres of the sister chromatids separate at the beginning of anaphase II.
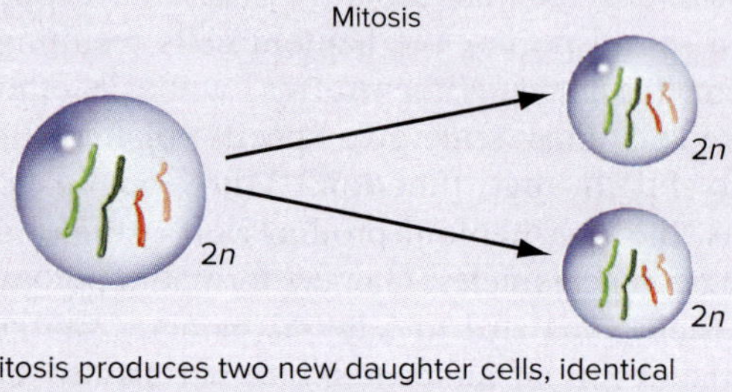 Mitosis produces two new daughter cells, identical to each other and the original cell. Mitosis is thus genetically conservative.	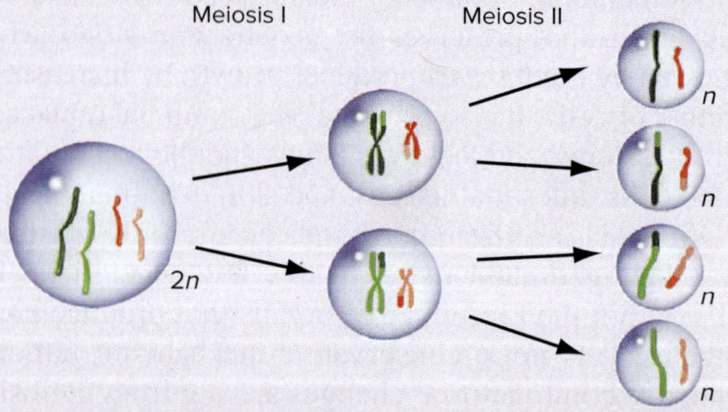 Meiosis produces four haploid cells, one (egg) or all (sperm) of which can become gametes. None of these is identical to each other or to the original cell, because meiosis results in combinatorial change.

Figure 4.18 In humans, egg formation begins in the fetal ovaries and arrests during the prophase of meiosis I. Fetal ovaries contain about 500,000 primary oocytes arrested in the diplotene substage of meiosis I. If the egg released during a menstrual cycle is fertilized, meiosis is completed. Only one of the three cells produced by meiosis serves as the functional gamete, or ovum.

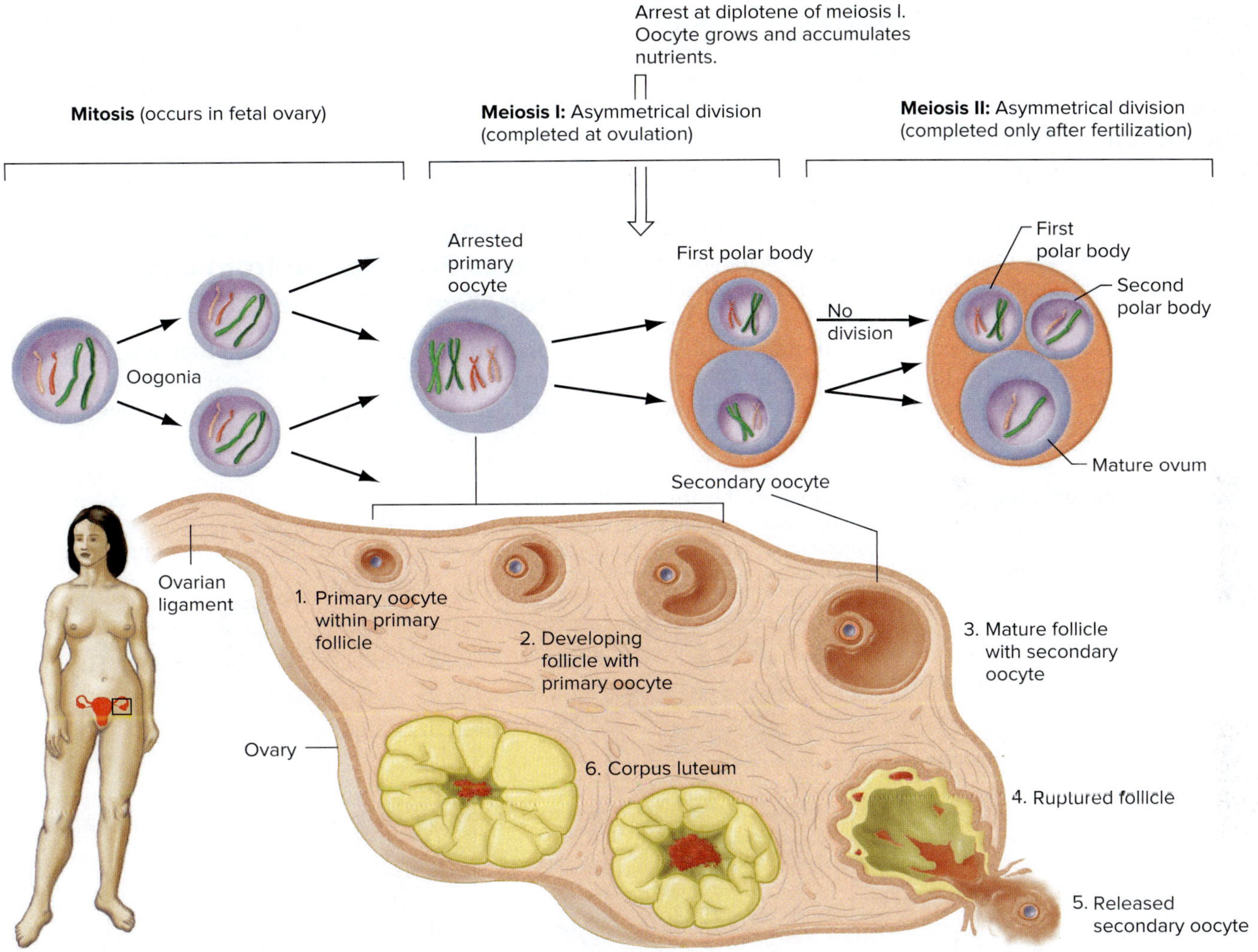

Oogenesis in Humans Produces One Ovum from Each Primary Oocyte

The end product of egg formation in humans is a large, nutrient-rich **ovum** whose stored resources can sustain the early embryo. The process, known as **oogenesis (Fig. 4.18)**, begins when diploid germ cells in the ovary, called **oogonia** (singular, *oogonium*), multiply rapidly by mitosis and produce a large number of **primary oocytes,** which then undergo meiosis.

For each primary oocyte, meiosis I results in the formation of two daughter cells that differ in size, so this division is asymmetric. The larger of these cells, the **secondary oocyte,** receives over 95% of the cytoplasm. The other small sister cell is known as the first **polar body.** During meiosis II, the secondary oocyte undergoes another asymmetrical division to produce a large haploid **ovum** and a small, haploid second polar body. The first polar body usually arrests its development. The two small polar bodies apparently serve no function and disintegrate, leaving one large haploid ovum as the functional gamete. Thus, only one of the three (or rarely, four) products of a single meiosis serves as a female gamete. A normal human ovum carries 22 autosomes and an X sex chromosome.

Oogenesis begins in the fetus. By six months after conception, the fetal ovaries are fully formed and contain about half a million primary oocytes arrested in the diplotene substage of prophase I. These cells, with their homologous chromosomes locked in synapsis, were thought for decades to be the only oocytes the female will produce. If so, a girl is born with all the oocytes she will ever possess. Remarkably, recent research has brought this long-held theory into question. Scientists have shown that germ-line precursor cells removed from adult ovaries can produce new eggs in a petri dish. However, it is not yet known whether these eggs are viable nor if these germ-line cells normally produce eggs in adults.

From the onset of puberty at about age 12, until menopause some 35–40 years later, most women release one primary oocyte each month (from alternate ovaries), amounting to roughly 480 oocytes released during the reproductive years. The remaining primary oocytes disintegrate during menopause. At ovulation, a released oocyte completes meiosis I and proceeds as far as the metaphase of meiosis II. If the oocyte is then fertilized, that is, penetrated by a sperm nucleus, it quickly completes meiosis II. The nuclear membranes of the sperm and ovum dissolve, allowing their chromosomes to form the single diploid nucleus of the zygote, and the zygote divides by mitosis to produce a functional embryo. In contrast, unfertilized oocytes exit the body during the menses stage of the menstrual cycle.

The long interval before completion of meiosis in oocytes released by women in their 30s, 40s, and 50s may contribute to the observed correlation between maternal age and meiotic segregational errors, including those that produce trisomies. Women in their mid-20s, for example, run a very small risk of trisomy 21; only 0.05% of children born to women of this age have Down syndrome. During the later childbearing years, however, the risk rises rapidly; at age 35, it is 0.9% of live births, and at age 45, it is 3%. You would not expect this age-related increase in risk if meiosis were completed before the mother's birth.

Spermatogenesis in Humans Produces Four Sperm from Each Primary Spermatocyte

The production of sperm, or **spermatogenesis (Fig. 4.19)**, begins in the male testes in germ cells known as **spermatogonia.** Mitotic divisions of the spermatogonia

Figure 4.19 Human sperm form continuously in the testes after puberty. Spermatogonia are located near the exterior of seminiferous tubules in a human testis. Once they divide to produce the primary spermatocytes, the subsequent stages of spermatogenesis—meiotic divisions in the spermatocytes and maturation of spermatids into sperm—occur successively closer to the middle of the tubule. Mature sperm are released into the central lumen of the tubule for ejaculation.

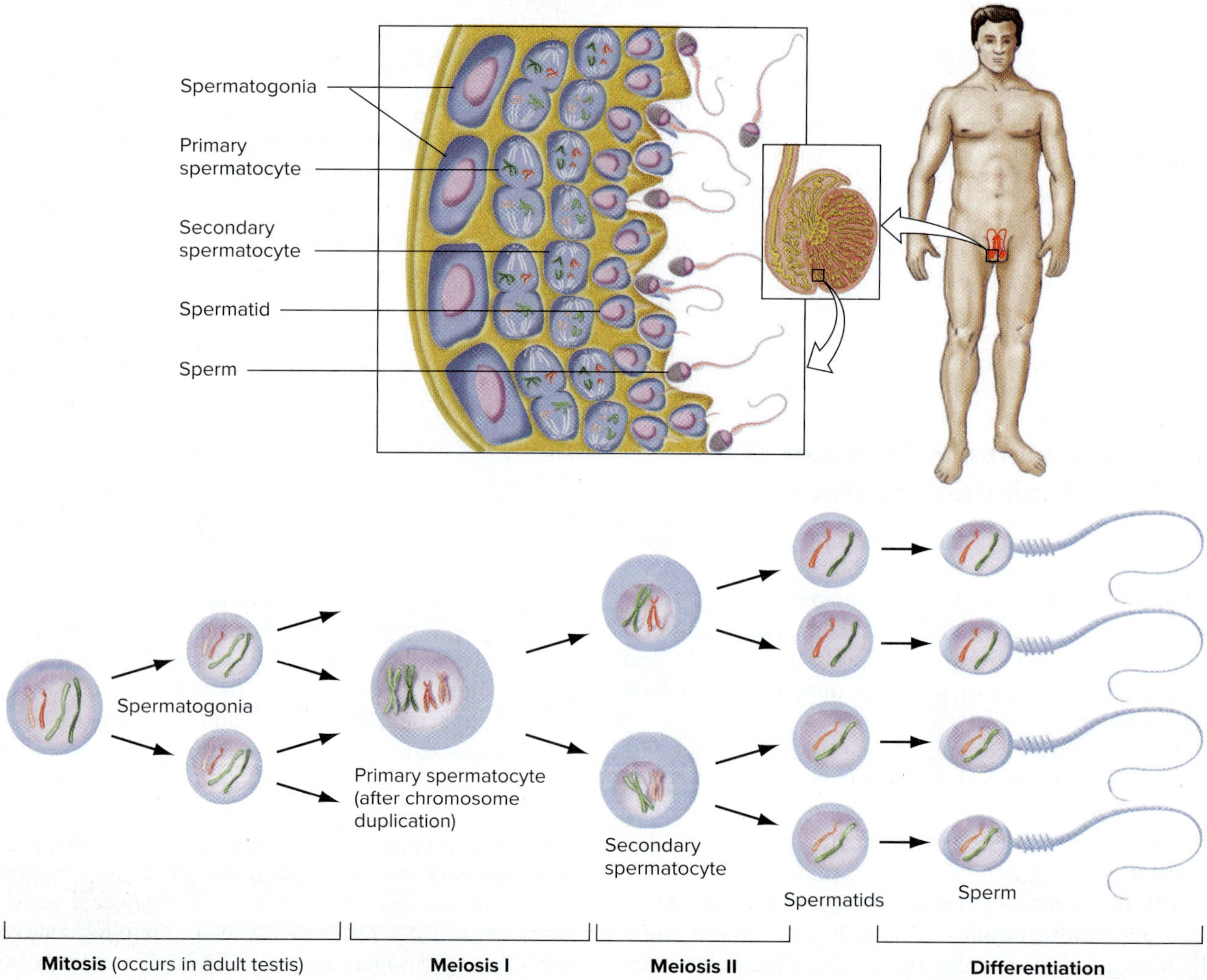

produce many diploid cells, the **primary spermatocytes.** Unlike primary oocytes, primary spermatocytes undergo a symmetrical meiosis I, producing two **secondary spermatocytes,** each of which undergoes a symmetrical meiosis II. At the conclusion of meiosis, each original primary spermatocyte thus yields four equivalent haploid **spermatids.** These spermatids then mature by developing a characteristic whiplike tail and by concentrating all their chromosomal material in a head, thereby becoming functional **sperm.** A human sperm, much smaller than the ovum it will fertilize, contains 22 autosomes and *either* an X *or* a Y sex chromosome.

The timing of sperm production differs radically from that of egg formation. The meiotic divisions allowing conversion of primary spermatocytes to spermatids begin only at puberty, but meiosis then continues throughout a man's life. The entire process of spermatogenesis takes about 48–60 days: 16–20 for meiosis I, 16–20 for meiosis II, and 16–20 for the maturation of spermatids into fully functional sperm. Within each testis after puberty, millions of sperm are always in production, and a single ejaculate can contain up to 300 million. Over a lifetime, a man can produce billions of sperm, almost equally divided between those bearing an X and those bearing a Y chromosome.

essential concepts

- Diploid *germ cell* precursors proliferate by mitosis and then undergo meiosis to produce haploid *gametes*.
- Human females are born with *oocytes* arrested in prophase of meiosis I. Meiosis resumes at *ovulation* but is not completed until fertilization. *Spermatogenesis* begins at puberty and continues through the lifetimes of human males.
- The two meiotic divisions of *oogenesis* are asymmetrical, so a *primary oocyte* results in a single egg. The two meiotic divisions of spermatogenesis are symmetrical, so a *primary spermatocyte* results in four *sperm*.
- All human oocytes contain a single X chromosome; human sperm contain either an X or a Y.

4.6 Validation of the Chromosome Theory

learning objectives

1. Describe the key events of meiosis that explain Mendel's first and second laws.
2. Infer from the results of crosses whether or not a trait is sex-linked.
3. Predict phenotypes associated with nondisjunction of sex chromosomes.

We have presented thus far two circumstantial lines of evidence in support of the chromosome theory of inheritance. First, the phenotype of sexual morphology is associated with the inheritance of particular chromosomes. Second, the events of mitosis, meiosis, and gametogenesis ensure a constant number of chromosomes in the somatic cells of all members of a species over time; one would expect the genetic material to exhibit this kind of stability even in organisms with very different modes of reproduction. Final acceptance of the chromosome theory depended on researchers going beyond the circumstantial evidence to a rigorous demonstration of two key points: (1) that the inheritance of genes corresponds with the inheritance of chromosomes in every detail, and (2) that the transmission of particular chromosomes coincides with the transmission of specific traits other than sex determination.

Mendel's Laws Correlate with Chromosome Behavior During Meiosis

Walter Sutton first outlined the chromosome theory of inheritance in 1902–1903, building on the theoretical ideas and experimental results of Theodor Boveri in Germany, E. B. Wilson in New York, and others. In a 1902 paper, Sutton speculated that "the association of paternal and maternal chromosomes in pairs and their subsequent separation during the reducing division (that is, meiosis I) . . . may constitute the physical basis of the Mendelian law of heredity." In 1903, he suggested that chromosomes carry Mendel's hereditary units for the following reasons:

1. Every cell contains two copies of each kind of chromosome, and two copies of each kind of gene.
2. The chromosome complement, like Mendel's genes, appears unchanged as it is transmitted from parents to offspring through generations.
3. During meiosis, homologous chromosomes pair and then separate to different gametes, just as the alternative alleles of each gene segregate to different gametes.
4. Maternal and paternal copies of each chromosome pair move to opposite spindle poles without regard to the assortment of any other homologous chromosome pair, just as the alternative alleles of unrelated genes assort independently.
5. At fertilization, an egg's set of chromosomes unites with a randomly encountered sperm's set of chromosomes, just as alleles obtained from one parent unite at random with those from the other parent.
6. In all cells derived from the fertilized egg, one-half of the chromosomes and one-half of the genes are of maternal origin, the other half of paternal origin.

The two parts of **Table 4.4** show the intimate relationship between the chromosome theory of inheritance and

TABLE 4.4 How the Chromosome Theory of Inheritance Explains Mendel's Laws

(a) The Law of Segregation

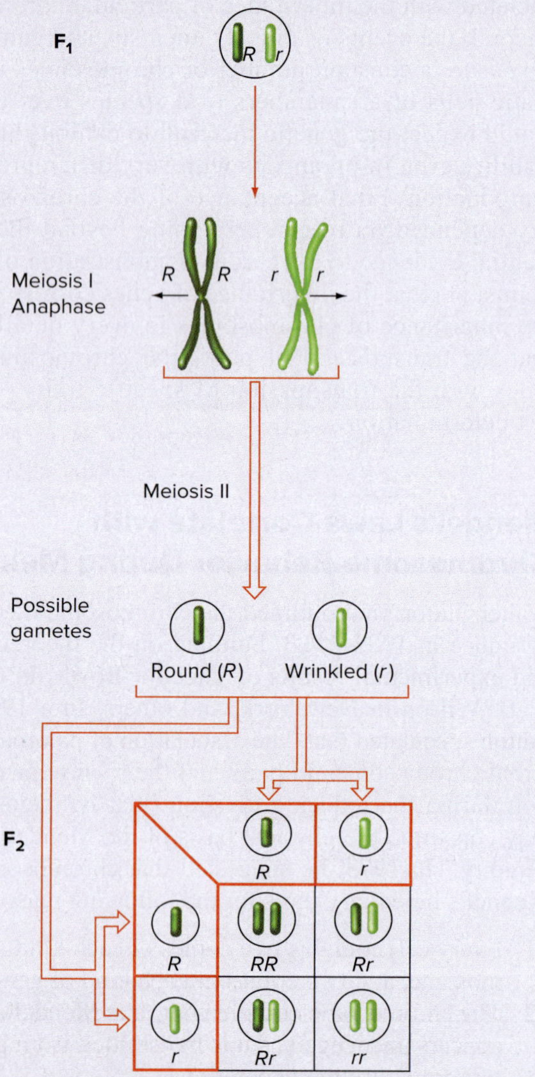

In an F_1 hybrid plant, the allele for round peas (*R*) is found on one chromosome, and the allele for wrinkled peas (*r*) is on the homologous chromosome. The pairing between the two homologous chromosomes during prophase through metaphase of meiosis I makes sure that the homologs will separate to opposite spindle poles during anaphase I. At the end of meiosis II, two types of gametes have been produced: half have *R*, and half have *r*, but no gametes have both alleles. Thus, the separation of homologous chromosomes at meiosis I corresponds to the segregation of alleles. As the Punnett square shows, fertilization of 50% *R* and 50% *r* eggs with the same proportion of *R* and *r* sperm leads to Mendel's 3:1 ratio in the F_2 generation.

(b) The Law of Independent Assortment

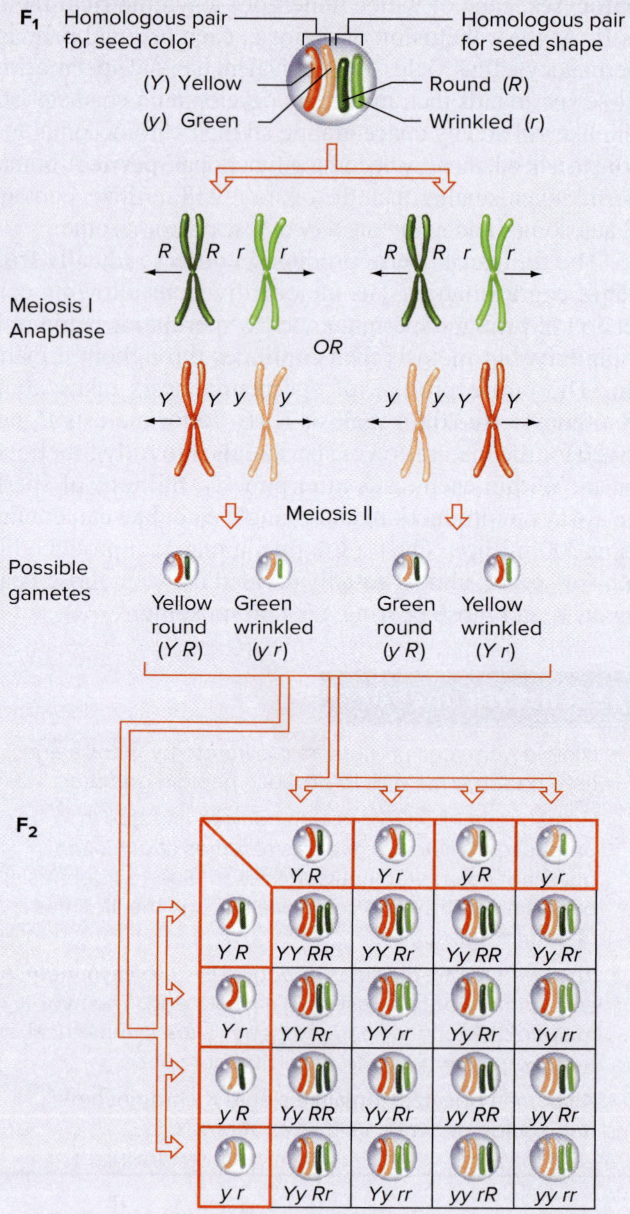

One pair of homologous chromosomes carries the gene for seed shape (alleles *R* and *r*). A second pair of homologous chromosomes carries the gene for seed color (alleles *Y* and *y*). Each homologous pair aligns at random at the metaphase plate during meiosis I, independently of the other homologous pair. Thus, two equally likely configurations are possible for the migration of any two chromosome pairs toward the poles during anaphase I. As a result, a dihybrid individual will generate four equally likely types of gametes with regard to the two traits in question. The Punnett square affirms that independent assortment of traits carried by nonhomologous chromosomes produces Mendel's 9:3:3:1 ratio.

Mendel's laws of segregation and independent assortment. If Mendel's genes for pea shape and pea color are assigned to different (that is, nonhomologous) chromosomes, the behavior of chromosomes can be seen to parallel the behavior of genes. Walter Sutton's observation of these parallels led him to propose that chromosomes and genes are physically connected in some manner. Meiosis ensures that each gamete will contain only a single chromatid of a bivalent and thus only a single allele of any gene on that chromatid (Table 4.4a). The independent behavior of two bivalents during meiosis means that the genes carried on different chromosomes will assort into gametes independently (Table 4.4b).

From a review of Fig. 4.17a, which follows two different chromosome pairs through the process of meiosis, you might wonder whether crossing-over abolishes the clear correspondence between Mendel's laws and the movement of chromosomes. The answer is no. Each chromatid of a homologous chromosome pair contains only one copy of a given gene, and only one chromatid from each pair of homologs is incorporated into each gamete. Because alternative alleles remain on different chromatids even after crossing-over has occurred, alternative alleles still segregate to different gametes as demanded by Mendel's first law.

Furthermore, because the orientations of nonhomologous chromosomes are completely random with respect to each other during both meiotic divisions, the genes on different chromosomes assort independently even if crossing-over occurs, as demanded by Mendel's second law. In Fig. 4.17a, you can see that without recombination, each of the two random alignments of the nonhomologous chromosomes results in the production of only two of the four gamete types: *AB* and *ab* for one orientation, and *Ab* and *aB* for the other orientation. With recombination, each of the alignments of alleles in Fig. 4.17a may in fact generate all four gamete types. (Imagine a crossover switching the positions of *A* and *a* nonsister chromatids in Fig. 4.17a). Thus, both the random alignment of nonhomologous chromosomes and crossing-over contribute to the phenomenon of independent assortment.

Specific Traits Are Transmitted with Specific Chromosomes

The fate of a theory depends on whether its predictions can be validated. Because genes determine traits, the prediction that chromosomes carry genes could be tested by breeding experiments that would show whether transmission of a specific chromosome coincides with transmission of a specific trait. Cytologists knew that one pair of chromosomes, the sex chromosomes, determines whether an individual is male or female. Would similar correlations exist for other traits?

A gene determining eye color on the *Drosophila* X chromosome

Thomas Hunt Morgan, an American experimental biologist with training in embryology, headed the research group whose findings eventually established a firm experimental base for the chromosome theory. Morgan chose to work with the fruit fly *Drosophila melanogaster* because it is extremely prolific and has a very short generation time, taking only 12 days to develop from a fertilized egg into a mature adult capable of producing hundreds of offspring. Morgan fed his flies mashed bananas and housed them in empty milk bottles capped with wads of cotton.

In 1910, a white-eyed male appeared among a large group of flies with brick-red eyes. A mutation had apparently altered a gene determining eye color, changing it from the normal wild-type allele specifying red to a new allele that produced white. When Morgan allowed the white-eyed male to mate with its red-eyed sisters, all the flies of the F_1 generation had red eyes; the red allele was clearly dominant to the white (**Fig. 4.20,** cross A).

Establishing a pattern of nomenclature for *Drosophila* geneticists, Morgan named the gene identified by the abnormal white eye color the *white* gene, for the mutation that revealed its existence. The normal wild-type allele of the *white* gene, abbreviated w^+, is for brick-red eyes, while the counterpart mutant *w* allele results in white eye color. The superscript + signifies the wild type. By writing the gene name and abbreviation in lowercase, Morgan symbolized that the mutant *w* allele is recessive to the wild-type w^+. (If a *Drosophila* mutation results in a dominant non-wild-type phenotype, the first letter of the gene name or of its abbreviation is capitalized; thus the mutation known as *Bar* eyes is dominant to the wild-type Bar^+ allele. (See the Appendix *Guidelines for Gene Nomenclature.*)

Morgan then crossed the red-eyed males of the F_1 generation with their red-eyed sisters (Fig. 4.20, cross B) and obtained an F_2 generation with the predicted 3:1 ratio of red to white eyes. But there was something askew in the pattern: Among the red-eyed offspring, there were two females for every one male, and all the white-eyed offspring were males. This result was surprisingly different from the equal transmission to both sexes of the Mendelian traits discussed in Chapters 2 and 3. In these fruit flies, the ratio of eye colors was not the same in male and female progeny.

By mating F_2 red-eyed females with their white-eyed brothers (Fig. 4.20, cross C), Morgan obtained some females with white eyes, which then allowed him to mate a white-eyed female with a red-eyed wild-type male (Fig. 4.20, cross D). The result was exclusively red-eyed daughters and white-eyed sons. The pattern seen in cross D is known as **crisscross inheritance** because the males inherit their eye color from their mothers, while the daughters inherit their eye color from their fathers. Note in Fig. 4.20 that the

Figure 4.20 A *Drosophila* eye color gene is located on the X chromosome. X-linkage explains the inheritance of alleles of the *white* gene in this series of crosses performed by Thomas Hunt Morgan. The progeny of crosses A, B, and C outlined with green dotted boxes are those used as the parents in the next cross of the series.

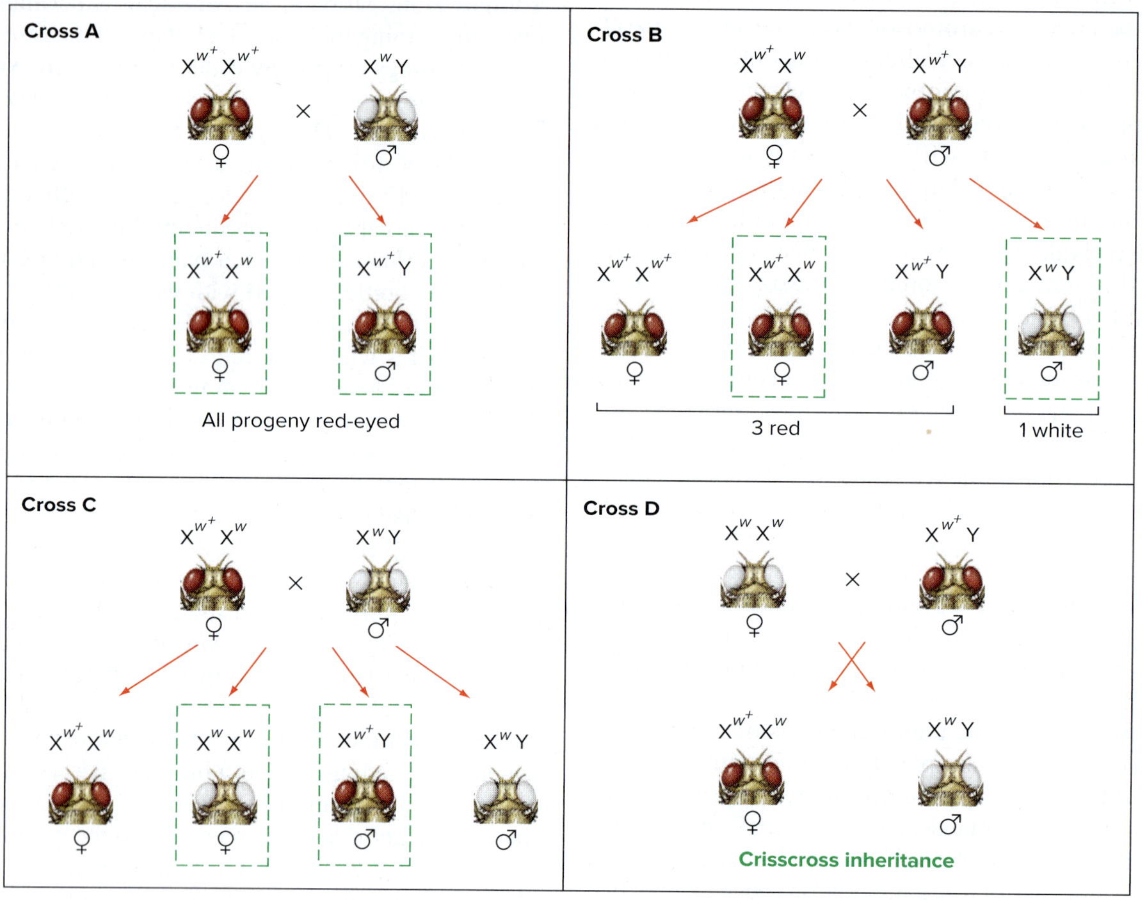

results of the reciprocal crosses red female × white male (cross A) and white female × red male (cross D) are not identical, again in contrast with Mendel's findings.

From the data, Morgan reasoned that the *white* gene for eye color is **X-linked,** that is, carried by the X chromosome. (Note that while symbols for genes and alleles are italicized, symbols for chromosomes are not.) The Y chromosome carries no allele of this gene for eye color. Males, therefore, have only one copy of the gene, which they inherit from their mother along with their only X chromosome; their Y chromosome must come from their father. Thus, males are **hemizygous** for this eye color gene, because their diploid cells have half the number of alleles carried by the female on her two X chromosomes.

If the single *white* gene on the X chromosome of a male is the wild-type w^+ allele, he will have red eyes and a genotype that can be written $X^{w^+}Y$. [Here we designate the chromosome (X or Y) together with the allele it carries, to emphasize that certain genes are X-linked.] In contrast to an $X^{w^+}Y$ male, a hemizygous X^wY male would have white eyes. Females with two X chromosomes can be one of three genotypes: $X^w X^w$ (white-eyed), $X^wX^{w^+}$ (red-eyed because w^+ is dominant to w), or $X^{w^+}X^{w^+}$ (red-eyed). As shown in Fig. 4.20, Morgan's assumption that the gene for eye color is X-linked explains the results of his breeding experiments. Crisscross inheritance, for example, occurs because the only X chromosome in sons of a white-eyed mother (X^wX^w) must carry the w allele, so the sons will be white-eyed. In contrast, because daughters of a red-eyed ($X^{w^+}Y$) father must receive a w^+-bearing X chromosome from their father, they should all have red eyes.

Validation of the chromosome theory from the analysis of nondisjunction

Although Morgan's work strongly supported the hypothesis that the gene for eye color lies on the X chromosome, he himself continued to question the validity of the chromosome theory until Calvin Bridges, one of his top students, found another key piece of evidence. Bridges repeated the cross Morgan had performed between white-eyed females and red-eyed males, but this time he did the experiment on a larger scale. As expected, the progeny of this cross consisted mostly of red-eyed females and white-eyed males. However, about 1 in every 2000 males had red eyes, and about the same small fraction of females had white eyes.

Figure 4.21 Nondisjunction: Rare mistakes in meiosis help confirm the chromosome theory. (a) Rare events of nondisjunction in an XX female produce XX and O eggs. The results of normal disjunction in the female are not shown. XO males are sterile because the missing Y chromosome is needed for male fertility in *Drosophila*. **(b)** In an XXY female, the three sex chromosomes can pair and segregate in two ways, producing progeny with unusual sex chromosome complements.

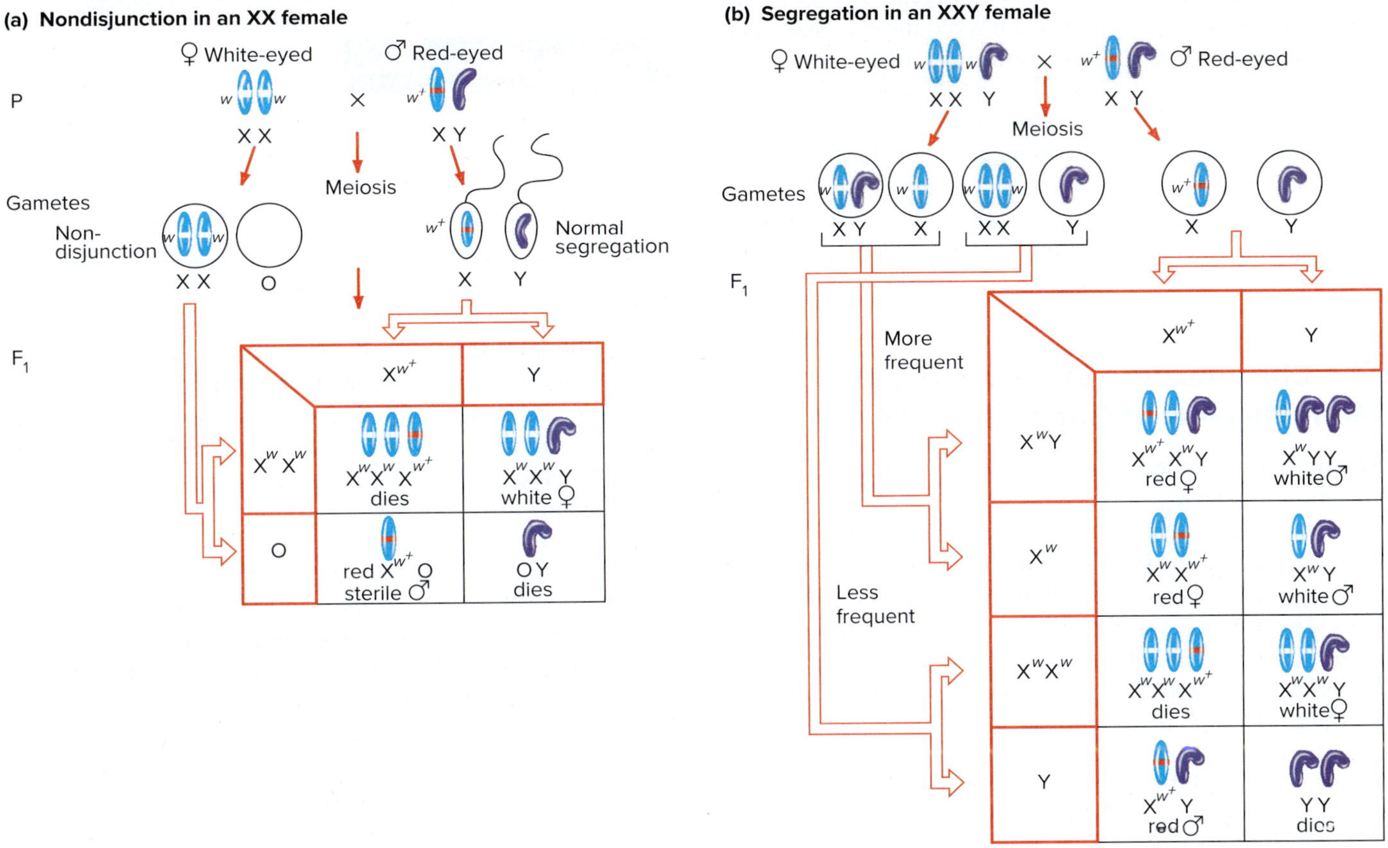

Bridges hypothesized that these exceptions arose through rare events in which the X chromosomes fail to separate during meiosis in females. He called such failures in chromosome segregation *nondisjunction*. Mistakes leading to nondisjunction can occur during either meiosis I or meiosis II, but in either case nondisjunction would result in some eggs with two X chromosomes and others with none. As **Fig. 4.21a** shows, fertilization of these chromosomally abnormal eggs could produce four types of zygotes: XXY (with two X chromosomes from the egg and a Y from the sperm), XXX (with two Xs from the egg and one X from the sperm), XO (with the lone sex chromosome from the sperm and no sex chromosome from the egg), and OY (with the only sex chromosome again coming from the sperm).

When Bridges examined the sex chromosomes of the rare white-eyed females produced in his large-scale cross, he found that they were indeed XXY individuals who must have received two X chromosomes and with them two *w* alleles from their white-eyed X^wX^w mothers. The exceptional red-eyed males emerging from the cross were XO; their eye color showed that they must have obtained their sole sex chromosome from their $X^{w^+}Y$ fathers. In this study then, transmission of the *white* gene alleles followed the predicted behavior of X chromosomes during rare meiotic mistakes, indicating that the X chromosome carries the gene for eye color. These results also suggested that zygotes with the two other abnormal sex chromosome karyotypes expected from nondisjunction in females (XXX and OY) die during embryonic development and thus produce no progeny.

Because XXY white-eyed females have three sex chromosomes rather than the normal two, Bridges reasoned they would produce four kinds of eggs: XY and X, or XX and Y (**Fig. 4.21b**). You can visualize the formation of these four kinds of eggs by imagining that when the three chromosomes pair and disjoin during meiosis, two chromosomes must go to one pole and one chromosome to the other. With this kind of segregation, only two results are possible: Either one X and the Y go to one pole and the second X to the other (yielding XY and X gametes), or the two Xs go to one pole and the Y to the other (yielding XX and Y gametes). The first of these two scenarios occurs more often because it comes about when the two similar X chromosomes pair with each other, ensuring that they will go to opposite poles during the first meiotic division. The second, less likely possibility happens only if the two X chromosomes fail to pair with each other.

Bridges next predicted that fertilization of these four kinds of eggs from an XXY female by normal sperm would generate an array of sex chromosome karyotypes associated with specific eye colors in the progeny. Bridges verified all his predictions when he analyzed the eye colors and sex chromosomes of a large number of offspring. For instance, he showed cytologically that all of the white-eyed females emerging from the cross in Fig. 4.21b had two X chromosomes and one Y chromosome, while one-half of the white-eyed males had a single X chromosome and two Y chromosomes. Bridges' painstaking observations provided compelling evidence that specific genes do in fact reside on specific chromosomes.

The Chromosome Theory Integrates Many Aspects of Gene Behavior

Mendel had assumed that genes are located in cells. The chromosome theory assigned the genes to a specific kind of structure within cells and explained alternative alleles as physically matching parts of homologous chromosomes. In so doing, the theory provided an explanation of Mendel's laws. The mechanism of meiosis ensures that the matching parts of homologous chromosomes will segregate to different gametes (except in rare instances of nondisjunction), accounting for the segregation of alleles predicted by Mendel's first law. Because each homologous chromosome pair aligns independently of all others at meiosis I, genes carried on different chromosomes will assort independently, as predicted by Mendel's second law.

The chromosome theory is also able to explain the creation of new alleles through *mutation,* a spontaneous change in a particular gene (that is, in a particular part of a chromosome). If a mutation occurs in the germ line, it can be transmitted to subsequent generations.

Finally, through mitotic cell divisions in the embryo and after birth, each cell in a multicellular organism receives the same chromosomes—and thus the same maternal and paternal alleles of each gene—as the zygote received from the egg and sperm at fertilization. In this way, an individual's genome—the chromosomes and genes he or she carries—remains constant throughout life.

essential concepts

- Segregation of homologous chromosomes into daughter cells at meiosis I explains Mendel's first law.
- Independent alignment of homologs with respect to each other and crossing-over of nonsister chromatids during meiosis I explain Mendel's second law.
- In organisms with XX/XY sex determination, males are *hemizygous* for X-linked genes, while females have two copies.

4.7 Sex-Linked and Sexually Dimorphic Traits in Humans

learning objectives

1. Determine from pedigree analysis whether human traits are X-linked or autosomal.
2. Explain how human cells compensate for the X-linked gene dosage difference in XX and XY nuclei.

A person unable to tell red from green would find it nearly impossible to distinguish the rose, scarlet, and magenta in the flowers of a garden bouquet from the delicately variegated greens in their foliage, or to complete a complex electrical circuit by fastening red-clad metallic wires to red ones and green to green. Such a person has most likely inherited some form of red-green color blindness, a recessive condition that runs in families and affects mostly males. Among Caucasians in North America and Europe, 8% of men but only 0.44% of women have this vision defect. **Figure 4.22** suggests to readers with normal color vision what people with red-green color blindness actually see.

In 1911, E. B. Wilson, a contributor to the chromosome theory of inheritance, combined family studies of the inheritance of color blindness with recent knowledge of the roles of the X and Y chromosomes in sex determination to make the first assignment of a human gene to a particular chromosome. The gene for red-green color blindness, he said, lies on the X because the condition usually passes from a maternal grandfather through an unaffected carrier mother to roughly 50% of the grandsons.

Several years after Wilson made this gene assignment, pedigree analysis established that various forms of hemophilia, or *bleeders disease* (in which the blood fails to clot properly), also result from mutations on the X chromosome that give rise to a relatively rare, recessive trait. In this context, rare means *infrequent in the population.* The family histories under review, including one following the descendants of Queen Victoria of England (**Fig. 4.23a**), showed that relatively rare X-linked traits appear more often in males than in females and often skip generations. The clues that suggest X-linked recessive inheritance in a pedigree are summarized in **Table 4.5.**

Unlike color blindness and hemophilia, some—although very few—of the known rare mutations on the X chromosome are dominant to the wild-type allele. With such dominant X-linked mutations, more females than males show the aberrant phenotype. This phenomenon occurs because all the daughters of an affected male but none of the sons will have the condition, while one-half the sons and one-half the daughters of an affected female will receive the dominant allele and therefore show the phenotype (see Table 4.5).

Figure 4.22 Red-green color blindness is an X-linked recessive trait in humans. How the world looks to a person with either normal color vision **(a)** or a kind of red-green color blindness known as deuteranopia **(b)**.

(both): Color deficit simulation courtesy of Vischeck (www.vischeck.com). Source image courtesy of NASA

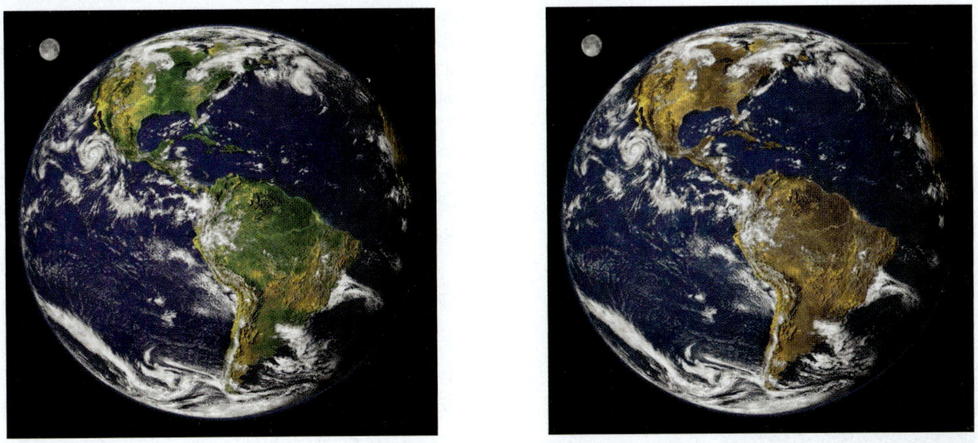

(a) (b)

Figure 4.23 X-linked traits may be recessive or dominant. **(a)** Pedigree showing inheritance of the recessive X-linked trait hemophilia in Queen Victoria's family. **(b)** Pedigree showing the inheritance of the dominant X-linked trait hypophosphatemia, commonly referred to as vitamin D–resistant rickets.

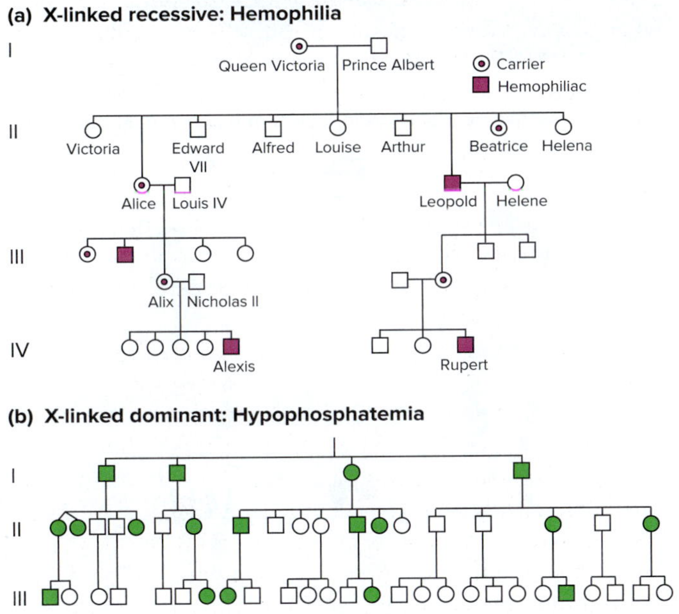

Vitamin D–resistant rickets, or hypophosphatemia, is an example of an X-linked dominant trait. **Figure 4.23b** presents the pedigree of a family affected by this disease.

In XX Human Females, One X Chromosome Is Inactivated

The XX and XY system of sex determination presents human cells with a curious problem that requires a solution

TABLE 4.5 Pedigree Patterns Suggesting Sex-Linked Inheritance

X-Linked Recessive Trait
1. The trait appears in more males than females because a female must receive two copies of the rare defective allele to display the phenotype, whereas a hemizygous male with only one copy will show it.
2. The mutation will never pass from father to son because sons receive only a Y chromosome from their father.
3. An affected male passes the X-linked mutation to all his daughters, who are thus carriers. Each son of these carrier females has a one-half chance to inherit the defective allele and thus the trait.
4. The trait often skips a generation as the mutation passes from grandfather through a carrier daughter to grandson.
5. The trait can appear in successive generations when a sister of an affected male is a carrier. If she is, each of her sons has a one-half chance of being affected.
6. With the rare affected (homozygous) female, all her sons will be affected and all her daughters will be carriers.
X-Linked Dominant Trait
1. More females than males show the aberrant trait.
2. The trait is seen in every generation as long as affected males have female children.
3. All the daughters but none of the sons of an affected male will be affected. This criterion is the most useful for distinguishing an X-linked dominant trait from an autosomal dominant trait.
4. The sons and daughters of an affected female each have a one-half chance of being affected.
5. For incompletely dominant X-linked traits, carrier females may show the trait in less extreme form than males with the defective allele.
Y-Linked Trait
1. The trait is seen only in males.
2. All male descendants of an affected man will exhibit the trait.
3. Not only do females not exhibit the trait, they also cannot transmit it.

Figure 4.24 Barr bodies are densely staining particles in XX cell nuclei. The arrow points to a Barr body in the nucleus of an XX cell treated with a DNA stain. The Barr body appears bright *white* in this negative image. Unlike the other chromosomes, the Barr body is highly condensed and attached to the nuclear envelope. XY cells have no Barr bodies.

a-b: From: Hong et al. (17 July 2001), "Identification of an autoimmune serum containing antibodies against the Barr body," *PNAS*, 98(15): 8703-8708, Fig 1A-B. © 2001 National Academy of Sciences, U.S.A.

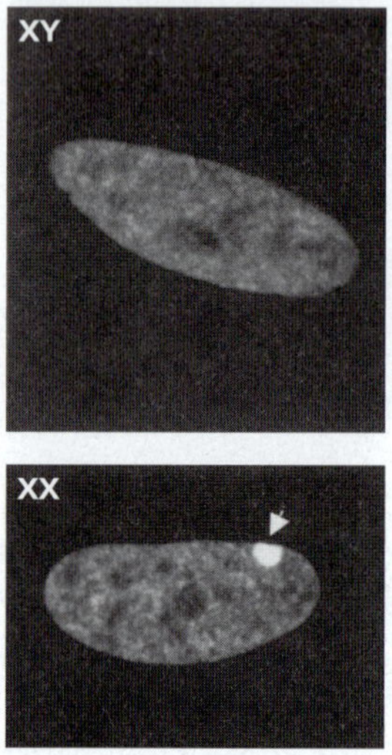

Figure 4.25 X chromosome dosage compensation makes human females a patchwork for X-linked gene expression. **(a)** Early in embryogenesis, each XX cell inactivates one randomly chosen X chromosome by condensing it into a Barr body (*black* oval). The same X chromosome remains a Barr body in all descendants of each cell. X^M = maternal X chromosome; X^P = paternal X chromosome. **(b)** The twins shown here are heterozygotes (*Dd*) for the X-linked recessive condition anhidrotic ectodermal dysplasia, which prevents sweat gland development. Patches of skin in *blue* lack sweat glands because the chromosome with the wild-type allele (*D*) is inactivated and the recessive *d* allele is nonfunctional.

(a) Perpetuation of X chromosome inactivation after cell divisions

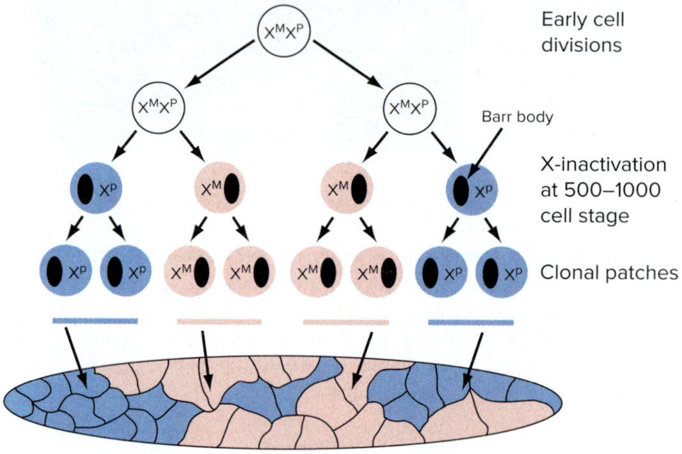

(b) X chromosome inactivation results in patchwork females

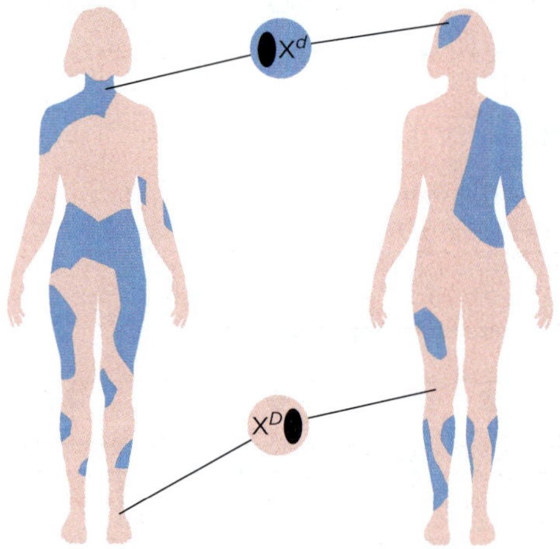

called **dosage compensation.** As mentioned earlier, the X chromosome contains about 1100 genes, and the proteins that they specify need to be present in the same amounts in male and female cells. To compensate for female cells having two copies of each X-linked gene and male cells having only one, XX cells inactivate one of their two X chromosomes. Almost all of the genes on the inactivated X chromosome are turned off, so no gene products can be made. X inactivation occurs at about two weeks after fertilization, when an XX human embryo is composed of only 500–1000 cells. At that time, each cell chooses one X chromosome at random to condense into a so-called **Barr body** and thereby inactivate it. Barr bodies, named after the cytologist Murray Barr who discovered them, appear as small, dark chromosomes in interphase cells treated with a DNA stain that allows chromosomes to be visible under a light microscope (**Fig. 4.24**).

Each embryonic cell "decides" independently which X chromosome will be inactivated—either the X inherited from the mother or the paternal X. Once the determination is made, it is clonally perpetuated so that all of the millions of cells descended by mitosis from a particular embryonic cell condense the same X chromosome to a Barr body (**Fig. 4.25a**). Human females are thus a patchwork of cells, some containing a maternally derived active X chromosome, and the others an active paternal X (**Fig. 4.25b**). The Fast Forward Box *Visualizing X Chromosome Inactivation in Transgenic Mice* explains how scientists have recently developed technology in mice to visualize, both outside and inside the body, the clonal patches of cells that express the genes on one X chromosome or the other.

The phenomenon of **X chromosome inactivation** may have interesting effects on the traits controlled by X-linked genes. When females are heterozygous at an X-linked gene, parts of their bodies are in effect hemizygous for one allele, and parts are hemizygous for the other allele in terms of gene function. Moreover, which body parts are functionally hemizygous for one allele or the other is random; even identical twins, who have identical alleles of all of their genes, will

FAST FORWARD

Visualizing X Chromosome Inactivation in Transgenic Mice

Scientists have recently used molecular techniques and transgenic technology (similar to that described in the earlier Fast Forward Box *Transgenic Mice Prove That* SRY *Is the Maleness Factor*) to visualize the pattern of X chromosome inactivation in mice. The researchers generated XX mice containing two different *transgenes* (in this case, genes from a different species). One of these transgenes was a jellyfish gene that specifies green fluorescent protein (GFP); the other was a gene from red coral that makes red fluorescent protein (RFP) **(Fig. A).**

In the XX mice, the GFP gene is located on the X chromosome from the mother, and the RFP gene resides on the X chromosome from the father. Clonal patches of cells are either green or red depending on which X chromosome was turned into a Barr body in the original cell that established the patch **(Fig. B).**

Different XX mice display different green and red patchwork patterns, providing a clear demonstration of the random nature of X chromosome inactivation. The patchwork patterns reflect the cellular memory of which X chromosome was inactivated in the founder cell for each clonal patch. Geneticists currently use these transgenic mice to decipher the genetic details of how cells "remember" which X to inactivate after each cell division.

Figure A Cells of transgenic mice glow either green or red in response to X chromosome inactivation. The mouse carries a green (GFP) transgene inserted in the maternal X chromosome (X^M), and a red (RFP) transgene in the paternal X chromosome (X^P). Cells in which X^P is inactivated (*top*) glow green; cells glow red (*bottom*) when X^M is inactivated.

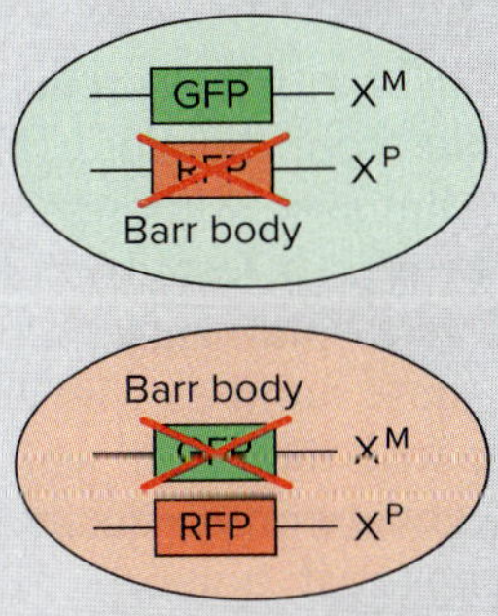

Figure B Heart cells of a transgenic mouse reveal a clonal patchwork of X inactivation. Patches of red or green cells represent cellular descendants of the founders that randomly inactivated one of their X chromosomes.

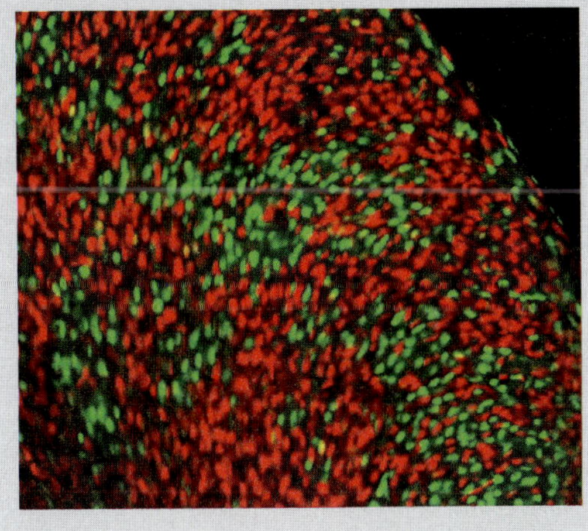

have a different pattern of X chromosome inactivation. In Fig. 4.25b, females heterozygous for the X-linked recessive trait anhidrotic epidermal dysplasia have patches of skin that lack sweat glands interspersed with patches of normal skin; the phenotype of a patch depends upon which X chromosome is inactivated. Each patch is a clone of skin cells derived from a single embryonic cell that made the decision to inactivate one of the X chromosomes. In a second example, women heterozygous for an X-linked recessive hemophilia allele are called *carriers* of the disease allele, even though they may have some symptoms of hemophilia. The severity of the condition depends on the particular random pattern of cells that inactivated the disease allele and cells that inactivated the normal allele. In Chapter 3, we discussed how chance events work through genes to affect phenotype; X inactivation is a perfect example of such an event.

Recall that the two tips of the X chromosome, the pseudoautosomal regions (PARs), contain genes also present at the tips of the Y chromosome (Fig. 4.8). In order to equalize the dosage of these genes in XX and XY cells, the PAR genes on the Barr body X chromosome escape inactivation. This feature of dosage compensation may explain at least in part why XXY males (Klinefelter syndrome) and XO females (Turner syndrome) have abnormal morphological features. Although one of the two X chromosomes in XXY males becomes a Barr body, Klinefelter males have three doses (rather than the normal two) of the genes in the PAR regions. The single X chromosome in XO cells does not become a Barr body, yet these cells have only one dose of the PAR genes (rather than two in XX females).

X chromosome inactivation is common to mammals, and we will present the molecular details of this process in later chapters. It is nonetheless important to realize that other organisms compensate for sex chromosome differences in alternative ways. Fruit flies, for example, hyperactivate the single X chromosome in XY (male) cells, so that most X chromosome genes produce twice as much protein product as each X chromosome in a female. The nematode *C. elegans,* in contrast, ratchets down the level of gene activity on each of the X chromosomes in XX hermaphrodites relative to the single X in XO males.

Figure 4.26 Male pattern baldness, a sex-influenced trait. **(a)** John Adams (1735–1826), second president of the United States, at about age 60. **(b)** John Quincy Adams (1767–1848), son of John Adams and the sixth president of the United States, at about the same age. The father-to-son transmission suggests that male pattern baldness in the Adams family is likely determined by an allele of an autosomal gene.

(a) **(b)**

Maleness and Male Fertility Are the Only Known Y-Linked Traits in Humans

Theoretically, phenotypes caused by mutations on the Y chromosome should also be identifiable by pedigree analysis. Such traits would pass from an affected father to all of his sons, and from them to all future male descendants. Females would neither exhibit nor transmit a Y-linked phenotype (see Table 4.5). However, besides the determination of maleness itself, as well as contributions to sperm formation and thus male fertility, no clear-cut Y-linked visible traits have turned up in humans. The paucity of known Y-linked traits reflects the fact that, as mentioned earlier, the small Y chromosome contains very few genes. Indeed, one would expect the Y chromosome to have only a limited effect on phenotype because normal XX females do perfectly well without it.

Autosomal Genes Contribute to Sexual Dimorphism

Not all genes that produce sexual dimorphism (differences in the two sexes) reside on the X or Y chromosomes. Some autosomal genes govern traits that appear in one sex but not the other, or traits that are expressed differently in the two sexes.

Sex-limited traits affect a structure or process that is found in one sex but not the other. Mutations in genes for sex-limited traits can influence only the phenotype of the sex that expresses those structures or processes. A vivid example of a sex-limited trait occurs in *Drosophila* males homozygous for an autosomal recessive mutation known as *stuck,* which affects the ability of mutant males to retract their penis and release the claspers by which they hold on to female genitalia during copulation. The mutant males have difficulty separating from females after mating. In extreme cases, both individuals die, forever caught in their embrace. Because females lack penises and claspers, homozygous *stuck* mutant females can mate normally.

Sex-influenced traits show up in both sexes, but the expression of such traits may differ between the two sexes because of hormonal differences. Pattern baldness, a condition in which hair is lost prematurely from the top of the head but not from the sides (**Fig. 4.26**), is a sex-influenced trait in humans. Although pattern baldness is a complex trait that can be affected by many genes, an autosomal gene appears to play an important role in certain families. Men in these families who are heterozygous for the balding allele lose their hair while still in their 20s, whereas heterozygous women do not show any significant hair loss. In contrast, homozygotes in both sexes become bald (though the onset of baldness in homozygous women is usually much later in life than in homozygous men). This sex-influenced trait is thus dominant in men, recessive in women.

Mutations in Sex Determination Pathway Genes Can Result in Intersexuality Disorders

We previously saw that the *SRY* gene on the Y chromosome is essential to maleness because it initiates testis development early in embryogenesis. But the functions of many genes are required for testis development, or for subsequent events that rely on hormones made in the testes for the development of sexual organs. Some of these genes are autosomal and some are X-linked; in either case, an XY individual with mutant alleles for any of these genes may have unusual intersexual phenotypes.

In one important example, XY people with nonfunctional mutant alleles of the X-linked *AR* gene specifying the androgen receptor have a disorder known as complete androgen insensitivity syndrome (CAIS). These

XY individuals have testes that make the hormone testosterone, but in the absence of the androgen receptor to which it binds, the testosterone has no effect. Without the androgen receptor, these people cannot develop male genitalia (penis and scrotum) nor male internal duct systems (the vas deferens, seminal vesicles, and ejaculatory ducts); instead, their external genitalia assume the default female state (labia and clitoris). However, the testes make another hormone that prevents the formation of female internal duct systems (including the fallopian tubes, uterus, and vagina). The result is that persons with CAIS are externally female but sterile because they lack the internal duct systems of either sex.

essential concepts

- *Sex-linked* (*X-linked*) traits show sex-specific inheritance patterns because sons always inherit their father's Y chromosome, while daughters always inherit their father's X chromosome.
- Random inactivation of either the maternal or paternal X chromosome in XX cells ensures that male and female mammalian cells express equivalent amounts of the proteins encoded by most X-linked genes.
- Mutations of genes—whether autosomal or X-linked—can have different effects in males and females.

WHAT'S NEXT

T. H. Morgan and his students, collectively known as the *Drosophila* group, acknowledged that Mendelian genetics could exist independently of chromosomes. "Why then, we are often asked, do you drag in the chromosomes? Our answer is that because the chromosomes furnish exactly the kind of mechanism that Mendelian laws call for, and since there is an ever-increasing body of information that points clearly to the chromosomes as the bearers of the Mendelian factors, it would be folly to close one's eyes to so patent a relation. Moreover, as biologists, we are interested in heredity not primarily as a mathematical formulation, but rather as a problem concerning the cell, the egg, and the sperm."

The *Drosophila* group went on to find several X-linked mutations in addition to white eyes. One made the body yellow instead of brown, another shortened the wings, yet another made bent instead of straight body bristles. These findings raised several compelling questions. First, if the genes for all of these traits are physically linked together on the X chromosome, does this linkage affect their ability to assort independently, and if so, how? Second, does each gene have an exact chromosomal address, and if so, does this specific location in any way affect its transmission? In Chapter 5 we describe how the *Drosophila* group and others analyzed the transmission patterns of genes on the same chromosome in terms of known chromosome movements during meiosis, and how they then used the information obtained to localize genes to specific chromosomal positions.

SOLVED PROBLEMS

I. In humans, chromosome 16 sometimes has a heavily stained area in the long arm near the centromere. This feature can be seen through the microscope but has no effect on the phenotype of the person carrying it. When such a "blob" exists on a particular copy of chromosome 16, it is a constant feature of that chromosome and is inherited.

A couple conceived a child, but the fetus had multiple abnormalities and was miscarried. When the chromosomes of the fetus were studied, it was discovered that it had three copies of chromosome 16 (it was *trisomic* for chromosome 16), and that two of the three chromosome 16s had large blobs. Both chromosome 16 homologs in the mother lacked blobs, but the father was heterozygous for blobs. Which parent experienced nondisjunction, and in which meiotic division did it occur?

Answer

This problem requires an understanding of nondisjunction during meiosis. When individual chromosomes contain some distinguishing feature that allows one homolog to be distinguished from another, it is possible to follow the path of the two homologs through meiosis. Because the fetus had two chromosome 16s with the blob, we can conclude that the extra chromosome came from the father (the only parent with a blobbed chromosome).

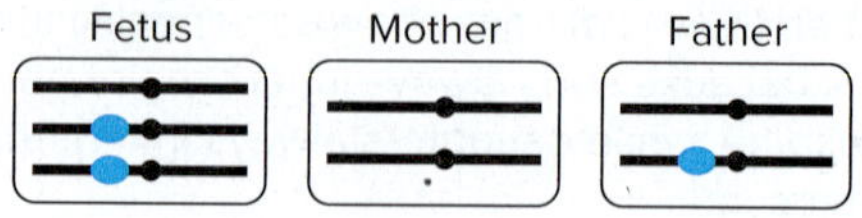

We know that nondisjunction must be involved because normal meiosis in the father would generate gametes with only a single chromosome 16 with the blob; as a result, the fetus could have only a single chromosome 16 with the blob.

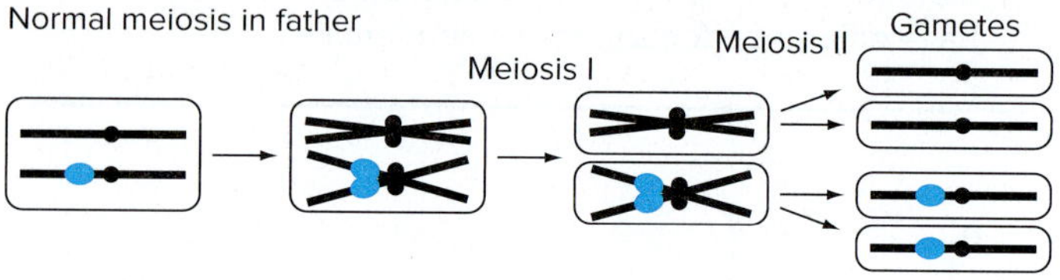

In which meiotic division did the nondisjunction occur? When nondisjunction occurs during meiosis I, homologs fail to segregate to opposite poles. If this occurred in the father, the chromosome with the blob and the normal chromosome 16 would segregate into the same cell (a secondary spermatocyte). After meiosis II, the gametes resulting from this cell would carry both types of chromosomes. If such sperm fertilized a normal egg, the zygote would have two copies of the normal chromosome 16 and one copy of the chromosome with a blob.

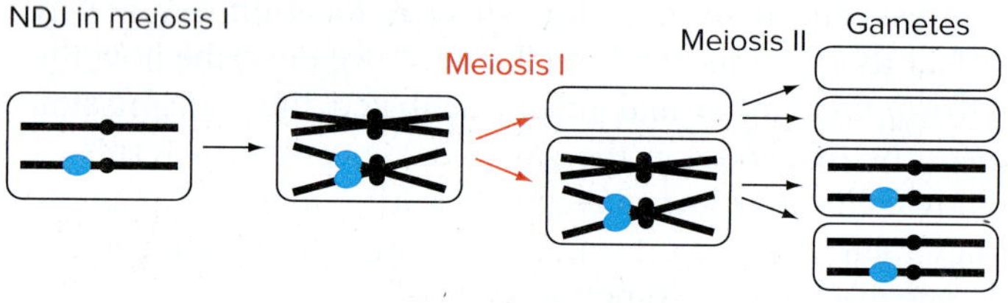

On the other hand, if nondisjunction occurred during meiosis II in the father in a secondary spermatocyte containing the blobbed chromosome 16, sperm with two copies of the blob-marked chromosome would be produced. After fertilization with a normal egg, the result would be a zygote of the type seen in this spontaneous abortion.

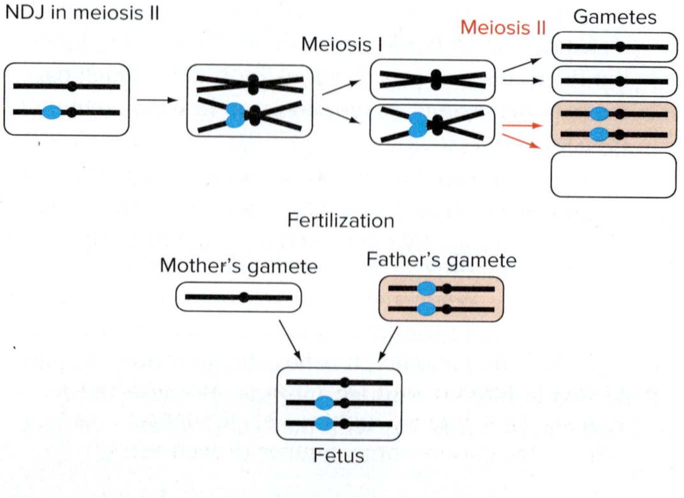

Therefore, the nondisjunction occurred in meiosis II in the father.

II. (a) What sex ratio would you expect among the offspring of a cross between a normal male mouse and a female mouse heterozygous for a recessive X-linked lethal gene? (b) What would be the expected sex ratio among the offspring of a cross between a normal hen and a rooster heterozygous for a recessive Z-linked lethal allele?

Answer

This problem deals with sex-linked inheritance and sex determination.

a. Mice have a sex determination system of XX = female and XY = male. A normal male mouse (X^RY) × a heterozygous female mouse (X^RX^r) would result in X^RX^R, X^RX^r, X^RY, and X^rY mice. The X^rY mice would die, so there would be a 2:1 ratio of females to males.

b. The sex determination system in birds is ZZ = male and ZW = female. A normal hen (Z^RW) × a heterozygous rooster (Z^RZ^r) would result in Z^RZ^R, Z^RZ^r, Z^RW, and Z^rW chickens. Because the Z^rW offspring do not live, the ratio of females to males would be 1:2.

III. A woman with normal color vision whose father was color-blind mates with a man with normal color vision.

a. What do you expect to see among their offspring?

b. What would you expect if it was the normal man's father who was color-blind?

Answer

This problem involves sex-linked inheritance.

a. The woman's father has a genotype of $X^{cb}Y$. Because the woman had to inherit an X from her father, she must have an X^{cb} chromosome, but because she has normal color vision, her other X chromosome must be X^{CB}. The man she mates with has normal color vision and therefore has an $X^{CB}Y$ genotype. Their children could with equal probability be $X^{CB}X^{CB}$ (normal female), $X^{CB}X^{cb}$ (carrier female), $X^{CB}Y$ (normal male), or $X^{cb}Y$ (color-blind male).

b. If the man with normal color vision had a color-blind father, the X^{cb} chromosome would not have been passed on to him, because a male does not inherit an X chromosome from his father. The man has the genotype $X^{CB}Y$ and cannot pass on the color-blind allele.

PROBLEMS

Vocabulary

1. Choose the best matching phrase in the right column for each of the terms in the left column.

a. meiosis	1. X and Y
b. gametes	2. chromosomes that do not differ between the sexes
c. karyotype	3. one of the two identical halves of a replicated chromosome
d. mitosis	4. microtubule organizing centers at the spindle poles
e. interphase	5. cells in the testes that undergo meiosis
f. syncytium	6. division of the cytoplasm
g. synapsis	7. haploid germ cells that unite at fertilization
h. sex chromosomes	8. an animal cell containing more than one nucleus
i. cytokinesis	9. pairing of homologous chromosomes
j. anaphase	10. one diploid cell gives rise to two diploid cells
k. chromatid	11. the array of chromosomes in a given cell
l. autosomes	12. the part of the cell cycle during which the chromosomes are not visible
m. centromere	13. one diploid cell gives rise to four haploid cells
n. centrosomes	14. cell produced by meiosis that does not become a gamete
o. polar body	15. the time during mitosis when sister chromatids separate
p. spermatocytes	16. site of the closest connection between sister chromatids

Section 4.1

2. Humans have 46 chromosomes in each somatic cell.
   a. How many chromosomes does a child receive from its father?
   b. How many autosomes and how many sex chromosomes are present in each somatic cell?
   c. How many chromosomes are present in a human ovum?
   d. How many sex chromosomes are present in a human ovum?

Section 4.2

3. The figure that follows shows the metaphase chromosomes of a male of a particular species. These chromosomes are prepared as they would be for a karyotype, but they have not yet been ordered in pairs of decreasing size.
   a. How many chromosomes are shown?
   b. How many chromatids are shown?
   c. How many centromeres are shown? (Count each sister centromere separately.)
   d. How many pairs of homologous chromosomes are shown?
   e. How many chromosomes on the figure are metacentric? Acrocentric?
   f. What is the likely mode of sex determination in this species? What would you predict to be different about the karyotype of a female in this species?

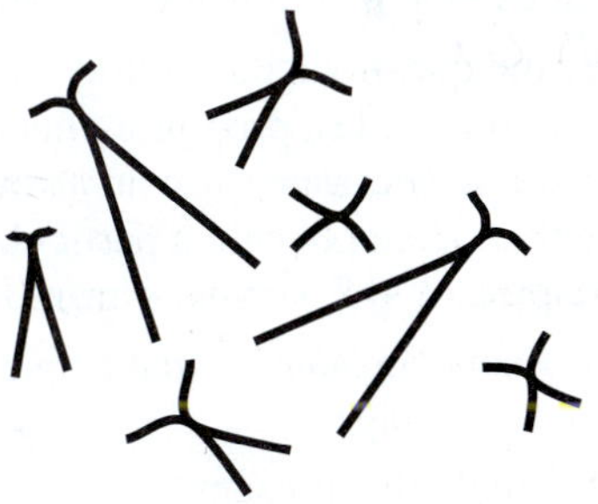

4. Human XX males who are sex-reversed because they have a mutant X chromosome like that shown in Fig. 4.7 often learn of their condition when they want to have children and discover that they are sterile. Can you explain why they are sterile?

5. Researchers discovered recently that the sole function of the SRY protein is to activate an autosomal gene called *Sox9* in the presumptive gonad (before it has "decided" to become a testis or an ovary).
   a. What would be the sex of an XY individual homozygous for nonfunctional mutant alleles of *Sox9?* Explain.
   b. Given your answer to part (a), why is *SRY*, rather than *Sox9*, considered the male determining factor? (*Hint:* What do you think would happen if you did an experiment like the one in the Fast Forward Box *Transgenic Mice Prove That* SRY *Is the Maleness Factor,* except that you used a *Sox9* transgene instead of *SRY?*)

Section 4.3

6. One oak tree cell with 14 chromosomes undergoes mitosis. How many daughter cells are formed, and what is the chromosome number in each cell?

7. Indicate which of the cells numbered i–v matches each of the following stages of mitosis:
   a. anaphase
   b. prophase
   c. metaphase
   d. G_2
   e. telophase/cytokinesis

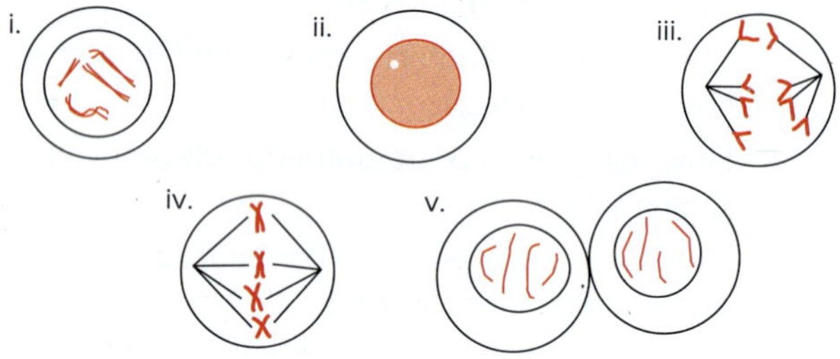

8. a. What are the four major stages of the cell cycle?
   b. Which stages are included in interphase?
   c. What events distinguish G_1, S, and G_2?
9. Answer the questions that follow for each stage of the cell cycle (G_1, S, G_2, prophase, metaphase, anaphase, telophase). If necessary, use an arrow to indicate a change that occurs during a particular cell cycle stage (for example, $1 \rightarrow 2$ or yes $\rightarrow$ no).
   a. How many chromatids make up each chromosome during this stage?
   b. Is the nucleolus present?
   c. Is the mitotic spindle organized?
   d. Is the nuclear membrane present?
10. Can you think of anything that would prevent mitosis from occurring in a cell whose genome is haploid?

Section 4.4

11. One oak tree cell with 14 chromosomes undergoes meiosis. How many cells will result from this process, and what is the chromosome number in each cell?
12. Which type(s) of cell division (mitosis, meiosis I, meiosis II) reduce(s) the chromosome number by half? Which type(s) of cell division can be classified as reductional? Which type(s) of cell division can be classified as equational?
13. Complete the following statements using as many of the following terms as are appropriate: mitosis, meiosis I (first meiotic division), meiosis II (second meiotic division), and none (not mitosis nor meiosis I nor meiosis II).
   a. The spindle apparatus is present in cells undergoing _____.
   b. Chromosome replication occurs just prior to _____.
   c. The cells resulting from _____ in a haploid cell have a ploidy of *n*.
   d. The cells resulting from _____ in a diploid cell have a ploidy of *n*.
   e. Homologous chromosome pairing regularly occurs during _____.
   f. Nonhomologous chromosome pairing regularly occurs during _____.
   g. Physical recombination leading to the production of recombinant progeny classes occurs during _____.
   h. The separation of sister centromeres occurs during _____.
   i. Nonsister chromatids are found in the same cell during _____.
14. The five cells shown in figures a–e are all from the same individual. For each cell, indicate whether it is in mitosis, meiosis I, or meiosis II. What stage of cell division is represented in each case? What is *n* in this organism?

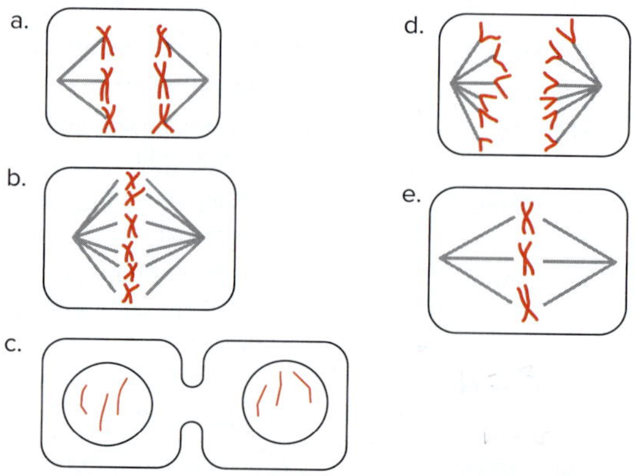

15. One of the first microscopic observations of chromosomes in cell division was published in 1905 by Nettie Stevens. Because it was hard to reproduce photographs at the time, she recorded these observations as *camera lucida* sketches. One such drawing, of a completely normal cell division in the mealworm *Tenebrio molitor,* is shown here. The techniques of the time were relatively unsophisticated by today's standards, and they did not allow her to resolve chromosomal structures that must have been present.

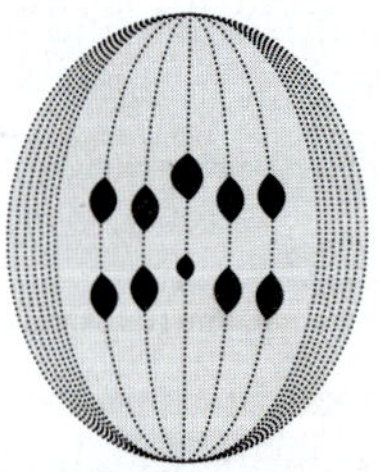

   a. Describe in as much detail as possible the kind of cell division and the stage of division depicted in the drawing.

b. What chromosomal structure(s) cannot be resolved in the drawing?

c. How many chromosomes are present in normal *Tenebrio molitor* gametes?

16. A person is simultaneously heterozygous for two autosomal genetic traits. One is a recessive condition for albinism (alleles *A* and *a*); this albinism gene is found near the centromere on the long arm of an acrocentric autosome. The other trait is the dominantly inherited Huntington disease (alleles *HD* and HD^+). The Huntington gene is located near the telomere of one of the arms of a metacentric autosome. Draw all copies of the two relevant chromosomes in this person as they would appear during metaphase of (a) mitosis, (b) meiosis I, and (c) meiosis II. In each figure, label the location on every chromatid of the alleles for these two genes, assuming that no recombination takes place.

17. Assuming (i) that the two chromosomes in every homologous pair carry different alleles of some genes, and (ii) that no crossing-over takes place, how many genetically different offspring could any one human couple potentially produce? Which of these two assumptions (i or ii) is more realistic?

18. In the moss *Polytrichum commune,* the haploid chromosome number is 7. A haploid male gamete fuses with a haploid female gamete to form a diploid cell that divides and develops into the multicellular *sporophyte*. Cells of the sporophyte then undergo meiosis to produce haploid cells called *spores*. What is the probability that an individual spore will contain a set of chromosomes all of which came from the male gamete? (Assume that no recombination occurs.)

19. Can you think of anything that would prevent meiosis from occurring in an organism whose genome is always haploid?

20. Sister chromatids are held together through metaphase of mitosis by complexes of *cohesin* proteins that form rubber band–like rings bundling the two sister chromatids. Cohesin rings are found both at centromeres and at many locations scattered along the length of the chromosomes. The rings are destroyed by protease enzymes at the beginning of mitotic anaphase, allowing the sister chromatids to separate.

    a. Cohesin complexes between sister chromatids are also responsible for keeping homologous chromosomes together until anaphase of meiosis I. With this point in mind, which of the two diagrams that follow (i or ii) properly represents the arrangement of chromatids during prophase through metaphase of meiosis I? Explain.

    b. What does your answer to part (a) allow you to infer about the nature of cohesin complexes at the centromere versus those along the chromosome arms? Suggest a molecular hypothesis to explain your inference.

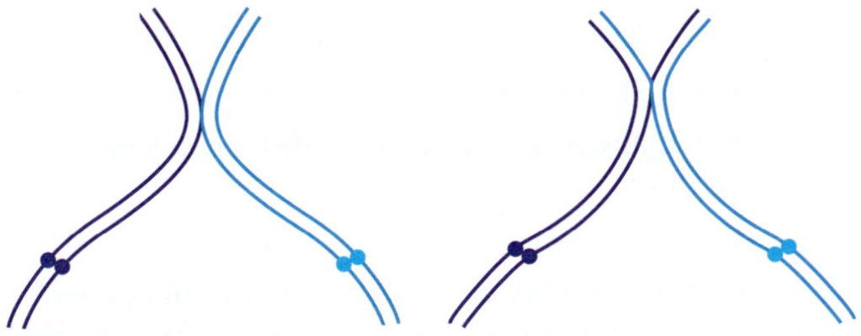

i ii

21. The pseudoautosomal regions (PARs) of the X and Y chromosomes enable the sex chromosomes to pair and synapse during meiosis in males. Given the location of the *SRY* gene near PAR1, can you propose a mechanism for how the mutant X and Y chromosomes in Fig. 4.7 (in which part of the X is on the Y, and part of the Y is on the X) may have arisen during meiosis?

22. Remarkably, the platypus has 10 sex chromosomes, the largest number found in any mammal. The female platypus has five pairs of different X chromosomes (X1–X5), while the male has X1–X5, and also five different Ys (Y1–Y5). During meiosis in the male, the five Xs always end up together in one gamete, and the five Ys always end up together in another gamete. To achieve this segregation, during prophase of meiosis I the sex chromosomes form a long chain, always in the order X1 Y1 X2 Y2 X3 Y3 X4 Y4 X5 Y5, in which the chromosomes are held together through pseudoautosomal regions (PARs).

    a. How many different PARs must exist to allow the formation of these chains? (*Hint:* To answer this question, try drawing the chain of chromosomes.)

    b. In terms of pairing ability of the PARs, explain the structural differences between the human and platypus sex chromosomes.

Section 4.5

23. Somatic cells of chimpanzees contain 48 chromosomes.

    How many chromatids and chromosomes are present at: (a) anaphase of mitosis, (b) anaphase I of meiosis, (c) anaphase II of meiosis, (d) G_1 prior to mitosis, (e) G_2 prior to mitosis, (f) G_1 prior to meiosis I, and (g) prophase of meiosis I?

    How many chromatids or chromosomes are present in: (h) an oogonial cell prior to S phase, (i) a spermatid, (j) a primary oocyte arrested prior to ovulation, (k) a secondary oocyte arrested prior to fertilization, (l) a second polar body, and (m) a chimpanzee sperm?

24. In humans:
    a. How many sperm develop from 100 primary spermatocytes?
    b. How many sperm develop from 100 secondary spermatocytes?
    c. How many sperm develop from 100 spermatids?
    d. How many ova develop from 100 primary oocytes?
    e. How many ova develop from 100 secondary oocytes?
    f. How many ova develop from 100 polar bodies?
25. Women sometimes develop benign tumors called *ovarian teratomas* or *dermoid cysts* in their ovaries. Such a tumor begins when a primary oocyte escapes from its prophase I arrest and finishes meiosis I within the ovary. (Normally meiosis I does not finish until the primary oocyte is expelled from the ovary upon ovulation.) The secondary oocyte then develops as if it were an embryo. Development is disorganized, however, and results in a tumor containing differentiated diploid tissues, including teeth, hair, bone, muscle, and nerve. If a dermoid cyst forms in a woman whose genotype is *Aa,* what are the possible genotypes of the cyst, assuming no recombination?
26. In a certain strain of turkeys, unfertilized eggs sometimes develop parthenogenetically to produce diploid offspring. (Females have ZW and males have ZZ sex chromosomes. Assume that WW cells are inviable.) What distribution of sexes would you expect to see among the parthenogenetic offspring according to each of the following models for how parthenogenesis occurs?
    a. The eggs develop from oogonial cells that never undergo meiosis.
    b. The eggs go all the way through meiosis and then duplicate their chromosomes to become diploid.
    c. The eggs go through meiosis I, and the chromatids separate to create diploidy.
    d. The egg goes all the way through meiosis and then fuses at random with one of its three polar bodies (this scenario assumes the first polar body goes through meiosis II).

Section 4.6

27. Imagine you have two pure-breeding lines of canaries, one with yellow feathers and the other with brown feathers. In crosses between these two strains, yellow female × brown male gives only brown sons and daughters, while brown female × yellow male gives only brown sons and yellow daughters. Propose a hypothesis to explain these results.
28. A system of sex determination known as *haplodiploidy* is found in honeybees. Females are diploid, and males (drones) are haploid. Male offspring result from the development of unfertilized eggs. Sperm are produced by mitosis in males and fertilize eggs in the females. Ivory eye is a recessive characteristic in honeybees; wild-type eyes are brown.
    a. What progeny would result from an ivory-eyed queen and a brown-eyed drone? Give both genotype and phenotype for progeny produced from fertilized and nonfertilized eggs.
    b. What would result from crossing a daughter from the mating in part (a) with a brown-eyed drone?
29. In *Drosophila,* the autosomal recessive *brown* eye color mutation displays interactions with both the X-linked recessive *vermilion* mutation and the autosomal recessive *scarlet* mutation. Flies homozygous for *brown* and simultaneously hemizygous or homozygous for *vermilion* have white eyes. Flies simultaneously homozygous for both the *brown* and *scarlet* mutations also have white eyes. Predict the F_1 and F_2 progeny of crossing the following true-breeding parents:
    a. vermilion females × brown males
    b. brown females × vermilion males
    c. scarlet females × brown males
    d. brown females × scarlet males
30. Barred feather pattern is a Z-linked dominant trait in chickens. What offspring would you expect from (a) the cross of a barred hen to a nonbarred rooster? (b) the cross of an F_1 rooster from part (a) to one of his sisters?
31. When Calvin Bridges observed a large number of offspring from a cross of white-eyed female *Drosophila* to red-eyed males, he found very rare white-eyed females and red-eyed males among the offspring. He was able to show that these exceptions resulted from nondisjunction, such that the white-eyed females had received two Xs from the egg and a Y from the sperm, while the red-eyed males had received no sex chromosome from the egg and an X from the sperm. What progeny would have arisen from these same kinds of nondisjunctional events if they had occurred in the male parent? What would their eye colors have been?
32. In a vial of *Drosophila,* a research student noticed several female flies (but no male flies) with *bag* wings each consisting of a large, liquid-filled blister instead of the usual smooth wing blade. When bag-winged females were crossed with wild-type males, 1/3 of the progeny were bag-winged females, 1/3 were normal-winged females, and 1/3 were normal-winged males. Explain these results.

33. In 1919, Calvin Bridges began studying an X-linked recessive mutation causing eosin-colored eyes in *Drosophila.* Within an otherwise true-breeding culture of eosin-eyed flies, he noticed rare variants that had much lighter cream-colored eyes. By intercrossing these variants, he was able to make a true-breeding cream-eyed stock. Bridges now crossed males from this cream-eyed stock with true-breeding wild-type females. All the F_1 progeny had red (wild-type) eyes. When F_1 flies were intercrossed, the F_2 progeny were 104 females with red eyes, 52 males with red eyes, 44 males with eosin eyes, and 14 males with cream eyes. Assume that these numbers represent an 8:4:3:1 ratio.

   a. Formulate a hypothesis to explain the F_1 and F_2 results, assigning phenotypes to all possible genotypes.

   b. What do you predict in the F_1 and F_2 generations if the parental cross is between true-breeding eosin-eyed males and true-breeding cream-eyed females?

   c. What do you predict in the F_1 and F_2 generations if the parental cross is between true-breeding eosin-eyed females and true-breeding cream-eyed males?

34. In *Drosophila,* a cross was made between a yellow-bodied male with vestigial (not fully developed) wings and a wild-type female (brown body). The F_1 generation consisted of wild-type males and wild-type females. F_1 males and females were crossed, and the F_2 progeny consisted of 16 yellow-bodied males with vestigial wings, 48 yellow-bodied males with normal wings, 15 males with brown bodies and vestigial wings, 49 wild-type males, 31 brown-bodied females with vestigial wings, and 97 wild-type females. Explain the inheritance of the two genes in question based on these results.

35. As we learned in this chapter, the *white* mutation of *Drosophila* studied by Thomas Hunt Morgan is X-linked and recessive to wild type. When true-breeding white-eyed males carrying this mutation were crossed with true-breeding purple-eyed females, all the F_1 progeny had wild-type (red) eyes. When the F_1 progeny were intercrossed, the F_2 progeny emerged in the ratio 3/8 wild-type females: 1/4 white-eyed males: 3/16 wild-type males: 1/8 purple-eyed females: 1/16 purple-eyed males.

   a. Formulate a hypothesis to explain the inheritance of these eye colors.

   b. Predict the F_1 and F_2 progeny if the parental cross was reversed (that is, if the parental cross was between true-breeding white-eyed females and true-breeding purple-eyed males).

Section 4.7

36. The following is a pedigree of a family in which a rare form of color blindness is found (filled-in symbols). Indicate as much as you can about the genotypes of all the individuals in the pedigree.

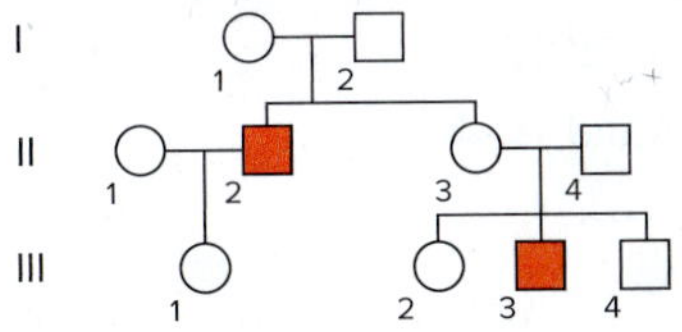

37. Each of the four pedigrees that follow represents a human family within which a genetic disease is segregating. Affected individuals are indicated by filled-in symbols. One of the diseases is transmitted as an autosomal recessive condition, one as an X-linked recessive, one as an autosomal dominant, and one as an X-linked dominant. Assume all four traits are rare in the population and completely penetrant.

   a. Indicate which pedigree represents which mode of inheritance, and explain how you know.

   b. For each pedigree, what would you tell the parents about the chance that their child (indicated by the hexagon shape) will have the condition?

Pedigree 1

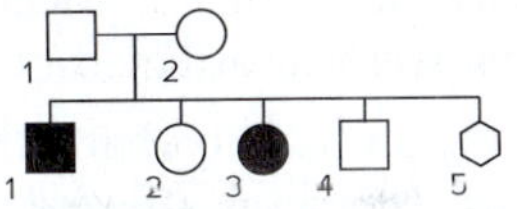

Pedigree 2

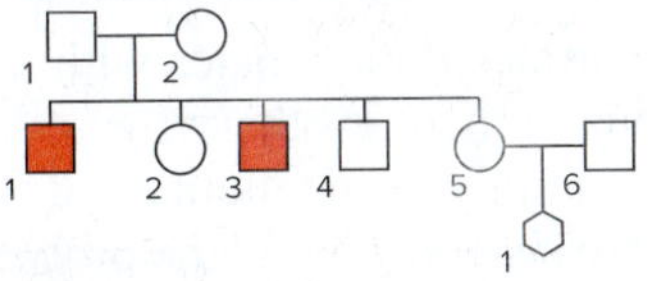

Pedigree 3

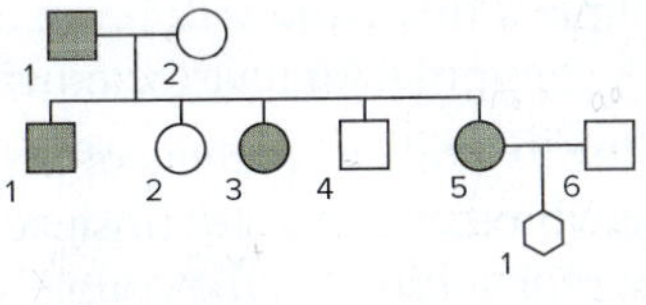

Pedigree 4

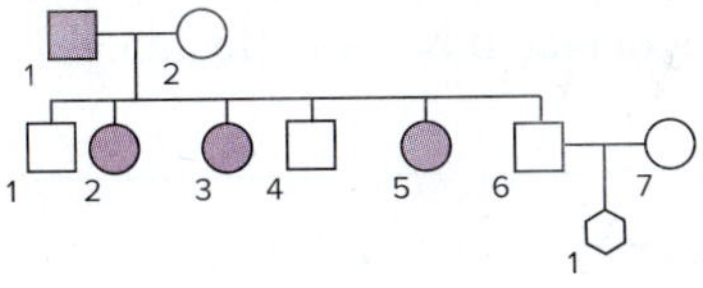

38. The pedigree that follows indicates the occurrence of albinism in a group of Hopi Indians, among whom the trait is unusually frequent. Assume that the trait is fully penetrant (all individuals with a genotype that could give rise to albinism will display this condition).

   a. Is albinism in this population caused by a recessive or a dominant allele?

   b. Is the gene sex-linked or autosomal?

What are the genotypes of the following individuals?

c. individual I-1

d. individual I-8

e. individual I-9

f. individual II-6

g. individual II-8

h. individual III-4

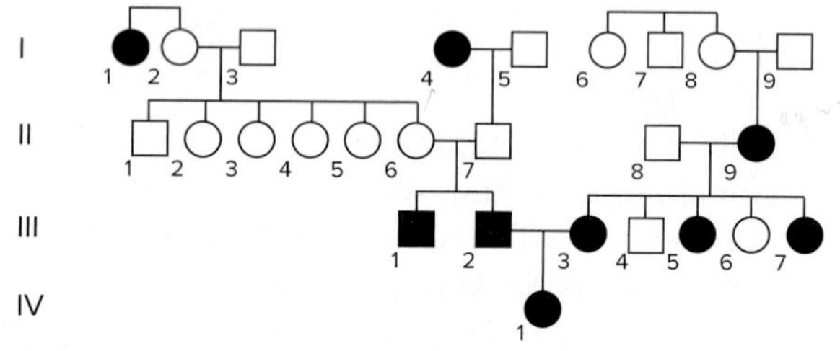

39. Duchenne muscular dystrophy (DMD) is caused by a relatively rare X-linked recessive allele. It results in progressive muscular wasting and usually leads to death before age 20. In this problem, an affected person is one with the severe form of DMD caused by hemizygosity or homozygosity for the disease allele.

    a. What is the probability that the first son of a woman whose brother is affected will be affected?

    b. What is the probability that the second son of a woman whose brother is affected will be affected, if her first son was affected?

    c. What is the probability that a child of an unaffected man whose brother is affected will be affected?

    d. An affected man mates with his unaffected first cousin; there is otherwise no history of DMD in this family. If the mothers of this man and his mate were sisters, what is the probability that the couple's first child will be an affected boy? An affected girl? An unaffected child?

    e. If the two related parents of the couple in part (d) were brother and sister (instead of sisters), what is the probability that the couple's first child will be an affected boy? An affected girl? An unaffected child?

40. The X-linked gene responsible for DMD encodes a protein called dystrophin that is required for muscle function. Dystrophin protein is not secreted—it remains in the cells that produce it. Given what you know about Barr body formation, do you think that females heterozygous for the recessive DMD disease allele could have the disease in some parts of their bodies and not others?

41. Males have hemophilia when they are hemizygous for a nonfunctional recessive mutant allele of the X-linked gene for clotting factor VIII. Factor VIII is normally secreted into the blood serum by cells in the bone marrow (and other specialized cells) that produce it.

    a. Do you think that females heterozygous for the hemophilia disease allele could have hemophilia in some parts of their bodies and not others?

    b. If such a female carrier of hemophilia suffered a cut, would her blood coagulate (form clots) faster, slower, or in about the same time as that of an individual homozygous for a normal allele of the factor VIII gene? Would the rate of clotting vary significantly among heterozygous females?

42. In the Fast Forward Box *Visualizing X Chromosome Inactivation in Transgenic Mice,* suppose the investigators had looked at the expression of green and red fluorescent protein in early mouse embryos, when the embryos have fewer than 500 cells. What patterns would they likely have observed? (Assume that the transgenes make gene product this early in development.)

43. The following pedigree shows five generations of a family that exhibits congenital hypertrichosis, a rare condition in which affected individuals are born with unusually abundant amounts of hair on their faces and upper bodies. The two small black dots in the pedigree indicate miscarriages.

    a. What can you conclude about the inheritance of hypertrichosis in this family, assuming complete penetrance of the trait?

    b. On what basis can you exclude other modes of inheritance?

    c. With how many fathers did III-2 and III-9 have children?

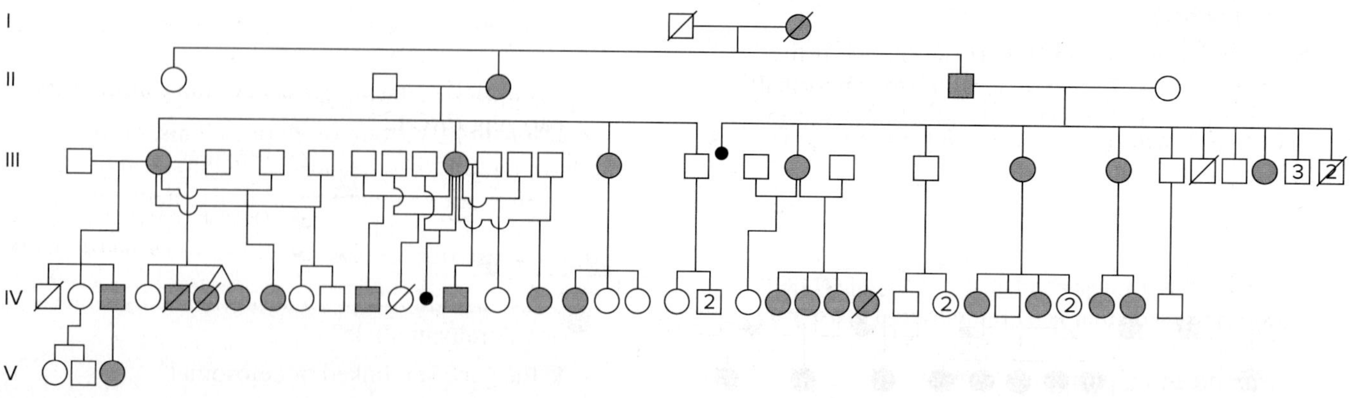

44. Consider the following pedigrees from human families containing a male with Klinefelter syndrome (a set of abnormalities seen in XXY individuals; indicated with shaded boxes). In each, *A* and *B* refer to codominant alleles of the X-linked *G6PD* gene. The *phenotypes* of each individual (A, B, or AB) are shown on the pedigree. Indicate if nondisjunction occurred in the mother or father of the son with Klinefelter syndrome for each of the three examples. Can you tell if the nondisjunction was in the first or second meiotic division?

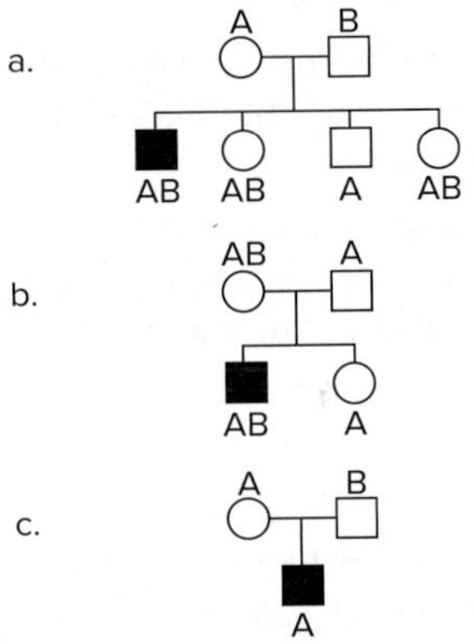

45. Several different antigens can be detected in blood tests. The following four traits were tested for each individual shown:

ABO type	(I^A and I^B codominant, i recessive)
Rh type	(Rh^+ dominant to Rh^-)
MN type	(*M* and *N* codominant)
$Xg^{(a)}$ type	[$Xg^{(a+)}$ dominant to $Xg^{(a-)}$]

All of these blood type genes are autosomal, except for $Xg^{(a)}$, which is X-linked.

Mother	AB	Rh^-	MN	$Xg^{(a+)}$
Daughter	A	Rh^+	MN	$Xg^{(a-)}$
Alleged father 1	AB	Rh^+	M	$Xg^{(a+)}$
Alleged father 2	A	Rh^-	N	$Xg^{(a-)}$
Alleged father 3	B	Rh^+	N	$Xg^{(a-)}$
Alleged father 4	O	Rh^-	MN	$Xg^{(a-)}$

a. Which, if any, of the alleged fathers could be the real father?

b. Would your answer to part (a) change if the daughter had Turner syndrome (the abnormal phenotype seen in XO individuals)? If so, how?

46. The ancestry of a white female tiger bred in a city zoo is depicted in the pedigree following part (e) of this problem. White tigers are indicated with unshaded symbols. (As you can see, there was considerable inbreeding in this lineage. For example, the white tiger Mohan was mated with his daughter.) In answering the following questions, assume that *white* is determined by allelic differences at a single gene and that the trait is fully penetrant. Explain your answers by citing the relevant information in the pedigree.

a. Could white coat color be caused by a Y-linked allele?

b. Could white coat color be caused by a dominant X-linked allele?

c. Could white coat color be caused by a dominant autosomal allele?

d. Could white coat color be caused by a recessive X-linked allele?

e. Could white coat color be caused by a recessive autosomal allele?

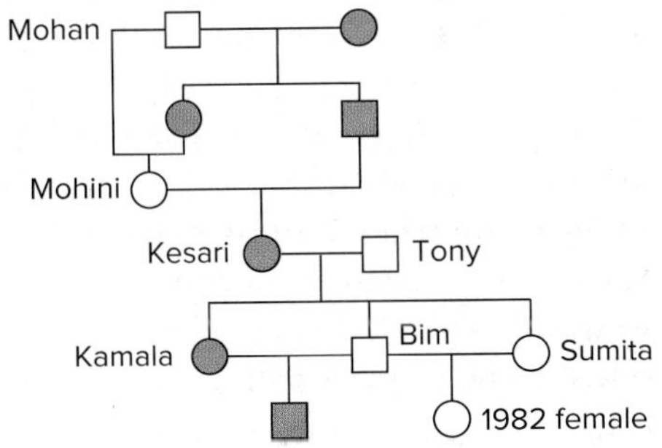

47. The pedigree that follows shows the inheritance of various types of cancer in a particular family. Molecular analyses (described in subsequent chapters) indicate that with one exception, the cancers occurring in the patients in this pedigree are associated with a rare mutation in a gene called *BRCA2*.

a. Which individual is the exceptional cancer patient whose disease is not associated with a *BRCA2* mutation?

b. Is the *BRCA2* mutation dominant or recessive to the normal *BRCA2* allele in terms of its cancer-causing effects?

c. Is the *BRCA2* gene likely to reside on the X chromosome, the Y chromosome, or an autosome? How definitive is your assignment of the chromosome carrying *BRCA2*?

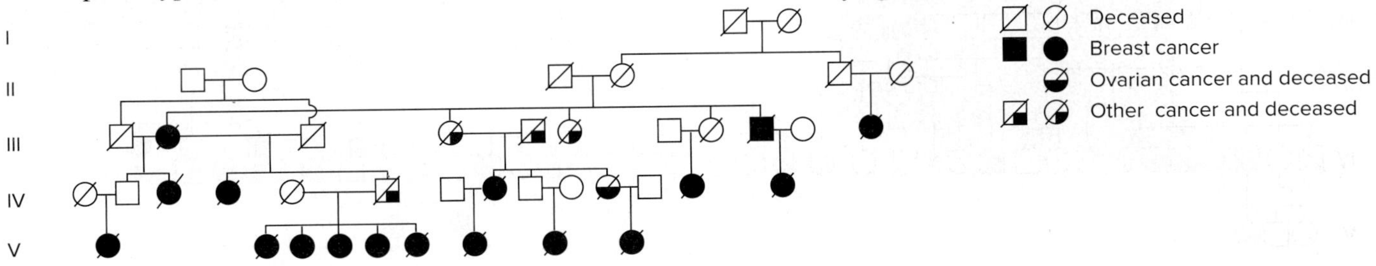

d. Is the penetrance of the cancer phenotype complete or incomplete?

e. Is the expressivity of the cancer phenotype unvarying or variable?

f. Are any of the cancer phenotypes associated with the *BRCA2* mutation sex-limited or sex-influenced?

g. How can you explain the absence of individuals diagnosed with cancer in generations I and II?

48. In 1995, doctors reported a Chinese family in which retinitis pigmentosa (progressive degeneration of the retina leading to blindness) affected only males. All six sons of affected males were affected, but all of the five daughters of affected males (and all of the children of these daughters) were unaffected.

a. What is the likelihood that this form of retinitis pigmentosa is due to an autosomal mutation showing complete dominance?

b. What other possibilities could explain the inheritance of retinitis pigmentosa in this family? Which of these possibilities do you think is most likely?

49. In cats, the dominant *O* allele of the X-linked orange gene is required to produce orange fur; the recessive *o* allele of this gene yields black fur. Tortoiseshell cats have coats with patches of orange fur alternating with patterns of black fur. Approximately 90% of all tortoiseshell cats are females.

a. Explain why tortoiseshell cats are nearly always female.

b. What types of crosses would be expected to produce female tortoiseshell cats?

c. Suggest a hypothesis to explain the origin of male tortoiseshell cats.

d. Calico cats (most of which are females) have patches of white, orange, and black fur. Suggest a hypothesis for the origin of calico cats.

50. In marsupials like the opposum or kangaroo, X inactivation selectively inactivates the paternal X chromosome.

a. Predict the possible coat colors of the progeny of both sexes if a female marsupial homozygous for a mutant allele of an X-linked coat color gene was mated with a male hemizygous for the alternative wild-type alleles of this gene.

b. Predict the possible coat colors of progeny of both sexes if a male marsupial hemizygous for an allele of an X-linked coat color gene was mated with a female homozygous for the alternative wild-type allele of this gene.

c. Why are the terms *recessive* and *dominant* not useful in describing the alleles of X-linked coat color genes in marsupials?

d. Why would marsupials heterozygous for two alleles of an X-linked coat color gene not have patches of fur of two different colors as did the tortoiseshell cats described in the previous problem?

51. The pedigree diagram below shows a family in which many individuals are affected by a disease called Leri-Weill dyschondrosteosis (LWD). People with LWD are short in stature due to leg bone deformities; arm bones are also malformed in some individuals. The mutant gene responsible for LWD was identified in 1998 as *SHOX*, a gene located in a pseudoautosomal region (PAR1) of the X and Y chromosomes.

a. Is the *SHOX* allele that causes LWD dominant or recessive? Explain. (*Note:* Sex reversal is not involved.)

b. Even though *SHOX* is located on the X chromosome, the pedigree is atypical for an X-linked allele. What features of the pedigree are incompatible with X-linkage?

c. For each affected individual in the pedigree, determine whether the *SHOX* disease allele is on the X or the Y.

d. Explain the inheritance pattern of the *SHOX* disease allele and the $SHOX^+$ normal allele in pedigree.

e. Diagram the crossover event that generated the Y chromosome in individual III-5. Your diagram should indicate the positions of the *SHOX* (disease) and $SHOX^+$ (normal) alleles on the X and Y chromosomes in the germ-line cells of individual II-3, and SRY^+ on Y.

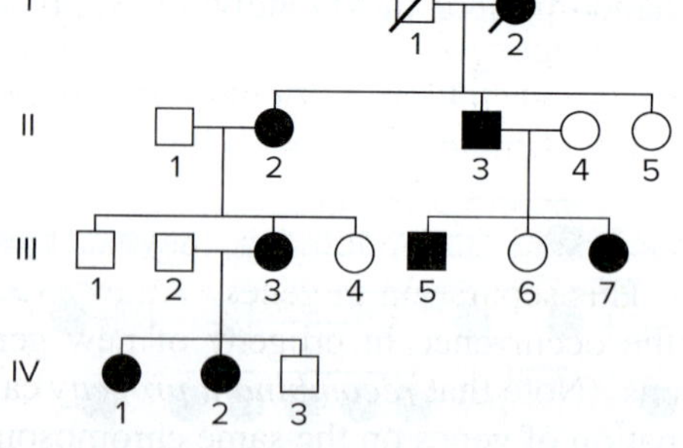

chapter **5**

Linkage, Recombination, and the Mapping of Genes on Chromosomes

Maps illustrate the spatial relationships of objects, such as the locations of subway stations along subway lines. Genetic maps portray the positions of genes along chromosomes.
© Rudy Von Briel/PhotoEdit

chapter outline

- 5.1 Gene Linkage and Recombination
- 5.2 Recombination: A Result of Crossing-Over During Meiosis
- 5.3 Mapping: Locating Genes Along a Chromosome
- 5.4 The Chi-Square Test and Linkage Analysis
- 5.5 Tetrad Analysis in Fungi
- 5.6 Mitotic Recombination and Genetic Mosaics

IN 1928, DOCTORS completed a four-generation pedigree tracing two known X-linked traits: red-green color blindness and hemophilia A (the more serious X-linked form of *bleeders disease*). The maternal grandfather of the family exhibited both traits, which means that his single X chromosome carried mutant alleles of the two corresponding genes. As expected, neither color blindness nor hemophilia showed up in his sons and daughters, but two grandsons and one great-grandson inherited both of the X-linked conditions (**Fig. 5.1a**). The fact that none of the descendants manifested one of the traits without the other suggests that the mutant alleles did not assort independently during meiosis. Instead they traveled together in the gametes forming one generation and then into the gametes forming the next generation, producing grandsons and great-grandsons with an X chromosome specifying both color blindness and hemophilia. Genes that travel together more often than not exhibit **genetic linkage.**

In contrast, another pedigree following color blindness and the slightly different B form of hemophilia, which also arises from a mutation on the X chromosome, revealed a different inheritance pattern. A grandfather with hemophilia B and color blindness had four grandsons, but only one of them exhibited both conditions. In this family, the genes for color blindness and hemophilia appeared to assort independently, producing in the male progeny all four possible combinations of the two traits—normal vision and normal blood clotting, color blindness and hemophilia, color blindness and normal clotting, and normal vision and hemophilia—in approximately equal frequencies (**Fig. 5.1b**). Thus, even though the mutant alleles of the two genes were on the same X chromosome in the grandfather, they had to separate to give rise to grandsons III-2 and III-3. This separation of genes on the same chromosome is the result of **recombination,** the occurrence in progeny of new gene combinations not seen in previous generations. (Note that *recombinant progeny* can result in either of two ways: from the recombination of genes on the same chromosome during gamete formation, discussed

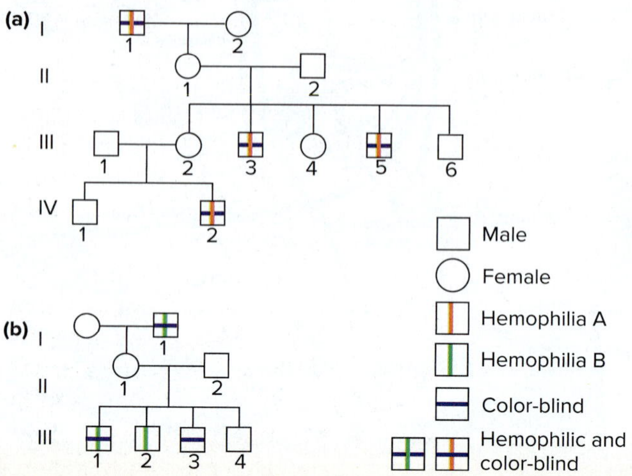

Figure 5.1 Pedigrees indicate that color blindness and two forms of hemophilia are X-linked traits. (a) Transmission of red-green color blindness and hemophilia A. The traits travel together through the pedigree, indicating their genetic linkage. **(b)** Transmission of red-green color blindness and hemophilia B. Even though both genes are X-linked, the mutant alleles are inherited together in only one of four grandsons in generation III. These two pedigrees suggest that the gene for color blindness is close to the hemophilia A gene but far away from the hemophilia B gene.

in this chapter, or from the independent assortment of genes on nonhomologous chromosomes, described in Chapter 4.)

Two important themes should be kept in mind as we follow the transmission of genes linked on the same chromosome. The first is that the farther apart two genes are, the greater is the probability of separation through recombination. Extrapolating from this general rule, you can see that the gene for hemophilia A is likely very close to the gene for red-green color blindness because, as Fig. 5.1a shows, the two rarely separate. By comparison, the gene for hemophilia B must lie far away from the color blindness gene because, as Fig. 5.1b indicates, new combinations of alleles of the two genes occur quite often. A second theme is that geneticists can use data about how often genes separate during transmission to map the genes' relative locations on a chromosome. Such mapping is a key to sorting out and tracking down the components of complex genetic networks; it is also crucial to geneticists' ability to isolate and characterize genes at the molecular level.

5.1 Gene Linkage and Recombination

learning objectives

1. Define linkage with respect to gene loci and chromosomes.
2. Differentiate between parental and recombinant gametes.
3. Conclude from ratios of progeny in a dihybrid cross whether two genes are linked.
4. Explain how a testcross can provide evidence for or against linkage.

If people have roughly 27,000 genes but only 23 pairs of chromosomes, most human chromosomes must carry hundreds, if not thousands, of genes. This is certainly true of the human X chromosome, which contains about 1100 protein-coding genes, as just described in Chapter 4. Recognition that many genes reside on each chromosome raises an important question. If genes on *different* chromosomes assort independently because nonhomologous chromosomes align independently on the spindle during meiosis I, how do genes on the *same* chromosome assort?

Some Genes on the Same Chromosome Do Not Assort Independently—Instead, They Are Linked

We begin our analysis with X-linked *Drosophila* genes because they were the first to be assigned to a specific chromosome. As we outline various crosses, remember that females carry two X chromosomes, and thus two alleles for each X-linked gene. Males, in contrast, have only a single X chromosome (from the female parent), and thus only a single allele for each of these genes.

We look first at two X-linked genes that determine a fruit fly's eye color and body color. These two genes are said to be **syntenic** because they are located on the same chromosome. The *white* gene was introduced in Chapter 4; you will recall that the dominant wild-type allele w^+ specifies red eyes, while the recessive mutant allele w confers white eyes. The alleles of the *yellow* body color gene are y^+ (the dominant wild-type allele for brown bodies) and y (the recessive mutant allele for yellow bodies). To avoid confusion, note that lowercase y and y^+ refer to alleles of the *yellow* gene, while capital Y refers to the Y chromosome (which does not carry genes for either eye or body color). You should also pay attention to the slash symbol (/), which is used to separate genes found on the two chromosomes of a pair (either the X and Y chromosomes as in this case, or a pair of X chromosomes or homologous autosomes). Thus

$w\ y$ / Y represents the genotype of a male with an X chromosome bearing w and y, as well as a Y chromosome; phenotypically this male has white eyes and a yellow body.

Detecting linkage by analyzing the gametes produced by a dihybrid

In a cross between a female with mutant white eyes and a wild-type brown body ($w\ y^+/w\ y^+$) and a male with wild-type red eyes and a mutant yellow body ($w^+\ y$ / Y), the F_1 offspring are divided evenly between brown-bodied females with normal red eyes ($w\ y^+/w^+\ y$) and brown-bodied males with mutant white eyes ($w\ y^+$/Y) (**Fig. 5.2**). Note that the male progeny look like their mother because their phenotype directly reflects the genotype of the single X chromosome they received from her. The same is not true for the F_1 females, who received w and y^+ on the X from their mother and $w^+\ y$ on the X from their father. These F_1 females are thus dihybrids. With two alleles for each X-linked gene, one derived from each parent, the dominance relations of each pair of alleles determine the female phenotype.

Figure 5.2 When genes are linked, parental combinations outnumber recombinant types. Doubly heterozygous $w\ y^+/\ w^+\ y$ F_1 females produce four types of male offspring. Sons that look like the father ($w^+\ y$ / Y) or mother ($w\ y^+$/ Y) of the F_1 females are parental types. Other sons (w^+y^+/ Y or $w\ y$ / Y) are recombinant types. For these closely-linked genes, many more parental types are produced than recombinant types.

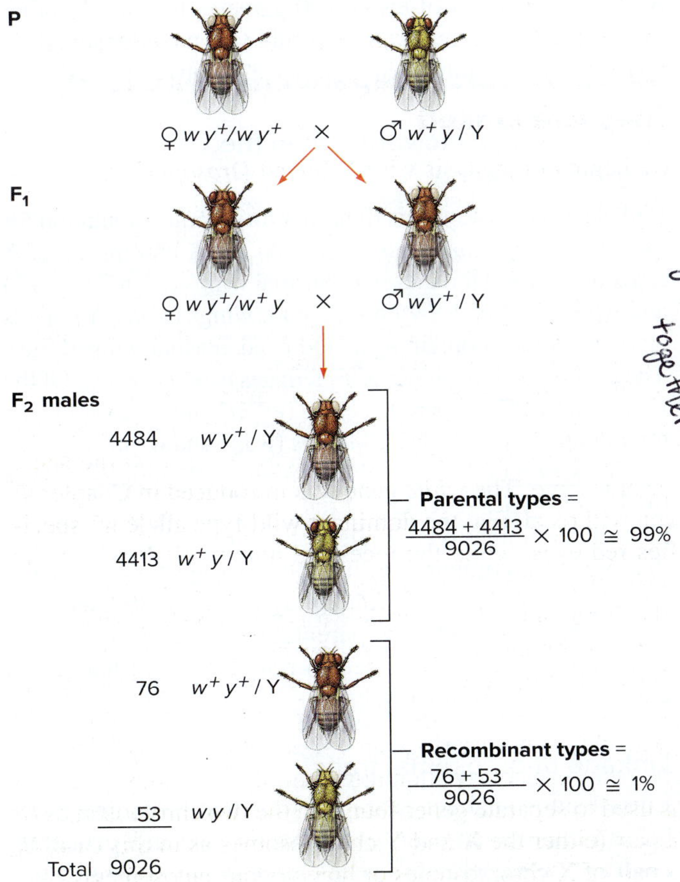

Now comes the significant cross for answering our question about the assortment of genes on the same chromosome. If these two *Drosophila* genes for eye and body color assort independently, as predicted by Mendel's second law, the dihybrid F_1 females should make four kinds of gametes, with four different combinations of genes on the X chromosome—$w\ y^+$, $w^+\ y$, $w^+\ y^+$, and $w\ y$. These four types of gametes should occur with equal frequency, that is, in a ratio of 1:1:1:1. If it happens this way, approximately half of the gametes will be of the two **parental types,** carrying either the $w\ y^+$ allele combination seen in the original female of the P generation or the w^+y allele combination seen in the original male of the P generation. The remaining half of the gametes will be of two **recombinant types,** in which reshuffling has produced either w^+y^+ or $w\ y$ allele combinations not seen in the P generation parents of the F_1 females.

We can see whether the 1:1:1:1 ratio of the four kinds of gametes actually materializes by counting the different types of male progeny in the F_2 generation, as these sons receive their only X-linked genes from their maternal gamete. The bottom part of Fig. 5.2 depicts the results of a breeding study that produced 9026 F_2 males. The relative numbers of the four X-linked gene combinations passed on by the dihybrid F_1 females' gametes reflect a significant departure from the 1:1:1:1 ratio expected of independent assortment. By far, the largest numbers of gametes carry the parental combinations $w\ y^+$ and w^+y. Of the total 9026 male flies counted, 8897, or almost 99%, had these genotypes. In contrast, the new combinations w^+y^+ and $w\ y$ made up little more than 1% of the total.

We can explain why the two genes fail to assort independently in one of two ways. The $w\ y^+$ and $w^+\ y$ combinations could be preferred because some intrinsic chemical affinity exists between these particular alleles. Alternatively, these combinations of alleles might show up most often because they are parental types. That is, the F_1 female inherited w and y^+ together from her P generation mother, and w^+ and y together from her P generation father; the F_1 female is then more likely to pass on these parental combinations of alleles, rather than the recombinant combinations, to her own progeny.

Linkage: A preponderance of parental classes of gametes

A second set of crosses involving the same genes but with a different arrangement of alleles explains why the dihybrid F_1 females do not produce a 1:1:1:1 ratio of the four possible types of gametes (see Cross Series B in **Fig. 5.3**). In this second set of crosses, the original parental generation consists of red-eyed, brown-bodied females ($w^+\ y^+$ / $w^+\ y^+$) and white-eyed, yellow-bodied males ($w\ y$ / Y), and the resultant F_1 females are all $w^+\ y^+$ / $w\ y$ dihybrids. To

Figure 5.3 Designations of *parental* and *recombinant* relate to past history. Figure 5.2 has been redrawn here as **Cross Series A** for easier comparison with **Cross Series B,** in which the dihybrid F_1 females received different allelic combinations of the *white* and *yellow* genes. Note that the parental and recombinant classes in the two cross series are the opposite of each other. The percentages of recombinant and parental types are nonetheless similar in both experiments, showing that the frequency of recombination is independent of the arrangement of alleles.

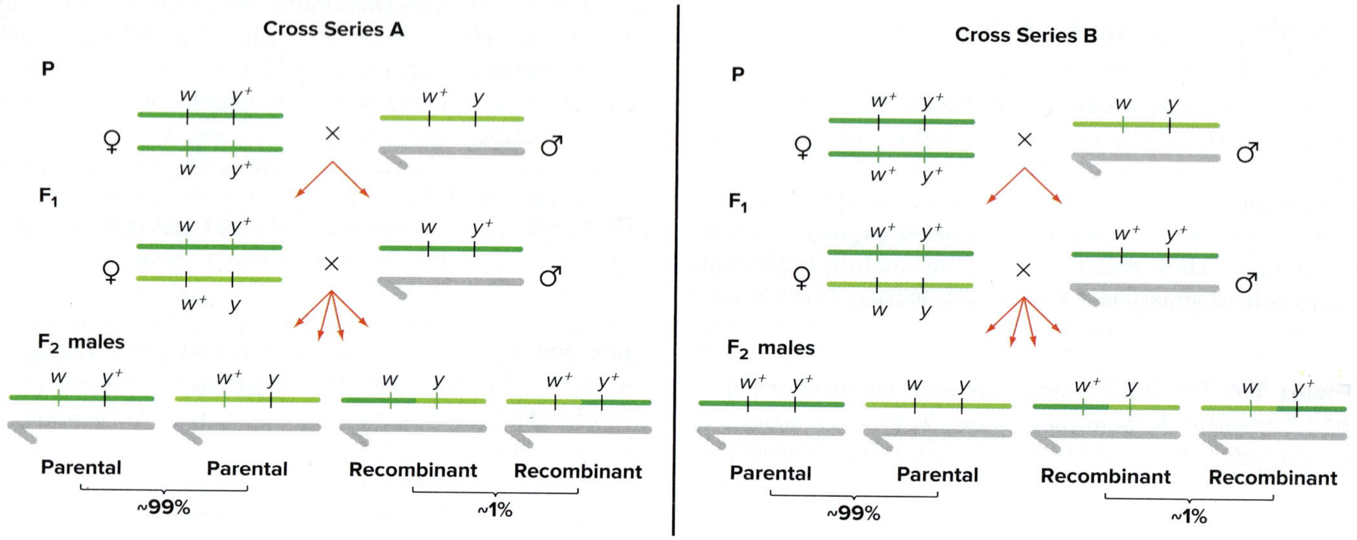

find out what kinds and ratios of gametes these F_1 females produce, we need to look at the telltale F_2 males.

This time, as Cross Series B in Fig. 5.3 shows, $w^+\ y\,/\,Y$ and $w\ y^+\,/\,Y$ are the recombinants that account for little more than 1% of the total, while $w\ y\,/\,Y$ and $w^+\ y^+\,/\,Y$ are the parental combinations, which again add up to almost 99%. You can see that no preferred association of w^+ and y or of y^+ and w exists in this cross. Instead, a comparison of the two experiments with these particular X chromosome genes demonstrates that the observed frequencies of the various types of progeny depend on how the arrangement of alleles in the F_1 females originated. We have redrawn Fig. 5.2 as Cross Series A in Fig. 5.3 so that you can make this comparison more directly. Note that in both experiments, it is the **parental classes**—the combinations originally present in the P generation—that show up most frequently in the F_2 generation. The reshuffled **recombinant classes** occur less often. It is important to appreciate that the designation of *parental* and *recombinant* gametes or progeny of a doubly heterozygous F_1 female is operational, that is, determined by the particular set of alleles she receives from each of her parents.

When genes assort independently, the numbers of parental and recombinant F_2 progeny are equal because a doubly heterozygous F_1 individual produces an equal number of all four types of gametes. By comparison, two genes are considered **linked** when the number of F_2 progeny with parental genotypes exceeds the number of F_2 progeny with recombinant genotypes. Instead of assorting independently, the genes behave as if they are connected to each other much of the time. The genes for eye and body color that reside on the X chromosome in *Drosophila* are an extreme illustration of the linkage concept. The two genes are so tightly coupled that the parental combinations of alleles—$w\ y^+$ and $w^+\ y$ (in Cross Series A of Fig. 5.3) or $w^+\ y^+$ and $w\ y$ (in Cross Series B)—are reshuffled to form recombinants in only 1 out of every 100 gametes formed. In other words, the two parental allele combinations of these tightly linked genes are inherited together 99 times out of 100.

Gene-pair-specific variation in the degree of linkage

Linkage is not always this tight. In *Drosophila,* a mutation for miniature wings (m) is also found on the X chromosome. A cross of red-eyed females with normal wings ($w^+\ m^+\,/\,w^+\ m^+$) and white-eyed males with miniature wings ($w\ m\,/\,Y$) yields an F_1 generation containing all red-eyed, normal-winged flies. The genotype of the dihybrid F_1 females is $w^+\ m^+\,/\,w\ m$. Of the F_2 males, 67.2% are parental types ($w^+\ m^+$ and $w\ m$), while the remaining 32.8% are recombinants ($w\ m^+$ and $w^+\ m$).

This preponderance of parental combinations among the F_2 genotypes reveals that the two genes are linked: The parental combinations of alleles travel together more often than not. But compared to the 99% linkage between the w and y genes for eye color and body color, the linkage of w to m is not that tight. The parental combinations for color and wing size are reshuffled in roughly 33 (instead of 1) out of every 100 gametes.

Linkage of autosomal traits

Linked autosomal genes are not inherited according to the 9:3:3:1 Mendelian ratio expected for two independently assorting, noninteracting genes, each with one completely dominant and one recessive allele. Mendel observed the 9:3:3:1 phenotypic ratio in the F_2 of his dihybrid crosses because the four possible gamete types (*A B*, *A b*, *A B*, and *a b*) were produced at equal frequency by both parents. Equal numbers of each of the four gamete types—independent assortment—means that each one of the 16 boxes in the Punnett square for the F_2 is an equally likely fertilization with a frequency of 1/16 (recall Fig. 2.15).

Had Mendel's two genes been linked, the phenotypic ratio in the F_2 would no longer have been 9:3:3:1 because the parental gametes would have been present at greater frequency than the recombinant gametes. **Figure 5.4** shows the consequences of linkage if the F_1 dihybrid individuals were both of genotype *A B / a b* : The 9/16 and 1/16 classes of F_2 would have increased at the expense of the two 3/16 classes. Conversely, if the alleles of the parents are configured differently (*A b / A b* × *a B / a B*) and the F_1 are therefore *A b / a B*, then the two 3/16 genotypic classes would increase at the expense of the 9/16 and 1/16 classes (*not shown*). Linkage thus undoes the basis of the 9:3:3:1 ratio. Unequal numbers of the four gamete types are produced, so each box of the Punnett square in Fig. 5.4 no longer represents an equally likely fertilization.

Figure 5.4 The 9:3:3:1 ratio is altered when genes *A* and *B* are linked. For linked genes, the F_2 genotypic classes produced most often by parental gametes increase in frequency at the expense of the other classes. In the *A B/a b* dihybrid cross shown here, the *A– B–* and *aa bb* classes in the F_2 will occur at higher frequencies, and the two other classes (*A– bb* and *aa B–*) at lower frequencies than predicted by the 9:3:3:1 ratios. Note that the blue colors and the relative sizes of the boxes in the Punnett square denote the frequencies at which particular genotypic classes will appear in the F_2 generation.

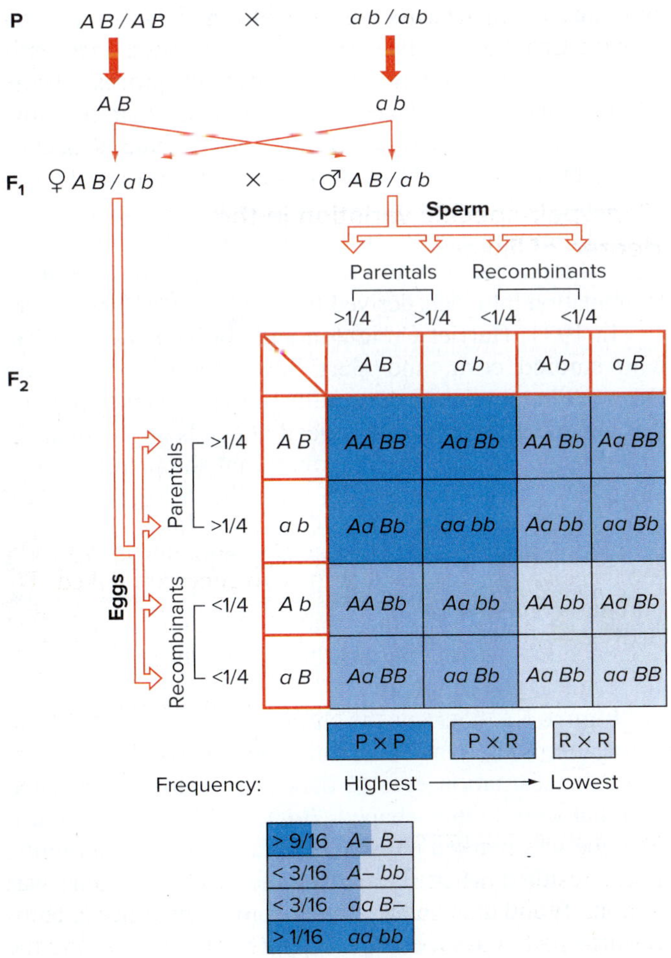

Testcrosses Simplify the Detection of Linkage

Early twentieth-century geneticists found it difficult to interpret crosses involving autosomal genes such as that shown in Fig. 5.4 because it was hard to trace which alleles came from which parent. For example, all the F_2 in Fig. 5.4 with genotype *A– B–* would have the same phenotype, but they could have arisen from fertilizations involving two parental gametes (*dark blue squares*), two recombinant gametes (*light blue squares*), or one of each kind (*medium blue squares*). However, by setting up testcrosses in which one parent was homozygous for the recessive alleles of both genes, as detailed in the next section, geneticists can easily analyze the gene combinations received in the gametes from the other, doubly heterozygous parent.

Fruit flies, for example, carry an autosomal gene for body color (in addition to the X-linked *y* gene); the wild type is once again brown, but a recessive mutation in this gene gives rise to black (*b*). A second gene on the same autosome helps determine the shape of a fruit fly's wing, with the wild type having straight edges and a recessive mutation (*c*) producing curves. **Figure 5.5** depicts a cross between

Figure 5.5 Autosomal genes can also exhibit linkage. A testcross shows that the recombination frequency for the body color (*b*) and wing shape (*c*) pair of *Drosophila* genes is 23%. Because parentals outnumber recombinants, the *b* and *c* genes are genetically linked and must be on the same autosome.

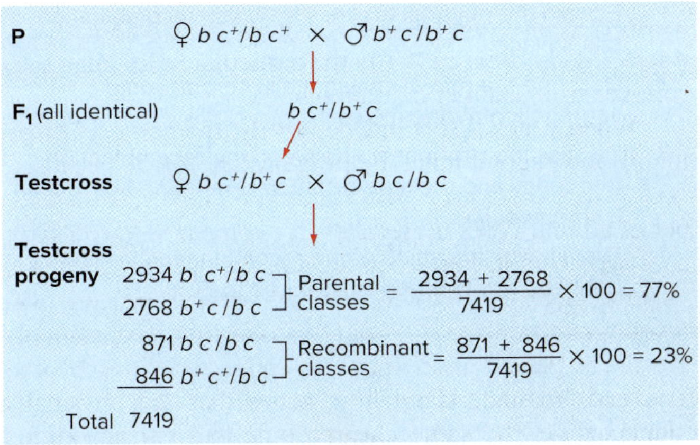

two pure-breeding strains: black-bodied females with straight wings ($b\ c^+ / b\ c^+$) and brown-bodied males with curved wings ($b^+\ c / b^+\ c$). All the F_1 progeny are double heterozygotes ($b\ c^+ / b^+\ c$) and are phenotypically wild type.

In a testcross of the F_1 females with $b\ c / b\ c$ males, all the offspring receive the recessive *b* and *c* alleles from their father. The phenotypes of the offspring thus indicate the kinds of gametes received from the mother. For example, a black fly with normal wings would be genotype $b\ c^+ / b\ c$; because we know it received the *b c* combination from its father, it must have received $b\ c^+$ from its mother. As Fig. 5.5 shows, roughly 77% of the testcross progeny in one experiment received parental gene combinations (that is, allelic combinations transmitted to the F_1 females by the gametes of each of her parents), while the remaining 23% were recombinants. Because the parental classes outnumbered the recombinant classes, we can conclude that the autosomal genes for black body and curved wings are linked.

essential concepts

- Genes on the same chromosome that do not assort independently are said to be *linked.*
- *Parental gametes* contain alleles inherited together from a single grandparent; *recombinant gametes* contain alleles inherited from different grandparents.
- The hallmark of linkage is that a dihybrid produces more parental gametes than recombinant gametes; as a result, the progeny ratio in a dihybrid cross involving linked genes is not 9:3:3:1.
- *Testcrosses* clarify linkage because each phenotypic class of progeny corresponds to each gamete type produced by the dihybrid parent.

5.2 Recombination: A Result of Crossing-Over During Meiosis

learning objectives

1. Explain the physical process by which recombination takes place.
2. Describe the role of chiasmata in chromosome segregation during meiosis.
3. Discuss the relationship between the recombination frequency and the map distance separating two loci on a chromosome.
4. Explain why the value of the recombination frequency between any two genes is limited to 50%.

It is easy to understand how genes that are physically connected on the same chromosome can be transmitted together and thus show genetic linkage. It is not as obvious why all linked genes always show some recombination in a sample population of sufficient size. Do the chromosomes participate in a physical process that gives rise to the reshuffling of linked genes that we call recombination? The answer to this question is of more than passing interest as it provides a basis for gauging relative distances between pairs of genes on a chromosome.

In 1909, the Belgian cytologist Frans Janssens described structures he had observed in the light microscope during prophase of the first meiotic division. He called these structures *chiasmata;* as described in Chapter 4, they seemed to represent regions in which nonsister chromatids of homologous chromosomes cross over each other (review Fig. 4.16). Making inferences from a combination of genetic and cytological data, Thomas Hunt Morgan suggested that the chiasmata observed through the light microscope were sites of chromosome breakage and exchange resulting in genetic recombination.

Reciprocal Exchanges Between Homologs Are the Physical Basis of Recombination

Morgan's idea that the physical breaking and rejoining of chromosomes during meiosis was the basis of genetic recombination seemed reasonable. But although Janssens's chiasmata could be interpreted as signs of the process, before 1930 no one had produced visible evidence that crossing-over between homologous chromosomes actually occurs. The identification of **physical markers,** or cytologically visible abnormalities that make it possible to keep track of specific chromosome parts from one generation to the next, enabled researchers to turn the logical deductions about recombination into facts derived from experimental evidence.

In 1931, Harriet Creighton and Barbara McClintock, who studied corn, and Curt Stern, who worked with *Drosophila,* published the results of experiments showing that genetic recombination indeed depends on the reciprocal exchange of parts between maternal and paternal chromosomes. Stern, for example, bred female flies with two different X chromosomes, each containing a distinct physical marker near one of the ends. These same females were also doubly heterozygous for two X-linked **genetic markers**—alleles of genes that could serve as points of reference in determining whether particular progeny were the result of recombination.

Figure 5.6 diagrams the chromosomes of these heterozygous females. One X chromosome carried mutations producing carnation eyes (a dark ruby color, abbreviated *car*) that were kidney-shaped (*Bar*); in addition, this chromosome was marked physically by a visible discontinuity, which resulted when the end of the X chromosome was broken off and attached to an autosome. The other X chromosome had wild-type alleles (+) for both the *car* and the

Figure 5.6 Evidence that recombination results from reciprocal exchanges between homologous chromosomes. Genetic recombination between the *car* and *Bar* genes on the *Drosophila* X chromosome is accompanied by the exchange of physical markers observable in the microscope. Note that this depiction of crossing-over is a simplification, as genetic recombination actually occurs after each chromosome has replicated into sister chromatids.

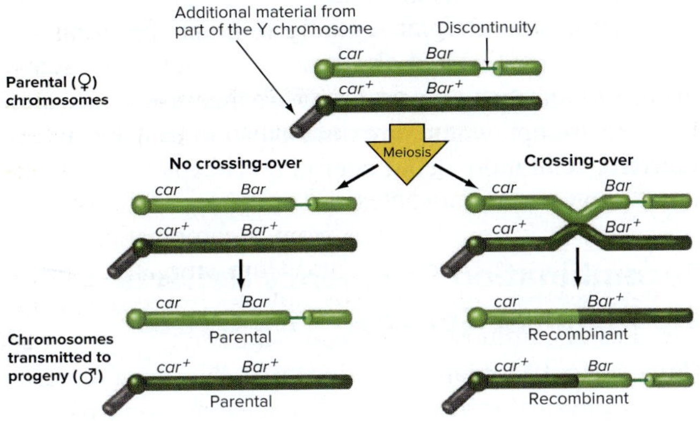

Bar genes, and its physical marker consisted of part of the Y chromosome that had become connected to the X-chromosome centromere.

Figure 5.6 illustrates how the chromosomes in these *car Bar* / car^+ Bar^+ females were transmitted to male progeny. According to the experimental results, all sons showing a phenotype determined by one or the other parental combination of genes (either *car Bar* or car^+ Bar^+) had an X chromosome that was structurally indistinguishable from one of the original X chromosomes in the mother. In recombinant sons, however, such as those that manifested carnation eye color and normal eye shape (*car* Bar^+ / Y), an identifiable exchange of the abnormal features marking the ends of the homologous X chromosomes accompanied the recombination of genes. The evidence thus tied an instance of genetic recombination to the crossing-over of specifically marked parts of particular chromosomes. This experiment demonstrated elegantly that genetic recombination is associated with the actual reciprocal exchange of segments between homologous chromosomes during meiosis.

Why Recombination?

In Chapter 4, we discussed one advantage that recombination provides for organisms on the earth measured over evolutionary time: Recombination contributes to genetic diversity by reshuffling the alleles of genes between homologous chromosomes (review Fig. 4.17). However, crossing-over also plays another, even more crucial role to ensure that chromosomes segregate properly when they are transmitted between parents and their progeny. As you will see, if recombination did not occur, nondisjunction during meiosis I would be a common, rather than a rare, occurrence, and species could never retain the same number of chromosomes in successive generations.

The issue is that proper chromosome segregation requires homologous chromosomes to be pulled to opposite spindle poles, which in turn requires the homologous chromosomes not only to pair with each other during prophase, but also to be linked to each other physically through metaphase until they separate at anaphase. The meiosis I spindle can exert tension on the chromosomes only if the homologs are pulled in opposite directions but remain joined by a physical link. If the tension did not exist, homologous chromosomes would not "know" that they were connected to opposite spindle poles. Without tension, both chromosomes therefore could often connect to fibers from the same spindle pole, and nondisjunction would occur. What then provides the physical link between homologous chromosomes until anaphase of meiosis I?

You might think from Fig. 4.16 that the synaptonemal complex or recombination nodules form the necessary link between homologous chromosomes. **Figure 5.7a** shows an actual fluorescence micrograph of these structures during the middle of prophase I (the pachytene substage). The synaptonemal complexes that help homologous chromosomes to pair with each other are shown in *red*. Although the DNA of the chromosomes is not illustrated in this figure, each red line represents a bivalent (tetrad) made of two homologous chromosomes, each of which has previously been duplicated into sister chromatids. Studded at intervals along the synaptonemal complex are recombination nodules that contain the enzymes responsible for the actual crossing-over; one of these proteins is stained in *green*. However, contrary to expectations, neither synaptonemal complexes nor recombination nodules can link homologous chromosomes until anaphase I begins, for the simple reason that these structures both disappear by the end of prophase I.

Figure 5.7b and **c** illustrate that the homologous chromosomes are still connected to each other even after the synaptonemal complexes and recombination nodules have dissolved. As discussed in Chapter 4, chiasmata mark the sites where recombination actually occurred earlier in prophase I (that is, where nonsister chromatids from homologous chromosomes exchanged places). However, crossing-over by itself is insufficient to keep the homologous chromosomes together. As seen in the artist's diagram in Fig. 5.7c, the physical linkage between homologous chromosomes involves molecular complexes called *cohesin* that make connections between sister chromatids soon after the chromosomes have replicated. Once a crossover takes place, it is cohesin complexes distal to the crossover point (that is, farther away from the centromere than the chiasmata) that keep the homologous chromosomes together at the metaphase plate and thus ensure proper chromosome segregation.

Cohesin plays many roles in the biology of chromosomes; for example, not only is it found along chromosomal

Figure 5.7 Recombination helps ensure proper chromosome segregation during meiosis I. (a) Mouse chromosomes during mid-prophase of meiosis I in a primary spermatocyte. A protein component of the synaptonemal complex is *red,* a component of recombination nodules is *green,* and a component of centromeres is *blue.* Each bivalent has at least one recombination nodule, although some are hard to see. **(b)** Mouse chromosomes in late prophase of meiosis I (diakinesis). Note that each bivalent has at least one chiasma (the *arrow* points to one example), indicating that crossing-over occurred earlier. **(c)** Artist's representation of cohesin complexes (*orange rings*) along the chromosome arms of a bivalent during metaphase of meiosis I; cohesin at the centromeres is *not shown.* Cohesin complexes distal to the crossover point keep sister chromatids together. *Arrows* pointing toward the poles indicate the forces that try to pull the homologous chromosomes apart.

a: © Dr. Paula Cohen & Dr. Miguel Angel Brieño-Enríquez, The Cohen Lab, Center for Reproductive Genomics, Cornell University, Ithaca, NY; b: © Dr. Paula Cohen & Dr. Kim Holloway, The Cohen Lab, Center for Reproductive Genomics, Cornell University, Ithaca, NY

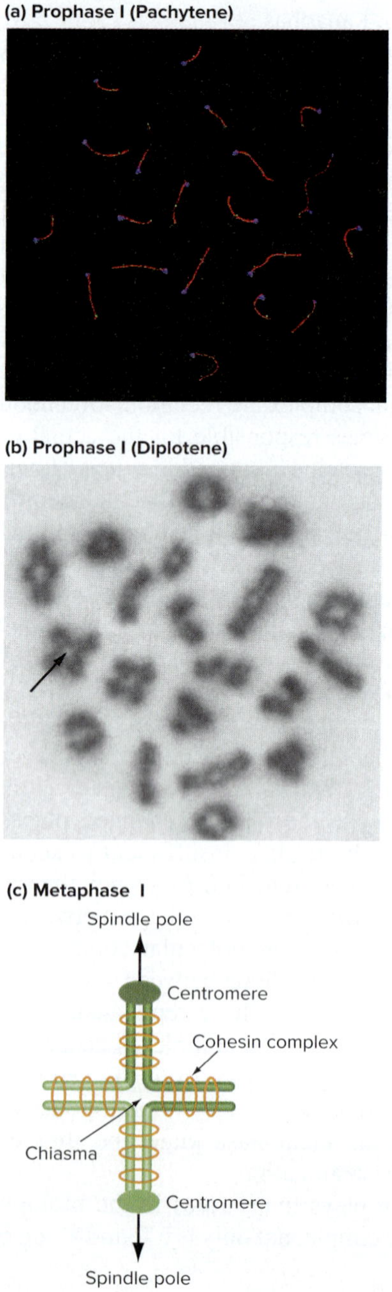

arms, but it also is a key component of the centromeres themselves. We will discuss in detail how cohesin connects sister chromatids during mitosis and meiosis in Chapter 12.

The importance of crossing-over to proper chromosome segregation is underlined by the fact that each bivalent in Figs. 5.7a and b has at least one recombination nodule or chiasma. In fact, a mechanism called *interference* that occurs in almost all sexually reproducing organisms helps ensure that each chromosome pair undergoes at least one crossover, thus preventing nondisjunction of any chromosome except when rare mistakes occur. We discuss interference in more detail later in this chapter.

Recombination Frequency Reflects the Distance Between Two Genes

Thomas Hunt Morgan's intuitions that chiasmata represent sites of physical crossing-over between chromosomes and that such crossing-over may result in recombination, led him to the following logical deduction: Different gene pairs exhibit different linkage frequencies because genes are arranged in a line along a chromosome. The closer together two genes are on the chromosome, the smaller their chance of being separated by an event that cuts and recombines the line of genes. To look at it another way, if we assume for the moment that chiasmata can form anywhere along a chromosome with equal likelihood, then the probability of a crossover occurring between two genes increases with the distance separating them. If this is so, the frequency of genetic recombination also must increase with the distance between genes.

To illustrate the point, imagine pinning to a wall a 10-inch piece of ribbon containing tiny black dots along its length and then repeatedly throwing a dart to see where you will cut the ribbon. You would find that practically every throw of the dart separates a dot at one end of the ribbon from a dot at the other end, while few if any throws separate any two particular dots positioned right next to each other.

Alfred H. Sturtevant, one of Morgan's students, took this idea one step further. He proposed that the percentage of total progeny that were recombinant types, the **recombination frequency (RF),** could be used as a gauge of the physical distance separating any two genes on the same chromosome. Sturtevant arbitrarily defined one RF percentage point as the unit of measure along a chromosome; later, another geneticist named the unit a **centimorgan (cM)** after T. H. Morgan. Mappers often refer to a centimorgan as a **map unit (m.u.)**. Although the two terms are interchangeable, researchers prefer one or the other, depending on their experimental organism. *Drosophila* geneticists, for example, use map units while human geneticists use centimorgans. In Sturtevant's system, 1% RF = 1 cM = 1 m.u.

Figure 5.8 Recombination frequencies are the basis of genetic maps. (a) 1.1% of the gametes produced by a female doubly heterozygous for the genes *w* and *y* are recombinant. The recombination frequency (RF) is thus 1.1%, and the genes are approximately 1.1 map units (m.u.) or 1.1 centimorgans (cM) apart. **(b)** The distance between the *w* and *m* genes is longer: 32.8 m.u. (or 32.8 cM).

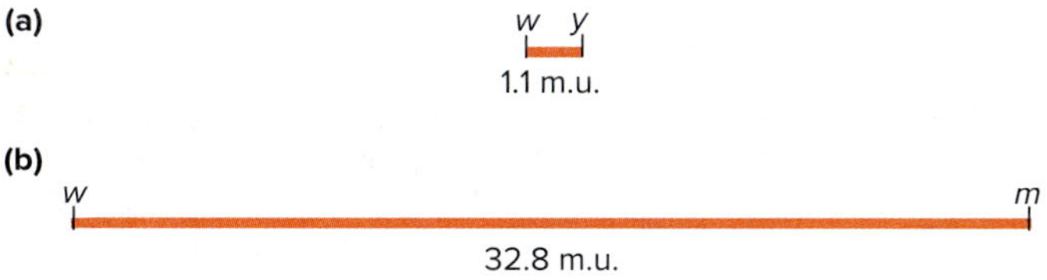

A review of the two pairs of X-linked *Drosophila* genes we analyzed earlier shows how Sturtevant's proposal works. Because the X-linked genes for eye color (*w*) and body color (*y*) recombine in 1.1% of F_2 progeny, they are 1.1 m.u. apart (**Fig. 5.8a**). In contrast, the X-linked genes for eye color (*w*) and wing size (*m*) have a recombination frequency of 32.8 and are therefore 32.8 m.u. apart (**Fig. 5.8b**).

It is easy to see why the fraction of recombinant gametes is a measure of the distance between two genes when you consider individual meioses that can occur in the germ-line cells of a dihybrid. If genes *A* and *B* are close together on a chromosome, meiosis will usually take place as shown in **Fig. 5.9a,** with no crossovers between the genes (an **NCO** meiosis). In a dihybrid (such as *A B* / *a b*), the outcome of an NCO meiosis will be four parental gametes (Fig. 5.9a). Occasionally, during a meiosis a single crossover will occur between the two genes (an **SCO** meiosis), resulting in two recombinant and two parental gametes (**Fig. 5.9b**). These crossovers will be rare because most of the length of the chromosome lies outside of the region between genes *A* and *B*. As the distance between the two genes increases, the frequency of SCO meioses increases, and the fraction of recombinant gametes increases.

You can also understand from Fig. 5.9 why the two types of parental gametes (*A B* and *a b* in this example) are produced at roughly equivalent frequencies, and why the two types of recombinant gametes (*A b* and *a B*) appear in approximately equal numbers with respect to each other also. Whenever two nonsister chromatids *do not* cross over between genes *A* and *B*, one of each type of parental chromosome is generated. Likewise, every time two nonsister chromatids *do* cross over between the two genes, both reciprocal recombinants are produced.

As a unit of measure, the map unit is simply an index of recombination probabilities assumed to reflect distances between genes. According to this index, the *y* and *w* genes are much closer together than the *m* and *w* genes. Geneticists have used this logic to map thousands of genetic markers to the chromosomes of *Drosophila,* building recombination maps step-by-step with closely linked markers.

Figure 5.9 Recombinant gametes are produced less frequently than parental gametes when two genes are linked. **(a)** Most meioses occur with no crossovers between tightly linked genes *A* and *B*, resulting in four parental gametes (shaded *orange*). **(b)** Occasionally, a single crossover occurs between *A* and *B* during meiosis, resulting in two recombinant gametes (shaded *blue*) and two parental gametes (*orange*).

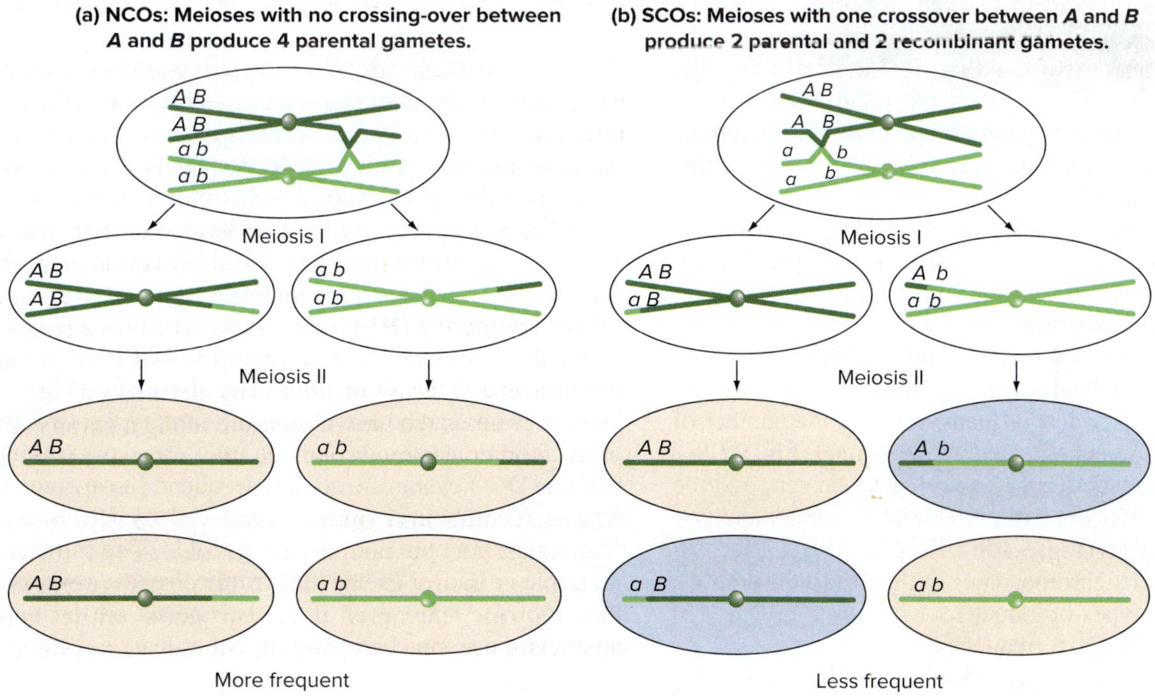

Figure 5.10 Why RF = 50% in a dihybrid for genes on nonhomologous chromosomes. (a) Nonhomologous chromosomes line up randomly with respect to one another so that meiosis results in all parental gametes (*left*) or all recombinant gametes (*right*) with equal frequency. **(b)** Meioses that occur with one crossover between one gene and the centromere generate all four gamete types with equal frequency.

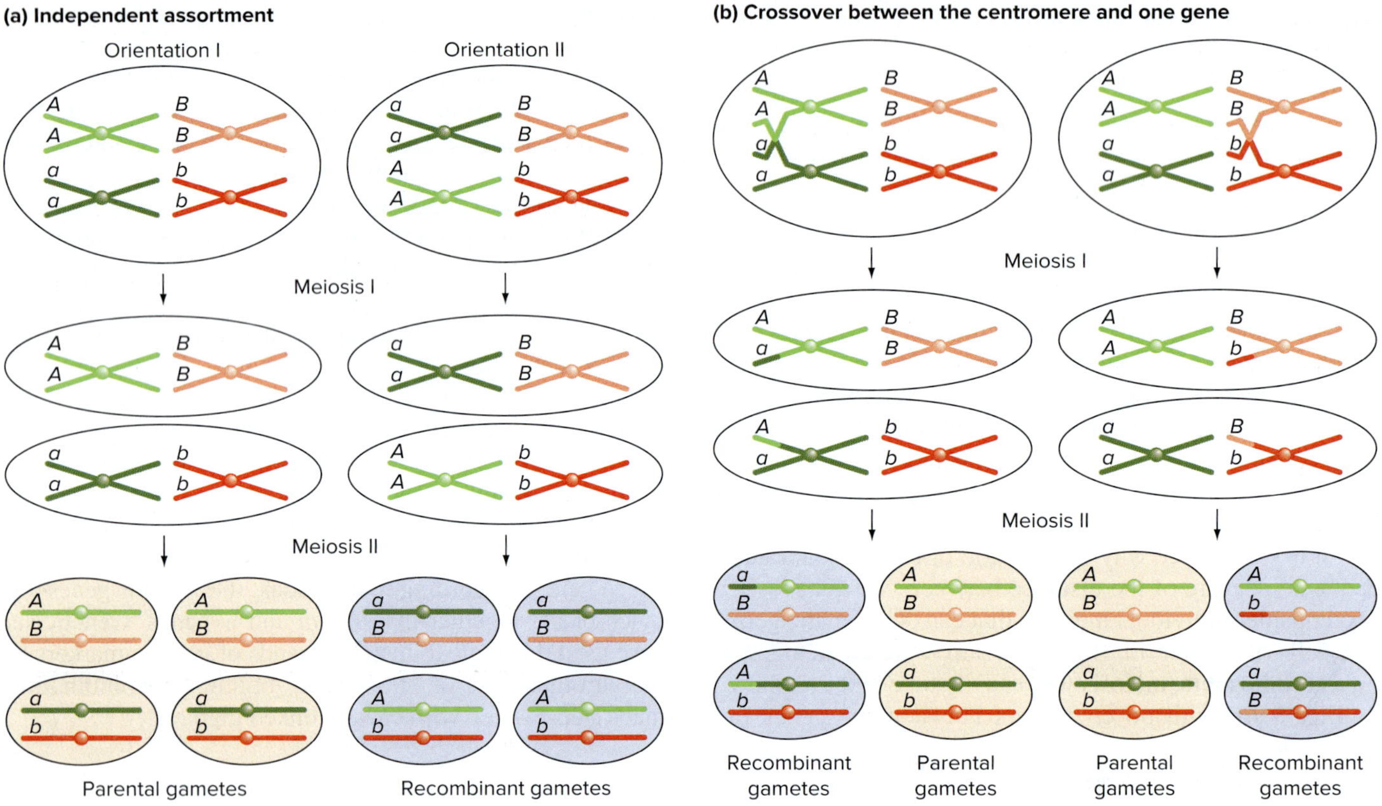

Recombination Frequencies Between Two Genes Never Exceed 50%

If the definition of linkage is that the proportion of recombinant classes is smaller than that of parental classes, a recombination frequency of less than 50% indicates linkage. But what can we conclude about the relative location of genes if roughly equal numbers of parental and recombinant progeny exist? And does it ever happen that recombinants are in the majority?

We already know one situation that can give rise to a recombination frequency of 50%. Genes located on different (that is, nonhomologous) chromosomes will obey Mendel's law of independent assortment for two reasons. First, the two chromosomes can line up on the spindle during meiosis I in either of two equally likely configurations. As a result, when summed over the products of many meioses, the number of parental and recombinant gametes will be equal (**Fig. 5.10a**). Second, if a crossover occurs between any one gene and the centromere, that meiosis will produce two parental and two recombinant gametes (**Fig. 5.10b**). Thus, if genes *A* and *B* are on nonhomologous chromosomes, a dihybrid will produce all four possible types of gametes (*A B, A b, a B,* and *a b*) with approximately equal frequency.

Importantly, experiments have established that genes located very far apart on the same chromosome also show recombination frequencies of approximately 50%. To understand why this is true, consider the different meioses that contribute to the pool of gametes counted in a testcross experiment. When two genes are very close together on the same chromosome, only two kinds of meioses are likely to occur: those with no crossovers (an NCO meiosis in Fig. 5.9a and **Fig. 5.11**) and rarer meioses with a single crossover (SCO in Fig. 5.9b and Fig. 5.11). An NCO meiosis yields only parental gametes, while an SCO event produces 50% recombinant gametes. These closely spaced genes are linked because some meioses (NCOs) do not make any recombinant-type gametes.

When the two genes are farther apart, SCOs become more frequent, and in some meioses, two crossovers between *A* and *B* (**DCO** meioses) can also occur (Fig. 5.11). DCO meioses can be one of four different types; two, three, or all four nonsister chromatids may cross over. Because the four DCO events are equally frequent, the average fraction of recombinant gametes produced by DCOs is 50% (see equation at the bottom of Fig. 5.11). The same is true for triple crossover events, quadruple crossover events, etc. (*not shown*). Thus, even if the two genes are far enough apart on the same chromosome that at least one crossover

Figure 5.11 Meioses possible in a dihybrid for genes on the same chromosome. Meioses with no crossovers (NCOs) between *A* and *B* result in all parental chromosomes. Meioses with an SCO between *A* and *B* result in half parental and half recombinant (R) chromosomes. Four kinds of DCOs are equally likely, and collectively they result in half parental and half recombinant gametes (*bottom*). In a 2-strand DCO event, one pair of nonsister chromatids undergoes two crossovers. In either of two types of 3-strand DCOs, a single chromatid of one homolog crosses over with each of the two chromatids of the other homolog. In a 4-strand DCO, each pair of nonsister chromatids recombines.

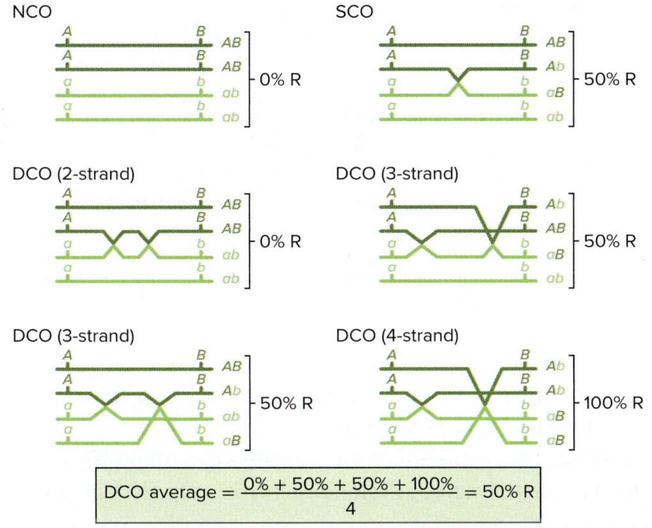

occurs between them in every meiosis, the pool of gametes produced by a dihybrid would be only 50% recombinant types. You can see now that if two genes on the same chromosome are so far apart that no meioses occur as NCOs, then the alleles of the genes assort independently, just as if they were on different chromosomes.

Even though crosses between two genes lying very far apart on a chromosome may show no linkage at all, you can demonstrate that they are on the same chromosome if you can tie each of the widely separated genes to one or more common intermediaries. **Table 5.1** summarizes the relationship between the relative locations of two genes and the presence or absence of linkage as measured by recombination frequencies.

TABLE 5.1 Properties of Linked Versus Unlinked Genes

Linked Genes	
Parentals > Recombinants (RF < 50%)	
Linked genes must be syntenic and sufficiently close together on the same chromosome so that they do not assort independently.	
Unlinked Genes	
Parentals = Recombinants (RF = 50%)	
Occurs either when two genes are on different chromosomes or when they are sufficiently far apart on the same chromosome that at least one crossover occurs between them in every meiosis.	

essential concepts

- *Recombination* occurs when chromatids of homologous chromosomes exchange parts during prophase of meiosis I.
- *Crossing-over* helps establish physical linkages between homologous chromosomes needed to prevent nondisjunction.
- The *recombination frequency* (*RF*) indicates how often two genes are transmitted together. For *linked* genes, RF < 50%.
- Two genes are *linked* if they are physically so close that some meioses occur with no crossovers between them.
- The maximum RF of 50% (independent assortment) occurs when two genes are on different chromosomes, or when two syntenic genes are so far apart that there is at least one crossover between them in every meiosis.

5.3 Mapping: Locating Genes Along a Chromosome

learning objectives

1. Establish relative gene positions using two-point cross data.
2. Refine genetic maps based on data from three-point testcrosses.
3. Explain how a genetic map (in map units) is related to actual physical distance (in base pairs of DNA).
4. Describe the relationship between linkage groups and chromosomes.

Maps are images of the relative positions of objects in space. Whether depicting the floor plan of New York's Metropolitan Museum of Art, the layout of the Roman Forum, or the location of cities served by the railways of Europe, maps turn measurements into patterns of spatial relationships that add a new level of meaning to the original data of distances. Maps that assign genes to specific locations on particular chromosomes called **loci** (singular **locus**) are no exception. By transforming genetic data into spatial arrangements, maps sharpen our ability to predict the inheritance patterns of specific traits.

Geneticists have been obsessed with mapping genes because knowing a gene's location gives scientists the ability to identify the segment of chromosomal DNA corresponding to a gene. In later chapters of this book, you will see how knowledge of a gene's location can be used to isolate its DNA, and furthermore, how molecular geneticists can use a gene's DNA to understand the gene's function.

We have seen that recombination frequency (RF) is a measure of the distance separating two genes along a chromosome. We now examine how data from many crosses following two and three genes at a time can be compiled and compared to generate accurate, comprehensive gene/chromosome maps.

Comparisons of Two-Point Crosses Establish Relative Gene Positions

In his senior undergraduate thesis, Morgan's student A. H. Sturtevant asked whether data obtained from a large number of two-point crosses (crosses tracing two genes at a time) would support the idea that genes form a definite linear series along a chromosome. Sturtevant began by looking at X-linked genes in *Drosophila.* **Figure 5.12a** lists his recombination data for several two-point crosses. Recall that the distance between two genes that yields 1% recombinant progeny—an RF of 1%—is 1 m.u.

As an example of Sturtevant's reasoning, consider the three genes *w, y,* and *m.* If these genes are arranged in a line (instead of a more complicated branched structure, for example), then one of them must be in the middle, flanked on either side by the other two. The greatest genetic distance should separate the two genes on the outside, and this value should roughly equal the sum of the distances separating the middle gene from each outside gene. The data Sturtevant obtained are consistent with this idea, implying that *w* lies between *y* and *m* (**Fig. 5.12b**). Note that the left-to-right orientation of this map was selected at random; the map in Fig. 5.12b would be equally correct if it portrayed *y* on the right and *m* on the left.

By following exactly the same procedure for each set of three genes considered in pairs, Sturtevant established a self-consistent order for all the genes he investigated on *Drosophila*'s X chromosome (**Fig. 5.12c;** once again, the left-to-right arrangement is an arbitrary choice). By checking the data for every combination of three genes, you can assure yourself that this ordering makes sense. The fact that the recombination data yield a simple linear map of gene position supports the idea that genes reside in a unique linear order along a chromosome.

Figure 5.12 Mapping genes by comparisons of two-point crosses. (a) Sturtevant's data for the distances between pairs of X-linked genes in *Drosophila.* **(b)** Because the distance between *y* and *m* is greater than the distance between *w* and *m,* the order of genes must be *y-w-m.* **(c)** and **(d)** Maps for five genes on the *Drosophila* X chromosome. The left-to-right orientation is arbitrary. Note that the numerical position of the *r* gene depends on how it is calculated. The best genetic maps are obtained by summing many small intervening distances as in (d).

(a)

Gene pair	RF
y-w	1.1
y-v	33.0
y-m	34.3
y-r	42.9
w-v	32.1
w-m	33.8
w-r	42.1
v-m	4.0
v-r	24.1
m-r	17.8

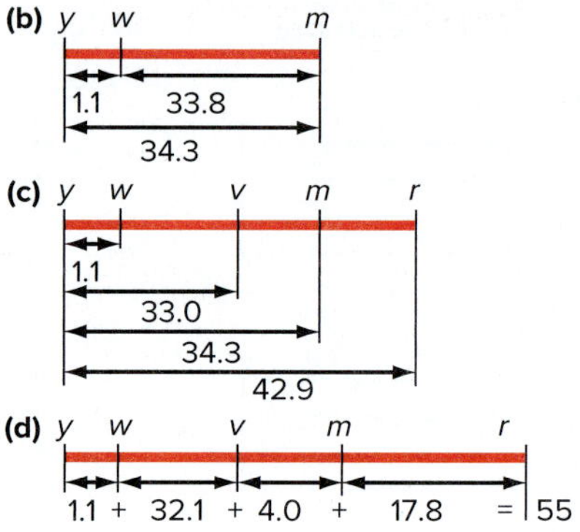

Limitations of two-point crosses

Though of great importance, the pairwise mapping of genes has several shortcomings that limit its usefulness. First, in crosses involving only two genes at a time, it may be difficult to determine gene order if some gene pairs lie very close together. For example, in mapping *y, w,* and *m,* 34.3 m.u. separate the outside genes *y* and *m,* while nearly as great a distance (33.8 m.u.) separates the middle *w* from the outside *m* (Fig. 5.12b). Before being able to conclude with any confidence that *y* and *m* are truly farther apart, that is, that the small difference between the values of 34.3 and 33.8 is not the result of sampling error, you would have to examine a very large number of flies and subject the data to a statistical test, such as the *chi-square test* that will be explained in the next section.

A second problem with Sturtevant's mapping procedure is that the actual distances in his map do not always add up, even approximately. As an example, suppose that the locus of the *y* gene at the far left of the map is regarded as position 0 (Fig. 5.12c). The *w* gene would then lie near position 1, and *m* would be located in the vicinity of 34 m.u. But what about the *r* gene, named for a mutation that produces rudimentary (very small) wings? Based solely on its distance from *y,* as inferred from the $y \leftrightarrow r$ data in Fig. 5.12a, we would place it at position 42.9 (Fig. 5.12c). However, if we calculate its position as the sum of all intervening distances inferred from the data in Fig. 5.12a, that is, as the sum of $y \leftrightarrow w$ plus $w \leftrightarrow v$ plus $v \leftrightarrow m$ plus $m \leftrightarrow r$, the locus of *r* becomes $1.1 + 32.1 + 4.0 + 17.8 = 55.0$ (**Fig. 5.12d**). What can explain this difference, and which of these two values is closer to the truth? Three-point crosses help provide some of the answers.

Three-Point Crosses Provide Faster and More Accurate Mapping

The simultaneous analysis of three markers makes it possible to obtain enough information to position the three genes in relation to each other from just one set of crosses. To describe this procedure, we look at three genes linked on one of *Drosophila*'s autosomes.

A homozygous female with mutations for vestigial wings (*vg*), black body (*b*), and purple eye color (*pr*) was mated to a wild-type male (**Fig. 5.13a**). All the triply heterozygous F_1 progeny, both male and female, had normal

Figure 5.13 Analyzing the results of a three-point cross. **(a)** Results from a three-point testcross of F_1 females simultaneously heterozygous for *vg, b,* and *pr.* **(b)** The gene in the middle must be *pr* because the longest distance is between the other two genes: *vg* and *b.* The most accurate map distances are calculated by summing shorter intervening distances, so 18.7 m.u. is a more accurate estimate of the genetic distance between *vg* and *b* than is 17.7 m.u.

(a) Three-point cross results

P ♀ $vg\ b\ pr\ /\ vg\ b\ pr$ × ♂ $vg^+\ b^+\ pr^+\ /\ vg^+\ b^+\ pr^+$

F_1 (all identical) $vg\ b\ pr\ /\ vg^+\ b^+\ pr^+$

Testcross ♀ $vg\ b\ pr\ /\ vg^+\ b^+\ pr^+$ × ♂ $vg\ b\ pr\ /\ vg\ b\ pr$

Testcross progeny			
	1779	$vg\ b\ pr$	Parental combinations for all three genes
	1654	$vg^+\ b^+\ pr^+$	
	252	$vg^+\ b\ pr$	Recombinants for *vg* relative to parental combinations for *b* and *pr*
	241	$vg\ b^+\ pr^+$	
	131	$vg^+\ b\ pr^+$	Recombinants for *b* relative to parental combinations for *vg* and *pr*
	118	$vg\ b^+\ pr$	
	13	$vg\ b\ pr^+$	Recombinants for *pr* relative to parental combinations for *vg* and *b*
	9	$vg^+\ b^+\ pr$	
	4197		

(b) Deduced genetic map

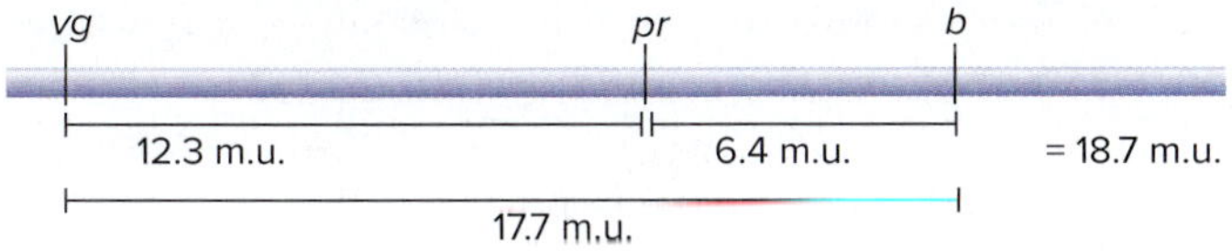

phenotypes, indicating that the mutations are autosomal recessive. In a testcross of the F_1 females with males having vestigial wings, black body, and purple eyes, the progeny were of eight different phenotypes reflecting eight different genotypes. The order in which the genes in each phenotypic class are listed in Fig. 5.13a is completely arbitrary. Thus, instead of *vg b pr,* one could write *b vg pr* or *vg pr b* to indicate the same genotype. Remember that at the outset we do not know the gene order; deducing it is the goal of the mapping study.

In analyzing the data, we look at two genes at a time (recall that the recombination frequency is always a function of a pair of genes). For the pair *vg* and *b,* the parental combinations are *vg b* and $vg^+\ b^+$; the recombinants are $vg\ b^+$ and $vg^+\ b$. To determine whether a particular class of progeny is parental or recombinant for *vg* and *b,* we do not care whether the flies are *pr* or pr^+. Thus, to the nearest tenth of a map unit, the $vg \leftrightarrow b$ distance, calculated as the percentage of recombinants in the total number of progeny, is

$$\frac{252 + 241 + 131 + 118}{4197} \times 100$$
$$= 17.7 \text{ m. u. } (vg \leftrightarrow b \text{ distance}).$$

Similarly, because recombinants for the *vg–pr* gene pair are $vg\ pr^+$ and $vg^+\ pr$, the interval between these two genes is

$$\frac{252 + 241 + 13 + 9}{4197} \times 100$$
$$= 12.3 \text{ m.u. } (vg \leftrightarrow pr \text{ distance})$$

while the distance separating the *b–pr* pair is

$$\frac{131 + 118 + 13 + 9}{4197} \times 100$$
$$= 6.4 \text{ m.u. } (b \leftrightarrow pr \text{ distance}).$$

These recombination frequencies show that *vg* and *b* are separated by the largest distance (17.7 m.u., as compared with 12.3 and 6.4) and must therefore be the outside genes, flanking *pr* in the middle (**Fig. 5.13b**). But as with the X-linked *y* and *r* genes analyzed by Sturtevant, the distance separating the outside *vg* and *b* genes (17.7) does not equal the sum of the two intervening distances (12.3 + 6.4 = 18.7). In the next section, we learn that the reason for this discrepancy is the rare occurrence of double crossovers.

Correction for double crossovers

Figure 5.14 depicts the homologous autosomes of the F_1 females that are heterozygous for the three genes *vg, pr,* and *b.* A close examination of the chromosomes reveals the kinds of crossovers that must have occurred to generate the classes and numbers of progeny observed. In this and subsequent figures, the chromosomes depicted are in late prophase/early metaphase of meiosis I, when there are four chromatids for each pair of homologous chromosomes. As we have suggested previously and demonstrate more rigorously in Chapter 6, prophase I is the stage at which recombination takes place. Note that *region 1* is the space between *vg* and *pr,* and the space between *pr* and *b* is *region 2.*

Recall that the progeny from the testcross performed earlier fall into eight groups (review Fig. 5.13). Flies in the two largest groups carry the same configurations of genes as did their grandparents in the P generation: *vg b pr* and $vg^+\ b^+\ pr^+$; they thus represent the parental classes (Fig. 5.14a). The next two groups—$vg^+\ b\ pr$ and $vg\ b^+\ pr^+$—are composed of recombinants that must be the reciprocal products of a crossover in region 1 between *vg* and *pr* (Fig. 5.14b). Similarly the two groups containing $vg^+\ b\ pr^+$ and $vg\ b^+\ pr$ flies must have resulted from recombination in region 2 between *pr* and *b* (Fig. 5.14c).

But what about the two smallest groups made up of rare $vg\ b\ pr^+$ and $vg^+\ b^+\ pr$ recombinants? What kinds of chromosome exchange could account for them? Most likely, they result from two different crossover events occurring simultaneously, one in region 1, the other in region 2 (Fig. 5.14d). The gametes produced by such double crossovers still have the parental configuration for the

Figure 5.14 Inferring the location of a crossover event. Once you establish the order of genes involved in a three-point cross, it is easy to determine which crossover events gave rise to particular recombinant gametes. Note that double crossovers are needed to generate gametes in which the gene in the middle has recombined relative to the parental combinations for the genes at the ends. Such events include the 2-strand DCO in part (d) as well as 3-strand DCOs (*not shown*).

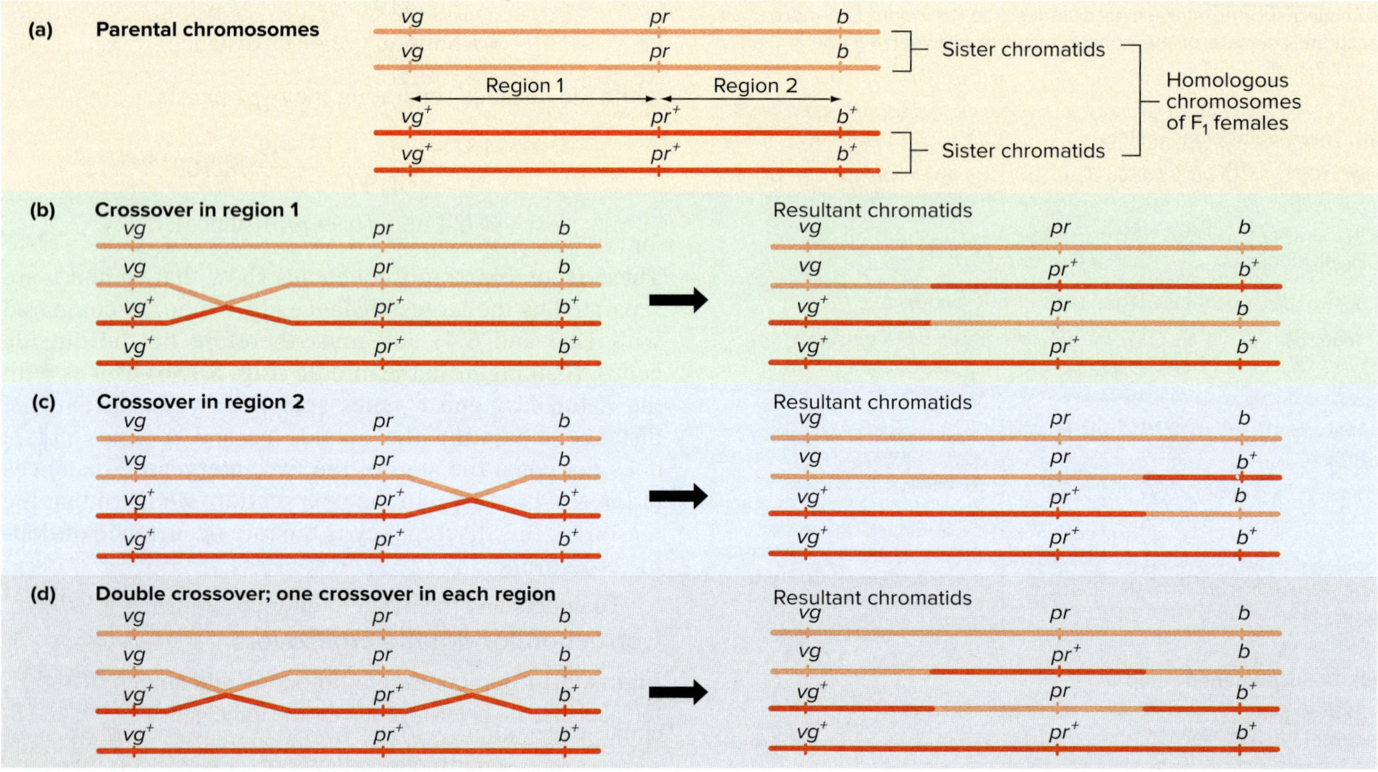

outside genes *vg* and *b,* even though not one but two exchanges must have occurred between them.

Because of the existence of double crossovers, the *vg* ↔ *b* distance of 17.7 m.u. calculated in the previous section does not reflect all of the recombination events producing the gametes that gave rise to the observed progeny. To correct for this oversight, it is necessary to adjust the recombination frequency by adding the double crossovers twice, because each individual in the double crossover groups is the result of two exchanges between *vg* and *b.* The corrected distance is

$$\frac{252 + 241 + 131 + 118 + 13 + 13 + 9 + 9}{4197} \times 100$$

$$= 18.7 \text{ m.u.}$$

This value makes sense because you have accounted for all of the crossovers that occur in region 1 as well as all of the crossovers in region 2. As a result, the corrected value of 18.7 m.u. for the distance between *vg* and *b* is now exactly the same as the sum of the distances between *vg* and *pr* (region 1) and between *pr* and *b* (region 2).

As previously discussed, when Sturtevant originally mapped several X-linked genes in *Drosophila* by two-point crosses, the locus of the rudimentary wings (*r*) gene was ambiguous. A two-point cross involving *y* and *r* gave a recombination frequency of 42.9, but the sum of all the intervening distances was 55.0 (review Fig. 5.12). This discrepancy occurred because the two-point cross ignored double crossovers that might have occurred in the large interval between the *y* and *r* genes. The data summing the smaller intervening distances accounted for at least some of these double crossovers by catching recombinations of gene pairs between *y* and *r.* Moreover, each smaller distance is less likely to encompass a double crossover than a larger distance, so each number for a smaller distance is inherently more accurate.

Note that even a three-point cross like the one for *vg, pr,* and *b* ignores the possibility of two recombination events taking place in, say, region 1. For greatest accuracy, it is always best to construct a map using many genes separated by relatively short distances.

Interference: Fewer double crossovers than expected

In a three-point cross following three linked genes, of the eight possible genotypic classes, the two parental classes contain the largest number of progeny, while the two double recombinant classes, resulting from double crossovers, are always the smallest (see Fig. 5.11). We can understand why double-crossover progeny are the rarest by looking at the probability of their occurrence. If an exchange in region 1 of a chromosome does not affect the probability of an exchange in region 2, the probability that both will occur simultaneously is the product of their separate probabilities (recall the product rule in Chapter 2). For example, if progeny resulting from recombination in region 1 alone account for 10% of the total progeny (that is, if region 1 is 10 m.u.) and progeny resulting from recombination in region 2 alone account for 20%, then the probability of a double crossover (one event in region 1, the second in region 2) is $0.10 \times 0.20 = 0.02$, or 2%. This makes sense because the likelihood of two rare events occurring simultaneously is even smaller than that of either rare event occurring alone.

If eight classes of progeny are obtained in a three-point cross, the two classes containing the fewest progeny must have arisen from double crossovers. The numerical frequencies of observed double crossovers, however, almost never coincide with expectations derived from the product rule. Let's look at the actual numbers from the cross we have been discussing. The probability of a single crossover between *vg* and *pr* is 0.123 (corresponding to 12.3 m.u.), and the probability of a single crossover between *pr* and *b* is 0.064 (6.4 m.u.). The product of these probabilities is

$$0.123 \times 0.064 = 0.0079 = 0.79\%.$$

But the observed proportion of double crossovers (see Fig. 5.11) was

$$\frac{13 + 9}{4197} \times 100 = 0.52\%.$$

The fact that the number of observed double crossovers is less than the number expected if the two exchanges are independent events suggests that the occurrence of one crossover reduces the likelihood that another crossover will occur in an adjacent part of the chromosome. This phenomenon—of crossovers not occurring independently—is called **chromosomal interference.**

As was shown in Fig. 5.7, interference likely exists to ensure that every pair of homologous chromosomes undergoes at least one crossover. It is crucial that every pair of homologous chromosomes sustain one or more crossovers because such events help the chromosomes orient properly at the metaphase plate during the first meiotic division. Indeed, homologous chromosome pairs without crossovers often segregate improperly. If only a limited number of crossovers can occur during each meiosis, and interference lowers the number of crossovers on large chromosomes, then the remaining possible crossovers are more likely to occur on small chromosomes. This increases the probability that at least one crossover will take place on every homologous pair. Though the molecular mechanism underlying interference is not yet clear, recent experiments suggest that interference is mediated by the synaptonemal complex.

Interference is not uniform and may vary even for different regions of the same chromosome. Investigators can obtain a quantitative measure of the amount of interference in different chromosomal intervals by first calculating a **coefficient of coincidence,** defined as the ratio between the actual frequency of double crossovers observed in an experiment and the number of double crossovers expected on the basis of independent probabilities.

$$\text{coefficient of coincidence} = \frac{\text{frequency observed}}{\text{frequency expected}}$$

For the three-point cross involving *vg, pr,* and *b,* the coefficient of coincidence is

$$\frac{0.52}{0.79} = 0.66.$$

The definition of interference itself is

$$\text{Interference} = 1 - \text{coefficient of coincidence}.$$

In this case, the interference is

$$1 - 0.66 = 0.34.$$

To understand the meaning of interference, it is helpful to contrast what happens when there is no interference with what happens when interference is complete. If interference is 0, the frequency of observed double crossovers equals expectations, and crossovers in adjacent regions of a chromosome occur independently of each other. If interference is complete (that is, if interference = 1), no double crossovers occur in the experimental progeny because one exchange effectively prevents another. As an example, in a particular three-point cross in mice, the recombination frequency for the pair of genes on the left (region 1) is 20, and for the pair of genes on the right (region 2), it is also 20. Without interference, the expected rate of double crossovers in this chromosomal interval is

$$0.20 \times 0.20 = 0.04\text{ , or }4\%,$$

but when investigators observed 1000 progeny of this cross, they found 0 double recombinants instead of the expected 40.

A method to determine the gene in the middle

The smallest of the eight possible classes of progeny in a three-point cross are the two that contain double recombinants generated by double crossovers. It is possible to use the composition of alleles in these double crossover classes to determine which of the three genes lies in the middle, even without calculating any recombination frequencies.

Consider again the progeny of a three-point testcross looking at the *vg, pr,* and *b* genes. The F_1 females are *vg pr b* / vg^+ pr^+ b^+. As Fig. 5.14d demonstrated, testcross progeny resulting from double crossovers in the trihybrid females of the F_1 generation received gametes from their mothers carrying the allelic combinations *vg* pr^+ *b* and vg^+

pr b^+. In these individuals, the alleles of the *vg* and *b* genes retain their parental associations (*vg b* and vg^+ b^+), while the *pr* gene has recombined with respect to both the other genes (*pr* b^+ and pr^+ *b*, *vg* pr^+ and vg^+ *pr*). The same is true in all three-point crosses: In those gametes formed by double crossovers, the gene whose alleles have recombined relative to the parental configurations of the other two genes must be the one in the middle.

Three-Point Crosses: A Comprehensive Example

The technique of looking at double recombinants to discover which gene has recombined with respect to both other genes allows immediate clarification of gene order even in otherwise difficult cases. Consider the three X-linked genes *y, w,* and *m* that Sturtevant located in his original mapping experiment (see Fig. 5.12). Because the distance between *y* and *m* (34.3 m.u.) appeared slightly larger than the distance separating *w* and *m* (33.8 m.u.), he concluded that *w* was the gene in the middle. But because of the small difference between the two numbers, his conclusion was subject to questions of statistical significance. If, however, we look at a three-point cross following *y, w,* and *m,* these questions disappear.

Figure 5.15 tabulates the classes and numbers of male progeny arising from females heterozygous for the *y, w,* and *m* genes. Because these male progeny receive their only X chromosome from their mothers, their phenotypes indicate directly the gametes produced by the heterozygous females. In each row of the figure's table, the genes appear in an arbitrary order that does not presuppose knowledge of the actual map. As you can see, the two classes of progeny listed at the top of the table outnumber the remaining six classes, which indicates that all three genes are linked to each other. Moreover, these largest groups, which are the parental classes, show that the two X chromosomes of the heterozygous females were w^+ y^+ *m* and *w y* m^+.

Figure 5.15 How three-point crosses verify Sturtevant's map of the *Drosophila* X chromosome. The parental classes correspond to the two X chromosomes in the F_1 female. The genotype of the double recombinant classes shows that *w* must be the gene in the middle.

F_1 ♀ $w^+ w\, y^+ y\, m^+ m$ × ♂ X / Y

Before data analysis, you do not know the gene order or allele combination on each chromosome.

Male progeny	2278	$w^+ y^+ m$ / Y	Parental class (noncrossover)
	2157	$w\ y\ m^+$ / Y	
	1203	$w\ y\ m$ / Y	Crossover in region 2 (between *w* and *m*)
	1092	$w^+ y^+ m^+$ / Y	
	49	$w^+ y\ m$ / Y	Crossover in region 1 (between *y* and *w*)
	41	$w\ y^+ m^+$ / Y	
	2	$w^+ y\ m^+$ / Y	Double crossovers
	1	$w\ y^+ m$ / Y	
	6823		

After data analysis, you can conclude that the gene order and allele combinations on the X chromosomes of the F_1 females were $y\,w\,m^+ / y^+ w^+ m$.

Among the male progeny in Fig. 5.15, the two smallest classes, representing the double crossovers, have X chromosomes carrying w^+ *y* m^+ and *w* y^+ *m* combinations, in which the *w* alleles are recombined relative to those of *y* and *m*. The *w* gene must therefore lie between *y* and *m*, verifying Sturtevant's original assessment.

To complete a map based on the *w y m* three-point cross, you can calculate the interval between *y* and *w* (region 1)

$$\frac{49 + 41 + 1 + 2}{6823} \times 100 = 1.3 \text{ m.u.}$$

as well as the interval between *w* and *m* (region 2)

$$\frac{1203 + 1092 + 2 + 1}{6823} \times 100 = 33.7 \text{ m.u.}$$

The genetic distance separating *y* and *m* is the sum of

$$1.3 + 33.7 = 35.0 \text{ m.u.}$$

Note that you could also calculate the distance between *y* and *m* directly by including double crossovers twice, to account for the total number of recombination events detected between these two genes.

$$\text{RF} = (1203 + 1092 + 49 + 41 + 2 + 2 + 1 + 1)/6823 \times 100 = 35.0 \text{ m.u.}$$

This method yields the same value as the sum of the two intervening distances (region 1 + region 2).

How Do Genetic Maps Correlate with Physical Reality?

Many types of experiments presented throughout this book show resoundingly that the *order of genes* revealed by recombination mapping always reflects the order of those same genes along the DNA molecule of a chromosome. In contrast, the actual *physical distances between genes*—that is, the amount of DNA separating them—does not always correspond linearly to genetic map distances.

Underestimation of physical distances between genes by recombination frequency

You have already seen that DCOs between two genes may go undetected in a testcross experiment, resulting in undercounting of the number of crossovers between the two genes, and therefore an underestimation of the distance between them. This is not much of a problem when two genes are close enough together that DCOs take place infrequently. However, as the distance between two genes increases, double and multiple crossovers occur often enough to affect the relationship between recombination frequency and map distance. This relationship cannot be linear because, as we have already seen, the RF for a two-point cross cannot exceed 50% regardless of how far apart two genes are on the same chromosome.

A second look at Fig. 5.11 makes it easy to see how DCOs result in underestimation of gene distances by RF. When genes *A* and *B* are close together, most meioses are NCOs, and the occasional meiosis is an SCO. Each SCO produces exactly two recombinant gametes, and so a perfect linear correspondence exists between the number of crossovers and the number of recombinant gametes (1 crossover: 2 recombinants). However, when genes *A* and *B* are farther apart, DCOs occur. Only one of the four equally frequent DCOs (4-strand) preserves the linear relationship between crossovers and recombinant gametes: Two crossovers occur in a 4-strand DCO and four recombinant gametes (2 crossovers: 4 recombinants). In contrast, the other three types of DCOs result in fewer than four recombinant gametes.

Figure 5.16 illustrates this discrepancy between the actual number of crossovers (*green* line) and observed RF (*purple* line) as a function of the amount of DNA separating the two genes. As you can see, the two graphs are nearly identical for distances of 5 m.u. or less. At genetic distances this small, the RF seen in a two-point cross is an accurate measure of the physical distance. The two graphs then diverge increasingly from each other, so the RF becomes a less precise estimate at genetic distances greater than 5 m.u.

Geneticists have developed mathematical equations called *mapping functions* to compensate for the inaccuracies inherent in relating recombination frequencies to physical distances. However, the corrections for large distances are at best imprecise because mapping functions are based on simplifying assumptions that are not completely true. Thus, the best way to create an accurate map is by summing many smaller intervals, locating widely separated genes through linkage to common intermediaries. Maps are subject to continual refinement as more and more newly discovered genes are included.

Figure 5.16 Recombination frequency underestimates crossover frequency. The lines representing the RF observed in a testcross (*purple*) and the crossover frequency (*green*) are nearly coincident when genes *A* and *B* are close together. As two genes become farther apart, the RF value increasingly underestimates the actual crossover frequency; eventually no NCOs will take place and the maximum RF of 50% will be observed. The *green* line represents crossover frequency/2 because every SCO produces 1/2 parental and 1/2 recombinant chromosomes.

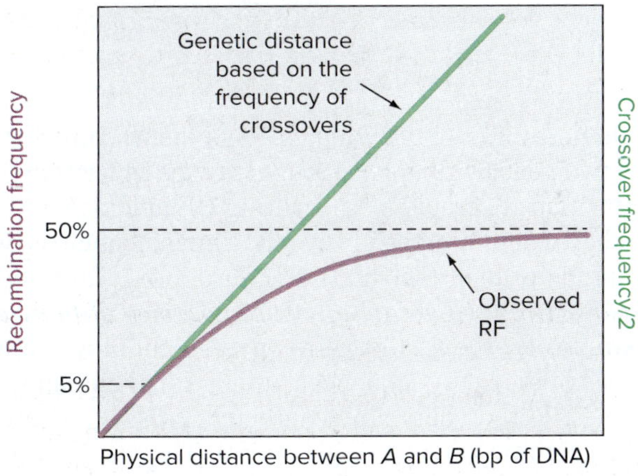

Nonuniform crossover frequencies

Although we have been assuming thus far that crossovers are just as likely to occur between any two base pairs along a chromosome, recombination is not in fact random. In human DNA, for example, most crossovers take place in so-called **recombination hotspots**—small regions of DNA where the frequency of recombination is much higher than average. As shown in **Fig. 5.17,** genes with hotspots between them (*A* and *B*) will be more distant from each other on a genetic map (measured in m.u.) than another gene pair (*B* and *C*) without a hotspot between them, even though the physical distance (measured in bp of DNA) separating each gene pair is the same. As hotspots are relatively frequent (a hotspot appears about every 50,000 bp in human chromosomes), recombination frequency is nevertheless a reasonable estimation of physical distance between most genes.

Frequencies of recombination may also differ from species to species. We know this because recent elucidation of the complete DNA sequences of several organisms' genomes has allowed investigators to compare the physical distances between genes with genetic map distances. They found that in humans, a map unit corresponds on average to

Figure 5.17 Recombination hotspots. Genes *A* and *B* are separated by the same number of base pairs as genes *B* and *C*. Because *A* and *B* flank a recombination hotspot, they appear much more distant from each other than do genes *B* and *C* on a genetic map.

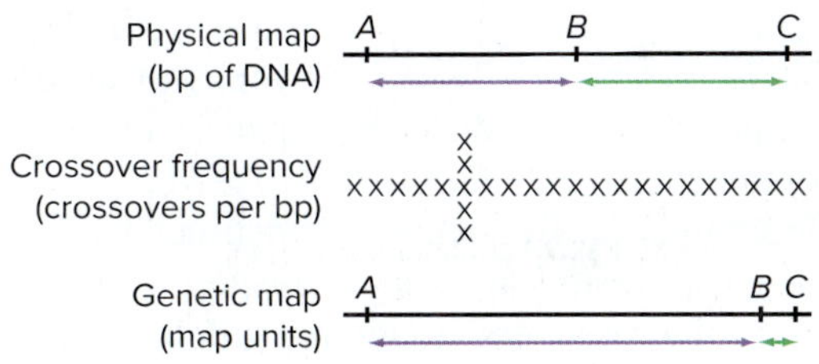

FAST FORWARD

Mapping the Crossovers that Generate the Chromosomes of Individual Human Sperm

Using DNA analysis technologies that will be described in Chapters 9 through 11, scientists can now examine the base pair sequences of the whole genomes of single sperm. In one such study, by comparing the DNA sequences in each of the pairs of homologous chromosomes in a man's somatic cells with those of individual sperm the same man produces, researchers could locate specific recombination events that occurred in the man's primary spermatocytes.

The homologous chromosomes any person inherits from his or her father or mother differ in about 1 out of every 1000 base pairs. The base pair differences in different genomes are called *SNPs* (pronounced *snips;* for *single nucleotide polymorphisms*). From comparisons of the genome sequences of many individuals, approximately 50,000,000 locations in the genome have been identified where SNPs commonly can occur. The different base pair sequences of SNPs are considered different alleles of the SNP locus (Fig. A). Researchers can zero in on SNP loci and determine which alleles of millions of SNPs are present in a genome.

In order to map recombination sites, first the scientists developed new technology to isolate individual chromosomes from diploid somatic cells. Once isolated, the SNP alleles of individual homologs could be determined (**Fig. A**).

Figure A DNA sequences of homologs reveal SNP loci. At a particular SNP locus, maternal (*M*) and paternal (*P*) homologs can have different alleles (for example, an A–T base pair or a G–C base pair).

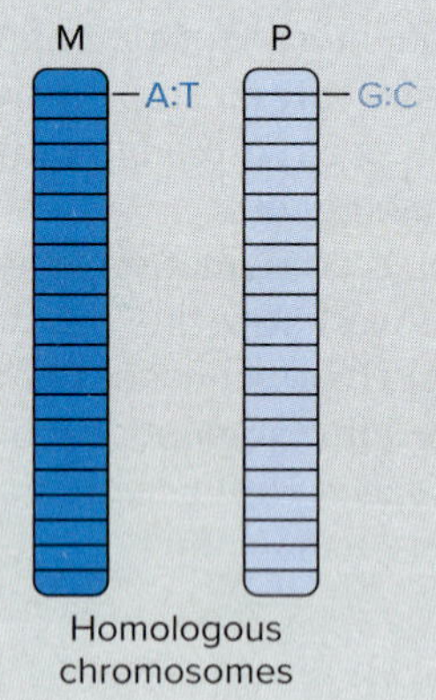

Next, the researchers determined which SNP alleles were present in individual sperm genomes. Then, by comparing the SNP alleles present on each chromosome in an individual sperm with the corresponding SNP alleles on each homolog of the man's somatic cells (prior to crossing over during meiosis), the locations of crossovers were revealed (**Fig. B**). By analyzing the crossovers in 91 different sperm, the researchers found that about 1 crossover per chromosome took place in each gamete, and crossover hotspots were detected.

The information obtained from this study and others like it is useful to scientists studying the biochemistry of recombination. In addition, you will see in later chapters of the book that the ability to determine the base pair sequence of individual chromosomes and individual gamete genomes has widespread applications in the study of mutation and human evolution.

Figure B Crossover map of a single sperm's autosomes. The autosomes (chromosomes *1–22*) of a sperm are depicted, where the *dark blue* and *light blue* regions correspond to the different homologs in the man's somatic cells (see Fig. A). Most chromosomes are the products of a single crossover; in this example, chromosomes 2 and 10 are exceptions.

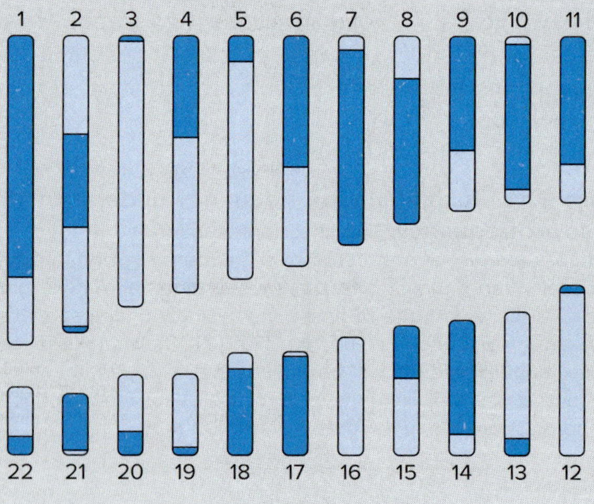

about 1 million base pairs. In yeast, however, where the frequency of recombination per length of DNA is much higher than in humans, one map unit is approximately 2500 base pairs. Thus, although map units are useful for estimating relative distances between the genes of an organism, 1% RF can reflect very different expanses of DNA in different organisms.

Recombination frequencies sometimes vary even between the two sexes of a single species. In humans, the frequency of crossovers is about twofold higher in the female germ line than in males. This fact means that the same two genes will appear roughly twice as far apart on a genetic map generated by measuring RF in female meiosis than they would if crossing-over during male meiosis were measured instead. The Fast Forward Box *Mapping the Crossovers that Generate the Chromosomes of Individual Human Sperm* explains how new technology allows analysis of the DNA sequences of individual human sperm genomes. Researchers now can detect crossovers directly in each chromosome of single sperm. The results of these analyses reveal that most chromosomes in human sperm have undergone only a single crossover.

Drosophila provides an extreme example: No recombination occurs during meiosis in males. If you review the examples already discussed in this chapter, you will discover that they all measure recombination among the progeny of doubly heterozygous *Drosophila* females. Problem 19 at the end of this chapter shows how geneticists can exploit the absence of recombination in *Drosophila* males to establish rapidly that genes far apart on the same chromosome are indeed syntenic.

Multiple-Factor Crosses Help Establish Linkage Groups

Genes chained together by linkage relationships are known collectively as a **linkage group.** When enough genes have been assigned to a particular chromosome, the terms *chromosome* and *linkage group* become synonymous. If you can demonstrate that gene *A* is linked to gene *B*, *B* to *C*, *C* to *D*, and *D* to *E*, you can conclude that all these genes are syntenic. When the genetic map of a genome becomes so dense that it is possible to show that any gene on a chromosome is linked to another gene on the same chromosome, the number of linkage groups equals the number of pairs of homologous chromosomes in the species. Humans have 23 linkage groups, mice have 20, and fruit flies have 4 (**Fig. 5.18**).

The total genetic distance along a chromosome, which is obtained by adding many short distances between genes, may be much more than 50 m.u. For example, the two long *Drosophila* autosomes are both slightly more than 100 m.u. in length (Fig. 5.18), while the longest human chromosome is approximately 270 m.u. Recall, however, that even with the longest chromosomes, *pairwise* crosses between genes

Figure 5.18 ***Drosophila melanogaster*** **has four linkage groups.** A genetic map of the fruit fly, showing the position of many genes affecting body morphology, including those used as examples in this chapter (*highlighted in bold*). Because so many *Drosophila* genes have been mapped, each of the four chromosomes can be represented as a single linkage group.

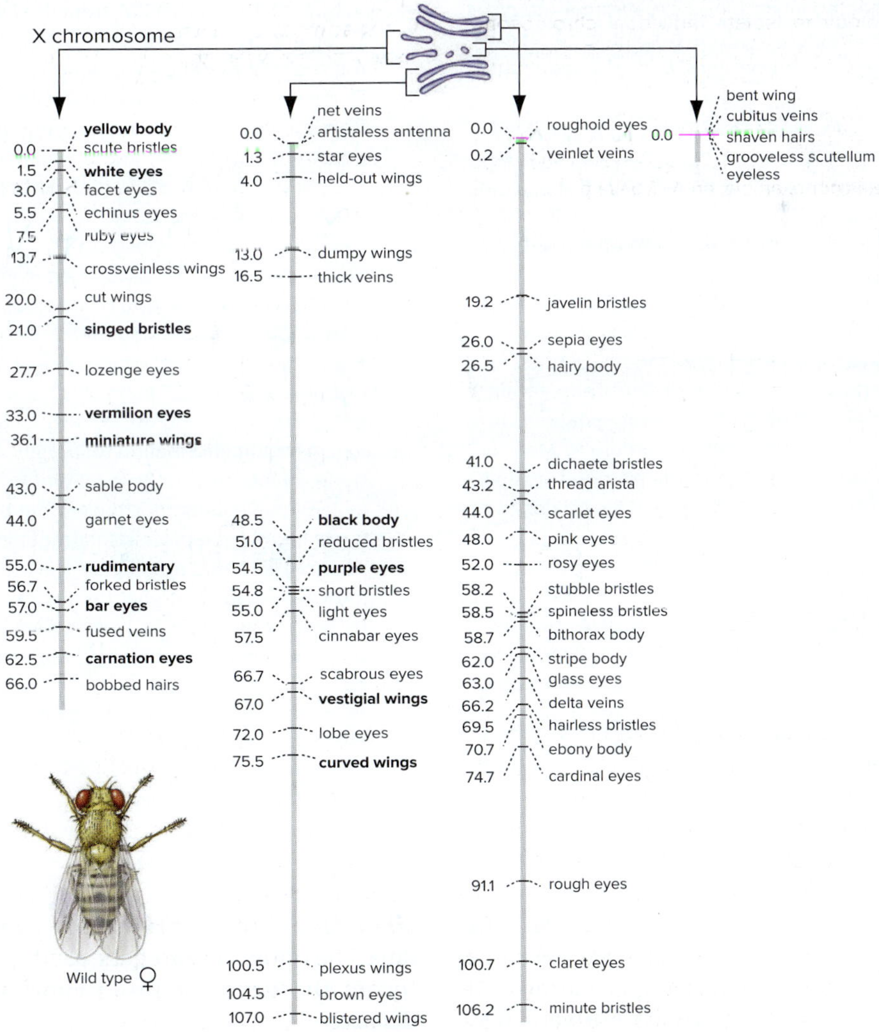

FAST FORWARD

GENE MAPPING MAY LEAD TO A CURE FOR CYSTIC FIBROSIS

For 40 years after the symptoms of cystic fibrosis were first described in 1938, no molecular clue—no visible chromosomal abnormality transmitted with the disease, no identifiable protein defect carried by affected individuals—suggested the genetic cause of the disorder. As a result, no effective treatment existed for the 1 in 2000 Caucasian Americans born with the disease, most of whom died before they were 30. In the 1980s, however, geneticists were able to combine recently invented techniques for looking directly at DNA with maps constructed by linkage analysis to pinpoint a precise chromosomal position, or *locus,* for the cystic fibrosis gene.

The mappers of the cystic fibrosis gene faced an overwhelming task. They were searching for a gene that encoded an unknown protein, a gene that had not yet even been assigned to a chromosome. It could lie anywhere among the 23 pairs of chromosomes in a human cell.

- A review of many family pedigrees confirmed that cystic fibrosis is most likely determined by a single gene (*CF*). Investigators collected white blood cells from 47 families with two or more affected children, obtaining genetic data from 106 patients, 94 parents, and 44 unaffected siblings.
- The researchers next tried to discover if any other trait is reliably transmitted with cystic fibrosis. Analyses of the easily obtainable serum enzyme paroxonase showed that its gene (*PON*) is indeed linked to *CF*. At first, this knowledge was not that helpful, because *PON* had not yet been assigned to a chromosome.
- Then, in the early 1980s, geneticists developed a large series of DNA markers, based on new techniques that enabled them to recognize variations in the genetic material. A **DNA marker** is a segment of DNA representing a specific locus that comes in identifiable variations. These allelic variations segregate according to Mendel's laws, which means it is possible to follow their transmission as you would any gene's. Chapter 11 explains the discovery and use of DNA markers in greater detail; for now, it is only important to know that they exist and can be identified.
- By 1986, linkage analyses of hundreds of DNA markers had shown that one marker called *D7S15,* and known to reside on the long arm of chromosome 7, is linked with both *PON* and *CF*. Researchers computed recombination frequencies and found that the distance from the DNA marker to *CF* was 15 cM; from the DNA marker to *PON,* 5 cM; and from *PON* to *CF,* 10 cM. They concluded that the order of the three loci was *D7S15-PON-CF* (**Fig. A**). Because *CF* could lie 15 cM in either of two directions from the DNA marker, the area under investigation was approximately 30 cM. And because the human genome consists of roughly 3000 cM, this step of linkage analysis narrowed the search to 1% of the human genome, in a small region of chromosome 7.
- Finally, investigators discovered linkage with several other markers on the long arm of chromosome 7, called *J3.11,* *βTR,* and *met*. Two of the markers turned out to be separated from *CF* by a distance of only 1 cM. It now became possible to place *CF* in band 31 of chromosome 7's long arm (band 7q31, Fig. A). By 1989, researchers had used this mapping information to identify and clone the *CF* gene on the basis of its location.
- In 1992, investigators showed that the *CF* gene specifies a cell membrane protein that regulates the flow of chloride ions into and out of cells (review Fig. 2.25). This knowledge has become the basis of new drug therapies to open up ion flow, as well as gene therapies to introduce normal copies of the *CF* gene into the cells of CF patients. Although only in the early stages of development, such gene therapy holds out hope of an eventual cure for cystic fibrosis.

Figure A **How molecular markers helped locate the gene for cystic fibrosis (*CF*).**

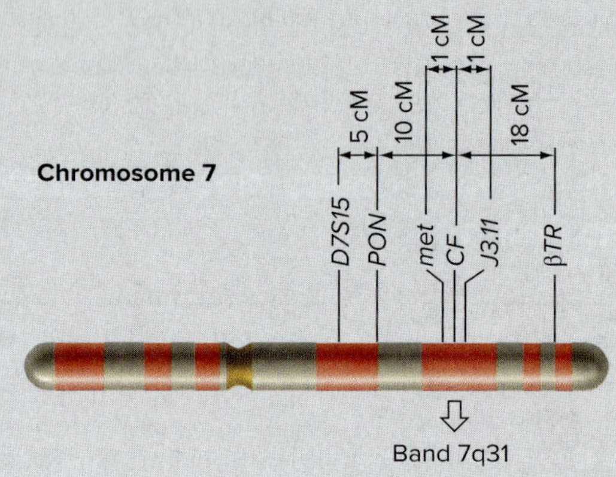

located at the two ends will not produce more than 50% recombinant progeny.

Linkage mapping has practical applications of great importance. For example, the Fast Forward Box *Gene Mapping May Lead to a Cure for Cystic Fibrosis* describes how researchers used linkage information to locate the gene for this important human hereditary disease.

essential concepts

- A series of *two-point crosses* can establish the order of linked genes and the distances between them through pairwise analysis of recombination frequencies.
- *Three-point testcrosses* can refine map distances and reveal the existence of crossover *interference,* a phenomenon that distributes among all chromosomes the limited number of crossovers that occur in each meiosis.
- Although *genetic maps* provide an accurate picture of gene order on a chromosome, the distances measured between genes can be misleading.
- Genes in a linkage group are by definition *syntenic.* With enough mapped genes, the entire chromosome becomes a single *linkage group.*

5.4 The Chi-Square Test and Linkage Analysis

learning objectives

1. Explain the purpose of the chi-square test.
2. Discuss the concept of the null hypothesis and its use in data analysis.
3. Evaluate the significance of experimental data based on the chi-square test.

How do you know from a particular experiment whether two genes assort independently or are genetically linked? At first glance, this question should pose no problem. Discriminating between the two possibilities involves straightforward calculations based on assumptions well supported by observations. For independently assorting genes, a dihybrid F_1 female produces four types of gametes in equal numbers, so one-half of the F_2 progeny are of the parental classes and the other half are of the recombinant classes. In contrast, for linked genes, the two types of parental classes by definition always outnumber the two types of recombinant classes in the F_2 generation.

The problem is that because real-world genetic transmission is based on chance events, in a particular study even unlinked, independently assorting genes can produce deviations from the 1:1:1:1 ratio, just as in 10 tosses of a coin, you may easily get 6 heads and 4 tails (rather than the predicted 5 and 5). Thus, if a breeding experiment analyzing the transmission of two genes shows a deviation from the equal ratios of parentals and recombinants expected of independent assortment, can we necessarily conclude the two genes are linked? Is it instead possible that the results represent a statistically acceptable chance fluctuation from the mean values expected of unlinked genes that assort independently? Such questions become more important in cases where linkage is not all that tight, so that even though the genes are linked, the percentage of recombinant classes approaches 50%.

The Chi-Square Test Evaluates the Significance of Differences Between Predicted and Observed Values

To answer these kinds of questions, statisticians have devised several different ways to quantify the likelihood that an experimentally observed deviation from the predictions of a particular hypothesis could have occurred solely by chance. One of these probabilistic methods is known as the **chi-square test** for *goodness of fit.* This test measures how well observed results conform to predicted ones, and it is designed to account for the fact that the size of an experimental population (the *sample size*) is an important component of statistical significance. To appreciate the role of sample size, let's return to the proverbial coin toss before examining the details of the chi-square test.

In 10 tosses of a coin, an outcome of 6 heads (60%) and 4 tails (40%) is not unexpected because of the effects of chance. However, with 1000 tosses of the coin, a result of 600 heads (60%) and 400 tails (40%) would intuitively be highly unlikely. In the first case, a change in the results of one coin toss would alter the expected 5:5 ratio to the observed 6:4 ratio. In the second case, 100 tosses would have to change from tails to heads to generate the stated deviation from the predicted 500:500 ratio. Chance events could reasonably, and even likely, cause one deviation from the predicted number, but not 100.

Two important concepts emerge from this simple example. First, a comparison of percentages or ratios alone will never allow you to determine whether or not *observed* data are significantly different from *predicted* values. Second, the absolute numbers obtained are important because they reflect the size of the experiment. The larger the sample size, the closer the observed percentages can be expected to match the values predicted by the experimental hypothesis, *if the hypothesis is correct.* The chi-square test is therefore always calculated with numbers—actual data—and not percentages or proportions.

The chi-square test cannot prove a hypothesis, but it can allow researchers to reject a hypothesis. For this reason, a crucial prerequisite of the chi-square test is the framing of a **null hypothesis:** a model that might possibly be refuted by the test and that leads to clear-cut numerical predictions. Although contemporary geneticists use the chi-square test to interpret many kinds of genetic experiments, they use it most often to discover whether data obtained from breeding experiments provide evidence for or against the hypothesis that two genes are linked. But one problem with the general hypothesis that *genes A and B are linked* is that no precise prediction exists for what to expect in terms of breeding data. The reason is that the frequency of recombinants, as we have seen, varies with each linked gene pair.

In contrast, the alternative hypothesis that *genes A and B are not linked* gives rise to a precise prediction: that alleles of different genes will assort independently and produce a 1:1:1:1 ratio of progeny types in a testcross. So, whenever a geneticist wants to determine whether two genes are linked, he or she actually tests whether the observed data are consistent with the null hypothesis of no linkage. If the chi-square test shows that the observed data differ significantly from those expected with independent assortment—that is, they differ enough not to be reasonably attributable to chance alone—then the researcher can reject the null hypothesis of no linkage and accept the alternative of linkage between the two genes.

The Tools of Genetics Box entitled *The Chi-Square Test for Goodness of Fit* presents the general protocol for this analysis. The final result of the calculations is the determination of the numerical probability—the ***p* value**—that a particular set of observed experimental results represents a chance deviation from the values predicted by a particular hypothesis. If the probability is high, it is likely that the hypothesis being tested explains the data, and the observed deviation from expected results is considered *insignificant.* If the probability is low, the observed deviation from expected results becomes *significant.* When this happens, it is unlikely that the hypothesis under consideration explains the data, and the hypothesis can be rejected.

It's important that you understand why the null hypothesis of no linkage (RF = 50%) is used, as opposed to a null hypothesis that assumes a particular degree of linkage (a particular RF < 50%). As stated earlier, a chi-square test can allow you to reject a null hypothesis, but not to prove it. This fact explains why geneticists test the null hypothesis that RF = 50 rather than a null hypothesis that the RF equals some specific number below 50, say 38, even though both models provide specific numerical predictions. If the deviations of the experimental values are *insignificant* (a high *p* value) relative to the hypothesis being tested, then the results could be consistent with that model, but they also could potentially be consistent with the other, untested hypothesis (RF = 50%) as well. Insignificant results are therefore not helpful. Suppose now that the deviations of the experimental values from the predictions of RF = 38 are *significant* (low *p* value), so that you could reject that hypothesis. This information would be similarly useless because you would not learn anything about the relative positions of the two genes other than that the map distance is not 38 m.u. Only one outcome is of real value: If you can reject the null hypothesis that the two genes are not linked (RF = 50%), then you have learned that they must be syntenic and are close enough together to be genetically linked.

Figure 5.19 Applying the chi-square test to determine if two genes are linked. The null hypothesis is that the two genes are unlinked. For Experiment 1, $p > 0.05$, so it is not possible to reject the null hypothesis. For Experiment 2, with a data set twice the size, $p < 0.05$, so most geneticists would reject the null hypothesis and conclude with greater than 95% confidence that the genes are linked.

Progeny Classes		Experiment 1			Experiment 2		
		O	E	$(O-E)^2/E$	O	E	$(O-E)^2/E$
Parentals	*A B*	**18**	12	36/12	**36**	24	144/24
	a b	**14**	12	4/12	**28**	24	16/24
Recombinants	*A b*	**7**	12	25/12	**14**	24	100/24
	a B	**9**	12	9/12	**18**	24	36/24
Total		**48**	48	74/12	**96**	96	269/24
				$\chi^2 = 6.17$			$\chi^2 = 12.3$
df = 3				$p > 0.10$			$p < 0.01$

Applying the Chi-Square Test to Linkage Analysis: An Example

Figure 5.19 depicts how the chi-square test could be applied to two sets of data obtained from testcross experiments asking whether genes *A* and *B* are linked. The columns labeled *O* for *observed* contain the actual data—the number of each of the four progeny types—from each experiment. In the first experiment, the total number of offspring is 48, so the *expected value* (*E*) for each progeny class, given the null hypothesis of no linkage, is simply 48/4 = 12. Now, for each progeny class, you square the deviation of the observed from the expected value, and divide the result by the expected value. Those calculations are presented in the column $(O\text{-}E)^2/E$. All four quotients are summed to obtain the value of chi square (χ^2). In experiment 1, $\chi^2 = 6.17$.

You next determine the **degrees of freedom (df)** for this experiment. Degrees of freedom is a mathematical concept that takes into consideration the number of independently varying parameters. In this example, the offspring fall into four classes. For three of the classes, the number of offspring can have any value, as long as their sum is no more than 48. However, once three of these values are fixed, the fourth value is also fixed, as the total in all four classes must equal 48. Therefore, the df with four classes is one less than the number of classes, or three. Next, you scan the chi-square table (see **Table 5.2**) for $\chi^2 = 6.17$ and df = 3. You find that the corresponding *p* value is greater than 0.10. From any *p* value greater than 0.05, you can conclude that it is not possible to reject the null hypothesis on the basis of this experiment, which means that this data set is not sufficient to demonstrate linkage between *A* and *B*.

If you use the same strategy to calculate a *p* value for the data observed in the second experiment, where there are a

TOOLS OF GENETICS

Blue DNA: © MedicalRF.com

The Chi-Square Test for Goodness of Fit

The general protocol for using the chi-square test for goodness of fit and evaluating its results can be stated in a series of steps. Two preparatory steps precede the actual chi-square calculation.

1. Use the data obtained from a breeding experiment to answer the following questions:
   a. What is the total number of offspring (events) analyzed?
   b. How many different classes of offspring (events) are present?
   c. In each class, what is the number of offspring (events) observed?
2. Calculate how many offspring (events) would be expected for each class if the null hypothesis (here, no linkage) were correct. To do so, multiply the fraction predicted by the null hypothesis (here, 1/4 of each possible progeny type) by the total number of offspring. You are now ready for the chi-square calculation.
3. To calculate chi square, begin with one class of offspring. Subtract the expected number from the observed number to obtain the deviation from the predicted value for the class. Square the result and divide this value by the expected number.
   Do this procedure for all classes and then sum the individual results. The final result is the chi-square (χ^2) value. This step is summarized by the equation:

   $$\chi^2 = \Sigma \frac{(\textit{Number observed} - \textit{Number expected})^2}{\textit{Number expected}}$$

   where Σ means *sum of all classes.*
4. Next, you consider the degrees of freedom (df). The df is a measure of the number of independently varying parameters in the experiment (see text). The value of degrees of freedom is one less than the number of classes. Thus, if N is the number of classes, then the degrees of freedom (df) = $N - 1$. If there are four classes, then there are 3 df.
5. Use the chi-square value together with the df to determine a p value: the probability that a deviation from the predicted numbers at least as large as that observed in the experiment would occur by chance. Although the p value is arrived at through a numerical analysis, geneticists routinely determine the value by a quick search through a table of critical χ^2 values for different degrees of freedom, such as Table 5.2.
6. Evaluate the significance of the p value. You can think of the p value as the probability that the null hypothesis is true. A value greater than 0.05 indicates that in more than 1 in 20 (or more than 5%) repetitions of an experiment of the same size, the observed deviation from predicted values could have been obtained by chance, even if the null hypothesis is actually true; the data are therefore *not significant* for rejecting the null hypothesis. Statisticians have arbitrarily selected the 0.05 p value as the boundary between rejecting or not rejecting the null hypothesis.
   A p value of less than 0.05 means that you can consider the deviation to be *significant,* and you can reject the null hypothesis.

TABLE 5.2 Critical Chi-Square Values

Degrees of Freedom	*p* Values						
	Cannot Reject the Null Hypothesis				Null Hypothesis Rejected		
	0.99	**0.90**	**0.50**	**0.10**	**0.05**	**0.01**	**0.001**
	χ^2 Values						
1	—	0.02	0.45	2.71	3.84	6.64	10.83
2	0.02	0.21	1.39	4.61	5.99	9.21	13.82
3	0.11	0.58	2.37	6.25	7.81	11.35	16.27
4	0.30	1.06	3.36	7.78	9.49	13.28	18.47
5	0.55	1.61	4.35	9.24	11.07	15.09	20.52

Note: χ^2 values that lie in the *yellow* region of this table allow you to reject the null hypothesis with > 95% confidence, and for recombination experiments, to postulate linkage.

total of 96 offspring, you find a p value less than 0.01 (Fig. 5.19). In this case, you can consider the difference between the observed and expected values to be significant. As a result, you can reject the null hypothesis of independent assortment and conclude it is likely that genes A and B are linked.

Statisticians have arbitrarily selected a p value of 0.05 as the boundary between significance and nonsignificance. Values lower than this indicate there would be fewer than 5 chances in 100 of obtaining the same results by random sampling if the null hypothesis were true. A p value of less than 0.05 thus suggests that the data show major deviations from predicted values significant enough to reject the null hypothesis with greater than 95% confidence. More conservative scientists often set the boundary of significance at $p = 0.01$, and they would therefore reject the null hypothesis only if their confidence was greater than 99%.

In contrast, p values greater than 0.01 or 0.05 do not necessarily mean that two genes are unlinked; it may mean only that the sample size is not large enough to provide an answer. With more data, the p value normally rises if the null hypothesis of no linkage is correct and falls if there is, in fact, linkage.

Note that in Fig. 5.19 all of the numbers in the second set of data are simply double the numbers in the first set, with the proportions remaining the same. Thus, by doubling the sample size from 48 to 96 individuals, it was possible to go from no significant difference to a significant difference between the observed and the expected values. In other words, the larger the sample size, the less the likelihood that a certain percentage deviation from expected results happened simply by chance. Bearing this in mind, you can see that it is not appropriate to use the chi-square test when analyzing small samples of less than 10. This issue creates a problem for human geneticists because human families produce only a small number of children. To achieve a reasonable sample size for linkage studies in humans, scientists often pool data from a large number of family pedigrees and use a different statistical analysis called the *Lod score* (see Chapter 11).

We emphasize again that the chi-square test *does not* prove linkage or its absence. What it *does* is provide a quantitative measure of the likelihood that the data from an experiment can be explained by a particular hypothesis. The chi-square analysis is thus a general statistical test for significance; it can be used with many different experimental designs and with hypotheses other than the absence of linkage. As long as it is possible to propose a null hypothesis that leads to a predicted set of values for a defined set of data classes, you can readily determine whether or not the observed data are consistent with the hypothesis.

When experiments lead to rejection of a null hypothesis, you may need to confirm an alternative. For instance, if you are testing whether two opposing traits result from the segregation of two alleles of a single gene, you would expect a testcross between an F_1 heterozygote and a recessive homozygote to produce a 1:1 ratio of the two traits in the offspring. If instead, you observe a ratio of 6:4 and the chi-square test produces a p value of 0.009, you can reject the null hypothesis (a 1:1 ratio). But you are still left with the question of what the absence of a 1:1 ratio means. Two alternatives exist: (1) Individuals with the two possible genotypes are not equally viable, *or* (2) more than one gene affects the trait. The chi-square test cannot tell you which possibility is correct, and you would have to study the matter further. The problems at the end of this chapter illustrate several applications of the chi-square test pertinent to genetics.

essential concepts

- A *null hypothesis* is a model that leads to a discrete numerical prediction.
- The *chi-square test for goodness of fit* helps determine whether two genes are linked by comparing differences between the numbers of progeny of different classes observed in an experiment, and the numbers of progeny of these classes expected from the null hypothesis that genes are unlinked and thus assort independently.
- The *probability value (p)* measures the likelihood that deviations from the predicted values have occurred by chance alone; the null hypothesis is rejected when $p < 0.05$.

5.5 Tetrad Analysis in Fungi

learning objectives

1. Explain the meaning of the term *tetrad* as applied to the asci produced by certain fungi.
2. Differentiate between parental ditype (PD), nonparental ditype (NPD), and tetratype (T).
3. Describe how the relative numbers of PDs and NPDs can be used to establish linkage.
4. Explain how ordered and unordered tetrad analysis can map the positions of genes and (for ordered tetrads) centromeres.

With *Drosophila,* mice, peas, people, and other diploid organisms, each individual represents only one of the four potential gametes generated by each parent in a single meiotic event. Thus, until now, our presentation of linkage, recombination, and mapping has depended on inferences derived from examining the phenotypes of diploid progeny resulting from random unions of meiotic products *en masse* (that is, among a large group). For such diploid organisms, we do not know which, if any, of the parents' other progeny arose from gametes created in the same meiosis. Because of this limitation, the analysis of random products of meiosis in diploid organisms must be based on statistical samplings of large populations.

In contrast, various species of fungi provide a unique opportunity for genetic analysis because they house all four haploid products of each meiosis in a sac called an **ascus** (plural, *asci*). These haploid cells, or **ascospores** (also known as *haplospores* or simply *spores*), can germinate and survive as viable haploid individuals that perpetuate themselves by mitosis. The phenotype of such haploid fungi is a direct representation of their genotype, without complications of dominance.

Figure 5.20 illustrates the life cycles of two fungal species that preserve their meiotic products in a sac. One,

Figure 5.20 The life cycles of the yeast *Saccharomyces cerevisiae* and the bread mold *Neurospora crassa*. Both *S. cerevisiae* and *N. crassa* have two mating types that can fuse to form diploid cells that undergo meiosis. **(a)** Yeast cells can grow vegetatively either as haploids or diploids. The products of meiosis in a diploid cell are four haploid ascospores that are arranged randomly in unordered yeast asci. **(b)** The diploid state in *Neurospora* exists only for a short period. Meiosis in *Neurospora* is followed by mitosis, to give eight haploid ascospores in the ascus. The ordered arrangement of spores in *Neurospora* asci reflects the geometry of the meiotic and mitotic spindles. The photographs showing a budding (mitotically dividing) yeast cell (*top*) and a yeast tetrad (*bottom*) in part (a) are at much higher magnification than the photograph displaying *Neurospora* asci in part (b).

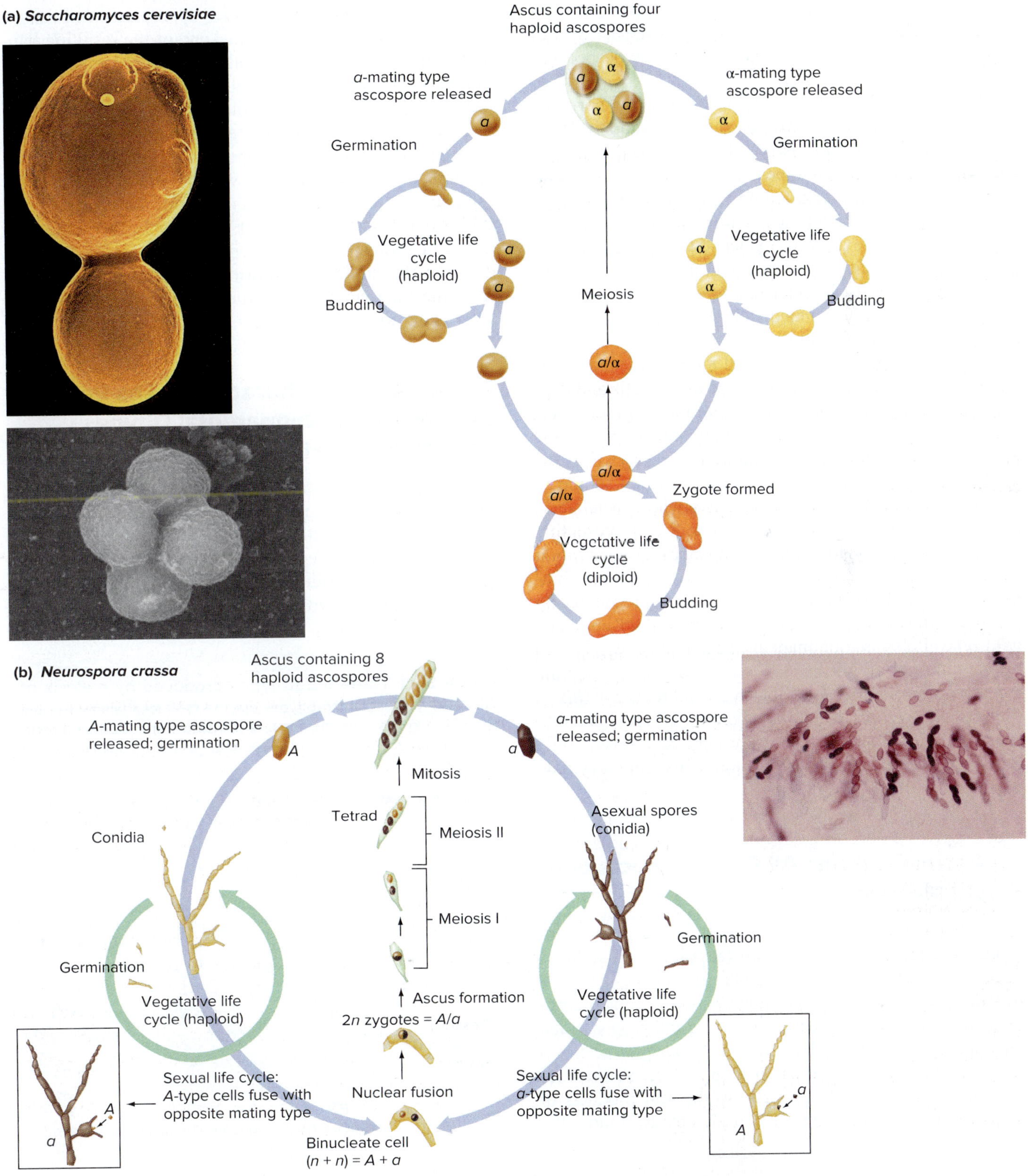

the normally unicellular baker's yeast (*Saccharomyces cerevisiae*), is sold in supermarkets and contributes to the texture, shape, and flavor of bread; it generates four ascospores with each meiosis. The other, *Neurospora crassa,* is a mold that renders the bread on which it grows inedible; it too generates four ascospores with each meiosis, but at the completion of meiosis, each of the four haploid ascospores immediately divides once by mitosis to yield four pairs, for a total of eight haploid cells. The two cells in each pair of *Neurospora* ascospores have the same genotype because they arose from mitosis.

Haploid cells of both yeast and *Neurospora* normally reproduce vegetatively (that is, asexually) by mitosis. However, sexual reproduction is possible because the haploid cells come in two mating types, and cells of opposite mating types can fuse to form a diploid zygote (Fig. 5.20). In yeast, these diploid cells are stable and can reproduce through successive mitotic cycles. Stress, such as that caused by a scarcity of essential nutrients, induces the diploid cells of yeast to enter meiosis. In bread mold, the diploid zygote instead immediately undergoes meiosis, so the diploid state is only transient.

Mutations in haploid yeast and mold affect many different traits, including the appearance of the cells and their ability to grow under particular conditions. For instance, yeast cells with the *his4* mutation are unable to grow in the absence of the amino acid histidine, while yeast with the *trp1* mutation cannot grow without an external source of the amino acid tryptophan. Geneticists who specialize in the study of yeast have devised a system of representing genes that is slightly different from the ones for *Drosophila* and mice. They use capital letters (*HIS4*) to designate dominant alleles and lowercase letters (*his4*) to represent recessives. For most of the yeast genes we will discuss, the wild-type alleles are dominant and may also be represented by the alternative shorthand +, while the symbol for the recessive alleles remains the lowercase abbreviation (*his4*). (See *Guidelines for Gene Nomenclature*.) Remember, however, that dominance or recessiveness is relevant only for diploid yeast cells, not for haploid cells that carry only one allele.

An Ascus Contains All Four Products of a Single Meiosis

The assemblage of four ascospores (or four pairs of ascospores) in a single ascus is called a **tetrad.** Note that this is a second meaning for the term *tetrad.* In Chapter 4, a tetrad was the four homologous chromatids—two in each chromosome of a bivalent—synapsed during the prophase and metaphase of meiosis I. Here, it is the four products of a single meiosis held together in a sac. Because the four chromatids of a bivalent give rise to the four products of meiosis, the two meanings of tetrad refer to almost the same things. Yeast make **unordered tetrads;** that is, the four meiotic products (the spores) are arranged at random within the ascus. *Neurospora crassa* produce **ordered tetrads,** with the four pairs, or eight haplospores, arranged in a line.

To analyze both unordered and ordered tetrads, researchers can release the spores of each ascus, induce the haploid cells to germinate under appropriate conditions, and then analyze the genetic makeup of the resulting haploid cultures. The data they collect in this way enable them to identify the four products of a single meiosis and compare them with the four products of many other distinct meioses. Ordered tetrads offer another possibility. With the aid of a dissecting microscope, investigators can recover the ascospores in the order in which they occur within the ascus and thereby obtain additional information that is useful for mapping. We look first at the analysis of randomly arranged spores in yeast tetrads as an example. We then describe the additional information that can be gleaned from the microanalysis of ordered tetrads, using *Neurospora* as our model organism.

Tetrads Can Be Characterized as Parental Ditypes (PDs), Nonparental Ditypes (NPDs), or Tetratypes (Ts)

When diploid yeast cells heterozygous for each of two genes are induced to undergo meiosis, three tetrad types can be produced whether the two genes are on the same chromosome or different chromosomes. Consider the cross in **Fig. 5.21,** in which haploid cells of opposite mating types (*a* versus *α*) and with alternate alleles of two genes

Figure 5.21 Three tetrad types produced by meiosis of dihybrid yeast. All three types may be produced whether or not genes *A* and *B* are on the same chromosome and whether or not they are linked. Parental spores are *orange* and recombinant spores are *blue*.

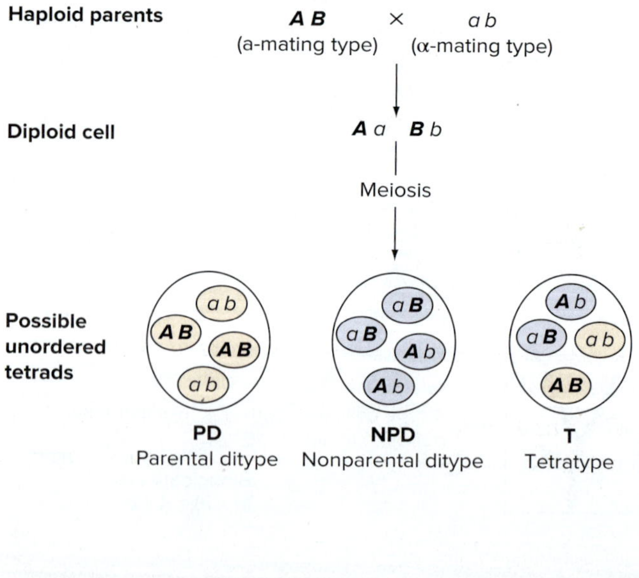

mate to form an *Aa Bb* diploid. One possibility is that all four spores in the resulting tetrad will have the parental configuration of alleles; such a tetrad is called a **parental ditype (PD).** The second kind of tetrad, called a **nonparental ditype (NPD),** contains four recombinant spores, two of each type. The final possibility is a **tetratype (T)** tetrad, which contains four different kinds of spores—two recombinants (one of each type) and two parentals (one of each type). Note that the spores in each yeast ascus are not arranged in any particular order (Fig. 5.21). The classification of a tetrad as PD, NPD, or T is based solely on the number of parental and recombinant spores in the ascus.

Recombination Frequencies May Be Determined by Counting the Number of Each Tetrad Type

In order to determine the RF between the two genes in Fig. 5.21, you could simply break open all the spore cases, pool the spores, and analyze them to determine which ones are parentals and which ones are recombinants. In this case, RF equals the number of recombinant spores divided by the total number of spores (parentals plus recombinants) counted.

Alternatively, as you determine the genotype of each spore, you could keep track of which spores came from the same ascus and count instead the number of each type of tetrad—PD, NPD, or T. When the latter method is used, the recombination frequency is simply the number of NPD-type tetrads (all of the spores in them are recombinant; Fig. 5.21) plus half the number of T-type tetrads (half of the spores in them are recombinant; Fig. 5.21) divided by the total number of tetrads counted:

$$RF = [NPD + 1/2\ (T)]/\text{total tetrads}.$$

Using either method, you will calculate the same value for RF. However, analyzing the products of a fungal cross as tetrads has several advantages. Some of these are technical; for example, in some fungi tetrad analysis enables you to determine the distance between genes and centromeres. But more importantly, analysis of tetrads enables you to develop a deeper appreciation for the events of meiosis. The best way to understand tetrad analysis is to examine how the different tetrad types are generated when the two genes in a dihybrid are on different chromosomes and when they are on the same chromosome.

Unlinked genes on different chromosomes: PDs = NPDs

What kinds of tetrads arise when diploid yeast cells heterozygous for two genes on different chromosomes are induced to undergo meiosis? Consider a mating between a haploid strain of yeast of mating type *a,* carrying the *his4* mutation and the wild-type allele of the *TRP1* gene, and a strain of the opposite mating type α that has the genotype *HIS4 trp1.* The resulting *a*/α diploid cells are *his4/HIS4; trp1/ TRP1,* as shown in **Fig. 5.22a.** [The semicolon (;) in the genotype separates genes on nonhomologous chromosomes.]

Three different types of meiosis can take place, each of which produces a different tetrad type. A PD tetrad will result from one of the two random alignments of homologous chromosomes during meiosis I (**Fig. 5.22b**). The equally likely alternative chromosome alignment yields an NPD tetrad (**Fig. 5.22c**). T tetrads are produced from a crossover between only one of the genes and the centromere (**Fig. 5.22d**).

These results reveal two important facts about the tetrads. First, because the meiosis events shown in panels (b) and (c) of Fig. 5.22 are equally likely, the number of PDs will equal the number of NPDs when the two genes are on different chromosomes. The production of Ts (Fig. 5.22d) does not affect the equality of PDs and NPDs because Ts are produced equally often with the two alternative chromosome alignments. Therefore, T meioses deplete PD and NPD production to the same extent.

Second, as expected for alleles of genes on different chromosomes, RF = 50%. The number of PDs, in which all of the spores are parentals, and the number of NPDs, in which all of the spores are recombinants, are equal. The only other tetrad type, T, contains half recombinants and half parentals.

Figure 5.22e displays the data from one experiment with *his4/HIS4; trp1/TRP1* diploids. (Bear in mind that the column headings of PD, NPD, and T refer to tetrads and not to individual haploid cells.) From these data, you can see that the number of PDs and NPDs are nearly the same. Chi-square analysis would indicate that the results do not differ significantly from the hypothesis that PD = NPD in this experiment.

Linked genes: PDs >> NPDs

You have just seen that if two genes in doubly heterozygous diploid yeast cells are on different chromosomes (unlinked), the number of PD tetrads will approximately equal the number of NPD tetrads. This will not be the outcome for linked genes. When yeast that are dihybrid for linked genes sporulate (that is, undergo meiosis), the number of PDs produced far exceeds the number of NPDs. By analyzing an actual cross involving linked genes, we can see how this statement follows from the events occurring during meiosis.

A haploid yeast strain containing the *arg3* and *ura2* mutations was mated to a wild-type *ARG3 URA2* haploid strain. When the resultant diploid was induced to sporulate,

Figure 5.22 How meiosis can generate three kinds of tetrads when two genes are on different chromosomes. **(a)** Parental cross. **(b)** and **(c)** In the absence of recombination, the two equally likely alternative arrangements of two pairs of chromosomes yield either PD or NPD tetrads. T tetrads are made only if either gene recombines with respect to its corresponding centromere, as in **(d).** Numerical data in **(e)** show that the number of PD tetrads ≈ the number of NPD tetrads when the two genes are unlinked. In (b)–(d), parental spores are *orange* and recombinant spores are *blue.*

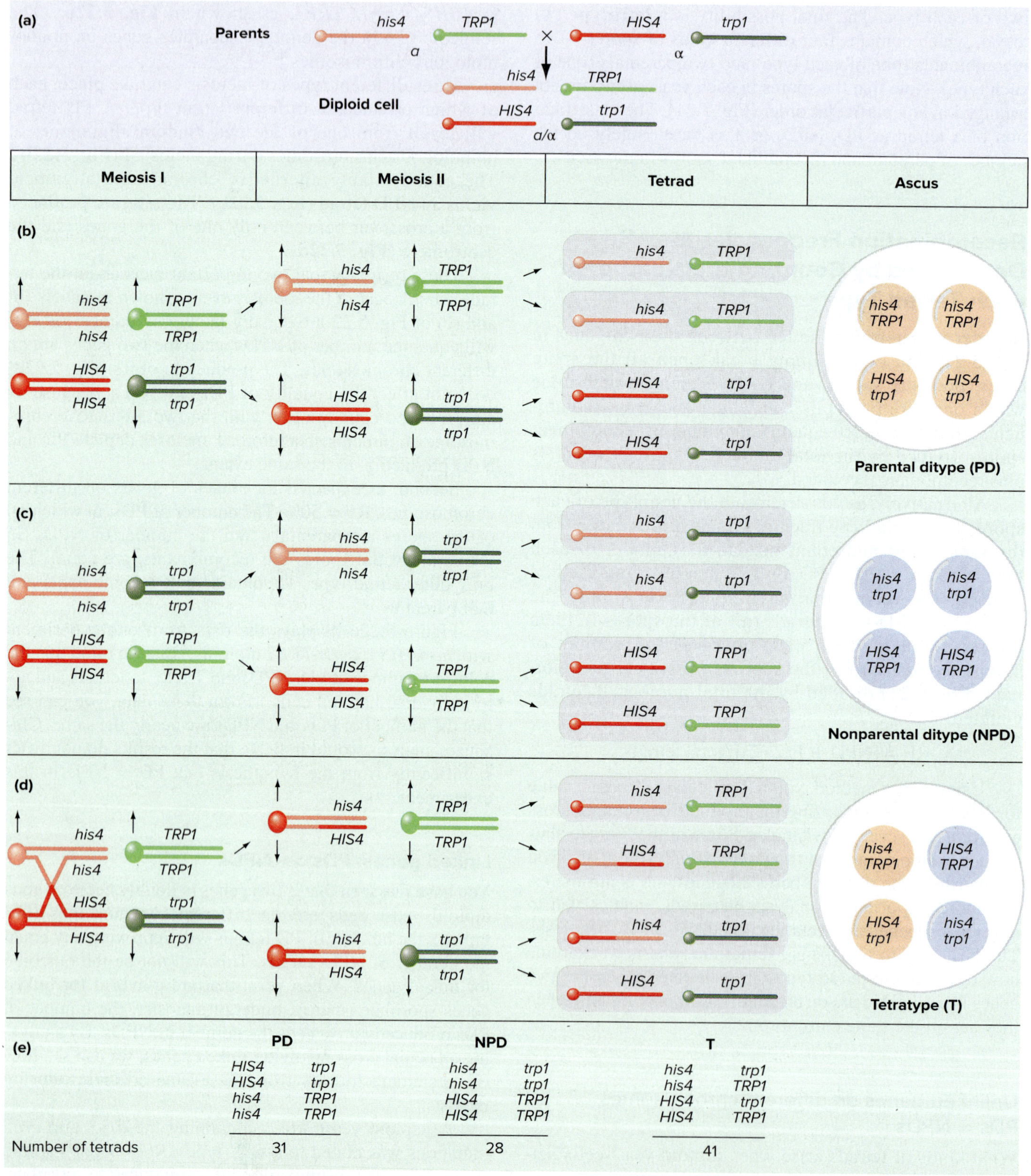

Figure 5.23 When genes are linked, PDs exceed NPDs.

P	*arg3 ura2* (a-mating type)	×	*ARG3 URA2* (α-mating type)
Diploid cell	*arg3 ura2* / *ARG3 URA2*		
	Meiosis		
Products of meiosis	**PD**	**NPD**	**T**
	arg3 ura2 *arg3 ura2* *ARG3 URA2* *ARG3 URA2*	*arg3 URA2* *arg3 URA2* *ARG3 ura2* *ARG3 ura2*	*arg3 ura2* *arg3 URA2* *ARG3 ura2* *ARG3 URA2*
Number of tetrads	127	3	70

Figure 5.24 **How crossovers between linked genes generate different tetrads.** **(a)** PDs arise when there is no crossing-over. **(b)** Single crossovers between the two genes yield tetratypes (Ts). **(c)** Double crossovers between linked genes can generate PD, T, or NPD tetrads, depending on which chromatids participate in the crossovers.

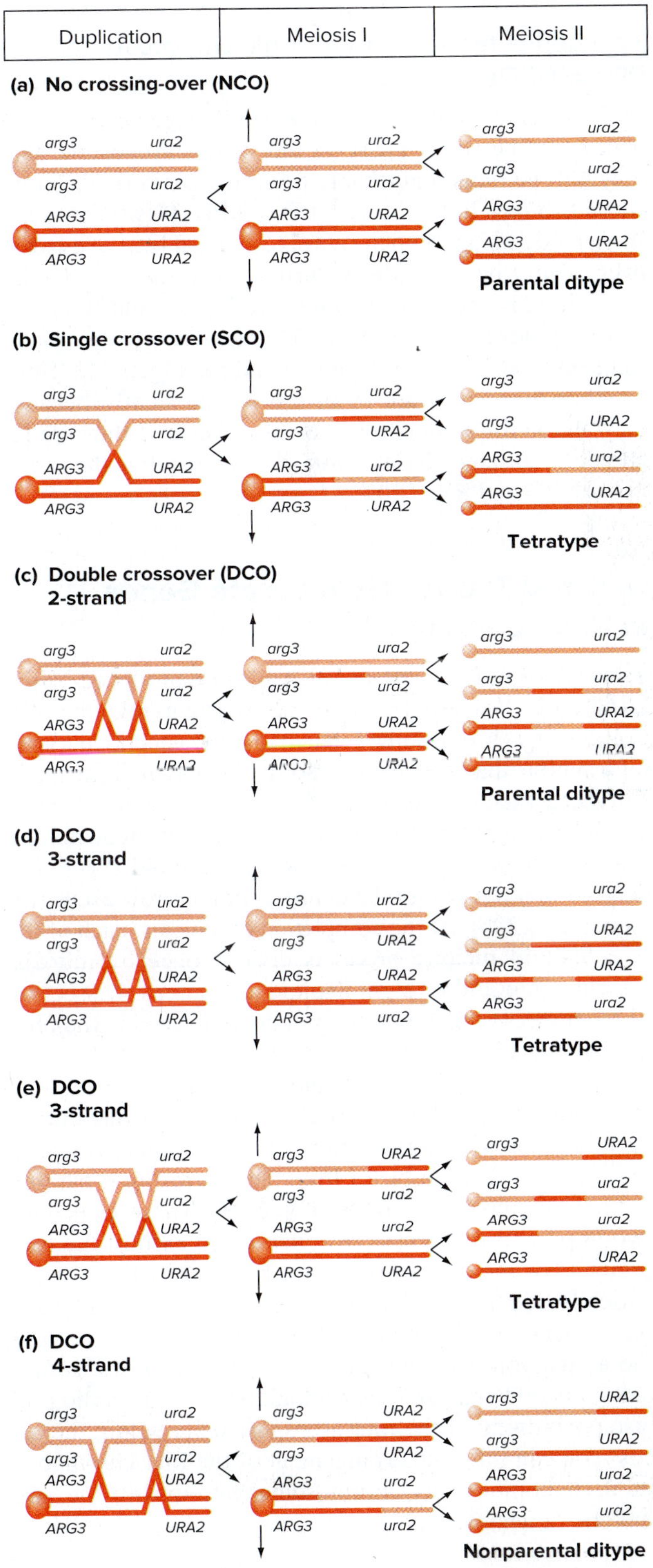

the 200 tetrads produced had the distribution shown in **Fig. 5.23.** As you can see, the 127 PD tetrads far outnumber the 3 NPD tetrads, suggesting that the two genes are linked.

Figure 5.24 shows how we can explain the particular kinds of tetrads observed in terms of the various types of meioses possible. If no crossing-over occurs between the two genes, the resulting tetrad will be PD (Fig. 5.24a). A single crossover between *ARG3* and *URA2* will generate a T tetrad (Fig. 5.24b). But what about double crossovers? As you saw earlier (Fig. 5.11), there are actually four different possibilities, depending on which chromatids participate, and each of the four should occur with equal frequency. A double crossover involving only two chromatids generates a PD tetrad (Fig. 5.24c). Three-strand double crossovers can occur in the two ways depicted in Fig. 5.24d and e; either way, a T tetrad results. Finally, if all four chromatids take part in the two crossovers (one crossover involves two strands and the other crossover, the other two strands), the resulting tetrad is NPD (Fig. 5.24f). Therefore, if two genes are linked, the only way to generate an NPD tetrad is through a four-strand double exchange. When two genes are close together on a chromosome, meioses with one of the four kinds of double crossovers are much rarer than those with no crossing-over or single crossovers, which produce PD and T tetrads, respectively. This explains why, if two genes are linked, the number of PDs must greatly exceed the number of NPDs.

If we calculate the RF from the data in Fig. 5.23 using the equation RF = [NPD + (1/2)T]/total tetrads, we find that

$$RF = [3 + (1/2)70]/200 \times 100 = 19 \text{ m.u.}$$

However, observation of Fig. 5.24 reveals that this equation for RF is not an accurate reflection of the actual number of crossover events when two genes are far enough apart that DCOs occur and NPDs appear. For example, the equation does not count any PDs in the numerator, even though DCO meioses generate some PDs (Fig. 5.24c).

Problem 44 at the end of the chapter helps you derive a corrected equation for RF. The corrected RF equation takes into account all of the DCO meioses that contribute to the tetrads resulting from a cross where PD >> NPD, but NPD is greater than zero.

Two genes far apart on a single chromosome: PDs = NPDs

In tetrad analysis, just as in *en masse* linkage analysis, two genes may be so far apart on the same chromosome that they will be indistinguishable from two genes on different chromosomes: In both cases, PD = NPD. If two genes are sufficiently far apart on the chromosome, at least one crossover occurs between them during every meiosis. Under such circumstances, no meioses are NCOs, and therefore all PD tetrads as well as all NPD tetrads come from equally frequent kinds of DCOs (events c and f in Fig. 5.24). Thus, whether two genes are assorting independently because they are on different chromosomes or because they are far apart on the same chromosome, the end result is the same: PD = NPD and RF = 50%.

Ordered Tetrads Help Locate Genes in Relation to the Centromere

Analyses of ordered tetrads, such as those produced by the bread mold *Neurospora crassa,* allow you to map the centromere of a chromosome relative to other genetic markers, information that you cannot normally obtain from unordered yeast tetrads. As described earlier, immediately after specialized haploid *Neurospora* cells of different mating types (*A* and *a*) fuse at fertilization, the diploid zygote undergoes meiosis within the confines of a narrow ascus (review Fig. 5.20b). At the completion of meiosis, each of the four haploid meiotic products divides once by mitosis, yielding an **octad** of eight haploid ascospores. Dissection of the ascus at this point allows one to determine the phenotype of each of the eight haploid cells.

The cross-sectional diameter of the ascus is so small that cells cannot slip past each other. Moreover, during each division after fertilization, the microtubule fibers of the spindle extend outward from the centrosomes parallel to the long axis of the ascus (**Fig. 5.25**). These facts have two important repercussions. First, when each of the four products of meiosis divides once by mitosis, the two genetically identical cells that result lie adjacent to each other. Because of this feature, starting from either end of the ascus, you can count the octad of ascospores as four cell pairs and analyze it as a tetrad. Second, from the precise positioning of the four ascospore pairs within the ascus, you can infer the arrangement of the four chromatids of each homologous chromosome pair during the two meiotic divisions.

Figure 5.25 How ordered tetrads form. Spindles form parallel to the long axis of the growing *Neurospora* ascus, and the cells cannot slide around each other. The order of ascospores thus reflects meiotic spindle geometry. After meiosis, each haploid cell undergoes mitosis, producing an eight-cell ascus (an octad). The octad consists of four pairs of cells; the two cells of each pair are genetically identical.

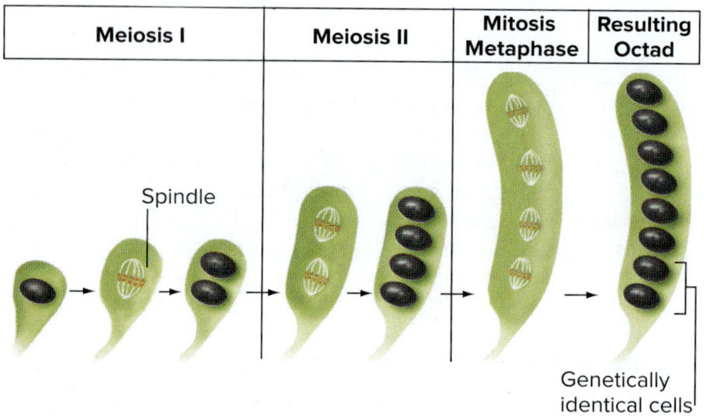

To understand the genetic consequences of the geometry of the ascospores, it is helpful to consider what kinds of tetrads you would expect from the segregation of two alleles of a single gene. (In the following discussion, you will see that *Neurospora* geneticists denote alleles with symbols similar to those used for *Drosophila,* as detailed in the *Guidelines for Gene Nomenclature.*) The mutant *white-spore* allele (*ws*) alters ascospore color from wild-type black to white. In the absence of recombination, the two alleles (ws^+ and *ws*) separate from each other at the first meiotic division because the centromeres separate at that stage. The second meiotic division and subsequent mitosis create asci in which the top four ascospores are of one genotype (for instance ws^+) and the bottom four of the other (*ws*). Whether the top four are ws^+ and the bottom four *ws,* or *vice versa,* depends on the random metaphase I orientation of the homologs that carry the gene relative to the long axis of the developing ascus.

The segregation of two alleles of a single gene at the first meiotic division is thus indicated by an ascus in which an imaginary line drawn between the fourth and the fifth ascospores of the octad cleanly separates haploid products bearing the two alleles. Such an ascus displays a **first-division (MI) segregation pattern** (**Fig. 5.26a**).

Suppose now that during meiosis I, a crossover occurs in a heterozygote between the *white-spore* gene and the centromere. As **Fig. 5.26b** illustrates, this can lead to four equally possible ascospore arrangements, each one depending on a particular orientation of the four chromatids during the two meiotic divisions. In all four cases, both ws^+ and *ws* spores are found on both sides of the imaginary line drawn between ascospores 4 and 5, because cells with only one kind of allele do not arise until the end of the second meiotic division. Octads carrying such configurations of

Figure 5.26 Two segregation patterns in ordered asci. **(a)** In the absence of a crossover between a gene and the centromere, the two alleles of a gene will separate at the first meiotic division. The result is an MI segregation pattern in which each allele appears in spores located on only one side of an imaginary line through the middle of the ascus. **(b)** A crossover between a gene and the centromere produces an MII segregation pattern in which both alleles appear on the same side of the middle line.

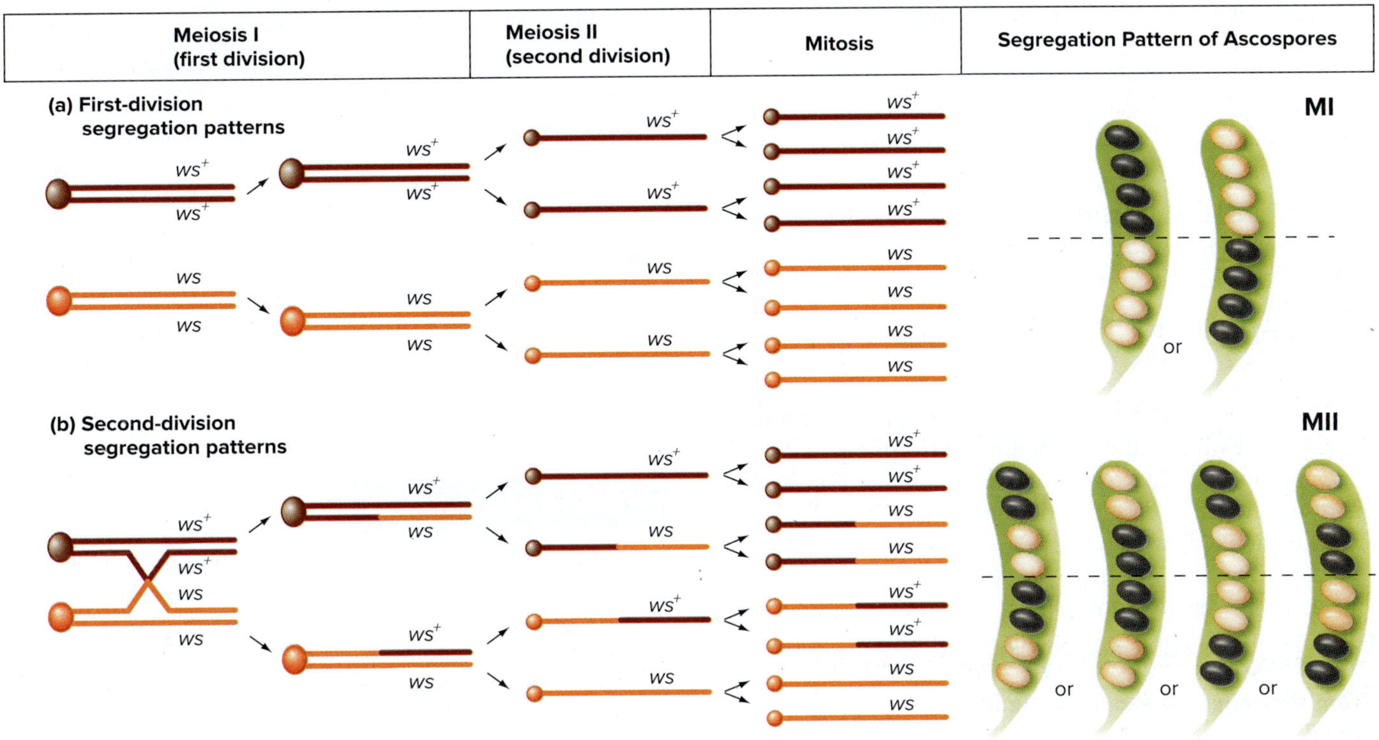

spores display a **second-division (MII) segregation pattern.**

Because MII patterns result from meioses in which a crossover has occurred between a gene and the centromere, the relative number of asci with this pattern can be used to determine the gene ↔ centromere distance. In an ascus showing MII segregation, one-half of the ascospores are derived from chromatids that have exchanged parts, while the remaining half arise from chromatids that have not participated in crossovers (Fig. 5.26b). To calculate the distance between a gene and the centromere, you therefore simply divide the percentage of MII octads by 2:

$$\text{gene} \leftrightarrow \text{centromere distance} = (1/2)\ \text{MII/total tetrads} \times 100.$$

Tetrad Analysis: A Numerical Example

In one experiment, a *thr*$^+$*arg*$^+$ wild-type strain of *Neurospora* was crossed with a *thr arg* double mutant. The *thr* mutants cannot grow in the absence of the amino acid threonine, while *arg* mutants cannot grow without a source of the amino acid arginine; cells carrying the wild-type alleles of both genes can grow in medium that contains neither amino acid. From this cross, 105 octads, considered here as tetrads, were obtained. These tetrads were classified in seven different groups—A, B, C, D, E, F, and G—as shown in **Fig. 5.27a.** For each of the two genes, we can now find the distance between the gene and the centromere.

Figure 5.27 Genetic mapping by ordered-tetrad analysis: An example. **(a)** In ordered-tetrad analysis, tetrad classes are defined not only as PD, NPD, or T but also according to whether they show an MI or MII segregation pattern. Each entry in this table represents a pair of adjacent, identical spores in the actual *Neurospora* octad. Red dots indicate the middle of the asci. **(b)** Genetic map derived from the data in part (a). Ordered-tetrad analysis allows determination of the centromere's position as well as distances between genes.

(a) A *Neurospora* cross

Tetrad group	A	B	C	D	E	F	G
Segregation pattern	*thr arg* *thr arg* • • • • *thr*$^+$*arg*$^+$ *thr*$^+$*arg*$^+$	*thr arg* *thr*$^+$*arg* • • • • *thr*$^+$*arg*$^+$ *thr arg*$^+$	*thr arg* *thr arg*$^+$ • • • • *thr*$^+$*arg* *thr*$^+$*arg*$^+$	*thr arg*$^+$ *thr*$^+$*arg* • • • • *thr*$^+$*arg*$^+$ *thr arg*	*thr arg*$^+$ *thr*$^+$*arg* • • • • *thr*$^+$*arg* *thr arg*$^+$	*thr arg*$^+$ *thr arg*$^+$ • • • • *thr*$^+$*arg* *thr*$^+$*arg*	*thr arg* *thr*$^+$*arg*$^+$ • • • • *thr*$^+$*arg*$^+$ *thr arg*
Total in group	72	16	11	2	2	1	1

(b) Corresponding genetic map

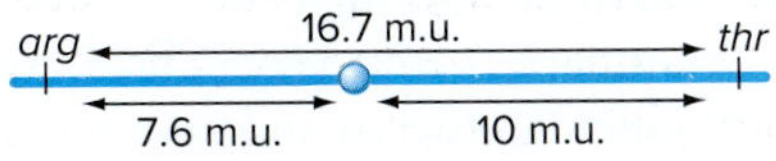

To do this for the *thr* gene, we count the number of tetrads with an MII pattern for that gene. Drawing an imaginary line through the middle of the tetrads, we see that those in groups B, D, E, and G are the result of MII segregations for *thr,* while the remainder show MI patterns. The centromere ↔ *thr* distance is thus:

Half the percentage of MII patterns =

$$\frac{(1/2)(16 + 2 + 2 + 1)}{105} \times 100 = 10 \text{ m.u.}$$

Similarly, the MII tetrads for the *arg* gene are in groups C, D, E, and G, so the distance between *arg* and its centromere is:

$$\frac{(1/2)(11 + 2 + 2 + 1)}{105} \times 100 = 7.6 \text{ m.u.}$$

To ascertain whether the *thr* and *arg* genes are linked, we need to evaluate the seven tetrad groups in a different way, looking at the combinations of alleles for the two genes to see if the tetrads in that group are PD, NPD, or T. We can then ask whether PD >> NPD. Referring again to Fig. 5.27a, we find that groups A and G are PD, because all the ascospores show parental combinations, while groups E and F, with four recombinant spores, are NPD. PD is thus 72 + 1 = 73, while NPD is 1 + 2 = 3. From these data, we can conclude that the two genes are linked.

What is the map distance between *thr* and *arg?* For this calculation, we need to find the numbers of T and NPD tetrads. Tetratypes are found in groups B, C, and D, and we already know that groups E and F carry NPDs. Using the same formula for map distances as the one previously used for yeast,

$$\text{RF} = \frac{\text{NPD} + 1/2\text{T}}{\text{Total tetrads}} \times 100$$

we get:

$$\text{RF} = \frac{3 + (1/2)(16 + 11 + 2)}{105} \times 100 = 16.7 \text{ m.u.}$$

Because the distance between *thr* and *arg* is larger than that separating either gene from the centromere, the centromere must lie between *thr* and *arg,* yielding the map in **Fig. 5.27b.** The distance between the two genes calculated by the preceding formula (16.7 m.u.) is smaller than the sum of the two gene ↔ centromere distances (10.0 + 7.6 = 17.6 m.u.) because the formula does not account for all of the double crossovers. As always, calculating map positions by adding shorter intervals produces the most accurate genetic maps. The gene ↔ centromere distances are shorter and are therefore more accurate than the *thr/arg* distance calculation in this example.

Table 5.3 summarizes the procedures for mapping genes in fungi producing ordered and unordered tetrads.

TABLE 5.3 Rules for Tetrad Analysis

For Ordered and Unordered Tetrads
Considering genes two at a time, assign tetrads as PD, NPD, or T.
If PD >> NPD, the two genes are linked.
If PD ≈ NPD, the two genes assort independently (they are unlinked).
The map distance between two genes if they are linked $= \frac{\text{NPD} + (1/2)\text{T}}{\text{Total tetrads}} \times 100$
For Ordered Tetrads Only
The map distance between a gene and the centromere $= \frac{(1/2)\text{MII}}{\text{Total tetrads}} \times 100$

essential concepts

- A *tetrad* is the group of four haploid spores within an ascus that results from a single meiosis in fungi.
- In a *parental ditype (PD),* a tetrad has four parental spores; in a *nonparental ditype (NPD),* a tetrad contains four recombinant spores; in a *tetratype (T),* an ascus contains two different parental spores and two different recombinant spores.
- When a dihybrid sporulates, if PD tetrads are equal to NPD tetrads, the genes in question are unlinked; when PDs greatly outnumber NPDs, the genes are linked.
- Analysis of *unordered tetrads* reveals linked genes and the map distance between them; analysis of *ordered tetrads* further allows determination of the distance between a gene and the centromere.

5.6 Mitotic Recombination and Genetic Mosaics

learning objectives

1. Explain how mitotic recombination leads to the mosaic condition termed twin spots.
2. Describe sectored colonies in yeast and their significance in evaluating mitotic recombination.

The recombination of genetic material is a critical feature of meiosis. It is thus not surprising that eukaryotic organisms express a variety of enzymes (to be described in Chapter 6) that specifically initiate meiotic recombination.

Recombination can also occur during mitosis. Unlike what happens in meiosis, however, mitotic crossovers are initiated by mistakes in chromosome replication or by chance exposures to radiation that break DNA molecules, rather than by a well-defined cellular program. As a result, mitotic recombination is a rare event, occurring no more often than once in a million somatic cell divisions. Nonetheless, the growth of a colony of yeast cells or the development of a complex multicellular organism involves so many cell divisions that geneticists can detect these rare mitotic events routinely.

Twin Spots Indicate Mosaicism Caused by Mitotic Recombination

In 1936, the *Drosophila* geneticist Curt Stern inferred the existence of mitotic recombination from observations of *twin spots* in fruit flies. **Twin spots** are adjacent islands of tissue that differ both from each other and from the tissue surrounding them. The distinctive patches arise from homozygous cells with a recessive phenotype growing amid a generally heterozygous cell population displaying the dominant phenotype. In *Drosophila,* the *yellow* (*y*) mutation changes body color from normal brown to yellow, while the *singed bristles* (*sn*) mutation causes body bristles to be short and curled rather than long and straight. Both of these genes are X-linked.

In his experiments, Stern examined *Drosophila* XX females of genotype $y\ sn^+\ /\ y^+\ sn$. These double heterozygotes were generally wild type in appearance, but Stern noticed that some flies carried patches of yellow body color, others had small areas of singed bristles, and still others displayed twin spots: adjacent patches of yellow cells and cells with singed bristles (**Fig. 5.28**). He assumed that mistakes in the mitotic divisions accompanying fly development could have led to these **mosaic** animals containing tissues of different genotypes. Individual yellow or singed patches could arise from chromosome loss or by mitotic nondisjunction. These errors in mitosis would yield XO cells containing only *y* (but not y^+) or *sn* (but not sn^+) alleles; such cells would show one of the recessive phenotypes.

The twin spots must have a different origin. Stern reasoned that they represented the reciprocal products of mitotic crossing-over between the *sn* gene and the centromere. The mechanism is as follows: During mitosis in a diploid cell, after chromosome duplication, homologous chromosomes occasionally—but rarely—pair up with each other. While the chromosomes are paired, nonsister chromatids can exchange parts by crossing-over. The pairing is transient, and the homologous chromosomes soon resume their independent positions on the mitotic metaphase plate. There, the two chromosomes can line up relative to each other in either of two ways (**Fig. 5.29a**). One of these orientations would yield two daughter cells that remain heterozygous for both genes and are thus indistinguishable from the surrounding wild-type cells. The other orientation, however, will generate two homozygous daughter cells: one $y\ sn^+/\ y\ sn^+$, the other $y^+sn\ /\ y^+sn$. Because the two daughter cells would lie next to each other, subsequent mitotic divisions would produce adjacent patches of *y* and *sn* tissue (that is, twin spots). Note that if crossing-over occurs between *sn* and *y,* single spots of yellow tissue can form, but a reciprocal singed spot cannot be generated in this fashion (**Fig. 5.29b**).

Figure 5.28 Twin spots: A form of genetic mosaicism. In a $y\ sn^+\ /\ y^+sn$ *Drosophila* female, most of the body is wild type, but aberrant patches showing either yellow color or singed bristles sometimes occur. In some cases, yellow and singed patches are adjacent to each other, a configuration known as *twin spots.*

Sectored Yeast Colonies Can Arise from Mitotic Recombination

Diploid yeast cells that are heterozygous for one or more genes exhibit mitotic recombination in the form of **sectors:** portions of a growing colony that have a different genotype than the remainder of the colony. If a diploid yeast cell of genotype *ADE2 / ade2* is placed on a petri plate, its mitotic descendents will grow into a colony. Usually, such colonies will appear white because the dominant wild-type *ADE2* allele specifies that color. However, many colonies will contain red sectors of diploid *ade2 / ade2* cells (**Fig. 5.30**). These cells are red because a block in the adenine biosynthesis pathway causes them to accumulate red pigment. The red sectors arose as a result of mitotic recombination between the *ADE2* gene and its centromere. (Homozygous *ADE2 / ADE2* cells will also be produced by the same event, but they cannot be distinguished from heterozygotes because both types of cells are white.)

The size of a red sector relative to the size of the colony as a whole indicates when mitotic recombination took place. If a red sector is relatively large, mitotic recombination

Figure 5.29 Mitotic crossing-over. **(a)** In a *y sn*$^+$ / *y*$^+$ *sn Drosophila* female, a mitotic crossover between the centromere and *sn* can produce two daughter cells, one homozygous for *y* and the other homozygous for *sn,* that can develop into adjacent aberrant patches (twin spots). This outcome depends on a particular distribution of chromatids at anaphase (*top*). If the chromatids are arranged in the equally likely opposite orientation, only phenotypically normal cells will result (*bottom*). **(b)** Crossovers between *sn* and *y* can generate single yellow patches. However, a single mitotic crossover in these females cannot produce a single singed spot if the *sn* gene is closer to the centromere than the *y* gene.

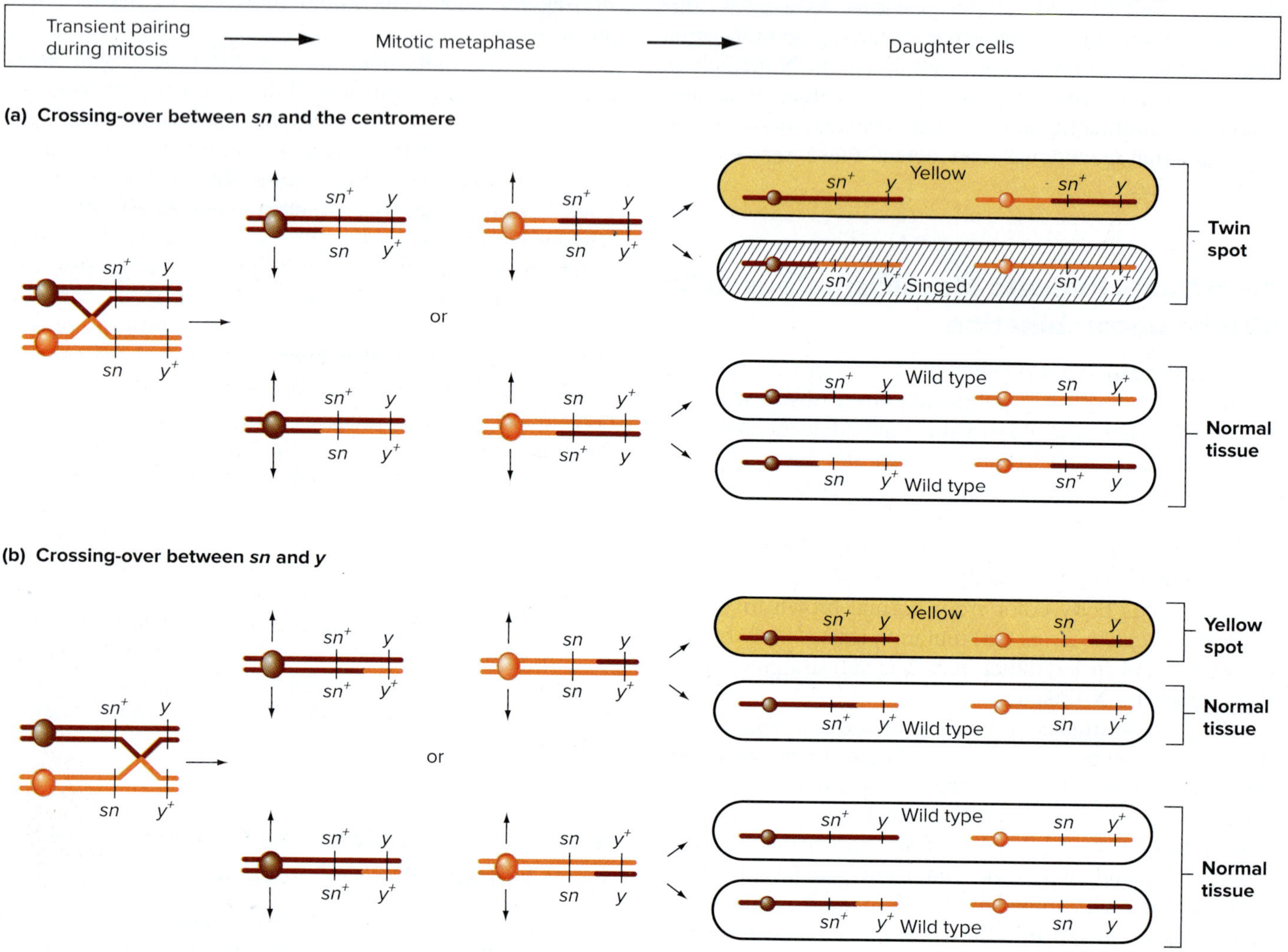

Figure 5.30 Mitotic recombination during the growth of diploid yeast colonies can create sectors. Arrows point to large, red *ade2 / ade2* sectors formed from *ADE2 / ade2* heterozygotes.

Image courtesy of B.A. Montelone, Ph.D. and T.R. Manney, Ph.D

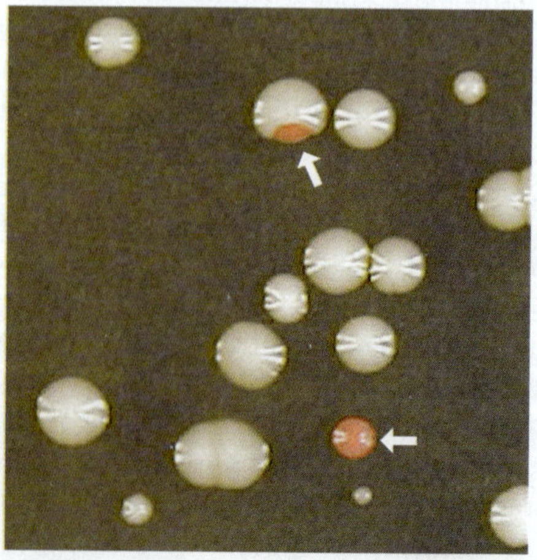

happened during a cell division early in the growth of the colony, giving the resulting daughter cells a long time to proliferate. If a red sector is small, the recombination happened later.

Mitotic Recombination Has Significant Consequences

Problem 51 at the end of this chapter illustrates how geneticists use mitotic recombination to obtain information about the locations of genes relative to each other and to the centromere. Mitotic crossing-over has also been of great value in the study of development because it can generate animals in which different cells have different genotypes (see Problem 52 and also Chapter 19). Finally, as the Genetics and Society Box *Mitotic Recombination and Cancer Formation* explains, mitotic recombination can have major repercussions for human health.

GENETICS AND SOCIETY

Crowd: © Image Source/Getty Images RF

Mitotic Recombination and Cancer Formation

In humans, some tumors, such as those found in retinoblastoma, may arise as a result of mitotic recombination. Recall from the discussion of penetrance and expressivity in Chapter 3 that retinoblastoma is a form of eye cancer. The retinoblastoma gene (*RB*) resides on chromosome 13, where the normal wild-type allele (RB^+) encodes a protein that regulates retinal growth and differentiation. Cells in the eye need at least one copy of the normal wild-type allele to maintain control over cell division. The normal, wild-type RB^+ allele is thus known as a *tumor-suppressor gene.*

People with a genetic predisposition to retinoblastoma are born with only one functional copy of the normal RB^+ allele; their second chromosome 13 carries either a nonfunctional RB^- allele or no *RB* gene at all. If a mutagen (such as radiation) or a mistake in gene replication or segregation destroys or removes the single remaining normal copy of the gene in a retinal cell in either eye, a retinoblastoma tumor will develop at that site. In one study of people with a genetic predisposition to retinoblastoma, cells taken from eye tumors were RB^- homozygotes, while white blood cells from the same people were RB^+/RB^- heterozygotes. As **Fig. A** shows, mitotic recombination between the *RB* gene and the centromere of the chromosome carrying the gene provides one mechanism by which a cell in an RB^+/RB^- individual could become RB^-/RB^-. Once a homozygous RB^- cell is generated, it can divide uncontrollably, leading to tumor formation.

Only 40% of retinoblastoma cases follow the preceding scenario. The other 60% occur in people who are born with two normal copies of the *RB* gene. In such people, it takes two mutational events to cause the cancer. The first of these must convert an RB^+ allele to RB^-, while the second could be a mitotic recombination producing daughter cells that become cancerous because they are homozygous for the newly mutant, nonfunctional allele.

The role of mitotic recombination in the formation of retinoblastoma helps explain the incomplete penetrance and variable expressivity of the disease. People born as RB^+/RB^- heterozygotes may or may not develop the condition (incomplete penetrance). If, as usually happens, they do, they may have tumors in one or both eyes (variable expressivity). It all depends on whether and in what cells of the body mitotic recombination (or some other "homozygosing" event that affects chromosome 13) occurs.

Figure A How mitotic crossing-over can contribute to cancer. Mitotic recombination during retinal growth in an RB^-/RB^+ heterozygote may produce an RB^-/RB^- daughter cell that lacks a functional retinoblastoma gene and thus divides out of control. The crossover must occur between the *RB* gene and its centromere. Only the arrangement of chromatids yielding this result is shown.

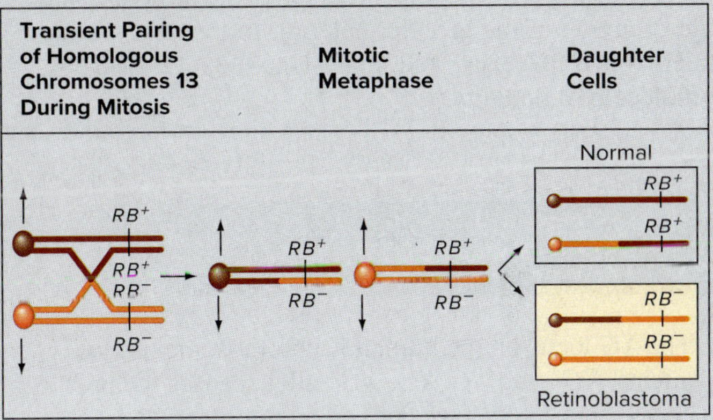

essential concepts

- *Twin spots* are a form of *genetic mosaicism;* these spots occur when *mitotic recombination* gives rise to two clones of cells having reciprocal mutant genotypes and phenotypes.
- Mitotic recombination can also produce *sectored colonies* in diploid yeast, in which part of a colony has a recognizable mutant phenotype.

WHAT'S NEXT

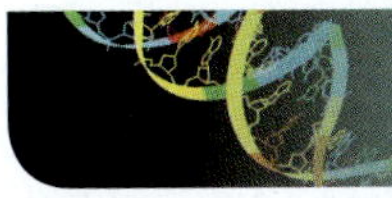

Medical geneticists have used their understanding of linkage, recombination, and mapping to make sense of the pedigrees shown at the beginning of this chapter (see Fig. 5.1). The X-linked gene for red-green color blindness must lie very close to the gene for hemophilia A because the two are tightly coupled. In fact, the genetic distance between the two genes is only 3 m.u. The sample size in Fig. 5.1a was so small that none of the individuals in the pedigree were recombinant types. In contrast, even though the hemophilia B locus is also on the X chromosome, it lies far enough away

DNA: © Design Pics/Bilderbuch RF

from the red-green color blindness locus that the two genes recombine often. The color blindness and hemophilia B genes may appear to be genetically unlinked in a small sample (as in Fig. 5.1b), but the actual recombination distance separating the two genes is about 36 m.u. Pedigrees pointing to two different forms of hemophilia, one very closely linked to color blindness, the other almost not linked at all, provided one of several indications that hemophilia is determined by more than one gene (**Fig. 5.31**).

Linkage and recombination are universal among life-forms and must therefore confer important advantages to living organisms. Geneticists reason that linkage provides the potential for transmitting favorable combinations of genes intact to successive generations, while recombination produces great flexibility in generating new combinations of alleles. Some new combinations may help a species adapt to changing environmental conditions, whereas the inheritance of successfully tested combinations can preserve what has worked in the past.

Thus far, this book has examined how genes and chromosomes are transmitted. As important and useful as this knowledge is, it tells us very little about the structure and mode of action of the genetic material. In the next section (Chapters 6–8), we carry our analysis to the level of DNA, the molecule of heredity.

Figure 5.31 A genetic map of part of the human X chromosome.

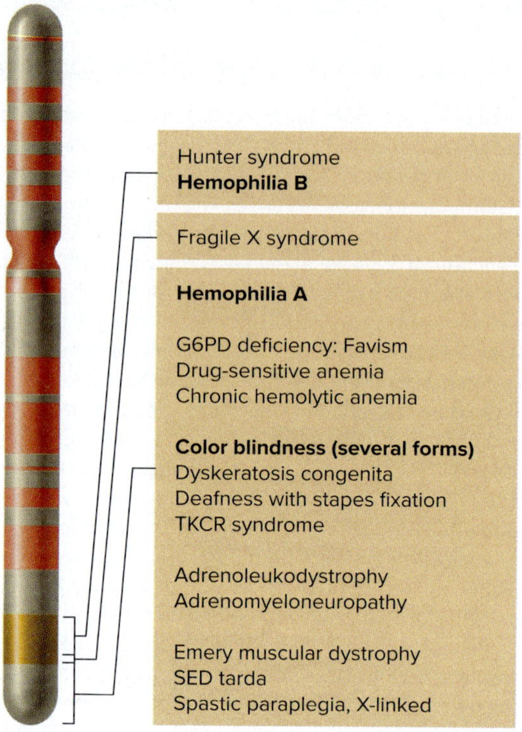

SOLVED PROBLEMS

I. The *XG* locus on the human X chromosome has two alleles, XG^+ and *XG*. The XG^+ allele causes the presence of the Xg surface antigen on red blood cells, while the recessive *XG* allele does not allow antigen to appear. The *XG* locus is 10 m.u. from the *STS* locus. The STS^+ allele produces normal activity of the enzyme steroid sulfatase, while the recessive *STS* allele results in the lack of steroid sulfatase activity and the disease ichthyosis (scaly skin). A man with ichthyosis and no Xg antigen has a normal daughter with Xg antigen. This daughter is expecting a child.

a. If the child is a son, what is the probability he will lack Xg antigen and have ichthyosis?

b. What is the probability that a son would have both the antigen and ichthyosis?

c. If the child is a son with ichthyosis, what is the probability he will have Xg antigen?

Answer

a. This problem requires an understanding of how linkage affects the proportions of gametes. First designate the genotype of the individual in which recombination during meiosis affects the transmission of alleles: in this problem, the daughter. The X chromosome she inherited from her father (who had icthyosis and no Xg antigen) must be *STS XG*. (No recombination could have separated the genes during meiosis in her father since he has only one X chromosome.) Because the daughter is normal and has the Xg antigen, her other X chromosome (inherited from her mother) must contain the STS^+ and XG^+ alleles. Her X chromosomes can be diagrammed as:

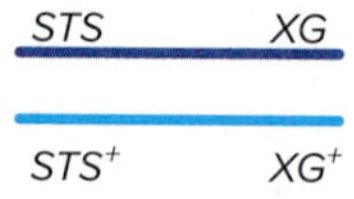

Because the *STS* and *XG* loci are 10 m.u. apart on the chromosome, the recombination frequency is 10%. Ninety percent of the gametes will be parental: *STS XG* or STS^+ XG^+ (45% of each type) and 10% will be recombinant: *STS* XG^+ or STS^+ *XG* (5% of each type). The phenotype of a son directly reflects the genotype of the X chromosome from his mother. Therefore, the probability that he will lack the Xg antigen and have icthyosis (genotype: *STS XG* / Y) is 45/100.

b. The probability that he will have the antigen and ichthyosis (genotype: *STS* XG^+/ Y) is 5/100.

c. Two classes of gametes exist that contain the ichthyosis allele: *STS XG* (45%) and *STS* XG^+ (5%). If the total number of gametes is 100, then 50 will have the *STS* allele. Of those gametes, 5 (or 10%) will have the XG^+ allele. Therefore a 1/10 probability exists that a son with the *STS* allele will have the Xg antigen.

II. *Drosophila* females of wild-type appearance but heterozygous for three autosomal genes are mated with males showing the three corresponding autosomal recessive traits: glassy eyes, coal-colored bodies, and striped thoraxes. One thousand (1000) progeny of this cross are distributed in the following phenotypic classes:

Wild type	27
Striped thorax	11
Coal body	484
Glassy eyes, coal body	8
Glassy eyes, striped thorax	441
Glassy eyes, coal body, striped thorax	29

a. Draw a genetic map based on these data.

b. Show the arrangement of alleles on the two homologous chromosomes in the parent females.

c. Normal-appearing males containing the same chromosomes as the parent females in the preceding cross are mated with females showing glassy eyes, coal-colored bodies, and striped thoraxes. Of 1000 progeny produced, indicate the numbers of the various phenotypic classes you would expect.

Answer

A logical, methodical way to approach a three-point cross is described here.

a. Designate the alleles:

t^+ = wild-type thorax	t = striped thorax
g^+ = wild-type eyes	g = glassy eyes
c^+ = wild-type body	c = coal-colored body

In solving a three-point cross, designate the types of events that gave rise to each group of individuals and the genotypes of the gametes obtained from their mother. [The paternal gametes contain only the recessive alleles of these genes ($t\ g\ c$). These alleles from the father allow the traits associated with the recessive maternal alleles to appear in the progeny.]

Progeny	Number	Type of event	Genotype
1. wild type	27	single crossover	$t^+\ g^+\ c^+$
2. striped thorax	11	single crossover	$t\ g^+\ c^+$
3. coal body	484	parental	$t^+\ g^+\ c$
4. glassy eyes, coal body	8	single crossover	$t^+\ g\ c$
5. glassy eyes, striped thorax	441	parental	$t\ g\ c^+$
6. glassy eyes, coal body, striped thorax	29	single crossover	$t\ g\ c$

Picking out the parental classes is easy. If all the other classes are rare, the two most abundant categories are those gene combinations that have not undergone recombination. Next, two sets of two phenotypes should exist, one set corresponding to a single crossover between the first and second genes and the other set to a single crossover between the second and third genes. Finally, there should be a pair of classes containing small numbers that result from double crossovers. In this example, no flies are found in the double crossover classes, which would have been the two missing phenotypic combinations: one is glassy eyes, and the other is coal body and striped thorax.

Look at the most abundant classes to determine which alleles were on each chromosome in the female heterozygous parent. One parental class had the phenotype of coal body (484 flies), so one chromosome in the female must have contained the t^+, g^+, and c alleles. (Notice that we cannot yet say in what order these alleles are located on the chromosome.) The other parental class was glassy eyes and striped thorax, corresponding to a chromosome with the t, g, and c^+ alleles.

To determine the order of the genes, compare the $t^+\ g\ c^+$ double crossover class (not seen in the data) with the most similar parental class ($t\ g\ c^+$). The alleles of g and c retain their parental associations ($g\ c^+$), while the t gene has recombined with respect to both other genes in the double recombinant class. Thus, the t gene is between g and c.

In order to complete the map, calculate the recombination frequencies between the center gene and each of the genes on the ends. For g and t, the nonparental combinations of alleles are in classes 2 and 4, so RF = (11 + 8)/1000 = 19/1000, or 1.9%. For t and c, classes 1 and 6 are nonparental, so RF = (27 + 29)/1000 = 56/1000, or 5.6%.

The genetic map is

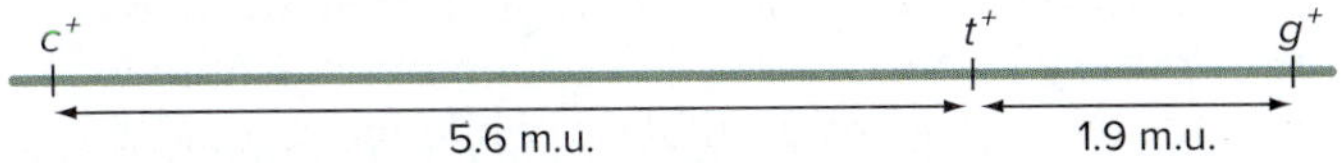

b. The alleles on each chromosome were already determined (c, g^+, t^+ and c^+, g, t). Now that the order of loci has also been determined, the arrangement of the alleles can be indicated as follows:

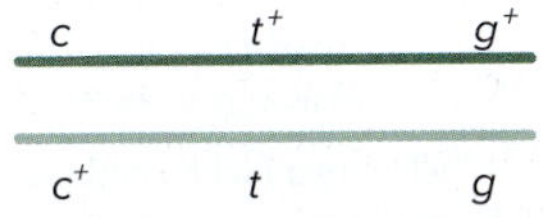

c. Males of the same genotype as the starting female ($c\ t^+\ g^+/\ c^+\ t\ g$) could produce only two types of gametes: parental types $c\ t^+\ g^+$ and $c^+\ t\ g$ because no recombination occurs in male fruit flies. The progeny expected from the mating with a homozygous recessive female are thus 500 coal body and 500 glassy-eyed, striped thorax flies.

III. The following *Neurospora* asci were obtained when a wild-type strain (ad^+ leu^+) was crossed to a double mutant strain that cannot grow in the absence of adenine or leucine (ad^- leu^-). Only one member of each spore pair produced by the final mitosis is shown because the two cells in a pair have the same genotype. Total asci = 120.

Spore pair	Ascus type				
1–2	ad^+ leu^+	ad^+ leu^-	ad^+ leu^+	ad^+ leu^-	ad^- leu^+
3–4	ad^+ leu^+	ad^+ leu^-	ad^+ leu^-	ad^- leu^+	ad^+ leu^+
5–6	ad^- leu^-	ad^- leu^+	ad^- leu^+	ad^- leu^-	ad^- leu^-
7–8	ad^- leu^-	ad^- leu^+	ad^- leu^-	ad^+ leu^+	ad^+ leu^-
# of asci	30	30	40	2	18

a. What genetic event causes the alleles of two genes to segregate to different cells at the second meiotic division, and when does this event occur?

b. Provide the best possible map for the two genes and the relevant centromere(s).

Answer

This problem requires an understanding of tetrad analysis and the process (meiosis) that produces the patterns seen in ordered asci.

a. A crossover between a gene and the centromere causes the segregation of alleles at the second meiotic division. The crossover event itself occurs during prophase of meiosis I.

b. Using ordered tetrads (or ordered octads) you can determine whether two genes are linked, the distance between two genes, and the distance between each gene and the centromere of the chromosome on which the gene is located. First designate the five classes of asci shown. The first class is a PD; the second is an NPD; the last three are Ts. Next determine if these genes are linked. The number of PD = number of NPD, so the genes are not linked. When genes are unlinked, the T asci are generated by a crossover between a gene and the centromere. Looking at the *leu* gene, there is an MII pattern of that gene in the third and fourth asci types. Therefore, the fraction of MII asci is:

$$\frac{40 + 2}{120} \times 100 = 35\%.$$

Because only half of the chromatids in the meioses that generated these T asci were involved in the crossover, the map distance between *leu* and the centromere is 35/2, or 17.5 m.u. Asci of the fourth and fifth types show an MII pattern for the *ad* gene:

$$\frac{2 + 18}{120} \times 100 = 16.6\%.$$

Dividing 16.6% by 2 gives the *ad* gene ↔ centromere map distance of 8.3 m.u. The map of these two genes is the following:

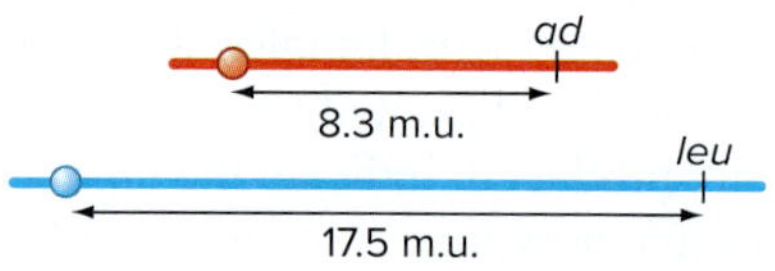

PROBLEMS

Vocabulary

1. Choose the phrase from the right column that best fits the term in the left column.

a. recombination	1. a statistical method for testing the fit between observed and expected results
b. linkage	2. an ascus containing spores of four different genotypes
c. chi-square test	3. one crossover along a chromosome makes a second nearby crossover less likely
d. chiasma	4. when two loci recombine in less than 50% of gametes
e. tetratype	5. the relative chromosomal location of a gene
f. locus	6. the ratio of observed double crossovers to expected double crossovers
g. coefficient of coincidence	7. individual composed of cells with different genotypes
h. interference	8. formation of new genetic combinations by exchange of parts between homologs
i. parental ditype	9. when the two alleles of a gene are segregated into different cells at the first meiotic division
j. ascospores	10. an ascus containing only two nonrecombinant kinds of spores
k. first-division segregation	11. structure formed at the spot where crossing-over occurs between homologs
l. mosaic	12. fungal spores contained in a sac

Section 5.1

2. a. A *Drosophila* male from a true-breeding stock with scabrous eyes was mated with a female from a true-breeding stock with javelin bristles. Both scabrous eyes and javelin bristles are autosomal recessive mutant traits. The F_1 progeny all had normal eyes and bristles. F_1 females from this cross were mated with males with both scabrous eyes and javelin bristles. Write all the possible phenotypic classes of the progeny that could be produced from

the cross of the F_1 females with the scabrous, javelin males, and indicate for each class whether it is a recombinant or parental type.

b. The cross in part (a) yielded the following progeny: 77 scabrous eyes and normal bristles; 76 wild type (normal eyes and bristles); 74 normal eyes and javelin bristles; and 73 scabrous eyes and javelin bristles. Are the genes governing these traits likely to be linked, or do they instead assort independently? Why?

c. Suppose you mated the F_1 females from the cross in part (a) to wild-type males. Why would this cross fail to inform you whether the two genes are linked?

d. Suppose you mated females from the true-breeding stock with javelin bristles to males with scabrous eyes and javelin bristles. Why would this cross fail to inform you whether the two genes are linked?

3. With modern molecular methods it is now possible to examine variants in DNA sequence from a very small amount of tissue like a hair follicle or even a single sperm. (See the Fast Forward Box *Mapping the Crossovers that Generate the Chromosomes of Individual Human Sperm.*) You can consider these variants to be alleles of a particular site on a chromosome (a *locus*; *loci* in plural). For example, AAAAAAA, AAACAAA, AAAGAAA, and AAATAAA at the same location (call it *B*) on homologous autosomes in different sperm might be called alleles *1, 2, 3,* and *4* of locus *B* (B^1, B^2, etc.). John's genotype for two loci *B* and *D* is B^1B^3 and D^1D^3. John's father was B^1B^2 and D^1D^4, while his mother was B^3B^3 and D^2D^3.

a. What is (are) the genotype(s) of the parental type sperm John could produce?

b. What is (are) the genotype(s) of the recombinant type sperm John could produce?

c. In a sample of 100 sperm, 51 of John's sperm were found to be B^1 and D^1, while the remaining 49 sperm were B^3 D^3. Can you conclude whether the *B* and *D* loci are linked, or whether they instead assort independently?

4. The Punnett square in Fig. 5.4 shows how Mendel's dihybrid cross results would have been altered had the two genes (*A* and *B*) been linked and had the P generation cross been *A B / A B* × *a b / a b*.

a. What would be the frequency of each F_2 phenotypic class if 80% of the gametes produced by the F_1s were parentals?

b. Answer part (a) assuming that the original, P generation cross was *A b / A b* × *a B / a B*.

Section 5.2

5. In mice, the dominant allele *Gs* of the X-linked gene *Greasy* produces shiny fur, while the recessive wild-type Gs^+ allele determines normal fur. The dominant allele *Bhd* of the X-linked *Broadhead* gene causes skeletal abnormalities including broad heads and snouts, while the recessive wild-type Bhd^+ allele yields normal skeletons. Female mice heterozygous for the two alleles of both genes were mated with wild-type males. Among 100 male progeny of this cross, 49 had shiny fur, 48 had skeletal abnormalities, 2 had shiny fur and skeletal abnormalities, and 1 was wild type.

a. Diagram the cross described and calculate the distance between the two genes.

b. What would have been the results if you had counted 100 female progeny of the cross?

6. In *Drosophila,* males from a true-breeding stock with raspberry-colored eyes were mated to females from a true-breeding stock with sable-colored bodies. In the F_1 generation, all the females had wild-type eye and body color, while all the males had wild-type eye color but sable-colored bodies. When F_1 males and females were mated, the F_2 generation was composed of 216 females with wild-type eyes and bodies, 223 females with wild-type eyes and sable bodies, 191 males with wild-type eyes and sable bodies, 188 males with raspberry eyes and wild-type bodies, 23 males with wild-type eyes and bodies, and 27 males with raspberry eyes and sable bodies. Explain these results by diagramming the crosses and calculating any relevant map distances.

7. If the *a* and *b* loci are 20 m.u. apart in humans and an *A B / a b* woman mates with an *a b / a b* man, what is the probability that their first child will be *A b / a b?*

8. *CC DD* and *cc dd* individuals were crossed to each other, and the F_1 generation was backcrossed to the *cc dd* parent. 997 *Cc Dd,* 999 *cc dd,* 1 *Cc dd,* and 3 *cc Dd* offspring resulted.

a. How far apart are the *c* and *d* loci?

b. What progeny and in what frequencies would you expect to result from testcrossing the F_1 generation from a *CC dd* × *cc DD* cross to *cc dd?*

c. In a typical meiosis, how many crossovers occur between genes *C* and *D?*

d. Assume that the *C* and *D* loci are on the same chromosome, but the offspring from the testcross described in part (b) were 498 *Cc Dd,* 502 *cc dd,* 504 *Cc dd,* and 496 *cc Dd*. How would your answer to part (c) change?

9. In mice, the autosomal locus coding for the β-globin chain of hemoglobin is 1 m.u. from the albino locus.

Assume for the moment that the same is true in humans. The disease sickle-cell anemia is the result of homozygosity for a particular mutation in the β-globin gene.

a. A son is born to an albino man and a woman with sickle-cell anemia. What kinds of gametes will the son form, and in what proportions?

b. A daughter is born to a normal man and a woman who has both albinism and sickle-cell anemia. What kinds of gametes will the daughter form, and in what proportions?

c. If the son in part (a) grows up and marries the daughter in part (b), what is the probability that a child of theirs will be an albino with sickle-cell anemia?

10. In a particular human family, John and his mother both have brachydactyly (a rare autosomal dominant allele causing short fingers). John's father has Huntington disease (another rare autosomal dominant allele). John's wife is phenotypically normal and is pregnant. Two-thirds of people who inherit the Huntington (*HD*) allele show symptoms by age 50, and John is 50 and has no symptoms. Brachydactyly is 90% penetrant.

a. What are the genotypes of John's parents?

b. What are the possible genotypes for John? How likely is John to have each of these genotypes?

c. What is the probability the child will express both brachydactyly and Huntington disease by age 50 if the two genes are unlinked?

d. How will your answer to part (c) change if instead these two loci are 20 m.u. apart?

11. Albino rabbits (lacking pigment) are homozygous for the recessive *c* allele (*C* allows pigment formation). Rabbits homozygous for the recessive *b* allele make brown pigment, while those with at least one copy of *B* make black pigment. True-breeding brown rabbits were crossed to albinos, which were also *BB*. F_1 rabbits, which were all black, were crossed to the double recessive (*bb cc*). The progeny obtained were 34 black, 66 brown, and 100 albino.

a. What phenotypic proportions would have been expected if the *b* and *c* loci were unlinked?

b. How far apart are the two loci?

12. In corn, the allele *A* allows the deposition of anthocyanin (blue) pigment in the kernels (seeds), while *aa* plants have yellow kernels. At a second gene, *W*– produces smooth kernels, while *ww* kernels are wrinkled. A plant with blue smooth kernels was crossed to a plant with yellow wrinkled kernels. The progeny consisted of 1447 blue smooth, 169 blue wrinkled, 186 yellow smooth, and 1510 yellow wrinkled.

a. Are the *a* and *w* loci linked? If so, how far apart are they?

b. What was the genotype of the blue smooth parent? Include the chromosome arrangement of alleles.

c. If a plant grown from a blue wrinkled progeny seed is crossed to a plant grown from a yellow smooth F_1 seed, what kinds of kernels would be expected, and in what proportions?

13. If the *a* and *b* loci are 40 cM apart and an *AA BB* individual and an *aa bb* individual mate:

a. What gametes will the F_1 individuals produce, and in what proportions? What phenotypic classes in what proportions are expected in the F_2 generation (assuming complete dominance for both genes)?

b. If the original cross was *AA bb* × *aa BB,* what gametic proportions would emerge from the F_1? What would be the result in the F_2 generation?

14. Write the number of *different kinds* of phenotypes, excluding sex, you would see among a large number of progeny from an F_1 mating between individuals of identical genotype that are heterozygous for one or two genes (that is, *Aa* or *Aa Bb*) as indicated. No gene interactions means that the phenotype determined by one gene is not influenced by the genotype of the other gene.

a. One gene; *A* completely dominant to *a.*

b. One gene; *A* and *a* codominant.

c. One gene; *A* incompletely dominant to *a.*

d. Two unlinked genes; no gene interactions; *A* completely dominant to *a,* and *B* completely dominant to *b.*

e. Two genes, 10 m.u. apart; no gene interactions; *A* completely dominant to *a,* and *B* completely dominant to *b.*

f. Two unlinked genes; no gene interactions; *A* and *a* codominant, and *B* incompletely dominant to *b.*

g. Two genes, 10 m.u. apart; *A* completely dominant to *a,* and *B* completely dominant to *b;* and with recessive epistasis between *aa* and the alleles of gene *B.*

h. Two unlinked duplicated genes (that is, *A* and *B* perform the same function); *A* and *B* completely dominant to *a* and *b,* respectively.

i. Two genes, 0 m.u. apart; no gene interactions; *A* completely dominant to *a,* and *B* completely dominant to *b.* (Two possible answers exist.)

15. A DNA variant has been found linked to a rare autosomal dominant disease in humans and can thus be used as a marker to follow inheritance of the disease allele. In an *informative family* (in which one parent is heterozygous for both the disease allele and the

DNA marker in a known chromosomal arrangement of alleles, and his or her mate does not have the same alleles of the DNA variant), the reliability of such a marker as a predictor of the disease in a fetus is related to the map distance between the DNA marker and the gene causing the disease.

Imagine that a man affected with the disease (genotype *Dd*) is heterozygous for the V^1 and V^2 forms of the DNA variant, with form V^1 on the same chromosome as the *D* allele and form V^2 on the same chromosome as *d*. His wife is V^3V^3 *dd,* where V^3 is another allele of the DNA marker. Typing of the fetus by amniocentesis reveals that the fetus has the V^2 and V^3 variants of the DNA marker. How likely is it that the fetus has inherited the disease allele *D* if the distance measured in a two-point cross between the *D* locus and the marker locus is (a) 0 m.u., (b) 1 m.u., (c) 5 m.u., (d) 10 m.u., (e) 50 m.u.?

16. Figure 5.7a shows chromosomes during prophase of meiosis I in a mouse primary spermatocyte.
    a. How would you know immediately that this figure shows male meiosis in a mouse and not in a human? (*Hint:* In mice, $n = 20$.)
    b. Are most mouse chromosomes metacentric or acrocentric? Explain.
    c. How many chromatids in total are represented in Fig. 5.7a?
    d. Where is the X-Y bivalent in Fig. 5.7a? (*Note:* Mouse sex chromosomes have only a single pseudoautosomal region, instead of two as in humans.) Diagram this bivalent, showing the X and Y chromosomes, the locations of the centromeres of these chromosomes, and the pseudoautosomal region.
    e. Explain the importance of the pseudoautosomal region(s) of the sex chromosomes of mammals for ensuring proper sex chromosome segregation during meiosis.
17. Figure 5.7b shows bivalents in mouse primary spermatocytes that have previously undergone recombination events, as indicated by the presence of a single chiasma. The artist's image in Fig. 5.7c depicts how cohesin complexes are involved in keeping the homologous chromosomes together within the bivalent. Explain using diagrams why the key cohesin complexes that connect the homologous chromosomes are those located distal to a chiasma (that is, farther away from the centromere) rather than those located proximal to a chiasma (that is, between the centromere and the chiasma).

Section 5.3

18. Cinnabar eyes (*cn*) and reduced bristles (*rd*) are autosomal recessive characters in *Drosophila.* A homozygous wild-type female was crossed to a reduced, cinnabar male, and the F_1 males were then crossed to the F_1 females to obtain the F_2. Of the 400 F_2 offspring obtained, 292 were wild type, 9 were cinnabar, 7 were reduced, and 92 were reduced and cinnabar. Explain these results and estimate the distance between the *cn* and *rd* loci.
19. In *Drosophila,* the autosomal recessive *dp* allele of the *dumpy* gene produces short, curved wings, while the autosomal recessive allele *bw* of the *brown* gene causes brown eyes. In a testcross using females heterozygous for both of these genes, the following results were obtained:

wild-type wings, wild-type eyes	178
wild-type wings, brown eyes	185
dumpy wings, wild-type eyes	172
dumpy wings, brown eyes	181

In a testcross using males heterozygous for both of these genes, a different set of results was obtained:

wild-type wings, wild-type eyes	247
dumpy wings, brown eyes	242

    a. What can you conclude from the first testcross?
    b. What can you conclude from the second testcross?
    c. How can you reconcile the data shown in parts (a) and (b)? Can you exploit the difference between these two sets of data to devise a general test for synteny in *Drosophila?*
    d. The genetic distance between *dumpy* and *brown* is 91.5 m.u. How could this value be measured?
20. From a series of two-point crosses, the following map distances were obtained for the syntenic genes *A, B, C,* and *D* in peas:

$B \leftrightarrow C$	23 m.u.
$A \leftrightarrow C$	15 m.u.
$C \leftrightarrow D$	14 m.u.
$A \leftrightarrow B$	12 m.u.
$B \leftrightarrow D$	11 m.u.
$A \leftrightarrow D$	1 m.u.

Chi-square analysis cannot reject the null hypothesis of no linkage for gene *E* with any of the other four genes.

    a. Draw a cross scheme that would allow you to determine the $B \leftrightarrow C$ map distance.
    b. Diagram the best genetic map that can be assembled from this data set.
    c. Explain any inconsistencies or unknown features in your map.
    d. What additional experiments would allow you to resolve these inconsistencies or ambiguities?

21. Map distances were determined for four different genes (*MAT, HIS4, THR4,* and *LEU2*) on chromosome III of the yeast *Saccharomyces cerevisiae:*

HIS4 ↔ *MAT*	37 cM
THR4 ↔ *LEU2*	35 cM
LEU2 ↔ *HIS4*	23 cM
MAT ↔ *LEU2*	16 cM
MAT ↔ *THR4*	20 cM

What is the order of genes on the chromosome?

22. In the tubular flowers of foxgloves, wild-type coloration is red while a mutation called *white* produces white flowers. Another mutation, called *peloria,* causes the flowers at the apex of the stem to be huge. Yet another mutation, called *dwarf,* affects stem length. You cross a white-flowered plant (otherwise phenotypically wild type) to a plant that is dwarf and peloria but has wild-type red flower color. All of the F_1 plants are tall with white, normal-sized flowers. You cross an F_1 plant back to the dwarf and peloria parent, and you see the 543 progeny shown in the chart. (Only mutant traits are noted.)

dwarf, peloria	172
white	162
dwarf, peloria, white	56
wild type	48
dwarf, white	51
peloria	43
dwarf	6
peloria, white	5

a. Which alleles are dominant?

b. What were the genotypes of the parents in the original cross?

c. Draw a map showing the linkage relationships of these three loci.

d. Do the data provide evidence for interference? If so, calculate the coefficient of coincidence and the interference value.

23. In *Drosophila,* the recessive allele *mb* of one gene causes missing bristles, the recessive allele *e* of a second gene causes ebony body color, and the recessive allele *k* of a third gene causes kidney-shaped eyes. (Dominant wild-type alleles of all three genes are indicated with a + superscript.) The three different P generation crosses in the table that follows were conducted, and then the resultant F_1 females from each cross were testcrossed to males that were homozygous for the recessive alleles of both genes in question. The phenotypes of the testcross offspring are tabulated. Determine the best genetic map explaining all the data.

Parental cross	Testcross offspring of F_1 females	
$mb^+\ mb^+,\ e^+\ e^+ \times$	normal bristles, normal body	117
$mb\ mb,\ e\ e$	normal bristles, ebony body	11
	missing bristles, normal body	15
	missing bristles, ebony body	107
$k^+\ k^+,\ e\ e \times k\ k,\ e^+\ e^+$	normal eyes, normal body	11
	normal eyes, ebony body	150
	kidney eyes, normal body	144
	kidney eyes, ebony body	7
$mb^+\ mb^+,\ k^+\ k^+ \times$	normal bristles, normal eyes	203
$mb\ mb,\ k\ k$	normal bristles, kidney eyes	11
	missing bristles, normal eyes	15
	missing bristles, kidney eyes	193

24. A snapdragon with pink petals, black anthers, and long stems was allowed to self-fertilize. From the resulting seeds, 650 adult plants were obtained. The phenotypes of these offspring are listed here.

78	red	long	tan
26	red	short	tan
44	red	long	black
15	red	short	black
39	pink	long	tan
13	pink	short	tan
204	pink	long	black
68	pink	short	black
5	white	long	tan
2	white	short	tan
117	white	long	black
39	white	short	black

a. Using *P* for one allele and *p* for the other, indicate how flower color is inherited.

b. What numbers of red : pink : white would have been expected among these 650 plants?

c. How are anther color and stem length inherited?

d. What was the genotype of the original plant?

e. Do any of the three genes show independent assortment?

f. For any genes that are linked, indicate the arrangements of the alleles on the homologous chromosomes in the original snapdragon, and estimate the distance between the genes.

25. In *Drosophila,* three autosomal genes have the following map:

a. Provide the data, in terms of the expected number of flies in the following phenotypic classes, when $a^+\ b^+\ c^+ / a\ b\ c$ females are crossed to $a\ b\ c / a\ b\ c$

males. Assume 1000 flies are counted and that no interference exists in this region.

a^+	b^+	c^+
a	b	c
a^+	b	c
a	b^+	c^+
a^+	b^+	c
a	b	c^+
a^+	b	c^+
a	b^+	c

b. If the cross was reversed, such that $a^+\ b^+\ c^+ / a\ b\ c$ males are crossed to $a\ b\ c / a\ b\ c$ females, how many flies would you expect in the same phenotypic classes?

26. *Drosophila* females heterozygous for each of three recessive autosomal mutations with independent phenotypic effects [thread antennae (*th*), hairy body (*h*), and scarlet eyes (*st*)] were testcrossed to males showing all three mutant phenotypes. The 1000 progeny of this testcross were

thread, hairy, scarlet	432
wild type	429
thread, hairy	37
thread, scarlet	35
hairy	34
scarlet	33

a. Show the arrangement of alleles on the relevant chromosomes in the triply heterozygous females.

b. Draw the best genetic map that explains these data.

c. Calculate any relevant interference values.

27. Male *Drosophila* expressing the autosomal recessive mutations *sc (scute), ec (echinus), cv (crossveinless),* and *b (black)* were crossed to phenotypically wild-type females, and the 3288 progeny listed were obtained. (Only mutant traits are noted.)

653	black, scute, echinus, crossveinless
670	scute, echinus, crossveinless
675	wild type
655	black
71	black, scute
73	scute
73	black, echinus, crossveinless
74	echinus, crossveinless
87	black, scute, echinus
84	scute, echinus
86	black, crossveinless
83	crossveinless
1	black, scute, crossveinless
1	scute, crossveinless
1	black, echinus
1	echinus

a. Diagram the genotype of the female parent.

b. Map these loci.

c. Do the data provide evidence of interference? Justify your answer with numbers.

28. a. In *Drosophila,* crosses between F_1 heterozygotes of the form $A\ b / a\ B$ always yield the same ratio of phenotypes in the F_2 progeny regardless of the distance between the two genes (assuming complete dominance for both autosomal genes). What is this ratio? Would this also be the case if the F_1 heterozygotes were $A\ B / a\ b$? (*Hint:* Remember that in *Drosophila,* recombination does not take place during spermatogenesis.)

b. If you intercrossed F_1 heterozygotes of the form $A\ b / a\ B$ in mice, the phenotypic ratio among the F_2 progeny would vary with the map distance between the two genes. Is there a simple way to estimate the map distance based on the frequencies of the F_2 phenotypes, assuming rates of recombination are equal in males and females? Could you estimate map distances in the same way if the mouse F_1 heterozygotes were $A\ B / a\ b$?

29. A true-breeding strain of Virginia tobacco has dominant alleles determining leaf morphology (*M*), leaf color (*C*), and leaf size (*S*). A Carolina strain is homozygous for the recessive alleles of these three genes. These genes are found on the same chromosome as follows:

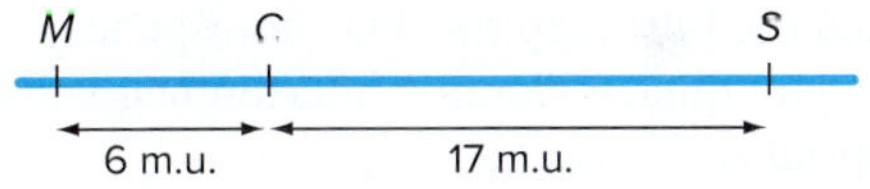

An F_1 hybrid between the two strains is now backcrossed to the Carolina strain. Assuming no interference:

a. What proportion of the backcross progeny will resemble the Virginia strain for all three traits?

b. What proportion of the backcross progeny will resemble the Carolina strain for all three traits?

c. What proportion of the backcross progeny will have the leaf morphology and leaf size of the Virginia strain but the leaf color of the Carolina strain?

d. What proportion of the backcross progeny will have the leaf morphology and leaf color of the Virginia strain but the leaf size of the Carolina strain?

30. In humans, the correlation between recombination frequency and length of DNA sequence is, on average, 1 million bp per 1% RF. During the process of mapping the Huntington disease gene (*HD*), it was found that *HD* was linked to a DNA marker called *G8* with an RF of 5%. Surprisingly, when the *HD* gene was finally identified, its physical distance from *G8* was found to be about 500,000 bp, instead of the expected 5 million bp. How can this observation be explained?

31. The following list of four *Drosophila* mutations indicates the symbol for the mutation, the name of the gene, and the mutant phenotype:

Allele symbol	Gene name	Mutant phenotype
dwp	*dwarp*	small body, warped wings
rmp	*rumpled*	deranged bristles
pld	*pallid*	pale wings
rv	*raven*	dark eyes and bodies

You perform the following crosses with the indicated results:

Cross #1: dwarp, rumpled females × pallid, raven males → dwarp, rumpled males and wild-type females

Cross #2: pallid, raven females × dwarp, rumpled males → pallid, raven males and wild-type females

F_1 females from cross #1 were crossed to males from a true-breeding *dwarp rumpled pallid raven* stock. The 1000 progeny obtained were as follows:

pallid	3
pallid, raven	428
pallid, raven, rumpled	48
pallid, rumpled	23
dwarp, raven	22
dwarp, raven, rumpled	2
dwarp, rumpled	427
dwarp	47

Indicate the best map for these four genes, including all relevant data. Calculate interference values where appropriate.

32. a. Explain in a qualitative sense how Fig. 5.7a suggests the existence of interference along mouse chromosomes.

    b. If you could examine many photographs similar to Fig. 5.7a, how would you apply statistics to provide evidence for the existence of interference during meiosis in mice? (No equations required; just outline the logic involved.)

33. The total length of the mouse genetic map is 1386 cM measured in males but 1817 cM measured in females. If Fig. 5.7a on shows a prophase I representative of all primary spermatocytes, how many recombination nodules would you expect to find in a representative prophase I primary oocyte?

34. The Fast Forward Box *Mapping the Crossovers that Generate the Chromosomes of Individual Human Sperm* describes how scientists used the whole genome DNA sequencing of ~100 single sperm from the same man to locate the crossovers involved in the production of each of these sperm. To determine the sequences of the parental chromosomes between which recombination occurred in the man's primary spermatocytes, these researchers isolated and then sequenced the DNA of individual chromosomes from the man's somatic cells. However, although this last step provided useful confirmation, it was not actually essential for determining the two DNA sequences for each chromosome the man inherited, one from his mother and one from his father. Instead, as you will demonstrate, this information can be inferred from the DNA sequences of the ~100 single sperm.

    a. Diagram three syntenic autosomal SNP loci for which the man is heterozygous, and suppose that the middle SNP locus is separated from each of those flanking it by 25 m.u. Use alleles of your choice for each of the three loci in your diagram.

    b. For simplicity, assume that interference = 1. Now, accounting only for these three loci, how many different types of sperm would you see, and in what frequencies?

    c. Given the results shown in the Fast Forward Box, explain why the assumption that interference = 1 is reasonable in this case.

    d. Explain how you would design a computer program that could reconstruct the DNA sequences of each chromosome in the somatic cells of a man, given the DNA sequences of 100 single sperm he produces.

Section 5.4

35. Do the data that Mendel obtained fit his hypotheses? For example, Mendel obtained 315 yellow round, 101 yellow wrinkled, 108 green round, and 32 green wrinkled seeds from the selfing of *Yy Rr* individuals (a total of 556). His hypotheses of segregation and independent assortment predict a 9:3:3:1 ratio in this case. Use the chi-square test to determine whether Mendel's data are significantly different from what he predicted. (The chi-square test did not exist in Mendel's day, so he was not able to test his own data for goodness of fit to his hypotheses.)

36. Two genes control color in corn snakes as follows: *O– B–* snakes are brown, *O– bb* are orange, *oo B–* are black, and *oo bb* are albino. An orange snake was mated to a black snake, and a large number of F_1 progeny were obtained, all of which were brown. When the F_1 snakes were mated to one another, they produced 100 brown offspring, 25 orange, 22 black, and 13 albino.

    a. What are the genotypes of the F_1 snakes?

    b. What proportions of the different colors would have been expected among the F_2 snakes if the two loci assort independently?

c. Do the observed results differ significantly from what was expected, assuming independent assortment is occurring?

d. What is the probability that differences this great between observed and expected values would happen by chance?

37. A mouse from a true-breeding population with normal gait was crossed to a mouse displaying an odd gait called *dancing*. The F_1 animals all showed normal gait.

a. If dancing is caused by homozygosity for the recessive allele of a single gene, what proportion of the F_2 mice should be dancers?

b. If mice must be homozygous for recessive alleles of each of two different genes to have the dancing phenotype, what proportion of the F_2 should be dancers if the two genes are unlinked? (Assume that all the mice in the population with normal gait were homozygous for dominant alleles.)

c. When the F_2 mice were obtained, 42 normal and 8 dancers were seen. Use the chi-square test to determine if these results better fit the one-gene model from part (a) or the two-gene model from part (b).

Section 5.5

38. *Neurospora* of genotype $a + c$ are crossed with *Neurospora* of genotype $+ b +$. (Here, + is shorthand for the wild-type allele.) The following tetrads are obtained (note that the genotype of the four spore pairs in an ascus are listed, rather than listing all eight spores):

$a + c$	$a\ b\ c$	$+ + c$	$+ b\ c$	$a\ b +$	$a + c$
$a + c$	$a\ b\ c$	$a + c$	$a\ b\ c$	$a\ b +$	$a\ b\ c$
$+ b +$	$+ + +$	$+ b +$	$+ + +$	$+ + c$	$+ + +$
$+ b +$	$+ + +$	$a\ b +$	$a + +$	$+ + c$	$+ b +$
137	141	26	25	2	3

a. In how many cells has meiosis occurred to yield these data?

b. Give the best genetic map to explain these results. Indicate all relevant genetic distances, both between genes and between each gene and the centromere.

c. Diagram a meiosis that could give rise to one of the three tetrads in the class at the far right in the list.

39. A cross was performed between one haploid strain of yeast with the genotype $a\,f\,g$ and another haploid strain with the genotype $\alpha\,f^+\,g^+$ (a and α are mating types). The resulting diploid was sporulated, and a random sample of 101 of the resulting haploid spores was analyzed. The following genotypic frequencies were seen:

α	f^+	g^+	31
a	f	g	29
a	f	g^+	14
α	f^+	g	13
a	f^+	g	6
α	f	g^+	6
a	f^+	g^+	1
α	f	g	1

a. Map the loci involved in the cross.

b. Assuming all three genes are on the same chromosome arm, is it possible that a particular ascus could contain an $\alpha\,f\,g$ spore but not an $a\,f^+\,g^+$ spore? If so, draw a meiosis that could generate such an ascus.

40. A cross was performed between a yeast strain that requires methionine and lysine for growth ($met^-\ lys^-$) and another yeast strain, which is $met^+\ lys^+$. One hundred asci were dissected, and colonies were grown from the four spores in each ascus. Cells from these colonies were tested for their ability to grow on petri plates containing either minimal medium (min), min + lysine (lys), min + methionine (met), or min + lys + met. The asci could be divided into two groups based on this analysis:

Group 1: In 89 asci, cells from two of the four spore colonies could grow on all four kinds of media, while the other two spore colonies could grow only on min + lys + met.

Group 2: In 11 asci, cells from one of the four spore colonies could grow on all four kinds of petri plates. Cells from a second one of the four spore colonies could grow only on min + lys plates and on min + lys + met plates. Cells from a third of the four spore colonies could only grow on min + met plates and on min + lys + met. Cells from the remaining colony could only grow on min + lys + met.

a. What are the genotypes of each of the spores within the two groups of asci?

b. Are the *lys* and *met* genes linked? If so, what is the map distance between them?

c. If you could extend this analysis to many more asci, you would eventually find some asci with a different pattern. For these asci, describe the phenotypes of the four spores. List these phenotypes as the ability of dissected spores to form colonies on the four kinds of petri plates.

41. Two crosses were made in *Neurospora* involving the mating type locus and either the *ad* or *p* genes. In both cases, the mating type locus (*A* or *a*) was one of the loci whose segregation was scored. One cross was $ad\ A \times ad^+\ a$ (cross i), and the other was $p\ A \times p^+\ a$ (cross ii). From cross i, 10 parental ditype,

9 nonparental ditype, and 1 tetratype asci were seen. From cross ii, the results were 24 parental ditype, 3 nonparental ditype, and 27 tetratype asci.

a. What are the linkage relationships between the mating type locus and the other two loci?

b. Although these two crosses were performed in *Neurospora,* you cannot use the data given to calculate centromere-to-gene distances for any of these genes. Why not?

42. Indicate the percentage of tetrads that would have 0, 1, 2, 3, or 4 viable spores after *Saccharomyces cerevisiae a* / α diploids of the following genotypes are sporulated:

a. A true-breeding wild-type strain (with no mutations in any gene essential for viability).

b. A strain heterozygous for a null (completely inactivating) mutation in a single essential gene.

For the remaining parts of this problem, consider crosses between yeast strains of the form *a* × *b,* where *a* and *b* are both temperature-sensitive mutations in different essential genes. The cross is conducted under permissive (low-temperature) conditions. Indicate the percentage of tetrads that would have 0, 1, 2, 3, or 4 viable spores subsequently measured under restrictive (high-temperature) conditions.

c. *a* and *b* are unlinked, and both are 0 m.u. from the centromere.

d. *a* and *b* are unlinked; *a* is 0 m.u. from the centromere, while *b* is 10 m.u. from the centromere.

e. *a* and *b* are 0 m.u. apart.

f. *a* and *b* are 10 m.u. apart. Assume all crossovers between *a* and *b* are SCOs (single crossovers).

g. In part (f), if a four-strand DCO (double crossover) occurred between *a* and *b,* how many of the spores in the resulting tetrad would be viable at high temperature?

43. The *a, b,* and *c* loci are all on different chromosomes in yeast. When $a\ b^+$ yeast were crossed to $a^+\ b$ yeast and the resultant tetrads analyzed, it was found that the number of nonparental ditype tetrads was equal to the number of parental ditypes, but there were no tetratype asci at all. On the other hand, many tetratype asci were seen in the tetrads formed after $a\ c^+$ was crossed with $a^+\ c$, and after $b\ c^+$ was crossed with $b^+\ c$. Explain these results.

44. This problem leads you through the derivation of a corrected equation for RF in yeast tetrad analysis that takes into account double crossover (DCO) meioses.

A yeast strain that cannot grow in the absence of the amino acid histidine (his^-) is mated with a yeast strain that cannot grow in the absence of the amino acid lysine (lys^-). Among the 400 unordered tetrads resulting from this mating, 233 were PD, 11 were NPD, and 156 were T.

a. What types of spores are in the PD, NPD, and T tetrads?

b. Are the *his* and *lys* genes linked? How do you know?

c. Using the simple equation RF = 100 × [NPD + (1/2)T]/total tetrads, calculate the distance in map units between the *his* and *lys* genes.

d. If you think about all the kinds of meiotic events that could occur (refer to Fig. 5.24), you can see that the calculation you did in part (c) may substantially underestimate RF. What kinds of meioses (NCO, SCO, or DCO) generated each of the tetrad types in this cross?

e. What incorrect assumptions does the simple RF equation you used in part (c) make about the meiotic events producing each type of tetrad? When could these assumptions nevertheless be correct?

f. Use your answers to part (d) to determine the number of NCO, SCO, and DCO meioses that generated the 400 tetrads.

g. Use your answers to part (f) to write a general equation that relates the number of DCO meioses to the number of the various tetrad types. Then write another general equation that computes the number of SCO meioses as a function of the number of the various tetrad types.

h. Based on your answer to part (f), calculate the average number of crossovers per meiosis (*m*) between *his* and *lys.*

i. Use your answer to (h) to write an equation for *m* in terms of NCO, SCO, and DCO meioses.

j. What is the relationship between RF and *m*?

k. Use your answer to part (j) to write a corrected equation for RF in terms of SCO, DCO, and NCO meioses.

l. Using your answer to part (g), rewrite the corrected RF equation from part (k) in terms of the numbers of the various tetrad types.

m. The equation you just wrote in part (l) is a corrected equation for RF that takes into account double crossovers that would otherwise have been missed. Use this improved formula to calculate a more accurate distance between the *his* and *lys* genes than the one you calculated in part (c).

45. a. In ordered tetrad analysis, what is the maximum RF that you could observe between a gene and the centromere? (*Hint:* What RF would you observe between a gene and the centromere if the gene were distant enough from the centromere so that at least one crossover, and often more than one

crossover, occurs between them in every meiosis?)

b. Can a gene and the centromere be unlinked? Explain.

c. Suppose in an ordered tetrad analysis you observe an RF of 30% between a gene and the centromere. Given your answer to part (a), do think that 30 m.u. is an accurate estimate of the distance between the gene and the centromere?

46. A research group has selected three independent Trp^- haploid strains of *Neurospora,* each of which cannot grow in the absence of the amino acid tryptophan. They first mated these three strains with a wild-type strain of opposite mating type, and then they analyzed the resultant octads. For all three matings, two of the four spore pairs in every octad could grow on minimal medium (that is, in the absence of tryptophan), while the other two spore pairs were unable to grow on this minimal medium.

a. What can you conclude from this result?

In the matings of mutant strains 1 and 2 with wild type, one of the two topmost pairs in some octads had spores that could grow on minimal medium while the other of the two topmost pairs in the same octads had spores that could not grow on minimal medium. In the mating of mutant strain 3 with wild type, either all the spores in the two topmost pairs could grow on minimal medium or all could not grow on minimal medium.

b. What can you conclude from this result?

The researchers next prepared two separate cultures of each mutant strain; one of these cultures was of mating type *A* and the other of mating type *a.* They mated these strains in pairwise fashion, dissected the resultant octads, and determined how many of the individual spores could grow on minimal medium. The results are shown here.

Mating	**% of octads with *x* number of spores viable on minimal medium**				
	$x = 0$	2	4	6	8
1×2	78	22	0	0	0
1×3	46	6	48	0	0
2×3	42	16	42	0	0

c. For each of the three matings in the table, how many of the 100 octads are PD? NPD? T?

d. Draw a genetic map explaining all of the preceding data. Assume that the sample sizes are sufficiently small that none of the octads are the result of double crossovers.

e. Although this problem describes crosses in *Neurospora,* it does not help in this case to present the matings in the table as ordered octads. Why not?

f. Why in this problem can you obtain gene ↔ centromere distances from the crosses in the table, even though the data are not presented as ordered octads?

Section 5.6

47. A single yeast cell placed on a solid agar will divide mitotically to produce a colony of about 10^7 cells. A haploid yeast cell that has a mutation in the *ade2* gene will produce a red colony; an $ade2^+$ colony will be white. Some of the colonies formed from diploid yeast cells with a genotype of $ade2^+ / ade2^-$ will contain sectors of red within a white colony.

a. How would you explain these sectors?

b. Although the white colonies are roughly the same size, the red sectors within some of the white colonies vary markedly in size. Why? Do you expect the majority of the red sectors to be relatively large or relatively small?

48. Figure 5.29 shows mitotic recombination leading to single spots or twin spots occurring in the G_2 stage of the cell cycle (after the chromosomes have replicated). However, because it usually is initiated by rare, random events of chromosome breakage, mitotic recombination can also take place in G_1, prior to S phase. Redraw Fig. 5.29 with mitotic recombination taking place in G_1 rather than G_2, and demonstrate why any such event could not yield single spots or twin spots.

49. A diploid strain of yeast has a wild-type phenotype but the following genotype:

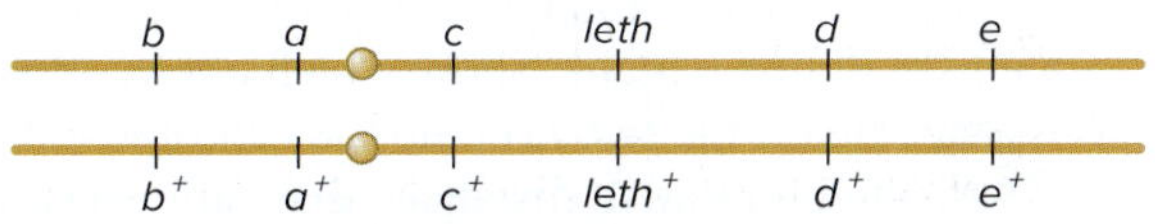

a, b, c, d, and *e* all represent recessive alleles that yield a visible phenotype, and *leth* represents a recessive lethal mutation. All genes are on the same chromosome, and *a* is very tightly linked to the centromere (indicated by a small circle). Which of the following phenotypes could be found in sectors resulting from mitotic recombination in this cell? (1) *a;* (2) *b;* (3) *c;* (4) *d;* (5) *e;* (6) *b e;* (7) *c d;* (8) *c d e;* (9) *d e;* (10) *a b.* Assume that double mitotic crossovers are too rare to be observed.

50. In *Drosophila,* the *yellow* (*y*) gene is near the telomere of the long arm of the acrocentric X chromosome, while the *singed* (*sn*) gene is located near the middle of the same X chromosome arm. On the wings of female flies of genotype $y\ sn / y^+\ sn^+$, you can very rarely find patches of yellow tissue within which a small subset of cells also have singed bristles.

a. How can you explain this phenomenon?

b. Would you find similar patches on the wings of females having the genotype $y^+\ sn / y\ sn^+$?

51. Neurofibromas are tumors of the skin that can arise when a skin cell that is originally $NF1^+/NF1^-$ loses the $NF1^+$ allele. This wild-type allele encodes a functional protein (called a *tumor suppressor*), while the $NF1^-$ allele encodes a nonfunctional protein.

    A patient of genotype $NF1^+/NF1^-$ has 20 independent tumors in different areas of the skin. Samples are taken of normal, noncancerous cells from this patient, as well as of cells from each of the 20 tumors. Extracts of these samples are analyzed by a technique called gel electrophoresis that can detect variant forms of four different proteins (A, B, C, and D) all encoded by genes that lie on the same autosome as *NF1*. Each protein has a slow (S) and a fast (F) form that are encoded by different alleles (for example, A^S and A^F). In the extract of normal tissue, slow and fast variants of all four proteins are found. In the extracts of the tumors, 12 had only the fast variants of proteins A and D but both the fast and slow variants of proteins B and C; 6 had only the fast variant of protein A but both the fast and slow variants of proteins B, C, and D; and the remaining 2 tumor extracts had only the fast variant of protein A, only the slow variant of protein B, the fast and slow variants of protein C, and only the fast variant of protein D.

    a. What kind of genetic event described in this chapter could cause all 20 tumors, assuming that all the tumors are produced by the same mechanism?
    b. Draw a genetic map describing these data, assuming that this small sample represents all the types of tumors that could be formed by the same mechanism in this patient. Show which alleles of which genes lie on the two homologous chromosomes. Indicate all relative distances that can be estimated. Note that *NF1* is one of the genes you can map in this way.
    c. Another mechanism that can lead to neurofibromas in this patient is a mitotic error producing cells with 45 rather than the normal 46 chromosomes. How can this mechanism cause tumors? How do you know, just from the results described, that none of these 20 tumors is formed by such mitotic errors?
    d. Can you think of any other type of error that could produce the results described?

52. Two important methods for understanding the genetic basis for development are mitotic crossing-over and the use of the gene from jellyfish called *GFP* (for green fluorescent protein) that makes these animals glow in the dark. By recombinant DNA techniques described later in the book, you can insert the jellyfish *GFP* gene anywhere into the genome of organisms like *Drosophila* or mice. Cells expressing this *GFP* gene will glow green in the microscope, while those without the *GFP* gene will not glow green.

    Mice homozygous for the recessive mutation *small cells* (*smc*) die as early embryos because their cells divide prematurely before they reach normal size.

    You want to design a mouse carrying one copy of the *GFP* gene and heterozygous for *smc* in which you could generate clones in adult mice by mitotic recombination. In this designer mouse, every cell in every clone that is not green would be homozygous for the *smc* mutation. The figure below shows a field of epithelial cells in the mouse you design. You will see some cells that are normal size and other cells that are small. You will also see cells of three different colors: blank, weakly glowing cells (*light green*), and brightly glowing cells (*dark green*). Most of the cells in the epithelium of this mouse are of normal size and weakly glowing. The epithelium also contains three clones of cells (1, 2, and 3) that have unusual appearances due to the occurrence of mitotic recombination.

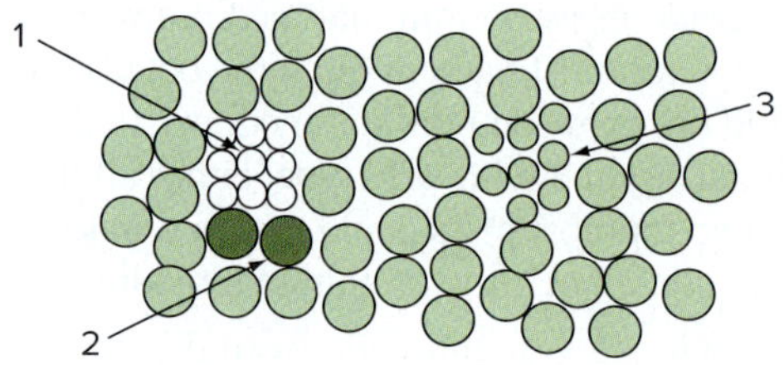

    a. Show the chromosomes and centromeres, the alleles smc^+ and *smc,* and GFP^+ (*GFP* gene present) and GFP^- (*GFP* gene absent) in your designer mouse. (As a reminder, this mouse will carry one copy of the *GFP* gene and will be heterozygous for *smc*. Every cell in every clone generated by mitotic recombination that is not green should be homozygous for the *smc* mutation.)
    b. Why do you need to use mitotic recombination to study the function of smc^+ in adult mice?
    c. Why do you see cells of three different colors?
    d. Why are clones 1 and 2 next to each other?
    e. On your map in part (a), place an arrow to show the position of a mitotic recombination event that could give rise to clones 1 and 2.
    f. Why do more cells exist in clone 1 than in clone 2?
    g. On your map in part (a), place an arrow to show the position of a mitotic crossover that could give rise to clone 3.

chapter 6

DNA Structure, Replication, and Recombination

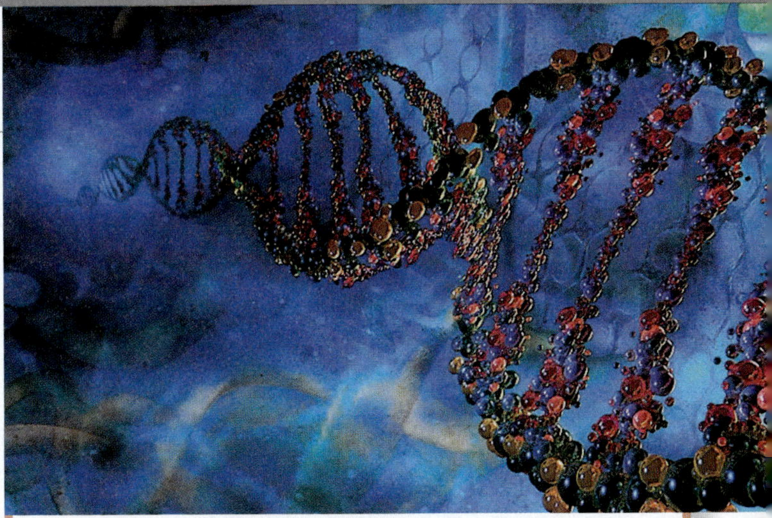

The double-helical structure of DNA provides an explanation for the transmission of genetic information from generation to generation over billions of years.

chapter outline

FOR NEARLY 4 BILLION YEARS, the double-stranded **DNA** molecule has served as the bearer of genetic information. It was present in the earliest single-celled organisms and in every other organism that has existed since. Over that long period of time, the *hardware*—the structure of the molecule itself—has not changed. In contrast, evolution has honed and vastly expanded the *software*—the programs of genetic information that the molecule stores, expresses, and transmits from one generation to the next.

Under special conditions of little or no oxygen, DNA can withstand a wide range of temperature, pressure, and humidity and remain relatively intact for hundreds, thousands, even tens of thousands of years. Molecular sleuths have retrieved the evidence: 38,000-year-old DNA from a Neanderthal skeleton (**Fig. 6.1**). Amazingly, this ancient DNA still carries readable sequences—shards of decipherable information that act as time machines for the viewing of genes in this long-vanished species. Comparisons with homologous DNA segments from living people make it possible to identify the precise mutations that have fueled evolution.

For example, comparisons of Neanderthal and human DNA have helped anthropologists settle a long-running debate about the genetic relationship of the two. The evidence shows that Neanderthals and our own species, *Homo sapiens,* last shared a common ancestor between 600,000 and 800,000 years ago. Neanderthal ancestors migrated to Europe about 400,000 years ago while our own ancestors remained in Africa. The two groups remained out of contact until 40,000 years ago, when *Homo sapiens* first arrived in Europe. Within a few millennia, the Neanderthals were extinct. However, their recently recovered DNA suggests that during the 10,000 years that Neanderthals shared Europe with *Homo sapiens*, some interbreeding took place; 1–4% of the genomes of modern non-Africans can be traced to Neanderthals.

Francis Crick, co-discoverer of DNA's double-helical structure and a leading twentieth-century theoretician of molecular biology, wrote that "almost all aspects of life are engineered at the molecular level, and without understanding molecules,

Figure 6.1 Ancient DNA still carries information. Molecular biologists have successfully extracted and determined the sequence of DNA from a 38,000-year-old Neanderthal skull. These findings attest to the chemical stability of DNA, the molecule of inheritance.

we can only have a very sketchy understanding of life itself." For this reason, we shift our perspective in this chapter to an examination of DNA, the molecule of which genes are made.

As we extend our analysis to the molecular level, bear in mind two important themes. First, DNA's genetic functions flow directly from its molecular structure—the way its atoms are arranged in space. Second, all of DNA's genetic functions depend on specialized proteins that interact with it and *read* the information it carries, because DNA itself is chemically inert. In fact, DNA's lack of chemical reactivity makes it an ideal physical container for long-term maintenance of genetic information in living organisms, as well as their nonliving remains.

6.1 Experimental Evidence for DNA as the Genetic Material

learning objectives

1. Describe the chemical components of DNA.
2. Summarize the methods that located DNA in chromosomes.
3. Explain how Avery and his colleagues demonstrated bacterial transformation, and explain the significance of this finding.
4. Describe the blender experiments of Hershey and Chase and what the results revealed about DNA's function.

At the beginning of the twentieth century, geneticists did not know that DNA was the genetic material. It took a cohesive pattern of results from experiments performed over more than 50 years to convince the scientific community that DNA is the molecule of heredity. We now present key pieces of the evidence.

Chemical Studies Locate DNA in Chromosomes

In 1869, Friedrich Miescher extracted a weakly acidic, phosphorus-rich material from the nuclei of human white blood cells and named it *nuclein.* It was unlike any chemical compound reported previously. Nuclein's major component turned out to be DNA, although it also contained some contaminants. The full chemical name of DNA is **deoxyribonucleic acid,** reflecting three characteristics of the substance: One of its constituents is a sugar known as deoxyribose; it is found mainly in cell nuclei; and it is acidic.

After purifying DNA from the nuclein by chemical means, researchers established that contains only four distinct chemical building blocks linked in a long chain (**Fig. 6.2**). The four individual components belong to a class of compounds known as **nucleotides;** the bonds joining one nucleotide to another are covalent **phosphodiester bonds;** and the linked chain of building block subunits is a type of **polymer.**

A procedure first reported in 1923 made it possible to discover where in the cell DNA resides. Named the *Feulgen reaction* after its designer, the procedure relies on a chemical which stains DNA red. In a preparation of stained cells, the chromosomes redden, while other areas of the cell remain relatively colorless. The reaction shows that DNA is localized almost exclusively within chromosomes.

The finding that DNA is a component of chromosomes does not itself prove that the molecule has anything to do with genes. Typical eukaryotic chromosomes also contain an even greater amount of protein by weight. Because proteins are built of 20 different amino acids, whereas DNA is made of only four different nucleotides, many researchers thought proteins had greater potential for diversity and were better suited to serve as the genetic material. These same scientists assumed that even though DNA was an important part of chromosome structure, it was too simple to contain the complexity of genes.

Bacterial Transformation Implicates DNA as the Genetic Material

Several studies eventually promoted the idea that DNA would be the chemical substance that carries genetic information. The most important of these used single-celled bacteria as experimental organisms. Bacteria carry their genetic material in a single circular chromosome that lies within the cell without being enclosed in a nuclear membrane. With

Figure 6.2 The chemical composition of DNA. A single strand of a DNA molecule consists of a chain of *nucleotide subunits* (*blue boxes*). Each nucleotide is made of the sugar **deoxyribose** (*tan pentagons*) connected to an inorganic **phosphate group** (*yellow circles*) and to one of four **nitrogenous bases** (*purple* or *green polygons*). The *phosphodiester bonds* that link the nucleotide subunits to each other attach the phosphate group of one nucleotide to the deoxyribose sugar of the preceding nucleotide.

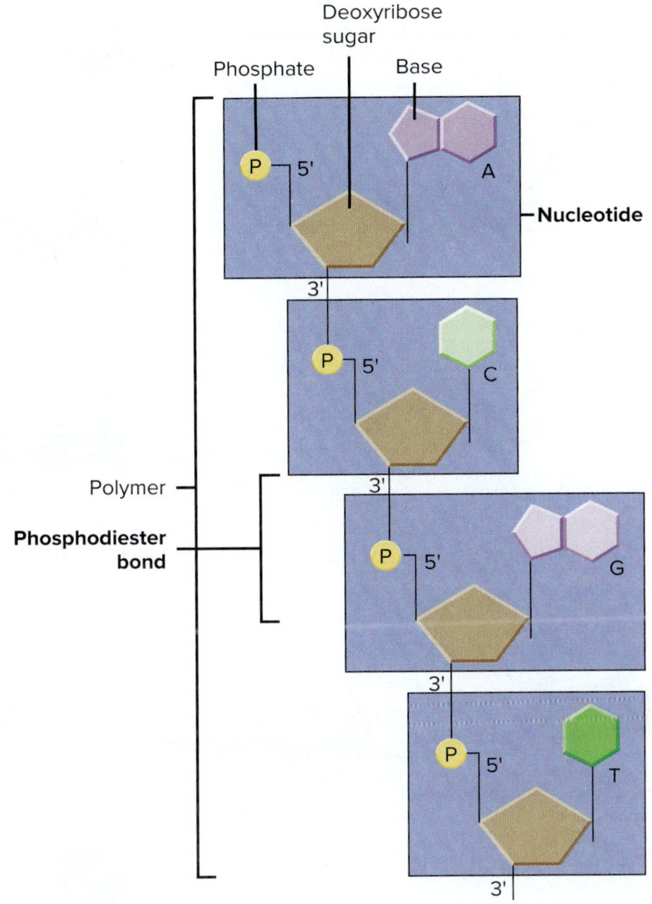

only one chromosome, bacteria do not undergo meiosis to produce germ cells, and they do not apportion their replicated chromosomes to daughter cells by mitosis; rather, they divide by a process known as *binary fission*. In spite of these obvious differences, at least some investigators in the first half of the twentieth century thought that the genetic material of bacteria might be the same as that found in eukaryotic organisms.

One prerequisite of genetic studies in bacteria, as with any species, is the detection of alternative forms of a trait among individuals in a population. In a 1923 study of *Streptococcus pneumoniae* bacteria grown in laboratory media, Frederick Griffith distinguished two bacterial forms: smooth (S) and rough (R). S is the wild type; a mutation in S gives rise to R. From observation and biochemical analysis, Griffith determined that S forms appear

Figure 6.3 Smooth (S) and rough (R) colonies of *S. pneumoniae*.

From: Arnold et al., "New associations with *Pseudomonas luteola* bacteremia: A veteran with a history of tick bites and a trauma patient with pneumonia," *The Internet Journal of Infectious Diseases,* 2005, 4(2): 1-5, Fig. 1 © & Courtesy of Dr. Forest Arnold, University of Louisville. Used with permission.

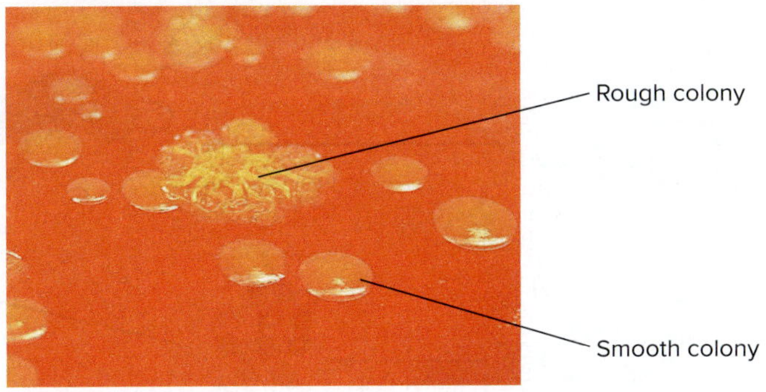

smooth because they synthesize a polysaccharide capsule that surrounds pairs of cells. R forms, which arise spontaneously as mutants of S, cannot make the capsular polysaccharide, and as a result, their colonies appear to have a rough surface (**Fig. 6.3**). We now know that the R form lacks an enzyme necessary for synthesis of the capsular polysaccharide. Because the polysaccharide capsule helps protect the bacteria from an animal's immune response, the S bacteria are virulent and kill most laboratory animals exposed to them (**Fig. 6.4a**); by contrast, the R forms fail to cause infection (**Fig. 6.4b**). In humans, the virulent S forms of *S. pneumoniae* can cause pneumonia.

The phenomenon of transformation

In 1928, Griffith published the astonishing finding that genetic information from dead bacterial cells could somehow be transmitted to live cells. He was working with two types of the *S. pneumoniae* bacteria—live R forms and heat killed S forms. Neither the heat-killed S forms nor the live R forms produced infection when injected into laboratory mice (Fig. 6.4b and **Fig. 6.4c**), but a mixture of the two killed the animals (**Fig. 6.4d**). Furthermore, bacteria recovered from the blood of the dead animals were living S forms (Fig. 6.4d).

The ability of a substance to change the genetic characteristics of an organism is known as **transformation.** Something from the heat-killed S bacteria must have transformed the living R bacteria into S. This transformation was permanent and most likely genetic, because all future generations of the bacteria grown in culture were the S form.

DNA as the active agent of transformation

By 1929, two other laboratories had repeated these results, and in 1931, investigators in Oswald T. Avery's laboratory found they could achieve transformation without using any animals at all, simply by growing R-form bacteria in

Figure 6.4 Transformation. (a) S bacteria are virulent and can cause lethal infections when injected into mice. **(b)** Injections of R mutants by themselves do not cause infections that kill mice. **(c)** Similarly, injections of heat-killed S bacteria do not cause lethal infections. **(d)** Lethal infection does result, however, from injections of live R bacteria mixed with heat-killed S strains; the blood of the dead host mouse contains living S-type bacteria.

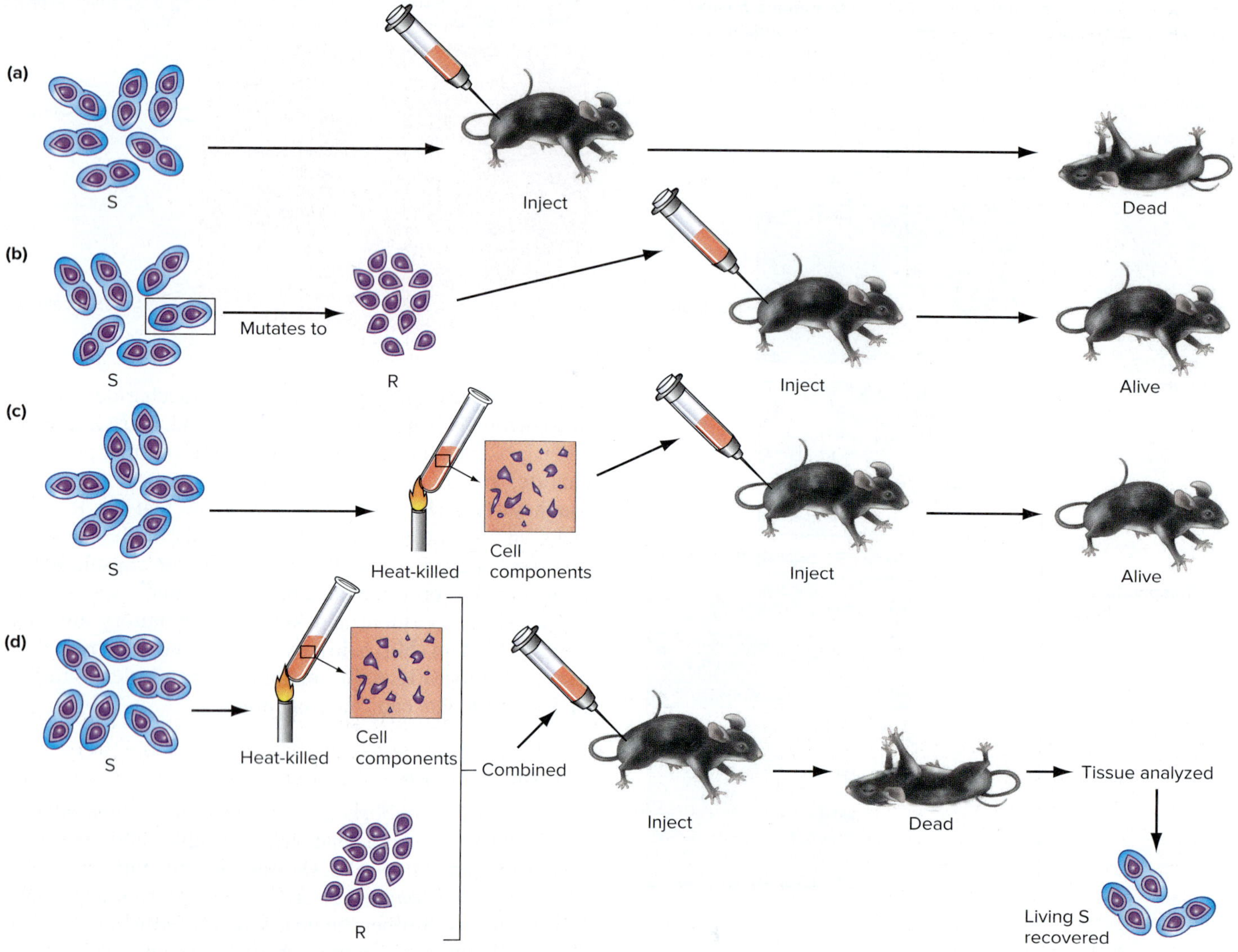

medium in the presence of components from dead S forms (**Fig. 6.5a**). Avery then embarked on a quest that would remain the focus of his work for almost 15 years: "Try to find in that complex mixture, the active principle!" In other words, try to identify the heritable substance in the bacterial extract that induces the transformation of harmless R bacteria into pathogenic S bacteria. Avery dubbed the substance he was searching for the *transforming principle* and spent many years trying to purify it sufficiently to be able to identify it unambiguously. He and his coworkers eventually prepared a tangible, active transforming principle. In the final part of their procedure, a long, whitish wisp materialized from ice-cold alcohol solution and wound around the glass stirring rod to form a fibrous wad of nearly pure transforming principle (**Fig. 6.5b**).

Once purified, the transforming principle had to be characterized. In 1944, Avery and two coworkers, Colin MacLeod and Maclyn McCarty, published the cumulative findings of experiments designed to determine the transforming principle's chemical composition (**Fig. 6.5c**). In these experiments, the purified transforming principle was active at the extraordinarily high dilution of 1 part in 600 million. Although the preparation was almost pure DNA, the investigators nevertheless exposed it to various enzymes to see if some molecule other than DNA might have caused the transformation. Enzymes that degraded RNA, protein, or polysaccharide had no effect on the transforming principle, but an enzyme that degrades DNA completely destroyed its activity. The tentative published conclusion was that the transforming principle appeared to be DNA. In a personal

Figure 6.5 The transforming principle is DNA. (a) Bacterial transformation occurs in culture medium containing the remnants of heat-killed S bacteria. Some *transforming principle* from the heat-killed S bacteria is taken up by the live R bacteria, converting (transforming) them into virulent S strains. **(b)** Purified DNA extracted from human white blood cells. **(c)** Chemical fractionation of the transforming principle. Treatment of purified DNA with a DNA-degrading enzyme destroys its ability to cause bacterial transformation, while treatment with enzymes that destroy other kinds of macromolecules has no effect on the transforming principle.

b: © Phanie/Science Source

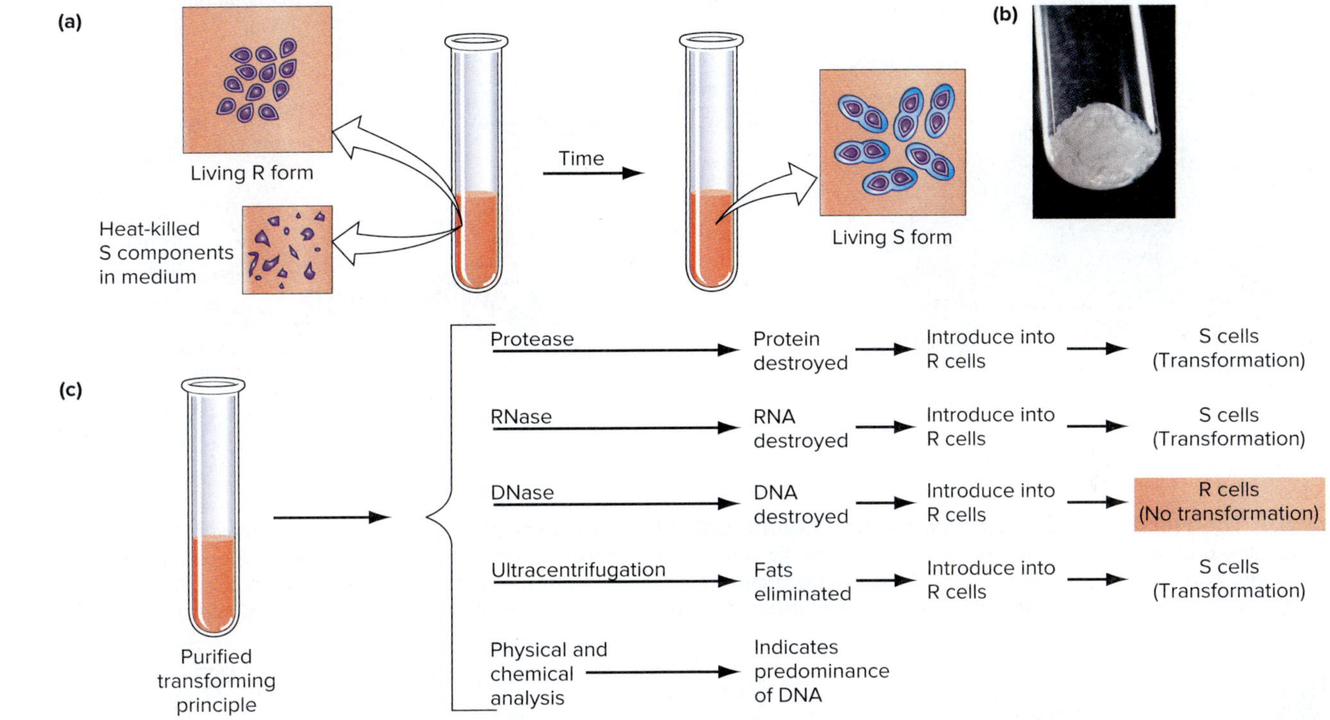

letter to his brother, Avery went one step further and confided that the transforming principle "may be a gene."

Despite the paper's abundance of concrete evidence, many within the scientific community still resisted the idea that DNA is the molecule of heredity. They argued that perhaps Avery's results reflected the activity of contaminants; or perhaps genetic transformation was not happening at all, and instead, the purified material somehow triggered a physiological switch that transformed bacterial phenotypes. Unconvinced for the moment, these scientists remained attached to the idea that proteins were the prime candidates for the genetic material.

DNA, Not Protein, Contains the Instructions for Virus Propagation

Not everyone shared this skepticism. Alfred Hershey and Martha Chase anticipated that they could assess the relative importance of DNA and protein in gene replication by infecting bacterial cells with viruses called **phages,** short for **bacteriophages** (literally *bacteria eaters*).

Viruses are the simplest of organisms. By structure and function, they fall somewhere between living cells capable of reproducing themselves and macromolecules such as proteins. Because viruses hijack the molecular machinery of their host cell to carry out growth and replication, they can be very small indeed and contain very few genes. For many kinds of phages, each particle consists of roughly equal weights of protein and DNA (**Fig. 6.6a**). These phage particles can reproduce themselves only after infecting a bacterial cell. Thirty minutes after infection, the cell bursts and hundreds of newly made phages spill out (**Fig. 6.6b**). The question is: What substance contains the information used to produce the new phage particles—DNA or protein?

With the invention of the electron microscope in 1939, it became possible to see individual phages, and surprisingly, electron micrographs revealed that the entire phage does not enter the bacterium it infects. Instead, a viral shell—called a *ghost*—remains attached to the outer surface of the bacterial cell wall. Because the empty phage coat remains outside the bacterial cell, one investigator likened phage particles to tiny syringes that bind to the cell surface and inject the material containing the information needed for viral replication into the host cell.

In their famous *Waring blender* experiment of 1952, Alfred Hershey and Martha Chase tested the idea that the ghost left on the cell wall is composed of protein, while the injected material consists of DNA (**Fig. 6.7**). A type of phage known as T2 served as their experimental system. Hershey and Chase grew two separate sets of T2 in bacteria

Figure 6.6 Experiments with viruses provide convincing evidence that genes are made of DNA. (a) and **(b)** Bacteriophage T2 structure and life cycle. The phage particle consists of DNA contained within a protein coat. The virus attaches to the bacterial host cell and injects its genes (the DNA) through the bacterial cell wall into the host cell cytoplasm. Inside the host cell, these genes direct the formation of new phage DNA and proteins, which assemble into progeny phages that are released into the environment when the cell bursts.

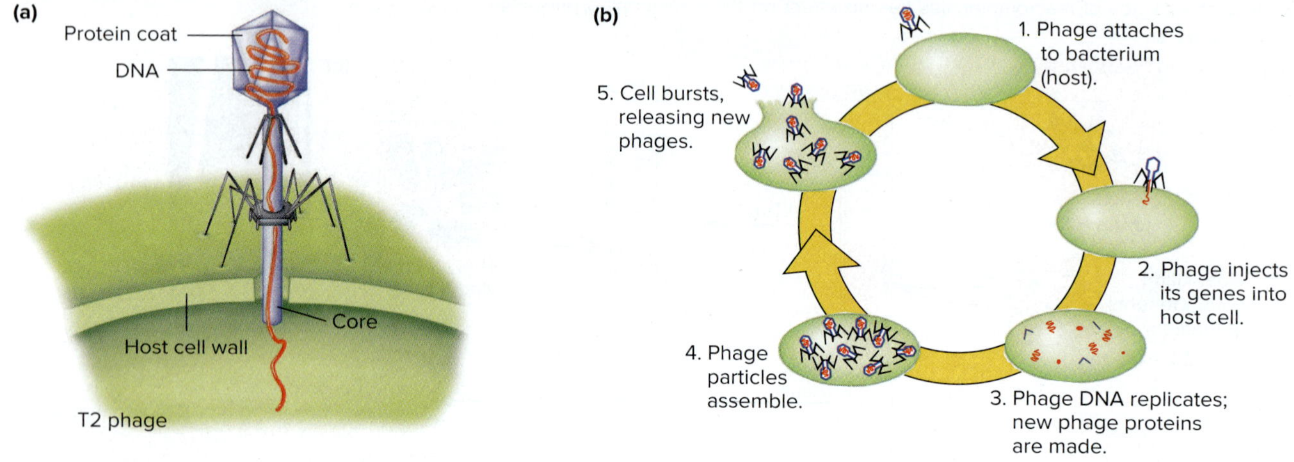

Figure 6.7 The Hershey-Chase Waring blender experiment. T2 bacteriophage particles either with ^{32}P-labeled DNA (*orange*) or with ^{35}S-labeled proteins (*purple*) were used to infect bacterial cells. After a short incubation, Hershey and Chase shook the cultures in a Waring blender and spun the samples in a centrifuge to separate the empty viral ghosts from the heavier infected cells. Most of the ^{35}S-labeled proteins remained with the ghosts, while most of the ^{32}P-labeled T2 DNA was found in the sedimented infected cells.

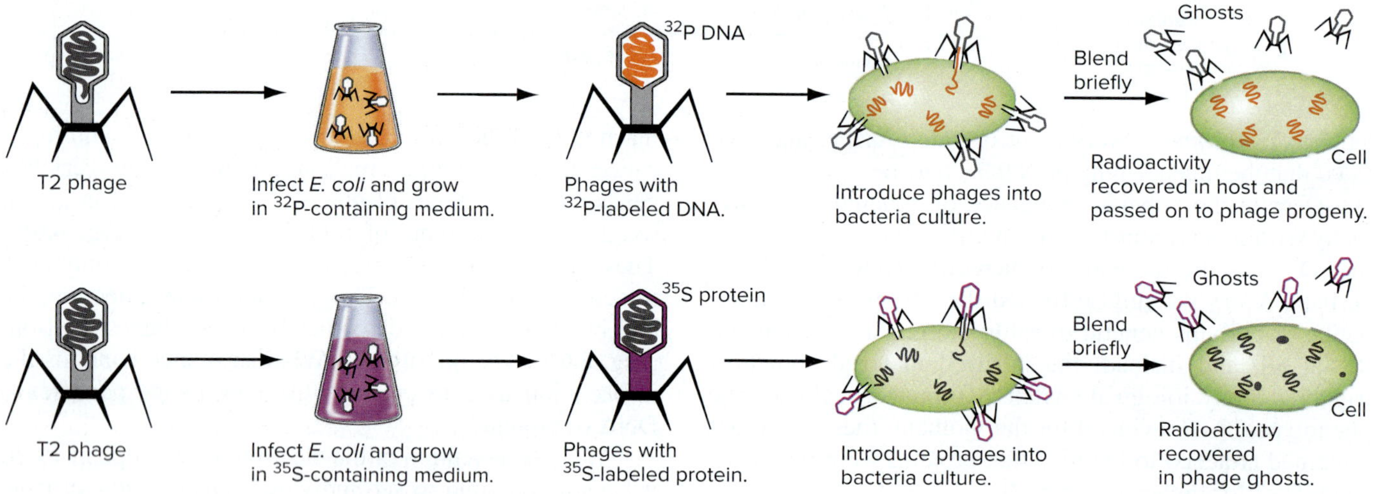

maintained in two different culture media, one infused with radioactively labeled phosphorus (^{32}P), the other with radioactively labeled sulfur (^{35}S). Because proteins incorporate sulfur but no phosphorus and DNA contains phosphorus but no sulfur, phages grown on ^{35}S would have radioactively labeled protein while particles grown on ^{32}P would have radioactive DNA. The radioactive tags would serve as markers for the location of each material when the phages infected fresh cultures of bacterial cells.

After exposing one fresh culture of bacteria to ^{32}P-labeled phages and another culture to ^{35}S-labeled phages, Hershey and Chase used a Waring blender to disrupt each one, effectively separating the viral ghosts from the bacteria harboring the viral genes. Centrifugation of the cultures then separated the heavier infected cells, which ended up in a pellet at the bottom of the tube, away from the lighter phage ghosts, which remained suspended in the supernatant solution. Most of the radioactive ^{32}P (in DNA) went to the pellet, while most of the radioactive ^{35}S (in protein) remained in the supernatant. This result confirmed that the extracellular ghosts were indeed mostly protein, while the injected viral material specifying production of more phages was mostly DNA. Bacteria containing the radio-labeled phage DNA behaved just as in a normal phage infection, producing and disgorging hundreds of progeny particles. From these observations, Hershey and Chase concluded that phage genes are made of DNA.

The Hershey-Chase experiment, although less rigorous than the Avery project, had an enormous impact. In the minds of many investigators, it confirmed Avery's results and extended them to viral particles. The spotlight was now on DNA.

essential concepts

- *DNA* is a polymer of *nucleotides* joined by *phosphodiester bonds*. Nucleotides are made of deoxyribose, phosphate, and one of four nitrogenous bases.
- DNA is localized almost exclusively in the chromosomes within the nucleus of a cell.
- Avery and his colleagues showed that a purified DNA preparation from S (virulent) bacteria could *transform* R (nonvirulent) bacteria into the S form; this result was strong evidence for DNA as the genetic material.
- Hershey and Chase grew T2 bacteriophages in the presence of either ^{35}S (labels proteins) or ^{32}P (labels DNA). They showed that it is the ^{32}P-tagged viral DNA that contains the genetic instructions to produce more virus particles.

6.2 The Watson and Crick Double Helix Model of DNA

learning objectives

1. Describe the key features of the Watson-Crick model for DNA structure.
2. Explain what is meant by the *antiparallel polarity* of the two strands of DNA within the double helix.
3. Distinguish the different structural forms of DNA from one another.

Under appropriate experimental conditions, purified molecules of DNA can align alongside each other in fibers to produce an ordered structure. And just as a crystal chandelier scatters light to produce a distinctive pattern on the wall, DNA fibers scatter X-rays to produce a characteristic *diffraction pattern* (**Fig. 6.8**). A knowledgeable X-ray crystallographer can interpret DNA's diffraction pattern to deduce certain aspects of the molecule's three-dimensional structure. When in the spring of 1951 the 23-year-old James Watson learned that DNA could project a diffraction pattern, he realized that it "must have a regular structure that could be solved in a straightforward fashion."

Figure 6.8 X-ray diffraction patterns reflect the helical structure of DNA. Photograph of an X-ray diffraction pattern produced by oriented DNA fibers, taken by Rosalind Franklin and Maurice Wilkins in late 1952. The crosswise pattern of X-ray reflections indicates that DNA is helical.

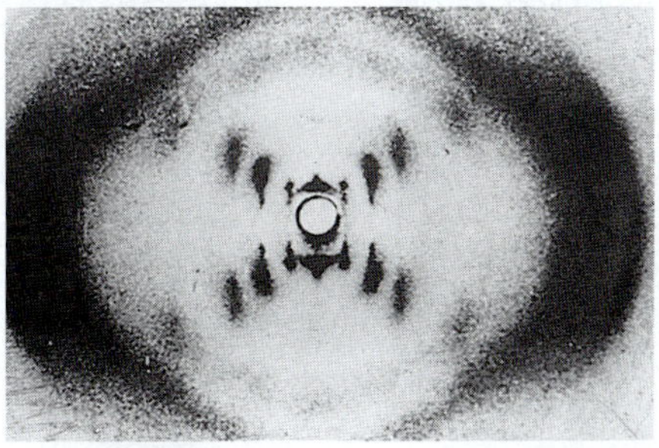

In this section, we analyze DNA's three-dimensional structure, looking first at significant details of the nucleotide building blocks, then at how those subunits are linked together in a polynucleotide chain, and finally, at how two chains associate to form a double helix.

Nucleotides Are the Building Blocks of DNA

DNA is a long polymer composed of subunits known as **nucleotides.** Each nucleotide consists of a *deoxyribose* sugar, a *phosphate*, and one of four *nitrogenous bases*. Detailed knowledge of these chemical constituents and the way they combine played an important role in Watson and Crick's model building.

The components of a nucleotide

Figure 6.9 depicts the chemical composition and structure of deoxyribose, phosphate, and the four nitrogenous bases; how these components come together to form a nucleotide; and how *phosphodiester bonds* link the nucleotides in a DNA chain. Each individual carbon or nitrogen atom in the central ring structure of a nitrogenous base is assigned a number: 1–9 for **purines,** and 1–6 for **pyrimidines.** The carbon atoms of the deoxyribose sugar are distinguished from atoms within the nucleotide base by the use of primed numbers from 1′ to 5′. Covalent attachment of a base to the 1′ carbon of deoxyribose forms a *nucleoside*. The addition of a phosphate group to the 5′ carbon forms a complete *nucleotide*.

Connecting nucleotides to form a DNA chain

As Fig. 6.9 shows, a DNA chain composed of many nucleotides has **polarity:** an overall direction. Phosphodiester

FEATURE FIGURE 6.9

A Detailed Look at DNA's Chemical Constituents

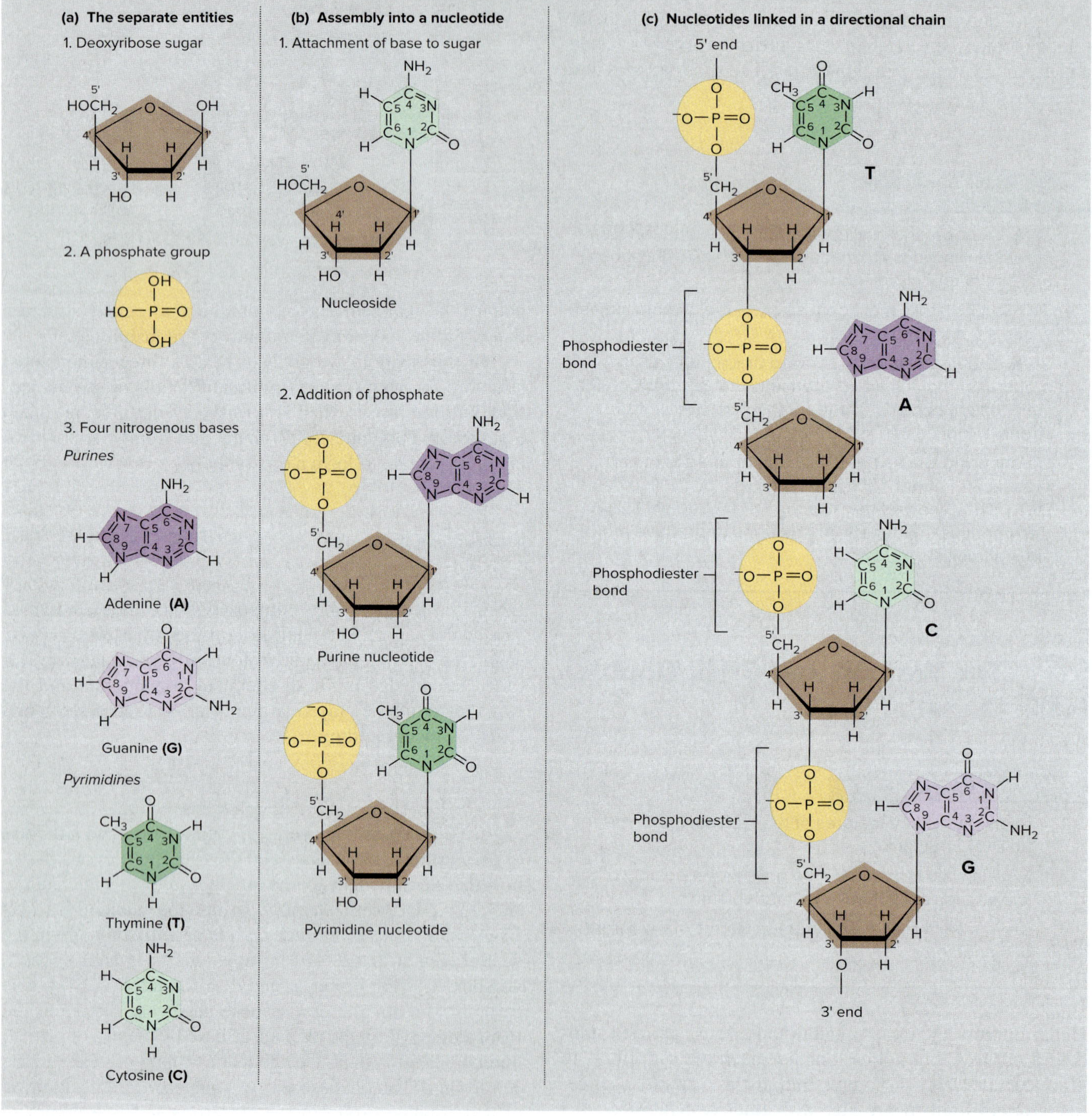

bonds always form a covalent link between the 3′ carbon of one nucleotide and the 5′ carbon of the following nucleotide. The consistent orientation of the nucleotide building blocks gives a chain overall direction, such that the two ends of a single chain are chemically distinct.

At the **5′ end,** the sugar of the terminal nucleotide has a free 5′ carbon atom, free in the sense that it is not linked to another nucleotide. Depending on how the DNA is synthesized or isolated, the 5′ carbon of the nucleotide at the 5′ end may carry either a hydroxyl or a phosphate group. At the other end of the chain—the **3′ end**—it is the 3′ carbon of the final nucleotide that is free. Along the chain between the two ends, this 5′-to-3′ polarity is conserved from nucleotide to nucleotide. By convention, a DNA chain is described in terms of its bases, written with the 5′-to-3′ direction going from left to right (unless otherwise noted). The chain depicted in Fig. 6.9c, for instance, would be 5′ TACG 3′.

DNA's information content

Information can be encoded only in a sequence of symbols whose order varies according to the message to be conveyed. Without this sequence variation, there is no potential for carrying information. Because DNA's *backbone* of alternating sugar and phosphate is chemically identical for every nucleotide in a DNA chain, the only difference between nucleotides is in the identity of the nitrogenous base. Thus, the genetic information in DNA must consist of variations in the sequence of the A, G, T, and C bases. The information constructed from the four-letter language of DNA bases is analogous to the information built from the 26-letter alphabet of English or French or Italian. Just as you can combine the 26 letters of the alphabet in different ways to generate the words of a book, so, too, different combinations of the four bases in very long sequences of nucleotides can encode the information for constructing an organism.

The DNA Helix Consists of Two Antiparallel Chains

Watson and Crick's discovery of the structure of the DNA molecule ranks with Darwin's theory of evolution by natural selection and Mendel's laws of inheritance in its contribution to our understanding of biological phenomena. The Watson-Crick structure, first embodied in a model that superficially resembled the Tinker Toys of preschool children, was based on an interpretation of all the chemical and physical data available at the time. Watson and Crick published their findings in the scientific journal *Nature* in April 1953.

Evidence from X-ray diffraction

The diffraction patterns of oriented DNA fibers do not, on their own, contain sufficient information to reveal structure. For instance, the number of diffraction spots, whose intensities and positions constitute the X-ray data (review Fig. 6.8), is considerably lower than the number of unknown coordinates of all the atoms in an oriented DNA molecule. Nevertheless, the photographs do reveal a wealth of structural information to the trained eye. Excellent X-ray images produced by Rosalind Franklin and Maurice Wilkins showed that the molecule is spiral-shaped, or helical; the spacing between repeating units along the axis of the helix is 3.4 Å (3.4×10^{-10} meters); the helix undergoes one complete turn every 34 Å; and the diameter of the molecule is 20 Å. This diameter is roughly twice the width of a single nucleotide, suggesting that a DNA molecule might be composed of two side-by-side DNA chains.

Complementary base pairing

If a DNA molecule contains two side-by-side chains of nucleotides, what forces hold these chains together? Erwin Chargaff provided an important clue with his data on the nucleotide composition of DNA from various species. Despite large variations in the relative amounts of the bases, the ratio of A to T is not significantly different from 1:1, and the ratio of G to C is also the same in every organism (**Table 6.1**). Watson grasped that the roughly 1:1 ratios of A to T and of G to C reflect a significant aspect of the molecule's inherent structure.

To explain Chargaff's ratios in terms of chemical affinities between A and T and between G and C, Watson made cardboard cutouts of the bases in the chemical forms they assume in a normal cellular environment. He then tried to match these up in various combinations, like pieces in a jigsaw puzzle. He knew that the particular arrangements of

TABLE 6.1 Chargaff's Data on Nucleotide Base Composition in the DNA of Various Organisms

	Percentage of Base in DNA				Ratios	
Organism	**A**	**T**	**G**	**C**	**A:T**	**G:C**
Escherichia coli	24.7	23.6	26.0	25.7	1.05	1.01
Saccharomyces cerevisiae	31.3	32.9	18.7	17.1	0.95	1.09
Caenorhabditis elegans	31.2	29.1	19.3	20.5	1.07	0.96
Drosophila melanogaster	27.3	27.6	22.5	22.5	0.99	1.00
Mus musculus	29.2	29.4	21.7	19.7	0.99	1.10
Homo sapiens	29.3	30.0	20.7	20.0	0.98	1.04

Note that even though the level of any one nucleotide is different in different organisms, the amount of A always approximately equals the amount of T, and the level of G is always similar to that of C. Moreover, the total amount of purines (A plus G) nearly always equals the total amount of pyrimidines (C plus T).

Figure 6.10 Complementary base pairing. An A on one strand can form two (noncovalent) hydrogen bonds with a T on the other strand. G on one strand can form three hydrogen bonds with a C on the other strand. The size and shape of A–T and of G–C base pairs are similar, allowing both to fill the same amount of space between the two backbones of the double helix.

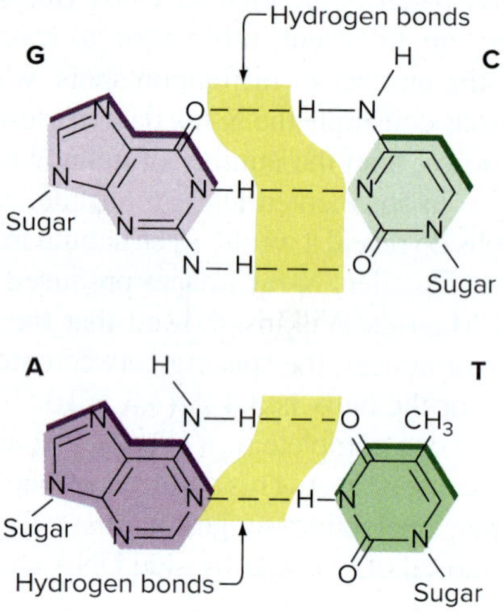

atoms on purines and pyrimidines play a crucial role in molecular interactions as they can participate in the formation of **hydrogen bonds:** weak electrostatic bonds that result in a partial sharing of hydrogen atoms between reacting groups (**Fig. 6.10**). Watson saw that A and T could be paired together such that two hydrogen bonds formed between them. If G and C were similarly paired, hydrogen bonds could also easily connect the nucleotides carrying these two bases. (Watson originally posited two hydrogen bonds between G and C, but there are actually three.) Remarkably, the two pairs—A–T and G–C—had essentially the same shape. This meant that the two pairs could fit in any order between two sugar-phosphate backbones without distorting the structure. This **complementary base pairing** also explained the Chargaff ratios—always equal amounts of A and T, and of G and C. Note that both of these base pairs consist of one purine and one pyrimidine.

Crick connected the chemical facts with the X-ray data, recognizing that because of the geometry of the base-sugar bonds in nucleotides, the orientation of the bases in Watson's pairing scheme could arise only if the bases were attached to backbones running in opposite directions. **Figure 6.11** illustrates and explains the model Watson and Crick proposed in April 1953: DNA as a double helix.

The Double Helix May Assume Alternative Forms

Watson and Crick arrived at the double helix model of DNA structure by building models, not by a direct structural determination from the data alone. And even though Watson has written that "a structure this pretty just had to exist," the beauty of the structure is not necessarily evidence of its correctness. At the time of its presentation, the strongest evidence for its correctness was its physical plausibility, its chemical and spatial compatibility with all available data, and its capacity for explaining many biological phenomena.

B DNA and Z DNA

The majority of naturally occurring DNA molecules have the configuration suggested by Watson and Crick. Such molecules are known as **B-form DNA;** they spiral to the right (**Fig. 6.12a**). DNA structure is, however, more polymorphic than originally assumed. One type, for example, contains nucleotide sequences that cause the DNA to assume what is known as a **Z form** in which the helix spirals to the left and the backbone takes on a zigzag shape (**Fig. 6.12b**). Researchers have observed many kinds of unusual non-B structures *in vitro* (in the test tube, literally *in glass*), and they speculate that some of these might occur at least transiently in living cells. There is some evidence, for instance, that Z DNA might exist in certain chromosomal regions *in vivo* (in the living organism). Whether the Z form and other unusual conformations have any biological role remains to be determined.

Linear and circular DNA

The nuclear chromosomes of all eukaryotic organisms are long, linear double helixes, but some smaller chromosomes are circular (**Fig. 6.13a** and **b**). These include the chromosomes of prokaryotic bacteria, the chromosomes of organelles such as the mitochondria and chloroplasts that are found inside eukaryotic cells, and the chromosomes of some viruses, including the papovaviruses that can cause cancers in animals and humans. Such circular chromosomes consist of covalently closed, double-stranded circular DNA molecules. Although neither strand of these circular double helixes has an end, the two strands are still antiparallel in polarity.

Single-stranded and double-stranded DNA

In some viruses, the genetic material consists of relatively small, single-stranded DNA molecules. Once inside a cell, the single strand serves as a *template* (pattern) for making a second strand, and the resulting double-stranded DNA then governs the production of more virus particles. Examples of viruses carrying single-stranded DNA are bacteriophages ϕX174 and M13, and mammalian parvoviruses, which are associated with fetal death and spontaneous abortion in humans. In both ϕX174 and M13, the single DNA strand is in the form of a covalently closed circle; in the parvoviruses, it is linear (**Fig. 6.13c** and **d**).

Alternative B and Z configurations; circularization of the molecule; and single strands that are converted to double

FEATURE FIGURE 6.11

The Double Helix Structure of DNA

(a) Watson and Crick took the known facts about DNA's chemical composition and its physical arrangement in space and constructed a wire-frame model that could explain the molecule's function.

(b) In the model, two DNA chains spiral around an axis with the sugar-phosphate backbones on the outside and flat pairs of bases meeting in the middle. One chain runs 5′ to 3′ upward, while the other runs in the opposite direction of 5′ to 3′ downward. In short, *the two chains are antiparallel.* The two chains wrap around each other once every 10 base pairs, or once every 34 Å. The result is a double helix that looks like a twisted ladder with the two spiraling structural members composed of sugar-phosphate backbones and the perpendicular rungs consisting of base pairs.

(a)

(c) In a space-filling representation of the model, the overall shape is that of a grooved cylinder with a diameter of 20 Å. The backbones spiral around the axis of the double helix like threads on a screw. Because two backbones exist, there are two threads, and these two threads are vertically displaced from each other. This displacement of the backbones generates two grooves, one (the **major groove)** much wider than the other (the **minor groove)**.

The two chains of the double helix are held together by hydrogen bonds between complementary base pairs, A–T and G–C. The spatial requirements of the double helix require that each base pair must consist of one small pyrimidine and one large purine, and even then, only for the particular pairings of A–T and G–C. In contrast, A–C and G–T pairs do not fit well and cannot easily form hydrogen bonds. Although any one nucleotide pair forms only two or three hydrogen bonds, the sum of these connections between successive base pairs in a long DNA molecule composed of thousands of nucleotides is a key to the molecule's great chemical stability.

(b)

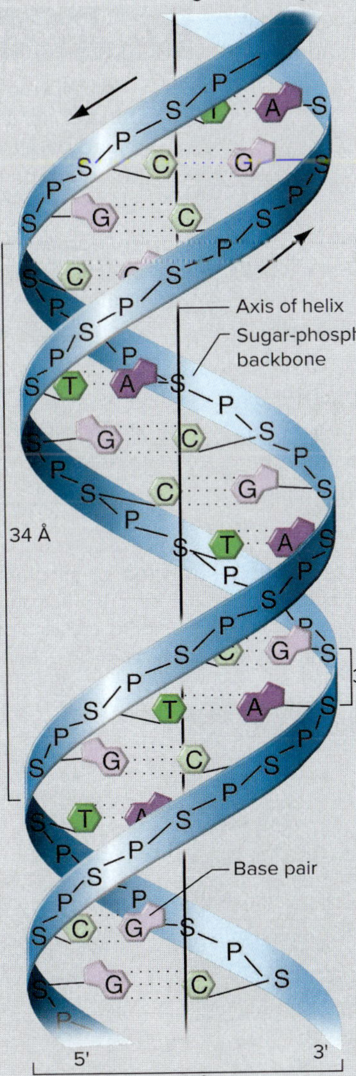

(c)

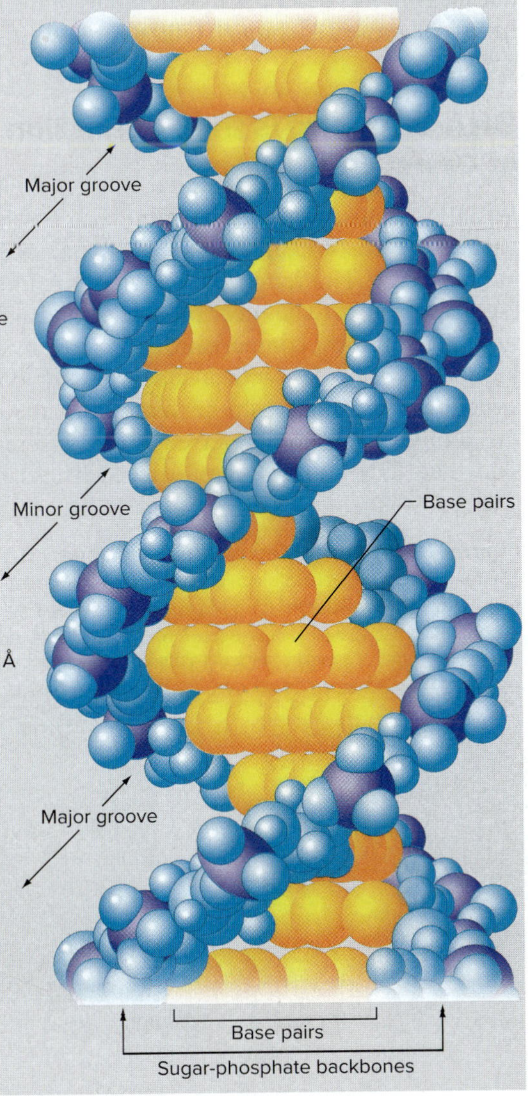

Figure 6.12 Z DNA is one variant of the double helix. **(a)** Typical Watson-Crick B-form DNA forms a right-handed helix with a smooth backbone. **(b)** Z-form DNA is left-handed and has an irregular backbone.

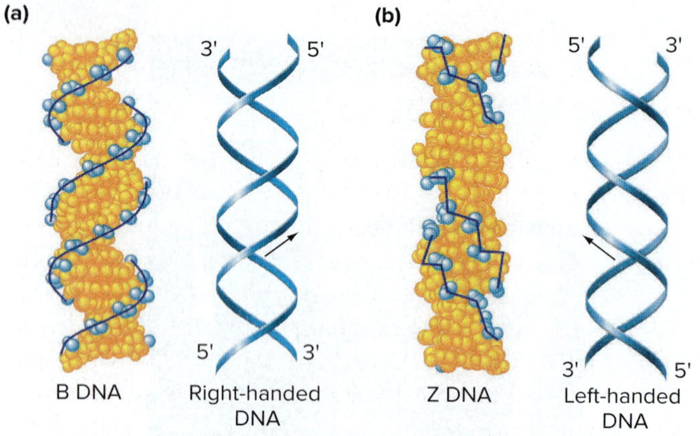

helixes before replication and expression—these are minor variations on the double-helical theme. Despite such experimentally determined departures of detail, the Watson-Crick double helix remains *the* model for thinking about DNA structure. This model describes those features of the molecule that have been preserved through billions of years of evolution.

DNA Structure Is the Foundation of Genetic Function

Without sophisticated computational tools for analyzing base sequence, one cannot distinguish bacterial DNA from human DNA. The reason is that all DNA molecules have the same general chemical properties and physical structure. Proteins, by comparison, are a much more diverse group of molecules with a much greater complexity of structure and function. In his account of the discovery of the double helix, Crick referred to this difference when he said that "DNA is, at bottom, a much less sophisticated molecule than a highly evolved protein and for this reason reveals its secrets more easily."

Four basic DNA secrets are embodied in the following four questions:

1. How does the molecule carry information?
2. How is that information copied for transmission to future generations?
3. What mechanisms allow the information to change?
4. How does DNA-encoded information govern the expression of phenotype?

The double-helical structure of DNA provides a potential answer to each of these questions, endowing the molecule with the capacity to carry out all the crucial functions required of the genetic material.

In the remainder of this chapter, we describe how DNA's structure enables it to carry genetic information, replicate that information with great fidelity, and reorganize the information through recombination. How the information changes through mutation and how the information determines phenotype are the subjects of Chapters 7 and 8.

essential concepts

- The DNA molecule is a *double helix* composed of two *antiparallel* strands, in each of which nucleotides are joined by phosphodiester bonds. Hydrogen bonding between the *complementary bases*—A with T, and G with C—holds the two strands together.
- *Antiparallel* means that one strand is oriented in the 5′-to-3′ direction, while the other, *complementary strand* is oriented in the 3′-to-5′ direction.
- Most eukaryotes have double-stranded, linear DNA, but prokaryotes, chloroplasts and mitochondria, and some viruses have double-stranded circular DNA. Certain other viruses contain a single-stranded DNA that can be linear or circular.

Figure 6.13 DNA molecules may be linear or circular, double-stranded or single-stranded. These electron micrographs of naturally occurring DNA molecules show: **(a)** a fragment of a long, linear, double-stranded human chromosome, **(b)** a circular double-stranded papovavirus chromosome, **(c)** a linear single-stranded parvovirus chromosome, and **(d)** circular single-stranded bacteriophage M13 chromosomes.

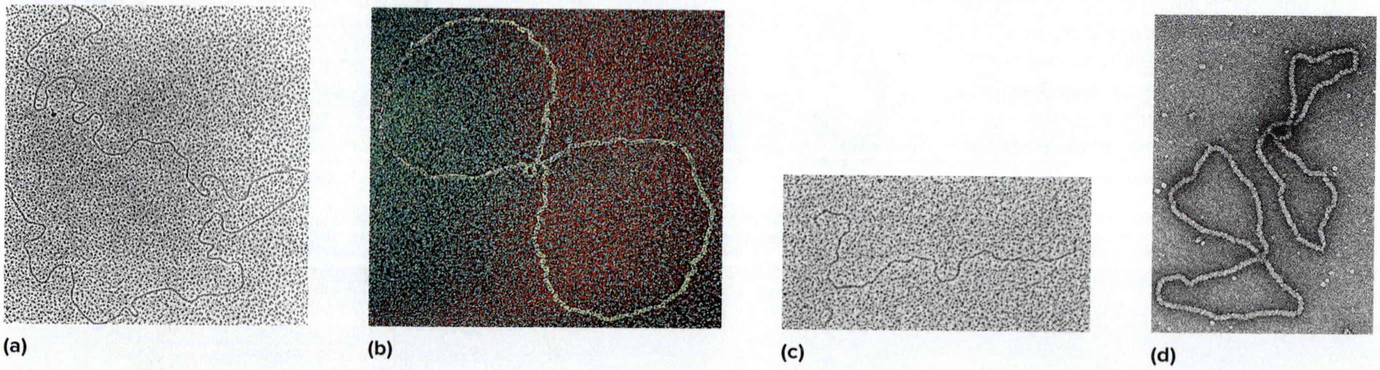

6.3 Genetic Information in Nucleotide Sequence

learning objectives

1. Explain how DNA stores complex information.
2. Compare the two ways in which the information in DNA may be accessed by proteins.
3. Describe the structural differences between DNA and RNA.

The information content of DNA resides in the sequence of its bases. The four bases in each chain are like the letters of an alphabet; they may follow each other in any order, and different sequences spell out different "words." Each word has its own meaning, that is, its own effect on phenotype. AGTCAT, for example, means one thing, while CTAGGT means another. Although DNA has only four different letters, or building blocks, the potential for different combinations and thus different sets of information in a long chain of nucleotides is staggering. Some human chromosomes, for example, are composed of chains that are 250 million nucleotides long; because the different bases may follow each other in any order, such chains could contain any one of $4^{250,000,000}$ (which translates to 1 followed by 150,515,000 zeros) potential nucleotide sequences.

Most Genetic Information Is Read from Unwound DNA Chains

The unwinding of a DNA molecule exposes a single sequence of bases on each of two strands (**Fig. 6.14**). Proteins read the information in a single DNA strand by synthesizing a stretch of RNA (a process called *transcription*) or DNA (a process called *replication*) complementary to a specific sequence.

Some Genetic Information Is Accessible Without Unwinding DNA

Some proteins can recognize and bind to specific base pair sequences in double-stranded DNA (**Fig. 6.15**). This information emerges mainly from differences between the four bases that appear in the major and minor grooves. Within the grooves, certain atoms at the periphery of the bases are exposed, and particularly in the major groove, these atoms may assume spatial patterns that provide chemical information. Proteins can access this information to sense the base sequence in a stretch of DNA without disassembling the double helix. Sequence-specific DNA-binding proteins include *transcription factors* that turn genes on and off (Chapters 16 and 17) as well as bacterial *restriction enzymes* that cut DNA at particular sites (Chapter 9).

In Some Viruses, RNA Is the Repository of Genetic Information

DNA carries the genetic information in all cellular forms of life and in many viruses. Prokaryotes such as *Escherichia coli* bacteria carry their DNA in a double-stranded, covalently closed circular chromosome. Eukaryotic cells package their DNA in double-stranded linear chromosomes. DNA viruses carry it in small molecules that are single- or double-stranded, circular, or linear.

By contrast, retroviruses, which include those that cause polio and AIDS, use **ribonucleic acid**, or **RNA** as their genetic material.

Figure 6.14 DNA stores information in the sequence of its bases. A partially unwound DNA double helix. Note that different structural information is available in the double-stranded and unwound regions of the molecule.

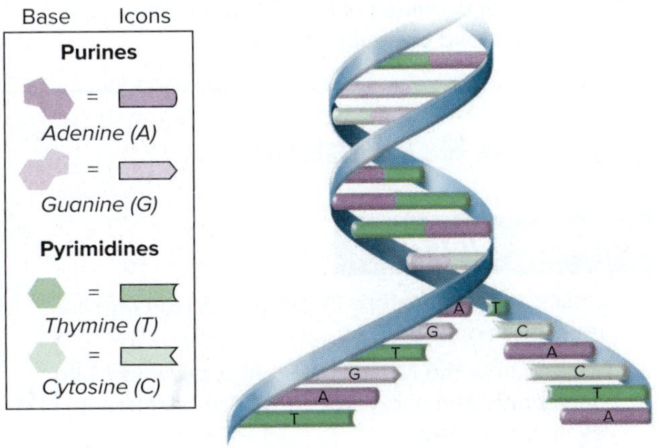

Figure 6.15 Proteins bind to specific sequences in DNA. Computer artwork of the *E. coli* catabolite gene activator protein (CAP) bound to DNA (*green* and *orange*). The structure of CAP is shown as a series of cylinders and ribbons. CAP can recognize specific sites in the major groove of double-helical DNA.
© Dr. Tim Evans/Science Source

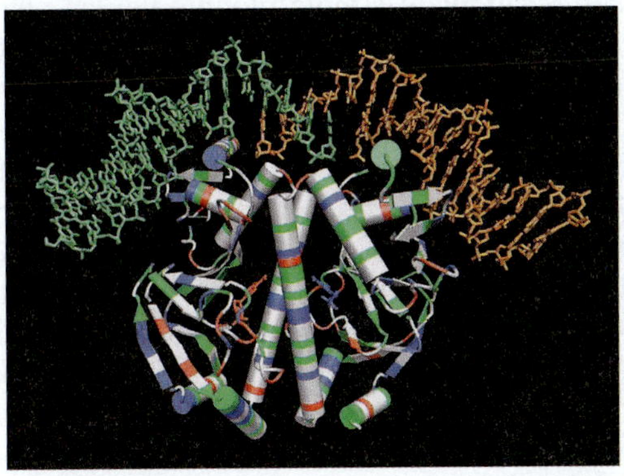

Figure 6.16 Differences in the chemical structure of DNA and RNA. Phosphodiester bonds join ribonucleotides into an RNA chain that differs from DNA in three ways (*bullets* in each column).

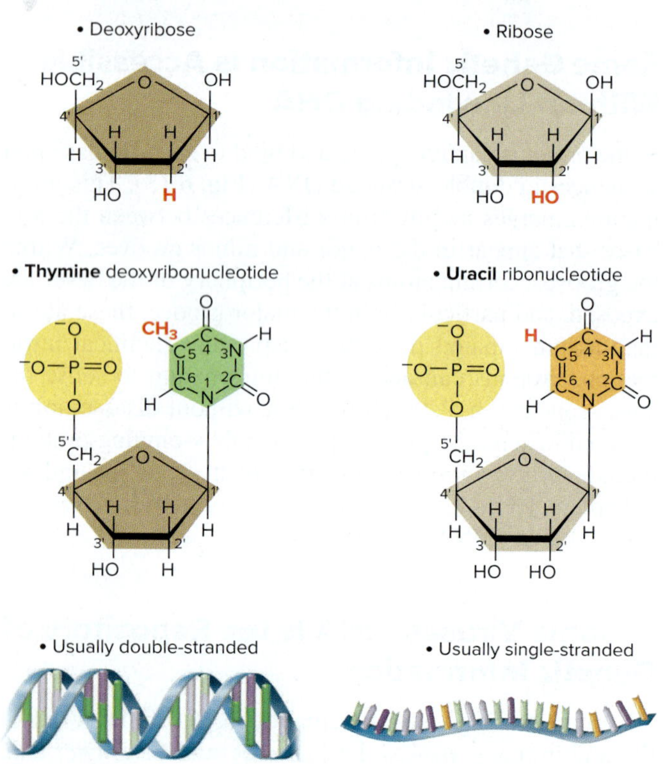

Figure 6.17 Complex folding patterns of RNA molecules. Most RNA molecules are single-stranded but are sufficiently flexible so that some regions can fold back and form base pairs with other parts of the same molecule.

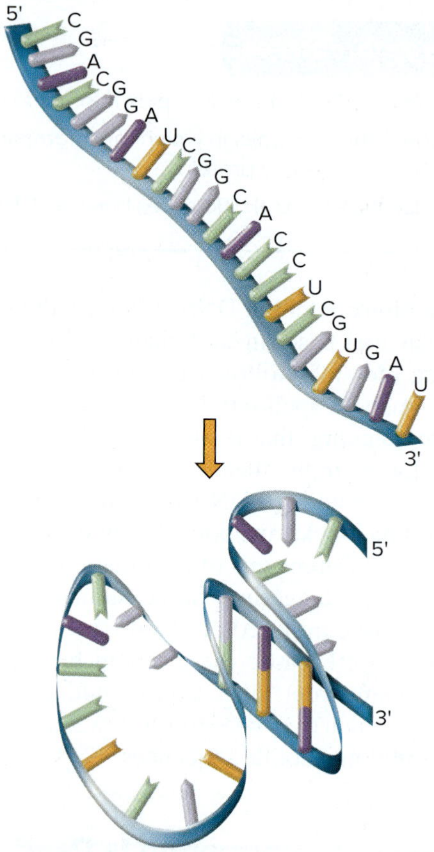

Differences between RNA and DNA

Three major chemical differences distinguish RNA from DNA (**Fig. 6.16**). First, RNA takes its name from the sugar **ribose**, which it incorporates instead of the deoxyribose found in DNA. Second, RNA contains the base uracil (U) instead of the base thymine (T); U, like T, base pairs with A. Finally, most RNA molecules are single-stranded and contain far fewer nucleotides than the very long DNA molecules found in nuclear chromosomes.

Within a single-stranded RNA molecule, folding can bring together two oppositely oriented regions that carry complementary nucleotide sequences to form a short, base-paired stretch within the molecule. This means that, compared to the relatively simple, double-helical shape of a DNA molecule, many RNAs have a complicated structure of short double-stranded segments interspersed with single-stranded loops (**Fig. 6.17**).

RNA has the same ability as DNA to carry information in the sequence of its bases, but it is much less stable than DNA. In addition to serving as the genetic material for an array of viruses, RNA fulfills several vital functions in all cells. For example, it participates in gene expression and protein synthesis, as presented in Chapter 8. RNA also plays a surprisingly significant role in DNA replication, which we now describe.

essential concepts

- DNA carries digital information in the sequence of its four *bases*.
- The *base sequence* of DNA can be read from a single, unwound strand during replication or transcription. In addition, specialized proteins can recognize and bind to short base sequences accessible in the *grooves* of double-stranded DNA.
- *RNA* contains ribose rather than deoxyribose and uracil (U) instead of thymine (T); it also is generally single-stranded instead of double-stranded.

6.4 DNA Replication

learning objectives

1. Describe the key steps in the semiconservative replication of DNA.
2. Explain how the Meselson-Stahl experiment with heavy nitrogen showed that DNA replication is semiconservative.

3. Summarize the key factors DNA polymerase requires to replicate DNA.
4. Outline the steps in the process of DNA replication and how they relate to the requirements of DNA polymerase.
5. Discuss three ways cells preserve the accuracy and integrity of the genetic information in DNA.

In one of the most famous understatements in the scientific literature, Watson and Crick wrote at the end of their 1953 paper proposing the double helix model: "It has not escaped our notice that the specific pairing we have postulated immediately suggests a possible copying mechanism for the genetic material." This copying, as we saw in Chapter 4, must precede the transmission of chromosomes from one generation to the next via meiosis, and it is also the basis of the chromosome duplication prior to each mitosis that allows two daughter cells to receive a complete copy of the genetic information in a progenitor cell.

Overview: Complementary Base Pairing Ensures Semiconservative Replication

In the process of replication postulated by Watson and Crick, the double helix unwinds to expose the bases in each strand of DNA. Each of the two separated strands then acts as a **template,** or molecular mold, for the synthesis of a new second strand (**Fig. 6.18**). The newly replicated strands form as complementary bases align opposite the exposed bases on the two parental strands. That is, an A at one position on the original strand signals the addition of a T at the corresponding position on the newly forming strand; a T on the original signifies the addition of an A; similarly, G calls for C, and C calls for G, as demanded by complementary base pairing.

Once the appropriate base has aligned opposite to and formed hydrogen bonds with its complement, enzymes join the base's nucleotide to the preceding nucleotide by a phosphodiester bond, eventually linking a whole new line of nucleotides into a continuous strand. This mechanism of DNA strand separation and complementary base pairing followed by the coupling of successive nucleotides yields two daughter double helixes that each contain one of the original DNA strands intact (this strand is *conserved*) and one completely new strand (**Fig. 6.19a**). For this reason, such a pattern of double helix duplication is called **semiconservative replication:** a copying in which one strand of each new double helix is conserved from the parent molecule and the other is newly synthesized.

Watson and Crick's proposal is not the only replication mechanism imaginable. **Figures 6.19b** and **c** illustrate two possible alternatives. With *conservative replication,* one of the two daughter double helixes would consist entirely of original DNA strands, while the other helix would consist of two newly synthesized strands. With *dispersive replication,* both daughter double helixes would carry blocks of original DNA interspersed with blocks of newly synthesized material. These alternatives are less satisfactory because they do not immediately suggest a mechanism for copying the information in the sequence of bases.

Figure 6.18 The model of DNA replication postulated by Watson and Crick. Unwinding of the double helix allows each of the two parental strands to serve as a template for the synthesis of a new strand by complementary base pairing. The end result: A single double helix is transformed into two identical daughter double helixes.

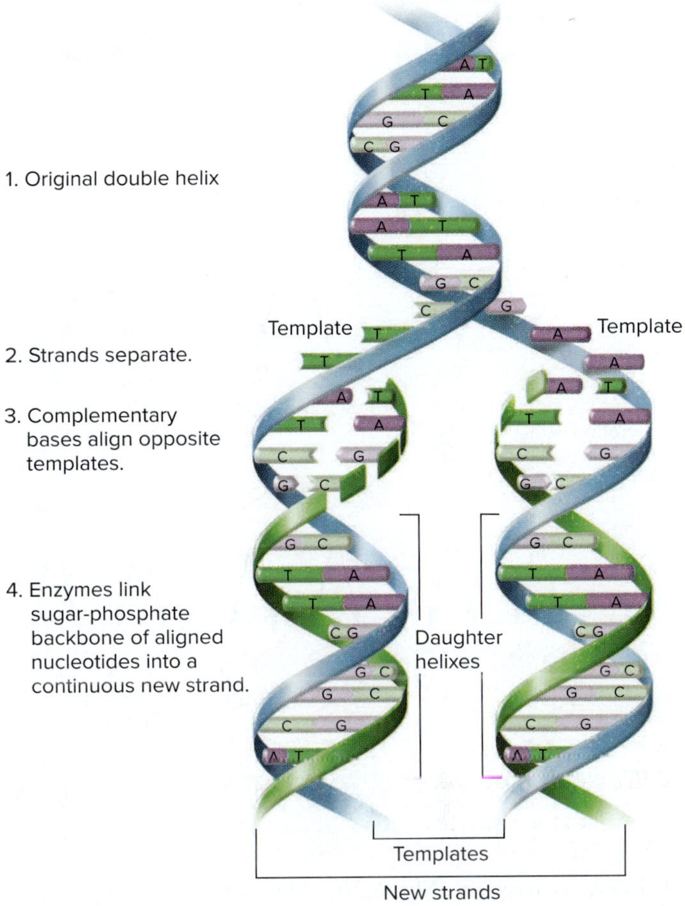

Experiments with Heavy Nitrogen Verify Semiconservative Replication

In 1958, Matthew Meselson and Franklin Stahl performed an experiment that confirmed the semiconservative nature

Figure 6.19 Three possible models of DNA replication. DNA from the original double helix is *blue;* newly made DNA is *magenta.* **(a)** Semiconservative replication (the Watson-Crick model). **(b)** Conservative replication: The parental double helix remains intact; both strands of one daughter double helix are newly synthesized. **(c)** Dispersive replication: Both strands of both daughter double helixes contain both original and newly synthesized material.

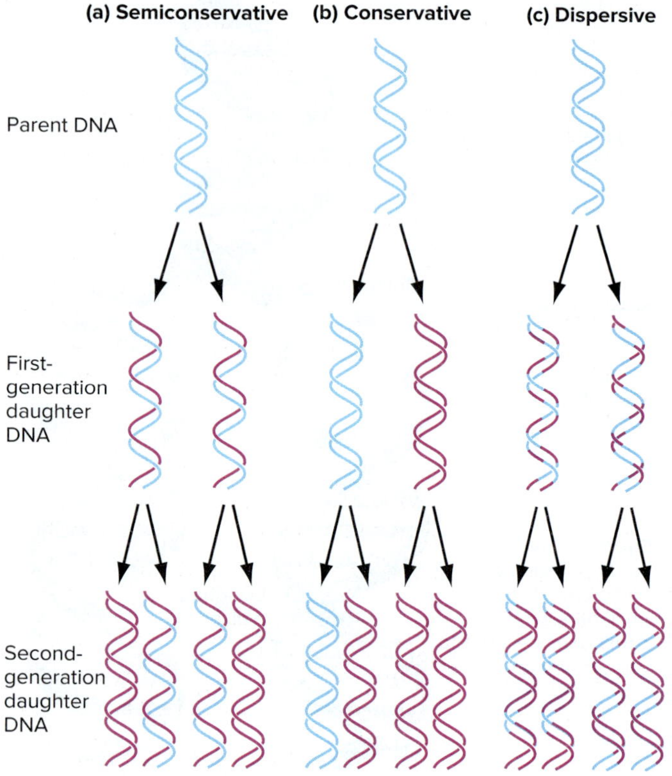

of DNA replication (**Fig. 6.20**). The experiment depended on being able to distinguish preexisting *parental* DNA from newly synthesized DNA. To accomplish this, Meselson and Stahl controlled the isotopic composition of the nucleotides incorporated in the newly forming strands by taking advantage of the fact that the purine and pyrimidine bases of DNA contain nitrogen atoms. They grew *E. coli* bacteria for many generations on media in which all the nitrogen was the normal isotope ^{14}N; these cultures served as a control. They grew other cultures of *E. coli* for many generations on media in which the only source of nitrogen was the heavy isotope ^{15}N. After several generations of growth on heavy-isotope medium, essentially all the nitrogen atoms in the DNA of these bacterial cells were labeled with (that is, contained) ^{15}N. The cells in some of these cultures were then transferred to new medium in which all the nitrogen was ^{14}N. Any DNA synthesized after the transfer would contain the lighter isotope.

Meselson and Stahl isolated DNA from cells grown in the different nitrogen-isotope cultures and then subjected these DNA samples to *equilibrium density gradient centrifugation,* an analytic technique they had just developed. In a test tube, they dissolved the DNA in a solution of the dense salt cesium chloride (CsCl) and spun these solutions at very high speed (about 50,000 revolutions per minute) in an ultracentrifuge. Over a period of two to three days, the centrifugal force (roughly 250,000 times the force of gravity) causes the formation of a stable gradient of CsCl concentrations, with the highest concentration, and thus the highest CsCl density, at the bottom of the tube. The DNA in the tube forms a sharply delineated band at the position where its own density equals that of the CsCl. Because DNA containing ^{15}N is denser than DNA containing ^{14}N, pure ^{15}N DNA will form a band lower, that is, closer to the bottom of the tube, than pure ^{14}N DNA (Fig. 6.20).

As Fig 6.20 shows, when cells with pure ^{15}N DNA were transferred into ^{14}N medium and allowed to divide once, DNA from the resultant first-generation cells formed a band at a density intermediate between that of pure ^{15}N DNA and that of pure ^{14}N DNA. A logical inference is that the DNA in these cells contains equal amounts of the two isotopes. This finding invalidates the conservative model, which predicts the appearance of bands reflecting only pure ^{14}N and pure ^{15}N with no intermediary band. In contrast, DNA extracted from second-generation cells that had undergone a second round of division in the ^{14}N medium produced two observable bands, one at the density corresponding to equal amounts of ^{15}N and ^{14}N, the other at the density of pure ^{14}N. This result invalidates the dispersive model, which predicts a single band between the two bands of the original generation.

Meselson and Stahl's observations are consistent only with semiconservative replication. In the first generation after transfer from the ^{15}N to the ^{14}N medium, one of the two strands in every daughter DNA molecule carries the heavy isotope label; the other, newly synthesized strand carries the lighter ^{14}N isotope. The band at a density intermediate between that of ^{15}N DNA and ^{14}N DNA represents this isotopic hybrid. In the second generation after transfer, half of the DNA molecules have one ^{15}N strand and one ^{14}N strand, while the remaining half carry two ^{14}N strands. The two observable bands—one at the hybrid position, the other at the pure ^{14}N position—reflect this mix. By confirming the predictions of semiconservative replication, the Meselson-Stahl experiment disproved the conservative and dispersive alternatives. We now know that the semiconservative replication of DNA is nearly universal.

Let's consider precisely how semiconservative replication relates to the structure of chromosomes in eukaryotic cells during the mitotic cell cycle (review Fig. 4.9). Early in interphase, each eukaryotic chromosome contains a single continuous linear double helix of DNA. Later, during the S-phase portion of interphase, the cell replicates the double helix semiconservatively; after this semiconservative replication, each chromosome is composed of two sister chromatids joined at their centromeres. Each sister chromatid is a double helix of DNA, with one strand of parental DNA and one strand of newly synthesized DNA. At the conclusion of

Figure 6.20 How the Meselson-Stahl experiment confirmed semiconservative replication. **(1)** *E. coli* cells were grown in heavy ^{15}N medium. **(2)** and **(3)** The cells were transferred to ^{14}N medium and allowed to divide either once or twice. When DNA from each of these cell preparations was centrifuged in a cesium chloride gradient, the density of the extracted DNA conformed to the predictions of the semiconservative mode of replication, as shown at the *bottom* of the figure, where *blue* indicates heavy DNA and *magenta* depicts light DNA. The results are inconsistent with the conservative and dispersive models for DNA replication (compare with Fig. 6.19).

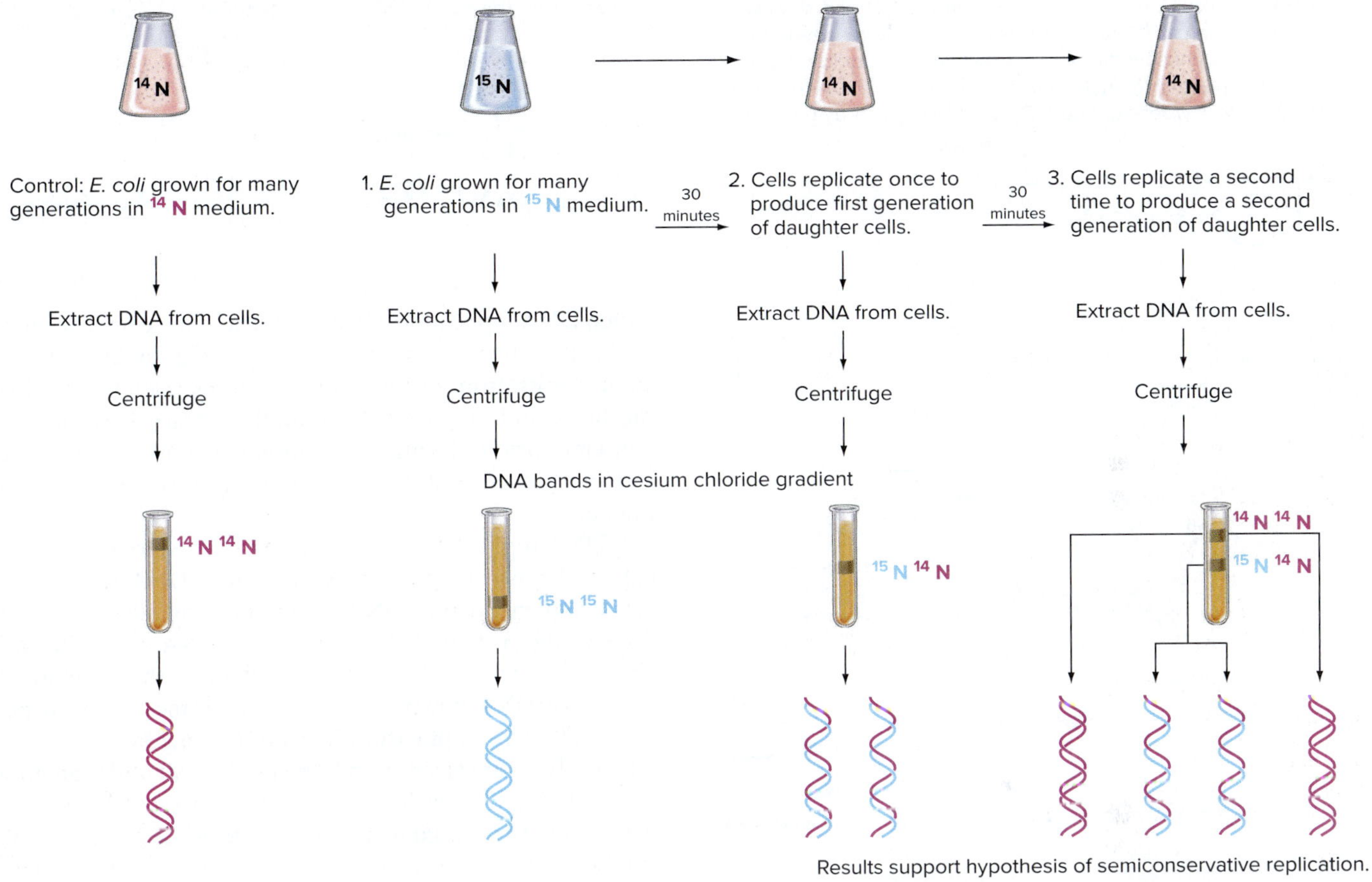

mitosis, each of the two daughter cells receives one sister chromatid from every chromosome in the cell. This process preserves chromosome number and identity during mitotic cell division because the two sister chromatids are identical in base sequence to each other and to the original parental chromosome.

DNA Polymerase Has Strict Operating Requirements

Watson and Crick's model for semiconservative replication is a simple concept to grasp, but the biochemical process through which it occurs is quite complex. Replication does not happen spontaneously any time a mixture of DNA and nucleotides is present. Rather, it occurs at a precise moment in the cell cycle, depends on a network of interacting regulatory elements, requires considerable input of energy, and involves a complex array of the cell's molecular machinery, including the key enzyme **DNA polymerase.** The salient details were deduced primarily in the laboratory of Arthur Kornberg, who won a Nobel prize for this work. The Kornberg group purified individual components of the replication machinery from *E. coli* bacteria. Remarkably, they were eventually able to elicit the reproduction of specific genetic information outside a living cell, in a test tube containing purified enzymes together with a DNA template, *primers* (defined below), and nucleotide triphosphates.

Although the biochemistry of DNA replication was elucidated for a single bacterial species, its essential features are conserved—just like the structure of DNA—within all organisms. The energy required to synthesize every DNA molecule found in nature comes from the high-energy phosphate bonds associated with the four *deoxyribonucleotide triphosphates* (dATP, dCTP, dGTP, and dTTP; or dNTP as a general term) that provide bases for incorporation into the growing DNA strand. As shown in **Fig. 6.21,** this conserved biochemical feature means that DNA synthesis can proceed only by adding nucleotides to the 3′ end of an existing polynucleotide. With energy released from severing the triphosphate arm of a dNTP substrate molecule, the DNA polymerase enzyme catalyzes the formation of a new phosphodiester bond. Once this bond is formed, the enzyme proceeds to join up the next nucleotide brought into position by complementary base pairing.

Figure 6.21 DNA synthesis proceeds in a 5′-to-3′ direction. The template strand is shown on the *right* in an antiparallel orientation to the new DNA strand under synthesis on the *left*. Here, a free molecule of dATP has formed hydrogen bonds with a complementary thymine base on the template strand. DNA polymerase (*yellow*) cleaves dATP between the first and second phosphate groups. This cleavage releases the energy needed to form a covalent phosphodiester bond between the terminal 3′-OH group on the preceding nucleotide to the first phosphate of the dATP substrate. Pyrophosphate (*PP_i*) is released as a by-product.

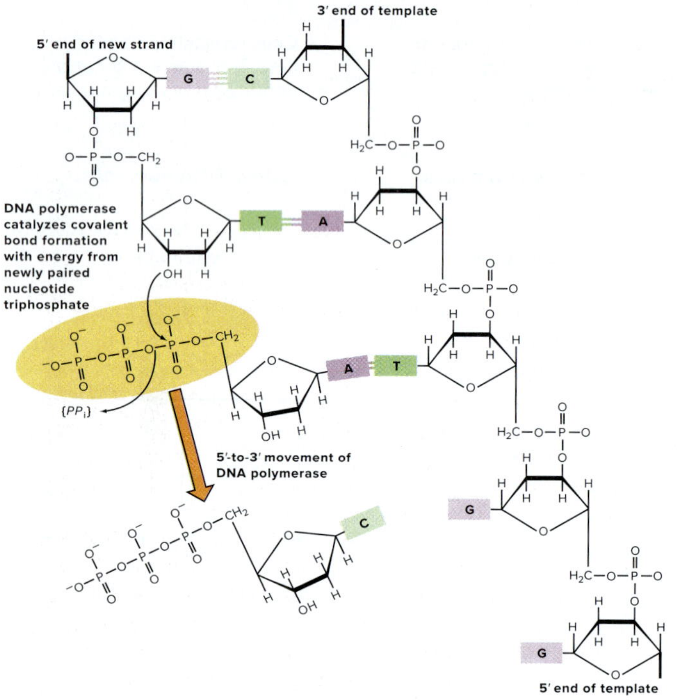

Figure 6.22 Requirements of DNA polymerase. To synthesize DNA, DNA polymerase requires a single-stranded DNA template, a primer that can be RNA or DNA, and free deoxyribonucleotide triphosphates (dNTPs). DNA polymerase adds nucleotides successively onto the 3′ end of the primer as instructed by the complementary nucleotides in the template.

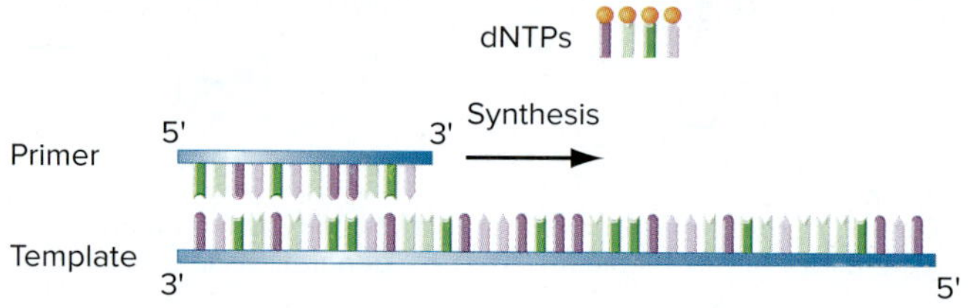

Many proteins in addition to DNA polymerase are required to replicate DNA. However, you will see below that the most important features of DNA replication reflect three strict requirements for DNA polymerase action (**Fig. 6.22**):

(1) The four dNTPs.

(2) A single-stranded template. Double-stranded DNA must be unwound, and DNA polymerase moves along the template strand in the 3′-to-5′ direction.

(3) A **primer** with a free 3′ hydroxyl group. DNA polymerase adds nucleotides successively to the 3′ end of the growing DNA chain. (That is, DNA polymerase synthesizes DNA only in the 5′-to-3′ direction.) However, DNA polymerase cannot establish the first link in a new chain. Polymerization therefore must start with a *primer,* a short, single-stranded molecule of DNA or RNA a few nucleotides long that base pairs with part of the template strand.

DNA Replication Is a Tightly Regulated, Complex Process

The formation of phosphodiester bonds by DNA polymerase is just one component of the highly coordinated process by which DNA replication occurs inside a living cell. The entire molecular mechanism, illustrated in **Fig. 6.23,** has two stages: **initiation,** during which proteins open up the double helix and prepare it for complementary base pairing, and **elongation,** during which proteins connect the correct sequence of nucleotides on both newly formed DNA double helixes.

DNA replication is complicated by the strict biochemical mechanism of polymerase function. DNA polymerase can lengthen existing DNA chains only by adding nucleotides to the 3′ hydroxyl group of the DNA strand, as was shown in Figs. 6.21 and 6.22. However, the antiparallel strands of DNA unwind progressively at the two Y-shaped areas called the **replication forks** in **Fig. 6.23a**. As a result, one newly synthesized strand (the **leading strand**) can grow continuously into each of the opening forks. But the other new strand (the **lagging strand**), made at the same fork but synthesized from the other template strand, can only be generated in pieces called **Okazaki fragments** as more and more template is unwound at the fork (**Fig. 6.23b**). These fragments must be joined together later in the process.

As Fig. 6.23 shows, DNA replication depends on the coordinated activity of many different proteins, including two different DNA polymerases called pol I and pol III (*pol* is short for polymerase). Pol III plays the major role in producing the new strands of complementary DNA, while pol I fills in the gaps between newly synthesized Okazaki segments. Other enzymes contribute to the initiation process: **DNA helicase** unwinds the double helix. A special group of *single-stranded DNA binding proteins* keep the DNA helix open. An enzyme called **primase** generates RNA primers to initiate DNA synthesis. During elongation, the *DNA ligase* enzyme welds together Okazaki fragments.

It took many years for biochemists and geneticists to discover how the tight collaboration of many proteins drives DNA replication. Today scientists think that programmed molecular interactions of this kind underlie many of the biochemical processes that occur in cells. In these processes, a group of proteins, each performing a specialized function, like the workers on an assembly line, cooperate in the manufacture of complex macromolecules.

FEATURE FIGURE 6.23

The Mechanism of DNA Replication

(a) Initiation: Preparing the double helix for use as a template. Initiation begins with the unwinding of the double helix at a particular short sequence of nucleotides known as the **origin of replication.** Several proteins bind to the origin, starting with the *initiator protein*. Initiator attracts **DNA helicase**, which unwinds the double helix. The opening up of a region of DNA generates two Y-shaped areas called **replication forks**, one at either end of the unwound area, the **replication bubble.** The single strands will serve as **templates** for synthesizing new strands of DNA.

An enzyme complex known as **DNA polymerase III** adds nucleotides to the 3′ end of a preexisting strand of nucleic acid. The requirement for an already existing chain means that something else must prime the about-to-be-constructed chain. In living cells, that something else is a short stretch of RNA called an **RNA primer,** synthesized by an enzyme called **primase.**

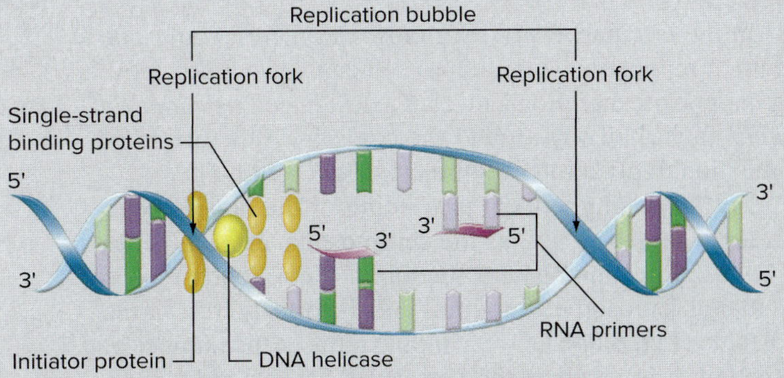

(b) Elongation: Connecting the correct sequence of nucleotides into a continuous new strand of DNA. Through complementary base pairing, the order of bases in the template strand specifies the order of bases in the newly forming strand. DNA polymerase III catalyzes the joining of a new nucleotide to the preceding nucleotide through the formation of a phosphodiester bond, a process known as **polymerization.** DNA polymerase III first joins the correctly paired nucleotide to the 3′ hydroxyl end of the RNA primer, and then it continues to add the appropriate nucleotides to the 3′ end of the growing chain. As a result, the DNA strand under construction grows in the 5′-to-3′ direction, while the DNA polymerase molecule actually moves along the antiparallel template strand in the 3′-to-5′ direction. (The three following figures diagram only the events occurring at the left replication fork.)

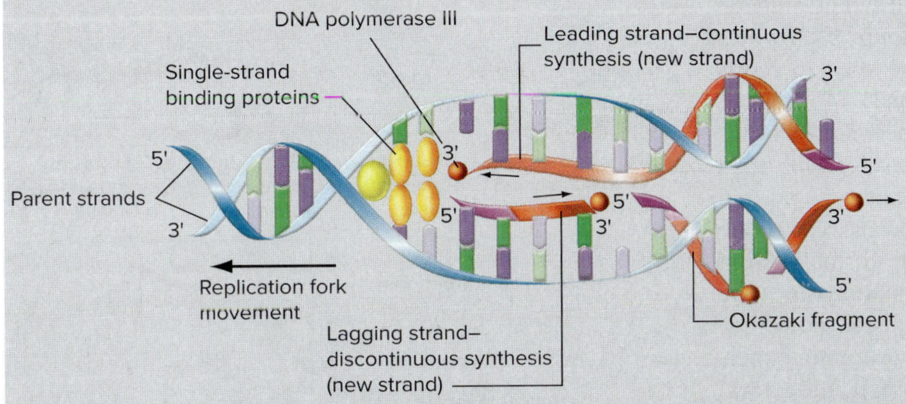

As DNA replication proceeds, helicase progressively unwinds the double helix. DNA polymerase III moves in the same direction as the fork to synthesize the **leading strand.** However, the synthesis of the second new DNA chain — the **lagging strand**— is problematic. The polarity of the lagging strand is opposite that of the leading strand, yet DNA polymerase functions only in the 5′-to-3′ direction. How can this work?

The answer is that the lagging strand is synthesized *discontinuously* in the normal 5′-to-3′ direction as small fragments of about 1000 bases called **Okazaki fragments** (after their discoverers, Reiji and Tuneko Okazaki).

Because DNA polymerase III can add nucleotides only to the 3′ end of an existing strand, each Okazaki fragment is initiated by a short RNA primer. Primase catalyzes the formation of the RNA primer for each upcoming Okazaki fragment as soon as the replication fork has progressed a sufficient distance along the DNA. DNA polymerase III then adds nucleotides to the primer, generating an Okazaki fragment that extends up to the 5′ end of the primer of the previously synthesized fragment.

Finally, **DNA polymerase I** and other enzymes replace the RNA primer of the previously made Okazaki fragment with DNA, and the enzyme **DNA ligase** covalently joins successive Okazaki fragments into a continuous strand of DNA.

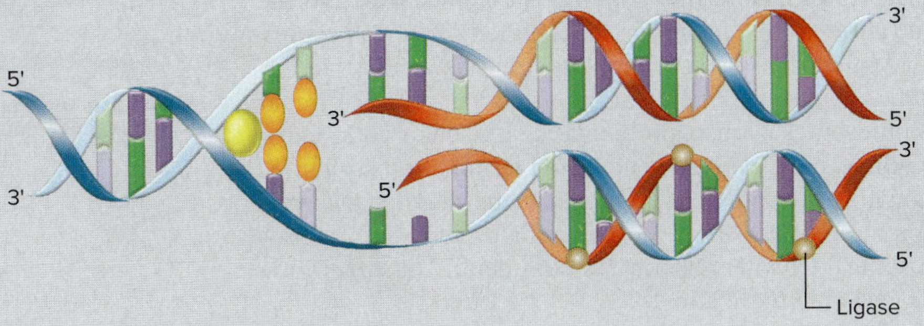

The unwinding of DNA beginning at the origin of replication produces two forks (Fig. 6.23a). As a result, replication is generally *bidirectional,* with the replication forks moving in opposite directions as unwinding proceeds. At each fork, polymerase copies both template strands, one in a continuous fashion, the other discontinuously as Okazaki fragments (Fig. 6.23b).

In the circular *E. coli* chromosome, there is only one origin of replication (**Fig. 6.24a**). When its two forks, moving in opposite directions, meet at a designated *termination region* about halfway around the circle from the origin of replication, replication is complete (**Figs. 6.24d–f**).

Not surprisingly, local unwinding of the double helix at a replication fork affects the chromosome as a whole. In *E. coli,* the unwinding of a section of a covalently closed circular chromosome overwinds and distorts the rest of the molecule (**Fig. 6.24b**). Overwinding reduces the number of helical turns to less than the 1-every-10.5-nucleotides characteristic of B-form DNA. The chromosome accommodates the strain of distortion by twisting back upon itself. You can envision the effect by imagining a coiled telephone cord that overwinds and bunches up with use. The additional twisting of the DNA molecule is called **supercoiling.** Movement of the replication fork causes more and more supercoiling.

This cumulative supercoiling, if left unchecked, would wind the chromosome up so tightly that it would impede the progress of the replication fork. A group of enzymes known as **DNA topoisomerases** helps relax the supercoils by nicking one or cutting both strands of the DNA—that is, cleaving the sugar-phosphate backbone between two adjoining nucleotides (**Fig. 6.24c**). Just as a telephone cord freed at the handset end can unwind and restore its normal coiling pattern, the DNA strands, after cleavage, can rotate relative to each other and thereby restore the normal coiling density of one helical turn per 10.5 nucleotide pairs. The activity of topoisomerases thus allows replication to proceed through the entire chromosome by preventing supercoils from accumulating in front of the replication fork. Replication of a circular double helix sometimes produces intertwined daughter molecules whose clean separation also depends on topoisomerase activity (Fig. 6.24e and f).

In the much larger, linear chromosomes of eukaryotic cells, bidirectional replication proceeds roughly as just described but from many origins of replication. The multiple origins ensure that copying is completed within the time allotted (that is, within the S period of the cell cycle). In addition, because the lagging strand is synthesized as Okazaki fragments, replication of the very ends of linear chromosomes is also problematic. But eukaryotic chromosomes have evolved specialized termination structures known as **telomeres,** which ensure the maintenance and accurate replication of the two ends of each linear chromosome. (Chapter 12 presents the details of eukaryotic chromosome replication.)

Figure 6.24 The bidirectional replication of a circular bacterial chromosome: An overview. **(a)** and **(b)** Replication proceeds in two directions from a single origin, creating two replication forks that move in opposite directions around the circle. Local unwinding of DNA at the replication forks creates supercoiled twists in the DNA in front of the replication fork. **(c)** The action of topoisomerase enzymes helps reduce this supercoiling. **(d)** and **(e)** When the two replication forks meet at the termination region, the entire chromosome has been copied. **(f)** Topoisomerase enzymes separate the two daughter chromosomes.

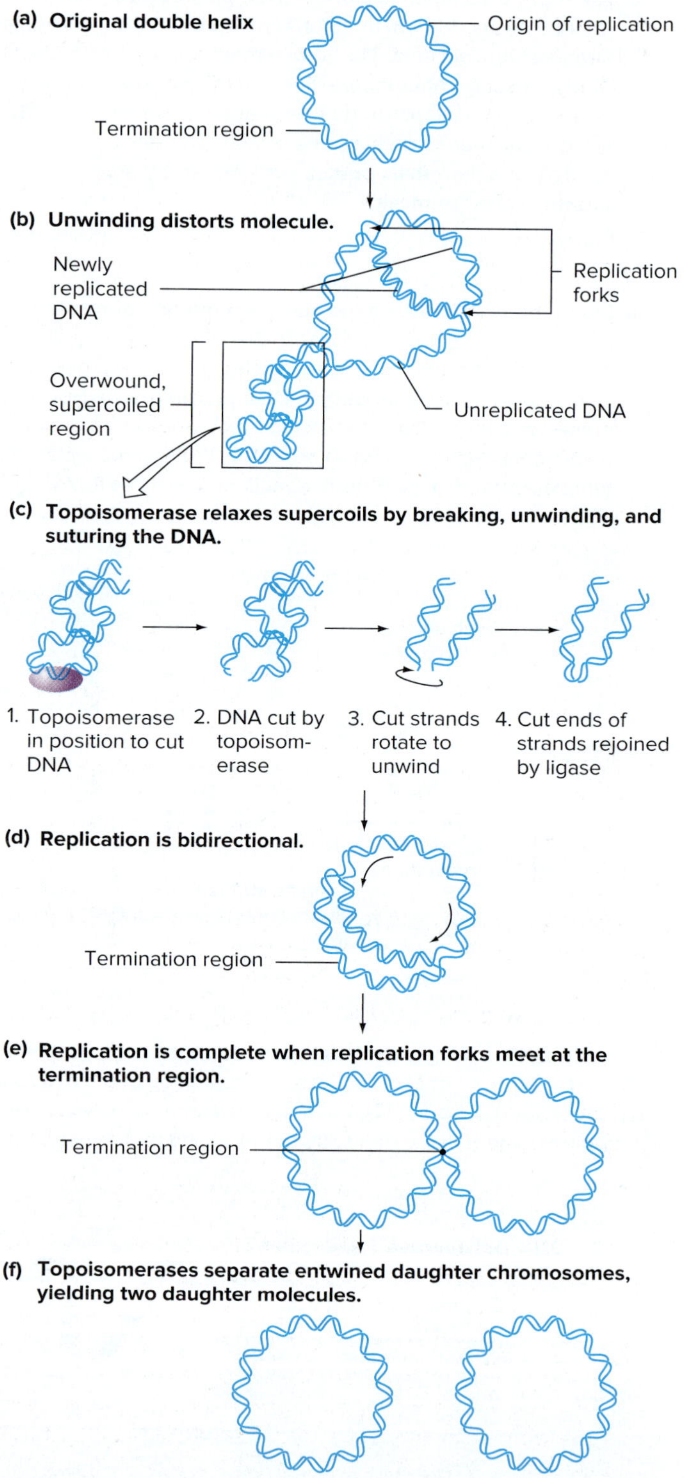

The Integrity of Genetic Information Must Be Preserved

DNA is the sole repository of the vast amount of information required to specify the structure and function of most organisms. In some species, this information may lie in storage for many years, or it may undergo replication many times before it is called on to generate progeny. During the time of storage and before gamete production, the organism must protect the integrity of the information, for even the most minor change can have disastrous consequences, such as causing severe genetic disease or even death. Each organism ensures the informational fidelity of its DNA in three important ways:

- *Redundancy.* Either strand of the double helix can specify the sequence of the other. This redundancy provides a basis for checking and repairing errors arising either from chemical alterations sustained during storage or from rare malfunctions of the replication machinery.
- *The remarkable precision of the cellular replication machinery.* Evolution has perfected the cellular machinery for DNA replication to the point where errors during copying are exceedingly rare. For example, DNA polymerase has acquired a proofreading ability to prevent unmatched nucleotides from joining a new strand of DNA; as a result, a free nucleotide is attached to a growing strand only if its base is correctly paired with its complement on the parent strand. We examine this proofreading mechanism in Chapter 7.
- *Enzymes that repair chemical damage to DNA.* The cell has an array of enzymes devoted to the repair of nearly every imaginable type of chemical damage. We describe how these enzymes carry out their corrections in Chapter 7.

All of these safeguards help ensure that the information content of DNA will be transmitted intact from generation to generation of cells and organisms. However, as we see next, new combinations of existing information arise naturally as a result of recombination.

essential concepts

- The DNA molecule is reproduced by *semiconservative replication;* the two DNA strands separate, and each acts a *template* for the synthesis of a new complementary strand.
- *DNA polymerase* synthesizes DNA in the 5′-to-3′ direction by adding nucleotides successively onto the 3′ end of a growing DNA chain.
- DNA polymerase requires: (i) a supply of the four *deoxyribonucleotide triphosphates,* (ii) a single-stranded DNA template, and (iii) a *primer* of either DNA or (in cells) RNA with a free 3′ hydroxyl group.
- At the DNA *replication fork*, DNA polymerase synthesizes one new strand (the *leading strand*) continuously, while the other (*lagging strand*) is synthesized as multiple *Okazaki fragments* that are then joined by DNA ligase.
- The integrity and accuracy of information in DNA is preserved by redundancy in the two strands, the precision of the enzymes synthesizing DNA, and the action of enzymes that repair damage to DNA.

6.5 Homologous Recombination at the DNA Level

learning objectives

1. Summarize the evidence from tetrad analysis confirming that recombination occurs at the four-strand stage and involves reciprocal exchange.
2. Explain how we know that DNA breaks and rejoins during recombination.
3. List the key steps of recombination at the molecular level.
4. Explain why recombination events do not always result in crossing-over.
5. Describe how mismatch repair of heteroduplex regions can lead to gene conversion in fungal tetrads.

Mutation, the ultimate source of all new alleles, is a rare phenomenon at any particular nucleotide pair on a chromosome. The most important mechanism for generating genomic diversity in sexually reproducing species is thus the production of new combinations of already existing alleles. This type of diversity increases the chances that at least some offspring of a mating pair will inherit a combination of alleles best suited for survival and reproduction in a changing environment.

New combinations of already existing alleles arise from two different types of meiotic events: (i) independent assortment, in which each pair of homologous chromosomes segregates free from the influence of other pairs, via random spindle attachment; and (ii) crossing-over, in which two homologous chromosomes exchange parts. Independent assortment can produce gametes carrying new allelic combinations of genes on different chromosomes, but for genes on the same chromosome, independent assortment alone will only conserve the existing combinations of alleles. Crossing-over, however, can generate new allelic combinations of linked genes. The evolution of crossing-over thus compensated for what would otherwise be a significant disadvantage of the linkage of the genes within chromosomes.

Historically, geneticists have used the term *recombination* to indicate the production of new combinations of alleles by any means, including independent assortment. But in the remainder of this chapter, we use **recombination** more narrowly to mean the generation of new allelic combinations through genetic exchange between homologous chromosomes. In this discussion, we refer to the products of crossing-over as **recombinants:** chromosomes that carry a mix of alleles derived from different homologs.

In eukaryotic organisms, recombination has an additional essential function beyond generating new combinations of alleles: It helps ensure proper chromosome segregation during meiosis. Chapter 4 has already described how crossovers, in combination with sister chromatid cohesion, keep homologous chromosomes together as bivalents during the period between prophase I and metaphase I. If homologs fail to recombine, they often cannot orient themselves toward opposite poles of the meiosis I spindle, resulting in nondisjunction.

As we examine recombination at the molecular level, we look first at experiments establishing the basic parameters of crossing-over. We will then describe the molecular details of a crossover event.

Tetrad Analysis Illustrates Key Aspects of Recombination

You saw in Chapter 5 that some fungi like yeast and *Neurospora* generate asci that contain in one sac all the products of individual meioses—that is, tetrads. Analysis of these tetrads allowed geneticists to infer basic information about recombination.

Evidence that recombination takes place at the four-strand stage

Recall that in tetrad analysis, the hallmark of linkage between two genes is that very few NPD tetrads are produced: The number of PDs, and also Ts, is always greater than the number of NPDs. This outcome makes sense because all SCO and some DCO meioses yield Ts, while only one-quarter of the rare DCOs yield NPDs (review Figs. 5.23 and 5.24).

The very low number of NPDs actually observed in crosses establishes that recombination occurs after the chromosomes have replicated, when four chromatids exist for each pair of homologs (**Fig. 6.25a**). If recombination instead took place before chromosome duplication, every single crossover event (that is, every SCO) would yield four recombinant chromatids and generate an NPD tetrad; such a model for recombination would be unable to account for any T tetrads (**Fig. 6.25b**).

Evidence that recombination is usually reciprocal

The discussion in Chapter 5 assumed that recombination is reciprocal, with nonsister chromatids from homologous chromosomes exchanging parts equally. That is, whatever is lost from one chromatid is gained from the other, and *vice versa*. We know that this assumption is legitimate because of the results of tetrad analysis.

Suppose you are following linked genes *A* and *B* in a cross between *A B* and *a b* strains of yeast. If the recombination that occurs during meiosis is reciprocal, every tetrad with recombinant progeny should contain equal numbers of both classes of recombinants. Observations have in general confirmed this prediction: Every T tetrad carries one *A b* and one *a B* spore, while every NPD tetrad contains two of each type of recombinant (recall Fig. 5.24). We can thus conclude that meiotic recombination is *almost* always reciprocal. We say *almost* always because as you will see later in this chapter, rare exceptions exist in which a tetrad cannot be classified PD, NPD, or T. These exceptional tetrads helped scientists to understand key features of the recombination mechanism.

DNA Molecules Break and Rejoin During Recombination

When viewed through the light microscope, recombinant chromosomes bearing physical markers appear to result from two homologous chromosomes breaking and exchanging parts as they rejoin (see Fig. 5.7). Because the recombined chromosomes, like all other chromosomes, are composed of

Figure 6.25 Recombination after chromosome replication. **(a)** Because recombination occurs after chromosomes have replicated, most tetrads containing recombinant spores are Ts. **(b)** A disproven model. If recombination occurred before chromosomes replicated and if two genes were linked, most tetrads containing recombinant spores would be NPDs instead of Ts.

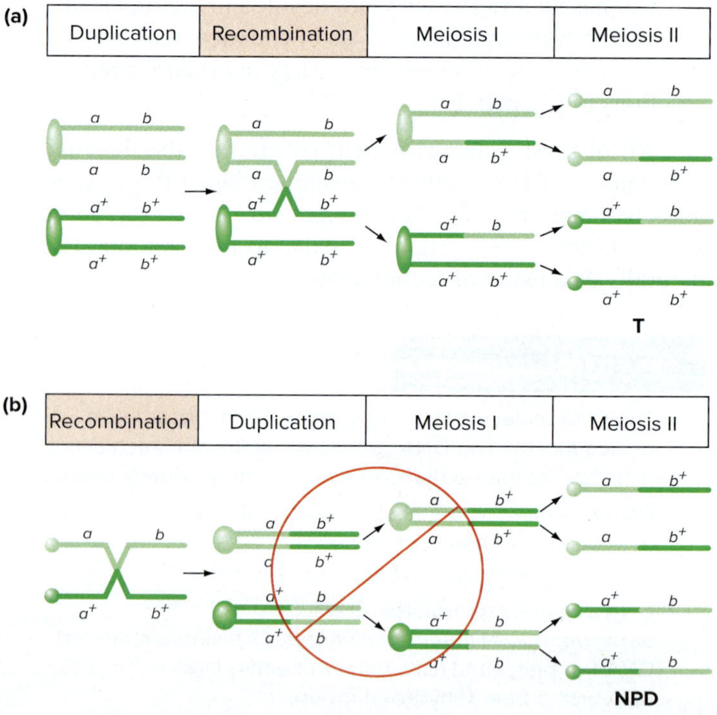

one long DNA molecule, a logical expectation is that they should show some physical signs of this breakage and rejoining at the molecular level. To evaluate this hypothesis, researchers chose a bacterial virus, *lambda,* as their model organism. Lambda had a distinct experimental advantage for this particular study: It is about half DNA by weight, so the density of the whole virus reflects the density of its DNA.

The experimental technique was similar in principle to the one Meselson and Stahl used to monitor a change in DNA density during DNA replication. In this case, however, the researchers (again Matthew Meselson but with a different collaborator, Jean Weigle) monitored DNA density to look at recombination. They used two strains of bacterial viruses that were genetically marked to keep track of recombination. They grew the wild-type strain (*A B*) in medium with heavy isotopes of carbon and nitrogen, and a strain with mutations in two genes (*a b*) in medium with the normal light isotopes of these atoms (**Fig. 6.26**). Meselson and Weigle then infected bacterial cells growing in normal (light isotope) medium with so many phages of each type that every cell was infected with both viral strains. After allowing time for the phages to replicate, recombine, and repackage their DNA into virus particles, the experimenters isolated the viruses released from the lysed cells and analyzed them on a density gradient.

It was important to the design of the experiment that both genes *A* and *B* were close to one end of the viral chromosome (Fig. 6.26). The idea was that some of the original phage chromosomes would undergo recombination before replicating in the light isotope medium. For example, some of the

Figure 6.26 DNA molecules break and rejoin during recombination. Meselson and Weigle infected *E. coli* cells with two different genetically marked strains of bacteriophage lambda previously grown in the presence of heavy (^{13}C and ^{15}N) or light (^{12}C and ^{14}N) isotopes. After growth on light medium, they spun the progeny bacteriophages on a CsCl density gradient. Some *A b* genetic recombinants (but almost no *a B* recombinants) had a heavy density almost the same as that of the *A B* parent.

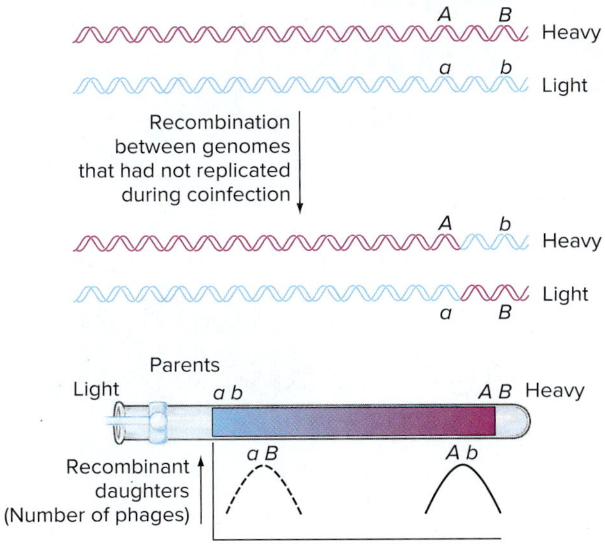

heavy *A B* chromosomes would recombine with light *a b* chromosomes. If crossing-over occurred through breakage and reunion of double-stranded DNA molecules, then some *A b* recombinant phages from the lysed cells should have a density almost as heavy as that of the parent phages that were *A B* (Fig. 6.26). In contrast, few if any recombinants of genotype *a B* should be composed of mostly heavy DNA.

Because the phages had replicated in light medium, recombinant phages could be found throughout the gradient. However, the key result was that a substantial proportion of the *A b* recombinants were indeed found near the heaviest density along with *A B* parent molecules. This result makes sense only if the *A b* chromosomes consisted mostly of double-helical heavy DNA, as expected for the kind of chromosomal breakage and reunion shown in Fig. 6.26.

Crossing-Over at the Molecular Level: A Model

Biochemical experiments performed mostly in yeast have informed our present understanding of the molecular mechanism for meiotic recombination. Researchers have found that the protein Spo11, which plays crucial roles in initiating meiotic recombination in yeast, is homologous to a protein essential for meiotic recombination in nematodes, plants, fruit flies, and mammals. This finding suggests that the mechanism of recombination presented in detail in **Fig. 6.27**—and known as the *double-strand-break repair model*—has been conserved throughout the evolution of eukaryotes.

In the figure, we focus on two nonsister chromatids, even though recombination takes place at the four-strand stage. Furthermore, we use the term *recombination event* to describe the molecular process initiated by Spo11, whether or not it results in crossing-over. As you are about to see, the molecular details of recombination events are such that crossing-over (reciprocal exchange of double-stranded DNA of nonsister chromosomes) results from the Spo11-mediated process only some of the time.

Initiating recombination

A meiotic recombination event begins when Spo11 makes a double-strand break in one of the four chromatids (Fig. 6.27, Step 1). Next, in a process called **resection,** an **exonuclease** (an enzyme that removes nucleotides from an end of a DNA molecule) degrades one strand of DNA from both sides of the cleavage, leaving 3′ single-stranded tails (Fig. 6.27, Step 2). In the next set of reactions, called **strand invasion,** one single-stranded tail displaces the corresponding strand on the nonsister chromatid (Fig. 6.27, Step 3). Strand invasion results in the formation of a **heteroduplex** region (from the Greek *hetero* meaning *other* or *different*) in which the DNA molecule is composed of one strand from each nonsister chromatid (Fig. 6.27, Step 3).

FEATURE FIGURE 6.27

A Model of Recombination at the Molecular Level

Step 1 Double-strand break formation. During meiotic prophase, **Spo11** protein makes a double-strand break on one of the chromatids by cleaving the phosphodiester bonds between adjacent nucleotides on both strands of the DNA. (Note that only the two nonsister chromatids undergoing recombination are shown.)

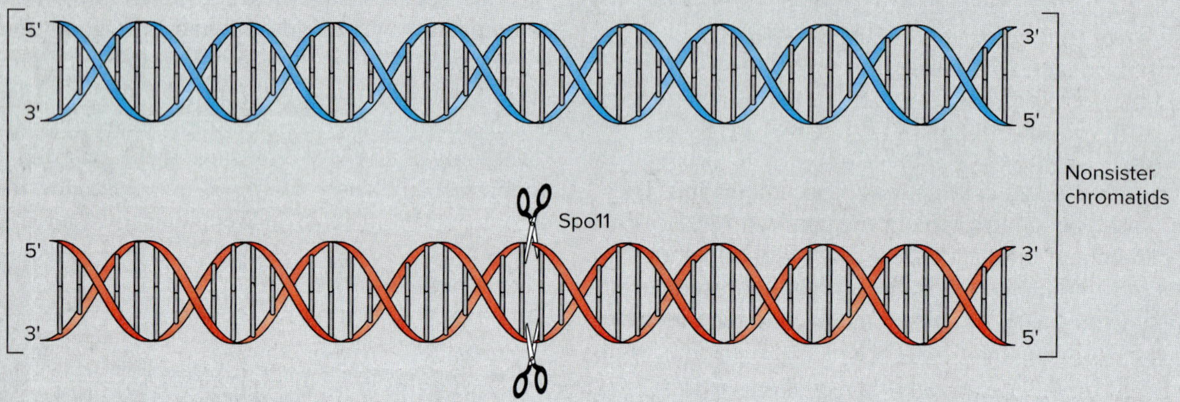

Step 2 Resection. An **exonuclease** degrades the 5′ ends on each side of the break to produce two 3′ single-stranded tails.

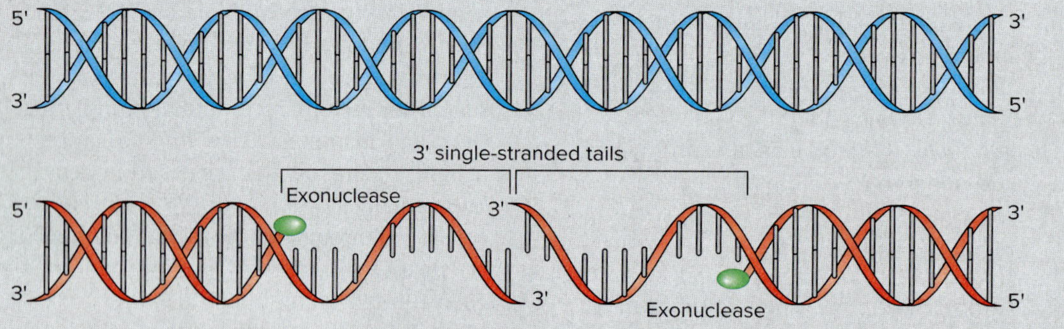

Step 3 First strand invasion. The protein Dmc1 (*orange* ovals) collaborates with other proteins (*not shown*) to help one of the tails invade and open up the other chromatid's double helix. Dmc1 then moves along the double helix, prying it open. The invading strand scans the base sequence in the momentarily unwound stretches of DNA duplex. As soon as it finds a complementary sequence of sufficient length, the two strands form a **heteroduplex** maintained by dozens of hydrogen bonds. The strand displaced by the invading tail forms a **D-loop** (for displacement loop), which is stabilized by binding of replication protein A (RPA) (*yellow* ovals).

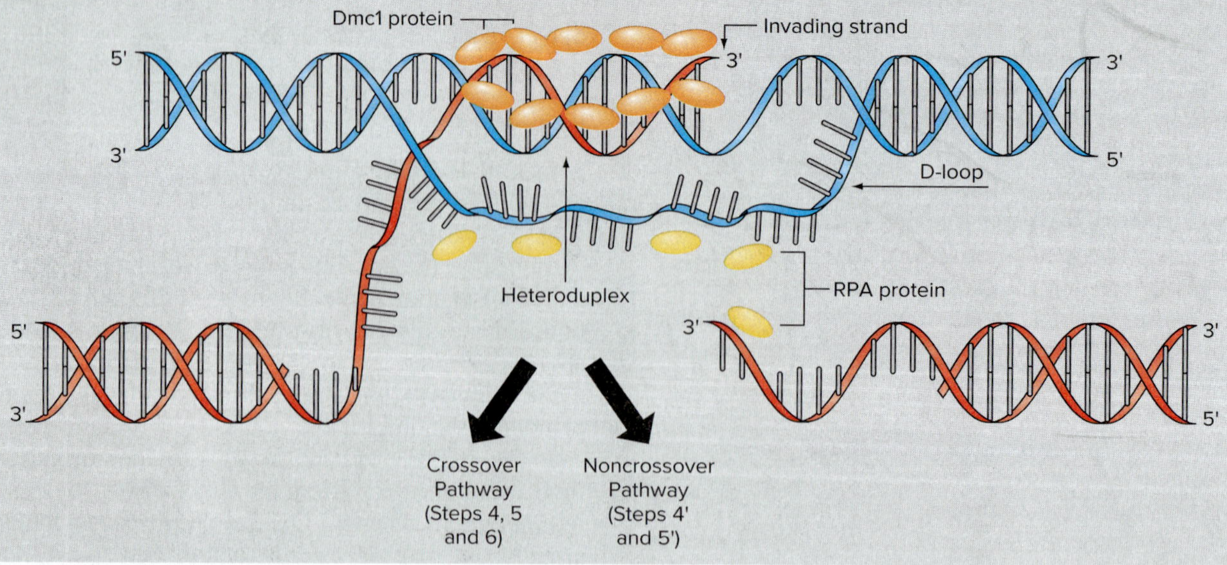

Crossover Pathway

Step 4 Formation of a double Holliday junction. New DNA added to the invading 3′ tail (*blue dots* at the top) enlarges the D-loop until the single-stranded bases on the displaced strand can form complementary base pairs with the 3′ tail on the nonsister chromatid. New DNA added to this latter tail (*blue dots* at the bottom) re-creates the DNA duplex on the bottom chromatid. At each side of the original break, the 3′ end of the newly synthesized DNA becomes adjacent to a 5′ end left after resection, and **DNA ligase** forms phosphodiester bonds to rejoin DNA strands without the loss or gain of nucleotides. The resulting X-shaped structures are called **Holliday junctions** after Robin Holliday, the scientist who first proposed their existence as a key intermediate in recombination.

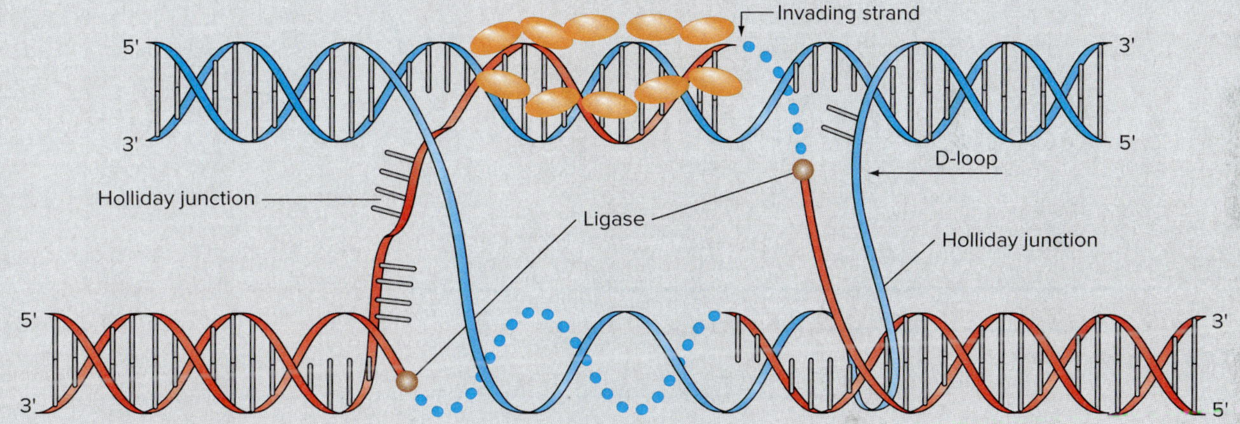

Step 5 Branch migration. The two invading strands tend to *zip up* by base pairing with the complementary strands of the parental double helixes they invade. The DNA double helixes unwind in front of this double zippering action, moving in the direction of the arrows in the figure, and two newly created heteroduplex molecules rewind behind it. Branch migration thus lengthens the heteroduplex region of both DNA molecules from tens of base pairs to hundreds or thousands.

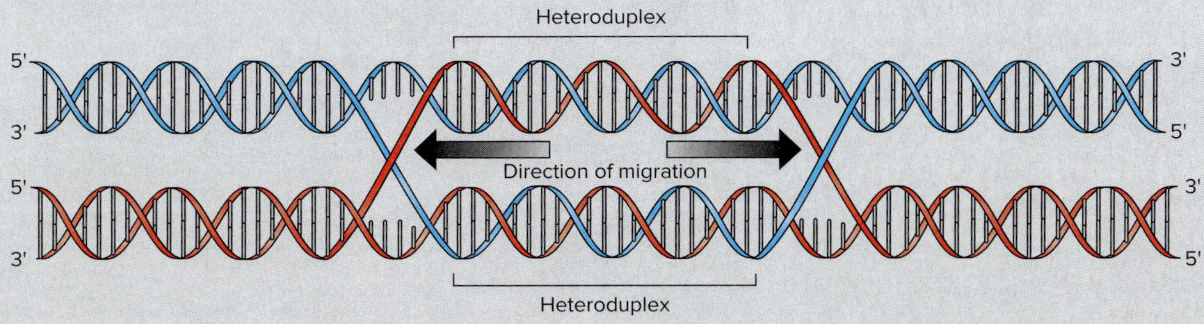

(*continued*)

FEATURE FIGURE 6.27 (*Continued*)

Step 6 Resolution of the double Holliday junction intermediate. The two interlocked nonsister chromatids must disengage. Separation is achieved by breakage of two DNA strands at each Holliday junction by an enzyme called **resolvase** (*not shown*); the strands are subsequently rejoined by DNA ligase (*not shown*). Different *blue* and *red* strands are cleaved at each junction; one junction is cleaved at the strands indicated by *yellow* arrows, and the strands indicated by the *green* arrows are cleaved at the other junction. At each junction, the strands are cut and rejoined so that *red* DNA connects to *blue* DNA and *vice versa*. Crossing-over results because each of the four strands is cut once and rejoined. Note in the diagram at the *bottom* that both of the recombinant chromatids have short heteroduplex regions.

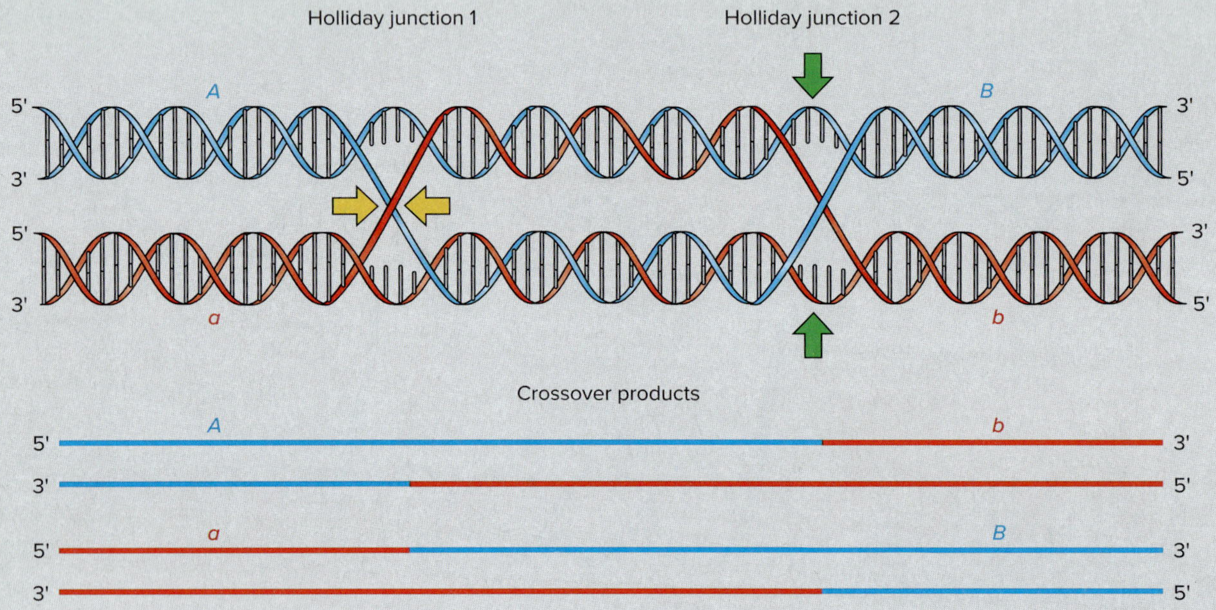

Noncrossover Pathway

Step 4′ Strand displacement and annealing. Just as in Step 4 of the Crossover Pathway, the invading strand (*arrow*) is first extended by DNA synthesis (*blue dots*) using the nonsister chromatid (*blue*) as a template. But next, an **anticrossover helicase** enzyme (*not shown*) disentangles the invading strand and the nonsister chromatid to yield the intermediates diagrammed.

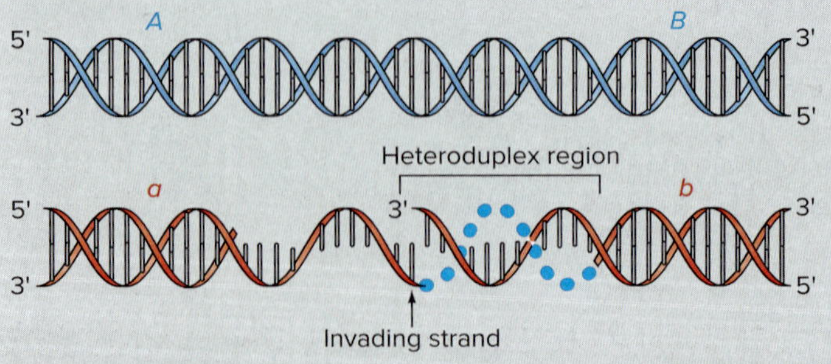

Step 5′ DNA synthesis and ligation. The remaining gap in the double-stranded DNA sequence is filled by DNA synthesis (*red dots*), and DNA ligase forms phosphodiester bonds to rejoin the DNA strands. The result is no crossover, but a heteroduplex region nonetheless remains in one chromatid.

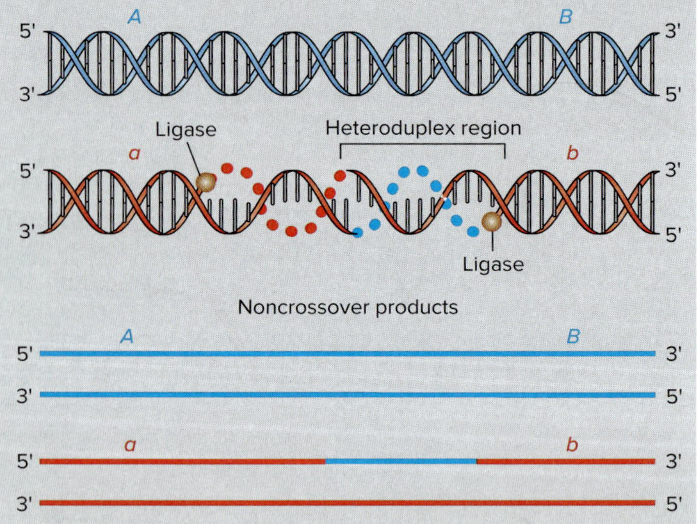

The molecular intermediate formed at the conclusion of Step 3 on Fig. 6.27 may have two alternative fates. One pathway, depicted in Steps 4 through 6, results in crossing-over. The second pathway, shown in Steps 4′ and 5′, does not yield a crossover, but one of the resultant chromatids has a heteroduplex region.

The crossover pathway

The strand displaced by strand invasion in Step 3 now forms a second heteroduplex with the other 3′ single-stranded tail (Fig. 6.27, Step 4). DNA synthesis to extend the two 3′ tails replaces the DNA that was degraded by the exonuclease, and **DNA ligase** reseals the DNA backbones (Fig. 6.27, Step 4). The result is that the two nonsister chromatids are interlocked at two **Holliday junctions** (Fig. 6.27, Step 5). The Holliday junctions move away from each other and thereby enlarge the heteroduplex between them—a process called **branch migration** (Fig. 6.27, Step 5).

Now, the two nonsister chromatids must be separated. The two chromatids disengage by the cutting and joining of two strands of DNA at each Holliday junction. As shown in Fig. 6.27 (Step 6), crossing-over (and recombination of flanking alleles) results when a different pair of DNA strands is cut and rejoined by **resolvase** and ligase enzymes at each junction. Because resolvase almost always cuts all four DNA strands, resolution of the double Holliday junctions usually results in crossing-over.

The noncrossover pathway

Recombination initiated by Spo11 can also result in no crossing-over through the action of an enzyme called **anticrossover helicase.** The helicase helps disentangle the invading strand from the nonsister chromatid, thus interrupting Holliday junction formation (Fig. 6.27, Step 4′). Note that although the end result of this pathway is no crossing-over (Step 5′), one of the resultant chromatids nonetheless contains a heteroduplex region.

Controlling where and when recombination occurs

Only cells undergoing meiosis express the Spo11 protein, which is responsible for a rate of meiotic recombination several orders of magnitude higher than that found in mitotically dividing cells. In yeast and humans, where meiotic double-strand breaks have been mapped, it is clear that Spo11 has a preference for cleavage of some genomic sequences over others, resulting in *hotspots* for crossing-over (recall Fig. 5.17).

Unlike meiotic cells, mitotic cells do not usually initiate recombination as part of the normal cell-cycle program; instead, recombination in mitotic cells is a consequence of environmental damage to the DNA. As you will see in Chapter 7, X-rays and ultraviolet light, for example, can cause either double-strand breaks or single-strand nicks. The cell's enzymatic machinery works to repair the damaged DNA site, and recombination is a side effect of this process.

A summary: Evidence for the current molecular model of homologous recombination

The double-strand-break repair model of meiotic recombination was proposed in 1983, well before the direct observation of any recombination intermediates. Scientists have now seen—at the molecular level—the formation of double-strand breaks, the resection of those breaks to produce 3′ single-strand tails, and double Holliday junction structures. The double-strand-break repair model has become established because it explains much of the data obtained from genetic and molecular studies as well as the six properties of recombination deduced from genetic experiments:

1. Homologs physically break, exchange parts, and rejoin. The Meselson-Weigle experiment with phage lambda provided key evidence for this key aspect of recombination (review Fig. 6.26).
2. Crossing-over occurs between nonsister chromatids after DNA replication. When yeast dihybrid for linked genes sporulate, the appearance of T tetrads and the rarity of NPDs make sense only if recombination happens at the four-strand, as opposed to the two-strand, stage (review Fig. 6.25).
3. Breakage and repair generate reciprocal products of recombination. Yeast and *Neurospora* tetrads are almost always NPD, PD, or T because the reciprocal recombinants are found in the same ascus.
4. Recombination events can occur anywhere along the DNA molecule. If enough progeny are counted, crossing-over can be observed between any pair of genes in a variety of different experimental organisms.
5. Precision in the exchange—no gain or loss of nucleotide pairs—prevents mutations from occurring during the process. Geneticists originally deduced the precision of crossing-over from observing that recombination usually does not cause mutations; today, we know this to be true from DNA sequence analysis.
6. **Gene conversion**—the physical change of one allele in a heterozygote into the other—sometimes occurs as a result of a recombination event. In the next section, you will see how gene conversion is explained by the formation of heteroduplexes during recombination events.

DNA Repair of Heteroduplexes Can Result in Gene Conversion

Yeast tetrad analysis allows us to see, just as Mendel predicted, that alleles segregate equally into gametes. Diploids heterozygous at a particular locus produce two spores containing one allele, and two spores containing the other allele (2:2 segregation)—most of the time. The opportunity to examine all four products of a single meiosis together in an ascus allowed the discovery that rarely, tetrads exhibit 3:1 or 1:3 segregation patterns, thereby breaking Mendel's first law. These rare tetrads are a consequence of heteroduplex formation during recombination; the phenomenon that produces these tetrads is called *gene conversion.*

Mismatched bases in heteroduplexes

The molecular model for recombination includes formation of a heteroduplex region, which occurs because the two strands of a recombined DNA molecule do not break and rejoin at the same location on the double helix. In addition, through branch migration, the heteroduplex region can be expanded to hundreds or even thousands of base pairs. The name *heteroduplex* applies not only because the two DNA strands came from different nonsister chromatids, but also because the base pairing of the strands may produce mismatches in which one or a few bases are not complementary. If the heteroduplex region is within a gene and the maternal and paternal alleles of the gene are different, gene conversion may result.

Figure 6.28 How gene conversion occurs. Alleles *B* and *b* differ by a single base pair; where *B* is G–C (*yellow*), *b* is T–A (*gray*). If gene *B* is within the heteroduplex region after a recombination event, repair of mismatched bases may convert *B* to *b* or *vice versa.* Gene conversion results when the bases changed by DNA repair (*black*) both originated from the same chromatid. Note that the *blue* and *red* lines are single DNA strands.

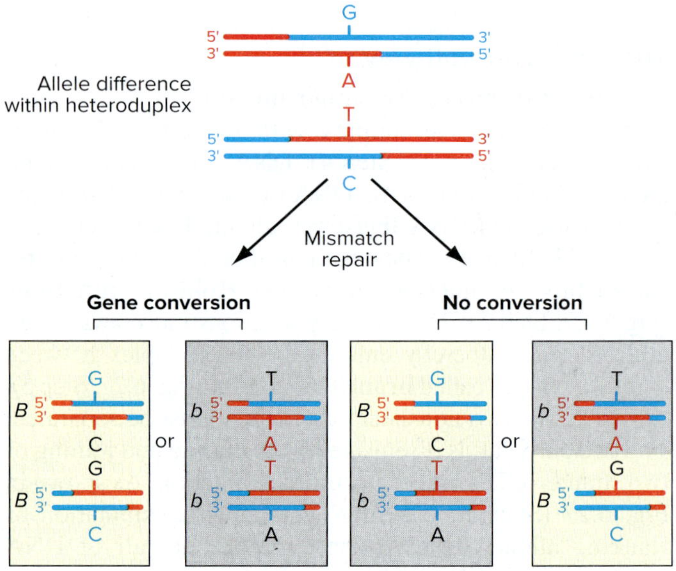

Gene conversion through mismatch repair

Mismatched heteroduplex molecules do not persist for long. The same DNA repair enzymes that operate to correct mismatches during replication (to be discussed in Chapter 7) also correct heteroduplexes during recombination. The outcome of the repair enzymes' work depends on which strands they alter. For example, the G–A mismatch in **Fig. 6.28** can become either G–*C* or *T*–A, and the T–C mismatch may be repaired to either *G*–C or T–*A* (*italics* indicate the altered base). Therefore, four possible repair outcomes exist for the two mismatches generated at a heteroduplex. Two of those outcomes—those in which both heteroduplexes are repaired to generate the same base pair—may result in gene conversion. Suppose that, as in Fig. 6.28, the base pair difference within a heteroduplex is the molecular difference between two alleles *B* and *b*. One nonsister chromatid started out with *B* and the other with *b*. The result of gene conversion is that both chromatids end up with the same allele—both are either *B* or *b*.

Gene conversion in yeast and *Neurospora* asci

Gene conversion is noticeable in yeast and *Neurospora* because all of the products of a single meiosis stay together in an ascus. Gene conversion may be detected as an unusual ascus that is neither PD, NPD, nor T. Recall from Fig. 5.22 that an *A B* /*a b* diploid yeast cell, where *A* and *B* are linked genes, could produce tetrads of any one of the three types. A key feature common to all three of these tetrad types is that the ratio of *A*:*a* or *B*:*b* alleles is always 2:2. However, a rare conversion of *b* to *B* results in a tetrad that is neither PD, NPD, nor T, because the ratio of *B*:*b* alleles is 3:1 (**Fig. 6.29**).

The idea that gene conversion is due to heteroduplex formation during a recombination event is supported by the observation that gene conversion is often associated with crossing-over of flanking alleles. For example, suppose during meiosis in an *A B C* / *a b c* diploid yeast, a recombination event occurs between gene *A* and gene *C* such that gene *B* is within the heteroduplex region (Fig. 6.29). Resolution of Holliday junctions on either side of gene *B* results in crossing-over—recombination between alleles of the flanking genes *A* and *C*. Subsequent DNA repair of the heteroduplex regions containing gene *B* can result in gene conversion, producing a tetrad that displays 3:1 segregation of *B*:*b* or *b*:*B* (**Fig. 6.29a**).

You should note that heteroduplexes resulting from recombination events that enter the noncrossover pathway can also generate tetrads with 3:1 segregation patterns. In such cases, gene conversion occurs but is not accompanied by recombination of the alleles of the flanking genes (**Fig. 6.29b**).

Figure 6.29 Detection of gene conversion in yeast tetrads. Throughout this figure, the *blue* and *red* lines represent single DNA strands. **(a)** Recombination during meiosis in an *A B C* / *a b c* diploid yeast cell generates heteroduplex regions in which each DNA strand has different alleles of gene *B*. Conversion of *b* to *B* by mismatch repair (*black B*s) results in an unusual tetrad with a 3:1 ratio of *B*:*b* alleles instead of 2:2. In this case, the recombination event resulted in crossing-over and thus recombination of the alleles of the flanking genes *A* and *C*. The tetrad is T with respect to *A* and *C*. **(b)** Here, the recombination event is resolved by the noncrossover pathway. Because crossing-over does not occur, the resulting tetrad is PD with respect to genes *A* and *C*. However, mismatch repair of the heteroduplex region converts *b* into *B*, so this tetrad also shows a 3:1 ratio of *B*:*b*.

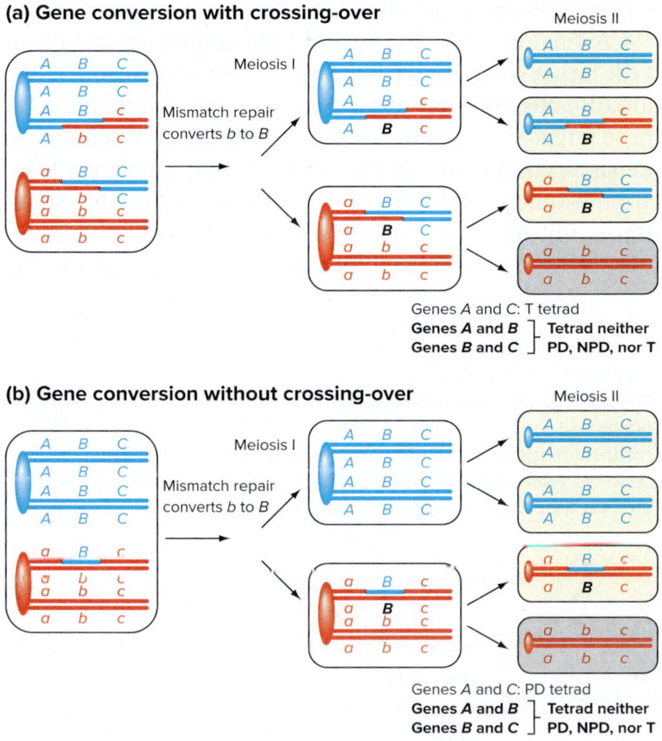

essential concepts

- In tetrad analysis, the existence of Ts and the very low number of NPDs observed establishes that recombination occurs after chromosome replication, when each pair of homologs contains four chromatids. T and NPD tetrads exhibit equal numbers of both classes of recombinants, indicating reciprocal exchange.
- The exchange of chromosome parts during recombination involves the breakage and rejoining of DNA molecules.
- At the molecular level, crossing-over during meiotic prophase entails the formation of *heteroduplex* DNAs between two *Holliday junctions* and resolution of the junctions by endonucleases and DNA ligase.
- Recombination events result in crossing-over only part of the time because helicases can disentangle the chromatids before Holliday junctions form.
- *Gene conversion,* a process whereby one allele in a heterozygote is physically changed into the other, provides evidence for heteroduplex formation during recombination events.

6.6 Site-Specific Recombination

learning objectives

1. Diagram the possible outcomes of site-specific recombination.
2. List the components that would have to be introduced to import site-specific recombination into a newly discovered organism.
3. Contrast the functions of Spo11 and Cas9, two enzymes that catalyze the formation of double-strand breaks.

Homologous recombination, as discussed in the previous section, begins with preexisting DNA molecules, breaks them apart, and then rejoins them to create new sequences of DNA. Natural selection then tests these new DNA molecules for their ability to help the organisms in which they are found to survive and reproduce in a changing environment. The more types of DNA molecules that are created in a population of organisms, the greater is the possibility that the population will continue in future generations. It is thus not surprising that homologous recombination can occur nearly at random at any of a very large number of sites in a genome, likely between any two adjacent pairs of nucleotides. In this way, homologous recombination helps to produce an enormous diversity in chromosome base sequences upon which natural selection can act.

Recombinase Enzymes Catalyze Recombination Between Specific DNA Sequences

In contrast with this type of nearly random homologous recombination, some organisms find it useful to have systems of **site-specific recombination** that promote the breakage and rejoining of DNA molecules at particular

DNA sequences. Site-specific recombination is crossing-over that occurs only between two specific DNA target sites that are usually less than 200 base pairs long. Site-specific recombination is much simpler at the molecular level than is the homologous recombination discussed in the previous section. In particular, in most systems of site-specific recombination, a single protein logically called a **recombinase** is sufficient to catalyze all the breakage and joining steps of the process. If you are curious, **Fig. 6.30** depicts the mode of action of one class of such recombinases.

The organisms that take advantage of site-specific recombination include certain kinds of bacteriophages that use this process for the **integration** (incorporation) of their small, circular genome into the chromosome of the host bacterium (**Fig. 6.31a**). In this way, the bacteriophage DNA "hitchhikes" along with the bacterial chromosome: When the host DNA replicates, so does the integrated bacteriophage genome.

Site-specific recombination is also important for the reverse process of **excision,** in which the DNA integrated between two target sites in a single chromosome is removed to create two independent DNA molecules (**Fig. 6.31b**). If a bacteriophage genome was previously integrated into the host chromosome, excision is crucial to allow the bacteriophage genome to extricate itself and then to become incorporated in the virus particle.

A third potential outcome of site-specific recombination systems is the **inversion** of a segment of DNA that is located between the two target sites (**Fig. 6.31c**). As you can imagine, such inversion could constitute a molecular switch between two configurations of the same chromosome. The in-between segment is oriented in one direction in one state and in the other direction in the other state.

A final mode of site-specific recombination can occur if the target site is found at the same position on each of two homologous chromosomes. Action of the recombinase on these target sites will result in the reshuffling of regions on nonsister chromatids, an outcome that leads to recombinant chromosomes (**Fig. 6.31d**). To our knowledge, this situation is not normally encountered in organisms that naturally use site-specific recombination. However, geneticists

Figure 6.30 One site-specific recombination mechanism. The Cre and Flp enzymes discussed in the text function as shown. The *red* and *blue* target DNA sequences are identical to each other but are represented in different colors for clarity. These targets are embedded in different DNA molecules (*black* and *gray dots*). The subunits of the recombinase tetramer are *yellow ovals;* this enzyme catalyzes all steps of the reaction. *Black triangles* are sites where recombinase cleaves single-stranded DNA. Note that resolution of the Holliday junction intermediate involves cleavage of the *blue* and *red* DNA strands that were not cleaved initially.

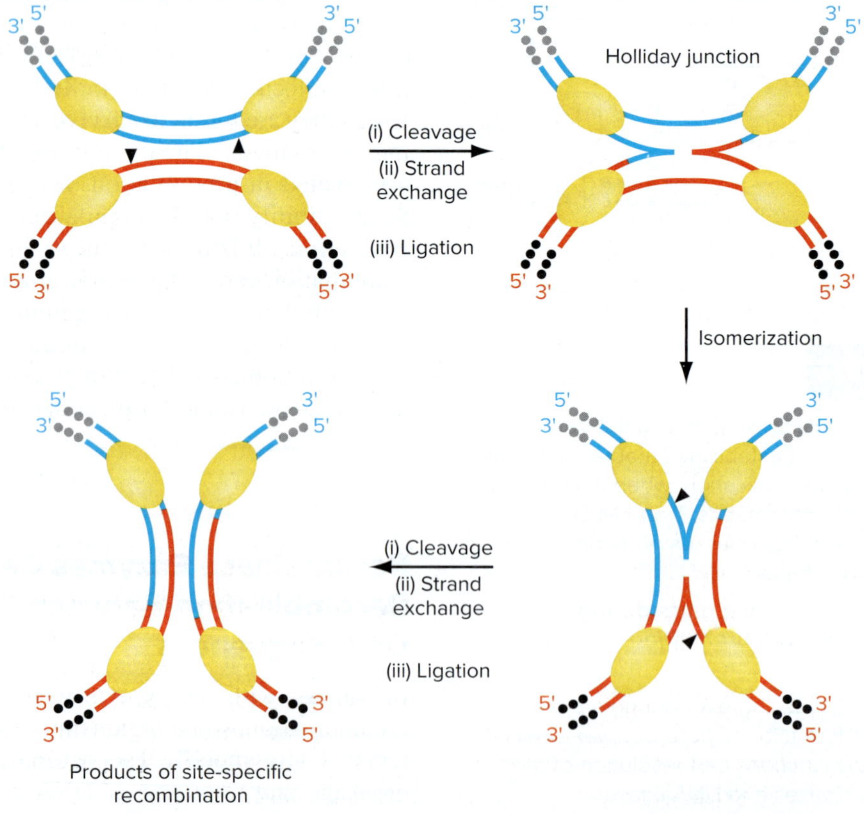

Figure 6.31 Possible outcomes of site-specific recombination. The *blue* and *red arrows* represent different identical target sites; the arrows can point in either of two directions because the target sites are asymmetric. The single *black* and *gray* lines in which the target sites are embedded are double-stranded DNA.

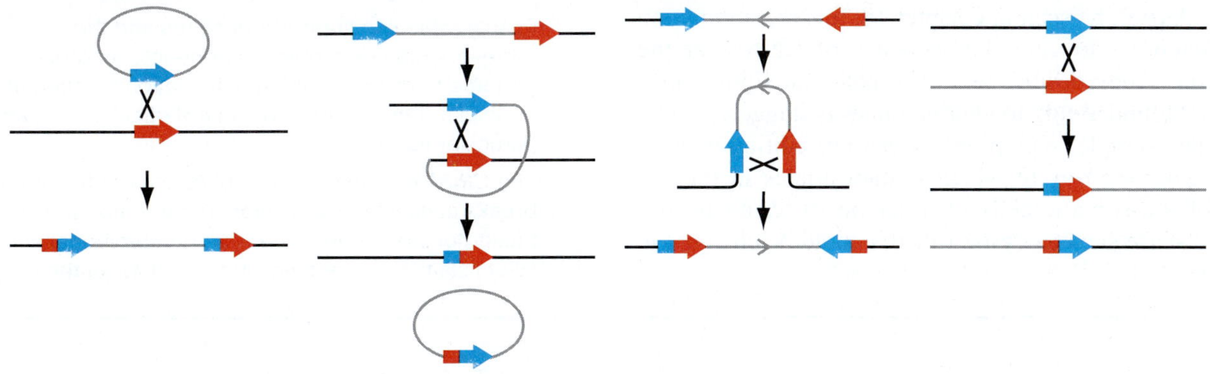

can create organisms with this arrangement of target sites that also make the recombinase protein. This technique is particularly useful in causing mitotic crossing-over to occur with high frequency at these defined locations.

Scientists can exploit the Flp/FRT and Cre/loxP site-specific recombination systems to turn genes on and off

Site-specific recombination is a property of only certain organisms, and its use in those organisms is usually restricted to a very specific process such as bacteriophage integration or excision. If site-specific recombination is not a general phenomenon like homologous recombination, why are we telling you about it? The answer is that geneticists can now export site-specific recombination to a wide variety of species, and these researchers have found such recombination to be incredibly useful. By adding target sequences to genomes, the geneticists can control precisely where in a genome recombination will take place. And by regulating the production of the recombinase enzyme, researchers can determine at what time and in what tissues the site-specific recombination occurs.

The later chapters of this book discuss two such systems of site-specific recombination: Flp recombinase/FRT sites (Flp/FRT), normally used for the replication of small circular DNAs (plasmids) in yeast cells; and Cre recombinase/loxP sites (Cre/loxP), needed for several stages in the life cycle of a type of bacteriophage called P1.

These feats of genetic engineering have several purposes. Using site-specific recombination, researchers can turn on or off the expression of a specific gene within an organism at a specific time or in a specific tissue. In addition, because site-specific recombination can occur with high efficiency in nearly all cell types, geneticists can use this method to induce mitotic recombination and thus reliably create clones of homozygous mutant cells within a heterozygous organism. By performing these manipulations, scientists can now ask important questions about the roles of particular genes in biological processes such as the development of a multicellular organism from a single cell, the fertilized egg; Chapters 18 and 19 will describe these issues in detail.

CRISPR-Cas9-induced recombination is a powerful tool for manipulating genomes

One important limitation in importing site-specific recombination to new organisms is that the target sequences need to be introduced into genomes, but in most cases researchers cannot direct those target sites into a preselected genomic region. Instead, the target sites become incorporated into random positions, and the scientists then search for a strain with the target site in the most advantageous location.

Remarkable methodologies developed very recently now allow researchers to alter genomes precisely in almost any way imaginable. One particularly exciting technology is based on small RNAs called *CRISPRs* and an enzyme called Cas9 that is produced in a few bacterial species. It is premature to describe this method in great detail so early in this book, but for the time being it is sufficient to tell you that a CRISPR can direct Cas9 to any specific DNA sequence in a complex genome. The importance is that Cas9 is an enzyme that produces double-strand breaks in DNA. As we saw in a previous section, the formation of a double-strand break (by Spo11) initiates the process of homologous recombination; in other words, double-strand breaks are *recombinogenic*.

Because CRISPR/Cas9 causes double- strand breaks at a specific genomic location determined by the sequence of the CRISPR RNA, researchers can now induce recombination to occur at high frequency at any specific location of the genome. As will be seen in Chapter 18, this recombination allows scientists to alter the sequence of DNA near the breakpoint in any desired way. The potential significance of this newfound ability to alter genomes is staggering. Just to cite one example, such pinpoint genome editing may allow for *gene therapy* in which mutant alleles in the genomes of the somatic cells of a person suffering from a genetic disease such as cystic fibrosis could be changed to wild-type alleles.

essential concepts

- *Site-specific recombination* is crossing-over between two short DNA target sites catalyzed by a *recombinase* enzyme.
- Researchers can import target sites and the corresponding recombinase gene into an organism's genome to promote site-specific recombination at a particular genomic location, at a particular time, and in a particular tissue.
- The *CRISPR/Cas9* system can induce double-strand breaks at almost any position in the genome. The fact that these double-strand breaks are recombinogenic allows scientists to edit genomes in the vicinity of the breakage.

WHAT'S NEXT

The Watson-Crick model for the structure of DNA, the single most important biological discovery of the twentieth century, clarified how the genetic material fulfills its primary functions of carrying and accurately reproducing information: Each DNA molecule carries one of a vast number of potential arrangements of the four nucleotide building blocks (A, T, G, and C). The model also suggested how base complementarity could provide a mechanism for faithful DNA replication. We have further seen how the structure of DNA enables the recombining of genetic information from maternal and paternal chromosomes.

Unlike its ability to carry information, DNA's capacities for replication and recombination are not solely properties of the DNA molecule itself. Rather they depend on the cell's complex enzymatic machinery. But even though they rely on the complicated orchestration of many different proteins, replication and recombination both occur with extremely high fidelity—normally not a single base pair is gained or lost. Occasionally, however, errors do occur, providing the genetic basis of evolution.

DNA copying or recombination errors that occur within genes sometimes produce dramatic changes in phenotype. How do mutations in genes arise? And how did we come to understand that different alleles of genes produce their phenotypic effects through the proteins that they specify?

We begin to answer these questions in Chapter 7. We first describe the molecular processes that lead to mutation. Next, you will see that scientists used mutations to determine what a gene actually is—a linear sequence of base pairs in DNA, and what a gene does—it encodes the information for producing a protein.

SOLVED PROBLEMS

5′ TAAGCGTAACCCGCTAA CGTATGCGAAC GGGTCCTATTAACGTGCGTACAC 3′
3′ ATTCGCATTGGGCGATT GCATACGCTTG CCCAGGATAATTGCACGCATGTG 5′

I. Imagine that the double-stranded DNA molecule shown was broken at the sites indicated by spaces in the sequence, and that before the breaks were repaired, the 11 base pair DNA fragment between the breaks was reversed (*inverted*). What would be the base sequence of the repaired molecule? Explain your reasoning.

Answer

To answer this question, you need to keep in mind the polarity of the DNA strands involved.

The top strand has the polarity left to right of 5′ to 3′. The reversed region must be rejoined with the same polarity. Label the polarity of the strands within the inverting region. To have a 5′-to-3′ polarity maintained on the top strand, the fragment that is reversed must also be flipped over, so the strand that was formerly on the bottom is now on top.

```
5′                                               3′
TAAGCGTAACCCGCTAAGTTCGCATACGGGGTCCTATTAACGTGCGTACAC
ATTCGCATTGGGCGATTCAAGCGTATGCCCCAGGATAATTGCACGCATGTG
3′                                               5′
```

II. A new virus has recently been discovered that infects human lymphocytes. The virus can be grown in the laboratory using cultured lymphocytes as host cells. Design an experiment using a radioactive label that would tell you if the virus contains DNA or RNA.

Answer

Use your knowledge of the differences between DNA and RNA to answer this question. RNA contains uracil instead of the thymine found in DNA. You could set up one culture in which you add radioactive uracil to the medium and a second one in which you add radioactive thymine to the culture. After the viruses have infected cells and produced more new viruses, collect the newly synthesized virus. Determine which culture produced radioactive viruses. If the virus contains RNA, the collected virus grown in medium containing radioactive uracil will be radioactive, but the virus grown in radioactive thymine will not be radioactive. If the virus contains DNA, the collected virus from the culture containing radioactive thymine will be radioactive, but the virus from the radioactive uracil culture will not.

(You might also consider using radioactively labeled ribose or deoxyribose to differentiate between an RNA- and DNA-containing virus. Technically this does not work as well because the radioactive sugars are processed by cells before they become incorporated into nucleic acid, thereby obscuring the results.)

III. If you expose a culture of human cells (for example, HeLa cells) to ^{3}H-thymidine during S phase, how would the radioactivity be distributed over a pair of homologous chromosomes at metaphase? Would the radioactivity be in (a) one chromatid of one homolog, (b) both chromatids of one homolog, (c) one chromatid each of both homologs, (d) both chromatids of both homologs, or (e) some other pattern? Choose the correct answer and explain your reasoning.

Answer

This problem requires application of your knowledge of the molecular structure and replication of DNA and how it relates to chromatids and homologs. DNA replication occurs during S phase, so the ^{3}H-thymidine would be incorporated into the new DNA strands. A chromatid is a replicated DNA molecule, and each new DNA molecule contains one new strand of DNA (semiconservative replication). The radioactivity would be in both chromatids of both homologs (answer d).

PROBLEMS

Vocabulary

1. For each of the terms in the left column, choose the best matching phrase in the right column.

a. transformation	1. the strand that is synthesized discontinuously during replication
b. bacteriophage	2. the sugar within the nucleotide subunits of DNA
c. pyrimidine	3. a nitrogenous base containing a double ring
d. deoxyribose	4. noncovalent bonds that hold the two strands of the double helix together
e. hydrogen bonds	5. Meselson and Stahl experiment
f. complementary bases	6. Griffith experiment
g. origin	7. structures at ends of eukaryotic chromosomes
h. Okazaki fragments	8. two nitrogenous bases that can pair via hydrogen bonds
i. purine	9. catalyzes site-specific recombination
j. topoisomerases	10. a nitrogenous base containing a single ring
k. semiconservative replication	11. a short sequence of bases where unwinding of the double helix for replication begins
l. lagging strand	12. a virus that infects bacteria
m. telomeres	13. short DNA fragments formed by discontinuous replication of one of the strands
n. recombinase	14. enzymes involved in controlling DNA supercoiling

Section 6.1

2. Griffith, in his 1928 experiments, demonstrated that bacterial strains could be genetically transformed. The evidence that DNA was the *transforming principle* responsible for this phenomenon came later. What was the key experiment that Avery, MacCleod, and McCarty performed to prove that DNA was responsible for the genetic change from rough cells into smooth cells?
3. During bacterial transformation, DNA that enters a cell is not an intact chromosome; instead it consists of randomly generated fragments of chromosomal DNA. In a transformation where the donor DNA was from a bacterial strain that was $a^+ b^+ c^+$ and the recipient was $a\ b\ c$, 55% of the cells that became a^+ were also transformed to c^+. But only 2% of the a^+ cells were b^+. Is gene b or c closer to gene a?
4. Nitrogen and carbon are more abundant in proteins than sulfur. Why did Hershey and Chase use radioactive sulfur instead of nitrogen and carbon to label the protein portion of their bacteriophages in their experiments to determine whether parental protein or parental DNA is necessary for progeny phage production?

Section 6.2

5. If 30% of the bases in human DNA are A, (a) what percentage are C? (b) What percentage are T? (c) What percentage are G?
6. Which of the following statements are true about double-stranded DNA?
   a. A + C = T + G
   b. A + G = C + T
   c. A + T = G + C
   d. A/G = C/T
   e. A/G = T/C
   f. (C + A) / (G + T) = 1
7. Imagine you have three test tubes containing identical solutions of purified, double-stranded human DNA. You expose the DNA in tube 1 to an agent that breaks the sugar-phosphate (phosphodiester) bonds. You expose the DNA in tube 2 to an agent that breaks the bonds that attach the bases to the sugars. You expose the DNA in tube 3 to an agent that breaks the hydrogen bonds. After treatment, how would the structures of the molecules in the three tubes differ?
8. What information about the structure of DNA was obtained from X-ray crystallographic data?
9. A portion of one DNA strand of the human gene responsible for cystic fibrosis is

   5′.....ATAGCAGAGCACCATTCTG.....3′

   Write the sequence of the corresponding region of the other DNA strand of this gene, noting the polarity. What do the dots before and after the given sequence represent?
10. When a double-stranded DNA molecule is exposed to high temperature, the two strands separate, and the molecule loses its helical form. We say the DNA has been *denatured.* (Denaturation also occurs when DNA is exposed to acid or alkaline solutions.)
    a. Regions of the DNA that contain many A–T base pairs are the first to become denatured as the temperature of a DNA solution is raised. Thinking about the chemical structure of the DNA molecule, why do you think the A–T-rich regions denature first?
    b. If the temperature is lowered, the original DNA strands can *reanneal,* or *renature.* In addition to the full double-stranded molecules, some molecules of the type shown here are seen when the molecules are examined under the electron microscope. How can you explain these structures?

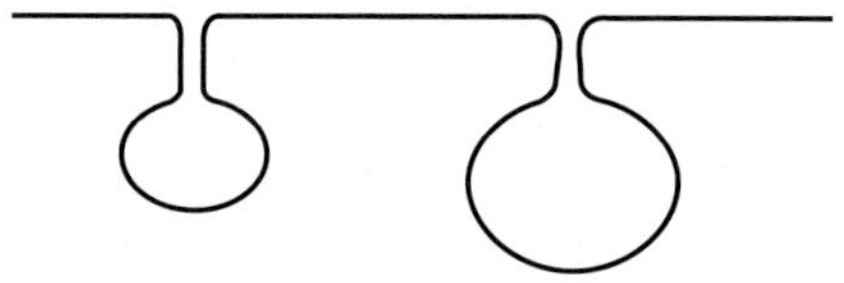

11. A particular virus with DNA as its genetic material has the following proportions of nucleotides: 20% A, 35% T, 25% G, and 20% C. How can you explain this result?

Section 6.3

12. The underlying structure of DNA is very simple, consisting of only four possible building blocks.
    a. How is it possible for DNA to carry complex genetic information if its structure is so simple?
    b. What are these building blocks? Can each block be subdivided into smaller units, and if so, what are they? What kinds of chemical bonds link the building blocks?
    c. How does the underlying structure of RNA differ from that of DNA?
13. An RNA virus that infects plant cells is copied into a DNA molecule after it enters the plant cell. What would be the sequence of bases in the first strand of DNA made complementary to the section of viral RNA shown here?

    5′ CCCUUGGAACUACAAAGCCGAGAUUAA 3′
14. Bacterial transformation and bacteriophage labeling experiments proved that DNA was the hereditary

material in bacteria and in DNA-containing viruses. Some viruses do not contain DNA but have RNA inside the phage particle. An example is the tobacco mosaic virus (TMV) that infects tobacco plants, causing lesions in the leaves.

Two different variants of TMV exist that have different forms of a particular protein in the virus particle that can be distinguished. It is possible to reconstitute TMV *in vitro* (in the test tube) by mixing purified proteins and RNA. The reconstituted virus can then be used to infect the host plant cells and produce a new generation of viruses. Design an experiment to show that RNA, rather than protein, acts as the hereditary material in TMV.

15. The CAP protein is shown bound to DNA in Fig. 6.15. CAP binds a specific sequence of base pairs in DNA (N = any base):

    5′ TGTGANNNNNNTCACA 3′
    3′ ACACTNNNNNNAGTGT 5′

    a. In a long double-stranded DNA molecule with random base sequence and an equal number of A–T and G–C base pairs, how many different kinds of DNA sequences could be bound by CAP?
    b. In the same DNA molecule, how frequently would a CAP binding site of any type be present? Of a particular type?
    c. CAP protein binds DNA as dimer; two identical CAP protein subunits bound to each other bind DNA. Can you detect a special feature of the DNA site that CAP binds that suggests that two identical protein subunits bind the DNA? (*Hint:* Try reading the sequence in the 5′-to-3′ direction on each strand.)
    d. CAP protein binds to the major groove of DNA. Do you expect that DNA helicase is required for CAP to bind DNA?

Section 6.4

16. In Meselson and Stahl's density shift experiments (diagrammed in Fig. 6.20), describe the results you would expect in each of the following situations:
    a. Conservative replication after two rounds of DNA synthesis on ^{14}N.
    b. Semiconservative replication after three rounds of DNA synthesis on ^{14}N.
    c. Dispersive replication after three rounds of DNA synthesis on ^{14}N.
    d. Conservative replication after three rounds of DNA synthesis on ^{14}N.
17. When Meselson and Stahl grew *E. coli* in ^{15}N medium for many generations and then transferred the cells to ^{14}N medium for one generation, they found that the bacterial DNA banded at a density intermediate between that of pure ^{15}N DNA and pure ^{14}N DNA following equilibrium density centrifugation. When they allowed the bacteria to replicate one additional time in ^{14}N medium, they observed that half of the DNA remained at the intermediate density, while the other half banded at the density of pure ^{14}N DNA. What would they have seen after an additional generation of growth in ^{14}N medium? After two additional generations?
18. If you expose human tissue culture cells (for example, HeLa cells) to ^{3}H-thymidine just as they enter S phase, then wash this material off the cells and let them go through a second S phase before looking at the chromosomes, how would you expect the ^{3}H to be distributed over a pair of homologous chromosomes? (Ignore the effect recombination could have on this outcome.) Would the radioactivity be in (a) one chromatid of one homolog, (b) both chromatids of one homolog, (c) one chromatid each of both homologs, (d) both chromatids of both homologs, or (e) some other pattern? Choose the correct answer and explain your reasoning. (This problem extends the analysis begun in Solved Problem III.)
19. Draw a replication bubble with both replication forks and label the origin of replication, the leading strands, lagging strands, and the 5′and 3′ ends of all strands shown in your diagram.
20. a. Do any strands of nucleic acid exist in nature in which part of the strand is DNA and part is RNA? If so, describe when such strands of nucleic acid are synthesized. Is the RNA component at the 5′ end or at the 3′ end?
    b. RNA primers in Okazaki fragments are usually very short, less than 10 nucleotides and sometimes as short at 2 nucleotides in length. What does this fact tell you about the *processivity* of the primase enzyme—that is, the relative ability of the enzyme to continue polymerization as opposed to dissociating from the template and from the molecule being synthesized? Which enzyme is likely to have a greater processivity, primase or DNA polymerase III?
21. As Fig. 6.21 shows, DNA polymerase cleaves the high-energy bonds between phosphate groups in nucleotide triphosphates (nucleotides in which three phosphate groups are attached to the 5′ carbon atom of the deoxyribose sugar). The enzyme uses this energy to catalyze the formation of a phosphodiester bond when incorporating new nucleotides into the growing chain.
    a. How does this information explain why DNA chains grow during replication in the 5′-to-3′ direction?
    b. The action of the enzyme DNA ligase in joining Okazaki fragments together is shown in Fig. 6.23.

Remember that these fragments are connected only after the RNA primers at their ends have been removed. Given this information, infer the type of chemical bond whose formation is catalyzed by DNA ligase and whether or not a source of energy will be required to promote this reaction. Explain why DNA ligase and not DNA polymerase is required to join Okazaki fragments.

22. The bases of one of the strands of DNA in a region where DNA replication begins are shown at the end of this problem. What is the sequence of the primer that is synthesized complementary to the bases in bold? (Indicate the 5′ and 3′ ends of the sequence.)

    5′ AGGCCTCG**AATTCGTATA**GCTTTCAGAAA 3′

23. Replicating structures in DNA can be observed in the electron microscope. Regions being replicated appear as bubbles.
    a. Assuming bidirectional replication, how many origins of replication are active in this DNA molecule?
    b. How many replication forks are present?
    c. Assuming that all replication forks move at the same speed, which origin of replication was activated last?

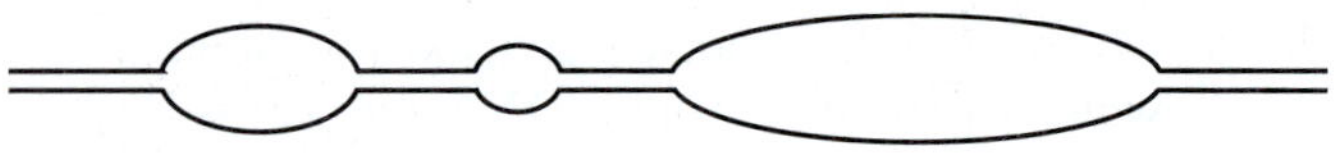

24. Indicate the role of each of the following in DNA replication: (a) topoisomerase, (b) helicase, (c) primase, and (d) ligase.
25. Draw a diagram of replication that is occurring at the end of a double-stranded linear chromosome. Show the leading and lagging strands with their primers. (Indicate the 5′ and 3′ ends of the strands.) What difficulty is encountered in producing copies of both DNA strands at the end of a chromosome?
26. Figure 6.18 depicts Watson and Crick's initial proposal for how the double-helical structure of DNA accounts for DNA replication. Based on our current knowledge, this figure contains a serious error due to oversimplification. Identify the problem with this figure.
27. Researchers have discovered that during replication of the circular DNA chromosome of the animal virus SV40, the two newly completed daughter double helixes are intertwined. What would have to happen for the circles to come apart?
28. A *DNA synthesizer* is a machine that uses automated chemical synthesis to generate short, single strands of DNA of any given sequence. You have used the machine to synthesize the following three DNA molecules:

    (DNA 1) 5′ CTACTACGGATCGGG 3′
    (DNA 2) 5′ CCAGTCCCGATCCGT 3′
    (DNA 3) 5′ AGTAGCCAGTGGGGAAAAACCCCACTGG 3′

    Now you add the DNA molecules either singly or in combination to reaction tubes containing DNA polymerase, dATP, dCTP, dGTP, and dTTP in a buffered solution that allows DNA polymerase to function. For each of the reaction tubes, indicate whether DNA polymerase will synthesize any new DNA molecules, and if so, write the sequence(s) of any such DNAs.
    a. DNA 1 plus DNA 3
    b. DNA 2 plus DNA 3
    c. DNA 1 plus DNA 2
    d. DNA 3 only

Section 6.5

29. Bacterial cells were coinfected with two types of bacteriophage lambda: One carried the c^+ allele and the other the c allele. After the cells lysed, progeny bacteriophage were collected. When a single such progeny bacteriophage was used to infect a new bacterial cell, it was observed in rare cases that some of the resulting phage progeny were c^+ and others were c. Explain this result.
30. A yeast strain with a mutant *spo11*$^-$ allele has been isolated. The mutant allele is nonfunctional; it makes no Spo11 protein. What do you suppose is the phenotype of this mutant strain?
31. Imagine that you have done a cross between two strains of yeast, one of which has the genotype *A B C* and the other *a b c*, where the letters refer to three rather closely linked genes in the order given. You examine many tetrads resulting from this cross, and you find two that do not contain the expected two *B* and two *b* spores. In tetrad I, the spores are *A B C, A B C, a B c,* and *a b c*. In tetrad II, the spores are *A B C, A b c, a b C,* and *a b c*. How have these unusual tetrads arisen?
32. The *Neurospora* octad shown came from a cross between a^+ and a^- strains.
    a. Is this an MI or an MII octad or neither? Explain.
    b. Diagram the production of this octad.
    c. Is it possible to observe evidence of heteroduplex formation in a *Neurospora* ascus even if gene

conversion did not occur during formation of the octad? Explain.

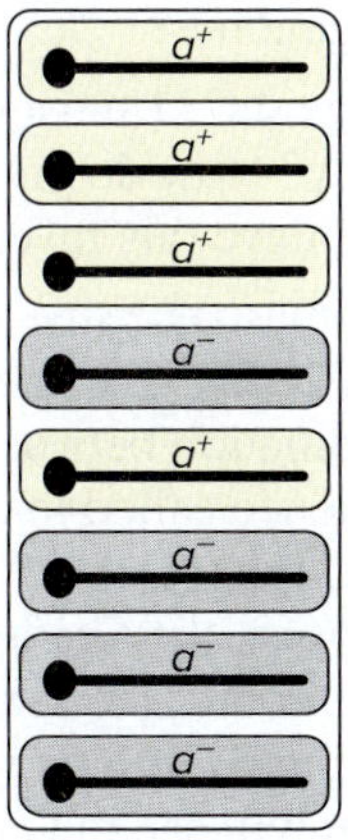

33. From a cross between $e^+ f^+ g^+$ and $e^- f^- g^-$ strains of *Neurospora*, recombination between these linked genes resulted in a few octads containing the following ordered set of spores:

$e^+ f^+ g^+$
$e^+ f^+ g^+$
$e^+ f^- g^+$
$e^+ f^- g^+$
$e^- f^- g^-$
$e^- f^- g^-$
$e^- f^- g^-$
$e^- f^- g^-$

a. Where was recombination initiated?

b. Did crossing-over occur between genes *e* and *g*? Explain.

c. Why do you end up with $2 f^+ : 6 f^-$ but $4 e^+ : 4 e^-$ and $4g^+ : 4g^-$?

d. Could you characterize these unusual octads as MI or MII for any of the three genes involved? Explain.

34. In Step 6 of Fig. 6.27, the resolvase enzyme almost always cuts all four strands of DNA in the double Holliday junction intermediate: both *blue* strands and both *red* strands. Another way of stating this fact is that the enzyme cuts the DNA at Holliday junctions 1 and 2 in different ways, represented by the *yellow arrows* at junction 1 and the *green arrows* at junction 2 in the figure. But rarely, the resolvase enzyme instead cuts the DNA at both Holliday junctions in the same way (*yellow arrows* at both junctions or *green arrows* at both junctions). In other words, at both junctions, the same *red* strand and the same *blue* strand are cut. What would be the outcome of this rare resolvase enzyme behavior?

Section 6.6

35. Figure 6.31 shows four potential outcomes of site-specific recombination that depend on the relative arrangement of the target sites for the recombinase enzyme. Could homologous recombination (in the absence of specific target sites and recombinase) also cause all of the same kinds of outcomes? If so, what is so different about how geneticists would use homologous recombination and site-specific recombination?

36. Each of the substrates for site-specific recombination listed (a–f) is also the product of site-specific recombination that occurred at a different one of these substrates. Match each substrate (a–f) with its product (a–f).

a. a linear DNA with two target sites in the same orientation

b. a linear DNA with two target sites in opposite orientations

c. a circular DNA with two target sites in the same orientation

d. a circular DNA with two target sites in opposite orientations

e. a circular DNA with one target site and a linear DNA with one target site

f. two circular DNAs each with one target site

37. Problem 52 in Chapter 5 discussed the use of mitotic recombination to study the function of a gene called *smc* during mouse development. The idea was that mitotic recombination in a cell of an smc^+/smc^- heterozygote could produce a clone of tissue derived from a daughter cell that was homozygous for the smc^- mutation. You could recognize the cells in this clone by the absence of green fluorescence from GFP.

a. Mitotic recombination is a rare event, making it difficult for researchers to find smc^-/smc clones to study. Explain why scientists might want to subject the same mice to X-rays to increase the frequency at which the desired clones could be found. At what stage of mouse development would the researchers expose the animals to X-rays?

b. An even more efficient way to induce mitotic recombination is to construct mice that include the Flp/FRT system. (Flp is a recombinase enzyme in yeast cells that promotes site-specific recombination between two identical copies of a particular 34 base pair long DNA sequence called an FRT site.) Assuming that these mice have a *transgene* that can specify the Flp recombinase protein, where would you put the FRT sites relative to the *smc* gene and to the *GFP* gene?

c. In part (b), what (in general terms) would you have to do with the Flp-encoding gene to make sure that mitotic recombination does not happen in every cell in the mouse? (This precaution is necessary because a mouse which has many *smc*$^-$/*smc*$^-$ homozygous cells could die as an embryo, before it reaches adulthood.)

38. Suppose that you could inject a wild-type mouse zygote with a specific CRISPR RNA and the Cas9 enzyme. The RNA directs the Cas9 enzyme to make a double-strand break within a gene that you think may be responsible for a heritable disease. Diagram in rough form how you might inject at the same time another nucleic acid molecule (here, a double-stranded DNA) to exploit homologous recombination so that you could convert the wild-type allele of the gene to a specific mutant allele.

39. ΦC31 is a type of bacteriophage that infects *Streptomyces* bacteria. One gene in the bacteriophage genome specifies a recombinase called *ΦC31 integrase* that works through a mechanism slightly different from that of the recombinase shown in Fig. 6.30. Most importantly, the two target DNA sequences are different from each other. One called *attP* is 39 base pairs and is found on the circular bacteriophage chromosome, while the other—*attB*—is 34 base pairs long and is located on the much larger circular bacterial chromosome. Excepting two base pairs roughly in the middle of both targets that are identical and at which recombination takes place, the DNA sequences of *attP* and *attB* are completely different from each other.
    a. Diagram the reaction that ΦC31 integrase performs. How could this reaction be important for the life cycle of the bacteriophage?
    b. Using the diagram you just drew, explain why ΦC31 integrase cannot reverse the reaction.
    c. Now consider how you might exploit this site-specific recombination to place genes from another species (a *transgene*) into the genome of an experimental organism like *Drosophila.* Assume you can make any DNA sequences you want and that you can introduce these DNA sequences into fruit fly germ-line cells by injection. Why is the irreversibility of the ΦC31 integrase–mediated reaction valuable for placing the transgene into the *Drosophila* genome?
    d. Bacteriophage ΦC31 must eventually reverse this reaction. Why? How do you think the bacteriophage can achieve this reversal?

40. Cre is a recombinase enzyme encoded by a gene in bacteriophage P1. The Cre enzyme promotes site-specific recombination between two copies of a 34 bp long DNA sequence called loxP that is derived from the same bacteriophage.

    Researchers use the Cre/loxP site-specific recombination system in order to make mice homozygous for a deletion of any particular gene and only in a specific tissue. You will see in Chapter 18 that scientists first make mice where a pair of loxP sites are configured in a particular way with respect to the gene to be deleted.
    a. Draw a diagram that shows the configuration of the loxP sites that would enable deletion of a gene by site-specific recombination.
    b. What else would the researchers need to introduce into the mouse genome in order to generate mice where only one tissue is homozygous for the gene deletion?
    c. Why do you think that scientists would want to generate mice like this?
    d. Unlike the DNA rearrangements at *attP* and *attB* sites catalyzed by ΦC31 integrase (described in Problem 39), the DNA rearrangements caused by Cre recombinase are reversible. Why?
    e. Why does the reversibility of Cre/loxP-mediated recombination not interfere with the use of the Cre/loxP system to generate mouse tissues with deletions?

41. Like Cre/loxP recombination, site-specific recombination mediated by the Flp/FRT system is reversible. Why doesn't this fact interfere with the experiment described in Problem 37?

chapter 7

Anatomy and Function of a Gene: Dissection Through Mutation

A scale played on a piano keyboard and a gene on a chromosome are both a series of simple, linear elements (keys or nucleotide pairs) that produce information. A wrong note or an altered nucleotide pair calls attention to the structure of the musical scale or the gene.

chapter outline

HUMAN CHROMOSOME 3 CONSISTS of approximately 220 million base pairs and carries about 1600 genes (**Fig. 7.1**). Somewhere on the long arm of the chromosome resides the gene for rhodopsin, a light-sensitive protein active in the rod cells of our retinas. The rhodopsin gene determines perception of low-intensity light. People who carry the normal, wild-type allele of the gene see well in a dimly lit room and on the road at night. One simple change—a mutation—in the rhodopsin gene, however, diminishes light perception just enough to lead to night blindness. Other alterations in the gene cause the destruction of rod cells, resulting in total blindness. Medical researchers have so far identified more than 30 mutations in the rhodopsin gene that affect vision in different ways.

The case of the rhodopsin gene illustrates some very basic questions. Which of the 220 million base pairs on chromosome 3 make up the rhodopsin gene? How are the base pairs that constitute this gene arranged along the chromosome? How can a single gene sustain so many mutations that lead to such divergent phenotypic effects? In this chapter, we describe the ingenious experiments performed by geneticists during the 1950s and 1960s as they examined the relationships among mutations, genes, chromosomes, and phenotypes in an effort to understand, at the molecular level, what genes are and how they function.

We can recognize three main themes from the elegant work of these investigators. The first is that mutations are heritable changes in base sequence that can affect phenotype. The second is that physically, a gene is usually a specific protein-encoding segment of DNA in a discrete region of a chromosome. (We now know that some genes encode various kinds of RNA that do not get translated into protein.) Third, a gene is not simply a bead on a string, changeable only as a whole and only in one way, as some had thought. Rather, genes are divisible, and each gene's subunits—the individual nucleotide pairs of DNA—can mutate independently and can recombine with each other.

Knowledge of what genes are and how they work deepens our understanding of Mendelian genetics by providing a biochemical explanation for how genotype

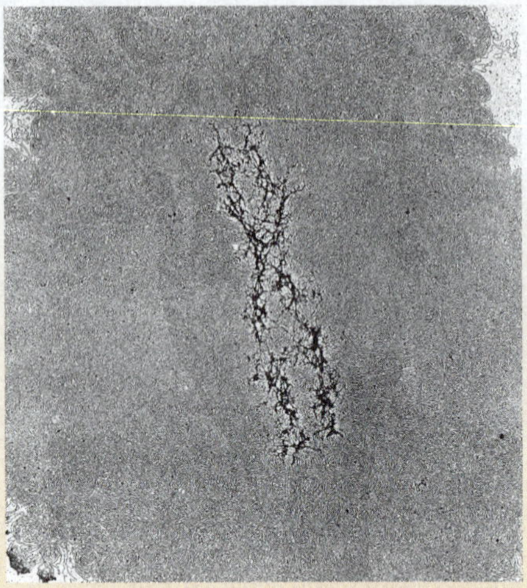

Figure 7.1 The DNA of each human chromosome contains hundreds to thousands of genes. The DNA of this human chromosome has been spread out and magnified 50,000×. No topological signs reveal where along the DNA the genes reside. The darker, chromosome-shaped structure in the middle is a scaffold of proteins to which the DNA is attached.

influences phenotype. One mutation in the rhodopsin gene, for example, causes the substitution of one particular amino acid for another in the construction of the rhodopsin protein. This single substitution changes the three-dimensional structure of rhodopsin and thus the protein's ability to absorb photons, ultimately altering a person's ability to perceive light.

7.1 Mutations: Primary Tools of Genetic Analysis

learning objectives

1. Distinguish between the effects of mutation in somatic and germ-line cells.
2. Describe four types of point mutations: transitions, transversions, deletions, and insertions.
3. Summarize the factors associated with differences in mutation rate.
4. Explain how the fluctuation test and replica plating have shown that mutations arise randomly and spontaneously.

We saw in Chapter 3 that genes with one common allele are monomorphic, while genes with several common alleles in natural populations are polymorphic. The term *wild-type allele* has a clear definition for monomorphic genes, where the allele found on the large majority of chromosomes in the population under consideration is wild type. In the case of polymorphic genes, the definition is less straightforward. Some geneticists consider all alleles with a frequency of greater than 1% to be wild type, while others describe the many alleles present at appreciable frequencies in the population as *common variants* and reserve *wild-type allele* for use only in connection with monomorphic genes.

Mutations Are Changes in DNA Base Sequences

A mutation that changes a wild-type allele of a gene (regardless of the definition) to a different allele is called a **forward mutation.** The resulting novel mutant allele can be either recessive or dominant to the original wild-type allele. Geneticists often diagram forward mutations as $A^+ \rightarrow a$ when the mutation is recessive to the wild-type allele, and as $b^+ \rightarrow B$ when the mutation is dominant to the wild-type. Mutations can also cause a novel mutant allele to revert back to wild type ($a \rightarrow A^+$, or $B \rightarrow b^+$) in a process known as **reverse mutation,** or **reversion.** In this chapter, we designate wild-type alleles, whether recessive or dominant to mutant alleles, with a plus sign ($^+$).

Mendel originally defined genes by the visible phenotypic effects—yellow or green, round or wrinkled—of their alternative alleles. In fact, the only way he knew that genes existed at all was because alternative alleles for seven particular pea genes had arisen through forward mutations. Mutations can occur in somatic cells or in germ-line cells. The mutations in Mendel's pea plants were heritable because they occurred in the germ-line cells of the plants and were thus transmitted through gametes. Close to a century later, knowledge of DNA structure clarified that such mutations are heritable changes in DNA base sequence. DNA thus carries the potential for genetic change in the same place it carries genetic information—the sequence of its bases.

Mutations May Be Classified by How They Change DNA

A **substitution** occurs when a base at a certain position in one strand of the DNA molecule is replaced by one of the other three bases (**Fig. 7.2a**); after DNA replication, a new base pair will appear in the daughter double helix. Substitutions can be subdivided into *transitions,* in which one purine (A or G) replaces the other purine or one pyrimidine

Figure 7.2 Point mutations classified by their effect on DNA.

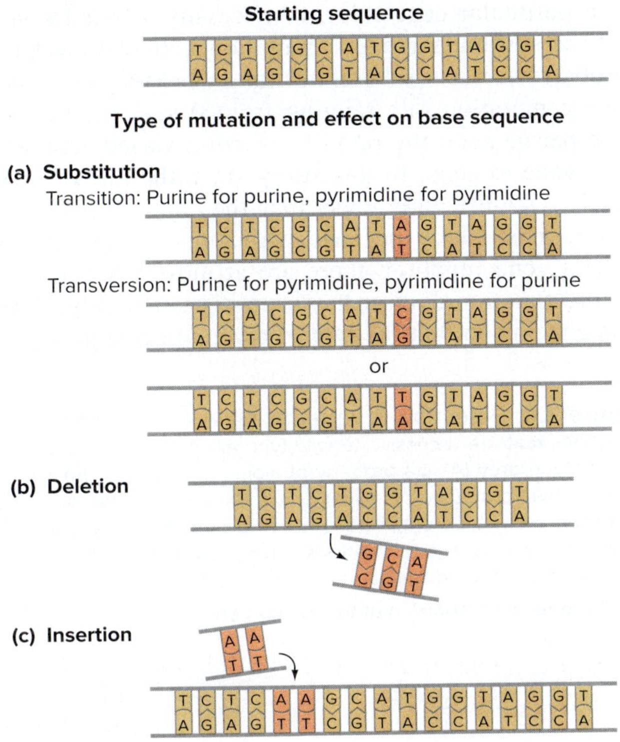

(C or T) replaces the other; and *transversions,* in which a purine changes to a pyrimidine, or *vice versa.*

Other types of mutations rearrange DNA sequences rather than just change the identity of a base pair. A **deletion** occurs when a block of one or more nucleotide pairs is lost from a DNA molecule; an **insertion** is just the reverse—the addition of one or more nucleotide pairs (**Figs. 7.2b** and **c**). Deletions and insertions can be as small as a single base pair or as large as megabases (that is, millions of base pairs). Large deletions and insertions are only some of the complex mutations that can reorganize genomes by changing either the order of genes along a chromosome, the number of genes in the genome, or even the number of chromosomes in an organism. We discuss all such chromosomal rearrangements, which affect many genes at a time, in Chapter 13. Here, we focus on **point mutations** (transitions, transversions, and small deletions or insertions) that affect one or just a few base pairs in the DNA and thus alter only one gene at a time.

Only a small fraction of the mutations in a genome actually alter the nucleotide sequences of genes in a way that affects gene function. By changing one allele to another, these mutations modify the structure or amount of a gene's protein product, and the modification in protein structure or amount can influence phenotype. Other mutations either alter genes in ways that do not affect their function or change the DNA between genes. We discuss mutations without observable phenotypic consequences in Chapter 11; such mutations are invaluable for mapping genes and tracking differences between individuals. In the remainder of this chapter, we focus on those mutations that have an impact on gene function and thereby influence phenotype.

Spontaneous Mutations Occur at a Very Low Rate

Mutations that modify gene function happen so infrequently that geneticists must examine a very large number of individuals from a formerly homogeneous population to detect the new phenotypes that reflect these mutations. In one ongoing study, dedicated investigators have monitored the coat colors of millions of specially bred mice and discovered that on average, a given gene mutates to a recessive allele in roughly 11 out of every 1 million gametes (**Fig. 7.3**). Studies of several other multicellular, eukaryotic organisms have yielded similar results: an average spontaneous rate of $2–12 \times 10^{-6}$ mutations per gene per gamete.

Looking at the mutation rate from a different perspective, you could ask how many mutations might exist in the

Figure 7.3 Rates of spontaneous mutation. (a) Wild-type (*left*) and mutant mouse coat colors: albino (*middle*), and brown (*right*). **(b)** Mutation rates from wild type to recessive mutant alleles for five coat color genes. Mice from highly inbred wild-type strains were mated with homozygotes for recessive coat color alleles. Progeny with mutant coat colors indicated the presence of recessive mutations in gametes produced by the inbred mice.

a (left): © imageBROKER/SuperStock RF; (middle, right): © Charles River Laboratories

(a)

(b)

Locus[a]	Number of gametes tested	Number of mutations	Mutation rate ($\times 10^{-6}$)
a^- *(albino)*	67,395	3	44.5
b^- *(brown)*	919,699	3	3.3
c^- *(nonagouti)*	150,391	5	33.2
d^- *(dilute)*	839,447	10	11.9
ln^- *(leaden)*	243,444	4	16.4
	2,220,376	25	11.2 (average)

[a] Mutation is from wild type to the recessive allele shown.

genes of an individual. To find out, you would simply multiply the rate of $2-12 \times 10^{-6}$ mutations per gene per gamete times 27,000, the current estimate of the number of genes in the human genome, to obtain an answer of between 0.05 and 0.30 mutations per haploid genome. This very rough calculation would mean that, on average, one new mutation affecting phenotype could arise in every 3 to 20 human gametes.

Different genes, different mutation rates

Although the average mutation rate per gene per gamete is $2-12 \times 10^{-6}$, this number masks considerable variation in the mutation rates for different genes. Experiments with many organisms show that mutation rates range from less than 10^{-9} to more than 10^{-3} per gene per gamete. Variation in the mutation rate of different genes within the same organism reflects differences in gene size (larger genes are larger targets that sustain more mutations) as well as differences in the susceptibility of particular genes to the various mechanisms that cause mutations.

Higher mutation rates in multicellular organisms than in bacteria

Estimates of the average mutation rates in bacteria range from 10^{-8} to 10^{-7} mutations per gene per cell division. Although the units here are slightly different than those used for multicellular eukaryotes (because bacteria do not produce gametes), the average rate of mutation in multicellular eukaryotes still appears to be considerably higher than that in bacteria. The main reason is that numerous cell divisions take place between the formation of a zygote and meiosis, so mutations that appear in a gamete may have actually occurred many cell generations before the gamete formed. In other words, more chances exist for mutations to accumulate.

Some scientists speculate that the diploid genomes of multicellular organisms allow them to tolerate relatively high rates of mutation in their gametes because a zygote would have to receive recessive mutations in the same gene from both gametes for any deleterious effects to occur. In contrast, a bacterium would be affected by just a single mutation that disrupted its only copy of the gene.

Gene function: Easy to disrupt, hard to restore

In the mouse coat color study, when researchers allowed brother and sister mice homozygous for a recessive mutant allele of one of the five mutant coat color genes to mate with each other, they could estimate the rate of reversion by examining the F_1 offspring (**Fig. 7.4a**). Any progeny expressing the dominant wild-type phenotype for a particular coat color, of necessity, carried a gene that had sustained a reverse mutation. Calculations based on observations of several million F_1 progeny revealed a reverse mutation rate ranging from 0 to 2.5×10^{-6} per gene per gamete; the rate of reversion varied somewhat from gene to gene. In this study, then, the rate of reversion was significantly lower than the rate of forward mutation, most likely because while many ways exist to disrupt gene function, there are at most a few ways to restore function once it has been disrupted (**Fig. 7.4b**). The conclusion that the rate of reversion is significantly

Figure 7.4 Detecting revertants. **(a)** Rare revertants of a^- mutations that are recessive to wild-type alleles (A^+) are detected as wild-type grey ($A^+ a^-$) progeny of albino ($a^- a^-$) mice. **(b)** The rate of forward mutation is usually much higher than the rate of reversion. Many different mutations can disrupt a gene's function, while at best only a few mutations can restore function to a previously inactivated gene.

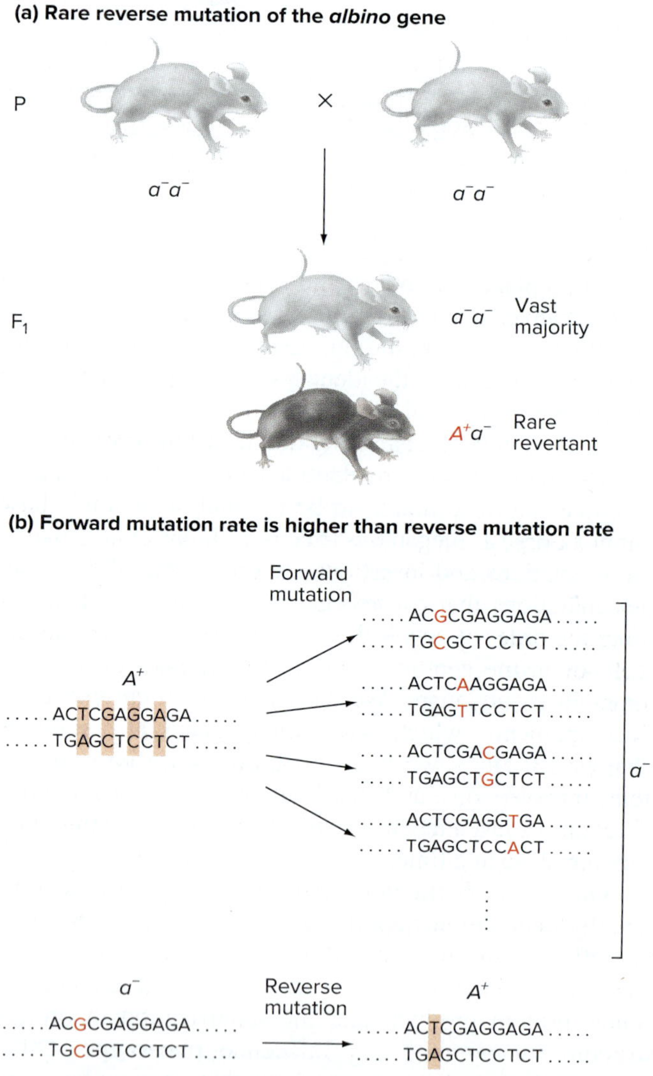

lower than the rate of forward mutation holds true for most types of mutations. In one extreme example, deletions of more than a few nucleotide pairs can never revert because DNA information that has disappeared from the genome cannot reappear spontaneously.

Higher mutation rate in human sperm than in human eggs

New technologies that will be explained in detail in later chapters have enabled researchers to determine the DNA sequence of the entire genome of thousands of people. By comparing the genome sequences of parents and their children, scientists have measured the human mutation rate with great precision. They found that the average value is about one mutation per hundred million base pairs (bp) per gamete (or 1×10^{-8}). Because the haploid human genome is about 3×10^9 bp, each gamete contains on average about 30 mutations, and each child contains about 60 mutations—that is, 60 base pairs that are different than those in either of their parents' genomes. You should note that this number includes all DNA changes, only very few of which influence phenotype.

Interestingly, most of these 60-odd new mutations in each human are obtained from the sperm rather than the egg. Advances in genome sequencing technology have recently made it possible to sequence the haploid genome contained in a single sperm. (See the Fast Forward Box *Crossovers Mapped in Chromosomes of Human Sperm* in Chapter 5.) By comparing the genome sequences of more than 100 individual sperm from the same person, the per-bp mutation rate was found to be $2\text{–}4 \times 10^{-8}$, which indicates that most of the new mutations seen in children come from the sperm rather than the egg.

The idea that sperm carry more mutations than oocytes makes sense. The reason is that more rounds of cell divisions are needed to produce human sperm than human eggs, presenting more opportunities for mutations to occur. Recall from Chapter 4 that human females are born with essentially all of the primary oocytes they will ever produce. It has been estimated that the germ-line cells of a female zygote need to undergo only 24 rounds of mitotic cell divisions to produce all of these oocytes. Male germ-line cells, on the other hand, undergo mitosis continually throughout life. Starting from a male zygote, the number of cell divisions to generate a sperm at age 13 is estimated to be 36. After that, about 23 rounds of mitotic divisions occur per year in the male germ line, meaning that at age 20, the cell lineage producing a given sperm has undergone 200 divisions; at age 30, 430; and at age 45, 770. Therefore, in humans, most new mutations found in the progeny come from the sperm rather than from the egg. Moreover, the older the father, the more mutations are likely to be found in his sperm.

Spontaneous Mutations Arise from Random Events

Because spontaneous mutations affecting a gene occur so infrequently, it is difficult to study the events that produce them. To overcome this problem, researchers turned to bacteria as the experimental organisms of choice. It is easy to grow many millions of individuals and then search rapidly through enormous populations to find the few that carry a novel mutation. In one study, investigators spread wild-type bacteria on the surface of agar containing sufficient nutrients for growth as well as a large amount of a bacteria-killing substance, such as an antibiotic or a bacteriophage. Although most of the bacterial cells died, a few showed resistance to the bactericidal substance and continued to grow and divide. The descendants of a single resistant bacterium, produced by many rounds of binary fission, formed a mound of genetically identical cells called a **colony.**

The few bactericide-resistant colonies that appeared presented a puzzle. Had the cells in the colonies somehow altered their internal biochemistry to produce a life-saving response to the antibiotic or bacteriophage? Or did they carry heritable mutations conferring resistance to the bactericide? And if they did carry mutations, did those mutations arise by chance from random spontaneous events that take place continuously, even in the absence of a bactericidal substance, or did they only arise in response to environmental signals (in this case, the addition of the bactericide)?

In 1943, Salvador Luria and Max Delbrück devised an experiment to examine the origin of bacterial resistance (**Fig. 7.5**). According to their reasoning, if bacteriophage-resistant colonies arise in direct response to infection by bacteriophages, separate suspensions of bacteria containing equal numbers of cells will generate similar, small numbers of resistant colonies when spread in separate petri plates on nutrient agar suffused with phages. By contrast, if resistance arises from mutations that occur spontaneously even when the phages are not present, then different liquid cultures, when spread on separate petri plates, will generate very different numbers of resistant colonies. The reason is that the mutation conferring resistance can, in theory, arise at any time during the growth of the culture. If the mutation occurs early, the cell in which it happens will produce many mutant progeny prior to petri plating; if it happens later, far fewer mutant progeny will be present when the time for plating arrives. After plating, these numerical differences will show up as fluctuations in the numbers of resistant colonies growing in the different petri plates.

The results of this **fluctuation test** were clear: Most plates supported zero to a few resistant colonies, but a few harbored hundreds of resistant colonies. From this

Figure 7.5 The Luria-Delbrück fluctuation experiment. **(a)** Hypothesis 1: If resistance arises only after exposure to a bactericide, all bacterial cultures of equal size should produce roughly the same number of resistant colonies. Hypothesis 2: If random mutations conferring resistance arise before exposure to bactericide, the number of resistant colonies in different cultures should vary (fluctuate) widely. **(b)** Actual results showing large fluctuations suggest that mutations in bacteria occur as spontaneous mistakes independent of exposure to a selective agent.

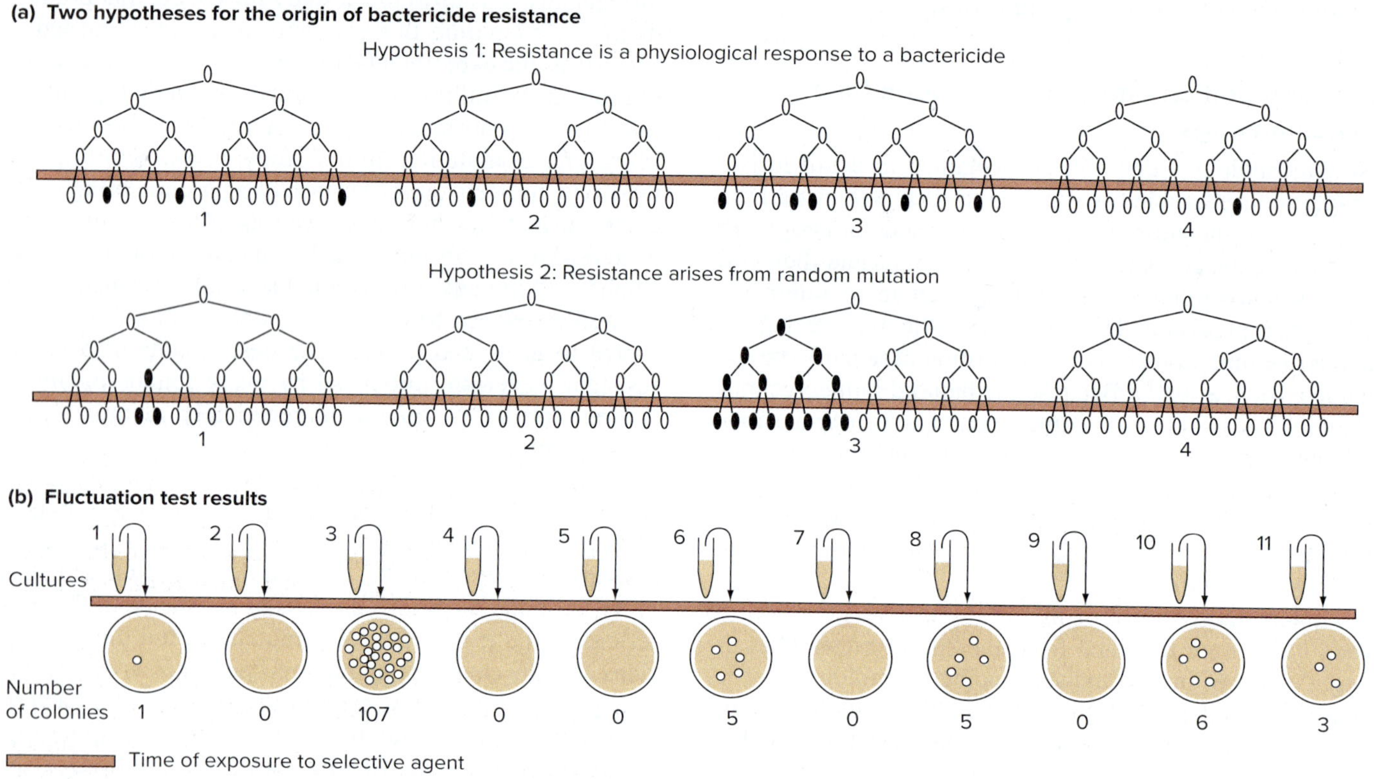

observation of a substantial fluctuation in the number of resistant colonies in different petri plates, Luria and Delbrück concluded that bacterial resistance arises from mutations that exist before exposure to bacteriophages. After exposure, however, the bactericide in the petri plate becomes a selective agent that kills off nonresistant cells, allowing only the preexisting resistant ones to survive. **Figure 7.6** illustrates how researchers used another technique, known as **replica plating,** to demonstrate even more directly that the mutations conferring bacterial resistance occur before the cells encounter the bactericide that selects for their resistance.

These key experiments showed that bacterial resistance to phages and other bactericides is the result of mutations, and these mutations do not arise in particular genes as a directed response to environmental change. Instead, mutations occur spontaneously as a result of random processes that can happen at any time and hit the genome at any place. Once such random changes occur, however, they usually remain stable. If the resistant mutants of the Luria-Delbrück experiment, for example, were grown for many generations in medium that did not contain bacteriophages, they would nevertheless remain resistant to this bactericidal virus.

We next describe some of the many kinds of random events that can cause mutations. We also discuss how cells cope with these events and minimize mutation creation.

essential concepts

- *Mutations* are heritable alterations in the base sequence of DNA.
- *Point mutations* change one or a few base pairs; they include *substitutions* (*transitions* and *transversions*) and small *insertions* and *deletions*.
- Spontaneous mutation rates are low and vary among different genes and organisms.
- The more cells divide, the more likely it is that mutations will accumulate in their genomes.
- Results of the *fluctuation test* and *replica plating* experiments showed that resistance mutations arise randomly in bacterial cells prior to bactericide exposure.

Figure 7.6 Replica plating verifies that bacterial resistance is the result of preexisting mutations. (a) Pressing a *master plate* onto a velvet surface transfers some cells from each bacterial colony onto the velvet. Pressing a *replica plate* onto the velvet then transfers some cells from each colony onto the replica plate. Investigators track which colonies on the master plate can grow on the replica plate (here, only penicillin-resistant ones). **(b)** Colonies on a master plate without penicillin are transferred sequentially to three replica plates with penicillin. Resistant colonies grow in the same positions on all three replicas, showing that some colonies on the master plate had multiple resistant cells even before exposure to the antibiotic.

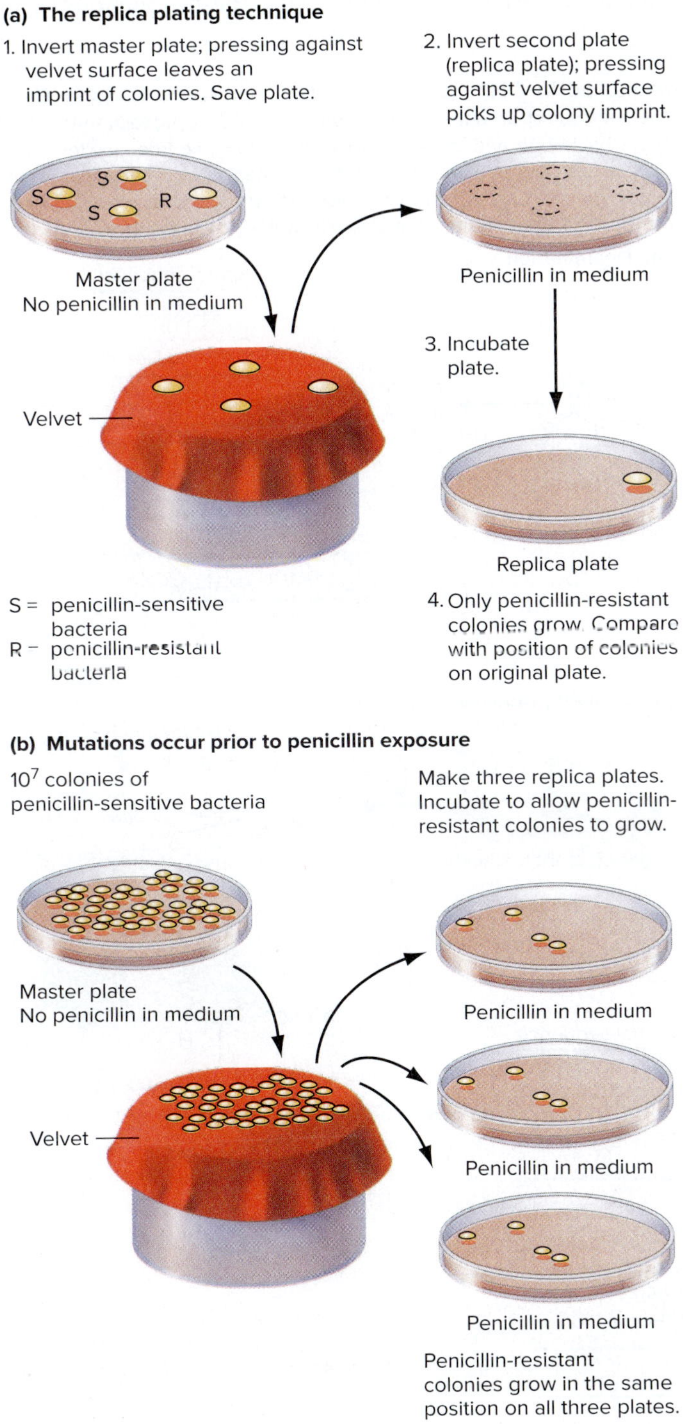

7.2 Molecular Mechanisms That Alter DNA Sequence

learning objectives

1. Outline natural processes that can produce mutations by damaging DNA.
2. Explain how errors in DNA replication can cause mutations.
3. Define *mutagen* and describe how mutagens are used in genetic research.
4. Describe how the Ames test can detect potential carcinogens.

The creation of a heritable mutation is the outcome of several competing processes: mutation, repair, and replication (**Fig. 7.7**). First, of course, a random event must occur to alter the DNA. Two different kinds of events initiate DNA changes: Either DNA can be damaged by chemical reactions or irradiation, or alternatively, mistakes can happen when DNA is copied during replication.

When DNA changes first occur, they are not yet actual mutations but only potential mutations. The reason is that most of them are quickly repaired by a variety of enzymatic systems within cells. These DNA repair machines are engaged in a continual race with DNA replication (Fig. 7.7). If repair of damaged DNA or misincorporated nucleotides occurs before the next round of DNA replication, then the sequence is corrected and no mutation will result. However, if the repair enzymes do not correct the problem before the next round of DNA replication, the mutation becomes established permanently in both strands of the double helix and a heritable mutation is the outcome.

In this section, we describe some of the most important mechanisms that can change DNA sequences. The subsequent

Figure 7.7 Point mutations result when DNA replication wins the race with DNA repair. An alteration that occurs in DNA is heritable only if DNA repair fails to reverse the change before the next round of DNA replication.

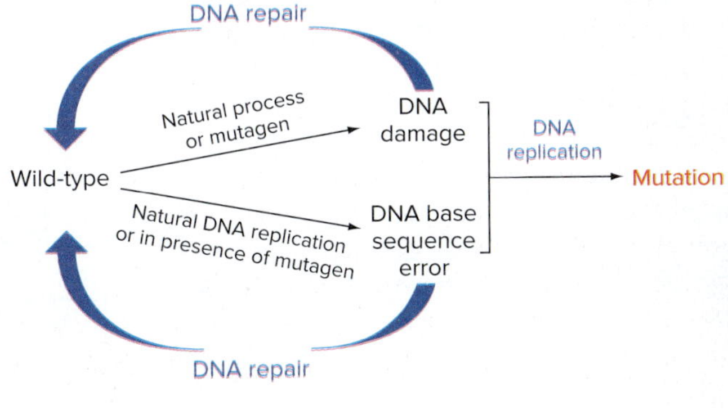

section of this chapter will discuss various biochemical pathways that biological systems have evolved to minimize the mutagenic consequences of these DNA alterations.

Natural Processes Cause Spontaneous Mutations Through DNA Damage

Chemical and physical assaults on DNA are quite frequent. Geneticists estimate, for example, that the hydrolysis of a purine base, A or G, from the deoxyribose-phosphate backbone occurs 1000 times an hour in every human cell. This kind of DNA alteration is called **depurination** (**Fig. 7.8a**). Because the resulting *apurinic site* cannot specify a complementary base, the DNA replication process introduces a random base opposite the apurinic site, causing a mutation in the newly synthesized complementary strand three-quarters of the time.

Another naturally occurring process that may modify DNA's information content is **deamination:** the removal of an amino ($-NH_2$) group. Deamination can change cytosine (C) to uracil (U), the nitrogenous base

Figure 7.8 How natural processes can change the information stored in DNA. (a) In depurination, the hydrolysis of A or G bases leaves a DNA strand with an unspecified base. **(b)** In deamination, the removal of an amino group from C initiates a process that causes a transition after DNA replication. **(c)** X-rays break the sugar-phosphate backbone and thereby split a DNA molecule into smaller pieces, which may be ligated back together improperly. **(d)** Ultraviolet (UV) radiation causes adjacent Ts to form dimers, which can disrupt the readout of genetic information. **(e)** Irradiation causes the formation of *free radicals* (such as oxygen molecules with an unpaired electron) that can alter individual bases. Here, the pairing of the altered base GO with A creates a transversion that changes a G–C base pair to T–A.

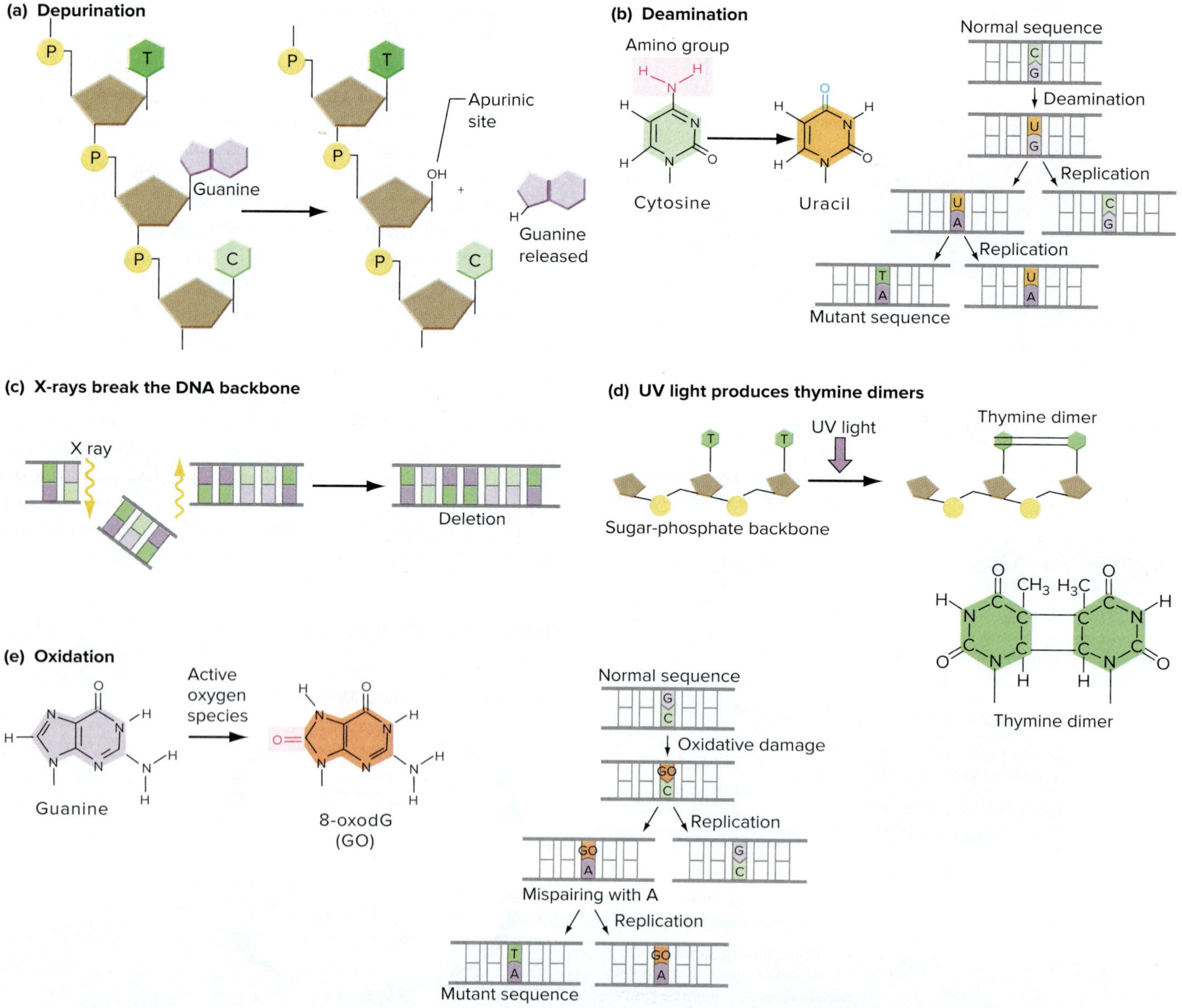

found in RNA but not in DNA. Because U pairs with A rather than G, deamination followed by replication may alter a C–G base pair to a T–A pair in future generations of DNA molecules (**Fig. 7.8b**); such a C–G to T–A change is a transition mutation.

Other assaults include naturally occurring radiation such as cosmic rays and X-rays, which break the sugar-phosphate backbone (**Fig. 7.8c**); ultraviolet light, which causes adjacent thymine residues to become chemically linked into **thymine dimers** (**Fig. 7.8d**); and oxidative damage to any of the four bases (**Fig. 7.8e**). If not repaired before DNA replication, all of these changes alter the information content of the DNA molecule permanently.

Mistakes in DNA Replication Also Cause Spontaneous Mutations

If the cellular machinery for some reason incorporates an incorrect base during replication, for instance, a C opposite an A instead of the expected T, then during the next replication cycle, one of the daughter DNAs will have the normal A–T base pair, while the other will have a mutant G–C. Careful measurements of the fidelity of replication *in vivo,* in both bacteria and human cells, show that such errors are exceedingly rare, occurring less than once in every 10^9 base pairs. That rate is equivalent to typing this entire book 1000 times while making only one typing error. Considering the complexities of helix unwinding, base pairing, and polymerization, this level of accuracy is amazing. How do cells avoid most DNA replication errors, and what kinds of mistakes occur nonetheless when DNA is copied?

The proofreading function of DNA polymerase

The replication machinery minimizes errors through successive stages of correction. In the test tube, DNA polymerases replicate DNA with an error rate of about one mistake in every 10^6 bases copied. This rate is about 1000-fold worse than that achieved by the cell. Even so, it is impressively low and is attained only because polymerase molecules provide, along with their polymerization function, a proofreading/editing function in the form of a nuclease that becomes active whenever the polymerase makes a mistake. This nuclease portion of the polymerase molecule, called the *3′-to-5′ exonuclease,* recognizes a mispaired base and excises it, allowing the polymerase to copy the nucleotide correctly on the next try (**Fig. 7.9**). Without its nuclease portion, DNA polymerase would have an error rate of one mistake in every 10^4 bases copied, so this editing function improves the fidelity of replication 100-fold.

DNA polymerase *in vivo* is part of a replication system including many other proteins that improve on the error rate collectively another 10-fold, bringing it to within about 100-fold of the fidelity attained by the cell. The 100-fold higher accuracy of the cell depends on a backup system called *methyl-directed mismatch repair* that notices and corrects residual errors in the newly replicated DNA. We present the details of this repair system later in the chapter when we describe the various ways in which cells attempt to correct mutations once they occur.

Figure 7.9 DNA polymerase's proofreading function. If DNA polymerase mistakenly adds an incorrect nucleotide at the 3′ end of the strand it is synthesizing, the enzyme's 3′-to-5′ exonuclease activity removes this nucleotide, giving the enzyme a second chance to add the correct nucleotide.

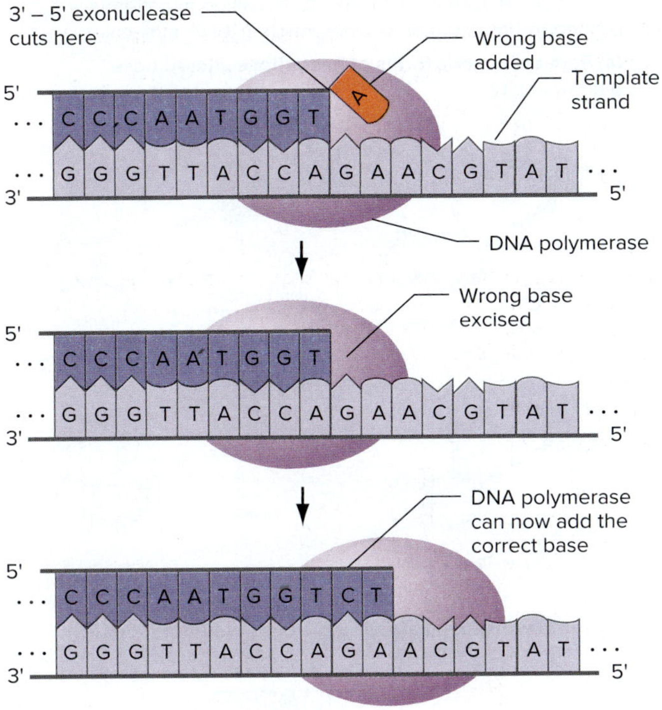

Base tautomerization

One reason why DNA polymerase may make mistakes is the **tautomerization** of bases. Each of the four bases has two **tautomers,** similar chemical forms that interconvert continually. The equilibrium between the tautomers is such that each base is almost always in the form in which A pairs with T and G pairs with C. However, if by chance a base in the template strand is in its rare tautomeric form when DNA polymerase arrives, the wrong base will be incorporated into the newly synthesized chain because the rare tautomers pair differently than do the normal forms (**Fig. 7.10a**). If the misincorporated nucleotide is not corrected by mismatch repair before the next round of replication, a point mutation results (**Fig. 7.10b**).

Figure 7.10 How base tautomerization causes mutation. **(a)** Rare tautomeric forms of the four bases have different pairing abilities than the usual base forms. **(b)** In its rare enol form, T causes DNA polymerase to insert a G in the complementary strand. If the mismatched T:G base pair is not repaired to T:A before the next round of replication, a T:A-to-C:G transition mutation is established in both strands of one daughter DNA molecule.

(a) Rare tautomeric forms of bases have altered base pairing ability.

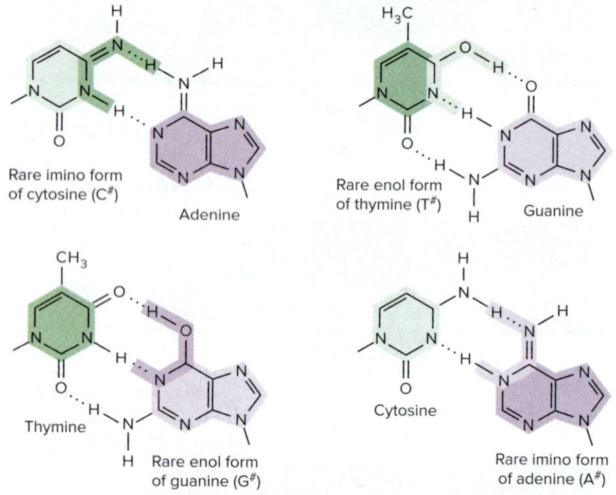

(b) Tautomerization causes single base pair mutations.

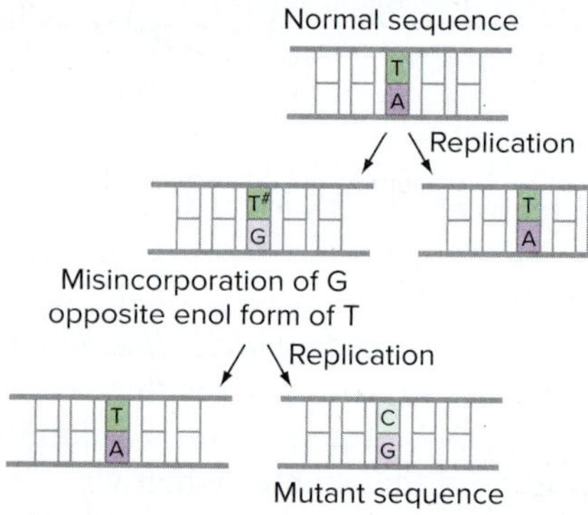

Unstable trinucleotide repeats

In 1992, molecular geneticists discovered a completely unexpected type of mutation in humans: the excessive amplification of a CGG base triplet normally repeated only a few to 50 times in succession. If, for example, a normal allele of a gene carries five consecutive repetitions of the base triplet CGG (that is, CGGCGGCGGCGGCGG on one strand), an abnormal allele could carry 200 repeats in a row. Repeats of several other trinucleotides—CAG, CTG, GCC, and GAA—can also be unstable, such that the number of repeats often increases or decreases in different somatic cells of a single individual. Instability can also occur during gamete production, resulting in changes in repeat number from one generation to the next.

Unstable trinucleotide repeats have now been found within about 20 different human genes, all associated with neurodegenerative diseases. In all cases, an expansion of the repeats beyond a certain number results in a disease-causing allele. The Fast Forward Box entitled *Trinucleotide Repeat Diseases: Huntington Disease and Fragile X Syndrome* explains that trinucleotide repeat diseases can be subdivided into two main groups according to the location of the repeats relative to the the part of the gene that specifies the protein product. One is exemplified by fragile X syndrome, the most common form of intellectual disability in boys; the other group is represented by Huntington disease, a neurological disorder discussed in Chapter 2.

A general feature of both groups of trinucleotide repeat diseases is that the more repeats at a particular location, the higher the probability that expansion and contraction will occur. Because larger repeat numbers mean more instability, some alleles with intermediate numbers of trinucleotide repeats behave as so-called **pre-mutation alleles (Fig. 17.11a)**. For example, in fragile X syndrome, individuals with pre-mutation alleles have a normal phenotype, but the expanded repeat number means that such pre-mutation alleles are highly likely to expand or contract during replication. Carriers of pre-mutation alleles thus have a high probability of giving new disease alleles (with an expanded number of repeats) to their children (**Fig. 7.11b**).

Figure 7.11 Inheritance of fragile X syndrome. **(a)** Wild-type, pre-mutation, and disease-causing alleles for fragile X syndrome differ in the number of CGG trinucleotide repeats. Disease alleles are nonfunctional. Pre-mutation alleles provide normal gene function, but they have a high probability of triplet repeat expansion during DNA replication in female germ-line cells. **(b)** Normal females heterozygous for pre-mutation alleles are likely to produce gametes with expanded triplet repeat numbers.

(a) Effect of (CGG) repeat number

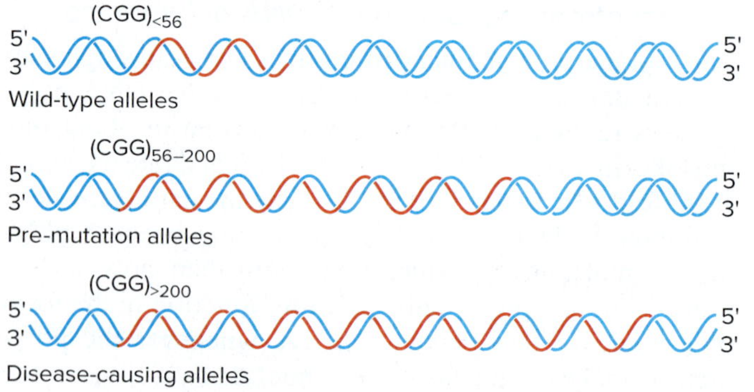

(b) A fragile X pedigree

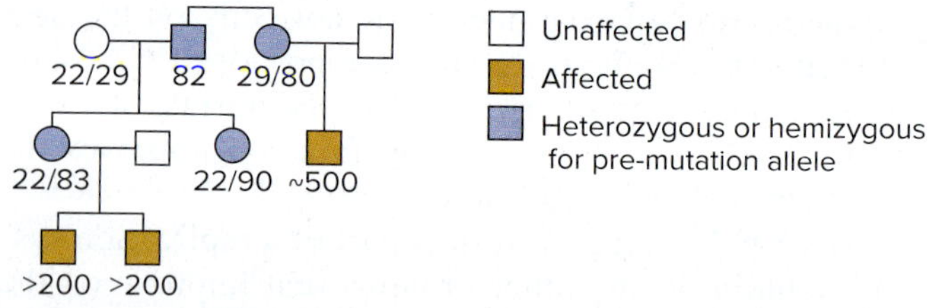

FAST FORWARD 

Sprinters: © Robert Michael/Corbis RF

Trinucleotide Repeat Diseases: Huntington Disease and Fragile X Syndrome

The approximately 20 known neurogenerative diseases caused by genes with unstable trinucleotide (triplet) repeats fall into two categories: **polyQ diseases** and **non-polyQ diseases** (where Q is the symbol for the amino acid glutamine). In polyQ disease genes, the repeated triplet is always CAG, while in non-polyQ disease genes, the trinucleotide repeat may be either CGG, CTG, GCC, or GAA. The two types of triplet repeat diseases are distinguished by the effect of the repeat sequence on gene function. In polyQ diseases, a disease allele with too many triplet repeats encodes an abnormal protein. Non-polyQ disease alleles encode either no protein or decreased protein amounts. The differences in the two classes of triplet repeat diseases are illustrated by the best-known example of each: Huntington disease, a polyQ disease; and fragile X syndrome, a non-polyQ disease.

Huntington disease affects about 1 in 10,000 people worldwide. The symptoms usually start at about 40 years of age and include muscle coordination difficulties, cognitive decline, and psychiatric problems. You saw in Chapter 2 that Huntington disease is inherited through autosomal dominant mutant alleles (*HD*). While normal HD^+ alleles have between 6 and 28 CAG repeats, *HD* disease alleles have an expanded repeat region that has 36 or more CAG repeats. The run of CAGs in the *HD* gene are in the *open reading frame*, or *ORF*, that contains the actual instructions to build a protein from its constituent *amino acids*.

Each CAG specifies that the amino acid glutamine (Q) should be added to the HD protein, so the normal protein has 6 to 28 Q amino acids in a row in its so-called *polyQ region* (**Fig. A**). An *HD* allele with 36 or more repeats encodes a mutant HD protein with an expanded polyQ region that is toxic to nerve cells. (Proteins encoded by pre-mutation alleles function normally but the alleles have an unstable repeat number.) Scientists do not yet understand the normal function of HD in nerve cells or the reason why the mutant HD protein is toxic.

PolyQ disease alleles like *HD* are called *gain-of-function* mutants because they specify proteins whose functions are qualitatively different from those of the corresponding wild-type protein. Typical for many gain-of-function mutants, polyQ disease alleles show dominant inheritance because the mutant polyQ proteins are toxic even in the presence of the normal proteins.

Non-polyQ diseases are exemplified by fragile X syndrome, a leading cause of inherited intellectual disability, affecting about 1 in 4000 males and 1 in 8000 females. The disease is caused by expansion of a CGG repeat region in an X-linked gene called *FMR-1* (for *fragile X mental retardation-1*).

The CGGs of *FMR-1* are located in a region of the gene outside of the ORF called the *5′ UTR* (**Fig. B**). Normal $FMR\text{-}1^+$ genes have between 6 and 55 CGG repeats; expansion of the repeat number to 200 CGGs or more results in an *FMR-1* disease allele that cannot produce the FMR-1 protein. Without FMR-1 protein, nerve cells cannot properly form connections called *synapses*.

A feature common to all non-polyQ diseases is that the triplets are located in a part of the gene outside of the ORF. The expanded repeats in non-polyQ disease genes generally prevent protein production, and so non-polyQ disease genes are *loss-of-function* alleles. Because females heterozygous for the disease allele have at least some disease symptoms most of the time, fragile X syndrome shows X-linked dominant inheritance with incomplete penetrance and variable expressivity. Other non-polyQ disease alleles may show either dominant or recessive inheritance patterns, depending on whether two doses or one dose of the normal gene product is required to avoid disease symptoms.

The triplet repeat diseases illustrate two fundamental principles regarding mutations. First, mutations may affect either the nature of the gene product (polyQ diseases) or the amount of the gene product (non-polyQ diseases). Second, certain DNA sequences can mutate at surprisingly high frequencies in special circumstances, as seen by pre-mutation alleles for either Huntington disease or fragile X syndrome. These two principles will be important themes in subsequent chapters.

Figure A Huntington disease: a polyQ repeat disease. The *HD* gene has a run of CAG repeats that specify glutamines (Qs) in the open reading frame (ORF). *HD* disease alleles direct the synthesis of a mutant, toxic HD protein with an expanded polyQ region. Pre-mutation alleles with an intermediate number of CAGs produce a normally functioning protein, but the allele is unstable.

PolyQ disease: Huntington disease

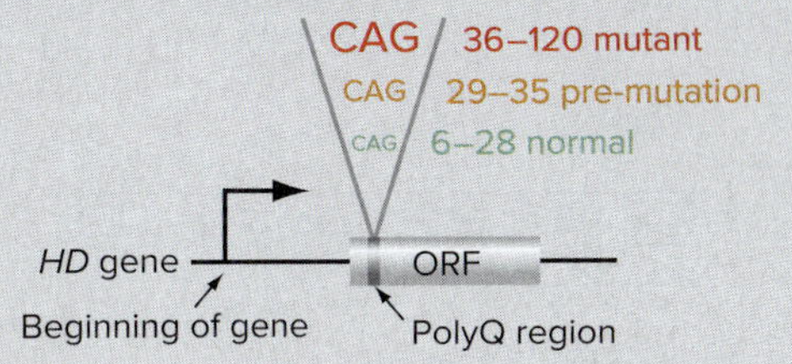

Figure B Fragile X syndrome: a non-polyQ repeat disease. The FMR-1 gene has a run of CGG repeats in the 5′ untranslated region (5′ UTR) outside the ORF. *FMR-1* disease alleles have an expanded repeat number, and this prevents synthesis of the gene's protein product. Pre-mutation alleles with an intermediate number of CGGs make normal amounts of protein, but these alleles are unstable.

Non-polyQ disease: fragile X syndrome

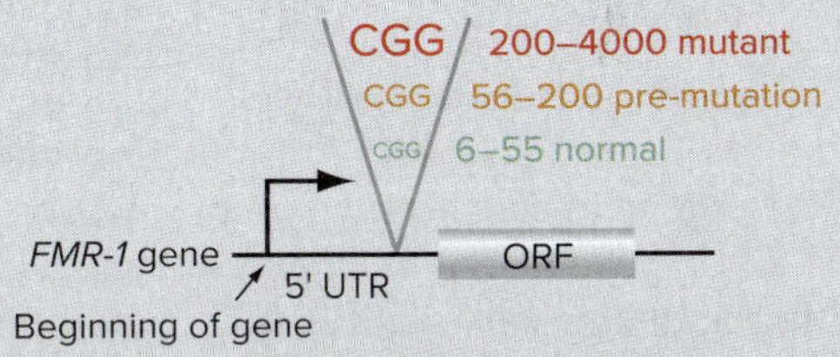

Figure 7.12 Expansion of trinucleotide repeats by slipped mispairing during DNA replication. **(a)** Pausing of DNA polymerase at repeat sequences during DNA replication allows slippage of the newly synthesized DNA strand (*blue*) relative to the template strand (*gray*). Because of the repeats, the slipped strand can still pair with the template, and DNA polymerization can continue. Another round of DNA replication will establish the additional repeats in double-stranded DNA. **(b)** Similarly, slippage of the template strand relative to the newly synthesized DNA strand can result in the deletion of repeats.

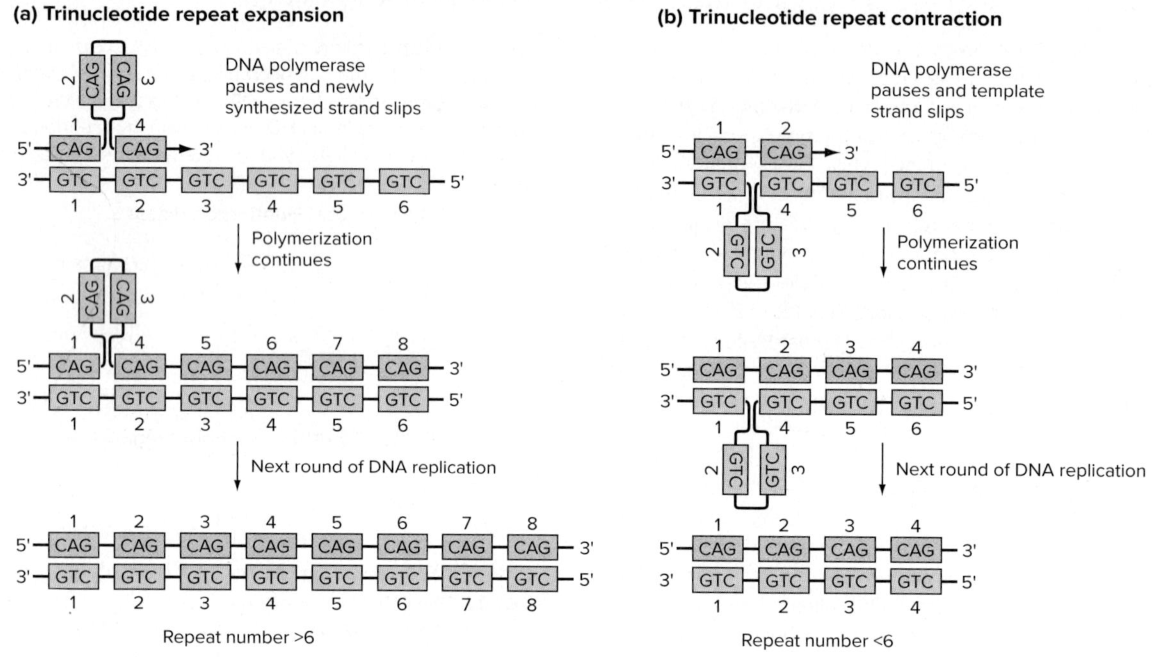

Researchers do not understand well a curious feature of trinucleotide repeat diseases: Pre-mutation alleles of particular genes tend to expand either in the male or female germ lines, but not both. For example, in Fig. 7.11b you can see that the alleles causing fragile X syndrome were inherited from mothers with pre-mutation alleles, but the repeat number does not expand appreciably in the sperm produced by a father with a pre-mutation allele. Strangely, for Huntington disease the situation is the opposite: Disease alleles almost always originate in the male, but not in the female, germ line.

A variety of biochemical mechanisms could be responsible for trinucleotide repeat expansion and contraction. One particularly well-characterized mechanism is **slipped mispairing** during DNA replication. DNA polymerase often pauses as it replicates through repeat regions, which allows one DNA strand (either the newly synthesized strand or the template strand) to slip relative to the other one (**Fig. 7.12**). Because the sequence contains repeats, the slipped strand and the other strand can pair out of register, forming a loop. After another round of DNA replication, this slipped mispairing can result in expansion or contraction of trinucleotide repeat number in both DNA strands.

Mutagens Induce Mutations

Mutations make genetic analysis possible, but most mutations appear spontaneously at such a low rate that researchers have looked for controlled ways to increase their occurrence. H. J. Muller, an original member of Thomas Hunt Morgan's *Drosophila* group, first showed that exposure to a dose of X-rays higher than the naturally occurring level increases the mutation rate in fruit flies (**Fig. 7.13**).

Muller exposed male *Drosophila* to increasingly large doses of X-rays and then mated these males with females that had one X chromosome containing an easy-to-recognize dominant mutation causing Bar eyes. This X chromosome (called a *Balancer*) also carried chromosomal rearrangements known as *inversions* that prevented it from crossing-over with other X chromosomes. (Chapter 13 explains the details of this phenomenon.) Some of the F_1 daughters of this mating were heterozygotes carrying a mutagenized X from their father and a *Bar*-marked X from their mother. If X-rays induced a recessive lethal mutation anywhere on the paternally derived X chromosome, then these F_1 females would be

Figure 7.13 Exposure to X-rays increases the mutation rate in *Drosophila*. F_1 females are constructed that have an irradiated paternal X chromosome (*red line*) and a *Bar*-marked *Balancer* maternal X chromosome (*wavy blue line*). These two chromosomes cannot recombine because the *Balancer* chromosome prevents crossing-over. Single F_1 females, each with a single X-ray-exposed X chromosome from their father, are then individually mated with wild-type males. If the paternal X chromosome in any one F_1 female has an X-ray-induced recessive lethal mutation (*m*), she can produce only Bar-eye sons (*left*). If the X chromosome has no such mutation, this F_1 female will produce both Bar-eye and non-Bar-eye sons (*right*).

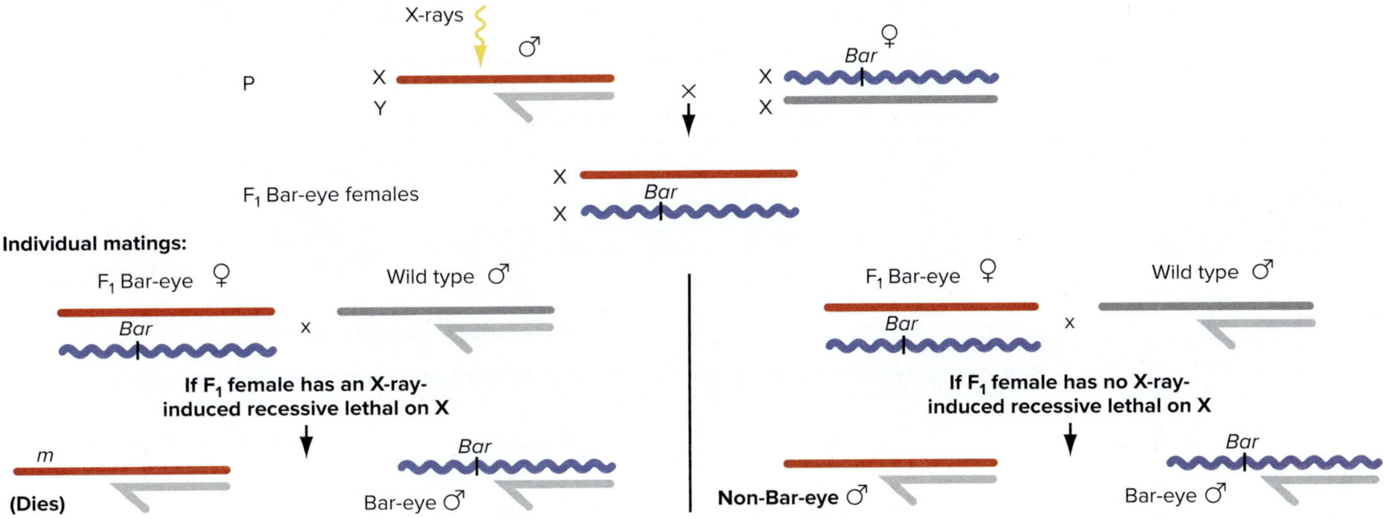

unable to produce non-Bar-eye sons. Thus, simply by noting the presence or absence of non-Bar-eye sons, Muller could establish whether a mutation had occurred in any of the more than 1000 genes on the X chromosome that are essential to *Drosophila* viability. He concluded that the greater the X-ray dose, the greater the frequency of recessive lethal mutations.

Any physical or chemical agent that raises the frequency of mutations above the spontaneous rate is called a **mutagen.** Researchers use many different mutagens to produce mutations for study. With the Watson-Crick model of DNA structure as a guide, they can understand the action of most mutagens at the molecular level. The X-rays used by Muller to induce mutations on the X chromosome, for example, can break the sugar-phosphate backbones of DNA strands, sometimes at the same position on the two strands of the double helix. Multiple double-strand breaks produce DNA fragmentation, and the improper stitching back together of the fragments can cause small deletions (review Fig. 7.8c) or large deletions and other rearrangements that will be discussed in Chapter 13.

Another molecular mechanism of mutagenesis involves mutagens known as **base analogs,** which are so similar in chemical structure to the normal nitrogenous bases that the replication machinery can incorporate them into DNA (**Fig. 7.14a**). Because a base analog may have tautomeric forms with pairing properties different from those of the base it replaces, the analog can cause base substitutions on the complementary strand synthesized in the next round of DNA replication.

Other chemical mutagens generate substitutions by directly altering a base's chemical structure and properties (**Fig. 7.14b**). Again, the effects of these changes become fixed in the genome when the altered base causes incorporation of an incorrect complementary base during a subsequent round of replication.

Yet another class of chemical mutagens consists of compounds known as **intercalators:** flat, planar molecules that can sandwich themselves between successive base pairs and disrupt the machinery for replication, generating deletions or insertions of a single base pair (**Fig. 7.14c**). The intercalator *proflavin* is often used in genetic research precisely for this reason.

Many Mutagens Are Carcinogens

Although only mutations that occur in the germ line can be passed on to the next generation, mutations in somatic cells can still have an impact on the well-being and survival of individuals. Somatic mutations in genes that help regulate the cell cycle may, for example, lead to cancer. For this reason, many mutagens act as *carcinogens* (cancer-causing agents).

Figure 7.14 How mutagens alter DNA. **(a)** Base analogs incorporated into DNA may pair aberrantly, allowing the addition of incorrect nucleotides to the opposite strand during replication. **(b)** Some mutagens alter the structure of bases such that they pair inappropriately in the next round of replication. **(c)** Intercalating agents are roughly the same size and shape as a base pair of the double helix. Their incorporation into DNA produces insertions or deletions of single base pairs.

Type of Mutagen	Chemical Action of Mutagen

(a) Replace a base:
Base analogs have a chemical structure almost identical to that of a DNA base.

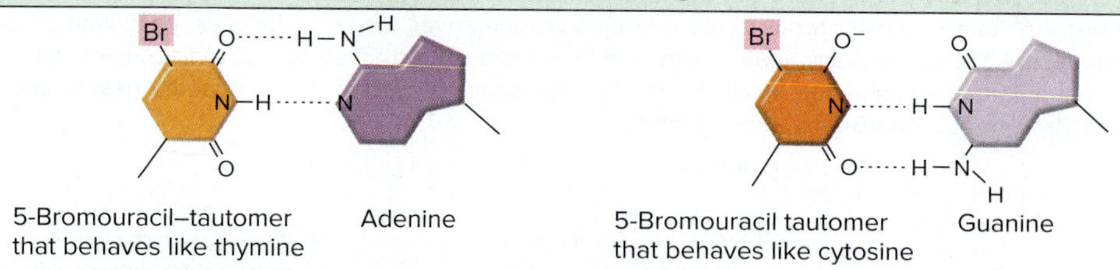

5-Bromouracil: One tautomer pairs with A; other tautomer pairs with G.

(b) Alter base structure and properties:
Hydroxylating agents: add a hydroxyl (–OH) group

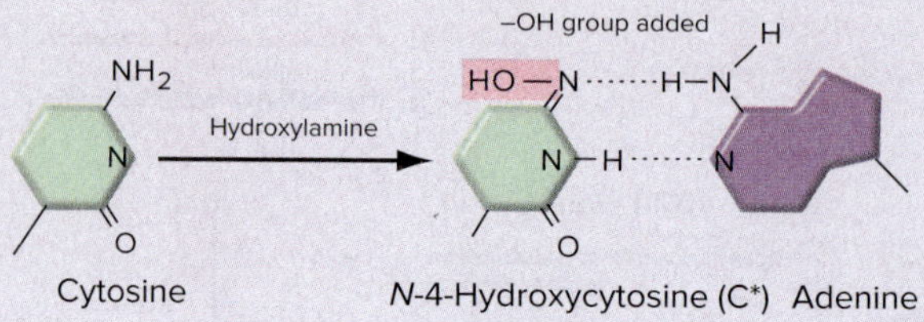

Hydroxylamine adds – OH to cytosine; with the – OH, hydroxylated C now pairs with A instead of G.

Alkylating agents: add ethyl ($–CH_2–CH_3$) or methyl ($–CH_3$) groups

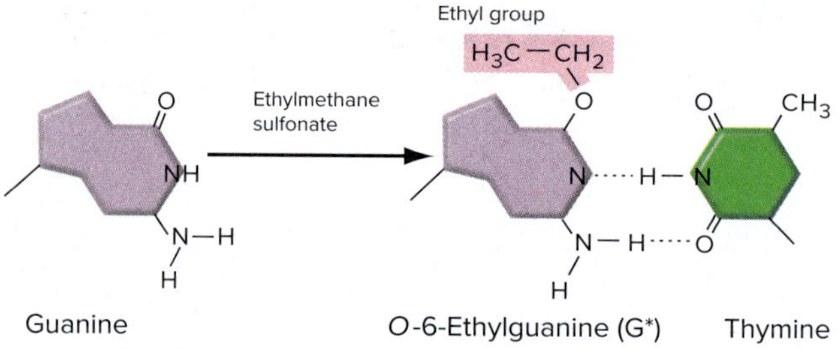

Ethylmethane sulfonate adds an ethyl group to guanine. Modified G pairs with T.

Deaminating agents: remove amine ($–NH_2$) groups

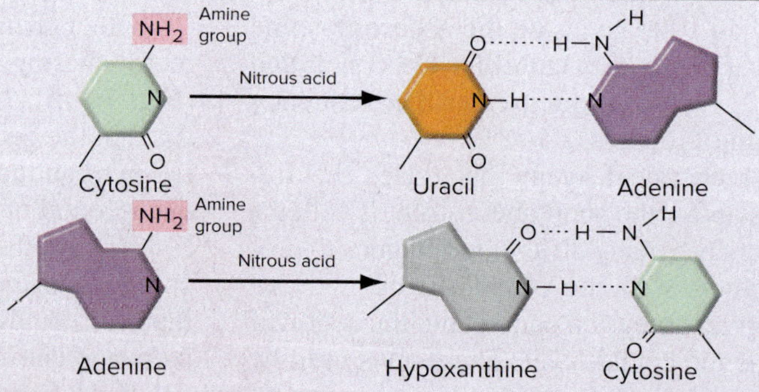

Nitrous acid modifies cytosine to uracil, which pairs with A instead of G; modifies adenine to hypoxanthine, a base that pairs with C instead of T.

(c) Insert between bases:
Intercalating agents

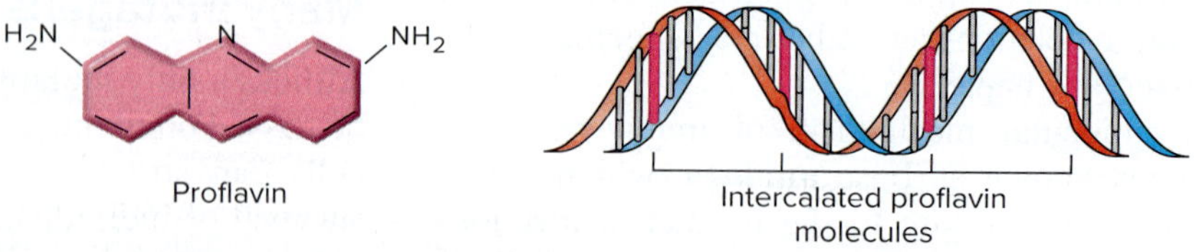

Proflavin intercalates into the double helix. During replication, the proflavin molecules cause single base insertions and deletions.

Figure 7.14 How mutagens alter DNA. (*Continued*)

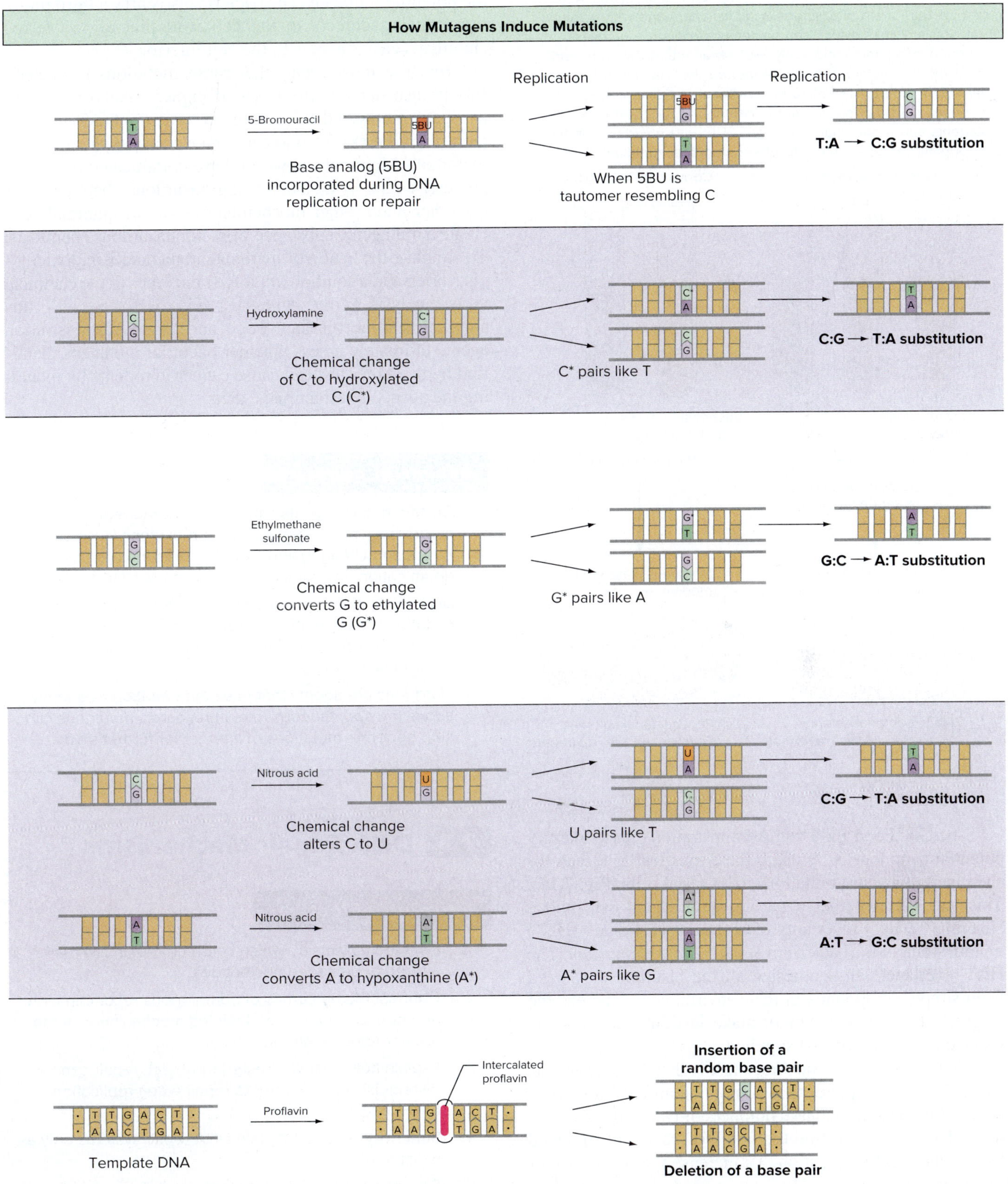

Figure 7.15 The Ames test identifies potential carcinogens. Investigators mix a compound to be tested with cells of a His^- strain of *Salmonella typhimurium* and with a solution of rat liver enzymes (which can sometimes convert a harmless compound into a mutagen). Only His^+ revertants grow on a petri plate without histidine. If this plate (*bottom left*) has more His^+ revertants than a control plate (also without histidine) containing unexposed cells (*bottom right*), the compound is considered mutagenic and a potential carcinogen. The rare revertants on the control plate represent the rate of spontaneous mutation.

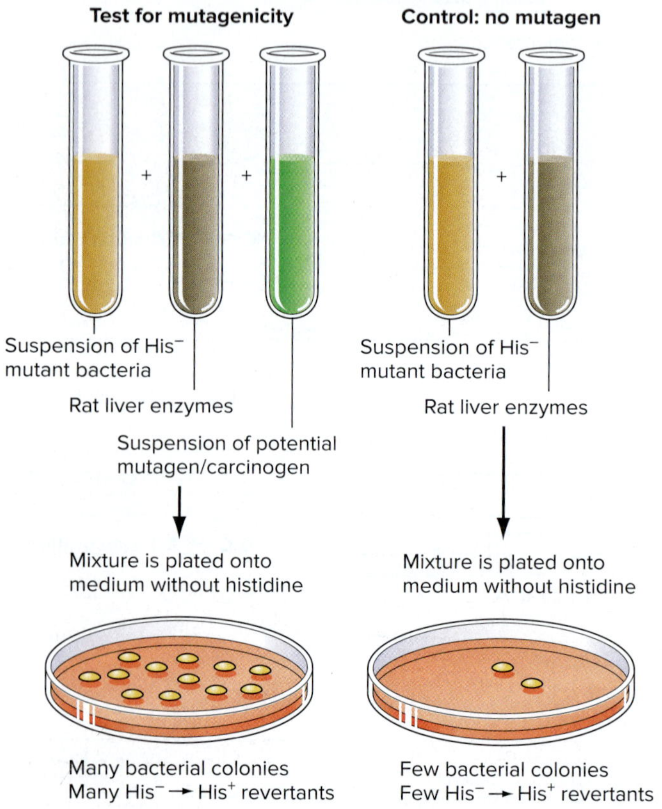

The U.S. Food and Drug Administration tries to identify potential carcinogens by using the **Ames test** to screen for chemicals that cause mutations in bacterial cells (**Fig. 7.15**). This test asks whether a particular chemical can induce Histidine$^+$ (His^+) revertants of a special Histidine$^-$ (His^-) mutant strain of the bacterium *Salmonella typhimurium*. The His^+ revertants can synthesize all the histidine they need from simple compounds in their environment, whereas the original His^- mutants cannot make histidine, so they can survive only if histidine is supplied.

The advantage of the Ames test is that only revertants can grow on petri plates that do not contain histidine, so it is possible to examine large numbers of cells from an originally His^- culture to find the rare His^+ revertants induced by the chemical in question. To increase the sensitivity of mutation detection, the His^- strain used in the Ames test system contains a second mutation that inactivates a DNA repair system (to be described in the next section) and thereby prevents the ready repair of mutations caused by the potential mutagen. The bacteria also carry a third mutation causing defects in the cell wall that allows tested chemicals easier access to the cell interior.

Because most agents that cause mutations in bacteria should also damage the DNA of higher eukaryotic organisms, any mutagen that increases the rate of mutation in bacteria might be expected to cause cancer in people and other mammals. Mammals, however, have complicated metabolic processes capable of inactivating hazardous chemicals. On the other hand, other biochemical events in mammals can create a mutagenic substance from nonhazardous chemicals. To simulate the action of mammalian metabolism, toxicologists often add a solution of rat liver enzymes to the chemical under analysis by the Ames test (Fig. 7.15). Because this simulation is not perfect, Food and Drug Administration agents ultimately assess whether bacterial mutagens identified by the Ames test can cause cancer in rodents by including the agents in test animals' diets.

essential concepts

- Certain natural agents can induce *spontaneous mutations*. These agents include radiations (such as X-rays and UV light) and chemical reactions (such as deamination and oxidation) that damage DNA.
- DNA replication errors are another source of spontaneous mutations. Many of these errors result from base tautomerization or expansions/contractions of trinucleotide repeats.
- *Mutagens* are agents that raise the mutation frequency above the spontaneous rate. In research, mutagens can help generate mutations of interest for further study.

7.3 DNA Repair Mechanisms

learning objectives

1. List mechanisms by which cells can repair DNA with altered or damaged nucleotides.
2. Contrast the outcomes of homologous recombination and nonhomologous end-joining mechanisms for the repair of double-strand breaks.
3. Explain how methyl-directed mismatch repair can distinguish which strand to repair when replication errors occur.
4. State why cells use certain DNA repair systems only as a last resort.
5. Describe the potential consequences for human health of mutations in genes that specify DNA repair factors.

Recall from Fig. 7.7 that if new DNA damage is repaired before DNA replication occurs, no mutation becomes established in the chromosomes. Cells have in fact evolved a variety of enzymatic systems that locate and repair damaged DNA and thereby dramatically minimize the occurrence of mutations. The combination of these repair systems must be extremely efficient, because the rates of spontaneous mutation observed for almost all genes are very low.

Some DNA Base Damage Can Be Reversed

Cells contain various enzyme systems that can reverse certain kinds of nucleotide alterations quickly and directly. For example, if methyl or ethyl groups are mistakenly added to guanine (as in Fig. 7.14b), *alkyltransferase* enzymes can remove them so as to re-create the original base.

In a second example, the enzyme *photolyase* recognizes the thymine dimers produced by exposure to ultraviolet light (review Fig. 7.8d) and reverses the damage by splitting the chemical linkage between the thymines. Interestingly, the photolyase enzyme works only in the presence of visible light. In carrying out its DNA repair tasks, it associates with a small molecule called a *chromophore* that absorbs light in the visible range of the spectrum; the enzyme then uses the energy captured by the chromophore to split thymine dimers. Because it does not function in the dark, the photolyase mechanism is called *light repair*, or *photorepair*.

Damaged Bases Can Be Removed and Replaced

Many repair systems use a general strategy of homology-dependent repair in which they first remove a small region from the DNA strand that contains the altered nucleotide, and then use the other strand as a template to resynthesize the region removed. This strategy makes use of one of the great advantages of the double-helical structure: If one strand sustains damage, cells can use complementary base pairing with the undamaged strand to re-create the original sequence.

Base excision repair

In this type of homology-dependent repair mechanism, enzymes called *DNA glycosylases* cleave an altered nitrogenous base from the sugar of its nucleotide, releasing the base and creating an apurinic or apyrimidinic (AP) site in the DNA chain (**Fig. 7.16**). Different glycosylase enzymes cleave specific damaged bases. Base excision repair is particularly important in the removal of uracil from DNA

Figure 7.16 Base excision repair removes damaged bases. Glycosylase enzymes (*light green* oval) remove aberrant bases [like uracil (*red*) formed by the deamination of cytosine], leaving an AP site. AP endonuclease (*purple* oval) cuts the sugar-phosphate backbone, creating a nick. Exonucleases extend the nick into a gap, which is filled in with the correct information (*dark green*) by DNA polymerase. DNA ligase reseals the corrected strand.

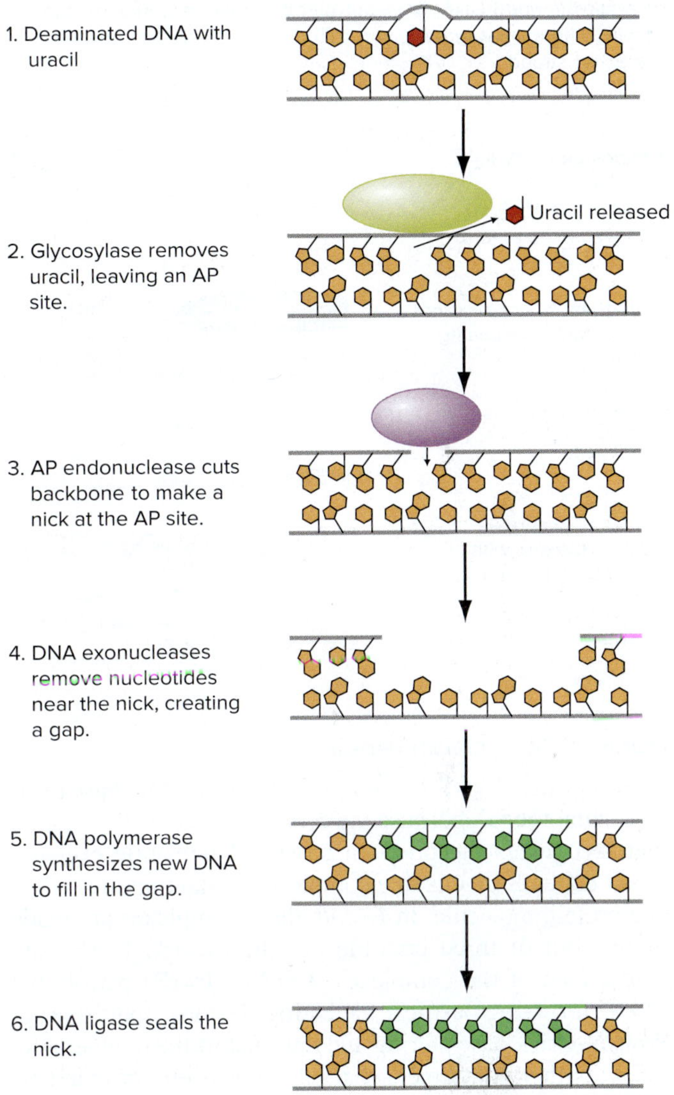

(recall that uracil often results from the natural deamination of cytosine; review Fig. 7.8b). In this repair process, after the enzyme uracil-DNA glycosylase has removed uracil from its sugar, leaving an AP site, the enzyme AP endonuclease makes a nick in the DNA backbone at the AP site. Other enzymes (known as *DNA exonucleases*) attack the nick and remove nucleotides from their vicinity to create a gap in the previously damaged strand. DNA polymerase fills in the gap by copying the undamaged strand, restoring the original nucleotide in the process. Finally, DNA ligase seals up the backbone of the newly repaired DNA strand.

Figure 7.17 Nucleotide excision repair corrects damaged nucleotides. A complex of the UvrA and UvrB proteins (*not shown*) scans DNA for distortions caused by DNA damage, such as thymine dimers. At the damaged site, UvrA dissociates from UvrB, allowing UvrB (*red*) to associate with UvrC (*blue*). These enzymes nick the DNA exactly four nucleotides to one side of the damage and seven nucleotides to the other side, releasing a small fragment of single-stranded DNA. DNA polymerases then resynthesize the missing information (*green*), and DNA ligase reseals the now-corrected strand.

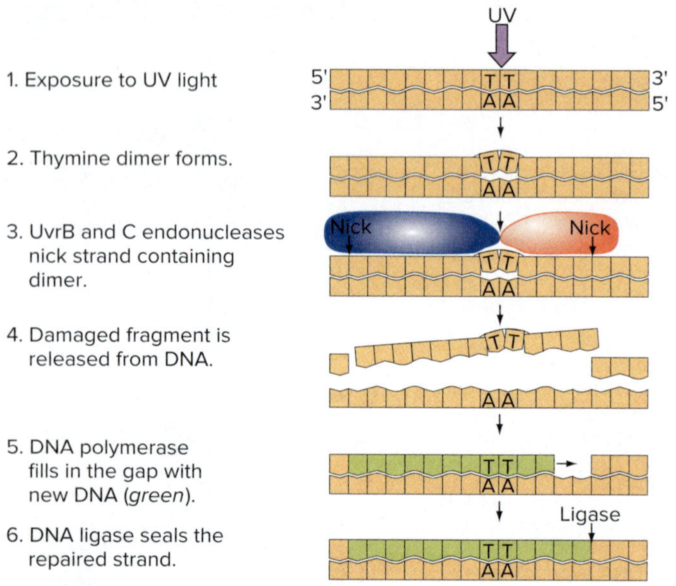

Nucleotide excision repair

This pathway (**Fig. 7.17**) removes alterations that base excision cannot repair because the cell lacks a DNA glycosylase that recognizes the problem base(s). Nucleotide excision repair depends on enzyme complexes containing more than one protein molecule. In *E. coli*, these complexes are made of two out of three possible proteins: UvrA, UvrB, and UvrC. One of the complexes (UvrA + UvrB) patrols the DNA for irregularities, detecting lesions that disrupt Watson-Crick base pairing and thus distort the double helix (such as thymine dimers that have not been corrected by photorepair). A second complex (UvrB + UvrC) cuts the damaged strand in two places that flank the damage. This double-cutting excises a short region of the damaged strand and leaves a gap that will be filled in by DNA polymerase and sealed with DNA ligase.

Two Important Mechanisms Can Repair Double-Strand Breaks

We have seen previously that X-rays can cause double-strand breaks, in which both strands of the double helix are broken at nearby sites (review Fig. 7.8c). Double-strand breaks represent a particularly dangerous kind of DNA lesion because if not repaired properly, such chromosomal breakages can lead not only to point mutations, but also to large deletions and other kinds of chromosomal rearrangements.

It is therefore not surprising that organisms have evolved at least two different ways of repairing double-strand breaks. One of these mechanisms, **homologous recombination (HR)**, uses complementary base pairing to repair breaks accurately with no loss or gain of nucleotides. The second pathway, called **nonhomologous end-joining (NHEJ)**, can bring together even DNA ends that were not previously adjacent to each other, and a few base pairs can be lost or added improperly in the process.

Both systems for the repair of double-strand breaks have great practical significance because they are fundamental to new, effective strategies for *genome editing* (altering an organism's genome in specific ways). We will describe these exciting methods for modifying genomes in Chapter 18, but it will be helpful for you to gain here some idea about how these repair mechanisms work.

Double-strand break repair via homologous recombination (HR)

You will recall that the first step of meiotic recombination is the formation of a double-strand break, and that through strand invasion, cells undergoing recombination eventually repair this double-strand break using the homologous chromosome as a template (review Fig. 6.27). Mitotic cells can employ much of the same enzymatic machinery for homologous recombination to repair double-strand breaks caused by X-ray exposure.

The HR system can use either a homologous chromosome, or more often a sister chromatid, as the template for repair. If the homologous chromosome serves as the template, repair of the break results in mitotic recombination. However, finding a homolog is inefficient, so repair through recombination usually occurs instead between sister chromatids during the G_2 phase of the cell cycle (that is, after the chromosomes have replicated). In this case, repair of the break does not produce mitotic recombination because the broken chromatid and the template chromatid base pair sequences are identical.

Repair of double-strand breaks by nonhomologous end-joining (NHEJ)

The NHEJ mechanism is an alternative to HR that is especially important for the repair of double-strand breaks formed during the G_1 phase of the cell cycle (that is, before a sister chromatid is available to serve as a template for homologous recombination). The proteins participating in NHEJ bind to DNA ends at the site of the breakage and protect the ends from nucleases. The NHEJ proteins also bridge the two ends, allowing them to be stitched together by the DNA ligase enzyme (**Fig. 7.18**).

Figure 7.18 Repair of double-strand breaks by nonhomologous end-joining. The proteins KU70, KU80, and PK_{CS} (in mammalian cells) bind to DNA ends, protect them from degradation, and bring them together so that DNA ligase can repair the phosphodiester backbone.

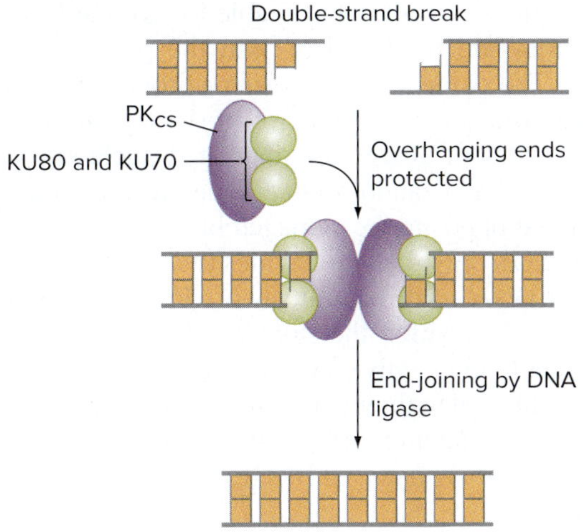

You should note that because NHEJ does not involve DNA homology, it can join together any DNA ends (other than telomeres, which are protected against this pathway), even if those ends were not adjacent to each other in the genome originally. Thus, if the genome suffers more than one double-strand break, NHEJ can potentially join the wrong ends together, causing chromosome rearrangements such as inversions or large deletions.

Another property of NHEJ is that although this mechanism is usually accurate, it can sometimes result in small changes to the DNA sequences where the broken ends are joined together. During the NHEJ process, DNA exonucleases and DNA polymerases can act at the broken ends, removing or adding a few base pairs to them before DNA ligase seals them together. Errors due to NHEJ are relatively infrequent, but they do occur and will become of considerable significance when we discuss genome editing techniques in Chapter 18 later in the book.

Mismatch Repair Corrects Errors in DNA Replication

DNA polymerase is remarkably accurate in copying DNA, but the DNA replication system still makes about 100 times more mistakes than most cells can tolerate. A backup repair system called **methyl-directed mismatch repair** corrects almost all of these errors (**Fig. 7.19**). Because mismatch repair is active only after DNA replication, this system needs to solve a difficult problem. Suppose that a G–C pair has been copied to produce two daughter molecules, one of which has the correct G–C base pair, the other an incorrect G–T. The mismatch repair system can easily recognize the incorrectly matched G–T base pair because the improper

Figure 7.19 In bacteria, methyl-directed mismatch repair corrects mistakes in replication. Parental strands are in *light blue* and newly synthesized strands are *magenta*. The MutS protein is *green*, MutL is *dark blue*, and MutH is *orange*. See text for details.

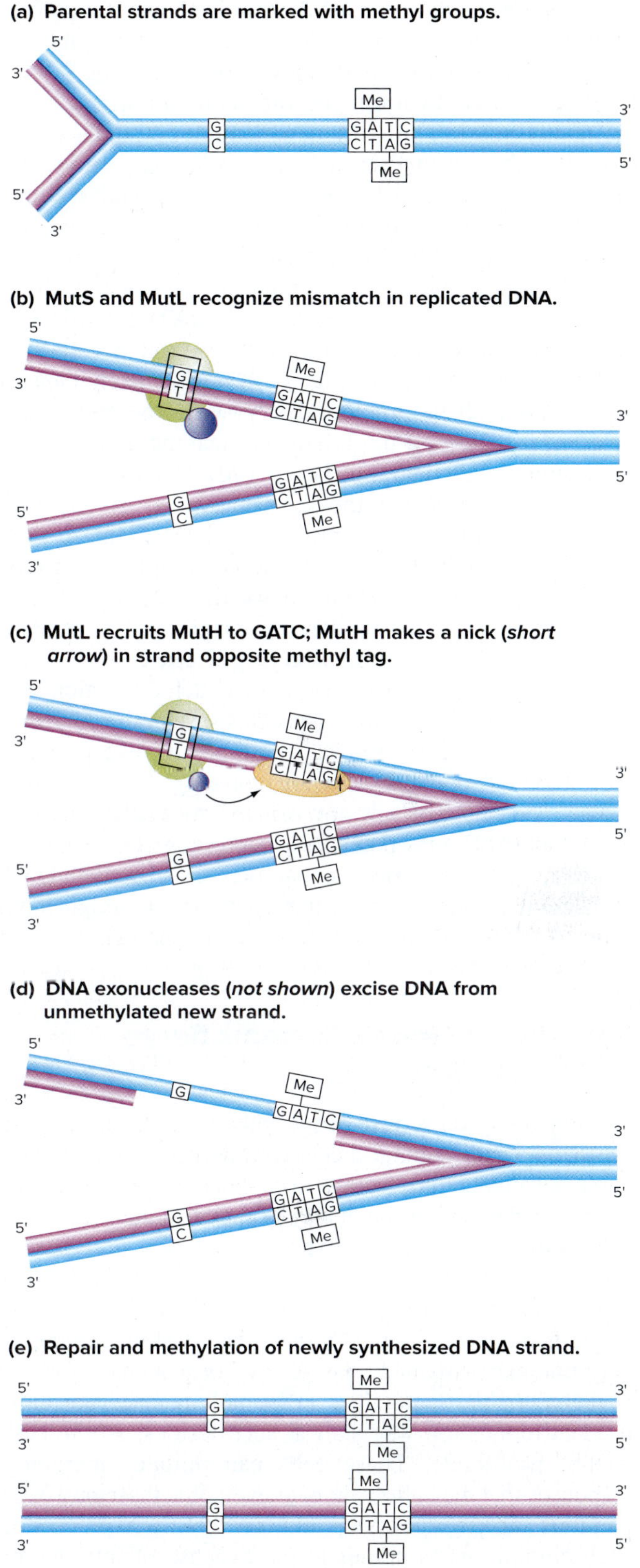

base pairing distorts the double helix, resulting in abnormal bulges and hollows. But how does the system know whether to correct the pair to a G–C or to an A–T?

Bacteria solve this problem by placing a distinguishing mark on the parental DNA strands at specific places: Everywhere the sequence GATC occurs, the enzyme adenine methylase puts a methyl group on the A (Fig. 7.19a). Shortly after replication, the old template strand bears the methyl mark, while the new daughter strand—which contains the wrong nucleotide—is as yet unmarked (Fig. 7.19b). A pair of proteins in *E. coli,* called MutL and MutS, detect and bind to the mismatched nucleotides. MutL and MutS direct another protein, MutH, to nick the newly synthesized strand of DNA at a position across from the nearest methylated GATC; MutH can discriminate the newly synthesized strand because its GATC is not methylated (Fig. 7.19c). DNA exonucleases then remove all the nucleotides between the nick and a position just beyond the mismatch, leaving a gap on the new, unmethylated strand (Fig. 7.19d). DNA polymerase can now resynthesize the information using the old, methylated strand as a template, and DNA ligase then seals up the repaired strand. With the completion of replication and repair, enzymes mark the new strand with methyl groups so that its parental origin will be evident in the next round of replication (Fig. 7.19e).

Eukaryotic cells also have a mismatch correction system, but we do not yet know how this system distinguishes templates from newly replicated strands. Unlike prokaryotes, GATCs in eukaryotes are not tagged with methyl groups, and eukaryotes do not seem to have a protein closely related to MutH. One potentially interesting clue is that the MutS and MutL proteins in eukaryotes associate with DNA replication factors; perhaps these interactions might help MutS and MutL identify the strand to be repaired.

Error-Prone Repair Systems Serve as Last Resorts

The repair systems just described are very accurate in repairing DNA damage because they can either replace damaged nucleotides with a complementary copy of the undamaged strand or ligate breaks back together. However, cells sometimes become exposed to levels or types of mutagens that they cannot handle with these high-fidelity repair systems. Strong doses of UV light, for example, might make more thymine dimers than the cell can mend. Any unrepaired damage has severe consequences for cell division; in particular, the DNA polymerases normally used in replication will stall at such lesions, so the cells cannot proliferate. These cells can initiate emergency responses that may allow them to overcome these problems and thus survive and divide, but their ability to proceed in such circumstances comes at the expense of introducing new mutations into the genome.

Using sloppy DNA polymerases

One type of emergency repair in bacteria, called the **SOS system** (after the Morse code distress signal), relies on error-prone (or *sloppy*) DNA polymerases. These sloppy DNA polymerases are not available for normal DNA replication; they are produced only in the presence of DNA damage. The damage-induced, error-prone DNA polymerases are attracted to replication forks that have become stalled at sites of unrepaired, damaged nucleotides. There the enzymes add random nucleotides to the strand being synthesized opposite the damaged bases.

The SOS polymerase enzymes thus allow the cell with damaged DNA to divide into two daughter cells, but because at each position the sloppy polymerases restore the proper nucleotide only one-quarter of the time, the genomes of these daughter cells carry new mutations. In bacteria, the mutagenic effect of many mutagens either depends on, or is enhanced by, the SOS system.

Sloppy repair of double-strand breaks

Another kind of emergency repair system, **microhomology-mediated end-joining (MMEJ)**, deals with dangerous double-strand breaks that have not been corrected by homologous recombination or NHEJ. The mechanism of MMEJ is similar to that of NHEJ (previously shown in Fig. 7.18), except in MMEJ the broken DNA ends are cut back on either side of the break (resected) by enzymes. The resection exposes small single-stranded regions of complementary DNA sequence (*microhomology*) on either side of the break that help in bringing the ends together.

Because nucleotides are removed at the sites of the double-stranded breaks during resection, MMEJ results in deletions of tens to hundreds of base pairs in the rejoined DNA. These deletions are longer than the small deletions of a few base pairs that sometimes result from NHEJ.

Mutations in Genes Encoding DNA Repair Proteins Impact Human Health

DNA repair mechanisms appear in some form in virtually all species. For example, humans have six proteins whose amino acids are about 25% identical with those of the *E. coli* mismatch repair protein MutS. DNA repair systems are thus very old and must have evolved soon after life emerged ~3.5 billion years ago. Some scientists think DNA repair became essential when plants started to deposit oxygen into the atmosphere, because oxygen helps form free radicals that can damage DNA.

The many known human hereditary diseases associated with the defective repair of DNA damage reveal how crucial these mechanisms are for survival. In one example, the cells of patients with *xeroderma pigmentosum* lack the ability to conduct nucleotide excision repair; these people

Figure 7.20 Skin lesions in a xeroderma pigmentosum patient. This heritable disease is caused by the lack of a critical enzyme in the nucleotide excision repair system.

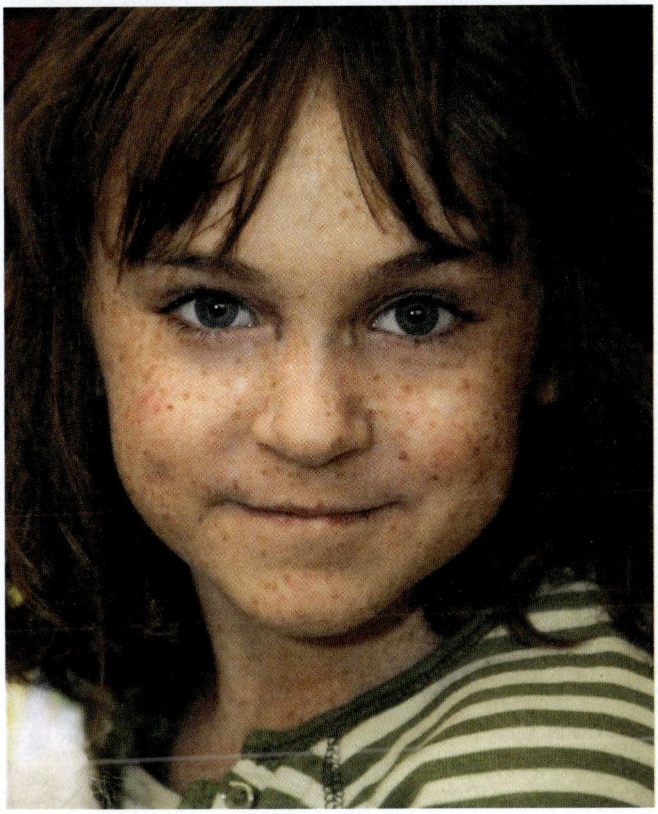

are homozygous for mutations in any one of seven genes encoding enzymes that normally function in this repair system. As a result, the thymine dimers caused by ultraviolet light cannot be removed efficiently. Unless these people avoid all exposure to sunlight, their skin cells begin to accumulate mutations that eventually lead to skin cancer (**Fig. 7.20**).

In another example, researchers have recently learned that hereditary forms of colorectal cancer in humans are associated with mutations in human genes that are closely related to the *E. coli* genes encoding the mismatch-repair proteins MutS and MutL. In yet another example, the breast cancer genes *BRCA1* and *BRCA2* (mutation of either of which is associated with a high risk of breast cancer in women) encode proteins that function in double-strand break repair via homologous recombination. Chapter 20 discusses the fascinating connections between DNA repair and cancer in more detail.

DNA Repair Cannot Be 100% Efficient

"The capacity to blunder slightly is the real marvel of DNA. Without this special attribute, we would still be anaerobic bacteria and there would be no music." In these two sentences, the eminent medical scientist and self-appointed "biology watcher" Lewis Thomas acknowledges that changes in DNA are behind the phenotypic variations that are the raw material on which natural selection has acted for billions of years to drive evolution.

As Dr. Thomas' poetic line suggests, the necessity for mutation is fundamental: Without mutations, life would have died out long ago because it could not have responded to changes in the environment. DNA repair processes must therefore walk a fine line. They must be efficient enough to protect genomes from the huge number of assaults on DNA that are always occurring, but the propagation of life requires some mutations to be transmitted to future generations.

essential concepts

- Cells have many different enzyme systems that minimize mutations by repairing DNA damage or replication errors.
- Double-strand breaks, which are particularly dangerous to genomes, can be repaired through *homologous recombination* (*HR*) or *nonhomologous end-joining* (*NHEJ*).
- Correction of DNA replication errors requires *mismatch repair* systems to choose the correct strand to change. Bacteria accomplish this task by marking parental strands with methyl groups.
- If normal repair mechanisms are overwhelmed by too much DNA damage, cells can then mobilize *error-prone DNA repair* systems.
- Mutations in genes specifying proteins that participate in DNA repair often lead to human diseases, including cancer.
- Mutations are the raw material of evolution. Although many mutations are harmful, rare mutations may confer a selective advantage.

7.4 What Mutations Tell Us About Gene Structure

learning objectives

1. Describe complementation testing and how its results distinguish mutations in a single gene from mutations in different genes.
2. Explain how Benzer's experimental results revealed that the *rII* region in bacteriophage T4 contains two genes, each composed of many nucleotide pairs.
3. Discuss how Benzer used deletions to map mutations in the *rII* region.

The science of genetics depends absolutely on mutations, because we can track genes in crosses only through the phenotypic effects of their mutant variants. In the 1950s

Figure 7.21 ***Drosophila*** **eye color mutations produce a variety of phenotypes.** Flies carrying different X-linked eye color mutations. From the *left:* ruby, white, and apricot; a wild-type eye is at the *far right.*
(all): © Science Source

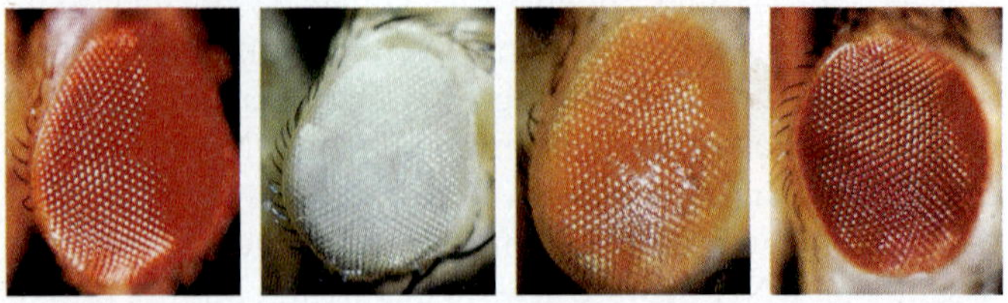

and 1960s, scientists realized they could also use mutations to learn how DNA sequences along a chromosome constitute individual genes. These investigators wanted to collect a large series of mutations in a single gene and analyze how these mutations were arranged with respect to each other. For this approach to be successful, they had to establish that various mutations were, in fact, in the same gene. This was not a trivial exercise, as illustrated by the following situation.

Early *Drosophila* geneticists identified a large number of X-linked recessive mutations affecting the normally red wild-type eye color (**Fig. 7.21**). The first of these to be discovered produced the famous white eyes studied by Morgan's group. Other mutations caused a whole palette of hues to appear in the eyes: darkened shades such as garnet and ruby; bright colors such as vermilion, cherry, and coral; and lighter pigmentations known as apricot, buff, and carnation. This wide variety of eye colors posed a puzzle: Were the mutations that caused them multiple alleles of a single gene, or did they affect more than one gene?

Complementation Testing Reveals Whether Two Mutations Are in a Single Gene or in Different Genes

Researchers commonly define a gene as a functional unit that directs the appearance of a molecular product that, in turn, contributes to a particular phenotype. They can use this definition to determine whether two mutations are in the same gene or in different genes.

If two homologous chromosomes in an individual each carry a mutation recessive to wild type, that individual will have a normal phenotype if the mutations are in different genes. Such a result is called **complementation.** The normal phenotype occurs because almost all recessive mutations disrupt a gene's function. The dominant wild-type alleles on each of the two homologs can make up for, or *complement,* the defect in the other chromosome by generating enough of both gene products to yield a normal phenotype (**Fig. 7.22a, left**).

In contrast, if the recessive mutations on the two homologous chromosomes are in the same gene, no wild-type allele of that gene exists in the individual, and neither mutated copy of the gene will be able to perform the normal function. As a result, no complementation will occur and no normal gene product will be made, so a mutant phenotype will appear (**Fig. 7.22a, right**). Ironically, a collection of mutations that do *not* complement each other is known as a **complementation group.** Geneticists often use *complementation group* as a synonym for *gene* because the mutations in a complementation group all affect the same unit of function, and thus, the same gene.

A simple test based on the idea of a gene as a unit of function can determine whether or not two recessive mutations are alleles of the same gene. You simply examine the phenotype of a heterozygous individual in which one homolog of a particular chromosome carries one of the mutations and the other homolog carries the other mutation. If the phenotype is wild-type, the mutations cannot be in the same gene. This technique is known as a **complementation test.** For example, because a female fruit fly simultaneously heterozygous for *garnet* and *ruby* ($garnet\ ruby^{+}/garnet^{+}\ ruby$) has wild-type brick-red eyes, it is possible to conclude that the mutations causing garnet and ruby colors complement each other and are therefore in different genes.

Complementation testing has, in fact, shown that garnet, ruby, vermilion, and carnation pigmentation are caused by mutations in separate genes. But chromosomes carrying mutations yielding white, cherry, coral, apricot, and buff phenotypes fail to complement each other. These mutations therefore constitute different alleles of a single gene. *Drosophila* geneticists named this gene the *white*, or *w*, gene after the first mutation observed; they designate the wild-type allele as w^{+} and the various mutations as w^{1} (the original white-eyed mutation discovered by T. H. Morgan, often simply designated as *w*), w^{cherry}, w^{coral}, $w^{apricot}$, and w^{buff}. As an example, the eyes of a $w^{1}\ /\ w^{apricot}$ female are a dilute apricot color; because the phenotype of this heterozygote is not wild-type, the two mutations are allelic. **Figure 7.22b** illustrates how researchers collate data from many complementation tests in a **complementation table.** Such a table helps visualize the relationships among a large group of mutants.

In *Drosophila*, mutations in the *w* gene map very close together in the same region of the X chromosome, while mutations in other sex-linked eye color genes lie elsewhere on the chromosome (**Fig. 7.22c**). This result suggests that genes are not disjointed entities with parts spread out from one end of a chromosome to another; each gene, in fact, occupies only a relatively small, discrete area of a chromosome. Studies defining genes at the molecular level have shown that most genes consist of 1000–20,000 contiguous base pairs (bp). In humans, among the shortest genes are the roughly 500 base pair–long genes that govern the production of histone proteins, while the longest gene so far identified is the *Duchenne muscular dystrophy* (*DMD*) gene, which has a length of more than 2 million nucleotide

Figure 7.22 Complementation testing of *Drosophila* eye color mutations. **(a)** A heterozygote has one mutation (m_1) on one chromosome and a different mutation (m_2) on its homolog. If the mutations are in different genes, the heterozygote will be wild type; the mutations complement each other (*left*). If both mutations affect the same gene, the phenotype will be mutant; the mutations do not complement each other (*right*). Complementation testing makes sense only when both mutations are recessive to wild type. **(b)** This complementation table reveals five complementation groups (five different genes) for eye color. A *plus* (+) indicates mutant combinations with wild-type eye color; these mutations complement and are thus in different genes. Several mutations fail to complement (–) and are thus alleles of one gene, *white*. **(c)** Recombination mapping shows that mutations in different genes are often far apart, while different mutations in the same gene are very close together.

(a) Complementation testing

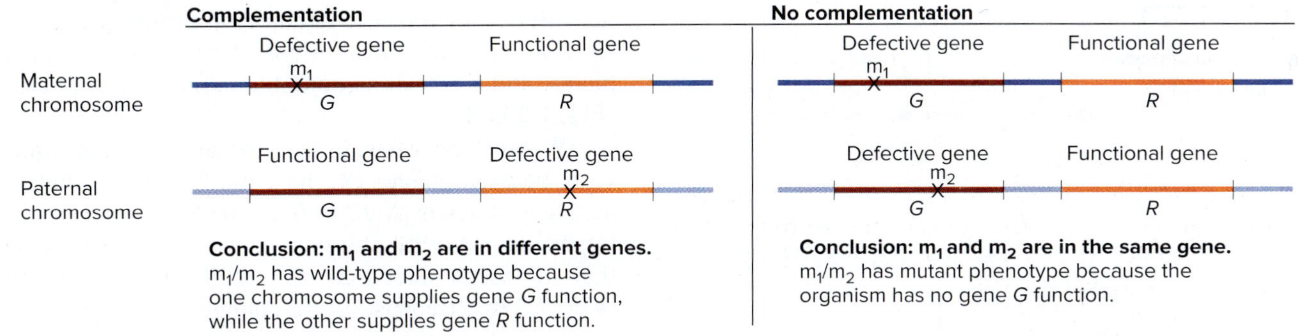

(b) A complementation table: X-linked eye color mutations in *Drosophila*

Mutation	*white*	*garnet*	*ruby*	*vermilion*	*cherry*	*coral*	*apricot*	*buff*	*carnation*
white	–	+	+	+	–	–	–	–	+
garnet		–	+	+	+	+	+	+	+
ruby			–	+	+	+	+	+	+
vermilion				–	+	+	+	+	+
cherry					–	–	–	–	+
coral						–	–	–	+
apricot							–	–	+
buff								–	+
carnation									–

(c) Genetic map: X-linked eye color mutations in *Drosophila*

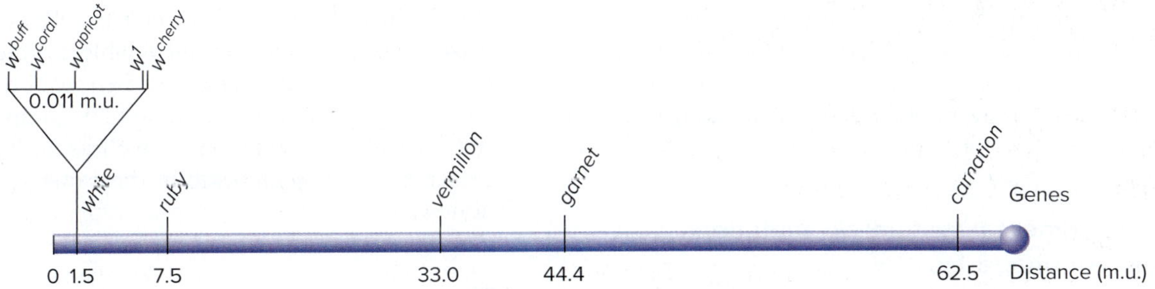

pairs. All known human genes fall somewhere between these extremes. To put these figures in perspective, an average human chromosome is approximately 130 million base pairs in length.

A Gene Is a Set of Nucleotide Pairs That Can Mutate Independently and Recombine with Each Other

Although complementation testing makes it possible to distinguish mutations in different genes from mutations in the same gene, it does not clarify how the structure of a gene can accommodate different mutations and how these different mutations can alter phenotype in different ways. Does each mutation change the whole gene at a single stroke in a particular way, or does it change only a specific part of a gene, while other mutations alter other parts?

In the late 1950s, the American geneticist Seymour Benzer used recombination analysis to show that two different mutations that did not complement each other and were therefore known to be in the same gene can in fact change different parts of that gene. He reasoned that if a gene is composed of separately mutable subunits, then it should be possible for recombination to occur within a gene, between these subunits. Therefore, crossovers

Figure 7.23 How recombination within a gene could generate a wild-type allele. Suppose a gene, indicated by the region between brackets, is composed of many sites that can mutate independently. Recombination between mutations m_1 and m_2 at different sites in the same gene produces a wild-type allele and a reciprocal allele containing both mutations.

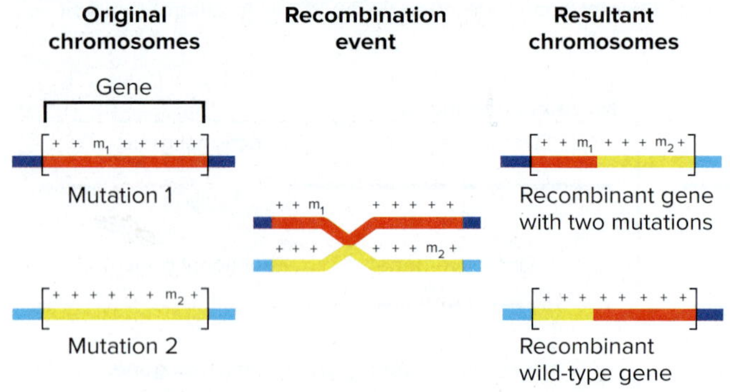

between homologous chromosomes carrying different mutations known to be in the same gene could in theory generate a wild-type allele (**Fig. 7.23**).

Because mutations affecting a single gene are likely to lie very close together, it is necessary to examine a very large number of progeny to observe even one crossover event between them. The resolution of the experimental system must thus be extremely high, allowing rapid detection of rare genetic events. For his experimental organism, Benzer chose bacteriophage T4, a virus that infects *E. coli* cells (**Fig. 7.24a.1**). Because each T4 phage that infects a bacterium generates 100 to 1000 progeny in less than an hour, Benzer could easily produce enough rare recombinants for his analysis (**Fig. 7.24a.2**). Moreover, by exploiting a peculiarity of certain T4 mutations, he devised conditions that allowed only recombinant phages, and not parental phages, to proliferate.

The experimental system: *rII*⁻ mutations of bacteriophage T4

Even though bacteriophages are too small to be seen without the aid of an electron microscope, a simple technique makes it possible to detect their presence with the unaided eye (**Fig. 7.24a.3**). To do this, researchers mix a population of bacteriophage particles with a much larger number of bacteria in molten agar and then pour this mixture onto a petri plate that already contains a bottom layer of nutrient agar. Uninfected bacterial cells grow throughout the top layer, forming an opalescent *lawn* of living bacteria. However, if a single phage infects a single bacterial cell somewhere on this lawn, the cell produces and releases progeny viral particles that infect adjacent bacteria, which, in turn, produce and release yet more phage progeny. With each release of virus particles, the bacterial host cell dies. The agar in the top layer prevents the phage particles from diffusing very far. Thus, several cycles of phage infection, replication, and release produce a circular cleared area in the lawn, called a **plaque,** devoid of living bacterial cells. The process of mixing phages with bacteria to produce a lawn and plaques on a petri plate is called *plating* phages.

Most plaques contain from 1 million to 10 million descendants of the single bacteriophage that originally infected a cell in that position on the petri plate. Sequential dilution of phage-containing solutions makes it possible to measure the number of phages in a particular plaque and arrive at a countable number of viral particles (**Fig. 7.24a.4**).

When Benzer first looked for genetic traits associated with bacteriophage T4, he found mutants that, when added to a lawn of *E. coli* B strain bacteria, produced larger plaques with sharper, more clearly rounded edges than those produced by the wild-type bacteriophages (**Fig. 7.24b**). Because these changes in plaque morphology result from the abnormally rapid lysis of the host bacteria, Benzer named the mutations *r* for *rapid lysis.* Many *r* mutations map to a region of the T4 chromosome known as the *rII* region; these are called rII^- mutations.

An additional property of rII^- mutations makes them ideal for the genetic **fine structure mapping** (the mapping of mutations within a gene) undertaken by Benzer. Wild-type rII^+ bacteriophages form plaques of normal shape and size on cells of both the *E. coli* B strain and a strain known as *E. coli* K(λ). The rII^- mutants, however, have an altered host range; they cannot form plaques with *E. coli* K(λ) cells, although as we have seen, they produce large, unusually distinct plaques with *E. coli* B cells (Fig. 7.24b). The reason that rII^- mutants are unable to infect cells of the K(λ) strain was not clear to Benzer, but this property allowed him to develop an extremely simple and effective test for rII^+ gene function, as well as an ingenious way to detect rare *intragenic* (within the same gene) recombination events.

The *rII* region has two genes

Before he could check whether two mutations in the same gene could recombine, Benzer had to be sure he was really looking at two mutations in a single gene. To verify this, he performed customized complementation tests tailored to two significant characteristics of bacteriophage T4: They are **monoploid** (that is, each phage carries a single T4 chromosome, so the phages have one copy of each of their genes), and they can replicate only in a host bacterium. Because T4 phages are monoploid, Benzer needed to ensure that two different T4 chromosomes entered the same bacterial cell in order to test for complementation between the mutations. In his complementation tests, he simultaneously infected *E. coli* K(λ) cells with two types of T4 chromosomes—one carried one rII^- mutation, the other carried a different rII^- mutation—and then looked for cell lysis

FEATURE FIGURE 7.24

How Benzer Analyzed the *rII* Genes of Bacteriophage T4

(a.1)

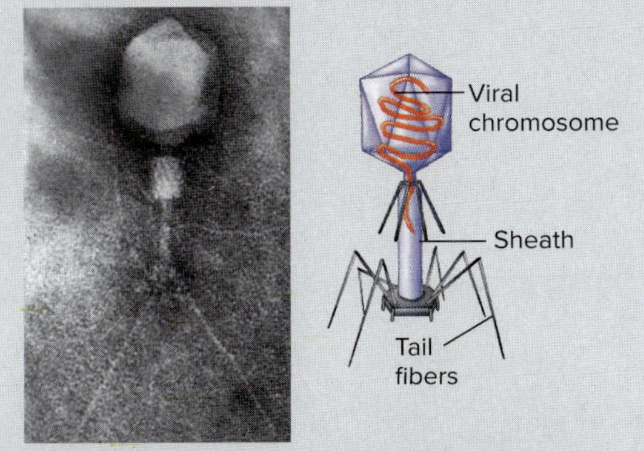

(a.2)

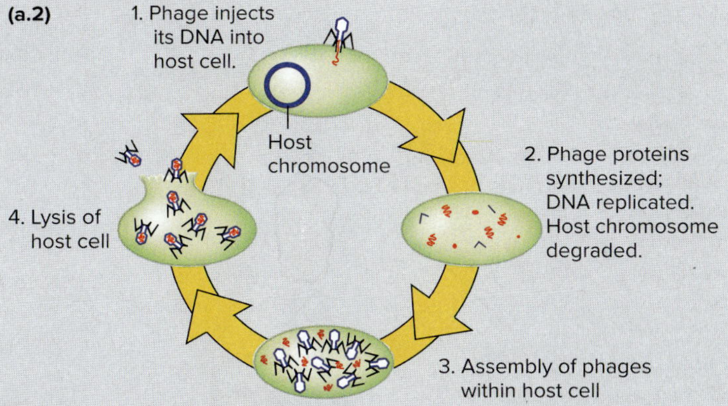

(a.3)

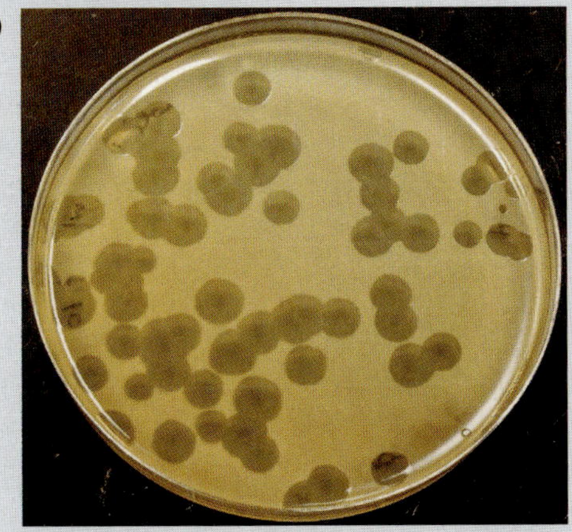

(a.4)

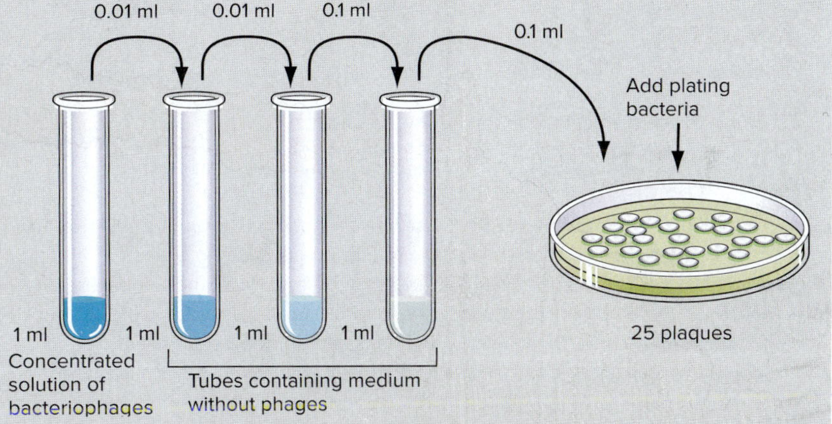

a.1: © Science Source; a.3: © McGraw-Hill Education/Lisa Burgess

(a) Working with bacteriophage T4

1. Bacteriophage T4 (at a magnification of approximately 100,000×) and in an artist's rendering. The viral chromosome is contained within a protein head. Other proteinaceous parts of the phage particle include the tail fibers, which help the phage attach to host cells, and the sheath, a conduit for injecting the phage chromosome into the host cell.

2. The lytic cycle of bacteriophage T4. A single phage particle infects a host cell; the phage DNA replicates and directs the synthesis of viral protein components using the machinery of the host cell; the new DNA and protein components assemble into new bacteriophage particles. Eventual lysis of the host cell releases up to 1000 progeny bacteriophages into the environment.

3. Clear plaques of bacteriophages in a lawn of bacterial cells. A mixture of bacteriophages and a large number of bacteria in molten agar are poured into a petri plate. Uninfected bacterial cells grow, producing an opalescent *lawn*. A bacterium infected by a single bacteriophage will lyse and release progeny bacteriophages, which infect adjacent bacteria. Several cycles of infection result in a *plaque*, a circular cleared area containing millions of genetically identical bacteriophages.

4. Counting bacteriophages by *serial dilution*. A small sample of a concentrated solution of bacteriophages is transferred to a test tube containing fresh medium, and a small sample of this dilution is transferred to another tube of fresh medium. Successive repeats of this process increase the degree of dilution. A sample of the final dilution, when mixed with bacteria in molten agar, yields a countable number of plaques from which it is possible to extrapolate the number of bacteriophages in the starting solution. The original 1 ml of solution in this illustration contained roughly 2.5×10^7 bacteriophages.

(b) Phenotypic properties of *rII*⁻ mutants of bacteriophage T4

1. *rII*$^-$ mutants, when plated on *E. coli* B cells, produce plaques that are larger and more distinct (with sharper edges) than plaques formed by *rII*$^+$ wild-type phages.

2. *rII*$^-$ mutants are particularly useful for finding rare recombination events because they have an altered host range. In contrast to *rII*$^+$ wild-type phages, *rII*$^-$ mutants cannot form plaques in lawns of *E. coli* strain K(λ)

(b.1)

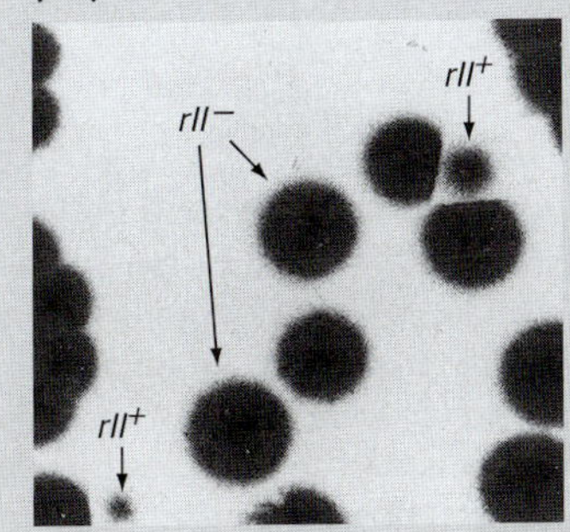

b.1: © Seymour Benzer

(b.2)

T4 strain	*E. coli* strain B	*E. coli* strain K(λ)
rII$^-$	Large, distinct	No plaques
rII$^+$	Small, fuzzy	Small, fuzzy

(Continued)

host bacteria. In a lysate with millions of *rII*$^-$ phages, even a single *rII*$^+$ recombinant phage can be identified because it can form a plaque on *E. coli* K(λ).

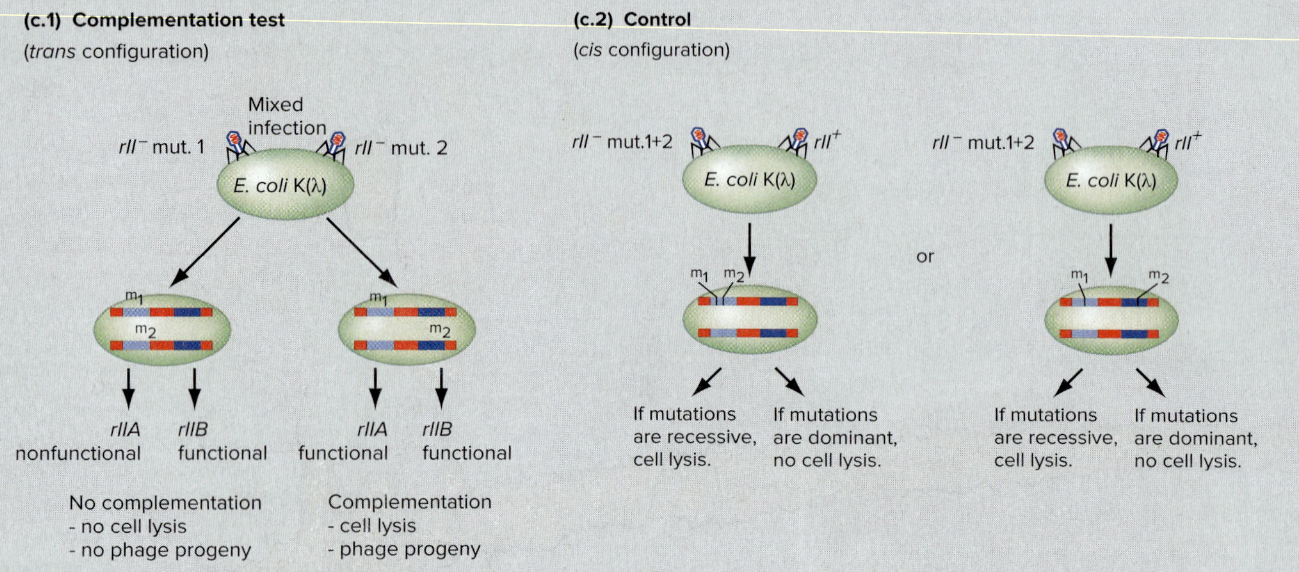

(c) A customized complementation test between *rII*$^-$ mutants of bacteriophage T4

1. *E. coli* K(λ) cells are infected simultaneously with an excess of two different *rII*$^-$ mutants (m_1 and m_2). Inside the cell, the two mutations will be in *trans;* that is, they lie on different chromosomes. If the two mutations are in the same gene, they will affect the same function and cannot complement each other, so no progeny phages will be produced. If the two mutations are in different genes (*rIIA* and *rIIB*), they will complement each other, leading to progeny phage production and cell lysis.

2. An important control for this complementation test is the simultaneous infection of *E. coli* K(λ) with a wild-type T4 and a T4 strain in which m_1 and m_2 pairs that fail to complement have been recombined onto the same chromosome—the two mutations will be in *cis*. Release of phage progeny shows that both mutations are recessive to wild type and that the mutations do not interact in a way that prevents the cells from producing progeny phages. Complementation tests are meaningful only if the two mutations tested are both recessive to wild type.

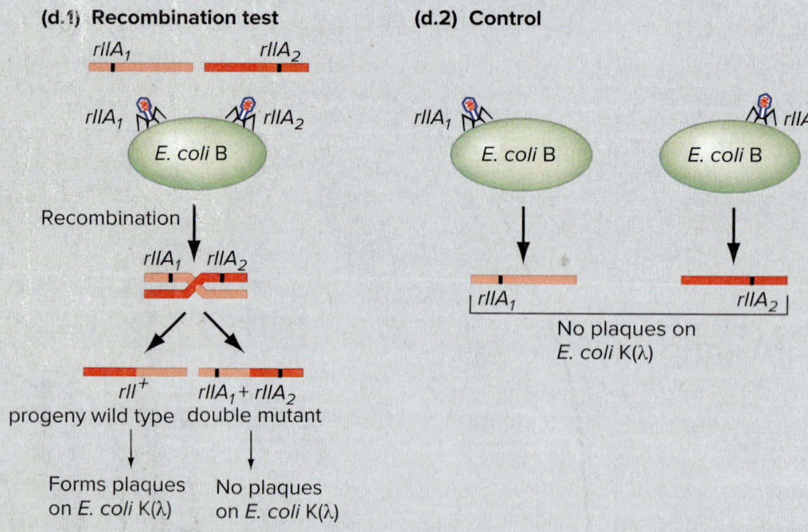

(d) Detecting recombination between two mutations in the same gene

1. *E. coli* B cells are infected with a large excess of two different *rIIA*$^-$ mutants (*rIIA*$_1$ and *rIIA*$_2$). If no recombination between the two *rIIA*$^-$ mutations takes place, all progeny phages will be *rII*$^-$. If recombination between the two mutations occurs, one of the products will be an *rII*$^+$ recombinant, while the reciprocal product will be a double mutant containing both *rIIA*$_1$ and *rIIA*$_2$. When the phage progeny subsequently infect *E. coli* K(λ) bacteria, only *rII*$^+$ recombinants will be able to form plaques.

2. As a control, *E. coli* B cells are infected with a large amount of only one kind of mutant (*rIIA*$_1$ or *rIIA*$_2$). The only *rII*$^+$ phages that can result are revertants of that mutation. Such revertants turn out to be extremely rare and can be ignored in most recombination experiments. Even if the two *rIIA*$^-$ mutations are in adjacent base pairs, the number of *rII*$^+$ recombinants obtained is more than 100 times higher than the number of *rII*$^+$ revertants the cells infected by a single mutant can produce.

(**Fig. 7.24c.1**). To ensure that the two kinds of phages would infect almost every bacterial cell, he added many more phages of each type than there were bacteria.

When tested by Benzer's method, if the two rII^- mutations were in different genes, they would complement each other: Each of the mutant T4 chromosomes would supply one wild-type rII^+ gene function, making up for the lack of that function in the other chromosome and resulting in lysis. On the other hand, if the two rII^- mutations were in the same gene, they would fail to complement: No plaques would appear because neither mutant chromosome would be able to supply the missing function.

Tests of many different pairs of rII^- mutations showed that they fall into two complementation groups: *rIIA* and *rIIB*. However, Benzer had to satisfy one final experimental requirement: For the complementation test to be meaningful, he had to make sure that pairs of rII^- mutations that failed to complement were each recessive to wild type and also did not interact with each other to produce an rII^- phenotype dominant to wild type. He checked these points by a control experiment in which he recombined pairs of $rIIA^-$ or $rIIB^-$ mutations onto the same chromosome (as described in the next section) and then simultaneously infected *E. coli* K(λ) with these double rII^- mutants and with wild-type phages (**Fig. 7.24c.2**). If the mutations were recessive and did not interact with each other, the cells would lyse, in which case the complementation test would be interpretable.

The significant distinction between the actual complementation test and the control experiment is in the placement of the two rII^- mutations. In the complementation test, one rII^- mutation is on one chromosome, while the other rII^- mutation is on the other chromosome (Fig. 7.24c.1); two mutations arranged in this way are said to be in the ***trans* configuration.** In the control experiment (Fig. 7.24c.2), the two mutations are on the same chromosome, in the so-called ***cis* configuration.** The complete test, including the complementation test and the control experiment, is known as a *cis-trans* test. In the complete experiment, two mutations that do not produce lysis in *trans* but do so when in *cis* are in the same complementation group. Benzer called any complementation group identified by the *cis-trans* test a **cistron,** and some geneticists still use the term *cistron* as a synonym for *gene*.

With the knowledge that the *rII* locus consists of two genes (*rIIA* and *rIIB*), Benzer could look for two mutations in the same gene and then see if they ever recombine to produce wild-type progeny.

Recombination between different mutations in a single gene

When Benzer infected *E. coli* B strain bacteria with a mixture of phages carrying different mutations in the same gene ($rIIA_1$ and $rIIA_2$, for example), he did observe the appearance of rare rII^+ progeny (**Fig. 7.24d.1**). He knew these wild-type progeny resulted from recombination and not from reverse mutations because the frequencies of the rII^+ phage particles he observed, even if rare, were much higher than the frequencies of rII^+ revertants seen among progeny produced by infecting B strain bacteria with either mutant alone (**Fig. 7.24d.2**).

These experiments were possible only because Benzer devised a **selection** for rare rII^+ recombinants. In a selection, conditions are such that the only survivors are the rare individuals you seek to identify. Benzer's selection condition for identifying rare rII^+ recombinant progeny was plating for plaques on *E. coli* K(λ). Benzer could assay a phage lysate containing tens of thousands of phage progeny on a single petri plate containing a lawn of *E. coli* K(λ). Because none of the rII^- phage in the lysate could form plaques, even a single rII^+ recombinant among them could be identified as a plaque.

On the basis of his observations with the *rII* genes, Benzer drew three conclusions about gene structure and function: (1) A gene consists of different parts that can each mutate; (2) recombination can occur between different mutable sites in the same gene; and (3) a gene performs its normal function only if all of its components are wild type. From what we now know about the molecular structure of DNA, this all makes perfect sense: The different mutable units are the base pairs that constitute the gene.

A Gene Is a Discrete Linear Set of Nucleotide Pairs

How are the multiple nucleotide pairs that make up a gene arranged—in a continuous row, or dispersed in precise patterns around the genome? And do the various mutations that affect gene function alter many different nucleotides, or only a small subset within each gene?

Using deletions to map mutations approximately

To answer these questions about the arrangement of nucleotides in a gene, Benzer eventually obtained thousands of spontaneous and mutagen-induced rII^- mutations that he needed to map with respect to each other.

To map the location of a thousand mutants through comparisons of all possible two-point crosses, Benzer would have had to set up a million ($10^3 \times 10^3$) matings. But by taking advantage of bacteriophage strains with large deletions, he could obtain the same information with far fewer crosses.

These large deletions are mutations that remove many contiguous nucleotide pairs along a DNA molecule. In crosses between bacteriophages carrying a mutation and bacteriophages carrying deletions of the corresponding region, no wild-type recombinant progeny can arise, because

Figure 7.25 Fine structure mapping of the bacteriophage T4 *rII* genes. (a) A phage cross between a point mutation and a deletion removing the DNA at the position of the mutation cannot yield wild-type recombinants. The same is true if two different deletion mutations overlap each other. **(b)** Large deletions divide the *rII* locus into regions; finer deletions divide each region into subsections. Point mutations, such as 271 (in *red* at *bottom*), map to region 3 if they do not recombine with deletions PT1, PB242, or A105 but do recombine with deletion 638 (*top*). Point mutations can be mapped to subsections of region 3 using other deletions (*middle*). Recombination tests map point mutations in the same subregion (*bottom*). Point mutations 201 and 155 cannot recombine to yield wild-type recombinants because they affect the same nucleotide pair. **(c)** Benzer's fine structure map. *Hotspots* are locations with many independent mutations that cannot recombine with each other.

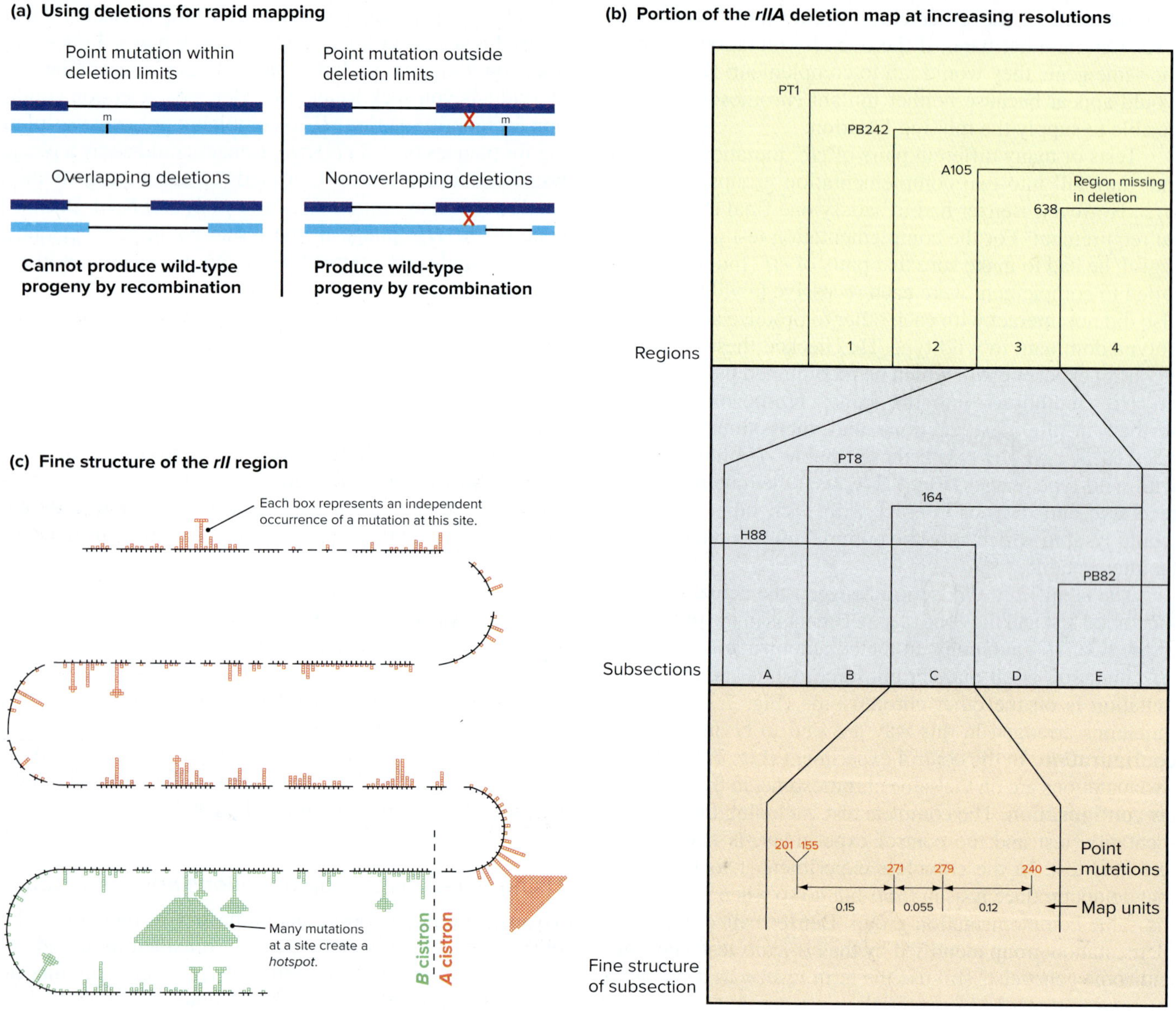

neither chromosome carries the proper information at the location of the mutation. However, if the mutation lies outside the region deleted from the homologous chromosome, wild-type progeny can appear (**Fig. 7.25a**). This is true whether the mutation is a point mutation affecting one or a few nucleotides, or is itself a large deletion. Crosses between any uncharacterized mutation and a known deletion thus immediately reveal whether the mutation resides in the region deleted from the other phage chromosome, providing a rapid way to find the general location of a mutation.

Using a series of overlapping deletions, Benzer divided the *rII* region into a series of relatively small regions, or *intervals*. He could then assign any point mutation to an interval by observing whether the point mutation recombined to give rII^+ progeny when crossed with the series of deletions (**Fig. 7.25b**).

Benzer mapped 1612 spontaneous point mutations and several deletions in the *rII* locus of bacteriophage T4 through recombination analysis. He first used recombination to determine the relationship between the deletions. He then found the approximate location of individual point mutations by observing which deletions could recombine with each point mutant to yield wild-type progeny.

Determining RF between rII^- mutations for precise mapping

Benzer next performed recombination tests to measure the genetic distance between pairs of point mutations he had found by deletion mapping to lie in the same small region of the chromosome. The distance between any two rII^- mutants could be measured simply by counting the number of rII^+ and total phages in an aliquot of lysate from a phage cross. The RF (in map units) is simply the number of rII^+ recombinants [plaques on *E. coli* K(λ)] divided by the total number of phages (plaques on *E. coli* B), multiplied by 2 to account for the rII^- double mutant reciprocal recombinants that must exist but cannot be detected easily:

$$\text{RF} = \frac{2[\text{number of plaques on } \textit{E. coli } \text{K}(\lambda)]}{(\text{number of plaques on } \textit{E. coli } \text{B})}$$

Benzer combined the results from deletion mapping and RF calculations to produce a map of the *fine structure* of the *rII* region (**Fig. 7.25c**). Note that all of the point mutations in the *rIIA* complementation group mapped to one side of the *rII* region, and all of the *rIIB* point mutations mapped to the other side.

How DNA nucleotides are organized into genes

Benzer knew that the genetic distances between all mapped genes in the T4 genome add up to about 1500 m.u. He also knew that the T4 chromosome constitutes about 169,000 bp of DNA, so he could calculate that for bacteriophage T4, 1 m.u. corresponds to 169,000/1500 = 113 bp. The lowest RF that he measured between any pair of rII^- mutants was 0.02 m.u., which would represent about 2 bp. Benzer thus inferred that a mutation can arise from the change of even a single nucleotide pair, and that recombination can occur between adjacent nucleotide pairs. From the observation that mutations within the *rII* region form a self-consistent, linear recombination map, he concluded that a gene is composed of a continuous linear sequence of nucleotide pairs within the DNA. And from observations that the positions of mutations in the *rIIA* gene did not overlap those of the *rIIB* gene, he determined that the nucleotide sequences composing those two genes are separate and distinct. A *gene* is thus a linear set of nucleotide pairs, located within a discrete region of a chromosome, that serves as a unit of function.

Hotspots of mutation

Some sites within a gene mutate spontaneously more often than others and as a result are known as **mutation hotspots** (Fig. 7.25c). The existence of hotspots suggests that certain nucleotides can be altered more readily than others. Treatment with mutagens also turns up hotspots, but because mutagens have specificities for particular nucleotides, the highly mutable sites that turn up with various mutagens are often at different positions in a gene than the hotspots resulting from spontaneous mutation.

Nucleotides are the same chemically whether they lie within a gene or in the DNA between genes. Furthermore, as Benzer's experiments imply, the molecular machinery responsible for mutation and recombination does not discriminate between those nucleotides that are *intragenic* (within a gene) and those that are *intergenic* (between genes). The main distinction between DNA within and DNA outside a gene is that the array of nucleotides composing a gene has evolved a function that determines phenotype. Next, we describe how geneticists discovered what that function is.

essential concepts

- A *complementation test* determines whether two different recessive mutations occur in the same gene or in different genes.
- At the DNA level, a gene is a linear sequence of nucleotide pairs in a discrete region of a chromosome that confers a specific unit of function.
- Recombination can occur between any two nucleotide pairs, whether they are within the same gene or not.

7.5 What Mutations Tell Us About Gene Function

learning objectives

1. Explain how the analysis of arginine auxotrophs implied that a single gene corresponds to a single enzyme.
2. Describe how missense mutations were used to show that genes determine the amino acid sequences of proteins.
3. Differentiate between primary, secondary, tertiary, and quaternary structures of proteins.

Mendel's experiments established that an individual gene can control a visible characteristic, but his laws do not explain how genes actually govern the appearance of traits.

Investigators working in the first half of the twentieth century studied carefully the biochemical changes caused by mutations in an effort to understand the genotype–phenotype connection.

In one of the first of these studies, conducted in 1902, the British physician Dr. Archibald Garrod showed that a human genetic disorder known as *alkaptonuria* is determined by the recessive allele of an autosomal gene. Garrod analyzed family pedigrees and performed biochemical analyses on family members with and without the trait. The urine of people with alkaptonuria turns black on exposure to air. Garrod found that a substance known as *homogentisic acid,* which blackens upon contact with oxygen, accumulates in the urine of alkaptonuria patients. Alkaptonuriacs excrete all of the homogentisic acid they ingest, while people without the condition excrete no homogentisic acid in their urine even after ingesting the substance.

From these observations, Garrod concluded that people with alkaptonuria are incapable of metabolizing homogentisic acid to the breakdown products generated by normal individuals (**Fig. 7.26**). Because many biochemical reactions within the cells of organisms are catalyzed by enzymes, Garrod hypothesized that lack of the enzyme that breaks down homogentisic acid is the cause of alkaptonuria. In the absence of this enzyme, homogentisic acid accumulates and causes the urine to turn black on contact with oxygen. He called this condition an *inborn error of metabolism.*

Figure 7.26 Alkaptonuria: An inborn error of metabolism. The biochemical pathway in humans that degrades phenylalanine and tyrosine via homogentisic acid (HA). In alkaptonuria patients, the enzyme HA hydroxylase is not functional, so it does not catalyze the conversion of HA to maleylacetoacetic acid. As a result, HA, which oxidizes to a black compound, accumulates in the urine.

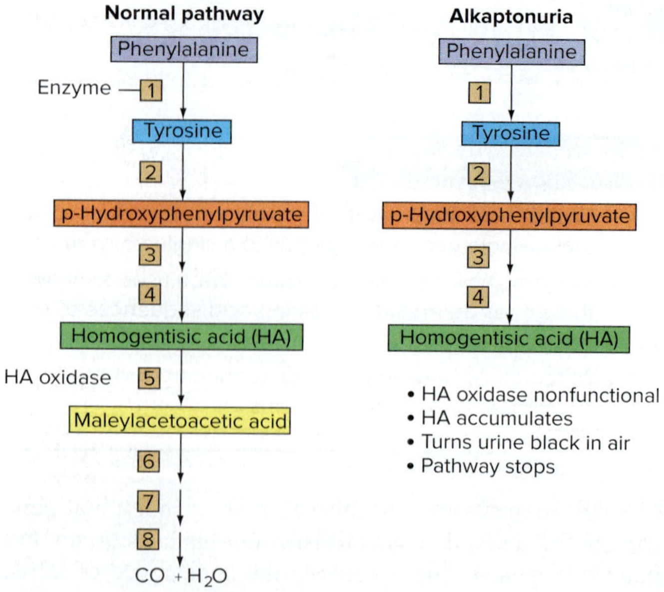

Garrod studied several other inborn errors of metabolism and suggested that all arose from mutations that prevented a particular gene from producing an enzyme required for a specific biochemical reaction. In today's terminology, the wild-type allele of the gene would allow production of functional enzyme (in the case of alkaptonuria, the enzyme is homogentisic acid oxidase), whereas the mutant allele would not. Because the single wild-type allele in heterozygotes generates sufficient enzyme to prevent the accumulation of homogentisic acid and thus the condition of alkaptonuria, the mutant allele is recessive.

A Gene Contains the Information for Producing a Specific Enzyme: The One Gene, One Enzyme Hypothesis

In the 1940s, George Beadle and Edward Tatum carried out a series of experiments on the bread mold *Neurospora crassa* (whose life cycle was described in Chapter 5) that demonstrated a direct relationship between genes and the enzymes that catalyze specific biochemical reactions. Their strategy was simple. They first isolated a number of mutations that disrupted the synthesis of the amino acid arginine, a compound needed for *Neurospora* growth. They next hypothesized that different mutations blocked different steps in a particular **biochemical pathway:** the orderly series of reactions that allows *Neurospora* to obtain simple molecules from the environment and convert them step-by-step into successively more complicated molecules culminating in the end product arginine.

Experimental evidence for *one gene, one enzyme*

Figure 7.27a illustrates the experiments Beadle and Tatum performed to test their hypothesis. They first obtained a set of mutagen-induced mutations that prevented *Neurospora* from synthesizing arginine. Cells with any one of these mutations were unable to make arginine and could therefore grow on a minimal medium containing salt and sugar only if it had been supplemented with arginine. A nutritional mutant microorganism that requires supplementation with substances not needed by wild-type strains is known as an **auxotroph.** The cells just mentioned were arginine auxotrophs. (In contrast, a cell that does not require the addition of a substance is a **prototroph** for that factor. In a more general meaning, *prototroph* refers to a wild-type cell that can grow on minimal medium alone.)

Recombination analyses located the auxotrophic arginine-blocking mutations in four distinct regions of the genome, and complementation tests showed that each of the four regions correlated with a different complementation group. On the basis of these results, Beadle and Tatum concluded that at least four genes support the biochemical

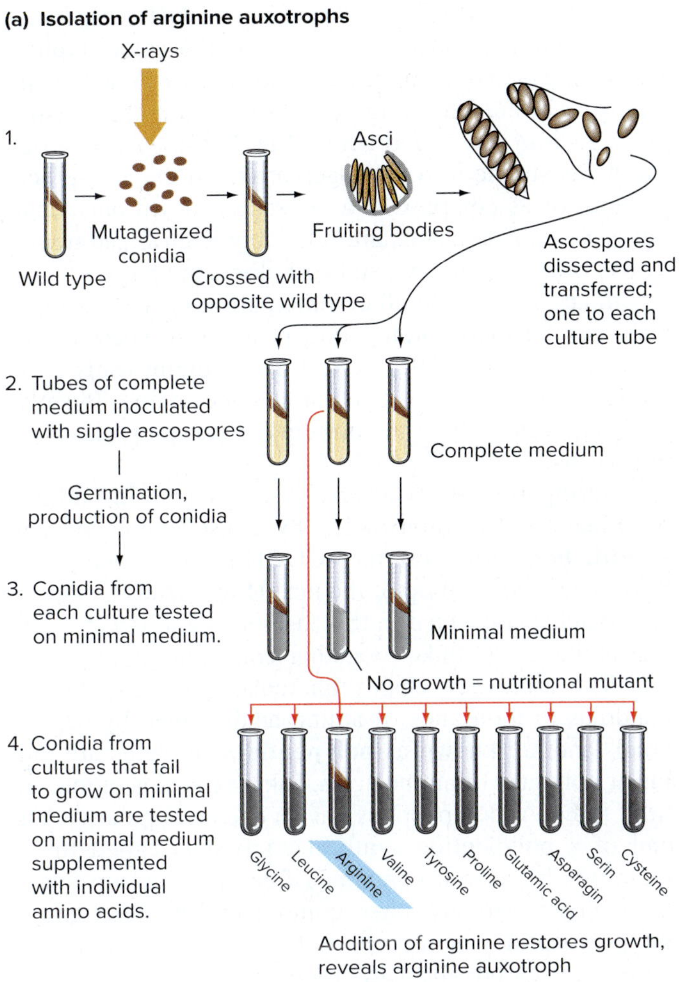

(b) Growth response if nutrient is added to minimal medium

Mutant strain	Supplements: Nothing	Ornithine	Citrulline	Arginino-succinate	Arginine
Wildtype: Arg^+	+	+	+	+	+
ARG-E$^-$	–	+	+	+	+
ARG-F$^-$	–	–	+	+	+
ARG-G$^-$	–	–	–	+	+
ARG-H$^-$	–	–	–	–	+

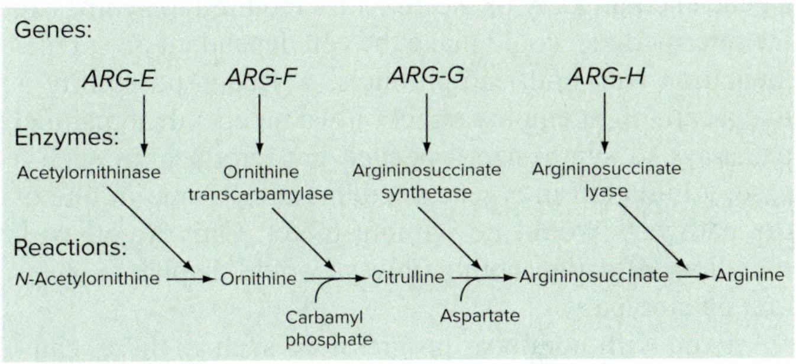

Figure 7.27 Experimental support for the *one gene, one enzyme* hypothesis. **(a)** Beadle and Tatum mated an X-ray-mutagenized strain of *Neurospora* with another strain, and they isolated haploid ascospores that grew on complete medium. Cultures that failed to grow on minimal medium were nutritional mutants. Nutritional mutants that could grow on minimal medium plus arginine were Arg^- auxotrophs. **(b)** The ability of wild-type and mutant strains to grow on minimal medium supplemented with intermediates in the arginine pathway. **(c)** Each of the four *ARG* genes specifies an enzyme needed to convert one intermediate to the next in the pathway.

pathway for arginine synthesis. They named the four genes *ARG-E, ARG-F, ARG-G,* and *ARG-H*.

They next asked whether any of the mutant *Neurospora* strains could grow in minimal medium supplemented with any of three known intermediates (ornithine, citrulline, and argininosuccinate) in the biochemical pathway leading to arginine, instead of with arginine itself. This test would identify *Neurospora* mutants able to convert the intermediate compound into arginine. Beadle and Tatum compiled a table describing which arginine auxotrophic mutants were able to grow on minimal medium supplemented with each of the intermediates (**Fig. 7.27b**).

Interpretation of results: Genes encode enzymes

On the basis of these results, Beadle and Tatum proposed a model of how *Neurospora* cells synthesize arginine (**Fig. 7.27c**). In the linear progression of biochemical reactions by which a cell constructs arginine from the constituents of minimal medium, each intermediate is both the product of one step and the substrate for the next. Each reaction in the precisely ordered sequence is catalyzed by a specific enzyme, and the presence of each enzyme depends on one of the four *ARG* genes.

A mutation in one gene blocks the pathway at a particular step because the cell lacks the corresponding enzyme and thus cannot make arginine on its own. Supplementing the medium with any intermediate that occurs beyond the blocked reaction restores growth to the mutant because the organism has all the enzymes required to convert the intermediate to arginine. Supplementation with an intermediate that occurs before the missing enzyme does not work because the mutant cell cannot convert the intermediate into arginine.

Each mutation abolishes the cell's ability to make an enzyme capable of catalyzing a certain reaction. By inference, then, each gene controls the synthesis or activity of an enzyme, or as stated by Beadle and Tatum: one gene, one enzyme. Of course, the gene and the enzyme are not the same thing; rather, the sequence of nucleotides in a gene contains information that somehow encodes the structure of an enzyme molecule.

Although the analysis of the arginine pathway studied by Beadle and Tatum was straightforward, studies of biochemical pathways are not always so easy to interpret. Some biochemical pathways are not linear progressions of stepwise reactions. For example, a branching pathway occurs if different enzymes act on the same intermediate to convert it into two different end products. If the cell requires both of these end products for growth, a mutation in a gene encoding any of the enzymes required to synthesize the intermediate would make the cell dependent on supplementation with both end products. A second possibility is that a cell might employ either of two independent, parallel pathways to synthesize a needed end product. In such a case, a mutation in a gene encoding an enzyme in one of the pathways would be without effect. Only a cell with mutations affecting both pathways would display an aberrant phenotype.

Even with nonlinear progressions such as these, careful genetic analysis can reveal the nature of the biochemical pathway on the basis of Beadle and Tatum's insight that genes specify proteins.

Genes Specify the Identity and Order of Amino Acids in Polypeptide Chains

Although the one gene, one enzyme hypothesis was a crucial advance in understanding how genes influence phenotype, it is an oversimplification. Not all genes govern the construction of enzymes active in biochemical pathways. Enzymes are only one class of the molecules known as *proteins,* and cells contain many other kinds of proteins. Among the other types are proteins that provide shape and rigidity to a cell, proteins that transport molecules in and out of cells, proteins that help fold DNA into chromosomes, and proteins that act as hormonal messengers. Genes direct the synthesis of all proteins, enzymes and nonenzymes alike. Moreover, as we see next, genes actually determine the construction of *polypeptides,* and because some proteins are composed of more than one type of polypeptide, more than one gene determines the construction of such proteins.

Proteins: Linear polymers of amino acids linked by peptide bonds

Proteins are polymers composed of building blocks known as **amino acids.** Cells use mainly 20 different amino acids to synthesize the proteins they need. All of these amino acids have certain basic features, encapsulated by the formula NH_2–CHR–COOH (**Fig. 7.28a**). The –COOH component, also known as *carboxylic acid,* is, as the name implies, acidic; the $–NH_2$ component, also known as an *amino group,* is basic. The R refers to side chains that distinguish each of the amino acids (**Fig. 7.28b**). An R group can be as simple as a hydrogen atom (in the amino acid glycine) or as complex as a benzene ring (in phenylalanine). Some side chains are relatively neutral and nonreactive, others are acidic, and still others are basic.

In addition to the 20 common amino acids, two rare ones can be incorporated into proteins in specific circumstances (**Fig. 7.28c**). A very few proteins (only 25 in humans) are known to contain selenocysteine. Pyrrolysine is present only in the proteins of certain prokaryotic organisms.

During protein synthesis, a cell's protein-building machinery links amino acids by constructing covalent **peptide bonds** that join the –COOH group of one amino acid to the $–NH_2$ group of the next (**Fig. 7.28d**). A pair of amino acids connected in this fashion is a *dipeptide;* several amino acids linked together constitute an *oligopeptide.* The amino acid chains that make up proteins contain hundreds to thousands of amino acids joined by peptide bonds and are known as **polypeptides.** Proteins are thus linear polymers of amino acids. Like the chains of nucleotides in DNA, polypeptides have a chemical polarity. The end of a polypeptide synthesized first is called the **N terminus** because it contains a free amino group that is not connected to any other amino acid. The other end of the polypeptide chain is the **C terminus** because it contains a free carboxylic acid group.

Mutations can alter amino acid sequences

Each protein is composed of a unique sequence of amino acids. The chemical properties that enable structural proteins to give a cell its shape, or allow enzymes to catalyze specific reactions, are a direct consequence of the identity, number, and linear order of amino acids in the protein.

If genes specify proteins, then at least some mutations could be changes in a gene that alter the normal sequence of amino acids in the protein specified by that gene. In the mid-1950s, Vernon Ingram began to establish what kinds of changes particular mutations cause in the corresponding protein. Using techniques that had just been developed for determining the sequence of amino acids in a protein, he compared the amino acid sequence of the normal adult form of hemoglobin (HbA) with that of hemoglobin in the bloodstream of people homozygous for the mutation that causes sickle-cell anemia (HbS). Remarkably, he found only a single amino acid difference

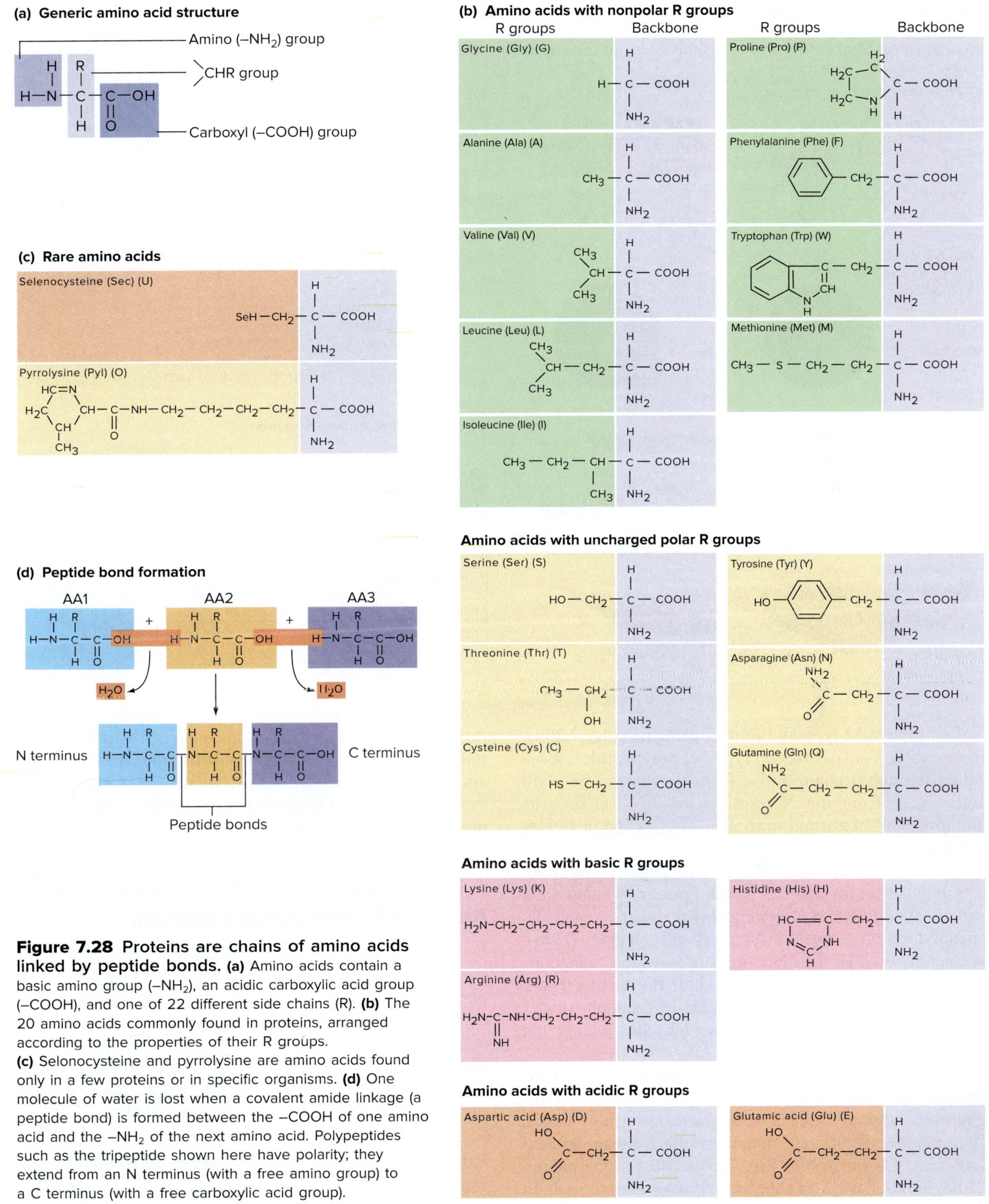

Figure 7.28 Proteins are chains of amino acids linked by peptide bonds. **(a)** Amino acids contain a basic amino group (–NH_2), an acidic carboxylic acid group (–COOH), and one of 22 different side chains (R). **(b)** The 20 amino acids commonly found in proteins, arranged according to the properties of their R groups. **(c)** Selonocysteine and pyrrolysine are amino acids found only in a few proteins or in specific organisms. **(d)** One molecule of water is lost when a covalent amide linkage (a peptide bond) is formed between the –COOH of one amino acid and the –NH_2 of the next amino acid. Polypeptides such as the tripeptide shown here have polarity; they extend from an N terminus (with a free amino group) to a C terminus (with a free carboxylic acid group).

Figure 7.29 The molecular basis of sickle-cell and other anemias. **(a)** Substitution of glutamic acid by valine at the sixth amino acid from the N terminus affects the three-dimensional structure of the β chain of hemoglobin. Hemoglobins incorporating the mutant β chain form aggregates that cause red blood cells to sickle. **(b)** Red blood cell sickling has many phenotypic effects. **(c)** Other mutations in the β chain gene also cause anemias.

(a) From mutation to phenotype

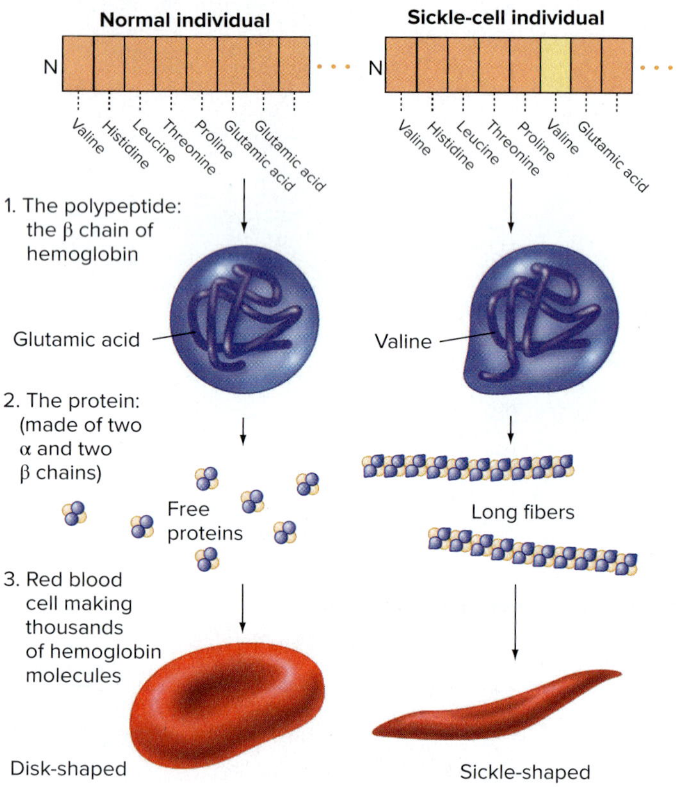

(b) Sickle-cell anemia is pleiotropic

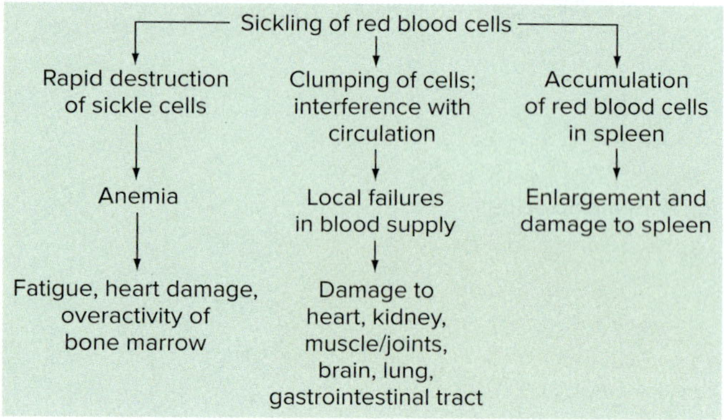

(c) β chain substitutions/variants

	Amino acid position									
	1	2	3 ···	6	7 ···	26···	63···	67···	125···	146
Normal (HbA)	Val	His	Leu	Glu	Glu	Glu	His	Val	Glu	His
HbS	Val	His	Leu	Val	Glu	Glu	His	Val	Glu	His
HbC	Val	His	Leu	Lys	Glu	Glu	His	Val	Glu	His
HbG San Jose	Val	His	Leu	Glu	Gly	Glu	His	Val	Glu	His
HbE	Val	His	Leu	Glu	Glu	Lys	His	Val	Glu	His
HbM Saskatoon	Val	His	Leu	Glu	Glu	Glu	Tyr	Val	Glu	His
Hb Zurich	Val	His	Leu	Glu	Glu	Glu	Arg	Val	Glu	His
HbM Milwaukee 1	Val	His	Leu	Glu	Glu	Glu	His	Glu	Glu	His
HbDβ Punjab	Val	His	Leu	Glu	Glu	Glu	His	Val	Gln	His

between the wild-type and mutant proteins (**Fig. 7.29a**). Hemoglobin consists of two types of polypeptides: a so-called α (alpha) chain and a β (beta) chain. The sixth amino acid from the N terminus of the β chain was glutamic acid in normal individuals but valine in sickle-cell patients.

Ingram thus established that a mutation substituting one amino acid for another had the power to change the structure and function of hemoglobin and thereby alter the phenotype from normal to sickle-cell anemia (**Fig. 7.29b**). We now know that the glutamic acid-to-valine change affects the solubility of hemoglobin within the red blood cell. At low concentrations of oxygen, the less soluble sickle-cell form of hemoglobin aggregates into long chains that deform the red blood cell (Fig. 7.29a).

Because people suffering from a variety of inherited anemias also have defective hemoglobin molecules, Ingram and other geneticists were able to determine how a large number of different mutations affect the amino acid sequence of hemoglobin (**Fig. 7.29c**). Most of the altered hemoglobins have a change in only one amino acid. In various patients with anemia, the alteration is generally in different amino acids, but occasionally, two independent mutations result in different substitutions for the same amino acid. Geneticists use the term **missense mutation** to describe a genetic alteration that causes the substitution of one amino acid for another.

A Protein's Amino Acid Sequence Dictates Its Three-Dimensional Structure

Despite the uniform nature of protein construction—a string of amino acids joined by peptide bonds—each polypeptide folds into a unique three-dimensional shape. Biochemists often distinguish between four levels of protein structure: *primary, secondary, tertiary,* and *quaternary.* The first three of these apply to any one polypeptide chain, while the quaternary level describes associations between multiple polypeptides within a protein complex.

Primary, secondary, and tertiary protein structures

The linear sequence of amino acids within a polypeptide is its **primary structure.** Each unique primary structure

Figure 7.30 Levels of polypeptide structure. **(a)** Covalent and noncovalent interactions determine the structure of a polypeptide. **(b)** A polypeptide's primary (1°) structure is its amino acid sequence. **(c)** Localized regions form secondary (2°) structures such as α helixes and β-pleated sheets. **(d)** The tertiary (3°) structure is the complete three-dimensional arrangement of a polypeptide. In this portrait of myoglobin, the iron-containing heme group, which carries oxygen, is *red,* while the polypeptide itself is *green.*

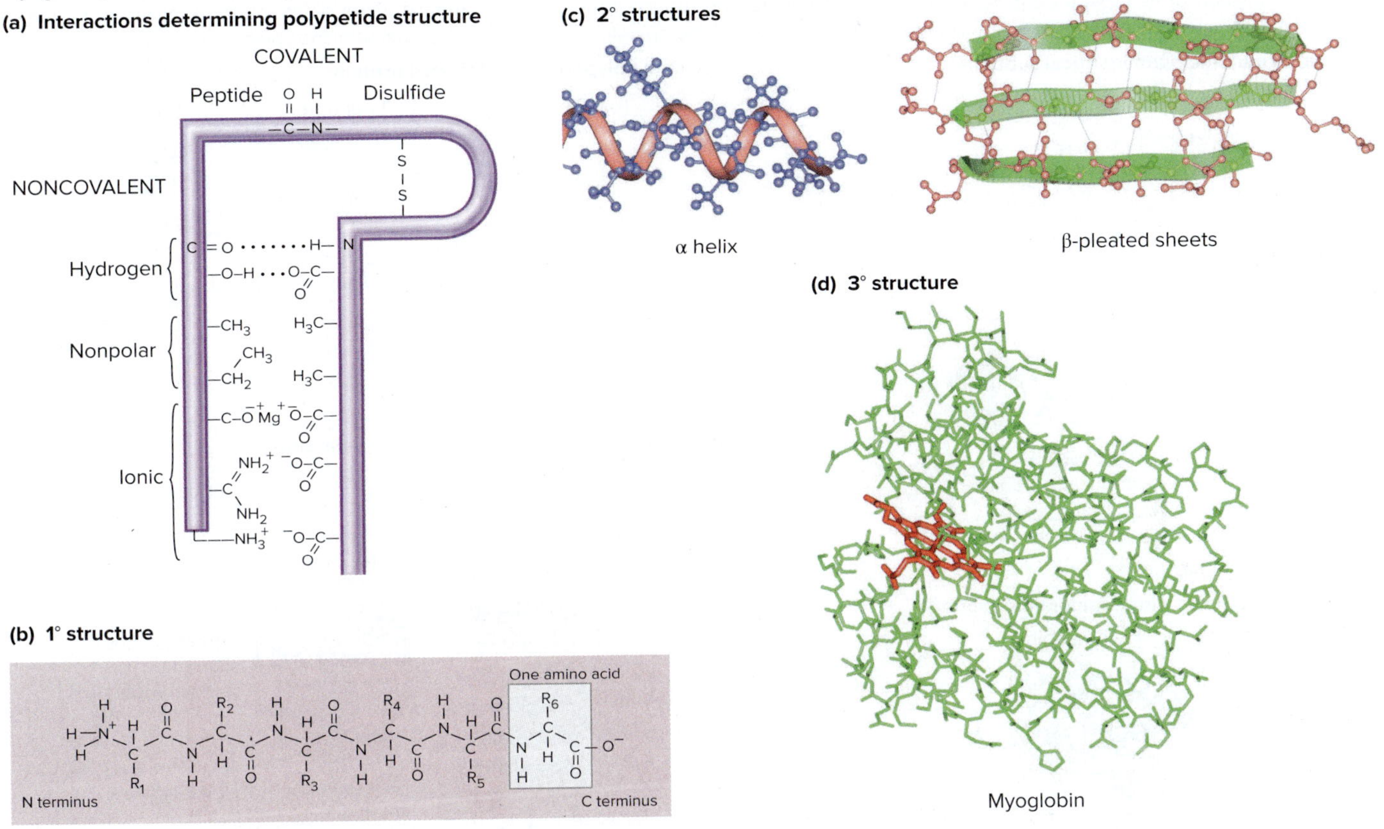

places constraints on how a chain can arrange itself in three-dimensional space. Because the R groups distinguishing the 22 amino acids have dissimilar chemical properties, some amino acids form hydrogen bonds or electrostatic bonds when brought into proximity with other amino acids. Nonpolar amino acids, for example, may become associated with each other by interactions that hide them from water in localized hydrophobic regions. As another example, two cysteine amino acids can form covalent disulfide bridges (–S–S–) through the oxidation of their –SH groups.

All of these interactions (**Fig. 7.30a**) help stabilize the polypeptide in a specific three-dimensional conformation. The primary structure (**Fig. 7.30b**) determines three-dimensional shape by generating **secondary structure:** localized regions with a characteristic geometry (**Fig. 7.30c**). Primary structure is also responsible for other folds and twists that together with the secondary structure produce the ultimate three-dimensional **tertiary structure** of the entire polypeptide (**Fig. 7.30d**). Normal tertiary structure—the way a long chain of amino acids naturally folds in three-dimensional space under physiological conditions—is known as a polypeptide's **native configuration.** Various forces, including hydrogen bonds, electrostatic bonds, hydrophobic interactions, and disulfide bridges help stabilize the native configuration.

It is worth repeating that primary structure—the sequence of amino acids in a polypeptide—directly determines secondary and tertiary structures. The information required for the chain to fold into its native configuration is inherent in its linear sequence of amino acids.

In one example of this principle, many proteins unfold, or become **denatured,** when exposed to urea and mercaptoethanol or to increasing heat or pH. These treatments disrupt the interactions that normally stabilize the secondary and tertiary structures. When conditions return to normal, some proteins spontaneously refold into their native configuration without help from other agents. No other information beyond the primary structure is needed to achieve the proper three-dimensional shape of such proteins.

You should know that some proteins are unable after denaturation to refold by themselves into their correct tertiary structure. The proper folding of these proteins requires other proteins called **chaperones** that help stabilize

Figure 7.31 Multimeric proteins. (a) β2 lens crystallin contains two copies of one kind of subunit; the two subunits are the product of a single gene. The peptide backbones of the two subunits are shown in different shades of *purple*. **(b)** Hemoglobin is composed of two different kinds of subunits, each encoded by a different gene. **(c)** Three distinct protein receptors for the immune-system molecules called interleukins (ILs; *purple*). All contain a common gamma (γ) chain (*yellow*), plus other receptor-specific polypeptides (*green*). A mutant γ chain blocks the function of all three receptors, leading to XSCID. **(d)** One α-tubulin (*red*) and one β-tubulin (*blue*) polypeptide associate to form a tubulin dimer. Many tubulin dimers form a single microtubule. The mitotic spindle is an assembly of many microtubules.

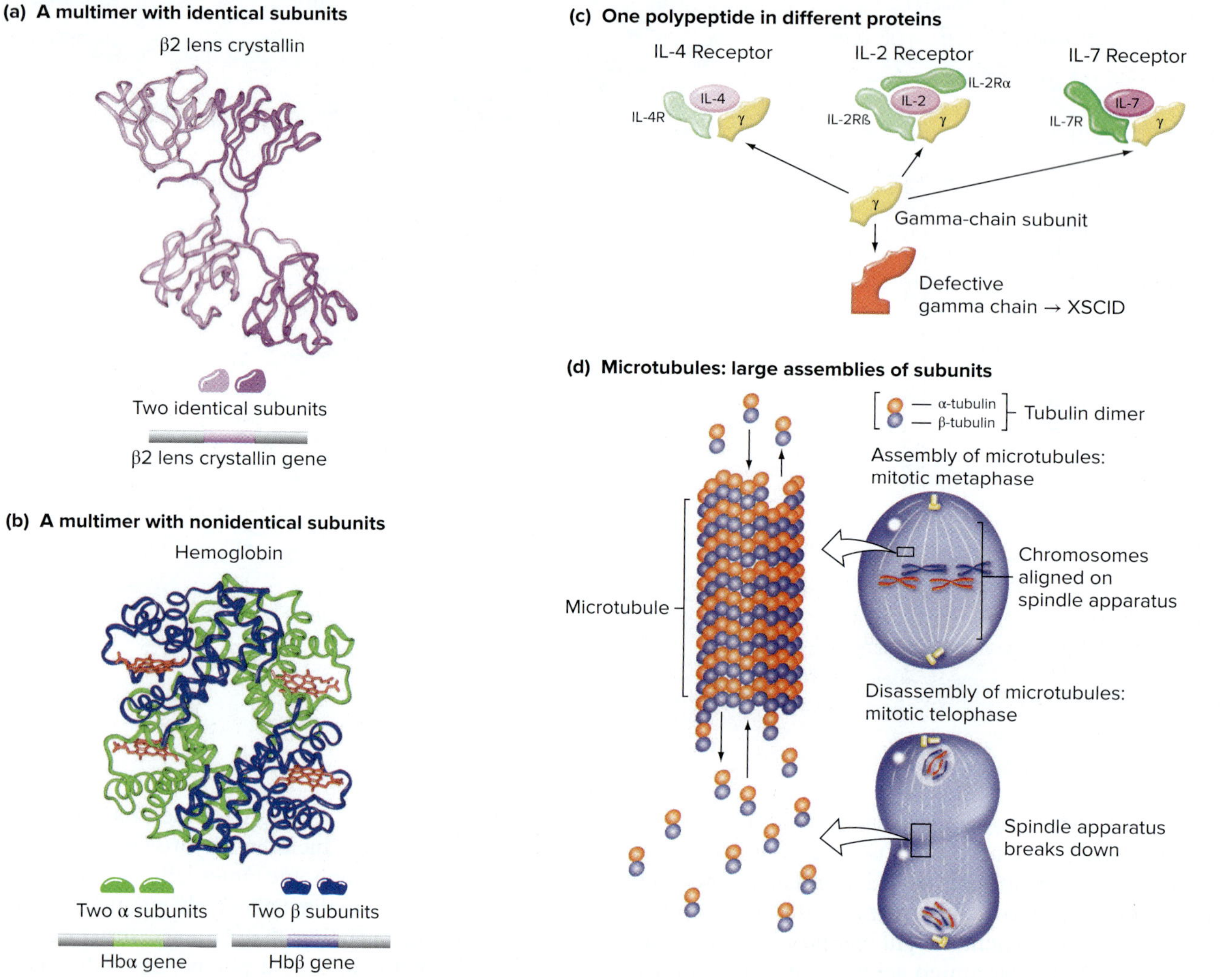

the native configuration. Because elevated temperatures cause protein unfolding, many chaperones are *heat shock proteins* that are made when organisms are exposed to high temperatures. These heat shock proteins protect cells from damage due to protein misfolding under high-temperature conditions. But even for proteins that need chaperones to achieve their native configurations, the amino acid sequence of the protein dictates the final three-dimensional structure.

Quaternary structure: Multimeric proteins

Certain proteins, such as the rhodopsin that promotes black-and-white vision, consist of a single polypeptide. Many others, however, such as the lens crystallin protein, which provides rigidity and transparency to the lenses of our eyes, or the hemoglobin molecule described earlier, are composed of two or more polypeptide chains that associate in a specific way (**Fig. 7.31a** and **b**). The individual polypeptides in an aggregate are known as **subunits,** and the complex of subunits is often referred to as a **multimer.** The three-dimensional configuration of subunits in a multimer is a complex protein's **quaternary structure.**

The same forces that stabilize the native form of a polypeptide (that is, hydrogen bonds, electrostatic bonds, hydrophobic interactions, and disulfide bridges) also contribute to the maintenance of quaternary structure. As Fig. 7.31a shows, in some multimers, the two or more interacting subunits are identical polypeptides. These identical chains are specified by one gene. In other multimers, by contrast, more than one kind of polypeptide makes up the protein (Fig. 7.31b). The different polypeptides in these multimers are specified by different genes.

Alterations in just one kind of subunit, caused by a mutation in a single gene, can affect the function of a multimer. The adult hemoglobin molecule, for example, consists of two α and two β subunits, with each type of subunit determined by a different gene—one for the α chain and one for the β chain. A mutation in the Hbβ gene resulting in an amino acid switch at position 6 in the β chain causes sickle-cell anemia.

Similarly, if several multimeric proteins share a common subunit, a single mutation in the gene encoding that subunit may affect all the proteins simultaneously. An example is an X-linked mutation in mice and humans that incapacitates several different proteins all known as *interleukin (IL) receptors*. Because all of these receptors are essential to the normal function of immune system cells that fight infection and generate immunity, this one mutation causes the life-threatening condition known as *X-linked severe combined immune deficiency* (*XSCID*; **Fig. 7.31c**).

The polypeptides of complex proteins can assemble into extremely large structures capable of changing with the needs of the cell. For example, the microtubules that make up the spindle during mitosis are gigantic assemblages of mainly two polypeptides: α-tubulin and β-tubulin (**Fig. 7.31d**). The cell can organize these subunits into very long hollow tubes that grow or shrink as needed at different stages of the cell cycle.

One gene, one polypeptide

Because more than one gene governs the production of some multimeric proteins and because not all proteins are enzymes, the *one gene, one enzyme* hypothesis is not broad enough to define gene function. A more accurate statement is *one gene, one polypeptide:* Each gene governs the construction of a particular polypeptide. As you will see in Chapter 8, even this reformulation does not encompass the function of all genes, as some genes in all organisms do not determine the construction of proteins; instead, they specify RNAs that are not translated into polypeptides.

Knowledge about the connection between genes and polypeptides enabled geneticists to analyze how different mutations in a single gene can produce different phenotypes. If each amino acid has a specific effect on the three-dimensional structure of a protein, then changing amino acids at different positions in a polypeptide chain can alter protein function in different ways. For example, most enzymes have an active site that carries out the enzymatic task, while other parts of the protein support the shape and position of that site. Mutations that change the identity of amino acids at the active site may have more serious consequences than those affecting amino acids outside the active site. Some kinds of amino acid substitutions, such as replacement of an amino acid having a basic side chain with an amino acid having an acidic side chain, would be more likely to compromise protein function than would substitutions that retain the chemical characteristics of the original amino acid.

Some mutations do not affect the amino acid composition of a protein but still generate an abnormal phenotype. As will be discussed in Chapter 8, such mutations change the amount of normal polypeptide produced by disrupting the biochemical processes responsible for decoding a gene into a polypeptide.

essential concepts

- Most genes specify the linear sequence of amino acids in a *polypeptide;* this sequence determines the polypeptide's three-dimensional structure and thus its function.
- A *missense mutation* changes the identity of a single amino acid in a polypeptide.
- *Multimeric proteins* include two or more polypeptides (*subunits*). If these subunits are different, they must be encoded by different genes.

7.6 A Comprehensive Example: Mutations That Affect Vision

learning objectives

1. Describe the functions of the four photoreceptor proteins in human vision.
2. Outline how the genes encoding the photoreceptors evolved through duplication and divergence of an ancestral gene.
3. Explain how mutations in the photoreceptor genes result in different vision defects.

Researchers first described anomalies of color perception in humans close to 200 years ago. Since that time, they have discovered a large number of mutations that modify human vision. By examining the phenotype associated with each mutation and then looking directly at the DNA alterations inherited with the mutation, they have learned a great deal about the genes influencing human visual perception and the function of the proteins they specify.

Using human subjects for vision studies has several advantages. First, people can recognize and describe variations in the way they see, from trivial differences in what the color red looks like, to not seeing any difference between red and green, to not seeing any color at all. Second, the highly developed science of *psychophysics* provides sensitive, noninvasive tests for accurately defining and comparing phenotypes. Finally, because inherited variations in the visual system rarely affect one's life span or ability to reproduce, mutations generating many of the new alleles that change visual perception remain in a population over time.

Cells of the Retina Contain Light-Sensitive Proteins

People perceive light through nerve cells (neurons) in the retina at the back of the eye (**Fig. 7.32a**). These neurons are of two types: *rods* and *cones.* The rods, which make up 95% of all light-receiving neurons, are stimulated by weak light over a range of wavelengths. At higher light intensities, the rods become saturated and no longer send meaningful information to the brain. This is when the cones take over, processing wavelengths of bright light that enable us to see color.

The cones come in three forms—one specializes in the reception of red light, a second in the reception of green, and a third in the reception of blue. For each photoreceptor cell, the act of reception consists of absorbing photons from light of a particular wavelength, transducing information about the number and energy of those photons to electrical signals, and transmitting the signals via the optic nerve to the brain. The brain integrates the information from the three types of cones and enables humans to discriminate more than 1 million colors.

Four related proteins with different light sensitivities

The protein that receives photons and triggers the processing of information in rod cells is *rhodopsin.* It consists of a single polypeptide chain containing 348 amino acids that snakes back and forth across the cell membrane (**Fig. 7.32b**). One lysine within the chain associates with *retinal,* a carotenoid pigment molecule that actually absorbs photons. The amino acids in the vicinity of the retinal constitute rhodopsin's active site; by positioning the retinal in a particular way, those amino acids determine its response to light. Each rod cell contains approximately 100 million molecules of rhodopsin in its specialized membrane. As you learned at the beginning of this chapter, the gene governing the production of rhodopsin is on chromosome 3.

The protein that receives and initiates the processing of photons in the blue cones is a relative of rhodopsin, also consisting of a single polypeptide chain containing 348 amino acids and also encompassing one molecule of retinal. Slightly less than half of the 348 amino acids in the

Figure 7.32 The cellular and molecular basis of vision. **(a)** Rod and cone cells in the retina carry membrane-bound photoreceptors. **(b)** The photoreceptor in rod cells is rhodopsin. The blue, green, and red receptor proteins in cone cells are related to rhodopsin. The *colored dots* are amino acids that differ between rhodopsin and the diagrammed protein. **(c)** One red photoreceptor gene and one to three green photoreceptor genes are clustered on the X chromosome. **(d)** The genes for rhodopsin and the three color receptors probably evolved from a primordial photoreceptor gene through three gene duplication events followed by divergence of the duplicated copies.

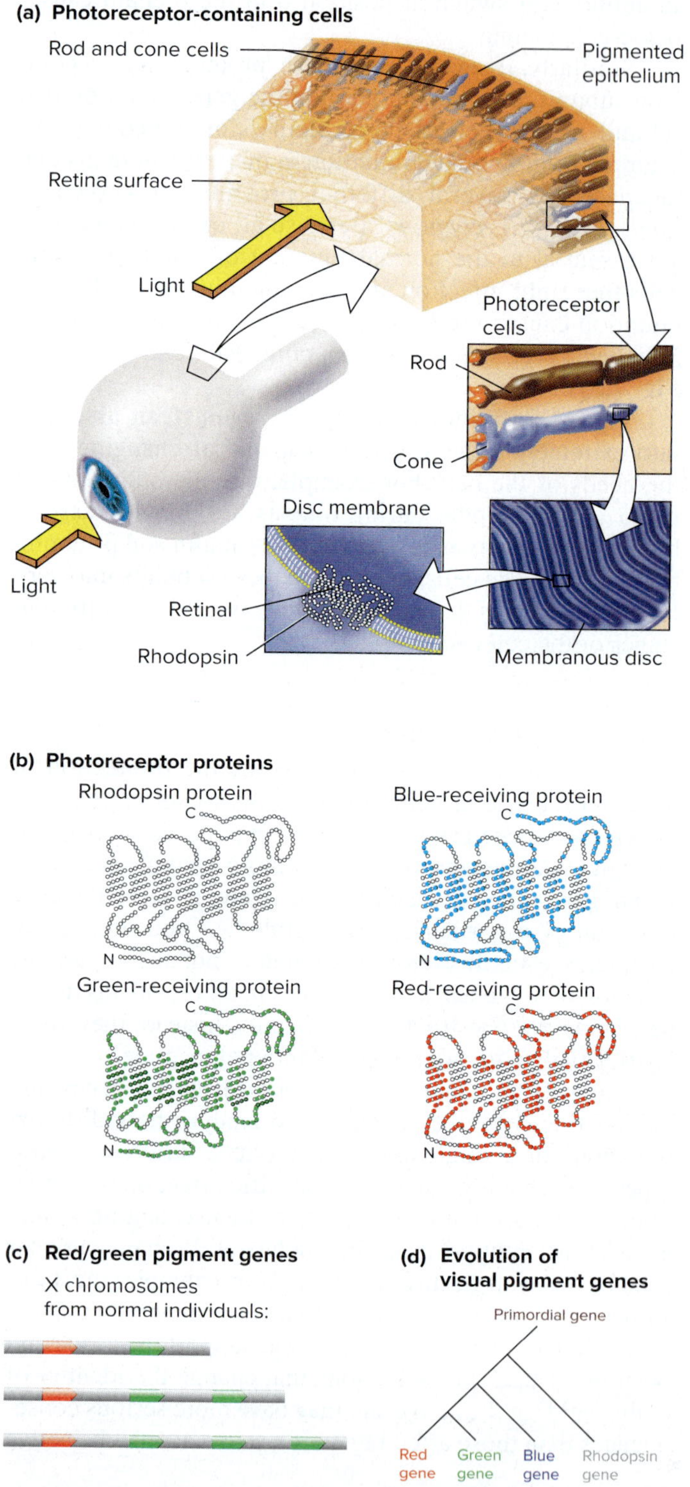

blue-receiving protein are the same as those found in rhodopsin; the rest are different and account for the specialized light-receiving ability of the protein (Fig. 7.32b). The gene for the blue protein is on chromosome 7.

Similarly related to rhodopsin are the red- and green-receiving proteins in the red and green cones. These are also single polypeptides associated with retinal and embedded in the cell membrane, although they are both slightly larger at 364 amino acids in length (Fig. 7.32b). Like the blue protein, the red and green proteins differ from rhodopsin in nearly half of their amino acids; they differ from each other in only 15 of their 364 amino acids. Even these small differences, however, are sufficient to differentiate the spectral sensitivities of red and green cone cells. The genes for the red and green proteins both reside on the X chromosome in a tandem head-to-tail arrangement. Most people have one red gene and one to three green genes on their X chromosomes (**Fig. 7.32c**).

Evolution of the rhodopsin gene family

The similarities in structure and function among rhodopsin and the three rhodopsin-related photoreceptor proteins suggest that the genes encoding these polypeptides arose by a series of gene duplication events in which the duplicated copies subsequently diverged through the accumulation of mutations. Many of the mutations that promoted the ability to see color must have provided selective advantages to their bearers.

Biologists can infer the evolutionary history of these duplications from the relatedness of the genes and protein products. The red and green genes are the most similar, differing by fewer than five nucleotides out of every hundred. This fact suggests they diverged from each other only in the relatively recent evolutionary past. The less pronounced amino acid similarity of the red or green proteins with the blue protein, and the even lower relatedness between rhodopsin and any color photoreceptor, reflect earlier duplication and divergence events (**Fig. 7.32d**).

How Mutations in the Rhodopsin Gene Family Affect the Way We See

Mutations in the genes encoding rhodopsin and the three color photoreceptor proteins can alter vision through many different mechanisms. These mutations range from point mutations that change the identity of a single amino acid in a single protein to larger aberrations that can increase or decrease the number of photoreceptor genes.

Mutations in the rhodopsin gene

At least 29 different single nucleotide substitutions in the rhodopsin gene cause an autosomal dominant vision disorder known as *retinitis pigmentosa* that begins with an early loss of rod function, followed by a slow progressive degeneration of the peripheral retina. **Figure 7.33a** shows the location of the amino acids affected by these mutations. These amino acid changes result in abnormal rhodopsin proteins that either do not fold properly or, once folded, are unstable. Although normal rhodopsin is an essential structural element of rod cell membranes, these nonfunctional mutant proteins are retained in the body of the cell, where they remain unavailable for insertion into the membrane. Rod cells that cannot incorporate enough rhodopsin into their membranes eventually die. Depending on how many rod cells die, partial or complete blindness ensues.

Other mutations in the rhodopsin gene cause the far less serious condition of night blindness (Fig. 7.33a). These mutations change the protein's amino acid sequence so that the threshold of stimulation required to trigger the vision cascade increases. With the changes, very dim light is no longer enough to initiate vision.

Figure 7.33 How mutations modulate light and color perception. (a) Amino acid substitutions (*black dots*) that disrupt rhodopsin's three-dimensional structure result in retinitis pigmentosa. Other substitutions diminishing rhodopsin's sensitivity to light cause night blindness. **(b)** Substitutions in the blue pigment can produce tritanopia (blue color blindness). **(c)** Red color blindness can result from particular mutations that destabilize the red photoreceptor. **(d)** Unequal crossing-over between the red and green genes can change gene number and create genes that specify hybrid photoreceptor proteins.

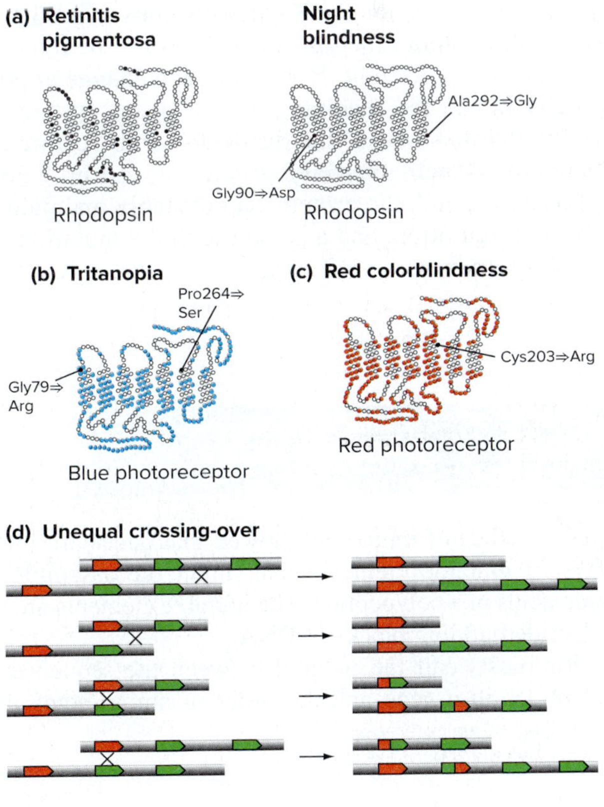

Figure 7.34 How the world looks to a person with tritanopia. Compare with Fig. 4.22.

Color deficit simulation courtesy of Vischeck (www.vischeck.com). Source image courtesy of NASA

Mutations in the cone cell pigment genes

Vision problems caused by mutations in the cone cell pigment genes are less severe than those caused by similar defects in the rod cell rhodopsin gene. Most likely, this difference occurs because the rods make up 95% of a person's light-receiving neurons, while the cones constitute only about 5%. Some mutations in the blue gene on chromosome 7 cause *tritanopia,* a defect in the ability to discriminate between colors that differ only in the amount of blue light they contain (**Figs. 7.33b** and **7.34**). Mutations in the red gene on the X chromosome can modify or abolish red protein function and as a result, the red cone cells' sensitivity to light. For example, a change at position 203 in the red-receiving protein from cysteine to arginine disrupts one of the disulfide bonds required to support the protein's tertiary structure (see **Fig. 7.33c**). Without that bond, the protein cannot stably maintain its native configuration, and a person with the mutation has red color blindness.

Unequal crossing-over between the red and green genes

People with normal color vision have a single red gene; some of these normal individuals also have a single adjacent green gene, while others have two or even three green genes. The red and green genes are 96% identical in DNA sequence; the different green genes, 99.9% identical.

Their proximity and high degree of homology make these genes unusually prone to an error in meiotic recombination called **unequal crossing-over.** When homologous chromosomes associate during meiosis, two closely related DNA sequences that are adjacent to each other, like the red and green photoreceptor genes, can pair with each other incorrectly. If recombination takes place between the mispaired sequences, photoreceptor genes may be deleted, added, or changed.

A variety of unequal recombination events produce DNA containing no red gene, no green gene, various combinations of green genes, or hybrid red-green genes (see **Fig. 7.33d**). These different DNA combinations account for the large majority of the known aberrations in red-green color perception, with the remaining abnormalities stemming from point mutations, as described earlier. Because the accurate perception of red and green depends on the differing ratios of red and green light processed, people with no red or no green gene perceive red and green as the same color (see Fig. 4.22).

essential concepts

- The vision pigments in humans consist of the protein rhodopsin in rods plus the blue-, red-, and green-sensitive photoreceptors in cones.
- The four genes of the rhodopsin family evolved from an ancestral photoreceptor gene by successive rounds of gene duplication and divergence.
- Mutations in the rhodopsin gene may disrupt rod function, leading to blindness. Mutations in cone cell photoreceptor genes are responsible for various forms of color blindness.

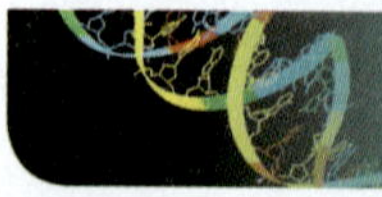

WHAT'S NEXT

Careful studies of mutations showed that genes are linear arrays of mutable elements that direct the assembly of amino acids in a polypeptide. The mutable elements are the nucleotide building blocks of DNA.

Biologists call the parallel between the sequence of nucleotides in a gene and the order of amino acids in a polypeptide **colinearity.** In Chapter 8, we explain how colinearity arises from base pairing, a genetic code, specific enzymes, and macromolecular assemblies like ribosomes that guide the flow of information from DNA through RNA to protein.

SOLVED PROBLEMS

I. Imagine that 10 independently isolated recessive lethal mutations (l^1, l^2, l^3, etc.) map to chromosome 7 in mice. You perform complementation testing by mating all pairwise combinations of heterozygotes bearing these lethal mutations, and you score the absence of complementation by examining pregnant females for dead fetuses. A + in the chart means that the two lethals complemented, and dead embryos were not found. A – indicates that dead embryos were found, at the rate of about one in four conceptions. (The crosses between heterozygous mice would be expected to yield the homozygous recessive showing the lethal phenotype in 1/4 of the embryos.) The lethal mutation in the parental heterozygotes for each cross are listed across the top and down the left side of the chart (that is, l^1 indicates a heterozygote in which one chromosome bears the l^1 mutation and the homologous chromosome is wild type).

	l^1	l^2	l^3	l^4	l^5	l^6	l^7	l^8	l^9	l^{10}
l^1	–	+	+	+	+	–	–	+	+	+
l^2		–	+	+	+	+	+	+	+	–
l^3			–	–	–	+	+	–	–	+
l^4				–	–	+	+	–	–	+
l^5					–	+	+	–	–	+
l^6						–	–	+	+	+
l^7							–	+	+	+
l^8								–	–	+
l^9									–	+
l^{10}										–

How many genes do the 10 lethal mutations represent? What are the complementation groups?

Answer

This problem involves the application of the complementation concept to a set of data. There are two ways to analyze these results. You can focus on the mutations that do complement each other, conclude that they are in different genes, and begin to create a list of mutations in separate genes. Alternatively, you can focus on mutations that do not complement each other and therefore are alleles of the same genes. The latter approach is more efficient when several mutations are involved. For example, l^1 does not complement l^6 and l^7. These three alleles are in one complementation group. l^2 does not complement l^{10}; they are in a second complementation group. l^3 does not complement l^4, l^5, l^8, or l^9, so they form a third complementation group. Three complementation groups exist. (Note also that for each mutant, the cross between individuals carrying the same alleles resulted in no complementation because homozygotes for the recessive lethal mutation were generated.) The three complementation groups consist of (1) l^1, l^6, l^7; (2) l^2, l^{10}; and (3) l^3, l^4, l^5, l^8, l^9.

II. W, X, and Y are the intermediates (in that order) in a biochemical pathway whose product is Z. Z^- mutants are found in five different complementation groups. *Z1* mutants will grow on Y or Z, but not W or X. *Z2* mutants will grow on X, Y, or Z. *Z3* mutants will only grow on Z. *Z4* mutants will grow on Y or Z. Finally, *Z5* mutants will grow on W, X, Y, or Z.

a. Order the five complementation groups in terms of the steps they block.

b. What does this genetic information reveal about the nature of the enzyme that carries out the conversion of X to Y?

Answer

This problem requires that you understand complementation and the connection between genes and enzymes in a biochemical pathway.

a. A biochemical pathway represents an ordered set of reactions that must occur to produce a product. This problem gives the order of intermediates in a pathway for producing product Z. The lack of any enzyme along the way will cause the phenotype of Z^-, but the block can occur at different places along the pathway. If the mutant grows when given an intermediate compound, the enzymatic (and hence gene) defect must be before production of that intermediate compound.

The *Z1* mutants that grow on Y or Z (but not on W or X) must have a defect in the enzyme that produces Y. *Z2* mutants have a defect prior to X; *Z3* mutants have a defect prior to Z; *Z4* mutants have a defect prior to Y; *Z5* mutants have a defect prior to W. The five complementation groups can be placed in order of activity within the biochemical pathway as follows:

$$\xrightarrow{Z5} W \xrightarrow{Z2} X \xrightarrow{Z1,\ Z4} Y \xrightarrow{Z3} Z$$

b. Mutants *Z1* and *Z4* affect the same step, but because they are in different complementation groups, we know they are in different genes. Mutations *Z1* and *Z4* are probably in genes that encode subunits of a multisubunit enzyme that carries out the conversion of X to Y. Alternatively, a currently unknown additional intermediate step between X and Y could exist.

PROBLEMS

Vocabulary

1. The following is a list of mutational changes. For each of the specific mutations described, indicate which of the terms in the right-hand column applies, either as a description of the mutation or as a possible cause. More than one term from the right column can apply to each statement in the left column.

    1. an A–T base pair in the wild-type gene is changed to a G–C pair
    2. an A–T base pair is changed to a T–A pair
    3. the sequence AAGCTTATCG is changed to AAGCTATCG
    4. the sequence CAGCAGCAGCAGCAGCAG is changed to CAGCAGCAGCAGCAGCAGCAGCAG
    5. the sequence AACGTTATCG is changed to AATGTTATCG
    6. the sequence AACGTCACACACACATCG is changed to AACGTCACATCG
    7. the sequence AAGCTTATCG is changed to AAGCTTTATCG

    a. transition
    b. base substitution
    c. transversion
    d. deletion
    e. insertion
    f. deamination
    g. X-ray irradiation
    h. intercalator
    i. slipped mispairing

Section 7.1

2. What explanations can account for the following pedigree of a very rare trait? Be as specific as possible. How might you be able to distinguish between these explanations?

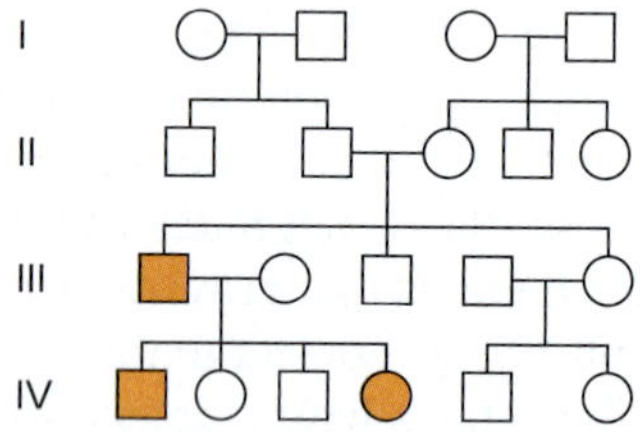

3. The DNA sequence of one strand of a gene from three independently isolated mutants is given here (5′ ends are at left). Using this information, what is the sequence of the wild-type gene in this region?

```
mutant 1   ACCGTAATCGACTGGTAAACTTTGCGCG
mutant 2   ACCGTAGTCGACCGGTAAACTTTGCGCG
mutant 3   ACCGTAGTCGACTGGTTAACTTTGCGCG
```

4. Among mammals, measurements of the rate of generation of autosomal recessive mutations have been made almost exclusively in mice, while many measurements of the rate of generation of dominant mutations have been made both in mice and in humans. What do you think is the reason for this difference?

5. Over a period of several years, a large hospital kept track of the number of births of babies displaying the trait achondroplasia. Achondroplasia is a very rare autosomal dominant condition resulting in dwarfism with abnormal body proportions. After 120,000 births, it was noted that 27 babies had been born with achondroplasia. One physician was interested in determining how many of these dwarf babies resulted from new mutations and whether the apparent mutation rate in this geographical area was higher than normal. He looked up the families of the 27 dwarf births and discovered that four of the dwarf babies had a dwarf parent. What is the apparent mutation rate of the achondroplasia gene in this population? Is it unusually high or low?

6. Suppose you wanted to study genes controlling the structure of bacterial cell surfaces. You decide to start by isolating bacterial mutants resistant to infection by a bacteriophage that binds to the cell surface. The selection procedure is simple: Spread cells from a culture of sensitive bacteria on a petri plate, expose them to a high concentration of phages, and pick the bacterial colonies that grow. To set up the selection you could (1) spread cells from a single liquid culture of sensitive bacteria on many different plates and pick every resistant colony; *or* (2) start many different cultures, each grown from a single colony of sensitive bacteria, spread one plate from each culture, and then pick a single mutant from each plate. Which method would ensure that you are isolating many independent mutations?

7. In a genetics lab, Kim and Maria infected a sample from an *E. coli* culture with a particular virulent bacteriophage. They noticed that most of the cells were lysed, but a few survived. The survival rate in their sample was about 1×10^{-4}. Kim was sure the bacteriophage induced the resistance in the cells, while Maria thought that resistant mutants probably already existed in the sample of cells they used. Earlier, for a different experiment, they had spread a dilute suspension of *E. coli* onto solid medium in a large petri dish, and, after seeing that about 10^5 colonies were growing up, they had replica-plated that plate onto three other plates. Kim and Maria decide to use these plates to test their theories. They pipette a suspension of the bacteriophage onto each of the three replica plates. What should they see if Kim is right? What should they see if Maria is right?

8. The results of the fluctuation test (Fig. 7.5) were interpreted to mean that different numbers of mutant bacteria preexisted in each of the 11 culture tubes because the mutations arose spontaneously at different times during the growth of each culture. However, another possibility is that the differences in the number of colonies on the plates are simply due to differences in the ability of the petri plates to support the growth of colonies. For example, perhaps the selective agent or the nutrients in the media were not evenly distributed in the molten agar poured into the petri dishes. What experiment could you do to determine whether or not differences in the petri plates were a factor in the experiment?

9. The following pedigree shows the inheritance of a completely penetrant, dominant trait called *amelogenesis imperfecta* that affects the structure and integrity of the teeth. DNA analysis of blood obtained from affected individuals III-1 and III-2 shows the presence of the same disease-causing mutation in one of the two copies of an autosomal gene called *ENAM* that is not seen in DNA from the blood of any of the parents in generation II. Explain this result, citing Fig. 4.19 and Fig. 7.5. Do you think this type of inheritance pattern is rare or common?

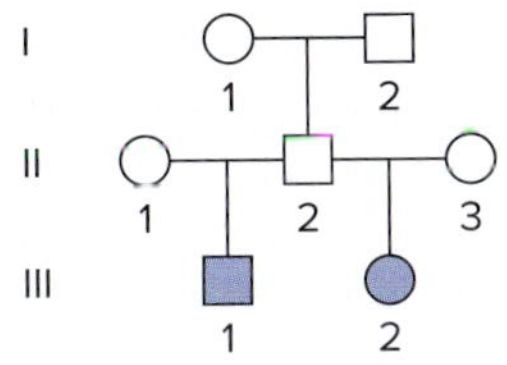

10. Autism is a neurological disorder thought to be caused by mutant alleles of one or more genes. Scientists had been wondering why the number of children diagnosed as autistic increased dramatically in a decade, from 1 in 500 in 2002 to 1 in 88 in 2012. Researchers now think that they might have found at least part of the answer: Men are fathering children at later and later ages. A paper published in the journal *Nature* in 2012 showed a correlation between paternal age and the incidence of autism; the age of the mother was not a factor. How does this observation provide a possible explanation for the apparent increase in the rate of autism?

11. Like the yellow Labrador retrievers featured in Chapter 3, golden retrievers are usually solid yellow. The golden retriever shown has an extremely rare black marking on its face. Explain the genetic basis for the appearance of this dog. Consider only the *E* and *B* genes listed in Table 3.3 (see also Figs. 3.12 and 3.13). (Also note that the black marking is not a mask.)

© Sally MacBurney

Section 7.2

12. Remember that *Balancer* chromosomes prevent the recovery of recombinant chromosomes between the *Balancer* and its normal homolog. Why was the *Balancer* X chromosome crucial to the design of Muller's experiment (Fig. 7.13)? (*Hint:* The best way to answer this question is to consider what the experimental results would have been without the *Balancer*.)

13. Figure 7.14 shows examples of base substitutions induced by the mutagens 5-bromouracil, hydroxylamine, ethylmethane sulfonate, and nitrous acid. Which of these mutagens cause transitions, and which cause transversions?

14. Figure 7.14a shows the mutagen 5-bromouracil (5-BU), which can resemble either T or C depending on its tautomeric state. The figure first shows 5-BU incorporated into DNA as the T-like tautomer, but then it changes its state to the C-like tautomer during the next round of DNA replication. The result was a T:A→C:G substitution. Suppose that the tautomeric states of 5-BU during the two rounds of replication were reversed. What kind of mutation would result?

15. So-called *two-way mutagens* can induce both a particular mutation and (when added subsequently to cells whose chromosomes carry this mutation) a reversion of the mutation that restores the original DNA sequence. In contrast, *one-way mutagens* can induce mutations but not exact reversions of these

mutations. Based on Fig. 7.14, which of the following mutagens can be classified as one-way and which as two-way?

a. 5-bromouracil

b. hydroxylamine

c. ethylmethane sulfonate

d. nitrous acid

e. proflavin

16. In 1967, J. B. Jenkins treated wild-type male *Drosophila* with the mutagen ethylmethane sulfonate (EMS) and mated them with females homozygous for a recessive mutation called *dumpy* that causes shortened wings. He found some F_1 progeny with two wild-type wings, some with two short wings, and some with one short wing and one wild-type wing. In a second cross, when he mated single F_1 flies with two short wings to *dumpy* homozygotes, he found, surprisingly, that only a fraction of these matings produced all short-winged progeny.

a. Explain these results in light of the mechanism of action of EMS shown in Fig. 7.14.

b. Should the short-winged progeny of the second cross have one or two short wings? Why?

17. When a particular mutagen identified by the Ames test is injected into mice, it causes the appearance of many tumors, showing that this substance is carcinogenic. When cells from these tumors are injected into other mice not exposed to the mutagen, almost all of the new mice develop tumors. However, when mice carrying mutagen-induced tumors are mated to unexposed mice, virtually all of the progeny are tumor free. Why can the tumor be transferred horizontally (by injecting cells) but not vertically (from one generation to the next)?

18. When the His^- *Salmonella* strain used in the Ames test is exposed to substance X, no His^+ revertants are seen. If, however, rat liver supernatant is added to the cells along with substance X, revertants do occur. Is substance X a potential carcinogen for human cells? Explain.

19. The Ames test uses the reversion rate (His^- to His^+) to test compounds for mutagenicity.

a. Is it possible that a known mutagen, like proflavin, would be unable to revert a particular His^- mutant used in the Ames test? How do you think that the Ames test is designed to deal with this issue?

b. Can you think of a way to use forward mutation (His^+ to His^-) to test a compound for mutagenicity? (*Hint:* Consider using the replica plating technique in Fig. 7.6.)

c. Given that the rate of forward mutation is so much higher than the rate of reversion, why does the Ames test use the reversion rate to test for mutagenicity?

Consult the Fast Forward Box *Trinucleotide Repeat Disease: Huntington Disease and Fragile X Syndrome* in considering the following two problems.

20. The mutant *FMR-1* allele that causes fragile X syndrome is considered to be X-linked dominant with incomplete penetrance and variable expressivity. Why do most females heterozygous for one mutant and one normal allele have at least some symptoms of the disease?

21. The physicist Stephen Hawking, famous for his theories about black holes, has lived past the age of 70 with amyotrophic lateral sclerosis (ALS), a paralyzing neurodegenerative disease that is usually fatal at a much younger age. Recently, geneticists discovered that a major cause of ALS is the unusual expansion of a hexanucleotide repeat (5′-GGGGCC-3′) that lies within a gene called *C9ORF72*, at a location outside of the gene's open reading frame (ORF). A single expanded allele is sufficient to cause ALS, but the reason the disease allele is dominant remains unclear. Some experimental results support the theory that the allele makes a toxic RNA containing the expanded repeat. If this theory is correct, in what ways is the mutant ALS-causing allele similar to the mutant allele that causes Huntington disease? In what ways is it similar to the mutant allele that causes fragile X syndrome?

Section 7.3

22. Aflatoxin B_1 is a highly mutagenic and carcinogenic compound produced by certain fungi that infect crops such as peanuts. Aflatoxin is a large, bulky molecule that chemically bonds to the base guanine (G) to form the aflatoxin-guanine *adduct* that is pictured below. (In the figure, the aflatoxin is *orange,* and the guanine base is *purple.*) This adduct distorts the DNA double helix and blocks replication.

a. What type(s) of DNA repair system is (are) most likely to be involved in repairing the damage caused by exposure of DNA to aflatoxin B_1?

b. Recent evidence suggests that the adduct of guanine and aflatoxin B_1 can attack the bond that connects it to deoxyribose; this liberates the adducted base, forming an apurinic site. How does this new information change your answer to part (a)?

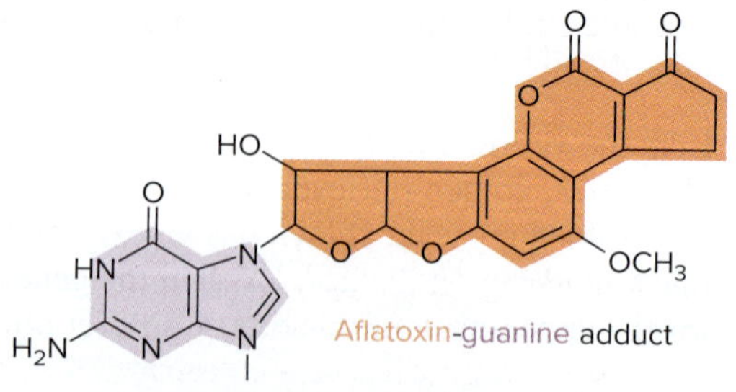

Aflatoxin-guanine adduct

23. In human DNA, 70% of cytosine residues that are followed by guanine (so-called *CpG dinucleotides,* where *p* indicates the phosphate in the phosphodiester bond between these two nucleotides) are methylated to form 5-methylcytosine. As shown in the following figure, if 5-methylcytosine should undergo spontaneous deamination, it becomes thymine.

5-methylcytosine Thymine

a. Methylated CpG dinucleotides are hotspots for point mutations in human DNA. Can you propose a hypothesis that explains why?

b. Making the simplifying assumptions that human DNA has an equal number of C–G and A–T base pairs, and that the human DNA sequence is random, how frequently in the human genome would you expect to find the base sequence CpG?

c. It turns out that, even after taking into account the actual GC content of human DNA (~42%), the frequency of CpG in human DNA is much lower than predicted by the calculation in part (b). Explain why this might be the case.

24. Bromodeoxyuridine (BrdU) is a synthetic nucleoside that can be incorporated into newly synthesized DNA in place of thymine (T). Researchers can add BrdU to medium in which cells are growing, and then they can detect its presence in chromosomes by staining with an antibody specific to BrdU.

a. The following figure at the *left* shows metaphase chromosomes isolated from somatic cells grown for some time in the presence of BrdU. Darker staining indicates more BrdU. For how many cell generations have these cells been growing in the presence of BrdU?

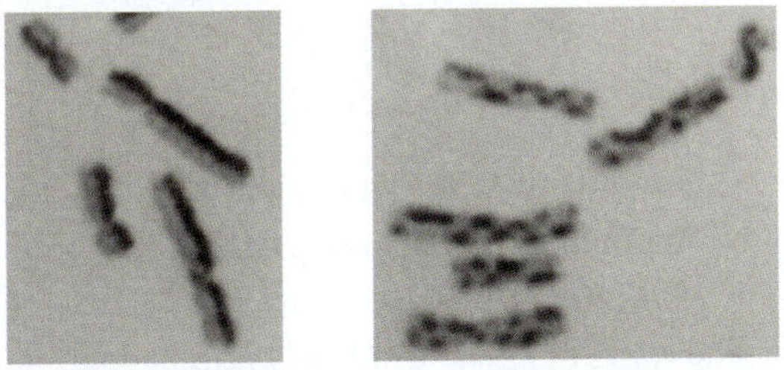

Normal chromosomes Harlequin chromosomes

© Maureen Sanz, Molloy College, Rockville Centre, NY

b. So-called *harlequin chromosomes* (similar to those at the *right* of the figure) can be isolated from cells grown in exactly the same way as those at the left, if the cells are exposed to X-rays a few hours before their chromosomes are prepared for analysis. The X-rays induce double-strand breaks in the DNA. What kind of repair process are you witnessing in these harlequin chromosomes?

c. At what point in the cell cycle did the repair process described in part (b) take place? Explain.

d. *Bloom syndrome* is an autosomal recessive disease in humans characterized by short stature, distinctive facial features, and an increased risk of cancer. The preceding figure at the *right* actually shows cells of Bloom syndrome patients grown in BrdU for the proper number of cell generations; almost all of these cells display harlequin chromosomes even though the cells were not subjected to X-rays. What do you think is the function of the protein specified by the wild-type allele of the Bloom syndrome gene (called BLM^+)?

Section 7.4

25. Albinism in animals is caused by recessive mutations in one of several autosomal genes required for synthesis of melanin, a chemical precursor for many skin and eye pigments. Albino animals are often confused with so-called *leucistic* animals that are white due to recessive mutations in a gene required in a different pathway, for example a pathway for development of the cells that produce all skin pigments. Suppose you have two white hummingbirds—a male and female—and they have mated. Assuming that all relevant mutations are rare, autosomal, and recessive to wild-type alleles, what would you expect their progeny to look like under the following conditions:

a. They are both albinos.

b. They are both leucistic.

c. One is albino and the other is leucistic.

26. a. In Figure 7.22b, what can you say about the phenotype(s) of the progeny indicated by a +? Explain.

b. What about the phenotypes of the progeny indicated by a – in the same figure? Explain.

27. Imagine that you caught a female albino mouse in your kitchen and decided to keep it for a pet. A few months later, while vacationing in Guam, you caught a male albino mouse and decided to take it home for some interesting genetic experiments. You wonder whether the two mice are both albino due to mutations in the same gene. What could you do to find out the answer to this question? Assume that both mutations are recessive.

28. Plant breeders studying genes influencing leaf shape in the plant *Arabidopsis thaliana* identified six independent recessive mutations that resulted in plants that had unusual leaves with serrated rather than smooth edges. The investigators started to perform complementation tests with these mutants, but some

of the tests could not be completed because of an accident in the greenhouse. The results of the complementation tests that could be finished are shown in the table that follows.

	1	2	3	4	5	6
1	−	+	−		+	
2		−				−
3			−	−		
4				−		
5					−	+
6						−

a. Exactly what experiment was done to fill in individual boxes in the table with a plus (+) or a minus (−) ? What does + represent? What does − represent? Why are some boxes in the table filled in *green*?

b. Assuming no complications, what do you expect for the results of the complementation tests that were not performed? That is, complete the table by placing a + or a − in each of the blank boxes.

c. How many genes are represented among this collection of mutants? Which mutations are in which genes?

29. In humans, albinism is normally inherited in an autosomal recessive fashion. Figure 3.24c in Chapter 3 shows a pedigree in which two albino parents have several children, none of whom is an albino.

a. Interpret this pedigree in terms of a complementation test.

b. It is very rare to find examples of human pedigrees such as Fig. 3.24c that in effect represent a complementation test. The reason is that most genetic conditions in humans are rare, so it is highly unlikely that unrelated people with the same condition would mate. In the absence of complementation testing, what kinds of experiments could be done to determine whether a particular human disease phenotype can be caused by mutations at more than one gene?

c. Complementation testing requires that the two mutations to be tested are both recessive to wild type. Suppose that two dominant mutations cause similar phenotypes. How could you establish whether these mutations affected the same gene or different genes?

30. a. Seymour Benzer's fine structure analysis of the *rII* region of bacteriophage T4 depended in large part on deletion analysis as shown in Fig. 7.25. But to perform such deletion analysis, Benzer had to know which *rII*$^-$ bacteriophage strains were deletions and which were point mutations. How do you think he was able to distinguish *rII*$^-$ deletions from point mutations?

b. Figure 7.25c shows Benzer's fine structure map of point mutations in the *rII* region. A key feature of this map is the existence of *hotspots,* which Benzer interpreted as nucleotide pairs that were particularly susceptible to mutation. How could Benzer say that all of the independent mutations in a hotspot were due to mutations of the same nucleotide pair?

31. a. You have a test tube containing 5 ml of a solution of bacteriophages, and you would like to estimate the number of bacteriophages in the tube. Assuming the tube actually contains a total of 15 billion bacteriophages, design a serial dilution experiment that would allow you to estimate this number. Ideally, the final plaque-containing plates you count should contain more than 10 and fewer than 1000 plaques.

b. When you count bacteriophages by the serial dilution method as in part (a), you are assuming a *plating efficiency* of 100%; that is, the number of plaques on the petri plate represents exactly the number of bacteriophages you mixed with the plating bacteria. Is there any way to test the possibility that only a certain percentage of bacteriophage particles can form plaques (so that the plating efficiency would be less than 100%)? Why is it fair to assume that any plaques are initiated by one rather than multiple bacteriophage particles?

32. You found five T4 *rII*$^-$ mutants that will not grow on *E. coli* K(λ). You mixed together all possible combinations of two mutants (as indicated in the following chart), added the mixtures to *E. coli* K(λ), and scored for the ability of the mixtures to grow and make plaques (indicated as a + in the chart).

	1	2	3	4	5
1	−	+	+	−	+
2		−	−	+	−
3			−	+	−
4				−	+
5					−

a. How many genes were identified by this analysis?

b. Which mutants belong to the same complementation groups?

33. The *rosy* (*ry*) gene of *Drosophila* encodes an enzyme called xanthine dehydrogenase. Flies homozygous for *ry* mutations exhibit a rosy eye color. Heterozygous females were made that had ry^{41} *Sb* on one homolog and *Ly* ry^{564} on the other homolog, where ry^{41} and ry^{564} are two independently isolated alleles of *ry*. *Ly* [*Lyra* (narrow) wings] and *Sb* [*Stubble* (short) bristles] are dominant mutant alleles of genes to the left

and right of *ry*, respectively. These females are now mated to males homozygous for ry^{41}. Out of 100,000 progeny, 8 have wild-type eyes, *Lyra* wings, and *Stubble* bristles, while the remainder have rosy eyes.

a. What is the order of these two *ry* mutations relative to the flanking genes *Ly* and *Sb*?

b. What is the genetic distance separating ry^{41} and ry^{564}?

34. Nine rII^- mutants of bacteriophage T4 were used in pairwise infections of *E. coli* K(λ) hosts. Six of the mutations in these phages are point mutations; the other three are deletions. The ability of the doubly infected cells to produce progeny phages in large numbers is scored in the following chart.

	1	2	3	4	5	6	7	8	9
1	–	–	+	+	–	–	–	+	+
2		–	+	+	–	–	–	+	+
3			–	–	+	–	+	–	–
4				–	+	–	+	–	–
5					–	–	–	+	+
6						–	–	–	–
7							–	+	+
8								–	–
9									–

The same nine mutants were then used in pairwise infections of *E. coli* B hosts. The production of progeny phages that can subsequently lyse *E. coli* K(λ) hosts is now scored. In the table, 0 means the progeny do not produce any plaques on *E. coli* K(λ) cells; – means that only a very few progeny phages produce plaques; and + means that many progeny produce plaques (more than 10 times as many as in the – cases).

	1	2	3	4	5	6	7	8	9
1	–	+	+	+	+	–	–	+	+
2		–	+	+	+	+	–	+	+
3			0	–	+	0	+	+	–
4				–	+	–	+	+	+
5					–	+	–	+	+
6						0	0	–	+
7							0	+	+
8								–	+
9									–

a. Which of the mutants are the three deletions? What criteria did you use to reach your conclusion?

b. If you know that mutation 9 is in the *rIIB* gene, draw the best genetic map possible to explain the data, including the positions of all point mutations and the extent of the three deletions.

c. One uncertainty should remain in your answer to part (b). How could you resolve this uncertainty?

35. In a haploid yeast strain, eight recessive mutations were found that resulted in a requirement for the amino acid lysine. All the mutations were found to revert at a frequency of about 1×10^{-6} except mutations 5 and 6, which did not revert. Matings were made between *a* and α cells carrying these mutations. The ability of the resultant diploid strains to grow on minimal medium in the absence of lysine is shown in the following chart (+ means growth and – means no growth.)

	1	2	3	4	5	6	7	8
1	–	+	+	+	+	–	+	–
2	+	–	+	+	+	+	+	+
3	+	+	–	–	–	–	–	+
4	+	+	–	–	–	–	–	+
5	+	+	–	–	–	–	–	+
6	–	+	–	–	–	–	–	–
7	+	+	–	–	–	–	–	+
8	–	+	+	+	+	–	+	–

a. How many complementation groups were revealed by these data? Which point mutations are found within which complementation groups?

The same diploid strains are now induced to undergo sporulation. The vast majority of resultant spores are auxotrophic; that is, they cannot form colonies when plated on minimal medium (without lysine). However, particular diploids can produce rare spores that do form colonies when plated on minimal medium (prototrophic spores). The following table shows whether (+) or not (–) any prototrophic spores are formed upon sporulation of the various diploid cells.

	1	2	3	4	5	6	7	8
1	–	+	+	+	+	–	+	+
2	+	–	+	+	+	+	+	+
3	+	+	–	+	–	+	+	+
4	+	+	+	–	–	–	+	+
5	+	+	–	–	–	–	+	+
6	–	+	+	–	–	–	+	+
7	+	+	+	+	+	+	–	+
8	+	+	+	+	+	+	+	–

b. When prototrophic spores occur during sporulation of the diploids just discussed, what ratio of auxotrophic to prototrophic spores would you generally expect to see in any tetrad containing such a prototrophic spore? Explain the ratio you expect.

c. Using the data from all parts of this question, draw the best map of the eight lysine auxotrophic mutations under study. Show the extent of any deletions involved, and indicate the boundaries of the various complementation groups.

36. In Problem 24, you learned that Bloom syndrome is an autosomal recessive disease characterized by the high frequency of harlequin chromosomes (as

detected after growth in BrdU). These chromosomes are caused by high levels of chromosome breakage followed by repair through homologous recombination. In some patients, every cell has many harlequin chromosomes. In other patients, the majority of cells have many harlequin chromosomes, but about 10% of the cells surprisingly have none.

a. What kinds of events produce the 10% of the cells in certain Bloom syndrome patients with no harlequin chromosomes? (*Hint:* Think about recombination.) What does the existence of these cells lacking harlequin chromosomes say about the alleles of the Bloom syndrome gene carried by these patients?

b. In what way do Bloom syndrome patients of both classes reflect the results of complementation tests?

c. Why does it make sense that the events you described in part (a) might occur in Bloom syndrome patients?

d. What is different about the events you described in part (a) from the events that give rise to harlequin chromosomes?

e. Could the events you described in part (a) occur during G_1 of the cell cycle? During G_2?

f. The events that give rise to the cells without harlequin chromosomes are very rare, occurring in less than one in a million cell divisions even in Bloom syndrome patients. Surprisingly, however, roughly 10% of the cells in certain patients lack harlequin chromosomes. How can these two statements be true simultaneously?

Section 7.5

37. The pathway for arginine biosynthesis in *Neurospora crassa* involves several enzymes that produce a series of intermediates.

ARG-E *ARG-F* *ARG-G* *ARG-H*

N-acetylornithine → ornithine → citrulline → argininosuccinate → arginine

a. If you did a cross between *ARG-E*$^-$ and *ARG-H*$^-$ *Neurospora* strains, what would be the distribution of Arg^+ and Arg^- spores within parental ditype and nonparental ditype asci? Give the spore types in the order in which they would appear in the ascus.

b. For each of the spores in your answer to part (a), what nutrients could you supply in the media to get spore growth?

38. In corn snakes, the wild-type color is brown. One autosomal recessive mutation causes the snake to be orange, and another causes the snake to be black. An orange snake was crossed to a black one, and the F_1 offspring were all brown. Assume that all relevant genes are unlinked.

a. Indicate what phenotypes and ratios you would expect in the F_2 generation of this cross if there is one pigment pathway, with orange and black being different intermediates on the way to brown.

b. Indicate what phenotypes and ratios you would expect in the F_2 generation if orange pigment is a product of one pathway, black pigment is the product of another pathway, and brown is the effect of mixing the two pigments in the skin of the snake.

39. In a certain species of flowering plants with a diploid genome, four enzymes are involved in the generation of flower color. The genes encoding these four enzymes are on different chromosomes. The biochemical pathway involved is as follows; the figure shows that either of two different enzymes is sufficient to convert a blue pigment into a purple pigment.

$$\text{white} \rightarrow \text{green} \rightarrow \text{blue} \overset{\rightarrow}{\underset{\rightarrow}{}} \text{purple}$$

A true-breeding green-flowered plant is mated with a true-breeding blue-flowered plant. All of the plants in the resultant F_1 generation have purple flowers. F_1 plants are allowed to self-fertilize, yielding an F_2 generation. Show genotypes for P, F_1, and F_2 plants, and indicate which genes specify which biochemical steps. Determine the fraction of F_2 plants with the following phenotypes: white flowers, green flowers, blue flowers, and purple flowers. Assume the green-flowered parent is mutant in only a single step of the pathway.

40. The intermediates A, B, C, D, E, and F all occur in the same biochemical pathway. G is the product of the pathway, and mutants 1 through 7 are all G^-, meaning that they cannot produce substance G. The following table shows which intermediates will promote growth in each of the mutants. Arrange the intermediates in order of their occurrence in the pathway, and indicate the step in the pathway at which each mutant strain is blocked. A + in the table indicates that the strain will grow if given that substance, an O means lack of growth.

	Supplements						
Mutant	**A**	**B**	**C**	**D**	**E**	**F**	**G**
1	+	+	+	+	+	O	+
2	O	O	O	O	O	O	+
3	O	+	+	O	+	O	+
4	O	+	O	O	+	O	+
5	+	+	+	O	+	O	+
6	+	+	+	+	+	+	+
7	O	O	O	O	+	O	+

41. In each of the following cross schemes, two true-breeding plant strains are crossed to make F_1 plants, all of which have purple flowers. The F_1 plants are

then self-fertilized to produce F_2 progeny as shown here.

Cross	Parents	F_1	F_2
1	blue × white	all purple	9 purple: 4 white: 3 blue
2	white × white	all purple	9 purple: 7 white
3	red × blue	all purple	9 purple: 3 red: 3 blue: 1 white
4	purple × purple	all purple	15 purple: 1 white

a. For each cross, explain the inheritance of flower color.

b. For each cross, show a possible biochemical pathway that could explain the data.

c. Which of these crosses is compatible with an underlying biochemical pathway involving only a single step that is catalyzed by an enzyme with two dissimilar subunits, both of which are required for enzyme activity?

d. For each of the four crosses, what would you expect in the F_1 and F_2 generations if all relevant genes were tightly linked?

42. The pathways for the biosynthesis of the amino acids glutamine (Gln) and proline (Pro) involve one or more common intermediates. Auxotrophic yeast mutants numbered 1–7 are isolated that require either glutamine or proline or both amino acids for their growth, as shown in the following table (+ means growth; – no growth). These mutants are also tested for their ability to grow on the intermediates A–E. What is the order of these intermediates in the glutamine and proline pathways, and at which point in the pathways is each mutant blocked?

Mutant	A	B	C	D	E	Gln	Pro	Gln + Pro
1	+	–	–	–	+	–	+	+
2	–	–	–	–	–	–	+	+
3	–	–	+	–	–	–	–	+
4	–	–	–	–	–	+	–	+
5	–	–	+	+	–	–	–	+
6	+	–	–	–	–	–	+	+
7	–	+	–	–	–	+	–	+

43. The following complementing *E. coli* mutants were tested for growth on four known precursors of thymine, A–D.

	Precursor/product				
Mutant	A	B	C	D	Thymine
9	+	–	+	–	+
10	–	–	+	–	+
14	+	+	+	–	+
18	+	+	+	+	+
21	–	–	–	–	+

a. Show a simple linear biosynthetic pathway of the four precursors and the end product, thymine. Indicate which step is blocked by each of the five mutations.

b. What precursor would accumulate in the following double mutants: 9 and 10? 10 and 14?

44. In 1952, an article in the *British Medical Journal* reported interesting differences in the behavior of blood plasma obtained from several people who suffered from X-linked recessive hemophilia. When mixed together, the cell-free blood plasma from certain combinations of individuals could form clots in the test tube. For example, the following table shows whether clots could form (+) or not (–) in various combinations of plasma from four people with hemophilia:

1 and 1	–	2 and 3	+
1 and 2	–	2 and 4	+
1 and 3	+	3 and 3	–
1 and 4	+	3 and 4	–
2 and 2	–	4 and 4	–

What do these data tell you about the inheritance of hemophilia in these individuals? Do these data allow you to exclude any models for the biochemical pathway governing blood clotting?

45. Mutations in an autosomal gene in humans cause a form of hemophilia called von Willebrand disease (vWD). This gene specifies a blood plasma protein cleverly called von Willebrand factor (vWF). vWF stabilizes factor VIII, a blood plasma protein specified by the wild-type hemophilia A gene. Factor VIII is needed to form blood clots. Thus, factor VIII is rapidly destroyed in the absence of vWF.

Which of the following might successfully be employed in the treatment of bleeding episodes in hemophiliac patients? Would the treatments work immediately or only after some delay needed for protein synthesis? Would the treatments have only a short-term or a prolonged effect? Assume that all mutations are null (that is, the mutations result in the complete absence of the protein encoded by the gene) and that the plasma is cell-free.

a. transfusion of plasma from normal blood into a vWD patient

b. transfusion of plasma from a vWD patient into a different vWD patient

c. transfusion of plasma from a hemophilia A patient into a vWD patient

d. transfusion of plasma from normal blood into a hemophilia A patient

e. transfusion of plasma from a vWD patient into a hemophilia A patient

f. transfusion of plasma from a hemophilia A patient into a different hemophilia A patient

g. injection of purified vWF into a vWD patient

h. injection of purified vWF into a hemophilia A patient

i. injection of purified factor VIII into a vWD patient

j. injection of purified factor VIII into a hemophilia A patient

46. Antibodies were made that recognize six proteins that are part of a complex inside the *Caenorhabditis elegans* one-cell embryo. The mother produces proteins that are believed to assemble stepwise into a structure in the egg, beginning at the embryo's inner surface. The antibodies were used to detect the protein location in embryos produced by mutant mothers [who are homozygous recessive for the gene(s) encoding each protein]. The *C. elegans* mothers are self-fertilizing hermaphrodites, so no wild-type copy of a gene will be introduced during fertilization.

In the following table, * means the protein was present and at the embryo surface, – means that the protein was not present, and + means that the protein was present but not at the embryo surface. Assume all mutations prevent production of the corresponding protein.

Mutant in gene for protein	Protein production and location					
	A	B	C	D	E	F
A	–	+	*	+	*	+
B	*	–	*	*	*	*
C	*	+	–	+	*	+
D	*	+	*	–	*	+
E	+	+	+	+	–	+
F	*	+	*	*	*	–

Complete the following figure, which shows the construction of the hypothetical protein complex, by writing the letter of the proper protein in each circle. The two proteins marked with arrowheads can assemble into the complex independently of each other, but both are needed for the addition of subsequent proteins to the complex.

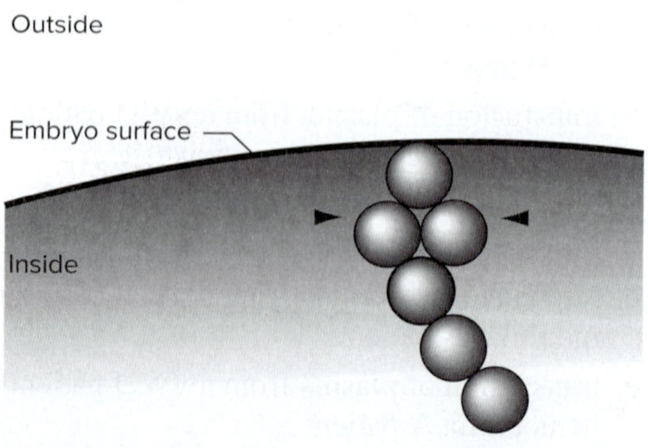

47. Adult hemoglobin is a multimeric protein with four polypeptides, two of which are α-globin and two of which are β-globin.

a. How many genes are needed to define the structure of the hemoglobin protein?

b. If a person is heterozygous for wild-type alleles and alleles that would yield amino acid substitution variants for both α-globin and β-globin, how many different kinds of hemoglobin protein would be found in the person's red blood cells and in what proportion? Assume all alleles are expressed at the same level.

48. Each complementation group (*ARG-E, ARG-F, ARG-G*, and *ARG-H*) in Fig. 7.27b can grow on a unique subset of supplements. Why were these four subsets the only ones observed? For example, why were no complementation groups observed that behaved like the four hypothetical ones shown in the following table? (Symbols as in Fig. 7.27b: + means growth, – means no growth.)

Hypothetical mutant strain	Supplements				
	Nothing	Ornithine	Citrulline	Arginino-succinate	Arginine
Wildtype: Arg^+	+	+	+	+	+
ARG-I⁻	–	+	–	+	+
ARG-J⁻	–	–	+	–	+
ARG-K⁻	–	+	–	–	+
ARG-L⁻	–	+	+	–	+

Section 7.6

49. In addition to the predominant adult hemoglobin, HbA, which contains two α-globin chains and two β-globin chains ($\alpha_2\beta_2$), there is a minor hemoglobin, HbA_2, composed of two α and two δ chains ($\alpha_2\delta_2$). The β- and δ-globin genes are arranged in tandem and are highly homologous. Draw the chromosomes that would result from an event of unequal crossing-over between the β and δ genes.

50. Most mammals, including New World primates such as marmosets (a kind of monkey), are *dichromats:* They have only two kinds of rhodopsin-related color receptors. Old World primates such as humans and gorillas are *trichromats* with three kinds of color receptors. Primates diverged from other mammals roughly 65 million years ago (Myr), while Old World and New World primates diverged from each other roughly 35 Myr.

a. Using this information, define on Fig. 7.32d the time spans of any events that can be dated.

b. Some New World monkeys have an autosomal color receptor gene and a single X-linked color receptor gene. The X-linked gene has three alleles, each of which specifies a photoreceptor that responds to light of a different wavelength (all three wavelengths are different from that recognized by the autosomal color receptor). How is color vision inherited in these monkeys?

c. About 95% of all light-receiving neurons in humans and other mammals are rod cells containing rhodopsin, a pigment that responds to low-level light of many wavelengths. The remaining 5% of light-receiving neurons are cone cells with pigments that respond to light of specific wavelengths of high intensity. What do these facts suggest about the lifestyle of the earliest mammals?

51. Humans are normally *trichromats;* we have three different types of retinal cones, each containing either a red, green, or blue rhodopsin-like photoreceptor protein. The reason is that most humans have genes for red and green photoreceptors on the X chromosome, and a blue photoreceptor gene on an autosome. Our brain integrates the information from each type of cone, making it possible for us to see about one million colors.

Some scientists think that rare people may be *tetrachromats,* that is, they have four different kinds of cones. Such people, if they exist, could potentially detect 100 million colors! For parts (a) and (b), assume that each X chromosome has one red and one green photoreceptor protein gene. For all parts, assume that mutant alleles can produce photoreceptors with altered spectral sensitivities.

a. Explain why scientists expect that many more females than males would be tetrachromats.

b. In X-linked, red/green color blindness, mutation of either the red or green photoreceptor gene results in a rhodopsin-like protein with altered spectral sensitivity. The mutant photoreceptor is sensitive to wavelengths in between the normal red and green photoreceptors. Why do scientists think that a woman with a son who is red/green color-blind is more likely to be a tetrachromat than a woman whose sons all have normal vision?

c. Suggest a scenario based on Fig. 7.33d that could explain how extremely rare males might be tetrachromats.

chapter 8

Gene Expression: The Flow of Information from DNA to RNA to Protein

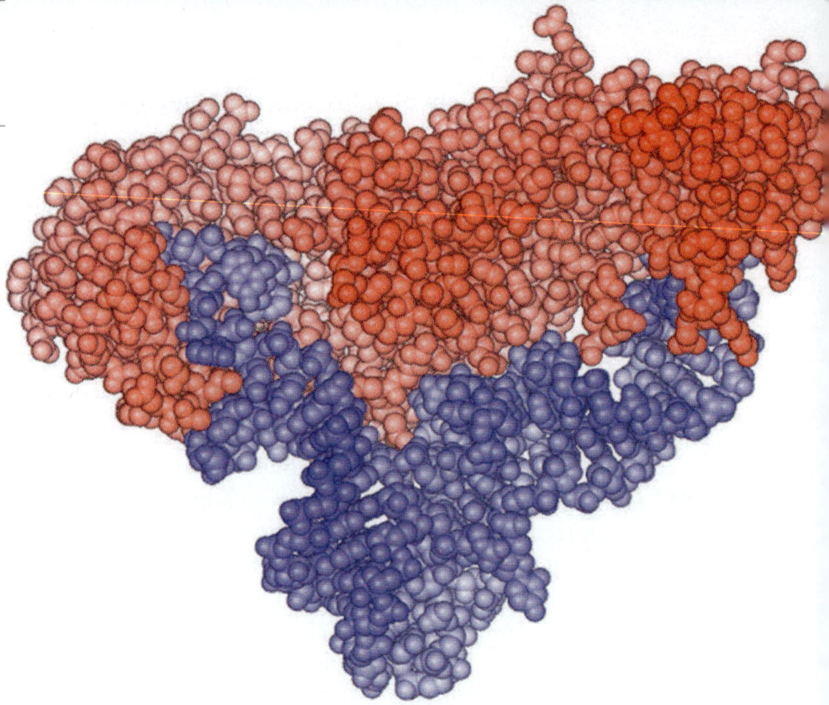

The ability of an aminoacyl-tRNA synthetase (red) *to recognize a particular tRNA* (blue) *and couple it to its corresponding amino acid* (not shown) *is central to the molecular machinery that converts the language of nucleic acids into the language of proteins.*

chapter outline

A DEDICATED EFFORT to determine the complete nucleotide sequence of the genome in a variety of organisms has been underway since 1990. This massive endeavor has been more successful than many scientists thought possible. By the time of this writing in 2016, the DNA sequence in the genomes of more than 8100 different species had already been deposited in databases, and sequencing projects for more than 35,000 additional species were in progress. With this sequence information in hand, geneticists can consult the genetic code—the cipher equating nucleotide sequence with amino acid sequence—to decide what parts of a genome are likely to be genes. As a result, modern geneticists can discover the number and amino acid sequences of all the polypeptides that determine phenotype. Knowledge of DNA sequence thus opens up powerful new possibilities for understanding an organism's growth and development at the molecular level.

In this chapter, we describe the cellular mechanisms that carry out **gene expression,** the means by which genetic information can be interpreted as phenotype. As intricate as some of the details may appear, the general scheme of gene expression is elegant and straightforward: *Within each cell, genetic information flows from DNA to RNA to protein.* This statement was set forward as the *Central Dogma* of molecular biology by Francis Crick in 1957. As Crick explained, "Once information has passed into protein, it cannot get out again."

The Central Dogma maintains that genetic information flows in two distinct stages (**Fig. 8.1**). The conversion of the information in DNA to its equivalent in RNA is known as **transcription.** The product of transcription is a **transcript:** a molecule of **messenger RNA (mRNA)** in prokaryotes, a molecule of RNA that undergoes processing to become an mRNA in eukaryotes.

In the second stage of gene expression, the cellular machinery decodes the sequence of nucleotides in mRNA into a sequence of amino acids—a **polypeptide**—by the process known as **translation.** It takes place on molecular workbenches called **ribosomes,** which are composed of proteins and **ribosomal RNAs (rRNAs).** Translation depends on the dictionary known as the **genetic code,** which defines each amino acid in terms of specific sequences of three nucleotides. Translation also requires **transfer RNAs (tRNAs),** small RNA adapter molecules that place specific amino acids at the correct position in a growing polypeptide chain.

The Central Dogma does not explain the behavior of all genes. As Crick himself realized, many genes are transcribed into RNAs that are never translated into proteins. You will see in this chapter that many nontranslated RNAs are critical to various steps of gene expression. The genes encoding rRNAs and tRNAs belong to this group.

Four general themes emerge from our discussion of gene expression. First, the pairing of complementary bases is key to the transfer of information from DNA to RNA, and from RNA to protein. Second, the polarities (directionality) of DNA, RNA, and polypeptides help guide the mechanisms of gene expression. Third, like DNA replication and recombination, gene expression requires an input of energy and the participation of specific proteins, RNAs, and macromolecular assemblies, such as ribosomes. Finally, mutations that change genetic information or obstruct the flow of its expression can have dramatic effects on phenotype.

Figure 8.1 Gene expression: The flow of genetic information from DNA via RNA to protein. In transcription, the enzyme RNA polymerase copies DNA to produce an RNA transcript. In translation, the cellular machinery uses instructions in mRNA to synthesize a polypeptide, following the rules of the genetic code.

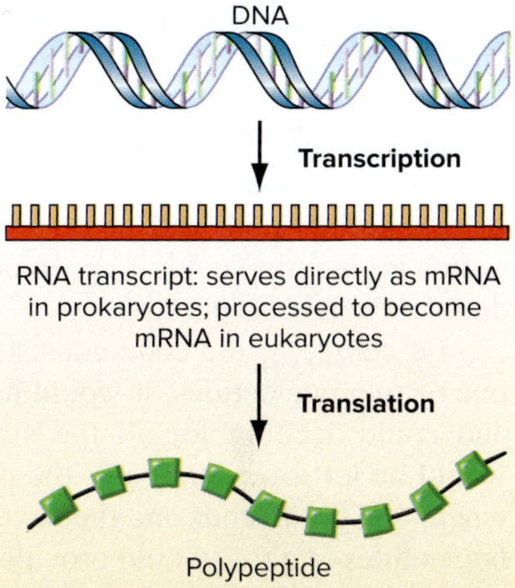

8.1 The Genetic Code

learning objectives

1. Explain the reasoning that established a sequence of three nucleotides (a triplet codon) as the basic unit of the code relating DNA to protein.
2. Summarize the evidence showing that the sequence of nucleotides in a gene is colinear with the sequence of amino acids in a protein.
3. Define *reading frame* and discuss its significance to the genetic code.
4. Describe experiments that determined which codons are associated with each amino acid and which are stop codons.
5. Explain how mutations were used to verify the genetic code.
6. Discuss evidence that the genetic code is almost universal, and cite some exceptions.

A code is a system of symbols that equates information in one language with information in another. A useful analogy for the genetic code is the Morse code, which uses dots and dashes to transmit messages over radio or telegraph wires. Various groupings of the dot-dash symbols represent the 26 letters of the English alphabet. And because many more letters exist than the two symbols (dot or dash), groups of one, two, three, or four dots or dashes in various combinations represent individual letters. For example, the symbol for C is dash dot dash dot (– · – ·), the symbol for O is dash dash dash (– – –), D is dash dot dot (– · ·), and E is a single dot (·). Because anywhere from one to four symbols specify each letter, the Morse code requires a symbol for *pause* (in practice, a short interval of time) to signify where one letter ends and the next begins.

Triplet Codons of Nucleotides Represent Individual Amino Acids

The language of nucleic acids is written in four nucleotides—A, G, C, and T in the DNA dialect; A, G, C, and U in the RNA dialect—while the language of proteins is written in amino acids. The first hurdle to be overcome in deciphering how sequences of nucleotides can determine the order of amino acids in a polypeptide is to determine how many amino acid "letters" exist.

Over lunch one day at a local pub, Watson and Crick produced the now accepted list of the 20 common amino acids that are encoded directly by DNA. They created the list by analyzing the known amino acid sequences of a variety of naturally occurring polypeptides. Amino acids present in only a small number of proteins or in only certain tissues or organisms did not qualify as standard building blocks; Crick and Watson correctly assumed that most such amino acids arise when proteins undergo chemical modification after their synthesis. By contrast, amino acids present in most proteins made the list. The question then became: How can four nucleotides encode 20 amino acids?

Like the Morse code, the four nucleotides encode 20 amino acids through specific groupings of A, G, C, and T (in DNA) or A, G, C, and U (in RNA). Researchers initially arrived at the number of letters per grouping by deductive reasoning and later confirmed their guess by experiment. They reasoned that if only one nucleotide represented an amino acid, information would exist for only four amino acids: A would encode one amino acid, G a second amino acid, and so on. If two nucleotides represented each amino acid, $4^2 = 16$ possible combinations of doublets would be possible.

Of course, if the code consisted of groups containing one *or* two nucleotides, it would have $4 + 16 = 20$ groups and could account for all the amino acids, but nothing would be left over to signify the pause required to denote where one group ends and the next begins. Groups of three nucleotides in a row would provide $4^3 = 64$ different triplet combinations, more than enough to code for all the amino acids. If the code consisted of doublets and triplets, a signal denoting a pause would once again be necessary. But a triplets-only code would require no symbol for *pause* if the mechanism for counting to three and distinguishing among successive triplets was very reliable.

Although this kind of reasoning generates a hypothesis, it does not prove it. As it turned out, however, the experiments described later in this chapter did indeed demonstrate that groups of three nucleotides represent all 20 amino acids. Each nucleotide triplet is called a **codon.** Each codon, designated by the bases defining its three nucleotides, specifies one amino acid. For example, GAA is a codon for glutamic acid (Glu), and GUU is a codon for valine (Val). Because the code comes into play only during the translation part of gene expression, that is, during the decoding of messenger RNA to polypeptide, geneticists usually present the code in the RNA dialect of A, G, C, and U, as depicted in **Fig. 8.2.** When speaking of genes, they can substitute T for U to show the same code in the DNA dialect.

If you knew the sequence of nucleotides in a gene or its transcript as well as the sequence of amino acids in the corresponding polypeptide, you could then deduce the genetic code without understanding how the underlying cellular machinery actually works. Although techniques for determining both nucleotide and amino acid sequence are available today, this was not true when researchers were trying to crack the genetic code in the 1950s and 1960s. At that time, they could establish a polypeptide's amino acid sequence, but not the nucleotide sequence of DNA or RNA. Because of their inability to read nucleotide sequence, scientists used an assortment of genetic and biochemical techniques

Figure 8.2 The genetic code: 61 codons represent the 20 amino acids, while 3 codons signify stop. To read the code, find the first letter in the *left column,* the second letter along the *top,* and the third letter in the *right column;* this reading corresponds to the 5′-to-3′ direction along the mRNA.

First letter	Second letter: U	Second letter: C	Second letter: A	Second letter: G	Third letter
U	UUU, UUC: Phe UUA, UUG: Leu	UCU, UCC, UCA, UCG: Ser	UAU, UAC: Tyr UAA Stop UAG Stop	UGU, UGC: Cys UGA Stop UGG Trp	U C A G
C	CUU, CUC, CUA, CUG: Leu	CCU, CCC, CCA, CCG: Pro	CAU, CAC: His CAA, CAG: Gln	CGU, CGC, CGA, CGG: Arg	U C A G
A	AUU, AUC, AUA: Ile AUG Met	ACU, ACC, ACA, ACG: Thr	AAU, AAC: Asn AAA, AAG: Lys	AGU, AGC: Ser AGA, AGG: Arg	U C A G
G	GUU, GUC, GUA, GUG: Val	GCU, GCC, GCA, GCG: Ala	GAU, GAC: Asp GAA, GAG: Glu	GGU, GGC, GGA, GGG: Gly	U C A G

to fathom the code. They began by examining how different mutations in a single gene affected the amino acid sequence of the gene's polypeptide product. In this way, they were able to use the abnormal (specific mutations) to understand the normal (the general relationship between genes and polypeptides).

A Gene's Nucleotide Sequence Is Colinear with the Amino Acid Sequence of the Encoded Polypeptide

As you know, DNA is a linear molecule with base pairs following one another down the intertwined chains. Proteins, by contrast, have complicated three-dimensional structures. Even so, if unfolded and stretched out from N terminus to C terminus, proteins have a one-dimensional, linear structure—a specific primary sequence of amino acids. If the information in a gene and its corresponding protein are colinear, the consecutive order of bases in the DNA from the beginning to the end of the gene would stipulate the consecutive order of amino acids from one end to the other of the outstretched protein.

In the 1960s, Charles Yanofsky was the first to compare maps of mutations within a gene to the particular amino acid substitutions that resulted. He began by generating a large number of Trp^- auxotrophic mutants in *E. coli* that carried mutations in the *trpA* gene for a subunit of the enzyme tryptophan synthase. He next made a fine structure recombination map of these mutations analogous to Benzer's fine structure map for the *rII* region of bacteriophage T4, which was discussed in Chapter 7. Yanofsky then purified and determined the amino acid sequences of the mutant tryptophan synthase subunits.

As **Fig. 8.3a** illustrates, Yanofsky's data showed that the order of mutations mapped within the DNA of the gene by recombination was indeed colinear with the positions of the amino acid substitutions occurring in the resulting mutant proteins. By carefully analyzing his results, Yanofsky deduced two other key features of the relationship between nucleotides and amino acids.

Evidence that a codon is composed of more than one nucleotide

Yanofsky observed that point mutations altering different nucleotide pairs may affect the same amino acid. In one example shown in Fig. 8.3a, mutation 23 changed the glycine (Gly) at position 211 of the wild-type polypeptide chain to arginine (Arg), while mutation 46 yielded glutamic acid (Glu) at the same position. In another example, mutation 78 changed the glycine at position 234 to cysteine (Cys), while mutation 58 produced aspartic acid (Asp) at the same position. These are all **missense mutations** that change a codon for one amino acid into a codon that specifies a different amino acid.

In both cases, Yanofsky found that recombination could occur between the two mutations that changed the identity of the same amino acid; such recombination would produce a wild-type tryptophan synthase gene (**Fig. 8.3b**). Because the smallest unit of recombination is the base pair, two mutations capable of recombination—in this case, in the same codon because they affect the same amino acid—must be in different (although nearby) nucleotides. Thus, a codon must contain more than one nucleotide.

Evidence that each nucleotide is part of only one codon

As Fig. 8.3a illustrates, each of the point mutations in the tryptophan synthase gene characterized by Yanofsky alters the identity of only a single amino acid. This is also true of the point mutations examined in many other genes, such as the human genes for rhodopsin and hemoglobin

Figure 8.3 Mutations in a gene are colinear with the sequence of amino acids in the encoded polypeptide. (a) The relationship between the genetic map of *E. coli's trpA* gene and the positions of amino acid substitutions in mutant tryptophan synthase proteins. **(b)** Codons must include two or more base pairs. When two mutant strains with different amino acids at the same position were crossed, recombination could produce a wild-type allele.

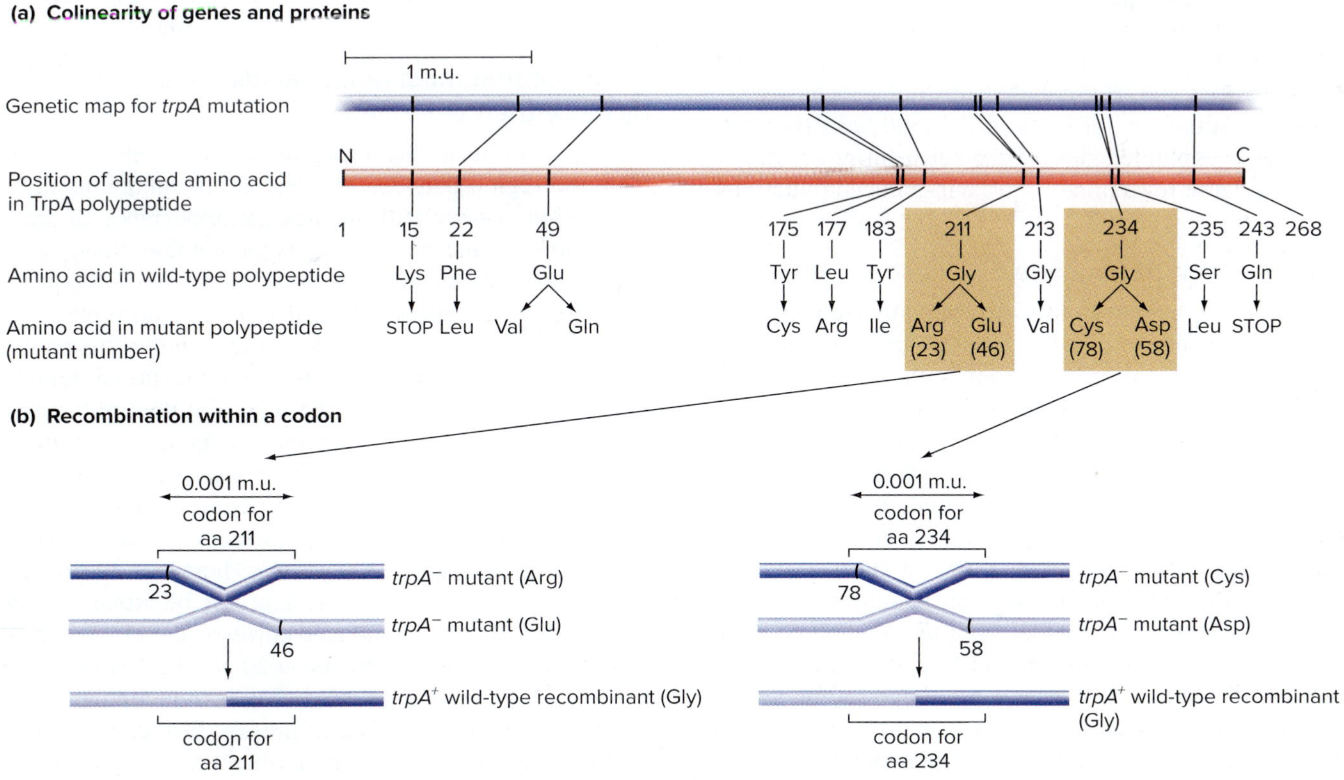

(see Chapter 7). Because point mutations that change only a single nucleotide pair affect only a single amino acid in a polypeptide, each nucleotide in a gene must influence the identity of only a single amino acid. In contrast, if a nucleotide were part of more than one codon, a mutation in that nucleotide would affect more than one amino acid.

Nonoverlapping Triplet Codons Are Set in a Reading Frame

Although the most efficient code to specify 20 amino acids requires three nucleotides per codon, more complicated scenarios are possible. But in 1955, Francis Crick and Sydney Brenner obtained convincing evidence for the triplet nature of the genetic code in studies of mutations in the bacteriophage T4 *rIIB* gene originally characterized by Seymour Benzer (Chapter 7). They induced the mutations with *proflavin,* an intercalating mutagen that can insert itself between the paired bases stacked in the center of the DNA molecule (recall Fig. 7.14c). Crick and Brenner's original assumption was that proflavin would act like other mutagens, causing single-base substitutions. If this were true, it would be possible to generate revertants through treatment with other mutagens that might restore the wild-type DNA sequence.

Surprisingly, genes with proflavin-induced mutations did not revert to wild-type upon treatment with other mutagens known to cause nucleotide substitutions. Only further exposure to proflavin caused proflavin-induced mutations to revert to wild-type. Crick and Brenner had to explain this observation before they could proceed with their phage experiments. With keen insight, they correctly guessed that proflavin does not cause base substitutions; instead, it causes insertions or deletions of a single base pair. This hypothesis explained why base-substituting mutagens could not cause reversion of proflavin-induced mutations.

Evidence for a triplet code

Crick and Brenner began their experiments with a particular proflavin-induced $rIIB^-$ mutation they called FC0. They next treated this mutant strain with more proflavin to isolate an $rIIB^+$ revertant (**Fig. 8.4a**). By recombining this revertant with wild-type bacteriophage T4, Crick and Brenner were able to show that the revertant's chromosome actually contained two different $rIIB^-$ mutations (**Fig. 8.4b**). One was the original FC0 mutation; the other was the newly induced FC7. Either mutation by itself yields a mutant phenotype, but their simultaneous occurrence in the same gene yielded an $rIIB^+$ phenotype. Crick and Brenner reasoned that if the first mutation was the addition of a single base pair, represented by the symbol (+), then the counteracting mutation must be the deletion of a base pair, represented as (−). The restoration of gene function by one mutation canceling another in the same gene is known as **intragenic suppression**.

Crick and Brenner supposed not only that each codon is a trio of nucleotides, but that each gene has a single starting point. This starting point establishes a **reading frame:** the sequential partitioning of nucleotides into groups of three to generate the correct order of amino acids in the resulting polypeptide chain (Fig. 8.4a). Changes that alter the grouping of nucleotides into codons are called **frameshift mutations;** they shift the reading frame for all codons beyond the point of insertion or deletion, almost always abolishing the function of the polypeptide product.

If codons are read in order from a fixed starting point, a deletion (−) can counterbalance an insertion (+) to restore the reading frame (Fig. 8.4a). Note that the gene would regain its wild-type activity only if the portion of the polypeptide encoded between the two mutations of opposite sign is not required for protein function, because in the double mutant, this region would have an improper amino acid sequence. Also, the incorrect amino acids must not prevent the protein from folding into a functional conformation.

Crick and Brenner realized that they could use + and − mutations in *rIIB* to test the hypothesis that codons were indeed nucleotide triplets. If codons are composed of three nucleotides, then combining two different $rIIB^-$ mutations of the same sign (+ + or − −) in the same gene should never lead to intragenic suppression (an $rIIB^+$ phenotype). Combinations of three + or three − mutations, however, should sometimes result in an $rIIB^+$ revertant. These predictions were exactly verified by the results (**Fig. 8.4c**).

Evidence that most amino acids are specified by more than one codon

As Fig. 8.4c illustrates, intragenic suppression occurs only if, in the region between two frameshift mutations of opposite sign, a gene still dictates the appearance of amino acids—even if these amino acids are not the same as those appearing in the normal protein. If the frameshifted part of the gene instead encodes instructions to stop protein synthesis by introducing a triplet that does not correspond to any amino acid, then production of a functional polypeptide will not be possible. The reason is that polypeptide synthesis would stop before the compensating mutation could re-establish the correct reading frame.

The fact that intragenic suppression occurs as often as it does suggests that the code includes more than one codon for some amino acids. Recall that there are 20 common amino acids but $4^3 = 64$ different combinations of three nucleotides. If each amino acid corresponded to only a single codon, there would be $64 - 20 = 44$ possible triplets not encoding an amino acid. These noncoding triplets would act as *stop* signals and prevent further polypeptide synthesis. In this scenario, more than half of all frameshift

Figure 8.4 Studies of frameshift mutations in the bacteriophage T4 *rIIB* gene showed that codons consist of three nucleotides. **(a)** Treatment with proflavin produces an *rIIB*⁻ frameshift mutation at one site (FC0) by insertion of a single nucleotide; the reading frame of all codons downstream of the insertion is shifted (*yellow*). A second proflavin exposure results in a second mutation (FC7), deletion of a single nucleotide within the same gene, which suppresses FC0 by restoring the proper reading frame (*green*). **(b)** When the revertant is crossed with a wild-type strain, crossing-over separates the two *rIIB*⁻ frameshift mutations (FC0 and FC7) onto separate DNA molecules. The reversion to an *rIIB*⁺ phenotype was thus the result of intragenic suppression. **(c)** When recombined onto a single DNA molecule, two addition (++) or two deletion (−−) mutations do not supply *rIIB*⁺ function, but three mutations of the same sign (+++ or −−−) restore the reading frame.

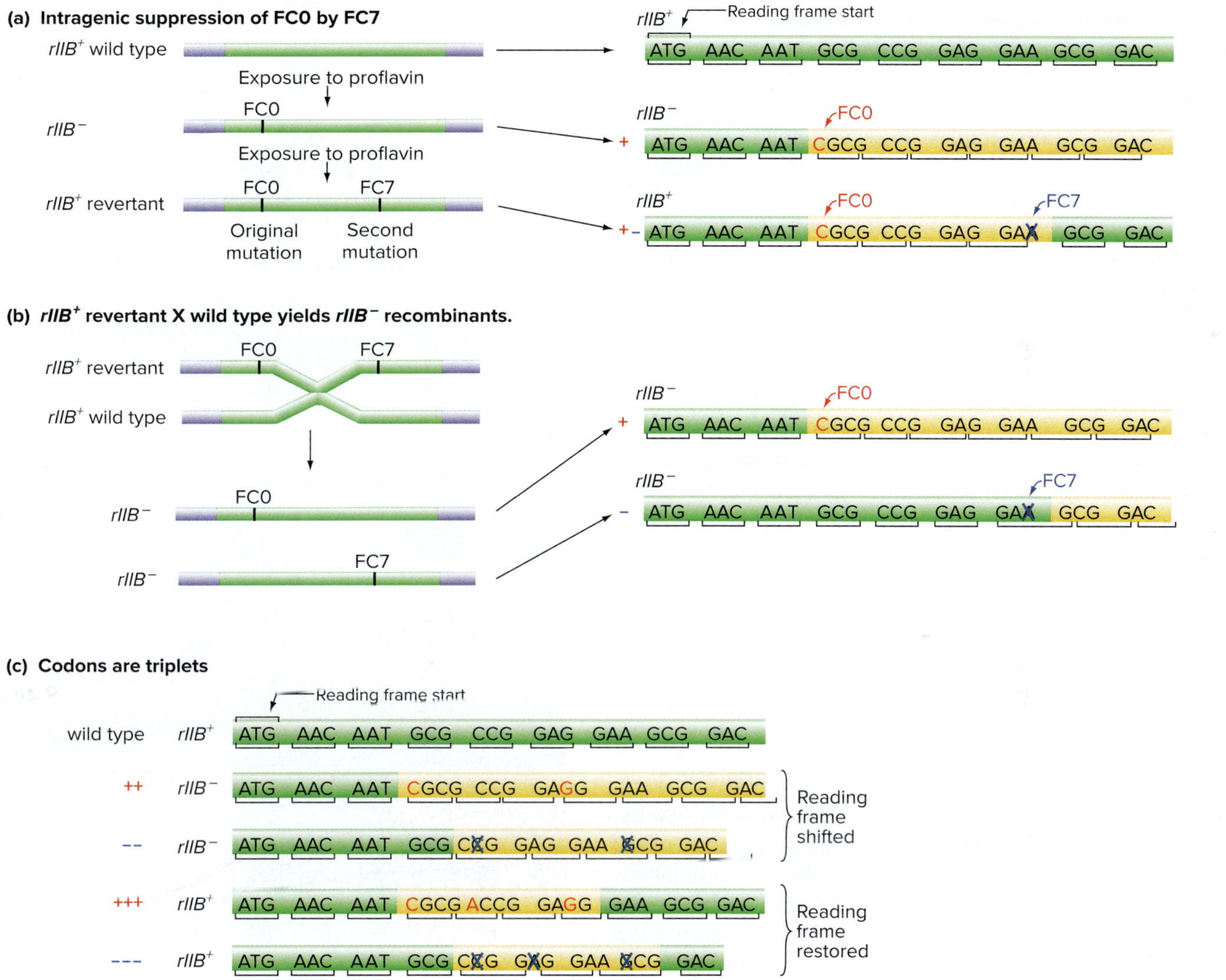

mutations (44/64) would cause protein synthesis to stop at the first codon after the mutation, and the chances of extending the protein would diminish exponentially with each additional amino acid. As a result, intragenic suppression would rarely occur. However, we have seen that many frameshift mutations of one sign can be offset by mutations of the other sign. The distances between these mutations, estimated by recombination frequencies, are in some cases large enough to code for more than 50 amino acids, which would be possible only if most of the 64 possible triplet codons specified amino acids. Thus, the data of Crick and Brenner provide strong support for the idea that the genetic code is **degenerate:** In most cases, two or more nucleotide triplets specify a single one of the 20 amino acids (see the genetic code in Fig. 8.2).

Cracking the Code: Which Codons Represent Which Amino Acids?

Although the genetic experiments just described allowed remarkably prescient insights about the nature of the genetic code, they did not establish a correspondence between specific codons and specific amino acids. The discovery of messenger RNA and the development of techniques for synthesizing simple messenger RNA molecules had to occur first so that researchers could manufacture simple proteins in the test tube.

The discovery of messenger RNAs

In the 1950s, researchers exposed eukaryotic cells to amino acids tagged with radioactivity and observed that protein synthesis incorporating the radioactive amino acids into polypeptides takes place in the cytoplasm, even though the genes for those polypeptides are sequestered in the cell nucleus. From this discovery, they deduced the existence of an intermediate molecule, made in the nucleus and capable of transporting DNA sequence information to the cytoplasm, where it can direct protein synthesis. RNA was a prime candidate for this intermediary information-carrying molecule.

Because of RNA's potential for base pairing with a strand of DNA, one could imagine the cellular machinery copying a strand of DNA into a complementary strand of RNA in a manner analogous to the DNA-to-DNA copying of DNA replication. Subsequent studies in eukaryotes using radioactive uracil, a base found only in RNA, showed that although the molecules are synthesized in the nucleus, at least some of them migrate to the cytoplasm. Among those RNA molecules that migrate to the cytoplasm are the messenger RNAs, or mRNAs, depicted in Fig. 8.1. They arise in the nucleus from the transcription of DNA sequence information and then move (after processing) to the cytoplasm, where they determine the order of amino acids during protein synthesis.

Using synthetic mRNAs and *in vitro* translation

Knowledge of mRNA served as the framework for two experimental breakthroughs that led to the deciphering of the genetic code. In the first, biochemists obtained cellular extracts that, with the addition of mRNA, synthesized polypeptides in a test tube. They called these extracts *in vitro translational systems*. The second breakthrough was the development of techniques enabling the synthesis of artificial mRNAs containing only a few codons of known composition. When added to *in vitro* translational systems, these simple, synthetic mRNAs directed the formation of simple polypeptides.

In 1961, Marshall Nirenberg and Heinrich Matthaei added a synthetic poly-U (5′. . . UUUUUUUUUUUU . . . 3′) mRNA to a cell-free translational system derived from *E. coli*. With the poly-U mRNA, phenylalanine (Phe) was the only amino acid incorporated into the resulting polypeptides (**Fig. 8.5a**). Because UUU is the only possible triplet in poly-U, UUU must be a codon for phenylalanine. In a similar fashion, Nirenberg and Matthaei showed that CCC encodes proline (Pro), AAA is a codon for lysine (Lys), and GGG encodes glycine (Gly) (**Fig. 8.5b**).

Figure 8.5 How geneticists used synthetic mRNAs to limit the coding possibilities. (a) Poly-U mRNA generates a poly-phenylalanine polypeptide. **(b)** Polydi-, polytri-, and polytetranucleotides encode simple polypeptides. Some synthetic mRNAs, such as poly-GUAA, contain stop codons in all three reading frames and thus specify the construction only of short peptides.

(a) Poly-U mRNA encodes polyphenylalanine.

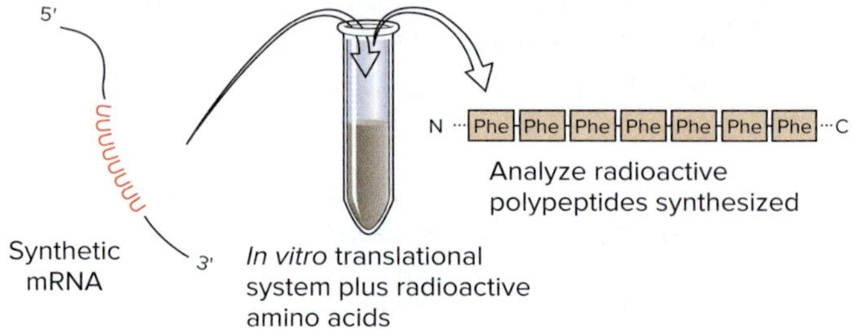

(b) Analyzing the coding possibilities.

Synthetic mRNA		Polypeptides synthesized
		Polypeptides with one amino acid
poly-U	UUUU ...	Phe-Phe-Phe ...
poly-C	CCCC ...	Pro-Pro-Pro ...
poly-A	AAAA ...	Lys-Lys-Lys ...
poly-G	GGGG ...	Gly-Gly-Gly ...
Repeating dinucleotides		Polypeptides with alternating amino acids
poly-UC	UCUCUC ...	Ser-Leu-Ser-Leu ...
poly-AG	AGAGAG ...	Arg-Glu-Arg-Glu ...
poly-UG	UGUGUG ...	Cys-Val-Cys-Val ...
poly-AC	ACACAC ...	Thr-His-Thr-His ...
Repeating trinucleotides		Three polypeptides each with one amino acid
poly-UUC	UUCUUCUUC ...	Phe-Phe.... and Ser-Ser.... and Leu-Leu....
poly-AAG	AAGAAGAAG ...	Lys-Lys.... and Arg-Arg.... and Glu-Glu....
poly-UUG	UUGUUGUUG ...	Leu-Leu.... and Cys-Cys.... and Val-Val....
poly-UAC	UACUACUAC ...	Tyr-Tyr.... and Thr-Thr.... and Leu-Leu....
Repeating tetranucleotides		Polypeptides with repeating units of four amino acids
poly-UAUC	UAUCUAUC ...	Tyr-Leu-Ser-Ile-Tyr-Leu-Ser-Ile...
poly-UUAC	UUACUUAC ...	Leu-Leu-Thr-Tyr-Leu-Leu-Thr-Tyr...
poly-GUAA	GUAAGUAA ...	none
poly-GAUA	GAUAGAUA ...	none

The chemist Har Gobind Khorana later made mRNAs with repeating dinucleotides, such as poly-UC (5′. . . UCUCUCUC . . . 3′), repeating trinucleotides, such as poly-UUC, and repeating tetranucleotides, such as poly-UAUC, and used them to direct the synthesis of slightly more complex polypeptides. As Fig. 8.5b shows, his results limited the coding possibilities, but some ambiguities remained. For example, poly-UC encodes the polypeptide N . . . Ser-Leu-Ser-Leu-Ser-Leu . . . C in which serine and leucine alternate with each other. Although the mRNA contains only two different codons (5′ UCU 3′ and 5′ CUC 3′), it is not obvious which corresponds to serine and which to leucine.

Nirenberg and Philip Leder resolved these ambiguities in 1965 with experiments in which they added short, synthetic mRNAs only three nucleotides in length to an *in vitro* translational system containing tRNAs attached to amino acids, where only one of the 20 amino acids was radioactive. They then poured through a filter a mixture of a synthetic mRNA and the translational system containing a tRNA-attached, radioactively labeled amino acid (**Fig. 8.6**). tRNAs carrying an amino acid normally go right through a filter. If, however, a tRNA carrying an amino acid binds to a ribosome, it will stick in the filter, because this larger complex of ribosome, amino-acid-carrying tRNA, and small mRNA cannot pass through.

Nirenberg and Leder used this approach to see which small mRNA caused the entrapment of which radioactively labeled amino acid. For example, they knew from Khorana's earlier work that CUC encoded either serine or leucine. When they added the synthetic triplet CUC to an *in vitro* system where the radioactive amino acid was serine, this tRNA-attached amino acid passed through the filter, and the filter thus emitted no radiation (Fig. 8.6). But when they added the same triplet to a system where the radioactive amino acid was leucine, the filter lit up with radioactivity, indicating that the radioactively tagged leucine attached to a tRNA had bound to the ribosome-mRNA complex and gotten stuck in the filter (Fig. 8.6). CUC thus encodes leucine, not serine. Nirenberg and Leder used this technique to determine most of the codon–amino acid correspondences shown in the genetic code table (see Fig. 8.2).

Figure 8.6 Cracking the genetic code with mini-mRNAs. Nirenberg and Leder added trinucleotides of known sequence, in combination with a mixture of amino acid–charged tRNAs where only one amino acid was radioactive, to an *in vitro* extract containing ribosomes. If the trinucleotide specified this amino acid, the radioactive charged tRNA formed a complex with the ribosomes that could be trapped on a filter. The experiments shown here indicate that the codon CUC specifies leucine, not serine.

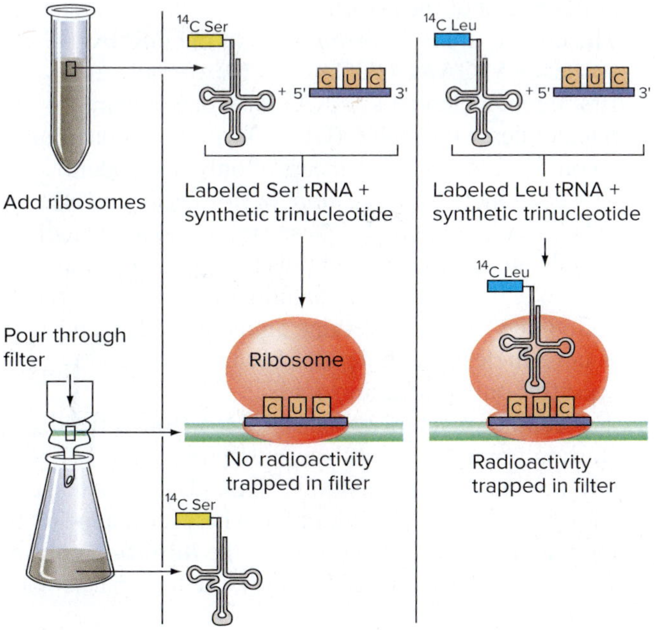

Polarities: 5′ to 3′ in mRNA corresponds to N to C in the polypeptide

In studies using synthetic mRNAs, when investigators added the six-nucleotide-long 5′ AAAUUU 3′ to an *in vitro* translational system, the product N Lys-Phe C emerged, but no N Phe-Lys C appeared. Because AAA is the codon for lysine and UUU is the codon for phenylalanine, this result means that the codon closest to the 5′ end of the mRNA encoded the amino acid closest to the N terminus of the corresponding polypeptide. Similarly, the codon nearest the 3′ end of the mRNA encoded the amino acid nearest the C terminus of the resulting polypeptide.

To understand how the polarities of the macromolecules participating in gene expression relate to each other, remember that although the gene is a segment of a DNA double helix, only one of the two strands serves as a template for the mRNA. This strand is known as the **template strand.** The other strand is the **RNA-like strand,** because it has the same polarity and sequence (written in the DNA dialect) as the RNA. Note that some scientists use the terms *sense strand* or *coding strand* as synonyms for the RNA-like strand; in these alternative nomenclatures, the template strand would be the *antisense strand* or the *noncoding strand.* **Figure 8.7** diagrams the respective polarities of a gene's DNA, the mRNA transcript of that DNA, and the resulting polypeptide.

Nonsense codons and polypeptide chain termination

Although most of the simple, repetitive RNAs synthesized by Khorana were very long and thus generated very long polypeptides, a few did not. These RNAs had signals that stopped construction of a polypeptide chain. As it turned out, three different triplets—UAA, UAG, and UGA—do not correspond to any of the amino acids. When these

Figure 8.7 Correlation of polarities in DNA, mRNA, and polypeptide. The template strand of DNA is complementary to both the RNA-like DNA strand and the mRNA. The 5′-to-3′ direction in an mRNA corresponds to the N terminus-to-C terminus direction in the polypeptide.

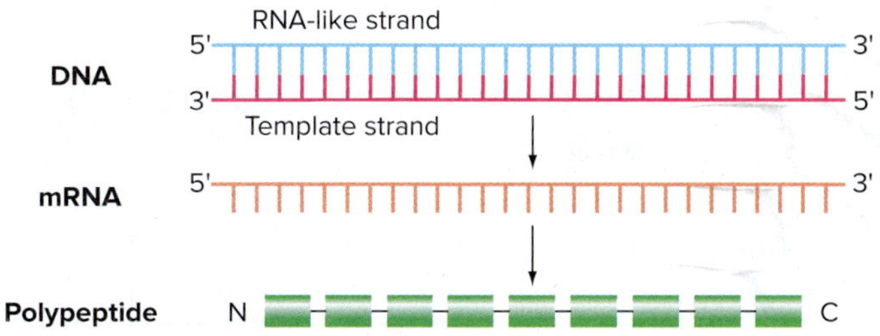

codons appear in frame, translation stops. As an example of how investigators established this fact, consider the case of poly-GUAA (review Fig. 8.5b). This mRNA will not generate a long polypeptide because in all possible reading frames, it contains the **stop codon** UAA.

Sydney Brenner helped establish the identities of the stop codons in an alternative way, through ingenious experiments involving point mutations in a T4 phage gene named *m*, encoding a protein component of the phage head capsule. As shown in **Fig. 8.8a,** Brenner determined that certain mutant alleles (m^1–m^6) encoded *truncated polypeptides* that were shorter than the wild-type M protein. Brenner found that the final amino acid at the C terminus in each of the truncated proteins would have been followed in the normal, full-length protein by an amino acid specified by a codon that differed from the triplet UAG by a single nucleotide. These data suggested that each *m* mutant had a point mutation that changed a codon for an amino acid into the stop codon UAG. Such a mutation is called a **nonsense mutation** because it changes a codon that signifies an amino acid (a *sense codon*) into one that does not (a *nonsense codon*). (It was not a coincidence that all of the truncation mutants had nonsense mutations where a codon was a changed to a particular stop codon—in this case UAG. Problem 56 at the end of this chapter explains why this was the case.)

Figure 8.8 Sydney Brenner's experiment showing that UAG is a stop signal. (a) The T4 phage m^+ gene encodes a polypeptide M whose amino acids are shown with *blue* circles. Mutant alleles m^1–m^6 direct synthesis of truncated M proteins (*black* circles). In the wild-type M protein, the amino acid that would follow the final amino acid in each truncated protein is encoded by a triplet that differs from UAG by a single nucleotide. **(b)** The genetic map positions of the m^1–m^6 mutations are colinear with the sizes of the corresponding truncated M proteins.

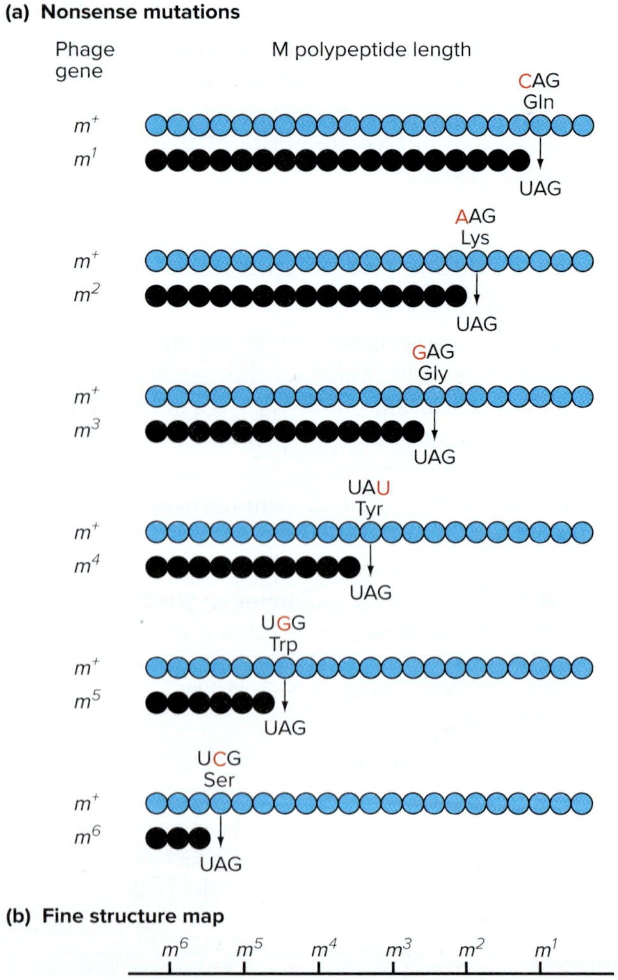

Brenner later established that a fine structure map of mutations m^1–m^6 corresponds in a linear manner to the size of the truncated polypeptide chains (**Fig. 8.8b**). It makes sense that the M protein encoded by m^6, for example, is shorter than that encoded by m^5 because the m^6 nonsense mutation is closer to the beginning of the reading frame than m^5.

Brenner also isolated analogous sets of nonsense mutations that defined UAA and UGA as stop codons. For historical reasons, researchers often refer to UAG as the *amber* codon, UAA as the *ochre* codon, and UGA as the *opal* codon. The historical basis of this nomenclature is the last name of one of the early investigators—Bernstein—which means *amber* in German; ochre and opal derive from their similarity with amber as semiprecious materials.

The Genetic Code: A Summary

The genetic code is a complete, unabridged dictionary equating the four-letter language of the nucleic acids with the 20-letter language of the proteins. The following list summarizes the code's main features:

1. *Triplet codons:* As written in Fig. 8.2, the code shows the 5′-to-3′ sequence of the three nucleotides in each mRNA codon; that is, the first nucleotide depicted is at the 5′ end of the codon.
2. The codons are *nonoverlapping.* In the mRNA sequence 5′ GAAGUUGAA 3′, for example, the first three nucleotides (GAA) form one codon; nucleotides 4 through 6 (GUU) form the second; and so on. Each nucleotide is part of only one codon.
3. The code includes three *stop,* or *nonsense, codons:* UAG, UAA, and UGA. These codons do not usually encode an amino acid and thus terminate translation.
4. The code is *degenerate,* meaning that more than one codon may specify the same amino acid. The code is nevertheless unambiguous because each codon specifies only one amino acid.
5. The cellular machinery scans mRNA from a fixed starting point that establishes a *reading frame.* As we will see later, the nucleotide triplet AUG, which specifies the amino acid methionine wherever it appears in the reading frame, also serves as the **initiation codon,** marking where in an mRNA the code for a particular polypeptide begins.

6. *Corresponding polarities* of codons and amino acids*:* Moving in the 5′-to-3′ direction along an mRNA, each successive codon is sequentially decoded into an amino acid, starting at the N terminus and moving toward the C terminus of the resulting polypeptide.
7. Mutations may modify the message encoded in a sequence of nucleotides in three ways. *Frameshift mutations* are nucleotide insertions or deletions that alter the genetic instructions for polypeptide construction by changing the reading frame. *Missense mutations* change a codon for one amino acid to a codon for a different amino acid. *Nonsense mutations* change a codon for an amino acid to a stop codon.

The Effects of Mutations on Polypeptides Helped Verify the Code

The experiments that first cracked the genetic code by assigning codons to amino acids were all *in vitro* studies using cell-free extracts and synthetic mRNAs. A logical question thus arose: Do living cells construct polypeptides according to the same rules? Early evidence that they do came from studies analyzing how mutations actually affect the amino acid composition of the polypeptides encoded by a gene. Most mutagens change a single nucleotide in a codon. As a result, most missense mutations that change the identity of a single amino acid should be single-nucleotide substitutions, and analyses of these substitutions should conform to the code. Yanofsky, for example, found two *trpA*$^-$ auxotrophic mutations in the *E. coli* tryptophan synthase gene that produced two different amino acids (arginine, or Arg, and glutamic acid, or Glu) at the same position—amino acid 211—in the polypeptide chain (**Fig. 8.9a**). According to the code, both of these mutations could have resulted from single-base changes in the GGA codon that normally inserts glycine (Gly) at position 211.

Even more informative were the *trpA*$^+$ revertants of these mutations subsequently isolated by Yanofsky. As Fig. 8.9a illustrates, single-base substitutions in the gene could also explain the amino acid changes in these revertants. Note that some of these substitutions restore Gly to position 211 of the polypeptide, while others place amino acids such as Ile, Thr, Ser, Ala, or Val at this site in the tryptophan synthase molecule. The substitution of these other amino acids for Gly at position 211 in the polypeptide chain is compatible with (that is, largely conserves) the enzyme's function.

Yanofsky obtained better evidence yet that cells use the genetic code *in vivo* by analyzing proflavin-induced frameshift mutations of the tryptophan synthase gene (**Fig. 8.9b**). He first treated populations of *E. coli* with proflavin to produce *trpA*$^-$ mutants. Subsequent treatment of these mutants with more proflavin generated some *trpA*$^+$ revertants among the progeny. The most likely explanation for the revertants was that their tryptophan synthase gene carried both a single-base-pair deletion and a single-base-pair insertion (− +). Upon determining the amino acid sequences of the tryptophan synthase enzymes made by the revertant strains, Yanofsky found that he could use the genetic code to predict the precise amino acid alterations that had occurred by assuming the revertants had a specific single-base-pair insertion and a specific single-base-pair deletion (Fig. 8.9b).

Yanofsky's results helped confirm not only amino acid codon assignments but other parameters of the code as well. His interpretations make sense only if codons do not overlap and are read from a fixed starting point, with no pauses or commas separating the adjacent triplets.

Figure 8.9 Experimental verification of the genetic code. **(a)** Single-base substitutions can explain the amino acid substitutions caused by *trpA*$^-$ mutations and *trpA*$^+$ reversions. **(b)** The genetic code predicts the amino acid alterations (*yellow*) that would arise from single-base-pair deletions and suppressing insertions.

(a) Altered amino acids in *trpA*$^-$ mutations and *trpA*$^+$ revertants

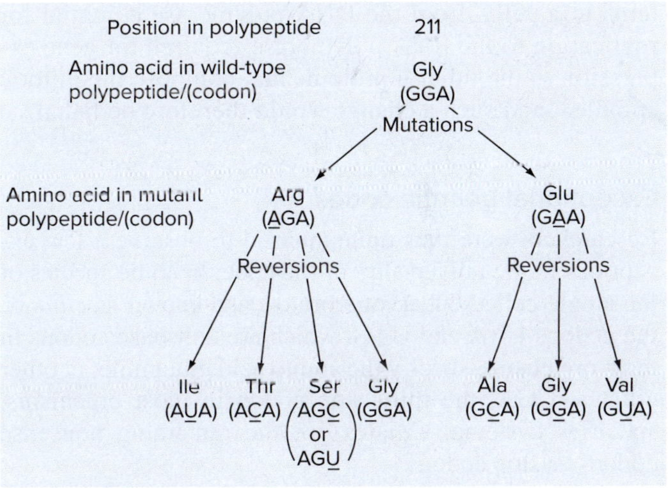

(b) Amino acid alterations that accompany intragenic suppression

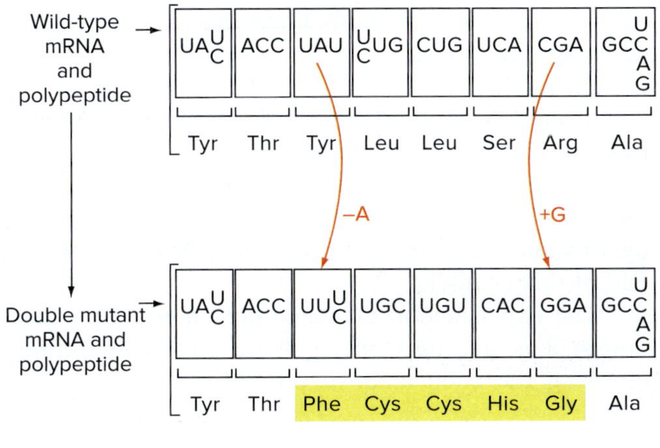

The Genetic Code Is Almost, but Not Quite, Universal

We now know that virtually all cells alive today use the same basic genetic code. One early indication of this uniformity was that a translational system derived from one organism could use the mRNA from another organism to convert genetic information to the encoded protein. Rabbit hemoglobin mRNA, for example, when injected into frog eggs or added to cell-free extracts from wheat germ, directs the synthesis of rabbit hemoglobin proteins. More recently, comparisons of DNA and protein sequences have revealed a perfect correspondence according to the genetic code between codons and amino acids in almost all organisms examined.

Conservation of the genetic code

The universality of the code is an indication that it evolved very early in the history of life. Once it emerged, the code remained constant over billions of years, in part because evolving organisms would have little tolerance for change. A single change in the genetic code could disrupt the production of hundreds or thousands of proteins in a cell—from the DNA polymerase essential for replication to the RNA polymerase required for gene expression to the tubulin proteins that compose the mitotic spindle—and such a change would therefore be lethal.

Exceptional genetic codes

Researchers were thus quite amazed to observe a few exceptions to the universality of the code. In some species of the single-celled eukaryotic protozoans known as *ciliates,* the codons UAA and UAG, which are nonsense codons in most organisms, specify the amino acid glutamine; in other ciliates, UGA, the third stop codon in most organisms, specifies cysteine. Ciliates use the remaining nonsense codons as stop codons.

Other systematic changes in the genetic code exist in mitochondria, the semiautonomous, self-reproducing organelles within eukaryotic cells that are the sites of ATP formation. Each mitochondrion has its own chromosomes and its own apparatus for gene expression (which we describe in detail in Chapter 15). In the mitochondria of yeast, for example, CUA specifies threonine instead of leucine. Yet another exception to the code is seen in certain prokaryotes that sometimes use the triplet UAG to specify insertion of the rare amino acid pyrrolysine (see Fig. 7.28c and also Problem 57 at the end of this chapter.)

The experimental evidence presented so far helped define a nearly universal genetic code. But although cracking the code made it possible to understand the broad outlines of information flow between gene and protein, these results did not explain exactly how the cellular machinery accomplishes gene expression. This question is our focus as we present in the next sections the details of transcription and translation.

essential concepts

- The nearly universal *genetic code* consists of 64 *codons,* each one composed of three nucleotides. Sixty-one codons specify amino acids, while three—UAA, UAG, UGA—are *stop codons.* The code is *degenerate* in that more than one codon can specify one amino acid.
- The codon AUG specifies methionine; it also serves as the *initiation codon* establishing the *reading frame* that groups nucleotides into successive, nonoverlapping codon triplets.
- *Missense mutations* change a codon so that it specifies a different amino acid; *frameshift mutations* alter the reading frame for all codons following the mutation; and *nonsense mutations* change a codon for an amino acid into a stop codon.

8.2 Transcription: From DNA to RNA

learning objectives

1. Describe the three stages of transcription: initiation, elongation, and termination.
2. Compare transcription initiation in prokaryotes and eukaryotes.
3. List three ways by which eukaryotes process mRNA after transcription.

Transcription is the process by which the polymerization of ribonucleotides, guided by complementary base pairing, produces an RNA transcript of a gene. The template for the RNA transcript is one strand of that portion of the DNA double helix that constitutes the gene.

RNA Polymerase Synthesizes a Single-Stranded RNA Copy of a Gene

Figure 8.10 depicts the basic components of transcription and illustrates key events in the process as it occurs in the bacterium *E. coli.* This figure divides transcription into

Figure 8.11 Control regions of bacterial and eukaryotic genes. Only the sequence of the RNA-like strand is shown; numbering starts at the first transcribed nucleotide (+1). **(a)** All promoters in *E. coli* share two different short stretches of nucleotides (*yellow*) essential for promoter recognition by RNA polymerase. The most common nucleotides in these short regions constitute the *consensus sequences* shown. **(b)** Eukaryotic genes transcribed by RNA pol II have a promoter, and also one or more distant DNA elements called *enhancers* (*orange*) that bind to protein factors aiding transcription.

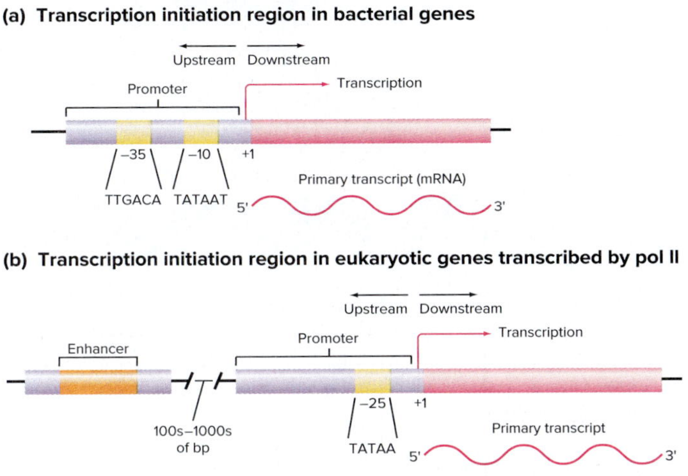

successive phases of *initiation, elongation,* and *termination.* The following four points are of particular importance:

1. The enzyme **RNA polymerase** catalyzes transcription.
2. DNA sequences near the beginning of genes, called **promoters,** signal RNA polymerase to begin transcription. Most bacterial gene promoters have almost identical nucleotide sequences in each of two short regions (**Fig. 8.11a**). These are the sites at which RNA polymerase makes particularly strong contact with the promoters.
3. RNA polymerase adds nucleotides to the growing RNA polymer in the 5′-to-3′ direction. The chemical mechanism of this nucleotide-adding reaction is similar to the formation of phosphodiester bonds between nucleotides during DNA replication (review Fig. 6.21), with one exception: Transcription uses ribonucleotide triphosphates (ATP, CTP, GTP, and UTP) instead of deoxyribonucleotide triphosphates. Hydrolysis of the high-energy bonds in each ribonucleotide triphosphate provides the energy needed for elongation.
4. Sequences in the RNA products, known as **terminators,** tell RNA polymerase where to stop transcription.

As you examine Fig. 8.10, bear in mind that a gene consists of two antiparallel strands of DNA, as mentioned earlier. One—the *RNA-like strand*—has the same polarity and sequence (except for T instead of U) as the emerging RNA transcript. The second—the *template strand*—has the opposite polarity and a complementary sequence that enables it to serve as the template for making the RNA transcript. When geneticists refer to the *sequence of a gene,* they usually mean the sequence of the RNA-like strand.

Transcription Initiation Varies Between Eukaryotes and Prokaryotes

Although the transcription of all genes in all organisms roughly follows the general scheme diagrammed in Fig. 8.10, prokaryotic and eukaryotic organisms vary in important details. In eukaryotes, promoters are more complicated than those in bacteria, and three different kinds of RNA polymerase exist that can transcribe different classes of genes. One of these is eukaryotic RNA polymerase II (pol II), which transcribes genes that encode proteins. **Figure 8.11b** illustrates the general structure of the DNA regions of eukaryotic genes that allow pol II to initiate transcription. A key difference with prokaryotes is that sequences called *enhancers* that can be thousands of base pairs away from the promoter are often also required for efficient transcription of eukaryotic genes.

Chapters 16 and 17 will describe how prokaryotic and eukaryotic cells can exploit these and other variations to control when, where, and at what level a given gene is expressed. Finally, the Genetics and Society Box *HIV and Reverse Transcription* describes how the AIDS virus uses an exceptional form of transcription, known as **reverse transcription,** to construct a double strand of DNA from an RNA template.

The result of transcription is a single strand of RNA known as a **primary transcript** (see Figs. 8.10 and 8.11). In prokaryotic organisms, the RNA produced by transcription is the actual messenger RNA that guides protein synthesis. In eukaryotic organisms, by contrast, most primary transcripts undergo **RNA processing** in the nucleus before they migrate to the cytoplasm to direct protein synthesis. As we see in the following section, this processing has played a fundamental role in the evolution of complex organisms.

In Eukaryotes, RNA Processing After Transcription Produces a Mature mRNA

Some RNA processing in eukaryotes modifies only the 5′ or 3′ ends of the primary transcript, leaving the information content of the rest of the mRNA untouched. Other processing deletes blocks of information from the middle of the primary transcript, so the content of the mature mRNA is

FEATURE FIGURE 8.10

Transcription in Bacterial Cells

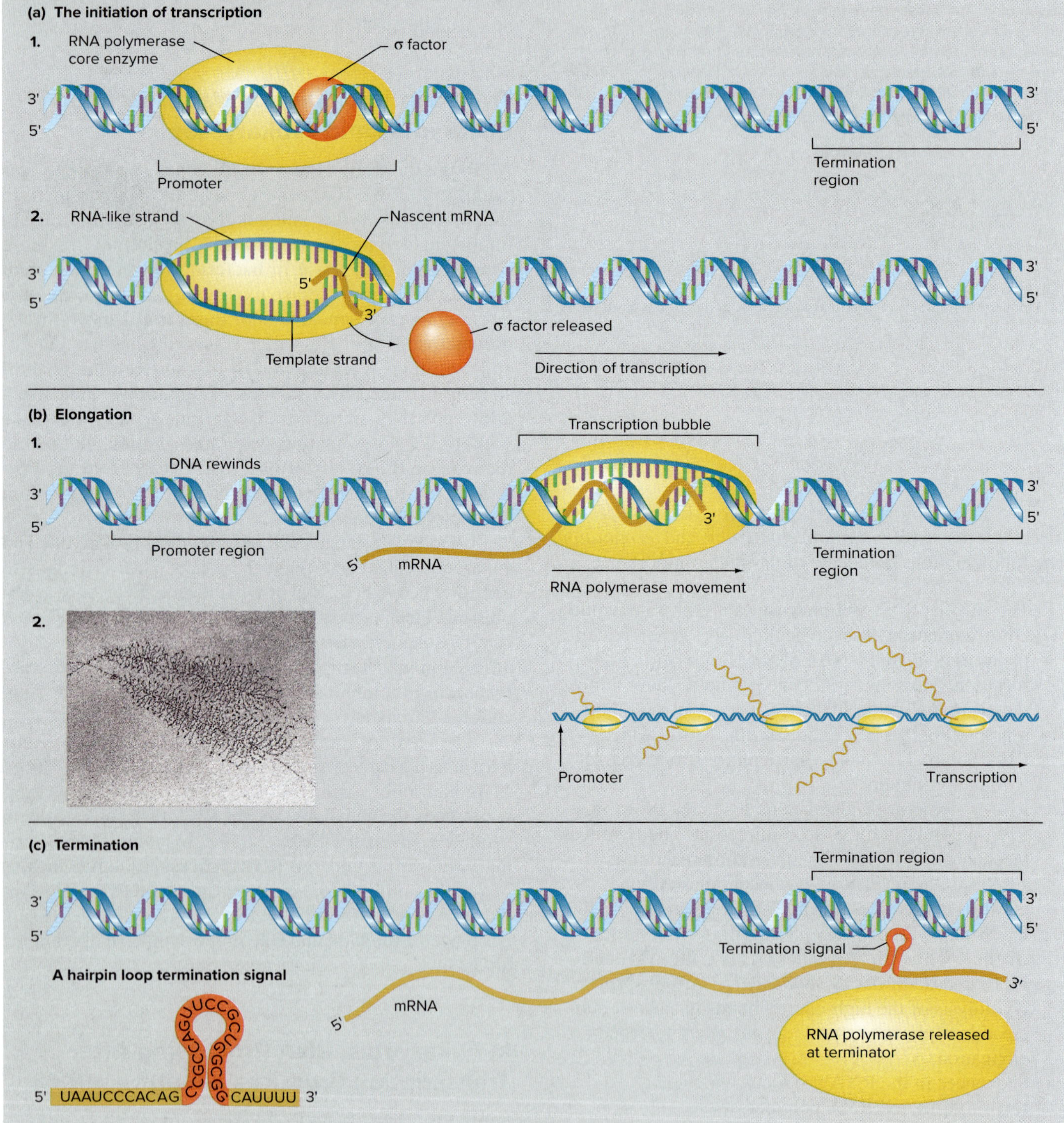

b.2: © Professor Oscar Miller/SPL/Science Source

(Continued)

FEATURE FIGURE 8.10 (*Continued*)

(a) The Initiation of Transcription

1. *RNA polymerase binds to double-stranded DNA at the beginning of the gene to be copied.* RNA polymerase recognizes and binds to **promoters,** specialized DNA sequences near the transcription start site. Although specific promoters vary substantially, all promoters in *E. coli* contain two characteristic short sequences of 6–10 base pairs (Fig. 8.11a). In bacteria, the complete RNA polymerase (the *holoenzyme*) consists of a *core enzyme,* plus a σ (*Sigma*) subunit involved only in initiation. The σ subunit reduces RNA polymerase's general affinity for DNA but simultaneously increases RNA polymerase's affinity for the promoter. As a result, the RNA polymerase holoenzyme can home in on a promoter and bind tightly to it, forming a *closed promoter complex.*

2. *After binding to the promoter, RNA polymerase unwinds part of the double helix, exposing unpaired bases on the template strand.* The complex formed between the RNA polymerase holoenzyme and an unwound promoter is called an *open promoter complex.* The enzyme identifies the template strand and chooses the two nucleotides to be copied. Guided by base pairing with these two nucleotides, RNA polymerase aligns the first two ribonucleotides at the 5′ end of the new RNA. The DNA transcribed as the 5′ end of the mRNA is the *5′ end of the gene.* RNA polymerase then forms a phosphodiester bond between the first two ribonucleotides. Soon thereafter, the RNA polymerase releases the σ subunit, marking the end of initiation.

(b) Elongation: Constructing an RNA Copy of the Gene

1. *When the σ subunit is released, RNA polymerase loses its enhanced affinity for the promoter sequence and regains its strong generalized affinity for any DNA.* These changes enable the core enzyme to leave the promoter yet remain bound to the gene. The core enzyme moves along the chromosome, unwinding the DNA to expose the next single-stranded region of the template. The enzyme extends the RNA by adding the correct ribonucleotide to the 3′ end of the growing chain. As the enzyme extends the mRNA in the 5′-to-3′ direction, it moves in the antiparallel 3′-to-5′ direction along the DNA template strand. RNA polymerase synthesizes RNA at an average speed of about 50 nucleotides per second.

The region of DNA unwound by RNA polymerase is the **transcription bubble.** Within the bubble, the nascent RNA chain remains base paired with the DNA template, forming a DNA–RNA hybrid. However, in those parts of the gene behind the bubble that have already been transcribed, the DNA double helix re-forms, displacing the RNA, which hangs out of the transcription complex as a single strand with a free 5′ end.

2. *Once an RNA polymerase has moved off the promoter, other RNA polymerase molecules can move in to initiate transcription.* If the promoter is very strong, that is, if it can attract RNA polymerase rapidly, many enzyme molecules can transcribe it simultaneously. Here we show an electron micrograph and an artist's interpretation of simultaneous transcription by several RNA polymerases. The promoter for this gene lies very close to where the shortest RNA is emerging from the DNA.

Geneticists often use the direction traveled by RNA polymerase as a reference when discussing gene structure. If, for example, you started at the 5′ end of a gene at point A and moved along the gene in the same direction as RNA polymerase to point B, you would be moving **downstream.** If, by contrast, you started at point B and moved in the opposite direction to point A, you would be traveling **upstream.**

(c) Termination: The End of Transcription

RNA sequences that signal the end of transcription are known as **terminators.** Two types of terminators exist: *intrinsic terminators,* which cause the RNA polymerase core enzyme to terminate transcription on its own, and *extrinsic terminators,* which require additional proteins—particularly a polypeptide known as *Rho*—to bring about termination. All terminators, whether intrinsic or extrinsic, are specific sequences in the mRNA that are transcribed from the gene. Terminators often form **hairpin loops** (also called **stem loops**) in which nucleotides within the mRNA pair with complementary nucleotides in the same molecule. Upon termination, RNA polymerase and a completed RNA chain are both released from the DNA.

related, but not identical, to the complete set of DNA nucleotide pairs in the original gene.

Adding a 5′ methylated cap and a 3′ poly-A tail

The nucleotide at the 5′ end of a eukaryotic mRNA is a G in reverse orientation from the rest of the molecule; it is connected through a triphosphate linkage to the first nucleotide in the primary transcript. This "backward G" is not transcribed from the DNA. Instead, a special *capping enzyme* adds it to the primary transcript after polymerization of the transcript's first few nucleotides. Enzymes known as *methyl transferases* then add methyl ($-CH_3$) groups to the backward G and to one or more of the

GENETICS AND SOCIETY

Crowd: © Image Source/Getty Images RF

HIV and Reverse Transcription

The AIDS-causing human immunodeficiency virus (HIV) is the most intensively analyzed virus in history. From laboratory and clinical studies spanning more than three decades, researchers have learned that each viral particle is a rough-edged sphere consisting of an outer envelope enclosing a protein matrix, which, in turn, surrounds a cut-off cone-shaped core (**Fig. A**). Within the core lies an enzyme-studded genome: two identical single strands of RNA associated with many molecules of an unusual DNA polymerase known as **reverse transcriptase.**

During infection, the AIDS virus binds to and injects its cone-shaped core into cells of the human immune system (**Fig. B**). The virus next uses reverse transcriptase to copy its RNA genome into double-stranded DNA molecules in the cytoplasm of the host cell. The double helixes then travel to the nucleus where another enzyme, called *integrase,* inserts them into a host chromosome. Once integrated into a host-cell chromosome, the viral genome can do one of two things. It can commandeer the host cell's protein synthesis machinery to make hundreds of new viral particles that bud off from the parent cell, taking with them part of the cell membrane. This process sometimes results in the host cell's death, which weakens the person's immune system. Alternatively, the HIV genome can lie latent inside the host chromosome, which then copies and transmits the viral genome to two new cells with each cell division.

The events of this life cycle make HIV a **retrovirus:** an RNA virus that after infecting a host cell copies its own single strands of RNA into double helixes of DNA, which a viral enzyme (integrase) then integrates into a host chromosome.

Reverse transcription, the foundation of the retroviral life cycle, is inconsistent with the one-way, DNA-to-RNA-to-protein flow of genetic information. Because it was so unexpected, the phenomenon of reverse transcription encountered great resistance in the scientific community when first reported by Howard Temin of the University of Wisconsin and David Baltimore, then of MIT. Now, however, it is an established fact. Reverse

Figure A Structure of the AIDS virus

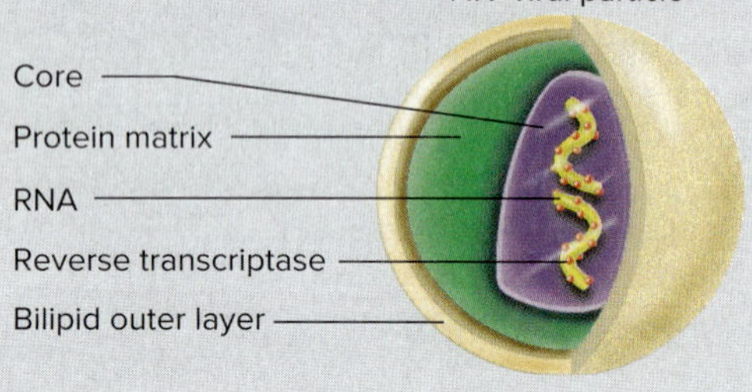

Figure B Life cycle of the AIDS virus

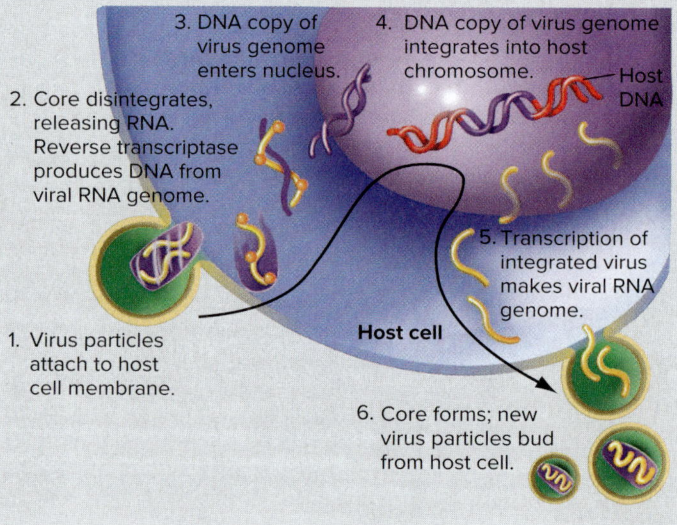

succeeding nucleotides in the RNA, forming a so-called **methylated cap (Fig. 8.12).**

Like the 5′ methylated cap, the 3′ end of most eukaryotic mRNAs is not encoded directly by the gene. In a large majority of eukaryotic mRNAs, the 3′ end consists of 100–200 As, referred to as a **poly-A tail (Fig. 8.13)**. Addition of the tail is a two-step process. First, a ribonuclease cleaves the primary transcript to form a new 3′ end; cleavage depends on the sequence AAUAAA, which is found in poly-A-containing mRNAs 11–30 nucleotides upstream of the position where the tail is added. Next, the enzyme *poly-A polymerase* adds As onto the 3′ end exposed by cleavage.

Unexpectedly, both the methylated cap and the poly-A tail are crucial for efficient translation of the mRNA into protein, even though neither helps specify an amino acid. Recent data indicate that particular eukaryotic *translation initiation factors* bind to the 5′ cap, while *poly-A binding protein* associates with the tail at the 3′ end of the mRNA. The interaction of these proteins in many cases shapes the mRNA molecule into a circle. This circularization both enhances the initial steps of translation and stabilizes the mRNA in the cytoplasm by increasing the length of time it can serve as a messenger.

Removing introns from the primary transcript by RNA splicing

Another kind of RNA processing became apparent in the late 1970s, after researchers had developed techniques that enabled them to analyze nucleotide sequences in both DNA

transcriptase is a remarkable DNA polymerase that can construct a DNA polymer from either an RNA or a DNA template.

In addition to its comprehensive copying abilities, reverse transcriptase has another feature not seen in most DNA polymerases: inaccuracy. As we saw in Chapter 7, normal DNA polymerases replicate DNA with an error rate of one mistake in every million nucleotides copied. Reverse transcriptase, however, introduces one mutation in every 5000 incorporated nucleotides.

HIV uses this capacity for mutation to gain a tactical advantage over the immune response of its host organism. Cells of the immune system seek to overcome an HIV invasion by multiplying in response to the proliferating viral particles (*virions*). The numbers are staggering. Each day of infection in every patient, from 100 million to a billion HIV particles are released from infected immune-system cells. As long as the immune system is strong enough to withstand the assault, it responds by producing as many as 2 billion new cells daily. Many of these new immune system cells produce antibodies targeted against proteins on the surface of the virus.

But just when an immune response wipes out those viral particles carrying the targeted protein, virions incorporating new forms of the protein resistant to the current immune response make their appearance. After many years of this complex chase, capture, and destruction by the immune system, the changeable virus outruns the host's immune response and gains the upper hand. Thus, the intrinsic infidelity of HIV's reverse transcriptase, by enhancing the virus's ability to compete in the evolutionary marketplace, increases its threat to human life and health.

This inherent mutability has undermined two potential therapeutic approaches toward the control of AIDS: drugs and vaccines. Some of the antiviral drugs approved in the United States for treatment of HIV infection—AZT (zidovudine), ddC (dideoxycytidine), and ddI (dideoxyinosine)—block viral replication by interfering with the action of reverse transcriptase. Each drug is similar to one of the four nucleotides, and when reverse transcriptase incorporates one of the drug molecules rather than a genuine nucleotide into a growing DNA polymer, the enzyme cannot extend the chain any further. However, the drugs are toxic at high doses and thus can be administered only at low doses that do not destroy all virions. Because of this limitation and the virus's high rate of mutation, mutant reverse transcriptases soon appear that work even in the presence of the drugs.

Similarly, researchers are having trouble developing effective vaccines. Even if a vaccine could generate a massive immune response against one, two, or even several HIV proteins, such a vaccine might be effective for only a short while—until enough mutations build up to make the virus resistant.

For these reasons, the AIDS virus will most likely not succumb entirely to drugs or vaccines that target proteins active at various stages of its life cycle. However, combinations of these therapeutic tools have nonetheless proven remarkably effective at prolonging an AIDS patient's life. In 2013, AIDS patients who received combination therapy had on average two-thirds of a normal life span. Newer drugs added to the cocktail include protease inhibitors that prevent the activity of enzymes needed to produce viral coat proteins, drugs that prevent viral entry into human cells, and inhibitors of the viral integrase protein.

A self-preserving capacity for mutation, perpetuated by reverse transcriptase, is surely one of the main reasons for HIV's success. Ironically, it may also provide a basis for its subjugation. Researchers are studying what happens when the virus increases its mutational load. If reverse transcriptase's error rate determines the size and integrity of the viral population in a host organism, greatly accelerated mutagenesis might push the virus beyond the error threshold that allows it to function. In other words, too much mutation might destroy the virus's infectivity, virulence, or capacity to reproduce. If geneticists could figure out how to make this happen, they might be able to give the human immune system the advantage it needs to overcome the virus.

and RNA. Using these techniques, which we describe in Chapter 9, they began to compare eukaryotic genes with the mRNAs derived from them. Their expectation was that just as in prokaryotes, the DNA nucleotide sequence of a gene's RNA-like strand would be identical to the RNA nucleotide sequence of the messenger RNA (with the exception of U replacing T in the RNA). Surprisingly, the investigators found that the DNA nucleotide sequences of many eukaryotic genes are much longer than their corresponding mRNAs. This fact suggested that RNA transcripts, in addition to receiving a methylated cap and a poly-A tail, undergo extensive internal processing.

An extreme example of the length difference between primary transcript and mRNA is seen in the human gene *DMD*, which encodes the protein Dystrophin. Abnormalities in the *DMD* gene underlie the genetic disorder Duchenne muscular dystrophy (DMD). The *DMD* gene is 2.5 million nucleotides—or 2500 kilobases (kb)—long, whereas the corresponding mRNA is roughly 14,000 nucleotides, or 14 kb, in length. Obviously the gene contains DNA sequences that are not present in the mature mRNA. Those regions of the gene that do end up in the mature mRNA are scattered throughout the 2500 kb of DNA.

Exons and introns Sequences found in both a gene's DNA and the mature messenger RNA are called **exons** (for *expressed regions*). The sequences found in the DNA of the gene but not in the mature mRNA are known as **introns** (for *intervening regions*). Introns interrupt, or separate, the exon sequences that actually end up in the mature mRNA.

Figure 8.12 Structure of the methylated cap at the 5′ end of eukaryotic mRNAs. Capping enzyme connects a backward G to the first nucleotide of the primary transcript through a triphosphate linkage. Methyl transferase enzymes then add methyl groups (*orange*) to this G and to one or two of the nucleotides first transcribed from the DNA template.

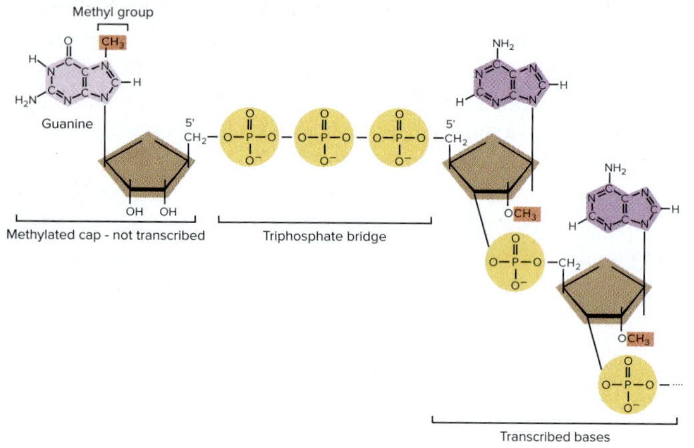

Figure 8.13 How RNA processing adds a tail to the 3′ end of eukaryotic mRNAs. A ribonuclease recognizes AAUAAA in a particular context of the primary transcript and cleaves the transcript 11–30 nucleotides downstream to create a new 3′ end. The enzyme poly-A polymerase then adds 100–200 As onto this new 3′ end.

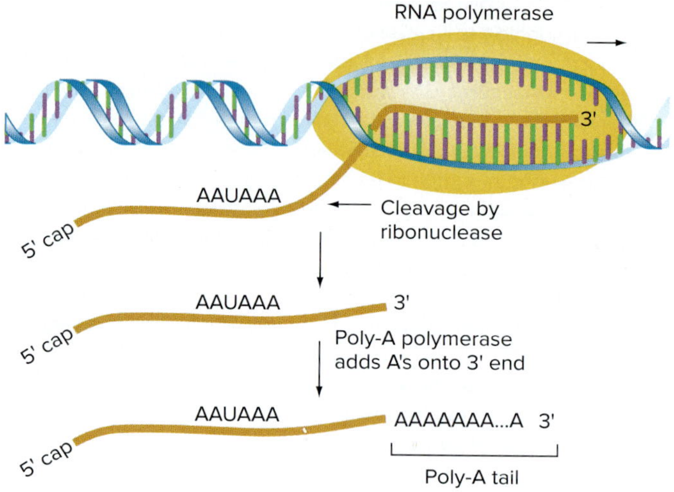

The gene for collagen (an abundant protein in connective tissue) shown in **Fig. 8.14** has two introns. By contrast, the *DMD* gene has more than 80 introns; the mean intron length is 35 kb, but one of its introns is an amazing 400 kb long. Other genes in humans generally have many fewer introns, while a few have none—and the introns range from 50 bp to over 100 kb. Exons, in contrast, vary in size from about 50 bp to a few kilobases; in the *DMD* gene, the mean exon length is 200 bp. The greater size variation seen in introns compared to exons reflects the fact that introns do not encode polypeptides and do not appear in mature mRNAs. As a result, fewer restrictions exist on the sizes and base sequences of introns.

Mature mRNAs must contain all the codons that are translated into amino acids, including the initiation and termination codons. In addition, mature mRNAs have sequences at their 5′ and 3′ ends that are not translated, but that nevertheless play important roles in regulating the efficiency of translation. These sequences, called the **5′ and 3′ untranslated regions (5′ and 3′ UTRs)**, are located just after the methylated cap and just before the poly-A tail, respectively (Fig. 8.14a). Excepting the cap and tail themselves, all of the sequences in a mature mRNA, including all the codons and both UTRs, must be transcribed from the gene's exons.

Introns can interrupt a gene at any location, even between the nucleotides making up a single codon. In such a case, the three nucleotides of the codon are present in two different (but successive) exons. You should also note that because introns can interrupt the 5′ and/or 3′ UTRs, the start codon is not always in the first exon, and neither is the stop codon always in the final exon.

How do cells make a mature mRNA from a gene whose coding sequences are interrupted by introns? The answer is that cells first make a primary transcript containing all of a gene's introns and exons, and then they remove the introns from the primary transcript by **RNA splicing,** the process that deletes introns and joins together successive exons to form a mature mRNA consisting only of exons (Fig. 8.14a). Because the first and last exons of the primary transcript become the 5′ and 3′ ends of the mRNA, while all intervening introns are spliced out, a gene must have one more exon than it does introns. To construct the mature mRNA, splicing must be remarkably precise. For example, if an intron lies within a codon, splicing must remove the intron and reconstitute the codon without disrupting the reading frame of the mRNA.

The mechanism of RNA splicing **Figure 8.15** illustrates how RNA splicing works. Three types of short sequences within the primary transcript—**splice donors, splice acceptors,** and **branch sites**—help ensure the specificity of splicing. These sites make it possible to sever the connections between an intron and the exons that precede and follow it, and then to join the formerly distant exons.

The mechanism of splicing involves two sequential cuts in the primary transcript. The first cut is at the splice-donor site, at the 5′ end of the intron. After this first cut, the new 5′ end of the intron attaches, via a novel 2′–5′ phosphodiester bond, to an A at the branch site located within the intron, forming a so-called *lariat*. The second cut is at the splice-acceptor site, at the 3′ end of the intron; this cut removes the intron. The discarded intron is degraded, and the precise splicing of adjacent exons completes the process of intron removal (Fig. 8.15).

SnRNPs and the spliceosome Splicing normally requires a complicated intranuclear machine called the **spliceosome,** which ensures that all of the splicing reactions take

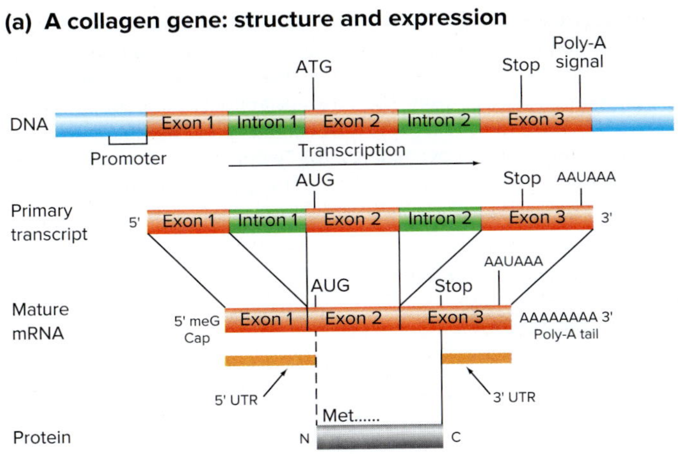

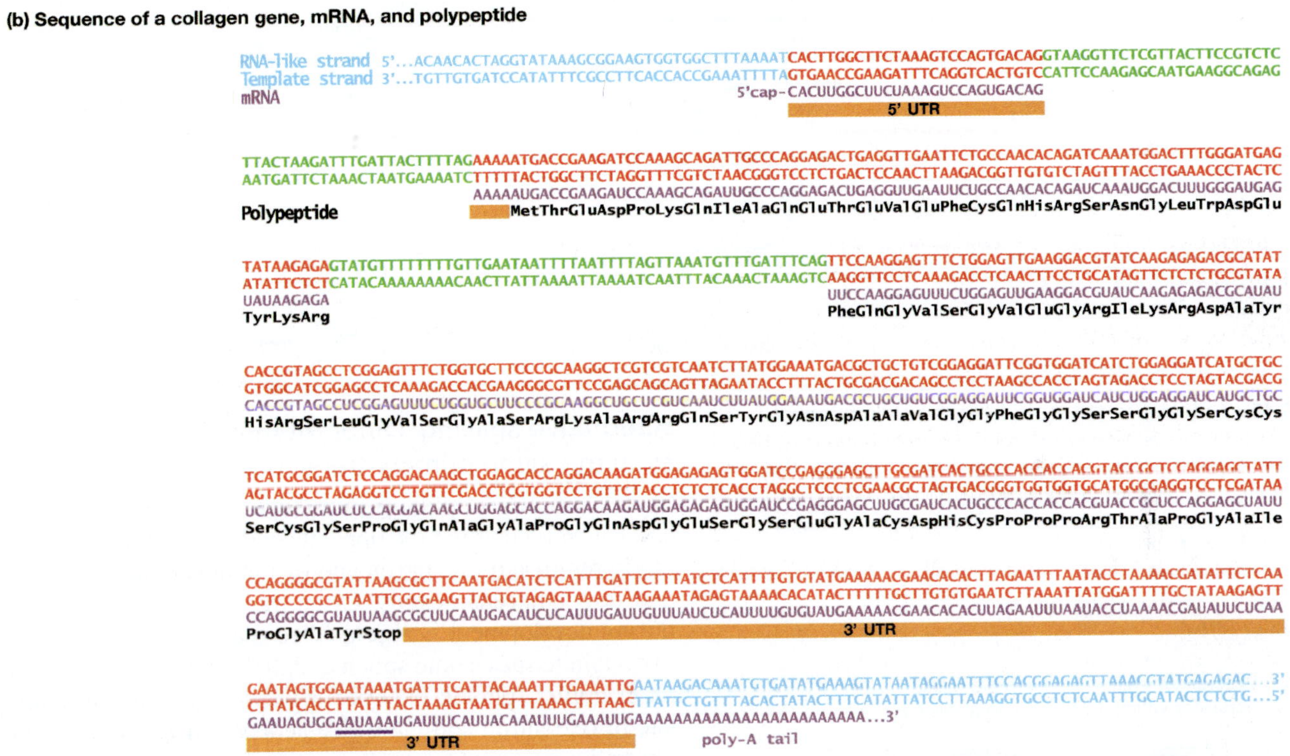

Figure 8.14 **Structure and expression of a typical eukaryotic gene.** **(a)** Schematic view of landmarks in a collagen gene and its products. Exons are shown in *red*, introns in *green*, and nontranscribed parts of the gene in *blue*. Mature mRNAs are processed from the primary transcript; introns are spliced out, a 5′ methyl-G cap is added, and a poly-A tail is added to the 3′ end. The 5′ untranslated region (5′ UTR) lies between the 5′ end and the start codon (AUG), and the 3′ untranslated region (3′ UTR) lies between the stop codon and the poly-A tail at the 3′ end of the mRNA (*orange bars*). **(b)** The same gene at the nucleotide level. Colors are the same as in part (a), except that the mature mRNA is shown in *purple* for emphasis. The AAUAAA poly-A addition signal in the mRNA is underlined. Introns can occur anywhere in the transcribed part of a gene, including within a codon or either of the UTRs.

place in concert (**Fig. 8.16**). The spliceosome consists of four subunits known as *small nuclear ribonucleoproteins*, or *snRNPs* (pronounced "snurps"). Each snRNP contains one or two *small nuclear RNAs (snRNAs)* 100–300 nucleotides long, associated with proteins in a discrete particle. Certain snRNAs can base pair with the splice donor and splice acceptor sequences in the primary transcript, so these snRNAs are particularly important in bringing together the two exons that flank an intron.

Given the complexities of spliceosome structure, it is remarkable that a few primary transcripts can splice themselves without the aid of a spliceosome or any additional factor. These rare primary transcripts function as **ribozymes:** RNA molecules that can act as enzymes and catalyze a specific biochemical reaction.

It might seem strange that eukaryotic genes incorporate DNA sequences that are spliced out of the mRNA before translation and thus do not encode amino acids. No one knows exactly why introns exist. One hypothesis proposes that introns make it possible to assemble genes from various exon building blocks that encode modules of protein function. This type of assembly would allow the shuffling of exons to make new genes, a process that appears to have played a key role in the evolution of complex organisms. The exon-as-module proposal is attractive because it is easy to understand the selective advantage of the potential for exon shuffling. Nevertheless, it remains a hypothesis without proof; introns may have become established through means that scientists have yet to imagine.

Figure 8.15 How RNA processing splices out introns and joins adjacent exons. Exons are in *red* and introns in *green.* **(a)** Three short sequences within the primary transcript are needed for splicing. (1) The splice-donor site occurs where the 3′ end of an exon abuts the 5′ end of an intron. In most splice-donor sites, a GU dinucleotide (*arrows*) that begins the intron is flanked on either side by a few purines (Pu; that is, A or G). (2) The splice-acceptor site is at the 3′ end of the intron where it joins with the next exon. The final nucleotides of the intron are always AG (*arrows*) usually preceded by 12–14 pyrimidines (Py; that is, C or U). (3) The branch site, which is located within the intron about 30 nucleotides upstream of the splice acceptor, must include an A (*arrow*) and is usually rich in pyrimidines. **(b)** Two sequential cuts, the first at the splice-donor site and the second at the splice-acceptor site, remove the intron, allowing precise splicing of adjacent exons.

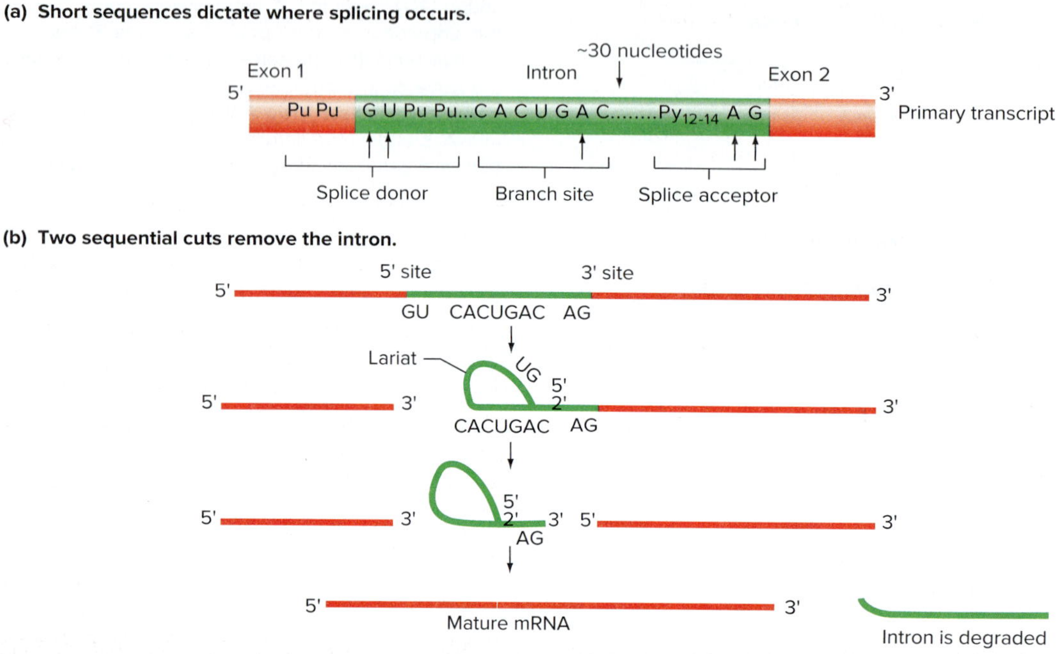

Figure 8.16 Splicing is catalyzed by the spliceosome. (*Top*) The spliceosome is assembled from four snRNP subunits, each of which contains one or two snRNAs and several proteins. (*Bottom*) Views of three spliceosomes in the electron microscope.

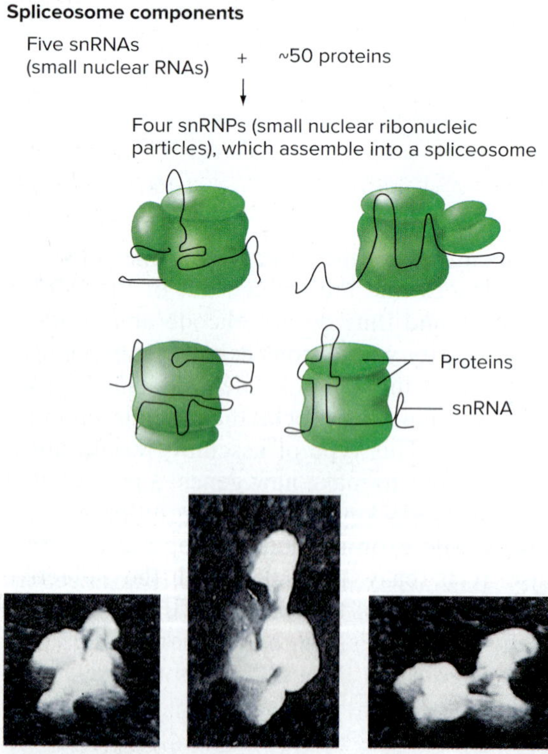

Alternative splicing: Different mRNAs from the same primary transcript

Sometimes RNA splicing joins together the splice donor and splice acceptor at the opposite ends of an intron, resulting in removal of the intron and fusion of two successive—and now adjacent—exons. Often, however, RNA splicing during development is regulated so that at certain times or in certain tissues, some splicing signals may be ignored. As an example, splicing may occur between the splice donor site of one intron and the splice acceptor site of a different intron downstream. Such **alternative splicing** produces different mRNA molecules that may encode related proteins with different—though partially overlapping—amino acid sequences and functions. In effect then, alternative splicing can tailor the nucleotide sequence of a primary transcript to produce more than one kind of polypeptide. Alternative splicing largely explains how the 27,000 genes in the human genome can encode the hundreds of thousands of different proteins estimated to exist in human cells.

In mammals, alternative splicing of the gene encoding the antibody heavy chain determines whether the antibody proteins become embedded in the membrane of the B lymphocyte that makes them or are instead secreted into the blood. The gene for antibody heavy chains has eight exons and seven introns; exon number 6 has a splice-donor site within it. To make the membrane-bound antibody, all exons except for the right-hand part of number 6 are joined to create an mRNA encoding a hydrophobic (water-hating, lipid-loving)

Figure 8.17 Different mRNAs can be produced from the same primary transcript. Alternative splicing of the primary transcript for the antibody heavy chain produces mRNAs that encode different kinds of antibody proteins.

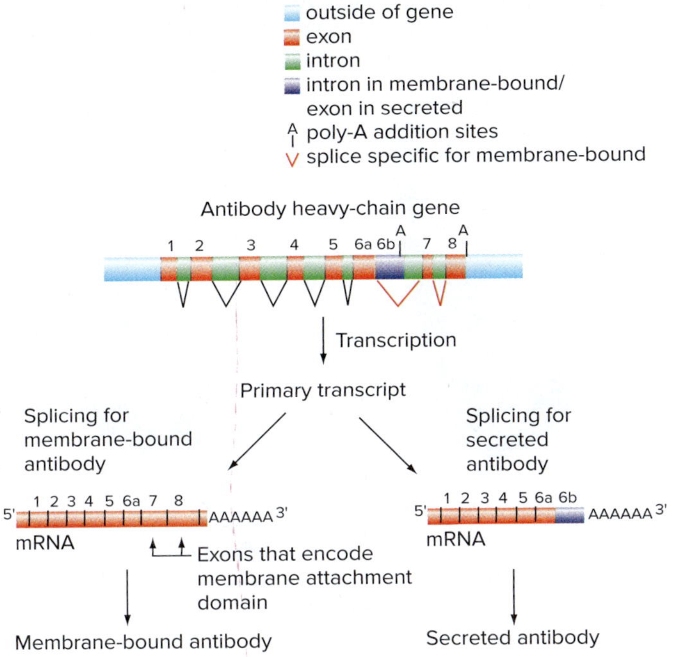

C terminus (**Fig. 8.17**). For the secreted antibody, only the first six exons (including the right part of 6) are spliced together to make an mRNA encoding a heavy chain with a hydrophilic (water-loving) C terminus. These two kinds of mRNAs formed by alternative splicing thus encode slightly different proteins that are directed to different parts of the body.

essential concepts

- *Transcription* is the process by which *RNA polymerase* synthesizes a single-stranded *primary transcript* from a DNA template.
- Transcription *initiation* requires a DNA sequence called the *promoter* that signals RNA polymerase to begin copying. In eukaryotes, initiation requires an additional DNA sequence called an *enhancer.*
- During transcription *elongation,* RNA polymerase adds nucleotides to the growing RNA strand in the 5′-to-3′ direction.
- A *terminator* in the RNA transcript tells RNA polymerase to cease transcription.
- In prokaryotes, the primary transcript is the *messenger RNA (mRNA).*
- In eukaryotes, *RNA processing* after transcription produces a mature mRNA; the RNA transcript is modified by the addition of a *5′ cap* and a *poly-A tail,* along with the excision of *introns* when *exons* are joined by *splicing.*
- Exons can be spliced together in alternative ways; *alternative splicing* produces different mRNA sequences and therefore different polypeptides from the same primary transcript.

8.3 Translation: From mRNA to Protein

learning objectives

1. Relate tRNA's structure to its function.
2. Describe the key steps of translation, indicating how each depends on the ribosome.
3. List three categories of posttranslational processing and provide examples of each.

Translation is the process by which the sequence of nucleotides in a messenger RNA directs the assembly of the correct sequence of amino acids in the corresponding polypeptide. Translation takes place on ribosomes that coordinate the movements of transfer RNAs carrying specific amino acids with the genetic instructions of an mRNA. As we examine the cell's translation machinery, we first describe the structure and function of tRNAs and ribosomes; and we then explain how these components interact during translation.

Transfer RNAs Mediate the Translation of mRNA Codons to Amino Acids

No obvious chemical similarity or affinity exists between the nucleotide triplets of mRNA codons and the amino acids they specify. Rather, **transfer RNAs (tRNAs)** serve as adapter molecules that mediate the transfer of information from nucleic acid to protein.

The structure of tRNA

Transfer RNAs are short, single-stranded RNA molecules 74–95 nucleotides in length. Several of the nucleotides in tRNAs contain chemically modified bases produced by enzymatic alterations of the principal A, G, C, and U nucleotides. Each tRNA carries one particular amino acid, and all cells must have at least one tRNA for each of the common 20 amino acids specified by the genetic code. The name of a tRNA reflects the amino acid it carries. For example, $tRNA^{Gly}$ carries the amino acid glycine.

As **Fig. 8.18** shows, it is possible to consider the structure of a tRNA molecule on three levels.

1. The nucleotide sequence of a tRNA constitutes the primary structure.
2. Short complementary regions within a tRNA's single strand can form base pairs with each other to create a characteristic cloverleaf shape; this is the tRNA's secondary structure.
3. Folding in three-dimensional space creates a tertiary structure that looks like a compact letter L.

Figure 8.18 tRNA structure. The nucleotide sequence of a tRNA (the primary structure) folds to form characteristic secondary and tertiary structures. The anticodon and the amino acid attachment site are at opposite ends of the L-shaped tertiary structure. Several unusual bases of the tRNA, indicated as I, ψ, UH_2, mI, m_2G, and mG, are enzymatically modified variants of A, G, C, and U.

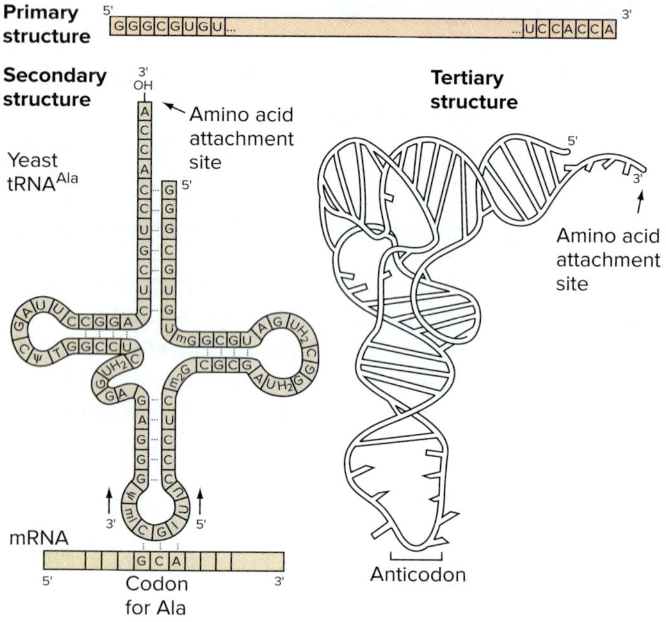

Figure 8.19 Aminoacyl-tRNA synthetases attach tRNAs to their corresponding amino acids. The aminoacyl-tRNA synthetase has recognition sites for an amino acid, the corresponding tRNA, and ATP. The synthetase first activates the amino acid, forming an AMP-amino acid. The enzyme then transfers the amino acid's carboxyl group from AMP to the hydroxyl (–OH) group of the ribose at the 3′ end of the tRNA, producing a charged tRNA.

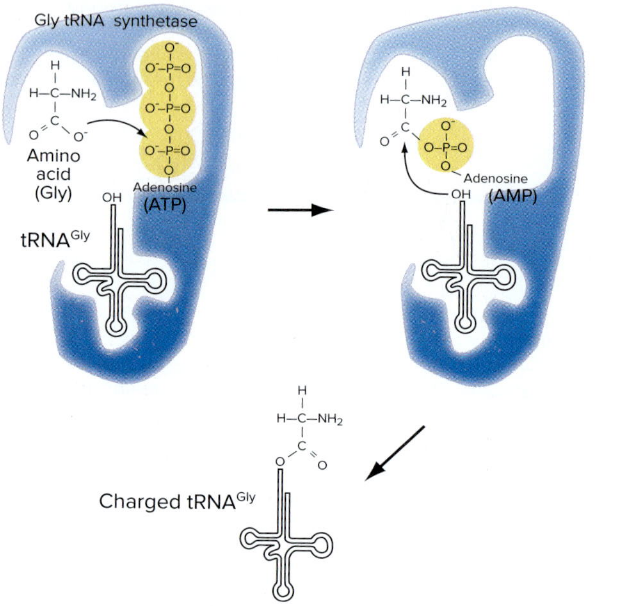

At one end of the L, the tRNA carries an **anticodon:** three nucleotides complementary to an mRNA codon specifying the amino acid carried by the tRNA (Fig. 8.18). The anticodon never forms base pairs with other regions of the tRNA; it is always available for base pairing with its complementary mRNA codon. As with other complementary base sequences, during pairing at the ribosome, the strands of anticodon and codon run antiparallel to each other. If, for example, the anticodon is 3′ CCU 5′, the complementary mRNA codon is 5′ GGA 3′, specifying the amino acid glycine. At the other end of the L, where the 5′ and 3′ ends of the tRNA strand are found (Fig. 8.18), the appropriate amino acid is attached to the tRNA's 3′ end.

Aminoacyl-tRNA synthetases: The molecular translators of the genetic code

Enzymes known as **aminoacyl-tRNA synthetases** connect the tRNA to the amino acid that corresponds to the anticodon. These enzymes are extraordinarily specific, recognizing unique features of a particular tRNA including the anticodon, while also recognizing the corresponding amino acid (see the opening figure of this chapter).

Aminoacyl-tRNA synthetases are, in fact, the only molecules that read the languages of both nucleic acid and protein. Normally, one aminoacyl-tRNA synthetase exists for each of the 20 common amino acids. Each synthetase functions with only one amino acid, but the enzyme may recognize several different tRNAs specific for that amino acid. **Figure 8.19** shows the two-step process that establishes the covalent bond between an amino acid and the 3′ end of its corresponding tRNA. A tRNA covalently coupled to its amino acid is called a **charged tRNA.** The bond between the amino acid and tRNA contains substantial energy that is later used to drive peptide bond formation.

The crucial role of base pairing between codon and anticodon

While attachment of the appropriate amino acid charges a tRNA, the amino acid itself does not play a significant role in determining where it becomes incorporated in a growing polypeptide chain. Instead, the specific interaction between a tRNA's anticodon and an mRNA's codon makes that decision. A simple experiment illustrates this point (**Fig. 8.20**). Researchers can subject a charged tRNA to chemical treatments that, without altering the structure of the tRNA, change the amino acid it carries. One treatment replaces the cysteine carried by $tRNA^{Cys}$ with alanine. When investigators then add the $tRNA^{Cys}$ charged with alanine to a cell-free translational system, the system incorporates alanine into the growing polypeptide wherever the mRNA contains a cysteine codon complementary to the anticodon of the $tRNA^{Cys}$.

Wobble: One tRNA, more than one codon

Although at least one kind of tRNA exists for each of the 20 common amino acids, cells do not necessarily carry tRNAs with anticodons complementary to all of the 61 possible

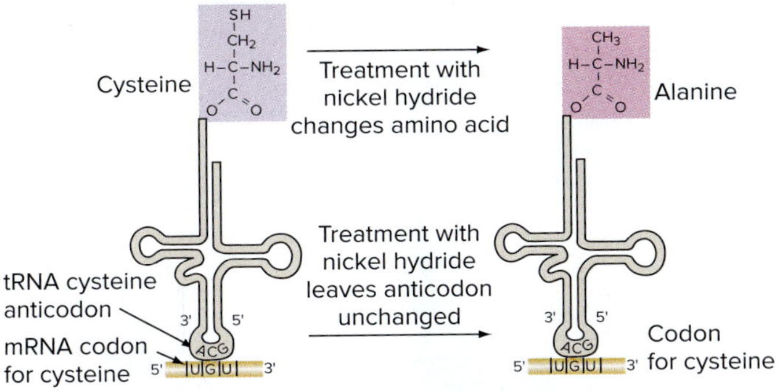

Figure 8.20 Base pairing between an mRNA codon and a tRNA anticodon determines which amino acid is added to a growing polypeptide. A tRNA with an anticodon for cysteine, but carrying the amino acid alanine, adds alanine whenever the mRNA codon for cysteine appears.

codon triplets in the genetic code. *E. coli,* for example, makes 79 different tRNAs containing 42 different anticodons. Although several of the 79 tRNAs in this collection obviously have the same anticodon, 61 – 42 = 19 of 61 potential anticodons are not represented. Thus 19 mRNA codons will not find a complementary anticodon in the *E. coli* collection of tRNAs. How can an organism construct proper polypeptides if some of the codons in its mRNAs cannot locate tRNAs with complementary anticodons?

The answer is that some tRNAs can recognize more than one codon for the amino acid with which they are charged. That is, the anticodons of these tRNAs can interact with more than one codon for the same amino acid, in keeping with the degenerate nature of the genetic code. Francis Crick spelled out a few of the rules that govern the promiscuous base pairing between codons and anticodons.

Crick reasoned first that the 3′ nucleotide in many codons adds nothing to the specificity of the codon. For example, 5′ GGU 3′, 5′ GGC 3′, 5′ GGA 3′, and 5′ GGG 3′ all encode glycine (review Fig. 8.2). It does not matter whether the 3′ nucleotide in the codon is U, C, A, or G as long as the first two letters are GG. The same is true for other amino acids encoded by four different codons, such as valine, where the first two bases must be GU, but the third base can be U, C, A, or G.

For amino acids specified by two different codons, the first two bases of the codon are, once again, always the same, while the third base must be either one of the two purines (A or G) or one of the two pyrimidines (U or C). Thus, 5′ CAA 3′ and 5′ CAG 3′ are both codons for glutamine; 5′ CAU 3′ and 5′ CAC 3′ are both codons for histidine. If Pu stands for either purine and Py stands for either pyrimidine, then CAPu represents the codons for glutamine, while CAPy represents the codons for histidine.

In fact, the 5′ nucleotide of a tRNA's anticodon can often pair with more than one kind of nucleotide in the 3′ position of an mRNA's codon. (Recall that after base pairing, the bases in the anticodon run antiparallel to the bases in the codon.) A single tRNA charged with a particular amino acid can thus recognize several or even all of the codons for that amino acid. This flexibility in base pairing between the 3′ nucleotide in the codon and the 5′ nucleotide in the anticodon is known as **wobble (Fig. 8.21a)**. The combination of normal base pairing at the first two positions of a codon with wobble at the third position clarifies why multiple codons for a single amino acid usually start with the same two letters.

An important aspect of wobble is the chemical modification of certain bases at the 5′ end of the anticodon (the *wobble position*) (**Fig. 8.21b** and **c**). An A in the wobble position of a tRNA is almost always modified to inosine

Figure 8.21 Wobble: Some tRNAs recognize more than one codon for the same amino acid. (a) The G at the 5′ end of the anticodon shown here can pair with either U or C at the 3′ end of the codon. **(b)** The table shows the pairing possibilities for nucleotides at the 5′ end of an anticodon (the wobble position). xo^5U only rarely pairs with C. k^2C occurs only in certain bacteria. **(c)** Chemical structures of the modified bases in anticodons.

(a)

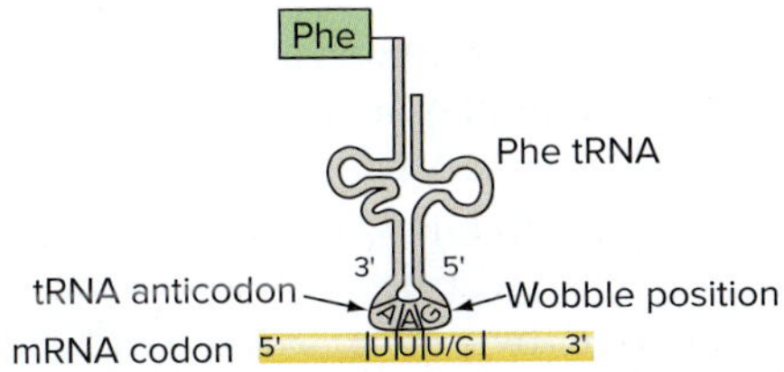

(b)

Wobble Rules		
5′ end of anticodon	→ can pair with →	3′ end of codon
G		U or C
C		G
I		U, C, or A
xm^5U		G
xm^5s^2U		A or G
xo^5U		A, G, U, or (C)
k^2C		A

(c) Modified bases in anticodon wobble position

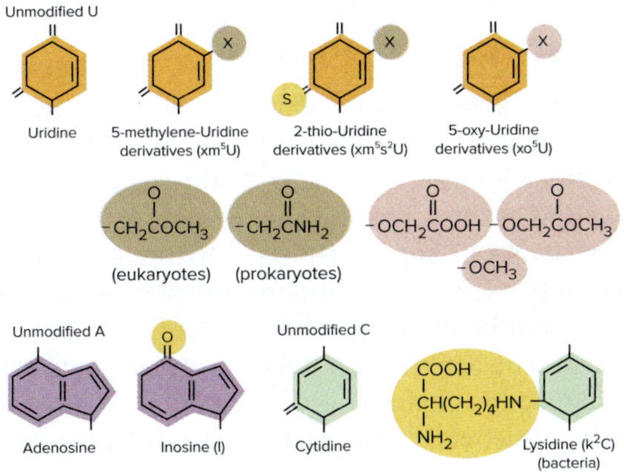

(I), and a U in the wobble position is always modified in one of three possible ways. By contrast, G in the anticodon wobble position is always unmodified, while modification of C occurs only in the tRNAs of some bacterial species. Wobble bases are modified by specific enzymes that act on the tRNA after it has been synthesized by transcription.

The wobble rules in Fig. 8.21c delimit the anticodon sequences and the wobble base modifications consistent with the genetic code. For example, methionine (Met) is specified by a single codon (5′ AUG 3′). As a result, Met-specific tRNAs must either have a C at the 5′ end of their anticodons (5′ CAU 3′) or a U that is modified to xm^5U, because these are the only nucleotides at that position that can base pair only with the G at the 3′ end of the Met codon. By contrast, a single isoleucine-specific tRNA with the modified nucleotide inosine (I) at the 5′ position of the anticodon can recognize all three codons (5′ AUU 3′, 5′ AUC 3′, and 5′ AUA 3′) for isoleucine.

A special tRNA for selenocysteine

Most mRNAs direct the synthesis of proteins containing only the 20 common amino acids. Exceptional mRNAs in bacteria and eukaryotes direct the synthesis of *selenoproteins,* which contain the amino acid selenocysteine (Sec), sometimes referred to as amino acid 21. Selenoproteins are rare; in humans, only 25 are known to exist.

As shown in **Fig. 8.22,** a dedicated selenocysteine tRNA ($tRNA^{Sec}$) is recognized by serine tRNA synthetase and charged with serine. Modification enzymes subsequently convert the Ser to Sec. The Sec-charged $tRNA^{Sec}$ interacts with 5′ UGA 3′ triplets found only in mRNAs that contain a special structure called the *Sec insertion sequence* (*SECIS*) element. The SECIS element is a region of the mRNA that forms a particular stem-loop (hairpin) structure through intramolecular complementary base pairing (Fig. 8.22). This stem loop prevents termination of polypeptide synthesis at the UGA triplet, which would otherwise act as a stop codon. The anticodon of the Sec-charged $tRNA^{Sec}$ binds to the UGA triplet in the mRNA, allowing the incorporation of Sec into the polypeptide product.

Figure 8.22 How rare proteins incorporate selenocysteine. **(a)** The serine carried by $tRNA^{Sec}$ with the anticodon 5′ UCA 3′ is modified to selenocysteine (Sec). The Sec-charged tRNA recognizes the triplet UGA only in rare mRNAs with a downstream SECIS element. The U in the wobble position of this tRNA is modified in an unusual manner (indicated as U^) and so it recognizes only A.

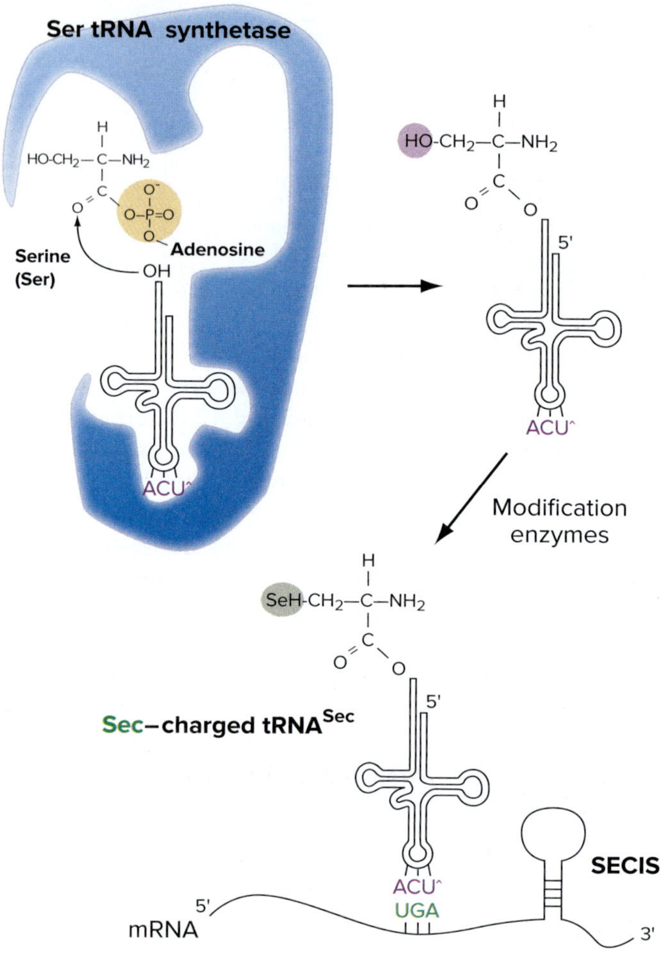

Ribosomes Are the Sites of Polypeptide Synthesis

Ribosomes facilitate polypeptide synthesis in various ways. First, they recognize mRNA features that signal the start of translation. Second, they help ensure accurate interpretation of the genetic code by stabilizing the interactions between tRNAs and mRNAs; without a ribosome, codon-anticodon recognition, mediated by only three base pairs, would be extremely weak. Third, ribosomes supply the enzymatic activity that links the amino acids in a growing polypeptide chain. Fourth, by moving 5′ to 3′ along an mRNA molecule, they expose the mRNA codons in sequence, ensuring the linear addition of amino acids. Finally, ribosomes help end polypeptide synthesis by dissociating both from the mRNA directing polypeptide construction and from the polypeptide product itself.

The structure of ribosomes

In *E. coli,* ribosomes consist of three different ribosomal RNAs (rRNAs) and 52 different ribosomal proteins (**Fig. 8.23a**). These components associate to form two different ribosomal subunits called the *30S subunit* and the *50S subunit.* (S designates a coefficient of sedimentation related to the size and shape of the subunit; the 30S subunit is smaller than the 50S subunit). Before translation begins, the two subunits exist as separate entities in the cytoplasm. Soon after the start of translation, they come together to reconstitute a complete ribosome. Eukaryotic ribosomes have more components than their prokaryotic counterparts, but they still consist of two dissociable subunits.

Figure 8.23 The ribosome: Site of polypeptide synthesis. **(a)** A ribosome has two subunits, each composed of rRNA and various proteins. **(b)** The small subunit initially binds to mRNA. The large subunit contributes the enzyme peptidyl transferase, which catalyzes the formation of peptide bonds. The two subunits together form the A, P, and E tRNA binding sites.

(a) A ribosome has two subunits composed of RNA and protein.

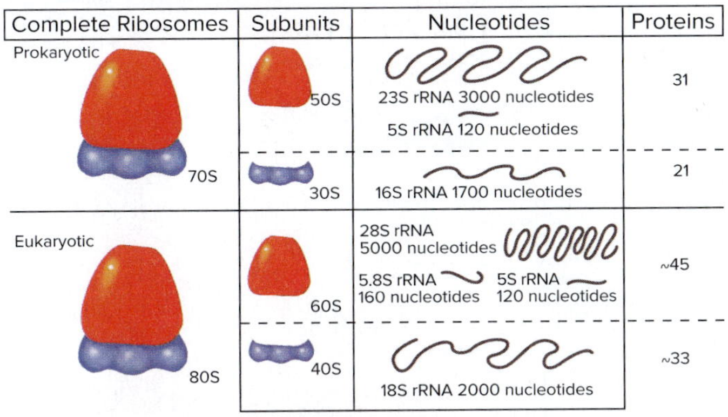

(b) Different parts of a ribosome have different functions.

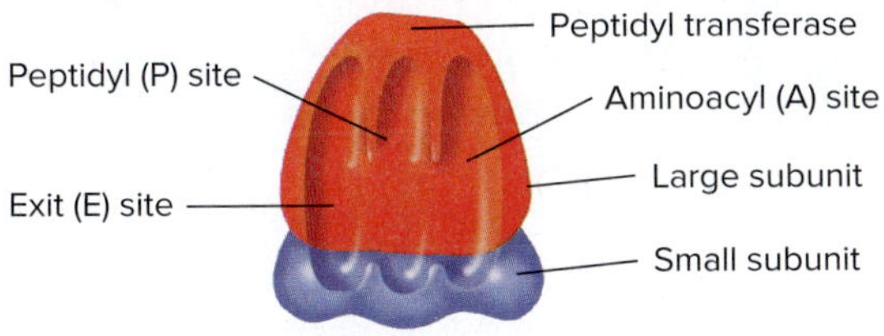

Figure 8.24 High resolution view of a bacterial ribosome in action. The large subunit is at the top; its 23S and 5S rRNA components are in *bright blue* and its various protein components in *bright green*. The small subunit is at the bottom, with the 16S rRNA in *gray* and its protein components in *aqua*. Two tRNA molecules are in *gold*, with the tRNA on the left at the E site and the tRNA to its right at the P site. The A site is occupied by a protein release factor shown in *pink*. A few of the nucleotides in the mRNA (*red*) can be seen near the bottoms of the tRNAs and release factor. This ribosome is acting during a stage of translation just prior to the ejection of the tRNA from the E site during the termination phase depicted in the left panel of Fig. 8.25c.

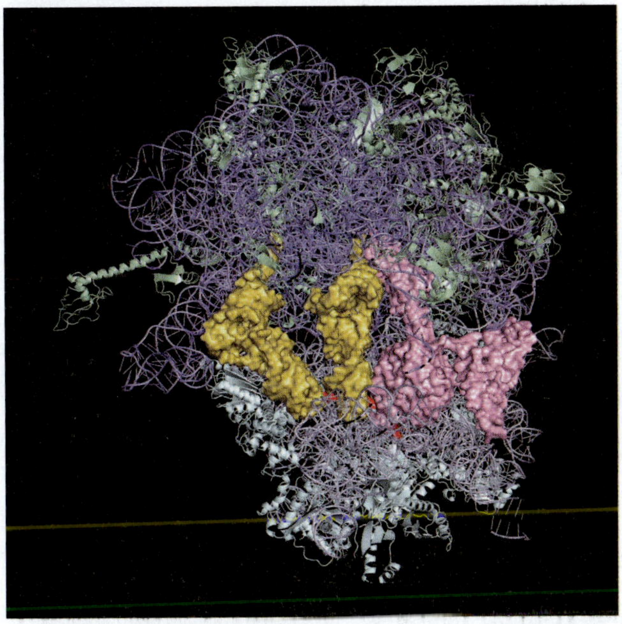

Functional domains of ribosomes

The small 30S subunit is the part of the ribosome that initially binds to mRNA. The larger 50S subunit contributes an enzyme known as **peptidyl transferase,** which catalyzes formation of the peptide bonds joining adjacent amino acids (**Fig. 8.23b**). Both the small and the large subunits contribute to three distinct tRNA binding areas known as the **aminoacyl** (or **A**) **site,** the **peptidyl** (or **P**) **site,** and the **exit** (or **E**) **site.** Finally, other regions of the ribosome distributed over the two subunits serve as points of contact for some of the additional proteins that play roles in translation.

Using X-ray crystallography and elegant techniques of electron microscopy, researchers have gained a remarkably detailed view of the complicated structure of the ribosome. **Figure 8.24** shows the interior of a ribosome nearing the completion of the translation of an mRNA; some parts of the ribosome that extend out of the page toward the reader were removed computationally so you can better see the tRNAs occupying the E site and P site.

With this illustration, you can see that the rRNAs occupy most of the space in the central part of the ribosome, while the various ribosomal proteins are studded around the exterior. Surprisingly, no proteins are found close to the region where the peptide bonds are formed during translation. This finding supports the conclusions of biochemical experiments that peptidyl transferase is actually a function of the large subunit's rRNA rather than any protein component of the ribosome. In other words, the rRNA acts as a *ribozyme* that joins amino acids together.

During translation, the ribosome associates briefly with various proteins that aid steps in the process. For example, **Fig. 8.24** shows that late in translation when the completed polypeptide is released from the ribosome, a protein called a *release factor* binds to the ribosome's A site. Remarkably, the release factor can associate with the A site because part of this protein folds in three-dimensional space in a way that mimics the structure of a tRNA.

Ribosomes and Charged tRNAs Collaborate to Translate mRNAs into Polypeptides

As was the case for transcription, translation consists of three phases: an **initiation** phase that sets the stage for polypeptide synthesis; **elongation,** during which amino acids are added to a growing polypeptide; and a **termination** phase that brings polypeptide synthesis to a halt and enables the ribosome to release a completed chain of amino acids. **Figure 8.25** illustrates the details of the process, focusing on translation as it occurs in bacterial cells. As

FEATURE FIGURE 8.25

Translation of mRNAs on Ribosomes

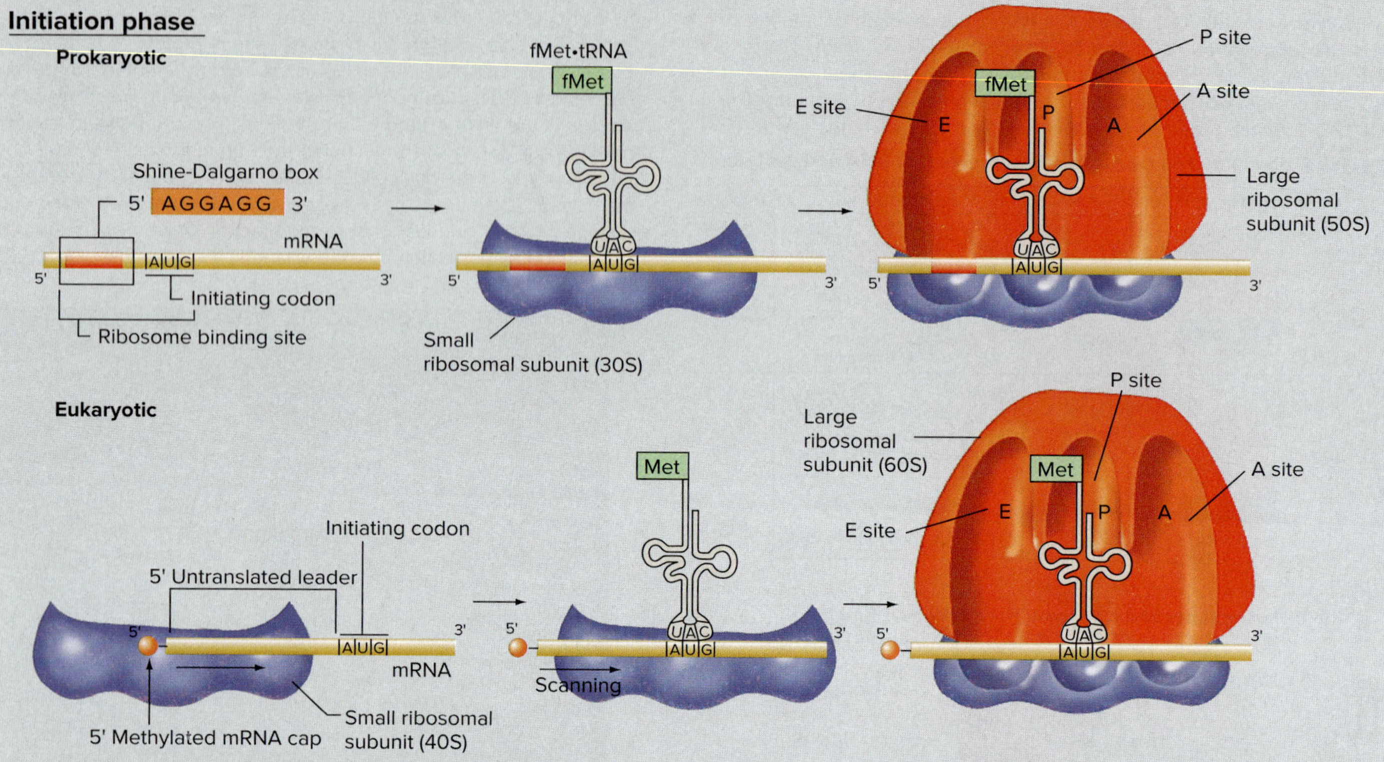

(a) Initiation: Setting the stage for polypeptide synthesis The first three nucleotides of an mRNA are not the first codon. Instead, a special signal indicates where along the mRNA translation should begin. In prokaryotes, this signal is the **ribosome binding site,** and it has two important elements. The first is a short sequence of six nucleotides–usually 5′. . . AGGAGG . . . 3′—named the **Shine-Dalgarno box.** The second element in an mRNA's ribosome binding site is the triplet 5′ AUG 3′, which serves as the initiation codon.

The 5′ CAU 3′ anticodon of a special initiator tRNA recognizes the AUG in the ribosome binding site. The initiator tRNA carries ***N*-formylmethionine (fMet),** a modified methionine whose amino end is blocked by a formyl group. A different tRNA that is charged with an unmodified methionine recognizes AUG codons located within an mRNA's reading frame; this tRNA cannot start translation.

During initiation, the 3′ end of the 16S rRNA in the 30S ribosomal subunit binds to the mRNA's Shine-Dalgarno box (*not shown*), the fMet tRNA binds to the mRNA's initiation codon, and a large 50S ribosomal subunit associates with the small subunit to round out the ribosome. At the end of initiation, the fMet tRNA sits in the P site of the completed ribosome. Proteins known as **initiation factors** (*not shown*) play a transient role in the initiation process.

In eukaryotes, the small ribosomal subunit binds first to the methylated cap at the 5′ end of the mature mRNA. The small subunit then migrates to the initiation site—usually the first AUG it encounters as it scans the mRNA in the 5′-to-3′ direction. The initiator tRNA in eukaryotes carries unmodified methionine (Met) instead of fMet.

(b) Elongation: The addition of amino acids to a growing polypeptide Proteins known as **elongation factors** (*not shown*) usher the appropriate tRNA into the A site of the ribosome. The anticodon of this charged tRNA must recognize the next codon in the mRNA. The ribosome simultaneously holds the initiating tRNA at its P site and the second tRNA at its A site so that peptidyl transferase can catalyze formation of a peptide bond between the amino acids carried by the two tRNAs. As a result, the tRNA at the A site now carries two amino acids. The N terminus of this dipeptide is fMet or Met; the C terminus is the second amino acid, whose carboxyl group remains covalently linked to its tRNA.

After the first peptide bond forms, the ribosome moves with the help of elongation factors, exposing the next mRNA codon. As the ribosome moves, the initiating tRNA, which no longer carries an amino acid, is transferred to the E site, and the other tRNA, which carries the dipeptide, shifts from the A site to the P site.

The empty A site now receives another tRNA, whose identity is determined by the next codon in the mRNA. The uncharged initiating tRNA is bumped off the E site and leaves the ribosome. Peptidyl transferase then catalyzes the formation of a second peptide bond, generating a chain of three amino acids connected at its C terminus to the tRNA currently in the A site. With each subsequent round of ribosome movement and peptide bond formation, the peptide chain grows one amino acid longer. Note that each tRNA moves from the A site to the P site to the E site (excepting the initiating tRNA, which first enters the P site).

Because the ribosome adds amino acids to the C terminus of the growing chain, polypeptide synthesis proceeds from the N terminus to the C terminus. As a result, the initial fMet in prokaryotes (Met in eukaryotes), will be the N-terminal amino

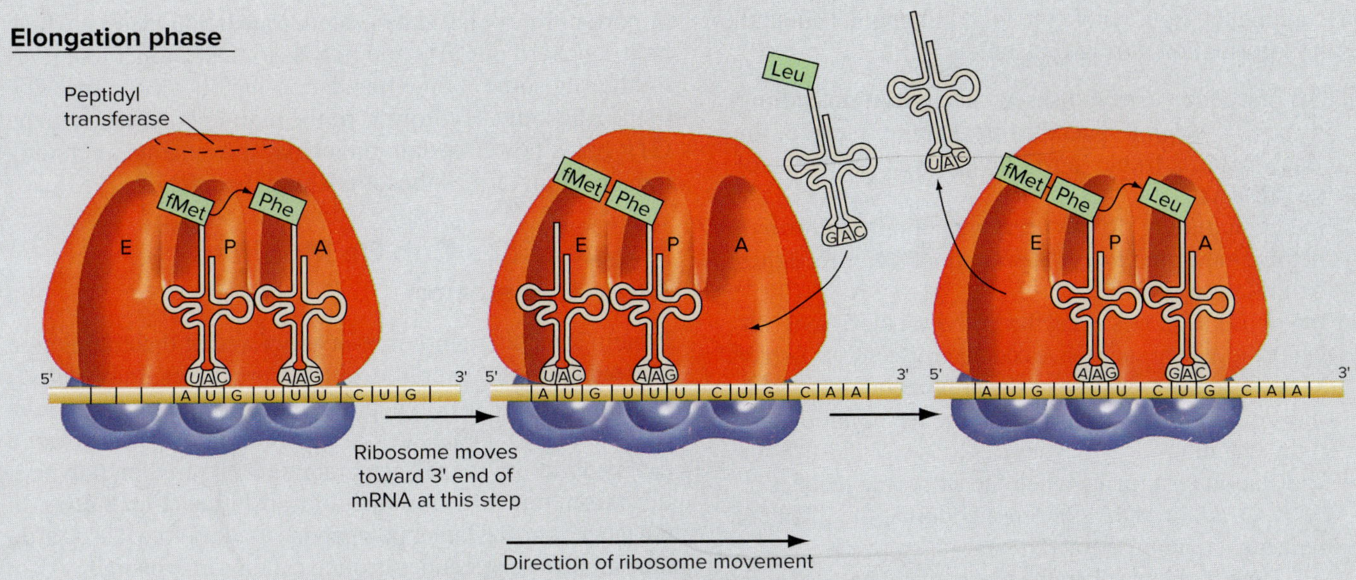

acid of all finished polypeptides prior to protein processing. Moreover, the ribosome must move along the mRNA in the 5′-to-3′ direction so that the polypeptide can grow in the N-to-C direction.

Once a ribosome has moved far enough away from the mRNA's ribosome binding site, that site becomes accessible to other ribosomes. In fact, several ribosomes can work on the same mRNA at one time. A complex of several ribosomes translating from the same mRNA is called a **polyribosome.** This complex allows the simultaneous synthesis of many copies of a polypeptide from a single mRNA.

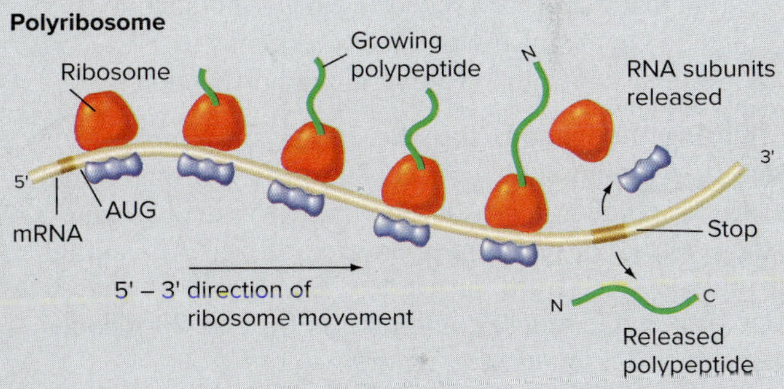

(c) Termination: The ribosome releases the completed polypeptide No normal tRNAs (except for $tRNA^{Sec}$) carry anticodons complementary to any of the three nonsense (stop) codons UAG, UAA, and UGA. Thus, when a nonsense codon moves into the ribosome's A site, no tRNAs can bind to that codon. Instead, proteins called **release factors** recognize the stop codons and halt polypeptide synthesis. The tRNA specifying the C-terminal amino acid releases the completed polypeptide, the same tRNA as well as the mRNA separate from the ribosome, and the ribosome dissociates into its large and small subunits.

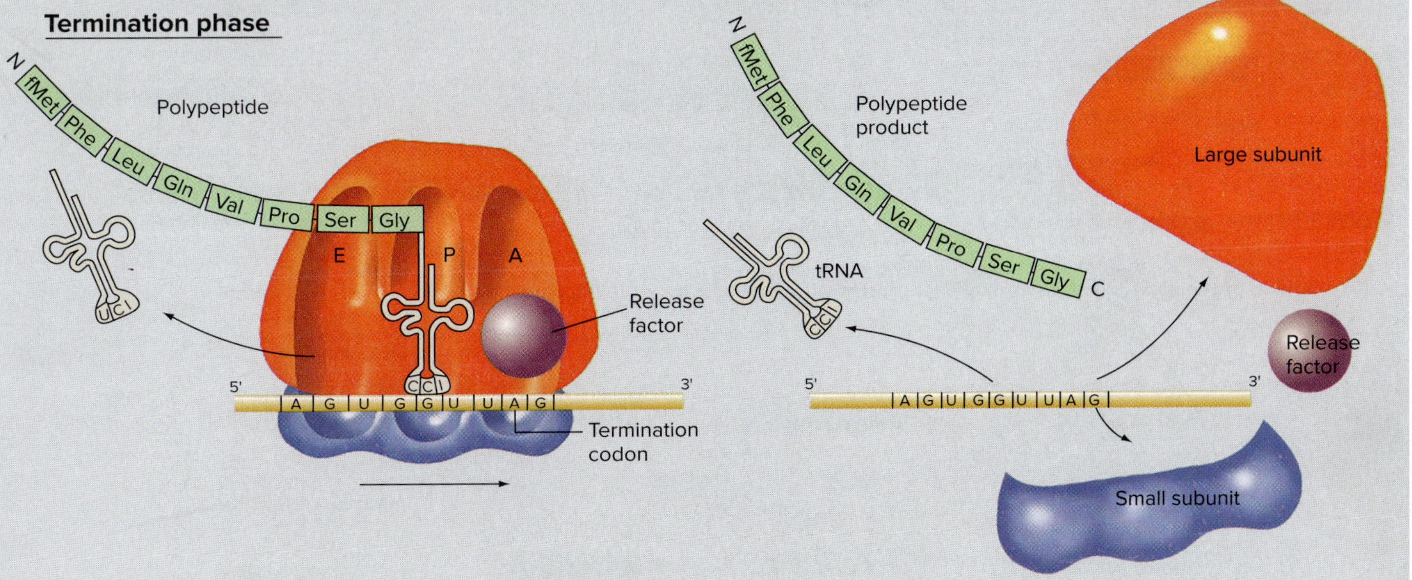

you examine the figure, note the following points about the flow of information during translation:

- The first codon to be translated—the **initiation codon**—is an AUG set in a special context at the 5′ end of the gene's reading frame (never precisely at the 5′ end of the mRNA).
- Special initiating tRNAs carrying a modified form of methionine called *formylmethionine* (*fMet*) recognize the initiation codon.
- The ribosome moves along the mRNA in the 5′-to-3′ direction, revealing successive codons in a stepwise fashion.
- At each step of translation, the polypeptide grows by the addition to its C terminus of the next amino acid in the chain.
- Translation terminates when the ribosome reaches a UAA, UAG, or UGA nonsense codon at the 3′ end of the gene's reading frame.

  These points explain the biochemical basis of colinearity, that is, the correspondence between the 5′-to-3′ direction in the mRNA and the N-terminus-to-C-terminus direction in the resulting polypeptide.

During elongation, the translation machinery adds about 2–15 amino acids per second to the growing chain. The speed is higher in prokaryotes and lower in eukaryotes. At these rates, construction of an average-size 300-amino-acid polypeptide (from an average-length mRNA that is about 1000 nucleotides) could take as little as 20 seconds or as long as 2.5 minutes.

Several details have been left out of Fig. 8.25 so that you can concentrate on the flow of information during translation. In particular, this figure does not depict the important roles played by protein translation factors, which help shepherd mRNAs and tRNAs to their proper locations on the ribosome. Some translation factors also carry GTP to the ribosome. Hydrolysis of the high-energy bonds in the GTP helps power certain molecular movements, including translocation of the ribosome along the mRNA.

Polypeptides Can Be Modified After Translation

Protein structure is not irrevocably fixed at the completion of translation. Several different processes may subsequently modify a polypeptide's structure. Cleavage may remove amino acids, such as the N-terminal fMet, from a polypeptide, or it may generate several smaller polypeptides from one larger product of translation (**Fig. 8.26a**). In the latter case, the larger polypeptide made before it is cleaved into smaller polypeptides is often called a **polyprotein.** Also, some proteins are synthesized in inactive forms called *zymogens* that are activated by enzymatic cleavage that removes an N-terminal *prosegment.*

Enzymatic addition of chemical constituents, such as phosphate groups, carbohydrates, fatty acids, or even other small peptides to specific amino acids may also modify a polypeptide after translation (**Fig. 8.26b**). Such changes to polypeptides are known as **posttranslational modifications.** Posttranslational changes to a protein can be very important: For example, the biochemical function of many enzymes depends directly on the addition (or sometimes removal) of phosphate groups. Posttranslational modifications may alter the way a protein folds, its ability to interact with other proteins, its stability, its activity, or its location in the cell.

Figure 8.26 Posttranslational processing can modify polypeptide structure. **(a)** Enzymatic cleavage processes many proteins into their mature forms. **(b)** Enyzmes add various functional groups to specific amino acids.

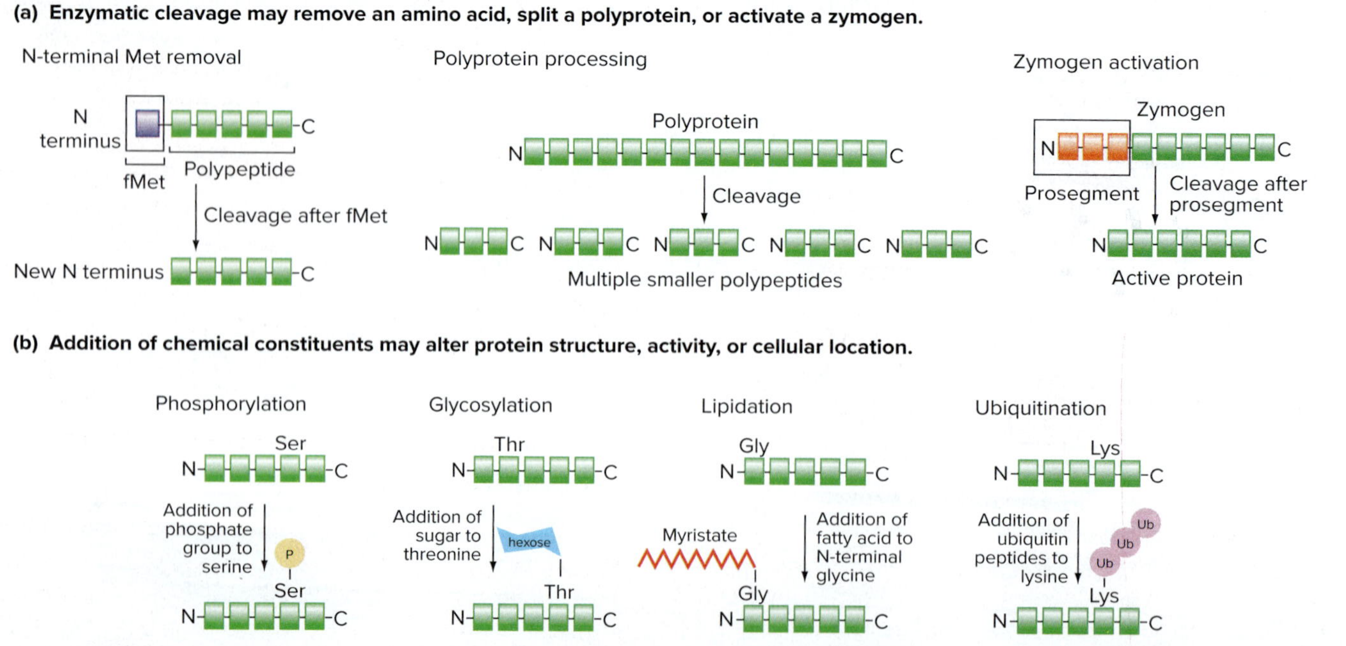

essential concepts

- *Translation* is the process by which *ribosomes* synthesize proteins according to the instructions in mRNAs. Ribosomes have specific binding sites for *tRNAs* (the A, P, and E sites) and supply *peptidyl transferase,* the ribozyme that forms peptide bonds between amino acids.
- *Transfer RNAs* (*tRNAs*) are the adapters that link mRNA codons to amino acids at the ribosome. *Aminoacyl-tRNA synthetases* connect the correct amino acids to their corresponding tRNAs.
- Each tRNA has an *anticodon* complementary to the mRNA codon specifying the particular amino acid. Because of *wobble,* a tRNA may recognize more than one codon.
- Translation *initiation* begins when a charged $tRNA^{fMet}$ (or $tRNA^{Met}$) binds the *start codon*, AUG, at the ribosomal P site.
- During *elongation,* the amino acid connected to the tRNA at the P site forms a peptide bond with the amino acid connected to the tRNA at the A site. The ribosome then moves in the 5′-to-3′ direction along the mRNA to the next codon.
- *Termination* occurs when the ribosome encounters an in-frame *stop codon* in the mRNA.
- *Posttranslational processing* enzymes may cleave a polypeptide or add chemical constituents to it.

8.4 Differences in Gene Expression Between Prokaryotes and Eukaryotes

learning objectives

1. Explain how the nuclear membrane affects gene expression in eukaryotes.
2. Discuss the function of enhancer sequences in eukaryotic transcription.
3. Describe the differences in translation initiation between prokaryotes and eukaryotes.
4. List the steps in mRNA formation that occur in eukaryotes but not in prokaryotes.

The processes of transcription and translation in eukaryotes and prokaryotes are similar in many ways but also are affected by certain differences, including (1) the presence of a nuclear membrane in eukaryotes, (2) eukaryotic-specific complexities in the mechanisms by which RNA polymerase recognizes promoters to start transcription, (3) variations in the way in which translation is initiated, and (4) the need for additional transcript processing in eukaryotes.

In Eukaryotes, the Nuclear Membrane Prevents the Coupling of Transcription and Translation

In *E. coli* and other prokaryotes, transcription takes place in an open intracellular space undivided by a nuclear membrane. Translation occurs in the same open space and is sometimes coupled directly with transcription (**Table 8.1**). This coupling is possible because transcription extends mRNAs in the same 5′-to-3′ direction as the ribosome moves along the mRNA. As a result, ribosomes can begin to translate a partial mRNA that the RNA polymerase is still in the process of transcribing from the DNA.

The coupling of transcription and translation has significant consequences for the regulation of gene expression in prokaryotes. For example, in an important regulatory mechanism called *attenuation,* which we describe in Chapter 16, the rate of translation of some mRNAs directly determines the rate at which the corresponding genes are transcribed into these mRNAs.

Such coupling cannot occur in eukaryotes because the nuclear envelope physically separates the sites of transcription and RNA processing in the nucleus from the site of translation in the cytoplasm. As a result, translation in eukaryotes can affect the rate at which genes are transcribed only in more indirect ways.

Distant Enhancer Sequences and Interactions with Chromatin Influence Eukaryotic Promoters

In eukaryotes, the promoters recognized by RNA polymerase to initiate transcription are affected by two situations not seen in prokaryotes (Table 8.1). First, as previously seen in Fig. 8.11, the stability of RNA polymerase's interaction with the promoter is often affected by enhancer sequences located far from the promoter. In prokaryotes, the DNA sequences that regulate transcription are all found much closer to the promoter. Second, eukaryotic chromosomes are tightly wound around *histone proteins* in a DNA/protein complex called *chromatin.* To be recognized by RNA polymerase, the promoter of a eukaryotic gene must first be unwound from chromatin. Interestingly, clearing the histones from the promoter is an important function of enhancers. (Histones and chromatin and their roles in transcription will be discussed in Chapters 12 and 17.)

Prokaryotes and Eukaryotes Initiate Translation Differently

In prokaryotes, translation begins at a ribosome binding site on the mRNA, which is defined by a short, characteristic sequence of nucleotides called a *Shine-Dalgarno box* adjacent to an initiating AUG codon (review Fig. 8.25a).

TABLE 8.1 Differences Between Prokaryotes and Eukaryotes in the Details of Gene Expression

	Prokaryotes	Eukaryotes
Overview	1. No nucleus. Transcription and translation take place in the same cellular compartments, and translation is often coupled to transcription. 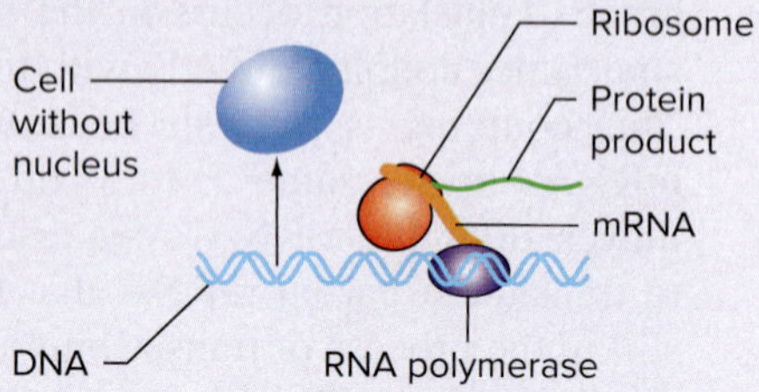	1. Nucleus separated from the cytoplasm by a nuclear membrane. Transcription takes place in the nucleus, while translation occurs in the cytoplasm. Direct coupling of transcription and translation is not possible. 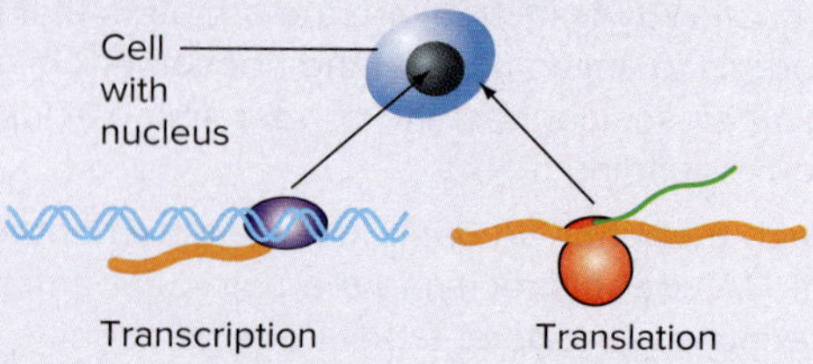
	2. Genes are not divided into exons and introns. 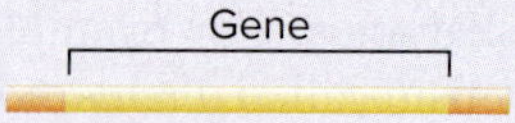	2. The DNA of a gene consists of exons separated by introns; the exons are defined by posttranscriptional splicing, which deletes the introns. 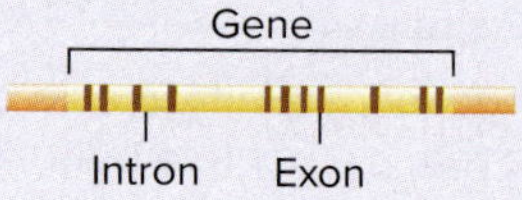
Transcription	1. One RNA polymerase consisting of five subunits.	1. Several kinds of RNA polymerase, each containing 10 or more subunits; different polymerases transcribe different genes.
	2. DNA sequences needed for transcription initiation are located close to the promoter.	2. Enhancer sequences far from the promoter are often needed for transcription initiation.
	3. Promoters are not wound up in chromatin. 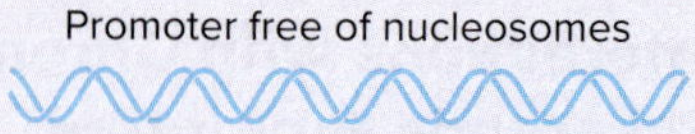	3. Transcription initiation requires promoters to be cleared of chromatin to allow access to RNA polymerase. 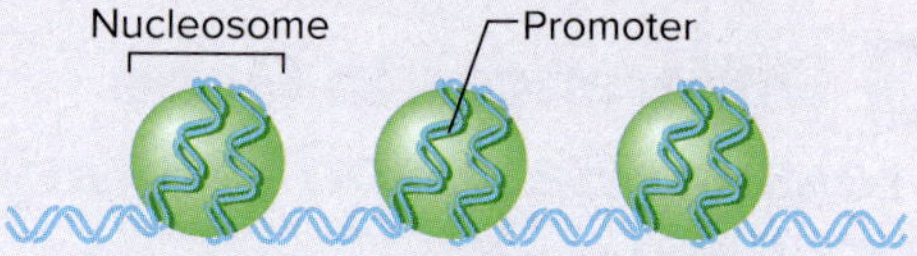
	4. Primary transcripts are the actual mRNAs; they have a triphosphate start at the 5′ end and no tail at the 3′ end. 	4. Primary transcripts undergo processing to produce mature mRNAs that have a methylated cap at the 5′ end and a poly-A tail at the 3′ end. 
Translation	1. Unique initiator tRNA carries formylmethionine.	1. Initiator tRNA carries methionine.
	2. mRNAs have multiple ribosome binding sites (RBSs) and can thus direct the synthesis of several different polypeptides. 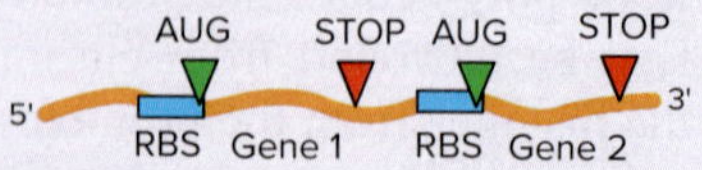	2. mRNAs have only one start site and can thus direct the synthesis of only one kind of polypeptide. 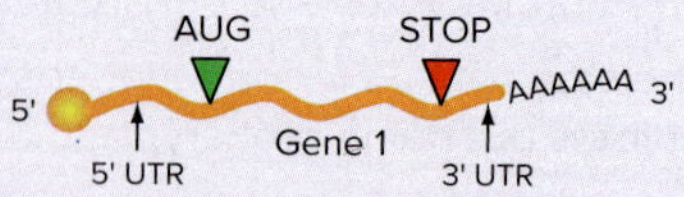
	3. Small ribosomal subunit immediately binds to the mRNA's ribosome binding site. 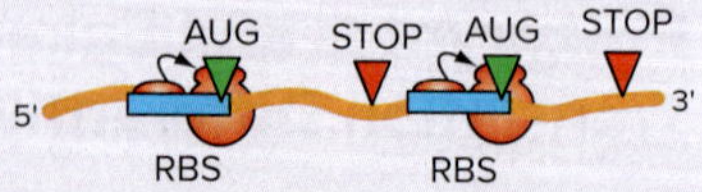	3. Small ribosomal subunit binds first to the methylated cap at the 5′ end of the mature mRNA and then scans the mRNA to find the ribosome binding site. 

There is nothing to prevent an mRNA from having more than one ribosome binding site, and in fact, many prokaryotic messages are **polycistronic:** They contain the information of several genes (sometimes referred to as *cistrons*), each of which can be translated independently starting at its own ribosome binding site (Table 8.1).

In eukaryotes, by contrast, the small ribosomal subunit first binds to the methylated cap at the 5′ end of the mature mRNA and then migrates through the 5′ UTR to the initiation site. This site is almost always the first AUG codon encountered by the ribosomal subunit as it moves along, or *scans,* the mRNA in the 5′-to-3′ direction (see Fig. 8.25a and Table 8.1). Because of this scanning mechanism, initiation in eukaryotes takes place at only a single site on the mRNA, and each mRNA is **monocistronic**—it contains the information for translating only a single kind of polypeptide.

Another difference between prokaryotic and eukaryotic translation is in the composition of the initiating tRNA. In prokaryotes, as already mentioned, this tRNA carries a modified form of methionine known as *N*-formylmethionine, while in eukaryotes, it carries an unmodified methionine (Table 8.1). Thus, immediately after translation, eukaryotic polypeptides all have Met (instead of fMet) at their N termini. Posttranslational cleavage in both prokaryotes and eukaryotes, however, often creates mature proteins that no longer have N-terminal fMet or Met (see Fig. 8.26a).

Eukaryotic mRNAs Require More Processing than Prokaryotic mRNAs

Table 8.1 reviews other important differences in gene structure and expression between prokaryotes and eukaryotes. In particular, introns interrupt eukaryotic, but not prokaryotic, genes such that the splicing of a primary transcript is necessary for eukaryotic gene expression. Other types of RNA processing that occur in eukaryotes but not prokaryotes add a methylated cap and a poly-A tail, respectively, to the 5′ and 3′ ends of the mRNAs.

essential concepts

- In prokaryotes, transcription and translation occur simultaneously. In eukaryotes, the nuclear membrane restricts transcription to the nucleus; mRNAs are translated only after transport into the cytoplasm.
- In eukaryotes, transcription initiation involves *enhancer sequences* located far from the promoter. In addition, the *chromatin* of eukaryotic chromosomes must be unwound to allow access by RNA polymerase.
- Prokaryotic mRNAs are *polycistronic* such that ribosomes can translate several different polypeptides from a single mRNA. Eukaryotes have *monocistronic* mRNAs that can be used to translate only a single protein.
- In prokaryotes, ribosomes bind to a sequence called the *Shine-Dalgarno box* adjacent to the AUG initiation codon. In eukaryotes, the small ribosome subunit binds at the 5′ cap and migrates until it encounters the initiation site.
- In prokaryotes, the primary transcripts are mRNAs immediately ready for translation. In eukaryotes, primary transcripts are processed prior to translation into mature mRNAs through the addition of 5′ caps and poly-A tails, as well as the removal of introns.

8.5 The Effects of Mutations on Gene Expression and Function

learning objectives

1. Compare silent mutations, missense mutations, nonsense mutations, and frameshift mutations in terms of how they alter a gene product.
2. Discuss mutations outside the coding sequence that could affect gene expression.
3. Explain why most loss-of-function alleles (hypomorphic or amorphic) are recessive to wild-type alleles, but some are incompletely dominant or dominant.
4. Contrast the actions of hypermorphic, neomorphic, and antimorphic gain-of-function alleles.
5. Give examples of mutations that can have global effects on gene expression.

We have seen that the information in DNA is the starting point of gene expression. The cell transcribes that information into mRNA and then translates the mRNA information into protein. Mutations that alter the nucleotide pairs of DNA can modify any of the steps or products of gene expression.

Mutations in a Gene's Coding Sequence May Alter the Gene Product

Because of the nature of the genetic code, mutations in a gene's amino acid–encoding exons generate a range of repercussions (**Fig. 8.27a**).

Silent mutations

One consequence of the code's degeneracy is that some mutations, known as **silent mutations,** can change a codon into a mutant codon that specifies exactly the same amino acid. The majority of silent mutations change the third nucleotide of a codon, the position at which most codons for the same amino acid differ. For example, a change from GCA to GCC in a codon would still yield alanine in the protein product. Because silent mutations do not alter the amino acid composition of the encoded polypeptide, such mutations usually affect neither gene expression nor phenotype.

Figure 8.27 **How mutations in a gene can affect its expression. (a)** Mutations in a gene's coding sequences. *Silent mutations* do not alter the protein's primary structure. *Missense mutations* replace one amino acid with another. *Nonsense mutations* shorten a polypeptide by replacing a codon with a stop signal. *Frameshift mutations* change the reading frame downstream of the addition or deletion. **(b)** Mutations outside the coding region can also disrupt gene expression.

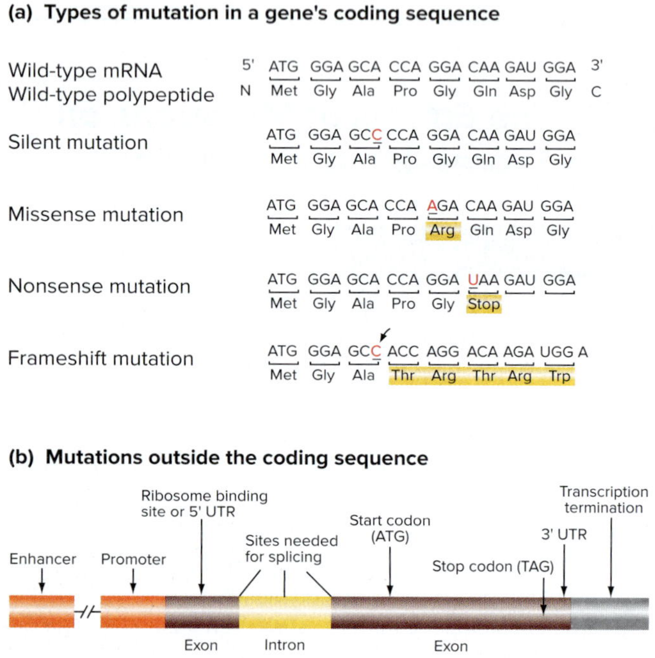

Missense mutations

Mutations that change a codon into a mutant codon that specifies a different amino acid are called **missense mutations.** If the substituted amino acid has chemical properties similar to the one it replaces, then this change may have little or no effect on protein function. Such mutations are **conservative substitutions.** For example, a mutation that alters a GAC codon for aspartic acid to a GAG codon for glutamic acid is a conservative substitution because both amino acids have acidic R groups.

By contrast, **nonconservative substitutions,** missense mutations that cause substitution of an amino acid with very different properties, are likely to have more noticeable consequences. A change of the same GAC codon for aspartic acid to GCC, a codon for alanine (an amino acid with an uncharged, nonpolar R group), is an example of a nonconservative substitution.

The effect on phenotype of any missense mutation is difficult to predict because it depends on how a particular amino acid substitution changes a protein's structure and function.

Nonsense mutations

Nonsense mutations change an amino acid–specifying codon to a premature stop codon. Nonsense mutations therefore result in the production of *truncated proteins* lacking all amino acids between the amino acid encoded by the mutant codon and the C terminus of the normal polypeptide. The mutant polypeptide will be unable to function if it requires the missing amino acids for its activity.

Frameshift mutations

Frameshift mutations result from the insertion or deletion of nucleotides within the coding sequence. As discussed earlier, if the number of extra or missing nucleotides is not divisible by 3, the insertion or deletion will skew the reading frame downstream of the mutation. As a result, frameshift mutations usually result in the formation of truncated proteins (because of the appearance of premature stop codons) with incorrect amino acids at their C termini.

Mutations Outside the Coding Sequence Can Also Alter Gene Expression

Mutations that produce a variant phenotype are not restricted to alterations in codons. Because gene expression depends on several signals other than the actual coding sequence, changes in any of these critical signals can disrupt the process (**Fig. 8.27b**).

We have seen that promoters and termination signals in the DNA of a gene instruct RNA polymerase to start and stop transcription. Changes in the sequence of a promoter that make it hard or impossible for RNA polymerase to associate with the promoter diminish or prevent transcription. Likewise, mutations in enhancers that disrupt them from being recognized by transcription factors also diminish the transcription of eukaryotic genes. Mutations in a termination signal can diminish the amount of mRNA produced and thus the amount of gene product.

In eukaryotes, most primary transcripts have splice-acceptor sites, splice-donor sites, and branch sites that allow splicing to join exons together with precision in the mature mRNA. Changes in any one of these sites can obstruct splicing. In some cases, the result will be the absence of mature mRNA and thus no polypeptide. In other cases, the splicing errors can yield aberrantly spliced mRNAs that encode altered forms of the protein.

Mature mRNAs have ribosome binding sites and in-frame stop codons indicating where translation should start and stop. Mutations affecting a ribosome binding site would lower the affinity of the mRNA for the small ribosomal subunit; such mutations are likely to diminish the efficiency of translation and thus the amount of polypeptide product. Mutations in a stop codon would produce longer-than-normal proteins that might be unstable or nonfunctional.

Most Mutations that Affect Gene Expression Reduce Gene Function

Mutations affect phenotype by changing either the amino acid sequence of a protein or the amount of the gene product produced. Any mutation inside or outside a coding region that reduces or abolishes protein activity in one of the many ways previously described is a **loss-of-function mutation.**

Figure 8.28 Why most loss-of-function mutant alleles are recessive to wild-type alleles. *Pink ellipses* represent amounts of an enzyme in *Drosophila* called xanthine dehydrogenase. Flies need only 10% of the enzyme produced in wild-type strains (A^+/A^+) to have normal eye color. Null allele a^1 and hypomorphic allele a^2 are recessive to wild-type because A^+/a^1 or A^+/a^2 heterozygotes have enough enzyme for normal eye color.

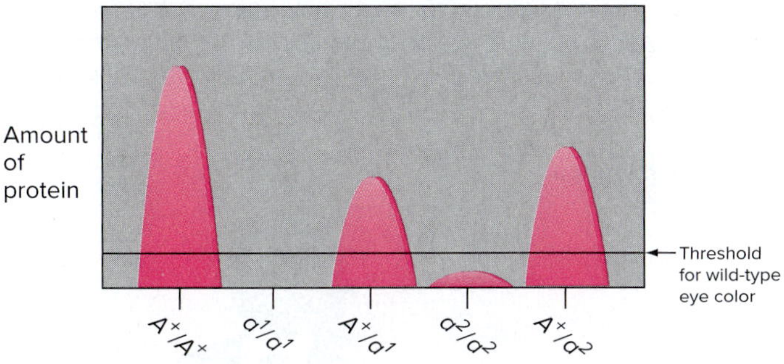

Recessive loss-of-function alleles

Loss-of-function alleles that block the function of a protein completely are called **null mutations,** or **amorphic mutations.** For protein-encoding genes, the mutations either prevent synthesis of the polypeptide or promote synthesis of a protein incapable of carrying out any function.

It is easy to understand why amorphic alleles are usually recessive to wild-type alleles. Consider an A^+/a^1 heterozygote, in which the wild-type A^+ allele generates functional protein, while the null a^1 allele does not (**Fig. 8.28**). If the amount of protein produced by the single A^+ allele (usually, though not always, half the amount produced in an A^+/A^+ cell) is above the threshold amount sufficient to fulfill the normal biochemical requirements of the cell, the phenotype of the A^+/a^1 heterozygote will be wild type. The vast majority of genes function in this way; A^+/A^+ cells actually make more than twice as much of the protein needed for the normal phenotype. Mendel's alleles for green pea color or wrinkled pea shape were likely null alleles and recessive to wild-type alleles for this reason. (Recall Fig. 2.20.)

A **hypomorphic mutation** is a loss-of-function allele that produces either less of the wild-type protein or a mutant protein that functions less effectively than the wild-type protein (a^2 in Fig. 8.28). Hypomorphic alleles are usually recessive to wild-type alleles for the same reason that amorphic alleles are usually recessive.

Incompletely dominant loss-of-function alleles

Some combinations of alleles generate phenotypes that vary continuously with the amount of functional gene product, giving rise to incomplete dominance. For example, loss-of-function mutations in a single pigment-producing gene can generate a red-to-white spectrum of flower colors, with the white resulting from the absence of an enzyme in a biochemical pathway (**Fig. 8.29**). Consider three alleles of the gene encoding enzyme R: R^+ specifies the wild-type amount of the enzyme; R^{50} generates half the normal amount of the same enzyme (or the full amount of an altered form that has half the normal level of activity); and R^0 is a null allele. R^+/R^0 heterozygotes produce pink flowers whose color is halfway between red and white because one-half the R^+/R^+ level of enzyme activity is not enough to generate a full red. Combining R^+ or R^0 with the R^{50} allele produces pigmentation intermediate between red and pink or between pink and white.

Figure 8.29 When a phenotype varies continuously with levels of protein function, incomplete dominance results.

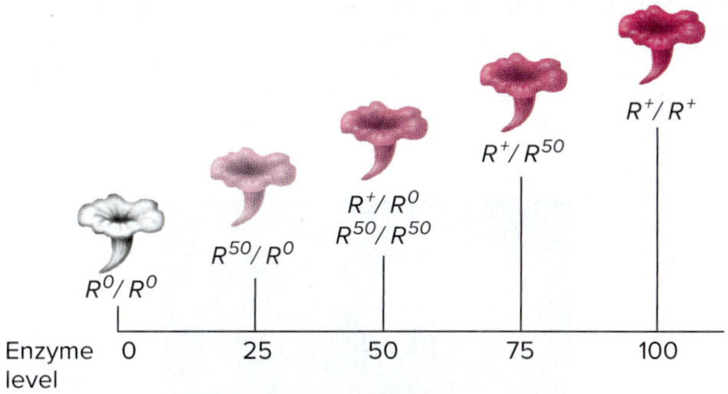

Unusual dominant loss-of-function alleles

With phenotypes exquisitely sensitive to the amount of functional protein produced, even a relatively small change of two-fold or less can cause a switch between distinct phenotypes. Therefore, a heterozygote for a loss-of-function mutation that generates less than the normal amount of functional gene product may look different from the wild-type organism. Geneticists use the term **haploinsufficiency** to describe relatively rare situations in which one wild-type allele does not provide enough of a gene product to avoid a mutant phenotype.

The number of haploinsufficient genes in humans is estimated to be about 800. One example of a human haploinsufficient gene is *GLI3,* which encodes a transcription factor important for the specification of digits. Heterozygosity for loss-of-function mutations in *GLI3* causes one form of *polydactyly*—the presence of extra fingers and toes (**Fig. 8.30**).

Unusual Gain-of-Function Alleles Are Almost Always Dominant

Because there are many ways to interfere with a gene's ability to make sufficient amounts of active protein, the large majority of mutations in most genes are loss-of-function alleles. However, rare mutations that either enhance a protein's function, confer a new activity on a protein, or express a protein at the wrong time or place act as **gain-of-function alleles.** Because a single such allele by itself usually produces a protein that can alter phenotype even in the presence of the normal protein, these unusual gain-of-function alleles are almost always dominant to wild-type alleles. Many dominant mutant alleles are lethal when homozygous.

Hypermorphic alleles

A **hypermorphic mutation** is one that generates either more normal protein product than the wild-type allele, or a

Figure 8.30 Haploinsufficiency: Some loss-of-function mutant alleles are dominant to wild-type alleles. The human *GLI3* gene is haploinsufficient. *GLI3/GLI3*$^{+}$ heterozygotes have extra fingers and toes, a condition known as *polydactyly*. One particular mutant *GLI3* allele is a nonsense mutation that changes codon 643 from arginine (R) to stop. (Wild-type GLI3 protein has 1580 amino acids.)

						643		
	V	T	K	K	Q	**R**	G	D
Normal allele (*GLI3*$^{+}$)	GTC	ACC	AAG	AAG	CAG	**C**GA	GGG	GAC
						↓		
Mutant allele (*GLI3*)	GTC	ACC	AAG	AAG	CAG	TGA	GGG	GAC
	V	T	K	K	**Q**	STOP		
					642			

more efficient mutant protein. For example, a hypermorphic mutation in the human *FGFR3* gene results in achondroplasia, the most common form of dwarfism (**Fig. 8.31a**). The *FGFR3* gene encodes a signaling protein (fibroblast growth factor receptor 3) that inhibits bone growth. FGFR3 protein is normally activated only when a small protein called FGF (fibroblast growth factor) binds to it (**Fig. 8.31b**). Most people with achondroplasia carry a mutant allele called *FGFR3*^{G480R}, which encodes an FGFR3 protein with arginine instead of the normal glycine at amino acid 480. This single amino acid change causes the mutant protein to become activated even in the absence of FGF. The mutant protein is thus a *constitutively active* receptor that is activated all the time (**Fig. 8.31c**). The hypermorphic allele (*FGFR3*^{G480R}) is dominant to the wild-type allele because the mutant protein remains active and continues to inhibit bone growth even if the normal protein is present.

Neomorphic alleles

A rare class of dominant gain-of-function alleles arises from **neomorphic mutations** that generate a novel phenotype. Some neomorphic alleles produce mutant proteins with a new function, while others cause genes to produce the normal protein but at an inappropriate time or place (**ectopic expression**).

Figure 8.31 Some hypermorphic alleles encode overactive proteins. (a) Achondroplasia, a form of dwarfism, is caused by a dominant hypermorphic mutant allele of the *FGFR3* gene, *FGFR3*^{G480R}. **(b)** The *FGFR3* gene encodes a dimeric transmembrane receptor protein that is normally activated only when it is bound to the small protein hormone FGF. The tyrosine kinase domain of one activated FGFR3 subunit adds phosphate groups (P in *yellow circles*) to the other subunit and *vice versa*. These phosphorylations initiate a signal that ultimately stops bone growth. **(c)** Mutant FGFR3^{G480R} protein is always activated, whether FGF is present or not, leading to improper bone development.

(a) Achondroplasia

(b) Normal FGFR **(c) Hypermorphic FGFR^{G480R}**

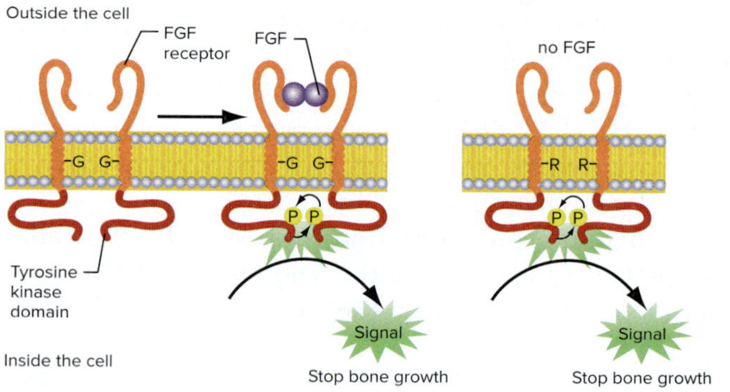

The dominant Huntington disease allele (*HD*) is an example of a neomorphic allele that makes a mutant protein. Recall from the Fast Forward Box in Chapter 7 (*Trinucleotide Repeat Diseases: Huntington Disease and Fragile X Syndrome*) that *HD*$^{+}$ is a polyQ-type trinucleotide repeat gene. Mutant HD proteins have an expanded run of glutamine (Q) amino acids, and for unknown reasons such mutant HD proteins cause neural degeneration. The *HD* disease allele is dominant to the normal allele because the presence of the normal HD protein (with fewer Qs) does not prevent the mutant HD protein from damaging nerve cells.

A striking example of a neomorphic allele that expresses a normal protein ectopically is the *Antp*^{Ns} mutant allele of the *Drosophila* gene *Antennapedia*. Flies that are *Antp*^{Ns}/*Antp*$^{+}$ heterozygotes have legs on their heads in place of antennae (**Fig. 8.32a**). The *Antp* gene encodes a protein that promotes leg development; accordingly, the wild-type allele *Antp*$^{+}$ is transcribed in tissues that will become the fly's legs. A mutation within the transcriptional control region of the gene instead causes the *Antp*^{Ns} allele to express normal protein in tissues destined to become the antennae (**Fig. 8.32b**). *Antp*^{Ns} is dominant because the *Antp*$^{+}$ allele does not prevent the ectopic expression of Antp protein in the cells normally destined to become antennae.

Antimorphic alleles

Some dominant mutant alleles of genes encode proteins that not only fail to provide the activity of the wild-type protein, but also prevent the normal protein from functioning. Such alleles are called **dominant-negative,** or **antimorphic,** alleles.

Figure 8.32 Neomorphic alleles can express a normal protein ectopically. (a) A neomorphic dominant mutation (*Antp*^{Ns}) in the fly *Antennapedia* gene produces flies with two legs growing out of the head (*right*); a normal fly head is shown at *left*. **(b)** *Antp*^{Ns} has a mutant transcriptional control region that results in ectopic expression of the leg-determining Antp protein in cells normally destined to become antennae.

a.1: © Eye of Science/Science Source; a.2: © Juergen Berger/Science Source

(a) The dominant neomorphic allele *Antp*^{Ns} causes antenna-to-leg transformation

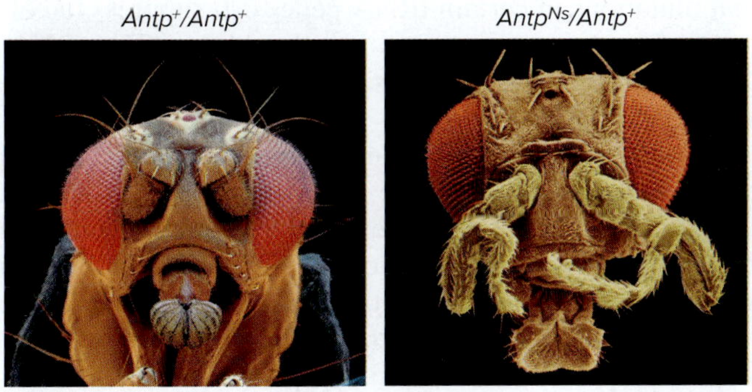

(b) *Antp*^{Ns} expresses Antp protein ectopically

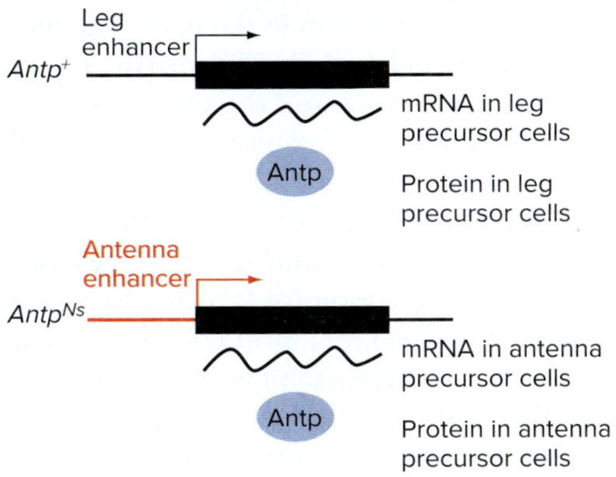

Consider, for example, a gene encoding a polypeptide that associates with three other identical polypeptides in a four-subunit enzyme. All four subunits are products of the same gene. If a dominant mutant allele *D* directs the synthesis of a *poison subunit* whose presence in the multimer—even as one subunit out of four—abolishes enzyme function, then active tetramers composed solely of functional wild-type d^{+} subunits are only 1/16 of all tetramers produced (**Fig. 8.33a**). As a result, total enzyme activity in D/d^{+} heterozygotes is far less than that seen in wild-type d^{+}/d^{+} homozygotes. The *Kinky* allele of the *Axin* gene in mice, which results in a malformed (kinky) tail, is an example of a dominant negative mutation with such a mechanism of action (**Fig. 8.33b**).

Figure 8.33 Why some dominant mutant alleles are antimorphic. (a) With proteins composed of four subunits encoded by a single gene, a dominant negative mutation may inactivate 15 out of every 16 multimers. **(b)** The *Kinky* allele of the mouse gene *Axin* (*Axin*^{Kinky}) is a dominant negative mutation that causes a kink in the tail. The Axin protein is a subunit of a protein complex; the protein encoded by *Axin*^{Kinky} prevents the complex from working.

b: © Tom Vasicek

(a) The antimorphic allele *D* encodes a mutant poison subunit

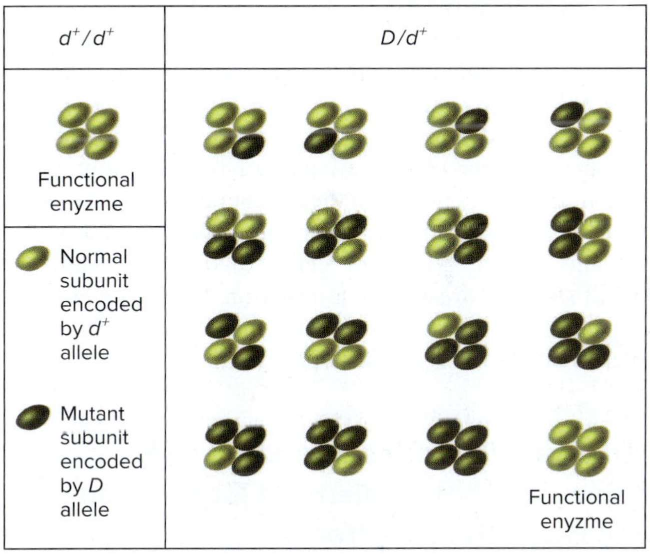

(b) Phenotypic effect of the poison subunit encoded by the antimorphic allele *Axin*^{Kinky}

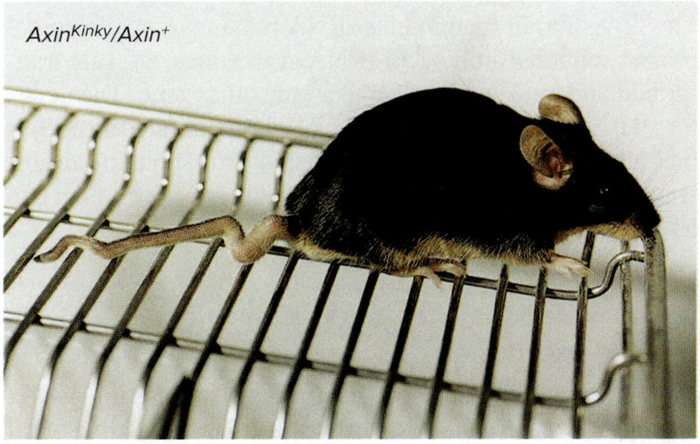

TABLE 8.2 Mutations Classified by Their Effects on Protein Function

	Loss-of-Function		Gain-of-Function		
Mutation Type	Amorphic (null)	Hypomorphic (leaky)	Hypermorphic	Neomorphic	Antimorphic (dominant negative)
Occurrence	Common	Common	Rare	Rare	Rare
Possible Dominance Relations	Usually recessive Can be incompletely dominant if phenotype varies continuously with gene product Can be dominant in cases of haploinsufficiency		Usually dominant or incompletely dominant	Usually dominant	Usually dominant or incompletely dominant

The Effects of a Mutation Can Be Difficult to Predict

As noted previously, most mutations constitute loss-of-function alleles. The reason is that many changes in amino acid sequence are likely to disrupt a protein's function, and because most alterations in gene regulatory sites, such as promoters, will make those sites less efficient. Nonetheless, rare mutations at almost any location in a gene can result in a gain of function.

Even when you know how a mutation affects gene function, you cannot always predict whether the mutant allele will be dominant or recessive to a wild-type allele (**Table 8.2**). Although most loss-of-function mutations are recessive and almost all gain-of-function mutations are dominant, exceptions to these generalizations do exist. The reason is that dominance relations between the wild-type and mutant alleles of genes in diploid organisms depend on how drastically a mutation influences protein production or activity, and how thoroughly phenotype depends on the normal wild-type level of the protein.

Mutations in Genes Encoding the Molecules that Implement Expression May Have Global Effects

Gene expression depends on an astonishing number and variety of proteins and RNAs, each encoded by a separate gene. The genes for all the proteins (RNA polymerases, ribosomal protein subunits, aminoacyl-tRNA synthetases, etc.) are transcribed and translated the same as any other gene. The genes for all the rRNAs, tRNAs, and snRNAs are **noncoding genes** that are transcribed but not translated. Mutations in almost any of these genes, whether protein-coding or noncoding, can have a dramatic effect on phenotype.

Lethal mutations affecting the machinery of gene expression

Loss-of-function mutations in the genes encoding molecules that implement gene expression, such as ribosomal proteins or rRNAs, are often lethal in homozygotes because such mutations adversely affect the synthesis of all proteins in a cell. Even a 50% reduction in the amount of some of the proteins or RNAs required for gene expression can have severe repercussions. In *Drosophila,* for example, null mutations in many of the genes encoding the various ribosomal proteins are lethal when homozygous. Due to haploinsufficiency, the same mutations in heterozygotes cause a dominant *Minute* phenotype, in which the slow growth of cells delays the fly's development.

Suppressor mutations in tRNA genes

If more than one gene encoded the same molecule with a role in gene expression, a mutation in one of these genes would not necessarily be lethal and might even be advantageous. Bacterial geneticists have found, for example, that mutations in certain tRNA genes can suppress the effects of nonsense mutations in other genes. The tRNA-gene mutations that have this effect give rise to **nonsense suppressor tRNAs.**

Consider, for instance, an otherwise wild-type *E. coli* population with an in-frame UAG nonsense mutation in the tryptophan synthase gene. All cells in this population make a truncated, nonfunctional form of the tryptophan synthase enzyme and are thus tryptophan auxotrophs (Trp^-) unable to synthesize tryptophan (**Fig. 8.34a**). Subsequent exposure of these auxotrophs to mutagens, however, generates some Trp^+ cells that carry two mutations: One is the original tryptophan synthase nonsense mutation, and the second is a mutation in the gene that encodes a tRNA for the amino acid tyrosine. Evidently, the mutation in the tRNA gene suppresses the effect of the nonsense mutation, restoring the function of the tryptophan synthase gene.

As **Fig. 8.34b** illustrates, the basis of this nonsense suppression is that the $tRNA^{Tyr}$ mutation changes an anticodon that recognizes the codon for tyrosine to an anticodon complementary to the UAG stop codon. The mutant tRNA can therefore insert tyrosine into the polypeptide at the position of the in-frame UAG nonsense mutation, allowing the cell to make at least some full-length enzyme.

Figure 8.34 Nonsense suppression. **(a)** A nonsense mutation that generates a stop codon causes production of a truncated, nonfunctional polypeptide. **(b)** A second, nonsense-suppressing mutation in a tRNA gene causes the addition of an amino acid in response to the stop codon, allowing production of a full-length polypeptide.

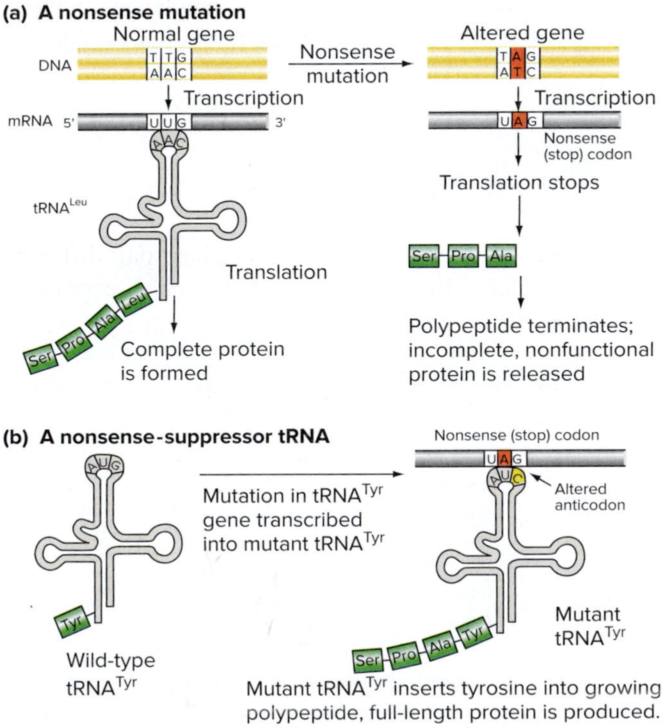

Similarly, mutations in the anticodons of other tRNA genes can suppress UGA or UAA nonsense mutations.

Cells with a nonsense-suppressing mutation in a tRNA gene can survive only if two conditions coexist with the mutation. First, the cell must have other tRNAs that recognize the same codon as the suppressing tRNA recognized before mutation altered its anticodon. Without such tRNAs, the cell has no way to insert the proper amino acid in response to that codon (in our example, the codon for tyrosine). Second, the suppressing tRNA must respond only inefficiently to the stop codons normally found at the ends of mRNA coding regions. If this were not the case, the suppressing tRNA would wreak havoc in the cell, producing a whole array of aberrant polypeptides that are longer than normal. One way cells guard against this possibility is to place two stop codons in a row at the ends of many genes. Because a suppressing tRNA's chance of inserting an amino acid at both of these codons is very low, only a small number of extended proteins arise.

essential concepts

- *Point mutations* in the coding sequences of a gene may modify the amino acid sequence of the polypeptide product.
- Mutations outside the coding sequences can modify gene expression by altering the amount, time, or place of protein production.
- *Loss-of-function mutations* reduce or abolish gene expression. Most loss-of-function alleles are recessive to wild-type alleles. When phenotype varies continuously with the amount of gene product, loss-of-function alleles are incompletely dominant. In *haploinsufficiency,* half the normal amount of gene product is not enough for a normal phenotype, so a loss-of-function mutant allele has dominant effects.
- Rare *gain-of-function mutations* cause increased protein production, synthesis of an altered protein, or production of the normal protein in the wrong context. Most gain-of-function alleles are dominant to wild-type alleles.
- Whether a mutation is recessive or dominant to wild type depends on how drastically the protein product is altered and how sensitive phenotype is to the abnormal gene function.
- Mutations in genes that encode molecules of the gene expression machinery are often lethal. Exceptions include mutations in tRNA genes that produce *nonsense suppressor tRNAs*.

WHAT'S NEXT

Our knowledge of gene expression enables us to redefine the concept of a gene. A gene is not simply the DNA that is transcribed into the mRNA codons specifying the amino acids of a particular polypeptide. Rather, *a gene is all the DNA sequences needed for expression of the gene as a polypeptide product.* A gene therefore includes the promoter sequences that govern where transcription begins and, at the opposite end, signals for the termination of transcription. A gene also must include sequences dictating where translation of the mRNA starts and stops. In addition to all of these features, eukaryotic genes contain introns that are spliced out of the primary transcript to make the mature mRNA. Because of introns, most eukaryotic genes are much larger than prokaryotic genes.

Even with introns, a single gene carries only a very small percentage of the nucleotide pairs in the chromosomes that make up a genome. In humans, the average gene is 16,000 nucleotide pairs in length. But the haploid human genome has roughly 3 billion (3,000,000,000) nucleotide pairs distributed among 23 chromosomes containing an average of 130 million nucleotide pairs apiece.

In Chapters 9, 10 and 11, we describe how researchers analyze the mass of genetic information in the chromosomes of a genome as they try to discover what parts of the DNA are genes and how those genes influence phenotype. They begin their analysis by breaking the DNA into pieces of manageable size, making many copies of those pieces to obtain enough material for study, and characterizing the pieces down to the level of nucleotide sequence. The scientists then try to reconstruct the DNA sequence of an entire genome by determining the spatial relationship between the many pieces. Finally, they use this knowledge to examine the genomic variations that make individuals unique.

SOLVED PROBLEMS

I. A geneticist examined the amino acid sequence of a particular protein in a variety of *E. coli* mutants. The amino acid in position 40 in the normal enzyme is glycine. The following table shows the substitutions the geneticist found at amino acid position 40 in six mutant forms of the enzyme.

mutant 1	cysteine
mutant 2	valine
mutant 3	serine
mutant 4	aspartic acid
mutant 5	arginine
mutant 6	alanine

Determine the nature of the base substitution that must have occurred in the DNA in each case. Which of these mutants would be capable of recombination with mutant 1 to form a wild-type gene?

Answer

To determine the base substitutions, use the genetic code table (see Fig. 8.2). The original amino acid was glycine, which can be encoded by GGU, GGC, GGA, or GGC. Mutant 1 results in a cysteine at position 40; Cys codons are either UGU or UGC. A change in the base pair in the DNA encoding the first position in the codon (a G–C to T–A transversion) must have occurred, and the original glycine codon must therefore have been either GGU or GGC. Valine (in mutant 2) is encoded by GUN (with N representing any one of the four bases), but assuming that the mutation is a single base change, the Val codon must be either GUU or GUC. The change must have been a G–C to T–A transversion in the DNA for the second position of the codon. To get from glycine to serine (mutant 3) with only one base change, the GGU or GGC would be changed to AGU or AGC, respectively. A transition occurred (G–C to A–T) at the first position. Aspartic acid (mutant 4) is encoded by GAU or GAC, so the DNA of mutant 4 is the result of a G–C to A–T transition at position 2. Arginine (mutant 5) is encoded by CGN, so the DNA of mutant 5 must have undergone a G–C to C–G transversion at position 1. Finally, alanine (mutant 6) is encoded by GCN, so the DNA of mutant 6 must have undergone a G–C to C–G transversion at position 2. Mutants 2, 4, and 6 affect a base pair different from that affected by mutant 1, so they could recombine with mutant 1.

In summary, the sequence of nucleotides on the RNA-like strand of the wild-type and mutant genes at this position must be:

wild type	5′ G G T/C 3′
mutant 1	5′ T G T/C 3′
mutant 2	5′ G T T/C 3′
mutant 3	5′ A G T/C 3′
mutant 4	5′ G A T/C 3′
mutant 5	5′ C G T/C 3′
mutant 6	5′ G C T/C 3′

II. The double-stranded circular DNA molecule that forms the genome of the SV40 tumor virus can be denatured into single-stranded DNA molecules. Because the base composition of the two strands differs, the strands can be separated on the basis of their density into two strands designated W(atson) and C(rick).

When each of the purified preparations of the single strands was mixed with mRNA from cells infected with the virus, hybrids were formed between the RNA and DNA. Closer analysis of these hybridizations showed that RNAs that hybridized with the W preparation were different from RNAs that hybridized with the C preparation. What does this tell you about the transcription templates for the different classes of RNAs?

Answer

An understanding of transcription and the polarity of DNA strands in the double helix are needed to answer this question. Some genes use one strand of the DNA as a template; others use the opposite strand as a template. Because of the different polarities of the DNA strands, one set of genes would be transcribed in a clockwise direction on the circular DNA (using, say, the W strand as the template), and the other set would be transcribed in a counterclockwise direction (with the C strand as template).

III. Geneticists interested in human hemoglobins have found a very large number of mutant forms. Some of these mutant proteins are of normal size (but have amino acid substitutions) while others are short, due to deletions or nonsense mutations. The first extra-long example was named Hb Constant Spring, in which the β-globin has all of its normal amino acids plus several extra amino acids attached after the normal C-terminal end of the protein.

a. What is the most plausible explanation for its origin?
b. Is it possible that Hb Constant Spring arose from failure to splice out an intron?
c. Estimate how many extra amino acids might be added to the C terminal end of the mutant protein.

Answer

An understanding of the principles of translation and RNA splicing are needed to answer this question.

a. Because an extension on the C-terminal end of the protein is present, the mutation probably affected the termination (nonsense) codon rather than affecting splicing of the RNA. This alteration could have been a base change or a frameshift or a deletion that altered or removed the termination codon. The information in the mRNA beyond the normal stop codon would be translated until another stop codon in the mRNA was reached.
b. A splicing defect could explain Hb Constant Spring only if an intron existed at a location just before the stop codon and the mutation prevented removal of the intron from the mature mRNA. In this case, the nucleotides in the intron would be read in frame as triplets until a stop codon was reached. The necessity for the presence of an intron in a specific location makes this scenario much less likely than that in part (a).
c. Regardless of whether the explanation for the Hb Constant Spring protein is a change to the termination codon or a change in splicing, you can estimate the number of amino acids added to the end of the protein by assuming they are encoded by a random DNA sequence. In the genetic code, 3 out of the 64 triplets (~1 in 21) are stop codons. So you would roughly estimate that on average you would add about 21 amino acids to the end of the mutant protein until the reading frame encountered a stop codon.

PROBLEMS

Vocabulary

1. For each of the terms in the left column, choose the best matching phrase in the right column.

a. codon	1. removing base sequences corresponding to introns from the primary transcript
b. colinearity	2. UAA, UGA, or UAG
c. reading frame	3. the strand of DNA that has the same base sequence as the primary transcript
d. frameshift mutation	4. a transfer RNA molecule to which the appropriate amino acid has been attached
e. degeneracy of the genetic code	5. a group of three mRNA bases signifying one amino acid
f. nonsense codon	6. most amino acids are not specified by a single codon
g. initiation codon	7. using the information in the nucleotide sequence of a strand of DNA to specify the nucleotide sequence of a strand of RNA
h. template strand	8. the grouping of mRNA bases in threes to be read as codons
i. RNA-like strand	9. AUG in a particular context
j. intron	10. the linear sequence of amino acids in the polypeptide corresponds to the linear sequence of nucleotide pairs in the gene
k. RNA splicing	11. produces different mature mRNAs from the same primary transcript
l. transcription	12. addition or deletion of a number of base pairs other than three into the coding sequence
m. translation	13. a sequence of base pairs within a gene that is not represented by any bases in the mature mRNA
n. alternative splicing	14. the strand of DNA having the base sequence complementary to that of the primary transcript
o. charged tRNA	15. using the information encoded in the nucleotide sequence of an mRNA molecule to specify the amino acid sequence of a polypeptide molecule
p. reverse transcription	16. copying RNA into DNA

Section 8.1

2. Match the hypothesis from the left column to the observation from the right column that gave rise to it.

a. existence of an intermediate messenger between DNA and protein	1. two mutations affecting the same amino acid can recombine to give wild type
b. the genetic code is nonoverlapping	2. one or two base deletions (or insertions) in a gene disrupt its function; three base deletions (or insertions) are often compatible with function
c. the codon is more than one nucleotide	3. artificial messages containing certain codons produce shorter proteins than messages not containing those codons
d. the genetic code is based on triplets of bases	4. protein synthesis occurs in the cytoplasm, while DNA resides in the nucleus
e. stop codons exist and terminate translation	5. artificial messages with different base sequences gave rise to different proteins in an *in vitro* translation system
f. the amino acid sequence of a protein depends on the base sequence of an mRNA	6. single base substitutions affect only one amino acid in the protein chain

3. How would the artificial mRNA

5′. . . GUGUGUGU . . . 3′

be read according to each of the following models for the genetic code?

a. two-base, not overlapping

b. two-base, overlapping

c. three-base, not overlapping

d. three-base, overlapping

e. four-base, not overlapping

4. An example of a portion of the T4 *rIIB* gene in which Crick and Brenner had recombined one + and one – mutation is shown here. (The RNA-like strand of the DNA is shown.)

```
wild type   5′ AAA AGT CCA TCA CTT AAT GCC 3′
mutant      5′ AAA GTC CAT CAC TTA ATG GCC 3′
```

a. Where are the + and – mutations in the mutant DNA?

b. The double mutant produces wild-type plaques. What alterations in amino acids occurred in this double mutant?

c. How can you explain the fact that amino acids are different in the double mutant than in the wild-type sequence, yet the phage has a wild-type phenotype?

5. Consider Crick and Brenner's experiments in Fig. 8.4, which showed that the genetic code is based on nucleotide triplets.

a. Crick and Brenner obtained FC7, an intragenic suppressor of FC0, that was a mutation in a second site in the *rIIB* gene near the FC0 mutation. Describe a different kind of mutation in the *rIIB* gene these researchers might have recovered by treating the FC0 mutant with proflavin and looking for restored $rIIB^+$ function.

b. How could Crick and Brenner tell the difference between the occurrence described in part (a) and an intragenic suppressor like FC7?

c. When FC7 was separated from FC0 by recombination, the result was two $rIIB^-$ mutant phages: One was FC7 and the other was FC0. How could they discriminate between the $rIIB^-$ recombinants that were FC7 and those that were FC0?

d. Explain how Crick and Brenner could obtain different deletion (–) or addition (+) mutations so as to make the various combinations such as ++, – –, +++, and – – – shown in Fig. 8.4c.

6. The $Hb\beta^S$ (sickle-cell) allele of the human β-globin gene changes the sixth amino acid in the β-globin chain from glutamic acid to valine. In $Hb\beta^C$, the sixth amino acid in β-globin is changed from glutamic acid to lysine. What would be the order of these two mutations within the map of the β-globin gene?

7. The following diagram describes the mRNA sequence of part of the *A* gene and the beginning of the *B* gene of phage ϕX174. In this phage, some genes are read in overlapping reading frames. For example, the code for the *A* gene is used for part of the *B* gene, but the reading frame is displaced by one base. Shown here is the single mRNA with the codons for proteins A and B indicated.

```
aa#    5  6  7  8  9  10  11 12 13 14 15 16
A      AlaLysGluTrpAsnAsnSerLeuLysThrLysLeu
mRNA   GCUAAAGAAUGGAACAACUCACUAAAAACCAAGCUG
B                MetGluGlnLeuThrLysAsnGlnAla
aa#               1  2  3  4  5  6  7  8  9
```

Given the following amino acid (aa) changes, indicate the base change that occurred in the mRNA and the consequences for the other protein sequence.

a. Asn at position 10 in protein A is changed to Tyr.

b. Leu at position 12 in protein A is changed to Pro.

c. Gln at position 8 in protein B is changed to Leu.

d. The occurrence of overlapping reading frames is very rare in nature. When it does occur, the extent of the overlap is not very long. Why do you think this is the case?

8. The amino acid sequence of part of a protein has been determined:

N . . . Gly Ala Pro Arg Lys . . . C

A mutation has been induced in the gene encoding this protein using the mutagen proflavin. The resulting mutant protein can be purified and its amino acid sequence determined. The amino acid sequence of the mutant protein is exactly the same as the amino acid sequence of the wild-type protein from the N terminus of the protein to the glycine in the preceding sequence. Starting with this glycine, the sequence of amino acids is changed to the following:

N . . . Gly His Gln Gly Lys . . . C

Using the amino acid sequences, one can determine the sequence of 14 nucleotides from the wild-type gene encoding this protein. What is this sequence?

9. The results shown in Fig. 8.5 may have struck you as incongruous because many synthetic RNAs that lacked AUG start codons (such as poly-U) were nonetheless translated into polypeptides *in vitro*. The reason this experiment was possible is that Marshall Nierenberg found that a high concentration of Mg^{2+} ions in the test tube, much higher than that found in cells, allows ribosomes to initiate translation at any position on an RNA molecule. Predict the outcomes of *in vitro* translation with each of the following synthetic mRNAs at both high and low Mg^{2+} concentrations:
   a. poly-UG (UGUGUG . . .)
   b. poly-CAUG (CAUGCAUGCAUG . . .)
   c. poly-GUAU (GUAUGUAUGUAU . . .)

10. Identify all the amino acid-specifying codons in the genetic code where a point mutation (a single base change) could generate a nonsense codon.

11. Before the technology existed to synthesize RNA molecules of defined sequence like those in Fig. 8.5, similar experiments were performed with synthetic mRNAs of undefined sequence. For example, RNAs consisting only of Us and Gs could be synthesized *in vitro*, but they would have random sequences. Suppose a pool of random-sequence RNAs was synthesized in a reaction mixture containing three times as much UTP as GTP, and that the resulting RNAs were translated *in vitro*.
   a. How many different codons exist in the RNAs?
   b. How many different amino acids would you find in the polypeptides synthesized?
   c. Why are your answers to (a) and (b) not the same?
   d. How often would you expect to find each of the codons in (a)?
   e. In what proportions would you expect to find each of the amino acids in the polypeptides?
   f. If you did this experiment—that is, synthesized random-sequence RNAs containing a 3:1 ratio of U:G, and quantified the amount of each amino acid in the polypeptides produced—prior to knowledge of the genetic code table, what would the results have told you?

12. A particular protein has the amino acid sequence

N . . . Ala-Pro-His-Trp-Arg-Lys-Gly-Val-Thr . . . C

within its primary structure. A geneticist studying mutations affecting this protein discovered that several of the mutants produced shortened protein molecules that terminated within this region. In one of them, the His became the terminal amino acid.
   a. What DNA single-base change(s) would cause the protein to terminate at the His residue?
   b. What other potential sites do you see in the DNA sequence encoding this protein where mutation of a single base pair would cause premature termination of translation?

13. How many possible *open reading frames* (frames without stop codons) exist that extend through the following sequence?

```
5′... CTTACAGTTTATTGATACGGAGAAGG...3′
3′... GAATGTCAAATAACTATGCCTCTTCC...5′
```

14. a. In Fig. 8.3, the physical map (the number of base pairs) is not exactly equivalent to the genetic map (in map units). Explain this apparent discrepancy.
   b. In Fig. 8.3, which region shows the highest frequency of recombination, and which the lowest?

15. Charles Yanofsky isolated many different *trpA*$^-$ mutants (Fig. 8.3).
   a. Explain how he could identify Trp$^-$ auxotrophs of *E. coli* using replica plating (recall Fig. 7.6).
   b. Assuming that the role of TrpA enzyme in the tryptophan biosynthesis pathway was known, explain how Yanofsky could have identified *trpA*$^-$ mutants among his Trp$^-$ auxotrophs. (*Hint:* Recall Beadle and Tatum's one gene, one enzyme experiments in Chapter 7.)

16. The sequence of a segment of mRNA, beginning with the initiation codon, is given here, along with the corresponding sequences from several mutant strains.

```
Normal    AUGACACAUCGAGGGGUGGUAAACCCUAAG...
Mutant 1  AUGACACAUCCAGGGGUGGUAAACCCUAAG...
Mutant 2  AUGACACAUCGAGGGUGGUAAACCCUAAG...
Mutant 3  AUGACGCAUCGAGGGGUGGUAAACCCUAAG...
Mutant 4  AUGACACAUCGAGGGGUUGGUAAACCCUAAG...
Mutant 5  AUGACACAUUGAGGGGUGGUAAACCCUAAG...
Mutant 6  AUGACAUUUACCACCCCUCGAUGCCCUAAG...
```

   a. Indicate the type of mutation present in each, and translate the mutated portion of the sequence into an amino acid sequence in each case.
   b. Which of the mutations could be reverted by treatment with EMS (ethylmethane sulfonate; see Fig. 7.14)? With proflavin?

17. You identify a proflavin-generated allele of a gene that produces a 110-amino acid polypeptide rather than the usual 157-amino acid protein. After subjecting this mutant allele to extensive proflavin mutagenesis, you are able to find a number of intragenic suppressors located in the part of the gene between the sequences encoding the N terminus of the protein and the original mutation, but no suppressors located in the region between the original mutation and the sequences encoding the usual C terminus of the protein. Why do you think this is the case?
18. Using recombinant DNA techniques (which will be described in Chapter 9), it is possible to take the DNA of a gene from any source and place it on a chromosome in the nucleus of a yeast cell. When you take the DNA for a human gene and put it into a yeast cell chromosome, the altered yeast cell can make the human protein. But when you remove the DNA for a gene normally present on yeast mitochondrial chromosomes and put it on a yeast chromosome in the nucleus, the yeast cell cannot synthesize the correct protein, even though the gene comes from the same organism. Explain. What would you need to do to ensure that such a yeast cell could make the correct protein?

Section 8.2

19. Describe the steps in transcription that require complementary base pairing.
20. Chapters 6 and 7 explained that mistakes made by DNA polymerase are corrected either by proofreading mechanisms during DNA replication or by DNA repair systems that operate after replication is complete. The overall rate of errors in DNA replication is about 1×10^{-10}, that is, one error in 10 million base pairs. RNA polymerase also has some proofreading capability, but the overall error rate for transcription is significantly higher (1×10^{-4}, or one error in each 10,000 nucleotides). Why can organisms tolerate higher error rates for transcription than for DNA replication?
21. The coding sequence for gene *F* is read from left to right on the accompanying figure. The coding sequence for gene *G* is read from right to left. Which strand of DNA (*top* or *bottom*) serves as the template for transcription of each gene?

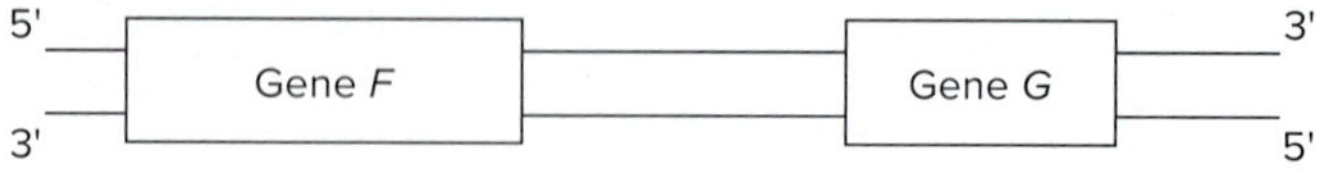

22. If you mixed the mRNA of a human gene with the genomic DNA for the same gene and allowed the RNA and DNA to form a hybrid molecule by base complementarity, what would you be likely to see in the electron microscope? Your figure should include hybridization involving both DNA strands (template and RNA-like) as well as the mRNA.
23. In studying normal and mutant forms of a particular human enzyme, a geneticist came across a particularly interesting mutant form of the enzyme. The normal enzyme is 227 amino acids long, but the mutant form was 312 amino acids long. The extra 85 amino acids occurred as a block in the middle of the normal sequence. The inserted amino acids do not correspond in any way to the normal protein sequence. What are possible explanations for this phenomenon? How would you distinguish among them?
24. The *Drosophila* gene *Dscam1* encodes proteins on the surface of nerve cells (neurons) that govern neuronal connections. Each neuron has on its surface a single Dscam1 protein of the tens of thousands that exist. The particular Dscam1 protein a neuron expresses is thought to tag the cell uniquely to determine the paths of the axons and dendrites it will grow. Eukaryotic genes are monocistronic. How then can a single *Dscam1* gene encode tens of thousands of different proteins?

Section 8.3

25. Describe the steps in translation that require complementary base pairing.
26. Locate as accurately as possible the listed items that are shown on the following figure. Some items are not shown. (a) 5′ end of DNA template strand; (b) 3′ end of mRNA; (c) ribosome; (d) promoter; (e) codon; (f) an amino acid; (g) DNA polymerase; (h) 5′ UTR; (i) centromere; (j) intron; (k) anticodon; (l) N terminus; (m) 5′ end of charged tRNA; (n) RNA polymerase; (o) 3′ end of uncharged tRNA; (p) a nucleotide; (q) mRNA cap; (r) peptide bond; (s) P site; (t) aminoacyl-tRNA synthetase; (u) hydrogen bond; (v) exon; (w) 5′ AUG 3′; (x) potential *wobble* interaction.

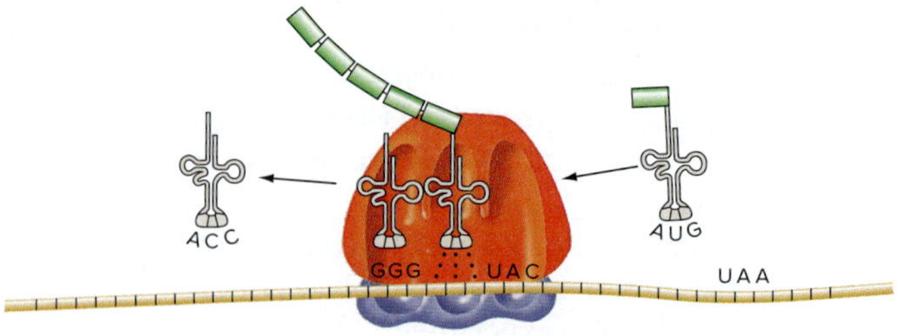

27. Concerning the figure for Problem 26:
    a. Which process is being represented?
    b. What is the next building block to be added to the growing chain in the figure? To what end of the

growing chain will this building block be added? How many building blocks will there be in the chain when it is completed?

c. What other building blocks have a known identity?

d. What details could you add to this figure that would be different in a eukaryotic cell versus a prokaryotic cell?

28. a. Can a tRNA exist that has the anticodon sequence 5′ IAA? If so, which amino acid would it carry?

b. Answer the same question for the anticodon sequence 5′ xm^5s^2UAA.

29. For parts (a) and (b) of Problem 28, consider the DNA sequence of the gene encoding the tRNA. What is the sequence of the RNA-like strand of the tRNA gene corresponding to the tRNA's anticodon? What is the sequence of the template strand of the gene for these same three nucleotides? Be sure to indicate polarities.

30. Remembering that the wobble base of the tRNA is the 5′ base of the anticodon:

a. In human tRNAs, what are the sequences of all possible anticodons that were originally transcribed with A in the wobble position? (Assume this A is always modified to I.)

b. In human tRNAs, what are the sequences of all possible anticodons that were originally transcribed with U in the wobble position? (*Note:* Any single type of tRNA with a U at the wobble position can be modified only in a single way.)

c. How might the wobble Us in each of the anticodons in (b) be modified and still be consistent with the genetic code?

d. What is the theoretical *minimal* number of different tRNA genes that must exist in the human genome? (Assume that xo^5U pairs with A, G, or U only.)

31. The human genome contains about 500 genes for tRNAs.

a. Do you think that each one of these tRNA genes has a different function?

b. Can you explain why the human genome might have evolved so as to house so many tRNA gen es?

32. The yeast gene encoding a protein found in the mitotic spindle was cloned by a laboratory studying mitosis. The gene encodes a protein of 477 amino acids.

a. What is the minimum length in nucleotides of the protein-coding part of this yeast gene?

b. A partial sequence of one DNA strand in an exon containing the middle of the coding region of the yeast gene is given here. What is the sequence of nucleotides of the mRNA in this region of the gene? Show the 5′ and 3′ directionality of your strand.

```
5′ GTAAGTTAACTTTCGACTAGTCCAGGGT 3′
```

c. What is the sequence of amino acids in this part of the yeast mitotic spindle protein?

33. The sequence of a complete eukaryotic gene encoding the small protein Met Tyr Arg Gly Ala is shown here. All of the written sequences on the template strand are transcribed into RNA.

```
5′ CCCCTATGCCCCCCTGGGGGAGGATCAAAACACTTACCTGTACATGGC 3′
3′ GGGGATACGGGGGGACCCCCTCCTAGTTTTGTGAATGGACATGTACCG 5′
```

a. Which strand is the template strand? In which direction (right-to-left or left-to-right) does RNA polymerase move along the template as it transcribes this gene?

b. What is the sequence of the nucleotides in the processed mRNA molecule for this gene? Indicate the 5′ and 3′ polarity of this mRNA.

c. A single base mutation in the gene results in synthesis of the peptide Met Tyr Thr. What is the sequence of nucleotides making up the mRNA produced by this mutant gene?

34. Arrange the following list of eukaryotic gene elements in the order in which they would appear in the genome and in the direction traveled by RNA polymerase along the gene. Assume the gene's single intron interrupts the open reading frame. Note that some of these names are abbreviated and thus do not distinguish between elements in DNA versus RNA. For example, *splice-donor site* is an abbreviation for *DNA sequences transcribed into the splice-donor site* because splicing takes place on the gene's RNA transcript, not on the gene itself. Geneticists often use this kind of shorthand for simplicity, even though it is imprecise. (a) splice-donor site; (b) 3′ UTR; (c) promoter; (d) stop codon; (e) nucleotide to which methylated cap is added; (f) initiation codon; (g) transcription terminator; (h) splice-acceptor site; (i) 5′ UTR; (j) poly-A addition site; (k) splice branch site.

35. Concerning the list of eukaryotic gene elements in Problem 34:

a. Which of the element names in the list are abbreviated? (That is, which of these elements actually occur in the gene's primary transcript or mRNA rather than in the gene itself?)

b. Which of the elements in the list are found partly or completely in the first exon of this gene (or in the RNA transcribed from this exon)? In the intron? In the second exon?

36. The human gene for ß2 lens crystallin has the components listed below. The numbers represent nucleotide pairs that make up the particular component. Assume for simplicity that no alternative splicing is involved.

5′ UTR	174
1st exon	119
1st intron	532
2nd exon	337
2nd intron	1431
3rd exon	208
3rd intron	380
4th exon	444
4th intron	99
5th exon	546
3′ UTR	715

Answer the following questions about the ß2 lens crystallin gene, primary transcript, and gene product. Questions asking *where* should be answered with one of the 11 components from the list or with *None*. Assume poly-A tails contain 150 As.

a. How large is the ß2 lens crystallin gene in bp (base pairs)?
b. How large is the primary transcript for ß2 lens crystallin in bases?
c. How large is the mature mRNA for ß2 lens crystallin in bases?
d. Where would you find the base pairs encoding the initiation codon?
e. Where would you find the base pairs encoding the stop codon?
f. Where would you find the base pairs encoding the 5′ cap?
g. Where would you find the base pairs that constitute the promoter?
h. Which intron interrupts the 3′ UTR?
i. Where would you find the sequences encoding the C terminus?
j. Where would you find the sequence encoding the poly-A tail?
k. How large is the coding region of the gene in bp (base pairs)?
l. How many amino acids are in the ß2 lens crystallin protein?
m. Which intron interrupts a codon?
n. Which intron is located between codons?
o. Where would you be likely to find the site specifying poly-A addition?

You find in lens-forming cells from several different people small amounts of a polypeptide that has the same N terminus as the normal ß lens crystallin, but it has a different C terminus. The polypeptide is 114 amino acids long, of which 94 are shared with the normal protein and 20 are unrelated *junk*. No mutation is involved in the production of this 114-amino acid protein.

p. Outline a hypothesis for the process that would produce this protein. Your hypothesis should explain why 94 amino acids are the same as in the normal ß2 lens crystallin.
q. Explain why you would expect that a polypeptide such as the 114-mer described above would on average have 20 amino acids of junk.

Section 8.4

37. In prokaryotes, a search for genes in a DNA sequence involves scanning the DNA sequence for long open reading frames (that is, reading frames uninterrupted by stop codons). What problem can you see with this approach in eukaryotes?

38. a. The genetic code table shown in Fig. 8.2 applies both to humans and to *E. coli*. Suppose that you have purified a piece of DNA from the human genome containing the entire gene encoding the hormone insulin. You now transform this piece of DNA into *E. coli*. Why can't *E. coli* cells containing the human insulin gene actually make insulin?
    b. Pharmaceutical companies have actually been able to obtain *E. coli* cells that make human insulin; such insulin can be purified from the bacterial cells and used to treat diabetic patients. How were the pharmaceutical companies able to create such bacterial factories for making insulin?

39. a. Very few if any eukaryotic genes contain tracts with more than 25 As or Ts in a row, yet almost all eukaryotic mRNAs have a tract with more than 100 As in a row. How is this possible?
    b. Scientists know the nucleotide sequences that direct the termination of bacterial gene transcription, but they generally have little idea about the nature of the nucleotide sequences that direct transcription termination in eukaryotic cells. Explain the basis of this statement.

40. Explain how differences in the initiation of translation dictate that eukaryotic mRNAs are monocistronic while prokaryotic mRNAs may be polycistronic.

Section 8.5

41. Do you think each of the following types of mutations would have very severe effects, mild effects, or no effect at all?
    a. Nonsense mutations occurring in the sequences encoding amino acids near the N terminus of the protein
    b. Nonsense mutations occurring in the sequences encoding amino acids near the C terminus of the protein
    c. Frameshift mutations occurring in the sequences encoding amino acids near the N terminus of the protein

d. Frameshift mutations occurring in the sequences encoding amino acids near the C terminus of the protein
e. Silent mutations
f. Conservative missense mutations
g. Nonconservative missense mutations affecting the active site of the protein
h. Nonconservative missense mutations not in the active site of the protein

42. Null mutations are valuable genetic resources because they allow a researcher to determine what happens to an organism in the complete absence of a particular protein. However, it is often not a trivial matter to determine whether a mutation represents the null state of the gene.
    a. Geneticists sometimes use the following test for the *nullness* of an allele in a diploid organism: If the abnormal phenotype seen in a homozygote for the allele is identical to that seen in a heterozygote (where one chromosome carries the allele in question and the homologous chromosome is known to be completely deleted for the gene) then the allele is null. What is the underlying rationale for this test? What limitations might there be in interpreting such a result?
    b. Can you think of other methods to determine whether an allele represents the null state of a particular gene?

43. The following is a list of mutations that have been discovered in a gene that has more than 60 exons and encodes a very large protein of 2532 amino acids. Indicate whether or not each mutation could cause a detectable change in the size or the amount of mRNA and/or a detectable change in the size or the amount of the protein product. (Detectable changes in size or amount must be greater than 1% of normal values.) What kind of change would you predict?
    a. Lys576Val (changes amino acid 576 from lysine into valine)
    b. Lys576Arg
    c. AAG576AAA (changes codon 576 from AAG to AAA)
    d. AAG576UAG
    e. Met1Arg (at least two possible scenarios exist for this mutation)
    f. promoter mutation
    g. one base pair insertion into codon 1841
    h. deletion of codon 779
    i. IVS18DS, G–A, + 1 (this mutation changes the first nucleotide in the eighteenth intron of the gene, causing exon 18 to be spliced to exon 20, thus skipping exon 19)
    j. deletion of the poly-A addition site
    k. G-to-A substitution in the 5′ UTR
    l. insertion of 1000 base pairs into the sixth intron (this particular insertion does not alter splicing)

44. Considering further the mutations described in Problem 43:
    a. Which of the mutations could be null mutations?
    b. Which of the mutations would be most likely to result in an allele that is recessive to a wild-type allele?
    c. Which of the mutations could result in an allele dominant to a wild-type allele? What mechanism(s) could explain this dominance?

45. Adermatoglyphia (described previously in Problem 18 in Chapter 3) is an extremely rare condition where people are born without fingerprints; only four families on earth are known to have this condition. The condition is inherited in an autosomal dominant fashion and is due to point mutations in a gene on chromosome 4 called *SMARCAD1*.

    The following figure shows that different point mutations—all near the 5′ end of the same intron of *SMARCAD1*—were found in each of the four families. All four mutations prevent the expression of a skin-specific transcript that uniquely contains exon 1, the first exon of this transcript; no other *SMARCAD1* mRNAs contain this exon. In the figure, the final three bases in the RNA-like strand of exon 1 are *shaded,* while the first five bases of intron 1 are *unshaded.*

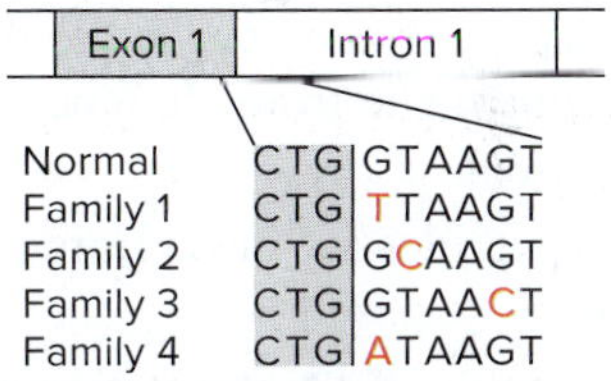

    a. No ATG sequence normally exists in exon 1 upstream of the sequence shown. Which part of the skin-specific mRNA corresponds to exon 1?
    b. What aspect of gene expression is likely to be affected most directly by these mutations?
    c. Are these mutations more likely to cause loss of function or gain of function?

46. Homozygosity for extremely rare mutations in a human gene called *SCN9A* cause complete insensitivity to pain (*congenital pain insensitivity* or *CPA*) and a total lack of the sense of smell (*anosmia*). The *SCN9A* gene encodes a sodium channel protein required for transmission of electrical signals from particular nerves in the body to the brain. The failure to feel pain is a dangerous condition as people cannot sense injuries.

    The *SCN9A* gene has 26 exons and encodes a 1977-amino acid polypeptide. Consanguineous matings in three different families have resulted in individuals with CPA/anosmia. In Family 1, a G-to-A transition in exon 15 results in a truncated protein that is

898 amino acids long; in Family 2, deletion of a single base results in a 766-amino acid polypeptide; and in Family 3, a C-to-G transversion in exon 10 yields a 458-amino acid protein.

a. Hypothesize as to how each of the three *SCN9A* mutations affects gene structure: Why are truncated proteins made in each case?

b. How would you classify the mutant alleles? Do these cause loss of function or gain of function? Are they amorphs, hypomorphs, hypermorphs, neomorphs, or antimorphs?

c. Explain in molecular terms why CPA/anosmia is a recessive condition.

47. You learned in Problem 21 in Chapter 7 that the neurodegenerative disease ALS can be caused by expansion of a hexanucleotide repeat region (5′-GGGGCC-3′) outside of the open reading frame (but within the first intron) of the gene called *C9ORF72.* While a normal *C9ORF72* allele has 2–23 copies of the hexanucleotide repeat unit, dominant disease-causing alleles have hundreds or even thousands of copies.

Researchers observed that the first intron of the *C9ORF72* disease allele is transcribed not only from the normal template strand of DNA, but also from the nontemplate strand. Even more unusual, both types of repeat-region transcripts are translated in all six reading frames in an AUG-independent manner—a process called *repeat-associated non-ATG translation,* or *RAN translation.* These discoveries led to the hypothesis that the proteins made from the repeats might contribute to ALS.

a. What polypeptides are made from the repeat-region transcripts?

b. According to the RAN translation hypothesis, why are disease-causing *C9ORF72* alleles dominant to normal alleles?

c. How would you classify the mutant alleles? Do they cause a loss of function or a gain of function? Are they amorphic, hypomorphic, hypermorphic, neomorphic, or antimorphic? (*Note:* More than one answer might be possible.)

48. When 1 million cells of a culture of haploid yeast carrying a *met*$^-$ auxotrophic mutation were plated on petri plates lacking methionine (Met), five colonies grew. You would expect cells in which the original *met*$^-$ mutation was reversed (by a base change back to the original sequence) would grow on the media lacking methionine, but some of these apparent reversions could be due to a mutation in a different gene that somehow suppresses the original *met*$^-$ mutations. How would you be able to determine if the mutations in your five colonies were due either to a precise reversion of the original *met*$^-$ mutation or to the generation of a suppressor mutation in a gene on another chromosome?

49. Why is a nonsense suppressor tRNA^{Tyr}, even though it has a mutant anticodon that cannot recognize a tyrosine codon, charged with tyrosine by Tyr tRNA synthetase? (*Hint:* Refer to Fig. 8.19.)

50. A mutant *B. adonis* bacterium has a nonsense suppressor tRNA that inserts glutamine (Gln) to match UAG (but not other nonsense) codons. [This species does not modify wobble position C residues to k^{2}C and does not have tRNA^{Pyl} (see Problem 57).]

a. What is the anticodon of the suppressing tRNA? Indicate the 5′ and 3′ ends.

b. What is the sequence of the template strand of the wild-type *tRNA^{Gln}* encoding gene that was altered to produce the suppressor, assuming that only a single-base-pair alteration was involved?

c. What is the *minimum* number of *tRNA^{Gln}* genes that could be present in a wild-type *B. adonis* cell? Describe the corresponding anticodons.

51. You are studying mutations in a bacterial gene that codes for an enzyme whose amino acid sequence is known. In the wild-type protein, proline is the fifth amino acid from the amino terminal end. In one of your mutants with nonfunctional enzyme, you find a serine at position number 5. You subject this mutant to further mutagenesis and recover three different strains. Strain A has a proline at position number 5 and acts just like a wild-type strain. Strain B has tryptophan at position number 5 and also acts like wild type. Strain C has no detectable enzyme function at any temperature, and you can't recover any protein that resembles the enzyme. You mutagenize strain C and recover a strain (C-1) that has enzyme function. The second mutation in C-1 that is responsible for the recovery of enzyme function does not map at the enzyme locus.

a. What is the nucleotide sequence in both strands of the wild-type gene at this location?

b. Why does strain B have a wild-type phenotype? Why does the original mutant with serine at position 5 lack function?

c. What is the nature of the mutation in strain C?

d. What is the second mutation that arose in C-1?

52. Another class of suppressor mutations, not described in the chapter, are mutations that suppress missense mutations.

a. Why would bacterial strains carrying such missense suppressor mutations generally grow more slowly than strains carrying nonsense suppressor mutations?

b. What other kinds of mutations can you imagine in genes encoding components needed for gene expression that would suppress a missense mutation in a protein-coding gene?

53. Yet another class of suppressor mutations not described in the chapter are mutations in tRNA genes that can suppress frameshift mutations. What would have to be true about a tRNA that could suppress a frameshift mutation involving the insertion of a single base pair?

54. At least one nonsense suppressing tRNA is known that can suppress more than one type of nonsense codon.
    a. What is the anticodon of such a suppressing tRNA?
    b. What stop codons would it suppress?
    c. Could this tRNA possibly also function as a missense suppressor?
    d. What are the amino acids most likely to be carried by this suppressing tRNA?

55. An investigator was interested in studying UAG nonsense suppressor mutations in bacteria. In one species of bacteria, she was able to select two different mutants of this type, one in a *tRNA*^{Tyr} gene and the other in a *tRNA*^{Gln} gene, but in a second species, she was not able to obtain any such nonsense suppressor mutations, even after very extensive effort. What could explain the difference between the two species?

56. Brenner's *m* mutant phages (m^1–m^6) described in Fig. 8.8 were suppressed when grown in suppressor (su^-) mutant bacteria; they produced full-length M proteins that functioned like wild-type M protein.
    a. What gene do you think was mutant in the su^- bacteria?
    b. When the m^- phages were propagated in the su^- bacterial strain, not all of the proteins made by the mutant *m* alleles were identical to wild-type M protein. How did some of them differ?

57. In certain bacterial species, pyrrolysine (Pyl), sometimes called amino acid 22, is incorporated into polypeptides through an unusual use of the genetic code: Pyl is specified by UAG triplets in the middle of the open reading frame of certain rare genes. These bacteria have a pyrrolysine tRNA synthetase that attaches Pyl to a tRNA with the anticodon 5′ CUA 3′ (see accompanying diagram). For Pyl to be incorporated into a protein, Pyl-charged tRNA^{Pyl} must arrive at the ribosome before translation is terminated.

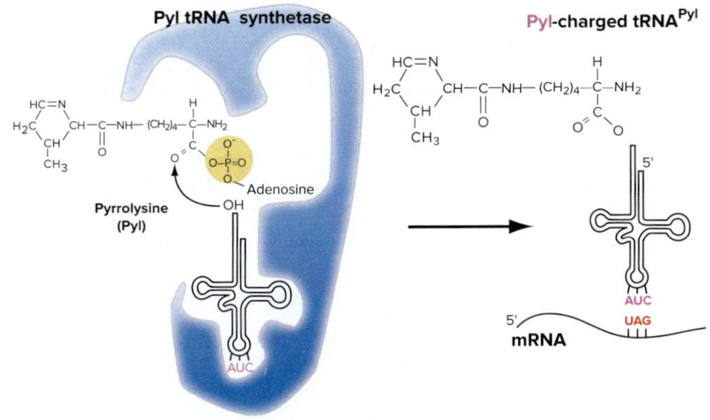

    a. Explain two ways in which the mechanism for Pyl specification differs from that of selenocysteine (Sec) incorporation.
    b. How is the mechanism for Pyl specification similar to nonsense suppression? (See Fig. 8.34.)

58. Canavanine is an amino acid similar to arginine (see accompanying figure) that is normally synthesized by some plants. Usually, in plants or animals that don't make canavanine, arginine aminoacyl-tRNA synthetase cannot distinguish between canavanine and arginine, and tRNA^{Arg} can be charged with canavanine. Incorporation of canvanine in proteins in place of arginine can cause misfolding and destroys protein structure and function.
    a. Can you think of a reason why a plant might have evolved the ability to make canavanine?
    b. How do you suppose plants that make canavanine escape its toxicity?
    c. A particular vining legume called *Dioclea megacarpa* makes canavanine and yet still has a single insect predator, a beetle, *Caryedes brasiliensis*. The beetle lays its eggs on the ripe fruit of the vine, and after hatching, the beetle larvae live in the fruit until they mature into adults. How do you suppose that the beetle evades canavanine toxicity?

Canavanine

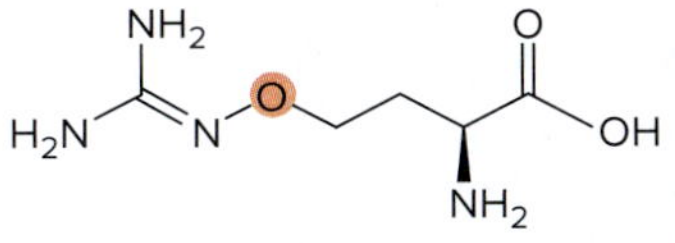

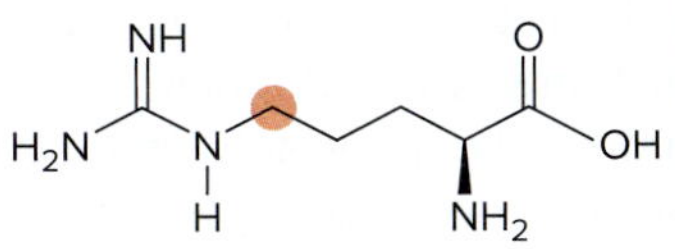

Arginine

Dioclea megacarpa

Caryedes brasiliensis

chapter **9**

Digital Analysis of DNA

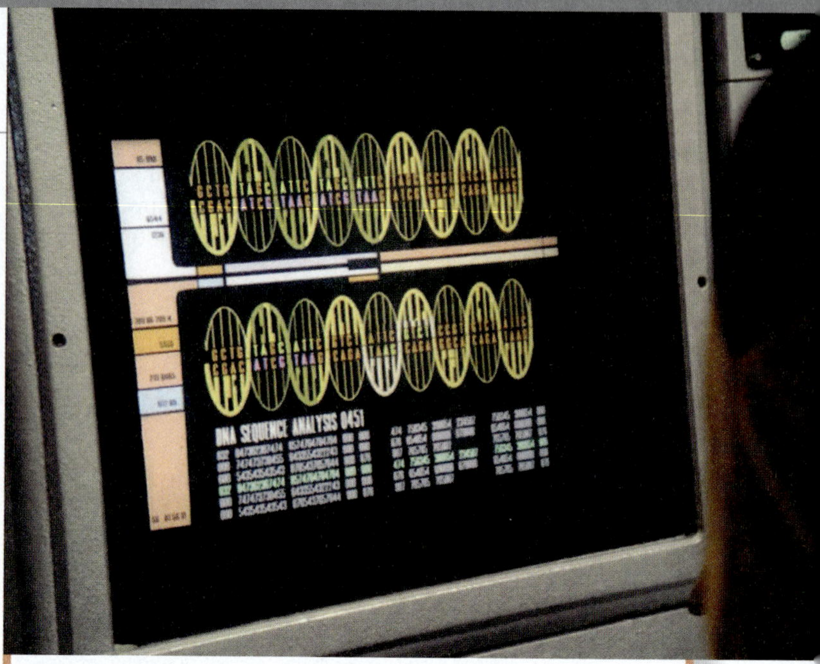

In 1989, when an episode of Star Trek: The Next Generation *featured this shot of DNA sequences on an imaginary computer screen, the ability to sequence the human genome appeared to be a distant dream in the realm of science fiction. Amazingly, the Human Genome Project, which began the following year, achieved this goal less than 15 years later.*

chapter outline

- 9.1 Fragmenting DNA
- 9.2 Cloning DNA Fragments
- 9.3 Sequencing DNA
- 9.4 Sequencing Genomes

SINCE THE MID-NINETEENTH CENTURY, three advances have transformed the field of genetics radically: Mendel's discovery of fundamental principles in the 1860s, Watson and Crick's elucidation of DNA structure in 1953, and the Human Genome Project from 1990 to the present. In this chapter and the next, we discuss the Human Genome Project and the field of **genomics** (the study of genomes) that it spawned.

The **Human Genome Project** was initiated to sequence and analyze the human genome in conjunction with the genomes of several model organisms. A **genome** is the total digital information contained within the DNA sequences of an organism's chromosomes. The haploid human genome contains a total of approximately 3 billion nucleotide pairs.

Prior to the inception of the Human Genome Project, the genome's enormous size caused many biologists to regard the objective of sequencing it as science fiction achievable only in the distant future. Nonetheless, some scientists could foresee the emergence of very fast and reliable automated (*high-throughput*) DNA sequencing methods as well as the computational tools necessary for capturing, storing, and analyzing the vast amounts of data involved. Persuaded by these arguments, agencies of the United States government agreed in 1990 to commit $3 billion over a projected 15-year period toward completion of the human genome sequence. Several international organizations also joined the enterprise.

Remarkably, investigators were able to determine a rough sequence of the human genome by February 2001. In this *draft,* the sequence had some gaps and did not yet have an appropriate level of accuracy (an error rate of 1/10,000 or less). An accurate sequence covering 97% of the genome was completed shortly thereafter in 2003, two years ahead of schedule. The early finish was prodded by the 1998 promise of Celera, a private company, to complete a draft of the genome in just three years at much lower cost, employing a novel sequencing strategy. The internationally supported genome effort reacted by moving its timetable ahead by several years.

The techniques and approaches developed by the public and private Human Genome Projects also catalyzed efforts to sequence the genomes of many species other than humans. By 2016, whole-genome sequences had been completed for more than 8100 distinct species, revolutionizing study in many areas such as microbiology and plant biology. The availability of genome sequences for these other organisms in turn has important benefits for our understanding of the human genome through the identification of genes and other DNA elements that are conserved across evolutionary lines.

In this chapter, we describe the methods that scientists developed to determine the sequence of the human genome. The general ideas behind genome sequencing are in fact not very complicated. Genomic researchers first fragment the genome into much smaller pieces, and then isolate and amplify (that is, *clone*) individual pieces by making so-called *recombinant DNA molecules.* Next, the scientists determine the DNA sequence of individual purified, bite-sized fragments of the genome. Finally, computer programs analyze the sequence of millions of these snippets to reconstruct the sequence of the whole genome from which the pieces originated.

9.1 Fragmenting DNA

learning objectives

1. Distinguish between digesting DNA with restriction enzymes and mechanical shearing of DNA.
2. Describe how certain restriction enzymes generate DNA fragments with sticky ends, while others generate blunt-ended fragments.
3. Calculate the average sizes and numbers of DNA fragments produced by digesting human genomic DNA with a given restriction enzyme.
4. Summarize the process by which gel electrophoresis separates DNA fragments.

Every intact diploid human cell, including the precursors of red blood cells, carries two nearly identical sets of 3 billion base pairs of information that, when unwound, extend 2 meters in length. This is much too much material and information to study as a whole. To reduce its complexity, researchers first cut the genome into bite-sized pieces that can be analyzed individually. One strategy to accomplish this goal is to use enzymes to cut the genome at specific DNA sequences; an alternative technique is to fragment the genome at random positions by shearing genomic DNA with mechanical forces. Both of these methods have their uses.

Restriction Enzymes Cut the Genome at Specific Sites

Researchers use *restriction enzymes* to cut the DNA released from the nuclei of cells at specific locations. These well-defined cuts generate fragments suitable for manipulation and characterization. A **restriction enzyme** recognizes a specific sequence of bases anywhere within the genome and then severs two phosphodiester bonds at that sequence, one in the sugar-phosphate backbone of each strand. The fragments generated by restriction enzymes are referred to as **restriction fragments,** and the act of cutting DNA is often called **digestion.**

Restriction enzymes originate in and can be purified from bacterial cells. As explained in the Tools of Genetics Box *Serendipity in Science: The Discovery of Restriction Enzymes,* these enzymes digest viral DNA to protect prokaryotic cells from viral infection. Bacteria shield their own genomes from digestion by these restriction enzymes through the selective addition of methyl groups ($-CH_3$) to the restriction recognition sites in their genomic DNA. In the test tube, restriction enzymes from bacteria recognize target sequences of four to eight base pairs (bp) in DNA isolated from any other organism and cut the DNA at or near these sites. **Table 9.1** lists the names, recognition sequences, and microbial origins of just 10 of the close to 300 commonly used restriction enzymes.

For the majority of these enzymes, the recognition site consists of four to six base pairs and exhibits a kind of palindromic symmetry in which the base sequences of each of the two DNA strands are identical when read in the 5′-to-3′ direction. Because of this fact, base pairs on either side of a central line of symmetry are mirror images of each other. Each enzyme always cuts at the same place relative to its specific recognition sequence, and most enzymes make their cuts in one of two ways: either straight through both DNA strands right at the line of symmetry to produce fragments with **blunt ends,** or displaced equally in opposite directions from the line of symmetry by one or more bases to generate fragments with single-stranded

TABLE 9.1 Ten Commonly Used Restriction Enzymes

Enzyme	Sequence of Recognition Site	Microbial Origin
*Taq*I	5' TCGA 3' 3' AGCT 5'	*Thermus aquaticus* YTI
*Rsa*I	5' GTAC 3' 3' CATG 5'	*Rhodopseudomonas sphaeroides*
*Sau*3AI	5' GATC 3' 3' CTAG 5'	*Staphylococcus aureus* 3A
*Eco*RI	5' GAATTC 3' 3' CTTAAG 5'	*Escherichia coli*
*Bam*HI	5' GGATCC 3' 3' CCTAGG 5'	*Bacillus amyloliquefaciens* H
*Hin*dIII	5' AAGCTT 3' 3' TTCGAA 5'	*Haemophilus influenzae*
*Kpn*I	5' GGTACC 3' 3' CCATGG 5'	*Klebsiella pneumoniae* OK8
*Cla*I	5' ATCGAT 3' 3' TAGCTA 5'	*Caryophanon latum*
*Bss*HII	5' GCGCGC 3' 3' CGCGCG 5'	*Bacillus stearothermophilus*
*Not*I	5' GCGGCCGC 3' 3' CGCCGGCG 5'	*Nocardia otitidiscaviarum*

ends (**Fig. 9.1**). Geneticists often refer to these protruding single strands as **sticky ends.** They are considered *sticky* because they are free to base pair with a complementary sequence from the DNA of any organism cut by the same restriction enzyme.

Restriction Enzymes with Longer Recognition Sites Produce Larger DNA Fragments

Researchers often need to produce DNA fragments of a particular length—larger ones to study the organization of a chromosomal region, smaller ones to examine a whole gene, and ones that are smaller still for DNA sequence analysis (that is, for the determination of the precise order of bases in a DNA fragment). To make these different-sized fragments, scientists can cut DNA with different restriction enzymes that recognize different sequences.

Figure 9.1 Restriction enzymes cut DNA molecules at specific locations to produce restriction fragments with either blunt or sticky ends. (a) The restriction enzyme *Rsa*I produces blunt-ended restriction fragments. **(b)** *Eco*RI produces sticky ends with a 5′ overhang. **(c)** *Kpn*I produces sticky ends with a 3′ overhang.

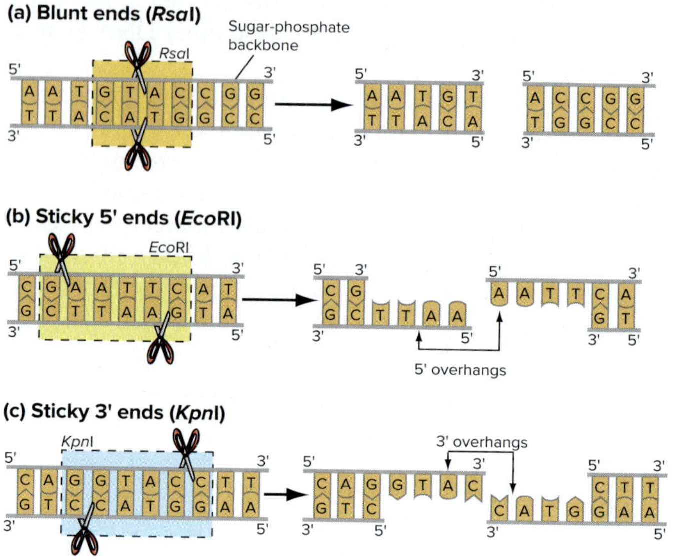

You can estimate the average length of the fragments that a particular restriction enzyme generates if you make two simplifying assumptions: first, that each of the four bases occurs in equal proportions such that a genome is composed of 25% A, 25% T, 25% G, and 25% C; and second, that the bases are distributed randomly in the DNA sequence. These assumptions enable us to estimate the average distance between recognition sites of any length by the general formula 4^n, where n is the number of bases in the site (**Fig. 9.2a**).

According to the 4^n formula, *Rsa*I, which recognizes the four-base-sequence 5′ GTAC 3′, will cut on average once every 4^4, or every 256 bp, creating fragments averaging 256 bp in length. By comparison, the enzyme *Eco*RI, which recognizes the six-base-sequence 5′ GAATTC 3′, will cut on average once every 4^6, or 4096 bp; because 1000 base pairs = 1 kilobase pair, researchers often round off this large number to roughly 4.1 kilobase pairs, abbreviated 4.1 kb. Similarly, an enzyme such as *Not*I, which recognizes the eight bases 5′ GCGGCCGC 3′, will cut on average every 4^8 bp, or every 65.5 kb. Note, however, that because the actual distances between restriction sites for any enzyme vary considerably, very few of the fragments produced by the three enzymes mentioned here will be precisely 256 bp, 4.1 kb, or 65.5 kb in length.

Once you know the average lengths of the fragments produced with a particular restriction enzyme, you can also estimate the number of the fragments that could be produced by treating a genome with that enzyme. For example, we have seen that the four-base cutter *Rsa*I cuts the

TOOLS OF GENETICS

Serendipity in Science: The Discovery of Restriction Enzymes

Most of the tools and techniques for cloning and analyzing DNA fragments emerged from studies of bacteria and the viruses that infect them. Molecular biologists had observed, for example, that viruses able to grow abundantly on one strain of bacteria grew poorly on a closely related strain. While examining reasons for this discrepancy, these scientists discovered restriction enzymes.

To follow the story, one must know that researchers compare rates of viral proliferation in terms of *plating efficiency:* the fraction of viral particles that enter and replicate inside host bacterial cells, causing the cells to lyse and release viral progeny. These progeny go on to infect neighboring cells, which in turn lyse and release further virus particles. When a petri dish is coated with a continuous *lawn* of bacterial cells, an active viral infection forms as a visibly cleared spot, or *plaque,* where bacteria have been eliminated (see Fig. 7.24). The plating efficiency of lambda virus grown on the *E. coli* C strain is nearly 1.0 (**Fig. A.1**). This means that 100 original virus particles will cause close to 100 plaques on a lawn of *E. coli* C bacteria.

The plating efficiency of the same virus grown on *E. coli* K12 is only 1 in 10^4, or 0.0001. The ability of a bacterial strain to prevent the replication of an infecting virus, in this case the growth of lambda on *E. coli* K12, is called **restriction.**

Restriction is rarely absolute. Although lambda virus grown on *E. coli* K12 produces almost no progeny (the viruses infect cells but can't replicate inside them), a few viral particles inside a few cells do manage to proliferate. If their progeny viruses are then tested on *E. coli* K12, the plating efficiency is nearly 1.0. The phenomenon in which growth on a restricting host modifies a virus so that succeeding generations grow more efficiently on that same host is called **modification.**

What mechanisms account for restriction and modification? Studies following viral DNA after bacterial infection found that during restriction, the viral DNA is broken into pieces and degraded (**Fig. A.2**). The enzyme responsible for the initial breakage was found to be an *endonuclease,* an enzyme that breaks phosphodiester bonds, usually making double-strand cuts at specific locations in the viral chromosome. Because this breakage restricts the biological activity of the viral DNA, researchers called the enzymes that accomplish it **restriction enzymes.**

Subsequent studies showed that the small percentage of viral DNA that escapes digestion and goes on to generate new viral particles has been modified by the addition of methyl groups during its replication in the host cell (**Fig. A.3**). Researchers named the enzymes that add methyl groups to specific DNA sequences **modification enzymes.**

Biologists have identified complementary restriction-modification systems in a wide variety of bacterial strains. Purification of the systems has yielded a mainstay of recombinant DNA technology: the battery of restriction enzymes used to cut DNA *in vitro* for cloning, mapping, and ligation (see Table 9.1).

This example of serendipity in science sheds light on the debate between administrators who distribute and oversee research funding and scientists who carry out the research. Researchers did not set out to find restriction enzymes; they could not have known these enzymes would be one of their finds. Rather, they sought to understand the mechanisms by which viruses infect and proliferate in bacteria. Along the way, they discovered restriction enzymes and how they work. The politicians and administrators in charge of allocating funds often want to direct research spending to urgent health or agricultural problems, while scientists often call for a broad distribution of funds to all projects investigating interesting biological phenomena. The validity of both views suggests the need for a balanced approach to the funding of research activities.

Figure A Operation of the restriction enzyme/modification system in nature. (1) *E. coli* strain C does not have a restriction enzyme/modification system and is susceptible to infection by the lambda phage. **(2)** In contrast, *E. coli* strain K12 generally resists infection by viral particles produced by a previous infection of *E. coli* C. Cells of *E. coli* K12 make the *Eco*RI restriction enzyme, which cuts the lambda DNA molecule before its genes can be expressed. **(3)** In rare K12 cells, modification enzymes add methyl groups (*me*) to lambda DNA, protecting it from the restriction enzymes. Modified lambda DNA can now replicate, and as the DNA methylation marks are copied during DNA replication, progeny viruses that readily form plaques on K12 bacteria are generated.

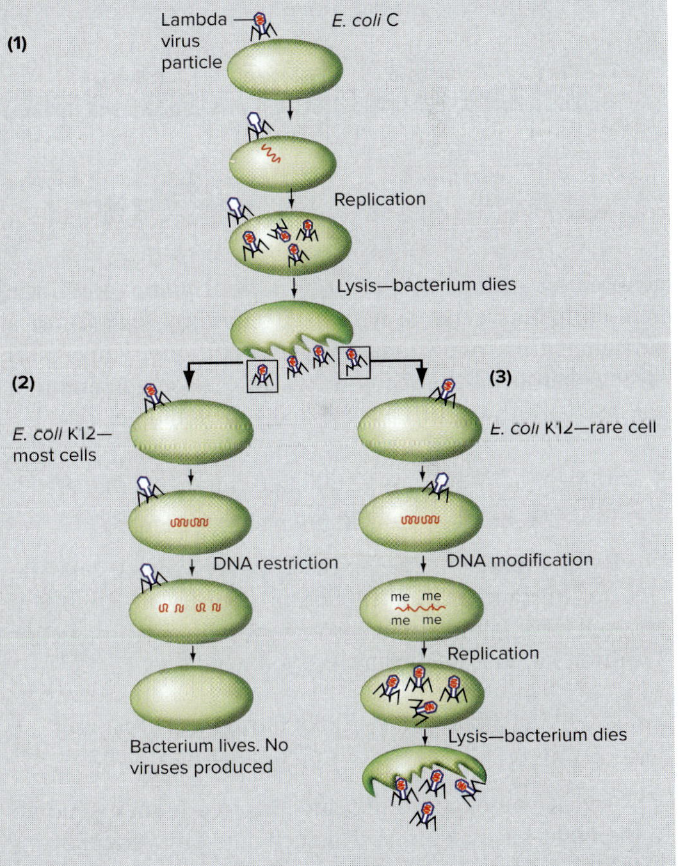

Figure 9.2 The number of base pairs in a recognition site determines the average size of the fragments produced. **(a)** *RsaI* recognizes and cuts at a 4 bp site, *Eco*RI cuts at a 6 bp site, and *NotI* cuts at an 8 bp site. **(b)** *RsaI*, *Eco*RI, and *NotI* restriction sites in a 200 kb region of human chromosome 11, followed by the names and locations of genes in this region. Numbered tick marks at the top are spaced 50 kb apart.

(a) Calculating Average Restriction Fragment Size

1. Probability that a four-base recognition site will be found at a given position in a genome =

   $1/4 \times 1/4 \times 1/4 \times 1/4 = 1/256$

2. Probability that a six-base recognition site will be found =

   $1/4 \times 1/4 \times 1/4 \times 1/4 \times 1/4 \times 1/4 = 1/4096$

3. Probability that an eight-base recognition site will be found =

   $(1/4)^8 = 1/65{,}536$

(b)

genome on average every 4^4 (256) bp. If you exposed the haploid human genome with its 3 billion bp to *Rsa*I for a sufficient time under appropriate conditions, you would ensure that all of the recognition sites in the genome that can be cleaved will be cleaved, and you would get

$$\frac{3{,}000{,}000{,}000 \text{ bp}}{\sim 256 \text{ bp}} = \begin{matrix}\sim 12{,}000{,}000 \text{ fragments that are} \\ \sim 256 \text{ bp in average length}\end{matrix}$$

By comparison, the six-base cutter *Eco*RI cuts the DNA on average every 4^6 (4096) bp, or every 4.1 kb. If you exposed the haploid human genome with its 3 billion bp, or 3 million kb, to *Eco*RI cleavage, you would get

$$\frac{3{,}000{,}000{,}000 \text{ bp}}{\sim 4100 \text{ bp}} = \begin{matrix}\sim 700{,}000 \text{ fragments that are} \\ \sim 4.1 \text{ kb in average length.}\end{matrix}$$

And if you exposed the same haploid human genome to the eight-base cutter *Not*I, which cuts on average every 4^8 (65,536) bp, or 65.5 kb, you would obtain

$$\frac{3{,}000{,}000{,}000 \text{ bp}}{\sim 65{,}500 \text{ bp}} = \begin{matrix}\sim 46{,}000 \text{ fragments that are} \\ \sim 65.5 \text{ kb in average length.}\end{matrix}$$

Figure 9.2b summarizes these relationships by depicting the results of cutting one small region of the human genome (containing only seven genes) with these three different restriction enzymes. Clearly, enzymes that recognize larger sites (like the 8 bp site for *Not*I) produce fewer fragments of larger average size than enzymes recognizing smaller sites (like the 4 bp sequence recognized by *Rsa*I).

Mechanical Shearing Forces Break DNA at Random Locations

As will be seen later in this chapter, some types of experiments require the random cutting of DNA so that in a given sample, different copies of the genome will be broken at different positions, rather than always at the same locations as with restriction enzymes. Random cutting of DNA can be achieved by subjecting the molecules to mechanical stress, such as passing the sample through very thin needles at high pressure, or by *sonication* (that is, the application of ultrasound energy). By pulling different parts of a DNA molecule in different directions, these mechanical forces can break phosphodiester bonds at random positions and thus fragment the DNA in the sample. Researchers can obtain fragments of various sizes by changing the amount of mechanical stress; for example, higher-energy ultrasound produces smaller fragments.

The ends of the DNA fragments produced by mechanical shearing are sometimes blunt, or they may have protruding single-stranded regions. If the latter, these single-stranded overhangs are not *sticky* in the same sense as the sticky ends produced by restriction enzymes because they are made up of random sequences and are thus not complementary with other overhangs. Molecular biologists have nonetheless developed elegant techniques that can convert any type of DNA end into any other type of end. All DNA fragments obtained by any procedure can thus ultimately be used in similar ways.

Gel Electrophoresis Separates DNA Fragments According to Size

To analyze the DNA in a sample, biologists employ a technique called **electrophoresis,** the movement of charged molecules in an electric field. Biologists use electrophoresis to separate many different types of molecules, for example, DNA of one length from DNA of other lengths, DNA from protein, or one kind of protein from another. In this discussion, we focus on its application to the separation in a gel of DNA fragments of varying length (**Fig. 9.3**).

To carry out such a separation, you place a solution of DNA molecules into indentations called *wells* at one end of a porous gel-like matrix. You then place the gel in a buffered aqueous solution and set up an electric field between bare wires at either end connected to a power supply. The electric field causes all charged molecules in the wells to migrate in the direction of the electrode having an opposite charge. Because all of the phosphate groups in the backbone of DNA carry a net negative charge in a solution near neutral pH, DNA molecules are pulled through the gel toward the wire with a positive charge.

Several variables determine the rate at which DNA molecules (or any other molecules) move during electrophoresis.

FEATURE FIGURE 9.3

Gel Electrophoresis

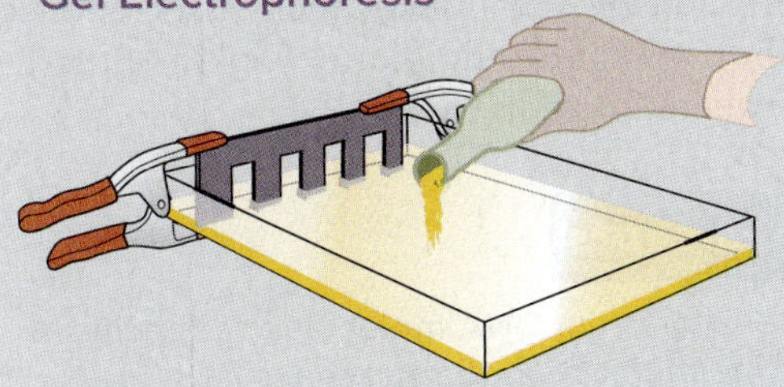

1. Pour heated molten agarose into a clear acrylic plate to which a comb has been attached with clamps. Allow the agarose to cool and harden.

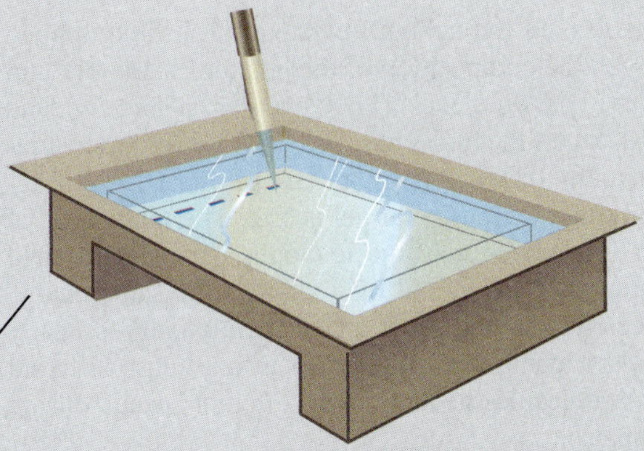

2. Remove the comb; shallow wells will be left in the gel. Remove the gel from the acrylic plate and transfer it to a tank containing a buffered solution. Use a micropipette to load a different DNA sample into each well of the gel. Each sample contains a blue dye to make it easier to see. One sample should contain DNA molecules of known length to serve as size markers.

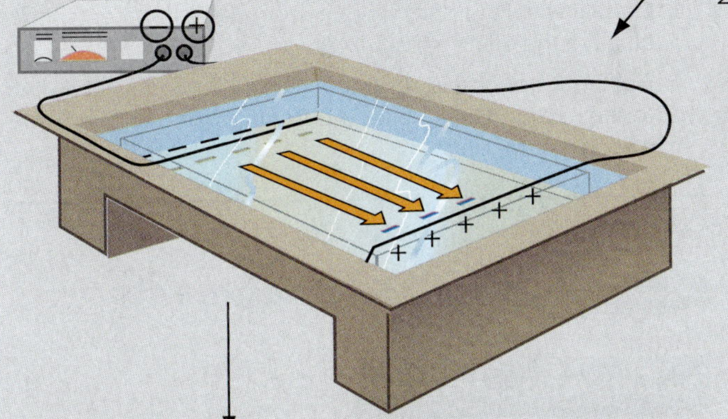

3. The tank contains electrode wires placed along each end of the gel. Attach these electrodes to a power supply. When you switch on the current, the negatively charged DNA molecules in each sample migrate toward the "+" end of the box, along the paths (*lanes*) shown by the orange arrows. Smaller DNA molecules will move faster toward the "+" end than larger DNA molecules.

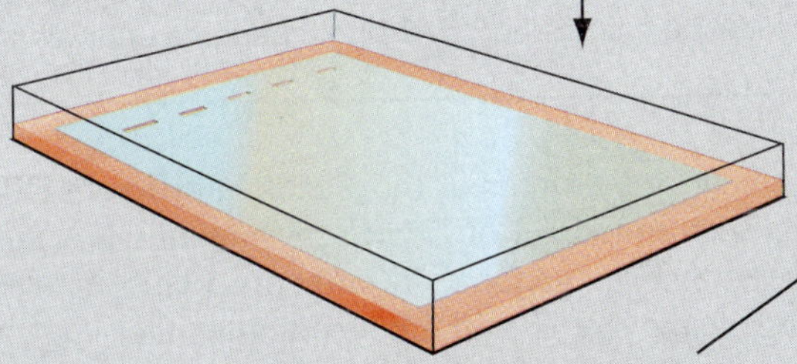

4. Remove the gel from the tank. Incubate in a solution containing ethidium bromide (which binds to DNA), then wash with water to remove excess dye from the gel.

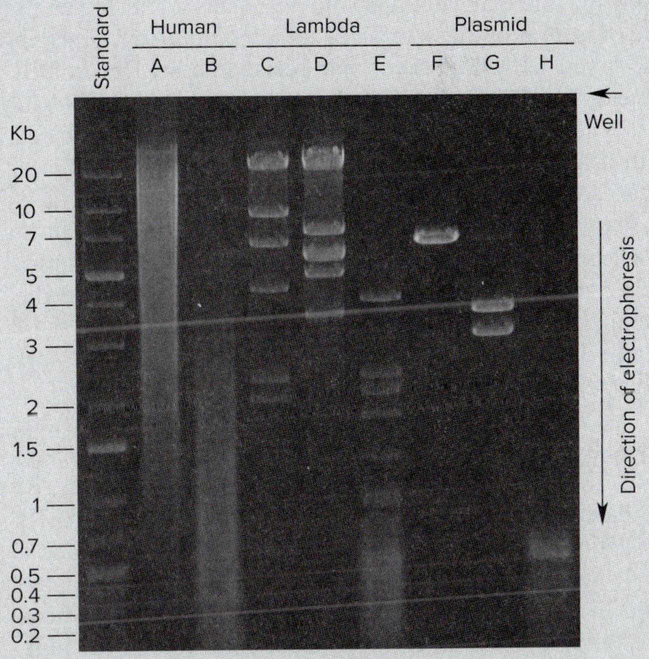

5. Expose the gel to ultraviolet (UV) light. DNA molecules will fluoresce as orange bands because the ethidium bromide bound to the DNA absorbs UV photons and gives off photons in the visible red range. You can estimate the size of the DNA molecules in the unknown samples by comparing their migration in the gel with that of the size markers (*standards*) in the lane at the left.

(5): © Lee Silver, Princeton University

Separating DNA molecules according to their size by agarose gel electrophoresis. To prepare an agarose gel with wells for samples, you pour the gel as shown in Step 1. You then transfer the gel to a tank containing a buffered solution with ions that allow current to flow, and load DNA samples in the wells (Step 2). You then connect the gel tank to a power supply and allow electrophoresis to run from 1 to 20 hours (depending on the DNA size and the voltage; Step 3). After incubating the gel with the fluorescent dye ethidium bromide (Step 4), you then expose the gel to UV light (Step 5). DNA molecules will appear as orange bands because they bind to the fluorescent dye.

Step 5 shows actual results from gel electrophoresis; because black-and-white film was used, DNA appears white rather than orange. The standard lane at left has DNA fragments of known sizes. Human genomic DNA was cut with *Eco*RI in lane A and with *Rsa*I in lane B. Smears containing hundreds of thousands of fragments are produced with an average size of about 4.1 kb for *Eco*RI and 256 bp for *Rsa*I. In C, D, and E, the chromosome of bacteriophage λ was cut with *Hin*dIII, *Eco*RI, and *Rsa*I, respectively. The sizes of the fragments in any one lane add up to 48.5 kb, the size of the viral genome. In F, G, and H plasmid DNA of total length 6.9 kb was cut with the same three enzymes. Note that the larger the genome analyzed, the more fragments are produced; moreover, the more bases in the restriction enzyme recognition site, the larger is the average size of the fragments produced.

These variables are the strength of the electric field applied across the gel, the composition of the gel, the charge per unit volume of the DNA molecule (known as *charge density*), and the physical size of the molecule. The only one of these variables that actually differs among any set of linear DNA fragments migrating in a particular gel is size. The reason is that all the DNA molecules are subjected to the same electric field and the same gel matrix, and they all have the same charge density (because the charge of all nucleotide pairs is nearly identical). As a result, only differences in size cause different linear DNA molecules to migrate at different speeds during electrophoresis.

The longer a molecule of linear DNA, the larger the volume it occupies as a random coil. The larger the volume a molecule occupies, the less likely it is to find a pore in the gel matrix big enough to squeeze through, and the more often it will bump into the matrix. And the more often the molecule bumps into the matrix, the lower its rate of migration (also referred to as its *mobility*). Thus, in any given period of electrophoresis, smaller DNAs will travel greater distances from the wells than larger DNAs.

When electrophoresis is completed, the gel is incubated with a fluorescent DNA-binding dye called *ethidium bromide*. After the unbound dye has been washed away, it is easy to visualize the DNA by placing the gel under an ultraviolet light, which causes the dye bound to the DNA fragments to glow in an orange color. You can determine the actual sizes of DNA molecules on gels by comparing their migration distances to those of known *marker fragments* subjected to electrophoresis in an adjacent lane of the gel.

Figure 9.3 (step 5) illustrates the types of results obtained by analyzing different DNA samples. If a genome is small, such as the 48.5 kb constituting the chromosome of the bacterial virus bacteriophage lambda (λ), then *Eco*RI restriction enzyme digestion of this DNA will produce a small number of discrete bands that can be discriminated from each other easily by gel electrophoresis and whose sizes total 48.5 kb. In contrast, electrophoresis of the hundreds of thousands of different fragments created when a sample of human genomic DNA is treated with the same enzyme will generate a smear centered around the average fragment size (about 4.1 kb for *Eco*RI, as previously discussed). Random breakage of DNA by mechanical forces will also produce a distribution of fragments whose average size will reflect the intensity of the shearing forces applied to the sample (*not shown*).

DNA molecules range in size from small fragments of fewer than 10 bp to whole human chromosomes that have an average length of 130,000,000 bp. No one sizing procedure has the capacity to separate molecules throughout this enormous range. To detect DNA molecules in different size ranges, researchers use a variety of protocols based mainly on two kinds of gels: *polyacrylamide* (formed by covalent bonding between acrylamide monomers), which is good for distinguishing smaller DNA fragments (less than 1 kb); and *agarose* (formed by the noncovalent association of agarose polymers), which is suitable for looking at larger fragments up to about 20 kb as in Fig. 9.3.

essential concepts

- *Restriction enzymes* cut DNA molecules at specific sequences; *mechanical shearing* breaks DNA at random locations.
- The longer the sequence recognized by a restriction enzyme, the fewer but larger will be the fragments the enzyme produces when cutting genomic DNA.
- Certain restriction enzymes can produce fragments that all have the same *sticky ends*.
- *Gel electrophoresis* separates DNA fragments according to their sizes. The smaller the fragment, the farther it will migrate in the gel.

9.2 Cloning DNA Fragments

learning objectives

1. Diagram the process by which restriction enzymes and DNA ligase are used to make recombinant DNA molecules.
2. Describe how scientists produce cellular clones of recombinant DNA molecules.
3. Contrast the use of plasmid vectors with that of BAC or YAC (bacterial or yeast artificial chromosome) vectors.
4. Explain why genomic DNA libraries require more colonies than are contained by a single genome equivalent.

The smear of hundreds of thousands of different DNA fragments seen in the gel of *Eco*RI-cut human DNA in Fig. 9.3 suggests that the genomes of animals, plants, and even microorganisms like *E. coli* are so complex that you can make sense of them only by looking at a small piece at a time. Ideally, you would like to purify just one of these fragments—a tiny bit of the genome—away from all the other fragments. You would then like to *amplify* this particular DNA fragment—that is, make many identical DNA copies of it. Amplification would allow you to obtain enough DNA to study, the most obvious type of analysis being to determine the sequence of nucleotides that constitute this particular fragment. If you could sequence separately each of the hundreds of thousands of fragments, you might ultimately be able to figure out the genome's entire DNA sequence.

The process that uses living cells both to isolate a single fragment of DNA from a complex mixture and to make many exact replicas of that fragment is called **molecular cloning.** This technique was central to the initial success of the Human Genome Project. Sophisticated methods have been developed recently to circumvent the need for molecular cloning in determining the sequence of genomes, and some of these techniques will be described in later chapters. However, molecular cloning remains today an essential component of many important approaches to the analysis and manipulation of DNA.

Molecular cloning consists of two basic steps. In the first, DNA fragments are inserted into specialized chromosome-like carriers called **cloning vectors,** which ensure the transport, replication, and purification of individual DNA inserts. In the second step, the combined vector-insert molecules are transported into living cells, and the cells make many copies of these molecules. Because all the copies of a given fragment are identical, the group of replicated DNA molecules is known as a **DNA clone.** DNA clones may be purified for immediate study or stored within cells or viruses as collections of clones known as *libraries* for future analysis. We now describe each step of molecular cloning.

Ligating Inserts to Vectors Produces Recombinant DNA Molecules

On their own, small fragments of human genomic DNA cannot reproduce themselves in a cell. To make replication possible, it is necessary to splice each fragment to a vector. Vectors must contain two kinds of specialized DNA sequences: one to provide a means of replication for the vector and the foreign DNA inserted into it, and the second to signal the vector's presence to an investigator by conferring a detectable property on the host cell. A vector must also have distinguishing physical traits, such as size or shape, by which it can be purified away from the host cell's genome. Several types of vectors are in use, and each one behaves as a minichromosome capable of accepting foreign DNA inserts and replicating independently of the host cell's genome. The cutting and ligating together of vector and inserted fragment—DNA from two different origins—creates a **recombinant DNA molecule.**

Sticky ends and base pairing

Two characteristics of sticky ends provide a basis for the efficient production of a vector-insert recombinant: First, the single-strand overhangs are available for base pairing. Second, no matter what the origin of the DNA (bacterial or human, for example), two sticky ends produced with the same enzyme are always *compatible,* that is, complementary in sequence.

To make recombinant DNA molecules, you simply cut the vector with the same restriction enzyme used to generate the fragment of genomic DNA, and then you mix the digested vector and genomic DNAs together in the presence of the enzyme DNA ligase (**Fig. 9.4**). The complementary sticky ends will form base pairs and the ligase will stabilize the molecule by forming phosphodiester bonds between adjacent nucleotides (one from the vector and one from the genomic DNA insert).

Laboratory tricks can increase the efficiency and general utility of molecular cloning. For example, certain procedures prevent two or more genomic fragments from joining with each other rather than with the vectors. Other methods minimize the chance that vector molecules can reseal themselves without including an insert of genomic DNA. Yet other manipulations can be performed to connect fragments of genomic DNA that do not have sticky ends to vectors. These techniques ensure that researchers can reliably produce the molecular clones they intend.

Choice of vectors

Available vectors differ from one another in biological properties, carrying capacity, and the type of host they can infect. Different types of vectors have different experimental uses.

The simplest vectors are small circles of double-stranded DNA known as **plasmids** that can gain admission to and replicate within many kinds of bacterial cells, independently of the bacterial chromosomes (Fig. 9.4a). The most useful plasmids contain a **polylinker,** which is a short, synthetic DNA sequence that contains a number of different restriction enzyme restriction sites (Fig. 9.4a). Each of these sites is found once in the polylinker but nowhere else in the plasmid vector. The polylinker provides flexibility in the choice of enzymes that can be used to digest the DNA containing the fragment of interest. Exposure to any one of these restriction enzymes opens up the vector at the corresponding recognition site, allowing the insertion of a foreign DNA fragment cut with the same enzyme, without at the same time splitting the plasmid into many pieces (Fig. 9.4b and c). Plasmid vectors can carry only relatively small foreign DNA fragments less than about 20 kb long.

Each plasmid vector carries an origin of replication and a gene for resistance to a specific antibiotic (Fig. 9.4a). The origin of replication enables the plasmid to replicate independently inside a bacterium. The gene for antibiotic resistance confers on the host cell the ability to survive in a medium containing a specific antibiotic; the resistance gene thereby enables experimenters to select for propagation only those bacterial cells that contain a plasmid. Antibiotic resistance genes and other vector genes that make it possible to pick out cells harboring a particular DNA molecule are called **selectable markers.** Plasmids fulfill the final requirement for vectors—ease of purification—because they can be purified away from the genomic DNA of the bacterial host by several techniques that take advantage of size and other differences.

The largest-capacity vectors are *artificial chromosomes:* recombinant DNA molecules that combine replication and segregation elements in such a way that they

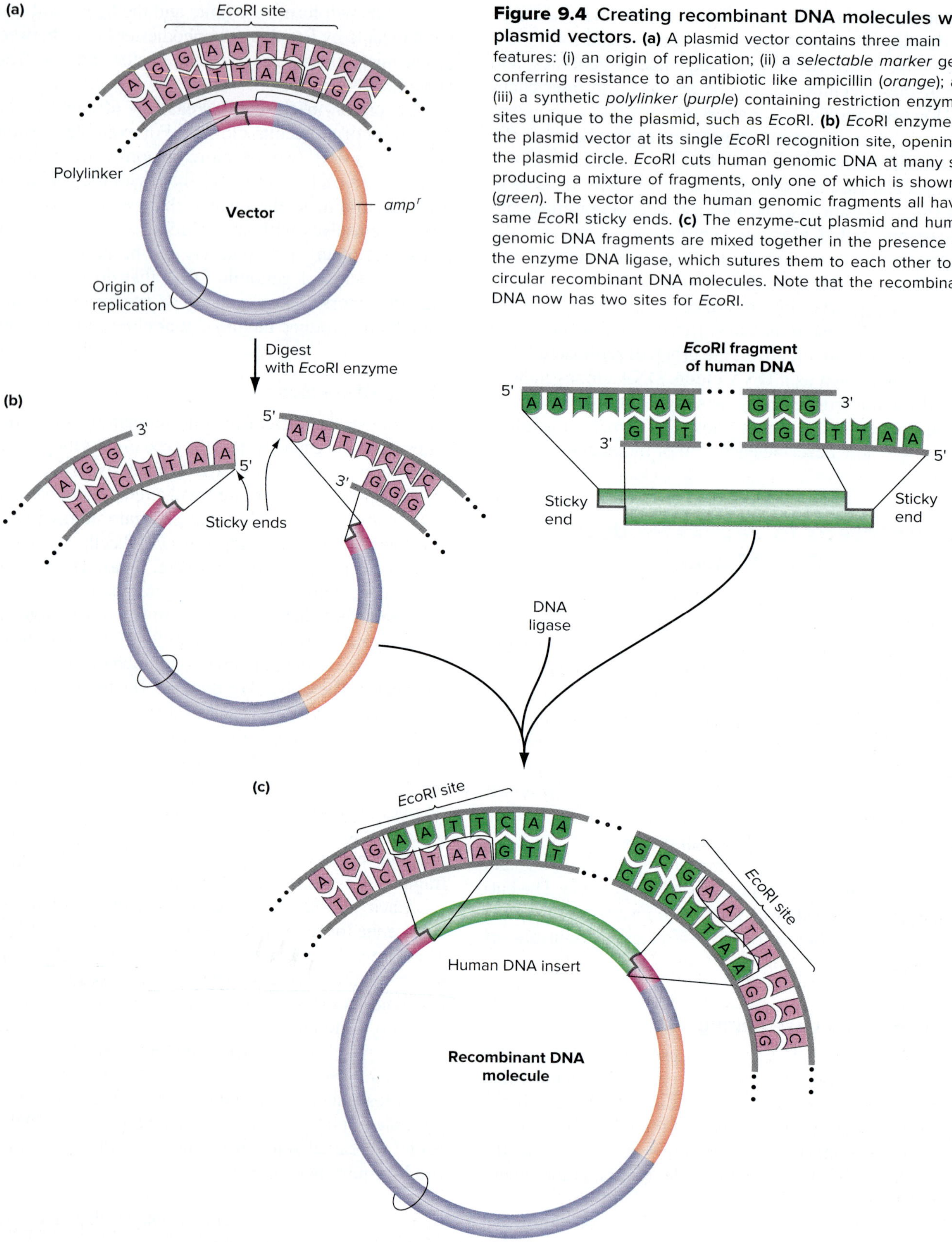

Figure 9.4 Creating recombinant DNA molecules with plasmid vectors. **(a)** A plasmid vector contains three main features: (i) an origin of replication; (ii) a *selectable marker* gene conferring resistance to an antibiotic like ampicillin (*orange*); and (iii) a synthetic *polylinker* (*purple*) containing restriction enzyme sites unique to the plasmid, such as *Eco*RI. **(b)** *Eco*RI enzyme cuts the plasmid vector at its single *Eco*RI recognition site, opening up the plasmid circle. *Eco*RI cuts human genomic DNA at many sites, producing a mixture of fragments, only one of which is shown (*green*). The vector and the human genomic fragments all have the same *Eco*RI sticky ends. **(c)** The enzyme-cut plasmid and human genomic DNA fragments are mixed together in the presence of the enzyme DNA ligase, which sutures them to each other to form circular recombinant DNA molecules. Note that the recombinant DNA now has two sites for *Eco*RI.

behave like normal chromosomes when introduced into a host cell. A bacterial artificial chromosome (BAC) can accommodate a DNA insert of 300 kb. Yeast artificial chromosomes (YACs) can incorporate even larger DNA inserts up to 2000 kb (2 Mb). In addition to their use in molecular cloning, YACs can help investigators analyze functional elements of chromosomes such as centromeres; we will thus discuss YACs in greater detail in Chapter 12 on The Eukaryotic Chromosome.

Host Cells Take Up and Amplify Recombinant DNA

Although each type of vector functions in a slightly different way and enters a specific kind of host, the general scheme of entering a host cell and taking advantage of the cellular environment to replicate itself is the same for all. **Figure 9.5** illustrates how scientists obtain *E. coli* cells that contain recombinant DNA molecules in which human DNA fragments were ligated into a plasmid vector. The procedure starts with vector and human genomic DNAs cut with the same restriction enzyme, which are then mixed together in the presence of DNA ligase to create hundreds of thousands of different recombinant DNAs, each with a different fragment of the human genome (Fig. 9.5a). Researchers must then introduce these molecules into *E. coli* such that each cell contains only a single type of recombinant DNA.

Transformation of host cells

Transformation, as you saw in Chapter 6, is the process by which a cell or organism takes up a foreign DNA molecule, changing the genetic characteristics of that cell or organism. What we now describe is similar to what Avery and his colleagues did in the transformation experiments that determined DNA was the molecule of heredity (recall Fig. 6.4), but the method outlined here is more efficient.

Recombinant DNA molecules are first added to a suspension of specially prepared *E. coli* that are sensitive to the antibiotic ampicillin. Under conditions favoring entry, such as suspension of the bacterial cells in a cold $CaCl_2$ solution or treatment of the solution with high-voltage electric shock (a technique known as *electroporation*), the plasmids will enter about 1 in 1000 cells (**Fig. 9.5b**). These protocols increase the permeability of the bacterial cell membrane, in essence punching temporary holes through which the DNA gains entry. The probability that any one plasmid will enter any one cell is so low (0.001) that the probability of simultaneous entry of two plasmids into a single cell is insignificant ($0.001 \times 0.001 = 0.000001$).

Identification and isolation of transformed cells

To identify the 0.1% of cells housing a plasmid, the bacteria-plasmid mixture is decanted onto a plate containing agar, nutrients, and ampicillin. Only cells transformed by a plasmid providing resistance to ampicillin will be able to grow and multiply in the presence of the antibiotic (Fig. 9.5b). The plasmid's origin of replication enables it to replicate in the bacterial cell independently of the bacterial chromosome; in fact, most plasmids replicate so well that a single bacterial

Figure 9.5 Cloning recombinant DNA molecules. **(a)** Recombinant DNA construction. Cutting genomic DNA with a restriction enzyme produces many fragments, each of which can form a different recombinant DNA molecule. **(b)** Obtaining clones of bacterial cells containing recombinant plasmids. Recombinant DNAs [from part (a)] are added to ampicillin-sensitive *E. coli* cells. Only cells transformed with recombinant plasmids (or more rarely with a religated vector lacking a foreign DNA insert) will grow on a petri plate containing ampicillin. Each colony on the plate contains millions of identical descendants from a single bacterial cell transformed with a single recombinant DNA molecule.

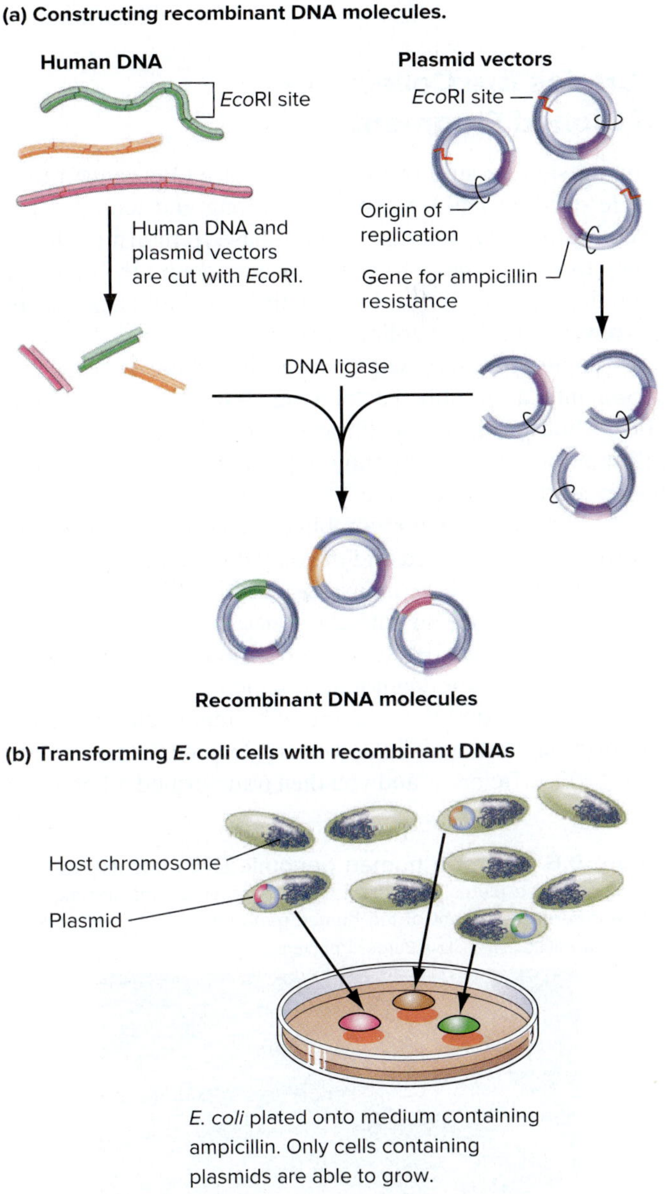

cell may end up with hundreds of identical copies of the same plasmid molecule.

Each viable plasmid-containing bacterial cell will multiply to produce a distinct spot on an agar plate, consisting of a colony of tens of millions of genetically identical cells. The colony as a whole is considered a **cellular clone.** Such clones can be identified when they have grown to about 1 mm in diameter. The millions of identical plasmid molecules contained within a colony together make up a **DNA clone** (Fig. 9.5b).

Libraries Are Collections of Cloned Fragments

Moving step by step from the DNA of any organism to a single purified DNA fragment is a long and tedious process. Fortunately, scientists do not have to return to step 1 every time they need to purify a new genomic fragment from the same organism. Instead, they can build a **genomic library:** a long-lived collection of cellular clones that contains copies of every sequence in the whole genome inserted into a suitable vector **(Fig. 9.6)**. Like traditional book libraries, genomic libraries store large amounts of information for retrieval upon request. They make it possible to start a new cloning project at an advanced stage, when the initial step of recombinant DNA construction has already been completed and the only difficult task left is to determine which of the many clones in the library contains the DNA sequence of interest. Once the correct cellular clone is identified, it can be amplified to yield a large amount of the desired genomic fragment.

If you digested the genome of a single cell with a restriction enzyme and ligated every fragment to a vector with 100% efficiency, and you then transformed all of these recombinant DNA molecules into host cells also with 100% efficiency, the resulting set of clones would represent the entire genome in a fragmented form. A hypothetical collection of cellular clones that includes one copy—and one copy only—of every sequence in the entire genome would be a single *complete genomic library.*

How many clones are present in this hypothetical library? If you started with the 3,000,000 kb of DNA from a haploid human sperm and reliably cut it into a series of 150 kb restriction fragments, you would generate 3,000,000/150 = 20,000 genomic fragments. If you placed each and every one of these fragments into BAC cloning vectors that were then transformed into *E. coli* host cells, you would create a perfect library of 20,000 clones that collectively carry every locus in the genome. The number of clones in this perfect library defines a **genomic equivalent.** To find the number of clones that constitute one genomic equivalent for any library, you simply divide the length of the genome (here, 3,000,000 kb) by the average size of the inserts carried by the library's vector (in this case, 150 kb).

In real life, it is impossible to obtain a perfect library. Each step of cloning is far from 100% efficient, and the DNA of a single cell does not supply sufficient raw material for the process. Researchers must thus harvest DNA from among the millions of cells in a particular tissue or organism. If you make a genomic library with this DNA by collecting only one genomic equivalent (20,000 clones for a human library in BAC vectors), then by chance some human DNA fragments will appear more than once, while others will not be present at all. Including four to five genomic equivalents produces an average of four to five clones for each region (locus) of the genome, and a 95% probability that any individual locus is present at least once.

Figure 9.6 Part of a human genomic DNA library. Each colony on these plates contains a different recombinant plasmid with a different fragment of the human genome.

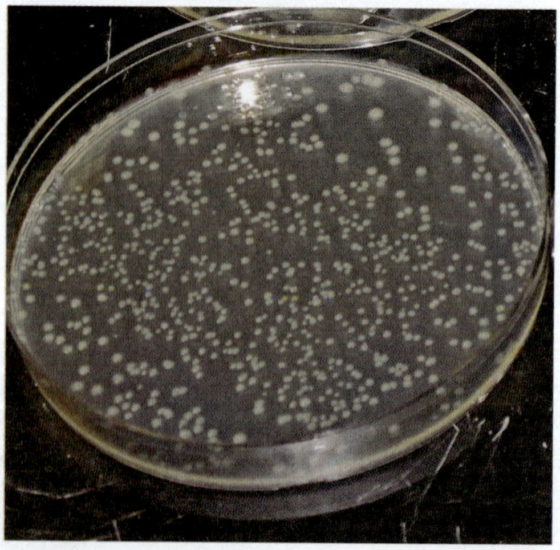

essential concepts

- To form *recombinant DNA molecules,* DNA ligase links together restriction-enzyme-cut *vector* and genomic DNA fragments with the same sticky ends.
- Vectors include an origin for DNA replication and a *selectable marker* such as a gene for antibiotic resistance. Plasmid vectors are used for small DNA inserts (<20 kb), while BAC or YAC vectors can carry much larger inserts.
- *Recombinant DNAs* are transformed into host cells. When a single transformed cell is grown into a *cellular clone,* each cell in the clone has the same recombinant DNA.
- *Genomic libraries* contain a species' entire genome as a collection of random DNA fragments; multiple *genome equivalents* are needed to ensure representation of all parts of the genome.

9.3 Sequencing DNA

learning objectives

1. Explain the roles of DNA polymerase, the template, and the primer in a Sanger sequencing reaction.
2. Describe the role of dideoxyribonucleotides in generating DNA fragments for analysis.
3. Interpret the fluorescent peaks obtained during a DNA sequencing run as a sequence of nucleotides with the proper polarity.

Looking at Fig. 9.6, you see petri plates containing thousands of separate colonies constituting part of a human genomic library. Each clone contains a different recombinant DNA molecule, each with a plasmid vector attached to a different fragment of human genomic DNA. Note that the colonies are scattered randomly around the plate, so their arrangement on the plate has no correspondence to their relative order in the genome. How can you then tell which colony contains which fragment of human DNA?

With the technology available today, the simplest way to answer this question is to sequence the human DNA insert in each clone. The DNA sequencing technology in widest current use is based on an original method developed in the mid-1970s by Fred Sanger, who won one of his two Nobel Prizes for this work (the other was awarded for his determination of the amino acid sequence of the hormone insulin). Sanger's methodology can be automated easily, providing the power needed to sequence the 3 billion nucleotides in the human genome.

Sanger Sequencing Depends on DNA Polymerase

Sanger based his technique on his knowledge of the way DNA is replicated in cells. You will recall from Chapter 6 that the enzyme DNA polymerase catalyzes DNA replication. As summarized in **Fig. 9.7a,** this enzyme's minimal requirements are: (1) a **template;** that is, a single strand of DNA to copy; (2) **deoxyribonucleotide triphosphates** (dATP, dCTP, dGTP, and dTTP) that are the basic building blocks for newly synthesized DNA; and (3) a **primer;** that is, a short single-stranded DNA molecule (an *oligonucleotide*) that is complementary to part of the template and that provides the free 3′ end to which DNA polymerase can attach new nucleotides.

To sequence DNA using Sanger's methods, you would need a template, part of whose sequence is known, but the remainder of which is unknown (because that is what you are trying to determine). One strand of the DNA of a recombinant plasmid could serve as such a template: You know the DNA sequence of the vector (which will be the same in all clones in the library) but not of the genomic DNA insert (which will vary in different clones) (Fig. 9.7a).

Next, you would need a short oligonucleotide primer complementary to the known sequence of the vector just adjacent to the unknown human DNA insert (Fig. 9.7a). Primers are made to order in *DNA synthesizers,* machines that can manufacture large quantities of any given DNA oligonucleotide up to 100 bases in length. The user simply types in the desired sequence of nucleotides into the computer controlling the DNA synthesizer, and the machine then strings those nucleotides together in the proper order using chemical reactions. You can design the primer because you already know the sequence of the vector, which was determined by alternative chemical techniques (not described here) that do not require prior knowledge.

Sanger sequencing allows the template and primer depicted in Fig. 9.7a to interact through the process of **hybridization:** the natural tendency of complementary single-stranded molecules of DNA or RNA to base pair and form double helixes. To make the template, you could simply amplify and purify double-stranded recombinant DNA from one particular clone, and then melt the DNA into single strands by raising the temperature so as to disrupt the hydrogen bonds that would otherwise keep the strands together. Although both strands of a DNA fragment are present in a typical DNA sample, only one is used as a template for sequencing. You then mix in large amounts of the previously made primer. As the temperature of the mixture is gradually lowered, hydrogen bonds will form between complementary nucleotides of the primer and the template strand of the recombinant DNA. The primer you make in the DNA synthesizer must be long enough to ensure that it will form a stable double-stranded region (that is, *anneal*) only with the one complementary sequence in the template; usually primers are between 17 and 25 bases long. The interaction of primer and template creates the substrate for the action of DNA polymerase (Fig. 9.7a).

Sanger Sequencing Generates Nested Sets of Single-Stranded DNA Fragments

To reveal the order of base pairs in an isolated DNA molecule, Sanger sequencing uses DNA polymerase to create a series of single-stranded fragments, where part of each fragment is complementary to the unknown portion of the DNA template under analysis (**Fig. 9.7b**). Each fragment differs in length by a single nucleotide from the preceding and succeeding fragments; the graduated set of fragments is known as a *nested array.* A critical feature of the fragments is that each one is distinguishable according to its terminal 3′ base. Thus, each fragment has two defining attributes—relative length and one of four possible terminating nucleotides.

The sequencing procedure to create the nested array begins by adding DNA polymerase to the annealed

Automated Sanger Sequencing

(a) Hybridization of template and primer. A cloned recombinant DNA is **denatured** (melted into single strands), using heat to break apart the hydrogen bonds joining the two strands. One of these strands serves as the **template.** The recombinant DNA is mixed with an **oligonucleotide primer** (previously made in a DNA synthesizer) whose sequence is complementary to about

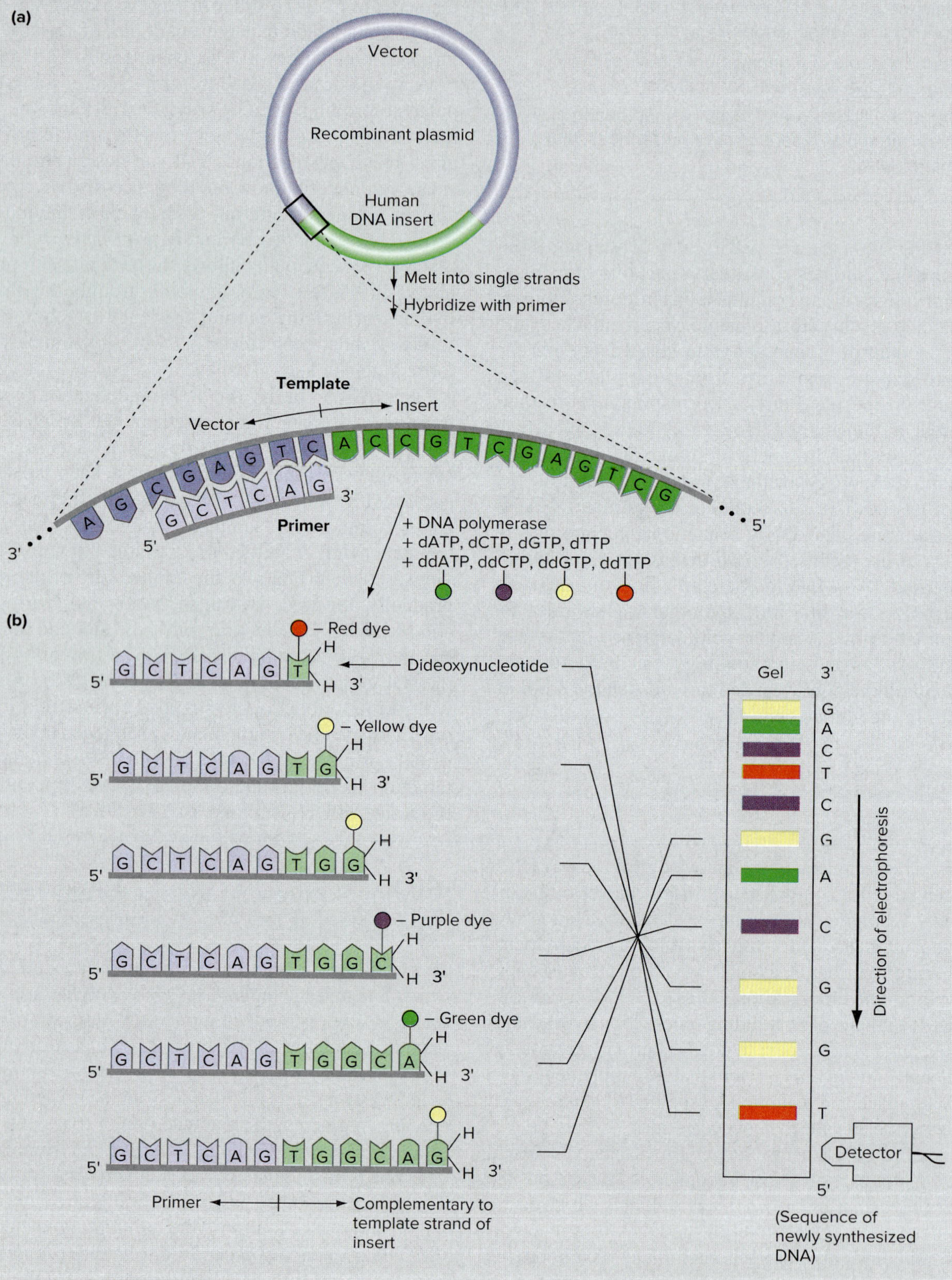

20 bases in the vector portion of the template strand. As the temperature is lowered, the template and primer **anneal (hybridize)** together. **(b)** Generating a nested set of polymerization products. The hybrid template-primer is now mixed with DNA polymerase, large amounts of the four **deoxynucleotide triphosphates (dNTPs),** and smaller amounts of the four **dideoxynucleotide triphosphates (ddNTPs).** Each ddNTP has a fluorescent tag of a different color. DNA polymerase synthesizes a new strand of DNA complementary to the template by adding nucleotides sequentially onto the 3′ end of the primer. Synthesis terminates when a dideoxynucleotide is added to the chain. The reaction generates a nested set of products, each with the same 5′ end but a different 3′ end. The terminating dideoxynucleotide at the 3′ end color-codes each product. After melting apart the newly synthesized DNA from the template, these products are electrophoresed on a special gel that can separate DNAs that differ in size by a single nucleotide. As each fragment moves past a laser beam, the color of the terminal base is detected and recorded. **(c)** Dideoxynucleotide structure. Because ddNTPs lack an -OH group on the 3′ carbon atom of deoxyribose, DNA polymerase cannot add any nucleotides onto a chain with a dideoxynucleotide at the 3′ end. **(d)** Analyzing nested sets of products on gels. **(e)** Image of a sequencing gel. Each lane displays the sequence obtained with a separate sample. **(f)** A DNA sequence trace from one gel lane. The raw data are displayed as peaks of four different colors. Here, yellow is pseudocolored as black for easier visualization. The base-calling software produces a text sequence of the newly synthesized DNA strand. It is possible to read almost 1000 bases from a single reaction.

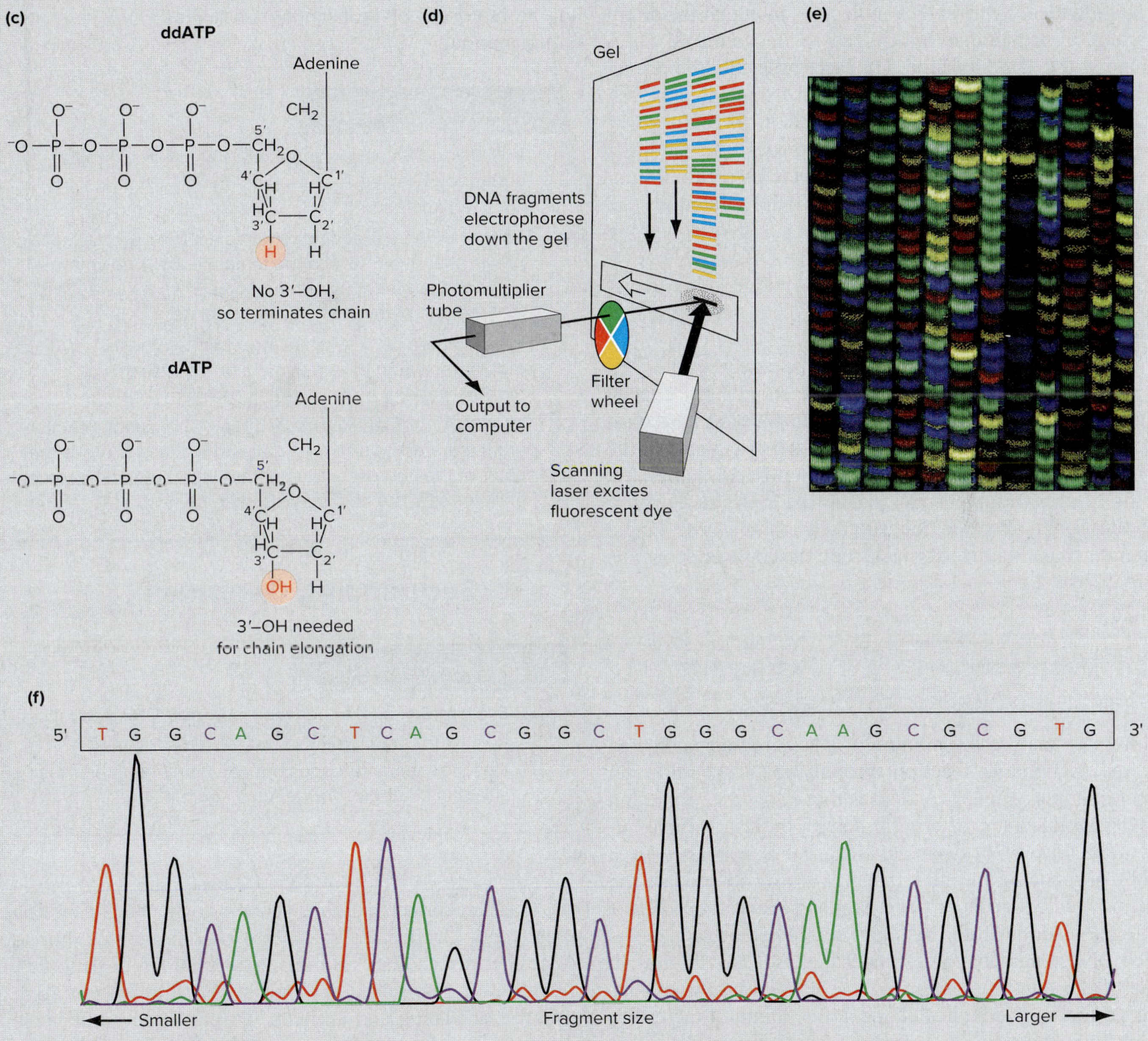

e: © Jean Claude Révy/ISM/Phototake; f: Courtesy of Joshua J. Filter, Cornell University, Ithaca, New York

template and primer, along with a carefully calibrated mixture of eight nucleotide triphosphates (Fig. 9.7b). Four of these are the normal deoxyribonucleotide triphosphates dATP, dCTP, dGTP, and dTTP. The other four are unusual and are added at lower concentrations: They are the **dideoxyribonucleotide triphosphates** (sometimes just called **dideoxynucleotides**) ddATP, ddCTP, ddGTP, and ddTTP (**Fig. 9.7c**). These dideoxynucleotides lack the 3′ hydroxyl group crucial for the formation of the phosphodiester bonds that extend the chain during DNA polymerization (review Fig. 6.21). Moreover, each dideoxynucleotide is labeled with a different color fluorescent dye; for example, ddATP can carry a dye that fluoresces in green, ddCTP has a purple dye, etc.).

The sequencing reaction tube contains billions of originally identical hybrid DNA molecules in which the oligonucleotide primer has hybridized to the template DNA strand at the same location. On each molecule, the primer supplies a free 3′ end for DNA chain extension by DNA polymerase. The polymerase adds nucleotides to the growing strand that are complementary to those of the sample's template strand. The addition of nucleotides continues until, by chance, a dideoxynucleotide is incorporated instead of a normal nucleotide. The absence of a 3′ hydroxyl group in the dideoxynucleotide prevents the DNA polymerase from forming a phosphodiester bond with any other nucleotide, ending the polymerization for that new strand of DNA (Fig. 9.7b).

When the reaction is completed, the newly synthesized strands are released from the template strands by denaturing the DNA at high temperature. The result is a nested set of fragments that all have the same 5′ end (the 5′ end of the primer) but different 3′ ends. The length and fluorescent color of each fragment making up the set is determined by the last nucleotide incorporated; that is, the single chain-terminating dideoxynucleotide the fragment contains (Fig. 9.7b).

The Fluorescence of DNA Fragments Reveals the Nucleotide Sequence

Biologists analyze the mixture of DNA fragments created by the sequencing reaction through polyacrylamide gel electrophoresis, under conditions that allow the separation of DNA molecules differing in length by just a single nucleotide (**Fig. 9.7b, d,** and **e**). The gel is examined by a DNA sequencing machine that has a laser to activate the dideoxynucleotide fluorescent tags and a sensitive detector that can distinguish the resultant colored fluorescence. As each DNA fragment passes under the laser, it will glow in one of the four fluorescent colors dictated by the dye attached to the dideoxynucleotide at the 3′ end of the chain. Each successive fluorescent signal represents a chain that is one nucleotide longer than the previous one.

The detector transmits information about the signals to a computer, which shows them as a series of different-colored peaks (**Fig. 9.7f**). The computers in DNA sequencing machines also have base-calling software that interprets the peaks as specific bases and that generates a digital file, called a **read,** of the sequence of As, Cs, Gs, and Ts comprising the newly synthesized DNA. Of course, this sequence is complementary to that of the template strand under analysis.

DNA sequencing machines available since the late 1990s can determine about 700–1000 bases from any single sample. These machines can also run hundreds of samples in parallel on separate gel lanes, each recorded with a separate fluorescence detector (Fig. 9.7d and e). Thus, a single machine running for a few hours can determine hundreds of thousands of bases of DNA sequence information.

essential concepts

- In *Sanger DNA sequencing*, the DNA molecule to be sequenced serves as a template for DNA synthesis by DNA polymerase.
- Sanger DNA sequencing requires a short oligonucleotide *primer* that *hybridizes* to the template. DNA polymerase extends the primer by adding (to its 3′ end) nucleotides that are complementary to the template.
- In automated DNA sequencing, chain synthesis terminates when DNA polymerase incorporates a *dideoxynucleotide* that has a fluorescent label.
- The DNA fragments made in the polymerization reaction are separated by size on a gel, and a detector reads the color of the fluorescent tag at the 3′ end of each fragment to determine the nucleotide sequence.

9.4 Sequencing Genomes

learning objectives

1. Explain why overlap between individual DNA sequences is required to reconstruct the sequence of a genome.
2. Describe the differences between the hierarchical and shotgun strategies for genome sequencing.

Genomes range from the 700,000 base pairs (700 kb) in the smallest known microbial genome, to more than 3 billion base pairs (3 gigabase pairs, or 3 Gb) distributed among the 23 chromosomes of humans, to even larger genomes. **Table 9.2** gives the genome sizes of representative microbes, plants, and animals. To put these numbers in perspective, the human genome is more than 700 times larger than that of *E. coli* and 45 times smaller than the

TABLE 9.2 Genome Comparisons

Organism: Type	Organism: Species	Number of Chromosomes[a]	Number of Genes[b]	Genome Size (Mb)
Bacterium	*Escherichia coli*	1	~4400	4.6[d]
Yeast	*Saccharomyces cerevisiae*	16	~6000	12.5
Worm	*Caenorhabditis elegans*	6	~22,000	100.3
Fly	*Drosophila melanogaster*	4	~17,000	122.7
Mustard weed	*Arabidopsis thaliana*	5	~28,000	135
Mouse	*Mus musculus*	20	~27,000	2,700
Human	*Homo sapiens*	23	~27,000	3,300
Lungfish	*Protopterus aethiopicus*	14	??	133,000
Canopy plant	*Paris japonica*	5[c]	??	152,400

[a]Haploid chromosome complement except where indicated.

[b]Includes non-protein-coding genes.

[c]This species is an octoploid; 5 is the *basic chromosome number* (see Chapter 13).

[d]*E. coli* genomes vary in size; 4.6 Mb is a representative length (see Chapter 14.)

genome of the plant *Paris japonica.* Thus, the information content of a genome is not necessarily proportional to the complexity of the organism.

The large size of some genomes, including the human genome, presents major challenges for their ultimate characterization and analysis. If any single DNA sequencing run can yield at most 1000 bases of information, then you might think you would need to obtain at least 3 million such sequences to determine the human genome's entire sequence. In fact this is a gross underestimate, because as we discussed previously, you would really need to examine at least five times this number of clones from a genomic library to ensure just a 95% chance that each portion of the genome would be represented once. How can you do so many DNA sequencing runs? And how can you deal with the immense amount of data you would obtain so that you could somehow figure out how these millions of small, 1000-base snippets are ordered with respect to each other in the intact genome?

The basic concept behind the method now used to sequence complex genomes, called the **whole-genome shotgun strategy,** is easy to explain: Determine DNA sequences, each about 1000 bases long, from both ends (*paired ends*) of random human genomic DNA inserts of millions of individual BAC (bacterial artificial chromosome) clones from a genomic library, and then look for overlaps between the sequences so that they can be assembled to reconstruct the sequence of the entire genome (**Fig. 9.8**). (*Shotgun* refers to the fact that the clones are chosen randomly for sequencing.) Ideally, in the case of the human genome, the ultimate output would be 24 linear strings of nucleotide sequences, one for each chromosome (the autosomes and the X and the Y).

When the Human Genome Project began in 1990, whole-genome shotgun strategies were thought to be impossible. One problem was recognized very early: As will be discussed later, genomes contain many kinds of repetitive DNA sequences, each of which can be located in many positions scattered throughout the genome. Many repeated sequences are longer than a typical sequence read of 1000 bp. This fact makes it impossible to assemble the genome from random reads. The reason is that unique sequences on one side of a long repeat from one particular genomic location cannot be present in the same read as unique sequences from the other side of the repeat (**Fig. 9.9a**). Scientists eventually realized that *paired-end sequencing* of random clones would make the whole-genome shotgun strategy possible. We will explain this method, but first we describe an alternative strategy researchers used before this conceptual breakthrough.

To get around the assembly problem presented by the existence of long repeats, the first genome scientists tried a divide-and-conquer method called a **hierarchical strategy** (Fig. 9.8). They first separated the genome into large chunks by cloning 200–300 kb fragments in BAC vectors, and then they applied strategies (not discussed here) to determine the order of the inserts with respect to each other in the original genome. Note that the genomic DNA fragments were generated using a method (such as sonication) that cleaves different copies of the genome at different locations, resulting in overlapping fragments (Fig. 9.8). These methods allowed the researchers to determine the smallest set of BAC clones with the least amount of overlap that could cover the entire genome (the so-called *minimal tiling path*). The scientists then determined the DNA sequence of the entire insert in each BAC clone of the minimal tiling path so as to reconstruct the genome

Figure 9.8 Strategies for genome sequencing. In the whole-genome shotgun (Celera) approach, BAC libraries are made from fragmented genomic DNA. Both ends of millions of clones are directly sequenced; a computer looks for overlaps between these sequences and assembles the sequence of the genome. The more systematic but less efficient hierarchical approach employs intermediate steps in which the BAC clones of a genomic library (with inserts of 200–300 kb) are characterized so as to determine the *minimum tiling path*. Each BAC clone in the path is fragmented into smaller pieces to make a mini-shotgun library, and the DNAs in the mini-libraries are sequenced. Computers reassemble the sequences of each BAC clone, and then look for the overlaps between BAC clones to reassemble the complete genome sequence.

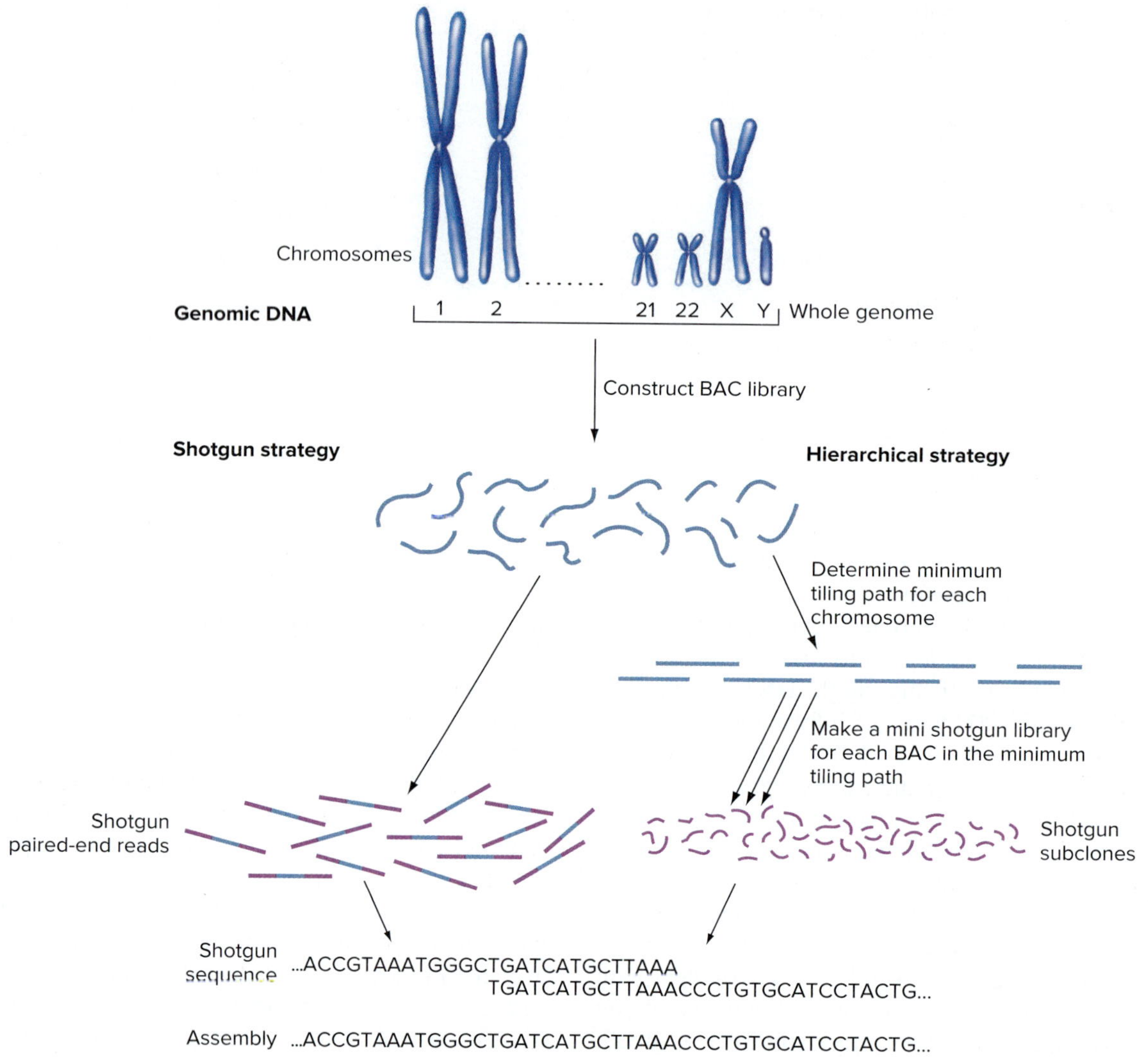

(Fig. 9.8). As most BAC clones would contain only a single copy of a particular repetitive element, assembling the genome one BAC clone at a time avoided the problem illustrated in Fig. 9.9a.

Although the hierarchical approach was ultimately successful, a private company called Celera astonished the scientific community by simultaneously undertaking and completing their own separate effort to sequence the human genome using the whole-genome shotgun strategy thought by many at the time to be hopeless. As mentioned earlier, the key to Celera's success was the idea of performing **paired-end sequencing;** that is, they obtained two sequence reads from each BAC clone, one from each end of the insert (**Fig. 9.9b** and **c**). Paired-end sequencing provided the Celera scientists with the information that the two sequences from each BAC clone must have originated from the same region of the genome and must have been 200–300 kb apart. As shown in Fig. 9.9b, this information allowed the scientists to properly align the unique sequences that flank repeat elements.

The whole-genome shotgun strategy has two important advantages over the hierarchical method. First, the shotgun approach does not require time-consuming mapping of the BAC clones to generate the tiling path. Second, the shotgun procedure can be highly automated. Celera invested in a huge facility containing hundreds of DNA sequencing machines fed by robots that first prepared DNA from the clones of genomic libraries, placed these

Figure 9.9 Whole-genome shotgun sequencing. **(a)** Repeated elements longer than a sequence read would prevent the assembly of shotgun sequences of the human genome because it is impossible to know which unique sequences (*green* or *orange*) flanking different copies of a repeat belong together. **(b)** The paired-end sequencing method. A recombinant BAC clone is melted into single strands and hybridized in one reaction with Primer 1 and in a second independent reaction with Primer 2. These primers correspond to the sequence of the vector just flanking the human DNA insert on either side. The two primers hybridize with opposite strands of the recombinant DNA and are oriented such that DNA polymerase will synthesize DNA corresponding to either end of the human DNA insert. [Note, as shown in part (c), that the *light green* portion of the insert is in fact much longer than the *dark green* portions sequenced.] **(c)** Paired-end sequencing allows correct assembly of genomes containing repeats because paired reads will include unique sequences on both sides of a repeat. In other words, the unique sequence reads (*green* or *orange boxes* at *right*) align, so the repeat-containing sequence reads (*green and black* or *orange and black boxes* at *left*) must align also.

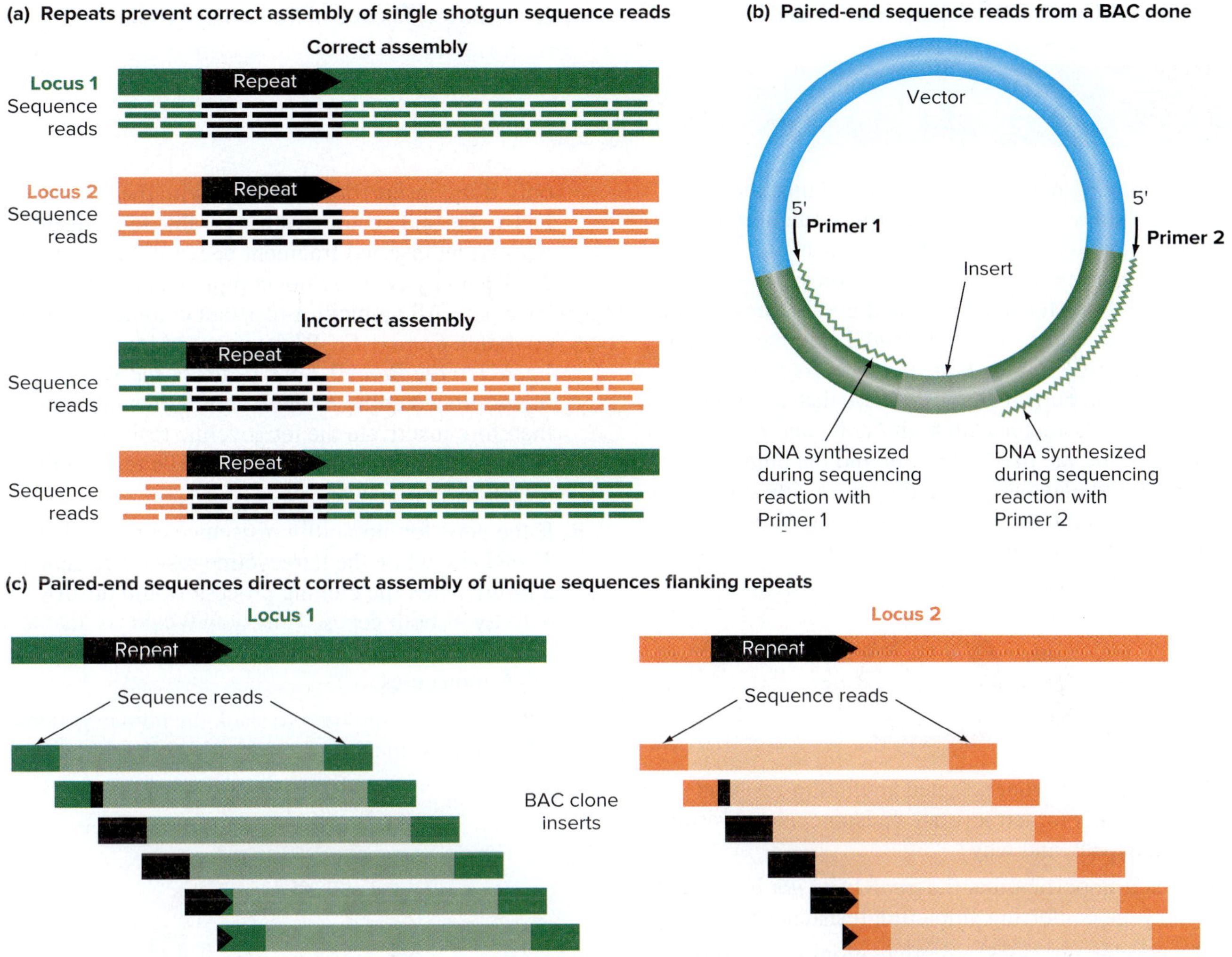

DNAs into sequencing reactions, and then loaded the reactions into the sequencing machines. This automation allowed Celera to obtain relatively cheaply the millions of DNA sequence reads required to provide about 10-fold genomic equivalent coverage. The DNA sequencing machines fed their data into a centralized supercomputer, whose complex software then assembled all these sequences into the chromosomal strings. The whole-genome shotgun approach had such large relative efficiencies that variants of this method have become the standard way to sequence genomes.

essential concepts

- The *whole-genome shotgun sequencing* strategy involves sequencing inserts of millions of clones selected randomly from libraries constructed with mechanically sheared genomic DNA to ensure overlap between fragments.
- Sequencing both ends of DNA inserts (*paired-end sequencing*) provides information useful for genome assembly.

WHAT'S NEXT

The Human Genome Project did not end with the determination of 3 billion base pairs of DNA sequence. An essential part of the project's task was to make sense of this vast amount of information. Where in all of these As, Cs, Gs, and Ts are the genes? What do the sequences of genes predict about the kinds of proteins and RNAs the genes encode and the possible functions of these proteins and RNAs? What kinds of DNA sequences make up other important features of chromosomes such as centromeres and telomeres? In the next chapter, we explain how scientists identified functional elements of the genome and how their findings revealed, at the level of DNA sequence, the architecture of the human genome.

SOLVED PROBLEMS

I. The following map of the plasmid cloning vector pBR322 shows the locations of the ampicillin (*amp*) and tetracycline (*tet*) resistance genes as well as two unique restriction enzyme recognition sites, one for *Eco*RI and one for *Bam*HI. You digested this plasmid vector with both *Eco*RI and *Bam*HI enzymes and purified the large *Eco*RI–*Bam*HI vector fragment. You also digested human genomic DNA that you want to insert into the vector with both *Eco*RI and *Bam*HI. After mixing the plasmid vector and the human genomic fragments together and ligating, you transformed an ampicillin-sensitive strain of *E. coli* and selected for ampicillin-resistant colonies.

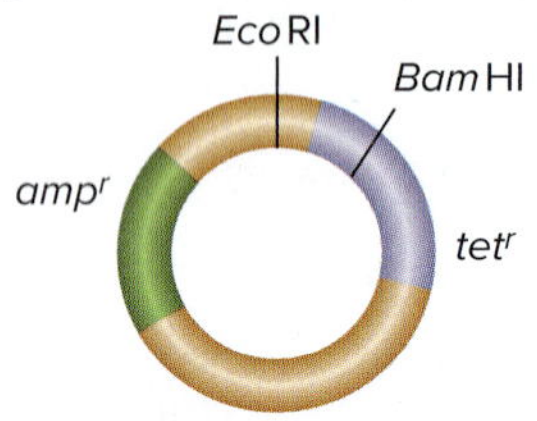

a. If you test all of your selected ampicillin-resistant transformants for tetracycline resistance, what result do you expect, and why?

b. Why is it important that the *Eco*RI site *not* be located in the gene conferring ampicillin resistance?

c. Diagram the positions and orientations of two oligonucleotide primers that you could use to sequence the two ends of the human DNA insert found in any recombinant DNA molecule made by this method.

d. Why would a library made in this fashion represent less than one genome equivalent?

Answer

This problem requires an understanding of vectors and the process of combining DNAs using sticky ends generated by restriction enzymes.

a. The plasmid must be circular to replicate in *E. coli,* and in this case, a circular molecule will be formed only if the insert fragment joins with the cut vector DNA. The cut vector will not be able to re-ligate without an inserted fragment because the *Bam*HI and *Eco*RI sticky ends are not complementary and cannot base pair. All ampicillin-resistant colonies therefore contain a *Bam*HI-*Eco*RI fragment of human DNA ligated to the *Bam*HI-*Eco*RI sites of the vector. Fragments cloned at the *Bam*HI-*Eco*RI sites interrupt and therefore inactivate the tetracycline resistance gene. All ampicillin-resistant clones will be tetracycline sensitive.

b. If the gene for ampicillin resistance contained an *Eco*RI site while the tetracycline resistance gene had a *Bam*HI site, the cloning process would destroy the activity of both genes. Thus, you would be unable to select bacterial cells transformed with recombinant DNA molecules.

c. The primers would have to flank the human genomic DNA inserts (*pink*) as shown in the following diagram.

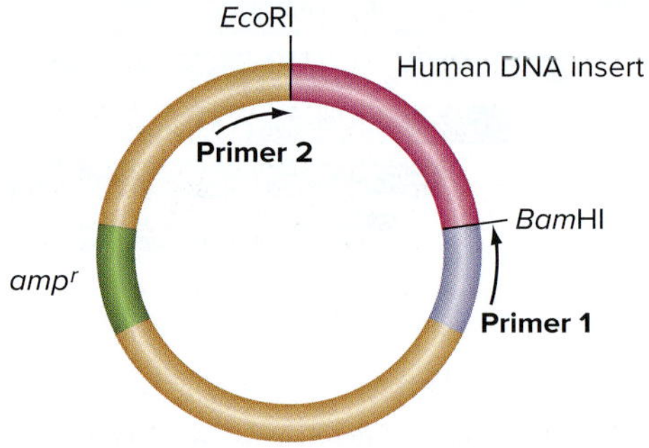

d. By chance, some regions of the human genome would have two *Eco*RI sites in a row without a *Bam*HI site in between. Other regions would have two *Bam*HI sites in a row without an *Eco*RI site in the middle. When such regions are cut with both enzymes, the resultant fragments would not have the two different kinds of sticky ends required for insertion into the vector cut with both enzymes. Thus, many regions of the human genome could not be cloned into the vector by this procedure.

II. In the course of sequencing a genome, a computer is trying to assemble the following six DNA sequences into *contigs* (that is, stretches of contiguous sequences that can be obtained from overlapping clones):

5' AGCAAATTACAGCAATATGAAGAGATC 3'
5' AAAATGCCCTAAAGGAAATGAGATTTT 3'
5' TGATCTCTTCATATTGCTGTAATTTGC 3'
5' TCCTTTTAAAAATCTCATTTCCTTTAG 3'
5' TACAGCAATATGAAGAGATCATACAGT 3'
5' AAATGCCCTAAAGGAAATGAGATTTTT 3'

a. How many contigs are represented by this set of DNA sequences, and what is the sequence of each contig?

b. Some of these sequences are complementary to each other in the region of overlap, while other sequences overlap but represent the same DNA strand. How is this possible?

c. If these sequences are all derived at random from the human genome, why would you actually expect them not to overlap with each other?

d. If you had enough sequence reads to cover all the base pairs in the human genome, how many contigs would there be?

Answer

a. Two contigs exist.

Sequences 1, 3, 5:

5' AGCAAATTACAGCAATATGAAGAGATCATACAGT 3'
3' TCGTTTAATGTCGTTATACTTCTCTAGTATGTCA 3'

Sequences 2, 4, 6:

5' TCCTTTTAAAAATCTCATTTCCTTTAGGGCATTTT 3'
3' AGGAAAATTTTTAGAGTAAAGGAAATCCCGTAAAA 5'

The orientation in which these fragments are written does not matter.

b. You are sequencing different molecules of DNA in clones that are overlapping. The clones are ligated into a vector in random orientations, and it's the orientation that determines which DNA strand is used as a template. Therefore, some of the sequences read the same strand but start in different places, and others read the complementary strand.

c. If you only had six short sequences from the entire 3 billion bp human genome, the chances would be vanishingly small that these few sequences all come at random from the same small region of the genome. Thus, six random sequences of the human genome should not overlap. Clearly, these sequences were not derived at random but were instead selected. (For example, these could be sequences all derived from a mini-library made from a particular human DNA insert in a BAC vector.)

d. Each human chromosome is a contig. A human male genome sequence would have 24 contigs, and a female genome 23 contigs.

Vocabulary

1. Match each of the terms in the left column to the best-fitting phrase from the right column.

a. oligonucleotide	1. gene in a vector that enables isolation of transformants
b. vector	2. a collection of the DNA fragments of a given species, inserted into a vector
c. sticky ends	3. synthetic DNA element in a cloning vector with unique restriction sites used for insertion of foreign DNA
d. recombinant DNA	4. stable binding of single-stranded DNA molecules to each other
e. ddNTPs	5. method for separating DNA molecules by size
f. genomic library	6. oligonucleotide extended by DNA polymerase during replication
g. genomic equivalent	7. contains genetic material from two different organisms
h. gel electrophoresis	8. the number of DNA fragments sufficient in aggregate length to contain the entire genome of a specified organism
i. selectable marker	9. short single-stranded sequences found at the ends of many restriction fragments
j. hybridization	10. a short DNA fragment that can be synthesized by a machine
k. primer	11. DNA chain-terminating subunits
l. polylinker	12. a DNA molecule used for transporting, replicating, and purifying a DNA fragment

When solving the problems in this chapter, unless instructed otherwise, make the simplifying assumptions that base pair sequences are random and that the number of A–T and G–C base pairs are equivalent.

Section 9.1

2. For each of the restriction enzymes listed below: (i) Approximately how many restriction fragments would result from digestion of the human genome (3×10^9 bases) with the enzyme? (ii) Estimate the average size of the pieces of the human genome produced by digestion with the enzyme. (iii) State whether the fragments of human DNA produced by digestion with the given restriction enzyme would have sticky ends with a 5′ overhang, sticky ends with a 3′ overhang, or blunt ends. (iv) If the enzyme produces sticky ends, would all the overhangs on all the ends produced on all fragments of the human genome with that enzyme be identical, or not? (The recognition sequence on one strand for each enzyme is given in parentheses, with the 5′ end written at the left. N means any of the four nucleotides; R is

any purine—that is, A or G; and Y is any pyrimidine—that is, C or T. ^ marks the site of cleavage.)

a. *Sau*3A (^GATC)

b. *Bam*HI (G^GATCC)

c. *Hpa*II (C^CGG)

d. *Sph*I (GCATG^C)

e. *Nae*I (GCC^GGC)

f. *Ban*I (G^GYRCC)

g. *BstY*I (R^GATCY)

h. *Bsl*I (CCNNNNN^NNGG)

i. *Sbf*I (CCTGCA^GG)

3. The calculations of the average restriction fragment size in Fig. 9.2 assume that DNA is composed equally of the four possible nucleotides. However, many genomes are somewhat enriched for certain nucleotides relative to others. As an example, the human genome is 29.6% A, 29.6% T, 20.4% C and 20.4% G. With this more accurate information in hand, re-estimate the average sizes of the pieces created by cleaving the human genome with enzymes (a–i) in the preceding Problem 9.2.

4. The DNA molecule whose entire sequence follows is digested to completion with the enzyme *Eco*RI (5′ G^AATTC 3′). How many molecules of DNA would result from this reaction? Write out the entire sequence(s) of the resultant DNA molecule(s), indicating all relevant 5′-to-3′ polarities. What about this problem appears unusual (though by no means impossible) in relationship to DNA made of random nucleotide sequences?

```
5' AGATGAATTCGCTGAAGAACCAAGAATTCGATT 3'
3' TCTACTTAAGCGACTTCTTGGTTCTTAAGCTAA 5'
```

5. Why do longer DNA molecules move more slowly than shorter ones during electrophoresis?

6. Agarose gels with different average pore sizes are needed to separate DNA molecules of different size classes. For example, optimal separation of 1100 bp and 1200 bp fragments would require a gel with a larger average pore size than optimal separation of 8500 bp and 8600 bp fragments. How do you think that scientists prepare gels of different average pore sizes? (*Hint:* Agarose gels are made in a manner similar to gelatin desserts such as JELL-O.)

7. The following picture shows the ethidium bromide–stained bands revealed by gel electrophoresis of two different DNA samples digested with two different restriction enzymes. One of the DNAs is human genomic DNA, the other is the small genome of a bacteriophage (bacterial virus) that infects *E. coli* cells. One of the restriction enzymes is *Eco*RI (5′ G^AATTC 3′); the other is *Hpa*II (5′ C^CGG 3′). For each of the four lanes on the gel (A–D), identify which of the two DNA samples was analyzed and which of the two restriction enzymes was used to digest that DNA. The arrow represents the direction of electrophoresis.

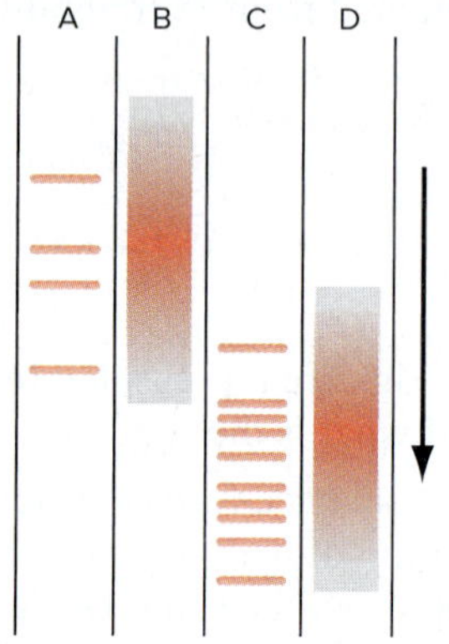

8. The linear bacteriophage λ genomic DNA has at each end a single-strand extension of 20 bases. (These are sticky ends but are not, in this case, produced by restriction enzyme digestion.) These sticky ends can be ligated to form a circular piece of λ DNA. In a series of separate tubes, either the linear or circular forms of the DNA are digested to completion with *Eco*RI, *Bam*HI, or a mixture of the two enzymes. The results are shown here.

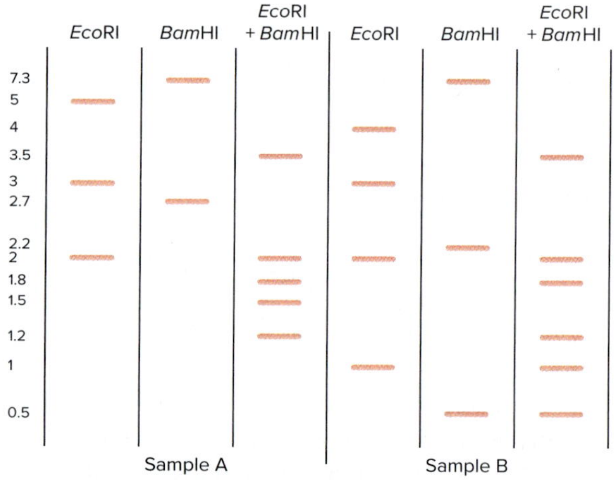

a. Which of the samples (A or B) represents the circular form of the DNA molecule? How do you know?

b. What is the total length of the linear form of the λ DNA molecule?

c. What is the total length of the circular form of the λ DNA molecule?

d. Draw diagrams of the circular and linear λ DNA molecules, showing the locations of the *Eco*RI and *Bam*HI sites.

9. Consider a *partial restriction digestion,* in which genomic DNA is exposed to a small, limiting amount of a restriction enzyme for a very short period of time.

a. Would the resultant fragments be longer or shorter or the same size as those produced by a complete digestion?

b. If you prepared genomic DNA from a tissue sample containing millions of cells, would the fragments produced by partial digestion of DNA from all of these cells be the same or different?

10. The text stated that molecular biologists have developed elegant techniques that can convert any type of DNA end into any other type of DNA end. In this problem, consider genomic DNA that is broken by mechanical shearing into random pieces. Some of the ends of these pieces are blunt, some have 5′-overhangs, and others have 3′-overhangs.
    a. Must the two ends of any one genomic DNA fragment be of the same type?
    b. Explain why the ends with 5′ or 3′ overhangs are not sticky.
    c. Researchers can convert ends with overhangs into blunt ends using either DNA polymerase (plus the four dNTPs), or nuclease S1, which degrades single-stranded regions of DNA but not double-stranded regions. Which kinds of ends with overhangs (5′ or 3′) could be converted into blunt ends using DNA polymerase? With S1 nuclease?

Section 9.2

11. a. What is the purpose of molecular cloning?
    b. What purpose do selectable markers serve in vectors?
    c. What is the purpose of the origin of replication in a plasmid vector?
    d. Why do cloning vectors have polylinkers?
12. Which of the enzymes from the following list would you need to make a recombinant DNA molecule? What is the function of those enzyme(s) in the process?
    a. DNA polymerase
    b. RNA polymerase
    c. A restriction enzyme
    d. DNA ligase
    e. An aminoacyl-tRNA synthetase
    f. Peptidyl transferase
    g. Reverse transcriptase
13. Is it possible that two different restriction enzymes could cut the human genome into exactly the same number of fragments and with exactly the same distribution of fragment sizes, yet the ends produced by the two enzymes could not be joined together by DNA ligase? Explain.
14. A plasmid vector pBS281 is cleaved by the enzyme *Bam*HI (5′ G^GATCC 3′), which recognizes only one site in the DNA molecule. Human DNA is digested with the enzyme *Mbo*I (5′ ^GATC 3′), which recognizes many sites in human DNA. These two digested DNAs are now ligated together. Consider only those molecules in which the pBS281 DNA has been joined with a fragment of human DNA. Answer the following questions concerning the junction between the two different kinds of DNA.
    a. What proportion of the junctions between pBS281 and all possible human DNA fragments can be cleaved with *Mbo*I?
    b. What proportion of the junctions between pBS281 and all possible human DNA fragments can be cleaved with *Bam*HI?
    c. What proportion of the junctions between pBS281 and all possible human DNA fragments can be cleaved with *Xor*II (5′ C^GATCG 3′)?
    d. What proportion of the junctions between pBS281 and all possible human DNA fragments can be cleaved with *Bst*YI (5′ R^GATCY 3′)? (R and Y stand for purine and pyrimidine, respectively.)
    e. What proportion of all possible junctions that can be cleaved with *Bam*HI will result from cases in which the cleavage site in human DNA was not a *Bam*HI site in the human chromosome?
15. A recombinant DNA molecule is constructed using a plasmid vector called pMBG36 that is 4271 bp long. The pMBG36 plasmid contains a so-called *polylinker* that has a single site for each of several restriction enzymes, including *Bam*HI (5′ G^GATCC 3′) and *Eco*RI (5′ G^AATTC 3′). The sequence of the polylinker region of pMBG36 is shown below; the dots indicate the large majority of the vector that is not shown. You now cut pMBG36 with *Eco*RI and insert into it a fragment of the DNA previously shown in Problem 4 that is also cut with *Eco*RI.

```
5'...CGGATCCCCTAAGATGAATTCCGCGCGCATCGGC...3'
3'...GCCTAGGGGATTCTACTTAAGGCGCGCGTAGCCG...5'
```

    a. Write out as much of the DNA sequence of the resultant recombinant DNA molecule as is possible. Two answers are possible; you need to show only one.
    b. Why are there two possible answers to part (a)?
    c. How many recognition sites for *Bam*HI will be found in the recombinant DNA molecule shown in your answer to part (a)?
    d. If you cut this recombinant DNA molecule with *Bam*HI and run the digest on a gel, how many bands would you see and how large would they be?
    e. How many recognition sites for *Eco*RI will be found in the recombinant DNA molecule shown in your answer to part (a)?
    f. If you cut this recombinant DNA molecule with *Eco*RI and run the digest on a gel, how many bands would you see and how large would they be?

16. Suppose you are using a plasmid cloning vector that has no *Eco*RI sites (5′ G^AATTC 3′) in its polylinker because the particular drug resistance gene your vector contains has an *Eco*RI site within it.
    a. How could you use the following two oligonucleotides (and ligase enzyme) to ligate an insert that is an *Eco*RI fragment into the *Bam*HI site (5′ G^GATCC 3′) in the polylinker of your vector?

       5′ GATCCGGGGGGGGGG 3′

       5′ AATTCCCCCCCCCCG 3′

    b. How many *Eco*RI sites will the recombinant DNA contain? How many *Bam*HI sites?
    c. In part (a), you used the two oligonucleotides to make a so-called *adapter*. Adapters can also be used to ligate blunt-ended inserts into vectors cut with sticky-ended enzymes. Design an adapter that would allow you to ligate blunt-ended inserts into the *Bam*HI site of your vector's polylinker. (*Note:* Two blunt-ended DNA fragments can be ligated together, although the reaction is much less efficient than sticky-end ligation.)
17. As a molecular biologist and horticulturist specializing in snapdragons, you have decided that you need to make a genomic library to characterize the flower color genes of snapdragons.
    a. How many genomic equivalents would you like to have represented in your library to be 95% confident of having a clone containing each gene in your library?
    b. How do you determine the number of independent clones that should be screened to guarantee this number of genomic equivalents?
18. Suppose you are constructing a human genomic library in BAC vectors where the human DNA fragments are on average 100,000 bp.
    a. What is the minimum number of different recombinant BACs you need to construct in order to have a greater than zero chance of having a complete library—meaning one in which the entire genome is represented?

    The simple statistical equation that follows allows you to determine the size that a genomic library needs to be (that is, the number of independent recombinant clones you need to make) for a given likelihood that the entire genome is represented in the library.

    $$N = \frac{\ln(1 - P)}{\ln(1 - f)}$$

    In the equation, N is the number of independent recombinant clones; P is the probability that any particular part of the genome is represented at least one time; f is the fraction of the genome in a single recombinant clone. (*Note: ln* is the natural log, sometimes written as log_e.)
    b. Calculate f for the genomic library described in part (a).
    c. How many different recombinant BAC clones would you need to have a 99% chance that a specific 100,000 bp region of the genome is represented? How many clones for a 99.9% chance?
    d. How many *genomic equivalents* correspond to each of your answers in part (c)?
    e. Suppose that after you ligated the human DNA inserts with the BAC vectors and transformed *E. coli* with the mixture, you find that you have only 30,000 drug-resistant colonies transformed with recombinant plasmids. What is the chance that any specific 100,000 bp region of the genome is represented in a recombinant plasmid?
    f. If you want to construct a complete human genomic library that contains the smallest number of independent recombinant clones possible, what is the key variable that you should adjust?

One difficulty in molecular cloning using plasmid vectors is that the restriction enzyme-digested vector can be resealed by DNA ligase without an insert of genomic DNA. The next two problems investigate methods to deal with this issue.

19. The *lacZ* gene from *E. coli* encodes the enzyme β-galactosidase, which can catalyze the conversion of a colorless compound called X-gal into a blue product. Molecular biologists have taken advantage of this property by constructing plasmid vectors that contain the *lacZ* gene with an *Eco*RI site in its middle (see figure that follows). After cutting this vector with the *Eco*RI enzyme, scientists ligate it together with *Eco*RI-digested human genomic DNA, transform the resultant molecules into ampicillin-sensitive *E. coli* cells, and plate these cells on petri plates containing ampicillin and X-gal. Some of the colonies growing on this plate are white in color, while others are blue. Why?

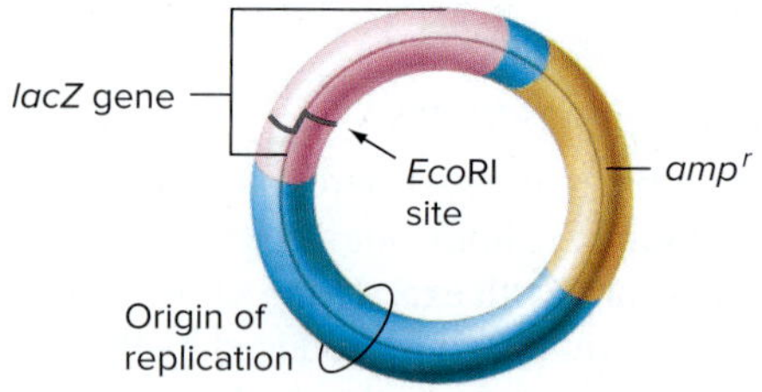

20. Your undergraduate research advisor has assigned you a task: Insert an *Eco*RI-digested fragment of frog DNA into the vector shown in Problem 19. Your advisor suggests that after you digest your plasmid with *Eco*RI, you should treat the plasmid with the enzyme alkaline phosphatase. This enzyme

removes phosphate groups that may be located at the 5′ ends of DNA strands. You will then add the fragment of frog DNA to the vector and join the two together with the enzyme DNA ligase.

You don't quite follow your advisor's reasoning, so you set up two ligations, one with plasmid that was treated with alkaline phosphatase and the other without such treatment. Otherwise, the ligation mixtures are identical. After the ligation reactions are completed, you transform *E. coli* with a small aliquot (portion) of each ligation and spread the cells on petri plates containing both ampicillin and X-gal. The next day, you observe 100 white colonies and one blue colony on the plate transformed with alkaline-phosphatase-treated plasmids, and 100 blue colonies and one white colony on the plate transformed with plasmids that had not been treated with alkaline phosphatase.

a. Explain the results seen on the two plates.

b. Why was your research advisor's suggestion a good one?

c. Why would you normally treat plasmid vectors with alkaline phosphatase, but not the DNA fragments you want to add to the vector?

Section 9.3

21. Which of the enzymes from the following list would you need to sequence DNA? What is the function of those enzyme(s) in the process?

    a. DNA polymerase

    b. RNA polymerase

    c. A restriction enzyme

    d. DNA ligase

    e. An aminoacyl-tRNA synthetase

    f. Peptidyl transferase

    g. Reverse transcriptase

22. You use the primer 5′ GCCTCGAATCGGGTACC 3′ to sequence part of the human DNA insert of a recombinant DNA molecule made with a plasmid vector. The result of the automated DNA sequence analysis is shown here. The height of the peaks is unimportant. (A = *green*; C = *purple*; G = *black*; T = *red*)

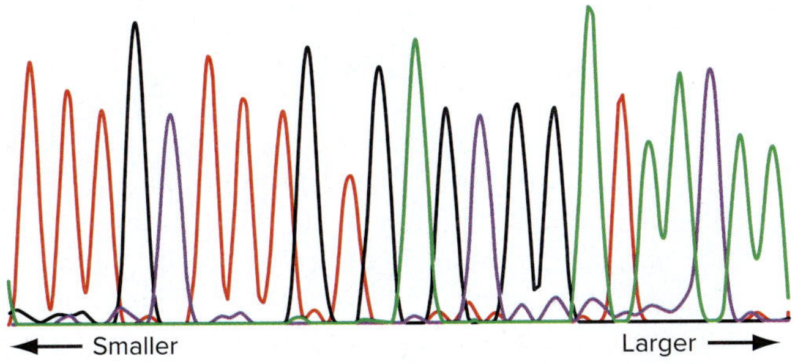

    a. Write the sequence of all the nucleotides of human DNA that you can determine. Indicate the 5′-to-3′ orientation of this sequence.

    b. Is the sequence you wrote in (a) part of the new DNA strand that was synthesized in the sequencing reaction or part of the template strand used in the sequencing reaction?

    c. How did you know how to design the primer you would need for the sequencing reaction? Diagram the recombinant DNA molecule to be sequenced, indicating the human and vector sequences, the position and orientation of the primer, and the position and orientation of the new DNA that would be synthesized during the sequencing reaction, using Fig. 9.7 as a guide.

    d. Show the *full* sequence of the smallest DNA molecule that would be synthesized in the sequencing reaction and that would contain dideoxyG (ddG). Indicate the 5′-to-3′ orientation of this molecule and the location of the ddG.

    e. How would the data differ from that shown if you accidentally left the dATP out of the reaction?

Section 9.4

23. a To make a genomic library useful for sequencing an entire genome, why would you ordinarily fragment the genomic DNA by mechanical shearing forces like sonication rather than by cutting the DNA with a restriction enzyme?

    b. Suppose that you wanted to make a genomic library to determine the complete sequence of a newly discovered organism's genome, but you did not have a sonicator readily available. Explain how you could nonetheless use two or more restriction enzymes to make libraries whose clones could be sequenced so that a computer could assemble the genomic sequence.

    c. Suppose you only had a single restriction enzyme available, and you want to make a single genomic library from which you could assemble the genomic sequence. How might you be able to achieve this goal? (*Hint:* See Problem 9.) To make this library, would it be preferable to use a restriction enzyme that recognizes a 4-base, 6-base, or 8-base sequence of DNA?

24. Problem 15 showed part of the sequence of the plasmid vector pMBG36. Suppose you make a genomic library by inserting *Eco*RI-digested fragments of a genome into *Eco*RI-digested pMBG36. Write out the sequences of two different primers you could use to sequence (in separate reactions) the two ends of all the clones in the library. These primers should be as long as possible based on the information given.

25. Eukaryotic genomes are replete with repetitive sequences that make genome assembly from sequence reads difficult. For example, sequences such as CTCTCTCTCT . . . (tandem repeats of the dinucleotide sequence CT) are found at many chromosomal locations, with variable numbers (n) of the CT repeating unit at each location. Scientists can assemble genomes despite these difficulties by using the paired-end sequencing strategy diagrammed in Fig. 9.9. In other words, they can make libraries with genomic inserts of defined size, and then sequence both ends of individual clones.

    Following are 12 DNA sequence reads from six cloned fragments analyzed in a genome project. 1A and 1B represent the two end reads from clone 1, 2A and 2B the two end reads from clone 2, etc. Clones 1–4 were obtained from a library in which the genomic inserts are about 2 kb long, while the inserts in clones 5 and 6 are about 4 kb long. All of these sequences have their 5′ ends at the left and their 3′ ends at the right. To simplify your analysis, assume that these sequences together represent two genomic locations (*loci;* singular *locus*), each of which contains a $(CT)_n$ repeat, and that each of the 12 sequences overlaps with one and only one other sequence.

    1A: CCGGGAACTCCTAGTGCCTGTGGCACGATCCTATCAAC
    1B: AGGACTCTCTCTCTCTCTCTCTCTCTCTCTCTCTCT
    2A: GTTTTTGAGAGAGAGAGAGAGAGAGAGAGACCTGGGGG
    2B: ACGTAGCTAGCTAACCGGTTAAGCGCGCATTACTTCAA
    3A: CTCTCTCTCTCTCTCTCTCTCTCAAAAACTATGGAAATTT
    3B: TAGTGATAGGTAACCCAGGTACTGCACCACCAGAAGTC
    4A: GGCCGGCCGTTGTTGACGCAATCATGAATTTAATGCCG
    4B: TCATGGGAGAGAGAGAGAGAGAGAGAGAGAGAGAGAGA
    5A: TAGTGCCTGTGGCACGATCCTATCAACTAACGACTGCT
    5B: AAGGAAAGGCCGGCCGTTGTTGACGCAATCATGAATTT
    6A: CAGCAGCTAGTGATAGGTAACCCAGGTACTGCACCACC
    6B: GGACTATACGTAGCTAGCTAACCGGTTAAGCGCGCATT

    a. Diagram the two loci, showing the locations of the repetitive DNA and the relative positions and orientations of the 12 DNA sequence reads.
    b. If possible, indicate how many copies of the CT repeating unit reside at either locus.
    c. Are the data compatible with the alternative hypothesis that these clones actually represent two alleles of a single locus that differ in the number of CT repeating units?

chapter **10**

Genome Annotation

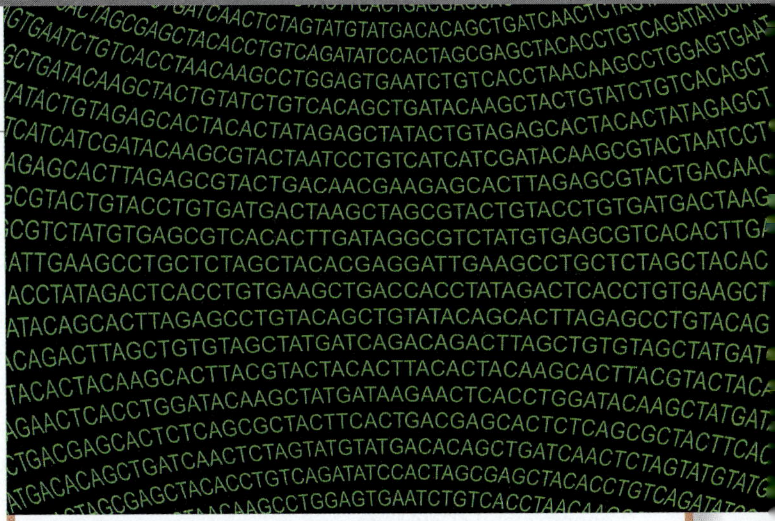

Once the sequence of a genome has been determined, researchers need to determine where functional elements such as genes reside within these billions of base pairs.

chapter outline

DETERMINATION OF THE 3 BILLION BASE PAIR SEQUENCE of the human genome, although an amazing achievement, was only the first step of the Human Genome Project. The nucleotide sequence by itself does not answer several key questions: Where are the genes? How many are there? What are their products? What lies in the genome other than genes? How are the genes and other genomic elements organized along the chromosomes? Without these answers, we cannot begin to understand how the human genotype (that is, the sequence) determines the complexity of the human phenotype.

In this chapter, we describe the **annotation** of the human genome; that is, the process of parsing out which sequences of DNA do which tasks. The process of annotation requires the compilation of data from diverse methods of investigation, including a variety of molecular experiments as well as complex computer algorithms to analyze the vast amount of data obtained. One lesson of this discussion is that the genomes of species other than humans provide important clues to the annotation of the human genome. We further describe some of the key findings from the Human Genome Project so that you can picture in a general fashion how these 3 billion nucleotides are organized.

A key feature of this chapter is an introduction to Internet-based resources, particularly a large database based at the National Center for Biotechnology Information (NCBI) at the U.S. National Institutes of Health, that you yourself can use to explore the human genome and other sequenced genomes. The chapter concludes with a comprehensive example illustrating how the sequence of the human genome has helped us to understand the nature of genetic diseases called *hemoglobinopathies* that disrupt our blood's ability to carry oxygen.

10.1 Finding the Genes in Genomes

learning objectives

1. Explain why a long open reading frame suggests the existence of a protein-coding exon.
2. Describe how scientists predict the location of genes by identifying sequences conserved in the genomes of widely divergent species.
3. Discuss the use of reverse transcriptase in the construction of a cDNA library.
4. Compare the information that can be obtained from genomic and cDNA libraries.

Genes are the key functional elements of genomes. In this section, we focus on methods to locate genes within genomic DNA sequences. You will see that information useful for the annotation of the genes within the human genome can be found in the sequence of the genome itself, the sequences of the genomes of species other than humans, and from the characterization of RNA molecules in human cells. These methods have successfully located and characterized more than 27,000 genes in the human genome, but in spite of all of these efforts, the task is still incomplete; some genes undoubtedly remain to be found.

Open Reading Frames (ORFs) Help Locate Protein-Coding Genes

One way to look specifically for regions that might correspond to the exons of protein-coding genes is to scan genomic DNA sequences for long **open reading frames (ORFs);** that is, stretches of nucleotides that have a reading frame of triplets uninterrupted by a stop codon. As you remember from Chapter 8's discussion of the genetic code, the four nucleotides can be arranged into $4^3 = 64$ possible triplets, of which three (TAA, TAG, and TGA written as DNA) signify stop. Thus, as a very rough estimate, if you looked at any random sequence of DNA starting at any one nucleotide, you would on average run into a stop codon after about $64/3 \approx 21$ triplets. If that nucleotide begins a reading frame that continues without a stop for significantly more than 21 triplets, there is a good chance that the DNA in this region is not a random set of nucleotides, but instead actually encodes amino acids within a protein (**Fig. 10.1**).

This method is useful but far from foolproof. Genomes are so large that regions that do not correspond to genes might rarely contain a long ORF by chance. On the other hand, because many genes in higher eukaryotes are interrupted by introns, some protein-coding exons are so small that they would not be identified as ORFs unless other information was available.

One type of additional information that could potentially aid computer programs in identifying genes is the fact that the splice acceptor and splice donor sites at intron/exon boundaries are composed of characteristic consensus sequences (review Fig. 8.15). Genome analysis programs can thus connect potential exons together and see if a long ORF suggestive of a gene would result.

Whole-Genome Comparisons Distinguish Genomic Elements Conserved by Natural Selection

The whole-genome shotgun approach to the sequencing of genomes described in Chapter 9 has been so successful that scientists have already deciphered the genomes of thousands of different species. Researchers can exploit this tremendous amount of information to look for regions of DNA that are similar in diverse organisms. Such regions usually, though not always, correspond to genes.

The justification for comparing genomes goes all the way back to Charles Darwin. Nearly a century before the DNA double helix was discovered, he proposed the evolution of species from now-extinct ancestors by a process of *descent with modification*. We now know that the actual entity undergoing descent with modification is the DNA sequence that defines an organism's genome. The modifications are random mutations that occur in DNA. **Natural selection** is the process whereby mutations that confer an advantage to the individuals carrying them will spread throughout a population, while deleterious mutations will disappear. The challenge is to trace such molecular evolution at the DNA level.

Figure 10.1 Open reading frames (ORFs). Any sequence of DNA can be read in any of six different reading frames (three from one strand, three from the other strand). Reading frames uninterrupted by stop codons (*red*) are ORFs. A long ORF suggests that the region may be part of a protein-coding exon. In this example, only one reading frame (Frame 5) is open.

Frame 1 → 5′ ...CCG ATG CTG AAT AGC GTA GAG GTT AGG TAA TCA TCA... 3′
Frame 2 → 5′ ... CGA TGC TGA ATA GCG TAG AGG TTA GGT AAT CAT CA... 3′
Frame 3 → 5′ ... GAT GCT GAA TAG CGT AGA GGT TAG GTA ATC ATC A... 3′

3′ ...GGC TAC GAC TTA TCG CAT CTC CAA TCC ATT AGT AGT... 5′ ← Frame 4
3′ ...GG CTA CGA CTT ATC GCA TCT CCA ATC CAT TAG TAG ... 5′ ← Frame 5
3′ ...G GCT ACG ACT TAT CGC ATC TCC AAT CCA TTA GTA ... 5′ ← Frame 6

Finding conserved DNA sequences

How can you tell whether DNA sequences from two sources are similar by chance or instead by common origin? As an example of a null hypothesis, consider a specific, but random, 50 bp sequence and calculate the probability that an independently derived DNA segment could be 100% identical, just by chance. The probability of the occurrence of any DNA sequence of length n is obtained simply by raising 0.25 (the chance occurrence of the same base at a particular position) to the 50th power (the number of independent chance events required): $(0.25)^{50} = 8 \times 10^{-31}$. This probability is very close to zero, which negates the null hypothesis and tells us that two perfectly matched 50 bp DNA sequences found in nature are almost certainly derived from the same ancestral sequence, rather than by chance.

A segment of DNA is said to be a **homolog** of a DNA segment in another species when the two show evidence of derivation from the same DNA sequence in a common ancestor. For perfectly matched sequences ~50 bp in length or longer, the evidence is clear. But evidence for homology of imperfectly matched DNA regions requires a more sophisticated statistical analysis, a task that is readily performed by specialized bioinformatics programs. When homologs of a DNA sequence are found in many different species, the sequence is said to be **conserved.**

The landscape of DNA sequence conservation

A traditional *phylogenetic tree,* like the one shown in **Fig. 10.2a,** was made by comparing genomic DNA sequences. The tree depicts the relatedness of multiple species to each other, with branch points that represent a series of nested common ancestors. When the human genome is compared as a whole with other representative vertebrate species, the percentage of sequence conservation is very high for chimps and monkeys, but it decreases as the elapsed time to a common ancestor increases (**Fig. 10.2b**). At a distance of over 400 million years, the fish genome contains only 2% of the DNA sequences present in the human genome. In contrast, when comparisons are restricted to human protein-coding sequences, conservation levels remain high—at more than 82%—throughout vertebrate evolution.

Mutations that disrupt the function of functional DNA sequences such as protein-coding regions may lessen the evolutionary fitness of the organism. As a result, functionally important sequences evolve more slowly than nonfunctional sequences, which do not contribute to phenotype. Unconstrained divergence of nonfunctional sequences would eventually eliminate all evidence of common ancestry. Thus, whole-genome comparisons can distinguish functional and nonfunctional DNA sequences by the degree of sequence conservation.

(a)

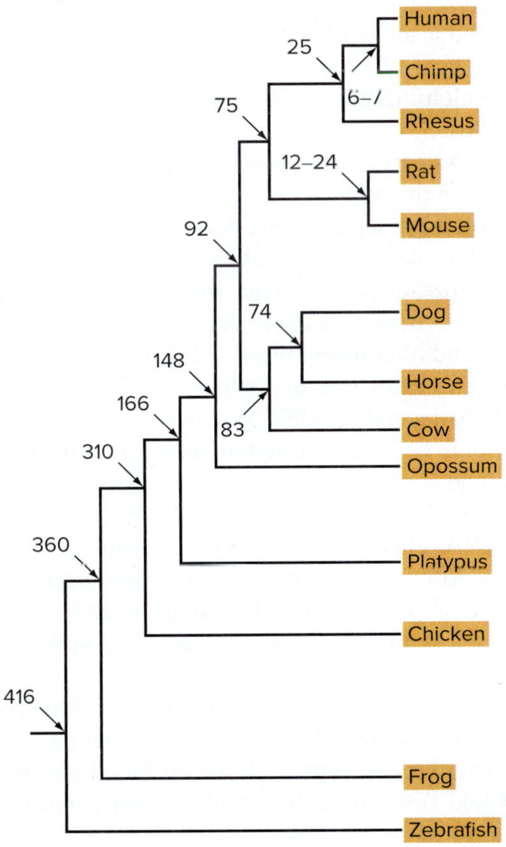

Figure 10.2 Species relatedness and genome conservation between *H. sapiens* and other vertebrates. **(a)** A phylogenetic tree showing branch points at which organisms diverged; the number at each branch point represents millions of years before the present. **(b)** Relatedness of the *H. sapiens* genome to that of other vertebrates. Column 1 shows the proportion of the complete human genome sequence found in the species being compared; column 2 indicates the proportions of human protein-coding sequences found in each vertebrate genome.

(b)

Scientific name	Common name	1	2
Homo sapiens	Human	100%	100%
Pan troglodytes	Chimp	93.9%	96.58%
Macaca mulatta	Rhesus	85.1%	96.31%
Rattus norvegicus	Rat	35.7%	94.47%
Mus musculus	Mouse	37.6%	95.36%
Canis familiaris	Dog	55.4%	95.18%
Equus caballus	Horse	58.8%	92.70%
Bos taurus	Cow	48.2%	94.78%
Monodelphis domestica	Opossum	11.1%	91.43%
Ornithorhynchus anatinus	Platypus	8.2%	86.43%
Gallus gallus	Chicken	3.8%	88.61%
Xenopus tropicalls	Frog	2.6%	87.44%
Danio rerio	Zebrafish	2.0%	82.38%

Figure 10.3 Homology map for a 100 kb region of the human genome. Regions in *black* are homologous between the human genome and the genome of the indicated species. Most DNA sequences conserved between humans and zebrafish are found in protein-coding exons. Some sequences outside of the exons are also constrained evolutionarily, suggesting that they may play functional roles that are currently unknown. (UTR: untranslated region; CDS: protein-coding sequence).

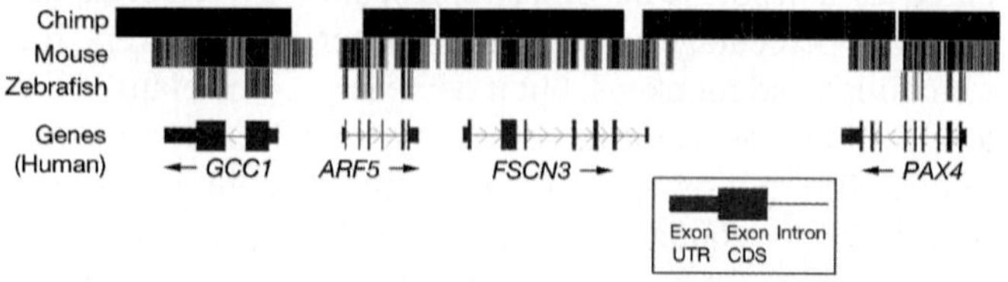

With a computerized genome visualization tool, it becomes possible to explore DNA sequence conservation directly along the genome, as well as across evolutionary time. An example of cross-species homology analysis is shown in **Fig. 10.3** for a 100 kb region containing four genes. The bottom row of the figure displays the locations and exon/intron structures of the four genes in the human genome. Above this row are homology maps for three representative vertebrate species; highly conserved DNA sequences are indicated with dark lines or blocks.

As anticipated from the close relationship between human and chimpanzee species, nearly complete conservation of human sequences exists across the entire region in a chimp genome. In other mammals, represented here by the mouse, conservation is also apparent across the entire region, but the pattern is choppy, indicating small regions of conservation interspersed with small, nonconserved regions.

As we move farther across the phylogenetic landscape to fish, we can distinguish sequences subject to evolutionary constraints more clearly from those that are not. Note in particular that large parts of the coding regions of three of the four genes are highly conserved in all the species examined (Fig. 10.3). This conservation suggests that the protein products of the three genes are crucial to the survival of all vertebrates. However, a homolog of the fourth gene is not found in zebrafish, indicating that its function is dispensable to fish. Regions of homology between the human and mouse or zebrafish genomes are much less frequent in introns, in the noncoding parts of exons (corresponding to the 5′ and 3′ UTRs of the genes), and in the spaces between genes.

Sequence conservation over long evolutionary periods, such as the time since humans last shared a common ancestor with mice or fish, therefore usually predicts the location of genes. However, exceptions do exist: Conserved DNA sequences can be observed rarely at locations outside of the coding regions. The fact that these features are so well conserved suggests strongly that they have a function that is subject to evolutionary constraints—even if in most cases we do not yet know what these functions may be. Scientists are actively exploring the potential roles of these conserved noncoding sequences; for example, some might represent enhancer elements (see Fig. 8.11) that help determine when and where nearby genes are transcribed into mRNA.

The Most Direct Method to Find Genes Is to Locate Transcribed Regions

Many genes encode proteins while some others, such as the genes for rRNAs and tRNAs, do not. However, all genes are transcribed into RNAs, even if some RNAs are not translated. If you knew the sequence of the RNA produced from a gene, it would be easy to find that gene in genomic DNA simply by looking for the DNA sequence complementary to the RNA. This approach in fact works well for RNAs that can be purified in large amounts like rRNAs (which can be isolated from other RNAs because they form part of the ribosome).

In contrast, most mRNAs are so relatively rare in cells that they cannot be purified readily. Moreover, although technologies for determining the nucleotide sequence of RNAs do exist, they are less widely available and much more difficult to perform than the methods available for sequencing DNA. As a result, the easiest way to study mRNAs is to copy them into DNA, to clone the resultant DNA molecules, and then to sequence these clones by the same methods already described for genomic DNA.

Making cDNA libraries

To produce DNA clones from mRNA sequences, researchers rely on a series of *in vitro* reactions that mimics part of the life cycle of viruses known as **retroviruses.** Retroviruses, which include among their ranks the HIV virus that causes AIDS, carry their genetic information in molecules of RNA. As part of their gene-transmission kit, retroviruses also contain the unusual enzyme known as **RNA-dependent DNA polymerase,** or simply **reverse transcriptase** (review the Genetics and Society Box in Chapter 8 entitled *HIV and Reverse Transcription*). After infecting a cell, a retrovirus uses reverse transcriptase to copy its single strand of RNA into a strand of **complementary DNA,** often abbreviated as **cDNA.** The reverse transcriptase, which can also function as a DNA-dependent DNA polymerase, then makes a second strand of DNA complementary to this first cDNA strand (and equivalent in sequence to the original RNA template). Finally, this double-stranded DNA copy of the retroviral RNA chromosome integrates into the host cell's genome. Although the designation *cDNA* originally meant a single strand of DNA complementary to an RNA molecule, it now refers to any DNA—single- or double-stranded—derived from an RNA template.

Let's see how you could use reverse transcriptase to make cDNA copies of all the mRNAs that are transcribed in a particular cell type such as red blood cell precursors. You would first isolate by simple chemical means the total population of RNA molecules in these cells **(Fig. 10.4a)**.

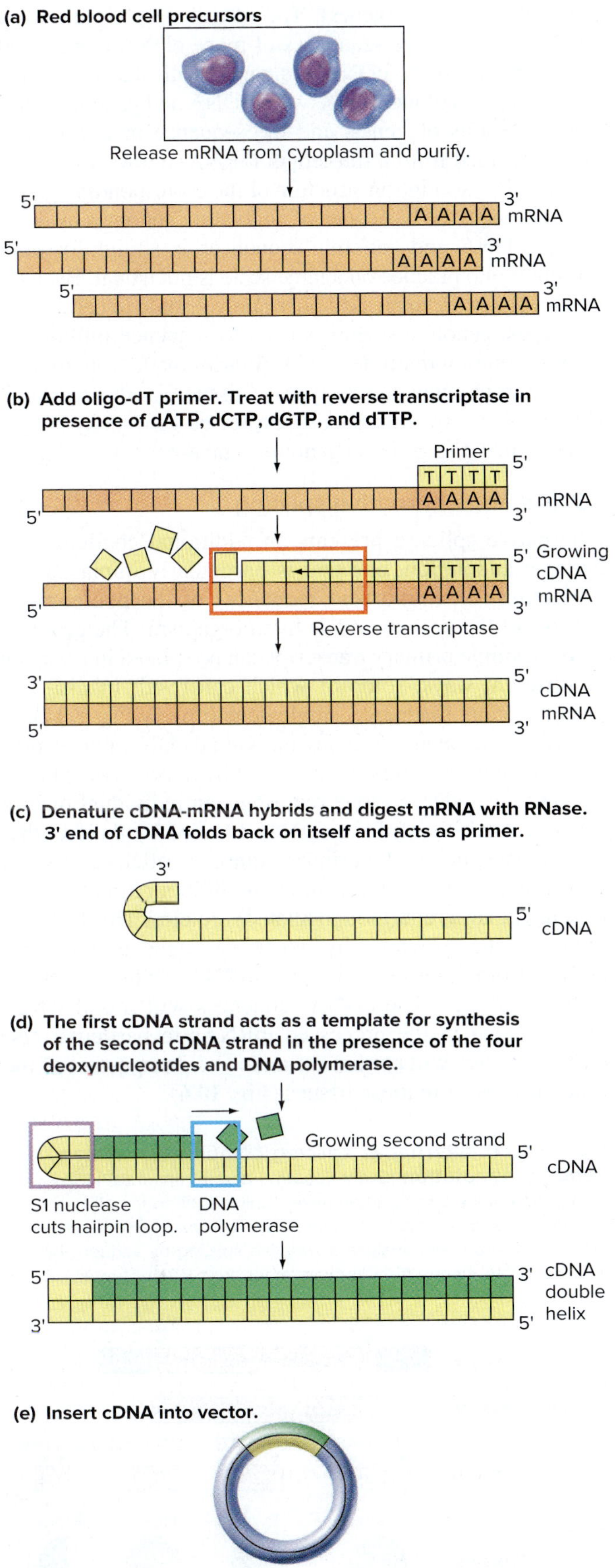

Figure 10.4 Converting RNA transcripts to cDNA. **(a)** Obtain mRNA from red blood cell precursors. **(b)** Create a hybrid cDNA-mRNA molecule using reverse transcriptase and oligo-dT primer. **(c)** Heat the mixture to separate mRNA and cDNA strands, and then eliminate the mRNA transcript. The 3′ ends of the cDNA strands bind by chance to complementary nucleotides within the same strand, forming a *hairpin loop* that can prime DNA polymerization. **(d)** Create a second cDNA strand complementary to the first. The enzyme S1 nuclease is used to cleave the hairpin loop. **(e)** Insert the newly created double-stranded DNA molecule into a vector for cloning.

Next, because mRNAs constitute only a small fraction of all the RNAs in the cell (1–5% depending on the cell type), it would be desirable to separate the mRNAs from the much more abundant rRNAs and tRNAs. This goal is possible because mRNAs in eukaryotic cells have poly-A tails at their 3′ ends. mRNAs will hybridize through their poly-A tails to the *oligo-dT* (single-stranded fragments of DNA containing about 20 Ts in a row). mRNA will thus bind to magnetic beads linked to oligo-dT, while other kinds of RNA will not. This interaction provides the basis for a separation technique (*not shown*) that will allow you to obtain a purified preparation of mRNA. The preparation will contain all of the mRNAs that are expressed in red blood cell precursors (Fig. 10.4a).

The addition of reverse transcriptase to this total mRNA—as well as ample amounts of the four deoxyribonucleotide triphosphates and primers to initiate synthesis—generates single-stranded cDNA bound to the mRNA template (**Fig. 10.4b**). The primers used in this reaction are also oligo-dT so as to initiate polymerization of the first cDNA strand from the 3′ ends of all mRNAs. After synthesis is finished, you can **denature** (separate) the mRNA-cDNA hybrids into single strands by heating the hybrids to high temperature. The addition of an RNase enzyme that digests the original RNA strands leaves intact single strands of cDNA (**Fig. 10.4c**). Most of these fold back on themselves at their 3′ ends to form transient hairpin loops that serve as primers for synthesis of the second DNA strand. Now the addition of DNA polymerase, in the presence of the requisite deoxyribonucleotide triphosphates, initiates the production of a second cDNA strand from the just-synthesized single-stranded cDNA template (**Fig. 10.4d**). The products are double-stranded cDNA molecules.

After using restriction enzymes and ligase to insert the double-stranded cDNA into a suitable vector (**Fig. 10.4e**) and then transforming the vector-insert recombinants into appropriate host cells, you would have a library of double-stranded cDNA fragments. The cDNA fragment in each individual clone will correspond to an mRNA molecule in the red blood cell precursors that served as your sample. It is important to note that this **cDNA library** includes only the exons from that part of the genome that these cells were actively transcribing for translation into protein. The clones in cDNA libraries do not contain introns because the mature mRNAs from which they were produced do not have introns. You should also understand that the cDNA library made from red blood cell precursors would contain many clones corresponding to mRNAs that are highly expressed in this tissue, but only a few clones that reflect genes that are expressed rarely.

Genomic versus cDNA libraries

Figure 10.5 compares genomic and cDNA libraries. The clones within genomic libraries represent all regions of DNA equally and show what the intact genome looks like in the region of each clone. The clones in cDNA libraries reveal which parts of the genome contain the information used in making proteins in specific tissues. The prevalence of the mRNAs for specific genes also gives some indication, though imperfect, of the relative amounts of the various proteins made in those cells.

As described previously, one of the main purposes of making cDNA libraries is to annotate genomes by finding transcribed regions (genes). The idea is very simple: You would determine the sequences of many cDNA clones, and then compare these cDNA sequences with that of the genome. Regions of identity between cDNA and genome represent the exons of genes, and the sequence of a complete cDNA (copied from a full-length mRNA) allows you to determine the exon/intron structure of the corresponding gene.

Although the basic idea of annotating genomes by comparing cDNA and genomic sequences is straightforward, putting it into practice on a large scale is not trivial. Because certain genes are expressed only rarely or only in certain tissue types, genomic scientists need to sequence millions of cDNA clones in multiple cDNA libraries made from mRNA derived from many diverse types of tissue. For this reason, it is likely that some infrequently expressed genes may not yet be recognized as genes in genome databases.

Figure 10.5 A comparison of genomic and cDNA libraries. Every tissue in a multicellular organism can generate the same genomic library, and the DNA fragments in that library collectively carry all the DNA of the genome. On average, the clones of a genomic library represent every locus an equal number of times. By contrast, different tissues in a multicellular organism generate different cDNA libraries. Clones of a cDNA library represent only the fraction of the genome that is transcribed in that tissue. The frequency with which particular fragments appear in a cDNA library is proportional to the level of the corresponding mRNA in that tissue.

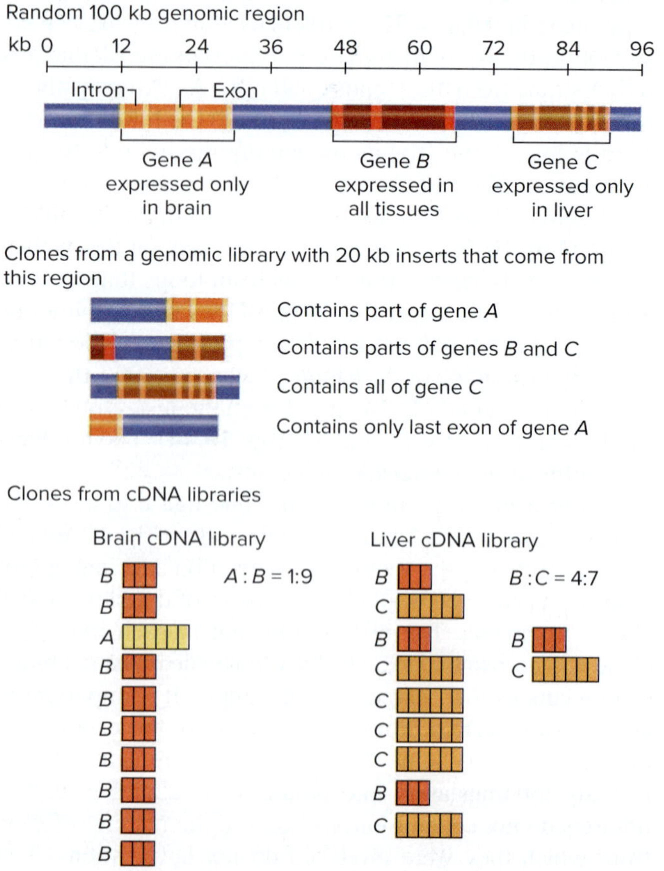

cDNAs and alternative splicing

Alternative splicing presents an additional challenge for genome annotation, one that is particularly important for predicting the amino acid sequences of the **proteome**—that is, all the proteins made in an organism. The problem is that a single primary transcript can be spliced in a variety of different ways, some of which can result in different proteins being made by a single gene (review Fig. 8.17).

The sequencing of many individual cDNA clones provides a solution to this issue caused by alternative splicing because each cDNA clone represents an individual mature mRNA. Analysis of these cDNAs is aided by the fact that alternative splicing of a primary transcript often occurs in a cell type–specific manner, allowing different kinds of cells to generate different (though related) proteins. This fact provides another reason why geneticists need to sequence cDNAs from libraries made using mRNAs from a variety of different tissues. The cDNA sequences will reveal which exons appear in the processed mRNAs in particular cell types, and thus will predict the amino acid sequences of the proteins present in those tissues (**Fig. 10.6**).

Figure 10.6 Alternative splicing complicates human genome annotation. Exons (*orange*) and introns (*red*) in the primary transcript can be alternatively spliced, often in cell type-specific ways; as a result, the same gene can express different proteins. Researchers analyze alternative splicing by sequencing multiple cDNA clones from libraries made from each of many different tissues.

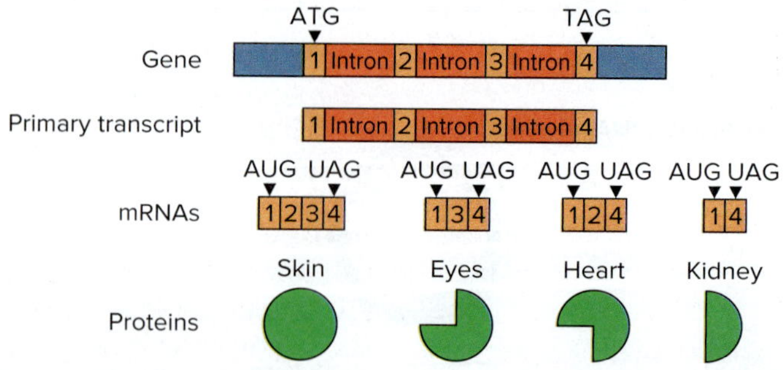

essential concepts

- Long *open reading frames (ORFs)* in genomic DNA usually identify protein-coding exons.
- DNA sequences *conserved* between the genomes of widely divergent species often correspond to the exons of genes.
- *Reverse transcriptase* produces *complementary DNA (cDNA)* from mRNA transcripts; cDNA clones thus represent only the exons of genes.
- cDNA sequences reveal how a primary transcript is spliced in a given cell type, and thus predict the amino acid sequence(s) of a gene's protein product(s) in that cell type.

10.2 Genome Architecture and Evolution

learning objectives

1. Discuss the arrangement of genes in genomes, including the number of genes, transcription direction, and gene density.
2. Explain how gene duplication and divergence lead to the formation of gene families and pseudogenes.
3. List three ways in which genomes can change over evolutionary time.
4. Describe how mechanisms at the DNA, RNA, and protein levels can produce complexity from a small number of genes.

The complete sequences of the human and other genomes have provided striking new insights into the organization and evolution of genomes. Our detailed knowledge of genomic sequences has changed profoundly the practice of biology. We now briefly describe some of the main lessons and surprises from these genome projects, focusing on the following three questions: How are the genes arranged in genomes? How do genes and genomes change during evolution? Lastly, how can genomes with a relatively small number of genes produce the vast complexity of phenotypes that results in living organisms, including humans?

The Arrangement of Genes in the Genome Is Not Uniform

A major shock to emerge from the completion of the human genome sequence was the discovery of only about 27,000 genes. Of these genes, roughly 19,000 encode proteins, while the remainder are transcribed into RNAs that are not translated, such as rRNAs and tRNAs that participate in translation, and the snRNAs that function in spliceosomes. These numbers are much lower than expected. Estimates made before the initiation of the Human Genome Project had suggested that 100,000 or more genes might exist. Many of these estimates anticipated that because humans have much greater biological complexity than simpler model organisms such as bacteria, yeasts, nematodes, and fruit flies, our genomes must have many more genes. Although the human genome indeed has more genes than these organisms, the difference in gene number is not nearly as great as had been thought (review Table 9.2). Mechanisms other than changes in gene number must therefore underly metazoan (multicellular animal) complexity.

The total length of genomes varies much more markedly over the course of evolution than the number of genes (compare especially the genomes of mammals such as mice and humans with model eukaryotes like worms and flies; Table 9.2). This generalization holds because the **exome,** the part of the genome corresponding to the exons of all known genes, constitutes only a small proportion of genomes, roughly 1.5–2.0% of the total in humans. The vast majority of DNA sequences instead are found in introns, in the spaces between genes (*intergenic regions*), in transposable elements that can move from one chromosomal position to another, and in structural features like centromeres and telomeres.

The tremendous variation seen in the size of genomes of different species is thus mostly due to expansions and contractions of noncoding DNA outside of the exome, rather than changes in gene number or gene sizes. For example, half or more of the human genome appears to be composed of transposable elements, often regarded as *selfish* or parasitic DNA that uses our genomes as a host for their own propagation. In a second example, the human genome also contains many simple repeating sequences (such as CGCGCGCG, etc.).

In this section, we focus on the small proportion of the genome corresponding to the genes, with an emphasis on those that encode proteins. Later chapters will describe in greater detail the nature of DNA sequences that constitute other chromosomal elements about which some information is available (centromeres, telomeres, and transposable elements). However, you should know that a large percentage of the human genome is "dark matter" whose presence we do not yet understand. Some of this DNA may have functions that are currently obscure, but much of it may in fact not have any function at all and may instead be the vestiges of chance events that occurred during evolution.

Random orientation of transcription of most genes

Students sometimes assume that all the genes on a chromosome are transcribed in the same direction, always using the same strand of double-stranded genomic DNA as the template. This assumption is absolutely incorrect. As previously shown in Fig. 10.3, neighboring genes can be transcribed either in the same or in opposite directions with respect to each other, and either toward the centromere or toward the telomere with respect to the chromosome as a whole. Gene maps such as that in Fig. 10.3 typically indicate the 5′-to-3′ direction of a gene's transcription (that is, the direction that RNA polymerase moves as it copies the gene into RNA) with an arrow.

Because neighboring genes can be transcribed in opposite directions, RNA polymerase uses the chromosome's *Watson* strand as the template for some genes, while for other genes, the template is the *Crick* strand. For most genes, the direction of transcription appears to be chosen at random, or at least no definitive patterns can yet be discerned. However, in a few exceptional genomic regions, such as those containing the hemoglobin genes that will be described later in this chapter, specific mechanisms of gene regulation require that neighboring genes have the same transcriptional orientation.

Variable gene density

On average, the density of genes in the human genome is slightly less than 1 gene in every 100 kb of DNA (27,000 genes in a 3,000,000 kb genome). However, this rough value obscures the fact that the packing of genes can be very different in various parts of the genome. Some regions on some chromosomes are gene-rich, with little space between densely packed genes. The most gene-rich region of the human genome is a 700 kb stretch of chromosome 6 that contains 60 genes encoding histocompatibility proteins with diverse functions (**Fig. 10.7**).

Other regions, called **gene deserts,** contain few or no genes. The largest known desert in the human genome is 5.1 Mb on chromosome 5 without a single identified gene. Some deserts are gene-poor because they contain so-called *big genes* whose nuclear transcript spans 500 kb or more of chromosomal DNA. The largest of the big genes in humans is the gene for dystrophin, which spans 2.3 Mb, most of which is composed of introns. Interestingly, because big genes can be expressed only through the production of enormous primary transcripts, their transcription cannot be completed in rapidly dividing cells. Many big genes are thus expressed only in neurons, which do not divide. It is possible that scientists have not yet detected some big genes, which might be transcribed into RNA only rarely.

A fundamental unanswered question is whether gene-rich and/or gene-poor regions have biological meaning. Is there a functional explanation for these variations in gene density, or do they instead reflect random fluctuations in evolutionary events that shape chromosomal architecture?

Genomes Undergo Evolutionary Change

Genomes continually undergo many different kinds of DNA sequence changes that provide the raw material for natural selection. In earlier chapters of this book, you have encountered some of the events that can alter the nucleotide sequence of pre-existing genes: In particular, environmental mutagens and mistakes in DNA replication can result in nucleotide substitutions. The accumulation of such point mutations within a gene can certainly change the gene's function over time. However, as we now describe, analysis of the genomes of humans and other species indicates that evolution can also create new genes and reorganize genomes by reshuffling blocks of DNA larger than a single nucleotide. Many kinds of processes promote genome plasticity over the evolutionary timeframe.

Alterations in domain architecture

Genome annotation has revealed that exons often encode discrete **protein domains,** each of which is a linear sequence of amino acids that folds up in three-dimensional space so as to act as a single functional unit. Genes with multiple exons often encode proteins with multiple domains analogous to the cars of a train. Each train is composed of many different cars, and each kind of car (engine, flat car, dining car, caboose) has a discrete function. Different trains may carry different combinations of cars and thus fulfill

Figure 10.7 Class III region of the human major histocompatibility complex. This densely packed 700 kb–long region contains 60 genes (*colored boxes*). Arrows below the genes indicate the direction of each gene's transcription; just as in Fig. 10.3, some genes are transcribed in one direction and others in the opposite direction.

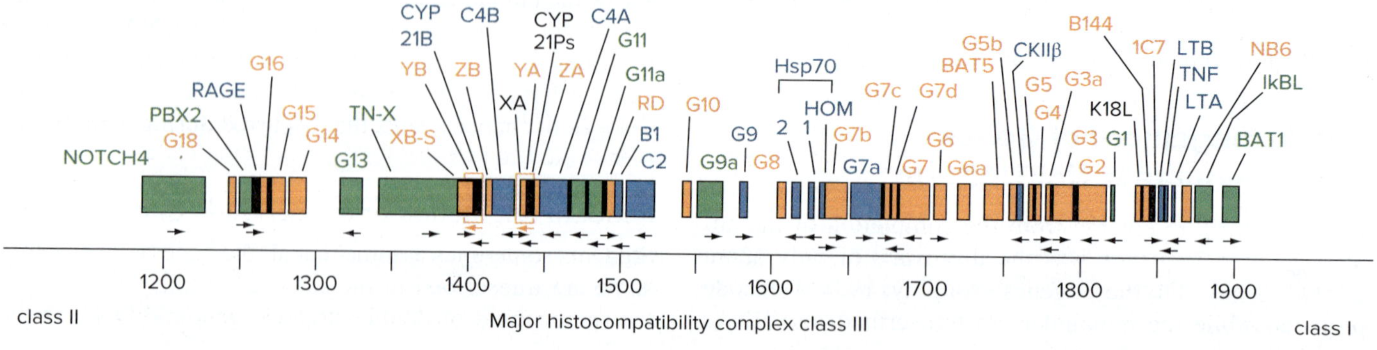

Figure 10.8 Domain architecture of transcription factors. Protein domains are indicated as colored icons labeled POZ, HD (homeodomain), etc. Horizontal lines connect domains found in the same protein. The numbers and types of transcription factors vary considerably between different species due to protein domain reshuffling during evolution. As an example, worms make about 143 different transcription factors containing POZ domains, of which three are shown, while fruit flies make about 93 POZ-containing proteins, of which two are shown. Some of the domains shown govern DNA binding, while others facilitate protein–protein interactions.

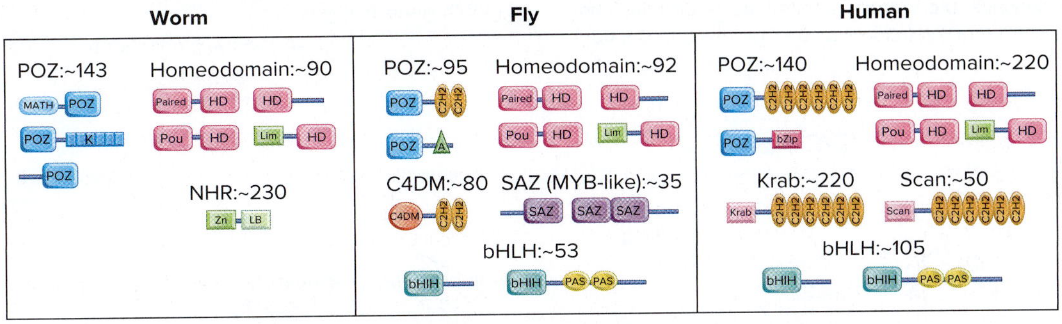

Figure 10.9 Homeodomain consensus sequence. The **consensus sequence** of amino acids shows the most commonly found amino acid at a given position within all known homeodomains in all organisms. Subsequent rows show matches to the consensus (*purple*) of homeodomains in nine *Drosophila* proteins that dictate key aspects of the animal's development. (These genes and proteins will be discussed in Chapter 19.)

Consensus	RRRKRTAYTRYQLLELEKEFHFNRYLTRRRRIELAHSLNLTERQVKIWFQNRRHKWKKEN
Ubx	RRRGRQTYTRYQTLELEKEFHTNHYLTRRRRIEMAHALSLTERQIKIWFQNRRMKLKKEI
Abd-A	RRRGRQTYTRFQTLELEKEFHFNHYLTRRRRIEIAHALSLTERQIKIWFQNRRMKLKKEL
Abd-B	VRKKRKPYSKFQTLELEKEFLFNAVSKQKRWILMRNAQSLTERVIKIWFQNRRMKNKKNS
lab	NNSGRTNFTNKQLTELEKEFHFNRYLTRRRRIEIANTLQLNETQVKIWFQNRRMKWKKEN
pb	PRRLRTAYTNTQLLELEKEFHTNKYLCRPRRIEIAASLDLTERQVKVWFQNRRMKHKRQT
Dfd	PKRQRTAYTLHQILELEKEFHYNRYLTRRRRIEIAHTLVLSERQIKIWFQNRRMKWKKDN
Scr	TKRQRTSYTRYQTLELEKEFHFNRYLTRRRRIEIAHALSLTERQIKIWFQNRRMKWKKEN
Antp	RKRGRQTYTRYQTLELEKEFHFNRYLTRRRRIEIAHALSLTERQIKIWFQNRRMKWKEIN

different purposes. Similarly, many genes are composed of multiple exons that encode discrete protein domains. The shuffling, addition, or deletion of exons during evolution can create new genes whose protein products have novel **domain architectures** (different numbers and kinds of domains in different orders) and thus can assume new roles in cells and organisms.

Figure 10.8 shows an example of the domains associated with various *transcription factors,* proteins that bind to regions of DNA such as enhancers that control the transcription of nearby genes. Exon shuffling over evolution has produced different transcription factors with differing domains that enable these proteins to recognize particular DNA sequences and also to interact uniquely with cofactors such as other proteins.

Biologists may guess at the function of a new protein (or the gene that encodes it) by analogy, if they find by computerized analysis that it contains a domain known to play a specific role in other proteins. As an example, many proteins that include a *homeodomain* (a particular DNA-binding motif) are transcription factors important for the development of multicellular organisms. Computer algorithms determine that a particular gene encodes a homeodomain-containing protein by comparing its putative amino acid sequence to those of known homeodomains and searching for similarity (**Fig. 10.9**).

The mechanism of RNA splicing facilitates this kind of exon rearrangement in eukaryotic genomes (and thus the creation of new genes) because the reshuffling does not have to be precise. Suppose, as shown in **Fig. 10.10,** that the exon of one gene plus its flanking introns is moved to a new location in the intron of a different gene. This exon can now be spliced together with the second gene's exons to make a single mRNA molecule, regardless of where within the introns these pieces of DNA were brought together.

Gene families

Gene families are groups of genes closely related in sequence and function; such gene families are abundant throughout genomes. Examples of gene families include the genes that encode the hemoglobins that allow us to transport oxygen in our blood (**Fig. 10.11**), the immunoglobins (antibodies) that help us ward off infections, and the olfactory receptors critical for our sense of smell.

Figure 10.10 Exon shuffling. Suppose two genes are broken in introns and joined together as shown. Transcription of the newly reshuffled gene will produce a primary transcript that can be spliced into a mature mRNA encoding a novel protein, regardless of where in the introns the breakages occurred. If the different exons encode different protein domains, the domain architecture of proteins can change over the course of evolution, as seen for the transcription factors in Fig. 10.8.

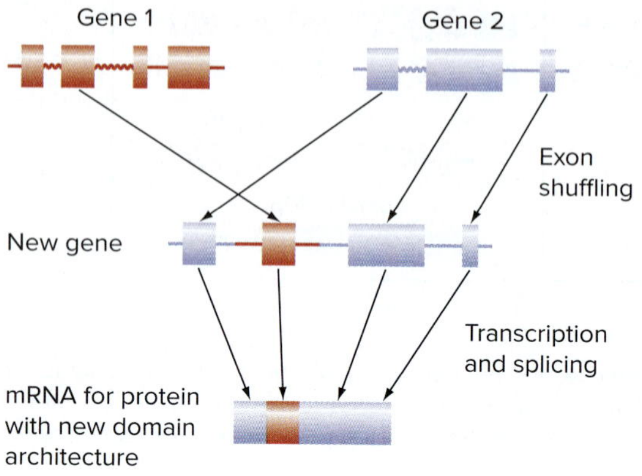

Figure 10.11 The genes for human hemoglobin polypeptides are located in two genomic clusters. **(a)** Schematic representation of the α-globin locus. The five functional genes are indicated with *purple* boxes, the two pseudogenes with *black* boxes. All of these genes are transcribed in the same direction (left-to-right on the map). The *red* box is the locus control region (LCR) described later in this chapter. **(b)** Schematic representation of the β-globin locus; this cluster has five functional genes (*green*) and one pseudogene (*brown*).

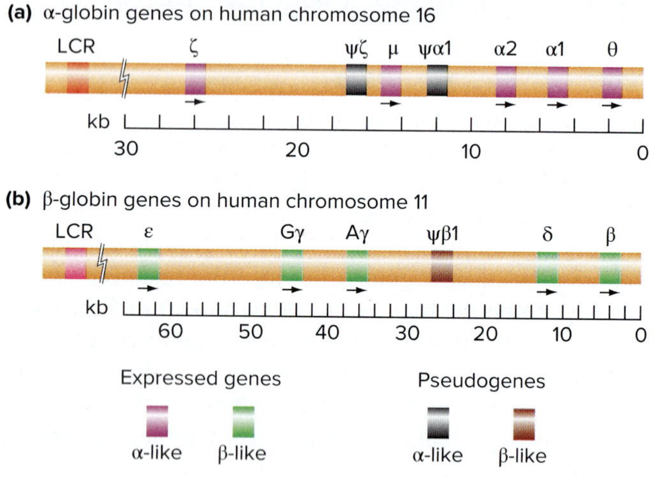

With the use of bioinformatics, researchers can see that each gene family evolved by a process of **duplication and divergence** from an ancestral gene. The two DNA sequence products of a duplication event, which start out identical, eventually diverge as they accumulate different mutations (**Fig. 10.12**). Additional rounds of duplication and divergence can further increase the number of related genes. For example, the human genome has ten functional members of the hemoglobin gene family, while the olfactory

Figure 10.12 Evolution of the globin gene family. Duplication of an ancestral gene, followed by divergence of the duplication products, established the α- and β-globin lineages. Further rounds of duplication and divergence within the separate lineages generated the genes and pseudogenes of the current-day globin gene family.

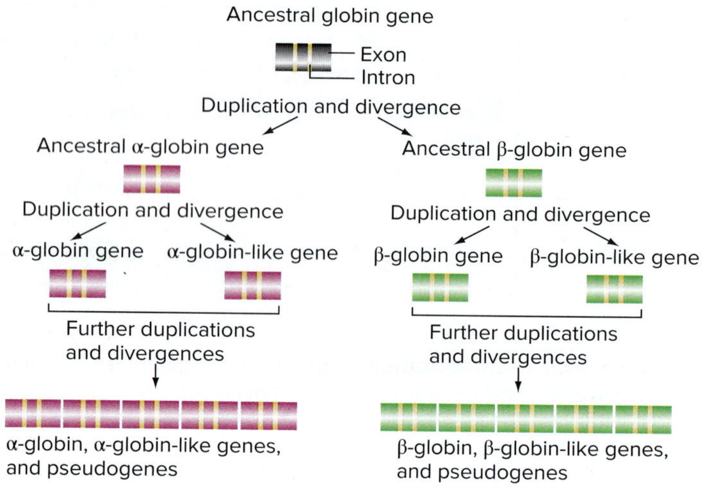

receptor family includes about 1000 genes. The duplication and divergence process is crucial for the creation of new raw material for evolution. Once a gene has duplicated, divergence allows either or both of the copies to assume new specialized but related functions, as long as one or both of the copies still fulfills the role of the original gene.

The genes in such families may be clustered together on one chromosome or dispersed on several chromosomes. In the case of the hemoglobin family, the α-globin gene cluster (also called the *α-globin locus*) on chromosome 16 contains five functional genes, while the β-globin cluster (*β-globin locus*) on chromosome 11 also has five genes (Fig. 10.11). The sequences of all the α-like genes are more similar to each other than they are to the β-like sequences, and *vice versa*. The β-like genes are exactly the same length, and the five β-like genes have two introns at exactly the same position; in fact, the α-like genes also have two introns at the same positions.

These comparisons suggest that all of the globin genes can be traced back to a single ancestral DNA sequence (Fig. 10.12). Hundreds of millions of years ago, this ancestral globin gene duplicated, and one copy moved to another chromosome. With time, one of the two copies gave rise to the α-lineage, the other to the β-lineage. Each lineage then underwent further duplications to generate the present array of α-like and β-like genes in humans. By comparing genomes of different organisms, it is possible to estimate when these various duplication events occurred. For example, the β-globin clusters of humans and chimpanzees have the same genes in the same order, but some other primates have one

Figure 10.13 Gene family nomenclature. *Orthologous genes* are separated by a speciation event. *Paralogous genes* are separated by a duplication event. *Homologous genes* are related to each other by descent from a common ancestral DNA sequence regardless of the mechanism of separation; all the genes shown in this figure are thus homologous.

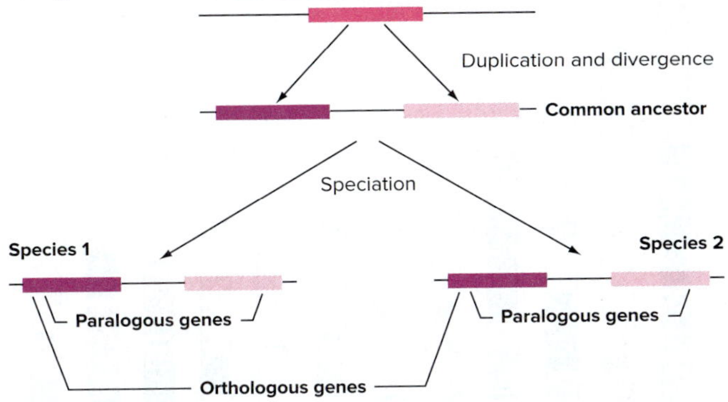

fewer β-like gene. Thus, the last gene duplication event in the β-globin cluster must have occurred in a common primate ancestor of humans and chimps.

The existence of gene families requires the definition of new terms to describe the relationship of the genes that compose them (**Fig. 10.13**). **Orthologous genes** are genes in two different species that arose from the same gene in the species' common ancestor; usually but not always, orthologous genes retain the same function. The genes for the ε-globin in humans and chimpanzees are orthologs because an ε gene already existed in their last common ancestor. By contrast, **paralogous genes** arise by duplication; this term is usually used to denote the different members of a gene family. Thus, the genes for δ-globin and ε-globin in the human β-globin locus (Fig. 10.11b) are close paralogs, and both are more distant paralogs of genes in the α-globin cluster. Finally, **homology** is a blanket term for all evolutionarily related sequences; all hemoglobin genes in all species are thus homologous, and all these genes share weaker homologies with myoglobin genes encoding more distantly related oxygen-carrying proteins in muscle tissues rather than red blood cells.

The duplications that gave rise to multiple functional hemoglobin genes also produced genes that eventually lost the ability to function. Molecular geneticists made this last deduction from data showing two additional α-like sequences within the α locus and one β-like sequence within the β locus that no longer have the capacity for proper expression (Fig. 10.11). The reading frames are interrupted by frameshifts, missense mutations, and nonsense codons, while regions needed to control the expression of the genes have lost key DNA signals. Sequences that look like, but do not function as, genes are known as **pseudogenes;** they occur in many gene families throughout all higher eukaryote genomes. Interestingly, the same pseudogene with almost all the same gene-inactivating mutations is found in the β cluster of both the human and chimpanzee genomes, indicating that the duplication giving rise to the pseudogene, as well as many of the mutations that disrupt its function, must have existed in a common primate ancestor of both species.

Because they serve no function, pseudogenes are subject to mutation without selection and thus accumulate mutations at a far faster pace than coding or regulatory sequences of a functional gene. Eventually, nearly all pseudogene sequences mutate past a boundary beyond which it is no longer possible to identify the functional genes from which they have been derived. Continuous mutation can thus turn a once functional sequence into an essentially random sequence of DNA.

De novo genes

Most annotated genes in any sequenced genome belong to gene families and are also homologs of genes that exist in many distantly related species. However, many genes discovered by genome sequencing appear either to lack homologs in any other species or to have homologs only in closely related species. For example, a few hundred genes in the human genome are human-specific. Genes without homologs are called **de novo genes.** The term *de novo* means *from new* in Latin.

De novo genes are young genes that evolved recently from ancestral intergenic sequences. Evidence exists for two different mechanisms of de novo gene evolution through mutation: Either transcribed intergenic regions gained an ATG and thus a short ORF (**Fig. 10.14a**), or small

Figure 10.14 Origins of de novo genes. Genes without homologs can arise either when **(a)** transcribed intergenic DNA mutates to generate an ATG and thus a small ORF, or when **(b)** a small ORF in intergenic DNA acquires transcriptional activation sequences.

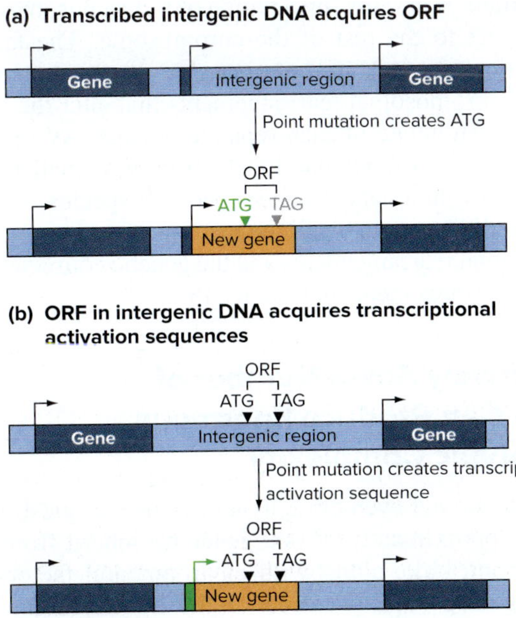

intergenic ORFs that were not originally transcribed acquired transcriptional regulatory sequences (**Fig. 10.14b**).

As expected of young genes that originated with short ORFs, de novo genes are smaller and simpler (they have fewer exons and less alternate splicing) than more ancient genes with homologs. Because de novo genes encode proteins that differ radically from others in the proteome, some scientists think that they may be particularly useful in facilitating the evolution of diverse morphologies.

Chromosomal rearrangements

The mouse and human genomes, which diverged 85 million years ago, exhibit striking similarities and differences not only in the sequences of individual genes, but also in the order of those genes on the chromosomes. The similarities are seen within each of approximately 180 homologous blocks of chromosomal sequence, ranging in size from 24 kb to 90.5 Mb—for an average of 17.6 Mb (**Fig. 10.15**). Within such blocks of linked loci, called **syntenic blocks,** the order of the genes is very similar in humans and mice. However, the orders of these blocks along the chromosomes are totally different in the two organisms. It is as if one genome had been cut into 180 pieces of varying size and then assembled randomly into the other genome.

Conserved synteny, in which the same two or more loci are linked in different species, also exists between the human and puffer fish genomes, which diverged more than 400 million years ago. In this case, though, the syntenic blocks are relatively small—averaging only about 250 kb in length.

The apparent cutting and reassembling of chromosomal blocks accompanying evolution are due to events that can occur within genomes called **chromosomal rearrangements.** For example, some rearrangements called *translocations* connect part of one chromosome to part of a different, nonhomologous chromosome. Other rearrangements called *inversions* flip a region of a chromosome 180° with respect to the rest of the chromosome. The farther back in time two species last shared a common ancestor, the more chromosomal rearrangements that alter the order of genes accumulate in each separate lineage. As a result, the average size of syntenic blocks becomes smaller with increasing evolutionary distances between species.

We will discuss in detail the mechanisms giving rise to chromosomal rearrangements and the genetic consequences of these rearrangements in Chapter 13.

Figure 10.15 Syntenic blocks in the human and mouse genomes. Human chromosomes, with segments containing at least two genes whose order is conserved in the mouse genome, appear as *color blocks.* Each color corresponds to a particular mouse chromosome as shown in the key.

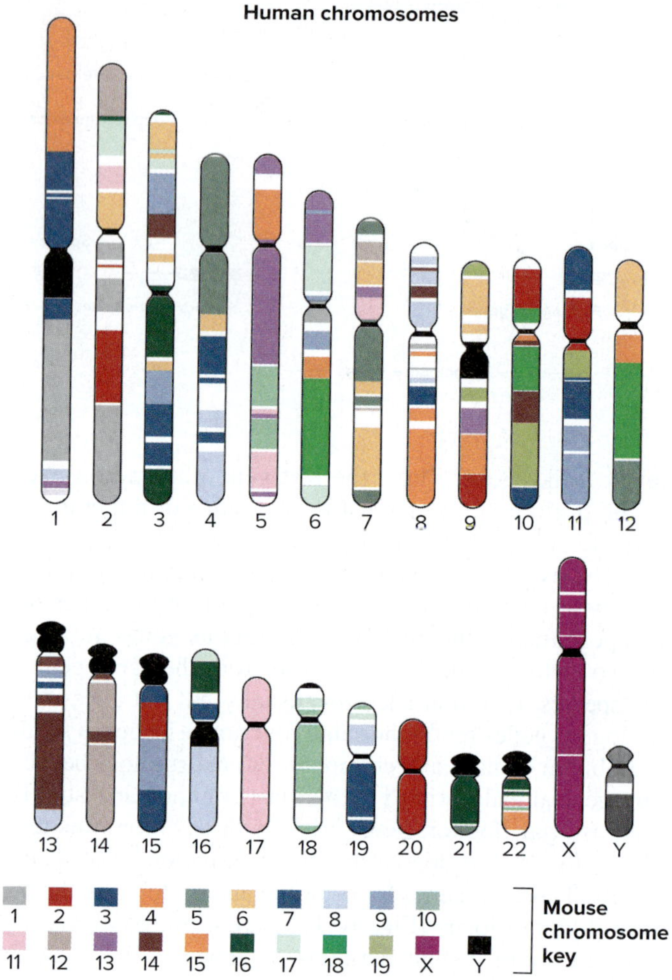

A Relatively Small Number of Genes Can Produce Enormous Phenotypic Complexity

Biologists are not even close to answering the question of how the approximately 27,000 genes we inherit from our parents contribute, along with environmental factors, to the staggering sophistication of the human organism. However, one important aspect of the answer must be that 27,000 does not even come close to the number of different proteins that our cells can form.

The diversity of proteins results in large part from combinatorial mechanisms that put together sequences of DNA or RNA in different ways in specific cells starting from the same inherited genome. *Combinatorial amplification* results from the potential for combining a set of basic elements in many different ways. A simple slot machine, for example, may contain three wheels, each carrying seven different symbols, but from its 21 basic elements (7 + 7 + 7), it can generate 343 different combinations (7^3, or 7 × 7 × 7). In biology, combinatorial amplification occurs at both the DNA and RNA levels.

A second contributor to protein diversification is the fact that proteins are subject to many molecular modifications after they are synthesized by translation. These modifications may alter protein structure and function.

Combinatorial strategies at the DNA level

T-cell receptor genes, among the best-studied examples of DNA-level combinatorial amplification, are encoded by a multiplicity of gene segments that become rearranged in one type of somatic cells—T cells—but not in the germ line or any other type of cell (**Fig. 10.16**). The human T-cell receptor family has 45 functional variable (V) gene segments, two functional diversity (D) gene segments, 11 functional joining (J) gene segments, and two almost identical constant (C) segments. In an individual T cell, any D element may first join to any J element by deletion of the intervening DNA. This joined D-J element may, in turn, join to any V element—once again by deletion of the intervening DNA—to generate a complete V-D-J exon. This combinatorial process can generate 990 different V-D-J exons ($45 \times 2 \times 11 = 990$), although in a given T cell, only one such functional rearrangement occurs. Thus, from 58 gene elements ($45 + 2 + 11$) within a single gene, a combinatorial joining mechanism can generate 990 different kinds of T-cell receptor proteins.

T-cell receptors are capable of interacting with foreign molecular structures, which are termed *antigens*. T cells are driven by contact with antigens to divide and expand their numbers 1000-fold or more. This antigen-triggered expansion by mitosis to a clone of genetically identical cells is a key part of every immune response. The particular combinatorial gene arrangements in a few of the original population of T cells by chance produce T-cell receptors fitted more precisely to a specific antigen. Binding with the antigen then triggers the clonal expansion of the cells that carry the tightly fitting receptors, amplifying the useful combinatorial information. The specificity and strength of the immune response thus increases over time because of the proliferation of T cells that have particular V-D-J rearrangements encoding the best receptors for the antigen to which the individual was exposed.

Figure 10.16 Gene for the human β T-cell receptor chain. In the germ line and in most somatic cells, the gene is composed of about 45 V elements, 2 D segments, about 11 J elements, and 2 nearly identical C (constant) regions. During T cell development, any D may join with any J. Subsequently, any V may join with any D-J. Finally, the rearranged V-D-J exon is spliced to a C exon. As a result of these sequential rearrangements of a single gene, different T cells can express one of almost 1000 different kinds of ß receptor chains.

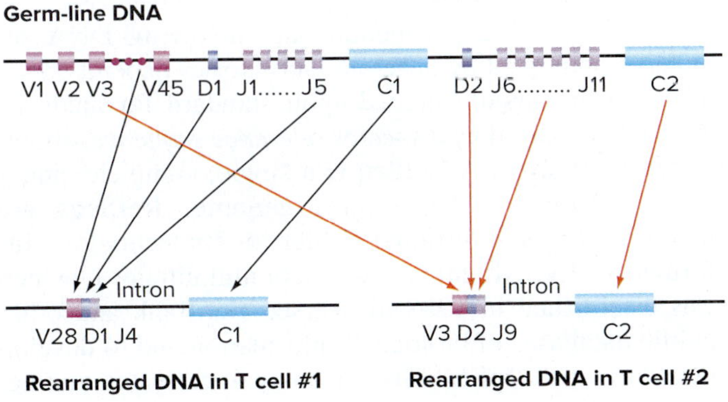

Combinatorial strategies at the RNA level

The splicing together of RNA exons in different orders—*alternative splicing*—is another way in which combinatorial strategies can increase information and generate diversity. Further diversity results from the initiation of transcription at distinct promoter regions, which creates transcripts with different numbers of exons.

The three neurexin genes, which encode proteins that help bind neurons together at synapses, illustrate both of these combinatorial RNA strategies (**Fig. 10.17**). Each neurexin gene contains two promoter regions (producing α- and β-class mRNAs) and five sites at which alternative splicing can occur. Together, these three genes can probably generate more than 2000 alternatively spliced forms of mRNA. Key questions include how many of the splice variants encode functionally distinct proteins (rather than proteins with the same function), and whether different variants represent different addresses for telling neurons where to go during embryonic development. By looking at the sequences of many cDNA clones, scientists have detected some splice variants that are specific for particular subsets of nerve tissue, suggesting the importance of this combinatorial strategy for nervous system organization.

Posttranslational modification of proteins

Human proteins may be modified by more than 400 different chemical reactions, each capable of altering the proteins' functions. Some examples of these posttranslational modifications were shown in Fig. 8.26, and they include reactions such as protein cleavage and protein phosphorylation. Thus, the typical human cell might have perhaps 50,000 different types of mRNAs (the primary transcripts of many genes are alternately spliced in a single cell type) but perhaps 1 million different proteins. Human cells can make more types of protein modifications than can cells of their simpler model-organism counterparts.

Figure 10.17 The organization of the human neurexin genes. The human genome has three genes encoding neurexin. Each gene has two promoters (α and β) to initiate mRNA synthesis and five sites at which alternative RNA splicing can occur. The *blue rectangles* indicate exons affected by alternative splicing. Numbers at the top of the figure designate individual exons.

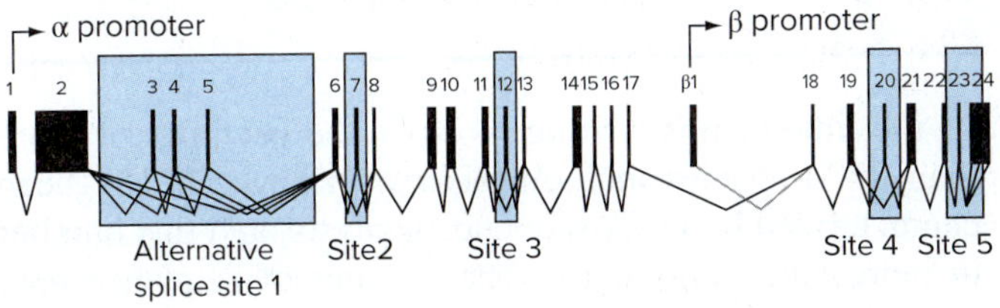

Genome Sequence Studies Affirm Evolution from a Common Ancestor

Comparisons of complete genomic sequences from nearly 10,000 species to date have resoundingly supported the ideas that began with Darwin and Mendel: All living organisms have similar genetic components for accomplishing basic cellular processes. This conclusion strongly supports the idea that we and other living organisms are all descendants of a single, fortuitous life-producing biochemistry. The similarity of basic genetic components also affirms that the analysis of appropriate biological systems in model organisms can provide fundamental insights into how the corresponding systems function in humans.

essential concepts

- Even the most complex genomes have surprisingly few genes (about 27,000 in the human genome).
- Gene density varies considerably within a genome, reflecting differences in intron size and in the spacing between genes.
- In most regions of the genome, the orientations of individual genes and thus the direction of gene transcription appears to be chosen at random.
- New genes can arise during evolution through: (i) *exon shuffling*, which can alter the domain structure of proteins; (ii) *duplication and divergence* that generates gene families; and (iii) *de novo mutations* in intergenic DNA sequences.
- *Combinatorial strategies* at the DNA level and RNA level, as well as *posttranslational modifications* of proteins, allow the production of highly diversified gene products even from a single gene.
- Genome comparisons affirm that all present life descended from a single common ancestor.

10.3 Bioinformatics: Information Technology and Genomes

learning objectives

1. Explain the relevance of a species RefSeq to bioinformatic studies.
2. Describe the uses of BLAST searches in comparative genomics.

At the time of this writing in 2016, the genomes of more than 8000 species including our own have already been characterized by DNA sequence analysis, and this number is increasing continually. The amount of sequence data available to researchers is staggering. Scientists must therefore rely on computers to store and help interpret this vast supply of information. The digital language used by computers for information storage and processing is ideally suited to handle the digital A, C, G, T code that exists naturally in genomes. These four values can be represented in two digits of binary code (00, 01, 10, and 11).

Keeping pace with the 1980s revolution in biological data generation fostered by the advent of automated DNA sequencing, a parallel revolution was occurring in information technology. The Internet came into existence along with personal computers that were linked together to establish rapid transmission of electronic data from one lab to another. It was a straightforward task to channel the output of DNA sequencer machines directly into electronic storage media, from which sequences were available for analysis and transmission to other scientists.

The **GenBank** database, established by the National Institutes of Health in 1982, still serves as the most widely used online repository of sequence data. The information is generated in molecular biology laboratories around the world, which deposit their sequences into GenBank electronically. From its establishment, the GenBank database has doubled in size roughly every 18 months, so that by 2016 it contained more than 300 billion annotated nucleotides of sequence information. One of the great powers of GenBank is that anyone in the world with an Internet connection can access this incredible storehouse of information easily.

Bioinformatics Provides Tools for Visualizing and Analyzing Genomes

Bioinformatics is the science of using computational methods—specialized software—to decipher the biological meaning of information contained within organismal systems. This section provides some examples of bioinformatics tools that can be accessed through any web browser to examine and interpret publicly available genome data.

The species RefSeq

Comparisons of experimental data involving DNA sequences generated by different laboratories depend on the use of a universally agreed-upon standard for analysis. This role is played by a *species reference sequence*, abbreviated as **RefSeq.** A RefSeq is a single, complete, annotated version of the species genome. RefSeqs are maintained by the National Center for Biotechnology Information (NCBI: http://www.ncbi.nlm.nih.gov), which was established in 1988 to oversee GenBank and other public databases of biological information and to develop bioinformatic applications for analyzing, systematizing, and disseminating the data. A RefSeq need not be derived from a single individual, and it need not contain the most common genetic variants found in species members. Rather, it is simply an arbitrary, but well-characterized, example against which all newly obtained sequences from that species can be compared.

Visualizing genes and genomes

Several web-based programs have been developed that allow a user to examine visual representations of genome data. One such program is the UCSC Genome Browser (https://genome.ucsc.edu/) that visualizes RefSeq genes and their associated annotations, showing features such as exon/intron structure and the location of protein-coding regions. Fig. 10.3 showed an example of the Genome Browser output, focusing on a 100 kb region of the human genome containing four genes. The transcription units are indicated at the bottom of the figure with large blue arrows that depict the extent of the gene, the direction of transcription, and each gene's exon/intron structure (exons represented as wider than the introns). Researchers can adjust their view of the browser to show many additional genomic features of interest, such as alternative splice variants, the location of repetitive DNA sequences, similarities with the genomes of other organisms, and the location of possible transcriptional regulatory elements.

BLAST Searches Automate the Finding of Homologous Sequences

Suppose that you have identified a gene, for example from the fruit fly *Drosophila,* that is of interest to you. You would like to know whether the human genome contains a homolog of this fly gene. One tool you could use is an NCBI program called BLAST (Basic Local Alignment Search Tool), which allows you to find nucleotide or amino acid sequences related to any given nucleotide or amino acid sequence. **Figure 10.18** displays a typical output of a BLAST search, in this case looking for human proteins that share similarity with a *Drosophila* protein of interest. The Query is the sequence you already know; here, the amino acid sequence of the *Drosophila* protein written in the one-letter code. The Subject is the homologous sequence found by the BLAST program; in this case, the related human protein. The row between the Query and the Subject indicates the conserved amino acids, with a + symbol denoting conservative amino acid replacements (missense substitutions in which an amino acid is replaced by a different amino acid with similar chemical properties).

To appreciate the power of bioinformatics programs such as the Genome Browser and BLAST search tool, you really need to access and use them yourself. Problems 23 and 24 at the end of this chapter involve some simple exercises that will place a few of these vast genomic databases at your disposal.

Figure 10.18 Output from a BLAST search. The program was asked to find a human protein related to a protein in *Drosophila*. The Query shows part of the sequence of the fly protein (from amino acids 688–720); the Subject (Sbjct) indicates the corresponding amino acids in the human protein found by the search. Some of these amino acids are identical in the fly and human proteins. Positions marked with a plus (+) are conservative substitutions in which the substituted amino acids have similar chemical properties. At some positions the amino acids are very different, suggesting that the identities of these particular amino acids are not crucial to protein function.

```
Query  688  GPLTASYKSDEIKHLIRALFQDTDWRAKAITQI  720
            GPL A++ S E+K LIRALFQ+T+ RA A+ +I
Sbjct  583  GPLAAAFSSSEVKALIRALFQNTERRAAALAKI  615
```

essential concepts

- *Bioinformatics* applications that are freely accessible online provide gateways for the exploration of genomic data.
- *Genome browsers* show the arrangement and structure of genes within RefSeq genomes.
- A *BLAST search* allows rapid, automated matching of particular DNA or amino acid sequences across multiple species for analysis of evolutionary relationships.

10.4 A Comprehensive Example: The Hemoglobin Genes

learning objectives

1. Discuss why it is advantageous for humans to produce different hemoglobins at different stages of development.
2. Explain how the clustering of hemoglobin genes impacts the cellular strategy to regulate their expression.
3. Predict the phenotypic severity of particular mutations in the α and β clusters.

The vivid red color of our blood arises from its life-sustaining ability to carry oxygen. This ability, in turn, derives from billions of red blood cells, each one packed with close to 280 million molecules of the protein pigment known as *hemoglobin* (**Fig. 10.19a**).

A normal adult hemoglobin molecule consists of four polypeptide chains—two alpha (α) and two beta (β) globins—each surrounding an iron-containing small molecular structure

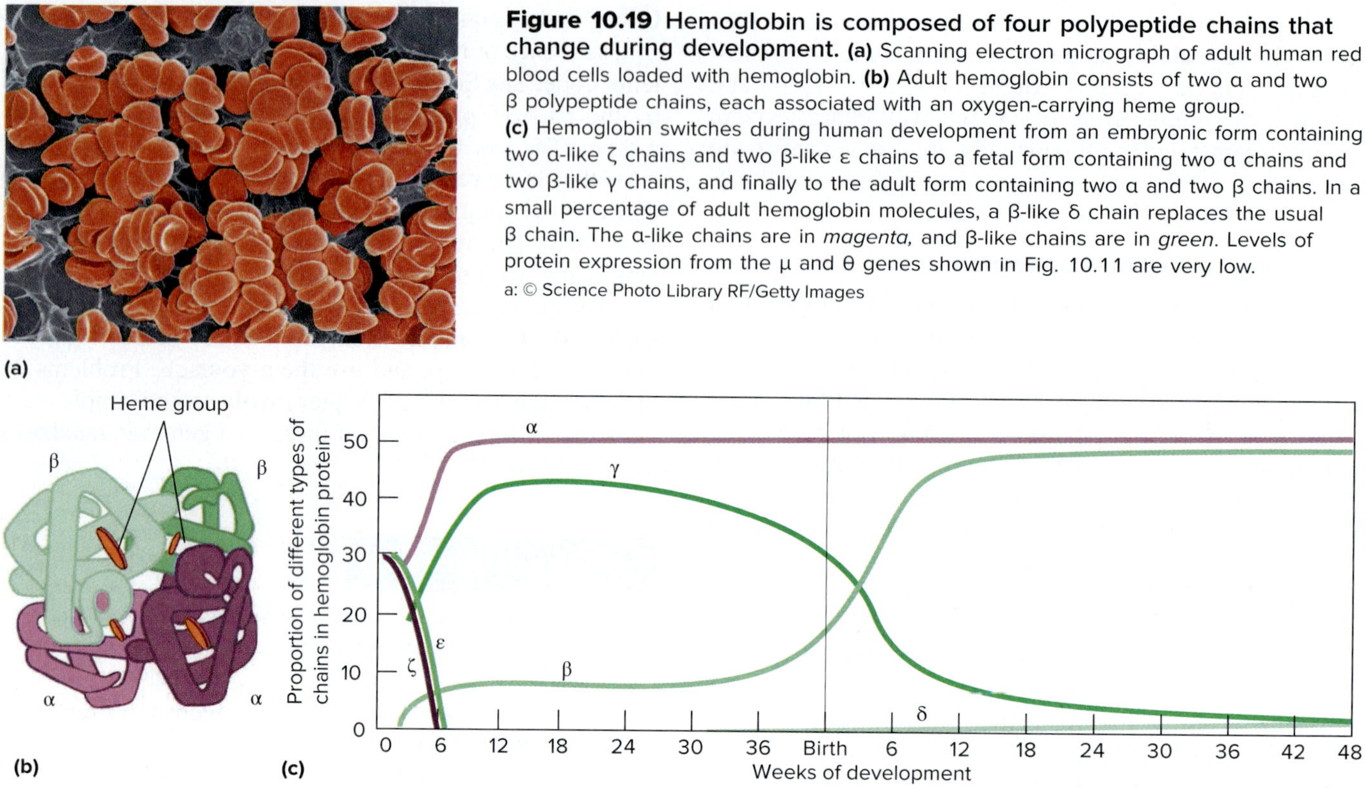

Figure 10.19 Hemoglobin is composed of four polypeptide chains that change during development. (a) Scanning electron micrograph of adult human red blood cells loaded with hemoglobin. **(b)** Adult hemoglobin consists of two α and two β polypeptide chains, each associated with an oxygen-carrying heme group. **(c)** Hemoglobin switches during human development from an embryonic form containing two α-like ζ chains and two β-like ε chains to a fetal form containing two α chains and two β-like γ chains, and finally to the adult form containing two α and two β chains. In a small percentage of adult hemoglobin molecules, a β-like δ chain replaces the usual β chain. The α-like chains are in *magenta,* and β-like chains are in *green*. Levels of protein expression from the μ and θ genes shown in Fig. 10.11 are very low.

known as a *heme group* (Fig. 10.19b). The iron atom within the heme sustains a reversible interaction with oxygen, binding it firmly enough to hold it on the trip from lungs to body tissue but loosely enough to release it where needed. The intricately folded α and β chains protect the iron-containing hemes from substances in the cell's interior. Each hemoglobin molecule can carry up to four oxygen atoms, one per heme, and these oxygenated hemes impart a scarlet hue to the pigment molecules and thus to the blood cells that carry them.

Different Hemoglobins Are Expressed at Different Developmental Stages

The genetically determined molecular composition of hemoglobin changes several times during human development, enabling the molecule to adapt its oxygen-transport function to the varying environments of the embryo, fetus, newborn, and adult (**Fig. 10.19c**). In the first five weeks after conception, the red blood cells carry *embryonic hemoglobin,* which consists of two α-like zeta (ζ) chains and two β-like epsilon (ε) chains. Thereafter, throughout the rest of gestation, the cells contain *fetal hemoglobin,* composed of two bona fide α chains and two β-like gamma (γ) chains. Then, shortly before birth, production of *adult hemoglobin,* composed of two α and two β chains, begins to climb. By the time an infant reaches three months of age, almost all of his or her hemoglobin is of the adult type.

Evolution of the various forms of hemoglobin maximized the delivery of oxygen to an individual's cells at different stages of development. The early embryo, which is not yet associated with a fully functional placenta, has the least access to oxygen in the maternal circulation. Both embryonic and fetal hemoglobin evolved to bind oxygen more tightly than adult hemoglobin does; they thus facilitate the transfer of maternal oxygen to the embryo or fetus. All the hemoglobins release their oxygen to cells, which have an even lower level of oxygen than any source of the gas. After birth, when oxygen is abundantly available in the lungs, adult hemoglobin, with its more relaxed kinetics of oxygen binding, allows for the most efficient delivery of the vital gas to other organs.

We have already seen that the hemoglobin genes occur in two clusters: the approximately 28 kb-long α-globin locus on chromosome 16 and the approximately 45 kb-long β-globin locus on chromosome 11 (review Fig. 10.11). As explained previously, the five functional genes plus two pseudogenes in the α-globin locus, and the five functional genes plus one pseudogene in the β-globin locus, all can be traced back to a single ancestral DNA sequence through multiple rounds of duplication and divergence.

Here, we show how the DNA sequence of these loci obtained from the Human Genome Project has provided fundamental insights into the mechanisms that change globin expression during normal development from embryonic to fetal to adult forms. Furthermore, the DNA sequence of these clusters reveals how various mutations give rise to a range of globin-related disorders. Hemoglobin disorders are the most common genetic diseases in the world and include *sickle-cell anemia,* which arises from an altered β chain, and *thalassemia,* which results from decreases in the amount of either α or β chain production.

The Order of the Hemoglobin Genes in the α and β Clusters Reflects the Timing of Their Expression

For the α-like chains, the temporal order of protein expression is ζ-globin during the first five weeks of embryonic life, followed by α-globin (encoded by both the α1 and α2 genes) during fetal and adult life. For the β-like chains, the order of protein production is ε-globin during the first five weeks of embryonic life; then γ-globin (encoded by the Aγ and Gγ genes) during fetal life; and finally, within a few months of birth, mostly β but also some δ chains (see Fig. 10.19c).

If you compare Figs. 10.11 and 10.19c, you will note that within each cluster, the order of globin genes on the chromosomes parallels the order of their expression during development. Furthermore, all the genes in the α-globin locus are oriented in the same direction relative to chromosome 16; that is, they all use the same strand of DNA as the template for transcription. The genes in the β-globin locus are also all oriented in the same direction, here relative to chromosome 11. The organization of these globin-gene-containing regions contrasts with most regions of the genome, where adjacent genes appear to be oriented randomly. These facts taken together suggest that whatever mechanism turns the globin genes on and off during different stages of development takes advantage of their relative positions and orientations.

We now understand what that mechanism is: Each of the two globin loci contains a **locus control region** (or **LCR**) at one end that controls sequential gene expression from that locus (Fig. 10.11). The LCR at each locus is a collection of regulatory elements called *enhancers* that are discussed in detail in Chapter 17. Through their interactions with proteins called *transcription factors*, the enhancers activate transcription of each gene in the right cells at the appropriate time.

One interesting consequence of the β-globin locus's organization and its control through the LCR is seen in a rare medical condition with a surprising prognosis. In some adults, the red blood cell precursors express neither the β nor the δ genes. Although this should be a lethal situation, these adults remain healthy. Sequence analysis of the β-globin locus from affected adults shows that they have certain deletions extending across the β and δ genes (compare **Fig. 10.20a** and **b**). Because of these specific deletions, the LCR can't switch, as it normally would near the time of birth, from γ-globin production to β- and δ-globin production. People with this rare condition, called *hereditary persistence of fetal hemoglobin,* continue to produce large enough amounts of fetal γ-globin throughout adulthood to maintain near-normal health.

Figure 10.20 Effects of deletions in the β-globin gene cluster. **(a)** Normal situation. The locus control region (LCR) sequentially turns on the transcription of the ε gene in embryos: the two γ genes during fetal development, and the β and δ genes in adults. **(b)** Loci with a deletion of the β and δ genes cannot switch gene expression at birth, so the γ fetal polypeptides are still produced in adults. This hereditary persistence of fetal hemoglobin is benign. **(c)** Deletion of the LCR prevents the expression of all genes in the cluster, causing severe β-thalassemia.

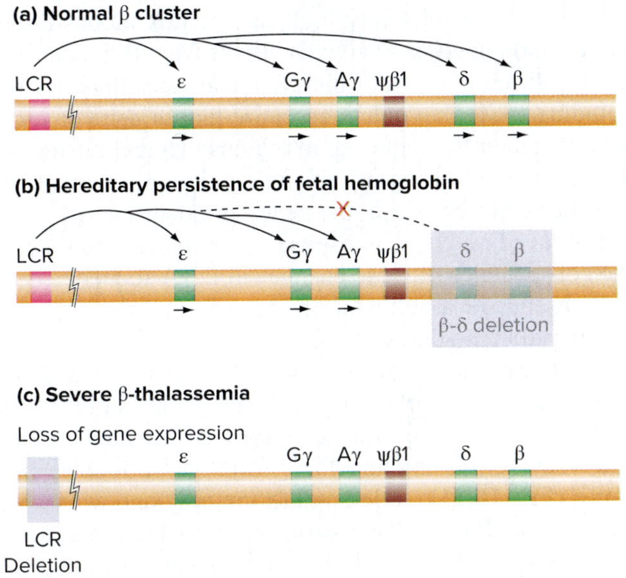

Globin-Related Diseases Result from a Variety of Mutations

By comparing DNA sequences from affected individuals with those from healthy individuals, researchers have learned that two general classes of disorders arise from alterations in the hemoglobin genes.

In one class, mutations change the amino acid sequence and thus the three-dimensional structure of the α- or β-globin chain. These structural changes result in an altered protein whose malfunction causes the destruction of red blood cells. Diseases of this type are known as *hemolytic anemias.* An example is sickle-cell anemia, caused by an A-to-T substitution in the sixth codon of the β-globin chain. This simple change in DNA sequence alters the sixth amino acid in the chain from glutamic acid to valine. Red blood cells carrying these altered molecules often have abnormal shapes that cause them to block blood vessels or be degraded (review Fig. 7.29).

The second major class of hemoglobin-related genetic diseases arises from DNA mutations that reduce or eliminate the production of one of the two globin polypeptides. The disease state resulting from such mutations is known as *thalassemia,* from the Greek words *thalassa* meaning *sea* and *emia* meaning *blood*; the name arose from the observation that a relatively high rate of this blood disease occurs among people who live near the Mediterranean Sea. Several different types of mutation can cause thalassemia, including those that delete an entire globin gene or locus; those that alter the sequence in regions that are outside of a globin gene but necessary for its regulation; or those that alter the sequence within the gene such that no protein can be produced, such as nonsense or frameshift mutations. The consequence of these changes in DNA sequence is the total absence or a deficient amount of one or the other of the normal hemoglobin chains.

Because each α-globin locus contains two α-globin genes (α1 and α2), normal people have four copies of this gene. Individuals carrying deletions within the α-globin locus may be missing anywhere from one to four of these copies (**Fig. 10.21**). A person lacking only one would be a heterozygote for the deletion of one of two α genes; a person missing all four would be a homozygote for deletions of both α genes. Given that α1 and α2 are expressed more or less equally beginning a few weeks after conception, the range of mutational possibilities explains the range of phenotypes seen in α-thalassemia. Individuals missing only one of four possible copies of the α genes are normal; those lacking two of the four have a mild anemia, and those without all four die before birth.

The fact that the α1 and α2 genes are expressed early in fetal life explains why the α-thalassemias are detrimental *in utero.* By contrast, β-thalassemia major, the disease occurring in people who are homozygotes for most deletions of the single β-globin gene, also usually results in death, but not until soon after birth. These individuals survive that long because the β-like protein δ-globin is expressed in the fetus (review Fig. 10.19c).

In some thalassemias, disease symptoms arise from mutations that alter the LCR found at one end of the α cluster or the LCR in the β cluster. Deletions of either LCR can produce severe thalassemias because none of the genes in the affected cluster are correctly transcribed (see **Fig. 10.20c** for an example). The fact that all the globin genes in such patients are intact while DNA off to the side of one of the clusters is missing was one of the first clues to the existence of LCRs.

Figure 10.21 Thalassemias associated with deletions of genes in the α-globin cluster. The fewer copies of the α genes that remain, the more severe the clinical symptoms of thalassemia.

Clinical condition	Genotype	Number of functional α genes	α-chain production
Normal	ζ α2 α1 αα/αα	4	100%
Silent carrier	αα/α–	3	75%
Heterozygous α-thalassemia—mild anemia	α–/α– or αα/– –	2	50%
HbH (β_4) disease—moderately severe anemia	α–/– –	1	25%
Homozygous α-thalassemia—lethal	– –/– –	0	0%

essential concepts

- Embryonic and fetal forms of hemoglobin bind oxygen more tightly than does the adult form, helping to ensure that the growing embryo/fetus receives sufficient oxygen from the mother's blood.
- The sequential expression of globin genes over the course of development is regulated by *locus control regions (LCRs)* in the α and β clusters.
- *Thalassemias* are blood diseases caused by mutations that eliminate or reduce the production of globin polypeptides from one of the clusters but not the other. These mutations can include deletions of specific genes or the LCR in either cluster.

WHAT'S NEXT

Determination of the nucleotide sequences of the human genome and the genomes of many other species constitutes an incredible milestone in our understanding of biology. We now know the fundamental blueprints for the lives of cells and organisms, and we have some idea of how differences in DNA lead to the emergence of different species.

However, in many ways the term "the human genome" lacks precision. People are not identical clones; instead, each of us has our own human genome that is closely related to that of all other humans but is also distinct and unique. It is the differences between the genomes of individuals that cause each of us to possess our own distinct and unique phenotype.

The sequence and even the annotation of one human genome is only the beginning. The human RefSeq provides a reference mark toward identifying and analyzing the differences between the genomes of many individuals so we can understand the genetic basis of phenotypic variation; for example, finding the nucleotide differences responsible for far-ranging and varied effects on human health. In Chapter 11, we describe how geneticists can now look at the genomes of many individuals to track genetic variation and to identify those differences in DNA sequence that underlie important traits.

SOLVED PROBLEMS

I. The following figure shows a screen shot from the UCSC Genome Browser, focusing on a region of the human genome encoding a gene called *MFAP3L*. (*Note: hg38* refers to version 38 of the human genome RefSeq.) If you do not remember how the browser represents the genome, refer to the key at the bottom of Fig. 10.3.

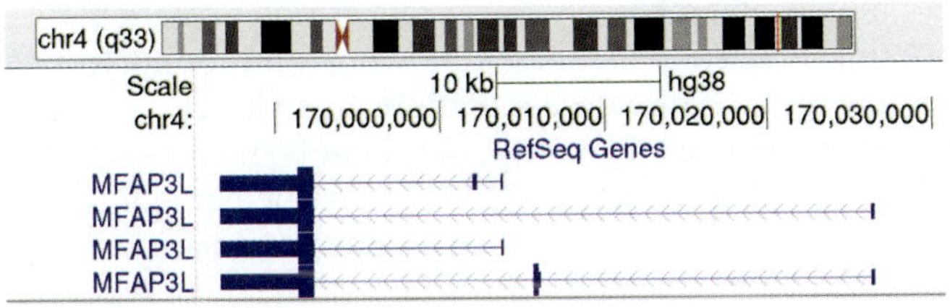

Source: University of California Genome Project, https://genome.ucsc.edu

a. Describe in approximate terms the genomic location of *MFAP3L*.

b. Is the gene transcribed in the direction from the centromere-to-telomere or from the telomere-to-centromere?

c. How many alternative splice forms of *MFAP3L* mRNA are indicated by the data?

d. How many different promoters for *MFAP3L* are suggested by the data?

e. How many different proteins does the *MFAP3L* gene appear to encode? Which alternatively spliced forms of the mRNA encode which proteins? Do the different forms vary at their N termini, their C termini, or somewhere in the middle? Estimate how many amino acids each of these proteins contains.

Answer

a. The gene is located on the long (*q*) arm of human chromosome 4; this position is denoted by the thin red vertical line on the chromosome representation (an *idiogram*) at the top of the figure. This location (in a band called 4q33) is roughly 170 million bp from the telomere of the small arm of chromosome 4 (from where the numbering begins); the total length of this chromosome is about 190 million bp.

b. The arrows within the introns of the gene show that the direction of transcription is from the telomere of *4q* toward the centromere of chromosome 4.

c. The data indicate four alternatively spliced forms of the mRNA. In the following parts, we list these as A to D from top to bottom.

d. The data suggest two promoters. One is roughly at position 170,037,000 and allows the transcription of a primary RNA alternatively spliced to produce mRNAs B and D. The other is roughly at position 170,013,000 and leads to the transcription of a primary RNA alternatively spliced to generate mRNAs A and C.

e. The data indicate that the *MFAP3* gene can encode two different but closely related proteins. mRNAs A, B, and C all encode the same protein; mRNA D a slightly larger protein that includes at its N terminus additional amino acids not found in the other protein. Otherwise these two proteins appear to be the same. The ORF that encodes the A B C protein form is about 880 bp long (a rough estimate); this corresponds to about (880/3 = 293 amino acids). The D protein is about 50 amino acids longer.

II. Two parents from Southeast Asia have a stillborn child with a lethal condition called *hydrops fetalis*. The parents themselves have α-thalassemia trait (mild anemia) and microcystosis (abnormally small red blood cells). Remember that humans have two essentially identical *Hbα* genes (*Hbα1* and *Hbα2*) and that the genes are autosomal, so normal humans have two copies of each (Fig. 10.21).

a. If this couple has many conceptions, what percentage of these conceptions is expected to result in hydrops fetalis?

b. Two other parents, who come from North Africa, also both have α-thalassemia trait, but a genetics counselor told them that none of their conceptions would result in hydrops fetalis. Explain how the genetic counselor's advice could be correct.

Answer

a. Both of these parents from Southeast Asia must have been heterozygotes with one normal copy of the *HbA* locus and the other copy deleted for both *Hbα1* and *Hbα2*. One fourth of their conceptions would lead to hydrops fetalis (homozygosity for the deletion), one half to α-thalassemia ($Hb\alpha 1^+\ Hb\alpha 2^+$ / - -), one fourth to a normal situation ($Hb\alpha 1^+\ Hb\alpha 2^+ / Hb\alpha 1^+\ Hb\alpha 2^+$).

b. These parents from North Africa both were homozygotes for a deletion of one of the two *Hbα* genes. Thus, all of their children would have α-thalassemia (two *Hbα* genes instead of the normal four), and none of their conceptions would lead to hydrops fetalis. Note that the genetic counselor could have given this advice only if the counselor knew from analysis of the parents' genomes what kind of defects were present in their *HbA* loci.

PROBLEMS

Vocabulary

1. Match each of the terms in the left column to the best-fitting phrase from the right column.

a. exome	1. a discrete part of a protein that provides a unit of function
b. de novo gene	2. a nonfunctional member of a gene family
c. gene desert	3. the joining together of exons in a gene in different combinations
d. pseudogene	4. most frequent residues, either nucleotide or amino acid, found at each position in a sequence alignment
e. syntenic block	5. set of genes related by processes of duplication and divergence
f. orthologs	6. chromosomal region with the same genes in the same order in two different species
g. natural selection	7. genes with sequence similarities in two different species that arose from a common ancestral gene
h. consensus sequence	8. genes that arose by duplication within a species
i. gene family	9. genomic DNA sequences containing exons
j. paralogs	10. gene-poor region of the genome
k. alternative RNA splicing	11. recently evolved from intergenic DNA sequences
l. protein domain	12. progressive elimination of individuals whose fitness is low and survival of individuals of high fitness

Section 10.1

2. List three independent techniques you could use to identify DNA sequences encoding human genes within a cloned genomic region.

3. Figure 10.2a has numbers indicating the approximate number of millions of years ago that species on separate branches of the tree last shared a common ancestor.

   a. About how many millions of years ago did humans last share a common ancestor with chimpanzees, mice, dogs, chickens, and frogs?

   These estimates for evolutionary events were obtained in part by comparing the genomic sequences of various current-day species. The basic supposition behind these estimates is that of a *molecular clock:* Differences in particular types of genomic sequences accumulate at a relatively linear rate during evolutionary time. Consider the three following kinds of nucleotide changes: (*i*) missense mutations in coding regions that alter amino acid identity; (*ii*) silent (synonymous) mutations that change a codon for a particular amino acid into a different codon for the same amino acid; and (*iii*) mutations in introns. Which of the three types of mutations would . . .

   b. . . . represent the slowest-ticking clock? (That is, which type of mutation would accumulate the least rapidly in genomes? *Hint:* See Fig. 10.3.)

   c. . . . you most likely use to estimate the divergence times of species that last shared a common ancestor more than 400 million years ago?

   d. . . . be most likely to vary in the rate at which they would accumulate in different genes?

4. Which of the enzymes from the following list would you need to make a cDNA library? What is the function of those enzyme(s) in the process?

   a. DNA polymerase
   b. RNA polymerase
   c. A restriction enzyme
   d. DNA ligase
   e. An aminoacyl-tRNA synthetase
   f. Peptidyl transferase
   g. Reverse transcriptase

5. One of the following sequences was obtained from a cloned piece of a genome that includes parts of two exons of a gene. The other sequence was obtained from the corresponding part of a cDNA clone representing the mRNA for this gene. (*Note:* For

simplicity, the intron is unrealistically short; some but not all sequence features needed for splicing are present.)

Sequence 1:
5′ TAGGTGAAAGAGTAGCCTAGAATCAGTTA 3′

Sequence 2:
5′ TAACTGATTTCTTTCACCTA 3′

a. Which sequence is the genomic fragment and which is the cDNA fragment?

b. Write the RNA-like strand of the genomic sequence and indicate the 5′ and 3′ ends. Draw vertical lines between the bases that are the exon/intron boundaries. (Refer to Fig. 8.15 for splice junction sequences.)

c. What sequence features needed for splicing are missing from this problem?

d. Assuming both exons are made only of protein-coding nucleotide sequences, what can you determine about the amino acid sequence of the protein product of the gene? (Indicate the N-to-C orientation.)

6. a. What sequence information about a gene is lacking in a cDNA library?

b. Can clones in a cDNA library contain 5′ UTR sequences? 3′ UTR sequences?

c. Would you be likely to find on average longer ORFs in cloned sequences from a genomic library or from a cDNA library? Explain.

7. Why do geneticists studying eukaryotic organisms often construct cDNA libraries, whereas geneticists studying bacteria almost never do? Why might bacterial geneticists have difficulties constructing cDNA libraries even if they wanted to?

8. Consider three different kinds of human libraries: a genomic library, a brain cDNA library, and a liver cDNA library.

a. Suppose that all three of these libraries are sufficiently large so as to represent all of the different human nucleotide sequences that the library could possibly include. Which of these libraries would then correspond to the largest fraction of the total human genome?

b. Would you expect any of these libraries not to overlap the others at all in terms of the sequences it contains? Explain.

c. How do these three libraries differ in terms of the starting material for constructing the clones in the library?

d. Why would you need to sequence many clones from many cDNA libraries to annotate a genome?

9. The human genome has been sequenced, but we still don't have an accurate count of the number of genes. Why not?

10. This problem investigates issues encountered in sequencing the inserts in cDNA libraries.

a. If you sequenced many clones individually, wouldn't you spend many of your resources inefficiently sequencing cDNAs for the same type of mRNA molecule over and over again? Explain. Does this apparently inefficient process provide any useful information beyond the sequences of individual mRNAs?

b. Suppose that you identified a clone with a cDNA insert that was 4 kb long. You could determine the entire sequence of the clone by shearing the DNA into small random fragments, cloning these fragments into a vector to make a mini-shotgun library, and then sequencing hundreds of these clones to allow the computer to assemble the full sequence of the 4 kb–long insert. However, this procedure would be inefficient.

An alternative that requires many fewer sequencing reactions is called *primer walking*. This technique involves the synthesis of additional oligonucleotide primers corresponding to cDNA sequences you have just obtained. Diagram how you would sequence the entire 4 kb–long cloned cDNA using primer walking, indicating the vector and insert, all primers that you would use, and all the sequences you would obtain. Assume that each sequence read is 1 kb.

11. For the sake of simplicity, Fig. 10.4 omitted one step of cDNA library construction. The figure implied that the last step of the process is the ligation of blunt-ended cDNAs into plasmid cloning vectors. Although such ligation reactions can occur, in reality they are highly inefficient. Instead, scientists convert blunt-ended cDNA molecules into sticky-ended molecules using *adapters,* and then they ligate the cDNAs into vectors with compatible sticky ends.

Adapters are short, partly double-stranded DNA molecules made by hybridization of two single-stranded oligonucleotides made in a DNA synthesizer. Suppose that the following two oligonucleotides were synthesized and then mixed together at high concentration and at a temperature that promotes hybridization of complementary DNA sequences:

5′ CCCCCG 3′
5′ AATTCGGGGG 3′

a. Draw the hybridized DNA molecules. These are the adapters.

b. Suppose you added the adapters and ligase enzyme to blunt-ended cDNAs at a very high molar ratio of adapters to cDNAs, so that each cDNA molecule is ligated to one adapter at each of its ends. Draw a picture of a resulting cDNA molecule.

c. The particular adapters discussed in this problem allow the cDNAs to be ligated efficiently into a vector treated with a commonly used restriction enzyme listed in Table 9.1. Name this restriction enzyme.

Section 10.2

12. Give two different reasons for the much higher ratio of total DNA to protein-encoding DNA in the human genome as compared to bacterial genomes.
13. Using a cDNA library, you isolated two different cDNA clones that have sequences indicating that they both correspond to mRNAs transcribed from the same nerve growth factor gene. The beginning and ending sequences of the clones are the same, but the middle sequence is different. How can you explain the different cDNAs?
14. The figure that follows shows part of a modified screen shot of part of the human genome as displayed on the UCSC Genome Browser. A through G map the sequences within individual cDNA clones to the genome sequence. Refer to the key in Fig. 10.3 if you need help in interpreting this diagram. Pay close attention to the vertical widths of the icons indicating exons.

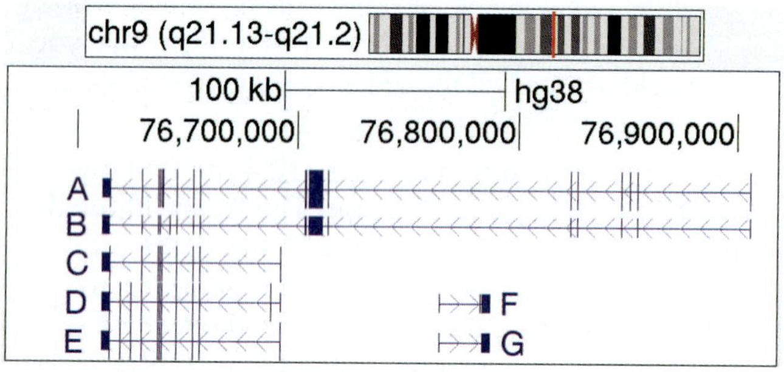

Source: University of California Genome Project, https://genome.ucsc.edu

a. How many annotated genes do you think are present in this region of the human genome?

b. For all annotated genes in this region, indicate whether they are transcribed in the direction from centromere to telomere or from telomere to centromere.

c. How many promoters are suggested by the data? Approximately where are these promoters located?

d. How many different proteins are encoded by the DNA sequences in this region?

e. What is unusual about this region of the human genome?

15. In Problem 14, cDNAs F and G could not be found in cDNA libraries (from any tissue) prepared using the method shown in Fig. 10.4. The reason is that the corresponding transcripts do not have poly-A tails.

a. Why is the lack of poly-A tails not surprising in light of your answer to part (d) of Problem 14?

b. Why does the lack of poly-A tails present a difficulty for the method diagrammed in Fig. 10.4?

c. Outline how you might adjust the protocol in Fig. 10.4 so as to find the cDNAs F and G annotated in the Genome Browser.

16. Fig. 10.10 presents a model for exon shuffling in which chromosomal fragments broken into introns can be restitched together to make novel genes that did not exist before. However, only a certain fraction of events like those shown on the figure could actually produce genes encoding proteins with functional domains from both original polypeptides. What is this fraction?
17. An interesting phenomenon found in vertebrate DNA is the existence of *pseudogenes,* nonfunctional copies of a gene found elsewhere in the genome. Some pseudogenes appear to have originated as double-stranded DNA copies of mature mRNA inserted into the chromosome; these copies later underwent mutations to make them into pseudogenes.

a. What sequence information might provide clues that the original source of some of these pseudogenes is cDNA copied in cells from mRNA and then inserted into the genome?

b. Would this mechanism of generating pseudogenes be more likely to have operated if the pseudogene was part of a gene family clustered in one region of the genome, or if it was instead part of a gene family whose members are scattered around the genome? Explain.

18. a. If you found a *zinc-finger domain* (which facilitates DNA binding) in a newly identified gene, what kinds of hypotheses could you make about the gene's function?

b. Suppose that this newly identified gene shares a high percentage of similarity throughout its length with a previously characterized gene in the same organism. What does this fact suggest about the origin of the two genes? Would you categorize these genes as being: (i) homologous, (ii) paralogous, or (iii) orthologous? (More than one answer may apply.)

19. You sequence the genomes of four different organisms and compare their sequences over a short region as shown below.

```
5′  AGGTATATAATTTGCG  3′
5′  CAATATAAAACCCTAC  3′
5′  GCGTATAAAAGAGCTA  3′
5′  TTATATATAAAGAAGT  3′
```

a. Determine the *consensus sequence* common to the four regions above.

b. Why would you want to define the consensus sequence? How would you decide whether the four sequences were worth comparing to define a consensus?

c. How could you use this general strategy for defining a consensus sequence to determine which amino acids of a protein are most crucial for its function?

20. In the human immune system, so-called *B cells* can make more than a billion different types of antibody molecules that protect us from infection. However, our genomes have only three genes that encode the polypeptides found in antibodies. What experiments could you perform to determine what kind of combinatorial events occur at the DNA level (V-D-J joining) and RNA level (alternative splicing) for any of these genes?

21. Chimpanzees have a set of hemoglobin genes very similar to the set in humans that was shown in Fig. 10.11. For example, the genomes of both species have $\alpha1$, $\alpha2$, β, $G\gamma$, $A\gamma$, δ, ε, and ζ genes.

a. Of the human and chimpanzee hemoglobin genes, which would be considered homologous? Which paralogous? Which orthologous?

b. When comparing genomes, geneticists would usually like to know which genes are the most likely to perform similar if not identical functions in different species. This determination can be somewhat complicated in the case of gene families. Would paralogous genes or orthologous genes be more likely to be functionally equivalent? Explain.

c. Which gene would have the greatest degree of nucleotide similarity to the human β gene: the chimpanzee β gene, or the human γ gene? Explain.

d. Rationalize the pattern of hemoglobin genes in the two species with the existence of duplication and divergence events among the hemoglobin genes depicted in Fig. 10.12.

22. Complete genome sequences indicate that the human genome has roughly 27,000 genes, while the worm (nematode) genome has about 22,000 genes. Explain how the human genome with only about 20% more genes can encode a creature enormously more complex than the worm.

Section 10.3

23. On your computer's browser, view the page accessed by the URL: http://genome.ucsc.edu/cgi-bin/hgGateway

In the Search Term box at the top, type *CFTR* (for the *CFTR* gene responsible for cystic fibrosis), then hit "GO." You will be directed to a window showing the organization of the *CFTR* gene on human chromosome 7. (If a list appears instead of a picture, click on the first link at the top of the list, and you will be directed to the proper window.) At the top of this window are control buttons that allow you to move your view to the left or right, zoom in (even to the level of the nucleotide sequence), zoom out, or (on the second row) jump to a different chromosomal position. Below these buttons is a diagram, called an *idiogram,* of the chromosome you are viewing, with a region in red indicating the particular region of the chromosome you are looking at. (You can also click on the idiogram to move around.)

a. How many exons are in the *CFTR* gene?

b. Is the *CFTR* gene located on the short arm (the *p* arm) or the long arm (the *q* arm) of human chromosome 7?

c. In which direction is the *CFTR* gene transcribed: toward the centromere, or away from the centromere?

Now zoom out the view by 10×.

d. What are the names of the genes that flank *CFTR* on either side? Are these genes transcribed from the same strand of chromosome 7 as *CFTR,* or from the other strand?

Now zoom out 100× until the entirety of chromosome 7 is visible.

e. What is the approximate size of chromosome 7 in Mb?

f. What is the approximate location of the centromere on human chromosome 7?

g. What is the significance of the RefSeq genes' appearing to pile up when you are viewing the whole chromosome?

24. On your computer's browser, view the page accessed by the URL: http://blast.ncbi.nlm.nih.gov/Blast.cgi

The heading "Basic BLAST" lists various programs that allow you to search for DNA or protein sequences related to any DNA or protein sequence query. For this problem, choose "nucleotide blast." On the window that comes up, make sure that the database "Human genomic + transcript" is selected (so that you will search through the RefSeq for humans and not any other species). Now, in the large box under "Enter Query Sequence" you should type in the nucleotide sequence from Fig. 9.7f. Then hit the blue "BLAST" button at the bottom and wait for the response from the NCBI computer, which may take up to a few minutes.

The response will come in several sections; you should migrate to the Descriptions. The "E value" column in the list is a statistical measure of the probability that a given sequence is related to your query by chance in the RefSeq database you are searching. The lower the E value (the closer to zero), the more significant is the match. For this exercise, look only at the first entry in the list, with the lowest E value.

a. What is the human gene that best matches the query sequence?

Now migrate down on the same page to the list of Alignments. Look only at the first entry, which corresponds to the first entry under the Descriptions.

b. Is the match exact or is it imperfect?

Go back to the nucleotide blast page. Using the same query sequence (from Fig. 9.7f), now choose "mouse genomic + transcript" as the database to search.

c. Is this gene conserved in the mouse? If so, how well conserved?

25. Use the UCSC Genome Browser (http://genome.ucsc.edu/cgi-bin/hgGateway) to identify the two major genes illustrated in the figure accompanying Problem 10.14.

Section 10.4

26. Certain individuals with mild forms of β-thalassemia produce, in addition to normal adult hemoglobin with two α chains and two β chains, lower levels of an unusual, so-called *Lepore hemoglobin* with two α chains and two chains in each of which the N-terminal half comes from a normal δ chain and the C-terminal half comes from a normal β chain. Certain other individuals who are asymptomatic produce a different, unusual *anti-Lepore hemoglobin* that contains two α chains and two chains in which the N-terminal half comes from a normal β chain and the C-terminal half comes from a normal δ chain.

a. Describe an event that could give rise to both Lepore and anti-Lepore hemoglobins.

b. Are the mildly thalassemic individuals with Lepore hemoglobin homozygotes or heterozygotes for the unusual allele?

c. Why might these mildly thalassemic people produce less Lepore hemoglobin than normal adult hemoglobin?

27. The α1 and α2 genes in humans are identical in their coding regions. With this fact and your answer to Problem 26 in mind, describe a mechanism that might frequently produce a deletion of one of these two genes.

28. The following figure shows an electrophoretic analysis of the hemoglobin proteins present in normal adults, newborns, and fetuses. Each band represents a complete hemoglobin protein with all of its subunits. The intensity of the band indicates the relative amount of that protein in the sample.

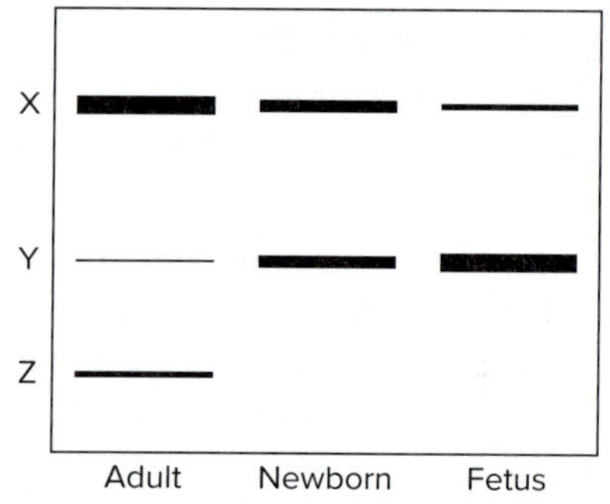

a. What is the subunit structure of the hemoglobin molecules marked X, Y, and Z?

b. Name one abnormal condition that should *increase* the percentage of hemoglobin in a newborn that is HbZ.

c. Name one abnormal condition that should *increase* the percentage of hemoglobin in an adult that is HbY.

chapter 11

Analyzing Genomic Variation

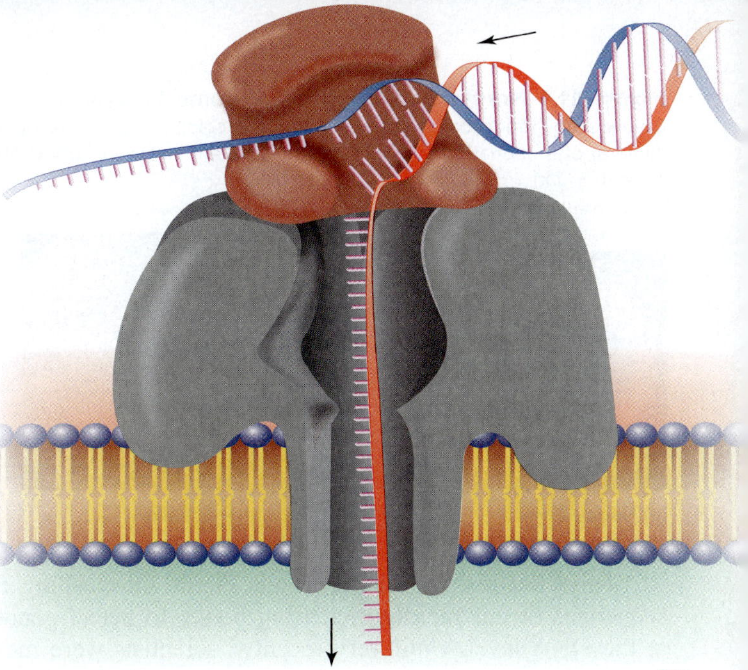

The era of whole-genome sequencing is being made possible by remarkable innovations. One novel technique for massively parallel DNA sequencing uses an enzyme (brown) *to thread a DNA strand through a narrow channel called a* micropore (gray). *The DNA in the channel limits (in a nucleotide-specific fashion) the flow of ions through the micropore. Recordings of the current flowing through the micropore as a function of time can be interpreted as the sequence of nucleotides.*

chapter outline

MILLIONS OF PEOPLE worldwide suffer from a group of conditions, including ulcerative colitis and Crohn's disease, featuring chronic inflammations of the digestive tract (**Fig. 11.1**). The patients' immune systems overreact to bacteria in the gut and begin to attack cells in the intestinal mucosa (gut lining). The symptoms are usually not life threatening, although flare-ups of intense abdominal pain, vomiting, diarrhea, nausea, and fatigue can completely disrupt a person's life. Variations in more than 100 genes have been found to predispose individuals to inflammatory bowel diseases (IBDs).

At the age of 2, Nic Volker acquired an IBD of unprecedented severity. His digestive tract developed lesions that extended through his body cavity all the way to the outside of his skin. As a result, fecal matter leached into his system, causing dangerous sepsis (systemic bacterial infections). Unfortunately, Nic's condition did not respond to the usual treatments for IBD, such as immunosuppressants or anti-inflammatory steroids. By the age of 4, Nic had already undergone more than 140 surgeries to resect parts of his digestive tract and heal the wounds in his skin; he weighed only 17 pounds (**Fig. 11.2**). Nic's long-term prognosis was obviously dire.

Nic's parents and doctors enlisted a team of human geneticists to attempt a novel approach for his desperate case: determining the sequence of all the protein-coding nucleotides in Nic's genome. Remarkably, in 2009 the research team found the mutation that caused Nic's disease. The mutation was located in a gene known to be associated with another inherited condition called X-linked lymphoproliferative disease (XLPD). The symptoms of XLPD were so different from Nic's that no one had ever guessed at the connection. Blood marrow transplantation was known to be effective for XLPD, so Nic's doctors decided to try this method, even though IBDs had never before been treated in this way. Within a few months of the transplant, Nic's health underwent an astonishing rebound, allowing him to live the normal life of a six-year-old (**Fig. 11.3**).

Figure 11.1 Normal and Crohn's syndrome bowels. Colonoscopy (endoscopic examination of the colon using a fiber optic camera) of a normal person (*left*) and of a patient suffering from Crohn's disease (*right*).

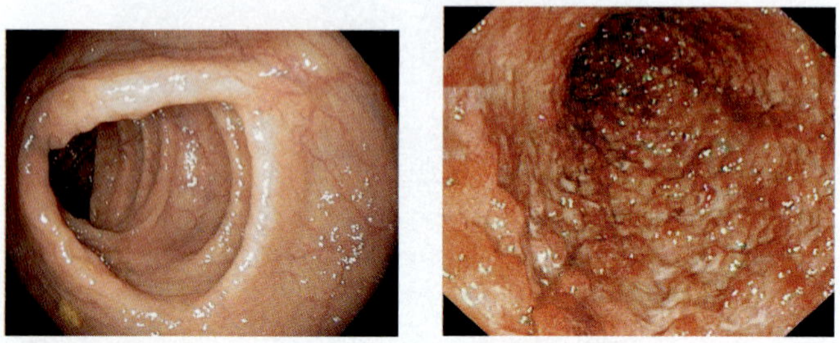

(a) Normal colon lining **(b)** Crohn's disease

Figure 11.2 Nic Volker at age 4.

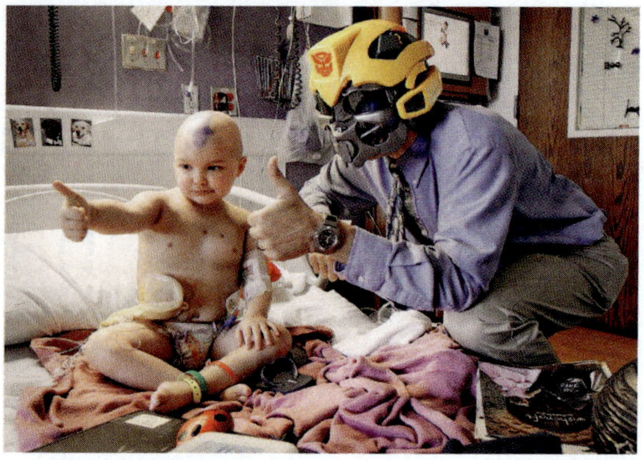

Figure 11.3 Nic Volker at age 6.

This case history illustrates one of the far-reaching medical consequences of our rapidly increasing power to detect genotype directly at the DNA level. Until very recently, scientists were more limited in their ability to look at human genomes. In the 1990s, researchers could examine an individual's genotype only a single gene at a time, and this was worthwhile only in the few cases where the disease-causing mutation was already identified. Even this limited amount of personal genetic information was often valuable in helping couples to make informed reproductive decisions.

By the turn of the twenty-first century, advances in genotyping allowed scientists to look at a much larger sample of the many nucleotide changes that differentiate one person's genome from another. For example, new methods including *DNA microarrays* allowed simultaneous examination of dozens or even millions of nucleotide variations at different positions, or **loci,** within individual genomes. As you will see, the ability to follow large numbers of nucleotide variations has many uses, even if these variations themselves have nothing to do with disease.

By 2013, scientists had the ability to genotype not just one or two or thousands of loci in a person's genome, but nearly all of the 6 billion nucleotides in a person's diploid genome. Costs are being driven down rapidly by new innovations in DNA sequencing technologies that will soon become a routine part of medical care. In this new and uncharted era of whole-genome DNA sequencing, the exponential increase in our knowledge of genome variations will provide details about people's genetic histories and destinies never before available.

11.1 Variation Among Genomes

learning objectives

1. Cite the approximate number of DNA polymorphisms that differentiate any two haploid human genomes.
2. Explain why most of these DNA polymorphisms are not responsible for the phenotypic differences between people.
3. Differentiate among different classes of DNA variants in terms of their structures, mechanisms of formation, and frequency in genomes.

There is no such thing as a wild-type human genome; instead, a staggering amount of variation exists between the genomes of any two people. With the advent of new technologies for whole-genome sequencing that will be described later in this chapter, the degree to which individual human genomes differ from each other is becoming apparent. Only a small minority of these DNA sequence variations is responsible for the phenotypic differences that characterize individuals. But even if certain DNA sequence differences have no effect on phenotypes, they are still highly useful as markers to track genes and chromosomes.

Figure 11.4 Comparison of three personal genomes. Single nucleotide substitutions in the genomes of J. Craig Venter, James D. Watson, and an anonymous Chinese man (YH), all relative to the human RefSeq. A substitution is counted once whether the individual is homozygous or heterozygous for that variant. Numbers of substitutions unique to each man's genome are in nonoverlapping portions of each circle. Variants not in the human RefSeq but shared by two of the three individuals are shown in the double overlap regions. The central three-way overlap indicates variants shared by all three men.

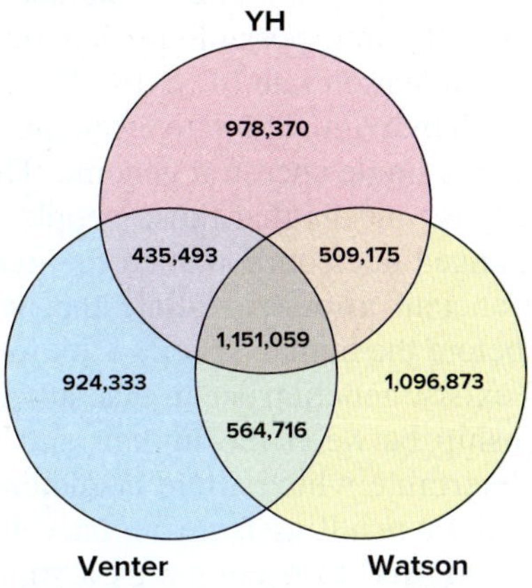

Extensive DNA Variation Distinguishes Individuals Within a Species

The genomes of James Watson, co-discoverer of the DNA double helix; J. Craig Venter, a pioneer of DNA sequencing; and an anonymous Chinese man reveal in total more than 5.6 million single nucleotide differences from the standard human genome (the GenBank RefSeq; see Chapter 10) (**Fig. 11.4**). Each man's diploid genome contains about 1 million unique **DNA polymorphisms** (that is, sequence differences) not shared by either of the other men, while the remaining approximately 2.6 million polymorphisms are shared in the genomes of two or in some cases all three of these individuals.

Not only does no single wild-type human genome sequence exist, there is even no such thing as a wild-type human genome length. Deletions, insertions, and duplications of DNA result in genome lengths that differ by as much as 1% in healthy people. For example, the genomes of Watson and Venter vary by small additions or subtractions of genetic material—insertions or deletions—at over 100,000 genomic sites.

Most DNA Polymorphisms Do Not Influence Phenotype

Some of the millions of DNA polymorphisms between the genomes of Watson and Venter must be responsible for the phenotypic differences that distinguish them as individuals. But in reality only a small fraction of these DNA sequence changes actually impacts phenotype. Only about 5000 of the millions of differences between these two people alter the amino acid sequences of proteins. This fact makes sense because:

(1) less than 2% of the human genome consists of codons within genes;
(2) even when they occur, many mutations of codons are silent (that is, they don't change the amino acid); and
(3) if a particular mutation is not silent and has deleterious effects, natural selection could often lead to its disappearance from the human population.

In addition to the approximately 5000 amino-acid-altering mutations, a few thousand other polymorphisms between these two genomes likely affect gene expression, for example the frequency of transcription or the efficiency of primary transcript splicing to produce mRNA. But even after accounting for these, we are left with the conclusion that the vast majority of sequence differences between genomes are **anonymous DNA polymorphisms** affecting neither the nature nor the amounts of any protein in the body. (You will see later that **nonanonymous DNA polymorphisms** do affect gene expression, and thus can affect phenotype.)

Figure 11.5 shows the actual distribution of polymorphisms that distinguish Watson and Venter from the human RefSeq within a 400 kb genomic region. This part of the genome includes the cystic fibrosis transmembrane receptor gene (*CFTR*), mutations in which cause cystic fibrosis, and two other genes. You can see that almost all of the

Figure 11.5 SNP distribution in a 400 kb region. This part of chromosome 7 (from base pairs 116,700,001 to 117,100,000) contains *CFTR* and two other genes. Vertical marks indicate locations at which a genome is either heterozygous or homozygous for a single nucleotide polymorphism (SNP) different from the human RefSeq. Two rows show SNPs that were read from the personal genomes of Watson and Venter. The third track compiles all SNPs from all human genomes analyzed that were deposited in the central SNP database as of 2009.

polymorphisms in both men's genomes, and indeed in all genomes, are located either between genes or in introns, consistent with the idea that most DNA changes do not alter phenotype.

The existence of such a vast number of anonymous DNA polymorphisms that distinguish different human genomes presents both challenges and opportunities for geneticists. The challenges are clear: How can we sort out the millions of polymorphisms in anyone's DNA to find the few that are relevant to traits such as genetic diseases? The opportunities lie in the fact that even if a polymorphism is anonymous, with no effect on phenotype, it still serves as a signpost in the genome, a **DNA marker.** You will see that researchers can use anonymous polymorphic loci to help locate the nearby mutations that are actually responsible for inherited diseases.

Genetic Variants Occur in Several Types

Geneticists usually place polymorphic DNA loci into one of the four categories shown in **Table 11.1** based on the number and kinds of nucleotide pairs involved. Although the borders between these classes are fuzzy and overlap to some extent, the categories help researchers describe what a particular genetic variant looks like. A useful generalization is that the smaller the number of nucleotide pairs involved in a given class of polymorphism, the more frequently variants of that class are found in the genome (Table 11.1).

Single nucleotide polymorphisms (SNPs)

Far and away the most common type of genetic variant is the class of **single nucleotide polymorphisms,** or **SNPs** (pronounced *snips*). SNPs are particular base positions in the genome where alternative letters of the DNA alphabet distinguish some people from others. SNPs account for the vast majority of the total variation that exists between human genomes, occurring on average once every 1000 bases in any pairwise comparison (Table 11.1). Chapter 7 discussed several of the many mechanisms that can give rise to SNPs, including rare mistakes in DNA replication, or exposure of the genome to mutagenic chemicals or radiation in the environment.

TABLE 11.1 Categories of Genetic Variants

	Size	Frequency (1 per. . . .)
SNP Single nucleotide polymorphism	1 bp	1 kb
DIP or **InDel** Insertion/deletion	1–100 bp	10 kb
SSR Simple sequence repeat	1–10 bp repeat unit	30 kb
CNV Copy number variant	10 bp–1 Mb	3 Mb

The right column shows how frequently on average you would find a polymorphism of the indicated class when comparing any two haploid human genomes.

Despite the existence of so many types of events that can create SNPs, the per-base spontaneous mutation rate is still less than one in 30 million (by some estimates as low as one in 100 million) per generation. This number is so low that most SNPs are biallelic in human populations, with only two of the four possible nucleotide pairs represented. The low mutation rate of SNPs allows researchers to trace back each individual SNP to a genomic change that occurred once in a single ancestral genome. The low mutation frequency also means that those people who did not inherit this changed nucleotide (called the **derived allele**) have a more ancient **ancestral allele** that was probably present long before the human species took form.

If a SNP exists, geneticists can take advantage of the close relationship between the human and chimpanzee genomes to determine which allele is ancient and which is the derived allele resulting from a relatively recent mutational event. **Figure 11.6** compares a small region of

Figure 11.6 Inferring the evolutionary history of SNPs. **(a)** A comparison of two human genomic sequences to the chimp RefSeq. Loci 1 and 2 are invariant in many sequenced chimp genomes. **(b)** *Cladogram* (diagram of evolutionary lineages). Locus 1 (*light blue*) differs between chimps and humans, but all humans have the same allele (G). The mutation causing the locus 1 difference must have occurred since the species diverged, either in the lineage leading to chimps or in that leading to humans; the allele at this position in the most recent common ancestor of the two species cannot be determined. At locus 2 (*red*), the C allele shared between chimps and some humans must be *ancestral,* while the T allele in other humans must be *derived* (that is, caused by a recent mutation specifically in the lineage of some humans).

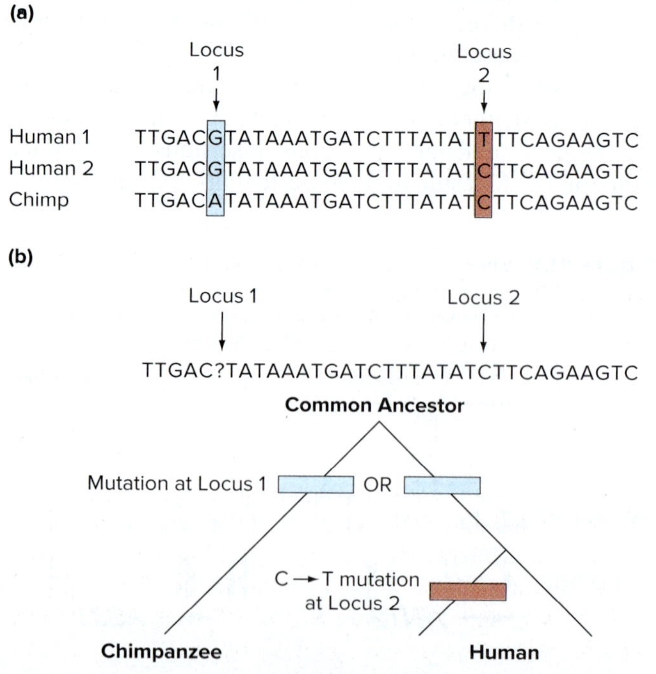

DNA in two different haploid human genomes and the chimp RefSeq genome. Two single-base changes have occurred in this small genomic region since the divergence of the two species. One is shared by all human genomes and is thus not polymorphic in humans. The second base change was from a C in the common chimp-human ancestor to a T (the derived allele) in a chromosome of the ancestor of some people but not others. This means that if you and a friend share a derived allele at an anonymous SNP locus, you both got that allele from the same ancestor who must have lived since the human and chimp lineages diverged from each other. The fact that every random pair of human beings on the planet shares many unlinked, derived SNP alleles indicates common ancestry for all people.

To date, the analysis of thousands of human genomes has led to the identification of more than 50 million SNPs that are catalogued in a SNP database (dbSNP) at the National Center for Biotechnology Information (NCBI). About 15 million of these SNPs are commonly found in human populations. The mutations giving rise to the derived allele of common SNPs must have occurred far enough back in human history to have been disseminated to a significant proportion of current-day people. The SNP database already includes a large fraction of all common SNPs in human populations. However, you should realize that mutational events in the very recent past also create rare SNPs that would be found in only one or a few people of the billions on the earth. Very few of these rare SNPs are yet accounted for in dbSNP because so few human genomes have been analyzed relative to the entire human population.

This brief discussion about the origin of SNPs suggests that genome sequencing provides powerful tools for understanding human ancestry. Chapter 21 presents some of the surprising findings about human history revealed by the analysis of DNA sequences from present-day humans and the fossilized remains of our primate ancestors.

Deletion-insertion polymorphisms (DIPs)

Short insertions or deletions of genetic material represent the second most common form of genetic variation in the human genome. These variants are referred to as **deletion-insertion polymorphisms** (**DIPs**) or **InDels.** While SNP loci occur about once per kilobase in the comparison of any two haploid genomes, DIPs are considerably rarer, occurring about once in every 10 kb of DNA (Table 11.1). DIPs range in length from one base pair to hundreds of base pairs, but their relative frequency declines steeply in relation to their length. As a result, DIPs involving only one or two nucleotides are the most common.

Several biochemical processes appear to contribute to the formation of DIPs. These include problems in DNA replication or recombination, and mistakes that occur when cells try to repair damage such as broken DNA strands.

You should remember that in protein-coding regions, DIP variants act as frameshift mutations unless the number of nucleotide pairs inserted or deleted is 3 or a multiple of 3.

Simple sequence repeats (SSRs)

The genomes of humans and higher eukaryotes are loaded with loci defined by **simple sequence repeats (SSRs)**, sometimes also termed **microsatellites.** SSR loci consist of sequences of one to a few bases that are repeated in tandem less than 10 to more than 100 times. Different alleles of an SSR locus have different numbers of repeating units. The most common repeating units are one-, two-, or three-base sequences. SSRs with larger repeating units are less frequent, and we employ here a relatively arbitrary cutoff in which the largest repeating unit of an SSR has 10 bases (Table 11.1; those with larger repeating units will be classified as *CNVs* below). Examples of SSRs are AAAAAAAAAAAAAAAA (a one-base repeat) or CACACACACACACACACACACA (a two-base repeating unit). SSRs of all types together account for about 3% of the total DNA in the human genome; an SSR locus can be found on average once in every 30 kb of human DNA.

As with all other polymorphisms, most SSRs occur outside the coding regions of genes and have no effect on phenotype. In contrast, SSR variations within genes can have profound phenotypic consequences. For example, we have already discussed in Chapter 7 (review the Fast Forward Box entitled *Trinucleotide Repeat Diseases: Huntington Disease and Fragile X Syndrome*) that long tracts of trinucleotide repeats are the molecular cause of several severe neurological conditions, including fragile X syndrome and Huntington disease.

SSRs arise spontaneously from rare, random events that initially produce a short repeated sequence with four to five repeat units. Once a short SSR mutates into existence, however, it can expand into a longer sequence by a form of faulty DNA replication called *slipped mispairing* or *stuttering*. Figure 7.12 showed in detail how this stuttering mechanism can change the number of repeat units at the SSR locus responsible for Huntington disease.

Because of events such as slipped mispairing, new alleles arise at SSR loci at an average rate of 10^{-3} per locus per gamete (that is, one in every thousand gametes). This frequency is much greater than the single nucleotide mutation rate of 10^{-9} and results in a large amount of SSR variation among unrelated individuals within a population. Unlike SNPs—which are biallelic and do not change after the mutational event that gave rise to them—SSRs are therefore highly polymorphic in the number of repeats they carry, often with 10 or more alleles distinguishable at a single SSR locus. The rate of SSR mutation is nonetheless low enough that changes usually do not occur within a few generations of even a large family. SSRs can thus serve as relatively stable, highly polymorphic DNA markers in linkage studies of many organisms, including humans.

Copy number variants (CNVs)

Individual human genomes also display DNA length polymorphisms involving more than just the few nucleotides characterizing SSRs and DIPs. Researchers were surprised to find that the genomes of many people showing no signs of any genetic disease carry a variable number of copies of large blocks of genetic material up to 1 Mb in length. This category of genetic variants is referred to as **copy number variants (CNVs)**. CNVs turn out to be quite common both in their distribution across the genome and in their frequency of occurrence within human populations (Table 11.1). Over 10,000 CNV loci have been identified in all human genomes examined to date, and pairwise comparison of any two genomes typically identifies differences at more than 1000 of these loci.

One of the most important mechanisms that can produce new alleles of a CNV locus is **unequal crossing-over** (**Fig. 11.7**). During meiosis I, tandem arrays of the repeating units on homologous chromosomes can pair out of register. If recombination takes place between mispaired repeating units, gametes are produced that have more or fewer copies of the repeating unit than the originals. Although mechanisms such as unequal crossing-over make CNV loci highly polymorphic, CNVs are still relatively stable when observed in families over a few generations: More than 99% of all CNV alleles in the current human population are thus derived from inheritance rather than new mutation.

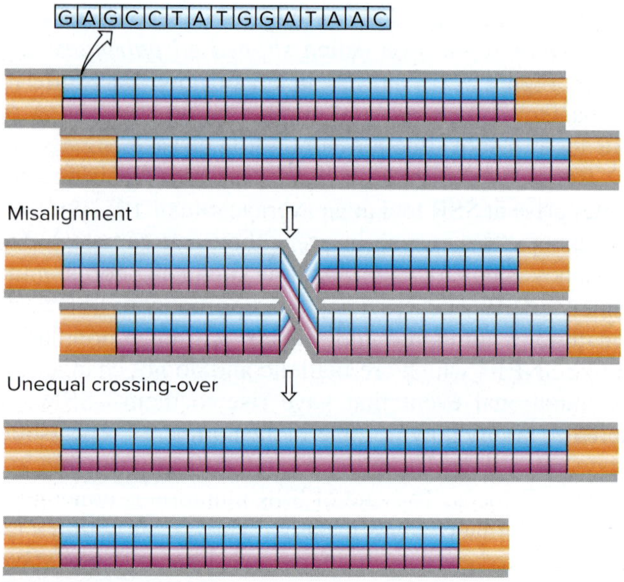

Figure 11.7 CNVs are highly polymorphic because of their potential for unequal crossing-over. CNVs are composed of tandem repeating units of identical or near identical sequences more than 10 bp long. (*Blue* and *purple* boxes are complementary strands of the repeating unit.) Misalignment and unequal crossing-over produce recombinant products—new alleles that have more or fewer repeating units than either parental allele.

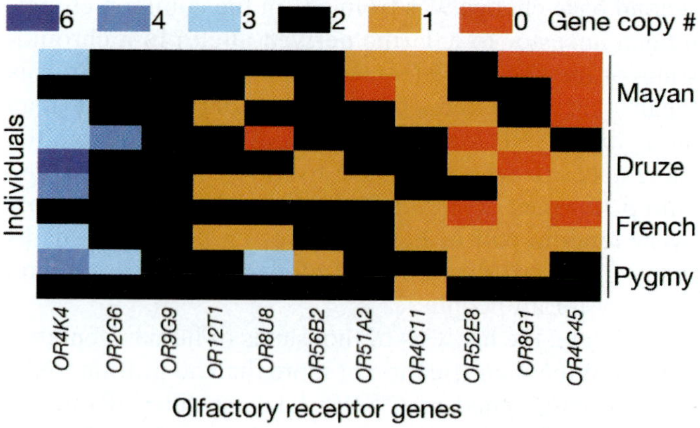

Figure 11.8 CNVs of olfactory receptor genes. Each row is a different person; each column a different olfactory receptor gene. The *colors* indicate the number of copies of a particular gene in a particular individual. Different humans vary substantially in copy numbers of olfactory receptor genes, accounting for much of the variation in people's ability to smell certain scents.

The olfactory receptor (*OR*) gene family, which encodes proteins that allow animals to smell a diverse array of odors, offers a fascinating example of variation in gene copy number. A typical mouse genome carries 1400 *OR* genes distributed at numerous chromosomal sites. But a keen sense of smell is no longer as important for human survival. As a result, *OR* genes can be lost without consequence, and people typically carry fewer than a thousand *OR* genes. However, individual humans vary widely around this mean. **Figure 11.8** shows the variation in copy number among 10 people at 11 representative *OR* loci. One gene, *OR4K4,* varies in copy number from two to six in different genomes, while five of the 11 genes are completely missing from some individuals. Some people can have hundreds more or hundreds fewer *OR* genes than others do, resulting in large differences in people's abilities to distinguish odors.

essential concepts

- When two or more alleles exist at a DNA locus, the locus is polymorphic, and the variations themselves are *DNA polymorphisms* (*DNA markers*). Most polymorphisms are *anonymous;* they have no effect on phenotype.
- *Single nucleotide polymorphisms,* or *SNPs,* are the most frequently found DNA polymorphisms. The low rate of SNP formation allows investigators to estimate when particular SNP-causing mutations occurred during evolution.
- Addition and subtraction of DNA sequences also cause genetic variations, including *DIPs, SSRs,* and *CNVs.* These variations can result from stuttering of DNA polymerase during replication of repeated sequences or from unequal crossing-over during meiosis.

11.2 Genotyping a Known Disease-Causing Mutation

learning objectives

1. Outline the steps by which the polymerase chain reaction (PCR) amplifies a specific region of a genome.
2. Describe how the sequencing or sizing of PCR products can elucidate genotypes.
3. Explain how PCR can be used to genotype fetuses *in utero* or embryos prior to implantation.

The ability to genotype individuals for genetic diseases provides information that can impact lives profoundly. If a person is diagnosed as having a genetic disease for which treatments are available, knowledge of the DNA genotype could save his or her life. Even if the condition cannot be treated, the genotyping of prospective parents, a fetus carried by a mother, or embryos created by *in vitro* fertilization allows families to make informed reproductive decisions.

The ability to determine whether a person is homozygous or heterozygous for a disease-causing allele of course presumes that scientists already know the precise change of nucleotides responsible for the disease. For certain diseases such as sickle-cell anemia, the identity of the disease-causing mutation is clear because we know how a particular protein (such as hemoglobin) is altered in the disease. But in most cases, the phenotype of the disease does not provide clear information about the disease gene. Later sections of this chapter will describe more general strategies to find mutations associated with various diseases. Once the mutation is identified, individuals can be genotyped by the methods we now discuss.

The Polymerase Chain Reaction (PCR) Amplifies Defined Regions of a Genome

Determining whether a person is homozygous or heterozygous for a disease-causing allele, or homozygous for a normal allele of the same gene, implies that you can isolate that gene from the person's genome and analyze the alleles by looking at the purified DNA. But genes are rare targets in complex genomes: The gene for the β chain of hemoglobin, for example, spans only about 1400 of the 3,000,000,000 nucleotide pairs in the haploid human genome. In 1985, Kary Mullis invented one of the most powerful techniques in molecular biology, called the **polymerase chain reaction (PCR)**, to deal with this problem of looking for the needle of a gene in the haystack of the genome. PCR is remarkably fast and efficient. Starting

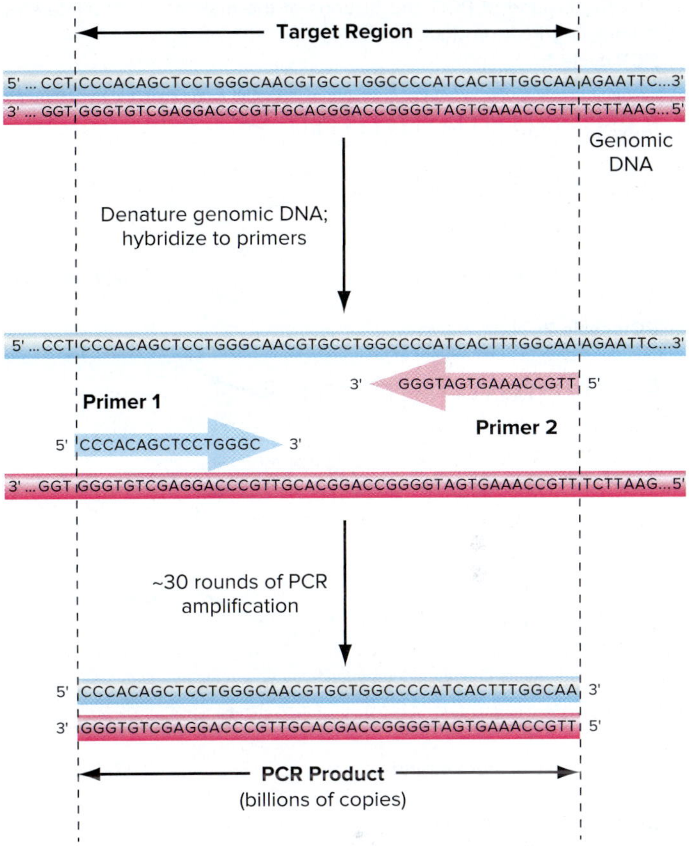

Figure 11.9 Overview of the polymerase chain reaction (PCR). The reaction amplifies a target region of genomic DNA defined by the 5′ ends of the two primers, making a PCR product. To help you understand the basis of PCR, the Watson and Crick strands of the same region of DNA are shown in different colors.

with minute amounts of DNA, such as that found in a single sperm or hair follicle, researchers can make a billion or more copies of a short, defined portion of that genome in just a few hours.

As shown in **Fig. 11.9,** PCR amplifies a *target region* of DNA. Two 16- to 30-base-long oligonucleotides, the *PCR primers,* define the ends of the target region. The investigator synthesizes these primers based on prior knowledge of the genome. One oligonucleotide is complementary to one strand of DNA at one end of the region; the other oligonucleotide is complementary to the other strand at the other end of the region. If these primers are drawn as 5′→3′ arrows to indicate their polarity, the arrows will point toward each other through the target region (Fig. 11.9). For DNA genotyping, the object of the PCR is to amplify sequences within the target region that may have different allelic forms.

The process of amplification is initiated by the hybridization of these synthetic oligonucleotides to one or more denatured template DNA molecules (melted into single strands) within the sample of genomic DNA to be analyzed (**Fig. 11.10**). The oligonucleotides act as primers that allow

Figure 11.10 PCR amplification of a target sequence. In the first PCR cycle of this example, a single molecule of double-stranded genomic DNA is heated to denature it into single strands. The temperature is then lowered to allow these single strands to hybridize with the two PCR primers. A heat-stable DNA polymerase now polymerizes new DNA onto the 3′ ends of each primer. In the second PCR cycle, the DNA in the reaction tube is denatured and then hybridized to the same primers. Both the original DNA strands and those made in the first cycle serve as templates in the second round. Each cycle of PCR thus doubles the amount of DNA in the target region. After the conclusion of the third round of PCR, the 5′ ends of the majority of template strands are defined by the 5′ ends of the primers, fixing the length of the accumulating PCR product.

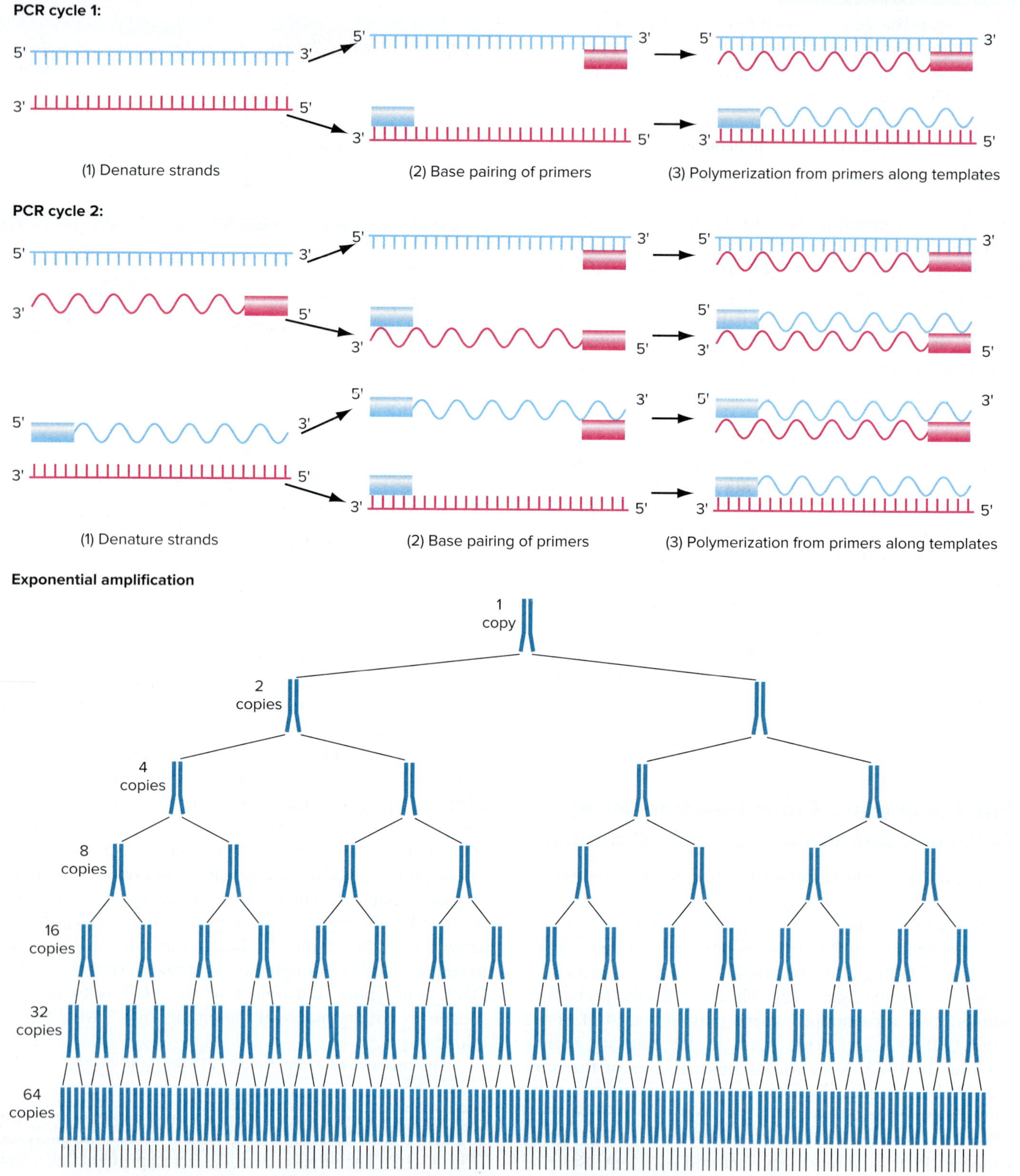

DNA polymerase to generate new strands of DNA complementary to both strands of genomic DNA between the primers; remember that DNA polymerase adds nucleotides sequentially onto the 3′ end of a primer.

After sufficient time has elapsed to allow copying of the target region, the reaction is heated in order to melt apart the original template strands of DNA from the newly synthesized strands. The reaction tube is allowed to cool, so that the starting DNA and the copies synthesized in the previous step become templates for further replication, using the oligonucleotides remaining in the tube as primers. Performing this same sequence of steps—denaturation into single strands, hybridization of primers, and polymerization by DNA polymerase—as an iterative loop results in an exponential increase in the number of copies of the target region with each step (Fig. 11.10). Repeating the cycle just 22 times would generate more than a million double-stranded copies of the target region; after 32 repetitions, the reaction tube would have over a billion copies of this part of the genome.

The iterative steps of the protocol can be automated in a PCR machine that heats up and cools down the sample according to a preprogrammed schedule. The reaction tubes placed into the machine contain enough nucleotide triphosphates and oligonucleotide primers to support these multiple rounds of DNA replication. Moreover, the tubes contain a special DNA polymerase from a bacterium that grows in hot springs. This DNA polymerase remains active after being subjected to the high temperatures used to melt apart the DNA strands at each round of the PCR protocol.

It is crucial for you to remember from Fig. 11.9 that the two priming oligonucleotides dictate the nature of the final PCR product. The ultimate PCR product is a double-stranded fragment of DNA that extends from the position of one primer's 5′ end to the position of the other primer's 5′ end. The primers must be complementary to opposite strands and have 5′-to-3′ polarities that point toward each other through the region of interest. In practice, PCR is inefficient if the primers are far apart, so the protocol generally cannot amplify DNA regions greater than 25 kb long.

PCR Products Are Genotyped by Sequencing or Sizing

For Mendelian genetic diseases caused by changes involving only one or a few nucleotides in a single gene, all of the information that distinguishes a normal allele from a mutant allele resides within a discrete region of the genome that can be encompassed by a PCR product. The differences between alleles can be recognized either by direct sequencing of PCR products, or in cases in which mutations add or subtract nucleotide pairs from the genome, by simply looking at the sizes of the PCR products. More complex polymorphisms such as copy number variants (CNVs) affect many nucleotide pairs that extend over regions much larger than can be amplified into a PCR product, so they must be analyzed by other methods that will be described later in the book.

Sequencing PCR products

As you may remember, the mutation $Hb\beta^S$ that causes sickle-cell anemia changes the identity of a single amino acid in the β chain of adult hemoglobin from glutamic acid to valine. The mutant allele is a single nucleotide substitution that changes an A to a T in the mRNA-like strand of the β-globin gene; the mutation is thus a single nucleotide polymorphism (SNP) (**Fig. 11.11a**). By genotyping the alleles of this SNP, we can identify people who will suffer from sickle-cell anemia or are carriers for this trait.

The process begins by PCR amplifying the locus from the person's genomic DNA using a pair of primers complementary to sequences on either side of the actual disease-causing mutation (Fig. 11.11a). Once the PCR product is made, its DNA sequence can be determined by the automated Sanger method shown previously in Fig. 9.7. Either one of the two PCR primers can serve as the primer for the sequencing reactions.

As **Fig. 11.11b** shows, the nucleotide substitution responsible for the disease shows up clearly in comparing the sequence obtained from the PCR products in $Hb\beta^S Hb\beta^S$ sickle-cell patients and $Hb\beta^A Hb\beta^A$ normal homozygotes. But importantly, both alleles are visible simultaneously in the sequence trace made from the PCR product generated using $Hb\beta^A Hb\beta^S$ heterozygous genomic DNA as the template. Genomic DNA prepared from somatic cells of the heterozygote contains both allelic variants. Because the primers hybridize equally well with the two homologous chromosomes (given that the sickle-cell mutation does not alter the genomic sequences complementary to the primers), about half the DNA molecules in the final PCR product will contain the mutant sequence and the other half the wild-type sequence. Heterozygosity for the disease-causing SNP is thus seen as a double peak showing both A and T in the DNA sequence trace.

The technique of sequencing PCR products amplified from genomic DNA is a straightforward way to determine one's genotype for any SNP. The same method can also be used to genotype other kinds of polymorphisms involving small numbers of nucleotides, such as small deletions/insertions (DIPs) or expansions/contractions of the numbers of repeats in simple sequence repeats (SSRs).

Size variation in PCR products

In some cases it is possible to genotype a polymorphism in a PCR product without actually sequencing it. Gel electrophoresis can easily distinguish small variations in the actual size of a locus caused by DIPs or SSRs, as illustrated in **Fig. 11.12.** Again, you begin by using a pair of primers

Figure 11.11 Detection of the sickle-cell mutation by sequencing of PCR products. (a) The mutation responsible for sickle-cell anemia is a SNP that changes a single amino acid in the hemoglobin β polypeptide from glutamic acid (E) to valine (V). This polymorphism is PCR amplified by primers complementary to flanking sequences that do not vary between alleles. **(b)** Sequencing of PCR products made from genomic DNA templates. Note that the sequence of the PCR product from a heterozygous carrier shows both the normal ($Hb\beta^A$) and mutant ($Hb\beta^S$) nucleotides at the position of the substitution (*black arrow*).

Courtesy of Joshua J. Filter, Cornell University, Ithaca, New York

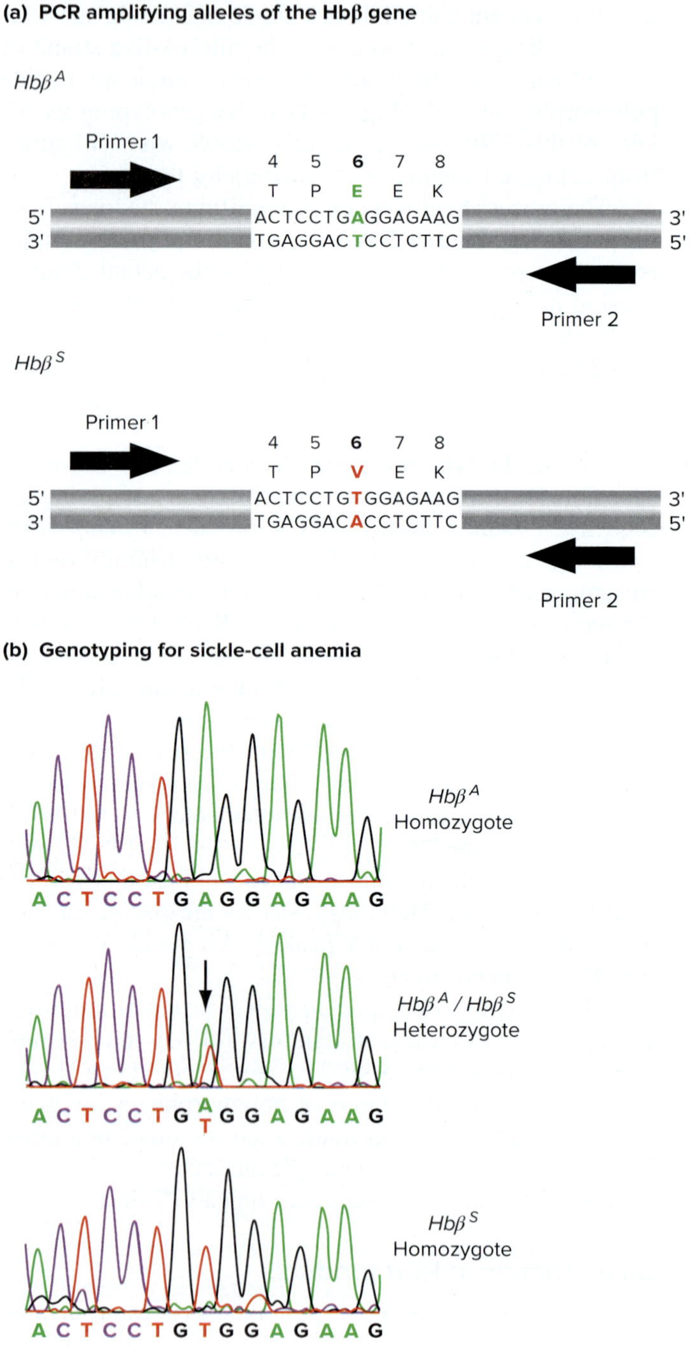

Figure 11.12 Detection of simple sequence repeat (SSR) polymorphisms by electrophoresis of PCR products. **(a)** SSR alleles differ in length. Left and right primers correspond to unique sequences that flank the SSR locus. **(b)** Genomic DNA is amplified by PCR with primers specific for the particular SSR locus. **(c)** Gel electrophoresis and ethidium bromide staining distinguish the alleles from each other. **(d)** SSRs are highly polymorphic with many different alleles present in a population, but each person has only two alleles of any given SSR locus.

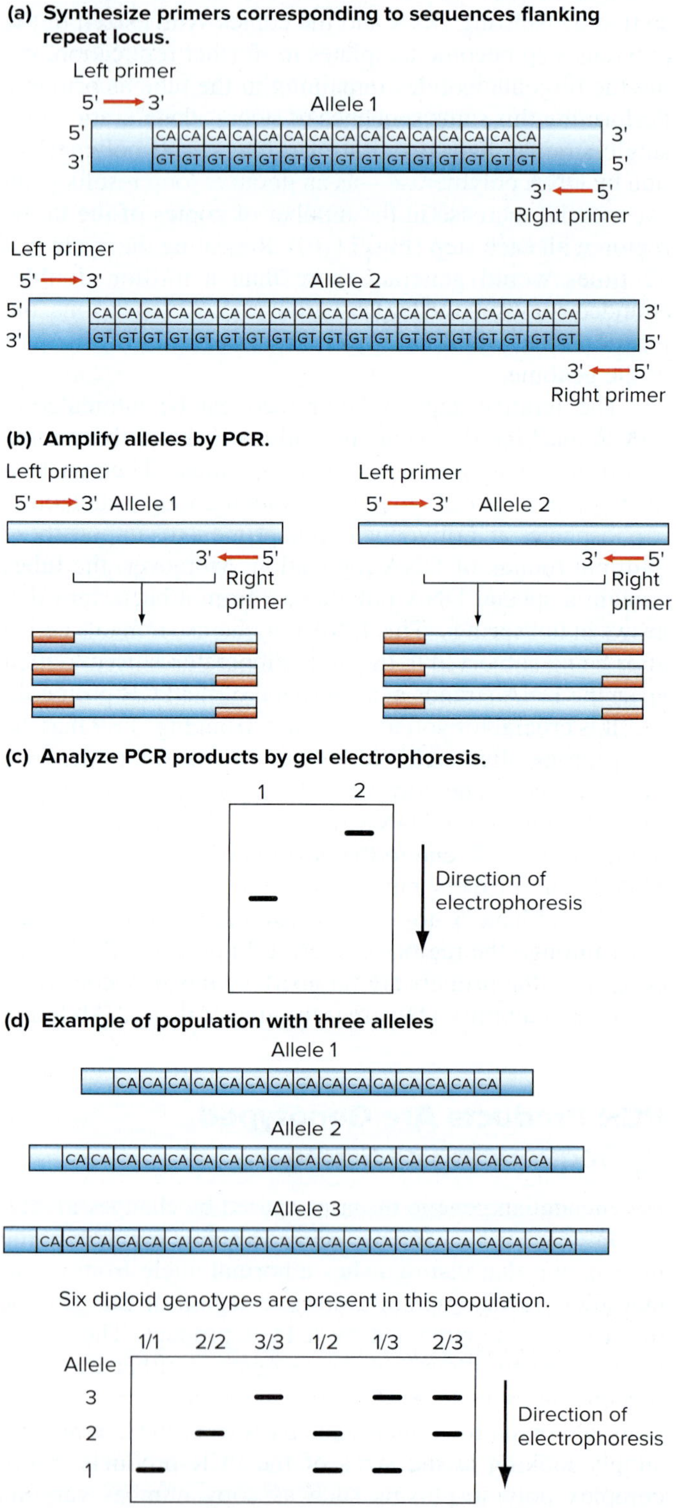

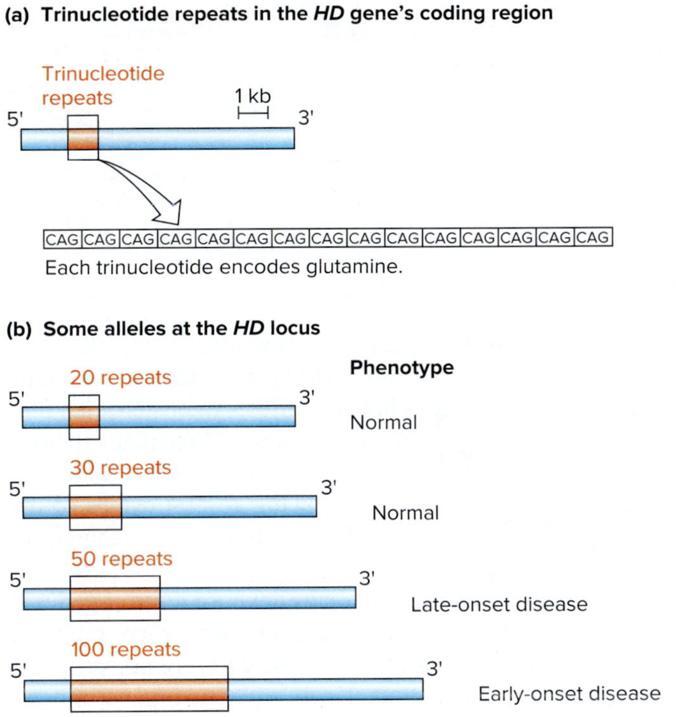

Figure 11.13 Mutations at the Huntington disease locus are caused by expansion of a trinucleotide repeat SSR in a coding region. **(a)** Near the 5′ end of the coding region is a repeating trinucleotide sequence that codes for a string of glutamines. **(b)** Different alleles at the *HD* locus have different numbers of repeating units. Normal alleles have 35 or fewer repeats. Dominant disease-causing alleles have 36 or more repeats; as the number of repeats increases, the onset of the disease is earlier.

complementary to sequences on either side of the actual length polymorphism in order to PCR amplify the locus from an individual's genomic DNA. But instead of sequencing, you then simply subject the PCR products to gel electrophoresis, which separates them according to their size. After staining with ethidium bromide, each allele appears as a specific band of DNA.

The ability to genotype size variants in this way is particularly important for the genetic diseases caused by trinucleotide repeat SSRs within genes. Recall from the Fast Forward Box in Chapter 7 (*Trinucleotide Repeat Diseases: Huntington Disease and Fragile X Syndrome*) that Huntington disease (HD) is one of the approximately 20 such trinucleotide repeat diseases in humans. HD is transmitted as an autosomal dominant mutation. Over 30,000 Americans currently show one or more symptoms of the disease—involuntary, jerky movements; unsteady gait; mood swings; personality changes; slurred speech; and impaired judgment. An additional 150,000 people in the United States have an affected parent, so they have a 50:50 chance of carrying and expressing the dominant condition themselves as they age. Although symptoms usually show up between the ages of 30 and 50, the first signs of the disease have appeared in people as young as 2 and as old as 83.

Normal alleles for the *HD* gene contain up to 35 tandem repeats of CAG, while disease-causing alleles carry 36 or more. The more repeats, the earlier the age of disease onset (**Fig. 11.13**). Those who inherit a disease allele invariably get the disease if they live long enough. Thus, although expressivity (in this case, the age of onset) is variable because it depends on the number of trinucleotide repeats, all disease alleles with 42 or more repeats are completely penetrant. Alleles with 36–41 CAGs are incompletely penetrant in that they cause disease in some people but no sign of the condition in others.

Some people with a family history of HD would like to know their genotype before deciding whether to have children. By amplifying the CAG-containing part of this person's *HD* genes (using primers that flank this SSR) and measuring the lengths of the resultant PCR product, geneticists can easily determine how many CAG repeats are found in each allele (Fig. 11.12). Other people from HD families elect not to undergo this procedure because definitive knowledge that they had a disease allele would have a devastating psychological impact.

Fetal and Embryonic Cells Can Be Genotyped

Suppose that a couple expecting a baby knows by genotyping their own genomes that they are both carriers for a highly deleterious recessive genetic disease such as cystic fibrosis, or that one of them has a dominant allele causing a late-onset condition like Huntington disease. Depending on their personal beliefs, some prospective parents would prefer to terminate the pregnancy if they knew the fetus had a disease-causing genotype.

Prenatal genetic diagnosis involves genotyping fetal cells by the methods we have just described. Physicians can isolate fetal cells by **amniocentesis,** a procedure in which some of the amniotic fluid surrounding the fetus in the mother's womb is extracted using a needle (see the Genetics and Society Box *Prenatal Genetic Diagnosis* in Chapter 4 for more details). The amniotic fluid contains some cells shed by the fetus. Geneticists PCR amplify the disease locus from genomic DNA prepared from these fetal cells and then analyze the PCR products by sequencing or sizing.

More recently, the dual success of technologies for *in vitro* fertilization and PCR has opened new options for reproductive decisions. It is now possible for couples to establish the genotype of embryos before they are placed in the mother's womb (**Fig. 11.14**). Such **preimplantation**

Figure 11.14 Preimplantation embryo diagnosis. Plucking one cell from an eight-cell embryo for the direct detection of genotype. The genomic DNA from the one cell is extracted and subjected to genotyping after PCR amplification. The remainder of the embryo survives and can be implanted into the mother's uterus.

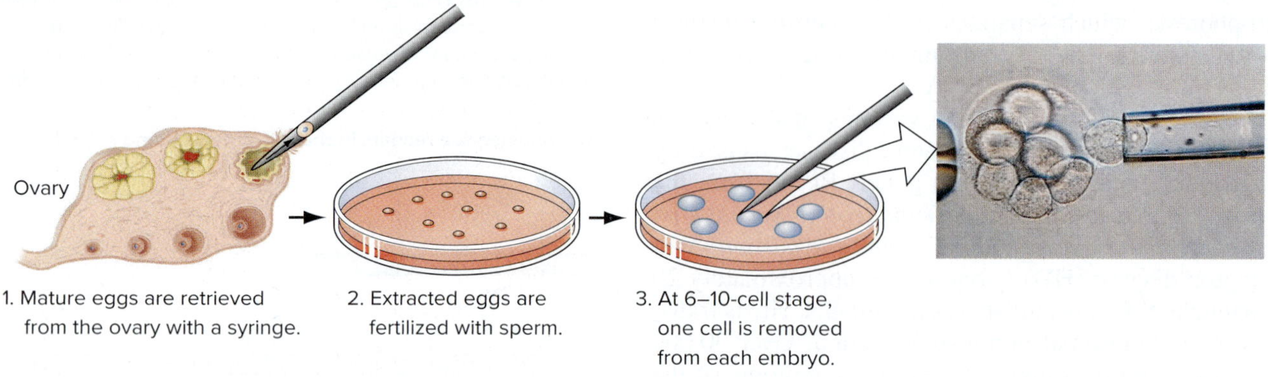

embryo diagnosis begins when a woman is injected with follicle stimulating hormone (FSH) to stimulate the maturation of about 10 eggs in her ovaries. An obstetrician then removes these eggs from the ovaries and fertilizes them *in vitro* with her partner's sperm. The fertilized eggs are then incubated for several days, allowing several cycles of mitotic division so as to produce early embryos containing 6–10 cells.

Specially trained technicians next use micropipettes to remove a single cell from each of these early embryos (Fig. 11.14). These early embryonic cells are not yet determined to become particular cell types or organs; indeed, embryos that split naturally at this stage can develop into healthy identical twins. Thus, the removal of a single cell does not harm the embryos nor prevent them from developing normally.

The technicians then prepare genomic DNA from the single cell obtained from each embryo and PCR amplify the specific region containing the site of the disease-causing mutation. They then analyze the PCR products by sequencing or sizing. In consultation with the physicians, the parents can select healthy embryos with genotypes that would not result in the disease (homozygous for the normal allele, or heterozygous for a recessive disease-causing allele). Usually two or three such embryos are placed into the mother's womb to improve the chances that at least one will implant properly into the uterus.

That embryos can be genotyped within a few days after fertilization by looking at the DNA from a single cell is conceptually and technically astonishing, but preimplantation embryo diagnosis has been used successfully in tens of thousands of pregnancies worldwide. The procedure is complex and expensive, costing thousands of dollars, but the information it provides can be invaluable to couples whose children would otherwise be at risk for serious genetic diseases.

essential concepts

- *PCR* amplifies specific regions of DNA defined by two oligonucleotide *primers.* Repeated cycles of synthesis increase exponentially the number of copies of the target DNA region.
- PCR product sequencing constitutes a simple method for genotyping many polymorphisms. Small insertions or deletions of DNA (DIPs and SSRs) can also be genotyped by examining PCR product sizes on gels.
- Fetuses can be genotyped *in utero* by obtaining fetal cells via *amniocentesis.* In *preimplantation diagnosis,* a single cell from an early embryo produced by *in vitro* fertilization is genotyped.

11.3 Sampling DNA Variation in a Genome

learning objectives

1. Explain why a relatively small number of SSR loci are sufficient to provide a DNA fingerprint of an individual.
2. Describe how a DNA microarray is constructed and how to genotype millions of loci on this microarray.

The vast majority of polymorphisms present in any genome do not cause disease or otherwise affect phenotypes. Several good reasons nonetheless exist for determining people's genotypes at these *anonymous loci.* Genotyping such loci allows people to be identified by their DNAs, which is highly useful for forensic purposes and for studying human evolution and history. Of additional practical significance, anonymous loci serve as molecular markers for specific regions of the genome. Even if these DNA markers do not themselves cause disease, scientists can follow their inheritance to locate hard-to-find genes responsible for genetic diseases and other phenotypes.

Forensic DNA Fingerprinting Examines Multiple SSR Loci

SSR loci are highly polymorphic: Many alleles that differ in the number of repeating units exist in the population, although any one person carries only two of these alleles for any given locus. The polymorphism of SSR loci makes them a powerful resource in identifying a person from his or her DNA.

The power comes from the possibility of examining multiple polymorphic SSR loci simultaneously. Suppose the likelihood that any two random people share exactly the same combination of two alleles of a particular SSR locus is 10% (0.1), and that the same is true for a second, independently segregating SSR locus. Using the product rule for independent events, the probability that two randomly chosen people will have the same alleles at the two SSR loci is $(0.1) \times (0.1) = 0.01$ or 1 in 100. Now consider 13 such SSR loci. The chance that two people will have the same combinations of alleles at all 13 positions in the genome is $(0.1)^{13}$, or one chance in 10 trillion. (This calculation is simplified but gives you a rough idea of these probabilities; Chapter 21 on population genetics will show you how to make such calculations precisely from actual data.) Given that the earth currently has only about 7 billion human inhabitants, you can see that a genotype for 13 unlinked, polymorphic SSR loci would serve as a **DNA fingerprint** unique to any one person, excepting identical twins.

A simple extension of a method we have already discussed to genotype a single SSR locus allows the simultaneous genotyping of multiple SSRs. Figure 11.12 showed that the PCR products amplified from an SSR locus will have different sizes reflecting the number of repeating units in each allele. To examine 13 SSRs at the same time, you would label the 13 pairs of PCR primers with dyes that fluoresce in different colors, and then combine all the primer pairs in the same PCR reaction tube. After gel electrophoresis, you could then identify the allelic variants for each SSR locus based on the fluorescent colors and sizes of the PCR products (**Fig. 11.15**).

In the United States, the Federal Bureau of Investigation (FBI) maintains a database called CODIS (Combined DNA Index System) that allows forensic laboratories throughout the country to share and compare DNA profiles. All these laboratories use the same 13 primer pairs to amplify the 13 SSR loci. The laboratories carefully catalog the sizes of the PCR products and submit the results to the

Figure 11.15 DNA fingerprinting. (a) Basis of DNA fingerprinting by *PCR multiplexing* (simultaneous analysis of PCR products from multiple loci). PCR primer pairs (ovals) amplify separate SSR loci, usually from nonhomologous chromosomes. The primer pairs are labeled with different fluorescent molecules (*blue* and *green* in this example for two loci). **(b)** Gel electrophoresis of multiplexed PCR products. This example shows the analysis in three people (A–C) of six loci, two of each color (*blue, green,* and *yellow*). The alleles for the two different loci of the same color (i.e., locus 1 and locus 2) differ sufficiently in their sizes that it is clear which allele belongs with which locus. *Red* bands in each lane are size standards.

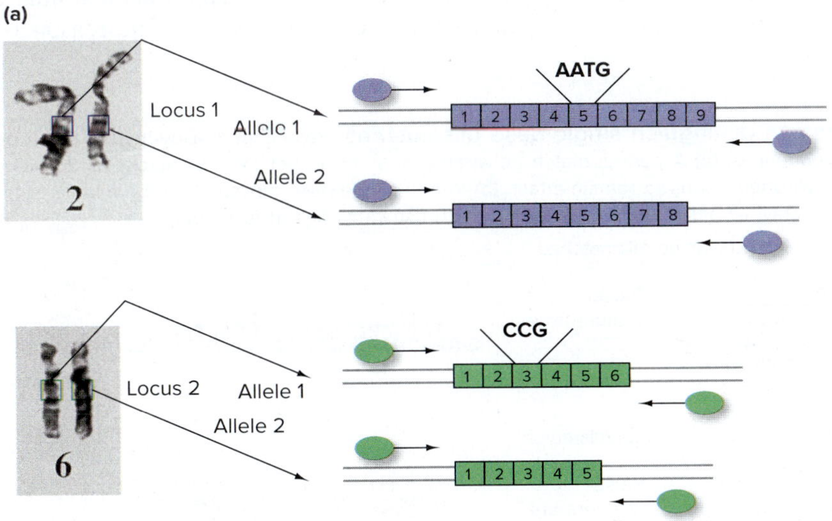

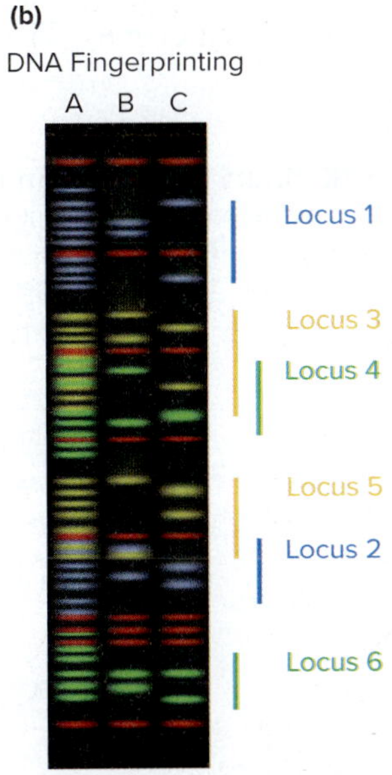

CODIS database. All 50 states mandate the collection of DNA fingerprint data for CODIS from felons convicted of certain crimes, such as sexual offenders; the database also includes profiles from missing persons.

As of 2016, CODIS has assisted more than 300,000 criminal investigations. Typically, forensic investigators use the database to match the DNA profile of evidence left at crime scenes with that of a felon. But DNA can also establish innocence: Suspects can be excluded if no match with crime scene evidence exists. In fact, a public policy organization called the Innocence Project has used DNA fingerprint evidence to help exonerate more than 300 people convicted of capital crimes, including several who were awaiting execution.

The power of DNA fingerprinting technology raises many concerns about privacy and possible discrimination in the collection of data. Just to give one interesting example, consider that siblings will share 50% of all SSR alleles; the same is true of parents and children. As a result, it is possible to identify the perpetrator of a crime not by a match to his or her own DNA, but instead by a partial match to the DNA of a close relative. This kind of familial DNA search was critical for the apprehension in 2010 of the major suspect for the "Grim Sleeper" serial killer in Los Angeles. The suspect's son had recently been convicted on a felony weapons charge, so his DNA was analyzed. The son's DNA profile partially matched DNA fingerprints from semen and saliva found on the Grim Sleeper's victims. Policemen followed the father, and one detective posing as a waiter obtained a partially eaten slice of pizza with the father's DNA. Stunningly, this DNA was a perfect match with the crime scene evidence. (The trial has just begun as of this writing in early 2016.) Should criminal investigators be allowed to conduct such familial searches given that family members of felons who have not committed any crime are in effect under lifelong genetic surveillance?

DNA fingerprints have many important uses beyond forensics for capital crimes. They now provide the most conclusive evidence in paternity suits, and DNA fingerprints can be used to identify human remains, as was the case for the victims of the World Trade Center disaster on September 11, 2001. The benefits of this technology are not restricted to DNA fingerprints of humans. Wildlife biologists study populations of endangered species by fingerprinting individual animals to increase the chance for success of captive breeding programs or to identify illegally poached animals. Owners of valuable domesticated animals such as show dogs, thoroughbred horses, or cattle can in some cases establish lineage through DNA fingerprints. In one fascinating if bizarre case from Argentina, scientists were enlisted to help apprehend a butcher who moonlighted as a cattle rustler. Meat hanging in the butcher's shop had the same DNA profile as a tissue sample that a rancher had taken from one of his cows before it was stolen.

DNA Microarrays Genotype Millions of SNPs

Nucleic acid hybridization, the ability of complementary single strands of DNA or RNA to come together to form double-stranded molecules, is the basis for many techniques in molecular biology. We have already discussed the importance of hybridizing oligonucleotide primers to DNA templates in Sanger DNA sequencing and in PCR. Both DNA sequencing and PCR assume that a perfect complementary match exists between all the nucleotides in the primers and templates. But what will happen if a mismatch exists between two single strands of nucleic acid?

Consider a 21-base oligonucleotide that hybridizes to a target strand that differs at a single base in the middle of the sequence (**Fig. 11.16**). The resulting double-stranded hybrid is significantly less stable than a similar hybrid in which all the nucleotides match. The reason is that the

Figure 11.16 Short hybridization probes can distinguish single-base mismatches. Researchers allow hybridization between a short 21-base *probe* and two different target sequences. **(a)** A perfect match between probe and target extends across all 21 bases. When the temperature rises, this hybrid has enough hydrogen bonds to remain intact. **(b)** With a single-base mismatch in the middle of the probe, the effective length of the probe-target hybrid is only 10 bases. When the temperature rises, this hybrid falls apart.

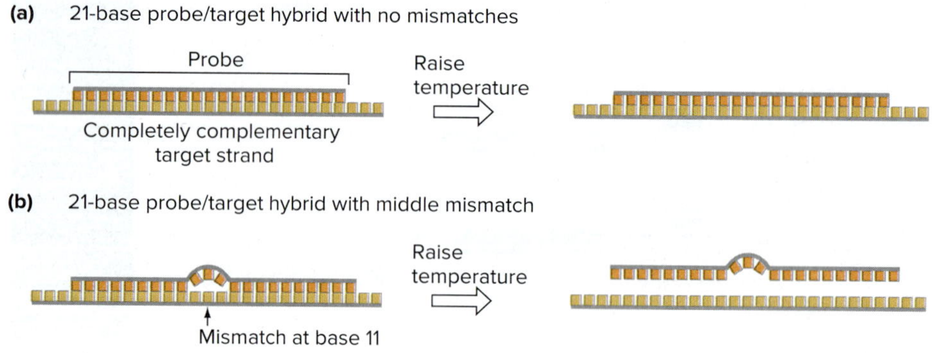

Figure 11.17 DNA microarrays. (a) Three identical DNA microarrays containing ASOs for two alleles of one SNP. (The ASOs are shown as 5 nucleotides long, but in practice they would be 20–40 nucleotides long.) A probe (with a *red* fluorescent tag) will hybridize only to perfectly complementary ASOs. The fluorescence intensity reflects the number of sequences in the genomic DNA that are complementary to the ASO. **(b)** Method to amplify genomic DNA for microarray analysis. Genomic DNA is cut with a restriction enzyme (RE), and the ends produced are ligated to a double-stranded oligonucleotide adapter. PCR then amplifies all genomic fragments using a single primer that hybridizes to part of the adapter. The resultant DNA fragments are denatured and fluorescently tagged (*red*). **(c)** A small region of a DNA microarray after hybridization with a genomic DNA probe.

c: Source: National Cancer Institute

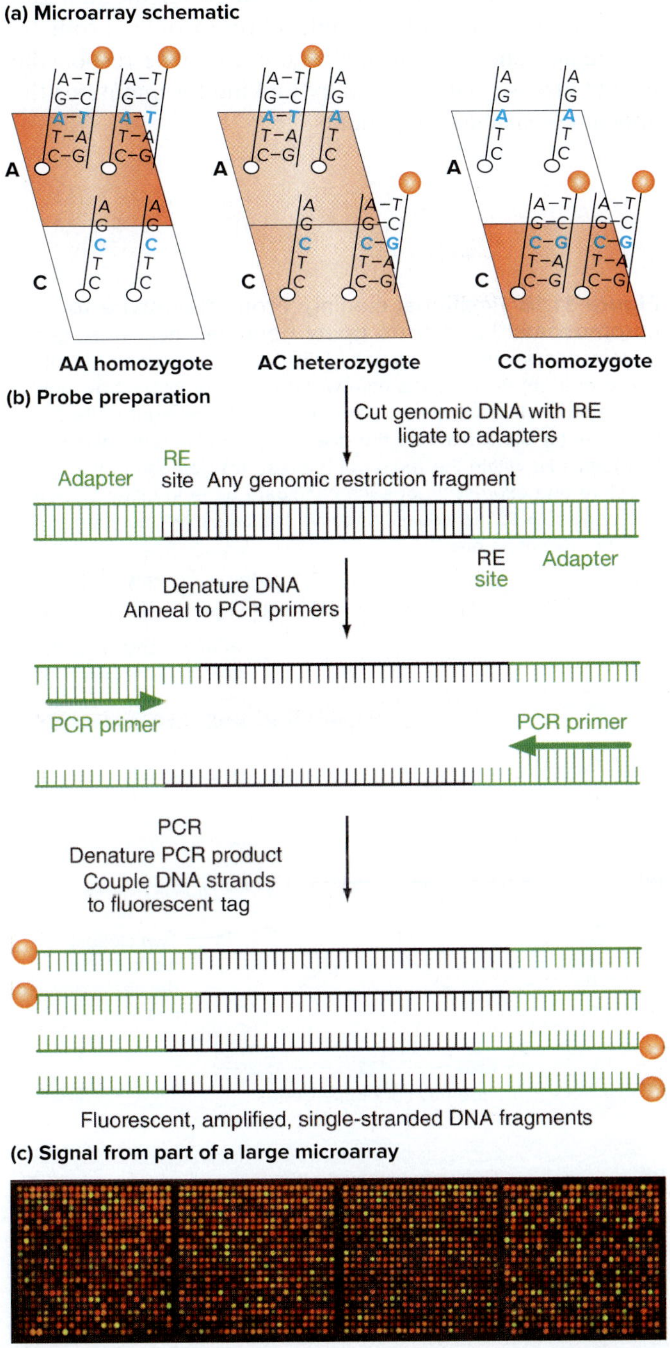

strength of the hydrogen-bond forces holding together the double helix depends on the longest stretch that does not contain any mismatches. When the two strands do not match exactly, there may not be enough weak hydrogen bonds in a row to hold them together. Thus, for small regions of up to about 40 bp, researchers can devise hybridization conditions (such as a particular temperature) under which the perfect hybrids will remain intact, while the less stable imperfect hybrids will not (Fig. 11.16).

Researchers can exploit the different stabilities of hybrid molecules for the genotyping of SNP loci. The idea, illustrated in **Fig. 11.17a,** is to attach to a solid support (such as a chip of silicon) short 20- to 40-base-long oligonucleotides that will hybridize under the right conditions to only one of the two alleles at a SNP locus. These oligonucleotides are logically called **allele-specific oligonucleotides,** or **ASOs.** The investigator now takes DNA from the genome to be analyzed and turns it into a **probe** by fragmenting the DNA, denaturing the fragments into single strands by heating, and attaching a fluorescent dye to these small pieces of single-stranded genomic DNA. The dye-carrying genomic DNA is now placed on the silicon chip (sometimes called a **DNA microarray**) and the temperature adjusted so that the genomic DNA probe will hybridize only to ASOs that match the probe perfectly. To visualize the hybridization signal, light is shined on the chip, and a detector records the amount of fluorescence emitted by each area containing a specific ASO. As you can see in Fig. 11.17a, the pattern of fluorescence allows straightforward determination of the genotype of the original genomic DNA for any SNP.

One additional feature of interest in Fig. 11.17a is that the intensity of the fluorescent signal over a particular ASO on the silicon chip is proportional to the number of copies of that allele in the genome. Fluorescence intensity can thus provide a way to monitor copy number variations (CNVs). We will discuss this use of ASO chips later in the book.

A single fluorescently-tagged genomic DNA molecule does not generate enough light to allow its detection on the microarray. Investigators must therefore amplify the genomic DNA so that many copies of each part of the genome can all be attached to fluorescent tags. **Figure 11.17b** illustrates one clever method for achieving this amplification. Researchers first digest the genome with a restriction enzyme that creates fragments with a sticky end, and then they ligate an *oligonucleotide adapter* to these restriction fragments. Part of the adapter can anneal to the overhang, while the rest of the oligonucleotide is complementary to a PCR primer. Fragments connected to adapters at the ends of both strands can now serve as templates for PCR amplification. In this way all parts of the genome could be amplified using a single adapter and a single PCR primer.

Rapid advances in DNA microarray manufacturing technology have led to the fabrication of chips capable of detecting SNP alleles at more than 4 million loci (**Fig. 11.17c**). At the time of this writing (2016), the cost of analyzing a sample of genomic DNA is only a few hundred dollars, which works out to a per-SNP genotyping cost that is a small fraction of a penny. The SNP loci analyzed on commercially available microarrays include all single-nucleotide variants known to be associated with genetic diseases, but most of the loci on the chip are common SNPs likely to be without phenotypic effect. The widespread occurrence of these particular anonymous SNPs makes them invaluable for locating the mutations that do cause diseases, as will be explained in the next section.

essential concepts

- In *DNA fingerprinting,* genotyping of multiple polymorphic loci such as SSRs provides enough information to identify individuals from their DNA.
- A *DNA microarray* contains *allele-specific oligonucleotides (ASOs)* for millions of SNP loci. Under the proper conditions, a *probe* made of fluorescently-labeled genomic DNA fragments binds only to complementary ASOs, allowing these loci to be genotyped.

11.4 Positional Cloning

learning objectives

1. Describe the process of positional cloning and how it allows mapping of disease-causing mutations.
2. Examine the limitations of pedigree analysis in providing the information needed for positional cloning.
3. Explain how a Lod score is obtained and what information it provides.
4. Discuss the consequences of allelic heterogeneity, compound heterozygosity, and locus heterogeneity.

Of the thousands of known human **disease genes** (genes whose mutant alleles cause a disease phenotype), scientists can identify only a small number based on the specifics of the abnormal condition. For example, sickle-cell anemia and thalassemias are diseases affecting red blood cells. About 97% of the dry weight of a red blood cell consists of hemoglobin, so researchers directed their attention to the genes encoding the polypeptides making up this oxygen-carrying protein as likely causes of these diseases. More often, it is difficult to make an educated guess about which protein is changed by a disease-causing mutation, so a different approach is needed.

Linkage Analysis with DNA Markers Gives Disease Genes an Approximate Chromosomal Address

A generally useful strategy to identify the defects causing hereditary diseases is called **positional cloning** (**Fig. 11.18**). The object is to obtain information about the unknown location of the disease gene by finding polymorphic loci to which the mutation is genetically linked. Because we know from the human genome sequence the exact position of each locus, discovering anonymous DNA polymorphisms closely linked to the disease gene allows researchers to focus their search for the mutation on a small region of a single chromosome. From the candidate genes within this region, the gene responsible for the disease can be found by looking for mutations that appear consistently in patients.

Figure 11.18 **Positional cloning: From phenotype to chromosomal location to guilty gene.** **(a)** The disease gene is located less than 50 map units (~50 Mb) away from any markers linked to it. **(b)** Researchers narrow the region of interest by looking for the most closely linked markers to the left and right of the mutation. **(c)** Candidates for the disease gene (different shades of *blue*) must lie within the region of interest. **(d)** Comparing the structure and expression of each candidate gene in many diseased and nondiseased individuals pinpoints the causative mutation and thus the disease gene.

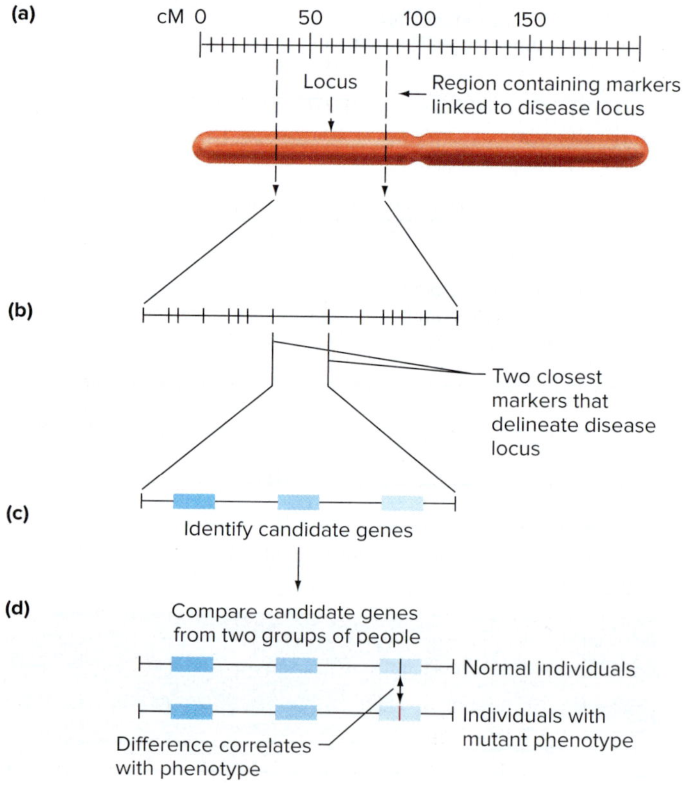

The strategy of positional cloning

You learned in Chapter 5 that two genes are genetically linked if they lie close together on the same chromosome. Genes are linked if parental types exceed recombinant types among the gametes. The frequency of recombination between the two loci provides a direct measure of the distance separating them, as recorded in centimorgans (cM), also called map units (m.u.), with 1 cM equal to a recombination frequency of 1%.

Positional cloning is a straightforward extension of this definition of linkage, with only two substantive differences. First, instead of tracking two genes by the traits they control, in positional cloning you track one locus by phenotype (noting which individuals in a pedigree are affected or unaffected), but you track the second locus by direct DNA genotyping of each person. Ultimately, you are looking for the same thing at both loci—a variation in DNA—but in different ways. The trait identifies the variant indirectly; the genotype shows this directly.

Second, instead of dealing with only two or three loci at a time (as in a two-point or three-point cross), you can use DNA microarrays to follow millions of anonymous loci in each person in the pedigree. You can think of this microarray/pedigree analysis most simply as the simultaneous performance of millions of two-point crosses, each one a test for linkage between an individual DNA marker and the disease locus. Discovery of a DNA marker that shows linkage to the disease locus is the first goal of positional cloning.

DNA microarrays are packed so densely with polymorphic loci that many of these loci must in fact be linked to any given Mendelian disease gene. In a two-point cross, two loci are linked when they are less than 50 cM apart. As a rough guide, 1 cM on average corresponds to 1 Mb (1 million basc pairs) in thc human genome, so linked loci must generally be within 50 Mb of each other (Fig. 11.18). If a microarray had only 1000 molecular markers spread out over the entire human genome, they would be on average 3 Mb apart. A disease-causing mutation would thus have to lie within 3 Mb of one polymorphic locus on thc microarray, and the two must be genetically linked. Modern DNA microarrays can analyze millions of polymorphisms simultaneously in a person's DNA, so this positional cloning method potentially could map disease genes even more precisely.

An example of positional cloning

Neurofibromatosis is a dominantly inherited, fully penetrant, autosomal condition that is rare but still affects more than 100,000 Americans. The disease causes nervous tissue to proliferate uncontrollably, forming tumorous bumps under the skin (**Fig. 11.19**). Although these tumors are usually benign, they can damage nerve cells and sometimes develop into malignant cancers. **Figure 11.20** documents one stage of a search by positional cloning for a mutation causing neurofibromatosis. The results allow us to ask whether a particular anonymous SNP locus (SNP1) is linked to the neurofibromatosis gene. SNP1 is located on chromosome 17 and has two alleles, G and T.

The pedigree (Fig. 11.20a) shows that the neurofibromatosis patient II-1 received from his affected mother I-1 both the dominant disease-causing allele *NF* and the G allele of SNP1. From his unaffected father, II-1 received the normal allele NF^+ of the neurofibromatosis gene and

Figure 11.19 Neurofibromatosis.

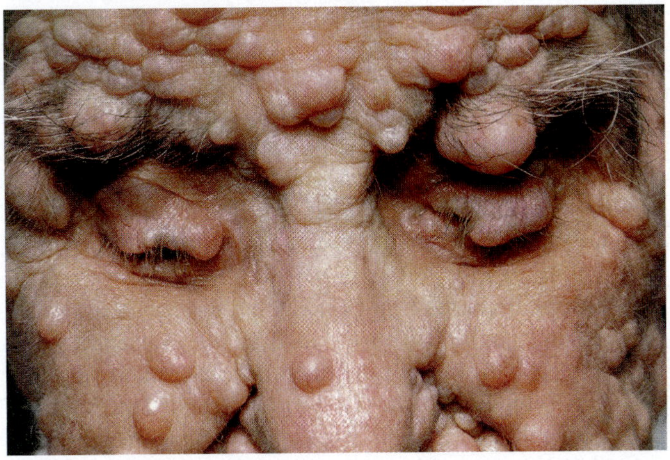

Figure 11.20 Positional cloning: an example. (a) Pedigree showing the inheritance of neurofibromatosis in a family. G and T refer to nucleotide alleles of the SNP1 locus. **(b)** Interpretation of the same pedigree, based on the hypothesis that the neurofibromatosis gene is linked to SNP1.

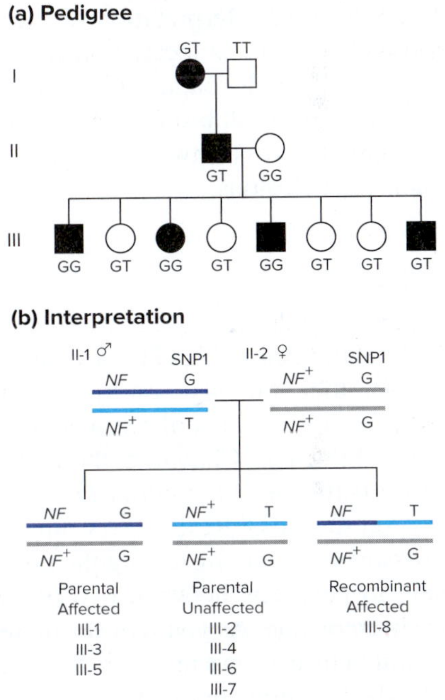

the T allele of the SNP. II-2, the unaffected mate of II-1, does not have the disease (so she is NF^+/NF^+), and the DNA analysis indicates that she is homozygous for the G allele of SNP1. The children in generation III of the pedigree are thus in effect the progeny of a testcross. Examining each child both for the presence or absence of the disease and the SNP genotype will reveal whether the child obtained a parental-type (nonrecombinant) or a recombinant-type sperm from the doubly heterozygous father II-1.

As Fig. 11.20a shows, seven out of eight progeny in generation III resulted from parental-type sperm: either the combination of *NF* and G seen in the grandmother I-1, or the combination of NF^+ and T in the grandfather I-2. Only one of the children in generation III was the product of a recombinant sperm (III-8, an affected son with *NF* and the T allele of the SNP). The data thus strongly suggest that the neurofibromatosis gene and this particular SNP are genetically linked, separated by a map distance $(1/8) \times 100 = 12.5$ cM.

A family size of eight children is large by today's standards, but this number of progeny is still not sufficiently large to achieve great statistical significance. Nonetheless, the most straightforward interpretation of the data is that the neurofibromatosis gene is located on chromosome 17, within a region that extends roughly 12.5 Mb to either side of the SNP1 locus. Fig. 11.20b shows graphically how you can easily interpret the cross between II-1 and II-2 based on these provisional conclusions.

Multipoint analysis

Microarray data include not just two-point information about linkage relationships between a disease gene and individual DNA loci, but also multipoint information about the behavior of millions of DNA loci with respect to each other. You saw in Chapter 5 the power of three-point crosses in fruit flies, but microarrays are like three-point crosses "on steroids." Researchers can now pinpoint particular crossovers occurring during the production of a single gamete to positions between two polymorphisms on the same chromosome. As you will see in Section 11.5, this information in turn allows researchers to pinpoint disease genes to short regions defined by mapped recombination events.

Positional Cloning Has Several Limitations

For traits expressed in plants or small animals, researchers can easily set up crosses that generate hundreds of progeny so as to allow accurate genetic mapping. But scientists do not direct human breeding, so not every mating between two people provides interpretable information about the relative positions of any two given loci. In addition, human family sizes are small, making it difficult to obtain sufficient data for precise mapping.

The phase problem

You should note that the calculation just presented for the map distance between the neurofibromatosis gene and SNP1 does not include one of the crosses shown in the pedigree: the mating of I-1 and I-2 (see Fig. 11.20). The reason is that we don't know the configuration of the alleles (or *phase*) in the affected grandmother I-1. It is possible that her mutant *NF* allele was on the same chromosome 17 as the G allele of the SNP, but it is also possible that the SNP allele on this *NF*-bearing chromosome was instead T. Because we cannot say with certainty whether II-1 (her affected son) results from a parental-type or recombinant-type egg, we did not consider this gamete in calculating the map distance.

The phase problem can be resolved in either of two circumstances. First, if you know the genotypes of two loci in both of a person's parents, sometimes you can determine which alleles came from one parent and which from the other. This is precisely how we knew that the phase in the double heterozygous father II-1 was *NF* G/NF^+ T (Fig. 11.20). So to determine the phase in I-1 by this method, we would need to have genotyping information about *her* affected parent. Second, if the two loci are sufficiently close together, you can infer the probable phase because the linked alleles should segregate with each other more often than not. In Fig. 11.20, where the parental classes considerably outnumber the recombinants in generation III, you could have inferred the likely arrangement of alleles in the doubly heterozygous II-1 even if you had no information about his parents.

Informative and noninformative crosses

Even if you know the phase in a doubly heterozygous parent, a mating may not provide any useful information about whether the two loci are linked. One example is seen in **Fig. 11.21a,** which again examines the mating between male II-1 with neurofibromatosis and his unaffected partner II-2. However, we are now looking at a different SNP locus (SNP2), where both parents are heterozygotes for the alleles A and C. If a child were also a heterozygote for A and C, you would not be able to tell which parent contributed the A and which the C, so you could not determine if the child was the result of a parental or recombinant gamete. But if the child were a homozygote for either SNP2 allele, say of genotype AA, then you know both the egg and sperm must have carried the A allele.

You should remember from Chapter 5 that the basic requirement for genetic mapping is that at least one parent must be a double heterozygote. **Figure 11.21b** emphasizes this crucial point by showing that if neither parent is a double heterozygote, the cross cannot be informative.

Figure 11.21 Some noninformative matings. (a) Even if you know that the *NF* gene and SNP2 are linked and you know the phase (arrangement of alleles) in the parents, the mating may or may not be informative. If the child is a CC or AA homozygote, the cross would be informative. If the child is an AC heterozygote, you don't know which allele came from which parent, so the cross would be noninformative. **(b)** If neither parent is a double heterozygote, it is impossible to perform linkage analysis. (SNP3 is a third SNP locus.)

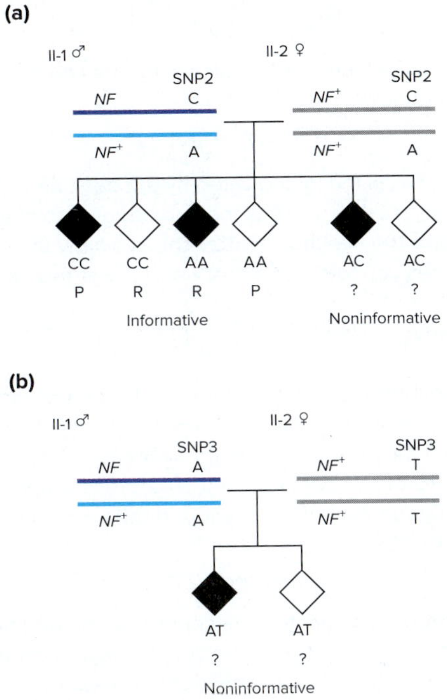

Even if a mating is noninformative for the linkage of a disease gene with a particular SNP locus, multipoint analysis on microarrays usually provides a way for scientists to overcome this constraint. This is because the microarray will likely contain other nearby SNPs that will be informative.

Obtaining sufficient pedigree data

With millions of polymorphic loci on a DNA chip, it should be possible in theory to map disease genes very accurately, even if certain crosses are uninformative for linkage of the disease gene to certain DNA markers. However, the resolution of positional cloning is always limited in practice by the number of people human geneticists can track in families in which the disease is segregating. If scientists have mapped a disease gene to within 1 cM of a DNA polymorphism, this value means they have examined the phenotypes (affected or unaffected) of at least 100 members of such families, and they have also genotyped on microarrays the DNA of all these people. (Remember that 1 cM means 1 recombinant gamete out of 100 total gametes.)

For this reason, positional cloning achieved its first successes for diseases that can be found in extended families with a large number of children. In 1984, the Huntington disease (*HD*) locus became the first human disease gene to be mapped successfully by positional cloning precisely because such a family was available. **Figure 11.22** shows the seven-generation, 65-member family pedigree used to demonstrate tight linkage between a DNA marker called *G8* and the *HD* locus.

The Lod Score Provides a Statistical Approach to Studying Linkage

Positional cloning is rarely as straightforward as it was for *HD*. Most human families have only a few children, and it is difficult to obtain DNA and phenotype information about multiple generations in a pedigree. For these reasons, human geneticists have developed a statistical tool called a **Lod score** (log of the odds). The purpose of the Lod score is reminiscent of that of the χ^2 statistic used in Chapter 5: to determine whether the data are sufficient to conclude with confidence whether a disease gene and a

Figure 11.22 A marker closely linked to the Huntington disease locus. Detection of linkage between the DNA marker *G8* and a locus responsible for Huntington disease (HD) was the first step in the cloning of the *HD* gene. The pedigree shows an extended Venezuelan family affected by HD. Alleles at the *G8* marker locus are indicated (A, B, C, and D), while affected individuals are indicated in *orange*. Cotransmission of marker alleles with the mutant and wild-type alleles at the *HD* locus is obvious.

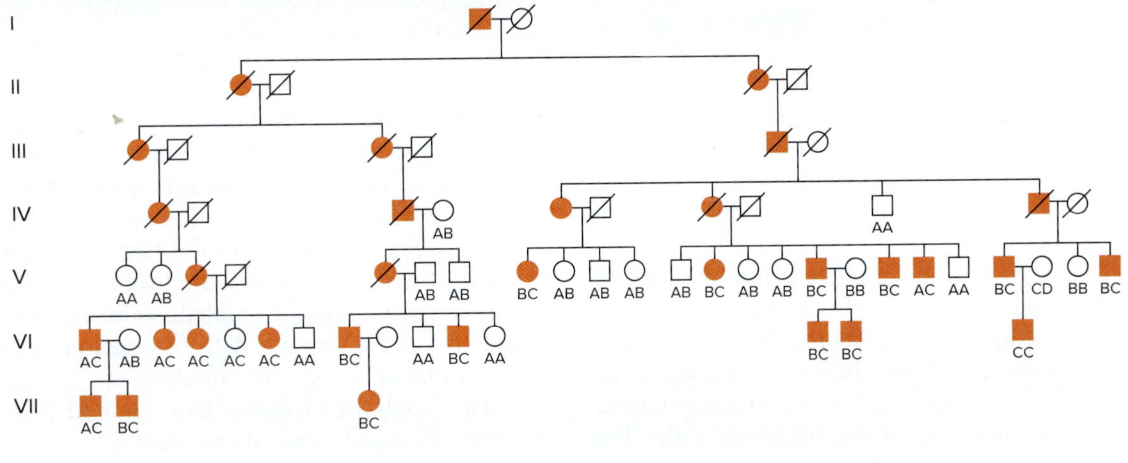

TOOLS OF GENETICS

The Lod Score Statistic

The Lod score is a mathematical answer to the question: How much more likely is it that the allele transmission pattern seen in a pedigree will occur if the loci are linked at a given recombination frequency (RF) less than 50%, than if they are not linked? The Lod score, as its name implies (log of the odds) is the logarithm of the ratio between these two probabilities:

$$\text{Lod} = \log\left[\frac{\text{P(obtaining observed results if loci are linked at a given RF)}}{\text{P(obtaining observed results if loci are unlinked)}}\right]$$

Here, we illustrate the Lod score calculation for the pedigree in Fig. 11.20a. The pedigree suggests that the *NF* gene is linked to a particular SNP on chromosome 17. The calculation will allow us to determine our degree of confidence in this preliminary conclusion.

1. **Tabulate which progeny are parental and which are recombinant.** In Fig. 11.20a, you can see that the first seven children in generation III have the parental (P) configuration of alleles and that only child III-8 has the recombinant (R) configuration. We'll abbreviate these data as *PPPPPPPR*.
2. **Calculate the Lod score denominator.** If two loci are unlinked, it is equally likely that any one child will be P or R (that is, the RF = 50%). The probability of P is thus ½, and the probability of R is also ½. The probability of obtaining children in the particular birth order *PPPPPPPR* if the *NF* gene and the SNP locus are unlinked is:

   $$P(RF_{50\%}) = \left(\frac{1}{2}\right)^8 = \frac{1}{256}$$

   You can see that a generalized formula for this part of the calculation is simply:

   $(½)^n$, where *n* is the total number of tabulated individuals.
3. **Calculate the Lod score numerator.** Loci could be linked if the RF is any value less than 50%, but the calculation requires us to assume an RF value. The pedigree in Fig. 11.20a indicates an RF of 1/8 = 12.5%, so we will use this as our best current estimate. With RF = 1/8, the expected frequency of P progeny is 7/8, and R progeny is 1/8. The probability of seven parentals and one recombinant in the particular birth order *PPPPPPPR* is:

   $$P(RF_{12.5\%}) = \left(\frac{7}{8}\right)^7\left(\frac{1}{8}\right)^1 \approx \frac{1}{20}$$

   A generalized formula for calculating the Lod score numerator is:

   $$(1 - RF_{obs})^{\#P} \times (RF_{obs})^{\#R}$$

   where RF_{obs} is the RF indicated by the data, #P is the number of parentals, and #R is the number of recombinants.
4. **Calculate the likelihood ratio.** This is simply the ratio of the values you found in steps 2 and 3. For this example,

   $$P(RF_{12.5\%})/P(RF_{50\%}) = \left(\frac{1}{20}\right)\Big/\left(\frac{1}{256}\right) = 12.8$$

   This likelihood ratio means that it is 12.8 times more likely that the *NF* gene and the SNP are linked with RF = 12.5% than that they are not linked (RF = 50%).
5. **Calculate the Lod score.** The Lod score is simply the base 10 logarithm of the likelihood ratio. For the example in Fig. 11.20a:

   $$\text{Lod score} = \log(12.8) = 1.1$$
6. **Interpret the Lod score.** The convention among human geneticists is that a Lod score ≥3 (that is, a likelihood ratio ≥1000) is required to be confident of linkage. The Lod score of 1.1 indicates that the data in Fig. 11.20a are insufficient to conclude that *NF* and SNP1 are linked.

Important points about Lod scores:

- The Lod score determined by assuming the precise RF implied by the data will always be the maximum Lod score obtainable for the data set.
- For a single pedigree, Lod scores can be calculated for any RF value less than 50%. A Lod score ≥3 indicates that the data obtained are compatible statistically with the particular distance (less than 50 m.u.) being tested.
- Likelihood ratios are converted into Lod scores because Lod scores calculated for the same RF value in different pedigrees may simply be added to see whether a Lod of 3 can be obtained.

DNA marker are genetically linked. The Lod score is used in human genetics instead of the χ^2 statistic because the Lod score better handles a small number of data points while allowing the data obtained from many different pedigrees to be combined.

The Lod score statistic is calculated from the ratio of two probabilities: the probability of obtaining a particular set of results in a pedigree if two loci are linked (assuming a particular RF value), and the chance of observing the same results if the loci are unlinked. The Lod score statistic is the base 10 logarithm (log) of this likelihood ratio. The convention adopted by human geneticists is that a Lod score greater than or equal to 3 indicates two loci are linked. A Lod score of 3 means that it is 1000 times more likely that the two loci are linked than that they are not (because 3 = log1000). The beauty of the Lod score statistic is that because it is a log function, the Lod scores from different pedigrees may simply be added, so researchers will know when they have enough data to conclude that a disease allele is linked to a specific marker.

The Tools of Genetics Box entitled *The Lod Score Statistic* illustrates how to calculate a Lod score, using as an example the pedigree in Fig. 11.20a. As you can see, the maximum Lod score for these data (which is obtained by assuming RF = 12.5%) is 1.1, indicating that this one pedigree is insufficient evidence for linkage of the SNP1 locus on chromosome 17 and the *NF* gene. However, if two additional pedigrees were available, each also with a Lod score of 1.1 (calculated for RF = 12.5%), the Lod score of the three pedigrees together would be 3.3, constituting strong evidence that the *NF* gene is on chromosome 17 and linked to SNP1.

Genetic Diseases Can Display Allelic or Locus Heterogeneity

Suppose that by positional cloning you have been successful in narrowing down the location of a disease gene to a 1 Mb long region between two polymorphic markers. In the human genome, the average gene density is about 1 gene per 100 kb of DNA. More than 10 genes might therefore lie in the 1 Mb region. How could you discriminate among these candidates to find the right one?

In some cases, it might be possible to find clues by looking for changes in patients in the amounts or sizes of the mRNA transcripts or the protein products of genes (see Problem 38 at the end of this chapter). But by far the most generally useful strategy is to use PCR to amplify DNA from all the candidate genes in all available patients, and then sequence all of these PCR products. If you found that the patients all had identifiable mutations in one of these candidate genes, particularly mutations that might affect the amino acid sequence of the gene's protein product, the evidence would be powerful for identifying that candidate as the actual disease gene (review Fig. 11.18d).

Some genetic conditions are always caused by the same single mutation in a single gene; we've already seen that all patients with sickle-cell anemia are homozygous for exactly the same base pair substitution in the gene encoding the β chain of hemoglobin. DNA sequencing would thus reveal the same mutation in the genomic DNA of all patients and carriers of sickle-cell anemia.

Allelic heterogeneity: Multiple mutant alleles in one gene

This simple scenario is not, however, always the case. Many other genetic diseases display **allelic heterogeneity,** meaning that they can be caused by a variety of different mutations in the same gene. An important example is cystic fibrosis, a recessive autosomal genetic condition inherited by 1 child in every 2500 born from two parents of European descent. Children with the disease have a variety of symptoms arising from abnormally viscous secretions in the lungs, pancreas, sweat glands, and several other tissues. Most cystic fibrosis patients die before the age of 30.

Positional cloning strategies allowed investigators to narrow their search for the causative gene to a 400 kb region between two DNA markers on chromosome 7 that contained only three candidate genes (previously shown in Fig. 11.5). One of these genes was *CFTR*, encoding the cystic fibrosis transmembrane receptor that allows chloride ions to pass through cell membranes; see Fig. 2.25). Significantly, both *CFTR* copies in all cystic fibrosis patients were found to contain mutations that would alter the amino acid sequence of the protein or would prevent normal amounts of the protein from being synthesized. Thus, as the name implies, *CFTR* is clearly the gene responsible for cystic fibrosis.

One mutation called ΔF508 (which removes the amino acid phenylalanine—F—from position 508 of the protein) accounts for about two-thirds of all mutant *CFTR* alleles worldwide. The remaining alleles consist of more than 1500 different rare mutations (**Fig. 11.23**). Many patients are thus so-called **compound heterozygotes** (sometimes known as ***trans*-heterozygotes**), in which one copy of chromosome 7 has one mutation in *CFTR* and the other copy of chromosome 7 has a different *CFTR* mutation. The disease results because neither chromosome 7 can encode a normal transmembrane receptor; in effect, the two different recessive *CFTR* mutations fail to complement each other.

The concept of allelic heterogeneity is central to understanding a drug that has very recently been developed to treat cystic fibrosis effectively, but only in a minority of patients. In 2012, the United States Food and Drug Administration approved the drug *ivacaftor* for patients who have one specific *CFTR* mutation called G551D (changing glycine at position 551 to aspartic acid). The mutant protein encoded by this allele assembles properly into the cell membrane, but the G551D protein is inefficient in

Figure 11.23 Allelic heterogeneity in the *CFTR* gene. Every cystic fibrosis patient has a mutated *CFTR* gene on both copies of chromosome 7. The diagram indicates the location of mutations at different positions in *CFTR* relative to the 24 exons of the gene. *Compound heterozygotes* are patients who have one copy of *CFTR* with one of these mutations, while the other copy of *CFTR* has a different mutation.

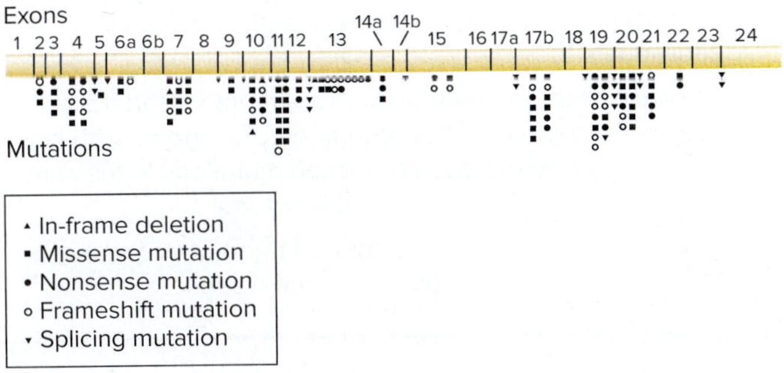

transporting chloride ions across the membrane. Ivacaftor interacts specifically at the cell surface with the G551D mutant CFTR protein, enhancing its ability to transport chloride. This treatment has been remarkably effective in preventing the symptoms of cystic fibrosis from developing in young children, but unfortunately G551D accounts for only about 4% of all mutant *CFTR* alleles in the human population.

More recently (in 2015), researchers developed a treatment for the much more prevalent ΔF508 mutation. This allele encodes a protein that cannot fold properly and thus is not inserted into cell membranes. The new drug, called *lumacaftor,* ameliorates the folding problem, resulting in an increase in the number of CFTR molecules in cell membranes. The mutant proteins are still partially defective in chloride ion transport. Remarkably, a combination pill containing lumacaftor and ivacaftor prevents the development of cystic fibrosis in many patients homozygous for ΔF508.

Locus heterogeneity: Mutations in different genes cause the same disease

In this chapter, we deal exclusively with Mendelian genetic diseases caused by mutations in a single gene, but you already know that many other conditions display **locus heterogeneity:** They are caused by mutations in one of two or more different genes. A previously discussed example of a heterogeneous condition is deafness (review Fig. 3.23). In confronting a new genetic disease, researchers must always be aware of the possibility of locus heterogeneity.

In *complex traits* (also called *quantitative traits*) such as high blood pressure, many different genes can influence the phenotype even in a single person. Chapter 22 outlines some methods geneticists can use to study the genetic basis of these complex traits.

essential concepts

- *Positional cloning* identifies DNA polymorphisms that are linked to disease genes.
- *Lod scores* allow statistical assessment of linkage when data are limited, as in human pedigrees.
- After a disease gene is mapped approximately, researchers sequence candidate genes in the region to identify one that is altered consistently in affected individuals.
- In *allelic heterogeneity,* a variety of mutations in a single gene cause disease. *Compound heterozygotes* with two different recessive loss-of-function mutations in the same gene may display the mutant phenotype.
- In *locus heterogeneity,* mutations in one of two or more different genes can cause the same disease.

11.5 The Era of Whole-Genome Sequencing

learning objectives

1. Describe a high-throughput, automated method by which millions of DNA templates may be sequenced simultaneously.
2. Summarize a sequence of investigative steps that can narrow the candidates for a disease-causing variant.
3. Explain how databases that catalog sequence variation in many people can facilitate the diagnosis of genetic diseases.

DNA microarrays with millions of SNPs sample only a small proportion of the variation between human genomes and can suggest only a disease gene's general chromosomal location. As we just saw, disease gene identification eventually requires DNA sequencing to correlate the disease phenotype with actual mutations. Suppose now that we could cheaply and accurately sequence all of the nucleotides—not just those of candidate genes—in an affected person's genome. The whole-genome sequence must somewhere include the causative mutation. Thus, unlike positional cloning, where the first goal is to find a molecular marker linked to the disease gene, the goal in the whole-genome approach is to find directly a DNA alteration that *is* the disease allele.

Startling developments are making the idea of routine and affordable whole-genome sequencing into a reality. Chapters 9 and 10 explained that the Human Genome Project, completed in 2003, sequenced the complete human genome at a cost of 3 billion dollars. Researchers have since invented imaginative new methods that have rapidly driven down the cost of DNA sequencing. In 2016, it is possible to sequence a person's whole genome (at a high coverage that will still miss some small regions) for about $2000, and the cost will undoubtedly fall under $1000 within a few years.

Whole-genome sequencing is still costly enough that researchers often economize by sequencing just that portion of the genome corresponding to the protein-coding exons. This is often informative because many, though far from all, disease-causing mutations alter the amino acid sequence of a protein. In **whole-exome sequencing,** investigators first enrich (by hybridization to cDNA sequences) for genomic DNA fragments that correspond to the exons of all genes, and then sequence these fragments. The **exome,** that is, the collection of all exons of all genes, constitutes less than 2% of whole-genome DNA, so whole-exome sequencing requires many fewer sequencing reads than whole-genome sequencing. When DNA sequencing becomes even cheaper, enriching for the exome will likely no longer have a significant cost advantage over whole-genome sequencing.

New Techniques Sequence Millions of Individual DNA Molecules in Parallel

The major technical advances that are making exome and genome sequencing fast and cheap enough for use in identifying disease genes permit millions of individual DNA molecules to be sequenced simultaneously. Many creative methods have been invented to perform so-called *high-throughput* or *massively parallel* sequencing.

Several of these high-throughput methods for sequencing human genomes are straightforward extensions of the Sanger sequencing by synthesis approach you already learned in Chapter 9, but three things are new. First, individual DNA molecules being synthesized by DNA polymerase are anchored in one place. Second, these methods control base addition temporally so that each base can be identified before the next one is added. Third, in some systems the sensitivity of detection is so high that a single molecule of DNA can be monitored without the need for cloning or PCR amplification steps. As shown in **Fig. 11.24,** the combination of these three innovations allows sequencing machines to record the successive addition of nucleotides to each of millions of growing DNA molecules in real time.

Figure 11.24 outlines only one of many ingenious technologies for inexpensive high-throughput sequencing currently under development. Other prototype systems are based on very different ideas; for example, the figure at the beginning of this chapter illustrates a novel method in which individual DNA molecules are threaded through small channels called *nanopores.* It is not clear which methods will become standard in the future as costs are steadily driven lower. But you should have no doubt that the era of whole-genome sequencing has already arrived.

Disease-Causing Mutations Are Hidden in a Sea of Variation

A patient's whole-exome or whole-genome sequence should include the sequence difference(s) responsible for

Figure 11.24 One method for high-throughput, single molecule DNA sequencing. (a) Millions of single-stranded genomic DNA fragments, to which poly-A has been enzymatically added at the 3′ end, are hybridized to oligo-dT molecules attached to the surface of a special microarray called a *flowcell.* **(b)** Using the genomic fragment as template and the oligo-dT as primer, DNA polymerase synthesizes new DNA containing nucleotides with colored, base-specific fluorescent tags. These nucleotides are also blocked at their 3′ ends so that only one nucleotide can be added at a time. **(c)** After a high-resolution camera photographs the fluorescence, chemicals applied to the flowcell remove the tag and the blocking group from the just-added nucleotide. **(d)** Each subsequent cycle begins by infusing the flowcell with a new dose of tagged nucleotides and polymerase, and then step (c) is iterated. The sequencing machine takes about 100 pictures that record a sequence of colored flashes at each of the millions of spots where a single DNA molecule is being synthesized. A computer rearranges the data into millions of short sequence reads of about 100 nucleotides and then assembles the genome sequence.

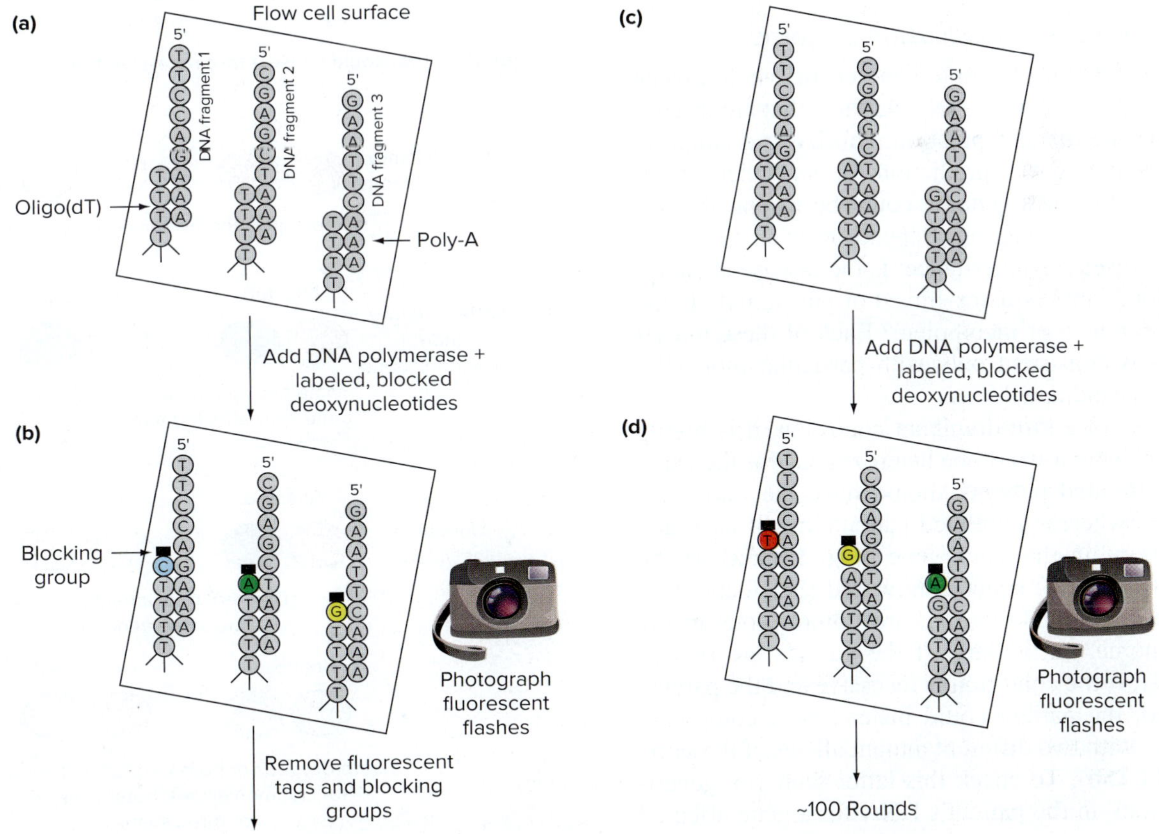

the genetic disease, but possession of this sequence information does not guarantee that geneticists will be able to identify the responsible mutation(s). One problem is technical: No genome sequence is 100% accurate or 100% complete. All sequencing methods have a low but real error rate in identifying nucleotides, and random sampling of DNA fragments will leave some regions of the genome unsequenced. These issues can be minimized by coverage of 10 or more genome equivalents, but they cannot be eliminated completely.

An even more fundamental problem is that the amount of variation among human genomes is huge. We saw at the beginning of this chapter that any person's genome differs at more than 3 million locations from the standard RefSeq human genome. How can we tell which of these millions of DNA polymorphisms causes a patient's disease? Our ability to deal with whole-genome sequences is still so limited that in many cases, the responsible mutation has yet to be identified. It should be lurking in the sequence, but it frustratingly remains hidden in front of our noses.

Despite these issues, investigators have been able to marshal the results of several types of data analysis, sometimes supported by inspired guesswork, to find an increasing number of disease genes. We focus in this section on the types of clues geneticists use to identify disease-causing mutations within whole-genome/exome sequences. However, it is crucial to keep in mind that these methods are not, at least not yet, always successful.

Clues from disease transmission patterns

The underlying logic of whole-genome or whole-exome sequencing requires that the DNA variants that are disease alleles will be rare in the population. This basic assumption allows scientists to make predictions about which of the variations in a patient's genome could be responsible for the disease. These predictions depend on what pedigrees say about the disease's inheritance: Is the disease allele recessive or dominant? Is it sex-linked or autosomal? Is the penetrance complete or incomplete? Each of these inheritance modes is consistent only with particular molecular genotypes at a candidate locus.

In the case of a rare dominant condition, it is highly likely that the patient would be heterozygous for the causative allele. Related patients should have the same rare mutant allele, whereas unrelated patients might have different mutations in the same gene (**Fig. 11.25a**). If the condition is recessive, geneticists would first focus their attention on rare mutations that are homozygous in the patient's genome, particularly if the parents are related even distantly. If the condition is recessive and the parents are unrelated, the patient could instead be a compound heterozygote, with two different mutant alleles of the same gene (**Fig. 11.25b**). To check this latter scenario, geneticists would look in the patient's DNA for a gene affected

Figure 11.25 SNP patterns consistent with inherited traits. Each oval represents a copy of a gene, so each person corresponds to two ovals. Common variants are in different shades of *gray*, while *orange, blue, yellow,* or *green* symbolize different rare variants in the same gene. **(a)** Variants that could cause a dominant trait. Within a family, affected individuals will be heterozygotes for the same rare variant. Unrelated affected people may be heterozygotes for different rare variants in the same gene. **(b)** Variants that could cause a recessive trait. In consanguineous families, affected individuals will be homozygotes for a single variant they inherit by descent from a recent common ancestor. The affected children of unrelated people most likely are compound heterozygotes who inherit different rare mutations in the same gene, one from each parent. In both (a) and (b), unaffected controls may or may not be related.

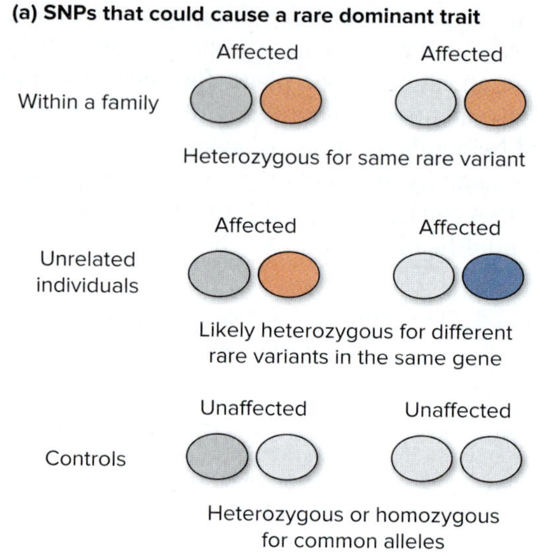

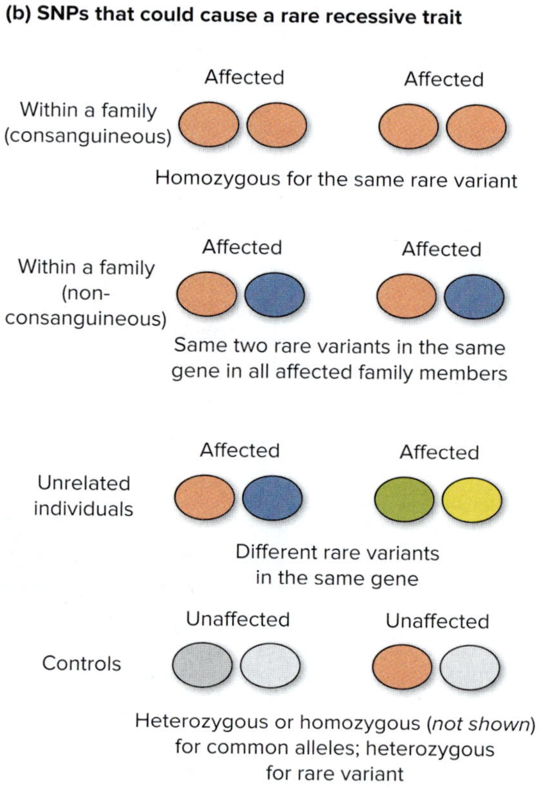

by two different mutations. Finally, if an inheritance pattern shows sex linkage, the search for candidate genes would be limited to the X chromosome; if autosomal, the X chromosome would be excluded.

DNA sequence information from the patient's relatives is particularly useful in winnowing down the list of candidate polymorphisms. As an example, SNP genotyping of relatives using microarrays, as discussed earlier in this chapter, could narrow the search to a region between two known SNPs. Positional cloning and whole-genome sequencing are thus not mutually exclusive approaches to disease gene identification; instead, they can provide complementary information. Better yet, though more expensive, would be comparisons of the patient's whole-genome or exome sequence with those of parents and/or siblings.

A recent case study illustrates the power of DNA sequence information from related individuals (**Fig. 11.26**). A brother and sister had Miller syndrome, a rare condition affecting development of the face and limbs, but neither parent was affected (Fig. 11.26a). These facts suggest (but don't prove) that Miller syndrome is a recessive autosomal condition, with the two children inheriting mutant alleles from both of their heterozygous, carrier parents. To find the Miller syndrome gene, researchers sequenced the entire genomes of both parents and both children; in fact, this was the first time in history that the genomes of all the members of a nuclear family were sequenced completely.

Figure 11.26 The first family with completely sequenced genomes. **(a)** Pedigree for Miller syndrome. **(b)** Map showing allele inheritance along chromosomes 16 and 17 in the affected children. In *identical* regions, the affected brother and sister share the same maternally and paternally derived alleles. In *nonidentical* regions, the siblings share no alleles. In *haploidentical maternal* regions, the siblings have the same allele from the mother but different alleles from the father. In *haploidentical paternal* regions, the brother and sister share a common allele from the father but have different alleles from the mother. If Miller syndrome is recessive, the responsible gene should lie in an identical region. This prediction was upheld when mutations in the *DHOD* gene on chromosome 16 were found to cause the disease.

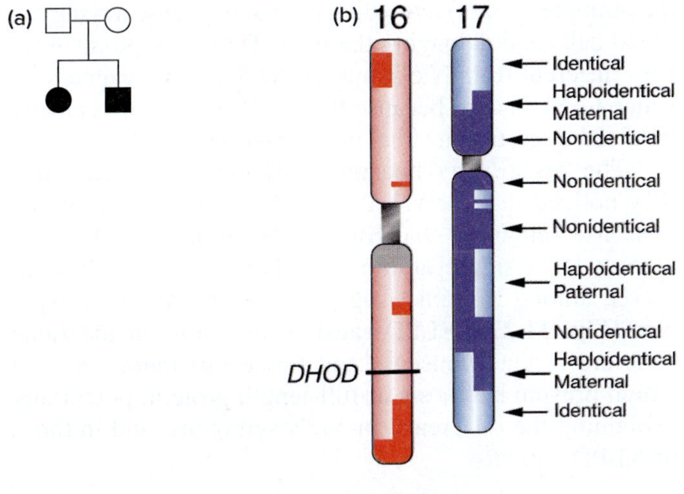

Figure 11.26b presents a graphical summary of part of this gigantic data set, showing the landscape of recombination that occurred on just two chromosomes during meiosis in the mother and father to produce the gametes resulting in the affected children. The hypothesis that Miller syndrome is recessive predicts that the responsible gene would lie in a region where the affected son and daughter share the same allele from the mother and also the same allele from the father (regions labeled *identical* in the figure). Geneticists studying the disease could thus focus their attention only on the approximately 25% of the genome where this was the case (Fig. 11.26b). We describe the outcome of this investigation later on in this chapter.

Clues from a variant's predicted effect on gene function

Researchers first try to look for disease-causing mutations in the protein coding regions of the exome because these parts of the genome are the easiest to look at: Coding regions constitute only a small fraction of the total genome, and alterations in the coding region are the most straightforward to interpret. In particular, investigators would search for rare polymorphisms that change the identity of an amino acid (that is, SNPs causing missense mutations) or alter the reading frame (SNPs, DIPs, or SSRs causing nonsense or frameshift mutations). Most nonsense/frameshift mutations and a subset of missense mutations will be *nonanonymous DNA polymorphisms* that affect phenotype through changes in protein function. In contrast, silent mutations that change a codon into a different codon for the same amino acid will not affect phenotype; these anonymous SNPs can therefore be discarded as candidates.

The assumption that a particular genetic disease results from a mutation in a protein-coding exon is nonetheless very uncertain. Some genetic diseases are caused not by alterations in amino acid sequence, but rather by the amount of a protein that the organism produces. Mutations that reside in regions of the genome outside of the exome could, for example, lower or prevent the transcription of a gene or the splicing of its primary transcript. Either case would lower the amount of the gene's protein product or even prevent its synthesis. Such mutations would never be found if researchers focused their attention only on the exome. And unfortunately, we still understand so little about the DNA sequences that regulate transcription or splicing that many such mutations will be overlooked even when the patient's whole-genome sequence is available.

Clues from previously determined genome sequences

Rare diseases are unlikely to be caused by variants common in the human population. As a result, variants that have been documented in databases as common are poor

TABLE 11.2 Finding the Disease-Causing Mutation in Nic Volker's Genome

Analysis Step	Candidate Variants Remaining
Sequence of Nic's exome	16,124
Filter for missense mutations changing an amino acid	7,157
Filter for novel variants not previously reported in databases	878
Filter for variants that are X-linked or display a recessive pattern in Nic's exome	136
Filter for variants that change evolutionarily conserved amino acids	35
Filter for variants in genes that are not frequently mutated in the general population	5
Filter for variants in genes known to be mutated in other genetic diseases of some relevance	1 (*XIAP*)

candidates for disease-causative mutations. In contrast, it is possible that different patients suffering from similar, or at least vaguely related symptoms, may share either the same rare mutation or they may have different rare mutations in the same gene.

Even genomes of other organisms can point genetic sleuths in the proper direction. For example, consider a gene that has been closely conserved during evolution so that some of the amino acids in the protein product are identical in diverse organisms, such as humans and fruit flies. Conservation suggests that these particular amino acids play crucial roles in the protein's function. If you now find a rare variant in a patient that changed such a highly conserved amino acid to a different one, this mutation would be a strong candidate because of the likelihood it would affect phenotype.

Pinpointing a Disease Gene Requires a Combination of Approaches

The beginning of this chapter introduced the case of Nic Volker, one of the first patients to be treated successfully for a genetic disease based on identification of the causative gene through sequencing of his personal exome. When Nic's DNA was characterized in 2009, sequencing was still relatively expensive, so investigators sequenced only Nic's exome, rather than his whole genome or the exomes/genomes of his parents or siblings. **Table 11.2** examines how geneticists analyzed Nic's exome, using many sources of information to narrow down the candidate variants to the single responsible mutation.

Nic's exome contained about 16,000 variants relative to the human standard RefSeq (Table 11.2). These variants were mostly SNPs, but some DIPs and SSRs were also found. Researchers started to winnow this list by excluding likely anonymous variants in the exons (that is, silent mutations in codons, or mutations in the sequences encoding the 5′ and 3′ UTRs of mRNAs). They next focused their attention on novel variants that had not been recorded in databases previously, in this way ignoring common variants known to exist in the genomes of normal individuals. The following step was to filter the list for mutations consistent with X-linked or recessive inheritance, as these were the most likely scenarios for a disease that affected Nic but not his parents or other relatives.

At this point, the list of good candidates for the cause of Nic's condition had been narrowed to 136 variants (Table 11.2). The researchers now took an evolutionary approach and asked whether any of these changes might have altered the identity of an amino acid that was tightly conserved in many diverse species. The next-to-last step of the analysis was to examine the remaining candidates for those that were in genes known from databases to be never or only infrequently inactivated (for example, by nonsense or frameshift mutations) in the general population. The idea behind this step was to ignore genes that are defective in many normal individuals and are thus unlikely to influence disease phenotypes.

Five candidate variants remained in the investigators' target list. The researchers realized that one of these candidate mutations was in a gene called *XIAP*. Other mutations in *XIAP* were known to cause a serious condition called *X-linked lymphoproliferative disease (XLPD),* in which the blood contains too many lymphocytes (white blood cells of the immune system), crowding out the oxygen-carrying red blood cells and damaging the liver. These symptoms were very different from Nic's, but XLPD had some vague relevance to Nic's case because the immune system is clearly involved in gastrointestinal inflammation.

The researchers became particularly excited when they noticed that the variation in Nic's *XIAP* gene was a missense mutation changing in the protein product the identity of a single amino acid that is completely conserved among humans, frogs, flies, and many other species (**Fig. 11.27**). XLPD-causing mutations in the same gene are instead nonsense and frameshift mutations that would prevent synthesis of full-length protein, potentially explaining the difference in Nic's symptoms and in those of XLPD patients.

Figure 11.27 A mutation in a conserved amino acid in the XIAP protein. Amino acids 195–211 of the XIAP protein in the one-letter code. Compared to the human RefSeq XIAP protein (second row), Nic Volker's XIAP (first row) has an amino acid substitution at position 203, from cysteine (C) to tyrosine (Y). In all other species examined, cysteine is found at this position, suggesting that the mutation in Nic's genome might alter XIAP function.

	195					200					205					210	
Nic's XIAP	G	D	Q	V	Q	C	F	C	**Y**	G	G	K	L	K	N	W	E
Human	G	D	Q	V	Q	C	F	C	C	G	G	K	L	K	N	W	E
Chimpanzee	G	D	Q	V	Q	C	F	C	C	G	G	K	L	K	N	W	E
Mouse	D	D	Q	V	Q	C	F	C	C	G	G	K	L	K	N	W	E
Dog	D	D	Q	V	Q	C	F	C	C	G	G	K	L	K	N	W	E
Cow	D	D	Q	V	Q	C	F	C	C	G	G	K	L	K	N	W	E
Chicken	D	D	Q	V	Q	A	F	C	C	G	G	K	L	K	N	W	E
Zebrafish	D	D	N	V	Q	C	F	C	C	G	G	G	L	S	G	W	E
Frog	R	D	H	V	K	C	F	H	C	D	G	G	L	R	N	W	E
Drosophila	L	D	H	V	K	C	V	W	C	N	G	V	I	A	K	W	E

Although this extensive bioinformatics analysis did not prove that the *XIAP* variation caused Nic's condition, it was an excellent candidate that fulfilled all the criteria examined in Table 11.2. What is more, if this identification was correct, Nic's disease might be *actionable,* meaning that something could be done to alleviate his condition. Maybe he could be treated by a method known to help XLPD patients: namely, a transplant into his bone marrow of umbilical cord blood from a newborn infant. This treatment would in theory provide Nic with a self-renewing source of stem cells that could continuously produce normal lymphocytes. Within a year of the transplantation procedure, Nic's condition improved remarkably (see Fig. 11.3).

The Study of Human Genetics Is an Ongoing Venture

Nic Volker is one of an increasing number of patients whose suffering has been ameliorated by information gleaned from whole-exome/genome sequencing. But the outcome of such studies is not always so favorable. For example, using bioinformatics filters similar to those employed in Nic's case, investigators were able to identify the mutations responsible for Miller syndrome that affected the brother and sister previously shown in Fig. 11.26. These siblings inherited one mutation in a gene called *DHOD* from their mother, and a different mutation in *DHOD* from their father; that is, they were compound heterozygotes. The Miller syndrome gene identification was unfortunately not actionable in terms of suggesting any kind of treatment. However, knowledge that *DHOD* is the Miller syndrome gene may nonetheless in the future help these patients to guide their reproductive decisions. If their partners do not carry a mutation in *DHOD*, none of their children would be affected by this condition.

In many other cases, the whole-genome sequence has not yet even allowed researchers to identify the responsible mutation. Perhaps the mutation lies in a poorly covered part of the sequence; perhaps the mutation lies outside of the protein-coding exome in sequences whose function has not yet been determined; perhaps the researchers made an incorrect (even though reasonable) assumption in one step of their bioinformatics analysis.

One lesson from Nic's story is the importance of databases and shared information. Critical steps in the identification of his disease-causing mutation depended on knowledge of variants from many other people's genomes, including unaffected controls and individuals affected by other genetic conditions sharing little or no phenotypic similarity with his own symptoms. The practice of human genetics is thus a giant bootstrapping operation: The more genomes that are sequenced, the more information is available to aid the analysis of all new genomes. The progress of human genetics therefore requires that databases be kept up to date and that their vast information be made accessible to all investigators using common methods of archiving. Allowing this degree of access, while preserving the confidentiality of the individuals whose genomes are cataloged, is a significant challenge for the future.

One of the most important databases for studies of human genetics is called **Online Mendelian Inheritance in Man** (**OMIM**; www.omim.org). OMIM is a catalog of human genes and the traits they control. The database lists the known variants in human genes that are associated with particular diseases or other traits and provides links to published research articles about these variants. Updated daily with new research findings, OMIM is an invaluable resource for researchers like those who figured out the genetic cause of Nic Volker's symptoms. This online database is so useful and easy to use that we encourage you to explore it on your own.

essential concepts

- High-throughput technologies allow parallel sequencing of millions of individual DNA molecules. These new methods are rapidly driving down the cost of whole-exome and whole-genome sequencing.
- Finding disease-causing mutations among the many DNA variations that distinguish individual genomes involves sequential filtering of information. These steps may involve deduction of likely transmission patterns, analysis of relatives' genomes, knowledge of similar genetic diseases, and predictions regarding a variant's effect on protein function.

WHAT'S NEXT

In this chapter and in Chapters 9 and 10, we have focused on the nucleotide content of genomes, particularly the 6 billion nucleotides organized into 46 chromosomes in each normal human diploid cell. In the next several chapters, we examine features of the chromosomes that allow these DNA sequences to function properly and to be transmitted from one generation to the next.

We begin by considering how in spite of the enormous complexity of DNA sequences, the DNA actually constitutes only about one-third of the total mass of a chromosome. The remainder of the chromosome is made of thousands of different types of proteins that help package and manage the information carried by DNA. These proteins have many roles. Certain proteins help compact the chromosomes to fit in the nucleus. Some proteins ensure that the chromosomal DNA is properly duplicated during each cell cycle, while others govern the distribution of chromosomes to daughter cells. Yet other proteins are responsible for regulating the availability of genes to the transcriptional machinery so that the genes can be expressed into proteins. In Chapter 12, we examine how proteins interact with DNA to generate the functional complexity of a chromosome.

SOLVED PROBLEMS

I. Genomic DNA from a woman's blood cells is PCR amplified by a single pair of primers representing a unique locus in the genome. The PCR products are then sequenced by the Sanger method, using one of the PCR primers as a sequencing primer. The following figure shows a trace of just part of the sequence read.

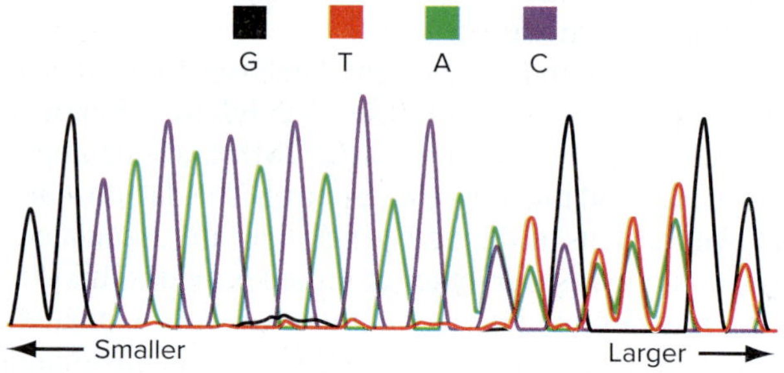

a. What kind of polymorphism is most likely represented?
b. With your answer to part (a) in mind, determine the woman's genotype at this locus. Indicate all nucleotides that can be read from both alleles and their 5′-to-3′ orientation.
c. What kind of molecular event was likely to have generated this polymorphism?
d. How would you know exactly where in the genome this locus is found?
e. What is another way in which you could analyze the PCR products to genotype this locus?
f. Suppose you wanted to genotype this locus based on single-molecule DNA sequencing of whole genomes as shown in Fig. 9.24. Would a single read suffice for genotyping the locus by this alternative method?

Answer

To solve this problem, you need to understand that PCR will simultaneously amplify both copies of a locus (one on the maternally derived chromosome and one on the paternally derived chromosome), as long as the primer can hybridize to both homologs as is usually the case. The DNA sequence trace has two nucleotides at several positions. This fact indicates that the woman must be a heterozygote and that the PCR is amplifying both alleles of the locus.

a. Notice that both alleles contain multiple repeats of the dinucleotide CA. The most likely explanation for the polymorphism is therefore that the locus contains an SSR polymorphism whose alleles have different numbers of CA repeats. One allele has six repeats; the second allele must have more CA units.

b. Writing out the first 14 nucleotides of both alleles is straightforward. If the assumption in part (a) is correct, then one allele should have more than six CA repeats. The trace shows evidence for two additional CA repeats in one allele at positions 15–18, for a total of eight CA repeats.

You can then determine the nucleotides beyond the repeats in the shorter allele by subtracting CACA from positions 15–18. The remaining peaks at these positions correspond to ATGT. Note that ATGT can also be found in the longer allele, but now at nucleotides 19–22, just past the two additional CACA repeats. You can determine the last four nucleotides in the shorter allele by subtracting ATGT from positions 19–22, revealing TAGG. The sequences of the two alleles of this SSR locus (indicating only one strand of DNA each) are thus:

Allele 1: 5′...GGCACACACACACAATGTTAGG...3′
Allele 2: 5′...GGCACACACACACACACAATGT...3′

c. The mechanism thought to be responsible for most SSR polymorphisms is stuttering of DNA polymerase during DNA replication.

d. You actually knew the location of this locus even before starting the experiment. This is because you design the PCR primers from knowledge of the entire human genome sequence.

e. The polymorphism involves a difference in the number of repeat units, and therefore the two alleles would produce PCR products that differ in length. You could genotype this locus by gel electrophoresis of the PCR products, as shown in Fig. 11.12.

f. Direct Sanger sequencing of a PCR product from genomic DNA produces a trace that includes both alleles. This is not true of single-molecule DNA sequencing techniques. You would require enough sequence runs from individual genomic DNA molecules to ensure that you could see both alleles if the person was a heterozygote.

II. It is difficult to obtain accurate recombination frequencies in humans because family sizes are small. An interesting way to circumvent this problem is to genotype individual sperm cells so as to obtain large data sets for linkage studies. The table that follows shows the genotype of four SNP loci from 20 single sperm that one man provided for this research. The genotypes were determined by microarray analysis of four SNP loci amplified from these samples by PCR. In the table, A, C, G, and T are the alleles of the SNPs (that is, the nucleotides on one strand) and a dash (–) means that no DNA corresponding to the locus was amplified from the sample.

SNP:	Locus 1	Locus 2	Locus 3	Locus 4
Sperm number				
1	G	C	—	T
2	G	A	G	C
3	G	C	G	C
4	G	C	G	T
5	G	A	—	C
6	G	C	—	T
7	G	A	G	C
8	G	A	—	C
9	G	A	G	T
10	G	C	—	T
11	G	C	—	T
12	G	C	G	T
13	G	A	—	C
14	G	A	—	C
15	G	C	G	T
16	G	A	G	C
17	G	C	—	T
18	G	A	—	T
19	G	A	G	C
20	G	C	G	T

a. Which SNP loci could be X-linked?

b. Which SNP loci could be on the Y chromosome?

c. Which SNP loci must be autosomal?

d. For any autosomal SNP loci, what is the sperm donor's genotype in somatic tissue?

e. Do any SNP loci appear to be linked to each other?

f. What is the distance between any two linked SNP loci?

Answer

SNP analysis by PCR is so sensitive that the single alleles present within individual sperm cells can be assayed, providing researchers with considerable information. You should also remember that a man's somatic cells have two copies of each autosome, one X chromosome, and one Y chromosome. Individual sperm will have one copy of each autosome and either an X or a Y chromosome.

a. For any X-linked SNP locus, half of the sperm cells will carry the same SNP allele, but the other half of the sperm would not have an X chromosome and would thus not yield a PCR product. Locus 3 shows this type of pattern.

b. Similarly, a locus on the Y chromosome would be found in only half the sperm, and all these sperm would have the same allele. Again, Locus 3 is a candidate for a Y-linked SNP. The data do not allow you to discriminate between an X or a Y chromosome location for Locus 3.

c. For an autosomal SNP locus, all the sperm will have one copy of the locus. Loci 1, 2, and 4 are thus autosomal.

d. If a man is homozygous for a single allele of an autosomal SNP locus, all the sperm he produces will have this one allele. If he is a heterozgote for two different alleles, approximately one-half of the sperm will have one allele and the other half of the samples will carry the other allele for the locus. The man's genotypes for the autosomal genes are SNP Locus 1: GG (homozygous); SNP Locus 2: CA (heterozygous); and SNP Locus 4: CT (heterozygous).

e. Alleles at linked loci will segregate together (end up in the same sperm) more than 50% of the time. This is true of the C allele of SNP Locus 2 and the T allele of Locus 4. The reciprocal alleles (A for Locus 2 and C for Locus 4) are also transmitted together more often than not. SNP Loci 2 and 4 are linked.

f. Sperm 3, 9, and 18 show evidence of recombination between alleles at Loci 2 and 4. Three out of 20, or 15%, of the sperm are recombinant. The distance between Loci 2 and 4 is therefore 15 cM.

PROBLEMS

Vocabulary

1. Choose the phrase from the right column that best fits the term in the left column.

a. DNA polymorphism	1. DNA element composed of short tandemly repeated sequences
b. phase	2. two different nucleotides appear at the same position in genomic DNA from different individuals
c. informative cross	3. arrangement of alleles of two linked genes in a diploid
d. ASO	4. location on a chromosome
e. SNP	5. a DNA sequence that occurs in two or more variant forms
f. DNA fingerprinting	6. a short oligonucleotide that will hybridize to only one allele at a chosen SNP locus
g. SSR	7. detection of genotype at a number of unlinked highly polymorphic loci
h. locus	8. allows identification of a gamete as recombinant or nonrecombinant
i. compound heterozygote	9. all exons in a genome
j. exome	10. individual with two different mutations in the same gene

Section 11.1

2. Would you characterize the pattern of inheritance of anonymous DNA polymorphisms as recessive, dominant, incompletely dominant, or codominant?
3. Would you be more likely to find single nucleotide polymorphisms (SNPs) in the protein-coding or in the noncoding DNA of the human genome?
4. A recent estimate of the rate of base substitutions at SNP loci is about 1×10^{-8} per nucleotide pair per gamete.
   a. Based on this estimate, about how many *de novo* mutations (that is, mutations not found in the genomes of your parents) are present in your own genome?
   b. Where and when did these *de novo* mutations in your genome most likely occur?
   c. It has been calculated that each sperm made in a 25-year-old man is the result on average of about 300 rounds of cell division, starting with the first mitotic division of the male zygote. In contrast, each mature oocyte found in a 5-month-old female human fetus is the result of about 25 rounds of division, starting with the first mitotic division of the female zygote. What bearing do these calculations have on the estimate of the rate of base substitutions in humans, and on your answer to part (b)?
5. If you examine Fig. 11.5 closely, you will note that in some regions, such as between nucleotides 116,870 K and 116,890 K, James Watson and Craig Venter share the same SNPs, and these regions are surrounded by others in which these two men do not share any SNPs. What does this fact say about the relationship between these two men, and how do you think this pattern of shared and unshared SNPs arose?
6. Approximately 50 million SNPs have thus far been recorded after the characterization of thousands of human genomes.
   a. About how many base pairs in the human genome are identical in these thousands of people?
   b. Do you think that many other SNPs exist among the human population? If so, why haven't they been found?
   c. Almost all of the SNP polymorphisms found to date are biallelic; that is, among all the genomes in the population studied to date, only two possible alleles can be found (for example, A and C). Provide a rough estimate for the number of triallelic SNP loci that could be found in the same group of humans (that is, the number of loci with three different alleles—for example, A, C, and T). At about how many loci would all four possible nucleotides be found among the human genomes studied to date?
7. Mutations at simple sequence repeat (SSR) loci occur at a frequency of 1×10^{-3} per locus per gamete, which is much higher than the rate of base substitutions at SNP loci (whose frequency is about 1×10^{-8} per nucleotide pair per gamete).
   a. What is the nature of SSR polymorphisms?
   b. By what mechanism are these SSR polymorphisms likely generated?
   c. Copy number variants (CNVs) also mutate at a relatively high frequency. Do these mutations occur by the same or a different mechanism than that generating SSRs?
   d. The SSR mutation rate is much higher than the mutation rate for new SNPs. Why then have geneticists recorded more than 50 million SNP loci but only about 100,000 SSR loci in human genomes?
8. Humans and gorillas last shared a common ancestor about 10 million years ago. Humans and chimps last shared a common ancestor about 6 million years ago. The table that follows shows the corresponding genomic region from two gorilla gametes, three chimpanzee gametes, and three human gametes.

a. Draw a cladogram similar to that in Fig. 11.6b to show the evolutionary relationships among these three species.

b. The data reveal six polymorphisms among these eight genomes, at positions 2 (A or G), 3 (A or T), 4 (G or T), 7 (C or T), 8 (C or T), and 9 (G or T). On your cladogram, indicate approximately when the mutations that produced each of these polymorphisms occurred. For each allele, state whether it is ancestral, derived, or that you can't tell.

c. Infer the sequences in (i) the last common ancestor of humans and chimpanzees, and (ii) the last common ancestor of all three species. Use a question mark (?) to represent any uncertainty.

Genome (haploid)	Gorilla	Chimpanzee	Human
1	CATGTCCTGA	CGAGTCCTGA	CAAGTCCTGA
2	CATGTCCTTA	CAAGTCCTGA	CAATTCCTGA
3		CAAGTCCCGA	CAATTCTTGA

9. In 2015, an international team of scientists assembled the complete genome sequences of two different woolly mammoths. Both specimens were discovered buried in the permafrost of Siberia, the coldest inhabited place on earth. Through radiocarbon dating, it was determined that one of the mammoths, found on Wrangel Island off the Siberian coast, died about 4000 years ago; the other mammoth, found in the town of Oimyakon, died about 45,000 years ago.

Analysis revealed that the genome sequences of these two animals differed significantly in the distribution of base pairs at which they are either homozygous or heterozygous. The Wrangel Island woolly mammoth had an extreme excess of *runs of homozygosity* (*ROHs*), regions in which the animal was homozygous for all of the base pairs. About 23.4% of the Wrangel Island animal's genome was composed of ROHs that were greater than 500 kb in length; some of these ROHs were in excess of 5 Mb long. In contrast, only 0.83% of the Oimyakon animal's genome consisted of ROHs longer than 500 kb.

a. Explain how polymorphisms are detected when sequencing a genome. How would researchers know, for any particular base pair, whether a genome is homozygous or heterozygous?

b. What does the extreme excess of ROHs in the Wrangel Island mammoth genome suggest about that animal's parents?

c. The Wrangel Island woolly mammoth is thought to have belonged to the last population on earth before the species went extinct about 4000 years ago. The answer to part (b) suggests one possible reason for the woolly mammoth's extinction. Explain.

Section 11.2

10. Using PCR, you want to amplify an approximately 1 kb exon of the human autosomal gene encoding the enzyme phenylalanine hydroxylase from the genomic DNA of a patient suffering from the autosomal recessive condition phenylketonuria (PKU).

a. Why might you wish to perform this PCR amplification in the first place, given that the sequence of the human genome has already been determined?

b. Calculate the number of template molecules that are present if you set up a PCR reaction using 1 nanogram (1×10^{-9} grams) of chromosomal DNA from blood cells as the template. Assume that each haploid genome contains only a single gene for phenylalanine hydroxylase and that the molecular weight of a base pair is 660 grams per mole. The haploid human genome contains 3×10^9 base pairs.

c. Calculate the number of PCR product molecules you will obtain if you perform 25 PCR cycles and the yield from each cycle is exactly twice that of the previous cycle. What would be the mass of these PCR products taken together?

11. Which of the following set(s) of primers a–d could you use to amplify the following target DNA sequence, which is part of the last protein-coding exon of the *CFTR* gene?

```
5′ GGCTAAGATCTGAATTTTCCGAG … TTGGGCAATAATGTAGCGCCTT 3′
3′ CCGATTCTAGACTTAAAAGGCTC … AACCCGTTATTACATCGCGGAA 5′
```

a. 5′ GGAAAATTCAGATCTTAG 3′;
5′ TGGGCAATAATGTAGCGC 3′

b. 5′ GCTAAGATCTGAATTTTC 3′;
3′ ACCCGTTATTACATCGCG 5′

c. 3′ GATTCTAGACTTAAAGGC 5′;
3′ ACCCGTTATTACATCGCG 5′

d. 5′ GCTAAGATCTGAATTTTC 3′;
5′ TGGGCAATAATGTAGCGC 3′

12. The previous problem raises several interesting questions about the design of PCR primers.

a. How can you be sure that the two 18-nucleotide-long primers you chose as your answer to Problem 11 will amplify only an exon of the *CFTR* gene, but no other region, from a sample of human genomic DNA?

b. Primers used in PCR are generally at least 16 nucleotides long. Why do you think the lower limit would be approximately 16?

c. Suppose one of the primers in your answer to Problem 11 had a mismatch with a single base in the genomic DNA of a particular individual. Would you be more likely to obtain a PCR product from this genomic DNA if the mismatch was at the 5′ end or at the 3′ end of the primer? Why?

13. You want to make a recombinant DNA in which a PCR product amplified from the human genome is inserted into a plasmid vector. The polylinker of this vector includes recognition sites for the enzymes *Eco*RI (5′ G^AATTC 3′) and *Bam*HI (5′ G^GATCC 3′). (The ^ symbolizes the cut site in the DNA.) PCR primers that could amplify the fragment of human DNA are: 5′ GCTACTTCGCGTATTCCA 3′ and 5′ CCCAAGTCCTAGCCGATA 3′.
    a. Describe in detail how these primers would need to be modified to create a fragment of the human genome flanked by *Eco*RI sticky ends so that this fragment could be cloned easily into the plasmid vector. You will need to consider the fact that most restriction enzymes, including *Eco*RI, cannot cut DNA if the restriction site is directly at the end of the DNA molecule; the restriction enzyme recognition site must be at least six base pairs distant from the end.
    b. Describe a potential feature of the PCR-amplified region of the human genome that could prevent you from using the strategy you described in part (a).
    c. Now describe how the primers must be modified to create a human DNA fragment with an *Eco*RI-compatible single-stranded overhang at one end and a *Bam*HI-compatible overhang at the other end. (Two possibilities exist; you need to describe only one. Assume that a *Bam*HI site also must be at least six base pairs from the end of the DNA.) Why might you want to make such a fragment?

14. You sequence a PCR product amplified from a person's genome, and you see a double peak such as that seen in Fig. 11.11b. Most of the time, this result indicates that the person is a heterozygote for a SNP at that position. But it is also possible that the result is due to a mistake in DNA replication during the PCR amplification, with DNA polymerase misincorporating the wrong nucleotide.
    a. If you saw an artifactual double peak in the sequence trace, did the mistake happen in the first few rounds of PCR amplification or in the last few rounds?
    b. *Whether or not* you see a double peak, is it more likely that a mistake would happen in the first few rounds of PCR amplification or in the last few rounds?
    c. Given that mistakes can happen during PCR amplification, what could you do to be sure of a person's genotype? Why would this degree of certainty be difficult to achieve if you were doing preimplantation genotyping of embryos?
    d. PCR relies on heat-stable DNA polymerases from thermophilic bacteria that grow in hot springs. The DNA polymerase originally used for PCR, from the bacterium *Thermus aquaticus,* lacks the 3′-to-5′ exonuclease found in other DNA polymerases such as that from *E. coli* (review Fig. 7.9). Why do scientists now most often use DNA polymerase from a different thermophilic bacterium (*Pyrococcus furiosa*) that does contain this exonuclease function?

15. Problem 8 shows three different sequences of the same autosome in human populations. These sequences are each from a single chromatid. You know this to be true because the PCR amplifications were from individual haploid gametes. If you wanted to obtain the same information by PCR amplification of genomic DNA from somatic cells, the problem would be somewhat more complicated because the starting cells are diploid. Each PCR product to be sequenced would thus actually be amplified from two homologous chromosomes. You could still verify the existence of the three different haploid sequences shown in Problem 8 by analyzing the somatic genomic DNA from as few as two people (if they happened to be the right people).
    a. Indicate the diploid genotypes of two people from whom you could identify these three different haploid sequences. Account for all 10 nucleotides in the sequences. Three possible correct answers exist; you need to show only one.
    b. If you PCR amplified DNA from somatic genomic DNA from a person with one particular genotype, you would not be able to conclude that their genome contains any of these three sequences. What is the genotype of this person? Explain why you could not reach this conclusion.

16. The trinucleotide repeat region of the Huntington disease locus (*HD*) in six individuals is amplified by PCR and analyzed by gel electrophoresis as shown in the following figure; the numbers to the right indicate the sizes of the PCR products in bp. Each person whose DNA was analyzed has one affected parent.
    a. Which individuals are most likely to be affected by Huntington disease, and in which of these people is the onset of the disease likely to be earliest?
    b. Which individuals are least likely to be affected by the disease?
    c. Consider the two PCR primers used to amplify the trinucleotide repeat region. If the 5′ end of one of these primers is located 70 nucleotides upstream of the first CAG repeat, what is the maximum distance downstream of the last CAG repeat at which the 5′ end of the other primer could be found? [Assume that the diagram shows the largest HD^+ allele possible (that is, 35 CAG repeats).]

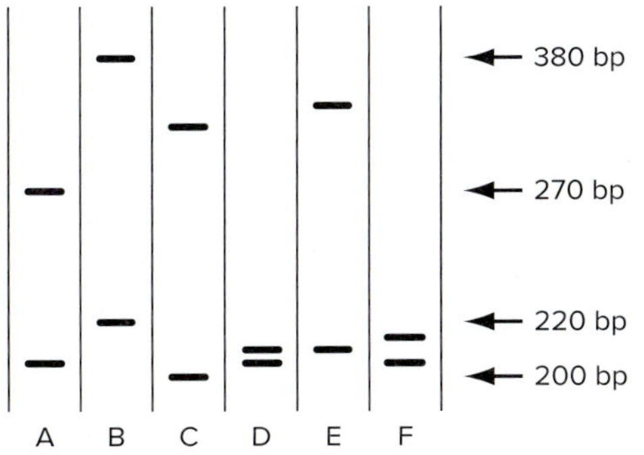

17. Sperm samples were taken from two men just beginning to show the effects of Huntington disease. Individual sperm from these samples were analyzed by PCR for the length of the trinucleotide repeat region in the *HD* gene. In the graphs that follow, the horizontal axes represent the number of CAG repeats in each sperm, and the vertical axes represent the fraction of total sperm of a particular size. The first graph shows the results for a man whose mutant *HD* allele (as measured in somatic cells) contained 62 CAG repeats; the man whose sperm were analyzed in the second graph had a mutant *HD* allele with 48 repeats.

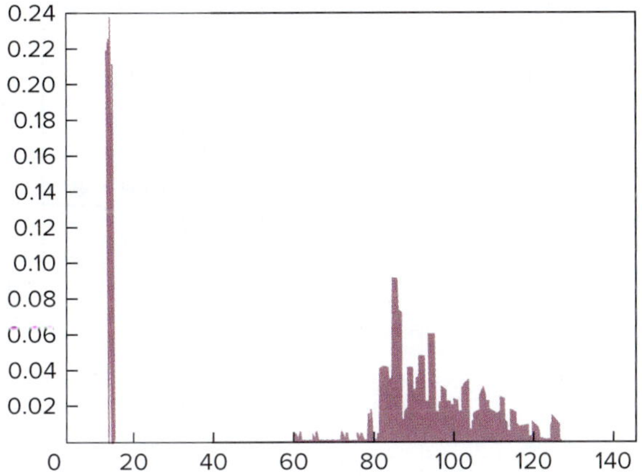

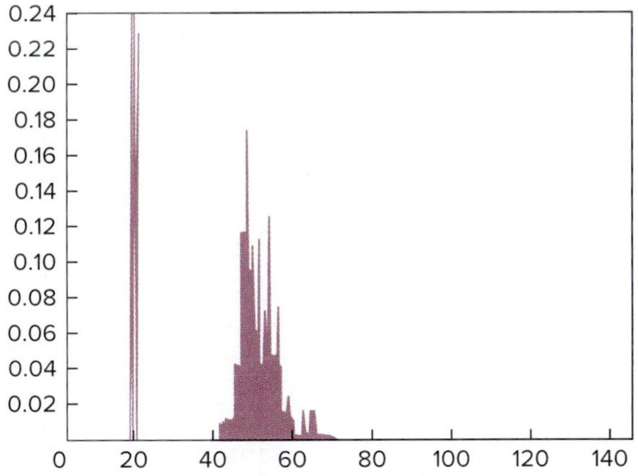

a. What is the approximate CAG repeat number in the HD^+ alleles from both patients?

b. Assuming that these results indicate a trend, what can you conclude about the processes that give rise to mutant *HD* alleles? In what kinds of cells do these processes take place?

c. How do these results explain why approximately 5–10% of Huntington disease patients have no family history of this condition?

d. Predict the results if you performed this same PCR analysis on single skin cells from each of these patients instead of single sperm.

Section 11.3

18. In 1993, the courts for the first time accepted plant DNA as evidence in a murder trial. The accused defendant owned a pickup truck in which the police found a few seed pods from a Palo Verde, the state tree of Arizona. The murdered woman was found abandoned in the Arizona desert. How could a prosecutor use DNA from the seed pods to build a strong case against the defendant?

19. a. It is possible to perform DNA fingerprinting with SNPs instead of SSRs as DNA markers, but in general you would need to examine more SNP markers than the 13 SSRs used in the CODIS database to be sure of a match. Explain why.

    b. DNA fingerprinting has been used to verify pedigrees of valuable animals such as show dogs, racing greyhounds, and thoroughbred horses. However, the technology is much harder to apply in these cases than it is in forensic applications for humans. In particular, many more DNA markers must be examined in domesticated animals to establish the identity or close familial relationship of two DNA samples. Why would you need to look at more polymorphic loci in these animals than you would in humans?

20. On July 17, 1918, Tsar Nicholas II; his wife the Tsarina Alix; their daughters Olga, Tatiana, Maria, and Anastasia; their son, the Tsarevitch (Crown Prince) Alexei; and four loyal retainers were murdered by Bolshevik revolutionaries. The bodies were not recovered for many years, fueling legends that Grand Duchess Anastasia had escaped, and allowing a woman named Anna Anderson to claim that she was Anastasia. In 1991 and in 2007, two mass graves with a total of nine skeletal remains were unearthed at Ekaterinburg in Russia's Ural Mountains. The table that follows presents partial DNA fingerprint analysis (using only five SSR loci and the sex chromosome marker Amel) of these skeletons. Entries separated by commas indicate alleles (number of repeating units).

    a. What is the most likely identification for each skeleton? (*Note:* You cannot differentiate among any of the daughters based on this information alone.)

Locus	A	B	C	D	E	F	G	H
Amel	XX	XX	XY	XX	XX	XX	XY	XX
D3S1358	17,18	17,18	14,17	16,17	16,18	13,15	14,16	17,18
D5S818	12,12	—	12,12	12,12	12,12	11,12	12,12	12,12
D13S317	11,11	11,11	—	11,11	11,11	10,14	11,12	11,11
D16S539	11,11	11,11	11,14	11,14	—	10,12	11,14	9,11
D8S1179	13,16	15,16	13,15	13,16	16,16	13,14	15,16	15,16

b. Three PCR reactions failed to yield PCR products. If the reactions had worked properly, what alleles would you expect to see in each case?

c. Are any of the daughters identical twins?

d. What kind of evidence could you obtain from the skeletons to differentiate among the daughters?

e. How do these DNA fingerprints repudiate the claims of Anna Anderson?

The DNA fingerprint data in the table are certainly consistent with the idea that some of these skeletons were members of the Tsar's family, but they do not prove the hypothesis. To investigate further, forensic scientists obtained blood samples from Prince Philip (the consort of Queen Elizabeth II of Great Britain) and compared his DNA fingerprint to those obtained from the skeletal remains in Russia. A family genealogy is provided here. The results validated that the Tsar's family was indeed interred in these graves.

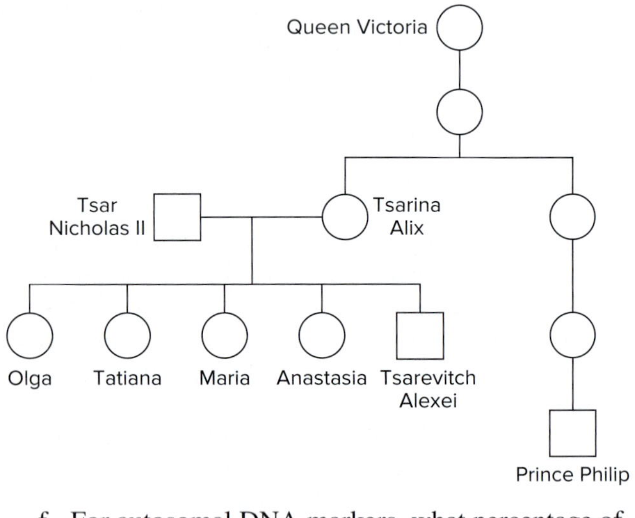

f. For autosomal DNA markers, what percentage of alleles in the Tsarina's skeleton should match with alleles in Prince Philip's genome?

g. For autosomal DNA markers, what percentage of alleles in the Tsarevitch's skeleton should match with alleles in Prince Philip's genome?

h. A question for genealogy *aficionados*: What is Prince Philip's relationship with the Tsarina?

21. The figure that follows shows DNA fingerprint analysis of the genomic DNA from semen associated with a rape (***) and from mouth swabs (somatic cells) of individuals 1–4. This analysis involves the PCR amplification of six SSR loci, each from a different (nonhomologous) chromosome. All PCR primers used are 20 nucleotides long; the primers for each locus have fluorescent tags in a locus-specific color. In the gel, some bands are thicker because relatively more of the corresponding PCR product was obtained. The figure has dots aligned on both sides

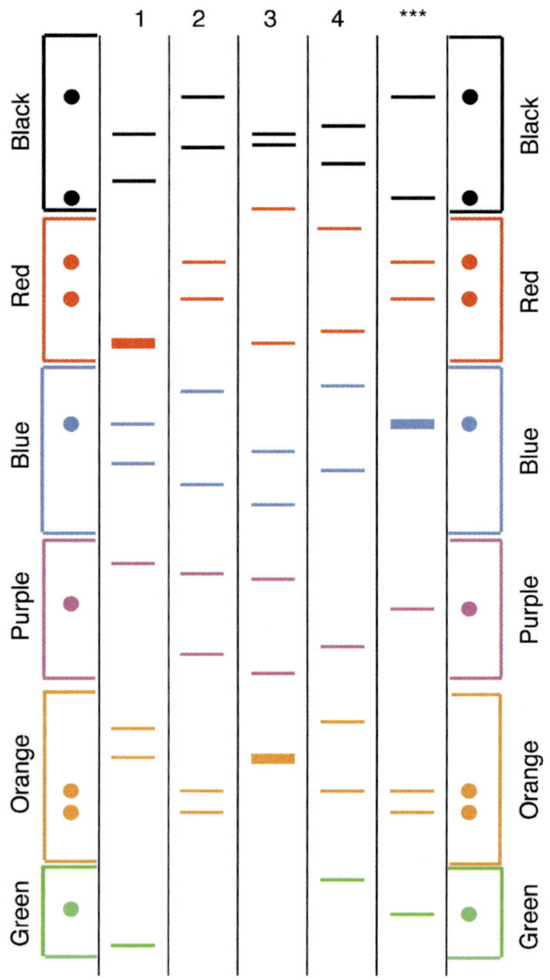

that you can use to find the critical bands, using the edge of a piece of paper as a guide.

a. Sperm are haploid, but the semen sample shows two different-sized PCR products for certain loci. How is this possible?

b. Is any locus on the X chromosome? If so, identify it.

c. Is any locus on the Y chromosome? If so, which one?

d. Explain why these results demonstrate that none of the four individuals is the rapist. What pattern would you expect by analyzing mouth swab DNA from the rapist?

e. Do these results nonetheless provide any information that could help catch the rapist? If so, be as specific as possible.

f. The two orange bands amplified by PCR from the semen are 200 and 212 bp long. How many tandem repeats of the SSR repeat unit are found in the two alleles of this locus in the rapist's genomic DNA? (Assume that the PCR products are the shortest possible and that the repeat unit for this locus is TCCG.)

22. Microarrays were used to determine the genotypes of seven embryos (made by *in vitro* fertilization) with regard to sickle-cell anemia. Each pair of squares in the accompanying figure represents two ASOs, one specific for the $Hb\beta^A$ allele (A) and the other for the $HB\beta^S$ allele (T), attached to a chip of silicon and hybridized with fluorescently labeled PCR product from a single cell from one of the embryos. The hybridizations were performed at three different temperatures (80°C, 60°C, and 40°C) as shown.

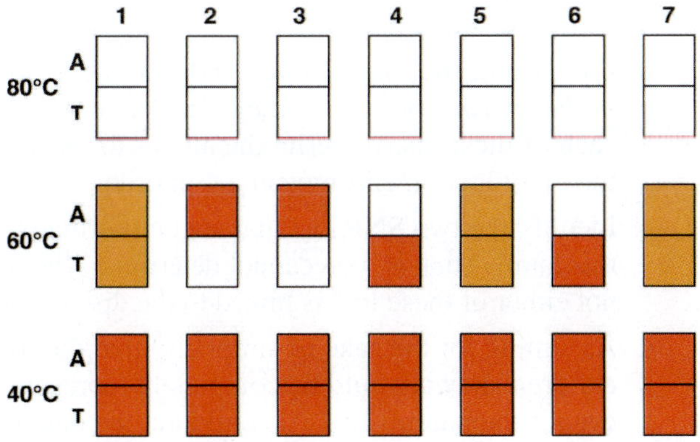

a. Why do you think the PCR step is needed for this microarray analysis?

b. Make a sketch of the location in genomic DNA of the PCR primers relative to the sickle-cell mutation. Indicate the 5′-to-3′ polarities of all DNA molecules involved.

c. Why is no hybridization seen at 80°C?

d. Why do you see strong hybridization of all genomic DNA probes to both ASOs at 40°C?

e. What are the genotypes of the seven embryos? Which of these embryos would you choose to implant into the mother's uterus to avoid the possibility that the child would have sickle-cell anemia?

23. A partial sequence of the wild-type $Hb\beta^A$ allele is shown here (the top strand is the RNA-like coding strand, and the location of the disease-causing mutation is underlined):

5′ ATGGTGCACCTGACTCCTGAGGAGAAGTCGCCG 3′
3′ TACCACGTGGACTGAGGACTCCTCTTCAGCGGC 5′

and the sickle-cell allele $HB\beta^S$ sequence is:

5′ ATGGTGCACCTGACTCCTGTGGAGAAGTCGCCG 3′
3′ TACCACGTGGACTGAGGACACCTCTTCAGCGGC 5′

Design two 21-nucleotide-long ASOs that could be attached to a silicon chip for the microarray analysis performed in Problem 22. Two possibilities exist for each ASO; you only need to show one possibility for each.

24. a. In Fig. 11.17b, PCR is performed to amplify genomic DNA for genotyping on microarrays. This PCR reaction requires only a single primer, but normally, PCR requires two primers. Why does only a single primer suffice in this case?

b. Again in Fig. 11.17b, genomic DNA was cut with a restriction enzyme before its PCR amplification. What kind of restriction enzyme would be most effective for this purpose: Would it create sticky ends or blunt ends? Would it recognize a site made of 4 bp, 6 bp, or 8 bp?

25. The following figure shows a partial microarray analysis for members of a nuclear family. The eight SNP loci examined are evenly spaced at about 10 Mb intervals on chromosome 4, and they are shown on the microarray in their actual order on this chromosome. For the time being, focus your attention only on the two parents and ignore whether they are affected or unaffected.

a. Write out the complete genotype for all the DNA markers in both parents.

b. The microarray data indicate that one SNP locus has three alleles in this family. Which one?

c. How would you know that these loci are in fact on chromosome 4 and are about 10 Mb apart?

d. About what percentage of the total length of chromosome 4 is present in the region between DNA markers 1 and 8? (Chromosome 4 is 191 Mb long; it is the fourth largest in the human genome.)

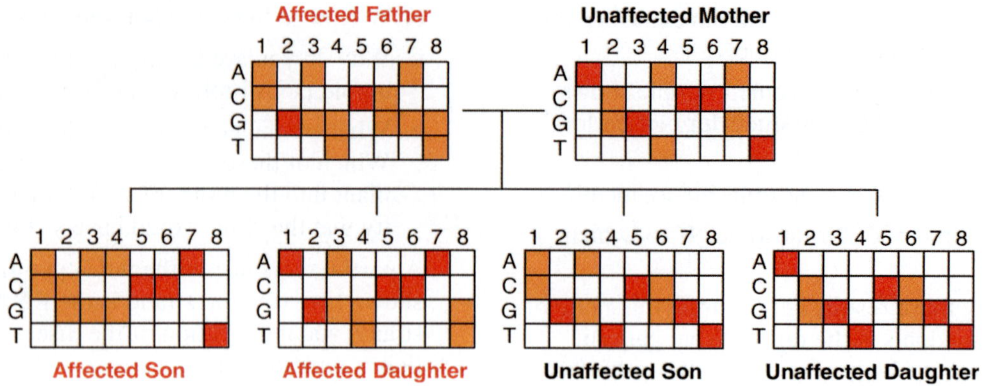

26. Scientists were surprised to discover recently that marmosets, a kind of monkey, are often *chimeric*—they have cells that originated from two different zygotes. Marmoset chimeras occur in fraternal twins. While in their mother's womb, these twins can share cells through the blood supply because their placentas are fused. Any organ or tissue, including the germ line, can be chimeric.

    To determine if a particular male fathered any of four baby marmosets from three different mothers, DNA fingerprinting of four SSR loci was performed using genomic DNA from hairs. In the accompanying diagram, the thickness of the bands correlates with the amount of DNA present. You can use the dots on both sides of the diagram, along with a ruler or paper edge, to help compare the positions of the bands.

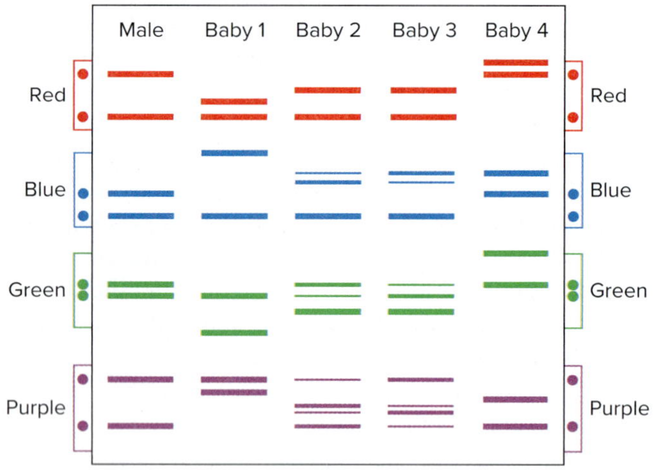

a. What is the maximum number of different alleles of any one SSR locus that the hair cells of any one marmoset could have?
b. Which of the four babies could have been fathered by the male whose DNA fingerprint is shown?
c. Identify the pair of chimeric fraternal twins.
d. Does any evidence exist for chimerism in the father of the twins? in the mother?
e. Explain how the biological mother of a baby marmoset can be its genetic uncle. (*Note:* This is not necessarily the situation for any of the baby marmosets in the figure.)

Section 11.4

27. The microarray shown in Problem 25 analyzes genomic DNA from a nuclear family in which the father, one son, and one daughter have rare, late onset polycystic kidney disease; while the mother, the second son, and the second daughter are unaffected. As stated in Problem 25, the eight SNP loci examined are evenly spaced at about 10 Mb intervals on chromosome 4, and they are shown on the microarray in their actual order on this chromosome.
    a. Is the allele responsible for the disease dominant or recessive with respect to wild type? Is the disease gene autosomal or X-linked?
    b. For each of the four siblings, indicate the genotype *of the sperm* from which they were created. For each of these sperm, write the alleles for each of the eight loci on chromosome 4 in order.
    c. Identify the two SNP loci that are uninformative in this family (that is, you cannot determine whether or not either of these loci is linked to the disease gene).
    d. Assuming for the sake of simplicity that the four children shown would be completely representative even if the parents had 100 children, the data in the figure indicate that one locus is unlinked to the disease gene. Which one?
    e. The microarray results indicate that during meiosis in the father, two different crossovers occurred in the region including the disease gene and the SNP loci that are genetically linked to it. Draw a map of chromosome 4 showing the locations of

the disease gene, the linked SNP loci, and the two crossovers. Your map should indicate any uncertainties in these positions.

f. Diagram the location and arrangement (phase) of all alleles of all genes that are on the two chromosomes in the father's diploid genome.

28. The figure that follows shows the pedigree of a family in which a completely penetrant, autosomal dominant disease is transmitted through three generations, together with microarray analysis of each individual for a biallelic SNP locus (the alleles are C and T).

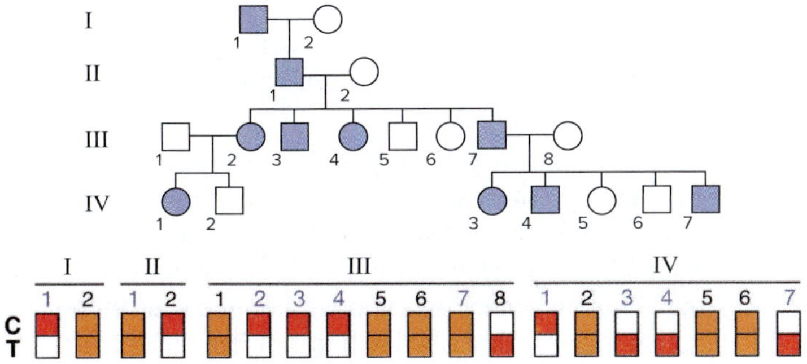

a. Do the data suggest the existence of genetic linkage between the SNP locus and the disease locus? If so, what is the estimated genetic distance between the two loci?

b. Calculate the maximum Lod score for linkage between the SNP and the disease locus for this pedigree. What does this value of the Lod score signify?

29. One of the difficulties faced by human geneticists is that matings are not performed with a scientific goal in mind, so pedigrees may not always provide desired information. As an example, consider the following matings (W, X, Y, and Z):

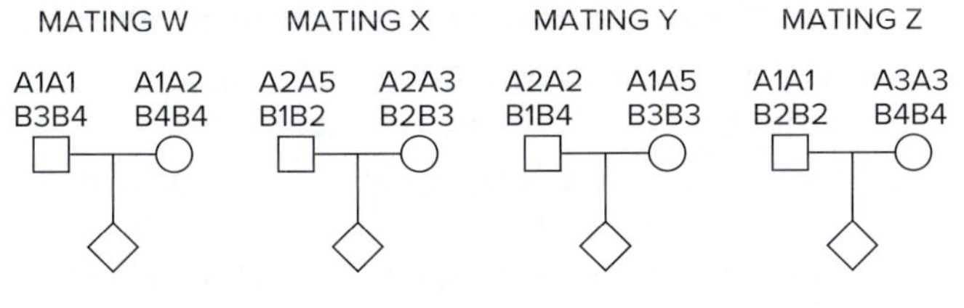

a. Which of these matings are informative and which noninformative for testing the linkage between anonymous loci A and B? (A1 and A2 are different alleles of locus A, B1 and B2 are different alleles of locus B, etc.) Explain your answer for each mating.

b. Is locus A more likely to be a SNP or an SSR? What about locus B? Explain.

30. Now consider a mating between consanguineous people involving a recessive genetic disease. The figures that follow show the genotypes of these two people at four SNP loci (1–4).

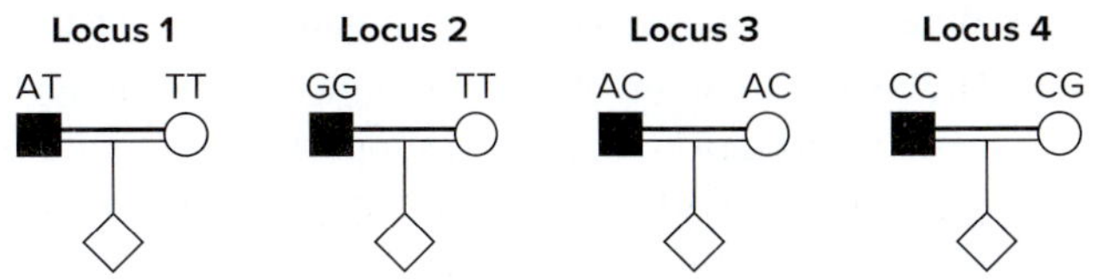

a. For which of these loci is it possible to obtain information about the linkage of the SNP to the disease gene? Explain your answer for each locus and describe any special conditions that may apply.

b. For any of the loci for which the mating is potentially informative, how would you tell whether the child is the product of a recombinant or nonrecombinant gamete? (That is, how could you solve the phase problem?) Be as specific as possible.

31. The pedigree shown in Fig. 11.22 was crucial to the identification of the Huntington disease gene *HD*, which is located on chromosome 4.

a. The data show that the DNA marker G8 is clearly linked to *HD*. For the large majority of the people in the pedigree with Huntington disease, which allele of G8 (A, B, C, or D) did they inherit, along with the dominant disease-causing allele of *HD*, on the copy of chromosome 4 from their affected parent?

b. How many people in the pedigree can you categorize absolutely as the product of parental or recombinant gametes from their affected parent, without making any assumptions at all (including the assumption of linkage)?

c. If you now make the assumption that G8 and *HD* are linked, how many of the people in this pedigree must be the product of a recombinant gamete from their affected parent?

d. Based solely on the data from this pedigree, what would be the best estimate for the map distance between G8 and *HD?*

e. Considering your answers to parts (b) through (d), calculate the maximum Lod score. The pedigree contains 47 people resulting from informative matings. (Note that $0^0 = 1$.) What does this Lod score signify?

32. You have identified a SNP marker that in one large family shows no recombination with the locus causing a rare hereditary autosomal dominant disease. Furthermore, you discover that all afflicted individuals in the family have a G base at this SNP on their

mutant chromosomes, while all wild-type chromosomes have a T base at this SNP. You would like to think that you have discovered the disease locus and the causative mutation but realize you need to consider other possibilities.

a. What is another possible interpretation of the results?
b. How would you go about obtaining additional genetic information that could support or eliminate your hypothesis that the base-pair difference is responsible for the disease?

Problems 33 and 34 show that you can make predictions about a child's genotype by genotyping linked markers even if you don't directly examine the disease-causing mutation. This method can be valuable for diseases showing high allelic heterogeneity if the linkage is extremely tight.

33. The pedigrees indicated here were obtained with three unrelated families whose members express the same disease caused by a completely penetrant dominant allele. The disease allele is linked at a distance of 10 cM from an SSR marker locus with three alleles numbered 1, 2, and 3. The SSR alleles present within each living person's genotype are indicated below the pedigree symbol. The phenotypes of the newly born labeled individuals—A, B, C, and D—are unknown.

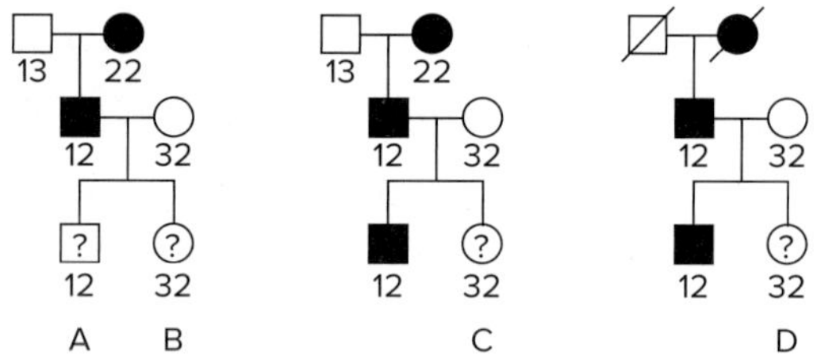

a. What is the probability of disease expression in each of these newborn babies?
b. Why would a human geneticist be unlikely to use this SSR marker for diagnosis of the genetic disease?

34 Approximately 3% of the population carries a mutant allele at the *CFTR* gene responsible for the recessive disease cystic fibrosis. A genetic counselor is examining a family in which both parents are known to be carriers for a *CFTR* mutation. Their first child was born with the disease, and the parents have come to the counselor to assess whether the new fetus inside the mother's womb is also diseased, is a carrier, or is homozygous wild type at the *CF* locus. DNA samples from each family member and the fetus are tested by PCR and gel electrophoresis for an SSR marker within one of the *CFTR* gene's introns. The following results are obtained:

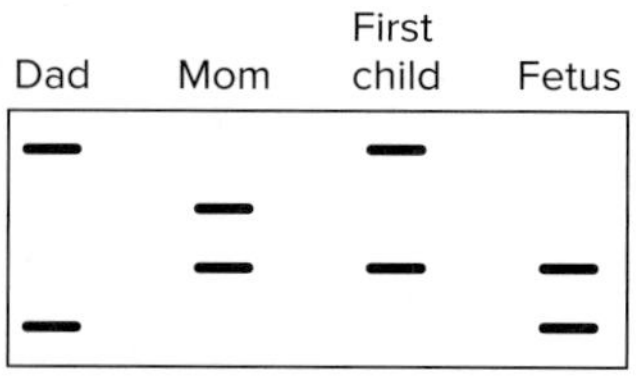

a. What is the probability that the child who will develop from this fetus will exhibit the disease?
b. When this child grows up, what is the probability that any one of her own children will be afflicted with the disease?
c. The cystic fibrosis gene displays extensive allelic heterogeneity: More than 1500 different mutations of the *CFTR* gene have been shown to be associated with cystic fibrosis worldwide. With this fact in mind, why might human geneticists choose to test the fetus in the indirect manner described in this problem rather than focusing directly on the mutations that actually caused the disease in the first child?

35. The drug *ivacaftor* has recently been developed to treat cystic fibrosis in children with the rare G551D mutant allele of *CFTR*.

a. Do you think that ivacaftor would be effective only in patients homozygous for the G551D mutation, or might it work as well in compound heterozygotes in which one copy of chromosome 7 had G551D and the other copy a different allele of *CFTR*, such as the more prevalent allele ΔF508? (The protein encoded by G551D folds up properly and inserts into the cell membrane, but is inefficient in chloride ion transport. Ivacaftor increases the efficiency of G551D's ion transport. The ΔF508 protein does not fold up properly and therefore does not get inserted into the cell membrane.)
b. Why do you think ivacaftor would be more effective in children than in older cystic fibrosis patients?
c. The scientists who developed ivacaftor had a model for cystic fibrosis: a line of cells that grow in culture and that are homozygous for G551D. These cells accumulate mucus at their surfaces that prevent cilia (tiny hairs on the outside of cells) from beating. Explain how the scientists could use this disease model to screen for drugs that would be effective against G551D-associated cystic fibrosis.
d. Ivacaftor is used in combination with an even newer drug called *lumacaftor* to treat individuals homozygous for the most common CF allele, ΔF508. Lumacaftor helps the ΔF508 mutant

CFTR protein fold properly so that it can insert in the cell membrane. Why do you think that neither lumacaftor nor ivacaftor alone are effective in treating ΔF508 homozygotes, while the combination of both drugs is effective?

Section 11.5

36. In the high-throughput DNA sequencing protocol shown in Fig. 11.24:
    a. What is the purpose of adding poly-A to fragments of single-stranded genomic DNA, and why is the poly-A added to the 3′ end of these fragments?
    b. Why at the end of every synthesis cycle do you need to remove the fluorescent tag on the incorporated nucleotide?
    c. Why do the incorporated nucleotides have a blocking group, and why does this blocking group need to be removed each cycle?
37. A researcher sequences the whole exome of a patient suffering from Usher syndrome, a rare autosomal recessive condition that is nonetheless the leading cause for simultaneous deafness and blindness. The exome sequence does not show homozygosity for any polymorphisms different from the human RefSeq.
    a. How could the researcher examine the data already gathered to try to find the disease gene, assuming the sequence is accurate?
    b. If the attempt described in part (a) was unsuccessful, the researcher might contemplate sequencing the patient's whole genome. What are the potential pitfalls of this strategy?
38. As explained in the text, the cause of many genetic diseases cannot yet be discerned by analyzing whole-exome/genome sequences. But in some of these seemingly intractable cases, important clues can be obtained by looking at mRNAs or proteins, rather than at the DNA.
    a. As you will see in more detail in later chapters, it is possible to use single-molecule methods to sequence cDNA copies of millions of mRNA molecules from any particular tissue cheaply. How could you sometimes use such information to find a disease gene? When would this information be noninformative?
    b. A technique called *Western blotting* allows you to examine any protein for which you have an antibody; it is possible to see differences in size or amount of that protein. How could you sometimes use such information to find a disease gene? When would this information be noninformative?
39. Figure 11.26 portrayed the analysis of Miller syndrome through the sequencing of four complete genomes: those of a brother and sister both affected by the disease, and of both their parents.
    a. Researchers made the assumption that Miller syndrome is a recessive trait. Could Miller syndrome instead be due to a dominant mutation? If so, what scenarios would make this possible?
    b. Why is it highly unlikely that Miller syndrome in this family is due to *de novo* mutations that occurred in the germ line of the mother, or of the father, or of both parents? Describe a scenario based on your understanding of cell divisions in human ovaries or testes (see Figs. 4.18 and 4.19) that make the *de novo* mutation hypothesis at least theoretically possible even if very unlikely.
    c. On Fig. 11.26b, indicate the location on chromosome 16 closest to the *DHOD* gene at which recombination took place during meiosis in one of the parents of the Miller syndrome patients. In which parent did this recombination occur?
    d. Do the number of crossovers you see in Fig. 11.26b fit previous estimates that in the human genome, 1 centiMorgan corresponds to about 1 Mb? Chromosome 16 is about 90 Mb long; chromosome 17 is about 81 Mb long.
    e. How could researchers use all the sequence data from this family to estimate the per-nucleotide rate of mutation in humans?
40. A research paper published in the summer of 2012 presented a method to obtain the whole-genome sequence of a fetus without any invasive procedure such as amniocentesis that could on rare occasions cause miscarriage. This new technique is based on the fact that some fetal cells leak into the mother's bloodstream and then break down, releasing their DNA. Assume that exactly 10% of the DNA fragments in the mother's blood serum come from the fetus, while the remaining 90% of the DNA fragments in the serum come from the mother's genome.

    The investigators collected cell-free DNA from a pregnant woman's bloodstream and subjected it to an advanced high-throughput sequencing method. The table at the end of this problem looks at seven unlinked loci; the number of *reads* of particular alleles (identified by Greek letters) are shown. You should assume for the sake of simplicity that all numerical differences are statistically significant (even though actual data are never this clean).
    a. Determine whether each locus is autosomal, X-linked, or Y-linked.
    b. Describe the diploid genomes of the mother and fetus by using Greek letters for the alleles, or a dash (–) if no Greek letter is appropriate.
    c. Is the fetus male or female?

d. At an eighth locus, 1500 reads of a single type of sequence were found. Provide a possible explanation for this result, being as specific as possible.

Locus		Sequences	Number of reads
1	α:	TCTTTGGTAAACGCAAG	1000
2	α:	GTACCGGAGGCAGCCTC	500
	β:	GTACCGGCGGCAGCCTC	500
3	α:	AGCCATTGCGGATCCGA	950
	β:	AGCTATTGCGGATCCGA	50
4	α:	GGGGCCTTATGATAAGG	50
5	α:	CAGTTCCTGGAGTTGTA	550
	β:	CAGTTCATGGAGTTGTA	450
6	α:	GCAGCCCGTGCTGTTAA	500
	β:	GCAGCCCGTGCTGTCAA	450
7	α:	CACTCAGTCCTACGGAC	500
	β:	CACTCGGTCCTACGGAC	450
	γ:	CACTCAGTCCTAAGGAC	50

41. Table 11.2 and Fig. 11.27 together portray the search for the mutation causing Nic Volker's severe inflammatory bowel disease. Neither of Nic's parents had the condition, so geneticists narrowed their investigation by focusing on rare variants that showed a recessive pattern and those on the X chromosome.
   a. For candidate variants on an autosome, would the researchers have looked only for variants for which Nic is homozygous? Explain.
   b. Apart from the recessive and X-linked hypotheses, do any other possible explanations exist for Nic's condition?
   c. The causative mutation was pinpointed by analyzing only Nic's exome, because at the time of these investigations, whole-genome or whole-exome sequencing was too expensive to perform on his parents. How could you determine inexpensively whether or not this mutation occurred *de novo* in the germ line of one of his parents (that is, during the formation of the particular egg or sperm that produced Nic)? Your answer should not involve whole-genome or whole-exome sequencing.
42. The human RefSeq of the entire first exon of a gene involved in Brugada syndrome (a cardiac disorder characterized by an abnormal electrocardiogram and an increased risk of sudden heart failure) is:

   5′ CAACGCTTAGGATGTGCGGAGCCT 3′

   The genomic DNA of four people (1–4), three of whom have the disorder, was subjected to single-molecule sequencing. The following sequences represent all those obtained from each person. Nucleotides different from the RefSeq are underlined.

   Individual 1:
   5′ CAACGCTTAGGATGTGCGGAGCCT 3′
   and
   5′ CAACGCTTAGGATGTGCGGAGACT 3′

   Individual 2:
   5′ CAACGCTTAGGATGTGAGGAGCCT 3′

   Individual 3:
   5′ CAACGCTTAGGATGTGCGGAGCCT 3′
   and
   5′ CAACGCTTAGGATGGCGGAGCCT 3′

   Individual 4:
   5′ CAACGCTTAGGATGTGCGGAGCCT 3′
   and
   5′ CAACGCTTAGGATGTGTGGAGCCT 3′

   a. The first exon of the RefSeq copy of this gene includes the start codon. Write as much of the amino acid sequence of the encoded protein as possible, indicating the N-to-C polarity.
   b. Are any of these individuals homozygotes? If so, which person and what allele?
   c. Is the inheritance of Brugada syndrome among these individuals dominant or recessive?
   d. Is Brugada syndrome associated with allelic heterogeneity?
   e. Are any of these individuals compound heterozygotes?
   f. Do the data show any evidence for locus heterogeneity?
   g. Which person has normal heart function?
   h. For each variant from the RefSeq, describe: (i) what the mutation does to the coding sequence; and (ii) whether the variation is a loss-of-function allele, a gain-of-function allele, or a wild-type allele.
   i. For each variant, indicate which of the following terms apply: null, hypomorphic, hypermorphic, nonsense, frameshift, missense, silent, SNP, DIP, SSR, anonymous.
   j. Is the function of this gene haploinsufficient? Explain.
43. Mutations in the *HPRT1* gene in humans result in at least two clinical syndromes. Consult OMIM (www.omim.org) by querying HPRT1; you will only need to look briefly at the top three hits (files #300322, 300323, and 308000).
   a. What is the full name of the HPRT1 enzyme?
   b. On which chromosome is the *HPRT1* gene located?

c. Mutations in *HPRT1* are associated with two different syndromes. What are these syndromes? For each, answer the following questions: (i) What are the symptoms associated with the syndrome? (ii) Is the mutant allele that causes the syndrome dominant, recessive, codominant, or incompletely dominant with respect to the normal allele, or do special conditions apply? (iii) Is the syndrome associated with a loss-of-function or a gain-of-function disease allele? (iv) Does the syndrome display allelic heterogeneity? (v) Does the syndrome display locus heterogeneity? (*Note:* You do not need to understand everything in the OMIM entries to answer these questions.)

44. We think of people as having two genomes (one from their mother and one from their father) that are identical in every somatic cell and in germ cell precursors. However, with every DNA replication and thus every cell division, a genome has an opportunity to mutate. When a mutation occurs in a single cell, all of that cell's mitotic descendants will thus have a mutation that the other cells in the body do not have. Humans are in this sense *mosaic*—we are a patchwork of cells with somewhat different genotypes. Explain how human mosaicism could complicate the process of pinpointing a recessive or dominant disease-causing mutation by positional cloning or by genome sequencing.

chapter 12

The Eukaryotic Chromosome

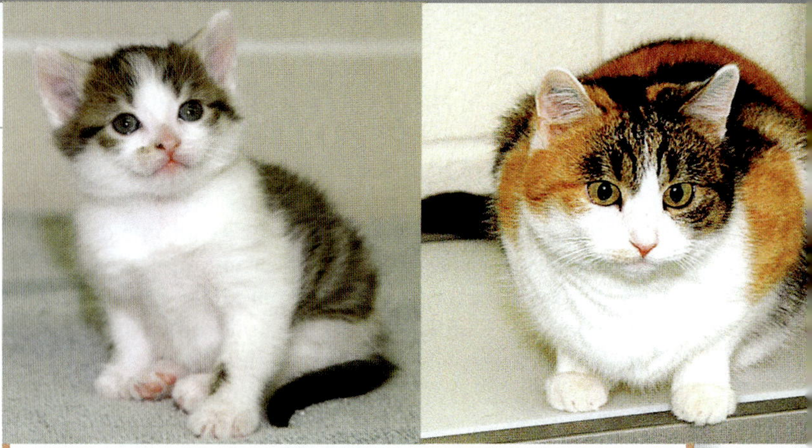

CC, the cloned cat (left), *and her nuclear donor Rainbow* (right) *share the same chromosomes yet have distinctive appearances.*
(left): © Texas A&M University/AP Photo
(right): © Alpha/ZUMAPRESS/Newscom

chapter outline

IN DECEMBER 2001, veterinarians delivered by caesarean section the world's first cloned pet, the kitten CC (a play on the words *Copy Cat* and *Carbon Copy*; accompanying figure). CC was the result of a nuclear transfer experiment in which the nucleus from a female calico cat named Rainbow (*right*) was injected into an egg (from an unnamed donor cat) whose own nucleus had been removed previously. The reconstituted egg was then implanted into the uterus of a surrogate mother. CC turned out to be a perfectly normal cat who has survived well past the age of 12 and has mothered three of her own normal kittens.

All of the cells you see in both CC and Rainbow have the same nuclear DNA because they are the mitotic descendants of a single cell: the zygote that became Rainbow. Yet even though the nuclear DNAs of all these cells are identical, all of the cells clearly are not. Some cells have differentiated into eyes, others into whiskers, yet others to skin, and so on. And despite the same DNA, CC and Rainbow are dissimilar in many phenotypes, ranging from the color patterns of their fur to the disposition of their characters.

The packaging of DNA into chromosomes underlies much of the biology you see on display in these photographs. Such packaging allows the billions of base pairs making up the cat genome to fit into a cell's nucleus, and the packaging is a prerequisite for the faithful copying and distribution of the genome through innumerable mitotic cell divisions. Some of the proteins on the chromosomes tell genes when to turn on and off, and are thus responsible for cell differentiation. Other macromolecules (proteins and RNAs) associated with chromosomes are responsible for X chromosome inactivation, which accounts for many differences seen in the coat color patterns of Rainbow and CC.

In this chapter, we examine the structure and function of the eukaryotic chromosome. One general theme emerges from our discussion. Chromosomes have a versatile, dynamic structure that supports their many functions in the packaging, replication, segregation, and expression of the information in a single long molecule of DNA.

12.1 Chromosomal DNA and Proteins

learning objectives

1. Diagram the DNA components of a chromosome, including the polarity of strands.
2. Contrast histone and nonhistone proteins in terms of structure and function.

When viewed under the light microscope, chromosomes change shape, character, and position as they pass through the cell cycle. During interphase, they look like tangled masses of spaghetti. By metaphase of mitosis, they appear as paired bars (the two sister chromatids) in the middle of the spindle apparatus. In this section, we describe what chromosomes are made of. Then, in succeeding sections of this chapter, we explain how these chromosomal components interact to produce the observed metamorphoses of structure.

Each Chromosome Is Composed of a Single Long Molecule of DNA

Researchers learned from physical analyses that each chromosome (or after replication, each chromatid) within a cell nucleus contains one long linear molecule of double-stranded DNA. In one early study, they placed chromosomal DNA between two cylinders, stretched the DNA by rotating one of the cylinders, and measured the DNA's rate of recoil. Shorter molecules recoil faster than longer ones. When the investigators applied this technique to the DNA in a *Drosophila* cell, the length of the longest DNA molecule measured corresponded to the amount of DNA in the largest chromosome. This chromosome must therefore contain a single linear molecule of DNA.

You have examined previously (in Chapter 11) general aspects of the way genes are organized along the DNA molecule making up a typical eukaryotic chromosome. Chromosomes also display another organizational feature of importance: Substantial stretches of noncoding repetitive DNA such as **simple sequence repeats** (*SSRs*) and **transposable elements** are concentrated in specific chromosomal regions, particularly at **centromeres** and **telomeres**. The repetitive sequences at these locations are crucial for several aspects of chromosome biology to be discussed in later sections of this chapter.

Chromosomes Contain Histone Proteins and Nonhistone Proteins

By itself, DNA does not have the ability to fold up small enough to fit in the cell nucleus. For sufficient compaction, it depends on interactions with two categories of proteins: *histones* and *nonhistone chromosomal proteins.* **Chromatin** is the generic term for any complex of DNA and protein found in a cell's nucleus. A chromosome is a piece of chromatin that contains (prior to S phase) a single DNA molecule and behaves as a unit during cell division.

Although chromatin is roughly one-third DNA, one-third histones, and one-third nonhistone proteins by weight, it also contains significant amounts of RNA. The roles of chromatin-associated RNAs are not well understood in general, but later in this chapter we will describe the function of a specific RNA molecule called *Xist* in controlling gene expression.

Histone proteins

Discovered in 1884, **histones** are relatively small proteins with a preponderance of the basic, positively charged amino acids lysine and arginine. The histones' strong positive charge enables them to bind to and neutralize the negatively charged DNA throughout the chromatin. Histones make up half of all chromatin protein by weight and are classified into five types of molecules: H1, H2A, H2B, H3, and H4. The last four types—H2A, H2B, H3, and H4—form the core of the most rudimentary DNA packaging unit—the **nucleosome**—and are therefore referred to as **core histones.** (We will examine the role of these histones in nucleosome structure momentarily.)

All five types of histones appear throughout the chromatin of nearly all diploid eukaryotic cells, and they are very similar in all eukaryotes. In the H4 proteins of pea plants and calves, for example, all but two of 102 amino acids in the polypeptide sequence are identical. That histones have changed so little throughout evolution underscores the importance of their contribution to chromatin structure.

Nonhistone proteins

The remaining half of the mass of protein in eukaryotic cell chromatin is not composed of histones. Rather, it consists of thousands of different kinds of **nonhistone chromosomal proteins.** The chromatin of a diploid genome contains from 200 to 2,000,000 molecules of each kind of nonhistone protein.

Not surprisingly, this large variety of proteins fulfills many different functions. Some nonhistone proteins play a purely structural role, helping to package DNA into more complex structures. The proteins that form the structural backbone, or *scaffold,* of the chromosome fall in this category (**Fig. 12.1**). Other nonhistone proteins, such as DNA polymerase, are active in replication. Still others are crucial for chromosome segregation: For example, the motor proteins of kinetochores help move chromosomes along

Figure 12.1 The chromosome scaffold. When this human chromosome was gently treated with detergents to remove the histones and some nonhistone proteins, a dark *scaffold* composed of the remaining nonhistone proteins became visible in the shape of the two sister chromatids. Loops of DNA freed by the detergent treatment surround the scaffold.

© Dr. Don Fawcett/J.R. Paulson, U.K. Laemmli/Science Source

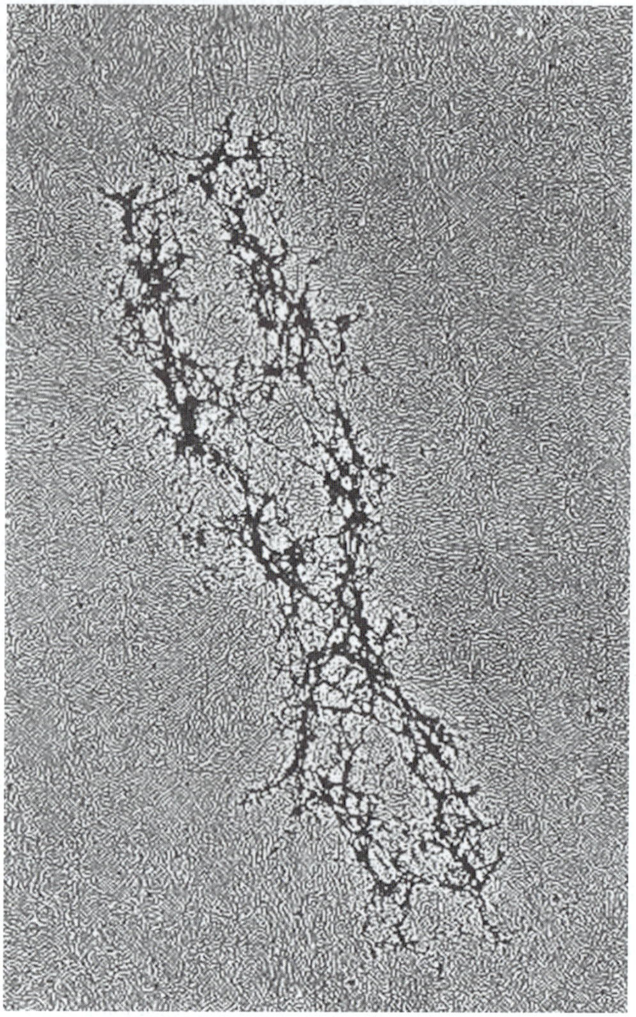

Figure 12.2 Kinetochore proteins. Some nonhistone proteins are required for chromosome movements along the spindle during cell division. Here, chromosomes are stained in *blue* and the nonhistone protein CENP-E in *red*. CENP-E molecules at the kinetochores help move sister chromatids toward the spindle poles during anaphase.

© Daniel A. Starr/University of Colorado

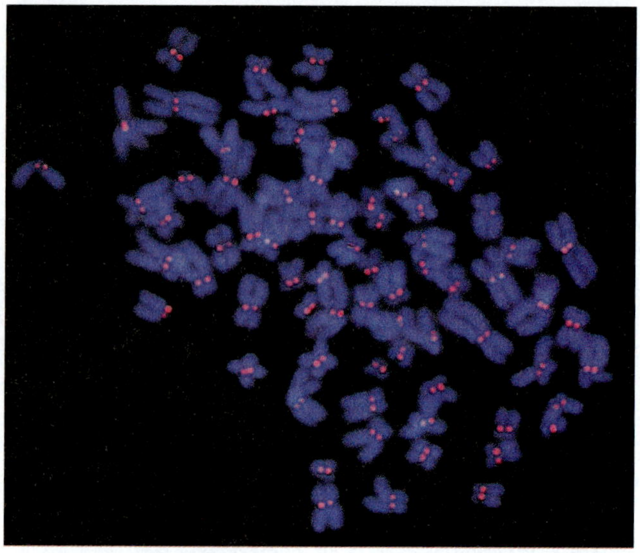

the spindle apparatus and thus expedite the transport of chromosomes from parent to daughter cells during mitosis and meiosis (**Fig. 12.2**).

By far the largest class of nonhistone proteins constitute those which foster or regulate transcription during gene expression. Mammals carry more than 5000 different proteins of this kind. By interacting with DNA, these proteins influence when, where, and how frequently genes are transcribed.

essential concepts

- Each eukaryotic chromosome contains (prior to its replication) a single molecule of linear, double-stranded DNA, without breaks or changes of polarity.
- The *histone proteins* help package DNA into *chromatin*; the *nonhistone proteins* function in chromosome structure, replication, and segregation, as well as in the control of gene expression.

12.2 Chromosome Structure and Compaction

learning objectives

1. Diagram the structure of a nucleosome.
2. Describe nucleosome supercoiling and its relationship to the radial loop–scaffold model of chromatin packaging.
3. Summarize the process of detecting G bands in a chromosome and how these bands are used in locating genes.
4. Describe FISH analysis and its application in finding specific DNA sequences in a chromosome.

TABLE 12.1 Levels of Chromosome Compaction

Mechanism	Status	What It Accomplishes
Nucleosome	Confirmed by crystal structure	Condenses naked DNA 7-fold to a 100 Å fiber
Supercoiling	Hypothetical model (although the 300 Å fiber predicted by the model has been seen in the electron microscope)	Causes additional 6-fold compaction, achieving a 40- to 50-fold condensation relative to naked DNA
Radial loop–scaffold	Hypothetical model (preliminary experimental support exists for this model)	Through progressive compaction of 300 Å fiber, condenses DNA to rodlike mitotic chromosomes that are 10,000 times more compact than naked DNA

Stretched out in a thin, straight thread, the DNA of a single human cell would be 6 feet (about 2 m) in length. This is, of course, much longer than the cell nucleus in which the genome must be contained; the diameter of the average human cell nucleus is only about 6 μm (6×10^{-6} m). Several levels of compaction enable the DNA to fit inside the nucleus (**Table 12.1**). First, the winding of DNA around histones forms small *nucleosomes*. Next, tight coiling gathers nucleosomes together into higher-order structures. Additional levels of compaction, which researchers do not yet understand, produce the metaphase chromosomes observable in the microscope.

The Nucleosome Is the Fundamental Unit of Chromosome Packaging

The electron micrograph of chromatin in **Fig. 12.3** shows long, nub-studded fibers bursting from the nucleus of a human blood cell. The nucleosomes resemble beads on a string, with the beads having a diameter of about 100 Å and the string a diameter of about 20 Å (1 Å = 10^{-10} m = 0.1 nm). The 20 Å string is DNA. **Figure 12.4a** illustrates how DNA wraps around histone cores to form the chromatin fiber's observed beads-on-a-string structure.

Each bead is a nucleosome containing roughly 160 bp of DNA wrapped around a core composed of eight histones—two each of H2A, H2B, H3, and H4, arranged as shown in Fig. 12.4a. The 160 bp of DNA wrap twice around this core histone octamer. An additional 40 or so base pairs form **linker DNA,** which connects one nucleosome with the next.

Figure 12.3 Nucleosomes. In the electron microscope, nucleosomes look like beads along a string of DNA.

© Dr. Barbara A. Hamkalo

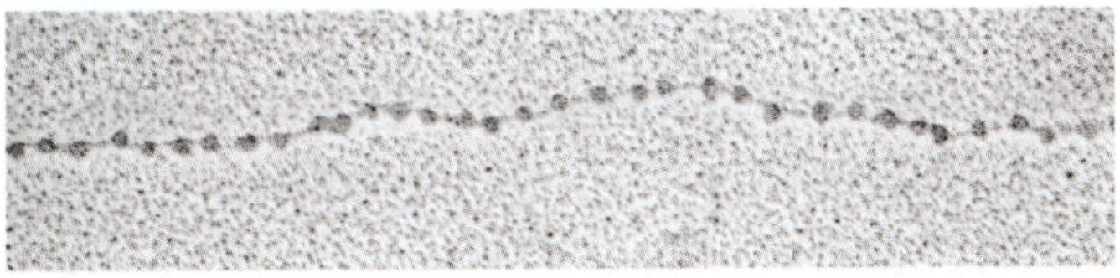

Figure 12.4 Nucleosome structure. (a) The DNA in each nucleosome wraps twice around a core made of two copies each of the histones H2A, H2B, H3, and H4. Histone H1 associates with the DNA as it enters and leaves the nucleosome. **(b)** Nucleosome structure as revealed by X-ray crystallography. The DNA (*orange*) bends sharply at several places as it wraps around the core histone octamer (*blue* and *turquoise*).

b: © Dr. Gerard J. Bunick, Oak Ridge National Laboratory

(a) Schematic diagram of a nucleosome

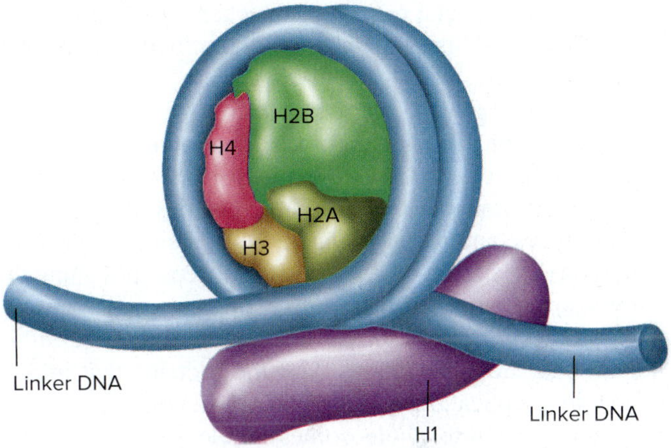

(b) Nucleosome structure at high resolution

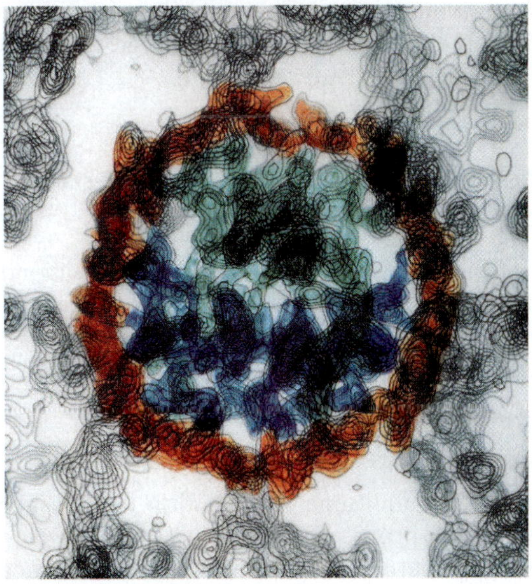

Histone H1 lies outside the core, associating with DNA entering and leaving the nucleosome (Fig. 12.4a). When investigators use specific chemical reagents to remove H1 from the chromatin, some DNA unwinds from each nucleosome, but the nucleosomes do not fall apart; about 140 bp remain wrapped around each core.

Scientists can crystallize the nucleosome cores and subject the crystals to X-ray diffraction analysis. The pictures have led to the model of nucleosome structure just described, and they also indicate that the DNA does not coil smoothly around the histone core (**Fig. 12.4b**). Instead, the DNA bends sharply at some positions and barely at all at others. Because the sharp bending may occur only with some DNA sequences and not others, base sequence helps dictate preferred nucleosome positions along the chromosome.

Duplication of the basic nucleosomal structure occurs in conjunction with DNA replication. Synthesis of the four basic histone proteins increases during S phase of the cell cycle so as to incorporate histones onto the newly replicated DNA. Additional proteins mediate the assembly of nucleosomes. Special regulatory mechanisms tightly coordinate DNA and histone synthesis so that both occur at the appropriate time.

The spacing and structure of nucleosomes correlate with genetic function. The nucleosomes of each chromosome are not evenly spaced, but they do have a particular arrangement along the chromatin. This arrangement varies among different cell types, and it can change even in a single cell when conditions are altered. The spacing of nucleosomes along the chromosome matters because DNA in the regions between nucleosomes is more readily available than the DNA within nucleosomes for interactions with the proteins that initiate expression, replication, and further compaction.

Packaging into nucleosomes condenses naked DNA about sevenfold. With this condensation, the 2 m of DNA in a diploid human genome shortens to approximately 0.25 m (a little less than a foot) in length. This is still much too long to fit in the nucleus of even the largest cell, and additional compaction is required.

Higher-Order Packaging Condenses Chromosomes Further

Many of the details of chromosomal condensation beyond the nucleosome remain unknown, but researchers have proposed several models to explain the different levels of compaction (see Table 12.1).

Nucleosome supercoiling

One model of additional compaction beyond nucleosomal winding proposes that the 100 Å nucleosomal chromatin supercoils into a 300 Å superhelix, achieving a further sixfold chromatin condensation. Support for this model comes in part from electron microscope images of 300 Å fibers that contain about six nucleosomes per turn (**Fig. 12.5a**). Whereas the 100 Å fiber is one nucleosome in width, the 300 Å fiber appears to be three beads wide. Histone H1 likely plays a special role in formation of the superhelix, because removal of some H1 causes the 300 Å to unwind to 100 Å, while adding back H1 reverses this process.

Although electron microscopists can actually see the 300 Å fiber, they do not know its exact structure. Higher levels of compaction are even less well understood.

The radial loop–scaffold model

This model proposes that several nonhistone proteins bind to chromatin every 60–100 kb and tether the super-helical, nucleosome-studded 300 Å fiber into structural loops (**Fig. 12.5b**). These proteins may gather the loops into daisylike rosettes and then compress the rosette centers into a compact bundle (**Fig. 12.5c**). This proposal, known as the **radial loop–scaffold model,** offers a simple explanation of chromosome packaging. Progressive levels of chromosome compaction involving nucleosome formation, nucleosome supercoiling into the 300 Å fiber, looping of the fibers, gathering of loops into rosettes, and rosette compression have the potential to give rise to the highly condensed, rodlike shapes we see as mitotic chromosomes.

Several pieces of biochemical and micrographic evidence support the radial loop–scaffold model. For example, metaphase chromosomes from which experimenters have extracted all the histones still maintain their familiar X-like shapes (see Fig. 12.1). The proteinaceous scaffold that remains includes proteins such as the helicase enzyme topoisomerase II, which maintains DNA winding (review Fig. 6.23), and protein complexes called **condensins** that help condense interphase chromosomes into metaphase chromosomes. Moreover, electron micrographs of mitotic chromosomes extracted in this way show loops of chromatin at the periphery of the chromosomes (**Fig. 12.6**), as predicted by the model.

Despite bits and pieces of experimental evidence, studies that directly confirm or reject the radial loop–scaffold model have not yet been completed. Thus, the loops and scaffold concept of higher-order chromatin packaging remains a hypothesis. The hypothetical status of this higher-order compaction model contrasts sharply with nucleosomes, which are entities that investigators have isolated, crystallized, and analyzed in detail.

Giemsa Staining Reveals Reproducible Chromosome Banding Patterns

We have just seen that different levels of packaging compact the DNA in human metaphase chromosomes

Figure 12.5 Models of higher-order chromosome packaging. **(a)** Artist's conception (*top*) and electron micrographs (*bottom*) contrasting the 100 Å fiber (*left*) with the 300 Å fiber (*right*). **(b)** The radial loop–scaffold model for higher levels of compaction. According to this model, the 300 Å fiber is first drawn into loops of 60–100 kb of DNA (*blue*) tethered by nonhistone scaffold proteins (*brown* and *orange*). **(c)** Additional nonhistone proteins might gather several loops together into daisylike rosettes and then compress the rosettes into bundles, forming metaphase chromosomes.

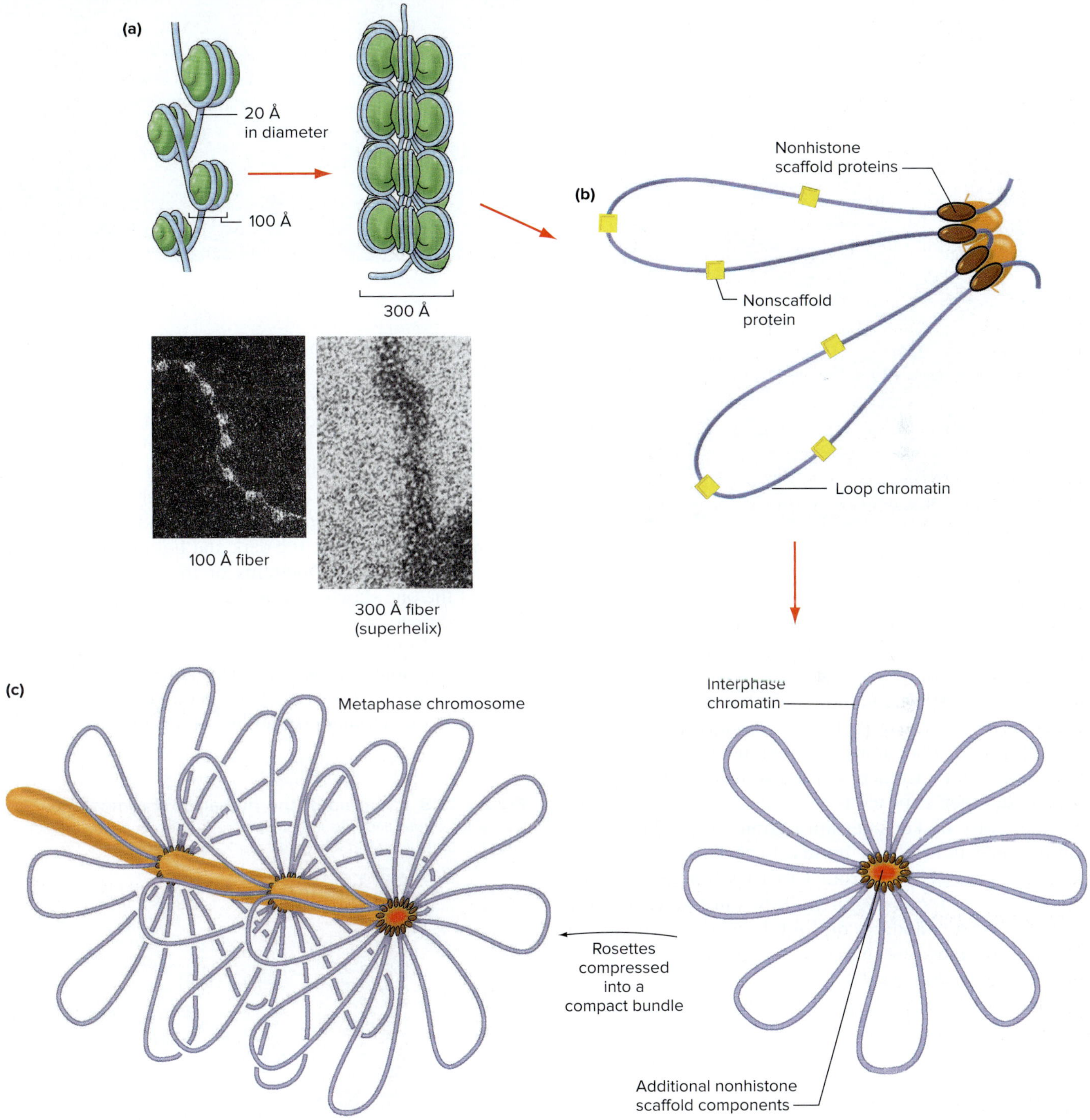

Figure 12.6 Experimental support for the radial loop–scaffold model. A close-up of part of the image in Fig. 12.1, this electron micrograph shows long DNA loops emanating from the protein scaffold at the *bottom* of the picture.

10,000-fold (see Table 12.1). With this amount of compaction, the centromere region and telomeres of each chromosome become visible. We have also seen (in Chapter 4) that various staining techniques reveal a characteristic banding pattern, size, and shape for each metaphase chromosome, establishing a karyotype. In *G-banding,* for instance, chromosomes are first gently heated and then exposed to Giemsa stain; this DNA dye preferentially darkens certain regions to produce alternating dark and light **G bands.** Each G band is a very large segment of DNA from 1 to 10 Mb in length, containing many loops. With high-resolution G-banding techniques, a standard diploid human karyotype of 46 chromosomes is seen to contain hundreds of dark and light G bands (**Fig. 12.7**).

The biochemical basis of banding is not yet understood. Most molecular geneticists think the bands produced by Giemsa staining probably reflect an uneven packaging of loops determined in some way by the spacing and density of short, repetitive DNA sequences along the chromosomes. Regardless of the underlying cause, every time a chromosome replicates, its banding pattern is faithfully reproduced. The fact that banding patterns are so highly reproducible from one generation to the next indicates they

Figure 12.7 Human chromosomes examined by high-resolution G-banding.

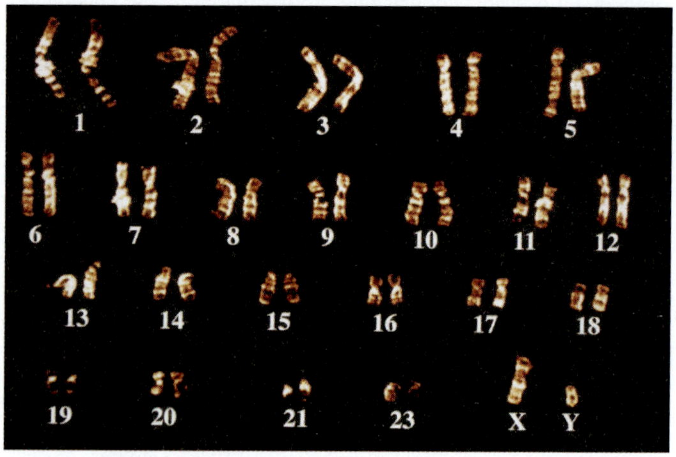

are an intrinsic property of each chromosome, initially determined by the DNA sequence itself.

The reproducibility of this pattern means that geneticists can designate the chromosomal location of a gene by describing its position in relation to the bands on the p (short; after the French *petit*) or q (long; for *queue*, the French word for *tail*) arm of a particular chromosome. For this purpose, the p and q arms are subdivided into regions, and within each region, the dark and light bands are numbered consecutively. Diagrams of the banding patterns, such as the one shown in **Fig. 12.8,** are called **idiograms.** As an example, the X-linked genes for color blindness reside at q27-qter, which indicates they are located on the X chromosome's long (q) arm somewhere between the beginning of the seventh band in the second region and the end of the telomere (terminus, or ter; Fig. 12.8).

Figure 12.8 Idiogram of the human X chromosome. Genes for color blindness in humans are located in a small region near the tip of the long arm (q) of the X chromosome.

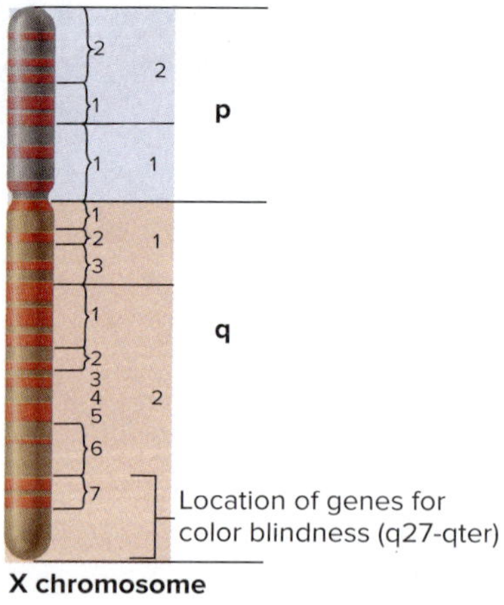

Fluorescent *In Situ* Hybridization (FISH) Helps Geneticists Characterize Genomes

Scientists are often faced with a problem of resolution when understanding the relationship of specific DNA sequences to the genome as a whole. Karyotypes show the G-banded chromosomes constituting the whole genome, but obviously at a resolution much lower than that of individual nucleotide pairs. In contrast, it is hard to step back from the mass of information derived from whole-genome sequencing to obtain a global view of genome organization. A technique called **fluorescent *in situ* hybridization (FISH)** provides a convenient bridge between the low resolution of a karyotype and the ultra-high resolution of a complete genomic sequence. In essence, FISH allows investigators to find the locations of specific DNA sequences with respect to the chromosomes in a karyotype.

As the name implies, the fundamental basis of FISH is inherent in nucleotide complementarity. Researchers first obtain cells in mitotic metaphase and then drop the cells onto a glass microscope slide. The cells are then subjected to treatments that successively burst the cells and nuclei open, spread the chromosomes apart, fix the chromosomes on the slide, and gently denature the chromosomal DNA. This latter denaturation step is performed in a way that preserves the overall chromosomal structure, even though the double helixes separate into single strands at numerous points. In a separate reaction, the researchers label a purified DNA sequence with a fluorescent tag, making a *DNA probe*. The probe DNA is denatured into single strands by heating, and the probe is then applied to the chromosomes on the plate. The probe will hybridize only with chromosomal regions that are complementary in nucleotide sequence, and the researchers can identify these regions by looking in a fluorescence microscope.

In one use of FISH, the probe is a short, defined sequence such as a cDNA clone. The results are fluorescent spots showing the location of the corresponding gene in the genome (**Fig. 12.9a**). Historically, this method was of considerable importance in verifying that the original draft of the Human Genome Project was properly assembled.

In a second variation of FISH called **spectral karyotyping (SKY)**, the probes are made from multiple DNAs that originated from positions scattered along the length of individual chromosomes. The probe for chromosome 1 is labeled with a mix of fluorescent tags (*dyes* or *fluors*) that together glow in one color, the probe for chromosome 2 is tagged with a different mix of fluors that yields a different color, and so forth, so that each of the 24 human chromosomes in a SKY karyotype can be recognized easily by its color (**Fig. 12.9b**). Chapter 13 will demonstrate that both of these FISH techniques remain important for characterizing chromosomal rearrangements such as deletions, duplications, inversions, and translocations that may cause genetic diseases.

Figure 12.9 Fluorescent *in situ* hybridization. (a) FISH analysis of a human cell's chromosomes. The *yellow* spots show where a probe made from a single gene hybridizes to the two sister chromatids on each of two homologous chromosomes. Because the two sister chromatids are extremely close together in this preparation, only one *yellow* spot appears on each homolog. **(b)** Spectral karyotyping (SKY). Probes made from DNA along the length of each chromosome are labeled with fluorescent tags of different colors.

a: © Patrick Landmann/Science Source; b: Courtesy Dr. Thomas Ried. Head, Cancer Genomic Section, Genetics Branch/CCR/NCI/NIH. Image created by Dr. Hesed Padilla-Nash

(a) FISH locates a human gene

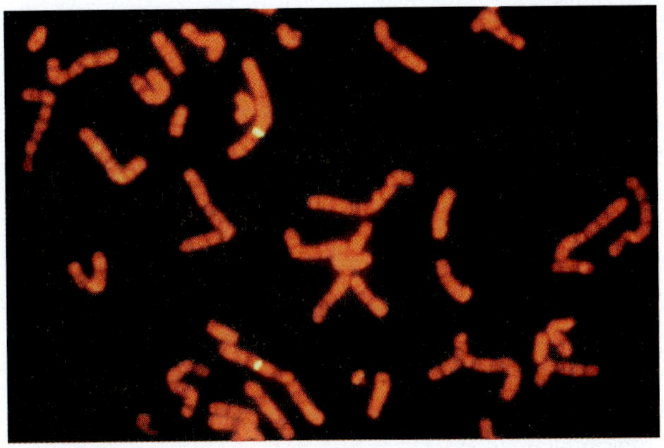

(b) SKY identifies human chromosomes

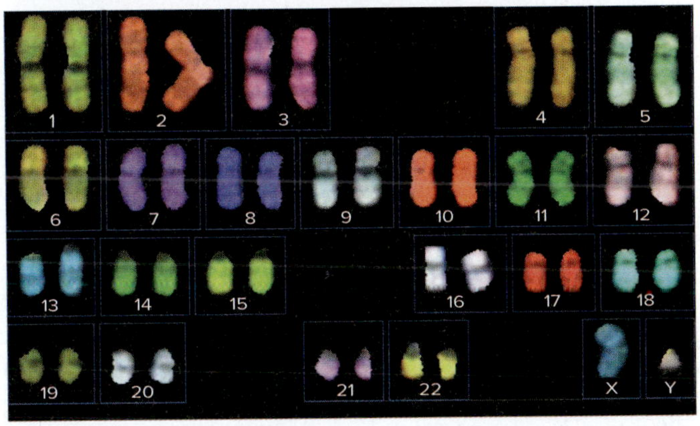

essential concepts

- In a *nucleosome*, DNA wraps twice around a core composed of histones H2A, H2B, H3, and H4. Histone H1 controls entry and exit of DNA.
- Models of higher-order compaction suggest that the nucleosomal fiber is supercoiled into a shorter but wider fiber. Nonhistone proteins anchor this fiber to the chromosome *scaffold*, forming loops that are then gathered together to condense DNA even more.
- Giemsa stain produces *G bands* in metaphase chromosomes. The pattern of G bands is highly specific and reproducible, allowing identification of chromosomes and gene locations.

- The *FISH* technique allows researchers to localize specific DNA sequences with respect to chromosomal bands or to visualize entire chromosomes to facilitate *karyotype* interpretation.

12.3 Chromosomal Packaging and Gene Expression

learning objectives

1. Describe how chromatin remodeling complexes allow gene expression to occur.
2. Differentiate between gene expression in heterochromatic and euchromatic regions.
3. Describe how scientists use position effect variegation to study the mechanisms underlying the formation of heterochromatin.
4. Outline how histone methylation and acetylation affect chromatin structure and gene expression.
5. Summarize the role of the *Xist* gene in X inactivation in mammalian cells.

The compaction of DNA into chromatin presents a problem for proteins that must recognize DNA sequences to carry out functions such as transcription, replication, and segregation. How do these proteins access particular nucleotide sequences that appear to be buried within complex chromatin structures? The answer is that chromatin structure is dynamic and can change to allow access of specific proteins when they need to act. These changes produce variations in chromatin structure necessary for different chromosomal functions.

In this section we focus on the relationship between chromatin structure and gene transcription. We first describe alterations in chromatin that allow RNA polymerase to recognize the promoters of genes and initiate transcription. We then discuss a type of chromatin structure called *heterochromatin* that is associated with chromosomal regions that are not transcribed. The formation of heterochromatin is the molecular basis for several important genetic phenomena, including X chromosome inactivation in mammals.

Transcription Requires Changes in Chromatin Structure and Nucleosome Position

An important generalization concerning gene expression in eukaryotes is that the less frequently a segment of DNA is transcribed, the more it is compacted. For example, cells express their genes mainly during interphase of the cell cycle when the chromosomes have decondensed, or decompacted. Little gene transcription occurs on highly compacted metaphase chromosomes. But even the relatively decompacted euchromatic interphase chromatin requires further unwinding to expose the DNA inside nucleosomes for transcription. Gene promoters are hidden from RNA polymerase and transcription factors when the promoter DNA is wrapped around the histone core of a nucleosome (**Fig. 12.10a**).

Studies of chromatin structure show that the promoters of most inactive genes are indeed wrapped in nucleosomes. In these studies, the positions of nucleosomes are investigated at the molecular level by treating chromatin with the enzyme DNase I, an enzyme that cleaves phosphodiester bonds in DNA. Sequences within nucleosomes are protected from DNase digestion, while chromosomal regions from which nucleosomes have been eliminated are recognizable by their hypersensitivity to DNase cleavage.

When a previously inactive gene prepares for transcription during a later step of cellular differentiation, the promoter region is observed to change from a DNase-resistant site to a **DNase hypersensitive site.** The reason is that transcription regulatory proteins (*transcription factors*) bind DNA at nearby *enhancers* and recruit proteins that reorganize the chromatin in the vicinity. In particular, these newly recruited proteins remove the promoter-blocking nucleosomes or reposition them in relation to the gene (**Fig. 12.10b**). One type of chromatin modulator consists of multisubunit **remodeling complexes** that use the energy of

Figure 12.10 Nucleosome packaging and gene expression. **(a)** Gene promoters wrapped around nucleosome cores are not accessible to RNA polymerase and transcription factors. **(b)** Chromatin remodeling complexes can expose promoters by placing them in nucleosome-free regions that are hypersensitive to DNase. **(c)** DNA in heterochromatin is so tightly packaged that it is transcriptionally inactive.

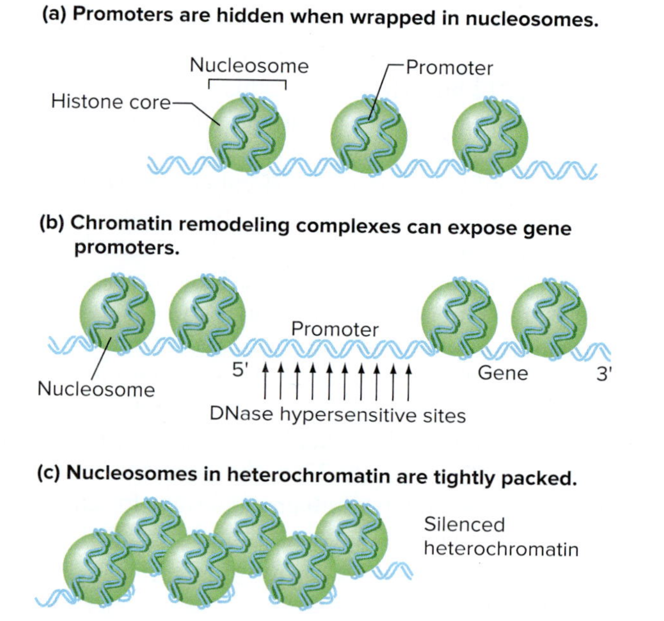

ATP hydrolysis to alter nucleosome positioning. Other chromatin modulators chemically modify the tails of the histones in the nucleosome core (as will be explained later). Chromatin changes accomplished by either mechanism expose the previously hidden promoter, allowing its recognition by RNA polymerase and thus facilitating the gene's transcriptional activation (Fig. 12.10b).

When differentiated cells divide, the transcriptional regulatory proteins that bind DNA and establish the chromatin structure are distributed into both daughter cells. After DNA replication, these proteins rebind DNA and reestablish the chromatin configuration that was present in the parent cell. As a result, differentiated cells have specific patterns of chromatin configuration and gene expression that persist after the cells divide by mitosis.

Most Genes in Heterochromatin Regions Are Silenced

One type of chromatin organization is widespread in genomes and is correlated with the strong suppression of gene expression. Chromatin organized in this way is visible even in the light microscope because it involves long stretches of DNA: In cells stained with certain DNA-binding chemicals, some chromosomal regions appear much darker than others. Geneticists call these darker regions **heterochromatin;** they refer to the contrasting lighter regions as **euchromatin.** The distinction between euchromatin and heterochromatin also appears in electron microscopy, where the heterochromatin appears much more condensed than the euchromatin. This observation reflects the much tighter packing of nucleosomes in heterochromatic regions (**Fig. 12.10c**).

Microscopists first identified dark-staining heterochromatin in the decondensed chromatin of interphase cells, where it tends to localize at the periphery of the nucleus. Even highly compacted metaphase chromosomes show the differential staining of heterochromatin versus euchromatin (**Fig. 12.11**). (This staining is distinct from, and should not be confused with, the G-banding in karyotyping described earlier.) Most of the heterochromatin in highly condensed chromosomes is found in regions flanking the centromere, but in some animals, heterochromatin forms in other regions of the chromosomes. In *Drosophila,* the entire Y chromosome, and in humans, most of the Y chromosome, is heterochromatic. Chromosomal regions that remain condensed in heterochromatin at most times in all cells are known as **constitutive heterochromatin.**

Figure 12.11 Constitutive heterochromatin. Human metaphase chromosomes were stained using a special technique that darkens the constitutive heterochromatin, most of which is in regions surrounding the centromeres.

Autoradiography (a method that detects radioactivity with photographic films) reveals that cells actively expressing genes incorporate radioactive RNA precursors into RNA almost exclusively in regions of euchromatin. This observation indicates that euchromatin contains most of the sites of transcription and thus almost all of the genes. By contrast, heterochromatin appears to be transcriptionally inactive for the most part; it is so tightly packaged that the enzymes required for transcription of the few genes it contains cannot access the correct DNA sequences (Fig. 12.10c).

A high proportion of the DNA located in regions of constitutive heterochromatin consists of long stretches of simple repetitive sequences like SSRs. Heterochromatic regions are also repositories for many *transposable elements*—segments of DNA that move around the genome. SSRs and transposable elements probably accumulate in constitutive heterochromatin because they are transcriptionally silenced there. Repetitive DNAs and transposable elements together constitute more than half of most genomes; their sequestration in transcriptionally inactive heterochromatin provides organisms with a way to minimize the effects of such *junk DNA* on normal cellular physiology.

We now discuss two specialized phenomena—position-effect variegation in *Drosophila* and Barr bodies in mammalian females—that illustrate clearly the correlation between loss of gene activity and heterochromatin formation. These phenomena also helped scientists investigate the biochemical differences between heterochromatin and euchromatin.

Heterochromatin Can Spread Along a Chromosome and Silence Nearby Euchromatic Genes

The *white*$^+$ (w^+) gene in *Drosophila* is normally located near the telomere of the X chromosome, in a region of relatively decondensed euchromatin. When a chromosomal rearrangement such as an inversion of a segment of DNA places the gene next to highly compacted heterochromatin near the centromere, the w^+ gene's expression may cease (**Fig. 12.12**). Such rearrangements silence w^+ gene expression in some cells and not others, producing **position-effect variegation (PEV).**

In flies carrying the wild-type w^+ allele, cells in the eye with an active w^+ gene are red, while cells with an inactive

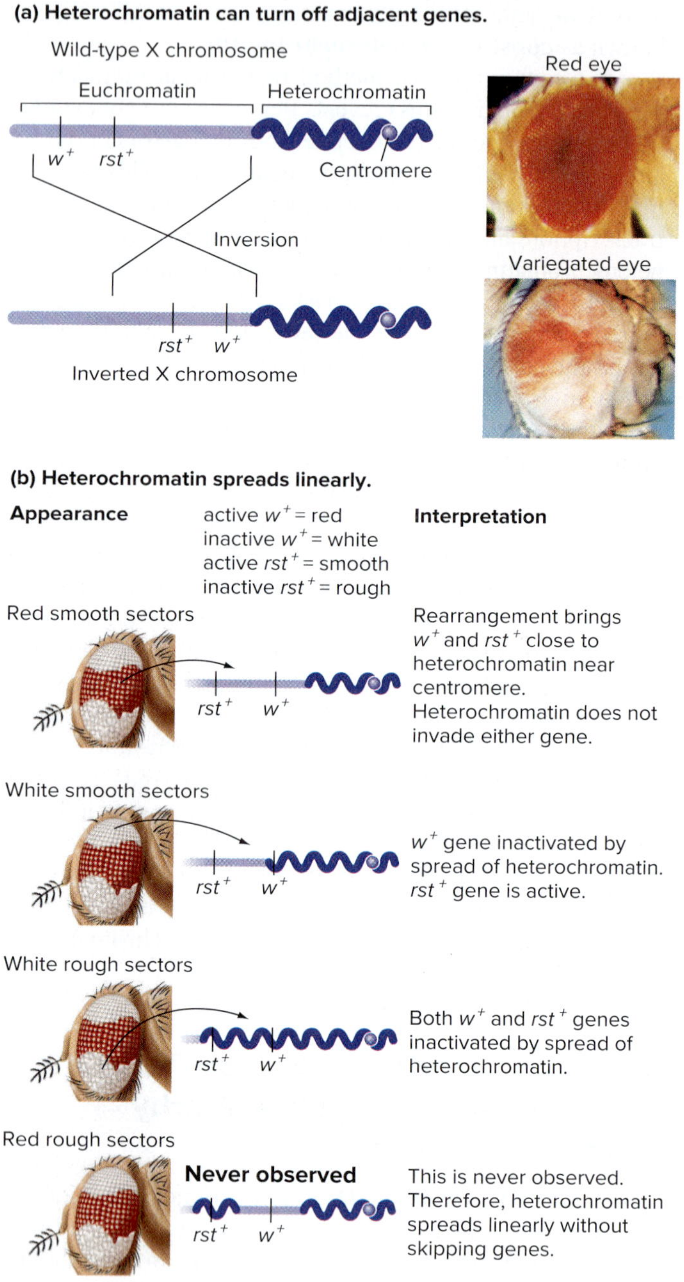

Figure 12.12 Position-effect variegation in *Drosophila*. **(a)** When the w^+ gene is brought near an area of heterochromatin by chromosomal inversion, the fly's eyes become variegated, with some red cells and some white cells. **(b)** A model for position-effect variegation postulates that heterochromatin can spread linearly from its normal location surrounding the centromere to nearby genes, causing their inactivation.

w^+ gene are white. Apparently, when normally euchromatic genes like w^+ come into the vicinity of heterochromatin, the heterochromatin can spread into the euchromatic regions, shutting off gene expression in those cells where the heterochromatic invasion takes place. In such a situation, the DNA of the gene has not been altered, but the relocation has altered the gene's packaging in some cells. The phenomenon of position-effect variegation thus reflects the existence of **facultative heterochromatin:** regions of chromosomes (or even whole chromosomes) that are heterochromatic in some cells and euchromatic in other cells of the same organism.

Position-effect variegation of red and white eye color in *Drosophila* produces eyes that are a mosaic of red and white patches (Fig. 12.12a). The numbers, positions, and sizes of the patches vary from eye to eye. Such variation suggests that the decision determining whether heterochromatin spreads to the w^+ gene in a particular cell is the result of a random process. Because patches composed of many adjacent cells have the same color, the decision must be made early in the development of the eye. Once made, the decision determining whether the *white* gene will be on or off is perpetuated by all of the cell's mitotic descendants. These descendants occupy a particular region of the eye, forming patches of red or white cells, respectively.

Scope of heterochromatin effects

One interesting property revealed by position-effect variegation is that heterochromatin can spread over more than 1000 kb of previously euchromatic chromatin. For example, some rearrangements that bring the w^+ gene near heterochromatin also place the *roughest*$^+$ gene in the same vicinity, although a little farther away from the centromeric heterochromatin (Fig. 12.12a). The wild-type *roughest*$^+$ (rst^+) gene normally produces a smooth eye surface. In flies carrying the rearrangements, some white-colored patches have smooth surfaces while others have rough surfaces (Fig. 12.12b). In the latter patches, the heterochromatin inactivated both the w^+ and the rst^+ gene. Red-colored, rough-surfaced patches never form, which means that the heterochromatin does not skip over genes as it spreads linearly along the chromosome.

If heterochromatin can spread, how is the boundary between heterochromatin and euchromatin normally formed? Research has identified DNA segments called **barrier elements** which block the spread of heterochromatin. The exact mechanism by which barrier elements work is unclear, but current models suggest these DNA elements recruit enzymes that modify histone proteins, as will be explained later in the chapter.

Using PEV to identify heterochromatin components

Scientists have taken advantage of the phenomenon of position-effect variegation in *Drosophila* to explore the molecules involved in heterochromatin formation. By looking for changes in the amount of variegation, researchers have obtained mutations that alter its efficiency. Enhancement of variegation (eyes more white) reflects gene inactivation in a larger number of cells; suppression of variegation (eyes more red) reflects gene inactivation in fewer cells.

The researchers later isolated several of the genes whose mutant alleles modified variegation, and they created antibodies against the mutant protein products of these genes. In this way, they discovered that some of the genes influencing heterochromatin formation encode proteins that localize selectively to the heterochromatin. As described in the next section, the identification of these proteins has provided important clues about the biochemical control of chromatin structure.

Heterochromatin and Euchromatin Have Different Histone Modifications

Several interdependent mechanisms govern the distinction between active (or potentially active) euchromatin and silenced heterochromatin. We focus here on one of the most important of these mechanisms, involving covalent modifications of the histones in the nucleosomes. Discussion of the interactions of chromatin modifier proteins with the transcription factors that regulate gene expression will be deferred until Chapter 17.

Histone tail modifications

The N-terminal regions of the four core histones—H2A, H2B, H3, and H4—form tails that extend outward from the nucleosome (**Fig. 12.13**). Enzymes can add several different kinds of chemical groups (among others, methyl groups, acetyl groups, phosphate groups, and ubiquitin; review Fig. 8.26) to various amino acids along these tails, while other enzymes can remove groups that were added previously. Such modifications of these **histone tails** can influence the packing of nucleosomes, and the modified tails can also serve as platforms to which chromatin modifier proteins can bind. The histone tails of a nucleosome core potentially could be modified in more than 100 ways (Fig. 12.13).

Figure 12.13 Histone tail modifications. The N-terminal tails of the core histone proteins extend outward from the nucleosome. Various amino acids in these tails are targets for modifications such as methylation and acetylation that can alter chromatin structure.

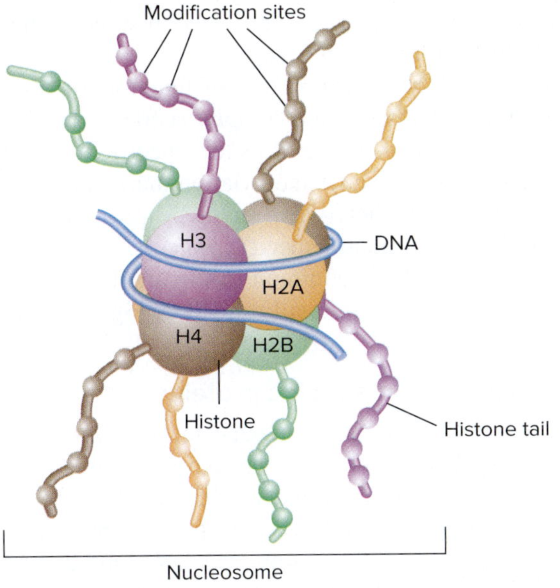

The best understood of these histone tail modifications are additions of acetyl groups to specific lysines (acetylation) and additions of methyl groups to specific lysines and arginines (methylation) (**Fig. 12.14**). Lysine acetylation, accomplished by a family of enzymes called *histone acetyl transferases (HATs),* opens chromatin by preventing the close packing of nucleosomes. Histone acetylation thus favors the expression of genes in euchromatic regions, as their promoters are now accessible to RNA polymerase and its associated proteins. Interestingly, the acetylated lysines on histone tails serve as binding sites for HAT enzymes, thus facilitating the spreading of histone acetylation to neighboring nucleosomes. *Histone deacetylase* enzymes (*HDACs*) that remove the acetyl groups reverse the process, resulting in closed chromatin and repressed transcription.

Figure 12.14 Histone tail modifications alter chromatin structure. In the heterochromatin (*orange*), K9-methylated (M) nucleosomes are bound by HP1 protein, which attracts HMTase enzymes that methylate adjacent nucleosomes and thus spread the inactive closed state. In euchromatic regions (*green*), HAT enzymes acetylate (Ac) core histones, resulting in open chromatin with nucleosomes sufficiently far apart so as to be accessible for transcription. *Barrier elements* that block the spread of heterochromatin are likely DNA sequences that attract HAT enzymes as shown, and/or demethylase enzymes that reverse K9 methylation (*not shown*).

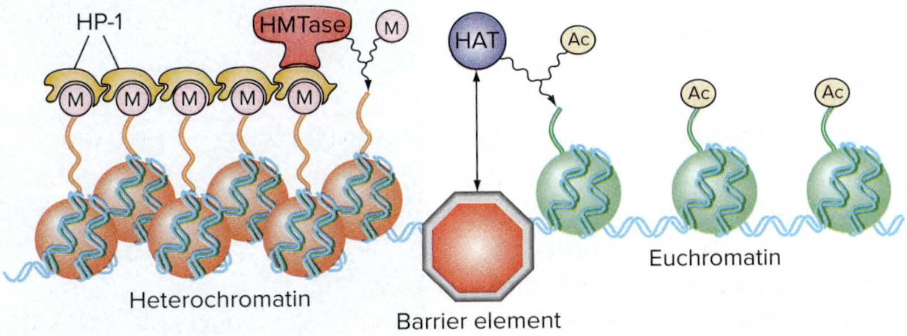

Methylation of histone tails is more complex, and may either close or open chromatin depending on the particular amino acid methylated. The enzymes that methylate histone tail amino acids are *histone methyltransferases* (*HMTases*), and enzymes that reverse histone methylation are called *histone demethylases.*

One of the genes whose loss-of-function mutant alleles act as PEV suppressors in *Drosophila* codes for an HMTase enzyme that adds methyl groups to a specific lysine (K9) in histone H3. This specific methylation marks the chromatin for assembly into heterochromatin by providing binding sites for heterochromatin-specific proteins. The methylation of histone H3 K9 is a common feature of chromosomal regions that are transcriptionally silenced (Fig. 12.14).

Heterochromatin-specific proteins

A different PEV suppressor gene in *Drosophila* encodes HP1, a key heterochromatin protein that binds to histone H3 tails containing methylated K9. The HP1 protein promotes chromatin compaction into heterochromatin in two ways (Fig. 12.14). First, it self-associates, and this helps to bring adjacent nucleosomes closer together. Second, HP1 binds other proteins, notably including the HMTase enzyme that adds methyl groups to the same K9 amino acid of histone H3. The recruited HMTase can methylate K9 on the histone H3s of adjacent nucleosomes. This autocatalytic effect provides at least a partial explanation for the linear spreading of heterochromatin observed in PEV.

As shown in Fig. 12.14, the choice between heterochromatic and euchromatic states involves competition between the enzymes that add and remove methyl and acetyl groups to various amino acids in the histone tails. Once the chromatin marks typical of euchromatin or heterochromatin predominate in a cluster of nucleosomes, they can spread to nearby nucleosomes until they reach barrier elements. The barriers that stop the spread of heterochromatin are probably DNA sequences that recruit HATs and histone demethylases, tilting the local competition toward the formation of euchromatin.

Heterochromatin Formation Inactivates an X Chromosome in Cells of Female Mammals

You will remember from Chapter 4 that mammals compensate for the difference in dosage of X-linked genes between males and females by randomly inactivating one of the two X chromosomes in each of the female's somatic cells. In some cells, the X inherited from the mother is inactivated; in others, it is the X inherited from the father (review Fig. 4.25). The inactive X chromosomes, or **Barr bodies**, are examples of facultative heterochromatin: An entire X chromosome becomes nearly completely heterochromatic in some cells, while other copies of this same X chromosome remain euchromatic in other cells. Most genes on an X chromosome are available for transcription only in cells where the chromosome is euchromatic. In contrast, only a few genes (mainly those in the pseudoautosomal regions) are available for transcription on an X chromosome that has become a heterochromatic Barr body.

Human X chromosomes contain a 450 kb region of DNA called the **X inactivation center (XIC)** that mediates dosage compensation (**Fig. 12.15a**). The role of the XIC

Figure 12.15 X chromosome inactivation. (a) One of the genes in the X inactivation center (*XIC*) is *Xist*, whose product is a nontranslated ncRNA. *Xist* is transcribed stably only from the inactive X chromosome. The *Xist* ncRNA binds to many sites on this chromosome; the *Xist* ncRNA then attracts histone-modifying proteins that silence the DNA. **(b)** Cells from an XX female mouse showing binding of *Xist* ncRNA (*red*) to the inactivated X chromosome (X_i; the Barr body) but not the active X chromosome (X_a). DNA is stained blue; the *Xist* genes on both X chromosomes are *yellow*.

(a) *Xist* is transcribed from the inactive X chromosome

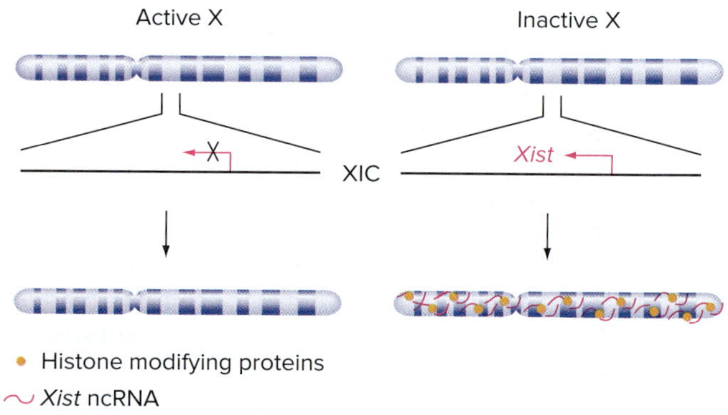

(b) *Xist* ncRNA binds the Barr body

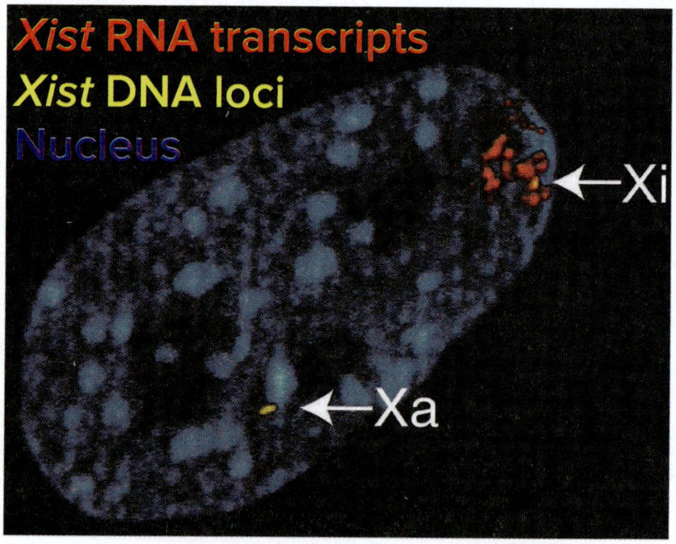

was established by experiments in which a copy of the XIC was transferred to an autosome, and remarkably, that autosome then became a Barr body. The most important known gene in the XIC is called ***Xist* (*X inactive specific transcript*)**. The *Xist* gene product is an unusually long (~17 kb) **noncoding RNA (ncRNA)** that, unlike most transcripts, never leaves the nucleus and is never translated into a protein.

Xist is transcribed stably only from the future inactive X chromosome, and it is the *Xist* ncRNA that triggers inactivation of the X chromosome from which it is transcribed. We know that *Xist* is a crucial gene that mediates inactivation because if *Xist* is deleted from one X chromosome, the other X always becomes the Barr body. Conversely, if a copy of *Xist* is added to an autosome, that autosome now becomes a heterochromatic Barr body. Researchers are currently exploring the possibility of adding *Xist* to one of the three copies of chromosome 21 in children with Down syndrome. In theory, this strategy could work as a kind of gene therapy to ameliorate the symptoms of Down syndrome, by turning off gene expression from the additional chromosome 21.

Studies using fluorescent molecular probes show that *Xist* ncRNA coats the X chromosome that produces it (**Fig. 12.15b**). The *Xist* ncRNA then recruits histone-modifying enzymes to this X chromosome. The enzymes methylate and deacetylate the histone tails in the nucleosomes of this chromosome, helping to condense it into a Barr body. Inactive X chromosomes also display other characteristics of silenced chromatin, such as the methylation of certain nucleotides in the DNA, as will be described in Chapter 17.

Although Fig. 12.15 shows some of the molecular details of X chromosome inactivation that have been worked out to date, several mysteries remain. Chapter 4 explains that in humans, about two weeks after fertilization, each of the 500–1000 cells in an XX female embryo decides independently to condense one of the two X chromosomes to a Barr body. It is unclear at present how individual cells choose which of the X chromosomes to inactivate. In addition, once an X chromosome is chosen to transcribe *Xist*, why does the *Xist* ncRNA bind only to that X chromosome?

Yet another area of active research involves the fact that X inactivation is inherited in somatic cells: Once the decision is made, it is clonally perpetuated so that all of the millions of cells descended by mitosis from a particular embryonic cell condense the same X chromosome to a Barr body. Scientists do not yet understand how the cell remembers which X chromosome to recondense into a Barr body following cell division.

essential concepts

- *Remodeling complexes* use the energy from ATP hydrolysis to change the position of nucleosomes, exposing promoters and allowing gene transcription.
- In regions of *heterochromatin*, promoters are tightly wrapped in nucleosomes, preventing transcription and thus silencing those genes.
- *Position-effect variegation* occurs when a chromosomal rearrangement places a gene next to the heterochromatic region near the centromere. If the heterochromatin spreads into the gene, then the gene will be silenced.
- *Methylation* of a specific lysine (K9) in histone H3 is a feature of silenced regions. In contrast, *acetylation* of lysines in histone tails opens chromatin and allows gene expression.
- The *Xist* gene is responsible for X inactivation. The *Xist noncoding RNA (ncRNA)* coats the X chromosome that produces it and recruits histone-modifying enzymes to inactivate the chromosome, which becomes a *Barr body*.

12.4 Replication of Eukaryotic Chromosomes

learning objectives

1. Explain how replication of the DNA in long eukaryotic chromosomes can occur in a short period of time.
2. Summarize the process by which nucleosomes are re-formed during replication.
3. Discuss the structure of telomeres and their roles in maintaining chromosome integrity.
4. Describe the action of telomerase and identify the cell types in which it continues to be synthesized.

Just as with transcription, replication of chromosomes requires reading of their DNA. Historically, scientists have been able to observe in the light microscope some aspects of the processes by which chromosomes replicate and become segregated into daughter cells. The mechanics at the molecular level, however, have been discovered only recently. In this section we review what is currently known about the replication of eukaryotic chromosomes.

In Chapter 6 we discussed in molecular detail how DNA polymerase cooperates with other factors to replicate DNA in bacterial cells. Although many aspects of DNA replication are similar in eukaryotes, these cells face additional challenges. First, eukaryotic cells have much more DNA than prokaryotic cells, all of which needs to be copied within the short span of a cell cycle. Second, the DNA replication machinery in eukaryotic cells must be able to operate even though the DNA is wrapped around nucleosomes. Finally, eukaryotic chromosomes are linear rather than circular, and as you will see, the ends of linear chromosomes are difficult to copy. We discuss in this section how eukaryotic cells deal with these special issues in chromosome replication.

Chromosomal DNA Replication Begins at Specific Origins

When DNA is copied, the enzyme DNA polymerase assembles a new string of nucleotides according to a DNA template, linking about 50 nucleotides per second in a typical human cell. At this rate and with only one origin of replication, it would take the polymerase about 800 hours, a little more than a month, to copy the 130 million base pairs in an average human chromosome. But the length of the cell cycle in actively dividing human tissues is much shorter, some 24 hours, and S phase (the period of DNA replication) occupies only about a third of this time. Eukaryotic chromosomes meet these time constraints by firing multiple origins of replication that can function simultaneously.

Most mammalian cells carry about 10,000 origins positioned strategically among the chromosomes. As you saw in Chapter 6, each origin of replication binds proteins that unwind the two strands of the double helix, separating them to produce two mirror-image replication forks. Replication then proceeds in two directions (bidirectionally), going one way at one fork and the opposite way at the other. As replication opens up a chromosome's DNA, a *replication bubble* becomes visible in the electron microscope, and with many origins, many bubbles appear (**Fig. 12.16**). These bubbles increase in size until adjacent bubbles run into each other, eventually allowing the entire chromosome to be copied.

The DNA running both ways from one origin of replication to the endpoints, where it merges with DNA from adjoining replication forks, is called a **replication unit,** or **replicon.** As yet unidentified controls tie the number of active origins to the length of S phase. In *Drosophila,* for example, early embryonic cells replicate their DNA in less than 10 minutes. To complete S phase in this short a time, their chromosomes use many more origins of replication than are active later in development when S phase is much longer. Thus, not all origins of replication are necessarily active during all the mitotic divisions that create an organism. The 10,000 origins of replication scattered throughout the chromatin of each mammalian cell nucleus are separated from each other by 30–300 kb of DNA, which suggests that at least one origin of replication exists per loop of chromatin.

Figure 12.16 Eukaryotic chromosomes have multiple origins of replication. Electron micrograph and diagrammatic interpretation, showing replicating DNA from a *Drosophila* embryo. Many origins of replication are active, creating multiple replicons.

© H. Kreigstein and D.S. Hogness, "Mechanism of DNA Replication in *Drosophila* Chromosomes: Structure of Replication Forks and Evidence of Bidirectionality", *PNAS,* 71(1974): 135-139

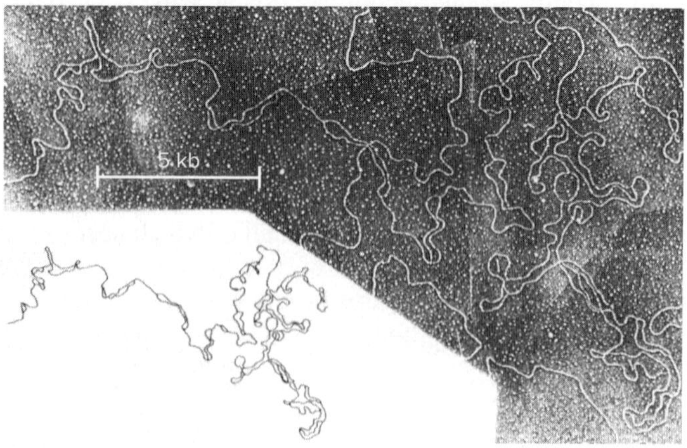

Origins of replication in yeast (known as *autonomously replicating sequences,* or *ARSs*) can be isolated by their ability to permit replication of plasmids in yeast cells. ARSs are capable of binding to the enzymes that initiate replication. Almost all ARSs contain a variant of the 11 bp AT-rich consensus sequence 5′ T/ATTTAYRTTTT/A 3′, where Y is either pyrimidine and R is either purine. Several other nearby sequence motifs also contribute to the function of ARSs as origins of replication. By digesting interphase chromatin with DNase I, an enzyme that fragments the chromatin only at points where the DNA is not protected by its association with a nucleosome, investigators have determined that origins of replication lie within accessible regions of DNA devoid of nucleosomes.

New Nucleosomes Must Be Formed During DNA Replication

DNA replication is only one step in chromosome duplication. The complex process also includes the synthesis and incorporation of histone and nonhistone proteins to regenerate nucleosomes and chromatin structure. While some aspects of the following model are controversial, researchers have obtained evidence that the process works something like this:

- The synthesis and transport of histones must be tightly coordinated with DNA synthesis because the nascent DNA becomes incorporated into nucleosomes within minutes of its formation.
- As DNA replication takes place, nucleosomes assemble rapidly on newly formed daughter DNA molecules. The new nucleosomes are a mixture of old (recycled) and newly formed histones, distributed randomly on the two daughter DNA molecules (**Fig. 12.17**).
- Some histone modifications may be retained at the replication fork to some extent through self-propagation because some modified histone tail amino acids attract modification enzymes (Fig. 12.14). However, even in these cases retention is inefficient. The reason is that only recycled histones can have modifications and about half the histones on newly replicated DNA are new. Moreover, histone modifications are labile, and so not all recycled histones retain their original marks.

Figure 12.17 Nucleosome creation after DNA replication. The replication fork disassembles the nucleosomes it encounters in the parental chromosome. New nucleosomes start assembling immediately after the replication fork passes. First, a tetramer containing two molecules each of histones H3 and H4 associates with DNA to form a half nucleosome, followed by two dimers each containing one molecule of H2A and one of H2B. The new nucleosomes can contain various random combinations of H3/H4 tetramers and H2A/H2B dimers that were either newly synthesized or that were previously found in parental nucleosomes.

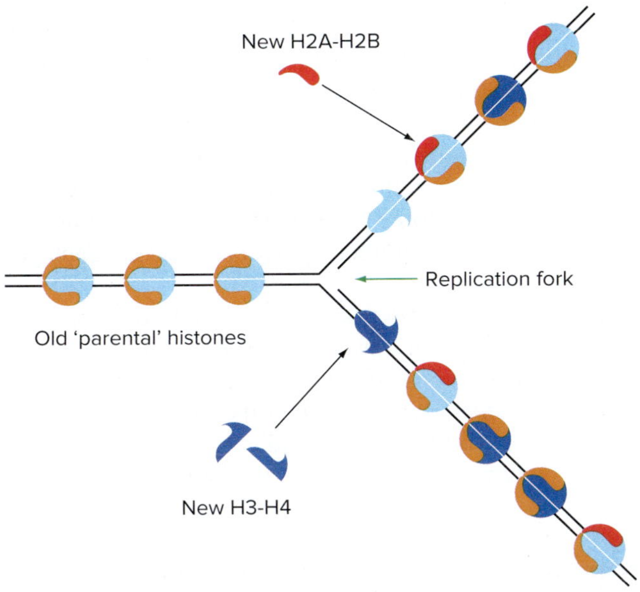

If no robust mechanism exists for copying histone marks during DNA replication, why is chromatin structure often recapitulated after differentiated cells undergo mitosis? The answer appears to be that just after replication, chromatin formed from new nucleosomes is open. A brief window of opportunity thus exists for transcription regulatory proteins in the nucleus to bind to the daughter DNA molecules. These proteins then recruit histone-modifying enzymes, re-creating the chromatin structure of the parental chromosome even though the histone modifications were briefly lost when the DNA was replicated.

Telomeres Protect the Ends of Linear Chromosomes and Allow Their Replication

The linear chromosomes of eukaryotic cells terminate at both ends in protective caps called **telomeres** (**Fig. 12.18**). Composed of special DNA sequences associated with specific proteins, these caps contain no genes but are crucial in preserving the structural integrity of each chromosome.

Telomeres and the replication of chromosome ends

The replication of the ends of linear chromosomes poses a difficult problem for cells. As you saw in Chapter 6, DNA polymerase, a key component of the replication machinery, functions only in the 5′-to-3′ direction: It can add nucleotides only to the 3′ end of an existing chain. With this constraint, the enzyme on its own cannot possibly replicate some of the nucleotides at the 5′ ends of the two DNA strands (one of which is in the telomere at one end of the chromosome, the other of which is in the telomere at the other end). As a result, DNA polymerase can reconstruct the 3′ end of each newly made DNA strand in a chromosome, but not the 5′ end (**Fig. 12.19**).

We saw in Chapter 5 that because DNA polymerase cannot start synthesis at the 5′ end of the new chain, it relies on RNA primers to do the job. These RNA primers are eventually removed. Thus, at the 5′ ends of linear chromosomes in eukaryotic cells, an RNA primer's length of nucleotides will be missing at the 5′ end of every new chromosomal strand every time the DNA is copied (Fig. 12.19). As a result, the chromosomes in successive generations of cells would become shorter and shorter, losing crucial genes as their DNA diminished.

Telomeres and an enzyme called **telomerase** provide a countermeasure to this limitation of DNA polymerase. Telomeres consist of particular repetitive DNA sequences that do not encode proteins. Human telomeres are composed of the base sequence 5′ TTAGGG 3′ (read on one strand) repeated 250 to 1500 times. The number of repeats varies with the cell type; sperm have the longest telomeres. The same TTAGGG sequence occurs in the telomeres of all mammals as well as in birds, reptiles, amphibians, bony

Figure 12.18 Telomeres protect the ends of eukaryotic chromosomes. Human telomeres light up in yellow upon FISH with probes that recognize the repeating base sequence TTAGGG.

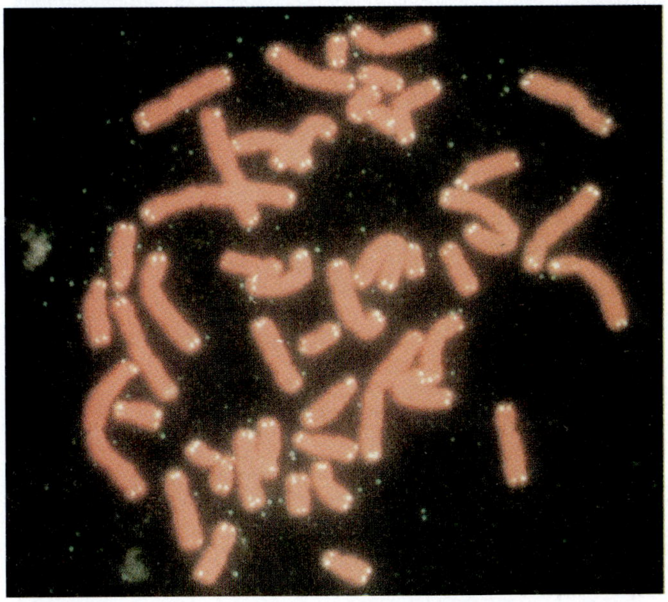

Figure 12.19 Replication at chromosome ends. Even if an RNA primer (*red*) at the 5′ end can begin synthesis of a new strand, a gap will remain when ribonucleases eventually remove the primer. DNA polymerase cannot fill this gap without a primer, so newly synthesized strands (*dark blue*) will be shorter than parental strands (*light blue*).

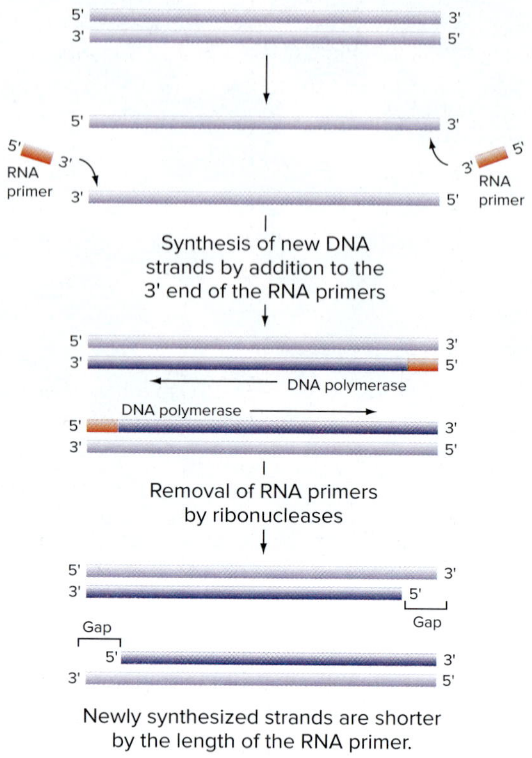

fish, and many plant species. Some much more distantly related organisms also have repeats in their telomeres but with slightly different sequences. For example, the telomeric repeat in the chromosomes of the ciliate *Tetrahymena* is TTGGGG. The close similarities of these repeated sequences across phyla suggests that they perform a vital function that emerged in the earliest stages of the evolutionary line leading to eukaryotic organisms.

Telomerase is an unusual enzyme consisting of protein in association with RNA. Because of this mix, it is called a *ribonucleoprotein.* The RNA portion of the enzyme contains 3′ AAUCCC 5′ repeats that are complementary to the 5′ TTAGGG 3′ repeats in telomeres, and they serve as a template for adding new TTAGGG repeats to the end of the telomere (**Fig. 12.20**). You can easily see how the addition of new repeats could counterbalance the loss of DNA that must occur when linear DNA molecules are copied.

Telomerase activity and cell proliferation

Some types of cells take advantage of the ability of telomerase to maintain the size of their chromosome ends upon cell division, but other cell types make very little if any telomerase, and in fact their chromosomes shorten during every cell generation. Most differentiated somatic cells in humans are of this latter type. Even though these cells have the telomerase gene in their genomes, they do not express this enzyme, so the telomeres shorten slightly with each cell division. After 30 to 50 cell generations, the chromosomes begin to lose essential genes from their ends; the cells start showing signs of senescence and then die. The lack of telomerase thus ensures that differentiated somatic cells have a finite life span.

In contrast, germ-line cells do express telomerase and thus maintain their chromosomal ends through repeated rounds of DNA replication without the loss of genes. (If this were not the case, our species would have died out long ago.) Apparently, germ-line cells have some kind of feedback mechanism that maintains the optimal number of repeats at the telomeres so that the chromosome ends neither shorten nor lengthen appreciably during each generation of humans.

Two kinds of somatic cells in our bodies can also make telomerase, and thus have the potential to reproduce for many generations, if not forever. One class of such cells are the *stem cells* that allow tissue renewal, such as the continual production of blood cells. The second class consists of tumor cells: somatic cells gone awry that can

Figure 12.20 How telomerase extends telomeres. The 3′ AAUCCC 5′ repeats of telomerase RNA (*red*) are complementary to the 5′ TTAGGG 3′ repeats of telomeres. Telomerase RNA thus serves as a template for adding TTAGGG repeats to telomere ends. After a new repeat is added, telomerase moves (translocates) to the newly synthesized end, allowing additional rounds of telomere elongation.

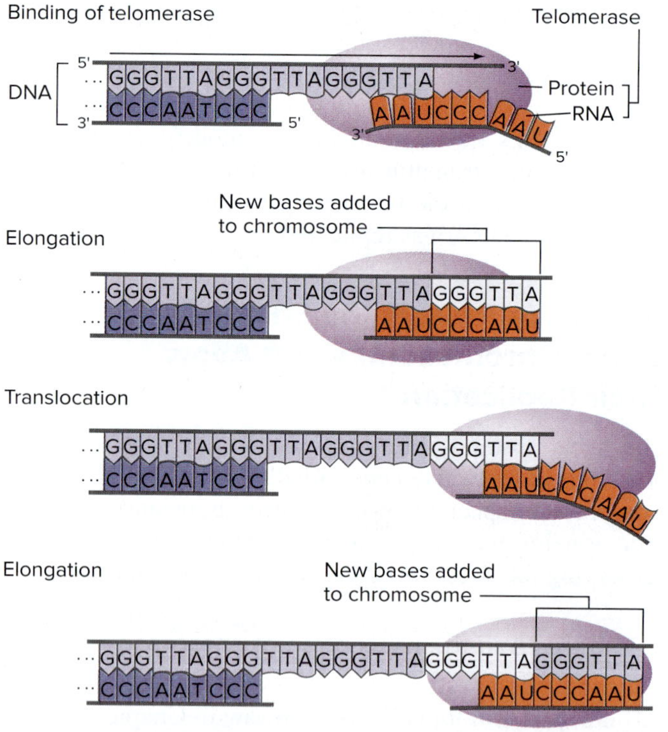

Figure 12.21 The shelterin complex protects telomeres. The proteins of the shelterin complex (colored shapes) bind to telomeres, folding the DNA ends (*gray*) so they can neither be attacked by nucleases nor subjected to nonhomologous end-joining.

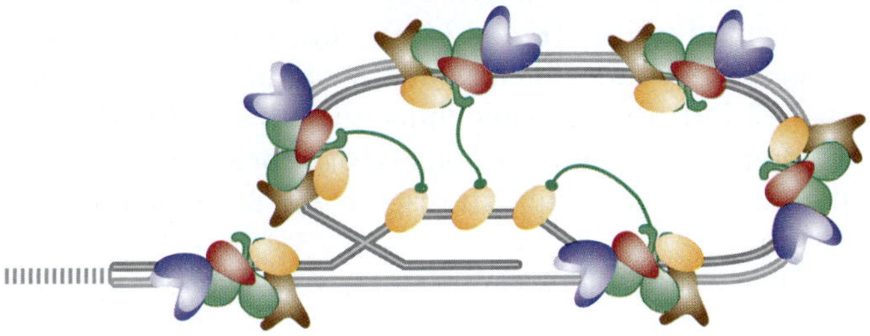

divide indefinitely, becoming seemingly immortal. Because high telomerase activity is a characteristic of many tumor cells, pharmaceutical companies are developing cancer treatment drugs that inhibit this important enzyme.

Telomeres and chromosome integrity

Chromosomes that are broken (for example, by exposure to X-rays) and detached from their telomeres pose many dangers for cells. One problem is that cells contain nuclease enzymes that can progressively degrade DNA inward from the broken ends. A second problem results from the enzyme systems responsible for nonhomologous end-joining (NHEJ) that were described in Fig. 7.18. If two different chromosomes are broken, these NHEJ enzymes will join the chromosomes together. The end-to-end fusion of broken chromosomes produces entities with two centromeres. During anaphase of mitosis, if the two centromeres are pulled in opposite directions, the DNA between them will rupture, resulting in broken chromosomes that segregate poorly and eventually disappear from the daughter cells. These examples demonstrate that the telomeres on normal, unbroken chromosomes have important functions in protecting the chromosomes and maintaining the correct genetic complement in cells.

The protective function of telomeres is due to proteins, different from telomerase, that also bind to the TTAGGG repeats at the very ends of a chromosome. These proteins form a complex called **shelterin** that folds up the telomeres into a structure that shields single-stranded TTAGGG sequences from nucleases and NHEJ enzymes (**Fig. 12.21**).

essential concepts

- In eukaryotic cells, DNA copying initiates nearly simultaneously at thousands of *origins of replication.*
- During replication, new nucleosomes assemble on the daughter DNA molecules. Regulatory proteins bind to the daughter DNAs and recruit histone-modifying enzymes to re-create the parental chromatin structure.
- Because DNA polymerase cannot copy the 5′ ends of linear DNA molecules, chromosomes will shorten every time they are replicated, leading to gene loss and cell death. To counteract this shortening, *telomeres* contain repeating sequences that can be extended by the enzyme *telomerase*.
- Most somatic cells do not express telomerase and thus have limited life spans. Cells that continue to express telomerase include germ-line cells, stem cells, and tumor cells.

12.5 Chromosome Segregation

learning objectives

1. Contrast the DNA sequences dictating centromere formation in yeast and in higher eukaryotes.
2. Describe spindle attachment during metaphase and the role tension on kinetochores plays in ensuring proper chromosome segregation.
3. Compare the behavior of cohesin complexes during mitosis, meiosis I, and meiosis II.

When cells divide during mitosis or meiosis II, the two chromatids of each replicated chromosome must separate from one another at anaphase and segregate such that each daughter cell receives one and only one chromatid from each chromosome. In contrast, during meiosis I, homologous chromosomes must pair and segregate such that each daughter cell receives one and only one chromosome from each homologous pair. The *centromeres* of eukaryotic chromosomes ensure this precise distribution during different kinds of cell division by serving as segregation centers. The centromeric region is where sister chromatids are most tightly bound together; in addition, structures at the centromeres called *kinetochores* are the locations at which chromosomes bind to spindle fibers (**Fig. 12.22**).

Figure 12.22 Centromeres. Cohesin proteins at centromeres (*yellow*) hold sister chromatids together. Centromeres also contain information for the construction of kinetochores (*orange*), the structures that allow the chromosomes to bind to spindle fibers.

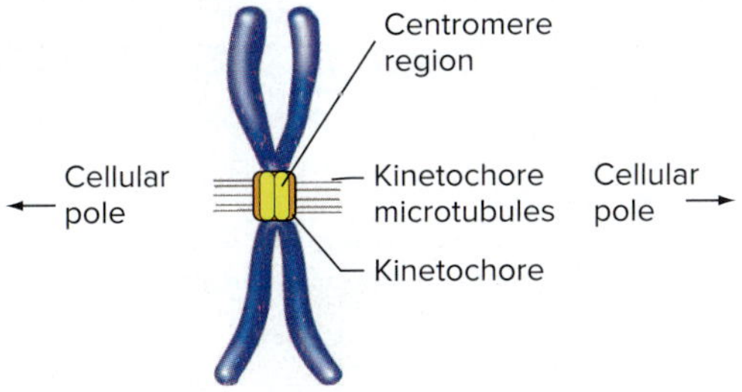

Special DNA Sequences Allow the Formation of Centromeres

In the yeast *S. cerevisiae,* centromeres consist of two highly conserved nucleotide sequences, each only 10–15 bp long, separated by approximately 90 bp of AT-rich DNA (**Fig. 12.23**). Evidently, a short stretch of roughly 120 nucleotides is sufficient to specify a centromere in this organism. The centromere sequences of different yeast chromosomes are so closely related that the centromere of one chromosome can substitute for that of another. This fact indicates that all centromeres play the same role in chromosome segregation, and they do not help distinguish one chromosome from another.

The centromeres of higher eukaryotic organisms are much larger and more complex than those of yeast. These more complicated centromeres are contained within blocks of certain repetitive, noncoding sequences known as **satellite DNAs.** Many different kinds of satellite DNA exist, each consisting of short sequences 5–300 bp long, repeated in tandem thousands or millions of times to form large arrays. The predominant human satellite at centromeres, α-satellite, is a noncoding sequence 171 bp in length; it is present in a block of tandem repeats extending over a megabase of DNA in the centromeric region of each chromosome. Various human centromeres also contain repetitive sequences unrelated to α-satellite; these sequences impart heterochromatic characteristics to centromeric regions, as was seen in Fig. 12.11.

Kinetochores Govern Attachment of Chromosomes to the Spindle

One of the important ways in which centromeres contribute to proper chromosome segregation is through the elaboration of **kinetochores:** specialized structures composed of DNA and proteins that are the sites at which chromosomes attach to the spindle fibers. Each of the simple kinetochores in yeast cells connect only with a single spindle fiber (Fig. 12.23), but the kinetochores in higher eukaryotes attach to many spindle microtubules (**Fig. 12.24**). Researchers think that these complex kinetochores are likely to consist of repeating structural subunits, with each subunit responsible for attachment to one fiber.

The kinetochore-forming DNA, such as the α-satellite in humans, has a different chromatin structure and different higher-order packaging than other chromosomal regions (Fig. 12.24). In this specialized chromatin, the normal histone H3 protein has been replaced by a histone variant called *CENP-A* in the nucleosome core. The CENP-A protein is very similar to histone H3 in its C-terminal region, but different from H3 in its N-terminal portion. Nucleosomes with this histone variant act as scaffolds to allow the assembly of many other proteins into kinetochores.

Figure 12.23 Yeast centromeric DNA sequences. Each yeast centromere has two short, conserved DNA elements and binds through kinetochore proteins (*not shown*) to a single spindle microtubule (*tan*), whose size is shown for comparison.

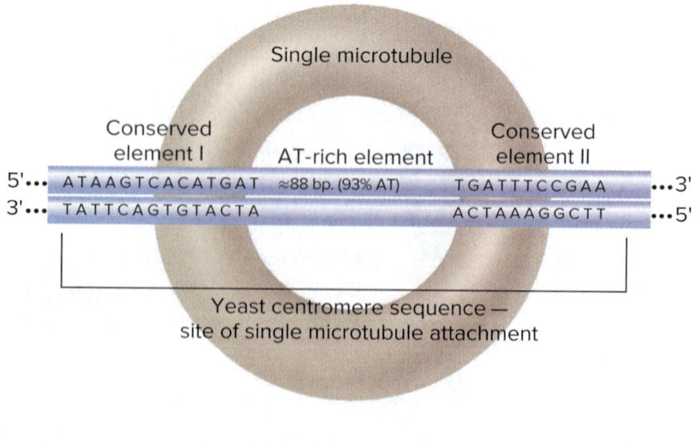

Figure 12.24 Kinetochores. In higher eukaryotes, centromeric DNA consists of repeated sequences organized into nucleosomes containing CENP-A, a variant of histone H3. Kinetochores containing dozens of proteins organize around these nucleosomes. Some of these proteins govern kinetochore assembly (*purple*), some bind microtubules (*yellow*); some are motors that move the chromosomes along the spindle (*red*); and some (*green*) act in a checkpoint that makes sure sister chromatids (mitosis, meiosis II) or homologous chromosomes (meiosis I) do not separate before all the chromosomes are properly attached to spindle fibers.

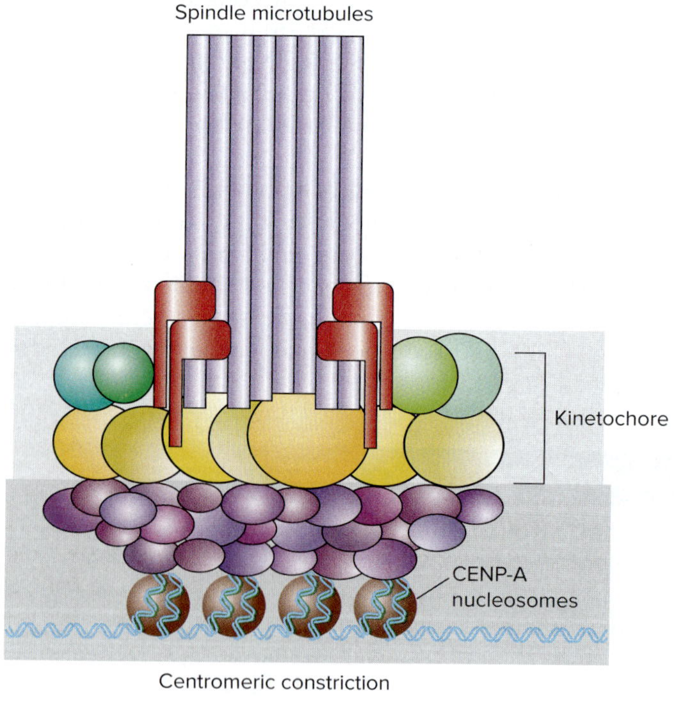

During mitosis, a kinetochore develops late in prophase on each sister chromatid, at the part of the centromere that faces one or the other cellular pole. By prometaphase, the kinetochores of the two sister chromatids attach to spindle fibers emanating from centrosomes at opposite poles of the cell. Some of the kinetochore proteins are motor proteins that exert pulling forces on chromatids toward the spindle pole to which they are attached.

At metaphase, tension arises when sister chromatids are pulled in opposite directions, yet are still held together at the centromere. Certain other kinetochore proteins monitor this tension to establish a cell cycle checkpoint: Only after all the kinetochores in the cell are under tension (meaning only after all chromosomes are properly attached to the spindle) do these proteins generate a molecular signal that allows the sister chromatids to disconnect from each other. The chromatids are then able to move to their respective poles during anaphase.

Cohesin Complexes Hold Sister Chromatids Together

A highly conserved, multisubunit protein complex called **cohesin** acts as the glue that holds sister chromatids together during mitosis and meiosis until segregation takes place. After the chromosomes replicate in S phase of the cell cycle, cohesin proteins associate with and hold sister chromatids together along the arms and in the centromere region. Cohesin encircles the two double helixes of the sister chromatids to keep them together. **Figure 12.25** shows one current (though controversial) model for the topological relationship between cohesin and the two molecules of DNA. Cohesin rings are scattered along the length of the chromosome, but they are found in particularly high concentrations in the vicinity of centromeric heterochromatin.

Figure 12.25 A molecular model for cohesin. The cohesin complex has protein subunits (*green, aqua,* and *purple*) that together surround the two sister chromatids (the two DNA molecules with strands in *blue* and *red*). Sister chromatids can separate when separase (*gold*) cleaves the *purple* cohesin subunit, freeing the two DNA molecules.

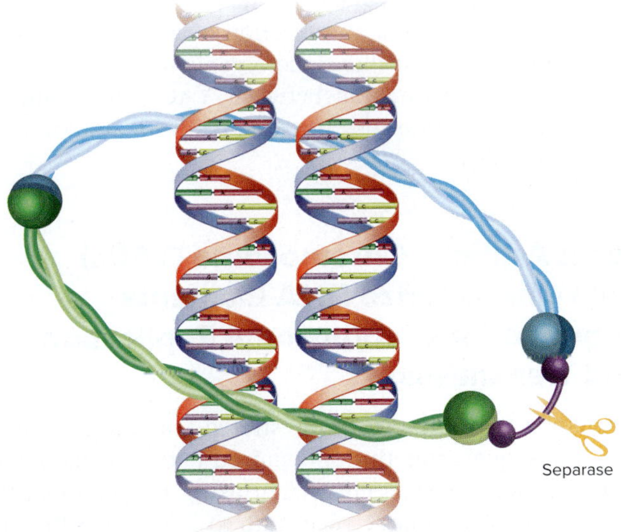

During metaphase of mitosis, the cohesin complexes resist the forces pulling the chromatids poleward at their kinetochores, generating tension across the chromosomes, as just mentioned. At anaphase a proteolytic enzyme called *separase* cleaves the cohesin complexes, allowing the sister chromatids to separate and move to opposite spindle poles (Fig. 12.25 and **Fig. 12.26a**). In support of this idea, dividing cells expressing a cohesin that cannot be cleaved by separase display many chromosome segregation errors.

In meiosis, a special problem arises: Sister chromatids must stay together during the entire first meiotic division but then separate during the second meiotic division. Cells solve this problem by making cohesin complexes with a meiosis-specific subunit (replacing the purple subunit in Fig. 12.25) that can interact with a protein called *shugoshin* (meaning *guardian spirit* in Japanese; **Fig. 12.26b**). Shugoshin protects the cohesin at the centromere from being cleaved by separase. Upon entering meiosis II, shugoshin is removed, and separase can now cleave the centromeric cohesin at anaphase II, allowing sister chromatids to segregate to opposite poles.

Interestingly, shugoshin does not protect cohesin along the arms of the sister chromatids; the molecular mechanism that discriminates between cohesin on the arms from that at the centromere is not yet known. Thus, at anaphase of meiosis I, the cohesin along the arms of the sister chromatids is cleaved, while that at the centromere is not. This fact is very important because, as you will discover in Problem 30 at the end of this chapter, cohesin along the arms is the glue keeping homologous chromosomes together while the spindle tries to pull them apart during metaphase of the first meiotic division. Cleavage of the cohesin on the arms is therefore essential to allow the homologous chromosomes to segregate to opposite spindle poles during anaphase of meiosis I.

essential concepts

- Short DNA sequences define yeast *centromeres*. In higher eukaryotes, centromeres are much more complex and contain repetitive DNA sequences.
- *Kinetochores* at the centromeres are sites where spindle fibers attach to the chromosomes. A cell cycle checkpoint ensures that chromatids do not separate until all kinetochores are properly connected to the spindle.
- *Cohesin* complexes hold chromatids together until the enzyme *separase* cleaves the cohesin at anaphase. In meiosis I, *shugoshin* protects cohesin at the centromere and thus keeps sister chromatids together; in meiosis II, shugoshin is removed, so separase can cleave cohesin at anaphase of this division.

Figure 12.26 Cohesin action in mitosis and meiosis. **(a)** During mitosis, cohesin holds sister chromatids together through metaphase. Cleavage of cohesin by separase releases sister chromatids so they can segregate at anaphase. **(b)** In anaphase of meiosis I, cohesin is cleaved from the chromatid arms but is protected at the centromeres by shugoshin. Cohesin remains at the centromere to hold sister chromatids together until anaphase II.

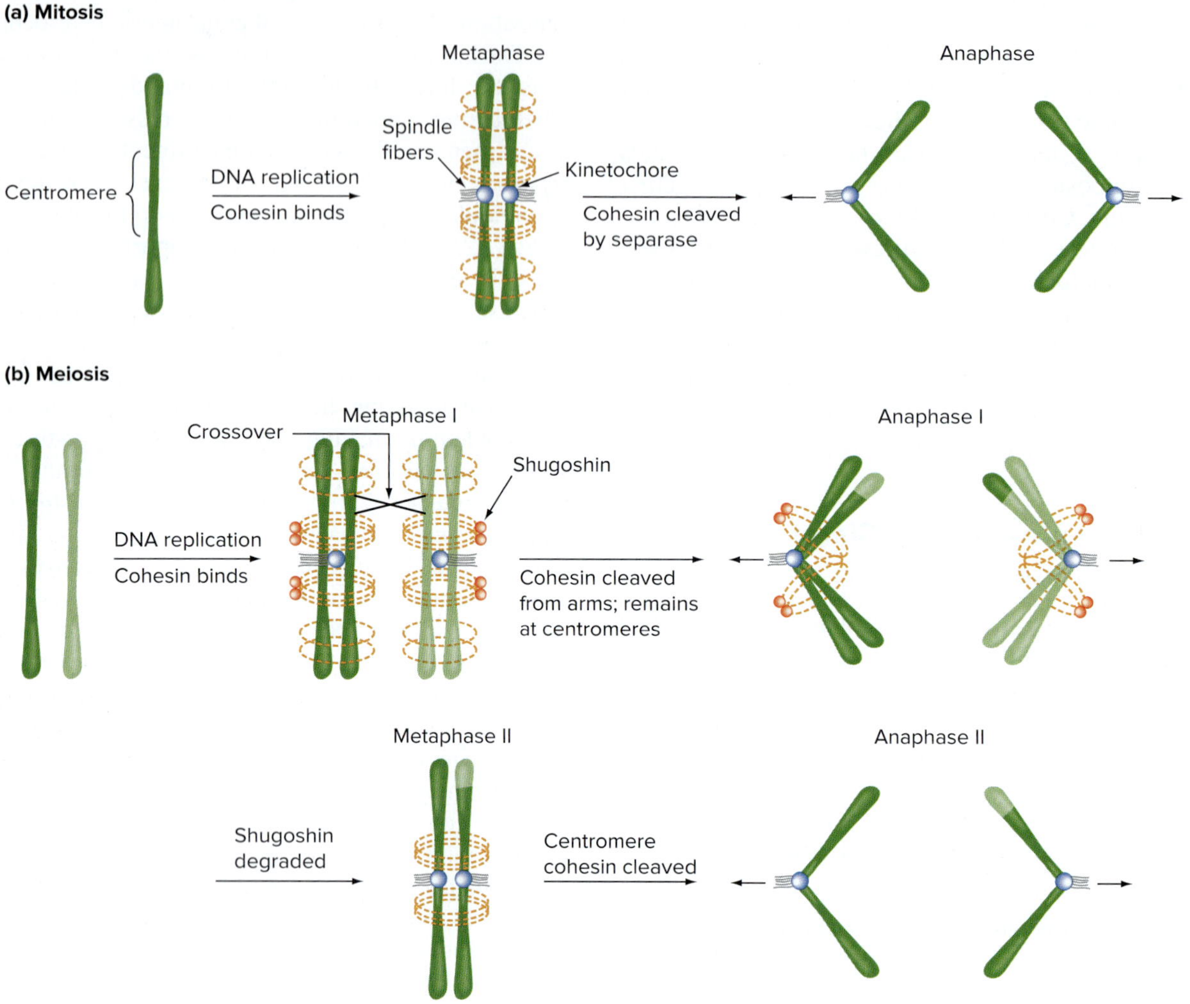

12.6 Artificial Chromosomes

learning objectives

1. List the elements that must be included in an artificial chromosome.
2. Discuss the reasons why scientists create artificial chromosomes.

In the 1980s, molecular geneticists constructed the first artificial eukaryotic chromosome, a **yeast artificial chromosome (YAC),** by combining in a single DNA molecule yeast (*Saccharomyces cerevisiae*) versions of the three key chromosomal elements described in this chapter—centromeres, telomeres, and origins of replication (**Fig. 12.27**). In 2014, a team of scientists at Johns Hopkins University reported the construction of a completely synthetic chromosome of this same organism. In this section, we discuss the distinction between YACs and synthetic chromosomes and discuss the uses of these two kinds of DNA molecules.

Yeast Artificial Chromosome (YACs) Help Characterize DNA Elements Required for Chromosome Replication and Transmission

In Chapter 9, we described the strategy of whole-genome shotgun sequencing through paired-end sequencing of cloned genome fragments. When using this method, the genome is cut into large pieces up to 2 Mb long that are

Figure 12.27 Yeast artificial chromosomes (YACs). To function effectively as artificial chromosomes, YAC vectors containing an ARS (yeast origin of replication), a centromere sequence, and telomeres at the ends need to be ligated (here through *Eco*RI sites) with inserts of foreign DNA more than 100 kb long (not drawn to scale). Yeast cells transformed with YACs can be identified if the vector has a selectable marker such as the TRP^{+} gene.

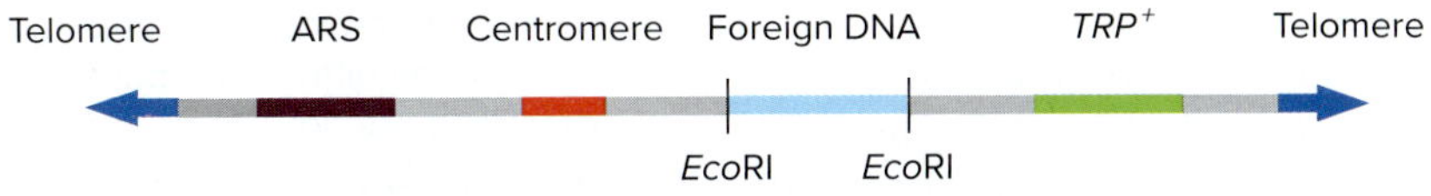

subsequently isolated, amplified, and characterized. Two types of vectors that allow such long DNA fragments to be cloned are BACs (bacterial artificial chromosomes) and YACs (yeast artificial chromosomes). We describe YACs in some detail here not only because they are useful for the cloning of large fragments of genomic DNA, but also because manipulation of these vector molecules has produced many insights into chromosome function.

Figure 12.27 illustrates how centromeres, telomeres, and origins of replication are combined to make a YAC. Once scientists join these elements using recombinant DNA technology, they transform the YAC into yeast cells, where it can be maintained as an autonomous chromosome. Researchers track the presence of the YAC by including a selectable marker, for example the TRP^{+} gene that allows an auxotrophic host cell to grow in the absence of the amino acid typtophan in the growth medium.

Manipulation of the YAC construction process was historically significant because it allowed geneticists to isolate and analyze the chromosomal regions corresponding to ARSs (yeast origins of replication) and centromeres. Plasmids containing only origins of replication but no centromere or telomere replicate but do not segregate properly. Plasmids with origins of replication and a centromere but no telomeres replicate and segregate fairly well if they are circular; if they are linear, they degrade and eventually become lost from the cell.

Interestingly, small DNA molecules carrying all three elements replicate and segregate as linear chromosomes independent of the normal yeast chromosomes, but they do not segregate accurately because they do not have enough DNA. YACs carrying inserts of 11,000 bp of random DNA mis-segregate in 50% of cell divisions. YACs containing 55,000 bp show segregation errors in 1.5% of cell divisions. With artificial chromosomes containing more than 100,000 bp inserts, the rate of segregation error falls to 0.3%. This error rate is still 200 times greater than that seen with natural yeast chromosomes of normal size, indicating that some subtle aspects of chromosome structure and function remain to be discovered.

Researchers have used the lessons from these YAC studies to develop human artificial chromosomes. They hope to use such chromosomes eventually to treat genetic diseases, the idea being they can serve as vectors for the transformation of human cells defective for a particular gene with a wild-type copy of that gene.

Synthetic Chromosomes May Help Define Minimal Genomes

Chromosomes undoubtedly contain many DNA sequences that are not required for the survival of the organism. These include transposable elements, repetitive DNAs, some of the DNA between genes, and many intronic sequences. In addition, only about 1000 of the approximately 6000 yeast genes have been found to be essential individually for yeast cell growth under laboratory conditions. In other words, deletion of any one of the remaining 5000 genes allows yeast to survive. But perhaps many of these 5000 nonessential genes are actually essential under specific growth conditions found in nature, or maybe a given gene becomes essential when a different gene is removed simultaneously. What then is the minimal set of DNA sequences that yeast cells actually need for their survival?

To answer this question, molecular biologists aim to create yeast cells with an entirely synthetic set of 16 chromosomes. In this approach, a **synthetic chromosome** differs from an artificial chromosome like a YAC in two ways. First, all the DNA in a synthetic chromosome is entirely man-made in DNA synthesizer machines; in contrast, a YAC is constructed by splicing together DNA elements that were originally in yeast chromosomes. Second, a synthetic chromosome would include (before any subsequent manipulation) all genes present on the corresponding yeast chromosome. In contrast, YAC vectors usually have a single protein-coding yeast gene—a selectable marker like TRP^{+} in Fig. 12.27.

Remarkably, in 2014, scientists reported the construction of Syn III, a synthetic version of yeast chromosome III, and the first synthetic eukaryotic chromosome. Even though Syn III lacks about 50,000 bp (mostly transposable elements and other intergenic sequences) of the normal (~317,000 bp) third chromosome, yeast cells containing Syn III instead of a normal third chromosome display normal appearance and growth. As you will see in Problem 37, Syn III also contains special short sequences that will in the future allow the researchers to make deletions of various regions of the chromosome to see which genes or gene combinations are essential or nonessential.

Anticipating that viable yeast with 16 minimized synthetic chromosomes can be obtained, these scientists are now busy creating similar synthetic versions of the other 15 yeast chromosomes. If the experiment is successful, not only will these researchers have created an entire genome by chemical synthesis, but the yeast they create will serve as the basis for future investigations aimed at determining the minimal yeast genome that can support life.

essential concepts

- Stable transmission of linear *yeast artificial chromosomes* (*YACs*) requires the inclusion of an origin of replication (ARS), a centromere, and telomeres; for unknown reasons, stably transmitted chromosomes must contain DNA in excess of 100 kb.
- *Artificial chromosomes* have been used as cloning vectors and in studies to identify the functional regions of natural chromosomes; they might be used in the future to treat genetic diseases in humans.
- A *synthetic chromosome* lacking most intergenic sequences functions normally. Such synthetic chromosomes may help to define a minimal yeast genome.

WHAT'S NEXT

Eukaryotic chromosomes manage the genetic information in DNA through a modular chromatin design whose flexibility allows shifts between different levels of organization and compaction. These reversible changes in chromatin structure reliably sustain a variety of chromosome functions, including packaging in nuclei, proper copying and segregation during cell division, and coordinating gene expression.

Although the faithful function, replication, and transmission of chromosomes underlie the perpetuation of life within each species, chromosomal changes do occur. We have already described two mechanisms of change: mutation of individual nucleotides (Chapter 7) and homologous recombination, which exchanges bases between homologs (Chapters 4, 5, and 6). In Chapter 13, we examine broader chromosomal rearrangements that produce different numbers of chromosomes, reshuffle genes between nonhomologous chromosomes, and reorganize the genes of a single chromosome. These large-scale modifications provide some of the important variations that fuel evolution.

SOLVED PROBLEMS

I. Scientists can construct YACs that range in size from 15 kb to 1 Mb. Based on DNA length, what level of chromosome compaction would you predict for a YAC of 50 kb compared with a YAC of 500 kb?

Answer

To answer this question, you need to apply information about the amount of DNA needed to achieve different levels of chromosome condensation.

The 500 kb YAC would probably be more condensed than the 50 kb YAC based on its larger size. The DNA of both YACs would be wound around histones to form the nucleosome structure (160 bp around the core histones plus 40 bp in linker region). That DNA would be further compacted into 300 Å fibers that contain six nucleosomes per turn. The 500 kb YAC would be compacted at a higher level of order, presumably in radial loops that occur every 60–100 kb in the chromosome, but the 50 kb YAC is not large enough to be packaged in this way.

DNA: © Design Pics/Bilderbuch RF

II. A protein called CBF1 was identified in yeast as a centromere-binding protein. The CBF1 protein is essential for proper chromosome segregation during cell division in yeast. You have identified a gene from the human genome that encodes a protein with some similarity in amino acid sequence with the yeast CBF1 protein.

a. How could you establish whether the protein encoded by the human gene is associated with human centromere regions? (Assume that you can make an antibody that can bind specifically to this protein.) Why would your test not be a FISH experiment?

b. Describe two methods to test if this human protein might actually participate in centromere function (as opposed to just being present at centromeres). For the first method, assume that you can easily produce mutations in any given gene in a human tissue culture cell or even in a whole mouse.

For the second method, you should use two different recombinant YACs. YAC-1 contains the yeast *CBF1*$^+$ gene and the yeast *URA3*$^+$ gene that allows *URA3*$^-$ yeast to grow in the absence of uracil. YAC-2 contains the wild type human *CBF1*-related gene and a *TRP*$^+$ gene that allows *TRP*$^-$ yeast cells to grow in the absence of the amino acid tryptophan. You will also want to use 5-FOA, a chemical closely related to the URA3 enzyme's normal substrate. Yeast that make the URA3 protein cannot grow in the presence of 5-FOA because the enzyme will convert 5-FOA to a lethal toxin.

Answer

This question requires an understanding of the structure and function of centromeres in ensuring proper chromosome segregation during cell division.

a. You would require a molecular probe that could specifically bind to the human protein related to yeast CBF1. An antibody that reacts with this human protein and that is labeled with a fluorescent tag would act as such a probe. You would purify large amounts of the human CBF1-related protein (Chapter 16 discusses how this can be easily accomplished using recombinant DNA technology), and then inject this protein into a rabbit or other experimental animal. The rabbit would develop antibodies against the protein (analogous to the way humans make antibodies against viral proteins in vaccines). You could obtain these antibodies from the rabbit's blood and label them with a fluorescent tag. Finally, you would place the tagged antibodies on a microscope slide on which human chromosomes are displayed. If the human CBF1-related protein is associated with human centromeres, you would see a fluorescent pattern similar to that seen in Fig. 12.2, which is the result of exactly this kind of experiment using antibodies against a different centromeric protein. This is not a FISH experiment because the probe is not a nucleic acid that hybridizes specifically to DNA sequences on a chromosome, but is instead an antibody that binds to a protein possibly associated with centromeric DNA sequences.

b. Two general lines of experimentation would allow you to ask if this human protein is involved in centromere function.

First, you could try to eliminate or reduce the function of the gene encoding this protein in human cells or cells of a related mammal like mice, and then determine if, when they divide, these cells mis-segregate their chromosomes at a high frequency. Chapter 18 discusses various ways to disrupt mammalian genes or interfere with their expression.

Second, you could determine if expression of the human protein can replace the function of the yeast *CBF1* gene. You would transform triple mutant yeast cells (*CBF1*$^-$ *URA3*$^-$ *TRP*$^-$) with YAC-1 (yeast *CBF1*$^+$ and *URA3*$^+$). The transformed yeast cells should be viable as long as they are grown in the presence of tryptophan. Next, you want to see if you can replace YAC-1 with YAC-2 (human *CBF1*$^+$ and *TRP*$^+$). You will select cells that are transformed with YAC-2 and also have lost YAC-1 by growing the yeast in the absence of tryptophan (selects for the *TRP*$^+$ gene in YAC-2) and in the presence of 5-FOA (selects against the *URA3*$^+$ gene in YAC-1). Yeast cells containing YAC-2 instead of YAC-1 will be obtained only if human CBF1 protein can substitute for the yeast *CBF1*$^+$ gene.

PROBLEMS

Vocabulary

1. For each of the terms in the left column, choose the best matching phrase in the right column.

a. telomere	1. protein complex that keeps sister chromatids together until anaphase
b. G bands	2. origin of replication in yeast
c. kinetochore	3. repetitive DNA found near the centromere in higher eukaryotes
d. nucleosome	4. specialized structure at the end of a linear chromosome
e. ARS	5. complexes of DNA, protein, and RNA in the eukaryotic nucleus
f. satellite DNA	6. small basic proteins that bind to DNA and form the core of the nucleosome
g. chromatin	7. complex of DNA and proteins where spindle fibers attach to a chromosome
h. cohesin	8. beadlike structure consisting of DNA wound around histone proteins
i. histones	9. protein complex that protects telomeres from degradation and end-to-end fusions
j. shelterin	10. regions of a chromosome that are distinguished by staining differences

Section 12.1

2. Many proteins other than histones are found associated with chromosomes. What roles do these nonhistone proteins play? Why do chromosomes have more different types of nonhistone than histone proteins?

Section 12.2

3. What difference exists between the compaction of chromosomes during metaphase and interphase? Give at least one reason why this difference may be necessary.
4. What is the role of the core histones in compaction as compared to the role of histone H1?
5. a. About how many molecules of histone H2A would be required in a typical human cell just after the completion of S phase, assuming an average nucleosome spacing of 200 bp?
   b. During what stage of the cell cycle is it most crucial to synthesize new histone proteins?
   c. The human genome contains 60 histone genes, with 10–15 genes of each type (H1, H2A, H2B, H3, and H4). Why do you think the genome contains multiple copies of each histone gene?
6. The enzyme *micrococcal nuclease* can cleave phosphodiester bonds on single- or double-stranded DNAs, but DNA that is bound to proteins is protected from digestion by micrococcal nuclease. When chromatin from eukaryotic cells is treated for a short period of time with micrococcal nuclease and then the DNA is extracted and analyzed by electrophoresis and ethidium bromide staining, the pattern shown in lane A on the following gel is found. Treatment for a longer time results in the pattern shown in lane B, and treatment for yet more time yields that shown in lane C. Interpret these results.

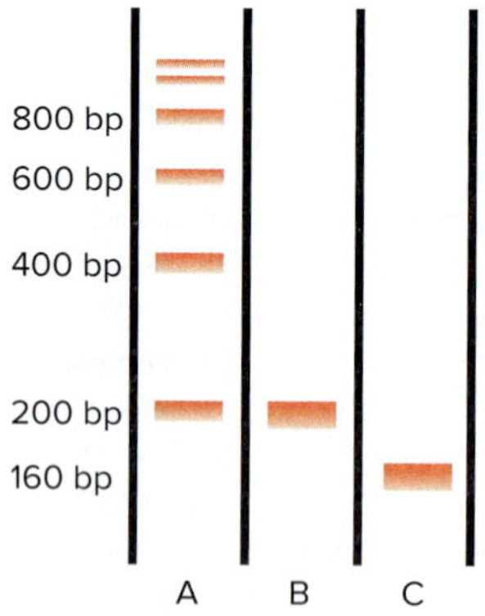

7. a. What letters are used to represent the short and long arms of human chromosomes?
   b. Sketch a schematic diagram of a hypothetical chromosome 3 that has 3 regions with 2 bands each on the short arm and 5 regions with 3 bands each on the long arm. Label the arms, regions, and bands and indicate a gene at position 3p32.
8. About 2000 G bands are visible in a high-resolution karyotype of the 3 billion base pairs in the haploid human genome. If the genome contains about 27,000 genes, about how many genes would be removed by a deletion of DNA that could be detected by karyotype analysis?
9. Suppose you performed a fluorescence *in situ* hybridization experiment (FISH) on chromosomes from a human cell using a probe corresponding to a gene located near (but not at) the telomere of the q arm of chromosome 4.
   a. On the following idiogram, which shows only chromosomes 1–5 contained in this diploid cell, indicate the location of all fluorescent signals.

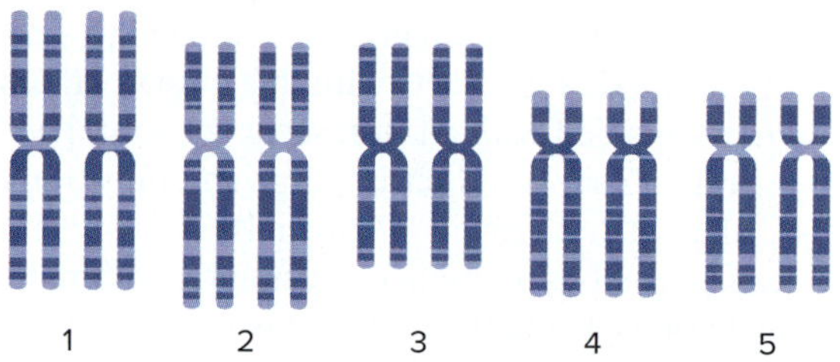

   b. Compare your idiogram with the result of the FISH experiment shown in Fig. 12.9a. Why are the chromosomes scattered in Fig. 12.9a, rather than being present in neatly arranged pairs of homologous chromosomes as in the idiogram? Do you think it is likely that the gene whose DNA was used as a probe in Fig. 12.9a is found on the q arm of human chromosome 21?

Section 12.3

10. Which of the following would be suggested by a DNase hypersensitive site?
    a. No transcription occurs in this region of the chromosome.
    b. The chromatin is in a more open state than a region without the hypersensitive site.
11. For each of the following pairs of chromatin types, which is the most condensed?
    a. 100 Å fiber or 300 Å fiber
    b. 300 Å fiber or DNA loops attached to a scaffold
    c. euchromatin or heterochromatin
    d. interphase chromosomes or metaphase chromosomes

12. Give examples of constitutive and facultative heterochromatin in:
    a. *Drosophila*
    b. humans
13. One histone modification that is seen consistently in many species is the addition of an acetyl group to the twelfth lysine in the H4 protein. If you were a geneticist working on yeast and had a clone of the H4 gene, what could you do to test whether the acetylation at this specific lysine was necessary for the functioning of chromatin?
14. Recently, scientists constructed a transgene that expresses a mutant form of *Drosophila* histone H3 in which lysine 27 in the histone tail was changed to methionine (H3K27M). Expression of the H3K27M transgene results in aberrant development of fruit flies because of inappropriate expression of many different genes. Explain this finding.
15. *Drosophila* geneticists have isolated many mutations that modify position-effect variegation. Dominant *suppressors of variegation* [*Su(var)*s] cause less frequent inactivation of genes brought near heterochromatin by chromosome rearrangements, while dominant *enhancers of variegation* [*E(var)*s] cause more frequent inactivation of such genes.
    a. What effects would each of these two kinds of mutations have on position-effect variegation of the *white* gene in *Drosophila* (that is, would the eyes be more red or more white)?
    b. Assuming that these *Su(var)* and *E(var)* mutations are loss-of-function (null) alleles in the corresponding genes, what kinds of proteins do you think these genes might encode?
16. On the following figures, genes *A* and *B* are on the X chromosome (*blue*) and both are subject to X inactivation, while genes *C* and *D* are on chromosome 17 (an autosome; *red*). *F* and *S* refer to alleles encoding fast- and slow-migrating forms of the corresponding proteins that can be discriminated by electrophoresis.

    For women 2 and 3 in the figures that follow, indicate all the possible forms of the four proteins that could be expressed in clones made from different individual somatic cells that already had one or more Barr bodies. As an example, some clones from normal woman 1 could express the A^F, B^F, C^F, C^S, D^F, and D^S proteins, while other clones could express the A^S, B^S, C^F, C^S, D^F, and D^S proteins. None of the clones from woman 1 should make both the slow and fast forms of proteins A or B. Woman 2 is a heterozygote for a deletion of the XIC. Woman 3 is a heterozygote for a *reciprocal translocation* in which parts of the X chromosome and chromosome 17 have exchanged places.

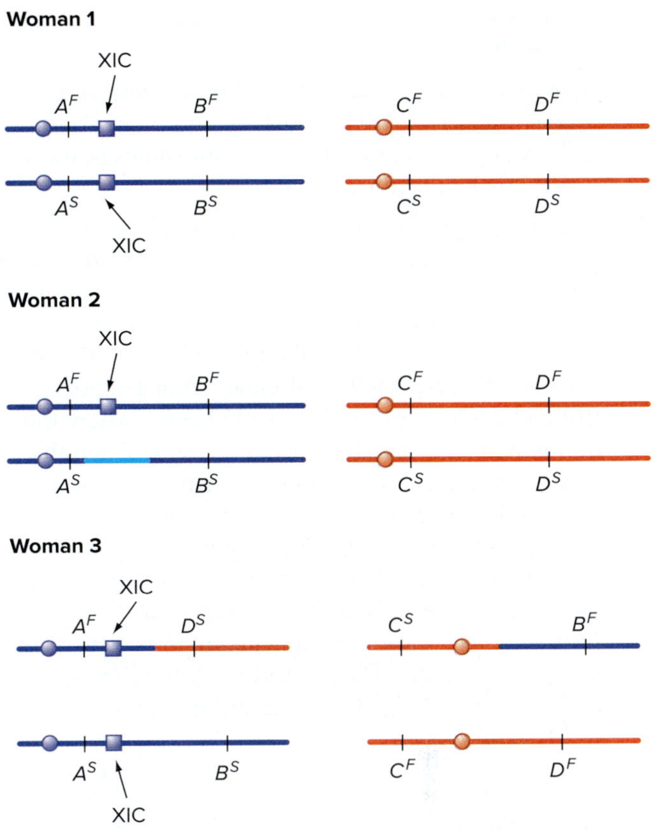

17. How one X chromosome is chosen to express *Xist* is unknown. One clue to the choice mechanism lies in another gene in the XIC called *Tsix*, which is transcribed only from the X chromosome that will remain active. *Tsix* overlaps *Xist* and is transcribed in the opposite direction, as shown in the following figure. *Tsix* produces a long noncoding RNA whose sequence is complementary (antisense) to that of *Xist*, explaining its name (*Xist* backwards is *Tsix*). *Tsix* expression from the future active X prevents *Xist* expression from that X chromosome.

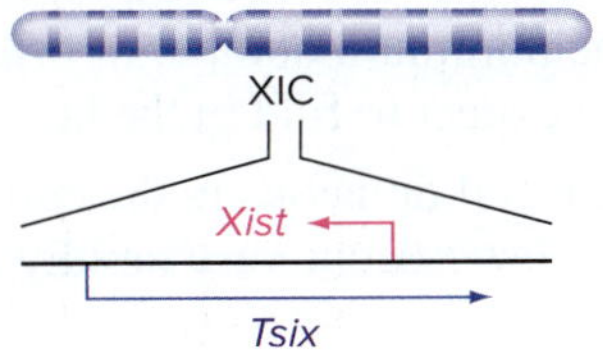

    a. Suppose a female mammal has one normal X chromosome and one X chromosome that has a mutation preventing expression of the *Tsix* RNA but allowing expression of the *Xist* RNA. In the cells of this female, which X chromosome would be more likely to become a Barr body?
    b. Formulate one or more hypotheses that might explain why *Tsix* transcription on an X chromosome might interfere with the expression of *Xist* from the

same X chromosome. Outline experiments to test your hypotheses.

c. Why does knowledge about the existence of the *Tsix* transcript still not solve the problem about how cells decide which X chromosomes to inactivate by heterochromatization?

18. The first page of this chapter displays photos of Rainbow and her clone, CC. Both cats are tabby (their colored fur has a mottled pattern) and both have white regions, mainly on their bellies and legs. However, a major difference exists in their appearances. Rainbow is a *calico tabby*; she has black and orange spots controlled by alleles of the X-linked *O* gene, where the dominant trait is black (*O*-) and the recessive trait is orange (*oo*). However, Rainbow's clone, CC, is a *black tabby;* she lacks the orange spots.

    CC was made by transferring into an enucleated oocyte a single nucleus from one of Rainbow's diploid somatic cells. That diploid oocyte, a mimic of a fertilized egg, was then implanted in the uterus of a surrogate mother. (The process of cloning by somatic cell nuclear transfer will be described in more detail in Chapter 18.)

    a. What is Rainbow's (and CC's) genotype for the *O* gene?

    b. Given that CC was made from a single one of Rainbow's somatic cell nuclei, explain the difference between CC's and Rainbow's appearances.

    c. Would every clone of Rainbow's be a black tabby like CC, or do other possibilities exist? Explain.

Section 12.4

19. The human genome contains about 3 billion base pairs. During the first cell division after fertilization of a human embryo, S phase is approximately three hours long. Assuming an average DNA polymerase rate of 50 nucleotides/second over the entire S phase, what is the minimum number of origins of replication you would expect to find in the human genome?

20. The mitotic cell divisions in the early embryo of *D. melanogaster* occur very rapidly (every eight minutes).

    a. If there were one bidirectional origin in the middle of each chromosome, how many nucleotides would DNA polymerase have to add per second to replicate all the DNA in the longest chromosome (66 Mb) during the eight-minute early embryonic cell cycles? (Assume that replication occurs during the entire cell division cycle.)

    b. In fact, many origins of replication are active on each chromosome during the early embryonic divisions and are spaced approximately 7 kb apart. Calculate the average rate (per second) with which DNA polymerase adds complementary nucleotides to a growing chain in the early *Drosophila* embryo, making the same assumption as in part (a).

21. In an experiment published in the journal *Cell* in 2014, Amnon Koren and Steven McCarroll isolated two populations of growing tissue culture cells from each of two unrelated people from different parts of the world. One population from each person consisted of millions of cells that were in G_1 of the cell cycle; the other population was a similar number of cells that were in S phase for various amounts of time. The scientists then performed high-throughput DNA sequencing on these cell populations.

    The two graphs that follow show the data for the two individuals. In each graph, the x-axis represents positions along a chromosome (here, chromosome 8), and the y-axis represents the ratio between the number of reads obtained for a given region of the genome from the S phase sample divided by the number of reads obtained for the same region from the G_1 sample. Each small purple dot is 2 kb along the chromosome; the black line is the moving average of the purple dots.

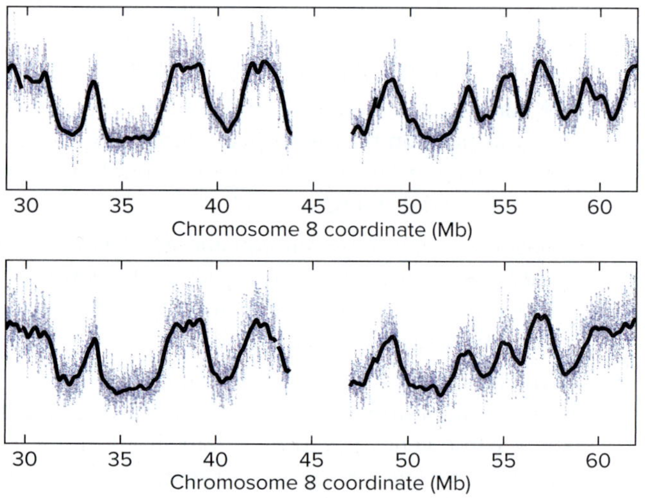

Source: Amnon Koren, Dept. of Molecular Biology and Genetics, Cornell University.

    a. At chromosomal coordinate 33 Mb, the y-axis value is much higher than at coordinate 35 Mb. What does this fact tell you about the timing of DNA replication at these two locations?

    b. Scientists still do not have a good idea about the nature of DNA sequences or chromatin structures that define origins of replication in human cells. If you were trying to locate such origins of replication, where would you look?

    c. Suppose you did a similar experiment using two populations with the same number of cells, one population in G_2 and the other in G_1. If you graphed the data in a similar fashion, with the y-axis representing the ratio of the number of reads from the G_2 sample divided by the number of reads from the G_1 sample, what would the plot look like?

d. The patterns for these two people are very nearly the same, even though they are completely unrelated. What does this fact suggest?

e. These scientists later reasoned that they could obtain the same kind of information from any person whose genome had been sequenced by high-throughput methods, without separating out populations of cells at different cell cycle stages. What would have to be true about the cells analyzed and the kinds of data available? Why would you want to look at this data from many different people?

22. a. What DNA sequences are found at the telomeres of human chromosomes?

b. What functions do the two telomere-associated complexes, telomerase and shelterin, fulfill at chromosome ends?

c. Where do you think that the RNA component of telomerase comes from?

23. a. Mice engineered to block expression of the gene encoding the telomerase protein age at a much faster rate than normal and have decreased life spans. When expression of the telomerase protein is turned back on in mice that are prematurely old, many negative effects of this aging are rapidly and dramatically reversed. Provide a possible explanation for these results.

b. The results of these mouse experiments have led some researchers to propose that treatments that could lead to overexpression of the telomerase gene might serve as a "fountain of youth" leading to reversal of aging in humans. Why do you think we should be very cautious about trying such treatments? Your argument should address why it might be advantageous to multicellular organisms for most of their somatic cells *not* to express telomerase.

24. a. In a fluorescent *in situ* hybridization (FISH) experiment, what would you see if you used DNA containing multiple copies of 3′ AATCCC 5′ as a probe? Show your results on the idiogram accompanying Problem 9.

b. What DNA sequence would you use as a probe to track one end of one particular chromosome in a FISH experiment? What would your results look like on the same idiogram?

25. If you are comparing the two telomeres in each entry in the following list, in which cases would you expect the two telomeres always to have exactly the same number of TTAGGG repeats?

a. One telomere at one end of a chromosome, one telomere at one end of a nonhomologous chromosome.

b. One telomere at one end of a chromosome, one telomere at the corresponding end of the homologous chromosome.

c. One telomere at one end of a chromosome, the other telomere at the other end of the same chromosome.

d. One telomere at one end of a chromatid, the other telomere at the corresponding position in the sister chromatid.

26. One hallmark of cancer cells is their ability to divide indefinitely, in contrast with most normal somatic cells that undergo senescence after 30 to 50 generations of divisions. We saw in this chapter that one reason for this difference is that many cancer cells express the telomerase enzyme that can mediate telomere lengthening.

Interestingly, about 15% of tumors do not express telomerase. Instead, they lengthen their telomeres by an alternative pathway. Tumor cells of this class appear to have telomeres that are highly heterogenous in length; some telomeres have many more TTAGGG repeats than others.

a. Diagram an event involving homologous recombination that would allow some telomeres in these cells to become longer. What feature(s) of telomeres make(s) such homologous recombination possible?

b. Does this recombination need to occur between homologous telomeres (that is, telomeres of the same arm of the same chromosome)? If such recombination could occur between nonhomologous telomeres, how might you detect it?

c. Almost all cells that undergo this alternative telomere lengthening pathway have *t-circles*: small molecules of circular DNA made up almost exclusively of telomeric sequences. Diagram how these circles might participate in telomere lengthening.

Section 12.5

27. a. What DNA sequences are commonly found at human centromeric regions?

b. What functions do the two centromere-associated complexes, cohesin and the kinetochore, play in chromosome mechanics?

28. On the graphs presented in Problem 21, no data is available for the region between coordinates 44 and 47 of chromosome 8. What kind of chromosome feature might this region represent? (*Hint:* It remains difficult for computer programs to assemble DNA sequences from regions containing large amounts of repetitive DNA.)

29. The Rec8 protein is a cohesin complex subunit that is normally made only during meiosis; it substitutes for the purple protein shown in the mitotic cohesin complex in Fig. 12.25. Rec8 is not cleaved during meiosis I, but it is cleaved early in meiosis II to allow sister chromatids to segregate during anaphase II.

Scientists hypothesized that a protein (shugoshin) protects the Rec8 protein from cleavage and degradation during meiosis I. To identify shugoshin, researchers first produced the Rec8 protein in mitotically dividing yeast cells. In these cells, Rec8 was cleaved during mitosis and the cells suffered no harmful effects. Researchers then expressed other, normally meiosis-specific proteins in cells expressing Rec8 mitotically. The scientists were able to identify shugoshin as a protein that protects Rec8 from degradation.

What effect do you think expressing shugoshin had on the mitotically dividing cells expressing Rec8? What phenotype would the cells show?

30. In the following diagram, each line represents a double-stranded DNA molecule.

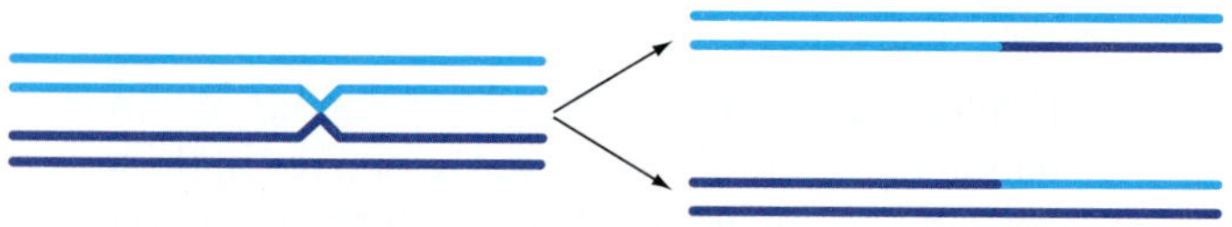

a. What type of cell division is being represented, and which stages of that cell division are shown in the two parts of the figure? What is the relationship between the lines drawn in the same shade of blue? Between lines drawn in different shades of blue?

b. Cohesin is added to chromosomes immediately after S phase. Cohesin complexes are concentrated at the centromeres, but some cohesin complexes are scattered along the chromosome arms. On a copy of the figure, indicate the distribution of cohesin complexes on the chromosomes. Distinguish between cohesin at the centromeres and cohesin along the arms. Your diagram should represent how cohesin might keep DNA molecules together.

c. Look carefully at your drawings. What keeps all the blue lines together at the metaphase plate of this kind of cell division, even though forces are pulling at the kinetochores in opposite directions?

d. Again looking carefully at your drawings, what can you conclude about the function of shugoshin during the type of cell division being depicted? What is the name of the enzyme whose function shugoshin prevents, and what does this enzyme do?

31. In the 1920s, Barbara McClintock, later a Nobel laureate for her discovery of transposable elements, examined the behavior of chromosomes in wheat cells that had been subjected to X-rays. She noticed that the X-rays produced chromosomal breaks during G1 phase, and that after subsequent chromosome replication in S phase, the broken ends of the two sister chromatids could join together to make a *fusion chromosome* larger than the original. Even later, during mitotic metaphase and early in mitotic anaphase, the joined sister chromatids would form an unusual *bridge* structure in which chromatin was stretched between the two spindle poles and could then eventually break. She called this phenomenon the breakage-fusion-bridge cycle. Each of photographs (a) and (b) that follow shows a cell in early mitotic anaphase that has two such chromatin bridges.

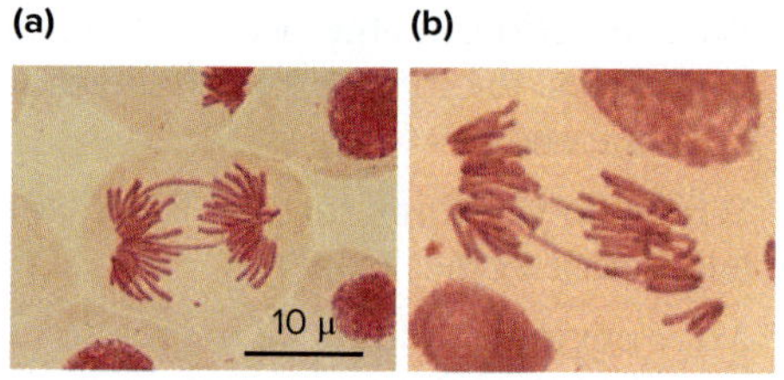

Courtesy Dr. Marin-Morales, São Paulo State University (UNESP), Ventura-Camargo BC, Maltempi PPP, Marin-Morales MA (2011), "The use of the cytogenetic to identify mechanisms of action of an azo dye in *Allium cepa* meristematic cells," *J Environment Analytic Toxicol,* 1:109. doi:10.4172/2161-0525.1000109

a. What ensures that the ends of normal chromosomes do not fuse together as do the ends of the sister chromatids after breakage?

b. The following figure shows a chromosome with genes *A–G;* the arrow indicates the location of X-ray-induced breakage. Draw the resulting bridge (that is, the large fused chromosome) as it would be seen in mitotic anaphase, and label all the genes and important chromosomal structures the bridge contains. Use arrows to show the forces exerted by the spindle apparatus on this bridge.

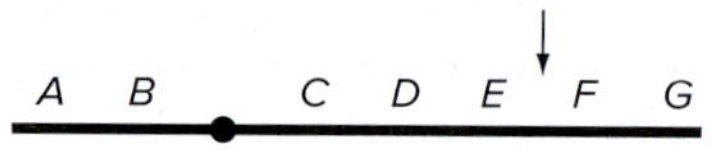

c. If the sister chromatids fuse, why *must* the fusion chromosome behave as a bridge during mitosis? [Think about the forces pulling on the bridge described in your answer to part (b).]

d. What is likely to happen to the bridge during mitotic anaphase? What then is likely to happen in the two daughter cells produced by the mitosis just described, and why? (*Hint:* McClintock's name for this phenomenon implies it is a cycle.)

32. Give at least one example of a chromosomal structure or function affected by the following mechanisms for modulating chromatin structure:
    a. Posttranslational changes of the normal histones found in the nucleosome
    b. Nucleosomes with variant histones encoded by special genes

33. *Cornelia de Lange syndrome* (*CdLS*) is a rare human disease caused by a dominant loss-of-function mutation in any one of at least five different genes, all of which encode components or regulators of the cohesin protein complex. People with CdLS have a wide range of morphological abnormalities, growth retardation, and mental impairment. Analysis of CdLS patients shows that in addition to chromosomal mis-segregation during cell division, their abnormal phenotype is likely due to widespread mis-regulation of gene expression during development. Cohesin may play a role in organizing chromatin loops necessary for proper regulation of transcription. (You will learn more about this topic in Chapter 17.)
    a. In different families, CdLS can show an autosomal dominant or X-linked dominant inheritance pattern. How is this possible?
    b. Explain how a loss-of-function allele in a gene encoding a cohesin protein could be dominant to its wild-type counterpart.
    c. CdLS is usually caused by new mutation in one parent's gamete. Why?

Section 12.6

34. a. Give at least three examples of types of mutations that would disrupt the process of mitotic chromosome segregation. That is, explain in what DNA structures or in genes encoding what kinds of proteins you would find these segregation-disrupting mutations.
    b. How could you use yeast artificial chromosomes (YACs) to find such mutations in *S. cerevisiae?*

35. A number of yeast-derived elements were added to the circular bacterial plasmid pBR322. Yeast that require uracil for growth (Ura^- cells) were transformed with these modified plasmids and Ura^+ colonies were selected by growth in media lacking uracil. For plasmids containing each of the elements listed in parts (a) to (c), indicate whether you expect the plasmid to integrate into a chromosome by recombination, or instead whether it is maintained separately as a plasmid. If the plasmid is maintained autonomously, is it stably inherited by all of the daughter cells of subsequent generations when you no longer select for Ura^+ cells (that is, when you grow the yeast in media containing uracil)?
    a. URA^+ gene
    b. URA^+ gene, ARS
    c. URA^+ gene, ARS, CEN (centromere)
    d. What would need to be added in order for these sequences to be maintained stably in yeast cells as a linear artificial chromosome?

36. A DNA fragment containing yeast centromere DNA was cloned into a TRP^+ ARS plasmid, YRp7, causing the plasmid to become mitotically very stable (that is, the plasmid was transmitted during mitotic divisions to each daughter cell). The assay for mitotic stability consists of growing a transformed cell without selection for the plasmid and determining the number of Trp^+ colonies remaining after 20 generations of growth under conditions that are not selective for the plasmid.

    To identify the region of the cloned fragment that contained centromere DNA, you cut the initial fragment into smaller pieces, reclone those pieces into YRp7, and test for mitotic stability. Based on the map that follows and results of the mitotic stability assay, where is the centromeric DNA located? (On the map, B, H, and S refer to recognition sites for three different restriction enzymes; the numbers on the map are restriction fragment sizes in kb.)

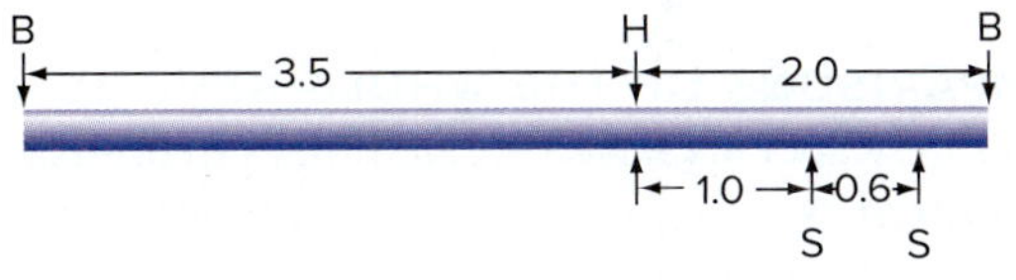

Results of Mitotic Stability Assay

Plasmid DNA	Percentage of Trp^+ colonies: after 20 generations
YRp7	0.9
YRp7 + 5.5 kb *Bam*HI (B)	68.1
YRp7 + 3.5 kb *Bam*HI-*Hin*dIII (H)	0.5
YRp7 + 2.0 kb *Bam*HI-*Hin*dIII	80.3
YRp7 + 0.6 kb *Sau*3A (S)	76.2
YRp7 + 1.0 kb *Hin*dIII-*Sau*3A	0.7

37. The completely synthetic yeast chromosome Syn III contains a loxP site in the 3′ UTR of every gene that is potentially nonessential to yeast survival. As you will recall from Chapter 6, loxP sites are targets of site-specific recombination. The researchers who constructed Syn III included these loxP sites as a way to "scramble" the chromosome, meaning that parts of the chromosome could easily be deleted or rearranged. The goal of these investigations is to drive the evolution of Syn III so as to define a minimal genome that can support the life of this organism. Outline the experiment the researchers would do to scramble Syn III in order to define a minimal genome.

chapter 13 Chromosomal Rearrangements and Changes in Chromosome Number

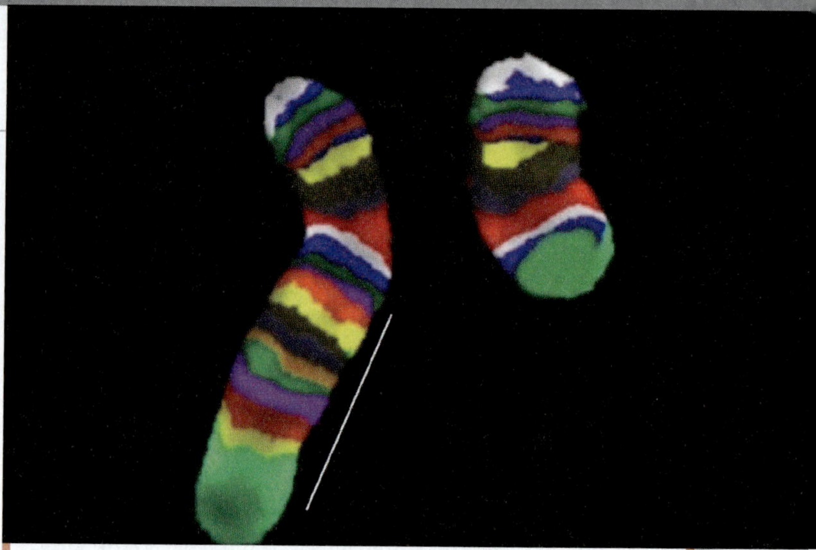

Multicolor banding, a FISH technique, generates a chromosome barcode *that enables detection of major chromosomal rearrangements. The different colors represent region-specific DNA probes labeled with different fluorescent tags. At left is a barcoded normal human chromosome 5. The chromosome 5 at right has a large deletion of the region indicated by the white line adjacent to the chromosome at left.*

chapter outline

COMPARISONS OF THE WHOLE-GENOME sequences from many species have revealed that **chromosome rearrangement** is a major feature of evolution. For example, each mouse chromosome consists of pieces of several different chromosomes found in humans, and *vice versa.* Focusing on just one case, mouse chromosome 1 contains large blocks of sequences found on human chromosomes 1, 2, 5, 6, 8, 13, and 18 (portrayed in different colors in **Fig. 13.1**). These blocks represent **syntenic segments** in which the identity, order, and transcriptional direction of the genes are almost exactly the same in the two genomes. In principle, scientists could reconstruct the mouse genome by breaking the human genome into 342 fragments, each an average length of about 16 Mb, and pasting these fragments together in a different order. Figure 13.1 illustrates this process in detail for mouse chromosome 1; Fig. 10.15 in Chapter 10 showed the syntenic relationships between the entire mouse and human genomes.

These findings contribute to our understanding of how complex life-forms evolved. Although mice and humans diverged from a common ancestor about 65 million years ago, the DNA sequences in many regions of the two genomes are very similar. It is thus possible to hypothesize that the mouse and human genomes evolved from the genome of a common ancestor through a series of approximately 300 reshaping events in which the chromosomes broke apart and the resulting fragments resealed end to end in novel ways. Some of these chromosomal rearrangements occurred in the lineage leading to mice; others in the diverged lineage leading to humans. Both nucleotide sequence differences and differences in genome organization therefore contribute to dissimilarities between the species.

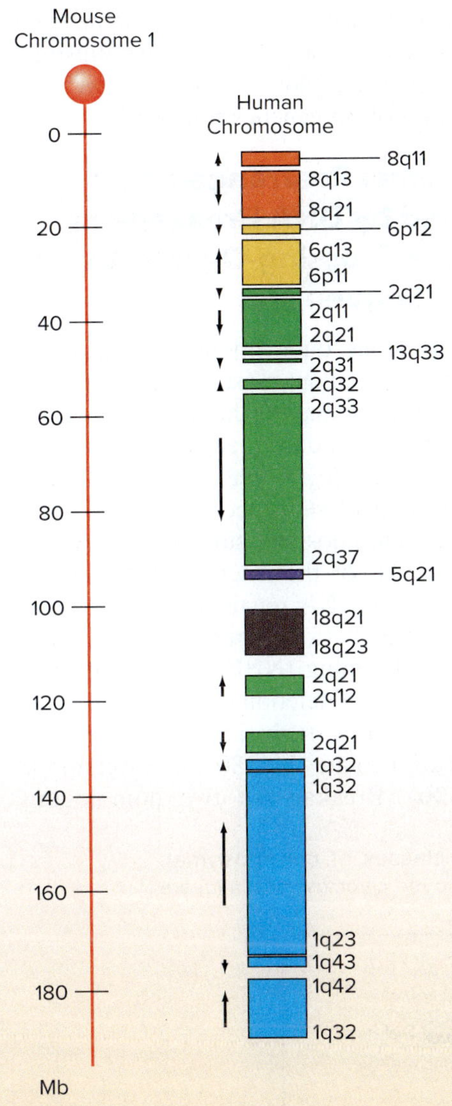

Figure 13.1 Comparing the mouse and human genomes. Mouse chromosome 1 contains large blocks of sequences found on human chromosomes 1, 2, 5, 6, 8, 13, and 18 (different *colors*). *Arrows* indicate the relative orientations of sequence blocks from the same human chromosome.

In this chapter, we examine two types of events that reshape genomes: (1) *rearrangements,* which reorganize the DNA sequences within one or more chromosomes, and (2) *changes in chromosome number* involving losses or gains of entire chromosomes or sets of chromosomes. Rearrangements and changes in chromosome number may affect gene activity or gene transmission by altering the position, order, or number of genes in a cell. Such alterations often, but not always, lead to a genetic imbalance that is harmful to the organism or its progeny.

We can identify two main themes underlying the phenomena of chromosomal changes. First, karyotypes generally remain constant within a species, not because rearrangements and changes in chromosome number occur infrequently (they are, in fact, quite common), but because the genetic instabilities and imbalances produced by such changes usually place individual cells or organisms and their progeny at a selective disadvantage. Second, despite selection against chromosomal variations, related species almost always have different karyotypes, with closely related species (such as chimpanzees and humans) diverging by only a few rearrangements and more distantly related species (such as mice and humans) diverging by a larger number of rearrangements. These observations suggest that significant correlation exists between karyotypic rearrangements and the evolution of new species.

13.1 Rearrangements of Chromosomal DNA

learning objectives

1. Summarize the four main classes of chromosomal rearrangements.
2. Explain the two major mechanisms by which chromosomal rearrangements take place.
3. Describe methods by which researchers can detect rearrangements.

All chromosomal rearrangements alter DNA sequence (**Table 13.1**). Some do so by removing or adding base pairs (**deletions** and **duplications** of particular chromosomal regions, respectively). Others relocate chromosomal regions without changing the number of base pairs they contain (**inversions,** which are half-circle rotations of a chromosomal region; and **reciprocal translocations,** in which two nonhomologous chromosomes exchange parts). This chapter focuses on heritable rearrangements that can be transmitted through the germ line from one generation to the next, but it also explains that the genomes of somatic cells can undergo changes in nucleotide number or order. For example, the Fast Forward Box *Programmed DNA Rearrangements and the Immune System* describes how the normal development of the human immune system depends on noninherited, programmed rearrangements of the genome in certain somatic cells.

In this section, we first explain how heritable chromosomal rearrangements come about and how scientists can track their presence. We then discuss how each kind of

TABLE 13.1 Major Classes of Chromosomal Rearrangements

Letters represent large chromosomal regions. Different (that is, nonhomologous) chromosomes are indicated as *red* and *blue*.

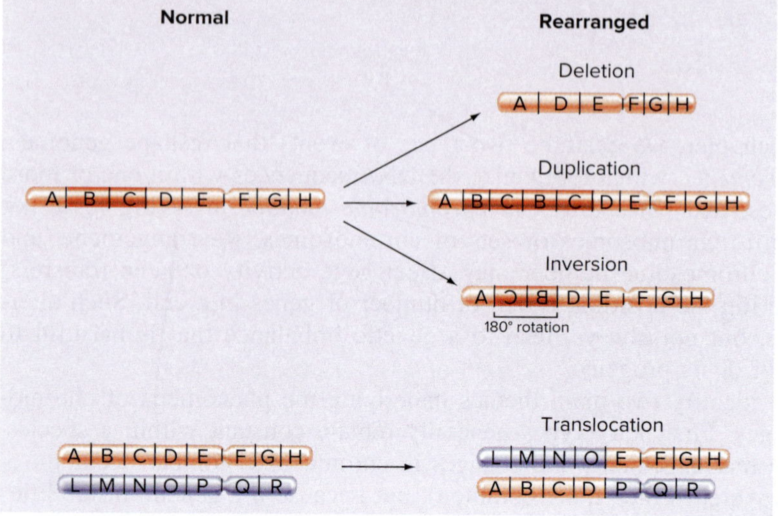

rearrangement can cause mutant phenotypes and affect chromosome behavior. Finally, we illustrate how geneticists can exploit the existence of rearrangements as tools in genetics research.

Chromosomal Rearrangements Are Caused by DNA Breakage or Illegitimate Crossing-Over Between Repeated Sequences

Deletions, inversions, duplications, and reciprocal translocations come about most often by either of two events: chromosomal breakage (produced, for example, by X-rays as described in Chapter 7) or aberrant, so-called *illegitimate* recombination at sites of repeated DNA sequences.

If a single chromosome suffers two double-strand breaks, loss of the fragment between the breaks followed by DNA repair that fuses the remaining broken ends [for example, through nonhomologous end-joining (NHEJ) as described in Chapter 7 results in a deletion (**Fig. 13.2a**). If the intervening fragment instead is stitched back into the chromosome, but only after it rotates by 180°, an inversion is generated (**Fig. 13.2b**). Breakage of two homologous

Figure 13.2 Chromosome breakage and subsequent DNA repair can result in all classes of chromosomal rearrangements. The chromosomes in (c) could be homologs or sister chromatids. The *blue* and *red* chromosomes in (d) are nonhomologous. Letters indicate chromosomal loci.

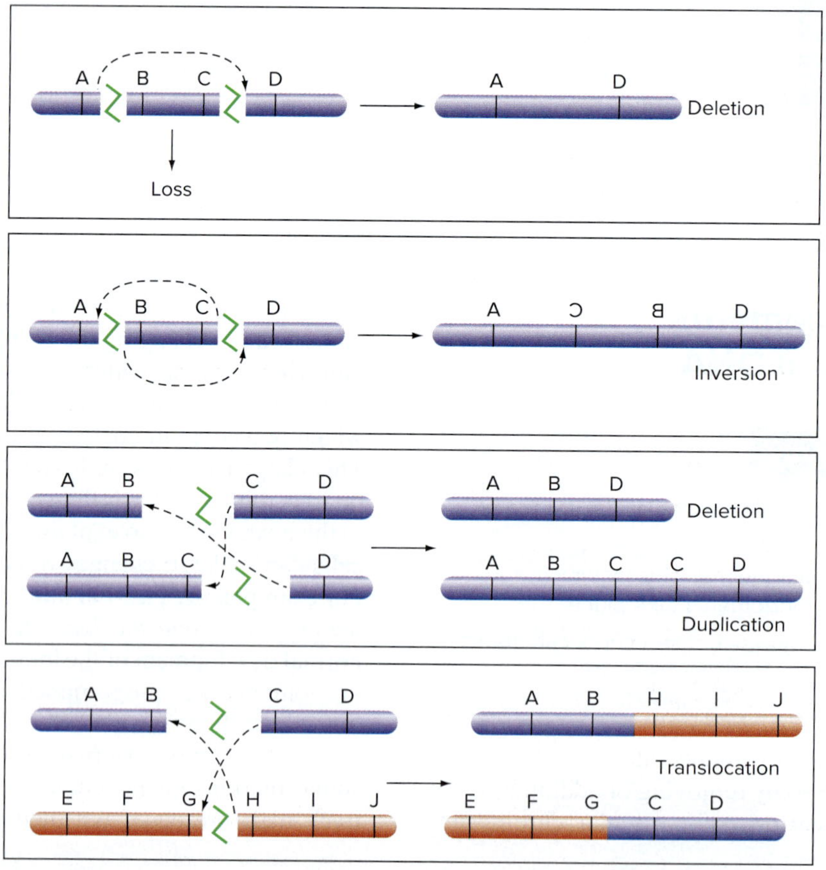

chromosomes or sister chromatids at different locations can result in a deletion and a duplication if the broken chromosome ends change places before they are brought together by DNA repair (**Fig. 13.2c**). Finally, breakage of two nonhomologous chromosomes can result in reciprocal translocation if the broken ends switch places before repair and fusion (**Fig. 13.2d**).

Rare, aberrant crossover events between repeated sequences on the same chromosome or on two different chromosomes can also generate rearrangments. The repeated sequences may be tandem repeats like simple sequence repeats (SSRs); recall that SSR loci with the same repeat sequences are located at several different places in the genome, where they provide sequence similarities that can be recognized by recombination enzymes. Alternatively, the repeats can be *transposable elements*. As will be explained later in the chapter, transposable elements are DNA sequences whose copies move from place to place. A single genome may accumulate hundreds of thousands of copies of such an element.

Crossovers between repeats of the same sequence at two locations on the same chromosome can result in a deletion if the repeats are in the same orientation (**Fig. 13.3a**), or an inversion if the repeats are in opposite orientations (**Fig. 13.3b**). If two homologous chromosomes misalign at repeated sequences and cross over, the result may be a deletion and a duplication (**Fig. 13.3c**). Crossovers at a repeated sequence on two nonhomologous chromosomes generate reciprocal translocations (**Fig. 13.3d**).

Figure 13.3 Aberrant crossing-over at repeated sequences can also produce rearrangements. The *green arrows* represent the repeated sequences and indicate their relative orientations. The repeated DNA sequences may be simple sequence repeats (SSRs) or transposable elements. The chromosomes in (c) could be homologs or sister chromatids. The *red* and *blue* chromosomes in (d) are nonhomologous.

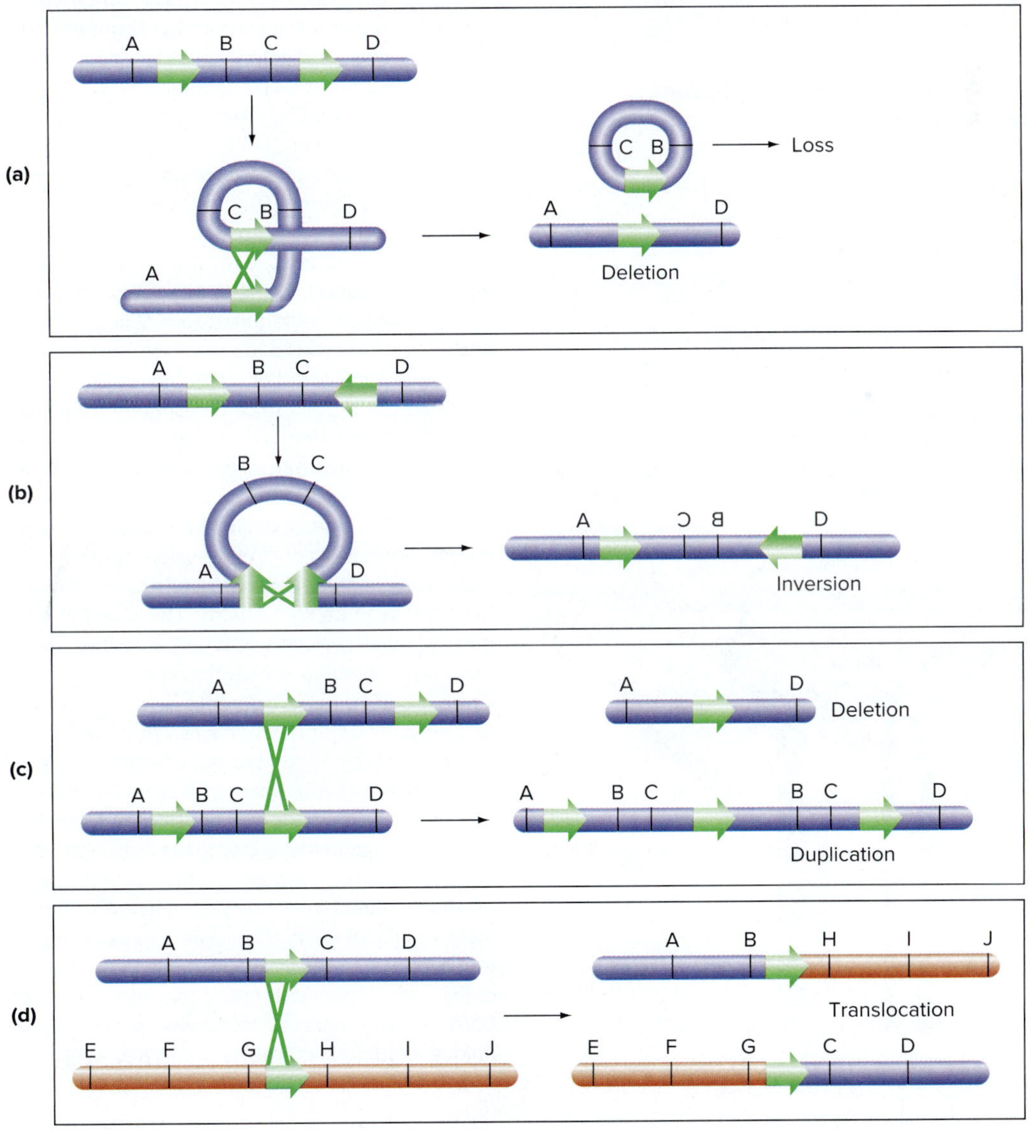

FAST FORWARD

Sprinters: © Robert Michael/Corbis RF

Programmed DNA Rearrangements and the Immune System

The human immune system is a marvel of specificity and diversity. It includes close to a trillion *B lymphocytes,* specialized white blood cells that make more than a billion different varieties of antibodies (also called *immunoglobulins,* or *Igs*). Each B cell, however, makes antibodies against only a single bacterial or viral protein (called an *antigen* in the context of the immune response). The binding of antibody to antigen helps the body attack and neutralize invading pathogens.

One intriguing question about antibody responses is: How can a genome containing only 20,000 to 30,000 ($2–3 \times 10^4$) genes encode a billion (10^9) different types of antibodies? The answer is that programmed gene rearrangements, in conjunction with somatic mutations and the diverse pairing of polypeptides of different sizes, can generate roughly a billion binding specificities from a much smaller number of genes. To understand the mechanism of this diversity, it is necessary to know how antibodies are constructed and how B cells come to express the antibody-encoding genes determining specific antigen-binding sites.

The Genetics of Antibody Formation Produce Specificity and Diversity

All antibody molecules consist of a single copy or multiple copies of the same basic molecular unit. Four polypeptides make up this unit: two identical light chains, and two identical heavy chains. Each light chain is paired with a heavy chain (**Fig. A**). Each light and each heavy chain has a constant (C) domain and a variable (V) domain. The C domain of the heavy chain determines whether the antibody falls into one of five major classes (designated IgM, IgG, IgE, IgD, and IgA), which influence where and how an antibody functions. For example, IgM antibodies form early in an immune response and are anchored in the B-cell membrane; IgG antibodies emerge later and are secreted into the blood serum. The C domains of the light and heavy chains are not involved in determining the specificity of antibodies. Instead, it is the V domains of light and heavy chains that come together to form the antigen-binding site, which defines an antibody's specificity.

Figure A How antibody specificity emerges from molecular structure. Two heavy chains and two light chains held together by disulfide (–S–S–) bonds form the basic unit of an antibody molecule. Both heavy and light chains have variable (V) domains (*yellow*) near their N termini, which associate to form the antigen-binding site. "Hypervariable" stretches of amino acids within the V domains vary extensively between antibody molecules. The remainder of each chain is composed of a C (constant) domain (*blue*); that of the heavy chain has several subdomains (C_{H1}, hinge, C_{H2}, and C_{H3}).

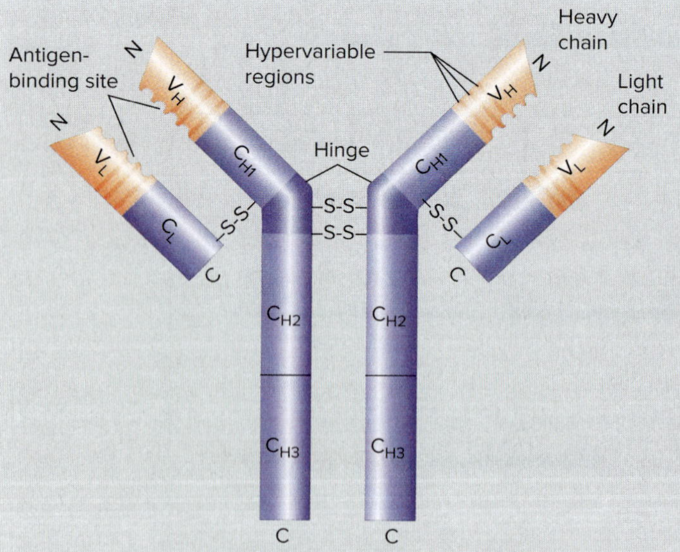

The DNA for all domains of the heavy chain resides on chromosome 14 (**Fig. B**). This heavy-chain gene region consists of more than 100 V-encoding segments, each preceded by a promoter, several D (for diversity) segments, several J (for joining) segments, and nine C-encoding segments preceded by an enhancer (a short DNA segment that aids in the initiation of transcription; see Chapter 17 for details). In all germ-line cells and in most somatic cells, including the cells destined to become B lymphocytes, these various gene segments lie far apart on the chromosome. During B-cell development, however, somatic rearrangements juxtapose random, individual V, D, and J segments together to form the particular variable region that will be transcribed. These rearrangements also place the newly formed variable region next to a C segment and its enhancer, and they further bring the promoter and enhancer into proximity, allowing transcription of the heavy-chain gene. RNA splicing removes the introns from the primary transcript, making a mature mRNA encoding a complete heavy-chain polypeptide.

The somatic rearrangements that shuffle the V, D, J, and C segments at random in each B cell permit the expression of one, and only one, specific heavy chain. Without the rearrangements, antibody gene expression cannot occur. Random somatic rearrangements also generate the actual genes that will be expressed as light chains. The somatic rearrangements allowing the expression of antibodies thus generate enormous diversity of binding sites through the random selection and recombination of gene elements.

Several other mechanisms add to this diversity. First, each gene's DNA elements are joined imprecisely, which occurs by cutting and splicing enzymes that sometimes trim DNA from or add nucleotides to the junctions of the segments they join. This imprecise joining helps create the *hypervariable regions* shown in Fig. A. Next, random somatic mutations in a rearranged gene's V region increase the variation of the antibody's V domain. Finally, in every B cell, two copies of a specific H chain that emerged from random DNA rearrangements combine with two copies of a specific L chain that also emerged from random DNA rearrangements to create molecules with a specific, unique binding site. The fact that any light chain can pair with

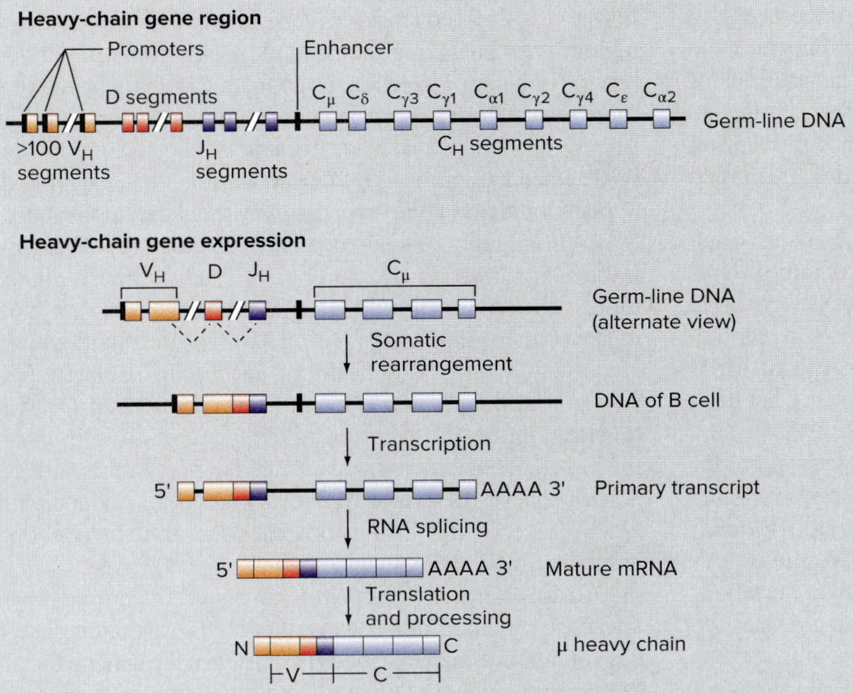

Figure B The heavy-chain gene region on chromosome 14. The DNA of germ-line cells (as well as all non-antibody-producing cells) contains more than 100 V_H segments, about 20 D segments, 6 J_H segments, and 9 C_H segments (*top row*). Each V_H and C_H segment is composed of two or more exons, as seen in the alternate view of the same DNA on the *second row*. In B cells, somatic rearrangements bring together random, individual V_H, D, and J_H segments. The primary transcript made from the newly constructed heavy-chain gene is subsequently spliced into a mature mRNA. The μ heavy chain translated from this mRNA is the type of heavy chain found in IgM antibodies. Later in B-cell development, other rearrangements *(not shown)* connect the same V-D-J variable region to other C_H segments such as C_δ, allowing the synthesis of other antibody classes.

any heavy chain increases exponentially the potential diversity of antibody types. For example, if there were 10^4 different light chains and 10^5 different heavy chains, there would be 10^9 possible combinations of the two.

Mistakes by the Enzymes that Carry Out Antibody Gene Rearrangements Can Lead to Cancer

RagI and RagII are enzymes that interact with DNA sequences in antibody genes to help catalyze the rearrangements just described. In carrying out their rearrangement activities, however, the enzymes sometimes make a mistake that results in a reciprocal translocation between human chromosomes 8 and 14. After this translocation, the enhancer of the chromosome 14 heavy-chain gene lies in the vicinity of the unrelated *c-myc* gene from chromosome 8. Under normal circumstances, *c-myc* generates a transcription factor that turns on other genes active in cell division, at the appropriate time and rate in the cell cycle. However, the translocated antibody-gene enhancer accelerates expression of *c-myc,* causing B cells containing the translocation to divide out of control. This uncontrolled B-cell division leads to a cancer known as *Burkitt lymphoma* (**Fig. C**).

Thus, although programmed gene rearrangements are necessary for the normal development of a healthy immune system, misfiring of the rearrangement mechanism can promote disease.

Figure C Misguided translocations can help cause Burkitt lymphoma. In DNA from a particular Burkitt lymphoma patient, a translocation brings transcription of the *c-myc* gene (*green*) under the control of the enhancer adjacent to C_μ. As a result, B cells produce abnormally high levels of the c-myc protein. Apparently, the RagI and RagII enzymes have mistakenly connected a J_H segment to the *c-myc* gene from chromosome 8, instead of to a D segment.

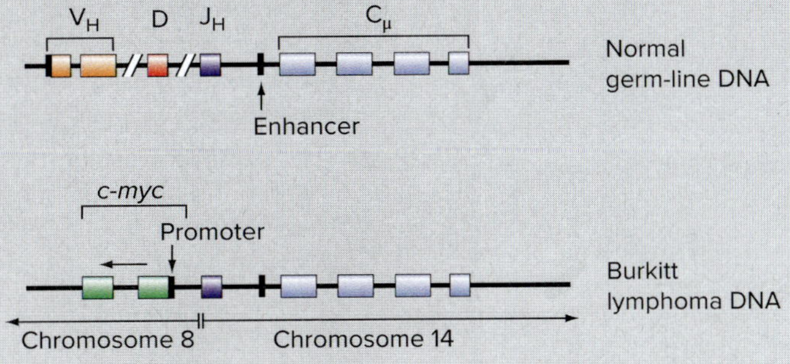

A Variety of Methods Can Detect Chromosomal Rearrangements

Geneticists require efficient tools to determine whether individual genomes contain chromosomal rearrangements, to define exactly which genes are rearranged, and to track the transmission of the rearrangements between generations of cells or individuals. Some techniques allow scientists to gain a general idea about the existence and rough location of a rearrangement, while others home in on the rearranged DNA at the molecular level.

Fluorescent *in situ* hybridization (FISH) techniques often provide the first hints about the presence and nature of rearrangements. Recall from Chapter 12 that in FISH, genomic DNA probes attached to fluorescent tags hybridize to complementary sequences on chromosomes. In one type of FISH analysis called spectral karyotyping (SKY), probes paint each chromosome a different color (see Fig. 12.9b). **Figure 13.4** shows a similar chromosome painting analysis with probes for just two different chromosomes that reveals a reciprocal translocation. In a more refined technique called *multicolor banding*, FISH probes specific for particular regions of chromosomes generate *chromosome barcodes*. The photograph at the beginning of this chapter shows how multicolor banding reveals the presence and nature of a deletion in a human chromosome.

Microarrays of genomic DNA can be used to detect deletions or duplications too small to be found by barcoding. Scientists prepare a probe from an individual's genomic DNA, and then hybridize the probe to a microarray of millions of fragments representing DNA throughout the genome (recall Fig. 11.17). The presence and approximate location of deletions or duplications are revealed by decreased or increased hybridization, respectively, to particular spots on the microarray.

Figure 13.4 Reciprocal translocation revealed by chromosome painting. This FISH karyotype is of a person heterozygous for a reciprocal translocation. The two reciprocally translocated chromosomes are stained both *red* and *green* (*arrows*). Two normal, nontranslocated chromosomes are stained entirely *red* or entirely *green* (*arrowheads*).

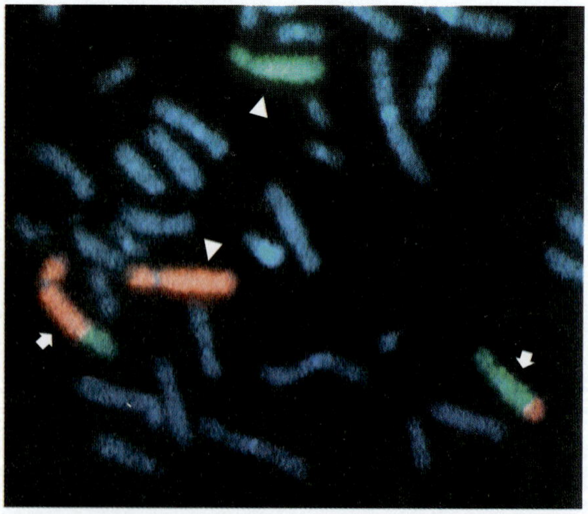

Genome sequencing can reveal chromosomal rearrangements at the ultimate level of resolution: the nucleotide pair. Fewer reads of sequences will be obtained from the deleted region when analyzing the genomic DNA from an individual heterozygous for a deletion, as compared with the reads of the same region from a normal genome without the deletion. Conversely, more sequence reads from the duplicated region would be detected in genomic DNA from an individual carrying that duplication than in wild-type DNA without the duplication.

Whole genome sequence analysis can provide detailed information beyond the number of reads because all rearrangement types will juxtapose DNA sequences that would not be connected normally (**Fig. 13.5**). The unusual sequences at these junctions will be found in certain reads from the whole genome sequence data. These juxtaposed DNA sequences define the precise **rearrangement breakpoints**—the base pairs at which the rearranged region begins and ends. Knowledge of these breakpoints is crucial for understanding which genes could be responsible for a mutant phenotype associated with the rearrangement.

Rearrangement breakpoints can also be determined by an alternative method that uses genomic DNA as a template for the polymerase chain reaction (PCR). Once the general location of a rearrangement is known from barcoding or microarrays, scientists can then design PCR primers that will amplify unique products from templates containing the rearranged chromosomes (Fig. 13.5). DNA sequencing of the amplification products will identify the precise rearrangement breakpoints.

PCR analysis is of particular value in providing an inexpensive and highly sensitive method for following the transmission of a known rearrangement. For example, you will see later in this chapter that a type of leukemia (a white blood cell cancer) is caused by a certain reciprocal translocation. Physicians can follow the success of chemotherapy designed to eradicate the leukemic cells by performing PCR analysis on samples of the patient's blood. Amplification of a product specific for one of the translocation breakpoints would show that at least a few cancerous cells have survived, indicating the need for more treatment.

essential concepts

- Chromosomal rearrangments include those that remove or add base pairs (*deletions* and *duplications*), and those that relocate DNA regions (*inversions* and *translocations*).
- Chromosome breakage followed by DNA repair, or illegitimate crossing-over in which chromosomes misalign, can bring about deletions, duplications, inversions, and translocations.
- Advanced FISH techniques can reveal chromosome rearrangements as differences in imaged bands (*barcoding*), while microarrays can detect very small deletions and duplications. DNA sequencing and PCR amplification can locate rearrangements down to the level of base pairs.

Figure 13.5 Breakpoints of chromosomal rearrangements can be detected by genome sequencing or PCR. Genome sequencing will detect rearrangement breakpoints (*thick black lines*) as unusually juxtaposed DNA sequences. Primer pairs for PCR (*half-arrows*) can be designed that will amplify from genomic DNA templates unique products that include rearrangement breakpoints. DNA sequencing of the amplification products will reveal the exact extents of the rearrangements.

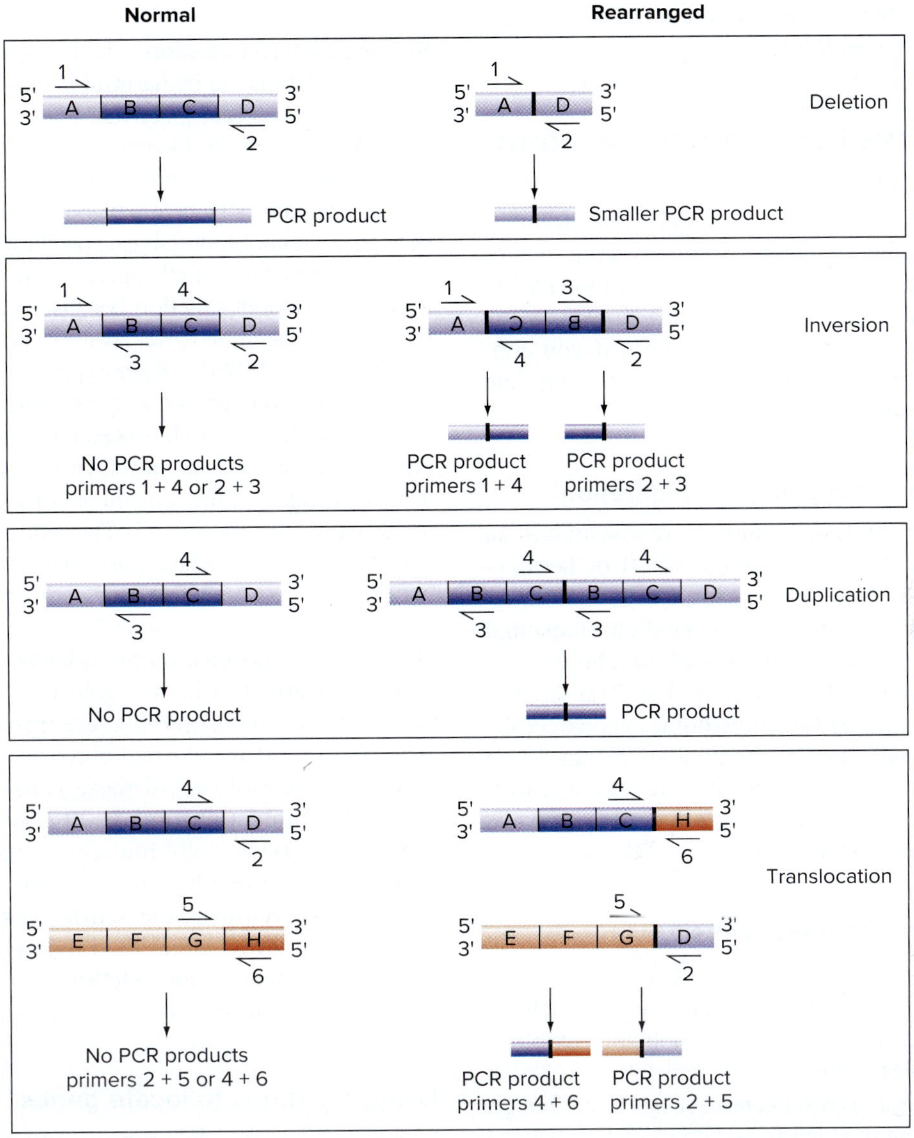

13.2 The Effects of Rearrangements

learning objectives

1. Describe the phenotypic consequences of deletions in homozygotes and heterozygotes.
2. Explain how researchers use deletions to locate genes.
3. Discuss the impacts of duplications on phenotype and on unequal crossing-over.
4. Distinguish between pericentric and paracentric inversions.
5. Explain why the breakpoints of inversions determine whether they have phenotypic effects.
6. Define *reciprocal translocation* and discuss when such rearrangements may have phenotypic consequences.
7. Summarize the effects of inversions and translocations on crossing-over and fertility.

Each of the four major classes of rearrangements can have an important impact on visible phenotype or even viability. The existence or severity of the effect usually depends on whether the individual is a homozygote or heterozygote for the rearranged chromosomes. In addition, these different types of changes to chromosomes can alter crossing-over as well as the fertility of individuals.

Deletions Remove DNA from the Genome

Small deletions of contiguous nucleotides, like those described in Chapter 7, often affect only one gene, whereas large deletions can generate chromosomes lacking tens or even hundreds of genes. We will use the symbol *Del* to designate a chromosome that has sustained a large deletion. In the following discussion, it will be helpful if you consider *Del* chromosomes as having amorphic (complete loss-of-function) alleles for the deleted genes.

Lethal effects of homozygosity for a deletion

Because many of the genes in a genome are essential to an individual's survival, homozygotes (*Del/Del*) or hemizygotes (*Del*/Y) for most deletion-bearing chromosomes do not survive. In rare cases where the deleted chromosomal region is devoid of genes essential for viability, however, a deletion hemi- or homozygote may survive. The smaller the deletion, the more likely it is to be homozygous viable. For example, *Drosophila* males hemizygous for an 80 kb deletion including the *white* (*w*) gene survive perfectly well in the laboratory; lacking the w^+ allele required for red eye pigmentation, they have white eyes.

Detrimental effects of heterozygosity for a deletion

Usually, an organism can survive with a chromosome deleted for more than a few genes only if the homologous chromosome has normal copies of the missing genes. Such a *Del*/+ individual is known as a *deletion heterozygote.* Even though all the genes are present in at least one copy, deletion heterozygotes can have mutant phenotypes for several reasons.

Haploinsufficiency Deletion heterozygotes sometimes have a mutant phenotype due to *haploinsufficiency;* that is, half of the normal **gene dosage** (the number of times a given gene is present in the genome) does not produce enough protein product for a normal phenotype. In some cases, the abnormal phenotype is due to lowered dosage of a single gene within the deletion. You saw previously in Chapter 8 that one form of polydactyly (extra fingers and toes) in humans is caused by heterozygosity for loss-of-function alleles for one particular gene (recall Fig. 8.30). Such effects of deletions are atypical because only about 800 genes in the human genome are haploinsufficient. However, lowered dosage of almost any gene is likely to have a small deleterious effect. Very large deletions that include many genes (such as the loss of half or more of a chromosome arm) are usually lethal even in heterozygotes because of the accumulated smaller effects of halving the dosage of many genes.

Vulnerability to mutation Another reason why heterozygosity for a deletion can be harmful is that cells become vulnerable to mutations that inactivate the one remaining copy of a gene. You saw in the Genetics and Society Box in Chapter 5 entitled *Mitotic Recombination and Cancer Formation* how people who are born heterozygous for loss-of-function mutations in the retinoblastoma gene on chromosome 13 (RB^-/RB^+) are predisposed to retinal cancer. Their cells are poised to lose all retinoblastoma function because many mechanisms exist that can change an RB^-/RB^+ heterozygous cell into one that is homozygous RB^-/RB^-. Karyotypes of normal, noncancerous tissues from many people with retinoblastoma reveal heterozygosity for deletions of chromosome 13 that include the retinoblastoma gene. Cells from retinal tumors of the same people have a mutation in the remaining copy of the *RB* gene on the nondeleted chromosome 13. The retinoblastoma gene is only one of many *tumor suppressor genes* whose role in the generation of cancers will be discussed in depth in Chapter 20.

Uncovering recessive mutant alleles A deletion heterozygote is, in effect, a hemizygote for genes on the normal, nondeleted chromosome that are missing from the deleted chromosome. If the normal chromosome carries a mutant recessive allele of one of these genes, the individual will exhibit the mutant phenotype. In *Drosophila,* for example, the *scarlet* (*st*) eye color mutation is recessive to wild type. However, an animal heterozygous for the *st* mutation and a deletion that removes the *scarlet* gene (*st/Del*) will have bright scarlet eyes, rather than wild-type, dark red eyes. In these circumstances, the deletion *uncovers* (that is, reveals) the phenotype of the recessive mutation (**Fig. 13.6a**).

Using deletions to locate genes

Geneticists can use deletions to find genes associated with abnormal phenotypes. The basic requirement is the availability of a recessive loss-of-function mutation *m* (that is, an amorphic or hypomorphic allele) that causes the phenotype. If the phenotype of an *m*/*Del* heterozygote is mutant (like that of *m/m*), the deletion has uncovered the mutated locus; at least part of the gene thus lies inside the region of deletion. In contrast, if the trait determined by the gene is wild type in these heterozygotes, the deletion has not uncovered the recessive allele, and all of the gene must lie outside the deleted region (**Fig. 13.6b**). You can consider this experiment as a complementation test between the mutation and the deletion: The uncovering of a recessive mutant phenotype demonstrates a lack of complementation because neither chromosome can supply wild-type gene function.

Figure 13.6 Gene mapping using deletion chromosomes. (a) Deletion chromosomes can *uncover* recessive mutations on the other homolog. A fly with the genotype *st/Del* displays the recessive scarlet eye color because the *Del* chromosome lacks an st^+ gene. **(b)** Heterozygotes for a deletion chromosome (*Del*) and a chromosome with a recessive *scarlet* mutation (*st*) will have scarlet eyes if the deletion includes the st^+ gene, and wild-type eyes if the st^+ gene is not deleted. The phenotypes of the five different deletion heterozygotes shown indicate that st^+ is located between the vertical dotted lines.

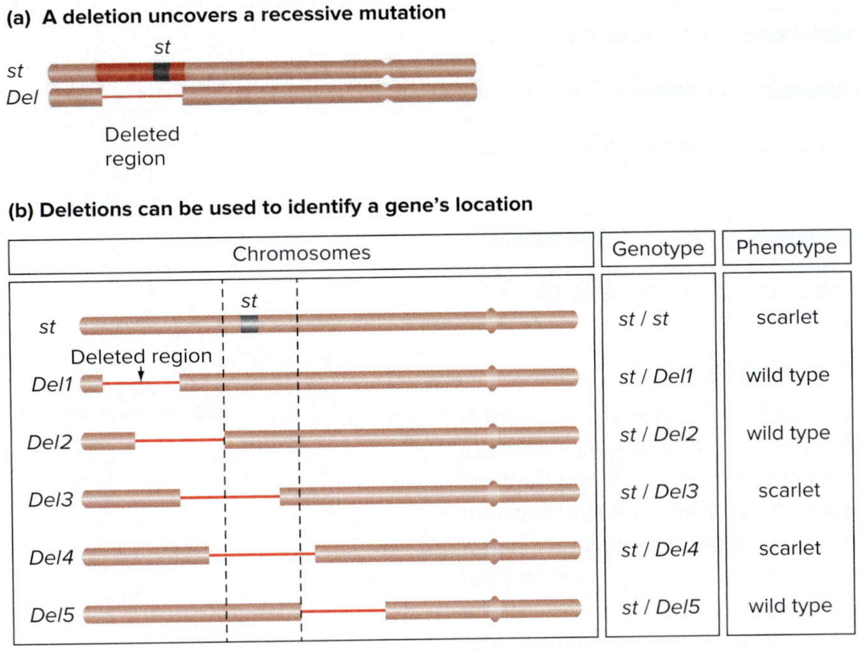

If several different, overlapping deletions in the same general region of the chromosome are available whose breakpoints are known at the DNA level (by PCR or whole-genome sequencing technologies as previously shown in Fig. 13.5), this approach of genetic mapping with *m*/*Del* heterozygotes can pinpoint the mutation quite precisely, often to only a single gene (Fig. 13.6b).

Effects of deletion heterozygosity on genetic map distances

Because recombination between maternal and paternal homologs can occur only at regions of similarity, map distances derived from genetic recombination frequencies in deletion heterozygotes will be aberrant. For example, no recombination is possible between genes *C, D,* and *E* in **Fig. 13.7** because the DNA in this region of the normal, nondeleted chromosome has nothing with which to recombine. In fact, during the pairing of homologs in prophase of meiosis I, the "orphaned" region of the nondeleted chromosome forms a **deletion loop**—an unpaired bulge of the normal chromosome that corresponds to the area deleted from the other homolog.

The progeny of a *Del*/+ heterozygote will always inherit the markers in a deletion loop as a unit (*C, D,* and *E* in Fig. 13.7). As a result, these genes cannot be separated by recombination, and the map distances between them, as determined by the phenotypic classes in the progeny of a *Del*/+ individual, will be zero. In addition, the genetic distance between loci on either side of the deletion (such as between markers *B* and *F* in Fig. 13.7) will be shorter than expected because fewer crossovers can occur between them.

Figure 13.7 Deletion loops form in the chromosomes of deletion heterozygotes. During prophase of meiosis I, the undeleted region of the normal chromosome has nothing with which to pair and forms a deletion loop. No recombination can occur within the deletion loop. Each line represents two chromatids.

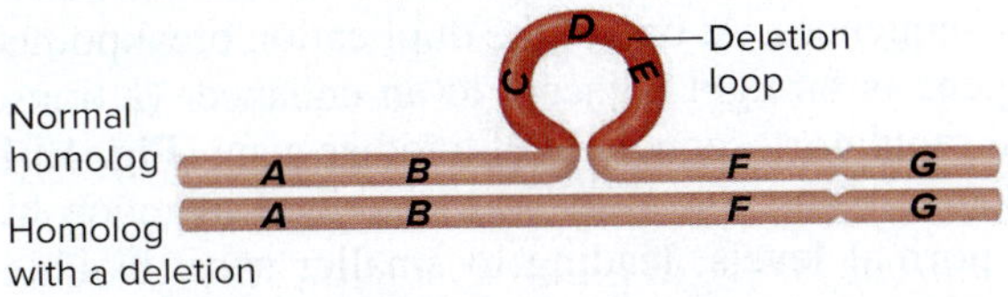

Duplication Chromosomes Have Extra Copies of Some Genes

The copies of the duplicated region can be arranged in different ways with respect to each other (**Fig. 13.8**). In **tandem duplications,** the repeated copies lie adjacent to each other, either in the same order or in reverse order. In **nontandem** (or *dispersed*) **duplications,** the copies of the region are not adjacent to each other and may lie far apart on the same chromosome or on different chromosomes. We use *Dp* as the symbol for a chromosome carrying a duplication.

Figure 13.8 Types of duplication chromosomes. In tandem duplications, the repeated regions lie adjacent to each other in the same or in reverse order. In nontandem duplications, the two copies of the same region are separated.

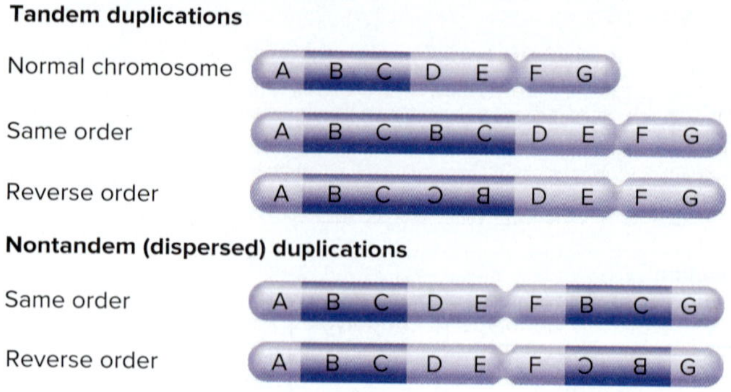

Figure 13.9 Phenotypic consequences of deletion and duplication heterozygosity. Deletion heterozygotes have only one copy of genes within the deletion, while duplication heterozygotes have three copies. Flies with one copy of the *Notch*$^+$ gene have notched wings. Flies with three copies of *Notch*$^+$ have wing vein pattern defects.

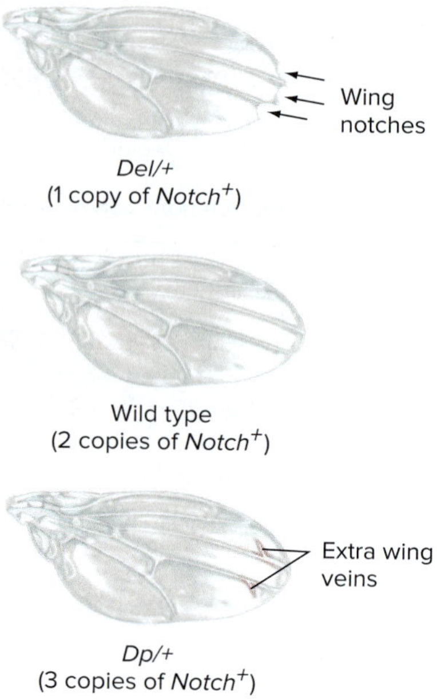

Phenotypic effects of duplications

Most duplications have no obvious phenotypic consequences, because an additional dose of most genes does not affect normal cellular or tissue physiology. Some duplications nevertheless do have phenotypic consequences for visible traits or for survival, and these abnormal phenotypes can occur for at least two reasons. First, certain phenotypes may be particularly sensitive to an increase in the number of copies of a particular gene or set of genes. Second, but more rarely, a gene near one of the borders of a duplication has altered expression because it is now found in a new chromosomal environment that does not exist in a wild-type chromosome.

Even duplication heterozygotes (*Dp/+*) may show unusual phenotypes. For example, *Drosophila* heterozygous for a duplication including the *Notch*$^+$ gene have abnormal wings that are caused specifically by the presence of the three copies of *Notch*$^+$ (**Fig. 13.9**). The *Notch*$^+$ gene is extraordinarily dosage sensitive; *Del/+* flies with only one copy of the *Notch*$^+$ gene have a different kind of wing abnormality, so this gene is also haploinsufficient (Fig. 13.9).

Organisms are not usually so sensitive to additional copies of a single gene; but just as for large deletions, imbalances for the many genes included in a large duplication have additive deleterious effects that jeopardize survival. In humans, a variety of disease syndromes are associated with heterozygosity for duplications of several megabases. Heterozygosity for even larger duplications (such as duplications of an entire chromosome arm) is most often lethal.

Unequal crossing-over between duplications

In individuals homozygous for a tandem duplication (*Dp/Dp*), homologs carrying the duplications occasionally pair out of register during meiosis. **Unequal crossing-over,** that is, recombination resulting from such out-of-register pairing, generates gametes containing increases to three and reciprocal decreases to one in the number of copies of the duplicated region.

In *Drosophila,* tandem duplication of a region of the X chromosome called 16A produces the Bar phenotype of kidney-shaped eyes (**Fig. 13.10a and b**). *Drosophila* females homozygous for the Bar eye duplication produce mostly Bar eye progeny. Some progeny, however, have wild-type eyes, whereas other progeny have double-Bar eyes that are even smaller than Bar eyes (Fig. 13.10a and b). The genetic explanation is that flies with wild-type eyes carry X chromosomes containing only one copy of the region in question, flies with Bar eyes have X chromosomes containing two copies of the region in tandem, and flies with double-Bar eyes have X chromosomes carrying three tandem copies (Fig. 13.10a and b). The wild-type and double-Bar-eyed progeny inherited X chromosomes produced by unequal crossing-over (**Fig. 13.10c**).

Unequal crossing-over in females homozygous for double-Bar chromosomes can yield progeny with even more extreme phenotypes associated with four or five copies of the duplicated region. Duplications in homozygotes thus allow for the expansion and contraction of the number of copies of a chromosomal region from one generation to the next.

Bar eyes are an example of a duplication-associated phenotype that is caused by the placement of a gene in a new chromosomal environment due to the reordering of DNA sequences. At one of the duplication breakpoints, the *Bar* gene is brought adjacent to an enhancer (a transcriptional regulatory sequence) of another gene (**Fig. 13.10d**). As a result, the *Bar* gene is transcribed at much higher than normal levels, leading to smaller eyes. Double-Bar

Figure 13.10 Unequal crossing-over can increase or decrease copy number. (a) Duplication of the 16A region of the *Drosophila* X chromosome causes Bar eyes; triplication causes double-Bar eyes. *Arrows* show the duplication breakpoint. **(b)** Bar eyes are narrower than wild type, and double-Bar eyes are even narrower. **(c)** Unequal pairing and crossing-over during meiosis in females homozygous for this duplication produce chromosomes that have either one copy of region 16A or three copies of 16A (causing the more abnormal double-Bar eyes). **(d)** The Bar eye is caused by abnormal juxtaposition of DNA sequences flanking the breakpoints such that a transcriptional enhancer from another gene causes overexpression of the *Bar* gene.

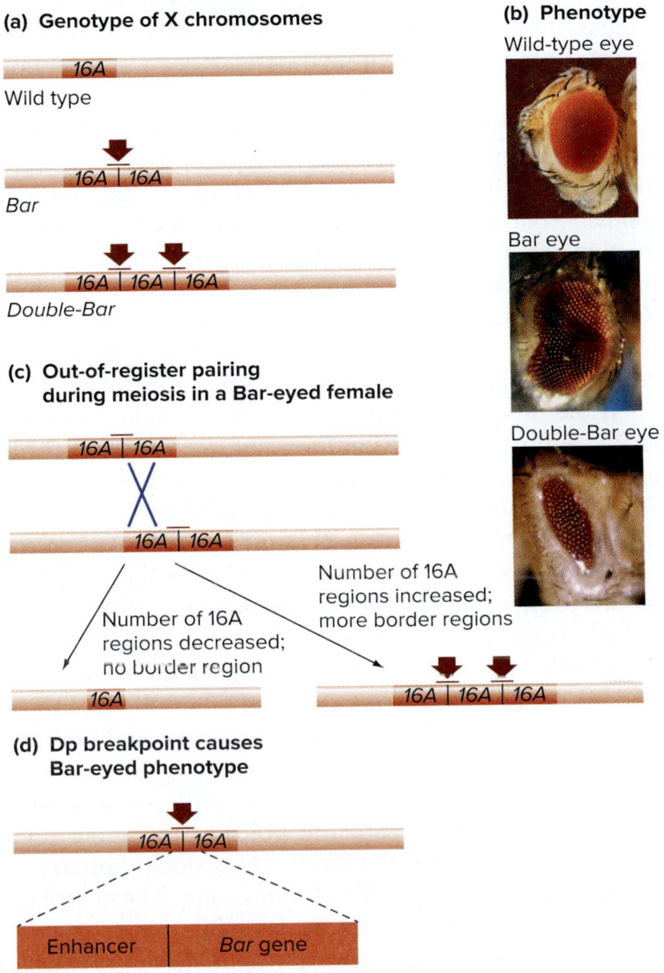

chromosomes with three copies of the 16A region have two copies of this breakpoint (that is, two copies of the abnormal enhancer-*Bar* gene fusion), even more of the mRNA, and even smaller eyes (Fig. 13.10a and b).

Inversions Reorganize the DNA Sequence of a Chromosome

Inversions are half-circle rotations of a region of a single chromosome. It is important to remember that because each DNA strand in a chromosome, even one with an inversion, must run continuously from its 5′ end to its 3′ end, the inverted part of the chromosome is not only rotated, but it is also flipped over with respect to the orientations of its two component strands (review Solved Problem I in Chapter 6).

Figure 13.11 Types of inversions. Inversions involve 180° rotations of a part of a chromosome. When the rotated segment includes the centromere, the inversion is *pericentric;* when the rotated segment does not include the centromere, the inversion is *paracentric.*

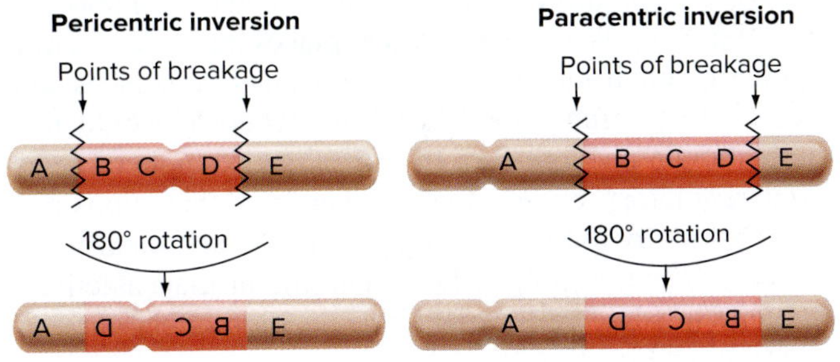

Geneticists distinguish between inversions that include the centromere, which are called **pericentric,** and inversions that exclude the centromere, which are **paracentric** (**Fig. 13.11**). As you will see later in this section, the location of the centromere relative to the inversion influences how an inversion-bearing chromosome behaves during meiotic cell divisions.

Phenotypic effects of inversions

Most inversions do not result in an abnormal phenotype, because even though they alter the order of genes along the chromosome, they do not add or remove DNA and therefore do not change the identity or number of genes. However, inversions can cause mutations in specific genes that span inversion breakpoints. As **Fig. 13.12** shows, if one inversion breakpoint lies within a gene, a loss-of-function mutation in that gene will occur. The inversion breaks the gene into two parts, relocating one part to a distant region of the chromosome, while leaving the other part at its original site. Such a split will disrupt the gene's function.

Figure 13.12 An inversion can affect phenotype if it disrupts a gene. Here, an inversion inactivates the *y* (*yellow*) gene by dividing it in two.

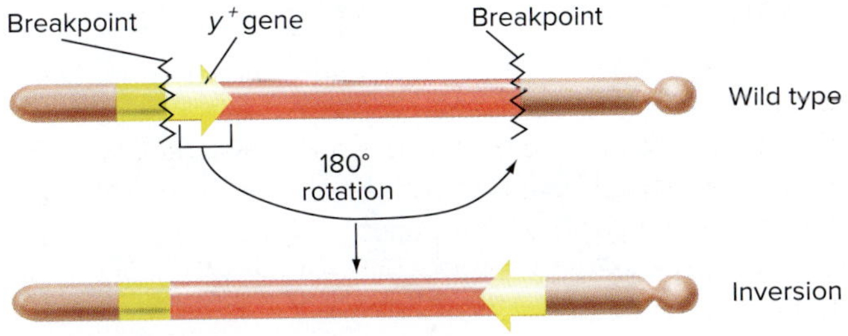

Inversions can also produce unusual phenotypes by moving genes residing near the inversion breakpoints to new chromosomal environments that alter their normal expression. For example, a mutation in the *Antennapedia* gene of *Drosophila* that transforms antennae into legs (review Fig. 8.32) is an inversion that places the gene in a new regulatory environment, next to enhancer sequences that cause it to be transcribed in tissues where it would normally remain unexpressed. In a different example of an altered regulatory environment, inversions that reposition genes normally found in a chromosome's euchromatin to a position near a region of heterochromatin may inactivate the gene in some cells, leading to position-effect variegation (review Fig. 12.12).

Figure 13.13 Inversion loops form in inversion heterozygotes. To maximize pairing during prophase of meiosis I in an inversion heterozygote (*In*/+), homologous regions form an inversion loop. (*Top*) Simplified diagram in which one line represents a pair of sister chromatids. (*Bottom*) Electron micrograph of an inversion loop during meiosis I in an *In*/+ mouse.

Courtesy and © Lorinda Anderson & Stephen Stack, Department of Biology, Colorado State University

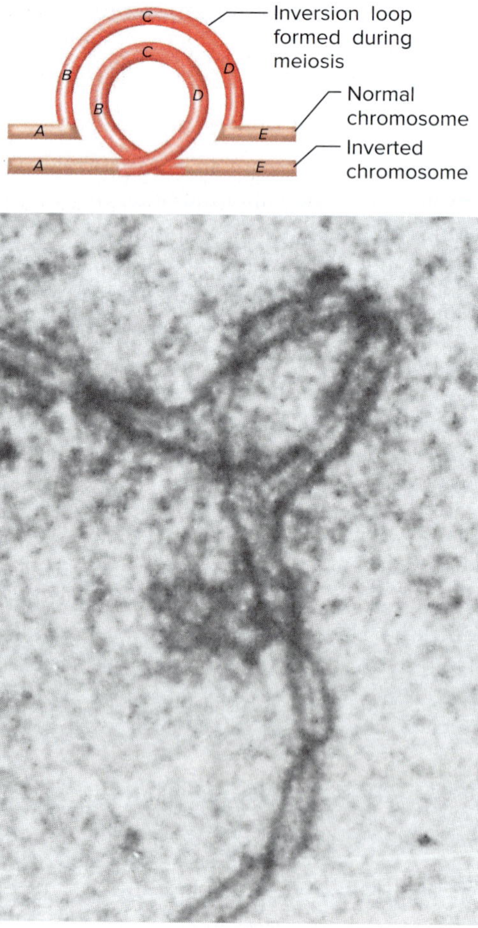

Inversion heterozygosity and reduced fertility

Individuals heterozygous for an inversion (*In*/+) are *inversion heterozygotes.* Because any specific inversion is rare in human populations, most people with an inversion chromosome are in fact inversion heterozygotes who inherited the inversion chromosome from only one parent. In such individuals, when the chromosome carrying the inversion pairs with its homolog at meiosis, formation of an **inversion loop** allows the tightest possible alignment of homologous regions. In an inversion loop, one chromosomal region rotates to conform to the similar region in the other homolog (**Fig. 13.13**). As we now discuss, crossing-over within an inversion loop produces aberrant recombinant chromatids whether the inversion is pericentric or paracentric.

If the inversion is pericentric and a single crossover occurs within the inversion loop, each recombinant chromatid will have a single centromere—the normal number—but will carry a duplication of one region and a deletion of a different region (**Fig. 13.14a**). Gametes carrying these recombinant chromatids will have an abnormal dosage of some genes. After fertilization, zygotes created by the union of these abnormal gametes with normal gametes are likely to die because of genetic imbalance.

If the inversion is paracentric and a single crossover occurs within the inversion loop, the recombinant chromatids will be unbalanced not only in gene dosage but also in centromere number (**Fig. 13.14b**). One crossover product will be an **acentric fragment** lacking a centromere, whereas the reciprocal crossover product will be a **dicentric chromatid** with two centromeres. Because the acentric fragment without a centromere cannot attach to the spindle apparatus during the first meiotic division, the cell cannot package it into either of the daughter nuclei; as a result, this chromosome is lost and will not be included in a gamete. By contrast, at anaphase of meiosis I, opposing spindle forces pull the dicentric chromatid toward both spindle poles at the same time with such strength that the dicentric chromatid breaks at a random position along the chromosome. These broken chromosome fragments are deleted for many of their genes. This loss of the acentric fragment and breakage of the dicentric chromatid results in genetically unbalanced gametes, which at fertilization will produce lethally unbalanced zygotes that cannot develop beyond the earliest stages of embryonic development. Consequently, no recombinant progeny resulting from a crossover in a paracentric inversion loop survive; any surviving progeny are nonrecombinants.

In summary, whether an inversion is pericentric or paracentric, crossing-over within the inversion loop of an inversion heterozygote has the same effect: formation of recombinant gametes that after fertilization prevent the zygote from developing.

You can see from Fig. 13.14 that every meiosis that occurs with a single crossover within the inversion loop results in two balanced gametes (one with a normal chromosome and the other containing the inversion) and two unbalanced gametes. The larger the inverted region, the more likely it is that a crossover will occur within the inversion loop, and the more unbalanced gametes will result. For this reason, inversion heterozygosity, especially when the inverted region is large, often results in reduced fertility.

Figure 13.14 Why inversion heterozygotes produce few if any recombinant progeny. Throughout this figure, each line represents one chromatid, and different shades of *green* indicate the two homologous chromosomes. **(a)** The chromatids formed by recombination within the inversion loop of a pericentric inversion heterozygote are genetically unbalanced. **(b)** The chromatids formed by recombination within the inversion loop of a paracentric inversion heterozygote are not only genetically unbalanced but also contain two centromeres or none, instead of the normal one.

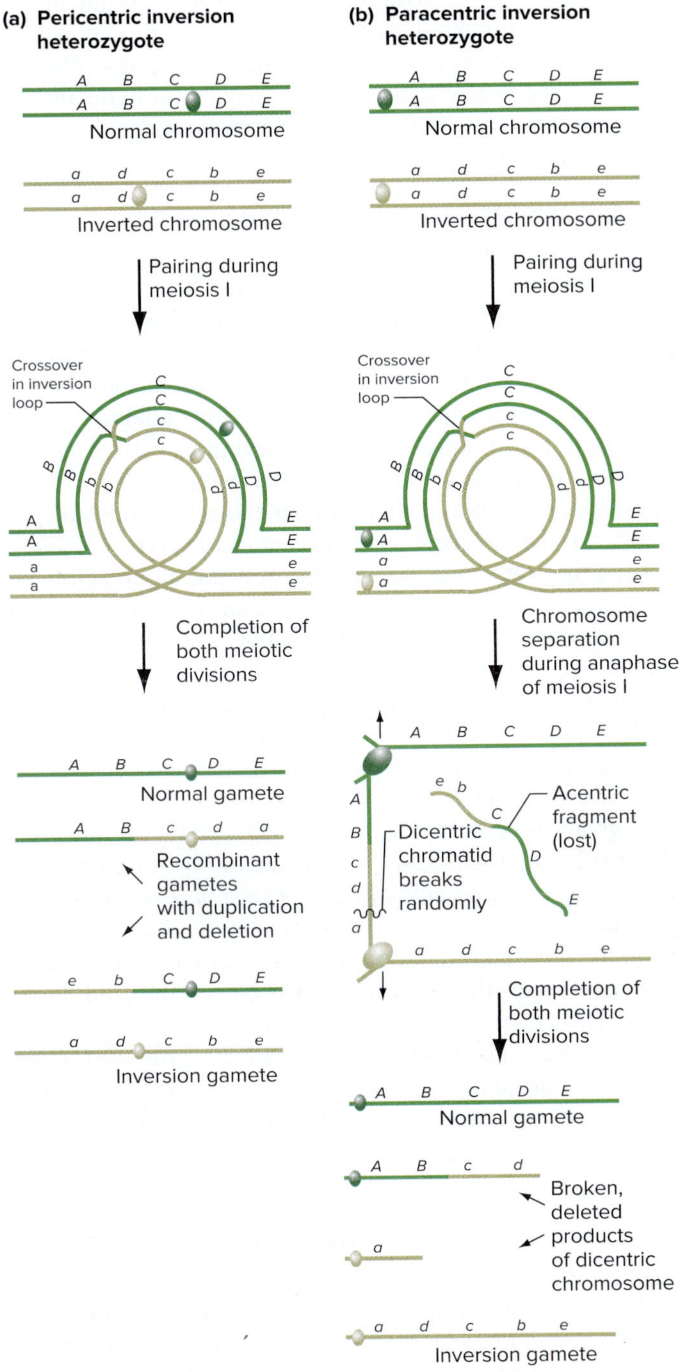

Inversion heterozygosity and crossover suppression

Because only gametes containing chromosomes that did not recombine within the inversion loop can yield viable progeny, inversions act as **crossover suppressors.** This does not mean that crossovers do not occur within inversion loops, but simply that few or no recombinants exist among the viable progeny of an inversion heterozygote.

Geneticists use crossover suppression to create ***Balancer* chromosomes,** which contain multiple, overlapping inversions (both pericentric and paracentric), as well as a marker mutation that produces a visible dominant phenotype (**Fig. 13.15**). The viable progeny of a *Balancer*/+ heterozygote will receive either the *Balancer* or the chromosome of normal order (+), but they cannot inherit a recombinant chromosome containing parts of both. Researchers can distinguish these two types of viable progeny by the presence or absence of the phenotype due to the dominant marker on the *Balancer* chromosome.

Figure 13.15 ***Balancer*** **chromosomes are useful tools for genetic analysis.** *Balancer* chromosomes carry both a dominant marker *D* and inversions (*brackets*) that prevent the recovery of recombinants between the *Balancer* and the chromosome carrying mutations of interest (m_1 and m_2). A parent heterozygous for the *Balancer* and an experimental chromosome will transmit either the *Balancer* or the double mutant chromosome, but not a recombinant chromosome, to its surviving progeny.

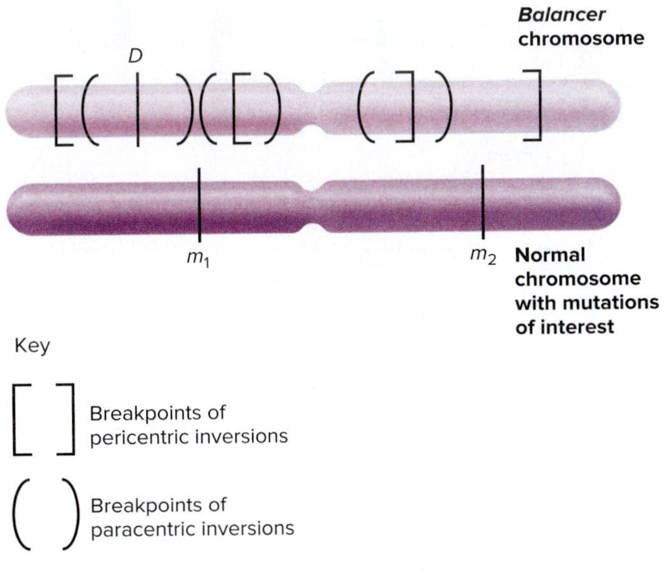

Geneticists often generate *Balancer* heterozygotes to ensure that a chromosome of normal order, along with any mutations of interest it may carry, is transmitted to the next generation unchanged by recombination. To help create genetic stocks, the marker in most *Balancer* chromosomes not only causes a dominant visible phenotype, but it also acts as a recessive lethal mutation that prevents the survival of *Balancer* chromosome homozygotes.

Figure 13.16 How a reciprocal translocation helps cause one kind of leukemia. (a) Uncontrolled divisions of large, dark-staining white blood cells in a leukemia patient (*right*) produce a higher ratio of white to red blood cells than that in a normal individual (*left*). **(b)** A reciprocal translocation between chromosomes 9 and 22 contributes to chronic myelogenous leukemia. This rearrangement makes an abnormal hybrid gene composed of part of the *c-abl* gene and part of the *bcr* gene. The hybrid gene encodes an abnormal fused protein that disrupts controls on cell division.

(a) Leukemia patients have too many white blood cells.

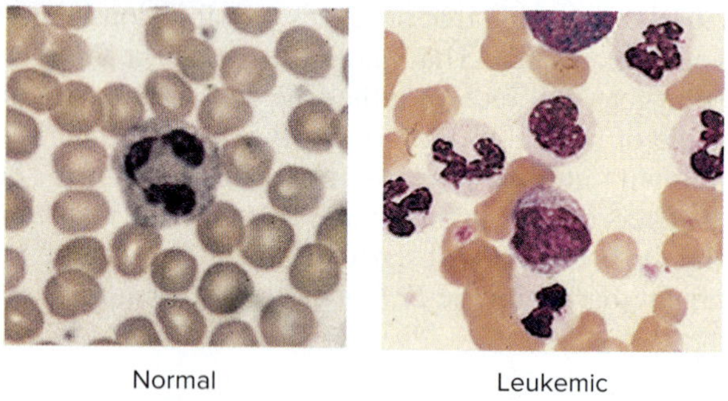

(b) The genetic basis for chronic myelogenous leukemia

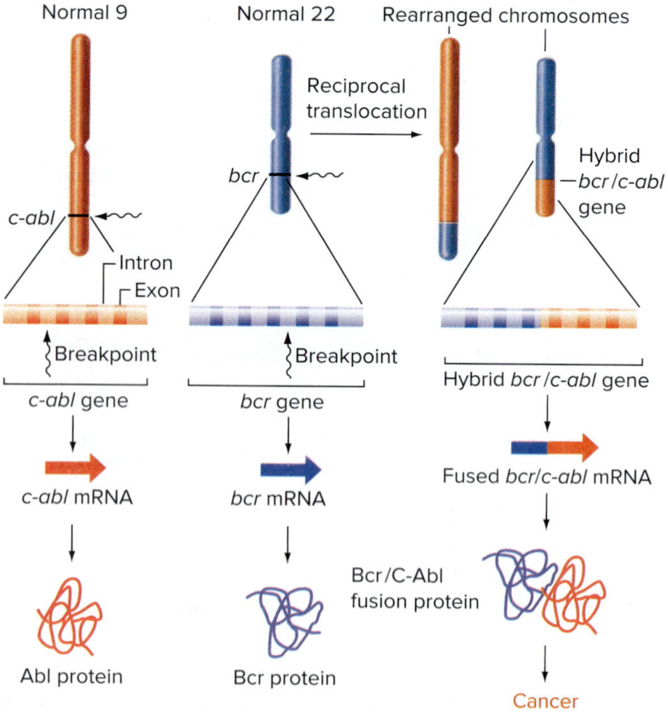

Translocations Attach Part of One Chromosome to Another Chromosome

Translocations are large-scale mutations in which part of one chromosome becomes attached to a nonhomologous chromosome. In this section we address exclusively the most common type of translocation, *reciprocal translocations* in which parts of two nonhomologous chromosomes exchange places, as previously shown in Table 13.1. Most individuals bearing reciprocal translocations are phenotypically normal because they have neither lost nor gained genetic material. As with inversions, however, if one of the translocation breakpoints occurs near or within a gene, that gene's function may change or be destroyed. Also like inversions, reciprocal translocations may result in decreased fertility, but as you will see, for different reasons.

Translocations that generate oncogenes

The potential effects of translocations on gene function are illustrated by the association of several kinds of cancer with translocations in somatic cells. In normal cells, genes known as *proto-oncogenes* control cell division. Translocations that relocate these genes can turn them into tumor-producing *oncogenes,* gain-of-function alleles whose protein products have an altered structure or level of expression that leads to runaway cell division.

As an example of this phenomenon, in almost all patients with *chronic myelogenous leukemia,* a type of cancer caused by overproduction of certain white blood cells, the leukemic cells have a reciprocal translocation between chromosomes 9 and 22 (**Fig. 13.16**). The breakpoint in chromosome 9 occurs within an intron of a proto-oncogene called *c-abl;* the breakpoint in chromosome 22 occurs within an intron of the *bcr* gene. After the translocation, parts of the two genes are adjacent to one another. During transcription, the RNA-producing machinery runs these two genes together, creating a long primary transcript. After splicing, the mRNA is translated into a fused protein in which 25 amino acids at the N terminus of the *c-abl*-determined protein are replaced by about 600 amino acids from the *bcr*-determined protein. The activity of this fused protein releases the normal controls on cell division, leading to leukemia.

The Fast Forward Box entitled *Programmed DNA Rearrangements and the Immune System* describes another example of a translocation-induced cancer called *Burkitt lymphoma.*

Diminished fertility in translocation heterozygotes

Translocations, like inversions, produce no significant genetic consequences in homozygotes if the breakpoints do not interfere with gene function. During meiosis in a translocation homozygote, chromosomes segregate normally

Figure 13.17 The meiotic segregation of reciprocal translocations. In all parts of this figure, each bar or line represents one chromatid. **(a)** In a translocation homozygote (*T/T*), chromosomes segregate normally during meiosis I. **(b)** In a translocation heterozygote (*T/+*), the four relevant chromosomes assume a *cruciform* (crosslike) configuration to maximize pairing. The alleles of genes on chromosomes in the original order (*N1* and *N2*) are shown in lowercase; the alleles of these genes on the translocated chromosomes (*T1* and *T2*) are in uppercase letters. **(c)** Three segregation patterns are possible in a translocation heterozygote. Only the alternate segregation pattern gives rise to balanced gametes. **(d)** This semisterile ear of corn comes from a plant heterozygous for a reciprocal translocation. It has fewer kernels than normal because unbalanced ovules are aborted.

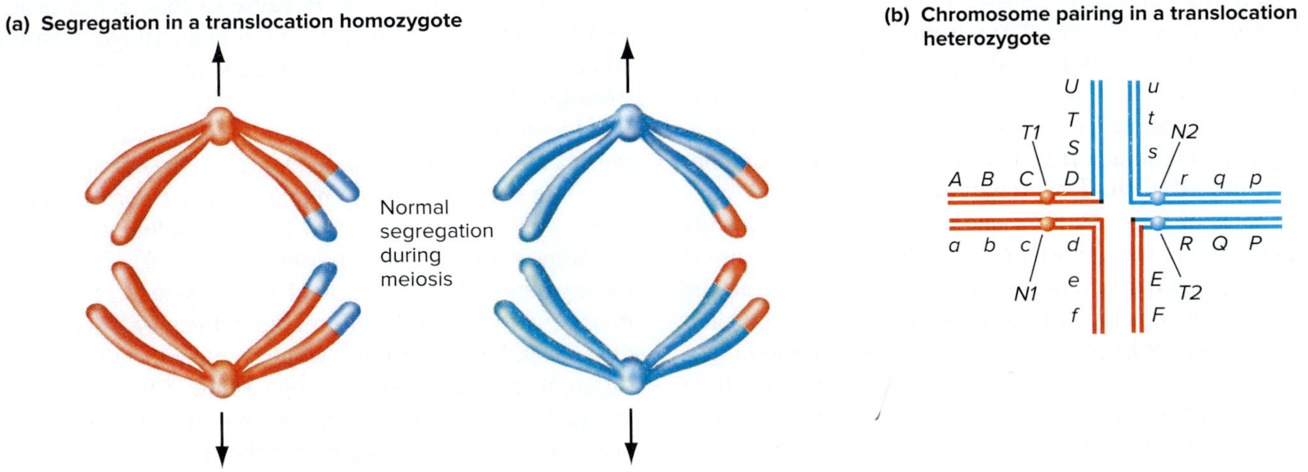

(c) Segregation in a translocation heterozygote

Segregation pattern	Alternate		Adjacent-1		Adjacent-2 (less frequent)	
	Balanced *N1* + *N2*	Balanced *T1* + *T2*	Unbalanced *T1* + *N2*	Unbalanced *N1* + *T2*	Unbalanced *N1* + *T1*	Unbalanced *N2* + *T2*
Gametes	*a b c d e f* / *p q r s t u*	*A B C D S T U* / *P Q R E F*	*A B C D S T U* / *p q r s t u*	*a b c d e f* / *P Q R E F*	*a b c d e f* / *A B C D S T U*	*p q r s t u* / *P Q R E F*
Type of progeny when mated with normal *abcdefpqrstu* homozygote	*abcdef* *pqrstu*	*ABCDEF* *PQRSTU*	None surviving	None surviving	None surviving	None surviving

(d) Semisterility in corn

according to Mendelian principles (**Fig. 13.17a**). Even though the genes have been rearranged, both haploid sets of chromosomes in the individual have the same rearrangement. As a result, all chromosomes will find a single partner with which to pair at meiosis, and there will be no deleterious consequences for the progeny.

In *translocation heterozygotes,* however, certain patterns of chromosome segregation during meiosis produce genetically unbalanced gametes that at fertilization become deleterious to the zygote. In a translocation heterozygote, the two haploid sets of chromosomes do not carry the same arrangement of genetic information. As a result, during prophase of the first meiotic division, the translocated chromosomes and their normal homologs assume a *cruciform* (crosslike) configuration in which four chromosomes, rather than the normal two, pair to achieve a maximum of synapsis between similar regions (**Fig. 13.17b**). To keep track of the four chromosomes participating in this cruciform structure, we denote the chromosomes carrying translocated material with a *T* and the chromosomes with a normal order of genes with an *N*. Chromosomes *N1* and *T1* have homologous centromeres found in wild type on chromosome 1; *N2* and *T2* have centromeres found in wild type on chromosome 2.

During anaphase of meiosis I, the mechanisms that attach the spindle to the chromosomes in this crosslike configuration usually ensure the disjunction of homologous centromeres, bringing homologous chromosomes to opposite spindle poles (that is, *T1* and *N1* go to opposite poles, as do *T2* and *N2*). Depending on the arrangement of the four chromosomes on the metaphase plate, this normal disjunction of homologs produces one of two equally likely patterns of segregation (**Fig. 13.17c**). In the **alternate segregation pattern,** the two translocation chromosomes (*T1* and *T2*) go to one pole, while the two normal chromosomes (*N1* and *N2*) move to the opposite pole. Both kinds of gametes resulting from this segregation (*T1, T2* and *N1, N2*) carry the correct haploid number of genes, and the zygotes formed by the union of these gametes with a normal gamete will be viable. By contrast, in the **adjacent-1 segregation pattern,** homologous centromeres disjoin so that *T1* and *N2* go to one pole, while *N1* and *T2* go to the opposite pole. As a result, each gamete contains a large duplication (of the region found in both the normal and the translocated chromosome in that gamete) and a correspondingly large deletion (of the region found in neither of the chromosomes in that gamete), which make them genetically unbalanced. Zygotes formed by the union of these gametes with a normal gamete are usually not viable.

Because of the unusual cruciform pairing configuration in translocation heterozygotes, nondisjunction of homologous centromeres occurs at a measurable but low rate. This nondisjunction produces an **adjacent-2 segregation pattern** in which the homologous centromeres *N1* and *T1* go to the same spindle pole, while the homologous centromeres *N2* and *T2* go to the other spindle pole (Fig. 13.17c). The resulting genetic imbalances are lethal after fertilization to the zygotes containing them.

Thus, of all the gametes generated by translocation heterozygotes, only those arising from alternate segregation, which account for slightly less than half the total, can produce viable progeny when crossed with individuals who do not carry the translocation. As a result, the fertility of most translocation heterozygotes, that is, their capacity for generating viable offspring, is diminished by at least 50%. This condition is known as **semisterility.** Corn plants illustrate the correlation between translocation heterozygosity and semisterility. The demise of genetically unbalanced ovules produces gaps in the ear where kernels would normally appear (**Fig. 13.17d**).

In humans, approximately 1 of every 500 individuals is heterozygous for some kind of translocation. While most such people are phenotypically normal, their fertility is diminished because many of the zygotes they produce abort spontaneously. As we have seen, this semisterility results from genetic imbalances associated with gametes formed by adjacent-1 or adjacent-2 segregation patterns. But such genetic imbalances are not inevitably lethal to the zygotes. If the duplicated or deleted regions are very small, the imbalanced gametes generated by these modes of segregation may produce normal children.

Pseudolinkage in translocation heterozygotes

The semisterility of translocation heterozygotes undermines the potential of genes on the two translocated chromosomes to assort independently. Mendel's second law requires that all gametes resulting from both possible metaphase alignments of two chromosomal pairs produce viable progeny. But as we have seen, in a translocation heterozygote, only the alternate segregation pattern yields viable progeny in outcrosses; the equally likely adjacent-1 pattern and the rare adjacent-2 pattern do not. Because of this, genes near the translocation breakpoints on the nonhomologous chromosomes participating in a reciprocal translocation exhibit **pseudolinkage:** They behave as if they are linked.

Figure 13.17c illustrates why pseudolinkage occurs in a translocation heterozygote. In the figure, lowercase *a b c d e f* represent the alleles of genes present on normal chromosome 1 (*N1*), and *p q r s t u* are the alleles of genes on a nonhomologous normal chromosome 2 (*N2*). The alleles of these genes on the translocated chromosomes *T1* and *T2* are in uppercase. In the absence of recombination, Mendel's law of independent assortment would predict that genes on two different chromosomes will appear in four types of gametes in equal frequencies; for example, *a p*, *A P*, *a P*, and *A p*. But alternate segregation, the only pattern that can give rise to viable progeny, produces (in the absence of crossing over) only *a p* and *A P* gametes. Thus, in translocation heterozygotes such as these, the genes on the two nonhomologous chromosomes act as if they are linked to each other.

Robertsonian translocations and Down syndrome

Robertsonian translocations arise from breaks at or near the centromeres of two acrocentric chromosomes (**Fig. 13.18**). The reciprocal exchange of broken parts generates one large metacentric chromosome and one very small chromosome containing few, if any, genes. This tiny chromosome may subsequently be lost from the organism. Robertsonian translocations are named after W. R. B. Robertson, who in 1911 was the first to suggest that during evolution, metacentric chromosomes may arise from the fusion of two acrocentrics.

An important example of this phenomenon is observed among individuals heterozygous for the Robertsonian translocation between chromosomes 14 and 21 shown in

Figure 13.18 Robertsonian translocations can reshape genomes. In a Robertsonian translocation, reciprocal exchanges between two acrocentric chromosomes generate a large metacentric chromosome and a very small chromosome. The latter may carry so few genes that it can be lost without ill effect.

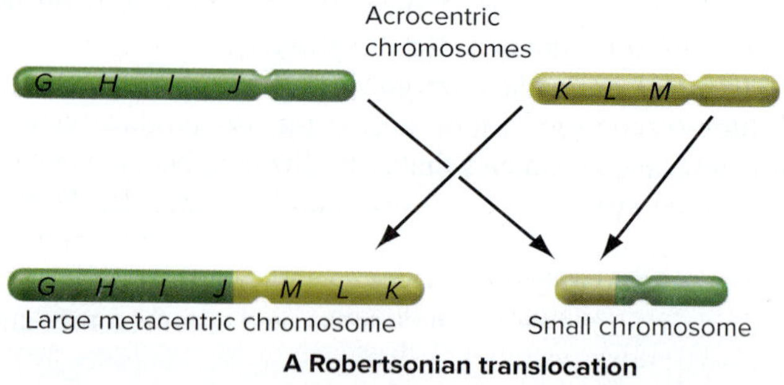

Figure 13.19 How translocation Down syndrome arises. In heterozygotes for a Robertsonian translocation involving chromosomes 14 and 21 (14q21q), adjacent-1 segregation can produce gametes with two copies of part of chromosome 21. If such a gamete unites with a normal gamete, the resulting zygote will have three copies of part of chromosome 21. Depending on which region of chromosome 21 is present in three copies, this tripling may cause Down syndrome. [In the original translocation heterozygote, the small, reciprocally translocated chromosome (14p21p) has been lost.]

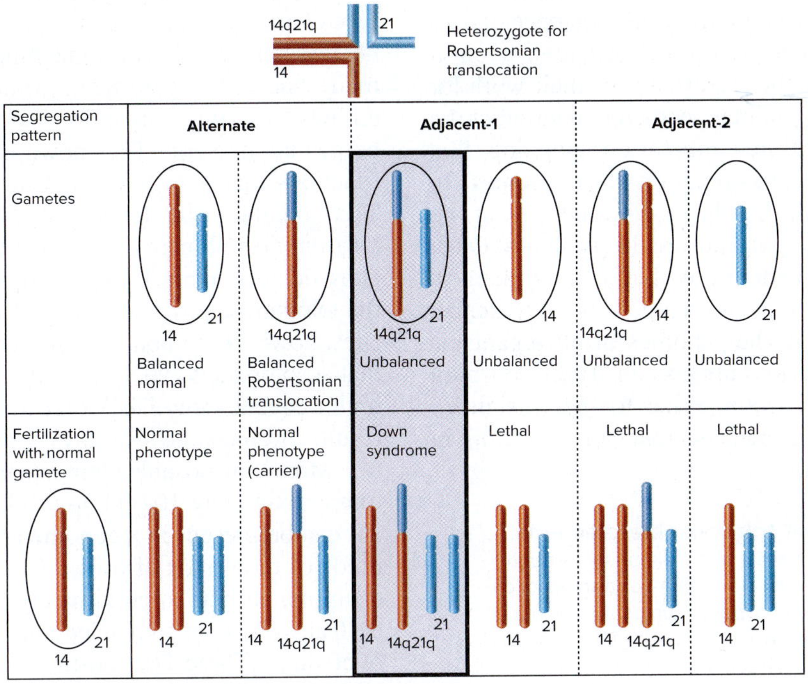

Fig. 13.19. These people are phenotypically normal but produce some gametes from the adjacent-1 segregation pattern that have two copies of most of chromosome 21: One copy is a normal chromosome 21, while the second copy is the Robertsonian translocation that joins chromosomes 14 and 21. At fertilization, if a gamete with the duplication unites with a normal gamete, the resulting child will have three copies of most of chromosome 21 and Down syndrome. About 4% of Down syndrome occurrences are due to Robertsonian translocations. You will learn later in this chapter that the majority of Down syndrome individuals instead have three normal, full-length copies of chromosome 21.

essential concepts

- *Deletions* remove DNA from a chromosome and cause mutant phenotypes mainly through effects on *gene dosage*.
- Deletion chromosomes are useful for locating genes.
- *Duplications* add DNA to a chromosome and cause mutant phenotypes either by increasing gene dosage or by changing the regulation of genes near a *breakpoint*.
- *Inversions* alter the order, but not the number, of genes on a chromosome.
- Inversions may affect the function of genes at or near a breakpoint by splitting a gene and disrupting its function, or by altering its regulation.
- Inversion heterozygotes have reduced fertility because crossing-over within the inverted region can generate *unbalanced gametes*.
- In *reciprocal translocations,* parts of two chromosomes trade places without the loss or gain of DNA. Translocations may modify the functions of genes at or near the translocation breakpoints.
- Heterozygosity for translocations reduces fertility because chromosome pairing and segregation during meiosis result in many unbalanced gametes.

13.3 Transposable Genetic Elements

learning objectives

1. Define *transposable element.*
2. Contrast the structures and mobilization mechanisms of retrotransposons and DNA transposons.
3. Discuss transposable elements in the human genome and their mobility.
4. Describe how transposable elements alter the genome.
5. Explain how cells prevent excess transposable element mobilization.

Another type of sequence rearrangement with a significant genomic impact is **transposition:** the movement of small segments of DNA—entities known as **transposable elements (TEs)**—from one position in the genome to another. Marcus Rhoades in the 1930s and Barbara McClintock in the 1950s first inferred the existence of TEs from intricate genetic studies of corn. The scientific community did not appreciate the importance of their work for many years because their findings seemed to contradict the conclusion from classical recombination mapping that genes are located at fixed positions on chromosomes. In addition, TEs are so small that they cannot be seen at the relatively low resolution of chromosomal karyotypes. Once the cloning of TE DNA made it possible to study them in detail, geneticists not only acknowledged their existence, but also discovered TEs in the genomes of all organisms analyzed, from bacteria to humans. In 1983, Barbara McClintock received the Nobel Prize for her insightful studies on movable genetic elements that cause mottling in corn kernels (**Fig. 13.20a**)

Figure 13.20 Discovery of transposable elements. **(a)** Barbara McClintock found transposable elements through experiments with maize (corn). **(b)** Movements of a transposon mottles corn kernels when the transposon jumps into or out of genes that influence pigmentation.

a: © Corbis; b: © Dr. Nina Fedoroff

(a) Barbara McClintock: Discoverer of transposable elements

(b) TEs cause mottling in corn.

Transposable Elements Are Classified According to How They Move

Any segment of DNA that evolves the ability to move from place to place within a genome is by definition a transposable element, regardless of its origin or function. TEs need not be sequences that do something for the organism; indeed, many scientists regard them primarily as "selfish" parasitic entities carrying only information that allows their self-perpetuation. Some TEs, however, appear to have evolved functions that help their host. In one interesting example, TEs maintain the length of *Drosophila* chromosomes. *Drosophila* telomeres, in contrast to those of most organisms, do not contain TTAGGG repeats that are extendable by the telomerase enzyme (review Fig. 12.20). Certain TEs in flies, however, combat the shortening of chromosome ends that accompanies every cycle of replication by jumping with high frequency into DNA very near chromosome ends. As a result, chromosome size stays relatively constant.

Most transposable elements in nature range from 50 bp to approximately 10,000 bp (10 kb) in length. A particular TE can be present in a genome anywhere from one to hundreds of thousands of times. *Drosophila melanogaster,* for example, harbors approximately 80 different types of TEs, each an average of 5 kb in length, and each present an average of 50 times. These TEs constitute $80 \times 50 \times 5 = 20{,}000$ kb, or roughly 12.5% of the 160,000 kb *Drosophila* genome.

FISH experiments using cloned TE DNA as probes illustrate that TEs can indeed transpose. For example, if you hybridized a probe made from a TE in flies called *copia* to two strains of *Drosophila* obtained from different geographical locations in the United States, you would see about 30 to 50 sites of hybridization in each strain's genome. Some of the sites would be the same in the two strains, but others would be different. In contrast, FISH performed with a probe from a normal gene would show labeling in the same single position in both strains. These observations suggest that since the time the strains were separated geographically, some of the *copia* sequences have moved around (transposed) in different ways in the two genomes, while normal genes have remained in fixed positions.

Classification of TEs on the basis of how they move around the genome distinguishes two groups. **Retrotransposons** transpose via reverse transcription of an RNA intermediate. The *Drosophila copia* elements just described are retrotransposons. **Transposons** (or **DNA transposons),** in contrast, move their DNA directly without the requirement of an RNA intermediate. The genetic elements discovered by Barbara McClintock in corn responsible for mottling the kernels are transposons (**Fig. 13.20b**).

Nearly Half of the Human Genome Consists of Transposable Elements

The human genome harbors more than 4 million transposable elements whose total length constitutes roughly 44%

TABLE 13.2 Transposable Elements in the Human Genome

LINES and SINES are poly-A type retrotransposons: LINES encode an RNA-binding protein and reverse transcriptase (the *ORF1* and *pol* genes) that enable their mobilization after pol II transcription. SINES, derived from pol III transcripts (such as tRNAs), rely on the LINE-encoded proteins to move after transcription by pol III. HERVs are LTR-type retrotransposons that, in addition to a *pol* gene, can include *gag* and *env* genes encoding retroviral coat proteins. DNA transposons in other organisms move due to the action of transposase enzyme on the inverted repeats at the ends of the transposon. Because of mutations in the genes they carry or in the end sequences needed for transposition, only a few LINEs and SINEs in the human genome are able to move; the HERVs and DNA transposons in the human genome are immobile relics.

Element	Structure	Length (kb)	Number	Genome fraction
Retrotransposons				
LINEs		6-8	1,000,000	20%
SINEs		<0.3	2,000,000	13%
HERVs		1-11	600,000	8%
DNA transposons				
		2-3	400,000	3%
		Total	4,000,000	44%

of the genome (**Table 13.2**). Most (about 90%) of the TEs in humans are retrotransposons. Transposable element movement in the human genome can be detected by comparing the human genome sequence with that of our closest relative, the chimpanzee, while more recent TE activity can be detected by comparing individual human genome sequences. These genome sequence studies reveal that since the time of our last common ancestor with the chimpanzee, no human DNA transposons have been mobile and only relatively few retrotransposons have moved. The low rate of TE mobilization in humans is due in part to the accumulation of mutations in TE DNA sequences, and in part to control mechanisms that minimize their movement.

Retrotransposons Move via RNA Intermediates

The transposition of a retrotransposon begins with its transcription by RNA polymerase into an RNA that encodes a reverse-transcriptase-like enzyme. This enzyme, like the reverse transcriptase made by the AIDS-causing virus HIV described in the Genetics and Society Box in Chapter 8 entitled *HIV and Reverse Transcription*, can copy RNA into a single strand of cDNA and then use that single DNA strand as a template for producing double-stranded cDNA. Many retrotransposons also encode polypeptides other than reverse transcriptase.

Some retrotransposons have a poly-A tail at the 3′ end of the RNA-like DNA strand, a configuration reminiscent of mRNA molecules. In humans, the two major types of poly-A-containing retrotransposons are called **LINEs (long interspersed elements)** and **SINEs (short interspersed elements)** (Table 13.2). Other retrotransposons end in *long terminal repeats* (*LTR*s), nucleotide sequences repeated in the same orientation at both ends of the element. The structure of this second type of retrotransposon is similar to the integrated DNA copies of *retroviruses* (RNA tumor viruses), suggesting that retroviruses evolved from this kind of retrotransposon, or *vice versa.* Human LTR-type retrotransposons are in fact now called **human endogenous retroviruses** (**HERVs**) (Table 13.2).

The structural parallels between retrotransposons, mRNAs, and retroviruses, as well as the fact that some retrotransposons have a gene that encodes a type of reverse transcriptase (the *pol* gene, Table 13.2), prompted investigators to ask whether retrotransposons move around the genome via an RNA intermediate. Experiments in yeast confirm that they do. In one study, a copy of the *Ty1* retrotransposon originally found on a yeast plasmid contained an intron in one of its genes; after transposition into the yeast chromosome, however, the intron was no longer there

Figure 13.21 Demonstration that retrotransposons move via an RNA intermediate. Researchers constructed a plasmid bearing a *Ty1* retrotransposon that contained an intron. After this plasmid was transformed into yeast cells, new insertions of *Ty1* into yeast genomic DNA were obtained. The newly inserted *Ty1* did not have the intron, which implies that transposition involves splicing of a primary transcript to form an intronless mRNA.

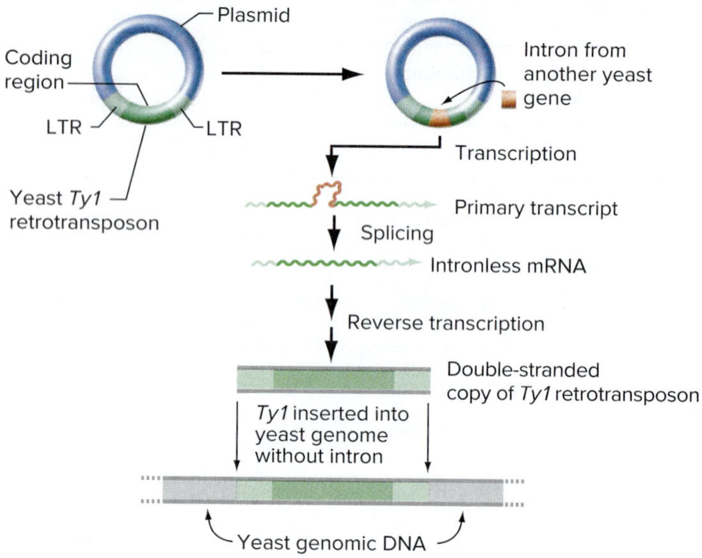

Figure 13.22 Mechanism of LTR-type retrotransposon movement. To mobilize, LTR-type retrotransposons rely on their LTRs and on the reverse transcriptase/endonuclease enzyme encoded by the *pol* gene. The reverse transcriptase synthesizes double-stranded retrotransposon cDNA. Insertion of this cDNA into a new genomic location (*blue*) involves a staggered endonuclease cleavage of the target site; polymerization to fill in the sticky ends produces two copies of the 5 bp target site.

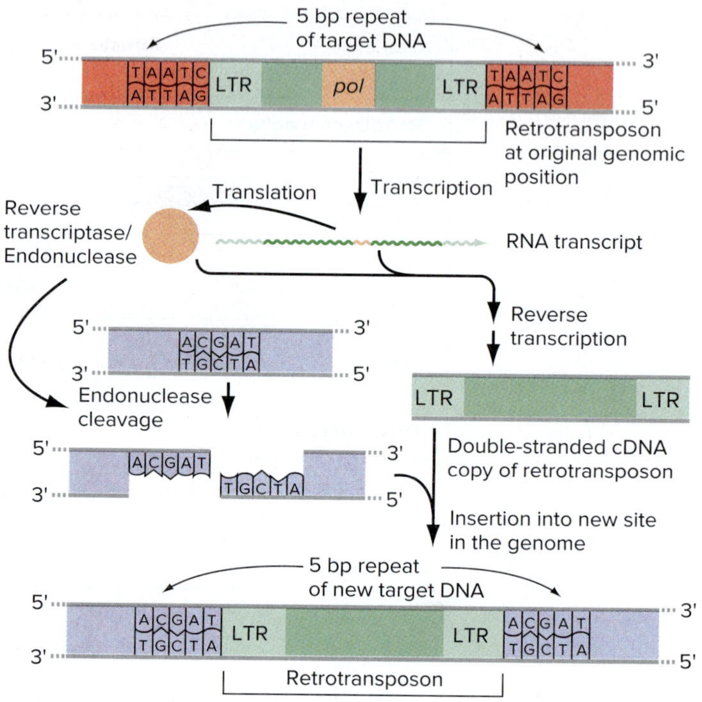

(**Fig. 13.21**). Because removal of introns occurs only during mRNA processing, researchers concluded that the *Ty1* element passed through an RNA intermediate during transposition from the plasmid to the chromosome.

The mechanisms by which poly-A-containing or LTR-containing retrotransposons move differ in detail but resemble each other in one important way—both mechanisms begin with transcription of the retrotransposon. **Figure 13.22** describes the mechanism by which LTR-containing retrotransposons (like yeast *Ty* elements, *Drosophila copia* elements, and the mobile ancestors of human HERVs) move around the genome. Translation of the *pol* gene in the retrotransposon transcript produces an enzyme with reverse transcriptase and endonuclease activity. The enzyme initiates the process of converting retrotransposon RNA into a double-stranded cDNA, and also cleaves a genomic DNA target site for insertion of the TE cDNA. Poly-A containing TEs (like human LINEs) mobilize through a more complex mechanism (*not shown*) that involves not only reverse transcriptase, but also another protein encoded by a different retrotransposon gene called *ORF1* (Table 13.2).

As Fig 13.22 illustrates, one outcome of transposition via an RNA intermediate is that the original copy of the retrotransposon remains in place while the new copy inserts in another location. With this mode of transmission, the number of copies potentially could increase rapidly with time. Organisms counteract the possibility of runaway retrotransposon proliferation through the evolution of elaborate mechanisms that limit retrotransposon mobilization.

Movement of DNA Transposons Is Catalyzed by Transposase Enzymes

A hallmark of transposons—TEs whose movement does not involve an RNA intermediate—is that their ends are inverted repeats of each other, that is, a sequence of base pairs at one end is present in mirror image at the other end (**Fig. 13.23a**). The inverted repeat is usually 10–200 bp long.

DNA between the transposon's inverted repeats commonly contains a gene encoding a **transposase,** a protein that catalyzes transposition through its recognition of those repeats. As Fig. 13.23a illustrates, the steps resulting in transposition include excision of the transposon from its original genomic position and integration into a new location.

The double-stranded break at the transposon's excision site is repaired in different ways in different cases. **Figure 13.23b** shows two of the possibilities. In *Drosophila,* after excision of a transposon known as a *P* element, DNA exonucleases first widen the resulting gap and then repair it (through double-strand-break repair, described in Chapter 7) using either a sister chromatid or a homologous chromosome as a template. If the template contains the *P* element and DNA replication is completely accurate, repair

Figure 13.23 DNA transposons: Structure and movement. (a) Most transposons contain inverted repeats at their ends (*light green; red arrows*) and encode a transposase enzyme that recognizes these inverted repeats. The transposase cuts at the borders between the transposon and adjacent genomic DNA, and it also helps the excised transposon integrate at a new site. **(b)** Transposase-catalyzed integration of *P* elements creates a duplication of 8 bp present at the new target site (*yellow*). A gap remains when transposons are excised from their original position. After exonucleases widen the gap, cells repair the gap using related DNA sequences as templates. Depending on whether the template contains or lacks a *P* element, the transposon will appear to remain at, or to be excised from, its original location.

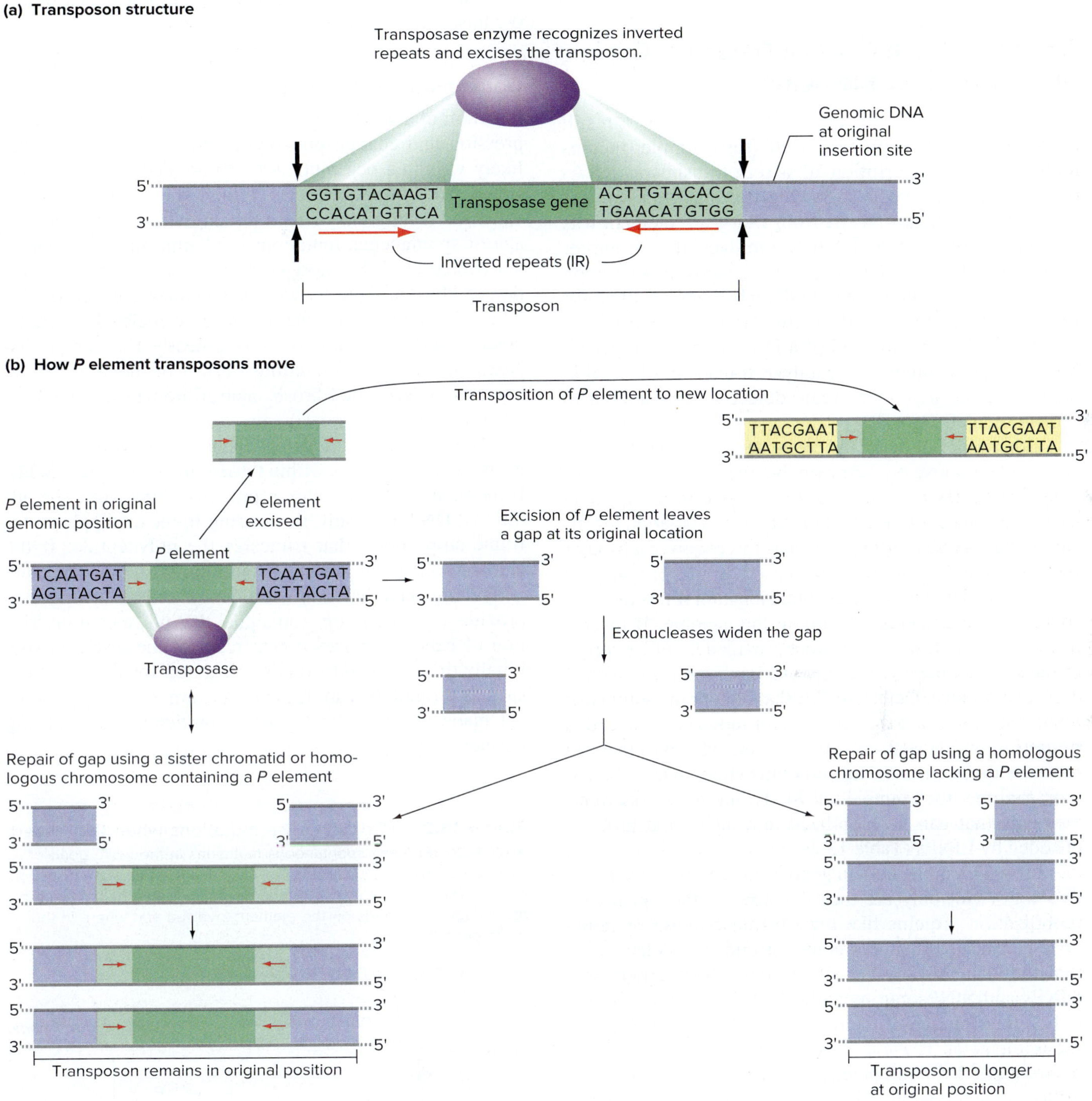

will restore a *P* element to the position from which it was excised; this will make it appear as if the *P* element remained at its original location during transposition (Fig. 13.23b, *bottom left*). If the template does not contain a *P* element, the transposon will be lost from the original site after transposition (Fig. 13.23b, *bottom right*).

Genomes Often Contain Defective Copies of Transposable Elements

Many copies of TEs sustain deletions either as a result of the transposition process itself (for example, incomplete reverse transcription of a retrotransposon RNA) or as a result of events following transposition (for example, faulty repair of a site from which a *P* element was earlier excised). If a deletion removes the promoter needed for transcription of a retrotransposon, that copy of the element cannot generate the RNA intermediate for future movements. If the deletion removes one of the inverted repeats at one end of a DNA transposon, transposase will be unable to catalyze transposition of that element. Such deletions create defective TEs unable to transpose again.

Other types of deletions create defective elements that are unable to move on their own, but they can move if nondefective copies of the element elsewhere in the genome supply the deleted function. For example, a deletion inactivating the reverse transcriptase gene in a retrotransposon or the transposase gene in a transposon would ground that copy of the element at one genomic location if it is the only source of the essential enzyme in the genome. If reverse transcriptase or transposase were provided by other copies of the same element in the genome, however, the defective copy could move. Defective TEs that require the activity of nondeleted copies of the same TE for movement are called **nonautonomous elements;** the nondeleted copies that can move by themselves are **autonomous elements.** In the human genome, for example, all SINEs are nonautonomous elements that can be mobilized only using the proteins encoded by LINEs (Table 13.2).

Most TEs in the human genome are relics, defective in two ways: Not only are they damaged in their genes for mobilization proteins like reverse transcriptase or transposase, but also their promoters or ends are defective so that they cannot move at all. As described earlier, only LINEs and SINEs can move in the human genome—and only a tiny fraction of them. The number of fully autonomous elements is even smaller; for example, scientists estimate that a diploid human genome has on average only 80 to 100 autonomous LINEs. Almost all of the 3 million LINE and SINE insertion points are the same in all people, while only about 8000 have mobilized during the course of human history, as evidenced by differing insertion points in different individuals.

Transposable Elements Can Disrupt Genes and Alter Genomes

Geneticists usually consider TEs to be segments of *selfish DNA* that exist for their own sake. However, the movement of TEs may have profound consequences for the organization and function of the genes and chromosomes of the organisms in which they are maintained.

Gene mutations caused by TEs

Insertion of a TE near or within a gene can affect gene expression and change phenotype. We now know that the likely wrinkled pea mutation first studied by Mendel resulted from insertion of a TE into the gene *SbeI* for a starch-branching enzyme. In *Drosophila,* a large percentage of spontaneous mutations, including the w^1 mutation discovered by T. H. Morgan in 1910, are caused by insertion of TEs. One way active TEs in humans are identified is when their movement generates disease alleles. Retrotransposon insertion mutations causing nearly 100 human diseases are known, including forms of hemophilia A, hemophilia B, cystic fibrosis, neurofibromatosis, and muscular dystrophy.

A TE's effect on a gene depends on what the element is and where it inserts within or near the gene (**Fig. 13.24**). If an element lands within a protein-coding exon, the additional DNA may shift the reading frame or supply an in-frame stop codon that truncates the polypeptide. If the element falls in an intron, it could diminish the efficiency of splicing. TEs that land within exons or introns may also provide a transcription stop signal that prevents transcription of gene sequences downstream of the insertion site. Finally, insertions into regions required for transcription, such as promoters, can influence the amount of gene product made in particular tissues at particular times during development.

Figure 13.24 TEs can cause mutations when they insert into a gene. Many spontaneous mutations in the *white* gene of *Drosophila* arise from insertions of TEs such as *P, copia, roo, pogo,* or *Doc.* The resultant eye color phenotype (indicated by the color in the *triangles*) depends on the element involved and where in the *white* gene it inserts.

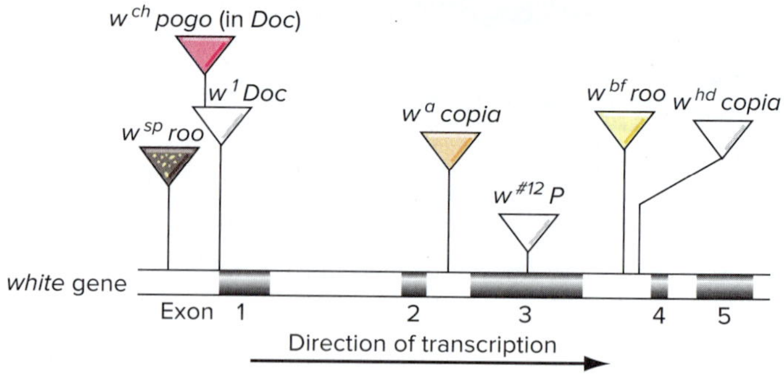

Instability of TE insertion mutations

Mutations caused by TE insertion are stable if an end of the element is damaged during mobilization, rendering the TE unmovable, or if elements encoding the mobilization proteins are no longer present in the genome. Otherwise, TE insertion mutations can be unstable: The TE can remobilize, usually resulting in reversion of the mutant allele to wild type.

An unstable mutation in a maize gene (gene *C*) caused by the insertion of a TE helped Barbara McClintock to discover transposable elements (**Fig. 13.25**). Gene *C* encodes a protein required for purple pigment in corn kernels. Kernels (corn plant embryos) were yellow when the *C* gene was inactivated by insertion of a nonautonomous transposon called *Ds* for *Dissociator*. In corn kernels that had autonomous copies of a transposon from the same family as *Ds*, called *Ac* for *Activator*, the mutation in gene *C* was unstable; that is, in some cells the nonautonomous *Ds* would hop out and gene *C* would regain function. Such kernels would be yellow with purple spots, which are clones of cells in which *Ds* had hopped out of the gene.

Chromosomal rearrangements caused by TEs

Retrotransposons and transposons can trigger spontaneous chromosomal rearrangements other than transpositions in several ways. As explained earlier (review Fig. 13.5), two copies of the same TE can pair with each other aberrantly and cross over to generate all kinds of chromosomal rearrangements. The duplication associated with the *Bar* mutation in *Drosophila* (recall Fig. 13.12) probably arose by unequal crossing-over at TEs flanking the duplicated region. Other chromosomal rearrangements result not from recombination between TEs, but rather from the process of transposition itself. Sometimes, mistakes can occur during transposition events that cause deletion or duplication of chromosomal material adjacent to the TE.

Figure 13.25 TE-associated alleles can be unstable. Cells in corn kernels are purple when gene *C* is active, while a *Ds* element insertion mutation that inactivates gene *C* produces yellow cells. Barbara McClintock observed that some yellow kernels had patches of purple cells (mottling). She explained the mottling as due to instability of the gene *C* mutation when an autonomous *Ac* element is present elsewhere in the genome. Later analysis by molecular biology showed that *Ac* elements provide transposase that can cause the nonautonomous *Ds* element insertion to jump out of gene *C*.

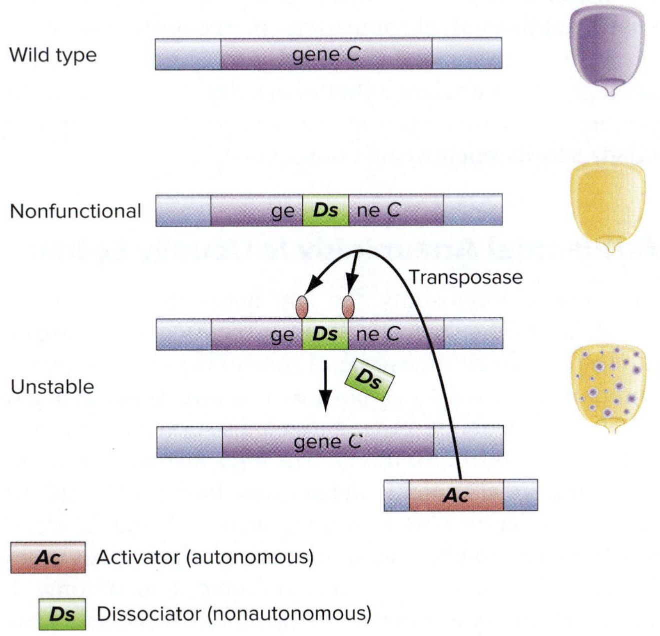

Figure 13.26 Transposons can move genes to new chromosomal locations. If two copies of a transposon are nearby on the same chromosome, transposase can recognize the outermost inverted repeats (IRs, *large arrows*), creating a composite transposon that allows intervening genes such as w^+ (*red*) to jump to new locations.

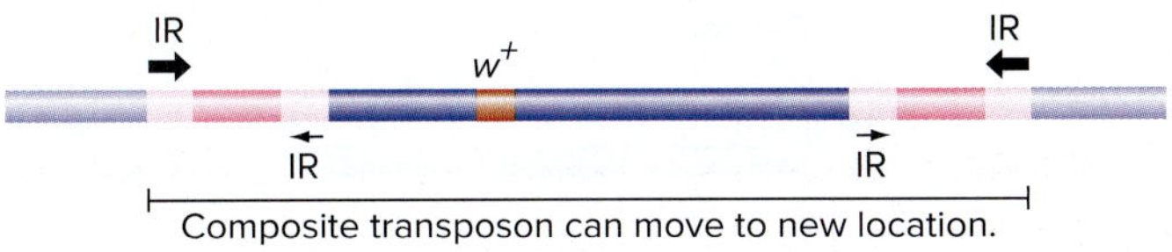

Gene relocation due to transposition

When two copies of a DNA transposon are found in nearby but not identical locations on the same chromosome, the inverted repeats at the outside ends of the two transposons (*bold arrows* in **Fig. 13.26**) are positioned with respect to each other just like the inverted repeats of a single transposon. If transposase acts on this pair of inverted repeats during transposition, it allows the entire region between them to move as one giant transposon, mobilizing and relocating any genes the region contains. In prokaryotes, the capacity of two TEs to relocate the intervening genes helps mediate the transfer of drug resistance between different strains or species of bacteria, as will be discussed in Chapter 14.

Several Mechanisms Limit Transposable Element Movement

We have just seen that the movement of TEs can mutate specific genes and rearrange chromosomes. Because frequent TE mobilization would wreak havoc in genomes, mechanisms have evolved to inhibit TE activity.

As an example, *P* element movement in *Drosophila* is limited by the production of both transposase and a repressor of transposition through alternative splicing of a *P* element RNA (**Fig. 13.27**). In germ-line cells, an intron (*yellow* in Fig. 13.27) in the primary transcript for transposase mRNA is sometimes spliced out and sometimes not. The mRNA with the intron removed encodes transposase, while the alternative splice form of the mRNA that retains

Figure 13.27 *P* element mobilization is regulated by mRNA splicing. The *P* element primary transcript is alternatively spliced. The intron between exons 3 and 4 (*yellow*) is sometimes spliced out and sometimes not. The mRNA that retains the intron encodes a repressor of transposition. Thus, in addition to transposase, cells produce repressor that limits *P* element mobilization.

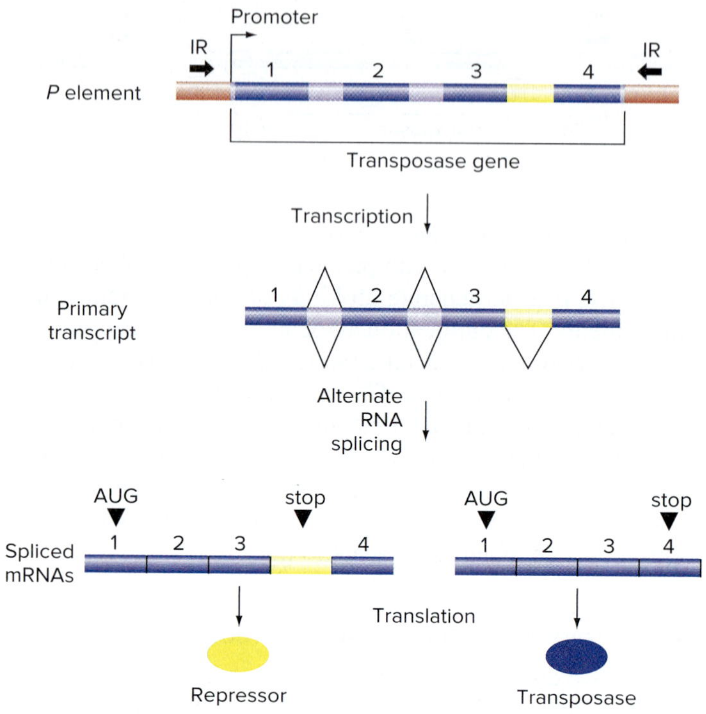

the intron encodes a smaller repressor polypeptide. In germ-line cells, repressor protein inhibits *P* element transposition by competing with transposase for binding to the transposon inverted repeats. Curiously, in somatic cells, the intron is never spliced out; only repressor is produced (not transposase), and so *P* elements do not mobilize in the soma.

A more general mechanism that inhibits germ-line activity of both DNA transposons (including *P* elements) and retrotransposons has been discovered recently and probably exists in all animals. This mechanism involves special small RNAs called **piRNAs (Piwi-interacting RNAs)** that block the transcription of TEs and translation of TE transcripts. We will defer detailed discussion of piRNAs until Chapter 17, which describes how several classes of small RNAs regulate gene expression in eukaryotes.

essential concepts

- *Transposable elements* are short segments of DNA, present in multiple locations, that move around the genome.
- *Retrotransposons* move via an RNA intermediate that is converted by *reverse transcriptase* into a cDNA subsequently inserted into the genome. In contrast, *DNA transposons* require *transposase* enzymes that recognize characteristic inverted repeats at the transposon ends.
- Nearly half the human genome is made up of transposable elements, only a few of which are currently able to move.
- Transposable elements alter the genome by disrupting genes, moving genes to new locations, or rearranging chromosomes.
- Mechanisms that inhibit TE activity include the production of transposition repressors through alternative splicing and the action of piRNAs that block TE transcription.

13.4 Aberrations in Chromosome Number: Aneuploidy

learning objectives

1. Define *aneuploidy, monosomy,* and *trisomy.*
2. Explain why autosomal aneuploidy is generally more deleterious than aneuploidy for sex chromosomes.
3. Describe how aneuploid and mosaic organisms arise.

We have seen that in peas, *Drosophila,* and humans, normal diploid individuals carry a $2n$ complement of chromosomes, where n is the number of chromosomes in the gametes. All the chromosomes in the haploid gametes of these diploid organisms differ from one another. Individuals whose chromosome number is not an exact multiple of the haploid number (n) for the species are **aneuploids** (**Table 13.3**). For example, in a normally diploid species, an individual lacking one chromosome is **monosomic** ($2n - 1$), whereas an individual having a single additional chromosome is **trisomic** ($2n + 1$). Monosomy, trisomy, and other forms of aneuploidy create a genetic imbalance that is usually deleterious to the organism. In this section we discuss how aneuploidy arises and its phenotypic consequences.

Autosomal Aneuploidy Is Usually Lethal

In humans, monosomy for any autosome is generally lethal, but medical geneticists have reported a few cases of monosomy for chromosome 21, one of the smallest human chromosomes. Although born with severe multiple abnormalities, these monosomic individuals survived for a short time beyond birth. Similarly, trisomies involving a human autosome are also highly deleterious. Individuals with trisomies for larger chromosomes, such as 1 and 2, are almost always aborted spontaneously early in pregnancy. Trisomy 18 causes Edwards syndrome, and trisomy 13 causes Patau syndrome; both phenotypes include gross

TABLE 13.3 Different Kinds of Aneuploidy in a Normally Diploid Organism

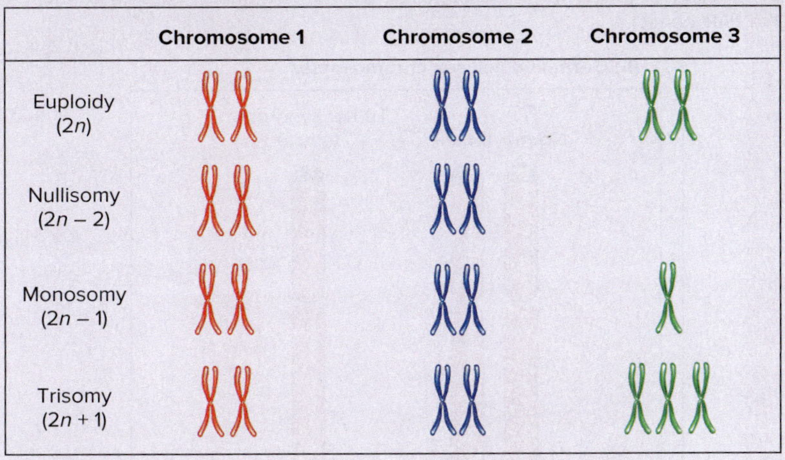

In this theoretical organism, $n = 3$.

developmental abnormalities that usually result in early death (**Table 13.4**).

The most frequently observed human autosomal trisomy, trisomy 21, results in *Down syndrome.* As one of the shortest human autosomes, chromosome 21 contains only about 1.5% of the DNA in the human genome. Although considerable phenotypic variation exists among Down syndrome individuals, traits such as intellectual disability and skeletal abnormalities are usually associated with the condition. Many Down syndrome babies die in their first year after birth from heart defects and increased susceptibility to infection.

Some people with Down syndrome have three copies of only part of, rather the entire, chromosome 21. For example, you saw earlier that one way people inherit a partial extra copy of chromosome 21 is through a Robertsonian translocation (review Fig. 13.19). Problem 38 at the end of the chapter discusses how such cases are being used to identify specific genes on chromosome 21 associated with the various aspects of Down syndrome.

TABLE 13.4 Aneuploidy in the Human Population

Chromosomes	Syndrome	Frequency at Birth
Autosomes		
Trisomic 21	Down	1/700
Trisomic 13	Patau	1/5000
Trisomic 18	Edwards	1/10,000
Sex chromosomes, females		
XO, monosomic	Turner	1/5000
XXX, trisomic XXXX, tetrasomic XXXXX, pentasomic		1/700
Sex chromosomes, males		
XYY, trisomic	Normal	1/10,000
XXY, trisomic XXYY, tetrasomic XXXY, tetrasomic XXXXY, pentasomic XXXXXY, hexasomic	Klinefelter	1/500

About 0.4% of all babies born have a detectable chromosomal abnormality that generates a detrimental phenotype.

Most Organisms Tolerate Aneuploidy for Sex Chromosomes

Although the X chromosome is one of the longest human chromosomes and contains 5% of the DNA in the genome, individuals with X chromosome aneuploidy, such as XXY males, XO females, and XXX females, survive quite well compared with aneuploids for the larger autosomes (Table 13.4). The explanation for this tolerance of X-chromosome aneuploidy is that X-chromosome inactivation equalizes the expression of most X-linked genes in individuals with different numbers of X chromosomes. We saw in Chapter 4 that in XX mammals, X-chromosome inactivation represses expression of most genes on one of the two X chromosomes; the genes that escape X inactivation are mainly in the pseudoautosomal regions (PARs) of the X and Y chromosomes (recall Fig. 4.8). In X-chromosome aneuploids with more than two X chromosomes, all but one X is inactivated in every cell. As a result, the amount of protein generated by most X-linked genes in X-chromosome aneuploids is the same as in normal XX or XY individuals.

Human X-chromosome aneuploidies are nonetheless not without consequence. XXY men have *Klinefelter syndrome,* and XO women have *Turner syndrome* (Table 13.4). The aneuploid individuals affected by these syndromes are usually infertile and display skeletal abnormalities, leading the XXY men to be unusually tall and long-limbed and the XO women to have unusually short stature.

The morphological abnormalities associated with Turner and Klinefelter syndromes are due at least in part to abnormal dosage of the 30 PAR genes in somatic cells. XO females have one copy fewer of each of these genes than normal females, while XXY men have one copy more of these genes than do normal males (**Fig. 13.28a**). One PAR gene called *SHOX* (*short stature homeobox*) encodes a protein important for bone development and is likely to play a leading role in the short stature of Turner females and the unusual tallness of Klinefelter males.

Figure 13.28 Why X-chromosome aneuploidy can affect morphology and cause sterility. (a) X-chromosome inactivation in somatic cells does not affect genes in the PARs (*green*) including *SHOX*. As a result, in XO Turner females PAR gene expression is half the normal level, and in XXY Klinefelter males the PAR genes are overexpressed. **(b)** Because all X chromosomes are active in the germ line, germ-line cells of XO females have half the normal dosage of all X-linked genes, while the germ-line cells of XXY males have twice the normal dosage of most X-linked genes (and three doses of the PAR genes).

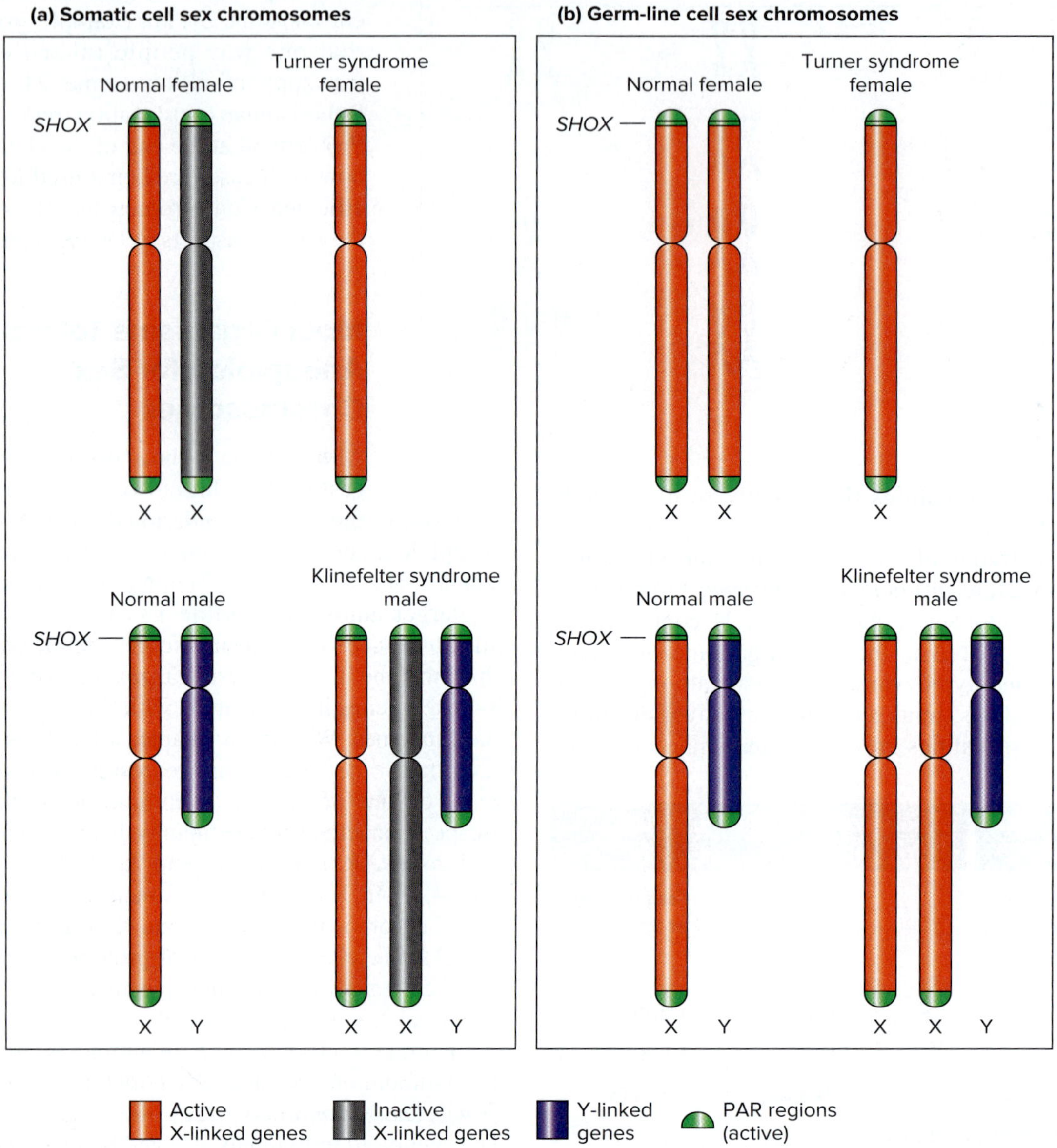

Infertility of individuals with Turner or Klinefelter syndrome is likely due to abnormal dosage in germ-line cells of X-linked genes outside the PAR regions (**Fig. 13.28b**). The reason is that germ-line cells undergo the reverse of X inactivation, that is, **X-chromosome reactivation.** In females, X reactivation normally occurs in the oogonia, the female germ-line cells that divide mitotically and whose daughters develop into the oocytes that subsequently undergo meiosis (review Fig. 4.18). Reactivation of the previously inactivated X chromosomes in the oogonia ensures that every mature ovum (the gamete) receives an active X. You should note that X reactivation is a necessary process: If it did not occur, half of a normal woman's eggs would contain an inactive X chromosome and would thus be incapable of supporting development after fertilization.

With X reactivation, oogonia in normal XX females have two functional doses of X chromosome genes, but the corresponding cells in XO Turner women have only one dose of the same genes and may thus undergo defective oogenesis (Fig. 13.28b). In XXY males, X reactivation in the spermatogonia, the male germ-line cells that divide mitotically and develop into the spermatocytes that undergo meiosis to produce sperm (review Fig. 4.19), results in twice the normal dose of X-linked genes (Fig. 13.28b). As a result, Klinefelter males usually make no sperm.

Aneuploidy Arises Through Meiotic Nondisjunction

Mistakes in chromosome segregation during meiosis produce aneuploids of different types, depending on when the mistakes occur. If homologous chromosomes do not separate (that is, do not disjoin) during the first meiotic division, two of the resulting haploid gametes will carry both homologs, and two will carry neither. Union of these gametes with normal gametes will produce aneuploid zygotes: half trisomic, half monosomic (**Fig. 13.29a**). By contrast, if **meiotic nondisjunction** occurs during meiosis II, only two of the four resulting gametes will be aneuploid (**Fig. 13.29b**).

Figure 13.29 Errors in meiosis cause aneuploidy. **(a)** For genes close to the centromere, if trisomic progeny inherit two different alleles (*A* and *a*) from one parent, the nondisjunction occurred in meiosis I. **(b)** If the two alleles inherited from one parent are the same (*A* and *A*; or *a* and *a*), the nondisjunction occurred during meiosis II.

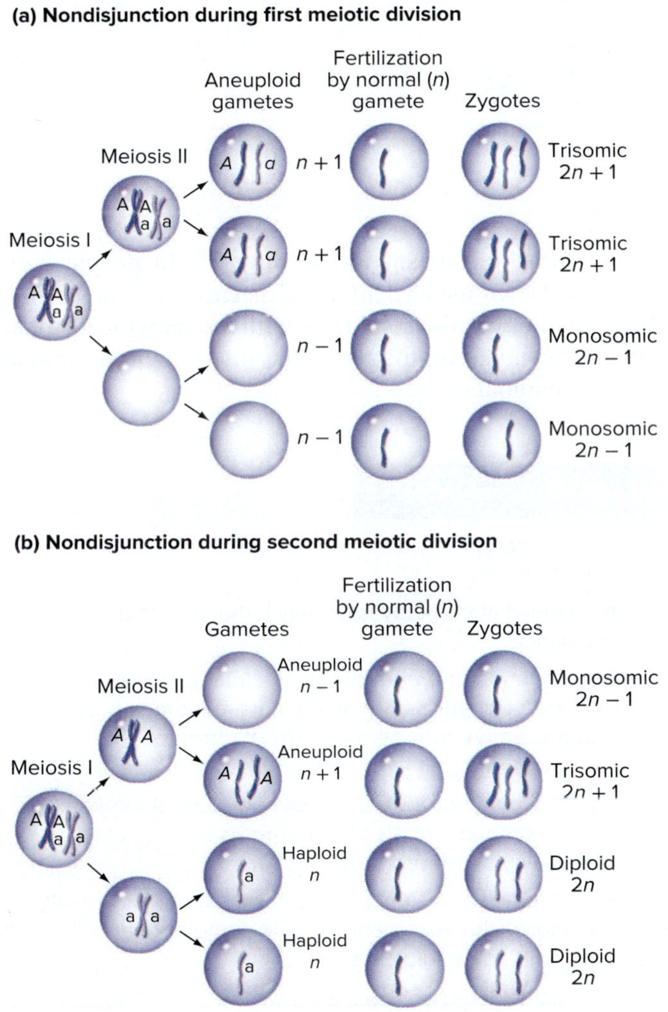

Abnormal $n + 1$ gametes resulting from nondisjunction in a cell that is heterozygous for alleles of genes close to the centromere on the nondisjoining chromosome will be heterozygous if the nondisjunction happens in the first meiotic division (Fig. 13.29a), but they will be homozygous if the nondisjunction takes place in the second meiotic division (Fig. 13.29b). It is possible to use this distinction to determine when and in which parent a particular nondisjunction occurred. The nondisjunction events that give rise to Down syndrome, for example, occur much more frequently in mothers (90%) than in fathers (10%). Curiously, in women, such nondisjunction events occur more often during the first meiotic division (about 75% of the time) than during the second. By contrast, when the nondisjunction event leading to Down syndrome takes place in men, the reverse is true.

Molecular studies have shown that many meiotic nondisjunction events in humans result from problems in meiotic recombination. By tracking DNA markers, clinical investigators can establish whether recombination took place anywhere along chromosome 21 during meioses that created $n + 1$ gametes. In many of the Down syndrome cases caused by nondisjunction during the first meiotic division in the mother, no recombination occurred between the homologous chromosome 21s in the defective meioses. This result makes sense because chiasmata, the structures associated with crossing-over, hold the maternal and paternal homologous chromosomes together in a bivalent at the metaphase plate of the first meiotic division (review Fig. 4.15). In the absence of recombination and thus of chiasmata, no mechanism exists to ensure that the maternal and paternal chromosomes will go to opposite poles at anaphase I. The increase in the frequency of Down syndrome children that is associated with increasing maternal age may therefore reflect a decline in the effectiveness of the mother's machinery for meiotic recombination.

Rare Mitotic Nondisjunction or Chromosome Loss Causes Mosaicism

As a zygote divides many times to become a fully formed organism, mistakes in chromosome segregation during the mitotic divisions accompanying this development may, in rare instances, augment or diminish the complement of chromosomes in certain cells. In **mitotic nondisjunction,** the failure of two sister chromatids to separate during mitotic anaphase generates reciprocal trisomic and monosomic daughter cells (**Fig. 13.30a**). Other types of mistakes, such as a lagging chromatid not pulled to either spindle pole at mitotic anaphase, result in a **chromosome loss** that produces one monosomic and one diploid daughter cell (**Fig. 13.30b**).

Figure 13.30 Mistakes during mitosis can generate clones of aneuploid cells. Mitotic nondisjunction **(a)** or mitotic chromosome loss **(b)** can create monosomic or trisomic cells that divide to produce aneuploid clones. **(c)** If an X chromosome is lost during the first mitotic division of an XX *Drosophila* zygote, one daughter cell will be XX (female), while the other will be XO (male). Such an embryo will grow into a gynandromorph. Here, the zygote was $w^+ \ m^+ \ / \ w \ m$, so the XX half of the fly (*left*) has red eyes and normal wings. Loss of the $w^+ \ m^+$ X chromosome gives the XO half of the fly (*right*) white eyes (w), miniature wings (m), and a male-specific sex comb.

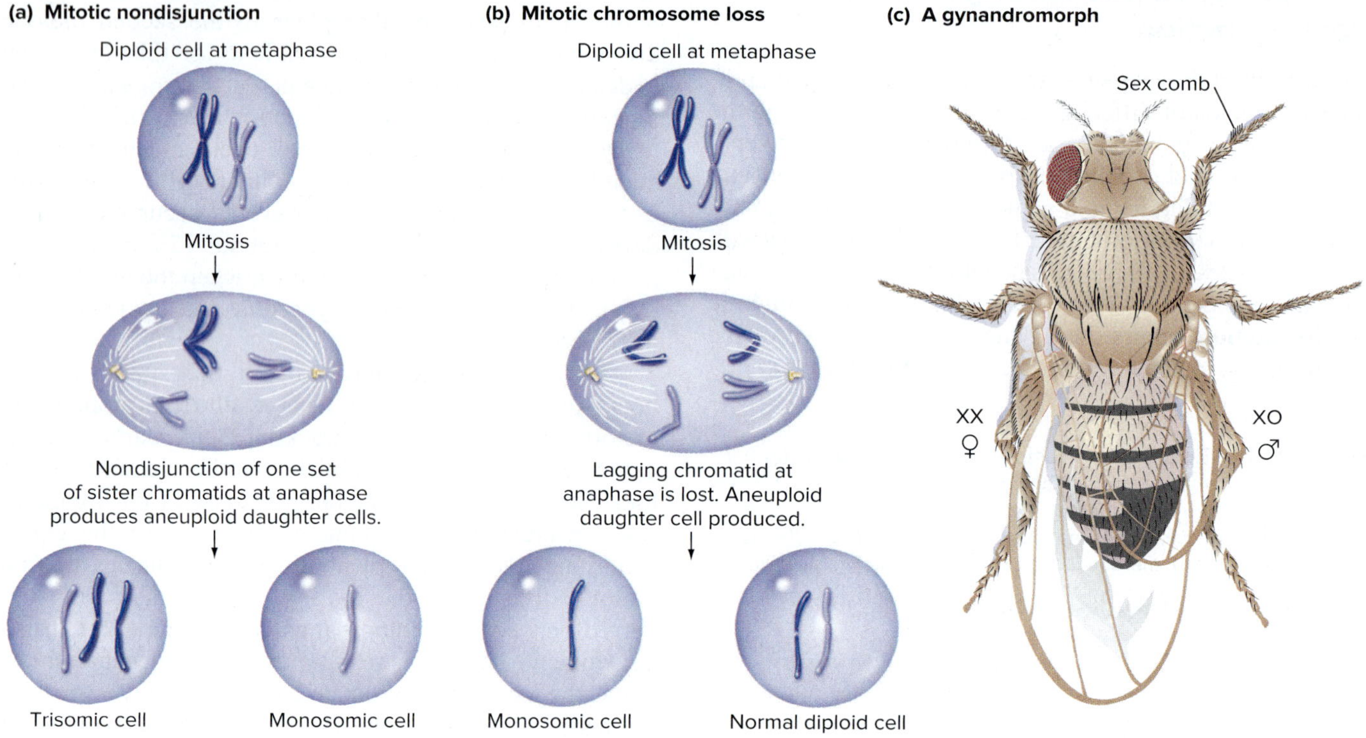

In a multicellular organism, aneuploid cells arising from either mitotic nondisjunction or chromosome loss may survive and undergo further rounds of cell division, producing clones of cells with an abnormal chromosome count. Nondisjunction or chromosome loss occurring early in development will generate larger aneuploid clones than the same events occurring later in development. The side-by-side existence of aneuploid and normal tissues results in a **mosaic** organism whose phenotype depends on what tissue bears the aneuploidy, the number of aneuploid cells, and the specific alleles of genes on the aneuploid chromosome. Many examples of mosaicism involve the sex chromosomes. If an XX *Drosophila* female loses one of the X chromosomes during the first mitotic division after fertilization, the result is a **gynandromorph** composed of equal parts male and female tissue (**Fig. 13.30c**).

Many Turner syndrome females are mosaics carrying some XX cells and some XO cells. These individuals began their development as XX zygotes, but with the loss of an X chromosome during the embryo's early mitotic divisions, they acquired a clone of XO cells. Similar mosaicism involving the autosomes also occurs. For example, physicians have recorded several cases of mild Down syndrome arising from mosaicism for trisomy 21. In people with Turner or Down mosaicism, the existence of some normal tissue ameliorates the condition, with the individual's phenotype depending on the particular distribution of diploid versus aneuploid cells.

essential concepts

- *Aneuploidy* is the loss or gain of one or more chromosomes.
- Autosomal aneuploidy is usually lethal due to genetic imbalance.
- Sex chromosome aneuploidy is usually well tolerated because only one X chromosome remains active and because the Y chromosome has few genes.
- Chromosome *nondisjunction* during either meiotic division creates unbalanced gametes and thus causes aneuploidy in the progeny.
- Rare mitotic nondisjunction or chromosome loss results in *mosaics;* that is, organisms with some normal cells and some aneuploid cells.

13.5 Variation in Number of Chromosome Sets: Euploidy

learning objectives

1. Differentiate between *x* and *n* as they apply to chromosome number in euploids and aneuploids.
2. Explain why organisms with an odd number of chromosome sets are usually sterile.
3. Compare autopolyploids with allopolyploids.
4. Discuss the ways in which plant breeders exploit the existence of monoploidy and polyploidy.

In contrast to aneuploids, **euploid** cells contain only complete sets of chromosomes. Most euploid species are diploid, but some euploid species are **polyploids** that carry three or more complete sets of chromosomes (**Table 13.5**). When speaking of polyploids, geneticists use the symbol x to indicate the **basic chromosome number,** that is, the number of different chromosomes that make up a single complete set. **Triploid** species, which have three complete sets of chromosomes, are then $3x$; **tetraploid** species with four complete sets of chromosomes are $4x$; and so forth.

For diploid species, x is identical to n—the number of chromosomes in the gametes—because each gamete contains a single complete set of chromosomes. This identity of $x = n$ does not, however, hold for polyploid species, as the following example illustrates. Commercially grown bread wheat has a total of 42 chromosomes: 6 nearly (but not wholly) identical sets each containing 7 different chromosomes. Bread wheat is thus a hexaploid with a basic number of $x = 7$ and $6x = 42$. But each triploid gamete has one-half the total number of chromosomes, so $n = 21$. Thus, for bread wheat, x and n are not the same.

TABLE 13.5 Variations of Euploidy

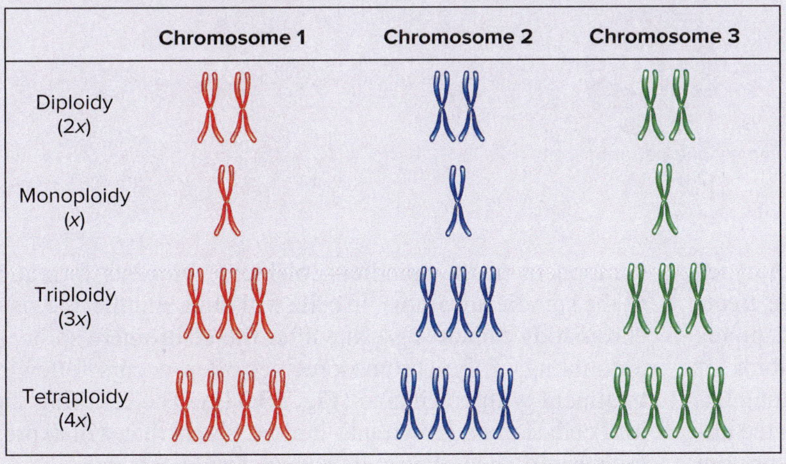

Another form of euploidy, in addition to polyploidy, exists in **monoploid** (x) organisms, which have only one set of chromosomes (Table 13.5).

Monoploidy and polyploidy are observed rarely in animals. Among the few examples of monoploidy are some species of ants and bees in which the males are monoploid, whereas the females are diploid. Males of these species develop by **parthenogenesis** from unfertilized eggs. These monoploid males produce gametes through a modified meiosis that in some unknown fashion ensures distribution of all the chromosomes to the same daughter cell during meiosis I; the sister chromatids then separate normally during meiosis II.

Polyploidy in animals exists normally only in species with unusual reproductive cycles, such as hermaphroditic earthworms, which carry both male and female reproductive organs, and parthenogenetically tetraploid species of goldfish. In *Drosophila,* it is possible, under special circumstances, to produce triploid and tetraploid females, but never males. In humans, polyploidy is always lethal, usually resulting in spontaneous abortion during the first trimester of pregnancy.

Monoploid Plants Are Useful to Plant Breeders

Botanists can produce monoploid plants experimentally from diploid species by special treatment of germ cells that have completed meiosis and would normally develop into pollen. (Note that monoploid plants obtained in this manner can also be considered haploids because $x = n$.) The treated cells divide into a mass of tissue known as an *embryoid.* Subsequent exposure to plant hormones enables the embryoid to develop into a plant (**Fig. 13.31a**). Monoploid plants may also arise from rare spontaneous events in a large natural population.

Most monoploid plants, no matter how they originate, are infertile. Because the chromosomes have no homologs with which to pair during meiosis I, they are distributed at random to the two spindle poles during this division. Rarely do all chromosomes go to the same pole, and if they do not, the resulting gametes are defective as they lack one or more chromosomes. The greater the number of chromosomes in the genome, the lower the likelihood of producing a gamete containing all of them.

Despite such gamete-generating problems, monoploid plants and tissues are of great value to plant breeders. Monoploids make it possible to visualize normally recessive traits directly, without crosses to achieve homozygosity. Plant researchers can introduce mutations into

Figure 13.31 The creation and use of monoploid plants. (a) Under certain conditions, haploid pollen grains can grow into haploid embryoids. When treated with plant hormones, haploid embryoids grow into monoploid plants. **(b)** Researchers select monoploid cells for recessive traits such as herbicide resistance. They then grow the selected cells into a resistant embryoid, which (with hormone treatment) eventually becomes a mature, resistant monoploid plant. Treatment with colchicine doubles the chromosome number, creating diploid cells that can be grown in culture with hormones to make a homozygous herbicide-resistant diploid plant. **(c)** Colchicine treatment prevents formation of the mitotic spindle and also blocks cytokinesis, generating cells with twice the number of chromosomes. *Blue, red,* and *green* colors denote nonhomologous chromosomes.

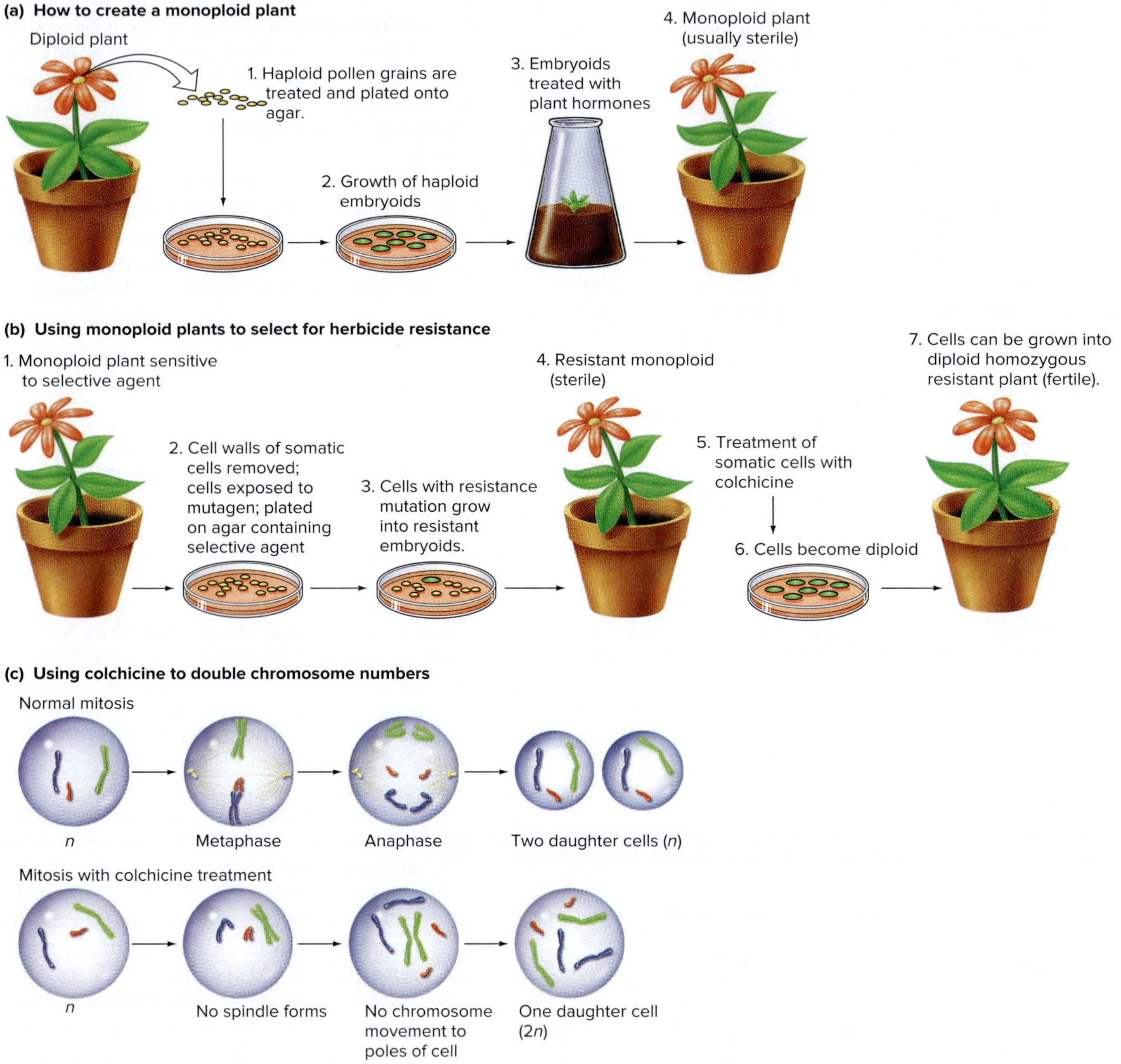

individual monoploid cells, select for desirable phenotypes such as resistance to herbicides, and use hormone treatments to grow the selected cells into monoploid plants (**Fig. 13.31b**). Breeders can then convert monoploids of their choice into homozygous diploid plants by treating tissue with *colchicine,* an alkaloid drug obtained from the autumn crocus. By binding to tubulin—the major protein component of the spindle—colchicine prevents formation of the spindle apparatus. In cells without a spindle, the sister chromatids cannot segregate after the centromere splits, so a doubling of the chromosome set often occurs following treatment with colchicine (**Fig. 13.31c**). The resulting diploid cells can be grown into diploid plants that will express the desired phenotype and produce fertile gametes.

Triploid Organisms Are Usually Infertile

Triploids ($3x$) result from the union of monoploid (x) and diploid ($2x$) gametes (**Fig. 13.32a**). The diploid gametes may be the products of meiosis in tetraploid ($4x$) germ cells, or they may be the products of rare spindle or cytokinesis failures during meiosis in a diploid.

Sexual reproduction in triploid organisms is extremely inefficient because meiosis produces mostly unbalanced gametes. During the first meiotic division in a triploid germ cell, three sets of chromosomes must segregate into two daughter cells. Regardless of how the chromosomes align in pairs, there is no way to ensure that the resulting gametes obtain a complete, balanced x or $2x$ complement of chromosomes. In most cases, at the end of anaphase I, two chromosomes of any one type move to one pole, while the remaining chromosome of the same type moves to the opposite pole. Because each set of three homologs decides independently of the other sets which pole gets two copies and which pole gets a single copy of that chromosome, the products of such a meiosis are almost always unbalanced, with two copies of some chromosomes and one copy of others (**Fig. 13.32b**). If x is large, the chance of obtaining any balanced gametes at all is remote. Thus, fertilization with gametes from triploid individuals does not usually produce viable offspring. However, if x is small, occasionally a meiosis will produce balanced gametes; by chance, two copies of each homolog will migrate to the same pole, and the remaining single copy of each homolog will migrate to the opposite pole.

Figure 13.32 The genetics of triploidy. **(a)** Production of a triploid ($3x = 6$) from fertilization of a monoploid gamete by a diploid gamete. Nonhomologous chromosomes are *blue* or *red.* **(b)** Meiosis in a triploid typically produces unbalanced (aneuploid) gametes because in meiosis I, one chromosome from each of the sets of three homologs segregates randomly into one or the other daughter cell.

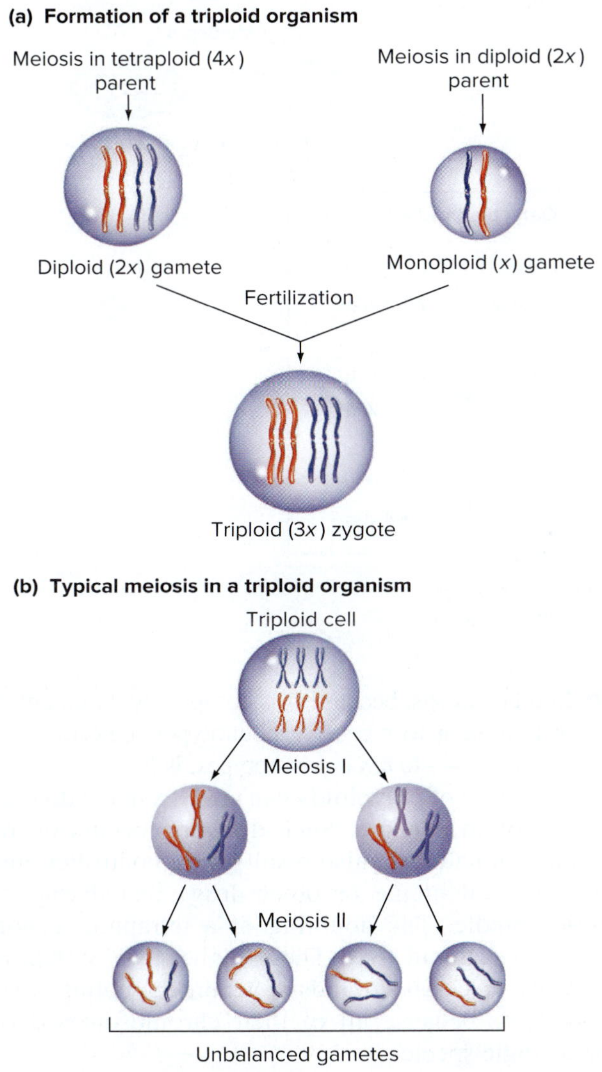

It is possible to propagate some triploid species, such as bananas and watermelons, through asexual reproduction. The fruits of triploid plants are seedless because the unbalanced gametes do not function properly in fertilization or, if fertilization occurs, the resultant zygote is so genetically unbalanced that it cannot develop. Either way, no seeds form. Like triploids, all polyploids with odd numbers of chromosome sets (such as $5x$ or $7x$) are sterile because they cannot reliably produce balanced gametes.

Polyploids with an Even Number of Chromosome Sets Can Become New Species

During mitosis, if the chromosomes in a diploid ($2x$) tissue fail to separate after replication, the resulting daughter cells will be tetraploid ($4x$; **Fig. 13.33a**). If such tetraploid cells arise in reproductive tissue, subsequent meioses will produce diploid gametes. Rare unions between diploid gametes produce tetraploid organisms. Self-fertilization of a newly created tetraploid organism will produce an entirely new species, because crosses between the tetraploid and the original diploid organism will produce infertile triploids (review Fig. 13.32). Tetraploids made in this fashion are **autopolyploids** (in this case, *autotetraploids*), a kind of polyploid that derives all its chromosome sets from the same species.

Maintenance of a tetraploid species depends on the production of gametes with balanced sets of chromosomes. Most successful tetraploids have evolved mechanisms ensuring that the four copies of each group of homologs pair two by two to form two **bivalents**—pairs of synapsed homologous chromosomes (**Fig. 13.33b**). Because the chromosomes in each bivalent become attached to opposite spindle poles during meiosis I, meiosis regularly produces gametes carrying two complete sets of chromosomes.

Tetraploids, with four copies of every gene, generate unusual Mendelian ratios. For example, even if only two alleles of a gene exist (say, A and a), five different genotypes are possible: $A\,A\,A\,A$, $A\,A\,A\,a$, $A\,A\,a\,a$, $A\,a\,a\,a$, and $a\,a\,a\,a$. If the

Figure 13.33 The genetics of tetraploidy. (a) Tetraploids arise from a failure of chromosomes to separate into two daughter cells during mitosis in a diploid. **(b)** In successful tetraploids, the pairing of chromosomes as bivalents generates genetically balanced gametes. **(c)** Gametes produced in an *A A a a* tetraploid heterozygous for two alleles of a centromere-linked gene, with orderly pairing of bivalents. The four chromosomes can pair to form two bivalents in three possible ways. For each pairing scheme, the chromosomes in the two pairs can assort in two different orientations. If all possibilities are equally likely, the expected genotype frequency in a population of gametes will be 1 (*A A*) : 4 (*A a*) : 1 (*a a*).

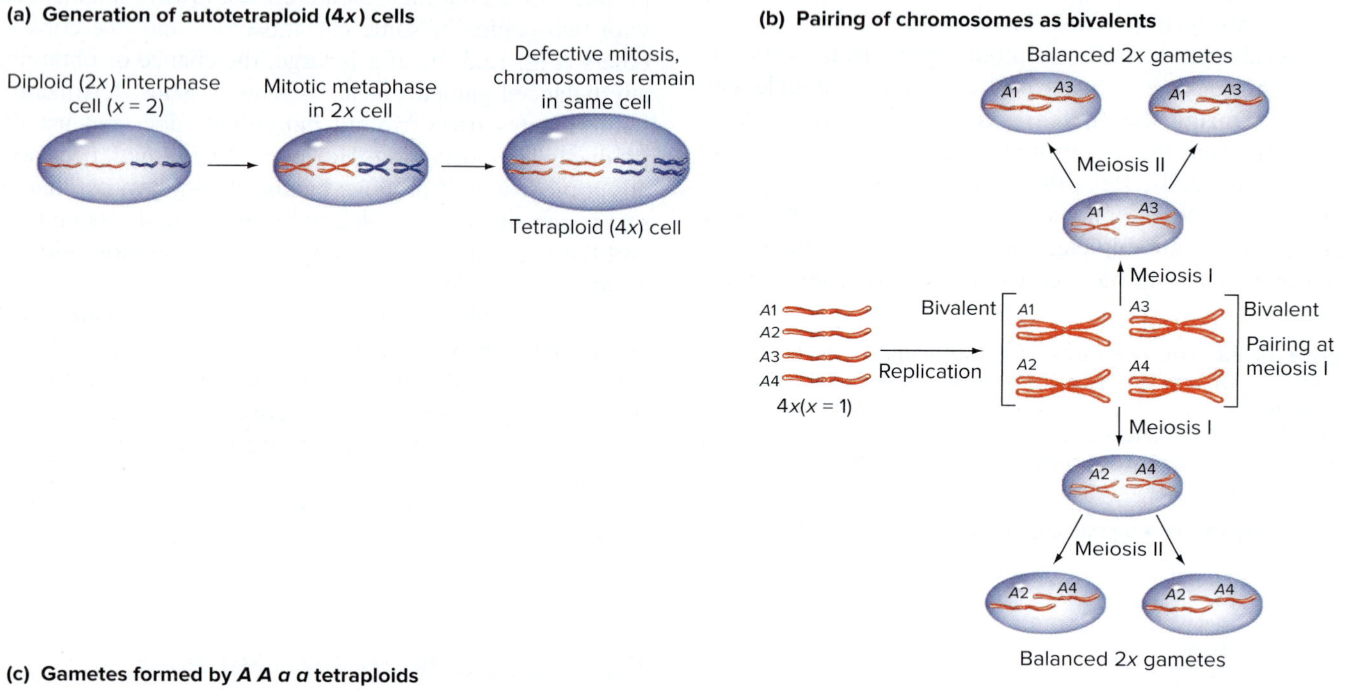

(c) Gametes formed by *A A a a* tetraploids

Chromosomes	Pairing and Alignment			Gametes Produced*		
1. *A* 2. *A* 3. *a* 4. *a*	1 ↑*A* / 2 ↓*A*; 3 ↑*a* / 4 ↓*a*	or	1 ↑*A* / 2 ↓*A*; 4 ↑*a* / 3 ↓*a*	1 + 3 *A a* 2 + 4 *A a*	or	1 + 4 *A a* 2 + 3 *A a*
	1 ↑*A* / 3 ↓*a*; 2 ↑*A* / 4 ↓*a*	or	1 ↑*A* / 3 ↓*a*; 4 ↑*a* / 2 ↓*A*	1 + 2 *A A* 3 + 4 *a a*	or	1 + 4 *A a* 2 + 3 *A a*
	1 ↑*A* / 4 ↓*a*; 2 ↑*A* / 3 ↓*a*	or	1 ↑*A* / 4 ↓*a*; 3 ↑*a* / 2 ↓*A*	1 + 2 *A A* 3 + 4 *a a*	or	1 + 3 *A a* 2 + 4 *A a*

*Assuming no crossovers between the centromere and gene *A*

Total:
2(*A A*) : 8(*A a*) : 2(*a a*)
= 1(*A A*) : 4(*A a*) : 1(*a a*)

phenotype depends on the dosage of *A*, then five phenotypes, each corresponding to one of the genotypes, will appear.

The segregation of alleles during meiosis in a tetraploid is similarly complex. Consider an *A A a a* heterozygote where the *A* allele is completely dominant. What are the chances of obtaining progeny with the recessive phenotype, generated by only the *a a a a* genotype? As **Fig. 13.33c** illustrates, if during meiosis I, the four chromosomes carrying the gene align at random in bivalents along the metaphase plate, the expected ratio of gametes is 2 (*A A*) : 8 (*A a*) : 2 (*a a*) = 1 (*A A*) : 4 (*A a*) : 1 (*a a*). The chance of obtaining *a a a a* progeny during self-fertilization is thus 1/6 × 1/6 = 1/36. In other words, because *A* is completely dominant, the ratio of dominant to recessive phenotypes, determined by the ratio of *A* – – – to *a a a a* genotypes, is 35:1.

New levels of polyploidy can arise from the doubling of a polyploid genome. Such doubling occurs on rare occasions in nature; it also results from controlled treatment with colchicine or other drugs that disrupt the mitotic spindle. The doubling of a tetraploid genome yields an octaploid (8*x*). These higher-level polyploids created by successive rounds of genome doubling are autopolyploids because all of their chromosomes derive from a single species.

Figure 13.34 Polyploid plants can be larger than their diploid counterparts. A comparison of octaploid (*left*) and diploid (*right*) strawberries.

Roughly one out of every three known species of flowering plants is a polyploid, and because polyploidy often increases plant size and vigor, many polyploid plants with edible parts have been selected for agricultural cultivation. Most commercially grown alfalfa, coffee, and peanuts are tetraploids ($4x$). MacIntosh apple and Bartlett pear trees that produce giant fruits are also tetraploids. Commercially grown strawberries are octaploids ($8x$) (**Fig. 13.34**).

The evolutionary success of polyploid plant species may stem from the fact that polyploidy, like gene duplication, provides additional copies of genes. While one copy continues to perform the original function, the others can evolve new functions that may be advantageous. As you have seen, however, the fertility of polyploid species requires an even number of chromosome sets.

Allopolyploids Are Hybrids with Complete Chromosome Sets from Two Different Species

Polyploidy can arise not only from chromosome doubling, but also from crosses between members of two different species, even if they have different numbers of chromosomes. Hybrids in which the chromosome sets come from two or more distinct, though related, species are known as **allopolyploids.**

Fertile allopolyploids arise only rarely because chromosomes from two different species usually differ in DNA sequence and number, so they cannot easily pair with each other. The resulting irregular segregation creates genetically unbalanced gametes such that the hybrid progeny will be sterile. Chromosomal doubling in germ cells, however, can restore fertility by creating a pairing partner for each chromosome. Allopolyploids produced in this manner are called **amphidiploids** if the two parental species were diploids; amphidiploids contain two diploid genomes, each one derived from a different parent. As the following illustrations show, it is hard to predict the characteristics of an amphidiploid or other allopolyploids.

A cross between diploid cabbages and diploid radishes, for example, leads to the production of amphidiploids known as *Raphanobrassica.* The gametes of both parental species contain 9 chromosomes; the sterile F_1 hybrids have 18 chromosomes, none of which has a homolog. Chromosome doubling in the germ cells after treatment with colchicine, followed by union of two of the resulting gametes, produces a new species: a fertile *Raphanobrassica* amphidiploid carrying 36 chromosomes—a full complement of 18 (9 pairs) derived from cabbages and a full complement of 18 (9 pairs) derived from radishes. Unfortunately, this amphidiploid has the roots of a cabbage plant and leaves resembling those of a radish, so it is not agriculturally useful.

By contrast, crosses between tetraploid (or hexaploid) wheat and diploid rye have led to the creation of several allopolyploid hybrids with agriculturally desirable traits from both species (**Fig. 13.35**). Some of the hybrids combine the high yields of wheat with rye's ability to adapt to unfavorable environments. Others combine wheat's high level of protein with rye's high level of lysine; wheat protein does not contain very much of this amino acid, an essential ingredient in the human diet. The various hybrids between wheat and rye form a new crop known as *Triticale.* Some Triticale strains produce nutritious grains that already appear in breads sold in health food stores. Plant breeders are currently assessing the usefulness of various Triticale strains for large-scale agriculture.

essential concepts

- *Euploid* organisms contain complete sets of chromosomes. *Monoploids* have only a single set of chromosomes, while *polyploids* have more than two sets.
- In polyploid organisms, x refers to the basic chromosome number that makes up a complete set, while n refers to the number of chromosomes in a gamete. For diploid organisms, ($2x$ somatic cells), $n = x$. For a hexaploid, ($6x$ somatic cells), $n = 3x$.
- Monoploids as well as polyploids containing odd numbers of chromosome sets are usually sterile because chromosomes cannot segregate during meiosis to produce balanced gametes.
- An *autopolyploid* derives all of its chromosomes from a single species; an *allopolyploid* has chromosomes from different species.
- Chromosome doubling due to mistakes in cell division produces genomes in which all chromosomes have a pairing partner and so the organism can be fertile.

Figure 13.35 Allopolyploids in agriculture. **(a)** Plant breeders crossed wheat with rye to create allopolyploid Triticale. Because this strain of wheat is tetraploid, x_1 (the number of chromosomes in the basic wheat set) is one-half n_1 (the number of chromosomes in a wheat gamete). For diploid rye, $n_2 = x_2$. Note that the F_1 hybrid between wheat and rye is sterile because the rye chromosomes have no pairing partners. Doubling of chromosome numbers by colchicine treatment of the F_1 hybrid allows regular pairing. **(b)** A comparison of wheat, rye, and Triticale grain stalks.

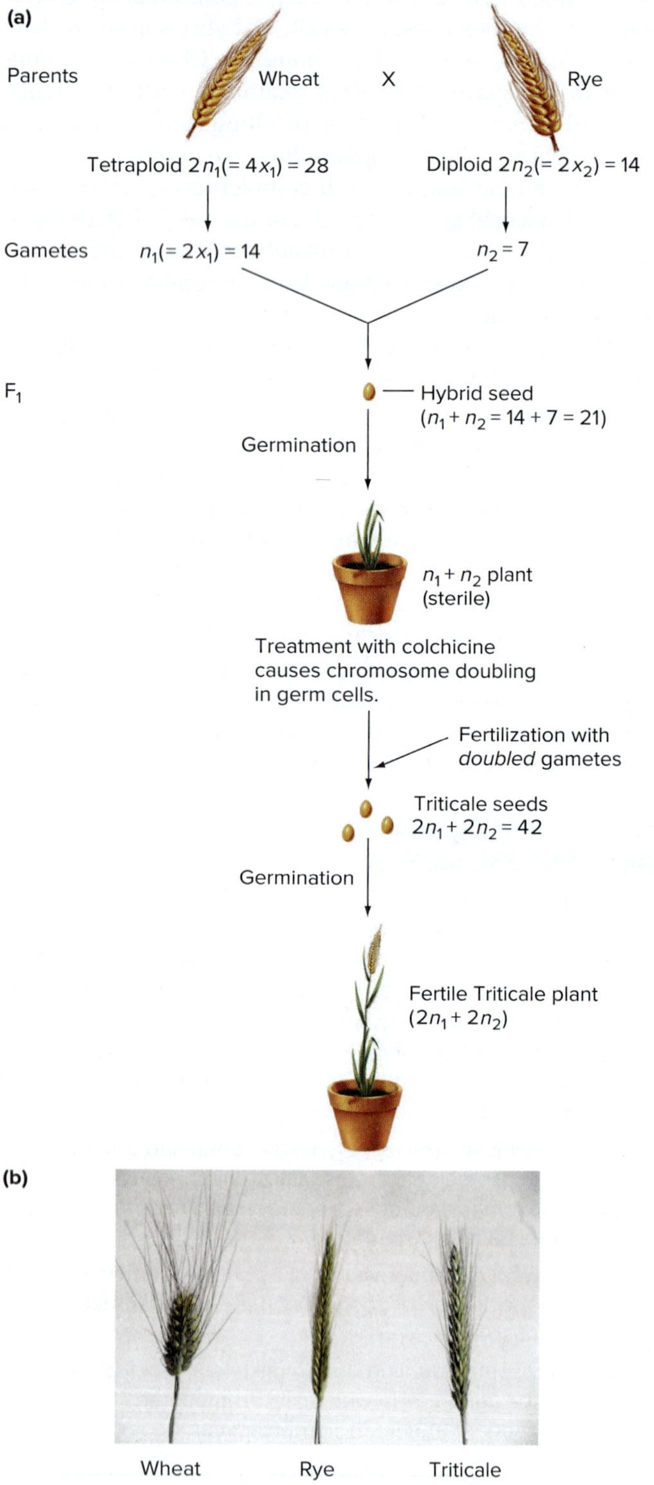

13.6 Genome Restructuring and Evolution

learning objectives

1. Describe how different types of chromosomal rearrangements alter gene expression patterns or generate new gene products.
2. Discuss how the extra gene copies generated by duplications help fuel evolution.
3. Explain why translocations contribute to speciation.

We saw at the beginning of this chapter that roughly 300 chromosomal rearrangements could reshape the human genome to a form that resembles the mouse genome. Direct DNA sequence comparison of the mouse and human genomes indicates that deletions, duplications, inversions, translocations, and transpositions have occurred in one or the other lineage since humans and mice began to diverge from a common ancestor 65 million years ago. We consider here three general ways in which the kinds of genomic restructuring discussed in this chapter might contribute to evolution.

Chromosome Rearrangements and Transposition Are Important Sources of Genome Variation

The process of evolution depends entirely on the existence of variation in DNA sequences. Variants that enable organisms to survive and reproduce in their particular environments will spread throughout the population, while variants with negative impacts on an organism's fitness will disappear. Without variation, there can be no natural selection and no evolution.

Although previous chapters have stressed the variation associated with single-base changes such as SNPs, it should be clear from this chapter that larger-scale rearrangements and the movement of TEs are also crucial contributors to DNA variation that can be acted upon by natural selection. Genes at or near rearrangement breakpoints may acquire new patterns of expression that increase or decrease their expression in particular tissues. In some cases, rearrangement breakpoints create new gene functions by the fusion of two previously separated genes (review Fig. 13.16b). The movement of a TE into a gene can alter gene expression or even disrupt a gene entirely to create an amorphic (null) mutation (as in Fig. 13.24); even gene inactivation could be advantageous in a new environment where the function of that gene might be detrimental.

Duplications Provide Extra Gene Copies That Can Acquire New Functions

An organism cannot normally tolerate mutations in a gene essential to its survival, but duplication would provide two copies of the gene. If one copy remained intact to perform the essential function, the other would be free to evolve a new function. The genomes of most higher plants and animals, in fact, contain many **gene families**—sets of closely related genes with slightly different functions—that most likely arose from a succession of gene duplication events. In vertebrates, some gene families have hundreds of members. Several examples of such gene families, such as the hemoglobin and taste receptor genes in humans, were discussed in Chapter 10.

If duplication of a few genes provides new raw material that can be acted on by divergence and natural selection to form gene families, you might imagine that the same must be true of duplications of whole genomes, which would yield additional copies of thousands of genes. Polyploidization has in fact been particularly important in the diversification of plant species. For example, plant biologists think that approximately 90 million years ago, an ancient diploid species with five different chromosomes ($x = 5$) underwent a whole genome duplication. The duplicated chromosomes began to diverge from each other and also underwent breakage and Robertsonian translocations. The result of all these events was the creation of a diploid species with 12 chromosomes ($x = 12$) that was an ancestor common to all present-day cereal grass species, including rice, wheat, barley, sorghum, and maize (corn).

Evidence for the proposed ancient whole genome duplication is seen in the genome sequences of present-day cereals such as rice (*Oryza sativa;* **Fig. 13.36**). Many blocks of homologous gene sequences are found on two different chromosomes, as indicated by the colored lines in the middle of the diagram. Within these blocks, the order of the genes is mostly conserved. However, the two copies of the genes are not identical: Some gene copies likely acquired new functions, while others have mutated beyond recognition and thus have become lost.

The general strategy of polyploidization to provide extra gene copies has been so successful in producing variation that the genomes of many plants show signs that they have undergone multiple rounds of genome duplications or triplication at different times in the past.

Figure 13.36 An ancient duplication is still visible in the present-day genome of *O. sativa* (rice). The ancient common ancestor to all cereal grasses is thought to have had five different chromosomes. After diploidization followed by chromosome rearrangements, the two copies of the original genome are now distributed among 12 chromosomes in *O. sativa,* called Os1 through Os12. The colored lines show homologous genes on different rice chromosomes that reveal this evolutionary history.

From: T. Thiel et al. (2009), "Evidence and evolutionary analysis of ancient whole-genome duplication in barley predating the divergence from rice," *BMC Evolutionary Biology,* 9:209, Fig. 1. © Thiel et al. Licensee BioMed Central Ltd. 2009. http://bmcevolbiol.biomedcentral.com/articles/10.1186/1471-2148-9-209

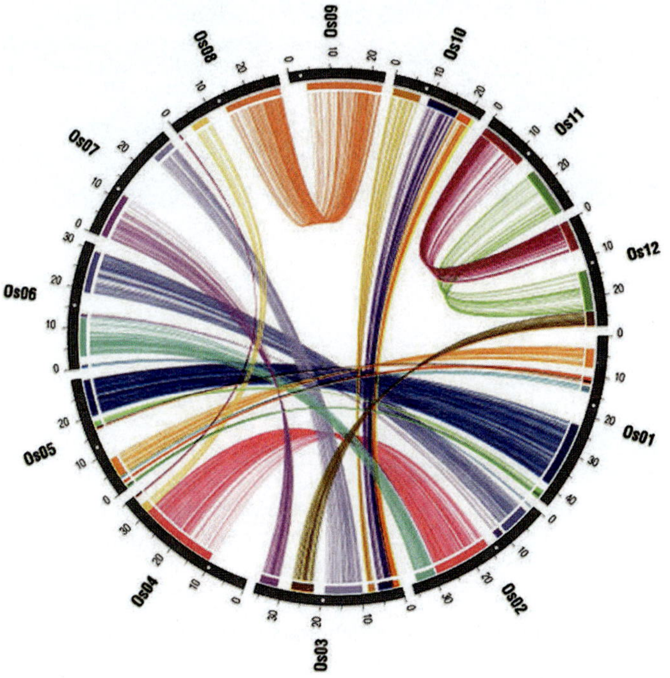

Translocations Contribute to Speciation

On the tiny volcanic island of Madeira off the coast of Portugal in the Atlantic Ocean, two populations of the common house mouse (*Mus musculus*) are becoming separate species because of translocations that have contributed to reproductive isolation. The mice live in a few narrow valleys separated by steep mountains. Geneticists have found that populations of mice on the two sides of these mountain barriers have very different sets of chromosomes because they have accumulated different sets of Robertsonian translocations (**Fig. 13.37**). Mice in one Madeira population, for example, have a diploid number ($2x$) of 22 chromosomes, whereas mice in a different population on the island have 24; for most house mice throughout the world, $2x = 40$.

The hybrid offspring of matings between individuals of these two mountain-separated populations are infertile because chromosomal complements that are so different cannot properly pair and segregate at meiosis. Thus, reproductive isolation has reinforced the already established geographical isolation, and the two populations are close to becoming two separate species. What is remarkable about this example of speciation is that mice were introduced into Madeira by Portuguese settlers only in the fifteenth century. This fact means that the varied and complicated sets of Robertsonian translocations that are contributing to speciation became fixed in the different populations in less than 600 years.

Figure 13.37 Rapid chromosomal evolution in mice on the island of Madeira. **(a)** Distribution of mouse populations with different sets of Robertsonian translocations (indicated by circles of different colors). **(b)** Karyotypes of female mice from two different populations. The karyotype I at the *top* is from the population shown with *blue dots* in part (a); the karyotype II at the *bottom* is from one of the populations indicated by *red dots*. Robertsonian translocations are identified by numbers separated by a comma. (For example, 2,19 is a Robertsonian translocation between chromosomes 2 and 19 of the standard mouse karyotype.)

a: Source: NASA Earth Observatory image created by Jesse Allen, using a digital elevation model from the Direcção Regional de Informação Geográfica e Ordenamento do Território (DRIGOT) of Madeira and the Telecommunication Advanced Networks for GMES Operations (TANGO). Special thanks to Pedro Soares and Antonio de la Cruz (European Union Satellite Centre); b: © Janice Britton-Davidian, Institut des Sciences de l'Evolution Montpellier

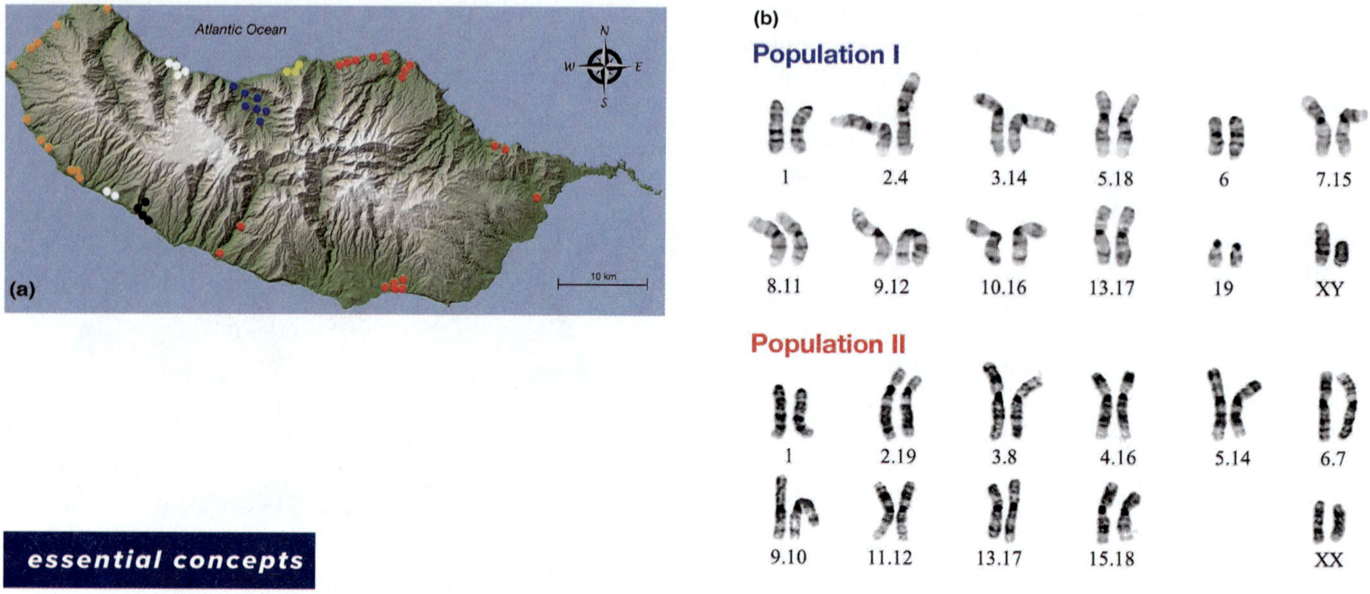

essential concepts

- Generation of new gene expression patterns or new gene products at rearrangement breakpoints provides variation that is subject to *natural selection*.
- Duplications produce extra copies of genes that can mutate independently of one another and adopt new functions.
- Translocations contribute to speciation because heterozygosity for translocations causes infertility; the most successful matings occur only between individuals who share the same chromosome complement.

WHAT'S NEXT

The detrimental consequences of most changes in chromosome organization and number cause considerable distress in humans (recall Table 13.4). Approximately 4 of every 1000 individuals has an abnormal phenotype associated with aberrant chromosome organization or number. Most of these abnormalities result from either aneuploidy for the X chromosome or trisomy 21. By comparison, about 10 people per 1000 suffer from a serious inherited disease caused by a single-gene mutation.

The incidence of chromosomal abnormalities among humans would be much larger were it not for the fact that many embryos or fetuses with abnormal karyotypes abort spontaneously in early pregnancy. Fully 15% to 20% of recognized pregnancies end with detectable spontaneous abortions; and half of the spontaneously aborted fetuses show chromosomal abnormalities, particularly trisomy, sex chromosome monosomy, and triploidy. These figures almost certainly underestimate the rate of spontaneous abortion caused by abnormal chromosomal variations, because embryos carrying aberrations for larger chromosomes, such as monosomy 2 or trisomy 5, may abort so early that the pregnancy goes unrecognized.

But despite all the negative effects of chromosomal rearrangements and changes in chromosome number, a few departures from normal genome organization survive to become instruments of evolution by natural selection.

As we see in the next chapter, chromosomal rearrangements occur in bacteria as well as in eukaryotic organisms. In bacteria, transposable elements catalyze many of the changes in chromosomal organization. Remarkably, the reshuffling of genes between different DNA molecules in the same cell helps speed the transfer of genetic information from one bacterial cell to another.

DNA: © Design Pics/Bilderbuch RF

SOLVED PROBLEMS

I. Male *Drosophila* from a true-breeding wild-type stock were irradiated with X-rays and then mated with females from a true-breeding stock carrying the following recessive mutations on the X chromosome: yellow body (*y*), crossveinless wings (*cv*), cut wings (*ct*), singed bristles (*sn*), and miniature wings (*m*). These markers are known to map in the order:

y - *cv* - *ct* - *sn* - *m*

Most of the female progeny of this cross were phenotypically wild type, but one female exhibited *ct* and *sn* phenotypes. When this exceptional *ct sn* female was mated with a male from the true-breeding wild-type stock, twice as many females as males appeared among the progeny.

a. What is the nature of the X-ray-induced mutation present in the exceptional female?

b. Draw the X chromosomes present in the exceptional *ct sn* female as they would appear during pairing in meiosis.

c. What phenotypic classes would you expect to see among the progeny produced by mating the exceptional *ct sn* female with a normal male from a true-breeding wild-type stock? List males and females separately.

Answer

To answer these questions, you need to think first about the effects of different types of chromosomal mutations in order to deduce the nature of the mutations. Then you can evaluate the consequences of the mutations on inheritance.

a. Two observations indicate that X-rays induced a deletion mutation. The fact that two recessive mutations are expressed phenotypically in the exceptional female suggests that a deletion was present on one of her X chromosomes that uncovered the two mutant alleles (*ct* and *sn*) on the other X chromosome.

Second, the finding that twice as many females as males were among the progeny of the exceptional female is also consistent with a deletion mutation. Males who inherit the deletion-bearing X chromosome from their exceptional mother will be inviable (because other essential genes are located in the region that is now deleted), but sons who inherit a nondeleted X chromosome will survive. On the other hand, all of the exceptional female's daughters will be viable: Even if they inherit a deleted X chromosome from their mother, they also receive a normal X chromosome from their father. As a result, the cross of the exceptional female with a wild-type male will produce half as many male progeny as females.

b. During pairing, the DNA in the normal (nondeleted) X chromosome will loop out because no homologous region exists in the deletion chromosome. In the simplified drawing of meiosis I that follows, each line represents both chromatids comprising each homolog.

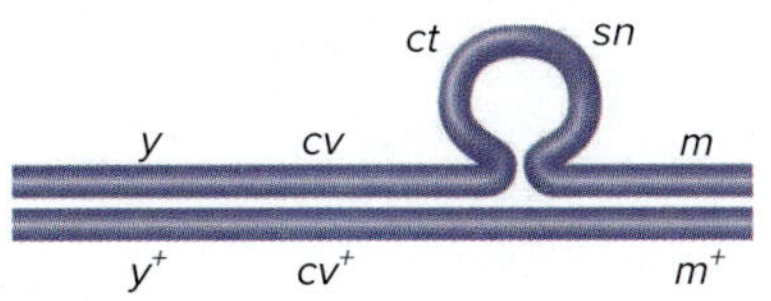

c. All daughters of the exceptional female will be wild type because the father contributes wild-type copies of all the genes. Each of the surviving sons must inherit a nondeleted X chromosome from the exceptional female. Some of these X chromosomes are produced from meioses in which no recombination occurred, but other X chromosomes are the products of recombination. Males can have any of the genotypes listed here and therefore the corresponding phenotypes. All classes contain the *ct sn* combination because no recombination between homologs is possible in this deleted region. Some of these genotypes require multiple crossovers during meiosis in the mother and will thus be relatively rare.

y	*cv*	*ct*	*sn*	*m*
+	+	*ct*	*sn*	+
+	*cv*	*ct*	*sn*	*m*
y	+	*ct*	*sn*	+
y	*cv*	*ct*	*sn*	+
+	+	*ct*	*sn*	*m*
+	*cv*	*ct*	*sn*	+
y	+	*ct*	*sn*	*m*

II. In maize trisomics, $n + 1$ pollen is not viable. If a dominant allele at the *B* locus produces purple color instead of the recessive phenotype bronze and a *B b b* trisomic plant is pollinated by a *B B b* plant, what proportion of the progeny produced will be trisomic and have a bronze phenotype?

Answer

To solve this problem, think about what is needed to produce trisomic bronze progeny: three *b* chromosomes in the zygote. The female parent would have to contribute two *b* alleles, because the $n + 1$ pollen from the male is not viable.

What kinds of gametes could be generated by the trisomic $B\ b\ b$ purple female parent, and in what proportion? To track all the possibilities, rewrite this genotype as $B\ b_1\ b_2$, even though b_1 and b_2 have identical effects on phenotype. In the trisomic female, the chromosomes carrying these alleles could pair as bivalents during the first meiotic division in three different ways, so that they would segregate to opposite poles: B with b_1, B with b_2, and b_1 with b_2. In all three cases, the remaining chromosome could move to either pole. To tabulate the possibilities as a branching diagram:

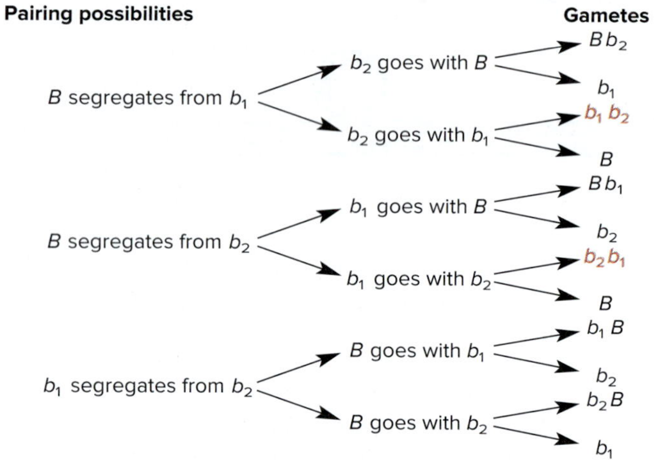

Of the 12 gamete classes produced by these different possible segregations, only the two classes written in red contain the two b alleles needed to generate the bronze ($b\ b\ b$) trisomic zygotes. The chance of obtaining such gametes is thus $2/12 = 1/6$.

Although segregation in the $B\ B\ b$ male parent is equally complicated, remember that males cannot produce viable $n + 1$ pollen. The only surviving gametes would thus be B and b, in a ratio (2/3 B and 1/3 b) that must reflect their relative prevalence in the male parent genome. The probability of obtaining trisomic bronze progeny from this cross is therefore the product of the individual probabilities of the appropriate $b\ b$ gametes from the female parent (1/6) and b pollen from the male parent (1/3): $1/6 \times 1/3 = 1/18$.

III. The figure at the beginning of this chapter shows a chromosomal deletion revealed by DNA barcoding.

Suppose the genome of a person heterozygous for the deletion was sequenced. Would the genome sequence data reveal both the the extent of the deleted region and the exact sequence of the deletion breakpoint? Consider separately the contributions of the two types of data that could be obtained: (i) the number of reads of any particular sequence, and (ii) the presence of novel sequences not seen in wild type.

Answer

To answer this question, you need to understand how genome sequencing data can reveal the breakpoints and extents of deletions.

Normal and deleted chromosome 5s are depicted in the diagram that follows; a large region (DEF) of the q arm (long arm) of the chromosome is missing in the deleted chromosome.

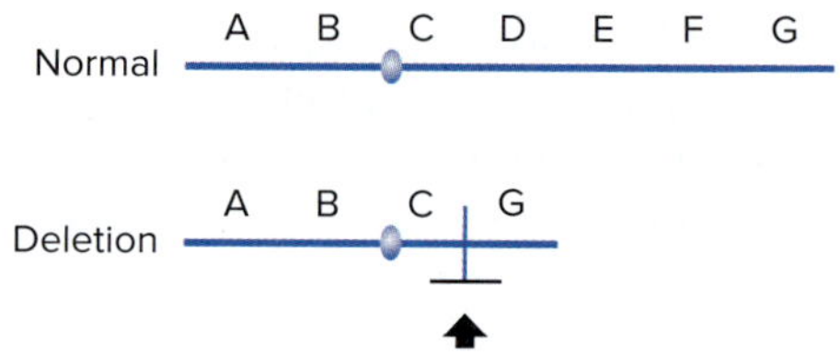

(i) Sequence reads from the deleted region (DEF) will be present at 50% the frequency of other sequence reads in the genome. (ii) In addition, genome sequencing of a deletion heterozygote will reveal the deletion breakpoint (*arrow*) because unique sequence reads will span it.

PROBLEMS

Vocabulary

1. For each of the terms in the left column, choose the best matching phrase in the right column.

a. reciprocal translocation	1. lacking one or more chromosomes or having one or more extra chromosomes
b. gynandromorph	2. movement of short DNA elements
c. pericentric	3. having more than two complete sets of chromosomes
d. paracentric	4. exact exchange of parts of two nonhomologous chromosomes
e. euploids	5. excluding the centromere
f. polyploidy	6. including the centromere
g. transposition	7. having complete sets of chromosomes
h. aneuploids	8. mosaic combination of male and female tissue

Section 13.1

2. Human chromosome 1 is a large, metacentric chromosome. A researcher decides to use multicolor banding to barcode chromosome 1 in a number of different human karyotypes. A map of a region near the telomere

of chromosome 1 is shown in the accompanying figure. Three of the barcode probe DNAs (A, B, and C) were from this region. Each was labeled with a different fluor and used for fluorescent *in situ* hybridization (FISH) to human mitotic metaphase chromosome squashes made with cells obtained from individuals with various genotypes. The breakpoints of chromosomal rearrangements in this region are indicated on the map. The black bars for deletions (*Del*) 1 and 2 represent DNA that is deleted. The breakpoints of inversions (*Inv*) 1 and 2 not shown in the figure are near but not at the centromere.

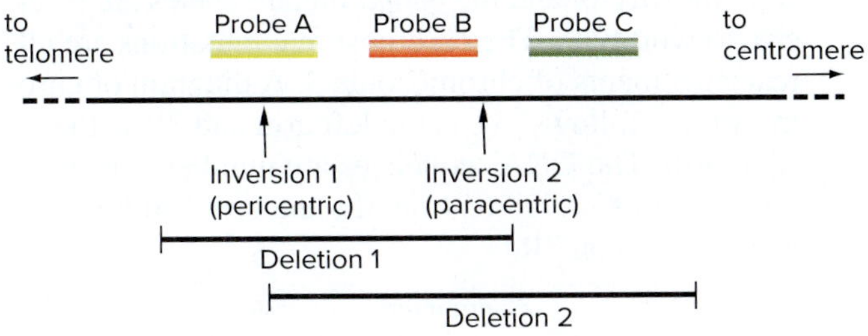

For each of the following genotypes, draw chromosome 1 as it would appear after FISH with all three probes. An example is shown in the following figure for hybridization of the yellow probe A to the two copies of chromosome 1 in wild type (+/+).

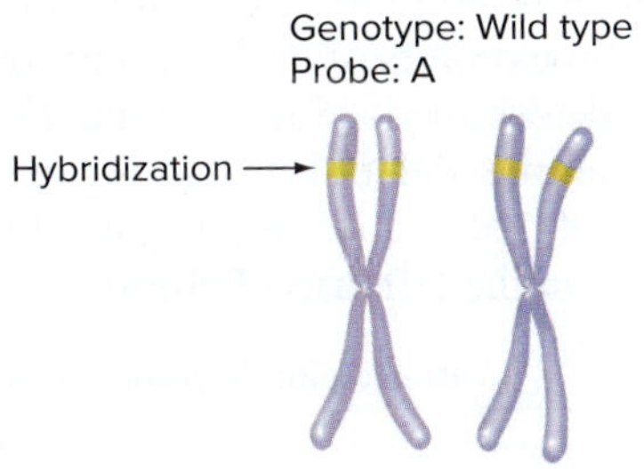

a. *Del1/Del2*
b. *Del1/+*
c. *Inv1/+*
d. *Inv2/+*
e. *Inv2/Inv2*

Section 13.2

3. For each of the following types of chromosomal aberrations, tell: (i) whether the chromosomes of an organism heterozygous for the aberration will form any type of loop during prophase I of meiosis; (ii) whether a chromosomal bridge can be formed during anaphase I in a heterozygote, and if so, under what conditions; (iii) whether an acentric fragment can be formed during anaphase I in a heterozygote, and if so, under what conditions; (iv) whether the aberration can suppress meiotic recombination; and (v) whether the two chromosomal breaks responsible for the aberration occur on the same side or on opposite sides of a single centromere, or if the two breaks occur on different chromosomes.
   a. Reciprocal translocation
   b. Paracentric inversion
   c. Small tandem duplication
   d. Robertsonian translocation
   e. Pericentric inversion
   f. Large deletion
4. For the following types of chromosomal rearrangements, would it theoretically ever be possible to obtain a perfect reversion of the rearrangement? If so, would such revertants be found only rarely, or would they be relatively common?
   a. A deletion of a region including five genes
   b. A tandem duplication of a region including five genes
   c. A pericentric inversion
   d. A Robertsonian translocation
   e. A mutation caused by a transposable element jumping into a protein-coding exon of a gene
5. One of the X chromosomes in a particular *Drosophila* female had a normal order of genes but carried recessive alleles of the genes for yellow body color (y), vermilion eye color (v), and forked bristles (f), as well as the dominant X-linked Bar eye mutation (B). Her other X chromosome carried the wild-type alleles of all four genes, but the region including y^+, v^+, and f^+ (but not B^+) was inverted with respect to the normal order of genes. This female was crossed to a wild-type male as diagrammed here.

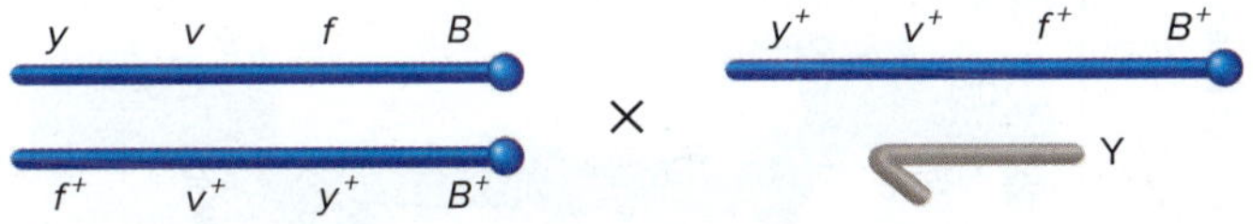

The cross produced the following male offspring:

y	v	f	B	48
y^+	v^+	f^+	B^+	45
y	v	f	B^+	11
y^+	v^+	f^+	B	8
y	v^+	f	B	1
y^+	v	f^+	B^+	1

   a. Why are there no male offspring with the allele combinations $y\ v\ f^+$, $y^+\ v^+\ f$, $y\ v^+\ f^+$, or $y^+\ v\ f$ (regardless of the allele of the *Bar* gene)?
   b. What kinds of crossovers produced the $y\ v\ f\ B^+$ and $v^+\ y^+\ f^+\ B$ offspring? Can you determine any genetic distances from these classes of progeny?
   c. What kinds of crossovers produced the $y^+\ v\ f^+\ B^+$ and $y\ v^+\ f\ B$ offspring?

6. A diploid strain of yeast was made by mating a haploid strain with a genotype w^-, x^-, y^-, and z^- with a haploid strain of opposite mating type that is wild type for these four genes. The diploid strain was phenotypically wild type. Four different X-ray-induced diploid mutants with the following phenotypes were produced from this diploid yeast strain. Assume a single new mutation is present in each strain.

Strain 1	w^-	x^+	y^-	z^+
Strain 2	w^+	x^-	y^-	z^-
Strain 3	w^-	x^+	y^-	z^-
Strain 4	w^-	x^+	y^+	z^+

When these mutant diploid strains of yeast go through meiosis, each ascus is found to contain only two viable haploid spores.

a. What kind of mutations were induced by X-rays to make the listed diploid strains?

b. Why did two spores in each ascus die?

c. Are any of the genes *w*, *x*, *y*, or *z* located on the same chromosome?

d. Give the order of the genes that are found on the same chromosome.

7. The two graphs that follow represent genomic sequencing data for two fruit flies with scarlet-colored eyes. Each fly is a heterozygote for a chromosome 3 with normal gene order and a recessive *scarlet* (*st*) mutation, and a rearranged chromosome 3; the rearrangements are different in the two flies. Each circle on the graph indicates the number of sequence reads (*y* coordinate) obtained for a 100 bp sequence starting at the position on chromosome 3 that is shown by the *x* coordinate.

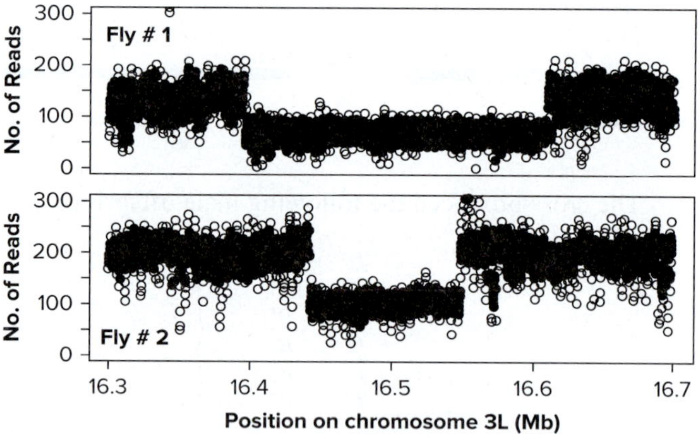

Courtesy of Tawny Cuykendall, Cornell University, Ithaca, New York

a. What is the nature of the rearranged chromosome in each fly? What are the approximate locations of the breakpoints defining these rearrangements? How could you find the exact nucleotides at the breakpoints?

b. Can you conclude that at least part of the *st* gene lies in a particular region of chromosome 3? If so, give the coordinates of that region defined as precisely as possible.

c. Is it possible that one or more protein-coding exons of the *st* gene lie outside of the region you defined in part (b)? Explain.

d. Does your ability to map the *st* gene on the basis of these data reflect a recombination experiment or a complementation experiment?

8. A series of chromosomal mutations in *Drosophila* were used to map the *javelin* gene, which affects bristle shape, and the *henna* gene, which affects eye pigmentation. Both the *javelin* and the *henna* mutant alleles are recessive to wild type. The chromosomal mutations are all rearrangements of chromosome 3. A diagram of chromosome 3 follows: 3L is the left arm and 3R is the right arm. The DNA sequences are numbered from bp #1 to bp #24,598,654 on 3L, and bp #1 to bp #27,929,329 on 3R.

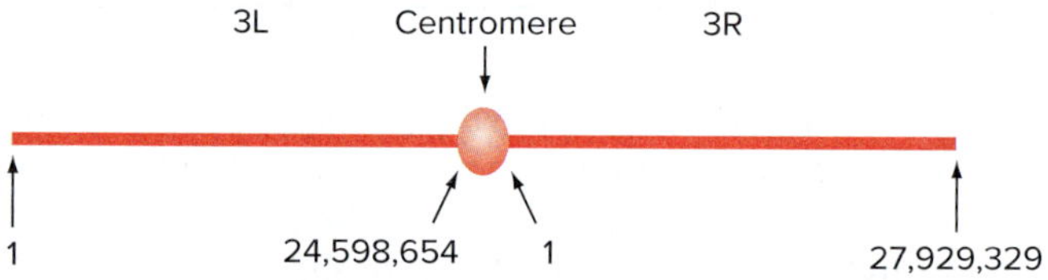

The chromosomal breakpoints for six chromosome 3 rearrangements are indicated in the following table. The exact breakpoints of each chromosomal mutation have been determined using PCR and DNA sequencing. For example, deletion A has one breakpoint just after bp #6,000,587, and another just after bp #6,902,063 on the left arm of chromosome 3.

		Breakpoints in chromosome 3L	
Deletions	A	6,000,587;	6,902,063
	B	6,703,444;	7,220,113
	C	6,880,255;	7,325,787
	D	6,984,866;	7,311,104
		Breakpoints in 3L; 3R	
Inversions	A	6,110,792 (3L);	40,272 (3R)
	B	6,520,488 (3L);	23,350 (3R)

Flies with a chromosome containing one of these six rearrangements (deletions or inversions) were mated to flies homozygous for both *javelin* and *henna*. The phenotypes of the heterozygous progeny (that is, *rearrangement/javelin, henna*) are shown here.

		Phenotypes of F_1 flies
Deletions	A	javelin, henna
	B	henna
	C	wild type
	D	wild type
Inversions	A	javelin
	B	wild type

a. Using these data, what can you conclude about the locations of the *javelin* and *henna* genes on chromosome 3?

b. For each chromosome rearrangement, draw on a diagram of a normal chromosome 3 the positions and 5′-to-3′ orientations of PCR primer pairs that could be used to amplify an approximately 100 bp region containing the rearrangement breakpoints.

9. Two wild-type fragments of human genomic DNA from the long arms of two nonhomologous acrocentric chromosomes are shown in the accompanying diagram. On both fragments, the centromere-to-telomere direction is left-to-right. The red lines numbered 1, 2, and 3 indicate the positions of breakpoints that could be caused by ionizing radiation. Certain combinations of these breaks could produce chromosomal rearrangements when the broken pieces are stitched back together by DNA repair systems.

   You will want to diagnose the presence of the rearrangements by using PCR. In every case that follows, one of the primers you will use for PCR is primer A, whose sequence is:

   5′ TCGATTCCGGAAAGCT 3′

   a. Which two of the breakpoints could yield a deletion?

   b. Write the sequence and polarity of a 16-base primer that, in conjunction with primer A, will allow you to diagnose the presence of a deletion in a patient's genomic DNA. The evidence should be positive, not negative. (That is, you will see a PCR product whose presence and/or size is specific to the deletion; the absence of a band is not informative.)

   c. Which two of the breakpoints could yield an inversion?

   d. Write the sequence and polarity of a 16-base primer that, in conjunction with primer A, will allow you to diagnose the presence of an inversion in a patient's genomic DNA. Your answer should show the primer that would yield the longest possible diagnostic PCR product. Your primer cannot cross any of the red lines.

   e. Which two of the breakpoints could yield a reciprocal translocation? (Two possible answers exist; you need only write one.)

   f. Write the sequence and polarity of a 16-base primer that, in conjunction with primer A, will allow you to diagnose by PCR the presence of a reciprocal translocation in a patient's genomic DNA. This reciprocal translocation can be stably inherited through meiosis without its loss or any chromosomal breakage. Again, the evidence should be positive, not negative.

   g. Is the translocation you described in parts (e) and (f) Robertsonian? Answer *yes* or *no,* and explain.

10. Indicate which of the four major classes of rearrangements (deletions, duplications, inversions, and translocations) are most likely to be associated with each of the following phenomena. In each case, explain the effect.

    a. semisterility
    b. lethality
    c. vulnerability to mutation
    d. altered genetic map
    e. haploinsufficiency
    f. neomorphic mutation
    g. hypermorphic mutation
    h. crossover suppression
    i. aneuploidy

11. The recessive, X-linked z^1 mutation of the *Drosophila* gene *zeste* (*z*) can produce a yellow (zeste) eye color only in flies that have two or more copies of the wild-type *white* (*w*) gene. Using this property, tandem duplications of the w^+ gene called w^{+R} were identified. Males with the genotype $y^+\ z^1\ w^{+R}\ spl^+ / Y$ thus have zeste eyes. These males were crossed to females with the genotype $y\ z^1\ w^{+R}\ spl\ /\ y^+\ z^1\ w^{+R}\ spl^+$. (These four genes are closely linked on the X chromosome, in the order given in the genotype, with the centromere to the right of all these genes: y = yellow bodies; y^+ = tan bodies; spl = split bristles; spl^+ = normal bristles.) Out of 81,540 male progeny of these females, the following exceptions were found:

Class A	2430 yellow bodies, zeste eyes, wild-type bristles
Class B	2394 tan bodies, zeste eyes, split bristles
Class C	23 yellow bodies, wild-type eyes, wild-type bristles
Class D	22 tan bodies, wild-type eyes, split bristles

    a. What were the phenotypes of the remainder of the 81,540 males from the first cross?

    b. What events gave rise to progeny of classes A and B?

    c. What events gave rise to progeny of classes C and D?

    d. On the basis of these experiments, what is the genetic distance between *y* and *spl?*

Figure for Problem 9

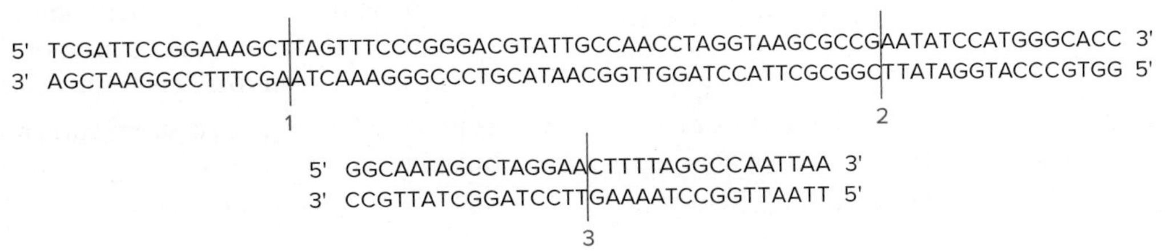

12. Genes *a* and *b* are 21 m.u. apart when mapped in highly inbred strain 1 of corn and 21 m.u. apart when mapped in highly inbred strain 2. But when the distance is mapped by testcrossing the F_1 progeny of a cross between strains 1 and 2, the two genes are only 1.5 m.u. apart. What arrangement of genes *a* and *b* and any potential rearrangement breakpoints could explain these results?

13. In the following group of figures, the *pink* lines indicate an area of a chromosome that is inverted relative to the normal (*black* line) order of genes. The diploid chromosome constitution of individuals 1–4 is shown. Match the individuals with the appropriate statement(s) that follow. More than one diagram may correspond to the following statements, and a diagram may be a correct answer for more than one question.

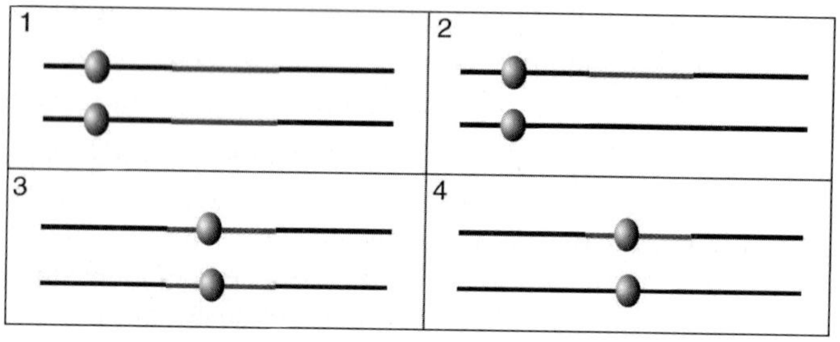

a. An inversion loop would form during meiosis I.

b. A single crossover involving the inverted region on one chromosome and the homologous region of the other chromosome would yield genetically unbalanced gametes.

c. A single crossover involving the inverted region on one chromosome and the homologous region of the other chromosome would yield an acentric fragment.

d. A single crossover involving the inverted region yields four viable gametes.

14. Three strains of *Drosophila* (Bravo, X-ray, and Zorro) are obtained that are homozygous for three variant forms of a particular chromosome. Karyotype analysis indicates that none of the strains is missing any part of any chromosome. When genetic mapping is performed in the Bravo strain, the following map is obtained (distances in map units).

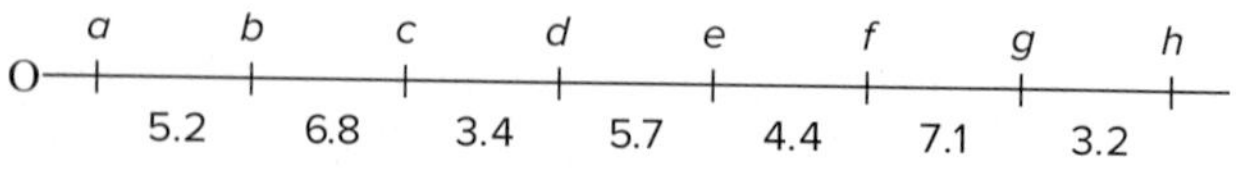

Bravo and X-ray flies are now mated to form Bravo/X-ray F_1 progeny, and Bravo flies are also mated with Zorro flies to form Bravo/Zorro F_1 progeny. In subsequent crosses, the following genetic distances were found to separate the various genes in the hybrids:

	Bravo/X-ray	**Bravo/Zorro**
a–b	5.2	5.2
b–c	6.8	0.7
c–d	0.2	<0.1
d–e	<0.1	<0.1
e–f	<0.1	<0.1
f–g	0.65	0.7
g–h	3.2	3.2

a. Make a map showing the relative order of genes *a* through *h* in the X-ray and Zorro strains. Do not show distances between genes.

b. In the original X-ray homozygotes, would the physical distance between genes *c* and *d* be greater than, less than, or approximately equal to the physical distance between these same genes in the original Bravo homozygotes?

c. In the original X-ray homozygotes, would the physical distance between genes *d* and *e* be greater than, less than, or approximately equal to the physical distance between these same genes in the original Bravo homozygotes?

15. Two yeast strains were mated and sporulated (allowed to carry out meiosis). One strain was a haploid with normal chromosomes and the linked genetic markers *ura3* (requires uracil for growth) and *arg9* (requires arginine for growth) surrounding their centromeres. The other strain was wild type for the two markers (*URA3* and *ARG9*) but had an inversion in this region of the chromosome as shown here in *pink:*

During meiosis, several different kinds of crossover events could occur. For each of the following events, give the genotype and phenotype of the resulting four haploid spores. Assume that any chromosomal deficiencies are lethal in haploid yeast. Do not consider crossovers between sister chromatids.

a. A single crossover outside the inverted region

b. A single crossover between *URA3* and the centromere

c. A double crossover involving the same two chromatids each time, where one crossover occurs between *URA3* and the centromere and the other occurs between *ARG9* and the centromere

16. Suppose a haploid yeast strain carrying two recessive linked markers *his4* and *leu2* was crossed with a strain that was wild type for *HIS4* and *LEU2* but had an

inversion of this region of the chromosome as shown here in *blue*.

Several different kinds of crossover events could occur during meiosis in the resulting diploid. For each of the following events, state the genotype and phenotype of the resulting four haploid spores. Do not consider crossover events between chromatids attached to the same centromere.

a. A single crossover between the markers *HIS4* and *LEU2*

b. A double crossover involving the same chromatids each time, where both crossovers occur between the markers *HIS4* and *LEU2*

c. A single crossover between the centromere and the beginning of the inverted region

17. In the mating between two haploid yeast strains depicted in Problem 16, describe a scenario that would result in a tetratype ascus in which all four spores are viable.

18. During ascus formation in *Neurospora,* any ascospore with a chromosomal deletion dies and appears white in color. How many of the eight ascospores in the ascus would be white if the octad came from a cross of a wild-type strain with a strain of the opposite mating type carrying:

a. a paracentric inversion, and no crossovers occurred between normal and inverted chromosomes?

b. a pericentric inversion, and a single crossover occurred in the inversion loop?

c. a paracentric inversion, and a single crossover occurred outside the inversion loop?

d. a reciprocal translocation, and an adjacent-1 segregation occurred with no crossovers between translocated chromosomes?

e. a reciprocal translocation, and alternate segregation occurred with no crossovers between translocated chromosomes?

f. a reciprocal translocation, and alternate segregation occurred with one crossover between translocated chromosomes (but not between the translocation breakpoint and the centromere of any chromosome)?

19. In the following figure, *black* and *pink* lines represent nonhomologous chromosomes. Which of the figures matches the descriptions (a)–(d) that follow? More than one diagram may correspond to any one statement, and a diagram may be a correct answer for more than one question.

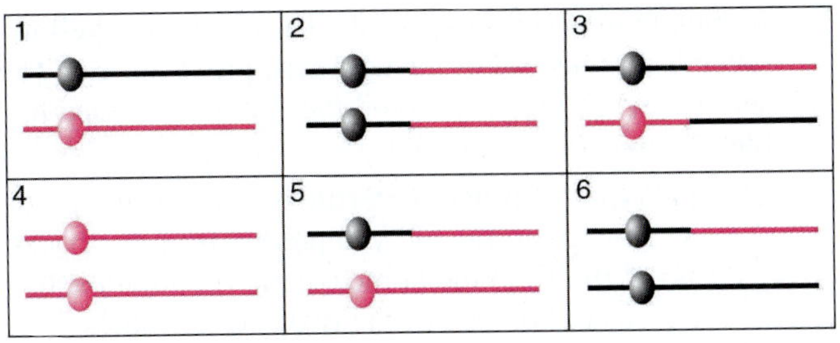

a. Gametes produced by a translocation heterozygote

b. Gametes that could not be produced by a translocation heterozygote

c. Genetically balanced gametes produced by a translocation heterozygote

d. Genetically imbalanced gametes that can be produced (at any frequency) by a translocation heterozygote

20. In *Drosophila,* the gene for cinnabar eye color is on chromosome 2, and the gene for scarlet eye color is on chromosome 3. A fly homozygous for both recessive *cinnabar* and *scarlet* alleles (*cn/cn; st/st*) is white-eyed.

a. If male flies (containing chromosomes with the normal gene order) heterozygous for *cn* and *st* alleles are crossed to white-eyed females homozygous for the *cn* and *st* alleles, what are the expected phenotypes and their frequencies in the progeny?

b. One unusual male heterozygous for *cn* and *st* alleles, when crossed to a white-eyed female, produced only wild-type and white-eyed progeny. Explain the likely chromosomal constitution of this male.

c. When the wild-type F_1 females from the cross with the unusual male were backcrossed to normal *cn/cn; st/st* males, the following results were obtained:

wild type	45%
cinnabar	5%
scarlet	5%
white	45%

Diagram a genetic event at metaphase I that could produce the rare cinnabar or scarlet flies among the progeny of the wild-type F_1 females.

21. Semisterility in corn, as seen by unfilled ears with gaps due to abortion of approximately half the ovules, is an indication that the strain is a translocation heterozygote. The chromosomes involved in the translocation can be identified by crossing the translocation heterozygote to a strain homozygous recessive for a gene on the chromosome being tested. The ratio of phenotypic classes produced from crossing semisterile F_1 progeny back to a homozygous recessive plant indicates whether the gene is on one of the chromosomes involved in the translocation. For example, a semisterile strain could be crossed to a strain homozygous for the

yg mutation on chromosome 9. (The mutant has yellow-green leaves instead of the wild-type green leaves.) The semisterile F_1 progeny would then be backcrossed to the homozygous *yg* mutant.

a. What types of progeny (fertile or semisterile, green or yellow-green) would you predict from the backcross of the F_1 to the homozygous *yg* mutant if the gene was not on one of the two chromosomes involved in the translocation?

b. What types of progeny (fertile or semisterile, green or yellow-green) would you predict from the backcross of the F_1 to the homozygous mutant if the *yg* gene is on one of the two chromosomes involved in the translocation?

c. If the *yg* gene is located on one of the chromosomes involved in the translocation, a few fertile, green progeny and a few semisterile, yellow-green progeny are produced. How could these relatively rare progeny classes arise? What genetic distance could you determine from the frequency of these rare progeny?

22. A promising biological method for insect control involves the release of insects that could interfere with the fertility of the normal resident insects. One approach is to introduce sterile males to compete with the resident fertile males for matings. A disadvantage of this strategy is that the irradiated sterile males are not very robust and can have problems competing with the fertile males. An alternate approach that is currently being tried is to release laboratory-reared insects that are homozygous for several translocations. Explain how this strategy will work. Be sure to mention which insects will be sterile.

23. A *Drosophila* male is heterozygous for a reciprocal translocation between an autosome and the Y chromosome. The part of the autosome now present on the Y chromosome contains the dominant mutation *Lyra* (shortened wings); the other (normal) copy of the same autosome is $Lyra^+$. This male is now mated with a true-breeding, wild-type female. What kinds of progeny would be obtained, and in what proportions?

24. a. Among the progeny of a self-fertilized semisterile corn plant heterozygous for a reciprocal translocation, what ratio do you expect for progeny plants with normal fertility versus those showing semisterility? In this problem, ignore the rare gametes produced by adjacent-2 segregation.

b. Among the progeny of a particular self-fertilized semisterile corn plant heterozygous for a reciprocal translocation, the ratio of fertile to semisterile plants was 1:4. How can you explain this deviation from your answer to part (a)?

25. Duchenne muscular dystrophy (DMD) is caused by a recessive mutant allele of an X-linked gene called *dystrophin*. Rarely, females have disease symptoms as severe as those in males hemizygous for the recessive allele. These females are heterozygous for X-autosome reciprocal translocations where the X chromosome breakage occurred in the middle of the *dystrophin* gene, breaking it into two pieces.

a. If it is equally likely for X chromosome inactivation to spread from either of the X chromosome inactivation centers (XICs; see Fig. 12.15) in the cells of this patient, what proportion of her cells would you expect to have normal function of the *dystrophin* gene?

b. Is it possible that some autosomal genes might be inactivated in some cells of this woman? Is it possible that some X-linked genes that are normally subject to X inactivation might be expressed from both X chromosomes in this woman? To answer these questions, draw a figure showing the normal and translocated chromosomes involved, and the locations of: centromeres, potentially inactivated autosomal genes, X chromosome genes potentially no longer subject to X inactivation, the *dystrophin* gene, and the XICs. Assume for simplicity that *dystrophin* and the XIC are located in the middle of the short and long arms of the X chromosome, respectively.

c. It is found that virtually none of the cells in this woman expresses any of the *dystrophin* gene product. Use your answers to parts (a) and (b) to explain this interesting observation. Consider what changes in gene expression might do to the survival and proliferation of cells.

d. Discuss how scientists might have used X:autosome reciprocal translocations to help map the XIC on the human X chromosome. Why does your answer to part (c) illustrate a potential pitfall of this approach?

26. WHIM syndrome is a disease of the immune system resulting in warts and frequent infections. The disease is caused by a dominant gain-of-function mutation in a gene on chromosome 2 called *CXCR4*. A 38-year-old woman suffering with WHIM syndrome her entire life was suddenly and mysteriously cured. Genome analysis of her blood precursor cells (stem cells) revealed that many of these cells had a chromosome 2 that had undergone *chromotripsis*—a rare (and poorly understood) process where a chromosome is "shattered" into small pieces that are subsequently stitched back together in random order, resulting in many deletions and inversions. Explain how chromotripsis of chromosome 2 in a blood stem cell could have cured the woman of WHIM syndrome.

Section 13.3

27. Explain how transposable elements can cause the movement of genes that are not part of the transposable element.

28. The *Drosophila* genome normally harbors about 40 *P* elements. Some of these DNA transposons are autonomous and some of which are nonautonomous. Review the structure of a *P* element (Fig. 13.27).

    a. Suppose one of the *P* elements suffered a deletion of one of its inverted repeats. Would this mutation affect the ability of the *P* element to move? Would it affect the ability of other *P* elements to move?

    b. Suppose one of the autonomous *P* elements suffered a mutation in the splice acceptor site for the intron drawn in *yellow* color in Fig. 13.27, so that now this intron cannot be spliced out of the primary transcript. Would this mutation affect the ability of the *P* element to move? Would it also affect other *P* elements?

    c. Answer part (b) again, assuming that the *P* element in question was the only autonomous *P* element in the genome.

    d. As illustrated in Fig. 13.27, *P* elements are not normally mobile in somatic cells. Describe a mutant *P* element that could mobilize in somatic cells. Would the mutation also affect the ability of other *P* elements to mobilize in the same somatic cells?

29. *Drosophila P* elements were discovered because of a phenomenon called *hybrid dysgenesis*—sterility of particular hybrid progeny. When scientists in the 1970s crossed their *D. melanogaster* laboratory strains to flies of the same species obtained from natural environments outside the lab, they observed a remarkable result: The progeny of the crosses were sterile, but only when outside males were crossed with lab strain females. Progeny resulting from crosses of outside females with lab males were perfectly normal.

    DNA analysis revealed that while the genomes of the outside flies contain *P* elements, the lab fly genomes have none. Apparently, *P* elements spread throughout the wild population of *D. melanogaster* after the capture of the originators of present-day laboratory strains over 100 years ago.

    a. The hybrid progeny are sterile because their germline cells have a high rate of mutation and chromosomal rearrangement (dysgenesis) caused by high rates of *P* element mobilization. Explain how *P* element movement can cause dysgenesis.

    b. Scientists first hypothesized that the deposition of *P* element-encoded repressor protein (see Fig. 13.27) in egg cytoplasm is behind the observation that dysgenic progeny result only from crosses of laboratory females with outside males, and not *vice versa.* Explain this hypothesis. Why do the *P* elements mobilize when the cross occurs in one direction but not the other? (You will see in Chapter 17 that this hypothesis is correct, but it accounts for only part of the story.)

    c. When males from certain outside strains are mated to lab females, the hybrid progeny are only partially sterile rather than completely sterile. Given this information, describe crosses that would allow you to isolate loss-of-function mutations in the X-linked *Drosophila* gene *yellow* that are caused by *P* element insertion. (These recessive mutant alleles will produce yellow rather than the wild-type tan body color.) At the molecular level, what do you think explains the difference between outside strains whose hybrid progeny are all sterile and outside strains whose progeny are only semisterile?

    d. In wild-type fruit flies, researchers can observe rare patches on the bodies that have yellow rather than tan color. Interestingly, the frequency of these yellow patches did not increase in the progeny of a cross between outside males and lab females. What property of hybrid dysgenesis does this result suggest?

30. Flies homozygous for mutant alleles of a *Drosophila* gene called *rough* have slightly malformed (rough) eyes, instead of normal smooth eyes. Two different strains of flies exist, each with a different *P* element-induced mutant *rough* allele. In each strain, the *P* element in the *rough* gene is the only *P* element in the fly genome.

    a. When homozygotes for one of the *rough* mutant alleles were bred for several generations, in each generation flies appeared either with wild-type eyes, or with much more severely rough eyes. Pure-breeding wild-type or severely rough-eyed lines could be generated from these unusual flies. Explain.

    b. When homozygotes for the other *rough* mutant allele were grown for many generations, neither wild-type flies nor flies with more severely rough eyes were ever seen. Explain.

Section 13.4

31. Fred and Mary have a child named Bob. The genomic DNAs of these three individuals are used as probes to ASO microarrays such as that shown in Fig. 11.17. The results with 10 SNP loci (1–10) located along chromosome 21 are shown in the figure that follows. *Yellow, orange,* and *red* indicate increasing intensities of hybridization.

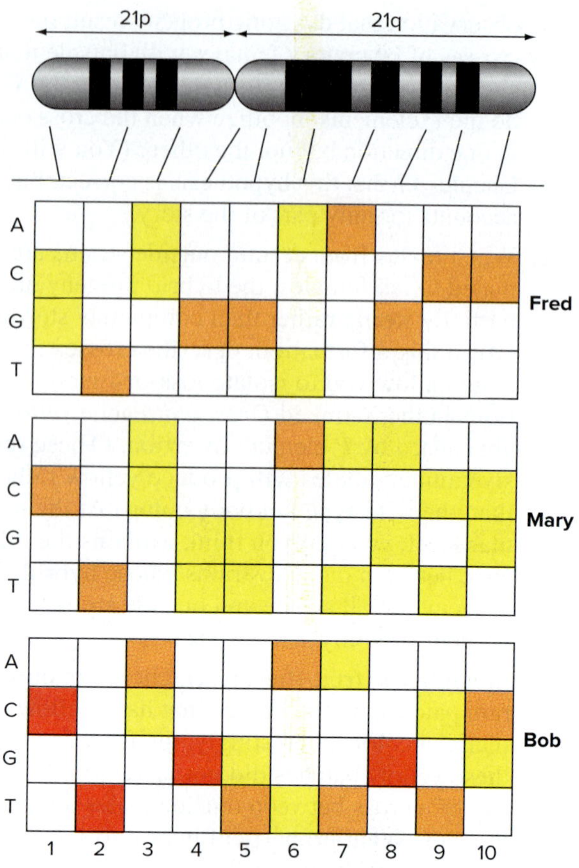

a. Is Bob likely to have Down syndrome? Explain.

b. Do the data provide evidence that a nondisjunction event occurred in one of the parents? If so, in which parent and during which meiotic division?

c. Do the data provide evidence that a recombination event occurred during meiosis in either parent? If so, in which parent, and describe the approximate location of the crossover.

d. Describe how this example illustrates why it is important when tracking nondisjunction events to look particularly at loci near the centromere.

32. *Uniparental disomy* is a rare phenomenon in which only one of the parents of a child with a recessive disorder is a carrier for that trait; the other parent is homozygous normal. By analyzing DNA polymorphisms, it is clear that the child received both mutant alleles from the carrier parent but did not receive any copy of the gene from the other parent.

a. Diagram at least two ways in which uniparental disomy could arise. (*Hint:* These mechanisms all require more than one error in cell division, explaining why uniparental disomy is so rare.) Is there any way to distinguish between these mechanisms to explain any particular case of uniparental disomy?

b. How might the phenomenon of uniparental disomy explain rare cases in which girls are affected with rare X-linked recessive disorders but have unaffected fathers, or other cases in which an X-linked recessive disorder is transmitted from father to son?

c. If you were a human geneticist and believed one of your patients had a disease syndrome caused by uniparental disomy, how could you establish that the cause was not instead mitotic recombination early in the patient's development from a zygote?

33. Among adults with Turner syndrome, it has been found that a very high proportion are genetic mosaics. These are of two types: In some individuals, the majority of cells are XO, but a minority of cells are XX. In other Turner individuals, the majority of cells are XO, but a minority of cells are XY. Explain how these two patterns of somatic mosaics could arise.

34. In *Neurospora, his2* mutants require the amino acid histidine for growth, and *lys4* mutants require the amino acid lysine. The two genes are on the same arm of the same chromosome, in the order

centromere - *his2* - *lys4*.

A *his2* mutant is mated with a *lys4* mutant. Draw all of the possible ordered asci that could result from meioses in which the following events occurred, accounting for the nutritional requirements for each ascospore. Ascospores without any copy of a chromosome will abort and die, turning white in the process.

a. A single crossover between the centromere and *his2*

b. A single crossover between *his2* and *lys4*

c. Nondisjunction during the first meiotic division

d. Nondisjunction during the second meiotic division

e. A single crossover between the centromere and *his2,* followed by nondisjunction during the first meiotic division

f. A single crossover between *his2* and *lys4,* followed by nondisjunction during the first meiotic division

35. Human geneticists interested in the effects of abnormalities in chromosome number often karyotype tissue obtained from spontaneous abortions. About 35% of these samples show autosomal trisomies, but only about 3% of the samples display autosomal monosomies. Based on the kinds of errors that can give rise to aneuploidy, would you expect that the frequencies of autosomal trisomy and autosomal monosomy should be more equal? Why or why not? If you think the frequencies should be more equal, how can you explain the large excess of trisomies as opposed to monosomies?

36. The incidence of Down syndrome will be very high (somewhat less than 50%) among the offspring of a parent with Down syndrome. Diagram meiosis in the Down syndrome parent to explain why progeny have such a high risk for chromosome 21 aneuploidy. Explain in addition why the incidence of Down syndrome among these children might be less than 50%.

37. The *Drosophila* chromosome 4 is extremely small; virtually no recombination occurs between genes on this chromosome. You have available three differently marked chromosome 4s: one has a recessive allele of the gene *eyeless* (*ey*), causing very small eyes; one has a recessive allele of the *cubitus interruptus* (*ci*) gene, which causes disruptions in the veins on the wings; and the third carries recessive alleles of both genes. *Drosophila* adults can survive with two or three, but not with one or four, copies of chromosome 4.

    a. How could you use these three chromosomes to find *Drosophila* mutants with defective meioses causing an elevated rate of nondisjunction?

    b. Would your technique allow you to discriminate nondisjunction occurring during the first meiotic division from nondisjunction occurring during the second meiotic division?

    c. What progeny types would you expect if a fly recognizably formed from a gamete produced by nondisjunction were testcrossed to a fly homozygous for a chromosome 4 carrying both *ey* and *ci?*

    d. Geneticists have isolated so-called *compound 4th chromosomes* in which two entire chromosome 4s are attached to the same centromere. How can such chromosomes be used to identify mutations causing increased meiotic nondisjunction? Are there any advantages relative to the method you described in part (a)?

38. Down syndrome is usually caused by having a complete extra copy of chromosome 21. The syndrome encompasses many different traits including low IQ, heart disease, characteristic facial features, gastrointestinal tract abnormalities, short stature, poor muscle tone, and increased risk of leukemia and dementia. Some people diagnosed with Down syndrome display only a subset of these anomalies. These people often have one normal chromosome 21 and one chromosome 21 containing a duplication; the region of chromosome 21 that is duplicated can differ among individuals.

    Some scientists think that one critical region of chromosome 21 exists (containing one or a few genes) that is responsible for Down syndrome. Researchers have used individuals with duplication Down syndrome to test this idea. Suppose that using genome sequencing, scientists determined which parts of chromosome 21 were present in three copies in eight individuals with duplication Down syndrome. A summary of the data is shown in the table that follows. An *X* indicates that the given individual is affected by the particular abnormality listed in the column heading.

Individual	Abnormalities						Duplicated region 21p 21q
	Low IQ	Facial features	GI tract	Short stature	Poor muscle tone	Heart disease	
1	X	X		X			
2	X			X			
3	X			X	X		
4				X		X	
5	X	X	X	X	X	X	
6	X	X	X		X	X	
7	X						
8	X	X					

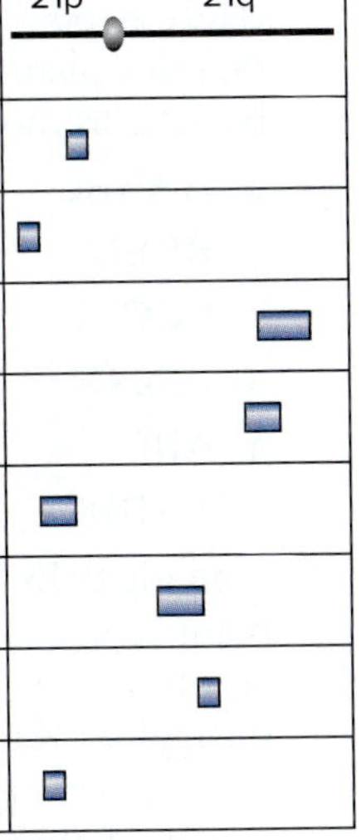

    a. Do the data support or refute the hypothesis that chromosome 21 contains a single Down syndrome critical region? Explain.

    b. Do the data support the idea that each Down syndrome abnormality is caused by overexpression of a different region of chromosome 21? Explain.

Section 13.5

39. Common red clover, *Trifolium pratense,* is a diploid with 14 chromosomes per somatic cell. What would be the somatic chromosome number of:

    a. a trisomic variant of this species?

    b. a monosomic variant of this species?

    c. a triploid variant of this species?

    d. an autotetraploid variant?

40. The numbers of chromosomes in the somatic cells of several oat varieties (*Avena* species) are: sand oats (*Avena strigosa*)—14; slender wild oats (*Avena barata*)—28; and cultivated oats (*Avena sativa*)—42.

    a. What is the basic chromosome number (x) in *Avena?*

    b. What is the ploidy for each of the different species?

    c. What is the number of chromosomes in the gametes produced by each of these oat varieties?

    d. What is the n number of chromosomes in each species?

41. Genomes A, B, and C all have basic chromosome numbers (x) of nine. These genomes were derived originally from plant species that had diverged from

each other sufficiently far back in the evolutionary past that the chromosomes from one genome can no longer pair with the chromosomes from any other genome. For plants with the following kinds of euploid chromosome complements, (i) state the number of chromosomes in the organism; (ii) provide terms that describe the individual's genetic makeup as accurately as possible; (iii) state whether or not it is likely that this plant will be fertile, and if so, give the number of chromosomes (n) in the gametes.

a. AABBC

b. BBBB

c. CCC

d. BBCC

e. ABC

f. AABBCC

42. Somatic cells in organisms of a particular diploid plant species normally have 14 chromosomes. The chromosomes in the gametes are numbered from 1 through 7. Rarely, zygotes are formed that contain more or fewer than 14 chromosomes. For each of the zygotes below, (i) state whether the chromosome complement is euploid or aneuploid; (ii) provide terms that describe the individual's genetic makeup as accurately as possible; and (iii) state whether or not the individual will likely develop through the embryonic stages to make an adult plant, and if so, whether or not this plant will be fertile.

a. 11 22 33 44 5 66 77

b. 111 22 33 44 555 66 77

c. 111 222 333 444 555 666 777

d. 1111 2222 3333 4444 5555 6666 7777

43. An allotetraploid species has a genome composed of two ancestral genomes, A and B, each of which have a basic chromosome number (x) of seven. In this species, the two copies of each chromosome of each ancestral genome pair only with each other during meiosis. Resistance to a pathogen that attacks the foliage of the plant is controlled by a dominant allele at the F locus. The recessive alleles F^a and F^b confer sensitivity to the pathogen, but the dominant resistance alleles present in the two genomes have slightly different effects. Plants with at least one F^A allele are resistant to races 1 and 2 of the pathogen regardless of the genotype in the B genome, and plants with at least one F^B allele are resistant to races 1 and 3 of the pathogen regardless of the genotype in the A genome. What proportion of the self-progeny of an $F^A\ F^a\ F^B\ F^b$ plant will be resistant to all three races of the pathogen?

44. You have haploid tobacco cells in culture and have made transgenic cells that are resistant to herbicide. What would you do to obtain a diploid cell line that could be used to generate a new fertile herbicide-resistant plant?

45. Chromosomes normally associate during meiosis I as *bivalents* (a pair of synapsed homologous chromosomes) because chromosome pairing involves the synapsis of the corresponding regions of two homologous chromosomes. However, Fig. 13.17b shows that in a heterozygote for a reciprocal translocation, chromosomes pair as *quadrivalents* (that is, four chromosomes are associated with each other). Quadrivalents can form in other ways: For example, in some autotetraploid species, chromosomes can pair as quadrivalents rather than as bivalents.

a. How could quadrivalents actually form in these autotetraploids, given that chromosomal regions synapse in pairs? To answer this question, diagram such a quadrivalent.

b. How can these autotetraploid species generate euploid gametes if the chromosomes pair as quadrivalents rather than bivalents?

c. Could quadrivalents form in an amphidiploid species? Discuss.

46. Using whole-genome sequencing, how could you distinguish between autopolyploids and allopolyploids?

47. Suppose you have an *AAaa* tetraploid plant and it undergoes self-fertilization. At least two copies of the dominant allele *A* are needed to obtain the dominant phenotype. At what frequency will progeny with the dominant phenotype appear?

48. Section 13.5 of this chapter discussed that *Raphanobrassica* is a hybrid species generated by crossing cabbages and radishes. How important was it to the viability and fertility of the hybrid plant that cabbages and radishes each have gametes with nine chromosomes ($x = 9$ in both species)?

49. Seedless watermelons that you find in the supermarket are triploids, where $x = 11$.

a. At what frequency are balanced gametes generated by triploid watermelons?

b. What is the probability that a particular seed in a triploid watermelon will be viable? (Recall that a viable seed is a euploid zygote.)

c. What is the ploidy of the viable seed in part (b)? More than one answer may apply.

50. The names of hybrid animals are usually themselves hybrids between the names of the species used to generate them, with the male gamete first. A cross between a male leopard and a female lion results in a *leopon;* the father of the *zonkey* in the picture that follows is a zebra and its mother is a donkey.

© Gary Neil Corbett/SuperStock RF

a. Not all animal hybrids are possible. For example, a cross between a dog and a bird, or an elephant and a koala, or a cat and a shark will not produce viable hybrid progeny. What do you think is the major factor determining whether or not a hybrid animal will be viable?

b. Some hybrid animals are fertile. What is the major factor that determines whether or not a viable animal hybrid will also be fertile?

51. While most animals cannot tolerate polyploidy, some mollusks, such as oysters, can. Although wild oysters are diploid, farmed oysters are often made triploid. The advantage of triploidy is that the oysters are sterile, so they expend the energy that would otherwise be used to make unappetizing gametes in the production of delicious muscle. These meaty triploid oysters are more valuable commercially than diploids.

   a. To make triploid oysters, researchers used cytochalasin B, a chemical that can inhibit cell division in either meiosis I or meiosis II during oogenesis. This drug does not interfere with chromosome replication. Explain how triploid oysters could be generated using cytochalasin B.

   b. Cytochalasin B is toxic to humans, so oyster farmers cannot use this drug. Can you think of an alternate method that might enable oyster farmers to produce large numbers of triploid oysters?

Section 13.6

52. What characteristic property of translocations causes scientists to believe that these rearrangements may be of importance in the emergence of new species?

53. In examining the genome of the rice (*Oryza sativa*) shown in Fig. 13.36:

   a. What is the evidence that the entire genome of an ancestral plant species in the lineage leading to rice underwent a duplication?

   b. If an ancestor of rice underwent a whole-genome duplication, why is rice a diploid and not a tetraploid species?

   c. Plant geneticists believe that the ancestral plant species in which the whole-genome duplication occurred had five chromosomes. Sequences from one of these chromosomes are mostly found in rice chromosomes Os02, 04, and 06; sequences from a second ancestral chromosome are now mostly found in rice chromosomes Os03, 07, and 10. Which of the remaining rice chromosomes have sequences that originated in the other three ancestral chromosomes?

   d. What evidence exists in Fig. 13.36 to support the idea that the genesis of chromosome Os12 involved a translocation that occurred in the rice lineage at some time after the whole-genome duplication event?

54. In the accompanying figure, the top and bottom lines represent regions of chromosomes 4 and 12 in the yeast *Saccharomyces cerevisiae* (*Scer 4* and *Scer 12*). Numbers refer to specific genes, and the *red arrows* represent the direction and extent of transcription. The middle line is the sequence of a region from chromosome 1 of a different, but related yeast species called *Klyuyveromyces waltii* (*Kwal* 1), with genes indicated in *light blue*. Similarities in DNA sequence are shown as lines joining chromosomes of the two species.

   a. What is the meaning of the two *K. waltii* genes filled in *dark purple?*

   b. Based on these data, formulate a hypothesis to explain the genesis of the part of the *S. cerevisiae* genome illustrated in the figure.

55. Two possible models have been proposed to explain the potential evolutionary advantage of gene duplications. In the first model, one of the two duplicated copies retains the same function as the

Figure for Problem 54

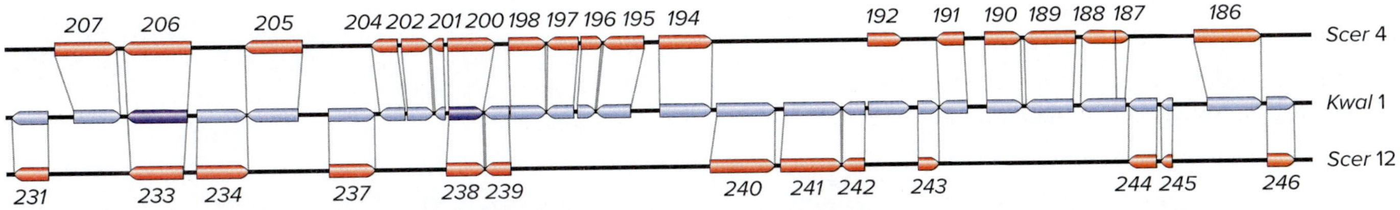

ancestral gene, leaving the other copy to diverge through mutation to fulfill a new biochemical function. In the second model, both copies can diverge rapidly from the ancestral gene, so that both can acquire new properties. Considering your answer to Problem 54, and given that both the *S. cerevisiae* and *K. waltii* genomes have been completely sequenced, how could you determine which of these two models better represents the course of evolution?

56. The accompanying figure shows idiograms of human chromosomes 1 and 2 and the corresponding chromosomes of the great apes. Although chromosome 1 is extremely similar in all four species, human chromosome 2's banding pattern resembles those of two different great ape chromosomes. Advance a hypothesis that explains the relationship between human chromosome 2 and the great ape chromosomes shown.

57. Some animals use mimcry in order to avoid predators. Female swallowtail butterflies of the species *Papilo ploytes*, for example, can mimic the color patterns of the toxic species *Pachilopta aristolochiae* (see accompanying figure). Normal and mimetic *Papilo ploytes* females can can coexist in a single population.

    Recently, scientists discovered that variants of a single autosomal gene called *dsx* control mimicry in swallowtails. The proteins encoded by the wild-type and mutant alleles of this gene differ at many amino acids, and these accumulated differences are thought to be responsible for the morphological differences in the normal females and the females that mimic other species.

    The *dsx* genes of the mimetic females are contained within an inversion. Explain how the presence of the inversion might have been crucial to the evolution of mimicry in swallowtails.

Figure for Problem 56

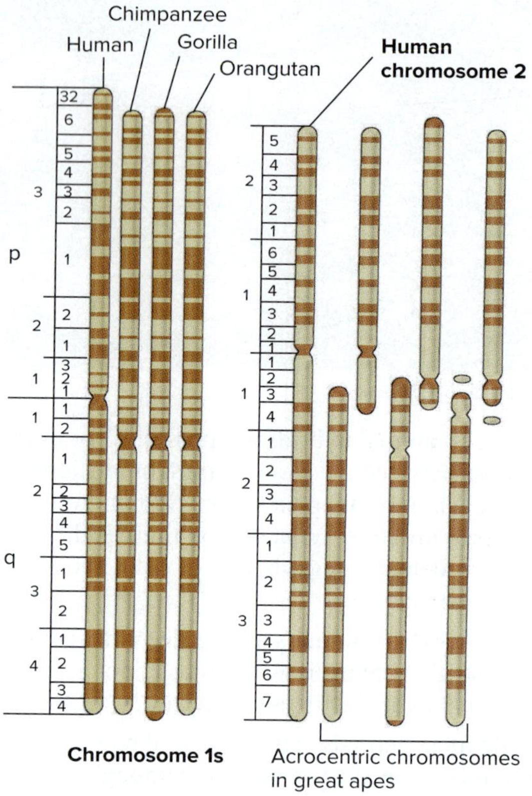

Normal male (*Papilo polytes*)

Normal female (*Papilo polytes*)

Mimetic female (*Papilo polytes*)

Toxic female (*Pachliopta aristolochiae*)

chapter **14**

Bacterial Genetics

Some species of bacteria can live in environments as hostile as hot springs (such as this beautiful pool in Yellowstone National Park, Wyoming). Colors in the pool other than the blue in the center are due to pigments in populations of bacteria that grow thickly around the edges of the water. Comparative genome analyses of bacteria that live in unusual environments will increase our understanding of the adaptations that allow survival in different niches.

chapter outline

GONORRHEA, A SEXUALLY transmitted infection of the urogenital tract in men and women, is caused by the bacterium *Neisseria gonorrhoeae*. The disease is rarely fatal, but it can lead to sterility in both sexes. Until the late 1970s, a few shots of penicillin were a certain cure for gonorrhea, but by 1995, more than 20% of *N. gonorrhoeae* bacteria isolated from patients worldwide were resistant to penicillin.

Geneticists now know that the agent of this alarming increase in antibiotic resistance was the transfer of DNA from one bacterium to another. According to epidemiologists, penicillin-resistant *N. gonorrhoeae* bacteria first appeared in Asia in the 1970s, in a patient receiving penicillin treatment for gonorrhea who was also fighting an infection caused by another species of bacteria—*Haemophilus influenzae*. Some of the patient's *H. influenzae* bacteria apparently carried a plasmid that contained a gene encoding penicillinase, an enzyme that destroys penicillin. When the doubly infected patient mounted a specific immune response to *H. influenzae* that degraded these cells, the broken bacteria released their plasmids. Some of the freed circles of DNA entered *N. gonorrhoeae* cells, transforming them to penicillin-resistant bacteria.

The transformed gonorrhea bacteria then multiplied, and successive exposures to penicillin selected for the resistant bacteria. As a result, the patient transmitted penicillin-resistant *N. gonorrhoeae* to subsequent sexual partners. Thus, while penicillin treatment does not create the genes for resistance, it accelerates the spread of those genes. Today in the United States, many *N. gonorrhoeae* are simultaneously resistant to penicillin and two other antibiotics—spectinomycin and tetracycline.

In this chapter, we focus first on the remarkable diversity of bacteria, and on how genome analysis has vastly increased our knowledge of the bacterial world. We next examine the mechanisms by which bacteria transfer genes between cells of the same species or even between cells of distantly related species, and how scientists exploit these mechanisms of gene transfer to map and identify bacterial genes with important functions. Finally, we explore how *N. gonorrhoeae* became resistant to multiple drugs, and what we as a society can do about the problem of multidrug-resistant pathogens.

One main theme can be found in our exploration of bacterial genetics: The DNA and genes of a single cell or a single species are neither unchangeable nor do they exist in complete isolation. Not only do DNA segments like transposable elements migrate within a genome, but DNA and genes are also capable of migration between different cells of one species and even between cells of different species in the same bacterial community. The transfer of genes between cells has been a key feature of the evolution of the microbial world.

14.1 The Enormous Diversity of Bacteria

learning objectives

1. List key features of prokaryotic cells.
2. Discuss how bacterial habitats influence bacterial metabolisms.
3. Summarize the properties that make certain bacteria pathogenic to humans.

Bacteria such as *N. gonorrhoeae* and *H. influenzae* constitute one of the three major evolutionary lineages of living organisms (**Fig. 14.1**). The two other lineages are the eukaryotes (organisms whose cells have nuclei encased in a membrane) and the *archaea,* organisms that tend to live in extreme conditions such as highly salty, hot, or anaerobic environments. Bacteria and archaea are both classified as *prokaryotes* because they lack the membrane-bound, true nucleus found in eukaryotes. Although they are both prokaryotes, bacteria and archaea are distinct from each other morphologically and biochemically. We confine this chapter to bacteria because they are more common laboratory organisms and thus better understood. However, many of the lessons learned from bacteria about genome organization and gene transfer between cells also have relevance to their distant cousins the archaea.

Figure 14.1 A family tree of living organisms. The three major evolutionary lineages of living organisms are bacteria and archaea (both prokaryotes) and eukaryotes. DNA sequencing results suggest that all living forms descended from a common prokaryotic ancestor that lived about 3.5 billion years ago.

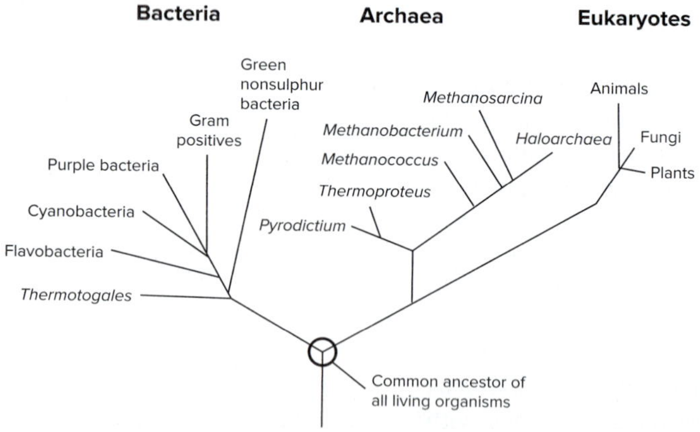

Understanding bacterial genetics is of profound importance to humans for many reasons, including the fact that we live so intimately with bacteria. An adult human carries around 30 to 50 trillion bacteria, which is about the same as the number of human cells. Most of these bacteria inhabit the intestines, but the skin, mouth, teeth, and respiratory tract are also homes for bacterial ecosystems. Bacteria aid human health in many ways, for example by synthesizing needed vitamins and helping to digest our food, but certain bacteria are also the causative agents of disease.

Bacteria Vary in Size and Shape

The smallest bacteria are about 200 nanometers (nm = billionths of a meter) in diameter. The largest are 500 micrometers (μm = millionths of a meter) in length, which makes them 10 billion times larger in volume and mass than the smallest bacterial cells. These large bacteria are visible without the aid of the microscope. The rod-shaped laboratory workhorse *Escherichia coli* (*E. coli*) is at the smaller end of this spectrum: It has a diameter of about 0.5 micrometers and is about 2 micrometers long (**Fig. 14.2**).

Although bacteria come in a variety of shapes and sizes, all lack a defined nuclear membrane as well as membrane-bounded organelles, such as the mitochondria and chloroplasts found in eukaryotic cells (**Fig. 14.3**). The single chromosome of a bacterium folds to form a dense **nucleoid body.** In most species of bacteria, the cell membrane is supported by a cell wall composed of carbohydrate and peptide polymers. Some bacteria have, in addition to the cell wall, a thick, mucus-like coating called a *capsule* that helps them resist attack by immune systems. Although

Figure 14.2 ***Escherichia coli.*** Scanning electron micrograph of *E. coli* (14,000×). Several of the cells are undergoing division by binary fission, as seen by constrictions near the middle.
© Mediscan/Alamy

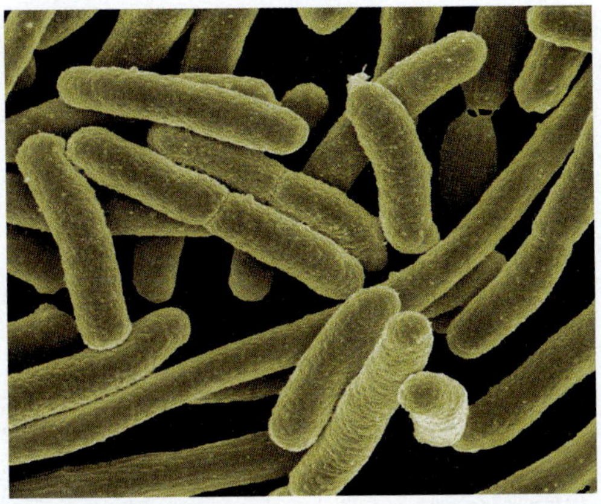

not visible in Fig. 14.2, many bacteria, including *E. coli,* have flagellae that propel them toward food or light (Fig. 14.3).

Bacteria Have Diverse Metabolisms

Bacteria have evolved to live in a wide variety of habitats. Some bacteria live independently on land, others float freely in aquatic environments, and still others live as parasites or symbionts inside other life-forms. The metabolism of bacterial species must be adapted to their particular environments. Some soil bacteria obtain the energy to fuel their growth from the chemical ammonia, while other, photosynthesizing bacteria obtain their energy from sunlight. Because of their metabolic diversity, bacteria play essential roles in many natural processes, such as the decomposition of materials essential for nutrient cycling. The balance of microorganisms is key to the success of these ecological processes, which help maintain the environment.

In the cycling of nitrogen, for example, decomposing bacteria break down plant and animal matter rich in nitrogen and produce ammonia (NH_3). Nitrifying bacteria then use this ammonia as a source of energy and release nitrate (NO_3^-), which some plants can use directly as a nitrogen source. Denitrifying bacteria convert the nitrate not used by plants to atmospheric nitrogen (N_2), while nitrogen-fixing bacteria, such as *Rhizobium,* that live in the roots of peas and other leguminous plants convert N_2 to ammonium (NH_4^+), which their host plants can use.

Figure 14.3 Structure of a bacterium. The bacterial chromosome (the nucleoid body) is in the cytoplasm, which is surrounded by a permeable cell membrane. A less permeable cell wall surrounds the membrane, and in many species, the entire bacterium is encased in a protective capsule.

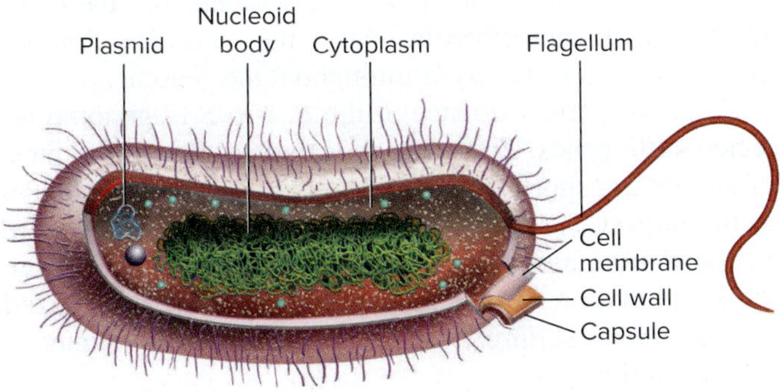

Some oceanic bacteria have the remarkable ability to eat oil. Species such as *Alcanivorax borkumensis*, which break down the hydrocarbons in oil to use as an energy source, are called *bioremediation bacteria* because they bloom after oil spills and assist in their clean-up. After the 2010 *Deepwater Horizon* disaster in the Gulf of Mexico, the population of oil-eating bacteria skyrocketed within the plumes of spilled oil. By sequencing the genomes of *Alcanivorax* species, scientists have identified some of the genes that give these bacteria their special oil-eating ability. Engineering of these genes may eventually lead to the creation of more voracious bacterial strains that would facilitate oil spill cleanup.

A Small Fraction of Bacteria Are Pathogens

A **pathogen** is an infectious agent that causes disease in the host organism. Most of the bacteria that inhabit the human body are either harmless or beneficial; only a few are pathogenic. Pathogenic bacteria have acquired genes that enable them to invade tissues and in some cases to produce **toxins,** proteins released by the bacteria that can destroy cell membranes or interfere with basic cellular functions in the host. For example, tetanus toxin (produced by *Clostridium tetani*) is a protease that inhibits communication between nerves and muscles, resulting in paralysis. In a second example of clinical importance, *Corynebacterium diphtheriae* produces diphtheria toxin, an enzyme that modifies a protein required for translation and thereby inhibits protein synthesis in human cells. Within the same bacterial species, some strains can be harmless while others are pathogens. You will see later in this chapter that a major reason for this intraspecies diversity is that bacteria transfer genes among themselves at a remarkable frequency.

essential concepts

- Bacteria are *prokaryotic cells*—they have no membrane-enclosed nucleus nor other membrane-bound organelles.
- Bacteria are capable of rapid evolution, and as a result these organisms vary enormously in size, shape, metabolism, and the habitats to which they have adapted.
- A typical human body carries 30 to 50 trillion bacteria; most of these are either harmless or helpful, but a few are *pathogenic* and thus cause disease.

14.2 Bacterial Genomes

learning objectives

1. Describe how genes are organized within a bacterial genome.
2. Differentiate between a species' core genome and its pangenome.
3. Discriminate between IS and Tn transposable elements in bacteria.
4. Describe how plasmids conferring multidrug resistance to bacteria may have evolved.
5. Explain how the study of metagenomics might yield practical benefits.

The essential component of a typical bacterial genome is the **bacterial chromosome:** a single molecule of double-helical DNA arranged in a circle (**Fig. 14.4**). This chromosome is 4–5 Mb long in most of the commonly studied species. The circular chromosome of *E. coli,* if broken at one point and laid out in a line, would form a DNA molecule 2.4 nm wide and 1.6 mm long, almost a thousand times longer than the *E. coli* cell in which it is found (**Fig. 14.5**). Inside the cell, the long, circular DNA molecule condenses by supercoiling and looping into a densely packed *nucleoid body* (Fig. 14.3).

During the bacterial cell cycle, each bacterium replicates its circular chromosome (review Fig. 6.24) and then divides by binary fission into two identical daughter cells, each with its own chromosome, generating two organisms from one. While the majority of bacteria contain a single circular chromosome, exceptions exist. Genomic analyses have shown that some bacteria, such as *Vibrio cholerae* (the cause of the disease cholera), carry two different circular chromosomes essential for viability. Certain other bacteria contain linear DNA molecules.

Figure 14.4 **The bacterial chromosome is a ring of double-stranded DNA.**

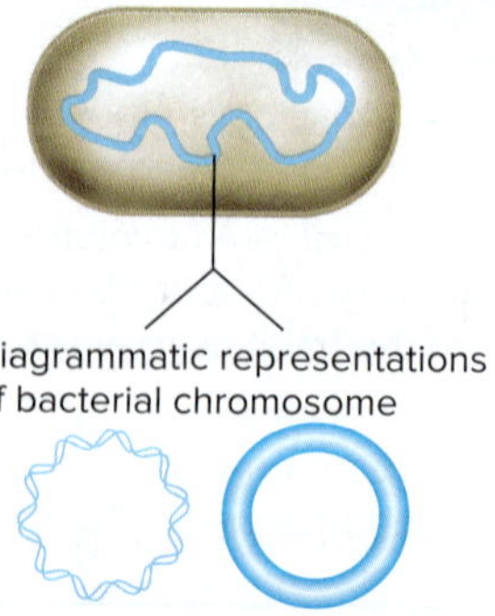

Figure 14.5 ***E. coli* chromosomal DNA.** An electron micrograph of an *E. coli* cell that has been lysed, allowing its chromosome to escape.

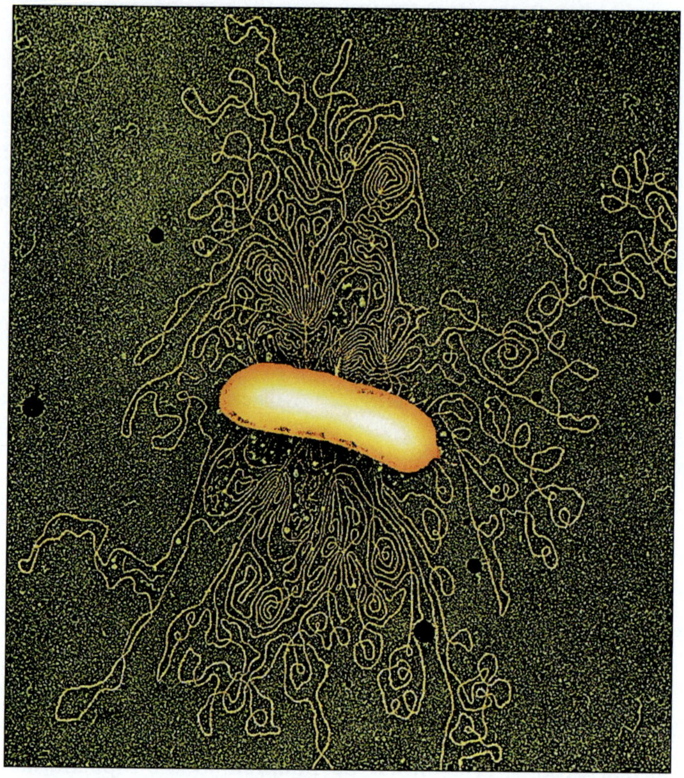

Genes Are Tightly Packed in Bacterial Genomes

In 1997, molecular geneticists completed sequencing the 4.6 million-base-pair genome of the *E. coli* strain known as K12 (**Fig. 14.6**). Close to 90% of *E. coli* DNA encodes proteins; on average, every kilobase (kb) of the chromosome contains one gene. This contrasts sharply with the human genome, in which less than 5% of the DNA encodes proteins and roughly one gene is present in every 100 kb. One reason for this discrepancy is that *E. coli* genes have no introns. In addition, little repeated DNA exists in bacteria, and intergenic regions tend to be very small. A glance at Fig. 14.6 will also show that some genes are transcribed from one strand of the bacterial DNA, while others are transcribed from the other strand. Genes transcribed in one or the other direction are interspersed with each other throughout the genome.

The complete sequence of the *E. coli* K12 genome revealed 4288 genes. The 427 genes that are thought to have a function transporting molecules into or out of cells make up the largest class. Other large classes include the genes for translation, amino acid biosynthesis, DNA replication, or recombination. About 20% of the genes are recognized only as *open reading frames* whose functions remain a mystery at this time.

Figure 14.6 Comparison of *E. coli* genomes. The outside two rings of the figure show the 4288 genes of *E. coli* K12 in many different colors. Genes transcribed from one strand of genomic DNA are on the outer ring, while genes transcribed in the opposite direction from the other strand are in the next ring. Each of the 16 inner rings shows which K12 genes are found in each of 16 other *E. coli* strains. *Red* color means that the K12 gene is present in the strain depicted in that ring; *blue* color means the K12 gene is absent, and *green* color means the other genome has a gene related to, but diverged from, the K12 gene. Note that many genes in the K12 strain are not found in one or more of the other strains; these genes are therefore not part of the *E. coli* core genome. Numbers indicate Mb (megabases) of DNA.

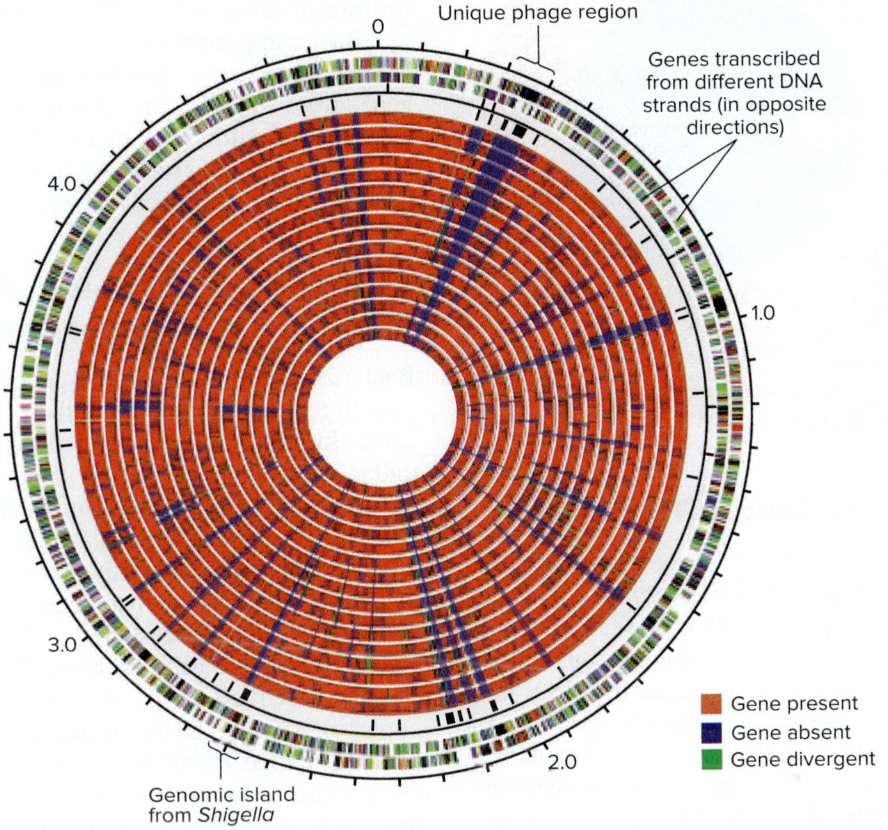

Another interesting feature of the *E. coli* K12 genome is the existence of remnants of bacteriophage genomes, including the *unique phage region* shown in Fig. 14.6. The presence of these sequences suggests an evolutionary history of these bacteria that included invasion by viruses on several occasions.

Individual *E. coli* Strains Contain Only a Subset of the *E. coli* Pangenome

As of this writing in 2016, the genomes of hundreds of different *E. coli* strains in addition to K12 have been sequenced. When scientists compared the *E. coli* genome sequences, they were astonished by the amount of variation they found (Fig. 14.6). All strains of *E. coli* have in common only about 1000 genes, meaning that about 80% of the *E. coli* genome is variable in different strains.

The approximately 1000 genes that are shared by all *E. coli* strains are called the **core genome.** In addition to the core genome, each strain also has genes that are either strain-specific or shared with only certain other strains. The core *E. coli* genome plus all of the other genes that are found in some strains but not others are collectively called the **pangenome** of the species (**Fig. 14.7**). So far, about 15,000 genes make up the *E. coli* pangenome, but this number is likely to grow larger as the genomes of additional strains are sequenced.

The core genome includes genes for metabolic functions that define the species *E. coli,* such as genes for the synthesis of lipids, cell walls and cell membranes, nucleotides, vitamins, and cofactors. While the core genes provide essential functions, the highly variable nature of the *E. coli* genome enables different strains to adapt to diverse environments. For example, different *E. coli* strains can break down different carbohydrates as sources of carbon.

The remarkable degree of genetic variation in *E. coli* is a general feature of bacterial species. These comparative genome studies provide evidence for a strikingly high incidence of gene transfer among bacteria of the same and different species.

Figure 14.7 Venn diagram of *E. coli* core and pangenomes.

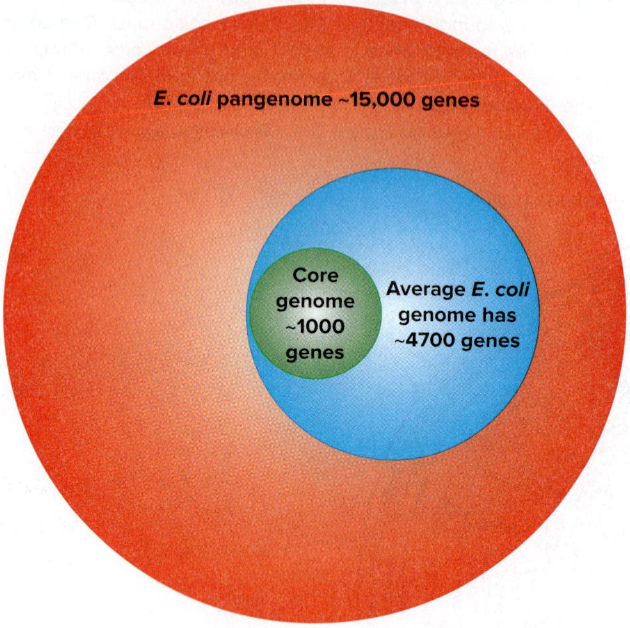

Bacterial Genomes Contain Transposons

DNA sequence analysis of bacterial genomes also revealed the positions of several small transposable elements called **insertion sequences (ISs)**. Researchers have identified several distinct elements ranging in length from 700 to 5000 bp; they named the elements IS1, IS2, IS3, and so forth, with the numbers designating the order of discovery. Like the ends of DNA transposons in eukaryotic cells (see Chapter 13), the ends of IS elements are inverted repeats of each other (**Fig. 14.8a**). Each IS element includes a gene encoding a transposase that initiates transposition by recognizing these mirror-image ends and cleaving the DNA there. Because IS elements can move to other sites on the bacterial chromosome when they transpose, their distribution varies in different strains of a single bacterial species. Like eukaryotic transposable elements, IS elements can cause mutations when they land within genes. Furthermore, recombination between two IS elements of the same class (for example, two IS1 elements) can rearrange bacterial genomes by causing deletion or inversion of DNA, as was previously seen in Chapter 13 for transposable elements in eukaryotes.

Bacterial genomes also have so-called *composite transposable elements,* or **Tn elements.** In addition to carrying a gene for transposase, Tn elements contain genes conferring resistance to antibiotics or toxic metals. One Tn element known as Tn10 consists of two IS10 elements flanking a gene encoding resistance to tetracycline (**Fig. 14.8b**).

Figure 14.8 Transposable elements in bacteria. (a) An IS element showing the inverted repeats (IR) at each end flanking the transposase gene. **(b)** The composite transposon Tn10, in which two slightly different IS10s (IS10L and IS10R) flank 7 kb of DNA, including a gene for tetracycline resistance. Because it is flanked by IS10 inverted repeats, Tn10 can be mobilized by the IS10 transposase.

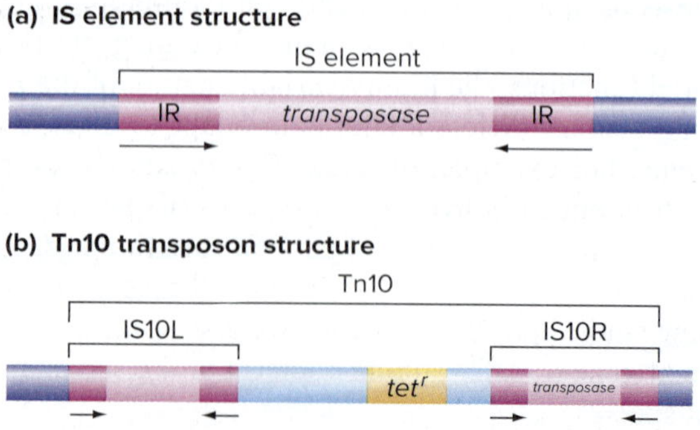

Plasmids Carry Genes in Addition to Those in the Bacterial Chromosome

Bacteria carry their essential genes—those necessary for growth and reproduction—in their large, circular chromosome. In addition, some bacteria carry genes not needed under normal conditions in smaller circles of double-stranded DNA known as **plasmids** (**Fig. 14.9**). Recall from Chapter 10 that bacterial plasmids have been genetically engineered by scientists for use as gene cloning vectors. Natural plasmids come in a range of sizes. The smallest are 1000 bp long; the largest are several megabases in length. Bacteria usually harbor no more than one extremely large plasmid, but they can house several or even hundreds of copies of smaller DNA circles. One important group of plasmids called *F episomes* allows the bacterial cells that carry them to make contact with another bacterium and transfer genes—both plasmid and bacterial—to the second cell. We describe this cell-to-cell mating, known as *conjugation,* in a later section of this chapter.

The genes carried by plasmids may benefit the host cell under certain conditions. For example, the plasmids in many bacterial species carry genes that protect their hosts against toxic metals such as mercury. The plasmids of various soil-inhabiting *Pseudomonas* species encode

Figure 14.9 Plasmids. Electron micrograph showing circular plasmid DNA molecules.

Figure 14.10 Resistance plasmids. Some plasmids contain multiple antibiotic resistance genes (shown in *yellow: cm^r* for chloramphenicol, *kan^r* for kanamycin, *str^r* for streptomycin, *su^r* for sulfonamide, *amp^r* for ampicillin, *hg^r* for mercury, and *tet^r* for tetracycline). Transposons (IS and Tn elements, shown in *light orange* and *red,* respectively) facilitate the movement of the antibiotic resistance genes onto the plasmid. Note that many antibiotic resistance genes are located between two IS1 elements, allowing them to transpose as a unit.

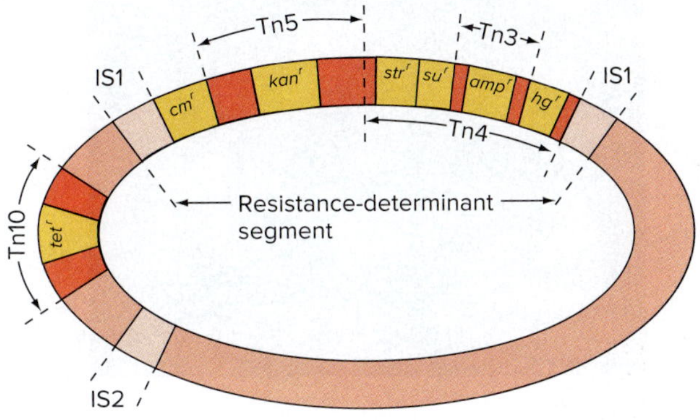

proteins that allow the bacteria to metabolize chemicals such as toluene, naphthalene, or petroleum products. In addition, many of the genes that contribute to pathogenicity reside in plasmids, such as the gene encoding the toxin produced by *Shigella dysenteriae,* the causative agent of dysentery. Genes specifying resistance to antibiotics are also often located on plasmids; the plasmid-determined resistance to multiple drugs was first discovered in *Shigella* in the 1970s. Multiple antibiotic resistance is often due to the presence of composite IS/Tn elements on a plasmid (**Fig. 14.10**).

As described later, plasmids can be transferred from one bacterium to another in nature, sometimes even across species. Plasmids thus have terrifying implications for medicine. If resistance plasmids are transferred to new strains of pathogenic bacteria, the new hosts can acquire resistance to many antibiotics in a single step. We encountered an example of this potential in the opening story on antibiotic-resistant gonorrhea.

Metagenomics Explores the Collective Genomes of Microbial Communities

Rarely (except in the laboratory) does a single bacterial species live in isolation; bacteria normally live in communities composed of hundreds or thousands of different species. Within these communities, various species of bacteria can influence each other, for example by exchanging metabolites or genes. In order to better understand the nature of these complex bacterial communities, scientists can now sequence the total DNA isolated from large populations of bacteria. From computer analysis of the sequence data researchers can determine all the species represented in the sample, the relative proportions of individuals of each species in the community, and the genes found in all of these species. The collective analysis of genomic DNA from a community of microbes sampled from their normal environment is called **metagenomics.**

One of the great powers of the metagenomics approach is that it allows the identification of new species of bacteria that could never before been detected because they do not grow in laboratory conditions. In 2016, the bacterial branch of the Tree of Life (Fig. 14.1) was increased by more than 1000 new bacterial species discovered by metagenomics to live in mud and meadow soils.

The U.S. National Institutes of Health has undertaken a metagenomic analysis of the bacteria that live on the human body—the **Human Microbiome Project.** In one of the first experiments in this project, bacterial samples were taken from different tissues like gut and skin on and in the bodies of 242 healthy individuals. This analysis revealed more than 10,000 different bacterial species in total, and as many as 1000 on a single individual. The most striking conclusion from this study so far is that individuals vary widely in the species of bacteria that they carry.

One future goal of the microbiome project is to track changes in the microbial communities over time and to correlate alterations in the bacterial metagenome with disease, diet, and drug treatments. For example, changes in the microbiome have been found to be associated with obesity: The populations of bacteria in the gut of some obese people are deficient in bacteria from a phylum called *Bacteroidetes,* while the microbiomes of these individuals are enriched for genes that are involved in carbohydrate and lipid metabolism. The Genetics and Society Box entitled *The Human Microbiome Project* discusses the progress of this endeavor in more detail.

Scientists also analyze the metagenomes of bacteria that live in extreme environments (*extremophiles*) because they harbor an abundance of genes for proteins that work under unusual conditions. These proteins can sometimes be useful in the laboratory. For example, *Taq* DNA polymerase, the enzyme used for PCR because it can withstand the hot temperatures that denature DNA, comes from the bacterial species *Thermus aquaticus*, first discovered in the hot springs in Yellowstone National Park (see the photograph at the beginning of this chapter).

Sequencing random DNA fragments from the metagenomes of these extremophile communities can reveal genes that confer unusual metabolic capabilities. In a commercially important example, alkaliphilic bacteria that grow well in conditions of high pH are sources of enzymes commonly added to laundry detergents. Various of these enzymes degrade proteins, lipids, or carbohydrates; the enzymes work efficiently in the alkaline environment of a soap bubble, where they inactivate chemicals or microorganisms that cause staining and odors.

GENETICS AND SOCIETY

The Human Microbiome Project

Established in 2008 and funded by the U.S. National Institutes of Health, The Human Microbiome Project (HMP, **Fig. A**) is one of several international consortia aiming to understand the complex relationship between our bodies and the trillions of microorganisms that inhabit them.

The HMP has already achieved its first goal of describing the diversity of the organisms that make up the human microbiome. Investigators analyzed the microbial metagenomes located at several different sites in the bodies of more than 250 people from around the globe. These studies focused on the sequence of the gene encoding the 16S rRNA of the ribosomal small subunit of these bacteria, because these sequences diverge substantially in different bacterial species and thus serve as markers for those species. The results showed that a single person can harbor up to 1000 different bacterial species, but people vary widely in the types of bacteria that make up their microbiome. Thus, it appears that worldwide more than 10,000 different bacterial species colonize human bodies. The researchers of the HMP have already sequenced the complete genomes of many of these kinds of bacteria.

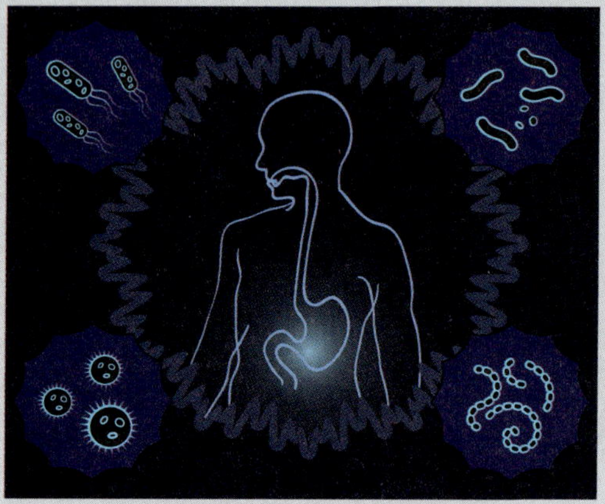

Figure A

The second phase of the HMP began in 2014, and is aimed ultimately at determining whether changes in the microbiome are the causes or effects of diseases or other important traits in humans. Diseases potentially linked to the microbiome include cancer, acne, psoriasis, diabetes, obesity, and inflammatory bowel disease; some investigators have suggested that the composition of microbiomes could also influence the mental health of their hosts. The first step in these studies will be to establish whether statistical correlations exist between specific kinds of microbial communities and disease states. As one example, one HMP phase II project currently underway is an analysis of vaginal host cells and microbes during pregnancy. Approximately 2000 pregnant women will be studied and their birth outcomes recorded. The goal of this project is to determine if changes in the microbiome correlate with premature birth or other complications of pregnancy.

Of course, the existence of any correlations found between microbiomes and disease does not prove cause or effect. But even if bacteria correlated with a disease state do not cause the disease, the existence of the correlation could be useful as a way to diagnose certain conditions. Nevertheless, the most exciting potential outcomes of the HMP will be results that point to bacteria within microbiomes as agents that contribute to complex diseases. Such bacteria would become obvious targets for therapeutics such as drugs that target proteins specifically made by these microorganisms.

How can researchers establish whether a statistical correlation between microbiomes and diseases reflects a cause or an effect? One method is to investigate in detail how the biological properties of the microbiome and the host might be changed by the interactions of bacteria and the humans they colonize. Thus, scientists will characterize whether and how the transcriptomes and proteomes of the bacteria and human cells are changed by bacterial colonization of human organs. These studies will further delve into *metabolomics* (characterizing metabolites in the human bloodstream).

A second and even more powerful method for establishing the cause and effect of microbiome changes is the use of *germ-free mice* raised in sterile environments. Surprisingly, germ-free mice can survive although they are not normal: they have altered immune systems, poor skin, and they need to eat more calories than do normal mice to maintain a normal body weight. Researchers can populate germ-free mice with a single bacterial species or a complex microbial community, and thus determine how microbiomes influence physiological states. Problem 8 at the end of this chapter will allow you to explore this approach by discussing an experiment recently performed with germ-free mice that asks if the microbiome plays a causal role in obesity.

If microbial communities indeed contribute to disease states in humans, then future treatments might aim to alter resident microbiomes. Thus, the flip side of the HMP is to investigate how human interventions might change bacterial communities. How effective are dietary changes or dietary additives such as probiotics in effecting long-lasting alterations in microbiomes? If acute infections are treated with antibiotics, how will bacterial communities change over time? Several HMP projects are already exploring these important questions.

essential concepts

- Genes in bacterial chromosomes are tightly packed, with no introns and very short intergenic regions.
- The *core genome* consists of genes that all members of a bacterial species share; the core genome plus all of the strain-specific genes is called the *pangenome* for the species.
- Transposons in bacterial genomes can carry genes, including those conferring drug resistance.
- *Plasmids* are circular DNA molecules that replicate autonomously and can be transferred between cells; they often contain genes for drug resistance or pathogenicity, or for specialized metabolic functions.
- *Metagenomics* is the analysis of the collective bacterial genome in a particular environment. Metagenomic studies may clarify the roles of bacteria in human health and may identify bacterial genes that have unique properties.

14.3 Bacteria as Experimental Organisms

learning objectives

1. Describe the features of bacteria, particularly *E. coli,* that make them useful laboratory organisms.
2. Describe different classes of bacterial mutants and the methods for identifying them.

The study of bacteria was crucial to the development of the science of genetics. From the 1940s to the 1970s, considered the era of classical bacterial genetics, virtually everything researchers learned about gene structure, gene expression, and gene regulation came from analyses of bacteria and the *bacteriophages* (bacterial viruses; often abbreviated as *phages*) that infect them. The advent of recombinant DNA technology in the 1970s and 1980s depended on an understanding of genes, chromosomes, and restriction enzymes in bacteria. Many recombinant DNA manipulations of genes from a variety of other organisms still rely on bacteria for the development and propagation of genetically engineered molecules.

Bacteria Grow in Liquid Cultures or on the Surface of Petri Plates

Researchers grow bacteria in liquid media (**Fig. 14.11a**) or on media solidified by agar in a plate, called a *petri dish* (**Fig. 14.11b**). In a liquid medium, the cells of commonly studied species, such as *E. coli*, grow to a concentration of 10^9 cells per milliliter within a day. On agar-solidified medium, a single bacterium will multiply to a visible colony containing 10^7 or 10^8 cells in less than one day. The ability to grow large numbers of cells quickly is one advantage that has made bacteria, especially *E. coli,* so attractive for genetic studies. It should be cautioned, however, that only a few bacterial species can be grown in culture in the laboratory; most species can be maintained only in their native environments.

Figure 14.11 Bacterial cultures. Bacteria can grow in the laboratory as a suspension in liquid medium **(a),** or as colonies on solid nutrient agar in a petri dish **(b).**

a: © Hank Morgan/Science Source; b: © Dr. Jeremy Burgess/SPL/Science Source

(a) (b)

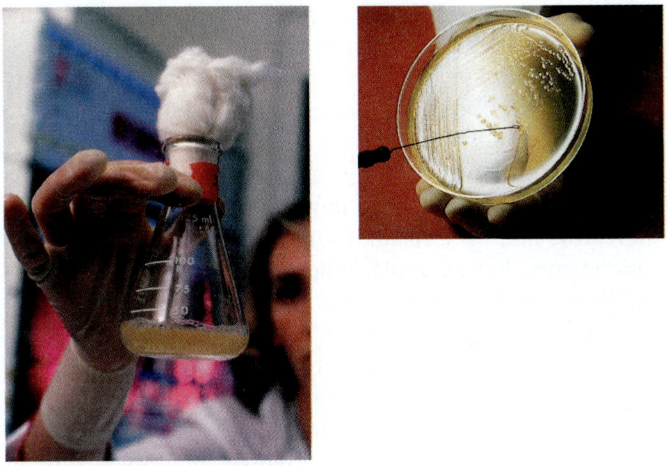

Genetic studies of bacteria require techniques to count these large numbers of cells and to isolate individual cells of interest. Researchers can use a solid medium to calculate the number of cells in a liquid culture. They begin with sequential dilutions (illustrated in Fig. 7.24) of cells in the liquid medium. They then spread a small sample of the diluted solutions on agar-medium plates and count the number of colonies that form. Although it is difficult to work with a single bacterial cell, the cells constituting a single colony contain the genetically identical descendants of the one bacterial cell that founded the colony.

E. coli Is a Versatile Model Organism

The most studied and best understood species of bacteria is *E. coli,* a common inhabitant of the intestines of warm-blooded animals. Many classical experiments and most modern recombinant DNA technologies have used *E. coli* as a model organism. *E. coli* cells can grow in the complete absence of oxygen—the condition found in the intestines—or in air. The *E. coli* strains studied in the laboratory are not pathogenic, but other strains of the species can cause a variety of intestinal diseases, most of them mild, a few life-threatening.

E. coli normally encodes all the enzymes it needs for amino acid and nucleotide biosynthesis. It is therefore a *prototrophic organism* that can grow in *minimal media,* which contain a single carbon and energy source, such as glucose, and inorganic salts to supply the other elements that compose bacterial cells. In a minimal medium, *E. coli* cells divide every hour, doubling their numbers 24 times a day. In a richer, more complex medium containing several sugars and amino acids, *E. coli* cells divide every 20 minutes to produce 72 generations per day. Two days of logarithmic growth at this latter rate, if unchecked by any limiting factor, would generate a mass of bacteria equal to the mass of the Earth.

The rapidity of bacterial multiplication makes it possible to grow an enormous number of cells in a relatively short time and, as a result, to obtain and examine rare genetic events. For example, wild-type *E. coli* cells are normally sensitive to the antibiotic streptomycin. By spreading a billion wild-type bacteria on an agar-medium plate containing streptomycin, it is possible to isolate a few extremely rare streptomycin-resistant mutants that have arisen by chance among the 10^9 cells. It is not as easy to find and examine such rare events with nonmicrobial organisms; in multicellular animals, this task is almost impossible.

Geneticists Identify Mutant Bacteria by the Presence or Phenotype of Colonies Under Specific Growth Conditions

Most bacterial genomes carry one copy of each gene and are therefore monoploid. The relationship between gene mutation and phenotypic variation is thus relatively straightforward; that is, in the absence of a second, wild-type allele for each gene, all mutations express their phenotype.

Bacteria are so small that the only practical way to examine them is in the colonies of cells they form on a petri dish. Within this constraint, it is still possible to identify many different kinds of mutations.

Classes of bacterial mutants

Bacterial mutants can be classified on the basis of the method used to identify them. Mutant classifications include:

1. Mutations affecting *colony morphology,* that is, whether a colony is large or small, shiny or dull, round or irregular.
2. Mutations conferring *resistance* to bactericidal agents such as antibiotics or bacteriophages.
3. Mutations that create *auxotrophs* unable to grow and reproduce on minimal medium. Auxotrophs cannot synthesize crucial complex compounds from simple materials.
4. Mutations affecting *catabolism,* the ability of cells to break down and use complicated chemicals in the environment. One example in *E. coli* is an inability to break down the complex sugar lactose because of mutations in the *lacZ* gene; such mutations are highly useful for investigating how gene expression is controlled, as will be described in Chapter 16.
5. Mutations in *essential genes* whose protein products are required for growth. Because a null mutation in an essential gene would prevent a colony from growing in any environment, bacteriologists must work with *conditional lethal mutations* such as *temperature-sensitive (ts) mutations:* hypomorphs that allow growth at low temperatures but not high ones.

Screens vs. selections

Bacteriologists use different techniques to isolate rare mutations. With mutations conferring resistance to a particular agent, researchers can do a straightforward **selection,** that is, establish conditions in which only the desired mutant will grow. For example, if wild-type bacteria are streaked on a petri dish containing the antibiotic streptomycin, the only colonies to appear will be streptomycin-resistant (Str^r). It is also possible to select for prototrophic revertants of strains carrying auxotrophic mutations by simply plating cells on minimal medium agar, which does not contain the compound auxotrophs require for growth.

Because the key characteristic of most of the other types of mutants just described is their inability to grow under particular conditions, it is not usually possible to select for them directly. Instead, researchers must identify these mutations by a genetic **screen:** an examination of each colony in a population for a particular phenotype. Scientists can, for example, use replica plating (review Fig. 7.6) to transfer cells from each colony growing on a minimal medium plate supplemented with methionine to a petri plate containing minimal medium without methionine. Failure of colonies to grow on the unsupplemented medium would indicate that the corresponding colony on the original plate is auxotrophic for methionine.

Spontaneous mutations in specific bacterial genes occur very rarely, in 1 in 10^6 to 1 in 10^8 cells, depending on the gene. Therefore, it would be virtually impossible to identify such rare mutations by screening for a particular phenotype in one million to one hundred million colonies. Treatment with mutagens increases the frequency with which a mutation in a gene appears in the population (review Fig. 7.14). Mutant screens in bacteria and mutant gene identification will be discussed further later in the chapter.

Designation of bacterial alleles

Researchers specify the genes of bacteria first by three lowercase, italicized letters that signify something about the function of the gene. For example, genes in which mutations result in the inability to synthesize the amino acid leucine are *leu* genes. In *E. coli,* there are four *leu* genes—*leuA, leuB, leuC,* and *leuD*—that correspond to the three enzymes (one constructed from two different polypeptides) needed for the synthesis of leucine from other compounds. A mutation in any one of the *leu* genes changes a bacterium into an auxotroph for leucine; that is, into a cell unable to synthesize leucine. Such a cell can grow only in media supplemented with leucine. Mutations in genes required for the breakdown of a sugar (for example, the *lacZ* gene) produce cells unable to grow in medium containing only that sugar (lactose) as a source of carbon. Other types of mutations give rise to antibiotic resistance; str^r is a mutation producing streptomycin resistance. To designate the alleles of genes present in wild-type bacteria (the genotype), researchers use a + superscript: leu^+, str^+, $lacZ^+$. To designate mutant alleles, they use a superscript –, as in $leuA^-$ and $lacZ^-$, or a superscript description, as in str^r.

The phenotype of a bacterium that is wild type or mutant for a particular gene is indicated by the three letters that designate the gene. However, these letters are written with an initial capital letter, no italics, and a superscript of minus, plus, or a one-letter abbreviation: Leu^- (requires leucine for growth); Lac^+ (grows on lactose); Str^r (is resistant to streptomycin). A Leu^- *E. coli* strain cannot multiply unless it grows in a medium containing leucine; a Lac^+ strain can grow if lactose replaces the usual glucose in the medium; a Str^r strain can grow in the presence of streptomycin.

essential concepts

- A single bacterial cell divides to form a *colony* of millions of genetically identical descendants.
- Features of bacteria that aid genetic research are monoploidy, which facilitates mutant identification, and rapid exponential growth, which allows recovery of rare mutants.
- Bacterial mutants can be identified in screens or selections. In a *screen,* individual colonies are tested for a particular phenotype. In a *selection,* only bacteria with the phenotype in question are recovered as colonies, enabling the identification of extremely rare mutants.

14.4 Gene Transfer in Bacteria

learning objectives

1. Compare the three mechanisms of gene transfer in bacteria: transformation, conjugation, and transduction.
2. Explain how each of these three methods of gene transfer can be used to map bacterial genes.
3. Discuss the role of horizontal gene transfer in the evolution of bacteria.

Gene transfer from one individual to another plays an important role in the evolution of new variants in nature. **Vertical gene transfer,** for example, occurs from one generation to the next and is particularly important in organisms utilizing sexual reproduction. By contrast, **horizontal gene transfer** means that the traits involved are not transferred by inheritance from parents to offspring; rather, they are introduced from unrelated individuals or from different species. Many cases of horizontal gene transfer have come to light through recent molecular and DNA sequencing analyses.

Comparative genomic analysis of many different genes in various bacterial species has revealed similarities of genes in species thought to be related only distantly. In addition, the existence of core genomes and pangenomes in individual species described earlier indicates that gene loss and gain occur frequently in *E. coli* and other species. The simplest explanation for these findings is that significant transfer of DNA between bacteria has occurred throughout evolution. A close examination of the known mechanisms of DNA transfer helps illuminate how horizontal gene transfer takes place. In addition, you will see that researchers can use the various methods of gene transfer to map genes and to construct useful bacterial strains.

Bacteria can transfer genes from one strain to another by three different mechanisms: *transformation, conjugation,* and *transduction* (**Fig. 14.12**). In all three mechanisms, one cell—the **donor**—provides the genetic material for transfer, while a second cell—the **recipient**—receives the material. In **transformation,** DNA from a donor is added to the bacterial growth medium and is then taken up from that medium by the recipient. In **conjugation,** the donor carries a special type of plasmid that allows it to come in contact with the recipient and transfer DNA directly. In **transduction,** the donor DNA is packaged within the protein coat of a bacteriophage and is transferred to the recipient when the phage particle infects it. The recipients of a gene transfer are known as **transformants, exconjugants,** or **transductants,** depending on the mechanism of DNA transfer that created them.

All bacterial gene transfer is asymmetrical by two criteria: First, transfer goes in only one direction, from donor to recipient. Second, most recipients receive 3% or less of a donor's DNA; only some exconjugants contain a greater percentage of donor material. Thus, the amount of donor DNA entering the recipient is small relative to the size of the recipient's chromosome, and the recipient retains most of its own DNA. We now examine each type of gene transfer in detail.

In Transformation, the Recipient Takes Up DNA that Alters Its Genotype

A few species of bacteria take up DNA fragments spontaneously from their surroundings in a process known as **natural transformation.** The large majority of bacterial

Figure 14.12 Gene transfer in bacteria: An overview. In this figure, and throughout this chapter, the donor's chromosome is *blue,* and the recipient's chromosome is *orange.* In transformation, fragments of donor DNA released into the medium enter the recipient cell. In conjugation, a specialized plasmid (shown in *red*) in the donor cell promotes contact with the recipient and initiates the transfer of DNA. In transduction, DNA from the donor cell is packaged into bacteriophage particles that can infect a recipient cell, transferring the donor DNA into the recipient.

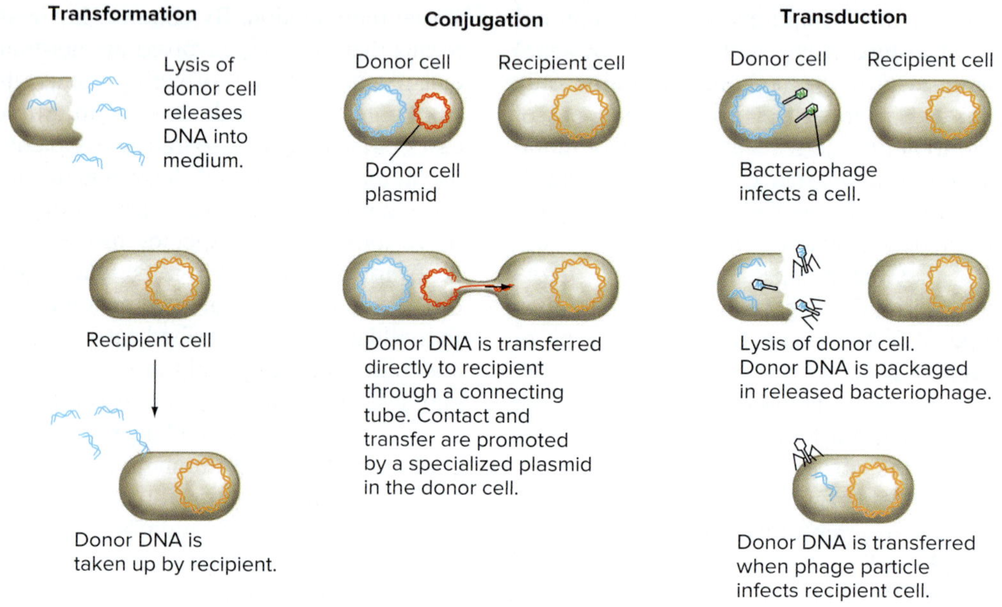

species, however, can take up DNA only after laboratory procedures make their cell walls and membranes permeable to DNA in a process known as **artificial transformation.**

Natural transformation

Researchers have studied several species of bacteria that undergo natural transformation, including *S. pneumoniae,* the pathogen in which transformation was discovered by Frederick Griffith and that causes pneumonia in humans (see Chapter 6); *B. subtilis,* a harmless soil bacterium; *H. influenzae,* a pathogen causing various diseases in humans; and *N. gonorrhoeae,* the microbial agent of gonorrhea.

In one study of natural transformation, investigators isolated *B. subtilis* bacteria with two mutations—*trpC*$^-$ and *hisB*$^-$—that made them Trp$^-$ His$^-$ double auxotrophs. These double auxotrophs served as the recipients in the study; wild-type cells (Trp$^+$ His$^+$) were the donors (**Fig. 14.13a**). The experimenters extracted and purified donor DNA and grew the *trpC*$^-$ *hisB*$^-$ recipients in a suitable medium until the cells became **competent,** that is, able to take up DNA from the medium.

When a recipient takes up DNA by natural transformation, only one strand of a fragment of donor DNA enters the cell, while the other strand is degraded (**Fig. 14.13b**). The entering strand recombines with the recipient chromosome, producing a transformant when the recipient cell divides.

To observe and count Trp$^+$ transformants, researchers spread newly transformed recipient cells onto petri dishes containing minimal medium with histidine. Recipient cells that do not take up donor DNA cannot grow on this medium because it lacks tryptophan, but the Trp$^+$ transformants can grow and be counted. To select for His$^+$ transformants, researchers poured the transformation mixture on minimal medium containing tryptophan, instead of histidine. In this study, the numbers of Trp$^+$ and His$^+$ transformants were equal. In conditions where *B. subtilis* bacteria become highly competent, 10^9 cells will produce approximately 10^5 Trp$^+$ transformants and 10^5 His$^+$ transformants.

To discover whether any of the Trp$^+$ transformants were also His$^+$, the researchers used sterile toothpicks to transfer colonies of Trp$^+$ transformants to a minimal medium containing neither tryptophan nor histidine. Forty of every 100 Trp$^+$ transferred colonies grew on this minimal medium, indicating that they were also His$^+$. Similarly, tests of the His$^+$ transformants showed that roughly 40% were also Trp$^+$. Thus, in 40% of the analyzed colonies, the *trpC*$^+$ and *hisB*$^+$ genes had been cotransformed. **Cotransformation** is the simultaneous transformation of two or more genes.

Because donor DNA replaces only a small percentage of the recipient's chromosome during transformation, it might seem surprising that the two *B. subtilis* genes are cotransformed with such high frequency. The explanation is that the *trpC* and *hisB* genes lie very close together on the chromosome and are thus genetically linked. The entire *B. subtilis* chromosome is approximately 4700 kb long. Only genes in the same chromosomal vicinity can be cotransformed; the closer together the genes lie, the more frequently they will be cotransformed. Therefore, although

Figure 14.13 Natural transformation in *B. subtilis.* **(a)** A wild-type donor and a $trpC^-$ $hisB^-$ double auxotroph recipient. Selection for Trp^+ and/or His^+ phenotypes identifies transformants. **(b)** Mechanism of natural transformation in *B. subtilis.* One strand of a fragment of donor DNA enters the recipient, while the other strand is degraded. The entering strand recombines with the recipient chromosome, resulting initially in a region of heteroduplex DNA in which one strand of the double helix comes from one parent, and the other strand from the other parent. When the recipient cell divides, one of the daughter cells is a transformant with donor genes.

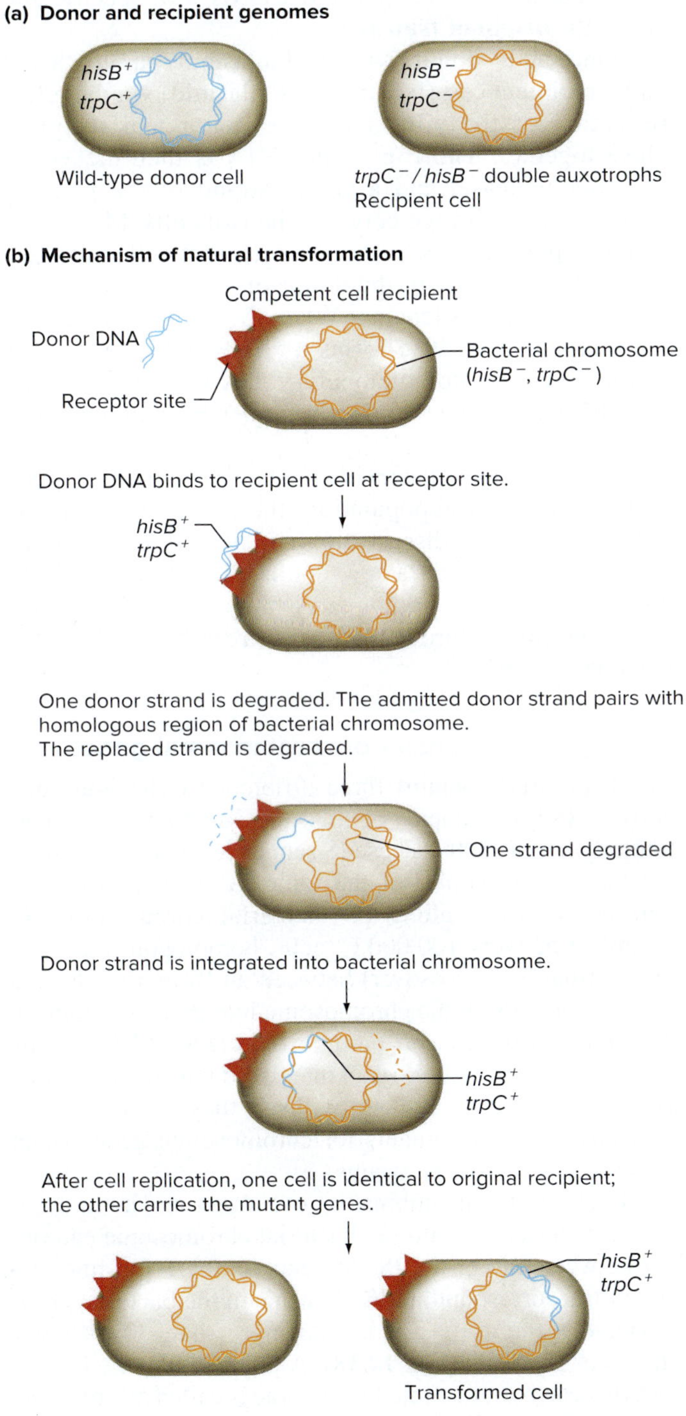

the donor chromosome is fragmented into small pieces of about 20 kb during its extraction for the transformation process, the wild-type $trpC^+$ and $hisB^+$ alleles are so close (about 7 kb apart) that they are often together in the same donor DNA molecule. By contrast, genes sufficiently far apart that they cannot appear together on a fragment of donor DNA will almost never be cotransformed, because transformation is so inefficient that recipient cells usually take up only a single DNA fragment.

Transformation also describes plasmid transfer; that is, if the donor DNA is a plasmid, recipient cells may take up an entire plasmid and acquire the characteristics conferred by the plasmid genes. Bacteriologists suspect that penicillin-resistant *N. gonorrhoeae,* described in the introduction to this chapter, originated through transformation by plasmids. The donors of the plasmids were *H. influenzae* cells disrupted by the immune defenses of a doubly infected patient. The plasmids carried the gene for penicillinase, and the recipient *N. gonorrhoeae* bacteria, transformed by the plasmids, acquired resistance to penicillin.

Artificial transformation

Although the study just described was a laboratory manipulation of natural transformation, researchers have devised many methods to transform bacterial species that do not undergo natural transformation. The existence of artificial transformation was crucial for the development of the gene-cloning technology described in Chapter 10. All the methods include treatments that damage the cell walls and membranes of recipient bacteria so that donor DNA can diffuse into the cells. With *E. coli,* the most common treatment is to suspend cells in a high concentration of calcium at cold temperature. Under these conditions, the cells become permeable to single- and even double-stranded DNA.

Another technique of artificial transformation is *electroporation,* in which researchers mix a suspension of recipient bacteria with donor DNA and then subject the mixture to a very brief high-voltage shock. The shock most likely causes holes to form in the cell membranes. With the proper shocking conditions, recipient cells take up the donor DNA efficiently. Transformation by electroporation works with most bacteria.

In Conjugation, a Donor Transfers DNA Directly to a Recipient

In the late 1940s, Joshua Lederberg and Edward Tatum analyzed two *E. coli* strains that were each multiple auxotrophs. They made the striking discovery that genes seemed to transfer from one type of *E. coli* cell to the other (**Fig. 14.14**). Neither strain could grow on a minimal medium. Strain A required supplementation with methionine and biotin (vitamin H); strain B required supplementation with threonine, leucine, and thiamine (vitamin B_1). Lederberg and Tatum

Figure 14.14 Conjugation. Neither of two multiple auxotrophic strains analyzed by Lederberg and Tatum formed colonies on minimal medium. When cells of the two strains were mixed, gene transfer produced some prototrophic cells that formed colonies on minimal medium.

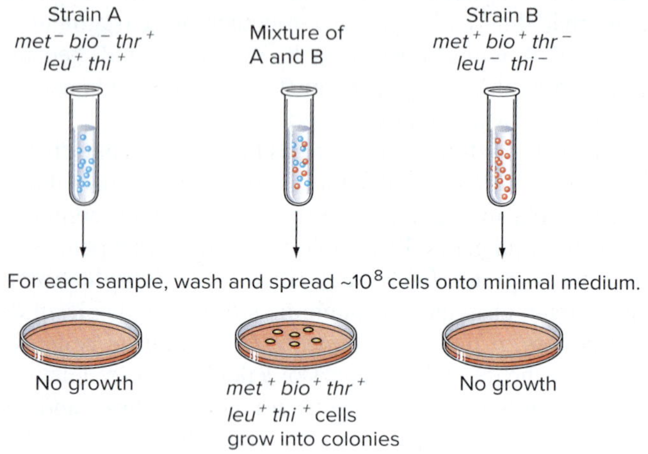

grew the two strains together on fully supplemented medium. When they then transferred a mixture of the two strains to minimal medium, about 1 in every 10^7 transferred cells proliferated to a visible colony. What were these colonies, and how they did they arise?

More than a decade of further experiments confirmed that Lederberg and Tatum had observed what became known as **bacterial conjugation** (**Fig. 14.15**): a one-way DNA transfer from donor to recipient that requires cell-to-cell contact and that is initiated by **conjugative plasmids** in donor strains. These plasmids can initiate conjugation because they carry genes that allow them to transfer themselves (and sometimes some of the donor's chromosome) to the recipient.

Figure 14.15 Bacterial cells conjugating.
© Eye of Science/Science Source

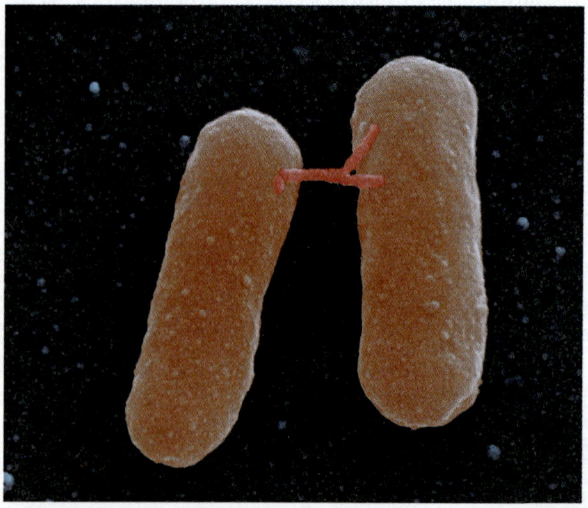

The F plasmid and conjugation

Figure 14.16 illustrates the type of bacterial conjugation initiated by the first conjugative plasmid to be discovered—the **F plasmid** of *E. coli.* Donor cells carrying an F plasmid are called F^+ cells; recipient cells without the plasmid are F^-. The F plasmid carries many genes required for the transfer of DNA, including genes for formation of an appendage, known as a *pilus,* by which a donor cell contacts a recipient cell, and a gene encoding an endonuclease enzyme that nicks the F plasmid's DNA at a specific site called the **origin of transfer**.

Once an F^+ donor (with the F plasmid) has contacted an F^- recipient cell (lacking the F plasmid) via the pilus, retraction of the pilus pulls the donor and recipient close together. The F plasmid DNA is then nicked by the endonuclease, and a single strand of the plasmid moves across a bridge between the two cells. Movement of the F plasmid DNA into the recipient cell is accompanied by synthesis in the donor of another copy of the DNA strand that is leaving. When the donor DNA enters the recipient cell, it re-forms a circle and the recipient synthesizes the complementary DNA strand. In this $F^+ \times F^-$ mating, the recipient becomes F^+, and the donor remains F^+.

By initiating and carrying out conjugation, the F plasmid acts in bacterial populations the way an agent of sexually transmitted disease acts in human populations. When introduced via a few donor bacteria into a large culture of cells that do not carry the plasmid, the F plasmid soon spreads throughout the entire culture, and all the cells become F^+.

Conjugational transfer of chromosomal genes

The F plasmid contains three different IS elements: one copy of IS2, two copies of IS3, and one copy of the particularly long IS1000. These IS sequences on the F plasmid are identical to copies of the same IS elements found at various positions along the bacterial chromosome. In roughly 1 of every 100,000 F^+ cells, homologous recombination (that is, a crossover) between an IS on the plasmid and the same IS on the chromosome integrates the entire F plasmid into the *E. coli* chromosome (**Fig. 14.17**). Cells whose chromosomes carry an integrated plasmid are called **Hfr bacteria** because, as we will see, they produce a high frequency of recombinants for chromosomal genes when they are mated with F^- strains.

Because the recombination event that results in the F plasmid's insertion into the bacterial chromosome can occur between any of the IS elements on the F plasmid and any of the corresponding IS elements in the bacterial chromosome, geneticists can isolate more than 30 different strains of Hfr cells (**Fig. 14.18**). A plasmid like the F plasmid that can integrate into the genome is called an **episome.** Various Hfr strains are distinguished by the location and

The F Plasmid and Conjugation

a. The F plasmid contains genes for synthesizing connections between donor and recipient cells. The F plasmid is a 100-kb-long circle of double-stranded DNA. F^+ cells generally have one copy of the plasmid. Researchers think of F^+ cells as male bacteria because the cells can transfer genes to other bacteria. About 35% of F plasmid DNA consists of genes that control plasmid transfer. Most of these genes encode polypeptides involved in the construction of the **F pilus** (plural, **pili**): a stiff, thin strand of protein that protrudes from the bacterial cell. Other regions of the plasmid carry ISs and genes for proteins involved in DNA replication.

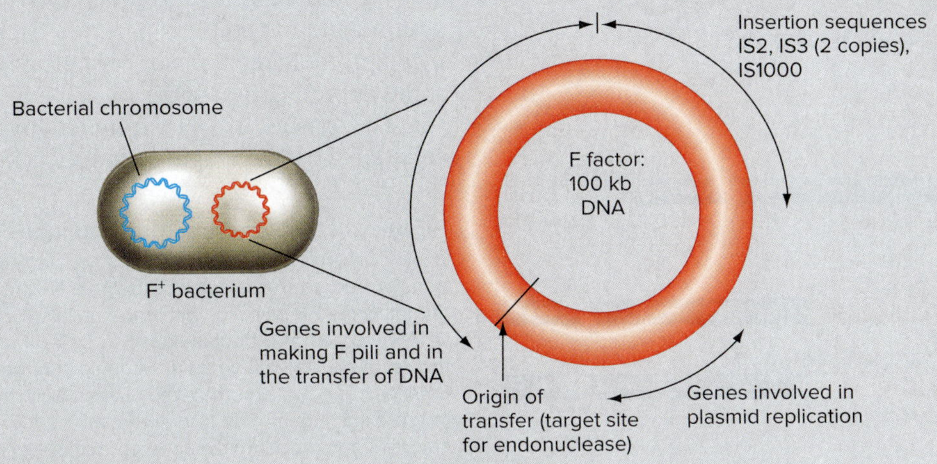

b. The process of conjugation.

1. **The pilus.** An average pilus is 1 μm in length, which is almost as long as the average *E. coli* cell. The distal tip of the pilus consists of a protein that binds specifically to the cell walls of F^- *E. coli* not carrying the F plasmid.
2. **Attachment to F^- cells (female bacteria).** Because they lack F factors, F^- cells cannot make pili. The pilus of an F^+ cell, on contact with an F^- cell, retracts into the F^+ cell, drawing the F^- cell closer. A narrow passageway forms through the two cell membranes.
3. **Gene transfer: A single strand of DNA travels from the male to the female cell.** An endonuclease cuts one strand of the F plasmid DNA at a specific site (the *origin of transfer*). The F^+ cell extrudes the cut strand through the passageway into the F^- cell. As it receives the single strand of F plasmid DNA, the F^- cell synthesizes a complementary strand. The formerly F^- cell contains a double-stranded F plasmid and is now an F^+ cell.
4. **In the original F^+ cell, newly synthesized DNA replaces the single strand transferred to the previously F^- cell.** When the two bacteria separate at the completion of DNA transfer and synthesis, they are both F^+.

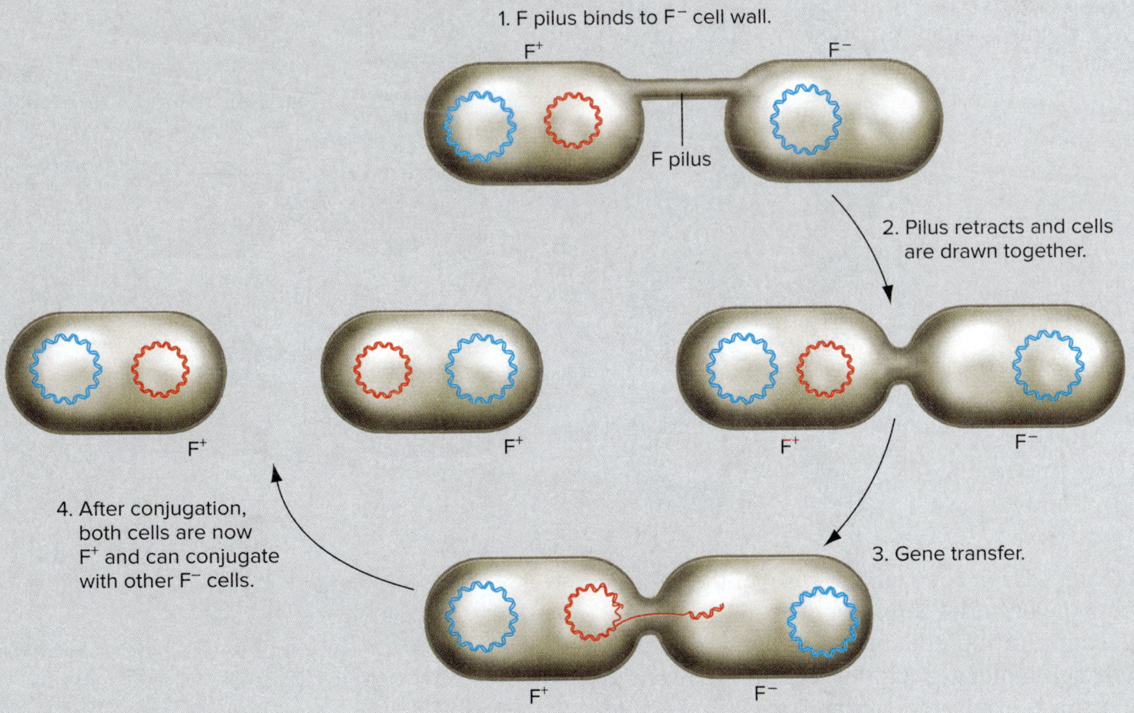

Figure 14.17 Genesis of an Hfr chromosome. In this figure, the bars represent both strands of DNA. Recombination between an IS on the F plasmid and the same kind of IS on the bacterial chromosome creates an Hfr chromosome. During conjugation, DNA will be transferred from the Hfr donor into the recipient starting with the origin of transfer (*arrow*), so bacterial genes will be transferred in the order *A B C*.

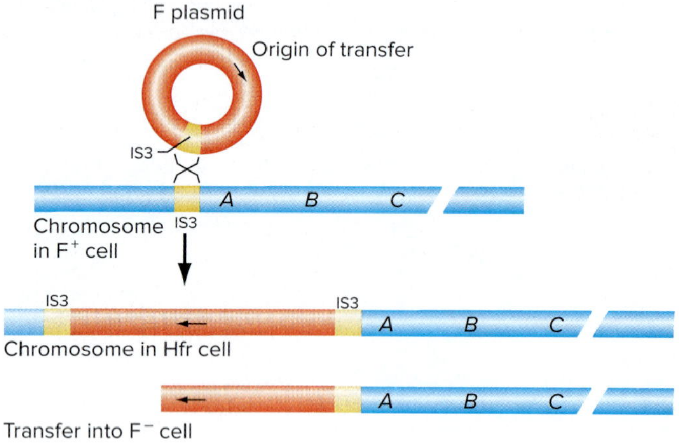

Figure 14.18 Different Hfr chromosomes. Recombination can occur between any IS on the F plasmid and any corresponding IS on the bacterial chromosome to create many different Hfr strains. Depending on the initial orientation of the IS element on the chromosome, recombination can produce some Hfr strains that transfer genes clockwise and others that transfer genes counterclockwise.

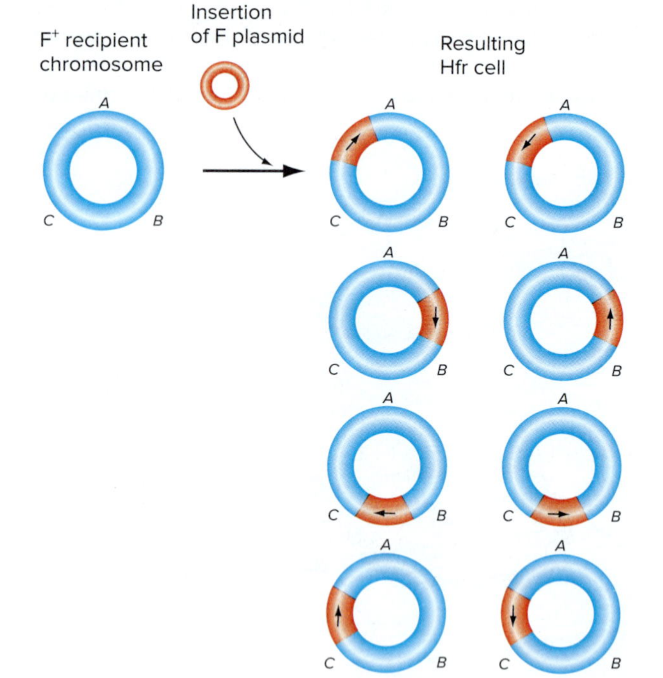

orientation (clockwise or counterclockwise) of the episome with respect to the bacterial chromosome.

During bacterial reproduction, the integrated plasmid of an Hfr cell replicates with the rest of the bacterial chromosome. As a result, the chromosomes in daughter cells produced by cell division contain an intact F plasmid at exactly the same location that the plasmid originally integrated into the chromosome of the parental cell. All progeny of an Hfr cell are thus identical, with the F plasmid inserted into the same chromosomal location and in the same orientation. The integrated F plasmid still has the capacity to initiate DNA transfer via conjugation, but now that it is part of a bacterial chromosome, it can promote the transfer of some of that donor chromosome as well (**Fig. 14.19**).

The transfer of DNA from an Hfr cell mated to an F$^-$ cell starts with a single-strand nick in the middle of the

Figure 14.19 Gene transfer between Hfr donors and F$^-$ recipients. In an Hfr × F$^-$ mating, single-stranded DNA is transferred into the recipient, starting with the origin of transfer on the integrated F plasmid. Within the recipient cell, this single-stranded DNA is copied into double-stranded DNA. If mating is interrupted, the recipient cell will contain a double-stranded linear fragment of DNA plus its own chromosome. Genes from the donor are retained in the exconjugant only if they recombine into the recipient's chromosome. Importantly, an even number of crossovers is required to ensure the recipient's chromosome remains circular so that the cell remains viable.

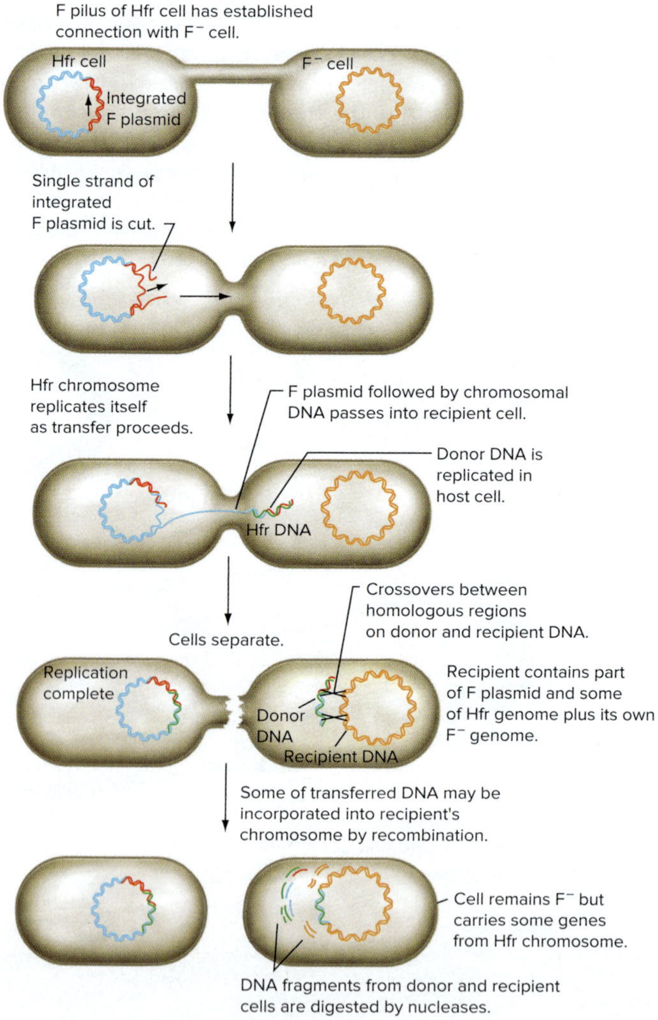

integrated F plasmid at the origin of transfer (**Fig. 14.20**). Once the donor DNA has been transferred to the recipient, recombination can occur between donor DNA and the chromosome in the recipient.

Mapping genes by gene transfer during conjugation

Because genes in an Hfr chromosome are transferred into the recipient in a consistent order, researchers realized that they could use Hfr $\times$ F^- crosses to map genes. In early studies of the *E. coli* genome, for example, Elie Wollman and Francois Jacob used an Hfr strain that was Str^s (streptomycin sensitive), Thr^+ (able to synthesize threonine), Azi^r (azide resistant), Ton^r (resistant to phage T1), Lac^+ (able to grow on lactose), Gal^+ (able to grow on galactose), and an F^- strain with alternate characteristics (Str^r, Thr^-, Azi^s, Ton^s, Lac^-, Gal^-). The use of these two strains enabled these investigators to isolate and analyze the exconjugants in which gene transfer and recombination had occurred.

Wollman and Jacob mixed the two strains in rich nonselective liquid medium to allow conjugation. Next, at 1-minute intervals, they agitated samples of the mating mixture in a kitchen blender to interrupt mating. (This is why the experiment is called the *interrupted-mating experiment*.) Samples of the terminated matings were spread onto petri plates containing streptomycin, which killed the original Hfr cells. The plates also lacked threonine to select against F^- cells that had not mated; any Thr^+ F^- cell must have received the *thr*$^+$ gene from the Hfr. The Str^r Thr^+ exconjugants that grew on the plates were replica plated to test for transfer of the four other markers from the Hfr into the F^- recipients (**Fig. 14.20a**).

Figure 14.20b shows the frequency of exconjugant colonies containing various alleles from the Hfr strain as a

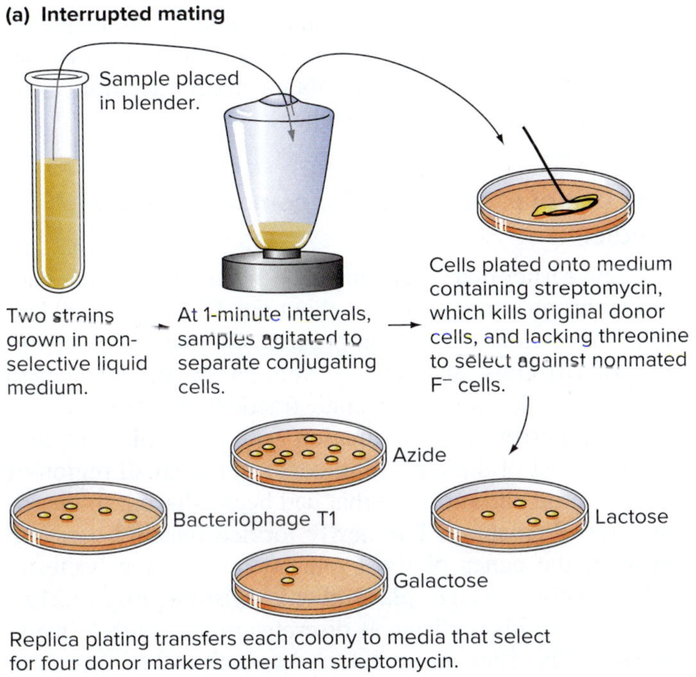

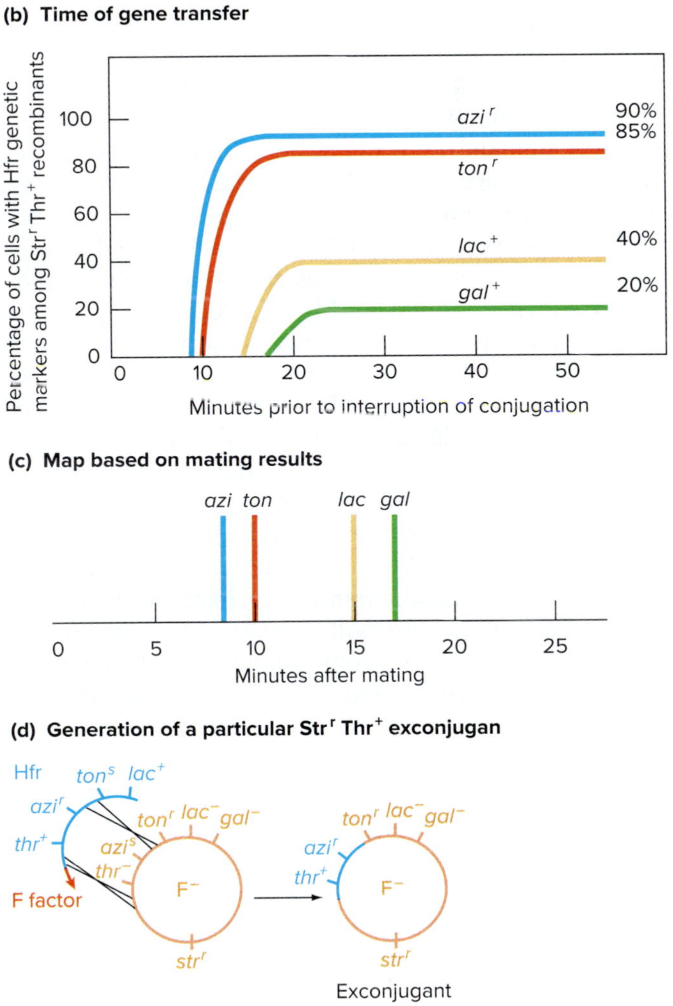

Figure 14.20 Interrupted-mating experiments. (a) Hfr (*thr*$^+$, *azi*^{r}, *ton*^{r}, *lac*$^+$, *gal*$^+$, *str*^{s}) and F^- (*thr*$^-$, *azi*^{s}, *ton*^{s}, *lac*$^-$, *gal*$^-$, *str*^{r}) cells were mixed to initiate mating. Samples were agitated at 1-minute intervals in a kitchen blender to disrupt gene transfer. Cells were plated onto a medium that contained streptomycin (to kill the Hfr donor cells) and that lacked threonine (to prevent the growth of F^- cells that had not mated). The genotypes of the exconjugants for other markers were established by replica plating. **(b)** Results of the interrupted-mating experiment. **(c)** Gene order established from the data, with positions determined by the time a donor gene first appears in the exconjugant. **(d)** Transfer of Hfr genes into F^- recipient. To appear in an exconjugant, the Hfr DNA must undergo two (or an even number) of crossovers with the F^- chromosome. As Thr^+ exconjugants are selected, two crossovers (*black lines*) must flank the *thr* gene. The position of the crossover to the right of *azi* (in this case between *azi* and *ton*) determines that both *thr*$^+$ and *azi*^{r} are transferred to the F^- chromosome.

function of time, from the beginning of conjugation to its interruption. After mating has proceeded for 8 minutes, a small fraction of the recombinants are Azi^r, but not one carries the other donor alleles. At about 10 minutes, some of the recombinants also have the donor's *ton^r* allele. By 15 minutes, the *lac$^+$* allele appears in the exconjugant colonies, and by 17 minutes, *gal$^+$* arrives. The percentages of recombinant colonies containing a particular gene from the Hfr increase with time until they reach a plateau characteristic of the gene. The first Hfr gene (*azi^r*) to enter the F$^-$ cell has the highest plateau (90%), and the last gene (*gal$^+$*) has the lowest (20%).

The presence of an integrated F episome in the Hfr strain explains the two characteristics of each transferred gene in the interrupted-mating experiment: the time at which transfer of the gene is first seen, and the plateau percentages of exconjugants that carry the donor gene. Each gene first appears in the exconjugants at a specific time because all Hfr donor cells carry the F factor at the same site and in the same orientation on their chromosomes, and transfer always initiates from the same spot in the F factor—the origin of transfer (recall Fig. 14.17). Thus, the time a gene first enters the recipient cell reflects the distance of that gene from the origin of transfer. This interpretation not only predicts the order of genes on the *E. coli* chromosome, it also makes it possible to map roughly the distances between the genes. The units of distance are defined as *minutes of chromosome transfer*, and each minute corresponds to approximately 47,000 bp (**Fig. 14.20c**).

The plateaus for the fraction of colonies carrying each transferred gene derive from the fact that new transfers are initiating continually and also from the inherent fragility of the conjugation bridges and the DNA being transferred. Transfer between Hfr donors and F$^-$ recipients can spontaneously abort if the cells separate or if the transferring chromosome breaks. As a result, not all exconjugants receive even early-arriving genes such as *azi^r*. In addition, the more time required to transfer a marker, the greater the chance that mating pairs will separate before the transfer occurs. In other words, conjugations were being interrupted naturally without the blender. As a result, the percentages of exconjugants receiving later-arriving genes became successively smaller.

As you can see from Fig. 14.20b, the order of genes could have been determined simply from the percentages of exconjugants bearing the different markers after 20 minutes of conjugation, without interrupting the mating at 1-minute intervals. This property of conjugation, where the distance of a gene from the F factor origin dictates the fraction of exconjugants with that marker, is called the *gradient of transfer*.

Once the genes from the Hfr donor have arrived in the F$^-$ recipient, stable replacement of the recipient with the donor alleles requires an even number of crossovers, two in the simplest case. **Figure 14.20d** shows how one particular class of stable exconjugants was created in the interrupted mating experiment. Because exconjugants were selected as being Trp$^+$, one of the two crossovers must occur between the origin of transfer and the *thr* gene. (In fact, the success of this experiment was based on Wollman and Jacob's prior knowledge that the *thr$^+$* marker was the first of all the markers to be transferred from this particular Hfr strain.) The second crossover must occur on the other side of the *thr* gene. For example, as demonstrated in Fig. 14.20d, cotransfer of *thr$^+$* and *azi^r* alleles (but not any other markers) from the Hfr to the F$^-$ recipient demands a second crossover between the *azi* and *ton* genes.

The problem set at the end of the chapter highlights two additional features of conjugation experiments. First, as seen in Problem 23, conjugation experiments using different Hfr donor strains with the F episome inserted with different orientations into various locations of the *E. coli* chromosome provided the first evidence that the bacterial chromosome is in fact circular. Second, Problems 20 and 21 emphasize that the formation of stable exconjugants requires an even number of crossovers. This fact provides scientists with a method to obtain highly accurate gene maps from conjugation experiments.

The use of F′ episomes for complementation studies

We saw earlier that insertion of the F plasmid into the bacterial chromosome occurs in about one F$^+$ cell per 100,000 to produce an Hfr cell. In approximately that same proportion of Hfr cells, an excision event causes the Hfr cell to revert to an F$^+$ cell. In a small fraction of these excision events, an error in recombination generates a plasmid containing most of the F plasmid genes plus a small region of the bacterial chromosome that had been adjacent to the integrated F episome. The newly formed plasmid carrying most of the genes of the F plasmid plus some bacterial DNA is known as an **F′ plasmid** or **F′ episome** (**Fig. 14.21a**).

F′ plasmids replicate as discrete circles of DNA inside *E. coli* cells. They are transferred to recipient (F$^-$) cells in the same way that F plasmids are transferred. The difference is that a few chromosomal genes will always be transferred as part of the F′ plasmid. The ability of the F′ episome to transfer chromosomal genes on a molecule that can replicate independently of the bacterial chromosome makes it a useful vehicle for complementation analysis.

Recall from Chapter 7 that complementation tests depend on cells that are diploid for the genes under analysis. Although bacteria are monoploid, F′ plasmids carrying bacterial genes can create specific regions of partial diploidy. For example, some F′ plasmids carry the *trp* genes that control the biosynthesis of tryptophan. An F′ plasmid carrying these genes is called an F′ *trp* plasmid.

To create partial diploids in bacterial cells using F′ plasmids, researchers must transfer the F′ into a strain

Figure 14.21 F′ episomes. (a) Rarely, when an F plasmid comes out of the bacterial chromosome, it takes some adjacent bacterial genomic DNA with it, generating an F′ episome. **(b)** Transfer of the F′ during conjugation can generate a *merodiploid.* **(c)** Merodiploids can be used for complementation analysis involving bacterial genes on the F′ episome. *trp*⁻ *x* and *trp*⁻ *y* are two independent *trp*⁻ mutations.

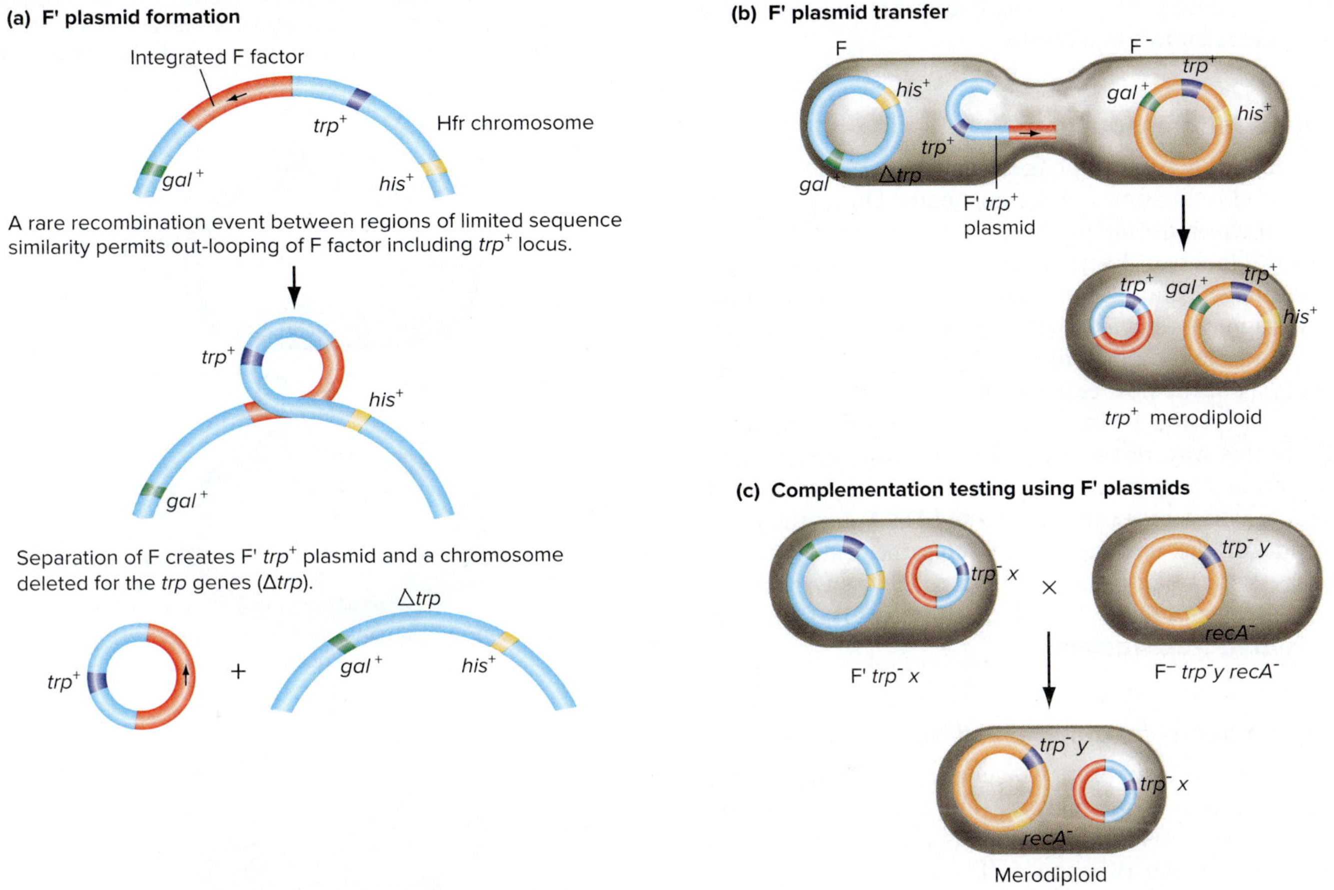

whose chromosome is not deleted for the genes carried by the F′ plasmid. This is accomplished by mating the F′-carrying cells with F⁻ cells (**Fig. 14.21b**). The exconjugants from these matings that contain two copies of some bacterial genes—one on the F′ and the other in the bacterial chromosome—are called **merodiploids.**

As an example of a complementation study using merodiploids, consider an analysis of mutations affecting tryptophan biosynthesis (**Fig. 14.21c**). All of these mutations map very close to each other. You must first construct a merodiploid by introducing an F′ plasmid carrying the entire *trp* region with a particular *trp*⁻ mutation (*trp*⁻ *x*) into a bacterial strain that carries a different *trp*⁻ mutation (*trp*⁻ *y*) in the chromosome. Growth of a particular merodiploid on minimal medium without tryptophan would indicate that the mutations complement each other and are thus in different genes. If the cells do not grow, the mutations must be in the same gene. Complementation studies using F′ *trp* merodiploids have shown that the *E. coli* genome has five different *trp* genes: *A*, *B*, *C*, *D*, and *E*, each one corresponding to one of the five enzymes required for the biosynthesis of tryptophan.

In Transduction, a Phage Transfers DNA from a Donor to a Recipient

The bacteriophages, or phages, that infect, multiply in, and kill various species of bacteria are distributed widely in nature. Most bacteria are susceptible to one or more such viruses. During infection, a virus particle may incorporate a piece of the bacterial chromosome and introduce this piece of bacterial DNA into other host cells during subsequent rounds of infection. The process by which viral particles transfer bacterial DNA from one host cell to another is known as *transduction.*

The lytic cycle of phage multiplication

When a bacteriophage injects its DNA into a bacterial cell, the phage DNA takes over the cell's protein synthesis and DNA replication machinery, forcing it to express the phage genes, produce phage proteins, and replicate the phage DNA (see Fig. 7.24). The newly produced phage proteins and DNA assemble into phage particles, after which the infected cell bursts, or *lyses,* releasing 100–200 new viral particles ready to infect other cells. The cycle resulting in cell lysis and release of progeny phage is called the **lytic cycle** of phage multiplication. The population of phage particles released from the host bacteria at the end of the lytic cycle is known as a **lysate.**

Generalized transduction

Many kinds of bacteriophages encode enzymes that destroy the chromosomes of the host cells. Digestion of the bacterial chromosome by these enzymes sometimes generates fragments of bacterial DNA about the same length as the phage genome, and these phage-length bacterial DNA fragments occasionally become incorporated into phage particles in place of the phage DNA (**Fig. 14.22**). After lysis of the host cell, the phage particles can attach to, and inject the DNA they carry into, other bacterial cells. In this way, phage transfer genes from the first bacterial strain (the donor) to a second strain (the recipient). Recombination between the injected DNA and the chromosome of the new host completes the transfer. This process, which can result in the transfer of any bacterial gene between related strains of bacteria, is known as **generalized transduction.**

Mapping genes by generalized transduction

As with cotransformation, two genes close together on the bacterial chromosome may be cotransduced. The frequency of **cotransduction** depends directly on the distance between the two genes: The closer they are, the more likely they are to appear on the same short DNA fragment and be packaged into the same transducing phage. Two genes that are farther apart than the length of DNA that can be packaged into a single phage particle can never be cotransduced. For bacteriophage P1, a phage often used for generalized transduction experiments with *E. coli,* the maximum separation allowing cotransduction is about 90 kb of DNA, which corresponds to about 2% of the bacterial chromosome.

Consider, for example, three genes—*thyA, lysA,* and *cysC*—that all map by interrupted-mating experiments to a similar region of the *E. coli* chromosome. Where do they lie in relation to one another? You can find out by using a P1 generalized transducing lysate from a wild-type strain to infect a *thyA*$^-$, *lysA*$^-$, *cysC*$^-$ strain and then selecting the transductants for either Thy$^+$ or Lys$^+$ phenotypes. After replica plating, you test each type of selected transductant for alleles of the two nonselected genes. As the phenotypic data in **Fig. 14.23a** indicate, *thyA* and *lysA* are close to each other but far from *cysC; lysA* and *cysC* are so far apart that they never appear in the same transducing phage particle; and finally, *thyA* and *cysC* are only rarely cotransduced. The order of the three genes therefore must be *lysA thyA cysC* (**Fig. 14.23b**).

Figure 14.22 Generalized transduction. The incorporation of random fragments of bacterial DNA from a donor into bacteriophage particles yields generalized transducing phages. When these phage particles infect a recipient, donor DNA is injected into the recipient cell. Recombination of donor DNA fragments with the recipient's chromosome yields transductants. An even number of crossovers is required.

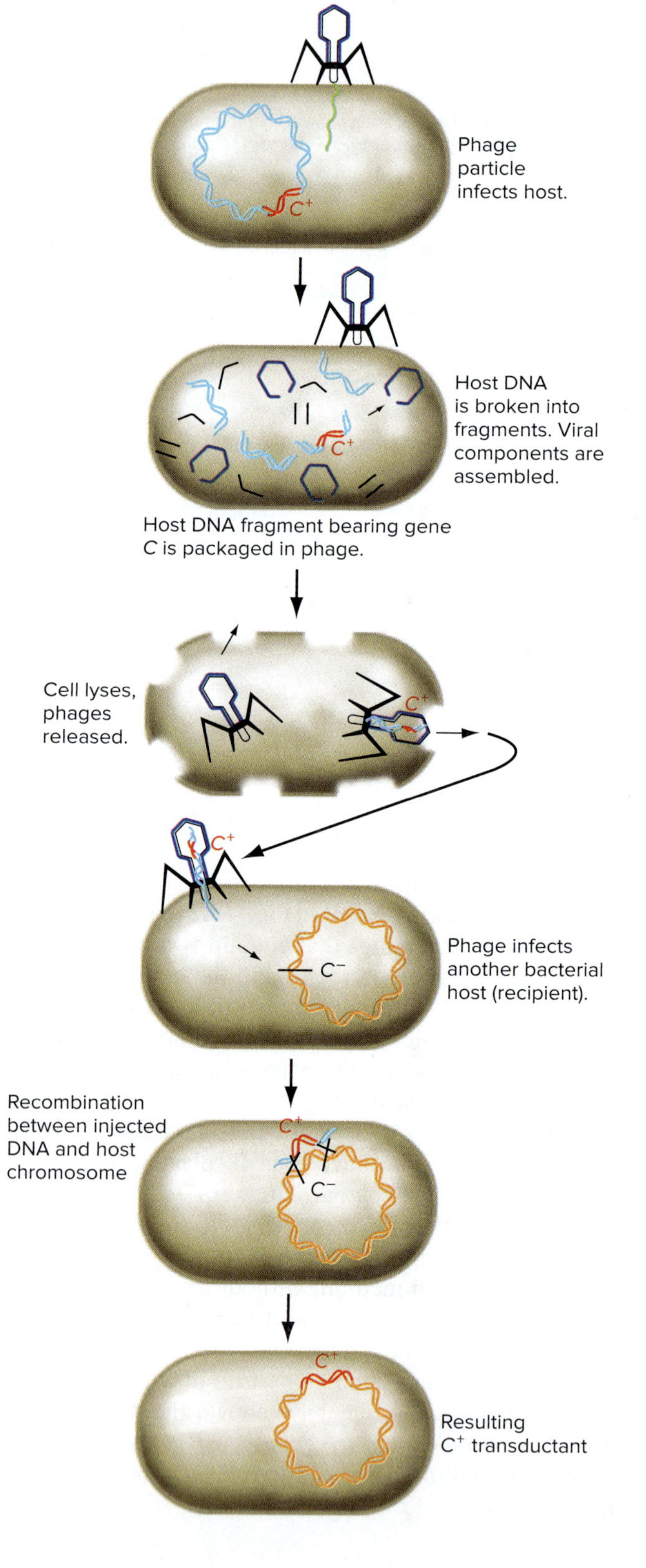

Figure 14.23 Mapping genes by cotransduction frequencies. **(a)** A P1 lysate of a *thyA*$^+$ *lysA*$^+$ *cysC*$^+$ donor infects a *thyA*$^-$ *lysA*$^-$ *cysC*$^-$ recipient. Either Thy$^+$ or Lys$^+$ cells are selected and then tested for the unselected markers. **(b)** Genetic map. The *thyA* and *cysC* genes were cotransduced at a low frequency, so they must be closer together than *lysA* and *cysC*, which were never cotransduced.

(a) Donor: *thyA*$^+$ *lysA*$^+$ *cysC*$^+$

↓ make P1 lysate; infect recipient

Recipient: *thyA*$^-$ *lysA*$^-$ *cysC*$^-$

Selected Marker	Unselected Marker
Thy$^+$	47% Lys$^+$; 2% Cys$^+$
Lys$^+$	50% Thy$^+$; 0% Cys$^+$

(b)

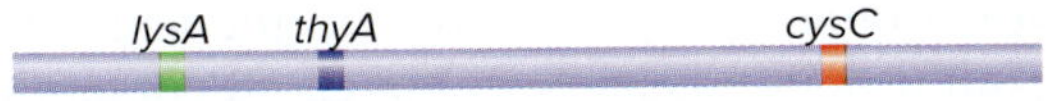

Temperate phages

The types of bacteriophages discussed so far are **virulent:** After infecting a host, they always enter the lytic cycle, multiplying rapidly and killing the cell. Other types of bacteriophages are **temperate:** Although they can enter the lytic cycle, they can also enter an alternative **lysogenic cycle,** during which their DNA integrates into the host genome and multiplies along with it, doing no harm to the host (**Fig. 14.24**). The integrated copy of the temperate bacteriophage DNA is called a **prophage,** and a bacterium containing a prophage is called a **lysogen.** Once integrated into the chromosome, the phage genome is a passive partner with the chromosomal DNA. The integrated prophage replicates along with the chromosome, but it does not produce the proteins that lead to production of more virus particles. The choice of lifestyle—lytic or lysogenic—occurs when a temperate phage injects its DNA into a bacterial cell, and it depends on environmental conditions. Normally when temperate phages inject their DNA into host cells, some of the cells undergo a lytic cycle, while others undergo a lysogenic cycle. One temperate phage commonly used in research is bacteriophage lambda (λ; **Fig. 14.25**).

Under certain conditions, it is possible to induce an integrated viral genome to excise from the chromosome, undergo replication, and form new viruses (**Fig. 14.26a**). In a small percentage of excisions, some of the bacterial genes

Figure 14.24 Lytic and lysogenic modes of reproduction. Cells infected with temperate bacteriophages enter either the lytic or lysogenic cycles. In the lytic cycle, phages reproduce by forming new bacteriophage particles that lyse the host cell and can infect new hosts. In the lysogenic cycle, the phage chromosome (*green*) becomes a prophage incorporated into the host chromosome (*orange*).

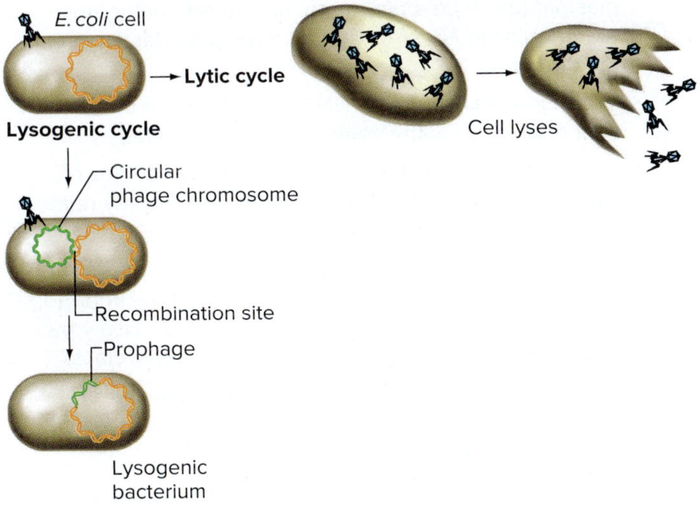

Figure 14.25 Bacteriophage lambda. Electron micrograph of two particles of a temperate phage, bacteriophage lambda (λ).

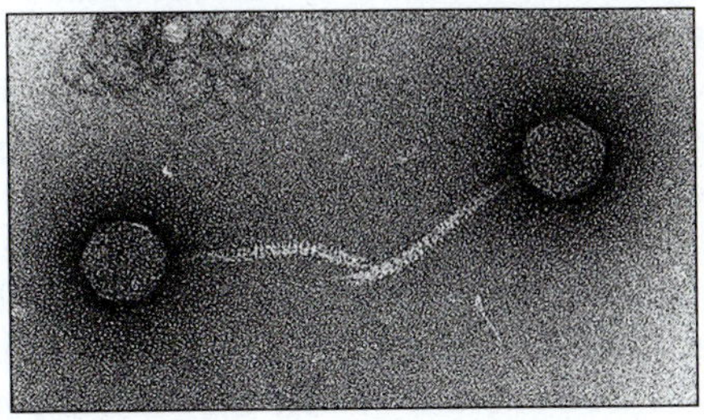

Figure 14.26 Lysogeny and excision. **(a)** Integration of phage λ DNA initiates the lysogenic cycle. Recombination between *att* sites on the phage and bacterial chromosomes allows prophage integration. **(b)** Errors in prophage excision produce specialized transducing phages. Normal excision produces circles containing only λ DNA. Illegitimate recombination between the prophage and bacterial chromosome causes inaccurate excision. The product is a DNA circle that lacks some phage genes but has acquired the adjacent *gal* genes.

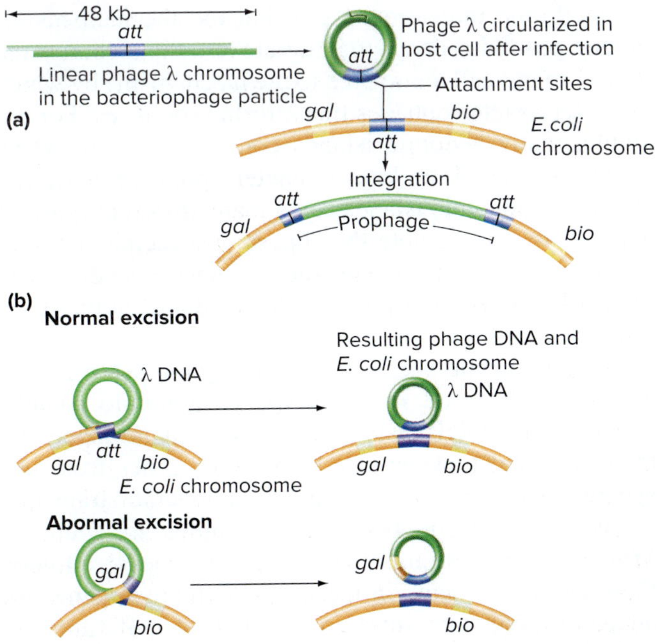

adjacent to the site where the bacteriophage integrated may be cut out along with the viral genome and be packaged as part of that genome. Viruses produced by the faulty excision of a lysogenic virus from the bacterial genome are called **specialized transducing phages (Fig. 14.26b).** During production of such phages, bacterial genes become passengers along with the viral DNA. When the specialized transducing phage then infects other cells, these few bacterial genes may be transferred into the infected cells. The phage-mediated transfer of a few bacterial genes is known as **specialized transduction.** Temperate phages are thought to be a significant vehicle for the horizontal transfer of genes from one bacterial strain to another or even from one species to another.

Horizontal Gene Transfer Has Significant Evolutionary and Medical Implications

The mechanisms of gene transfer just described (transformation, conjugation, and transduction) occur in many bacterial species. The widespread evidence of horizontal gene transfer indicates that these mechanisms are crucial for rapid adaptation of bacteria to a changing environment.

Horizontal gene transfer between different bacterial species is behind the presence in many bacterial genomes of large segments of DNA (10–200 kb in size), called **genomic islands,** whose properties suggest that they originated from transfer of foreign DNA into a bacterial cell. One such genomic island in the *E. coli* K12 map was shown earlier in Fig. 14.6. This island was apparently obtained from the genome of a *Shigella* species; note that it is not seen in the other strains of *E. coli* depicted in the figure. Some indications exist that the mechanisms allowing genomic islands to integrate into the chromosome of a recipient cell are related to the mechanisms by which temperate bacteriophages like λ form prophages. For example, many genomic islands contain genes encoding enzymes related to known bacteriophage integration enzymes. Genomic islands carry many different types of genes that can promote the fitness of a recipient bacterium in a new environment, such as genes encoding new metabolic enzymes or proteins that mediate antibiotic resistance.

In pathogenic bacteria, the pathogenic determinants are often clustered in a subtype of genomic islands, called **pathogenicity islands.** With such an arrangement, the horizontal transfer of a package of genes from one species to another can turn a nonpathogenic strain into a pathogenic strain. Important examples are found in *Vibrio cholerae* strains that cause the disease cholera. Pathogenicity islands in these strains include genes for an enterotoxin that interferes with host cell function,

Figure 14.27 Pathogenicity island. Pathogenicity islands within bacterial genomes can contain many genes involved in causing disease.

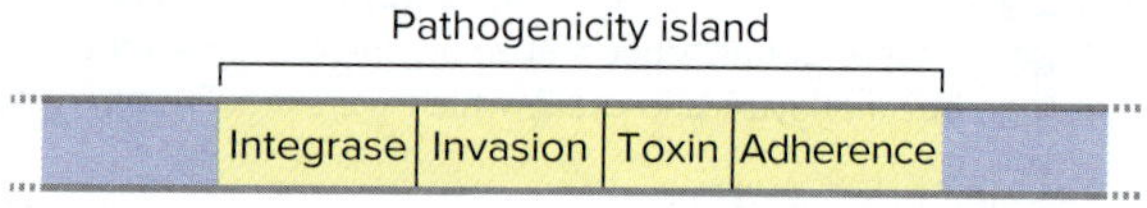

for invasion proteins that allow the bacteria to make their way through mucus of the intestinal tract, for proteins that allow the bacteria to adhere to host cells, for phage-related integrases, and many more **(Fig. 14.27).** Epidemics of cholera are caused by specific strains of *V. cholerae,* and genomic analysis of several of these disease strains reveals variation in the genes present in the pathogenicity islands, although all contain the toxin gene. The severity of an epidemic depends on the genes present in the strain.

Pathogenicity islands called **integrative and conjugative elements (ICEs)** are cause for particular concern. In addition to the characteristics of other pathogenicity islands, ICEs contain features of conjugative plasmids like the F episome. As a result, ICE elements encode the machinery needed for conjugation, including genes that mediate the connection between two cells and transfer the DNA. Conjugation initiated by ICEs is usually promiscuous, allowing ready transfer of pathogenicity island DNA between many different species.

essential concepts

- *Horizontal gene transfer* between bacteria occurs through three mechanisms: transformation, conjugation, and transduction.
- In *transformation*, donor DNA in the growth medium enters a recipient cell.
- *Conjugation* depends on direct cell-to-cell contact between a donor F^+ carrying either a conjugative plasmid (the *F plasmid*) or an integrated conjugative element (as in *Hfr strains*), and a recipient lacking such an element (F^-).
- In *transduction*, bacterial DNA packaged into the protein coat of a phage is the vehicle for gene transfer.
- For genes that are close together, the frequencies of *cotransformation* or *cotransduction* are inversely related to the distance between the genes.
- In an Hfr × F^- conjugation, genes can be mapped roughly by the time at which different alleles from the Hfr donors first appear in F^- *exconjugants*, and more precisely by counting the exconjugants of each phenotypic class.
- Bacteria evolve rapidly due to horizontal transfer of genes, including packets of genes called *pathogenicity islands*, between bacterial species.

14.5 Using Genetics to Study Bacterial Life

learning objectives

1. Explain how to identify mutant genes molecularly by transformation with recombinant plasmids.
2. Discuss the use of transposons as mutagens in bacteria.
3. Describe how to generate specific mutations in any *E. coli* gene by gene targeting.

One of the primary goals of bacterial genetics is the molecular identification of genes whose products have important functions for bacterial life. In this way, researchers can study various aspects of bacterial metabolism such as the biosynthesis of amino acids or nucleotides, the resistance or sensitivity of bacteria to agents such as antibiotics or bacteriophages, the pathogenesis caused by certain bacteria, or bacterial behavior.

Scientists can connect genotype and phenotype in either of two ways. First, they can find a mutation that affects the property of interest and then identify the gene affected by the mutation. Alternatively, researchers can start with a known gene suspected of involvement with the process, then make a mutation in the gene, and finally ask whether the mutation causes an aberrant phenotype related to the process being studied. We describe in this section several efficient techniques that geneticists are now using to identify important bacterial genes.

Recombinant Plasmid Libraries Simplify Gene Identification

Bacterial genomic DNA libraries serve as generally useful resources for gene identification. As an example, suppose a scientist has identified a mutagen-induced arginine auxotroph (*arg*$^-$) and wants to identify the mutant gene. The *arg*$^-$ bacteria can be transformed with a genomic library in which fragments of a wild-type *E. coli* strain were cloned into a plasmid vector marked by an ampicillin resistance gene. Investigators would look for clones producing transformants that are ampicillin resistant and that can grow without an arginine supplement (**Fig. 14.28**). Such a colony contains a plasmid from the library that "rescues" the Arg$^-$ mutant phenotype to Arg$^+$, and therefore should contain a wild-type copy of the mutant gene. This procedure would lead to rapid identification of the gene of interest because its sequence would be found in all clones that rescue the arginine auxotrophy.

To verify the gene identification, the corresponding gene in the *arg*$^-$ *E. coli* chromosome can be amplified by PCR and its sequence analyzed. If the correct gene has been identified, the copy in the *arg*$^-$ genome should have an inactivating mutation.

Figure 14.28 Identifying mutant genes by plasmid library transformation. In this example, mutant auxotrophic bacterial cells (*arg*$^-$) are transformed with a recombinant library made from wild-type *E. coli* genomic DNA. The plasmid purified from a colony that grows on minimal medium without an arginine supplement contains the *arg*$^+$ allele of the gene that is mutant in the auxotrophs.

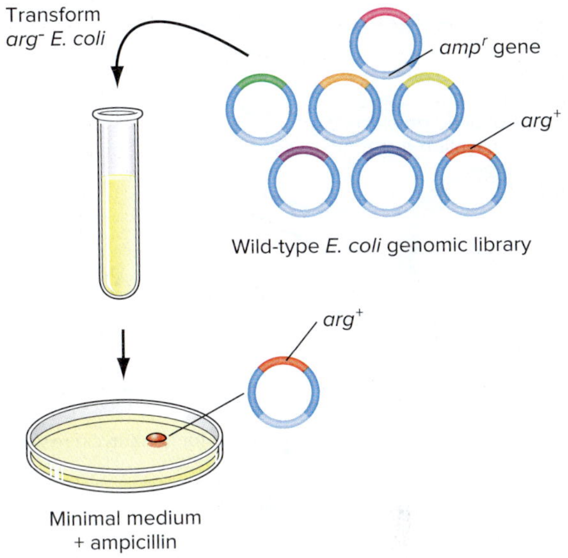

Transposons Can Be Used as Gene-Tagging Mutagens

As you saw in Chapter 13, transposable elements can cause mutations when they move and land in genes. The advantage of transposons over other mutagens is that the transposon serves as a molecular tag to help researchers identify the mutant gene rapidly.

Geneticists have cleverly engineered a DNA transposon from fruit flies, called *Mariner,* to serve as a gene-tagging mutagen in bacteria such as *E. coli.* The bacteria are transformed with a plasmid containing two genes: a kanamycin resistance (*kan*^{r}) gene flanked by *Mariner* element inverted repeats, and also a gene for *Mariner* transposase, which recognizes the inverted repeats to catalyze movement of the engineered transposon containing *kan*^{r}. The plasmid has no origin of replication, and so for the *kan*^{r} gene to be retained by cells during cell division, it must transpose from the plasmid to the *E. coli* chromosome. Cells in which transposition has occurred are selected by spreading the transformed bacteria on petri plates containing kanamycin; each kanamycin-resistant (Kan^r) colony contains a transposon at a different location in the *E. coli* chromosome. A researcher can screen the resulting colonies for a mutant phenotype of interest.

Figure 14.29 Transposons as mutagens. A genetically engineered fly *Mariner* transposon carrying a gene for kanamycin resistance (*green*) can hop from a plasmid containing a transposase gene into the *E. coli* chromosome. Growth on medium containing kanamycin selects for cells whose genomes contain a randomly integrated Mariner element. The *black arrows* are inverted repeats recognized by the *Mariner* transposase.

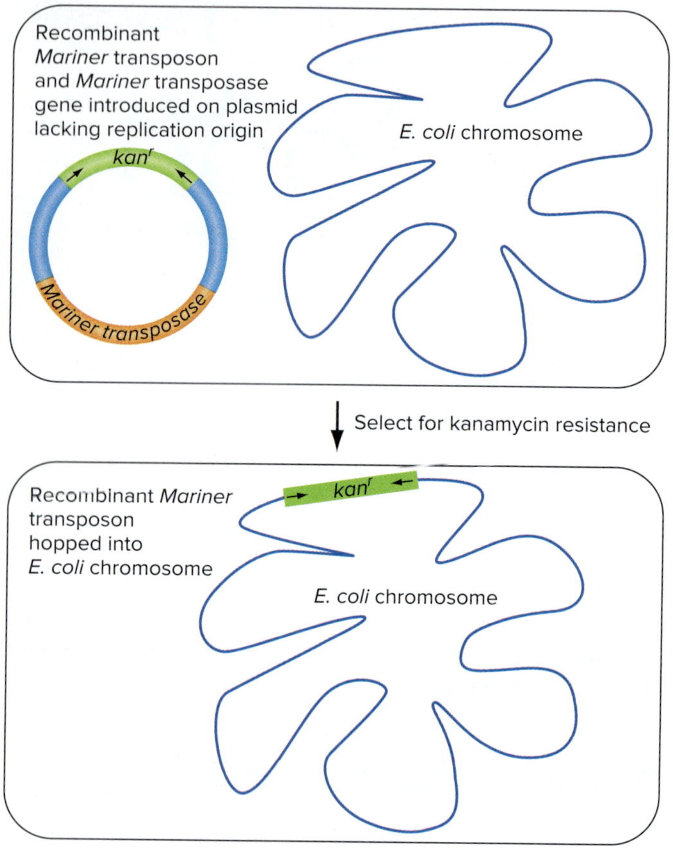

Figure 14.30 Inverse PCR identifies genes with transposon insertions. DNA from a bacterial genome with a transposon insertion (*green*) into a gene *X* (*purple*) is cut with a restriction enzyme (RE) that recognizes a site in the transposon. The fragments produced are circularized by DNA ligase; one circle contains some transposon DNA along with adjacent genomic DNA up to the next RE site. A pair of PCR primers within the transposon (*purple arrows*) each amplify one strand of the circle (*squiggly lines*) containing part of gene *X*. Sequencing of the resultant PCR product reveals the gene into which the transposon has jumped.

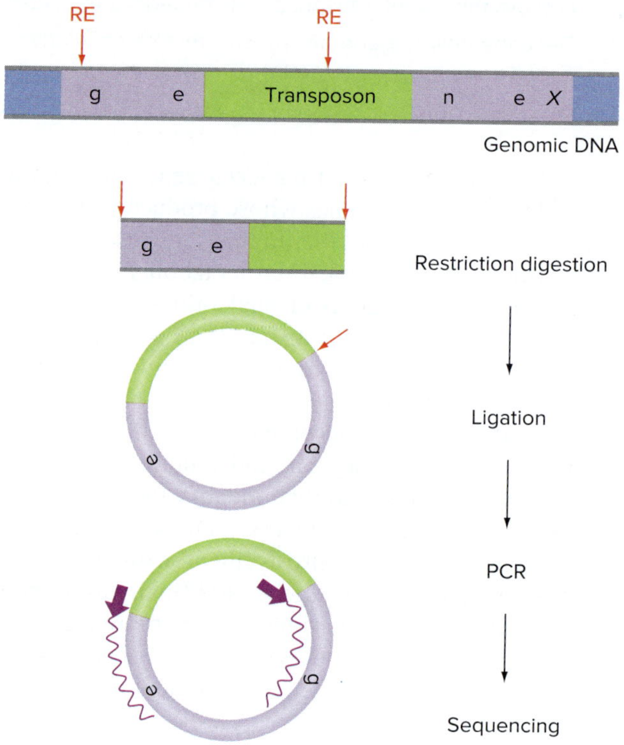

The use of the eukaryotic *Mariner* transposon has two advantages over bacterial transposons. First, because the source of *Mariner* transposase was lost when the plasmid failed to replicate, the transposon cannot mobilize again, and the mutation it causes will remain stable. Second, because each Kan^r bacterial colony selected will contain only the single transposon that moved, and because the *E. coli* genome has no DNA sequences related to *Mariner,* it is easy to find the gene that was disrupted by the transposon. **Figure 14.30** illustrates one method called *inverse PCR* that can readily identify the genomic DNA sequences adjacent to the tagging transposon.

Gene Targeting Provides a Way to Mutagenize Specific Genes

The analysis of bacterial genome sequences has led to the identification of many genes whose functions are not yet known. One approach to determining the functions of such genes is to make null mutations in them using recombinant DNA techniques and homologous recombination. This approach, shown in **Fig. 14.31,** is known as **gene targeting.**

To make a null mutation of gene *X*, researchers introduce into bacterial cells a linear DNA fragment constructed *in vitro* in which 50 or more bps of the 5′ and 3′ ends of gene *X* flank a drug resistance gene (Fig. 14.30). The drug resistance gene will be retained in dividing bacterial cells only if it is incorporated into the bacterial chromosome by homologous recombination at both ends of the fragment. These homologous recombination events will replace the wild-type gene *X* with the drug resistance marker, generating a null mutation (Fig. 14.31). Adding antibiotic to the medium selects for colonies in which the integration occurred. These cells can then be analyzed to reveal a mutant phenotype.

Figure 14.31 Gene targeting. Linear DNA fragments (generated using PCR) introduced into *E. coli* undergo recombination with bacterial chromosome sequences homologous to the fragment's free ends. Here, incorporation of the DNA fragment replaces most of gene *X* with an ampicillin resistance gene, generating gene *X* null mutants. Cells that have undergone gene *X* replacement can be selected by growing on medium containing ampicillin.

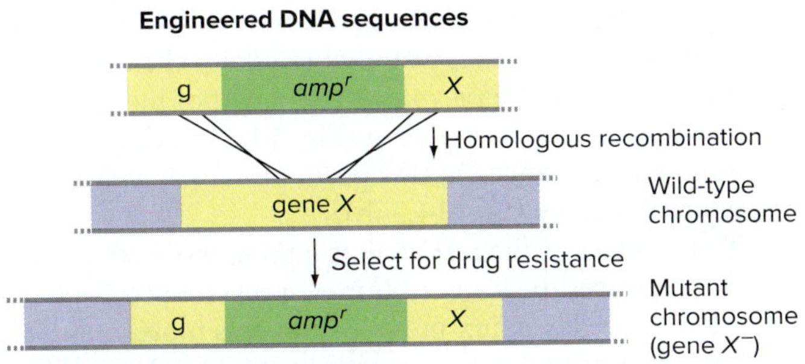

essential concepts

- To identify mutant genes, a wild-type bacterial genomic library in plasmids is used to transform the mutant bacterial strain. A transformed bacterium in which the mutant phenotype is *rescued* to wild type likely harbors a plasmid containing a wild-type copy of the corresponding gene.
- Transposons are useful as mutagens because they act as molecular tags for genomic DNA sequences that can be identified rapidly by inverse PCR.
- In *gene targeting,* homologous recombination between the bacterial chromosome and a linear DNA construct synthesized *in vitro* can generate a null mutation in any gene.

14.6 A Comprehensive Example: How *N. gonorrhoeae* Became Resistant to Penicillin

learning objectives

1. Explain how penicillin kills bacteria.
2. Describe mechanisms of penicillin resistance and how *N. gonorrhoeae* has become resistant.
3. Discuss potential solutions to the worldwide problem of drug-resistant pathogens.

As discussed at the beginning of this chapter, the sexually transmitted bacterium *Neisseria gonorrhoeae* has become resistant to many antibiotics, including penicillin. Gonorrhea is one of the most prevalent sexually transmitted bacterial infections worldwide, and it is in danger of becoming untreatable with currently available drugs. Here we will explore how antibiotics kill bacteria, focusing on penicillin's action on *N. gonorrhoeae.* We will then examine how bacteria develop resistance to drugs, using penicillin resistance in *N. gonorrhoeae* as an example. Increased understanding of the mechanisms of drug resistance will be needed to help avert the impending crisis of multidrug-resistant bacteria.

Penicillin Interferes with Synthesis of the Bacterial Cell Wall

Surrounding their permeable cell membrane, bacteria have a less permeable cell wall (review Fig. 14.3). Because the bacterial cytoplasm contains many solutes, without the cell wall, bacteria would take in so much water through osmosis that they would burst. A large component of the cell wall is *peptidoglycan,* a substance made of two sugar molecules: N-acetylglucosamine (NAG) and N-acetylmuramic acid (NAM). Bacteria synthesize long chains of alternating NAG and NAM molecules that are cross-linked by short peptides attached to the NAMs by an enzyme called *transpeptidase* (**Fig. 14.32**). Penicillin prevents cross-linking by binding transpeptidase and inhibiting its enzymatic activity; for this reason, transpeptidase is also known as *penicillin-binding protein (PBP).*

As bacterial cells grow and divide, the cell wall is remodeled continually. When penicillin is present, the bacteria cannot rebuild their cell walls after cell division, so the cells die. Because human cells do not have peptidoglycan, the lethal effect of penicillin is specific to bacteria.

N. gonorrhoeae Become Resistant to Penicillin Through Multiple Mechanisms

Bacteria that are resistant to the lethal effects of antibiotics sometimes have the ability to inactivate the drug molecules directly. Alternatively, the physiology of the bacteria can be altered so as to block access of the drug to its target in the cell. Bacteria gain these capabilities for drug resistance

Figure 14.32 Peptidoglycan. A major component of the bacterial cell wall is peptidoglycan, in which chains of sugars are cross-linked with peptides. NAG (*green*) is N-acetylglucosamine, and NAM (*purple*) is N-acetylmuramic acid.

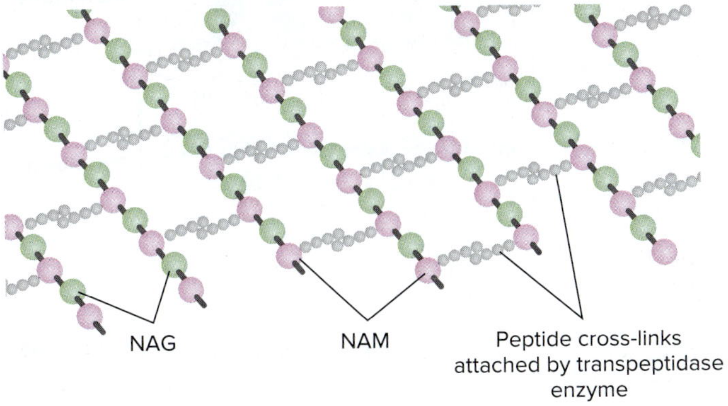

Figure 14.33 Penicillinase action. The *pen*^{r} gene encodes penicillinase, an enzyme that cleaves the β-lactam ring of penicillin and thus inactivates the drug.

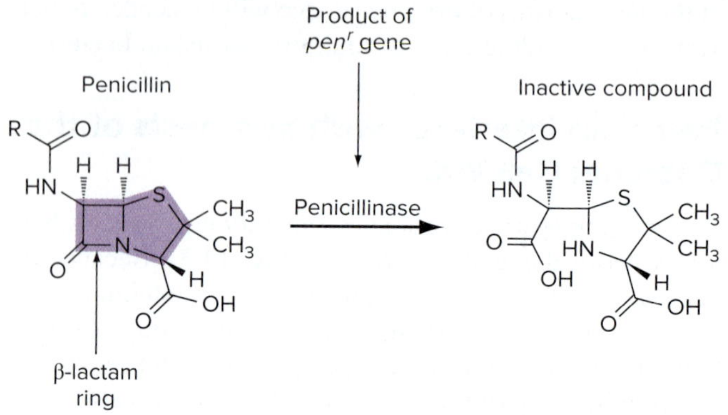

through spontaneous mutation of chromosomal genes or through the transfer of genes from other bacteria by transformation, conjugation, or transduction.

At the beginning of the chapter, you saw that one way *N. gonorrhoeae* became resistant to penicillin was by acquiring a plasmid from *H. influenzae* that carries a penicillin resistance gene. The penicillin resistance gene (*pen*^{r}) encodes an enzyme called *penicillinase,* which cleaves the β-lactam ring of the penicillin molecule, rendering the drug inactive (**Fig. 14.33**).

Another way that *N. gonorrhoeae* has become penicillin resistant is through mutation of several different chromosomal genes, including *penA, penB,* and *mtr;* the more of these gene mutations the bacterial strain has, the more resistant to penicillin it becomes.

- PBP (transpeptidase), the main target of penicillin (**Fig. 14.34a**), is encoded by the *penA* gene. A *penA* missense mutation decreases PBP's affinity for penicillin (**Fig. 14.34b**).
- The *penB* gene encodes a *porin,* a protein in the outer membrane of the cell wall that regulates the entry of molecules into the *periplasm*—the part of the cell wall that includes peptidoglycan (Fig. 14.34a). Specific amino acid changes in this porin protein decrease penicillin entry (Fig. 14.34b).
- The product of the *mtr* gene is a protein called MtrR that represses the transcription of genes for polypeptides making up the efflux pump that expels molecules like penicillin from the periplasm (Fig. 14.34a). Loss of *mtr* activity by mutation results in an increase in the number of efflux pumps and thus less penicillin inside the bacterial cell (Fig. 14.34b).

What Should We Do About the Problem of Drug Resistance?

The use of antibiotics results in selection for drug-resistant pathogenic strains, eventually decreasing or even eliminating the effectiveness of drugs that have saved countless lives. One important way to slow down this process is to decrease the selection pressure on the pathogens by reducing

Figure 14.34 *N. gonorrhoeae* gene mutations that cause penicillin resistance. (a) In gram-negative bacteria like *N. gonorrhoeae,* the cell wall consists of an outer membrane and a periplasmic space containing peptidoglycans. Porin proteins allow the influx of penicillin (*red circles*) into the periplasmic space, while efflux pump proteins pump penicillin out. Penicillin-binding proteins (PBPs) catalyze the formation of cross-links between peptidoglycans. Penicillin inhibits the enzymatic activity of PBPs. **(b)** How three gene mutations contribute to penicillin resistance.

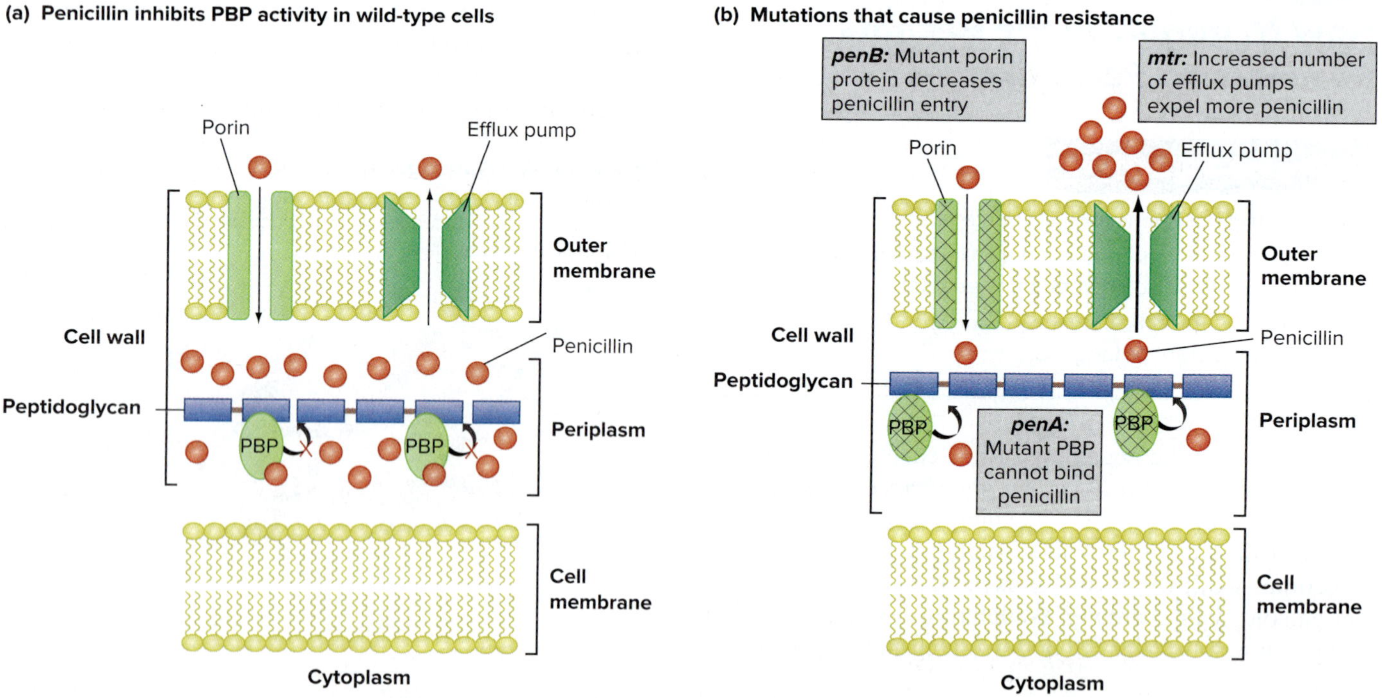

antibiotic use. However, this strategy alone cannot be a long-term solution because many infections will ultimately require antibiotic treatment.

Unquestionably, scientists and pharmaceutical companies need to develop new antimicrobials and new ways to make the ones we have more effective. Novel classes of antibiotics that target many different kinds of molecules in bacterial cells must be discovered. Metagenomic analysis of microorganism populations isolated from many different environments presents a new and exciting means of prospecting for new types of antibiotics.

Other imaginative approaches for dealing with the problem of drug resistance are currently under development. One intriguing idea is to develop drugs that self-destruct over time, so that less antibiotic accumulates in the environment. A different avenue of research is exploring chemicals that either block or circumvent bacterial resistance mechanisms. For example, scientists are trying to find agents that inhibit efflux pump activity, which would increase susceptibility to a broad range of antibiotics. And in Chapter 16, you will see that the results of research into a phenomenon called *quorum sensing,* a mechanism bacteria use to communicate with one another, has suggested to scientists new ideas for creating antibiotics.

Finding novel antibiotics is expensive and difficult, and as a result, drug companies often direct their research dollars elsewhere. Public funding of antibiotic research may be one answer to solving the growing problem of drug resistance.

essential concepts

- Some resistant *N. gonorrhoeae* strains have acquired a plasmid carrying the gene for penicillinase, an enzyme that destroys penicillin; other resistant strains have accumulated mutations that prevent the build-up of penicillin in the cell.
- Reduced use of antibiotics can slow the generation of drug-resistant strains. Scientists are also developing new antibiotics with different chemical structures or that target different molecules within the pathogen.

WHAT'S NEXT

The study of bacterial genetics underscores the unity of genetic phenomena in all types of living organisms. Double-stranded DNA serves as the genetic material in bacteria as it does in the nuclear genome of eukaryotes. However, we have also seen a remarkable diversity of mechanistic detail in biological processes. Although bacteria do not produce gametes that fuse to become zygotes, they can exchange genes between different strains through transformation, conjugation, and transduction. These three modes of gene transfer increase the potential for the evolution of prokaryotic genetic material. Indeed, the pangenome of some bacterial species may be larger than the human genome.

We learned in this chapter of the intimate relationship between humans and bacteria. On a cell number basis, we humans are equally bacterial and eukaryotic. Remarkable as this fact is, it actually understates the extent of this relationship. Not only do we carry around pounds of bacteria in our intestines and elsewhere, but each of our eukaryotic cells houses vital organelles that have prokaryotic origins.

Biologists think that *mitochondria,* the cellular organelles that produce energy for metabolic processes, and *chloroplasts,* the photosynthetic organelles of plant cells, are descendants of bacteria that fused with the earliest nucleated cells. Mitochondria are similar in size and shape to today's aerobic bacteria and have their own DNA, which replicates independently of the cell's nuclear genetic material. Chloroplasts are similar in shape and size to certain cyanobacteria, and they too have self-replicating DNA that carries bacteria-like genes. Based in part on these observations, the **endosymbiont theory** proposes that chloroplasts and mitochondria originated when free-living bacteria were engulfed by primitive nucleated cells. Chapter 15 examines the phenomenon of *organellar inheritance*—how genes outside of an organism's nuclear genome can affect phenotype and can be inherited in a non-Mendelian manner.

SOLVED PROBLEMS

I. Using bacteriophage P22, you performed a three-factor cross in *Salmonella typhimurium.* The cross was between an Arg^- Leu^- His^- recipient bacterium and bacteriophage P22, which was grown on an Arg^+ Leu^+ and His^+ strain. You selected for 1000 Arg^+ transductants and tested them on several selective media by replica plating. You obtained the following results:

Arg^+ Leu^- His^-	585
Arg^+ Leu^- His^+	300
Arg^+ Leu^+ His^+	114
Arg^+ Leu^+ His^-	1

a. What are the cotransduction frequencies of *arg* with either *his* or *leu?*

b. Is *arg* closer to *his* than it is to *leu,* or is *arg* closer to *leu* than it is to *his?* What are the relative map distances?

c. Why can you not obtain an accurate cotransduction frequency for *his* and *leu* from the data provided?

d. What is the order of the three markers?

e. Draw the crossovers that would produce the class with only one transductant.

f. Estimate the proportion of the 4.9 Mb *S. typhimurium* genome that contains these three genes. Assume that bacteriophage P22 can package the same amount of DNA as the *E. coli* virus P1.

Answer

a. Cotransduction frequency is the percentage of cells that received two markers. For *arg* and *his,* the cotransductants are the Arg^+ Leu^- His^+ cells (300) and Arg^+ Leu^+ His^+ cells (114). The cotransduction frequency of *arg* and *his* is 414/1000 = 41.4%. The cotransduction frequency of *arg* and *leu* is 114 + 1 or 115/1000 = 11.5%.

b. Because the cotransduction frequency of *arg* and *his* is larger than that of *arg* and *leu, arg* is closer to *his* than it is to *leu.* The map distance between *arg* and *leu* is about 41.4/11.5 = 3.6 times longer than that between *arg* and *his.*

c. All the transductants were selected for being Arg^+, so you could not detect the Arg^- Leu^+ His^+ transductants that must have been produced in the experiment.

d. and e. The answer to part (b) above is compatible with two possible gene orders: either *arg-his-leu* or *his-arg-leu*. To discriminate between these possibilities, you need to consider the crossovers between the linear fragments from the donor (which were transduced into the recipient by the bacteriophage) and the circular recipient chromosome that could form stable transductants of each class. A single crossover (or any odd number of crossovers) would yield a large linear chromosome that could not be replicated in bacteria, so this would be lethal (no colonies would form). On the other hand, two crossovers, or any even number of crossovers, would successfully replace part of the recipient chromosome with DNA from the donor.

Note that the smallest class of transductants by far is Arg^+ Leu^+, which implies that recovery of this class requires four rather than two crossovers. Thus, the order of the genes is *arg-his-leu.* The following figure shows the four crossovers involved. You should make diagrams for yourself to show that all three of the other classes of transductants can be obtained with only two crossovers given this gene order, explaining why these classes would be larger.

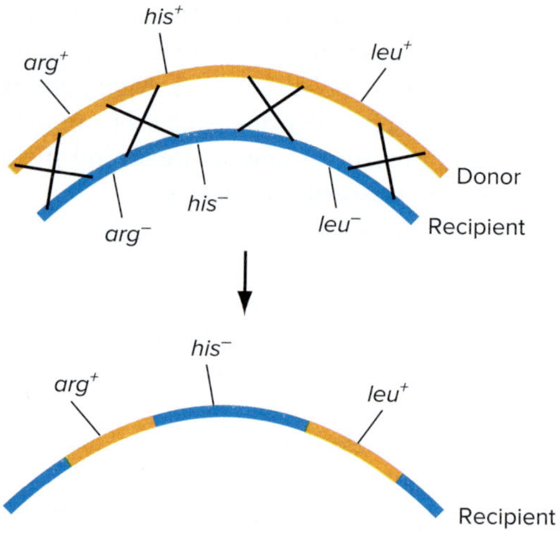

f. Bacteriophage P22 can package genomic DNA fragments that are roughly 90 kb long. The genes *arg, his,* and *leu* can be contransduced, so the distance between the outside genes (*arg* and *leu*) must be less than 90 kb. This region represents less than 90/4900 × 100 = 1.8% of the *S. typhimurium* genome; these genes are very close together on the genome map.

II. While perusing the *E. coli* K12 genome sequence, you come across a gene with no known function. The amino acid sequence of the gene's protein product shows weak similarities with known *porins,* proteins that cross a cellular membrane to let molecules such as amino acid or sugar nutrients (or drugs like penicillin) pass through. Some porins are nonspecific and let any solute up to a certain size transit into the cell. Other porins are specific and allow the transit of certain sugars but not others. What genetic experiments could you do to try to determine whether this new gene has a specific function in allowing bacterial cells to scavenge the sugar maltose from the environment? Describe scenarios that might complicate your experimental approach.

Answer

You could attempt to generate a null mutation in the gene by gene targeting (see Fig. 14.31). Using recombinant DNA technology, you would create a DNA construct in which a drug resistance gene is flanked on one side by sequences at the 5′ end of the gene in question, and on the other side by sequences from the 3′ end of the gene. You would introduce these linear DNA fragments into wild-type *E. coli* and select for colonies containing the drug resistance gene by growing cells on agar media containing the drug in petri plates. Colonies that appear on the plates have a null mutation in the gene; they descend from a bacterium that incorporated, by homologous recombination, the drug resistance gene in place of the gene of interest.

In order to learn about the function of the gene, you would use replica plating to examine the phenotype of these colonies. In particular, if your hypothesis is that the gene encodes a porin that allows maltose to enter the cell, your prediction would be that in contrast with wild type, these mutant cells would grow very poorly and make small or no colonies on petri plates in which the only sugar nutrient was maltose. These same mutant cells would make normal colonies on media containing other kinds of sugars that could be used to supply cells with energy. A potential complication is that if the gene is vital for the bacteria to survive or replicate under any conditions, you will not recover any drug-resistant colonies because bacteria with null mutations in the gene will not live.

PROBLEMS

Vocabulary

1. Choose the phrase from the right column that best fits the term in the left column.

a. transformation	1. requires supplements in medium for growth
b. conjugation	2. a method for mutagenizing genes in bacterial genomes
c. transduction	3. small circular DNA molecule that can integrate into the chromosome
d. lytic cycle	4. the core genes that define a bacterial species plus all of the genes unique to individual strains
e. lysogeny	5. transfer of DNA requiring direct physical contact
f. episome	6. integration of phage DNA into the chromosome
g. auxotroph	7. infection by phages in which lysis of cells releases new virus particles
h. pangenome	8. transfer of naked DNA
i. gene targeting	9. transfer of DNA between bacteria via virus particles

Section 14.1

2. The unicellular, rod-shaped bacterium *E. coli* is ~2 μm long and 0.8 μm wide, and has a genome consisting of a single 4.6 Mb circular DNA molecule. The unicellular archaean *Methanosarcina acetivorans* is spherical (coccus-shaped) with a diameter of 3 μm and has a 5.7 Mb circular genome. The unicellular eukaryote *Saccharomyces cerevisiae* is roughly spherical, with a diameter of 5–10 μm. It has a haploid genome of 12 Mb divided among 16 linear chromosomes. Given these descriptions, how could you determine whether a new, uncharacterized microorganism was a bacterium, an archaean, or a eukaryote?

Section 14.2

3. Now that the sequence of the entire *E. coli* K12 strain genome (roughly 5 Mb) is known, you can determine exactly where a cloned fragment of DNA came from in the genome by sequencing a few bases and matching that data with genomic information.
   a. About how many nucleotides of sequence information would you need to determine exactly where a fragment is from?
   b. If you had purified a protein from *E. coli* cells, roughly how many amino acids of that protein would you need to know to establish which gene encoded the protein?
   c. You determine 100 nucleotides of sequence of genomic DNA from a different *E. coli* strain, but you cannot find a match in the *E. coli* K12 genome sequence. How is this possible?

4. Bacterial genomes such as that of *E. coli* typically have only a single origin of replication, from which replication proceeds bidirectionally. Pol III, the DNA polymerase responsible for replicating the *E. coli* chromosome, synthesizes DNA at a rate of about 1000 nucleotides per second.
   a. From this information, estimate the minimum generation time of *E. coli*.
   b. Under optimal conditions, *E. coli* have been observed to divide in as little as 17 minutes. Speculate how this might be possible, given your answer to part (a).

5. List at least three features of eukaryotic genomes that are not found in bacterial genomes.

6. Describe a mechanism by which a gene could move from the bacterial genome to a plasmid in the same cell, or *vice versa*.

7. High salt concentrations tend to cause protein aggregation. Suggest a way to identify proteins normally

expressed in particular bacterial species that can retain their solubility despite high salt conditions.

8. Recently, scientists tested the possibility that human gut bacteria may play a role in determining body weight. The study subjects were four sets of twins (one set of identical twins and three sets of fraternal twins), where one twin was of normal weight and the other was obese. Samples of their gut bacteria were collected and transplanted into bacteria-free mice. Mice with the different bacterial transplants were all fed the same diet and monitored over the course of about one month. For each of the four twin pairs, the mice with the bacteria from the obese twin gained significantly more weight and fat than the mice transplanted with the bacteria from the normal twin.
   a. What would you conclude about the relationship between the human gut microbiome and body weight?
   b. Why were twins used in the study?
   c. Do the results of this study mean that human genes (genes in the nuclei of human cells) do not play a role in body weight and fat content? Explain.
   d. Mice are *coprophagic,* meaning that they eat feces. How could you test whether a certain bacterial species associated with leanness or obesity could successfully invade the gut microbiome of an animal in which that bacterial species was not previously found?
   e. One problem with using bacteria-free mice in experiments such as this is that the mouse gut is not equivalent to the human gut as a bacterial host: Different bacterial species thrive in mice and humans. Explain how this fact could affect the experiment discussed in this problem.

9. A recent metagenomic study analyzed the microorganisms present on surfaces within the entire subway system of New York City. The researchers found hundreds of bacterial species in the subway, most of them nonpathogenic. Interestingly, almost half of all DNA found in the subway matches no known organism.
   a. The scientists found that different subway stations had characteristic microbiomes. How might this observation be useful to the police?
   b. Because the majority of the subway DNA that could be identified was bacterial, the researchers presume that most of the DNA fragments that could not be matched to a known organism are bacterial. Why do you think that so many bacterial species are unknown to us? What feature of these unknown bacteria might prevent us from studying them?

Section 14.3

10. Linezolid is a new type of antibiotic that inhibits protein synthesis in several bacterial species by binding to the 50S subunit of the ribosome and inhibiting its ability to participate in the formation of translational initiation complexes. Physicians are particularly interested in this antibiotic for treating pneumonia caused by penicillin-resistant *Streptococcus pneumoniae* (also called *pneumococci*). To explore the mechanisms by which pneumococci can develop resistance to linezolid, you first want to identify linezolid-resistant strains. Next, using one of these strains as starting material, you want to identify derivatives of these mutants that are no longer tolerant of linezolid.
    a. Outline the techniques you would use to identify linezolid-resistant mutant pneumococci and linezolid-sensitive derivatives of these mutants. In each case, would your techniques involve direct selection, screening, replica plating, treating with mutagens, or testing for a visible phenotype?
    b. Suggest possible mutations that could be responsible for the two kinds of phenotypes you will identify. What types of events in the bacterial cells would be altered by the mutations? Do you think that these mutations would be loss-of-function or gain-of-function? Explain.

11. A liquid culture of *E. coli* at a concentration of 2×10^8 cells/ml was diluted serially, as shown in the following diagram, and 0.1 ml of cells from the final two test tubes were spread on agar plates containing rich medium. How many colonies do you expect to grow on each of the two plates?

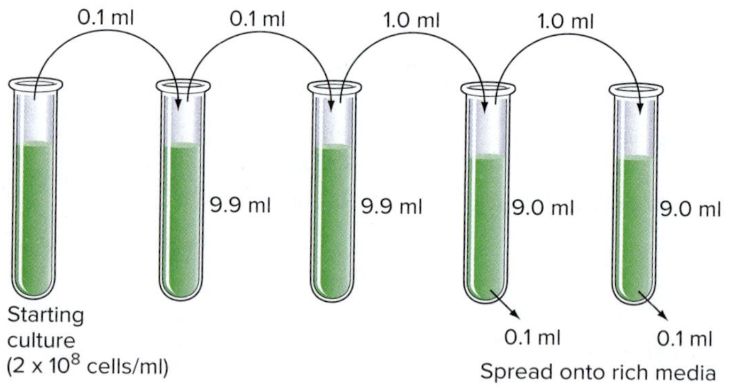

12. Pick out the medium (i, ii, iii, or iv) onto which you would spread cells from a Lac^- Met^- *E. coli* culture to:
    a. select for Lac^+ cells
    b. screen for Lac^+ cells
    c. select for Met^+ cells
       i. minimal medium + glucose + methionine
       ii. minimal medium + glucose (no methionine)
       iii. rich medium + X-gal
       iv. minimal medium + lactose + methionine

Section 14.4

13. This problem concerns Fig. 14.14, which illustrates the experiment performed by Lederberg and Tatum that first indicated the existence of bacterial conjugation.
    a. Strain A had mutations in two genes, while strain B had three mutations. The reason is that Lederberg and Tatum wanted to ensure that the phenomenon they were examining did not involve reversion of mutations. Explain the logic behind this aspect of their experimental design, assuming that the rate of reversion of a single gene is one in 10 million (1 in 10^7) cells. How did these investigators know that the cells they found after mixing the two cultures were indeed not due to reversion?
    b. The experiment shown in Fig. 14.14 did not inform the investigators which strain was the donor and which was the recipient. Describe a way in which they could modify this experiment to answer this question.
14. In two isolates (one is resistant to ampicillin and the other is sensitive to ampicillin) of a new bacterium, you found that genes encoding ampicillin resistance are being transferred into the sensitive strain.
    a. How would you know that gene transfer is taking place?
    b. To determine if the gene transfer is transformation or transduction, you treat the mixed culture of cells with DNase. Why would this treatment distinguish between these two modes of gene transfer? Describe the results predicted if the gene transfer is transformation versus transduction.
    c. To determine if the gene transfer involves transformation, conjugation, or transduction, you separate the ampicillin-resistant and ampicillin-sensitive strains by a membrane with pores that are smaller than the size of a bacterium, but larger than the sizes of bacteriophage or DNA fragments. If gene transfer is still observed, what mechanisms are possibly involved and which are excluded?
15. *E. coli* cells usually have only one copy of the F plasmid per cell. You have isolated a cell in which a mutation increases the copy number of F to three to four per cell. How could you distinguish between the possibility that the copy number change was due to a mutation in the F plasmid versus a mutation in a chromosomal gene?
16. In *E. coli,* the genes *purC* and *pyrB* are located halfway around the chromosome from each other. These genes are never cotransformed. Why not?
17. DNA sequencing of the entire *H. influenzae* genome was completed in 1995. When DNA from the nonpathogenic strain *H. influenzae Rd* was compared to that of the pathogenic *b* strain, eight genes of the fimbrial gene cluster (located between the *purE* and *pepN* genes) involved in adhesion of bacteria to host cells were completely missing from the nonpathogenic strain. What effect would this deletion have on cotransformation of *purE* and *pepN* genes using DNA isolated from the nonpathogenic versus the pathogenic strain?
18. Genes encoding toxins are often located on plasmids. A recent outbreak has just occurred in which a bacterium that is usually nonpathogenic is producing a toxin. Plasmid DNA can be isolated from this newly pathogenic bacterial strain and separated from the chromosomal DNA. To establish whether the plasmid DNA contains a gene encoding the toxin, you could determine the sequence of the entire plasmid and search for a sequence that looks like other toxin genes previously identified. An easier way exists to determine whether the plasmid DNA carries the gene(s) for the toxin; this strategy does not involve DNA sequence analysis. Describe this easier method.
19. a. You want to perform an interrupted-mating mapping with an *E. coli* Hfr strain that is Pyr^+, Met^+, Xyl^+, Tyr^+, Arg^+, His^+, Mal^+, and Str^s. Describe an appropriate bacterial strain to be used as the other partner in this mating.
    b. In an Hfr × F^- cross, the *pyrE* gene enters the recipient in 5 minutes, but at this time point there are no exconjugants that are Met^+, Xyl^+, Tyr^+, Arg^+, His^+, or Mal^+. The mating is now allowed to proceed for 30 minutes and Pyr^+ exconjugants are selected. Of the Pyr^+ cells, 32% are Met^+, 94% are Xyl^+, 7% are Tyr^+, 59% are Arg^+, 0% are His^+, and 71% are Mal^+. What can you conclude about the order of the genes?

Problems 20–23 require you to diagram recombination events that can replace specific genes on the chromosome of a recipient cell with copies of those genes introduced from a donor cell. As seen in the solution to Solved Problem I, only an even number of crossovers can produce viable recombinant chromosomes. Gene mapping is simplified if you remember that progeny classes that result from four crossovers are found much less often than progeny classes that require two crossovers.

20. In Problem 19, do you think that most of the Pyr^+ Arg^+ exconjugants are also Xyl^+ and Mal^+, or not? Explain your answer by considering the recombination events that would be required to generate colonies that are Pyr^+ Arg^+ Xyl^+ Mal^+ and those required to make Pyr^+ Arg^+ Xyl^- Mal^- colonies.
21. One issue with interrupted-mating experiments such as that in Problem 19 is that gene order may be ambiguous if the genes are close together. Another shortcoming is that such experiments do not provide

accurate map distances. The reason is that researchers select for the first Hfr marker transferred into the recipient, but the recovery of F^- exconjugants with a later Hfr marker is complex, depending both on transfer of the marker into the cell and on crossovers that transfer the marker into the recipient chromosome.

To make more accurate maps, bacterial geneticists often do Hfr × F^- crosses in a different way: They select for exconjugants that contain a late Hfr marker, and then screen for the presence of the earlier markers. This method ensures that all of the markers have entered the F^- cell, so relative gene distances are now accounted for solely by crossover frequencies. Furthermore, gene order is clarified by considering the crossovers responsible for each class of exconjugants.

As an example, suppose you performed the same cross as in Problem 19, but you selected for Arg^+ exconjugants, and then screened them for the earlier Hfr markers Mal^+ Xyl^+ and Pyr^+. You obtained the following data:

Exconjugant type	Number of exconjugants
Arg^+ Mal^+ Xyl^+ Pyr^+	80
Arg^+ Mal^+ Xyl^+ Pyr^-	40
Arg^+ Mal^+ Xyl^- Pyr^-	20
Arg^+ Mal^- Xyl^- Pyr^-	20
Arg^+ Mal^- Xyl^+ Pyr^-	1
Arg^+ Mal^- Xyl^- Pyr^+	1

a. Explain why four of the exconjugant types are much more frequent than the other two.

b. What can you conclude about the relative distances between the four genes?

c. The data allow you to estimate one other relevant genetic distance. Explain.

22. Suppose you have two Hfr strains of *E. coli* (HfrA and HfrB), derived from a fully prototrophic streptomycin-sensitive (wild-type) F^- strain. In separate experiments you allow these two Hfr strains to conjugate with an F^- recipient strain (Rcp) that is streptomycin resistant and auxotrophic for glycine (Gly^-), lysine (Lys^-), nicotinic acid (Nic^-), phenylalanine (Phe^-), tyrosine (Tyr^-), and uracil (Ura^-). By using an interrupted-mating protocol you determine the earliest time after mating at which each of the markers can be detected in the streptomycin-resistant recipient strain, as shown here.

	Gly^+	Lys^+	Nic^+	Phe^+	Tyr^+	Ura^+
HfrA × Rcp	3	*	8	3	3	3
HfrB × Rcp	8	3.	13	8	8	8

(The * indicates that no Lys^+ cells were recovered in the 60 minutes of the experiment.)

a. Draw the best map you can from these data, showing the relative locations of the markers and the origins of transfer in strains HfrA and HfrB. Show distances where possible.

b. To resolve ambiguities in the preceding map, you studied cotransduction of the markers by the generalized transducing phage P1. You grew phage P1 on strain HfrB and then used the lysate to infect strain Rcp. You selected 1000 Phe^+ clones and tested them for the presence of unselected markers, with the following results:

Number of transductants	Phenotype Gly	Lys	Nic	Phe	Tyr	Ura
600	–	–	–	+	–	–
300	–	–	–	+	–	+
100	–	–	–	+	+	+

a. Draw the order of the genes as best you can based on the preceding cotransduction data.

b. Suppose you wanted to use generalized transduction to map the *gly* gene relative to at least some of the other markers. How would you modify the cotransduction experiment just described to increase your chances of success? Describe the composition of the medium you would use.

23. Starting with an F^- strain that was prototrophic (that is, had no auxotrophic mutations) and Str^s, several independent Hfr strains were isolated. These Hfr strains were mated to an F^- strain that was Str^r Arg^- Cys^- His^- Ilv^- Lys^- Met^- Nic^- Pab^- Pyr^- Trp^-. Interrupted-mating experiments showed that the Hfr strains transferred the wild-type alleles in the order listed in the following table as a function of time. The time of entry for the markers within parentheses could not be distinguished from one another.

Hfr strain	Order of transfer →
HfrA	*pab ilv met arg nic (trp pyr cys) his lys*
HfrB	*(trp pyr cys) nic arg met ilv pab lys his*
HfrC	*his lys pab ilv met arg nic (trp pyr cys)*
HfrD	*arg met ilv pab lys his (trp pyr cys) nic*
HfrE	*his (trp pyr cys) nic arg met ilv pab lys*

a. From these data, derive a map of the relative position of these markers on the bacterial chromosome. Indicate with labeled arrows the position and orientation of the integrated F plasmid for each Hfr strain.

b. To determine the relative order of the *trp, pyr,* and *cys* markers and the distances between them, HfrB was mated with the F^- strain long enough to allow transfer of the *nic* marker, after which Trp^+ recombinants were selected. The unselected markers *pyr* and *cys* were then scored in the Trp^+ recombinants, yielding the following results:

Number of recombinants	Trp	Pyr	Cys
790	+	+	+
145	+	+	−
60	+	−	+
5	+	−	−

Draw a map of the *trp, pyr,* and *cys* markers relative to each other. (Note that you cannot determine the order relative to the *nic* or *his* genes using these data.) Express map distances between adjacent genes as the frequency of crossing-over between them.

24. You can carry out matings between an Hfr and F^- strain by mixing the two cell types in a small patch on a plate and then replica plating to selective medium. This methodology was used to screen hundreds of different cells for a recombination-deficient $recA^-$ mutant. Why is this an assay for RecA function? Would you be screening for a $recA^-$ mutation in the F^- or Hfr strain using this protocol? Explain.

25. Genome sequences show that some pathogenic bacteria contain virulence genes that promote disease next to genes that originally came from bacteriophage. Why does this result suggest horizontal gene transfer, and what would the mechanism of transfer have been?

26. Generalized and specialized transduction both involve bacteriophages. What are the differences between these two types of transduction?

Section 14.5

27. This problem highlights some useful variations of the gene identification by plasmid transformation procedure shown in Fig. 14.28.

    a. Suppose you have obtained a new bacterial mutant strain with a phenotype of interest. To determine the affected gene, you sequence the entire genome of the mutant strain and compare it with that of a wild-type strain. One of the differences found is a nonsense mutation that seems to be a good candidate. How would you use a plasmid library to verify that this nonsense mutation is responsible for the mutant phenotype?

    b. Figure 14.28 showed how plasmid libraries could be used to identify genes with loss-of-function mutations that are responsible for a given aberrant phenotype. How could you use a plasmid library to identify a gene affected by a gain-of-function mutation?

28. A researcher has a Trp^- auxotrophic strain of *E. coli* with a mutation in a single gene. To identify that mutant gene, she uses a genomic library made from a wild-type version of that same strain to find plasmids that rescue the mutant phenotype. The result is surprising. She recovers 10 plasmids that provide a Trp^+ phenotype, but six of the plasmids contain gene *X,* while the other four contain gene *Y.* Our scientist has encountered a phenomenon called *multicopy suppression,* related to the fact that plasmids are usually present in several copies per bacterium. Because the genes in the plasmids are present in more than their usual single copy in the bacterial chromosome, more than the usual amount of Protein X or Protein Y is being produced from the plasmids. Sometimes, overexpression of one protein can rescue the mutant phenotype caused by loss of a different protein. Suggest at least two ways that our scientist could determine which of the two genes, gene *X* or gene *Y,* actually corresponds to the mutant gene causing the Trp^- phenotype.

29. *Streptococcus parasanguis* is a bacterial species that initiates dental plaque formation by adhering to teeth. To investigate ways to eliminate plaque, researchers constructed a plasmid, depicted in the figure shown, to mutagenize *S. parasanguis.* The key features of this plasmid include $repA^{ts}$ (a temperature-sensitive origin of replication), kan^r (a gene for resistance to the antibiotic kanamycin), and the transposon *IS256.* This transposon contains the erm^r gene for resistance to the antibiotic erythromycin. *IS256* transposes in *S. parasanguis* thanks to a gene encoding a transposase enzyme that moves all DNA sequences located between the transposon's inverted repeats (IRs).

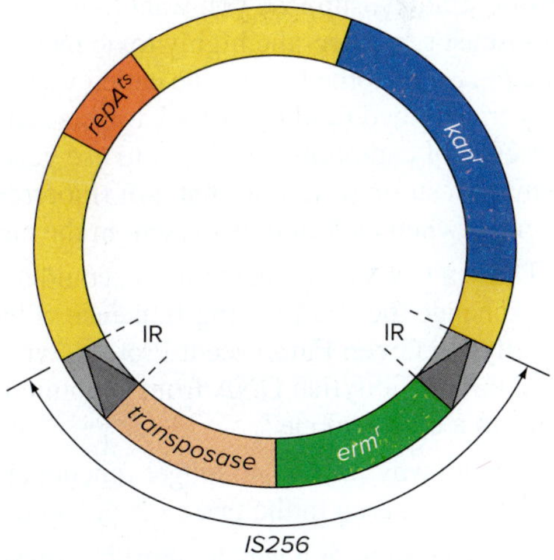

    a. How could the researchers use this plasmid as a mutagen? Consider how they could get the transposon into the bacteria, and how they could identify strains that had new insertions of *IS256* into *S. parasanguis* genes. Your answer should explain why the plasmid has two different antibiotic resistance genes as well as a temperature-sensitive origin of replication.

b. Why would the researchers use this plasmid as a mutagen?

c. If the investigators found a mutant strain of *S. parasanguis* that was defective in plaque formation, how could they identify the affected gene?

30. The sequence at one end of one strand of the *Drosophila* transposon *Mariner* is shown below (dots indicate sequences within the transposon):

5′ TTAGTTTGGCAAATATCTCCCTTCCGCCTTTTTGATCTTATGT... 3′

You obtain a mutant bacterial strain tagged with an engineered *Mariner* transposon, cut the genomic DNA from this strain with the restriction enzyme *Mbo*I (whose recognition site is ^GATC), and circularize the resultant DNA fragments by diluting the restriction enzyme digest and adding DNA ligase.

a. Design two 17 bp PCR primers that you could use to identify (by inverse PCR) the gene into which the transposon inserted.

b. What DNA sequence will be amplified from the circularized fragments of the mutant genome? Show the extent of this DNA sequence on a map of the genome of the mutant strain, indicating the locations of the transposon insertion and any relevant sites for the enzyme *Mbo*I.

31. Scientists can use gene targeting not just to knock out genes (as was shown in Fig. 14.31), but also to introduce nonbacterial genes into bacterial chromosomes. One such gene in wide use is a gene from jellyfish encoding Green Fluorescent Protein. In one example of this strategy, suppose you want to make *E. coli* into a biosensor to detect the highly toxic metal cadmium. The *E. coli* genome has a gene called *yodA* that is only transcribed (and its mRNA translated) in the presence of cadmium. You want to use gene targeting to make a strain of *E. coli* that will fluoresce brightly in green when cadmium is present in the environment.

a. Draw a DNA construct that you could use to exchange the *yodA* coding sequence with that for jellyfish Green Fluorescent Protein. Would you obtain the jellyfish DNA from genomic DNA or from a cDNA clone?

b. Explain why *yodA* is no longer functional in bacteria that glow green in the presence of cadmium.

c. Can you think of a way to alter the approach so that *yodA* might remain functional?

Section 14.6

32. Scientists who study amino acid biosynthesis pathways want to isolate auxotrophic bacteria. A technique called *penicillin enrichment* makes this task easier. This procedure starts by exposing a liquid culture of wild-type (prototrophic) bacteria growing in rich (complete) medium to a chemical mutagen. After this treatment, the cells are centrifuged to remove the liquid and the mutagen. The pellet of cells at the bottom of the centrifuge tube is now resuspended in medium that lacks one amino acid (in this example, cysteine) but contains penicillin. Subsequently, the bacteria are poured onto a filter that concentrates them and allows them to be washed free of the penicillin. The living bacteria retained on the filter are highly enriched for cysteine auxotrophs.

a. Given what you know about the action of pencillin, explain why this enrichment occurs.

b. Penicillin enrichment is not a selection, because the drug does not kill 100% of the prototrophs. The cells on the filter thus need to be screened for cysteine auxotrophy. How would the scientists perform this screen?

c. If the starting strain contained a pen^r gene on a plasmid, would this scheme still enrich for auxotrophs? Explain.

33. Suppose that you could obtain radioactively labeled penicillin. How could you use this compound to distinguish whether a penicillin-resistant bacterium harbors a gene encoding penicillinase or whether the bacterium has acquired a mutation in *penA, penB,* or *mtr?*

34. Scientists are using metagenomics to tackle one of the most significant problems affecting human beings: the resistance of many pathogenic bacteria to currently available antibiotics. One aspect of solving this problem is developing different bactericidal drugs. To do so, researchers are taking advantage of the discovery that several bacterial species synthesize toxins that enable them to prey on other bacterial species. Remarkably, these scientists have discovered that the enzymes that work in synthesis pathways to make such toxins often have particular structures and therefore characteristic patterns within their amino acid sequences.

Describe how metagenomic analysis of the microbiomes of soil, the ocean, or the human body could enable researchers to discover new antibiotics that could be effective on human pathogens.

35. Some scientists are trying to engineer bacteriophage to treat bacterial infections in humans when the infections do not respond to chemical antibiotics.

a. What possible advantages might phage therapy have over antibiotic therapy?

b. Describe potential difficulties that would need to be overcome for phage therapy to succeed.

c. How might researchers best confront the issue that bacterial cells could become resistant to bacteriophage just as they could to antibiotics?

chapter 15

Organellar Inheritance

Inflorescence of a titan arum (Amorphophallus titanum) *plant. Described by some as the world's smelliest plant, titan arum releases odorous molecules that attract pollinating insects. Mitochondria play a key role in generating this noxious signal. (See* What's Next *at the end of the chapter for details.)*

chapter outline

JUST NINE YEARS after the rediscovery of Mendel's laws, the plant geneticist Carl Correns reported a perplexing phenomenon that challenged one of Mendel's basic assumptions. In a 1909 paper, Correns described the results of reciprocal crosses analyzing the inheritance of leaf color in flowering plants known as four-o'clocks. Most four-o'clocks have green leaves, while those of other individuals are *variegated*, with some leaves or parts of leaves being green and others white (**Fig. 15.1**).

Fertilization of eggs from a plant with variegated leaves by pollen from a green-leafed plant produced variegated offspring. Surprisingly, the reciprocal cross—in which the leaves of the mother plant were green and those of the father variegated—did not lead to the same outcome; instead, all of the progeny from this cross displayed green foliage. From these results, it appeared that offspring inherit their form of the variegation trait from the mother only. This type of transmission, known as **maternal inheritance,** challenged Mendel's assumption that maternal and paternal gametes contribute equally to inheritance. Geneticists thus said that the trait in question exhibited **non-Mendelian inheritance.**

We now know that the non-Mendelian transmission of the leaf color trait in four-o'clocks is due to the fact that the genes that control leaf color do not reside on the chromosomes in the nucleus. Instead, the genes that dictate this trait are found on the genomes of nonnuclear organelles called *chloroplasts*. These organelles have their own genomes, in the form of small circular chromosomes called *chloroplast DNA* (*cpDNA*). In four-o'clocks,

Figure 15.1 Four-o'clocks. The first known example of non-Mendelian inheritance was of leaf variegation in four-o'clocks.

a plant receives chloroplasts and cpDNA only from the egg; the pollen does not contribute any chloroplasts or cpDNA to the embryonic plant.

Chloroplasts are not found in animals, yet both plants and animals have other kinds of organelles called *mitochondria* that also possess their own genomes of *mitochondrial DNA (mtDNA)*. Because mtDNAs are separate from the chromosomes in the nucleus and are transmitted independently of them, traits controlled by genes on mtDNAs also display non-Mendelian inheritance.

Three major themes surface in our detailed discussion of the genes and genomes of mitochondria and chloroplasts. First, unlike the rules governing transmission of nuclear genes, the rules for transmission of organellar genomes can vary in different organisms. In many organisms like four-o'clocks and humans, only the mother passes on organelles to the next generation; but in some species the organelles are inherited only from the father, and in yet other species, from both parents. Second, the maintenance and function of organelles requires cooperation between the organellar genome and the genome in the nucleus of the same cell. Finally, the genomes and biochemical processes of organelles have more similarity to those in bacteria than to those in other parts of the eukaryotic cell. These observations formed the basis of the *endosymbiont theory* which proposes that organelles are the evolutionary remnants of bacteria that had a symbiotic relationship with the ancient precursors of the earliest eukaryotes.

15.1 Mitochondria and Their Genomes

learning objectives

1. Describe the structure and function of a typical mitochondrion.
2. List ways in which mitochondrial genomes vary among different species.
3. Summarize RNA editing in mitochondria.
4. Discuss exceptions found in mitochondria to the universal genetic code.

Mitochondria, membrane-bounded organelles found in most eukaryotic cells, convert energy derived from glucose and other nutrient molecules into ATP. Researchers have shown that the mitochondria have their own DNA, separate from the nuclear genome. The mitochondrial genome encodes some, but not all, of the proteins needed for energy conversion. The remainder are encoded by the nuclear genome and imported into the organelle.

Each eukaryotic cell houses many mitochondria (**Fig. 15.2a**), with the exact number depending on the energy requirements of the cell as well as the chance distribution of mitochondria during cell division. In humans, nerve, muscle, and liver cells each carry more than a thousand mitochondria; human oocytes have about 100,000 mitochondria.

Mitochondria are not static structures. The mitochondria in a cell can grow, fuse with each other, or divide. Roughly speaking, mitochondria double in size, replicate their mtDNA, and then divide in half in each cell generation. When the cell divides, the mitochondria are distributed randomly and passively to the daughter cells. These mitochondrial processes are largely independent of what occurs elsewhere in the cell, resulting in large variations in the number of mitochondria and mtDNA molecules in individual cells.

Mitochondria Produce ATP

Figure 15.2b reveals the organelle's basic structure: The unwrinkled outer membrane surrounds an inner membrane bent into wrinkles called *cristae*. The inner membrane in turn surrounds an area called the *matrix*.

Mitochondria produce packets of energy (ATP) in two stages. In the first, enzymes in the matrix catalyze the *Krebs cycle,* which metabolizes pyruvate (the product of the breakdown of glucose in the cytoplasm) to generate the high-energy electron carriers NADH and $FADH_2$. In the second

Figure 15.2 Anatomy of a mitochondrion. (a) False color micrograph of mitochondrion (62,800X). **(b)** The organization and structure of a single mitochondrion. Regions entirely enclosed within the inner membrane are known as the *matrix* (*blue*). The matrix contains the mitochondrial DNA and enzymes of the Krebs cycle. Inner-membrane foldings are called *cristae*. A single crista is magnified to show how enzymes of the electron transport chain carry out oxidative phosphorylation.

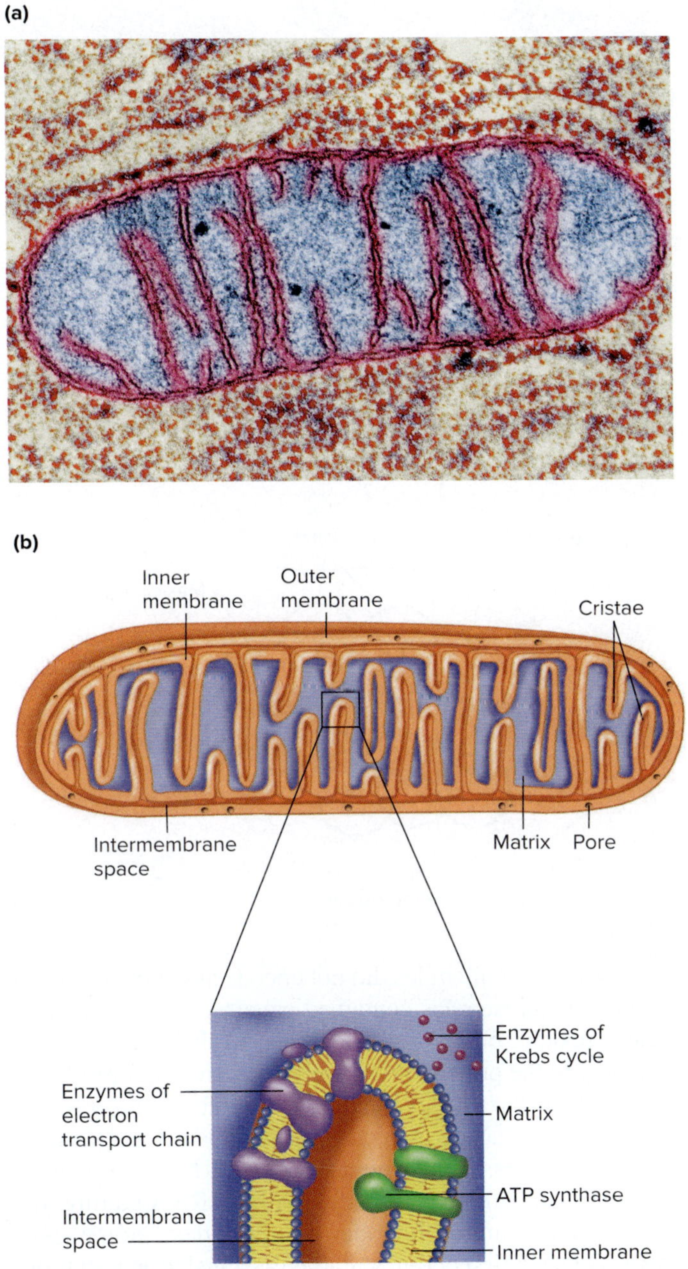

stage, a series of multisubunit enzyme complexes embedded in the inner mitochondrial membrane harness this energy in a process called **oxidative phosphorylation.** Some of the enzyme complexes form an electron transport chain that transfers the electrons from NADH and $FADH_2$ to the ultimate electron acceptor, oxygen. The energy released from these steps of electron transport is used to pump protons out of the matrix into the space between the inner and outer membranes, creating an electrical potential across the inner membrane. The protons then flow back into the matrix through an enzyme complex called *ATP synthase,* which is embedded in the cristae. ATP synthase uses the energy released by this flow of protons to phosphorylate ADP and thus form ATP.

Mitochondrial Genomes Vary Among Species

A single mitochondrion usually contains several copies of its genome within the matrix; the number of copies in an organelle can vary depending on the energy needs of the cell, but it is usually between 2 and 10. We will first describe the human mitochondrial genome as an example, but as you will see, mitochondrial genomes can vary astonishingly in size and form in different species.

Human mitochondrial DNA

The 16.5 kb human mitochondrial genome, which is only about 1/100,000th the length of the haploid genome in a human gamete, is a circular DNA molecule that carries 37 genes (**Fig. 15.3**). Thirteen of these genes encode polypeptide subunits of the protein complexes that make up the oxidative

Figure 15.3 The human mitochondrial genome. The 37 genes in human mtDNA are shown as follows: *green* for genes encoding cytochrome oxidase proteins; *red* for genes encoding ATPase subunit proteins; *yellow* for genes encoding NADH complex proteins; *tan* for cytochrome complex protein genes; and *purple* for ribosomal protein or ribosomal RNA genes. Each tRNA gene is indicated by a *black* ball and stick. Genes on the outer and inner circles are transcribed in opposite directions.

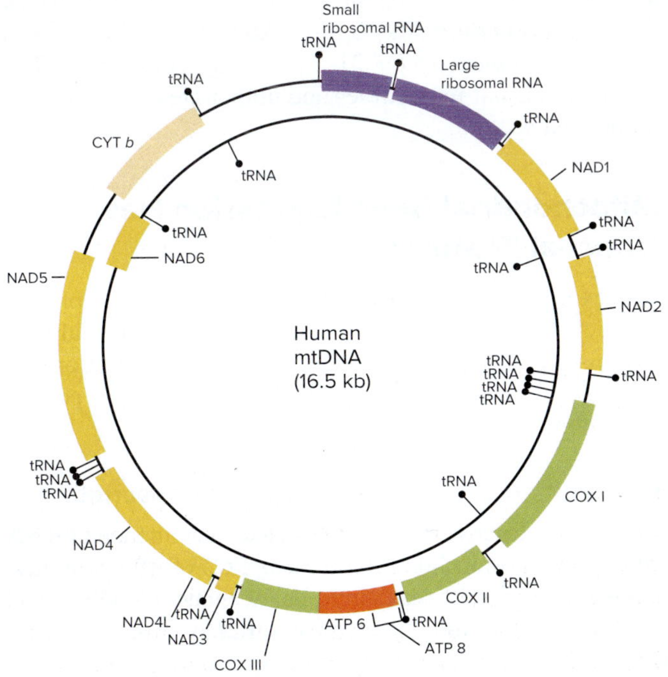

phosphorylation apparatus. The mitochondrial genome also encodes 22 different tRNA genes and two genes for the large and small rRNAs found in mitochondrial ribosomes.

A significant feature of the human mitochondrial genome is the compactness of its gene arrangement. Adjacent genes either abut each other or in a few cases, even slightly overlap. With virtually no nucleotides between them and no introns within them, the genes are tightly packaged.

Mitochondrial genomes

The size and gene content of mitochondrial DNA vary from organism to organism. The mtDNAs in the malaria parasite, *Plasmodium falciparum,* are only 6 kb in length; those in the muskmelon, *Cucumis melo,* are a giant 2400 kb long. These mtDNA size differences do not necessarily reflect comparable differences in gene content. Although the large mtDNAs of higher plants do contain more genes than the smaller mtDNAs of other organisms, the 75 kb mtDNA of baker's yeast encodes fewer proteins of the respiratory chain than does the 16.5 kb mtDNA of humans. The larger size of yeast mitochondria is due in part to the existence of introns in yeast mitochondrial genes and in part to large spacers between the genes.

Even the shape of mtDNAs varies considerably. Biochemical analyses and mapping studies have shown that the mtDNAs of humans and other animals are circular. However, the mtDNAs of most fungi and plants are linear. The difficulty in isolating unbroken mtDNA molecules from some organisms has made it hard to be certain of the shape of their mtDNA *in vivo.*

Protozoan parasites of the genera *Trypanosoma* and *Leishmania* exhibit mtDNAs that have a highly unusual organization. These single-celled eukaryotic organisms carry a single mitochondrion known as a **kinetoplast.** Within this structure, the mtDNA exists as a large network of 10–25,000 *minicircles* 0.5–2.5 kb in length interlocked with 50–100 *maxicircles* 21–31 kb long (**Fig. 15.4**). The differing roles of maxicircles and minicircles are described in the next section.

Figure 15.4 Kinetoplast DNA network. In certain protozoan parasites, the single mitochondrion, or *kinetoplast,* contains a large interlocking network of mini- and maxi-circles of DNA. **(a)** Electron micrograph of part of a kinetoplast. The arrow points to a single maxicircle. **(b)** Diagram illustrating how these circular DNA molecules interlock (*catenate*) with each other.

a: Electron micrograph by Dr. Stephen Hajduk/University of Alabama at Birmingham

(a)

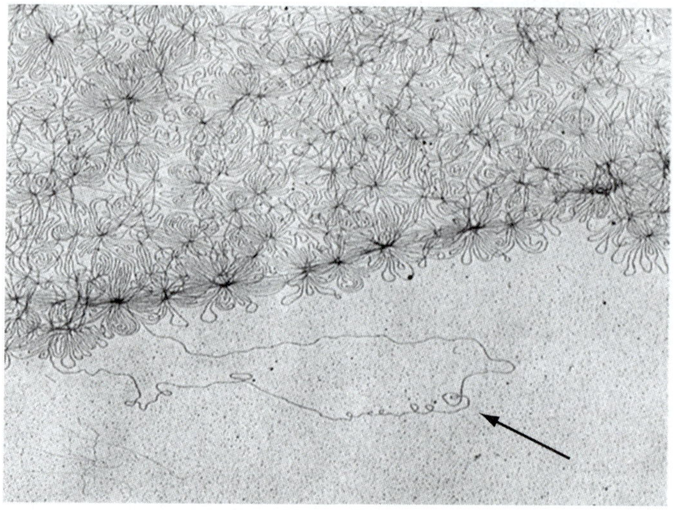

(b)

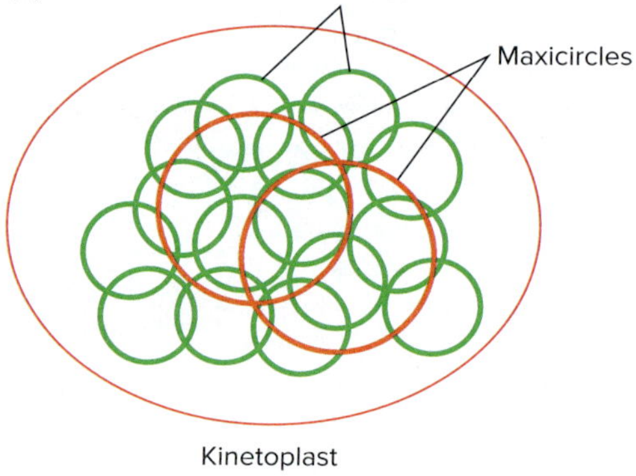

Mitochondrial Gene Expression Has Unusual Features

The expression of mitochondrial genes into their protein products has unique features. These include a special type of transcript processing called *RNA editing.* In addition, the translational machinery of mitochondria requires certain exceptions to the universal genetic code.

RNA editing of mitochondrial gene transcripts

Researchers discovered the unexpected phenomenon of RNA editing in the mitochondrion (kinetoplast) of trypanosomes. DNA sequencing of maxicircle DNA, minicircle DNA, and cDNAs copied from kinetoplast mRNAs revealed major surprises. The minicircles did not encode any protein-coding genes. The maxicircles contained gene sequences that were clearly related, but far from identical to the cDNAs corresponding to the protein-coding mRNAs. The latter result implied that the maxicircles are transcribed into precursor RNAs (pre-mRNAs) that are then changed into mature mRNAs.

The process that converts pre-mRNAs to mature mRNAs is called **RNA editing.** Without RNA editing, the pre-mRNAs cannot encode polypeptides. Some pre-mRNAs lack a first codon suitable for translation initiation; others lack a stop codon for the termination of translation. RNA editing creates both types of sites, as well as many new codons within the genes.

In trypanosomes, the RNA editing machinery adds or deletes uracils to convert pre-mRNAs into mature mRNAs. As **Fig. 15.5** shows, uracil editing occurs in stages in which enzymes organized into a structure called an *editosome* use

Figure 15.5 RNA editing in trypanosomes. A portion of a pre-mRNA sequence is shown at the *top*. This pre-mRNA forms a double-stranded hybrid with a guide RNA through both standard Watson-Crick A–U and G–C base pairing, as well as atypical G–U base pairing. Unpaired G and A bases within the guide RNA initiate the insertion of Us within the pre-mRNA sequence (*blue*), while unpaired Us in the pre-mRNA are deleted (*red*), bringing about the final edited mRNA.

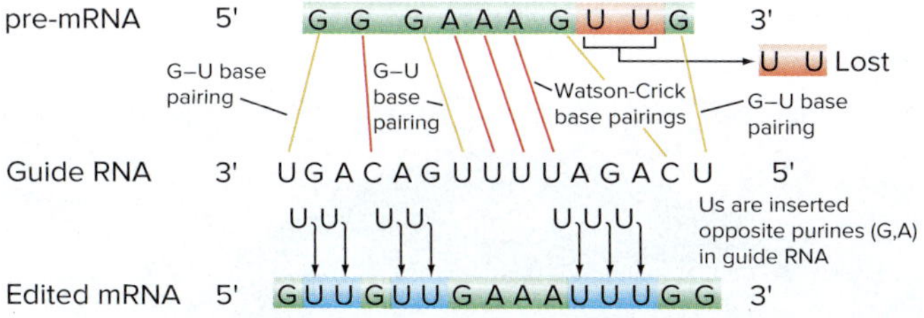

an RNA template as a guide for correcting the pre-mRNA. The guide RNAs are transcribed from short stretches of DNA on minicircles, explaining why kinetoplasts have minicircles as well as maxicircles.

RNA editing is an unusual phenomenon, but it is not limited to the kinetoplasts of trypanosomes. The mitochondrial transcripts of the slime mold *Physarum* undergo an RNA editing process in which cytosines are added. In plant mitochondria and chloroplasts, a different kind of editing occurs in which cytosines in the pre-mRNAs are changed to uracils in the mature mRNAs. The mechanisms underlying these other forms of RNA editing are not yet understood.

Mitochondrial exceptions to the universal genetic code

Mitochondria have their own distinct translational apparatus, as suggested by the fact that mtDNAs carry their own rRNA and tRNA genes (Fig. 15.3). Mitochondrial translation is quite unlike the cytoplasmic translation of mRNAs transcribed from nuclear genes in eukaryotes; in fact, many aspects of the mitochondrial translational system resemble details of translation in prokaryotes. For example, as in bacteria, *N*-formyl methionine and $tRNA^{fMet}$ initiate translation in mitochondria. Moreover, drugs that inhibit bacterial translation, such as chloramphenicol and erythromycin, which have no effect on eukaryotic cytoplasmic protein synthesis, are potent inhibitors of mitochondrial protein synthesis.

We stated in Chapter 8 that the genetic code is almost, but not quite, universal. Many exceptions to the "universal" code involve mitochondria. For example, in human mtDNA, five kinds of triplets are used differently than they would be in the nucleus (**Table 15.1**). No single mitochondrial genetic code functions in all organisms, and the mitochondria of higher plants use the universal code. The genetic codes of some mitochondria therefore probably diverged from the universal code by a series of mutations occurring some time after the organelles became established components of eukaryotic cells.

TABLE 15.1 Differences Between the Universal and Human Mitochondrial Genetic Codes

Triplet	Universal	mtDNA
UGA	Stop	Trp
AGG	Arg	Stop
AGA	Arg	Stop
AUA	Ile	Met
AUU	Ile	Ile-elongation Met-initiation

essential concepts

- Mitochondria generate ATP through the oxidation of pyruvate, the product of glycolysis. ATP generation involves electron transport chains and the enzyme ATP synthase, which are embedded in the mitochondrion's inner membrane.
- Mitochondrial genomes in different organisms vary greatly according to the length of the genome, whether the mtDNA is linear or circular, and whether or not introns are present.
- Pre-mRNAs in mitochondria are converted to mature mRNAs through *RNA editing,* which makes translation of the transcripts possible.
- Translation in mitochondria has many similarities to that of bacteria. Although some mitochondria use the universal DNA code, in other organisms some triplets are used differently in mitochondria than they are in nuclear DNA.

15.2 Chloroplasts and Their Genomes

learning objectives

1. Describe the structure and function of a typical chloroplast.
2. Contrast the variation in chloroplast genomes among species with that of mitochondrial genomes.
3. Describe the process by which transgenic chloroplasts can be produced.

Chloroplasts capture solar energy and store it in the chemical bonds of carbohydrates through the process of **photosynthesis.** Every time a bird takes flight, a person speaks, a worm turns, or a flower unfurls, the organism's cells are using energy that was captured originally from sunlight by chloroplasts and is then released through the functions of mitochondria. In corn, one of the many crop plants adept at carrying out photosynthesis, each leaf cell contains 40 to 50 chloroplasts (**Fig. 15.6a**), and each square millimeter of leaf surface carries more than 500,000 of these organelles.

Chloroplasts Are the Sites of Photosynthesis

Figure 15.6b illustrates the structure of a chloroplast. Embedded in the membranes of internal structures called *thylakoids* are the light-absorbing pigment chlorophyll and light-absorbing proteins, as well as proteins of the photosynthetic electron transport system. During the light-trapping phase of photosynthesis, the energy of photons of light from the sun boosts electrons in chlorophyll to higher energy levels. The energized electrons are then conveyed to an electron transport system that uses the energy to convert water to oxygen and protons.

Photosynthetic electron transport forms NADPH and drives the synthesis of ATP via an ATP synthase similar to the one in mitochondria. During the second, sugar-building phase of photosynthesis, enzymes of the *Calvin cycle* use that ATP and NADPH to convert atmospheric carbon dioxide into carbohydrates. The energy stored in the bonds of these nutrient molecules fuels the activities of both the plants that make them and the animals that eat the plants.

Chloroplast Genomes Are Relatively Uniform

Chloroplasts exist in plants and algae. The genomes they carry are much more uniform in size than the genomes of mitochondria. Although chloroplast DNAs (cpDNAs) range in size from 120 to 217 kb, most are between 120 and 160 kb long. cpDNA contains many more genes than mtDNA. Like the genes of bacteria and human mtDNA, these genes are closely packed, with relatively few nucleotides between adjacent coding sequences. Like the genes of yeast (but not human) mtDNA, they contain introns. Although some are circular, many cpDNAs exist as linear and branched forms. Like mitochondria, chloroplasts contain more than one copy of their genome—usually 15–20 copies.

The chloroplast genome of the liverwort *M. polymorpha*, the first cpDNA to be sequenced completely, is depicted in **Fig. 15.7.** The cpDNA-encoded proteins include many of the molecules that carry out photosynthetic electron transport and other aspects of photosynthesis, as well as RNA polymerase, translation factors, ribosomal proteins, and other molecules active in chloroplast gene expression. The RNA polymerase of chloroplasts is similar to the multisubunit

Figure 15.6 Anatomy of a chloroplast. **(a)** Electron micrograph of an isolated chloroplast from a tobacco leaf cell (*Nicotiana tabacum*) (11,000X). **(b)** Internal organization. A chloroplast has an outer and an inner membrane. The space within the inner membrane—containing the chloroplast DNA and photosynthetic enzymes—is called the *stroma*. In the stroma are vesicles called *thylakoids* stacked in columns termed *grana*. Photosynthesis takes place on the surface of thylakoids.

(a)

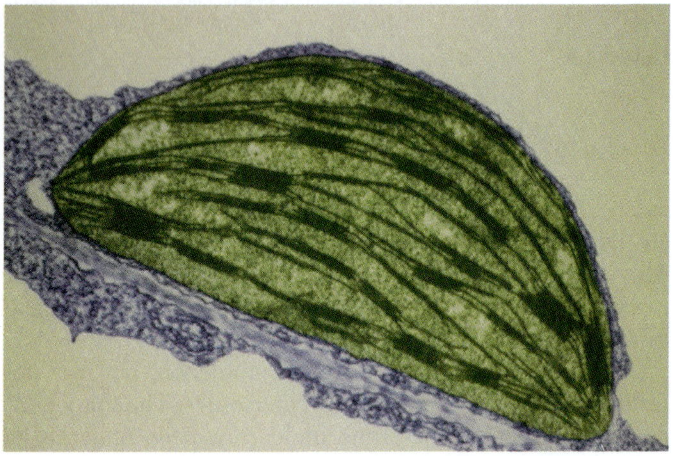

(b)

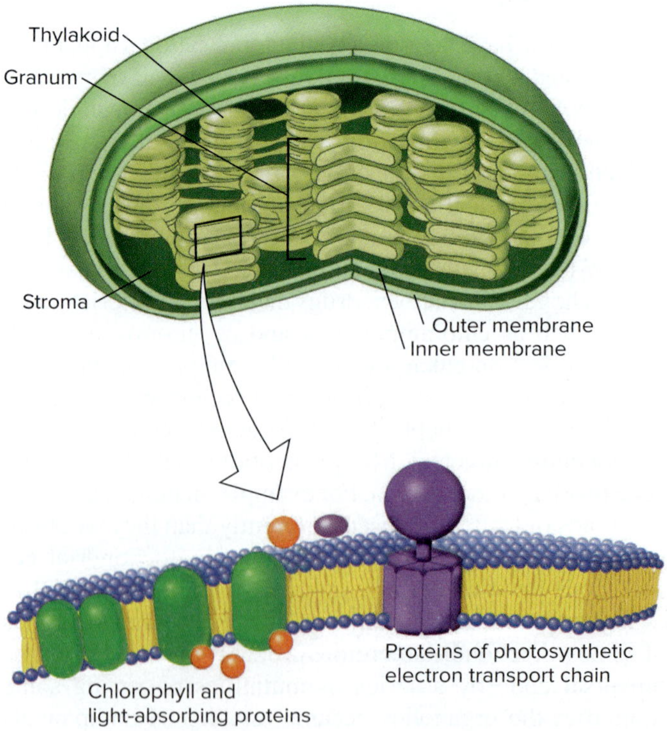

Figure 15.7 Chloroplast genome of the liverwort *M. polymorpha*. The relative locations and symbols of some of the 128 genes are indicated. Genes are color-coded according to function.

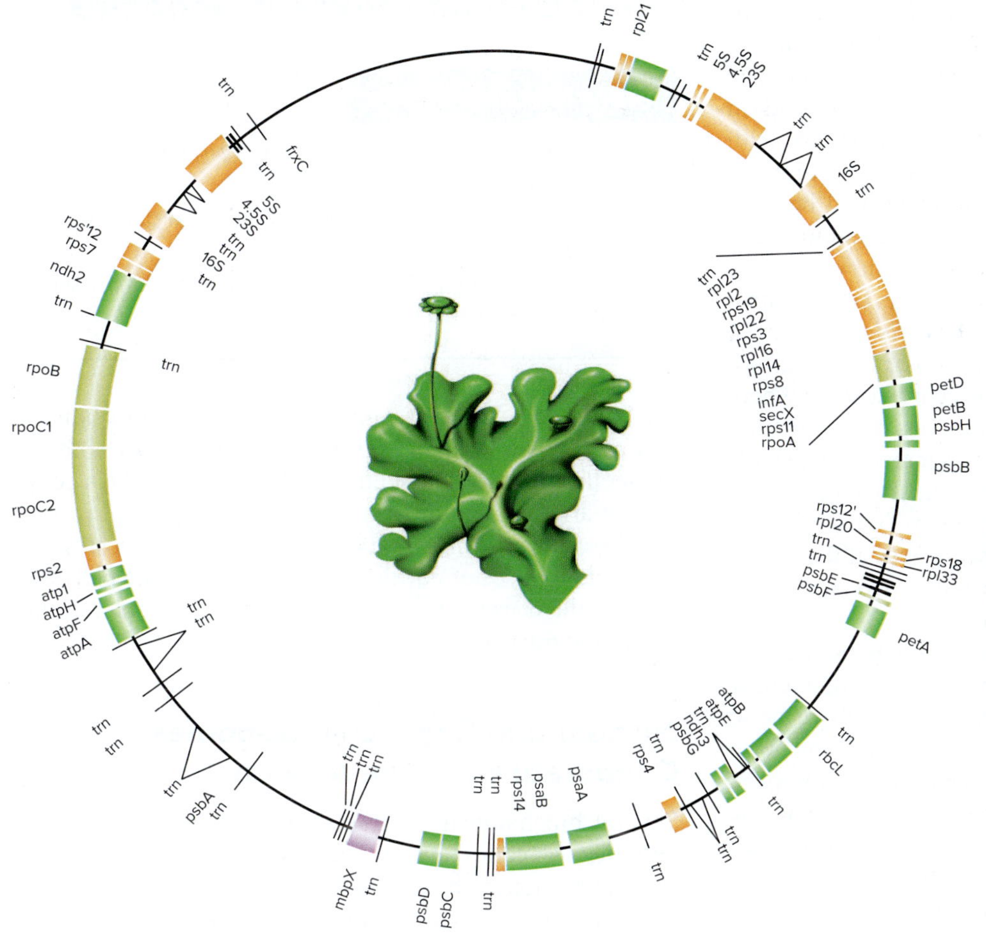

	Translation
rps	30S ribosomal proteins
rpi	50S ribosomal subunit proteins
trn	tRNAs
4.5S, 5S 16S, 23S	rRNAs
infA	initiation factor
secX	50S ribosomal protein
	Photosynthesis and electron transport
rbc	ribulose bisphosphate carboxylase
psa	photosystem 1
psb	photosystem 2
pet	cytochrome *b/f* complex
atp	ATP synthase
frx	iron-sulfur proteins
ndh	NAD(P)H oxidoreductase
	Transcription
rpo	RNA polymerase
	Miscellaneous
mbpx	permease

bacterial RNA polymerases. Drugs that inhibit bacterial translation, such as chloramphenicol and streptomycin, inhibit translation in chloroplasts, as they do in mitochondria.

Scientists Can Produce Transgenic Chloroplasts

In the early days of recombinant DNA technology, researchers studying organelles were frustrated by an inability to transfer cloned genes and mutated DNA fragments into organellar genomes. Development of the *gene gun* and a gene delivery method known as **biolistic transformation** in the late 1980s solved the problem (**Fig. 15.8**). This technique has been particularly important for investigations of chloroplast genomes, as we describe here. Scientists have also been successful in transforming the mitochondria of the yeast *S. cerevisiae* with exogenous DNA using the same method, but the stable transformation of mitochondria in multicellular organisms remains an elusive goal.

Figure 15.8 A gene gun. The gun is used to propel DNA-coated beads into plant cells, thus enabling chloroplast transformation.

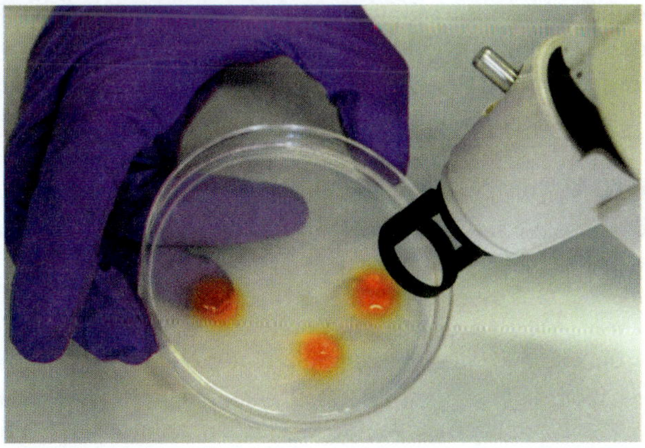

The basic idea is to coat small (1 μm) metal particles with DNA and then shoot these DNA-carrying "bullets" at cells (Fig. 15.8) or leaves. The DNAs shot into the plant cells can enter the chloroplast and integrate into a specific location in cpDNA via homologous recombination. A drug resistance gene in the introduced DNA allows for selection of transformed cells that can be grown into transplastomic plants containing transgenic chloroplasts. The gene for spectinomycin resistance is typically used in biolistic transformation. Spectinomycin interferes with translation of chloroplast gene mRNAs, and therefore chloroplasts that do not contain a transgene will be nonfunctional. Plant cells with nontransformed chloroplasts that survive drug selection would be white and weak.

To insert a *transgene* into cpDNA, the DNAs introduced into the plant cells are *gene-targeting* constructs, where cloned cpDNA sequences within the construct determine the location in the endogenous chromosome where the transgene will integrate via homologous recombination (review Fig. 14.31). Typically, the transgene is alongside a spectinomycin resistance gene, and the two genes are flanked by cpDNA sequences for targeting. Transgene integration occurs when these cpDNA sequences at the 5′ and 3′ ends of the construct undergo crossing-over with their counterparts in the chloroplast genome.

Transformation of the chloroplast genome has considerable potential for altering the properties of commercially important crop plants. For example, one goal might be to produce herbicide-resistant plants. A major advantage to introducing an herbicide resistance gene into chloroplast DNA instead of nuclear DNA is that foreign DNA in the chloroplasts will be inherited maternally, but not through the pollen. The risk that introduced genes will spread to neighboring plant populations is therefore low.

Just as in bacteria (Fig. 14.31), gene-targeting in chloroplasts also provides a way to determine the function of ORFs—open reading frames—for which no function has yet been assigned. To explore the function of an ORF, a DNA molecule is constructed that contains a spectinomycin-resistance gene cloned within the ORF. This DNA integrates into the chloroplast genome and replaces the wild-type ORF with the mutant ORF. Researchers have used this protocol to identify chloroplast genes encoding novel subunits of photosynthetic enzymes in several plant species.

essential concepts

- In chloroplasts, sunlight activates an electron transport chain that produces ATP and NADPH. These high-energy molecules are used subsequently to convert carbon dioxide and water into carbohydrates.
- Chloroplast genomes generally contain more genes than those of mitochondria; the products of chloroplast genes are needed for photosynthesis and gene expression within the organelle.
- *Biolistic transformation,* by which DNA is shot into cells on microscopic metal particles, has enabled production of plant lines that contain *transgenic chloroplasts.*

15.3 The Relationship Between Organellar and Nuclear Genomes

learning objectives

1. Describe the cooperation between organellar and nuclear genomes.
2. Summarize the endosymbiont theory of organelle origin.
3. Explain the implications of gene transfer from organellar genomes to the nucleus.

The maintenance and assembly of functional mitochondria and chloroplasts depend on gene products from both the organelles themselves and the nuclear genome. This cooperative arrangement did not happen overnight, but instead developed over evolutionary time. Evidence indicates that the ancient ancestors of these organelles and the cells that contain them were free-living organisms that entered into symbiotic relationships.

Nuclear and Organellar Genomes Cooperate with One Another

Several biochemical processes require components from both the organelles and the nucleus. As one example, cytochrome *c* oxidase, the terminal protein of the mitochondrial electron transport chain, in most organisms is composed of seven subunits. Three of these are encoded by mitochondrial genes whose mRNAs are translated on mitochondrial ribosomes. The remaining four are encoded by nuclear genes whose messages are translated on ribosomes in the cytoplasm; these proteins must then be imported into mitochondria.

In all organisms, nuclear genes encode the majority of the proteins needed for gene expression in mitochondria and chloroplasts. For example, although mitochondrial genomes carry the rRNA genes, nuclear genomes carry the genes for most (in yeast and plants) or all (in animals) of the proteins in the mitochondrial ribosome. Because mitochondria and chloroplasts do not carry genes for all the proteins they need to function and reproduce, these organelles must be provisioned constantly with molecules imported from other parts of the cell. Mitochondria and chloroplasts thus cannot exist independently of the cells in which they are found.

Mitochondria and Chloroplasts Originated from Bacteria

Chloroplasts are remarkably similar in size and shape to certain photosynthetic bacteria alive today. And while it is difficult to generalize due to the huge diversity of mitochondrial genetic systems, at least some mitochondria resemble certain present-day aerobic bacteria. These likenesses suggest that mitochondria and chloroplasts started out as free-living bacteria that merged with the ancestors of modern eukaryotic cells to form a cellular community in which host and guest benefited from the group arrangement.

The endosymbiont theory

In the 1970s, Lynn Margulis was one of the first biologists to propose that mitochondria and chloroplasts originated when ancient precursors of eukaryotic cells established a symbiotic relationship with, and ultimately engulfed, certain bacteria. The primitive cells carrying a mitochondrion-like or chloroplast-like bacterial cell would have gained an edge in the fierce competition for energy production and eventually evolved into complex eukaryotes. So much evidence now supports this hypothesis that it is generally accepted as the **endosymbiont theory.**

The molecular evidence for the endosymbiont theory includes the following facts:

1. Both mitochondria and chloroplasts have their own DNA, which replicates independently of the nuclear genome.
2. Like the DNA of bacteria, mtDNA and cpDNA are not organized into nucleosomes by histones.
3. Mitochondrial gene expression uses *N*-formyl methionine and $tRNA^{fMet}$ in translation.
4. Inhibitors of bacterial translation, such as chloramphenicol and erythromycin, block mitochondrial and chloroplast translation but have no effect on eukaryotic protein synthesis in the cytoplasm.
5. Comparisons of organelle and bacterial rRNA gene sequences suggest that mitochondrial genomes derive from a common ancestor of present-day gram-negative nonsulfur purple bacteria, while chloroplast genomes derive from cyanobacteria (formerly referred to as blue-green algae).

Scientists estimate that the endosymbiotic event(s) giving rise to mitochondria occurred as long as 2 billion years ago, while the endosymbiosis resulting in chloroplasts happened perhaps 500 million years later. These events are so ancient that the exact processes that were involved are understood only dimly. Some scientists theorize that instead of an early primitive eukaryotic cell engulfing a bacterium, the first eukaryotic cell might have emerged from a symbiosis between archaea and bacteria.

Gene transfer between organelle and nucleus

In the time since the original endosymbiotic events that gave rise to mitochondria and chloroplasts, some genes likely moved from the organellar genome to that of the nucleus. We have seen, for example, that some of the genes required for oxidative phosphorylation and photosynthesis reside in the nuclear genome; they may have been transferred there from the organellar genome. The same is likely true for the genes encoding organelle ribosomal proteins.

The idea that genes can move from the organelle to the nucleus has important implications. First, once copies of such genes are incorporated into nuclear chromosomes, the copy in the organelle would become redundant and then could be lost. If the gene was originally necessary for independent growth of the endosymbiotic bacterium, then the proto-organelle could no longer survive outside of the host cell. Second, different evolutionary lineages of eukaryotes could have moved different subsets of organellar genes to the nucleus, resulting in some of the enormous diversity of current-day organellar genomes.

Researchers have some understanding of the mechanisms by which genes transfer between an organelle and the nucleus. In many plants, the mitochondrial genome encodes the *COXII* gene of the mitochondrial electron transport chain. In other plants, the nuclear DNA encodes that same gene. And in several plant species where the nuclear *COXII* gene is functional, the mtDNA still contains a recognizable, but nonfunctional, copy of the gene (that is, a *COXII* pseudogene). Remarkably, the mtDNA gene contains an intron, while the nuclear gene does not. Geneticists have interpreted this finding to mean that the *COXII* gene transferred from mtDNA to nuclear DNA via an RNA intermediate using reverse transcriptase. The intron would have been spliced out of the *COXII* transcript, and when the mRNA was copied into DNA by reverse transcriptase and integrated into a chromosome in the nucleus, the resulting nuclear gene also would have had no intron. Other mechanisms for the transfer of DNA between organelles and the nucleus that do not involve an RNA intermediate also appear to exist.

essential concepts

- Cooperation between an organelle and the nucleus is required because many proteins needed for organelle functions are encoded by nuclear DNA, while others are specified by organellar DNA.
- The *endosymbiont theory* states that mitochondria and chloroplasts evolved from symbiotic relationships established between bacteria-like cells and the precursors of eukaryotic cells that engulfed them.
- Transfer of genes from the organellar genome to the nuclear genome and subsequent loss of the original organellar genes made organelles unable to live outside their host cells.

15.4 Non-Mendelian Inheritance of Mitochondria and Chloroplasts

learning objectives

1. Describe experimental approaches that demonstrated organelles are often maternally inherited.
2. Explain how mixtures of different cpDNAs in the same organism account for variegation in plants.
3. Describe genetic studies which showed that yeast mitochondria are inherited biparentally.

Mutations in organelle genes often produce readily detectable whole-organism phenotypes because the altered proteins and RNAs they encode disrupt the production of cellular energy. For example, mtDNA mutations can cause colonies of unicellular organisms to grow more slowly, or tissues of multicellular organisms to be unusually weak. Mutations in cpDNA can incapacitate proteins essential for the production of chlorophyll, the green pigment required for photosynthesis, and can thus change the color of plant leaves. An alternative way to track mutations in organellar genomes is by following DNA polymorphisms directly through sequencing.

Modes of organelle transmission vary wildly among different species. One of the questions that geneticists ask of any given species is whether the progeny of a cross obtain their organelles from both parents (**biparental inheritance**), or from just one parent (**uniparental inheritance**), which can be either maternal (if all the organelles come from the mother) or paternal. All of these possibilities are seen in nature.

In Many Organisms, Organelles and Their DNA Are Inherited from One Parent

For many eukaryotic species, and particularly in animals, organelles are inherited uniparentally. When uniparental inheritance occurs, progeny most often inherit their organelles from the maternal parent (**maternal inheritance**), but many exceptions exist. For example, in bananas inheritance of the chloroplast genome is maternal while that of mitochondrial genomes is paternal. The situation is reversed in alfalfa, while in sequoia trees, both chloroplast and mitochondrial DNAs are paternally inherited. We focus here on one of the classic examples of the more common maternal pattern.

Maternal inheritance of *Neurospora* mtDNA mutants

In Chapter 5 we discussed the genetics of the bread mold *Neurospora crassa*. Recall that *Neurospora* colonies are haploid (1*n*) and are of two different mating types, each of which can generate specialized cells capable of mating (fusion) with cells of the opposite mating type (review Fig. 5.19b). Each mating type can generate both male and female mating cells, where the male cells are considerably smaller than the female cells. After mating of one type's male cell with the other type's female cell, the nuclei fuse to form a diploid (2*n*) cell that undergoes meiosis followed by mitosis to generate an octad of haploid spores. If the original haploid cells had different alleles of a gene on a nuclear chromosome, Mendelian inheritance dictates that half the spores in the octad would contain one allele, and the other half the other allele (4:4 segregation).

In 1952, Mary and Herschel Mitchell isolated a mutant *Neurospora* strain they called *poky* that exhibited a slow growth phenotype. They crossed the *poky* mutant strain of one mating type, which was also wild type for the gene ad^+ (can synthesize adenine), to a strain of the other mating type with normal growth ($poky^+$) that was also ad^- (requires an adenine supplement). The nuclear gene markers ad^+ and ad^- segregated 4:4 as expected. Surprisingly, however, all of the spores were either uniformly $poky^+$ or $poky^-$; the segregation of the slow growth phenotype was therefore 8:0. Moreover, the spore phenotypes were always identical to that of the colony that provided the female mating cell (**Fig. 15.9**). The *poky* trait thus exhibits *non-Mendelian inheritance* because maternal and paternal gametes do not make equal contributions to the phenotypes of the progeny.

The explanation is that $poky^-$ is a mutation in mtDNA. The *poky* gene encodes a mitochondrial ribosomal RNA, and growth is slow in $poky^-$ mutants because translation in mitochondria is inefficient. The larger female mating cell provides all of the mitochondria to the cytoplasm of the transient diploid cell and to all eight spores. In *Neurospora,* mitochondrial inheritance is thus uniparental and maternal.

Mechanisms leading to maternal inheritance of organelles

Differences in gamete size help explain maternal inheritance in species such as *Neurospora crassa* in which the male gamete is much smaller than the female gamete. As a result, the zygote receives a very large number of maternal organelles and at most a very small number of paternal organelles. The same situation explains uniparental, maternal inheritance in many multicellular organisms, including that of chloroplasts in four o'clocks and of mitochondria in humans, examples that will be explored in the next section.

However, zygote size is not the only mechanism leading to maternal inheritance of organelles. In some organisms, paternal organelles are actively excluded or destroyed. In some plants, the early divisions of the zygote distribute most or all of the paternal organellar genomes to cells that

Figure 15.9 Maternal inheritance of the Poky phenotype in *Neurospora*. **(a)** In the octads from a mating of a Poky ad^+ (*blue* cytoplasm, *red* nuclei) female gamete and a Normal ad^- (*brown* cytoplasm, *yellow* nuclei) male gamete, half the spores are ad^+ and half are ad^-. All of the spores are Poky because this phenotype is controlled by a mitochondrial gene, and all the mitochondria are supplied by the larger female gamete. **(b)** Diploids from the reciprocal cross also give rise to half ad^+ and half ad^- spores. However, all the spores are Normal (not Poky).

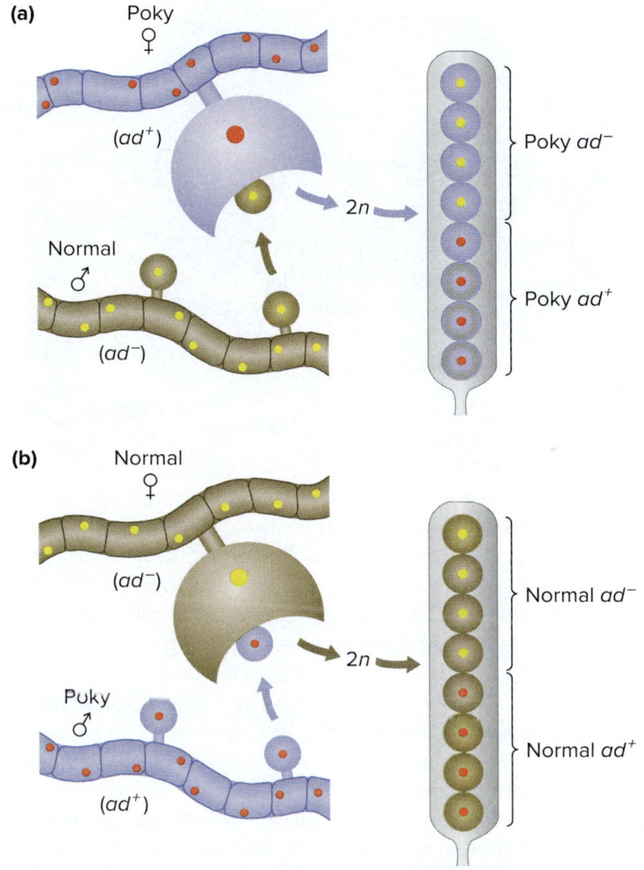

are not destined to become part of the embryo. In certain animals, details of fertilization prevent a paternal cell from contributing its organelles to the zygote. For example, in the prevertebrate chordates called tunicates, fertilization allows only the sperm nucleus to enter the egg, while it physically excludes the paternal mitochondria. In yet another mechanism that occurs in many animals, the zygote destroys the paternal organelles after fertilization.

The Genetics and Society Box, *Mitochondrial DNA Tests as Evidence of Kinship in Argentine Courts,* describes how a human rights organization in Argentina used mtDNA sequences as the legal basis for reuniting kidnapped children with their biological families. The maternal inheritance of mitochondria in humans makes it possible to compare and match the DNA of a grandmother and a grandchild.

Because it occurs in only a relatively small number of species, the basis of uniparental paternal inheritance of organelles is not yet well understood.

Variants of Organellar Genomes Segregate During Cell Division

A single eukaryotic cell may harbor thousands of mitochondria, and a single plant cell may have dozens of chloroplasts. Moreover, each of these organelles may contain multiple copies of the organellar genome. These facts have important consequences for the inheritance of traits determined by genes on the organelle chromosomes.

Variegation in *Mirabilis jalapa*

At the beginning of the chapter, we mentioned landmark studies performed by the plant geneticist Carl Correns on the inheritance pattern of color variegation in the plant *Mirabilis jalapa* (four-o'clocks). Variegated plants typically have branches that are variegated with some parts green and some white, and also branches that are solid green or solid white (**Fig. 15.10**). Correns performed all nine possible pairwise crosses between male

Figure 15.10 Variegated four-o'clock plant. Variegated plants typically have variegated, solid green, and solid white branches (sectors). The main shoot is usually variegated.

GENETICS AND SOCIETY

Crowd: © Image Source/Getty Images RF

Mitochondrial DNA Tests as Evidence of Kinship in Argentine Courts

Between 1976 and 1983, the military dictatorship of Argentina kidnapped, incarcerated, and killed more than 10,000 university students, teachers, union members, and others who did not support the regime. Many very young children disappeared along with the young adults, and close to 120 babies that were born to women in detention centers. In 1977, the grandmothers of some of these infants and toddlers held vigils in the main square of Buenos Aires to bear witness and inform others about the disappearance of their children and grandchildren (**Fig. A**). They soon formed a human rights group—the *Grandmothers of the Plaza de Mayo.*

The grandmothers' goal was to locate the more than 200 grandchildren they suspected were still alive and to reunite them with their biological families. To this end, they gathered information from eyewitnesses, such as midwives and former jailers, and set up a network to monitor the papers of children entering kindergarten. They also contacted organizations outside the country, including the American Association for the Advancement of Science (AAAS).

The grandmothers asked AAAS to help provide genetic analyses that would stand up in court. By the time a democracy had replaced the military regime and the grandmothers could argue their legal cases before an impartial court, children abducted at age 2 or 3 or born in 1976 were 7–10 years old. Although the external features of the children had changed, their genes—relating them unequivocally to their biological families—had not. The grandmothers, who had educated themselves about the potential of genetic tests, sought help with the details of obtaining and analyzing such tests. Starting in 1983, the courts agreed to accept their test results as proof of kinship.

In 1983, the best way to confirm or exclude the relatedness of two or more individuals was to compare proteins called human lymphocyte antigens (HLAs). People carry a unique set of HLA markers on their white blood cells, or lymphocytes, and these markers are diverse enough to form a kind of molecular fingerprint. HLA analyses can be carried out even if a child's parents are no longer alive, because for each HLA marker, a child inherits one allele from the maternal grandparents and one from the paternal grandparents. Statistical analyses can establish the probability that a child shares genes with a set of grandparents. In some cases, HLA analysis was sufficient to make a very strong case that a tested child belonged to the family claiming him or her. But in other cases, a reliable match could not be accomplished through HLA typing.

Figure A Grandmothers of the Plaza de Mayo (1977).
© Horacio Villalobos/Corbis

The AAAS put the grandmothers in touch with Mary Claire King, then at the University of California. King and two colleagues—C. Orrego and A. C. Wilson—developed an mtDNA test based on the then new techniques of PCR amplification and direct sequencing of a highly variable noncoding region of the mitochondrial genome. The maternal inheritance and the lack of recombination of mtDNAs mean that as long as a single maternal relative is available for matching, the approach can resolve cases of disputed relatedness. The extremely polymorphic noncoding region makes it possible to identify grandchildren through a direct match with the mtDNA of only one person—their maternal grandmother, or their mother's sister or brother—rather than through statistical calculations.

To validate their approach, King and colleagues amplified sequences from three children and their three maternal grandmothers without knowing who was related to whom. The mtDNA test unambiguously matched the children with their grandmothers. Thus, after 1989, the grandmothers included mtDNA data in their archives.

Today, the grandchildren—the children of *Los Desaparecidos* (the Disappeared)—have reached adulthood and attained legal independence. Although most of their grandmothers have died, the grandchildren may still discover their biological identity and establish what happened to their families through the mtDNA data the grandmothers left behind.

(pollen) and female (ova) gametes from flowers growing in each type of branch: variegated, green, or white. The progeny phenotypes always resembled those of the source of the female gamete (**Table 15.2**) because the color of the plant cells is controlled by maternally inherited cpDNA.

The reason for the variegated phenotype is that variegated four-o'clocks have two kinds of chloroplasts: wild-type and mutant. The mutant cpDNAs have a defective allele of a gene required for synthesis of the green pigment chlorophyll; cells without chlorophyll are white. A cell or organism with more than one genotype of an organellar

TABLE 15.2 Results of Correns's *Mirabilis* Crosses

Plant Branch Providing the Egg	Plant Branch Providing the Pollen	Progeny Phenotype
Green	Green	Green
Green	White	Green
Green	Variegated	Green
White	Green	White
White	White	White
White	Variegated	White
Variegated	Green	Green or White or Variegated
Variegated	White	Green or White or Variegated
Variegated	Variegated	Green or White or Variegated

genome is said to be **heteroplasmic (Fig. 15.11)**. The variegated plant considered as a whole is heteroplasmic because it came from a heteroplasmic egg—one that contained both wild-type and mutant chloroplasts. The plant also has **homoplasmic** cells—cells with only one type of cpDNA (Fig. 15.11). For example, the solid white areas of the plant are homoplasmic for mutant chloroplasts. Why are some cells heteroplasmic and others homoplasmic?

Figure 15.11 Cytoplasmic segregation of chloroplasts. Variegated plants contain both wild-type (*green*) and mutant (*white*) chloroplasts. Homoplasmic cells (cpDNAs all mutant or all wild type) can be generated when heteroplasmic cells (with mutant and wild-type cpDNAs) divide. Unequal distribution of chloroplasts may occur and by chance generate a daughter cell with only one type of cpDNA.

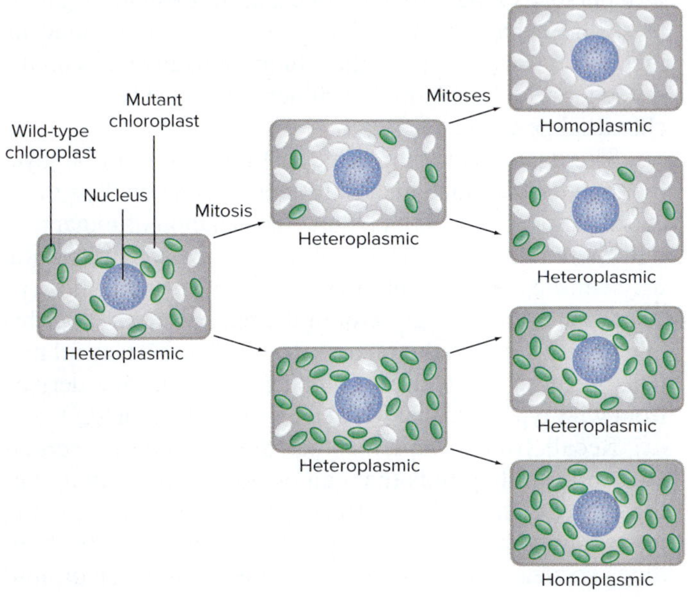

Cytoplasmic segregation of organellar genomes

When a cell undergoes mitosis, approximately half of the chloroplasts end up in each daughter cell. But this distribution is not precise, so the two daughters of a heteroplasmic cell do not receive exactly the same proportions of wild-type and mutant chloroplasts. It is easy to see in Fig. 15.11 that after several cell divisions, a homoplasmic descendant cell containing only one type of cpDNA could arise by such random **cytoplasmic segregation**.

Once a cell becomes homoplasmic, it cannot become heteroplasmic again (except by new mutation), and so all of its descendants from that point on are homoplasmic. Chance cytoplasmic segregation of chloroplasts explains at least in part how a plant that is heteroplasmic for wild-type and mutant chloroplasts could have a mixture of heteroplasmic, homoplasmic wild-type, and homoplasmic mutant cells.

The relationship of cytoplasmic segregation and variegation

Variegated plants usually have a variegated main shoot with patches of green and white tissue, and also branches that are solid green or solid white (Fig. 15.10). In the variegated regions, the green patches contain mainly heteroplasmic cells (**Fig. 15.12**). During mitosis, cells homoplasmic for mutant chloroplasts can arise, and they establish the white patches and also white branches. Cells homoplasmic for wild-type chloroplasts also arise and establish the solid green branches.

In four-o'clocks, heteroplasmic cells are green because the amount of chlorophyll even in a small number of wild-type chloroplasts is sufficient for green color. This phenomenon, where a particular fraction of wild-type organelles is sufficient for the normal phenotype, is called the **threshold effect.** The precise fraction of wild-type organelles needed to avoid a mutant phenotype will depend on the particular gene and mutation.

Now we can understand the results of Correns's crosses (Table 15.2). Female gametes from flowers on white branches are homoplasmic for mutant chloroplasts; they always give rise to solid white plants (which ultimately die because they cannot photosynthesize). Flowers from green branches give rise to eggs homoplasmic for wild-type chloroplasts and therefore solid green (nonvariegating) progeny. Finally, flowers from variegated branches can have any one of the three egg types (Fig. 15.12). The plant depicted in Fig. 15.10 came from a heteroplasmic egg of a flower from a variegated branch.

Heteroplasmy of individual organelles

Each chloroplast or mitochondrion may have several copies of its genome, so an individual organelle can itself be either heteroplasmic or homoplasmic for wild-type

Figure 15.12 Three egg types in variegated four-o'clocks. In variegated branches, the green sectors are composed mainly of heteroplasmic cells. Cytoplasmic segregation gives rise to some homoplasmic mutant cells that form the white areas, and also some homoplasmic wild-type cells in the green areas. Like the somatic cells, eggs in variegated branches can be homoplasmic (for either wild-type or mutant chloroplasts) or heteroplasmic. All eggs and somatic cells in solid green or solid white branches are homoplasmic for wild-type or mutant chloroplasts, respectively.

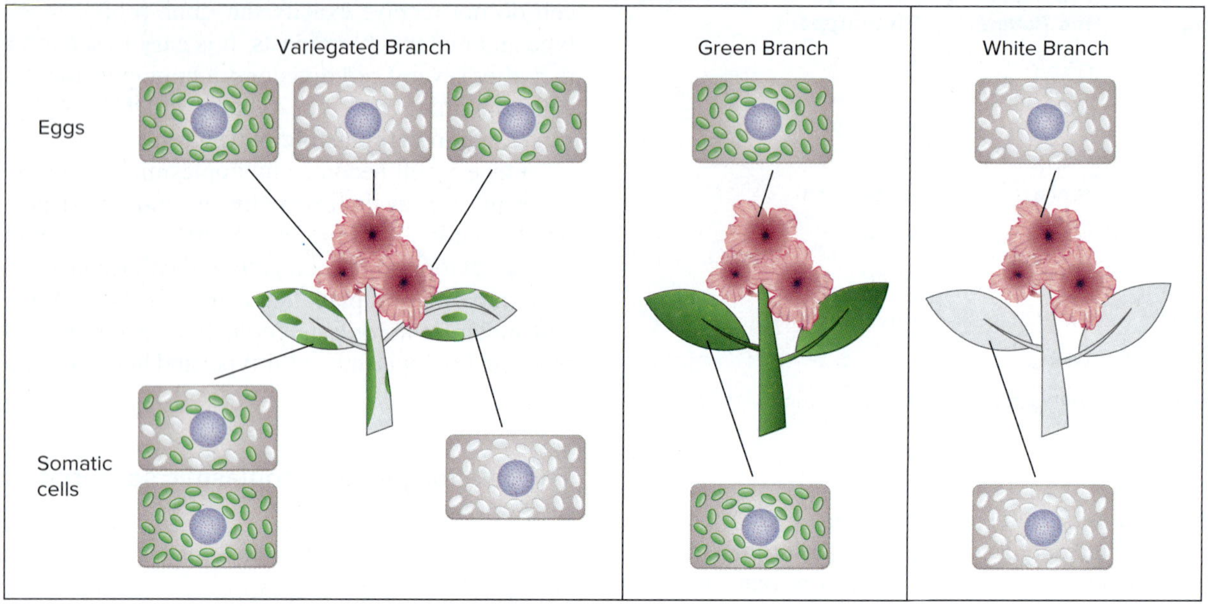

versus mutant DNAs. Two kinds of events can lead to cytoplasmic segregation of the genomes within an originally heteroplasmic organelle. First, when an organelle divides, distribution of genome copies to the daughter organelles is random, and therefore subject to chance cytoplasmic segregation of one kind of cpDNA. Second, not all DNA molecules within the organelle undergo replication, and which ones do seems to be random. As a result, some genomes replicate many times, while others do not replicate at all.

Whether or not an individual organelle will function normally (wild type) or not (mutant) is potentially subject to threshold effects. Therefore, just like the cells they inhabit, the phenotype of an individual organelle—whether it is functionally wild-type or mutant—is affected by its relative fractions of wild-type and mutant genome copies.

Some Organisms Exhibit Biparental Inheritance of Organellar Genomes

Although uniparental inheritance of organelles is the norm among most metazoans and plants, certain single-celled yeasts and some plants inherit their organellar genomes from both parents—that is, in a biparental fashion. In this section, we look at the example of the budding yeast *Saccharomyces cerevisiae*.

Studying mitochondrial gene inheritance in yeast

Drugs that inhibit mRNA translation in bacteria, such as chloramphenicol and erythromycin, are potent inhibitors of mitochondrial (but not normal cytoplasmic) protein synthesis. This fact was discovered in the early 1960s, when researchers found that chloramphenicol inhibits the growth of wild-type yeast on media containing a nonfermentable source of carbon (glycerol or ethanol); but the drug does not inhibit yeast growth on medium containing glucose, a fermentable carbon source. Because fermentation generates ATP anaerobically, independent of mitochondria, these scientists concluded that chloramphenicol acts on the mitochondrial translation machinery, encoded by the mitochondrial genome.

These investigators realized that growing yeast on glycerol or ethanol, which makes the cells depend on their mitochondria for growth, could allow the isolation of mutants in mitochondrial genes. The procedure was straightforward: they selected for mutants that, in contrast with wild-type yeast, could grow on glycerol in the presence of a drug that inhibits mitochondrial protein synthesis. The first useful mutants were resistant to chloramphenicol (C^r); they derived from wild-type cells that were sensitive to the drug (C^s).

Recall from Chapter 5 that *Saccharomyces cerevisiae* cells are haploids that can be either one of two mating types: a or α (Fig. 5.19a). Cells of opposite mating types can fuse to form a diploid, which grows by budding, so yeast crosses can generate cultures of diploid cells. Yeast is an *isogamous species* because the gametes (the haploid cells of opposite mating type) are of similar size and morphology.

Biparental inheritance and cytoplasmic segregation of mitochondria in yeast

To analyze the inheritance pattern of mtDNA in yeast, researchers mixed C^r and C^s parental cells of opposite mating types, and isolated diploids which they could select by taking advantage of nuclear auxotrophic markers in the parental strains. They allowed the diploid cells to divide through several generations of vegetative growth, and then scored individual diploid progeny for the C^r or C^s phenotype by replica plating the cells onto petri plates containing glycerol medium with chloramphenicol. The investigators found that some diploids were C^r and others were C^s (**Fig. 15.13**). This result shows that both parents transmitted organelles to the progeny.

Figure 15.13 Biparental mitochondrial inheritance in *Saccharomyces cerevisiae*. Resistance or sensitivity to chloramphenicol is controlled by a mitochondrial gene in yeast. A cross between chloramphenicol resistant (C^r) and chloramphenicol sensitive (C^s) haploid yeast of opposite mating types shows that diploids inherit mitochondria from both haploid cells. These mtDNA variants subsequently undergo cytoplasmic segregation when the diploid cells divide by mitosis.

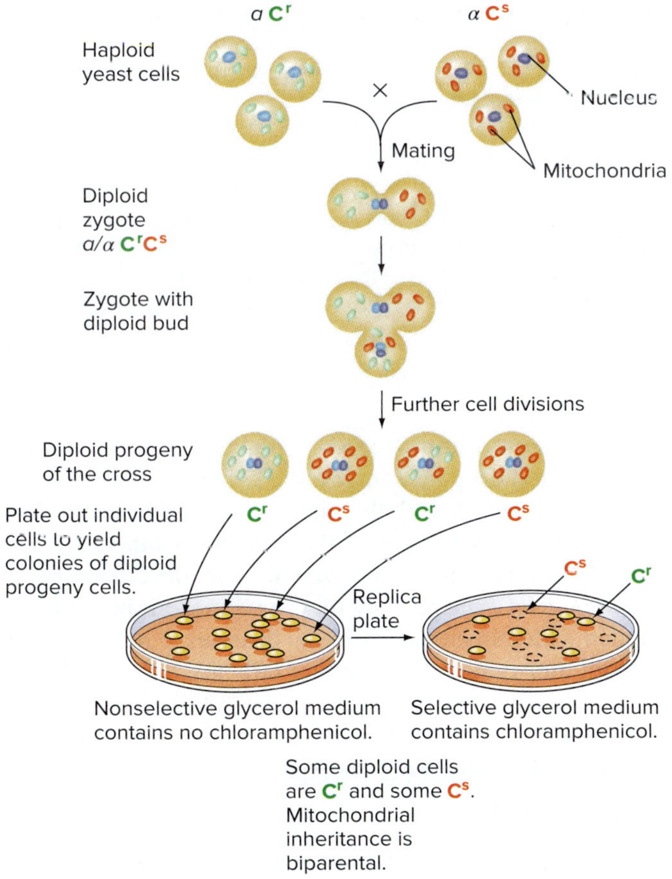

The experiment also shows that organelles from the parental cells of different mating types segregated during the mitotic divisions of vegetative growth. Immediately after mating, the diploids would contain both kinds of mitochondria, and the diploid cells would all be phenotypically C^r because that allele is dominant. But after several rounds of division, some cells contained only C^s mitochondria and were thus phenotypically C^s. Clearly, the C^s and C^r mtDNAs underwent random cytoplasmic segregation (Fig. 15.13).

One aspect of *S. cerevisiae* biology makes this process of cytoplasmic segregation of mtDNAs so rapid that it is often achieved after only a few rounds of mitosis. You will recall that this yeast divides mitotically through a process of budding that creates a small bud from a larger mother cell (review Fig. 5.19a). Because of this inequality in size, only a few mtDNA molecules are transferred into the newly formed bud, making it more likely that the bud will contain mostly or only one type of mtDNA.

essential concepts

- In many organisms, transmission of organelle-encoded traits is uniparental and maternal. The phenotype of all the progeny resembles that of the mother.
- *Maternal inheritance* of organelles is often a consequence of the larger size of the female gamete relative to the male gametes. Additional mechanisms can also destroy or exclude organelles that originate in the male gamete.
- Mitochondria are inherited biparentally in some organisms like yeast, in which the two gametes are of similar size.
- Cells may carry a mixture of organellar DNAs. The organellar genomes of these *heteroplasmic cells* can segregate from each other during several rounds of mitosis, generating *homoplasmic cells* with only one type of organellar DNA. Cytoplasmic segregation can cause *variegation*—patches of tissue with different phenotypes.

15.5 Mutant Mitochondria and Human Disease

learning objectives

1. Recognize mitochondrial diseases in human pedigrees.
2. Explain the effect of heteroplasmy on the manifestation of mitochondrial diseases.
3. Discuss why some scientists think a relationship exists between mitochondria and aging.
4. Describe how oocyte nuclear transfer can be used to prevent the transmission of mitochondrial disease.

FAST FORWARD

Sprinters: © Robert Michael/Corbis RF

Mitochondrial Eve

The strictly maternal inheritance of mitochondria provides a unique opportunity to make inferences about our evolutionary history. Through such studies, the geneticist Allan C. Wilson and his colleagues reached the startling conclusion that the mitochondrial DNAs of all humans alive today trace back through maternal lineages to the mtDNA of a single woman who lived in Africa some 200,000 years ago.

The carrier of this ancestral mtDNA, dubbed *Mitochondrial Eve,* probably lived in a population of 10,000 to 50,000 people who interbred and lived together in the same area.

Although all humans alive today can trace their mitochondria to a single female who lived a long time ago, our mitochondrial DNA is not all alike. In fact, it is these differences in the base sequences of our mitochondrial genomes, caused by random mutations that have occurred over the past 200,000 years, that enabled these scientists to trace the path of the mitochondria back to Mitochondrial Eve.

Wilson and his coworkers analyzed mitochondrial DNA sequences of about 250 individuals representing all of the inhabited continents of the world. The researchers found more sequence differences among different African individuals than among Asians or Europeans. Because mutations occur over time, the African population must have had the longest time to accumulate variations. The conclusion is straightforward: Modern humans originated in Africa.

The researchers could estimate the probable date of Mitochondrial Eve's existence by assuming that the rate at which mutations accumulate in mtDNA is relatively constant. The scientists knew that chimpanzees and humans diverged from a common ancestor approximately 5 million years ago and that human mtDNA differs from that of chimpanzees in about 15% of the genome. Adjusting these data to account for multiple substitutions at the same base pair, they estimated that the mtDNA of humans and chimpanzees has been diverging at an average rate of about 13.8% per million years.

To determine approximately when Mitochondrial Eve lived, Wilson and his colleagues considered that the greatest amount of human mtDNA variation represents about 2.8% of the organellar DNA. The researchers then simply divided this number by the rate determined from the comparison of chimp and human genomes to estimate the last time the most divergent present-day humans last shared a common maternal ancestor:

$$\frac{2.8\%\ \text{bp changes}}{13.8\%\ \text{bp changes/million years}} = 0.20\ \text{million years}$$

$$= 200{,}000\ \text{years ago}$$

Although some controversy has existed over the statistical methods and assumptions that formed the basis of this analysis, most geneticists now agree with the conclusion that the women carrying our ancestral mtDNA lived roughly 200,000 years ago in sub-Saharan Africa.

Who was Mitochondrial Eve? The answer is that she is simply the one woman to whom all the mtDNA present in humans alive today can be traced. **Figure A** illustrates the concept.

You can see that despite her name, Mitochondrial Eve was certainly not the first woman; many other women were alive at the same time, and many other women existed before her. But the mtDNAs from these other women were not passed on to present-day humans. During the last 200,000 years, either these women or their surviving matrilineal descendants in later generations had no children or they only had sons. You can visualize this concept in Fig. A in the colored lineages that end before the current generation.

A similar ancestry trace determined that the man who carried the most recent common ancestor of all present-day Y chromosomes—so-called *Y chromosome Adam*—also lived in sub-Saharan Africa, but most likely at a different time (about 200,000 to 300,000 years ago). In Chapter 21, you will learn in more detail how scientists compare the DNA sequences of individuals to make inferences about human evolutionary history.

Just as in all chromosomes, mutations occur in human mtDNAs. As a result, present-day humans have various polymorphisms that distinguish their mitochondrial genomes. Geneticists can analyze the sequences of human mtDNAs to explore the evolutionary history of our species. The fact that mitochondrial genomes are maternally inherited provides a unique perspective to these evolutionary studies. The Fast Forward Box *Mitochondrial Eve* explains how comparisons of mtDNA sequences suggest that all humans alive today descend from a woman in Africa who lived about 200,000 years ago.

Certain polymorphisms in mtDNAs have profound phenotypic consequences. In particular, several debilitating diseases of the human nervous system are caused by mutations in the mitochondrial genome. These disease genes are passed from mothers to daughters and sons, and from affected daughters to granddaughters and grandsons, and so on down through the maternal line. Because of heteroplasmy, the symptoms of these diseases can vary enormously among family members. The development of *mitochondrial gene therapy* may soon make it possible for women with mitochondrial-based diseases to have children without transmitting their mutant mtDNA.

In LHON, Affected People Are Usually Homoplasmic

Leber's hereditary optic neuropathy, or LHON, is a disease in which flaws in the mitochondrial electron

Figure A Most recent common ancestor (MRCA) of mtDNA. The mtDNA in all present-day people (*generation 6*) can be traced to a single woman (*black* in *generation 1*)—the carrier of the MRCA of mtDNA. The mtDNA of the other females shown in earlier generations *(colors other than black)* were lost sometime prior to generation 5. Because mtDNA transmission is maternal, the mtDNA variants carried by males do not influence the distribution of mtDNAs in the next generation. Once all the females in the population carry the MRCA variant, all the males in the next generation must do so as well.

Source: C. Rottensteiner, http://en.wikipedia.org/wiki/File:MtDNA-MRCA-generations-Evolution.svg

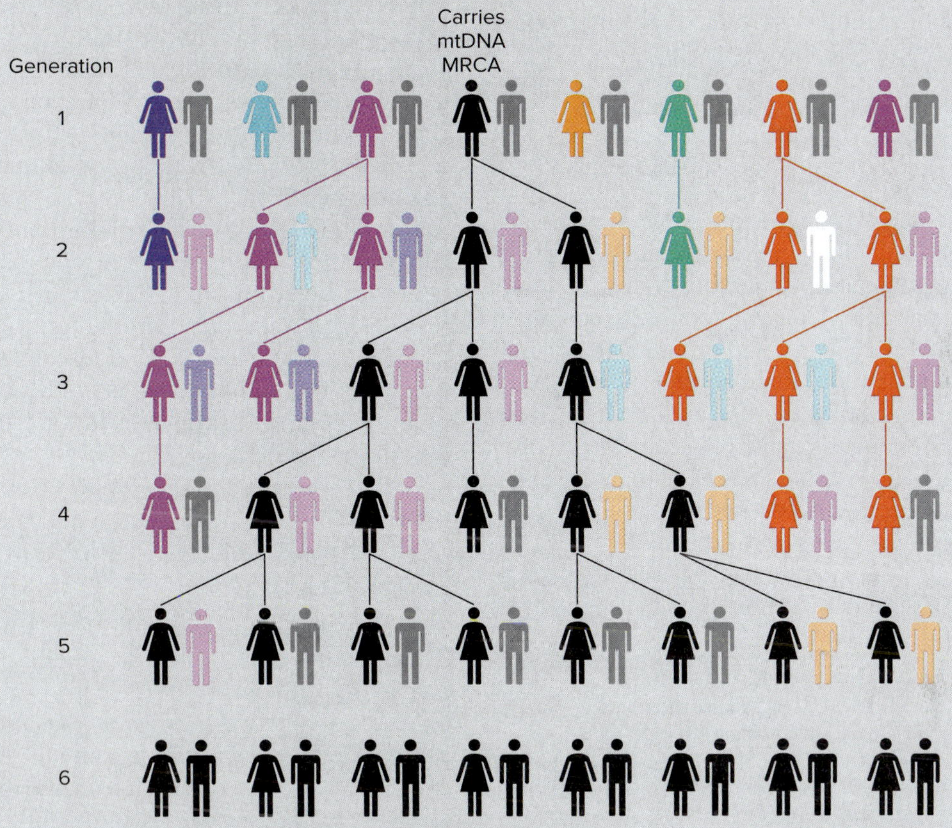

transport chain lead to optic nerve degeneration and blindness. Family pedigrees show that LHON passes only from mother to offspring (**Fig. 15.14**). LHON is caused by hypomorphic mutations in any one of three mitochondrial genes—*ND1*, *ND4*, or *ND6*—each of which encodes a different subunit of the enzyme NADH dehydrogenase, the first enzyme in the electron transport pathway previously shown in Fig. 15.2b. Diminished electron flow down the respiratory transport chain reduces the mitochondrion's production of ATP, causing a gradual decline in cell function and ultimately cell death. Optic nerve cells have a relatively high requirement for energy, so the genetic defect affects vision before it affects other physiological systems.

Figure 15.14 Characteristic pedigree for mitochondrial disease. All offspring of diseased mothers, but none of the children of diseased fathers, are affected. Although some LHON pedigrees show this idealized inheritance pattern, the trait can show incomplete penetrance and variable expressivity because some LHON patients are heteroplasmic, and because the trait can be influenced by various alleles of nuclear genes.

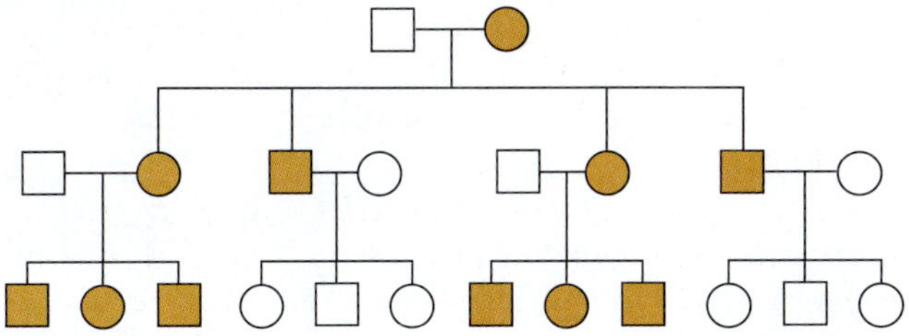

Because the disease-causing alleles are weak mutants, in most people affected by LHON the optic nerve cells are homoplasmic for the disease mutation (all the mitochondria are mutant); usually this means that the person as a whole is homoplasmic for mutant mitochondria. For this reason, LHON often (but not always) shows the simplest possible inheritance pattern for a mitochondrial disease (Fig. 15.14): All the mtDNAs in the ova of every affected female are mutant, so all of their progeny—both males and females—are homoplasmic for mutant mitochondria. LHON pedigrees are not always as clear-cut as that shown in Fig. 15.14; in some families the inheritance is incompletely penetrant, likely due to heteroplasmy.

In MERRF, Affected Individuals Are Heteroplasmic

People with a rare inherited condition known as myoclonic epilepsy and ragged red fiber disease (MERRF) have a range of symptoms: uncontrolled jerking (the myoclonic epilepsy part of the condition), muscle weakness, deafness, heart problems, kidney problems, and progressive dementia. Affected individuals often have an unusual "ragged" staining pattern in regions of their skeletal muscles, which explains part of the condition's name (**Fig. 15.15**).

MERRF is caused by loss-of-function mutations in mitochondrial tRNA genes; 90% of people with MERRF carry a mutation in $tRNA^{Lys}$. Because these mutations affect the translation of all mitochondrial mRNAs, they have a major deleterious effect on ATP production. People with these mutant mtDNAs are always heteroplasmic for mutant and wild-type mitochondria because cells homoplasmic for such mutant DNAs would die. Heteroplasmy results in wide variations in the penetrance and expressivity of disease symptoms.

Figure 15.15 Muscle fiber cross section in a MERRF patient. Mitochondria are stained in *red*, cell cytoplasm is *blue*. In muscle fibers of this heteroplasmic individual that by chance have a higher proportion of mutant mtDNAs, the mitochondria function poorly. To try to compensate, these *ragged red fibers* make more mitochondria (and thus display more red staining). In spite of this compensation, ragged red fibers generate less energy and are thus less robust than normal fibers.

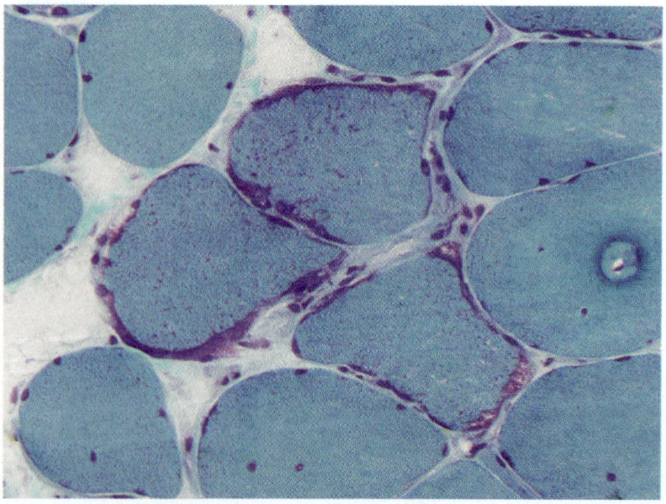

Variation in the proportion of mutant mitochondria in eggs

As the pedigree in **Fig. 15.16** shows, family members inherit MERRF from their mothers; none of the offspring of affected males exhibit disease symptoms. The family history also reveals individual variations in the severity of symptoms, which correlate roughly with their overall fraction of mutant mitochondria. Note in Fig. 15.16 that the fraction of mutant mitochondria in mothers and their progeny are not always the same, and neither is the proportion of mutant mitochondria the same among siblings. The reason is that a woman whose cells are heteroplasmic for the MERRF mutation produces eggs that vary in the relative proportions of mtDNAs with wild-type and mutant $tRNA^{Lys}$ genes. The precise combination depends on the random partitioning of mitochondria during the mitotic divisions that gave rise to the germ line.

Variation in the tissue distribution of mutant mitochondria and phenotypes

In each person heteroplasmic for MERRF-causing mitochondria, the ratio of mutant to wild-type mtDNA varies considerably from tissue to tissue due to random cytoplasmic segregation during mitosis. And because each tissue has its own energy requirements, even the same ratio can affect different tissues to varying extents (**Fig. 15.17**). Muscle and nerve cells have the highest energy needs of all types of cells and therefore depend the most on oxidative phosphorylation. Chance segregation of mutant mtDNAs to these tissues generates the defining features of MERRF.

Figure 15.16 Maternal inheritance of MERRF. Although this pedigree shows maternal transmission as expected for mitochondrial mutations, the overall fraction of mutant mtDNA in the cells of heteroplasmic individuals varies (see *bottom*) and corresponds with the severity of the condition (indicated by different color coding).

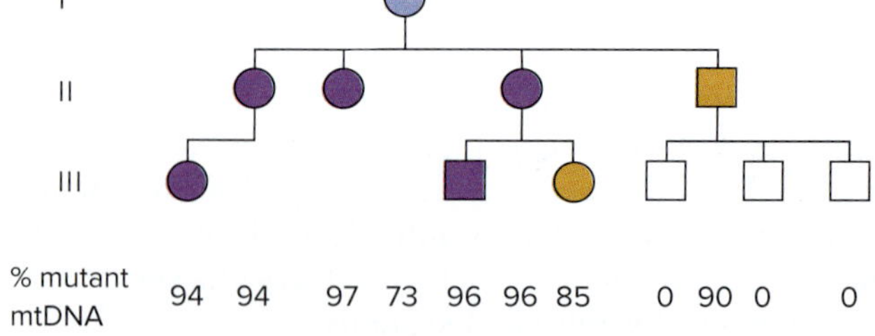

Figure 15.17 MERRF symptoms and the ratio of mutant to wild-type mtDNAs. The proportion of mutant mitochondria (*blue*) and the nature of tissues that are affected together determine the severity of MERRF symptoms. Tissues with higher energy requirements (for example, brain) are least tolerant of mutant mitochondria. Tissues with low energy requirements (for example, skin) are affected only when the proportion of wild-type mitochondria (*purple*) is greatly reduced.

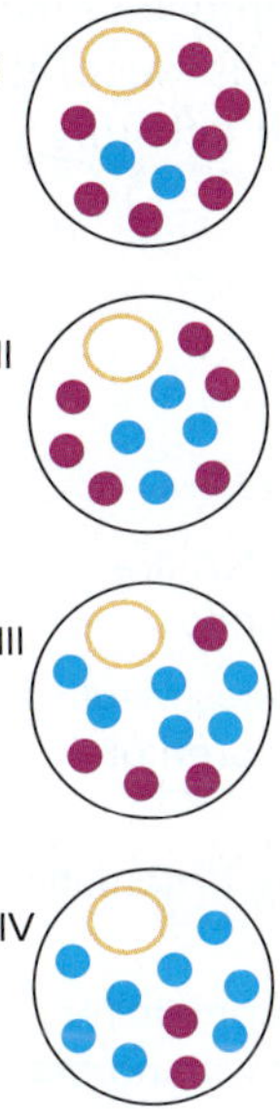

Individual mtDNA Genotypes		Tissues Affected: Brain	Heart	Skeletal Muscle: Type I	Skeletal Muscle: Type II	Skin
I	20% mutant mtDNAs	+	–	–	–	–
II	40% mutant mtDNAs	+	+/–	–	–	–
III	60% mutant mtDNAs	+	+	+	–	–
IV	80% mutant mtDNAs	+	+	+	+/–	+/–

Mitochondrial Mutations May Affect Human Aging

Some mutations in mtDNA are inherited through the germ line, while others arise sporadically in somatic cells as a result of random events, such as exposure to radiation or chemical mutagens. In fact, in humans the rate of somatic mutations is much higher in mitochondrial DNA than in nuclear DNA. One reason is that the mitochondrial oxidative phosphorylation system generates high levels of DNA-damaging free radicals within the organelle.

Some researchers focusing on the genetics of aging think that the accumulation of mtDNA mutations over a person's lifetime results in an age-related decline in oxidative phosphorylation. This decline, in turn, accounts for some of the symptoms of aging, such as decreases in heart and brain function. One piece of evidence supporting this hypothesis is that the brain cells of people showing symptoms of Alzheimer's disease (AD) have an abnormally low energy metabolism. Intriguingly, 20% to 35% of the mitochondria in the brain cells of most AD patients carry mutations in two of their three cytochrome *c* oxidase genes, which could impair the brain's energy metabolism. Further research will be necessary to discover whether mitochondrial damage indeed makes a significant contribution to the aging process.

Oocyte Nuclear Transplantation Can Sidestep Transmission of Mitochondrial Disease

Mitochondrial disease is relatively rare, yet every year in the United States alone, several thousand babies are born with a disease caused by mtDNA mutation. New technology may soon make it possible for an affected mother to avoid transmitting the mitochondrial disease to her children.

The idea behind this technology, referred to as **mitochondrial gene therapy,** is to remove the nuclei from donor eggs that have normal mitochondria and replace each of these nuclei with a nucleus from an egg obtained from the prospective mother with the mitochondrial disease (**Fig. 15.18**). The resulting eggs, which have nuclei

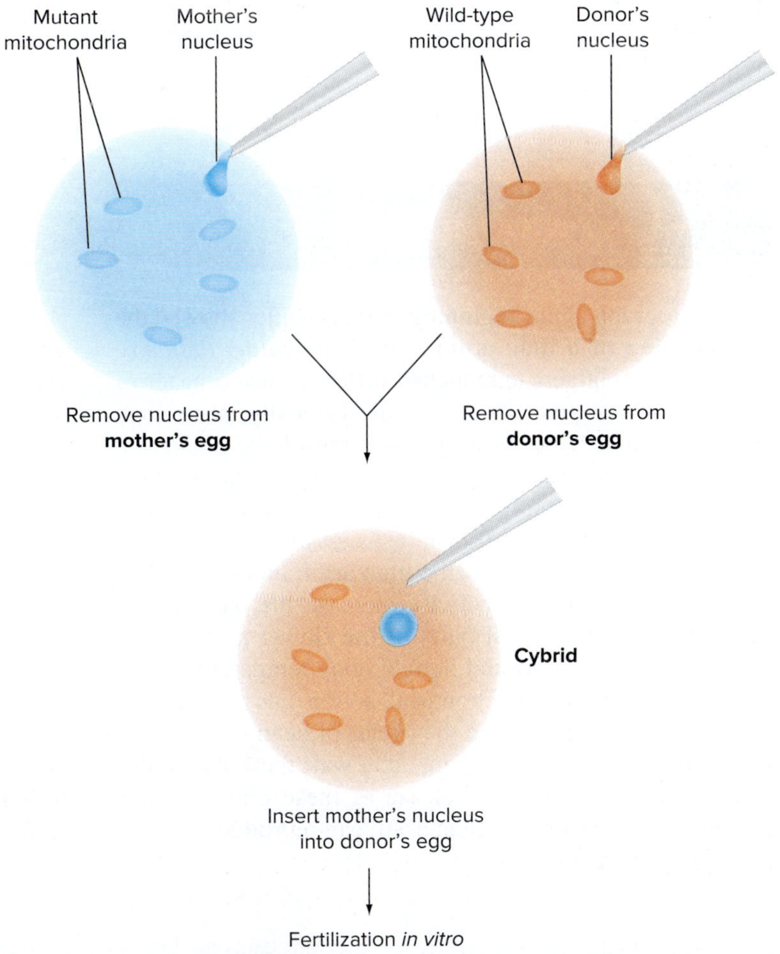

Figure 15.18 Mitochondrial gene therapy. Ooctye nuclear transfer from an egg with mutant mitochondria to an egg with normal mitochondria creates a *cybrid* egg. This technology could potentially prevent transmission of mitochondrial disease.

Figure 15.19 Mito and Tracker: successful oocyte nuclear transfer in primates. The two rhesus monkeys shown are products of oocyte nuclear transfer and *in vitro* fertilization. Their cells contain nuclear DNA from both the father and the oocyte nuclear donor, while their mtDNA is derived only from the enucleated oocytes of the cytoplasm donor.

from one source and mitochondria from a different source, are called **cybrids** (for cytoplasmic hybrids).

After *in vitro* fertilization, the zygotes can be implanted in a womb. The resulting child would have the nuclear genes of his or her parents, and the normal mtDNA of the egg donor. A crucial aspect of this procedure is the clean transfer of the mother's nucleus unaccompanied by any mitochondria—otherwise the resulting zygote would be heteroplasmic.

In 2009, a team of researchers transferred the nuclei of rhesus monkey eggs into enucleated donor eggs and fertilized these manipulated eggs *in vitro* with rhesus monkey sperm. The zygotes were implanted into a surrogate mother's womb. Apparently normal fraternal twins, Mito and Tracker, were born (**Fig. 15.19**), and molecular analysis showed they had mtDNA only from the cytoplasmic donor.

The success of the monkey experiment led the government of the United Kingdom in 2015 to approve clinical trials of mitochondrial gene therapy in humans. As of this writing in 2016, the U.S. government is still debating the ethics of this new technology.

essential concepts

- Diseases caused by mutation in human mtDNA are recognized by a pattern of maternal inheritance.
- Heteroplasmy for mitochondrial disease alleles results in phenotypic differences even among the children of a single affected mother.
- Accumulation of mtDNA mutations over time in somatic cells may play a role in aging.
- Transfer of an oocyte nucleus from an egg containing mutant mtDNA to an enucleated donor egg with normal mtDNA may help people to avoid transmitting mitochondrial diseases.

WHAT'S NEXT

The photograph at the beginning of this chapter showed the rare opening of a titan arum (*Amorphophallus titanum*), which has the largest unbranched inflorescence (cluster of flowers) of any plant in the world. Typically, more than 10 years elapse between each blooming of the plant, and the inflorescence lasts only about two days. Titan arum is famous for its stench; due to the characteristic smell of rotting flesh, it is known colloquially as *corpse plant.*

During the few days of the titan arum bloom, cyclical bursts of mitochondrial activity increase the temperature inside the bloom, so that parts of the flower periodically become hotter than normal human body temperature. The heat generated by mitochondrial action causes odorous molecules to volatilize, and the cycles of heating and cooling produce convection currents that waft these molecules away from the plant. The release of these odors attracts pollinators, ensuring that titan arum can reproduce.

Even though the same nuclear and mitochondrial genes are present in all cells of the titan arum, the use of these genes varies in ways that matter to the organism. The expression of certain genes in some cells but not others ensures that various cells develop into different structures in the plant, such as roots or flowers. The expression of genes can also vary over time, causing phenomena like the cyclical heating and cooling of the titan arum in bloom.

In the next two chapters, we will explore the molecular mechanisms by which organisms control the expression of their genes—the process of *gene regulation.* In Chapter 16, we will discuss mechanisms of gene regulation in bacteria. Then, in Chapter 17, we explore the regulation of nuclear genes in eukaryotes. We will see that gene regulation controls not only the metabolic activities of all cells, but the differentiation of cells in multicellular organisms.

SOLVED PROBLEMS

I. In the early 1970s, Igor Dawid and Antonie Blackler conducted classical experiments that first showed directly the maternal inheritance of mtDNA in vertebrates. Their studies used crosses between two closely related species of frogs, *Xenopus laevis* and *Xenopus borealis*, which have mtDNAs that vary in many nucleotides. The techniques they used at the time were not sensitive enough to detect small amounts of paternal DNA. What techniques that are highly sensitive to small amounts of DNA could be used today? How could you use these techniques to determine if paternal mitochondrial DNA was present in the progeny of the interspecies cross?

Answer

The polymerase chain reaction (PCR) is a sensitive technique that detects very small amounts of DNA. Oligonucleotide primers that are specific for each of the mitochondrial DNAs in each of the two different species could be used to determine if paternal DNA is present in the offspring from the interspecies cross. You would take, for example, DNA from several tadpoles resulting from a cross between *X. laevis* females and *X. borealis* males and employ the PCR test. You would expect that only *X. laevis*-specific PCR products would be obtained if this is purely maternal inheritance. As a control, you would look at tadpoles from a cross between *X. laevis* males and *X. borealis* females; here you would expect only *X. borealis*-specific PCR products.

An alternative approach is high-throughput DNA sequencing. If you purified total DNA from tadpoles and then sequenced random fragments, you would determine the sequence not only of the nuclear genome, but also mitochondrial genomes. If you performed enough sequencing reactions, you would be able to detect whether even small amounts of paternal mtDNA were present in the progeny.

II. a. Does the following pedigree suggest mitochondrial inheritance? Why or why not?

b. What other mode(s) of inheritance is (are) consistent with these data?

c. How could you distinguish between mitochondrial inheritance and any other possibility that is consistent with the data?

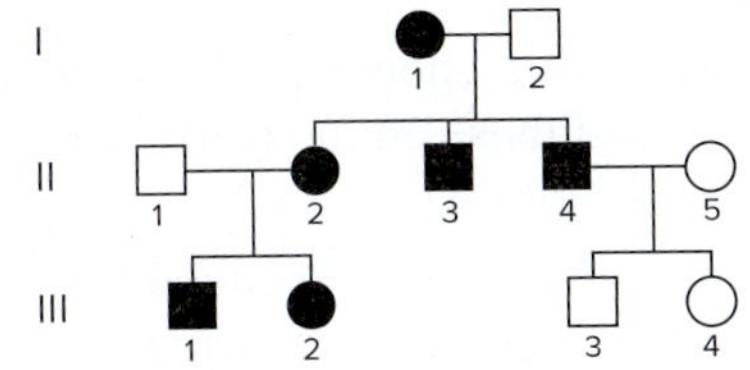

Answer

a. The data presented in this pedigree are consistent with mitochondrial inheritance because the trait is transmitted by females; the affected males in this family did not transmit the trait; and all of the females' progeny have the trait.

b. This inheritance pattern is also consistent with transmission of an autosomal dominant trait. According to this hypothesis, individuals I-1 and II-2 passed on the dominant allele to all children, but II-4 did not pass on the dominant allele to either child.

c. The symptoms exhibited by affected individuals provide one clue. If the disease is mitochondrial, you would expect to see symptoms similar to those in patients known to have mitochondrial diseases such as LHON or MERRF. However, such evidence is not conclusive because some diseases that affect mitochondria result from mutations in nuclear genes. You would ultimately need to examine the mtDNA in the affected individuals to see if any alterations exist that could disrupt the structure or expression of genes in the mitochondrial genome.

PROBLEMS

Vocabulary

1. Choose the phrase from the right column that best fits the term in the left column.

a. cytoplasmic segregation	1. transmission of genes through maternal gamete only
b. heteroplasmic	2. cell that has mtDNAs or cpDNAs all of one genotype
c. homoplasmic	3. having gametes of similar size
d. maternal inheritance	4. a cell with a mixture of different mtDNAs generates a daughter cell with only one kind
e. uniparental inheritance	5. a specific fraction of wild-type organellar DNAs is required for a wild-type phenotype
f. isogamous	6. cell with mtDNAs or cpDNAs with different genotypes
g. threshold effect	7. transmission of genes through either a maternal or a paternal gamete, but not both

Section 15.1

2. Assuming human cells have on average 1000 mitochondria, what percentage by weight of the total DNA isolated from human tissue would be mtDNA?

3. *Reverse translation* is a term given to the process of deducing the DNA sequence that could encode a particular protein. If you had the amino acid sequence Trp His Ile Met:
   a. What human nuclear DNA sequence could have encoded these amino acids? (Include all possible variations.)
   b. What human mitochondrial DNA sequence could have encoded these amino acids? (Include all possible variations.)

4. The human nuclear genome encodes tRNAs with 32 different anticodons (excluding $tRNA^{Sec}$ that was described in Fig. 8.22). The mitochondrial genome encodes only 22 different tRNAs that are sufficient to translate all mitochondrial mRNAs. The differences in the nuclear and mitochondrial genetic codes (see Table 15.1) are not great enough to explain the difference in the numbers of tRNAs needed in each case. How can the difference be explained? (*Hint:* Think about the wobble rules shown in Fig. 8.21b.)

5. The human mitochondrial genome includes no genes for tRNA synthetases.
   a. How are mitochondrial tRNAs charged with amino acids?
   b. Given your answer to part (a), explain how AUA can specify Met in mitochondria but Ile in the nucleus.

6. How do you know if the halibut you purchased at the supermarket is really halibut? To identify the source of a biological sample, scientists PCR amplify and then sequence a region of DNA known to vary between species. For animals, this DNA region is a 648–base pair portion of the mitochondrial *cytochrome oxidase I* gene. The sequence of this mtDNA region acts as a so-called *DNA barcode* because a database exists that contains the sequences of this mtDNA region that are unique for hundreds of thousands of animal species.
   a. Why do you think that a region of mitochondrial DNA is used for barcoding animals, as opposed to a region of nuclear genomic DNA?
   b. A single pair of PCR primers can be used to barcode any species of fish. Explain how this is possible.
   c. List criteria that scientists would have considered when determining which mitochondrial DNA sequence to use for barcoding animals.

Section 15.2

7. Is each of these statements true of chloroplast or mitochondrial genomes, both, or neither?
   a. contain tRNA genes
   b. encode proteins that participate in electron transport pathways
   c. all genes necessary for function of the organelle are present
   d. vary greatly in size from organism to organism

8. Suppose you are characterizing the DNA of a diploid plant species that had never been analyzed previously. You purify all the DNA that can be isolated from a seedling, and subject this DNA to high-throughput sequencing involving millions of reads of random DNA fragments.
   a. If you obtained on average 100 reads of a given single-copy nuclear DNA sequence, about how many reads would you obtain for mtDNA? For cpDNA? (Assume each mitochondrion has 10 genome copies and that each choloroplast has 20 genome copies. Assume also that the average cell of this plant species has 1000 mitochondria and 50 chloroplasts.)
   b. Beyond the number of reads, what other criteria would allow you to conclude whether a particular read was of nuclear DNA, mtDNA, or cpDNA?

9. An example of a gene-targeting DNA plasmid vector for insertion of a transgene into chloroplast DNA by biolistic transformation is shown in the following diagram. The plasmid DNA can be prepared in large quantities in *E. coli* before being shot into plant cells with a gene gun. Match the component of the construct with its function. (RE1, RE2, and RE3 are different restriction enzyme recognition sites unique to the plasmid vector.)

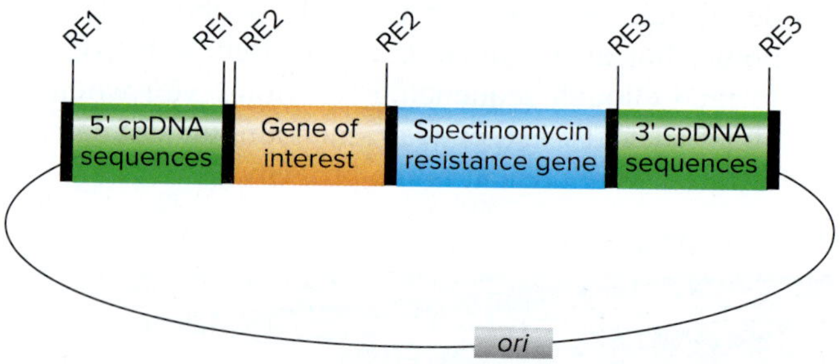

a. spectinomycin resistance gene	1. DNA that mediates integration
b. cpDNA sequences	2. gene used to select chloroplast transformants
c. RE2	3. sequence for plasmid replication in *E. coli*
d. *ori*	4. sites at which targeting chloroplast DNA can be inserted into the vector
e. RE1 and RE3	5. site at which transgene can be inserted into the vector

Section 15.3

10. Which of the following characteristics of chloroplasts and/or mitochondria make them seem more similar to bacterial cells than to eukaryotic cells?
    a. Translation is sensitive to chloramphenicol and erythromycin.
    b. Alternate codons are used in mitochondria genes.
    c. Introns are present in organelle genes.
    d. DNA in organelles is not arranged in nucleosomes.
11. The *Saccharomyces cerevisiae* nuclear gene *ARG8* encodes an enzyme that catalyzes a key step in biosynthesis of the amino acid arginine. This protein is normally synthesized on cytoplasmic ribosomes, but then is transported into mitochondria, where the enzyme conducts its functions. In 1996, T. D. Fox and his colleagues constructed a strain of yeast in which a gene encoding the Arg8 protein was itself moved into mitochondria, where functional protein could be synthesized on mitochondrial ribosomes.
    a. How could these investigators move the *ARG8* gene from the nucleus into the mitochondria, while permitting the synthesis of active enzyme? In what ways would the investigators need to alter the *ARG8* gene to allow it to function in the mitochondria instead of in the nucleus?
    b. Why might these researchers have wished to move the *ARG8* gene into mitochondria in the first place?

Section 15.4

12. The so-called *hypervariable regions* (HV1 and HV2) of the human mitochondrial genome are sometimes used in forensic analysis. They are two noncoding regions of the mitochondrial genome, each approximately 300 bp, that flank the origin of replication; the function of these DNA sequences is not well understood. However, these two regions of mtDNA show the most variation (SNPs and InDels) among different people. The DNA within HV1 and HV2 accumulates mutations at ten times the rate of DNA sequences in the nuclear genome.
    a. Under what circumstances would human mtDNA be preferable over nuclear DNA for identifying individuals?
    b. What are the disadvantages of using mtDNA, relative to nuclear DNA, in order to identify individuals?
13. Suppose a new mutation arises in a mitochondrial genome. Explain what would have to happen in order for the mutation to express itself phenotypically.
14. Describe at least two ways in which the contribution of mitochondrial genomes from male parents is prevented in the offspring of different species.
15. Why are severe mitochondrial or chloroplast gene mutations usually found in heteroplasmic cells instead of homoplasmic cells?
16. Suppose you are examining a newly found plant species, and you want to determine whether the inheritance of mtDNA is maternal, paternal, or biparental. You find that in the population two variants of mtDNA exist that can be distinguished by size differences of PCR amplification products made with a particular pair of primers. You first perform PCR analysis on DNA isolated from a leaf on each of two individual plants. Then you cross eggs from plant 1 with pollen from plant 2, obtain four seedlings, and perform PCR analysis with DNA from each of the whole seedlings. The results are shown below.

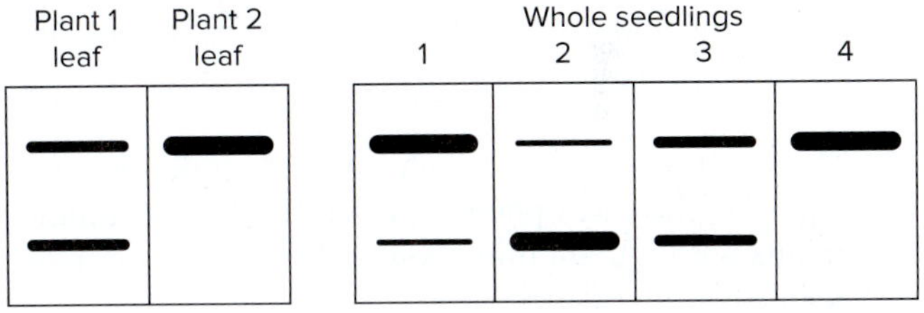

    a. Assuming plant 2 is homoplasmic, do these results exclude any of the three possible models for the inheritance of mtDNA in this species (maternal, paternal, or biparental)?
    b. What experiment could you perform to distinguish between the models that remain?
    c. What experiment(s) could you perform to check your assumption that plant 2 is in fact homoplasmic? Why is such an experiment necessary to make a conclusion about the mode of inheritance of mtDNA in this species?
    d. Explain the differing proportions of the two forms of mtDNA in the four seedlings. Be as specific as possible.
17. A form of male sterility in corn is inherited maternally. Marcus Rhoades first described this *cytoplasmic male sterility* by crossing female gametes from a male sterile plant with pollen from a male fertile plant. The resulting progeny plants were male sterile.
    a. Diagram the cross, using different colors and shapes to distinguish between nuclear (lines) and cytoplasmic (circles) genomes from the male sterile (one color) and male fertile (another color) strains.

b. Female gametes from the male sterile progeny were backcrossed with pollen from the same male fertile parent of the first cross. The process was repeated many times. Diagram the next two generations including possible crossover events.

c. What was the purpose of the series of backcrosses? [*Hint:* Look at your answer to part (b) and think about what is happening to the nuclear genome.] Why could Rhoades interpret these results as a demonstration of cytoplasmic male sterility?

18. Plant breeders have long appreciated the phenomenon called *hybrid vigor* or *heterosis,* in which hybrids formed between two inbred strains have increased vigor and crop yield relative to the two parental strains. Starting in the 1930s, seed companies exploited the cytoplasmic male sterility (CMS) phenomenon in corn that was described in Problem 17 so that they could cheaply produce hybrid corn seed to sell to farmers. This type of CMS is caused by mutant mitochondrial genomes that prevent pollen formation.

    a. How would CMS aid seed companies in producing hybrid corn seed?

    Dominant *Rf* alleles of a nuclear gene called *Restorer* suppress the CMS phenotype, so that *Rf*-containing plants with mutant mitochondrial genomes are male fertile.

    b. Describe a cross generating hybrid corn seed that would grow into fertile (self-fertilizing) plants. (Farmers planting hybrid seed want fertile plants because corn kernels result from fertilized ovules.)

    c. One of the historical challenges in the commercialization of hybrid corn produced through CMS was the maintenance of strains with CMS mitochondria: How could the seed companies keep producing male sterile corn plants if these plants never themselves produced pollen? Suggest a strategy by which the seed companies could continue to obtain male sterile plants every breeding season.

    d. Are there any potential disadvantages to the use of hybrid corn? If so, what issues might arise?

19. A mutant haploid strain of *Saccharomyces cerevisiae* (yeast) called *cox2-1* was found that was unable to grow on media containing glycerol as the sole source of carbon and energy. (Glycerol is a nonfermentable substrate for yeast.) This strain could, however, grow on the fermentable substrate glucose. Researchers discovered that *cox2-1* cells lack a mitochondrial protein called cytochrome *c* oxidase.

    a. Explain why *cox2-1* cells can grow on medium containing glucose but not on glycerol medium.

    b. When *cox2-1* was crossed with a wild-type yeast strain and the resultant diploid cells were allowed to grow mitotically, it was found that about half the diploid clones were able to grow on glycerol, while the other half could not. The diploid clones that could grow on glycerol were induced to sporulate, and they yielded tetrads with four spores that were all able to grow on glycerol medium. In all of these tetrads, two of the haploid progeny were of mating type a and two of mating type α. The diploids that could not grow on glycerol could not sporulate. What do the results of the mating say about the location of the *cox2-1* mutation?

    c. A different mutant strain of yeast called *pet111-1* is also unable to grow on glycerol medium but still can grow on glucose medium. These mutant cells similarly lacked the cytochrome *c* oxidase. When *pet111-1* was crossed with a wild-type haploid strain of the opposite mating type, the resultant diploids were able to grow on glycerol and yielded asci that all showed a 2:2 segregation of haploid cells that could or could not grow on glycerol. Explain these results in light of your answer to part (b).

20. In the late 1940s, the French researcher Boris Ephrussi discovered one of the first examples of non-Mendelian inheritance in the yeast *Saccharomyces cerevisiae.* He found mutants that were not able to respire (that is, were unable to do oxidative phosphorylation). The mutant cells formed small (*petite* in French) colonies when grown on petri plates containing the sugar glucose as a carbon source. These *petite* cells cannot grow at all if the only carbon source is nonfermentable, like glycerol or ethanol. In contrast, wild-type (*grande*) cells grow well on both kinds of petri plates.

    Ephrussi found that in all crosses between *grande* and *petite* haploid strains of yeast, all four spores in every ascus were *grande;* that is, the ratio of *grande: petite* was always 4:0, showing uniparental inheritance.

    a. Explain the phenotypes of *grande* and *petite* yeast cells (that is, their ability to grow on the two types of media) in terms of the ways the cells extract energy from their environment (oxidative phosphorylation and fermentation).

    b. Figure 15.13 shows that mitochondrial inheritance in *S. cerevisiae* is biparental. The difference between *petite* and *grande* strains of yeast turns out to be solely a property of mtDNA, but in this case the inheritance is uniparental. How can all of these seemingly paradoxical statements be true simultaneously? (*Hint: S. cerevisiae* cells do not require mtDNA to perform fermentation.)

Section 15.5

21. What characteristics in a human pedigree suggest a mitochondrial location for a mutation affecting the trait?

22. The first person in the family represented by the pedigree shown here who exhibited symptoms of the mitochondrial disease MERFF was II-2.

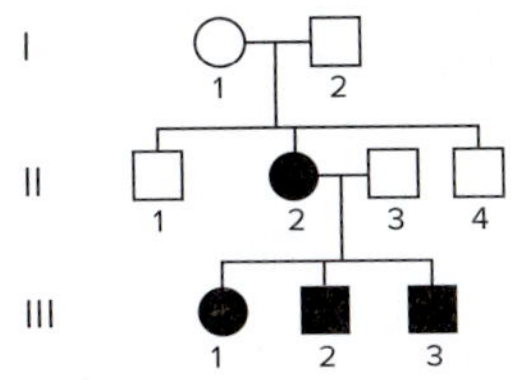

a. What are two possible explanations of why the mother I-1 was unaffected but daughter II-2 was affected?

b. How could you differentiate between the two possible explanations?

23. In 1988, neurologists in Australia reported the existence of identical twins who had developed myoclonic epilepsy in their teens. One twin remained only mildly affected by this condition, but the other twin later developed other symptoms of full-blown MERRF, including deafness, ragged red fibers, and ataxia (loss of the ability to control muscles). Explain the phenotypic dissimilarity in these identical twins.

24. If you were a genetic counselor and had a patient with MERRF who wanted to have a child, what kind of advice could you give about the chances the child would also have the disease? Are there any tests you could suggest that could be performed prenatally to determine if a fetus would be affected by MERRF?

25. Kearns-Sayre syndrome (KSS), Pearson syndrome, and progressive external opthalmoplegia (PEO) are rare diseases in which up to 7.6 kb of the mitochondrial genome is deleted. KSS affects the central nervous system, skeletal muscle, and heart; patients often die in young adulthood. Pearson syndrome is characterized by severe anemia and pancreatic dysfunction. The condition is usually fatal during infancy, but the few survivors often develop the symptoms of KSS. PEO patients have ptosis (drooping eyelids) and weakness in the limbs, but they have normal life spans.

a. How can you explain the variation in tissues affected and severity of symptoms in patients with these three conditions, given that they all bear large deletions of mtDNA? (Assume that the size of the deletion does not contribute to phenotypic differences.)

b. Assuming that mtDNA begins its replication from a single origin, what can you conclude from these diseases about the location of this replication origin?

c. Although these syndromes are due to mtDNA deletions, they are not usually maternally inherited but instead arise as a new mutation in an individual. For example, mothers with PEO usually do not transmit this trait to their offspring. Propose an explanation for this surprising finding.

26. Many clinically relevant mitochondrial diseases are caused by mutations in mitochondrial genes affecting tRNAs. For example, one form of MELAS (mitochondrial myopathy, encephalopathy, lactic acidosis and stroke-like episodes) is caused by a point mutation in the gene encoding the mitochondrial $tRNA^{Leu}$ whose anticodon recognizes the codons 5′ UUA and 5′ UUG. The mutation makes the aminoacylation of this tRNA inefficient.

a. The rate of synthesis of most mitochondrial proteins is either unaffected or slightly decreased in MELAS cells, but one mitochondrial protein called NAD6 is synthesized at only 10% of the normal rate. How is it possible that the translation of this single mitochondrial protein might be affected specifically?

b. Why might the decreased translation of this one protein be responsible for the pathological condition?

c. Researchers are currently investigating ways to treat the symptoms of MELAS patients. One strategy involves a change to a nuclear gene. What nuclear gene might the investigators be targeting? (Assume that you can make any desired change to nuclear genes; we will describe methods to alter genomes in Chapter 18.)

27. Leigh syndrome is characterized by psychomotor regression: that is, the progressive loss of mental and movement abilities. Patients also suffer from lactic acidosis, a condition in which mitochondrial respiration is deficient, so their tissues metabolize glucose anaerobically, leading to the buildup of lactate. Some patients with Leigh syndrome have a mutation in the mitochondrial gene *MT-CO3,* which encodes a subunit of the electron transport complex cytochrome *c* oxidase. Other patients diagnosed with Leigh syndrome have a loss-of-function mutation in the nuclear gene *SURF1,* which encodes a factor needed for the assembly of this same enzyme complex.

a. How can the same symptoms result from mutations in a mitochondrial gene and from mutations in a nuclear gene?

b. Does the pedigree shown provide enough evidence to discriminate between the possibilities that Leigh syndrome in this family is due to a mutation in a mitochondrial gene or in a nuclear gene? (Remember that many diseases, whether resulting from nuclear or mtDNA mutations, are incompletely penetrant or show variable expressivity.)

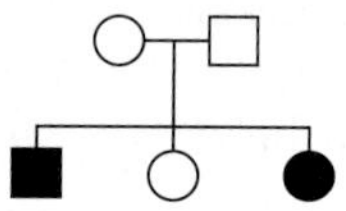

c. In what practical way would it be helpful to members of this family to discriminate between these two possible modes of inheritance?

28. All mutations in mitochondrial genes ultimately affect (whether directly or indirectly) the key function of mitochondria, which is to make ATP. Why then do mutations in different genes cause different diseases, with specific symptoms? (*Note*: The answer to this question is not known, but your speculations will help you think about the material in this chapter.)
29. How could researchers have determined that the rhesus monkeys Mito and Tracker (see Fig. 15.19) were devoid of the mitochondrial DNA from their nuclear donor mother?

chapter 16

Gene Regulation in Prokaryotes

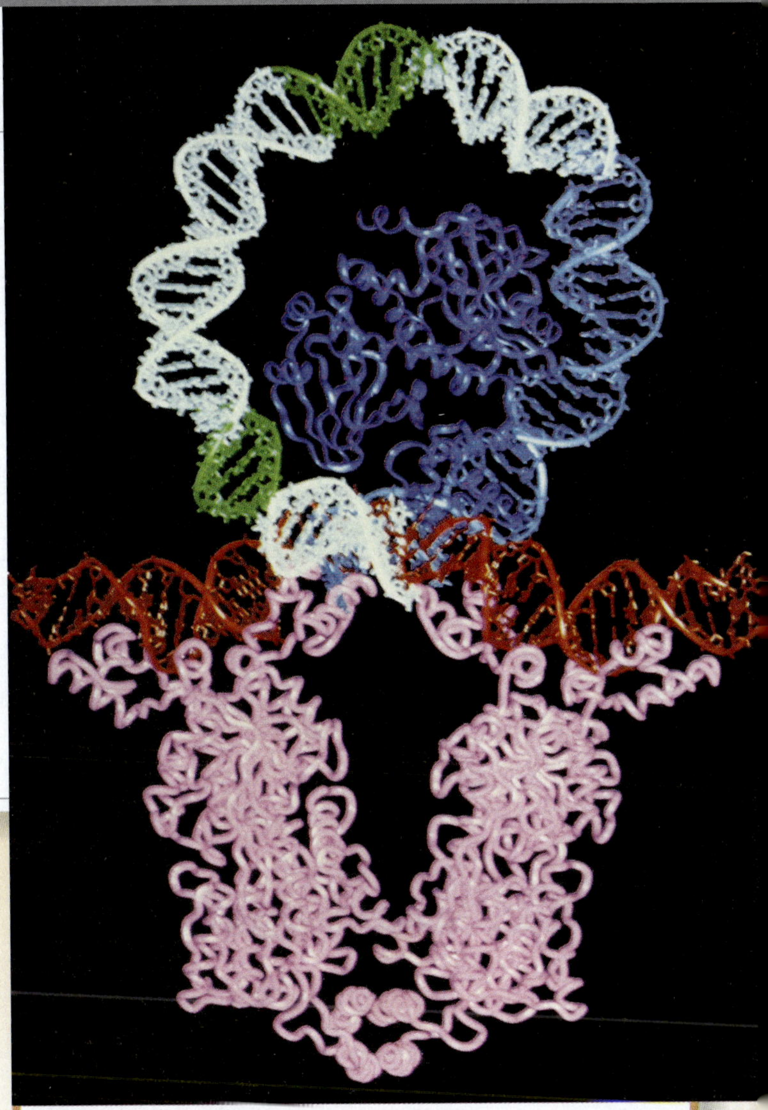

Lac repressor protein (violet) *binds to specific sites in DNA to turn off expression of the lac operon in* E. coli. *Lac repressor is a tetramer with two subunits binding to each of two operator sites, causing a loop* (blue *and* green) *to form in the DNA. Each operator site* (red) *exhibits rotational symmetry, so the two subunits forming a dimer are oriented oppositely on the chromosome. This model also shows where the CRP protein* (dark blue) *binds to* lac *DNA.*

BACTERIA USE A communication system called *quorum sensing* to adjust their behavior according to their population density. Quorum sensing was first observed in *Vibrio fischeri*, a bioluminescent bacterium that lives in the light-producing organ of the Hawaiian bobtail squid (**Fig. 16.1**). By matching the moonlight that would otherwise highlight their dark silhouette in the water, the bioluminescence camouflages the squid from predators. Scientists observed that the bacteria produce light only when they reach a certain density in the squid's light organ. This behavior makes energetic sense: A single bacterium could not produce enough light to illuminate the squid, and light production requires the bacteria to use energy.

How do the *V. fischeri* bacteria "know" when their population in the squid's light organ is sufficient to produce light? Researchers curious about this fascinating biological question used the powerful techniques of bacterial genetics to dissect the bioluminescence pathway. You will see in this chapter that *V. fischeri* synthesize and release into their environment regulatory factors that control the transcription of genes encoding light-making proteins. Only at high population density is the level of regulatory molecules high enough to activate transcription of the bioluminescence genes.

The story of bioluminescent *V. fischeri* illustrates two key aspects of the life of a unicellular prokaryote: First, these organisms are always in direct contact with their external environment; and second, bacteria need to be able to respond to changes in that environment by altering gene expression. The coordinated control of gene expression in

chapter outline

V. fischeri is an example of **prokaryotic gene regulation,** the subject of this chapter. Prokaryotes regulate gene expression by activating, increasing, diminishing, or preventing the transcription of specific genes and/or translation of the mRNAs made from these genes.

One overarching theme emerges from our discussion. To adapt and survive in a constantly changing world, bacteria must be able to tune the expression of many genes in a coordinated way so that the cells respond appropriately to many different environments and do not waste energy by making unneeded proteins.

Figure 16.1 Luminescent bacteria protect squid from predators. **(a)** *Vibrio fischeri* bacteria generate light. **(b)** These bacteria inhabit the light organ of Hawaiian bobtail squid.

a: Image of GFP-labeled *V. fischeri* cells, provided by the *Vibrio fischeri* Genome Project courtesy of L. Sycuro and E.G. Ruby. Individual cells are approximately 0.7 by 1.5 microns; b: © & Courtesy of Mattias Ormsestad and Eric Roettinger/Kahi Kai

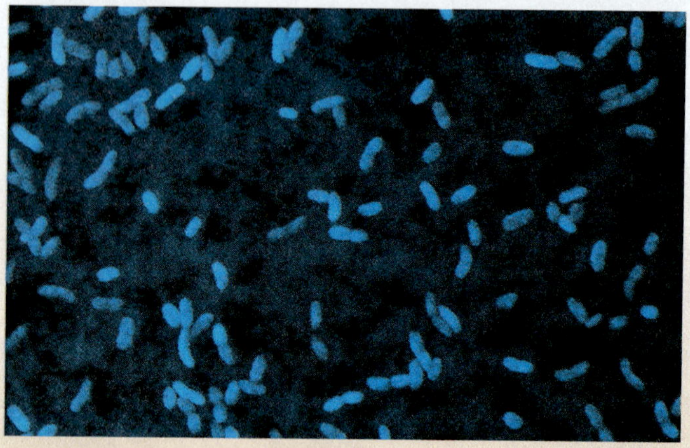

(a) Bioluminescent *Vibrio fischeri*

(b) Hawaiian bobtail squid

16.1 The Elements of Prokaryotic Gene Expression

learning objectives

1. Describe the function in prokaryotic transcription of the RNA polymerase core enzyme, sigma (σ) factor, and rho (ρ) factor.
2. Explain why an mRNA molecule can be transcribed and translated simultaneously in prokaryotes.
3. List the steps in gene expression that are potentially subject to regulation.

We saw in Chapter 8 that *gene expression* is the production of RNAs and proteins according to instructions encoded in DNA. During gene expression, the information in DNA is transcribed into RNA, and the RNA message is translated into a string of amino acids. Both transcription and translation provide opportunities for cells to regulate synthesis of the RNAs and proteins that they need in the amounts that they need them.

RNA Polymerase Is the Key Enzyme for Transcription

To begin the process of gene expression in prokaryotes, RNA polymerase transcribes a gene's DNA into RNA. RNA polymerase participates in all three phases of transcription: initiation, elongation, and termination. The details of RNA polymerase's function in all three steps of transcription were presented previously in Fig. 8.10, but we briefly review the most salient points here.

You will recall that initiation requires a special subunit of RNA polymerase—the sigma (σ) subunit—in addition to the other subunits that make up the core enzyme. When bound to the core enzyme, the σ subunit recognizes and binds specific DNA sequences at the *promoter.* When bound to the promoter, the complete *RNA polymerase holoenzyme*—core enzyme plus σ—functions as a complex that both initiates transcription by unwinding the DNA and begins polymerization of bases complementary to the DNA template strand.

The switch from initiation to elongation requires the movement of RNA polymerase away from the promoter and the release of σ. Elongation continues until the RNA polymerase encounters a signal in the RNA sequence that triggers termination. Two types of termination signals are found in prokaryotes: Rho dependent and Rho independent. In Rho-dependent termination, a protein factor called Rho (ρ) is a helicase enzyme that unwinds the mRNA from the DNA template, and helps dissociate RNA polymerase from the template. In Rho-independent termination, a sequence of bases in the RNA forms a secondary structure, known as a *hairpin loop,* that serves as a signal for the release of RNA polymerase from the completed RNA (review Fig. 8.10).

Translation in Prokaryotes Begins Before Transcription Ends

No membrane encloses the bacterial chromosome, so translation of the RNA message into a polypeptide can begin while mRNA is still being transcribed. Ribosomes bind to special initiation sites at the 5′ end of the reading frame (the *ribosome binding site*) while transcription of downstream regions of the RNA is still in progress. Signals for the initiation and termination of translation are distinct from signals for the initiation and termination of transcription. Figure 8.25 reviews how ribosomes, tRNAs, and translation factors mediate mRNA translation to produce a polypeptide that grows from its N terminus to its C terminus, according to instructions embodied in the sequence of mRNA codons.

Another unique feature of prokaryotic gene expression is that ribosomes can initiate translation at several positions along a single mRNA. Many bacterial mRNAs are thus *polycistronic;* that is, they contain open reading frames for several different proteins. As you will see, this fact has important consequences for gene regulation.

Regulation of Gene Expression Can Occur at Many Steps

Many levels of control determine the amount of a particular polypeptide in a bacterial cell at any one time. Some controls affect an aspect of transcription, such as the binding of RNA polymerase to the promoter, the shift from transcriptional initiation to elongation, or the release of the mRNA at the termination of transcription. Other controls are posttranscriptional and determine the stability of the mRNA after its synthesis, the efficiency with which ribosomes recognize the various translational initiation sites along the mRNA, or the stability or activity of the polypeptide product.

As we see next, a crucial step in the regulation of many bacterial genes is the binding of RNA polymerase to DNA at the promoter. Later in the chapter, we will discuss how the termination of transcription as well as posttranscriptional mechanisms also play important roles in regulating the expression of many bacterial genes.

essential concepts

- *RNA polymerase* is the crucial enzyme for prokaryotic transcription. *Sigma (σ) factor* allows the enzyme to recognize promoters, while *Rho (ρ) protein* terminates the transcription of some genes.
- No membrane encloses the bacterial chromosome, so translation of mRNA into a polypeptide can begin while transcription is taking place.
- Many bacterial mRNAs are *polycistronic;* each open reading frame in the transcript has its own ribosome binding site.
- Regulation of prokaryotic gene expression can occur at many different levels: transcription initiation, elongation, or termination; mRNA stability; translation initiation; or protein stability or activity.

16.2 Regulation of Transcription Initiation via DNA-Binding Proteins

learning objectives

1. Compare the regulation requirements of catabolic pathways to those of anabolic pathways.
2. Outline the operon theory using the *lac* operon as an example.
3. Discuss genetic evidence that *lacI* encodes an allosteric repressor protein that binds the operator DNA.
4. Explain why it is advantageous for transcriptional regulatory proteins to be multimeric and for their binding sites to be clustered.
5. Compare the actions of positive and negative regulatory proteins at promoters.
6. Explain how repressor proteins can be central to the regulation of both catabolic and anabolic operons.

The initiation of transcription is the first step in gene expression, so it makes sense that one of the fundamental modes of regulating the expression of many genes involves the binding of regulatory proteins to DNA targets at or near promoters to control transcription. The DNA binding of these regulatory proteins either inhibits or enhances the effectiveness of RNA polymerase in initiating transcription.

In our discussion, we consider the inhibition of RNA polymerase activity as *negative regulation* and the enhancement of RNA polymerase activity as *positive regulation.*

Catabolic and Anabolic Pathways Require Different Types of Regulation

Researchers first delineated basic principles of gene regulation through studies of various metabolic pathways in *Escherichia coli.* Many of these pathways are **catabolic pathways** in which complicated molecules are broken down for the use of the cell; examples of catabolic pathways are those that break down sugars to provide cells with energy and carbon atoms. Other pathways in the cells are **anabolic pathways** that allow cells to construct end product molecules they need, such as amino acids and nucleotides, from simpler constituents.

The underlying logic cells must follow to regulate catabolic and anabolic pathways is entirely different. Catabolic pathways demand **inducible regulation:** This means the pathway should be turned on—that is, induced—only when the complex molecules to be broken down (*catabolites*) are present in the cell's environment. The cell would waste resources in synthesizing the enzymes needed to break down a particular sugar if that sugar was not available to the cell. In contrast, anabolic pathways require **repressible regulation:** This means the pathway should be turned on only when the cell does not have enough of the needed *end product,* such as a specific amino acid. If the end product is present in sufficient quantities, the pathway should be turned off—repressed—so the cell does not waste resources trying to make molecules that it already has.

In the rest of this section, we focus our attention first on the inducible regulation of one particular catabolic pathway in *E. coli:* the pathway that allows these bacteria to use the sugar lactose as a source of carbon and energy. Many of the lessons learned from this story will also be useful when we turn later to a discussion of the repressible regulation of one particular anabolic pathway involved in the synthesis of the essential amino acid tryptophan.

E. coli's Utilization of Lactose Provides a Model System of Gene Regulation

Proliferating *E. coli* can use any one of several sugars as a source of carbon and energy. One of these is *lactose,* a complex sugar composed of two monosaccharides: glucose and galactose. A membrane protein, Lac permease, transports lactose in the medium into the *E. coli* cell. There, the enzyme β-galactosidase splits the lactose into galactose and glucose (**Fig. 16.2**). Note that this is a catabolic pathway that breaks down lactose into simpler subcomponents.

Figure 16.2 Lactose utilization in an *E. coli* cell. Lactose passes through the membranes of the cell via an opening formed by the Lac permease protein. Inside the cell, β-galactosidase splits lactose into galactose and glucose.

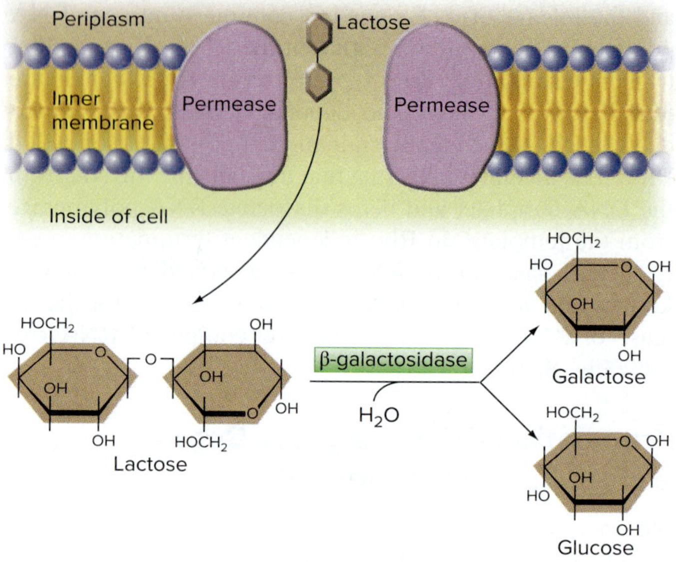

Induction of coordinated gene expression by lactose

The two proteins Lac permease and β-galactosidase, both required for lactose utilization, are present at very low levels in cells grown without lactose. The cell has no need for either of these proteins if lactose is not present. But soon after lactose is added to the bacterial medium, the production of these proteins increases 1000-fold. The process by which a specific molecule stimulates synthesis of a given protein is known as **induction.** The molecule responsible for stimulating production of the protein is called the **inducer.** In the regulatory system under consideration, lactose modified to a derivative known as *allolactose* is the inducer of the genes for lactose utilization.

How lactose in the medium induces the simultaneous expression of the proteins required for its utilization was the subject of a major research effort in the 1950s and 1960s—a period some refer to as the golden era of bacterial genetics.

Research advantages of the lactose system in *E. coli*

Lactose utilization in *E. coli* was a wise choice as a model for studying gene regulation. The possibility of culturing large numbers of the bacteria made it easy to isolate rare mutants. Once isolated, the mutations responsible for the altered phenotypes could be located by mapping techniques. Another advantage was that the lactose utilization genes are not essential for survival, because the bacteria can grow using other sugars as a carbon source. In addition,

Figure 16.3 *E. coli* **colonies on plates containing X-gal.** Colonies that express β-galactosidase are blue; those that do not are white.

© anyaivanova/Getty Images RF

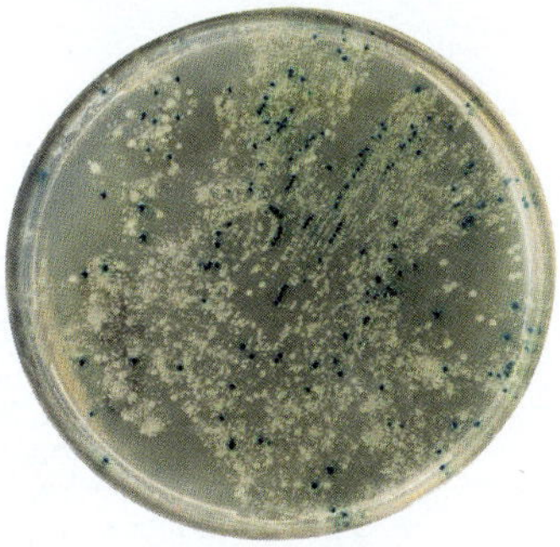

Figure 16.4 Jacques Monod played a key role in discovering the principles of gene regulation.

© Bettmann/UPI/Corbis

as just mentioned, a striking 1000-fold difference exists between lactose utilization protein levels in induced and uninduced cells. This latter fact makes it easy to see the difference between the mutant and wild-type states, and it also allows the identification of mutants that have partial—not just all-or-none—effects.

The ability to measure levels of expression was crucial for many of these experiments. To this end, chemists synthesized compounds other than lactose, such as *o*-nitrophenyl-galactoside (ONPG), that could be split by β-galactosidase into products that were easy to assay. One product of ONPG splitting has a yellow color, whose intensity is proportional to the amount of product made and thus reflects the level of activity of the β-galactosidase enzyme. A spectrophotometer can easily measure the amount of cleaved yellow product in a sample. Another substrate of the β-galactosidase enzyme that produces a color change upon cleavage is *X-gal,* whose cleavage produces a blue substance. As will be described later in this chapter, researchers find it valuable to add X-gal to media in petri plates because a bacterial colony growing on the plate will turn blue if the cells in the colony are expressing β-galactosidase (**Fig. 16.3**).

The Operon Theory Explains How a Single Substance Can Regulate Several Clustered Genes

Jacques Monod (**Fig. 16.4**), a man of diverse interests, was a catalyst for research on the regulation of lactose utilization. A political activist and a chief of French Resistance operations during World War II, he was also a fine musician and esteemed writer on the philosophy of science. Monod led a research effort centered at the Pasteur Institute in Paris, where scientists from around the world came to study enzyme induction. Results from many genetic studies led Monod and his close collaborator François Jacob to propose a model of gene regulation known as the **operon theory,** which suggested that a single signal can simultaneously regulate the expression of several genes that are clustered together on a chromosome and are involved in the same process. They reasoned that because these genes form a cluster, they can be transcribed together into a single mRNA, and thus anything that regulates the transcription of this mRNA will affect all the genes in the cluster. Clusters of genes regulated in this way are called **operons.** We first summarize the theory itself and then describe key experiments that influenced Jacob and Monod's thinking.

Figure 16.5 presents the molecular players in the theory and how they interact to achieve the coordinated regulation of the genes for lactose utilization. As shown, three so-called *structural genes* (*lacZ, lacY,* and *lacA*) encoding proteins needed for lactose utilization, together with two regulatory elements—the promoter (*P*) and the operator (*o*)—make up the ***lac* operon:** a single DNA unit enabling the simultaneous regulation of the three structural genes in response to environmental changes. Molecules that interact with the operon include the *repressor,* which binds to the operon's operator, and the *inducer* (allolactose), which when present binds to the repressor and prevents it from binding to the operator. The repressor is an **allosteric protein**—a protein that undergoes a reversible change in conformation when bound to another molecule (in this case, the inducer allolactose).

Jacob and Monod's theory was remarkable because the authors were working with an abstract sense of the molecules in the bacterial cell: The Watson-Crick model of DNA structure was only eight years old, mRNA had only recently been identified, and the details of transcription had not yet been described. In 1961, the details of information flow from DNA to RNA to protein were still being established, and knowledge of proteins' roles in the cell was limited. For example, although Monod was a biochemist with a special interest in allostery and its effects, the repressor itself was a purely conceptual construct. At the time of publication, the repressor had not yet been isolated, and it was unknown whether it was RNA or protein or some

FEATURE FIGURE 16.5

The Lactose Operon in *E. coli*

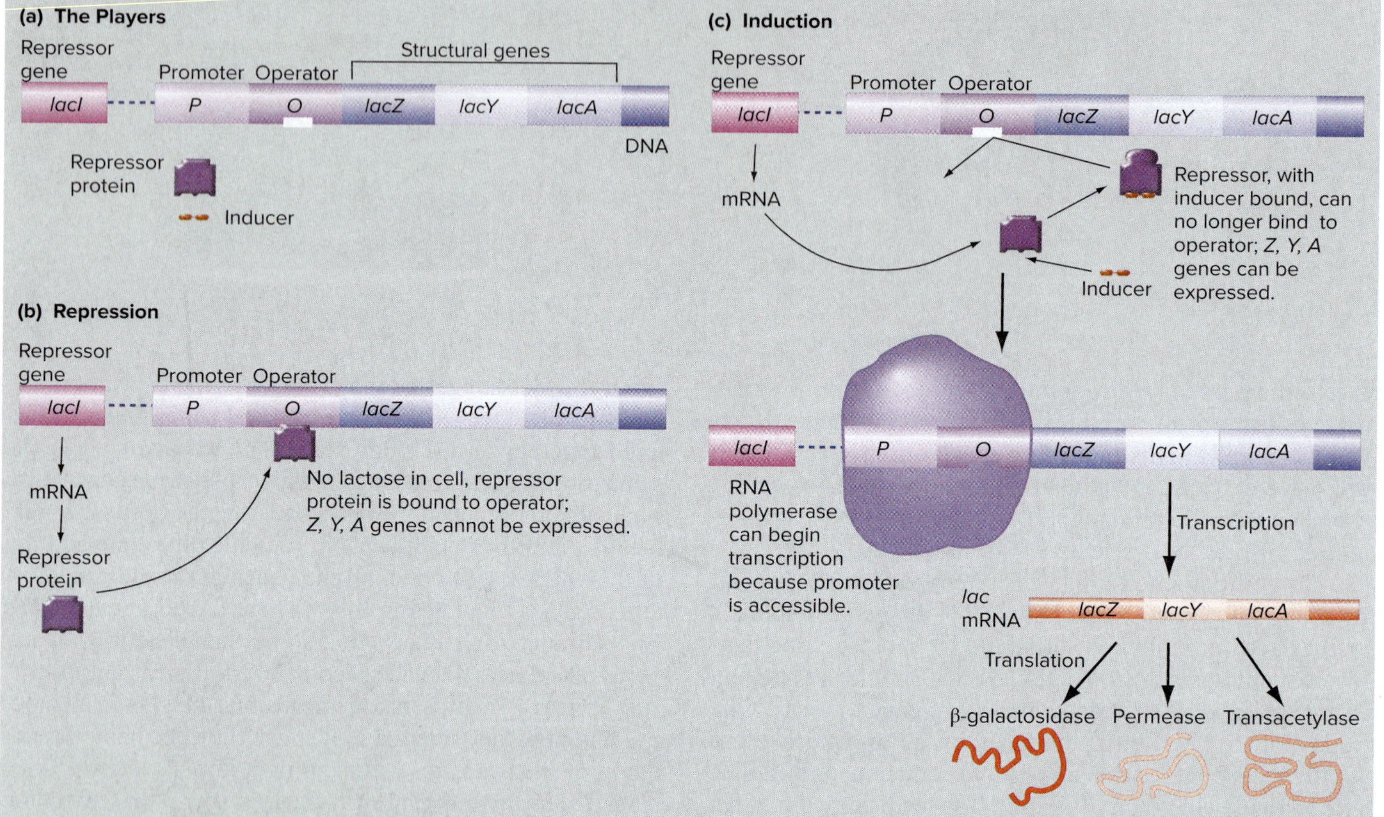

a. The Players

The coordination of various elements enables bacteria to use lactose in an energy-efficient way. These elements include:

1. A closely linked cluster of three structural genes—*lacZ, lacY,* and *lacA*—that encode the enzymes active in splitting lactose into glucose and galactose.
2. A promoter site (*P*), from which RNA polymerase initiates transcription of a polycistronic mRNA. The promoter acts in *cis,* affecting the expression of only downstream structural *lac* genes on the same DNA molecule.
3. A *cis*-acting DNA operator site (*O*) lying very near the *lac* operon promoter on the same DNA molecule.
   The three structural genes together with the promoter and the operator constitute the *lac operon.*
4. A *trans*-acting repressor protein that can bind to the operator. The repressor is encoded by the *lacI* gene, which is separate from the operon and is unregulated. Once made, the repressor diffuses through the cytoplasm and binds with its target.
5. An inducer that prevents the repressor's binding to the operator. Although early experimenters thought lactose was the inducer, we now know that the inducer is actually allolactose, a molecule derived from lactose.

How the Players Interact to Regulate the Lactose-Utilization Genes

b. Repression

In the absence of lactose, the repressor protein binds to the DNA of the operator, and this binding prevents transcription. The repressor thus serves as a negative regulatory element.

c. Induction

1. When lactose is present, the inducer allolactose binds to the repressor. This binding changes the shape of the repressor, making it unable to bind to the operator.
2. With the release of the repressor from the operator, RNA polymerase gains access to the *lac* operon promoter and initiates transcription of the three lactose-utilization genes into a single polycistronic mRNA.

other molecule. Jacob and Monod thus required a major leap of imagination in order to propose their theory.

We now know that a key concept of the theory—that proteins bind to DNA to regulate gene expression—holds true for the positive as well as the negative regulation of the *lac* operon. The binding of proteins to DNA is also central to the control of many prokaryotic genes outside the *lac* operon, including the inducible regulation of other catabolic genes and the repressible regulation of anabolic genes, and to eukaryotic genes as well.

Genetic Analysis Led Jacob and Monod to the Operon Hypothesis

On the way to developing the operon theory of gene regulation, Monod and his collaborators isolated many different mutations that either prevented the cells from utilizing lactose or that allowed the cells to synthesize the enzymes needed to break down lactose all the time, whether lactose was present in the environment or not.

Complementation and mapping analyses of Lac⁻ mutants

Lac⁻ mutants are bacterial cells unable to utilize lactose. Using complementation analysis of a large number of Lac⁻ mutants, the researchers showed that the cells' inability to break down lactose resulted from mutations in two genes: *lacZ,* which encodes β-galactosidase, and *lacY,* which encodes Lac permease. They also discovered a third *lac* gene, *lacA,* which encodes a transacetylase enzyme that adds an acetyl (CH_3CO) group to lactose and other sugars. Genetic mapping showed that the three genes appear on the bacterial chromosome in a tightly linked cluster, in the order *lacZ-lacY-lacA* (Fig. 16.5). Because LacA protein is not required for the breakdown of lactose, most studies of lactose utilization do not follow the *lacA* gene.

Evidence for a repressor protein: the effect of *lacI⁻* mutations

Loss-of-function mutations in a gene called *lacI,* located near but not within the *lac* operon (Fig. 16.5), produce **constitutive mutants** that synthesize β-galactosidase and Lac permease even in the absence of lactose. Constitutive mutants synthesize certain enzymes all the time, irrespective of environmental conditions. The existence of these constitutive mutants suggested that *lacI* encodes a negative regulator, or **repressor.** Cells would need such a repressor to prevent expression of *lacY* and *lacZ* in the absence of inducer. In constitutive mutants, however, a mutation in the *lacI* gene generates a defect in the repressor protein that prevents it from carrying out this negative regulatory function.

Figure 16.6 The PaJaMo experiment. DNA carrying *lacI*$^+$ and *lacZ*$^+$ genes is introduced (by conjugation) into a *lacI*$^-$ *lacZ*$^-$ cell. In the recipient, β-galactosidase is synthesized from the introduced *lacZ*$^+$ gene initially, but as repressor (made from the introduced *lacI*$^+$) accumulates, the synthesis of β-galactosidase stops. If inducer is added (*dotted line*), the synthesis of β-galactosidase resumes.

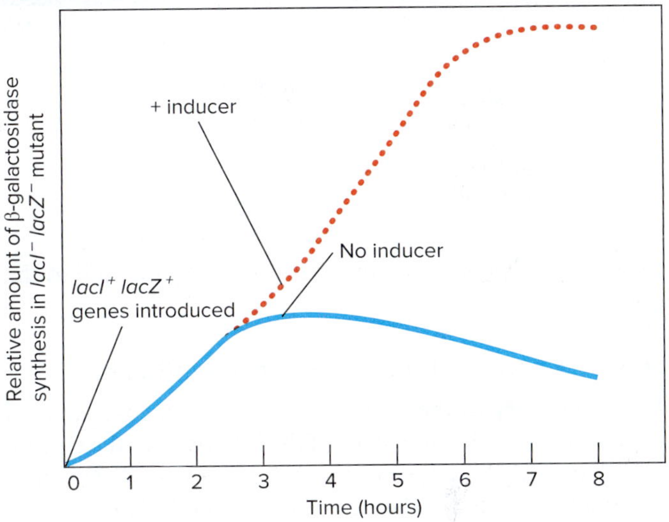

The historic PaJaMo experiment—named after Arthur Pardee (a third collaborator), Jacob, and Monod—provided further evidence that *lacI* indeed encodes this hypothetical negative regulator of the *lac* genes. Matings in which the chromosomal DNA of an Hfr donor cell is transferred into an F⁻ recipient cell served as the basis of the PaJaMo study. The researchers transferred the *lacI*$^+$ and *lacZ*$^+$ alleles into a bacterial cell devoid of LacI and LacZ proteins in a medium containing no lactose (**Fig. 16.6**). Shortly after the transfer of the *lacI*$^+$ and *lacZ*$^+$ genes, the researchers detected synthesis of β-galactosidase. Within about an hour, however, this synthesis stopped.

Pardee, Jacob, and Monod interpreted these results as follows: When the donor DNA is first transferred to the recipient, no repressor (LacI protein) is in the recipient cell's cytoplasm because the recipient cell's chromosome is *lacI*$^-$. In the absence of repressor, the *lacZ*$^+$ gene is expressed. Over time, the recipient cell begins to make the Lac repressor protein from the *lacI*$^+$ gene introduced by the mating, so expression then becomes repressed.

On the basis of these experiments, Monod and company proposed that the repressor protein prevents further transcription of *lacZ* by binding to a hypothetical **operator site:** a DNA sequence near the promoter of the lactose-utilization genes. They suggested that the binding of repressor to this operator site blocks the promoter, and that this binding occurs only when lactose is not present in the medium (Fig. 16.5). They further predicted that although some *lacI*$^-$ alleles would be null mutations that could not make any protein, other *lacI*$^-$ mutations would make a

Figure 16.7 Repressor mutant (*lacI*⁻). In some *lacI*⁻ mutants, the repressor cannot bind to the operator site and therefore cannot repress the operon. Other *lacI*⁻ mutants are null mutants that make no protein (*not shown*).

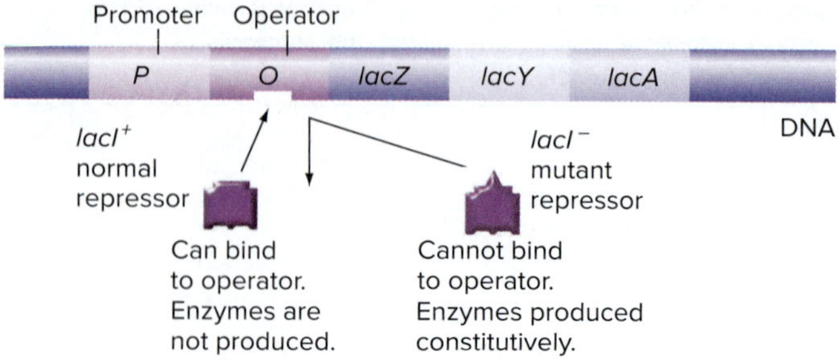

form of the repressor protein that was unable to bind to the operator (**Fig. 16.7**).

How the inducer triggers enzyme synthesis

In the final step of the PaJaMo experiment, the researchers added lactose—the precursor of the inducer—to the culture medium. With this addition, the synthesis of β-galactosidase resumed (Fig. 16.6). Their interpretation of this result was that the inducer binds to the wild-type repressor. This binding changes the shape of the repressor protein so that it can no longer bind to DNA. When the inducer is removed from the environment, the repressor, free of inducer, reverts to its DNA-bindable shape (Fig. 16.5). The binding of inducer to repressor thus causes an allosteric effect that abolishes the repressor's ability to bind the operator. In this sequence of events, the inducer is an *effector* that releases repression without itself binding to the DNA.

Operator mutants

While *lacI*⁻ mutations in the repressor gene can erase repressor activity, mutations that alter the specific nucleotide sequence of the operator recognized by the repressor can have the same effect (**Fig. 16.8**). When mutations change the nucleotide sequence of the operator, the repressor is unable to recognize and bind to the site; the resulting phenotype is the *constitutive synthesis* of the lactose-utilization proteins. Researchers have isolated constitutive mutants whose genetic defects map to the *lac* operator site, which is adjacent to the *lacZ* gene. They call the constitutive operator DNA alterations o^c mutations.

Figure 16.8 Operator mutants. The repressor cannot recognize the altered DNA sequence in the *lac* o^c mutant site, so it cannot bind and repress the operon.

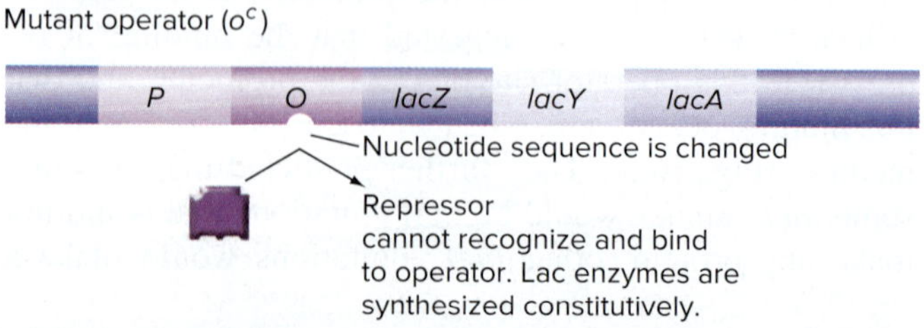

Figure 16.9 Superrepressor mutant *lacI*ˢ. In superrepressor mutants, LacIˢ binds to the operator but cannot bind to the inducer, so the repressor cannot be removed from the operator, and genes are continually repressed.

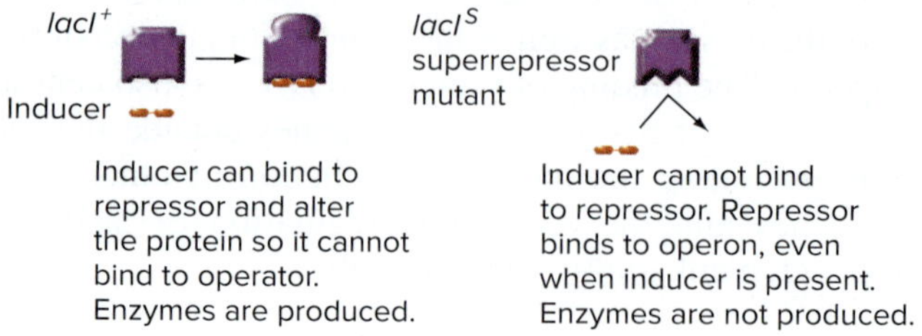

Superrepressor (*lacI*ˢ) mutations

If binding of the inducer to the repressor protein prevents the repressor from binding to the operator, what outcome would you predict for mutations that prevent the repressor from interacting with the inducer? Clearly, you would expect that such mutations would result in cells that could not turn on the operon, even when inducer was added to the medium. Researchers eventually isolated such noninducible mutations in the repressor gene and designated them *lacI*ˢ, or superrepressor mutations (**Fig. 16.9**). The *lacI*ˢ mutants, although they cannot bind inducer, can still bind to DNA and repress transcription of the operon. This repressed state is independent of the presence or absence of lactose or allolactose.

Proteins act in *trans*, DNA sites act in *cis*

The phenotypes of the bacteria with mutations in the *lac* operon and *lacI* gene are summarized in **Table 16.1.** One of the key findings shown in this table is that two different kinds of mutations allow expression of the *lac* operon even when the inducer is not present: these are the constitutive operator (o^c) mutants (genotype 4 in Table 16.1) and the constitutive *lacI*⁻ mutants (genotype 2 in Table 16.1). How can you distinguish these mutations from each other, considering that both prevent repression? The answer is found in a *cis/trans* test.

Elements that act in ***trans*** can diffuse through the cytoplasm and act at target DNA sites on any DNA molecule in the cell. Elements that act in ***cis*** can influence the expression only of adjacent genes on the same DNA molecule. Studies of *merodiploids* (partial diploids) in which a second copy of the *lac* genes was introduced helped distinguish mutations in the operator site (o^c), which act in *cis,* from mutations in *lacI,* which encodes a protein that acts in *trans.*

The merodiploids were made using F′ plasmids that carry a few chromosomal bacterial genes. When F′ (*lac*) plasmids are present in a bacterium, the cell has two copies of the region containing both the lactose-utilization genes and *lacI*—one on the plasmid and one on the bacterial chromosome. Using F′ (*lac*) plasmids, Monod's group

TABLE 16.1 Summary of key *lac* operon phenotypes

No. Genotype	*lacZ* Activity		*lacY* Activity		Conclusions
	Without inducer	With inducer	Without inducer	With inducer	
(1) $I^+o^+Z^+Y^+$	–	+	–	+	*lac* operon is inducible
(2) $I^-o^+Z^+Y^+$	+	+	+	+	*I* encodes repressor
(3) $I^So^+Z^+Y^+$	–	–	–	–	I^S encodes superrepressor
(4) $I^+o^cZ^+Y^+$	+	+	+	+	o^+ is a DNA site that binds repressor
(5) $I^-o^+Z^+Y^-$/F′ ($I^+o^+Z^-Y^+$)	–	+	–	+	I^+ is dominant to I^-; repressor acts in *trans*
(6) $I^+o^+Z^+Y^-$/F′ ($I^So^+Z^-Y^+$)	–	–	–	–	I^S is dominant to I^+; superrepressor acts in *trans*
(7) $I^+o^cZ^+Y^-$/F′ ($I^+o^+Z^-Y^+$)	+	+	–	+	o^+ and o^c act in *cis*

could create bacterial strains with diverse combinations of regulatory (o^c and *lacI*) mutations and mutations in enyzme-encoding structural genes (*lacZ* and *lacY*). The phenotypes of these partially diploid cells allowed Monod and his collaborators to determine whether particular constitutive mutations were in the genes that produce diffusible, *trans*-acting proteins or at *cis*-acting DNA sites that affect only genes on the same molecule. Genotypes 5–7 in Table 16.1 summarize the results of key merodiploid experiments.

$lacI^+$: dominant to $lacI^-$ in *trans* In one experiment, Monod and colleagues started with a $lacI^-$ $lacZ^+$ $lacY^-$ bacterial strain that was constitutive for β-galactosidase production because it could not synthesize repressor (**Fig. 16.10** and genotype 5 in Table 16.1). The introduction of an F′ ($lacI^+$ $lacZ^-$ $lacY^+$) plasmid into this strain created a merodiploid that was phenotypically wild type with respect to both β-galactosidase and permease expression: Both $lacZ^+$ and $lacY^+$ were repressible in the absence of lactose and inducible in its presence. The wild-type phenotype of the merodiploid indicated that $lacI^+$ is dominant to $lacI^-$. Moreover, the inducibility of not only $lacY^+$ (on the plasmid with $lacI^+$), but also $lacZ^+$ (on the bacterial chromosome), meant that LacI protein produced from the $lacI^+$ gene on the plasmid can bind to the operator on its own chromosome and also to the operator on the bacterial chromosome. Thus, the product of the *lacI* gene is a *trans*-acting protein able to diffuse inside the cell and bind to any operator site it encounters, regardless of the operator's chromosomal location.

$lacI^s$: dominant to $lacI^+$ in *trans* In a second experiment, the introduction of a $lacI^s$ plasmid into a $lacI^+$ strain of bacteria that was originally both repressible and inducible created bacteria that were still repressible but were no longer inducible (**Fig. 16.11** and genotype 6 in Table 16.1). This effect occurred because the mutant LacI^s repressor,

Figure 16.10 ***$LacI^+$* protein acts in *trans*.** Repressor protein, made from the $lacI^+$ gene on the plasmid, can diffuse in the cytoplasm and bind to the operator on the chromosome as well as to the operator on the plasmid. (Genotype No. 5 in Table 16.1.)

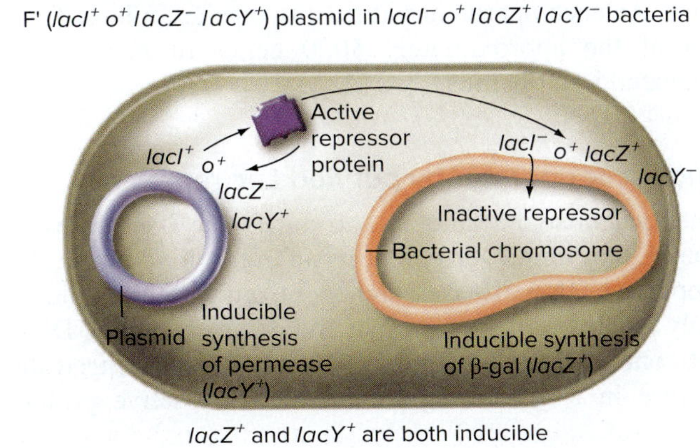

Figure 16.11 **LacI^s protein acts in *trans*.** The superrepressor encoded by $lacI^s$ on the plasmid diffuses and binds to operators on both the plasmid and the chromosome to repress the *lac* operon, even if the inducer is present. (Genotype No. 6 in Table 16.1.)

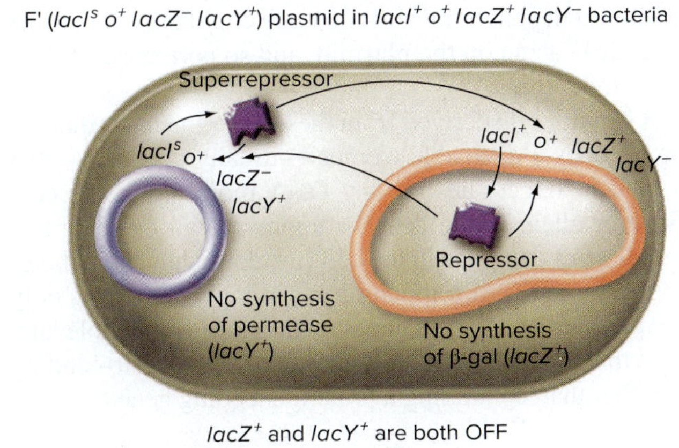

Figure 16.12 o^c and o^+ act in *cis*. The o^c constitutive mutation and the o^+ wild-type operator each affect only the operon of which they are a part. In this cell, only the chromosomal copy of the operon will be transcribed constitutively. (Genotype No. 7 in Table 16.1.)

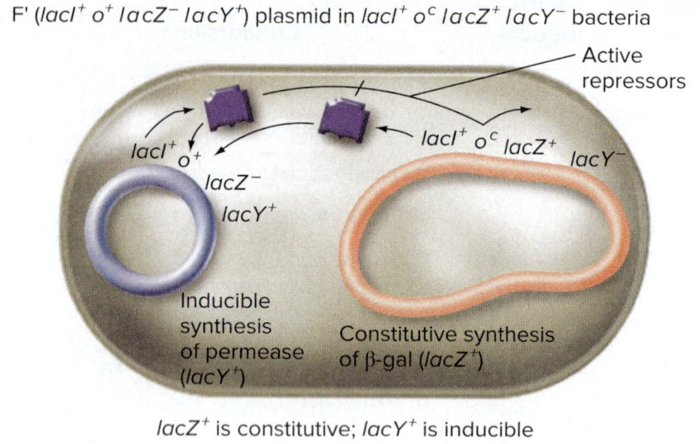

while still able to bind to the operator, could no longer bind inducer. The allele encoding the noninducible superrepressor was dominant to the wild-type repressor allele because after a while, the mutant repressor, unable to bind inducer, occupied all the operator sites and blocked all *lac* gene transcription in the cell.

o^c and o^+: action in *cis* In a third set of experiments, the researchers began with *lacI*$^+$ o^c *lacZ*$^+$ *lacY*$^-$ bacteria that were constitutive for β-galactosidase synthesis because the wild-type repressor they produced could not bind to the altered operator (**Fig. 16.12** and genotype 7 in Table 16.1). Introduction of an F′ (*lacI*$^+$ o^+ *lacZ*$^-$ *lacY*$^+$) plasmid did not change this state of affairs—the cells remained constitutive for β-galactosidase production, although they now were inducible for permease. The explanation is that the o^+ operator on the plasmid had no effect on the *lacZ*$^+$ gene on the chromosome DNA because the operator DNA acts only in *cis*. Because it was able to influence gene expression only of the *lacZ*$^-$ and *lacY*$^+$ genes on its own DNA molecule, the wild-type operator on the plasmid could not override the mutant chromosomal operator to allow repression of genes on the bacterial chromosome. Conversely, the o^c mutant operator on the bacterial chromosome could not act in *trans* on the *lacY*$^+$ gene on the plasmid, and so permease synthesis was inducible.

A general rule derived from these experiments is that if a gene encodes a diffusible element—usually a protein—that can bind to target sites on any DNA molecule in the cell, whichever allele of the gene is dominant will override any other allele of that gene in the cell (the dominant allele therefore acts in *trans*). If a mutation is *cis*-acting, it affects only the expression of adjacent genes on the same DNA molecule; it does this by altering a DNA site, such as a protein-binding site, rather than by altering a protein-encoding gene.

Biochemical Experiments Support the Operon Hypothesis

With the development in the 1970s of cloning, DNA sequencing, and techniques for analyzing protein–DNA interactions, researchers increased their ability to isolate specific macromolecules, determine their structures, and analyze the relationships between molecules. These studies verified the basic tenets of the Jacob-Monod operon theory, fleshed out the molecular details of the *lac* operon, and revealed how these lessons could be applied to many other examples of gene regulation at the level of transcriptional initiation.

Coordinate expression of the *lacZYA* genes as a single polycistronic mRNA

You will recall that Monod and coworkers showed that the repression and alternatively the induction of *lacZ, lacY,* and *lacA* occur in unison. To explain this coordinate expression of the genes for lactose utilization, they proposed that the three genes are transcribed as part of the same polycistronic mRNA, as you saw previously in Fig. 16.5. The results shown in Table 16.1 all fit well with this idea, although these findings do not by themselves prove that the three coding sequences are together on the same transcript.

Biochemical studies showed eventually that RNA polymerase indeed initiates transcription of the tightly linked *lac* gene cluster from a single promoter. During transcription, the polymerase produces a single polycistronic mRNA containing the *lac* gene information in the order 5′-*lacZ-lacY-lacA*-3′. As a result, mutations in the promoter (which must be located just upstream of *lacZ*) affect the transcription of all three genes. You should remember that for each of the genes to be translated from this polycistronic mRNA in bacterial cells, each open reading frame must be preceded by its own independent ribosome binding site.

The clustering of genes with similar functions into operons is a simple and efficient way to achieve *coordinate gene expression.* It is thus not at all surprising that many operons have evolved in bacterial genomes. As an example, the *E. coli* genome has roughly 400 verified operons. As each operon has at least two genes, a large fraction of the approximately 5000 genes in *E. coli* are organized into operons.

Binding of operator by purified Lac repressor

When scientists became able to purify the Lac repressor protein, they could verify that it can in fact bind physically to operator DNA. The researchers mixed together radioactively labeled repressor protein and a bacterial virus DNA that contained the *lac* operon. When they centrifuged the mixture in a glycerol gradient, the radioactive protein

Figure 16.13 The Lac repressor binds to operator DNA. A radioactive tag is attached to the Lac repressor protein so it can be followed in the experiment. **(a)** When repressor protein from *lacI*$^+$ cells was purified and mixed with DNA containing the *lac* operator (on a bacterial virus chromosome), the protein cosedimented with the DNA. **(b)** When wild-type repressor was mixed with DNA containing a mutant operator site, no radioactivity sedimented with the DNA.

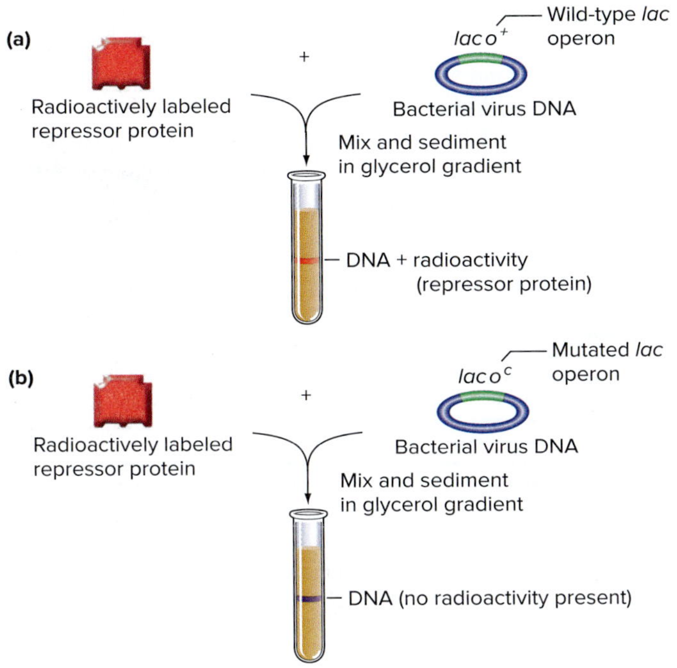

cosedimented with the DNA (**Fig. 16.13a**). If the viral DNA contained a *lac* operon that had an o^c mutation, the protein did not cosediment with the DNA, because it could not bind to the altered operator site (**Fig. 16.13b**).

Domains of the Lac repressor protein

The purified repressor protein is a dimer of two identical *lacI*-encoded subunits; in some situations, two dimers of Lac repressor can associate to form a tetramer. Importantly, each subunit contains three distinct domains (**Fig. 16.14**). One of these regions binds to the inducer, while a second domain recognizes and binds to DNA at operator sites. (Note that we use the term *domains* for the functional parts of proteins but the term *sites* for the DNA sequences with which a protein's DNA-binding domain interacts.)

When researchers investigated the molecular nature of mutations that affected repressor functions, their results made complete sense with this picture of the repressor subunits. *lacI*$^-$ mutations encoding proteins that could not bind to the operator affected amino acids in the protein's DNA-binding domain. In contrast, *lacI*^{s} superrepressor mutations, encoding proteins that could not be induced, were clustered in codons for amino acids in the inducer-binding domain.

The third domain of a Lac repressor subunit, which is found at the C-terminal region of the polypeptide

Figure 16.14 Domains of Lac repressor protein. X-ray crystallographic data provide a model of Lac repressor structure that shows a region to which operator DNA binds and another region to which inducer binds. A third domain near the C terminus allows subunits to multimerize into dimers and tetramers.

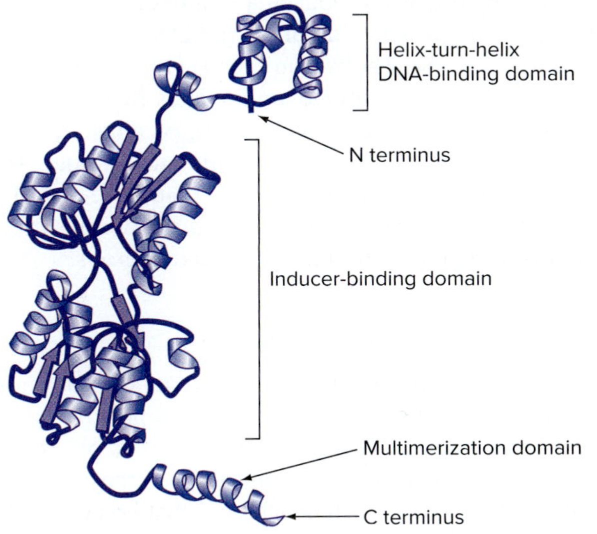

(Fig. 16.14), interacts with the same domain of other subunits to allow formation of the dimeric and tetrameric proteins. The significance of this multimerization will be discussed later in this section.

Helix-turn-helix proteins

X-ray crystallographic studies revealed that the DNA-binding domain of Lac repressor subunits has a characteristic three-dimensional structure: Two α-helical regions are separated by a turn (Fig. 16.14). This **helix-turn-helix (HTH) motif** in the protein fits well into the major groove of the DNA (**Fig. 16.15**).

Figure 16.15 Helix-turn-helix motifs recognize specific DNA sequences. A protein motif that has the shape of a helix-turn-helix (*helixes shown here inside a cylindrical shape*) fits into the major groove of the DNA helix. Specific amino acids within the helical regions of the protein recognize a particular base sequence in the DNA.

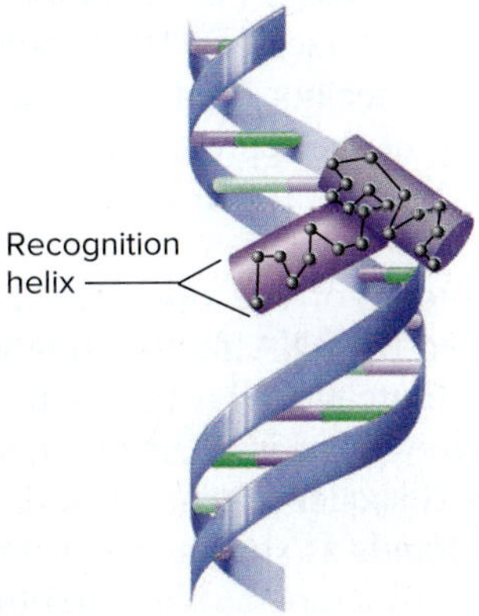

The HTH motif is found in hundreds of DNA-binding proteins, not only in bacteria but also in eukaryotic cells. The structural similarity of these proteins suggests that they evolved from a common ancestral gene whose duplication and divergence produced a family of transcriptional repressor proteins with similar overall DNA-binding structures. However, the α helixes in each HTH-containing transcription factor carry unique amino acids that recognize a specific DNA sequence of nucleotides. As a result, various HTH-containing proteins can bind to unique DNA sequences. In bacteria, this means that different HTH-containing transcription factors interact with different operators to regulate different genes and operons.

Nature of the operator sequences

In the 1970s, geneticists studying gene regulation developed new *in vitro* techniques to determine where regulatory proteins bind to the DNA. Purified proteins that bind to fragments of DNA protect the region to which they bind from digestion by enzymes such as DNase I that break the phosphodiester bonds between nucleotides. If a sample of DNA, labeled at one end of one strand and bound by a purified protein, is digested partially with DNase I, the enzyme will cleave any given phosphodiester bond in at least some DNA molecules in the sample, except for those phosphodiester bonds that are in regions protected by the bound protein. Gel electrophoresis of the DNA and *autoradiography* (exposure of the gel to radiation-sensitive film) reveal bands at positions corresponding to the cleavage between each base, except in the region where bound protein protected the DNA. Portions of the gel without bands are thus *footprints,* indicating the nucleotides of the DNA fragment that were protected by the DNA-binding protein (**Fig. 16.16**).

Figure 16.16 DNase footprints show where proteins bind to DNA. A piece of DNA is labeled at one end with radioactivity (*red asterisk*). Partial digestion with DNase I produces a series of fragments. The enzyme cannot digest the DNA at sites covered by a bound protein. Gel electrophoresis of the digested DNA shows which fragments were not generated and thus indicates where the protein binds.

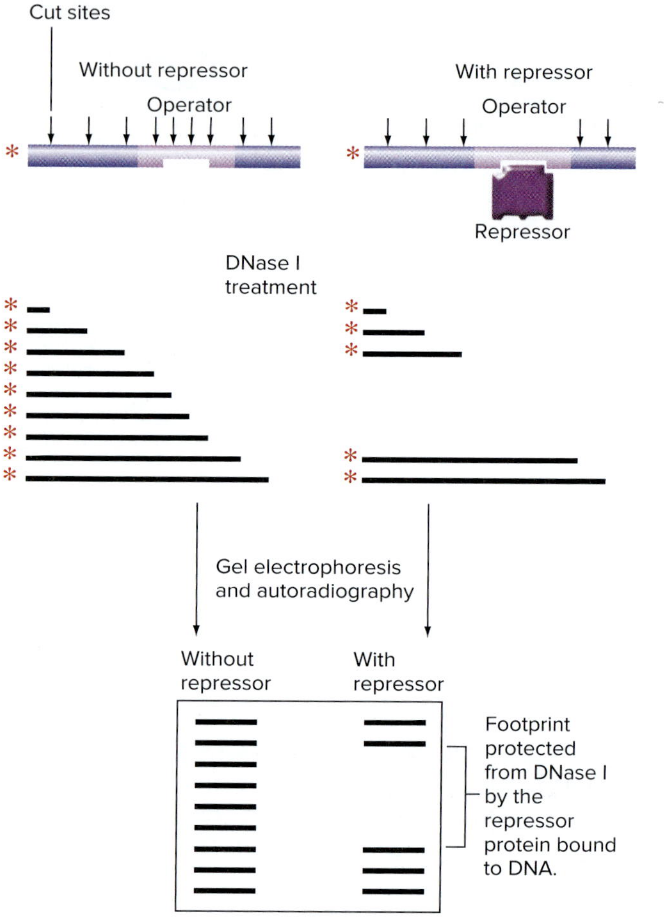

Overlap of operator and promoter DNA footprinting experiments showed that some of the nucleotides in the *lac* operon operator are also part of the *lac* operon promoter (that is, where RNA polymerase first binds to the gene) (**Fig. 16.17**). In fact, some of the nucleotides in the operator are actually transcribed into mRNA when the operon is turned on. These observations mean that if the operator site is occupied by the repressor (as would happen in a normal cell in the absence of lactose), RNA polymerase cannot recognize the promoter nor bind to it. These findings explain why binding of the repressor to the operator blocks the expression of the *lac* operon genes.

Rotational symmetry of the operator The *lac* operator sequences revealed by DNA footprinting experiments using Lac repressor protein display the interesting property of *rotational symmetry.* That is, their two DNA strands have an almost identical sequence when read in the 5′-to-3′ direction on both strands as shown in Fig. 16.17, where palindromic regions of the operator are highlighted in purple.

This rotational symmetry makes sense when you consider that one form of the Lac repressor protein is a dimer. One of the subunits of the dimer forms tight contacts with the bases making up one-half of the operator sequence. The other subunit of the Lac repressor dimer faces in the opposite direction, and it associates with the other, rotationally symmetrical half of the operator sequence.

Rotational symmetry is seen not only in the *lac* operator, but also in many different DNA sequences to which transcription factors of all types bind. This kind of symmetry reflects the fact that many transcription factors are assembled from identical or similar subunits.

Cooperativity in binding

As mentioned earlier, two dimers of the Lac repressor protein can associate to form a tetramer. DNA footprinting experiments performed with tetrameric Lac repressor, and

Figure 16.17 Regulatory protein binding sites overlap. The Lac repressor bound to the operator prevents RNA polymerase from binding to the promoter. The binding sites for RNA polymerase and repressor (determined by DNase digestion experiments) show that there is overlap between the operator and promoter. Note the rotational symmetry of nucleotide sequences in the operator (*purple*) and also in the binding site for CRP-cAMP (*green*) centered on the *black circles*.

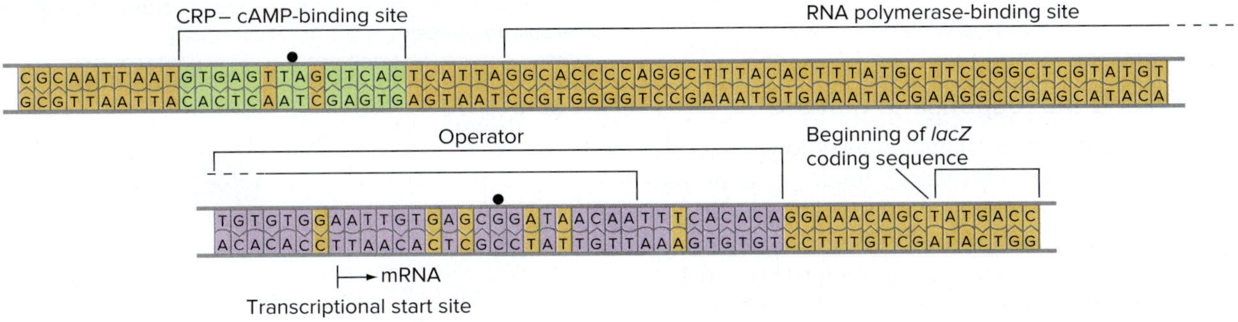

with larger fragments of *lac* operon DNA that include more sequences upstream of the promoter, revealed that the Lac repressor protein can actually bind to three sites in the vicinity of the operon. One of these sites is called o_1 (o for operator; this is the site originally identified by o^c mutations and shown in Fig. 16.17); the other sites (not shown in the figure) are o_2 and o_3.

Site o_1 has the strongest binding affinity for the repressor, and one of the dimers making up the tetramer always binds to the rotationally symmetrical sequences at this site. The other dimer within the tetramer binds to either o_2 or o_3 (**Fig. 16.18**). Mutations in *either* o_2 *or* o_3 thus have very little effect on repression. By contrast, mutations in *both* o_2 *and* o_3 make repression 50 times less effective. The conclusion is that for maximal repression, two dimers (and thus all four of the tetrameric repressor's subunits) must bind DNA simultaneously.

The o_1 site is located roughly 400 bases away from o_2 and 100 bases away from o_3. These distances are sufficiently large so that a tetrameric repressor molecule can bind simultaneously to o_1 and either o_2 or o_3 only if a loop of DNA forms between operator sites (Fig. 16.18). Binding at four recognition sequences (two in each of two operator sites) increases the stability of the protein–DNA interactions so much that it can compensate for any energy required to form the loop. In fact, the DNA binding of the Lac repressor is so efficient that only 10 repressor tetramers per cell are sufficient to maintain repression in the absence of lactose.

Figure 16.18 Lac repressor tetramer binds to two sites. The Lac repressor is a tetramer of identical subunits. Two of the subunits bind as a dimer to the sequence in one operator site (o_1), and the other two subunits bind to a second operator (either o_2 or o_3).

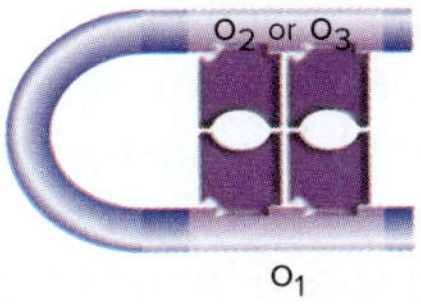

The figure shown at the beginning of this chapter shows the actual structure of the repressor tetramer bound to *lac* operon DNA; the looping required for this binding is clearly visible. You should examine this figure in some detail. See if you can identify the two dimers of repressor, the location of the multimerization domains on the repressor subunits, and the four helix-loop-helix domains binding to DNA. This figure demonstrates why each of the operator sites exhibits rotational symmetry: The two DNA-binding domains in a dimer are oriented in opposite directions with respect to the bacterial chromosome.

Similar strategies are used by most regulatory proteins that bind to DNA whether in prokaryotes or eukaryotes. Because many DNA-binding proteins are multimers, each subunit of which has a DNA-binding domain, an assembled transcription factor often has multiple DNA-binding domains. If the sites to which a multimeric protein can bind are clustered in a gene's regulatory region, many contacts can be established between the protein and the regulatory region. By increasing the stability of protein–DNA interactions, these multiple binding domains collectively produce the strength of binding necessary to maintain the control of transcription.

The *lac* Operon Is Also Regulated by Positive Control

Lactose is only one of the sugars bacterial cells can use as sources of energy and carbon atoms. In fact, if given a choice, most bacteria prefer glucose. *E. coli* grown in medium containing both glucose and lactose, for example, will deplete the glucose before gearing up to use lactose. While the glucose is present, the cells do not turn on expression of the Lac proteins even if lactose is present.

Why doesn't lactose act as an inducer under these conditions? The reason is that transcriptional initiation at the *lac* operon is a complex event. In addition to the release of repression, initiation depends on a positive regulator protein that assists RNA polymerase in the start-up of transcription. Without this assist, the polymerase does not open up the double helix efficiently. As we see next, the presence of glucose blocks the function of this positive regulator indirectly.

CRP protein and the response to glucose

Inside bacterial cells, the small nucleotide known as *cAMP (cyclic adenosine monophosphate)* binds to a protein called *cAMP receptor protein,* or *CRP.* The binding of cAMP to CRP enables CRP to bind to DNA in the regulatory region of the *lac* operon near the promoter (Fig. 16.17). When bound to DNA, CRP helps recruit RNA polymerase to the promoter by making physical contacts with the polymerase enzyme; in essence, then, the DNA binding of CRP increases the ability of RNA polymerase to transcribe the *lac* genes (**Fig. 16.19**). CRP thus functions as a positive regulator that enhances the transcriptional activity of RNA polymerase at the *lac* promoter, while cAMP is an **effector** whose binding to CRP enables CRP to bind to DNA near the promoter and carry out its regulatory function. An *effector* is a small molecule that binds to an allosteric protein or an RNA molecule and causes a conformational change. (An inducer such as allolactose is, like cAMP, an effector that activates gene expression; we will shortly look at other effectors that turn down gene expression at other operons.)

The CRP protein associated with cAMP binds to DNA as a dimer (Fig. 16.19). Just as is the case with the *lac* operator, the DNA sequence to which the CRP–cAMP complex binds has rotational symmetry (you can see this in the actual sequence on Fig. 16.17). Thus, CRP-binding sites consist of two recognition sequences pointing in opposite directions, each able to bind one subunit of the CRP dimer. This example again stresses the importance both of the multimerization of DNA-binding protein subunits and of the clustering of their corresponding binding sites in the vicinity of promoters.

Figure 16.19 Positive regulation of the *lac* operon. High-level expression of the *lac* operon requires that a positive regulator, the CRP–cAMP complex, be bound to a site near the promoter. The complex is a dimer, and the site to which it is bound contains rotationally symmetrical DNA sequences. The CRP–cAMP complex contacts RNA polymerase directly to help initiate transcription.

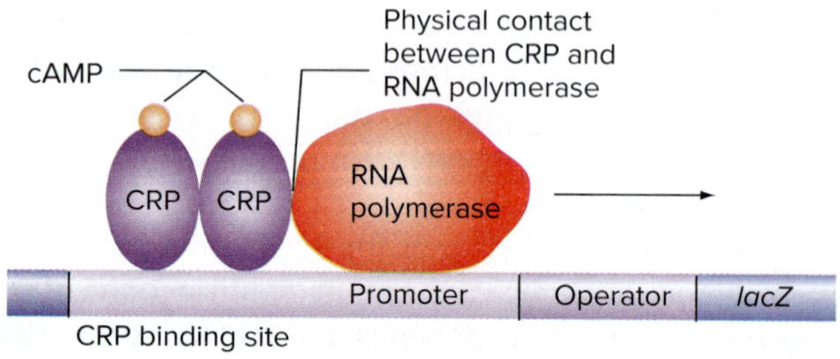

Figure 16.20 Catabolite repression. Glucose controls the activity of the enzyme adenyl cyclase that synthesizes cAMP. When glucose is high, adenyl cyclase activity and thus cAMP levels are low, so the positive regulator CRP does not bind to the *lac* operon. As a result, transcription of the *lac* operon is low when glucose is available to the cell.

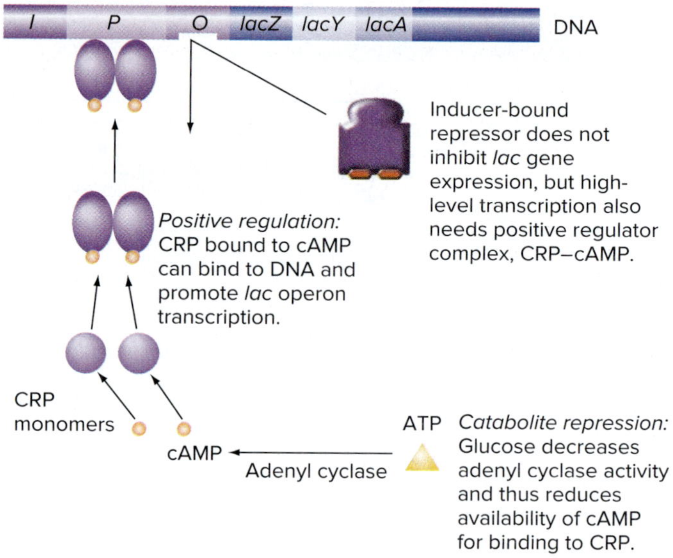

Glucose controls the amount of cAMP in the cell indirectly by decreasing the activity of *adenyl cyclase,* the enzyme that converts ATP into cAMP (**Fig. 16.20**). Thus, when glucose is present, the level of cAMP remains low; when glucose is absent, cAMP synthesis increases. As a result, when glucose is present in the culture medium, there is little cAMP available to bind to CRP and therefore little induction of the *lac* operon, even if lactose is present in the culture medium. The overall effect of glucose in preventing *lac* gene transcription is known as **catabolite repression,** because the presence of a preferred catabolite (glucose) represses transcription of the operon.

Global regulation by CRP protein

In addition to functioning as a positive regulator of the *lac* operon, the CRP–cAMP complex increases transcription in several other catabolic gene systems, including the *gal* operon (whose protein products help break down the sugar galactose) and the *ara* operon (contributing to the breakdown of the sugar arabinose). As you would expect, these other catabolic operons are also sensitive to the presence of glucose, exhibiting a low level of expression when glucose is present and cAMP is in short supply. Mutations in the gene encoding CRP that alter the DNA-binding domain of the protein reduce transcription of the *lac* operon and other catabolic operons. The binding of the CRP–cAMP complex to many operons is an example of a global regulatory strategy in response to a limited substrate (such as glucose) in the environment.

Positive and negative regulators

You can see that having operons rely on the activity of two different regulators—Lac repressor and CRP in the case of the *lac* operon—that respond to different environmental cues increases the range of gene regulation. *E. coli* cells can adjust their gene expression exquisitely to benefit from the particular mix of sugar carbon sources or other nutrients available to them at any given time.

Although positive and negative regulators obviously have opposite effects on transcription, you should keep in mind that most regulators of both types work through their effects on RNA polymerase. As we have seen, many negative regulators, such as the Lac repressor, prevent initiation by blocking the functional binding of RNA polymerase to the promoter. Positive regulators, by contrast, usually act by establishing a physical contact with RNA polymerase that attracts RNA polymerase to the promoter or that keeps RNA polymerase bound to the DNA longer so that it initiates transcription more often.

Repressor/Effector Interaction Enables Repressible Regulation of Transcription Initiation

In bacteria, the multiple genes of both catabolic and anabolic pathways are often clustered together and coregulated in operons. We have seen that the catabolic *lac* operon responds to the presence of lactose by inducing expression of the *lac* genes. By contrast, anabolic operons respond to the presence of the pathway's end product by shutting down expression of the structural genes whose protein products manufacture the end product. In other words, anabolic pathways require repressible regulation.

Several anabolic bacterial operons are involved in the production of amino acids. A well-studied example is the *E. coli* tryptophan (*trp*) operon, a group of five genes—*trpE, trpD, trpC, trpB,* and *trpA*—required for biosynthesis of the amino acid tryptophan (**Fig. 16.21**). As you would expect, maximal expression of the *trp* genes occurs when tryptophan is absent from the growth medium.

The *trp* operon is controlled at the level of transcription initiation through the action of a repressor protein that is the product of the *trpR* gene. Tryptophan functions an an effector for the TrpR repressor, an allosteric protein. Binding of tryptophan to TrpR protein causes TrpR to change its shape, thus enabling it to bind the operator and inhibit transcription of the *trp* operon genes. Mutations in the *trpR* gene that change either the protein's tryptophan-binding domain or its DNA-binding domain destroy the TrpR repressor's ability to associate with DNA, and both types of mutations therefore result in constitutive expression of the *trp* genes even when tryptophan is present in the growth medium.

It is remarkable that control of both catabolic and anabolic pathways depends on the same kind of regulatory molecule: a repressor protein that binds to operator sequences at or near the promoter, preventing RNA polymerase from recognizing the promoter. The difference is simple but essential: In the presence of the inducer allolactose, the Lac repressor *cannot* bind to the *lac* operator; while the TrpR repressor *can* bind to the operator only when the effector tryptophan is present. Again, all of this makes sense when thinking of the cellular logic behind catabolic and anabolic pathways. Catabolic pathways need to be turned on only when the starting substrate (lactose for the *lac* operon) is present, but catabolic pathways need to be turned off when the end product of the pathway (tryptophan for the *trp* operon) is present.

Although TrpR-mediated repression is a crucial first step in regulating expression of the *trp* operon, we will see in a later section of this chapter that other regulatory mechanisms, which occur subsequent to the initiation of transcription, can fine-tune the control of this operon.

Figure 16.21 Tryptophan acts as an effector. (a) When tryptophan is available, it binds to the TrpR repressor, causing TrpR to change shape so that it can bind to the operator of the *trp* operon and repress transcription. **(b)** When tryptophan is not available, the repressor cannot bind to the operator, and the tryptophan biosynthetic genes are expressed. The leader and the attenuation site will be described in the next section.

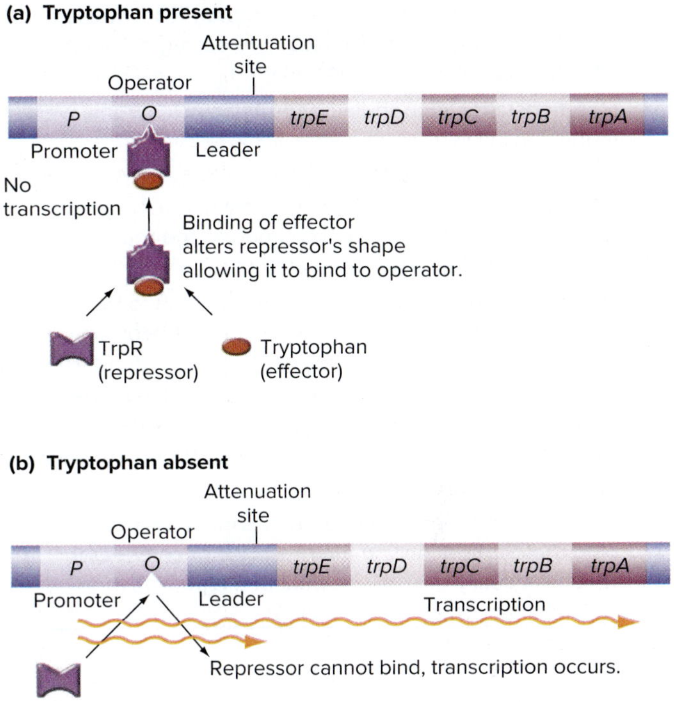

essential concepts

- *Catabolic pathways* are induced when the molecule to be broken down is present; *anabolic pathways* are repressed when the molecule that is the end product is present.
- In the *lac* operon, binding of the Lac *repressor* protein to the DNA *operator* prevents transcription. The *inducer* allolactose binds to the repressor, releasing it from the operator and allowing transcription.
- Mutations affecting regulatory proteins act in *trans*, while mutations in DNA-binding sites act in *cis*.
- Lac repressor protein has a helix-turn-helix DNA-binding domain and an inducer-binding domain. A third domain governs multimerization, which enables the repressor to bind tightly to clustered operators.
- The cAMP regulatory protein (CRP) can bind to the *lac* promoter region to allow maximum transcription only in the absence of glucose, when cAMP levels are high.
- *Negative regulators* like the Lac repressor inhibit RNA polymerase from binding to the promoter; *positive regulators* like CRP–cAMP encourage RNA polymerase's binding to the promoter.
- In anabolic (biosynthetic) operons, repressor proteins bind to operators only in the presence of an *effector*, which is usually the end product of the pathway.

16.3 RNA-Mediated Mechanisms of Gene Regulation

learning objectives

1. Explain how RNA leader devices can regulate gene expression in response to environmental conditions.
2. List two different ways in which *trans*-acting small RNAs (sRNAs) regulate the expression of target genes.
3. Describe the relationship between the promoters for sense and antisense RNAs of a gene.

We have just seen that many bacterial genes, like those in the *lac* and *trp* operons, are regulated primarily at the level of transcription initation by the action of *trans*-acting regulatory proteins such as repressors or CRP that bind to *cis*-acting DNA sites like operators or CRP-binding sites. Allostery of the regulatory proteins (that is, changes in their shapes that occur when they bind to small molecules like allolactose or tryptophan) enables genes to be turned on or off in response to the cellular environment.

Bacteria have evolved many other exquisite and important mechanisms for adjusting gene expression in response to ever-changing conditions. Many of these mechanisms function after transcription is initiated; for example, some regulate the termination of transcription, while others control the translation of mRNA transcripts. A remarkable finding in recent years is that many mechanisms that control gene expression in bacterial cells depend on RNA molecules that also exhibit allosteric changes in shape. In some cases, the control features are part of the transcript itself and therefore act in *cis;* in other cases, one RNA molecule can act in *trans* to influence the expression of a different RNA.

Diverse RNA Leader Devices Act in *cis* to Regulate Gene Expression

All bacterial mRNAs begin with an untranslated region called the 5′ UTR, or **RNA leader sequence.** Through complementary base-pairing, many RNA leaders form secondary structures called **stem loops** (or **hairpin loops**). The stem loops can terminate transcription of the rest of the mRNA prematurely, or they can prevent translation by blocking access of the mRNA to the ribosome binding site. These RNA leaders are allosteric in that they can alter their stem-loop structures and thus their function in response to a wide variety of environmental cues.

Attenuators

The first RNA leader mechanism discovered involves fine-tuning of the response of the *trp* operon to the amount of tryptophan available to the cell. You will remember from Fig. 16.21 that maximal expression of the *trp* genes occurs when tryptophan is absent from the growth medium: The TrpR repressor protein cannot bind to the operator because the effector tryptophan is lacking. The TrpR-mediated repression of the *trp* operon is indeed a crucial first step, but it is not the only regulatory mechanism controlling expression of the *trp* genes in *E. coli.*

If TrpR binding to the repressor were the only event of importance, you would expect *trpR*$^-$ mutants to show constitutive expression of their *trp* genes. With or without tryptophan in the medium, if there is no repressor to bind at the operator, RNA polymerase should have uninterrupted access to the *trp* promoter. Surprisingly, the actual experiments showed that the *trp* genes of *trpR*$^-$ mutants are not completely de-repressed (that is, turned on maximally) when tryptophan is present in the growth medium. The removal of tryptophan from the medium caused expression of the *trp* genes to increase a further threefold. Apparently, tryptophan can affect the expression of the *trp* operon by some kind of additional mechanism that does not involve the TrpR repressor protein.

In a series of elegant experiments, Charles Yanofsky and coworkers found that this repressor-independent change in *trp* operon expression involves the production of alternative *trp* operon transcripts. Sometimes initiation at the promoter leads to transcription of a truncated mRNA

Figure 16.22 An attenuator in the tryptophan operon of *E. coli.* **(a)** Stem loops form by complementary base pairing in the *trp* leader RNA. Two different conformations are possible: One of these includes a *terminator* that signals RNA polymerase to stop polymerization, while the other is an *antiterminator* that allows transcription of the operon's structural genes. The leader also contains a small open reading frame with two UGG codons for tryptophan. **(b)** When tryptophan is present, the ribosome follows quickly along the transcript, and the terminator forms. **(c)** If tryptophan is absent, the ribosome stalls at the Trp codons, allowing formation of the antiterminator.

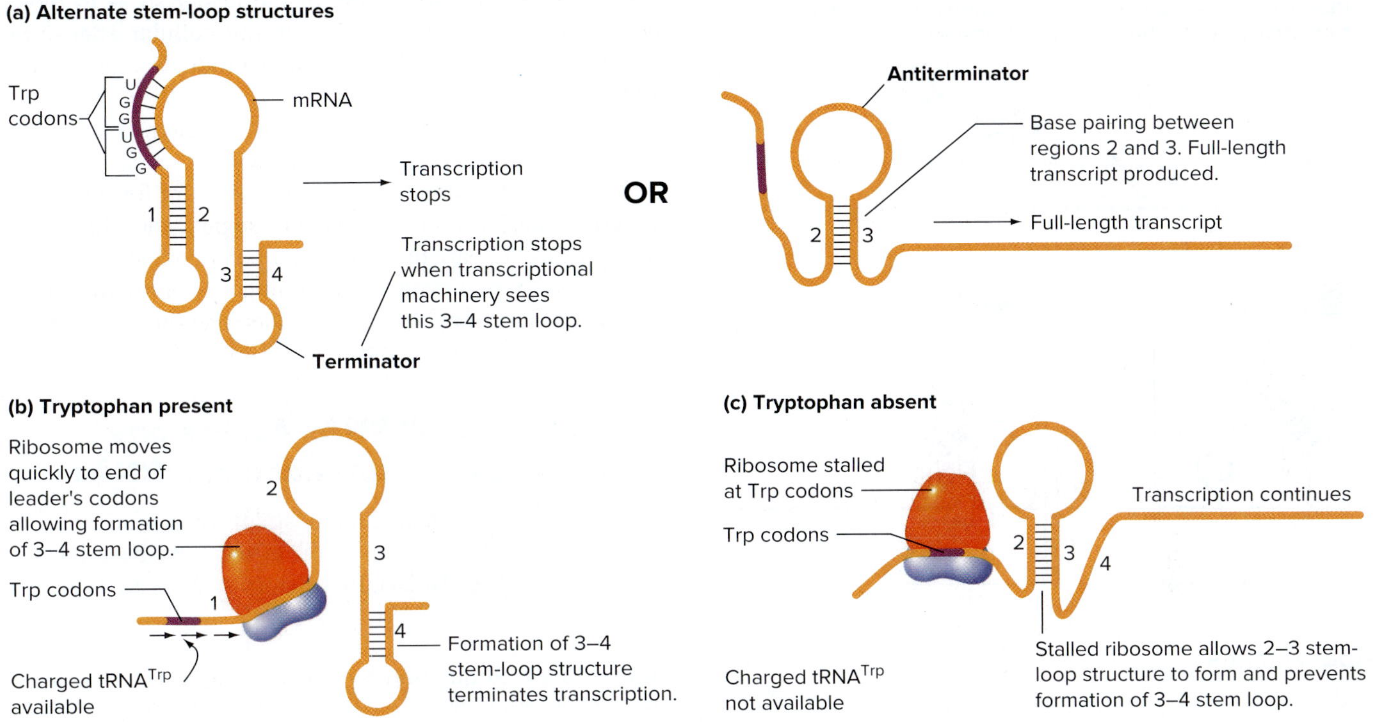

about 140 bases long containing only the RNA leader but none of the structural genes. At other times, transcription continues beyond the end of the leader sequence to produce a full operon-length transcript (Fig. 16.21b). In analyzing why some mRNAs terminate before they can transcribe the structural *trp* genes, while others do not, the researchers discovered **attenuation:** control of gene expression by RNA leader–mediated premature termination of transcription that involves unusual translation of part of the leader sequence. Whether or not transcription terminates depends on how the translation machinery interacts with a portion of the RNA leader called the **attenuator.**

The *trp* RNA leader can fold into two different stable conformations, each one based on the complementarity of bases in the same molecule of RNA (**Fig. 16.22a**). The first conformation contains two stem-loop structures: Region 1 makes a stem loop with region 2, while region 3 associates with region 4. The 3–4 stem-loop configuration is called a **terminator** because when it forms in the *trp* operon transcript, RNA polymerase contacts it and stops transcription, producing a short, *attenuated* RNA. The alternative RNA structure, called the **antiterminator,** forms by base pairing between regions 2 and 3. In this conformation, the leader RNA cannot form the terminator (because region 3 is no longer available to pair with region 4), and as a result, the transcription machinery continues to produce a full-length mRNA that includes the *trp* structural gene sequences.

Early translation of a short portion of the RNA leader (while transcription of the rest of the leader is still taking place) determines which of the two alternative RNA structures forms. That key portion of the RNA leader includes a short open reading frame containing 14 codons, two of which are Trp codons (Fig. 16.22a). When tryptophan is present, the ribosome moves quickly past the Trp codons in the RNA leader and proceeds to the end of the leader's codons, allowing formation of the terminator (**Fig. 16.22b**). In the absence of tryptophan, the ribosome stalls at the two Trp codons in the RNA leader because of the lack of charged $tRNA^{Trp}$ in the cell. The antiterminator is then able to form, which prevents formation of the terminator (**Fig. 16.22c**). As a result, transcription proceeds through the leader into the structural genes.

Why has such a complex system evolved in the regulation of the *trp* operon and other biosynthetic pathways? Whereas the TrpR repressor shuts off transcription in the presence of tryptophan and allows it in the amino acid's absence, the attenuation mechanism provides a way to fine-tune this on/off switch. It allows the cell to avoid expending energy synthesizing gene products unnecessarily; the cell senses the level of charged $tRNA^{Trp}$ and adjusts the level of *trp* mRNA accordingly. In *E. coli,* similar attenuation mechanisms also exist for several other amino acid biosynthetic operons, including histidine, phenylalanine, threonine, and leucine.

Figure 16.23 Riboswitches. The leader in the mRNA of many bacterial genes and operons is a *riboswitch* that controls expression in response to an effector molecule. A riboswitch contains an *aptamer* that binds to an effector, altering conformation, and an *expression platform* that responds to the conformational change.

(a) Riboswitch that regulates transcription elongation **(b) Riboswitch that regulates translation**

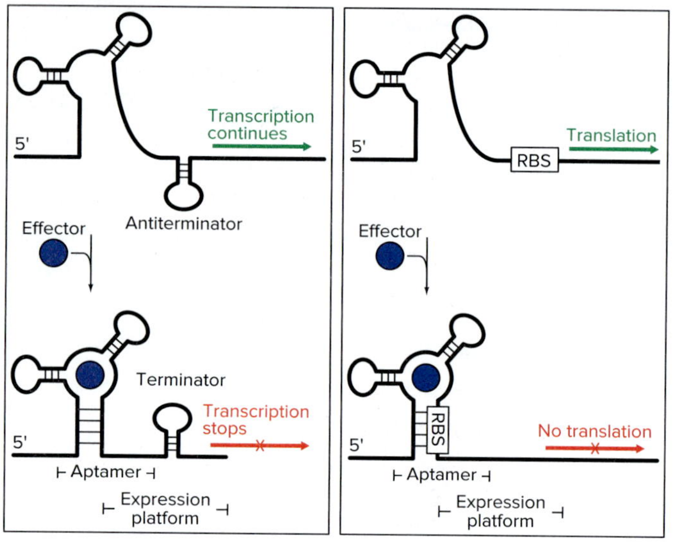

Riboswitches

In the years since Yanofsky and his colleagues elucidated the molecular details of attenuation, it has become increasingly clear that RNA-based mechanisms do much more than fine-tune the expression of certain operons. In fact, shifts between different conformations of RNA molecules play the primary role in the regulation of many genes. One widespread mechanism involves **riboswitches:** allosteric RNA leaders that bind small molecule effectors to control gene expression (**Fig. 16.23**).

Whereas in attenuation the level of an amino acid like tryptophan governs the conformation of the leader indirectly (through participation of tRNAs and ribosomes), leaders that act as riboswitches have a region called the *aptamer* that binds a particular effector directly. Riboswitches also have a second region, called the *expression platform,* which controls gene expression by altering its stem-loop structures in response to the aptamer configuration.

In some riboswitches, the expression platform controls the termination of transcription (**Fig. 16.23a**). For example, one of the simplest riboswitches controls transcription in response to guanine. When the aptamer is bound to guanine, the most stable conformation of the expression platform forms a terminator. The expression platform switches to form an antiterminator when the aptamer is not bound to the nucleotide base guanine. In this way, the aptamer is a sensor for guanine levels; the riboswitch turns off the expression of genes that participate in guanine synthesis when guanine levels are high, and turns those genes on when guanine levels are low.

In other riboswitches, the expression platform regulates translation by blocking or unblocking the ribosome binding site (**Fig. 16.23b**). In an anabolic pathway, binding of the effector (in this case the end product) to the aptamer would shift the conformation of the leader so as to block the ribosome binding site, conserving cellular energy by preventing the synthesis of protein products that are not currently needed by the cell.

Seventeen different riboswitch aptamers have been identified thus far in the *E. coli* genome, each of which binds to specific effectors such as cofactors for enzymes, nucleotide derivatives, amino acids, sugars, and Mg^{2+} ions. Because any one aptamer type can be connected to different expression platforms in different genes or operons, bacterial cells can coordinate complicated responses to diverse changes in the environment.

Regulatory Small RNAs Act in *trans* to Regulate the Translation of mRNAs

Bacterial genomes encode many small RNA molecules, or **sRNAs,** that regulate translation in *trans* by base pairing with mRNAs (**Fig. 16.24**). Regulatory sRNAs are typically

Figure 16.24 Regulation by *trans*-acting sRNAs. (a) Base-pairing of an sRNA with the leader sequence can inhibit translation of an mRNA by hiding the ribosome binding site (*RBS*). **(b)** An sRNA can facilitate mRNA translation through base pairing interactions with the leader that prevent the formation of a stem loop that occludes the ribosome binding site.

(a) Negative regulation of translation by sRNAs

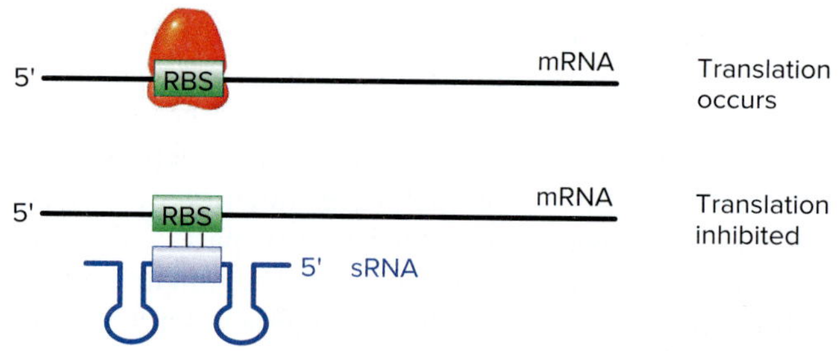

(b) Positive regulation of translation by sRNAs

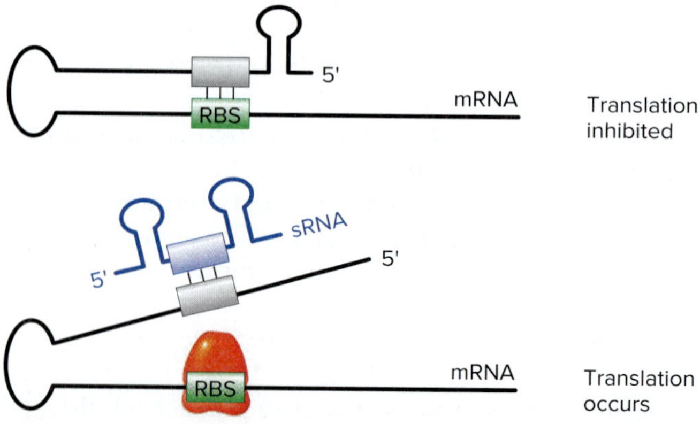

50–400 nt long, and their sequences contain a region complementary to several different mRNA targets. Most sRNAs are repressive, meaning that they inhibit translation of their target mRNAs by base pairing with the ribosome binding site (**Fig. 16.24a**). Some sRNAs, however, activate translation of their target mRNAs by disrupting the formation of a stem-loop structure in the leader of the mRNA that would otherwise block the ribosome binding site (**Fig. 16.24b**). Another way in which certain sRNAs can influence the expression of particular target genes is by promoting the degradation of the mRNA: The double-stranded RNA region resulting from sRNA binding to the mRNA causes the mRNA to be degraded by ribonuclease enzymes (*not shown*).

You should note that most sRNAs do not bind to effectors, so the sRNA usually cannot respond to environmental changes directly. Instead, the transcription or stability of an sRNA is often controlled through other regulatory molecules, such as transcription factors that interact with the sRNA gene promoter so as to increase or decrease the cellular concentration of the sRNA. This generalization means that sRNAs often act as intermediaries in *regulatory cascades* in which one regulator influences the expression of a different regulator. You will see examples of such regulatory circuits later in the chapter.

Genes Can Also Be Regulated by Antisense RNAs

The regulatory sRNAs just described are encoded by genes that can be far removed from the genes encoding their mRNA targets. In contrast, some bacterial genes are regulated by RNAs that are complementary in sequence to the mRNA because their transcription template is the opposite strand of DNA. These regulatory RNAs are called **antisense RNAs;** the mRNAs they regulate are *sense RNAs* (**Fig. 16.25**). Antisense RNAs range in size from 10–1000 nt, and may be complementary to the entire mRNA encoded on the opposite DNA strand, or they may overlap only part of it. Note that the promoters for the production of sense mRNAs and antisense regulatory RNAs are located on opposite sides of the coding region (Fig. 16.25).

Some antisense RNAs function like the *trans*-acting sRNAs: They inhibit translation by base pairing with the sense mRNA and blocking the ribosome binding site, similar to what you saw previously in Fig. 16.24. In other cases, the double-stranded RNA formed by base pairing between the sense mRNA and antisense RNA can be degraded by ribonucleases. In yet other cases, it is not the antisense RNA itself, but the act of transcribing it that inhibits expression of the sense gene; that is, antisense transcription can interfere with initiation of transcription of the sense gene.

Often, when the sense mRNA encodes a protein that is toxic to the cell at high concentrations, the antisense RNA is transcribed constitutively to ensure that the mRNA for the toxic product stays at low levels. At other genes, the antisense promoter is controlled by elements of a regulatory cascade that adjusts transcription of the antisense RNAs (and thus, indirectly, the sense RNAs) according to environmental conditions.

Figure 16.25 Antisense RNAs inhibit gene expression. Antisense transcripts that overlap all or part of the sequence of some sense mRNAs are transcribed from nearby antisense promoters.

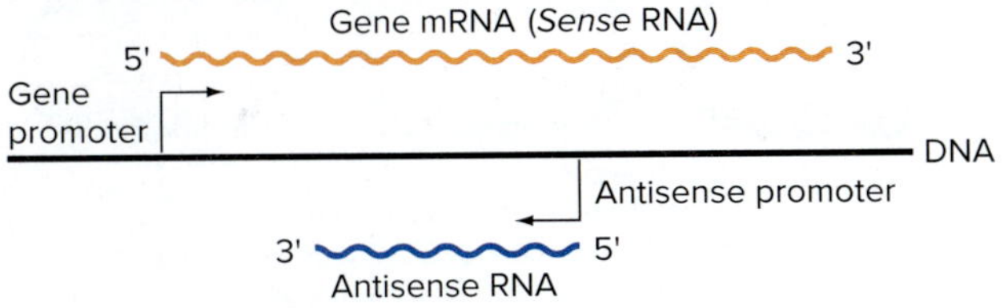

essential concepts

- Prokaryotic *RNA leaders* can act in *cis* as regulators of transcription or translation by folding into alternate *stem loops* in response to environmental conditions.
- *Small RNAs (sRNAs)* regulate the translation of mRNAs in *trans* through complementary base pairing that can hide or expose the ribosome binding site.
- *Antisense RNAs,* transcribed from the opposite DNA strand, can regulate some genes by decreasing sense mRNA translation, stability, or transcription.

16.4 Discovering and Manipulating Bacterial Gene Regulatory Mechanisms

learning objectives

1. Describe the ways in which scientists employ *lacZ* reporter genes to study gene regulation.
2. Explain how regulatory regions of the *lac* operon can be used in the production of pharmaceutical proteins.
3. Discuss RNA-Seq and its application in the study of bacterial responses to heat shock.
4. List at least two ways in which computerized analysis of transcriptomes aids genetic research.

Suppose that you are a geneticist interested in understanding bacterial responses to environmental changes, such as how bacteria survive when the ambient temperature rises, how pathogenic bacteria evade the body's defense mechanisms, or how the bacteria that exist in symbiotic relationships with plants are triggered to fix nitrogen (converting gaseous N_2 to nitrogen-containing molecules useful to the plant). What genes are turned on or off by a given change in the environment? Which of the general mechanisms described in this chapter, or what novel mechanisms not yet found, are involved in the gene regulation?

In this section, we describe some of the methods currently being used to answer these kinds of questions. We also describe how scientists and genetic engineers employ their understanding of bacterial gene regulation for the practical application, crucial for the pharmaceutical industry, of synthesizing large amounts of important protein drugs.

lacZ Reporter Genes Help Reveal the Regulation of Other Genes

We have seen earlier that one of the key advantages of the *lac* operon is the existence of simple assays to measure the amount of β-galactosidase enzyme in a sample: Enzyme function can turn one colorless substrate into a soluble yellow compound, while β-galactosidase can also change the colorless compound X-gal into an insoluble blue precipitate. Because it is so easy to measure β-galactosidase levels, the *lacZ* gene that encodes it can serve as a **reporter gene** to detect the amount of transcription that occurs in response to any specific regulatory element.

For this purpose, researchers use recombinant DNA methods to create DNA molecules in which the coding region of *lacZ* is fused to *cis*-acting regulatory regions (including promoters and operators) of any other gene (gene *X*). The synthetic reporter gene cloned in a plasmid can then be introduced by transformation into bacterial cells. In cells with this **fusion gene,** conditions that normally induce the expression of gene *X* will generate β-galactosidase (**Fig. 16.26**).

Figure 16.26 A *lacZ* reporter gene. The *lacZ* coding sequences can be fused to a regulatory region of gene *X*. Expression of β-galactosidase will depend on signals impacting the regulatory region to which *lacZ* is fused.

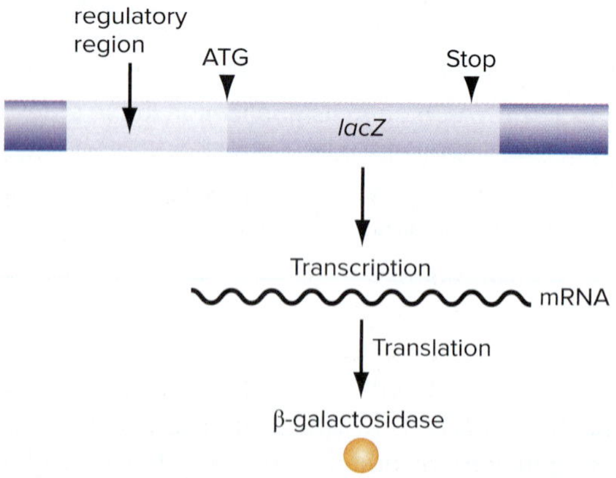

The scientists can then assess the activity of the gene *X* regulatory elements by monitoring the amount of β-galactosidase the cell produces.

The availability of *lacZ* reporter genes makes it possible to identify the DNA sites necessary for regulation as well as the genes and signals involved in that regulation. For example, you could mutate gene *X*'s control region *in vitro* and then transform the altered gene *X*–*lacZ* fusion molecule into bacterial cells to identify *cis*-acting sites important for the regulation. In an alternative protocol, you could identify *trans*-acting genes encoding molecules that interact with DNA sequences in the gene *X* control region by mutagenizing bacteria harboring a reporter fusion construct. In both procedures, you would ultimately look for changes in the level of *lacZ* expression as measured by blue or white colony color on agar media containing X-gal.

Reporter fusions also make it possible to identify genes that are regulated by a given environmental stimulus. To this end, researchers can use transposons to insert the *lacZ* gene, without its regulatory region, at various sites around the bacterial chromosome (**Fig. 16.27**). This *lacZ* gene cannot by itself make β-galactosidase, because it lacks

Figure 16.27 Using a promoterless *lacZ* gene. Transposition of the *lacZ* gene without its promoter creates a collection of *E. coli* cells with insertions at random chromosomal positions. If *lacZ* integrates within a gene in the orientation of transcription, *lacZ* expression will be controlled by that gene's regulatory region. Researchers screen the library of clones to identify genes regulated by a common signal.

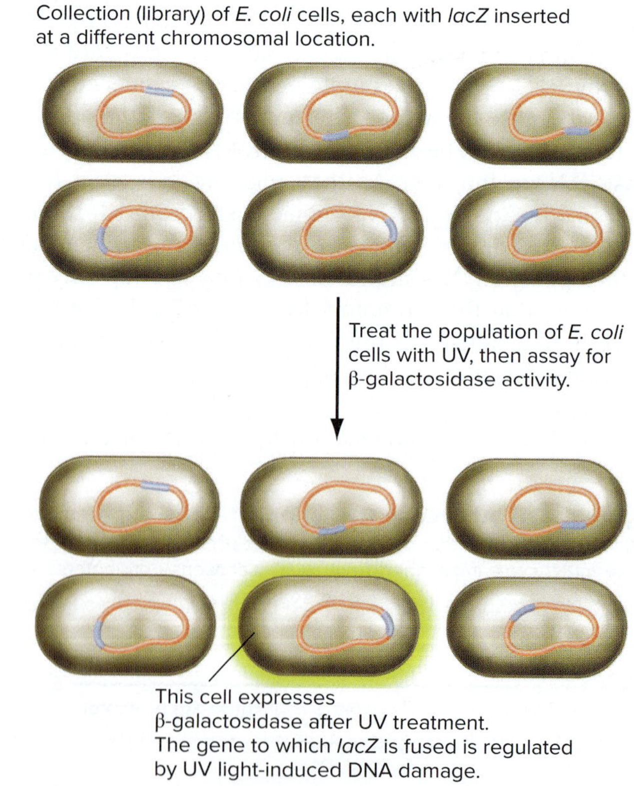

a promoter. However, in some cells, the promoterless *lacZ* reporter gene becomes inserted adjacent to the promoters and regulatory regions of other genes. If conditions are right to turn on the expression of a gene into which the reporter has inserted, the cell will produce β-galactosidase. Using this method, researchers identified a set of genes activated by exposure to DNA-damaging agents like UV light (Fig. 16.27).

lac Operon Regulatory Sequences Help Produce Protein Drugs in Bacteria

When constructing a reporter gene, the crucial part of the *lac* operon is the *lacZ* coding sequence. However, geneticists can exploit other parts of the *lac* operon, namely the regulatory region that allows high levels of transcription, for a very important practical application: making bacteria into mini-factories for the production of large amounts of medically important proteins. These include hormones like insulin for diabetics, clotting factors for hemophiliacs, and polypeptide antigens that can be used as vaccines.

The basic idea is to construct recombinant plasmids in which the *lac* operon control region DNA is fused to the open reading frame encoding the protein to be expressed. *E. coli* cells transformed with such a recombinant plasmid will produce large amounts of the protein drug when lactose is added to the medium (**Fig. 16.28**). The ability to control expression of the recombinant gene is particularly important if the foreign protein has deleterious effects on the growth of *E. coli*. The culture can be grown to a high density of cells in the absence of the inducer, after which the addition of lactose turns on protein production to high levels.

RNA-Seq Is a General Tool for Characterizing Transcriptomes and Their Regulation

A **transcriptome** is the base sequence of every transcript that a cell produces under a particular set of conditions. Transcriptome sequencing requires the production of a cDNA library and subsequent sequencing of many of these cDNAs; the entire procedure is called **RNA-Seq** or sometimes **cDNA deep sequencing.** Current RNA-Seq technology allows a researcher to obtain about one billion sequence reads of cDNA sequence, each read about 150 nucleotides long, in a single experiment. The more copies of a particular mRNA in a cell, the more times cDNAs corresponding to that RNA will be sequenced. Deep sequencing thus enables a researcher to quantify the relative levels of individual mRNAs as the proportion of total cDNA reads that represent the particular mRNA in question.

Figure 16.28 Making *E. coli* into a factory for protein production. **(a)** The *lac* regulatory region can be fused to gene *X* to control the expression of genes. **(b)** In this example, a cDNA encoding human growth hormone is cloned next to the *lac* control region and transformed into *E. coli.* Conditions that induce *lac* transcription will cause expression of growth hormone that can be purified from the cells.

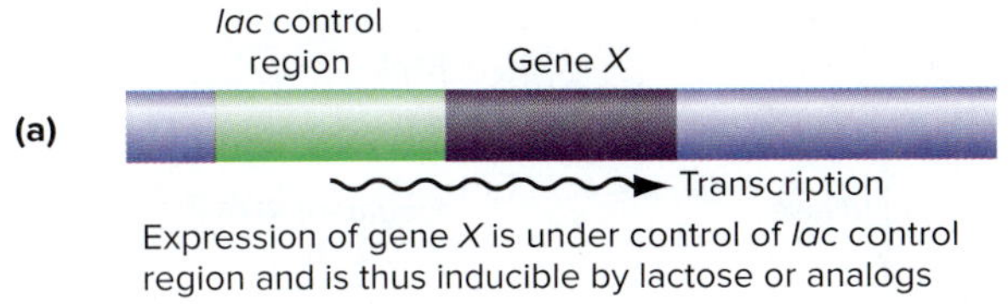

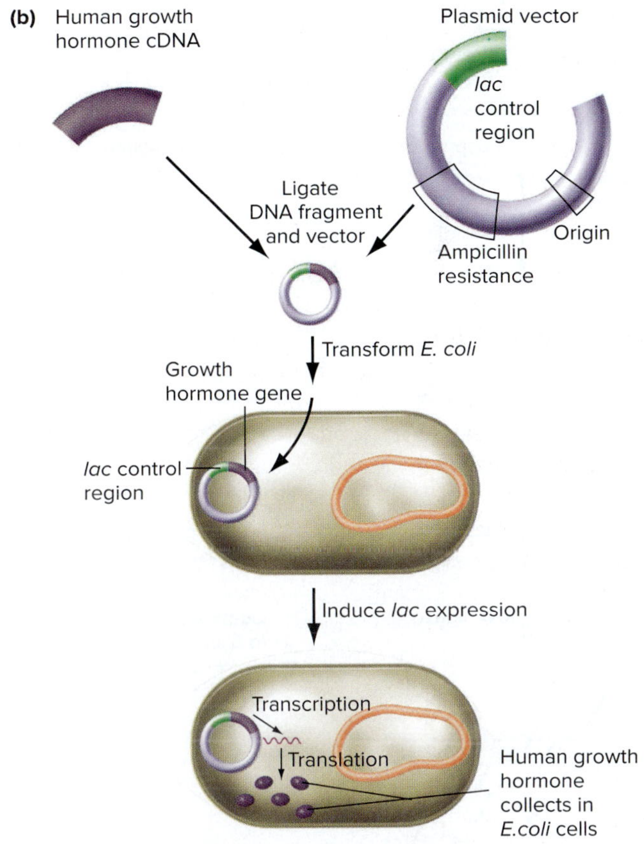

By comparing the transcriptomes of a population of bacterial cells grown in different environments, investigators can determine which of all of the genes in the genome have their expression turned up and which are turned down when the conditions are changed. In an alternative use of this methodology, investigators can compare the transcriptome of a wild-type strain with that of a mutant strain of the same species that is defective for a regulator gene or regulatory DNA sequence, in order to determine which genes in the genome depend most on that regulatory feature.

Figure 16.29 Construction of a directional cDNA library for RNA-Seq. Bacterial RNAs are fragmented with RNase, and the resultant 5′ ends are dephosphorylated enzymatically. RNA ligase then connects a short synthetic single-stranded RNA sequence (*Adapter A*) specifically to the 3′ ends. [The adapter cannot attach to the 5′ ends of the RNA fragments because no phosphate group (*yellow circle*) is present there.] After rephosphorylation of the 5′ ends with a kinase enzyme, a second adapter with a different base sequence (*Adapter B*) is ligated specifically to the phosphorylated 5′ ends of the RNA fragments. On the cDNAs synthesized from these templates, the two adapter sequences A and B mark the 3′ and 5′ ends of the original RNAs, respectively.

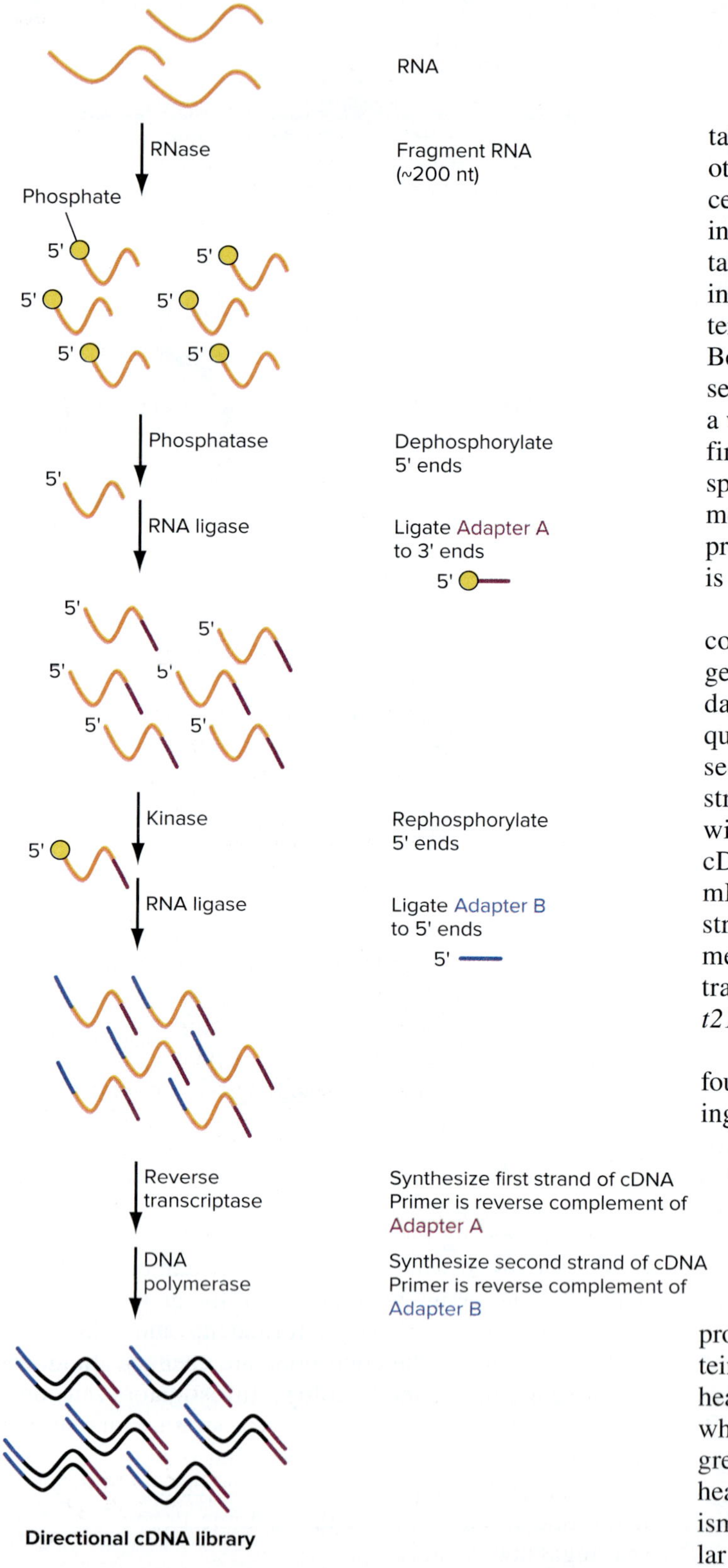

In contrast with mRNAs in eukaryotic cells, poly-A tails are not characteristic features of mRNAs in prokaryotic organisms. As a result, cDNA libraries for bacterial cells must be constructed in a different manner from those in eukaryotic cells, which depend on the existence of poly-A tails as was shown in Fig. 10.4. In place of the poly-A tail, investigators use RNA ligase to connect fragments of bacterial RNAs with adapter oligonucleotides (**Fig. 16.29**). Because some genes are transcribed in both sense and antisense directions, the cDNA library must be constructed in a way that indicates which strand of the cDNA was copied first, thereby tracking which cDNA strand sequence corresponds to the RNA and which one is the reverse complement. A cDNA library constructed in a manner that preserves this information, such as that shown in Fig.16.29, is called a *directional cDNA library.*

To analyze the data from an RNA-Seq experiment, a computer lines up each cDNA sequence read with the genome sequence of the bacterium. An example of such data for one small segment of a bacterial genome sequence is shown in **Fig. 16.30.** The data reveal which sequences in the genome are transcribed and which DNA strand is used as the template. In addition, the frequency with which each nucleotide in the genome was found in a cDNA read indicates the relative levels of the different mRNAs present. The results shown in Fig. 16.30 demonstrate that under the particular conditions of the experiment, genes such as *ybhC, ybhE,* and *galM* are strongly transcribed, while other genes such as *hutC, hutI,* and *t2110* are not expressed.

One important example of the RNA-Seq method is found in studies of the heat shock response in *E. coli* resulting from exposure of the cells to extremely high temperatures (up to 45°C). RNA-Seq has found that high temperatures specifically induce the transcription of a suite of genes encoding specialized *heat-shock proteins* that allow cells to survive. These heat-shock proteins counteract the tendency of high temperatures to cause the denaturation and aggregation of most of the other proteins in the cell. Some of the induced heat-shock proteins recognize and degrade aberrant proteins, while other heat-shock proteins act as so-called *molecular chaperones* which help refold other proteins and also prevent their aggregation. *E. coli*'s induction of the proteins that combat heat shock is a highly conserved stress response. Organisms as different as bacteria, flies, and plants induce similar proteins in response to high temperatures.

Figure 16.30 RNA-Seq results. At bottom is a map of part of the *S. typhimurium* genome: Numbers indicate base pairs and genes are shown as *colored arrows* pointing in the direction of transcription. Adjacent genes in the same color are likely in the same operon. At top, the RNA-Seq data are plotted. The *green line* shows the number of times each base in the genomic sequence was read in a cDNA sequence made from an RNA transcribed from right to left. The *purple line* indicates the reads per base from cDNAs transcribed from left to right.

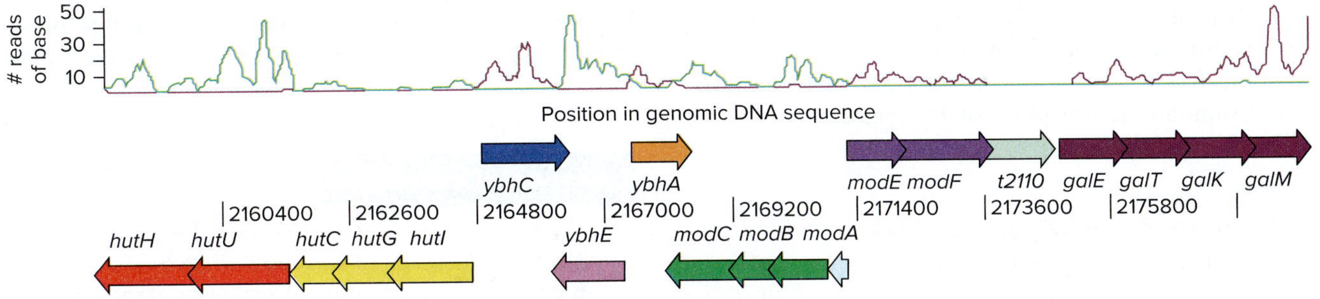

Computer Analysis Reveals Many Aspects of Gene Regulation

A major goal in the postgenomic era is to identify the complete set of *trans*-acting factors (proteins and RNAs) that regulate transcription in an organism, all of the *cis*-acting elements to which these factors bind, and ultimately to catalog how the various factors and elements interact so as to regulate each gene in the genome. This goal is of course extremely ambitious and far from being realized, but the pace of progress is increasing rapidly and these studies have already provided information of value both to our basic understanding of how cells work and to practical applications of that knowledge. The progress depends largely on computerized analysis of bacterial genomes and transcriptomes. We discuss briefly here a few of the ways in which computers can help interpret these gigantic data sets.

Operon discovery

A key issue in uncovering the regulatory machinery in bacteria is to identify operons. It is easy to correlate genomic DNA sequence with potential open reading frames, but how can you find genes that are cotranscribed on a single mRNA? One way is to examine transcriptome information gained by RNA-Seq and related methods to determine the base sequences of cDNAs corresponding to all of the transcripts made by a bacterial species. Computer programs can identify cotranscribed genes in an operon simply as one transcript containing several open reading frames. Computer experts have furthermore developed algorithms that search for several closely spaced genes adjacent to a known promoter, as well as for transcription termination signals.

These predictions can be assessed further by comparative species analysis. If two very different species have homologous sets of genes that lie adjacent to each other in the two genomes, it is likely that the genes remain together because they are all cotranscribed as an operon.

Discovery of transcription factors and their binding sites

New *trans*-acting regulatory proteins can be identified by searching the genome for sequences that could encode known DNA binding motifs, such as the helix-turn-helix region of the Lac repressor protein. Computers can in some cases find the *cis*-acting sites bound by the *trans*-acting regulatory proteins by looking at the DNA sequences of the suites of genes that exhibit similar regulation in response to a given environmental condition. You might expect, for example, that if several genes or operons were regulated by the same repressor protein, you would find similar or related operator DNA sequences just upstream of those genes or operons.

One interesting use of this kind of methodology is seen in the analysis of the global regulatory response of *E. coli* cells to high temperatures, which was introduced earlier. You will remember that high temperatures induce transcription of the heat-shock genes, whose products protect cells from the deleterious consequences of this environmental condition. Examination of the DNA sequences just upstream of the heat-shock gene coding region showed that the promoters of the heat-shock genes have significant differences from the promoters of the everyday *housekeeping genes* expressed at normal temperature. The reason turns out to be that alternative sigma factors allow RNA polymerase to recognize the two different classes of promoters (**Fig. 16.31**). The normal housekeeping sigma factor, σ^{70}, is active in the cell under normal physiological conditions, but it becomes denatured and therefore is

Figure 16.31 Sigma factor recognition sequences. The promoters of *housekeeping genes* expressed at normal temperatures are recognized by σ^{70}. The promoters of genes encoding heat-shock proteins have sequences that are recognized by the heat-resistant sigma factor σ^{32}. (The N indicates that any base can be found at this position.)

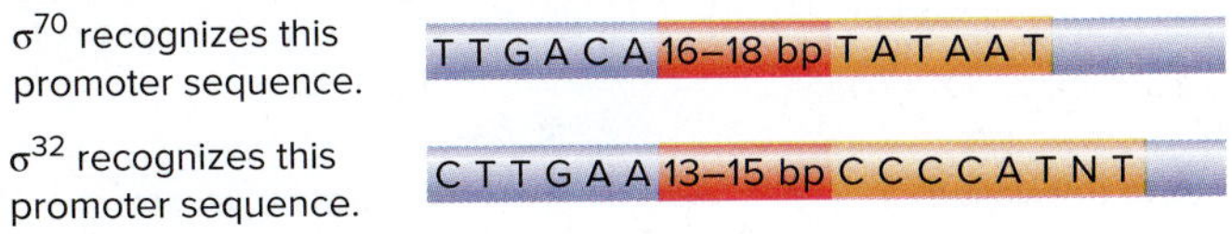

inactivated at high temperatures. By contrast, the alternative σ^{32} can function at high temperatures; it also recognizes different promoter sequences than those recognized by σ^{70}.

Many genes induced by heat shock contain nucleotide sequences in their promoters that are recognized by σ^{32}. The σ^{32} factor mediates the heat-shock response by binding to the core RNA polymerase, thereby allowing the polymerase to initiate transcription of the genes encoding the heat-shock proteins.

Identification of new regulatory RNAs

Regulatory RNAs can be difficult to find. Although RNA leader devices such as riboswitches often have similar secondary structure features, little or no base sequence conservation exists that computers could identify from the genome sequence alone. However, judicious examination of the transcriptome can supply useful clues for the discovery of RNA leader devices, sRNAs, and antisense RNAs. RNA leaders would be located at the 5′ ends of transcripts connected to the open reading frames of genes. sRNAs should have short regions displaying at least limited complementarity to portions of mRNA transcripts (review Fig. 16.24). These regions of complementarity should be even more pronounced in the case of antisense RNAs (review Fig. 16.25), and the inverse polarities of the sense and antisense transcripts should be revealed by the directionalities of the corresponding cDNAs.

Computerized analysis of the mRNA encoding the alternative sigma factor σ^{32} revealed the existence of a specialized kind of RNA leader mechanism called an **RNA thermometer,** which can be regarded as a rudimentary riboswitch. Such RNA thermometers inhibit translation at low temperatures by forming stem-loop structures that involve the ribsome binding site in base pairing interactions, preventing mRNA translation. At higher temperatures, the stem-loop structure becomes unstable and unzips, freeing the ribosome binding site (**Fig. 16.32**). This mechanism makes perfect sense for the σ^{32} mRNA. You will remember that σ^{32} is needed only at high temperature, when it allows RNA polymerase to transcribe the heat-shock genes. The RNA thermometer in the leader of the σ^{32} mRNA prevents this sigma factor (and therefore the heat-shock proteins) from being made at low temperature, when their production could be damaging to the cells.

Figure 16.32 An RNA thermometer. Some bacterial mRNAs contain leader sequences that regulate translation in response to temperature. Stem-loop structures that occlude the ribosome binding site form at low temperatures. At high temperatures, the stem-loop unzips and the mRNA can be translated.

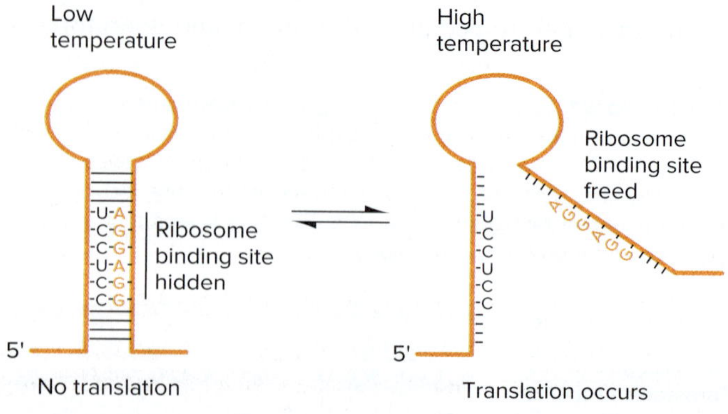

essential concepts

- *Reporter genes* constructed with *lacZ* coding sequences allow the detection of transcription via production of β-galactosidase. Insertion of *lacZ* without regulatory sequences into random genomic positions can pinpoint promoters active under specific conditions.
- *Gene fusions* that combine the *lac* regulatory region with other genes' coding sequences allow the production of medically useful protein products in bacteria.
- *RNA-Seq* is a procedure for sequencing *transcriptomes* and determining the relative levels of the transcription of all genes under different environmental conditions.
- Computerized analysis facilitates the discovery of operons, transcription factors and their binding sites, and regulatory RNAs.

16.5 A Comprehensive Example: Control of Bioluminescence by Quorum Sensing

learning objectives

1. Explain how scientists used *E. coli* to identify the *V. fischeri* bioluminescence genes.
2. Describe the molecular mechanism by which quorum sensing controls bioluminescence in *V. fischeri.*
3. Discuss the possible advantages of quorum sensing proteins as antibiotic targets.

In the beginning of the chapter, we described the remarkable phenomenon whereby *V. fischeri* bacteria not only make light, but do so only when their population is sufficiently large for the light to be visible at night to potential predators of the squid that they inhabit. The control of bioluminescence in *V. fischeri* was the first characterized example of **quorum sensing,** a communication mechanism now known to be used universally by bacteria to regulate gene expression in response to population density. We describe here some of the key experiments that uncovered how quorum sensing works at the levels of genes and gene products.

Recombinant *Vibrio fischeri* Genes Make *E. coli* Bioluminescent

Unlike *V. fischeri*, *E. coli* do not normally fluoresce. Scientists used this fact to isolate the *V. fischeri* bioluminescence genes. The idea was to identify a region of the *V. fischeri* genome that, via gene transfer, could confer to *E. coli* the ability to produce light. The researchers constructed a *V. fischeri* genomic library in plasmids and used it to transform *E. coli*. Remarkably, a few *E. coli* colonies became bioluminescent. Even more remarkably, these light-producing bacteria began to glow only when the culture reached a high density of cells. Therefore, the recombinant plasmid in the *E. coli* cells contained all the *V. fischeri* bioluminescence genes, and all the regulatory elements needed for quorum sensing. This plasmid contained only 9 kb of *V. fischeri* DNA.

To determine how many bioluminescence genes were contained in the cloned 9 kb *V. fischeri* genomic DNA fragment, the researchers grew the bioluminescent *E. coli* strain in the presence of a chemical mutagen and screened the colonies for mutants that could no longer produce light. The scientists then sorted the mutations in the plasmid's *V. fischeri* genes into complementation groups. To do this, they transformed *E. coli* with pairs of different plasmids, each containing the 9 kb fragment but each carrying a different, independently isolated mutation that prevented bioluminescence. If the doubly transformed cells produced light, then the two recombinant plasmids complemented, meaning that each plasmid had a mutation in a different *V. fischeri* gene. Colonies did not produce light if the two plasmids failed to complement because they had mutations in the same *V. fischeri* gene.

This analysis revealed seven complementation groups—seven genes—in the *V. fischeri* DNA fragment. DNA sequencing and other molecular experiments showed that the seven genes are transcribed from two divergent promoters. One promoter generates the *luxR* transcript, which produces a single protein (LuxR). A second promoter is for transcription of a polycistronic mRNA that generates the proteins LuxI, LuxC, LuxD, LuxA, LuxB, and LuxE (**Fig. 16.33a**).

The genes *luxA* and *luxB* encode subunits of the enzyme Luciferase, which catalyzes light production. The genes *luxC, luxD,* and *luxE* encode proteins that synthesize and recycle Luciferase substrates and cofactors.

LuxI and LuxR Are the Quorum-Sensing Proteins in *Vibrio fischeri*

The 9 kb genomic DNA fragment also contains the two genes that mediate quorum sensing in *V. fischeri: luxI* and *luxR* (Fig. 16.33a). LuxR protein is a transcriptional activator protein called Receptor, and it is needed for transcription of *luxICDABE* mRNA. LuxI protein is a Synthase enzyme that generates a molecule called an *autoinducer.* The autoinducer is an effector—it binds to LuxR and enables the Receptor to bind DNA.

Figure 16.33 Quorum sensing controls *Vibrio fischeri* bioluminescence. (a) The *luxR* and *luxICDABE* transcripts are required for *V. fischeri* to generate light. The genes *luxR* and *luxI* encode the quorum-sensing proteins (Receptor and Synthase). The other *lux* genes encode bioluminescence proteins, including the subunits of Luciferase, LuxA and LuxB. **(b)** LuxI generates *autoinducer,* a molecule that the cell releases into the environment and which can also reenter the cell. When its levels reach a threshold, autoinducer binds LuxR protein, which can then bind the promoter of the bioluminescence operon and activate its transcription.

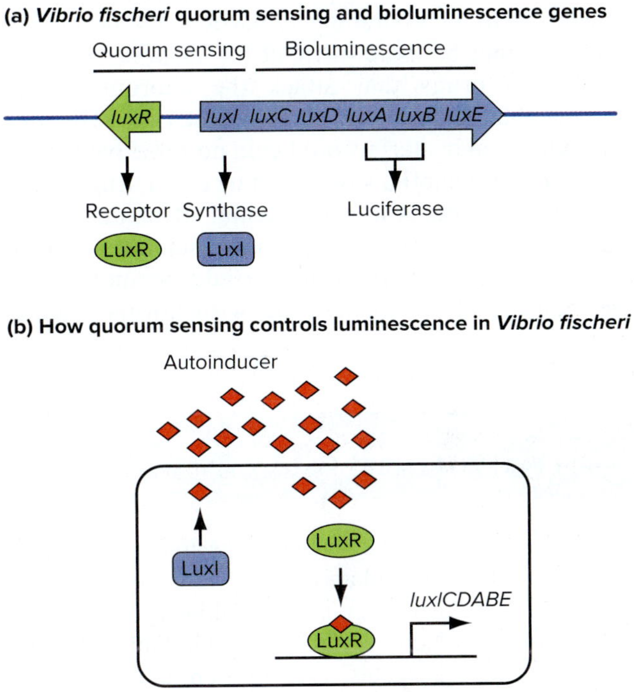

The autoinducer is released into the extracellular environment, and it can also reenter cells. When the bacterial population is dense, the cell cytoplasm is filled with autoinducer, which binds LuxR. As a result, the bioluminescence genes are expressed at high levels and the bacteria produce light (**Fig. 16.33b**).

When only a few bacteria are present, the autoinducer concentration is too low for much of it to have reentered cells. The bioluminescence genes are transcribed only at low levels, and no light is produced. Problem 43 at the end of this chapter illustrates how scientists discovered the regulatory mechanism of quorum sensing by fusing a *lacZ* reporter gene to sequences of *V. fischeri* DNA.

Quorum Sensing Suggests a New Approach to Developing Antibiotics

Quorum sensing is not limited to bioluminescent bacteria; in fact, almost all bacterial species have quorum-sensing pathways. For example, some pathogenic bacteria use quorum sensing to coordinate the release of toxins for maximum effect on the host. The discovery of quorum sensing suggested to scientists that the introduction of an agent that interferes with quorum sensing might prevent pathogenic bacteria from releasing toxins.

Drugs that target the quorum-sensing mechanism would have several potential advantages for helping physicians deal with the growing problem of antibiotic-resistant pathogens. This new strategy is particularly promising because it might be more difficult for bacteria to evolve resistance to drugs that attack the quorum-sensing machinery than to antibiotics that kill the cells. The idea is that individual bacteria would gain no selective advantage or survival benefits from resistance to a drug targeting quorum sensing. Another potential advantage is that the quorum-sensing proteins do not resemble any proteins in humans, so such drugs would be unlikely to cause side effects by interfering with normal human biochemistry. It remains to be seen if this interesting idea will become the basis of new treatments for diseases caused by bacterial pathogens.

essential concepts

- The genes for bioluminescence and *quorum sensing* were identified by transforming *E. coli* cells with *V. fischeri* DNA and looking for colonies that glowed.
- *V. fischeri* uses quorum sensing so that the cells make light only when their population is dense. The basis of quorum sensing is that only large groups of cells accumulate levels of autoinducer sufficient to activate transcription of the bioluminescence operon to high levels.
- Quorum-sensing proteins may be excellent targets for future antimicrobial drugs because immunity would provide no selective advantage to individual cells.

WHAT'S NEXT

Most of the methods that bacteria use to regulate their genes are available to eukaryotes as well. For example, both types of organisms can use diffusible regulatory proteins to increase or decrease transcription initiation. In both prokaryotes and eukaryotes, transcription and translation are also regulated after the intiation step by sRNAs.

Several features unique to eukaryotes nonetheless dictate that the mechanisms these organisms use to regulate gene expression cannot all be the same as those used in prokaryotes. In eukaryotes, transcription in the nucleus is physically separated by the nuclear membrane from the sites of translation on the ribosomes in the cytoplasm. Thus, eukaryotes cannot employ mechanisms such as attenuation that depend on the coupling between transcription and translation. Another contrast with the situation in bacteria is that eukaryotic mRNAs must be spliced, modified at their 5′ and 3′ ends, and transported from the nucleus into the cytoplasm where they are translated. In addition, eukaryotic chromosomes are wound up in chromatin. You will see in Chapter 17 that all of these processes, as well as multicellularity itself, necessitate and provide additional avenues for the regulation of gene expression in eukaryotic organisms.

SOLVED PROBLEMS

I. In the galactose operon of *E. coli,* a repressor, encoded by the *galR* gene, binds to an operator site, *galo,* to regulate the expression of three structural genes, *galE, galT,* and *galK*. Expression is induced by the presence of galactose in the media. For each of the strains listed, would the cell show constitutive, inducible, or no expression of each of the structural genes? (Assume that $galR^-$ is a loss-of-function mutation.)

a. $galR^- galo^+ galE^+ galT^+ galK^+$

b. $galR^+ galo^c\ galE^+ galT^+ galK^+$

c. $galR^- galo^+ galE^+ galT^+ galK^- /$
$galR^+ galo^+ galE^- galT^+ galK^+$

d. $galR^- galo^c\ galE^+ galT^+ galK^- /$
$galR^+ galo^+ galE^- galT^+ galK^+$

Answer

This problem requires an understanding of how regulatory sites and the proteins that bind to these sites behave. Focus first on the wild-type copies of the structural genes, and ask how their expression would be influenced by the particular regulatory elements on the same chromosome. For parts (c) and (d), you also need to consider whether expression of a structural gene would be impacted by regulatory elements on the other chromosome.

a. The *galR* gene encodes a repressor, so the lack of GalR protein would lead to constitutive expression of the *galE, T,* and *K* genes.

b. The *galo^c* mutation is an operator site mutation. By analogy with the *lac* operon, the designation *galo^c* indicates that repressor cannot bind, so constitutive expression of *galE, T,* and *K* would result.

c. The first copy of the operon listed has a *galR*⁻ mutation, but the other copy is wild type for the *galR* gene. The wild-type allele produces a repressor that can act in *trans* on both copies of the operon, overriding the effect of the *galR*⁻ mutation. Overall, the three *gal* genes will be inducible. The GalE protein will be made from the first copy of the operon, the GalK protein from the second, and the GalT protein from both copies.

d. The first copy of the operon contains a *galo^c* mutation, leading to constitutive synthesis of *galE* and *galT.* The other copy has a wild-type operator, so it is inducible, but neither operator has effects on the other copy of the operon. The net result is constitutive *galE* and *galT* expression and inducible *galK* expression.

II. Recall that in studies of the *trp* operon in *E. coli,* the existence of attenuation became apparent because in *trpR*⁻ mutants that lacked TrpR repressor protein, expression of the operon was threefold higher in media without tryptophan than in media with this amino acid.

a. Researchers have generated a strain in which the RNA leader is deleted from the *trp* operon. If such a deletion strain is also *trpR*⁻, would expression of the *trp* operon be altered if tryptophan levels changed? Explain.

b. A different strain that is *trpR*⁻ also has a point mutation that alters a base in region 4 of the RNA leader (see Fig. 16.22). How would this point mutation affect expression of the operon? Describe a second site mutation in the RNA leader that could reverse this effect.

c. Suppose the A of the AUG codon in the *trp* RNA leader was changed to C. What effect would this mutation have on the regulation of the operon if the strain was also *trpR*⁻?

d. What would happen if the strain without the *trp* operon leader in part (a) was also *trpR*⁺?

Answer

To answer these questions, remember that attenuation involves alternative configurations of the leader sequence in the *trp* operon mRNA. One of these configurations forms a transcription *terminator* that blocks expression of the structural genes, while the alternative configuration is an *antiterminator* that allows transcription of the remainder of the operon. If tryptophan is present, ribosomes can translate the two Trp codons in the small ORF contained in the leader, and this leads to formation of the terminator. If tryptophan is absent, ribosomes stall at these codons, favoring formation of the antiterminator.

a. If the leader is completely absent, no mechanism exists to terminate transcription of the operon prematurely. Furthermore, if the strain is also *trpR*⁻, tryptophan would have no other way of influencing the expression of the operon. As a result, the structural genes of the operon would be expressed at similar high constitutive levels in the presence and the absence of tryptophan.

b. If a nucleotide in region 4 was altered, it could no longer base pair with another nucleotide in the terminator stem loop. This point mutation would destabilize the terminator, so if tryptophan was present, the operon would be expressed at a higher level than in wild type. In the absence of tryptophan, the RNA leader mutation would have no effect (the expression levels would be as high as in wild type) because the terminator would not normally form under these conditions. If the leader had a second mutation such that the two mutant nucleotides could base pair with each other, the terminator stem loop would likely function normally.

c. The ribosome would be unable to translate the Trp codons in the leader if the AUG was missing. The leader would then form its most stable configuration that would have the greatest amount of intramolecular base pairing. In this most stable configuration, region 1 pairs with region 2 and region 3 pairs with region 4 to make the terminator stem loop. The expected result is that expression of the *trp* operon would be low both in the presence and in the absence of tryptophan.

d. If TrpR protein is available, but the RNA leader of the *trp* operon is missing, tryptophan could control the operon's expression by acting as an effector for TrpR. Transcription of the operon would therefore be high in the absence of tryptophan and low in the presence of tryptophan. It is also likely that this strain would have slightly higher levels of expression in the absence of tryptophan than would a wild-type strain that still had the *trp* operon leader. The reason is that the leader in the wild-type case could sometimes form the terminator configuration, but this would be impossible if the leader was completely missing.

PROBLEMS

Vocabulary

1. For each of the terms in the left column, choose the best matching phrase in the right column.

a. induction	1. glucose prevents expression of catabolic operons
b. repressor	2. protein or RNA undergoes a reversible conformational change
c. operator	3. regulates translation of mRNAs in *trans*
d. allostery	4. RNA leader that regulates gene expression in response to a small molecule or ion
e. operon	5. site to which repressor binds
f. catabolite repression	6. termination of transcription elongation in response to translation
g. reporter gene	7. group of genes transcribed into one mRNA
h. attenuation	8. negative regulator
i. sRNA	9. a fusion of the regulatory region of one gene to the coding region of another gene whose product is assayed readily
j. riboswitch	10. stimulation of protein synthesis by a specific molecule

Section 16.1

2. The following statement occurs early in this chapter: ". . . a crucial step in the regulation of many bacterial genes is the binding of RNA polymerase to DNA at the promoter." Why might it be advantageous for bacteria to regulate the expression of their genes at this particular step?

3. One of the main lessons of this chapter is that several bacterial genes are often transcribed from a single promoter into a large multigene (polycistronic) transcript. The region of DNA containing the set of genes that are cotranscribed, along with all of the regulatory elements that control the expression of these genes, is called an *operon*.

   a. Which of the mechanisms in the following list could explain differences in the levels of the mRNAs for different operons?

   b. Which of the mechanisms in the following list could explain differences in the levels of the protein products of different genes in the same operon?

      i. Different promoters might have different DNA sequences.

      ii. Different promoters might be recognized by different types of RNA polymerase.

      iii. The secondary structures of mRNAs might differ so as to influence the rate at which they are degraded by ribonucleases.

      iv. In an operon, some genes are farther away from the promoter than other genes.

      v. The translational initiation sequences at the beginning of different open reading frames in an operon might result in different efficiencies of translation.

      vi. Proteins encoded by different genes in an operon might have different stabilities.

4. All mutations that abolish function of the Rho termination protein in *E. coli* are conditional mutations; no cells with null mutations of the Rho-encoding gene have ever been isolated. What does this tell you about the *rho* gene and its product?

Section 16.2

5. The figure at the beginning of this chapter shows the binding of both a Lac repressor tetramer and a CRP-cAMP complex to the regulatory region of the *lac* operon.

   a. What is the key feature of a regulatory protein such as the Lac repressor or CRP that allows it to regulate specifically the genes or operons it is supposed to control?

   b. On the figure, show the positions of the following components: (i) A Lac repressor monomer; (ii) a Lac repressor dimer; (iii) all four DNA binding domains of the Lac repressor tetramer; (iv) a single helix-turn-helix motif; (v) the o_1 part and either the o_2 or o_3 parts of the *lac* operator (assume the operon would be transcribed from right to left on the figure); (vi) the multimerization domains of the four Lac repressor monomers; (vii) an inducer-interacting domain; (viii) the CRP-cAMP complex; and (ix) a DNA loop.

   c. What is the physical basis for the formation of the DNA loop shown in the figure?

   d. On the figure, show the position of two axes of symmetry in the sequence of DNA. How do you know, only on the basis of the figure and without prior information about the precise DNA sequence, that these two axes of symmetry are likely to be present in the DNA and that the sequences around these axes are rotationally symmetrical?

6. The promoter of an operon is the site to which RNA polymerase binds to begin transcription. Certain base changes in the promoter result in a mutant site to which RNA polymerase cannot bind. Would you expect mutations in the promoter that prevent binding of RNA polymerase to act in *trans* on another copy of the operon on a plasmid in the cell, or only in *cis* on the copy immediately adjacent to the mutated site?

7. You are studying an operon containing three genes that are cotranscribed in the order *hupF, hupH,* and *hupG.* Diagram the mRNA for this operon, showing the location of the 5′ and 3′ ends, all open reading frames, translational start sites, stop codons, transcription termination signals, and any regions that might be in the mRNA but do not serve any of these functions.

8. You have isolated a protein that binds to DNA in the region upstream of the promoter sequence of the *sys* gene. If this protein is a positive regulator, which of the following would be true?
   a. Loss-of-function mutations in the gene encoding the DNA-binding protein would cause constitutive expression of *sys*.
   b. Loss-of-function mutations in the gene encoding the DNA-binding protein would result in little or no expression of *sys*.

9. You have isolated two different mutants (*reg1* and *reg2*) causing constitutive expression of the *emu* operon (*emu1 emu2*). One mutant contains a defect in a DNA-binding site, and the other has a loss-of-function defect in the gene encoding a protein that binds to the site.
   a. Is the DNA-binding protein a positive or negative regulator of gene expression?
   b. To determine which mutant has a defect in the site and which one has a mutation in the binding protein, you decide to do an analysis using F′ plasmids. Assuming you can assay levels of the Emu1 and Emu2 proteins, what results do you predict for the two strains (i and ii; see descriptions below) if *reg2* encodes the regulatory protein and *reg1* is the regulatory site?
      i. F' $(reg1^- reg2^+ emu1^- emu2^+)/$ $reg1^+ reg2^+ emu1^+ emu2^-$
      ii. F' $(reg1^+ reg2^- emu1^- emu2^+)/$ $reg1^+ reg2^+ emu1^+ emu2^-$
   c. What results do you predict for the two strains (i and ii) if *reg1* encodes the regulatory protein and *reg2* is the regulatory site?

10. Bacteriophage λ, after infecting a cell, can integrate into the chromosome of the cell if the repressor protein, cI, binds to and shuts down phage transcription immediately. (A strain containing a bacteriophage DNA integrated into the chromosome is called a *lysogen.*) The alternative fate is the production of many more viruses and lysis of the cell. In a mating, a donor strain that is a lysogen was crossed with a lysogenic recipient cell, and no phages were produced. However, when the lysogen donor strain transferred its DNA to a nonlysogenic recipient cell, the recipient cell burst, releasing a new generation of phages.
    a. Why did the mating with a nonlysogenic recipient result in phage growth and release, but the infection of a lysogenic recipient did not?
    b. Explain how this phenomenon relates to the PaJaMo experiment in Fig. 16.6.
    c. Explain how this phenomenon relates to hybrid dysgenesis, described in Problem 29 of Chapter 13.

11. Mutants were isolated in which the constitutive phenotype of a missense *lacI* mutation was suppressed. That is, the operon was now inducible. These suppressor mutations mapped to the operon, not to the *lacI* gene. What could these mutations be?

12. Suppose you have six strains of *E. coli*. One is wild type, and each of the other five has a single one of the following mutations: $lacZ^-$, $lacY^-$, $lacI^-$, o^c, and $lacI^S$. For each of these six strains, describe the phenotype you would observe using the following assays. [*Notes:* (1) IPTG is a colorless synthetic molecule that acts as an inducer of *lac* operon expression but cannot serve as a carbon source for bacterial growth because it cannot be cleaved by β-galactosidase; (2) X-gal cannot serve as a carbon source for growth; (3) *E. coli* requires active lactose permease (the product of *lacY*) to allow lactose, X-gal, or IPTG into the cells.]
    a. Growth on medium in which the only carbon source was lactose.
    b. Colony color in medium containing glycerol as the only carbon source, X-gal, and IPTG.
    c. Colony color in medium containing glycerol as the only carbon source and X-gal, but no IPTG.
    d. Colony color in medium containing high levels of glucose as the only carbon source, X-gal, and IPTG.
    e. Colony color in medium containing high levels of glucose as the only carbon source and X-gal, but no IPTG.

13. The previous problem raises some interesting issues:
    a. In most experiments using the *lac* operon, researchers use the synthetic inducer IPTG to turn on operon expression, instead of lactose or allolactose. What do you think is the advantage of using IPTG?
    b. Scientists were originally puzzled by what they termed the *lactose paradox.* To turn on expression of the *lac* operon, an inducer (whether IPTG or lactose/allolactose) needs to be able to get into the cell. Import of this inducer requires the presence of the Lac permease enzyme in the cell membrane (Fig. 16.2). But if the *lac* operon is being repressed prior to addition of the inducer, no Lac permease should be present, so no inducer could be imported, and induction could never occur. Yet induction obviously does occur; how might this be possible?

14. For each of the *E. coli* strains containing the *lac* operon alleles listed, indicate whether the strain is inducible, constitutive, or unable to express β-galactosidase and permease.
    a. $I^+ o^+ Z^- Y^+/I^+ o^c Z^+ Y^+$
    b. $I^+ o^+ Z^+ Y^+/I^- o^c Z^+ Y^-$
    c. $I^+ o^+ Z^- Y^+/I^- o^c Z^+ Y^-$
    d. $I^- P^- o^+ Z^+ Y^-/I^+ P^+ o^c Z^- Y^+$
    e. $I^s o^+ Z^+ Y^+/I^- o^+ Z^+ Y^-$
15. For each of the following growth conditions, what proteins would be bound to *lac* operon DNA? (List the proteins, but do not include RNA polymerase.)
    a. glucose
    b. glucose + lactose
    c. lactose
16. For each of the following mutant *E. coli* strains, plot a 30-minute time course of concentration of β-galactosidase, permease, and acetylase enzymes grown under the following conditions: For the first 10 minutes, no lactose is present; at 10 minutes, lactose becomes the sole carbon source. Plot concentration on the y-axis, time on the x-axis. (Don't worry about the exact units for each protein on the y-axis.)
    a. $I^- P^+ o^+ Z^+ Y^+ A^+/I^+ P^+ o^+ Z^- Y^+ A^+$
    b. $I^- P^+ o^c Z^+ Y^+ A^-/I^+ P^+ o^+ Z^- Y^+ A^+$
    c. $I^s P^+ o^+ Z^+ Y^+ A^+/I^- P^+ o^+ Z^- Y^+ A^+$
    d. $I^- P^- o^+ Z^+ Y^+ A^+/I^- P^+ o^c Z^+ Y^- A^+$
    e. $I^- P^+ o^+ Z^- Y^+ A^+/I^- P^- o^c Z^+ Y^- A^+$
17. Maltose utilization in *E. coli* requires the proteins encoded by genes in three different operons. One operon includes the genes *malE, malF,* and *malG;* the second includes *malK* and *lamB;* and the genes in the third operon are *malP* and *malQ*. The MalT protein is a positive regulator that controls the expression of all three operons; expression of the *malT* gene itself is catabolite sensitive.
    a. What phenotype would you expect to result from a loss-of-function mutation in the *malT* gene?
    b. Do you expect the three maltose operons to contain binding sites for CRP (cAMP receptor protein)? Why or why not?

    In order to infect *E. coli,* bacteriophage λ binds to the maltose transport protein LamB (also known as the λ receptor protein) that is found in the outer membrane of the bacterial cell. The synthesis of LamB is induced by maltose in the medium via expression of the MalT protein, as described above.

    c. List the culture conditions under which wild-type *E. coli* cells would be sensitive to infection by bacteriophage λ.
    d. *E. coli* cells that are resistant to infection by bacteriophage λ have been isolated. List the types of mutations in the maltose *regulon* (the set of all genes regulated by maltose) that λ-resistant mutants could contain.
18. Seven *E. coli* mutants were isolated. The activity of the enzyme β-galactosidase produced by cells containing each mutation alone or in combination with other mutations was measured when the cells were grown in medium with different carbon sources.

	Glycerol	**Lactose**	**Lactose + Glucose**
Wild type	0	1000	10
Mutant 1	0	10	10
Mutant 2	0	10	10
Mutant 3	0	0	0
Mutant 4	0	0	0
Mutant 5	1000	1000	10
Mutant 6	1000	1000	10
Mutant 7	0	1000	10
F′ *lac* from mutant 1/ mutant 3	0	1000	10
F′ *lac* from mutant 2/ mutant 3	0	10	10
Mutants 3 + 7	0	1000	10
Mutants 4 + 7	0	0	0
Mutants 5 + 7	0	1000	10
Mutants 6 + 7	1000	1000	10

    Assume that each of the seven mutations is one and only one of the genetic lesions in the following list. Identify the type of alteration each mutation represents.
    a. superrepressor
    b. operator deletion
    c. nonsense (amber) suppressor tRNA gene (assume that the suppressor tRNA is 100% efficient in suppressing amber mutations)
    d. defective CRP–cAMP binding site
    e. nonsense (amber) mutation in the β-galactosidase gene
    f. nonsense (amber) mutation in the repressor gene
    g. defective *crp* gene (encoding the CRP protein)
19. Cells containing missense mutations in the *crp* gene (encoding the positive regulator CRP) are Lac^-, Mal^-, Gal^-, etc. To find cells with suppressors of the *crp* mutation (that is, cells with the *crp* mutation that behave as if they are crp^+), cells were screened to find those that were both Lac^+ and Mal^+.
    a. What types of suppressor mutations would you expect to obtain using this screen compared with a screen for Lac^+ only?
    b. All suppressors isolated were mutant in the gene for the α-subunit of RNA polymerase. What hypothesis could you propose based on this analysis?

20. Six strains of *E. coli* (mutants 1–6) that had one of the following mutations (i–vi) affecting the *lac* operon were isolated.
    i. deletion of *lacY*
    ii. o^c mutation
    iii. missense mutation in *lacZ*
    iv. inversion of the *lac* operon (but not an inversion of the *lacI* gene)
    v. superrepressor mutation
    vi. inversion of *lacZ, Y,* and *A* but not *lacI, P, o*
    a. Which of these mutations would prevent the strain from utilizing lactose?
    b. The entire *lac* operon (including the *lacI* gene and its promoter) from each of the six *E. coli* strains was cloned into a plasmid vector containing an ampicillin resistance gene. Each recombinant plasmid was transformed into each of the six strains to create partial diploids. In analysis of these strains, mutant 1 was found to carry a deletion of *lacY*, so this strain corresponds to mutation i in the list above. Which of the other types of mutations would be expected to complement mutant 1 in these partial diploids so as to allow lactose utilization?
    c. In part (b), each strain was plated on ampicillin media in which lactose was the only carbon source. (Ampicillin was included to ensure maintenance of the plasmid.) Growth of the transformants is scored below (a + sign indicates growth, a – sign means no growth). Synthesis of β-galactosidase and permease are both required for growth on this medium. Results of this merodiploid analysis are shown here. Which mutant bacterial strain (1–6) contained each of the alterations (i–vi) listed previously?

	1	2	3	4	5	6
1	–	+	–	+	–	+
2	+	–	–	+	–	+
3	–	–	–	+	–	+
4	+	+	+	+	–	+
5	–	–	–	–	–	+
6	+	+	+	+	+	+

21. a. The original constitutive operator mutations in the *lac* operon were all base changes in o_1. Why do you think mutations in o_2 or o_3 were not isolated in these screens?
    b. Explain how a mutagen that causes small insertions could produce an o^c mutation.
    c. Would a strain with one of the o^c mutations described in part (b) and also a $lacI^S$ mutation be able to make β-galactosidase either in the presence or absence of inducer? Explain.

22. In an effort to determine the location of an operator site for a negatively regulated gene, you have made a series of deletions within the regulatory region. The extent of each deletion is shown by the line underneath the sequence, and the resulting expression from the operon (*i* = inducible; *c* = constitutive; – = no expression) is also indicated.

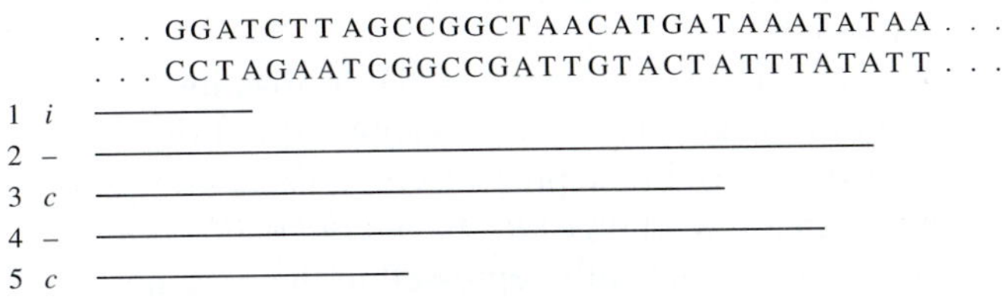

    a. What can you conclude from these data about the location of the operator site?
    b. Why do you think deletions 2 and 4 show no expression of the gene?

23. Figure 16.17 shows that in the *lac* operon, both the operator (o_1) and the binding site for CRP–cAMP show rotational symmetry. This is not true of the promoter (the RNA polymerase binding site) as a whole. Why do you think the promoter does not exhibit rotational symmetry?

24. The footprinting experiment described in Fig. 16.16 depended on having a fragment of double-stranded DNA that was labeled with radioactivity at one end of one strand.
    a. How would you make such a labeled fragment of DNA? Outline the steps you would perform, starting with two PCR primers and some genomic DNA. You also have available a kinase enzyme that adds phosphate groups to the 5′ ends of DNA strands, radioactive ATP, and a restriction enzyme of your choice. You could also use a phosphatase enzyme that removes phosphate groups from the 5′ ends of DNA strands.
    b. Why does the footprinting experiment require a fragment of DNA labeled only on one end of one strand? In other words, how would the results of the experiment differ from those shown in Fig. 16.16 if the DNA was labeled on both strands at their 5′ ends, or if the DNA was labeled with radioactive phosphate along its entire length at every phosphodiester bond?

Section 16.3

25. Why is the *trp* attenuation mechanism unique to prokaryotes?

26. a. How many ribosomes are required (at a minimum) for the translation of *trpE* and *trpC* from a single transcript of the *trp* operon?
    b. How would you expect deletion of the two tryptophan codons in the RNA leader to affect the expression of the *trpE* and *trpC* genes?

27. The following is a sequence of the leader region of the *his* operon mRNA in *Salmonella typhimurium*. What bases in this sequence could cause a ribosome to pause when histidine is limiting (that is, when there is very little of it) in the medium?

5′ AUGACACGCGUUCAAUUUAAACACCACCAUCAUCACCAUCA
UCCUGACUAGUCUUUCAGGC 3′

28. For each of the *E. coli* strains that follow, indicate the effect of the genotype on the expression of the *trpE* and *trpC* genes in the presence or absence of tryptophan. [In the wild type (R^+ P^+ o^+ att^+ $trpE^+$ $trpC^+$), *trpC* and *trpE* are fully repressed in the presence of tryptophan and are fully expressed in the absence of tryptophan.]

R = repressor gene; R^n product cannot bind tryptophan; R^- product cannot bind operator

o = operator for the *trp* operon; o^- cannot bind repressor

att = attenuator; att^- is a deletion of the attenuator

P = promoter; P^- is a deletion of the *trp* operon promoter

$trpE^-$ and $trpC^-$ are null (loss-of-function) mutations

a. R^+ P^- o^+ att^+ $trpE^+$ $trpC^+$
b. R^- P^+ o^+ att^+ $trpE^+$ $trpC^+$
c. R^n P^+ o^+ att^+ $trpE^+$ $trpC^+$
d. R^- P^+ o^+ att^- $trpE^+$ $trpC^+$
e. R^+ P^+ o^- att^+ $trpE^+$ $trpC^-$/R^- P^+ o^+ att^+ $trpE^-$ $trpC^+$
f. R^+ P^- o^+ att^+ $trpE^+$ $trpC^-$/R^- P^+ o^+ att^+ $trpE^-$ $trpC^+$
g. R^+ P^+ o^- att^- $trpE^+$ $trpC^-$/R^- P^+ o^- att^+ $trpE^-$ $trpC^+$

29. One mechanism by which antisense RNAs act as negative regulators of gene expression is by base pairing with the ribosome binding site on the sense mRNA to block translation. In a second, alternative mechanism, the act of transcribing an antisense RNA can somehow prevent RNA polymerase from recognizing the sense promoter for the same gene. Design an experimental approach that would enable you to distinguish between these two modes of action at a specific gene. (*Hint:* What would be the outcome in each case if high levels of the antisense RNA were transcribed from a gene on a plasmid?)

30. For each element in the list that follows, indicate what kind of molecule it is (DNA, RNA, protein, small molecule), whether it acts as a positive or negative regulator, what stage of gene expression it affects, and whether it acts in *cis* or in *trans*. (In its most general sense, the term *cis* describes elements that affect the function of the molecule of which it is a part, while *trans* describes elements on one molecule that affect the function of a different molecule.)

a. Lac repressor
b. *lac* operator
c. CRP
d. CRP-binding site
e. Trp repressor
f. charged $tRNA^{Trp}$ (in terms of its function at the *trp* operon)
g. the antiterminator at the *trp* operon
h. a terminator in the expression platform of a riboswitch
i. an sRNA that blocks mRNA translation

31. Among the structurally simplest riboswitches are the two so-called *purine riboswitches,* one of which responds to guanine, and the other to adenine. The accompanying diagram shows a guanine riboswitch. Base-pairing between free guanine and a particular cytosine residue within the aptamer determines the riboswitch conformation.

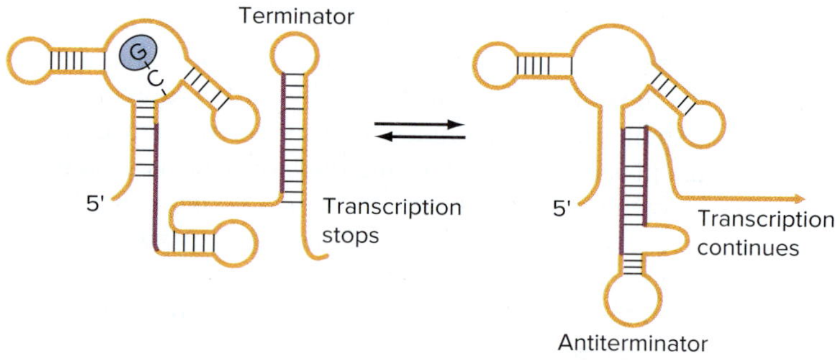

a. What condition in the cellular environment would favor each of the riboswitch configurations?
b. In *B. subtilis,* guanine riboswitches are located in 5 different transcription units containing 17 different genes. Based on the diagram and your answer to part (a), what biological processes do you think these 17 genes might be involved in? Explain your reasoning.

32. Great variation exists in the mechanisms by which RNAs can mediate gene regulation. In one recently discovered example shown in the following diagram, the genes *CsrA* and *CsrB* are global regulators of suites of target genes that are involved in the use of carbon atoms. The product of *CsrA* is the CsrA protein, which binds to the ribosome binding site (RBS) of target gene mRNA, preventing target gene expression. The product of the *CsrB* gene is the CsrB RNA, which contains 22 binding sites for CsrA protein. CsrB RNA can thus compete with target mRNAs for CsrA protein binding. In the presence of high CsrB

RNA concentrations, CsrA protein cannot bind to mRNA binding sites, so expression of the target genes is turned on.

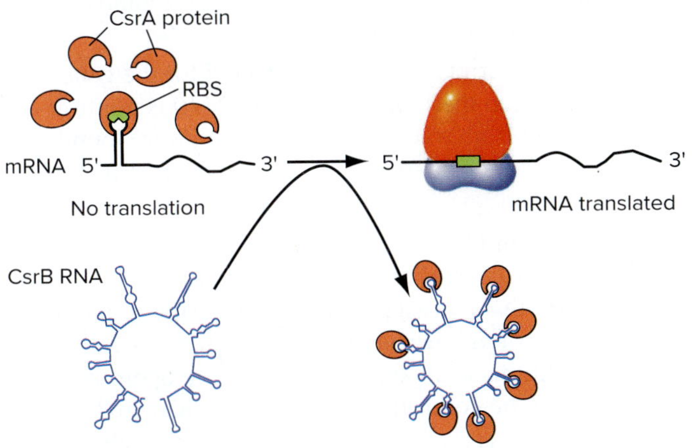

a. For the *CsrA* and *CsrB* genes, indicate what kind of molecule the gene product is (DNA, RNA, protein, small molecule), whether it acts as a positive or negative regulator, what stage of gene expression it affects, and whether it acts in *cis* or in *trans*. (It will be interesting to compare your answers to those for Problem 30.)

b. CsrA/CsrB regulate glycogen biosynthesis and breakdown; glycogen is a polymer of glucose and a major source of stored energy in the human body. CsrA/CsrB are negative regulators of glycogen biosynthesis and positive regulators of glycogen breakdown.

To what environmental factor do you think that the CsrA/CsrB system is most likely to respond? Suggest a possible way that this system might be repressible, and then suggest a different hypothesis for how this system might be inducible. (Assume in both cases that CsrB expression is modulated.) Which of your hypotheses would be most consistent with the target genes being involved in glycogen biosynthesis (an anabolic pathway), and which is most consistent with glycogen breakdown (a catabolic pathway)? Explain.

Section 16.4

33. Many genes whose expression is turned on by DNA damage have been isolated. Loss-of-function mutations in the *lexA* gene lead to the expression of many of these genes, even when there has been no DNA damage. Would you hypothesize that LexA protein is a positive or a negative regulator? Why?

34. In 2005, Frederick Blattner and his colleagues found that *E. coli* cells have a global transcriptional program that helps them forage for better sources of carbon. Many genes, including genes needed for bacterial motility, are turned on in response to poorer carbon sources so that the bacteria can search for better nutrition. You now want to search for genes that regulate this response. How could you use *lacZ* fusions to try to identify such regulatory genes?

35. The *E. coli* MalT protein is a positive regulator of several *mal* operons, which are induced in the presence of the sugar maltose. The gene that encodes MalT was identified in a screen for mutants causing constitutive expression of *mal* operons; the operons were transcribed even in the absence of maltose. The screen involved a *lacZ* transcriptional fusion reporter gene in which the regulatory region of a maltose-inducible operon was fused to the coding sequences of *lacZ*.

a. Bacteria with a *lacZ*$^-$ mutation are transformed with the reporter gene and spread on petri plates containing the β-galactosidase substrate X-gal. What color would the colonies be if the plates also contained maltose? What if the plates had X-gal but no maltose?

b. In the screen, scientists mutagenized the *lacZ*$^-$ bacteria before transforming them with the reporter gene, and then spread the transformed bacteria on plates with X-gal and no maltose. All of the colonies were white except for one colony that was blue. At this stage of the analysis, researchers could not establish whether the gene mutant in the blue colony encoded a positive or a negative regulator of *mal* operons.

Suppose first that the gene encoded a positive regulator. (i) How could the wild-type protein respond to maltose? (ii) How would the mutation affect protein function? (iii) Describe the likely nature of the mutation in the gene at the molecular level.

Now answer these same three questions for the hypothesis in which the gene encoded a negative regulator (a repressor) of *mal* operon expression.

c. How do you think the scientists figured out that MalT was a positive regulator and not a repressor? (*Hint:* Recall Fig. 14.28. Think about what would happen in each case if the researchers attempted to identify the *malT* mutant using a plasmid library made from the genome of a wild-type strain versus a plasmid library made from the genome of the mutant strain.)

36. Erythropoietin is a human protein hormone that stimulates the production of red blood cells. Imagine that you are a researcher for a pharmaceutical company, and you want to make this hormone in bacteria so it can be used to treat patients with anemias. You will create a recombinant DNA molecule that has the following elements, some of whose importance will be explained later in the problem: (i) Coding sequences

for human erythropoeitin. (ii) Regulatory sequences of the *lac* operon. (iii) Sequences encoding the *E. coli* maltose binding protein (MBP). (iv) Sequences encoding a series of five amino acids (DDDDK in the one-letter code). The pharmaceutical company's engineers will transform a recombinant plasmid with these sequences into *E. coli* and induce the expression of a *tagged* fusion protein N MBP-DDDDK-erythropoietin C.

a. Diagram the recombinant plasmid, indicating the order of these four components and how they are arranged with respect to the plasmid vector.
b. Which one of the four elements encodes the ribosome binding site for the mRNA that could make this fusion protein?
c. Which of these four elements must be placed in the same reading frame with respect to each other?
d. Would you obtain the erythropoietin coding sequences from a human genomic DNA clone or from a human cDNA clone? Explain.
e. What compound would you use to induce expression of the fusion protein? Would it be best to add this compound to the medium before you seeded it with *E. coli* cells, or after the population of cells had grown to high density? Explain.
f. Cells that express the fusion protein also contain many other *E. coli* proteins. For pharmaceutical use, it is important to purify drugs away from contaminants. Given that MBP binds tightly to the sugar maltose, and that maltose can be attached to an insoluble resin, explain how you would purify the fusion protein away from all other *E. coli* proteins.
g. For pharmaceutical use, the human erythropoietin must not be attached to any other amino acid sequences. A protease called *enterokinase* cleaves proteins just C-terminal to DDDDK. Explain how you would use enterokinase to separate erythropoietin away from the rest of the fusion protein and then to purify the desired pharmaceutical.

37. To find genes that are turned on or off in response to changes in osmolarity (the total concentration of solutes in solution), you grow a culture of *E. coli* in a medium with high osmolarity and another culture in a medium with low osmolarity. You now perform RNA-Seq analysis on each culture. It is possible that osmotic changes may induce a general stress response that may be seen with other stresses as well (for example, heat shock). How could you distinguish the genes that might be involved in a general stress response from those that are specific for the osmolarity change? (*Note:* You will need to grow additional bacterial cultures.)

38. The following questions concern Fig. 16.30:
    a. How many genes are depicted in this figure? How many operons? What is the average gene density in the region shown? Is this value representative of most bacterial genomes?
    b. No transcripts at all were detected for the gene *t2110*. Do these data prove the *t2110* is nonfunctional? If no transcripts of the gene were found, how could scientists assign a direction to its transcription?
    c. What kind of evidence in the figure would suggest the existence of an attenuation or riboswitch mechanism that causes premature transcriptional termination of an operon? Under the environmental conditions analyzed, do any of these operons appear to be controlled by such a mechanism?
    d. The data shown in Fig. 16.30 do not provide any evidence that any of the genes or operons depicted are regulated by an antisense transcript. What would the data look like if an antisense mechanism was involved in controlling a gene or operon?
    e. Although the *galM* gene is depicted as part of the operon that contains *galETK,* it is possible that *galM* is actually transcribed from a separate promoter. What evidence in the figure suggests this possibility?
    f. Could the data shown in Fig. 16.30 reveal the existence of a regulatory mechanism in which an sRNA occludes a ribosome binding site?

39. In many bacterial species, regulatory sRNAs have been identified by transcriptome sequencing (RNA-Seq). How do researchers know that the small RNA species identified by cDNA sequencing are regulatory sRNAs rather than fragments of longer mRNAs?

40. Many bacterial genes involved in amino acid biosynthesis are regulated by RNA leaders that respond, indirectly, to the level of a specific amino acid. Like attenuators or some riboswitches, these allosteric RNA leaders, called *T-box leaders,* can form either terminators or antiterminators. The name *T-box* refers to a 14-nucleotide sequence present in all of these RNA leaders; a 5′ UGGU 3′ sequence within the T-box is complementary to the conserved 3′ end of tRNAs (5′ ACCA 3′).

    The first T-box RNA device was discovered in the *B. subtilis tyrS* gene, which encodes tyrosyl-tRNA synthetase, the enzyme that charges $tRNA^{Tyr}$ with tyrosine. As shown in the following diagram, the T-box leader can bind to the anticodon, and at the same time, to the 3′ end of the same uncharged $tRNA^{Tyr}$ (*left*). When uncharged $tRNA^{Tyr}$ is bound, an antiterminator forms in the leader; otherwise, a terminator forms (*right*).

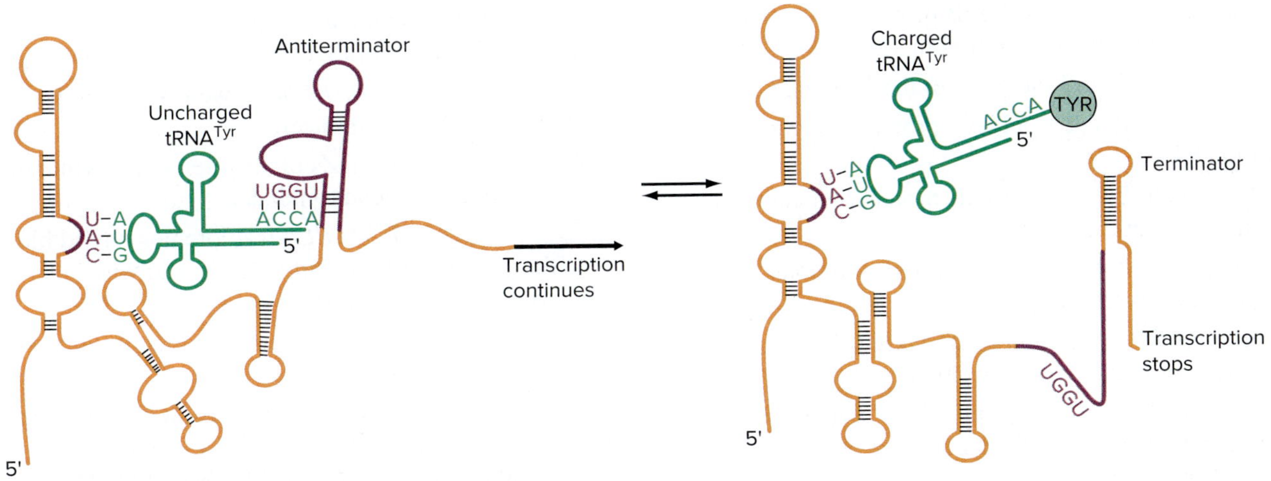

a. Most of the tRNA synthetase genes in *B. subtilis* are regulated by T-box RNA leaders that respond to specific uncharged tRNAs. Explain the logic of this regulation.

b. How could you alter the base sequence of the *B. subtilis tyrS* T-box leader so that it might respond to uncharged $tRNA^{Phe}$ instead of uncharged $tRNA^{Tyr}$?

c. The T-box is responsible for nearly all of the regulation of *tyrS* gene expression. What do you predict would happen to *tyrS* gene expression if the the 5′ UAC 3′ in the RNA leader that interacts with the $tRNA^{Tyr}$ anticodon was changed to to 5′ CUA 3′? Explain.

d. Key experimental support for the idea that the T-box RNA leader binds tRNAs was the finding that normal function could be restored to the mutant T-box described in part (c) by a particular mutation in a gene different from *tyrS*. What specific bacterial gene mutation would render the mutant T-box RNA leader functional again?

Results of experiments involving a *tyrS-lacZ* reporter gene indicated that the *tyrS* T-box leader responds directly to relative levels of charged versus uncharged tRNAs rather than to the availability of tyrosine. These experiments involved expression of mutant $tRNA^{Tyr}$ species. These mutations were outside of the anticodon or the four base pairs at the 3′ end, yet they prevented the tRNA from being charged by tRNA synthetase.

e. An otherwise wild-type *B. subtilis* strain contains a *tyrS-lacZ* reporter transgene that includes the T-box RNA leader. Compare the expression of β-galactosidase in this strain when tyrosine levels are low as opposed to when tyrosine levels are high.

f. Suppose now that the $tRNA^{Tyr}$ gene in the strain described in part (e) was replaced with the mutant $tRNA^{Tyr}$ gene described earlier with mutations that prevented the tRNA from being charged. Compare the expression of β-galactosidase in this new strain when tyrosine levels are low as opposed to when tyrosine levels are high. Explain how this experiment distinguished the two hypotheses.

g. T-box regulators have been identified in many other bacterial species using computer algorithms. What do you think the computer programs searched for?

Section 16.5

41. Describe how RNA-Seq analysis could have been used to discover the components of the *V. fischeri* quorum-sensing pathway. Would any of the components of the pathway shown in Fig. 16.33 have escaped detection? Explain.

42. The researchers who investigated bioluminescence and quorum sensing found that *E. coli* transformed with a plasmid containing a 9 kb fragment of *V. fischeri* DNA could glow when the cell population was dense. They mutagenized these *E. coli* cells and isolated many mutations that mapped to the 9 kb fragment and prevented the cells from glowing. They then performed complementation testing on these mutants by transforming *E. coli* cells simultaneously with two plasmids, each containing the 9 kb fragment with one of the mutations. To ensure the *E. coli* cells were transformed with both plasmids, one of the two plasmids had a gene conferring resistance to ampicillin, while the other plasmid had a gene conferring resistance to tetracycline, and cells were selected on petri plates that had both antibiotics.

a. Construct a 9 × 9 complementation table for the nine mutations list that follows, using + to indicate cells that would glow and – to indicate cells that would remain dark. (You only need to fill in half the table.)

- Mutation 1: Encodes a LuxA protein that cannot bind a substrate for the luciferase enzyme
- Mutation 2: Encodes a LuxA protein that cannot associate with the LuxB protein

- Mutation 3: Encodes a LuxB protein that cannot associate with the LuxA protein
- Mutation 4: A null mutation in the *luxI* gene
- Mutation 5: Encodes a LuxR protein that cannot bind DNA
- Mutation 6: Encodes a LuxR protein that cannot bind to the autoinducer
- Mutation 7: A mutation in the *luxR* promoter that prevents transcription
- Mutation 8: A mutation in the *luxICDABE* promoter that prevents transcription
- Mutation 9: A mutation in the *luxICDABE* promoter region that blocks binding of the LuxR protein

b. How many complementation groups exist among these nine mutations?

c. Is your answer to part (b) also the number of different genes? Explain.

43. A key experiment in understanding the molecular mechanism of quorum sensing involved the use of two transcriptional fusion reporter genes, each made within the 9 kb fragment of *V. fischeri* DNA described in Problem 42. In one reporter (*luxR/lacZ*), the *luxR* regulatory region drives *lacZ* transcription (that is, the *luxR* coding sequences are replaced by those of *lacZ*). In the other reporter (*luxICDABE/lacZ*), the *luxICDABE* operon regulatory sequences drive *lacZ* expression (that is, the operon structural genes are replaced by *lacZ* coding sequences). Fig. 16.33 shows the structure of the *luxR* and *luxICDABE* region of *V. fischeri*. *E. coli* colonies (chromosomally *lacZ*$^-$) containing either reporter (*luxR/lacZ* or *luxICDABE/lacZ*) were white. When purified autoinducer molecules were added to the media, the *luxR/lacZ* colonies remained white, but the *luxICDABE/lacZ* colonies turned blue over time.

a. Explain why *luxR/lacZ* colonies were white in the absence of the autoinducer.

b. Explain why *luxICDABE/lacZ* were white in the absence of the autoinducer.

c. Explain why *luxR/lacZ* colonies remained white in the presence of the autoinducer.

d. Explain why *luxICDABE/lacZ* colonies turned blue in the presence of the autoinducer, and why this reaction was time-dependent.

e. What do these results suggest about the transcription of the *luxR* gene?

44. Quorum sensing controls the expression of virulence in many pathogenic bacteria. Usually, pathogens express toxins in response to receptor activation by ligand binding at high cell density. *V. cholerae* (the causative agent of cholera) does the opposite; its virulence genes are expressed only at low cell density because its quorum-sensing receptor is repressed by ligand binding. The unusual "reversed" mechanism for activating virulence genes in *V. cholerae* has suggested to scientists a simple idea for generating a new kind of antibiotic for the treatment of cholera. Explain.

45. Scientists are currently screening a *chemical library* of small molecules for inhibitors of bioluminescence in response to high cell density in *V. fischeri.* The small molecules were chosen for their potential ability to bind LuxR.

a. How could a molecule that binds LuxR prevent bioluminescence?

b. Hundreds of different bacterial species use quorum sensing mechanisms similar to that of *V. fischeri* and encode proteins similar to LuxI and LuxR. In light of this information, why do you think scientists want to identify the molecule described in part (a)?

chapter 17

Gene Regulation in Eukaryotes

Drosophila melanogaster *male* (bottom) *courting a female.*

chapter outline

WHEN A *DROSOPHILA* male courts a female, he sings an ancient song and dances a ritual dance. He taps his prospective partner's abdomen with his foreleg and then performs his song by stretching out his wings and vibrating them at a set frequency. When the song is over, he begins to follow the female. If she is receptive, the male will lick her genitals with his proboscis, curl his abdomen and mount the female, and copulate with her for about 20 minutes.

Mutations in a gene called *fruitless* cause behavioral changes that prevent the male from mating properly. For example, males with some hypomorphic mutant alleles cannot distinguish females from males, and they court males and females indiscriminately. The *fruitless* gene encodes male- and female-specific proteins through sex-specific mRNA splicing; a primary transcript produced in both sexes is spliced differently in males and females.

The sex-specific Fruitless proteins are themselves regulatory proteins that control the transcription of many different target genes. The male-specific version of Fruitless protein is present in only a few thousand of the approximately 100,000 neurons that make up the male *Drosophila's* nervous system. These Fruitless-expressing cells include motor neurons that control wing and leg movements; sensory neurons for odors, taste, sights and sounds; and brain cells. By controlling target gene transcription in these cells, the male-specific Fruitless protein governs mating behaviors such as the ritualized song and dance described above.

In this chapter, we see that **eukaryotic gene regulation**—the control of gene expression in the cells of eukaryotes—depends on an array of interacting regulatory elements that turn genes on and off in the right places at the right times. In multicellular eukaryotes, gene regulation controls not only the elaboration of sex-related characteristics and behaviors but also the differentiation of tissues and organs. Regulation can take place at many levels of gene expression. For example, the case of *Drosophila* courtship just described involves both gene regulation at the level of transcription initiation, and posttranscriptional gene regulation at the level of transcript splicing.

In contrast to the theme of environmental adaptation found in the unicellular prokaryotes, the theme in multicellular eukaryotes is the initial establishment of particular gene expression programs in different cell types and the subsequent maintenance and modification of these programs. Although many of the basic principles we have already learned about gene regulation in prokaryotes are also relevant to eukaryotes, the molecular interactions governing gene regulation in eukaryotes are more complex and have several unique features.

17.1 Overview of Eukaryotic Gene Regulation

learning objectives

1. List steps in the process of gene expression that potentially could be regulated.
2. Cite key differences between eukaryotes and prokaryotes that impact gene regulation.

As you explore the intricacies of eukaryotic gene regulation, bear in mind the key similarities and differences between eukaryotes and prokaryotes. In both, transcriptional regulation occurs mainly through the attachment of DNA-binding proteins to specific DNA sequences that are in the vicinity of the transcription unit itself. However, additional levels of complexity are both possible and necessary for controlling expression in eukaryotes for several reasons:

- Chromatin structure often makes DNA unavailable to the transcription machinery.
- Additional RNA processing events occur.
- Transcription occurs in the nucleus, but translation takes place in the cytoplasm.
- Gene regulation needs to control cellular differentiation into hundreds of specialized cell types.

As in prokaryotes, gene expression in eukaryotes can be regulated at the time of transcriptional initiation, when RNA polymerase starts to make a primary transcript. Important decisions concerning the amount of gene product in the cell are indeed often made at this point. However, many steps in the process of gene expression leading to an active gene product exist beyond transcription initiation (**Fig. 17.1**). Transcript processing (including splicing), export of mRNA from the nucleus, translatability and stability of the message, localization of the protein product in specific organelles in the cell, and modifications to the protein that alter its function or stability are all activities that can be regulated and that affect the amount of final active product.

Figure 17.1 Gene expression in eukaryotes. Gene expression first involves transcription and mRNA processing in the nucleus. The mRNA is transported to the cytoplasm, where it is translated into a protein. Posttranslational modifications such as phosphorylation (*P*) can affect the activity, stability, or localization of the protein product. Nontranscribed DNA, *orange;* exons, *green;* introns, *blue.*

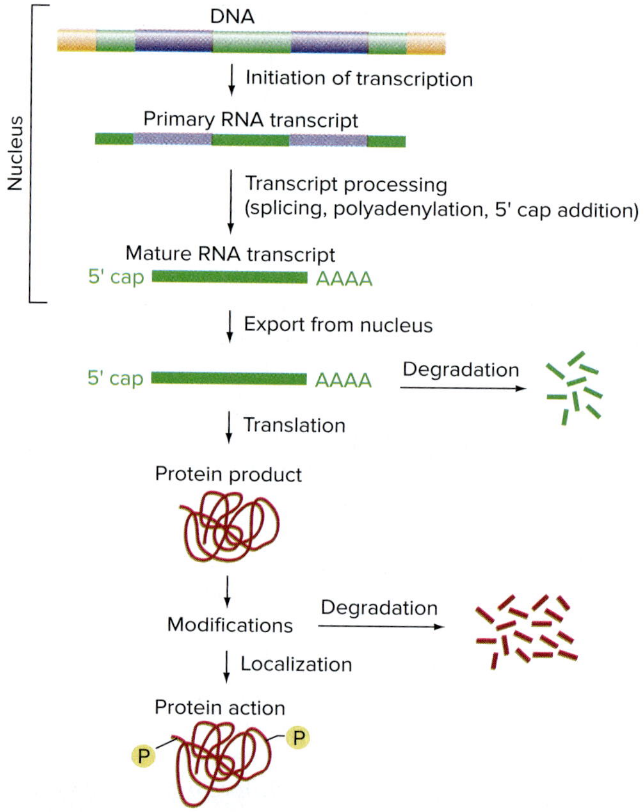

essential concepts

- Control of eukaryotic gene expression occurs at many levels; these include transcription initiation, transcript processing, mRNA stability, mRNA translation, protein modifications, and protein stability.
- The mechanisms of gene regulation in eukaryotes are more complex than those in prokaryotes because eukaryotes have chromatin, eukaryotic transcripts require more processing, and transcripts are exported from the nucleus to the cytoplasm for translation. In multicellular eukaryotes, complex gene regulation directs the development of numerous cell types.

17.2 Control of Transcription Initiation Through Enhancers

learning objectives

1. Describe the *cis*-acting elements that control tissue-specific transcription of eukaryotic protein-encoding genes.
2. Compare the structures and functions of transcriptional activators and repressors.
3. Explain how enhancers work and how they are identified.
4. Describe the function of insulators and how scientists can locate them.
5. Discuss how computer programs and the ChIP-Seq technique can identify transcription factors and their target sites.

Three types of RNA polymerases transcribe genes in eukaryotes. RNA polymerase I (pol I) transcribes genes that encode the major RNA components of ribosomes (rRNAs). RNA polymerase II (pol II) transcribes genes that encode all proteins. RNA polymerase III (pol III) transcribes genes that encode the tRNAs as well as certain other small RNA molecules. We focus in this chapter on the major transcription activity that produces proteins: pol II transcription.

Some of the genes expressed by pol II, like those for ribosomal proteins, are so-called *housekeeping genes* that are transcribed in all cell types nearly all of the time. Other pol II–transcribed genes, like those for the polypeptides making up hemoglobin in red blood cells, are expressed only in one or a few types of cells. We focus first on mechanisms for regulation of genes transcribed in a cell-type-specific manner. Later in the chapter, we will address a mechanism involving DNA methylation that is key for the constitutive transcription of housekeeping genes and is also of importance for the regulation of certain cell-type specific genes.

Promoters and Enhancers Are the Major *cis*-Acting Regulatory Elements

Although each of the regulatory regions of the thousands of pol II–transcribed genes in a eukaryotic genome is unique, genes that are transcribed in a cell-type-specific manner all contain two kinds of essential DNA sequences. The first of these is the **promoter,** which is always very close to the gene's protein-coding region. Promoters usually contain a **TATA box** (or *initiation box*), consisting of roughly seven nucleotides of the sequence T–A–T–A–(A or T)–A–(A or T), located just upstream of the transcription initiation site. As it attracts RNA polymerase only weakly on its own (without an enhancer), the TATA box allows a low, so-called *basal level* of transcription.

Figure 17.2 ***cis*-acting elements of a gene.** *cis*-acting regulatory elements are regions of DNA sequence located on the same DNA molecule as the gene they control. *Promoters* are typically adjacent to the start of transcription. *Enhancers* that regulate expression can sometimes lie thousands of base pairs away from a gene, either upstream of the promoter (*above*), or downstream of the promoter (*below*). Enhancers may even reside in one of the gene's introns (*not shown*).

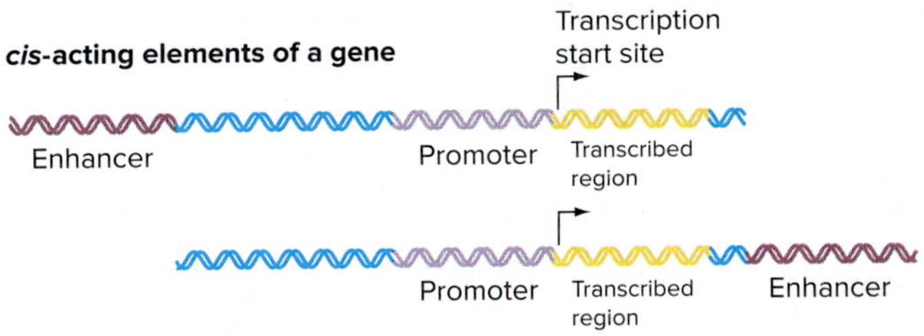

The second type of DNA sequence element important for transcription in eukaryotes is called an **enhancer:** a regulatory site that can be quite distant—up to tens of thousands of nucleotides away—from the promoter. In specific cell types, binding of proteins to enhancers can augment or repress basal levels of transcription. Enhancers may be located either 5′ or 3′ to the transcription start site (some are even found in introns) (**Fig. 17.2**), and a single gene can have one enhancer or several. As described in more detail below, a single enhancer may have multiple binding sites for different transcription factors.

Scientists often use a gene from jellyfish that encodes green fluorescent protein (GFP) as a reporter to identify enhancers in eukaryotes, similar to the way the *lacZ* gene was used as a reporter in bacteria (recall Fig. 16.26). The idea is to construct a recombinant DNA molecule in which the putative regulatory sequence of the gene of interest is fused to the *GFP* gene coding sequence (**Fig. 17.3a**). This recombinant DNA is then used to generate a *transgenic organism* whose genome harbors the fusion. (You will learn details of transgenic technology in Chapter 18.) If the DNA fragment contains an enhancer that directs transcription in a particular type of tissue, only then will the reporter express detectable levels of GFP in that tissue. GFP fluoresces green when it is exposed to light of a particular wavelength. Thus, a tissue-specific enhancer can be detected if that tissue glows when the transgenic animal is illuminated with light of that wavelength (**Fig. 17.3b**).

Researchers can make mutations in the control region of the GFP reporter construct and then make transgenic animals with the mutant DNA. If GFP fluorescence is no longer seen in the same tissue, then the mutation must have affected the tissue-specific enhancer. In this way, scientists can define the DNA sequences constituting particular enhancers. One interesting outcome of such experiments is the finding that enhancers can still function if their orientations or positions relative to the promoter (upstream or downstream) are changed. The reason for this result will be discussed later in this section.

Figure 17.3 Identifying enhancers with GFP reporters. **(a)** An enhancer, such as the eye-specific one shown here, can be found by fusing different fragments of DNA from the vicinity of a gene to an *enhancerless* reporter for jellyfish green fluorescent protein (GFP). When introduced into an organism's genome, only reporters fused to the enhancer will produce large amounts of GFP in the correct tissue. **(b)** A mouse containing a GFP reporter transgene containing an eye-specific mouse gene enhancer.

(b): © & Courtesy John H. Wilson, Baylor College of Medicine

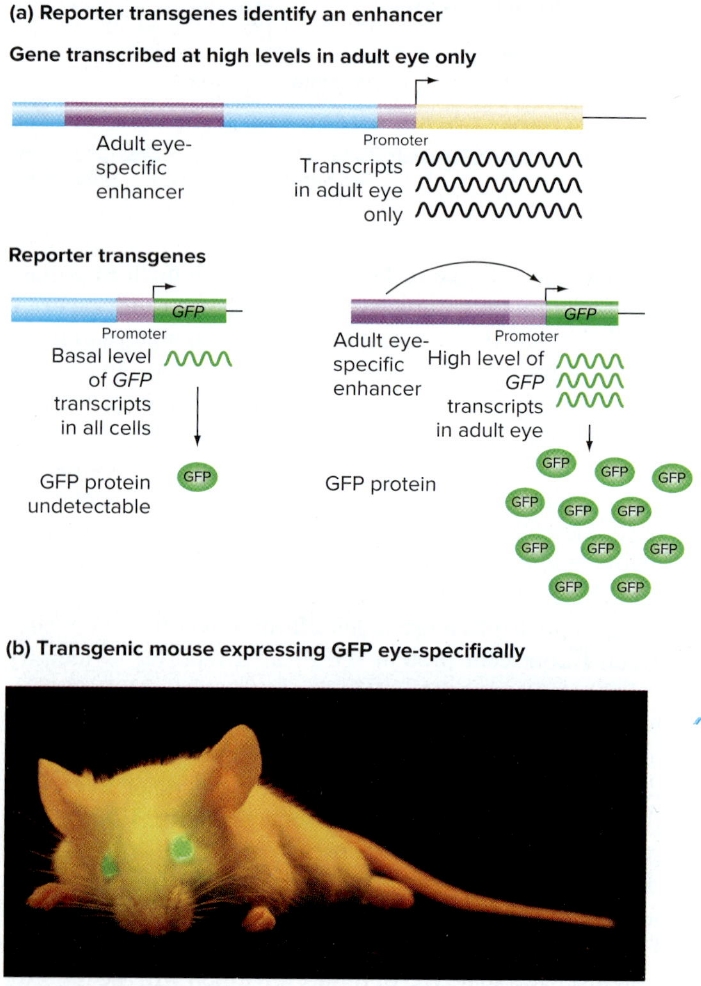

Proteins Act in *trans* to Control Transcription Initiation

The binding of proteins to a gene's promoter and enhancer (or enhancers) controls the frequency of transcriptional initiation (**Fig. 17.4**). Different types of proteins bind to each of the *cis*-acting regulatory regions: *Basal factors* bind to the promoter, while *activators* and *repressors* bind to the enhancers. In this book, we use the term **transcription factors** to describe all sequence-specific DNA-binding proteins that influence transcription (whether they are basal factors, activators, or repressors). Once the transcription factors bind to the DNA, they recruit additional proteins to the gene that can also influence transcription.

Figure 17.4 ***trans*-acting factors.** Transcription factors are *trans*-acting proteins that interact with *cis*-acting elements (the promoter and enhancer) directly through DNA binding. Other *trans*-acting proteins bind DNA indirectly through interactions with transcription factors.

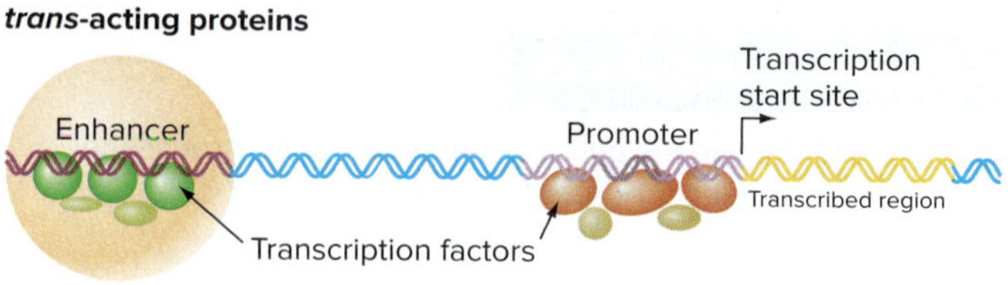

Figure 17.5 Basal factors bind to the promoters of protein-encoding genes. Schematic representation of the binding of the TATA box–binding protein (TBP) to the promoter DNA, the binding of TBP-associated factors (TAFs) to TBP, and the binding of RNA polymerase (pol II) to these basal factors. Once RNA polymerase associates with the promoter, it can begin low-level (*basal*) transcription of the gene.

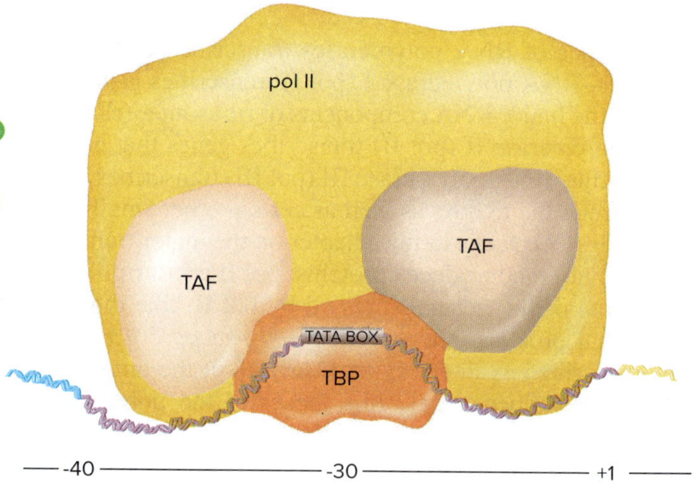

Basal factors

Basal factors assist the binding of RNA polymerase II to the promoter. The key component of the basal factor complex that forms on most promoters is the *TATA box–binding protein,* or *TBP.* This protein recruits other proteins called *TBP-associated factors,* or *TAFs,* to the promoter in an ordered pathway of assembly (**Fig. 17.5**). Once the basal factor complex has formed, RNA polymerase can initiate a low level of transcription (basal transcription). The primary sequences and three-dimensional structures of the basal factors are highly conserved in all eukaryotes, from yeast to humans.

Mediator

Transcription of many eukaryotic genes requires a multisubunit complex called *Mediator* that contains more than 20 proteins. Mediator does not bind DNA directly but instead serves as a bridge between the RNA pol II complex at the promoter and activator or repressor proteins bound at the enhancer (**Fig. 17.6a**).

Figure 17.6 What transcriptional activators do. **(a)** Activators associated with enhancer DNA bind basal factors (in some cases, indirectly through Mediator) to stabilize the interaction of pol II with promoter DNA. **(b)** Activator proteins can also recruit *coactivator proteins*—for example, enzymes that modify histone tails so as to clear the promoter DNA of nucleosomes. If the enhancer and promoter are far apart, DNA looping is required for either mechanism of activator function.

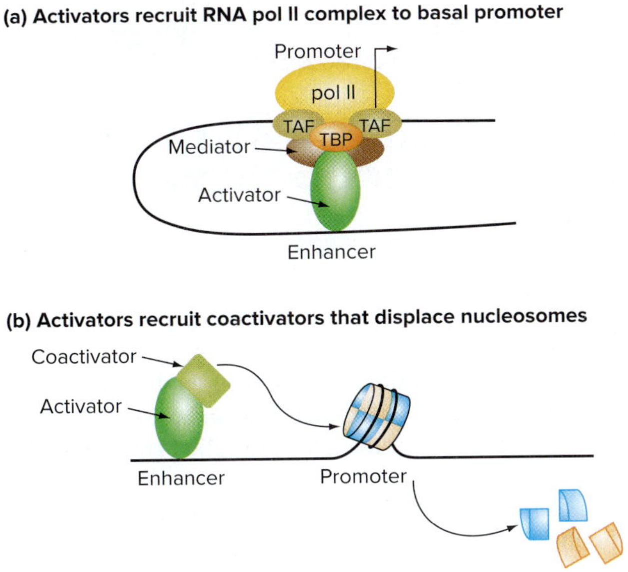

Activators

Although similar basal factor complexes bind to all the promoters of the tens of thousands of genes in eukaryotic genomes, not all genes are transcribed in all cell types. The enormous range of transcriptional regulation occurs through the binding of different transcription factors to distinct enhancer elements associated with different genes. When bound to an enhancer element, transcriptional **activators** increase transcription above the basal levels that occur by action of the promoter alone. Activators can interact directly or indirectly with basal factors at the promoter in a three-dimensional protein/DNA complex to cause an increase in transcriptional activity.

Because enhancers may be far removed from promoters, the DNA between the enhancer and the promoter loops as a consequence of interactions between basal factors, Mediator, and activators (or repressors) (Fig. 17.6a). Long stretches of DNA can be quite flexible explaining why enhancer sequences can function if moved to different positions relative to the promoter.

What activators do At the mechanistic level, transcriptional activator proteins bind their target sites on DNA and increase RNA synthesis by doing one or both of the following:

1. Activators help recruit the basal factors and pol II to core promoter sequences by interacting directly with the components of this complex, as shown in Fig. 17.6a. This function is similar to that of most transcriptional activators in bacteria; recall that positive regulators in bacteria like CRP stabilize the binding of RNA polymerase to the promoter.
2. Activators recruit **coactivators;** these are proteins that open local chromatin structure to allow gene transcription (**Fig. 17.6b**). Recall from Chapter 12 that chromosomal DNA is wound around histone proteins in nucleosomes. Promoter DNA that is covered with nucleosomes is inaccessible to basal factors (Fig. 12.10). Thus, in order for a gene to be transcribed, the promoter DNA must be free of nucleosomes.

   You will also recall that two types of coactivators open chromatin: Histone acetyl transferase (HAT) enzymes and chromatin remodeling complexes. HATs acetylate particular amino acid residues of histone tails, while chromatin remodelers contain ATPase subunits and use the energy from ATP hydrolysis to displace nucleosomes from the promoter. Activator proteins can help attract either type of coactivator to the promoter.

How do activators bind enhancers in the first place if the enhancer DNA is covered with nucleosomes? Scientists think that sometimes activators bind enhancer DNA just after DNA replication, before the nucleosomes are assembled. Also, some activators may be able to bind enhancer DNA even when it is wound up in nucleosomes.

Structures of activators Transcriptional activator proteins must bind to enhancer DNA in a sequence-specific way, and after binding, they must be able to interact with other proteins (a basal factor or a coactivator) to activate transcription. Two structural domains within the activator protein—the *DNA-binding domain* and the *activation domain*—mediate these two biochemical functions (**Fig. 17.7**).

Transcription factors belong to several protein families that share similar DNA-binding domains. Two well-characterized DNA-binding structures are *helix-turn-helix* domains and *zinc fingers* (**Fig. 17.8**). The zinc finger motif is found mainly in eukaryotic proteins, but helix-turn-helix factors are also present in prokaryotes; for example, the Lac

Figure 17.7 Modular structure of activators and repressors. Some activators and some repressors can bind DNA as monomers (*left*), while others bind as dimers (*right*). Activation and repression domains interact with coactivators/corepressors, Mediator, or the basal promoter complex (*not shown*).

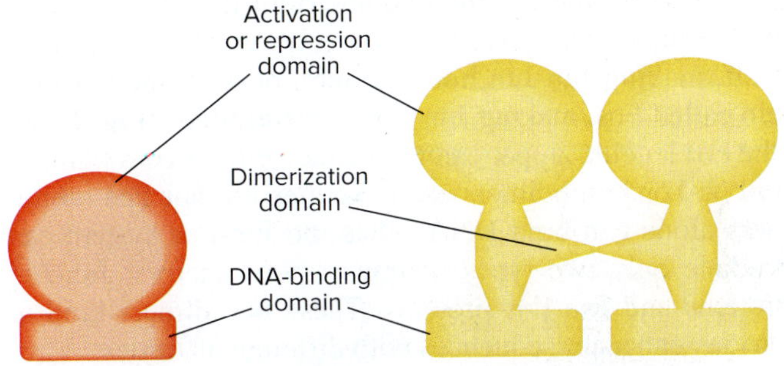

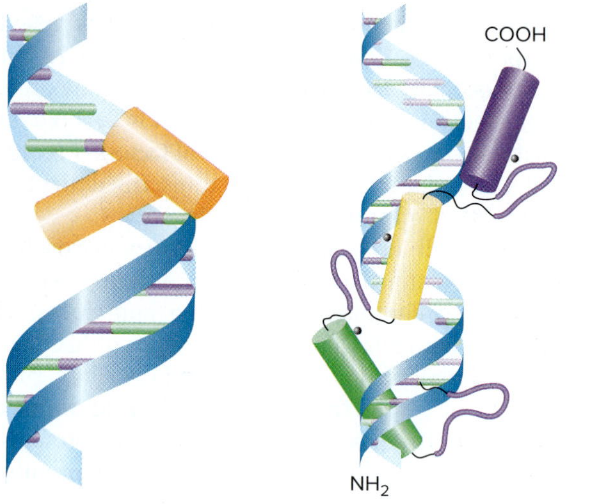

Figure 17.8 Activator protein families. Two common DNA-binding motifs found in activators are the helix-turn-helix and the zinc finger. The colored cylinders represent helical regions of the DNA-binding domains.

repressor is a helix-turn-helix protein. (Recall Fig. 16.15.) DNA-binding domains fit within the major groove of DNA. Subtle differences in the amino acid sequences among activators of the same family can allow these proteins to recognize specific DNA sequences at different enhancer elements.

The activation domains of transcription factors are less well characterized than the DNA-binding domains, and may be less structured. The amino acid sequences of activation domains depend on whether the activator interacts with the basal complex or with one or more coactivators.

Many activator polypeptides also have a third domain—the *dimerization domain* (Fig. 17.7)—that enables them to interact with other copies of the same polypeptide or with other transcription factor subunits to form multimeric proteins, as was the case for several regulatory proteins in bacteria such as the Lac repressor (review Fig. 16.18). One structural motif in many dimerization domains is a *leucine zipper,* a helix with leucines at regular intervals. The ability of two leucine zipper proteins to interlock depends on the specific amino acids that lie between the leucines.

Among the best-characterized transcription factors with a leucine zipper is Jun, a protein important for cellular proliferation and other processes such as the loss and regeneration of the uterine endometrium during the mammalian menstrual cycle. Jun can form dimers either with itself, making Jun-Jun **homodimers,** or with another protein called Fos, making Jun-Fos **heterodimers** (**Fig. 17.9**). The Fos leucine zipper cannot interact with its own kind, so no Fos-Fos homodimers exist. Neither Jun nor Fos monomers alone can bind DNA. Thus the Jun-Fos system can produce only two types of transcription factors: Jun-Jun proteins and Jun-Fos proteins. These two dimers bind to various enhancer sequences with different affinities.

Figure 17.9 Jun-Jun and Jun-Fos dimers. Homodimers contain two identical polypeptides, whereas heterodimers contain two different polypeptides. The leucine zipper motifs in two subunits interact tightly with each other, allowing dimerization to occur. So-called *basic DNA-binding domains* are characteristic of most leucine zipper–containing transcription factors.

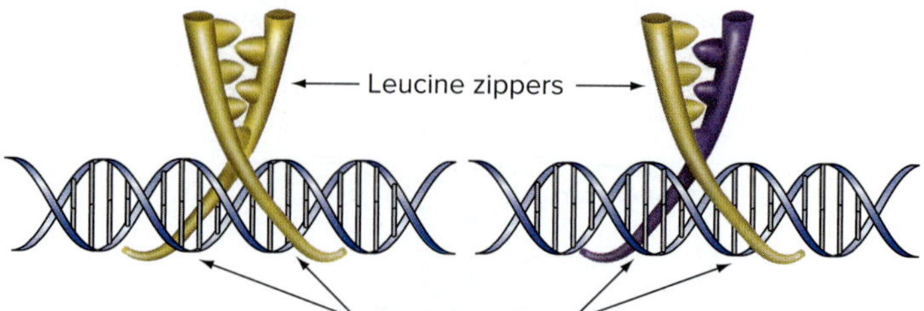

Repressors

Eukaryotic transcription factors that bind specific DNA sites near a gene, such as enhancers, and prevent the initiation of transcription of the gene are called **repressors.**

What repressors do In prokaryotes, negative regulators generally work by physically blocking RNA polymerase from binding the promoter. In eukaryotes, the primary function of repressors is different: Eukaryotic repressors generally recruit **corepressor** proteins to enhancers. Corepressor proteins on their own cannot bind to DNA, so they can associate with enhancers only if a repressor with which they can associate is already there. Corepressors can have either of two functions:

1. Some corepressors can interact directly with the RNA pol II basal complex and prevent it from binding the promoter (**Fig. 17.10a**).
2. Other corepressors are enzymes that modify histone tail amino acids, resulting in closed chromatin (**Fig. 17.10b**). Recall from Chapter 12 that such enzymes are histone deacetylases (HDACs) and histone methyl transferases (HMTs).

Structures of repressors Repressor structures are similar to those of activators: Repressors have DNA-binding motifs, repression domains for interacting with corepressors, and some have dimerization domains (Fig. 17.9). In fact, certain transcription factors can act as either activators or repressors, depending on context. For example, the *Drosophila* transcription factor called Dorsal is an activator when bound to the enhancers of some target genes and a repressor when bound to the enhancers of other genes. How can one protein have two opposing functions in transcription? Dorsal is intrinsically an activator that binds coactivators. However, at some enhancers, interactions with another protein causes Dorsal to recruit instead a corepressor called Groucho (**Fig. 17.11**).

Figure 17.10 What transcriptional repressors do. Repressor proteins bind DNA at enhancers and recruit corepressors. **(a)** Some corepressor proteins directly prevent the basal pol II complex from binding the promoter. **(b)** Other corepressors are histone-modifying enzymes that close the chromatin at the promoter and prevent transcription.

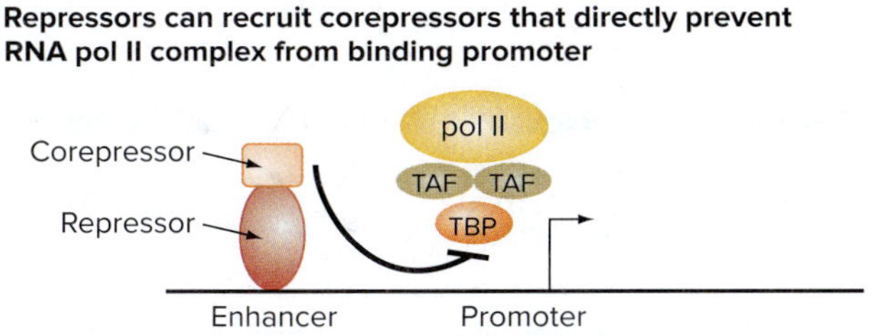

Figure 17.11 A single transcription factor can be both an activator and a repressor. The *Drosophila* Dorsal protein acts as an activator of gene *1*. At gene *2*, another transcription factor (*gray*) bound to enhancer 2 causes Dorsal to act as a repressor by helping it recruit a corepressor called Groucho.

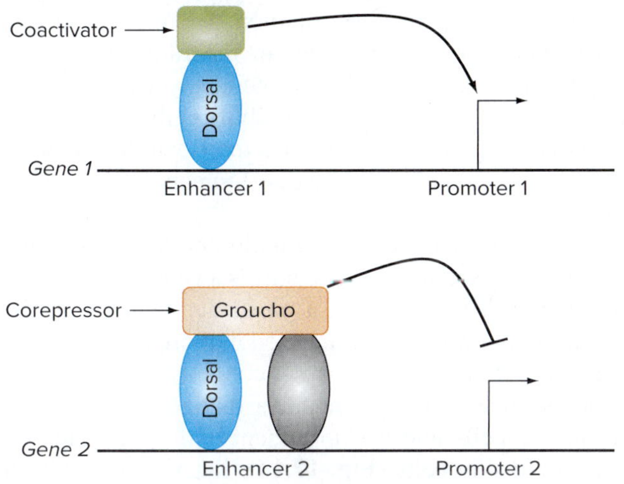

Indirect repression

Many regulatory proteins—called **indirect repressors**—prevent transcription initiation indirectly, not by recruiting corepressors, but instead by interfering with the function of activators. In one such mechanism, some repressors can compete with activators for access to an enhancer because the binding sites of the repressor and the activator overlap (**Fig. 17.12a**). In another form of indirect repression called *quenching,* a protein can bind the activation domain of an activator bound to an enhancer and thereby prevent the activator from functioning (**Fig. 17.12b**). Some indirect repressors bind to activators and hold them in the cytoplasm.

Figure 17.12 Indirect repression. **(a)** A repressor (*orange icon*) may act both directly and indirectly by sharing with an activator (A; *green icon*) a common or overlapping binding site within an enhancer; the repressor can out-compete the activator for enhancer binding. **(b)** An indirect repressor can bind an activator and hide its activation domain. **(c)** An activator can be sequestered in the cytoplasm when bound to an indirect repressor. **(d)** Formation of nonfunctional indirect repressor/activator heterodimers can prevent the formation of functional activator homodimers.

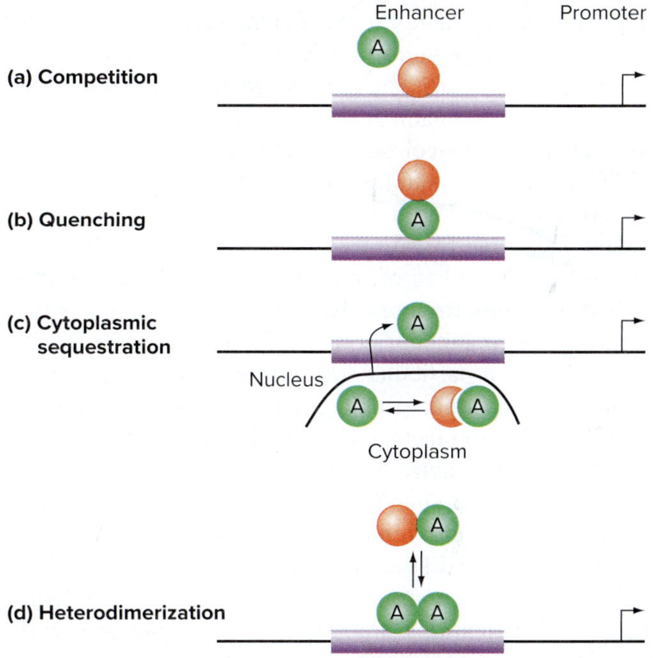

Posttranslational modification of the indirect repressor causes it to release the activator, which can then enter the nucleus and bind its target enhancers (**Fig. 17.12c**). Finally, some indirect repressors can form heterodimers with activators. If only activator homodimers can bind DNA, the indirect repressors can titrate the activators so that few homodimers are able to form (**Fig. 17.12d**).

Histone modifications

As already described in **Chapter 12,** correlations exist between gene transcription and various covalent modifications that can be added to specific amino acids in the N-terminal tails of the histone proteins in nucleosomes. For transcription, the most important of these modifications are acetylation and methylation—the addition of acetyl groups or methyl groups, respectively.

Acetylation of particular lysine amino acids histone tails by enzymes called **histone acetyl transferases (HATs)** favors gene expression by helping to clear promoters of nucleosomes. Thus, many transcription factor coactivators are HATs. Histone acetylation helps open chromatin in two ways. First, lysine acetylation reduces the positive charge on the histones and thus lessens the electrostatic interaction of nucleosomes with negatively charged

DNA. Second, histone tails with acetylated lysines act as landing pads that can recruit certain proteins to nucleosomes. These can be HATs that acetylate other histone tail lysines within the same or adjacent nucleosomes, or *DNA remodeling proteins* that use the energy from ATP hydrolysis to remove histones from DNA.

The methylation of certain lysines (or arginines) in histone tails by **histone methyltransferases (HMTases)** can help either to activate or to repress transcription, depending on the particular proteins that the methylated site recruits to the nucleosome. Particular methylated amino acids bind factors that open chromatin at specific genes, while certain others bind proteins that close chromatin regionally. Thus, some HMTases are coactivators and others are corepressors.

Histone acetylation and methylation are dynamic processes. These modifications can be added rapidly by HATs or HMTases, or taken off rapidly by **histone deacetylases** or **histone demethylases**. But the transcriptional regulation of genes is inherently dynamic. You will see that the interaction of a particular transcription factor with an enhancer can respond to changes in cellular history (what other transcription factors are available at a particular time) or in the cellular environment (for example, the presence of allosteric effectors such as steroid hormones).

It is important to remember that histone modifying enzymes cannot bind specific DNA sites in the genome on their own. All gene-specific or genomic region-specific histone modifications are initiated by sequence-specific DNA-binding proteins (transcription factors). In addition, as of this writing in 2016, no mechanism is known whereby histone modifications are copied at the DNA replication fork. Thus, modified histones function downstream of the transcription factors that regulate gene expression.

Identifying transcription factors

Scientists often use GFP reporter genes to verify that a transcription factor suspected of regulating the expression of a target gene indeed plays such a role. This experimental approach has two requirements. First, researchers must construct an organism with a transgene containing the enhancer of the target gene fused to a GFP reporter (a promoter plus *GFP* coding sequences). As we saw in Fig. 17.3, in an otherwise wild-type animal, GFP will be expressed at high levels in the appropriate cells.

The second requirement is that mutations in the gene encoding the putative transcription factor must be available. If the animal with the GFP reporter also has a loss-of-function mutation in a gene encoding a putative activator, the reporter will no longer be expressed if the activator in fact interacts with that particular enhancer (**Fig. 17.13**). By contrast, loss-of-function mutations in a gene encoding a repressor may yield higher levels of GFP fluorescence than normal and/or cause the expression of GFP in inappropriate cells (*not shown*).

Figure 17.13 Identifying transcriptional activators. **(a)** Activators bind to an enhancer fused to a reporter gene and enable transcription, detected as GFP expression. **(b)** A mutation in a gene encoding an activator can reduce transcription of the reporter, leading to loss of GFP expression.

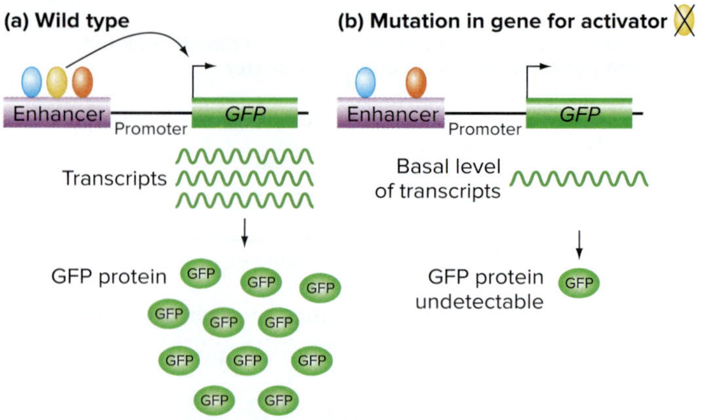

Enhancers Integrate Cellular Information to Control Gene Transcription

In complex multicellular organisms, a large percentage of genes are devoted to transcriptional regulation. Of the estimated 27,000 genes in the human genome, scientists estimate that more than 2000 of them encode transcription factors. Many proteins can regulate the transcription of any one gene, and each regulatory protein may act on many different genes. The number of possible combinations of regulators is staggering and provides the flexibility important for differentiation of cells and development in multicellular eukaryotes.

An enhancer element is usually defined operationally as a *cis*-acting sequence that controls a gene's expression in a particular type of cell at a particular moment in time. Because many genes are expressed in more than one tissue, a gene's regulatory region may contain several enhancer elements, such as an eye enhancer that activates transcription in eye cells and a skin enhancer that activates transcription in skin cells (**Fig. 17.14**). Each enhancer in turn has one or more binding sites with varying affinities for each of several different activators and repressors. At any moment, dozens of activators and repressors in a cell may compete for these binding sites; coactivators and corepressors may also compete with each other for binding to different activators or repressors. The biochemical integration of all this information from an enhancer guides the cell not just to a binary decision of whether a gene should be turned on or off, but in fact helps the cell to fine-tune a gene's transcriptional activation or repression to a level optimal for the cell's role in the organism.

In yeast, a unicellular eukaryote, genes are are regulated by simple elements similar to enhancers called **upstream activating sequences (UASs).** A UAS binds

Figure 17.14 Enhancers govern tissue- and temporal-specific gene expression. A hypothetical gene includes enhancers for expression in eyes or skin. Sequences in the enhancers are recognized by transcriptional activators (*green icons*) or repressors (*orange icons*) that may be present in some tissues but not others. Certain transcription factors can be regulated allosterically (for example, by hormone binding) or posttranscriptionally (for example, by phosphorylation). The enhancer integrates information from the binding of all the transcription factors to determine whether the target gene is transcribed in the tissue, at what level and at what cell cycle stage.

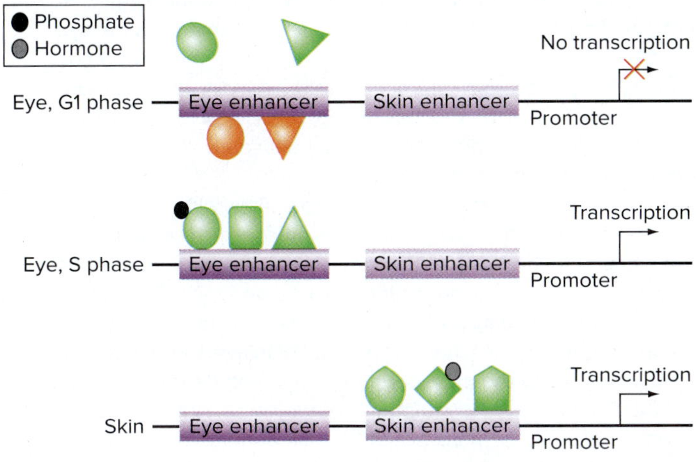

Figure 17.15 Steroid hormone receptors. The DNA-binding domains of steroid hormone receptors are in an inactive conformation until allosteric changes, caused by the binding of a hormone molecule to another domain, allow the DNA-binding domain to recognize the enhancer.

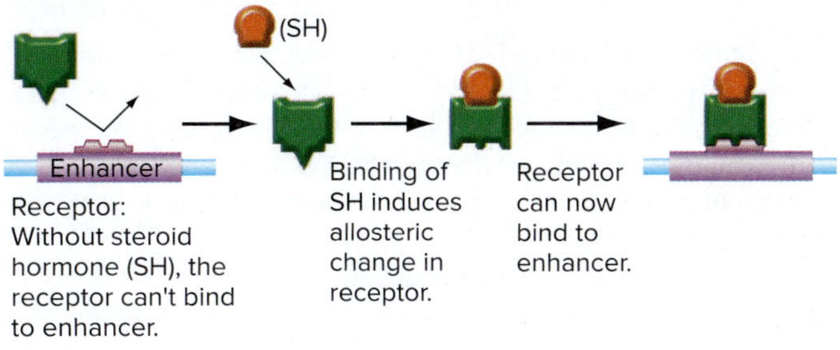

multiple copies of a single transcription factor. The Tools of Genetics Box entitled *The Gal4/UAS$_G$ Binary Expression System* describes how a particular UAS has been exploited to express any cloned gene in virtually any expression pattern in a transgenic organism.

Within a multicellular organism, all cells have the same nuclear DNA, so a gene's enhancer elements are present in all types of cells. The cell-type specificity of transcription depends therefore on changes that occur during the course of development to the constellation of transcription factors that interact with these enhancers (Fig. 17.14). We discuss here three of several ways in which the set of functional *trans*-acting proteins can change over time.

Allosteric interactions

Eukaryotic transcription factor activity can be regulated through allosteric interactions with effectors, similar to what you saw in Chapter 16 for the binding of allolactose to the Lac repressor and tryptophan to the TrpR repressor. Steroid hormone receptor transcription factors constitute an important example of allostery in eukaryotes: Hormone binding causes a shape change in the receptor protein that greatly increases the affinity of its DNA-binding domain for its target enhancer (**Fig. 17.15**).

In humans, the steroid hormones testosterone and dihydrotestosterone work precisely through this mechanism. At puberty in males, the testes generate high levels of these androgen steroids, which bind to the androgen receptor in target cells like those in facial hair follicles. Modulation of androgen receptor transcription factors by steroids controls gene activity in the target cells, leading to the development of male secondary sexual characteristics like facial hair in post-pubescent boys. This example clearly shows that each cell of a multicellular eukaryote must constantly modify its program of gene activity in response to ever-changing signals from elsewhere in the body.

Transcription factor modifications

Transcription factor proteins can be modified after they are synthesized by the covalent addition of any of several different chemical groups, as was previously described in Fig. 8.26b. One of the most important of these modifications is *phosphorylation,* the addition of a phosphate group to a protein by action of an enzyme called a *kinase.* Phosphorylations can either activate or deactivate a transcription factor in any of a number of ways: by influencing movement of the factor into the nucleus, the factor's DNA-binding properties, its ability to multimerize, or its ability to interact with other proteins, including coactivators or corepressors.

Cells often rely on phosphorylations to control events that must occur rapidly, such as responses to changes in the environment or transitions between states in the cell cycle. In Chapter 20, you will see how the phosphorylation of a particular transcription factor called p53 plays an important role in protecting organisms from cancer.

Transcription factor cascades

As Fig. 17.14 illustrated, not all transcription factors are made in all cells at all times. Clearly, if a factor is not present in a given cell, it will be unable to influence the initiation of the transcription of any target genes. In other words, the availability of various transcription factors is crucial to a cell's determination of which genes will be transcribed, and at what levels.

TOOLS OF GENETICS

Blue DNA: © MedicalRF.com

The Gal4/UAS_G Binary Gene Expression System

In yeast, a unicellular eukaryote, DNA elements called *upstream activating sequences* (*UAS*s) serve as simple enhancers that bind multiple copies of a single transcription factor. For example, *UAS_G*, which regulates several genes involved in galactose metabolism, has four binding sites for an activator protein called *Gal4*.

Researchers have exploited the simple nature of the Gal4–*UAS_G* interaction to develop an experimental system whereby a transgenic model organism can express any given gene in any of thousands of tissue-specific patterns. **Figure A** illustrates how this system works in *Drosophila*.

Researchers first clone the gene's cDNA downstream of a cloned *UAS_G* and a promoter, and they introduce this transgene construct into the fly genome. When flies carrying a *UAS_G-cDNA* transgene are crossed to flies containing a transgene that expresses Gal4, the progeny carrying both transgenes express the cDNA specifically in those cells making Gal4 protein.

The transgenes expressing Gal4 protein are maintained in a collection of fly lines called *drivers*. Each driver line contains a copy of a *promoter-Gal4* fusion gene in which an enhancerless fly promoter is fused to a cDNA encoding Gal4. Scientists generated thousands of lines, each containing a different insertion of the *promoter-Gal4* transgene into random locations in the fly genome. Enhancers near the site of the *promoter-Gal4* insertion determine the pattern of Gal4 expression. Thousands of lines are available, each of which expresses Gal4 in a different tissue-specific pattern.

The Gal4/UAS_G binary system has many applications; we briefly discuss here one interesting example that uses the system to kill (or *ablate*) specific classes of cells. In this application, the cDNA placed downstream of *UAS_G* encodes Reaper, a protein that activates the process of *apoptosis*, or *programmed cell death*. Cells that express Reaper will ultimately die and be removed from the animal.

Researchers crossed *UAS_G-reaper* flies with strains containing drivers that express Gal4 only in certain specific classes of neurons in the brain. In this way, the scientists generated different animals that lacked different neuron types. They found that females lacking particular neurons that normally make a specific kind of ion channel protein displayed enhanced responsiveness to the male courtship song and dance described at the beginning of this chapter. In other words, although normal females may reject many potential male partners before choosing a mate, these mutant flies would mate with just about any male. These results show the importance of those specific neurons and that kind of ion channel in this female mating behavior.

Figure A Gene expression using Gal4/UAS_G. The *Gal4* and *UAS_G* transgenes are in different genomic locations. The *P* element ends are used to integrate the transgenes into the genome (explained in detail in Chapter 18.)

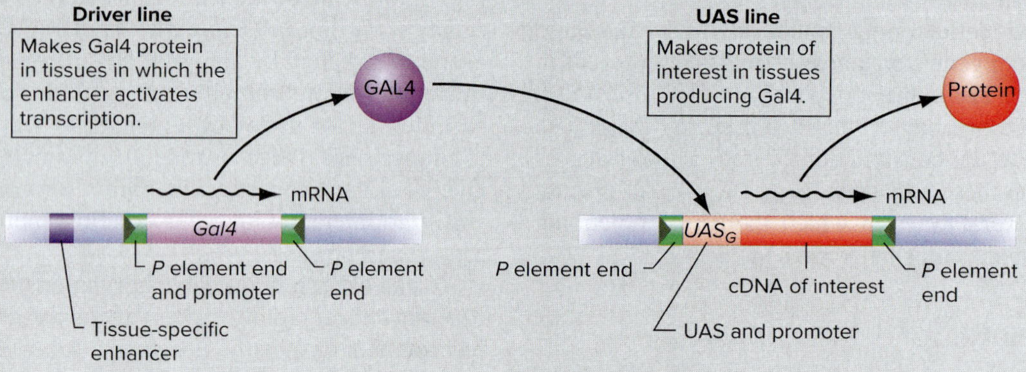

A transcription factor is, like any protein, a gene product. The expression of a gene encoding a transcription factor is thus subject to control by other transcription factors, implying that cascades of transcription factor expression must occur. One set of factors turns on or represses another set of factors, which in turn controls the expression of yet other transcription factors. You will see in Chapter 19 that such *transcription factor cascades* are crucial to the biochemical mechanisms that control the development of multicellular eukaryotes.

Insulators Organize DNA to Control Enhancer/Promoter Interactions

As mentioned above, an enhancer may be located upstream or downstream of the promoter that it regulates and in either orientation with respect to the promoter. These facts pose a conceptual problem: How does an enhancer "know" which of the two genes it inevitably sits between is the right one? And because enhancers may work at

Figure 17.16 Identifying insulators. An enhancer placed between the promoters of two reporter genes will activate transcription of both unless an insulator sequence is located between the enhancer and one of the promoters. RFP is *red fluorescent protein*; the *RFP* gene was cloned from a red fluorescent coral. With the construct at the *top*, cells in a transgenic organism would fluoresce both red and green (that is, in yellow); with the construct at the *bottom*, the same cells would fluoresce only in green.

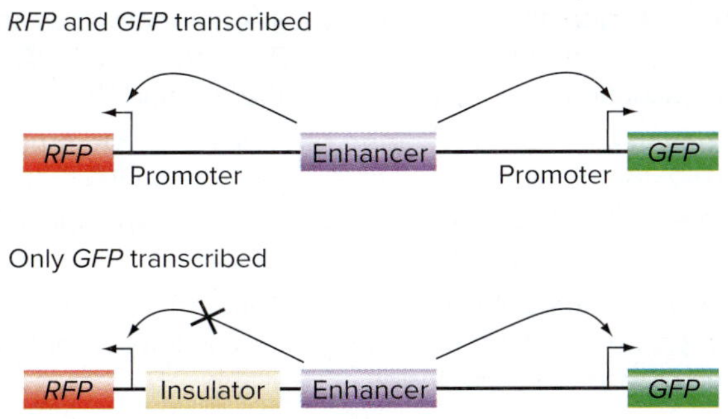

Figure 17.17 How insulators work. Insulators organize genomic DNA into loops, while enhancers activate transcription (A = activator) only from promoters within the same loop. Interactions between CTCF proteins (*yellow*) bound to different insulators facilitates the formation of a DNA loop in the region between the insulators.

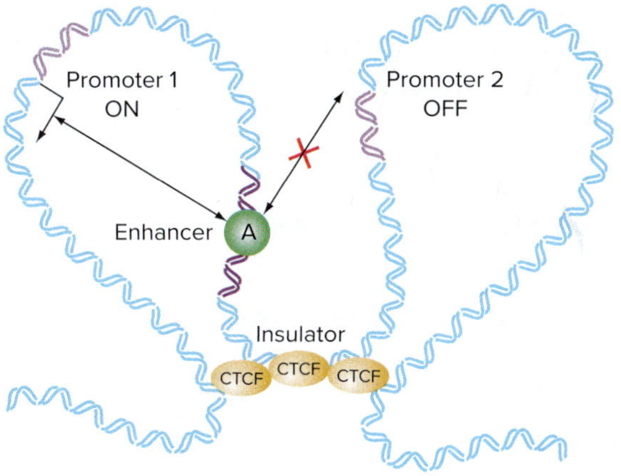

great distances from the promoters that they regulate, what prevents any enhancer on a chromosome from influencing any promoter for any gene anywhere on that chromosome? The answer is that DNA elements called **insulators** organize chromatin so that enhancers have access only to particular promoters.

How scientists identify insulators

Insulators are characterized as DNA elements located between a promoter and an enhancer that block the enhancer from activating transcription from that promoter. To identify insulators, researchers insert suspect DNA sequences between an enhancer and a promoter of a reporter gene in a recombinant construct inserted into a genome. If reporter gene expression is blocked, then the DNA sequence is deemed an insulator (**Fig. 17.16**).

How insulators work

Human insulators bind a protein called *CTCF (CCCCTC-binding factor)*. Connections between CTCF proteins bound to different insulators facilitate the formation of DNA loops. A promoter and an enhancer will be in separate loops and cannot interact with each other if an insulator lies between them (**Fig. 17.17**).

Recent research has revealed that insulator functions can be much more complex than simply blocking enhancers. Some developmental regulator genes have several enhancers separated by insulators. DNA loop formation at these genes is dynamic, and the insulators may deliver specific enhancers to particular promoters in response to signals that change as the organism develops.

New Methods Provide Global Views of *cis*- and *trans*-Acting Transcriptional Regulators

The recent emergence of **bioinformatics**—a field of science in which biology, computer science, and information technology merge to form a single discipline—promises to facilitate the understanding of complex transcriptional programs. As an example, computer programs virtually translate coding sequences in cDNA clones and putative open reading frames in genomes into the amino acid sequences of proteins. The computers then search for signatures within these amino acid sequences, such as helix-loop-helix or zinc finger motifs, that would indicate the proteins are transcription factors. In this way, researchers can find *trans*-acting transcription factors encoded by an organism's genome.

To search for possible *cis*-acting transcriptional regulatory sites such as enhancers, computers compare genomic sequences of closely related species. As was shown previously in Fig. 10.3, nucleotide sequences tend to be poorly conserved outside of coding regions, so any such conserved sequences that are found are strong candidates for having roles in important processes such as gene regulation.

Chromatin immunoprecipitation-sequencing (ChIP-Seq) is a powerful new technology for finding all the target genes of a particular transcription factor within the entire genome of a particular type of cell (**Fig. 17.18**). Scientists first isolate chromatin from nuclei of the cells being studied, chemically cross-link the DNA and protein components of the chromatin and then fragment the DNA in the chromatin. Researchers next add microscopic beads coated with an antibody that binds specifically to

Figure 17.18 ChIP-Seq. Antibodies (Y-shaped molecules) against a specific transcription factor (*blue oval*) are used to purify the protein bound to its target gene DNA sites (by attaching the antibody to microscopic beads, *not shown*). Sequencing of the DNA fragments within the purified protein–DNA complexes identifies the genes the transcription factor regulates.

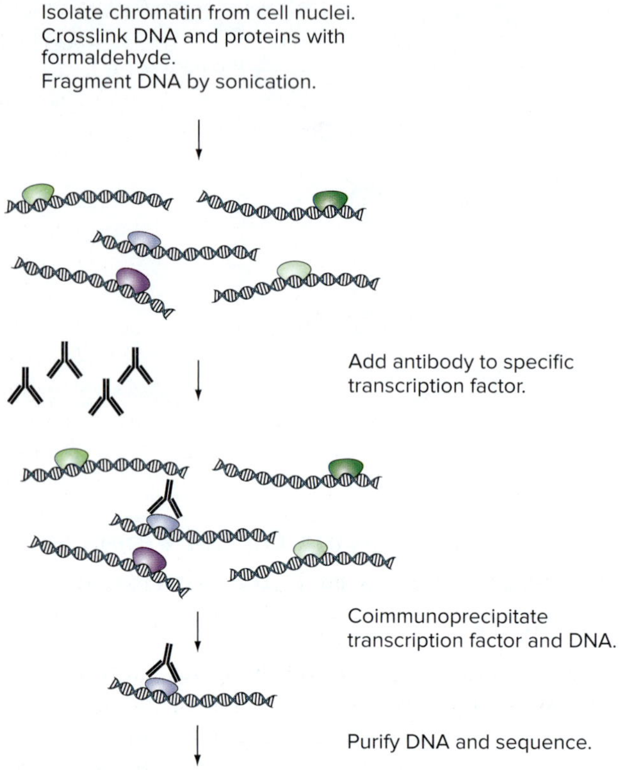

the transcription factor of interest. The only protein–DNA complexes that will stick to the beads are those containing the transcription factor cross-linked to the enhancers with which it interacts. These complexes can be washed free of other, nonspecific chromatin pieces. Purification, through antibody binding, of specific proteins bound to other proteins or nucleic acids is called *co-immunoprecipitation.* The scientists sequence the DNA fragments in these purified complexes so as to identify the genes targeted by the particular transcription factor in the type of cell being analyzed.

As a follow-up to the Human Genome Project, a consortium of hundreds scientists is attempting to map all of the *cis*-acting regulatory elements (promoters, enhancers, and insulators) in the human genome. The project, called ENCODE (the Encyclopedia of DNA Elements), uses computer technology and biochemical experiments such as ChIP-Seq. The enormous ENCODE database can be especially useful for scientists searching for disease-causing mutations located in enhancers far away from the exons of the gene they regulate.

essential concepts

- *Enhancers* are DNA sequences which may be distant from a gene's promoter and act in particular cell types to increase or decrease the amount of transcription relative to a basal level.
- *Transcription factors* are *trans*-acting proteins that include *basal factors* that bind the promoter, and *activators* and *repressors* that bind enhancers. Once bound to DNA, transcription factors can recruit other proteins to the gene.
- Enhancers can have binding sites for many different activators and repressors; this property of enhancers enables them to impart temporal and cell type specificity to gene transcription.
- *Insulators* are DNA elements that organize chromatin into loops; an enhancer and a promoter can interact only if they are in the same loop.
- *Bioinformatics* enables genome-wide searches for new transcription factors and their binding sites. *ChIP-Seq* uses specific antibodies to identify genes regulated by transcription factors of interest.

17.3 Epigenetics

learning objectives

1. Describe gene regulation by CpG islands.
2. Discuss how genomic imprinting can be inferred from inheritance patterns in human pedigrees.
3. Define an *epigenetic phenomenon.*
4. Explain the relationship between DNA methylation and genomic imprinting.

In the preceding section, we discussed how the binding of transcription factors to enhancers ensures that many genes are expressed only in particular tissues at specific times during development. A second method by which cells can regulate transcription initiation is through the control of **DNA methylation:** a biochemical modification of DNA in which a methyl (CH_3) group is added to the fifth carbon of the cytosine base in a 5′ CpG 3′ dinucleotide pair on one strand of the double helix (**Fig. 17.19a**). (The *p* in CpG stands for phosphate.) Enzymes called **DNA methyl transferases (DNMTs)** catalyze the methylation of cytosines in CpG dinucleotides.

DNA methylation is particularly important to the control of expression of housekeeping genes in vertebrates, though it also helps regulate some cell-type-specific genes. In the human genome, about 70% of the C residues in CpG dinucleotides are methylated. You will see that because DNA methylation affects gene transcription, and because methylation patterns are copied during DNA replication,

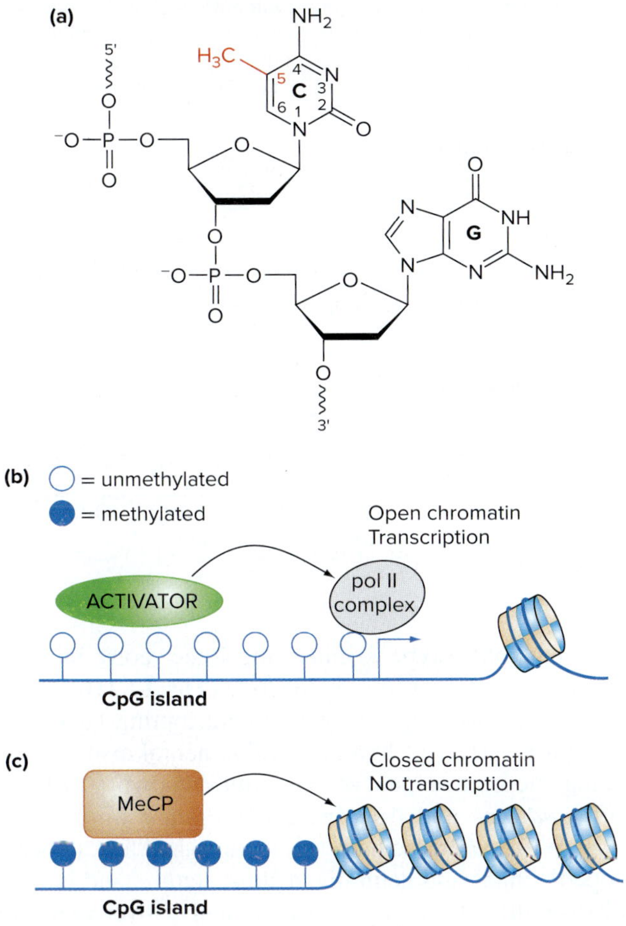

Figure 17.19 DNA methylation and CpG islands. **(a)** Chemical structure of a CpG dinucleotide where the C is methylated (*red*). **(b** and **c)** Transcription of some genes is controlled by *CpG islands* (sequences rich in CpG residues), which contain binding sites for activators. **(b)** Bound activators prevent methylation of the C residues. **(c)** If activators are no longer present, the CpG island becomes methylated; repressors called *methyl-CpG binding proteins (MeCPs)* bind methylated sites and close chromatin.

DNA methylation can alter gene expression heritably without changing the base sequence of DNA—and thus causes a so-called **epigenetic phenomenon.** Methylation is key to an epigenetic phenomenon seen in mammals (including humans) that is called *genomic imprinting*.

Invertebrate animals and unicellular eukaryotes have little or no DNA methylation, while the worm *C. elegans* and yeasts have none. The information in this section may thus not be relevant to all eukaryotic organisms, but it is very important to human genetics.

DNA Methylation at CpG Islands Silences Gene Expression

CpG islands are DNA sequences that may be a few hundred or a few thousand bp long, and within which the frequency of CpG dinucleotides is much higher than that of the rest of the genome. However, unlike the CpG dinucleotides in the rest of the mammalian genome, the C residues in CpG islands are usually unmethylated. When the CpG islands in the vicinity of a gene's promoter are unmethylated, the chromatin is *open* and the gene is transcriptionally active. Methylation of the CpG islands *closes* the chromatin and represses transcription (**Fig. 17.19b** and **c**).

The reason that CpG islands are usually unmethylated is that the proteins that activate transcription by binding to CpG islands prevent DNMTs from methylating these islands (Fig. 17.19b). These transcriptional activators will be found in many cell types if the target gene is a housekeeping gene expressed in most cells.

If the activators are absent, the CpG island becomes methylated. The gene cannot be transcribed because repressors called *methyl-CpG-binding proteins* (*MeCPs*) bind to methylated CpG islands and close the chromatin structure (Fig. 17.19c). Repression of genes by DNA methylation is often long-term because the methylation pattern is maintained through numerous cell divisions; long-term repression through DNA methylation is called **silencing.** DNA methylation patterns are copied during DNA replication by a special DNMT present at the replication fork that recognizes hemi-methylated DNA (DNA methylated on only one strand, in this case the parental strand) and methylates the newly synthesized DNA strand (**Fig. 17.20**).

The potential importance of gene regulation through DNA methylation in humans was revealed by the discovery of the mutant gene responsible for Rett syndrome, a developmental disorder of the brain that results in seizures and mental and physical impairment. The disease shows X-linked dominant inheritance, and it is caused by loss-of-function mutations in an X-linked gene called *MeCP2* that encodes a methyl-CpG-binding protein.

Sex-Specific DNA Methylation Is Responsible for Genomic Imprinting

A major tenet of Mendelian genetics is that the parental origin of an allele—whether it comes from the mother or the father—does not affect its function in the F_1 generation. For the vast majority of genes in plants and animals, this

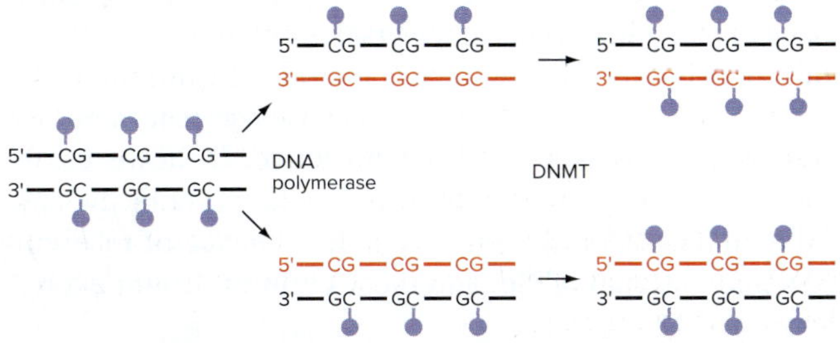

Figure 17.20 Cytosine methylation is perpetuated during DNA replication. A dedicated DNA methyltransferase (DNMT) functions at the DNA replication fork; the pattern of cytosine methylation (*blue circles*) on the template strand is replicated on the newly synthesized strand of DNA (*red*).

principle still holds true. Surprisingly, however, geneticists have found that some genes in mammals are exceptional and do not obey this general rule.

The unusual phenomenon in which the expression of an allele depends on the parent that transmits it is known as **genomic imprinting.** In genomic imprinting, the copy of a gene an individual inherits from one parent is transcriptionally inactive, while the copy inherited from the other parent is active. The term *imprinting* signifies that whatever silences the maternal or paternal copy of an imprinted gene is not a change in the nucleotide sequence of DNA. Instead, as you will see later in this section, the "whatever" is sex-specific methylation of certain DNA sequences called **imprinting control regions (ICRs).**

Only about 100 of the approximately 27,000 genes in the human genome exhibit imprinting. This number of imprinted genes was estimated by RNA-Seq experiments (review Figs. 16.29 and 16.30) that could distinguish transcripts of a gene from the two homologs in a heterozygous individual. About half of these 100 are *paternally imprinted genes,* meaning that the allele inherited from the father is not expressed, while the allele from the mother is transcribed. For *maternally imprinted genes,* the allele inherited from the mother is not transcribed, and all of the mRNA for this gene is made from the paternal allele. It may be helpful for you to track this nomenclature by equating the term *imprinted* with *silenced.*

Figure 17.21 Genomic imprinting and human disease. **(a)** A typical pedigree for a disease associated with deletion of a paternally imprinted autosomal gene. Fathers can pass the deletion to their sons or daughters who are unaffected (*dots* in pedigree symbols indicate unaffected carriers of the deletion); mothers can pass the deletion and the disease (*yellow* shading) to their children. **(b)** A typical pedigree for a disease associated with deletion of a maternally imprinted gene. Here, it is the mothers who can pass the deletion to their sons and daughters without effect (*dots*); fathers can pass the deletion and the disease to their sons and daughters who will be affected (*purple*). Both pedigrees also apply for inheritance of a recessive loss-of-function mutation of the imprinted gene instead of a deletion.

(a) Paternal imprinting

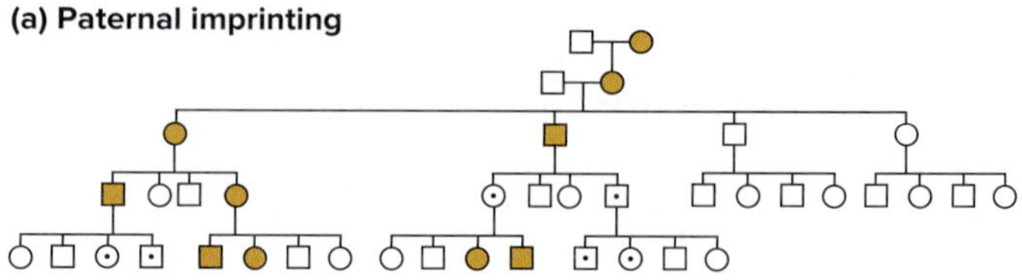

(b) Maternal imprinting

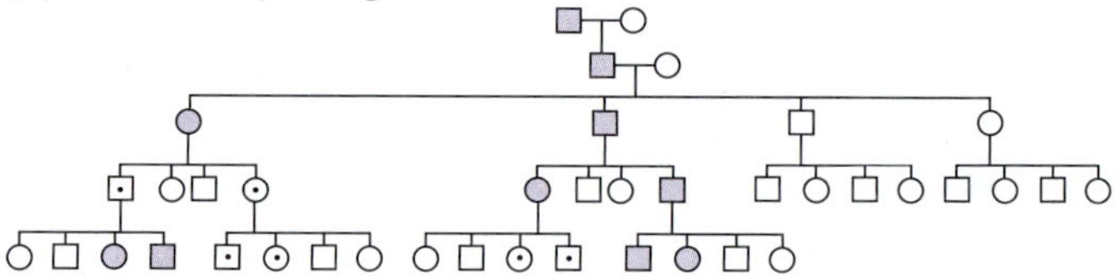

Imprinting and human disease

The existence of genomic imprinting was first inferred well before the RNA-Seq technique was developed, when clinical geneticists in the 1980s observed pedigrees in which the sex of the parent carrying the mutant allele determined whether the child would manifest the disease. These kinds of pedigree patterns were particularly clear in certain rare cases where the condition was caused by a deletion that removed an imprinted gene, because the inheritance of the deletion as well as the disease could be followed in karyotypes.

As seen in **Fig. 17.21a,** a deletion of a paternally imprinted gene could pass without effect from a father to any child, because the child's wild-type maternal allele would be expressed. However, if a woman was heterozygous for the same deletion, 50% of her children would receive the deletion from her. All of these heterozygous children would have the mutant phenotype because the one intact copy of the gene they inherited from their father would be inactive; no gene product would be made, and this would produce the aberrant phenotype. Conversely, deletion of a maternally imprinted gene could pass unnoticed from mother to daughter for many generations because the paternally derived gene copy would always be active. If, however, the deletion passed from a man to his children, both the sons and daughters would each have a 50% chance of receiving a deleted paternal allele, and those children would express the mutant phenotype because the intact copy inherited from their mothers would be inactive (**Fig. 17.21b**).

Evidence for imprinting as a contributing factor now exists for a variety of human developmental disorders, including the related pair of conditions known as *Prader-Willi syndrome* and *Angelman syndrome.* Children with Prader-Willi syndrome have small hands and feet, underdeveloped gonads and genitalia, a short stature, and intellectual disability; they are also compulsive overeaters and obese. Children affected by Angelman syndrome have red cheeks, a large jaw, and a large mouth with a prominent tongue; they also show severe intellectual and motor disability. Both syndromes are often associated with small deletions in the q11–13 region of chromosome 15. When the deletion is inherited from the father, the child develops Prader-Willi syndrome; when the same deletion comes from the mother, the child has Angelman syndrome.

The explanation for this phenomenon is that at least two genes in the region of these deletions are differently imprinted. One gene is maternally imprinted; children receiving a deleted chromosome from their father and a wild-type (nondeleted) chromosome with an imprinted copy of this gene from their mother exhibit Prader-Willi syndrome because the imprinted, wild-type gene is inactivated. In the case of Angelman syndrome, a different gene in the same region is paternally imprinted; children receiving a deleted chromosome from their mother and a normal, imprinted gene from their father develop this syndrome.

Imprinting as an epigenetic phenomenon

Genes may be modified in a manner that does not change the base pair sequence of the DNA but nevertheless affects gene transcription in a heritable manner. As mentioned earlier, modifications to genes that alter gene expression without changing the base pair sequence and that are inherited directly through cell divisions are called *epigenetic changes.* The type of epigenetic change responsible for genomic imprinting is sex-specific DNA methylation of CpG dinucleotides found in specific ICRs (imprinting control regions) that are located near the 100-odd imprinted genes.

Imprints are maintained when somatic cells divide by mitosis because the pattern of methylation can be transmitted during DNA replication. The presence of a methyl group on one strand of a newly synthesized double helix signals DNMT methylase enzymes to add a methyl group to the other strand (review Fig. 17.20). Sex-specific methylation of imprinted loci thus generally remains in the somatic cells throughout the life of the individual.

Note, however, that the pedigrees shown in Fig. 17.21 require that the patterns of DNA methylation must be reset during meiosis before being passed on to the next generation. If this were not true, the imprinting would not be sex-specific. **Figure 17.22** shows that the methylations are erased (removed) in the germ-line cells, and sex-specific methylation marks are then generated during each passage of the gene through the germ line into the next generation. Some genes are methylated in the maternal germ line; others receive methylation marks in the paternal germ line. For each gene subject to this effect, imprinting occurs in either the maternal or paternal line, never in both. The molecular differences in the male and female germ line that result in different patterns of methylation are unknown.

Figure 17.22 Genomic methylation marks are reset during meiosis. Maternally methylated genes are shown in *red* and paternally methylated genes in *black*. In germ-line cells, somatic cell methylation marks are erased, and new sex-specific methylation marks are established.

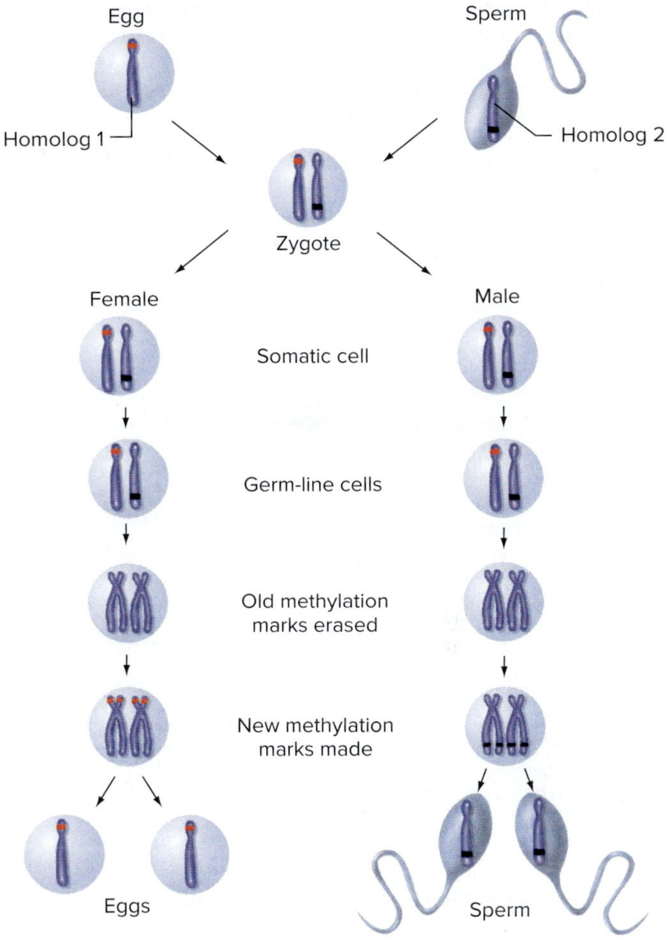

How imprinting works

DNA methylation at ICRs controls the transcription of nearby genes. In contrast with methylation at CpG islands, which always represses transcription, methylation at ICRs can turn imprinted genes either on or off. Biochemical studies have uncovered two ways in which ICR methylation can influence gene expression.

Insulator mechanism Here, the ICR contains an insulator whose function is controlled by DNA methylation. An example of this mechanism for ICR function is seen in the maternally imprinted mouse gene *Igf2* (for *insulin-like growth factor 2*). Imprinting at the *Igf2* locus works through methylation of an insulator that lies between the *Igf2* promoter and its enhancer (**Fig. 17.23a**). The nonmethylated insulator on the maternal chromosome is functional—it binds CTCF, which as we saw earlier, is a protein whose association with insulators forms loops in chromatin. As a result, the enhancer on the maternal chromosome cannot activate transcription of the *Igf2* gene because it is not in the same loop as the gene's promoter.

On the paternal chromosome, by contrast, the insulator is methylated, which prevents it from binding CTCF. Without a functional insulator, the enhancer activates transcription from the *Igf2* promoter because these two elements are now in the same loop. Note in this case that even though it is the paternal chromosome that is methylated, it is the maternal allele that is silenced (that is, *Igf2* is maternally imprinted).

Noncoding RNA (ncRNA) mechanism In the vicinity of some imprinted genes, the ICR encodes an ncRNA whose transcription is controlled by a CpG island. The paternally imprinted *insulin growth factor receptor 2* gene (*Igfr2*), which encodes the receptor for Igf2, provides an example of this imprinting mechanism (**Fig. 17.23b**). On the paternal chromosome, an ncRNA called *Air* is transcribed from a promoter within an intron of *Igfr2* but in the opposite direction

Figure 17.23 Genomic imprinting mechanisms.
(a) Maternal imprinting of *Igf2* is controlled by methylation of an insulator located between the *Igf2* enhancer and promoter. On the maternal homolog, the insulator is unmethylated and is therefore functional (it binds CTCF). On the paternal homolog, the insulator is methylated and does not function. **(b)** Paternal imprinting of *Igfr2* depends on methylation of a CpG island that controls transcription of the *Air* ncRNA; when *Air* is transcribed, *Igfr2* is not expressed. The CpG island on the maternal homolog is methylated, silencing *Air* transcription and allowing *Igfr2* expression. The paternal *Igfr2* allele is silenced because the CpG island is unmethylated and *Air* is transcribed.

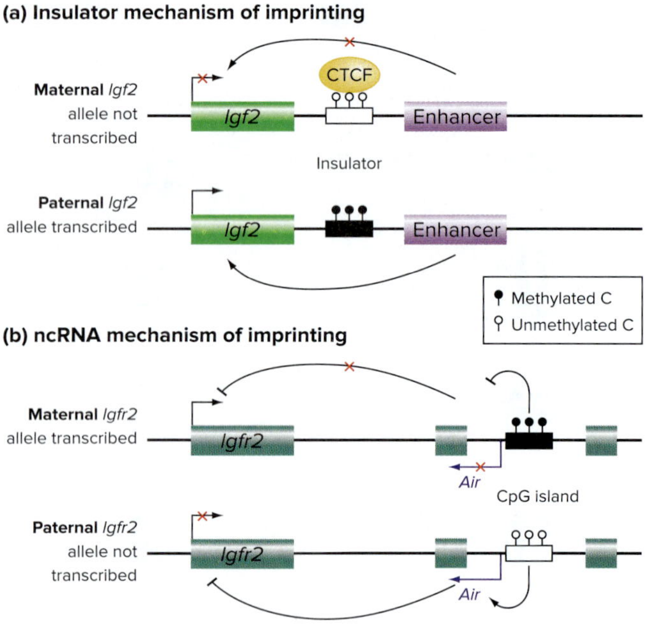

from *Igfr2*. The Air ncRNA is thus an antisense transcript that suppresses the expression of *Igfr2*. On the maternal chromosome, a CpG island that controls *Air* transcription is methylated, silencing transcription of *Air* and thus permitting expression of *Igfr2*. It is not clear how the antisense *Air* suppresses *Igfr2;* perhaps the act of *Air* transcription itself somehow interferes with transcription of *Igfr2,* or interaction of *Air* and *Igfr2* transcripts could lead to the latter's destruction by *RNA interference* (described later).

Why imprinting?

The answer to this question is not known, but several hypotheses have been proposed. One interesting idea concerns the facts that imprinting occurs only in placental mammals and that most imprinted genes, like *Igf2* and *Igfr2,* control prenatal growth. This so-called *parental conflict hypothesis* imagines that because a fetus growing in the womb uses tremendous maternal resources, it is in the mother's interest for her baby to be small so as to balance her own needs with those of the child. Conversely, the father's only interest is for his babies to be large and therefore more robust. According to the parental conflict theory, imprinting may be nature's way of playing out this struggle in the womb. For example, *Igf2* encodes a ligand that promotes growth, and it is maternally imprinted, while *Igfr2* encodes a receptor for the ligand that represses growth and is paternally imprinted. Although the parental conflict hypothesis is compelling on its surface, many biologists think that it is overly simplistic, and they have very different ideas about the origins of genomic imprinting.

Can Environmentally Acquired Traits Be Inherited in Mammals?

One of the liveliest and most controversial areas of current genetic research asks whether a trait acquired by a multicellular organism through environmental influences can be passed on to that individual's progeny. It is clear that the environment can influence gene expression in eukaryotes. For example, chemicals introduced into the environment can alter gene expression patterns by modifying the proteins and RNAs that regulate transcription or translation, by modifying histone tails or by modifying DNA through methylation. Can the effects of these changes, as opposed to DNA mutations, be transmitted between generations?

Intergenerational inheritance of an acquired trait in mice

A famous example of the inheritance of an environmentally acquired trait concerns the A^Y allele of the *agouti* gene. Recall from Fig. 3.8 that while normal mice (*AA*) are gray, A^YA heterozygotes are yellow. The A^Y allele is a gain-of-function mutation that leads to the exclusive production of the yellow pigment pheomelanin and its deposition in hairs (**Fig. 17.24a**). The gray mouse at the right in **Fig. 17.24b** (F_1) inherited the A^Y allele from its A^YA heterozygous mother at left (P♀), but its coat is not yellow. Why doesn't the dominant yellow color show up?

The reason is that the A^YA mother was fed a diet rich in methyl donors (garlic and beets). In her germ-line cells, the unusually high concentration of methyl groups caused cytosine methylation within a regulatory region specific to the mutant A^Y allele, which silenced its transcription (**Fig. 17.24c**). Because methylation marks are copied during DNA replication (Fig. 17.20), all of the A^Y alleles in the somatic cells of the F_1 mouse were methylated and silenced, explaining the wild-type (gray) appearance of the A^YA progeny. Thus, what a female mouse eats can affect the phenotype of her progeny.

Figure 17.24 shows a clear example of *intergenerational epigenetic inheritance* of an acquired trait, that is, the passage through gametes of an altered gene expression state for a single generation, in the absence of the environmental factor that induced it. But what about the F_2 and the F_3?

Figure 17.24 Intergenerational inheritance of an acquired trait. **(a)** Normal (*AA*) mice appear gray because they have black hairs with a yellow stripe. The hairs of yellow mice (A^YA) are completely yellow. **(b)** The A^Y allele of the yellow female who ate a diet rich in methyl donors was silenced in her gametes and in the somatic cells of her progeny. Methylation of the A^Y allele is erased in the germ line of the F_1 progeny, so the yellow phenotype is expressed in the F_2 (*not shown*). **(c)** The A^Y mutation is caused by insertion of a transposable element; methylation of the transposable element silences the allele.

(b): Courtesy Randy L. Jirtle, Ph.D. Originally published in B. Weinhold, "Color by Soy: Genistein Linked to Epigenetic Effects," *Environ Health Perspect*. 2006 Apr., 114(4): A240. Environews, Science Selections

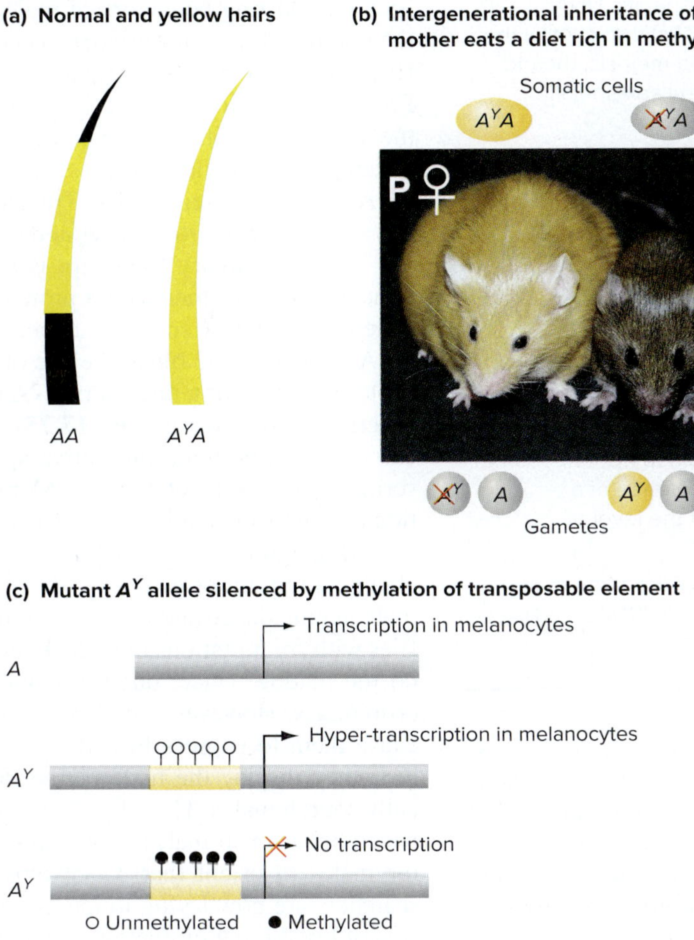

Transgenerational epigenetic inheritance in mammals?

The normal-looking (gray) A^YA F_1 mouse in Fig. 17.24b was fed a normal diet, and its progeny (F_2) that inherited the A^Y allele had yellow coats, as did the subsequent F_3 progeny that inherited the A^Y allele (*not shown*). The A^Y allele was no longer silenced in the F_2 or the F_3, showing that *transgenerational epigenetic inheritance* of the acquired trait did not occur. In other words, environmentally altered gene expression was not inherited stably through multiple generations in the absence of the environmental factor.

The absence of transgenerational epigenetic inheritance of A^Y silencing should not be surprising. You saw earlier (Fig. 17.22) that in mammalian germ lines, DNA methylation marks are normally removed and sex-specific ones are added. However, observations in *Drosophila* and the worm *C. elegans* indicate that transgenerational inheritance of environmentally influenced traits does normally occur in these organisms. These mechanisms involve small RNA modulators of gene expression called *piRNAs* that will be described later in this chapter. If this phenomenon does exist in humans even though it has not yet been detected, we would need to consider that our actions, as well as our genes, might affect the traits we pass on to future generations.

essential concepts

- Certain repressors bind methylated *CpG islands,* blocking transcription activators. Cells maintain this repression through many cellular generations because they copy CpG methylation patterns during DNA replication.
- The expression patterns of about 100 human genes depend on whether they were inherited from the male or female parent. *Paternally imprinted genes* are silenced

when inherited from the father, while *maternally imprinted genes* are silenced when inherited from the mother.

- *Epigenetic phenomena*, such as *imprinting,* are caused by changes in DNA that alter gene expression without changing base-pair sequence and that are heritable during cell division.
- Genomic imprinting results from sex-specific DNA methylation of *cis*-acting elements (ICRs) that control the expression of particular genes. During meiosis, the old methylation marks are erased and new sex-specific methylation patterns are established.

17.4 Regulation After Transcription

learning objectives

1. Explain how the primary transcript of a single eukaryotic gene can produce different proteins.
2. Describe results that could be obtained from ribosome profiling that would indicate the existence of a regulatory mechanism operating at the level of translational initiation.
3. Contrast the origins and functions of the three main categories of small regulatory RNAs (miRNAs, siRNAs, and piRNAs).

Gene regulation can take place at any point in the process of gene expression. Thus far we have discussed mechanisms that influence the frequency of transcription initiation, but many other systems exist that regulate posttranscriptional events. These include the splicing, stability, and localization of mRNAs; the translation of these mRNAs into proteins; and the stability, localization, and modifications of the protein products of these mRNAs. It is impossible to discuss all of these mechanisms in a single chapter, so we focus here on a few of the key decision points.

Sequence-Specific RNA Binding Proteins Can Regulate RNA Splicing

The genomes of eukaryotic cells have many fewer genes than the number of different proteins expressed in those cells. One of the ways cells can generate more than one type of protein from a single gene is through **alternative splicing:** that is, the splicing of primary transcripts into distinct mRNAs that produce different proteins (review Fig. 8.17).

The spliceosomes that assemble at the splice junction sites of primary transcripts can contain more than 100 proteins. The spliceosome proteins carry out different functions, including RNA cleavage and ligation to join exons together. Some of the spliceosome components are crucial for determining which exons become spliced together. These include spliceosome proteins that recognize specific RNA sequences in the primary transcript to either facilitate or prevent the use of particular splice junction sequences, so they are crucial for gene regulation through alternative splicing.

We mentioned at the beginning of this chapter that sex-specific courting behaviors of male *Drosophila* are under the control of the *fruitless* (*fru*) gene. In their brains, male flies produce a male-specific form of the *fru* gene product, Fru-M, a zinc-finger transcription factor. The synthesis of the Fru-M protein only in males requires male-specific splicing that depends on the absence of a female-specific RNA-binding protein called Transformer (Tra). (We will discuss later why Tra is generated only in females.)

The *fru* primary transcript is made in both sexes. In females, Tra (together with a protein present in both sexes called Tra2) binds specific sequences in the *fru* primary RNA; Tra and Tra2 block the use of a particular splice acceptor site, resulting in a *fru* mRNA that produces a female specific Fru-F protein (**Fig. 17.25**). In males, whose cells carry no Tra protein, alternative splicing of the *fru* transcript generates a related Fru-M protein with 101 additional amino acids at its N terminus (Fig. 17.25).

Although Fru-F appears not to have a function, Fru-M elicits a program of gene expression that controls both the male mating dance and its orientation toward females. Male flies with *fru* mutations that block production of Fru-M still do the mating dance, but court males and females indiscriminately. However, female flies with *fru* mutations that cause them to express Fru-M acquire male sexual behaviors; they display the male dating dance and also specifically court females. Thus, Fru-M is redundant for the mating dance behavior in males, and Fru-M is required absolutely for males to orient that dance only toward females. Researchers are now trying to identify the transcriptional targets of Fru-M that ultimately dictate these behaviors.

Figure 17.25 Sex-specific splicing of the primary *fru* transcript. Splicing of *fru* RNA in the absence of Tra protein (in males) produces an mRNA that is translated into Fru-M protein. In females, Tra protein (with Tra2) blocks the use of one exon, causing the *fru* transcript to be spliced so as to encode Fru-F.

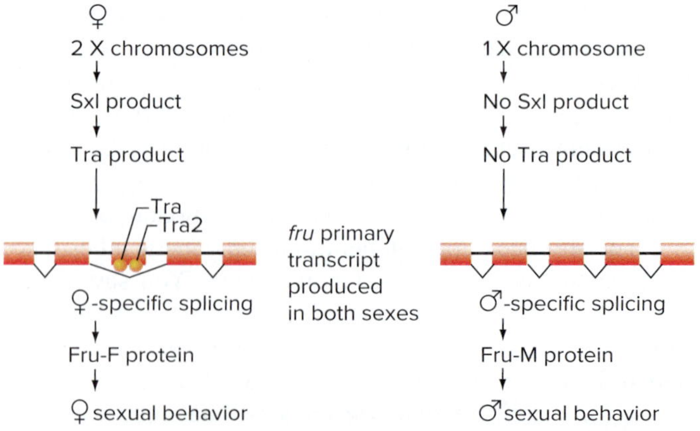

Several Mechanisms Regulate mRNA Translation

Control of translation often happens at the initiation of this process. You will remember from Fig. 8.25 that in eukaryotes, translation begins when the small subunit of the ribosome binds the cap at the 5′ end of the mRNA. The ribosome then scans the mRNA in the 5′-to-3′ direction to find the first AUG, which serves as the initiation codon that specifies Met at the N-terminus of the protein product.

The small subunit of the ribosome does not see the 5′ cap alone, but instead it recognizes a more complicated structure built around the cap (**Fig. 17.26**). A complex of three initiation factors called eIF4A, eIF4E, and eIF4G binds to the 5′ cap. The eIF4G protein in this complex then interacts with *poly-A binding protein* (*PABP*) at the poly-A tail. As a result, the mRNA circularizes, and it is this circularized structure with complexed initiation factors that the small ribosomal subunit recognizes to initiate translation.

In this section we first describe two of the eukaryotic translational regulation mechanisms that control assembly of initiation factors at the 5′ cap. We then discuss an interesting mode of translation regulation that is a property of mRNAs that have so-called *decoy AUGs* upstream of the actual AUG. The presence of these decoy AUGs causes the ribosome to initiate the translation of small peptides instead of the normal polypeptide product of the mRNA.

Regulating assembly of the translation initiation complex

The fact the ribosome recognizes the 5′ cap of the mRNA in the context of the complex shown in Fig. 17.26 enables the regulation of translation through the control of complex assembly.

Regulating mRNA translation in response to nutrients An important pathway that allows eukaryotic cells to respond appropriately to extracellular stimuli depends on a protein called *eIF4E binding protein 1* (*4E-BP1*). As its name implies, this protein can bind to the initiation factor eIF4E, and this binding blocks the assembly of the remainder of the initiation complex on the 5′ cap (**Fig. 17.27**).

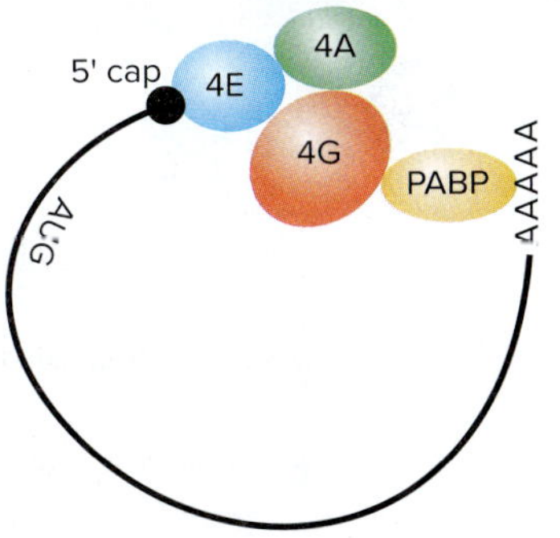

Figure 17.26 Eukaryotic translation initiation complex. The small ribosomal subunit recognizes the mRNA 5′ cap (*black circle*) in the context of an RNA/protein complex. In the complex are eIF4A (*4A*), eIF4E (*4E*), eIF4G (*4G*), and poly-A binding protein (*PABP*).

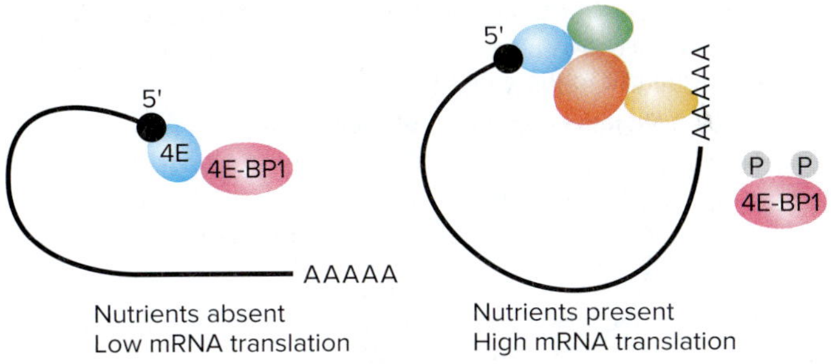

Figure 17.27 Regulation of translation by 4E-BP1. *Left:* Unphosphorylated 4E-BP1 (*pink*) prevents ribosome binding to the mRNA 5′ cap (*black circle*) by binding to eIF4E (*4E*). *Right:* When phosphorylated (*P*), 4E-BP1 no longer binds eIF4E, and the translation initiation complex can assemble.

The presence of nutrients and growth factors in the extracellular environment activates kinase enzymes that add phosphate groups to 4E-BP1. When 4E-BP1 becomes phosphorylated, it can no longer bind to eIF-4E, so the initiation complex can assemble at the 5′ caps of mRNAs (Fig. 17.27). When cells need to grow and make large amounts of many proteins, this mechanism ensures that the global frequency of mRNA translation increases significantly.

Circadian control of mRNA translation The circularization of mRNAs through the association of initiation factors at the 5′ cap and PABP at the 3′ poly-A tail is the physical basis of a mechanism for regulating translational initiation indirectly through control of poly-A tail length. Longer tails attract PABP more efficiently than shorter tails do, and thus the longer the poly-A tail, the more translation. Sequence-specific RNA-binding proteins bind sites on the mRNA and recruit enzymes that can add As to the tail or remove them (**Fig. 17.28**).

Intriguingly, this mechanism is partially responsible for *circadian oscillations* in the amounts of certain proteins at different times of day. Although the amounts of mRNA for these proteins remain constant, the size of the poly-A tails and thus the efficiency of mRNA translation and the amounts of these proteins peak late at night and early in the morning.

Upstream ORFs

Certain mRNAs have one or more *upstream open reading frames* (*uORFs*) that begin with *decoy AUGs* and encode small peptides that have no function. If ribosomes translate a uORF, translation of the major ORF in the mRNA is inhibited. Thus, the choice between translating a uORF and the main ORF is a potential point at which the gene expression can be regulated. Because uORFs are found in about half of all transcripts in human cells, this mechanism for blocking translation is likely to be widespread.

Figure 17.28 Translational control through poly-A tail length. Longer poly-A tails bind PABP more efficiently, and thus the translation initiation complex forms more efficiently (see Fig. 17.26). **(a)** Translation is negatively regulated through recruitment of a deadenylase enzyme (*purple*) by a sequence-specific RNA-binding protein (*gray oval*). **(b)** Translation is positively regulated when a sequence-specific RNA binding protein (*gray rectangle*) recruits poly-A polymerase to the mRNA.

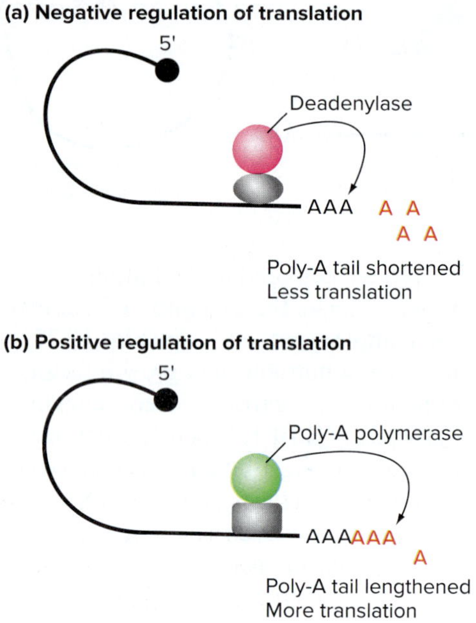

An example of translational regulation by uORFs is seen in the *Drosophila* sex differentiation pathway that will be explained in detail at the end of this chapter. For the time being, you only need to know that a protein called Sex lethal (Sxl) is expressed only in XX flies, and it is needed for the development of XX flies as females. On the other hand, if females make a different, normally male-specific protein called Msl-2, they will die. Sxl protein ensures that Msl-2 protein is not synthesized in females through several mechanisms; in one of these, Sxl protein blocks the *msl-2* mRNA from being translated (**Fig. 17.29**). Sxl protein binds to the *msl-2* mRNA at a specific binding site, and this binding promotes the translation of a uORF, preventing translation of the ORF encoding the Msl-2 protein and thus allowing these XX animals to survive.

Figure 17.29 Translational regulation by a decoy ORF. To prevent translation of *msl-2* mRNA in *Drosophila* females, Sxl protein (*green*) binds to the mRNA and causes the ribosome (*blue*) to translate the uORF. Sxl protein thus prevents translation of the main ORF that encodes Msl-2 protein.

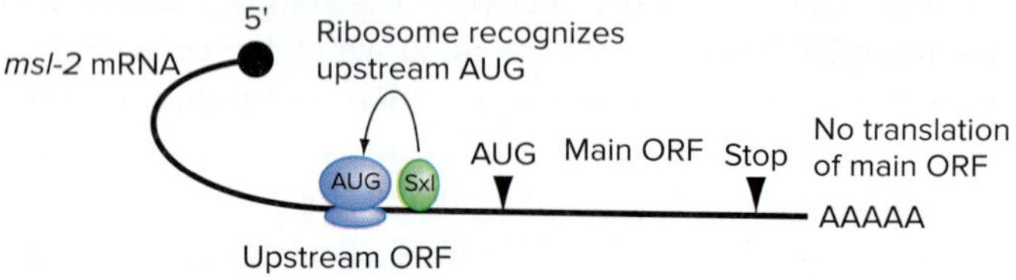

Ribosome Profiling Measures Translation Efficiency

Because gene expression can be regulated at the level of translation, the amount of a gene's protein product does not always correlate directly with the amount of mRNA. Scientists thus require methods that allow them to determine how efficiently mRNAs are being translated. **Ribosome profiling** is a new technology that allows researchers to observe the positions of ribosomes on all the mRNAs in any given type of cell or tissue.

The first part of the procedure is similar conceptually to DNase footprinting (recall Fig. 16.16), in which DNA-binding sites of a transcription factor protein can be identified because the bound protein protects that DNA from DNase I digestion. Ribosome profiling identifies the locations on mRNAs where ribosomes are bound because the ribosomes protect these RNA sequences from RNase digestion.

The procedure for ribosome profiling is shown in **Fig. 17.30**. After mRNA–ribosome complexes are purified and digested with RNase, protected RNA fragments (~30 nt long) are converted to single-stranded cDNAs in a manner that preserves information about the polarity of the RNA (*not shown*). These cDNAs are then PCR-amplified and subjected to a high-throughput technique that allows the sequencing of hundreds of millions or even billions of randomly chosen cDNA fragments.

Figure 17.30 Ribosome profiling. Purification of mRNA (*orange*)-ribosome (*blue*) complexes and *footprinting* of the RNase-resistant regions on the mRNA provides information about how often different mRNAs are translated and the positions of ribosomes on the mRNAs.

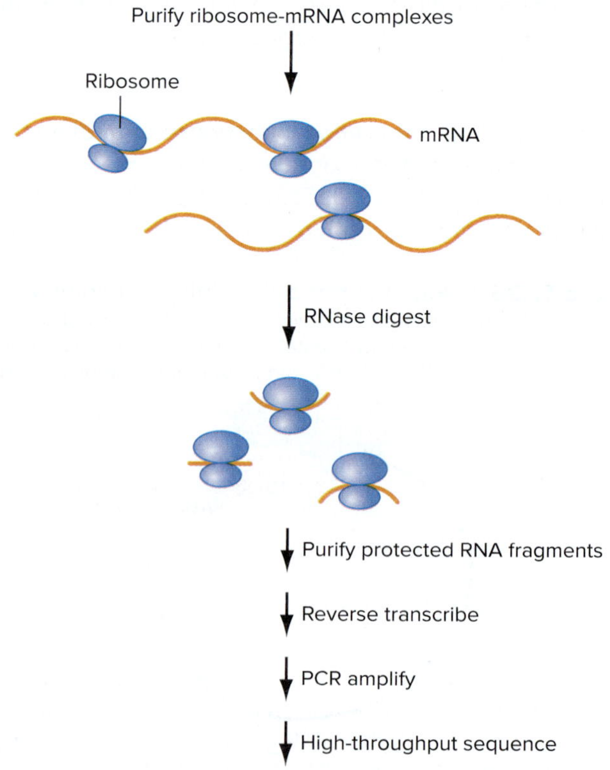

TABLE 17.1 Small RNAs in Eukaryotes

	Targets	Effects
miRNAs (micro-RNAs)	• mRNAs	• Block mRNA translation • Destabilize mRNAs
siRNAs (small interfering RNAs)	• mRNAs • Nascent transcripts of chromosomal regions destined to become heterochromatin	• Block translation/Destabilize mRNAs • Recruit histone-modifying enzymes to DNA, resulting in heterochromatin formation
piRNAs (Piwi-interacting RNAs)	• Transposable element transcripts • Transposable element promoters	• Degradation of transposable element mRNA • Facilitate histone modifications that inhibit transposable element transcription

The number of times sequences of a particular mRNA are obtained reveals how often that mRNA species was being translated at the time the tissue sample was prepared. In addition, the results provide a snapshot of the positions of the ribosomes along any given mRNA, revealing where translation starts, stops, or pauses. When researchers perform ribosome profiling using cells grown under different conditions, and then compare the findings with those obtained from RNA-Seq (which, as we have seen, indicates levels of mRNA transcripts), the analysis can indicate the existence of global regulatory mechanisms that alter translation efficiency.

Small RNAs Regulate mRNA Stability and Translation

In the first five years of the twenty-first century, new types of gene regulators were discovered in the form of small, specialized RNAs that prevent the expression of specific genes through complementary base pairing (**Table 17.1**). Three classes of small regulatory RNAs have now been identified: **micro-RNAs (miRNAs)**, **small interfering RNAs (siRNAs)**, and **Piwi-interacting RNAs (piRNAs)**. Each small RNA class is generated through a distinct pathway, leading to the production of single-stranded RNAs of slightly different lengths but always within the range of 21–30 nucleotides.

To exert their functions, each small RNA class forms ribonucleoprotein complexes with distinct members of the Argonaute/Piwi protein family. The small RNA in each complex serves to guide the complex to particular nucleic acid targets that have perfect or partial complementarity with the small RNA. All three classes of small RNAs regulate gene activity at the posttranscriptional level through the modulation of RNA stability and/or translation; siRNAs and piRNAs also act at the transcriptional level by affecting chromatin structure.

miRNAs

In animals, one of the most abundant classes of small RNAs is composed of the micro-RNAs (miRNAs). As will be seen shortly, miRNAs are usually negative regulators of target mRNAs, resulting in the destruction of these mRNAs or prevention of their translation.

The human genome has close to 1000 genes encoding miRNAs. These genes are transcribed by RNA polymerase II into long primary transcripts called *pri-miRNAs* that contain one or more miRNA sequences in the form of mostly double-stranded stem loops (**Fig. 17.31**). The pri-miRNAs undergo processing to form the active miRNAs, which are short and single-stranded. **Figure 17.32** diagrams this multistep process, which is aided by two ribonuclease enzymes called *Drosha* and *Dicer.* During the process, the miRNA sequences are transported out of the nucleus (where they were transcribed) into the cytoplasm (where they will act). Furthermore, the miRNAs become incorporated into ribonucleoprotein complexes called *miRNA-induced silencing complexes (miRISCs);* each miRISC contains a particular member of the Argonaute protein family.

The ribonucleoprotein complexes (miRISCs) containing miRNAs mediate diverse functions depending on the particular Argonaute protein they possess, and on the extent of sequence complementarity between the miRNA in the complex (called the *guide*) and the target sequences in mRNA 3′ UTRs. A miRISC whose guide miRNA has perfect complementarity with the target RNA causes mRNA cleavage (**Fig. 17.33a**). With less complementarity, the mechanism is usually inhibition of translation (**Fig. 17.33b**), although exactly how miRISCs regulate translational activity is not yet understood.

Figure 17.31 Micro-RNA-containing genes. Primary miRNA transcripts (pri-miRNAs) can contain one or several miRNAs. Some of these primary transcripts do not encode proteins **(a),** but in other cases miRNAs can be processed from the introns of protein-coding transcripts **(b).**

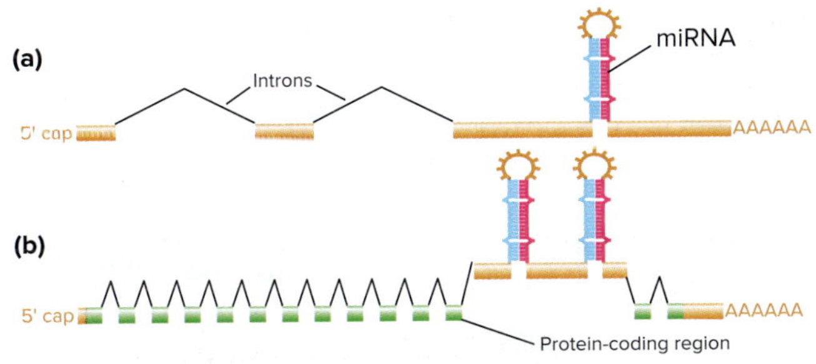

Figure 17.32 miRNA processing. Immediately after transcription, pri-miRNAs are recognized by the nuclear enzyme Drosha, which crops out pre-miRNA stem-loop structures from the larger RNA. The pre-miRNAs undergo active transport from the nucleus into the cytoplasm, where the enzyme Dicer recognizes them. Dicer reduces the pre-miRNA into a short-lived miRNA*:miRNA duplex, which is released and picked up by a RISC. The RISC becomes a functional and highly specific miRISC by eliminating the *blue* miRNA* strand that is partially complementary to the *red* miRNA that will serve as the guide in the miRISC.

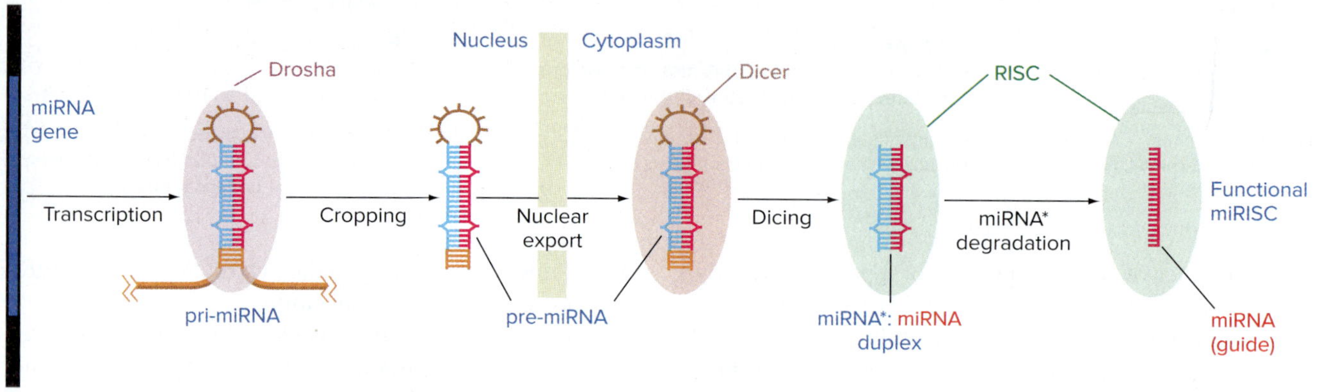

Because exact complementarity between guide and target RNAs is not required, each type of miRNA ultimately can control several different mRNAs (about 10 on average). As a result, scientists estimate that about half of all human genes are controlled by miRNAs. Moreover, each miRNA gene is transcribed according to its own temporal and spatial pattern during the development of a multicellular organism, so any single miRNA can influence gene expression in different ways in different tissues.

siRNAs

The small-interfering RNA (siRNA) pathway has many similarities with that just described for miRNAs. A key difference is the source of the small RNA. Instead of resulting from the processing of a long, single-stranded transcript, as was the case with miRNAs, siRNAs result from the processing (also by Dicer) of double-stranded RNAs (dsRNAs). These dsRNAs are produced originally by transcription of both strands either of an endogenous DNA sequence in the genome or an exogenous source such as a virus. Processing of the dsRNAs produces single-stranded RNAs that form ribonucleoprotein complexes with Argonaute proteins. Using the single-stranded RNA as a guide, these complexes can interfere with the expression of a gene containing the complementary sequence by mechanisms previously shown for miRNA-containing complexes in Fig. 17.33. The siRNA pathway may also protect the cell from invading viruses by destroying viral mRNAs.

Researchers have exploited the siRNA pathway to selectively shut off the expression of specific genes in order to evaluate their function. The idea is to introduce dsRNA

Figure 17.33 How miRISCs interfere with gene expression. The miRISC can down-regulate target genes in two different ways. **(a)** If the miRNA and its target mRNA contain perfectly complementary sequences, miRISC cleaves the mRNA. The two cleavage products are no longer protected from RNase and are rapidly degraded. **(b)** If the miRNA and its target mRNA have only partial complementarity, translation of the mRNA is inhibited by an unknown mechanism.

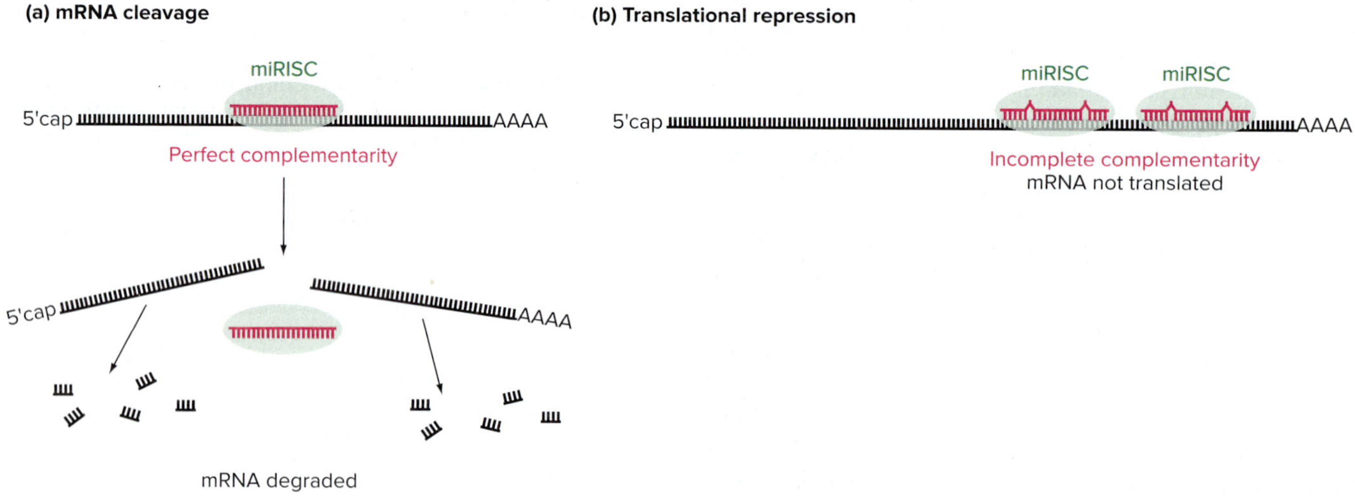

corresponding to a particular gene into the cell or organism to shut off or knock down the expression of the endogenous gene in the genome; this technique is called **RNA interference.** The processing pathway for siRNA will convert the double-stranded RNA into a single-stranded siRNA. Then, within the context of an Argonaute-containing complex, the siRNA will hybridize with the complementary mRNA transcript for the gene and mediate the destruction of that mRNA. In this way, scientists can turn down the expression of any gene of interest and investigate any possible phenotypic consequences of this loss of function. In Chapter 19, we will explore the use of RNA interference to investigate many biological processes.

Another important role of siRNAs is in the formation of heterochromatin. Recall from Chapter 12 that heterochromatin formation involves modifications of histone tails that facilitate binding of a protein called HP1 (review Fig. 12.14). A chromosomal region destined to become heterochromatin is first transcribed bidirectionally, and the resulting long double-stranded RNA is processed by Dicer into small double-stranded RNAs. A complex similar to RISC called the *RNA-induced transcriptional silencing* (RITS) complex incorporates one strand of these duplexes and uses this siRNA as a guide to bind its complement in a nascent transcript being transcribed from the DNA destined to become heterochromatin. The RITS complex brings histone-modifying enyzmes to the DNA; the result is HP1 binding, closed chromatin, and the inactivation of transcription.

piRNAs

You will recall from previous chapters that the genomes of eukaryotic organisms contain many transposable elements (TEs) that propagate themselves by mobilization and transposition. The organisms harboring these TEs must limit TE movement to prevent their genomes from being destroyed by rapid mutation and rearrangement. One important mechanism by which organisms can minimize TE mobilization is through the action of small Piwi-interacting RNAs (piRNAs). These piRNAs block both the transcription of TEs in the genome and the translation of the TE mRNAs that do get transcribed. Without the synthesis of enzymes like transposase (for DNA transposons) or reverse transcriptase (for retrotransposons), the TEs cannot move.

piRNAs are generated by cleavage of long RNAs transcribed from piRNA gene clusters located throughout the genome, each of which encodes between 10 and 1000 piRNAs. After processing, piRNAs are loaded onto complexes containing Piwi proteins (one subfamily of Argonaute proteins), and the piRNA guides the Piwi complexes mainly to TE DNA or TE transcripts. Piwi complexes at TE DNA facilitate histone modifications that interfere with TE transcription, while Piwi complexes bound to TE transcripts degrade the TE RNAs.

Many details of the piRNA pathway remain to be worked out. Intriguingly, recent observations in organisms such as *Drosophila* and the worm *C. elegans* indicate that piRNAs mediate certain specialized phenomena in which phenotype is altered in response to a change in the environment, and this change can be inherited transgenerationally even after the environment returns to its initial conditions. For example, if a new environmental condition stimulates the synthesis of a piRNA that silences a target gene, this gene can be silenced in descendants of the organism that have been maintained for many generations in the absence of the initial stimulus. One possible reason is that the mechanism of piRNA biogenesis is somehow self-perpetuating. As you might imagine, this fascinating phenomenon has made the piRNA pathway a focus of intensive investigation in many laboratories.

essential concepts

- In eukaryotes, *alternative splicing* can produce different proteins from a single transcript. Sequence-specific RNA-binding proteins can inhibit or promote the use of particular splice-junction sequences.
- Because mechanisms exist to regulate mRNA translation, levels of protein synthesis measured by *ribosome profiling* do not always correlate precisely with mRNA levels measured by RNA-seq.
- Three classes of small RNAs regulate mRNA stability, translation, or transcription through complementary base pairing: *miRNAs*, *siRNAs*, and *piRNAs*. These small RNAs act as guides to bring protein complexes to particular target mRNAs (leading to mRNA destruction or preventing translation) or to DNA sequences near promoters (blocking transcription or promoting heterochromatization).

17.5 A Comprehensive Example: Sex Determination in *Drosophila*

learning objectives

1. Explain how the *Sxl* promoter "counts" the number of X chromosomes in *Drosophila*.
2. Describe the cascade of RNA splicing events initiated by the Sxl protein that results in female morphology and behavior.
3. Discuss the role of transcriptional regulation in *Drosophila* sex determination.

Male and female *Drosophila* exhibit many sex-specific differences in morphology, biochemistry, behavior, and function of the germ line (**Fig. 17.34**). Through decades of work, researchers concluded that in *Drosophila*, it is the number of X chromosomes, not the presence of the Y, that

Figure 17.34 Sex-specific traits in *Drosophila*. Objects or traits shown in *blue* are specific to males. Objects or traits shown in *red* are specific to females. Objects or traits shown in *green* are found in different forms in the two sexes.

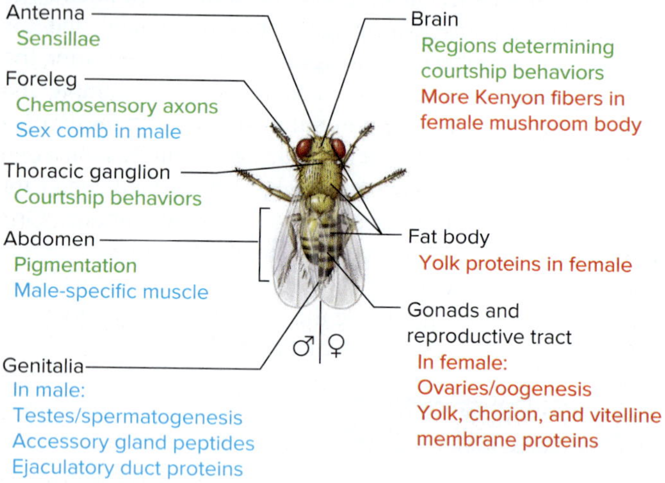

determines sex, and that sex determination first occurs through transcriptional regulation of the *Sxl* gene. Transcription of *Sxl* in XX (and not XY) animals initiates a cascade of events that influences sex through three independent pathways: One determines whether the flies look and act like males or females; another determines whether germ cells develop as eggs or sperm; and a third produces dosage compensation by doubling the frequency of transcription of X-linked genes in males.

We focus here mainly on the first-mentioned pathway: the determination of somatic sexual characteristics. An understanding of this pathway emerged from analyses of mutations that affect particular sexual characteristics in one sex or the other. For example, as we saw at the beginning of the chapter, XY flies carrying mutations in the *fruitless* gene (*fru*) exhibit aberrant male courtship behavior, whereas XX flies with the same *fru* mutations appear to behave as normal females.

Table 17.2 shows that mutations in other genes also affect the two sexes differently. Clarification of how these mutations influence somatic sex determination came from a combination of genetic experiments (studying, for example, whether one mutation is epistatic to another) and molecular biology experiments (in which investigators cloned and analyzed mutant and normal genes). Through such studies, *Drosophila* geneticists dissected various stages of sex determination to delineate the following complex regulatory network.

TABLE 17.2 *Drosophila* Mutations that Affect the Two Sexes Differently

Mutation	Phenotype of XY	Phenotype of XX
Sex lethal (*Sxl*)	Male	Dead
transformer (*tra*)	Male	Male (sterile)
doublesex (*dsx*)	Intersex	Intersex
intersex (*ix*)	Male	Intersex
fruitless (*fru*)	Male with aberrant courtship behavior	Female

All mutant alleles are loss-of-function and recessive to wild type.

The Number of X Chromosomes Determines Sex in *Drosophila*

In Chapter 4, you saw that in both humans and flies, XY animals are male and XX are female. However, the underlying molecular mechanism of sex determination is different in humans and flies. In humans the key to maleness is the presence of the *SRY* gene on the Y chromosome; femaleness is the default state in the absence of *SRY*. In flies, maleness is the default state brought about by the presence of only one X chromosome instead of two as in females. The reason is that two X chromosomes are required to activate transcription of the *Sxl* gene in early *Drosophila* embryogenesis.

Counting of X chromosomes by the *Sxl* promoter

In early embryogenesis (before sex determination and dosage compensation have taken place), XX cells transcribe *Sxl* from the *establishment promoter* (P_e). Transcription from P_e depends on four transcriptional activator proteins: Scute, Runt, SisA, and Upd (**Fig. 17.35a**). Because the genes for these activators are on the X chromosome, XX embryos have twice as much of these four activators as XY embryos. Only in cells with two X chromosomes is the concentration of activators sufficient for *Sxl* transcription to occur.

The action of the Sxl protein in females

Sxl is an RNA-binding protein that controls the alternative splicing of specific RNA targets, including its own RNA (**Fig. 17.35b**). As embryogenesis progresses, the transcription factors that activate *Sxl* transcription from P_e disappear, and *Sxl* is transcribed instead from the *maintenance promoter* (P_m). In males, splicing of the primary *Sxl* transcript produced from the maintenance promoter generates an RNA that includes an exon (exon 3) containing a stop codon in its reading frame. As a result, this RNA in males is not productive—it does not generate any functional Sxl protein.

In females, however, the Sxl protein previously produced by transcription from the establishment promoter P_e influences the splicing of the primary transcript initiated at the maintenance promoter P_m. When the earlier-made Sxl protein binds to the later-transcribed RNA, this binding alters splicing so that exon 3 is no longer part of the final mRNA. Without exon 3, the mRNA can be translated to make more Sxl protein. Thus, a small amount of Sxl

Figure 17.35 ***Sxl*** **expression only in XX** ***Drosophila.*** **(a)** In the early female—but not the male—embryo, transcriptional activators encoded by X-linked genes are present in concentrations sufficient to initiate transcription from the P_e promoter. This mRNA, whose first two exons are *E1* and *4,* encodes Sxl protein. **(b)** Later in development, transcriptional activators produced equally in XX and XY animals activate *Sxl* transcription from the P_m promoter. *L1, 2, 3,* and *4* denote the first few exons. When Sxl protein is present (in females), it binds the *Sxl* primary transcript to make a spliced mRNA that can be translated into more Sxl protein. The result is a feedback loop that maintains Sxl protein in females but not in males.

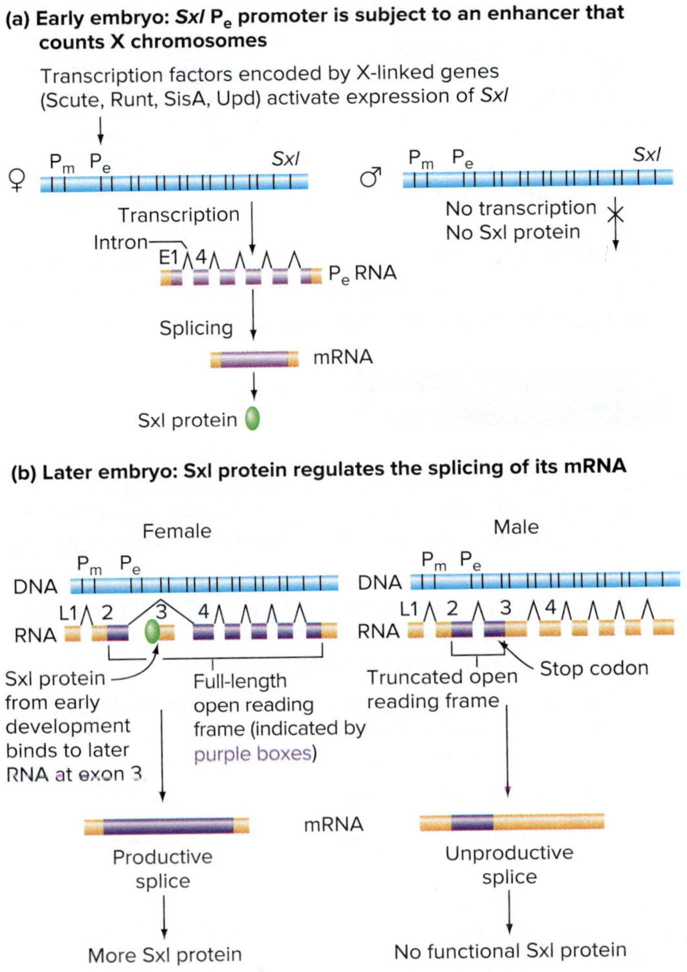

protein synthesized very early in development establishes a positive feedback loop that ensures more synthesis of Sxl protein later in female development.

The effects of *Sxl* mutations

Recessive *Sxl* mutations that produce nonfunctional gene products have no effect in XY males, but they are lethal in homozygous mutant XX females (see Table 17.2). The reason is that males, which do not normally express the *Sxl* gene, do not miss its functional product, but females, which depend on the Sxl protein for sex determination, do.

The lethality of females with loss-of-function mutations in *Sxl* is due to aberrant expression of certain dosage-compensation genes that normally increase (specifically in males) transcription of the genes on the X chromosome. Fig. 17.29 previously showed an example: Sxl protein in normal females prevents the translation of the mRNA for one of these dosage-compensation genes, *msl-2.* But females with mutations in *Sxl* incorrectly make Msl-2 protein (as well as other dosage-compensation factors), and this causes each X-linked gene to be transcribed at twice the frequency at which it is transcribed in normal females. Because females have two X chromosomes rather than the one X in males, hypertranscription of the genes from the two X chromosomes proves lethal.

Sxl Protein Triggers a Cascade of Splicing

The Sxl protein influences the splicing of RNAs transcribed not only from its own gene, but also from other genes. Among these is the *transformer* (*tra*) gene. In the presence of the Sxl protein (as in normal females), the *tra* primary transcript undergoes productive splicing that produces an mRNA translatable to a functional protein. In the absence of Sxl protein (as in normal males), the splicing of the *tra* transcript results in a nonfunctional protein (**Fig. 17.36a**).

The cascade continues. You saw in Fig. 17.25 that Tra protein (and the non-sex-specific Tra2) control the splicing of *fru* primary RNA so that the transcription factor Fru-M is produced only in males; Fru-M controls male sexual behaviors. The Tra and Tra2 proteins also influence the splicing of the *doublesex* (*dsx*) gene's primary transcript.

Figure 17.36 A cascade of alternate splicing. (a) Sxl protein alters the splicing of *tra* RNA; female transcripts produce functional Tra protein, while male transcripts cannot. **(b)** Tra protein, in turn, alters the splicing pattern of *dsx* RNA; a different Dsx product results in males (Dsx-M) and in females (Dsx-F).

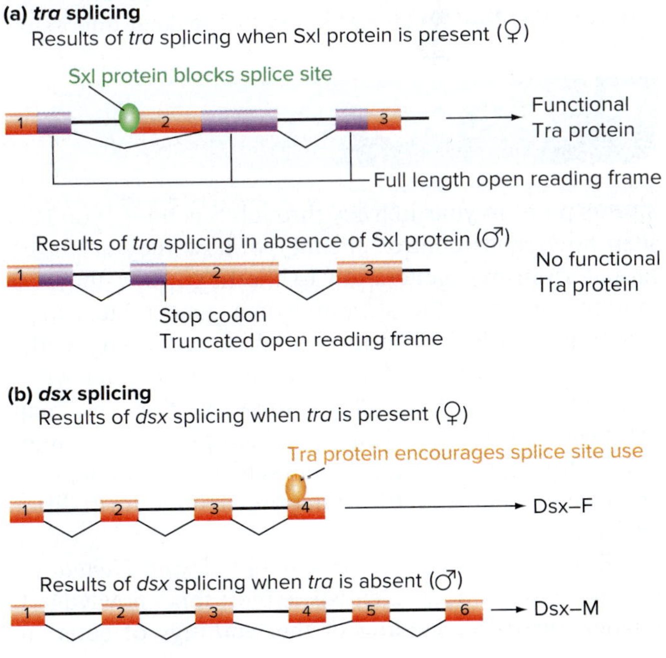

This splicing pathway results in the production of a female-specific Dsx protein called Dsx-F. In males, where there is no Tra protein, the splicing of the *dsx* primary transcript produces the related, but different, Dsx-M protein (**Fig. 17.36b**). The N-terminal parts of the Dsx-F and Dsx-M proteins are the same, but the C-terminal parts of the proteins are different.

Dsx-F and Dsx-M Proteins Control Development of Somatic Sexual Characteristics

Although both Dsx-F and Dsx-M function as transcription factors, they have opposite effects. In conjunction with the protein encoded by the *intersex* (*ix*) gene, Dsx-F primarily represses the transcription of genes whose expression would generate the somatic sexual characteristics of males. However, it also activates the transcription of genes that promote somatic femaleness. Dsx-M, which works independently of the Intersex protein, does the opposite; it is primarily a transcriptional activator of maleness genes, and it also represses femaleness genes.

Interestingly, the two Dsx proteins can bind to the same enhancer elements, but their binding produces opposite outcomes (**Fig. 17.37**). For example, both bind to an enhancer upstream of the promoter for the *yp1* gene, which encodes a yolk protein; females make this protein in their fat body organs and then transfer it to developing eggs. The binding of Dsx-F stimulates transcription of the *yp1* gene in females; the binding of Dsx-M to the same enhancer helps inactivate *yp1* transcription in males.

Mutations in *dsx* affect both sexes because in both males and females, the production of Dsx proteins represses certain genes specific to the development of the opposite sex. Null mutations in *dsx* that make it impossible to produce either functional Dsx-F or Dsx-M result in intersexual individuals that cannot repress either certain male-specific or certain female-specific genes (Table 17.2).

Figure 17.37 Male- and female-specific forms of Dsx protein. At the *yp1* gene enhancer, Dsx-F acts as a transcriptional activator, whereas Dsx-M acts as a transcriptional repressor.

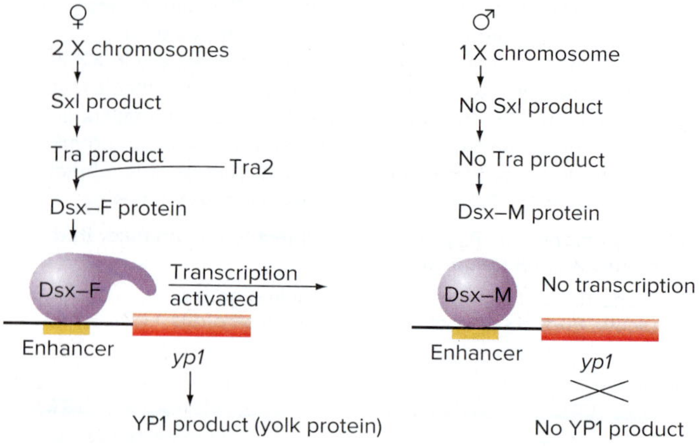

essential concepts

- The *Sxl* gene is the master regulator of sex determination in *Drosophila*. Early transcription of *Sxl* depends on activator proteins encoded by X-linked genes; only in XX cells is the concentration of activators high enough for *Sxl* transcription.
- In females, the Sxl protein initiates a splicing factor cascade that culminates in the synthesis of a *dsx* mRNA that encodes Dsx-F protein. In males without Sxl protein, alternative splicing results in a *dsx* mRNA that makes *Dsx*-M.
- The Dsx-F and Dsx-M proteins are transcription factors that have opposite effects on the expression of genes whose products influence female- and male-specific morphologies and behaviors.

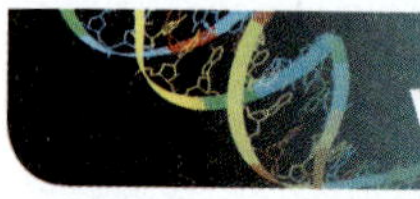

WHAT'S NEXT

At this point in your journey through this book, you have seen how genes control phenotype, how they are transmitted from one generation to the next, how they can mutate, and how the structure of a gene relates to its function. You have also learned about technologies that allow scientists to analyze individual genes and whole genomes at the molecular level. In the previous two chapters, we discussed how genes and their products are regulated, allowing a single-celled organism to respond to its environment and a multicellular organism to form different organs.

In the next section of this book—*Using Genetics*—we will explore how scientists exploit this knowledge to further our understanding of the workings of cells and whole organisms. In Chapter 18, you will see that genes can be inserted into or removed from the genomes of model organisms at will—any gene, indeed any base pair, in the genome can be changed at the whim of a molecular geneticist. Genome manipulation is the basis for gene therapies that hold promise for curing some human diseases. In Chapter 19, we will explore how scientists use the genetics of model organisms to dissect biological pathways. In particular, the analysis of mutants that develop aberrantly from a fertilized egg to a multicellular organism has helped uncover many details of this remarkable process. Finally, in Chapter 20, you will see how new technologies for studying genomes are revolutionizing our understanding and treatment of cancer, the most important of all genetic diseases.

SOLVED PROBLEMS

I. The retinoic acid receptor (RAR) is a transcription factor that is similar to steroid hormone receptors. The substance (*ligand*) that binds to this receptor is retinoic acid. One of the genes whose transcription is activated by retinoic acid binding to the receptor is *myoD*. The diagram that follows shows a schematic view of the RAR proteins produced by genes into which one of two different 12-base double-stranded oligonucleotides had been inserted in the ORF. The insertion site (a–m) associated with each mutant protein is indicated with the appropriate letter on the polypeptide map. For constructs encoding proteins a–e, oligonucleotide 1 (5′ TTAATTAATTAA 3′ read off either strand) was inserted into the *RAR* gene. For constructs encoding proteins f–m, oligonucleotide 2 (5′ CCGGCCGGCCGG 3′) was inserted into the gene.

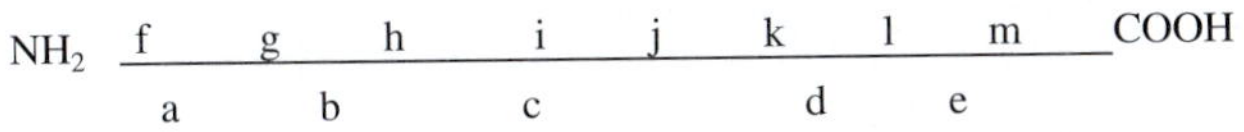

The wild-type RAR protein can both bind DNA and activate transcription weakly in the absence of retinoic acid (RA) and strongly in RA's presence. Each mutant protein was tested for its ability to bind RA and DNA and to activate transcription of the *myoD* gene in the presence and absence of RA. Results are tabulated as follows:

	RA	DNA binding		Transcriptional activation	
Mutant	binding	−RA	+RA	−RA	+RA
a	–	–	–	–	–
b	–	–	–	–	–
c	–	–	–	–	–
d	–	+	+	+	+
e	+++	+	+++	+	+++
f	+++	+	+++	+	+++
g	+++	+	+++	–	–
h	+++	+	+++	–	–
i	+++	–	–	–	–
j	+++	–	–	–	–
k	–	+	+	+	+
l	–	+	+	+	+
m	+++	+	+++	+	+++

a. What is the effect of inserting oligonucleotide 1 anywhere in the ORF?

b. What are the possible effects of inserting oligonucleotide 2 anywhere in the ORF?

c. Indicate the three protein domains of RAR on a copy of the preceding drawing. Note that the three domains are separate—they do not overlap.

Answer

This question involves the concept of domains within proteins and the use of the genetic code to understand the effects of oligonucleotide insertions.

a. Oligonucleotide 1 contains a stop codon in any of its three reading frames. This means oligonucleotide 1 will cause termination of translation of the protein in either orientation, wherever it is inserted.

b. Oligonucleotide 2 does not contain any stop codons and it is 12 nucleotides long. Thus, oligonucleotide 2 will only add amino acids to the protein, and it will not change the reading frame of the protein. Insertion of the oligonucleotide can disrupt the function of a domain in which it inserts, although this will not necessarily be the case.

c. Looking at the data overall, notice that all mutants that are defective in RA binding are also defective in both DNA binding and transcriptional activation. This result makes sense because binding to RA causes the RAR protein to change shape so that its DNA-binding domain is available.

Inserts a, b, and c using oligonucleotide 1, which truncates the protein at the site of insertion, are defective in all three activities, meaning that at least part of the RA-binding domain must be C-terminal to c. DNA binding and transcription activation are both seen in mutant d, so these two activities must lie closer to the N terminus than d, and at least part of the RA-binding domain must to C-terminal to d. As truncation at e has no effect on RA binding, the RA-binding domain must be completely N-terminal to e.

Using oligonucleotide 2 transcriptional activation was disrupted by insertions at sites g and h, indicating that this region is part of the activation domain; i and j insertions disrupted DNA binding, and k and l insertions disrupted RA binding. The minimal endpoints of the domains of the RAR protein as determined from these data are summarized in the following schematic.

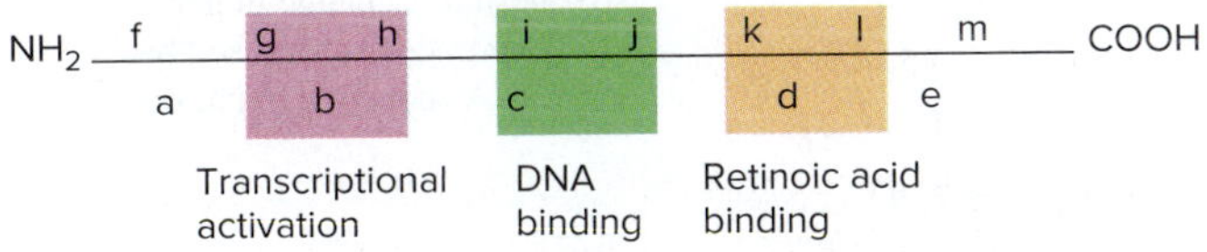

II. Assume that the disease illustrated with the following pedigree is due to the phenotypic manifestation of a rare recessive allele of an autosomal gene that is paternally imprinted. What would you predict is the genotype of individuals (a) I-1, (b) II-1, and (c) III-2?

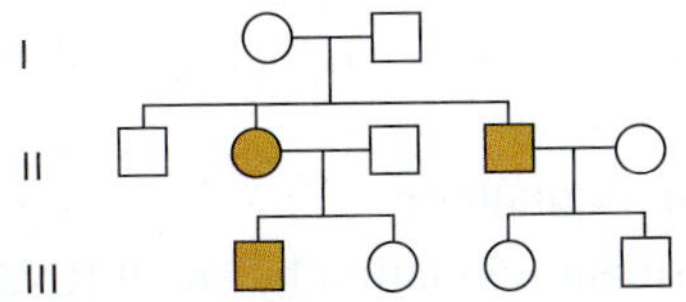

Answer

This question requires you to understand how imprinting influences phenotype, and how imprints are reset in the germ line. Alleles of paternally imprinted genes are expressed only if they are inherited from the mother. Neither parent in generation I displays the disease, but two out of their three children do. At first glance, either parent could have the mutant allele of the gene. Because neither suffers from the disease, each parent in generation I could have received an inactivated mutant allele from their father and a normal, transcriptionally active allele from their mother. However, further consideration shows that it cannot be the male parent (I-2) who is the source of the mutant allele. He can provide only transcriptionally inactive alleles to generation II. The two generation II children showing the disease phenotype must have received the mutant allele from their mother, because hers is the only allele that they express. In the answers that follow, *A* is the normal allele and *a* is the disease allele.

a. The genotype of I-1 is *Aa*. The a allele is inactive and was inherited from her father.

b. The genotype of II-1 is *AA*. He must have inherited the normal allele from his mother, as this is the only allele he expresses. We assume that his father is *AA* because disease alleles are rare, and no data in the pedigree forces us to conclude that II-1 or his father is other than *AA*.

c. The genotype of III-2 is *AA*. She must have inherited *A* from her mother (who is *Aa*) because this is the only allele III-2 expresses and she is unaffected. Because disease alleles are rare, the most likely assumption is that III-2's father is *AA*.

PROBLEMS

Vocabulary

1. For each of the terms in the left column, choose the best matching phrase in the right column.

a. basal factors	1. organizes enhancer/promoter interactions
b. repressors	2. pattern of expression depends on which parent transmitted the allele
c. CpG	3. activates gene transcription temporal- and tissue-specifically
d. imprinting	4. site of DNA methylation
e. miRNA	5. identifies DNA-binding sites of transcription factors
f. coactivators	6. bind to enhancers
g. epigenetic effect	7. bind to promoters
h. insulator	8. bind to activators
i. enhancer	9. prevents or reduces gene expression posttranscriptionally
j. ChIP-Seq	10. heritable change in gene expression not caused by DNA sequence mutation

Section 17.1

2. For each of the following types of gene regulation, indicate whether it occurs in eukaryotes only, in prokaryotes only, or in both prokaryotes and eukaryotes.
   a. differential splicing
   b. positive regulation
   c. chromatin compaction
   d. attenuation of transcription through translation of the RNA leader
   e. negative regulation
   f. translational regulation by small RNAs

3. List five events other than transcription initiation that can affect the type or amount of active protein produced in a eukaryotic cell.

Section 17.2

4. Which eukaryotic RNA polymerase (RNA pol I, pol II, or pol III) transcribes which genes?
   a. tRNAs
   b. mRNAs
   c. rRNAs
   d. miRNAs

5. As shown in the following diagram, a single nucleotide difference in a hair follicle enhancer of a human gene called *KITLG* contributes to the trait of hair color. People with an A–T base pair in the enhancer tend to have dark hair, while people with a G–C base pair at the same position tend to have blond hair. The base pair difference affects the level of *KITLG* transcription: The blond-associated allele is transcribed only 80% as frequently as the dark hair-associated allele. Explain how a single base pair difference in an enhancer sequence can have this effect.

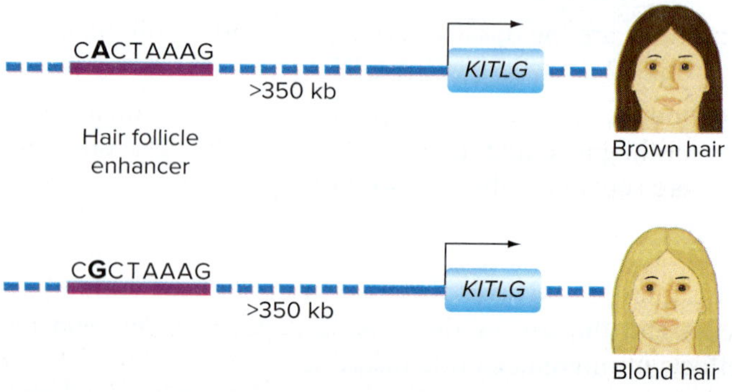

6. You have synthesized an *enhancerless* GFP reporter gene in which the jellyfish *GFP* cDNA is placed downstream of a basal promoter that functions in mice. You will now fuse this enhancerless reporter to the three types of sequences listed below (x–z).

   a. Which of the three types of sequences would you use for which of the three listed purposes (i–iii)? In each case, explain how the particular fusion would address the particular use.

      Types of sequences fused to the reporter:

      x. random mouse genome sequences

      y. known mouse kidney-specific enhancer

      z. fragments of genomic DNA surrounding the transcribed part of a mouse gene

      Uses:

      i. to identify a gene's enhancer(s)

      ii. to express GFP tissue-specifically

      iii. to identify genes expressed in neurons

   b. Which of the sequences (x–z) would you fuse to a particular mouse gene of interest in order to express the protein product of the gene *ectopically*, that is, in a tissue in which the gene is not usually expressed? Why might you want to do this experiment in the first place?

7. You isolated a gene expressed in differentiated neurons in mice. You then fused various fragments of the gene (shown as dark lines in the following figure) to a *GFP* reporter lacking either enhancers or a promoter. The resulting clones were introduced into neurons in tissue culture, and the level of GFP expression was monitored by looking for green fluorescence. From the results that follow, which region contains the promoter and which contains a neuronal enhancer?

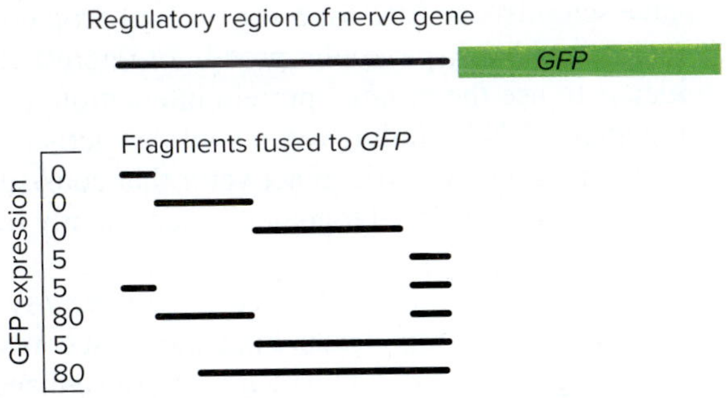

8. Yeast genes have *cis*-acting elements upstream of their promoters, similar to enhancers, called *upstream activating sequences* or *UASs*. Several target genes involved in galactose utilization are regulated by one type of UAS called UAS_G, which has four binding sites for an activator called GAL4. Two target genes regulated by UAS_G are *GAL7* and *GAL10*. The GAL80 protein is an indirect repressor of *GAL7* and *GAL10* transcription: At UAS_G, GAL80 binds to GAL4 protein and blocks GAL4's activation domain. In the presence of galactose, GAL80 no longer binds GAL4.

   In which gene(s) (*GAL4* and/or *GAL80*) should you be able to isolate mutations that allow the constitutive expression of the target genes *GAL7* and *GAL10* in the absence of galactose? In each case, what characteristics of the protein would the mutation disrupt?

9. A single UAS_G regulates the expression of three genes, all of which are adjacent: *GAL7* and *GAL10* as described in Problem 8, and also *GAL1*.

   a. Would you expect these genes to be transcribed into individual transcripts, or to be cotranscribed as one mRNA? Explain.

   b. How could you determine experimentally whether each gene is transcribed separately or instead that the three are cotranscribed into a single mRNA?

   c. *GAL1* and *GAL10* are not only adjacent to each other, but also are transcribed divergently with a single UAS_G between them. Describe experiments using *GFP* and *RFP* transgenes that would allow you to determine which of the four GAL4 binding sites in this UAS_G element is (are) important for the transcription of *GAL1* and/or *GAL10*.

10. MyoD is a transcriptional activator that turns on the expression of several muscle-specific genes in human cells. The Id gene product inhibits MyoD action.

    a. One possibility is that the Id protein directly represses the expression of these muscle-specific genes. Explain how Id would function if it were a repressor.

    b. Another possibility is that Id inhibits muscle-specific gene transcription indirectly, by preventing MyoD function. Explain how Id could function as an indirect repressor.

    c. Suppose you know the amino acid sequence of the Id protein. How might this information support the hypothesis in part (a) or in part (b)?

11. a. Assume that two transcription factors are required for expression of the blue pigmentation genes in pansies. (Without the pigment, the flowers are white.) What phenotypic ratios would you expect from crossing strains heterozygous for wild-type and recessive amorphic alleles for each of the genes encoding these transcription factors?

    b. Now assume that either transcription factor is sufficient to get blue color. What phenotypic ratios would you expect from crossing the same two heterozygous strains?

12. a. You want to create a genetic construct that will express GFP in *Drosophila*. In addition to the *GFP* coding sequence, what DNA element(s) must you include in order to express this protein in flies if the construct were integrated into the *Drosophila* genome? Where should such DNA element(s) be located? How would you ensure that GFP is expressed only in certain tissues of the fly, such as the wing?

    b. Suppose you insert the *GFP* coding region plus all of the DNA elements required by the answer to part (a)—except the enhancer—between inverted repeats found at the ends of a particular transposable element. Because all of the DNA sequences located between these inverted repeats can move from place to place in the *Drosophila* genome, you can generate many different fly strains, each with the construct integrated at a different genomic location. You now examine animals from each strain for GFP fluorescence. Animals from different strains show different patterns: some glow green in the eyes, others in legs, some show no green fluorescence, etc. Explain these results and describe a potential use for this experiment.

13. In Problem 12, you identified a genomic region that is likely to behave as an eye-specific enhancer. What experiments could you perform to verify that these DNA sequences indeed share all the characteristics of an enhancer and to determine the precise boundaries of the enhancer in the genomic DNA?

14. A graduate student came up with the following idea for identifying insulators in the *Drosophila* genome: Perform an experiment like the one described in Problem 12, but instead of using an *enhancerless* construct, use one that contains an enhancer, and screen for lines that do not express GFP.

    a. What is wrong with this experimental design?

    b. Can you think of a different experiment that you could perform to identify insulators?

15. Myc is a transcription factor that regulates cell proliferation; mutations in the *myc* gene contribute to many cases of the cancer Burkitt lymphoma. Initial experiments on Myc were puzzling. The Myc protein contains both a leucine zipper dimerization domain and a specialized type of helix-loop-helix DNA-binding domain called a bHLH motif, but purified Myc can neither homodimerize nor bind to DNA efficiently.

    Discovery of the Max and Mad proteins helped resolve this dilemma. Like Myc, Max and Mad each contain a bHLH motif and a leucine zipper, but neither Max nor Mad homodimerize readily nor bind DNA with high affinity. However, Myc-Max and Mad-Max heterodimers do readily form and bind DNA; in fact, they bind the same sites on the enhancers of the same target genes. Myc contains an activation domain, while Mad contains a repression domain, and Max contains neither.

    The *max* gene is expressed in all cells at all times. In contrast, *mad* is expressed in resting cells (in the G_0 phase of the cell cycle), while *myc* is not transcribed in resting cells but starts to be expressed when cells are about to divide (at the transition from G_1 to S phase). Mad and Myc proteins are unstable relative to Max protein; when expression of *mad* or *myc* ceases, Mad or Myc proteins soon disappear.

    a. Do you think target genes with enhancers containing binding sites for Myc-Max encode proteins that would arrest the cell cycle or that would drive the cell cycle forward? What about genes with enhancers containing binding sites for Mad-Max? Explain your answers.

    b. Draw diagrams similar to those seen in Fig. 17.14 to show the control region for a target gene, the proteins binding to the enhancer, and whether or not transcription is taking place in (i) resting cells, and (ii) cells that are about to divide.

    c. Provide a concise summary of how these three proteins can regulate cell proliferation.

    d. Would cancer-causing mutations in *myc* be loss-of-function or gain-of-function mutations?

16. Genes in both prokaryotes and eukaryotes are regulated by activators and repressors.

    a. Compare and contrast the mechanism of function of a prokaryotic repressor (for example, Lac repressor) with a typical eukaryotic repressor protein (a direct repressor).

    b. Compare and contrast the mechanism of function of a prokaryotic activator (for example, CAP) with a typical eukaryotic activator protein.

17. The modular nature of eukaryotic activator proteins gave scientists an idea for a way to find proteins that interact with any particular protein of interest. The idea is to use the protein–protein interaction to bring together a DNA-binding region with an activation region, creating an artificial activator that consists of two polypeptides held together noncovalently by the interaction.

    The method is called the *yeast two-hybrid system*, and it has three components. First, the yeast contains a reporter gene construct in which UAS_G (an enhancer-like sequence that binds the activator Gal4 as described in Problem 8) drives the expression of an *E. coli lacZ* reporter (encoding the enzyme ß-galactosidase) from a yeast promoter. Second, the yeast also expresses a fusion protein in which the DNA-binding domain of Gal4 is fused to the protein of interest; this fusion protein is called the *bait.* The third component is a cDNA library made in plasmids, where each cDNA is fused in frame to the activation domain of Gal4, and these can be expressed in yeast cells as *prey* fusion proteins.

How could you use a yeast strain containing the first two components, along with the plasmid cDNA expression library described, to identify prey proteins that bind to the bait protein? How is this procedure relevant to the goal of finding proteins that might interact with each other in the cell?

18. Lysine 4 of histone H3 (H3K4) is methylated in the nucleosomes of many transcriptionally active genes. Suppose you want to determine all the places in the human genome where nucleosomes contain methylated H3K4.
    a. Starting with an antibody that specifically binds only to the tails of histone H3s that have K4 methylation, what kind of experiment would you perform? Outline the major steps of this experiment.
    b. Do you think that you would get the same results if your starting material was skin cells in one experiment and blood precursor cells in a second experiment? Explain.
    c. Describe a follow-up experiment that could determine if your data from part (a) are consistent with the idea that H3K4 methylation marks appear only at transcriptionally active genes.

19. J.T. Lis and collaborators have developed an experimental protocol called PRO-Seq that pinpoints all the positions in a genome at which *transcriptionally engaged* RNA polymerase (that is, enzyme molecules that are actively in the process of synthesizing RNA) is located at a specific time. The accompanying figure shows the results of a PRO-Seq analysis of one region of the human genome, in the vicinity of the *Dnaj4* gene. Vertical *red lines* show the location of transcriptionally engaged RNA polymerase that is moving along the DNA in the left-to-right direction, while vertical *blue lines* show transcriptionally engaged RNA polymerase in the opposite direction. The length of a line indicates the frequency at which active polymerase is found at that position along the genome in the sample.

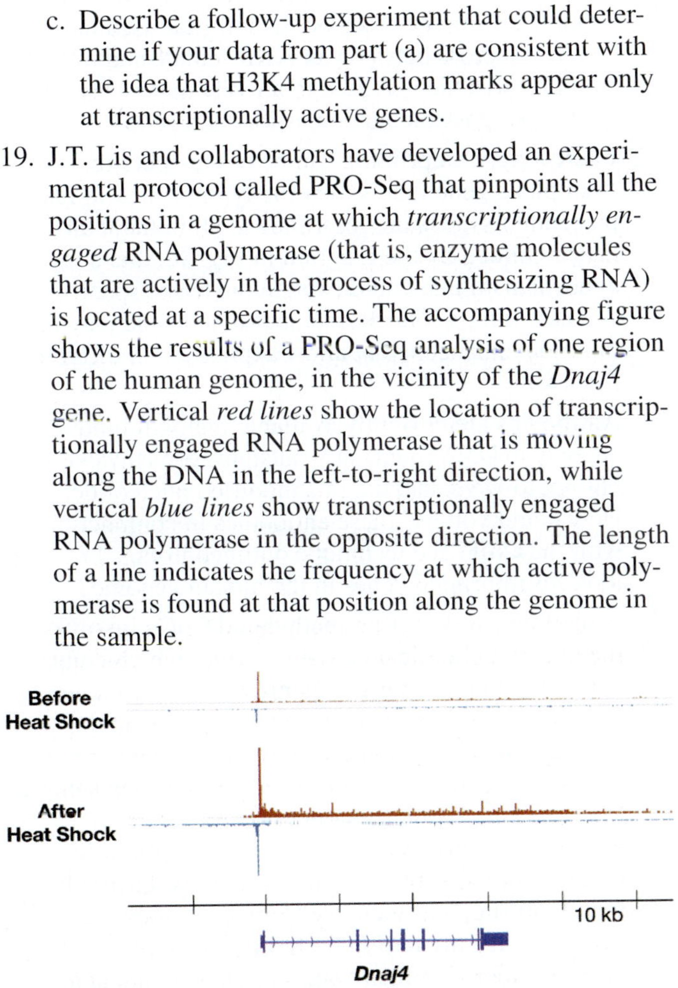

Figure source: Digbijay Mahat and John T. Lis, Cornell University, Ithaca, NY

The two samples were taken from the same culture of human cells grown on a petri plate. The culture was grown under normal conditions and then sampled (Before Heat Shock); then the culture was grown at high temperature and sampled one hour later (After Heat Shock).

    a. Transcriptionally engaged RNA polymerase is not uniformly distributed along the gene. What does this fact signify?
    b. The data, along with your answer to part (a), together suggest that the binding of RNA polymerase to the promoter is not the only rate-limiting step in the transcription of the *Dnaj4* gene, whether before or after heat shock. What other step is involved? Which step appears to be most directly regulated by heat shock?
    c. When RNA polymerase binds to the promoter of this gene, does it know which strand of DNA to use as the template?
    d. Where does the transcription of the *DNAj4* gene end? Do the data clearly show the existence of a single, well-defined transcription stop site?
    e. How do you think that the PRO-Seq method can localize transcriptionally engaged RNA polymerase specifically as opposed to RNA polymerase that might bind to DNA but is not catalyzing transcription?

Section 17.3

20. *Hydatiform moles* are growths of undifferentiated tissues that form within the uterus during an abnormal *molar pregnancy.* These moles are usually made up of XX diploid cells, although some can be XY diploids. Surprisingly, all of the DNA in the nuclei of the cells in the mole is paternal in origin. Most hydatiform moles are benign, but because they sometimes can develop into cancers, these moles should be removed surgically when they are detected.
    a. What kinds of events could lead to the generation of a hydatiform mole?
    b. Hydatiform moles are diploid cells with the normal numbers of genes and chromosomes. Why do you think they develop as undifferentiated tissues rather than as normal embryos?

21. Prader-Willi syndrome is caused by a mutation in an autosomal maternally imprinted gene. Label the following statements as true or false, assuming that the trait is 100% penetrant.
    a. Sons of affected males have a 50% chance of showing the syndrome.
    b. Daughters of affected males have a 50% chance of showing the syndrome.

c. Sons of affected females have a 50% chance of showing the syndrome.

d. Daughters of affected females have a 50% chance of showing the syndrome.

22. The human *IGF2* gene is autosomal and maternally imprinted. Copies of the gene received from the mother are not expressed, but copies received from the father are expressed. You have found two alleles of this gene that encode two different forms of the IGF2 protein distinguishable by gel electrophoresis. One allele encodes a 60K (Kilodalton) blood protein; the other allele encodes a 50K blood protein. In an analysis of blood proteins from a couple named Bill and Joan, you find only the 60K protein in Joan's blood and only the 50K protein in Bill's blood. You then look at their children: Jill is producing only the 50K protein, while Bill Jr. is producing only the 60K protein.

a. With these data alone, what can you say about the *IGF2* genotype of Bill Sr. and Joan?

b. Bill Jr. and a woman named Sara have two children, Pat and Tim. Pat produces only the 60K protein and Tim produces only the 50K protein. With the accumulated data, what can you now say about the genotypes of Joan and Bill Sr.?

23. Follow the expression of a paternally imprinted gene through three generations. Indicate whether the copy of the gene from the male in generation I in the accompanying diagram is expressed in the germ cells and somatic cells of the individuals listed.

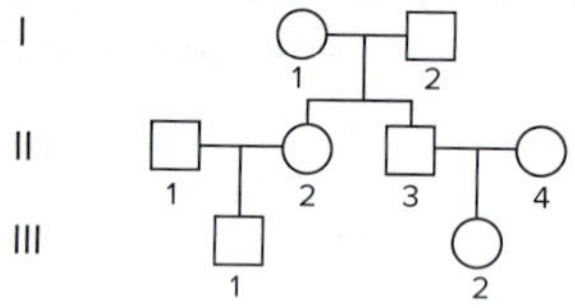

a. generation I male (I-2): germ cells

b. generation II daughter (II-2): somatic cells

c. generation II daughter (II-2): germ cells

d. generation II son (II-3): somatic cells

e. generation II son (II-3): germ cells

f. generation III grandson (III-1): somatic cells

g. generation III grandson (III-1): germ cells

24. Reciprocal crosses were performed using two inbred strains of mice, AKR and PWD, that have different alleles of many polymorphic loci. In each of the two crosses, placental tissue was isolated whose origin was strictly from the fetus (this can be separated by dissection from placental tissue originating from the mother). RNA was prepared from the fetal placental tissue and then subjected to *deep sequencing* (that is, RNA-Seq). Because of the polymorphisms, investigators could compare the number of reads of mRNAs for specific genes that were transcribed from maternal or paternal alleles, as shown in the following figure. (The x-axis shows the percentage of reads for the given mRNA that correspond to the AKR allele of that gene.)

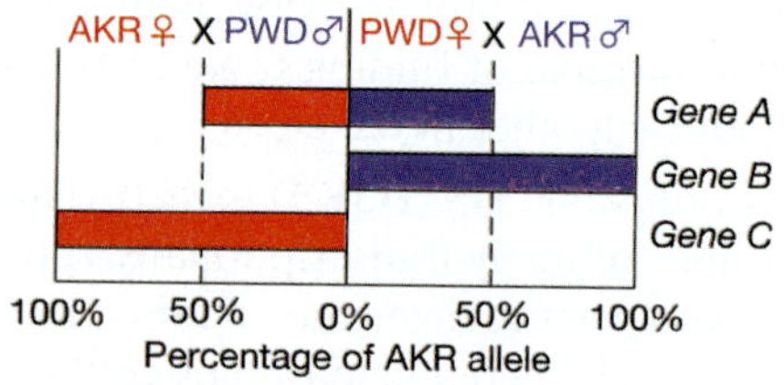

a. Which of the genes (A, B, or C) is maternally imprinted? Which is paternally imprinted? Which is not imprinted?

b. Why was it important to perform reciprocal crosses to determine whether any of the genes were imprinted?

c. Using the same type of diagram that indicates the percentage of AKR alleles, diagram the expected results for these same three genes if a female F_1 mouse from the cross on the left (that is, a daughter of a cross between an AKR female and a PWD male) was then crossed to a PWD male. Describe the two possible outcomes for each gene.

25. Interestingly, imprinting can be tissue-specific. For example, a gene that is maternally imprinted in fetal placental tissue is not imprinted at all in the fetal heart. Guided by the diagram in Fig. 17.23a, suggest a mechanism that could explain the tissue specificity of imprinting. (*Hint:* Remember that a gene may have multiple enhancers that allow expression in different tissues.)

26. Antibodies are currently available that will bind specifically to DNA fragments containing 5-methylcytosine but not to DNA lacking this modified nucleotide. How could you use these antibodies in conjunction with the ChIP-Seq technique outlined in Fig. 17.18 to look for imprinted genes in the human genome?

27. A method for detecting methylated CpGs involves the use of a chemical called *bisulfite,* which converts cytosine to uracil but leaves methylated cytosine untouched. You want to know whether a particular CpG dinucleotide at one location in the genome is methylated on one or both strands in a tissue sample. The genomic sequence containing this CpG is: 5'...TCCAT**CG**CTGCA...3'. You take genomic DNA from the sample tissue, treat it exhaustively with bisulfite, and then use flanking primers to PCR-amplify the region including this CpG dinucleotide. You then want to Sanger sequence (see Fig. 9.7) the amplified PCR product.

a. After you treat genomic DNA with bisulfite, the two DNA strands will melt into single strands. Why?

b. Your answer to part (a) introduces a potential complication, because if you do not account for this result of bisulfite treatment, the PCR primers will not amplify the DNA. What special considerations would be necessary when you design your PCR primers for this experiment? Could one pair of PCR primers amplify both strands of DNA?

c. What sequence would you see if you amplified the DNA strand shown and the CpG was methylated? If it was not methylated?

d. Using the bisulfite method, can you tell if this CpG dinucleotide in the tissue sample is hemimethylated (methylated on one strand) or methylated on both strands? Explain.

28. Honeybees (*Apis mellifera*) provide a striking example of environmental effects on eukaryotic gene expression and thus phenotype. Fertile queens and sterile workers are both female bees with the same diploid genomes. However, their morphologies and behaviors are different.

Source: State of New Jersey, Department of Environmental Protection.

It turns out that the diet of the larval (*larvae* are young developing insects) females controls their development as either workers or queens. When the hive needs new queens, a few female larvae are fed *royal jelly,* a substance secreted by worker bees, instead of the normal diet of nectar and pollen.

Investigators determined that many genes are more highly CpG-methylated in the larval-stage workers than in the larval-stage queens.

a. Describe an experiment you could do to determine if DNA methylation correlates with differences in gene expression in workers and queens.

b. Researchers were able to mimic the effect of royal jelly by knocking down the expression in female larvae of a DNA methyltransferase enzyme called *DNMT3*. (The knockdown was accomplished through a technique called *RNA interference* that is described in a later problem; the details are unimportant here.) Remarkably, when fed a normal diet, most of these larvae developed as queens. Based on the description of the transgenerational silencing of the A^Y allele of the mouse *agouti* gene presented in Fig. 17.24, suggest a simple hypothesis that accounts for the different effects on female honeybee larval development of normal versus royal jelly diets.

29. Consider the experiment in Fig. 17.24, where the A^Y allele was silenced in the F_1 progeny of a female mouse who ate a diet rich in methyl donors (beets and garlic). To be convinced that transgenerational epigenetic inheritance had occurred, in how many generations of progeny not fed the methyl-donor-rich diet would you need to observe silencing of the A^Y allele? (Assume that female progeny in which A^Y was silenced are always used to produce the next generation.) Would your answer change if the original mouse fed the beets and garlic was male, and male progeny in which A^Y was silenced were always used to produce the next generation? Explain.

Section 17.4

30. A protein or RNA that regulates gene expression in *trans*—either at the level of transcription, RNA splicing, or translation—*must* have specificity for one target gene or a group of target genes. Explain how specificity is achieved in each case.

31. a. How can a single eukaryotic gene give rise to several different types of mRNA molecules?

b. Excluding the possible rare polycistronic message, how can a single mRNA molecule in a eukaryotic cell produce proteins with different activities?

32. The *hunchback* gene, a gene necessary for proper patterning of the *Drosophila* embryo, is translationally regulated. The position of the coding region within the transcript is known. How could you determine if the sequences within the 5′ UTR or 3′ UTR, or both, are necessary for proper regulation of the mRNA's translation?

33. You know that the mRNA and protein produced by a particular gene are present in brain, liver, and fat cells, but you detect an enzymatic activity associated with this protein only in fat cells. Provide a possible explanation for this phenomenon.

34. You are studying a transgenic mouse strain that expresses a *GFP* reporter gene under the control of *cis*-acting regulatory elements that normally control a gene needed for the early development of mice. Previous evidence from transcriptome sequencing (RNA-seq) indicates that mRNA for the gene of interest

can be identified between days 8.5 and 10.5 of gestation. In your strain, GFP fluorescence can be seen from about day 8.75 until at least day 12.

a. Explain the discrepancy between mRNA and protein expression.

b. Would you expect GFP protein expression to indicate more accurately the normal onset of activity for this gene or the normal cessation of this gene's activity? Explain.

35. By searching a human genomic database, you have found a gene that encodes a protein with weak homology to Argonaute, a factor present in complexes that bind to certain miRNAs and mediate their ability to regulate the stability or translation of target mRNAs (review Figs. 17.32 and 17.33).

a. How would you determine which specific miRNAs might be associated with the new protein you discovered? (Think about how you might use a variation of the ChIP-Seq technique described in Fig. 17.18 to explore this question.)

b. If a mouse could be obtained that was homozygous for a null mutation of a gene almost identical to the human gene you found, how could you use this mutant mouse to ask what mRNAs might be targeted by the miRNA-RISC complexes containing your Argonaute-like protein?

36. Scientists have exploited the siRNA pathway to perform a technique called *RNA interference*—a means to knock down the expression of a specific gene without having to make mutations in it. The idea is to introduce dsRNA corresponding to the target gene into an organism; the dsRNA is then processed into an siRNA that leads to the degradation of the target gene's mRNA. One clever method for delivery of the dsRNA to some organisms (the nematode *C. elegans*, for example) is to feed them bacteria transformed with a recombinant plasmid that expresses dsRNA.

a. Draw a gene construct that, when expressed from a plasmid in bacteria, could be used to knock down by RNA interference the expression of gene *X* of *C. elegans.*

b. How can you test if gene *X* expression is obliterated in worms that have eaten the bacteria transformed with a plasmid containing your construct?

c. Do you think that only gene *X* expression will be affected in these worms? Explain.

37. Persimmons (*Diospyros lotus*) are *dioecious plants,* meaning that male (XY) and female (XX) flowers exist on separate plants. Male persimmon flowers have rudimentary, nonfunctional carpels (the female sex organ that includes the ovary), while female flowers have stamens and anthers (male sex organs), but no pollen is produced. Remarkably, sex determination in *D. lotus* is controlled by a Y-linked miRNA gene called *OGI* (*male tree* in Japanese), whose target mRNA is encoded by an autosomal gene called *MeGI* (*female tree*).

a. What features of the *OGI* transcript base sequence might have suggested to the investigators that *OGI* encodes a miRNA?

b. RNA-Seq experiments detected high levels of *MeGI* mRNA in female sex organs but low levels in male sex organs. Suggest a molecular mechanism by which the *OGI* miRNA could regulate *MeGI* gene expression.

Integrating transgenes into the *D. lotus* genome is difficult. Therefore, to examine the function of *MeGI,* the scientists constructed a transgene that overexpresses *MeGI* and used it to transform *Arabidopsis thaliana,* a plant species more amenable to such studies. *Arabidiposis* have so-called *perfect flowers*—individual flowers have both male and female structures. *MeGI* overexpression in the transformants sterilized the anthers, but the carpels were unaffected and functional.

c. What do you predict would happen if the *Arabidopsis* transformants also contained a transgene that overexpressed *OGI?*

d. The MeGI protein is a homeobox transcription factor. What do the results with transformed *Arabidopsis* suggest is one role of this protein in persimmon sex determination?

e. Speculate about possible mechanisms that could account for the rudimentary development of the nonfunctional carpels in male persimmon flowers. Could any of these mechanisms involve *OGI* and MeGI?

38. *Drosophila* females homozygous for loss-of-function mutations in the gene *aubergine* are sterile. RNA-Seq experiments show that in the ovaries of these females, the levels of RNAs for many kinds of transposable elements are more than 10× higher than in wild-type ovaries. The *aubergine* gene encodes a Piwi-family protein.

a. Why do you think these females are sterile?

b. Piwi proteins interact with piRNAs that are transcribed from piRNA gene clusters. Given that the levels of many kinds of TEs are elevated in mutant ovaries, what kinds of DNA sequences do you think are located in these clusters?

c. Many investigators think of piRNAs as a kind of defensive mechanism that protects organisms from the effects of new transposable elements that might be introduced into genomes, for example from other species. Explain.

d. In what way might piRNA systems be beneficial to transposable elements?

e. Problem 29 in Chapter 13 described a phenomenon called *hybrid dysgenesis* that occurs when males from so-called *P strains,* whose genomes have *P* element transposons, mate with females from *M strains,* whose genomes lack *P* elements. In the germ lines of the hybrid progeny, frequencies of mutations and chromosomal rearrangements are elevated and can lead to sterility. How might the piRNA system be involved in hybrid dysgenesis?

39. The text has discussed the RNA-Seq technique, which quantifies the amounts of mRNAs present in a given cell type at a given time. We have also described the method of ribosome profiling, which quantifies the levels of mRNA sequences that are being translated actively in a cell population at a given time. But even when both kinds of experiments are performed, the data do not tell us accurately the relative amounts of the protein products of these mRNAs in the cells at that time. Why not? What elements of gene expression do these techniques account for and what elements do they ignore?

Section 17.5

40. Researchers know that Fru-M controls male sexual behavior in *Drosophila* because inappropriate Fru-M expression in females causes them to behave like males: Such females display male behaviors that are oriented toward other females.

a. Describe a *fru* mutation that could cause the expression of Fru-M in females.

b. Describe a transgene construct that scientists could generate and insert into *Drosophila* females that would have the same effect as the mutant you described in (a).

41. The *Drosophila* gene *Sex lethal* (*Sxl*) is deserving of its name. Certain alleles have no effect on XY animals but cause XX animals to die early in development. Other alleles have no effect on XX animals but cause XY animals to die early in development. Thus, some *Sxl* alleles are lethal to females, while others are lethal to males.

a. Would you expect a null mutation in *Sxl* to cause lethality in males or in females? What about a constitutively active *Sxl* mutation?

b. Why do *Sxl* alleles of either type cause lethality in a specific sex?

The gene transformer (*tra*) gets its name from *sexual transformation,* as some *tra* alleles can change XX animals into morphological males, while other *tra* alleles can change XY animals into morphological females.

c. Which of these sex transformations would be caused by null alleles of *tra* and which would be caused by constitutively active alleles of *tra?*

d. In contrast with *Sxl,* null *tra* mutations do not cause lethality either in XX or in XY animals. However, the Sxl protein regulates the production of the Tra protein. Why then do all *tra* mutant animals survive?

e. Predict the consequences of null mutations in *tra-2* on XX and XY animals. (Recall that *tra-2* encodes a protein, expressed in both sexes, that is required for Tra function.)

f. XY males carrying loss-of-function mutations in the *fruitless* (*fru*) gene display aberrant courtship behavior. Would you predict that either XX or XY animals with wild-type alleles of *fru* but loss-of-function mutations of *tra* would also court abnormally?

42. Figure 17.29 shows that the Sxl protein binds to the mRNA of the *msl-2* gene, inhibiting translation of the mRNA's proper reading frame. The MSL-2 protein is a transcription factor that binds to the X chromosome in XY males to double the level of X-linked gene transcription, thus equalizing X-linked gene expression in XY males and XX females.

a. In which sex, XY males or XX females, would the Sxl protein bind to the *msl-2* mRNA?

b. As discussed in Problem 41, some *Sxl* alleles are lethal to females and others are lethal to males. Is the function of Sxl in regulating the synthesis of Msl-2 protein sufficient to explain the sex-specific lethality caused by both kinds of alleles?

c. Predict the effect of loss-of-function mutations in *msl-2* on male and female fertility and viability.

chapter **18**

Manipulating the Genomes of Eukaryotes

A statue in front of the Institute of Cytology and Genetics in Novosibirsk, Russia pays homage to the laboratory mouse.

chapter outline

UNTIL RECENTLY, CHILDREN born with poor vision due to a genetic disease called Leber congenital amaurosis (LCA) were destined to become completely blind by early adulthood. Now, for many of these children, the success of gene therapy trials provides hope not only for a halt to the retinal degeneration characteristic of the disease, but even for restoration of normal sight.

One form of LCA is caused by homozygosity for a recessive loss-of-function allele of a gene called *RPE65*. This gene encodes a protein found in the retinal pigment epithelium (a cell layer just beneath the retina) that is crucial for the function of photoreceptors. The RPE65 enzyme functions in the visual cycle—the process by which the retina detects light. LCA patients lose sensitivity to light, which eventually results in a reduction in the amount of brain cortex devoted to visual processing (**Fig. 18.1**).

Gene therapy is the manipulation of genes—adding DNA to the genome or altering the DNA of a gene—in order to cure a disease. The experimental gene therapy strategy for this form of LCA was simple: Scientists delivered normal copies of the *RPE65* gene to the retinal pigment epithelium cells of patients, simply by injecting DNA packaged in viral particles through the eye into these cells. Since the first results of *RPE65* gene therapy clinical trials were reported in 2008, more than 30 patients have undergone the procedure, and almost all of them have had their vision restored at least in part; several are no longer considered legally blind.

In this chapter, you will learn about two general strategies for altering genomes: creation of *transgenic organisms* and *targeted mutagenesis*. Development of these exciting technologies has relied on knowledge of the natural processes by which DNA can move within a genome, can be transferred between individuals and between species, and can be protected from alteration or degradation. The overarching theme of

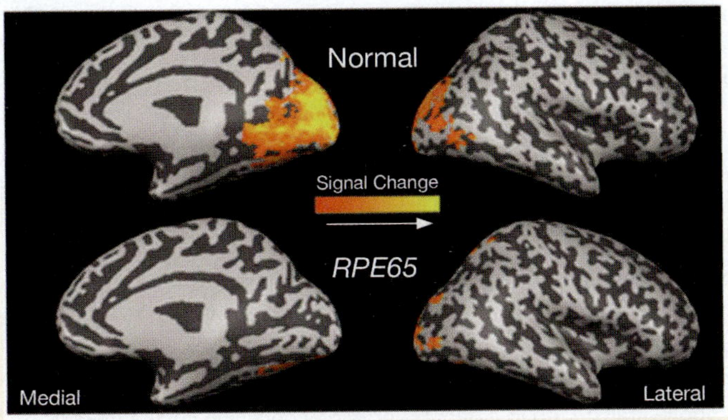

Figure 18.1 Activation of the brain cortex in response to light. The cortexes of the brain of a normal dog (*top*) and a dog with Leber congenital amaurosis (LCA) due to mutations in the *RPE65* gene (*bottom*) are mapped in shades of *gray*. The *right* and *left* images for each dog are views from different angles. *Yellow* and *orange* signals indicate the amplitude of cortical activation in response to a controlled amount of light shined on the eyes of these animals. Much less of the cortical region of the LCA dog participates in visual processing.
Aguirre GK, Komáromy AM, Cideciyan AV, Brainard DH, Aleman TS, et al. (2007), "Canine and Human Visual Cortex Intact and Responsive Despite Early Retinal Blindness from RPE65 Mutation," *PLoS Med,* 4(6): e230. doi:10.1371/journal.pmed.0040230

this chapter is that by using recombinant DNA technology, scientists can harness these natural processes to develop creative and powerful methods to alter genomes—not only in order to treat disease, but also to improve the production of medicines and food crops and to enhance modern biological research.

18.1 Creating Transgenic Organisms

learning objectives

1. Summarize how scientists create transgenic mice with pronuclear injection.
2. Describe how *P* elements are used to produce transgenic *Drosophila*.
3. Explain how researchers use the Ti plasmid from *Agrobacterium* to insert genes into plant genomes.

The genome of a **transgenic organism** contains a gene from another individual of the same species or of a different species; such a gene is called a **transgene.** In this section we will discuss some of the ways researchers can make transgenic organisms. Then in the next section we will examine just a few of the possible uses of transgenic technology, which are limited only by the scientist's imagination.

Scientists Exploit Natural Gene Transfer Mechanisms to Create Transgenic Organisms

Transgenes can be made *in vitro* using the types of recombinant DNA techniques described in Chapter 9. But to make transgenic eukaryotes, researchers need to introduce the transgene DNA into one or more cells. This goal can be accomplished in various ways depending on the organism. Some unicellular eukaryotes like the yeast *S. cerevisiae* can be subjected to treatments that disrupt the cell wall, and DNA can then enter the cells in a fashion very similar to the artificial transformation of *E. coli* (see Fig. 9.5b). For many other organisms, the most efficient means of transferring DNA into cells involves injection of a DNA solution directly into a cell or embryo (**Fig. 18.2**).

Figure 18.2 Injecting transgenes into cells. (a) An investigator is injecting DNA into one of the two pronuclei present in a mouse embryo soon after fertilization. **(b)** DNA is being injected into the posterior end of an early *Drosophila* embryo that is a single syncytial cell with many nuclei.
a: © Martin Oeggerli/Science Source; b: © Solvin Zankl

(a) Injecting a transgene into a recently fertilized mouse egg

(b) Injecting a transgene into a *Drosophila* embryo

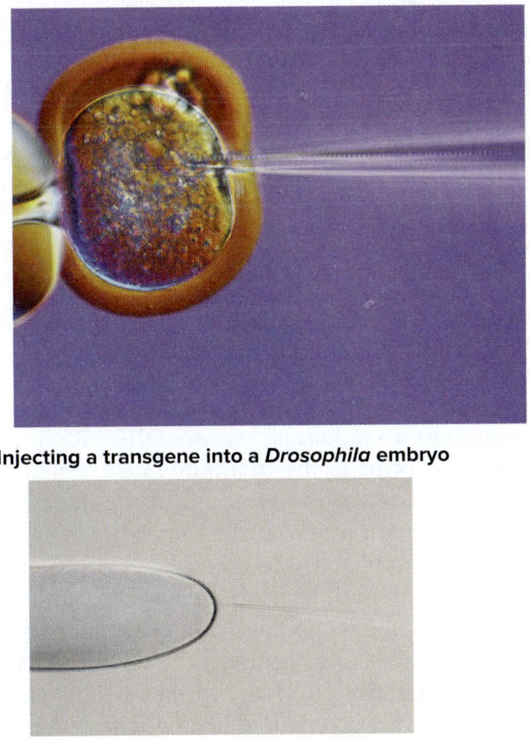

In yet other cases, the transgenic DNA can be incorporated into a virus particle or even a bacterium that can then infect a cell (described later).

Once introduced into a cell, the transgene has to be replicated and maintained as the cell divides. In most cases, these goals are accomplished by integrating the transgene into a random location in the genome of the host cell. However, in some species, the transgenes can be maintained outside of the host chromosomes, either as an extrachromosomal array (in *C. elegans*) or as a plasmid (in yeast). Finally, in order for the transgene to be propagated between generations of a multicellular organism, it is crucial that cells containing the transgene have the ability to develop eventually into gametes. In animals, this requirement means that the transgene must be incorporated into a germ-line cell. In contrast, in plants, almost any cell can carry the transgene because entire plants can be regenerated from isolated cells.

We describe here methods to create transgenic mice, flies, and plants that illustrate many of these points. These techniques are in large part based on our knowledge of natural gene transfer mechanisms.

DNA Injection into Pronuclei Generates Transgenic Mice

A fertilized mouse egg (zygote) contains two haploid **pronuclei**—one maternal and one paternal. The pronuclei come close together, their nuclear membranes break down, and the maternal- and paternal-derived chromosomes intermingle so that their sister chromatids separate on the same spindle for the first mitotic division. At the conclusion of this mitosis, each cell of the two-cell embryo has a single diploid nucleus.

To create a transgenic mouse carrying a foreign DNA sequence integrated into one of its chromosomes by **pronuclear injection,** as shown in **Fig. 18.3,** a researcher mates a male and female mouse and harvests the just-fertilized eggs from the female's reproductive tract. The investigator then injects linear copies of the foreign DNA into either one of the pronuclei of the fertilized egg (see also Fig. 18.2a). The injected, fertilized egg is then implanted into the oviduct of a pseudo-pregnant female, where it can continue its development as an embryo.

Roughly 25% to 50% of the time, the injected DNA will integrate into a random chromosomal location. Integration can occur prior to the first mitosis, in which case the transgene will appear in every cell of the adult body. Alternatively, integration may occur somewhat later, after the embryo has completed one or two cell divisions; in such cases, the mouse will be a mosaic of cells, some with the transgene and some without it. As long as the transgene is present in germ-line cells, the transgene will be transmitted to the next generation. A mouse formed from a gamete containing a transgene can then be mated with other mice to establish stable lines of transgenic animals.

Figure 18.3 Making transgenic mice by pronuclear injection.

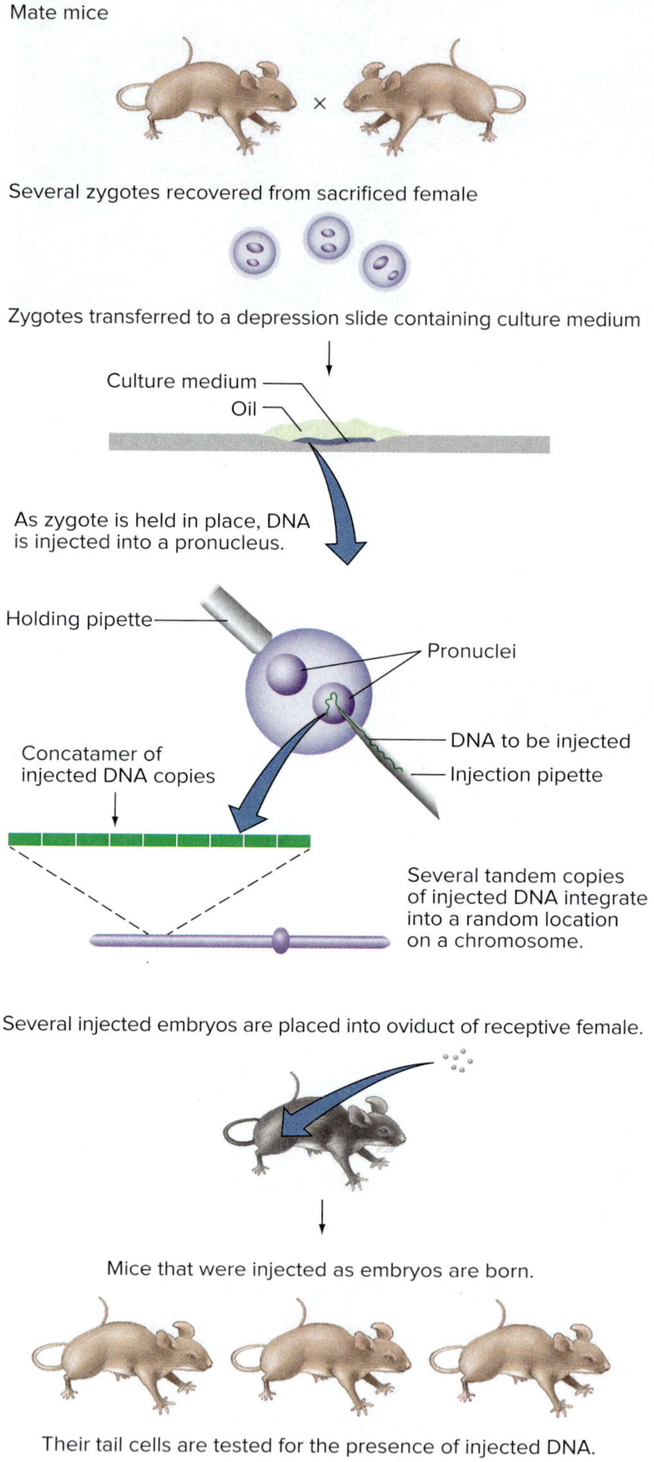

The exact mechanism for random transgene integration is unknown, but it clearly depends on DNA repair enzymes that seek out and repair broken ends of DNA, probably similar to those involved in nonhomologous end-joining (NHEJ; review Fig. 7.18). Usually, many tandem copies of the transgene become integrated together into the same random site in the genome (Fig. 18.3).

Recombinant *P* Elements Can Transform *Drosophila*

P elements are a class of DNA transposon in *Drosophila* (recall Fig. 13.23b). Autonomous *P* elements contain a gene for transposase protein, and the transposon ends are inverted repeats. Transposase binds the inverted repeats, "cuts" the transposon out of the genome, and "pastes" it into a new location. *Drosophila* geneticists use *P* elements as *vectors* (vehicles) for transfer of genes into germ-line cells—a process called ***P* element transformation.**

P element vectors are plasmids that contain the *P* element ends but not the transposase gene (**Fig. 18.4**). Using recombinant DNA techniques, scientists replace the transposase gene with the transgene and a **marker gene;** the marker gene enables detection of transgenic flies. A widely used marker gene is the wild-type *white* gene (w^+), which confers normal red eye color to flies with mutations in their endogenous *white* genes (w^-).

Figure 18.4 shows a common procedure for generating transgenic flies with *P* element vectors. Investigators inject two plasmids into w^- embryos at an early stage of development, when at most several hundred nuclei are present (see also Fig. 18.2b). One plasmid was made by cloning the transgene into the vector; this plasmid now contains the transgene and the w^+ marker gene, both located within the *P* element inverted repeats. The other plasmid, called the *helper plasmid,* contains the transposase gene but no *P* element inverted repeats (Fig. 18.4).

When these two plasmids are injected into embryos, transposase protein produced by the helper plasmid can mobilize the recombinant *P* element, cutting it out of the other plasmid and pasting the element into a random site in a *Drosophila* host chromosome (Fig. 18.4).

After the injected embryos mature into adults, researchers cross each adult to w^- flies. If a recombinant *P* element integrates into a chromosome of a germ-line precursor cell, some gametes produced by the injected animal will carry a chromosome with the recombinant *P* element. Investigators can recognize transgenic progeny (flies containing the recombinant *P* element) because they will have red (w^+) eyes. These red-eyed flies can be used in cross schemes to establish stable lines of transgenic flies. A recombinant *P* element containing a transgene will not subsequently mobilize and move around the genome in flies of this stable line because laboratory strains of *Drosophila* do not contain *P* elements, so no transposase will be present.

Figure 18.4 Constructing transgenic *Drosophila* by *P* element transformation. A transgene (*Gene*) is first ligated into a vector containing the *white*$^+$ gene (w^+) located between *P* element inverted repeats (*green*). Researchers inject this plasmid, along with a helper plasmid containing the *P* element transposase gene, into w^- host embryos where transposition occurs in some germ-line cells. When adults with these germ cells are mated with w^- flies, some progeny will have red eyes and an integrated transgene.

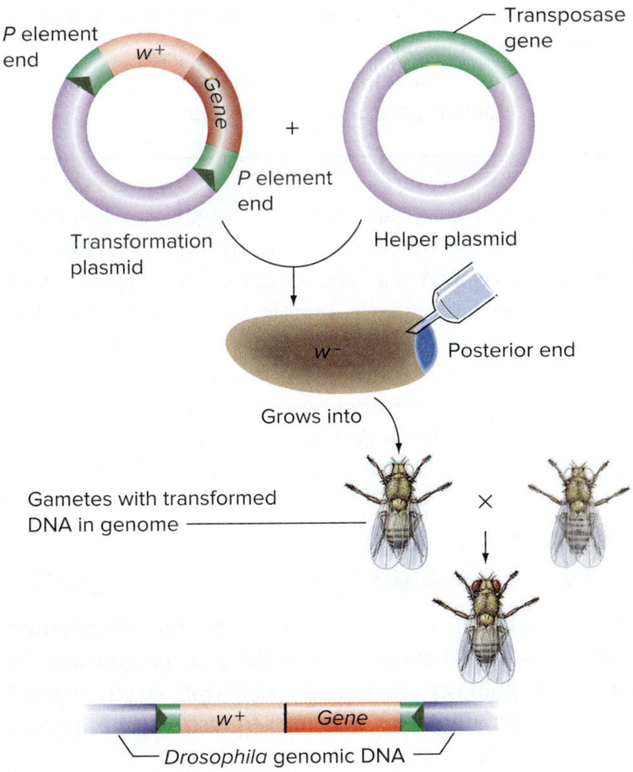

Agrobacterium Ti Plasmid Vectors Accomplish Plant Transgenesis

A vector derived from the tumor-inducing (Ti) plasmid of the bacterium *Agrobacterium tumefaciens* is the basis for an efficient method for introducing transgenes into plants—***Agrobacterium*-mediated T-DNA transfer (Fig. 18.5)**. *A. tumefaciens* bacteria infect plant cells; during infection the bacteria can transfer DNA into the host cell in a process reminiscent of conjugation in bacteria because it involves formation of a pilus connecting the *A. tumefaciens* donor and the plant cell recipient.

The transferred DNA is called the **T-DNA,** which is a portion of the Ti plasmid DNA present in *A. tumefaciens.* The T-DNA integrates into the plant cell genome. Because the T-DNA contains a gene that causes cell overgrowth, the descendants of the T-DNA–containing cell form a tumor called a *crown gall.* T-DNA transfer depends on 25-base pair sequences at each end of the T-DNA called the left and right borders (LB and RB), and on several proteins encoded by the *vir* genes normally present on the Ti plasmid.

You can see that the integration of T-DNA into the host chromosome is in many ways analogous to the mobilization of a DNA transposon, with the *vir* gene proteins being *trans*-acting enzymes that work on the LB and RB sequences at the border. The method for using T-DNA to transfer genes into plant genomes thus has some underlying similarities with the *P* element procedure just outlined for *Drosophila.*

Figure 18.5 Transgenic plants produced using a T-DNA plasmid vector. Researchers infect plants with *Agrobacterium tumefaciens* bacteria containing two plasmid constructs. A T-DNA plasmid contains a transgene (*Gene*) and marker gene that confers resistance to an herbicide, both within the T-DNA ends LB and RB. A helper plasmid contains the *vir* genes, required for T-DNA transfer to a plant cell. Upon infection, the recombinant T-DNA integrates into the host plant genome in both somatic cells and eggs. Investigators select for single cells or seeds with a transgene insertion by growing cells or seeds in the presence of herbicide. They then grow the selected cell or seedling into a whole transgenic plant.

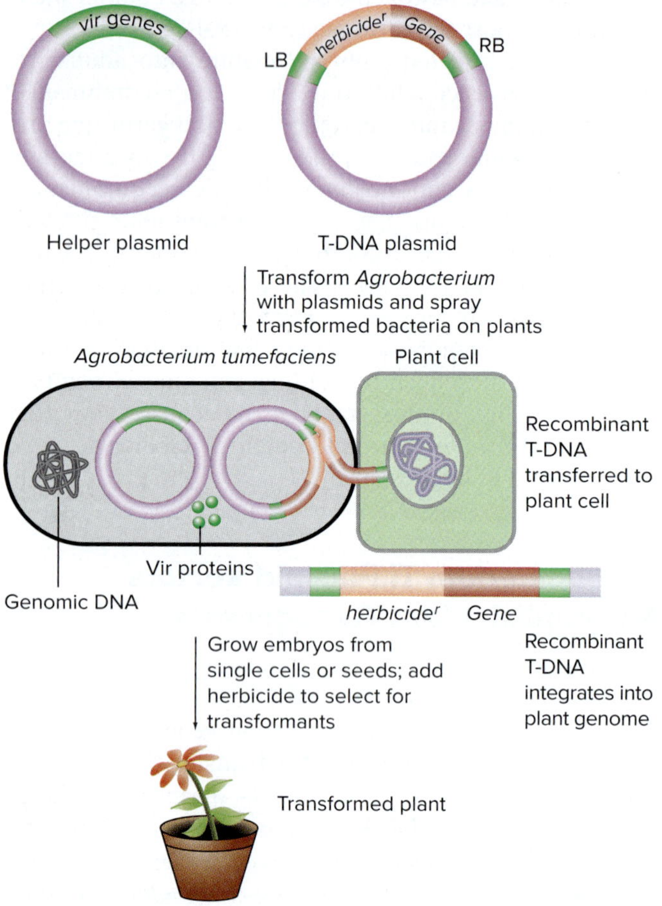

Researchers transform *A. tumefaciens* with two different plasmids (Fig. 18.5). One is a helper plasmid that contains the *vir* genes but no border sequences. The other plasmid is the T-DNA vector engineered to contain the gene to be transferred and a marker gene (often a gene that confers resistance to an herbicide), both located between the LB and RB sequences. To start the infection, investigators spray the transformed *A. tumefaciens* onto whole plants or plant cells. They next grow individual infected plant cells in culture or seeds in soil to generate embryonic plants, and they select embryos or seedlings transformed with the recombinant T-DNA by adding herbicide to the growth medium (Fig. 18.5).

These examples of methods used for constructing transgenic organisms show how scientists can take advantage of natural processes to alter genomes. Researchers in essence have "hijacked" the process by which *A. tumefaciens* makes crown galls to introduce foreign DNA into plants. Naturally occurring enzymatic processes, whether those used for DNA repair or for mobilizing transposons or T-DNA, are thus the basis for integrating foreign DNA into host chromosomes.

essential concepts

- Transgenic mice are produced by injecting foreign DNA into a pronucleus of a fertilized egg.
- Transformation of *Drosophila* relies on the construction of transgenes inserted into *P* element transposon vectors.
- Researchers make transgenic plants by infecting plant cells with *Agrobacterium* containing a Ti (tumor-inducing) plasmid engineered to contain the transgene.
- These methods of creating transgenic organisms result in the integration of transgenes at random locations in the host genome.

18.2 Uses of Transgenic Organisms

learning objectives

1. Describe how transgenes can clarify which gene causes a mutant phenotype.
2. Summarize the use of transgene reporter constructs in gene expression studies.
3. Discuss examples of how transgenic organisms serve to produce proteins needed for human health.
4. List examples of GM organisms and discuss the pros and cons of their production.
5. Explain the use of transgenic animals to model gain-of-function genetic diseases in humans.

Our ability to generate transgenic organisms has had a major impact on biological research and is also increasingly important for several aspects of daily life. Studies with transgenic model organisms enable researchers to understand better the functions of particular genes and their regulation and to model certain human diseases in animals. In addition, scientists have engineered transgenic plants and animals to produce drugs and (more controversially) better agricultural products, and even glowing pets (**Fig. 18.6**).

Transgenes Assign Genes to Phenotypes

In many genetic investigations, the available information may not allow scientists to pinpoint the gene responsible for a particular phenotype. The construction of transgenic organisms often allows investigators to resolve ambiguities.

Figure 18.6 Glofish®. Transgenic zebrafish (*Danio rerio*) that express different-colored variants of GFP and RFP were the first genetically modified pets.

As an example, suppose a geneticist interested in how the *Drosophila* eye develops isolates a mutant fly strain homozygous for a recessive mutation (m^-) that results in malformed eyes (**Fig. 18.7a**). Molecular analysis reveals that the mutation is a small deletion that removes the 5′ portions of two different genes (**Fig. 18.7b**). Is it the loss of gene *A* or the loss of gene *B* that accounts for the eye defects?

You could answer this question by creating recombinant *P* element constructs containing either gene *A* or gene *B* wild-type genomic DNA (**Fig. 18.7c**). You would then test the ability of each transgene to restore a normal eye. For example, if homozygous m^-/m^- flies carrying a wild-type gene *A* transgene have malformed eyes, but m^-/m^- flies carrying a wild-type gene *B* transgene have normal eyes, you would conclude that the loss of gene *B* is the cause of the mutant phenotype; in other words, m = gene *B*.

You saw previously in Chapter 4 an important historical example of similar logic. You will recall that an XX mouse containing an *SRY* transgene developed as a male (Fast Forward Box entitled *Transgenic Mice Prove that SRY Is the Maleness Factor*). This experiment exemplifies another way that transgenic technology can be used to understand the function of a particular gene: Here, the expression of the *SRY* gene in an unusual context (in an organism with two X chromosomes and no Y) showed that *SRY* controls maleness in mammals like mice and humans.

Transgenes Are Key Tools for Analyzing Gene Expression

In Chapter 17, we described how scientists use reporter constructs containing foreign genes whose protein products are easy to detect (such as the jellyfish gene for GFP—green fluorescent protein—or *E. coli*'s *lacZ* gene for the enzyme β-galactosidase) to study many aspects of the regulation of gene expression in eukaryotes. Such reporter constructs help researchers identify enhancers that dictate the transcription of a gene in specific tissues at particular times in development (review Fig. 17.3). Reporter constructs are also valuable in finding genes that encode transcription factors encoding with the enhancers (review Fig. 17.13). Here, we remind you that the function of these reporter constructs can be monitored only when they are introduced into eukaryotic organisms as transgenes.

Figure 18.7 Using *Drosophila* transgenes to link a mutant phenotype to a gene. (a) Scanning electron micrographs show that loss of *m* gene activity results in malformed fly eyes. **(b)** The m^- mutation is a deletion of parts of two adjacent genes: gene *A* and gene *B*. **(c)** Flies containing the gene *A* transgene (*left*) that are also m^-/m^- still have malformed eyes, while m^-/m^- flies containing the gene *B* transgene (*right*) have wild-type eyes. Therefore, the malformed eyes are due to the loss of gene *B*, not to the loss of gene *A*. (The plasmid vector is described in Fig. 18.4.)

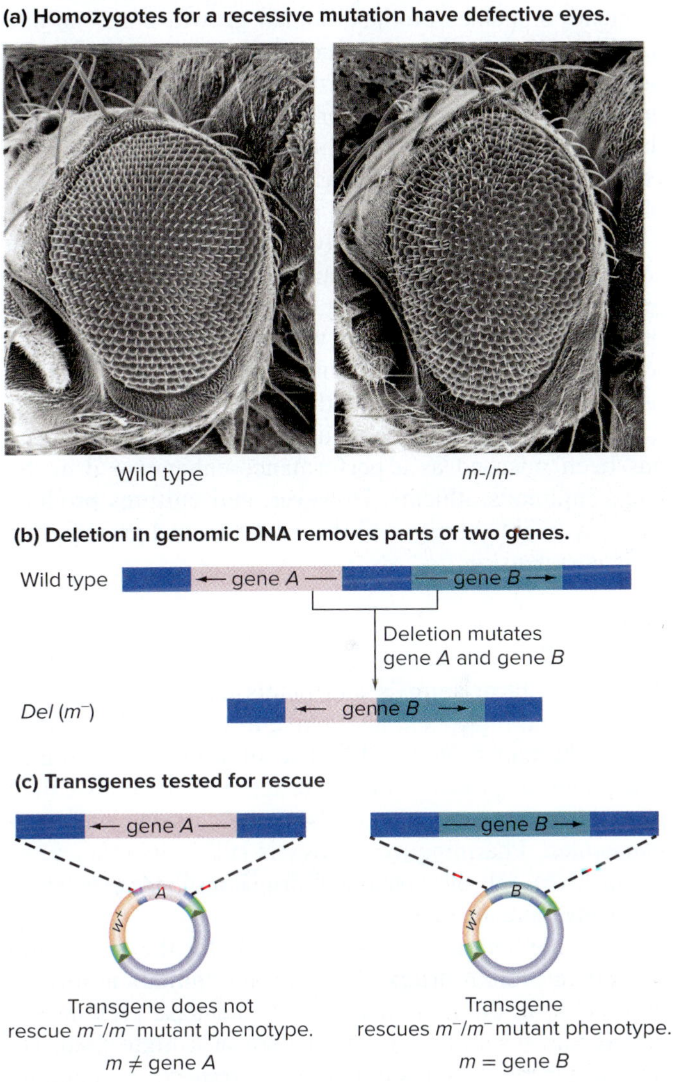

Transgenic Cells and Organisms Serve as Protein Factories

In Chapter 16, you saw that some human proteins used as drugs can be produced in bacteria transformed with fusion gene constructs. In these constructs, the coding sequences for the human protein were placed under the control of bacterial promoters and Shine-Dalgarno sequences that ensure high levels of gene expression (review Fig. 16.28). Pharmaceutical companies produce human growth hormone and insulin in this way. However, not all human proteins can be produced in a functional form in bacteria. Bacteria are unable to perform many important posttranslational operations, including proper folding or cleavage of certain polypeptides, or modifications such as glycosylation and phosphorylation.

To circumvent such problems, drug companies can sometimes use transgenic mammalian or plant cells that grow suspended in liquid culture. Several pharmaceutical proteins are produced in this way. One is factor VIII protein, the blood-clotting factor that is deficient in some people with hemophilia. Another is erythropoietin (EPO), a hormone that stimulates red blood cell production that has been misused as a performance-enhancing drug by some infamous athletes. However, cell cultures produce only low yields of recombinant proteins, and growing the cells is expensive.

Pharming

Transgenic farm animals and plants can provide a cost-effective and high-yield alternative to cell culture for producing human proteins. The use of transgenic animals and plants to produce protein drugs is sometimes called **pharming,** a combination of the words *farming* and *pharmaceutical.* Pharming technology is still in its infancy; so far (in 2016), only one "pharmed" drug is available to patients, but many more are in development.

The method used most commonly for the production of human protein drugs in transgenic animals is protein expression in the mammary glands, because proteins secreted into the milk can be purified at a high yield. By pronuclear injection (as in Fig. 18.3), transgenes encoding human proteins have been transferred to goats, pigs, sheep, and rabbits. In 2009, the United States Food and Drug Administration (FDA) approved the first human protein drug produced in the milk of a transgenic animal: the blood factor antithrombin III. The goats that produce this drug were transformed with a fusion gene in which the regulatory sequences of a goat gene normally expressed in the mammary gland were fused with the coding region of the human gene that encodes antithrombin III, a blood plasma factor that inhibits coagulation (**Fig. 18.8**). People with only one functional copy of the gene for antithrombin III tend to develop blood clots, particularly after surgery or childbirth; the drug is currently approved for patients with this genetic condition of *venous thromboembolic disease.*

Figure 18.8 A gene construct that produces a human anticoagulant protein in the milk of transgenic goats.

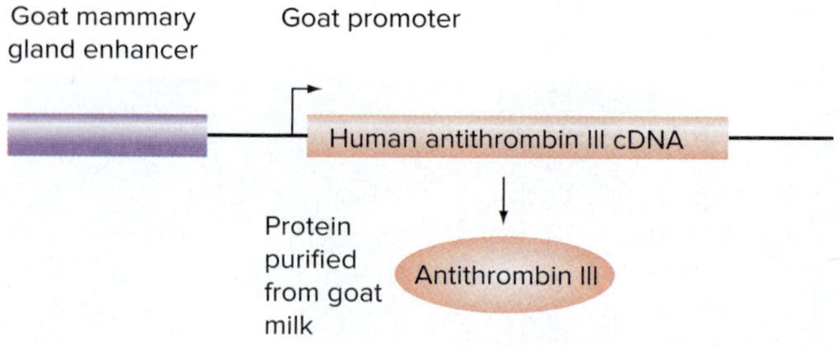

Individual transgenic animals produced by pronuclear injection will have variable numbers of transgene copies, and the transgene array will be present at different random genomic locations. These variations result in large differences in the human protein yield among individual injected animals. One way to enhance the value of a rare, high-producing animal is by **reproductive cloning:** using somatic cell nuclei of transgenic adults to generate other animals with the identical genomes. Not surprisingly, the same pharmaceutical companies that are developing the technology to produce drugs in transgenic animals are funding the development of animal cloning technology. The Tools of Genetics Box entitled *Cloning by Somatic Cell Nuclear Transfer* describes the most commonly used reproductive cloning technology.

Vaccine production in transgenic plants

Like transgenic animals, plants carrying transgenes can be used for the production of human protein drugs. Transgenic plants have particular advantages for making **vaccines,** antigens of a disease-causing agent that stimulate an immune response to that particular foreign substance. Vaccine proteins produced by transgenic crop plants such as tobacco, sunflower, spinach, potatoes, rice, soybeans, corn, or tomatoes could be stored in the leaves or seeds. The plants could simply be eaten to protect individuals from the pathogen. Edible vaccines could be especially advantageous for less-developed countries: No refrigeration is required for seed transport, plants could be grown on site, and no needles, syringes, or medical professionals would be necessary.

Despite the theoretical promise of producing vaccines in transgenic plants, trials to date have had only partial success, and many problems need to be overcome before any of these vaccines can be marketed. One major difficulty is controlling the dose of the antigen: Individual plants can vary in the amount of antigen they produce, and too little antigen will result in an ineffective vaccine. In addition, vaccines that are eaten require higher antigen doses than

TOOLS OF GENETICS

Cloning by Somatic Cell Nuclear Transfer

In Chapter 12, you were introduced to CC, the world's first cloned cat. *Cloning* in this sense refers to *reproductive cloning,* in which the genome of a single somatic cell from one individual now becomes the genome of every somatic cell in a different individual.

Researchers create reproductive clones through a protocol known as **somatic cell nuclear transfer.** Scientists take the diploid nucleus of a somatic cell from one individual and insert it into an egg cell whose own nucleus has been removed **(Fig. A)**. After several days of growth, the researchers implant the manipulated embryo into the uterus of a surrogate mother. After the embryo develops to term, a cloned animal is born. The cloned cat at the bottom of Fig. A could be thought of as having three different mothers: the somatic nuclear donor, the oocyte donor, and the surrogate who provided the womb. It is also possible to clone male animals if the somatic cell nucleus comes from a male.

Even though all of the nuclear chromosomes in all of the cells of the clone are derived only from the somatic nuclear donor, the cloned animal and this donor are not perfectly identical in all respects, for several reasons: (1) The mitochondrial genomes of the clone come from the oocyte donor, not the nuclear donor. (2) For female clones, the pattern of X chromosome inactivation in the clone and in the mother will not be the same because the decision of which X to inactivate is made randomly in individual cells early in the animal's development. (3) The uterine environment in the surrogate is not exactly the same as in the womb of the donor's mother.

The work leading to the cloning of CC was funded by a biotechnology company called Genetic Savings & Clone, whose mission was to provide commercial cloning services for pet owners who might want to replicate their animals after their deaths. The company was not ultimately successful. Few people could afford the high costs of the cloning procedure, and furthermore, some ill-informed clients were disappointed to find that the clone they received was not in fact exactly the pet they knew.

Good reasons nonetheless exist for the cloning of certain animals. Research on cloned animals enables scientists to better understand basic processes such as gene imprinting. Drug companies are investing in reproductive cloning technology with an eye toward being able to generate large numbers of high-producing transgenic animals. In fact, well before the cloning of CC, the first animal ever to be cloned from an adult cell was a sheep called Dolly in 1996. Dolly was cloned by scientists in Scotland, in part with funding from a pharmaceutical company. Before Dolly died in 2003, she gave birth to five progeny who live on. Finally, several endangered species have been cloned for the purpose of their preservation.

Figure A Cloning a cat by somatic cell nuclear transfer.

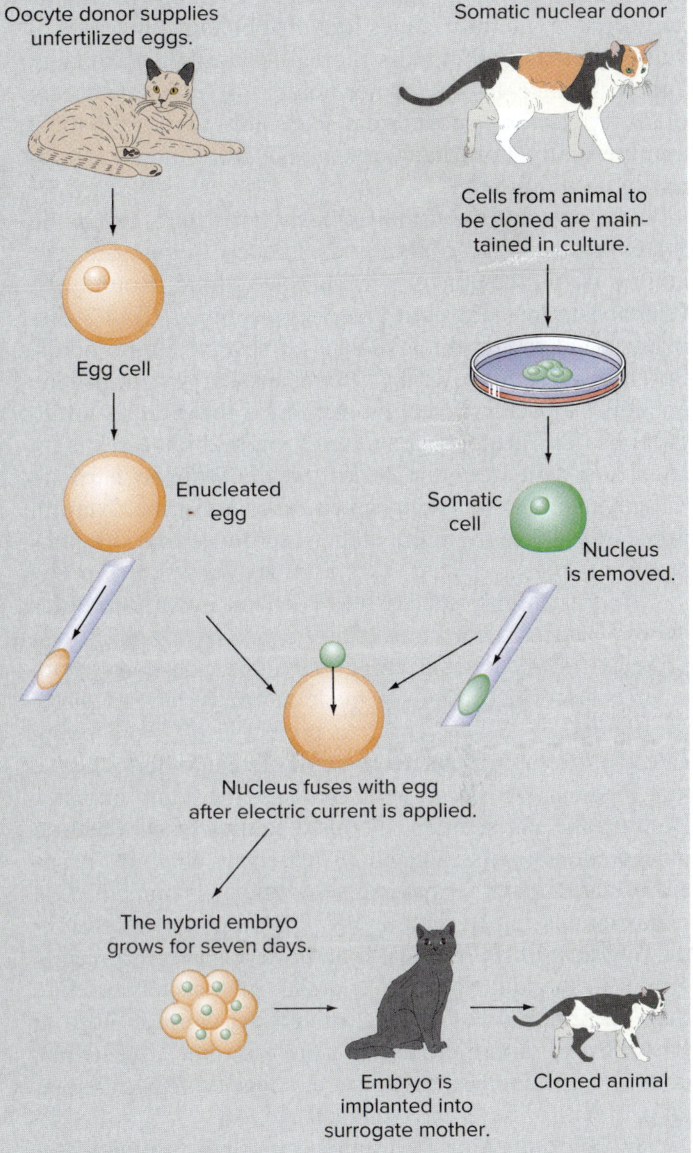

those that are injected. Even if the scientific problems can be overcome, drug companies will encounter many regulatory hurdles before making these plant-produced vaccines available to humans. Because the regulations are less strict, considerable recent attention has been placed instead on feeding transgenic vaccine-making plants to domestic animals, so as to protect them from various diseases caused by pathogenic organisms.

GM Organisms Are Used Widely in Modern Agriculture

As of 2016, more than 100 different transgenic plant varieties with improved traits have been created and grown by farmers. Such **GM (genetically modified) crops** now exist for many species. The improvements conferred by the transgenes include enhanced nutritional value; increased shelf life; increased yield or plant size; and resistance to stress, herbicides, or infestations by plant viruses or insects. We discuss here two of the most commercially important transgenic crops that are currently in wide use.

More than 90% of the soybeans grown in the United States are transgenic plants resistant to glyphosate, the active ingredient in the herbicide called Roundup®. Glyphosate interferes with an enzyme called EPSPS that plants need for the biosynthesis of several amino acids. So-called Roundup Ready® soybeans carry a transgene encoding a bacterial version of EPSPS that is resistant to glyphosate. Farmers spray fields of herbicide-resistant soybeans with Roundup to kill weeds with no harm to the soybeans, thus saving much labor and time. Natural processes in the environment then rapidly degrade glyphosate itself.

Another highly successful GM plant is corn that produces a natural, organic insecticide called Bt protein, which protects the plant from being eaten by corn-borer moth caterpillars. This protein is made naturally by the bacterium *Bacillus thuringiensis* to protect itself from being eaten by the caterpillars. Bt protein is lethal to insect larvae that ingest it, but not to other animals, including humans. Because the engineered corn manufactures its own natural insecticide, farmers can avoid using costly chemical pesticides that damage farmworkers and the environment. GM plants expressing Bt protein were grown commercially for the first time in 1996; at least one-third of all corn currently grown in the United States contains *Bt* transgenes. More than 10 billion acres of land around the world is used to grow Bt-expressing crops, not only corn but also canola, cotton, corn, papaya, potato, rice, soybean, squash, sugar beet, tomato, wheat, and eggplant.

In 2015, the first GM animal—Atlantic salmon produced by AquaBounty Technologies®—was approved for human consumption by the U.S. Food and Drug Administration. Atlantic salmon normally take three years to grow to their full size of about 9 pounds; their growth hormone gene is shut off during the coldest months when food is scarce, and so they grow only about eight months of the year. The GM salmon, which contain a growth hormone transgene that is expressed year-round, achieve their full weight in half that time.

GM crops are grown in the United States and in more than 25 other countries, where these organisms are regarded as important tools to help limit environmental problems caused by large-scale agriculture and to meet the food requirements of an increasing world population. However, other countries restrict or even completely bar the importation of GM organisms. Few scientists now believe that the ingestion of food made from GM organisms poses direct dangers to humans. However, some objections to GM organisms must be considered. The widespread use of GM crops may disrupt the lives of farmers and farm communities, and it places considerable power in a small number of transnational agribusinesses. Some potential environmental consequences may also exist, such as the unwanted transmission of traits from GM organisms to other species in the wild. These issues are likely to remain contentious in the coming years.

Transgenic Animals Model Human Gain-of-Function Genetic Diseases

Animal models of human genetic diseases have for decades been an important tool for scientists trying to understand disease biochemistry so as to design and test new drugs and other treatments. The idea of an animal model for a monogenic human disease is simple—to generate an animal with a corresponding mutation and a similar disease phenotype.

You should note that because transgenes are added to otherwise wild-type genomes, transgenic animals made by the techniques just described can serve as models only for dominant, gain-of-function mutations. (We discuss animal models for diseases caused by loss-of-function mutations in a subsequent section of this chapter.)

For many reasons, mice are the animals used most often to model human genetic diseases. Mice are mammals, and similar versions of most human genes are present in their genome. In addition, mice are small and relatively economical laboratory animals. But for the study of human neurological disorders, unfortunately, mice cannot replicate the complex effects of some gene mutations on brain functions and behavior. Instead, scientists have recently begun to model human diseases in transgenic laboratory monkeys—rhesus macaques.

The first transgenic primate model for a human neurological disorder was for Huntington disease. You will recall that Huntington disease is caused by a dominant allele of the *HD* gene with an expanded number of CAG trinucleotide repeats within the coding region. (Review the Fast Forward Box in Chapter 7 entitled *Trinucleotide Repeat Diseases: Huntington Disease and Fragile X Syndrome.*) The mutant allele encodes a form of the protein product (Huntingtin) that has more than the normal number of glutamine residues in its so-called polyQ region. Rhesus monkeys that model Huntington disease carry a transgene containing a mutant copy of the *HD* gene with an expanded CAG repeat region. These monkeys show disease symptoms similar to those of people with Huntington disease, helping scientists to understand the disorder and to develop more effective therapies.

Experiments with primates raise substantial ethical concerns for many people, so the future of primate models for human genetic diseases is unclear. As of this writing in 2016, the United States National Institutes of Health is in the process of phasing out most, though not all, invasive research on primate species.

essential concepts

- A wild-type transgene can be inserted into an embryo homozygous for a recessive mutant allele. If the normal phenotype is restored, then the transgene identifies the gene that was mutated.
- The creation of reporter constructs allows easy detection of when and in which tissues a gene is turned on or turned off in eukaryotes.
- Transgenic organisms produce medically important human proteins including insulin, blood clotting factors, and erythropoietin; transgenic crop plants can potentially make ingestible vaccines.
- GM soybeans are resistant to the weed killer glyphosate. Many crops, such as corn, soybean, canola, and cotton have been genetically modified to express Bt protein which discourages insect predation.
- Adding a transgene that carries a disease-causing, gain-of-function allele to a nonhuman animal model allows researchers to observe disease progression and to test possible therapeutic interventions.

18.3 Targeted Mutagenesis

learning objectives

1. Describe how ES cells are used to generate knockout mice.
2. Explain why an investigator might want to create a conditional knockout mouse.
3. Discuss how scientists employ a bacteriophage site-specific recombination system to generate knockin mice.
4. Describe CRISPR/Cas9 and how it is used to modify genomes.

In the previous section, you saw that genes can be transferred easily into random locations in the genomes of many animals and plants. Here we will explore more advanced technology that enables scientists to change specific genes in virtually any way desired—that is, **targeted mutagenesis.** A researcher needs only to know the DNA sequence of a gene in order to alter it; now that the genome sequences of all model organisms normally used in the laboratory have been determined, any gene in these species can be mutated at will.

We focus here mostly on methods to alter specific genes in mice, which are the animal of choice for many studies relevant to human biology. However, at the end of this section we describe an exciting new technique just coming into widespread use that is applicable to many different species.

Knockout Mice Have Loss-of-Function Mutations in Specific Genes

Homologous recombination provides a way for DNA sequences to zero in on specific regions of a genome. In fact, in Chapter 14 you have seen already that gene transfer by means of homologous recombination can make mutations in specific bacterial genes—a process called **gene targeting** (recall Fig. 14.31). In gene targeting, scientists mutagenize a specific gene *in vitro,* and then introduce the mutant DNA into bacterial cells. Homologous recombination then replaces the normal copy of the gene in the bacterial genome with the mutant copy. Although homologous recombination events are rare, investigators can grow large numbers of bacteria easily and then identify rare cells containing targeted mutations by selecting for a drug resistance marker present within the transferred DNA. Gene targeting in single-celled eukaryotes such as the yeast *S. cerevisiae* by the same method is also quite routine.

Mouse geneticists use mouse **embryonic stem cells (ES cells)** to surmount two main obstacles for gene targeting in multicellular organisms. First, for a chromosome containing a targeted gene to be transmitted to progeny, gene targeting has to occur in germ-line cells. Second, given the low efficiency of homologous recombination, investigators need to screen through a large number of germ-line cells to obtain one with the desired mutation. Mouse ES cells grow in a culture dish, so just as is done with bacteria or yeast, investigators can select rare cells containing a targeted mutation. A crucial aspect of this procedure is that the ES cells with targeted chromosomes can be moved from a cell culture dish to a developing embryo, where they can contribute to all different cell types, including germ-line cells.

Gene targeting in ES cells to generate knockout mice

Mouse ES cells are undifferentiated cells derived from the *inner cell mass* of early-stage embryos called *blastocysts* (**Fig. 18.9**). These ES cells are not yet committed to

Figure 18.9 Constructing knockout mice.

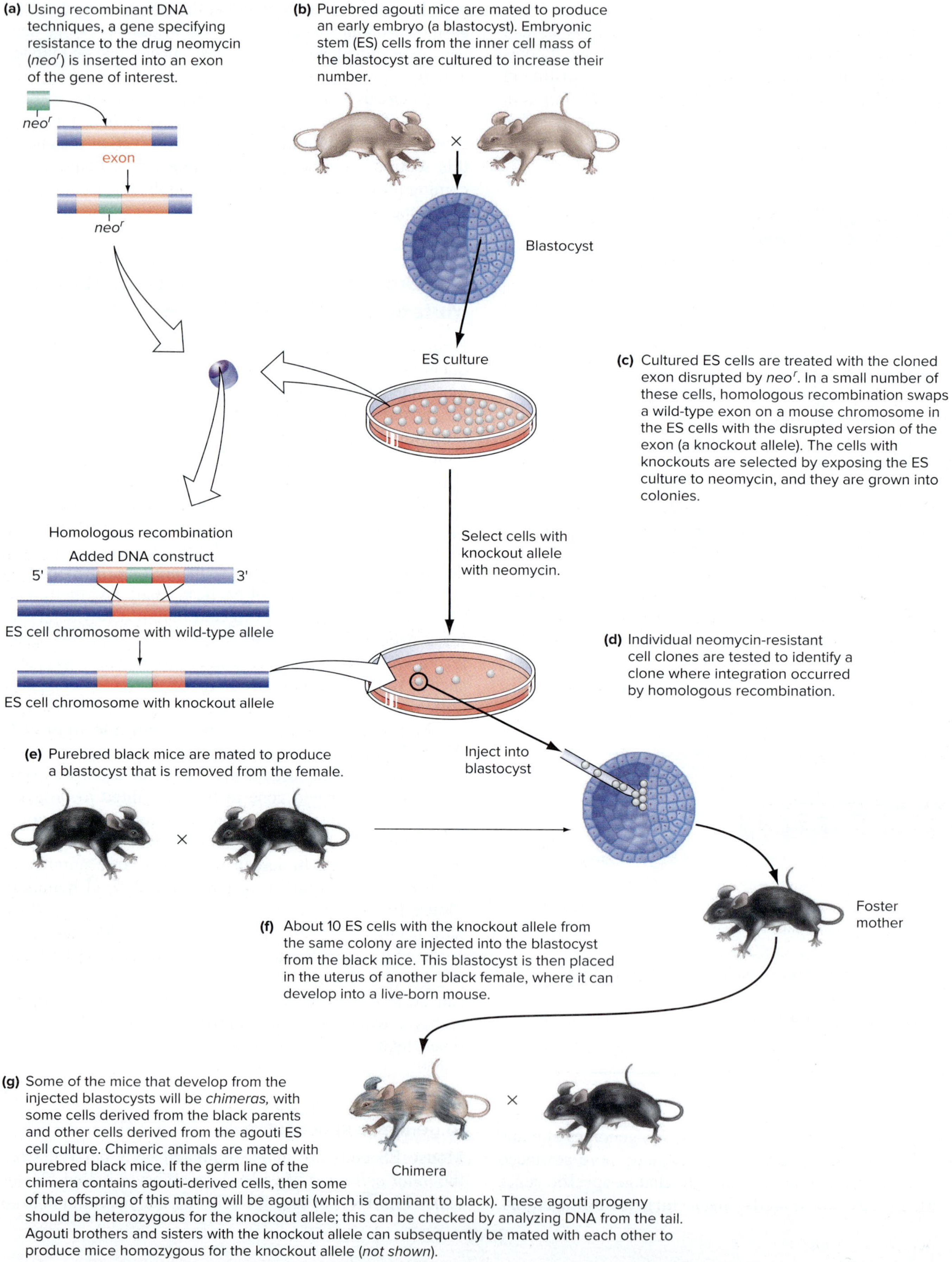

carrying out a pattern of gene expression characteristic of a particular type of cell, such as skin, bone, or blood. Because they are undifferentiated, mouse ES cells can divide for many generations in petri plates containing special culture medium. Most important, ES cells growing in culture are **totipotent,** meaning that they retain the ability to become any cell type—including germ-line cells and gametes—when exposed to the appropriate signals in a developing embryo.

Figure 18.9 presents the details of gene targeting with cultured ES cells. Researchers create DNA constructs in which a specific gene is mutagenized by the insertion of a drug resistance marker, and then they add this DNA into culture medium in which the ES cells are growing. Some of the cells will take up the DNA from the medium. After some time, the investigators add the drug to the culture medium; only the rare cells that have incorporated the exogenous DNA into their genomes survive and divide. The scientists then use PCR to identify lines derived from those surviving cells whose chromosomes acquired the drug resistance gene through homologous recombination; these cells are heterozygous for a loss-of-function mutation in the targeted gene.

The next steps of the gene targeting protocol take advantage of the ability of ES cells to be moved into an early embryo and to develop into any type of cell. The ES cells with the desired mutation are injected into host blastocysts, which are then implanted in surrogate mothers (Fig. 18.9). The mice that are born are called **chimeras,** meaning that they are made up of cells from two different organisms: The cells in these animals derived from ES cells are heterozygous for a mutation in the targeted gene, while the host cells are homozygous for wild-type alleles of this locus. Geneticists mate the chimeric mice to wild-type mice to generate nonmosaic heterozygous progeny. These progeny are called **knockout mice** because they contain a chromosome with an amorphic (knocked out) allele of the targeted gene. The geneticists then cross heterozygous knockout mice to each other to generate homozygous mutants.

Uses of mouse knockouts

The first knockout mouse was created in 1989, and eight years later the three scientists who developed this technology were awarded a Nobel Prize. As nearly every human gene has a counterpart in the mouse that has the same or a similar function, knockout mice are useful for studies of a variety of human diseases caused by loss of gene function.

One of the first monogenic diseases modeled in a knockout mouse was cystic fibrosis. Recall that the *CF* gene encodes CFTR, a membrane protein that regulates the flow of chloride ions into and out of lung cells (review Fig. 2.25). Using *CF* knockout mice, researchers discovered that cystic fibrosis results from mucus buildup in the lungs, caused at least in part by failure to clear bacteria from lung cells. Scientists are currently testing new experimental therapies for cystic fibrosis on *CF* knockout mice.

In a more general sense, knockout mice are invaluable for helping researchers understand the function of any gene in a mammalian organism. Simply put, a geneticist would generate homozygous knockout mice completely lacking the function of any given gene and then observe the effects on the organism's phenotype. In recognition of the importance of this approach, the U.S. National Institutes of Health funds the Knockout Mouse Project, a collaboration between academic, government, and industrial laboratories with the goal of creating, for every gene in the mouse genome, an ES cell line containing a knockout mutation.

Conditional Knockout Mice Reveal Functions of Essential Genes

For some genes, it is impossible to generate homozygous knockout mice. These so-called *essential genes* may be required for early stages of development of the animal, or for some process crucial to the viability of all cells. To investigate what the product of an essential gene does in the organism, researchers can use gene targeting to create mosaic individuals in which most cells are homozygous for wild-type alleles of the gene, and only certain cells are homozygous for mutant alleles. In an alternative application of the same kind of technology, the entire animal can remain homozygous for the wild-type allele until it reaches adulthood, after which the investigator can direct that some or even all of its cells become homozygous for the mutant allele.

Key to these strategies is the idea that scientists add DNA sequences to the targeted gene that, under certain conditions, cause an exon to be eliminated from the genomic DNA. Mice with genes that can be inactivated specifically when investigators alter the environmental conditions are **conditional knockout mice.** To achieve the conditional deletion of a gene's exon, genetic engineers exploit a naturally occurring site-specific recombination system that is important for the growth of bacteriophage P1 when it infects *E. coli* cells.

Cre protein and loxP sites

During its normal process of DNA replication, bacteriophage P1 produces large DNA circles containing many copies of the P1 genome. To generate single genomes that can be packaged into individual phage particles, an enzyme called *Cre recombinase* causes crossing-over to occur between two specific DNA sequences on the large circles called *loxP sites* **(Fig. 18.10)**. (Review Section 6.6 on

Figure 18.10 The Cre/loxP recombination system. Bacteriophage P1 generates single copies of its genome by site-specific recombination between 34-base-pair loxP sites mediated by Cre recombinase protein.

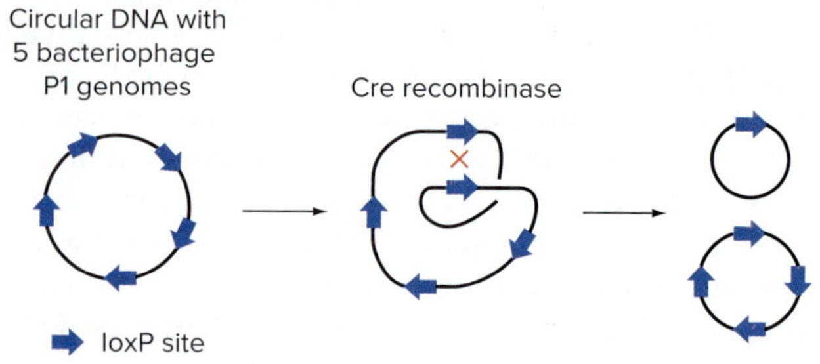

site-specific recombination for more details.) How do scientists harness the **Cre/loxP recombination system** to generate deletion mutations in a particular mouse gene in a specific tissue of the animal and/or at a specific time?

Making conditional knockouts

The first step is to use recombinant DNA technology to engineer a gene-targeting construct like the one shown in **Fig. 18.11a.** The construct contains two introns and an exon of the gene to be conditionally knocked out. One intron has a loxP site within it, while the other intron contains a drug resistance gene flanked by two loxP sites. The exon in the construct must be necessary for the function of the gene; in the case of protein-coding genes, the exon chosen usually contains part of the open reading frame (ORF).

In the second step, researchers use the procedure we just described in Fig. 18.9 to generate ES cells in which sequences in the targeting construct replace the corresponding DNA sequences in the endogenous wild-type gene in the mouse genome. The special ES cells used in these experiments were previously modified to contain a transgene that expresses Cre protein in an inducible manner; for example, the transgene could have a mouse promoter that is activated by heat shock fused to the coding sequences for Cre protein.

The third step is to remove the drug resistance gene. Growing the ES cells at high temperature turns on Cre expression, causing recombination to occur between a pair of loxP sites. One possible outcome is the chromosome shown at the bottom of Fig. 18.11a; the exon (or gene) is said to be **floxed**—*flanked by loxP sites.*

Now that ES cells heterozygous for the floxed gene have been obtained, researchers inject them into host blastocysts and generate heterozygous mice, just as described earlier in Fig. 18.9. Finally, geneticists perform a series of crosses to generate a mouse that is homozygous for the floxed gene and also carries a transgene that expresses Cre at a specific time or in a specific tissue. For convenience, we discuss here one example in which the *cre* transgene is transcribed only in the eye (**Fig. 18.11b**).

Because the loxP sites are placed in introns at positions that do not interfere with RNA splicing, the floxed gene functions normally. However, in eye cells, where Cre is expressed, Cre-mediated crossover between the loxP sites removes an exon from both copies of the targeted gene (Fig. 18.11b). Thus, the eye cells—and only the eye cells—are homozygous for a knockout of the gene.

You can see why floxed alleles are called conditional knockouts: These alleles function normally in all tissues, except in those cells in which Cre is made and deletes sequences from the gene.

Figure 18.11 Conditional knockout mice. (a) A multistep process replaces part of a gene in ES cells with a *floxed* exon flanked by loxP sites. These ES cells are incorporated into the mouse germ line. **(b)** In this example, transgenic mice homozygous for the floxed gene also contain a transgene that expresses Cre only in the eye. Only in eye cells but not elsewhere in the body, Cre-mediated site-specific recombination at the loxP sites removes the exon from both copies of the floxed transgene.

(a) Floxing a mouse gene

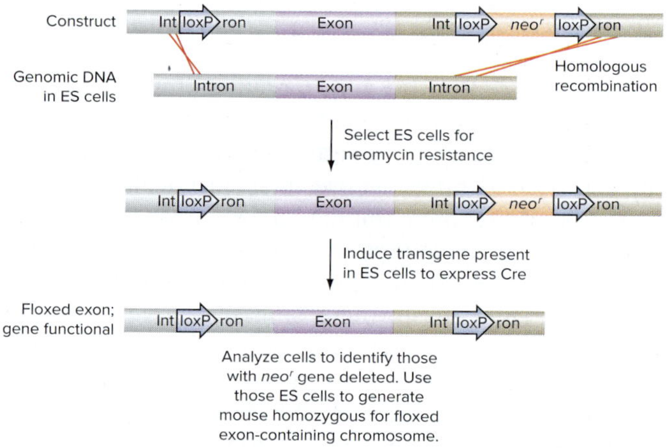

(b) Conditional knockouts with floxed genes.

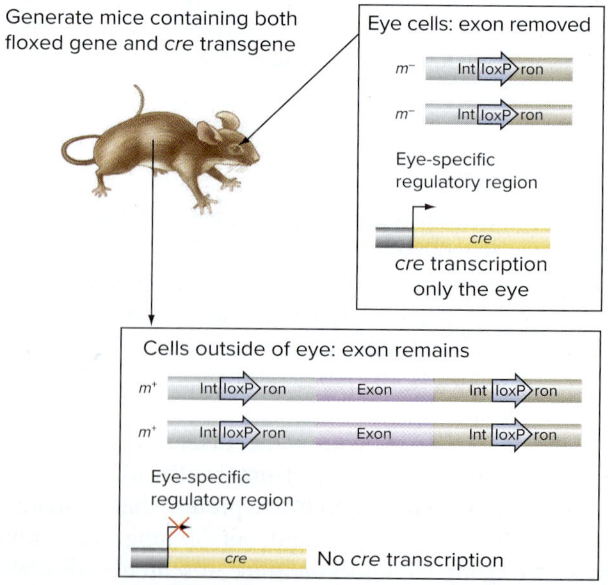

Using conditional knockouts

What kind of information can conditional knockout mice provide to scientists? As just one example, suppose that the targeted gene in Fig. 18.11 is required for the viability of all cells. If this is the case, it is likely that after Cre expression the animal with the floxed allele will be eyeless. Alternatively, the targeted gene may be essential only in tissues outside of the eye, in which case the eyes (and indeed the whole animal) will be normal. Yet another possibility is that the gene is required in the eye for a specific function, such as forming the retina; in this case, Cre expression in homozygotes for the floxed gene would result in malformed retinas.

Interestingly, the juxtaposition of knockout and wild-type cells in mosaic animals can allow scientists to determine whether a gene product expressed in one cell can affect the function of a neighboring cell. This topic—so-called *mosaic analysis*—will be discussed in Chapter 19.

Knockins Introduce Specific Mutations into Specific Genes

Scientists can use ES cell technology not only to destroy the function of a mouse gene, but also to alter genes in other, more specific, ways. For example, an investigator may want to create a mouse that has a particular missense mutation that changes one amino acid in a protein to a different amino acid. **Figure 18.12** illustrates that researchers can easily modify the techniques we have just described for creating conditional knockout mice so as to change a given gene in any way desired. The new DNA introduced, in this case an exon with an altered DNA sequence, is said to be *knocked in*. **Knockin mice** homozygous for the mutant gene will produce only the mutant form of the protein, although mice heterozygous for the knocked-in mutation may also be valuable if the mutation has dominant effects.

This ability to replace a gene in the mouse genome with an allele engineered to have any change of interest is important for the creation of mouse models for certain genetic diseases in humans. Many inherited diseases are associated not with a completely amorphic allele of a gene, but rather with missense mutations in the codons for specific amino acids that might have hypomorphic, hypermorphic, or neomorphic effects.

One such example concerns mouse models for achrondroplasia, or dwarfism. Recall from Chapter 8 that short-limb dwarfism in humans is caused by a missense mutation in the human *FGFR3* gene, resulting in a gain-of-function, constitutively active FGF receptor protein (Fig. 8.31). Researchers have used knockin technology to engineer mice with exactly the same point mutation in their homologous *FGFR3* gene. Remarkably, this mutant gene produces in mice what appears to be the same dominant dwarfism phenotype seen in human achrondroplasia (**Fig. 18.13**).

Figure 18.12 Knockins to make specific mutations in a gene. Genetic engineers can use the construct shown with a point mutation indicated by the *red* asterisk (*) to replace the corresponding sequences in ES cells that contain an inducible Cre transgene. After removal of the drug resistance gene by Cre expression, the loxP site that remains in the intron will not interfere with splicing.

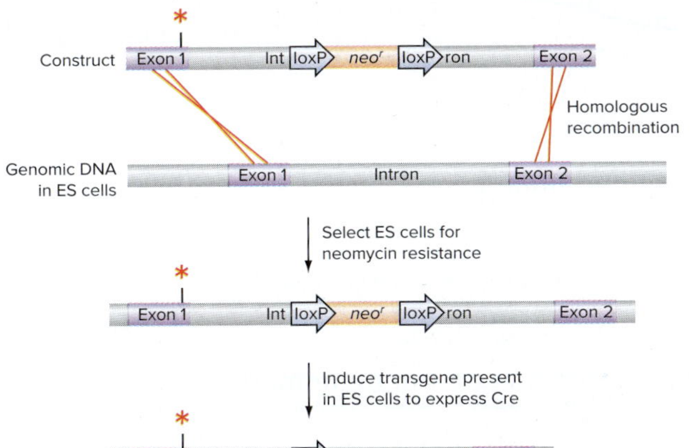

CRISPR/Cas9 Allows Targeted Genome Editing in Any Organism

Using the ES cell technology just described, a researcher can remove, add, or change any DNA sequence in the mouse genome in almost any manner. Until recently, only the mouse genome could be altered with this kind of precision. Several newly developed **gene editing** technologies now allow scientists to alter the genome of virtually any organism so as to make knockouts or knockins even without the use of ES cells. This new technology enables scientists to create

Figure 18.13 Achondroplastic dwarfism in the mouse. The dwarf mouse at the *right* is heterozygous for an *FGFR3* allele with the same amino acid sequence as that causing achondroplasia in humans. A wild-type littermate is at the *left*.

From: Wang et al. (1999), "A mouse model for Achondroplasia produced by targeting fibroblast growth factor receptor 3," *PNAS*, 96(8): 4455-4460. © 1999 National Academy of Sciences, U.S.A.

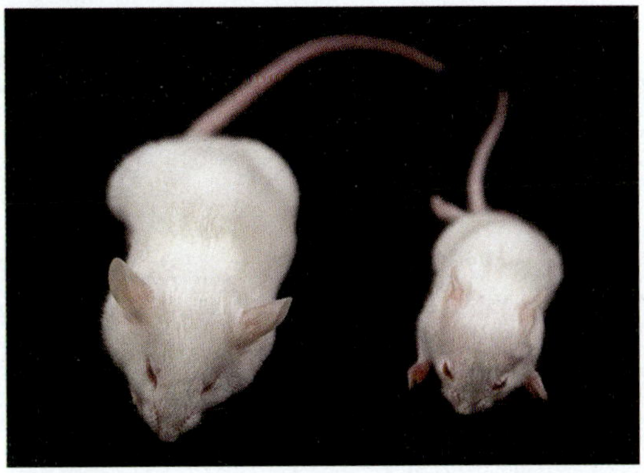

mutant mice more efficiently. Of wider importance, researchers can apply the same tools in animals other than mice, or even in cultured cells, opening up many possibilities for the study of gene function and to establish new models for human diseases.

In all of these technologies, either a protein or an RNA molecule serves as a guide that brings a DNA-cleaving enzyme to a specific genomic location. DNA repair of the break can then result in a point mutation (a base pair change, or insertion or deletion of one or a few pairs) or a knockin of specific DNA sequences. We describe here the newest and most efficient genome editing system, called *CRISPR/Cas9.*

CRISPR is an acronym for *clustered regularly interspaced short palindromic repeats*. Many bacterial genomes contain a CRISPR region, which functions as an antiviral immune system. CRISPR immunity also depends on endonucleases called **Cas proteins** (CRISPR-associated proteins) encoded by the bacterial genome; these enzymes can make double-stranded breaks in DNA. The Tools of Genetics Box entitled *How Bacteria Vaccinate Themselves Against Viral Infections with CRISPR/Cas9* describes in detail how bacteria use this mechanism to ward off infection by bacteriophages. The attention of the scientific community became focused on CRISPR/Cas9 when researchers realized they could adapt this system for use in any organism.

The genetically engineered **CRISPR/Cas9 system** has two components. The first is an investigator-designed, single-stranded RNA called *sgRNA* (*single guide RNA*). At the 5′ end of the sgRNA is a 20 bp sequence that is complementary in sequence to a target site of interest in the genome to be altered. The 3′ end of the sgRNA binds specifically to the Cas9 protein. (As an aside, the 5′ and 3′ regions of the sgRNA correspond respectively to the *crRNA* and *tracrRNA* in the Tools of Genetics Box.) The second component is a Cas9 polypeptide that has been altered so that it includes a short stretch of amino acids that constitute a *nuclear localization signal*, allowing the protein to be imported into the nucleus where it can act on DNA.

In the nucleus, Cas9/sgRNA complexes seek out and bind to their designated genomic DNA target. The Cas9 enzyme within the complexes then makes a double-strand break in the target DNA (**Fig. 18.14**). Repair of the break by nonhomologous end-joining (NHEJ; review Fig. 7.18) often results in a small insertion or a deletion of a few base pairs at the break. Such a mutation can knockout the function of a gene, for example if it corresponds to a frameshift mutation in an open reading frame.

Alternatively, if DNA molecules corresponding to the DNA flanking the break are introduced into cells at the same time as the Cas9/sgRNA, double-strand break repair by homologous recombination can incorporate that DNA into the genome at the break site, generating a knockin (Fig. 18.14). Double-strand breaks are *recombinogenic*

Figure 18.14 Genome editing by CRISPR/Cas9. The sgRNA sequence is designed to bring the Cas9 endonuclease to a specific target in the genome. Repair after Cas9 cleavage can result in a knockout or a knockin, depending on whether or not a DNA fragment suitable for homologous recombination is introduced. NHEJ: nonhomologous end-joining.

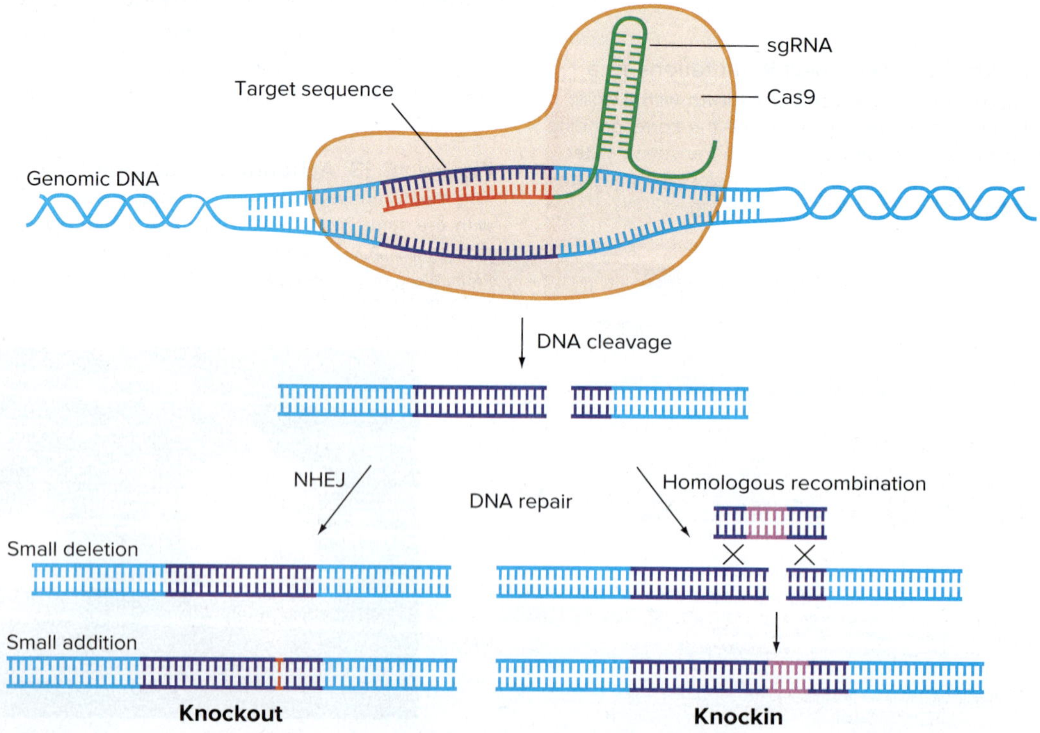

TOOLS OF GENETICS

How Bacteria Vaccinate Themselves Against Viral Infections with CRISPR/Cas9

Researchers discovered clustered sequence repeats (CRISPRs) in bacterial genomes as early as 1987. When in 2005 some of these sequences were found to originate from bacteriophage genomes, several astute scientists speculated that CRISPRs might mediate a viral immunity system in bacteria. These ideas were largely ignored for several more years until the mechanism of resistance became clarified. And finally, in 2012–2013, the so-called *CRISPR craze* reached its full bloom when researchers including Feng Zhang, Jennifer Doudna, and Emmanuelle Charpentier developed methods to adapt this viral immunity system to engineer genomes in bacterial cells and in eukaryotic organisms.

At the *CRISPR* locus of bacterial genomes, short direct repeats are interrupted at regular intervals by unique spacer sequences (**Fig. A**). The spacer sequences are fragments of bacteriophage genomes captured by the host cell and integrated into the host genome by the action of two bacterially encoded Cas proteins (Cas1 and Cas2). The repeats within the *CRISPR* arrays are added by these endonucleolytic enzymes during the capture and integration process.

Viral immunity results from steps that begin with transcription of the *CRISPR* array into long RNA molecules called *pre-crRNAs* that are processed into short (24–48 nt) so-called *CRISPR RNAs* (*crRNAs*). In the bacterial species

Figure A The *CRISPR/Cas9* locus vaccinates bacteria against viruses.

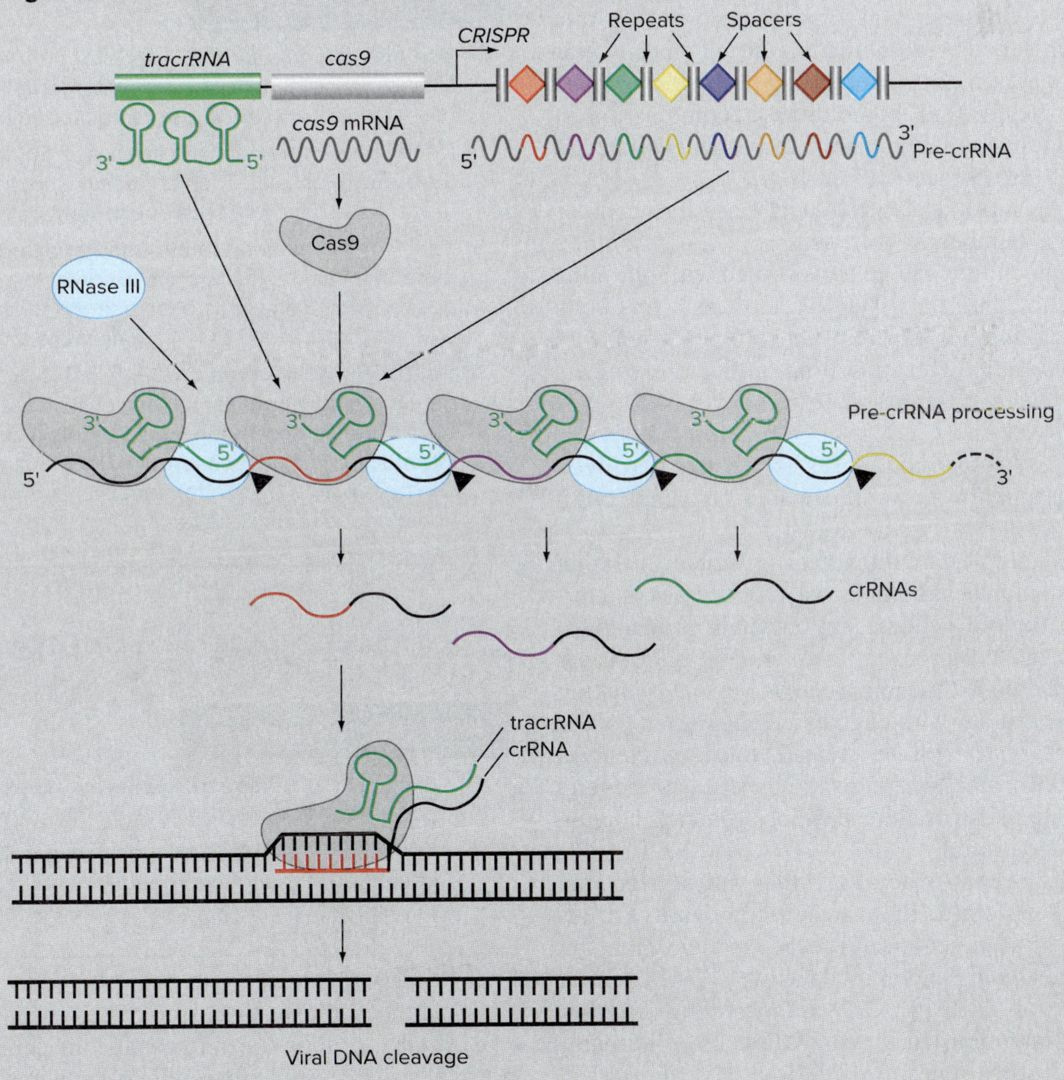

Streptococcus pyogenes, this chopping-up of the large RNA requires another small RNA called *tracrRNA* (*trans-acting CRISPR RNA*) that is transcribed from a gene in the host genome (Fig. A). The tracrRNA forms complementary base pairs with the repeat sequences in the pre-crRNA. These double-stranded RNA regions become substrates for the endonuclease RNase III; cleavage at these locations produces the short crRNAs complexed with Cas9 protein.

When an invading virus injects its double-stranded DNA chromosome into the host cell, a specific crRNA, the tracrRNA, and the Cas9 enzyme cooperate to cleave the viral genome (Fig. A). The 5′ end of the crRNA base pairs with its target DNA in the bacteriophage chromosome, while the 3′ end of the crRNA base pairs with tracrRNA to form a stem loop that binds Cas9. Thus, the crRNA and tracrRNA together bring Cas9 enzyme to the target sequence in the viral genome. Cas9 cleaves the phage chromosome, preventing infection.

For editing the chromosomes of eukaryotic organisms, researchers found that they could join the crRNA and tracrRNA together in a single RNA molecule (an sgRNA) that brings Cas9 to the site of interest in the eukaryotic genome (Fig. 18.14).

It is worthwhile to note that two of the major technologies for manipulating DNA molecules—the use of restriction enzymes for DNA cloning and CRISPR/Cas9 for genome editing—emerged from studies of bacterial immunity mechanisms. These examples of the ways bacterial cells degrade viral chromosomes demonstrate the importance of basic science research: Topics that might have seemed obscure at the outset can have immense practical applications once they are understood.

(that is, they stimulate recombination); as you will recall from Chapter 6, double-strand break formation is in fact the first step in genetic recombination during meiosis. For this reason, knockin formation using CRISPR/Cas9 is much more efficient than homologous recombination in cases where double-strand breaks in the target genome are not introduced (as was seen in Fig. 18.9).

Researchers introduce sgRNA and Cas9 into cells or organisms in a number of different ways. For example, cells growing in culture can be transformed with plasmids containing genes that make sgRNA and Cas9, and cells with modified genomes can be used to clone organisms (refer to the Tools of Genetics Box entitled *Cloning by Somatic Cell Nuclear Transfer*). Alternatively, scientists can inject sgRNA and Cas9 genes (DNA) or transcripts (RNA) or sgRNA/Cas9 complexes into newly fertilized eggs that subsequently grow into whole organisms with modified genomes.

These technologies are remarkably efficient, such that researchers can mutate simultaneously both copies of a target gene in a diploid genome, immediately producing a mutant homozygote. Moreover, if two or more sgRNAs are introduced at the same time into a cell or egg that is making Cas9, a double mutant with engineered changes in two or more different genes will be created. Such efficiencies eliminate tedious and time-consuming steps of genetic crosses to create plants or animals whose genomes contain multiple modified genes.

In the few years since scientists introduced the CRISPR/Cas9 system to the research community, it has been used to edit the genomes of every model organism, and many other plants and animals as well. One exciting potential use for CRISPR/Cas9 is correcting disease-causing gene mutations in the somatic cells of humans. *Gene therapy,* using genes to cure disease, is the subject we discuss next.

essential concepts

- Use of *embryonic stem (ES) cells* for *gene targeting* allows mouse geneticists to mutate germ-line cells in order to create stable lines of *knockout* mutant mice.
- *Conditional knockout* mice are useful for analyzing the functions of essential genes because such mice can be made mosaic for wild-type and mutant cells.
- *Knockin* mice generated through gene targeting carry a gene that has been altered in any way the researcher desires, including single base pair missense mutations that may correspond to human hereditary diseases.
- Genetically engineered *CRISPR/Cas9* can alter the genome of any organism at a specific location designated by the researcher. The key step of this technology is the introduction at that location of a double-strand break that can be repaired by nonhomologous end-joining or by homologous recombination.

18.4 Human Gene Therapy

learning objectives

1. Explain how therapeutic genes may be delivered to patients.
2. Describe problems associated with the use of viral vectors to introduce therapeutic genes into the cells of patients.

In all of the examples discussed to this point in the chapter, the ultimate goal has been to change the genomes of germ-line cells so that stable lines of experimental organisms can be created. However, changing a human gene in a manner that allows the changed gene to be transmitted to the next generation presents many ethical difficulties and is currently considered unethical in most of the world. The Genetics and Society Box entitled *Should We Alter Human Germ-Line Genomes?* discusses some of these issues.

Medical scientists developing methods for **gene therapy**—the use of DNA to cure disease—instead focus on altering the genomes of somatic cells in patients. As of 2016, more than 2200 gene-therapy clinical trials have been conducted. Although gene therapy in humans is still experimental, recently some promising successes have been achieved.

Different Diseases Require Different Gene Therapies

The idea of gene therapy is straightforward: Introduce into a patient's somatic cells a **therapeutic gene**—a gene whose expression will fight disease. However, no single strategy of gene therapy can work for all diseases.

Choosing a therapeutic gene

The molecular nature of the particular disease dictates what type of gene could be employed for therapy. For a disease caused by a loss of gene function, such as cystic fibrosis, the therapeutic gene would simply be a wild-type copy of the gene whose function was lacking. A different strategy is required to combat gain-of-function conditions such as Huntington disease, in which expression of a mutant protein (or in other cases overexpression of the normal protein) causes the aberrant phenotype. In such cases, the therapeutic gene would need somehow to inactivate the disease gene or its protein product.

Finally, for diseases with complex genetic origins such as cancer, the most generally useful strategy might be more indirect. As an example, a cancer gene therapy might target a gene or protein needed for a biochemical pathway leading to cell proliferation, even if that gene or protein was normal in the patient's cancerous cells.

Methods of therapeutic gene delivery

After a medical researcher chooses what therapeutic DNA to use, the next issue is to decide how to deliver this DNA into the somatic cells where they would do the most good. For example, it makes sense to introduce the DNA into retinal cells to treat congenital blindness, into lung cells to aid cystic fibrosis patients, or into blood cell precursors to treat anemias.

If doctors can access the tissue easily but the cells cannot be removed from the patient's body, the technique of choice is ***in vivo* gene therapy,** in which genes are delivered directly to somatic cells. The therapeutic gene could be injected into retinal cells, for example, or inhaled into the lungs. These types of delivery methods are easy to perform, but they can introduce the DNA into only a fraction of the affected cells. For other diseases in which the affected cells can be removed from the body, more potent ***ex vivo* gene therapy** can be used. Here, researchers remove tissues such as blood precursor cells from the patient's bone marrow, expose the cells to large amounts of the therapy gene while the cells grow briefly in culture, and then return the cells to the patient's body.

For either *in vivo* or *ex vivo* approaches, the foreign DNA needs to be packaged in some kind of vector so that it can be taken up by enough cells for successful therapy. Most gene therapy trials have been performed with either of two types of viral vectors: **retroviral vectors** or **adeno-associated viral vectors (AAV vectors).** In both cases, specialized cell lines package the therapeutic DNA into virus-like particles (**Fig. 18.15**). These cell lines carry

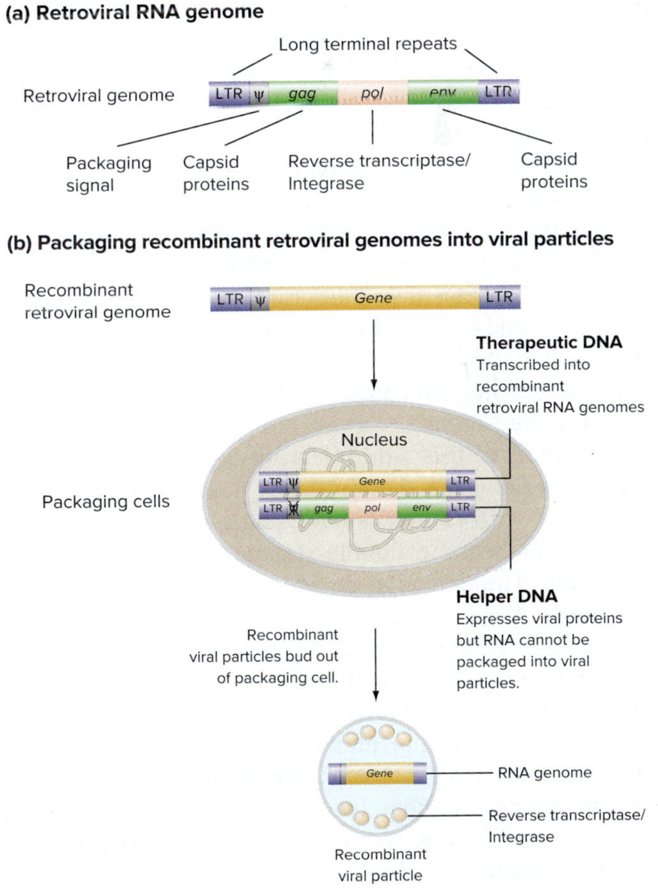

Figure 18.15 Packaging recombinant DNAs into retrovirus particles. (a) Normal retroviral RNA genome. LTRs, long terminal repeats needed to integrate the viral cDNA into the host chromosome; Ψ, *cis*-acting sequence needed to package retroviral RNA into a capsid. **(b)** Making recombinant retroviruses. In the recombinant retroviral genome, the therapeutic gene replaces the *gag, pol,* and *env* genes. In a *packaging cell,* a defective *helper* viral genome supplies *gag, pol,* and *env,* but the helper RNA lacks Ψ, so it cannot be packaged into viral particles. These cells produce virus-like particles containing the therapeutic gene, but no viruses that can replicate by themselves within cells.

Figure 18.16 Gene therapy with viral vectors.
(a) Recombinant retroviral genomes integrate into the patients' genome. **(b)** Recombinant AAV genomes usually remain extra-chromosomal.

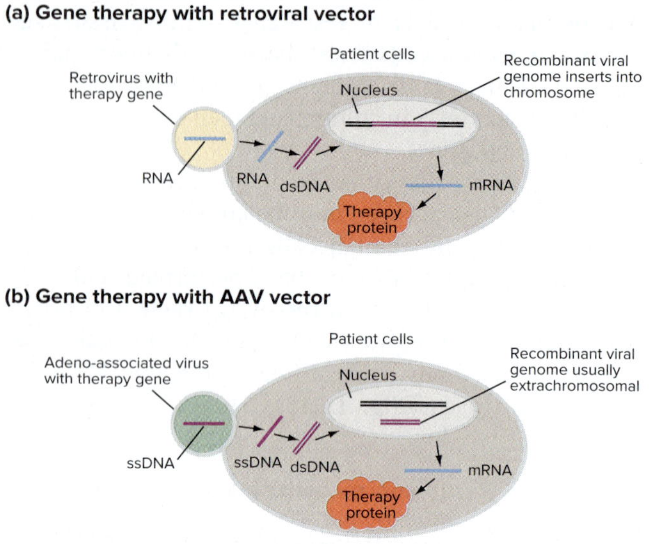

helper DNA that expresses all the proteins that make up the viral particles, but the cell lines by themselves do not make active viruses because transcripts from the helper DNA lack sequences that would allow them to be packaged into viral particles. The therapeutic DNA is cloned into modified, defective viral genomes that lack the viral genes but have the packaging sequences. Thus, when the cells are transfected with the cloned therapeutic DNA, it is packaged into virus-like particles that can infect cells but that lack almost the entire viral genome.

Once made, researchers infect the patient's cells with the recombinant virus-like particles, whether *in vivo* or *ex vivo*. In the case of recombinant retroviruses with RNA genomes, the therapeutic gene integrates into the patient's genome, where it is then transcribed; the mRNA produces the therapeutic protein (**Fig. 18.16a**). One problem associated with retroviral vectors is that their integration can result in genome mutation and in some cases can cause cancer in the patient.

Because of this side effect observed with retroviral vectors, other gene therapy trials have been performed with AAV vectors. AAVs have single-stranded DNA genomes that become double-stranded after infection of host cells. The defective double-stranded AAV genomes containing the therapeutic gene usually do not integrate into the host cell chromosomes, though they can still produce the therapy protein (**Fig. 18.16b**). Although the use of AAV vectors alleviates serious problems caused by vector integration into human chromosomes, the extrachromosomal AAV DNA is eventually degraded. This means that AAV-mediated gene therapy needs to be repeated periodically to provide a constant supply of the therapy protein.

Human Gene Therapy Holds Promise for the Future

The first encouraging results with gene therapy came in 2000, when several people with X-linked severe combined immune deficiency (SCID-X1) were cured substantially. SCID-X1 is commonly known as *bubble boy disease* because children with the disease have no functional immune system and are forced to live in sterile environments to avoid infections.

The cause of SCID-X1 is loss-of-function mutation of a gene called *IL2-RG,* which normally encodes a protein that promotes the growth of several different kinds of immune system cells. Researchers employed an *ex vivo* approach, in which they obtained immune system precursor cells from the patients' bone marrows and then infected these cells with recombinant retroviruses containing a wild-type copy of *IL2-RG.* Patients administered the altered cells regained immune system function and were able to resist infection. Of nine children treated at that time, eight are still alive and have successfully resisted many infections. However, four of the patients eventually developed leukemia because the retroviral vector had inserted adjacent to a gene involved in cell proliferation; one of these children has succumbed to the cancer.

Because of the problems sometimes caused by retroviral insertion, most recent attempts at gene therapy employ AAV vectors. At the beginning of the chapter, we discussed how a form of congenital blindness has been partially cured by gene therapy. Doctors injected recombinant AAV vectors containing a normal copy of the *RPE65* gene missing from the patients' genomes into their retinal epithelial cells. In the majority of cases, the patients regained at least some sight and suffered no ill effects from the gene therapy. These people may at some time in the future require additional treatments when the introduced extrachromosomal DNA gets degraded or lost.

Gene therapy is still experimental; you can see that many technical problems need to be surmounted for gene therapy to become standard medical practice. Nonetheless, results to date are sufficiently encouraging so that researchers around the world are testing new ideas to treat more conditions with gene therapies. For example, for diseases caused by gain-of-function mutations in a patient's genome, scientists are examining the use of synthetic therapy genes that express small interfering RNAs (siRNAs). As was discussed in Chapter 17, the presence of an siRNA in a cell could potentially block the translation, or cause the degradation, of the mRNA transcribed from the mutant gene (review Fig. 17.33). A particularly intriguing future possibility that has generated much excitement among gene therapy researchers is the use of genome editing, such as CRISPR/Cas9 technology, to repair mutant alleles, whether gain-of-function or loss-of-function, in human cells.

GENETICS AND SOCIETY

© Image Source/Getty Images RF

Should We Alter Human Germ-Line Genomes?

In April 2015, Chinese scientists reported the use of CRISPR/Cas9 to correct ß-globin gene mutations—the cause of the disease ß-thalassemia—in human embryos. Although these embryos were never placed in a womb, this publication opened a firestorm of controversy because some descendants of embryonic cells eventually will become sperm or eggs that could be passed down to future generations. In other words, these studies demonstrated forcefully that gene editing technology is becoming powerful enough that humans will soon be able to change their own evolutionary destiny.

In response to this report, the governments of the United States, the United Kingdom, and China organized an international summit on human gene editing, held in Washington, D.C. in December of 2015 and attended by more than 500 scientists, ethicists, and legal experts from 20 countries. The strong consensus of the summit was that genome editing of human embryos intended for pregnancy is premature because its safety cannot be ensured, but the participants were divided as to whether the goal of altering human germ lines is ethical or desirable. As of this writing in 2016, the British and Chinese governments are likely to continue to fund research involving genome editing of human embryos not destined for pregnancy, but in the United States only private agencies fund such investigations.

Gene editing of somatic cells to cure the symptoms of disease is relatively noncontroversial, but altering germ-line genomes raises many issues. Some of these issues are technical. For example, CRISPR/Cas9 technology is powerful, but it can cause unwanted off-target effects that alter sequences elsewhere in the genome. The consequences of these off-target mutations when transmitted over many generations are unpredictable; this is why the international summit concluded the method is premature.

But even if the technologies can be perfected, should we ever employ them to alter human germ lines in eggs, sperm, or embryos? Some people believe the entire idea is unethical because decisions made now will impact our descendants without their consent. Because it is conceivable that genome modifications can be made eventually that will enhance traits like intelligence, some people argue that germ-line editing technologies will inevitably lead to a further stratification of society: Likely, only wealthy individuals would be able to afford to have "designer children" with these enhanced characteristics. But on the other side of the issue, some scientists argue that if gene editing can be shown to be safe, without off-target effects, it would be unethical *not* to use this technology, at least to eradicate disease if not to improve human traits.

Genome editing methods are advancing so rapidly that these issues will soon go beyond interesting theoretical debates to the point where they have real impact on people's lives and those of future generations. If mankind will intentionally alter its own evolution, we had better be sure that the vast potential implications of these decisions have been thoroughly considered.

essential concepts

- *Therapeutic genes* can be delivered in recombinant viral vectors to somatic cells of patients either *in vivo* or *ex vivo*.
- *Retroviral vectors* insert therapy genes into human chromosomes, but this method can result in gene mutation and cancer.
- DNA introduced in *adenoviral vectors* remains extrachromosomal, necessitating periodic repeats of the therapy.
- Scientists are gearing up to use genome editing methods such as CRISPR/Cas9 to repair mutant genes in human somatic cells.

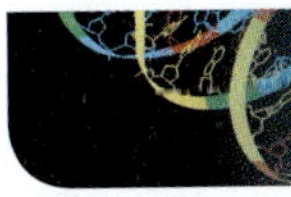

WHAT'S NEXT

Manipulation of the genome is the basis for many of the experimental strategies we will describe in Chapter 19, where we discuss how genetic analysis has been a crucial tool in unraveling the biochemical pathways of development—the process by which a single-celled zygote becomes a complex multicellular organism. Transgenic technology is key to cloning the genes identified in mutant screens that are crucial for regulating development, and also to manipulating these genes in order to understand their precise functions in the organism.

DNA: © Design Pics/Bilderbuch RF

SOLVED PROBLEMS

If you wanted to make a mouse model for any of the following human genetic conditions (a–d), indicate which of the following types of mice (i–vi) would be useful to your studies. If more than one answer applies, state which type of mouse would most successfully mimic the human disease: (i) transgenic mouse overexpressing a normal mouse protein; (ii) transgenic mouse expressing normal amounts of a mutant human protein; (iii) transgenic mouse expressing a dominant negative form of a protein; (iv) a knockout mouse; (v) a conditional knockout mouse; and (vi) a knockin mouse in which the normal allele is replaced with a mutant allele that is at least partially functional. In all cases, the transgene or the gene that is knocked out or knocked in is a form of the gene responsible for the disease in question.

a. Marfan syndrome (a dominant disease caused by haploinsufficiency for the *FBN1* gene);

b. A dominantly inherited autoinflammatory disease caused by a hypermorphic missense mutation in the gene *PLCG2;*

c. A deletion of the *amelogenin* gene that results in a recessive X-linked condition in which patients have much reduced tooth enamel;

d. An inherited form of deafness due to homozygosity for a recessive mutation that prevents expression of an ear-specific alternative splice of the primary transcript for the gene *TRIOBP*. Other splice forms of this gene are expressed in, and necessary for, the growth of all cells.

Answer

a. Although Marfan syndrome is a dominantly inherited disease, it is due to a haploinsufficient loss-of-function mutation. Transgenic mice cannot provide a loss of function, except in the case of transgenic mice expressing a dominant negative form of the protein. However, the latter would be very tricky to use, as you would somehow need to obtain a transgenic mouse that had exactly half of the function of the protein in question. The simplest approach would be to construct a knockout mouse for the Marfan syndrome gene (iv). A conditional knockout may not be needed because a mouse heterozygous for a simple knockout might survive and have symptoms similar to those of human Marfan patients.

b. For this gain-of-function condition, it is possible that a transgenic mouse of classes (i) or (ii) might have appropriate disease symptoms. However, the closest situation to the human disease would be a mouse that is heterozygous for a knockin allele in which the mouse *PLCG2* gene was replaced by the mutant human gene or by a mouse gene carrying the analogous mutation (vi).

c. This disease condition, called *amelogenesis imperfecta,* is caused by hemizygosity or homozygosity for a null allele of the *amelogenin* gene. It is possible that this situation could be mimicked by a transgene expressing a dominant negative version of the protein (iii), but this protein would have to be extremely efficient at disrupting the protein expressed by the normal alleles in the genome. The best mouse model would be a homozygous knockout for the *amelogenin* gene (iv).

d. This condition is caused by a loss-of-function mutation of a gene whose general activity is needed in all cells, but the protein translated from one alternative splicing product plays a specific role in hearing. A simple knockout would not work because a mouse homozygous for this knockout would likely die before it was born. A mouse with a conditional knockout that would remove gene function in the ear might not be completely appropriate because this mutation would prevent the expression of all alternative mRNA splice products, including those needed for the survival of all cells. The best choice would thus be a mouse homozygous for a knockin (vi) that re-creates the same mutation found in deaf patients.

PROBLEMS

Vocabulary

1. Match each of the terms in the left column to the best-fitting phrase in the right column.

a. transgene	1. genetically engineered viral genome that transfers a therapy gene
b. pronuclear injection	2. contains additional or altered DNA through gene targeting
c. floxed gene	3. useful for making conditional knockouts
d. T-DNA	4. can develop into any cell type
e. AAV vector	5. plant or animal that carries a transgene
f. packaging cells	6. causes crossovers at loxP sites
g. Cas9	7. gene transferred by a scientist into an organism's genome

h. knockout mouse	8. vector of bacterial origin used for constructing transgenic plants
i. knockin mouse	9. method of DNA transfer used for many vertebrates
j. Cre recombinase	10. endonuclease used for genome editing along with sgRNA
k. ES cells	11. loss-of-function mutant through gene targeting
l. GM organism	12. generate viral particles for gene therapy

Sections 18.1 and 18.2

2. Mice are usually gray, but a mouse geneticist has a pure-breeding white-furred strain that is homozygous for a recessive mutation. Molecular analysis shows that the mutation, represented by an asterisk in the following diagram, is a missense mutation in an exon common to three alternative splice forms of a gene expressed in hair follicles.

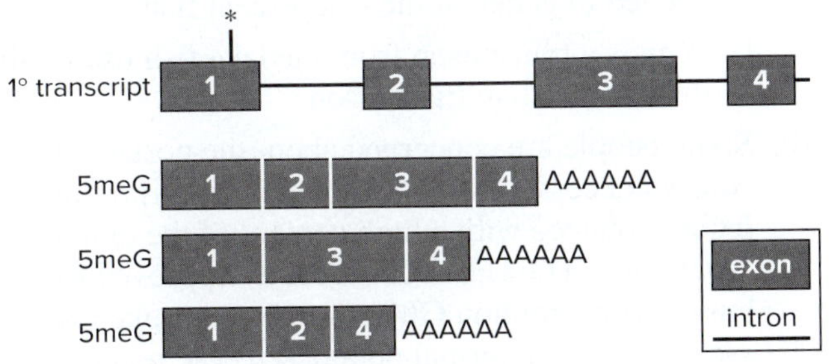

   a. Suggest an experimental approach using transgenic animals that you might employ to determine which splice form(s), when mutant, cause the white fur phenotype.
   b. Diagram the gene construct(s) you would create for your experiment.
   c. A potential problem with this approach concerns the amounts of mRNA that might be transcribed from the transgenes. Explain.

3. Sometimes, genes transferred into the mouse genome by pronuclear injection disrupt a gene at the (random) site of integration, resulting in a mutation. In one such case, investigators identified a recessive mutation that causes limb deformity in transgenic mice.
   a. The mutant phenotype could be due to the insertion of the transgene in a particular chromosomal site, or a chance point mutation that arose somewhere in the mouse genome different from the integration site. How could you distinguish between these two possibilities?
   b. The mutation in this example was in fact caused by the insertion of the transgene. How could you use this transgene insertion as a tag for identifying the mutant gene responsible for the aberrant limb phenotype?
   c. The insertion mutation was mapped to chromosome 2 of mice in a region where a recessive mutation called *limb deformity* (*ld*) had been identified previously. Mice carrying this mutation are available from a major mouse research laboratory. How could you tell if the *ld* mutation was in the same gene as the transgene insertion mutation?

4. In mice, a group of so-called *Hox* genes encode transcription factors that control the patterning of the animal's vertebral column. For example, the cervical vertebrae (labeled C1 and C2 below) express both HoxA1 and HoxD4 proteins, while the occipital bones at the base of the animal's skull (labeled E below) express HoxA1 only.

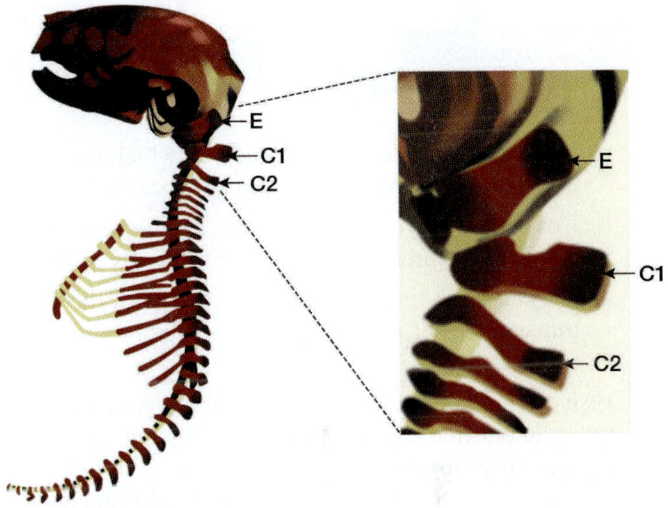

   Scientists hypothesized that the expression of HoxD4, controlled at the level of transcription initiation, is what makes C1 and C2 develop differently from the occipital bones.
   a. Describe a gene construct you could introduce by pronuclear injection that you could use to test this hypothesis.
   b. If the hypothesis is correct, what result would you expect in the transgenic mice?

5. The fly eyes shown in Fig. 18.7 are malformed because they lack a functional copy of a gene called *fat facets* that is required for eye development. The human genome encodes a protein called Usp9 that is similar in amino acid sequence to the fly Fat facets (Faf) protein. Likewise, the mouse genome encodes a protein called Fam that is similar to Faf and Usp9.
   a. How could you determine if human Usp9 can substitute for the *Drosophila* Faf protein?
   b. How could you determine if, like Faf in flies, the mouse Fam protein is required for mouse eye development?

6. This problem concerns a technique called *enhancer trapping* which scientists first developed in *Drosophila.* The purpose is to find enhancers in the genome and thereby to identify genes that are active in particular cell types. In enhancer trapping, thousands of fly lines are created, each of which has a single copy of the transgene shown here integrated into a random location in the genome via *P* element–mediated gene transfer. Note that the transgene has a promoter but no enhancer. (The thick horizontal arrows indicate the inverted repeats at the ends of *P* elements.)

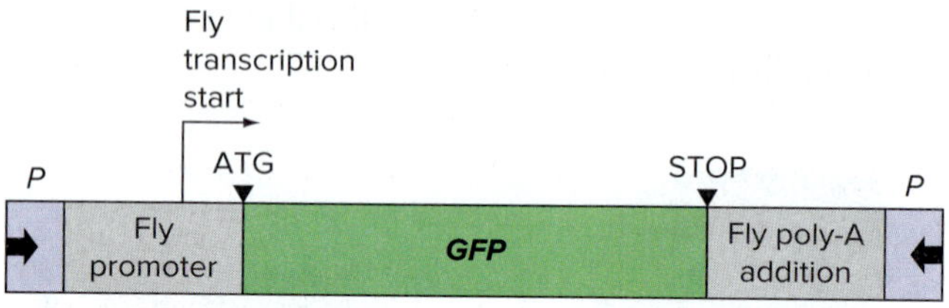

   a. How could you identify fly lines in which the transgene integrated next to an enhancer?
   b. Describe how you could use this technique to identify genes expressed specifically in the fly wing.
   c. Do you think that homozygotes for any of the transgene insertions might have a mutant phenotype? Explain.
7. Fish and other organisms that live in the Arctic express an antifreeze protein that prevents subzero temperatures from damaging their cells. Scientists have generated transgenic strawberries and tomatoes that express the fish antifreeze protein. However, neither crop has been commercialized.
   a. Explain the steps involved in generating the transgenic crop plants. What enables these plants to make a fish protein?
   b. The reason neither crop is on the market is that the antifreeze protein did not produce the desired effect: protection of the fruits from damage in a freeze. Provide a possible hypothesis explaining why the fish antifreeze protein might not work in plants.
8. a. Describe two ways you could potentially make a transgene that would inhibit the function of a specific gene in a transgenic organism. (*Hint:* For one of these techniques, recall the discussion of RNA interference in Chapter 17.)
   b. Discuss how you could use either of these methods to construct a mouse model for a recessive human genetic condition associated with a loss of function, such as cystic fibrosis.
9. Figure 18.6 shows a picture of Glofish®, transgenic zebrafish (*Danio rerio*) that express GFP, RFP, and some of their derivatives. Glofish® were the first GM pets. Genes can be transferred into the zebrafish genome using a transposable element called Tol2 from medaka fish (*Oryzias latipes*). The Tol2 transposon serves as a vector for gene transfer in a manner similar to *Drosophila P* elements as shown in Fig. 18.4. To use the Tol2 vector, researchers inject two-cell stage zebrafish embryos with two DNAs: One DNA expresses the Tol2 transposase gene, while the other recombinant transposon contains the transgene. If the transposon hops into a chromosome of a germ-line cell, a stable line of transgenic fish can be established.

   Scientists have synthesized several different genes encoding derivatives of GFP and RFP. The derivatives have amino acid changes that alter their color, so that fish glowing green, red, orange, blue, or purple are available in many pet stores.
   a. Diagram the two recombinant DNA constructs needed to generate the fluorescent fish.
   b. Why is a transposon from medaka fish used rather than a zebrafish transposon?
10. Some people are concerned about the possible health consequences of eating acrylamide, a by-product of frying potatoes with a high content of the amino acid asparagine. The U.S. FDA recently approved for human consumption GM potatoes that make less asparagine than normal potatoes. These potatoes express *RNA interference* (*RNAi*) transgenes that lower the amounts of two different enzymes in the asparagine synthesis pathway, called StAs1 and StAs2.
    a. Diagram the transgenes that could be used to reduce the expression of the *StAs1* and *StAs2* genes and explain how they work. (Recall from Chapter 17 that RNAi is initiated by double-stranded RNAs.)
    b. Why is it important that the RNAi transgene lower the levels of *StAs1* and *StAs2* gene expression but do not eliminate expression of either gene entirely?
    c. Scientists have recently made transgenic potatoes expressing an RNAi insecticide that kills the Colorado potato beetle, a pest of this crop. Speculate about the kind of transgene that might have been involved and the advantages of this approach.

Section 18.3

11. The goal of the Knockout Mouse Project is to generate a set of ES cell lines, each with a knockout mutation in a single gene, that collectively contains a mutation in every gene in the mouse genome.
    a. Do you think that it will be possible, for every gene, to generate a heterozygous knockout ES cell line? Explain.

b. Do you think that for every heterozygous knockout ES cell line, it will be possible to generate a heterozygous knockout mouse? Explain.

c. In fact, investigators failed in their attempts to generate ES cell knockouts for the *Fam* gene described in Problem 5. How could these researchers use ES cell technology to determine whether Fam is required for eye development in the mouse? Diagram a construct that the researchers could introduce into ES cells to explore this issue.

d. Describe various outcomes that might be obtained from the experiment in part (c) and what you could conclude in each case.

12. Gene targeting in yeast is performed just as in bacteria (review Fig. 14.31). Because the sequence of the yeast genome is available, researchers can easily use gene targeting to create a yeast strain with a knockout mutation in any gene.

a. Draw a gene targeting construct that you could introduce into yeast in order to generate a strain with a knockout of the histone subunit *H2A* gene. Your construct should contain a gene that confers resistance to kanamycin (*kan*^{r}).

b. List the steps that you would perform to generate the mutant yeast.

c. Recall that yeast can grow as diploids or haploids. Would you perform the gene targeting in a haploid or a diploid strain? Explain.

d. Suppose you knock out the *H2A* gene in a diploid strain to generate a heterozygote, sporulate the diploid, and find that all four haploid spores are viable (they produce haploid colonies). What would you conclude? Is this the result that you expect for the *H2A* gene? How can you explain the result?

13. Recall that constructs used for *floxing* a gene contain, within one of the gene's introns, two loxP sites flanking a gene for neomycin resistance (Fig. 18.11a). A loxP site is only 34 base pairs long, as shown in the following figure.

Explain how you could use PCR to generate a neomycin resistance gene flanked by loxP sites, starting with a plasmid containing a *neo*^{r} gene. If you had the intron of the target gene cloned in a plasmid vector, how could you insert your PCR product into the intron?

14. a. Which genome manipulation technique would you use to create a mouse model for the human disease Leber congenital amaurosis (LCA), described at the very beginning of the chapter? Explain.

b. Which procedure would you use to generate a mouse model for fragile X syndrome, a trinucleotide repeat disease caused by a mutant allele of the *FMR-1* gene? (Review the Fast Forward Box in Chapter 7 entitled *Trinucleotide Repeat Diseases: Huntington Disease and Fragile X Syndrome*.) Explain.

c. Which procedure would you use to create a mouse model for Huntington disease?

15. a. Diagram a knockin construct that could have been used to create the mouse model for achondroplasia shown in Fig. 18.13.

b. Explain how CRISPR/Cas9 could have been used to produce a mouse model for this condition.

16. In Problem 6, you learned about the *enhancer trapping* method used in *Drosophila* research. Reporter transgenes are integrated randomly into the fly genome, and their expression identifies enhancers and thereby genes expressed in particular patterns.

In mice, a similar technique for gene identification is called *gene trapping*. In gene trapping, an exon encoding a reporter protein such as GFP (see accompanying figure) is integrated into random locations in ES cell chromosomes. (When DNA constructs are introduced into mouse ES cells, most recombination events are into random sites; homologous recombination is relatively rare.) ES cell lines with randomly integrated transgenes are used to generate mice.

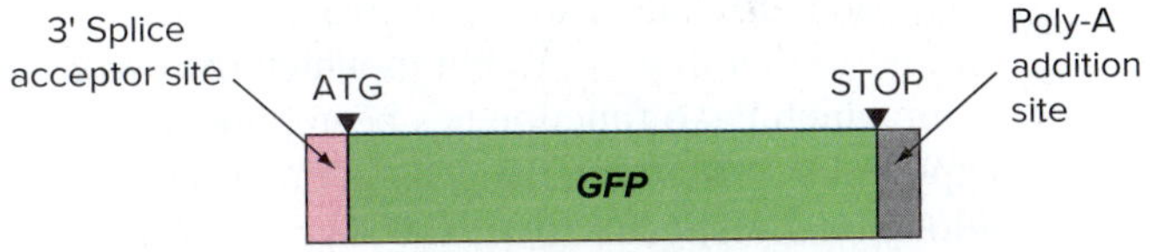

a. Where in the genome must a gene trapping construct integrate in order for ES cells to express GFP?

b. Will all of the mice with transgenes integrated the way that you describe in part (a) express GFP? Explain.

c. Both enhancer trapping and gene trapping identify genes that are expressed in particular patterns. How does the information obtained about the genes differ using the two techniques?

d. Which of the two methods is more likely to create a mutant allele of a gene near the integration site?

17. In tumor cells obtained from patients with Burkitt lymphoma, a cancer of the immune system's B cells, the *myc* gene often appears close to one of the breakpoints of a reciprocal translocation between chromosomes 8 and 14. In this translocated position, *myc* is

expressed at a higher-than-normal level. Scientists hypothesize that Myc protein overexpression in B cells contributes to lymphoma formation.

a. Explain how transgenic mice produced using pronuclear injection could be used to test this hypothesis. (Assume that you previously cloned a gene regulatory region that is active specifically in B cells throughout the life of the mouse.)

b. Suppose you wanted to overexpress Myc only in the immune cells of mice, starting at one week of age. To restrict Myc transcription spatially, you will use same promoter described in part (a). To restrict Myc transcription temporally, you will use a *cre* transgene whose expression is controlled by heat shock (*hs-cre*). Describe the mouse you would create to accomplish this goal and how you would generate this mouse. (*Hint:* What would happen during Cre-mediated recombination between two loxP sites in the same DNA molecule and in opposite orientations with respect to each other?)

18. The transcription factor Pax6 is required continually during the life of a mouse (or a human) for the development and maintenance of the retina. Homozygous *Pax6* knockout mice die soon after birth because Pax6 protein is also required in essential organs, such as the pancreas.

a. In order to study the role of Pax6 in eye development, a researcher wants to generate a mouse that expresses Pax6 everywhere except in its eyes. Describe how you could construct such a mouse.

b. Suppose the scientist wants to create a mouse similar to that in part (a), but in which the eye cells from which Pax6 function has been removed now express GFP. Marking the cells in this way will allow the investigator to see the shapes of the $Pax6^{-}$ eye cells more easily than if they did not express GFP. Diagram a *Pax6* gene construct that would enable her to do this experiment.

19. Mouse models for human genetic diseases are potentially powerful tools to help geneticists understand the cause of the aberrant phenotypes and develop new therapeutic measures. However, such mice are not always as useful to investigators as it might seem at first glance. Suppose that you have a mouse knockout model for a human disease caused by homozygosity for a null allele of a gene. Discuss how the following situations might complicate investigations of the human disease based on this mouse model.

a. Mice have a shorter life span than humans.

b. Mice homozygous for certain knockout mutations die *in utero.*

c. Mouse genomes may have additional copies of the gene whose mutation causes the disease in humans.

d. Mice from different inbred lines homozygous for the same gene knockout vary in the penetrance and expressivity of the phenotype.

e. Manipulations to create the knockout mouse, such as the presence of a drug resistance gene that allows the selection of cells containing the knockout (see Fig. 18.9), can disrupt not only the targeted gene, but also the expression of other, nearby genes.

20. One way to determine where inside a cell a protein (protein X) normally localizes is to generate a reporter gene construct containing: (i) the gene *X* regulatory region and coding sequences, and (ii) coding sequences for GFP fused in frame to the 3′ end of the gene *X* coding sequences just before the stop codon. A mouse containing such a transgene will express a hybrid protein X-GFP only in those cells in which gene *X* is normally expressed.

a. The gene *X-GFP* fusion gene described could be generated by knocking in *GFP* coding sequences instead of by random insertion of a transgene. Diagram the knockin construct you could use for this purpose.

b. What might the advantage be of the knockin strategy versus the transgene strategy?

21. In Problem 5 in Chapter 17, you saw that a SNP located in a hair follicle enhancer of the human *KITLG* gene is associated with the difference between blond and dark hair. People with blond hair have a high probability of having the A allele, while dark-haired people tend to have the G allele of the SNP.

a. The information given does not prove that the A allele of the SNP causes blond hair. Explain.

b. The hair follicle enhancer DNA sequences of the human and mouse *KITLG* genes are extremely similar; the mouse enhancer normally has the G allele of the SNP. Design a knockin experiment in mice that could test whether or not the SNP allele associated with blondness in humans actually contributes to the blond hair phenotype. Diagram any construct you would use to set up your experiment. (Assume that the A and G alleles are incompletely dominant.)

22. Scientists now routinely use CRISPR/Cas9 to make defined deletions of a gene that can remove several kb of DNA from the genome. This method is possible even in cells defective in homologous recombination, as long as the cells can still perform nonhomologous end-joining (NHEJ).

a. How could researchers make such deletions?

b. A GM animal that may be approved for human consumption by the time this book is published is a "super-muscly" pig made by inactivation of the *myostatin* gene. During normal development, Myostatin protein prevents the overgrowth of muscles. Given your answer to part (a), how could the super-muscly pig have been generated?

23. Geneticists are currently considering using technologies described in this chapter to *de-extinct* the woolly mammoth, a species that disappeared roughly 4000 years ago.
    a. Frozen specimens of woolly mammoths have been found in the Siberian tundra. If intact, living cells could be obtained from these samples, how could you attempt to bring back these long-extinct animals using these cells and oocytes from Asian elephants, a closely related species?
    b. Scientists have determined the genome sequence of the mammoth from frozen samples. These researchers are now trying to understand the adaptations that allowed these creatures to survive extreme cold. For example, mammoths had much thicker hair than do any elephants. How in theory could you use CRISPR/Cas9 to investigate the genetic basis of the difference in hair thickness between mammoths and elephants?
    c. How (again in theory) might it be possible to extend the CRISPR/Cas9 technique to de-extinct the mammoth? What kinds of technical challenges are involved in this approach? Why do you think some people consider the idea of de-extinction to be unethical?
24. a. Figures 18.9 and 18.12 demonstrated methods to produce mouse knockouts and knockins, respectively. CRISPR/Cas9 can make the same knockouts and knockins, and most mouse geneticists would now choose to use this new technology instead of the other methods. Explain why , respectively CRISPR/Cas9 is an easier and more efficient way to perform targeted mutagenesis in mice.
    b. How could you use the CRISPR/Cas9 technique to obtain a conditional knockout of a gene's function only in a specific tissue of a multicellular organism?
25. Nonhomologous end-joining (NHEJ) of a double-strand break almost always results in perfect resealing of the DNA lesion, without the loss or gain of nucleotide pairs. Yet CRISPR/Cas9, which produces double-strand breaks, is a highly efficient method of making small deletions or insertions at the targeted site. How can you resolve this apparent contradiction?
26. One problem that researchers sometimes encounter when editing genomes with CRISPR/Cas9 is that one or more loci other than the intended target can be recognized by Cas9/sgRNA and cleaved. Part of the reason is that single base pair mismatches between the target site and the sgRNA in the 5′-most half of the 20 bp DNA/RNA hybrid do not prevent Cas9 cleavage of the target site. How could scientists use bioinformatics to avoid such *off-target effects?*
27. Researchers at the University of California at San Diego have designed a strategy, alternatively called the *mutagenic chain reaction* (*MCR*) or *gene drive,* that can introduce rapidly a designed mutation into almost all of the chromosomes within an entire interbreeding population. Their idea was surprisingly simple, and it depends on plasmids such as that in the diagram that follows. In these MCR constructs, genes that can express high levels of Cas9 protein (*gray*) and an sgRNA (*green*) for a particular target in the genome are flanked by sequences that surround the target site in the genome (*blue*).

    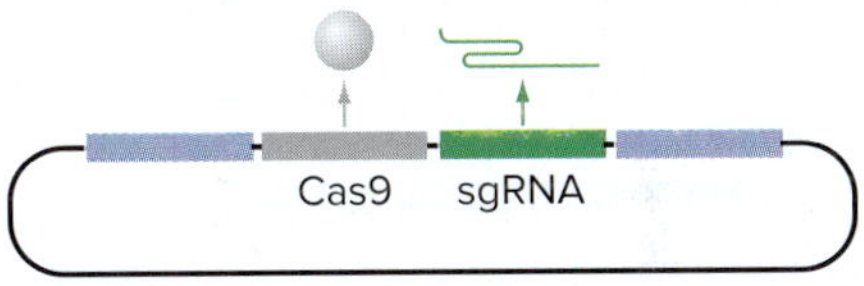

    Suppose you make a recombinant MCR plasmid in which the plasmid contains sequences for the X-linked *yellow* body color gene in *Drosophila.* The *Cas9* gene and sg*RNA* genes in the plasmid replace a protein-coding exon of the *yellow* gene that is needed for *yellow* gene function. The sg*RNA* is specific for a site within the wild-type *yellow* gene.
    a. The researchers injected this plasmid into a wild-type male embryo, where it became incorporated into some germ-line cells by homologous recombination. The sperm that developed from these germ-line cells fertilized a wild-type egg. The females that developed from these fertilized eggs were yellow bodied, which was surprising because loss-of-function alleles of *yellow* are recessive to wild-type alleles. Explain (include diagrams) the genesis of these yellow-bodied females. (*Hint:* Think about the name *mutagenic chain reaction.*)
    b. When a single such yellow-bodied female was introduced into a population of 100 wild-type flies, within a couple of generations almost every fly had yellow bodies. Explain this result.
    c. Researchers are now trying to use the gene drive system to prevent *Anopheles stephensi* mosquitoes from spreading malaria, a disease caused by a protozoan called *Plasmodium* that parasitizes mosquitoes and humans. This use of the technology is based on the availability of DNA sequences, from mice resistant to malaria, that encode an antibody that interrupts the *Plasmodium* life cycle. Describe how this system would work to control the spread of malaria.

d. In 2016, an expert panel convened by the National Academies of Science, Engineering, and Medicine released a report that cautioned against the release to the environment of *Anopheles* mosquitoes engineered as in part (c). Why was this panel so concerned about using MCR to control malaria?

Problems 28 and 29 are about CRISPR/Cas9 and relate to the following diagram, which shows the complex between an sgRNA and the genomic site it targets so that Cas9 can cut the genomic DNA at the positions shown. You should note the so-called *protospacer adjacent motif* or *PAM* site located adjacent to the target site. The PAM site (5′-NGG-3′ on one strand, N is any base) must be present in the position indicated for cleavage to occur. One way to think about this situation is that the sgRNA brings the Cas9 enzyme to the adjacent PAM site to initiate cleavage.

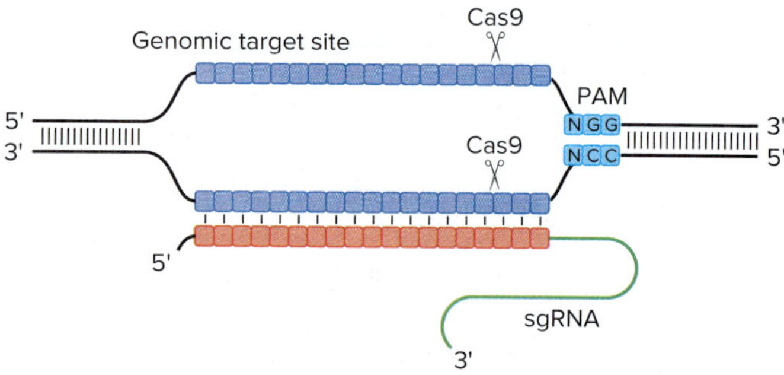

28. As was discussed in the Tools of Genetics box entitled *How Bacteria Vaccinate Themselves Against Viral Infections with CRISPR/Cas9,* the CRISPR/Cas9 system evolved in bacteria to provide a form of immunity against bacteriophage infection.
    a. Part of an sgRNA corresponds to a crRNA while the other part corresponds to the tracrRNA. (The utility of the CRISPR/Cas9 system for genome engineering is largely due to the fact that these two components are brought together into one RNA molecule.) On the preceding figure, which part of the sgRNA corresponds to the crRNA and which part to the tracrRNA?
    b. The CRISPR locus in the genome of a bacterium such as *Streptococcus pyogenes* does not contain any PAM sites. Why is this fact crucial? How does the PAM site function in bacterial immunity if it is not in the CRISPR locus?
29. F. Port and S. Bullock at the University of Cambridge (UK) designed the elegant plasmid vector *pCFD3* for the expression of sgRNAs in *Drosophila*. The following figure shows a part of this vector. The *orange* sequences are part of a strong promoter (transcription from this promoter starts at the **G** in bold—which must be present—and goes from left to right). The *purple* sequences are a portion of the tracrRNA component of the sgRNA. After cutting the *pCFD3* plasmid with the restriction enzyme *Bbs*I (whose recognition site is also shown in the following figure), you will replace the *blue* sequences in the figure with sequences that will allow the expression of an sgRNA that targets a *Drosophila* gene called *NiPp1*.

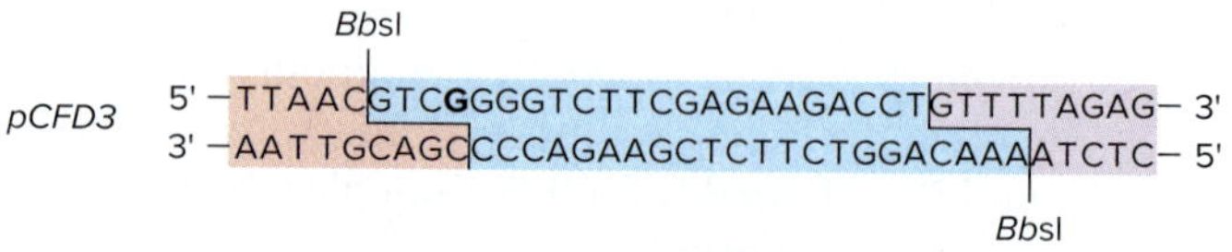

*Bbs*I recognition site 5' GAAGACNN^ 3'
3' CTTCTGNNNNNN^ 5'

The last part of the jigsaw puzzle you will need is the following sequence, which shows part of the *NiPp1* gene including the triplet corresponding to the start codon. Capital letters are in the gene's first exon with the coding region in *blue*; lowercase letters are in the first intron. The NiPp1 protein is 383 amino acids long. Your assignment is to generate a knockout allele of this gene by inducing Cas9 to produce a double-strand break into the gene that will be repaired imprecisely by nonhomologous end-joining (NHEJ).

Met Thr Asn Ser Tyr Asp Ile His Ser
5'...GTTAAAAGTATGACTAACAGCTACGACATACACAGTTGgtgagtttggcatc... 3'

a. Identify the two PAM sites in this sequence. Which of these PAM sites would you want to use in order to produce a null allele of the *NiPp1* gene? Why would you prefer this site?
b. If you targeted Cas9 to the proper location in the *NiPp1* gene, and the resultant double-strand break was repaired imprecisely by NHEJ (so that a few—usually ≤6 bp are deleted or added at that location), about what percentage of the imprecisely repaired genes could you say with confidence would be null alleles? Explain.
c. Diagram the *pCFD3* vector after it has been cut with the *Bbs*I enzyme. Don't worry about the small *blue* fragment that will be removed; the emphasis here is to show the 5′-overhangs that will be made.
d. Design two 24-nt DNA oligonucleotides that you could anneal together and clone into *Bbs*I-cut *pCFD3* vector so that the recombinant plasmid could express an sgRNA useful for making null mutations in the *NiPp1* gene.
e. Show exactly where Cas9 would cut the *NiPp1* gene.
f. Briefly outline what you would do with your recombinant plasmid to make a null mutation in the fly *NiPp1* gene.

g. Briefly outline how you would modify this technique to generate a knockin allele in which the first amino acid in the NiPp1 protein after the initiating Met (that is, Thr) would be changed to Ala.

30. On Fig 18.14, locate the PAM site and identify the 5′ and 3′ ends of the sgRNA.

Section 18.4

31. In contrast with the genomic manipulations of animals and plants described in this chapter, human gene therapy is directed specifically at altering the genomes of somatic cells rather than germ-line cells. Why couldn't or wouldn't medical scientists try to alter the genome of human germ-line cells?

32. a. Compare the means by which retroviral and AAV vectors deliver therapeutic genes to human cells.

    b. Explain the advantages and disadvantages of each of the two viral vectors.

33. Recall that Leber congenital amaurosis (LCA), a form of congenital blindness in humans, can be caused by homozygosity for recessive mutations in the *RPE65* gene. Recently, a rare dominant mutation in *RPE65* has been implicated as one cause of an eye disease called *retinitis pigmentosa,* which is characterized by retinal degeneration that can progress to blindness. The dominant *RPE65* mutation is a missense mutation causing amino acid 447 in the polypeptide to change from Asp to Glu. Little is known about the nature of the mutant protein.

    a. Do you think that the dominant allele is more likely a loss-of-function or a gain-of-function mutation? Explain.

    b. As described in this chapter, gene therapy for LCA has been at least partially successful. Do you think that the same kind of gene therapy can be used for patients with retinitis pigmentosa caused by the dominant mutant allele of *RPE65?* Explain.

34. One potential strategy for gene therapy to correct the effects of dominant gain-of-function mutations is to express small interfering RNAs (siRNAs) that will cause degradation of mRNA produced by the dominant mutant allele. The siRNAs can be delivered to patient cells by a synthetic gene encoding a *hairpin RNA* that will be processed by Dicer and used to target the RISC complex to the mutant allele mRNA (review Figs. 17.32 and 17.33).

    a. A major problem with this gene therapy strategy is designing an siRNA that will prevent the expression of the mutant allele specifically and not the normal allele. Explain why this is a problem.

    b. Specificity for the mutant allele is a particular problem in designing an siRNA therapy gene for Huntington disease. Explain this issue and suggest a possible solution that would allow you to use RNAi.

    c. Another potential strategy to correct the mutant Huntington disease allele is to cut the DNA in the middle of the repeat tract. Exonucleases present in the nucleus would then degrade many of the repeats prior to repair by NHEJ. Could you use CRISPR/Cas9 technology to correct Huntington disease mutations with this approach? (*Hint:* See Problems 28 and 29.)

35. Recently, scientists have used a mouse model for Duchenne muscular dystrophy (called the *mdx* mouse) to test whether Cas9 and an sgRNA could be an effective therapy for this disease. The cause of muscular dystrophy is homozygosity or hemizygosity for loss-of-function mutations in the X-linked *Dmd* gene. The *mdx* mouse has a nonsense mutation in exon 23 (of 79 exons) of *Dmd.* Researchers tested a technique called *exon skipping;* their idea was to use CRISPR/Cas9 to delete exon 23 in the mutant *Dmd* gene in the *mdx* mouse. AAV vectors with genes that express Cas9 and two different sgRNAs were injected into the muscles of adult mice. In about 10% of the muscle cells, exon 23 was deleted, and functional dystrophin protein was detected. Some muscle function was restored, although only to a small extent.

    a. Draw a diagram of exon 23 and the introns that flank it. On your diagram, draw the locations where the sgRNAs would hybridize with the genomic DNA. Explain how the deletion of exon 23 would take place.

    b. Design a PCR assay to determine if exon 23 is deleted from the genomic DNA of cell clones. Where would the PCR primers hybridize, and how would you be able to tell if the exon was deleted?

    c. Skipping exon 23 restored dystrophin protein function at least partially. What does this say about the amino acids encoded by exon 23?

    d. Considering your answer to part (c), would the exon-skipping strategy work for all *Dmd* point mutations that cause muscular dystrophy?

    e. What must be true about exons 22, 23, and 24 that would allow the researchers to consider this exon-skipping strategy? (*Hint:* See Problems 28 and 29.)

chapter 19

The Genetic Analysis of Development

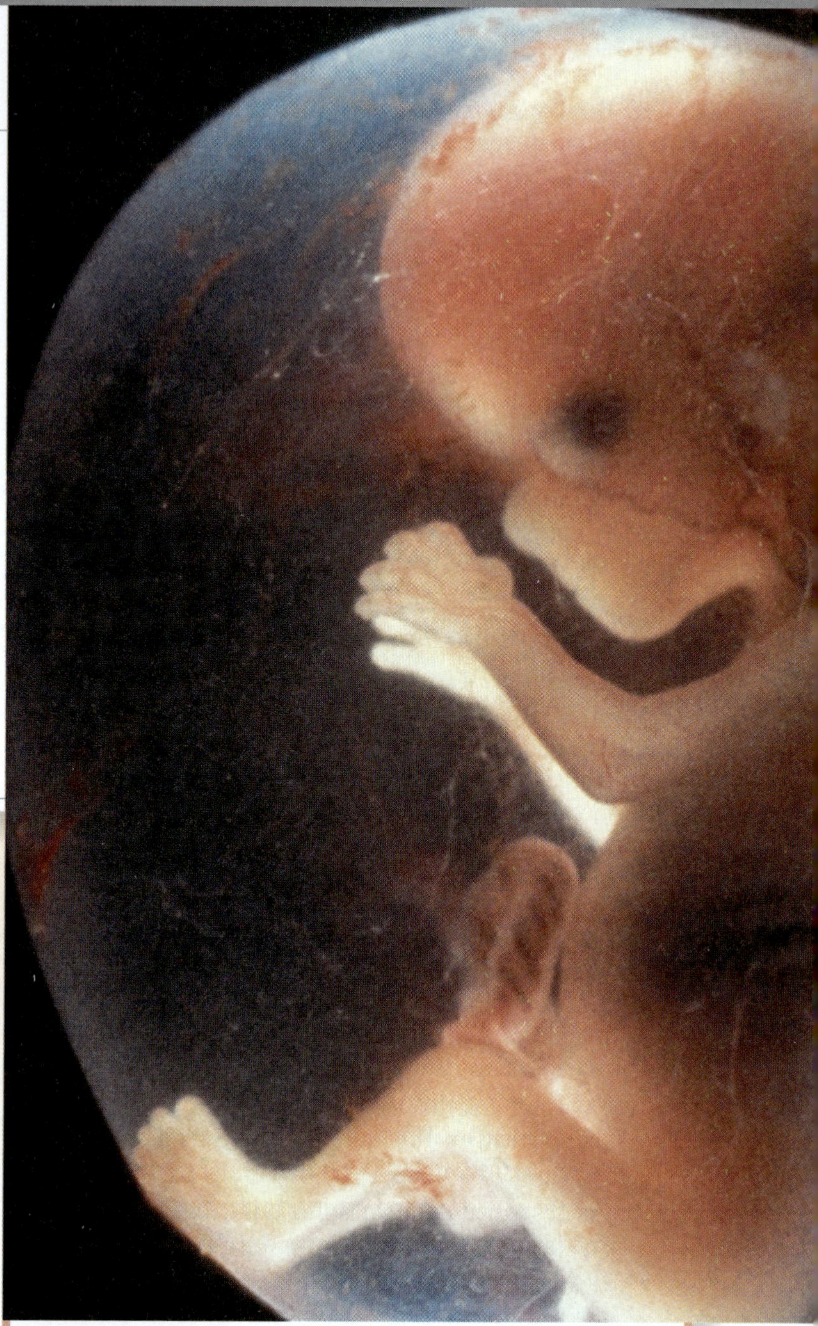

A human fetus three months after fertilization.

chapter outline

THE UNION OF A HUMAN sperm and egg (**Fig. 19.1**) initiates the amazing process of development in which a single cell—the fertilized egg—divides by mitosis into trillions of genetically identical cells. These cells differentiate from each other during embryonic development to form hundreds of different cell types. Cells of various types assemble into wondrously complex yet carefully structured systems of organs, including two eyes, a heart, two lungs, and an intricate nervous system. Within a period of three months, the human embryo develops into a fetus (see opening photograph) whose form anticipates that of the baby who will be born six months later. At birth, the baby is already capable of crying, breathing, and eating; and the infant's development does not stop there. New cells form and differentiate throughout a person's growth, maturation, and even senescence.

Biologists now accept that genes direct the cellular behaviors underlying development, but as recently as the 1940s, this idea was controversial. Many embryologists could not understand how cells with identical chromosome sets, and thus the same genes, could form so many different types of cells if genes were the major determinants of development. As we now know, the answer to this riddle is very simple: Not all genes are "turned on" in all tissues. Cells regulate the expression of their genes so that each gene's protein product appears only when and where it is needed. Two central challenges for scientists studying development are to identify which genes are crucial for the development of particular cell types or organs, and to figure out how these genes work together to ensure that each is expressed at the right time, in the right place, and in the right amount.

Developmental geneticists are scientists who use genetics as a tool to study how the fertilized egg of a multicellular organism becomes an adult. Like other geneticists, they analyze mutations—in this case, mutations that produce developmental abnormalities. An understanding of such mutations helps clarify how wild-type alleles of genes control cell growth, cell communication, and the emergence of specialized cells, tissues, and organs. Significant ethical and practical limitations prevent most studies of developmental genetics in humans. As a result, most modern developmental geneticists study mutations that affect the development of model organisms such as *Drosophila*.

In this chapter we present an overview of the experimental strategies scientists have used to examine the fundamental question of developmental biology: How do the single cells of a fertilized egg, or *zygote,* differentiate into hundreds of cell types? We can discern two key themes in our exploration of **developmental genetics.** One is that, surprisingly, many genes that control development have been highly conserved through evolution. Thus, the study of a process in *Drosophila* can shed light on events that occur during the development of other animals, including humans. A second theme is that genes involved in key developmental decisions often function in *hierarchies,* in which the product of one gene controls the expression of the next gene. This hierarchical pattern ensures that cells in multicellular organisms can develop into successively more specialized types.

Figure 19.1 Fertilization. Fertilization of an egg by a sperm creates a zygote that undergoes many rounds of division and cell differentiation to produce a fetus.

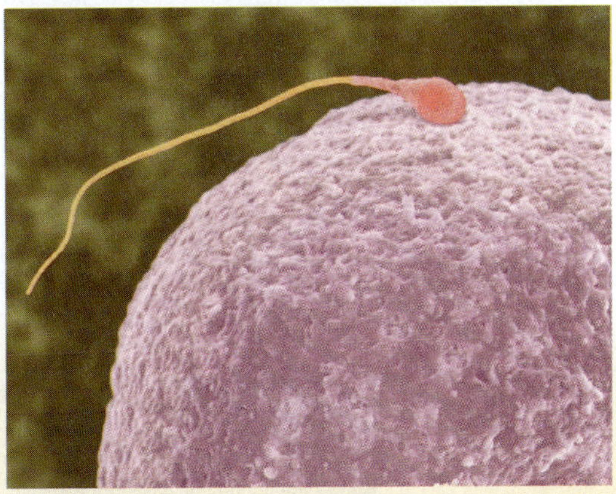

19.1 Model Organisms: Prototypes for Developmental Genetics

learning objectives

1. Explain why geneticists use model organisms to study development.
2. Cite evidence demonstrating that all living organisms are related, yet unique.

Developmental geneticists have concentrated their research efforts on a small number of organisms that include (but are not limited to):

- the yeast *Saccharomyces cerevisiae*
- the plant *Arabidopsis thaliana*
- the fruit fly *Drosophila melanogaster*
- the nematode (roundworm) *Caenorhabditis elegans*
- the zebrafish *Danio rerio*
- the mouse *Mus musculus*

These model organisms are easy to cultivate, and they produce large numbers of progeny rapidly. Geneticists can thus find rare mutations and study their behavior through successive generations. Each organism has attracted a dedicated cadre of researchers who share information, mutants, and other materials. Each model organism's genome has been sequenced completely, making it much easier for geneticists to identify genes whose mutant alleles have phenotypic effects on the organism's development.

All Living Forms Are Related...

Biologists have come to realize that life-forms are related on many levels. For example, the cells of all eukaryotic organisms have many structural features in common, such as a nucleus and mitochondria. The metabolic pathways by which cells make or degrade organic molecules are virtually identical in all living organisms, and almost all cells use the same genetic code to synthesize proteins. The relatedness of organisms is even visible in the amino acid sequences of individual proteins. As one example, over roughly two billion years, evolution has conserved the sequence of the histone protein H4, so the H4 proteins of widely divergent species are identical at all but a few amino acids. Most other proteins are not as invariant as H4, but nonetheless, scientists can often trace the evolutionary descent of a protein through the amino acid similarities of its homologs in various species.

Many basic strategies of development are conserved in multicellular eukaryotes, even in organisms with body plans that look quite different. A graphic example is seen in studies of the genetic control of eye development in fruit

Figure 19.2 The *eyeless/Pax-6/AN* gene is crucial for eye development. **(a)** Homozygosity for hypomorphic or null mutations in the *eyeless* gene reduce the size of eyes or completely abolish them in adult flies. **(b)** Homozygosity for loss-of-function mutations in the homologous mouse *Pax-6* gene have similar effects. **(c)** Heterozygosity for *AN* loss-of-function mutations disrupts iris development in humans.

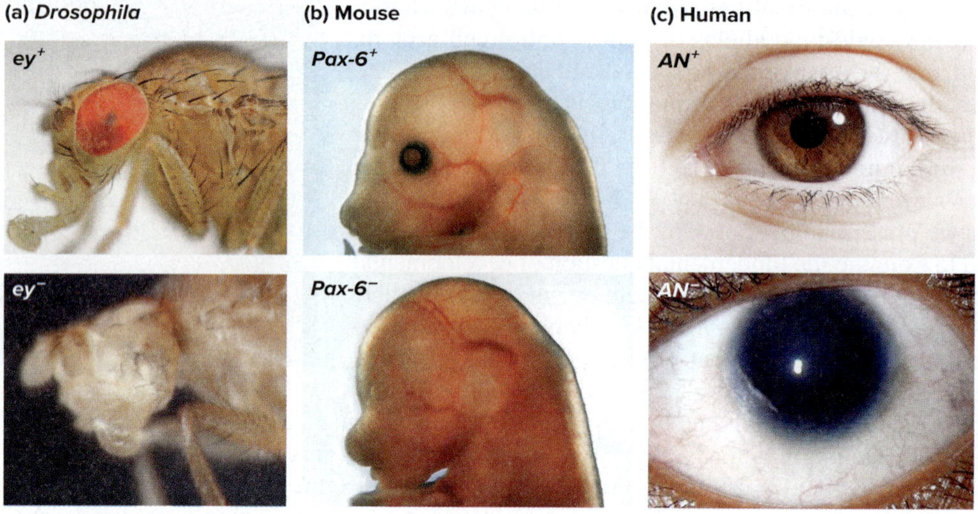

flies, mice, and humans. *Drosophila* homozygous for mutations in the *eyeless* (*ey*) gene have either no eyes at all or, at best, very small eyes (**Fig. 19.2a**). Mutations in the *Pax-6* gene in mice have a similar effect (**Fig. 19.2b**). Humans heterozygous for loss-of-function muations in the *Aniridia* (*AN*) gene lack irises (**Fig. 19.2c**).

When researchers cloned the *ey*, *Pax-6*, and *AN* genes, they found that the amino acid sequences of all three encoded proteins were closely related. This result was surprising because the eyes of vertebrates and insects are so dissimilar: Insect eyes are composed of many facets called *ommatidia,* whereas the vertebrate eye is a single camera-like organ. Biologists had thus long assumed that the two types of eyes evolved independently. However, the homology of *ey*, *Pax-6*, and *AN* suggests instead that the eyes of insects and vertebrates evolved from a single prototypical light-sensing organ whose development required a gene ancestral to *ey* and its mouse and human homologs.

...Yet All Species Are Unique

Although the conservation of developmental pathways makes it tempting to conclude that humans are simply large fruit flies, this is obviously not true. Evolution is not only conservative, but it is also innovative. Organisms sometimes use disparate strategies to accomplish the same developmental goal.

One example is the difference between the two-cell embryos that form in *C. elegans* and humans upon completion of the first mitotic division in the zygote. If one of the two cells is removed or destroyed in a *C. elegans* embryo at this stage, a complete nematode cannot develop. Because each of the two cells has already received a different set of molecular instructions to guide development, the descendants of one of the cells can differentiate into only certain cell types, and the descendants of the other cell into other types. The situation is very different in humans: If the two embryonic cells are separated from each other, two complete individuals (identical twins) will develop.

An intrinsic difference exists, therefore, in the way worm and human embryos develop at very early stages. As soon as the *C. elegans* zygote undergoes cell division, each daughter cell has already been assigned a specific fate; this pattern of development is often called *mosaic determination*. In contrast, the cells of an early human embryo can alter or regulate their fates according to the environment, for example, to make up for missing cells; this is called *regulative determination*.

The difference between mosaic and regulative determination is only one of countless examples of the myriad strategies for the development of multicellular organisms created during the course of evolution. It is impossible to describe this enormous diversity in the confines of a single chapter. Instead, we have chosen to center the discussion of this chapter on only two aspects of the development of *Drosophila melanogaster*: how the structure of the eye is determined (Sections 19.2–9.4), and how the body is divided into a series of segments along its length (Section 19.5). These studies have revealed some of the general mechanisms that guide the development of different organs in many different species. But more importantly, we hope that you will gain an appreciation for the ways in which various kinds of genetic analysis can help scientists gain insights about specific questions in developmental biology.

essential concepts

- Scientists use model organisms because they are easy to grow, produce large numbers of offspring, and have relatively short generation times.
- Evidence of the relatedness of life-forms can be found in conserved gene sequences that play similar roles in many different animal groups; for example, the *ey, Pax-6,* and *AN* genes that control eye development in *Drosophila,* mice, and humans, respectively.
- Evolution has created many different solutions to the problem of developing a complex multicellular organism from a single fertilized egg.

19.2 Mutagenesis Screens

learning objectives

1. Discuss how a primary mutant screen can identify genes required for a particular developmental process.
2. Explain the limitations of screens for mutants with morphological phenotypes, and discuss how sensitized modifier screens can help overcome these issues.
3. Describe how researchers use RNA interference to identify pleiotropic genes that function in a particular developmental pathway.

The geneticist's approach to understanding the nature of a developmental process is simply to ask: What are the genes required for the process? The first experimental steps toward answering that question almost always involve a **mutant screen.** Researchers examine a large number of mutagenized organisms and identify rare individuals with a phenotype of interest, such as a specific defect in eye development. Developmental geneticists have performed thousands of mutant screens in model organisms, leading to an intricate (but still incomplete) understanding of the mechanisms that guide plant and animal development.

Genetic Screens Identify Genes Required for Specific Developmental Processes

Developmental geneticists have performed intensive studies of the molecular mechanisms that guide development of the fruit fly compound eye (**Fig. 19.3a**). One reason for this choice is that mutant phenotypes are easy to analyze simply by looking at eyes in a microscope. The compound eye contains about 800 identical ommatidia, or facets, each composed of a small number of cells that assemble stepwise in precisely the same order in every ommatidium. In a

Figure 19.3 Cell signaling during *Drosophila* eye development. **(a)** Scanning electron micrograph of an adult *Drosophila* eye, showing the individual facets, or *ommatidia.* **(b)** Eight photoreceptor cells are recruited sequentially into each ommatidium. *Blue* indicates the commitment of cells to a photoreceptor fate. The sev^+ and $boss^+$ gene products help specify R7. **(c)** Light micrographs of sections through the retina. R7 is missing in every ommatidium of sev^- or $boss^-$ eyes. (Note that R8 is present but unseen because it lies beneath R7 and thus beneath the plane of the section.) **(d)** The Boss protein is a ligand expressed on the surface of R8. Binding of Boss to the Sev receptor on the surface of the R7 precursor cell activates a signaling pathway that results in the R7 fate.

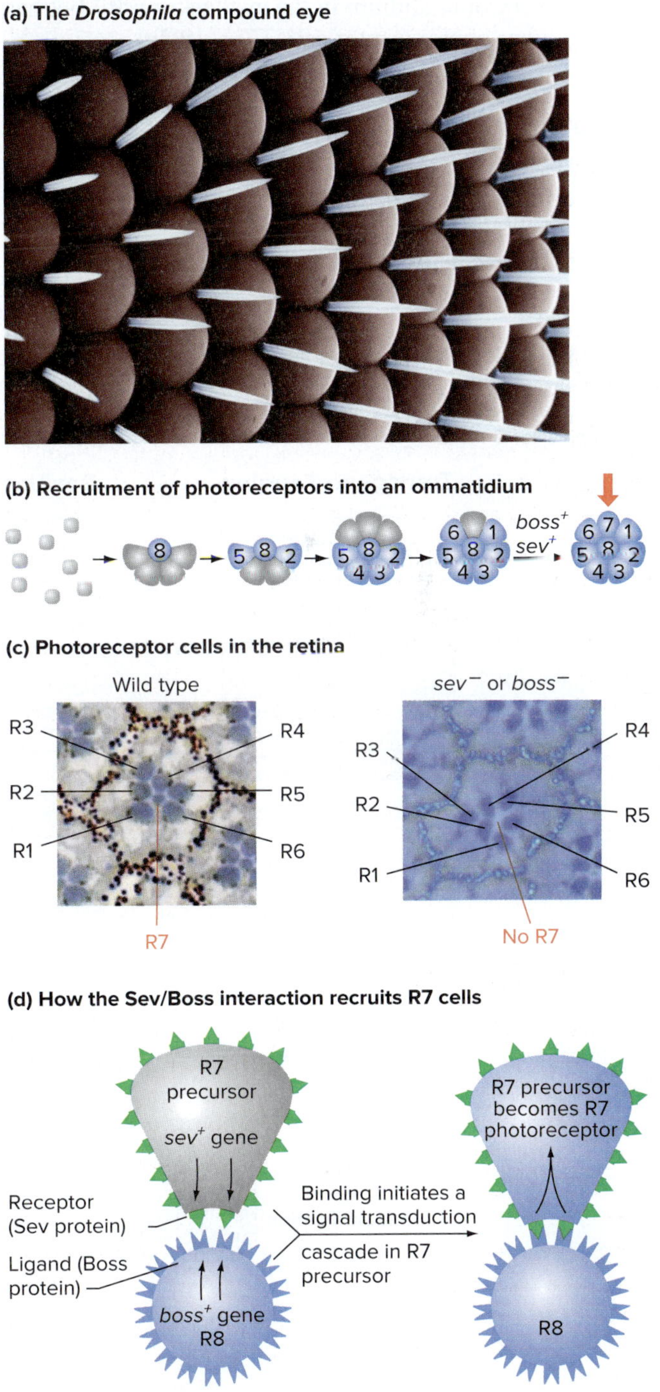

mutant where the assembly process is disrupted, the defect will be iterated 800 times in each eye, which makes it simple to recognize the nature of the mutant phenotype.

The first cells to assemble in each ommatidium are the eight photoreceptors, and the last of these cells to be recruited into a facet is a photoreceptor cell named *R7* (**Fig. 19.3b**). R7 is unique in that it contains rhodopsin proteins that enable flies to detect UV light; this ability is important for the fly's survival in the real world outside of the laboratory. Researchers conducted screens for *Drosophila* mutants with defective eyes, and the scientists focused their attention on several strains that displayed a very specific defect, in which every ommatidium in the eye lacks an R7 but has all of the other photoreceptor cells (**Fig. 19.3c**).

Separating mutations into complementation groups

After a screen identifies interesting mutants, the next step is to determine the number of different mutant genes represented in the collection. This goal is achieved by sorting the mutations into *complementation groups,* each containing different mutant alleles of the same gene. When researchers conducted pairwise crosses between many recessive, loss-of-function mutations that caused the specific loss of the R7 cell in fly ommatidia, they found that the mutations resolved into only two complementation groups (genes) they called *sevenless* (*sev*) and *bride-of-sevenless* (*boss*) (Fig. 19.3b and c).

Mutant gene identification

The availability of the genome sequence for each model organism usually makes it possible for researchers to identify the genes corresponding to mutant phenotypes in a matter of a few months. The method chosen for this purpose depends on the mutagen used to generate the mutants and also on the resources available to the investigator (**Fig. 19.4**). For example, the *sev* and *boss* strains had point mutations in which chemical mutagens altered a single nucleotide pair; these mutations were mapped using chromosomes containing deletions with molecularly defined breakpoints as described in Chapter 13 (Fig. 13.6).

Figure 19.4 Identifying genes responsible for mutant phenotypes in model organisms. A mutation induced by the insertion of a transposable element or transgene can be identified by its proximity to the inserted DNA. Scientists can find point mutations causing mutant phenotypes by genetic mapping (positional cloning) or by sequencing the whole genome of the mutant animal and looking for key polymorphisms. Gene assignments by any of these techniques need to be verified as shown at the bottom of the figure.

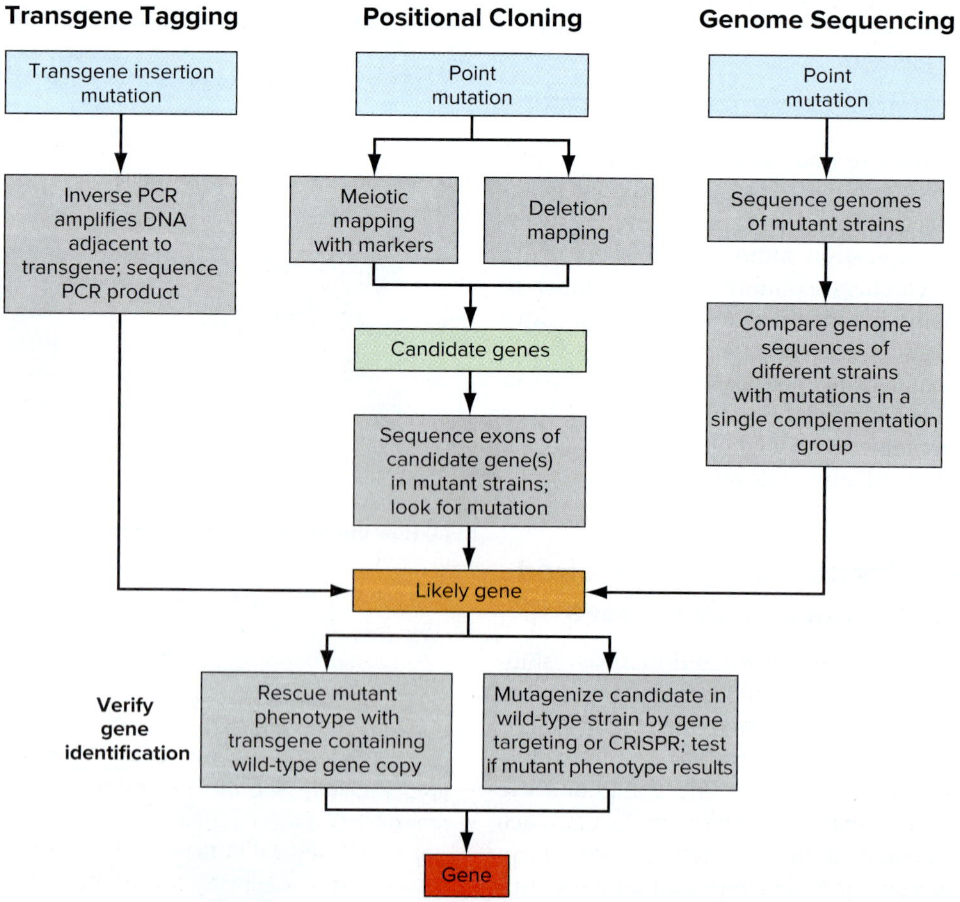

Other genes important for eye development have been identified by *P* element transposon tagging and inverse PCR (review Fig. 14.30). Finally, because the cost of whole-genome sequencing is plummeting rapidly, researchers are increasingly able to identify mutant genes in model organisms using genome sequencing and analysis techniques similar to those described in Chapter 11 for finding human disease genes.

Scientists can sometimes be misled into identifying the wrong gene as the cause of a mutant phenotype. It is thus important to verify the gene assignment (Fig. 19.4). One method of verification is to add back a transgenic copy of the wild-type presumptive gene to the mutant organism, to see if a wild-type phenotype results. Another method made possible by gene targeting or new techniques such as CRISPR/Cas9 (see Fig. 18.14) is to mutagenize the candidate gene in an otherwise wild-type organism to establish whether the mutant phenotype is recapitulated.

Clues from the nature of the encoded protein

Once researchers have identified a gene important for a developmental process, the amino acid sequence of the protein encoded by the gene can often provide key information about the molecular nature of that process. In the case of R7 cell determination in the ommatidia of *Drosophila* eyes, the proteins encoded by the *sev* and *boss* genes both had hydrophobic domains, suggesting that both proteins crossed through the cell membrane so that they would be found at the surface of particular cells.

Together with information derived from other methods to be described, scientists were able to formulate a simple hypothesis about how these two proteins function in recruiting R7 cells to ommatidia. Sev is a transmembrane receptor protein present on the surface of R7 precursor cells, while Boss is a transmembrane ligand present on the R8 surface. When an R7 precursor contacts R8, Boss binds Sev. This contact initiates a **signal transduction cascade** (a series of molecular events often including phosphorylations) within the R7 precursor cell, resulting in the expression of genes that determine R7 cell fate (**Fig. 19.3d**).

Primary Mutant Screens Can Miss Key Genes

Screens for loss-of-function mutants with specific morphological defects can identify genes dedicated to a particular *developmental pathway* in which the products of many genes cooperate to produce a particular outcome, such as the specification of R7 cells in ommatidia. However, for almost any process, mutations in genes that encode important pathway components will be impossible to recover with such primary screens. Here we discuss the reasons why and the alternative approaches that geneticists can use to try to find these missing components.

The problem of pleiotropic or redundant genes

Figure 19.3d indicates that the interaction of Sev and Boss at the surface of cells initiates a series of events that ultimately influences the expression of genes in the nucleus of the R7 cell. Why didn't the screen for the absence of R7 reveal mutants in any of the other genes whose products are involved in the signaling cascade? The answer is that these genes were missed in the screen because they are **pleiotropic**—that is, they are required for more than one developmental pathway, not only for R7 specification. In particular, the function of many of these genes is required for *Drosophila* viability: If the gene function is lacking, the organism dies before you can even look at its eyes.

Mutations in **redundant genes** will also be missed in screens for morphological phenotypes like R7 cell fate determination. If two genes perform the same function, the loss of either one will not result in a mutant phenotype. The existence of pleiotropic and redundant genes is a major limitation on the success of genetic screens. Researchers can sometimes overcome these problems through genetic tricks, some of which we describe next.

Dominant modifiers of sensitized mutant phenotypes

Scientists often try to identify pleiotropic genes involved in a developmental pathway by conducting a **modifier screen.** The idea is that heterozygous mutation of a pleiotropic gene can modify the phenotype caused by hypomorphic mutation of a different gene that has a dedicated role in a particular developmental pathway. This approach is most likely to be successful if the hypomorphic mutation produces a *sensitized phenotype* that could be affected by small changes in the levels of other proteins that function in the same pathway.

Investigators tried this approach to find pleiotropic genes they speculated might also participate in the determination of R7 cell fate. To identify these putative genes, the researchers devised a genetic background with a highly sensitive mutant eye phenotype caused by a *sev* hypomorphic mutation. This allele encodes a Sevenless protein with compromised activity; during eye development in these flies, only about half the ommatidia recruit R7. Because the Sevenless protein in these mutants just barely functions, reducing the level of a protein that aids Sevenless function could disable the R7 recruitment pathway completely, resulting in all the ommatidia lacking R7 (resembling a *sev* null phenotype). On the other hand, reducing the amount of a protein that antagonizes Sevenless function could make the disabled signaling pathway more robust, resulting in successful recruitment of R7 in all of the facets (resembling a normal phenotype).

Figure 19.5 illustrates the screen for modifiers of the sensitized *sev* mutant phenotype. Scientists created flies homozygous for the hypomorphic *sev* mutation (sev^{hypo} in

Figure 19.5 Dominant modifiers of the *sev*⁻ mutant phenotype. The eyes of sev^- hypomorphs have both phenotypically wild-type and mutant ommatidia. Dominant suppressor mutations cause all ommatidia to appear wild type (R7 present), while dominant enhancers cause them all to appear mutant (R7 absent). In an otherwise wild-type cell, the suppressor or enhancer mutations have no effect on eye morphology (box at *bottom right*).

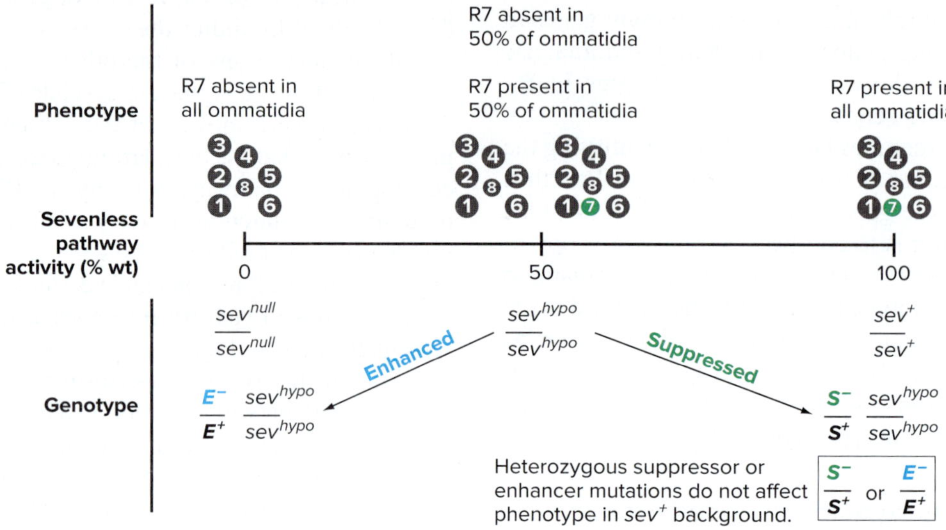

the figure) and heterozygous for random, mutagen-induced mutations. Some of these random mutations create null alleles in genes, so animals heterozygous for such mutations have only half the amount of the corresponding gene product. The investigators then looked for eyes different from those of homozygous *sev* hypomorphs. Mutations that cause the eyes of *sev* hypomorphs to appear more mutant are called *dominant enhancers* (E^-); mutations that cause *sev* hypomorphic eyes to appear more like wild type are called *dominant suppressors* (S^-). In an otherwise wild-type background, the heterozygous enhancer and suppressor mutations do not cause a mutant phenotype. Only in the sensitized background (that is, in *sev* hypomorphs) do the mutations have an effect on eye morphology.

Among the mutants identified in this screen were several components of the signaling pathway shown in **Fig. 19.6.** Ras^- and Sos^- mutations behave as dominant enhancers of *sev* hypomorphs, indicating that the Ras and Sos proteins help Sevenless recruit R7 cells into ommatidia. In contrast, loss-of-function mutations in a different gene called *Gap* act as dominant suppressors of the sev^{hypo} phenotype; the researchers thus deduced that the Gap protein antagonizes the action of Sevenless in promoting the R7 cell fate.

None of these mutants was identified in the original screen for homozygotes whose eyes lack R7 because all three genes (*Ras*, *Sos*, and *Gap*) are pleiotropic. The proteins encoded by these genes relay signals emanating not only from the Sevenless transmembrane receptor, but also from many other transmembrane receptors that are essential for cell-to-cell signaling during early embryonic development. As a result, homozygous Ras^-, Sos^-, or Gap^- mutants die early during embryogenesis, long before the adult eye forms.

Using transgenes in modifier screens

Researchers assumed that Sev, Boss, Ras, Sos, and Gap were not the only factors participating in the R7 cell specification pathway. To find these other suspected proteins,

Figure 19.6 The Sevenless signaling pathway. Some of the proteins in the signaling cascade activated when Boss binds to Sevenless (Sev) at the cell surface. The ultimate outcome is that a protein called MAPK enters the nucleus, where it activates the transcription factor Pnt and represses the transcription factor Yan. A program of transcription that causes R7 specification results.

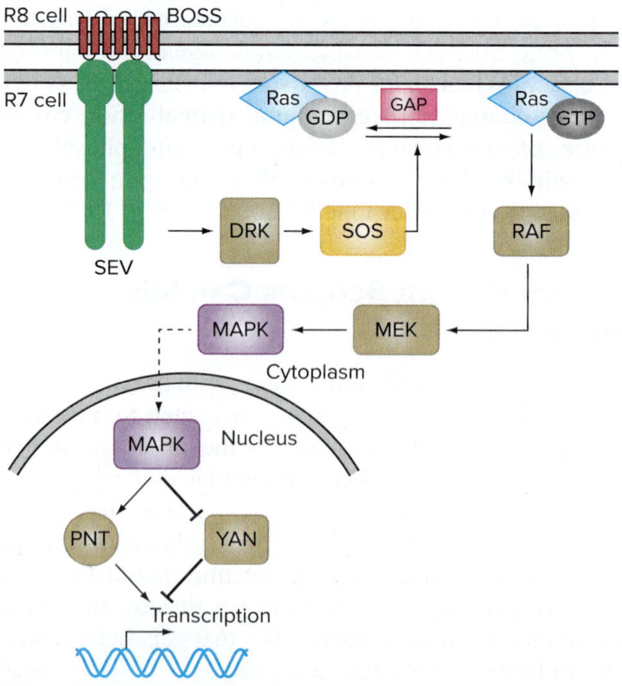

the investigators wanted to conduct other modifier screens using sensitized phenotypes associated, for example, with mutations in *Ras*. However, as just mentioned, the *Ras* gene is pleiotropic, and so animals homozygous for hypomorphic Ras^{-} mutations or heterozygous for hypermorphic *Ras* alleles die before they develop eyes. To circumvent this problem, the investigators constructed a transgene that creates a hypermorphic *Ras* mutant phenotype, but only in eye tissue.

In this transgene, the transcription regulatory region of *sev* was fused to the coding region of a hypermorphic *Ras* allele, Ras^{G12V} (**Fig. 19.7a**). When transformed into flies, the *sev* regulatory region causes Ras^{G12V} to be expressed only in the eye, and only in certain cells in the eye. These are the five precursor cells associated with each developing ommatidium that normally express Sevenless and thus are competent to become R7s (**Fig. 19.7b**). In wild-type eyes, only one of these five precursor cells contacts R8 (which has Boss on its surface) and becomes R7. The other four cells do not become photoreceptors and instead develop into non-neuronal cells that secrete the lens.

However, in flies with the transgene, the situation is different because the amino acid change in Ras^{G12V} causes this mutant protein to function constitutively, meaning that it is active even in cells in which Sevenless is not bound by Boss.

Furthermore, the amino acid change in the Ras^{G12V} protein causes this mutant protein to be constitutively functional even when Boss has not activated Sevenless. The result is that the Sevenless signaling pathway shown in Fig. 19.6 is always active in all five of the ommatidial precursor cells that express the transgene (Fig. 19.7b). Even though only one of those five cells contacts R8 (which has Boss on its surface), all five of these cells become R7s (**Fig. 19.7c**). In flies carrying the *sev-*Ras^{G12V} transgene, recruitment of extra R7 cells causes morphological aberrations visible on the outside of the eye (**Fig. 19.7d**).

Because Ras^{G12V} is expressed from the transgene only at low levels, the eye defects of the transgenic flies are sensitive to modification by heterozygous loss-of-function mutations in other genes. Mutant screens for modifiers of this eye phenotype identified several different signaling proteins in the Ras pathway in the eye, helping to fill in the scheme shown in Fig. 19.6.

Figure 19.7 Use of a transgene expressing hypermorphic Ras in a modifier screen. **(a)** Flies containing the transgene shown express Ras^{G12V} in five R7 precursor cells. **(b)** In wild-type animals, of the five R7 precursors (*red*) that express Sev protein, the four that do not contact R8 become nonneural cone cells that secrete the lens. **(c)** All five R7 precursor cells expressing Ras^{G12V} become R7s. **(d)** Flies containing the transgene in (a) have abnormal eye morphology that is made worse by dominant enhancer mutations (E^{-}/E^{+}) and more normal by dominant suppressor mutations (S^{-}/S^{+}).
(all photos): Courtesy of Andrew Tomlinson, Columbia University Medical Center

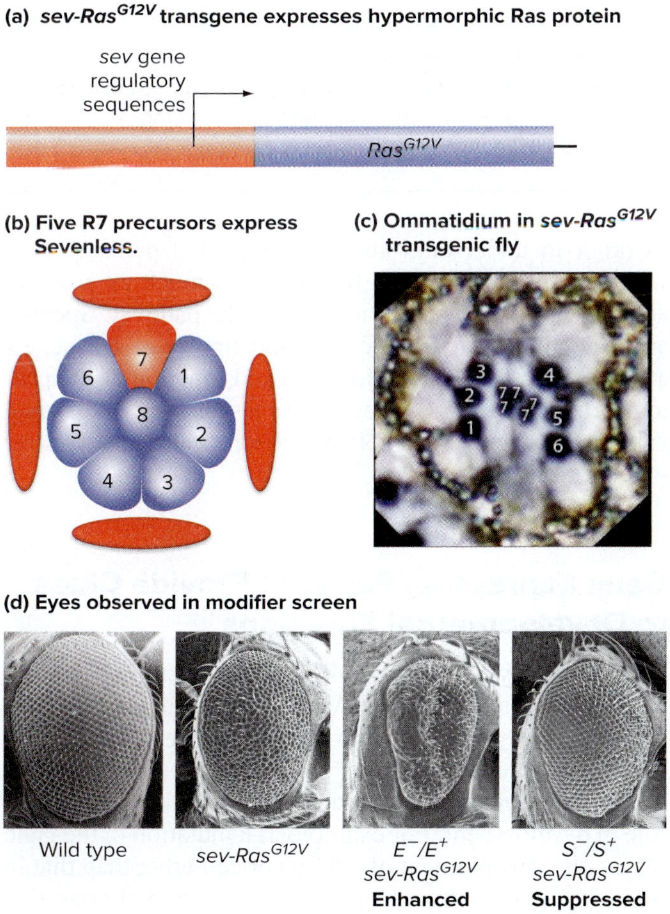

Stock Repositories and Genome Sequences Allow Systematic Screening of Mutations

Mutagens increase the frequency of mutations, but they alter the genome at random. As a result, traditional mutagenesis screens are inefficient and incomplete: In any collection of mutants, some genes will be represented by many similar mutations, while the collection will lack mutations in many other genes. The resources available to modern geneticists permit more systematic approaches to finding genes involved in developmental processes.

For the model organisms mentioned at the beginning of this chapter, centralized centers maintain collections of thousands of stocks, each with a mutation in a specific gene. These collections represent many (and for some organisms, all) of the known genes in the genome. Researchers can obtain these mutant strains from the stock centers, and thus they can screen each known gene one-by-one for effects on specific phenotypes.

For some model species such as *Drosophila* or mice, classic loss-of-function mutations are not yet available for all genes. An alternative way to screen individual genes for functions in a specific process is to use *RNA interference*.

Figure 19.8 Synthesis of double-stranded RNA (dsRNA) for RNA interference (RNAi) screens. Transcription occurs from both strands of a cDNA cloned between two promoters. Complementary RNA transcripts anneal with each other to make dsRNA that causes the degradation of the corresponding mRNA in cells.

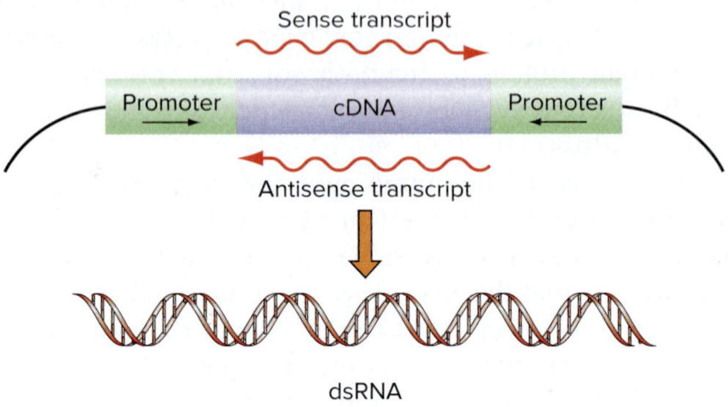

You will recall from Chapter 17 that eukaryotic cells have cellular machinery that causes double-stranded RNAs (dsRNAs) to trigger the specific degradation of mRNAs of complementary sequence (review Figs. 17.32 and 17.33). To employ this RNAi strategy, researchers can synthesize a dsRNA *in vitro* and then deliver it into the cells of a developing organism. Investigators working with *C. elegans* can deliver the dsRNA into cells in a simple and elegant way: They simply feed larvae with *E. coli* cells that contain a plasmid with a construct like the one shown in **Fig. 19.8.** RNA polymerase within the *E. coli* cells containing such a plasmid will synthesize the desired dsRNA, which is then taken up by *C. elegans* larval cells as the bacteria are digested in the worm's gut.

An alternative and more general method to perform genetic screens with RNAi is to construct transgenes that enable the transgenic organism to synthesize the dsRNA. Researchers make transgenes similar to those shown in Fig. 19.8, but the promoters included in the transgenic construct are those for the species under investigation rather than for *E. coli*. If the transcriptional regulatory regions included in the transgene have tissue-specific enhancers, the dsRNA will be expressed only in those cell types. The RNAi approach provides a convenient way to avoid the issue of pleiotropy: Researchers can create an animal in which only the cells in the tissue being studied lack the function of a specific gene, while all the other cells in the organism retain wild-type function of that gene.

One shortcoming of RNAi screens is that the efficiency of knockdown can vary for different genes, and full elimination of expression is rarely achieved. However, new genome editing tools such as CRISPR/Cas9 are so efficient that in the near future it is likely that collections of null alleles for every gene in the genome of model organisms will be established.

essential concepts

- Researchers can identify genes required for a developmental process by performing a *mutant screen* in which strains with random loss-of-function mutations are examined for a particular phenotype
- Mutations in pleiotropic or redundant genes are often not recovered in mutant screens. Scientists can sometimes circumvent this problem by searching for dominant modifiers of a *sensitized mutant phenotype*.
- In *RNA interference* (*RNAi*), the introduction of a double-stranded RNA molecule into cells causes degradation of an mRNA of complementary sequence. Investigators can induce gene-specific RNAi in particular tissues to test for functions of pleiotropic genes.

19.3 Determining Where and When Genes Act

learning objectives

1. Summarize methods for monitoring the mRNA and protein products of specific genes.
2. Discuss how genetic mosaics can help determine the focus of action of a gene.
3. Explain how researchers use temperature-sensitive alleles to determine when genes act during development.

In order to understand any developmental pathway as a whole, investigators must first learn as much as possible about each of the genes that make up the pathway. Specifically, details about the location and timing of the gene's expression, as well as the location and function of the protein product within particular tissues or even individual cells, help scientists establish a theoretical framework to guide further analysis.

Gene Expression Patterns Provide Clues to Developmental Functions

A wide variety of methods allow scientists to monitor the expression of specific mRNAs or proteins in whole organisms, or in sections of tissues on microscope slides. Defining the tissues in which a gene is expressed can help researchers formulate hypotheses concerning the gene's role in development. For example, if a mutation in the gene affects the development of a tissue or cell other than that in which the gene is transcribed, you might hypothesize that the gene encodes a signaling molecule. In one such case, the *boss* gene is transcribed in R8 cells, but the ligand it encodes affects the fate of adjacent R7 cell precursors that have Sev receptors (recall Fig. 19.3).

Tracking mRNA expression

One way to determine where and when a gene's transcripts accumulate in an organism is to perform an **RNA *in situ* hybridization** experiment. To do this, you first label cDNA sequences corresponding to the gene's mRNA. Next you use the labeled cDNA as a probe for mRNA on preparations of thinly sectioned tissues, or in some cases whole mount organisms or tissues. Signals where the probe is retained indicate cells containing the gene's mRNA. We will present the results of an RNA *in situ* hybridization experiment later in this chapter.

An alternative approach to determining which cells transcribe a particular mRNA at high levels is to determine the sequences of all the mRNAs found in the cell. We have previously described this kind of *deep sequencing* (or *RNA-Seq*) strategy in Chapter 16 (see Fig. 16.29 and Fig.16.30). In essence, researchers make cDNAs corresponding to the mRNAs in tissue samples and then obtain sequence data for millions of these cDNAs. The frequency of finding a particular cDNA indicates the abundance of that mRNA in that tissue. Modern RNA-Seq techniques are so efficient and sensitive that investigators can characterize the mRNAs present in a single isolated cell.

Tracking protein expression

It is often easier technically to evaluate the tissues in which a gene is expressed by following the gene's protein product rather than by looking at the gene's mRNA. The protein approach also in many cases provides a more accurate view of gene expression because cells have controls that regulate the translation of certain mRNAs; thus, a tissue might have a great deal of a particular mRNA but only a small amount of the corresponding protein. Finally, the intracellular localization of a protein often provides clues to its function. For example, concentration of a protein in the nucleus would be consistent with a role as a transcription factor.

One way to monitor a protein involves the generation of antibodies that bind very tightly to a part of the protein. We described previously in Chapter 16 the use of recombinant DNA techniques to use *E. coli* cells as mini-factories to make large amounts of any given polypeptide (review Fig. 16.28). Researchers can purify this *E. coli*–made protein and inject it into rabbits or other animals. The proteins act as *immunogens* in the bloodstream of the injected animal, whose immune system will synthesize antibodies against the foreign protein. Scientists obtain these antibodies and label them with a fluorescent tag, allowing the investigators to track the tagged antibodies as they bind to the protein of interest in preparations of tissues and cells. **Figure 19.9** shows a particularly colorful example of this approach.

Figure 19.9 Using antibodies to monitor protein localization. A developing *Drosophila* eye is labeled with antibodies against several proteins, each expressed in specific photoreceptor cells. Each antibody is tagged with a dye that fluoresces in a particular color.

Courtesy Helen McNeill, Lunenfeld-Tanenbaum Research Institute Mount Sinai Hospital, Toronto

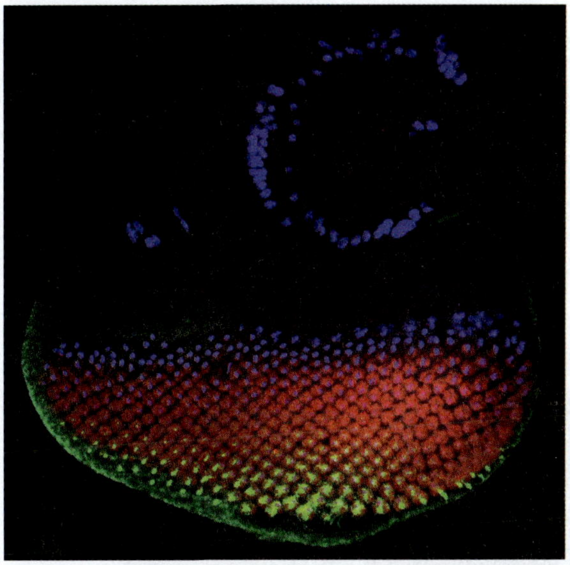

An alternative way to track a protein is to construct a **fusion gene** encoding a protein tagged with jellyfish green fluorescent protein (GFP). Researchers synthesize an open reading frame that encodes not only the entire protein of interest, but also, at the protein's N or C terminus, the amino acids constituting GFP (**Fig. 19.10a**). The construct also contains the promoter and enhancer sequences needed for proper expression of the gene. When this recombinant gene is introduced into the genome (for example, by *P* element–mediated transformation in flies), the organism will make the GFP **fusion protein** in the same places and times it makes the normal untagged protein. Investigators can keep track of the fusion protein by following GFP fluorescence (**Fig. 19.10b**). A major advantage of this approach is that researchers can use it to follow the GFP-tagged protein in living cells or animals; for technical reasons, tagged antibodies cannot be tracked in live material.

Mosaic Analysis Can Determine the Focus of Gene Action

Genetic mosaics are individuals composed of cells of more than one genotype. Studies of mosaics allow developmental geneticists to address an important question: In what particular cells must a gene be expressed for a developmental process to occur? Those cells are considered the

Figure 19.10 Visualizing proteins tagged with GFP. **(a)** The fusion gene shown encodes a fusion protein that contains GFP at its C terminus. **(b)** This fly contains a transgene that expresses a GFP-tagged fusion protein only in its eyes.
b: Courtesy of Malcolm J. Fraser, Jr., University of Notre Dame

(a) Tagging a protein with GFP

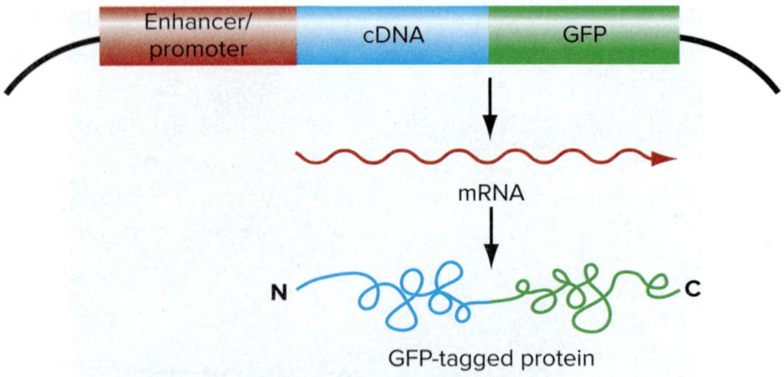

(b) Fly carrying transgene expressing GFP in eyes

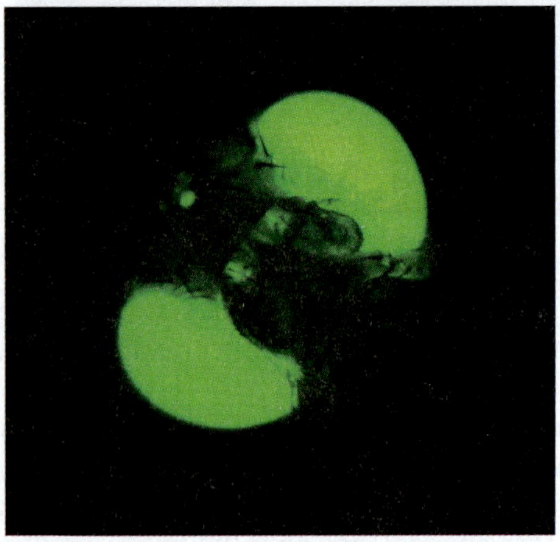

focus of action of the gene. As discussed in Chapter 18, investigators can use several different techniques to generate mosaic experimental organisms, depending on the species.

In fruit flies, mosaics are most often created by site-specific recombination between nonsister chromatids during mitosis using the **FLP/FRT recombination system.** Recombination is enabled by a recombinase enzyme from yeast called FLP that catalyzes reciprocal crossing-over at specific recombination targets called FRTs (recall Fig. 6.31). Investigators cloned the *FLP* gene downstream of, and therefore under the control of, a heat-shock promoter whose expression is turned on at elevated temperatures, and they introduced this construct into *Drosophila* using a *P* element vector. Researchers also inserted FRT sites into *Drosophila* chromosomes, and they collected strains in each of which an FRT is inserted close to the centromere on a particular chromosome arm. If FRTs are present at the same location on both homologs, FLP recombinase (whose synthesis is induced by heat shock) catalyzes recombination between them. In a cell heterozygous for a mutant and a wild-type allele, mitotic recombination can result in a homozygous mutant daughter cell which gives rise to a clone of homozygous mutant descendant cells.

Scientists used **mosaic analysis** in the fly eye to ask: In which cells do the *sevenless* and *boss* genes need to be expressed in order for R7 to be recruited to ommatidia? To answer this question for the *sevenless* gene, the researchers generated mosaics with marked (white) cells that were mutant for both *white* and *sevenless* in an eye where the other cells (red) carried at least one dominant wild-type allele of each gene (**Fig. 19.11a**). Some facets contained a mix of red and white photoreceptor cells. But in every facet with an R7 cell, the R7 cell was red (that is, it had both the w^+

Figure 19.11 Mosaic analysis determines the focus of action of Sevenless protein. **(a)** FLP-mediated site-specific recombination at FRT sites results in a clone of *white* cells that do not express Sevenless protein (w^- sev^-) in a background of *red* cells that do express Sevenless (w^+ sev^+). **(b)** In mosaic ommatidia that are phenotypically wild-type (contain an R7), the R7 cell is always w^+ sev^+; the *sev* genotypes of the other seven photoreceptor cells are irrelevant. Therefore, for R7 determination, Sevenless protein must be expressed in R7.

(a) Generating w^- sev^- clones

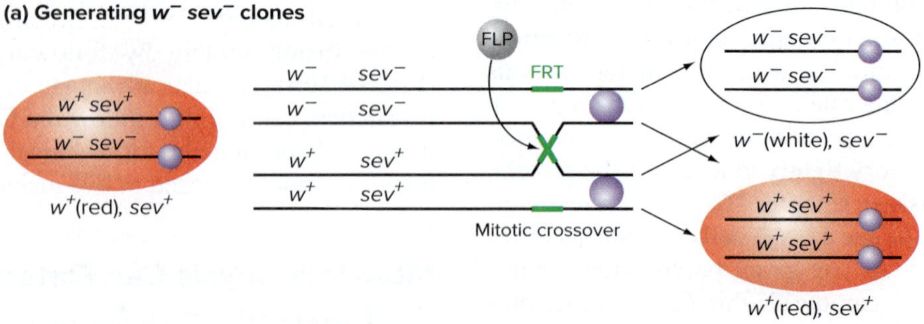

(b) Examples of mosaic phenotypically wild-type ($R7^+$) ommatidia

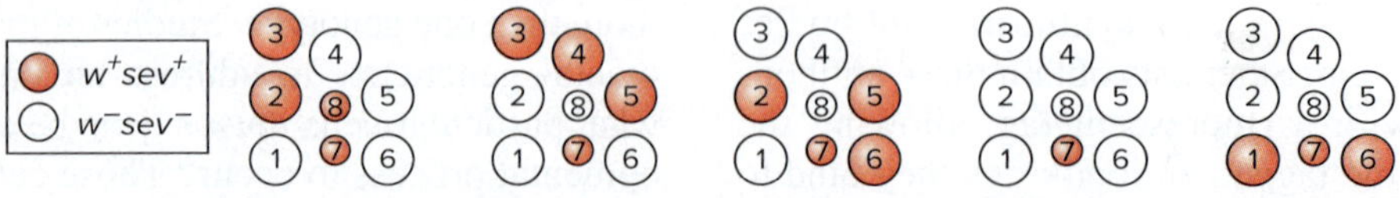

and *sev*$^+$ genes), even if all the other photoreceptor cells in the same ommatidium were white (homozygous *w*$^-$ and *sev*$^-$) (**Fig. 19.11b**). No ommatidia in which the R7 cell was white were ever seen. These results indicate that the requirement for the *sev*$^+$ gene product is in the R7 cell itself. To develop correctly, an R7 precursor cell must make the Sevenless protein, and it does not matter whether the protein is made in any other photoreceptors in the same ommatidium. The Sevenless protein affects only the cells in which it is made.

Temperature-Sensitive Alleles Help Determine the Time of Gene Action

Many of the events crucial for an organism's development occur in a sequence whose relative timing matters. For example, the action of a particular protein may be important only during a narrow window of time. If the protein interacts with other molecules or structures, then the presence of the protein will have no effect until those molecules or structures are present. Conversely, the expression of a protein that helps a cell make a developmental decision will have no effect after that decision has been made.

Researchers often use **temperature-sensitive (ts) mutations** as a tool to determine when the function of a particular protein product is important for a developmental process. These alleles produce a protein that functions at a lower *permissive temperature* but that fails to function at a higher *restrictive temperature*. In contrast, the wild-type protein functions at both temperatures. In one experiment using this approach, illustrated in **Fig. 19.12,** investigators used ts alleles of the *sevenless* gene to establish that the Sevenless receptor protein is needed during a period of several hours within a presumptive R7 cell in order for it to adopt an R7 fate. If Sev protein is not activated by Boss continually during this span of time, then the cell will commit to an alternative fate (such as becoming a lens-secreting cone cell as was seen in Fig. 19.7).

Figure 19.12 Using temperature-sensitive alleles for time-of-function analysis. *sev*^{ts} larvae were grown at 18°C (the permissive temperature) for various lengths of time. Wild-type ommatidia (containing R7) appeared in adult eyes only when the larvae were at 18°C for at least 12 hours after Sev protein first appears.

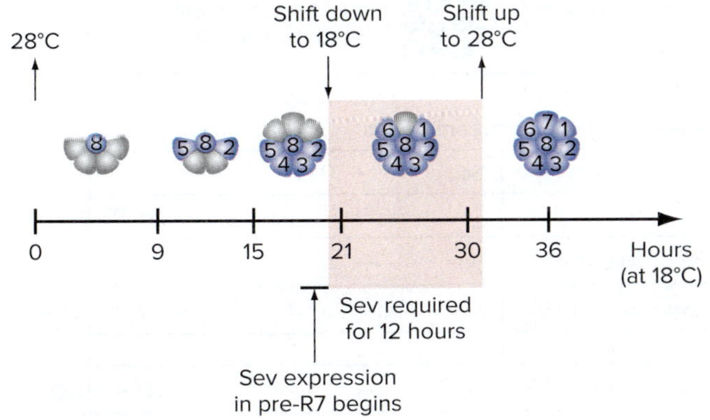

You can see that the results of several types of experiments have converged to provide a clear view of the function of the Sevenless receptor protein in specifying the R7 cell. Using antibodies to Sevenless and Boss proteins, researchers observed that several cells produce Sevenless protein, but only one of them—the presumptive R7 cell—ever contacts R8. The focus of action of the *sev* gene is the presumptive R7 cell (Fig. 19.11). And at the surface of this cell, the Sevenless receptor must be activated by the Boss ligand from the R8 cell continually for many hours to ensure that the presumptive R7 cell will indeed assume an R7 cell fate (Fig. 19.12).

essential concepts

- Tracking the timing and location of mRNA expression can be accomplished by *RNA in situ hybridization* or by tissue-specific RNA-Seq. Protein expression can be monitored by using specific antibodies or by engineering a gene that encodes a GFP-tagged protein.
- *Mosaic tissues* can indicate which cells constitute a gene's *focus of action*—that is, the cells or tissues in which the gene's expression is required.
- *Temperature-shift experiments* to inhibit or allow the function of a temperature-sensitive allele enable researchers to determine when the gene product is required for a developmental process.

19.4 Ordering Genes in a Pathway

learning objectives

1. Explain how to determine the order of genes in a pathway by monitoring gene expression.
2. Discuss how epistatic interactions between mutant alleles of two genes can sometimes help determine gene order in a pathway.

Up until now, you have seen how to use genetics to identify genes in a developmental pathway, and how to determine where and when the individual genes act and in which cells they need to be expressed. Sometimes mutants can also be analyzed to help determine the order of gene action in a pathway; that is, which genes or gene products activate or inactivate which other genes or gene products. Here we present two different approaches to ordering gene functions.

The Effects of One Gene on the Expression of Another Can Reveal the Order of Action

Once you have defined the tissue distribution and intracellular location of one gene's mRNA or protein, you can ask how mutations in a different gene affect this distribution or localization. For example, a transcription factor called Prospero is essential for the development of the R7 photoreceptor in the *Drosophila* eye. Prospero expression depends on activation of the Sevenless pathway. In sev^- null mutants, Prospero is absent from the R7 precursor (**Fig. 19.13**). When Ras is activated inappropriately in R7 precursors in flies with a *sev-Ras*^{G12V} transgene, Prospero is detected in all five R7 precursors (Fig. 19.13). In contrast, neither Sevenless nor Ras expression is affected by mutations in the *prospero* gene (*not pictured*). Therefore, Sevenless pathway activation is upstream of (here, turns on the expression of) *prospero* in the R7 cell.

The general principle demonstrated in this example is that if either loss or overactivity of gene *a* function affects the expression of gene *b*, and not the converse, then gene *a* likely functions upstream of gene *b* in a pathway that regulates gene *b* expression.

Double Mutant Phenotypes Can Help to Determine the Order of Gene Action

Another way that researchers can establish the gene order in a pathway is by making **double mutants.** The idea is to start with two mutations, each in a different gene (gene *a* and gene *b*). The two genes work in the same pathway, although the mutant phenotypes associated with each of them is different. The investigators construct an organism with both mutations and ask whether one mutation is **epistatic** to the other. In other words, does the a^- b^- double mutant have either the a^- phenotype or the b^- phenotype? If a clear epistasis interaction is observed, the result can indicate which gene is upstream of the other in the pathway.

Figure 19.13 A mutation in one gene can affect the expression of another gene. Prospero protein (*pink*) is normally expressed in R7 of developing ommatidia. In developing sev^- eyes, the cell that would have become R7 instead becomes a lens-secreting cone cell (*), and no Prospero is detected in it. In developing transgenic eyes that express constitutively active Ras^{G12V} in five potential R7 precursors in each ommatidium, these five cells express Prospero.

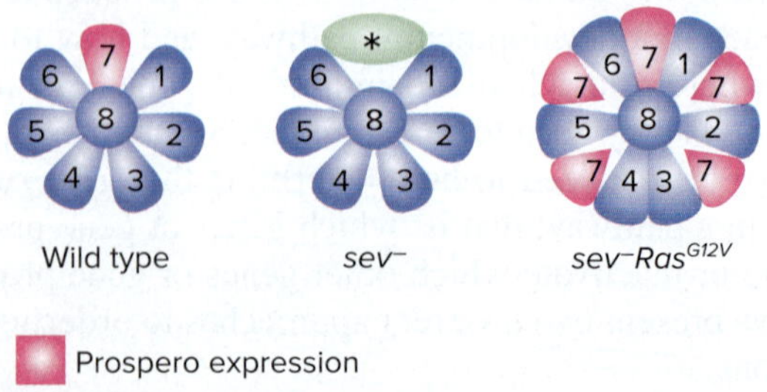

Epistasis analysis is particularly useful in investigating *switch/regulation pathways* that play important roles in development. In such a pathway, a signal controls a switch that can be either ON or OFF in any particular cell. If the switch is ON, it initiates a cascade of events where genes in a linear pathway are regulated (either activated or inactivated) in turn, leading to a change in the cell's fate. The outcome of the pathway is thus one of two developmental states, depending on whether the switch is ON or OFF.

Consider the example in **Fig. 19.14,** which shows that R7 cell fate is determined by a switch/regulation pathway.

Figure 19.14 Epistasis analysis. (a) R7 cell specification depends on a switch/regulation pathway where the switch is Sevenless and the signal is Boss. In the R7 precursor cell that both expresses Sevenless and touches R8, the switch is ON (Sevenless is activated), which initiates a signaling cascade that results in R7 development. **(b)** Double mutants for sev^- and *sev-Ras*^{G12V} display the *sev-Ras*^{G12V} mutant phenotype. **(c)** This epistatic interaction is consistent only with the action of Sevenless upstream of Ras (*shaded row*) but not the other way around (*bottom row*). *Green* indicates the protein is active; *red* represents inactive protein.

(a) Sevenless signaling is a switch/regulation pathway

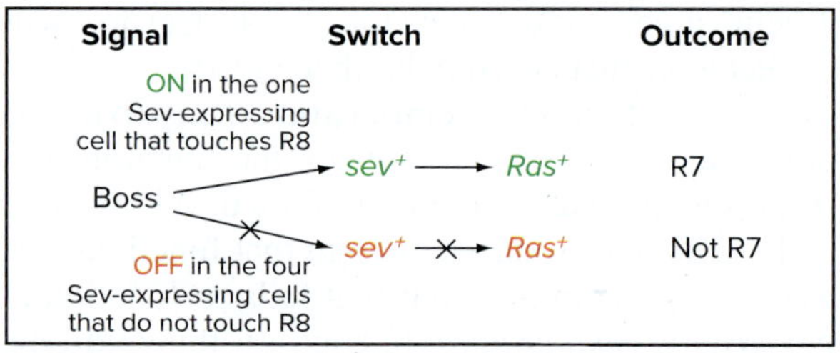

(b) *Ras*^{G12V} is epistatic to *sev*$^-$

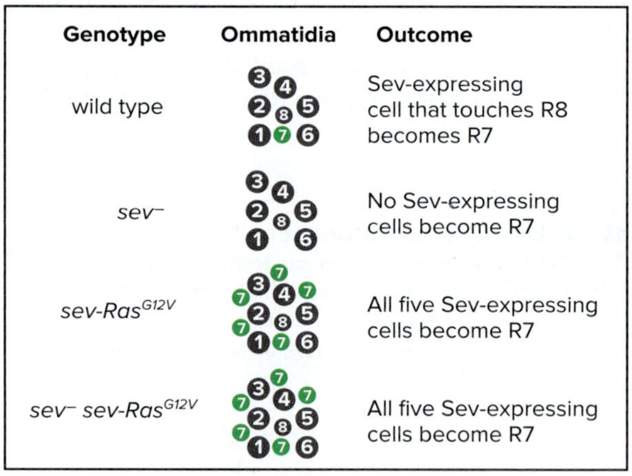

(c) *sev*$^+$ functions upstream of *Ras*$^+$

Wild type	Sev ⟶ Ras ⟶ R7	
sev$^-$	Sev ✕⟶ Ras No R7	
sev-Ras^{G12V}	Sev ⟶ Ras^{G12V} ⟶ R7	
sev$^-$ *sev-Ras*^{G12V}	Sev ✕⟶ Ras^{G12V} ⟶ R7	**YES**
OR		
sev$^-$ *sev-Ras*^{G12V}	Ras^{G12V} ⟶ Sev No R7	**NO**

The signal, Boss, flips the switch, Sevenless, ON in the one cell that both expresses Sevenless and touches R8. The switch is OFF in the four other Sevenless-expressing cells that are not adjacent to R8, because the Sevenless receptor in these cells never comes into contact with Boss. The outcome of the pathway is either R7 development (switch ON) or not (switch OFF) (**Fig. 19.14a**).

Recall that in *Drosophila* sev^- null mutants, ommatidia lack the R7 photoreceptor cell. Flies containing a transgene ($sev\text{-}Ras^{G12V}$) express a constitutively active Ras protein in the five R7 precursor cells that produce the Sevenless receptor; as a result, all five precursor cells become R7s (recall Fig. 19.7b). Double mutants that contain $sev\text{-}Ras^{G12V}$ and that are also sev^- have eyes in which all five precursor cells become R7s (**Fig. 19.14b**). Therefore, $sev\text{-}Ras^{G12V}$ is epistatic to sev^-, meaning that the double mutant has the same phenotype as a fly with only the $sev\text{-}Ras^{G12V}$ mutation.

The epistatic interaction observed is consistent with the model, based on biochemical experiments, that Ras is downstream of Sevenless in the signaling pathway (review Fig. 19.6). Although in wild type Sevenless protein (after binding to Boss) activates Ras, a constitutively active mutant *Ras* is epistatic to a sev^- null mutation because activated Ras does not need Sevenless protein in order to generate an R7 cell (**Fig. 19.14c**). If activated Ras still required Sevenless in order to signal, the opposite epistasis interaction would have been observed. This incorrect model predicts that the sev^- $sev\text{-}Ras^{G12V}$ double mutants would have the sev^- phenotype because signaling would be blocked in the double mutants, and no R7 would develop (Fig. 19.4c, *bottom row*).

Epistasis analysis can be informative, but to interpret the results several conditions must be met. For example, the phenotypes seen in the two single mutants must differ from each other, and the mutant alleles under consideration must be either null or constitutive. Even when these conditions are fulfilled, conclusions about gene order in a pathway made from epistasis analysis should always be viewed as hypotheses that require verification with biochemical experiments.

essential concepts

- In a developmental pathway, if the expression of gene *a* is altered in a gene *b* mutant, but gene *b* expression is unaffected in a gene *a* mutant, then gene *b* must function upstream of gene *a*.
- In a *switch/regulation pathway,* the phenotype of double mutants can reveal epistatic interactions and help determine the order of action of the two genes.

19.5 A Comprehensive Example: Body Plan Development in *Drosophila*

learning objectives

1. Describe key events in early *Drosophila* development that occur before segmentation is first apparent.
2. Define the term *maternal effect genes,* and explain why the protein products of some of these genes are called *morphogens.*
3. Summarize the hierarchy of zygotic segmentation genes.
4. Discuss the functions of the homeotic genes in specifying segment identity.

Studies on the genetic control of the basic body plan of *Drosophila* have revolutionized our understanding of development. Here, we focus on the aspect of this work that explains how the fly's body becomes differentiated and specialized along the *anterior/posterior (AP) axis,* the line running from the animal's head to its tail.

The research we describe was based on the observation that a fertilized *Drosophila* egg becomes subdivided into an embryo with several clearly defined segments, each of which eventually has a specific appearance and function. Some segments become parts of the head, others parts of the thorax, and still others, parts of the abdomen. Scientists designed experiments to answer two fundamental questions about segmentation. First, how does the animal develop the proper number of body segments? And second, how does each body segment know what kinds of structures it should form? Results showed that very early in development, before transcription even begins, **maternal effect genes,** whose products are deposited by the mother into her eggs, help establish regional differences in the embryo that lead to the proper segment number. Next, the action of a large group of genes, called the **zygotic segmentation genes,** expressed by the zygote's own genome, subdivides the body into an array of essentially identical body segments. Later in development, the expression of the **homeotic genes,** a set of genes that encode transcription factors, assigns a unique identity to each body segment.

Drosophila Embryos Become Divided into Segments

To understand how the maternal effect, segmentation, and homeotic genes function, it is helpful to consider some of the major events that take place in the first few hours of *Drosophila* development (**Fig. 19.15**). The egg is fertilized as it is being laid, and the meiotic divisions of the oocyte

Figure 19.15 Early *Drosophila* development: From fertilization to cellular blastoderm. The zygotic nucleus undergoes 13 rapid mitotic divisions in a single syncytium. A few nuclei at the posterior end become the germ-line *pole cells*. At the *syncytial blastoderm* stage, the egg surface is covered by a monolayer of nuclei. At the end of the thirteenth division cycle, cell membranes enclose the nuclei at the cortex into separate cells to produce a *cellular blastoderm*.

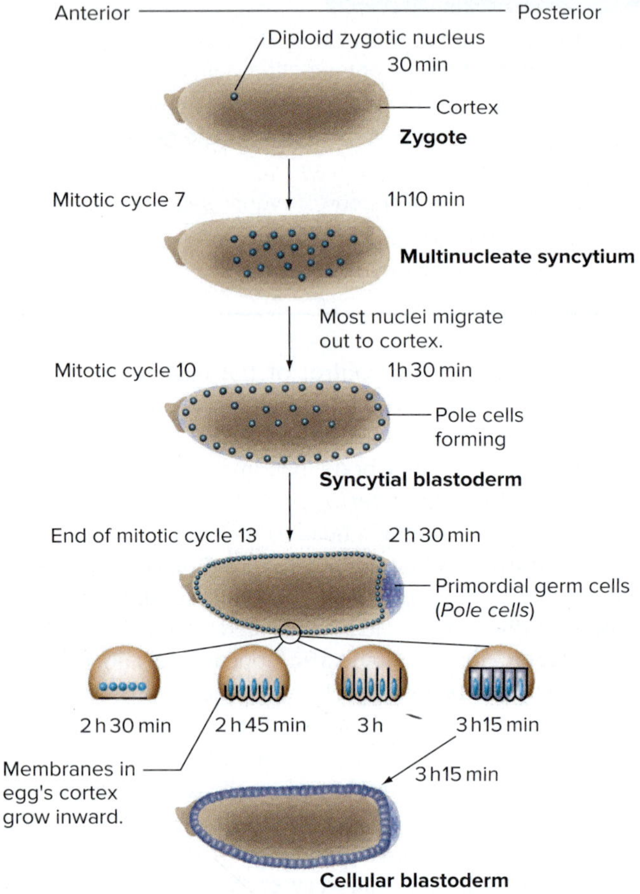

nucleus, which had arrested previously in the metaphase of meiosis I, resume at this time. After fusion of the haploid male and female pronuclei, the diploid zygotic nucleus of the embryo undergoes 13 rounds of nuclear division at an extraordinarily rapid rate, with the average time of mitotic cycles 2 through 9 being only 8.5 minutes.

Nuclear division in early *Drosophila* embryos, unlike most mitoses, is not accompanied by cell division, so the early embryo becomes a multinucleate *syncytium*. During the first eight division cycles, the multiple nuclei are centrally located in the egg. During the ninth division, most of the nuclei migrate out to the *cortex*—just under the surface of the embryo—to produce the **syncytial blastoderm.** During the tenth division, nuclei at the posterior pole of the egg are enclosed in membranes that invaginate from the egg cell membrane to form the first embryonic cells; these *pole cells* are the primordial germ cells. At the end of the thirteenth division cycle, about 6000 nuclei are present at the egg cortex.

Figure 19.16 *Drosophila* development after formation of the cellular blastoderm. **(a)** Scanning electron micrograph of a cellular blastoderm. Individual cells are visible at the embryo's periphery, as are the larger pole cells at the posterior end (*arrow*). **(b)** A ventral view of the furrows that form during gastrulation, roughly 4 hours after fertilization: vf, ventral furrow; cf, cephalic furrow. **(c)** By 10 hours after fertilization, the embryo is subdivided into segments. Ma, Mx, and Lb are the three head segments. CL, PC, O, and D are nonsegmented regions of the head. The three thoracic segments (T1, T2, and T3) are the prothorax, the mesothorax, and the metathorax, respectively, whereas the abdominal segments are labeled A1–A8.

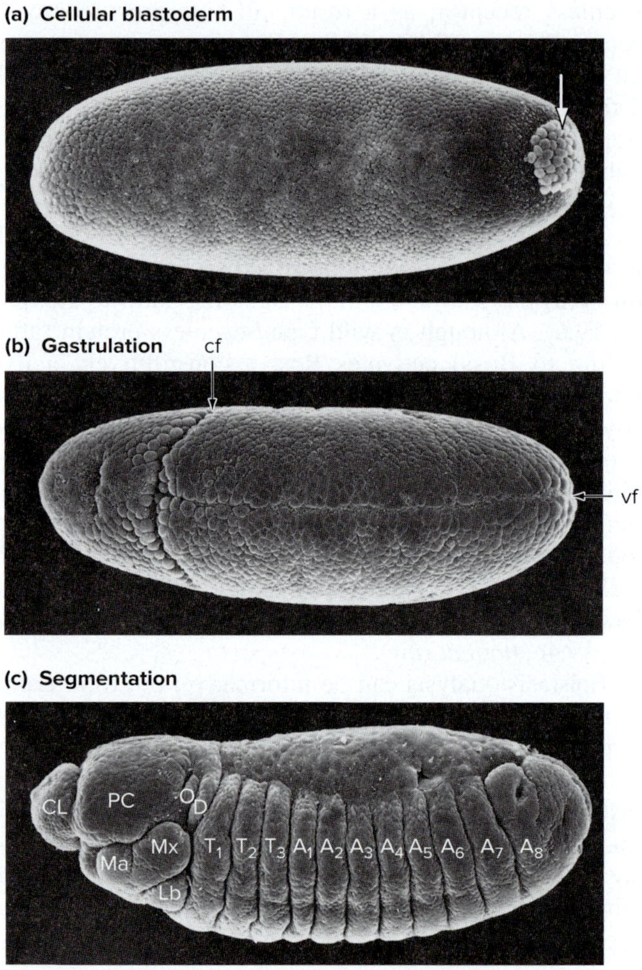

During the interphase of the fourteenth cycle, membranes in the egg's cortex grow inward between these nuclei, creating an epithelial layer called the **cellular blastoderm** (Fig. 19.15). The embryo completes formation of the cellular blastoderm about 3 hours after fertilization. At the cellular blastoderm stage, no regional differences in cell shape or size are apparent, with the exception of the pole cells at the posterior end (**Fig. 19.16a**). Molecular studies nonetheless reveal that most segmentation and homeotic genes function during or even before the cellular blastoderm stage.

Immediately after cellularization, an infolding process called **gastrulation** begins to establish the embryonic cell layers. The *mesoderm* forms by invagination of a band of midventral cells that extends most of the length of the embryo. This infolding (the *ventral furrow;* **Fig. 19.16b**) produces an internal tube; the tube cells divide and migrate to produce a mesodermal layer. The *endoderm* forms by distinct invaginations anterior and posterior to the ventral furrow; one of these invaginations is the *cephalic furrow* seen in Fig. 19.16b. The cells of the endodermal infoldings migrate over the yolk to produce the gut. Finally, the nervous system arises from neuroblasts located laterally on each side of the ventral *ectoderm.*

The first visible signs of segmentation are periodic bulges in the mesoderm, which appear about 40 minutes after gastrulation begins. Within a few hours of gastrulation, the embryo is divided into clear-cut body segments that will become the three head segments, three thoracic segments, and eight major abdominal segments of the larva (**Fig. 19.16c**). Even though the animal eventually undergoes metamorphosis to become an adult fly, the same basic body plan is conserved in the adult stage (**Fig. 19.17**).

Maternal Effect Genes Help Specify Segment Number

Very little transcription of genes occurs in the embryonic nuclei between the time of fertilization and the end of the 13 rapid syncytial divisions. Because of this near (but not total) absence of transcription, developmental biologists suspected that formation of the basic body plan initially requires *maternally supplied components* deposited by the mother into the egg during oogenesis. How could they identify the genes encoding these maternally supplied components? Christiane Nüsslein-Volhard and Eric Wieschaus realized that the embryonic phenotype determined by such genes does not depend on the embryo's own genotype; rather, it is determined by the genotype of the mother. They devised genetic screens to identify recessive mutations in maternal genes that influence embryonic development. These recessive mutations are called **maternal effect mutations.**

To carry out their screens, Nüsslein-Volhard and Wieschaus established individual balanced stocks for thousands of mutagen-treated chromosomes, and they then examined the embryos obtained from homozygous mutant mothers. They focused their attention on stocks in which homozygous mutant females were sterile, because they anticipated that the absence of maternally supplied components needed for the earliest stages of development would result in embryos so defective that they could never grow into adults (**Fig. 19.18**). Through these large-scale

Figure 19.17 Segment identity is preserved throughout development. The identities of embryonic segments (*left*) are preserved through the larval stages and are also retained through metamorphosis into the adult (*right*).

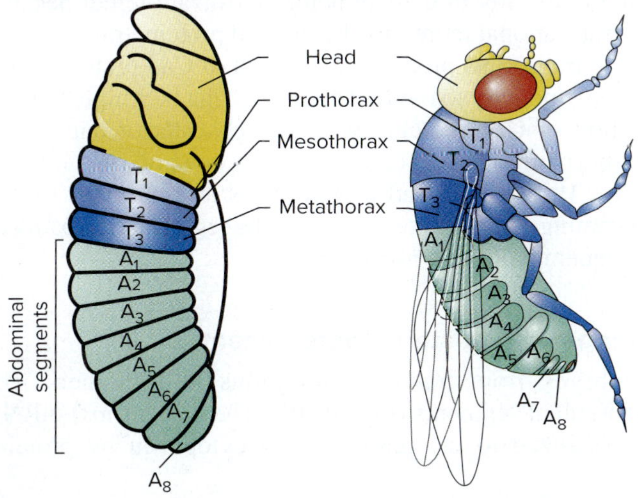

Figure 19.18 Mutagenesis screens for maternal effect and zygotic lethal mutations. How Nüsslein-Volhard and Wieschaus found recessive mutations on *Drosophila* chromosome 2 that affect segmentation is shown. Mutagenized 2nd chromosomes were recovered in F_1 males who also had a *Balancer* chromosome (*squiggly line*) with the dominant *Cy* marker. These F_1 males were crossed individually to *Balancer/D* females. (*D* is a different dominant marker.) Curly-winged F_2 progeny were intercrossed to generate homozygous mutants (m^-/m^-) in the F_3. If these animals died as embryos with segmentation defects, the 2nd chromosome carried interesting zygotic lethal mutations. Viable homozygous F_3 females were crossed to wild-type males to identify female sterile mutations affecting segmentation of the F_4 progeny.

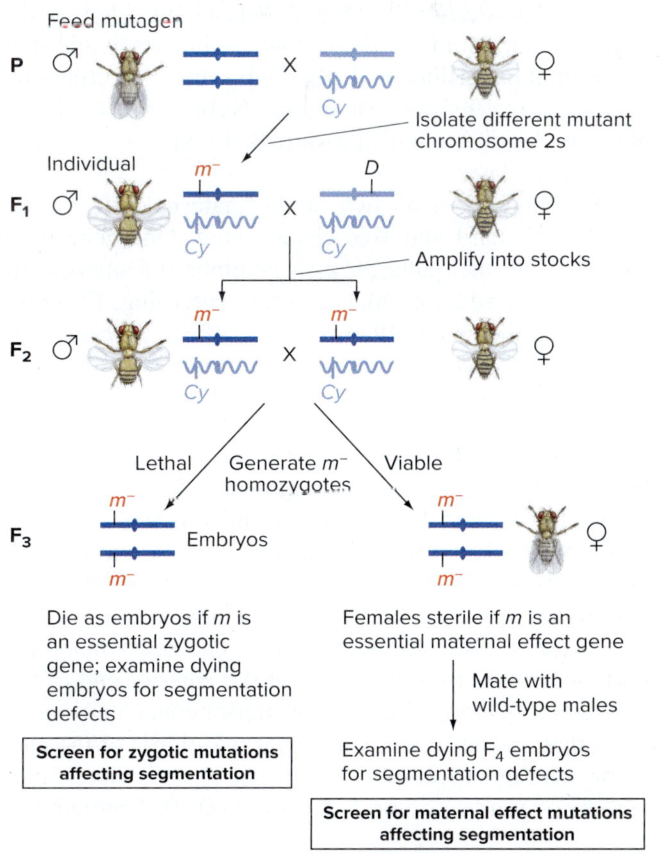

Figure 19.19 Maternal effect mutants affecting AP axis patterning. In embryos of wild-type mothers (*left*), the body between the head (anterior) and the tail (posterior) is divided into segments with specific patterns of hairs called *denticles* on their ventral surfaces. Embryos of mothers that fail to deposit Bcd protein in their eggs (*bcd*$^-$; *middle*) lack anterior segments, while embryos of mothers that fail to deposit Nos protein (*nos*$^-$; *right*) lack abdominal segments. A, anterior; P, posterior; D, dorsal; V ventral.
Courtesy of Christiane Nuesslein-Volhard, Max Planck Institute for Developmental Biology

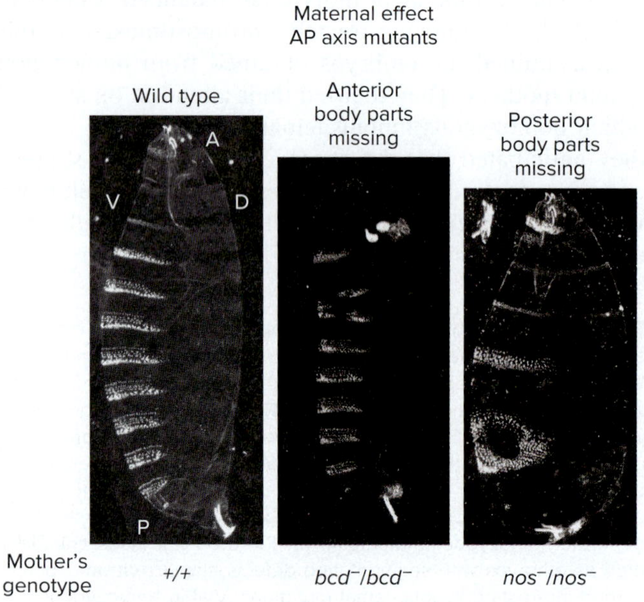

Figure 19.20 Bicoid is the anterior morphogen. (a) The *bicoid* mRNA (visualized by *in situ* hybridization in *purple*) concentrates at the embryo's anterior tip. **(b)** The Bicoid (Bcd) protein (seen by *green* antibody staining) is distributed in a gradient: high at the anterior end and trailing off toward the posterior. Bcd is a transcription factor that accumulates in the nuclei of the syncytial blastoderm embryo.
a: © Steve Small, New York University; b: © David Kosman and John Reinitz

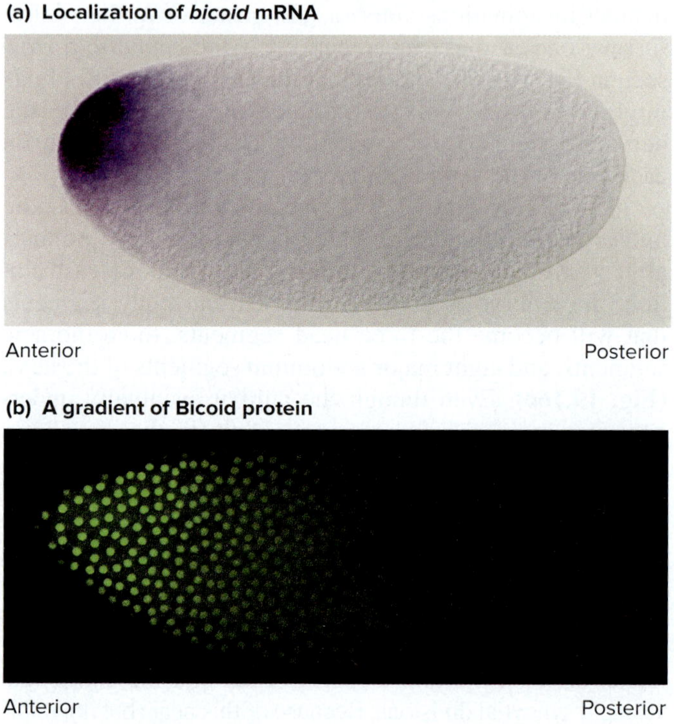

screens, Nüsslein-Volhard and Wieschaus identified a large number of *maternal effect genes* that are required for the normal patterning of the body. For this and other contributions, in 1995 they shared the Nobel Prize for Physiology or Medicine with Edward B. Lewis—whose work we describe later.

We focus here on two of the maternal effect genes Nüsslein-Volhard and Wieschaus found. One gene is required for normal patterning of the embryo's anterior; the other is required for normal posterior patterning. These two genes, together with allied maternal effect genes, initiate the process that determines segment number.

Bicoid: The anterior morphogen

Embryos from mothers homozygous for null alleles of the *bicoid* (*bcd*) gene lack all head and thoracic structures (**Fig. 19.19**). The protein product of *bcd* is a transcription factor whose mRNA is localized at the anterior tip of the egg cytoplasm (**Fig. 19.20a**). Translation of the *bcd* transcripts takes place after fertilization. The newly made Bcd protein diffuses from its source at the anterior end to produce a high-to-low, anterior-to-posterior concentration gradient that extends over the anterior two-thirds of the early embryo (**Fig. 19.20b**). This gradient determines most aspects of head and thorax development. Bcd protein thus functions as a **morphogen**—a substance that defines different cell fates in a concentration-dependent manner.

The Bcd protein itself works in two ways: as a transcription factor that helps control the transcription of genes farther down the regulatory pathway (discussed later), and as a translational repressor. The target of its repressor activity is the transcript of the *caudal* (*cad*) gene, which also encodes a transcription factor. The *cad* mRNAs are distributed uniformly in the egg before fertilization, but because of translational repression by the Bcd protein, translation of these transcripts produces a gradient of Cad protein that is complementary to the Bcd gradient. That is, a high concentration of Cad protein exists at the posterior end of the embryo and lower concentrations toward the anterior (**Fig. 19.21**). The Cad protein plays an important role in activating genes expressed later in the segmentation pathway to generate posterior structures.

Nanos: The posterior determinant

Embryos from *nanos* homozygous null mothers lack abdominal segments (Fig. 19.19). The *nanos* (*nos*) mRNA is localized to the posterior egg cytoplasm by proteins

Figure 19.21 mRNA and protein products of maternal effect genes within the early embryo. *Top*: In the oocyte prior to fertilization, *bicoid* mRNA is concentrated near the anterior tip and *nanos* mRNA at the posterior tip, while *hunchback* and *caudal* mRNAs are distributed uniformly. *Bottom*: In early cleavage stage embryos, the Bicoid and Hunchback proteins are found in concentration gradients high at the anterior and lower toward the posterior (A-to-P), whereas the Nanos and Caudal proteins are distributed in opposite (P-to-A) gradients.

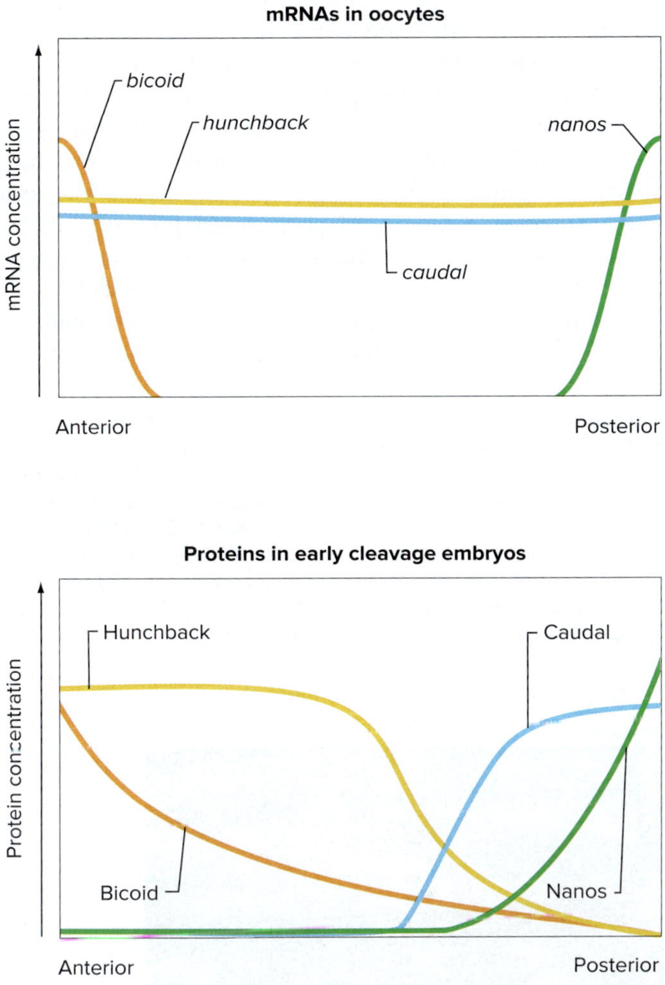

encoded by other maternal genes. Like *bcd* mRNAs, *nos* transcripts are translated during the early nuclear division stages. After translation, diffusion produces a posterior-to-anterior Nos protein concentration gradient (Fig. 19.21).

The Nos protein, unlike the Bcd protein, is not a transcription factor; rather, the Nos protein functions only as a translational repressor. Its target is the maternally supplied transcript of the *hunchback* (*hb*) gene, which is deposited in the egg during oogenesis and is distributed uniformly before fertilization (Fig. 19.21). However, unlike the Bcd gradient, the Nos gradient is not important to Nos function; what is important is that high levels of Nos are present at the embryo's posterior.

For development to occur properly, Hb protein (another transcription factor) must be absent from the posterior of the embryo. The Nos protein, which is present at high concentrations at the posterior, represses the translation of *hb* maternal mRNA and thus eliminates Hb protein at the embryo's posterior pole (Fig. 19.21).

Zygotic Genes Determine Segment Number and Polarity

The maternally determined Bcd and Cad protein gradients control the spatial expression of zygotic segmentation genes. Unlike the products of maternal genes, whose mRNAs are placed in the egg during oogenesis, the products of zygotic genes are transcribed and translated from DNA in the nuclei of embryonic cells descended from the original zygotic nucleus. The expression of zygotic segmentation genes begins in the syncytial blastoderm stage, a few division cycles before cellularization.

Most of the zygotic segmentation genes were identified in a second mutant screen also carried out in the late 1970s by Christiane Nüsslein-Volhard and Eric Wieschaus. In this screen, the *Drosophila* geneticists placed individual mutagenized chromosomes into balanced stocks and then examined homozygous mutant embryos from these stocks for defects in the segmentation pattern of the embryo (see Fig. 19.18). These embryos were so aberrant that they were unable to grow into adults; thus, the mutations causing these defects would be classified as *recessive lethals.*

After screening several thousand such stocks for each of the *Drosophila* chromosomes, Nüsslein-Volhard and Wieschaus identified three classes of zygotic segmentation genes: **gap genes** (9 different genes); **pair-rule genes** (8 genes); and **segment polarity genes** (~17 genes). These three classes of zygotic genes fit into a hierarchy of gene expression.

Gap genes

The gap genes are the first zygotic segmentation genes to be transcribed. Embryos homozygous for mutations in the gap genes show a "gap" in the segmentation pattern caused by an absence of particular segments that correspond to the position at which each gene is expressed (**Fig. 19.22a** and **b**).

How do the maternal transcription factor gradients ensure that the various gap genes are expressed in their broad zones at the proper position in the embryo? Part of the answer is that the binding sites in the enhancers of the gap genes have different affinities for the maternal transcription factors. For example, some gap genes are activated by Bcd protein (the anterior morphogen). Gap genes such as *hb* with low-affinity Bcd protein-binding sites are activated only in the most anterior regions, where the concentration of Bcd is at its highest; by contrast, genes with high-affinity sites have an activation range extending farther toward the posterior pole.

Another part of the answer is that the gap genes themselves encode transcription factors that can influence the expression of other gap genes. The *Krüppel* (*Kr*) gap gene, for example, appears to be turned off by high amounts of Hb protein at the anterior end of its band of expression, activated within its expression band by Bcd protein in conjunction with lower levels of Hb protein, and turned off at the posterior end of its expression zone by the products of the *knirps* (*kni*) gap gene (Fig. 19.22b). (Note that the *hb* gene is usually classified as a gap gene, despite the maternal supply of some *hb* RNA, because scientists have detected a role in patterning only for the protein translated from the transcripts of zygotic nuclei.)

Figure 19.22 Gap genes. (a) Zones of expression of four gap genes [*hunchback (hb), Kruppel (Kr), knirps (kni),* and *giant (gt)*] in late syncytial blastoderm embryos, as visualized with fluorescently stained antibodies. **(b)** Zones of gap gene expression in the embryo are shown. Mutation of a particular gap gene results in the loss of segments corresponding to the zone of expression of that gap gene.

a: © David Kosman and John Reinitz

(a) Zones of gap gene expression

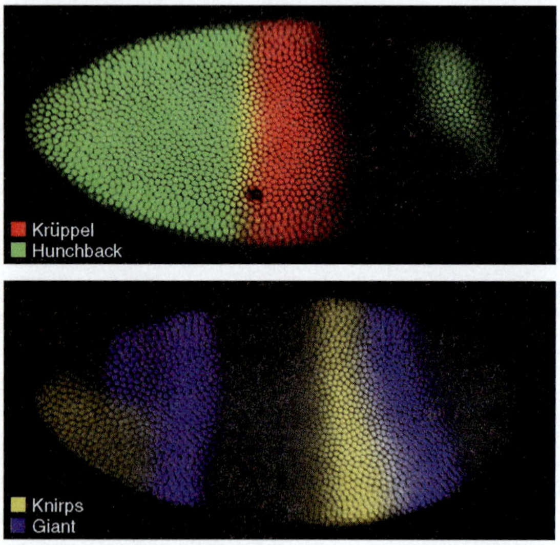

(b) Gap genes: a summary

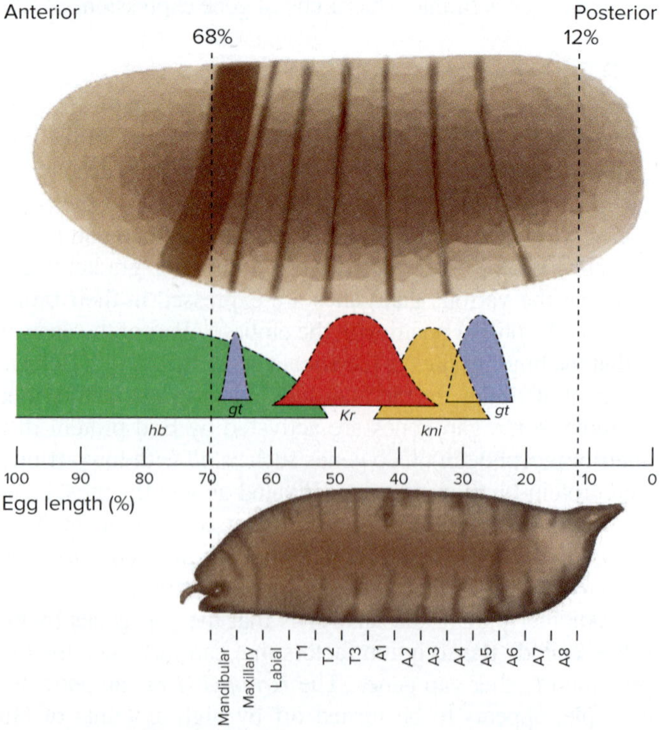

Pair-rule genes

After the gap genes have divided the body axis into rough, generalized regions, activation of the pair-rule genes generates more sharply defined sections. These genes encode transcription factors that are expressed in seven stripes in preblastoderm and blastoderm embryos (**Fig. 19.23a**). The stripes have a two-segment periodicity; that is, one stripe exists for every two segments. Mutations in pair-rule genes cause the deletion of similar pattern elements from every alternate segment. For example, larvae mutant for *fushi tarazu* (*segment deficient* in Japanese) lack parts of abdominal segments A1, A3, A5, and A7; mutations in *even-skipped* cause the loss of even-numbered abdominal segments.

Two classes of pair-rule genes exist: primary and secondary. The striped expression pattern of the three primary

Figure 19.23 Pair-rule genes. (a) Expression zones of the protein encoded by the pair-rule genes *fushi tarazu* (*ftz*) and *even-skipped* (*eve*) in the cellular blastoderm. Each gene is expressed in seven stripes. Eve stripe 2 is the second green stripe from the *left*. **(b)** The formation of Eve stripe 2 requires activation of *eve* transcription by the Bcd and Hb proteins, and the absence of repression by the Gt and Kr proteins. **(c)** The enhancer between 800 and 1500 bp upstream of the *eve* gene that directs the Eve second stripe contains multiple binding sites for the four proteins shown in part (b).

a: © David Kosman and John Reinitz

(a) Distribution of pair-rule gene products

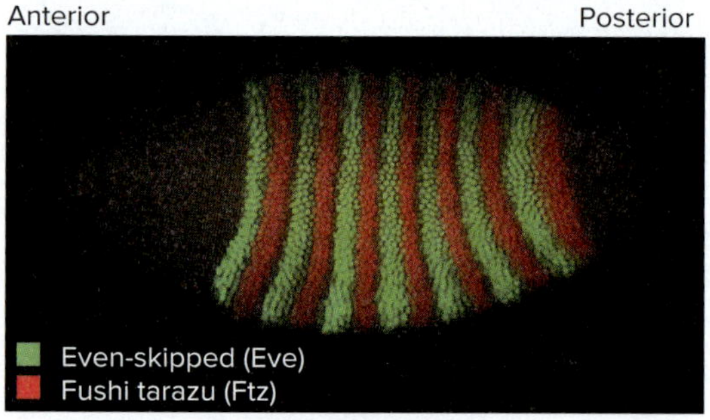

(b) Proteins regulating *eve* transcription

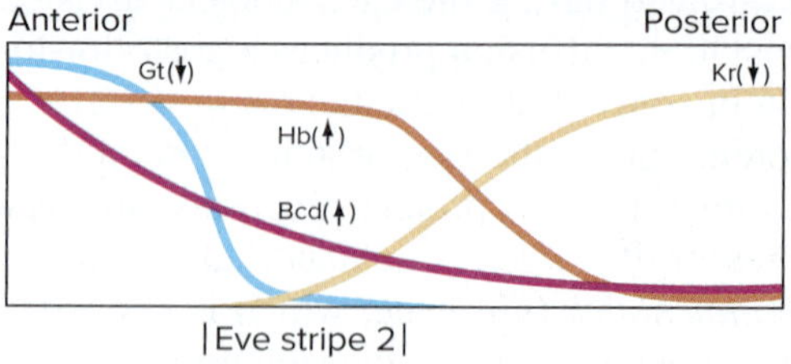

(c) Upstream regulatory region of *eve*

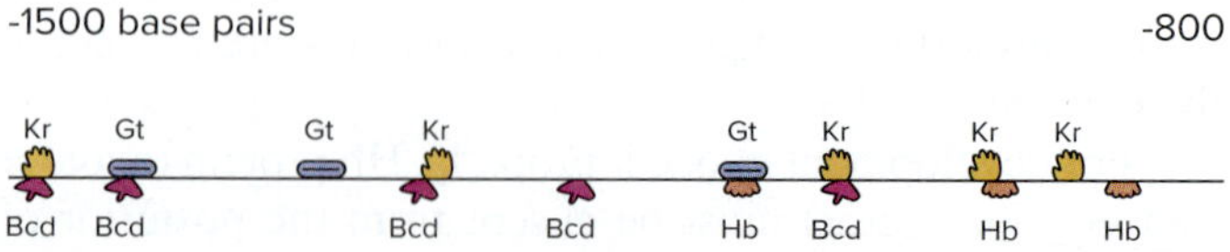

pair-rule genes depends on the transcription factors encoded by the maternal genes and the zygotic gap genes. Specific elements within the upstream regulatory region of each pair-rule gene drive the expression of that pair-rule gene within a particular stripe. For example, as Fig. **19.23b** and **c** show, the DNA region responsible for driving the expression of *even-skipped* (*eve*) in the second stripe contains multiple binding sites for the Bcd protein and the proteins encoded by the gap genes *Krüppel, giant* (*gt*), and *hb*. The transcription of *eve* in this stripe of the embryo is activated by Bcd and Hb, while it is repressed by Gt and Kr. Only in the stripe 2 region are Gt and Kr levels low enough, and Bcd and Hb levels high enough, to allow activation of the enhancer driving *eve* expression.

In contrast with the primary pair-rule genes, transcription of the five pair-rule genes of the secondary class is controlled by the transcription factors encoded by other pair-rule genes.

Segment polarity genes

Many segment polarity genes are expressed in stripes that repeat with a single segment periodicity; that is, one stripe is present in each of 14 segments (**Fig. 19.24a**). Mutations in segment polarity genes cause deletion of part of each segment, often accompanied by mirror-image duplication of the remaining parts. The segment polarity genes thus function to determine certain patterns that are repeated in each segment.

The regulatory system that directs the expression of segment polarity genes in a single stripe per segment is quite complex. In general, the transcription factors encoded by pair-rule genes initiate the pattern by directly regulating certain segment polarity genes. Interactions between various cell polarity genes then maintain this periodicity later in development. Significantly, the activation of segment polarity genes occurs after cellularization of the embryo is complete, so the diffusion of transcription factors within

(a) Distribution of Engrailed protein

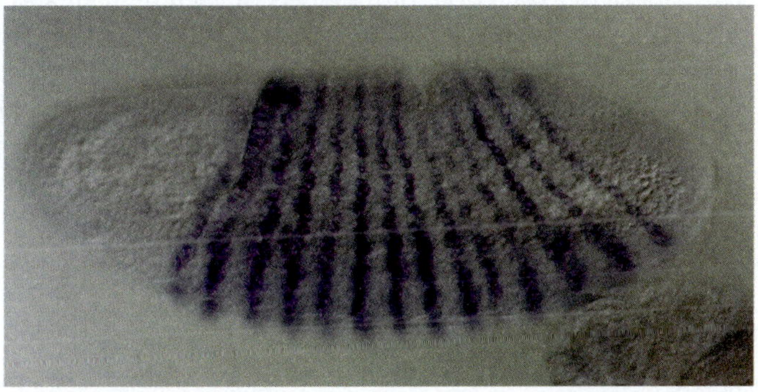

(b) Segment polarity genes establish compartment borders.

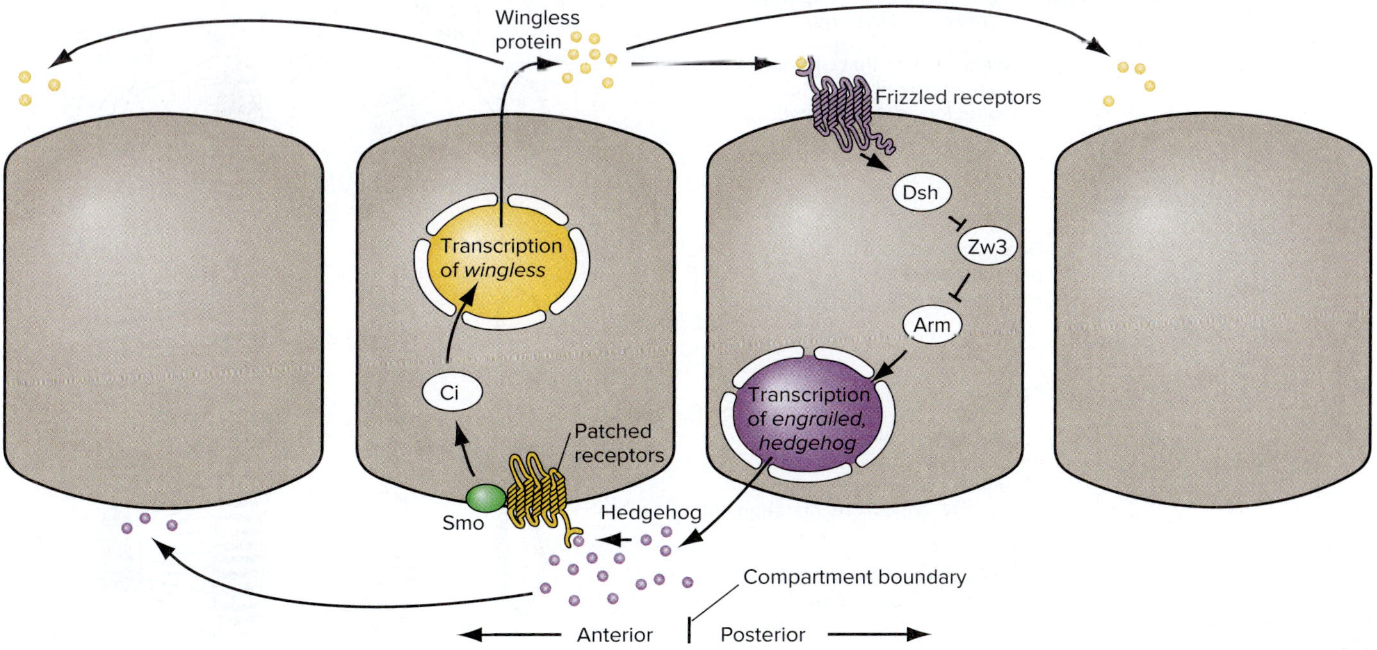

Figure 19.24 Segment polarity genes. (a) Wild-type embryos express the segment polarity gene *engrailed* in 14 stripes. **(b)** The border between a segment's posterior and anterior halves, or *compartments,* is governed by the *engrailed* (*en*), *wingless* (*wg*), and *hedgehog* (*hh*) segment polarity genes. Cells in posterior compartments express *en*. En protein activates transcription of *hh*, which encodes a secreted ligand. Binding of Hh protein to the Patched receptor in the adjacent anterior cell initiates a signal transduction pathway (through the Smo and Ci proteins) leading to *wg* transcription. Wg is a secreted protein that binds to a receptor in the posterior cell, encoded by *frizzled*. Binding of Wg to the Frizzled receptor initiates a different signal transduction pathway (including the Dsh, Zw3, and Arm proteins) that stimulates transcription of *en* and *hh*. The result is a reciprocal loop stabilizing the alternate fates of adjacent cells at the border (the compartment boundary).

a: © Steve Small, New York University

the syncytium ceases to play a role. Instead, intrasegmental patterning is determined mostly by the diffusion of secreted proteins between cells.

Two of the segment polarity genes, *hedgehog* (*hh*) and *wingless* (*wg*), encode secreted proteins. These proteins, together with the transcription factor encoded by the *engrailed* (*en*) segment polarity gene, are responsible for many aspects of segmental patterning (**Fig. 19.24b**). A key component of this control is that a one-cell-wide stripe of cells secreting the Wg protein is adjacent to a stripe of cells expressing the En protein and secreting the Hh protein. The interface of these two cell types is a self-reinforcing, reciprocal loop. The Wg protein secreted by the more anterior of the two adjacent stripes of cells is required for the continued expression of *hh* and *en* in the adjacent posterior stripe. The Hh protein secreted by the more posterior stripe of cells maintains the expression of *wg* in the anterior stripe. The interface between En- and Hh-expressing cells and the Wg-expressing cells forms the *compartment boundary.* Cells within the anterior and posterior compartments of a segment do not intermingle.

Gradients of Wg and Hh proteins made from these adjacent stripes of cells control many aspects of patterning in the remainder of the segment. The products of both *wg* and *hh* are morphogens; that is, responding cells adopt different fates depending on the concentration of Wg or Hh protein to which they are exposed.

Other segment polarity genes encode proteins involved in **signal transduction pathways** initiated by the binding of Wg and Hh proteins to receptors on cell surfaces. Signal transduction pathways enable a signal received from a receptor on the cell's surface to be converted to a final intracellular regulatory response, usually the activation or repression of particular target genes. The signal transduction pathways initiated by the Wg and Hh proteins determine the ability of cells in portions of each segment to differentiate into the particular cell types characteristic of those locations.

Homologs of the segment polarity genes are key players in many important patterning events in vertebrates. For example, the chicken *sonic hedgehog* gene (related to fly *hh*) is crucial for the initiation of the left-right asymmetry in the early chicken embryo as well as for the processes that determine the number and polarity of digits produced by the limb buds. The mammalian homolog of *sonic hedgehog* has the same conserved functions as in chickens.

Summary of segment number specification

The pattern of expression for members of each segmentation gene class is controlled either by genes higher in the hierarchy or by members of the same class, never by genes of a lower class (**Fig. 19.25**). In this regulatory cascade, the maternal genes control the gap and pair-rule genes, the gap genes

Figure 19.25 The genetic hierarchy leading to segmentation in *Drosophila*. **(a)** Genes in successively lower parts of the hierarchy are expressed in narrower bands within the embryos. **(b)** Mutations in segmentation genes cause the loss of segments that correspond to regions where the gene is expressed (*yellow*). The denticle bands (*dark brown*) are features that help researchers identify the segments.

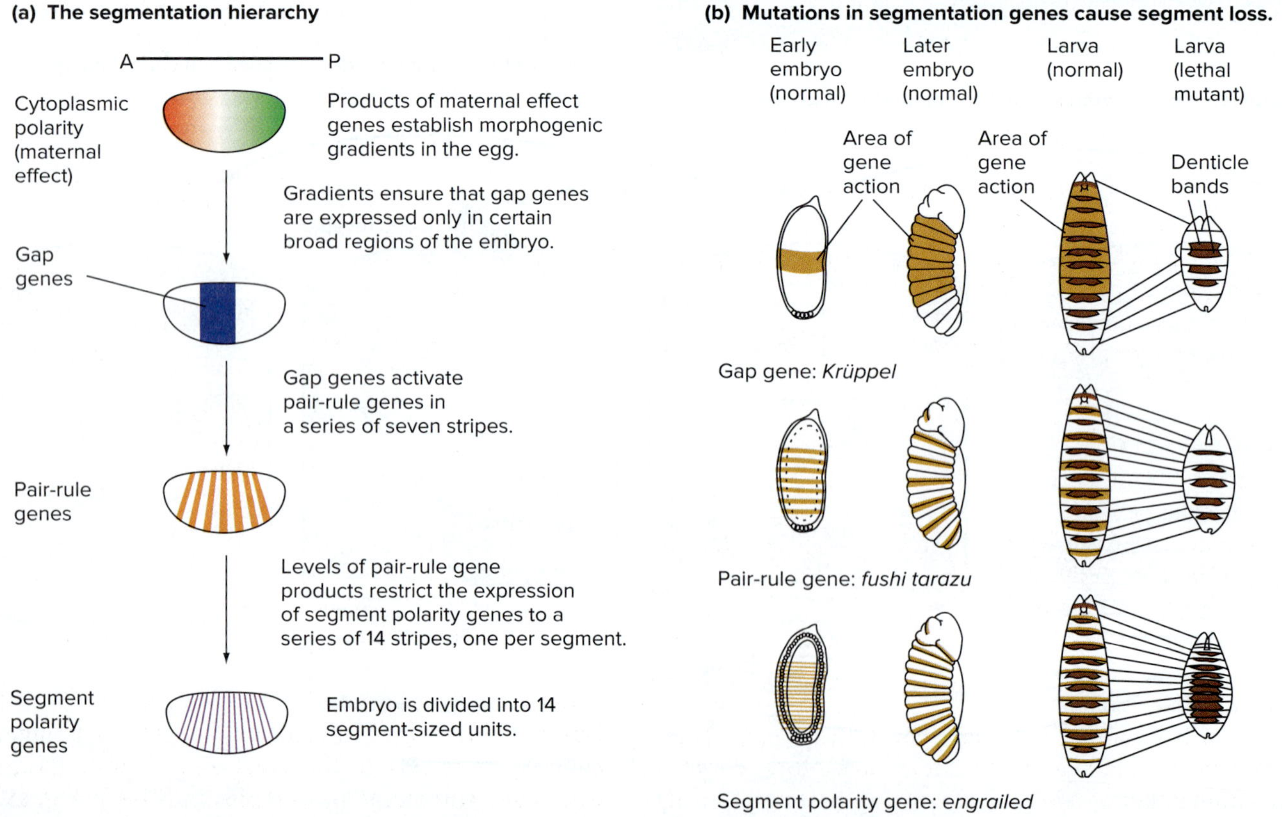

control themselves and the pair-rule genes, and the pair-rule genes control themselves and the segment polarity genes. The expression of genes in successively lower parts of the hierarchy is increasingly restricted spatially within the embryo.

The cellular blastoderm looks from the outside like a uniform layer of cells (as seen in Fig. 19.16a), but the coordinated action of the segmentation genes has actually already divided the embryo into segment primordia. A few hours after gastrulation, these primordia become distinguishable as clear-cut segments (Fig. 19.16c).

Homeotic Genes Specify Parasegment Identity

After the segmentation genes have subdivided the body into a precise number of segments, the homeotic genes help assign a unique identity to each segment. Each homeotic gene controls a unit called a **parasegment (PS)**, which is the posterior compartment of one segment and the anterior compartment of the segment just posterior to it. Homeotic genes do this by functioning as "master regulators" that control the transcription of batteries of genes responsible for the development of segment-specific structures. The homeotic genes themselves are regulated by the gap, pair-rule, and segment polarity genes so that at the cellular blastoderm stage, or shortly thereafter, each homeotic gene becomes expressed within a specific subset of parasegments. Most homeotic genes then remain active throughout development, functioning continuously to direct proper segmental specialization.

Mutations in homeotic genes, referred to as **homeotic mutations,** cause particular whole parasegments or specific individual compartments to develop as if they were located elsewhere in the body. Because some of the homeotic mutant phenotypes are quite spectacular, researchers noticed them very early in *Drosophila* research. In 1915, for example, Calvin Bridges found a mutant he called *bithorax* (*bx*). In homozygotes for this mutation, the anterior compartment of the third thoracic segment (T3) develops like the anterior compartment of the second thoracic segment (T2); in other words, this mutation transforms part of T3 into the corresponding part of T2. The *bx* mutant phenotype is dramatic, as anterior T3 normally produces only small club-shaped balancer organs called *halteres,* whereas anterior T2 produces the wings (**Fig. 19.26**).

Another homeotic mutation is *postbithorax* (*pbx*), which affects only the posterior compartment of T3, causing its transformation into the posterior compartment of T2 (Fig. 19.26). (Note that in this context, *Drosophila* geneticists use the term *transformation* to mean a change of body form.) In the *bx pbx* double mutant, all of T3 develops as T2 to produce the now famous four-winged fly shown in Chapter 1 (Fig. 1.9).

In the last half of the twentieth century, researchers isolated many other homeotic mutations, most of which map within either of two gene clusters. Mutations affecting segments in the abdomen and posterior thorax lie within a cluster known as the **bithorax complex (BX-C);** mutations affecting segments in the head and anterior thorax lie within the **Antennapedia complex (ANT-C) (Fig. 19.27).**

Figure 19.26 Homeotic transformations. In animals homozygous for the mutation *bithorax* (*bx*), the anterior compartment of T3 (the third thoracic segment that makes the haltere) is transformed into the anterior compartment of T2 (the second thoracic segment that makes the wing). The mutation *postbithorax* (*pbx*) transforms the posterior compartment of T3 into the posterior compartment of T2.

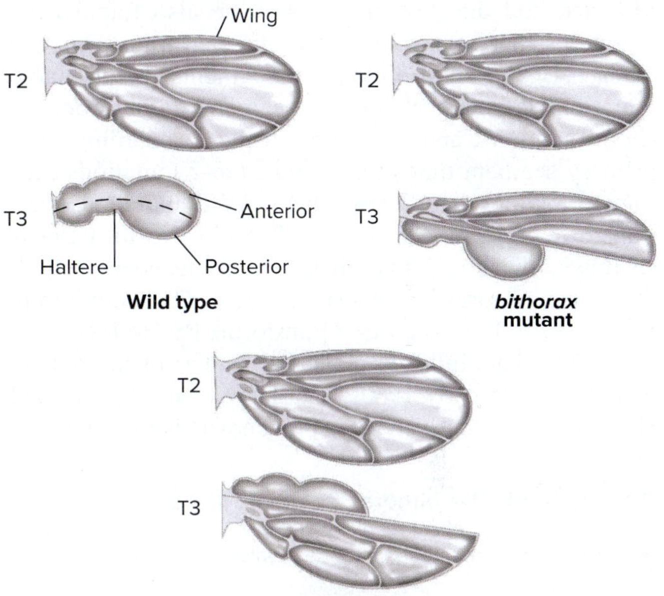

Figure 19.27 Homeotic selector genes. Two clusters of genes on *Drosophila's* chromosome 3—the Antennapedia complex and the bithorax complex—determine most aspects of segment identity. Interestingly, the order of genes in these complexes is the same as the order of the parasegments each gene controls.

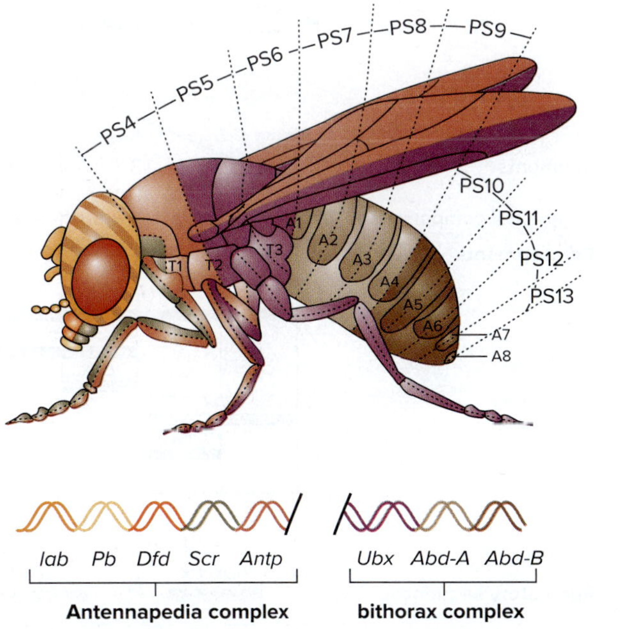

The bithorax complex

Edward B. Lewis shared the 1995 Nobel Prize for Physiology or Medicine with Christiane Nüsslein-Volhard and Eric Wieschaus for his extensive genetic studies of the BX-C. In his work, Lewis isolated BX-C mutations that, like *bx* and *pbx,* affected the posterior thorax; he also found BX-C mutations that caused anteriorly directed transformations of each of the eight abdominal segments. Lewis named mutations affecting abdominal segments *infra-abdominal* (*iab*) mutations, and he numbered these according to the primary segment they affect. Thus, *iab-2* mutations cause transformations of A2 toward A1, *iab-3* mutations cause transformations of A3 toward A2, and so forth. Later researchers discerned that these mutations were actually transforming parasegments (Fig. 19.27). Thus, *iab-2* transforms PS7 to PS6, and *iab-3* transforms PS8 to PS7.

Researchers initiated molecular studies of the bithorax complex in the early 1980s, and in 15 years, they not only characterized extensively all of the genes and mutations in the BX-C at the molecular level but also completed the sequencing of the entire 315 kb region. **Figure 19.28** summarizes the structure of the complex. A remarkable feature of the BX-C is that mutations map in the same order on the chromosome as the anterior/posterior order of the parasegments each mutation affects. Thus, *bx* mutations, which affect posterior PS5, lie near the left end of the complex, whereas *pbx* mutations, which affect anterior PS6, lie immediately to their right. In turn, *iab-2,* which affects PS7, is to the right of *pbx* but to the left of the PS8-determining *iab-3.*

Because the *bx, pbx,* and *iab* elements are mutable independently, Lewis thought that each was a separate gene. However, the molecular characterization of the region revealed that the BX-C actually contains only three protein-coding genes: *Ultrabithorax* (*Ubx*), which controls the identities of PS5 and PS6; *Abdominal-A* (*Abd-A*), which controls the identities of PS7–PS9; and *Abdominal-B* (*Abd-B*), which controls the identities of PS10–PS13 (Fig. 19.28).

Figure 19.28 The bithorax complex. The complex contains three homeotic genes: *Ubx*, *Abd-A*, and *Abd-B*. Many homeotic mutations such as *bx* and *pbx* affect regulatory regions that influence the transcription of one gene in a particular parasegment or compartment. Note that the order of both the genes and the regulatory regions corresponds to the anterior-to-posterior order of segments in the animal.

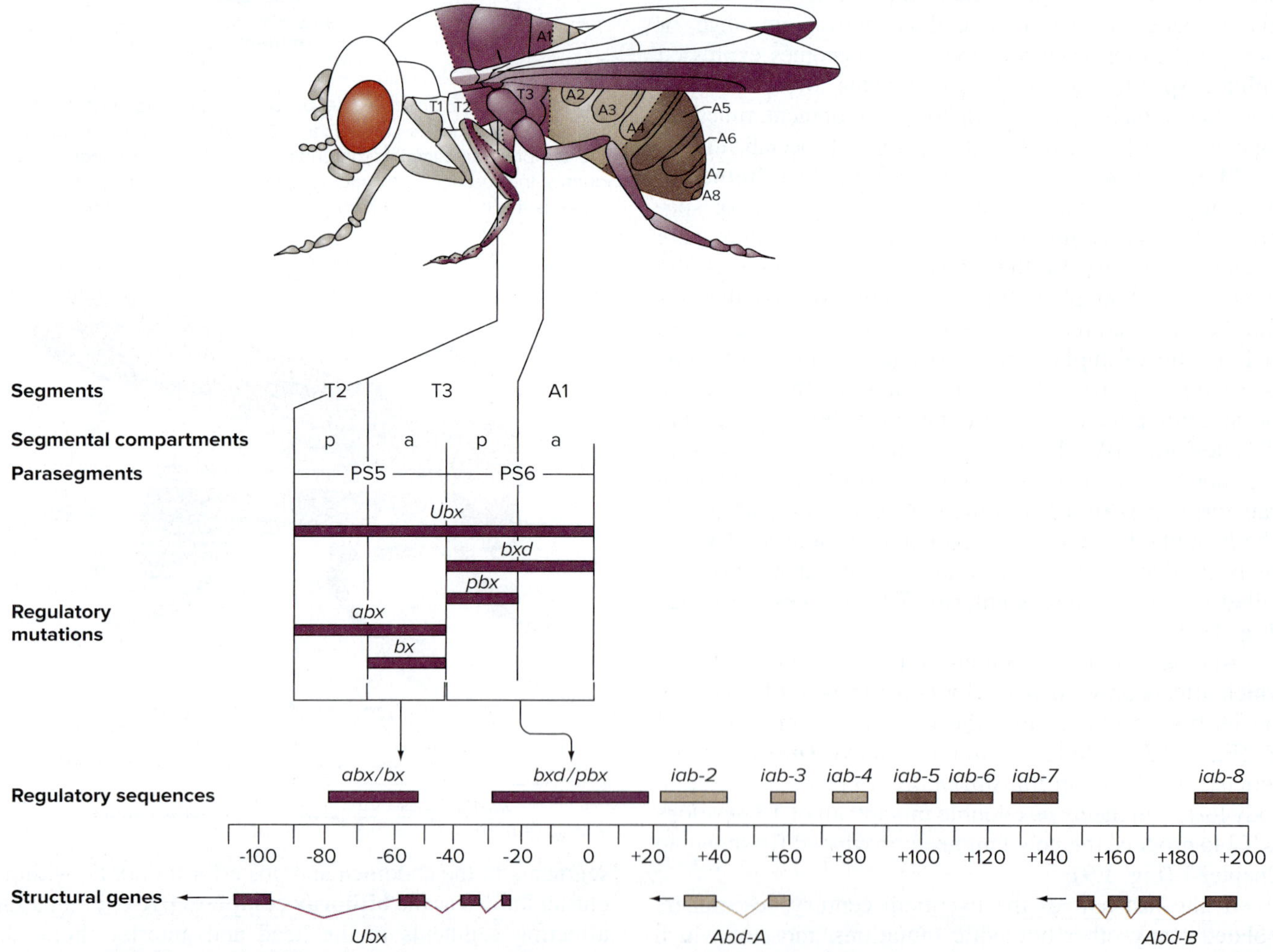

Figure 19.29 Expression patterns of bithorax complex genes. Expression of each BX-C gene begins at the position along the AP axis where it is first required. The segment and parasegment identities controlled by each gene are indicated by color: *Ubx* (*purple*); *Abd-A* (*light brown*); *Abd-B* (*dark brown*). (PS14 is not controlled by the BX-C.) For each gene, darker shades of its color indicate overall higher levels of its protein.

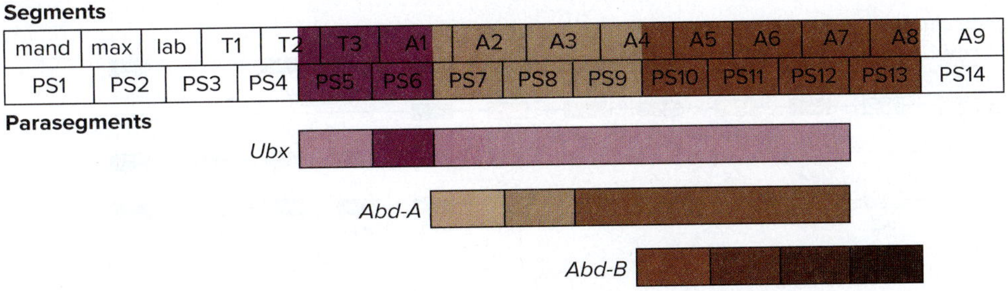

The expression patterns of these genes are consistent with their roles (**Fig. 19.29**). Along the AP axis, *Ubx* expression begins in PS5, *Abd-A* expression begins in PS7, and *Abd-B* expression begins in PS10. The *bx, pbx,* and *iab* mutations studied by Lewis affect large *cis*-regulatory regions that control the intricate spatial and temporal expression of these genes within specific segments.

The Antennapedia complex

Genetic studies in the early 1980s showed that a second homeotic gene cluster, the Antennapedia complex (ANT-C), specifies the identities of segments in the head and anterior thorax of *Drosophila.* Three of the five homeotic genes of the ANT-C—*labial* (*lab*), *proboscipedia* (*pb*), and *Deformed* (*Dfd*)—control the head parasegments PS1 and PS2. *Sex combs reduced* (*Scr*) controls head and thoracic parasegments (PS2 and PS3), and *Antennapedia* (*Antp*) controls the thoracic parasegment PS4. (Figure 19.16c shows these head and thoracic segments, whereas Fig. 19.27 illustrates the order of the homeotic genes in the ANT-C.) As with the BX-C, the order of genes in the ANT-C is the same (with the exception of *Pb*) as the order of parasegments each controls.

The homeodomain in development and evolution

As researchers started to characterize the genes of the ANT-C and the BX-C at the molecular level, they were surprised to find that all of these genes contained some closely related (although not identical) DNA sequences. Similar sequences were also found in many other genes important for development, such as *bicoid* and *eyeless,* which are located outside the homeotic gene complexes. The region of sequence homology, called the **homeobox,** is about 180 bp long and is located in the protein-coding part of each gene. The 60 amino acids encoded by the homeobox constitute the **homeodomain,** a region of each protein that can bind to DNA (**Fig. 19.30**). We now know that almost all proteins containing homeodomains are transcription factors in which the homeodomain is responsible for the sequence-specific binding of the proteins to the *cis*-acting control sites of the genes they regulate.

Figure 19.30 The homeodomain: A DNA-binding motif found in many transcription factors that regulate development. Amino acids within the homeodomain (*yellow*) interact with specific sequences in a DNA double helix (*red* and blue).
Modeled by Thomas R. Bürglin, based on the data of Otting et al. (1990), "Protein—DNA contacts in the structure of a homeodomain—DNA complex determined by nuclear magnetic resonance spectroscopy in solution," *EMBO J,* 9(10): 3085-3092

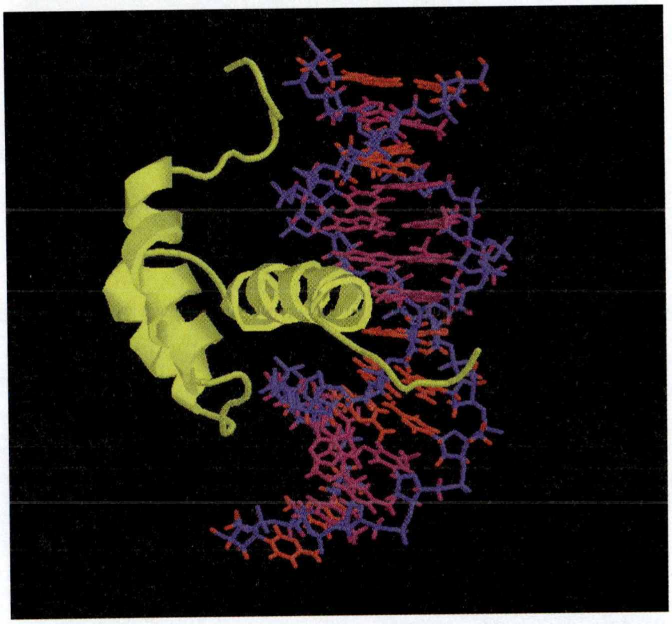

The discovery of the homeobox was one of the most important advances in the history of developmental biology because it allowed scientists to identify by homology many other genes with roles in the development of *Drosophila* and other organisms. In the late 1980s and 1990s, the biological community was astonished to learn that the mouse and human genomes contain clustered homeobox genes called ***Hox* genes** with clear homologies to the ANT-C and BX-C genes in *Drosophila* (**Fig. 19.31**). Remarkably, in all mammals studied to date, the genes within these clusters are arranged in a linear order that reflects their expression in particular regions along the spine of developing mammalian embryos (Fig. 19.31). In other words, these gene clusters in mice and humans are arranged in the genome and are

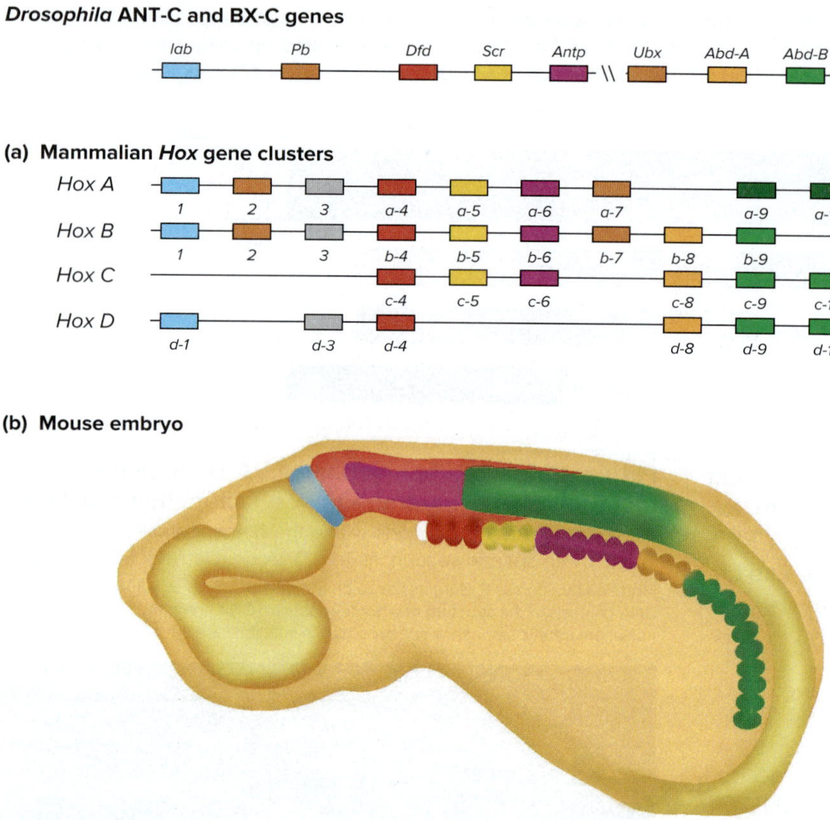

Figure 19.31 The mammalian *Hox* genes are organized into four clusters. **(a)** Mammalian genomes contain multiple homologs of each of the ANT-C and BX-C homeobox genes in *Drosophila*. **(b)** Just as in *Drosophila*, the mammalian (mouse) *Hox* genes in each cluster are arranged in the order in which they are expressed along the AP axis of the embryo. The *colored disks* represent *somites*, precursors of the vertebrae and other structures. The colors represent the *Hox* gene expressed in that tissue.

regulated along the anterior/posterior axis in almost exactly the same way as in the fly ANT-C and BX-C complexes.

As it turns out, all animal genomes, even those of sponges, the most primitive animals, contain *Hox* genes, so these genes are ancient and have played important roles in the developmental patterns of all animals. Generally, the more complex the body plan, the more *Hox* genes: Humans and other mammals have four *Hox* clusters that together contain 38 *Hox* genes (see Fig. 19.31). In just one demonstration that *Hox* genes mediate the developmental fate of specific regions in the bodies of animals other than *Drosophila*, malformation of the digits in humans, in a condition called *synpolydactyly*, is caused by mutations in *HoxD13*, one of these 38 *Hox* genes (**Fig. 19.32**).

Figure 19.32 Synpolydactyly caused by mutations in the human *HoxD13* gene.

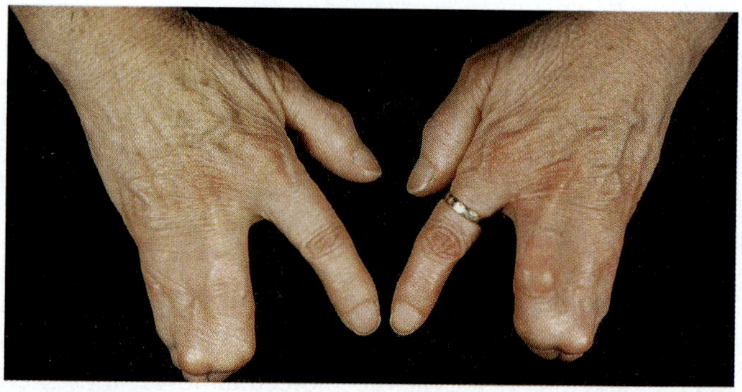

essential concepts

- By performing *mutant screens* and analyzing the mutants, *Drosophila* geneticists have identified a *gene hierarchy* that determines body patterning (segmentation) along the AP axis of the animal.
- *Maternal effect genes* are expressed from the maternal genome, and their products are deposited in the egg. Some of these gene products affect segmentation of the embryo in a concentration-dependent manner and are thus termed *morphogens*.
- In *Drosophila*, three classes of *zygotic segmentation genes* (*gap genes*, *pair-rule genes*, and *segment polarity genes*) are expressed in a hierarchy that eventually subdivides the embryo into 14 identical segments. Many of these genes are transcription factors that control the expression of genes in the same class or the next lower class in the hierarchy.
- *Drosophila homeotic genes* encode homeobox transcription factors that are expressed in different segments, providing each segment with its individual identity. *Homeobox genes* that pattern the AP axis of *Drosophila* are conserved in all animals, where they function in body plan patterning.

WHAT'S NEXT

The development of a single-celled zygote into a complex multicellular organism depends on the careful coordination of cell division. In general, the mitotic divisions in early embryos occur rapidly to supply the developing organism with a large pool of undifferentiated cells that can subsequently differentiate into the multitude of cell types that form various organ systems. After the embryo has grown into an adult organism, most somatic cells in animals (excepting stem cells, whose division is needed to replenish lost cells) divide much less often to ensure that one tissue does not overgrow others.

Rarely, these controls on cell division break down, leading to the unregulated cell growth we call a *cancer*. In Chapter 20, we discuss the evidence showing that cancer represents many different diseases caused by the accumulation of mutations in various genes whose protein products either promote or inhibit cell growth. Scientists are increasingly able to identify the key mutations that cause the particular type of cancer found in each patient. These exciting new discoveries hold the promise that more effective treatments can be devised because they can now be targeted to individual cancers.

SOLVED PROBLEMS

I. Using recombinant DNA technology, scientists can create transgenes that express genes *ectopically*—at an abnormal place or time. The flies shown here, with eye tissue on a wing or a leg, contain a transgene in which the cDNA of the *eyeless* gene of *Drosophila* (review Fig. 19.2a) is fused to the control region of a heat-shock gene whose transcription in any tissue is turned on by higher-than-normal temperatures. Eyeless protein is usually expressed only in the developing eye, but animals bearing this fusion gene that were grown at high temperature made the Eyeless protein throughout their bodies.

a. What can you conclude about *eyeless* gene function?

b. Ectopic eyes also arise when the mouse *Pax-6* or the human *Aniridia* gene is expressed in *Drosophila* under the control of the same heat-shock gene promoter. What does this result signify?

c. Sometimes in ectopic expression experiments, we wish to express a gene in one specific tissue rather than in the whole organism. Why? How could you express *eyeless* only in the legs of the fly?

Answer

a. The *eyeless* gene encodes a master regulator that causes the cells that express it to become eyes.

b. This result means that both the amino acid sequence and the actual function of this master switch have been conserved throughout animal evolution.

c. Global ectopic gene expression can often be lethal to the organism. To express *eyeless* only in legs, you would fuse the *eyeless* cDNA to a characterized gene regulatory region that activates transcription leg-specifically.

II. The *Gal4/UAS system* enables an investigator to express a gene in any one of thousands of patterns using only one cDNA transgene. Gal4 is a yeast transcription factor that can activate transcription from genes that carry a yeast enhancer called UAS_G. (Review the Tools of Genetics Box in Chapter 17 entitled *The Gal4/UAS_G Binary Gene Expression System.*)

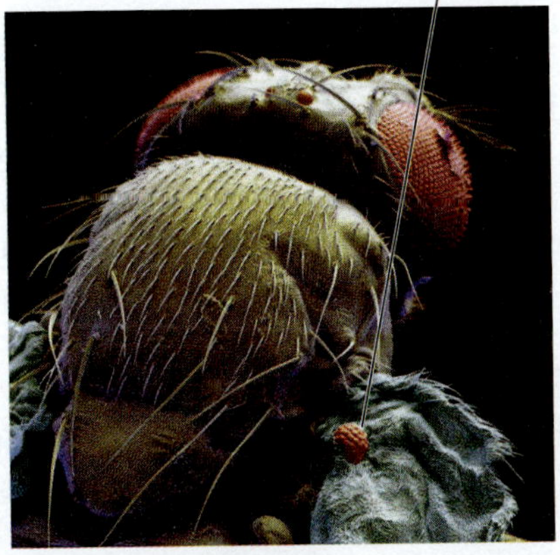

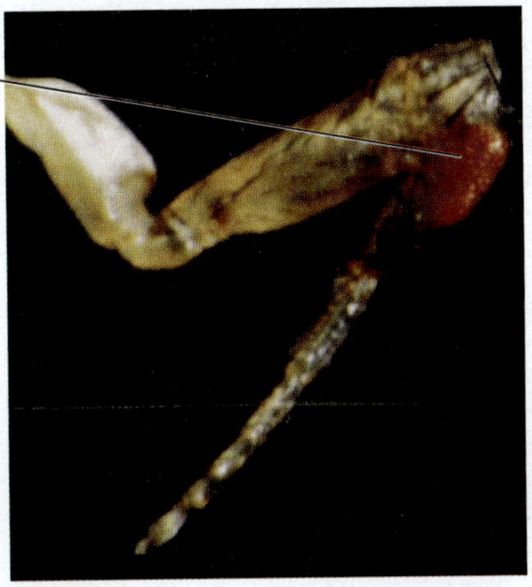

How could you use the Gal4/UAS system to express Eyeless protein specifically in legs? Explain the advantage of using the Gal4/UAS system instead of the type of fusion gene generated in part (c) of Solved Problem I.

Answer

First, you would transform flies with a *P* element construct in which UAS_G was placed upstream of *eyeless* cDNA. You would then cross the UAS_G-*eyeless* flies to a driver line that expresses Gal4 protein in cells destined to become legs.

The advantage of using the binary system is that one UAS_G-*eyeless* construct can be used with a variety of different Gal4 drivers that are already available and that are specific to different tissues. Thus, by crossing flies with the UAS_G-*eyeless* construct to strains with different drivers, you could make flies that make the Eyeless protein ectopically, in any given tissue, whether legs or wings or even in internal organs. In addition, scientists would not need to define particular gene regulatory regions in order to achieve tissue-specific gene expression.

PROBLEMS

Vocabulary

1. Match each of the terms in the left column to the best-fitting phrase from the right column.

a. epistatic interaction	1. divide the body into identical units (segments)
b. regulative determination	2. initiated by binding of ligand to receptor
c. modifier screen	3. individuals with cells of more than one genotype
d. RNAi	4. the fate of early embryonic cells can be altered by the environment
e. ectopic expression	5. assign identity to body segments
f. homeodomain	6. substance whose concentration determines cell fates
g. green fluorescent protein	7. suppression of gene expression by double-stranded RNA
h. genetic mosaics	8. method for identifying pleiotropic genes
i. segmentation genes	9. a DNA-binding motif found in certain transcription factors
j. homeotic genes	10. encode proteins that accumulate in unfertilized eggs and are needed for embryo development
k. morphogen	11. double mutant has phenotype of one of the two mutants
l. maternal effect genes	12. a gene is turned on in an inappropriate tissue or at the wrong time
m. signal transduction pathways	13. a tag used to follow proteins in living cells

Section 19.1

2. a. If you were interested in the role of a particular gene in the embryonic development of the human heart, why would you probably study this role in a model organism, and which model organism(s) would you choose?

   b. If you were interested in finding genes that might be required for human heart development, which model organism(s) would you choose? Describe two different experimental approaches you could use.

3. Early *C. elegans* embryos display mosaic determination, whereas early mouse embryos exhibit regulative determination. Predict the results you would expect if the following treatments were performed on four-cell embryos of each of these two species (assuming these manipulations could actually be performed):

   a. A laser is used to destroy one of the four cells (this technique is called *laser ablation*).

   b. The four cells of the embryo are separated from each other and allowed to develop.

   c. The cells from two different four-celled embryos are joined together to make an eight-celled embryo.

4. Hypomorphic mutations in the *wingless* gene of *Drosophila* result in animals lacking wings.

   a. Starting with a set of *wingless* mutations, how could researchers have identified the *wingless* gene in the *Drosophila* genome sequence?

   b. Part of the amino acid sequence encoded by the ORF of the *wingless* gene is:

   (N)...EAGRAHVQAEMRQECKCHGMSGSCTVKTCWMRL...(C)

   Perform a **protein blast** at the following website to ask whether the human genome has a gene related to the fly *wingless* gene: https://blast.ncbi.nlm.nih.gov/Blast.cgi?PROGRAM=blastp&PAGE_TYPE=BlastSearch&LINK_LOC=blasthome

   Enter **Homo sapiens (taxid:9606)** as the organism. Leave all the other settings in their default state, and hit the **blue BLAST button** at the bottom of the page. The results of the database search will appear in a few minutes. What do the results of the search tell you about the existence of human genes homologous to the fly *wingless* gene?

Section 19.2

5. Flies homozygous for recessive null mutations in the *sevenless* (*sev*) or *bride-of-sevenless* (*boss*) genes

have the same mutant phenotype: Every ommatidium (facet) in their eyes lacks photoreceptor cell 7 (R7). The R7 cells enable flies to detect UV light.

a. Given that flies normally move toward light, suggest a screening method that would enable you to identify mutations in additional genes required for R7 determination.

b. Would you be able to recover mutations in every gene required for R7 development with your method? Explain.

c. How could you tell whether any of the new mutations you found in your screen are alleles of *sev* or *boss?*

d. Suppose you found one recessive mutant allele of a gene not previously known to be involved in eye development. How could you use this allele in a new mutagenesis screen to find additional alleles of this gene? Why might you want additional mutant alleles to study the process?

Problem 6 concerns a *Drosophila* gene called *rugose* (*rg*). Adult flies homozygous for recessive mutations in this gene have rough eyes in which the regular pattern of ommatidia (facets) is disrupted. The disruption of the ommatidial pattern is caused by the absence of one or more so-called *cone cells* from ommatidia; in the wild type, each ommatidium has four cone cells.

6. In 1932, H. J. Muller suggested a genetic test to determine whether a particular mutation whose phenotypic effects are recessive to wild type is a null (amorphic) allele or is instead a hypomorphic allele of a gene. Muller's test was to compare the phenotype of homozygotes for the recessive mutant alleles to the phenotype of a heterozygote in which one chromosome carries the recessive mutation in question and the homologous chromosome carries a deletion for a region including the gene.

   In a study using Muller's test, investigators examined two recessive, loss-of-function mutant alleles of *rugose* named rg^{41} and $rg^{\gamma 3}$. The eye morphologies displayed by flies of several genotypes are indicated in the following table. *Df(1)JC70* is a large deletion that removes *rugose* and several genes to either side of it.

Genotype	Eye surface	Cone cells per ommatidium
wild type	smooth	4
rg^{41}/rg^{41}	mildly rough	2–3
rg^{41}/*Df(1)JC70*	moderately rough	1–2
$rg^{\gamma 3}/rg^{\gamma 3}$	very rough	0–1
$rg^{\gamma 3}$/*Df(1)JC70*	very rough	0–1

a. Which allele (rg^{41} or $rg^{\gamma 3}$) is stronger (that is, which causes the more severe mutant phenotype)?

b. Which allele directs the production of higher levels of functional Rugose protein?

c. How would Muller's test discriminate between a null allele and a hypomorphic allele? Suggest a theoretical explanation for Muller's test. Based on the results shown in the table, is either of these two mutations likely to be a null allele of *rugose?* If so, which one?

d. Explain why an investigator would want to know whether a particular *rg* allele was amorphic or hypomorphic.

e. Suppose that a hypermorphic *rg* allele exists (rg^{hyper}) that causes rough eyes due to an excess of cone cells. Could you use Muller's genetic method to determine that the dominant allele is hypermorphic? Explain.

f. Suppose an antimorphic *rg* allele exists (rg^{anti}). Can you think of a way to determine if a dominant mutation is antimorphic? (*Hint:* Assume that in addition to the chromosome with a deletion that deletes *rg*, a chromosome with a duplication that includes the wild-type *rg* gene is available.)

Problems 7 and 8 concern a recombinant DNA construct called *myo-2::GFP* that *C. elegans* developmental geneticists have transformed into worms. Worms containing this construct express green fluorescent protein (GFP) in their pharynx, an organ located between the mouth and the gut that grinds up the bacteria *C. elegans* eats as a food source (see photo). In the *myo-2::GFP* construct, the open reading frame for jellyfish GFP was ligated downstream of the promoter and enhancer for *myo-2,* a gene that is expressed specifically in the muscle cells of the pharynx.

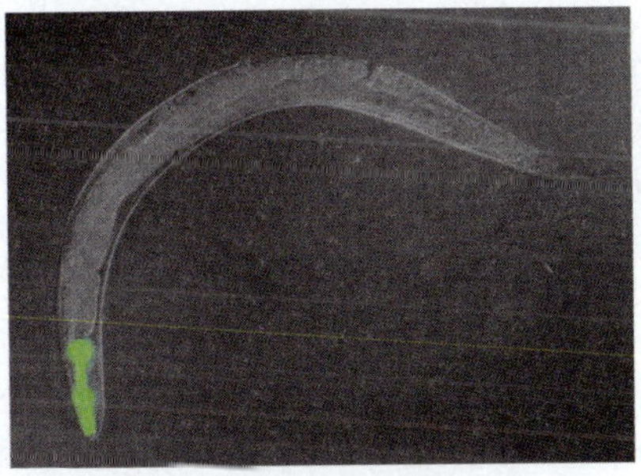

Image courtesy of John M. Kemner

7. a. Explain how you could use worms transformed with *myo-2::GFP* to find mutations that disrupt the structure of the pharynx. How would the presence of the transgene facilitate the mutant screen?

   b. Nematodes homozygous for loss-of-function mutations in a gene called *pha-4* have no detectable pharyngeal structures. How could you use *myo-2::GFP* to determine if *pha-4* is a master regulatory gene that directs development of the pharynx in a manner similar to the way *Pax-6/eyeless* controls eye development? (*Hint:* Review Solved Problem I.)

8. Suppose you want to determine whether a particular gene *X* is important for specification of the pharynx, but mutations in this same gene disrupt embryonic development well before pharyngeal structures appear. How could you use *myo-2::GFP*, the *myo-2* promoter, the DNA sequence of gene *X*, and your knowledge of RNA interference (RNAi) to generate worms that lack gene *X* expression in the pharynx but express gene *X* in all other tissues in which it is expressed in wild-type *C. elegans?*

9. Sevenless is an unusual receptor protein in that it is required only in one nonessential cell type—R7 precursor cells. That is, sev^- flies (null mutants) lack R7 cells in their eyes, yet they are fully viable and fertile under laboratory conditions. In contrast, the epidermal growth factor receptor (EGFR) is required in a variety of cell types beginning early in development. Because of the pleiotropic function of EGFR, flies that lack this protein die during embryogenesis.
   a. The Ras/MAPK pathway that relays the Sevenless signal to the nucleus (Fig. 19.6) also operates downstream of EGFR. Explain why the screen for modifiers of *sev* mutations used to identify components of the Ras/MAPK pathway (Fig. 19.5) would not have worked with *Egfr* mutations.
   b. Design a modifier screen that would identify mutations in genes in the *Egfr* pathway. (*Hint:* Use a transgene.)

10. Suppose that you generated flies containing a transgene ($sev\text{-}Ras^{S17N}$) that uses wild-type *sevenless* regulatory sequences to drive the expression of coding sequences encoding a dominant negative form of the Ras protein.
    a. Would you expect these flies to have a mutant phenotype? Explain.
    b. Suppose you used these transgenic flies in a modifier screen. For each protein in Fig. 19.6, state whether loss-of-function mutations in the corresponding gene could have been identified as enhancers or suppressors of the mutant phenotype.
    c. Answer part (b) if the flies contained a $sev\text{-}Ras^{G12V}$ (hypermorphic Ras) transgene instead.

11. *Drosophila* researchers have collected many strains that carry a single recombinant *P* element containing a wild-type *white* gene (a $P[w^+]$ transgene) inserted into a known genomic location. These strains can be used to map the location of any mutant gene in the fly genome.

    Investigators performed a testcross to map a recessive mutation *rough* (*ro*), which causes rough eyes, relative to a $P[w^+]$ element on chromosome 3. Females heterozygous for the $P[w^+]$ on one chromosome 3 and a ro^- mutation on the other, homologous chromosome 3 were crossed to ro^-/ro^- males, and the progeny in the following list were obtained. In both the parents and the progeny, the endogenous *white* gene is nonfunctional—the flies have red eyes only if they contain the $P[w^+]$ transgene.

145	red, smooth (wild-type) eyes
152	white, rough eyes
2	white, smooth eyes
1	red, rough eyes

    a. Are *ro* and the $P[w^+]$ linked? If so, how many map units separate them?
    b. The data in part (a) do not indicate on which side of the $P[w^+]$ (toward the centromere or telomere) the *ro* gene is located. How could the experiment be modified to reveal that information?
    c. Suppose you map the *ro* mutation to a genomic region between two different $P[w^+]$ elements that are 5000 bp apart. Describe some experimental approaches that would allow you to identify the *ro* gene at the molecular level.
    d. How could you use the DNA sequence of the *ro* gene to determine the function of the protein it encodes?

12. As an alternative to random mutagenesis, scientists can screen for mutant phenotypes by knocking down individual gene functions systematically using RNAi.
    a. Suggest ways to construct transgenes that in flies would express RNAi to knock down a gene.
    b. How could you perform a mutant screen for fly genes required for wing development using RNAi? How could this screen avoid the problem of pleiotropy?

13. A *C. elegans* (nematode) gene called *par-1* helps to determine the AP axis of the animal early in development. Scientists determined that *par-1* is pleiotropic—it also has a later function in forming the vulva of the adult animal. How could researchers circumvent the lethality of $par\text{-}1^-$ mutants to observe the later function of the *par-1* gene? (*Hint: C. elegans* larvae can eat bacteria expressing RNAi for any gene.)

Section 19.3

14. The molecular identity of the fruit fly *rugose* gene described in Problem 6 has been established. As a result, cDNA clones corresponding to the *rugose* gene mRNA and antibodies that recognize the Rugose protein are now available.
    a. How could you use these reagents to determine in which tissues the *rugose* gene is expressed? Does the expression of the gene in any particular tissue establish that the Rugose protein plays an essential function there?
    b. How could these same materials also help you to determine if a newly discovered recessive allele of *rugose* is a null or a hypomorphic mutation? If a new allele is dominant to wild type, could the mRNA

clone or the antibody help you decide if the allele is antimorphic, hypermorphic, or neomorphic?

15. To determine the focus of action of $boss^+$, researchers performed a mosaic experiment like the one shown in Fig. 19.11): Ommatida mosaic for w^+ $boss^+$ and w^- $boss^-$ photoreceptors were generated by mitotic recombination using FLP/FRT.

    a. Describe the appearance of the mosaic ommatidia (which cells are w^+ and which are w^-) that have a wild-type phenotype (R7 present) and that have a $boss^-$ mutant phenotype (R7 absent). What is the focus of action of the $boss^+$ gene?

    b. Although the *boss* gene is located on chromosome 3, it was still possible to use the w^+ gene as a marker for $boss^+$ cells. This was achieved by using flies in which the endogenous *white* gene on the X chromosome was mutant and that carried a $P[w^+]$ transgene on chromosome 3. Diagram the chromosomes used, including in your diagram the positions of the FRTs, the $P[w^+]$, $boss^-$ and $boss^+$ alleles, and the mitotic crossover. Show also how the chromosomes segregate into daughter cells to generate the mosaics.

16. Suppose a particular gene is required for early development and also later for development of a particular tissue, such as the adult nervous system. By generating a homozygous mutant clone in that tissue of a heterozygote, researchers can circumvent the lethality that would result if the entire animal is homozygous for a loss-of-function mutation in that gene.

    A technique called *MARCM (Mosaic Analysis with a Repressible Cell Marker)* was developed to enable *Drosophila* geneticists to generate homozygous mutant cell clones that are marked by the presence of a reporter protein such as GFP. Marker expression enables the investigator to observe clearly the mutant phenotype within a clone of mutant cells. This technique relies on a yeast protein called Gal80 that is a negative regulator of the Gal4 protein described previously in Solved Problem II. Gal80 binds to Gal4 and prevents it from activating transcription. The idea of MARCM is that Gal4/UAS_G-driven GFP expression is blocked by Gal80 throughout the fly, except within the homozygous mutant clone where the Gal80-expressing transgene is lost by mitotic recombination.

    a. Diagram the chromosomes and the mitotic crossover that generate a homozygous m^- mutant clone marked by GFP expression.

    b. How could you restrict the clones to the adult nervous system?

17. Researchers have exploited *Minute* mutations in order to study the phenotypes associated with recessive lethal mutations (l^-) that decrease the rate of cell division and thus make only very tiny homozygous mutant clones that are difficult to analyze. Many different strains of *Drosophila* carry dominant loss-of-function *Minute* (*M*) mutations in a variety of genes encoding ribosomal protein subunits. The *M* genes are haploinsufficient; flies with only one wild-type M^+ gene copy have a slower pace of cell division, and thus prolonged development and subtle morphological abnormalities.

    To circumvent the tiny clone problem, researchers generate GFP-marked homozygous l^-/l^- clones that are also M^+/M^+, in flies that are l^-/l^+ and M^-/M^+. The loss of the *Minute* mutation only in cells within the clone gives the l^-/l^- cells a growth advantage over their neighbors, enabling the mutant clone to grow large enough to study. Diagram chromosomes that could be used to generate such clones.

18. Some ts alleles are temperature sensitive during protein synthesis: If translation occurs at the restrictive temperature, the newly forming protein cannot fold correctly. Other ts alleles are temperature sensitive for activity: When the temperature is raised, the existing, properly folded protein unfolds and can no longer perform its function. Which kind of ts allele is better for temperature shift experiments (like the one in Fig. 19.12) aimed at determining when a protein functions? Explain your answer.

19. The following figure shows the temperature-shift analysis of *C. elegans* embryos from mothers homozygous for a temperature-sensitive allele of the *zyg-9* gene, which helps determine the polarity of the early embryo. Each dot represents an individual embryo subjected to a short (5-minute) pulse of high temperature. *Green* dots indicate that the embryo ultimately survives and becomes a normal worm, whereas *red* dots represent animals with abnormal polarity that ultimately die.

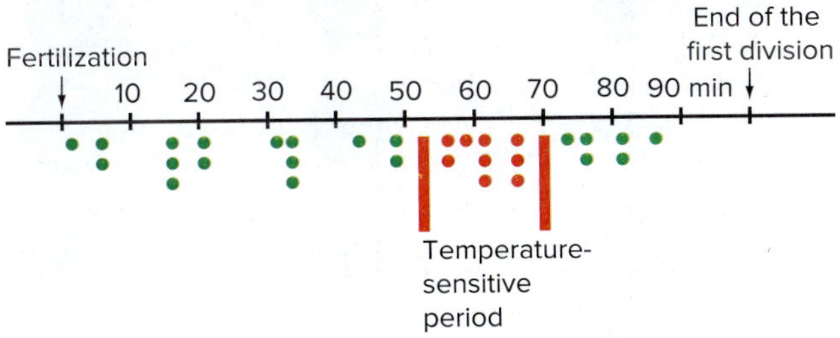

    a. At what time after fertilization is the *zyg-9* gene required for normal development of *C. elegans?*

    b. The same results were obtained whether the sperm used in the fertilization had wild-type or mutant alleles of *zyg-9*. What does this fact say about the source of the Zyg-9 protein that is present during the narrow window of time when it is required?

20. A temperature-sensitive allele of the gene encoding the Notch protein (N^{ts}) helped researchers understand the many roles of this protein in fly eye development. Notch is a transmembrane receptor that, when bound to a ligand, relays a signal to the nucleus. In one

experiment, wild-type and N^{ts} homozygous developing eyes were allowed to grow in larvae for several hours at permissive temperature, and then the temperature was shifted to the restrictive temperature. After 4 hours, the eyes were dissected from the larvae and the photoreceptors were labeled with an antibody to a protein expressed in all photoreceptors (*blue cells* in the figure that follows are labeled with antibody). The *black dots* represent ommatidia at more advanced stages of development that are not shown in the figure.

Eye development occurs in a structure called the *eye imaginal disc* present in the larva. Ommatidia develop behind an indentation called the *morphogenetic furrow* (*mf* in the diagram). The furrow forms at the posterior of the disc and moves anteriorly; every 2 hours, a new row of ommatidia initiates development posterior to the furrow, while the rows behind that row mature successively to the next stages of assembly. (Only one ommatidium is shown in the diagram, rather than an entire row.) Therefore, in a single eye disc, ommatidia at all stages of development are present. As you saw in Fig. 19.3, the first cells to join the ommatidium are the photoreceptors, R1–R8, and they do so in a particular order.

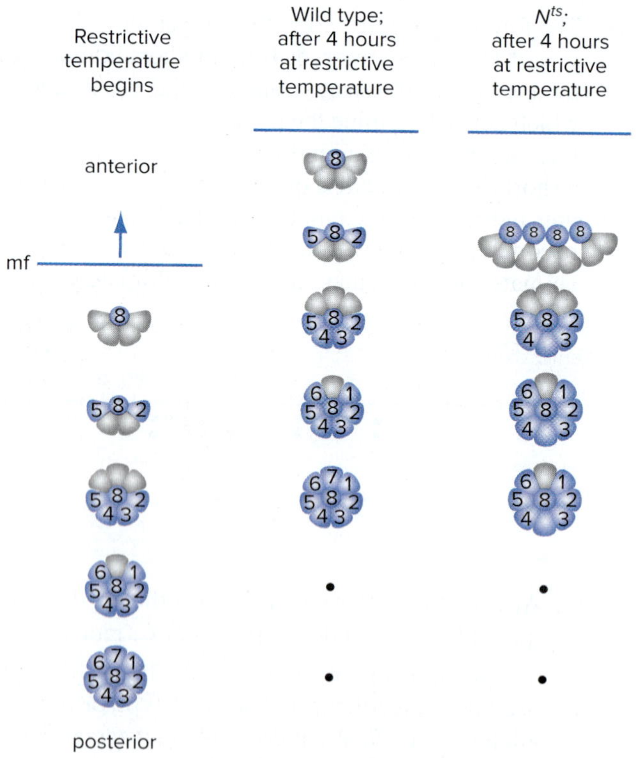

Describe the different roles of the Notch protein at different stages of ommatidial assembly.

21. Hypomorphic alleles of a pleiotropic gene essential for early development can sometimes provide enough gene activity for an organism to survive early development. In such cases, the mutant phenotype can reveal later functions of the gene.

Mice homozygous for null alleles of *Fgf8* (*fibroblast growth factor 8*) die during early embryogenesis, obscuring a later role for Fgf8 in setting up left/right asymmetry of organs. However, mice homozygous for hypomorphic *Fgf8* alleles, develop much further and reveal that Fgf8 protein is a determinant of leftness. In *Fgf8* mutants, many organs such as the heart are left/right reversed as shown in the following figure.

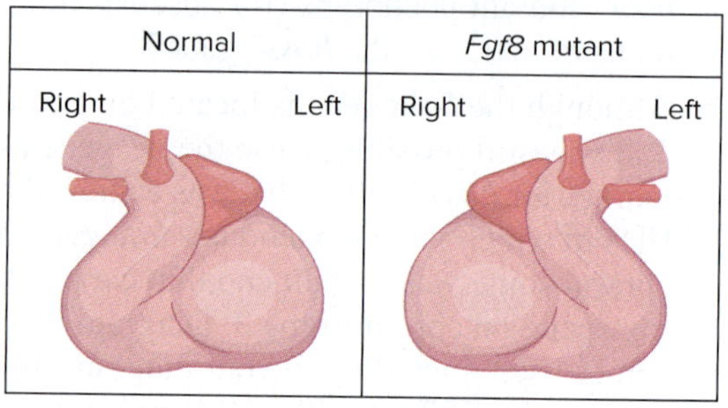

How might these *Fgf8* hypomorphic mutant alleles been generated? (More than one answer is possible.)

22. In addition to the maternal effect genes that establish anterior/posterior polarity in the *Drosophila* embryo (like *bicoid* and *nanos*), other maternal effect genes, including *dorsal, pelle,* and *Toll,* independently determine dorsal/ventral polarity. The *dorsal* gene encodes a transcription factor (Dorsal), originally deposited in the egg cytoplasm, that determines ventralness in a concentration-dependent manner. As shown in the following figure (*Wild type*), a gradient of Dorsal nuclear localization exists in early embryos: Cells whose nuclei have the highest Dorsal protein concentration become the ventral-most cells, cells whose nuclei have no Dorsal protein are dorsal-most, and lateral cells "learn" their positions and fates through the particular intermediate Dorsal protein levels in their nuclei. The figure shows sections through blastoderm embryos, where D = dorsal and V = ventral.

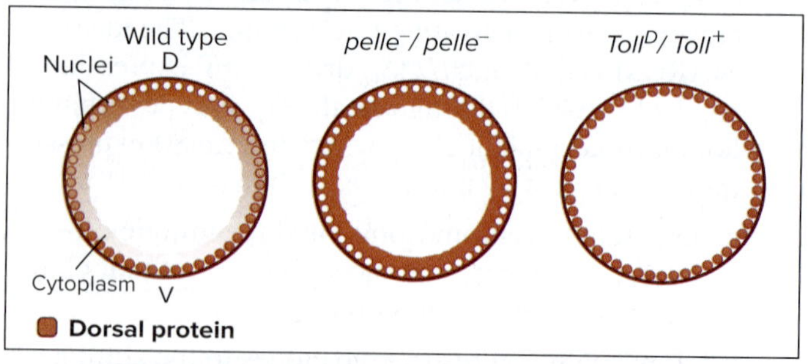

Embryos from mothers homozygous for *dorsal* null mutations are dorsalized—every cell along the dorsal/ventral axis "thinks" it is the ventral-most cell. Embryos from mothers homozygous for loss-of-function *pelle* mutations or heterozygous for gain-of-function (constitutively active) $Toll^D$ mutations show altered patterns of Dorsal protein localization, as shown in the figure. Pelle and Toll expression are unaltered in *dorsal* loss-of-function mutants (*not shown*).

a. Describe the alterations in Dorsal protein localization in embryos produced by $pelle^-/pelle^-$ or $Toll^D/Toll^+$ mutant mothers.

b. Based on the information given, order the *dorsal, pelle,* and *Toll* genes in a pathway.

c. The two kinds of embryos described in part (a) die just before hatching. Describe their morphological mutant phenotypes.

Section 19.4

23. The *yan* gene encodes a transcription factor that represses R7 development in the five cells that express Sevenless and are therefore competent to become R7. In *yan* null mutants, all five cells in each ommatidium can become R7 (just as in *sev-*Ras^{G12V} transgenic flies; see Fig. 19.7b). The eyes of sev^- yan^- double null mutants display the yan^- mutant phenotype.

 a. Describe the epistasis relationship between sev^- and yan^- mutants.

 b. Based on what you know about the roles of Sevenless and Yan proteins in the signaling pathway (Fig. 19.6), is this the epistatic interaction you would have expected? Explain.

 c. Epistasis analysis can be performed only if two mutants in genes in the same pathway have different mutant phenotypes. How is it possible for two different null mutations in the R7 signaling pathway, *sev* and *yan*, to elicit different mutant phenotypes?

 d. The constitutively active Ras^{G12V} mutation is epistatic to the sev^- null mutation, and so is a yan^- null mutation; yet both Ras and Yan function downstream of Sevenless in the signaling pathway. Considering your answer to part (c), explain why this makes sense.

 e. Suppose that a researcher identifies flies with a dominant constitutively active mutant allele of *yan* (yan^D) that expresses *yan* even in cells where the Sevenless pathway is activated. Predict the mutant phenotype of the flies. Would you expect yan^D to be epistatic to sev^-? to Ras^{G12V}? Explain.

24. Recall from Chapter 17 that in *Drosophila,* sex determination depends on the number of X chromosomes: XX flies are female, and XY flies are male. The elaboration of female morphology depends on a cascade of gene regulation initiated by the expression of the *Sxl* gene in XX embryos (Fig. 17.35 and Fig. 17.36). Sxl causes splicing of the *tra* gene transcript in a manner that allows translation of Tra protein. Tra protein in turn causes female-specific splicing of the *dsx* mRNA and expression of Dsx-F protein, a transcription factor. Dsx-F activates the transcription of female morphology genes and represses male morphology gene expression. The *ix* gene product, Ix protein, is required for Dsx-F to repress the expression of male morphology genes.

 a. *Drosophila* sex determination is a switch/regulation pathway. Define the signal and the switch in this pathway. In what cells is the switch ON? OFF?

 b. Describe the tra^- and ix^- mutant phenotypes in XX flies and XY flies.

 c. Predict the mutant phenotype of tra^- ix^- null double mutants. Which mutation would you expect to be epistatic?

 d. In the two examples discussed previously concerning epistasis analysis in the study of eye development—sev^- and Ras^{G12V} described in the text, and sev^- and yan^- in Problem 23—the epistatic mutation defined the downstream gene in the pathway. Can you explain why that is not the case in this example for the sex determination pathway?

Section 19.5

25. a. Explain the difference between maternal inheritance of organelle DNAs and maternal effect inheritance.

 b. How do the inheritance patterns of phenotypes caused by mitochondrial genes differ from those caused by maternal effect genes?

26. In the 1920s, Arthur Boycott, working with the snail *Limnaea peregra,* discovered the first maternal effect phenotype. Like most gastropods, the shell and internal organs of *Limnaea peregra* normally coil to the right—a phenotype described as *dextral.* Boycott found a mutant whose body was mirror-image symmetrical to the wild type: The shell and internal body of the mutant coiled to the left—a phenotype called *sinistral.*

Courtesy of Bill Frank, www.jaxshells.org

More recent studies have shown that sinistrality and dextrality are controlled by an autosomal maternal effect gene called *s;* the dominant s^+ allele causes dextrality, and the recessive, loss-of-function s^- allele causes sinistrality.

a. *Limnaea* are either hermaphrodites or males; the hermaphrodites can self-fertilize or they can mate with males. Suppose that a pure-breeding line of sinistral *Limnaea* hermaphrodites are crossed with males from a pure-breeding dextral line. What will be the phenotype of the progeny?

b. Answer part (a) for a cross performed in the opposite direction: dextral hermaphrodites crossed with sinistral males.

c. Describe the progeny phenotypes if the F_1 from part (a) self-fertilize. Do the same for the F_1 from part (b).

d. Suppose you now self the F_2 from parts (a) and (b). What results would you expect in the F_3 in each case?

27. The *Drosophila* mutant screen shown on the right-hand side of Fig. 19.18 was limited; it could identify mutations only in maternal effect genes whose functions were not required in female tissues other than the eggs they produced.

a. What aspect of the screen imposed this limitation?

b. How could you determine if any of the mutations identified are in genes required in males?

c. Why was the *Balancer* chromosome needed in the screen?

28. Some genes are required both zygotically and maternally. One experimental approach to studying such genes relies on the existence of ovo^D, a dominant female sterile mutation of the *ovo* gene, which is located near the middle of the acrocentric *Drosophila* X chromosome. Females that are ovo^D/ovo^+ are sterile; ovo^D-containing germ-line cells cannot produce eggs.

a. Mutations in gene *X* are recessive lethals, so homozygotes for these mutations do not develop into adults. Explain how researchers could use the ovo^D mutation in a mitotic recombination experiment to determine (i) whether or not females might supply the RNA or protein product of gene *X* to the eggs they make in their ovaries, and (ii) whether this maternally supplied product is needed for proper development of their progeny. Where in the genome would gene *X* need to be located for this approach to work?

b. The ovo^D mutant gene has been cloned, so genomic DNA for this mutant gene is available. How could you use this cloned DNA to determine whether any embryonic lethal mutation located anywhere in the genome was an allele of a maternal effect gene?

c. Regardless of its chromosomal location, how could you distinguish in such an experiment whether the gene in question was a maternal effect gene, as opposed to a gene whose product is needed for oogenesis in the female?

d. How can fly strains containing ovo^D mutations be maintained if females carrying these mutations are sterile?

29. How would a human with a mutation in a maternal effect gene most likely be recognized?

30. One important demonstration that Bicoid is an anterior determinant came from injection experiments analogous to those performed by early embryologists. These experiments involve the introduction, by direct injection into the egg, of components such as cytoplasm from an egg or mRNA that is synthesized *in vitro*. Describe injection experiments that would demonstrate that Bicoid is an anterior determinant.

31. The *hunchback* gene contains a 5′ transcriptional regulatory region, a 5′ UTR, a structural region (the coding sequences), and a 3′ UTR.

a. What important sequences required to control *hunchback* gene expression are found in the transcriptional regulatory region of *hunchback?*

b. What sequence elements that encode specific protein domains are found in the structural region of *hunchback?*

c. Another important kind of sequence is located in the 3′ UTR of the *hunchback* mRNA. What might this sequence do?

32. In flies developing from eggs laid by a $nanos^-$ mother, development of the abdomen is inhibited. Flies developing from eggs that have no maternally supplied *hunchback* mRNA are normal. Flies developing from eggs laid by a $nanos^-$ mother that also have no maternally supplied *hunchback* mRNA are normal. If too much Hunchback protein accumulates in the posterior of the egg, abdominal development is prevented.

a. What do these findings say about the function of the Nanos protein and of the *hunchback* maternally supplied mRNA?

b. What do these findings say about the efficiency of biological processes that are subject to evolution?

33. Wild-type embryos and mutant embryos lacking the gap gene *knirps* (*kni*) are treated with fluorescent antibodies at the syncytial blastoderm stage to examine the distributions of the Hunchback and Krüppel proteins. The results are shown schematically on the following figure.

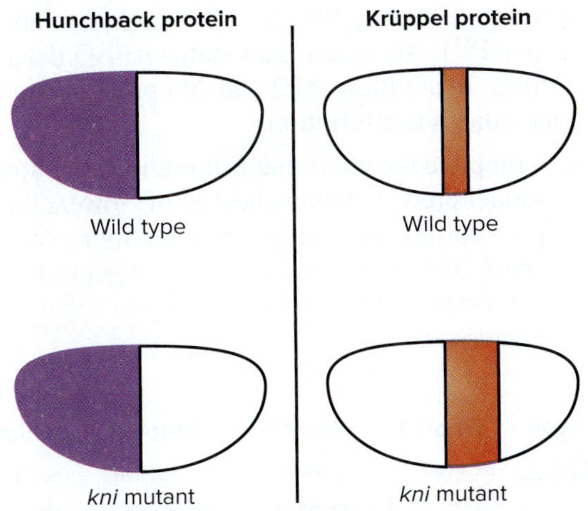

a. Based on these results, what can you conclude about the relationships among these three genes?

b. Would the pattern of Hunchback protein in embryos from a *nanos*$^-$ mutant mother differ from that shown? If yes, describe the difference and explain why. If not, explain why not.

34. The *Drosophila even-skipped* (*eve*) gene has four different enhancers that control its transcription in a pattern of seven stripes. (One of them—the stripe 2 enhancer—is shown in Fig. 19.23c.)

    a. Why is the stripe 2 enhancer active only in cells corresponding to stripe 2?

    b. The other three enhancers each work in the cells corresponding to two different (nonadjacent) stripes. How is this possible?

    c. Secondary pair-rule genes like *ftz* have only one enhancer that is active in all seven stripes. How is this possible? Describe an experiment that would have led scientists to this conclusion.

    d. Would you expect the segment polarity gene *engrailed* to have two enhancers or fourteen enhancers, one active in each stripe?

35. In *Drosophila* with loss-of-function mutations affecting the *Ubx* gene, transformations of body segments are always in the anterior direction. That is, in *bx* mutants, the anterior compartment of T3 is transformed into the anterior compartment of T2, whereas in *pbx* mutants, the posterior compartment of T3 is transformed into the posterior compartment of T2 (Fig. 19.26). In wild type, the *Ubx* gene itself is expressed in posterior T2–anterior A7 (PS5–PS12) and most strongly in posterior T3–anterior A1 (PS6). (See Fig. 19.29.)

    a. The *Abd-B* gene is transcribed in the abdominal parasegments 10–13. Assuming the mode of function of *Abd-B* is the same as that of *Ubx*, what is the likely consequence of homozygosity for a null allele of *Abd-B* (that is, what segmental or parasegmental transformations would you expect to see)?

    b. Because *Abd-A* is expressed in parasegments 7–12, all three genes of the BX-C (*Ubx, Abd-A*, and *Abd-B*) are transcribed in parasegments 10–12 (see Fig. 19.29). Why then are the abdominal parasegments 10, 11, and 12 morphologically distinguishable?

    c. What parasegment transformations would you expect to see in an animal deleted for all three genes of the BX-C (*Ubx, Abd-A*, and *Abd-B*)?

    d. Certain *Contrabithorax* mutations in the BX-C cause transformations of wing to haltere. Propose an explanation for this phenotype based on the transcription of the *Ubx* gene in particular parasegments. Do you anticipate that *Contrabithorax* mutations would be dominant or recessive to wild type? Explain.

36. It is crucial to the development of *Drosophila* that the *Hox* genes must not only be on where they should be expressed, but also that they must be off where they should not be expressed. Thus, in addition to enhancers, the *Hox* genes are controlled by *silencers*—elements like enhancers that bind only repressors. This fact was discovered by zygotic mutations in about 30 so-called *Polycomb group* (*PcG*) genes. Embryos homozygous for loss-of-function mutations in any one of the *PcG* genes have a similar mutant phenotype: As seen in the accompanying figure, parasegments whose identities are controlled by *BX-C* genes are all transformed into the most posterior abdominal parasegment (PS13):

    a. What types of proteins do you think that the *PcG* genes encode?

    b. Explain the mutant phenotype of the embryos lacking PcG proteins.

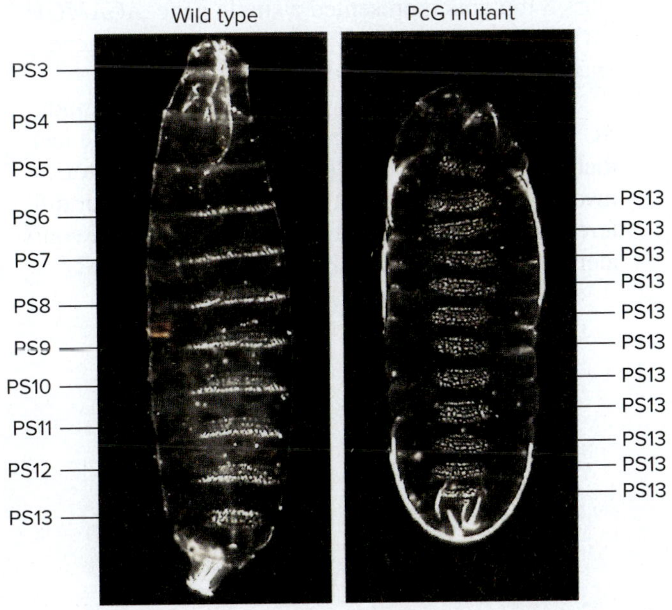

37. In the plant *Arabidopsis thaliana*, every flower is constructed of four concentric whorls of modified leaves. The first whorl (whorl 1) consists of four green leaf-like sepals, whorl 2 is composed of four white petals, whorl 3 is made of six stamens bearing the pollen that houses the male gametes (sperm), and whorl 4 contains the two carpels, within which lie the ovules that hold the female gametes (eggs). As shown in the diagram that follows, a shorthand description of the wild-type flower pattern is: sepal, petal, stamen, carpel.

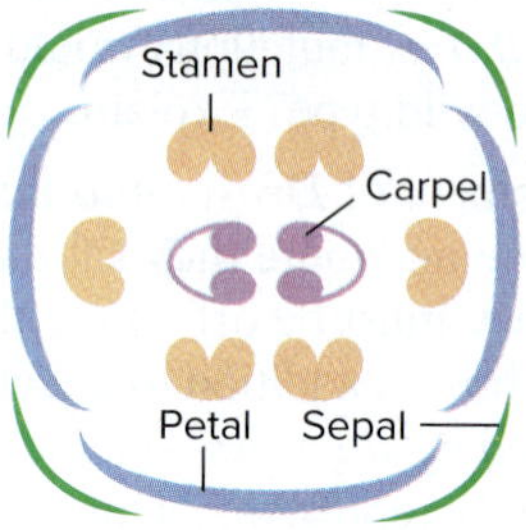

Source: USDA/Peggy Greb, photographer

Scientists wanted to understand how this pattern of whorls arises. They generated *Arabidopsis* strains homozygous for randomly induced mutations and screened them for mutant flowers with an abnormal order or selection of floral organs. The interesting mutants identified fell into three phenotypic classes: (1) carpel, stamen, stamen, carpel; (2) sepal, sepal, carpel, carpel; (3) sepal, petal, petal, sepal.

The investigators found that all of the class 1 mutants were alleles of the same gene which they called *APETELA2* (*AP2*). Class 2 mutants were alleles of either one of two genes, which were named *APETELA3* (*AP3*) and *PISTILLATA* (*PI*). Finally, the class 3 mutants represented a single gene, *AGAMOUS* (*AG*). Molecular analysis showed that all four genes encode transcription factors.

Based on the phenotypes of *AP2, AP3, PI,* and *AG* mutants (all of them are null alleles) and the fact that all four gene products are transcription factors, investigators formulated the following model for differentiation of the four flower whorls from four equivalent whorl precursor cell groups: AP2 protein determines sepals; AP2 + AP3 + PI determines petals; AP3 + PI + AG determines stamens; AG determines carpels. In addition, AP2 and AG proteins repress each other's transcription.

a. Complete the chart that follows to show how the model predicts the phenotype of flowers homozygous for mutations in each of the four genes. In the chart, the colored boxes represent genes that are expressed; the white boxes are genes that are not expressed.

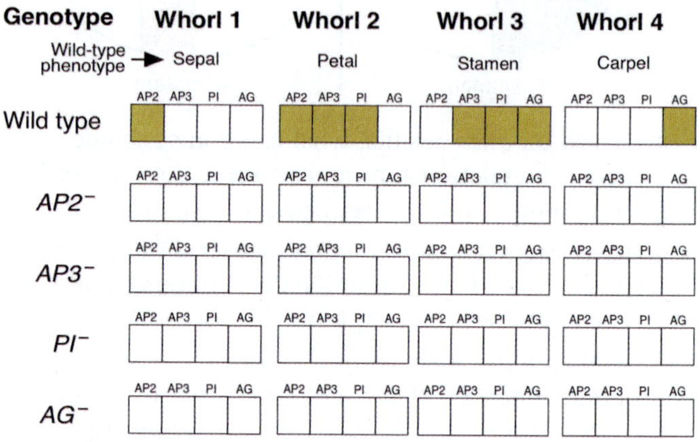

b. Scientists tested the flower patterning model with RNA *in situ* hybridization experiments using cDNA probes for each of the four genes. The goal was to see whether each gene's mRNA was expressed in the precursor cells for each of the four whorls. What results would you predict with each probe on wild-type flowers? on *AP2* mutants? on *AG* mutants?

c. Another way the researchers tested their model for flower patterning was by making double mutants. What phenotypes does the model predict for each of the six double mutant combinations? [*Hint:* It will be helpful to expand the chart shown in part (a) for each possible double mutant.]

d. Are the roles of the four genes described in this problem more similar to those of the segmentation genes or to those of the homeotic genes from *Drosophila* described in the text?

chapter 20

The Genetics of Cancer

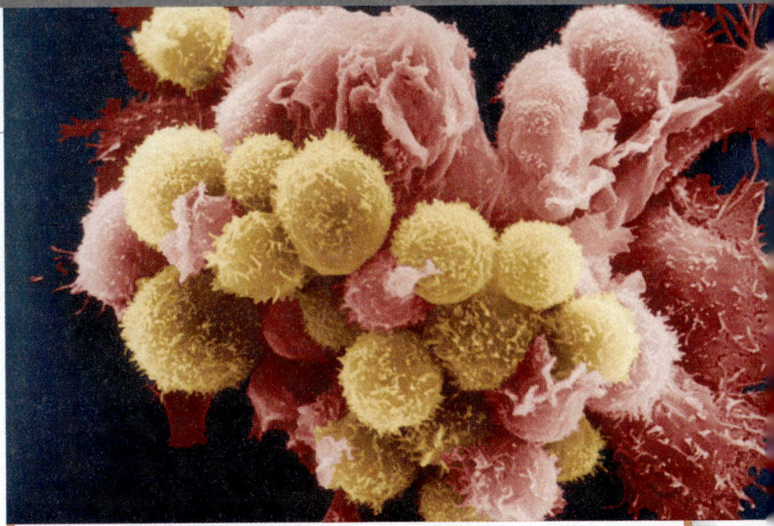

Killer cells of the immune system (yellow) *attacking a large cancer cell* (pink).

chapter outline

MUTATIONS THAT OCCUR in the germ line are transmitted to successive generations through the gametes. These germ-line mutations are the causes of inherited diseases and are the substrates for evolution by natural selection. But what about mutations in somatic cells that are not transmitted to the next generation? Are they of any interest to geneticists? In this chapter, you will see that somatic mutations can indeed affect organisms profoundly, particularly because they are the underlying causes of cancer.

Cancer is not a single disease, but is instead a bewildering variety of different diseases all characterized by the uncontrolled growth and division of cells. Two reasons exist for this enormous variety. First, cancer-causing somatic mutations can occur in cells of different tissues, producing lung cancers, skin cancers, breast cancers, and so on (**Fig. 20.1**). Different cell types vary in their phenotypic responses to the same cancer-causing mutations. And second, mutations in a variety of different genes encoding different proteins that regulate cell division can help produce cancer. Thus, two lung cancers may actually be very different diseases because they result from diverse sets of mutations.

The recognition that any cancer might represent a unique situation increases the challenges for medical science. However, in the coming age of inexpensive, massively parallel nucleic acid sequencing, cancer treatments will become more personalized and thus more effective. The idea is simple in concept: Comparing the whole-genome sequence of cells from a tumor with that of normal cells from the same patient may pinpoint the distinct mutations causing this particular cancer. Such information could then help physicians choose already available drugs, or design new drugs, that would specifically target the defective protein or biochemical pathway in the patient's unique cancer.

This chapter is organized around the unifying theme that cancer is actually a diverse set of genetic diseases that share some common features. In all cancers, somatic cells accumulate mutations in a variety of genes, many of whose products function in normal cells either as "accelerators" or "brakes" for the processes that make cells divide. The cancer-causing mutations either step down on the accelerators or

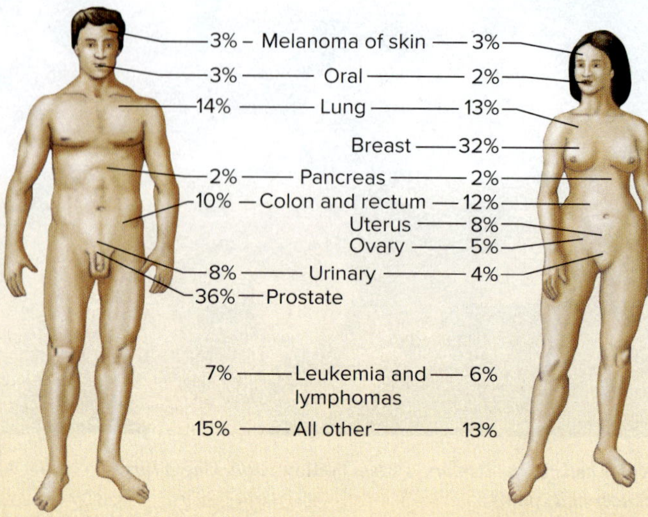

Figure 20.1 Percentages of cancers in the United States that occur at different sites in the body. Cancers affecting the same tissue may have very different origins and effects; for example, breast cancers in different patients often have unique properties.

let up on the brakes, so the cancer cells divide in a rapid and uncontrolled manner. The proteins encoded by yet other cancer-associated genes normally help safeguard the cell's DNA; mutations in these genes contribute to cancer by increasing the rate of mutation in accelerator or brake genes. The particular set of mutations in a tumor cell dictates not only its particular cancerous properties, but also the susceptibility of the tumor to particular drugs. The insights of the new science of *personalized cancer genetics* are thus increasingly guiding researchers to treatments tailored for individual cancers.

20.1 Characteristics of Cancer Cells

learning objectives

1. Discuss the behaviors of cancer cells that may differentiate their growth from that of normal cells.
2. Explain how telomerase expression contributes to the immortality of cancer cells.
3. List the types of changes in the human genome that can help lead to cancer.
4. Describe how tumors can grow to large sizes and also spread to other parts of the body.

Tumor cells have many properties that distinguish them from normal cells. However, not all cancer cells manifest the same suite of phenotypic changes.

Cancerous Cells Evade Normal Controls on Cell Growth

To form the human body from a single cell—the fertilized zygote—requires millions of rounds of cell divisions. These divisions are subject to controls ensuring that cells are distributed correctly into discrete organs and tissues. Many controls on cell division involve signals relayed between cells. For example, most cells will divide only when they encounter growth factors such as hormones made elsewhere in the body. In another example, most cells will stop dividing when they come into physical contact with other cells; this property is called **contact inhibition.** Yet other controls on cell division actually involve cell death. In particular, most cells are instructed to die through a process called **apoptosis** (or **programmed cell death**) that is activated when the genomic DNA becomes too damaged.

Figure 20.2 outlines ways in which cancer cells can evade these controls. Some cancer cells make their own division-stimulating signals (**autocrine stimulation**), or they can lose contact inhibition. Many cancer cells do not die when their DNA is damaged in ways that would cause apoptosis in normal cells.

Cancer Cells Often Acquire a Potential for Immortality

With a few exceptions (such as rare stem cells), normal somatic cells taken out of the body either do not grow readily in culture, or they undergo a limited number of rounds of divisions (20–50) and then die. In contrast, tumor cells not only grow well in culture, but many of them can divide indefinitely (**Fig. 20.3**). These characteristics make it relatively easy for scientists to develop useful lines of cancerous cells that can be distributed throughout the world for study.

One reason for tumor cells' immortality has to do with expression of the enzyme *telomerase.* As you learned in Chapter 12, telomerase prevents telomeres from shrinking when chromosomes replicate. Most normal somatic cells do not make telomerase, so with every cell cycle their chromosomes lose DNA sequences from their ends. After several rounds of division, essential genes are lost, so the cells age and die. In contrast, most cancer cells express telomerase, a feature they share with germ-line cells and some rare stem cells. The telomerase enzyme regenerates new copies of the repeating sequences at telomeres, allowing cancer cell telomeres to remain approximately the same length over many cycles of cell division.

Figure 20.2 Phenotypic changes causing uncontrolled growth of cancer cells. **(a)** Many tumor cells can divide in the absence of external growth signals required for the proliferation of normal cells. **(b)** Normal cells stop growing when they contact each other; as a result, they form a monolayer one cell thick when grown in culture. Tumor cells lose this contact inhibition and climb over each other to form piles many cells thick called *transformed foci*. **(c)** When normal cells are damaged, they die by a process called *programmed cell death*, or *apoptosis*. Many cancer cells fail to die when damaged to the same extent.

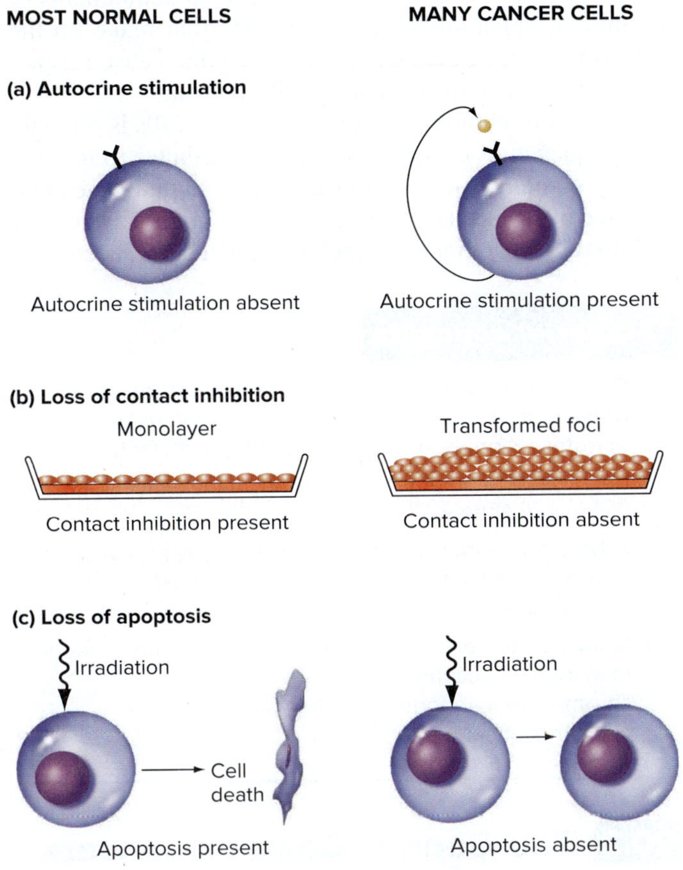

Figure 20.3 Many cancer cells are immortal. Most normal somatic cells spontaneously stop growing after a specific number of cell divisions (*red* line). Tumor cells, by contrast, can divide indefinitely (*blue* line). One reason for this difference is that cancer cells often express the enzyme telomerase, while normal cells do not.

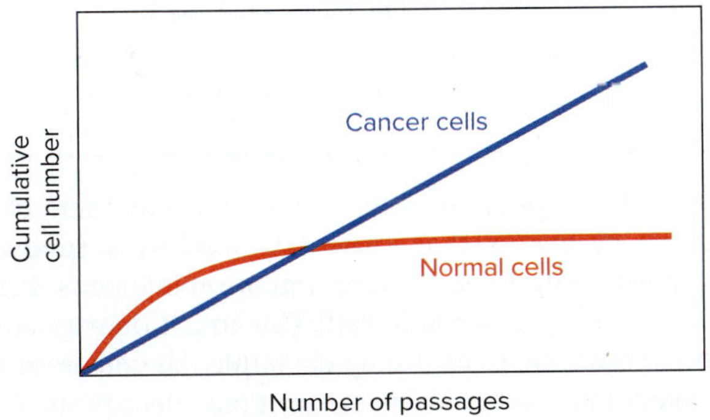

Some Changes Alter a Tumor's Interactions with the Body

The rapid growth of tumors can choke the function of nearby tissues. Tumor growth requires many nutrients, but nutrient delivery to the cancer cells is limited by the local blood supply. Some tumors evade this potential restriction to their growth by secreting substances that cause blood vessels to grow toward them. The growth of blood cells in response to secreted signals is called **angiogenesis** (**Fig. 20.4a**). The new vessels serve as supply lines through which the tumor can tap new nutrient sources.

The lethality of many cancers involves their spread to many locations within the body and thus the disruption of many different tissues. Normal cells stay restricted to a single area by membrane barriers surrounding the tissue or organ. But tumor cells often acquire the ability to break through such membranes (**Fig. 20.4b**). The cancerous cells can then travel through the bloodstream to colonize distant tissues in a process called **metastasis.** The new blood vessels generated by angiogenesis help provide escape routes for metastasizing cells.

A different way in which cancer cells influence their interactions with the body is that they can evade **immune surveillance.** The human immune system usually recognizes cancer cells as foreign and attacks them, thereby helping to eliminate tumors even before they are large enough to be detected (see the photograph at the beginning of this chapter). Successful tumor cells, however, somehow develop the ability to mask themselves from the immune system, so they can bypass this protective mechanism.

Figure 20.4 Changes that enable a tumor to grow and to invade distant tissues. **(a)** Tumor cells secrete substances that cause blood vessels to grow toward them, providing new supplies of nutrients for the tumor. **(b)** Some cancerous cells can break through membranes that define particular tissues and then travel through the bloodstream to colonize distant tissues.

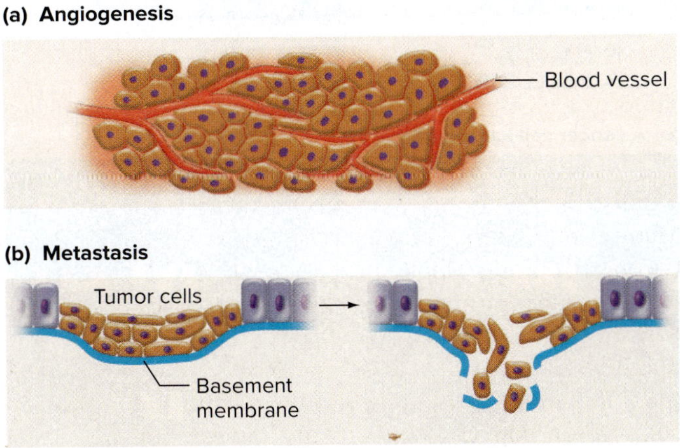

Most Cancer Cells Have Unstable Genomes

The rate of mutation in many cancer cells is much higher than in normal cells of the same person. One major reason is that cancer cells are often defective either in the enzymatic systems that repair DNA damage caused by external agents such as radiation or ultraviolet light, or in the mismatch repair systems that correct mistakes in nucleotide incorporation during DNA replication (**Fig. 20.5a**).

Figure 20.5 Genomic instability in cancer cells. (a) In many cancer cells, enzymatic systems for repairing DNA damage are disrupted. As a result, the cells have enormously increased rates of mutation. **(b)** Cancer cells often have karyotypes showing abnormal numbers of chromosomes and many chromosomal rearrangements such as translocations. These karyotypic changes result from defective DNA damage and chromosome segregation mechanisms. (Compare this spectral karyotype with that of the normal cell shown in Fig. 12.9b.)

b: Jonathan Landry et al., "The genomic and transcriptomic landscape of a HeLa cell line," *G3: Genes, Genomes, Genetics*, March 11, 2013, Fig. 3A reprinted courtesy Genetics Society of America

(a) Cancer mutations disrupt DNA repair

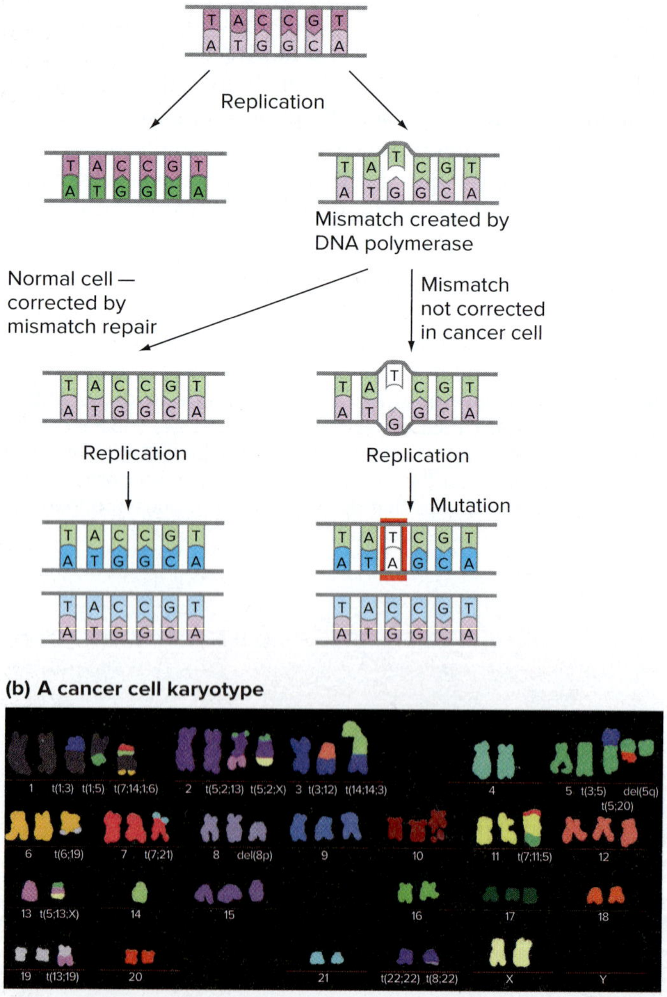

Another manifestation of **genomic instability** is that many cancer cells have major chromosomal aberrations (**Fig. 20.5b**). Tumor-cell karyotypes often display gross abnormalities, such as chromosomal rearrangements (deletions, duplications, inversions, and translocations), aneuploidy, and even polyploidy. Chromosome rearrangements are, like point mutations, consequences of problems in the DNA replication and repair machineries. Aneuploidies and polyploidies can result from defects in components of the mitotic apparatus (such as proteins that make up the centrosomes, kinetochores, or spindles) that cause misdistribution of chromosomes upon cell division.

The high mutation rate in some cancer cells is not only a consequence of cancer, but it also contributes to the disease by speeding the occurrence of gene mutations responsible for other cancer phenotypes. Later in the chapter, we will discuss what kinds of genes are mutated in cancer cells.

essential concepts

- Cancer cells do not respond to controls that limit the division of normal cells, such as *contact inhibition, apoptosis* signals, or the absence of growth factors.
- Because of telomerase expression, many cancer cells can divide indefinitely.
- Advanced tumors can stimulate *angiogenesis*, evade the body's normal immune defenses, and *metastasize* to other tissues.
- Many cancer cells exhibit high *genomic instability,* leading to increased accumulation of new mutations and chromosomal abnormalities that cause the other hallmarks of cancer.

20.2 The Genetic Basis of Cancers

learning objectives

1. Describe the current model for the genesis of cancer.
2. Summarize the evidence that cancer is a genetic disease, requiring the acquisition of mutations in several key genes within a somatic cell lineage.
3. Explain why some individuals have a genetic predisposition to certain kinds of cancers.
4. Analyze the feedback loop between cell proliferation and genomic instability in cancer cells.

Figure 20.6 summarizes our current understanding of the sequence of events that ultimately lead to a cancerous **malignant cell.** First, a rare mutation happens in the genome of a single somatic cell. This mutation may confer a growth advantage, or it may decouple the cell from normal constraints on cell division, or it may disrupt the DNA

Figure 20.6 A model for the genetic basis of cancer. Cancer is thought to arise by successive mutations to key genes within a clone of proliferating cells.

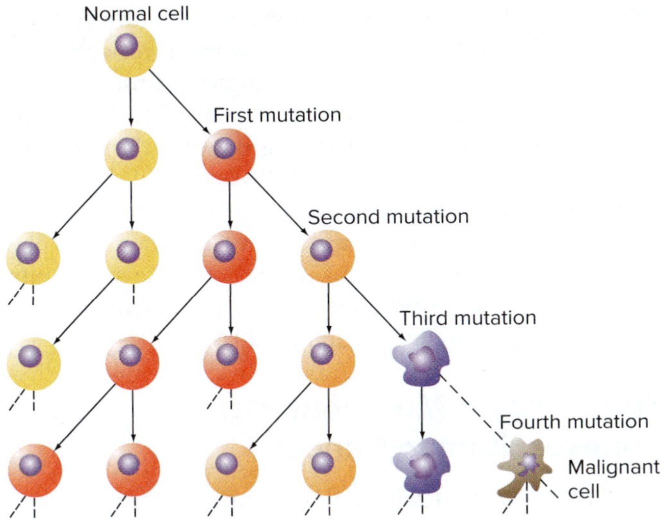

Figure 20.7 Evidence for the clonal origin of tumors. Because of X chromosome inactivation, each cell in a female expresses only one allele of a polymorphic X-linked gene. A patch of normal tissue will usually contain some cells that express one allele, and others that express the other allele. All the cells of a tumor express the same one allele, demonstrating that the tumor arose from a single cell. Electrophoresis distinguishes the two allelic forms of the gene's protein product.

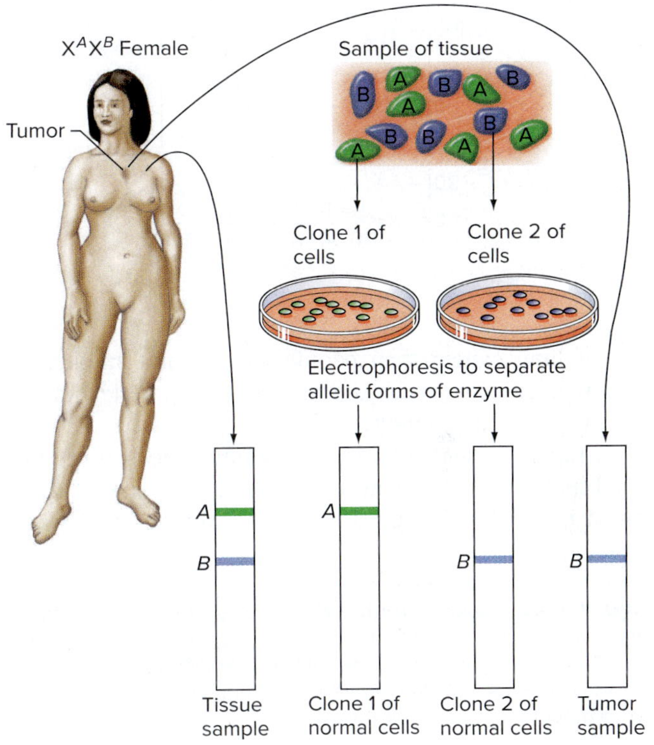

repair machinery so as to increase the rate of mutation in the cell's genome. A second mutation of one of these types then occurs in one of the mitotic descendants of this cell. Other mutations that confer additional cancerous properties successively occur in later descendant cells. Once a cell within this lineage has acquired enough such mutations, it generates a clone of proliferating cancer cells that grow and divide so rapidly that they overwhelm the surrounding normal tissue. The cancer can progress over time as some of the cells in this clone acquire yet other mutations that increase the tumor's malignancy.

We now discuss several lines of evidence support the hypothesis outlined in Fig. 20.6 that cancer results from the accumulation of mutations in a cell lineage to produce the progenitor of a clone of cancer cells.

Cancer Involves the Proliferation of a Clone of Cells

Examination of cells from women heterozygous for X-linked alleles provides evidence that cancer originates in a single somatic cell (**Fig. 20.7**). You will recall from previous chapters that X inactivation in female mammals results in the expression of only one of the two alleles of an X-linked gene in any given cell. Samples of a normal somatic tissue will almost always include some cells in which the maternal X was inactivated, and other cells in which the paternal X was inactivated. The reason is that most somatic tissues are constructed from many clones of cells descended from individual early embryonic cells that randomly inactivated one of the two X chromosomes. Once made, this decision is perpetuated through many rounds of cell division.

In contrast with normal tissue, tumors in females invariably express only one allele of an X-linked gene. This finding suggests that the cells of each tumor are the clonal descendants of a single somatic cell that had already inactivated one of its X chromosomes and then sustained a rare mutation to initiate the cancer-causing process.

Cancer Is Generally a Disease of Old Age

The incidence of cancer rises dramatically with age (**Fig. 20.8**). The prevalence of cancer in older people supports the ideas that cancer develops over time and that it involves the accumulation of many mutations in the clonal descendants of a somatic cell.

Environmental Mutagens Increase the Likelihood of Cancer

Epidemiological investigations have shown that numerous environmental agents increase cancer rates; most of these agents are mutagens, as established by the Ames test described in Fig. 7.15. One such mutagen is cigarette smoke. People who smoke for many years have a higher

Figure 20.8 Incidence of cancer with age. The incidence of most cancers shows a dramatic increase with age, consistent with the accumulation of mutations in somatic cells.

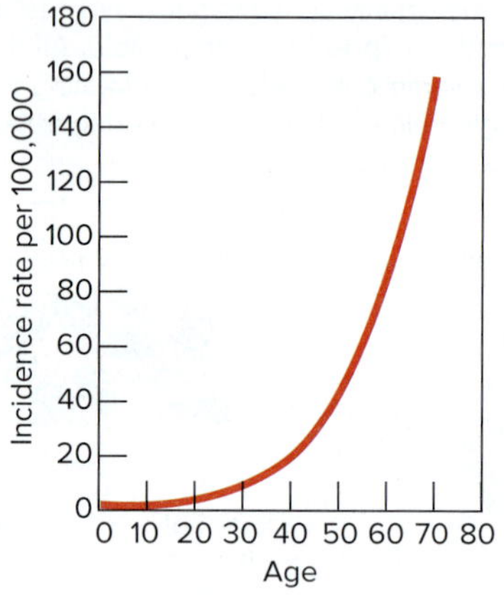

risk of lung cancer than people who do not smoke, and their risk increases with the number of cigarettes and the length of time they smoke.

Graphs of the death rates from lung cancer in America as a function of historical time illustrate the risk of smoking (**Fig. 20.9**). Lung cancer rates among men began increasing dramatically in the 1940s but among women only in the 1960s, reflecting the fact that smoking became prevalent among men before it did among women. Since 1990, rates of lung cancer have dropped significantly for men, and they have leveled off for women, reflecting the effectiveness of public health campaigns against smoking.

Figure 20.9 Carcinogenic effects of smoking. Lung cancer death rates in the United States began increasing rapidly for men in the 1940s and for women in the 1960s. This difference reflects the fact that cigarette smoking became prevalent among men about 20 years before it did among women.

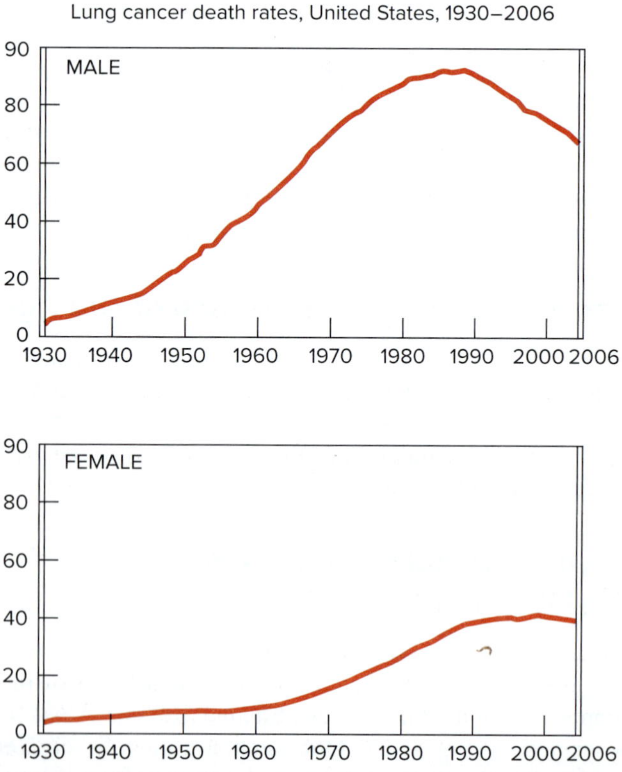

Rates are per 100,000 and are age-adjusted to the 1970 U.S. census population.

An interesting aspect of the data in Fig. 20.9 is the existence of a two-decade-long lag between the time men began frequent smoking in the 1920s and the dramatic rise of lung cancer in the 1940s. Delays between exposure to a mutagen and the occurrence of cancer are a common phenomenon. Such observations suggest that several mutations must occur to produce cancer, consistent with the model presented previously in Fig. 20.6.

Some Known Mutations Increase Predisposition to Cancer

The frequency with which first-degree relatives, such as sisters and brothers or even identical twins, have the same type of cancer is low for most forms of the disease. If one sibling or twin gets a cancer, the other usually does not. The mutations that produce these cancers are not inherited through the germ line in a dominant or recessive pattern; rather, they arise sporadically as a result of random mutations in somatic cells.

Important exceptions to this rule exist. In some families, a specific type of cancer recurs in many members, indicating the inheritance of a predisposition through the germ line. Retinoblastoma, described in the Genetics and Society Box in Chapter 5 (entitled *Mitotic Recombination and Cancer Formation*) as associated with mutant alleles of a gene called *RB,* is one example of this type of cancer.

Interestingly, some (but by no means all) of the genes involved in inherited forms of cancer encode DNA repair proteins. For example, women who inherit mutant alleles of the *BRCA1* or *BRCA2* genes have a highly elevated risk for breast and ovarian cancer. The wild-type BRCA1 and BRCA2 proteins are components of the machinery for repairing double-strand breaks in DNA.

We will explore these familial cancers in more detail later in this chapter, but at present you should understand at a simple level why the inheritance of mutations in particular genes such as *RB, BRCA1,* or *BRCA2* predisposes individuals to cancer. The reason is that the cells in these people already have a head start along the process of cancer progression. The person's cells already have a mutation in one cancer-related gene, so their somatic cells will need to accumulate one mutation fewer to become cancerous.

The existence of such inherited predispositions clearly shows that cancers are fundamentally genetic diseases. The fact that individuals inheriting the mutant alleles are usually not born with cancers but instead develop them over time emphasizes again that changes in more than one gene are required for cells to acquire a cancer phenotype.

Tumor Genome Sequencing Reveals Multiple Mutations in Cancer Cells

Recent advances in human genome sequencing technology have made it possible to sequence the genomes of both cancerous and normal somatic cells from individual patients. Comparisons of these genome sequences reveal many mutations in the cancer cells that are not found in normal cells. Individual cancers vary widely in the number of mutations their genomes have accumulated, from hundreds, to hundreds of thousands.

The large majority of mutations in cancer genomes are **passenger mutations** that occur due to the increased mutation rate of cancer cells but do not contribute to the disease. Only a few of the mutations in a cancer cell genome are **driver mutations** that cause cancer phenotypes. As an example, a typical colon cancer has thousands of mutations not found in normal cells of the same patient. Only 60 to 70 of these mutations alter the open reading frames of genes. And of these, only 3 to 10 of these are likely to be driver mutations.

You should understand two key points concerning these driver mutations. First, a single driver mutation is by itself insufficient to cause cancer; instead, several driver mutations in different genes must accumulate. And second, underlining each cancer's unique nature, cancers of the same tissue (such as colon) from different patients accumulate different sets of driver mutations.

We will describe later in the chapter the molecular nature of various driver mutations, but at this stage you can guess that some of them disrupt the function of genes encoding DNA repair proteins. Cells with defective DNA repair have increased rates of mutation and other genetic instabilities, making it more likely that their genomes will accumulate driver mutations in other genes.

Feedback Between Cell Proliferation and Genomic Instability Underlies Tumor Progression

Cancer cells are always changing, acquiring more and more properties that distinguish them from normal cells. Thus, the phenomenon of **tumor progression:** Over time, tumors tend to grow faster and become more invasive of other tissues. This process is accelerated by mutations in certain genes (such as those encoding DNA repair enzymes) that cause genomic instability and thus increase the rate of mutation in other genes. A small fraction of the mutations and chromosomal abnormalities that accumulate in tumor cells can increase the rate of proliferation, allow the cells to spread to other parts of the body, promote angiogenesis, and help the cancer cells evade immune surveillance.

Increased rates of cell proliferation and genomic instability are tied together intimately through a feedback loop that is devastating for the patient. An increase in proliferation alone, without other changes, generates benign growths that are not life threatening and can be removed by surgery. The real danger of increased proliferation is that it provides a large clone of cells within which further mutations can occur, and these further mutations may lead to even faster proliferation and eventual malignancy. The more cells that exist in a clone, the more likely that rare mutations will occur in the clone—which already has the potential for propagating them rapidly.

essential concepts

- Cancers result from the accumulation of mutations in certain critical genes (*driver mutations*) in a mitotically dividing clone of somatic cells.
- Support for this model of cancer genesis comes from observations that cancer rates are increased by the advancing age of the patient, by exposure to mutagens, and by rare inherited mutations.
- Increased cell proliferation and genomic instability form a feedback loop that leads to the acquisition of additional cancer-promoting properties.

20.3 How Cell Division Is Normally Controlled

learning objectives

1. Describe the action of the following molecular players in signal transduction pathways: growth factors, growth factor receptors, kinase cascades, and transcription factors.
2. Explain the action of cyclin-dependent kinases in regulation of the cell cycle.
3. Summarize how cell-cycle checkpoints ensure genomic stability, using the p53 protein as an example.

To understand how the accumulation of somatic mutations can lead to cancer, it is first necessary to gain some knowledge of the molecular basis of cell-cycle regulation in normal cells. The reason is that the mutations involved in cancer progression often occur in genes whose protein products control normal cell proliferation. Some of these proteins provide signals telling cells when to divide, others allow progression from one stage of the cell cycle to the next stage when all is well, and yet others cause the cellular machinery to slow down when damage to the genome requires repair. A comprehensive view of the molecular mechanisms governing cell division is beyond the scope of this book, but we discuss here some highlights pertinent to the remainder of the chapter.

Growth Factors Initiate Cell Division Through Signal Transduction Cascades

Cells must signal each other to ensure coordinated function of the body as a whole. These signals tell individual cells whether to divide, metabolize (that is, make the molecules they are programmed to make), or die. The molecular signals that influence cell proliferation and division are called **growth factors.** We focus in this chapter on **mitogens:** growth factors that stimulate cell proliferation. However, you should know that other growth factors inhibit proliferation, and yet others can have either effect, depending on the cellular environment.

When growth factors come into contact with proteins named **growth factor receptors** found on the surface of a cell, a series of biochemical reactions—a so-called **signal transduction cascade**—occurs within that cell. The signal transduction cascade eventually reaches the nucleus, where it activates genes that encode **transcription factors.** The transcription factors in turn regulate sets of genes whose protein products are the molecules that ultimately cause cells to divide or stop dividing.

Growth factors

The two basic types of signals initiating cell division are extracellular signals and cell-bound signals.

Extracellular signals in the form of steroids, peptides, and proteins act over long or short distances and are collectively known as *hormones* (**Fig. 20.10a**). The thyroid-stimulating hormone (TSH) produced by the brain's pituitary gland, for example, travels through the bloodstream to the thyroid gland, where it stimulates cells to produce another hormone, thyroxine, which in turn influences metabolic rate.

Cell-bound signals, such as the histocompatibility proteins that distinguish an individual's cells from all foreign cells and molecules, require direct contact between cells for signal transmission (**Fig. 20.10b**). Cells of the immune system communicate via cell-bound signals about the presence of viral particles, bacteria, and toxins.

Figure 20.10 Extracellular signals can be delivered by diffusion or by cell-to-cell contact. (a) The pituitary gland produces thyroid-stimulating hormone (TSH) that moves through the circulation to the thyroid gland. Cells in the thyroid produce another hormone, thyroxine, that acts on many cells throughout the body. **(b)** A killer T cell recognizes its virus-infected target cell by direct cell-to-cell contact. Receptors on the T cell bind to complexes made between viral antigens and histocompatibility proteins on the infected cell's surface.

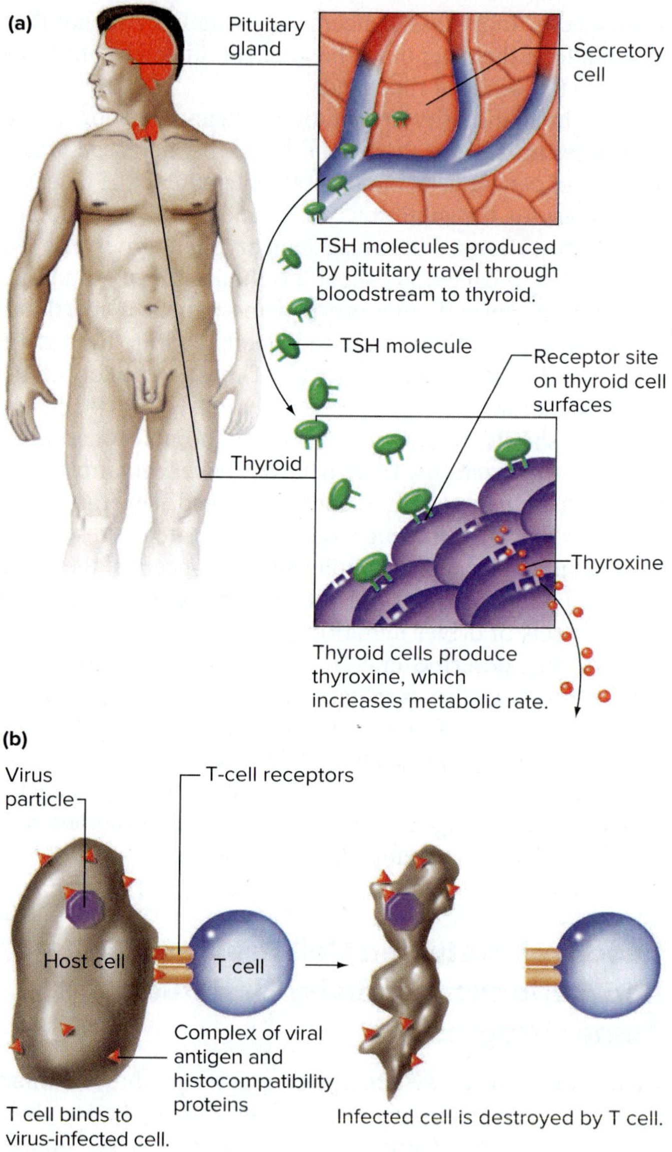

Binding of growth factors to receptors

Most growth factors, whether extracellular hormones or cell-bound signals, deliver their message by binding to specific receptors embedded in the membrane of the receiving cell—that is, the cell which will be prompted to grow or to stop growing (**Fig. 20.11a**). The growth factor receptors are proteins that have three parts: a signal-binding site outside the cell, a transmembrane segment that passes through the cell membrane, and an intracellular domain that relays the signal (that is, the binding of growth factor) to other proteins inside the cell's cytoplasm.

Signal transduction cascades

The cytoplasmic proteins responsible for relaying the signal inside the cell are known as **signal transducers.** Typically, several different signal transducers are linked in a

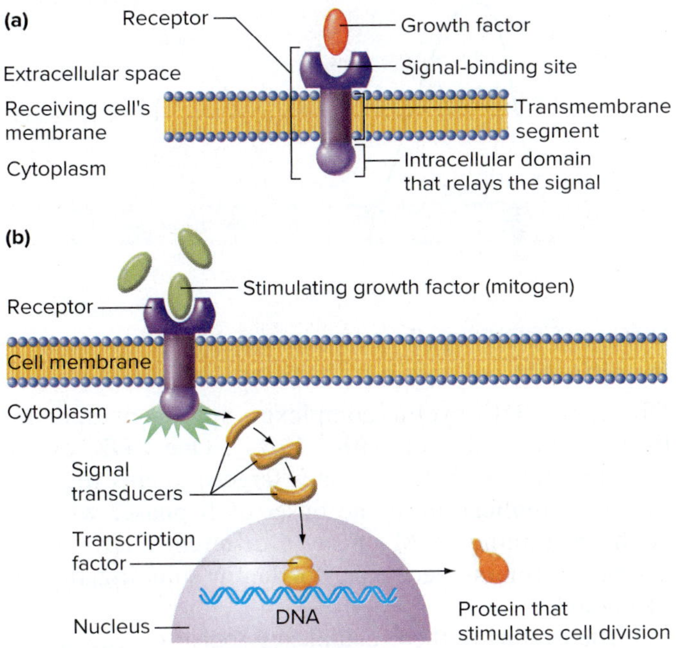

Figure 20.11 Binding of growth factors to receptors initiates signal transduction cascades. (a) A growth factor binds to the extracellular domain of a receptor embedded in the cell membrane. **(b)** This binding transmits a signal to the intracellular domain of the receptor, which in turn interacts with other signaling molecules. At the end of the signal transduction cascade are transcription factors that can turn on or off the expression of genes. The end result is stimulation of cell division.

biochemical cascade in which each molecule transmits the receptor's binding-of-growth-factor signal by activating or inhibiting another molecule (**Fig. 20.11b**).

One of many examples of a signal transduction system includes the Ras protein that was introduced in Chapter 19. Ras is a molecular switch that exists in two forms: an inactive form in which it is bound to guanosine diphosphate (Ras–GDP), and an active form in which it is bound to guanosine triphosphate (Ras–GTP). Once a growth factor activates a receptor, the intracellular domain of the receptor flips the Ras switch to on by exchanging GDP for GTP (**Fig. 20.12**). Next, Ras–GTP activates a series of three enzymes called **protein kinases** that can add phosphate groups to other proteins. The first kinase in the chain activates the second kinase, which in turn activates the third kinase. The trio turned on by Ras–GTP is known as a *MAP (Mitogen-activated protein) kinase cascade.*

Transcriptional regulation of cell-cycle genes

The final components in most growth-factor-initiated signal transduction cascades are transcription factors in the nucleus that activate or repress the expression of specific genes whose products either promote or inhibit cell proliferation (Fig. 20.11b and Fig. 20.12). In Ras-mediated signaling, the link between transcription factors and the signal transduction cascade is that after it is phosphorylated, the final MAP kinase in the chain translocates from the cytoplasm to the nucleus. In the nucleus, the MAP kinase phosphorylates specific transcription factors that influence the expression of the cell-cycle genes. Important classes of cell-cycle genes activated transcriptionally in response to growth factors are the *cyclins* and *cyclin-dependent kinases* described in the next section.

Figure 20.12 A Ras-mediated signal transduction cascade. The Ras protein is an intracellular signaling molecule that is induced to exchange a bound GDP (inactive) for a bound GTP (active) when a growth factor binds to the cellular receptor with which Ras interacts.

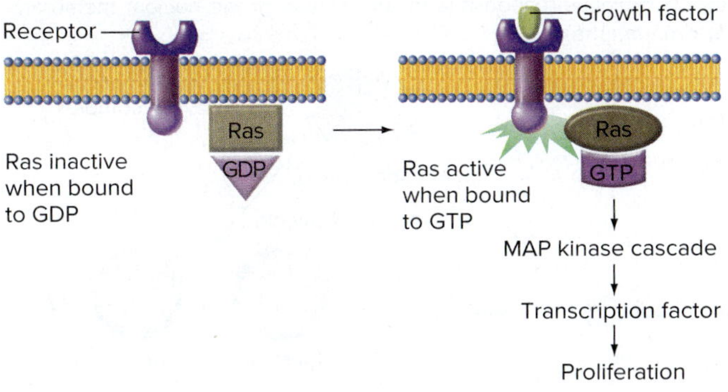

Cyclins and Cyclin-Dependent Kinases Are the Essential Drivers of Cell-Cycle Events

Cell division, as you learned in Chapter 4, requires the duplication of chromosomes as well as the precise partitioning of the duplicated chromosomes to two daughter cells. The cell orchestrates these events as the four successive stages of the cell cycle: G_1, S, G_2, and M (review Fig. 4.9). In brief, G_1 is the gap period between the end of mitosis and the DNA synthesis that precedes the next mitosis. During G_1, the cell grows in size, imports materials to the nucleus, and prepares in other ways for DNA replication. S is the period of DNA synthesis, or replication. G_2 is the gap between DNA synthesis and mitosis. During G_2, the cell prepares for division. M, the stage of mitosis, includes the breakdown of the nuclear membrane, the condensation of the chromosomes, their attachment to the mitotic spindle, and the segregation of chromosomes to the two poles. At the completion of mitosis, the cell divides by cytokinesis.

A family of protein kinases, called **cyclin-dependent kinases (CDKs),** are central controlling elements that guide the transitions from one cell-cycle stage to the next. As the name implies, CDKs function only after associating with proteins called **cyclins.** CDK–cyclin complexes work by phosphorylating hundreds of target proteins that perform specific functions at specific times in the cell

Figure 20.13 Cyclin-dependent kinases (CDKs) control the cell cycle by phosphorylating other proteins. (a) A CDK combines with a cyclin and acquires the capacity to phosphorylate target proteins. Phosphorylation of a protein can either inactivate or activate it. **(b)** CDK phosphorylation of proteins called lamins is responsible for dissolution of the nuclear membrane at prometaphase.

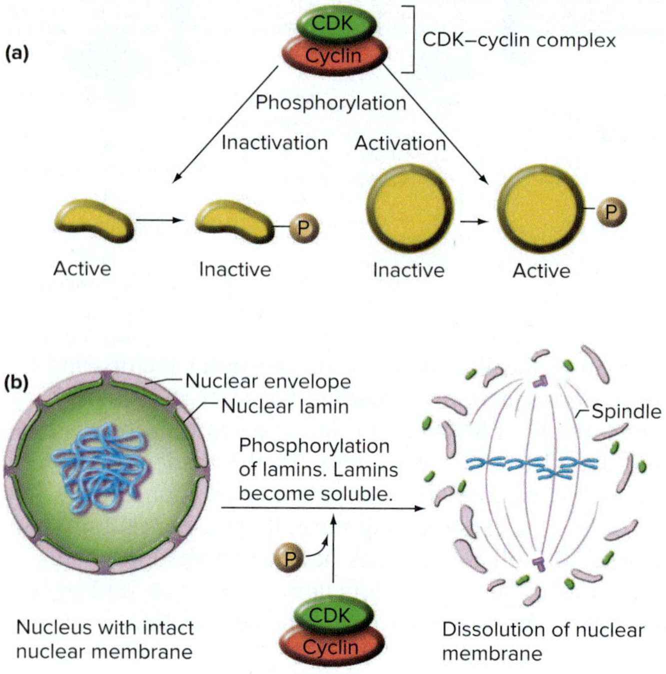

cycle. These phosphorylations can activate some target proteins or inactivate others (**Fig. 20.13a**). The cyclin portion of a CDK–cyclin complex specifies which set of proteins a particular CDK phosphorylates, while the CDK portion of the complex then performs the actual phosphorylations.

As an example, consider the action of one CDK–cyclin complex on the *nuclear lamins,* proteins that underlie the inner surface of the nuclear membrane and provide structural support for the nucleus (**Fig. 20.13b**). During most of the cell cycle, the lamins form an insoluble structural matrix. At mitosis, however, the lamins become soluble, allowing dissolution of the nuclear membrane. Lamin solubility requires phosphorylation; mutant lamins that resist phosphorylation do not become soluble at mitosis. A particular CDK–cyclin complex that becomes activated only at the beginning of prometaphase catalyzes lamin phosphorylation. Thus, a critical mitotic event—dissolution of the nuclear membrane—is triggered by CDK phosphorylation of nuclear lamins.

The Tools of Genetics Box entitled *Analysis of Cell-Cycle Mutants in Yeast* discusses how the study of yeast cells defective for cell division provided much of the evidence that the highly conserved CDK–cyclin complexes are key controlling agents for cell-cycle progression in all eukaryotes.

Figure 20.14 Different CDK–cyclin complexes govern different cell-cycle transitions. Complexes made of various cyclin-dependent kinases and cyclins appear during specific cell cycle stages and prompt the cell to proceed to the next stage. The regular appearance and disappearance of these complexes is due to the synthesis of particular cyclins at particular times, as well as the destruction of cyclins when they are no longer needed.

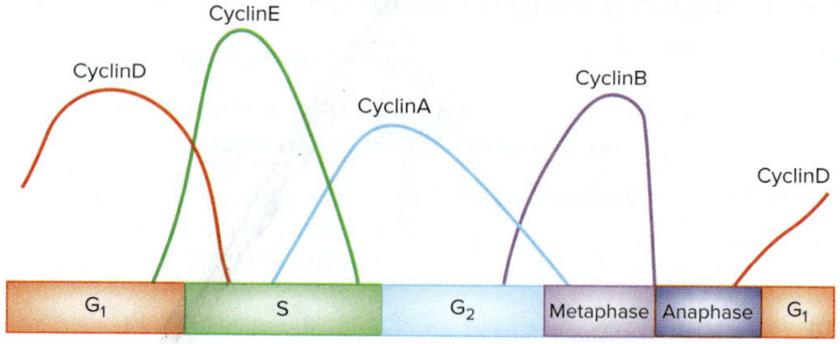

Different CDK–cyclin complexes for different cell-cycle transitions

Different CDK–cyclin complexes appear at specific times in the cell cycle (**Fig. 20.14**). One CDK–cyclin complex, for example, activates target proteins required for DNA replication at the onset of S phase; whereas at the beginning of M phase, a different CDK–cyclin activates proteins necessary for lamin solubilization as just noted.

The cyclins in these complexes appear on cue at particular stages of the cell cycle. After they associate with the appropriate CDKs and point out the proper protein targets, they then disappear to make way for the succeeding set of cyclins. The cycle of precisely timed cyclin appearances and disappearances is the result of two mechanisms: gene regulation that turns on and off the synthesis of particular cyclins, and regulated protein degradation that removes the cyclins once they are no longer needed. Cyclin degradation is one of the most interesting features of cell-cycle regulation because the molecular machinery responsible for this degradation is itself activated by the CDK–cyclin. Thus, the cell cycle has an intrinsic ratchetlike mechanism, ensuring that activation of one phase later leads to the irreversible end of that phase.

Cell-Cycle Checkpoints Ensure Genomic Stability

Damage to a cell's genome can cause serious problems. It is therefore not surprising that elaborate mechanisms have evolved to give cells time to repair DNA damage or correct potential segregation errors. These additional controls are called **checkpoints** because they check for the integrity of the genome before allowing the cell to continue to the next phase of the cell cycle.

We discussed the existence of several cell-cycle checkpoints in Chapter 4. For example, you may recall that during

TOOLS OF GENETICS

Analysis of Cell-Cycle Mutants in Yeast

The budding yeast *Saccharomyces cerevisiae* has been instrumental in identifying the genes that control cell division. Three properties make *S. cerevisiae* particularly useful for studying the cell cycle. First, it can grow as either a haploid or a diploid. Recessive mutations can be identified in the haploid cells, and then diploid cells can be constructed containing two different mutations, allowing scientists to determine the number of complementation groups defined by the mutations.

Second, the morphology of a yeast cell indicates its cell-cycle status. Toward the end of G_1, a new daughter cell arises as a bud on the surface of the mother cell. As the mother cell progresses through the division cycle, the bud grows in size; it is small during S phase and large during mitosis. Bud size thus serves as a marker of progress through the cell cycle. A normal, asynchronously cycling population of growing yeast cells contains nonbudding cells as well as cells with buds of all sizes.

Finally, researchers can isolate *S. cerevisiae* with mutations in essential cell-cycle genes, even though such mutants are lethal. The trick is to find cell-cycle-defective mutants that are temperature sensitive. At the low permissive temperature, the protein needed for cell division functions normally so the cells can be grown. A shift up to the higher restrictive temperature causes the temperature-sensitive protein in the mutant population to become nonfunctional; researchers can then study the consequences to the cell cycle of its loss.

Researchers have focused their attention on temperature-sensitive mutants that, when grown at the permissive temperature, make populations with unbudded cells and cells with buds of all sizes (**Fig. A**), but when the same population is then shifted to the restrictive temperature, the cells come to have a uniform appearance. In the mutant population shown in **Fig. B,** for example, all cells have a single large bud, indicating that the cells are arrested in mitosis. Other cell-cycle mutants would arrest with different but also uniform morphologies, for example, with all unbudded cells. Thus, mutants that acquire a uniform bud-related morphology at the restrictive temperature are each defective at a particular stage of the cell cycle.

Figure A A cell-cycle mutant of yeast grown at permissive temperature. These cells display buds of all sizes, including some with small buds (e.g., *yellow arrow*) and some with large buds (e.g., *red arrow*).

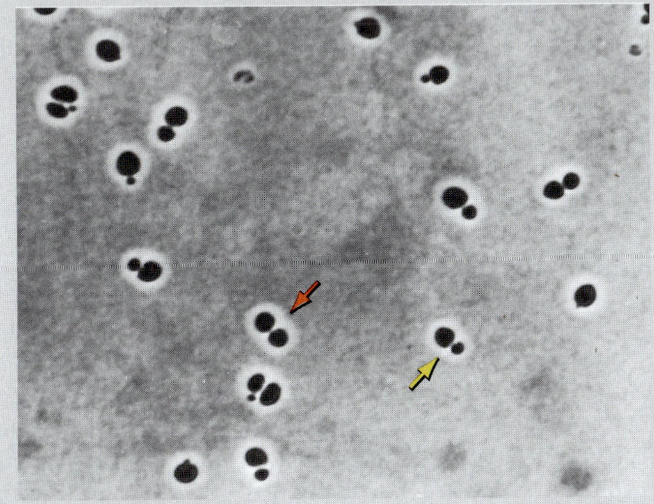

Figure B A cell-cycle mutant of yeast grown at restrictive temperature. The same cells as in Fig. A photographed after incubation at the restrictive temperature. These cells have all arrested with a large bud. (Other cell-cycle mutants may arrest with buds of different sizes or with no bud at all.) Cells that were early in the cell cycle (with small buds in Fig. A) at the time of the temperature shift arrest in the first cell cycle (*yellow arrow*). Cells that were later in the cell cycle (with larger buds in Fig. A) at the time of the shift finish the first cell cycle and arrest in the second, producing clumps with two large-budded cells (*red arrow*).

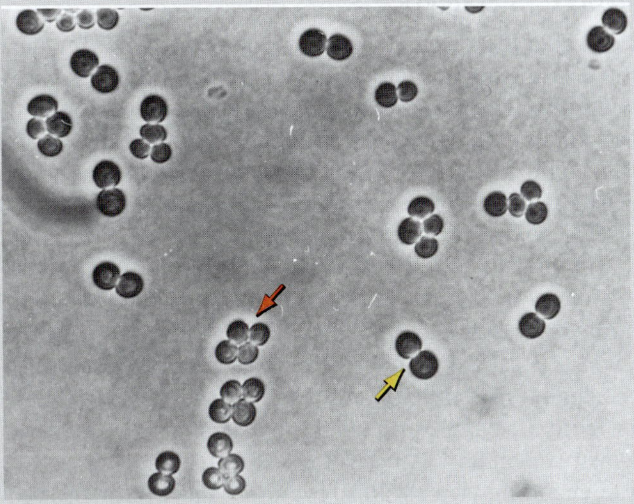

Yeast geneticists have identified over 100 cell-cycle genes in this manner, and the significance of many of these became apparent when geneticists identified related genes in other organisms. They found, for example, that one of the yeast cell-cycle genes encodes a cyclin-dependent kinase called CDK1. Homologous genes were found in the fission yeast *Schizosaccharomyces pombe,* the African clawed frog *Xenopus laevis,* and in humans. In the two yeast species (*S. cerevisiae* and *S. pombe*), CDK1 controls a step that commits cells to initiating a new round of division. The frog CDK1 protein participates in a CDK–cyclin complex that controls the rapid early divisions in this organism.

Remarkably, gene swapping experiments showed that the budding yeast and the fission yeast genes for CDK1 can replace one another in either organism, demonstrating that they encode proteins that carry out the same activity. The *Xenopus* and human CDK1-encoding genes can also function in either kind of yeast. Thus, in four very different organisms, genes that seem to be the central controlling elements of the cell cycle encode highly conserved, functionally homologous protein kinases.

mitosis (M phase), one checkpoint oversees formation of the mitotic spindle and proper engagement of all pairs of sister chromatids. If even a single chromosome fails to attach to the spindle, the cell does not initiate sister chromatid separation or anaphase chromosome movement until the correct attachment is made. This spindle assembly checkpoint is important for preventing aneuploidy among the daughter cells of a mitotic division.

In this section, we focus on the molecular mechanism of a different system, the *G_1-to-S checkpoint,* which ensures that cells will not perform the G_1-to-S transition if they have damaged DNA.

The G_1-to-S checkpoint

When radiation or chemical mutagens damage DNA during G_1, DNA replication is postponed. This postponement allows time for DNA repair before the cell proceeds to DNA synthesis; otherwise, replication of the unrepaired DNA could have many dangerous consequences. Before we discuss the checkpoint process, you first need to understand a few key molecular aspects of the way that cells normally transit from G_1 phase to S phase.

Quiescent cells in G_1 are stimulated to enter S by the binding of growth factors to their corresponding receptors. As we have seen, this interaction on the cell surface initiates a signal transduction cascade that activates transcription factors, which turn on the expression of a suite of downstream genes. Among these downstream genes are those encoding two cyclins, cyclin D and cyclin E. As a result, the first CDK–cyclin complexes to appear during G_1 in human cells are CDK4–cyclinD and CDK2–cyclinE (**Fig. 20.15**).

These two CDK–cyclin complexes initiate the transition to S in large part by phosphorylating the protein product of the retinoblastoma (*RB*) gene. The unphosphorylated Rb protein inhibits a transcription factor called E2F, but phosphorylated Rb can no longer inhibit E2F. When the brakes on E2F are released, it can now turn on the expression of various genes needed for DNA synthesis. The chain of events is complicated, but you can see that in the overall scheme, mitogen activation of a growth factor receptor provokes a cell to progress from G_1 to S because the enzymes needed for DNA replication can now be produced.

Now that you know the basis for the G_1-to-S transition, we can discuss the checkpoint that ensures this transition does not occur when cells have damaged DNA. In mammals, cells exposed to ionizing radiation or UV light during G_1 delay entry into S phase by activating the *p53 pathway* (**Fig. 20.16**).

The protein p53 is a transcription factor that participates in the G_1-to-S checkpoint in several ways (Fig. 20.16). First, p53 turns on transcription of the CDK inhibitor known as p21. The p21 protein binds to CDK–cyclin complexes and inhibits their activity; specifically, p21 prevents entry into S by inhibiting the activity of CDK4–cyclinD complexes. The second function of p53 is to turn on the expression of several genes encoding DNA repair enzymes.

Figure 20.15 How CDK–cyclin complexes mediate the G_1-to-S phase transition. CDK4 complexed to cyclinD, and CDK2 complexed to cyclinE, phosphorylate the Rb protein, causing it to dissociate from, and thus activate, the E2F transcription factor. E2F stimulates the transcription of many genes needed for DNA replication, including that for cyclinA. When cells enter S phase, cyclinD is destroyed, while CDK2–cyclinA complexes are formed and then activate DNA replication.

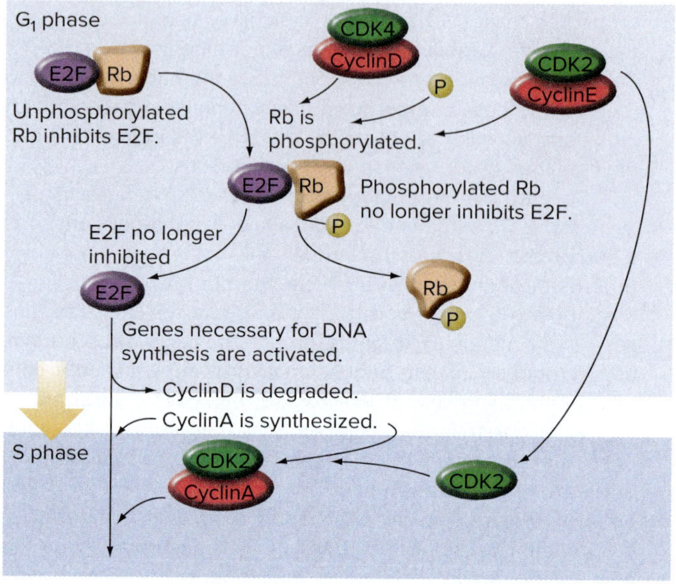

Figure 20.16 The p53 pathway is part of the G1-to-S checkpoint. DNA damage activates the p53 transcription factor, which in turn induces expression of the *p21* gene. The p21 protein inhibits CDK–cyclin complexes, resulting in a G_1 phase arrest. Activated p53 protein also induces the expression of many DNA repair and apoptosis genes.

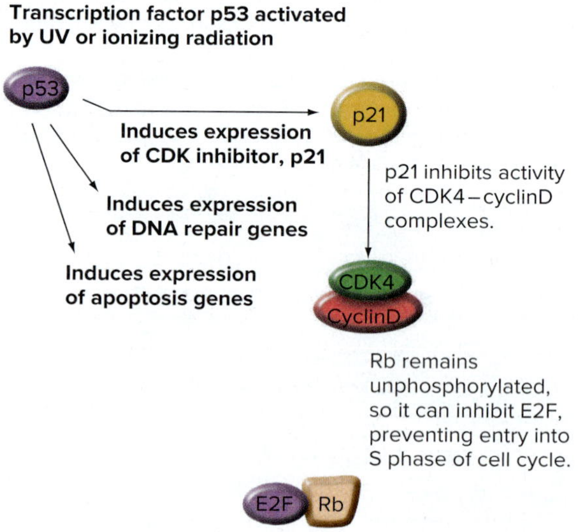

Figure 20.17 Apoptosis. If a normal cell's DNA is too highly damaged for repair, the cell will undergo programmed cell death (apoptosis). The DNA is degraded, the nucleus condenses, and the damaged cells emit signals attracting phagocytic cells to "eat up" the dying cell. Many apoptotic proteins are expressed only when the p53 transcription factor is active (see Fig. 20.16).

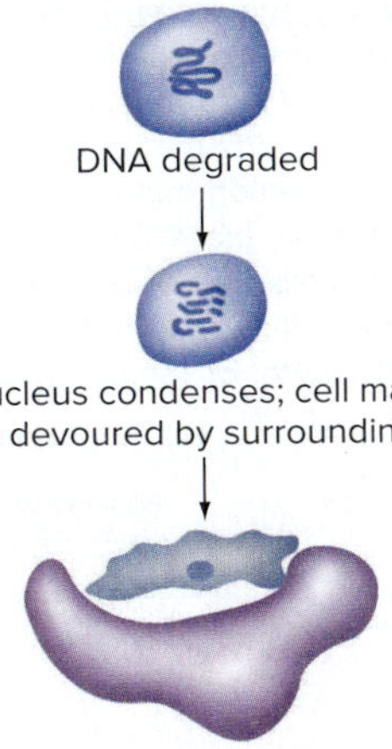

While the cell is arrested in the G_1 phase by p21, these enzymes can repair the DNA damage. Once this is accomplished, the p53 pathway is turned off and the cell can proceed into S phase.

The p53 pathway and apoptosis

If the DNA damage is great enough, wild-type cells producing p53 not only arrest in G_1, but they also commit suicide in a process known as *programmed cell death,* or *apoptosis.* During apoptosis, the cellular DNA is degraded, and the nucleus condenses. The dying cell sends out signals that attract phagocytic cells to devour and destroy it (**Fig. 20.17**). The p53 transcription factor, if activated sufficiently by massive DNA damage, turns on the expression of genes whose protein products participate in various aspects of apoptosis. This function constitutes a third way in which p53 participates in the G_1-to-S checkpoint (Fig. 20.16).

It makes sense for multicellular organisms to have a mechanism to eliminate cells that have sustained too much chromosomal damage, because the survival and reproduction of such cells could help generate cancers. The proteins involved in apoptosis are thus conserved in multicellular animals from roundworms to flies to humans.

The necessity of checkpoints

Checkpoints are not essential for cell division *per se,* but they are needed for the well-being of the organism. Experiments in mice and other animals demonstrate that mutant cells with a defective checkpoint are viable and divide at a normal rate. These mutant cells, however, are much more vulnerable to DNA damage than normal cells.

A common feature of cells with defective checkpoints is the instability of their genomes. Take for example the G_1-to-S checkpoint. Single-strand nicks resulting from oxidative or other types of DNA damage occur quite often. A cell normally repairs such nicks during G_1 before it enters S phase. If the G_1-to-S checkpoint coordinating this repair fails to function, however, the copying of single-strand breaks during replication would produce double-strand breaks that could lead to point mutations and chromosome rearrangements such as translocations.

Another manifestation of chromosome instability in cells with a defective G_1-to-S checkpoint is a propensity for **gene amplification:** an increase from the normal two copies to hundreds of copies of a gene. This amplification is often visible under the microscope, appearing either as an enlarged area within a chromosome known as a **homogeneously staining region (HSR)** that contains many tandem repeats of a gene, or as small chromosome-like bodies (called **double minutes** because of their small size) that lack centromeres and telomeres (**Fig. 20.18**). Normal human cells do not generate gene amplification in culture, but *p53* mutant cells exhibit high rates of such amplification.

Figure 20.18 Gene amplification in cells with defective G_1-to-S checkpoints. Tumor cells with defective checkpoints often exhibit amplified regions of DNA. These can appear as homogeneously staining regions (HSRs) containing many tandem repeats of one or more genes (*left*); or as double minutes, which are small pieces of extrachromosomal DNA (*right*).

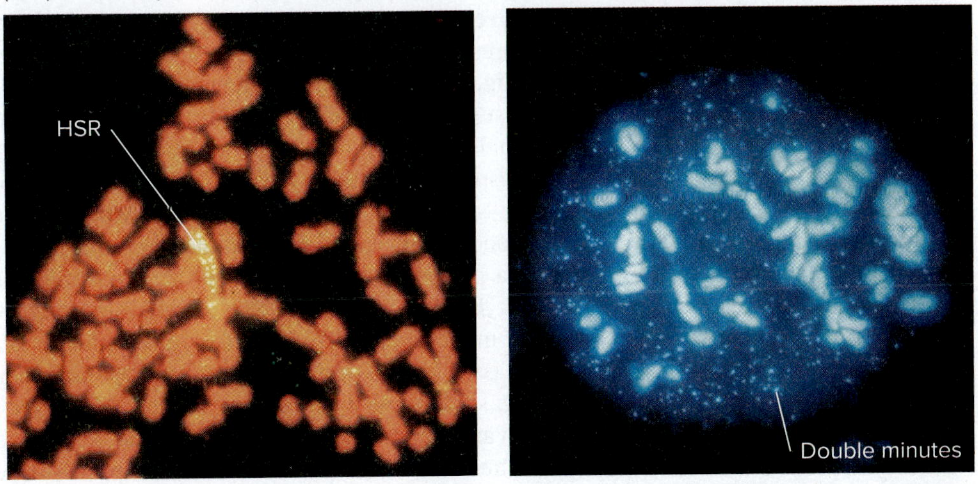

essential concepts

- Cell division is initiated by the binding of growth factors to receptor proteins on the cell surface. This binding initiates *signal transduction cascades* that activate transcription factors controlling the expression of cell-cycle genes.
- The formation and subsequent degradation of different *CDK–cyclin complexes* drive transitions between cell-cycle stages.
- The G_1-to-S *checkpoint* prevents cells from replicating their chromosomes if the DNA is damaged. The p53 protein is a key component of this checkpoint.

20.4 How Mutations Cause Cancer Phenotypes

learning objectives

1. Distinguish between proto-oncogenes and tumor-suppressor genes.
2. Explain how mutations in tumor-suppressor genes can be recessive at the level of the cell but cause dominantly inherited predispositions to cancer.

The mutant alleles that lead to cancer are popularly referred to as *cancer genes,* but this term is a misnomer because all cancer genes are, in fact, mutant alleles of normal genes. Scientists have already found more than 100 genes in which mutations can fuel the progression to cancer.

Cancer genes and their associated mutations can be subdivided into two important classes. Mutant alleles that act in a dominant fashion to promote cancer are called **oncogenes;** in a diploid cell, one mutant oncogenic allele is sufficient to help cause a cancer-related phenotype (**Fig. 20.19a**). Mutant alleles that act recessively to promote cancer are mutations in **tumor-suppressor genes;** in a diploid cell, both copies of a tumor-suppressor gene must be mutant to make the cell abnormal (**Fig. 20.19b**). The distinction between oncogenes and tumor-suppressor genes can perhaps best be understood in analogy with the pedals—the accelerator and the brake, respectively—that govern the movement of an automobile.

The normal, nonmutant allele of an oncogene is called a **proto-oncogene,** and it often encodes a protein needed for cell-cycle progression (Fig. 20.19a). In this way, a proto-oncogene is like the accelerator that moves a car forward. The oncogenic mutation increases gene expression or enhances the gene product's efficiency, so the cell will undergo more mitotic cell cycles in a given amount of time. You can thus think of the oncogenic mutation as the equivalent of pressing harder on the accelerator. A single oncogenic mutation thus has a dominant, gain-of-function effect in making the cell divide more frequently (or acquire other malignant characteristics).

Figure 20.19 Two kinds of cancer-producing mutations. **(a)** Proto-oncogenes (*light green DNA*) often encode proteins (*light green circles*) that promote cell proliferation. A dominant gain-of-function mutation in a wild-type proto-oncogene generates a cancer-causing oncogene (*dark green DNA*). Mutant oncogenic alleles are hypermorphs or neomorphs that produce either proteins with abnormal activity (*dark green stars*) or excessive amounts of normal protein (*not shown*). **(b)** Wild-type tumor-suppressor genes (*red DNA*) encode proteins (*red octagons*) that restrict cell proliferation or DNA repair proteins that guard against genome instability. Recessive loss-of-function mutations in tumor-suppressor genes (amorphs or hypomorphs; *black DNA*) exhibit no mutant phenotype in heterozygous cells that also have a wild-type allele. However, when a second mutation inactivates the wild-type allele, the cell proliferates out of control or rapidly accumulates mutations.

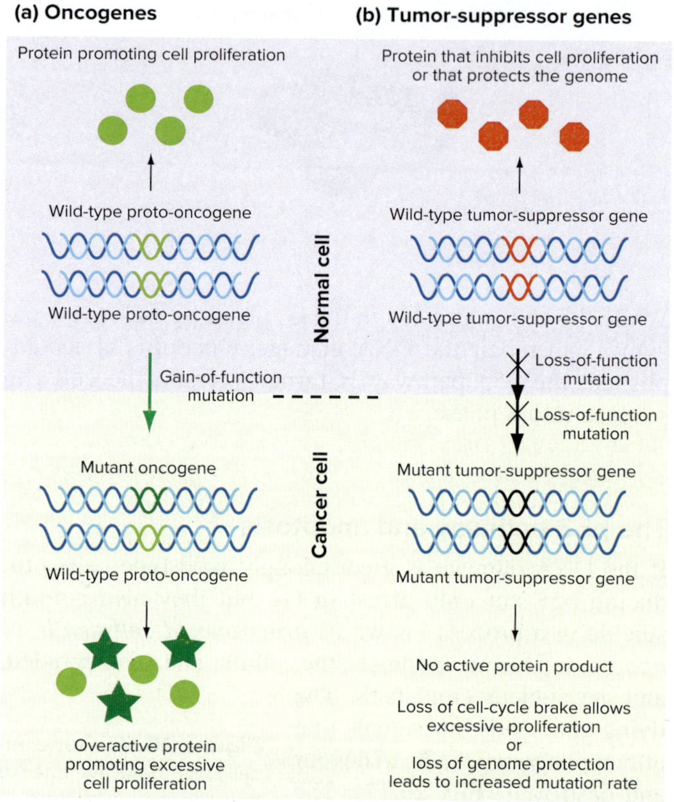

In contrast, each normal copy of a tumor-suppressor gene encodes a protein that either slows cell division down or guards against genome instability, so the analogy for a diploid cell would be a car with two redundant brakes. The mutations in tumor-suppressor genes that can contribute to cancer are loss-of-function mutations that are the equivalent of removing feet from the brakes (Fig. 20.19b). If one foot is removed, one brake is still in place and the car will not move; nothing happens until both feet are taken off both brakes. Mutations in tumor-suppressor genes are recessive at the cellular level because both normal copies of the tumor-suppressor gene must be inactivated to produce an effect. When neither allele is functional, then the cell will either proliferate faster, accumulate mutations at a faster rate, or both.

Most cancers require the accumulation of multiple mutations both in oncogenes and in tumor-suppressor genes. The cell will divide in an increasingly uncontrolled fashion the harder the accelerator is pushed down and the fewer the brakes that are engaged. Oncogene and tumor-suppressor gene mutations are called the *drivers* of cancer as they initiate the vicious cycle of proliferation and mutation that leads to malignancy.

Oncogenic Mutations Accelerate the Cell Cycle

How can we find candidates for oncogenes? What kinds of mutations convert a proto-oncogene into an oncogene? What do the proto-oncogenes do in normal cells? How do oncogenic mutations actually contribute to the induction of cancer? We address here recent research on these important questions.

Finding oncogenes through association with tumor viruses

Tumor viruses are useful tools for studying cancer-causing genes, first because these viruses carry very few genes themselves, and second because they infect and change cultured cells to tumor cells, which makes it possible to study them *in vitro*. Many of the viruses that generate tumors in animals are retroviruses whose RNA genome, upon infecting a cell, is copied to cDNA, which then integrates into the host chromosome. (Review the Genetics and Society Box in Chapter 8 entitled *HIV and Reverse Transcription*.) Occasionally by chance, the site into which the viral cDNA integrates is located near a proto-oncogene in the host cell genome. Rare deletion events can subsequently bring the viral cDNA and the proto-oncogene even closer together (**Fig. 20.20a**). Later, when the cDNA is transcribed from the host chromosome, the viral RNAs that result can pick up copies of host proto-oncogenes (**Fig. 20.20b**).

RNA tumor viruses can contribute to cancer in at least three ways. First, when viruses carrying proto-oncogenes are propagated, gain-of-function mutations can occur in the proto-oncogene to convert it into an oncogene. When a virus carrying the oncogene infects a new host cell, the oncogene is expressed, and its aberrant protein product causes the new host cell to proliferate abnormally. Second, proto-oncogenes or oncogenes incorporated into viral genomes are placed under the transcriptional control of powerful promoters and enhancers on the viral chromosome. As a result, the proto-oncogene or oncogene is transcribed at an unusually high rate, leading to excessive synthesis of the protein product (Fig. 20.20b). Third, when viral cDNA integrates adjacent to the proto-oncogene in a host chromosome (as in Fig. 20.20a), the same strong promoters and enhancers elevate transcription of the proto-oncogene, again resulting in excessive amounts of the proto-oncogene-encoded protein.

Figure 20.20 Oncogenes and retroviruses. (a) Rarely, double-stranded viral cDNA integrates adjacent to a proto-oncogene in the host genome. Powerful promoters in the cDNA lead to overexpression of the proto-oncogene. **(b)** Examples of highly oncogenic retroviruses. Occasionally, an adjacent proto-oncogene (such as *src*, *myb*, *fes*, or *rel*) can become "captured"—that is, incorporated into the viral genome so as to become an oncogene. U3, R, U5, *gag*, *pol* (the gene for reverse transcriptase), and *env* are general features of retroviral chromosomes, including that of the weakly oncogenic virus without an oncogene pictured in the *top row*.

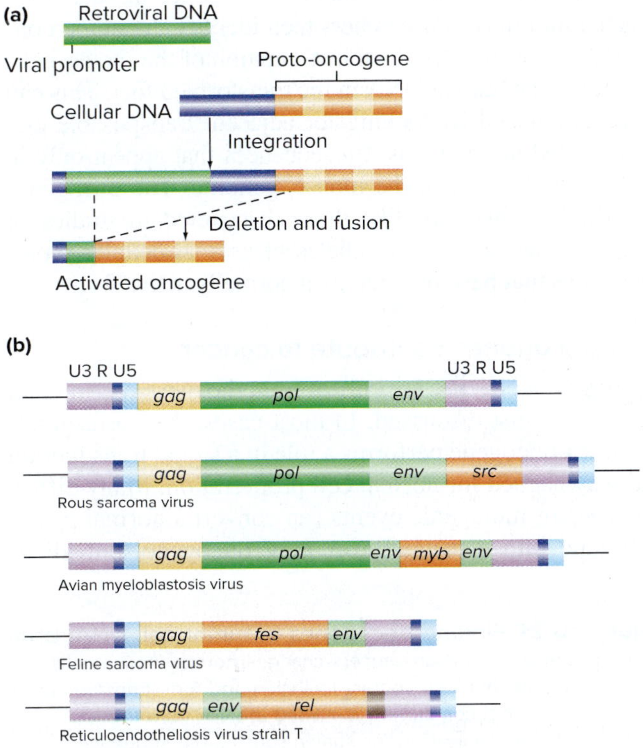

The analysis of retroviruses that cause tumors in various animal species led to the discovery of many known oncogenes. Scientists simply look for oncogenes carried in the small viral genomes, or they look at the genomes of tumors induced by retroviruses for genes next to viral cDNA integration sites.

Finding oncogenes through cell transformation assays

Scientists have also used a second, very different strategy to identify other oncogenes. The researchers isolate DNA from tumor cells and expose noncancerous cells in culture to this tumor DNA. If the noncancerous cells pick up a fragment of tumor DNA that contains a complete oncogene, some of them will be transformed into cells that are capable of producing tumors. This effect occurs because the oncogenes are dominant to the normal proto-oncogenes present in the genomes of the original noncancerous cells.

Figure 20.21 illustrates how this method has been used to identify oncogenes in human cancers. Scientists isolate genomic DNA from human tumors, fragment the DNA into gene-sized pieces, and then add the DNA to noncancerous mouse cells. Some of these cells take up an oncogene, lose contact inhibition, and form *transformed foci* in petri plates. Cells from the transformed foci often produce tumors when injected into mice. Researchers then identify the human oncogene responsible for the transformation of the mouse cells by finding the human DNA in the transformed foci. This can be accomplished by looking for adjacent transposable elements (SINEs) known as *Alu* sequences that appear only in the human, but not in the mouse, genome. The oncogenes identified in this way, like those discovered in studies of tumor viruses, are mutant alleles of normal cellular proto-oncogenes that have mutated to abnormally active forms.

How oncogenes contribute to cancer

Dozens of oncogenes are known from the two experimental approaches just discussed. In most cases, the corresponding proto-oncogene performs a role in a signal transduction pathway needed for normal cell proliferation. Many different types of mutagenic events can convert a normal proto-oncogene into a cancer-promoting oncogene, but all of these mutations have some kind of gain-of-function, dominant (hypermorphic or neomorphic) effect.

We discuss here three important examples that illustrate these points.

Ras Several oncogenic alleles of the *Ras* gene are point mutants encoding Ras proteins that are always (that is, constitutively) in the GTP-activated form. One such mutant allele, Ras^{G12V}, was discussed in Chapter 19 (Fig. 19.7). A cell carrying a constitutively active *Ras* oncogene generates its own *autocrine signals* to divide whether or not a growth factor is present (**Fig. 20.22a**).

c-Abl Chapter 13 on chromosomal rearrangements described how the genome of cancerous cells in many patients suffering from chronic myelogenous leukemia contains a translocation between chromosomes 9 and 22 that fuses a gene named *c-abl* with another gene called *bcr* (see

Figure 20.21 Finding oncogenes in tumor cell DNA. DNA isolated from some human cancers changes normal mouse cells into cancer cells that lose contact inhibition and form transformed foci. Human oncogenes can be identified in DNA from the transformed cells by looking for human-specific *Alu* sequences adjacent to the oncogenes.

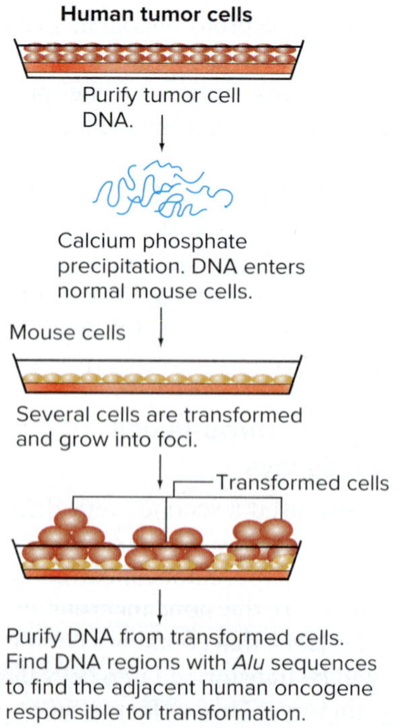

Figure 20.22 Examples of oncogenic mutations. (a) Certain point mutations in the *Ras* gene encode an altered Ras protein locked in the GTP-activated form. **(b)** In breast cancer cells (*right*) but not in normal cells (*left*), the *Her2* gene is amplified on multiple double-minute chromosomes. In this FISH analysis, *red* spots indicate chromosome 17 centromeres (two exist in normal and most tumor cells), while *black* spots represent *Her2* genes (two in normal cells, many in the tumor cells).

b: From: H. Nitta et al. (2008), "Development of automated brightfield double *In Situ* hybridization (BDISH) application for *HER2* gene and chromosome 17 centromere (CEN 17) for breast carcinomas and an assay performance comparison to manual dual color *HER2* fluorescence *In Situ* hybridization (FISH)," *Diagnostic Pathology*, 3:41, Fig. 3A-B. © Nitta et al. Licensee BioMed Central Ltd. 2008. http://diagnosticpathology.biomedcentral.com/articles/10.1186/1746-1596-3-41

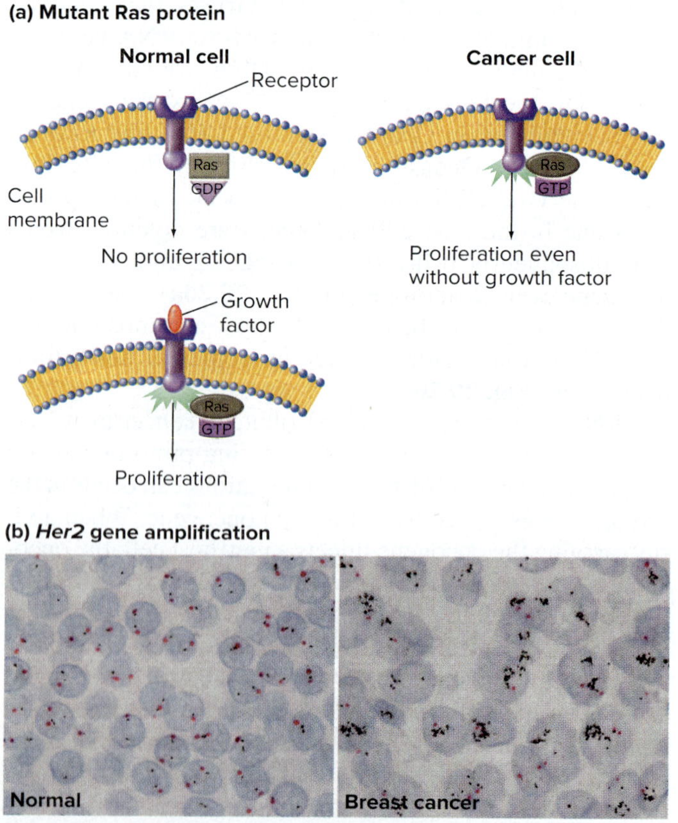

Fig. 13.16). The fused gene is an oncogene that encodes a hybrid protein. The c-Abl part of the protein is a *protein tyrosine kinase,* which adds phosphate groups to tyrosine amino acids in other proteins. This enzyme participates in certain growth-factor-induced signal transduction pathways.

Normal cells closely regulate the activity of the c-Abl protein, blocking its function most of the time but activating it in response to stimulation by growth factors in the environment. By contrast, the fused protein encoded by *bcr/c-abl* in cells carrying the translocation is not susceptible to regulation. Similar to oncogenic Ras proteins, the Bcr/c-Abl protein is always active, even in the absence of growth factors.

Her2 About 20% of all breast cancers overexpress human epidermal growth factor receptor 2 (Her2). As the name implies, this protein is a member of the growth factor receptor family, but it is currently called an *orphan receptor* because the growth factor that activates it is not yet known. Cells with too many copies of the Her2 protein in their membranes activate signal transduction pathways constitutively and thereby cause inappropriate divisions.

In some Her2-positive cells, overexpression of the protein reflects tremendous amplification of the number of copies of the *Her2* gene. As shown in **Fig. 20.22b,** the additional copies of *Her2* are usually found on tiny, double-minute chromosomes. This finding suggests strongly that a mitotic ancestor of the cancer cell must have first accumulated mutations in genes encoding DNA repair enzymes or components of a DNA damage checkpoint like p53. Such mutations would allow the amplification of certain regions of the genome, and by chance one of these regions included the *Her2* gene.

These Her2-positive breast cancers are particularly aggressive, but as we will see in a later section, drugs targeted at the Her2 protein itself can treat many such cancers effectively.

Mutations in Tumor-Suppressor Genes Release the Cell-Cycle Brakes or Destabilize the Genome

Some mutant tumor-suppressor genes are recessive alleles of genes whose normal alleles help put cell division on hold, for example in terminally differentiated cells or in cells with DNA damage. One wild-type copy of these genes apparently produces enough protein to regulate cell division. It is only when both wild-type copies are lost that a restraint (brake) on proliferation is released (review Fig. 20.19b). Other tumor-suppressor genes encode DNA repair proteins. Loss of both normal alleles can increase the mutation rate and produce driver mutations in other cancer genes.

Finding tumor-suppressor genes through analysis of pedigrees and genomes

Researchers have identified dozens of tumor-suppressor genes through the genomic analysis of families with an inherited predisposition to specific types of cancer, or through the analysis of specific chromosomal regions that are deleted reproducibly in certain tumor types.

Retinoblastoma provides an example of such an identification process. As cancers of the color-perceiving cone cells in the retina, retinoblastoma tumors are easy to diagnose and remove before they become invasive (**Fig. 20.23a**). Retinoblastoma is one of several cancers that can be inherited in a dominant fashion in human families (**Fig. 20.23b**). Roughly half the children of a parent with retinoblastoma develop the disease.

Karyotypes of normal, noncancerous tissues from many people suffering from retinoblastoma reveal heterozygosity for deletions in the long arm of chromosome 13; that is, the patients carry one normal and one partially deleted copy of 13q. Cancerous retinal cells from some of these same patients are homozygous for the same chromosome 13 deletions that are heterozygous in the noncancerous cells.

Figure 20.23 Mutations in tumor-suppressor genes are recessive at the cellular level but dominant at the organismal level. **(a)** A child with a retinoblastoma tumor in one eye. **(b)** A pedigree indicating that inheritance of a single RB^- allele dominantly predisposes an individual to retinoblastomas. **(c)** In an RB^+/RB^- person, loss of the remaining RB^+ from the genome of a single retinal cell can lead to a retinoblastoma. Such *second hits* are rare, but the retina contains so many cells that such an event likely will occur in one or more of them.

a: Source: National Cancer Institute

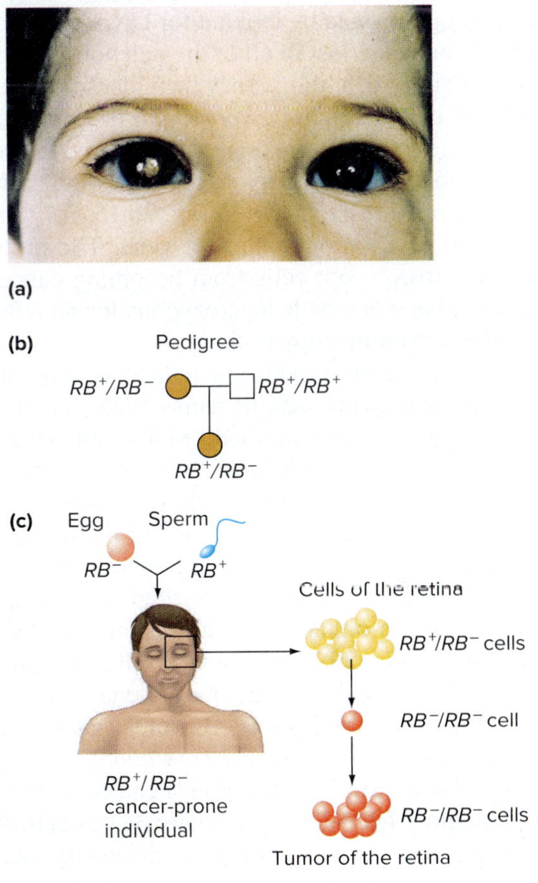

Figure 20.24 Events causing loss-of-heterozygosity in somatic cells of RB^+/RB^- individuals. RB^+ is inherited through the germ line as an autosomal recessive mutation. Subsequent changes to the RB^+ allele during somatic divisions generate a clone of cells homozygous or hemizygous for the nonfunctional RB^- allele. *Uniparental disomy* means that the two copies of a chromosome in a cell were obtained from one parent. Two events are needed to produce a cell with uniparental disomy; for example, the first event could be loss of one chromosome, and the second event could be duplication of the remaining chromosome.

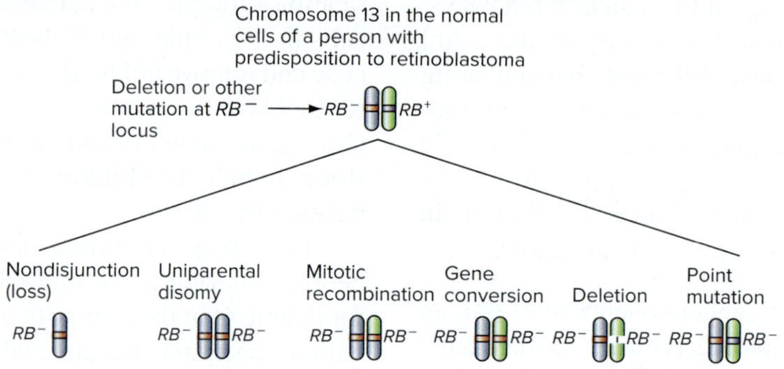

Although the deletions vary in size and position from patient to patient, they all remove band 13q14.

These observations indicate that band 13q14 includes a gene whose removal contributes to the development of retinoblastoma. *RB* is the symbol for this gene. The heterozygous cells in a patient's normal tissues carry one copy of the gene's wild-type allele (RB^+), and this one copy prevents the cells from becoming cancerous. Tumor cells homozygous for the deletion, however, do not carry any copies of RB^+, and without it, they begin to divide out of control.

Geneticists used their understanding of retinoblastoma inheritance to find the *RB* gene. They located the general region carrying the gene by looking for DNA sequences in band 13q14 that were lost in all of the deletions associated with the hereditary condition. They then identified the specific gene by characterizing a very small deletion that affected only one transcriptional unit—the *RB* gene itself. It encodes one of many proteins involved in regulating the cell cycle, as previously shown in Fig. 20.15. *RB* thus fits our definition of a tumor-suppressor gene: The protein it determines helps prevent cells from becoming cancerous. Cancer can arise when cells heterozygous for an *RB* deletion lose the remaining copy of the gene.

This picture of the genetics of retinoblastoma raises a perplexing question: How can the retinoblastoma trait be inherited in a dominant fashion if a deletion of the *RB* gene is recessive to the wild-type RB^+ allele? The answer is that *RB* deletions are dominant at the level of the organism because of the strong likelihood that in at least one of the hundreds of thousands of retinal cells heterozygous for the deletion, a subsequent genetic event will disable the single remaining RB^+ allele, resulting in a mutant cell with no functional tumor-suppressor gene (**Fig. 20.23c**). This one cell then multiplies out of control, eventually generating a clone of cancerous cells (see the Genetics and Society Box of Chapter 5 entitled *Mitotic Recombination and Cancer Formation*).

Many different kinds of rare events can occur to knock out the remaining RB^+ allele in a heterozygous cell. All of these events produce a **loss-of-heterozygosity** because they change an RB^-/RB^+ cell into one that is RB^-/RB^- (**Fig. 20.24**). The RB^+ copy could itself become deleted or suffer an inactivating point mutation. The chromosome carrying RB^+ could be lost by nondisjunction. Mitotic recombination or gene conversion could substitute the RB^+ allele with an RB^- mutation or deletion from the homologous chromosome inherited from the affected parent.

The hypothesis that cancer requires two independent *hits* that knock out the function of both alleles of a tumor-suppressor gene is supported strongly by the fact that sporadic, noninherited forms of cancers like retinoblastoma exist. Sporadic retinoblastomas generally occur later in life than inherited retinoblastomas. In addition, most patients with sporadic retinoblastomas have only a single tumor in one eye, whereas many individuals with inherited forms have multiple tumors that can affect both eyes. Sporadic retinoblastomas are rare (and take longer to develop than the inherited disease) because they require two successive, independent hits in a clone of cells. In contrast, cells in people with an inherited RB^- mutation require only one additional hit to lose function of this tumor-suppressor gene. Why the loss of *RB* affects eyes more than the loss of other tumor suppressors does is not understood. However, exposure to UV light could be one reason why RB^-/RB^- heterozygotes are likely to develop retinoblastoma.

The loss of both copies of RB^+ is a critical driver step that can initiate the vicious feedback loop leading to cancer. Within this loop, subsequent genetic changes occur that create oncogenes or disrupt the function of other tumor-suppressor genes, increasing the oncogenic potential of cells in the clone.

How tumor-suppressor mutations contribute to cancer

The normal copies of tumor-suppressor genes encode proteins that have three types of interrelated functions:

- Some tumor-suppressor gene products are proteins that are an inherent part of the basic cell proliferation

machinery, though they act in a negative fashion (that is, to slow cell division).
- Certain tumor-suppressor gene products serve as components of cell-cycle checkpoints.
- Other wild-type tumor-suppressor genes encode enzymes that are involved in DNA damage repair or that promote apoptosis when DNA damage is too great.

The following examples illustrate some of the molecular mechanisms leading from loss-of-function mutations in tumor-suppressor genes to increased predispositions to cancer.

Rb Figure 20.15 showed that unphosphorylated, wild-type Rb protein delays cell-cycle progression into S phase through its inhibition of the E2F transcription factor. In RB^-/RB^- cells that have no Rb function, E2F cannot be inhibited. Thus, the cells progress into S phase before they are ready or in the absence of growth factors.

p53 We saw previously that the p53 protein is essential for the G_1-to-S checkpoint (Fig. 20.16). The p53 protein, and thus the checkpoint, is activated in wild-type cells by stressful conditions such as the presence of single-stranded breaks (nicks) in DNA. Activation of p53 induces the expression of suites of genes whose products (1) block the action of CDK–cyclin complexes; (2) help repair DNA damage, or (3) promote apoptosis when all else fails.

In $p53^-/p53^-$ cells, none of these events occur (**Fig. 20.25**). Cells with DNA damage continue into S phase. Single-stranded breaks become converted into double-stranded breaks when the chromosomes replicate, leading to the generation of many chromosomal rearrangements. Cells with these damaged genomes do not die by apoptosis as they should.

BRCA1 and BRCA2 *BRCA1* and *BRCA2* (*breast cancer 1* and *breast cancer 2*) encode protein components of the machinery for repairing double-strand breaks in DNA. Women heterozygous for either one of these mutations have a much higher than average risk of developing breast or ovarian cancer in their lifetimes. If a cell loses its remaining normal copy of *BRCA1* or *BRCA2,* the mutation rate increases, eventually producing new driver mutations which can lead to malignancy. Scientists do not yet understand why the loss of this particular DNA repair system affects breast and ovarian cells more than other tissues.

Figure 20.25 Loss of p53 function leads to genomic instability. In cells without the p53 transcription factor that also sustain single-stranded breaks in their DNA: (1) p21 is not induced, so the cells progress to S phase. (2) Genes for DNA repair enzymes are not transcribed. Thus, upon DNA replication, the single-stranded nicks are converted to double-strand breaks, which in turn cause mutations and chromosomal instability. (3) Apoptotic genes are not induced, so the cells with DNA damage do not die as they should (*not shown*). (Compare with Fig. 20.16.)

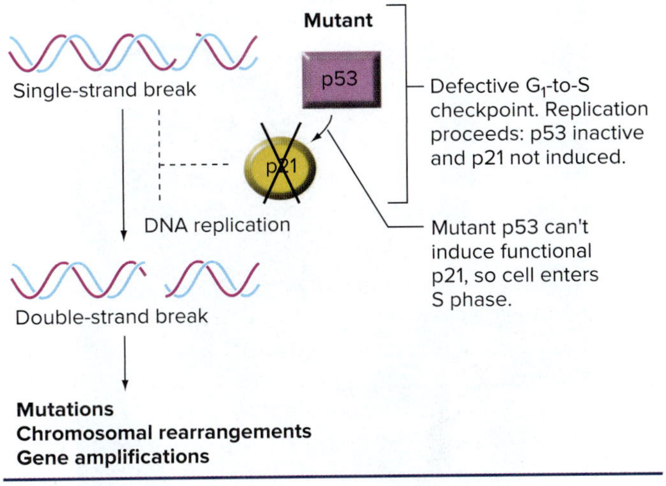

essential concepts

- *Proto-oncogenes* usually encode proteins that participate in growth-factor-dependent signal transduction pathways. *Oncogenes* are dominant, gain-of-function mutant alleles of proto-oncogenes that result in activation of these pathways even in the absence of the growth factor.
- The normal alleles of *tumor-suppressor genes* encode proteins that act as brakes on the cell cycle or that maintain genome stability. Loss-of-function mutations in tumor-suppressor genes are recessive at the level of the cell because both normal copies must be lost to promote cancer. Individuals with a mutant copy of a tumor-suppressor gene have a dominantly inherited predisposition to cancer because of the high probability that the remaining wild-type copy of the tumor-suppressor gene will be lost in at least one cell.

20.5 Personalized Cancer Treatment

learning objectives

1. Explain how drugs such as Gleevec® and Herceptin® target specific products of oncogenes.
2. Discuss the challenges to producing anticancer drugs that counteract the effects of mutations in tumor-suppressor genes.
3. Summarize current approaches to tailoring treatments for individual cancers.

Cancer has always afflicted humans; **Fig. 20.26** shows that prostate cancer has been found in a mummy preserved from ancient Egypt. Ancient papyrus scrolls reveal that as early as 1500 B.C., Egyptian physicians not only recognized cancer, but also tried to excise tumors by surgery or by plunging hot pokers into them. Surgery remains today a major method for

Figure 20.26 Prostate cancer detected by X-ray tomography in a 2000-year-old mummy.
Siemens Press Picture, © Siemens AG, Munich/Berlin.

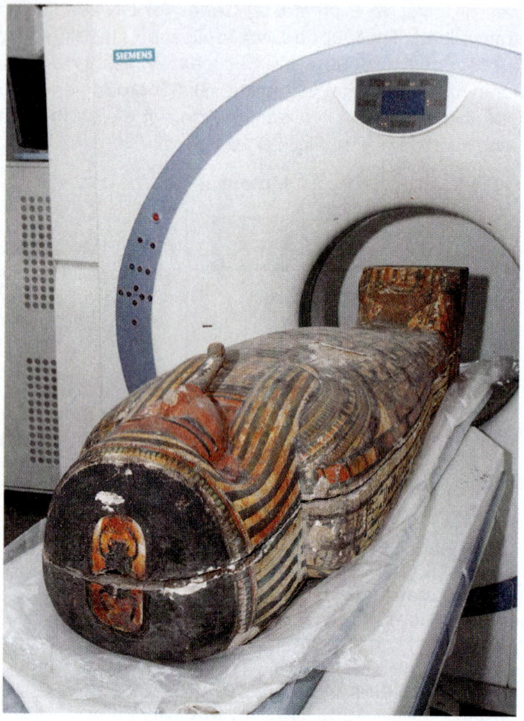

treating cancers, while the modern equivalent of the hot poker to kill cancer cells is radiation treatment. These techniques can be effective, but they have major limitations. Some tumors are located in parts of the body too dangerous to reach. All the cancerous cells may not be killed or removed, particularly if the tumor has already metastasized to other tissues.

In the 1940s, medical scientists began to explore **chemotherapy;** that is, treating patients with drugs that kill cancer cells. These chemicals were directed against biochemical pathways needed by proliferating cells. Some chemotherapies disrupt DNA replication, while others prevent the formation of the mitotic spindle. This approach has had notable successes, but a major problem exists because our bodies continually require cell renewal. Normally dividing cells might be only slightly less susceptible than rapidly dividing cancer cells to a chemotherapy agent. Chemotherapy thus requires a delicate balance of doses that kill cancer cells but not the patient. Even when treatments are successful, the side effects on the patient can be severe.

Most New Antitumor Drugs Target the Products of Oncogenes

Our increasing knowledge of the molecular basis for cancer suggests new, more specific therapeutic approaches with fewer side effects. The changes that convert proto-oncogenes into oncogenes mean that cancer cells will have unique molecular properties that might be targeted by drugs. The protein products of oncogenes are either different from those of the corresponding proto-oncogenes, or alternatively, the oncogenes express more of the same protein. Biochemists can thus design drugs that bind specifically to oncogenic driver proteins, inactivating them and thus slowing cancer cell proliferation, and/or mobilizing the body's immune defenses to kill cells that (over)express the proteins. We describe here drugs that illustrate these two strategies.

Oncogenic protein inactivation: Gleevec®

As mentioned earlier, a fused *bcr/c-abl* gene is an important cause of some forms of chronic myelogenous leukemia, because the abnormal protein tyrosine kinase it encodes is always active even when no growth factors are present. Pharmaceutical companies have developed a drug called Gleevec® that specifically inhibits the enzymatic activity of the protein tyrosine kinase part of the fusion protein (the part encoded by *c-abl*). For the kinase to function, it must bind to ATP, from which it obtains a phosphate group to transfer to other proteins. Gleevec® was designed to fit into the ATP binding site of the Bcr/c-Abl kinase and prevent ATP from binding there, thus inactivating the oncogenic protein (**Fig. 20.27**).

In clinical trials, 98% of participants whose cancers carried the fusion gene experienced a complete disappearance of leukemic blood cells and the return of normal white cells. Of these patients, only 5% died from leukemia five years after the trial began. This drug is now the standard treatment for chronic myelogenous leukemia and is a model for new types of cancer treatments that home in on cancer cells without hurting healthy ones.

Monoclonal antibodies to growth factor receptors: Herceptin®

A substantial fraction of breast tumors overexpress the Her2 growth factor receptor on the cell surface. The location of part of the Her2 protein outside of the cell made possible a clever strategy to attack such cancers. Scientists have

Figure 20.27 Gleevec®: Mechanism of action. The product of the *bcr/c-abl* fusion gene is a protein tyrosine kinase that transfers a phosphate group from ATP to many substrate proteins. Gleevec® binds tightly to the enzyme's ATP binding site, preventing the kinase function.

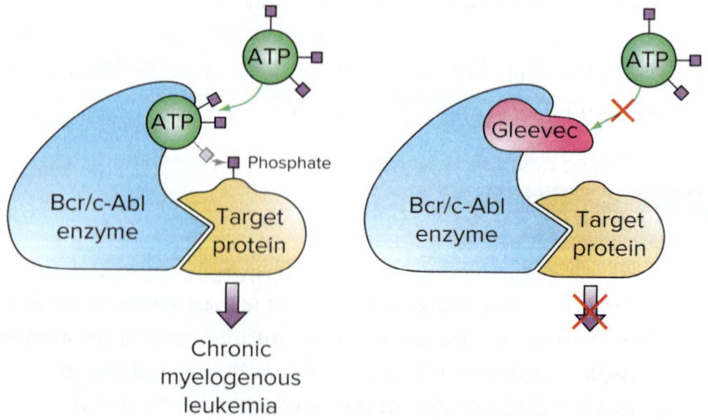

Figure 20.28 Herceptin®: Possible modes of action. This drug is a monoclonal antibody that binds to the Her2 protein on the surfaces of certain cancer cells. Binding of Herceptin® to Her2 prevents receptor subunits from dimerizing, thus blocking the growth factor–initiated signaling cascade. Binding of Herceptin® to Her2 also directs killer T cells to the cancer cells.

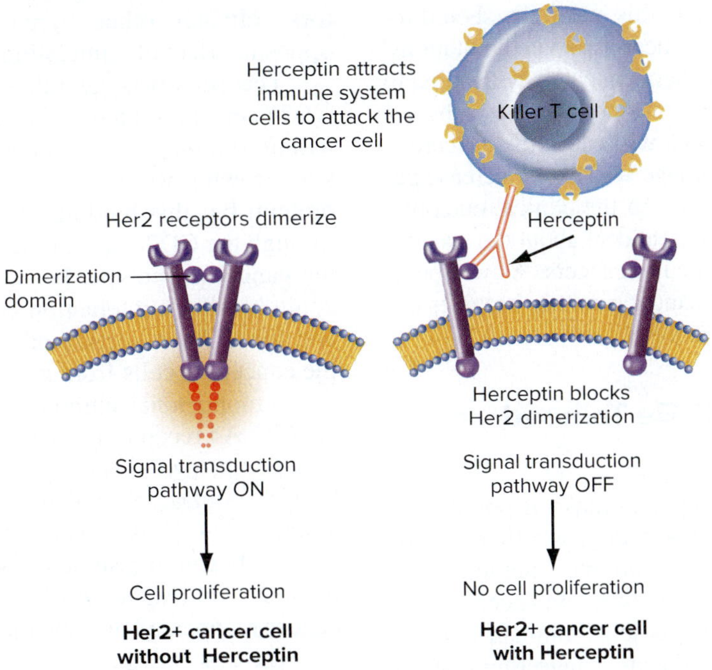

developed a monoclonal antibody called Herceptin® that binds tightly to the extracellular part of the Her2 protein.

The binding of Herceptin® to Her2 presumably has two effects (**Fig. 20.28**). First, Her2 receptors bound to Herceptin® cannot start the signaling cascade normally initiated by Her2's interaction with its corresponding growth factor. As a result, the Herceptin®-treated Her2-positive cells do not proliferate well. Second, the binding of Herceptin® to Her2 receptors mobilizes specialized immune system cells called *killer T cells* to target the cancer cells and eventually destroy them.

Herceptin® has been found in large-scale clinical trials to have considerable effectiveness against Her2-positive breast cancers, reducing the likelihood of cancer reoccurrence by 25% to 50% after surgery or standard chemotherapy. One reason the drug is not even more effective is that cancerous cells can acquire new mutations fairly easily that make them resistant to Herceptin®—for example, mutations that inappropriately activate later steps in the same signal transduction pathway.

Some Chemotherapies Target Cells Without Functional Tumor-Suppressor Genes

Drug development focuses more often on oncogenes than on tumor-suppressor genes. The reason is that cancer cells lack both functional copies of one or more tumor-suppressor genes and are thus missing factors needed for DNA repair or mitotic checkpoints. No protein thus exists that a drug can directly target. Furthermore, scientists have not yet devised efficient means of gene replacement to add wild-type copies of the tumor-suppressor gene back into the cancerous cells that do not have them.

As described previously, however: (1) Many tumor-suppressor gene products are involved in DNA repair mechanisms and DNA damage checkpoints; and (2) in many cells, excessive DNA damage induces programmed cell death (apoptosis). These facts suggested a paradoxical idea: Many cancer cells should be ultra-sensitive to molecules that increase DNA damage, because the tumor cells cannot efficiently repair that damage and would therefore be more likely to die by apoptosis.

This hypothesis seems to hold true, at least for certain cancers. For example, *PARP inhibitors* block the action of an enzyme called poly ADP ribose polymerase (PARP). These inhibitors are effective in treating many breast cancers with loss-of-function mutations in the tumor-suppressor gene *BRCA1*. The PARP enzyme is needed to heal nicks (single-strand breaks) in DNA. If the nicks are not repaired, they turn into double-strand breaks when DNA is replicated during S phase. The BRCA1 protein helps repair double-strand breaks.

BRCA1-negative breast cancer cells die more readily than normal cells when treated with PARP inhibitors because the drug will produce single-strand breaks that are converted into double-strand breaks in the highly proliferative cancer cells. These double-strand breaks will accumulate in cells that lack the BRCA1 protein. If the genomes of these BRCA1-negative cancer cells have enough double-strand breaks, they will self-destruct by apoptosis.

You probably realize already that for some tumors, PARP inhibitors might be exactly the wrong treatment. These drugs increase genomic instability because they prevent cells from repairing single-strand breaks, and this instability should result in the accumulation of new cancer-associated mutations. The key issue is whether the biochemical pathways leading to apoptosis can take place in the tumor cells. If the pathways still operate, excess DNA damage will cause the cancerous cells to die. If not, the DNA damage will likely increase genomic instability and perhaps worsen the clinical outcome. Physicians thus need to know a great deal about the specific mutations that characterize particular cancers. As we see in the next section, sequencing of cancer genomes provides exactly this kind of crucial information.

Future Treatments Will Be Tailored to Individual Cancers

For some cancers, fairly simple tests make it possible to predict, although somewhat imperfectly, whether the tumors might be treatable with available chemotherapy drugs. For example, cytogenetic analysis of cancer cells in patients with chronic myelogenous leukemia can indicate the presence of a fused *bcr/c-abl* gene, suggesting that the cancer will likely respond to Gleevec®. Similarly, if analysis of a breast tumor shows that cells have excessive amounts of Her2 protein on their surfaces, then physicians might treat the patient with Herceptin®.

Whole-genome sequencing of tumor cells: a case study

Such simple tests unfortunately often turn up negative for many cancer patients. What to do next? This is where whole-genome sequencing comes in. If you could determine the entire DNA sequence of the genome of the cells in a tumor, and compare it to the complete genomic sequence of normal cells from the same patient, you might be able to pinpoint one or more of the driver mutations in oncogenes and/or tumor-suppressor genes responsible for the cancer. This knowledge in turn might inform medical scientists of possible ways to treat the specific kind of cancer in the patient. We describe here a case study that shows the promise and potential pitfalls of cancer genome sequencing.

One of the first clinical tests of this approach involved a woman named Beth McDaniel with an advanced case of Sézary syndrome, a rare T cell lymphoma (a cancer of specific immune system cells) that did not respond to standard treatments. Scientists performed whole-genome sequencing of normal cells from Mrs. McDaniel's saliva and cancerous cells from the tumor and found roughly 18,000 differences. Most of these were unlikely to have any effects on cell proliferation. However, one mutation revealed a promising candidate. In the cancer genome, two genes—*CTLA4* and *CD28*—had fused together. The products of both genes are receptor proteins found on the surfaces of immune cells. The binding of ligands to CTL4A receptors activates signal transduction pathways that suppress cell growth and division, while the binding of different ligands to CD28 receptors activates other signaling pathways that have the opposite effect of stimulating cells to proliferate.

The scientists hypothesized that the fused *CTLA4/CD28* gene might have reversed the normal patterns of signaling. Inhibitory signals that would tell normal T cells to stop growing would bind to the CTL4A part of the fused protein, but this binding would activate cell proliferation through the CD28 part of the protein and thus contribute to the cancer phenotype. If so, then perhaps treatments that could block the interaction between the inhibitory ligands and the CTLA4 section of the fused receptor would stop the cancerous cells from growing.

A monoclonal antibody called Yervoy® directed against the CTLA4 receptor had previously been developed to treat melanomas (skin cancers). Doctors administered Yervoy® to Mrs. McDaniel, and the initial results were stunning. Within a few weeks, many signs of the cancer disappeared, and she began to resume aspects of her normal life. Unfortunately, the respite was brief, and a few weeks later, the cancer reappeared and she died soon after.

The costs of cancer genome sequencing are plummeting rapidly, and so this kind of personalized cancer study is becoming increasingly available to patients. In several recent cases, the effects of drug treatments suggested by the identification of key mutations in the cancer cells have proven to be more long-lasting, but as explained in a subsequent section, even in these cases recurrence of the cancer remains a serious threat.

Finding driver mutations and druggable targets

Given that cancers usually involve loss-of-function mutations in tumor-suppressor genes that disrupt DNA repair mechanisms, it is not surprising that cancer cells accumulate thousands or even tens of thousands of mutations that differentiate their genomes from those of normal cells in the same person. How can researchers pinpoint the particular driver mutations most responsible for the cancer phenotype?

Some critical mutations may be immediately obvious from cancer genome sequences, while others remain hard to find among the thousands of possibilities. Mutations that affect the amino acid sequences of known oncogenes and tumor-suppressor genes are the most obvious candidates. As of this writing in 2016, 47 oncogenes and 70 tumor-suppressor genes have been discovered. However, because these lists are undoubtedly incomplete, mutations (missense, nonsense, frameshift, rearrangements) that alter the coding sequences of any gene must be considered, even if the gene's function is not yet known.

Regulatory mutations that increase the transcription of oncogenes or decrease that of tumor-suppressor genes are also clearly of potential importance, but are as yet difficult to find in the whole-genome sequence of a cancer cell. The ENCODE project (see Chapter 17), which mapped regulatory elements such as enhancers and insulators in the human genome, has in some cases enabled researchers to pinpoint regulatory mutations that alter cancer gene transcription. More generally useful at the present time is the deep sequencing of cDNAs (that is, RNA-Seq), which can help find cancer-related genes whose expression is affected by regulatory mutations, even if the mutations themselves are not identified.

Whole-genome sequencing is not a panacea that will lead to immediate cures for all cancers. The genomes of cancer cells obtained from Steve Jobs, the founder of the Apple computer company, and Christopher Hitchens, a famous author and journalist, did not reveal any clues leading to effective treatments for the diseases to which they succumbed. In these cases, researchers did not recognize the driver mutations in the genomes of their cancers. But even when mutations of interest are found in cancer genomes, many of them are not *druggable,* meaning that pharmaceutical strategies to counteract the effects of the mutations have not yet been developed.

Cancer remission and recurrence

Beth McDaniel's case study indicates that cancers in remission can reappear, and sometimes very quickly. Some cancer cells may evade treatments, whether surgeries or new targeted chemotherapies. If even a few such cells survive, their proliferative potential is so great that the cancer can reemerge.

Recent studies that have looked at the whole-genome sequences of cells derived from different regions of the same tumor have suggested one important reason for cancer recurrence. Some tumors are heterogeneous, in that various cells within the tumor have somewhat different genomes. Certain mutations in oncogenes or tumor-suppressor genes are common to all the cells in the tumor, while some are not.

These findings make sense in terms of the hypothesis that cancers involve an accumulation of mutations in a clone of cells. The common mutations are those that occurred early in the mitotic proliferation of the clone, while the mutations specific to subpopulations of cancer cells occurred more recently in the past. Clearly, the most effective cancer treatments would be directed against the common mutations, but it is difficult to tell which mutations are common without sequencing the genomes of many cells throughout the tumor.

The cancer landscape and future horizons

Several large-scale projects, including the Cancer Genome Atlas funded by the National Institutes of Health (NIH), are currently characterizing the whole genomes/exomes of thousands of cancers. These studies have found recurrent patterns of mutation that subdivide cancers into groups of probable clinical relevance. As one example, a study published in 2012 classifies breast cancers into four groups that have very different properties. However, the results of these cancer genome projects also underscore how each tumor genome encompasses a unique set of mutations.

Figure 20.29 presents a summary of some of the results obtained from characterizing the genomes of 178 lung

Figure 20.29 Analysis of 178 lung squamous cell carcinomas. Each horizontal *row* represents a tumor from a different patient. The colored rectangles indicate base substitutions in particular genes (*columns*) with various effects on the gene's coding region or on splicing (*colors*). The percentage of patients with mutations in a given gene is at the bottom (e.g. 81% have mutations in the *TP53* gene encoding p53). All the genes shown are tumor-suppressor genes, except for *PIK3CA* and *NFE2L2* (oncogenes), *NOTCH1* (surprisingly an oncogene in some tumors and a tumor-suppressor gene in others), and *HLA-A* (which helps regulate immune surveillance of tumors). *Not listed* are rare mutations in known cancer-related genes found in only one or two of these 178 tumors.

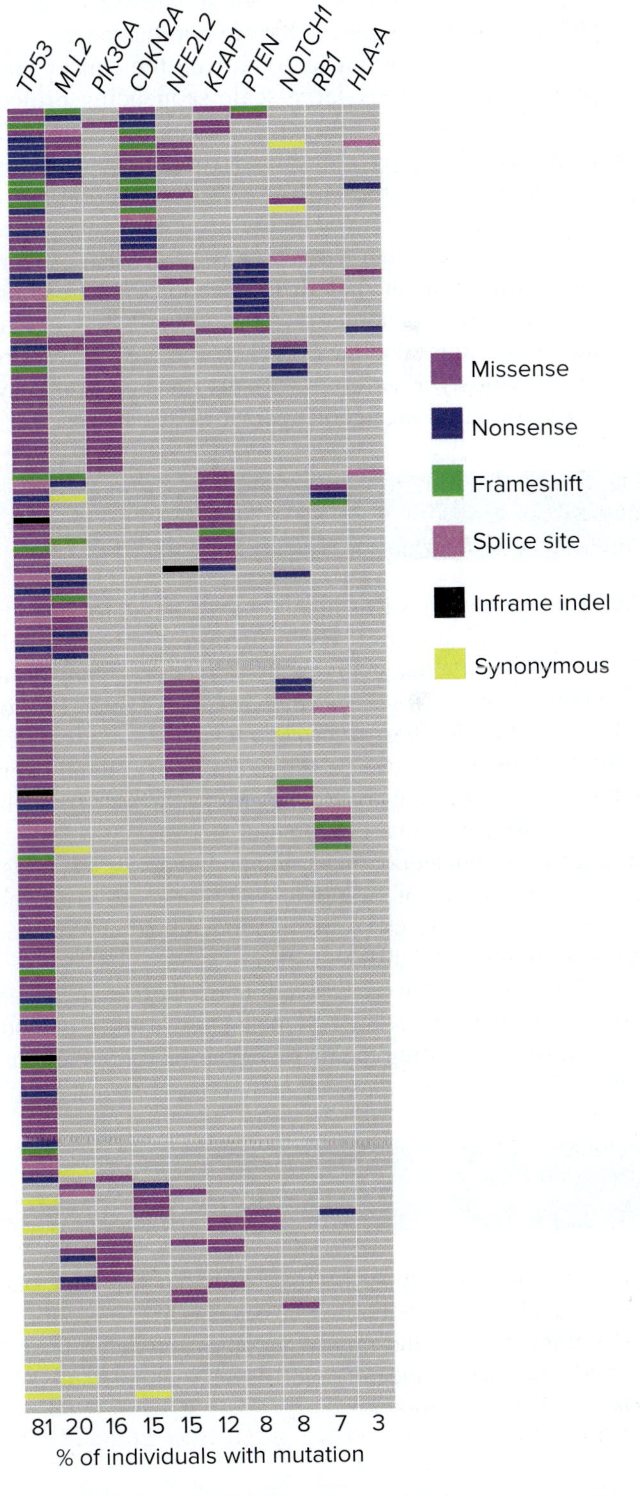

squamous cell carcinomas with respect to matched normal DNA from the same patients. Some genes, like the *p53* tumor-suppressor gene, are mutant in a high proportion, but not all, of these cancers. Mutations in certain other tumor-suppressor genes and certain oncogenes show up at relatively high frequencies among these cancers. But Fig. 20.29 tells only part of the story. Mutations in other cancer-related genes not shown in the table appear in only a few of the 178 cases. Furthermore, the genomes of many squamous cell carcinomas harbor chromosomal rearrangements, including deletions of tumor-suppressor genes and amplifications of oncogenes, that are not illustrated.

One of the important lessons from the cancer landscape revealed by these large-scale sequencing projects is that specific cancers affecting different tissues may share more underlying mutational similarities than two cancers of the same tissue. For example, the pattern of mutation in certain breast cancers resembles many ovarian cancers more than other types of breast cancers. These findings mean that drugs designed to treat one type of cancer might sometimes be very effective against cancer of a different organ. In fact, the case study presented previously illustrates precisely this point. Yervoy® was developed as an agent against skin cancers, yet it had dramatic (if temporary) effects against Mrs. McDaniel's T cell lymphoma.

This is hopeful news for the future. The number of oncogenes and tumor-suppressor genes is finite, and mutations in a relative handful (some dozens) of genes are apparently responsible for many kinds of cancer that were thought to be disparate. Targeted chemotherapies like Gleevec®, Herceptin®, and Yervoy®, if they are used appropriately in patients selected by the profile of mutations in their tumors, might thus be much more useful than previously imagined.

essential concepts

- Several anticancer drugs target mutant proteins encoded by oncogenes.
- Other drugs exploit the fact that many cancer cells are particularly sensitive to DNA damage because of the absence of checkpoints or DNA repair proteins.
- Whole-genome sequencing of cancer cells helps pinpoint potential treatments directed against the particular set of mutations in individual tumors.

WHAT'S NEXT

The Tools of Genetics Box in this chapter demonstrated that the gene encoding CDK1, one of the most important cell-cycle control proteins, is highly conserved in organisms as different as yeast, frogs, and humans. In Chapter 19, you saw that the *eyeless/Pax-6/Aniridia* gene, whose product is a master regulator of eye development, is also tightly conserved in many diverse taxonomic groups. In both cases, the evidence for gene conservation was not restricted only to similarities in nucleotide pairs and amino acids, but it also included the results of swapping experiments showing that a gene of one species can substitute functionally for the corresponding gene of a different species.

Despite the tight conservation of these two genes, yeast, frogs, and humans are undoubtedly very different kinds of organisms exhibiting highly divergent morphologies and behaviors. During the course of evolution, some genes must have changed much more than did *Cdk1* or *eyeless/Pax-6/Aniridia.* Evolution creates and then preserves genetic solutions to the problems or opportunities organisms find in their environments, but evolution also tinkers with these solutions to produce novel outcomes.

Evolution occurs not at the level of individual genes in individual organisms, but rather in the context of populations that change over many generations and phenotypes determined by the interactions of many genes. To understand the evolutionary forces propelling the genetic changes that ultimately create different species, we thus shift our focus in the next section of the book from investigating the activity of one gene in one organism to the study of gene transmission in whole populations over long periods of time (Chapter 21) and to the analysis of complex traits governed by multiple genes (Chapter 22).

SOLVED PROBLEMS

I. A common way to verify that particular genes are oncogenes or tumor-suppressor genes is to use genetic engineering to alter mouse genomes and determine whether the engineered mice develop tumors and die of cancer. The two figures that follow show the results of such experiments. In experiment 1, mice have a single transgenic copy of gene *A* or gene *B*, or a single transgenic copy of both genes. In experiment 2, mice are homozygous or heterozygous for knockouts of gene *C*.

a. Of genes *A, B,* and *C,* which ones are oncogenes and which are tumor-suppressor genes? Explain your reasoning.

b. What can you infer from the result that mice doubly transgenic for *A* and *B* are particularly susceptible to tumors?

c. Why do the mice that are heterozygous for the *C* gene knockout display lower survival than the C^+/C^+ controls, but much better survival than homozygotes for the knockout?

d. How can homozygous knockout mice survive past birth?

e. What would you predict about the progeny of two mice, both of which carried one copy of the *A* transgene? (*Note:* Female mice become sexually mature at about six weeks after birth, males at eight weeks.)

f. What would you predict about the progeny of two mice, both of which were heterozygous for the *C* gene knockout?

Experiment 1

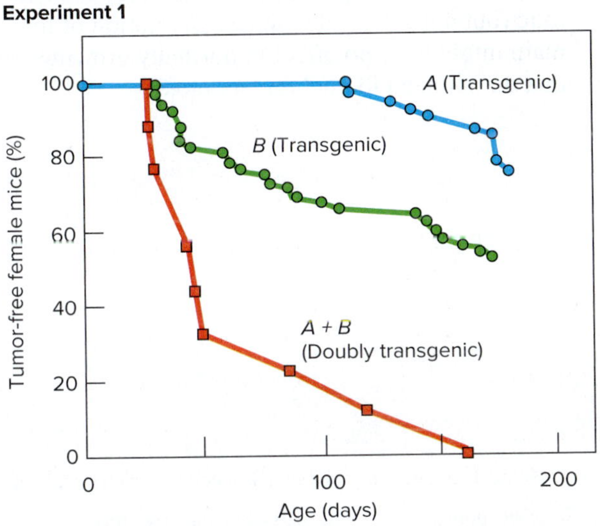

Experiment 2

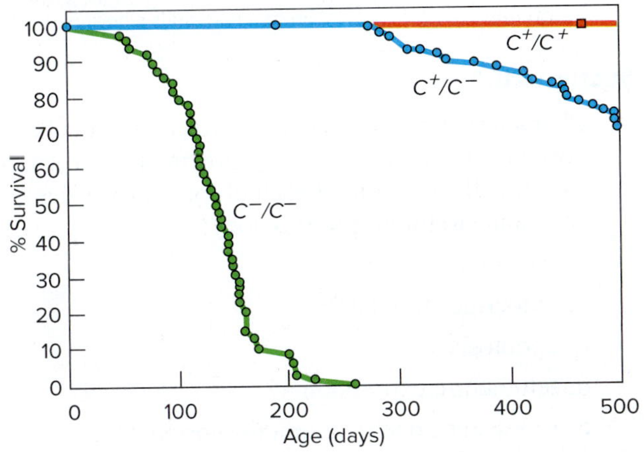

Answer

Answering this question requires you to know that the mutations that convert proto-oncogenes into oncogenes have dominant effects in promoting cancer, while cancer-associated mutations in tumor-suppressor genes are recessive at the cellular level but dominant at the level of the organism.

a. Genes *A* and *B* are oncogenes. A single transgenic copy of either gene promotes tumor growth, so the transgenes have dominant effects. Gene *C* is a tumor-suppressor gene. A predisposition to cancer is particularly obvious when both wild-type copies of gene *C* are knocked out.

b. The fact that the two transgenic oncogenes have synergistic effects is consistent with the idea that the generation of cancer involves the accumulation of mutations in more than one gene.

c. Recall the *two-hit* idea that cancerous properties arise only when both copies of a tumor-suppressor gene become inactivated. Homozygotes for the knockout would have no functional copies and therefore would develop cancer relatively early in life. Heterozygotes for the knockout would be born with one remaining copy of C^+, but these heterozygotes would eventually begin to die of cancer after a second hit inactivates that copy in one or more cells. Note that this latter result indicates that inheritance of a single knockout allele predisposes to cancer in a dominant fashion at the level of the organism.

d. Homozygous knockout mice survive past birth because mutations must accumulate in several oncogenes and tumor-suppressor genes to produce cancer, not just in this one tumor-suppressor gene. Note that this tumor-suppressor gene cannot be necessary for the survival of cells.

e. One-quarter of the progeny would not inherit any copy of the transgene, so they would develop no or very few tumors over their normal life spans. One-half the progeny inherit one copy of the transgene, so they would develop tumors at about the same rate as their parents (the blue line in experiment 1). One-quarter of the progeny would be homozygous for the transgene. These animals might show the same rate of tumorigenesis as their parents, or the rate might be faster, depending on the transgene's molecular nature.

f. One-quarter of the progeny would not inherit any copy of the knockout, so they would survive normally. One-half would inherit one copy of the knockout, so their survivability should resemble the blue line in experiment 2. The remaining one-quarter would be homozygotes for the transgene, so they would die at a rate similar to that shown by the green line in experiment 2.

II. The text described several oncogenic viruses with RNA genomes (retroviruses). Some cancer-causing viruses instead have DNA genomes; one of these viruses is called *SV40.* Mouse tissue culture cells infected with the SV40 virus lose normal growth control and become transformed. If transformed cells are transferred into mice, they grow into tumors. The SV40 protein responsible for this transformation is called *T antigen,* and it has been found to associate with the cellular protein p53. If the *p53* gene fused to a high-level-expression (strong) promoter is transfected into tissue culture cells, the cells are no longer transformed by infection with SV40.

a. Propose a hypothesis to explain how the high expression of *p53* saves the cells from transformation by T antigen.

b. You have decided to examine the functional domains of the p53 protein by mutagenizing the cDNA, fusing it to the strong promoter, and transfecting it into cells. Each mutation alters an amino acid that is crucial for the activity of one functional domain of the p53 protein. Results are shown in the following table. How would you explain the effects of mutations 1 and 2 on p53 function?

c. What is the effect of mutation 3 on p53 function?

	Morphology	
p53 construct	**Noninfected cells**	**SV40-infected cells**
None	Normal	Transformed
Wild type	Normal	Normal
Mutation 1	Normal	Normal
Mutation 2	Normal	Normal
Mutation 3	Normal	Transformed

Answer

a. p53 protein is inactivated when it is bound by T antigen. By supplying excess p53 from the strong promoter, there is now enough p53 protein in the cell to bind to all the T antigen and still have enough unbound p53 to regulate the cell cycle. Thus, the effect of the T antigen is minimized.

b. Mutants 1 and 2 are expressing large amounts of altered p53 protein, but they no longer are affected by T antigen. Mutations 1 and 2 could therefore be loss-of-function mutations that block the ability of p53 to bind T antigen. Such mutant p53 proteins could still regulate the cell cycle normally even in T antigen's presence. Another possibility is that mutant proteins 1 and 2 cannot function in cell cycle control but can bind sufficient T antigen to allow the endogenous p53 to function.

c. Mutant 3 cannot rescue cells from the oncogenic effect of T antigen, so this mutation must affect a functional domain other than those that bind T antigen or that allow p53 to serve as a transcription factor for cell-cycle genes. As one of several possibilities, p53 has a domain that specifically helps it regulate apoptotic (but not cell-cycle) genes. Disruption of this domain might have no effect in normally growing cells that do not need to undergo apoptosis.

PROBLEMS

Vocabulary

1. For each of the terms in the left column, choose the best matching phrase in the right column.

a. mitogenic growth factor	1. mutations in these genes are dominant for cancer formation
b. tumor-suppressor genes	2. programmed cell death
c. cyclin-dependent protein kinases	3. series of steps by which a message is transmitted
d. apoptosis	4. proteins that are active cyclically during the cell cycle
e. oncogenes	5. control progress in the cell cycle in response to DNA damage
f. growth factor receptor	6. mutations in these genes are recessive at the cellular level for cancer formation
g. signal transduction	7. signals a cell to leave G_0 and enter G_1
h. checkpoints	8. cell-cycle enzymes that phosphorylate proteins
i. cyclins	9. protein that binds a hormone

Section 20.1

2. Characterize the differences between tumor cells and normal cells in terms of the following properties. In cancer cells, how might each of these properties contribute to tumor progression?

a. contact inhibition

b. autocrine stimulation

c. apoptosis

d. telomerase expression

e. senescence due to telomere shortening

f. genomic stability
g. angiogenesis
h. metastasis
i. susceptibility to immune surveillance

Section 20.2

3. The incidence of colon cancer in the United States is 30 times higher than it is in India. Differences in diet and/or genetic differences between the two populations may contribute to these statistics. How would you assess the role of each of these factors?
4. Some germ-line mutations predispose individuals to cancer, yet often environmental factors (chemicals, exposure to radiation) are considered major risks for developing cancer. Do these views of the cause of cancer conflict, or can they be reconciled?
5. A carcinogenic compound is placed on the skin of inbred laboratory mice. In many of these mice, skin tumors develop at the site of exposure, but only months after the chemical is no longer detectable. Why don't all the mice develop tumors, and why don't the tumors appear much sooner?
6. You have decided to study genetic factors associated with colon cancer. An extended family from Morocco in which the disease presents itself in a large percentage of family members at a very early age has come to your attention. (The pedigree is shown in the accompanying diagram.) In this family, individuals either get colon cancer before the age of 16, or they don't get it at all.

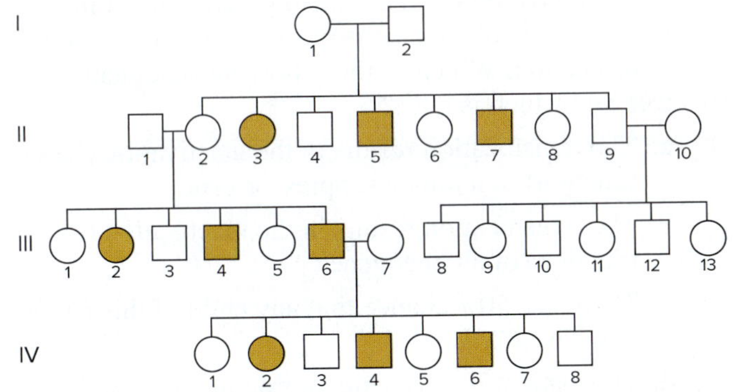

   a. Based on the information you have been given, what evidence, if any, suggests an inherited contribution to the development of this disease?
   b. You decide to take a medical history of all of the 33 people indicated in the pedigree and discover that a very large percentage drink a special coffee on a daily basis, while the others do not. The only ones who don't drink the coffee are individuals numbered I-1, II-2, II-4, II-9, III-7, III-13, IV-1, and IV-3. Could the drinking of this special coffee possibly play a role in colon cancer? Explain your answer.
7. B cells are specialized blood cells that secrete antibodies. Normally, human blood has millions of different types of B cells making millions of different kinds of antibody molecules. This variety occurs because, as described in the Fast Forward Box in Chapter 13 entitled *Programmed DNA Rearrangements and the Immune System*, antibody genes undergo rearrangements in the precursors of B cells. Individual B cell precursors rearrange their antibody genes in different ways.

   In the blood of patients with cancers called *B cell lymphomas,* almost all of the antibody molecules are all of one type, but this single type of antibody is different in different lymphoma patients.
   a. Based on this information, provide a brief description of the genesis of B cell lymphomas, focusing on the cells that are overproliferating.
   b. How does the nature of B cell lymphomas provide support for the clonal theory of cancer shown in Fig. 20.6?

Section 20.3

8. Molecules outside and inside the cell regulate the cell cycle, making it start or stop.
   a. What is an example of an external molecule that regulates the cell cycle?
   b. What is an example of a molecule inside the cell that is involved in cell-cycle regulation?
9. Put the following steps in the correct ordered sequence.
   a. kinase cascade
   b. activation of a transcription factor
   c. hormone binds transmembrane receptor
   d. expression of target genes in the nucleus
   e. Ras molecular switch
10. a. Would you expect a cell to continue or to stop dividing at a nonpermissive high temperature if it is a temperature-sensitive *Ras* mutant whose protein product is fixed in the GTP-bound form at nonpermissive temperature?
    b. What would you expect if you had a temperature-sensitive mutant in which the Ras protein stays in the GDP-bound form at high temperature?
11. Two different protein complexes called *SCF* and *APC* covalently add a small polypeptide called *ubiquitin* to cyclin proteins. The addition of ubiquitin to a protein targets that protein to be degraded by another protein complex called the *proteasome*. The SCF complex is activated during S phase, and the APC complex is activated during M phase.

a. To which cyclins (A, B, D, or E in Fig. 20.14) do you think SCF couples ubiquitin? What about the APC?

b. How might cells activate SCF and APC only at the correct times?

12. One of the hallmarks of mitotic anaphase is the separation of sister chromatids. A protein complex called *cohesin* holds sister chromatids together, as described in Fig. 12.25. Based on the answer you have just given for Problem 11, propose a mechanism that would allow sister chromatids to separate during anaphase. How might your proposed mechanism also explain the checkpoint operating in M phase that prevents sister chromatid separation until all the chromosomes have connected properly to the mitotic spindle?

13. Concerning the Tools of Genetics Box *Analysis of Cell-Cycle Mutants in Yeast:*

a. Describe how you would use replica plating of mutagenized, haploid yeast cells to identify temperature-sensitive (*ts*) mutations in essential genes needed for yeast growth and survival.

b. Among the many *ts* mutations you found in part (a), how would you distinguish mutations in genes needed for cell-cycle progression from those in genes needed for other aspects of the life of yeasts?

c. If you had a large collection of yeast cell-cycle mutants, how would you determine which of the mutations are in the same gene and which are in different genes?

d. Figures A and B in the Tools of Genetics Box show a culture of a single yeast *ts* cell-cycle mutant. The two figures show the same petri plate of cells examined at different times: Fig. A before the shift to restrictive temperature, and Fig. B after the temperature shift. Cells with small buds in Fig. A arrest as a single large-budded cell in Fig. B (the *yellow* arrows point to an example). In contrast, cells with large buds in Fig. A arrest as two large-budded cells in Fig. B (*red* arrows). What do these observations tell you about when during the cell cycle the protein product of the gene in question normally functions?

e. Describe in detail an experiment to show that the human gene for the cyclin-dependent kinase CDK1 can replace the function of the homologous gene in yeast.

Section 20.4

14. Are genome and karyotype instabilities consequences or causes of cancer?

15. Which one of the following events is unlikely to be associated with cancer?

a. mutation of a cellular proto-oncogene in a normal diploid cell

b. a chromosomal translocation with a breakpoint near a cellular proto-oncogene

c. deletion of a cellular proto-oncogene

d. mitotic nondisjunction in a cell carrying a deletion of a tumor-suppressor gene

e. incorporation of a cellular oncogene into a retrovirus chromosome

16. Why don't all loss-of-function mutations that are recessive at the cellular level behave as dominants at the organismal level? Is this property restricted to tumor-suppressor gene mutations?

17. *Chromothripsis* is a rare phenomenon, first discovered in cancer cells, where a single chromosome "shatters" into many fragments and is reassembled in a rearranged form by the DNA repair machinery. (The underlying mechanism causing the shattering is not understood.) Approximately 2% of cancers contain cells with a shattered chromosome. Explain how chromothripsis could contribute to cancer.

18. The chromosome 9/22 translocation associated with CML (chronic myelogenous leukemia) is called the *Philadelphia chromosome* after the city in which its cancer association was first discovered in 1960. People with CML do not inherit this translocation—it occurs in somatic cells. Why do you think that this particular translocation that fuses the *bcr* and *abl* genes happens independently in the somatic cells of many different people?

19. A female patient 19 years old, whose symptoms are anemia and internal bleeding due to a massive buildup of leukemic white blood cells, is diagnosed with chronic myelogenous leukemia (CML). Karyotype analysis shows that the leukemic cells of this patient are heterozygous for a reciprocal translocation involving chromosomes 9 and 22. However, none of the normal, nonleukemic cells of this patient contain the translocation. Which of the following statements is true and which is false?

a. The translocation results in the inactivation (loss of function) of a tumor-suppressor gene.

b. The translocation results in the inactivation (loss of function) of an oncogene.

c. There is a 50% chance that any child of this patient will have CML.

d. This patient is a somatic mosaic in terms of the karyotype.

e. DNA extracted from leukemic cells of this patient, if taken up by normal mouse tissue culture cells, could potentially transform the mouse cells into cells capable of causing tumors.

f. The normal function of the affected tumor-suppressor gene or proto-oncogene at the translocation breakpoint could potentially block the function of the cyclin proteins that drive the cell cycle forward.

g. Two rare events must have occurred to disrupt both copies of the tumor-suppressor gene or proto-oncogene at the translocation breakpoint in the leukemic cells.

h. A possible treatment of the leukemia would involve a drug that would turn on the expression of the tumor-suppressor gene or oncogene at the translocation breakpoint in the leukemic cells.

20. Describe a molecular test to determine if chemotherapy given to the patient described in Problem 19 would be completely successful. That is, devise a method to make sure that the patient's blood would be free of leukemic cells. Be as specific as possible in describing the reagents you would need for the test, how you would perform it, and what the different results would show.

21. A generic signaling cascade is shown in the following figure. A growth factor (GF) binds to a growth factor receptor, activating the kinase function of an intracellular domain of the growth factor receptor. One substrate of the growth factor receptor kinase is another kinase, kinase A, that has enzymatic activity only when it is itself phosphorylated by the GF receptor kinase. Activated kinase A adds phosphate to a transcription factor. When it is unphosphorylated, the transcription factor is inactive and stays in the cytoplasm. When it is phosphorylated by kinase A, the transcription factor moves into the nucleus and helps turn on the transcription of a *mitosis factor* gene whose product stimulates cells to divide.

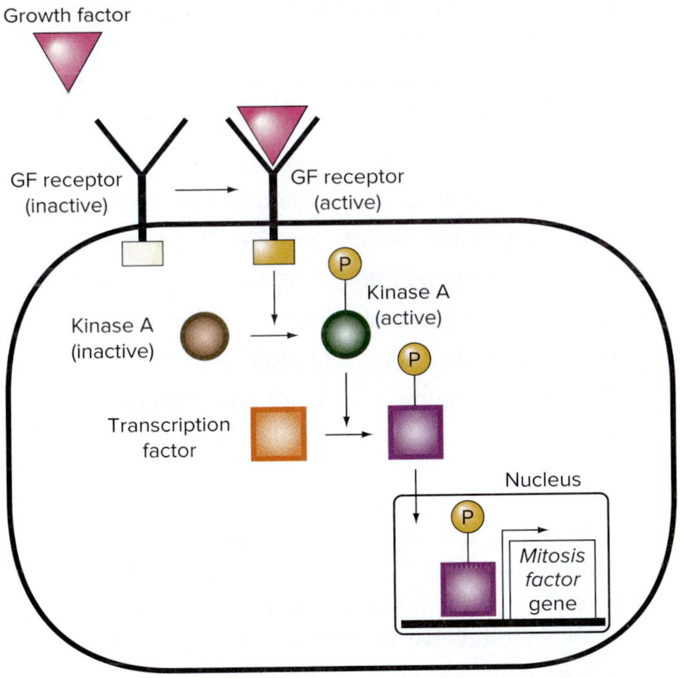

a. The following list contains the names of the genes encoding the corresponding proteins. Which of these could potentially act as a proto-oncogene? Which might be a tumor-suppressor gene?
   i. *growth factor*
   ii. *growth factor receptor*
   iii. *kinase A*
   iv. *transcription factor*
   v. *mitosis factor*

Though it is not pictured, the cell in the figure also has a *phosphatase,* an enzyme that removes phosphates from proteins—in this case, from the transcription factor. This phosphatase is itself regulated by kinase A.

b. What would you expect to be the effect when kinase A adds a phosphate group to the phosphatase? Would this activate the phosphatase enzyme or inhibit it? Explain.

c. Is the *phosphatase* gene likely to be a proto-oncogene or a tumor-suppressor gene or neither?

d. Several mutations are listed below. For each, indicate whether the mutation would lead to excessive cell growth or decreased cell growth if the cell were either homozygous for the mutation, or heterozygous for the mutation and a wild-type allele. Assume that 50% of the normal activity of all these genes is sufficient for normal cell growth.
   i. A null mutation in the *phosphatase* gene
   ii. A null mutation in the *transcription factor* gene
   iii. A null mutation in the *kinase A* gene
   iv. A null mutation in the *growth factor receptor* gene
   v. A mutation that causes production of a constitutively active growth factor receptor whose kinase function is active even in the absence of the growth factor
   vi. A mutation that causes production of a constitutively active kinase A
   vii. A reciprocal translocation that places the *transcription factor* gene downstream of a strong enhancer
   viii. A mutation that prevents phosphorylation of the phosphatase enzyme
   ix. A mutation that causes the production of a phosphatase that acts as if it is always phosphorylated

22. Neurofibromatosis type 1 (NF1; also known as von Recklinghausen disease) is an inherited dominant disorder. The phenotype (see Fig. 11.19) usually involves the production of many skin neurofibromas (benign tumors of the fibrous cells that cover the nerves).

a. Is it likely that *NF1* is a tumor-suppressor gene or an oncogene?

b. Are the *NF1* neurofibromatosis-causing mutations that are inherited by affected children from affected parents likely to be loss-of-function or gain-of-function mutations?

c. Neurofibromin, the protein product of *NF1*, has been found to be associated with the Ras protein.

Ras is involved in the transduction of extracellular signals from growth factors. The active form of Ras (the form initiating the signal transduction cascade causing proliferation) is complexed with GTP; the inactive form of Ras is complexed with GDP. Would the wild-type neurofibromin protein favor the formation of Ras–GTP or Ras–GDP?

d. Which of the following events in a normal cell from an individual inheriting a neurofibromatosis-causing allele could cause the descendants of that cell to grow into a neurofibroma?

   i. A second point mutation in the allele of *NF1* inherited from the affected parent
   ii. A point mutation in the allele of *NF1* inherited from the normal parent
   iii. A large deletion that removes the *NF1* gene from the chromosome inherited from the affected parent
   iv. A large deletion that removes the *NF1* gene from the chromosome inherited from the normal parent
   v. Mitotic chromosomal nondisjunction or chromosome loss
   vi. Mitotic recombination in the region between the *NF1* gene and the centromere of the chromosome carrying *NF1*
   vii. Mitotic recombination in the region between the *NF1* gene and the telomere of the chromosome carrying *NF1*

e. A much rarer form of neurofibromatosis exists called *segmental neurofibromatosis.* In this form of the disease, neither parent of the patient has any clinical sign of the disease. The tumors in the patient are restricted to one part of the body, like the right leg. Suggest an explanation for the genesis of segmental neurofibromatosis that clarifies why it is restricted to one part of the body.

23. Families with germ-line *BRCA1* or *BRCA2* loss-of-function mutations usually display *Hereditary Breast and Ovarian Cancer (HBOC).* The accompanying diagram shows a *BRCA2* pedigree.

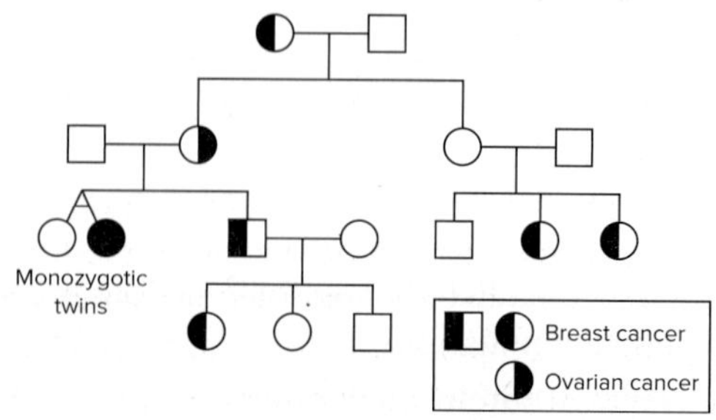

a. *BRCA1* and *BRCA2* are tumor-suppressor genes; cells homozygous for loss-of-function mutations in either gene can become cancerous due to a loss of DNA repair machinery. Explain why the pedigree shows a dominant inheritance pattern of HBOC.

b. Can you tell from the pedigree diagram whether or not HBOC is completely penetrant?

c. Does HBOC show varying expressivity?

d. Explain your answers to parts (b) and (c) in terms of current models for the origins of cancers.

24. The text explained that retroviruses can cause cancer. Some viruses with DNA genomes can also cause cancer. For example, herpes papilloma virus (HPV) causes cervical cancer. The HPV genome encodes a protein called E6 that interferes with p53 function, and another protein called E7 that inhibits the function of Rb protein. Explain how HPV causes cancer. Are the viral E6 and E7 protein functions more similar to oncogenes or tumor suppressors?

Section 20.5

25. *Hepatocellular carcinoma* is the most frequent form of liver cancer. In a patient with heritable hepatocellular carcinoma, formation of the tumor was associated with eight genetic alterations affecting two different oncogenes and three different tumor-suppressor genes. These alterations are:

   i. Mitotic recombination
   ii. A deletion of a chromosomal region
   iii. Trisomy
   iv. A duplication of a chromosomal region
   v. Uniparental disomy (see Fig. 20.24)
   vi A point mutation
   vii. Another point mutation
   viii. Yet another point mutation

For parts a–c below, supply all possible correct answers from the preceding list. Remember that the majority of point mutations are loss-of-function mutations.

a. Which of the mutations from the preceding list is likely to affect a proto-oncogene?

b. Which of the mutations from the preceding list is likely to involve a tumor-suppressor gene?

c. Which of the mutations from the preceding list involves *copy-neutral loss-of-heterozygosity* (that is, a loss-of-heterozygosity in which the genomes of the cancerous cells still have two copies of the gene in question, whether or not those copies are functional)?

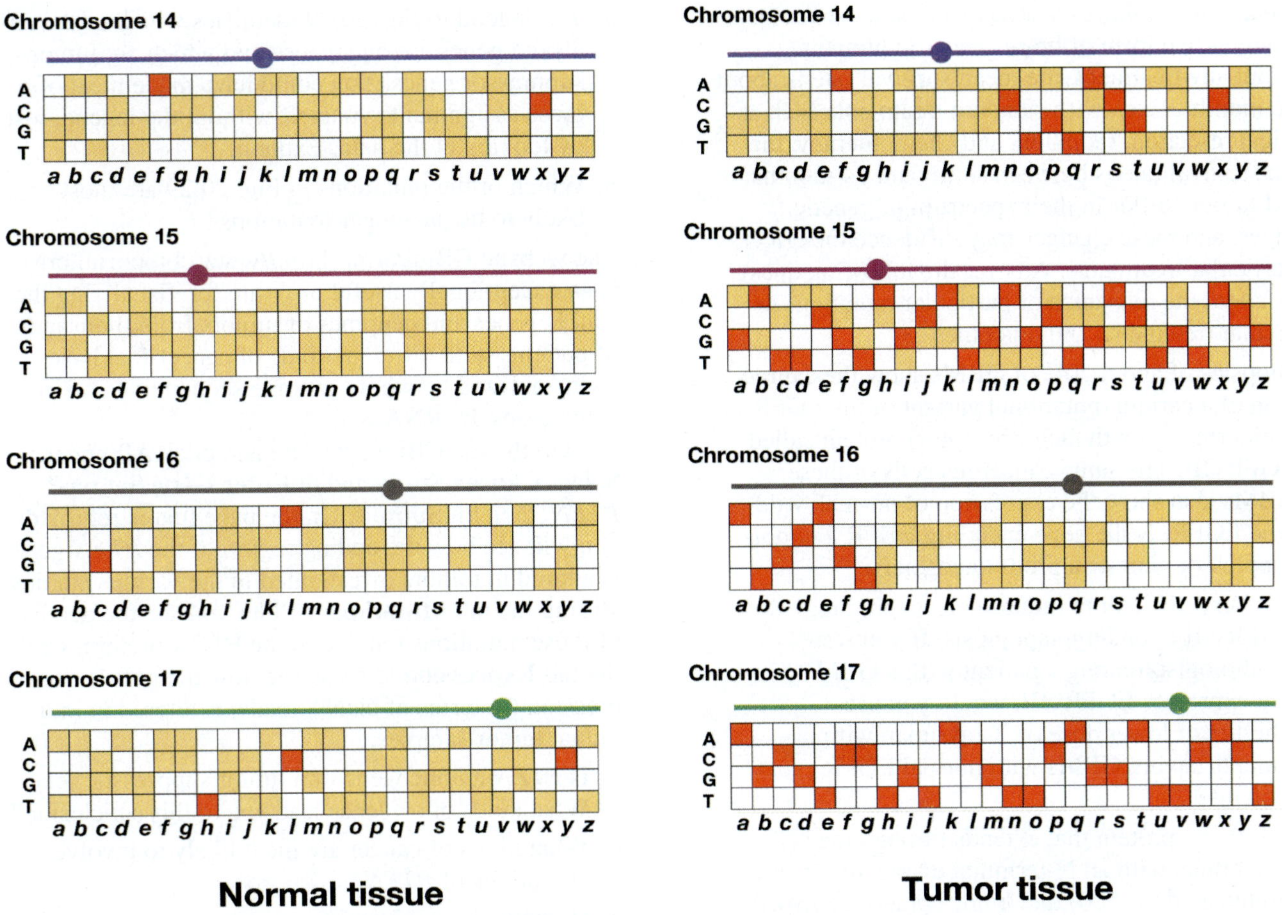

Genomic DNA is prepared from normal white blood cells and from a biopsy of the tumor in this patient. These genomic DNAs are prepared as fluorescent probes that are each hybridized to an ASO microarray of polymorphisms in the human genome (review Figs. 11.16 and 11.17). The results for SNPs *a–z* on chromosomes 14, 15, 16, and 17 are shown in the accompanying figure. *Red* and *orange* represent different levels of fluorescence.

d. Based on the microarray data, provide the most accurate localization of the first five types of genetic alterations in the list (i–v). For example, if an alteration involves markers *a–e* of chromosome 15, write 15*a–e*.

e. As precisely as possible, indicate the location of the mitotic recombination event involved in the genesis of this cancer.

f. If these data allow you to map any of the three cancer-promoting point mutations, provide the most accurate mutation location(s) possible.

g. Of all the genetic alterations i–viii, for which one do you see clear-cut evidence that the mutation or other event was inherited from a parent of the patient?

h. For a tumor-suppressor gene to play a role in cancer, normally both of the copies in the tumor cells must be nonfunctional. For each of the three tumor-suppressor genes contributing to the cancer in this patient, provide a scenario explaining which two *hits* (i–viii in the list, with vi–viii equivalent) could be responsible, the order in which the hits must have occurred, and whether the hits in question could be inherited or could have occurred somatically.

26. Suppose that instead of microarrays, you analyzed the normal and cancerous tissue from the patient described in Problem 25 by whole-genome sequencing. What evidence would you find in the whole-genome sequence data for the existence and location of the eight genetic alterations on the list?

27. Suppose that you also analyzed the normal and cancerous tissue from the patient described in Problems 25 and 26 by *deep sequencing* of the mRNAs in the cancerous and normal cells of the patient (RNA-Seq). What evidence could you obtain for the involvement in tumorigenesis of a novel oncogene or tumor-suppressor gene that would be very difficult to find from the whole-genome sequence?

28. *Glioblastoma multiforme (GBM)* is the most common and aggressive form of brain cancer in humans. Without any treatment, the mean survival rate is about three months. Even with standard treatments such as surgical resection, radiation, and chemotherapy, the mean survival rate is between seven and 14 months. GBM tumors differ in their spectrum of genetic changes, and these changes may influence the effect of particular treatments. Answer the following questions about the relevance of particular mutations to particular treatments and outcomes.
   a. Biopsies of about 20% of GBMs show the expression of a certain mutational variant of the EGFR (epidermal growth factor receptor) protein called EGFRvIII. The same cancerous cells of these GBMs also show the expression of normal, wild-type EGFR. Is the gene encoding EGFR a tumor-suppressor gene or a proto-oncogene?
   b. It is very difficult to induce cells expressing EGFRvIII to undergo apoptosis. If you were a radiologist treating a patient with a GBM that expresses EGFRvIII, would you use a higher or lower dose of X-rays than with patients having GBMs with normal EGFR proteins?
   c. EGFR is a protein that extends through the cell membrane, with an N-terminal extracellular part (amino acids 1–500) that binds epidermal growth factor (EGF) and an intracellular C-terminal kinase part (amino acids 501–1000) that is normally activated when EGF binds to EGFR. The active kinase phosphorylates (adds phosphate groups to) other proteins, setting off a signal transduction cascade that promotes cell growth and division. EGFRvIII is a deletion that removes amino acids 6 through 273 of the EGFR protein. How might this mutant protein contribute to cancer?
   d. Iressa™ is a drug that blocks the kinase activity of EGFR. Would you expect Iressa™ to be a potential treatment for GBMs expressing EGFRvIII, or would you instead anticipate the drug would make the tumors grow faster?
   e. Cisplatin is a platinum compound that binds to DNA and damages it, eventually leading to cell death. *ERCC1* is a gene that encodes a DNA repair protein. GBMs are found that show much higher levels of transcription of *ERCC1* than normal. Would you treat patients having such GBMs with higher or lower doses of cisplatin than patients whose tumors have normal amounts of *ERCC1* mRNAs?

29. a. The legend to Fig. 20.29 identifies which of the analyzed genes are oncogenes and which are tumor-suppressor genes. You could have made most of these assignments yourself without the legend, just by looking at the data. Explain.
   b. Which of the mutations in Fig. 20.29 are most likely to be passenger mutations?

30. The website CBioPortal (http://www.cbioportal.org) is an exceptionally useful program for visualizing the cancer genes and genomes of tumors from thousands of patients with different kinds of cancer that have been analyzed by whole genome sequencing and in some cases, by RNA-Seq.

    Go the the CBioPortal site and click **All** under **Select Cancer Study** and in **Enter Gene Set** type *PTEN*, then hit **Submit**. On the page that is returned you will see how the coding region of the *PTEN* gene is altered in tumors investigated in the various studies. Hitting the tab **Mutations** will let you see the details of these mutations relative to the PTEN protein, while the tab **Expression** lets you see how the gene's expression (in terms of cDNA reads) is altered in individual tumor samples.
   a. Is *PTEN* an oncogene or a tumor suppressor gene? What kinds of evidence lead you to this conclusion?
   b. What kinds of cancer are most likely to involve alterations of *PTEN*?
   c. How would you identify patients whose tumor cells are particularly likely to have a somatic mutation in the *PTEN* gene that is outside of the coding region but nonetheless contributes to cancer by affecting the gene's regulation?

Now return to the CBioPortal home page. Again, select **All** under **Select Cancer Study**, but this time type *ERBB2* under **Enter Gene Set** and then hit **Submit**.

   d. Is *ERBB2* an oncogene or a tumor-suppressor gene? What kinds of evidence lead you to this conclusion?
   e. Are any kinds of listed mutations in the *ERBB2* gene almost certainly passenger mutations as opposed to driver mutations? What does it mean to be a passenger mutation?
   f. If you were looking for regulatory mutations in the *ERBB2* gene that are not in the coding sequence but that contribute to cancer, what attributes would you look for under the **Expression** tag?
   g. In comparing your results with the *PTEN* and *ERBB2* genes, how informative are missense mutations in these genes with respect to possible contributions of such mutations to cancer phenotypes?

chapter 21

Variation and Selection in Populations

Populations are groups of families.

"Milk is for babies. When you grow up you have to drink beer." Arnold Schwarzenegger

chapter outline

MILK IS ADVERTISED in the United States as an excellent source of protein and calcium. However, some adolescents and adults have uncomfortable symptoms soon after drinking milk, including diarrhea, nausea, and abdominal cramps. These unpleasant reactions are often signs of a deficiency for the enzyme lactase in the lining of the small intestine.

In our primate relatives and in other mammals, lactase is expressed only in newborns so that they can obtain nutrition from their mother's breast milk by breaking down the sugar lactose. Transcription of the lactase-encoding gene shuts down after weaning in these species. From their primate ancestors, early humans inherited *lactase* genes that behave the same way. Modern humans in eastern Asia, parts of Africa, and of indigenous heritage in North and South America and Australia still have *lactase* genes that are turned off in later childhood. These people are thus lactose intolerant as adults, and their diets include little or no milk or other dairy products.

So why can many people in other ethnic groups drink milk as adults? The reason is that chance mutations in human populations in both Europe and in pastoral regions of sub-Saharan Africa led to the independent occurrence of *lactase* alleles that remain expressed during adulthood. In populations that raised domesticated cattle, the ability for adults to derive nutrition from milk likely provided a survival and/or reproductive advantage to individuals with these mutations (**Fig. 21.1a**). This advantage led to a rapid increase in the frequencies of variants that continue *lactase* expression into adulthood, so the frequency of these alleles is now over 90% in European and central sub-Saharan African populations (**Fig. 21.1b**).

Population genetics is the study of the transmission between generations of genetic variations such as those that determine lactose tolerance or intolerance in *populations:* groups of families of the same species living in the same time and place. The logic of population genetics is a simple extension of what we have seen previously in this book, where we followed genetic variants as they passed from parent to offspring in a single generation of a cross. What is new in population genetics is that we are following variants as they change in frequency over time and space within collections of

Figure 21.1 Worldwide variation in lactose tolerance among human populations. (a) Rock paintings from the Tassili n'Ajjer mountain range in the Sahara Desert of Africa, depicting early humans herding domesticated cattle during a period (3000–1900 BC) when the region was a savannah rather than today's desert. **(b)** Present-day populations shown in *green* have high percentages of lactose-tolerant individuals; for populations shown in *red*, most people are lactose intolerant. The earliest human populations were lactose intolerant. Lactose tolerance results from either of two independent mutations that provided a selective advantage to individuals who could gain nutrition from the milk from domesticated animals. One of these mutations occurred in Europe or western Asia, while the second was in sub-Saharan Africa (*green circles*).

(a) Cattle domestication early in human history

(b) Variation in lactose intolerance among human populations

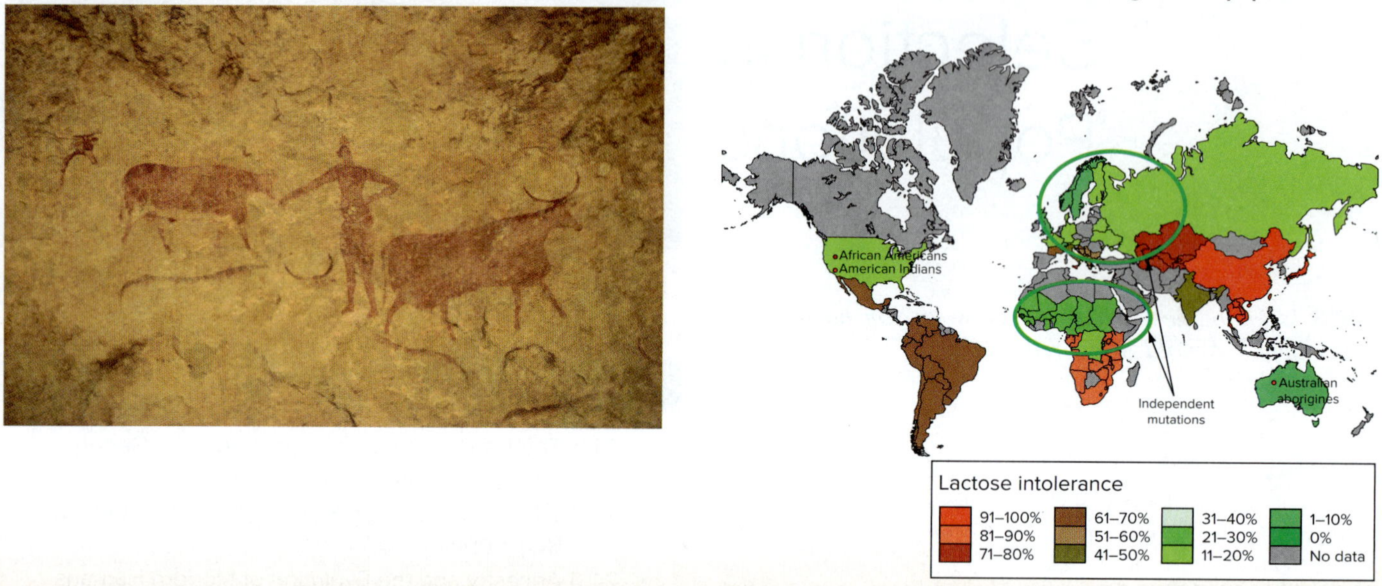

individuals who may be related to each other only distantly. *Population geneticists* formulate models of how allele and genotype frequencies can change, and then they compare the resulting predictions to patterns of variation found in real populations.

Two themes underlie the material presented in this chapter. First, because populations are simply groups of families, population geneticists extend Mendel's basic principles of trait transmission in families to make predictions about changes in genetic variation in populations. Second, population genetics is not simply a collection of abstractions. Instead, it is an important tool that allows scientists to gain insights into biological function, evolutionary mechanisms, and even human history.

21.1 The Hardy-Weinberg Law: Predicting Genetic Variation in "Ideal" Populations

learning objectives

1. Calculate the allele frequencies from the genotype frequencies in a given population.
2. Given the allele frequencies, determine the genotypic proportions predicted for a locus in a population at Hardy-Weinberg equilibrium.
3. Describe how forensic scientists determine the likelihood that the DNA profile of a random person in the population would match that of a sample found at a crime scene.

Geoffrey H. Hardy and Wilhelm Weinberg realized as early as 1908 that the easiest way to understand the forces shaping the patterns of genetic variation within and between populations in nature is to start with a simple, randomly mating "ideal" population that fulfills several assumptions. With this imaginary population as a starting point, geneticists could then examine what happens when each of the assumptions is violated in turn. In this way, scientists can build realistic models of how genetic variations change in time and space in natural populations.

Population Geneticists Measure Frequencies that Describe Populations

To follow Hardy and Weinberg's reasoning, we first need to define some terms that will provide us with clear-cut measures of variation. A **population** is a group of individuals of a single species living in the same time and place; if the organism is sexual, the individuals in the population must be capable of mating among themselves. Examples would be all the white-tailed deer on Angel Island in San Francisco Bay in 1990 or all the rock cod at the mouth of the bay. The sum total of all alleles carried in all members of a population is that population's **gene pool.** Individuals carry at most two allelic copies of each gene. Thus, considering an autosomal gene in a population of *N* individuals, the gene pool is made up of 2*N* allelic copies of the gene. We hardly ever examine the alleles in every individual in a population, but instead we take a **sample:** a finite number of individuals used to make inferences about the population as a whole. Usually, scientists sample individuals randomly without consideration of their genotype or phenotype.

The term **allele** describes a variant at a specific locus, gene, region, or nucleotide position of the genome. If the sample reveals only one allele at a nucleotide position, that site is **monomorphic;** if more than one allele or variant at a site exists, that site is **polymorphic.** Our sample of individuals also allows us to determine the relative frequencies of different variants that are present (or *segregating*) in the population at any moment in time.

Genotype and phenotype frequencies

Genotype frequency is the proportion of total individuals in a population that carry a particular genotype. To determine genotype frequencies, you simply count the number of individuals of each genotype and divide by the total number of individuals in the population (**Fig. 21.2**). For recessive traits such as blue eyes, it is not possible to distinguish between homozygotes for dark eye alleles and heterozygous individuals containing alleles for both dark and blue eyes; both genotypes give rise to individuals with dark eyes. Thus, for such genes, the only way to determine genotype frequencies directly is to use a molecular assay that distinguishes between the different eye color alleles at the DNA level.

Figure 21.2 Describing genotype frequencies and allele frequencies in a population.

Genotype	*AA*	*Aa*	*aa*	Total
Number of individuals	12	4	4	20
Genotype frequency	0.6	0.2	0.2	1.0

Allele	*A*	*a*	Total
Number of chromosomes	28	12	40
Allele frequency	0.7	0.3	1.0

Consider a sample of 20 humans of which 16 individuals have dark eyes, meaning that they are either homozygotes for the *A*, or dominant dark eye allele, or they are heterozygous for *A* and the recessive blue eye allele *a*. The other four people have blue eyes and thus are *aa* homozygotes (Fig. 21.2). The sample has two phenotypes—dark-eyed and blue-eyed individuals—in relative proportions **(phenotype frequencies)** of 16/20 = 80% and 4/20 = 20%, respectively. Suppose that DNA analyses show that of these 20 individuals, 12 are of genotype *AA*, 4 are *Aa*, and 4 are *aa*. In this sample, the *AA* genotype frequency is 12/20 = 0.6; the *Aa* genotype frequency is 4/20 = 0.2; and the *aa* genotype frequency is also 0.2. Note that these three frequencies (0.6 + 0.2 + 0.2) sum to 1, the total of all genotypes at this locus in the sample and, by inference, in the whole population.

Allele frequency

The **allele frequency** is the proportion of gene copies in a population that are of a given allele type. Because each individual in a population has two copies of each chromosome, the total number of gene copies (the gene pool) is two times the number of individuals. Thus, for our hypothetical population of 20 people, 40 copies of an autosomal gene would exist. Of course, both homozygotes and heterozygotes contribute to the frequency of an allele. But homozygotes contribute to the frequency of a particular allele twice, while heterozygotes contribute only once (Fig. 21.2).

To find the frequencies of *A* and *a*, you first use the number of people with each genotype to compute the number of *A* and *a* alleles.

For the *A* alleles,

$$12\ AA \rightarrow 24 \text{ copies of } A$$

$$4\ Aa \rightarrow 4 \text{ copies of } A$$

$$4\ aa \rightarrow 0 \text{ copies of } A$$

The sample contains 24 + 4 + 0 = 28 copies of the *A* allele. Similarly, for the *a* alleles,

$$12\ AA \rightarrow 0 \text{ copies of } a$$

$$4\ Aa \rightarrow 4 \text{ copies of } a$$

$$4\ aa \rightarrow 8 \text{ copies of } a$$

The people in the sample together have 0 + 4 + 8 = 12 copies of the *a* allele.

Next, you add the 28 *A* alleles to the 12 *a* alleles to find the total number of chromosome copies sampled (40); as expected, this is twice the number of people in the sample. Finally, you divide the number of copies of each allele by the total number of gene copies to find the proportion, or frequency, for each allele.

For the A allele, the allele frequency is $28/40 = 0.7$.

For the a allele, the allele frequency is $12/40 = 0.3$.

Note that here again, the frequencies sum to 1.0, representing all the alleles in the gene pool.

An equivalent way to calculate the frequency of any allele is to use the following formulas based on genotype frequencies:

$$\text{Frequency of allele } A = \text{frequency of } AA + (1/2) \text{ frequency of } Aa \quad (21.1)$$

$$\text{Frequency of allele } a = \text{frequency of } aa + (1/2) \text{ frequency of } Aa$$

Substituting the genotype frequencies at the top of Fig. 21.2 into these formulas will yield the same allele frequencies shown at the bottom of the figure that were determined from counting individual gene copies.

The Hardy-Weinberg Law Correlates Allele and Genotype Frequencies

Prior to the twentieth century, when the blending of traits was still considered to be the basis of inheritance, many people thought that recessive traits such as naturally blond or red hair would be destined for extinction in human populations over time: The information for blonde or red hair color would be lost gradually when people with lighter hair color mated with people with darker hair. In 1908, the British mathematician Geoffrey H. Hardy explored this claim in light of the newly rediscovered principles of Mendelian segregation. Hardy showed that if certain assumptions were met, allele frequencies, genotype frequencies, and phenotype frequencies would instead remain constant over time and between generations. A German physician, Wilhelm Weinberg, reached the same conclusion in the same year.

Assumptions of the Hardy-Weinberg law

The simplifying assumptions that allowed Hardy and Weinberg to formulate this principle involve both the nature of the population and the nature of the genetic variation under investigation:

1. The population is composed of a very large number of diploid individuals that, for mathematical simplicity, is assumed to be infinite.
2. An individual's genotype at the locus of interest has no influence on his or her choice of a mate—that is, mating is random.
3. No new mutations appear in the gene pool.
4. No migration of individuals takes place into or out of the population.
5. Different genotypes at the locus of interest have no impact on **fitness**—the ability to survive to reproductive age and transmit genes to the next generation.

As we shall see, if a population is a reasonably close fit to the preceding assumptions for the locus of interest, then allele frequencies will not change over time, and only genotype frequencies may change briefly to reach proportions predicted by the frequencies of the alleles that make up each genotype. The population is then in **Hardy-Weinberg equilibrium (HWE).** Here, equilibrium means that the allele and genotype frequencies for that locus will not change unless one of the assumptions just listed is violated.

Of course, no actual population is a perfect fit to these assumptions for an ideal population. All populations are finite; alternative genotypes can make a difference in mating; mutations occur constantly; migration into and out of a population can be common; and many genotypes of interest, such as those that cause diseases, affect the ability to survive or reproduce. Nevertheless, even when many of the assumptions do not apply, the Hardy-Weinberg equilibrium is still remarkably robust at providing estimates of genotype and phenotype frequencies in real populations over a limited number of breeding generations. Furthermore, the discovery of frequencies that are inconsistent with the Hardy-Weinberg equilibrium can sometimes provide scientists with insight into special biological properties of particular genes and populations.

Predicting genotype and phenotype frequencies from one generation to the next

In diploid sexual populations, allele frequencies are transformed by the mating system into genotype frequencies. The law of segregation and also random mating in such a population have two important consequences.

First, haploid gametes are produced by the diploid adults of one generation according to Mendel's law of segregation so that each allele of the diploid adult appears in half the gametes. If the likelihood of producing a gamete does not depend on the gamete's genotype, then the allele frequencies in the adults of the population should be the same as the allele frequencies in all the gametes these adults would produce. For example, if p is the frequency of allele A and q is the frequency of allele a in the adults, p and q will also be the frequencies of the two alleles among all the gametes made by this generation of adults.

Second, if random mating occurs and the population is large, the allele frequencies in the gametes can be used to predict the genotype frequencies in the zygotes of the next generation. We can see how this happens using a special type of Punnett square, which provides a systematic means

Figure 21.3 Predicting the genotype frequencies of the offspring produced as a result of random mating within a population.

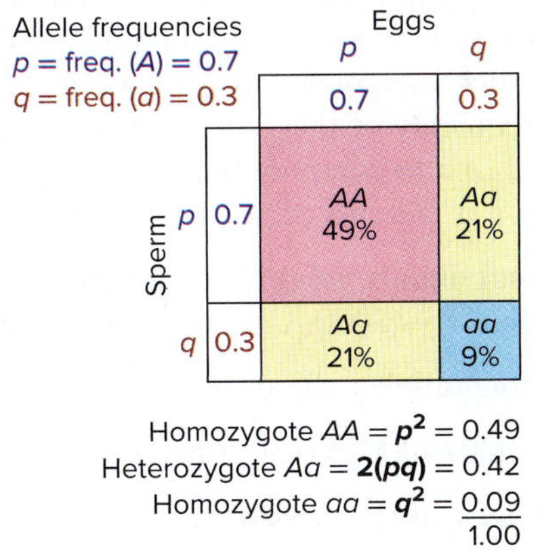

of considering all possible combinations of uniting gametes (**Fig. 21.3**).

If *A*-carrying sperm fertilize *A*-carrying eggs, *AA* zygotes will be formed. Because the genotype of a sperm is independent of the genotype of the egg it fertilizes, we can apply the product rule so as to multiply the frequency of *A* sperm (p) by the frequency of *A* eggs (also p) to find the frequency of *AA* zygotes: $p \times p = p^2$. Similarly, the frequency of *aa* zygotes among the progeny, which must result from fertilization of *a*-carrying eggs (whose frequency is q) by *a*-carrying sperm (whose frequency is also q), is the product of $q \times q = q^2$. Finally, *Aa* zygotes result either from the fertilization of *A* eggs by *a* sperm, with a frequency of $p \times q = pq$, or from the fertilization of *a* eggs by *A* sperm, also occurring at a frequency of $q \times p = pq$. The total frequency of *Aa* zygotes is thus $pq + pq = 2pq$.

The resemblance of the Hardy-Weinberg square shown in Fig. 21.3 to the Punnett squares that we first encountered in the visual representation of formal genetics in Chapter 2 is not a coincidence. The similarity of these diagrams results from the fact that populations are simply groups of families. In the original Punnett square for crosses between heterozygotes (Fig. 2.11), the top and left sides were divided into two equal sectors representing the equal frequencies of each genetically distinct class of sperm or egg produced by two individual parents.

If we replace the gametes from the two single parents with the male and female gametes produced by the population as a whole, then Fig. 21.3 is a metaphorical representation of the sperm produced by all breeding males along the left side, and of eggs produced by all breeding females along the top. In other words, random mating among the different genotypes in the population is equivalent to the random combination of the gametes produced by all the individuals in the population. But you should also notice an important difference from the classical Punnett squares: The sizes of the sectors are not necessarily equal; instead, the proportions of gametes bearing the two alleles correspond to whatever the frequencies of the two alleles are in the population being considered.

The key finding from the analysis shown in Fig. 21.3 is that the genotype frequencies of zygotes arising in a large, randomly mating population of sexually reproducing diploid organisms that satisfies all the Hardy-Weinberg assumptions are p^2 for *AA*, $2pq$ for *Aa*, and q^2 for *aa*. These genotype frequencies are known as the **Hardy-Weinberg proportions.** Because these genotype frequencies represent the totality of genotypes in the population, they must sum to 1. Hardy-Weinberg proportions can thus be expressed as the following binomial equation:

$$p^2 + 2pq + q^2 = 1 \qquad (21.2)$$

You should note that $p^2 + 2pq + q^2 = (p + q)^2$; this makes sense because $(p + q)$ represents all the sperm and also all of the eggs produced by all the adults. The set-up of the Hardy-Weinberg square illustrates precisely this point. Thinking about the pools of sperm and eggs in a population in this way allows us to extend the Hardy-Weinberg prediction in Eq. (21.2) to genes that have more than two alleles. Recall from Fig. 3.5 that the ABO blood groups are determined by three alleles of a single gene: I^A, I^B, and i. If we call the frequencies of these three alleles in the population p, q, and r, respectively, the predicted Hardy-Weinberg equilibrium genotype frequencies would then be $(p + q + r)^2 = p^2 + q^2 + r^2 + 2pq + 2pr + 2qr = 1$ for the genotypes I^AI^A, I^BI^B, ii, I^AI^B, I^Ai, and I^Bi in that order.

Populations with genotype frequencies in Hardy-Weinberg proportions will be in equilibrium, meaning that the allele and genotype frequencies will remain unchanged over the generations, as long as the assumptions remain valid. You can use the rules for computing allele frequencies from genotype frequencies [Eq. (21.1)] to compute algebraically the allele frequencies of the next generation and show that they do not change. From the Hardy-Weinberg equation for a gene with two alleles, you know that p^2 of the individuals are *AA*, all of whose alleles are *A*, and $2pq$ of the individuals are *Aa*, half of whose alleles are *A*. Thus, remembering that $p + q = 1$ (and therefore $q = 1 - p$), the frequency of allele *A* in the progeny of the original generation will be:

$$\begin{aligned} p^2 + (1/2)[2p(1-p)] &= p^2 + p(1-p) \\ &= p^2 + p - p^2 = p \end{aligned}$$

Similarly, $p = 1 - q$, and the frequency of the *a* allele in the next-generation population is:

$$\begin{aligned} q^2 + (1/2)[2q(1-q)] &= q^2 + q(1-q) \\ &= q^2 + q - q^2 = q \end{aligned}$$

As you can see, the allele frequencies among the progeny are the same as those among the parents; they remain p and q.

Figure 21.4 For any set of allele frequencies, only one set of genotype frequencies results in Hardy-Weinberg equilibrium.

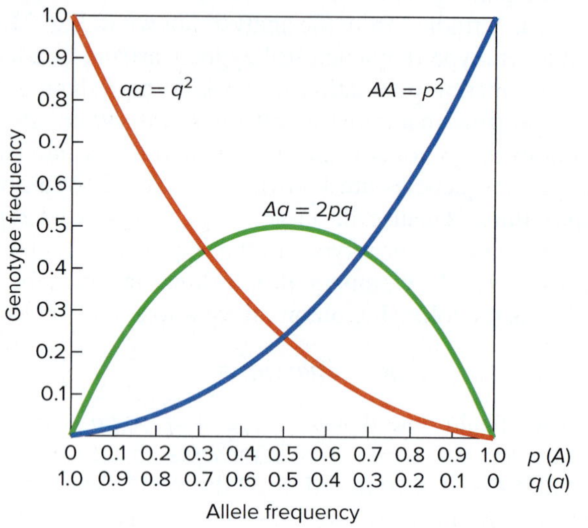

The potential for equilibrium provides the answer to the question of the extinction of recessive traits like hair color in blonds and redheads. In the absence of a fitness difference or other major deviations from Hardy-Weinberg assumptions, neither of these phenotypes will go extinct simply because their phenotypes are caused by recessive alleles. Instead, the frequencies of the alleles and genotypes that produce these hair colors will stay constant over time.

As illustrated in **Fig. 21.4,** one and only one HWE exists for a specific set of allele frequencies p and q, but different values of p and q imply different HWEs. Figure 21.4 plots the frequencies of the three genotypes for the full range of allele frequencies from $p = 0$ to 1.0 (and thus $q = 1.0$ to 0, since $q = 1 - p$). As an example, if $p = 0.8$ and $q = 0.2$, then at HWE the *AA* genotype is the most frequent of the three possible genotypes: The frequency of *AA* homozygotes $= p^2 = 0.64$, while the frequency of *Aa* heterozygotes is $2pq = 0.32$, and that of *aa* homozygotes is only $q^2 = 0.04$.

Notice in Fig. 21.4 that when q is small, most of the *a* alleles are carried by heterozygotes. This point will become important when we consider rare, recessive harmful alleles that cause human genetic disease later in this chapter. Also of interest, note that the frequency of heterozygotes is highest (50%) when $p = q = 0.5$.

The power of random mating in shaping genotype frequencies

Even if a population is not currently at HWE, one generation of random mating can be sufficient to establish equilibrium. Consider the following extreme example: Suppose that an island is colonized simultaneously by 1000 reproductive adults (equally divided among males and females) of a single species of insect from each of two different mainland locations. The 1000 colonizers from one location all are homozygous *AA*; those from the other location all happen to be homozygous *aa*. The initial genotype frequencies in the newly established population of adults are thus 50% *AA* homozygotes and 50% *aa* homozygotes. To predict the frequency of the genotypes in the next generation, you need first to compute the allele frequencies in the new population of parents on the island.

1000 *AA* individuals, no *Aa* individuals, and 1000 *aa* individuals → 2000 *A* alleles; 2000 *a* alleles.

Out of 4000 total alleles, the frequency of the *A* allele is

$$2000/4000 = 0.50; \text{ thus } p = 0.50,$$

and the frequency of the *a* allele is

$$2000/4000 = 0.50; \text{ thus } q = 0.50.$$

Because mating between all the adults is random, you can consider the pools of both sperm and eggs as reflecting the allele frequencies. As was seen in Fig. 21.3, this way of viewing the gamete pools predicts that the next generation of insects on this island will be distributed in Hardy-Weinberg proportions:

$$p^2 + 2pq + q^2 = 1$$
$$(0.50)^2 + 2(0.50 \times 0.50) + (0.50)^2 = 1$$
$$0.25 + 0.50 + 0.25 = 1$$

Supposing 1000 progeny are produced in this next generation, there will be;

$1000 \times 0.25 = 250$ *AA* individuals

$1000 \times 0.50 = 500$ *Aa* individuals

$1000 \times 0.25 = 250$ *aa* individuals

The lesson is clear: A population that is initially stratified because of its founding by individuals from two or more distinct populations having different allele frequencies for an autosomal locus will reshuffle to the Hardy-Weinberg genotype proportions in a single generation of random mating.

Have the allele frequencies also changed as the population goes from the first generation of colonizers to the second generation (the colonizers' progeny)? The answer is no. Using Eq. (20.1), you can calculate the allele frequencies in the progeny generation from the numbers of each genotype among the 1000 individuals present, thus:

$$p = \text{the frequency of the } A \text{ allele} = \frac{2(250)+500}{2(1000)} = \frac{1000}{2000} = 0.5$$

$$q = \text{the frequency of the } a \text{ allele} = \frac{2(250)+500}{2(1000)} = \frac{1000}{2000} = 0.5$$

These frequencies are the same as those in the previous generation. Thus, even though the genotype frequencies have changed dramatically from the first generation to the next, the allele frequencies have not. Note that this is true of both the dominant and the recessive alleles. The *conservation of allele proportions* principle holds from each generation to the next, as long as the population is sufficiently large and mates at random, alleles are not lost by mutation or selection, and alleles are not gained by mutation or immigration.

You have just seen that before HWE has been reached, populations with the same allele frequencies do not necessarily have the same genotype or phenotype frequencies. The reason is that a single allele can exist in homozygous or heterozygous genotypes, but a recessive trait is expressed only in homozygotes. In a different extreme hypothetical example, a population could have a blue-eyed allele frequency of 0.5 without actually having any people with blue eyes, if everyone in the population is a heterozygote. Even in this second example, the Hardy-Weinberg equilibrium tells us that with random mating and an autosomal gene, the Hardy-Weinberg genotype frequencies described by p^2, $2pq$, and q^2 will appear in the very next generation. Accordingly, 25% of the individuals in this next generation will have blue eyes.

The Hardy-Weinberg equilibrium and X-linked genes

Sex-linked genes, such as those on the X chromosome of humans, take several additional generations to reach HWE if allele frequencies differ between males and females and genotype frequencies are not initially in Hardy-Weinberg proportions (**Fig. 21.5**). The requirement for these extra generations is a consequence of the fact that males have only one X chromosome while females have two. Thus, the allele frequency in males in the next generation is equal to the allele frequency in females of the present generation because males receive their X from their mother. The allele frequency in females of the next generation will be equal to the average frequency in males and females of the present generation because females receive one X from each parent.

Figure 21.5 X-linked genes may require several generations of random mating to achieve equilibrium. The model population shown is established from *AA* females and *a*Y males. When HWE is established after several generations, the allele frequencies in males and females will be equal to each other; this value will be two-thirds of the initial allele frequency in females plus one-third of the initial frequency in males.

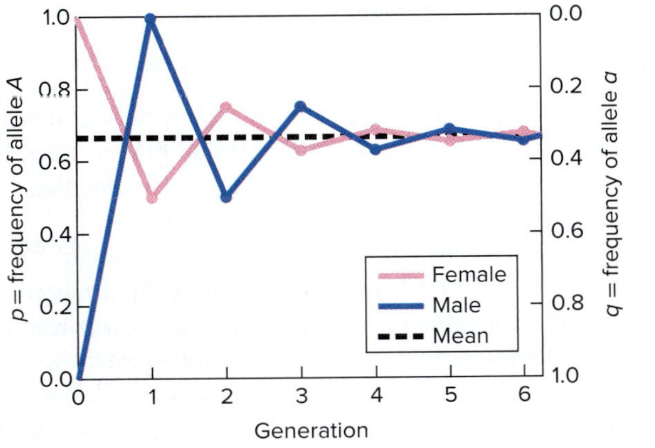

You can understand the consequences of these facts by imagining a new population introduced on a different desert island that is composed of 100 *a*Y males and 100 *AA* females. If the frequency of allele *A* is *p*, then initially $p_{male} = 0$ but $p_{female} = 1.0$. As seen in Fig. 21.5, in the next generation, the allele frequencies will be $p_{male} = 1.0$ but $p_{female} = (1.0 + 0)/2 = 0.5$. The frequencies in the two sexes will continue to change for several more generations before approaching HWE. More than six generations of random mating would be required to obtain Hardy-Weinberg proportions of all three possible genotypes in females; you should trace this out for yourself. At HWE, the allele frequencies in males and females will be equal to each other, and for either allele will be 2/3 the initial allele frequency in females plus 1/3 the initial frequency in males, so at HWE in this example, $p_{male} = p_{female} = 0.67$ (which we will now just call *p*).

Because the frequencies of alleles *A* and *a* in both the males and females of the population at HWE are respectively *p* and *q*, then the frequencies of *A*Y and *a*Y males will be simply *p* and *q*, respectively, because the Y chromosome has no copy of the gene. And at HWE, we expect to have three kinds of females (*AA*, *Aa*, and *aa*) in frequencies p^2, $2pq$, and q^2, just as was the case for autosomal loci.

The fact that allele frequencies are the same in males and females while genotypic frequencies differ underlies the common observation that many more males are red-green color blind than females. If we call the allele for color blindness *a*, then the frequency of color-blind males (*a*Y) will be equal to *q*, but the frequency of color-blind females (*aa*) will be q^2. When *q* is less than 1, then q^2 will always be less than *q*. For example, the frequency of color blindness among males in the United States is about 7%. The value of *q* is thus 0.07, while the value of q^2 is $(0.07)^2 = 0.0049$, or 0.49%.

For your convenience, **Table 21.1** summarizes the genotype frequencies at HWE for the scenarios discussed thus far.

TABLE 21.1 Genotype Frequencies at Hardy-Weinberg Equilibrium

	Females	Males
Autosomal gene with two alleles:		
AA	p^2	p^2
Aa	$2pq$	$2pq$
aa	q^2	q^2
X-linked gene with two alleles:		
AA	p^2	
Aa	$2pq$	
aa	q^2	
*A*Y		p
*A*Y		q
Autosomal gene with three alleles:		
A^1A^1	p^2	p^2
A^2A^2	q^2	q^2
A^3A^3	r^2	r^2
A^1A^2	$2pq$	$2pq$
A^1A^3	$2pr$	$2pr$
A^2A^3	$2qr$	$2qr$

Many Loci in Human Populations Are Near Hardy-Weinberg Proportions

Once a population is known to be in Hardy-Weinberg equilibrium, investigators can easily predict allele frequencies from genotype frequencies, and genotype and phenotype frequencies from allele frequencies. Random mating may seem at first thought to be a particularly unrealistic assumption for humans, whose mating choices are influenced by many factors, including geographical proximity, cultural norms, and the traits of possible mates. Yet from the point of view of population genetics, what matters is mating randomly with respect to the specific genotypes under study. As we will explore more in the next chapter, the connections between genotype and phenotype are so complex that we rarely select mates on the basis of any one genotype being investigated. For this reason, studies of most genetic loci in humans in fact reveal a surprisingly good fit to Hardy-Weinberg predictions, particularly when looking at anonymous loci that do not influence phenotype. This simple observation has contributed to the usefulness of DNA variation in solving crimes and in the identification of human remains in tragedies such as mass disasters and plane crashes.

Using Hardy-Weinberg to analyze DNA fingerprints

The Hardy-Weinberg equilibrium is crucial to the interpretation of DNA fingerprinting evidence for forensic investigations, a technique that was discussed in Chapter 11 (review Fig. 11.15). Suppose that a perfect match has been obtained between the blood under a murder victim's fingernails and a particular suspect. Population genetics allows forensic scientists to answer with numerical precision the question: How much more likely is it that the DNA found on the victim is the suspect's DNA, than that the DNA came from another (random) person?

As described in Chapter 11, the most useful DNA markers for forensic analysis are polymorphic anonymous loci that are highly variable in human populations. To enable comparisons of DNA samples, law enforcement agencies in the United States and many other countries focus on 13 unlinked, simple sequence repeat loci (SSRs) found throughout the human genome. The results obtained are deposited in a database called the Combined DNA Indexing System (CODIS; **Fig. 21.6**). From a practical perspective, criminal investigators and juries need to be able to conclude that a perfect match at all 13 CODIS loci between the suspect and the sample under the victim's fingernails is not simply a chance match. In other words, what is the likelihood that this particular 13-locus genotype could be found among the pool of potential suspects constituting all the people in the population?

The SSRs chosen for CODIS were selected because they are unlinked to each other and have been found from surveys of thousands of individuals to be highly variable and in HWE. The critical data from these surveys consist of the allele frequencies for the 13 loci. From these allele frequencies, investigators can use the Hardy-Weinberg equation to calculate the likelihood of a match of the diploid genotype for any one locus. Because the CODIS loci are unlinked and alleles at each locus are statistically independent of alleles at the other loci, forensic scientists can simply multiply together the expected genotype frequencies for each separate locus to obtain the expected frequency of any observed 13-locus genotype.

Figure 21.6 Chromosomal locations of the 13 CODIS core autosomal SSR loci. *AMEL* (amelogenin) is an additional marker that allows sex determination of a sample because it always amplifies different-sized PCR products from the X and Y chromosomes.

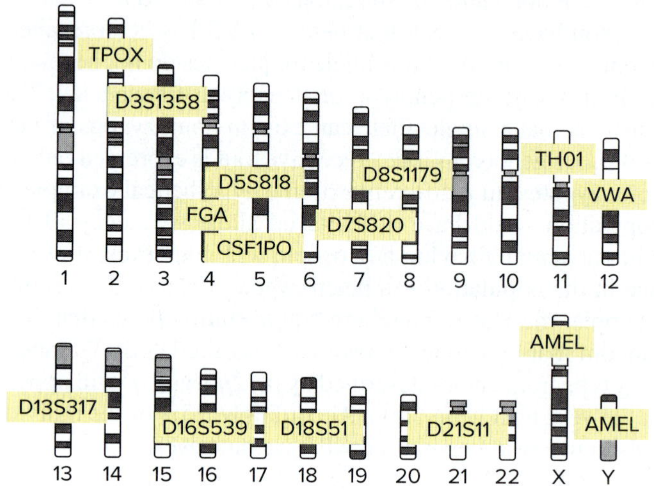

Calculating the match probability

Suppose that the DNA profile common to both the crime scene sample and the suspect was heterozygous for two alleles of one CODIS locus, with allele frequencies of 0.05 and 0.03, respectively, and homozygous for an allele of a second locus, whose frequency is 0.04. Even though more than 20 alleles for each CODIS SSR can be distinguished, the genotype frequency for any particular heterozygous genotype is still 2 × frequency of allele 1 × frequency of allele 2, while the frequency for a homozygous genotype is the square of the allele frequency. Thus, the **match probability** for a random person in the population having the particular two-locus genotype seen in the crime scene sample would be $2(0.05)(0.03) \times (0.04)^2 = 0.0000048 = 4.8 \times 10^{-6}$. In other words, the frequency of that particular two-locus genotype is .0000048 or 1/208,334 individuals.

The match probability becomes very small for any given 13-locus CODIS genotype because the frequency of each allele of any one of these polymorphic loci is low. Suppose the 13-locus match probability for the DNA sample under the fingernails of the crime victim is 7.7×10^{-15}, which is in the range of those typically found in forensic investigations. We would compare the likelihood of obtaining a perfect match if the DNA is from the suspect, which is 100% or 1, against the likelihood of obtaining a match if the DNA were from a different, random person, which is 7.7×10^{-15}. Thus, it is $1/(7.7 \times 10^{-15}) = 1.3 \times 10^{14}$ times more likely that the DNA is from the suspect than that it is from a different, random person. Put another way, only one in 130 trillion people would be expected to have the suspect's particular set of CODIS alleles. Considering that the world's total population (in 2016) is about 7.4×10^9 (~7.4 billion) individuals, the DNA fingerprinting results would constitute compelling evidence connecting the suspect (or his or her identical twin) with the DNA under the victim's fingernails.

essential concepts

- Given the *genotype frequencies* for a particular population, you can calculate the *allele frequencies* by adding the frequency of homozygotes for a given allele with one-half the frequency of heterozygotes for that allele.
- In a population at *Hardy-Weinberg equilibrium,* the allele frequencies and the genotype frequencies remain constant from one generation to the next. For an autosomal gene with two alleles, the genotype frequencies at equilibrium are distributed according to the formula $p^2 + 2pq + q^2 = 1$, where p and q are the allele frequencies.
- To determine the *match probability* for a DNA profile, forensic scientists multiply together the genotype frequencies for each independent locus, as calculated from the known allele frequencies using the Hardy-Weinberg proportions.

21.2 What Causes Allele Frequencies to Change in Real Populations?

learning objectives

1. Explain why the Hardy-Weinberg model is more accurate in predicting allele and genotype frequencies in the short run than in the long run.
2. Discuss how the finite size of populations means that new mutations eventually will be either lost or fixed.
3. Describe scenarios by which natural selection can promote the loss, spread, or maintenance of an allele in a population.
4. Explain why the frequency of alleles for insecticide resistance in mosquito populations decreased when the insecticide was no longer sprayed.

As **Fig. 21.7a** shows, different populations of the human species have dramatically different proportions of individuals with blue eye color. These differences reflect variations in the allele frequencies of a SNP upstream of the *OCA2* gene that is responsible for this recessive phenotype (**Fig. 21.7b**). Many other loci in the human genome, such as the one responsible for lactose tolerance/intolerance, also display geographical differences in allele frequency (review Fig. 21.1b). What do these observed differences in allele frequencies between populations suggest about the applicability of the Hardy-Weinberg ideal model assumptions to real populations?

Figure 21.7 Phenotype and allele frequencies at one of the genes associated with eye color in human populations. **(a)** Geographical differences in the proportions of European populations having blue eyes. **(b)** Pie diagrams depict frequencies of alleles G (*blue*) and A (*brown*) at the SNP locus *rs12913832*. This polymorphism is located in an enhancer of the *OCA2* gene, whose protein product is involved in the production of the pigment melanin.

(a) Frequencies of blue eyes **(b) Frequencies of SNP *rs12913832* allele**

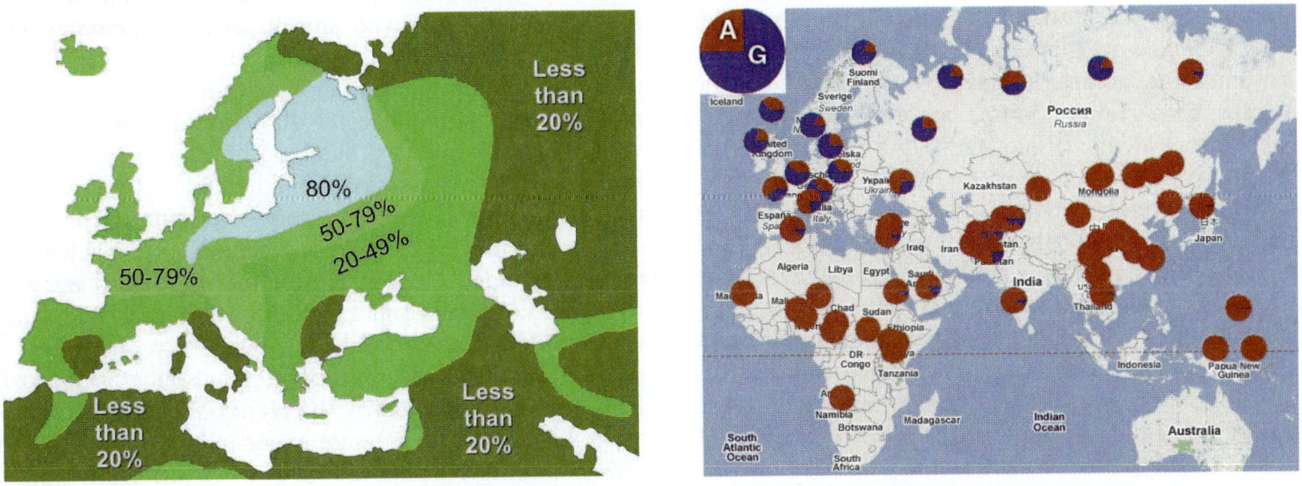

Hardy-Weinberg Provides a Starting Point for Modeling Actual Populations

In natural populations, conditions always deviate at least slightly from the Hardy-Weinberg assumptions. New mutations do appear occasionally at every locus, no population is infinite in size, small groups of individuals sometimes migrate from the main group to become founders of new populations, separate populations can merge together, individuals do not always mate at random, and different genotypes do generate differences in rates of survival and reproduction. These exceptions to Hardy-Weinberg conditions change the genetic makeup of populations over time and are thus essential for the evolution of living forms.

Even though these deviations from the Hardy-Weinberg assumptions always exist, the Hardy-Weinberg equation nonetheless provides remarkably good estimates of allele, genotype, and phenotype frequencies *over the short run,* that is, through one or a few breeding generations of large populations. Over the long run, the realities of natural populations mean that the Hardy-Weinberg equation by itself cannot predict how genetic frequencies change over the course of many generations. The Hardy-Weinberg viewpoint nonetheless serves a useful role in providing the foundation for mathematical models that incorporate factors responsible for deviations from equilibrium conditions, allowing population geneticists to model successfully the dynamics of actual populations.

In Finite Populations, Chance Plays a Crucial Role

Hardy and Weinberg derived their famous equation by extending Mendel's first law of segregation. You should remember that Mendel's first law does not determine which allele a heterozygous parent will transmit to a particular child. Instead, it tells us that allele inheritance is like a flip of a coin: A child can receive heads or tails (that is, either allele of an autosomal gene from a heterozygous parent) with an equal probability. Mendel's law *does* predict the approximate proportion of a large cohort of offspring that will inherit a particular allele; the larger the cohort, the more accurate the prediction.

The Hardy-Weinberg equation is based on the idea that the frequencies of alleles that are transmitted to the next generation are the same as those in the parental generation. In other words, alleles are contributed to the pool of gametes and end up in zygotes in exactly the same frequencies as they are present in the parental genotypes. But this idea is valid only as the size of a population, and thus the number of gametes contributed to the next generation, approaches infinity. (In a similar fashion, you cannot predict that any finite number of coin tosses will produce exactly one-half heads and one-half tails.) Because no population is infinite, no population truly abides by the Hardy-Weinberg conditions for equilibrium.

Computer simulation of chance in populations

To simulate long-term allele frequency changes in a finite population, the input of allele frequencies for the Hardy-Weinberg equation must take into account the effect of random chance on choosing the gametes used in each generation. Researchers model the effects of these so-called *sampling errors* using a **Monte Carlo simulation,** a computer program that uses a random-number generator to choose an outcome for each probabilistic event. The Monte Carlo program begins with a population having a defined number of individuals of each homozygous and heterozygous class. As we will see shortly, the size of the population is a key variable. The program sets up matings between individuals chosen by the random-number generator. If a chosen parent is a heterozygote, the program also flips a coin (metaphorically) to decide which allele will be transmitted to a child.

Monte Carlo simulations usually adjust the birth rate so that the population size remains constant between generations, and they do not allow mating between different generations. Thus, once the first-generation parents are eliminated, their children become the progenitors for the subsequent generation. The program continues this process for as many generations as the investigator requests. After the data are recorded, the computer simulates a new run starting with the same initial conditions. With a sufficient number of independent Monte Carlo simulations, researchers can get a sense of what outcomes are possible and with what probabilities.

Genetic drift

In the example shown in **Fig. 21.8a,** the computer ran six Monte Carlo simulations initialized with populations of only 10 individuals who are all set to be heterozygotes. Each population thus has $2 \times 10 = 20$ total gene copies (two in each individual), and each allele (A or a) occurs initially with a frequency of 0.5. Because of the way in which this example is structured, the Monte Carlo simulation in this first generation is mathematically equivalent to the results obtained from tossing a coin 20 times. As you can see in Fig. 21.8a, in the first generation the actual simulations yielded frequencies for the A allele that ranged from 0.25 (5 heads and 15 tails) to 0.65 (13 heads and 7 tails), with an average of 0.48.

Although the average of populations is not too far from the 0.5 predicted by the Hardy-Weinberg equation, the values for each separate experiment guide each individual simulated population down a different path of

Figure 21.8 Monte Carlo modeling of population drift in populations of different sizes. **(a)** Population size = 10. **(b)** Population size = 500. In both cases the initial condition is equal numbers of the two alleles of the locus, with no selection. Each colored line represents a different run of the simulation.

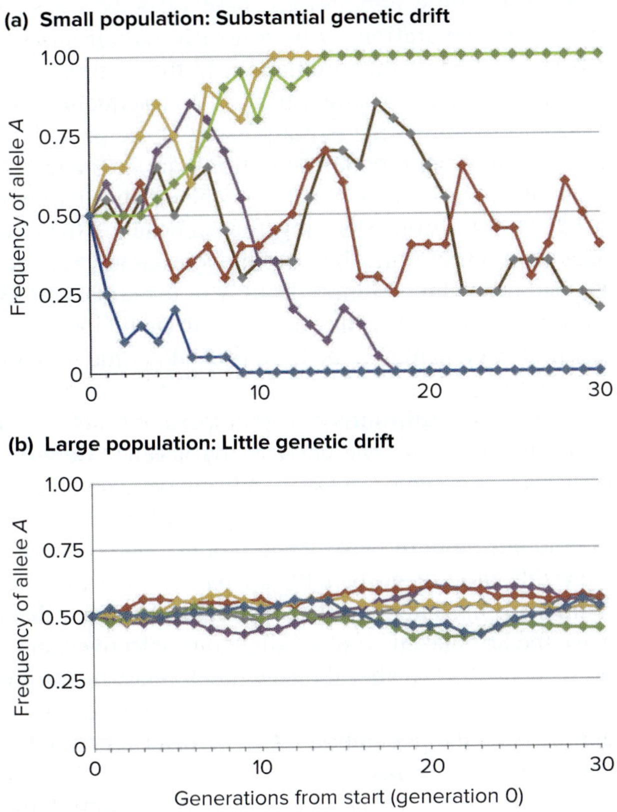

genetic drift, which is a change in allele frequencies as a consequence of the randomness of inheritance due to sampling error from one generation to the next. Genetic drift occurs because the allele frequency in any given generation provides the basis for the possible allele frequencies in the next generation. So, for example, it turns out that if one allele has already drifted to a high frequency, a 50% chance exists that it will go even higher in the following generation.

In four of the six simulated runs shown in Fig. 21.8a, genetic drift culminated in the loss, or **extinction,** of one or the other of the two original alleles by generation 18. In each of these instances, the cumulative effects of changes in allele frequency from one generation to the next caused extinction of one allele, and **fixation** of the remaining allele. Geneticists consider a population to be *fixed* at a locus when only one allele has survived and all individuals are homozygous for this allele. At this point, no further changes in allele frequency can occur (in the absence of migration or mutation).

Population size has a dramatic effect on allele frequency dynamics, as seen by comparing the experiments just seen in Fig. 21.8a with six simulations of populations that are initially all heterozygous but that have 500 members rather than 10 (**Fig. 21.8b**). If we follow the lines representing each of the latter populations, we can see that single generation changes in allele frequency are always relatively small. Because these changes in allele frequency are small, the traditional Hardy-Weinberg equation provides good estimates of allele and genotype frequencies in large populations over the course of a few generations. But a series of small changes can still add up to large consequences over the long run, so each of these populations will eventually become fixed for one allele or the other many generations later than those shown in Fig. 21.8b.

Founder effects and population bottlenecks

When populations become very small, genetic drift can be accelerated by two related processes: *founder effects* and *population bottlenecks*. **Founder effects** occur when a few individuals separate from a larger population and establish a new one that is isolated from the original (**Fig. 21.9**). The small number of founder individuals in the new population carries only a fraction of the gene copies from the original population. By chance sampling error, allele frequencies among the founders can be different than in the original population from which they came; some alleles may even fail to be transferred to the new population. If the population stays small for a period of time after founding, then additional genetic drift will change allele frequencies rapidly, further exacerbating the founder effect.

About 200 individuals emigrated from Germany in the early eighteenth century to form the Amish community in eastern Pennsylvania. Because this founding population was completely cut off from Europe and people married only within the group, it was subjected to genetic drift. Today, the Amish in the United States number more than 150,000 in total, but the population exhibits a much higher incidence of manic-depressive illness than does the Amish population remaining in Germany, most likely because several founders carried alleles producing this condition.

Plant and animal populations are frequently subjected to **population bottlenecks,** which occur when a large proportion of individuals perish, often as a consequence of environmental disturbances or disease. The surviving individuals are essentially equivalent to a founder population, and as long as the population size stays small, genetic drift will be accelerated further. In Ashkenazi Jews, high frequencies of certain *BRCA1* and *BRCA2* mutations that lead to an elevated risk of breast and ovarian cancer are thought to have resulted from a population bottleneck or founder effect that occurred 600 to 800 years ago.

Figure 21.9 Founder effects and bottlenecks can alter allele frequencies rapidly. Because of sampling error, the frequencies of alleles among the small number of founders of a population on an island can be very different from those in the population from which the founders came. One consequence is loss of genetic variation; note that on the island, the *yellow* allele is no longer present. In addition, as long as the population on the island remains small, genetic drift accelerates.

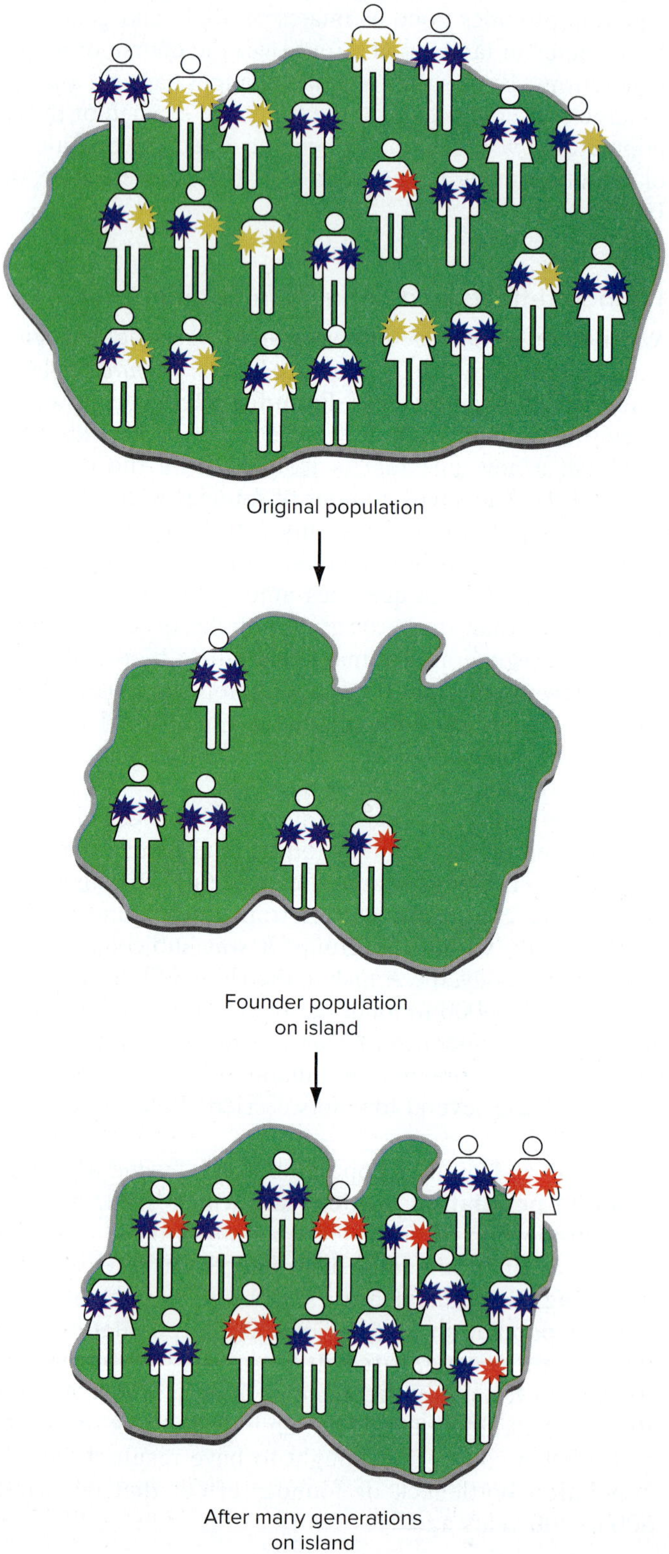

Mutations Introduce New Genetic Variation

While genetic drift leads eventually to the loss of genetic variability in finite populations, new variants arise continually as new mutations or by **migration,** their introduction from a neighboring population. In the context of this discussion, a *mutation* is a variant DNA sequence in an individual's genome that is not present in the genomes of either parent. Spontaneous mutations are in general so infrequent at individual genes (on the order of once in 100,000 to 1 million offspring for each gene) that their impact on changing the allele frequencies at any one gene in a population can be ignored safely in the short run. But mutations are of course the source of all new alleles at all loci, which means that some mutations do come to matter in the long term.

Deleterious mutations disrupt important functions like the activity of an enzyme encoded by a gene. **Beneficial mutations** that provide a selective advantage to an organism or population are in contrast relatively rare. But many mutations belong to a third class: They produce polymorphisms that provide little or no benefit nor harm to the organism. Mutations of this latter class are **neutral mutations;** that is, because the original and mutant alleles are selectively equivalent, their fates—whether the new mutation or the original DNA sequence will be maintained or will eventually disappear from the population—will be determined only by stochastic, chance events.

In a population where a new, selectively neutral allele occurs by mutation, Monte Carlo simulations show that the average number of generations to fixation is roughly equal to twice the total number of gene copies in breeding individuals ($4N$, where N is the number of diploid breeding individuals in the population). For a population of size 10, the average fixation time is 40 generations; for a population of 500, it is 2000 generations; and for a breeding population of 200 million people, it would be 800 million generations, or 16 billion years (assuming a birth-to-birth generation time of 20 years).

The chance that any specific new mutation will go to fixation is reduced as populations get larger. However, more mutations occur in large populations. It turns out that these two factors exactly counterbalance each other for selectively neutral variants so that the average rate of change in DNA sequences over time will be simply equal to the rate of input by mutation (measured as mutations per locus per generation). This rate is independent of the size of the population.

Mutation rates appear to be relatively constant over time for many organisms. The implication is that neutral genetic drift alone leads to a time-dependent accumulation of DNA differences, at a roughly constant rate, between genetically isolated populations. This fact provides evolutionary geneticists with a **molecular clock,** meaning that they can predict how long in the past different types of organisms diverged from a common ancestor by examining how different the DNA sequences of these organisms are from each other. We will discuss the molecular clock concept more fully later in the chapter.

Natural Selection Acts on Differences in Fitness to Alter Allele Frequencies

For many traits, including inherited diseases, genotype does influence survival and the ability to reproduce, contrary to the Hardy-Weinberg assumptions. Thus, in real populations, not all individuals survive to adulthood, and some probability always exists that an individual will not live long enough to reproduce. As a result, the genotype frequencies of real populations change as their individual members mature from zygotes to adults.

Fitness and natural selection

To population geneticists, an individual's relative ability to survive and transmit its genes to the next generation is its **fitness.** Although fitness is an attribute associated with each genotype, it cannot be measured within a single individual; the reason is that each animal with a particular genotype survives and reproduces in a manner greatly affected by chance circumstances. However, by considering all the individuals of a particular genotype together as a group, it becomes possible to measure the relative fitness for that genotype. Thus, for population geneticists, fitness is a statistical measurement only. Nevertheless, differences in fitness can have a profound effect on the allele frequencies within a population.

Fitness has two components: *viability* and *reproductive success.* The fitness of individuals with variations that help them survive and reproduce in a changing environment is relatively high; the fitness of individuals without those adaptive variations is relatively low. In nature, the process that progressively eliminates individuals whose fitness is lower and chooses individuals of higher fitness to survive and become the parents of the next generation is known as **natural selection.**

Field studies show that natural selection acts on traits in all natural populations. A straightforward example is the fur coloration of pocket mice living in the deserts of New Mexico. In areas where the soil and rocks are light colored, the pocket mice have light-colored fur, while darkly colored pocket mice are found in areas with black volcanic rock. Studies have revealed that predators can more readily find and consume pocket mice whose color does not match that of the soil and rock on which they live (**Fig. 21.10**). This process leads to strong changes in the underlying genetic variation affecting coat color between populations living on these two soil types. In this section, we consider how selection alters allele and genotype frequencies from those predicted by the Hardy-Weinberg model.

Figure 21.10 Genetic variation can lead to dramatic differences in selective fitness. Pocket mice from New Mexico living on sandy soils or very black volcanic rock have diverged by natural selection to match their substrate (*top row*). When on the wrong substrate given their fur color, they are readily detected (and eaten) by predators (*bottom row*).

From: Nachman et al. (2003), "The genetic basis of adaptive melanism in pocket mice," *PNAS*, 100: 5268–5273. © 2003 National Academy of Sciences, U.S.A.

Adding selection to Hardy-Weinberg predictions

We can see how to apply the Hardy-Weinberg equation in populations undergoing selection with an analysis of a particular gene in a population of zygotes that begins in Hardy-Weinberg proportions. In this population, the genotype frequencies *AA*, *Aa*, and *aa* are p^2, $2pq$, and q^2, respectively. For simplicity, we will assume that the two components of fitness—viability and success at reproduction—both depend on genotype in the same way. If we define the **relative fitness (*W*)** of each of the three genotypes as W_{AA}, W_{Aa}, and W_{aa}, respectively, the relative frequencies of the three genotypes at adulthood are p^2W_{AA}, $2pqW_{Aa}$, and q^2W_{aa} (**Fig. 21.11**). Usually, geneticists assign values to W_{AA}, W_{Aa}, and W_{aa} by calling the largest of these numbers 1.0, so the relative fitness of a less fit genotype would be less than 1.0.

The fitness-modified Hardy-Weinberg equation is most useful when the relative fitnesses are *normalized* so that each term in the equation represents an actual rather than a relative genotype frequency. Normalization is accomplished by a two-step calculation. First, we set the sum of the terms in the modified equation to a new variable, designated $\overline{W}$, or the *mean fitness,* which represents

Figure 21.11 Changes in allele frequencies caused by selection. To calculate the genotype frequencies after selection (in adults), first multiply the zygote genotype frequencies formed through random mating by their relative fitness values. Next, normalize these terms through division by the mean fitness $\overline{W}$. Finally, calculate the allele frequencies of the gametes produced by the offspring generation from the adult offspring genotype frequencies using Eq. (21.1).

Parental Gametes

Allele	*A*	*a*
Frequency	p	q

↓

Offspring Generation

Genotypes	*AA*	*Aa*	*aa*
Zygote frequency	p^2	$2pq$	q^2
Relative fitness	W_{AA}	W_{Aa}	W_{aa}
Relative frequency after selection	p^2W_{AA}	$2pqW_{Aa}$	q^2W_{aa}
Normalized adult frequency	$\frac{p^2W_{AA}}{\overline{W}}$	$\frac{2pqW_{Aa}}{\overline{W}}$	$\frac{q^2W_{aa}}{\overline{W}}$

↓

Allele Frequencies in Offspring Generation

Allele	*A*	*a*
Frequency	$p' = \frac{(p^2W_{AA} + pqW_{Aa})}{\overline{W}}$	$q' = \frac{(q^2W_{aa} + pqW_{Aa})}{\overline{W}}$

Normalization factor $\overline{W} = p^2W_{AA} + 2pqW_{Aa} + q^2W_{aa}$

the sum of the relative contributions of each genotype to the next generation.

$$p^2W_{AA} + 2pqW_{Aa} + q^2W_{aa} = \overline{W} \qquad (21.3a)$$

When genotype fitnesses differ, $\overline{W}$ is less than 1 because not all gametes in the initial generation will contribute to the next generation.

To continue with the process of normalization, we next divide each side of Eq. (21.3a) by $\overline{W}$ so that the new equation for adults in the offspring generation becomes

$$\frac{p^2W_{AA}}{\overline{W}} + \frac{2pqW_{Aa}}{\overline{W}} + \frac{q^2W_{aa}}{\overline{W}} = 1 \qquad (21.3b)$$

Each term in this normalized equation represents the actual frequency that each genotype will assume in the next generation following the one used for the original calculation.

One important outcome of Eqs. (21.3a and 21.3b) is the prediction that in the presence of selection, the mean fitness of the population, $\overline{W}$, will change from one generation to the next in a way that pushes its value toward 1 over time. For example, a value of $\overline{W}$ less than 1 means that not all individuals in the initial generation will contribute to the next generation; the mean fitness of the original population is lower than what it could be if all genotypes present were of the optimal fitness. However, selection will result in an increase in the frequency of favored genotypes in the next generation, so the mean fitness of the population (that is, $\overline{W}$) in the next generation would be higher (closer to 1) than it was in the previous one.

To see how these equations allow us to calculate the effects of selection on allele frequencies over time, let's use the variables p' and q' to represent the frequencies of the *A* and *a* alleles in this next generation (that is, the offspring generation in Fig. 21.11). Among the gametes produced by the original population, the frequency of allele *a* will be the result of contributions of *a* alleles from both *Aa* and *aa* adults relative to the number of individuals in the entire adult population. If q' represents the frequency of the *a* allele in the next-generation adults, then

$$q' = \frac{q^2W_{aa} + \frac{1}{2}2pqW_{aa}}{\overline{W}} = \frac{q(qW_{aa} + pW_{Aa})}{\overline{W}} \qquad (21.4)$$

Thus, in one generation of selection, the allele frequency of *a* has changed from q to q'.

It is often useful to know the change in allele frequency over one generation of selection. We can estimate this change as $\Delta q = q' - q$. Substituting Eq (21.4) for q' yields (after some algebra):

$$\Delta q = \frac{pq[q(W_{aa} - W_{Aa}) - p(W_{AA} - W_{Aa})]}{\overline{W}} \qquad (21.5)$$

Equation (21.5) shows that selection can cause the frequency of an allele to change from one generation to the next, and this change depends both on the frequencies of the two alleles and on the relative fitnesses of the three genotypes. Note that if the fitnesses of all genotypes are the same, as in populations at Hardy-Weinberg equilibrium, then $\Delta q = 0$. In other words, if no genotype-related differences in fitness exist, there is no possibility of selection, and allele frequencies will be subjected only to genetic drift, as we described earlier.

We can use Eq. (21.5) to examine how the deleterious effects of a recessive genetic disease such as cystic fibrosis can affect the frequency of the mutant allele (call it *d*) over time in a population. If the disease decreases fitness by decreasing the probability of surviving to adulthood, then the fitnesses of *DD* and *Dd* individuals are the same, while the fitness of *dd* individuals is reduced. Because only the relative values of the fitnesses are important, we will set the values of $W_{DD} = 1$, $W_{Dd} = 1$, and $1 \geq W_{dd} \geq 0$.

For a genotype with deleterious effects, the fitness W_{dd} can vary from just less than 1 (minimal selection against

Figure 21.12 Decrease in the frequency of a recessive lethal allele over time. The *dotted line* represents the mathematical prediction. The *blue line* represents the actual data obtained from one experiment with an autosomal recessive lethal allele.

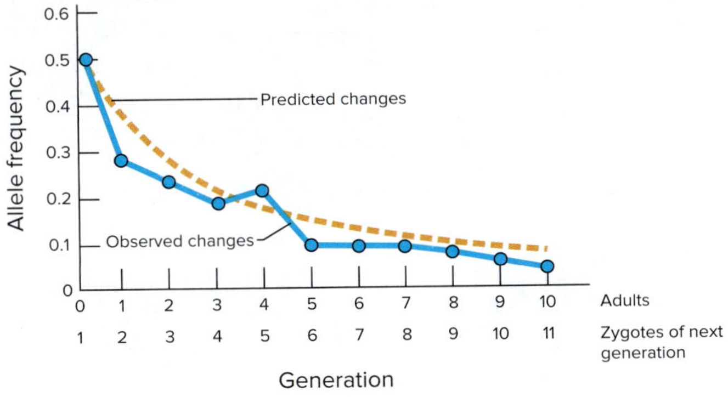

Figure 21.13 Natural selection together with genetic drift. Each colored line represents an independent Monte Carlo simulation of a population of $N = 500$ individuals in which a new mutant allele with a slight dominant fitness advantage appears at time 0. In three of the simulations, the mutant allele goes extinct in <100 generations because of genetic drift when the allele frequency is very low. The few simulations that escape from the loss of the advantageous mutation move inevitably to its fixation.

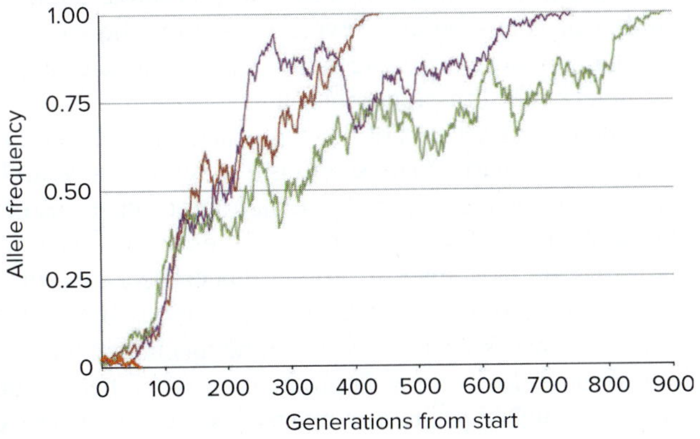

dd) to 0 (*dd* is lethal, so no *dd* individuals survive to adulthood). If selection against *dd* homozygotes exists, Δq is always negative, and the frequency of the *d* allele decreases with each generation.

A key feature of Eq. (21.5) when W_{dd} is less than 1 is its prediction that the rate at which q decreases over time diminishes as q becomes smaller. This prediction emerges because Δq varies with q^2, and because q is always less than 1, $q^2 < q$.

To understand this effect, consider the special case of a lethal recessive disease for which $W_{dd} = 0$. The dotted line in **Fig. 21.12** shows the decrease in allele frequency predicted by Eq. (21.5), starting from an initial allele frequency of 0.5. The decrease in allele frequency is rapid at first and then slows. After 10 generations, the predicted frequency of the recessive disease allele is still nearly 10%, even though the homozygous recessive genotype is lethal. The solid line in Fig. 21.12 plots actual data for the decrease in frequency of an autosomal recessive lethal allele in a large experimental population of *Drosophila melanogaster;* the predicted and observed changes in allele frequency match quite closely.

Why does selection become less effective as the frequency of a recessive lethal allele moves closer to zero? The answer is that when q is small, individuals homozygous for the disease allele (at a frequency of q^2) are rare because most copies of the *d* allele occur in *Dd* heterozygotes (at a frequency of $2pq$) who do not experience negative selection. In mathematical terms, the ratio of q^2 to q decreases exponentially for all values of q less than 1. Over successive generations, then, the allele frequency q should continue to decline, albeit more and more slowly over time as q moves closer and closer to a value of zero.

Natural selection in finite populations

Modifying the Hardy-Weinberg equation with relative fitnesses overcomes one limitation of the original equation: the assumption that all possible genotypes are equal in fitness. But the analytical solution of this modified equation to determine Δq still suffers from depending on the assumption of an infinite population. However, we can use the modified Hardy-Weinberg equation to develop Monte Carlo simulations that explore the impact of natural selection on finite populations.

As an example, let's consider a population of 500 individuals in which 499 are homozygous initially for the *b* allele, and one is heterozygous with a *B* mutation on one chromosome that provides a slight dominant advantage in survival described by the following relative fitness values: $W_{BB} = 1.0$, $W_{Bb} = 1.0$, and $W_{bb} = 0.98$. These conditions can be modeled with a Monte Carlo approach that randomly eliminates 2% of *bb* individuals created in each generation, and replaces them with offspring from a new mating of the parental generation.

Figure 21.13 shows the results of six simulations of this population model. Notice first that in three of these, the new *B* allele never takes off, going extinct within 65 generations. But in the populations where the *B* allele increases in frequency to about 0.10, it inevitably moves toward fixation.

This example illustrates two important points concerning the impact that a new mutant allele with a small, yet realistic, fitness advantage can have on a population. First, even though the novel allele provides a selective advantage, it will often go extinct due to chance events of reproduction in the initial generations. But second, if the advantageous allele reaches a threshold frequency level that ensures its survival, its frequency will always increase all the way to fixation eventually, even if the small fitness advantage is imperceptible at the individual level.

The impact of natural selection on humans

Beginning about 60,000 to 80,000 years ago, when people migrated out of the East African region in which *H. sapiens* originated (as will be discussed later in this chapter), founder populations encountered environmental conditions in Europe and Asia that were distinct from those in Africa. As a result, the relative fitnesses of alternative alleles at a number of genes became reversed. Among the most obvious changes were differences in allele frequencies at genes that determine skin pigmentation.

The ultraviolet rays of the sun provide people with benefits as well as harm. One benefit is the catalysis of vitamin D production; the harm is in the induction of mutations in our skin that can lead to skin cancer. Closer to the equator, the sun's rays are most intense. Alleles that cause a darkening of the skin are advantageous in tropical regions because they protect against skin cancer while allowing enough ultraviolet light through for vitamin D production. At higher latitudes, where the sun's rays are less intense, skin cancer is less of a problem, and alleles that lighten the skin allow enough UV penetration for sufficient vitamin D production.

As was described in Chapter 3, skin pigmentation is a complex quantitative trait determined by alleles at many genes, but about a half dozen are most influential. One fascinating question concerning our history as a species is whether European and Asian populations derived lighter skin pigmentation from a common ancestral population, or whether the trait evolved separately on the two continents. A mixed answer has been obtained by surveying allele frequencies at multiple pigmentation loci in populations indigenous to different geographical locations around the Old World.

KITLG is among the few genes with a prominent role in skin pigmentation. As you can see in **Fig. 21.14a,** Europeans and Asians share a common SNP variant of *KITLG* responsible for a reduction in pigmentation, suggesting that they derived it from a common ancestor who lived after humans migrated from Africa into the Arabian Peninsula and prior to the separation of populations heading northwest and northeast. In contrast, Europeans and Asians independently accumulated variants with roles in skin pigmentation at two other loci (*SLC24A5* and *MC1R;* **Fig. 21.14b** and **c**). Thus, although the same selective pressure of reduced sunlight existed in both populations, the selection acted on different mutations that occurred at different times in human history.

Figure 21.14 Geographical distribution of allele frequencies at skin pigmentation loci.

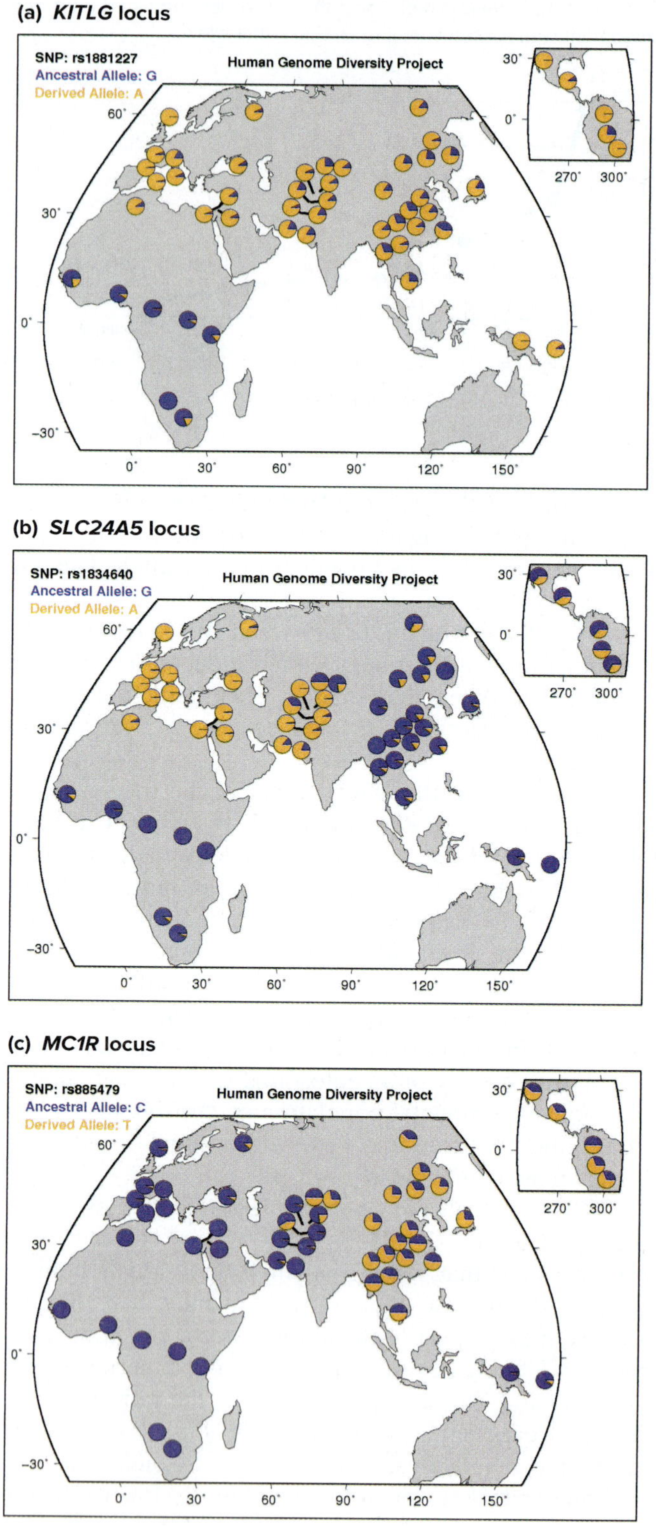

Another example of recent strong natural selection changing allele frequencies in different human populations is lactase persistence, which we introduced at the beginning of this chapter (review Fig. 21.1). Here, selection was brought about not by exposure to different environments, but rather through the development by humans of agriculture and the domestication of cattle that provided milk as a source of nutrition. The chance occurrence of mutations in regions upstream of the gene encoding the enzyme for milk sugar (lactose) digestion eliminated the turning off of gene expression past weaning that takes place in all other mammals. People who could digest milk as adolescents and adults apparently survived better and/or had more offspring when food was scarce, leading to a fitness advantage in certain parts of the world for the lactase persistence mutations.

Balancing Selective Forces Can Maintain Alleles in a Population

Sickle-cell anemia, which includes episodes of severe pain, serious anemia, and a probability of early death, is a recessive condition resulting from two copies of the sickle-cell allele at the β-globin gene ($Hb\beta$). It is thus surprising that the disease allele has not disappeared from several African populations, where it seems to have existed for a very long time.

One clue to the maintenance of the sickle-cell allele $Hb\beta^S$ in human populations lies in the observation that its allele frequency is highest in regions of Africa in which malaria is endemic (**Fig. 21.15**). A second clue is that heterozygotes for the normal and sickle-cell alleles ($Hb\beta^A$ $Hb\beta^S$) are resistant to malaria. This resistance is due, in part, to the fact that red blood cells containing a sickle-cell allele break open after being infected by the malaria parasite, destroying the parasite as well as the red blood cell itself. By contrast, in red blood cells with two normal hemoglobin alleles, the malaria parasite thrives. Thus, individuals of genotype $Hb\beta^A$ $Hb\beta^S$ have a **heterozygote advantage** in malaria-infested regions over either type of homozygote: The carriers are less susceptible to malaria than are $Hb\beta^A$ $Hb\beta^A$ homozygotes, and less susceptible to anemia than are $Hb\beta^S$ $Hb\beta^S$ homozygotes. Heterozygote advantage is one of several processes leading to **balancing selection** that actively maintains genetic polymorphisms.

To understand heterozygote advantage mathematically, assume that $Hb\beta^A$ $Hb\beta^S$ heterozygotes have the maximum relative fitness of 1, while the relative fitness for the $Hb\beta^A$ $Hb\beta^A$ homozygotes is W_{AA}, and the relative fitness for $Hb\beta^S$ $Hb\beta^S$ homozygotes is W_{aa}. (To simplify the following equations we are temporarily renaming the $Hb\beta^A$ allele as A whose frequency is p, and renaming the $Hb\beta^S$ allele as a whose frequency is q.) Selection will maintain both alleles in the population only if $\Delta q = 0$ for some value of q between 0 and 1. The q value at which $\Delta q = 0$ is known as the allele's *equilibrium frequency*. This value occurs when the term inside the brackets of Eq. (21.5) is 0, that is, when

$$[q(W_{aa} - W_{Aa}) - p(W_{AA} - W_{Aa})] = 0 \qquad (21.6)$$

Substituting $1 - q$ for p and solving Eq. (21.6) for q reveals that the equilibrium frequency of $Hb\beta^S$ represented by q_e is reached when

$$q_e = \frac{W_{AA} \quad W_{Aa}}{(W_{aa} - W_{Aa}) + (W_{AA} - W_{Aa})} \qquad (21.7)$$

Thus, to find the equilibrium frequency, that is, the value of q at which $\Delta q = 0$ such that both alleles persist in the population, you need know only the relative fitnesses for the two homozygotes, because W_{Aa} was set to 1.0.

On the other hand, if you know the equilibrium frequency and the relative fitness of one of the homozygotes, you can use Eq. (21.7) to estimate the relative fitness of the other homozygote. For example, we can assume that the African populations in which sickle-cell anemia is prevalent are roughly at equilibrium with regard to natural selection acting on the β-globin gene. Several field studies have revealed that the average frequency of the $Hb\beta^S$ allele in tropical populations is 0.17, so we will take this number as the equilibrium frequency q_e. Because the heterozygote $Hb\beta^A$ $Hb\beta^S$ has the highest fitness in areas with malaria, we will assign $W_{Aa} = 1$. Further, if you assume that $Hb\beta^S$ $Hb\beta^S$ homozygotes

Figure 21.15 High frequency of the sickle-cell allele $Hb\beta^S$ in regions of Africa where malaria is prevalent. **(a)** Geographical distribution of $Hb\beta^S$. **(b)** Geographical distribution of the malaria-causing parasite *Plasmodium falciparum*. **(c)** $Hb\beta^A$ $Hb\beta^S$ heterozygotes have decreased susceptibility to malaria, and thus have a selective advantage in areas with malaria relative to both $Hb\beta^A Hb\beta^A$ homozygotes who are fully susceptible to malaria and to $Hb\beta^S Hb\beta^S$ homozygotes who suffer from sickle cell anemia.

(a) Distribution of $Hb\beta^S$

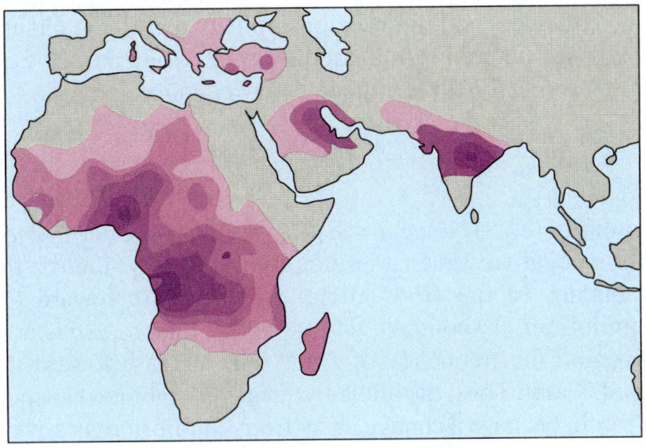

(b) Distribution of malaria

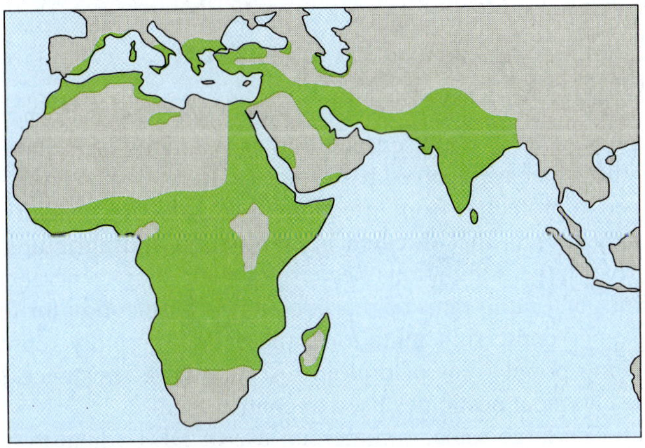

(c) $Hb\beta$ genotype fitnesses

Genotype:	$Hb\beta^A Hb\beta^A$	$Hb\beta^A Hb\beta^S$	$Hb\beta^S Hb\beta^S$
Relative fitness:	0.8	1.0	0

Equilibrium frequency of $Hb\beta^S = 0.17$
predicted (and observed) in areas with malaria

never reproduce, as was essentially true before medical advances enabled the survival of children expressing the sickle-cell trait, then $W_{aa} = 0$. With $W_{aa} = 0$, and $W_{Aa} = 1$, Eq. (21.7) can be rearranged to provide the following estimate of the relative fitness of the wild-type genotype W_{AA} given q_e:

$$W_{AA} = \frac{1 - 2q_e}{1 - q_e} = \frac{1 - 2(0.17)}{1 - 0.17} = 0.8$$

To understand the relationship between q, the change in q, and the equilibrium frequency q_e, you can use Eqs. (21.5) and (21.7) to formulate a new equation for Δq:

$$\Delta q = \frac{-pq[(1 - W_{AA}) + (1 - W_{aa})](q - q_e)}{\overline{W}} \qquad (21.8)$$

From Eq. (21.8), you can see that when q is greater than q_e, Δq is negative. Under these circumstances, q (that is, the frequency of the $Hb\beta^S$ allele) will decrease toward the equilibrium. By contrast, when q is less than q_e, Δq is positive and the frequency of $Hb\beta^S$ will increase toward the equilibrium. Thus, the allele frequency is stabilized at equilibrium, because a change away from equilibrium is always followed by a change back toward it.

A Comprehensive Example: Human Behavior Can Affect Evolution of Insect Pests

In Chapter 14, we discussed how populations of bacterial pathogens have evolved resistance to drugs humans developed to protect us from infections. Like infectious bacteria, many insects that threaten human health and agriculture spawn large populations because of their short generation times and rapid rates of reproduction. Via selection for resistance-conferring mutations, these large, rapidly reproducing populations of diploid insects evolve resistance to the chemical pesticides used to control them.

The large-scale, commercial use of DDT (dichlorodiphenyltrichloroethane) and other synthetic organic insecticides, begun in the 1940s, was at first highly successful at reducing crop destruction by agricultural pests, such as the boll weevil, and insect vectors of disease, such as the mosquitoes that transmit malaria and yellow fever. Within a few years, however, resistance to these insecticides was detectable in the targeted insect populations. In fact, resistance to every known insecticide has evolved within 10 years of its commercial introduction. Because different populations within a species can become resistant independently of other populations, insecticide resistance likely developed many separate times in many insect species since the introduction of insecticides.

Genetic studies show that insecticide resistance can result from mutations in several different genes. DDT, for example, is a nerve toxin in insects because it binds to a sodium channel protein and therefore disrupts the protein's function in nerve transmission. Some insects develop DDT resistance through recessive mutations in the channel-encoding gene that produce a channel protein that binds DDT poorly. Houseflies and certain mosquito species develop resistance to DDT through dominant mutations in other genes that encode enzymes that detoxify DDT, rendering it harmless to the insect. In some cases, these dominant alleles are mutations in gene regulatory regions that lead to the overexpression of a detoxifying enzyme.

Both recessive and dominant mutations causing DDT resistance are of concern for the control of insect pests, but we focus here on dominant mutations, which as we have seen can spread very rapidly through populations. Consider, for example, the dominant mutation R (for insecticide resistance), which occurs initially at low frequency in a population. Soon after the mutation appears, most of the R alleles are in Rr heterozygotes (in which r is the wild-type susceptibility allele). With the application of insecticide, strong selection favoring Rr heterozygotes will rapidly increase the frequency of the resistance allele in the population.

A field study of the use of DDT in Bangkok, Thailand, to control *Aedes aegypti* mosquitoes, the carriers of yellow fever, illustrates the rapid evolution of resistance. Spraying of the insecticide began in 1964 and was very effective in controlling the mosquitoes. Within a year, however, dominant DDT-resistant mutant alleles (R) emerged and rapidly increased in frequency. By mid-1967, the frequency of resistant RR homozygotes was nearly 100% (**Fig. 21.16**).

Because DDT became ineffective in reducing mosquito populations in Bangkok due to the near fixation of the DDT-resistant allele, the insecticide spraying program was stopped. The response of the mosquito population to the cessation of spraying was intriguing: The frequency of the R allele decreased rapidly, and by 1969, RR genotypes had virtually disappeared (Fig. 21.16).

The precipitous decline of the R allele suggests that in the absence of DDT, the RR genotype produces a lower

Figure 21.16 **How genotype frequencies among populations of *A. aegypti* mosquitoes changed in response to insecticide application.** Results observed after the insecticide DDT was used in a suburb of Bangkok, Thailand, beginning in 1964 and ending in 1968.

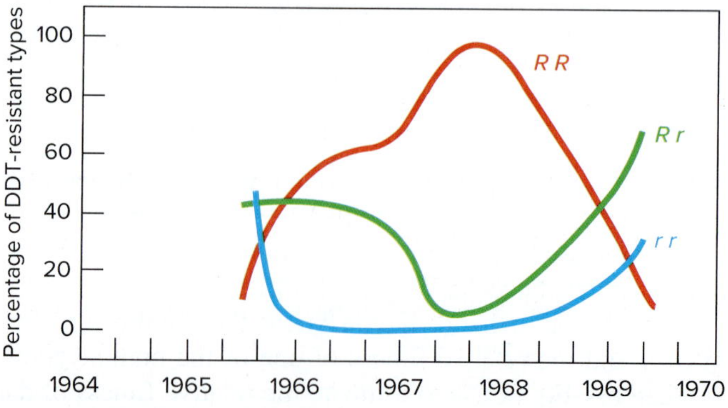

fitness than the *rr* genotype. In other words, the homozygous resistance genotype imposes a **fitness cost** on individuals such that in the absence of insecticide, resistance is subject to a negative selection that decreases the frequency of *R* in the population. This dependence of the fitness of individual genotypes on their environment is quite similar to the heterozygous advantage seen for humans carrying the mutation causing sickle-cell anemia in parts of the world in which malaria is endemic.

essential concepts

- Although the Hardy-Weinberg equation almost always provides accurate estimates of allele and genotype frequencies over the course of a few generations, it fails in the long run because real populations do not conform to the Hardy-Weinberg assumptions.
- In small populations, *genetic drift* due to random sampling of finite gamete pools can alter the frequency of an allele rapidly, until it eventually becomes either lost or *fixed*.
- Because genotypes may not display equal *fitness, natural selection* may increase or decrease allele frequencies in populations over time.
- Relative to an alternative genotype, the same genotype can be more fit in one environment, but less fit in a different environment. For example, heterozygotes for the sickle-cell mutation are resistant to malaria, explaining the high frequency of this allele in tropical populations of humans.
- The fitness benefits to insects of insecticide resistance often come with *fitness costs;* resistance allele frequencies thus change rapidly when humans apply or stop applying insecticides.

21.3 Ancestry and the Evolution of Modern Humans

learning objectives

1. Distinguish between an individual's biological and genetic ancestries.
2. Summarize the evidence supporting the origin of modern humans in Africa.
3. Explain how DNA sequencing has clarified the relationships of ancient and modern human lineages.

At the beginning of this chapter, we saw that human populations in different regions of the globe vary greatly in the frequency of alleles dictating lactose tolerance or intolerance. These current-day variations in allele frequency reflect many processes that occurred during the course of human history. The earliest humans almost certainly were all lactose intolerant, as are other primates. At least twice in human history, and in different geographical locations, mutations occurred that allowed expression of the lactase enzyme to continue through adulthood. The frequency of the mutant alleles increased rapidly in populations that raised dairy animals because of the selective advantage afforded by obtaining sustenance from milk products. Genetic drift in small populations of dairy herders may also have contributed to the spread of these alleles. The mutant alleles were then introduced into other populations by migration of individuals carrying the alleles.

This example illustrates that the existence of specific DNA variants, and the frequencies at which these variants are represented in different current-day human populations, serve as molecular fossils that can provide scientists with insights into the events that shaped human history. We explore here how population genetics provides important tools for *anthropology,* the study of humankind.

Shared Alleles Denote Common Genetic Ancestry

Individual people have two kinds of ancestors: **biological ancestors** and **genetic ancestors.** Biological ancestry is simply a description of who begat whom: You have two biological parents, four grandparents, eight great grandparents, and so forth (**Fig. 21.17a** and **b**). Any individual alive today could potentially have 2^k biological ancestors *k* generations ago, assuming that the ancestors in any one generation were unrelated. Thus, 20 generations ago (about 400 years) you could have had over 1 million biological ancestors, and 30 generations ago (about 600 years) more than 1 billion. This latter number is much higher than the number of humans thought to have been on the earth at that point in history (during the Middle Ages), the reason being that some ancestors in previous generations must have been related. But nonetheless, you still had a very large number of biological ancestors not so very far back in the past. In fact, you are almost certainly related to someone famous, and also to someone infamous.

The most recent common ancestor (MRCA)

Genetic ancestry refers to the actual inheritance of segments of the genome from biological ancestors (**Fig. 21.17c**). Comparison of Fig. 21.17b and c shows that for diploid regions of the human genome, we have many more biological ancestors than genetic ancestors. The reason is that we each have two parents, each of whom has two alleles at a diploid autosomal locus, but we inherit only one allele from each parent. As we just saw, the number of biological ancestors for a given person increases exponentially in every past generation. But in any single

Figure 21.17 Biological and genetic ancestries. (a) The great grandparents of Dion and Ana came from four different regions of the world. **(b)** Tracing the Y chromosomal, autosomal, and mitochondrial (Mt) contributions from biological ancestors to Dion and Ana. **(c)** Y chromosome DNA variation tracks the paternal lineage, while the mtDNA traces the maternal lineage. The autosomes (chromosomes 1–22) undergo recombination, so the ancestry of individual segments must be traced separately.

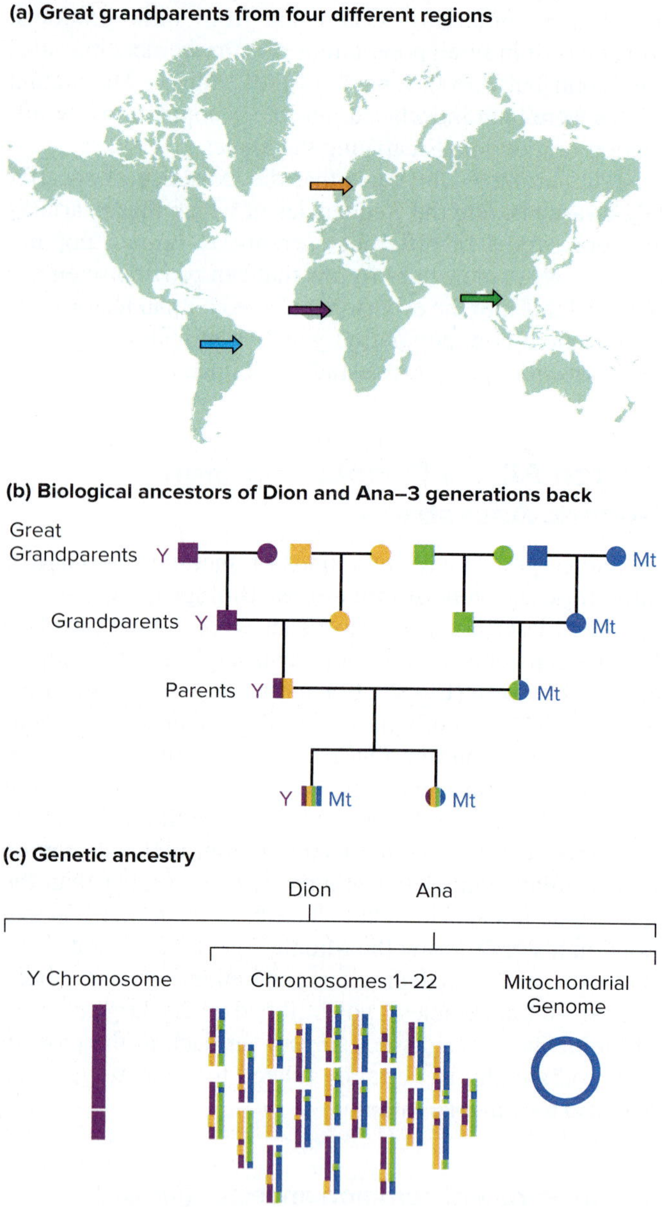

Figure 21.18 Tracing genetic ancestry. (a) For a single genomic locus, the genetic lineages of five people alive today coalesce to just one ancestral allele, the most recent common ancestor (MRCA) for that region of the genome. **(b)** Analysis of shared mutations allows scientists to trace lineages back to the MRCA.

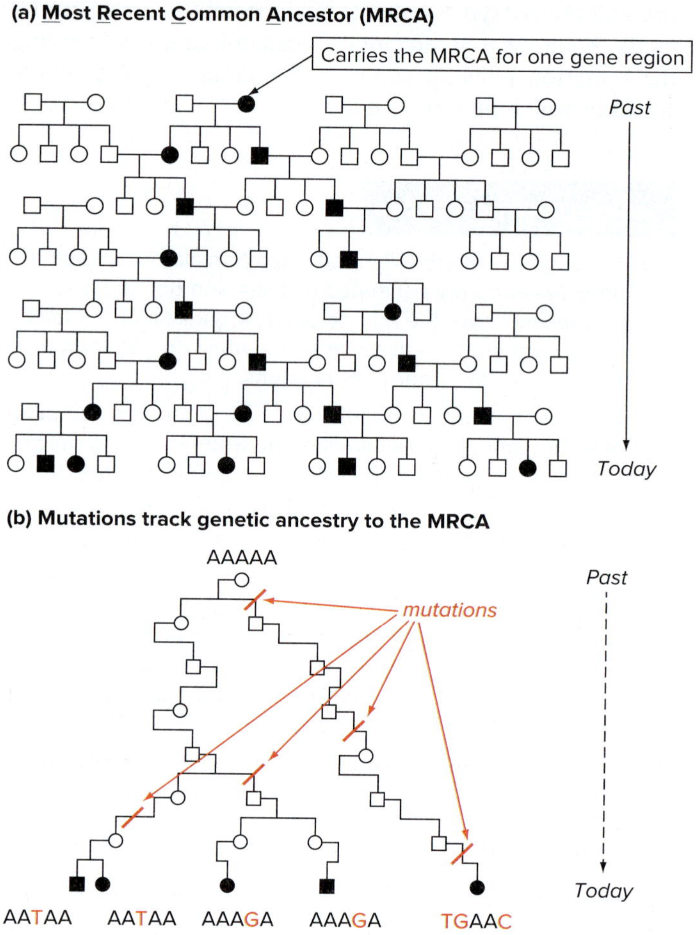

generation in the past, only two genetic ancestors exist for any given locus in that person's genome.

A highly useful term for describing genetic ancestry is the **MRCA,** or **most recent common ancestor** for a specific genomic region that is shared by two or more given people (**Fig. 21.18a**). The MRCA describes the most recent sequence of a DNA locus that existed in a past generation that was passed on in an unbroken line to two or more current individuals. It is important to understand that the MRCA is a past sequence of a region of DNA and *not* a particular individual in the past, although some common ancestor of the current individuals must have harbored the MRCA in his or her genome. In Fig. 21.18a, for example, the individual carrying the most recent common ancestor for an allele of a particular autosomal region in the five people alive today highlighted in *black* at the bottom of the figure existed seven generations ago.

Because of recombination, different regions of the genomes of two current-day relatives may have MRCAs that were present in different common ancestors. In fact, across their genomes, two relatives have MRCAs that originated in many individuals in the past. This statement reflects the fact that any given person inherits genomic bits and pieces from many (but not all) of his or her many biological ancestors (see Fig. 21.17c).

For any specific region of the genome, the MRCA for *all humans* is the most recent allele from which *all* current-day people have obtained DNA sequences in an unbroken line of descent. The MRCA for that particular locus was carried by an individual who coexisted with other people who also contributed some of their genes only to some (or to no) current-day humans.

In contrast to the recombining diploid portions of our genome, mitochondrial DNA (mtDNA) is passed directly from mothers to their offspring with no contribution from the fathers (Fig. 21.17b and c). Similarly, excluding the small PAR regions shared by the X and Y chromosomes, the DNA on Y chromosomes is passed directly from fathers to sons: Daughters do not inherit a Y chromosome (Fig. 21.17b and c). Each of us thus has just one genetic ancestor each generation for mtDNA or Y-chromosomal DNA. These sequences cannot have come from the same person because mtDNA is maternally derived and Y DNA is paternally derived. Due to a lack of recombination, the entire mtDNA and almost the entire Y chromosome each have only single MRCAs for all humans.

Lines of genetic ancestry revealed by mutations

Looking at the history of human populations through the lens of genetic ancestry is a powerful method for interpreting DNA sequence variation in present-day people. The different allele lineages *coalesce* to the MRCA as we go backwards in time (Fig. 21.18a). The MRCA thus provides a starting point for analysis, in the form of an ancestral allele whose descendant sequences are found in all people now on the earth.

Even though an unbroken line of descent connects the MRCA with sequences found in all modern-day people, this does not mean that all MRCA-related sequences are identical. The reason is that mutations can occur by chance along the lineages that connect each modern-day person to the MRCA, leaving trails in our genes of our genetic ancestors in the generations that separate us from the MRCA (**Fig. 21.18b**).

Researchers determine these lines of descent by analyzing mutations shared by present-day individuals. Because mutations accumulate over time, the longer the ancestry branch length between alleles in two individuals alive today, the more different are their DNA sequences. You can see the implications of this fact in Fig. 21.18b, where, as we have seen, alleles in five individuals alive at present trace back to a single MRCA just seven generations ago. The DNA sequence of that MRCA was AAAAA. As that sequence was passed down over generations, germ-line mutations could occur in it, creating novel descendant alleles. Note in Fig. 21.18b how the alleles of individuals closely related in the family history (such as the siblings at the bottom left) are identical in sequence (AATAA), but they differ at four of the five nucleotide positions from the individual at the bottom right (TGAAC), whose allele comes from a different family lineage that is nonetheless still derived from the MRCA at the top.

In Chapter 11, we saw how this picture of descent with modification allows population geneticists to interpret DNA sequence data obtained from many people in many different human populations (review Fig. 11.6). If a SNP allele is found both in some present-day humans and in some present-day members of other primate species such as chimpanzees, the allele is *ancestral* and must have been inherited directly from a common ancestor of humans and chimps without mutational modification. By contrast, for a *derived* SNP allele found in some human populations but not in other human groups nor in chimps, the mutational event must have occurred at some generation after the MRCA in a specific human sublineage. Human populations who share more alleles must have separated from each other more recently than populations whose DNA sequences are more divergent.

As was suggested in Fig. 21.17, tracing variations in mtDNA and Y-chromosome sequences is particularly valuable because these sequences provide clear estimates of matrilineal and patrilineal genetic ancestries, respectively. Our autosomes also provide ancestry information, but the data require more complex analysis because of diploid inheritance and recombination.

Modern Humans Originated in Africa

The rapid growth of inexpensive DNA sequencing technologies that allow data to be collected from easily obtained samples such as saliva has brought into view dramatically the genetic ancestries and relationships of contemporary humans.

Mitochondrial Eve and Y Chromosome Adam

The first insights came from the study of the matrilineally inherited mtDNA molecule, as was discussed previously in the Fast Forward Box in Chapter 15 entitled *Mitochondrial Eve*. These investigations established that people representing populations in sub-Saharan Africa, southeast Asia, Europe, aboriginal Australians, and aboriginal New Guineans all shared a most recent common ancestral mtDNA from a female—*Mitochondrial Eve*—who lived no more than 200,000 years ago. The estimate of time came from a calibration of the rate of accumulation of mutations in mtDNA from samples with independent archeological or geological estimates of times since divergence. This rate has been found to be remarkably constant for many lineages and genomes, and it thus constitutes a *molecular clock*.

Scientists further concluded that Mitochondrial Eve lived in Africa. The evidence for this statement is that African populations show much greater mtDNA sequence diversity than do populations in other parts of the world. That is, the branch point at which the lineages of the most different current-day Africans diverged occurred longer ago in history than the comparable branch point for any human populations found on other continents.

Studies of the patrilineally inherited Y chromosome reached a similar conclusion. The lineages leading to all present-day human Y chromosomes coalesce to an MRCA that must have been carried by a man—*Y chromosome Adam*—who lived 200,000 to 300,000 years ago. African populations display much greater diversity in Y chromosome sequences than is found within or even between populations from other parts of the world. These observations indicate that Y chromosome Adam was also African. Although it is possible that Mitochondrial Eve and Y chromosome Adam were contemporaneous, their lives were more likely separated by tens of thousands of years.

Tracking human population of the earth

These studies of mtDNA and Y chromosomal DNA variation, coupled with more recent surveys showing that the MRCAs for autosomal regions were also carried by individuals who lived in Africa, have revealed a remarkably consistent picture of human origins and the spread of humanity around the globe (**Fig. 21.19**). Modern humans all originated from a sub-Saharan African population that was established roughly 200,000 years ago. This population subsequently dispersed throughout Africa. Then, no later than 60,000 years ago, a subgroup of Africans left the continent and dispersed along a southern Asian route, followed by a more recent dispersal from Africa that settled in the Middle East. From these initial groups in Asia and the Middle East, people then spread out further to colonize the globe in several waves of migration.

The reason that populations outside of Africa have less genetic diversity than do African populations is that non-Africans all share a more recent common ancestor for all genomic regions than do Africans. Another way to think of this phenomenon is as a kind of founder effect, in which the subgroup that left Africa had only a subset of the genetic diversity that was found in Africa 60,000 to 80,000 years ago (**Fig. 21.20**).

DNA sequencing surveys reveal many striking and interesting stories about human ancestries. For example, one Y chromosome lineage that originated in Mongolia approximately 1000 years ago is now found in 8% of men throughout an extensive part of Asia, from the Caspian Sea to the Pacific Ocean; this lineage constitutes roughly 0.5% of the world's total male population today. Such rapid dispersal through such a large region could not have occurred by chance. Rather, it appears that the dissemination of this lineage accompanied the establishment of a massive land empire led by Genghis Khan and his male relatives. They slaughtered the males they encountered and fathered many children, whose descendants are now spread widely across southern Asia.

Figure 21.19 How modern humans populated the earth. A subgroup of humans left their ancestral home in Africa and moved across the Red Sea into southern Asia and then into Australia; subsequently other members of the subgroup migrated from Africa into the Middle East. People then spread to other parts of the globe via the major routes of migration shown. This map incorporates not only genetic data but also anthropological, cultural, language, and paleontological information.

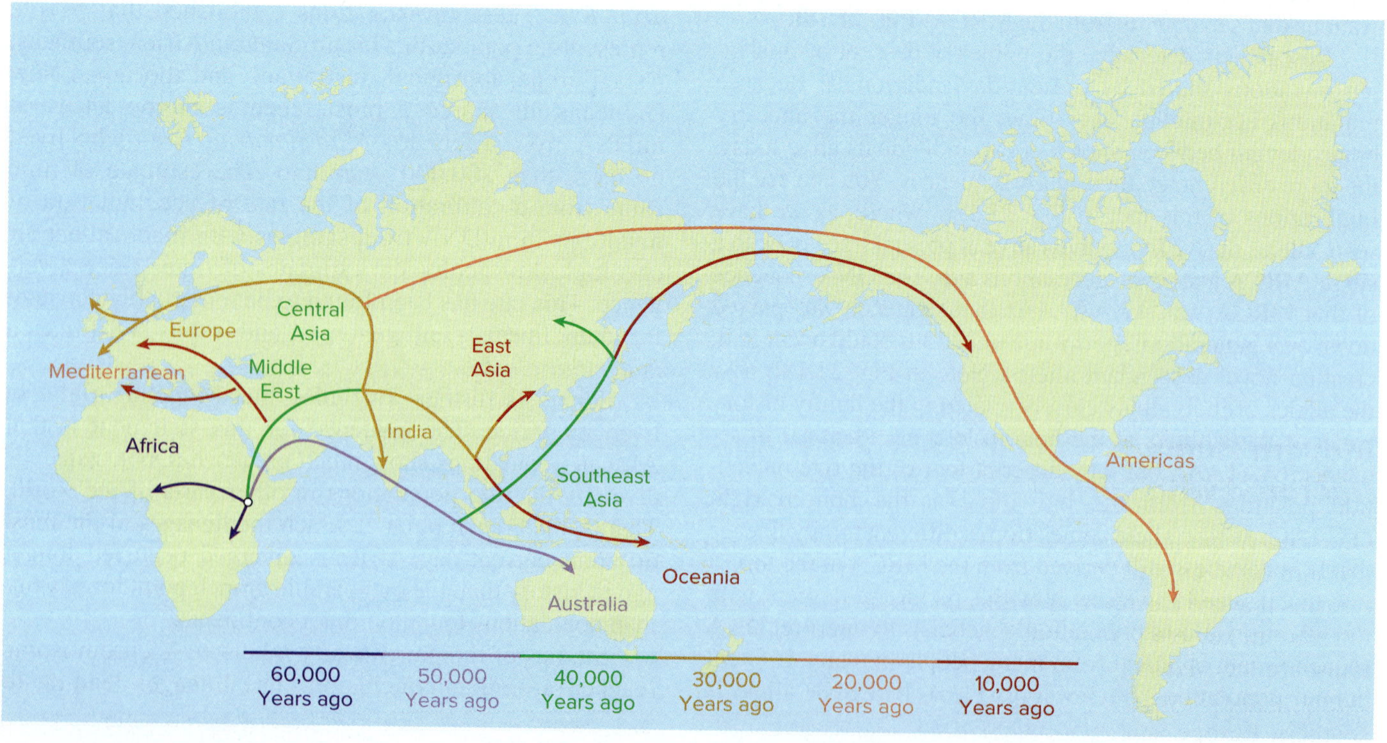

Figure 21.20 Why the genetic diversity of humans in Africa is greater than that elsewhere in the world. Human migration out of Africa no later than 60,000 years ago was initiated by a small founder population that contained only a fraction of the genetic diversity present on the African continent at that time. *Colors* represent different alleles of a given genomic region. (*Homo sapiens sapiens* is the subspecies of humans alive today.)

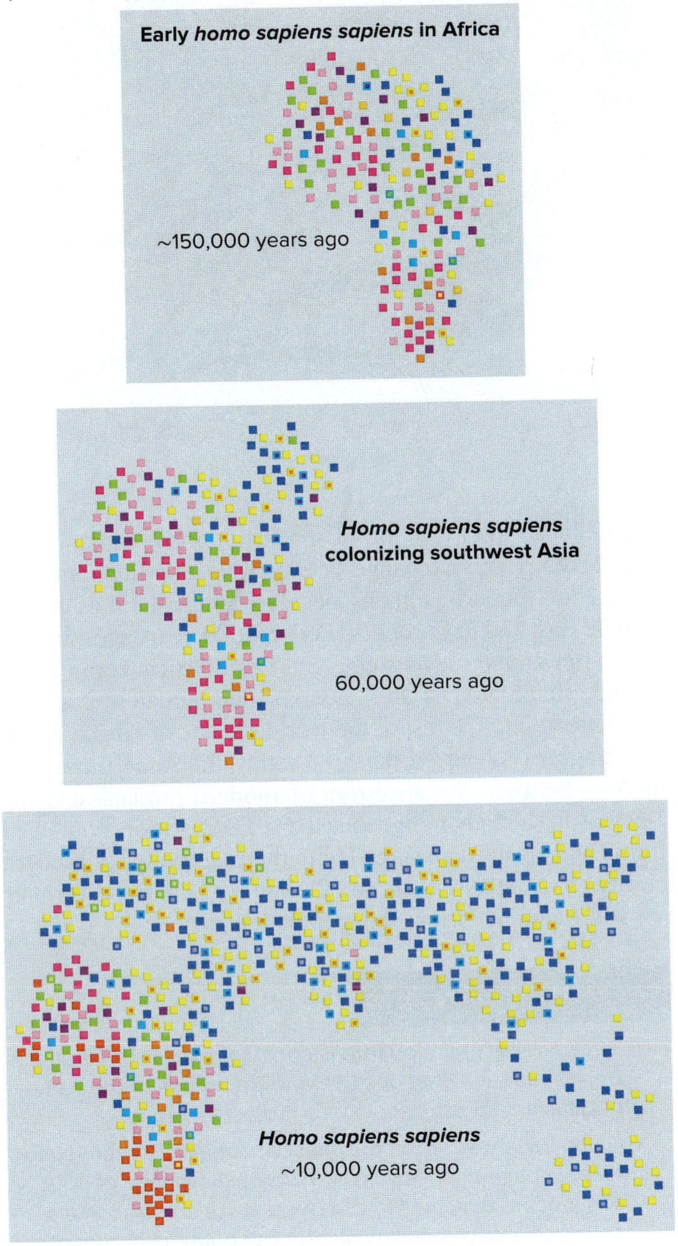

Modern Humans in Europe and Asia Interbred with Other Hominins

One long-running issue in anthropology has been the relationship between anatomically modern humans and fossil remains that are clearly humanlike *hominins* but also quite different in key elements of morphology. Of particular interest are the *Neanderthals,* whose fossils, dating to as recently as 30,000 years ago, have been found in caves across southern Europe and into central Asia. These specimens showed a very rugged physique with large-browed and heavily boned skulls (**Fig. 21.21a**). Were Neanderthals a completely separate lineage of the genus *Homo* that died out? Or, given that they coexisted, possibly for 10,000 or so years, with other populations more clearly related to our own, is it conceivable that Neanderthals bred with individuals from these other populations, so that some Neanderthal genes are found in current-day humans?

Figure 21.21 Neanderthals: Archaic humans. (a) Comparisons of full skeletons of a Neanderthal (*left*) and a modern human (*right*). **(b)** Artist's reconstruction of a Neanderthal face.

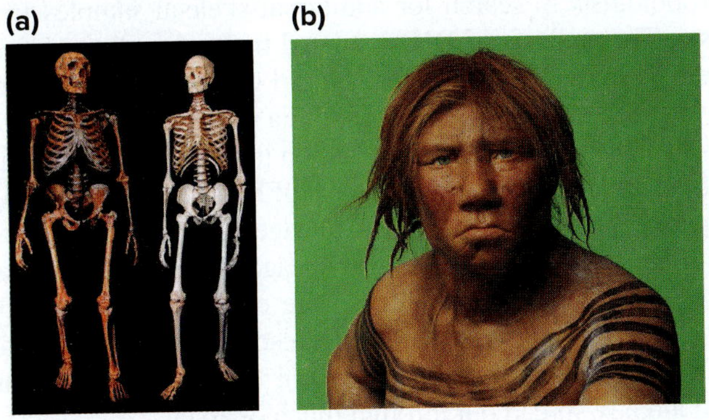

Neanderthal and Denisovan genomic DNA sequences

Recently, scientists have been able to address questions concerning the relationship between *Homo sapiens* and Neanderthals by comparing genome sequences of current-day humans with those obtained from particularly well-preserved Neanderthal skeletons such as that shown in Fig. 21.21a. Remarkably, researchers have, for example, been able to sequence the full nuclear genome from a fragment of a Neanderthal femur (a leg bone) found in a cave in Croatia and dated to approximately 38,000 years ago. Several lines of evidence, including the finding that the sequence divergence between Neanderthal DNA and that of any present-day human is several-fold higher than that between any two present-day humans, resulted in the estimate that the hominin lineages leading to Neanderthals and *Homo sapiens* diverged between 500,000 and 800,000 years ago.

The DNA sequence obtained from this Neanderthal femur and from the bones of several additional Neanderthal individuals allows scientists to describe certain morphological characteristics of Neanderthals. In the reconstruction shown in **Fig. 21.21b,** the hair of this Neanderthal man is reddish and the skin color light. These traits are not a figment of the artist's imagination, but are instead specific predictions from the sequences of the *melanocortin 1 receptor* gene (*MC1R*). Several Neanderthal specimens have a variant of this gene that is not found in modern humans, but that is nonetheless similar to an allele in some modern humans that encodes an MC1R protein that functions inefficiently.

People with this allele of the *MC1R* gene have a higher-than-average probability of having red hair and fair skin.

The success of studying ancient samples of Neanderthal bone fragments obtained from caves in Europe motivated anthropologists to search for additional skeletal samples from Asia. The analysis of DNA isolated from a small bone fragment from the tip of a finger of a girl who lived and died in a cave in Siberia just north of Mongolia revealed an unexpected result. The pattern of DNA variation indicated that she represented a previously unknown type of hominin whose lineage separated from the Neanderthal lineage perhaps 600,000 years ago and from the modern human lineage roughly 800,000 years ago. Individuals in this lineage are called *Denisovans;* the fossil record indicates that the Denisovans died out roughly 30,000 years ago, during a period in which modern humans had already spread out through much of Asia.

Hominin interbreeding and human history

Given that modern humans coexisted with Neanderthals and Denisovans possibly for thousands of years, did our ancestors interbreed with these other hominins at least to some degree? Did the Neanderthals and Denisovans mate with each other before their lineages died out? Comparisons of DNA sequences from many modern humans, various nonhominin primates such as chimpanzees, multiple Neanderthals, and now a few Denisovans allowed investigators to conclude that all of these types of interbreeding between hominin groups in fact occurred.

A simple statistical approach allows population geneticists to estimate ancestral interbreeding between genetically distinct lineages. They compare, for example, the sequence variants found at individual positions in the genomes of two different humans (H_1 and H_2) to those of a Neanderthal (*N*) and a chimpanzee (*C*). The scientists then focus their attention on sites where the Neanderthal does not match the chimpanzee [that is, derived variants that could not have been inherited from a common ancestor of all three lineages (*H*, *N*, and *C*); this idea was previously illustrated in Fig. 11.6]. If H_1 and H_2 differ significantly in the percentage of matches to the Neanderthal sequences at such sites, then some present-day humans but not others inherited variants from Neanderthals.

Strikingly, roughly 2% of the variants found in Neanderthals but not chimpanzees are found in the genomes of contemporary humans living outside of Africa, while people indigenous to Africa show few to none of the Neanderthal-specific variants. Application of this test to Denisovan sequences shows a different pattern in which Denisovan ancestry in contemporary human populations is restricted largely to southeastern Asia and the south Pacific (Oceania).

These insights into past interbreeding as well as inferences of the relatedness of DNA sequences of contemporary and ancient human lineages (using the logic of *descent with modification* illustrated in Fig. 21.18b) suggest a hypothetical timetable for the evolution of modern humans illustrated in **Fig. 21.22.** Gene exchange between Neanderthals and the lineage of modern humans occurred several times in the Middle East and in Europe 30,000–85,000 years ago, starting soon after *Homo sapiens* emerged from Africa. From there, Neanderthal DNA variants accompanied human migrants as they spread into the rest of the world outside of Africa. Interbreeding also occurred during roughly the same timeframe in Asia between the ancestors of modern humans and the Denisovans already indigenous there. DNA variants obtained from Denisovans then spread with migrant groups of modern humans to Southeast Asia and Oceania, where these variants are still found in current-day populations.

Figure 21.22 Hypothetical evolutionary relationships of modern humans, Neanderthals, and Denisovans inferred from their genomic DNA sequences. Overlaps between branches of different colors indicate gene transfer through interbreeding of different hominins. *kya* = thousand years ago.

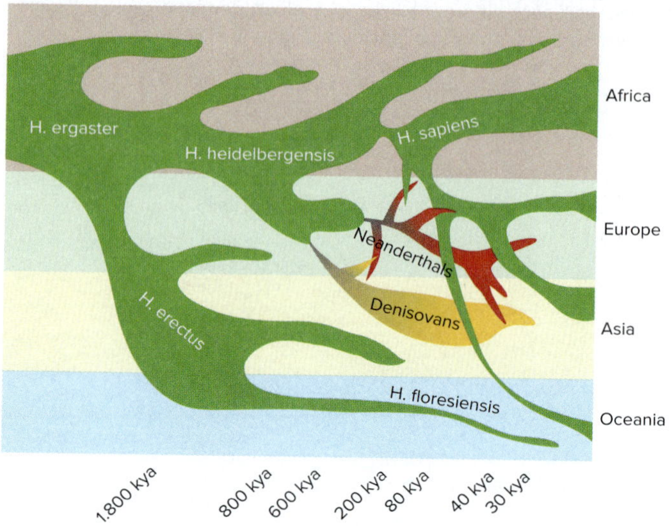

essential concepts

- Because the mutation rate is constant over time, neutral DNA sequence changes between genomes of the same or different species serve as a *molecular clock.*
- Going backwards in time, genetic lineages coalesce to *most recent common ancestors* (*MRCAs*). Examination of DNA sequences using the logic of *descent with modifications* allows scientists to make inferences about the history of lineages.
- The fact that the diversity of DNA sequences among Africans is greater than that found among non-Africans suggests that modern humans first evolved in Africa, and that a subset of Africans later emigrated to populate the rest of the earth.
- Certain, but not all, populations of modern humans share alleles with Neanderthals and Denisovans, indicating some degree of interbreeding between these lineages after the emergence of *Homo sapiens* from Africa.

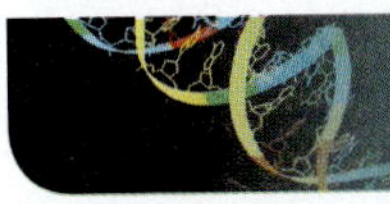

SOLVED PROBLEMS

I. A population called the *founder generation,* consisting of 2000 *AA* individuals, 2000 *Aa* individuals, and 6000 *aa* individuals is established on a remote island. Mating within this population occurs at random, the three genotypes are selectively neutral, and mutations occur at a negligible rate.

a. What are the frequencies of alleles *A* and *a* in the founder generation?

b. Is the founder generation at Hardy-Weinberg equilibrium?

c. What is the frequency of the *A* allele in the second generation (that is, the generation subsequent to the founder generation)?

d. What are the frequencies for the *AA*, *Aa*, and *aa* genotypes in the second generation?

e. Is the second generation at Hardy-Weinberg equilibrium?

f. What are the frequencies for the *AA*, *Aa*, and *aa* genotypes in the third generation?

Answer

This question requires calculation of allele and genotype frequencies and an understanding of the Hardy-Weinberg equilibrium principle.

a. To calculate allele frequencies, count the total alleles represented in individuals with each genotype and divide by the total number of alleles.

Number of individuals	Number of *A* alleles	Number of *a* alleles
2000 *AA*	4000	0
2000 *Aa*	2000	2000
6000 *aa*	0	12,000
Total	6000	14,000

The frequency of the *A* allele (p) = 6000/20,000 = 0.3.

The frequency of the *a* allele (q) = 14,000/20,000 = 0.7.

b. If a population is at Hardy-Weinberg equilibrium, the genotype frequencies are p^2, $2pq$, and q^2. We calculated in part (a) that $p = 0.3$ and $q = 0.7$ in this population. Therefore,

$$p^2 = (0.3)^2 = 0.09$$

$$2pq = 2(0.3)(0.7) = 0.42$$

$$q^2 = (0.7)^2 = 0.49$$

For a population of 10,000 individuals, the number of individuals with each genotype, if the population were at equilibrium and the allele frequencies were $p = 0.3$ and $q = 0.7$, would be *AA*, 900; *Aa*, 4200; and *aa*, 4900. Therefore, the founder population described is not at equilibrium.

c. Given the conditions of random mating, selectively neutral alleles, and no new mutations, allele frequencies do not change from one generation to the next; $f(A) = p = 0.3$, and $f(a) = q = 0.7$.

d. The genotype frequencies for the second generation would be those calculated for part (b) because in one generation the population will go to equilibrium. $AA = p^2 = 0.09$; $Aa = 2pq = 0.42$; and $aa = q^2 = 0.49$.

e. Yes, in one generation a population not at equilibrium will go to equilibrium if mating is random and no selection or significant mutation exists.

f. The genotype frequencies will be the same in the third generation as in the second generation.

II. Two alleles have been found at the X-linked phosphoglucomutase gene (*Pgm*) in a *Drosophila persimilis* population in California. The frequency of the Pgm^A allele is 0.25, while the frequency of the Pgm^B allele is 0.75. Assuming the population is at Hardy-Weinberg equilibrium, what are the expected genotype frequencies in males and females?

Answer

This problem requires application of the concept of allele and genotype frequencies to X-linked genes. For X-linked genes, males (XY) have only one copy of the X chromosome, so the genotype frequency is equal to the allele frequency. Therefore, $p = 0.25$ and $q = 0.75$.

The frequency of male flies with genotype $X^{PgmA}Y$ is 0.25; the frequency of males with genotype $X^{PgmB}Y$ is 0.75. Three genotypes exist for females: $X^{PgmA}\,X^{PgmA}$, $X^{PgmA}\,X^{PgmB}$, and $X^{PgmB}\,X^{PgmB}$ corresponding to p^2, $2pq$, and q^2. The frequencies of female flies with these three genotypes are $(0.25)^2$, $2(0.25)(0.75)$, and $(0.75)^2$; or 0.0625, 0.375, and 0.5625, respectively.

III. Three different genes (*red, blue,* and *green*) each have two alleles; one recessive allele for each gene has deleterious effects. The relative fitness of the recessive homozygotes for one of these genes is 0.9, for another it is 0.8, and for the third it is 0.7. The following graph depicts changes in the frequencies of these alleles in populations of infinite size over time; in each case, the frequency of the allele in question (q) is 0.7 at the beginning of the experiment.

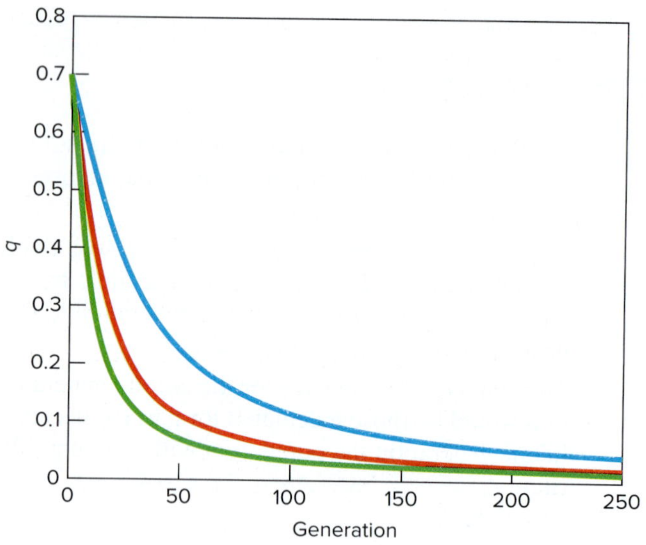

a. For each gene (identified by the fitness of the recessive homozygotes), calculate Δq between the parental generation and the first generation of progeny (q = frequency of the recessive deleterious allele). Assume that the relative fitnesses of heterozygotes and of homozygotes for the dominant allele are 1.0. Then calculate q' (the frequency of the recessive allele in the first generation of progeny). You do not yet need to know which gene corresponds to which color in the graph.

b. Which of the three genes (*blue, red,* or *green*) is the one for which the relative fitness of the recessive homozygote is 0.9? 0.8? 0.7?

c. Briefly explain why Δq will become a smaller negative number in each successive generation for all of these three genes.

d. Briefly explain why q would never go to 0 in any of these populations of infinite size.

e. Alleles do disappear from real populations, but not in the populations examined in the graph. How can this be?

Answer

a. For one of the genes (which we will call gene A for the time being), $W_{AA} = 1.0$; $W_{Aa} = 1.0$; $W_{aa} = 0.9$. For gene B, $W_{BB} = 1.0$; $W_{Bb} = 1.0$; $W_{bb} = 0.8$. For the remaining gene C, $W_{CC} = 1.0$; $W_{Cc} = 1.0$; $W_{cc} = 0.7$. (That is, the deleterious effects involved are all completely recessive.) To make the desired calculations, you would plug in these fitness values into Eq. (21.5). For gene A, $\Delta q = -0.0154$ and $q' = 0.684$. For gene B, $\Delta q = -0.0326$ and $q' = 0.667$. For gene C, $\Delta q = -0.0517$ and $q' = 0.648$.

b. For the *green* gene, the change in q is steeper than for the other genes, meaning that Δq must be the highest negative number in any given generation. Thus, the recessive homozygote fitness for the *green* gene must be 0.7 (this is gene C), that for the *red* gene (B) is 0.8, and that for the *blue* gene (A) must be 0.9.

c. The rate of change decreases for each gene in every successive generation because each generation has a successively smaller proportion of homozygotes for the recessive allele who would be subject to selection.

d. The populations of infinite size would always have heterozygous individuals who would retain the deleterious allele but would not be selected against.

e. Real populations are not infinite in size. Thus, genetic drift would eventually cause the deleterious allele to be lost from the population.

PROBLEMS

Vocabulary

1. Choose the best matching phrase in the right column for each of the terms in the left column.

a. fitness	1. the genotype with the highest fitness is the heterozygote
b. gene pool	2. chance fluctuations in allele frequency
c. fitness cost	3. mutations accumulate at a relatively constant rate
d. allele frequencies	4. $p = 1.0$
e. heterozygote advantage	5. ability to survive and reproduce
f. equilibrium frequency	6. p and q
g. genetic drift	7. event that drastically lowers N
h. molecular clock	8. the advantage of a particular genotype in one situation is a disadvantage in another situation
i. population bottleneck	9. frequency of an allele at which $\Delta q = 0$
j. hominins	10. collection of alleles carried by all members of a population
k. fixation	11. *Homo sapiens,* Neanderthals, and Denisovans

Section 21.1

2. When an allele is dominant, why does it not always increase in frequency to produce the phenotype proportion of 3:1 (3/4 dominant : 1/4 recessive individuals) in a population?
3. A population with an allele frequency (p) of 0.5 and a genotype frequency (p^2) of 0.25 is at equilibrium. How can you explain the fact that a population with an allele frequency (p) of 0.1 and a genotype frequency (p^2) of 0.01 is also at equilibrium?
4. In a certain population of frogs, 120 are green, 60 are brownish green, and 20 are brown. The allele for brown is denoted G^B, while that for green is G^G, and these two alleles are incompletely dominant.
   a. What are the genotype frequencies in the population?
   b. What are the allele frequencies of G^B and G^G in this population?
   c. What are the expected frequencies of the genotypes if the population is at Hardy-Weinberg equilibrium?
5. Which of the following populations are at Hardy-Weinberg equilibrium?

Population	*AA*	*Aa*	*aa*
I	0.25	0.50	0.25
II	0.10	0.74	0.16
III	0.64	0.27	0.09
IV	0.46	0.50	0.04
V	0.81	0.18	0.01

6. A dominant mutation in *Drosophila* called *Delta* causes changes in wing morphology in *Delta*/+ heterozygotes. Homozygosity for this mutation (*Delta* / *Delta*) is lethal prior to the adult stage. In a population of 150 flies, it was determined that 60 had normal wings and 90 had abnormal wings.
   a. What are the allele frequencies in this population?
   b. Using the allele frequencies calculated in part (a), how many total zygotes must be produced by this population in order for you to count 160 viable adults in the next generation?
   c. Given that there is random mating, no migration, and no mutation, and ignoring the effects of genetic drift, what are the expected numbers of the different genotypes in the next generation if 160 viable offspring of the population in part (a) are counted?
   d. Is this next generation at Hardy-Weinberg equilibrium? Why or why not?
7. A large, random mating population is started with the following proportion of individuals for the indicated blood types:

   0.5 MM
   0.2 MN
   0.3 NN

   This blood type gene is autosomal, and the *M* and *N* alleles are codominant.
   a. Is this population at Hardy-Weinberg equilibrium?
   b. What will be the allele and genotype frequencies after one generation under the conditions assumed for Hardy-Weinberg equilibrium?
   c. What will be the allele and genotype frequencies after two generations under the conditions assumed for Hardy-Weinberg equilibrium?
8. A gene called Q has two alleles, Q^F and Q^G, that encode alternative forms of a red blood cell protein that allows blood group typing. A different, independently segregating gene called R has two alleles, R^C and R^D, permitting a different kind of blood group typing. A random, representative population of football fans was examined, and on the basis of their blood typing, the following distribution of genotypes was inferred (all genotypes were distributed equally between males and females):

$Q^FQ^F\ R^CR^C$	202
$Q^FQ^G\ R^CR^C$	101
$Q^GQ^G\ R^CR^C$	101
$Q^FQ^F\ R^CR^D$	372
$Q^FQ^G\ R^CR^D$	186
$Q^GQ^G\ R^CR^D$	186
$Q^FQ^F\ R^DR^D$	166
$Q^FQ^G\ R^DR^D$	83
$Q^GQ^G\ R^DR^D$	83

   This sample contains 1480 fans in total.
   a. Is the population at Hardy-Weinberg equilibrium with respect to either or both of the Q and R genes?
   b. After one generation of random mating within this group, what fraction of the *next* generation of football fans will be Q^FQ^F (independent of their R genotype)?
   c. After one generation of random mating, what fraction of the *next* generation of football fans will be R^CR^C (independent of their Q genotype)?
   d. What is the chance that the first child of a Q^FQ^G R^CR^D female and a Q^FQ^F R^CR^D male will be a Q^FQ^G R^DR^D male?
9. Alkaptonuria is a recessive autosomal genetic disorder associated with darkening of the urine. In the

United States, approximately 1 out of every 250,000 people has alkaptonuria.

a. Assuming Hardy-Weinberg equilibrium, estimate the frequency of the allele responsible for this trait.

b. What proportion of people in the U.S. population are carriers for this trait? In this population, what is the ratio of carriers to individuals affected by alkaptonuria?

c. If a woman without alkaptonuria had a child with this trait with one husband then remarried, what is the chance that a child produced by her second marriage would have alkaptonuria?

d. Alkaptonuria is a relatively benign condition, so there is little selective advantage to individuals with any genotype; as a result, your assumption of Hardy-Weinberg equilibrium in part (a) is reasonable. Could you also use the assumption of Hardy-Weinberg equilibrium to estimate the allele frequencies and carrier frequencies of more severe recessive autosomal conditions such as cystic fibrosis? Explain.

10. Two hypothetical lizard populations found on opposite sides of a mountain in the Arizonan desert have two alleles (A^F, A^S) of a single gene A with the following three genotype frequencies:

	A^FA^F	A^FA^S	A^SA^S
Population 1	38	44	18
Population 2	0	80	20

a. What is the allele frequency of A^F in the two populations?

b. Do either of the two populations appear to be at Hardy-Weinberg equilibrium?

c. A huge flood opened a canyon in the mountain range separating populations 1 and 2. They were then able to migrate such that the two populations, which were of equal size, mixed completely and mated at random. What are the frequencies of the three genotypes (A^FA^F, A^FA^S, and A^SA^S) in the next generation of the single new population of lizards?

11. It is the year 1998, and the men and women sailors (in equal numbers) on the American ship the *Medischol Bounty* have mutinied in the South Pacific and settled on the island of Bali Hai, where they have come into contact with the local Polynesian population. Of the 400 sailors that come ashore on the island, 324 have MM blood type, 4 have the NN blood type, and 72 have the MN blood type. Already on the island are 600 Polynesians between the ages of 19 and 23. In the Polynesian population, the allele frequency of the *M* allele is 0.06, and the allele frequency of the *N* allele is 0.94. No other people come to the island over the next 10 years.

a. What is the allele frequency of the *N* allele in the sailor population that mutinied?

b. It is the year 2008, and 1000 children have been born on the island of Bali Hai. If the mixed population of 1000 young people on the island in 1998 mated randomly and the different blood group phenotypes had no effect on viability, how many of the 1000 children would you expect to have MN blood type?

c. In fact, 50 children have MM blood type, 850 have MN blood type, and 100 have NN blood type. What is the observed frequency of the *N* allele among the children?

12. a. Alleles of genes on the X chromosome can also be at equilibrium, but the equilibrium frequencies under the Hardy-Weinberg assumptions must be calculated separately for the two sexes. For a gene with two alleles *A* and *a* at frequencies of p and q, respectively, write expressions that describe the equilibrium frequencies for all the genotypes in men and women.

b. Approximately 1 in 10,000 males in the United States is afflicted with hemophilia, an X-linked recessive condition. If you assume that the population is at Hardy-Weinberg equilibrium, what proportion of American females would be hemophiliacs? About how many female hemophiliacs would you expect to find among the 170 million women living in the United States? (Assume that all females with hemophilia are homozygous for the disease allele.)

13. In 1927, the ophthalmologist George Waaler tested 9049 schoolboys in Oslo, Norway, for red-green color blindness and found 8324 of them to be normal and 725 to be color blind. He also tested 9072 schoolgirls and found 9032 that had normal color vision while 40 were color blind.

a. Assuming that the same sex-linked recessive allele c causes all forms of red-green color blindness, calculate the allele frequencies of c and C (the allele for normal vision) from the data for the schoolboys. (*Hint:* Refer to your answer to Problem 12a.)

b. Does Waaler's sample demonstrate Hardy-Weinberg equilibrium for alleles of this gene? Explain your answer by describing observations that are either consistent or inconsistent with this hypothesis.

On closer analysis of these schoolchildren, Waaler found that there was actually more than one c allele causing color blindness in his sample: one kind for the *prot* type (c^p) and one for the *deuter* type (c^d). (Protanopia and deuteranopia are slightly different forms of red-green color blindness.) Importantly, some of the apparently normal females in Waaler's studies were probably of genotype c^p/c^d. Through further analysis of the 40 color-blind females, he found that 3 were prot (c^p/c^p), and 37 were deuter (c^d/c^d).

c. Based on this new information, what are the frequencies of the c^p, c^d, and C alleles in the population examined by Waaler? Calculate these values as if the frequencies obey the Hardy-Weinberg equilibrium. (*Note:* Again, refer to your answer to Problem 12a.)

d. Calculate the frequencies of all genotypes expected among men and women if the population is at equilibrium.

e. Do these results make it more likely or less likely that the population in Oslo is indeed at equilibrium for red-green color blindness? Explain your reasoning.

14. The equation $p^2 + 2pq + q^2 = 1$ representing the Hardy-Weinberg proportions examines genes with only two alleles in a population.

a. Derive a similar equation describing the equilibrium proportions of genotypes for a gene with three alleles. [*Hint:* Remember that the Hardy-Weinberg equation can be written as the binomial expansion $(p + q)^2$.]

b. A single gene with three alleles (I^A, I^B, and i) is responsible for the ABO blood groups. Individuals with blood type A can be either $I^A I^A$ or $I^A i$; those with blood type B can be either $I^B I^B$ or $I^B i$; people with AB blood are $I^A I^B$, and type O individuals are ii. Among Armenians, the frequency of I^A is 0.360, the frequency of I^B is 0.104, and the frequency of i is 0.536. Calculate the frequencies of individuals in this population with the four possible blood types, assuming Hardy-Weinberg equilibrium.

In Problems 15–17, you will see that because mating between individuals within populations at Hardy-Weinberg equilibrium is random, it is possible to predict *mating frequencies:* that is, the proportion of all matings in the population between individuals of particular genotypes or phenotypes.

15. A gene has two alleles A (frequency $= p$) and a (frequency $= q$). If a population is at Hardy-Weinberg equilibrium, develop mathematical expressions in terms of p and q that predict the following mating frequencies:

a. Between two AA homozygotes

b. Between two aa homozygotes

c. Between two Aa heterozygotes

d. Between an AA homozygote and an aa homozygote

e. Between an AA homozyote and an Aa heterozygote

f. Between an aa homozygote and an Aa heterozygote

Considering your answers to parts (a)–(f):

g. Do the six possibilities listed account for all possible matings? How would you know whether this is true mathematically? Demonstrate this latter point by setting p equal to an arbitrary number between 0 and 1 such as 0.2.

h. Can you develop a simple, general rule for calculating the mating frequencies between individuals of the same genotype versus the mating frequencies between individuals of different genotypes?

i. If the population is equally divided between males and females, what proportion of all matings will be between an AA male and an AA female? Between an AA male and an Aa female? Between an AA male and an aa female?

16. Some people can taste the bitter compound phenylthiocarbamide while others cannot. This trait is governed by a single autosomal gene; the allele for tasting is completely dominant with respect to the allele for nontasting. Among 1707 Hawaiians tested for the ability to taste, 1326 tasters were found. Assuming that the population is at Hardy-Weinberg equilibrium for this gene and that mating is purely random:

a. What are the allele frequencies for the tasting allele T [= (p)] and for the nontasting allele t [= (q)]?

b. What are the genotype frequencies in the population?

c. Of all the matings in the population, what proportion will be between two nontasters?

d. Of all the matings in the population, what proportion will be between a taster and a nontaster?

e. Of all the matings in the population, what proportion will be between a taster male and a nontaster female?

f. What proportion of all of the progeny produced by all matings between a taster male and a nontaster female will be nontasters?

g. Of all the matings in the population, what proportion will be between two tasters?

17. *Androgenetic alopecia* (pattern baldness) is a complex trait in humans governed by several genes, but suppose a human population exists in which a single autosomal allele determines pattern baldness. This allele is dominant in males and recessive in females. The population is in Hardy-Weinberg equilibrium, and 51% of the men are bald.

a. What is the allele frequency of the baldness allele among males?

b. What is the allele frequency of the baldness allele among females?

c. What percentage of the women in this population will exhibit pattern baldness?

d. Assuming random mating, what proportion of all matings should be between a bald man and a nonbald woman?

e. What percentage of the bald men in the population are heterozygotes?

f. If a nonbald couple produces a bald son, what is the probability that their next son will be bald?

g. A woman with androgenetic alopecia has a daughter, but nothing is known about the father. What is the probability that the daughter will be bald?

18. The following figure shows the FBI-style analysis of the genomic DNA of 10 people (1–10), and also of hair found at a crime scene left by the murderer [***]. This analysis involves the PCR amplification of SSR loci, each from a different (nonhomologous) chromosome. The PCR primers are for each SSR locus are labeled with a unique fluorescent molecule. Some bands are thicker because relatively more of the corresponding PCR product was obtained. The figure has dots aligned on both sides to help you find the crucial bands; it will help to use a straight-edge as a guide. The numbers at right are the total number of copies of the SSR locus among the population of 11 samples.

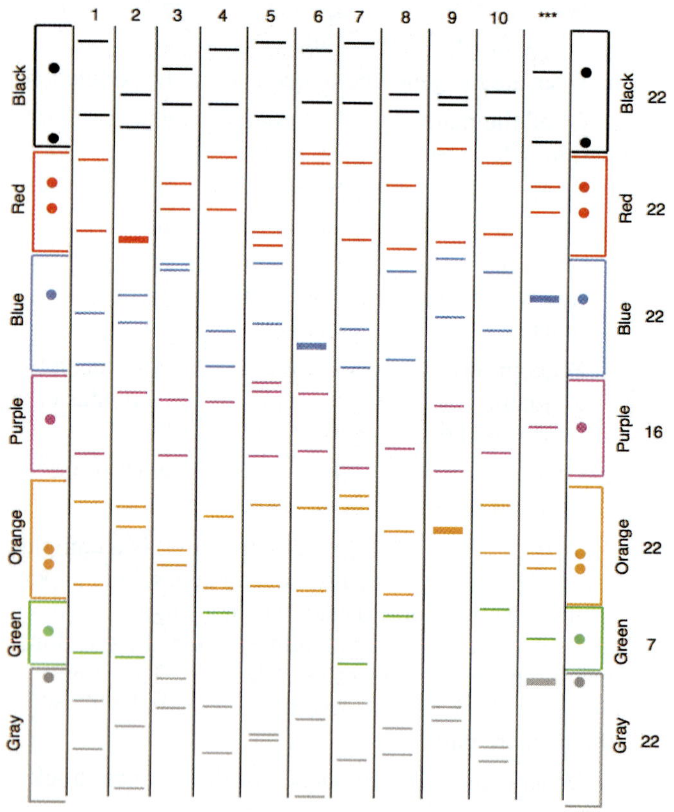

a. Are any of individuals 1–10 the murderer? If so, which one?

b. Are any of the loci on the X chromosome? If so, identify this (these) locus (loci) by color.

c. Are any of the loci on the Y chromosome? If so, identify this (these) locus (loci) by color.

d. Are any of individuals 1–10 probable relatives of the murderer? If so, identify this person and describe the degree of relationship to the criminal.

e. One of individuals 1–10 is aneuploid. Write the identity of this person, their sex (M or F), the type of aneuploidy involved, whether the nondisjunction occurred in the mother or father of this person or you can't tell, and whether the nondisjunction occurred during meiosis I or meiosis II or that you can't tell.

f. What is the probability that any random *male* in the United States would share the same genotype as the murderer (the match probability)? Assume that all 11 DNA samples analyzed in the diagram are together representative of the U.S. population as a whole. Show what numbers you would multiply to do this calculation.

g. Explain why the assumption in part (f) that the sample is representative cannot be completely accurate. How might the inaccuracy of this assumption affect the match probability?

Section 21.2

19. Why is the elimination of a fully recessive deleterious allele by natural selection difficult in a large population and less so in a small population?

20. Tristan da Cunha is a group of small islands in the middle of the Atlantic Ocean. In 1814, a group of 15 British colonists founded a settlement on these islands. In 1885, 15 of the 19 males on the island were lost in a shipwreck. In the late 1960s, four cases of retinitis pigmentosa, which progressively leads to blindness, were found among the 240 descendants of these settlers remaining on the island. The frequency of retinitis pigmentosa in Britain is about 1 in 6000. Explain the high incidence of this disease on Tristan da Cunha relative to that seen in Britain.

21. Small population size causes genetic drift because of chance sampling of different alleles from one generation to the next. We can predict how much genetic drift occurs for a given population size using binomial sampling statistics. With a population of size N, we can estimate that 95% of the time the allele frequency (p) in the next generation will be within

the confidence interval of $p \pm 1.96\left(\sqrt{\frac{p(1-p)}{2N}}\right)$,

where $\frac{p(1-p)}{2N}$ is an estimate of the statistical variance in allele frequencies from one generation to the next with random sampling of $2N$ alleles each generation.

a. What is the confidence interval for $p = 0.5$ when $N = 100{,}000$?

b. What is the confidence interval for $p = 0.5$ when $N = 10$?

c. How are the results in parts (a) and (b) related to the consequences of a population bottleneck?

22. Three basic predictions underlie genetic drift in populations: (1) As long as the population size is finite, some level of genetic drift will occur; thus, without new mutations, all variation will drift either to fixation or to loss. (2) Drift happens faster in small populations than in large populations. (3) The probability that an allele is fixed (goes to a frequency of 1.0) is equal to its initial frequency (p) in the population, while its probability of loss from the population due to drift is equal to $1 - p$. Given these three predictions:

a. What is the allele frequency of a new autosomal mutation immediately after it occurs in a diploid population of size $N = 100{,}000$?

b. What is the allele frequency of a new autosomal mutation immediately after it occurs in a diploid population of size $N = 10$?

c. In which population does the new mutation have a higher probability of going to fixation by chance with genetic drift?

23. A mouse mutation with incomplete dominance (t = *tailless*) causes short tails in heterozygotes (t^+/t). The same mutation acts as a recessive lethal that causes homozygotes (t/t) to die *in utero*. In a population consisting of 150 mice, 60 are t^+/t^+ and 90 are heterozygotes.

a. What are the allele frequencies in this population?

b. Given that there is random mating among mice, no migration, and no mutation, and ignoring the effects of random genetic drift, what are the expected numbers of the different genotypes in this next generation if 200 offspring are born?

c. Two populations (called Pop 1 and Pop 2) of mice come into contact and interbreed randomly. These populations initially are composed of the following numbers of wild-type (t^+/t^+) homozygotes and tailless (t^+/t) heterozygotes:

	Pop 1	Pop 2
Wild type	16	48
Tailless	48	36

What are the frequencies of the two genotypes in the next generation?

24. In *Drosophila,* the vestigial wings recessive allele, *vg*, causes the wings to be very small. A geneticist crossed some true-breeding wild-type males to some vestigial virgin females. The male and female F_1 flies were wild type. He then allowed the F_1 flies to mate with one another and found that 1/4 of the male and female F_2 flies had vestigial wings. He dumped the vestigial F_2 flies into a morgue and allowed the wild-type F_2 flies to mate and produce an F_3 generation.

a. Give the genotype and allele frequencies among the wild-type F_2 flies.

b. What will be the frequencies of wild-type and vestigial flies in the F_3?

c. Assuming the geneticist repeated the selection against the vestigial F_3 flies (that is, he dumped them in a morgue and allowed the wild-type F_3 flies to mate at random), what will be the frequency of the wild-type and mutant alleles in the F_4 generation?

d. Now the geneticist lets all of the F_4 flies mate at random (that is, both wild-type and vestigial flies mate). What will be the frequencies of wild-type and vestigial F_5 flies?

25. In a population of infinite size, three loci A, B, and C have two alleles each. Alleles A^1, B^1, and C^1 are found in 1% of the population at a particular moment in time, and each has beneficial effects on the organisms' fitnesses as compared to the other allele of that locus (A^2, B^2, and C^2, respectively). The relative fitnesses of the three possible genotypes at each of these loci is:

$W_{A^1A^1}$	$W_{A^1A^2}$	$W_{A^2A^2}$
1.00	0.99	0.99
$W_{B^1B^1}$	$W_{B^1B^2}$	$W_{B^2B^2}$
1.00	1.00	0.90
$W_{C^1C^1}$	$W_{C^1C^2}$	$W_{C^2C^2}$
1.00	1.00	0.99

The frequencies of alleles A^1, B^1, and C^1 over thousands of generations is shown in the following graph:

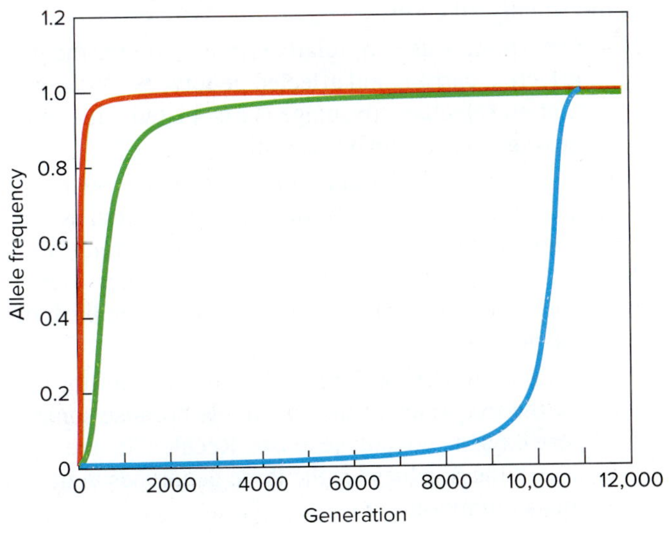

a. Which line (*blue, red,* or *green*) corresponds to A^1? B^1? C^1?

b. Why does the allele represented by the *red* line go to fixation more quickly than that represented by the *green* line?

c. Why does the allele represented by the *blue* line go to fixation more slowly than the alleles represented by either the *red* or *green* lines?

d. Suppose the population only had 1000 individuals. Discuss how this change in population size might affect the shapes of the three lines.

26. You have identified an autosomal gene that contributes to tail size in male guppies, with a dominant allele *B* for large tails and a recessive allele *b* for small tails. Female guppies of all genotypes have similar tail sizes. You know that female guppies usually mate with males with the largest tails, but the effects of population density and the ratio of the sexes on this preference have not been studied. You therefore place an equal number of males in three tanks. In tank 1, the number of females is twice the number of males. In tank 2, the numbers of males and females are equal. In tank 3, half as many females as males are present. After mating, you find the following proportions of small-tailed males among the progeny: tank 1, 16%; tank 2, 25%; tank 3, 30%.

    a. In your original population (before the animals were placed in the three tanks), 25% of the males have small tails. Assuming that the allele frequencies in males and females are the same, calculate the frequencies of *B* and *b* in your original population.

    b. Calculate Δq for each tank.

    c. If $W_{BB} = 1.0$, what is W_{Bb} for each tank?

    d. If $W_{BB} = 1.0$, is W_{bb} less than, equal to, or greater than 1.0 for each tank?

27. In Europe, the frequency of the CF^- allele causing the recessive autosomal disease cystic fibrosis is about 0.04. Cystic fibrosis causes death before reproduction in virtually all cases.

    a. Determine values of relative fitness (*W*) for the unaffected, carrier, and affected genotypes. Assume that no selective advantage is associated with heterozygosity for the disease allele.

    b. Given your answer to part (a), determine the average (mean) fitness at birth of the population as a whole with respect to the cystic fibrosis trait ($\overline{W}$) and the expected change in allele frequency over one generation (Δq) when measured at the birth of the next generation.

    c. Suppose the European population is at equilibrium for the frequency of the CF^- allele because some heterozygote advantage exists. Recalculate the relative fitness values for the three genotypes under this assumption.

    d. The CFTR protein encoded by the *CF* gene is a chloride ion channel. People suffering from cholera have diarrhea that pumps water and chloride ions out of the small intestine. Use these facts to explain why a heterozygote advantage might in fact exist for the *CF* gene.

28. An allele of the *G6PD* gene acts in a recessive manner to cause sensitivity to fava beans, resulting in a hemolytic reaction (lysis of red blood cells) after ingestion of the beans. The same allele also confers dominant resistance to malaria. The heterozygote has an advantage in a region where malaria is prevalent. Will the equilibrium frequency (q_e) be the same for an African and a North American country? What factors affect q_e?

29. Explain why evolutionary biologists monitor selectively neutral polymorphisms as molecular clocks.

30. Tiny foxes live on the Channel Islands off the coast of Southern California; the adults weigh less than 3 lbs. These so-called *island foxes* (*Urocyon littoralis*) derived from the mainland gray fox (*Urocyon cinereoargenteus*). Analysis of genome sequences revealed that unlike the mainland foxes, the foxes on a single island have shockingly little genetic diversity.

    a. The genome of only one fox from each island was sequenced. How would the lack of genetic diversity be evident in a single genome sequence?

    b. The populations of the foxes on each island are small. How might the low diversity have occurred?

    c. Why is low genetic diversity thought to lead to species extinction?

    d. Hypothesize as to why the Channel Island foxes are thriving without human assistance despite their lack of genome sequence diversity.

Section 21.3

31. What is the most straightforward evidence at the molecular level in support of the idea that modern humans first appeared in Africa?

32. In March 2013, the *American Journal of Human Genetics* published a report that an African-American man who submitted his genome for commercial genealogical analysis had a Y chromosome whose sequence was very different from that of other Y chromosomes that had been characterized previously. The investigators then found that certain males among the Mbo (an ethnic group in Cameroon) shared many of the polymorphisms first found in this African-American man. How do you think these findings would have altered estimates of when a man carrying the MRCA for the human Y chromosome would have lived on the earth?

33. If you go back 40 generations into your biological ancestry:
    a. How many ancestors are you predicted to have?
    b. How could you reconcile that prediction with the fact that the world's population of humans is now roughly 7 billion people?
34. In Fig. 21.17, to what part of the world does Dion's mitochondrial DNA recently trace? Do his Y chromosome and autosomal chromosomes 1–22 also trace back recently to this same region of the world? Why or why not?
35. Predict the DNA sequences at the four *nodes* (branching points) on the cladogram in Fig. 21.18b.
36. A cladogram (not drawn to scale) for the taxonomic family *Hominidae* is shown here. The numbers 1–10 represent evolutionary lineages or events. The letters A–F represent entries from the following list:

    Homo neanderthalensis
    Pan troglodytes (chimpanzees)
    Homo sapiens (African Bantu)
    Homo sapiens (European Danish)
    Homo sapiens (Native American Hopi)
    Homo sapiens (Asian Uighurs)

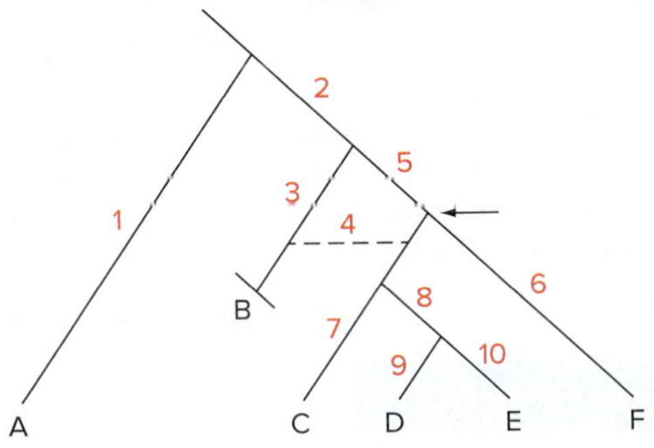

    a. Match the entries in the preceding list with an appropriate letter from the cladogram. Two of the groups in the list are equivalent on this diagram; either possibility is correct.
    b. One evolutionary divergence is indicated with a small arrow. Describe this divergence and estimate how many years ago it occurred based on Fig. 21.19.
    c. Six SNPs (α, β, γ, δ, ε, and ζ) are sequenced in several individuals in each of the six groups; the allele frequencies are given in the table that follows. At the bottom of each column in the table, write a number from 1 to 10 (corresponding to a *red number* on the figure) that indicates where along the cladogram a mutation occurred that changed the allele in the common ancestor of all humans and chimps to a derived allele. One blank can be filled by either of two numbers; you only need to show one. Also indicate (in the last row at the bottom) the identity of the allele found in the common ancestor, or write *can't tell*.

	SNP α	**SNP β**	**SNP γ**	**SNP δ**	**SNP ε**	**SNP ζ**
H. neanderthalensis	A 1.0	C 0.4 A 0.6	C 1.0	T 1.0	T 1.0	T 1.0
P. troglodytes	G 1.0	A 1.0	C 1.0	T 1.0	G 1.0	T 1.0
African Bantu	G 1.0	A 1.0	C 1.0	A 0.7 T 0.3	T 1.0	C 1.0
European Danes	G 1.0	C 0.25 A 0.75	C 1.0	T 1.0	T 1.0	C 1.0
Native American (Hopi)	G 1.0	C 0.8 A 0.25	A 1.0	T 1.0	T 1.0	C 1.0
Asian Uighurs	G 1.0	C 0.5 A 0.5	A 0.6 C 0.4	T 1.0	T 1.0	C 1.0
Number (1–10)						
Ancestral allele						

37. As noted in Fig. 21.22, humans now living in Oceania (e.g, Melanesia, Micronesia, Polynesia, and Australia) represent an early offshoot in the spread of humans around the world from an origin in Africa.
    a. Given these population sizes and histories, would you expect to see more, or less, genetic variation in population samples of humans from Oceania compared to Africa? Explain.
    b. Explain why variants found in Denisovan DNA are found only in modern humans living in Southeast Asia and Oceania but nowhere else on earth.
38. As of this writing in 2016, no Neanderthal-derived Y chromosome nor mitochondrial DNA sequences has ever been found in a modern human. Propose two alternative explanations.

chapter 22

The Genetics of Complex Traits

Artificial selection by dog breeders has led to dramatic differences in size among dog breeds. Population geneticists have found that only six genes are responsible for more than half the variation in sizes among dog breeds.

chapter outline

- 22.1 Heritability: Genetic Versus Environmental Influences on Complex Traits
- 22.2 Mapping Quantitative Trait Loci (QTLs)

TODAY (IN 2016), the cost of sequencing a whole human genome is roughly $1000. At such prices, how worthwhile would it be to you to obtain this information for your own genome or for that of a fetus conceived by you and your partner?

The answer will be different for different people, but for everyone, a major component in weighing the costs and benefits is the degree to which genomic sequence data can be interpreted as predictions about specific traits. We have already seen that whole-genome sequences will reveal with near certainty whether an individual is a carrier or will be afflicted by many Mendelian conditions such as sickle-cell anemia or cystic fibrosis, where the trait is governed by alleles of a single gene and the penetrance is essentially complete.

However, the thousands of dollars you spend on your (or your child's) whole-genome sequence will provide, at least in the near future, almost no clue about many other traits such as intelligence or personality. The reason is that these are **complex traits** influenced by many factors, including multiple genes, interactions between alleles of different genes, variations in the environment, interactions between genes and the environment, and chance events.

The height of adult humans is one such complex trait. Tall parents tend to have tall children, suggesting a genetic contribution to height. Scientists have recently established that hundreds of genes influence human height, and many of these genes have not yet been identified. Except in special cases such as the mutation causing achondroplasia (dwarfism), the contribution of any one particular polymorphism to height is so small as to have virtually no predictive power.

Another reason why genotypic information cannot easily anticipate adult height is that a key environmental factor—nutrition—has a strong influence on this trait.

Figure 22.1 shows that in many different populations in Europe, average height increased dramatically during the twentieth century. This period of time is so short that it is improbable that allele frequencies underwent significant changes; instead, the changes almost certainly represent improvements in diet.

In this chapter, we begin our discussion of complex traits by considering how scientists try to distinguish between the contributions of genes and those of the environment to a given trait. We then focus on two different methods to identify the specific genes, usually referred to as **quantitative trait loci (QTLs),** that contribute to these complex traits. Parsing out the many factors responsible for such phenotypes will be an increasingly important goal of genetics in the future, for both theoretical and practical reasons. Research into complex traits will help us understand the causes of, and help develop treatments for, complicated degenerative conditions such as arthritis and coronary disease. Improvement of agriculturally valuable plants and animals will require us to track the factors that control complex traits including yield, drought and heat resistance, and nutritional value.

A central theme of this chapter is that studies of complex traits require scientists to compare individuals within large populations. The conclusions from complex trait analysis therefore apply only to the specific population living in the specific environment currently under investigation, and they cannot be generalized to other populations or conditions.

Figure 22.1 Adult male height by birth cohort for various European populations.

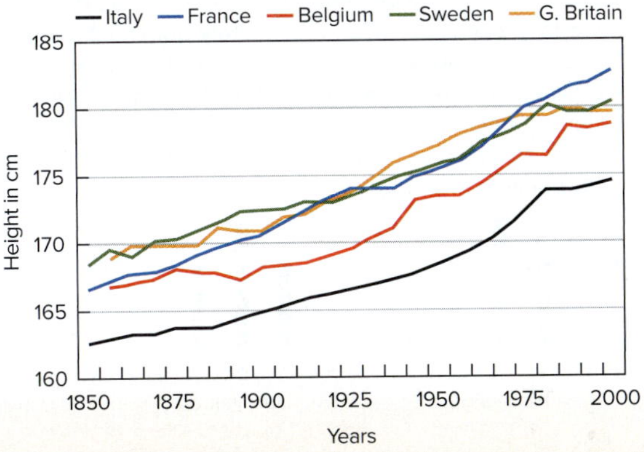

22.1 Heritability: Genetic Versus Environmental Influences on Complex Traits

learning objectives

1. Summarize the meaning of *phenotypic variance*.
2. Outline experiments that would allow you to distinguish genetic and environmental influences on phenotypic variance.
3. Explain the term *heritability* and why it applies to populations and not to individuals.
4. Describe how scientists estimate heritability in wild animal populations by comparing parents and their progeny, or in human populations by studying monozygotic and dizygotic twins.
5. Diagram how plant breeders use *truncation selection* to improve agricultural crops.

Many complex traits are **quantitative traits** for which the phenotype can be measured over a range of numbers called **phenotypic values** or **trait values.** Many such traits show a roughly bell-shaped **normal distribution** of phenotypic values in populations. Human height provides a good example (**Fig. 22.2**). A few individuals are either very tall or very short, but most people have heights clustered around the average for the group. The apparently continuous distribution of height in this bell-shaped distribution is shaped by the contributions of both genes and the environment. (The word *environment* here includes chance events.)

As we saw in Chapter 3, the more genes involved in the expression of a trait, the more possibilities exist for phenotypic variation in any given environment and the more the potential phenotypes resemble a normal distribution (review Fig. 3.28). Furthermore, rather than being identical, the phenotypes exhibited by individuals of any one genotype are actually distributed in narrower bell-shaped curves centered around the average phenotype for that genotype.

Figure 22.2 Distribution of human height for members of a genetics class at the University of Notre Dame. Note the roughly bell-shaped distribution, with most individuals of intermediate height.

Figure 22.3 How genetic and environmental influences on a quantitative trait can interact to produce normal distributions. As was demonstrated in Fig. 3.28, genetic variation at two loci, each with two incompletely dominant alleles with equal effects on phenotype, can produce five phenotypic classes with the trait values shown at the *left*. Each of these trait values becomes blurred by environmental influences (*center*) to yield a continuous bell-shaped distribution of trait values among all the individuals in the population (*right*).

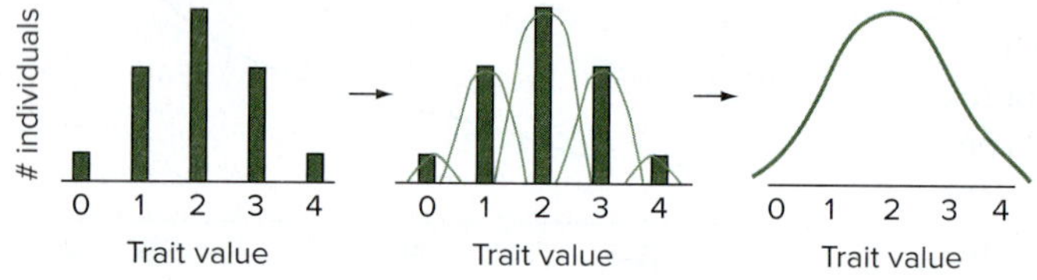

The reason is that individuals with the same genotype are subjected to slightly different microenvironments (**Fig. 22.3**). For the population as a whole, the effects of the environment superimposed on variation in just a few genes can therefore easily approximate a bell-shaped curve of phenotypic values.

How do scientists evaluate distributions of phenotypes to estimate the *heritability* of a complex trait—that is, the contributions of genes (as opposed to the environment) to the phenotypic differences observed in a particular population? As you will see, the answers to this question are of more than theoretical interest.

Variance Is a Statistical Measure of the Amount of Variation in a Population

To estimate heritability, scientists first need to obtain a numerical description for the curve, usually bell-shaped, of the trait's distribution in the population under study. Researchers track the amount of variation by comparing the phenotypic value for each individual in the population to the average phenotypic value for the population as a whole. Statistically, the result of this analysis is termed the **total phenotype variance (V_P),** and it is calculated as the average squared difference between each individual trait value and the mean.

As an example, let's consider the trait of stem length in a population of dandelions, a common weedy plant in North America (**Fig. 22.4a**). The stem lengths in the population exhibit a bell-shaped normal distribution. As **Fig. 22.4b** shows, you find the *mean stem length* by summing the values of all stem lengths and dividing by the number of stems. You then find the *variance in stem length* (V_P for this trait) by expressing each stem length as plus or minus deviations from the mean, squaring those deviations, summing the squares, and dividing the sum by the number of stems measured. The variance provides a mathematical description of this distribution; the narrower the curve relative to the peak, the lower the value of the variance.

Figure 22.4 Phenotypic variance. (a) The familiar dandelion (*Taraxacum* sp.). **(b)** Variance is the average squared difference between each individual value and the mean. On the plot, the square root of the variance is shown so as to be in the same units as stem length. (Statistical theory tells us that 95% of the observations under the bell-shaped curve will be within the interval of the mean minus 1.96 times the square root of the variance, and the mean plus 1.96 times the square root of the variance.)

(a) The dandelion

(b) Calculating the mean and variance of stem length

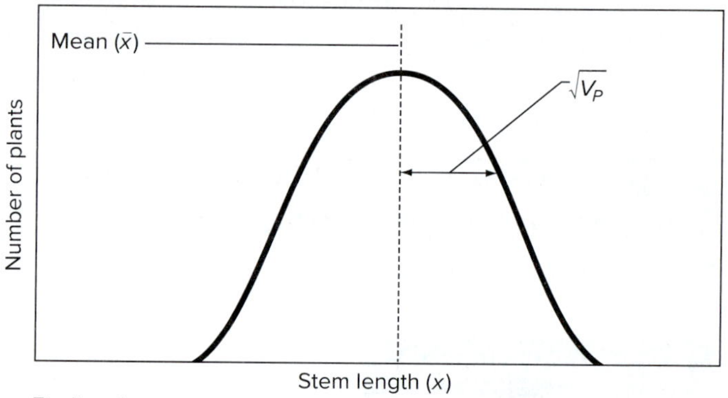

Finding the mean:

Let x_i = the stem length of the plant i in a population of N plants. The mean of stem length, $\bar{x}$, for the population is defined as

$$\bar{x} = \frac{\sum_{i=1}^{N} x_i}{N}$$

Finding the variance:

The variance V_P of stem length for the population is defined as

$$V_P = \frac{\sum_{i=1}^{N} (x_i - \bar{x})^2}{N}$$

Once the total variance of the trait is determined, scientists can then begin to ask what fraction of this variance is due to differences in the genes carried by individual organisms and what part is due to differences in the microenvironments to which these individuals are subjected.

Genetic Variance Can Be Separated from Environmental Variance

To distinguish environmental from genetic effects on phenotypic variation, you need to vary one of the two, say the environment, while controlling for the other one (that is, while holding the genetic contribution steady). This particular experiment is easy to accomplish in the case of dandelions because most dandelion seeds arise from mitotic, rather than meiotic, divisions. As a result, all the seeds from a single plant are genetically identical. You could begin by planting genetically identical seeds on a grassy hillside and allowing them to grow undisturbed until they flower. You then measure the length of the stem of each flowering plant and determine the mean and variance of the distribution of values for this trait in this dandelion population (**Fig. 22.5a**).

Because all members of this population are genetically identical (if we ignore rare mutations), any observed variation in stem length among individuals should be a consequence of environmental variations, such as different amounts of water and sunlight at different locations on the hillside. When represented as a variance from the mean, these observed environmentally determined differences in stem length are called the **environmental variance (V_E).**

To examine the impact of genetic differences on stem length, you take seeds from many different dandelion plants produced in many different locations, and you plant them in a controlled greenhouse (**Fig. 22.5b**). Because you are raising genetically diverse plants in a relatively uniform environment, the observed variation in stem length is mostly the result of genetic differences promoting **genetic variance (V_G).**

Now, to illustrate the total impact on phenotype of variation in both genes and environment, you take the seeds of many different plants from many different locations (and thus with different genetic variants) and grow them on a hillside with a variety of microenvironments (**Fig. 22.5c**). For the population of dandelions that grow up from these genetically diverse seeds, the total phenotype variance (V_P) in stem length will be the sum of the genetic variance (V_G) and the environmental variance (V_E):

$$V_P = V_G + V_E \qquad (22.1)$$

Note that the bell-shaped curve for dandelion stem length in Fig 22.5c is broader than either that for genetically identical individuals on a hillside (Fig. 22.5a) or for genetically variable individuals grown in a controlled greenhouse (Fig. 22.5b). For natural populations of dandelions, both genetic variation among individuals and variation in the environmental conditions experienced by each plant contribute to the total phenotypic variation.

Figure 22.5 Environmental and genetic components of phenotypic variance. (a) The phenotypic variance of genetically identical plants grown in a variable environment like a hillside field is all due to the environmental variance V_E. **(b)** The phenotypic variance of genetically diverse plants grown in a uniform environment such as a controlled greenhouse is all due to the genetic variance V_G. **(c)** For natural populations grown in diverse natural environments, the total phenotypic variance V_P is the sum of the genetic and environmental variance components ($V_G + V_E$).

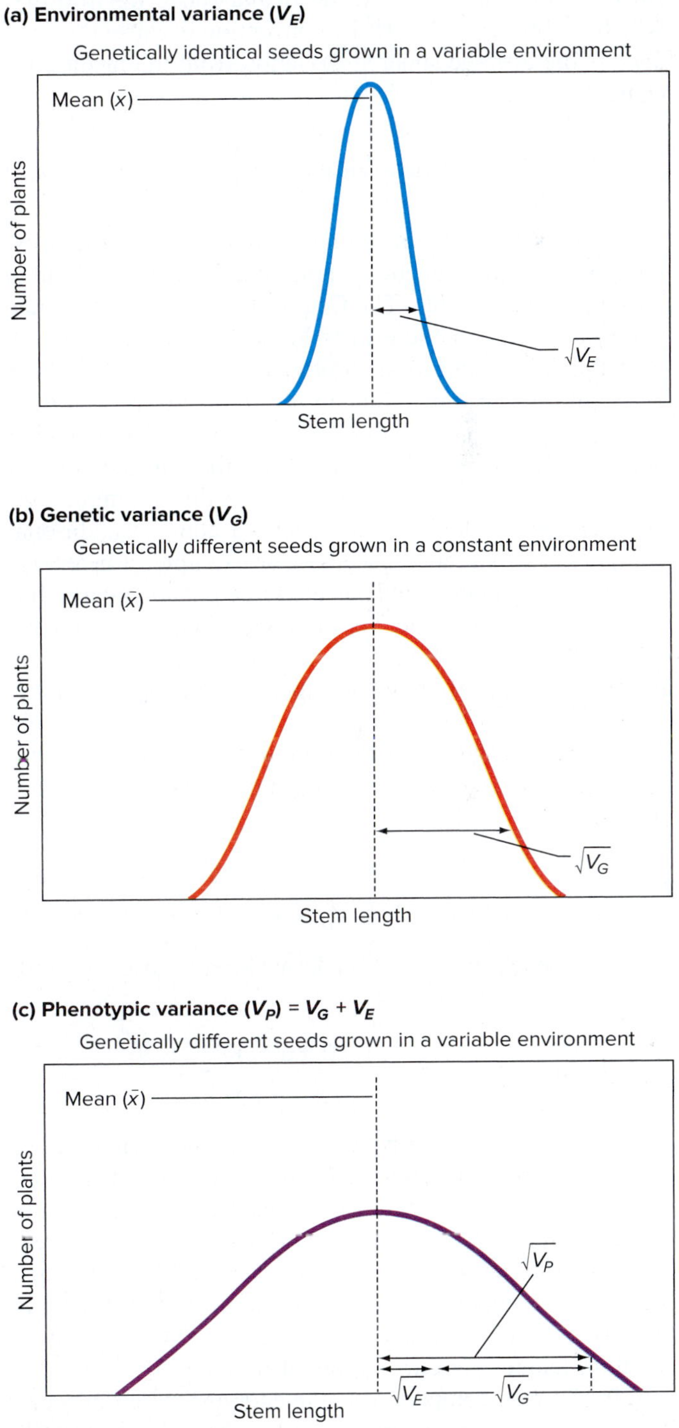

Heritability Is the Proportion of Phenotypic Variance Due to Genetic Variance

Geneticists define the **heritability** of a phenotypic trait as the proportion of total phenotypic variance (V_P) ascribable to genetic variation alone (V_G). Formally, heritability defined in this way is called the **broad-sense heritability,** which is abbreviated as $\boldsymbol{H^2}$ by convention because the variance is the average squared deviation from the mean for a characteristic:

$$\text{Broad-sense } H^2 = \frac{V_G}{V_P} \qquad (22.2)$$

To understand the meaning of broad-sense heritability, it helps to see that genetic variance itself can be subdivided into three components. The first of these is V_A, or variance due to additive genetic effects V_A (for example, if allele *a* contributes one unit to the trait value, *A* two units, *b* three units, and *B* four units, then an *Aa Bb* heterozygote would have $1 + 2 + 3 + 4 = 10$ units of trait value). The two other contributions to genetic variance are the variance due to dominance effects (V_D) and the variance due to interactions between genetic loci (V_I). Dominance adds an additional component to variance because, for example, a heterozygote for a dominant allele has the same phenotype as a homozygote for that allele, yet these individuals do not have the same genotype. Similarly, nonadditive interactions between alleles at different loci (for example, epistasis) can cause an allele of one gene to have different phenotypic values depending on the alleles present at the second gene. The total genetic variance is the sum of its three components:

$$V_G = V_A + V_D + V_I, \text{ thus} \qquad (22.3)$$

$$V_P = V_A + V_D + V_I + V_E \qquad (22.4)$$

Using these identities, we can define broad-sense heritability more precisely as:

$$\text{Broad-sense } H^2 = \frac{V_A + V_D + V_I}{V_A + V_D + V_I + V_E} = \frac{V_G}{V_P} \qquad (22.5)$$

You will see in the next section that the heritability of particular traits can be estimated in studies of genetic relatives. Broad-sense heritability is typically measured only in studies that compare identical twins to each other. The reason is that identical twins share the same alleles at all loci, so all three components of genetic variance are the same in both twins. This fact means that the V_G measured in twin studies includes not only V_A, but also V_D and V_I.

In contrast, comparisons of phenotypic values between parents and offspring cannot measure broad-sense heritability because only one of the alleles at any individual locus is shared between any one offspring and any one parent, and because the combinations of alleles among different loci also differ. The allele of each locus that is shared between a parent and an offspring instead must be considered as a genetic factor that acts in a simple, additive fashion. Comparisons of parents and offspring represent only the additive (V_A) component of overall genetic variance (V_G) because V_I and V_D are randomized over the population being analyzed, and so their effects cannot be measured. The heritability value estimated in studies of parents and offspring is called **narrow-sense heritability ($\boldsymbol{h^2}$)**—the proportion of total variation due specifically to variance of the additive genetic component:

$$\text{Narrow-sense } h^2 = \frac{V_A}{V_A + V_D + V_I + V_E} = \frac{V_A}{V_P} \qquad (22.6)$$

Because the additive component of the genetic effects most precisely predicts the range of phenotypic values expected among the progeny of crosses, plant and animal breeders often calculate the narrow-sense heritability of traits of interest. As you will see, the narrow-sense heritability is also important to breeders for a second reason: It dictates how strongly a particular trait will respond to selection for a trait value. Although for the sake of accuracy we will differentiate between H^2 and h^2 in the discussion that follows, the distinction is relatively minor for most of the purposes of this chapter.

At one extreme, a heritability of 0 (whether in the broad or narrow sense) means that no heritable variation influences the trait in the population, and that all of the observed phenotypic variation is due to environmental effects. At the other extreme, a heritability of 1 means that all of the observed phenotypic variation in the population is due to genetic variation and none is due to environmental effects.

A crucial fact to keep in mind is that heritability in either sense is a property of a specific population, not an individual. Thus, a statement that the heritability of a trait such as susceptibility to alcoholism is 0.4 does not imply that 40% of a particular person's susceptibility is due to genes and 60% to the environment. Instead, this value indicates that 40% of the phenotypic variation in this trait observed in the population can be attributed to genetic differences between the individuals in this particular population.

Because the amounts of genetic, environmental, and phenotypic variation may differ among traits, among populations, and among different environments, the heritability of a trait is always defined for a specific population and a specific set of environmental conditions. The heritability of any trait could thus differ for different populations with genetic variants and/or environments that are not the same.

The analysis of human height again provides a good example. When measured in a prosperous population with modern standards of food production, human height shows a very high heritability, greater than 0.9; that is, 90% of the variation in height seen in this population is due to allele differences. In contrast, in a poor country where not everyone gets adequate nutrition, the heritability of height would be much lower. The explanation comes from the fact that a person's genome determines their maximum height potential. If their nutrition is sufficient, they will reach this potential; further intake of food will make no difference. However, in underdeveloped countries, great differences exist between individuals in the amount of nourishment they can consume. This environmental difference will express itself as an increase in the environmental component of height variance.

Heritability Studies Examine Phenotypic Variation in Genetic Relatives

Relatives (such as parents and offspring) share genes, so we expect them to be similar phenotypically to the degree that their genes are related and the trait is heritable. If genetic similarity contributes to phenotypic similarity for a trait, it is logical to expect that a pair of close genetic relatives will be more similar phenotypically than a pair of individuals chosen at random from the population. Thus, by comparing the phenotypic variation among a well-defined set of genetic relatives with the phenotypic variation of the entire population over some range of environments, it is possible to estimate the heritability of a trait.

We first need to define quantitatively the **genetic relatedness** of two individuals as the average fraction of alleles at all genetic loci that the individuals share because they inherited them from a common ancestor. For an autosomal gene, the genetic relatedness of a parent and a child would be 0.5, because for any given locus, one of the two alleles in the child's genome came from that parent. In other words, a parent and a child share half of their alleles. The genetic relatedness of two siblings is also 0.5, because if you assume that one sibling received allele A^1 from an A^1A^2 heterozygous parent, the probability that the second sibling received the same allele is 0.5. Extending this same kind of analysis, we can see that an aunt and a niece have a genetic relatedness of 0.25, while that between first cousins is 0.125.

Although heritability could in theory be determined using even distant relatives, it makes sense to study first the closest relatives possible: individuals for whom genetic relatedness is 0.5 or higher. The reason is simply that the contributions of genes to a trait will be most obvious in cases where the most alleles are shared. In the remainder of this section, we discuss two approaches to estimating heritability in close relatives. The first of these compares phenotypic values in parents and progeny, focusing on a famous field study involving birds first studied by Darwin; the second compares traits in human twins.

Estimating heritability by comparing parents and progeny

Figure 22.6 illustrates the implication of heritability with respect to the mean phenotypic values of the parents, their offspring, and the population as a whole. Narrow-sense heritability can be estimated by quantifying how closely progeny phenotypes resemble parental phenotypes. If the mean phenotypic value of the progeny is always similar to that of the population as a whole, regardless of the the parents' phenotypes, then the heritability is 0 because none of the phenotypic variance is due to additive genetic effects of the alleles inherited from the parents. However, if the progeny deviate from the population average by 25% as much as their parents do, then the heritability is 0.25. If the progeny deviate from the population mean as much as their parents do, then the heritability is 1.0 because the alleles inherited from the parents closely predict the phenotypes of the offspring.

The finches observed by Darwin in the Galápagos Islands (often referred to as *Darwin's finches*) provide an example of a population for which geneticists have measured the heritability of a quantitative trait under natural conditions in the field by comparing the phenotypes of parents and offspring. Scientists studied the medium ground finch, *Geospiza fortis,* on the island of Daphne Major by banding many of

Figure 22.6 Determining heritability by parent/progeny comparisons. The phenotypic distribution of the parent generation population is shown in *tan* with the mean of the two specific parents chosen to breed indicated near the right of the distribution. The *purple line* below indicates the range of the offspring phenotypes. If $h^2 = 1$, then the mean of the offspring will equal the mean of the chosen parents because the alleles inherited from the parents dictate completely the phenotypes of the progeny. If $h^2 = 0.25$, then the mean of the offspring is one-quarter of the difference between the mean of the selected parents and the mean of the total parental generation. If $h^2 = 0$, then the mean of the offspring will be the same as the mean of the total parental population; in such cases, the genotypes of the parents do not influence the phenotypes of the progeny at all.

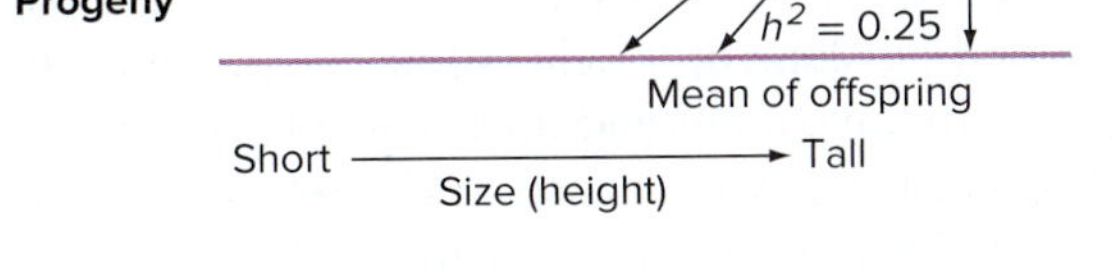

Figure 22.7 Measuring the heritability of bill depth in populations of Darwin's finches. (a) *Geospiza fortis* is a seed-eating bird endemic to the Galápagos Islands. **(b)** A scatter plot relating beak size of individual offspring and their *midparent value* (the average of the parents' beak depth) in 1976 and in 1978 is shown. Heritability is estimated as the *correlation coefficient,* which is the slope of the *line of correlation* (*blue* and *red*). The mean beak depth increased by natural selection associated with a drought in 1977 that caused a shift in available seeds from small, soft seeds to larger, harder seeds. The correlation coefficient is, however, almost the same both years, indicating a constant high heritability that is relatively independent of the environmental change. **(c)** Plots if heritability were close to 1.0. **(d)** Plots if heritability were 0.0. In panels (c) and (d), the graphs to the *right* show the phenotypic value averages for four of the mating pairs and their progeny represented in the same format as Fig. 22.6; each arrow in the plot at the *right* corresponds to one dot in the plot at the *left.*

(a) © Ralph Lee Hopkins/Getty Images RF

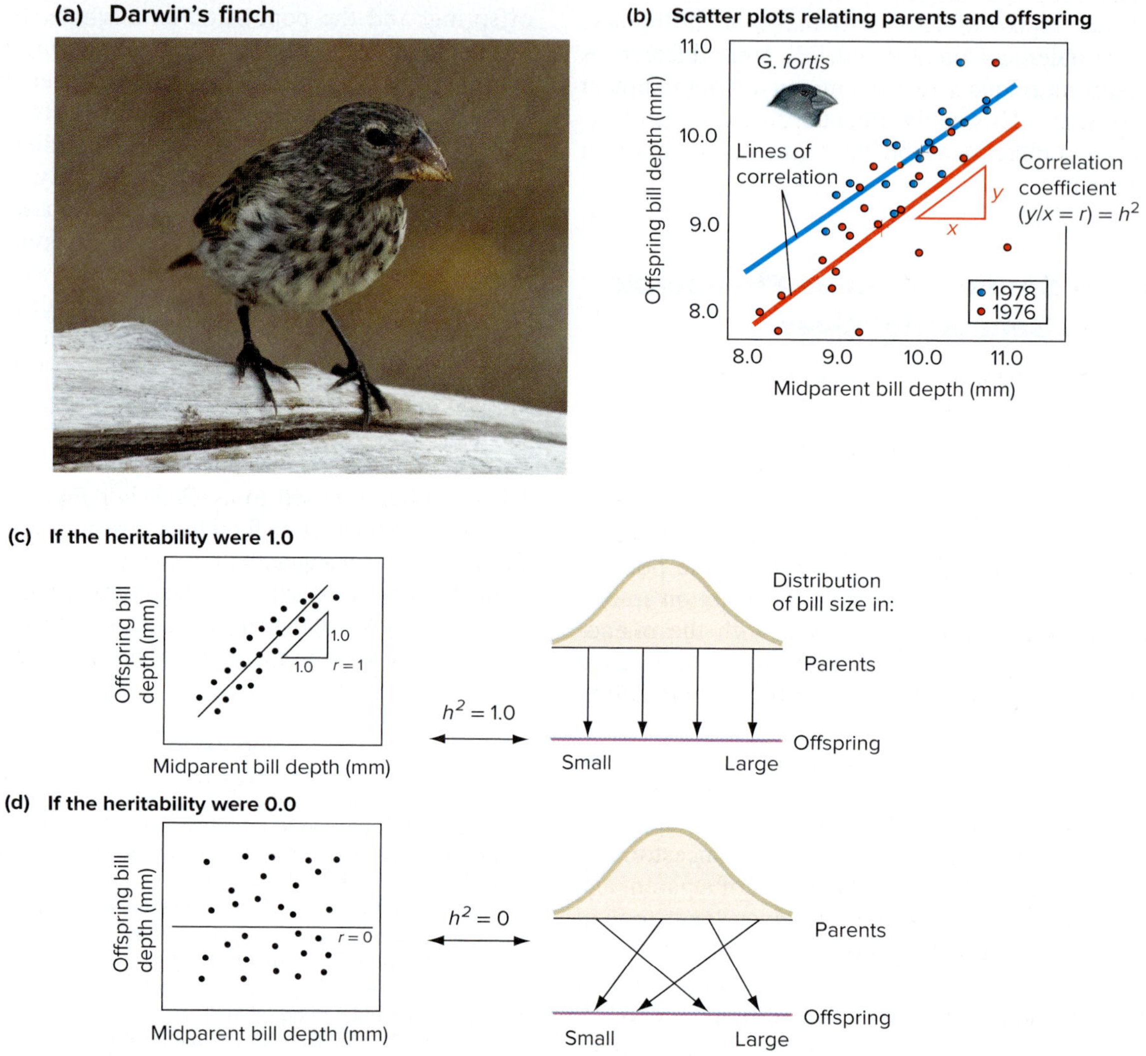

the individual birds in the population (**Fig. 22.7a**). The researchers then measured the depth of the bill for the mother, father, and offspring in each nest on the island and calculated whether the bill depth of the offspring was statistically correlated with the average bill depth of the mother and father (the *midparent value*; **Fig. 22.7b**).

The diagram in Fig. 22.7b is a *scatter plot* of the trait values of the parents and their progeny. The red and blue lines are the **lines of correlation** (also called the *best-fit line*) for the two data sets. The scatter plots show a clear positive correlation (the slope of the line of correlation is a positive number) between parents and offspring; parents with deeper bills had offspring with deeper bills, while parents with smaller bill depth had offspring with smaller bill depth.

For reasons that will become clear momentarily, when a positive correlation exists, the slope of the line of correlation relating offspring to midparent value—called the **correlation coefficient (*r*)**—is an estimate of the narrow-sense heritability for bill depth. In the figure, the heritability of bill depth, as represented by the slope of the line correlation relating midparent bill depth to offspring bill depth is 0.82. This means that roughly 82% of the variation in bill depth in this population of Darwin's finches is attributable to additive genetic variation among individuals in the population; the other 18% results from variation in the environment and nonadditive genetic effects.

In **Fig. 22.7c–d,** we examine the extreme cases to illustrate why the correlation coefficient (*r,* the slope of the line of correlation) provides an estimate of the narrow-sense heritability h^2. Suppose first that the environment had no influence at all on the trait (Fig. 22.7c). In such a case, the slope of the line of correlation relating bill depth in parents with bill depth in offspring would be 1.0 (Fig. 22.7c; *left*). At the *right* of this figure, we show the results for four mating pairs displayed in the same fashion previously used in Fig. 22.6. You can see that both of these representations are essentially equivalent.

Now consider a population in which the bill depth for parents and their offspring is, on average, no more or less similar than the bill depths for any pair of individuals chosen from the population at random (Fig. 22.7d). In such a population, no correlation exists between the bill depth trait in parents and in offspring, and a plot of midparent and offspring bill depths produces a "cloud" of points with no correlation between midparent and offspring bill depths (Fig. 22.7d, *left*). The *right* panel of Fig. 22.7d shows an equivalent, alternative representation of this same case. You can see that analyzing data that tracks the phenotypic values of parents, their progeny, and the population as a whole by either of the two methods used in Fig. 22.7 allows researchers to estimate heritability.

From these examples, you might conclude that phenotypic similarity among genetically related individuals provides evidence for the heritability of a trait. However, conversion of the phenotypic similarity among genetic relatives to a measure of heritability depends on a crucial assumption: that the distribution of genetic relatives is random with respect to environmental conditions experienced by the population. In the finch example, we assumed that parents and their offspring do not experience environments that are any more similar than the environments of unrelated individuals.

In nature, however, genetic relatives may violate this assumption by inhabiting similar environments. With finches, for example, all offspring produced by a mother and father during a breeding season normally hatch and grow in a single nest where they receive food from their parents. Because bill depth affects a finch's capacity to forage for food, the amount of feeding in a nest may cause progeny bill depth to correlate partially with parental bill depth for reasons quite distinct from genetic similarities.

Researchers studying Darwin's finches cannot easily deal with the issue that genetic relationship and environment could be intertwined. The reason is that the Galápagos islands are a national park of Ecuador, so researchers are forbidden to manipulate the birds in any way other than banding them for identification and measuring their physical attributes like beak sizes. But scientists studying domesticated species can try to reduce the confounding issue of environment by removing the eggs from the nest of one pair of parents and placing them randomly in nests built by other parents in the population; this random relocation of eggs is called **cross-fostering.** In heritability studies of animals that receive parental care, cross-fostering helps randomize environmental conditions.

Using twins to estimate complex trait heritability in humans

The heritabilities of human traits have been estimated by comparing the trait values of close genetic relatives, usually mother and child, father and child, or siblings. In these studies, heritability is the correlation coefficient (r) divided by the genetic relatedness of the individuals compared. For parent and child or for siblings, the relatedness is 0.5, so $h^2 = r/0.5 = 2r$. This method is problematic in that it overestimates heritabilities.

In human societies, family members share not only similar genes but also similar physical and cultural environments. Thus, phenotypic similarity between genetic relatives may result either from genetic similarities or from similar environments, or most often, from both. How can you distinguish the effects of genetic similarity from the effects of a shared environment?

The existence of **monozygotic twins** (**MZ** or *identical* twins) in placental mammals like humans provides researchers with a powerful tool for heritability studies. MZ twins result from a split in the zygote after fertilization. They are genetically identical because they come from a single sperm and a single egg; they share all alleles at all loci (**Fig. 22.8a**). Phenotypic differences between identical twins therefore cannot be due to genetic variation.

Twins can be used to investigate the heritability of either *quantitative complex traits* in which phenotypic values vary over a continuous range, or *discrete complex traits* that have only two phenotypic values: you either have the trait or you do not. Because the kinds of data generated in studies of quantitative and discrete complex traits are inherently different, we discuss these methods separately.

Figure 22.8 Different types of twins. (a) MZ twins have a genetic relatedness of 1.0. **(b)** The genetic relatedness of DZ twins is 0.5.

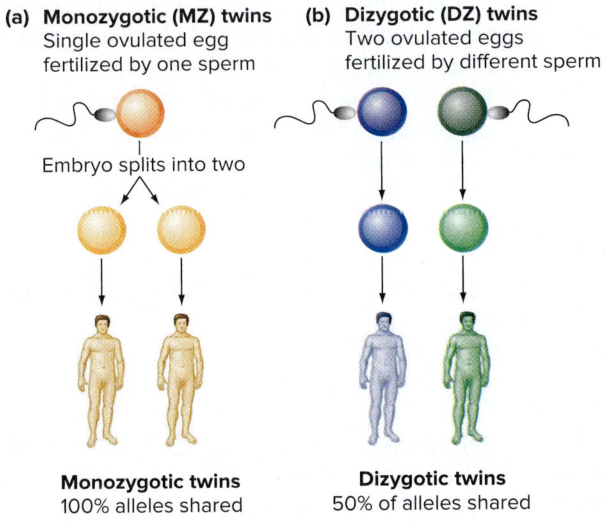

Fig. 22.9 Estimating heritability using MZ twins. **(a)** A scatter plot comparing the trait values of pairs of MZ twins adopted by different families is shown. The correlation coefficient (*r*) is the slope of the line of correlation (*orange*); *r* is an estimate of heritability (H^2). **(b)** Scatterplots of the trait values of MZ twins (*red*) or DZ twins (*blue*) for the same trait. Heritability can be estimated as: *heritability* = 2 ($r_{MZ} - r_{DZ}$).

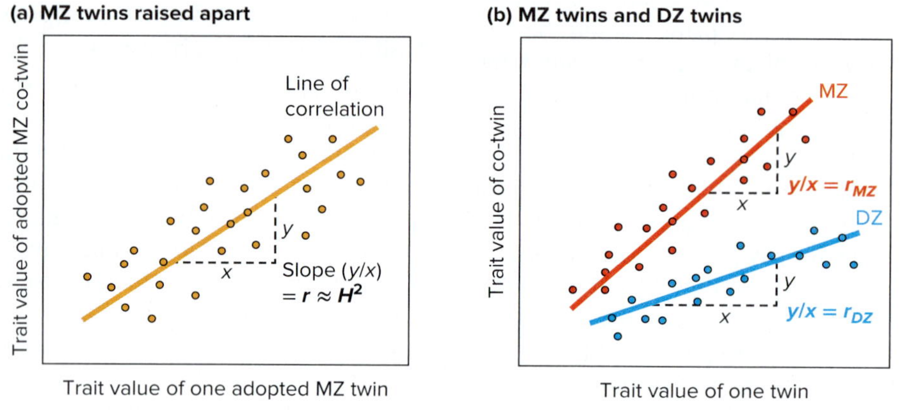

Twin studies of quantitative complex traits

Occasionally, MZ twins are subjected to an inadvertent cross-fostering experiment: They are given up for adoption shortly after birth and raised in different families. In such a pair of identical twins, any phenotypic similarity must in theory be the result of genetic similarity. Researchers can thus estimate the fraction of the phenotypic variance controlled by genes (the broad-sense heritability H^2) from a scatter plot of the trait values for different pairs of adopted MZ twins (**Fig. 22.9a**). The slope of the line of correlation (the correlation coefficient) estimates heritability: $H^2 = r$.

Only relatively few pairs of MZ twins have been raised separately, making these studies difficult to perform and limiting the data sets. But MZ twins can be used in a second way to estimate heritability by comparing their phenotypic differences with those of **dizygotic (DZ) twins.** Such DZ twins are the result of different sperm from a single father fertilizing two different maternal eggs (**Fig. 22.8b**). Just like siblings born at separate times, DZ twins share on average 50% of their alleles at all loci. Comparing the trait values for different pairs of MZ twins (whose genetic relatedness is 1.0) with the trait values for different pairs of DZ twins (whose genetic relatedness is 0.5) can help distinguish the relative contributions of genes and the environment.

Fig. 22.9b demonstrates this method of estimating heritability by comparisons of MZ and DZ twins. In essence, you compare the slopes of the lines of correlation (the correlation coefficients) for the two types of twins. The correlation coefficient for MZ twins (r_{MZ}) should always be greater than that for DZ twins (r_{DZ}) because MZ twins share twice as many genes (100%) as do DZ twins (50%). Assuming that the environmental effects encountered by MZ and DZ twins are equivalent, the greater the difference in slopes on the scatter plots, the greater is the heritability. In mathematical terms:

$$heritability = \frac{V_G}{V_P} = 2\,(r_{MZ} - r_{DZ}) \qquad (22.7)$$

To understand why the difference between r_{MZ} and r_{DZ} estimates heritability, you should realize that all the phenotypic variation between MZ twins is due to environmental factors, while the phenotypic variation between DZ twins has both genetic and environmental components. The factor of 2 in the equation is necessary because DZ twins share only half of their genes. Problem 8 at the end of the chapter encourages you to explore Eq. 22.7 in more detail.

Table 22.1 lists heritabilities for quantitative human traits estimated from twin studies. Although it is useful to gain some idea of the relative heritabilities of these traits, you should exercise care in interpreting these numbers. As explained earlier, heritability estimates can be applied only to the specific population being studied at the specific time it is being studied. In addition, the investigations listed in Table 22.1 are subject to complications that often can lead to overestimates of the actual heritability. For example, when different families adopt MZ twins, those families are likely to be more similar than a pair of families chosen at random. The two families might themselves be related, or adoption agencies might place the twins in two families of

TABLE 22.1 Heritability estimates from twin studies of quantitative traits

Trait	Heritability*
Height	0.68 – 0.90
Body mass index	0.64 – 0.84
Birth weight	0.64 – 0.84
Brain frontal lobe volume	0.90 – 0.95
Exercise participation	0.48 – 0.71
Dietary patterns	0.41 – 0.48

*Heritability estimates are specific to a population at a particular time.

similar higher socioeconomic backgrounds to ensure the children's financial security.

Twin studies of discrete complex traits

Some complex traits in humans are *discrete*, with only two possible phenotypic states, in contrast with continuous traits that have a range of possible phenotypic values. Many discrete traits are complex in the sense that they are controlled by multiple genes and the environment; examples include whether or not you have a heart attack, get cancer, or become an alcoholic. Twin studies are also useful for estimating the heritability of discrete traits. Because of the nature of the data obtained, these studies cannot measure correlation coefficients, but instead they compare the trait **concordance**—the frequency with which the other twin has the trait in question when one of them of does—between MZ twins and DZ twins.

Consider a trait in which the differences in phenotype among individuals in the population arise entirely from differences in the environment experienced by each individual, that is, a trait for which the heritability is 0.0 (**Fig. 22.10a**). For this trait, you would expect the likelihood that MZ twins share the same phenotype to be the same as that for DZ twins. In fact, the likelihood of trait sharing would also be the same for genetically unrelated siblings who had been adopted into the same family. If one child expresses a trait for which the heritability is 0.0, then the only factor that influences the chance a sibling will show the same phenotype is the probability that the range of environments investigated can produce the phenotype; the degree of genetic relationship between the children has no effect.

Now consider a trait for which differences in phenotypes among individuals in a population arise entirely from genetic differences, that is, a trait for which the heritability is 1.0 (**Fig. 22.10b**). Because MZ twins are genetically identical, they are expected to show 100% *concordance* in expression: if one expresses the trait, the other does as well. The concordance of trait expression between unrelated individuals varies based on the commonality of the trait. That is, the more common the trait in the population, the greater the chance that two unrelated people will have that phenotype (Fig. 22.10b). Regardless of the commonality of the trait, DZ twins, because they share half of their alleles, will always display greater concordance than genetically unrelated individuals, but less than the 100% concordance of MZ twins when heritability = 1.

Table 22.2 shows the concordances measured in twin studies for some complex discrete traits in humans. The higher the ratio of MZ concordance to DZ concordance, the higher the heritability of the trait. Note that in all cases the MZ concordance is less than 1.0. This result tells us that the environment influences all of these traits to some degree. The values of MZ and DZ concordance can be used to estimate heritability as follows:

$$heritability = \frac{\text{MZ concordance} - \text{DZ concordance}}{1 - \text{DZ concordance}} \qquad (22.8)$$

Considering the two extreme cases will help you appreciate Eq. 22.8. If all of the variance in the trait is

Figure 22.10 Probability that a second child will express a dominant trait that is expressed by a first child.

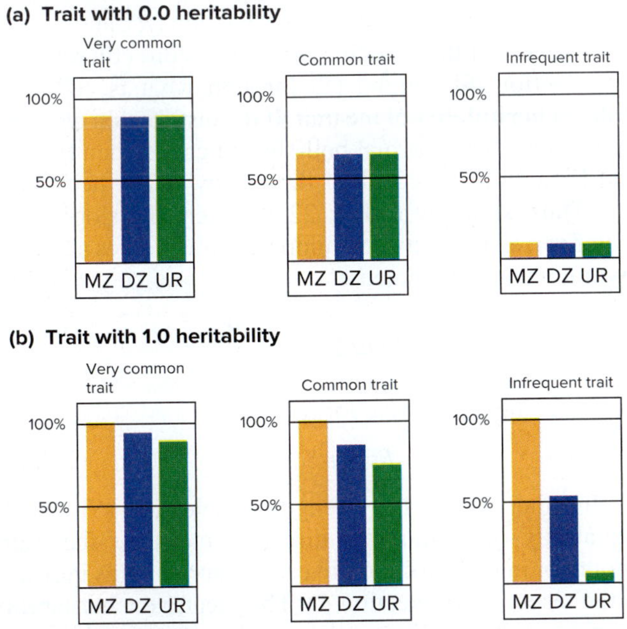

MZ = monozygotic twins, DZ = dizygotic twins,
UR = siblings unrelated genetically due to adoption

TABLE 22.2 MZ and DZ twin concordance for complex discrete traits

Trait	Concordance* MZ twins	DZ twins
Type 1 diabetes	0.43	0.074
Type 2 diabetes	0.34	0.16
Schizophrenia	0.41	0.053
Autism spectrum	0.94	0.47
Alzheimer's disease	0.32	0.087
Parkinson's disease	0.16	0.11
Multiple sclerosis	0.25	0.054
Crohn's disease	0.38	0.02
Colorectal cancer	0.11	0.05
Breast cancer	0.13	0.09
Prostate cancer	0.18	0.03

* Concordance values are specific to a population at a particular time.

controlled by environment and none by genes, the concordances of MZ and DZ twins would be expected to be equal, as we are assuming that the effects of their growing up together in the same families are equivalent. Therefore, the numerator of Eq. 22.8 is 0, and heritability = 0.

At the other extreme, where all of the phenotypic variation is due to genes, the MZ concordance will be 1. The DZ concordance will always be a fraction of 1, but this fraction depends on the rarity of the trait, whether the trait is controlled by dominant or recessive alleles, and how many genes control the trait. Suppose that we apply Eq. 22.8 to the simplest possible situation of a Mendelian trait controlled by a single, rare, 100% penetrant, dominant allele. In this simplified case, MZ concordance would be 1, DZ concordance would be 0.5, and Eq. 22.8 yields heritability = 1 as expected.

A Trait's Heritability Determines Its Potential for Evolution

We saw in Chapter 21 how the selection of preexisting mutations generates evolutionary change. Because the heritability of a complex trait is a measure of the genetic component of its variation, heritability quantifies the potential for selection and thus the potential for evolution from one generation to the next. A trait with high heritability has a large potential for evolution via selection, whether this selection is natural or artificial. This idea is important for understanding how phenotypes evolve over time.

The role of heritability in evolution also has considerable practical significance for breeding programs to improve agriculturally important plant and animal species. If the heritability of a trait is low, such a program would have little chance of success: It would make more sense either to alter the environment in which the crop or herd was raised, or to search out other representatives of the same species in remote areas of the globe to increase the genetic component of variation.

Figure 22.11 illustrates how a plant breeder can exploit situations of high heritability to improve a crop by artificial selection. In this example, the breeder wants to select for larger edible beans, and he will employ a simple but powerful strategy called **truncation selection.** The essence of this method is that he would plant beans with trait values above a certain cutoff (in this case, beans of a chosen minimal size) to produce the next generation.

In Fig. 22.11, S represents the *selection differential,* measured as the difference between the average trait of the selected parents and the average trait value in the entire parental population (that is, both breeding and nonbreeding individuals). Among the offspring of the selected parents, the average trait value will be higher than the average trait value in the entire parental generation. In the figure, R represents this difference. Used in this way, R signifies the *response to selection,* that is, the amount of evolution, or change in mean trait value, resulting from the selection applied by the breeder.

Figure 22.11 The strategy of truncation selection. Narrow-sense heritability h^2 predicts the response to selection on a quantitative trait. The total parental generation has a bell-shaped distribution for the phenotypic trait, but only individuals above a certain size are allowed to breed (*dark shading*). The difference in the mean phenotypic value between the selected group and the total population is S, the *selection differential.* If h^2 heritability is high, the offspring of those selected parents will have a mean phenotypic value that is also larger than the mean of the original population; the difference between these values is the *response to selection R.*

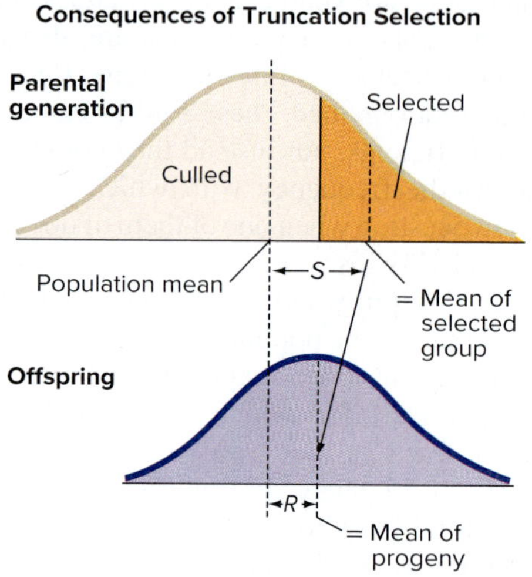

The significance of heritability to the breeder comes from the fact that the response to selection (R), and thus the effectiveness of the breeding program, is directly related to the selection differential (S) through what is called the **realized heritability** of the trait. If the distribution of phenotypic values fits a normal bell-shaped curve, then the realized heritability is equal to the narrow-sense heritability (h^2). Thus, similarly to Fig. 22.7b where the slope of the line of correlation between midparent trait values and progeny trait values estimated h^2:

$$\text{realized } h^2 = \frac{R}{S}$$

Rearranging to solve for R:

$$R = h^2 S \qquad (22.9)$$

In other words, the strength of selection (S) and the heritability of a trait (h^2) directly determine the trait's amount or rate of evolution in each generation (as indicated by the response to selection R). The greater the heritability, the greater the likelihood that the breeding program will improve the crop.

essential concepts

- The narrower the distribution around the mean of a *normal distribution*, the lower the *variance*.
- The *total phenotypic variance* of a complex trait in a population has both genetic and environmental components.
- *Heritability* describes the proportion of total phenotypic variation due to genetic variation in a specific population in a given environment.
- Heritability can be estimated by examining the degree to which traits are more similar in close relatives than in less related individuals.
- The heritability of a trait determines its potential for evolution.

22.2 Mapping Quantitative Trait Loci (QTLs)

learning objectives

1. Explain how researchers identify QTLs by establishing backcross lines and then fine-mapping the QTLs with introgressions.
2. Discuss the importance of linkage disequilibrium (LD) to association mapping of QTLs in humans.

Two main approaches exist to mapping **quantitative trait loci (QTLs)**, the genes that contribute to quantitative traits. The first approach, which we call here *direct QTL mapping,* requires researchers to conduct crosses between individuals with different phenotypic values for the trait of interest; this technique can be used only for species that can be bred through controlled crosses. The second approach, termed *association mapping,* takes advantage of past events that occurred in previous generations of populations. Association mapping can be applied even to species such as humans in which controlled breeding experiments cannot be performed.

As you will see, the idea underlying both approaches is the same, and it relies on recombination to produce individuals with different genetic compositions. In direct QTL mapping, recombination occurs during several generations of a series of crosses controlled by the researcher. By contrast, in association mapping, the recombination already happened during the history of a randomly breeding population. In both cases, investigators ultimately test for statistical correlations between markers in different regions of the genome and particular phenotypic values for the trait of interest. These correlations can often pinpoint mutations responsible for variations in the phenotype.

Researchers Map QTLs by Analyzing Recombinants Obtained Through Breeding Programs

To conduct **direct QTL mapping** in experimental organisms, investigators cross individuals showing two extreme phenotypic values for the trait of interest (such as size; very large and very small) and examine the joint segregation of trait value and genetic markers distributed throughout the genome of the organism. Markers showing a strong correlation between their presence/absence and the trait value (size in this example) are likely to be linked to one or more genes that influence the trait.

Researchers track the presence of particular genomic regions in the original two parents by examining DNA sequence variants (SNPs, InDels, or SSRs) chosen because they vary in the mapping population and are distributed throughout the genome. As the costs of gene chip and DNA sequencing methodologies continue to drop, researchers can screen more and more markers, increasing the resolution of the resulting QTL maps.

Identification of QTLs by rough mapping

Some of the most important applications of direct QTL mapping involve agricultural plants and animals. If researchers can find markers that correlate with QTLs for a commercially valuable trait, then they can develop useful strains that maximize the expression of that trait by finding recombinants that contain particular combinations of alleles of several QTLs. In such applications, the researchers do not necessarily have to identify the particular genes that contribute to the trait value. Instead, they just need to find polymorphisms linked closely enough to the QTLs so that DNA analysis will identify the strains most likely to have the most desirable phenotypes.

To illustrate the general method for identifying QTLs, we examine a landmark study of tomato fruit size conducted in the late 1990s. More than two thousand years of domesticated breeding has resulted in the tomatoes used in today's cuisines. Some domesticated tomatoes have fruits that are hundreds of times larger than those of their wild ancestors, which were originally from Mexico (**Fig. 22.12**). The size increase leading to today's large domesticated tomato occurred through the accumulation of mutations in many different genes over thousands of generations of selection.

To identify the relevant genes, researchers started with two closely related species, easily interbred, that exhibited opposite extreme trait values for fruit size: *Solanum lycopersicum* (large) and *Solanum pennellii* (small) (Fig. 22.12). Each of the two starting strains was inbred for several generations so that each line became homozygous for essentially every allele of every one of its genes. Globally homozygous strains such as these, in which all individuals have the identical alleles at all of the genes, are referred to

Figure 22.12 Size variation between domesticated tomatoes and their wild relatives. Domestic tomato fruit, *Solanum lycopersicum* (*left*) and fruit of three wild tomato species, *S. pimpinellifolium*, *S. habrochaites*, and *S. pennellii*.
Courtesy of Brad Townsley, UC Davis

as **isogenic lines.** Like Mendel's strains that were pure breeding (homozygous) for the alleles affecting the trait in question, isogenic *small* or *large* strains are homozygous for each of the QTLs affecting tomato size. As you will see, isogenic starting strains are crucial to the success of the experiment because they simplify the later analysis. The researchers crossed the large and small isogenic strains to produce the F_1, which were medium-sized tomatoes (**Fig. 22.13a**).

The investigators next backcrossed F_1 plants to the large isogenic parent strain (*S. lycopersicum*) to produce the BC_1 (backcross 1) generation. Because several different genes control tomato size and the small and large parents have different alleles of many of them, individual plants of the BC_1 generation display a wide range of sizes (Fig. 22.13a). The researchers then weighed and genotyped hundreds of individual BC_1 tomatoes to answer the following question: Do *p* (*pennellii*) or *l* (*lycopersicum*) alleles of any given marker locus correlate with tomato size?

Because the starting strains of *S. pennellii* and *S. lycopersicum* were isogenic, we can represent them as being *p/p* or *l/l* for every gene or marker locus at which the two strains were different. The F_1 are thus genetically identical and heterozygous (*p/l*) at every one of these loci, while each BC_1 tomato is either homozygous *l/l* or heterozygous *p/l*. You can see at this point the advantage of beginning with isogenic strains: The BC_1 progeny will inherit different single known alleles of molecular markers specific to *S. pennellii* or *S. lycopersicum* that can be tracked easily.

The researchers then calculated the mean weights of the tomatoes homozygous and heterozygous for each marker. In the cases of most of these marker loci, the mean weights of the homozygotes and the heterozygotes were the same. But for markers linked to the relevant QTLs, the mean weights were different (**Fig. 22.13b**). (To determine

Figure 22.13 Strategy for genetic mapping of quantitative trait loci (QTLs) influencing complex traits. **(a)** Isogenic *S. pennellii* (*p/p*) and *S. lycopersicum* (*l/l*) parents were crossed. The resulting F_1 generation was then backcrossed to isogenic *S. lycopersicum*, creating the BC_1 generation. **(b)** BC_1 individuals were weighed and genotyped. If a significant difference in trait value (here, the weight) exists between heterozygotes with the *S. pennellii* and *S. lycopersicum* alleles of a marker and homozygotes for the *S. lycopersicum* allele, the marker is linked to a QTL influencing the phenotype.

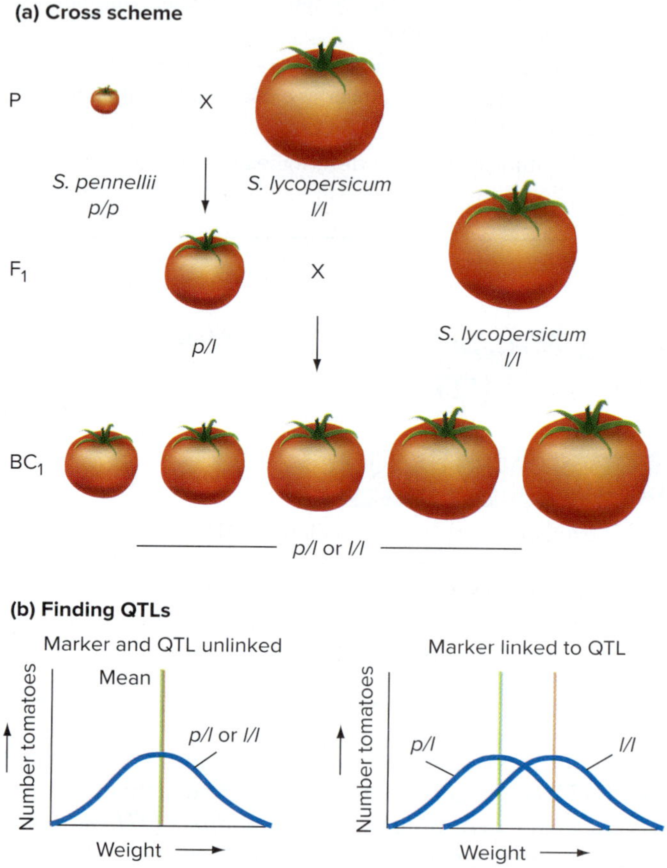

if a calculated difference was significant, the scientists used a version of the Lod score mapping statistic described in the Tools of Genetics Box in Chapter 11.) Using this method, the investigators discovered 28 QTLs that influence tomato size.

As an example of the direct QTL mapping technique, suppose the starting strain of *S. pennellii* was homozygous for allele A^1 of a particular marker, the isogenic *S. lycopersicum* strain was homozygous for allele A^2, and the heterozygous *p/l* (A^1/A^2) BC_1 tomatoes are significantly smaller than the *l/l* homozygous (A^2/A^2) BC_1 tomatoes. The marker is thus linked to a QTL for tomato size. Plant breeders can then predict that tomatoes with the A^2 allele will likely be larger than those with the A^1 allele of this marker. Such predictions do not require this marker to be the polymorphism responsible for the trait, only that the marker differences be linked to the responsible polymorphism.

Fine-mapping of QTLs with nearly isogenic (congenic) lines

As you have just seen, rough mapping does not identify the causal gene for the QTL, but instead establishes a chromosomal segment in which the gene could lie, and whose boundaries are defined by the nearest linked molecular markers. In most such studies, this region is between 1 and 10 cM long, and it could include over 100 genes. Although successful breeding programs may not require scientists to find the causal gene, good reasons often exist for extending the research to accomplish this goal.

In the study just described, the QTL with the largest effect on size was called *fw2.2* (*fruit weight 2.2*); the *S. lycopersicum* alleles of *fw2.2* may increase fruit weight by up to 30%. In order to begin to understand what molecular factors govern tomato size, investigators wanted to identify the *fw2.2* causal gene through a process called *fine-mapping.*

The first step in this fine-mapping procedure is to conduct a long series of backcrosses and intercrosses, starting with the BC_1 progeny like those discussed in Fig. 22.13, to generate strains called **NILs (nearly isogenic lines)** or **congenic lines.** In the example shown in **Fig. 22.14a,** the congenic lines were essentially *S. lycopersicum*, except each line had a different small region of the *S. pennellii* genome, called an **introgression,** in the region of the QTL.

The second step is to measure the trait value of each NIL and also to genotype it for molecular markers to determine the precise boundaries of each *S. pennellii* introgression. The idea is that the *pennellii* genomic region that all of the small tomato (*pennellii* phenotype) congenic lines share and all the large tomato (*lycopersicum* phenotype) congenic lines lack must contain the causal *fw2.2* gene. If this region is sufficiently small (which will be the case when a number of independent introgressions are characterized), researchers can then identify the causal gene by analyzing the DNA sequence of that region in the *S. pennellii* and *S. lycopersicum* genomes to look for polymorphisms that distinguish the two starting isogenic strains (**Fig. 22.14b**).

The final step of the fine-mapping process is to validate the gene assignment, which in this case was done by phenotypic rescue. Researchers cloned the suspect *fw2.2* gene from the *pennellii* genome and introduced it as a transgene into *lycopersicum*. The result, shown in **Fig. 22.14c,** was that the transgene made the *lycopersicum* tomatoes substantially smaller. This phenotypic rescue test was possible because the *pennellii fw2.2* allele (causing smaller fruit size) is dominant to *lycopersicum fw2.2* allele (causing larger fruit size).

New technologies introduced in the last few years allow alternative means to validate suspect QTL genes; for example, techniques such as the use of CRISPR/Cas9 (described in Fig. 18.14) allow investigators to change one allele into another directly to verify if this change alters the trait under consideration.

Figure 22.14 Fine-mapping of QTLs through generation and analysis of nearly isogenic (congenic) lines. (a) Cross scheme for making nearly isogenic lines (NILs) whose genomes are almost entirely from *S. lycopersicum* (*unshaded*) but that are homozygous for small regions of *S. pennellii* DNA (introgressions; *black*). **(b)** Researchers map QTLs by comparing the trait values of NILS that have overlapping introgressions. **(c)** Using transgenes to verify the assignment of a QTL to a particular gene.

c: Courtesy of Stephen D. Tanksley, Cornell University

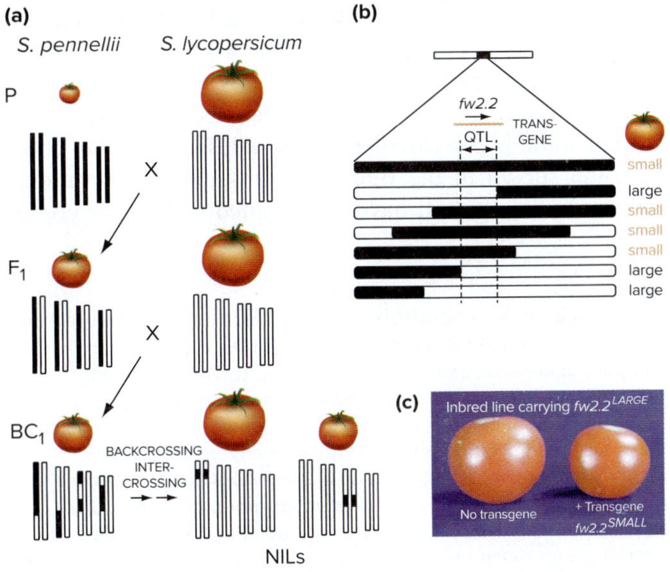

The *fw2.2* gene was found to encode a protein that is a negative regulator of cell division. In fact, the product of this gene is related to a tumor-suppressor protein that when mutated in humans causes the uncontrolled cell growth of cancer (see Chapter 20). Furthermore, the recessive *S. lycopersicum* allele of *fw2.2* was found to be hypomorphic—that is, it has reduced function relative to the *S. pennellii* allele. These results help explain why *S. lycopersicum* tomatoes are smaller than those of *S. pennellii;* reduced function of a negative regulator of cell growth should result in larger fruit.

Researchers worked for 10 years throughout the 1990s to identify the causal gene in the *fw2.2* QTL. Now, with high-density marker maps, microarray tools for genotyping, and low-cost, high-throughput DNA sequencing, the timeline is significantly shorter. Hundreds of QTL genes controlling a wide variety of traits have been identified in yeast, maize, *Arabidopsis*, rice, *Drosophila*, mice, pigs, and cattle.

Association Mapping Can Identify QTLs in Populations

The standard QTL mapping methods just described require controlled matings of phenotypically different individuals; but for many organisms including humans, such experiments are neither practical nor ethical. In addition, the number of recombination events that occur in the experimental crosses performed limits the resolution of standard QTL mapping.

Fortunately, nature already provides an alternative way to map QTLs: Geneticists can use a method called *association mapping* to take advantage of past recombination events that occurred in the ancestors of present-day individuals. Association mapping is really just an extension of linkage mapping (**Fig. 22.15a**), in which recombination occurs not over just a single generation, but instead accumulates over many generations.

As shown in **Fig. 22.15b,** the idea behind association mapping is quite simple. Suppose a new mutation that affects a trait took place many generations ago. This mutation (*red* in Fig. 22.15) would have occurred on a particular chromosome in a particular individual, like the *blue* chromosome shown in the figure. In subsequent generations, this chromosome will have recombined with other copies of this chromosome (*green, yellow,* and *purple*) that had different variants of polymorphic markers scattered over the chromosome's length. Through many rounds of recombination, all chromosomes in the present-day population are patchworks of the chromosomal types that were present in the population's ancestors.

In **association mapping,** scientists test large numbers of present-day individuals for genetic variants to find those variants that correlate statistically with differences in phenotype. For example, if the trait is a condition such as coronary artery disease, then the goal is to find a marker whose frequency in a population of patients is significantly greater than that in a population of nondiseased controls. If the researchers find variants strongly correlated with the condition, these markers must be closely linked to QTLs influencing whether people will develop coronary artery disease.

Figure 22.15 Association mapping is a simple extension of linkage mapping with a longer time frame. (a) In linkage analysis, geneticists examine recombinants to find correlations between specific genomic regions and discrete traits. **(b)** In association mapping, the recombinants already exist in a population because recombination has occurred in many generations of ancestors. Researchers identify a QTL by its linkage to marker alleles whose frequencies in the affected group of current-day individuals are greater than their corresponding frequencies in the control group. Such marker alleles would have been present on the *blue* chromosome when the disease-causing mutation (*red*) occurred nearby on this same chromosome.

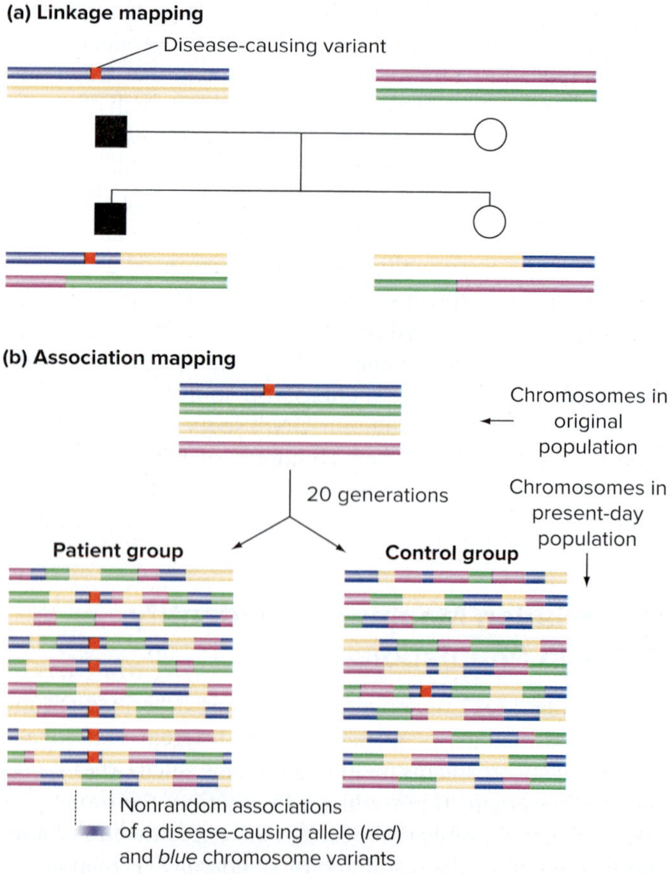

Linkage disequilibrium: Statistical correlations between variations at two loci

To understand how geneticists perform association mapping, we need first to examine how variants at different sites across the genome tend to be organized with respect to each other in natural populations. The basic question is whether alternative variants at one site are randomly associated with variants at other sites, or instead whether they are statistically correlated.

For example, in a hypothetical population, nucleotide position 300500 on chromosome 1 has two variants (A and T) in equal frequency, while nucleotide position 300600 on the same chromosome has two variants (G and C), also in equal frequency. With free recombination between sites, we would expect to have four different two-site haploid types (*haplotypes*) among gametes produced by parents in this population: A–G, A–C, T–G, and T–C in frequencies of 25% each. In such a case, the presence of an A at site 1 does not provide any information about the variant at site 2, which is equally likely to be a G or a C. In such a case, no correlations exist between the identities of the alleles at the two sites, and we would then say that variation at these two sites is in **linkage equilibrium.** A **haplotype,** then, is simply a particular collection of alleles at nearby variant loci.

An extreme departure from random association would be where only two haplotypes are present, for example A–G and T–C, each in a frequency of 0.5. In the latter case, a perfect positive correlation exists between variants at the two sites: When we find an A at site 1 then we can predict with certainty that we will see a G at the second site. Variants T and C are also perfectly positively correlated with each other. When the variants of two loci are correlated, then we say the variation is in **linkage disequilibrium (LD).**

LD measurements range from 0 (for no correlation as in the case of linkage equilibrium) to 1 (indicating a perfect correlation between variation at two sites). As we compare sites further and further apart along a chromosome, with greater genetic map distances between them, LD gradually decays to 0 because more possibilities exist for genetic recombination to disrupt allele associations. Geneticists often illustrate this concept by a plot of pairwise LD values which has the SNPs listed across the top of a *triangle diagram* whose component elements are diamonds. The shade of color of each diamond indicates the strength of LD between the SNPs compared. **Figure 22.16** shows through an analogy how such plots are made and interpreted; in the figure, the colors indicate the distances between six cities along the west coast of the United States rather than LD values.

Rates of recombination in humans average roughly 1×10^{-8} crossover events per generation between adjacent base pairs. However, recombination tends to be clustered into *hotspots* of genetic exchange. This discontinuity of recombination, as well as the randomness of when and where recombination occurs over evolutionary time, leads to discontinuities in LD across chromosomes for human populations. These discontinuities can be tracked by the presence of **LD blocks** (or **haplotype blocks**) on triangle diagrams (**Fig. 22.17**). The borders between LD blocks likely correspond to recombination hotspots.

Figure 22.16 Analogy for linkage disequilibrium plots: matrix of physical distances between cities in California. Different colors indicate increasing distance. Turning the matrix in (**a**) on its side so that the cities are listed linearly across the top (as in **b**) results in a *triangle diagram* analogous to that shown in Fig. 22.17 for linkage disequilibrium (LD) among SNPs.

(a) Distances between cities

	Los Angeles	Bakersfield	Fresno	San Jose	San Francisco
San Jose					42
Fresno				124	162
Bakersfield			104	211	270
Los Angeles		101	206	307	348
San Diego	112	212	316	418	459

<150 | 151-300 | 301-400 | >400

(b) Triangle diagram of intercity distances

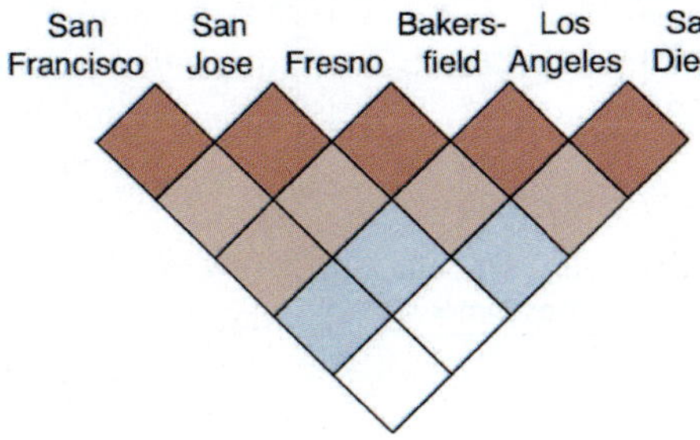

Figure 22.17 Clusters (blocks) of LD found among SNPs across the human genome. A segment of the genome is expanded in the second row with the locations of nine SNPs indicated. At the bottom is a *triangle diagram* with colors indicating the strength of the statistical correlation (LD) for pairwise comparisons of the indicated SNPs (progressively weaker correlations are in *dark brown, light brown, blue,* and *white*, respectively).

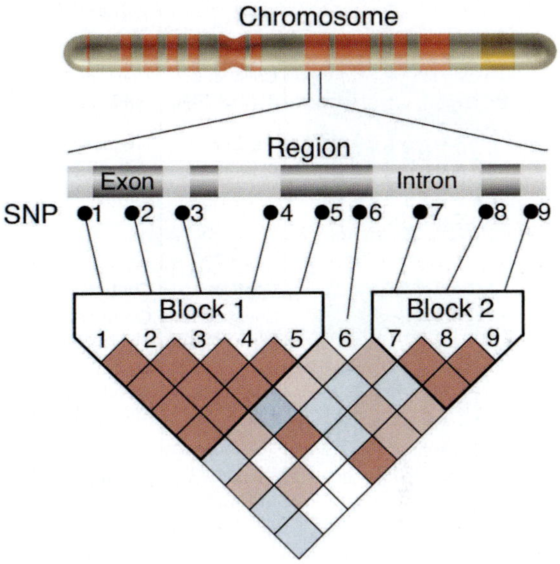

GWAS: Using LD to map genes in populations

The presence of LD in genomes makes it possible for scientists to assay variation at random polymorphisms at different locations across chromosomes and then test for statistical correlations between the variants and the trait of interest. When carried out on a genome-wide scale, such a survey is called a **genome-wide association study (GWAS).** Ideally, you would want to assay every DNA variant in the genome to discover the most likely causal variant(s) for the difference in phenotype. However, genotyping to this resolution would require whole-genome sequencing, and as of this writing in 2016, the cost of high-quality full-genome sequencing is still too high for studies involving the tens of thousands of individuals needed. Thus, researchers performing GWASs to date usually use DNA microarrays (see Fig. 11.17) that assay millions of common SNPs tagging different regions of the genome.

The chi-square test for independence To carry out a GWAS analysis, scientists assess the frequencies of variants distributed across genomes in large samples of *Cases* (individuals with the trait) and *Controls* (individuals that do not have the trait). Then they use the chi-square test to locate variants whose frequencies are positively correlated with the trait of interest (**Fig. 22.18**). In other words, to identify a QTL, the chi-square test for a tightly linked variant would need to show a significant difference between the frequencies for that variant in the patient and control populations; the null hypothesis is that no difference exists.

Figure 22.18 Comparing Cases and Controls in a GWAS of a human disease. The presence of a QTL is indicated if the frequency of a particular SNP allele is higher among patients with the disease than among controls. The chi-square test for independence (see Fig. 22.19) establishes the significance of this difference for each SNP. In this example, *SNP1* is linked to a QTL that contributes to risk for the disease, while *SNP2* is not.

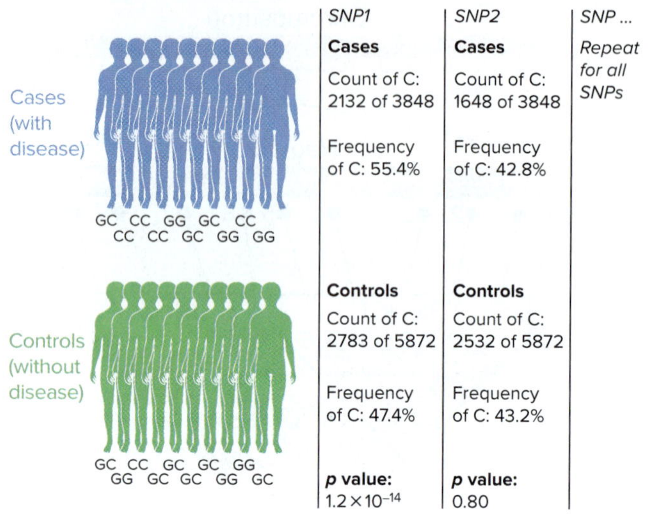

We cannot use the method described in the Chapter 5 Tools of Genetics Box entitled *The Chi-Square Test for Goodness of Fit* to assess the linkage between a SNP and a human trait because we don't know what progeny frequencies to expect based on the null hypothesis of no linkage. One reason why is that we don't know the frequencies of any pair of alternate SNP alleles in the population, and a second reason is that we don't know the phase of the SNP and disease alleles in the parents of the sampled populations. Instead, we can test the data against the null hypothesis that the distributions of the alternate SNP alleles are the same in the Cases and Controls. The statistical method required is called the *chi-square test for independence.*

To illustrate how this test works, let's use the example of *SNP1* in Fig. 22.18. We need to ask whether the difference between the frequencies of allele C in the Cases and Controls is statistically significant. The null hypothesis is that the occurrence of the disease and allele C are *independent*; in other words, that the distribution of the two alleles of *SNP1* (G and C) is random with respect to the two populations.

To calculate the χ^2 value, we use the data in Fig. 22.18 to construct a *contingency table* **(Fig. 22.19).** First, we enter the Observed values (O), which are simply the number of times that each *SNP1* allele (C or G) was observed in Cases and Controls. We also enter the total numbers of Cases, Controls, G alleles, and C alleles. Next, given the null hypothesis of independence, we calculate the Expected (*E*) values for each of the four observed classes. For example, if the C allele of *SNP1* is distributed the same way in the Cases and Controls, the E value for allele C in the Cases is simply the frequency of the occurrence of C in all of the genomes observed (4915/9720) multiplied by the total number of Cases (3848) (Fig. 22.19, *bottom left blue rectangle*). The same logic allows us to calculate the other three E values in the contingency table (Fig. 22.19).

For each of the four classes, $\chi^2 = (O - E)^2 / E$, the same as in the test for the goodness of fit. We then add the four individual χ^2 values to obtain the χ^2 statistic for the entire data set. Note that in the contingency table, only one degree of freedom exists (df = 1). The reason is that when one value for E is set, the the other three are determined.

For *SNP1*, the χ^2 value is 59.6 (Fig. 22.19), which means (with df = 1) that $p = 1.2 \times 10^{14}$. (Recall from Chapter 5 that *p* values for given χ^2 and df values can be obtained from tables or algorithms available on the Internet.) This *p* value means that if allele C of *SNP1* is not in fact associated with the disease, you would expect to find the observed results in only about 1 in 83 trillion [$1/(1.2 \times 10^{-14}) = 8.3 \times 10^{15}$] experiments like the one in Fig. 22.18.

The Tools of Genetics Box entitled *The Chi-Square Test for Independence* outlines the process of constructing a contingency table and interpreting the χ^2 results.

Figure 22.19 A chi-square test for independence contingency table. The table shown is to determine if the apparent association between the C allele of *SNP1* and heart disease (see Fig. 22.18) is statistically significant. O = observed; E = Expected; $\chi^2 = (O\text{-}E)^2/E$; df = degrees of freedom; *p* = probability. See the text for details.

SNP1	**Cases**	**Controls**	
G	**O = 1716** E = [(4805/9720)] × 3848 = **1902** $\chi^2 = [(1716 - 1902)^2/1902] =$ **18.2**	**O = 3089** E = [(4805/9720)] × 5872 = **2903** $\chi^2 = [(3089 - 2903)^2/2903] =$ **11.9**	**Total G = 4805**
C	**O = 2132** E = [(4915/9720)] × 3848 = **1946** $\chi^2 = [(2132 - 1946)^2/1946] =$ **17.8**	**O = 2783** E = [(4915/9720)] × 5872 = **2969** $\chi^2 = [(2783 - 2969)^2/2969] =$ **11.7**	**Total C = 4915**
	Total Cases = 3848	**Total Controls = 5872**	**Total genomes = 9720**

$\chi^2 = 18.2 + 11.9 + 17.8 + 11.7 = \mathbf{59.6}$; **df = 1;** $p = \mathbf{1.2 \times 10^{-14}}$

TOOLS OF GENETICS

The Chi-Square Test for Independence

The chi-square test for independence is used when asking if a particular SNP allele in the human genome is associated with a disease. This version of the chi-square test is useful in several other situations in genetic analysis where, based on a null hypothesis, expected frequencies of individuals cannot be predicted using the simple logic of Mendelian transmission genetics.

In performing a chi-square test for independence, the first step is to construct a *contingency table,* which enables you to calculate the chi-square value for the data set. Using the example of association of a SNP allele with a human trait, we outline below how to construct a contingency table and how to evaluate the results.

In the example that follows, two alleles of a particular SNP locus, G and C, are distributed unevenly in the genomes of the *Cases* (individuals with the trait in question) and the *Controls* (individuals who do not have the trait). Suppose that G is present more frequently in the Cases, and C is present more frequently in the Controls. The question we ask is: Is the association of G with the trait statistically significant?

Note that because in this example we are counting alleles (not genotypes), *each Observed number in the contingency table represents a genome, not a person.* Thus, the number of Cases or Controls means the number of genomes. For an autosomal SNP, the number of individual people who are either Cases or Controls is half the number of genomes.

1. **Draw a contingency table.** In our example, the table has four columns and four rows (**Fig. A**). Label the central two columns Cases and Controls, and the central two rows G and C.
2. **Fill in the Observed values (O).** Using your data, fill in numerical values for O in each of the four central boxes.
3. **Add the Observed values across the rows and down the columns.** These sums are the numerical values for Total G, Total C, Total Cases, and Total Controls.
4. **Assign a numerical value to the Total genomes box.** This box contains the sum of the rows above (Total G + Total C), which equals the sum of the two rows across (Total Cases + Total Controls).
5. **Formulate a null hypothesis.** In this example, the null hypothesis is that no association exists between the G allele and the phenotype.
6. **Calculate the Expected values (E).** Based on the assumption that the null hypothesis is true, for each of the four central boxes in the table, calculate E as shown in Fig. A.
7. **Calculate the χ^2 value for the data set.** First, calculate χ^2 for each of the four Observations as shown in Fig. A. The sum of each of these four χ^2 values is the total χ^2 statistic for the data set.
8. **Determine the degrees of freedom (df).** By examining the E value calculations in Fig. A, you can see that if one E value is set, the other three are determined. Therefore, df = 1.
9. **Find the probability (*p*).** Using a table on the Internet, find the value for *p* that corresponds to the χ^2 and df values. The *p* value is the probability that a deviation from the predicted numbers at least as large as that observed in the experiment would occur by chance.
10. **Evaluate the significance of the *p* value.** For an experiment with a single SNP, by convention the *p* value cutoff is 0.05. That is, the null hypothesis that the SNP allele in question and the occurrence of the disease are independent can be rejected only if $p < 0.05$. However, in GWAS experiments using many SNPs, the *p* value cutoff is 0.05 / (the number of SNPs examined). The data indicate a significant association of the SNP and the disease only if the *p* value is lower than this number.

Figure A How to construct a contingency table.

SNP allele	Cases	Controls	
G	O_{CaG} = Cases with G $E_{CaG} = (O_{CaG}/\text{Total individuals}) \times \text{Total Cases}$ $\chi^2_{CaG} = (E_{CaG} - O_{CaG})^2/E_{CaG}$	O_{CoG} = Controls with G $E_{CoG} = (O_{CoG}/\text{Total individuals}) \times \text{Total Cases}$ $\chi^2_{CoG} = (E_{CoG} - O_{CoG})^2/E_{CoG}$	Total G = $O_{CaG} + O_{CoG}$
C	O_{CaC} = Cases with C $E_{CaC} = (O_{CaC}/\text{Total individuals}) \times \text{Total Cases}$ $\chi^2_{CaC} = (E_{CaC} - O_{CaC})^2/E_{CaC}$	O_{CoC} = Controls with C $E_{CoC} = (O_{CoC}/\text{Total individuals}) \times \text{Total Cases}$ $\chi^2_{CoC} = (E_{CoC} - O_{CoC})^2/E_{CoC}$	Total C = $O_{CaC} + O_{CoC}$
	Total Cases = $O_{CaG} + O_{CaC}$	Total Controls = $O_{CoG} + O_{CoC}$	Total genomes = $O_{CaG} + O_{CaC} + O_{CoG} + O_{CoC}$

Total $\chi^2 = \chi^2_{CaG} + \chi^2_{CaC} + \chi^2_{CoG} + \chi^2_{CoC}$

Manhattan plots In the context of a GWAS experiment, the interpretation of p values is complicated by the fact that millions of SNPs are being examined, leading to a problem of false positives. With this many loci, a large number of SNPs may appear to differ in their frequencies between affected cases and controls even if these differences reflect only chance sampling errors. Recall that the typical cutoff for a single chi-square test to be considered statistically significant is a p value less than 0.05, meaning that we would observe deviation as large or larger from the expected equal proportions between cases and controls by chance 5% of the time. Thus, if we repeat the chi-square test for all 1 million SNPs in a genome-wide study, we will expect that $1{,}000{,}000 \times 0.05 = 50{,}000$ of them appear to be associated with that trait only by chance.

To deal with this false positive problem, researchers do not consider individual tests in GWAS analysis to be statistically significant unless the probability of observed deviation from the expectations of the null hypothesis (that is, the p value in the chi-square test) is less than 0.05/(1,000,000 SNPs) $= 5 \times 10^{-8}$. This condition necessitates very large sample sizes of Cases and Controls. And in our example (Figs. 22.18 and 22.19), we can conclude that the association of *SNP1* with the disease is statistically significant.

In 2007, one of the first large-scale GWAS explorations of human diseases examined nearly 1 million SNPs in approximately 2000 patients and 3000 control individuals for each of seven complex diseases, including coronary artery disease. The most significant associations have a very small probability of being observed by chance (such as 1×10^{-14}). Therefore, researchers plotted the $-\log_{10}$ (p value) [for example, $-\log_{10}(1 \times 10^{-14}) = 14$] for the chi-square tests of statistical significance for the differences between the frequency of each individual SNP among the Cases (for example, people with heart disease) versus the Controls (no heart disease) (**Fig. 22.20**). This way of representing the data is often called a **Manhattan plot** as it looks somewhat like the skyline of Manhattan in New York City. The rare peaks of $-\log_{10}(5 \times 10^{-8}) = 7.7$ or greater indicate the presence of QTLs for this disease in this study of 1 million SNPs.

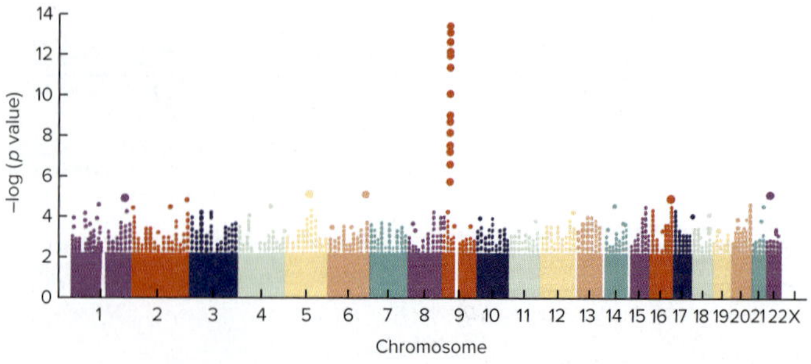

Figure 22.20 Manhattan plot for a GWAS of coronary artery disease in humans. Peak of red dots on chromosome 9 represent a cluster of SNPs with high −log p values, indicating highly significant statistical evidence for an association with heart disease.

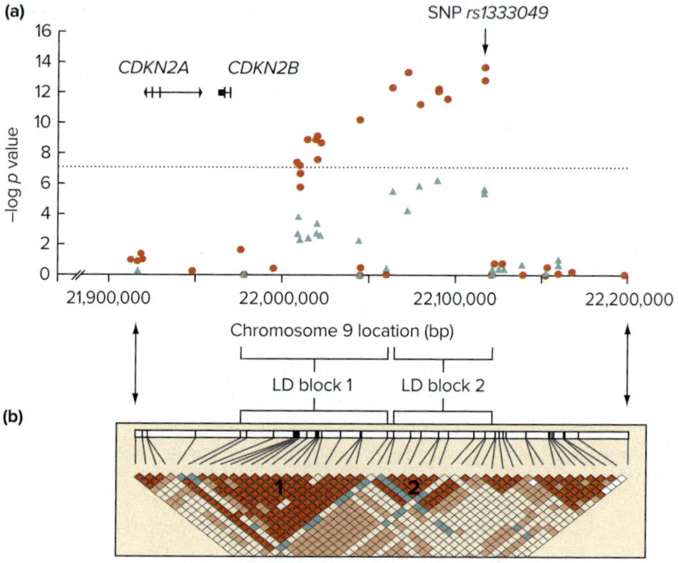

Figure 22.21 Associations with coronary artery disease of SNPs in a 300 kb region of human chromosome 9. **(a)** A blown-up scale of the Manhattan plot from Fig. 22.20 shows that multiple SNPs between nucleotides 22,010,000 and 22,120,000 display strong statistical associations with the disease in two different GWAS surveys (*red dots and blue triangles*). A cutoff for statistical significance in studies of about 500,000 SNPs is $-\log_{10}$ (p value) = 7 (*dotted line*). The two annotated genes (*CDKN2A* and *CDKN2B*) in the region are located at positions that do not show significant disease associations. **(b)** Triangle diagram of LD among the SNPs shown in (a). Several blocks of strong LD (*dark brown* = maximum LD, *light brown* = significant LD, *blue* = weakly significant LD, and *white* = no significant LD) are apparent. Multiple SNPs in LD blocks 1 and 2 such as *rs1333049* have strong associations with coronary heart disease.

If we zoom in on the peak of red dots on chromosome 9, we find a cluster of significantly associated SNPs (**Fig. 22.21a**). A follow-up study of German patients with myocardial infarctions (heart attacks) also showed strong associations between SNPs in this region and the occurrence of heart attacks before the age of 60 (Fig. 22.21a). *SNP1* in Figs. 22.18 and 22.19, whose association with heart disease is very strong, is actually one of the SNPs from this latter study (it is called *rs1333049* in Fig. 22.21a).

Odds ratios It is important to note that *rs1333049* is only one of several SNPs with strong disease associations that are clustered together in two adjacent blocks of LD (**Fig. 22.21b**). Because of the linkage disequilibrium between alleles of these polymorphisms, we cannot say which if any of them is the causative mutation for coronary disease. In fact, the region may contain more than one causative mutation. Very little recombination has occurred within either of these LD blocks and only marginally more between them, so we cannot discriminate where the actual mutation(s) that increase(s) the risk of disease lie in this region.

Furthermore, because no obvious genes have been found to be located within these blocks of LD (Fig. 22.21a), the nature of the genomic region has provided researchers with no clear-cut clues about the specific location of the disease-causing variant(s).

However, even if we cannot identify the relevant mutation(s), the lack of recombination in the region also means that SNP alleles with strong disease associations can help physicians identify individuals who might have an elevated likelihood of coronary artery disease. We can quantify the average increase in risk associated with a particular SNP allele by calculating the **allelic odds ratio (Fig. 22.22a)**. In the case of SNP *rs1333049,* one of the SNPs with the strongest disease association in this region, the presence of a C in the genome confers a 1.38 times greater risk of having coronary artery disease than the presence of a G.

We can also calculate a **genotypic odds ratio**—the increased risk associated with a particular SNP genotype—by comparing the risk allele homozygote or the heterozygote genotype frequencies between Cases and Controls to those for the nonrisk allele homozygote (**Fig. 22.22b**). It is important to appreciate that even a person with the genotype of greatest risk is not guaranteed to have a heart attack before the age of 60; the chances of such an event are simply higher than for people of other genotypes.

QTLs Reflect Population Histories

A QTL is by definition a genetic variant that contributes in a detectable way to a particular quantitative trait in a particular population. This definition poses two major challenges to QTL detection. First, the number of QTLs depends on the amount of variation in the population, and the amount of variation in turn reflects the history of the population. If, for example, you were examining an inbred population of genetically identical plants, you would never be able to find a QTL influencing a trait such as viability in stress conditions like drought, because no genetic variation exists in the population. A large number of genes must exist that encode proteins which could potentially influence a plant's survival during a drought, but none of these would constitute a QTL in this population because all of the plants have the same alleles of all of these genes. In other words, the set of QTLs that could in theory be found by experiment depends on the existence of mutations that occurred previously in the population.

A second challenge to searches for QTLs is that our ability to find any QTL depends both on the frequency of that particular QTL allele in the population and on the strength of its contribution to the trait. If the allele is rare, statistical significance indicating the presence of the QTL might be obtained only in very large-scale studies that include hundreds of thousands of individuals. And matters become further clouded when many different QTLs for a trait exist in the population, each of which contributes only a small amount to the trait.

Striking contrasts that have been observed between GWAS analyses in humans and in domesticated animals such as dogs and horses illustrate how QTLs reflect population histories.

Figure 22.22 Calculations of association *p* values and allelic odds ratios. (a) The *allelic odds ratio* indicates the magnitude of risk conferred by the tested SNP (*SNP1* in Figs. 22.18 and 22.19). The ratio of 1.38 in this study means that individuals having the C allele of this SNP have a 1.38 times greater risk of having the disease than people having the G allele. **(b)** The *genotypic odds ratio* indicates the magnitude of risk conferred by particular genotypes at the tested SNP.

(a) Calculating the allelic odds ratio

Odds of C allele given Case status =
Freq of C in Case/Freq of G in Case
Odds of C allele given Control status =
Freq of C in Control/Freq of G in Control

$$\text{Odds Ratio} = \frac{\text{Case C/Case G}}{\text{Control C/Control G}} = \frac{2132/1716}{2783/3089} = 1.38$$

(b) Calculating the genotypic odds ratio

	CC *N* (%)	CG *N* (%)	GG *N* (%)	χ^2 (2 df)	*p* value
Cases	586 (30.5)	960 (49.9)	378 (19.6)	59.6	1×10^{-13}
Controls	676 (23.0)	1431 (48.7)	829 (28.2)		

Homozygote (CC) odds ratio (relative to GG) =
(586/378)/(676/829) = 1.91
Heterozygote (CG) odds ratio (relative to GG) =
(960/378)/(1431/829) = 1.47

Effects of the recent human population explosion

Human height provides an instructive example of the challenges facing geneticists studying many complex traits and diseases in humans. The heritability (h^2) of human height is over 80% in most populations, yet the first small GWAS surveys found no major height genes in humans (excepting obviously rare and unusual states such as dwarfism). A much more massive, recent GWAS investigation of more than 180,000 people found 180 QTLs that influence height. These genes are, as expected, enriched for those involved in growth-related processes that influence adult height. However, each of these QTLs makes only a very small contribution to the phenotypic value, and all of these 180 QTLs put together explain only about 10% of the total variation in adult human height. In other words, the population of humans on the earth must have genomes containing many other mutations not yet detected as QTLs that can also contribute to height.

In a similar fashion, significant GWAS "hits" have been found for an increasing number of medically important traits (**Fig. 22.23**). However, as was seen for height, the individual genes that were identified contributed in most cases only a very small amount to the total variation in the trait or susceptibility to the disease.

Figure 22.23 Genomic distribution across the 23 human chromosomes of published statistically significant GWAS associations for 17 medically important traits. For all associations shown, $p \leq 5 \times 10^{-8}$. The diagram was updated at the end of 2015. Numbers next to the categories in the key are the number of different studies.

Source: Welter, D., MacArthur, J., Morales, J., Burdett, T., Hall, P., Junkins, H., Klemm, A., Flicek, P., Manolio, T., Hindorff , L., and Parkinson, H. (2014) The NHGRI GWAS Catalog, a curated resource of SNP-trait associations Nucl. Acids Res. 42 (D1): D1001–D1006 (www.ebi.ac.uk/fgpt/gwas & www.genome.gov/gwastudies.)

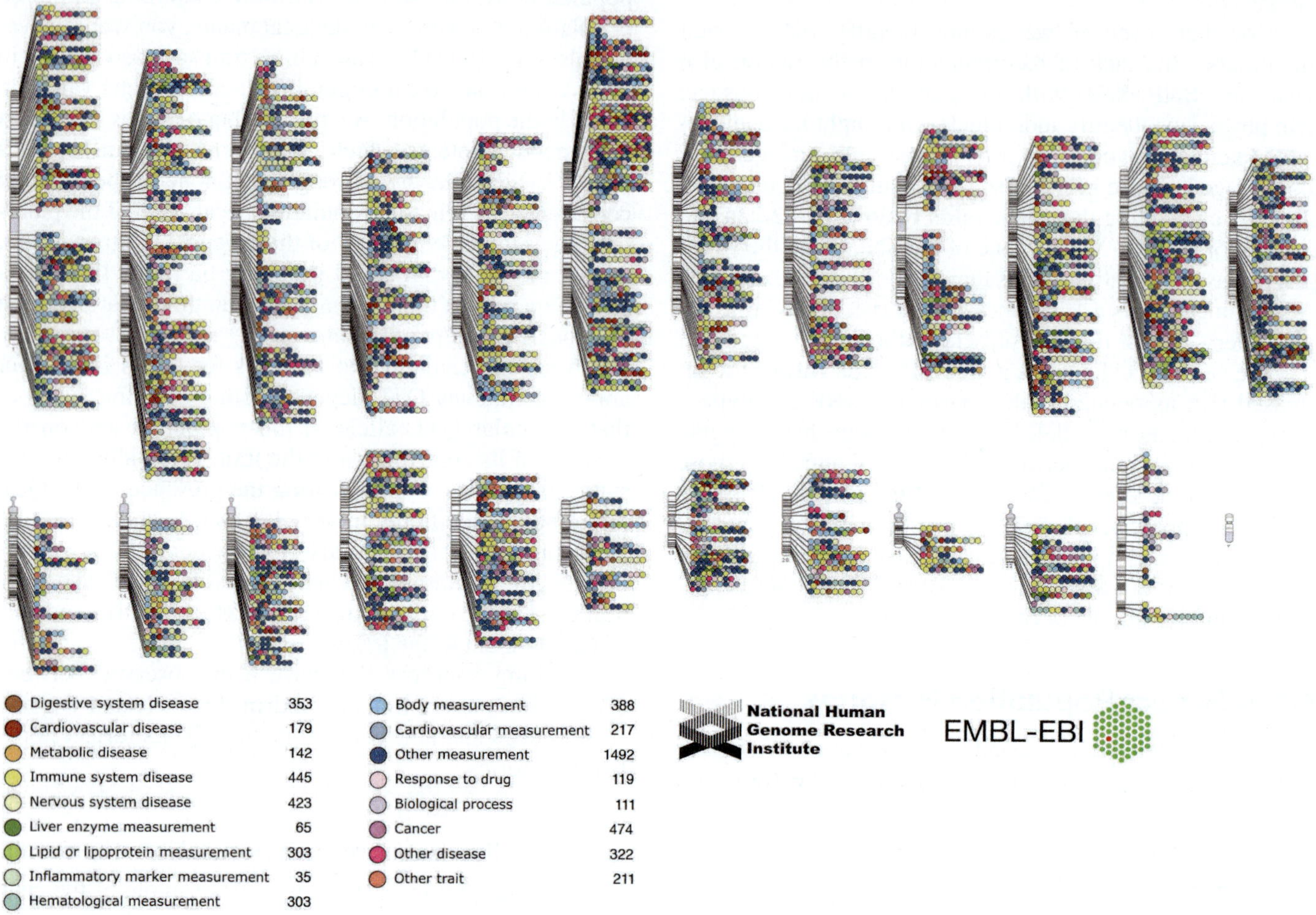

Results such as these have been a disappointment to geneticists who anticipated that GWAS analyses would be able to explain much more of the total trait variation in human populations. These scientists had pinned their hopes on the idea that common diseases would be due to common variants, so GWASs would provide much more information about the genetic basis of important traits like disease susceptibility.

Why has this hope thus far not panned out for GWASs of humans? The answer is in large part based on the recent history of humans on the earth. In particular, the human population has exploded since the onset of the Industrial Revolution in about 1750 (**Fig. 22.24**). Furthermore, as you already know, the more gametes, the greater the chance that new mutations can arise. These facts mean that many of the mutations in the current human population occurred only within the last few hundred years. Such recent mutations have not yet had a chance to spread widely around the

Figure 22.24 The population of humanity on earth. The population explosion since the beginning of the Industrial Revolution has shaped the genetic variation underlying many complex traits.

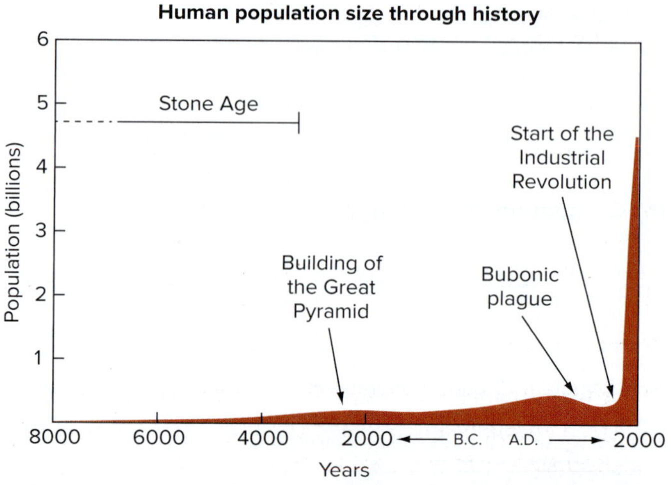

globe. As a result of the human population explosion, complex traits will often be governed by a large variety of mutations in many genes. These QTLs will be difficult to find because most of these recent mutations will be present in the genomes of only a very small fraction of the people on earth. Furthermore, given the large number of genes influencing most quantitative traits, most individual QTLs could have only small effects on the phenotype.

The effects of artificial selection during animal domestication

Domesticated animals show a strikingly different pattern of inheritance for quantitative traits such as adult height than that seen for such traits in humans. Dogs are particularly interesting because a huge range of morphologically and diverse breeds exists today (see the photo at the beginning of this chapter). Studies comparing SNP frequencies genome-wide between large and small breeds of dogs revealed that a single gene, *Insulin-like growth factor 1* (*IGF1*), explains 10–15% of the variation in dog size. Remarkably, only six genes (including *IGF1*) explain 50% of the observed variance in body size in dogs. Similar studies examining large and small breeds of horses found that only four genes explain 83% of size variation in these domesticated animals. In addition, researchers have found that only a small number of loci (usually four or fewer) explains most of the variation in dogs for other morphological traits as diverse as average body shape and ear erectness.

How can we explain the striking contrast in the number of major genes underlying quantitative traits between domesticated animals (where only a few QTLs are typically involved) and humans (where the QTLs for many traits number in the hundreds)? The answer again involves the history of these populations. Humans have domesticated dogs and horses during a relatively short period of time, beginning roughly 15,000 years ago for dogs and perhaps 6000 years ago for horses. In the course of domestication, humans imposed very strong artificial selection for desired traits. Only a small population of dogs underwent domestication, and so only a few variants were favored by this artificial selection.

The future of complex trait analysis in humans

Clearly, the fact that complex traits in humans are influenced by a large number of genetic variants, each of which may be found only in a small number of people and may have only small effects, complicates our understanding of these traits. The solutions to these problems will require the acquisition of huge amounts of data. For example, efforts are now being mounted to obtain the whole-genome sequences of tens of thousands of individuals from diverse parts of the world. Furthermore, researchers are measuring in detail many complex traits of these people so that they can be placed into Case and Control groups for association mapping.

New methods will need to be developed to interpret this vast amount of information. One important goal will be to evaluate how combinations of variants (rather than individual variants considered one at a time) can contribute to specific traits. A second new approach to GWAS analyses will increase the statistical sensitivity of the method by asking if different rare nonanonymous SNPs *within the same gene* (rather than individual anonymous SNPs as at present) correlate with disease. The success of all of these endeavors by the next generation of geneticists will be crucial to our ability to use the personal genomic information that is becoming cheaper and cheaper to acquire as the basis for the preventive and diagnostic medicine of the future.

essential concepts

- *QTL mapping* extends the basic cross concept by using more genetic markers and statistical tests for correlations between alleles and phenotypic values.
- *Genome-wide association studies (GWASs)* take advantage of recombination events that occurred in the ancestors of present-day populations. Researchers compare allele frequencies in Case and Control subpopulations to test the association of specific variants with the trait.
- Variation in complex traits reflects the history of the population. In domesticated animals, recent strong artificial selection means that polymorphisms at just a few loci are responsible for much of the variation observed. In humans, the recent population explosion means that many loci each make small contributions to complex trait variation.
- Although it is difficult to analyze the genetic underpinnings of complex traits in humans, this information will be essential for future progress in preventive and diagnostic medicine.

SOLVED PROBLEMS

I. A plant breeder is interested in selecting for larger-sized edible beans. Beans from the upper range of the bean size distribution (that is, in the *dark orange* region of the distribution at the *top* of the following figure) were chosen as parents for the next generation. The distribution of bean sizes among the progeny of

DNA: © Design Pics/Bilderbuch RF

these selected parents is shown in *purple* at the *bottom* of the figure. The numbers above the distributions indicate the mean weights of beans of the entire parental generation (403.5), of the beans selected for producing the next generation (691.7), and of the beans in the progeny generation (609.1).

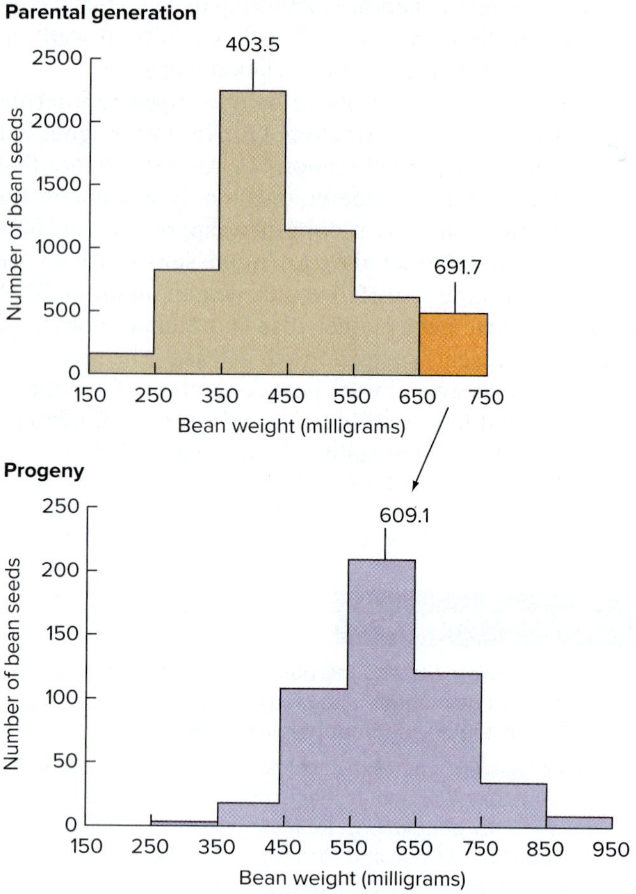

a. For this experiment, what is the selection differential (S)?

b. What is the response to selection (R) in this experiment?

c. Calculate the heritability for bean weight based on these data.

d. Is the heritability you just calculated in part (c) an estimate of narrow-sense or broad-sense heritability? Explain.

e. How successful was the breeder in applying artificial selection? Could the breeder have improved this experiment to obtain even larger beans in the same growing period?

Answer

a. The selection differential $S = 691.7$ mg $- 403.5$ mg $= 288.2$ mg. This value is the difference between the mean weight of the beans from the parental generation that were selected for breeding and the mean weight of the entire parental generation (including all plants whether or not they were selected for breeding).

b. The response to selection $R = 609.1 - 403.5 = 205.6$ mg. This value is the difference between the mean weight of the progeny and the mean weight of the entire parental generation.

c. The heritability from these data is calculated as $h^2 = R/S = 205.6/288.2 = 0.713$.

d. Heritability calculated as R/S is formally a measure of *realized heritability*, which is equal to the narrow-sense heritability when the phenotypic values are normally distributed. The estimate is of narrow and not broad-sense heritability because only additive genetic variation is measured. The offspring beans share only one of the two alleles found in a parental bean, and in the offspring that allele is paired with a different allele than it was in either parent. Thus, dominance and gene interaction effects that could influence phenotype are different in the parents and offspring. Broad-sense heritability would include such dominance and interaction genetic effects as well as the additive effects.

e. This breeding program was very efficient: The breeder started with a parental population whose mean bean weight was 403.5 mg, and in a single generation obtained progeny whose mean weight was 609.1 mg. The success of this program reflects the high selection differential and the high heritability involved. It is possible that the breeder could have obtained progeny beans of even higher mean weight by selecting a very few parents at the absolute high end of the range, but then fewer progeny would be available to harvest.

II. Systemic lupus erythematosus (SLE) is a chronic autoimmune disease that can lead to inflammation and tissue damage in any part of the body and even death. A large-scale GWAS analysis of this condition was performed; the study involved the genotyping of more than 500,000 SNP loci in more than 1000 Cases (patients) and 3000 Controls of European ancestry. The data revealed significant associations with five genetic loci, whose names and positions are shown in the Manhattan plot that follows.

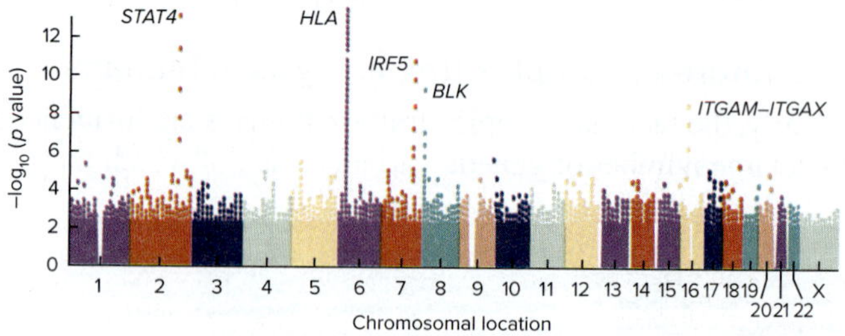

a. Are any of these five loci genetically linked?

b. Do the heights of these five peaks reflect the relative risks that a person with a particular genotype for the corresponding locus would develop the disease?

c. In the Manhattan plot, it appears that a SNP with a $-\log_{10}(p \text{ value})$ of approximately 3 exists at every chromosomal location in the genome. Explain why this impression is misleading. In addition, explain how the appearance of the Manhattan plot relates to the *p* value threshold that would indicate that the difference between the frequency of the allele in the patient population and in the control population is significant.

d. Given the accomplishments of this study in identifying loci associated with SLE, why might it still be worthwhile to conduct similar studies in other populations?

Answer

a. These five loci are all unlinked because they are on different chromosomes. *STAT4* is on chromosome 2; *HLA* on chromosome 6; *IRF5* on chromosome 7; *BLK* on chromosome 8; and *ITGAM-ITGAX* on chromosome 16.

b. No, the heights of the peaks shows the significance of the association of the disease with individual SNPs. The relative risk is only one component of the significance of the association; another major component of the significance is the allele frequency. Thus, a SNP allele closely linked to a rare mutation which might have a very high impact on the trait (that is, a high risk ratio) might not register as a significant association if only a small percentage of the patient group had that particular mutation.

c. If you broadened the *x*-axis of the Manhattan plot to show all 500,000 SNPs examined, the $-\log_{10}(p \text{ value})$ of most would be close to zero, as was seen in Fig. 22.21. Because the *x*-axis in the Manhattan plot is compressed, your attention focuses on the relatively small fraction of SNPs where random sampling error makes the *p* value about 0.001.

The Manhattan plot shows that due to random sampling error, thousands of the 500,000 SNPs have a $-\log_{10}(p \text{ value})$ higher than 3, hundreds have a $-\log_{10}(p \text{ value})$ higher than 4, and dozens have a $-\log_{10}(p \text{ value})$ higher than 5. This fact provides a graphic illustration of why most researchers would place the *p* value threshold of significance for this experiment at $(0.05 / 500{,}000) = 1 \times 10^{-7}$, or a $-\log_{10}(p \text{ value}) = 7$.

d. The QTLs behind this complex trait may differ in populations with different genetic histories. The results of this study of individuals with European descent may not be particularly informative for other groups. Furthermore, GWAS investigations of other populations may pinpoint genes other than these four that might contribute to the trait, and the study of these other genes may help researchers better understand the biology underlying the SLE disease.

PROBLEMS

Vocabulary

1. Choose the best matching phrase in the right column for each of the terms in the left column.

a. isogenic lines	1. fraternal
b. QTL	2. blocks of association between variants at different loci
c. response to selection	3. proportion of total phenotypic variance attributed to genetic variance
d. association mapping	4. homozygous for all genomic regions
e. MZ twins	5. genes contributing to complex traits
f. DZ twins	6. identical
g. congenic lines	7. measure of evolution
h. linkage disequilibrium	8. 0.5 for siblings
i. heritability	9. takes advantage of recombination over the course of a population's history
j. genetic relatedness	10. contain introgressions

Section 22.1

2. Suppose you grew genetically identical dandelion seeds in an environmentally controlled environment like a greenhouse.
   a. Sketch the distribution of stem length phenotypic classes you would expect to see relative to the distributions shown in Figs. 22.5a and 22.5b.
   b. Why would you expect to see a phenotypic variance greater than zero for these genetically identical dandelions grown in a constant environment?
   c. How could you use this information to refine the estimate of genetic variance shown originally in Fig. 22.5b?

3. How can each of the following be used in determining the role of genetic and/or environmental factors in phenotypic variation in different organisms?
   a. genetic clones
   b. human monozygotic versus dizygotic twins
   c. cross-fostering

4. Two different groups of scientists studying a rare trait in ground squirrels report very different heritabilities. What factors influencing heritability values make it possible for both conclusions to be correct?
5. Which of the following statements would be true of a human trait that has high heritability in a population of one country?
   a. The phenotypic difference within monozygotic twin pairs would be about the same as the phenotypic differences among members of dizygotic twin pairs.
   b. Very little phenotypic variation exists between monozygotic twins but high variability exists between dizygotic twins.
   c. The trait would have the same heritability in a population of another country.
6. Studies have indicated that for pairs of twins raised in the same family, the environmental similarity for monozygotic (MZ) twins is not significantly different from the environmental similarity for fraternal (dizygotic or DZ) twins. Why is this fact important for calculations of heritability?
7. A study published in 1937 examined the average differences between pairs of twins [either monozygotic (MZ) or dizygotic (DZ)] and pairs of siblings for three different quantitative traits: height, weight, and intelligence quotient (IQ) as measured by the Stanford-Binet test. (The concept of IQ is extremely controversial as it is unclear to what extent IQ tests measure native intelligence, but for this problem, consider IQ to be a measurable trait even if its significance is unknown.) Some of the MZ twins were raised together in the same household (RT), while other MZ twins were raised apart in different families (RA). The results of this study, shown as average differences, are as follows:

	MZ(RT)	MZ(RA)	DZ	Siblings
Height	1.7 cm	1.8 cm	4.4 cm	4.5 cm
Weight	1.86 kg	4.49 kg	4.54 kg	4.72 kg
IQ	5.9	8.2	9.9	9.8

   a. Which of these three traits appears to have the highest heritability? The lowest heritability? (*Note:* You do not have enough data to calculate numerical values for heritability. You are meant to think about this in general quantitative terms.)
   b. The Centers for Disease Control and Prevention (CDC) of the National Institutes of Health recently reported that in the United States during the period 1960–2002, the average weight of a 15-year-old boy increased from 135.5 pounds (61.46 kg) to 150.3 pounds (68.17 kg). During the same period, the average height of a 15-year-old boy increased from 67.5 inches (171.5 cm) to 68.4 inches (173.7 cm). How well do these statistics match your estimates of relative heritabilities from part (a)?
8. This problem is about Eq. 22.7, which relates heritability to the difference between MZ and DZ twins' correlation coefficients for a quantitative trait: heritability $= 2(r_{MZ} - r_{DZ})$.
   a. Explain why Eq. 22.7 is true for the extreme case where all phenotypic variation is due to genes.
   b. Do the same for the case where no phenotypic variation is due to genes.
9. Table 22.2 lists concordance values for MZ and DZ twins with respect to a number of discrete complex traits.
   a. How do you know at a glance that the heritabilities of all of these traits is less than 1.0?
   b. To estimate the heritabilities of these traits, would you use Eq. 22.7 or Eq. 22.8? Explain why.
   c. Calculate heritability estimates for each trait using the data in Table 22.2. For which trait is the contribution of genetic variance to the total phenotypic variance more than any other in the table?
10. In 1959, the Russian geneticist Dmitri Belyaev began an experiment in which he bred silver foxes for tameness. He started with wild foxes, and selected for reproduction the progeny who showed the least fear of, and least aggression toward, humans. By 1999, after 30 generations of inbreeding and outbreeding, Belyaev and his colleagues obtained a population of foxes that behaved essentially like domesticated dogs; they wagged their tails and licked their human caretakers.
   a. What does this experiment say about the heritability of docile behavior in foxes? Explain.
   b. Remarkably, as shown in the photographs here, along with the behavioral changes came morphological changes, including spotted coats, floppy ears, and curly tails.

© Vincent J. Musi/National Geographic/Getty Images

Wild silver fox

© Lee H. Rentz/Photoshot/Newscom

Tame silver foxes

© Michael Goldberg, Cornell University, Ithaca, NY

A similar *domestication phenotype* has been observed in other animals bred and selected in a similar manner. Do you think that this outcome necessarily means that the same genes control both the behavior and the appearance? Explain.

11. Two traits with similar phenotypic variance exist in a population. One trait has two major genes and six minor loci that influence the phenotypic value, and the second trait has 12 minor loci and no major genes affecting the phenotypic value. Does this information tell you which trait you should expect to respond most consistently to selection? Explain why it does or does not.

Section 22.2

12. Two alleles at one locus produce three distinct phenotypes. Two alleles of two genes lead to five distinct phenotypes. Two alleles of six genes lead to 13 distinct phenotypes. (These statements assume that the alleles at any one locus are codominant or incompletely dominant and that each gene makes an equal contribution to the phenotype.)
    a. Derive a formula to express this relationship. (Let n equal the number of genes.)
    b. Each of the most extreme phenotypes for a trait determined by two alleles at one locus are found in a proportion of 1/4 in the F_2 generation. If two alleles of two genes determine the trait, each extreme phenotype will be present in the F_2 as 1/16 of the population.

       In common wheat (*Triticum aestivum*), kernel color varies from red to white and the genes controlling the color act additively, that is, alleles for each gene are incompletely dominant and each gene contributes equally to the color. A true-breeding red variety is crossed to a true-breeding white variety, and 1/256 of the F_2 have red kernels and 1/256 have white kernels. How many genes control kernel color in this cross?

13. In a certain plant, leaf size is determined by four genes whose alleles assort independently and act additively. Thus, alleles A, B, C, and D each add 4 cm to leaf length and alleles A', B', C', and D' each add 2 cm to leaf length. Therefore, an $AA\ BB\ CC\ DD$ plant has leaves 32 cm long and an $A'A'\ B'B'\ C'C'\ D'D'$ plant has leaves 16 cm long.
    a. If true-breeding plants with leaves 32 cm long are crossed to true-breeding plants with leaves 16 cm long, the F_1 will have leaves 24 cm long and the genotype $AA'\ BB'\ CC'\ DD'$. List all possible leaf lengths and their expected frequencies in the F_2 generation produced from these F_1 plants. (*Hint:* Recall Problem 45e in Chapter 2.)
    b. Now assume that in a randomly mating population the following allele frequencies occur:

       frequency of $A = 0.9$
       frequency of $A' = 0.1$
       frequency of $B = 0.9$
       frequency of $B' = 0.1$
       frequency of $C = 0.1$
       frequency of $C' = 0.9$
       frequency of $D = 0.5$
       frequency of $D' = 0.5$

       Calculate separately the expected frequency in this population of the three possible genotypes for each of the four genes.
    c. What proportion of the plants in the population described in part (b) will have leaves that are 32 cm long?

14. Compare and contrast the use of SNP genotyping: (i) in the positional cloning of Mendelian disease genes, (ii) in direct QTL mapping, and (iii) in GWAS.

15. Explain the similarities and differences between the procedures used for the coarse-scale mapping and fine-scale mapping of QTLs through controlled crosses, as was accomplished for QTLs involved with tomato size.

16. In Fig. 22.14c, the *fw2.2* causal gene was validated by introducing the *S. pennellii* version of the gene into *S. lycopersicum*. Why was the experiment done this way, rather than the other way around?

17. Among the most prevalent pathologies that afflict human beings is heart disease, which can have a severe impact on quality of life and can even result in premature death. While heart disease mostly afflicts those who are older, 1% or 2% of people in their 30s, and even in their 20s, suffer from this disease. Genetic and environmental components of this disease exist. What strategy might you use to choose families to participate in a GWAS of heart disease–causing genes? Explain your reasoning.

18. Human geneticists have found the Finnish population to be very useful for studies of a variety of conditions. The population is small; Finns have extensive church records documenting lineages; and few people have migrated into Finland. The frequency of some recessive disorders is higher in the Finnish population than elsewhere in the world, and diseases such as PKU and cystic fibrosis that are common elsewhere do not occur in the Finnish population.
    a. How would a population geneticist explain these variations in disease occurrence?
    b. The Finnish population is also a source of information for the study of quantitative traits. The genetic basis of schizophrenia is one question that can be explored in this population. What advantage(s) and disadvantage(s) can you imagine for studying complex traits based on the Finnish population structure?

19. Canavan disease, caused by homozygosity for a recessive allele, is a severe neurodegenerative syndrome usually resulting in death by the age of 18 months. The frequency of Canavan disease is particularly high in Jewish populations. In an effort to map the gene causing this condition, researchers looked at 10 SNPs (1–10) spaced at roughly 100 kb distances along chromosome 17 in five affected Jewish patients (Cases) and four unaffected Jewish individuals (Controls). In the accompanying table, each row depicts a single haplotype. (Every individual is diploid and therefore has two haplotypes, although only one is shown in the table.) G, C, A, and T represent the actual nucleotide at the indicated SNP location.

Case	SNP1	SNP2	SNP3	SNP4	SNP5	SNP6	SNP7	SNP8	SNP9	SNP10
1	G	T	G	T	T	T	C	A	G	T
2	A	T	G	T	T	T	C	A	G	T
3	G	T	G	T	T	T	C	A	G	C
4	A	A	G	T	T	T	C	T	C	C
5	G	A	G	C	C	T	G	A	C	C
Control										
6	A	A	G	T	T	T	C	A	G	T
7	G	T	G	G	C	T	G	A	G	T
8	A	T	C	T	C	G	C	T	C	C
9	G	T	C	G	T	G	G	A	C	T

a. Does the disease-causing mutation appear to be in linkage disequilibrium with any of the SNP alleles? If so, which ones?

b. Where is the most likely location for the Canavan disease gene? About how long is the region to which you can ascribe the gene?

c. How many independent mutations of the Canavan gene are suggested by these data?

d. Suppose that individuals 2–9 are Ashkenazic (whose ancestors lived in the Rhine river basin of Germany and France after the Jews were expelled from Judea in 70 A.D.) while individual 1 is Sephardic (a non-Ashkenazic Jew). Would these facts provide any information about the history of the mutations causing Canavan disease?

e. For mapping genes by haplotype association, why is it often helpful to focus on certain subpopulations? Does this strategy have any disadvantages?

f. Human chromosome 17 is an autosome, so each person has two copies of each region along the chromosome. With this in mind, explain the practical difficulty in determining haplotypes. (*Hint:* Consider heterozygosity.) In light of this difficulty, how could the researchers determine any individual haplotype, such as any of those shown in the table?

20. In GWAS analysis, because of the existence of *LD blocks* (or *haplotype blocks*), it is not necessary to genotype a person for every one of the 50 million known SNPs. Haplotype blocks are stretches of DNA containing particular SNP variants that tend to be inherited together (as a block) because recombination within the region is rare. In the accompanying figure, three different SNP loci are shown at *top* (SNP10, SNP11, and SNP12), each with two alleles among the world's population of humans.

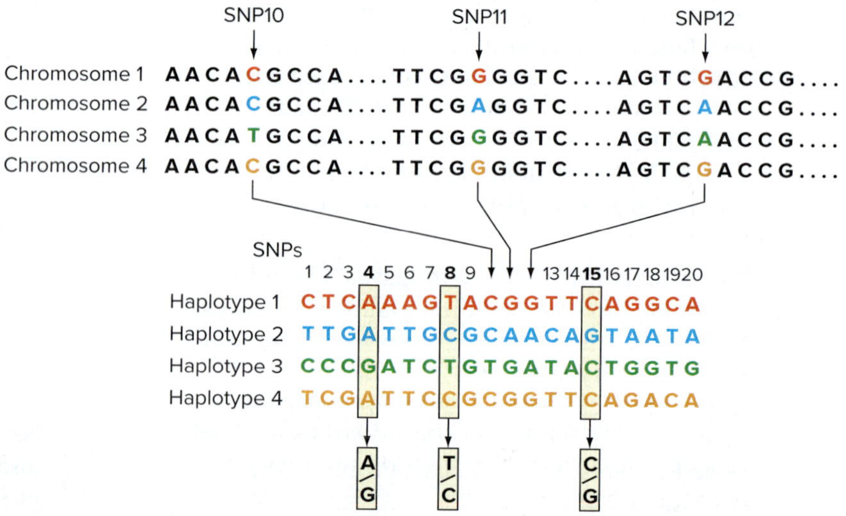

Only four of all the possible combinations of these SNP alleles are found in human genomes, as shown in the four chromosome types pictured. These three SNPs are part of a larger block of 20 SNPs that are usually inherited in one of the four configurations, or *haplotypes,* shown. Because these 20 SNPs are inherited as haplotype blocks, genotyping any individual for the three so-called *Tag SNPs* (SNP4, SNP8, and SNP15 shown in bold) should be sufficient to predict that individual's alleles for the other 17 SNPs.

a. How many configurations of the three SNPs shown at the top of the diagram (SNPs 10, 11, and 12) are theoretically possible?

b. How many haplotype variants are theoretically possible considering all 20 SNPs in the haplotype block, and assuming that each of them has two possible alleles?

c. Given that humans are diploid, every individual has two copies of every (autosomal) haplotype block, one on each homolog. Does heterozygosity for the haplotype blocks interfere with genotyping individuals using the Tag SNPs shown in the diagram? Explain.

d. In part (c), you saw that the three Tag SNPs shown in the diagram are sufficient to type any individual for this particular haplotype block. Is this the only set of three Tag SNPs that could be used?

21. In Fig. 22.15:

a. Why do some chromosomes in the disease group not carry the disease-causing variant shown in *red?*

b. Why do some chromosomes in the control group carry the variant shown in *red?*

c. Discuss how scientists would evaluate data from association mapping studies to find QTLs that contribute to the disease. What would researchers look for? And at the bottom of the figure, why are the nonrandom associations of the disease specifically discussed in terms of the *blue* alleles?

d. Suppose that researchers identified two different regions of the genome with statistically significant associations of SNPs with the disease in question. In Region 1, SNPs extending over 200 kb of DNA showed such associations. In Region 2, the region containing disease-associated SNPs was 2 Mb long. If you assume that each of these two regions has only one disease-causing variant, which of the two mutations was likely to have occurred earlier in human history? Explain.

e. Given your answer to part (d), explain why the length of the region containing disease-associated SNPs is nonetheless not a perfect indicator of the time in human history at which the disease-causing mutation occurred.

22. Consider the triangle diagram shown in Fig. 22.17.

a. What feature of the genome is likely to be located between the two LD blocks that allows scientists to visualize them as separate blocks?

b. Even though the figure analyzes eight different SNPs, genotyping just two of these SNPs would allow you to predict the genotype of almost everyone in the population. Explain why this limited genotyping has predictive value.

c. When researchers obtain the data that enables them to build triangle diagrams, do they typically genotype common SNPs or rare SNPs? Explain.

d. Considering your answer to part (c), why are human population geneticists interested in obtaining complete whole-genome sequences, as opposed to genotyping only the small subset of predictive SNPs?

23. In Fig. 22.18:

a. Why are some patients of genotype GG, some of genotype GC, and others of genotype CC? That is, why don't all the patients have the same genotype? Discuss at least two reasons.

b. How many people participated in this study? How many of them were patients (Cases)? Controls?

24. You conduct a Case/Control study comparing the frequency of a single SNP (with alleles C and T) among individuals who have developed high blood pressure as compared with control individuals of a similar age who show no sign of this condition. You obtain the following results:

	Cases	Controls
C	1025	725
T	902	922

a. Does a statistically significant association exist between this SNP and the risk of developing high blood pressure, if this is the only SNP that you test?

b. Would this association still be significant if this was one of a million SNPs that you tested in your study? Explain.

c. Which allele (C or T) is associated with a higher risk of developing high blood pressure?

d. Calculate the allelic odds ratio for the risk allele. What does the allelic odds ratio mean about a person's chance of developing high blood pressure?

25. In Fig. 22.21, the association of the SNP *rs1333049* with coronary artery disease has a $-\log_{10}(p \text{ value})$ greater than the cutoff of 7 in one GWAS but not in a second such study. Assuming that this SNP is indeed associated with the condition, provide two possible reasons that the $-\log_{10}(p \text{ value})$ in the second study was less than the cutoff.

26. ALS (amyotrophic lateral sclerosis) is a rare, fatal neurodegenerative disease that is genetically complex. During the last several years, using GWAS analyses and other methods, 11 different genes have been identified that are thought to be connected with ALS. The most recently discovered of these genes, *TBK1*, was identified through analysis of the whole exome sequences of several thousand Cases and Controls. Recall from Chapter 11 that the exome is the approximately 1% of the human genome that corresponds to exons. Each person's exome sequence was evaluated on a gene-by-gene basis as to whether or not a SNP variant likely to alter gene function was present. The *TBK1* gene was not identified in previous GWAS analyses that genotyped similar numbers of individuals for common SNPs. Cite possible explanations for the different outcomes of these two experiments.
27. Through GWAS explorations, scientists have identified several SNPs linked to obesity in people who live in the United States. One of these SNPs was within a gene called *FTO*. Interestingly, a common *FTO* variant is associated with obesity, but only in people born after 1945. Moreover, the later the birth year, the higher the risk for obesity associated with this variant of *FTO*. Why would a genetic risk factor for obesity vary by birth year?
28. In domesticated dogs, size has a high heritability, and the trait is determined by only a small number of genes. In contrast, genetic variation at more than 180 QTLs explains only a very small proportion of the high heritability for height in humans. What could explain the *missing heritability* in humans, and how could you test your hypothesized explanations?
29. Suppose a GWAS investigation found a particular LD block to be associated with people's preference for chocolate or vanilla ice cream. How could you identify the specific gene within this region of the genome whose alleles help determine this preference?
30. In 2008, *Time* magazine named as its invention of the year the development of personal genomics services by a company named 23andMe. Customers sent saliva samples to the company, which then genotyped approximately one million SNPs located across the genome, and communicated the data online to the customer along with what was claimed to be a "for education use only" assessment of potential risk for a variety of traits.

    However, on November 22, 2013, the United States Food and Drug Administration (FDA) ordered 23andMe to stop marketing its personal genomics services because the accuracy of its SNP genotyping and risk predictions had not been validated as sufficiently accurate for medical use. The FDA was concerned that people might make serious medical decisions based on information from a test that was not clinically approved. Some elements of this ban were relieved in 2015 and 2017, but DNA testing services are still restrained from offering their customers all the pre-ban predictions. Defining these limits remains a contentious and unresolved issue.

    a. Can the information you would obtain from this personal genomics service tell you whether or not you have a Mendelian genetic disease? Explain.
    b. Can the information you would obtain from this personal genomics service inform you about your likelihood of having a disease that is a complex trait? Explain.
    c. In December 2013, a reporter for *The New York Times* reported that she sent samples of her own DNA to three different companies (one of which was 23andMe), but the three companies provided very different estimates of her risk for a variety of complex traits. What were the likely causes of the differences in these estimates?
    d. Do you think new scientific developments will help resolve these issues in the near future?

Guidelines for Gene Nomenclature

Inconsistencies within the various branches of genetics exist on some nomenclature in part because scientists working with different model organisms or in different fields each made their own rules. The following guidelines can be applied to all chapters in this book.

General Rules

- Names of genes are in italics (*lacZ, CDC28*).
- Names of proteins are in regular (Roman) type with an initial uppercase letter (LacZ, Cdc28).
- Chromosomes: Sex chromosomes are represented by a capital letter in Roman type (X, Y). Autosomes are designated by a cardinal number (1, 2, 21, 22).
- Symbols for genes on different chromosomes are separated by semicolons (*y; bw; st*). Homozygosity is assumed unless heterozygosity is indicated.
- In a diploid, symbols for different alleles of genes on homologous chromosomes are separated by a slash ($th^{1}\ st^{2} / th^{2}\ st^{1}$).

Specific Rules for Different Organisms

- **Bacteria:** Lowercase italics for genes (*lac, ara*), with the addition of a capital letter to designate a specific gene in a pathway or operon (*lacZ, lacA, araB*); numbers (not superscript) for specific alleles (*trpC2, hisB2*); superscript plus (+) for wild-type alleles and superscript minus (–) for mutant alleles ($lacZ^{+}$, $lacZ^{-}$).

  Bacterial phenotypes are designated in regular (Roman) type with an abbreviation starting with a capital letter, followed by a descriptive superscript (Lac^{-} cannot grow on medium containing only lactose as a carbon source; Arg^{+} can grow on medium without an arginine supplement; Str^{r} is streptomycin resistant).
- **Yeast:** All uppercase for wild-type alleles (*CDC28*), and all lowercase for mutant alleles, with a cardinal number indicating the specific mutation (*cdc28-1*).
- ***Arabidopsis:*** All uppercase for wild-type alleles (*LEAFY,* abbreviated *LFY*); all lowercase for mutant alleles (*lfy*).
- ***C. elegans:*** Lowercase italics for genes and wild-type alleles (*dpy-10*); mutant alleles in parentheses following the gene name [*dpy-10(e128)*]
- ***Drosophila:*** Many genes are named for the mutant phenotype that first revealed them. If the mutation that first revealed the gene is dominant, the gene designation has an initial cap (*Deformed,* abbreviated *Dfd*)*;* if the mutation that first revealed the gene is recessive, the gene designation is all lowercase (*white,* abbreviated *w; wingless,* or *wg*). Gene names often describe the mutant phenotype caused by the allele that first revealed the gene.
- **Mice:** Genes are named for the protein they encode and designated by an initial uppercase letter followed by any mix of letters or numbers (for example, *Tcp1* designates the gene for T complex protein 1); alleles are designated by superscripts. For multiple wild-type alleles, the superscripts may be *a, b, c,* or *1, 2, 3* ($Tcp1^{a}$, $Tcp1^{b}$); some wild-type alleles are indicated by a superscript plus (Kit^{+}). For mutant alleles, the superscripts often describe the resulting phenotype (the Kit^{W} mutant allele causes white spotting).
- **Zebrafish:** Gene names are all lowercase (*pdgfra*); mutant alleles have a letter and a number, where the letter denotes where the allele was generated or discovered, and the numbers count the alleles from that place ($pdgfra^{b1059}$). Sometimes wild-type alleles are indicated by a superscript plus (+) and mutant alleles by a superscript minus (–) . Homozygous or heterozygous genotypes indicate the gene name once ($pdgfra^{b1059/b1059}$ or $pdgfra^{b1059/+}$).
- **Humans:** All uppercase letters for the names of genes or of mutant alleles of these genes (*HD, CF*); wild-type alleles are designated by a superscript plus (HD^{+}, CF^{+}). Sometimes mutant alleles are designated with a superscript minus (–) to stress that these are loss-of-function alleles.

Glossary

2*n* the number of chromosomes in a normal diploid cell.

3′ end the final nucleotide of an RNA or DNA molecule.

5′ and 3′ untranslated regions (UTRs) the sequences located just after the methylated cap and just before the poly-A tail, respectively; encoded by exons and do not include codons.

5′ end the first nucleotide of an RNA or DNA molecule.

A

acentric fragment a chromatid fragment lacking a centromere; usually the result of a crossover in an inversion loop.

acrocentric describes a chromosome in which the centromere is close to one end.

activator a type of transcription factor that binds to specific DNA sequences within enhancer elements and increases the level of transcription from a nearby promoter.

adeno-associated viral (AAV) vector a gene therapy vehicle whose recombinant single-stranded DNA genome contains a therapeutic gene that does not integrate into patient chromosomes.

adjacent-1 segregation pattern one of two patterns of segregation resulting from the normal disjunction of homologs during meiosis I. Homologous centromeres disjoin so that one translocation chromosome and one normal chromosome go to each pole.

adjacent-2 segregation pattern a pattern of segregation resulting from a nondisjunction in which homologous centromeres go to the same pole during meiosis I.

***Agrobacterium*-mediated T-DNA transfer** a method for generating transgenic plants where bacteria containing a recombinant plasmid infects plants and the part of the plasmid containing the transgene integrates into the plant genome.

allele frequency the proportion of all copies of a gene in a population that are of a given allele type.

allele-specific oligonucleotides (ASOs) short oligonucleotides that hybridize with only one of a pair of alleles distinguished by a single base difference.

alleles alternative forms of a single gene.

allelic heterogeneity a situation where a genetic disease is caused by a variety of different mutations in the same gene.

allelic odds ratio the average increase in disease risk conferred by a particular SNP allele.

allopolyploids polyploid hybrids in which the chromosome sets come from two or more distinct, though related, species.

allosteric proteins proteins that undergo reversible changes in conformation when bound to another molecule (an effector).

alternate segregation pattern one of two patterns of segregation resulting from the normal disjunction of homologs during meiosis I. Two translocation chromosomes go to one pole, while two normal chromosomes move to the opposite pole, resulting in balanced gametes.

alternative splicing the production of different mature mRNAs from the same primary transcript by joining different combinations of exons.

Ames test screens for chemicals that cause mutations in bacterial cells.

amino acids the building blocks of proteins.

aminoacyl (A) site the site on a ribosome to which a charged tRNA first binds.

aminoacyl-tRNA synthetases enzymes that catalyze the attachment of tRNAs to their corresponding amino acids, forming charged tRNAs.

amniocentesis a medical procedure in which a sample of amniotic fluid is taken from a pregnant woman to determine the karyotype of an unborn baby.

amorphic mutations mutations that abolish the function of a gene product encoded by the wild-type allele. Such mutations either prevent synthesis of both the protein and RNA (this is the strict definition used by some geneticists exclusively) or promote the synthesis of a protein or RNA that is incapable of carrying out any function. Synonymous with *null mutations*.

amphidiploid a special kind of allopolyploid produced by the mating of two different diploid parents; contains two diploid genomes, each one derived from a parent of a different species.

anabolic pathways biochemical pathways for the synthesis of complex molecules from simpler ones.

anaphase the stage of mitosis in which the connection of sister chromatids is severed, allowing the chromatids to be pulled to opposite spindle poles.

anaphase I the phase of meiosis I during which the chiasmata joining homologous chromosomes dissolve, allowing maternal and paternal homologs to move toward opposite spindle poles; the centromeres do not separate, so that the chromosomes moving toward the poles each consist of two chromatids.

anaphase II the phase of meiosis II when the dismantling of cohesin complexes allows sister chromatids to move to opposite spindle poles.

ancestral allele the allele carried by the last common ancestor of two species.

aneuploid describes a cell or an organism whose chromosome sets are not all complete.

angiogenesis the growth of blood vessels toward cells.

anneal base pairing between a probe and and its complementary DNA sequence. Synonym for *hybridize*.

annotation the analysis of a genome to determine gene locations in the DNA sequence and gene functions.

anonymous DNA polymorphisms differences in genomic DNA sequence with no effect on gene function.

Antennapedia complex (ANT-C) in *Drosophila,* a region containing several homeotic genes specifying the identity of parasegments in the head and anterior thorax.

anticodon a group of three nucleotides on transfer RNA (tRNA) molecules that recognize codons on the mRNA by complementary base pairing and wobble.

anticrossover helicase an enzyme that helps disentangle the invading strand from the nonsister chromatid, thus interrupting Holliday junction formation and preventing crossing-over.

antimorphic mutations (or alleles) block the activity of wild-type alleles of the same gene, causing a loss of function even in heterozygotes. (Synonymous with *dominant-negative mutations.*)

antisense RNAs regulatory RNAs that are complementary in sequence to the mRNAs they regulate because they are transcribed using the opposite strand of DNA as a template. Antisense RNAs can block transcription or translation of their target mRNAs.

antiterminator a stem-loop structure in an RNA that prevents formation of a terminator.

apoptosis programmed cell death; a process in which DNA is degraded and the nucleus condenses; the cell may then be devoured by neighboring cells or phagocytes.

artificial selection the purposeful control of mating by choice of parents for the next generation.

artificial transformation a process to transfer genes from one bacterial strain to another, using laboratory procedures to weaken cell walls and make membranes permeable to DNA.

ascospores in some fungi, the haploid cells that result from meiosis. Also known as *haplospores.*

ascus saclike structure in some varieties of fungi that houses all four haploid products of meiosis.

association mapping identification of QTLs through analysis of the correlations of phenotypic values with molecular markers.

astral microtubules short, unstable microtubules that extend out from a centrosome toward the cell's periphery to stabilize the mitotic spindle.

attenuation a type of gene regulation in which transcription of a gene terminates in the RNA leader sequence before a complete mRNA transcript is made; involves interaction between stem loops in the RNA leader and the translation machinery.

attenuator part of the RNA leader of some bacterial genes that forms alternate stem-loop structures that depend on interaction with the translation machinery. In one conformation, a stem loop (the *terminator*) terminates transcription; in the alternate conformation, a different stem loop (the *antiterminator*) forms and transcription continues.

autocrine stimulation a process by which many tumor cells make and respond to their own signals to divide.

autonomous elements intact transposable elements that can move from place to place in the genome by themselves.

autopolyploids polyploids that derive all of their chromosome sets from the same species.

autosomes chromosomes not involved in sex determination. The diploid human genome consists of 46 chromosomes: 22 pairs of autosomes, and 1 pair of sex chromosomes (the X and Y chromosomes).

auxotroph a mutant microorganism that can grow on minimal medium only if it has been supplemented with one or more nutrients not required by wild-type strains.

B

B-form DNA the most common form of DNA, in which the double helix spirals to the right.

bacterial chromosome a bacterial genome; usually a single circular molecule of double-helical DNA.

bacterial conjugation one of the mechanisms by which bacteria transfer genes from one strain to another; in this case, the donor carries a special type of plasmid that allows it to transfer DNA directly when it comes in contact with the recipient. The recipient is known as an *exconjugant.*

bacteriophage a virus for which the natural host is a bacterial cell; literally *bacteria eaters.*

***Balancer* chromosome** a special chromosome created for use in genetic manipulations; prevents recovery of crossover chromosomes, thereby maintaining chromosomes containing multiple mutations. As *Balancer* chromosomes cause lethality or infertility when present in two copies, they are also useful for maintaining heterozygous stocks of chromosomes bearing lethal mutations.

balancing selection selection that actively maintains multiple alleles of a gene in a population.

Barr body an inactive X chromosome observable at interphase as a darkly stained heterochromatin mass.

barrier elements DNA elements that block the spread of heterochromatin.

basal factor a protein (a transcription factor) that can bind directly to the DNA of all promoters in a genome; also other proteins that bind the transcription factor(s) at the promoter and help to recruit RNA polymerase. A complex containing basal factors at a promoter is required for the initiation of transcription by RNA polymerase.

base analogs mutagens that are so similar in chemical structure to the normal nitrogenous bases in DNA that the replication mechanism can incorporate them into DNA in place of the normal bases. Their incorporation can cause base substitutions on the complementary strand synthesized in the next round of DNA replication.

bases components of nucleotides. In DNA, the four bases are guanine, cytosine, adenine and thymine; in RNA, uracil substitutes for thymine. (Also called *nitrogenous bases.*)

basic chromosome number (*x*) the number of different chromosomes that make up a single complete set.

beneficial mutations rare mutations that provide a selective advantage to an organism or population.

biochemical pathway an orderly series of chemical reactions within a cell in which molecules are converted stepwise into a final product.

bioinformatics the science of using computational methods—specialized software—to decipher the biological meaning of information contained within organisms.

biolistic transformation the use of a gene gun to generate transgenic chloroplasts.

biological ancestors the individuals from whom any person inherited his or her genes; parents, grandparents, great grandparents, etc.

biparental inheritance the inheritance of organelles from both parents. Occurs in single-celled yeast and some plants.

bithorax complex (BX-C) in *Drosophila,* a cluster of homeotic genes that control the identity of parasegments in the abdomen and posterior thorax.

bivalent a pair of synapsed homologous chromosomes during prophase of meiosis I.

blunt end an end of a double-stranded DNA molecule that has no 5′ or 3′ overhang.

branch migration a process during recombination whereby Holliday junctions move away from each other and thereby enlarge the heteroduplex region between them.

branch sites special base sequences of RNA nucleotides within an intron that helps form the *lariat* intermediate required for RNA splicing.

branched-line diagram a system for listing the expected results of multihybrid crosses.

broad-sense heritability (H^2) a heritability measurement that includes all three components of genetic variance (V_G): additive effects (V_A), dominance effects (V_D), and the effects of interactions between alleles at different loci (V_I). Typically measured in studies of identical twins.

C

C terminus the end of the polypeptide chain that contains a free carboxylic acid group.

carriers heterozygous individuals of normal phenotype that have a recessive allele for a trait.

Cas proteins CRISPR-associated proteins. These are bacterial endonucleases integral to the CRISPR immunity system. Cas proteins can cleave viral DNA, leading to its degradation.

catabolic pathways metabolic pathways by which complex molecules are broken down.

catabolite repression the repression of transcription in sugar-metabolizing operons like the *lac* operon when glucose or another preferred catabolite is present.

cDNA a DNA molecule synthesized using mRNA as a template; a double-stranded DNA representation of an mRNA.

cDNA deep sequencing method for analysis of the transcriptome of an organism in which millions of cDNAs are sequenced. Also called *RNA-Seq.*

cDNA library a large collection of cDNA clones that are representative of the mRNAs expressed by a particular cell type, tissue, organ, or organism.

cell cycle the repeating pattern of cell growth, replication of genetic material, and mitosis.

cell-bound signals signals requiring direct contact between cells for transmission.

cell-free fetal DNA analysis a procedure where DNA in the blood of a pregnant woman, which contains DNA released from broken fetal cells that leaked into her bloodstream, is analyzed in order to genotype the fetus.

cellular blastoderm in *Drosophila* embryos, the one-cell-deep epithelial layer resulting from cellularization of the syncytial blastoderm.

cellular clone a group of cells that all descend from a single progenitor cell.

centimorgan (cM) a unit of measure of recombination frequency. One cM is equal to a 1% chance that a marker at one genetic locus will be separated from a marker at a second locus due to crossing-over in a single generation. (Synonymous with *map unit.*)

centrioles short cylindrical structures that help organize microtubules. Two centrioles at right angles to each other form the core of a centrosome. Each centrosome serves as a pole of the mitotic spindle.

centromere a specialized chromosome region at which sister chromatids are connected and to which spindle fibers attach during cell division.

centrosomes microtubule organizing centers at the poles of the spindle apparatus.

chaperone a protein that helps other proteins fold into their native tertiary structure.

charged tRNA a tRNA molecule to which the corresponding amino acid has been attached by an aminoacyl-tRNA synthetase.

checkpoints mechanisms that prevent cells from continuing to the next phase of the cell cycle until a previous step has been completed successfully, thus safeguarding genomic integrity.

chemotherapy treatment of patients with drugs that kill cancer cells.

chi-square (χ^2) test a statistical test to determine the probability that an observed deviation from an expected outcome occurs solely by chance.

chiasmata observable regions in which nonsister chromatids of homologous chromosomes cross over.

chimera an embryo or animal composed of cells from two or more different organisms.

ChIP-Seq chromatin immunoprecipitation-sequencing; a method for finding the target genes of a particular transcription factor that involves using an antibody to that transcription factor to purify that protein bound to DNA.

chloroplasts plant organelles that capture solar energy and store it in the chemical bonds of carbohydrates through the process of photosynthesis.

chromatid one of two copies of a chromosome that exist immediately after DNA replication.

chromatin the complexes of DNA and protein found in a cell's nucleus that form chromosomes.

chromosomal interference the phenomenon of crossovers not occurring independently.

chromosome loss a mechanism causing aneuploidy in which a particular chromatid or chromosome fails to become incorporated into either daughter cell during cell division.

chromosome rearrangement a change in the order of base sequences along one or more chromosomes.

chromosome theory of inheritance the idea that chromosomes are the carriers of genes.

chromosomes the self-replicating DNA/protein complexes in the nucleus that contain genes.

cis describes the action of a DNA site or an RNA molecule that acts only on the DNA or RNA to which it is connected physically.

***cis* configuration** on the same DNA molecule.

cistron a term sometimes used as a synonym for complementation group or gene. The word originates from Benzer's *cis-trans* test that defined a gene as a complementation group.

cloning vector vehicles for introducing foreign DNA into host cells, where that DNA can be reproduced in large quantities; a DNA molecule into which another DNA fragment of appropriate size can be integrated without loss of the vector's capacity for replication.

coactivator a protein that binds to a transcriptional activator and plays a role in increasing levels of transcription.

codominant describes the relationship between two alleles of a gene where the heterozygote has the traits of both homozygotes.

codon a nucleotide triplet that represents a particular amino acid to be inserted in a specific position in the growing amino acid chain during translation. Codons are in both the mRNA and the DNA from which the RNA is transcribed.

coefficient of coincidence the ratio between the actual frequency of double crossovers observed in an experiment and the number of double crossovers expected on the basis of independent probabilities.

cohesin a multisubunit protein complex that associates with sister chromatids in eukaryotic cells and holds the chromatids together until anaphase; can be found at both the centromere and along the chromosome arms.

colinearity the parallel between the sequence of nucleotides in a gene and the order of amino acids in a polypeptide.

colony a mound of genetically identical cells that all descend from a single cell.

common variants high-frequency alleles of a polymorphic gene.

competent cells cells treated so that they are able to take up DNA from the medium.

complementarity the property of DNA whereby the base sequences of the two strands in the double helix are reverse complements of one another; A is opposite T, and G is opposite C.

complementary base pairing during DNA replication, base pairing in which a complementary strand aligns opposite the exposed bases on the parent strand to create the nucleotide sequence of the new strand of DNA; hydrogen bonding between A-T and G-C that holds the two antiparallel strands of the DNA double helix together; can also enable formation of RNA:DNA or RNA:RNA double strands. (A pairs with U in RNA.)

complementary base pairs A-T and G-C held together by hydrogen bonds in a DNA double helix.

complementation the process in which heterozygosity for loss-of-function mutant recessive alleles for two different genes produces a normal phenotype.

complementation group a collection of mutations that do not complement each other. Often used synonymously with *gene*.

complementation table a method of collating data that helps visualize the relationship (which pairs of mutants complement) among a large group of mutants.

complementation test method of discovering whether two mutations are in the same or separate genes. Two mutant strains with the same mutant phenotype are crossed. If the progeny are all wild type (complementation occurred), the strains had mutations in different genes; if the progeny are all mutant (no complementation occurred), the strains had mutations in the same gene.

complex traits traits controlled by multiple genes; complex traits can also be influenced by interactions between alleles of different genes, variations in the environment, and interactions between genes and the environment.

compound heterozygotes (*trans*-heterozygotes) heterozygotes for two different mutant alleles of the same gene.

concordance describes the extent to which two different individuals share a discrete trait. In twin studies, the frequency with which the co-twin has the trait in question when one of them does.

condensation cellular process of chromatin compaction that results in the visible emergence of individual chromosomes.

condensin a multisubunit complex of proteins in eukaryotic cells that compacts chromosomes during mitosis.

conditional knockouts transgenic mice that have a *floxed* gene; the gene can be made nonfunctional through partial deletion in specific tissues.

conditional lethal an allele that is lethal only under certain conditions.

congenic lines strains that are genetically identical (isogenic) except for small regions of their genomes. Also called *NILs*.

conjugation a mechanism by which bacteria transfer genes from one strain to another; the donor carries a special type of plasmid that allows it to transfer DNA directly when it comes in contact with the recipient. The recipient is known as an *exconjugant*.

conjugative plasmids plasmids that initiate conjugation because they carry the genes that allow the donor to transfer genes to the recipient.

consanguineous mating mating between genetic relatives sharing a recent common ancestor.

consensus sequence a sequence of amino acids in proteins (or bases in DNA or RNA) that is most commonly found within all known protein domains (or regions of DNA or RNA) with the same particular function.

conservative substitutions mutations that substitute an amino acid in a protein with a different amino acid having similar chemical properties.

conserved (DNA sequence) describes a DNA sequence that has homologs in many different species.

constitutive heterochromatin chromosomal regions that remain condensed in heterochromatin at most times in all cells.

constitutive mutants strains that synthesize certain enzymes all the time, irrespective of environmental conditions.

contact inhibition a cell-to-cell signaling mechanism that normally limits cell growth to a monolayer in culture.

continuous trait an inherited trait that exhibits many intermediate forms; a trait determined by alleles of many different genes whose interaction with each other and possibly also the environment produces the phenotype. Also called a *quantitative trait*.

contractile ring transitory organelle composed of actin microfilaments aligned around the circumference of dividing animal cell's equator; contraction of the filaments pinches the cell in two.

copy number variant (CNV) a category of genetic variation arising from large regions (from 10 bp up to 1 Mb) of duplication or deletion.

core genome genes shared by all bacterial strains of a given species.

core histones proteins that form the core of the nucleosome: H2A, H2B, H3, and H4.

corepressor a protein that binds to a repressor and helps to prevent transcription.

correlation coefficient (*r*) the slope of the line of correlation.

cotransduction transfer of different bacterial genes together in one phage by transduction.

cotransformation the simultaneous transformation of two or more genes.

CpG island a regulatory element of some eukaryotic genes which is unusually rich in 5′ CpG 3′ dinucleotides. Methylation of the C residues in the CpG island can silence gene transcription.

Cre/loxP recombination system the use in transgenic organisms of Cre recombinase protein from P1 phage and the loxP DNA site that it binds to perform site-specific recombination; typically used in mice to delete parts of genes only in specific tissues. Similar to the FLP/FRT recombination system.

CRISPR acronym for *clustered regularly interspaced short palindromic repeats*. A region in many bacterial genomes that confers immunity to viral infection. The CRISPR mechanism has been exploited by biotechnologists for genome editing of higher organisms.

CRISPR/Cas9 system a genetically engineered version of the immunity system of *Streptococcus pyogenes* that is used for genome editing. An investigator-designed sgRNA brings Cas9 endonuclease to a target site in the genome. Cas9 makes double-stranded breaks in the DNA that, through DNA repair, can result in knockouts or knockins.

crisscross inheritance an inheritance pattern in which males inherit a trait from their mothers, while daughters inherit the trait from their fathers.

cross-fertilization brushing the pollen from one plant onto the female organ of another plant.

cross-fostering the random relocation of offspring to the care of other parents, typically done in animal studies to randomize the effects of environment on outcome.

crossing-over during meiosis, the breaking of one maternal and one paternal chromatid, resulting in the exchange of corresponding sections of DNA and the rejoining of the chromosomes. This process can result in the exchange of alleles between chromosomes.

crossover suppressors inversion chromosomes when heterozygous with normal homologs, because no viable recombinant progeny are possible.

cybrid a *cytoplasmic hybrid*. A cell that has nuclear DNA from one source and mitochondrial DNA from another.

cyclin-dependent kinases (CDKs) protein kinases that depend on proteins known as cyclins for the targeting of their activity to a specific substrate. CDKs regulate the transition from G_1 to S and from G_2 to M through phosphorylations that activate or inactivate target proteins.

cyclins a family of proteins that combine with cyclin-dependent kinases and thereby determine the substrate specificity of the kinases. By directing kinases to specific substrates, the cyclins help regulate passage of the cell through the cell cycle. Concentrations of the various cyclins rise and fall throughout the cell cycle.

cytokinesis the final stage of cell division, which begins during anaphase but is not completed until after telophase. In this stage, the daughter nuclei emerging at the end of telophase are packaged into two separate daughter cells.

cytoplasmic segregation chance distribution of all of one type of organelle DNA in a heteroplasmic cell into a single daughter cell during mitosis.

D

DCO a meiosis where two crossovers occur between a given gene pair.

de novo genes genes lacking homologs except in closely related species, thought to have evolved anew from intergenic DNA sequences.

deamination the removal of an amino (—NH_2) group from normal DNA.

degenerate describes the ability of elements that are structurally different to perform the same function, as is the case in the genetic code: Several different codons can specify the same amino acid, and a single tRNA anticodon can recognize several different codons.

degrees of freedom (df) the number of independently varying parameters in an experiment.

deleterious mutations mutations that disrupt important gene functions.

deletion the loss of a block of one or more nucleotide pairs from a DNA molecule.

deletion loop an unpaired bulge of a normal chromosome that corresponds to an area deleted from its paired homolog.

denaturation (denature, denatured) the disruption of hydrogen bonds within a macromolecule that normally uses hydrogen bonds to maintain its structure and function. Hydrogen bonds can be disrupted by heat, extreme conditions of pH, or exposure to chemicals such as urea. When normally soluble proteins are denatured, they unfold and expose their nonpolar amino acids, which can cause them to become insoluble. When DNA is denatured, double-stranded molecules break apart into two separate strands.

deoxynucleotide triphosphates the building blocks of DNA: dATP, dGTP, dCTP, and dTTP. Each contains a deoxyribose molecule, one of the four nitrogenous bases, and three phosphates. Synonym for *deoxyribonucleotide triphosphates.*

deoxyribonucleic acid (DNA) the molecule of heredity that encodes genetic information.

deoxyribonucleotide triphosphates (dNTPs) the building blocks of DNA: dATP, dGTP, dCTP, and dTTP. Each contains a deoxyribose molecule, one of the four nitrogenous bases, and three phosphates.

deoxyribose a molecule similar to ribose, except that the 2′ carbon has a hydrogen rather than a hydroxyl group.

depurination a DNA alteration in which a purine base, either A or G, is hydrolyzed from the deoxyribose-phosphate backbone.

derived allele an allele that arises through mutation.

developmental genetics the use of genetics to study how the fertilized egg of a multicellular organism becomes an adult.

diakinesis the substage of prophase I during which chromosomes condense to the point where each tetrad consists of four separate chromatids; at the end of this substage, the nuclear envelope breaks down and the microtubules of the spindle apparatus begin to form.

dicentric chromatid a chromatid with two centromeres.

dideoxynucleotide triphosphates dideoxyribonucleotides containing three phosphates: ddATP, ddGTP, ddCTP, ddTTP. Synonym for *dideoxyribonucleotide triphosphates.*

dideoxynucleotides nucleotide analogs lacking the 3′-hydroxyl group that is crucial for the formation of phosphodiester bonds. Dideoxynucleotides are key components of the most common methods of DNA sequencing.

dideoxyribonucleotide triphosphates (ddNTPs) dideoxyribonucleotides containing three phosphates: ddATP, ddGTP, ddCTP, ddTTP.

digestion the enzymatic process by which a complex biological molecule (DNA, RNA, protein, or complex carbohydrate) is broken down into smaller components.

dihybrid an individual that is heterozygous at two different genes.

diploid describes cells carrying two matching sets of chromosomes; symbolized as $2x$.

diplotene the substage of prophase I during which regions of homologous chromosomes are separated slightly, but the aligned homologous chromosomes of each bivalent remain tightly merged at chiasmata.

direct QTL mapping identifying quantitative trait loci (QTLs) by controlled crosses.

discrete trait an inherited trait that clearly exhibits an either/or status (that is, purple versus white flowers). Synonymous with *discontinuous trait.*

disease genes genes whose mutant alleles cause human genetic diseases.

dizygotic (DZ) twins two different embryos, each derived from a different zygote, developing together in one womb; also called *fraternal twins.*

DNA deoxyribonucleic acid; the molecule of heredity that encodes genetic information.

DNA clone a purified sample containing a large number of identical DNA molecules.

DNA fingerprint the multilocus pattern produced by the detection of SSR loci.

DNA helicase an enzyme that unwinds the double helix.

DNA ligase an enzyme that forms phosphodiester bonds between DNA fragments.

DNA marker an identifiable physical location on a chromosome with DNA sequence variants whose inheritance can be monitored.

DNA methyl transferases (DNMTs) enzymes that catalyze the addition of methyl groups to bases of DNA.

DNA methylation enzymatic addition of methyl groups to bases in DNA. In the human genome, cytosine residues in a 5′ CpG 3′ dinucleotide are often methylated.

DNA microarray a collection of oligonucleotides attached to a solid surface.

DNA polymerase a complex enzyme that forms a new DNA strand during replication by adding nucleotides reverse complementary to a template, one by one, to the 3′ end of a growing strand.

DNA polymerase I also known as *pol I;* the enzyme that replaces RNA primers with DNA during DNA replication.

DNA polymerase III also known as *pol III;* complex enzyme that plays the major role in synthesizing a new DNA strand during replication.

DNA polymorphisms two or more alleles at a locus. The sequence variations of a DNA polymorphism can occur at any position on a chromosome and may (*nonanonymous polymorphism*), or may not (*anonymous polymorphism*), have an effect on phenotype.

DNA topoisomerases a group of enzymes that help relax supercoiling of the DNA helix by nicking one or both strands to allow the strands to rotate relative to each other.

DNA transposons or simply *transposons;* segments of DNA that move from place to place within the genome without an RNA intermediate, sometimes causing a change in gene function when they insert in a new chromosomal location.

DNase hypersensitive sites sites on DNA that contain few, if any, nucleosomes; these sites are susceptible to cleavage by DNase enzymes.

domain architecture the number and order of a protein's functional domains.

dominance series the dominance relations of all possible pairs of alleles arranged in a linear order.

dominant epistasis a phenomenon where the effects of a dominant allele at one gene hide the effects of alleles at another gene.

dominant trait the trait that appears in the F_1 hybrids (heterozygotes) resulting from a mating between pure-breeding parental strains showing antagonistic phenotypes.

dominant-negative mutations (or alleles) mutant alleles that block the activity of wild-type alleles of the same gene, causing a loss of function even in heterozygotes. (Synonymous with *antimorphic mutations.*)

donor in gene transfer in bacteria, the cell that provides the DNA to the recipient.

dosage compensation mechanism that equalizes the levels of X-linked gene expression independent of the number of copies of the X chromosome; in mammals, the dosage compensation mechanism is X chromosome inactivation.

double minutes small chromosome-like bodies lacking centromeres and telomeres produced by gene amplification.

double mutants organisms whose genomes contain mutations in two different genes.

downstream the direction traveled by RNA polymerase as it moves from the promoter to the terminator; toward the 3′ end of a gene.

driver mutations DNA alterations in cancerous cells that contribute to the cancer phenotype.

duplication a chromosomal rearrangement where the number of copies of a particular chromosomal region is increased.

duplication and divergence the process that creates gene families. A duplicate of a gene is generated, freeing one of the two gene copies to accumulate mutations and possibly acquire a new function.

E

ectopic expression gene expression that occurs outside the cell type, tissue, or time where or when the gene is normally expressed.

effector a small molecule that binds to an allosteric protein or RNA and causes a conformational change.

elongation phase of DNA replication, transcription, or translation when nucleotides or amino acids are added successively to a growing macromolecule.

elongation factors proteins that aid in the elongation phase of translation.

embryonic stem cells (ES cells) cultured embryonic cells that continue to divide without differentiating and are capable of becoming any cell type.

endosymbiont theory the proposal that chloroplasts and mitochondria originated when free-living bacteria were engulfed by primitive nucleated cells.

enhancers *cis*-acting elements that regulate transcription from nearby promoters. Enhancers function by acting as binding sites for transcription factors and are responsible for the spatiotemporal specificity of transcription.

environmental variance (V_E) in a population, the deviation of a phenotypic value from the mean attributed to the influence of external, noninheritable factors.

epigenetic phenomenon a heritable alteration in gene expression not due to mutation in base pair sequence.

episomes plasmids, like the F plasmid, that can integrate into the host genome.

epistasis a gene interaction in which the effects of alleles at one gene hide the effects of alleles at another gene.

equational division cell division that does not reduce the number of chromosomes, but instead distributes sister chromatids to the two daughter cells. Mitosis and meiosis II are both equational divisions.

euchromatin a loosely condensed chromosomal region in which many genes are transcriptionally active.

eukaryotic gene regulation the control of gene expression in the cells of eukaryotes.

euploid describes cells containing only complete sets of chromosomes.

evolution change in the heritable traits of biological populations over successive generations.

***ex vivo* gene therapy** a procedure where a patient's somatic cells are removed, a therapeutic gene is delivered to the cells in culture, and the cells are reintroduced into the patient.

excision removal of a segment of DNA from a larger DNA molecule.

exconjugants recipient cells resulting from gene transfer in which donor cells carrying specialized plasmids establish contact with and transfer DNA to the recipients.

exit (E) site one of three transfer RNA binding sites in ribosomes. The E site is occupied by tRNAs during the period just after their disconnection from amino acids by the action of peptidyl transferase and just before the release of the tRNAs from the ribosome.

exome the portion of a genome corresponding to exons; in humans, the exome is ~2% of the genome.

exons sequences found both in a gene's DNA and in the corresponding mature mRNA.

exonuclease enzyme that removes nucleotides from an end of a DNA molecule.

expressivity the degree or intensity with which a particular genotype is expressed as a phenotype.

extinction loss of an allele from a population.

extracellular signals signals that can diffuse from cell to cell or by cell-to-cell contact; they can be steroids, peptides, or proteins.

F

F pilus hollow protein tube that protrudes from an F^+, Hfr, or F' bacterial cell and binds to the cell wall of an F^- cell. Retraction of the pilus into the F^+ cell draws the two cells close together in preparation for gene transfer.

F plasmid a conjugative plasmid that carries many genes required for the transfer of DNA. Cells carrying an F plasmid are called *F^+ cells;* cells without the plasmid are called *F^- cells.*

F′ plasmids F plasmid variants that carry most F plasmid genes plus some bacterial genomic DNA; particularly useful in genetic complementation studies.

facultative heterochromatin regions of chromosomes (or even whole chromosomes) that are heterochromatic in some cells and euchromatic in other cells of the same organisms.

fine structure mapping recombination mapping of mutations in the same gene.

first filial (F_1) generation progeny of the parental generation in a controlled series of crosses.

first-division (MI) segregation pattern a tetrad in which the arrangement of ascospores indicates that the two alleles of a gene segregated from each other in the first meiotic division.

FISH fluorescence *in situ* hybridization; a physical mapping approach where fluorescent tags are used to detect hybridization of nucleic acid probes with chromosomes.

fitness the relative advantage or disadvantage in reproduction that a particular genotype provides to members of a population in comparison to alternative genotypes at the same locus.

fitness cost the effect of a deleterious allele—it is selected against.

fixation a process whereby a single allele of a locus becomes the only one in a population.

floxed describes a gene in a transgenic organism in which an exon is flanked by loxP sites; the gene can be knocked out conditionally in cells that express Cre recombinase.

FLP/FRT recombination system the use in transgenic organisms of FLP recombinase enzyme from yeast and the FRT DNA site that it binds to perform site-specific recombination. One use in flies is to cause mitotic recombination for mosaic analysis. Similar to the *Cre/loxP recombination system.*

fluctuation test the Luria-Delbrück experiment to determine the origin of bacterial resistance. Fluctuations in the numbers of resistant colonies growing in different petri plates showed that resistance is not caused by exposure to bactericides.

focus of action the cells in which a gene must be active to allow the animal to develop and function normally.

forward mutation a mutation that changes a wild-type allele of a gene to a different allele.

founder effect a variation of genetic drift, occurring when a few individuals separate from a larger population and establish a new one that is isolated from the original population, resulting in altered allele frequencies in the new population.

frameshift mutations insertions or deletions of base pairs that alter the grouping of nucleotides into codons.

fusion gene a gene made of up of parts of two or more different genes.

fusion protein a protein encoded by open reading frames from more than one gene.

G

G bands alternating dark and light segments (1-10 Mb) of a chromosome after staining with Giemsa dye.

G_1 phase the stage of the cell cycle from the birth of a new cell until the onset of chromosome replication at S phase.

G_2 phase the stage of the cell cycle from the completion of chromosome replication until the onset of cell division.

gain-of-function mutations (or alleles) rare mutations that alter a gene's function rather than disable it.

gametes specialized cells (eggs and sperm) that carry genes between generations.

gametogenesis the formation of gametes.

gap genes in *Drosophila,* these are the first zygotic segmentation genes to be transcribed; they encode transcription factors. Embryos homozygous for mutations in the gap genes show a "gap" in the segmentation pattern caused by the absence of the segments in which the gap gene is normally expressed.

gastrulation folding of the cell sheet early in embryo formation; usually occurs immediately after the blastula stage of development.

gel electrophoresis a process used to separate DNA fragments, RNA molecules, or polypeptides according to their size. Electrophoresis is accomplished by passing an electrical current through agarose or polyacrylamide gels. In response to the current, molecules migrate through the gel, and at different rates that depend on their sizes.

GenBank the NIH database of annotated DNA sequences.

gene a specific segment of DNA in a discrete region of a chromosome that serves as a unit of function by encoding a particular RNA or protein.

gene amplification an increase from the ormal two copies to hundreds of copies of a gene; often due to mutations in p53, which disrupt the G_1-to-S checkpoint.

gene conversion in a heterozygote, change in the base sequence of one allele to that of the other allele as a result of heteroduplex formation and mismatch repair during recombination.

gene desert a gene-poor region of the genome.

gene dosage the number of times a given gene is present in the cell nucleus.

gene editing describes a variety of technologies, including CRISPR/Cas9, that allows the creation of knockout and knockin animals without the use of ES cells.

gene expression the process by which a gene's information is converted into RNA and then (for protein-coding genes) into a polypeptide.

gene family a set of closely related genes with slightly different functions that most likely arose from a succession of gene duplication events.

gene superfamily a large set of related genes that is divisible into smaller sets, or families, with the genes in each family being more closely related to each other than to other members of the superfamily. The genes that compose a superfamily can reside at different chromosomal locations. The families of genes encoding the globins and the *Hox* genes are examples of gene superfamilies.

gene targeting method for inserting DNA into a genome that relies on homologous recombination; the DNA is targeted for insertion into a specific place in the genome by sequence similarity.

gene therapy manipulation of the genome in order to cure a disease.

generalized transduction a type of transduction (gene transfer mediated by bacteriophages) that can result in the transfer of any bacterial gene between related strains of bacteria.

genetic ancestors genomic DNA segments (alleles) inherited by an individual from biological ancestors.

genetic background all of the alleles of genes in a organism's genome; the set of unknown modifier genes that influence the action of the known genes that control specific aspects of phenotype.

genetic code the sequence of nucleotides, coded in triplets (codons) along the mRNA, that determines the sequence of amino acids in proteins.

genetic drift unpredictable, chance fluctuations in allele frequency due to random sampling as opposed to natural selection.

genetic linkage or simply *linkage*; the phenomenon where particular alleles of genes tend to travel together during vertical or horizontal gene transfer.

genetic markers genes identifiable through phenotypic variants that can serve as points of reference in determining whether particular progeny are the result of recombination.

genetic mosaic an organism containing tissues of different genotypes.

genetic relatedness the average fraction of common alleles at all gene loci that individuals share because they inherited them from a common ancestor.

genetic variance (V_G) in a population, the deviation of a phenotypic value from the mean attributable to inheritable factors.

genetics the science of heredity.

genome the sum total of genetic information in a particular cell or organism.

genome-wide association study (GWAS) analysis of the entire genome of many individuals in a population to identify SNP alleles that correlate with a specific phenotype; such SNP loci are linked to QTLs for that trait.

genomic equivalent the number of different DNA clones—with inserts of a particular size—that would be required to carry a single copy of every sequence in a particular genome.

genomic imprinting the phenomenon in which a gene's expression depends on the parent that transmits it; caused by sex-specific DNA methylation.

genomic instability in cancer cells, the accumulation of point mutations and chromosomal rearrangements that lead to an abnormal karyotype.

genomic islands large DNA segments transferred from one bacterial species to another.

genomic library a collection of DNA clones that together carry a representative copy of every DNA sequence in the genome of a particular organism.

genomics the study of whole genomes.

genotype the alleles present in an individual.

genotype frequency proportion of total individuals in a population that are of a particular genotype.

genotypic class a grouping defined by a set of related genotypes that will produce a particular phenotype. The term is most useful in describing progeny of dihybrid or multihybrid crosses involving complete dominance; for example, in a cross between *Aa Bb* individuals, the genotypic classes are *A- B-*, *A- bb*, *aa B-*, and *aa bb*.

genotypic odds ratio the average increase in disease risk conferred by either heterozygosity or homozygosity for a particuar SNP allele.

germ cells specialized cells that incorporate into the reproductive organs, where they ultimately undergo meiosis, thereby producing gametes that transmit genes to the next generation.

germ line all the germ cells in a sexually reproducing organism. In animals, the germ line is set aside from the somatic cells during embryonic development. The germ cells in the germ line divide by mitosis to produce a collection of specialized diploid cells that then divide by meiosis to produce haploid cells, or gametes. The germ line thus includes the precursors of the gametes such as oogonia, spermatogonia, primary and secondary oocytes, and primary and secondary spermatocytes as well as the gametes.

GM crops genetically modified crops; agricultural plants that contain transgenes.

growth factor receptors proteins on the cell surface that bind growth factors and trigger signal transduction cascades.

growth factors extracellular hormones and cell-bound ligands that stimulate or inhibit cell proliferation.

gynandromorph a rare genetic mosaic with some male tissue and some female tissue, usually in equal amounts.

H

hairpin loop a structure formed when a single strand of RNA folds back on itself because of complementary base pairing between different regions in the same molecule. Also called a *stem loop*.

haploid describes cells, such as the gametes of diploids, that contain one set of chromosomes.

haploinsufficiency a property of a gene in diploids where two wild-type (functional) gene copies are required; an individual heterozygous for a wild-type allele and a loss-of-function allele of a haploinsufficient gene shows an abnormal phenotype because the level of gene activity is not enough to produce a normal phenotype.

haplotype a particular combination of linked alleles.

haplotype blocks stretches of genomic DNA within which clusters of alleles of different loci display linkage disequilibrium. Such blocks of DNA sequence are flanked by recombination hotspots. Also called *LD blocks.*

Hardy-Weinberg equilibrium (HWE) describes a population in which the allele and genotype frequencies do not change from generation to generation; the state of an ideal population that obeys the assumptions of the Hardy-Weinberg law.

Hardy-Weinberg proportions the genotype frequencies for a particular locus in a population in Hardy-Weinberg equilibrium.

helix-turn-helix a transcription factor DNA-binding domain; for example, the Lac repressor is a helix-turn-helix protein, and the homeodomain of Hox proteins is a helix-turn-helix domain.

hemizygous describes the genotype for genes present in only one copy in an otherwise diploid organism, such as X-linked genes in a male.

heredity the way genes transmit physiological, physical, and behavioral traits from parents to offspring.

heritability the proportion of total phenotype variance in a population ascribable to genetic variance.

HERVs human endogenous retroviruses; retrotransposons in the human genome that resemble retroviruses structurally.

heterochromatin highly condensed chromosomal regions within which genes are usually transcriptionally inactive.

heterodimers protein complexes consisting of two different polypeptides.

heteroduplex a region of double-stranded DNA in which the two strands have nonidentical (though similar) sequences. Heteroduplex regions are often formed as intermediates during crossing-over.

heterogametic sex the sex of a species in which the two sex chromosomes are dissimilar; e.g., males are the heterogametic sex in humans because they have an X and a Y chromosome.

heterogeneous trait a phenotype caused by a mutation in any one of a number of different genes.

heteroplasmic genomic makeup of a cell's organelles characterized by a mixture of organellar genomes.

heterozygote individual with two different *alleles* for a given *gene* or *locus.*

heterozygote advantage the situation in which heterozygotes have a higher fitness than either homozygote.

heterozygous describes a genotype in which the two copies of a gene are different alleles.

Hfr bacteria that produce a high frequency of recombinants for chromosomal genes in mating experiments because their chromosomes contain an integrated F plasmid.

hierarchical strategy a method for genome sequencing where a small set of genomic library clones are first ordered to generate a minimal tiling path, and then genomic DNA in each recombinant clone is sequenced.

histone acetyl transferases (HATs) enzymes that acetylate histone tail lysines resulting in open chromatin.

histone deacetylase enzyme that removes acetyl groups from histone tail lysines thereby closing chromatin.

histone demethylases enzymes that remove methyl groups from histone tail lysines, thereby affecting chromatin structure.

histone methyl transferases (HMTases) enzymes that methylate histone tail lysines and arginines, thereby affecting chromatin stucture.

histone tails the N termini of histone proteins whose amino acid residues are enzymatically modified to affect local chromatin structure.

histones small DNA-binding proteins with a preponderance of the basic, positively charged amino acids lysine and arginine. Histones are the fundamental protein components of nucleosomes.

Holliday junctions interlocked regions of two nonsister chromatids in recombination intermediates.

homeobox in homeotic genes that encode homeodomain-containing transcription factors, the region of DNA homology, usually 180 bp in length, that encodes the homeodomain. The homeodomain is the DNA-binding region.

homeodomain a conserved DNA-binding region of transcription factors encoded by the homeobox of homeotic genes.

homeotic gene a gene that gives originally identical groups of cells individual identities during development.

homeotic mutation a mutation that causes cells to misinterpret their position in the body pattern and become normal organs in inappropriate positions.

homodimers protein complexes consisting of two identical polypeptides.

homogametic sex the sex of a species in which the two sex chromosomes are identical; in humans, females are the homogametic sex because they have two X chromosomes.

homogeneously stained region (HSR) a region of a chromosome that contains many tandem repeats of a gene due to gene amplification and is visible under a microscope as an enlarged area.

homologous chromosomes (homologs) a pair of chromosomes containing the same linear gene sequence, each derived from one parent.

homologous recombination or *crossing-over;* during meiosis, the breaking and rejoining of nonsister chromatids, resulting in the exchange of corresponding sections of DNA. Also during repair of broken chromosomes, sister chromatids can cross over.

homologs homologous chromosomes; also refers to genes or regulatory DNA sequences that are similar in different species because of descent from a common ancestral sequence.

homology describes DNA or amino acid sequences that are similar because they derive from the same ancestral sequence.

homoplasmic describes the genomic makeup of a cell's organelles characterized by a single type of organellar DNA.

homozygote an individual with identical alleles for a given gene or locus.

homozygous describes a genotype in which the two copies of the gene that determine a particular trait are the same allele.

horizontal gene transfer the introduction and incorporation of DNA from an unrelated individual or from a different species.

***Hox* genes** a gene superfamily in *Drosophila* and humans encoding homeobox transcription factors. The *Hox* genes pattern flies and people along the AP body axis during development.

Human Genome Project the initiative to determine the complete sequence of the human genome and to analyze this information.

Human Microbiome Project the initiative to identify all species of microorganisms that are symbionts with humans, and to correlate differences in microorganism populations with phenotypic differences and disease states.

hybridization (hybridize) base pairing between a probe and its complementary DNA sequence.

hybrids offspring of genetically dissimilar parents; often used as synonym for *heterozygotes.*

hydrogen bonds weak electrostatic bonds that result in a partial sharing of hydrogen atoms between reacting groups.

hypermorphic mutation (or allele) a mutant allele that generates either more protein than the wild-type allele or a more efficient protein.

hypomorphic mutation (or allele) an allele that produces either less of a wild-type protein or a mutant protein with a weak but detectable function.

I

idiogram a black-and-white diagram of the chromosomes converted from the light and dark bands (G bands) observed under the microscope.

immune surveillance a process by which the human immune system recognizes and attacks foreign bodies.

imprinting control regions (ICRs) large regions of DNA containing genes that regulate sex-specific methylation of nearby genes.

***in vivo* gene therapy** delivery of a therapeutic gene directly to somatic cells.

incomplete dominance the relationship between two alleles of a gene where the heterozygote has a phenotype intermediate between that of the two homozygotes.

InDel (or DIP) a genomic DNA polymorphism caused by insertion or deletion; in humans, they occur once in every 10 kb of DNA.

independent assortment the random distribution of alleles of different genes during gamete formation.

indirect repressor a protein that interferes with the function of an activator without necessarily binding DNA.

inducer a small molecule that causes transcription from a gene or set of genes.

inducible regulation gene control where transcription occurs only in the presence of an inducer.

induction the process by which a signal causes expression of a gene or set of genes.

initiation the first phase of DNA replication, transcription, or translation needed to set the stage for the addition of nucleotide or amino acid building blocks during elongation.

initiation codon the nucleotide triplet that marks the precise spot in the nucleotide sequence of an mRNA where the code for a particular polypeptide begins: AUG.

initiation factors a term usually applied to proteins that help promote the association of ribosomes, mRNA, and initiating tRNA during the first phase of translation.

insertion the addition to a DNA molecule of one or more nucleotide pairs.

insertion sequences (ISs) small bacterial transposons that do not contain selectable markers.

insulator a transcriptional regulation element in eukaryotes that blocks interaction between enhancers on one side of it with promoters on the other side.

integration insertion of one DNA molecule into another.

integrative and conjugative elements (ICEs) pathogenicity islands that also include genes for proteins that mediate conjugation.

intercalators a class of chemical mutagens composed of flat, planar molecules that can sandwich themselves between successive base pairs, disrupt the machinery of DNA replication, and thereby cause mutation.

interkinesis the brief interphase between meiosis I and meiosis II.

interphase the period in the cell cycle between divisions.

intragenic suppression the restoration of gene function by one mutation canceling the effects of another mutation in the same gene.

introgression a small genomic region from one species that is present in a different species.

introns the sequences found in a gene's DNA that are spliced out of the primary transcript and are therefore not found in the mature mRNA.

inversion a 180-degree rotation of a segment of a chromosome relative to the remainder of the chromosome.

inversion loop a DNA loop formed in a chromosome of an inversion heterozygote when the inverted region or its noninverted counterpart rotates to pair with the similar region in its homolog.

isogenic lines globally homozygous strains in which every individual has the identical genotype due to inbreeding.

K

karyotype the visual description of the complete set of chromosomes in one cell of an organism; usually presented as a photomicrograph with the chromosomes arranged in a standard format showing the number, size, and shape of each chromosome type.

kinetochore a specialized chromosomal structure composed of DNA and proteins that is the site at which chromosomes attach to the spindle fibers.

kinetochore microtubules microtubules of the mitotic spindle that extend between a centrosome and the kinetochore of a chromatid. Chromosomes move along the kinetochore microtubules during cell division.

kinetoplast a network of circular DNAs inside a mitochondrion including genome copies (maxicircles) and DNA encoding guide RNAs for RNA editing (minicircles).

knockin mice mice in which a gene has been altered by targeted mutagenesis. The alteration can be a point mutation or a large insertion of DNA.

knockout mice mice homozygous for an induced mutation in a targeted gene; the mutation destroys (knocks out) the function of the gene.

L

***lac* operon** a single DNA unit in *E. coli,* composed of the *lacZ, lacY,*and *lacA* genes together with the promoter (P) and operator site (o), that enables the simultaneous regulation of the three structural genes in response to environmental changes.

lagging strand during replication, the DNA strand replicated discontinuously, 5′ to 3′ away from the Y-shaped replication fork, as small Okazaki fragments that are ultimately joined into a continuous strand.

late-onset describes a genetic trait in which symptoms are not present at birth, but manifest themselves later in life.

law of independent assortment Mendel's second law, which states that alleles of different genes segregate into gametes independently of each other.

law of segregation Mendel's first law, which states that the two alleles for each trait separate (segregate) during gamete formation and then unite at random, one from each parent, at fertilization.

LD blocks stretches of genomic DNA within which clusters of alleles of separate loci display linkage disequilibrium. Such blocks of DNA sequence are flanked by recombination hotspots. Also called *haplotype blocks.*

leading strand during replication, the DNA strand replicated continuously 5′ to 3′ toward the unwinding Y-shaped replication fork.

leptotene the first definable substage of prophase I during which the long, thin, already duplicated chromosomes begin to thicken.

line of correlation the *best-fit line* relating phenotypic values of offspring to the midparent value; its slope estimates the heritability of the trait.

LINEs long interspersed elements; one of the two major classes of retrotransposons. Structurally, LINEs resemble mRNAs in mammals.

linkage disequilibrium (LD) when particular alleles at separate loci (such as marker alleles and disease alleles) are associated with each other at a significantly higher frequency than would be expected by chance.

linkage equilibrium when particular alleles at separate loci on the same chromosome associate with one another randomly.

linkage group a group of genes chained together by linkage relationships.

linked describes genes whose alleles are inherited together more often than not; linked genes are usually located close together on the same chromosome.

linker DNA a stretch of ~40 base pairs of DNA that connect one nucleosome with the next.

locus a designated location on a chromosome; sometimes refers to a gene.

locus control region (LCR) a *cis*-acting regulatory element (a collection of enhancers) that regulates transcription from individual genes within a gene complex such as the α-globin complex.

locus heterogeneity describes a trait where mutations in any one of two or more genes results in the same mutant phenotype.

Lod score log of odds; statistic used to analyze linkage data in humans. The statistic determines, for a particular data set, how much more likely it is that two loci are linked with a given RF value than that they are unlinked.

loss-of-function mutation (or allele) a mutation that either reduces or abolishes the activity of a gene.

loss-of-heterozygosity a phenomenon in which one of a number of rare events can cause a cell originally heterozygous for a tumor-suppressor gene mutation to become homozygous.

lysate a population of phage particles released from the host bacteria at the end of the lytic cycle.

lysogen a bacterial cell that carries a prophage.

lysogenic cycle the integration of a bacteriophage into the bacterial host genome as a *prophage,* which does no immediate harm to the cell.

lytic cycle the bacterial cycle of phage-infected cells resulting in cell lysis and release of progeny phage.

M

M phase the part of the cell cycle where mitosis and cytokinesis take place.

major groove in a space-filling representation of the DNA double helix model, the wider of the two grooves resulting from the vertical displacement of the two backbone threads.

malignant cell a cancerous cell.

Manhattan plot the display of data from a GWAS that resembles the Manhattan skyline. For each SNP in the genome (*x*-axis), the negative log of the probability that the association of the SNP and the trait in question would be observed by chance is on the *y*-axis.

map unit (m.u.) a unit of measure of recombination frequency. One m.u. is equal to a 1% chance that a marker at one genetic locus will be separated from a marker at a second locus due to crossing-over in a single generation. Synonymous with *centimorgan.*

marker an identifiable physical location on a chromosome, whose inheritance can be monitored. Markers can be genes or any segment of DNA with variant forms that can be followed.

marker gene a gene inserted into a genome along with a transgene whose expression indicates successful incorporation of the transgene.

match probability the chance that two different people have the same diploid genotype for a particular set of molecular markers, such as SSRs.

maternal effect genes genes encoding maternal components (those supplied to the egg by the mother) that enable the development of her progeny.

maternal effect mutations mutations in maternal effect genes. Progeny of mothers carrying such mutations have a mutant phenotype—the mutant mothers themselves do not.

maternal inheritance the transfer of organellar genomes to progeny in the egg cytoplasm.

meiosis the process of two consecutive cell divisions in the progenitors of gametes. In the first division, pairs of homologous chromosomes segregate into two different daughter cells; in the second division, the chromatids of each homolog segregate into two different daughter cells. A diploid ($2x$) cell produces four haploid ($1x$) gametes by meiosis.

meiosis I (division I of meiosis) when the parent nucleus divides to form two daughter nuclei; during meiosis I, the previously replicated homologous chromosomes segregate into different daughter cells.

meiosis II (division II of meiosis) when each of the two daughter nuclei resulting from meiosis I divides to produce two nuclei; the chromatids of each homolog segregate into different daughter cells.

meiotic nondisjunction failures in chromosome segregation during meiosis, when either chromatids or homologs do not separate properly.

merodiploids partial diploids in which two copies of some genes exist.

messenger RNA (mRNA) RNA that serves as a template for protein synthesis.

metacentric describes a chromosome whose centromere is at or near the middle.

metagenomics the collective analysis of genomic DNA from natural communities of microbes.

metaphase the stage in mitosis or meiosis during which the chromosomes are aligned along the equatorial plane of the cell.

metaphase I the phase of meiosis I when the kinetochores of homologous chromosomes attach to microtubules from opposite spindle poles, positioning the bivalents at the equator of the spindle apparatus.

metaphase II the second phase of meiosis II during which kinetochores of sister chromatids attach to microtubule fibers emanating from opposite poles of the spindle apparatus. Two characteristics distinguish metaphase II from its counterpart in mitosis: (1) the number of chromosomes is one-half that in mitotic metaphase of the same species, and (2) in most chromosomes, the two sister chromatids are no longer identical because of the recombination through crossing-over that occurred in meiosis I.

metaphase plate the imaginary equator of the cell toward which chromosomes move during metaphase.

metastasis the colonization of distant tissue by cancer cells that travel through the bloodstream.

methyl-directed mismatch repair a DNA repair mechanism that corrects mistakes in replication, discriminating between newly synthesized and parental DNA by the methyl groups on the parental strand.

methylated cap the structure at the 5′ end of eukaryotic mRNA formed by the action of capping enzyme and methyl transferase; crucial for efficient translation of the mRNA into protein.

micro-RNA (miRNA) an RNA molecule 21–24 bases long that is encoded in the genome of an organism and, through complementary base pairing, targets specific mRNAs for destruction or blocks their translation.

microhomology-mediated end-joining (MMEJ) a DNA double-strand break repair process similar to NHEJ except that DNA ends are resected, which results in small deletions in rejoined DNA.

microsatellite an *SSR locus*. The term derives from the fact that genomic DNA fragments containing SSR loci separate from fragments without repeat sequences in density gradient centrifugation experiments.

migration the movement of individuals between populations.

minor groove in a space-filling representation of the DNA double helix model, the narrower of the two grooves resulting from the vertical displacement of the two backbone threads.

missense mutation a mutation in a gene that changes a codon for one amino acid to a codon that specifies a different amino acid.

mitochondria organelles that convert energy derived from nutrient molecules into ATP; mitochondria have their own genomes.

mitochondrial gene therapy transfer of the nucleus from an oocyte with mutant mitochondria to an enucleated oocyte with normal mitochondria.

mitogen a growth factor that stimulates cell proliferation.

mitosis the process of division that produces daughter cells that are genetically identical to each other and to the parent cell.

mitotic nondisjunction the failure of two sister chromatids to separate during mitotic anaphase. In a diploid, this generates reciprocal trisomic and monosomic daughter cells.

mitotic spindle a structure composed of three types of microtubules (kinetochore microtubules, polar microtubules, and astral microtubules). The mitotic spindle provides a framework for the movement of chromosomes during cell division.

modification the phenomenon in which growth on a restricting host changes a phage so that succeeding generations grow more efficiently on that same host.

modification enzymes enzymes that add methyl groups to specific DNA sequences, preventing the action of specific restriction enzymes on that DNA.

modifier genes genes that produce a subtle, secondary effect on phenotype. No formal distinction exists between major and modifier genes, rather it is a continuum of degrees of influence.

modifier screen a mutant screen performed in the background of mutations in a specific gene which cause a mutant phenotype. A researcher looks for changes in that mutant phenotype that are due to an additional mutation in a different gene; the second mutation can identify a second gene that works in the same pathway as the first gene.

molecular clock the hypothesis stating that one can assume a constant rate of amino acid or nucleotide substitution as a means of determining the genealogies of organisms.

molecular cloning the process by which a single DNA fragment is purified from a complex mixture of DNA molecules and then amplified into a large number of identical copies.

monocistronic mRNA mRNA containing the coding region of only one gene.

monohybrid crosses crosses between parents that differ in only one trait.

monohybrids individuals having two different alleles for a single trait.

monomorphic a gene with only one wild-type allele.

monoploid describing cells, nuclei, or organisms that have a single set of unpaired chromosomes. For diploid organisms, monoploid and haploid are synonymous.

monosomic describes an individual lacking one chromosome from the diploid number for the species.

monozygotic (MZ) twins genetically identical twins that derive from the splitting of an embryo from a single zygote into two separate embryos; also called *identical twins.*

Monte Carlo simulation a computer simulation that uses a random number generator to choose an outcome for probabilistic events in a dynamic system defined by predetermined rules of probability; allows multiple runs of generational sequences.

morphogens substances that define different cell fates in a concentration-dependent manner.

mosaic an organism containing cells of different genotypes.

mosaic analysis the observation of mosaic tissues in order to determine the focus of action of a gene.

most recent common ancestor (MRCA) the most recent occurrence of a particular segment of genomic DNA (an allele) that a given group of individuals all inherited, even if in mutated form; the most recently shared genetic ancestor for a particular segment of genomic DNA.

multifactorial traits traits determined by several factors, including multiple genes interacting with each other and one or more genes interacting with the environment.

multihybrid crosses crosses between parents heterozygous at three or more genes.

multimer a protein made from more than one polypeptide; each polypeptide in the multimeric protein is called a subunit.

mutagen any physical or chemical agent that raises the frequency of mutations above the spontaneous rate.

mutant allele (1) an allele, or DNA variant, whose frequency is less than 1% in a population; (2) an allele that causes a phenotype seen only rarely in a population.

mutant screen a process where researchers examine a large number of mutagenized organisms and identify rare individuals with a mutant phenotype of interest.

mutation hotspots sites within a gene that mutate more frequently than others, either spontaneously or after treatment with a particular mutagen.

mutations heritable alterations in DNA sequence.

N

n the number of chromosomes in gametes.

N terminus the end of a polypeptide chain that contains a free amino group that is not connected to any other amino acid.

***N*-formylmethionine (fMet)** a modified methionine whose amino end is blocked by a formyl group; fMet is carried by a specialized tRNA that functions only at a ribosome's translation initiation site.

narrow-sense heritability (h^2) heritability measurement that includes only the additive component (V_A) of genetic variance (V_G).

natural selection in nature, the process that progressively eliminates individuals whose fitness is low and chooses individuals of high fitness to survive and become the parents of the next generation.

natural transformation a process by which a few species of bacteria transfer genes from one strain to another by spontaneously accepting DNA from their surroundings.

NCO a meiosis that occurs with no crossovers between a particular gene pair.

nearly isogenic lines (NILs) strains that are genetically identical (isogenic) except for a small region of their genomes; also called *congenic lines.*

neomorphic mutations (or alleles) rare mutations that produce a novel phenotype due to production of a protein with a new function or due to ectopic expression of the protein.

neutral mutations (or alleles) alleles that provide no selective advantage or disadvantage as compared with the original allele.

nitrogenous bases components of nucleotides. In DNA, the four different bases are guanine, cytosine, adenine, and thymine; in RNA, uracil substitutes for thymine.

non-Mendelian inheritance a pattern of inheritance that does not follow Mendel's laws and does not produce Mendelian ratios among the progeny of various crosses.

non-polyQ diseases human single-gene neurological diseases caused by germ-line expansion of trinucleotide repeats. The expanded trinucleotide region affects gene expression although it is outside the open reading frame.

nonanonymous DNA polymorphisms differences in genomic DNA sequence that have an effect on gene function.

nonautonomous elements defective transposable elements that cannot mobilize unless the genome contains nondefective autonomous elements that can supply necessary functions such as tranposase enzymes.

noncoding genes genes that are transcribed but not translated.

noncoding RNA (ncRNA) a transcript that lacks an open reading frame and functions as an RNA molecule.

nonconservative substitutions mutations that substitute an amino acid in a protein with a different amino acid with dissimilar chemical properties.

nondisjunction failures in chromosome segregation during cell division, when either chromatids or homologs do not separate properly.

nonhistone proteins chromatin constituents other than histones with a wide variety of functions.

nonhomologous chromosomes chromosomes in a single genome that are not homologs—they do not have similar DNA sequences and do not pair during meiosis.

nonhomologous end-joining (NHEJ) mechanism for stitching back together ends formed by double-strand breaks. NHEJ relies on proteins that bind to the ends of the broken DNA strands and bring them close together.

nonparental ditype (NPD) a fungal tetrad containing four recombinant spores.

nonsense mutation a mutation in which a codon for an amino acid is changed to a stop codon, resulting in the formation of a truncated protein.

nonsense suppressor tRNAs mutant tRNAs containing anticodons that can recognize stop codons, thus suppressing the effects of nonsense mutations by inserting an amino acid into a polypeptide in spite of the stop codon.

nontandem duplications chromosomal aberrations where an additional copy of a genomic region is present and is not adjacent to the original copy; the duplicated region can be on the same chromosome or on a different chromosome from the original copy.

normal distribution a set of data points that scatter around a central mean (average) without left or right bias; an equal number of data points fall below and above the mean. Normally distributed data can be represented by a *bell curve* graph in which the x-axis is the phenotypic value and the y-axis is the number of individuals in the population who show a particular phenotypic value. The phenotypic values of quantitative traits are often normally distributed.

nuclear envelope two membranes that surround the nucleus of a eukaryotic cell.

nucleic acid hybridization formation of double-stranded molecules through complementary base pairing of single strands of DNA or RNA.

nucleoid body a folded bacterial chromosome.

nucleolus a large, spherical organelle visible in the nucleus of interphase eukaryotic cells with a light microscope; the site of ribosome biosynthesis.

nucleotide a subunit of DNA or RNA consisting of a nitrogenous base (adenine, guanine, thymine, or cytosine in DNA; adenine, guanine, uracil, or cytosine in RNA), a phosphate group, and a sugar (deoxyribose in DNA; ribose in RNA).

null hypothesis a statistical hypothesis to be tested and either accepted or rejected in favor of an alternative.

null mutations (or alleles) mutations that abolish the function of a gene product encoded by the wild-type allele. Such mutations either prevent synthesis of both the protein and RNA (this is the strict definition used by some geneticists exclusively) or promote the synthesis of a protein or RNA that is incapable of carrying out any function. Synonymous with *amorphic mutations*.

O

octads asci produced in *Neurospora* containing eight spores because of a round of mitosis that occurs after meiosis is completed.

Okazaki fragments during DNA replication, small fragments of about 1000 bases that are joined after synthesis to form the lagging strand.

oligonucleotide primer a short single-stranded DNA molecule that hybridizes to a DNA or RNA template; its 3′ end is extended by DNA polymerase or reverse transcriptase.

oncogene a dominant gain-of-function mutant allele found in cancer cells; mutant alleles of proto-oncogenes whose normal cellular roles are to promote proliferation.

Online Mendelian Inheritance in Man (OMIM) a database located on the Internet (www.omim.org) of human genes and the traits that they control. The database includes known variants of genes and the diseases or other traits that they cause.

oogenesis formation of the female gametes (eggs).

oogonia diploid germ cells in the ovary.

open reading frames (ORFs) DNA sequences with long stretches of codons in the same reading frame uninterrupted by stop codons.

operator site a short DNA sequence near a promoter that can be recognized by a repressor protein; binding of repressor to the operator blocks transcription of the gene.

operon a unit of DNA composed of two or more genes transcribed as a polycistronic mRNA under the control of a single promoter and operator.

operon theory the model that explains the repression and induction of genes in *E. coli.*

ordered tetrads tetrads in fungi such as *Neurospora* in which the order of ascospores in the ascus reflects the geometry of the meiotic divisions.

origin of replication short sequence of nucleotides where DNA replication initiates.

origin of transfer during bacterial conjugation, the spot on the F plasmid where replicative transfer of DNA from the donor to the recipient cell initiates.

orthologous genes genes with sequence similarities in two different species that arose from the same gene in the two species' common ancestor.

ovum haploid female germ cell (the egg).

oxidative phosphorylation a metabolic pathway carried out by mitochondria in which the energy released by oxidation of nutrients is used to generate ATP.

P

***P* element transformation** a method for generating transgenic *Drosophila* in which a recombinant *P* element containing a transgene and a marker gene is injected into embryos; the recombinant transposon integrates into a chromosome of a germ-line cell.

***p* value** the numerical probability that a particular set of observed experimental results represents a chance deviation from the values predicted by a particular hypothesis.

pachytene the substage of prophase I that begins at the completion of synapsis and includes crossing-over.

pair-rule genes in *Drosophila,* these zygotic segmentation genes encode transcription factors that are expressed in seven stripes in embryos with a two-segment periodicity; that is, there is one stripe for every two segments. Mutations in pair-rule genes cause the deletion of similar pattern elements from every alternate segment, corresponding to the regions where the gene is normally expressed.

paired-end sequencing a strategy for determining the base pair sequence of whole genomes in which ~1000 bp at both ends of single BAC clones are sequenced. Knowing that the two sequence reads are connected on a single BAC insert allows genome assembly despite the presence of repeated elements.

pangenome the core genome of a bacterial species plus all genes found in some strains but not others.

paracentric inversions chromosomal inversions that exclude the centromere.

paralogous genes that arise by duplication within the same species, often within the same chromosome; paralogous genes often constitute a gene family.

parasegments units of the *Drosophila* body along the AP axis whose identities are controlled by homeotic genes.

parental (P) generation individuals whose progeny in subsequent generations will be studied for specific traits.

parental classes combinations of alleles present in the original parental generation.

parental ditype (PD) a tetrad that contains four parental class haploid cells.

parental types phenotypes that reflect a previously existing parental combination of alleles that is retained during gamete formation.

parthenogenesis reproduction in which offspring are produced by an unfertilized female; common in ants, bees, wasps, and certain species of fish and lizards.

passenger mutations DNA alterations in cancerous cells that occur due to the increased mutation rate of cancer cells but do not contribute to the cancer phenotype.

pathogen an agent, for example a microorganism, that causes disease in its host.

pathogenicity islands segments of DNA in disease-causing bacteria that encode several genes involved in pathogenesis. Pathogenicity islands appear to have been transferred into the bacteria by horizontal gene transfer from a different species.

pedigree an orderly diagram of a family's relevant genetic features, extending through as many generations as possible.

penetrance in a population, the fraction of individuals with a particular genotype that show the associated phenotype.

peptide bond a covalent bond that joins amino acids during protein synthesis.

peptidyl (P) site the site on a ribosome to which the initiating tRNA first binds and at which the tRNAs carrying the growing polypeptide are located during elongation.

peptidyl transferase the enzymatic activity of the ribosome responsible for forming peptide bonds between successive amino acids.

pericentric inversions chromosomal inversions that include the centromere.

permissive condition an environmental condition that allows the survival of an individual with a conditional lethal allele.

phage short for bacteriophage; a virus for which the natural host is a bacterial cell; literally *bacteria eaters.*

pharming the use of transgenic animals and plants to produce protein drugs.

phenocopy a change in phenotype arising from environmental agents that mimics the effects of a mutation in a gene. Phenocopies are not heritable because they do not result from a change in a gene.

phenotype an observable characteristic.

phenotype frequency the proportion of individuals in a population that have a particular phenotype.

phenotypic value a number that describes a specific intensity of expression of a quantitative trait; also called a *trait value.*

phosphodiester bonds covalent bonds joining one nucleotide to another that form the backbone of DNA.

photosynthesis a metabolic pathway in chloroplasts in which light energy is stored in the chemical bonds of carbohydrates.

physical markers cytologically visible abnormalities that make it possible to keep track of specific chromosome parts from one generation to the next.

piRNAs (Piwi-interacting RNAs) small RNAs generated by processing long RNAs transcribed from the human genome. PiRNAs recruit Piwi protein complexes to transposable element sites in genomic DNA and prevent TE mobilization.

plaque a clear area on a bacterial lawn, devoid of living bacterial cells, containing the genetically identical descendants of a single bacteriophage.

plasmids small circles of double-stranded DNA that can replicate in bacterial cells independently of the bacterial chromosome; commonly used as cloning vectors.

pleiotropic describes a gene that functions in several different pathways.

pleiotropy phenomenon in which a single gene determines a number of distinct and seemingly unrelated characteristics.

point mutation a mutation of one or a few base pairs, or a small deletion or insertion in DNA.

polar body a cell produced by meiosis I or meiosis II during oogenesis that does not become the primary or secondary oocyte.

polar microtubules microtubules that originate in centrosomes and are directed toward the middle of the cell; polar microtubules that arise from opposite centrosomes interdigitate near the cell's equator and push the spindle poles apart during anaphase.

polarity the property of having distinct ends.

poly-A tail the 3′ end of eukaryotic mRNA, consisting of 100–200 A residues, that stabilizes the mRNA and increases the efficiency of translation initiation.

polycistronic mRNA an mRNA that contains more than one protein-coding region; often the transcriptional product of an operon in bacterial cells.

polygenic describes a trait controlled by multiple genes.

polylinker a synthetic DNA sequence in a cloning vector containing several different restriction enzyme recognition sites that can be used for insertion of a DNA molecule.

polymer a linked chain of repeating subunits that form a larger molecule; DNA is a type of polymer.

polymerase chain reaction (PCR) a fast and inexpensive method of replicating a DNA sequence when short sequences at each end are known; based on a reiterative loop that amplifies the products of each previous round of replication.

polymerization the linkage of subunits to form a multisubunit chain. For example, in DNA replication, the polymerization of nucleotides occurs through the formation of phosphodiester bonds by DNA polymerase.

polymorphic describes a locus with two or more distinct alleles in a population.

polypeptides polymers containing hundreds to thousands of amino acids joined in a linear fashion by peptide bonds.

polyploids euploid species that carry three or more complete sets of chromosomes.

polyprotein a polypeptide produced by translation that can subsequently be cleaved by protease enzymes into two or more separate proteins.

polyQ diseases human neurological diseases caused by germ-line expansion of CAG repeats. The CAG repeats are codons specifying a run of glutamine (Q) amino acids within triunucleotide repeat genes.

polyribosome a structure formed by the simultaneous translation of a single mRNA molecule by multiple ribosomes.

population a group of interbreeding individuals of the same species that inhabit the same space at the same time.

population bottleneck a phenomenon where a large proportion of individuals in a population die, often because of environmental disturbances, resulting in survivors who essentially are equivalent to a founder population.

population genetics the scientific discipline concerned with genetic events that occur in whole populations.

position-effect variegation variable expression of a gene in a population of cells, caused by the gene's location near highly compacted heterochromatin.

positional cloning the process by which researchers use linkage analysis to obtain the clone of a gene.

posttranslational modifications changes such as phosphorylation that occur to a polypeptide after translation has been completed.

pre-mutation allele a normally-functioning allele of an unstable trinucleotide repeat gene with a repeat region expanded to the point where further expansion to a disease allele occurs at high frequency.

preimplantation embryo diagnosis a process by which human eggs are fertilized *in vitro* and genotyped for particular disease alleles prior to implantation in a uterus.

prenatal genetic diagnosis genotyping fetal cells to determine alleles of disease loci.

primary oocytes/spermatocytes germ-line cells in which meiosis I occurs.

primary structure the linear sequence of amino acids within a polypeptide.

primary transcript the single strand of RNA resulting from transcription.

primase the enzyme that synthesizes RNA primers during DNA replication.

primer a short, preexisting DNA oligonucleotide or RNA molecule to which nucleotides can be added by DNA polymerase.

probe an oligonucleotide labeled with a radioactive isotope or fluorescent dye and used to identify complementary sequences by means of hybridization.

product rule the probability of two or more independent events occurring together is the product of the probabilities that each event will occur by itself.

programmed cell death (PCD) apoptosis; a process in which DNA is degraded and the nucleus condenses; the cell may then be devoured by neighboring cells or phagocytes.

prokaryotic gene regulation control of gene expression in a bacterial cell via mechanisms to increase or decrease the transcription or translation of specific genes or groups of genes.

prometaphase the stage of mitosis or meiosis just after the breakdown of the nuclear envelope, when the chromosomes connect to the spindle apparatus and begin to move toward the metaphase plate.

promoters gene sequences near the transcription start site that attract RNA polymerase.

pronuclear injection a method for generating transgenic mammals in which DNA is injected into a pronucleus after fertilization.

pronuclei (pronucleus) the sperm and egg nuclei present in the same cytoplasm after fertilization in mammals.

prophage a phage genome integrated into the bacterial host genome.

prophase the phase of the cell cycle marked by the emergence of the individual chromosomes from the undifferentiated mass of chromatin, indicating the beginning of mitosis.

prophase I the longest, most complex phase of meiosis consisting of several substages.

prophase II the first phase of meiosis II; if the chromosomes decondensed during interkinesis, they recondense. At the end of prophase II, the nuclear envelope breaks down and the spindle apparatus re-forms.

protein domains discrete functional regions of a protein.

protein kinases enzymes that add phosphate groups to their protein substrates.

proteins large polymers composed of hundreds to thousands of amino acids strung together in a specific linear order. Proteins are required for the structure, function, and regulation of the body's cells, tissues, and organs.

proteome the complete set of proteins encoded by a genome.

proto-oncogene a gene that can mutate into an oncogene—an allele that causes a cell to become cancerous.

prototroph a microorganism (usually wild type) that can grow on minimal medium.

pseudoautosomal regions (PARs) homologous regions at both ends of the X and Y chromosomes.

pseudogene a nonfunctioning gene; the result of duplication and divergence events in which one copy of an originally functioning gene has undergone mutations such that it no longer functions.

pseudolinkage a characteristic of a heterozygote for a reciprocal translocation, in which genes that flank a translocation breakpoint behave as if they are linked even though they originated on nonhomologous chromosomes.

Punnett square a simple method for visualizing the fertilization events possible in a given cross.

pure-breeding (lines) families of organisms that produce offspring with specific parental traits that remain constant from generation to generation.

purines a chemical group that includes the nitrogenous bases adenine and guanine.

pyrimidines a chemical group that includes the nitrogenous bases cytosine, thymine, and uracil.

Q

quantitative trait inherited trait that exhibits many intermediate forms; determined by alleles of many different genes whose interactions with each other and the environment produce a phenotype. Synonymous with *continuous trait.*

quantitative trait loci (QTLs) genes that control the expression of continuous traits.

quaternary structure the three-dimensional configuration of subunits in a multimeric protein.

quorum sensing a communication system whereby bacteria detect their population density.

R

radial loop–scaffold model the model of looping and gathering of DNA by nonhistone proteins that results in high compaction of chromosomes at mitosis.

read in a single DNA sequencing run, a digital file of the sequence of As, Cs, Gs, and Ts constituting the newly synthesized DNA.

reading frame the partitioning of groups of three nucleotides from a fixed starting point such that the sequential interpretation of each succeeding triplet codon generates the order of amino acids in the resulting polypeptide chain.

realized heritability the heritability value measured as a response to selection.

rearrangement breakpoints the abnormal juxtapositions of DNA sequences in a chromosome caused by rearrangement; identifies the precise base pairs at which rearranged chromosome segments begin and end.

recessive epistasis a gene interaction in which the effects of recessive alleles at one gene hide the effects of alleles at another gene.

recessive lethal allele an allele that prevents the birth or survival of homozygotes, though heterozygotes carrying the allele survive.

recessive trait the trait that remains hidden in the F_1 hybrids (heterozygotes) resulting from a mating between pure-breeding parental strains showing antagonistic phenotypes; the recessive trait usually reappears in the second filial (F_2) generation.

recipient during gene transfer in bacteria, the cell that receives the DNA.

reciprocal crosses crosses performed in two directions, with the traits in the males and females reversed relative to the other, thereby controlling whether a particular trait is transmitted by the male or female gamete.

reciprocal recessive epistasis an interaction between alleles of two different genes where the homozygous recessive genotype of each gene prevents the phenotypic expression of (is epistsatic to) the dominant allele of the other gene.

reciprocal translocation a chromosomal rearrangement that results when two breaks, one in each of two chromosomes, yield DNA fragments that switch places and become attached to the other chromosome.

recombinant classes reshuffled combinations of alleles that were not present in the parental generation.

recombinant DNA molecules a combination of DNA molecules with parts having different origins that were joined using recombinant DNA technologies.

recombinant types phenotypes reflecting a new combination of genes that occurs during gamete formation.

recombinants (1) chromosomes that carry a mix of alleles derived from different homologous chromosomes; (2) gametes with combinations of alleles not inherited from the same parent; (3) recombinant plasmids.

recombinase an enzyme that performs site-specific recombination.

recombination the process by which offspring derive a combination of alleles different from that of either parent; the generation of new allelic combinations. In higher organisms, this can occur by independent assortment and by crossing-over.

recombination frequency (RF) the percentage of recombinant progeny;

can be used as an indication of the physical distance separating any two loci on a chromosome.

recombination hotspots small DNA regions where the frequency of recombination is much higher than average.

recombination nodules structures that appear during pachytene of prophase I. An exchange of parts between nonsister chromatids occurs at recombination nodules.

reduction(al) division cell division that reduces the number of chromosomes, usually by segregating homologous chromosomes to two daughter cells. Meiosis I is a reductional division.

redundant gene action a phenomenon where dominant, functional alleles of either one or the other of two genes is required in a pathway.

redundant genes genes whose products serve the same function in a pathway; a mutant phenotype is observed only if both gene products are absent.

RefSeq (species reference sequence) an annotated, complete genome sequence of a species.

relative fitness (*W*) the average number of surviving progeny of a particular genotype compared with the average number of surviving progeny of competing genotypes after a single generation.

release factors proteins that recognize stop codons and help end translation.

remodeling complex a complex of proteins that removes promoter-blocking nucleosomes or repositions them in relation to the gene, helping to prepare a gene for transcription.

replica plating a process whereby colonies on a master plate are picked up on velvet and then transferred to media in other petri plates to test for phenotype.

replication bubble unwound area of the original DNA double helix during replication.

replication fork the Y-shaped area consisting of the two unwound DNA strands branching out into unpaired (but complementary) single strands during replication.

replicon (replication unit) the DNA running both ways from one origin of replication to the endpoints where it merges with DNA from adjoining replication forks.

reporter gene a protein-coding region of a gene incorporated into a recombinant DNA molecule along with putative DNA regulatory elements. After transformation of bacteria or incorporation of the reporter gene into the genome of an organism, the reporter gene "reports" the activity of the putative regulatory elements by expressing the protein.

repressible regulation gene control where transcription occurs only in the absence of a repressor.

repressor a type of transcription factor that can bind to specific *cis*-acting elements and thereby diminish or prevent transcription. Repressors bind operators in prokaryotes and enhancers (or silencers) in eukaryotes.

reproductive cloning the creation of a cloned embryo by insertion of the nucleus of a somatic cell from one individual into an egg cell whose nucleus has been removed. The hybrid egg is stimulated to begin embryonic cell divisions, and the resulting cloned embryo is transplanted into the uterus of a foster mother and allowed to develop to term.

resection during homologous recombination, the process whereby single-stranded 5′ ends of DNA are produced by an exonuclease at the site of double-stranded cleavage.

resolvase the enzyme that breaks and joins DNA strands at Holliday junctions to separate nonsister chromatids during crossing-over.

restriction the bacterial capacity for limiting viral growth.

restriction enzymes bacterial proteins that recognize specific, short nucleotide sequences and cleave the DNA backbone at those sites.

restriction fragments DNA fragments generated by the action of restriction enzymes.

restrictive condition an environmental condition that prevents the survival of an individual with a conditional lethal allele.

retrotransposons genetic elements that mobilize via reverse transcription of an RNA intermediate. One class of transposable elements.

retroviral vectors gene therapy vectors that are partial retrovirus genomes containing a therapeutic gene; the therapeutic gene becomes stably integrated into the genomes of somatic cells.

retroviruses viruses that hold their genetic information in a single strand of RNA and carry the enzyme reverse transcriptase to convert that RNA into DNA within a host cell.

reverse mutation a mutation that causes a mutant to change back to wild type. Synonymous with *reversion.*

reverse transcriptase a retroviral RNA-dependent DNA polymerase that synthesizes DNA strands complementary to an RNA template.

reverse transcription the process by which reverse transcriptase synthesizes DNA strands complementary to an RNA template. The product of reverse transcription is a cDNA molecule.

reversion a mutation that causes a mutant to change back to wild type. Synonymous with *reverse mutation.*

ribonucleic acid (RNA) a polymer of ribonucleotides found in the nucleus and cytoplasm of cells; it plays an important role in protein synthesis. Several classes of RNA molecules exist, including messenger RNA (mRNA), transfer RNA (tRNA), ribosomal RNA (rRNA), miRNAs, piRNAs, and siRNAs.

ribose the 5-carbon sugar found in RNA.

ribosomal RNAs (rRNAs) RNA components of ribosomes, which are composed of both rRNAs and proteins.

ribosome binding site region on prokaryotic mRNAs containing both an initiation codon and a Shine-Dalgarno box; ribosomes bind to these elements to start translation.

ribosome profiling a technique for *footprinting* ribosomes on mRNAs. High-throughput sequencing of mRNA fragments protected by ribosomes can provide information about translational regulation by revealing the kinds of different mRNAs being translated and the frequency of their translation. In addition, by providing a snapshot of the positions of ribosomes on mRNAs, translation start and stop sites and pause sites can be identified.

ribosomes cytoplasmic structures composed of ribosomal RNA and protein; the sites of protein synthesis.

riboswitches allosteric RNA leaders that bind small molecule effectors to control gene expression.

ribozymes RNA molecules that can act as enzymes to catalyze specific chemical reactions.

RNA ribonucleic acid; a polymer of ribonucleotides. Different RNAs play important roles in protein synthesis and regulation of gene expression.

RNA editing specific alteration of the base sequence carried within an RNA molecule after transcription is completed.

RNA *in situ* hybridization an experimental approach to determining the expression pattern of a particular mRNA in the context of an entire organism or tissue. Labeled cDNA sequences corresponding to the gene's mRNA are used as a probe for mRNA on preparations of thinly sectioned tissues, or in some cases whole organisms or tissues. Signals where the probe is retained through hybridization indicate cells containing the gene's mRNA.

RNA interference (RNAi) the sequence-specific modulation of eukaryotic gene expression by a 21–24 nucleotide-long siRNA molecule. Through complementary base pairing with their target mRNAs, the siRNAs bring the RISC complex to the mRNA, resulting in its degradation.

RNA leader sequence the 5′ untranslated region (5′ UTR) of an mRNA.

RNA polymerases enzymes that transcribe a DNA sequence into an RNA transcript. Eukaryotes have three types of RNA polymerases called pol I, pol II, and pol III that are responsible for transcribing different classes of genes.

RNA primer a short stretch of RNA that initiates DNA synthesis during DNA replication.

RNA processing the modification of primary transcripts in eukaryotes that converts them into mRNAs. Modifications include splicing of exons (removal of introns), and addition of a poly-A tail to the 3′ end and a methylated cap at the 5′ end.

RNA splicing the process that deletes introns and joins together adjacent exons to form a mature mRNA consisting of only exons.

RNA thermometer an RNA leader that regulates translation in response to temperature through a stem-loop structure whose stability is temperature-dependent.

RNA-dependent DNA polymerase an enzyme that synthesizes DNA strands complementary to an RNA template.

RNA-like strand the strand of a double-helical DNA molecule that has the same nucleotide sequence as an mRNA (except for the substitution of T for U) and that is complementary to the template strand.

RNA-Seq a method for analysis of the transcriptome of an organism in which millions of cDNAs are sequenced. Also called *cDNA deep sequencing*.

Robertsonian translocation a translocation arising from breaks at or near the centromeres of two acrocentric chromosomes. The reciprocal exchange of broken parts generates one large metacentric chromosome and one very small chromosome.

S

S phase the stage of the cell cycle when chromosome replication occurs.

sample in population genetics, a finite number of individuals used to make inferences about the population as a whole.

satellite DNAs blocks of repetitive, noncoding sequences, usually around centromeres; these blocks have a different chromatin structure and different higher-order packaging than other chromosomal regions. The term derives from the fact that genomic DNA fragments containing repeated sequences separate from fragments without repeat sequences in density gradient centrifugation experiments.

SCO a meiosis that occurs with a single crossover between a particular gene pair.

screen a process whereby researchers examine a large number of organisms and identify rare individuals with a mutant phenotype of interest.

second filial (F_2) generation progeny resulting from self-crosses or intercrosses between individuals of the F_1 generation in a series of controlled matings.

second-division (MII) segregation pattern a fungal tetrad in which the arrangement of ascospores indicates that the two alleles of a gene segregated from each other in the second meiotic division.

secondary oocytes/spermatocytes germ-line cells in which meiosis II occurs.

secondary structure a localized region of a polypeptide chain with a characteristic geometry, such as an α-helix or β-pleated sheet.

sectors portions of a growing colony of microorganisms that have a different genotype than the remainder of the colony.

segment polarity genes in *Drosophila,* a group of zygotic segmentation genes expressed in fourteen stripes that are repeated with a single segment periodicity along the AP axis. Mutations in these genes cause deletion of that part of each segment corresponding to the cells in which the gene is expressed, often accompanied by mirror-image duplication of the remaining parts. The segment polarity genes encode cell communication proteins and determine certain patterns that are repeated in each segment.

segregation separation of alleles during gamete formation, in which one allele of each gene goes to each gamete.

selectable markers vector genes that make it possible to identify cells harboring a recombinant DNA molecule.

selection a process that establishes conditions in which only the desired mutant will grow; in population genetics, a process whereby alleles that confer the highest fitness to an organism become most frequent in the population.

self-fertilization (selfing) fertilization in which both egg and pollen come from the same plant or animal.

semiconservative replication a mechanism of DNA replication in which each single strand of the parent double helix serves as template for synthesis of its complement. The result is two daughter double helixes that each contain one of the original DNA strands intact (conserved) and one completely new strand.

semisterility a condition in which the capacity to generate viable offspring is diminished by at least half.

sex chromosomes in humans, the X and Y chromosomes, which determine the sex of an individual.

sex reversal the phenomenon whereby males are XX or females are XY.

sex-influenced traits traits that can appear in both sexes but are expressed differently in each sex due to hormonal differences.

sex-limited traits traits involving a structure or process that is found in one sex but not the other.

shelterin a protein complex that binds to telomeres and protects them from enzymatic activities.

Shine-Dalgarno box a sequence of six nucleotides in mRNA that is one of two elements constituting a ribosome binding site. (The other element is the initiation codon.)

signal transducers cytoplasmic proteins that relay signals inside the cell.

signal transduction pathway (cascade) a form of molecular communication in which the binding of proteins to receptors on cell surfaces constitutes a signal that is converted through a series of intermediate steps to a final intracellular regulatory response, usually the activation or repression of target gene transcription.

silencing long-term repression of transcription through DNA methylation.

silent mutations mutations without effects on phenotype; usually point mutations that change one of the three bases in a codon but, because of the degeneracy of the genetic code, do not change the identity of the specified amino acid.

simple sequence repeat (SSR) genomic locus that consists of one or a few bases repeated in tandem up to 100 times. Different alleles have different repeat numbers. The human genome contains ~100,000 SSR loci. Also called a *microsatellite* locus.

SINEs short interspersed elements; one of the two major classes of retrotransposons in mammals; their structures resemble mRNAs.

sister chromatids the two identical copies of a chromosome that exist immediately after DNA replication. Sister chromatids are held together by protein complexes called cohesins.

site-specific recombination crossing-over that occurs between two specific short DNA sequences, due to the action of specific enzymes called *recombinases.*

SKY (spectral karyotyping) a type of fluorescent *in situ* hybridization that results in labeling of each of the 24 human chromosomes as a different color.

slipped mispairing slippage of one DNA strand relative to the other during DNA replication through repeat regions. This process, also called *stuttering*, can result in expansion or contraction of SSR repeat numbers.

small interfering RNAs (siRNAs) short (21–24 nt) RNAs, generated by processing of double-stranded RNAs, that destroy specific RNA targets through complementary base pairing. In this, process called *RNA interference,* siRNAs guide the RISC complex to the mRNA target. Through RITS complex binding, siRNAs also function in heterochromatization of specific DNA regions.

SNP (single nucleotide polymorphism) a single nucleotide locus with two naturally existing alleles defined by a single base pair substitution. SNP loci are useful as DNA-based markers for genetic linkage analysis.

somatic cell nuclear transfer the method for reproductive cloning; the nucleus of a somatic cell replaces the nucleus of an oocyte, which is subsequently fertilized *in vitro,* and the zygote is introduced into the womb of a surrogate mother.

somatic cells cells in an organism other than gametes and their precursors.

SOS system an emergency repair system in bacteria that relies on error-prone DNA polymerases; these special SOS polymerases allow cells with damaged DNA to divide, but the daughter cells carry many new mutations.

specialized transducing phages bacteriophage carrying mainly phage DNA but also one or a few of the bacterial genes that lie near the site of prophage insertion. They can transfer these genes to another bacterium in the process known as specialized transduction.

specialized transduction bacteriophage-mediated transfer of a few bacterial genes located next to the prophage in the bacterial chromosome.

sperm a male gamete produced by meiosis. (In humans, sperm are haploid.)

spermatids haploid cells produced at the end of meiosis that will mature into sperm.

spermatogenesis the production of sperm.

spermatogonia diploid germ cells in the testes.

splice acceptors nucleotide sequences in a primary transcript at the border between an intron and the downstream exon that follows it; required for RNA splicing.

splice donors nucleotide sequences in a primary transcript at the border between an intron and the upstream exon that precedes it; required for RNA splicing.

spliceosome a complex intranuclear machine that performs RNA splicing.

sRNAs small RNA molecules that regulate translation in *trans* by base pairing with sites on mRNAs that can hide or expose the ribosome binding site.

SRY a Y-linked gene (*sex determining region of Y*) that determines maleness in mammals.

stem loop a structure formed when a single strand of RNA folds back on itself due to complementary base pairing between different regions in the same molecule. Also called a *hairpin loop.*

sticky ends the result achieved after digestion by a restriction enzyme that breaks the phosphodiester bonds on the two strands of a DNA molecule at offset locations. The resulting double-stranded DNA molecule has a single protruding strand at each end that is usually one to four bases in length.

stop codons triplets that do not correspond to an amino acid and instead signal termination of transcription: UAA, UGA, UAG.

strand invasion during recombination, one single-stranded DNA displaces the corresponding strand on the nonsister chromatid.

substitution a mutation that occurs when a base pair in a DNA molecule is replaced by one of the other three base pairs.

subunit a single polypeptide that is a constituent of a multimeric protein.

sum rule the probability that any of two or more mutually exclusive events will occur is the sum of their individual probabilities.

supercoiling additional twisting of the DNA molecule caused by movement of the replication fork during unwinding.

synapsis the process during which homologous chromosomes become aligned and zipped together; occurs during zygotene of prophase I.

synaptonemal complex the structure that helps align homologous chromosomes during prophase of meiosis I.

syncytial blastoderm in *Drosophila* embryos, the developmental stage when most of the syncytial nuclei have migrated out to the cortex of the embryo.

syncytium an animal cell with two or more nuclei.

syndrome a group of symptoms that appear together consistently.

syntenic describes the relationship of two or more loci located on the same chromosome.

syntenic blocks blocks of linked loci within a genome.

syntenic segments in the comparison of two genomes, large blocks of DNA sequences in which the identity, order, and transcriptional direction of the genes are almost exactly the same.

synthetic chromosome a chromosome whose DNA component is made using a DNA synthesis machine.

T

T-DNA the part of the Ti plasmid of *Agrobacterium tumefaciens* that integrates into the host plant genome. T-DNA is used as a vector for generating transgenic plants.

tandem duplications repeats of a chromosomal region that lie adjacent to each other, either in the same order or in reverse order.

targeted mutagenesis technology that enables scientists to alter any particular base pairs in the genome; includes gene targeting and CRISPR/Cas9.

TATA box in some eukaryotic genes, a promoter sequence ~25 nt upstream of the transcription start site that is bound by TBP, a transcription factor that brings the basal complex and RNA pol II to the promoter.

tautomerization interconversion of bases between two similar forms, or *tautomers*.

tautomers similar chemical forms of molecules, such as bases in DNA, that interconvert.

telomerase an enzyme crucial to the successful replication of telomeres at chromosome ends.

telomeres specialized terminal structures on eukaryotic chromosomes that ensure the maintenance and accurate replication of the two ends of each linear chromosome.

telophase the final stage of mitosis in which the daughter chromosomes reach the opposite poles of the cell and nuclei re-form.

telophase I the phase of meiosis I when nuclear membranes form around the chromosomes that have moved to the poles; each incipient daughter nucleus contains one-half the number of chromosomes in the original parent cell nucleus, but each of these chromosomes consists of two sister chromatids held together by cohesin protein complexes.

telophase II the final phase of meiosis II during which membranes form around each of the four daughter nuclei, and cytokinesis places each nucleus in a separate cell.

temperate bacteriophages phages that can enter either the lytic cycle or the alternative lysogenic cycle.

temperature-sensitive (ts) mutations gene mutations—usually missense mutations—that render the allele temperature sensitive.

template a strand of DNA or RNA that is used as a model by DNA or RNA polymerase or by reverse transcriptase for the creation of a new complementary strand of DNA or RNA.

template strand the strand of the double helix that is complementary to both the RNA-like DNA strand and the mRNA. The DNA strand used as a template for transcription.

termination the phase of translation that brings polypeptide synthesis to a halt.

terminators sequences in the RNA transcripts that cause RNA polymerase to stop transcription; stem-loop structures in an RNA leader sequence that stop transcription of the RNA downstream.

tertiary structure the ultimate three-dimensional shape of a polypeptide.

testcross a cross used to determine the genotype of an individual showing a dominant phenotype by mating with an individual showing the recessive phenotype.

tetrad (1) in some fungi, the assemblage of four ascospores (resulting from meiosis) in a single ascus. (2) a pair of synapsed homologous chromosomes during prophase of meiosis I, also known as a bivalent.

tetraploid describes cells or organisms with four complete sets of chromosomes.

tetratype (T) a fungal ascus that carries four kinds of spores, or haploid cells: two different parental types and two different recombinant types.

therapeutic gene a cloned gene introduced into patient's somatic cells whose product is meant to cure a disease.

threshold effect a phenomenon where a particular fraction of wild-type organellar DNAs is sufficient for the normal phenotype.

thymine dimers covalent linkage between adjacent thymine residues in DNA that can cause mutation.

Tn element a bacterial transposon carrying transposase and drug resistance genes flanked by ISs.

total phenotype variance (V_P) a measure of the amount of variation in a population, calculated as the average squared difference between each individual's trait value and the mean trait value for the population; the sum of the genetic variance (V_G) and the environmental variance (V_E).

totipotent describes a cell state during early embryonic development in which the cells have not yet differentiated and retain the ability to produce every type of cell found in the developing embryo and adult animal.

toxin a poison of plant or animal origin that causes disease at low concentrations.

trait value a number that describes a specific phenotype of a quantitative trait; also called a *phenotypic value.*

trans describes the action of a protein or RNA that can bind to target sites on any DNA or RNA in the cell.

***trans* configuration** on different DNA molecules.

transcript an RNA molecule that is the product of transcription.

transcription the conversion of DNA-encoded information to its RNA-encoded equivalent.

transcription bubble the region of DNA unwound by RNA polymerase.

transcription factor a protein whose DNA sequence-specific binding to a *cis*-control element regulates the timing, location, or level of a particular gene's transcription. Functional categories include basal factors, activators and repressors.

transcriptome the population of mRNAs expressed in a single cell, cell type, or organism.

transductants cells resulting from gene transfer mediated by bacteriophages.

transduction one of the mechanisms by which bacteria transfer genes from one strain to another; donor DNA is packaged within the protein coat of a bacteriophage and transferred to the recipient when the phage particle infects it. Recipient cells are known as transductants.

transfer RNA (tRNA) a small RNA adapter molecule that, through complementary base pairing with codons in mRNA, places a specific amino acid at the correct position in a growing polypeptide chain at the ribosome.

transformants cells that have received donor DNA.

transformation one of the mechanisms by which bacteria transfer genes from one strain to another; occurs when DNA from a donor is added to the bacterial growth medium and is then taken up from the medium by the recipient. The recipient cell is known as a transformant.

transgene any piece of foreign DNA that researchers have inserted into the genome of an organism.

transgenic organisms individuals carrying a transgene.

translation the process in which the codons carried by mRNA direct the synthesis of polypeptides from amino acids according to the genetic code.

translocation a rearrangement that occurs when part of a chromosome moves to a different chromosome.

transplastomic plant a plant containing transgenic chloroplasts.

transposable elements (TEs) DNA segments that move about in the genome, regardless of mechanism.

transposase a protein encoded by a DNA transposon that binds to the transposon's inverted repeats and catalyzes mobilization.

transposition the movement of transposable elements from one position in the genome to another.

transposons segments of DNA that move from place to place within the genome without an RNA intermediate. Synonymous with *DNA transposons*.

triploid describes cells or organisms with three complete sets of chromosomes.

trisomic describes individuals having one extra chromosome in addition to the normal diploid set.

truncation selection a form of artificial selection used by plant breeders in which only individuals with phenotypic values above or below a particular cutoff are bred to produce the next generation.

tumor progression the phenomenon where, over time, cancerous tumors grow faster and become more invasive.

tumor-suppressor genes genes whose loss-of-function mutant alleles contribute to cells becoming cancerous.

twin spots adjacent patches of tissue that are phenotypically distinct from each other and from the surrounding tissue; can be produced as a result of mitotic recombination.

U

unequal crossing-over erroneous recombination caused by misalignment of homologs where one homologous chromosome ends up with a duplication, while the other homolog sustains a deletion.

uniparental inheritance transmission of organelle genes via one parent. Most species transmit mitochondrial DNA and chloroplast DNA through the mother.

unordered tetrads tetrads in yeast in which the four ascospores are arranged in the ascus randomly.

unstable trinucleotide repeats SSR loci with a 3 bp repeat unit. In the very rare instances when these are within a gene, expansion of the repeat number during DNA replication results in disease alleles.

upstream movement opposite the direction RNA follows when moving along a gene; toward the 5′ end of a gene.

upstream activating sequence (UAS) a *cis*-control element in yeast similar to an enhancer; unlike enhancers, UASs cannot function downstream of promoters and their orientation with respect to the promoter matters.

V

vaccines antigens of a disease-causing agent that stimulate an immune response to that particular foreign substance.

vertical gene transfer the passage of genes from one generation to the next, particularly during sexual reproduction.

virulent bacteriophages phages that always enter the lytic cycle, multiply rapidly, and kill the host.

W

whole-exome sequencing sequencing of only the genomic DNA corresponding to exons.

whole-genome shotgun strategy a method for determining an entire genome sequence where random overlapping genomic DNA fragments are sequenced, and the sequences are assembled by a computer until the entire genome sequence is complete.

wild-type allele (1) an allele, or DNA variant, whose frequency is more than 1% in a population; (2) an allele that dictates the most frequently observed phenotype in a population. Wild-type alleles are often designated by a superscript plus sign ($^+$).

wobble the ability of the 5′-most base of an anticodon to pair with more than one type of base at the 3′ ends of codons; helps explain the degeneracy of the genetic code.

X

X chromosome inactivation in mammals, a mechanism of dosage compensation in which all X chromosomes in a cellular genome but one are inactivated at an early stage of development through the formation of heterochromatic Barr bodies.

X inactivation center (XIC) a region of the X chromosome (~450 kb) that mediates dosage compensation.

X chromosome reactivation in mammals, a mechanism by which X chromosomes that were inactivated become reactivated in oogonia so that the haploid cells in the germ line all have an active X chromosome.

X-linked carried by the X chromosome.

Xist (X inactive specific transcript) ncRNA transcribed from the XIC of the X chromosome that becomes a Barr body; *Xist* ncRNA binds to the X chromosome and mediates X chromosome inactivation.

Y

yeast artificial chromosome (YAC) a vector used to clone DNA fragments up to 400 kb in length. A YAC is constructed from telomeric, centromeric, and origin-of-replication sequences needed for replication in yeast cells.

Z

Z-form DNA DNA in which the nucleotide sequences cause the structure to assume a zigzag shape due to the helixes spiraling to the left. The significance of this variation on DNA structure is unknown.

zygote the cell formed by the fertilization of the egg by the sperm during sexual reproduction; in humans, eggs and sperm are haploid, and zygotes are diploid.

zygotene the substage of prophase I when homologous chromosomes become zipped together in synapsis.

zygotic segmentation genes in *Drosophila,* a group of genes transcribed from the genome of the zygote that divide the organism into an array of identical segments. The three subgroups of zygotic segmentation genes are: gap genes, pair-rule genes, and segment-polarity genes.

Index

Note: Page numbers followed by *f* and *t* indicate figures and tables, respectively.

A

B

D

E

F

G

H

I

M

P

Q

R

S

T

U

V

W

X

Y

Z